Question Bank in

Electronics and Communication Engineering

DR. B.R. GUPTA
VANDANA SINGHAL

Question Bank in

Electronics And Communication Engineering

With more than 12,000 Objective Types Quesitons and Answers

Dr. B.R. GUPTA
B.Sc. Engg. (Hons.); M.Sc. Engg. (Distinction)
Ph.D., Senior Member IEEE (USA)
MIE (India)
Professor of Electrical Engg. (Retd.)
Punjab Engineering College
CHANDIGARH

VANDANA SINGHAL
B.E. (Hons.) Gold Medalist
M.E. (Distinction)

S.K. KATARIA & SONS®
Educational Publishers
4885/109, Prakash Mahal, Dr. Subhash Bhargav Lane,
Opposite Delhi Medical Association, Daryaganj, New Delhi-110002 (INDIA)
Phone: +91-11-43551243; *Mobile:* +91-9871775858
e-mail: katariabook@gmail.com; katariabooks@yahoo.com
Website: www.skkatariaandsons.com
(e-books, subscriptions and bulk buying also available on our website)

Published by:

S.K. Kataria & Sons®
4885/109, Prakash Mahal, Dr. Subhash Bhargav Lane,
Opposite Delhi Medical Association, Daryaganj, New Delhi-110002 (INDIA)
Phone: +91-11-43551243; *Mobile:* +91-9871775858
e-mail: katariabook@gmail.com; katariabooks@yahoo.com
Website: www.skkatariaandsons.com

Send all correspondence to : **M/s. S.K. Kataria & Sons,** New Delhi
Email: katariabook@gmail.com

First Edition: 2002; Third Edition: 2012; Reprint: 2024

Laser Typesetting by : S.N. Computers, Shahdara

Printed at : J.S. Printing Press

Preface to The Third Edition

Many public sector undertakings and private organisations recruit engineers through written tests consisting of Objective Type Questions. This book has been designed to meet the needs of electronics, communication, instrumentations and computer engineering students so that they can prepare and face these tests confidently. It includes the syllabii of a large number of such organisations.

Each chapter has a brief theory pertaining to that chapter. It is followed by **multiple choice questions, true-false questions** and **fill in the blanks questions**. Most of the questions included in the book have been taken from question papers of competitive examinations.

It is hoped that students will find the book useful. Suggestions for improvement of the book are welcome.

AUTHORS

Contents

CHAPTER 1

ELECTRICAL AND ELECTRONIC MATERIALS AND COMPONENTS

1.1. CONDUCTING MATERIALS

1.1.1. Resistivity

Metals are good conductors of electricity. The resistance of a conductor is given by the equation.

$$R = \frac{\rho l}{a} \quad ...(1.1)$$

where R = resistance in ohms

ρ = specific resistance or resistivity, ohms–m

l = length, m

a = area of cross-section, m^2

The specific resistance for some conducting materials is given in Table 1.1.

TABLE 1.1. Specific resistance at 20°C

Material	ρ (*ohms – m*)	*Material*	ρ (*ohms – m*)
Advance	48.9×10^{-8}	Manganin	45×10^{-8}
Aluminium	2.83×10^{-8}	Nichrome	109.7×10^{-8}
Brass (annealed)	7×10^{-8}	Nickel	7.81×10^{-8}
Constantan	49×10^{-8}	Platinum	10×10^{-8}
Copper (annealed)	1.724×10^{-8}	Silver	1.64×10^{-8}
Copper (hard drawn)	1.78×10^{-8}	Steel (hard)	45.7×10^{-8}
Gold	2.44×10^{-8}	Tungsten	5.52×10^{-8}
Iron (commercial)	12×10^{-8}	Zinc	6.22×10^{-8}
Lead	22×10^{-8}		

1.1.2. Temperature Coefficient of Resistance

The resistance of all pure metals increases with temperature. Thus

$$R_{t_2} = R_{t_0} [1 + \alpha_0 (t_2 - t_0)] \quad ...(1.2)$$

where R_{t_2} = resistance at t_2°C

R_{t_0} = resistance at t_0°C

α_0 = temperature coefficient of resistance at t_0°C

The temperature coefficient α is also a function of temperature. Thus

$$\alpha_2 = \frac{\alpha_1}{1 + \alpha_1 (t_2 - t_1)} \quad ...(1.3)$$

where α_2 is temperature coefficient of resistance of t_2°C.

α_1 is temperature coefficient of resistance at t_1°C.

Temperature coefficient of some materials at 20°C is given in Table 1.2.

TABLE 1.2. Temperature Coefficient at 20°C

Material	α	*Material*	α
Advance	0.00005	Manganin	0.000006
Aluminium	0.0039	Nichrome	0.0004
Brass	0.002	Nickel	0.0062
Carbon	– 0.0005	Silver	0.0038
Constantan	0.000008	Tungsten	0.0045
Copper	0.00393	Zinc	0.0037

1.1.3. Ohm's Law

$$J = \sigma E \quad ...(1.4)$$

Where J = current density, A/m^2

σ = Conductivity, $\frac{1}{\text{ohm} - \text{m}}$

E = Electric field strength, $\frac{\text{Volts}}{\text{m}}$

The usual form of Eq. (1.4) is $I = \frac{V}{R}$

Conductivity σ is reciprocal of specific resistance ρ.

1.1.4. Drift Velocity, Mobility

The steady average velocity of electron in an applied field E is called drift velocity of the electron. It is very small, less than 1 mm/sec.

Mobility is the average drift velocity per unit field. Thus

$$\text{Drift Velocity} = -\left(\frac{e^{\tau}}{m}\right)E \quad ...(1.5)$$

where E = electric field strength, volts/m

e = electric charge = 1.602×10^{-19} C

m = electronic mass = 9.109×10^{-31} kg

τ = relaxation time of electron

The ratio $\frac{e}{m}$ is 1.759×10^{11}

$$\text{Conductivity } \sigma = \frac{1}{\rho} = ne^2 \frac{\tau}{m} \quad ...(1.6)$$

where n is the number of electrons per m^3.

1.1.5. Super Conductor

In some materials resistance becomes zero when the material is cooled to a very low temperature. This temperature is given in Table 1.3.

TABLE 1.3. Superconducting Elements

Element	*Temp K*	*Element*	*Temp K*
Niobium	9.17	Mercury	4.15
Lead	7.23	Tin	3.72
Vanadium	5.03	Indium	3.40
Tantalum	4.39	Aluminium	1.17

1.1.6. Varistor (Non-Linear Resistor)

It is a voltage dependent metal oxide material, *e.g.*, silicon carbide, zinc oxide etc. whose resistance decreases sharply with increase in voltage. Its v–i relationship is

$$i = kv^n \quad ...(1.7)$$

where i is instantaneous current, v is instantaneous voltage, k is a constant whose value depends on geometry of resistance. The value of exponent n depends on the material. For silicon carbide n lies in the range of 2 to 5. for zinc oxide n is in the range of 25 to 50. Greater the value of n, higher is non-linearity. These materials are used in lightning arresters (also called surge arresters).

1.1.7. Thermistor

It is a resistor having negative temperature coefficient α. It is manufactured from manganese, metal and cobalt oxides with suitable binding material. It is used in temperature alarm circuits of electric machines and for temperature measurement and control.

1.1.8. Photo Conductor (Photo Resistor)

It is semiconductor whose resistance decreases with increase in light intensity. It is also called LDR (light dependent resistor). It is used in photographic exposure meters and automatic control of street lights.

1.1.9. Fermi Level

The free electrons in a metal can accept a series of discrete energy levels. At OK all energy levels below a certain value W_F are filled while all those above W_F are empty. W_F is called fermi level. Table 1.4 gives the value of conductivity, fermi level, mean free paths λ and relaxation time τ for some monovalent metals.

TABLE 1.4. Conductivity σ, Fermi Level W_F, Mean Free Path λ and Relaxation time (τ) of some monovalent metals

Metal	σ (10^8 ohm^{-1} m^{-1})	W_F (eV)	λ (angstroms)	τ (10^{-14} secs)
Li	0.12	4.7	110	0.9
Na	0.23	3.1	350	3.1
K	0.19	2.1	370	4.4
Cu	0.64	7.0	420	2.7
Ag	0.68	5.5	570	4.1

1.1.10 Heat Developed in a Conductor

$$W = JE = \sigma E^2$$

where W = Heat developed, watts per m^3 per second

J = current density, A/m^2

σ = conductivity, ohm^{-1} m^{-1}

E = Electric field intensity, volts/m

1.2. DIELECTRIC MATERIALS

Materials which do not have free electrons cannot conduct. They are called dielectric materials or insulating materials. Their resistivity at room temperature is very high (ρ for rubber is $5 \times 10^{12}\ \Omega - \text{m}$).

1.2.1. Solid Dielectrics

Glass, mica, paper, porcelain are solid dielectrics. According to temperature limits they are classified into 7 categories Y. A, E, B, F, H and C. Their temperature limits and examples are given in Table 1.5.

TABLE 1.5. Temperature Limits of Solid Dielectrics

Class	*Temperature*	*Examples*
Y	90°C	Cotton, silk, wood, paper, fibre without impregnation. They are hygroscopic and greatly affected by moisture.
A	105°C	Black tape, empire cloth, sleeve etc.
E	120°C	PVC, synthetic plastic, synthetic resin, enamel
B	130°C	Combination of mica, glass, fibre (mica tape etc.)
F	155° C	Mica sheets, Micanite with cement, glass fibre, dried and impregnated paper.
H	180°C	Silica based inorganic and organic compounds, glass fibre etc.
C	above 180°C	Mica, procelain, glass, quartz etc.

1.2.2. Dielectric Strength

The minimum electrostatic field which causes breakdown of dielectric is called dielectric or breakdown strength. Its values for some dielectrics are given in Table 1.6.

TABLE 1.6. Dielectric Strength

Dielectric	*Dielectric Strength* $\frac{kV}{mm}$	*Dielectric*	*Dielectric Strength* $\frac{kV}{mm}$
Air	3	Paper	4 – 10
Asbestos	10	Paper (kraft, dry)	20 – 30
Bakelite	10 – 28	Paraffin Wax	8
Bitumin (Vulcanised)	14	Porcelain	7 – 20
Cotton	3 – 4	Rubber	10 – 25
Empire cloth	10 – 20	Shellac	520
Fibre	6	Varnished cambric	32
Glass	10 – 40	Vinyl (plastic)	16
Mica	50 – 200		

1.2.3. Permittivity, Dielectric Constant

Electric permittivity specifies the ease with which electric flux is permitted to pass through a dielectric material. It is the ratio of electric flux density D and electric field intensity E. Thus

$$D = \epsilon E \qquad ...(1.9)$$

where D = Electric flux density, coulomb/m^2

E = Electric field intensity, V/m

ϵ = Permittivity, F/m

Permittivity of a material is written as

$$\epsilon = \epsilon_r, \epsilon_0$$

where ϵ_r = Relative permittivity or Dielectric constant

ϵ_0 = Permittivity of free space (= 8.85×10^{-12} F/m)

ϵ_r for some materials given in Table 1.7.

TABLE 1.7. Relative Permittivity (Dielectric Constant)

Dielectric	ϵ_r	*Dielectric*	ϵ_r
Vacuum	1	Paper	2 – 4
Air	1.006	Paraffin	2 – 2.5
Bakelite	4.5 – 5.5	Plastic	2 - 4.5
Ceramic (low loss)	6 – 20	Porcelain	6
Ceramic (high ϵ_r)	> 1000	Rubber	2 – 3
Fibre	2 – 5	Shellac	2.5 – 3.5
Glass	5 – 100	Transformer oil	4
Mica	3 – 7	Wood	2.5 – 7
Oxide films	5 – 20		

1.2.4. Insulation Resistance of Cable (Fig 1.1)

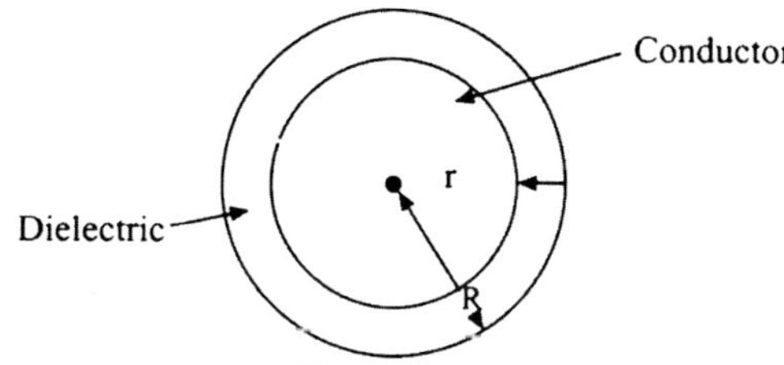

Fig. 1.1.

$$\text{Insulation resistance} = \frac{\rho}{2\pi l} \ln \frac{R}{r} \qquad ...(1.10)$$

ρ = resistivity, ohm m

l = length of cable, m

R = outside radius, m

r = core radius, m

1.2.5. Capacitance of Parallel Plate Capacitor

$$C = \frac{\epsilon A}{d} \qquad ...(1.11)$$

where C = capacitance, F

ϵ = Permittivity, F/m = $\epsilon_r \epsilon_0$.

A = Area of each plate, m^2

d = distance between plates, m

1.2.6. Capacitor with Composite Dielectric (Fig. 1.2)

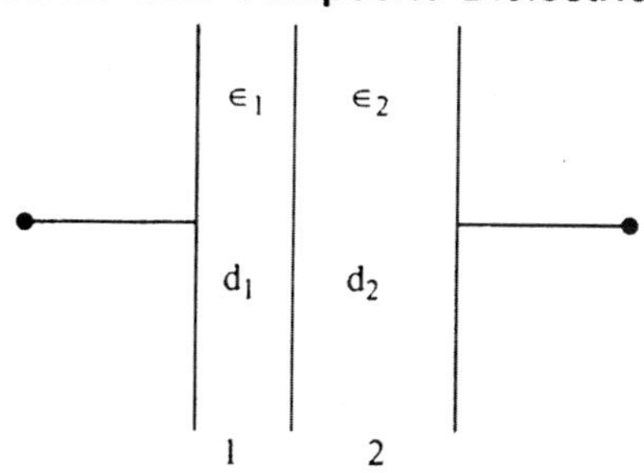

Fig. 1.2.

$$C_1 = \frac{\epsilon_1 \epsilon_0 A}{d_1} \qquad ...(1.12)$$

$$C_2 = \frac{\epsilon_2 \epsilon_0 A}{d_2} \qquad ...(1.13)$$

$$C = \frac{C_1 C_2}{C_1 + C_2} \qquad ...(1.14)$$

1.2.7. Cylindrical Capacitor (Fig. 1.3)

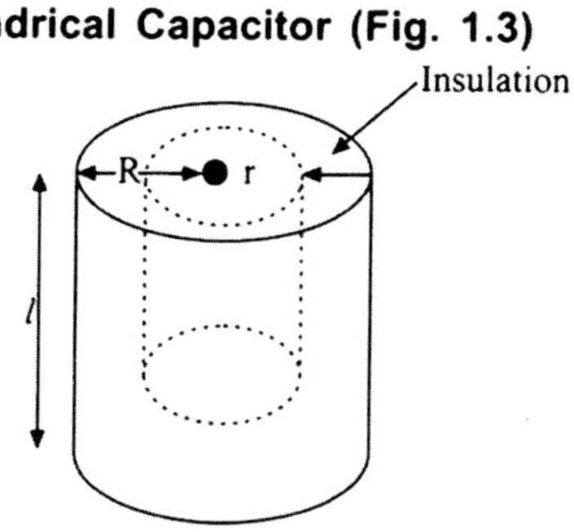

Fig. 1.3.

$$C = \frac{2\pi l \in_0 \in_r}{\ln\left(\frac{R}{r}\right)} \quad ...(1.15)$$

1.2.8. Energy stored in Capacitor

$$\text{Energy} = \frac{1}{2} CV^2 \quad ...(1.16)$$

1.2.9. Types of Capacitor

Air Capacitors, Paper Capacitor, Plastic Film Capacitor, Mica Capacitor, Ceramic Capacitor, Electrolytic Capacitor.

1.2.10. Polarization

A dielectric subjected to homogenous field carries a dipole moment P per unit volume.

$$P = (\in_r - 1)\, E \quad ...(1.17)$$

The dipole moment per unit volume is called polarization. Its units are coulomb m^{-2}.

The polarization in a polyatomic gas consists of orientation polarization (P_0), electronic polarization (P_e) and ionic polarization (P_i). Total polarization is sum of these quantities.

$$P_e + P_i = N(\alpha_e + \alpha_i)\, E$$

where N = number of atoms/m^3

α_e = electronic polarizabilily, farads/m^2

α_i = ionic polarizabilily, farads/m^2

$$P_0 = \frac{N\mu_p^2 E}{3kT} \quad ...(1.18)$$

μ_p = permanent dipole moment

T = absolute temperature, K

k = Boltzmann's constant

= 1.38×10^{-23} Joule/K

= 8.62×10^{-5} eV/K

$$\text{Thus, } \in_0(\in_r - 1) = N(\alpha_e + \alpha_i + \mu_p^2/3kT) \quad ...(1.19)$$

1.2.11. Elemental and Ionic Solid Dielectrics

Elemental dielectrics are built up from only one kind of atoms e.g. diamond, phosphorous etc. These materials do not have permanent dipoles ($P_i = P_0 = 0$). Thus for elemental solid dielectrics

$$P = N\alpha_e E_i \quad ...(1.20)$$

where N is number of atoms/m^3 and E_i is internal field.

Applied field E and internal field E_i are related as

$$E_i = E + \frac{\gamma}{\in_0} P \quad ...(1.21)$$

where P is dipole moment per unit volume and γ is called internal field constant.

In ionic crystals without permanent dipoles, the total polarization is

$$P = P_e + P_i \quad ...(1.22)$$

Alkali hides, *e.g.*, LiF, NaF, Nal etc. are examples of ionic crystals without permanent dipole moment. In solids containing permanent dipoles (*e.g.*, liquid HCl) all the three polarizations are present.

1.2.12. Ferro Electric Materials

Some dielectrics exhibit hysteresis effect. They are called ferroelectric materials. In these materials polarization is not a unique function of field strength. The graph of polarization versus E is similar to hysteresis loop of magnetic materials. Rochelle salt, dihydrogen phosphate, barium titanate are some such materials.

1.2.13. Spontaneous Polarization

Ferro electric materials exhibit the property of spontaneous polarization. In the ferro electric region the dielectric constant is not a constant but is a function of electric field strength E. For such materials we have an expression

$$\in_0(\in_r - 1) = \frac{dP}{dE} \quad ...(1.23)$$

The spontaneous polarization vanishes above a temperature called ferro electric Curie temperature θ_f. Above θ_f, the dielectric constant varies with temperature as per Curie-Weiss law, *i.e.*,

$$\in = \frac{C}{(T - \theta)} \quad ...(1.24)$$

where C is a constant and θ is a characteristic temperature, a few degrees lower than θ_f.

Polarization of a material exhibiting spontaneous polarization may be written as

$$P = \frac{N\alpha E}{1 - \left(\frac{N\alpha\gamma}{\in_0}\right)} \quad ...(1.25)$$

where N is the number of ferroelectric cubic units and α is total polarizabilily. If $\frac{N\alpha\gamma}{\in_0} = 1$, spontaneous polarization (*i.e.*, finite value of P even if E is zero) can occur.

1.2.14 Piezo Electricity

Dielectric polarization causes mechanical distortion in many dielectric materials. This effect is called electrostriction.

Some materials become polarized when a mechanical stress is applied. They are called piezo electric. These materials permit conversion of mechanical energy into electrical energy and vice versa.

1.2.15 Complex Dielectric Constant

Complex dielectric constant of non-dipolar solids can be expressed as

$$\in_r^* = \in_r' - j\in_r'' \quad ...(1.26)$$

For frequencies up to microwave frequencies $\in_r$ is real. Above these frequencies imaginary part of dielectric constant exists.

1.2.16 Dipolar Relaxation

Polarization can be expressed as a function of time, *i.e.*,

$$P_0(t) = P_0 \in_r^{-t/\tau} \quad ...(1.27)$$

where $P_0(t)$ is polarization at time t, P_0 is polarization at $t = 0$ and τ is relaxation time.

1.2.17 Dielectric Loss

The energy dissipated in dielectric medium is called dielectric loss. The current in a capacitor built up from an imperfect dielectric leads the voltage by an angle ϕ (and not 90°) as shown in Fig. 1.4. $\angle\delta$ is called loss angle of dielectric.

P_d, *i.e.*, Dielectric loss can be written as

$$P_d = VI \cos\phi = VI \sin\delta$$

or $$P_d = \omega CV^2 \sin\delta \quad ...(1.28)$$

Since δ is very small, $P_d = \omega CV^2 \tan\delta$

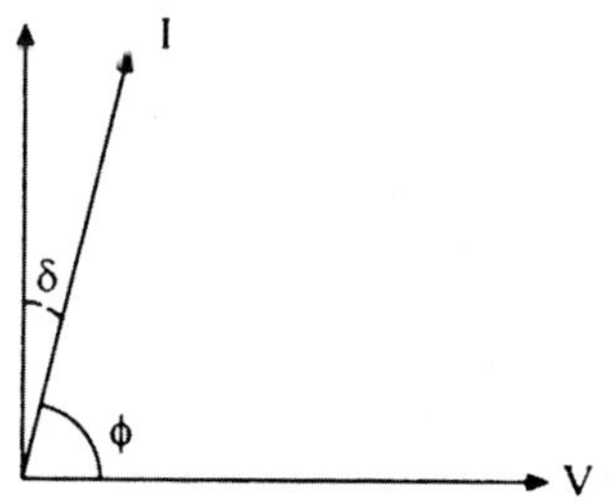

Fig. 1.4.

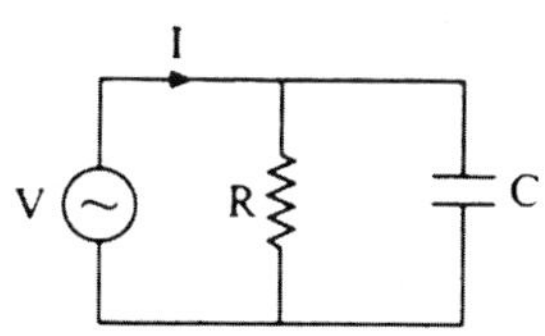

Fig. 1.5.

An imperfect capacitor is represented as an ideal (or perfect) capacitor C is parallel with a resistance R (Fig. 1.5). This resistance R contributes to dielectric loss. It can be shown that

$$\tan\delta = \frac{\epsilon_r''}{\epsilon_r'} \quad ...(1.29)$$

1.3. MAGNETIC MATERIALS

1.3.1. Magnetic Dipole Moment

$$M = (\mu_r - 1) H \quad ...(1.30)$$

where M is magnetic dipole moment per unit volume

μ_r is relative permeability.

H is magnetic field strength

$$B = \mu_0(H + M) \quad ...(1.31)$$

where B is flux density, Wb/m^2.

On the basis of permanent atomic magnetic dipoles the classification is given in Table 1.8.

TABLE 1.8. Classification of Materials

Material	*Permanent dipoles*	*Interaction between neighbouring dipoles*	*Example*
Diamagnetic	No	–	Copper, silver
Paramagnetic	Yes	Negligible	Aluminium, Platinum
Ferromagnetic	Yes	Parallel Orientation	Fe, Ni, CO
Anti-ferromagnetic	Yes	Anti-parallel orientation of equal moments	MnO
Ferrimagnetic	Yes	Anti-parallel orientation of unequal moments	Ferrite

1.3.2. Permanent Magnetic Dipoles

The angular momentum of an atom is contributed by orbital angular momentum of electrons, electron spin angular momentum and nuclear spin angular momentum.

1.3.3. Curie Temperature, Curie-Weiss Law

Above a temperature called ferromagnetic curie temperature θ_f a ferromagnetic material behaves as a paramagnetic material. Susceptibility ψ is given by Curie-Weiss law as

$$\psi = \mu_r - 1 = \frac{C}{T - \theta} \quad ...(1.32)$$

where C is called Curie constant, T is absolute temperature and θ is paramagnetic Curie temperature. θ is somewhat higher than θ_f. Some values are given in Table 1.9.

TABLE 1.9. Curie Temperature

	Fe	Co	Ni
θ_f	1043	1393	631
θ	1093	1428	650

when $T < \theta_f$, ferromagnetic materials exhibit the hysteresis phenomenon.

When $T = \theta$, spontaneous magnetisation may occur (*i.e.*, finite M for H = 0).

1.3.4. Ferromagnetic Domains

A virgin specimen of iron has a number of regions or domains (about 10^{-6} m size) each domain having spontaneous magnetisation. The directions of spontaneous magnetisation varies from domain to domain resulting in net zero magnetisation.

1.3.5. Neel Temperature

In anti-ferromagnetic materials susceptibility has a maximum value at a temperature known as Neel temperature T_N. Thus we can write the following equations for susceptibility.

Paramagnetic	Ferromagnetic	Anti-ferromagnetic	
$\psi = \frac{C}{T}$	$\psi = \frac{C}{T - \theta}$	$\psi = \frac{C}{T + \theta}$	...(1.33)
	For $T > \theta_f$	For $T > T_N$	

1.4. SEMICONDUCTORS

Silicon and germanium are semiconducting materials. Their resistivity is much lower than that of insulators but much higher as compared to that of conductors. The resistivity decreases with increase in temperature. Some compounds, *e.g.*, metallic oxides, metallic sulphides also possess semiconducting properties.

Both silicon and Germanium have 4 valence electrons each. At 0K they do not have any free electronic and hence no conductivity. At higher temperatures some valence electrons get thermally excited and break the covalent bonds. As an electron breaks a covalent bond it leaves a hole (*i.e.*, positive charge) behind. The total current is partly due to holes and partly due to electrons. Silicon and germanium are called intrinsic semiconductors.

The conducting property can be greatly enhanced by adding small amount of impurity. Such semiconductors are called extrinsic semiconductors. When a pentavalent impurity (*e.g.*, antimony, arsenic) is added it becomes *n* type semiconductor. When a trivalent impurity (indium, boron, gallium) is added we get a *p* type semiconductor.

The measurement of Hall coefficient of an extrinsic semiconductor provides us with the information about the sign of charge carriers (electrons or holes) and thus the type of semiconductor and the density of charge carriers.

OBJECTIVE TYPE QUESTIONS

A. Choose the most appropriate alternative

1. For making a capacitor it is better to have a dielectric having
   (*a*) high permittivity (*b*) low permittivity
   (*c*) permittivity same as that of air
   (*d*) permittivity which is neither high nor low
2. A dielectric material must be
   (*a*) insulator (*b*) good conductor
   (*c*) resistor (*d*) semi conductor
3. Which of the following is the poorest conductor of electricity?
   (*a*) aluminium (*b*) nichrome
   (*c*) copper (*d*) all the above
4. Nichrome is used for
   (*a*) overlead line wires (*b*) heater coils
   (*c*) lamp filaments (*d*) all the above
5. A magnet does not attract
   (*a*) nickel (*b*) iron
   (*c*) copper (*d*) iron and copper
6. Tesla is a unit of
   (*a*) flux (*b*) field strength
   (*c*) flux density (*d*) mmf
7. A permeable substance is one
   (*a*) which is a strong magnet
   (*b*) which is a weak magnet
   (*c*) which is a good conductor
   (*d*) through which magnetic lines of force can pass easily
8. The two conductors of a transmission line carry equal current I in opposite direction. The force on each conductor is
   (*a*) proportional to I
   (*b*) proportional to I^2
   (*c*) proportional to distance between the lines
   (*d*) proportional to I^3
9. A piece of iron is placed in a magnetic field
   (*a*) the magnetic field will not be affected
   (*b*) the magnetic lines of force will bend to pass through the piece
   (*c*) the magnetic lines of force will bend away from the piece
   (*d*) any of above depending on type of iron
10. The materials having low retentivity are suitable for making
    (*a*) temporary magnets
    (*b*) weak magnets
    (*c*) permanent magnets
    (*d*) both temporary and permanent magnets
11. A magnetic field exists around
    (*a*) moving charges (*b*) static charges
    (*c*) iron (*d*) copper
12. A material which is slightly repelled by a magnetic field is known as
    (*a*) diamagnetic material
    (*b*) non-magnetic material
    (*c*) ferromagnetic material
    (*d*) paramagnetic material
13. Fleming's left hand rule is used to find
    (*a*) direction of force on a current carrying conductor
    (*b*) direction of flux in solenoid
    (*c*) direction of magnetic field due to a current carrying conductor
    (*d*) direction of induced emf
14. If a piece of wood is placed in a magnetic field
    (*a*) the magnetic field will not be affected
    (*b*) the magnetic lines of force will bend towards the piece
    (*c*) the magnetic lines of force will bend away from the piece
    (*d*) the flux may increase or decrease
15. The unit for mmf is
    (*a*) A (*b*) Wb
    (*c*) T (*d*) V
16. The direction of force on a conductor carrying current and lying in a magnetic field can be found by using
    (*a*) Fleming's left hand rule
    (*b*) Fleming's right hand rule
    (*c*) Cork screw role (*d*) any of above.

17. The relative permeability of a ferromagnetic material is
(*a*) 1 (*b*) less than 1
(*c*) about 10 (*d*) 1000 or more

18. The iron core of a coil is removed so that it becomes an air cored coil. Its inductance will
(*a*) increase (*b*) decrease
(*c*) remain the same (*d*) may increase or decrease

19. In a specimen of ferro magnetic material, as the flux density is increased from zero to 2.5 T, μ_r will
(*a*) increase (*b*) decrease
(*c*) first decrease then increase
(*d*) first increase then decrease

20. In a specimen of silicon steel, a field strength of 200 A/m causes a flux density of 0.5 T. In a specimen of cast iron, the same H would cause B to be
(*a*) 0.5 T (*b*) less than 0.5 T
(*c*) more than 0.5 T (*d*) 0.5 T or more or less

21. Permeability is analogous to
(*a*) resistivity (*b*) retentively
(*c*) conductivity (*d*) coercivity

22. An iron specimen has μ_r = 1000 when B = 0.3 T. If B = 0.35 T, μ_r
(*a*) will be 1000 (*b*) will be less than 1000
(*c*) will be more than 1000
(*d*) may be more or less than 1000

23. Some typical examples of macroscopic quantities are
(*a*) conductivity (*b*) μ_r
(*c*) dielectric constant (*d*) all the above

24. Macroscopic quantities of a given material
(*a*) are absolutely constant
(*b*) are a function of temperature
(*c*) are a function of temperature, pressure, field
(*d*) are a function of temperature, pressure, field etc.

25. The charge per proton is
(*a*) 1.6×10^{-19} C
(*b*) 2.6×10^{-19} C
(*c*) 0.6×10^{-19} C
(*d*) 1.6×10^{-13} C

26. The mass of proton and electron are
(*a*) 1.672×10^{-26} kg and 9.107×10^{-31} kg respectively
(*b*) 9.107×10^{-26} kg and 9.107×10^{-31} kg respectively
(*c*) 9.107×10^{-29} kg and 9.107×10^{-31} kg respectively
(*d*) 9.107×10^{-30} kg and 9.107×10^{-31} kg respectively

27. In terms of classical representation of electrons
(*a*) an atom is essentially empty
(*b*) an atom is essentially full
(*c*) an atom may be essentially empty or full depending on the material
(*d*) an atom may be essentially empty or full depending on temperature

28. If *v* is velocity of electron, ϵ_0 is absolute permittivity, *m* is mass of electron, *e* is charge an electron and *r* is the radius of orbit of electron in hydrogen atom, the stability of orbit requires the following equation to be satisfied
(*a*) $\frac{mv}{r} = \frac{e^2}{4\pi\epsilon_0 r^2}$ (*b*) $\frac{mv^2}{r} = \frac{e^2}{4\pi\epsilon_0 r^2}$
(*c*) $\frac{mv^2}{r} = \frac{e^2\epsilon_0}{4\pi r^2}$ (*d*) $\frac{mv^2}{r} = \frac{e^2}{\epsilon_0 r^2}$

29. In its lowest state, the electron is bound to the field of proton by an energy equal to
(*a*) 100.5 eV (*b*) 51.2 eV
(*c*) 13.6 eV (*d*) 0.2 eV

30. If *m, v, r* are the mass, velocity and radius of orbit of electron and *h* is planck's constant, quantum condition for electron's orbit is
(*a*) $\frac{mv}{r} = \frac{nh}{2\pi}$ (*b*) $mvr = \frac{nh}{2\pi}$
(*c*) $mvr = \frac{2\pi n}{h}$ (*d*) $mvr = \frac{h}{2n\pi}$

31. The transition of electron from energy level W_1 to W_2, is associated with emission or absorption of electro magnetic radiation of frequency *f* such that
(*a*) $hf = |W_1 - W_2|$ (*b*) $|W_1 - W_2|f = h$
(*c*) $hf = 0.5|W_1 - W_2|$ (*d*) $(hf)^2 = |W_1 - W_2|$

32. Schrodinger wave equation is a
(*a*) linear equation
(*b*) differential equation
(*c*) partial differential equation
(*d*) none of above

33. When schrodinger equation is applied to the motion of electron in the field of protons, the physically acceptable solutions for the wave function exist for quantum number which are
(*a*) 2 (*b*) 3
(*c*) 4 (*d*) 5

34. The quantum numbers associated with electron motion are designed as
(*a*) *a, b, c* (*b*) *h, k, l*
(*c*) *n, m, l* (*d*) *h, n, l*

35. In the wave mechanical theory, the maximum of the charge distribution in the ground state occurs for a distance from the nucleus equal to
(*a*) first Bohr radius
(*b*) second Bohr radius
(*c*) third Bohr radius
(*d*) either first or second Bohr radius

36. For a given value *n* of principal quantum number the number of possible electronic states are
(*a*) n (*b*) n^2
(*c*) n^3 (*d*) $\sqrt{n}$

37. An electron shell corresponding to principal quantum number *n* can contain a maximum of
(*a*) n^2 electrons (*b*) $2n^2$ electrons
(*c*) $2n$ electrons (*d*) n^3 electron

38. On the basis of the type of chemical bonds, solids may be classified into
(*a*) Ionic crystals
(*b*) Ionic and valence crystals
(*c*) Ionic crystal, valence crystals and metals
(*d*) Ionic crystals, valence crystals, metals and vender wall crystals

39. In a valence crystal, the valence electrons
(*a*) are not shared
(*b*) are shared between neighbouring only atoms
(*c*) are shared by all atoms
(*d*) any of above

40. In a metal, the valence electrons
(*a*) are not shared
(*b*) are shared only between neighbouring atoms
(*c*) are shared by all atoms
(*d*) either (*b*) or (*c*)

41. As regards dielectric constant $\in_r$
(*a*) the units are farads/m
(*b*) the units are newton/coulomb
(*c*) it is dimensionless
(*d*) the units are farads

42. The electric dipole moment of a neutral system of point charges $Q_1, Q_2...$ is $\sum_i Q_i r_i$. The charges Q_i
(*a*) must be all positive
(*b*) must be all negative
(*c*) must be such that $\sum_i Q_i = 0$
(*d*) either (*a*) or (*b*)

43. The units for electric dipole moment are
(*a*) coulombs (*b*) coulomb-metre
(*c*) coulom/metre (*d*) coulomb/m^2

44. From a homogenous and isotropic dielectric material of dielectric constant $\in_r$, a small volume element is taken out creating a cavity. The field can be kept homogenous if D_i (*i.e.*, electric flux density inside the cavity) and D_0 (*i.e.*, electric flux density outside the cavity) are such that
(*a*) $D_0 = D_i$ (*b*) $D_i = \frac{D_0}{\in_r}$
(*c*) $D_0 = \frac{D_i}{\in_r}$ (*d*) either (*a*) or (*b*)

45. A dielectric of relative permittivity $\in_r$ is subjected to a homogenous electric field E. The dipole moment P per unit volume is given by
(*a*) P = constant
(*b*) $P = \in_0 E$
(*c*) $P = \in_0 (\in_r - 1) E$
(*d*) $P = \in_r E$

46. An atom of a rare gas is placed is an electric field E. Then
(*a*) the nucleus will be shifted in the direction of E by an amount which is proportional to E
(*b*) the nucleus will not be shifted
(*c*) the nucleus will be shifted in the direction of E by an amount which is independent of E
(*d*) The nucleus will be shifted in a direction opposite to that of E

47. The units of polarization of dielectric are
(*a*) $\frac{C}{m^2}$ (*b*) $\frac{C}{m}$
(*c*) $C - m$ (*d*) $\frac{C}{m^3}$

48. If the radius of electron cloud around the nucleus of a rare gas atom is R, the electronic polarizability of the atom is proportional to
(*a*) R (*b*) R^2
(*c*) R^3 (*d*) $\frac{1}{R}$

49. When the atom of rare gas is placed in an electric field the nucleus shifts from the centre of electron cloud by an amount *x*. If the radius of electron cloud is R, then
(*a*) $x << R$ (*b*) $x \simeq 0.5\,R$
(*c*) $x \simeq 0.75\,R$ (*d*) $x \simeq R$

50. A rare gas has N atoms per m^3. If electronic polarizability of single atom of this gas is α_e, then
(*a*) $\in_r = N\alpha_e$ (*b*) $\in_0 \in_r = N\alpha_e$
(*c*) $\in_0 (\in_r - 1) = N\alpha_e$ (*d*) $\in_0 (\in_r - 2) = N\alpha_e$

51. As the atoms of rare gases become larger, electronic polarizability
(*a*) increases
(*b*) decreases
(*c*) remains the same
(*d*) may increase or decrease

52. In a polyatomic gas, the types of polarization which exist together are
(*a*) electronic and orientational
(*b*) electronic and ionic
(*c*) ionic and orientational
(*d*) electronic, orientational and ionic

53. If μ_p is permanent dipole moment and T is absolute temperature, the orientation polarization of polyatomic gases is
(*a*) proportional to $\mu_p^2 T$ (*b*) proportional to $\frac{\mu_p^2}{T}$
(*c*) proportional to $\mu_p T$ (*d*) proportional to $\frac{\mu_p}{T^2}$

54. Electric dipole moment is usually expressed in debye units. 1 debye unit equals
(*a*) 3.33×10^{-30} C – m
(*b*) 3.33×10^{-30} C – m^2
(*c*) 3.30×10^{-10} C – m
(*d*) 3.33×10^{-10} C – m^2

55. For a polyatomic gas, the equation relating dielectric constant $\in_r$ with molecular properties is

(a) $\in_0 (\in_r - 1) = N\left(\alpha_e + \alpha_i + \frac{\mu_p^2}{3kT}\right)$

(b) $\in_0 (\in_r - 1) = N\left(\alpha_e + \alpha_i + \frac{\mu_p^2}{3kT}\right)T^2$

(c) $\in_0 (\in_r - 1) = N\left(\frac{\alpha_e}{T} + \alpha_i + \mu_p^2\right)$

(d) $\in_0 (\in_r - 1) = N\left(\alpha_e + \frac{\alpha_i}{T} + \frac{\mu_p^2}{3kT}\right)$

56. When an electric field E is applied to solid and liquid insulating materials, the internal field E_i acting at the location of atom is such that
(a) $E_i = E$
(b) $E_i > E$
(c) $E_i < E$
(d) E_i may be equal to or less than E

57. In solid insulator materials having only one kind of atom, the types of polarization which exist are
(a) electronic only
(b) electronic and ionic
(c) electronic, ionic and permanent
(d) electronic and permanent

58. Ferro electric materials are those which
(a) cannot be polarized
(b) have a permanent polarization
(c) have α_e equal to zero
(d) have $\mu_p = 0$

59. The dielectric constant of ferroelectric materials can be expressed by the equation

(a) $\in_0 (\in_r - 1) = \frac{dP}{dE}$ (b) $\in_0 (\in_r - 1) = \frac{dE}{dP}$

(c) $\in_0 (\in_r - 1) = \left(\frac{dP}{dE}\right)^2$ (d) $\in_0 (\in_r - 1) = \left(\frac{dP}{dE}\right)^{1.5}$

60. A particle of charge Q coulombs is bound elastically to an equilibrium position with a force constant of f newton per metre. The polarizability of the system is

(a) $\frac{Q}{f}$ (b) $\frac{Q^2}{f}$

(c) $\frac{Q}{f^2}$ (d) Qf

61. An atom of polarizability α is placed in a homogenous electric field E. The energy stored in polarized atom is
(a) αE^2 (b) $\alpha^2 E$
(c) $\frac{1}{2}\alpha E^2$ (d) $\frac{1}{2}\alpha E$

62. The capacitance of an insulated conducting sphere of radius R in vacuum is

(a) $2\pi \in_0 R$ (b) $4\pi \in_0 R$

(c) $4\pi \in_0 R^2$ (d) $4\pi \frac{\in_0}{R}$

63. Two coaxial metallic cylinders have radii R_1 and R_2. The space between them is filled with a dielectric having dielectric constant $\in_r$. The capacitance per metre length is

(a) $\pi \in_0 \in_r / \ln (R_2/R_1)$ (b) $2\pi \in_0 \in_r / \ln (R_2/R_1)$

(c) $\pi \ln (R_2/R_1)/\in_0 \in_r$ (d) $2\pi \in_0 \in_r /(R_2/R_1)$

64. If m is the mass of electron cloud, the equation of motion of electron cloud in the presence of an alternating electric field is of the form

(a) $m\frac{d^2x}{dt^2} = -ax - 2b\frac{dx}{dt} - e\,E_0 \cos \omega t$

(b) $m\frac{d^2x}{dt^2} = -ax - 2b\frac{dx}{dt} - E_0 \cos \omega t$

(c) $m\frac{d^2x}{dt^2} = -ax - 2b\frac{dx}{dt} - e^2 E_0$

(d) $m\frac{d^2x}{dt^2} - eE_0 \cos \omega t = -ax - 2b\frac{dx}{dt}$

65. For dielectrics in alternating field, polarizability α_e is a complex quantity. The imaginary part of α_e is zero for
(a) $\omega = 0$
(b) $\omega \to \infty$
(c) $\omega = 0$ and $\omega \to \infty$
(d) ω = natural frequency ω_0

66. A liquid has N permanent dipoles μ_p per unit volume and has been subjected to a dc field for a long time such that orientation polarization is P_0. At $t = 0$ the field is suddenly removed, then polarization
(a) will remain constant at P_0
(b) will become zero at $t = 0$
(c) will decay to zero with a time constant
(d) either (b) or (c)

67. As the viscosity of a liquid increases, the relaxation time
(a) remains constant
(b) increases
(c) decreases
(d) first decreases then increases

68. A dielectric is subjected to alternating field. The dielectric losses are proportional to
(a) real part of dielectric constant
(b) imaginary part of dielectric constant
(c) both real and imaginary parts of dielectric constant
(d) either real or imaginary part of dielectric constant

69. If $\in_r'$ and $\in_r''$ are the real and imaginary parts of complex dielectric constant $\in_r$, and δ is the loss angle, then (*a*) $\tan\delta = \frac{\epsilon'_r}{\epsilon''_r}$ (*b*) $\tan\delta = \frac{\epsilon''_r}{\epsilon'_r}$

(*c*) $\tan\delta = \epsilon'_r\epsilon''_r$ (*d*) $\tan\delta = 1-\epsilon'_r$

70. A parallel plate capacitor has area of plate equal to 1 m^2 and distance between plates is also 1 m. The lossy dielectric has a complex dielectric constant $\epsilon_r = \epsilon'_r - j\epsilon''_r$. If the capacitor is represented by R and C is parallel, then

(*a*) $R = \frac{1}{\epsilon_0\epsilon''_r\,\omega}$ and $C = \epsilon_0\epsilon'_r$

(*b*) $R = \frac{\omega}{\epsilon_0\epsilon'_r}$ and $C = \epsilon_0\epsilon''_r$

(*c*) $R = \frac{1}{\omega\epsilon_0\epsilon'_r}$ and $C = \epsilon_0\epsilon''_r$

(*d*) $R = \frac{1}{\omega\epsilon_0\epsilon'_r\epsilon''_r}$ and $C = \epsilon_0\epsilon'_r$

71. A parallel plate capacitor having $C = 200 \times 10^{-12}$ F has lossy dielectric having $\tan\delta = 4 \times 10^{-4}$. At an ac voltage of $f = 10^6$ Hz and peak value 2 V, the dielectric losses are

(*a*) 10^{-8} W (*b*) 10^{-6} W
(*c*) 10^{-4} W (*d*) 10^{-2} W

72. As regards ϵ_0 and ϵ_r (absolute and relative primitively).

(*a*) both ϵ_0 and ϵ_r can be interpreted in terms of atomic properties of medium
(*b*) only ϵ_r can be interpreted in terms of atomic properties of medium
(*c*) only ϵ_0 can be interpreted in terms of atomic properties
(*d*) ϵ_r has no physical significance

73. The units of μ_0 and μ_r are

(*a*) H/m for both
(*b*) H/m for μ_r, and no units for μ_0
(*c*) H/m for μ_0 and no units for μ_r
(*d*) Wb/m for μ_0 and no units for μ_r

74. A wire is carrying current I. The field intensity H at a distance *r* from the wire

(*a*) is directed towards wire
(*b*) is directed away from wire
(*c*) is tangential everywhere to a circle of radius *r*
(*d*) none of above

75. The core of a solenoid is made of material of relative permeability μ_r. A small cavity area *d*A and length *dl* is cut in the core. If it is desired that flux density in the cavity should be equal lo B_0,

(*a*) the fields inside and outside the cavity must be equal
(*b*) the field inside the cavity must be made stronger than field outside the cavity
(*c*) the field inside the cavity must be made weaker than the field outside the cavity
(*d*) none of above

76. In a homogenous magnetic field H of a solenoid, the core carries a magnetic dipole moment M per unit volume equal to

(*a*) $\mu_0 H$ (*b*) $\mu_r H$
(*c*) $(\mu_r - 1)$ H (*d*) $\mu_0\mu_r H$

77. For a core having a relative permeability μ_r, magnetic dipole moment per unit volume M and field strength H, the flux density B is

(*a*) $B = \mu_r (H + M)$
(*b*) $B = \mu_0 (H + M)$
(*c*) $B = \mu_0 \mu_r (H + M)$
(*d*) $B = \mu_0 \mu_r (H - M)$

78. An electron (charge – *e*) is rotating in a circular orbit of radius R, around a proton with charge +*e* at a constant angular velocity ω. The orbital magnetic dipole moment is

(*a*) $0.5\, e\omega R^2$ (*b*) $e\omega R^2$
(*c*) $2\, e\omega R^2$ (*d*) $0.5\, e\omega R^3$

79. In terms of Planck's constant *h*, the orbital angular momentum of an electron is

(*a*) h (*b*) $\frac{h}{2\pi}$

(*c*) $2\pi h$ (*d*) $\frac{h}{\pi}$

80. An electron is moving around a proton, in a circle of radius R. If flux density in the region is increased from zero to B, the force on the electron is

(*a*) $\frac{eR}{2}(dB/dt)$ (*b*) $eR\,(dB/dt)$

(*c*) eRB (*d*) $\frac{dR}{2}B$

81. An electron is moving in a circular orbit of radius R at angular velocity ω_0 in a field free space. If a magnetic field B is created, the velocity of electron ω will be given by the equation

(*a*) $\omega^2 = \omega_0^2 + \frac{eB}{m}\omega$ (*b*) $\omega^2 = \omega_0^2 - \frac{eB}{m}\omega$

(*c*) $\omega^2 = \omega_0^2 + \frac{2eB}{m}\omega$ (*d*) $\omega = \omega_0 + \frac{eB}{m}\omega^2$

82. Material which lack permanent magnetic dipoles are known as

(*a*) paramagnetic (*b*) diamagnetic
(*c*) ferromagnetic (*d*) ferrimagnetic

83. In paramagnetic materials

(*a*) permanent magnetic dipoles exist but the interaction between neighbouring dipoles is negligible
(*b*) permanent magnetic dipoles do not exist
(*c*) permanent magnetic dipoles exist and the interaction between neighbouring dipoles is very strong
(*d*) permanent magnetic dipole moment may or may not exist

84. As regards ferromagnetic curie temperature θ_f and paramagnetic curie temperature θ

(*a*) $Q_f > \theta$
(*b*) $\theta > Q_f$
(*c*) θ equal to or less than Q_f
(*d*) θ may be equal to less than or more θ_f

85. For iron, the ferromagnetic curie temperature is
(*a*) about 20 K (*b*) about 100°C
(*c*) about 20°C (*d*) about 1000 K

86. When temperature is above ferromagnetic curie temperature, a ferromagnetic material
(*a*) becomes a paramagnetic material
(*b*) becomes a very strong ferromagnetic material
(*c*) becomes diamagnetic material
(*d*) none of above

87. A material has N spins per m^3. If N_P are the dipoles per m^3 parallel to field and N_a are the dipoles per m^3 anti-parallel to field, the magnetisation is
(*a*) $M = \frac{(N_p - N_a)\, eh}{4\pi m}$ (*b*) $M = \frac{(N_p + N_a)\, eh}{4\pi m}$
(*c*) $M = \frac{(N_p - N_a)^2\, eh}{4\pi m}$ (*d*) $\frac{2\,(N_p - N_a)\, eh}{m}$

88. In a ferromagnetic material the internal field seen by a dipole is
(*a*) equal to applied field
(*b*) equal to applied field minus a contribution from neighbouring dipoles
(*c*) equal to applied field plus a contribution from neighbouring dipoles
(*d*) equal to or less than applied field

89. The fact that domains exhibit spontaneous magnetisation is
(*a*) first hypothesis is of Weiss
(*b*) second hypothesis is of Weiss
(*c*) Curie-Weis law
(*d*) Boltzmann's statistics

90. Magnetic hysteresis phenomenon is explained by
(*a*) motion of domain walls
(*b*) motion of domain walls and domain rotation
(*c*) domain rotation
(*d*) none of above

91. In anti-ferromagnetic materials, the graph of susceptibility versus temperature shows a sharp maximum at a certain temperature. This temperature is known as
(*a*) Curie temperature
(*b*) Neel temperature
(*c*) Weiss temperature
(*d*) Bitter temperature

92. A piece of Copper and another piece of Germanium are cooled from 30°C to 80 K. The resistance of
(*a*) copper decreases and germanium increases
(*b*) both decreases
(*c*) both increases
(*d*) copper increases and germanium decreases

93. Ferromagnetic materials
(*a*) show spontaneous magnetisation below a certain temperature
(*b*) have resistivity pretty high as compared to that of iron
(*c*) can be used even at microwave frequencies
(*d*) all the above

94. An electron in a homogenous magnetic field and having velocity perpendicular to the direction of field moves
(*a*) perpendicular to the field
(*b*) along the field
(*c*) along a straight line inclined at 45° to field
(*d*) in a circular path

95. A conductor carries 10 A current in $+x$ direction and is lying in a field having a flux density 0.2 T directed along $+z$ direction. The magnitude of force per metre length and direction are
(*a*) 2N, $+y$ direction
(*b*) 2N, $-y$ direction
(*c*) 0.02 N, $+y$ direction
(*d*) 0.02 N, $-y$ direction

96. The law $J = \sigma E$ where J is current density, σ is electrical-conductivity and E is field strength is
(*a*) Ohm's law (*b*) Gauss law
(*c*) Ampere's law (*d*) Biot savart law

97. A metal has a large number of free electrons. If e is electronic charge and m is mass of electron and an electric field E is applied along (say) x direction, the electrons acquire a velocity having a component v_x is x direction. Then
(*a*) $\left(\frac{\partial}{\partial t} v_x\right) = \frac{e}{m} E$ (*b*) $\left(\frac{\partial}{\partial t} v_x\right) = \frac{m}{e} E$
(*c*) $\left(\frac{\partial}{\partial t} v_x\right) = \frac{-e}{m} E$ (*d*) $\left(\frac{\partial}{\partial t} v_x\right) = \frac{e}{Em}$

98. A specimen of metal has n valence electrons per m^3. If v_x is the average velocity of these electrons, each having a charge e, the current density J is
(*a*) $J = ne\, v_x$ (*b*) $J = -\,ne\, v_x$
(*c*) $J = \frac{-ne}{v_x}$ (*d*) $J = \frac{-ne}{v_x}$

99. In a metal the drift velocity of electrons in the presence of a field E is
(*a*) proportional to E (*b*) proportional to $\sqrt{E}$
(*c*) inversely proportional to E
(*d*) proportional to E^2

100. If a metal specimen has n valence electrons per m^3 each having charge e and mass m and τ is the relaxations time, the conductivity is
(*a*) $\sigma = \frac{ne^2 m}{\tau}$ (*b*) $\sigma = \frac{ne^2\tau}{m}$
(*c*) $\sigma = \frac{nem}{\tau}$ (*d*) $\sigma = \frac{ne^2\tau}{m}$

101. An electron in a metal suffers collision and if the probability of it being scattered over an angle $d\theta$ is

$2\pi\ P(\theta)\sin\theta\ d\theta$ where $P(\theta)$ is an arbitrary function of scattering angle θ, then

(*a*) $\int_{\theta=0}^{\pi} 2\pi P(\theta)\sin\theta\, d\theta = 0$

(*b*) $\int_{\theta=0}^{\pi} 2\pi P(\theta)\sin\theta\, d\theta = 2$

(*c*) $\int_{\theta=0}^{\pi} 2\pi P(\theta)\sin\theta\, d\theta = 1$

(*d*) $\int_{\theta=0}^{\pi} 2\pi P(\theta)\sin\theta\, d\theta = -1$

102. If τ is relaxation time and τ_c is average time between collisions, then in an isotropic metal
(*a*) $\tau = \tau_c$ (*b*) $\tau < \tau_c$
(*c*) $\tau_c < \tau$ (*d*) $T = 0.01\ \tau_e$

103. If W_F is the fermilevel, then in a metal at T = 0
(*a*) all energy levels above W_F filled, all those below W_F are empty
(*b*) all energy levels below W_F are filled, all those above W_F are empty
(*c*) some energy levels, both below and above W_F are filled
(*d*) all energy levels below and above W_F are filled

104. If k is Boltzman's constant and T is absolute temp then at room temp.
(*a*) $kT \sim 0.025$ eV (*b*) $kT \sim 0.5$ eV
(*c*) $kT \sim 1$ eV (*d*) $kT \sim 100$ eV

105. At T = 0, the resistivity of a perfect single crystal of a metal is
(*a*) 0 (*b*) low
(*c*) high (*d*) finite

106. The heat developed per m^3 per second (W) in a conductor having, a conductivity σ and a field E is
(*a*) $W = \sigma E^2$ (*b*) $W = \sigma E$
(*c*) $W = \frac{E^2}{\sigma}$ (*d*) $W = (\sigma E)^{0.5}$

107. In metals the thermal conductivity K and electrical conductivity are related as $(K/\sigma T) = L$. L is known as
(*a*) Lattice constant (*b*) Lorenz number
(*c*) Lanevin function (*d*) Larmor number

108. The temperature at which some materials become superconductors is called
(*a*) transition temperature (*b*) Curie temperature
(*c*) Neel temperature (*d*) Onnes temperature

109. The elements which become superconducting at extremely low temperatures are
(*a*) good conductors at room temperature
(*b*) ferromagnetic at room temperature
(*c*) relatively poor conductors at room temperature
(*d*) insulators at room temperature

110. When a material becomes a superconductor, its resistivity becomes
(*a*) very small
(*b*) zero
(*c*) about 10% of normal value
(*d*) about 20% of normal value

111. When a material is in superconducting state and a magnetic field is applied
(*a*) the superconducting property is not affected
(*b*) the superconducting properly may be affected and resistivity may increase from zero to a low value
(*c*) the superconducting property may be destroyed
(*d*) the material may become an insulator

112. The application of magnetic field to superconducting material
(*a*) increases transition temperature
(*b*) decreases transition temperature
(*c*) has no effect on transition temperature
(*d*) may increase or decrease the transition temperature

113. For a superconductor, the relative permeability is
(*a*) zero (*b*) high
(*c*) low (*h*) either low or high

114. The relaxation time of electrons in a metal is of the order of
(*a*) 10^{-6} secs (*b*) 10^{-10} secs
(*c*) 10^{-14} sees (*d*) 10^{-20} secs

115. A copper specimen has resistivity equal to 1.8×10^{-8} Ohm – m at room temperature. At a temperature of 700 K, the resistivity is likely to be
(*a*) $1.6 \times 10^{-8}\ \Omega - m$ (*b*) $2 \times 10^{-8}\ \Omega - m$
(*c*) $3 \times 10^{-8}\ \Omega - m$ (*d*) $6 \times 10^{-8}\ \Omega - m$

116. The rule that resistivity of conductors can be separated into two parts viz temperature independent and temperature dependent is known as
(*a*) Matthiessen's rule (*b*) Debye rule
(*c*) Curie rule (*d*) Onne's rule

117. Above a certain temperature the specific heat of a metal becomes constant. This temperature is called
(*a*) Debye temperature (*b*) Curie temperature
(*c*) Neel temperature (*d*) transition temperature

118. Resistivity of copper is lower than that of nickel. When a small percentage of copper is added to nickel, the resistivity
(*a*) increases
(*b*) decreases
(*c*) may increase or decrease
(*d*) remains the same

119. Debye temperature of metals is
(*a*) about 10 K (*b*) about 200 K
(*c*) about 600 K (*d*) about 2 K

120. In a metal the probability of a state 0.1 eV below Fermi level being occupied
(*a*) is more than the probability of state 0.1 eV above Fermi level being occupied
(*b*) is the same as the probability of a state 0.1 eV above Fermi level being occupied
(*c*) is less than the probability of a state 0.1 eV above Fermi level being occupied

(*d*) may be equal to or more or less than the probability of a state 0.1 eV above Fermi level being occupied

121. The relation between thermal conductivity K (W/m/°C), heat flow density Q (watts/m^2) and $\frac{dT}{dx}$ (°C per metre) is

(*a*) $Q = K(dT/dx)$ (*b*) $Q = -K(dT/dx)$
(*c*) $K = Q(dT/dx)$ (*d*) $dT/dx = -QK$

122. The law that $\frac{K}{\sigma T}$ (where K is thermal conductivity, σ is electrical conductivity and T is absolute temperature) for all metals is equal to a universal constant is known as

(*a*) Lorenz law
(*b*) Wiedemann Franz law
(*c*) Curie-Weiss law
(*d*) Kittel law

123. The temperature coefficient of resistivity of semi-conductors is

(*a*) positive
(*b*) negative
(*c*) may be positive or negative
(*d*) very low

124. At a very low temperature a semiconductor

(*a*) becomes a conductor
(*b*) becomes an insulator
(*c*) may become a conductor or an insulator
(*d*) will remain a semiconductor or become a conductor or an insulator

125. Semiconducting materials are all

(*a*) elements
(*b*) compounds
(*c*) elements or compounds
(*d*) elements or mixtures

126. Silicon and germanium are elements in the

(*a*) second column of periodic table
(*b*) third column of periodic table
(*c*) fourth column of periodic table
(*d*) fifth column of periodic table

127. In an intrinsic semiconductor at a given temperature T

(*a*) the number of electron hole pairs generated per second due to thermal motion are equal to the number of electron hole pairs lost per second due to recombination
(*b*) the number of electron hole pairs generated per second due to thermal motion is always less than the number of electron hole pairs lost per second due to recombination
(*c*) the number of electron hole pairs generated per second due to thermal motion is always more than the number of electron hole pairs lost per second due to recombination
(*d*) any of above

128. As the temperature of semiconductor is increased

(*a*) the average number of free charge carriers decreases
(*b*) the average number of free charge carriers increases
(*c*) the average number of free charge carriers remains the same
(*d*) the average number of feree charge carriers may increase or decrease

129. The number of protons in a copper atom are

(*a*) 2 (*b*) 9
(*c*) 11 (*d*) 29

130. The core of an atom consists of

(*a*) nucleus
(*b*) nucleus and all orbits
(*c*) nucleus and inner orbits
(*d*) none of above

131. A copper atom is neutral. Its core has a net charge of

(*a*) 0 (*b*) + 1
(*c*) – 1 (*d*) + 2

132. If the valence electron is separated from a copper atom, the remaining part of atom has a net charge of

(*a*) 1 (*b*) –1
(*c*) 0 (*d*) + 3

133. The amount of time between the creation and recombination of a free electron and hole is called

(*a*) relaxation time (*b*) life time
(*c*) Lorenz time (*d*) collision time

134. The attraction between the nucleus and valence electron of copper atom is

(*a*) zero (*b*) weak
(*c*) strong (*d*) either zero or strong

135. The number of valence electrons in silicon atom is

(*a*) 4 (*b*) 2
(*c*) 1 (*d*) zero

136. The number of valence electrons in germanium atom is

(*a*) 4 (*b*) 2
(*c*) 1 (*d*) zero

137. The number of protons in a silicon atom are

(*a*) 4 (*b*) 8
(*c*) 12 (*d*) 14

138. The number of tightly bound electrons in silicon atom are

(*a*) 14 (*b*) 10
(*c*) 6 (*d*) 4

139. The number of protons in germanium atom is

(*a*) 4 (*b*) 16
(*c*) 28 (*d*) 32

140. Silicon and germanium atoms combine into orderly array of atoms located in repetitive geometric pattern. This pattern is called

(*a*) crystal (*b*) covalent bond
(*c*) molecule (*d*) valence bond

141. The resistivity of intrinsic semiconductors, at room temperature, is of the order of

(*a*) 1 Ω – m (*b*) 1000 Ω – m
(c) 10^6 Ω – m (d) 10^{-3} Ω – m

142. Even at room temperature, an intrinsic semiconductor has some holes. These holes are due to

(*a*) electrons (*b*) doping

(*c*) thermal energy (*d*) free electrons

143. The merger of hole and free electron is called
(*a*) covalent bonding (*b*) recombination
(*c*) thermal energy (*d*) free electrons

144. In a pure sample of silicon
(*a*) the number of holes and free electrons is always equal
(*b*) the number of free electrons is more than the number of holes
(*c*) the number of free electrons and holes is equal at low temperature
(*d*) the number of free electrons is less than number of holes

145. In a conductor the current flow is due to
(*a*) holes (*b*) electrons
(*c*) ions (*d*) electrons and ions

146. The current flow in a semiconductor is due to
(*a*) holes (*b*) electrons
(*c*) holes and electrons (*d*) holes, electrons and ions

147. When a potential is applied across an intrinsic semi-conductor, holes flow
(*a*) towards positive terminal
(*b*) away from positive terminal
(*c*) in the external circuit only
(*d*) none of above

148. As the temperature of a pure silicon specimen is increased
(*a*) the number of free electrons increases
(*b*) the number of holes decreases
(*c*) the number of holes increases
(*d*) the number of free electrons and holes increases

149. The number of valence electrons in trivalent impurity is
(*a*) 5 (*b*) 4
(*c*) 3 (*d*) 1

150. When a trivalent impurity is added, the semiconductor becomes
(*a*) *p* type (*b*) *n* type
(*c*) intrinsic (*d*) neutral

151. The number of valence electrons in pentavalent impurity is
(*a*) 5 (*b*) 4
(*c*) 3 (*d*) 1

152. When a pentavalent impurity is added, the semiconductor becomes
(*a*) *p* type (*b*) *n* type
(*c*) intrinsic (*d*) neutral

153. The number of valence electrons in donor impurity are
(*a*) 5 (*b*) 4
(*c*) 3 (*d*) 1

154. The number of valence electrons in acceptor impurity are
(*a*) 5 (*b*) 4
(*c*) 3 (*d*) 1

155. In which of the following semiconductor are holes minority carriers
(*a*) extrinsic (*b*) *p* type
(*c*) *n* type (*d*) intrinsic

156. In which of the following semiconductor are holes majority carriers
(*a*) intrinsic (*b*) extrinsic
(*c*) *p* type (*d*) *n* type

157. Silicon and germanium have diamond structure. The number of atoms in a volume equal to a^3 where a is the edge of elementary cube are
(*a*) 32 (*b*) 16
(*c*) 8 (*d*) 4

158. The measurement of Hall coefficient of a semiconductor with one type of charge carriers gives the information about
(*a*) sign of charge carrier
(*b*) density of charge carrier
(*c*) both sign and density of charge carrier
(*d*) none of above

159. The probability of recombination of electrons and holes in a semiconductor is proportional to
(*a*) density of electrons
(*b*) density of holes
(*c*) density of electrons and holes
(*d*) none of above

160. At room temperature the number of conduction electrons in an intrinsic semiconductor
(*a*) is almost equal to the number of valence electrons
(*b*) is about half the number of valence electrons
(*c*) is small as compared to the number of valence electrons
(*d*) is a very small fraction of the number of valence electrons

161. Specific resistance of silver is about
(*a*) $0.16 \times 10^{-8}\ \Omega - \text{m}$
(*b*) $1.64 \times 10^{-8}\ \Omega - \text{m}$
(*c*) $16 \times 10^{-8}\ \Omega - \text{m}$
(*d*) $164 \times 10^{-8}\ \Omega - \text{m}$

162. If temperature is increased, the temperature coefficient of a metal
(*a*) increases
(*b*) decreases
(*c*) may increase or decrease
(*d*) remains constant

163. One mil is equal to
(*a*) 0.1 mm (*b*) 0.01 mm
(*c*) $\frac{1}{100}$ inch (*d*) $\frac{1}{1000}$ inch

164. A circular mil is the area of cross-section of a wire
(*a*) having a diameter equal to 1 mil
(*b*) having a radius equal to 1 mil
(*c*) having circumference equal to I mil
(*d*) none of above

165. The temperature coefficient of a thermistor
(*a*) is positive
(*b*) is negative
(*c*) zero
(*d*) may be positive or negative depending on its composition

166. Thermistor is used in
(*a*) temperature measurement
(*b*) temperature control
(*c*) high temperature alarm circuits of devices
(*d*) all the above

167. The resistance of a photoconductor
(*a*) increases with increase in light intensity
(*b*) decreases with increase in light intensity
(*c*) may increase or decrease, with increase in light intensity
(*d*) remains constant

168. A photo conductor is a
(*a*) conductor (*b*) semiconductor
(*c*) insulator (*d*) superconductor

169. Which of the following is used in automatic control of street lights?
(*a*) thermistor (*b*) photoconductor
(*c*) transistor (*d*) varistor

170. Which of the following is used in temperature alarm circuits?
(*a*) thermistor (*b*) photoconductor
(*c*) transistor (*d*) varistor

171. Which of the following is used in lightning arresters?
(*a*) thermistor (*b*) photoconductor
(*c*) transistor (*d*) varistor

172. The $v - i$ characteristic shown in Fig. 1.6 is for

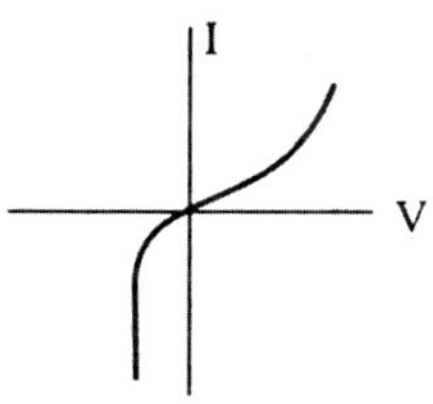

Fig. 1.6.

(*a*) thermistor (*b*) photoconductor
(*c*) varistor (*d*) none of above

173. At a temperature of about 1 K, Aluminium is a
(*a*) conductor (*b*) insulator
(*c*) photoconductor (*d*) superconductor

174. The resistance of a superconductor is
(*a*) low (*b*) very low
(*c*) very very low (*d*) zero

175. Two materials having temperature coefficients of 0.004 and 0.0004 respectively are joined in series. The overall temperature coefficient is likely to be
(*a*) 0.08 (*b*) 0.04
(*c*) 0.001 (*d*) 0.0001

176. Nichrome is used for
(*a*) overhead line wires (*b*) heater coils
(*c*) lamp filaments (*d*) all the above

177. If the diameter of a wire is doubled, its current carrying capacity becomes
(*a*) one-fourth (*b*) half
(*c*) twice (*d*) four times

178. The colour code on a carbon resistor is yellow-violet-orange. The value of resistance is
(*a*) 17000 Ω (*b*) 27000 Ω
(*c*) 37000 Ω (*d*) 47000 Ω

179. In carbon resistor colour code, if the third band is of silver. It indicates
(*a*) that three zeros are to be added to indicate the resistance value
(*b*) multiplier is 0.01
(*c*) multiplier is 0.1
(*d*) tolerance is ± 5%

180. In carbon resistor colour code, the third band is red. It indicates
(*a*) two zeros
(*b*) three zeros
(*c*) multiplier is 0.01
(*d*) multiplier is 0.1

181. The colour code on a carbon resistor is brown-black-red-gold. The value of resistance is
(*a*) 100 Ω (*b*) 1000 Ω
(*c*) 1000 Ω ± 5% (*d*) 1000 Ω ± 10%

182. The $v - i$ relation for an arc is $v = 44 + \frac{30}{i}$. If $v = 50$ V, $i =$
(*a*) 10 A (*b*) 5 A
(*c*) 2 A (*d*) 0.5 A

183. The colour code on a carbon resistor is green-blue-yellow. The tolerance is
(*a*) ± 5% (*b*) ± 10%
(*c*) ± 20% (*d*) ± 30%

184. The colour code on a carbon resistor is red-red-black-silver, The value of resistor is
(*a*) 22000 Ω (*b*) 2200 Ω
(*c*) 22 ± 5% Ω (*d*) 22 ± 10% Ω

185. The classes of solid dielectrics are
(*a*) A, B, C, D, E, F, H (*b*) A E B F I J, C D
(*c*) Y A E B F H C (*d*) A E B F H C

186. Which solid dielectric has highest temperature limit?
(*a*) E (*b*) H
(*c*) C (*d*) F

187. The maximum temperature limit of class B insulation is
(*a*) 90°C (*b*) 105°C
(*c*) 120°C (*d*) 130°C

188. The class of insulation of impregnated paper is
(*a*) A (*b*) B
(*c*) C (*d*) F

189. Transformer oil is
(*a*) refined vegetable oil
(*b*) refined mineral oil
(*c*) mixture of vegetable and mineral oil
(*d*) none of above

190. If the operating temperature increases by 10°C, the rate of deterioration of solid dielectric becomes
(*a*) double (*b*) one and a half time
(*c*) 3 times (*d*) 4 times

191. Resistivity of rubber is about
(*a*) 10^{-2} Ω – m (*b*) 10^{2} Ω – m
(*c*) 5×10^{8} Ω – m (*d*) 5×10^{12} Ω – m

192. Which capacitor cum store higher amount of energy?
(*a*) air capacitor (*b*) paper capacitor
(*c*) mica capacitor (*d*) plastic film capacitor

193. Hydrogen is used in
(*a*) large size transformers
(*b*) MCB
(*c*) large size generators
(*d*) circuit breakers

194. Sulphur hexa fluoride is used in
(*a*) transformers (*b*) generators
(*c*) relays (*d*) circuit breakers

195. As per temperature limits, the number of categories of solid dielectrics are
(*a*) 7 (*b*) 6
(*c*) 8 (*d*) 5

196. The phenomenon of dielectric hysteresis occurs with
(*a*) dc (*b*) ac
(*c*) both ac and dc (*d*) none of above

197. The units for intrinsic strength of dielectric are
(*a*) kV (*b*) kV/mm
(*c*) A (*d*) A/m

198. Which of these has the highest dielectric strength?
(*a*) mica (*b*) air
(*c*) cotton (*d*) rubber.

199. Dielectric constant of transformer oil is about
(*a*) 1 (*b*) 4
(*c*) 10 (*d*) 20

200. Which of these has lowest dielectric strength
(*a*) dry air (*b*) wet air
(*c*) cotton (*d*) paper

201. For a multiplate capacitor having *n* plates, A as area of each plate, the capacitance is proportional to
(*a*) nA (*b*) $(n - 1)$A
(*c*) $(n + 1)$A (*d*) $\frac{nA}{2}$

202. An electrolytic capacitor can be used for
(*a*) ac only (*b*) dc only
(*c*) both ac and dc (*d*) high frequency ac

203. A sheet of bakelite is inserted between the plates of an air capacitor. The capacitance will
(*a*) increase (*b*) decrease
(*c*) remain the same (*d*) may increase or decrease

204. A capacitor can be represented by a capacitance and resistance in parallel. For a good capacitcr, this parallel resistance should be
(*a*) negligible (*b*) low
(*c*) high (*d*) very high

205. The capacitance of a capacitor is not affected by
(*a*) area of plates (*b*) thickness of plates
(*c*) ϵ_r (*d*) distance between plates

206. The units for ϵ_r are
(*a*) Farads (*b*) Farads/m
(*c*) $\frac{kV}{mn}$ (*d*) no units

207. Figure 1.7 shows a composite capacitor. If C_A, C_B, C_C are the capacitances of capacitors A, B, C the overall capacitance C is
(*a*) $C = C_A + C_C + C_B$ (*b*) $C = (C_A + C_B)C_C$
(*c*) $C = C_A + \frac{C_B + C_C}{C_B + C_C}$ (*d*) $\frac{1}{C} = \frac{1}{C_A} + \frac{1}{C_B} + \frac{1}{C_C}$

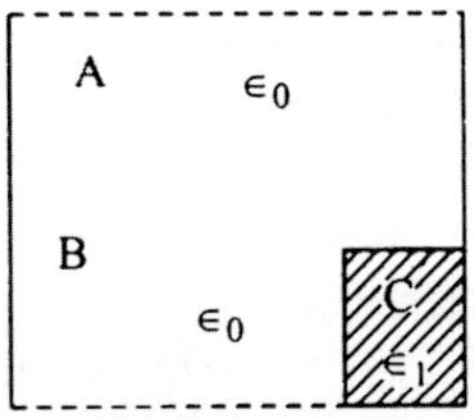

Fig. 1.7.

208. A good dielectric should have
(*a*) low losses
(*b*) good heat conductivity
(*c*) high intrinsic strength
(*d*) all the above

209. Which element exhibits the property of inertia?
(*a*) resistance
(*b*) capacitance
(*c*) inductance
(*d*) both resistance and inductance

210. The core of a coil has a length of 10 cm. The self-inductance is 8 mH. If the core length is doubled all other quantities remaining the same, the self-inductance will be
(*a*) 32 mH (*b*) 16 mH
(*c*) 8 mH (*d*) 4 mH

211. A 200 turn coil has a self-inductance of 10 mH. If the number of turns is doubled, all other quantities remaining the same the self-inductance will be
(*a*) 2.5 mH (*b*) 5 mH
(*c*) 20 mH (*d*) 40 mH

212. Automobile ignition systems use the principle of
(*a*) resistance (*b*) self-inductance
(*c*) mutual inductance (*d*) capacitance

213. Which inductor is linear?
(*a*) air cored (*b*) cast iron cored
(*c*) sheet steel cored (*d*) iron alloy cored

214. Which inductor is used at high frequencies?
(*a*) cast iron cored (*b*) sheet steel cored
(*c*) iron alloy cored (*d*) ceramic cored

215. The hysteresis loss is
(*a*) proportional to frequency
(*b*) independent of frequency
(*c*) proportional to $(\text{frequency})^2$
(*d*) proportional to $\frac{1}{\text{frequency}}$

216. Eddy current loss is
(*a*) proportional of frequency
(*b*) independent of frequency
(*c*) proportional to $(\text{frequency})^2$
(*d*) proportion of $(\text{frequency})^3$

217. If B_{max} is the maximum flux density, eddy current loss is proportional to
(*a*) B_{max} (*b*) $B^{1.6}_{max}$
(*c*) B^2_{max} (*d*) B^3_{max}

218. If B_{max} is the maximum flux density, hysteresis loss is proportional to
(*a*) B_{max} (*b*) $B^{1.6}_{max}$
(*c*) B^2_{max} (*d*) B^3_{max}

219. The relationship of hysteresis loss to maximum flux density was first determined by
(*a*) C. Steinmetz (*b*) G.R. Kirchhoff
(*c*) Mr. Laplace (*d*) Mr. Ampere

220. Inductors used at V.H.F. are made using
(*a*) air core (*b*) cast iron core
(*c*) sheet steel core (*d*) thin film technique

221. As per Matthiessen's rule the total resistivity ρ of a conductor can be written as $\rho = \rho_i + aT$ where ρ_i, and a are constants and T is absolute temperature. At very low temperatures
(*a*) $aT = 0$ (*b*) $aT > p_i$
(*c*) $aT \sim \rho_i$ (*d*) $aT < \rho_i$

222. Which of the following are true for relative dielectric constant?
1. It is dimension less
2. It is equal lo 1 for vacuum
3. It is equal to 1 for all substances

(*a*) 1 and 2 only (*b*) 2 and 3 only
(*c*) 1 and 3 only (*d*) 1, 2 and 3

223. In a coaxial cable braided copper is used for
(*a*) conductor (*b*) shield
(*c*) dielectric (*d*) jacket

224. The orientation polarization is
1. directly proportional to temperature
2. inversely proportional to temperature
3. inversely proportional to square of permanent dipole moment
4. directly proportional to square of permanent dipole moment

Which of the above are correct?
(*a*) 1 and 3 only (*b*) 1 and 3 only
(*c*) 1 and 4 only (*d*) 2 and 4 only

225. Electrical conductivity of metals expressed in $ohm^{-1} m^{-1}$ is of the order of
(*a*) 10^{10} (*b*) 10^5
(*c*) 10^{-4} (*d*) 10^{-6}

226. Match List I with List II and give correct answer using the given codes

List I	List II
A. Mu Metal	1. very high magnetic permeability
B. Samrium/cobalt	2. very high magnetic remanence
C. Saphire	3. very high thermal conductivity
D. Ferrites	4. very stable magnetic permeability

Codes:

	A	B	C	D
(*a*)	1	2	3	4
(*b*)	3	4	1	2
(*c*)	2	3	4	1
(*d*)	4	1	2	3

227. The temperature below which certain materials are anti-ferromagnetic and above which they are paramagnetic is called
(*a*) Weiss temperature
(*b*) Curie temperature
(*c*) Neel temperature
(*d*) none of above

228. *n* type semiconductors
(*a*) are negatively charged
(*b*) are produced when indium as impurity is added to germanium
(*c*) are produced phosphorus as impurity is added to silicon
(*d*) none of above

229. The number of turns in a coil is doubled. If initial time constant is T, the new time constant is
(*a*) T (*b*) 2T
(*c*) 4T (*d*) $\frac{T}{2}$

230. In Fermi Dirac statistics the probability of electron occupation of an energy level equal to Fermi level is
(*a*) 0 (*b*) 0.25
(*c*) 0.5 (*d*) 1

231. The most important primary force of attraction in the formation of solids is
(*a*) electromagnetic (*b*) electrostatic
(*c*) quantum-electrostatic
(*d*) quantum electromagnetic

232. Electrical contact materials used in switches, brushes, relay must possess
(*a*) high thermal conductivity and high melting point
(*b*) low thermal conductivity and low melting point
(*c*) high thermal conductivity and low melting point
(*d*) low thermal conductivity and high melting point

233. If a sample of germanium and a sample of silicon have the same impurity density and are kept at room temperature
(*a*) both will have equal value of resistivity
(*b*) both will have equal negative resistivity
(*c*) resistivity of germanium will be higher than that of silicon
(*d*) resistivity of silicon will be higher than that of germanium

234. Match List I with List II and give correct answer using the given codes

List I (Electron spin)	List II (Phenomenon)
A. ↑↑↑↑	1. Anti-ferromagnetism
B. ↑↓↑↓	2. Ferrimagnetism
C. ↑↓↑↑	3. Ferromagnetism

Codes:

	A	B	C
(*a*)	1	2	3
(*b*)	3	2	1
(*c*)	2	3	1
(*d*)	3	1	2

235. By inserting a plate of dielectric between the plates of a parallel plate capacitor, the energy stored in the capacitor is increased five terms. The dielectric constant of material is

(*a*) $\frac{1}{25}$ (*b*) $\frac{1}{5}$

(*c*) 5 (*d*) 25

236. For small size high frequency coils, the most common core material is

(*a*) air (*b*) ferrite

(*c*) powdered iron (*d*) steel

237. The actual equivalent circuit of an air dielectric variable capacitor is shown in Fig. 1.8. The equivalent circuit for this capacitor at high frequencies only is given by

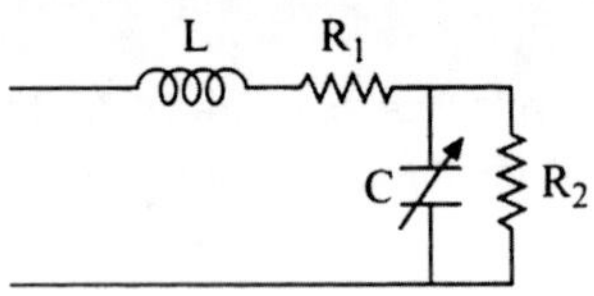

Fig. 1.8.

(*a*) 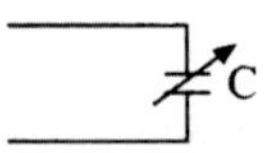

(*b*) 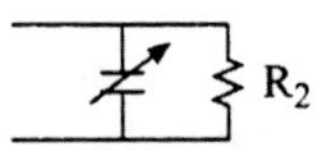

(*c*) 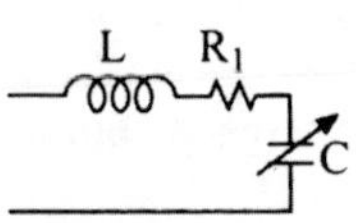

(*d*) 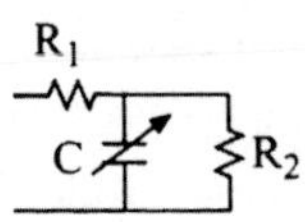

238. The apparent capacitance at a high angular frequency ω of an air cored capacitor having capacitance C and inductance L is given by

(*a*) $C_{app} = \frac{C}{1+\omega^2 LC}$ (*b*) $C_{app} = \frac{C}{1+\omega^2 LC}$

(*c*) $C_{app} = \frac{L}{1+\omega^2 LC}$ (*d*) $C_{app} = \frac{L}{1+\omega^2 LC}$

239. High frequency inductors and capacitors are commonly plated with silver. The main purpose of this is to

(*a*) reduce dc resistance

(*b*) reduce ac resistance

(*c*) increase dc resistance

(*d*) increase ac resistance

240. If a capacitor C having selectivity factor Q is scaled down in all its dimensions by a factor K without in any way altering the dielectric material, the new capacitor will have capacitance and selectivity as

(*a*) $C, \frac{Q}{K}$ (*b*) $\frac{C}{K}, Q$

(*c*) $\frac{C}{K}, \frac{Q}{K}$ (*d*) C, Q

241. The residual resistivity of a binary alloy at 0K is

(*a*) sum of residual resistivities of component metals

(*b*) difference of residual resistivities of component metals

(*c*) product of residual resistivities of component metals

(*d*) dependent on the concentration of minor component in the alloy

242. If the temperature of an extrinsic semiconductor is increased so that the intrinsic carrier concentration is doubled, then

(*a*) the majority carrier density doubles

(*b*) the minority carrier density is doubled

(*c*) the minority carrier density becomes 4 times the original value

(*d*) both the majority and minority carrier densities double

243. When a ferromagnetic substance is magnetised there are small changes in dimensions. This phenomenon is called

(*a*) hysteresis (*b*) magnetostriction

(*c*) diamagnetism (*d*) dipolar relaxation

244. At room temperature the current is intrinsic semiconductor is due to

(*a*) holes (*b*) electrons

(*c*) ions (*d*) holes and electrons

245. If a capacitor is represented by *r* and *c* is series (Fig. 1.9). The equivalent series resistance *r* in terms of power factor and operating frequency *f* is given by

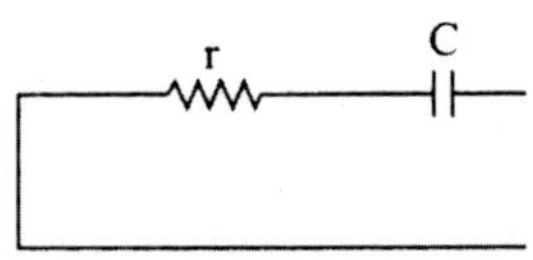

Fig. 1.9.

(*a*) $\frac{\text{Power factor}}{2\pi f C}$ (*b*) $\frac{2\pi f C}{\text{Power factor}}$

(*c*) $2\pi f C$ (Power factor) (*d*) $\frac{1}{2\pi f C \text{ (Power factor)}}$

246. Consider the following statements

1. An iron cored choke is non-linear and passive device
2. A carbon resistor kept in sunlight is a time invariant and passive device
3. A dry cell is a time varying and active device
4. An air capacitor is time invariant and passive device

Of these statements

(*a*) 1, 2, 3 and 4 are correct

(*b*) 1, 2, 3 are correct

(*c*) 1, 2, 4 are correct

(*d*) 2 and 4 are correct

247. Match List I and List II and give correct answer using the given codes

List I	**List II**
A. Conductor	1. 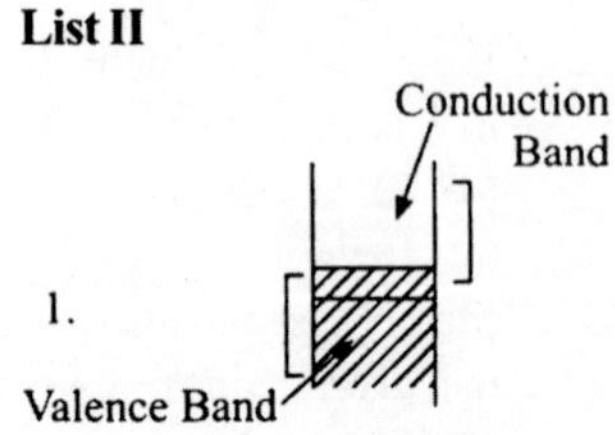
B. Semiconductor	2. 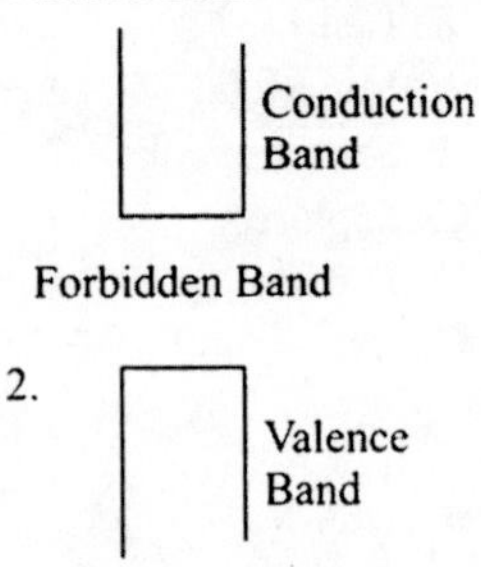

C. Insulator	3. 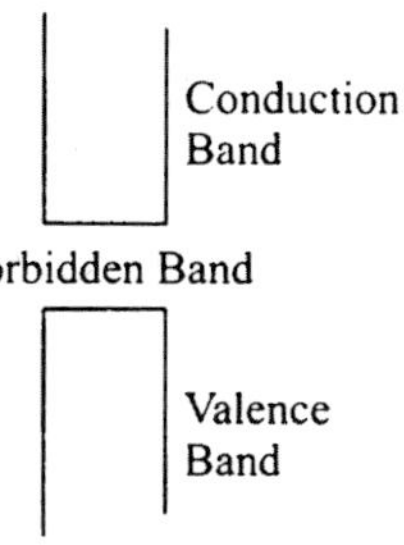

Codes:

	A	B	C
(*a*)	1	3	2
(*b*)	3	1	2
(*c*)	1	2	3
(*d*)	2	3	1

248. For a photoconductor with equal electron and hole mobilities and perfect ohmic contacts at the ends, an increase in intensity of optical illumination results in
(*a*) a change in o.c. voltage
(*b*) a change in s.c. current
(*c*) decrease in resistance
(*d*) increase in resistance

249. A sample of *n* type semiconductor has an electron density of $6.25 \times 10^{18}/cm^3$ at 300 K. If intrinsic concentration of carriers in this sample is $2.5 \times 10^{13}/cm^3$ at this temperature, the hole density is
(*a*) $10^6/cm^3$ (*b*) $10^8/cm^3$
(*c*) $10^{10}/cm^3$ (*d*) $10^{12}/cm^3$

250. A coil of wire has inductive impedance. At high frequencies the impedance will be represented by

(*a*) 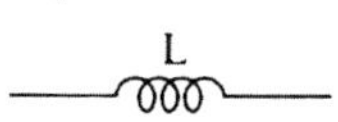

(*b*) 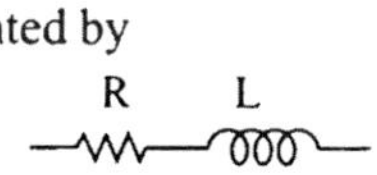

(*c*) 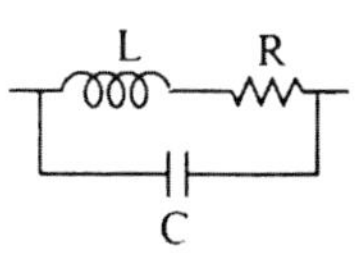

(*d*) 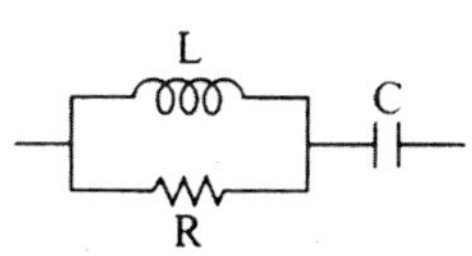

251. A parallel plate capacitor has its length, width and separation doubled. It fringing effects are neglected, to keep the capacitance same, the dielectric constant must be
(*a*) halved (*b*) kept the same
(*c*) doubled (*d*) made 4 times

252. Hall effect is observed in a specimen (metal or semiconductor) when it is carrying current and is placed in a magnetic field. The resultant electric field inside the specimen is
(*a*) normal to both current and magnetic field
(*b*) in the same direction as current
(*c*) in a direction antiparallel to magnetic field
(*d*) none of above

253. In *p* type semiconductor the conductivity due to holes is (e = charge on hole, μ_p = hole mobility, p = hole concentration)
(*a*) $\frac{pe}{\mu_p}$ (*b*) $\frac{\mu_p}{pe}$
(*c*) $pe\mu_p$ (*d*) $\frac{1}{pe\mu_p}$

254. The minority carrier life time and diffusion constant in a semiconducting material are 100 μs and 100 cm^2/s respectively. The diffusion length of carriers is
(*a*) 0.1 cm (*b*) 0.01 cm
(*c*) 0.0141 cm (*d*) 1 cm

255. The conductivity of intrinsic semiconductor is given by
(*a*) $en_i(\mu_p - \mu_n)$ (*b*) $en_i(\mu_p + \mu_n)$
(*c*) $n_i(\mu_p + \mu_n)$ (*d*) $n_i(\mu_p \mu_n)$

256. If q is charge on an electron, R is radius of electron orbit and ω is angular velocity of electron rotation, the magnetic dipole moment of electron in orbit is
(*a*) $0.5\,q\omega R$ (*b*) $0.5\,q^2\omega^2R^2$
(*c*) $0.5\,q\omega R^2$ (*d*) $0.5\,q\omega^2R$

257. The dipole moment per unit volume P as a function of E for an insulator is
(*a*) $P = \epsilon_0 E(\epsilon_r)$ (*b*) $P = \epsilon_0 E$
(*c*) $P = \epsilon_0(\epsilon_r - 1)E$ (*d*) $P = \frac{E}{\epsilon(\epsilon_r - 1)}$

258. The wavelength of light emitted by Ga As laser is 8670×10^{-10} m. If $h = 6.626 \times 10^{-34}$ Js, velocity of light $= 3 \times 10^8$ m/s and $eV = 1.602 \times 10^{-19}$ J, the energy gap in Ga As is
(*a*) 0.18 eV (*b*) 0.7 eV
(*c*) 1.43 eV (*d*) 2.39 eV

259. A single layer air cored coil has *n* turns and inductance L. If a new coil is formed by having 2*n* turns while keeping the coil length and diameter the same, the new inductance is
(*a*) 0.5 L (*b*) L
(*c*) 2 L (*d*) 4 L

260. A parallel plate capacitor has area of plate A and plate separation *d*. Its capacitance is C. A metallic plate P of area A and negligible thickness is added as shown in Fig. 1.10. The new value of capacitance is

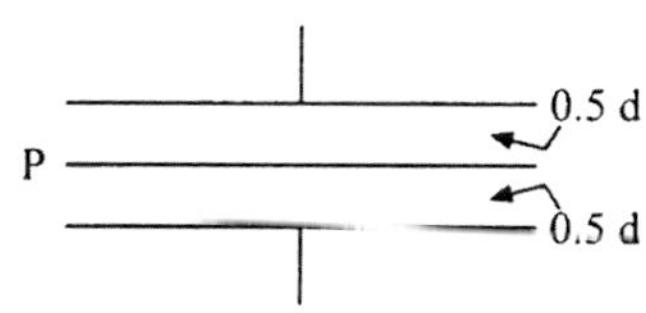

Fig. 1.10.

(*a*) C (*b*) 2C
(*c*) 0.5 C (*d*) 0.25 C

261. If ρ is resistivity, *l* is length of wire, *a* is area of cross-section, ω is angular frequency, F is form factor, *n* is number of turns. *d* is coil diameter and *r* is resistance of a coil, match List I with List II and give correct answer using the given codes

List I	**List II**
A. Value of inductance	1. $\rho\frac{l}{a}$
B. Power factor	2. $\frac{\omega L}{r}$
C. Q factor	3. $\frac{r}{\omega L}$
D. Value of resistance	4. Fn^2d

Codes:

	A	B	C	D
(a)	1	2	3	4
(b)	4	3	2	1
(c)	4	2	3	1
(d)	1	3	2	4

262. With increase in temperature, the electrical conductivity would
(a) increase in metals as well as in intrinsic semiconductor
(b) increase in metals but decrease in intrinsic semiconductor
(c) decrease in metals but increase in intrinsic semiconductor
(d) decrease in metals as well as in intrinsic semiconductor

263. Ferrites have
(a) a low copper loss
(b) low eddy current loss
(c) low resistivity
(d) higher specific gravity compared to iron

264. As per Curie-Weiss law the magnetic susceptibility of a material varies as
(a) T^{-2} (b) $\frac{1}{T}$
(c) T (d) T^2

265. When subjected to alternating stresses, an insulating material is characterised by complex dielectric constant $\epsilon'_r - j\epsilon''_r$. In such materials the dielectric losses under alternating stress is proportional
(a) ϵ''_r (b) both ϵ'_r and ϵ''_r
(c) ϵ'_r (d) $(\epsilon''_r)^2$

266. Yttrium-iron garnet ($Y_2Fe_5O_{12}$) is a soft magnetic material suitable for applications involving
(a) dc (b) 50 Hz ac
(c) ac of a few kHz
(d) ac of a few hundred MHz

267. An air cored inductance is a
(a) linear circuit element because its reactance varies linearly with voltage at a fixed frequency
(b) linear circuit element because its current varies linearly with voltage at a fixed frequency
(c) non-linear circuit element due to saturation of core
(d) non-linear circuit element because $v = \frac{L\,di}{dt}$

268. Consider the following statement: In case of a superconductor
1. B = 0 2. μ_r is high
3. diamagnetism is high
4. transition temperature varies with isotopic mass
Of the above statements
(a) 1 and 2 are correct (b) 2, 3, 4 are correct
(c) 1, 3, 4 are correct (d) 2, 4 are correct

269. The resistance of a metallic wire would
(a) increase as *f* increase (b) decrease as *f* increase
(c) remains the same at all frequencies
(d) increase with *f* up to a certain frequency and then decrease

270. The carrier mobility in a semiconductor is 0.4 m^2/Vs. Its diffusion constant al 300 K will be (in m^2/s)
(a) 0.40 (b) 0.16
(c) 0.04 (d) 0.01

271. In a piezo electric crystal, the application of mechanical force will cause
(a) plastic deformation of crystal
(b) magnetic dipoles in the crystal
(c) electrical polarization in the crystal
(d) shift in Fermi level

272. Consider the following materials
1. Nickel, 2. Silver, 3. Oxygen, 4. Aluminium
The correct sequence in the order of decreasing permeability is
(a) 1, 4, 3, 1 (b) 1, 4, 2, 3
(c) 4, 1, 3, 2 (d) 4, 1, 2, 3

273. A vacuum parallel plate capacitor is charged. The field between the plates is 2×10^4 V/m. If the space between the plates is filled with a material having $\epsilon_r = 10$, then the value of field in the dielectric will be
(a) 2×10^3 V/m (b) $\frac{2}{\sqrt{10}} 10^4$ V/m
(c) $2\sqrt{10} \times 10^4$ V/m (d) 2×10^5 V/m

274. Some ceramic materials become superconducting
(a) below liquid helium temperature
(b) between liquid helium and liquid nitrogen temperature
(c) above liquid nitrogen but below room temperature
(d) at room temperature

275. The real part of complex dielectric constant and tan δ for a dielectric are 2.1 and 5×10^{-4} at 100 Hz respectively. The imaginary part of dielectric constant at 100 Hz is
(a) 1.05×10^{-3} (b) 2.1×10^{-3}
(c) 5×10^{-3} (d) 1.05×10^{-2}

276. Which of the following are donor impurities
1. Gold. 2. Phosphorous 3. Boron, 4. Antimony, 5. Arsenic 6. Indium
Select the correct answer using the following codes
(a) 1, 2, 3 (b) 1, 2, 4, 6
(c) 3, 4, 5, 6 (d) 2, 4, 5

277. The ohm's law for conduction in metals is
(a) $J = \sigma E$ (b) $J = \frac{E}{\sigma}$
(c) $J \propto \sigma E$ (d) $J \propto \frac{E}{\sigma}$

278. The magnetic moment in units of Bohr magneton of a ferrous ion in any ferrite is
(a) zero (b) 2
(c) 4 (d) 6

279. For a permanent magnetic material
(a) the residual induction and coercive field should be large
(b) the residual induction and coercive field should be small
(c) area of hysteresis loop should be small
(d) initial relative permeability should be high

280. The most important set of specifications of transformer oil include
(a) dielectric strength and viscosity
(b) dielectric strength and flash point

(c) viscosity and flash point
(d) flash point and viscosity

281. For a superconductor, magnetisation M in a field H is
(a) extremely small (b) – H
(c) – 1 (d) zero

282. If a small amount of Cu is added to Ni conductor, then
(a) resistivity of Ni will decrease at all temperatures because Cu is a better conductor than Ni
(b) residual resistivity of Ni at low temperature will increase as Cu atoms act as defect centres
(c) resistivity of Ni will increase at all temperatures as Cu destroys the periodicity of Ni and acts as defects
(d) resistivity of Ni remains the same

283. Match List I with List II and select the correct answer using the given codes

List I (Characteristics)	List II (Decrease)
A. Paramagnetic	1. All dipoles are aligned is one preferred direction
B. Ferromagnetic	2. Half the dipoles are aligned in opposite direction and have equal magnitude
C. Anti-ferromagnetic	3. Half the dipoles (with equal magnitudes) are aligned in opposite direction to the other half having equal but lower magnitude
D. Ferrimagnetic	4. All dipoles have equal magnitude but are randomly oriented

Codes:

	A	B	C	D
(a)	4	3	2	1
(b)	4	1	2	3
(c)	2	1	4	3
(d)	2	3	4	1

284. For an insulating material, dielectric strength and dielectric losses should be
(a) high and high (b) low and high
(c) high and low (d) low and low

285. Match List I (optical devices) with List II (electrical optical characteristics) and give correct answer as per the given codes

List I	List II
A. Laser	1. emits monochromatic light of low intensity
B. Solar cell	2. consumes electrical power due to incident light
C. Photo diode	3. delivers power to a load
D. LED	4. emits monochromatic light of high intensity

	A	B	C	D
(a)	4	3	1	2
(b)	3	4	2	1
(c)	4	3	2	1
(d)	3	4	1	2

286. Consider the following statement
If an electric field is applied to an n type semiconductor bar, the electrons and holes move in opposite directions due to their opposite charges. The net current is
1. both due to electrons and holes with electrons as majority carriers
2. sum of hole and electron currents
3. difference between electron and hole currents

Which of above statements are correct
(a) 1 only (b) 1 and 2
(c) 2 only (d) 3 only

287. Alumina is a
(a) dielectric (b) ceramic
(c) semiconductor (d) conductor
[IES 2010]

288. In ionic crystals electrical conductivity is
(a) very high (b) depends on material
(c) depends on temperature
(d) practically zero **[IES 2010]**

289. Which of the following is not a ferromagnetic material?
(a) cobalt (b) iron
(c) nickel (d) bismuth **[IES 2010]**

290. Which of the following statements are true for diamond structure?
(1) Coordination number is 4
(2) Packing fraction is 0.34
(3) Copper crystallizes into diamond structure
(4) Lattice is FCC
(a) 1 only (b) 1, 2 and 4
(c) 2 and 3 (d) 1, 2, 3, 4 **[IES 2009]**

291. Which of the following statements are true for a good conductor of electricity?
(1) its conductivity decreases with increasing temperature
(2) number of free electrons is about 10^{28} m^{-3}
(3) its conductivity decreases with addition of impurities
(4) it is a good conductor of heat also
(a) 1, 2, 3, 4 (b) 1 only
(c) 2 and 3 only (d) 3 and 4 only **[IES 2009]**

292. Which of the following is not an insulator?
(a) diamond (b) graphite
(c) bakelite (d) lucite **[IES 2009]**

293. Consider the following statements
The conductivity of a metal has negative temperature coefficient since
(1) the electron concentration increases with temperature
(2) electron mobility decreases with temperature
(3) electron lattice scattering rate increases with temperature

Which of the above statements are correct?
(a) 1 only (b) 1 and 2
(c) 2 and 3 (d) 3 only **[IES 2009]**

294. Which of the following is not a conducting material
(a) copper (b) tungsten
(c) germanium (d) platinum **[IES 2009]**

295. What is the chemical bonding in silicon semiconductor?
(*a*) metallic (*b*) ionic
(*c*) covalent (*d*) vanderwalls **[IES 2009]**

296. Which of the following is a trivalent material?
(*a*) antimony (*b*) phosphrous
(*c*) arsenic (*d*) feron **[IES 2009]**

297. The fuse material used in electrical energy must have one of the following properties **[IES 2009]**
(*a*) high resistivity (*b*) low conductivity
(*c*) high melting point (*d*) low melting point

298. Manganin, an alloy of copper and manganese is used in
(*a*) soldering material (*b*) heating elements
(*c*) ballast resistance (*d*) standard resistance **[IES 2009]**

299. Hysteresis loss in a transformer working at 220 V and a frequency of 50 Hz is 100 W. When the transformer is working at 220 V and 100 Hz. What is the hysteresis loss?
(*a*) 50 W (*b*) 100 W
(*c*) 200 W (*d*) 400 W` **[IES 2009]**

300. Which of the following is used for making permanent magnets?
(*a*) steel (*b*) carbon
(*c*) carbon-steel (*d*) graphite **[IEEE 2009]**

301. The following properties are associated with ferroelectric materials.
1. Its susceptability is negative **[IES 2009]**
2. Susceptability is expressed as

$$\psi = \frac{C}{T - T_C}$$ where C is Curie constant and T_C is Curie temperature

3. It has permanent dipoles oriented randomly

Which of these statements are correct?
(*a*) 1 only (*b*) 1 and 3
(*c*) 2 only (*d*) 1, 2 and 3

302. The materials not having negative temperature coefficient of resistivity are
(*a*) metals (*b*) semiconductors
(*c*) insulators (*d*) none of above **[IES 2009]**

303. Effective Q of the equivalent circuit of quartz crystal is of the order of
(*a*) 200 (*b*) 2000
(*c*) 20000 (*d*) 200000 **[IES 2009]**

304. In a material the fermi level is located between the centre of forbidden bond and conductor bond. The material is
(*a*) *p* type semiconductor
(*b*) *n* type semiconductor
(*c*) intrinsic semiconductor
(*d*) insulator **[IES 2009]**

305. The transition temperature of superconducting material is changed by which of the following **[IES 2008]**
(*a*) electric field (*b*) magnetic field
(*c*) mechanical stress (*d*) none of these

306. Ceramic resonators use which one of the following?
(*a*) barium titanate (*b*) silicon
(*c*) piezo electric quartz crystal
(*d*) zirconium titanate **[IES 2007]**

307. An air capacitor made of square plates 10 cm × 10 cm has capacitance C. If plates are reduced to 2.5 cm × 2.5 cm the new value of capacitance is
(*a*) C/4 (*b*) C/8
(*c*) C/16 (*d*) C/36 **[IES 2007]**

308. Velocity of light in a particular medium is 10^8 m/s. What is the relative permittivity of medium?
(*a*) 1.732 (*b*) 3
(*c*) 9 (*d*) 0.333 **[IES 2007]**

309. What does quality factor of a dielectric mean?
(*a*) it is related to permittivity of material
(*b*) it is related to breakdown strength of material
(*c*) it is related to resistivity of material
(*d*) it is related to the ratio of maximum stored energy and average power loss in the dielectric **[IES 2007]**

310. In Curie law for paramagnetic materials how is susceptibility ψ related to absolute temperature T.
(*a*) $\psi \propto T$ (*b*) $\psi \propto \frac{1}{T}$
(*c*) $\psi \propto T^2$ (*d*) $\psi \propto \frac{1}{T^2}$ **[IES 2007]**

311. Why are askerals not used as transformer or capacitor insulation?
(*a*) they have low dielectric strength
(*b*) their density is high
(*c*) they decompose easily giving out toxic gases
(*d*) they are highly inflammable **[IES 2006]**

312. A solid specimen of dielectric has relative permittivity $\epsilon_r = 4.2$ and loss angle tan $\delta = 0.001$ at 50 Hz. It is subjected to 50 kv/cm. What is the heat generated in the specimen due to dielectric loss?
(*a*) 291 W/cm^3 (*b*) 5.82 mW/cm^3
(*c*) 19.5 W/cm^3 (*d*) 0.291 mW/cm^3 **[IES 2006]**

313. The dielectric constant of an insulating material is given by $\epsilon = \epsilon' - j\epsilon''$. If a parallel plate capacitor of plate area A and separation *d* is filled with this material, what would be the loss factor? **[IES 2006]**
(*a*) $\tan^{-1}[A\epsilon'/d\epsilon'']$ (*b*) $A\epsilon'/d\epsilon''$
(*c*) ϵ''/ϵ' (*d*) $\tan^{-1}[d\epsilon''/A\epsilon']$

314. Piezo electric effect is the polarization of a dielectric under the influence of
(*a*) light (*b*) mechanical stress
(*c*) electrical stress (*d*) heat **[IES 2006]**

315. Consider the following statements with regard to FCC structure

(1) number of nearest neighbor atoms is 12

(2) packing efficiency is 0.74

(3) there is an atom at the body centre of unit cell

Which of the above are correct?

(*a*) 1, 2, 3 (*b*) 1, 2

(*c*) 2, 3 (*d*) 1, 3 **[IES 2005]**

316. Magnetic cores required for RF application should have

(*a*) high hysteresis and eddy current loss

(*b*) high hysteresis and low eddy current loss

(*c*) low hysteresis and high eddy current loss

(*d*) low hysteresis and eddy current loss **[IES 2005]**

317. Which of the following correctly represents the variation of critical magnetic field with temperature for a superconductor? **[IES 2005]**

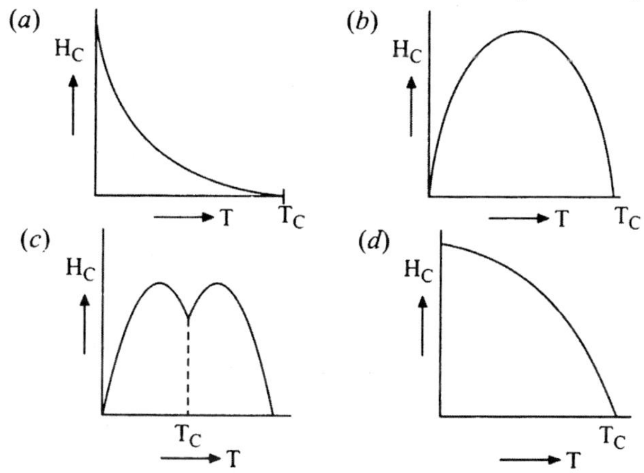

Fig. 1.11.

318. Ferrites are the materials which have

(*a*) low permeability and low dielectric loss

(*b*) low permeability and high dielectric loss

(*c*) high permeability and low dielectric loss

(*d*) high permeability and high dielectric loss **[IES 2005]**

319. Which of the following correctly represents *v-i* characteristics of thermister? **[IES 2005]**

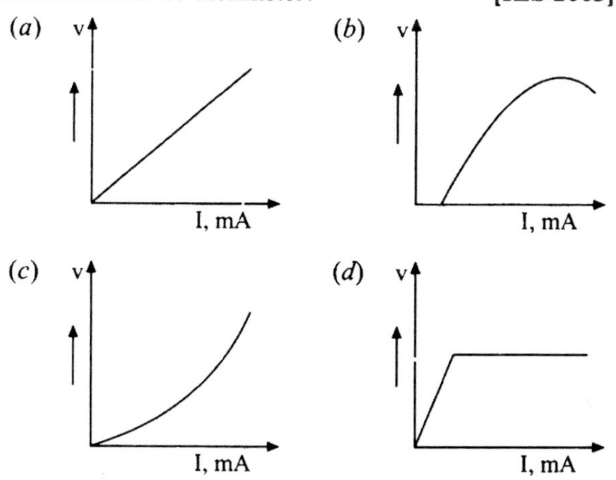

Fig. 1.12.

320. A thin resistive film having length *l*, width *w* and thickness *t* is used as a resistors (R_1) for conducting current in direction of *l*. It is connected as a resistance R_2 for conducting current in the direction of *w*, then R_1/R_2 is

(*a*) l/w (*b*) w/l

(*c*) l^2/w^2 (*d*) w^2/l^2 **[IES 2005]**

321. The relative dielectric constant of Al_2O_3 is 8. The dielectric constant of free space is 8.854×10^{-12} F/m. What is the capacitance of a layer of Al_2O_3 which is 0.5×10^{-6} m thick and 1000 mm^2 of surface area?

(*a*) 7.08×10^{-7} F (*b*) 1.42×10^{-7} F

(*c*) 2.83×10^{-7} F (*d*) 0.92×10^{-7} F **[IES 2005]**

322. How can eddy current loss of transformer be minimized

1. by increasing resistivity of core
2. by using laminated sheets with insulator coating
3. by using highly insulating non-magnetic core
4. by using paramagnetic material as core

Which of the following are correct?

(*a*) 1, 2 (*b*) 2, 3

(*c*) 1, 3 (*d*) 1, 2, 4 **[IES 2005]**

323. A copper wire is 1 m long and has a uniform cross-section of 0.1 mm^2. Its resistance at room temperature 0.171 Ω. What is the resistivity of material?

(*a*) 1.71×10^{-6} ohm-m (*b*) 1.71×10^{-7} ohm-m

(*c*) 1.71×10^{-8} ohm-m (*d*) 1.71×10^{-9} ohm-m **[IES 2004]**

324. The material which has the property of becoming electrically polarized is response to applied mechanical stress is termed as **[IES 2004]**

(*a*) ferroelectric (*b*) piezoelectric

(*c*) optoelectronic (*d*) superconducting

325. The electric displacement D in a material is expressed as $D = \epsilon E + ds$ where ϵ is permitivity, E is electric field, *s* is strain and *d* is characteristic constant of the material. What is this material? **[IES 2004]**

(*a*) metal (*b*) antiferromagnetic

(*c*) semiconductor (*d*) piezoelectric

326. A conductor carries a current of 4 A. If magnitude of charge on electron is 1.6×10^{-19} coulomb, the number of electrons which flows past the cross-section per second is

(*a*) 2.5×10^{19} (*b*) 1.6×10^{19}

(*c*) 6.4×10^{19} (*d*) 0.4×10^{19} **[IES 2003]**

327. The correct sequence of the following materials in the increasing order of conductivity is

(*a*) silicon doped with boron – silver – aluminum intrinsic silicon

(*b*) intrinsic silicon – aluminium – silver – silicon doped with boron

(*c*) aluminium – intrinsic silicon-silicon doped with boron – silver

(*d*) intrinsic silicon-silicon doped with boron – aluminium silver **[IES 2003]**

328. The correct sequence of following materials in the increasing order of susceptibility is

(*a*) diamagnetic – ferromagnetic – paramagnetic superconductor

(*b*) ferromagnetic – paramagnetic – diamagnetic superconductor

(*c*) paramagnetic – diamagnetic – superconductor ferromagnetic

(*d*) superconductor – diamagnetic – paramagnetic – ferromagnetic **[IES 2003]**

329. Consider the following statements

A semiconductor to be used in optoelectronic device should have

1. direct energy band gap
2. indirect energy band gap
3. any value of forbidden energy band gap
4. right value of band gap corresponding to light wavelength

Which of the above statements are correct?

(*a*) 1 (*b*) 1 and 4

(*c*) 2 and 3 (*d*) 2 and 4 **[IES 2003]**

330. An LED made using Ga As emits radiation in

(*a*) visible region (*b*) ultra-violet region

(*c*) infrared region (*d*) microwave region

[IES 2003]

331. A resistance thermometer has a temperature coefficient of resistance of 10^{-3}/°C. At 0°C resistance is 1 ohm. At what temperature is resistance 1.1 ohms?

(*a*) 10°C (*b*) 100°C

(*c*) 120°C (*d*) –10°C **[IES 2003]**

332. When BCC iron is heated it changes to FCC resulting in

(*a*) contraction in volume

(*b*) increase in volume

(*c*) no change in volume

(*d*) crack in material **[IES 2002]**

333. When copper is added to silver in small quantity so as to form an alloy, the resistivity of alloy a

(*a*) equal to that of copper

(*b*) equal to that of silver

(*c*) greater than resistivity of copper

(*d*) in between that of silver and copper

334. When a certain percentage of silicon is added to iron the resistivity of iron increases by a factor of 5. If a transformer core is made out of iron as compared to previous one when silicon in not added, eddy current loss could be

(*a*) 77% (*b*) 80%

(*c*) 83% (*d*) 86% **[IES 2003]**

335. Match list 1 (materials) and list II (applications of materials) and selected correct answer using the codes

List I	List II
A. Aluminium	1. Current carrying spring
B. Phosphor bronz	2. Heating element
C. Carbon	3. Commutator brush
D. nichrome	4. Telephone cords and trolley wires

	A	B	C	D
(*a*)	4	3	1	2
(*b*)	2	1	3	4
(*c*)	4	1	3	2
(*d*)	2	3	1	4

[IES 2002]

336. Which one is an essential component of electromechanical relays

(*a*) graphite rod (*b*) LED

(*c*) electromagnet (*d*) MOSFET **[IES 2002]**

B. The following questions have two statements Assertion A and Reason R. Examine the statements and give correct answer as per following codes.

Codes:

(*a*) both A and R are true and R is correct explanation of A

(*b*) both A and R are true but R is not correct explanation of A

(*c*) A is true but R is false

(*d*) A is false but R is true

337. A. The resistance of a copper specimen changes by about 22% when temperature is increased from room temperature to 700°C.

R. The physical properties of a material are determined mostly by the extent of impurities.

338. A. Copper is a good conductor of electricity.

R. Copper has a face centred cubic lattice.

339. A. Atomic number of sodium is 11.

R. Sodium has a body centred cubic lattice.

340. A. Both sodium and ferrous are ferromagnetic.

R. Both sodium and ferrous have body centred cubic structure.

341. A. The regular or irregular stacking of atoms has important effect on properties of materials.

R. The arrangement of atoms in a given material can be studied using X-rays.

342. A. An amorphous material is obtained when mobility of atoms is inhibited during solidification.

R. Crystalline state is a natural state for most solids.

343. A. Copper is a monatomic solid.

R. In a metal the valence electrons belong to all atoms.

344. A. Diamond and corborudum are both very hard materials.

R. Valence bond can be extremely strong.

345. A. Ionic solids, in general, have no electrical conductivity.

R. In ionic solids, electrons are tightly bound to the nucleus.

346. A. Helium, Argon and Neon are gaseous at room temperature.

R. The atoms in Helium, Argon and Neon are chemically extremely inactive.

347. A. Relative permittivity is determined by the atomic structure of the material.

R. Absolute permittivity is determined by the atomic structure of the material.

348. A. $\in_0$ has the dimensions of farad per meter.

R. $D = \in_0 \in_r E$.

349. A. $\in_r$ can be determined experimentally by measuring the capacitance of a parallel plate capacitor with and without the dielectric.

R. Capacitance of a parallel plate Capacitor

$= \frac{\in_0 \in_r A}{d}$.

350. A. The equation $D = \in_0 \in_r E$ is applicable only to isotropic materials.

R. In poly crystalline materials, the directional effects are absent.

351. A. For a dielectric $P = \in_0(\in_r - 1) E$.

R. Dipole moment per unit volume is called polarization of dielectric.

352. A. Electronic polarizability increases as atoms become larger.

R. For rare gas $\in_0(\in_r - 1) = N\alpha_e$ where N is number of atoms/m^3 and α_e is electronic polarizabilily of material.

353. A. Dielectric constant is independent of temperature unless the temperature is extremely high.

R. For normal temperatures the number of atoms per unit volume is constant.

354. A. Electronic and Ionic polarization in a poly atomic gas are independent of temperature.

R. The orientation polarization is independent of temperature.

355. A. In Fig. 1.13, the intercept of the line on *y*-axis is equal to $N(\alpha_e + \alpha_i)$ where N is number of atoms/m^3, α_i is electronic polarizability and α_e is ionic polarizability.

R. α_e and α_i are independent of temperature.

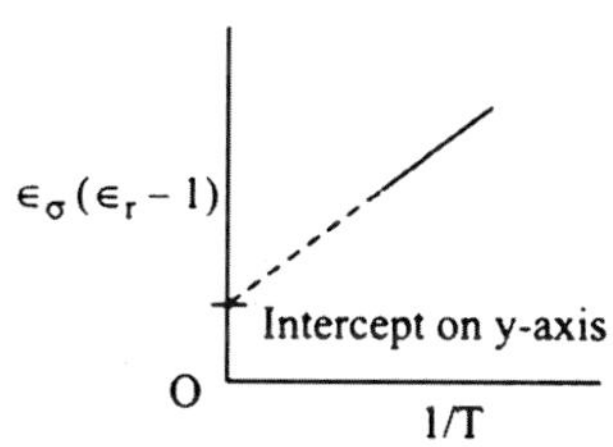

Fig. 1.13.

356. A. From the slope of the line in Fig 1.13 we can find permanent dipole moment of the molecules.

R. Total polarization of a polyatomic gas is given by $P = N(\alpha_e + \alpha_i + \mu_p^2/3kT) E$.

357. A. CO_2 has no resultant dipole moment but CO has a dipole moment.

R. The structure of CO_2 is $0 = C = 0$.

358. A. For elemental dielectrics $P = N\alpha_e E_i$ where N is number of atoms/m^3, P is polarization, α_i is electronic polarization and E_i is internal field.

R. In elemental dielectrics there are no permanent dipoles or ions.

359. A. In ionic dielectrics $P = P_c + P_i$.

R. In ionic crystals, permanent electric dipole moment is zero.

360. A. The optical dielectric constant (*i.e.*, $\in_r$ at optical frequencies) of ionic dielectrics is less than static dielectric constant.

R. At optical frequencies $P_i = 0$.

361. A. If difference between static and optical dielectric constant is high, the mechanical compressibility of that material is also high.

R. For soft ions, ionic polarization per unit field is higher than that of hard ions.

362. A. In some materials $\in_r$ is independent of temperature while in others $\in_r$ depends on temperature.

R. If permanent dipoles are absent, $\in_r$ varies with temperature.

363. A. Ferroelectric materials have spontaneous polarization.

R. Above curie temperature $\in = \frac{C}{(T - \theta)}$ for ferroelectric materials.

364. A. Rochelle salt and Barium titanate are both ferroelectric.

R. Ferroelectric materials exhibit hysteresis effect.

365. A. In many materials dielectric polarization causes mechanical distortion.

R. If dielectric polarization causes mechanical distortion then mechanical distortion must cause dielectric polarization.

366. A. If an atom of polarizabilily α is placed in homogenous field E, the energy stored is $0.5 \alpha E^2$.

R. Capacitance of an isolated conducting sphere R in vacuum is $4\pi\in_0 R$ farads.

367. A. In imperfect capacitors the current does not lead the applied ac voltage by 90°.

R. When subjected to ac fields the dielectric constant can be expressed as $\in'_r - \in''_r$.

368. A. In liquids and glassy substances, dielectric losses occur.

R. The orientation polarization in many substances is frequency dependent.

369. A. capacitor with imperfect dielectric can be represented by a capacitance in parallel with resistance.

R. For imperfect dielectrics, dielectric constant has real and imaginary parts.

370. A. In a perfect capacitor the current density is given by $\omega \in_0 E_0 \in'_r \cos(\omega t + 90°)$ where $\in'_r$ is real part of dielectric constant.

R. In a perfect capacitor dielectric losses are zero.

371. A. Dielectric losses are proportional to tan δ.

R. $\tan\delta = \frac{\in''_r}{\in'_r}$ where $\in'_r$ and $\in''_r$ denote the real and imaginary parts of $\in_r$.

372. A. In a parallel plate capacitor having area of plate 1 m^2 and plate separation 1 m, the capacitance C and resistance R are given by $C = \in_0 \in'_r$ and $R = \frac{1}{\in_0 \in''_0 \omega}$, where ω is frequency, $\in'_r$ and $\in''_r$ are real and imaginary parts of dielectric constant.

R. The imaginary part of dielectric constant is responsible for dielectric losses.

373. A. For a solenoid having N turns,

$B = \frac{\mu_0 \mu_r NI}{L}$.

R. For all magnetic materials B increases linearly with I.

374. A. For a magnetic material $M = (\mu_r - 1)H$ where M is magnetic dipole moment/m³, H is field intensity A/m and μ_r is relative permeability.
R. μ_r can have any value from 0 to hundreds of thousands.

375. A. If an electron having charge e is revolving in an orbit of radius R and with angular velocity ω, the magnetic dipole moment of orbit is $0.5\, e\omega R^2$.
R. Orbital angular momentum and magnetic dipole moment are always equal.

376. A. Orbital magnetic moment of an electron in an atom is of the order of 1 Bohr magneton.
R. 1 Bohr magneton = 9.27×10^{-24} A/m².

377. A. When an electron is revolving in its orbit and magnetic field is also present, the angular frequency of electron will be affected by the magnetic field.
B. In the case of electron revolving in its orbit in the presence of magnetic field, the orbital magnetic dipole moment is not affected by the presence of magnetic field.

378. A. For Diamond $\mu_r = 1$.
R. Diamond is a diamagnetic material.

379. A. The angular momentum of an atom is due to three contributions *viz* orbital, electron spin and nuclear spin.
R. The nuclear spin magnetic moment is much more than electron spin magnetic moment.

380. A. Magnetic susceptibility of diamagnetic materials is much less than that of paramagnetic materials.
R. For diamagnetic as well as paramagnetic materials μ_r is nearly equal to 1.

381. A. Ferromagnetic materials possess ferromagnetic properties only when temperature is less than paramagnetic curie temperature.
R. For a ferromagnetic material

$$\psi = \frac{C}{T - \theta} \text{ for } T > \theta_f.$$

382. A. Magnetism can be explained by the theory of magnetic domains.
R. The existence of magnetic domains is evident from the Bitter powder patterns.

383. A. Ferrites are very useful at very high frequencies.
R. Ferrites have high permeability and high resistivity.

384. A. Magnetic susceptibility of copper is -0.5×10^{-5}.
R. Copper is diamagnetic material.

385. A. In a conductor current density remains constant in time as long as E is constant.
R. In a conductor $J = \sigma E$.

386. A. An electron having charge $-e$ and mass m and lying in an electric field E is subjected to an acceleration $-(e/m)$ E.
R. If there are n electrons/m³ moving with velocity v_x and each electron has a charge $-e$, the current density $J = -n\, e v_x$.

387. A. Drift velocity of electrons is proportional to the applied electric field *E*.
R. The average drift velocity per unit field is called mobility.

388. A. Relaxation time and mean time between collisions are equal when scattering is isotropic.
R. Mean free path of an electron λ, velocity of electron v and mean time between collisions τ_c are related as $\lambda = v T_c$.

389. A. If W_F is fermi energy, v is velocity of electron and m is mass of electron, then

$$\frac{1}{2} m v^2 = W_F.$$

R. Conductivity of a metal depends on temperature and impurity concentration.

390. A. Resistivity of a perfect single crystal of a metal goes to zero as T approaches zero.
R. Scattering of electrons is due to deviations from perfect periodicity of lattice.

391. A. Power loss in a conductor of resistance $R = I^2R$.
R. When a conductor is carrying current with current density J as a result of applied field E, then heat devloped/m³/second = JE.

392. A. Metals are better heal conductors than insulators.
R. If $\frac{dT}{dx}$ is temperature gradient, Q is heat flow density and *K* is thermal conductivity, then $Q = -K\,(dT/dx)$.

393. A. If E is available energy state eV and E_F is fermi level, then probability of energy state E being occupied = 0.5.
R. $f(E) = \dfrac{1}{1 + e^{(E - E_F)/kT}}$.

394. A. For most metals, fermi energy E_F is less than 10 eV.
R. Fermi level is a metal is given by $E_F = 3.64 \times 10^{-19}\,(n)^{2/3}$ where n is number of free electrons/m³ of metal.

395. A. At a temperature of 1000 K, thermionic emission current is about 0.1 A/cm² of surface.
R. Thermionic emission current is given by

$$I_{th} = SA_0\, T^2 e^{-E_W/KT}.$$

396. A. When cathode temperature is increased from 2500 K to 2550 K, thermionic emission current may increase by about 50%.
R. Thermionic emission current ∝ $T^2\, e^{-E_W/kT}$.

397. A. In an intrinsic semiconductor $J = (\mu_n \times \mu_p) e n_i$.
R. Intrinsic charge concentration n_f at temperature T is given by

$$n_i^2 = A_0 T^3\, e^{-EG_0/kT}.$$

398. A. Intrinsic resistivity of silicon is lower than that of germanium.
R. Magnitude of free electron concentration in germanium is more than that of silicon.

399. A. Acceptor extrinsic semiconductor has higher conductivity then intrinsic semiconductor.
R. Addition of p-type impurity produces an allowable discrete energy level just above valence band.

400. A. Holes are majority carriers in p type semiconductor.
R. In p type semiconductor the electrons produced by thermal agitation recombine with holes.

401. A. Silicon is less sensitive to changes in temperature than germanium.
R. Cut in voltage in silicon is less than that in germanium.

402. A. Silicon is the most favoured semiconductor material.
R. PIV for silicon diode is more than that for germanium diode.

403. A. Hall effect is used to determine whether the semiconductor is *p* or *n* type.
R. Under the influence of field, holes and electrons move in opposite direction.

404. A. Electron concentration *n* hole concentration *p* and charge concentration n_i are related as $n_i^2 = np$.
R. *p* type semiconductor is obtained by adding trivalent impurity to intrinsic semiconductor.

405. A. Doping level in extrinsic semiconductors is very small.
R. Additional of impurity in the ratio of 1 part in 10^8 parts increases the number of charge carriers by about 20.

C. State whether the following are True or False

406. In a hydrogen atom, an electron moves in the field of a proton. The stability of orbit requires equilibrium between coulomb's force and centrifugal force.

407. As per Bohr's postulate, only those circular orbits are stable for which the angular momentum is equal to an integer times $\frac{h}{2\pi}$.

408. Schrodinger wave equation is partial differential equation.

409. If *n* denotes principal quantum number, the angular quantum number $l = 0, 1 \ldots (n-1)$.

410. The magnetic quantum number determines the component of angular momentum along a prescribed direction.

411. In atomic physics a state with $l = 0$ is called *p* state.

412. The chemical properties of atoms are determined by all the electrons and their configuration.

413. In a true valence crystal, the bonding between the atoms is accomplished by the sharing of valence electrons.

414. In metals the valence electron wave functions are strongly perturbed by the presence of neighbouring atoms.

415. In vander walls crystals, there exists a high degree of stability of the outer electron shell.

416. At 0K, silicon and germanium behave like mono atomic crystals.

417. In crystalline solids, atoms are stacked in a regular manner.

418. A characteristic property of crystal is its periodicity of structure.

419. In body centred cubic structure comers and centre of the cube are occupied by identical atoms.

420. We can know the arrangement of atom in a material by studying electron diffraction pattern.

421. The circumference of an electron orbit is equal to integer times the wavelength of electron.

422. ϵ_0 has no other physical meaning than that of a fundamental conversion factor.

423. ϵ_r is determined by the atomic structure of the dielectric.

424. The equation $D = \epsilon_0 \epsilon_r E$ refers only to isotropic materials.

425. When a dielectric is subjected to an electric field E, each volume element of dielectric can be considered to be carrying an electric dipole moment.

426. A homogenous dielectric material is subjected to an electric field E. If a small cavity is made inside the material, the electric field intensity inside and outside the cavity are different.

427. In a homogenous dielectric subjected to an electric field E, the dipole moment per unit volume is $\epsilon_0(\epsilon_r - 1)$ E.

428. A rare gas having N atoms/m^3 is subjected to an electric field E. The electric dipole moment per unit volume is proportional to N and E.

429. If the number of atoms per unit volume of a monotonic gas changes, the dielectric constant still remains constant.

430. In a rare gas, the molecules of some atoms have an excess positive or negative charge. The electric field tends to shift positive ions relative to negative ions. This is known as ionic polarization.

431. The orientation polarization of a polyatomic gas is inversely proportional to absolute temperature.

432. Electric dipole moment is expressed in debye unit.

433. The dielectrics having only one kind of atoms are known as elemental dielectrics.

434. In elemental dielectrics the orientation and ionic polarization are absent.

435. In molecules of ionic crystals, orientational polarization is zero.

436. If there are no permanent dipoles in a dielectric material, the dielectric constant is independent of temperature.

437. There is no hysteresis phenomenon is any dielectric material.

438. Spontaneous polarization occurs in ferro electric materials.

439. Above ferro electric curie temperature, spontaneous polarization in ferro electric materials becomes stronger.

440. Curie-Weiss law is $\epsilon = \frac{C}{(T - \theta)}$.

441. Rochelle salt is a ferro electric material.

442. Piezo electricity means the slight change in dimensions of a specimen when an electric field is applied.

443. The materials which become polarized on the application of mechanical stress are called piezoelectric.

444. When an atom of polarizability α is placed in an electric field E, the energy stored in atom is $0.5\,\alpha E^2$.

445. Dielectric constant, under the condition of alternating fields, is a complex quantity.

446. When a dielectric material is subjected to alternating field the absorption of energy by the material from the field is given by the imaginary pan of polarizabilily.

447. If $\epsilon_r = \epsilon_r' - j\epsilon_r''$, $\tan\delta = \frac{\epsilon_r''}{\epsilon_r'}$.

448. A leaky dielectric has $\epsilon_r = \epsilon_r - j\epsilon_r''$. If a field E is applied, the current component in phase with E is proportional to ϵ_r'.

449. A leaky capacitor can be represented by a capacitance C and resistance R in parallel.

450. The dielectric losses do not depend on frequency.

451. Magnetisation of a material is the magnetic dipole moment per m^3.

452. If B, H and M denote flux density, field intensity and magnetisation, $B = \mu_0(H + M)$.

453. $M = (\mu_r - 1)$ H.

454. Orbital magnetic dipole moment is due to the motion of electron in its orbit around nucleus.

455. 1 Bohr magneton $= \frac{eh}{(4\pi m)}$.
456. If e is the charge an electron, R is the radius of its orbit and ω is the angular velocity of electron, the magnetic dipole moment μ_m of the orbit is $|\mu_m| = e\omega R^2$.
457. The induced magnetic diploic moment has a direction opposite to the applied magnetic field.
458. If ω_0 is the angular frequency of rotation of electron in the absence of magnetic field, the angular frequency ω, when a magnetic field B is acting is $\omega \sim \omega_0 + (e/2m)B$.
459. The induced magnetic dipole moment is independent of the permanent magnetic dipole moment.
460. Diamagnetic materials do not have permanent magnetic dipoles.
461. Paramagnetic materials have permanent magnetic dipoles but interaction between neighbouring dipoles is negligible.
462. Anti-ferromagnetic materials do not have permanent dipoles.
463. In ferromagnetic materials, the orientation between neighbouring permanent dipoles is parallel.
464. Ferrimagnetic materials have anti-parallel orientation of equal moments of permanent dipoles.
465. Diamond is a paramagnetic material.
466. For a diamagnetic material $\mu_r \sim 1$.
467. An electron for which $l = 0$ has zero orbital magnetic dipole moment.
468. The spin angular momentum along a given direction is either $\frac{+h}{4\pi}$ or $\frac{-h}{4\pi}$.
469. For a paramagnetic material susceptiblity increases with increasing temperature.
470. When temperature is more than ferromagnetic curie temperature, the behaviour of a ferromagnetic material is similar to that of paramagnetic material.
471. The hysteresis phenomenon, in ferromagnetic materials, exists at all temperatures.
472. Spontaneous magnetisation is characteristic of ferromagnetic materials.
473. Carbon steel has a residual flux density of 1 Wb/m^2.
474. In a virgin specimen of iron, the direction of spontaneous magnetisation varies from domain to domain.
475. Hysteresis curve is associated with motion of domain walls.
476. Bitter powder patterns prove the existence of domains.
477. When an electron moves at velocity v in combined electric and magnetic fields E and H, the Lorentz force F is $F = -e(E + v \times B)$.
478. Iron has a body centred cubic structure.
479. In metals the average drift velocity of electron per unit field is called mobility.
480. If v is the velocity of electron, τ_c is the average time between collisions, the mean free path $\lambda = v\tau_c$.
481. Resistivity of metals consists of two parts, one part constant and the other temperature dependent.
482. Semiconductors have a positive temperature coefficient of resistivity.
483. Many compounds also exhibit semiconducting properties.
484. In intrinsic semiconductors, the number of free electrons is equal to the number of mobile holes.
485. A vacant electronic state moves in the same direction as would a positive charge carrier.
486. When two conductors carry currents in the same direction the force is repulsive.
487. Wood is diamagnetic material.
488. For magnetic shielding we need a ferromagnetic material.
489. Lines of magnetic field attract each other.
490. The field, pattern of a solenoid is similar to that of bar magnet.
491. To produce magnetic flux, *mmf* is necessary.
492. Lines of magnetic field cannot intersect.
493. Nickel is a non-magnetic material.
494. Soft iron is suitable for making temporary magnets.
495. All magnetic materials exhibit the phenomenon of saturation.

D. Fill in the Blanks with suitable words

496. In metals the bonding between atom is ....
497. Ionic bonding exists in some ....
498. At very low temperature, semiconducting materials behave as ....
499. The reciprocal of resistance is ....
500. As temperature increases the temperature coefficient of metals ....
501. In carbon resistor code the fourth band indicates ....
502. At a temperature of 7 K lead is a ....
503. Zinc oxide is a .... resistance.
504. Dielectric materials have high ....
505. In solid dielectrics the rate of deterioration .... for every 10 – 12°C rise is temperature.
506. Dielectric constant of a dielectric should be ....
507. The units for permittivity are ....
508. Insulation resistance of a cable in inversely proportional to its ....
509. The capacitance of a parallel plate capacitor is directly proportional to ....
510. The current flowing through a capacitor is called ....
511. For a capacitor $C = (2)(.............)/V^2$.
512. A capacitor should have ....
513. The losses is mica capacitors are ....
514. All other parameters remaining the same a multiplate capacitor has .... capacitance than a capacitor with two plates.
515. The direction of magnetic filed due to a current carrying conductor is given by ....
516. Lines of magnetic field .... each other.
517. Lines of magnetic field cannot ....
518. The field pattern of a .... is similar to that of a bar magnet.
519. Diamagnetic materials exhibit a slight .... to the magnetic flux lines.
520. The material used for making electromagnets has to be ....
521. Soft iron is suitable for making .... magnet.
522. To produce magnetic flux it is necessary to have an ....
523. The ....emission is by modifying the potential energy barrier through an intense electric field.
524. Intrinsic semiconductor doped with .... impurity results is p type semiconductor.

525. Drift velocity of electrons and holes is proportional to ....

526. The number of free .... and .... in a material are equal.

527. Atoms with five valence electrons are called ....

528. p type material has more .... than ....

529. When a $p - n$ junction is reverse biased the width of .... increases.

530. An important characteristic property of a crystal is the .... of the structure.

531. The structure of copper is ....

532. The structure of Sodium and Potassium is ....

533. The arrangement of atom in a material can be studied by ....

534. Glass is an .... material.

535. Elemental dielectrics donot have ....

536. In a dielectric the electric dipole moment per unit volume is called ....

537. As atoms become larger .... increases.

538. The three types of polarization in a polyatomic gas are called ....

539. Out of the three polarizations in a polyatomic gas only .... depends at temperature.

540. A characteristic property of ferro electric material is ....

541. .... materials becomes polarized on the application of a mechanical stress.

542. When subjected to an ac field, the dielectric constant of a dielectric material can be written as a ....

543. For a good dielectric loss angle δ is ....

544. The quantity $(\mu_r - 1)$ is called ....

545. From magnetic point of view silicon and germanium are ....

546. The three contributions to angular momentum of an atom are called ....

547. For a paramagnetic material $\psi =$ ....

548. The magnetic susceptibility is negative for .... materials.

549. The temperature at which a material becomes superconducting is called ....

550. The average drift velocity of an electron per unit field is called ....

ANSWERS WITH HINTS AND EXPLANATION

1. (*a*) Capacitance is directly proportional to permittivity.

2. (*a*) Any material other than insulator will short-circuit the plates.

3. (*b*) See Table 1.2

4. (*b*) Because heater coil must be made of high resistivity material.

5. (*c*) Only copper is non-magnetic.

6. (*c*) **7.** (*d*) **8.** (*b*) **9.** (*b*)

10. (*a*) **11.** (*a*) **12.** (*a*) **13.** (*a*)

14. (*a*) Wood is non-magnetic

15. (*a*) **16.** (*a*) **17.** (*d*)

18. (*b*) Because magnetic flux will decrease.

19. (*d*)

20. (*b*) Because cast iron has lower permeability than silicon steel.

21. (*c*) **22.** (*d*) **23.** (*d*) **24.** (*d*)

25. (*a*) Because charge on proton is positive but numerically equal to that on electron.

26. (*a*) **27.** (*a*)

28. (*b*) Centrifugal force is equal to $\frac{mv^2}{r}$ and coulomb force is equal to $\frac{e^2}{4\pi\epsilon_0 r^2}$. For stability they must be equal.

29. (*c*) 13.6 eV is required to ionize hydrogen atom.

30. (*b*) For stable circular orbit angular momentum is equal to integer times $\frac{h}{2\pi}$.

31. (*a*) **32.** (*c*)

33. (*b*) There are three principal quantum numbers.

34. (*c*)

35. (*a*) The total charge associated with an electron moving about nucleus in the ground state of hydrogen atom is given by $\int_{r=0}^{\infty} 4\pi^2 r\rho(r)\, dr$. This integrand has a maximum value at r = radius of first Bohr orbit.

36. (*b*) The total number of states corresponding to a given value of n is equal to

$$\sum_{l=0}^{l=n-1} (2l+1) = 1+3+...+[2(n-1)+1] = n^2.$$

37. (*b*) The electron shell corresponding to the principal quantum number can contain no more than $2n^2$ electrons.

38. (*d*)

39. (*b*) In true valence crystals the bonding between the atoms is accomplished by the sharing of valence electrons. This sharing can only be between neighbouring atoms only.

40. (*c*) In a metal the valence electrons belong to all atoms.

41. (*c*) ϵ_r is only a number and does not have any units.

42. (*c*) The system is neutral and hence sum is zero.

43. (*b*) It is product of charge and distance.

44. (*b*) Since the field is homogenous $D_i = \frac{D_0}{\epsilon_r}$.

45. (*c*)

46. (*a*) For equilibrium nucleus is shifted in the direction of E by an amount which is proportional to E.

47. (*a*) Polarization $P = \epsilon_0(\epsilon_r - 1)$ E. Therefore units of Polarization $= \frac{F}{m} \times \frac{V}{m} = \frac{C}{m^2}$.

48. (*c*) Since $(\epsilon_r - 1) = 4\pi NR^3$, electronic polarizability $\propto R^3$.

49. (*a*) **50.** (*c*) **51.** (*c*)

52. (*d*) All the three types of polarization exist in polyatomic gases.

53. (*b*) Orientation polarization = $N\mu^2 pE/3kT$.
54. (*a*) 55. (*a*)
56. (*b*) The actual field seen by an atom is larger than applied field.
57. (*a*) In elemental dielectrics only electronic polarization exists.
58. (*b*) Ferroelectric materials have permanent polarization.
59. (*a*) 60. (*b*) 61. (*c*) 62. (*b*)
63. (*b*) 64. (*a*)
65. (*c*) Imaginary part of α_e is zero for $\omega = 0$ and $\omega = \infty$.
66. (*c*)
67. (*b*) Relaxation time increases with viscosity.
68. (*b*) Absorption of energy is proportional to imaginary part of complex dielectric constant.
69. (*b*) If $\in''_r = 0, \delta = 0$.
70. (*a*)
71. (*b*) Dielectric losses = $V^2 \omega C \tan \delta$.
72. (*b*)
73. (*c*) μ_r is only a numeric.
74. (*c*) 75. (*b*) 76. (*c*) 77. (*b*)
78. (*a*) 79. (*b*) 80. (*a*) 81. (*a*)
82. (*b*) Diamagnetic materials have no permanent magnetic diploes.
83. (*a*) In paramagnetic materials interaction between neighbouring dipoles is negligible.
84. (*b*) 85. (*d*)
86. (*a*)Above ferromagnetic curie temperature, ferromagnetic material behaves as a paramagnetic material.
87. (*a*) 88. (*c*) 89. (*b*) 90. (*b*)
91. (*b*)
92. (*a*) As temperature is decreased, resistance of conductors decreases and resistance of semiconductors increases.
93. (*d*) 94. (*d*) 95. (*b*)
96. (*a*) It is another form of Ohm's law.
97. (*c*) 98. (*b*)
99. (*a*) Drift velocity is proportional to field strength.
100. (*b*) 101. (*c*) 102. (*a*) 103. (*b*)
104. (*a*)
105. (*a*) At absolute zero temperature resistivity of metals is zero.
106. (*a*) 107. (*b*) 108. (*a*) 109. (*c*)
110. (*b*) A superconductor has zero resistance.
111. (*c*) 112. (*b*) 113. (*a*) 114. (*c*)
115. (*d*) 116. (*a*) 117. (*a*) 118. (*a*)
119. (*b*) 120. (*a*) 121. (*b*) 122. (*a*)
123. (*b*) Resistance of semiconductors decreases with increase in temperature.
124. (*b*) Semiconductor becomes insulator at very low temperature.
125. (*c*) Many compounds also show semiconducting properties.
126. (*c*) 127. (*a*)
128. (*b*)As temperature increases, conductivity of semiconductors increases.
129. (*d*) 130. (*c*) 131. (*b*) 132. (*a*)
133. (*b*)
134. (*b*) The valence electron, in copper atom, can be easily detached from nucleus.
135. (*a*) 136. (*a*) 137. (*d*)
138. (*b*) Silicon has 14 electrons and 4 valence electrons. The 10 electrons in the inner orbits are tightly bound.
139. (*d*) 140. (*a*) 141. (*a*) 142. (*c*)
143. (*b*) 144. (*a*) 145. (*b*) 146. (*c*)
147. (*b*) Holes are positively charged and move away from positive terminal.
148. (*d*)
149. (*c*) Trivalent means 3 valence electrons.
150. (*a*)
151. (*a*) Pentavalent means 5 valence electrons.
152. (*b*)
153. (*a*) Donor impurity has 5 valence electrons (while silicon has 4 valence electrons). Hence donor impurity can donate electrons.
154. (*c*) 155. (*c*) 156. (*c*) 157. (*c*)
158. (*c*) 159. (*c*) 160. (*d*)
161. (*b*) See Table 1.1.
162. (*b*) See Eq. (1.3).
163. (*d*) 164. (*a*) 165. (*b*) 166. (*d*)
167. (*b*) 168. (*b*) 169. (*b*) 170. (*a*)
171. (*d*) 172. (*c*) 173. (*d*) 174. (*d*)
175. (*c*) The overall temperature coefficient will be less than 0.004 and more than 0.0004.
176. (*b*)
177. (*d*) Current carrying capacity depends on area of cross-section.
178. (*d*) 179. (*b*) 180. (*a*) 181. (*c*)
182. (*d*) $i = \frac{30}{(50-44)}$.
183. (*d*) 184. (*d*) 185. (*c*) See Table 1.5.
186. (*c*) 187. (*d*) 188. (*a*) 189. (*b*)
190. (*a*) 191. (*d*)
192. (*c*) Energy stored is proportional to capacitance and capacitance is proportional to permittivity.
193. (*c*) Modern large size generators have hydrogen cooling.
194. (*d*) Sulphur hexafluoride is used in circuit breakers for arc quenching and insulation.
195. (*a*) See Table 1.5.
196. (*b*) Hysteresis phenomenon occurs when direction of electric field is changed. This can happen only with ac.
197. (*b*) Dielectric strength is expressed in volts (or kV) per unit distance (cm or mm).
198. (*a*) See Table 1.6.
199. (*b*)
200. (*b*) Cotton and paper have higher dielectric strength than air. Moisture decreases dielectric strength.
201. (*b*) If the number of plates is *n*, the number of capacitors is $(n-1)$.
202. (*b*)
203. (*a*) Bakelite has higher permittivity than air.
204. (*d*) The dielectric of a good capacitor should have very high resistivity.

205. (*b*) Capacitance depends on area, distance and $\in_r$ but not on thickness of plates.

206. (*d*) $\in_r$ is only a numeric. Hence no units.

207. (*c*) B and C are in series and A is in parallel with this combination.

208. (*d*)

209. (*c*) Inductance opposes rise and decay of current. Hence it has the property of inertia.

210. (*d*) Inductance is inversely proportional to the core length.

211. (*d*) Inductance is proportional to square of number of turns.

212. (*c*) A high voltage is induced due to mutual inductance because the number of turns in secondary coil is very high.

213. (*a*) For magnetic materials, permeability is not constant.

214. (*d*) Cast iron, sheet steel and iron alloy have very high hysteresis and eddy current losses at high frequencies.

215. (*a*) **216.** (*c*) **217.** (*c*) **218.** (*b*)

219. (*a*)

220. (*d*) Thin film technique ensure low hysteresis and eddy current loss.

221. (*d*) At very low temperatures $a\,T \to 0$. Since resistivity has a finite value even at low temperature, $a\,T < \rho_r$.

222. (*a*) $\in_r$ has no units and equal to 1 for vaccum. For other substances it has different value.

223. (*b*) Braided material is used for shield.

224. (*d*)

225. (*b*) Conductivity is reciprocal of resistivity. Since resistivity of metals is around 10^{-7} ohm-m, conductivity is about 10^7 ohm^{-1} m^{-1}. So 10^5 is the nearest value.

226. (*a*) **227.** (*c*)

228. (*c*) *n* type semiconductors are produced by adding pentavalent material like phosphorous to silicon.

229. (*b*) When N is doubled, resistance becomes twice and inductance becomes four times. Since time constant is $\frac{L}{R}$, time constant becomes twice.

230. (*c*) **231.** (*b*)

232. (*a*) Heat generated at contact should be dissipated quickly and contact should not melt.

233. (*d*) Resistivity of silicon is more than that of germanium.

234. (*d*)

235. (*c*) Capacitance is directly proportional to dielectric constant.

236. (*a*) **237.** (*b*)

238. (*b*) The combination can be represented as L and C in series.

$$X = \frac{1}{\omega C} - \omega L = \frac{1-\omega^2 LC}{\omega C}. \text{ Therefore } C_{app.} = \frac{C}{1-\omega^2 LC}$$

239. (*b*)

240. (*b*) Since the dimension of plates is scaled down by K, the area decreases by a factor of K^2, the distance between plates decreases by a factor of K. Therefore capacitance changes to $\frac{C}{K}$. The selectivity will not be affected.

241. (*d*)

242. (*d*) As the intrinsic carrier concentration is doubled, the concentration of both electrons and holes is doubled.

243. (*b*)

244. (*d*) Both electrons and holes contribute to flow of current in semiconductors.

245. (*a*) Power factor $= \frac{r}{\sqrt{r^2 + \frac{1}{\omega^2 C^2}}} \simeq r(2\pi f C)$

or $r = \frac{\text{Power factor}}{2\pi f C}$

246. (*a*) Since all the statements are correct.

247. (*a*) The bands overlap in conductors. The forbidden gap in small in semiconductors and large in insulators.

248. (*c*) It is the characteristic of a photo conductor that its resistance decreases with increase in illumination.

249. (*b*)

250. (*a*) At high frequencies the inductive reactance is much higher than resistance. The capacitive effect is negligible.

251. (*a*) $C = \frac{\in_0 \in_r A}{d}$. When dimensions are changed, A becomes four times and *d* becomes twice. Since $\in_0$ is constant, $\in_r$ must be halved to keep C constant.

252. (*a*) **253.** (*c*)

254. (*a*) Diffusion length = (Carrier life time × Diffusion constant)$^{0.5}$ = $(100 \times 10^{-6} \times 100)^{05}$ = 0.1 cm.

255. (*b*) For intrinsic semiconductor
Conductivity = electron charge × number of charge carriers × sum of mobilities of holes and electrons.

256. (*c*) **257.** (*c*) **258.** (*c*)

259. (*d*) Inductance is proportional to square of number of turns.

260. (*a*) The new configuration has two capacitors in series and the capacitance of each is 2C.

261. (*b*) Resistance $= \frac{\rho l}{a}, Q = \frac{\omega L}{r}, \text{p.f.} = \frac{r}{Z} \simeq \frac{r}{\omega L}$, inductance = F$n^2 d$.

262. (*c*) Increase in temperature causes resistivity of metals to increase and resistivity of intrinsic semiconductors to decrease. Conductivity is reciprocal of resistivity.

263. (*b*) Resistivity of ferrites is very high as compared to that of iron. Hence eddy currents are much less.

264. (*b*)

265. (*a*) The imaginary part of $\in_r$ gives rise to absorption of energy.

266. (*a*) Yttrium iron garnet has very low eddy current loss. Hence it is used at frequencies in MHz range.

267. (*a*) Because of air core, inductance is constant.

268. (*c*)

269. (*a*) As *f* increases the resistance increases due to skin effect.

270. (*d*)

271. (*c*) Piezo electric materials permit conversion of mechanical force into electrical energy.

272. (*b*)

273. (*a*) $D \propto \epsilon_r E$. The flux density is the same. Since ϵ_r has become 10 times, E decreases by 10 times.

274. (*a*)

275. (*a*) $\tan \delta = \frac{\epsilon_r''}{\epsilon_r'}$.

276. (*d*) Donor impurities have five valence electrons.

277. (*a*) **278.** (*c*)

279. (*a*) Materials suitable for permanent magnet have large hysteresis loop. Hence residual and coercive field are large.

280. (*c*) **281.** (*b*) **282.** (*c*) **283.** (*b*)

284. (*c*) An insulating material should have high dielectric strength and low losses.

285. (*c*)

286. (*b*) The net current is sum of currents due to holes and electrons. In *n* type materials, electrons are majority carriers.

287. (*b*) Alumina is a ceramic material having high resistivity. It can withstand high temperature.

288. (*d*) Ionic crystals are insulating materials.

289. (*d*) Iron, nickel, cobalt are ferromagnetic.

290. (*b*) Copper does not crystallize into diamond structure. Other statements are true.

291. (*a*) In conductors number of free electrons is very high. Their resistivity increases (and hence, conductivity decreases) with increasing temperature. When impurities are added resistivity increases (and hence conductivity decreases).

292. (*b*) Graphite is a conductor.

293. (*c*) In metals the electron concentration is constant.

294. (*c*) Germanium is a well known semiconductor.

295. (*c*) Silicon has 4 valence electrons. Each atom shares its four valence electrons with neighbouring atoms. This bonding is called covalent.

296. (*d*) Antimony, phosphorus and arsenic are pentavalent only boron is trivalent.

297. (*d*) The fuse material must have low melting point so that it easily melts when there is a short-circuit.

298. (*d*) Mangamin has very low temperature coefficient (0.000006). Therefore, it is very suitable for standard resistances.

299. (*c*) Hysteresis loss is proportional to frequency.

300. (*c*) Carbon steel has very high retentivity.

301. (*c*) Susceptibilty is positive. Permanent dipoles are oriented in the same direction.

302. (*a*) In metal temperature coefficient of resistivity is positive.

303. (*b*) **304.** (*b*)

305. (*b*) The change in transition temperature is dependent on magnetic field.

$$H_C = H_0[1 - (T/T_C)^2]$$

where T_C = transition temperature
H_0 = critical field at 0K

306. (*c*) piezoelectric quartz crystals are very commonly used as oscillator. They have good frequency stability.

307. (*c*) Capacitance $= \frac{\in A}{d}$

Area is reduced form 100 cm^2 to 6.25 cm^2. Therefore capactance is also reduced to C/16.

308. (*c*) $\frac{c}{v} = \sqrt{\epsilon_r}$

or $\frac{3\times10^8}{10^8} = \sqrt{\epsilon_r}$ or $\epsilon_r = 9$

309. (*d*) Quality of a dielectric refers to the power loss. Lower the power loss better is quality of dielectric. Thus quality factor is the ratio of maximum stored energy to the average power loss.

310. (*b*) As per Curie law susceptability is inversely proportional to temperature.

311. (*c*) Askerals are very harmful for environment. They decompose and emit toxic gases.

312. (*d*) Dielectric loss $= \frac{E^2 f \epsilon_r \tan\delta}{1.8\times10^{12}}$ W/cm^3

$$= \frac{(50\times10^3)^2(50)(4.2)(0.001)}{1.8\times10^{12}}$$

$= 0.291$ m W/cm^3

313. (*c*) Loss factor $= \frac{\epsilon''}{\epsilon'}$

314. (*b*)

315. (*b*) In FCC structure there is an atom at the centre of each face and at every corner.

316. (*d*) Hysteresis loss is proportional to frequency. Eddy current loss is proprotional to (frequency)^2. It is necessary that core for RF applications have both these losses low.

317. (*d*) In superconductors the variation of critical field H_C with temperature T is given by

$H_C = H_0[1 - T/T_C)^2]$

where H_0 is critical field at 0K and T_C is transition temperature. Above T_C superconducting materials go from superconducting state to normal state.

318. (*c*) Ferrites must have high permeability and low dielectric loss.

319. (*b*) Thermistor is non-linear resistance. Its temperature coefficient is negative. Voltage across it increases with increasing current and then starts decreasing.

320. (*c*) $R_1 = \frac{\rho l}{w.t}$ and $R_2 = \frac{\rho w}{l.t}$

$$\frac{R_1}{R_2} = \frac{l^2}{w^2}$$

321. (*b*) $C = \frac{\epsilon_0 \epsilon_r A}{d} = \frac{8.854\times10^{-12}\times8\times1000\times10^{-6}}{0.5\times10^{-6}}$

$= 1.42 \times 10^{-7}$ F.

322. (*a*) Transformer cores are made of ferromagnetic material (silicon steel) having high resistivity and laminated sheet.

323. (*c*) $R = \frac{\rho l}{a}$ or $\rho = \frac{aR}{l_{-8}} = \frac{0.1 \times 10^{-6} \times 0.171}{1}$
$= 1.71 \times 10^{-8}$ ohm–m

324. (*b*) Piezoelectric materials become polarized when mechanical stress is applied.

325. (*d*)

326. (*a*) Number of electrons $= \frac{\text{Current}}{\text{Charge on electron}}$
$= \frac{4}{1.6 \times 10^{-19}} = 2.5 \times 10^{19}$.

327. (*d*) Silver has highest conductivity Aluminium has higher conductivity than semiconductor. Doping increase conductivity.

328. (*d*)

Material	Susceptability
Superconductor	about – 1
diamagnetic	about – 10^{-6}
paramagnetic	about – 10^{5}
ferromagnetic	very high

329. (*b*) In direct gap material, electron releases energy in the form of light when it jumps from higher energy state to lower energy state. The band gap must have right value corresponding L to wave length of light. In indirect band gap material energy is released in the form of heat.

330. (*c*) Band gap of Ga As = 1.43 eV $= \frac{hc}{\lambda}$

or $\lambda = \frac{hc}{1.43 \times 1.6 \times 10^{-19}} = \frac{6.626 \times 10^{-34} \times 3 \times 10^{8}}{1.43 \times 1.6 \times 10^{-19}}$
$= 8.688 \times 10^{-7}$ m

This wavelength is in infrared region.

331. (*b*) Temperature $= \frac{1.1 - 1}{10^{-3}} = 100°C$

332. (*a*) For BCC iron $a\sqrt{3} = 4r$

For FCC iron $a\sqrt{2} = 4r$

where a = side of cube, r is radius of atom

Hence volume decreases.

333. (*c*) Resistivity of alloy is more than both silver and copper.

334. (*b*) New eddy current loss $= \frac{5\rho - \rho}{5\rho} = 80\%$.

335. (*c*) Heating elements are made of nichrome. Commutator brush is made of carbon and so on.

336. (*c*) Every electromechanical relay has an electromagnet.

337. (*b*) The increase of resistance with temperature is governed by temperature coefficient of resistance.

338. (*b*) Conductivity does not depend on lattice structure.

339. (*b*) Atomic number is not dependent on lattice structure.

340. (*d*)

341. (*b*) Both the statements are correct but not related.

342. (*b*) Both the statements are correct but not related.

343. (*b*) Both the statements are correct but not related.

344. (*a*) Valence bond results in hardness.

345. (*a*) If electrons are tightly bound, conductivity is not there.

346. (*a*) **347.** (*c*)

348. (*a*) The dimensions of ϵ_0 can be derived from the expression $D = \epsilon_0 \epsilon_r E$.

349. (*a*) **350.** (*b*)

351. (*b*) The two statements are correct but independent of each other.

352. (*b*)

353. (*a*) For normal temperatures the dielectric constant depends on the number of atoms per unit volume.

354. (*c*) Orientation polarization is inversely proportional to temperature.

355. (*a*) The intercept on *y*-axis equals N ($\alpha_e + \alpha_i$). This intercept is constant because α_e and α_i are independent of temperature.

356. (*a*) The slope of the line in Fig. 1.13 is $\frac{N\mu_p^2}{3k}$. Since the total polarization is as given by the expression. We can find the permanent dipole moment of the molecules from the slope.

357. (*a*)

358. (*a*) In elemental dielectrics $P_i = P_o = 0$

359. (*a*) In ionic dielectrics $\mu_p = \Sigma_i e_i r_i$ and $P_o = 0$. Therefore $P = P_e + P_i$.

360. (*a*) **361.** (*a*)

362. (*c*) R is false.

363. (*b*) Both A and R are correct. Spontaneous polarization vanishes above ferroelectric curie temperature.

364. (*b*) Both the materials exhibit ferroelectric properties. Hysteresis effect is one of the features of ferroelectric materials.

365. (*c*) R is not correct.

366. (*b*) Both A and R are independent statements.

367. (*a*) Imaginary part of ϵ_r gives rise to absorption of energy by the material from alternating field. Hence current does not lead the voltage by 90°.

368. (*a*)

369. (*a*) Imaginary part of dielectric constant is responsible for dielectric losses.

370. (*a*) If dielectric losses are zero, the current in capacitor leads the applied voltage by 90°. Also imaginary part of ε_r is zero.

371. (*a*) **372.** (*a*)

373. (*c*) Increase of B with I is not linear. Hence R is false.

374. (*c*) μ_r is always more than I. It cannot be 0 or less than I.

375. (*c*) Orbital angular momentum and magnetic dipole moment are not equal.

376. (*b*) Both A and R are independent.

377. (*c*) Orbital magnetic moment is equal to ($-e/2m$).

378. (*a*) For diamagnetic materials $\mu_r = 1$.

379. (*c*) Nuclear spin is of the same order as electron spin.

380. (*a*) Magnetic susceptibility = ($\mu_r - 1$) H. For diamagnetic and paramagnetic materials magnetic susceptibility is very small and $\mu_r = 1$.

381. (*a*)
382. (*b*) Both A and R are independent but correct.
383. (*a*) Both A and R are correct but independent.
384. (*a*) If magnetic susceptibility is very low μ_r is nearly equal to 1 and the material is diamagnetic.
385. (*a*) In conductors constant E means constant current because conductivity is constant.
386. (*b*) Both the statements are correct but independent.
387. (*b*) 388. (*b*) 389. (*b*) 390. (*b*)
391. (*a*) Heat developed depends on power loss.
392. (*b*) 393. (*a*) If $E = E_F$, $f(E) = 0.5$.
394. (*a*)
395. (*a*) If different values are substituted in the equation for I_{th}, the value of emission current comes out to L be about 0.1 A/cm^2.
396. (*a*)
397. (*b*) Both the statements are correct but independent.
398. (*d*) 399. (*a*) 400. (*a*)
401. (*c*) Cut in voltage of silicon is higher than that in germanium.
402. (*a*) Higher PIV for silicon is a definite advantage.
403. (*a*) 404. (*b*) 405. (*a*)
406. (T) Orbit is stable only if the two forces are equal and apposite.
407. (T) It is a quantum condition postulated by Bohr.
408. (T)
409. (T) Angular quantum number determines the angular momentum of electron.
410. (T)
411. (F) The state with $l = 0$ is called *s* state.
412. (F) Chemical properties are governed by valence electrons only.
413. (T) Sharing of valence electrons is a characteristic of true valence crystal.
414. (T) In a metal valence electrons are shared by all atoms.
415. (T) This is the reason for chemical inactivity in these crystals.
416. (F) 417. (T) 418. (T) 419. (T)
420. (T) 421. (T) 422. (T) 423. (T)
424. (T) 425. (T)
426. (F) The electric field intensity inside and outside the cavity are equal.
427. (T) 428. (T)
429. (F) The dielectric constant changes as it is proportional to the number of atoms per unit volume.
430. (T) 431. (T) 432. (T) 433. (T)
434. (T) 435. (T) 436. (T)
437. (F) Hysteresis phenomenon exists in dialectic materials.
438. (T)
439. (F) Spontaneous polarization vanishes above ferroelectric curie temperature.
440. (T) 441. (T) 442. (T) 443. (T)
444. (T) 445. (T) 446. (T) 447. (T)
448. (T) 449. (T)
450. (F) Dielectric losses vary with frequency.
451. (T) 452. (T) 453. (T) 454. (T)
455. (T)
456. (F) $|\mu_m| = 0.5\, e\, \omega\, r^2$
457. (T) 458. (T) 459. (T) 460. (T)
461. (T)
462. (F) Anti-ferromagnetic materials have permanent dispoles but anti-parallel orientation of equal moments.
463. (T) 464. (T)
465. (F) Diamond is diamagnetic.
466. (T) 467. (T) 468. (T)
469. (F) Susceptibility $\propto \dfrac{1}{T}$.
470. (T)
471. (F) Hysteresis phenomenon exists below ferromagnetic curie temperature.
472. (T) 473. (T) 474. (T) 475. (T)
476. (T) 477. (T) 478. (T) 479. (T)
480. (T) 481. (T)
482. (F) In semiconductors the temperature coefficient of resistivity is negative.
483. (T) 484. (T) 485. (T)
486. (F) The force is attractive.
487. (F) 488. (T)
489. (T) Lines of magnetic field repel each other
490. (T) 491. (T) 492. (T)
493. (F) Nickel is ferromagnetic material.
494. (T) 495. (T)
496. metallic 497. insulator
498. insulator 499. conductance
500. decreases 501. tolerance
502. superconductor 503. non-linear
504. specific resistance 505. doubles
506. high 507. F/m
508. length 509. permittivity
510. displacement current 511. Energy
512. low losses 513. low
514. higher 515. right hand rule
516. repel 517. intersect
518. solenoid 519. opposition
520. ferromagnetic 521. temporary
522. mmf 523. field
524. trivalent 525. electric field strength
526. electrons, holes 527. pentavalent
528. holes, free electrons 529. depletion layer
530. periodicity 531. face centred cubic
532. body centred cubic 533. X-ray
534. amorphous 535. electric dipolemoment
536. polarization 537. polarizability
538. orientation polarization, electronic polarization, ionic polarization
539. orientation polarization
540. spontaneous polarization
541. piezo electric 542. complex quantity
543. very small 544. magnetic susceptibility
545. diamagnetic materials
546. orbital, electron spin and nuclear spain
547. C/T 548. diamagnetic material
549. temperature 550. mobility.

PHYSICAL ELECTRONICS, ELECTRONIC DEVICES AND ICs

CHAPTER 2

2.1. ELECTRONIC EMISSION IN METALS

2.1.1. Energy Levels

An atom consists of a positively charged nucleus which contains all its mass. Surrounding this nucleus are negatively charged electrons moving about in closed orbits. Each orbit is characterized by a discrete energy level. When electrons are in states corresponding to these discrete energies they do not emit radiations and are in stationary or non-radiating states. A stationary state is determined by the condition that the angular momentum of electon in this state is an integral multiple of $\frac{h}{2\pi}$ where h is Planck's constant (= 6.63×10^{-34} Js).

Thus $$mvr = n\frac{h}{2\pi} \quad ...(2.1)$$

If we substitute $n = 1, 2, 3,...$ in Eq. (2.1) we get stable orbits with redii $r_1, r_2, r_3,...$

The lower the energy level, the more tightly bound is the electron. In transition from one state (corresponding to energy W_2) to another stationary state (corresponding to energy W_1) radiation is emitted. The frequency of this radiant energy is

$$f = \frac{(W_2 - W_1)}{h} \quad ...(2.2)$$

The distribution of electrons over a range of allowed energy levels at thermal equilibrium in given by the following equation known as Fermi-Dirac distribution function, *i.e.*,

$$f(E) = \frac{1}{1 + e^{(E - E_F)/kT}} \quad ...(2.3)$$

In Eq. (2.3) E_F is Fermi level in eV, k is Boltzmann's constant (= 0.8625×10^{-4} eV/K), T is temperature (Kelvin) and E is available energy state (eV).

Fermi-Dirac distribution function is shown in Fig. 2.1.

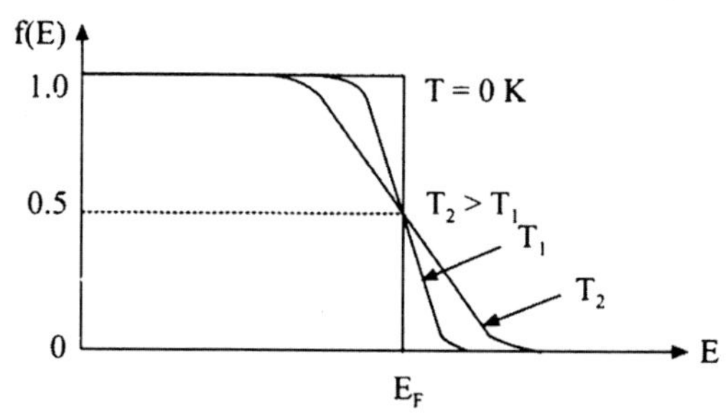

Fig. 2.1. Fermi-Dirac distribution function.

2.1.2. Thermionic Emission

At sufficiently high temperatures some electrons acquire enough kinetic energy to escape from metal surface. This is called thermionic emission. The current due to thermionic emission is given by Richardson–Dushman equation, *i.e.*,

$$I_{th} = SA_0\, T^2\, e^{-E_w/kT} \quad ...(2.4)$$

where I_{th} is current due to thermionic emission (A), S is surface area of cathode (m^2), A_0 is constant having dimension $A/m^2/K^2$, T is absolute temperature (K), E_w is work function eV and k is Boltzmann's constant (eV/K)

2.1.3. Secondary Emission

When a moving particle (say electron) strikes a solid surface, it may impart sufficient energy to the electron in the solid to escape from the surface of solid. This is called secondary emission.

2.1.4. Field Emission

When a very high electric field is created at metal surface, electrons may be emitted from the surface.

The current density due to field emission in given by

$$J = A_f\, E^2\, e^{-(b_f/E)} \quad ...(2.5)$$

where J = density of field emission current, b_f is a constant which depends on work function of the metal of cathode.

2.1.5. Photoelectric Emission

When a proton of light impinges on a metallic surface, it may transfer enough energy to the electrons at the surface of the metal to escape from the surface. This is called photoelectric emission. As per Einstein's equation, the velocity of emitted electron in given by

$$\frac{1}{2}\, mv^2 \le hf - U_w$$

where f is frequency of light falling on the metallic surface having work function U_w joules (corresponding to E_w electron volts).

2.2. SEMICONDUCTOR PHYSICS

2.2.1. Energy Bands in Solids

Every solid has a characteristic band structure. The wide range of electrical properties are due to these different band structures.

In a conductor the conduction and valence bands overlap resulting in low resistively (of the order of 10^{-7} Ω–m). In insulators a large forbidden gap of several electron volts exists between conduction band and valence bond resulting in high resistivity (of the order of 10^{10} to 10^{16} Ω-m). In semiconductors the forbidden gap between valence and conduction bands is small (about 1 eV). Germanium and silicon are the most important semiconductors.

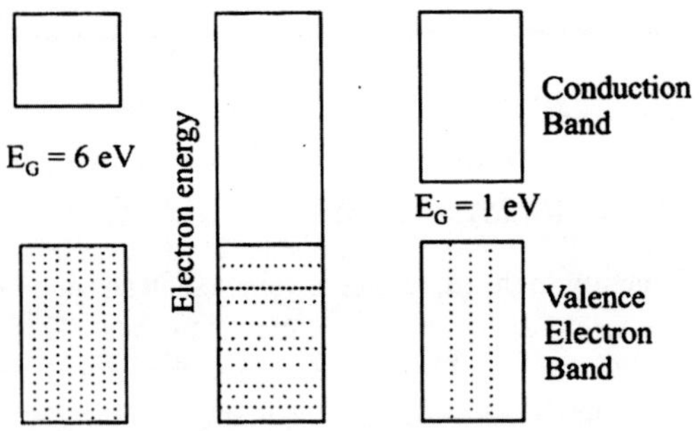

Fig. 2.2. Energy bands in solids.

2.2.2. Fermi Level

The distribution of free electrons in a solid can be written as

$$dn_e = P_e dE \qquad ...(2.7)$$

where dn_e is the number of free electrons per m^3 whose energies lie between E and E + dE (*i.e.*, in energy interval dE) and P_e is the density of electrons in this interval. P_e can be written as

$$P_e = f(E)\,N(E) \qquad ...(2.8)$$

where N(E) is the density of states (*i.e.*, number of states per eV per m^3) in the conduction band and f(E) is the probability that a quantum state with energy E is occupied by electron. f(E) is given by Eq. (2.3). For most metals E_F is less than 10 eV. For tungsten E_F = 8.95 eV. Moreover,

$$N(E) = \gamma E^{0.5} \qquad ...(2.9)$$

where $\gamma = \frac{4\pi}{h^3}(2\,m)^{1.5}\,(1.6 \times 10^{-19})^{1.5} = 6.82 \times 10^{27}$

2.2.3. Instrinsic Semiconductors

Germanium atom has 32 electrons out of which 28 are tightly bound to the nucleus and 4 are valence electrons. Silicon has 14 electrons out of which 10 are tightly bound and 4 are valence electrons.

Thus both germanium and silicon have 4 valence electrons. At 0K there no free electrons and hence no conductivity. At higher temperatures some valence electrons get thermally excited and break the covalent bond. The energy required to break the covalent bond is 0.72 eV (at 300 K) for germanium and 1.1 eV (at 300 K) for silicon. As an electron breaks the covalent bond and leaves the vicinity of an aton, it leaves a positive charge behind. This positive charge in called 'hole'. The total current has two components, one due to electrons and the other due to holes. Current density J is given by

$$J = (n\mu_n + p\mu_p)\,eE = \sigma E \qquad ...(2.10)$$

where n is magnitude of free electron concentration, p is magnitude of hole concentration, σ is conductivity, E is electric field, μ_p is mobility of hole and μ_n is mobility of electron. For intrinsic semiconductors $n = p = n_i$. Therefore

$$J = (\mu_n + \mu_p)\,en_iE \qquad ...(2.11)$$

The intrinsic charge concentration n_i varies with absolute temperature T as per the equation

$$n_i^2 = A_0\,T^3 e^{-E_{Go}/kT} \qquad ...(2.12)$$

where E_{Go} is the forbidden energy gap in eV at 0K.

The conductivity of germanium increases by about 6% per degree increase in temperature. In silicon this increase is about 8% per degree increase in temperature. Some constants for Germanium and Silicon are given in Table 2.1.

TABLE 2.1. Properties of Germanium and Silicon

Property	*Germanium*	*Silicon*
Atomic weight	72.6	2.8.1
Atomic number	32	24
Density (g/cm^3)	5.32	2.33
Relative permittivity	16	11.9
Number of atoms/cm^3	4.4×10^{22}	5×10^{22}
E_{GO} at 0 K, eV	0.785	1.21
E_G at 300 K, eV	0.72	1.12
n_i at 300 K, cm^{-3}	2.5×10^{13}	1.45×10^{10}
Intrinsic resistivity at 300 K	45 Ωcm	230000 Ωcm
μ_n at 300 K	3800 cm^2/V-sec	1500 cm^2/V-sec.
μ_p at 300 K	1800 cm^2/V-sec	475 cm^2/V-sec.

2.2.4. Extrinsic Semiconductor, Donor and Acceptor Impurities

A semiconductor to which an impurity is added is called extrinsic semiconductor and the process of adding impurity is called doping. Both silicon and germanium are tetravalent. To increase conductivity either a pentavalent (*i.e.*, one having 5 valence electrons) or trivalent (*i.e.*, one having 3 valent electrons) is added. A pentavalent impurity is called domer or n type impurity. A trivalent impurity is called acciptor or p type impurity. Examples of pentavalent impurity are antimony, arsenic etc. On the other hand, indium, boron, gallium, aluminium are trivalent impurities. The extent of doping is very small say 1 part in 10^8 parts. However the increase in conductivity by addition of minute impurities is about 20 times.

In p type semiconductor holes are majority carries and electrons are minority carriers. In n type semiconductor electrons are majority carries and holes are minority carriers.

Diffusion hole current density J_p is

$$J_p = -e\,D_p \frac{dp}{dx} \qquad ...(2.13)$$

Diffusion electron current density J_n is

$$J_n = e\,D_n \frac{dn}{dx} \qquad ...(2.14)$$

D_p is diffusion constant for holes (and has units of m^2/s). D_n is diffusion constant for electron (units m^2/s) Diffusion constants D_p, D_n and mobility μ_p, μ_n are related as

$$\frac{D_p}{\mu_p} = \frac{D_n}{\mu_n} = V_T \qquad \text{...(2.15)}$$

and $$V_T - \frac{kT}{e} = \frac{T}{11600} \qquad \text{...(2.16)}$$

At $$T = 300, \; V_T = \frac{300}{11600}$$

and $$\mu = 38.67 \, D$$

2.2.5. Hall Effect

Hall voltage $$E_H = Bvt = \frac{BJt}{\rho} = \frac{BI}{\rho d} \qquad \text{...(2.17)}$$

Hall coefficient $$K_H = \frac{1}{\rho} = \frac{dE_H}{BI} \qquad \text{...(2.18)}$$

2.3. P-N JUNCTION DIODE

When a P-N junction is formed, a depletion layer or space charge region or transition region is created around the junction (Fig. 2.3). The equilibrium potential across the deplection layer is called barrier potential. Its value is 0.3 V for germanium diode and 0.7 V for silicon diode at 25°C.

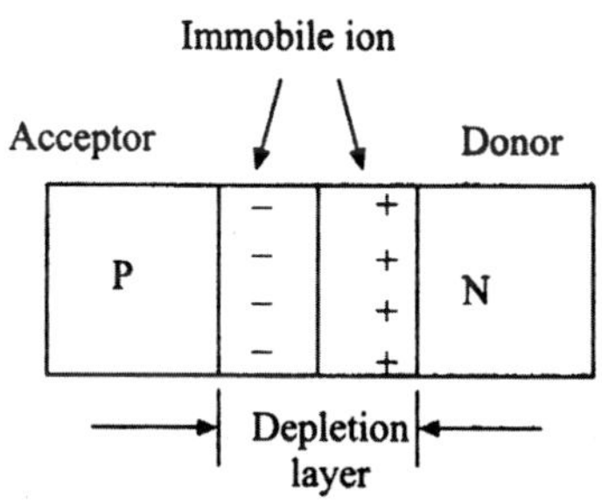

Fig. 2.3. P-N junction and Deplection layer

When a forward bias V is applied (Fig. 2.4*a*) the width of depletion layer is decreased. The current is given by

$$i = I_0 \, (\in^{V/\eta V_T} - 1) \qquad \text{...(2.19)}$$

where I_0 is saturation current. V is applied voltage, $V_T = \frac{T}{11600}$, T is absolute temperature, $\eta = 1$ for germanium and 2 for silicon for low levels of current and 1 for silicon for higher levels of current.

When a reverse bias is applied, the width of duplication layer increases (Fig. 2.4*b*) and current is very small.

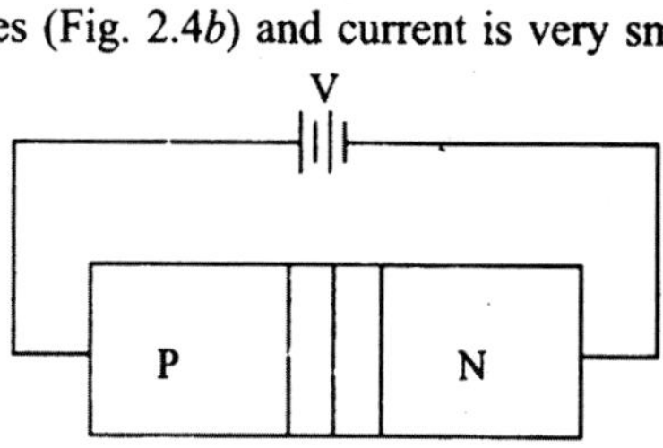

(*a*)

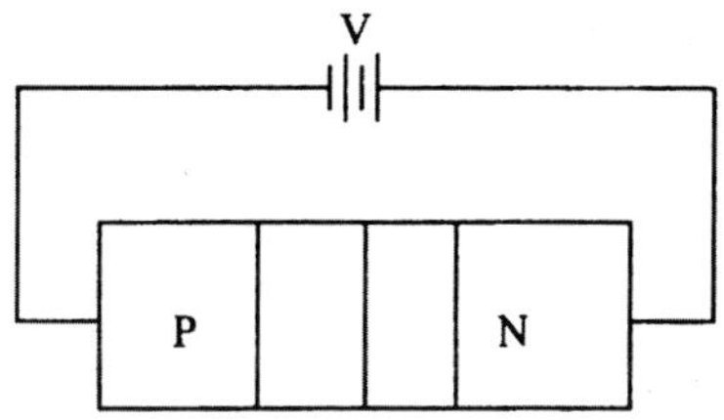

(*b*)

Fig. 2.4. P-N junction with (*a*) forward bias (*b*) reverse bias.

Figure 2.5 shows the *v-i* characteristics of semiconductor diode. The ratio *v*/*i* is called static resistance of diode. In forward direction the static resistance is only a few ohms. In reverse direction the static resistance is very high (a few mega ohms). The ratio $\frac{dv}{di}$ is called dynamic resistance.

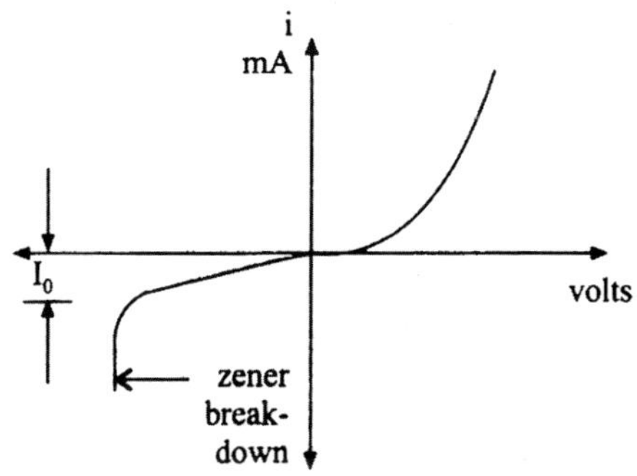

Fig. 2.5. *v-i* characteristics of P-N junction diode.

$$\text{Dynamic resistance } \frac{dv}{di} = r_e = \frac{\eta V_T}{i} \qquad \text{...(2.20)}$$

When reverse bias is applied, the increase in depletion layer can be considered as a capacitive effect. Incremental capacitance is

$$C_T = \left|\frac{dQ}{dV}\right| = \frac{\in A}{W} \qquad \text{...(2.21)}$$

where A is area of P-N junction, W is thickness of depletion layer and $\in$ is permittivity of semiconductor material. Typical value of C_T is 20 *p*F for a reverse voltage of 4 V.

2.4. BIPOLAR JUNCTION TRANSISTOR (BJT)

A BJT consists of three doped elements. It can be P-N-P or N-P-N. These are shown in Figs. 2.6 and 2.7.

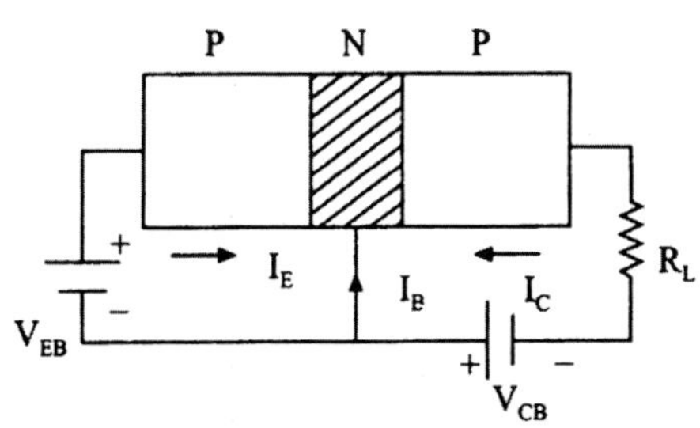

(*a*)

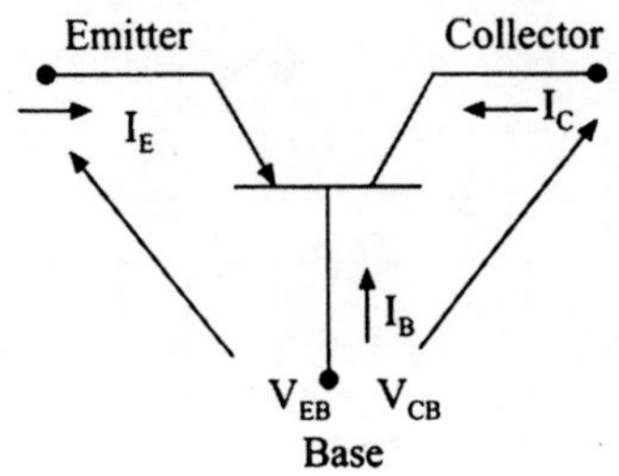

Fig. 2.6. P–N–P junction transistor
(*a*) Circuit, (*b*) Schematic representation (common base).

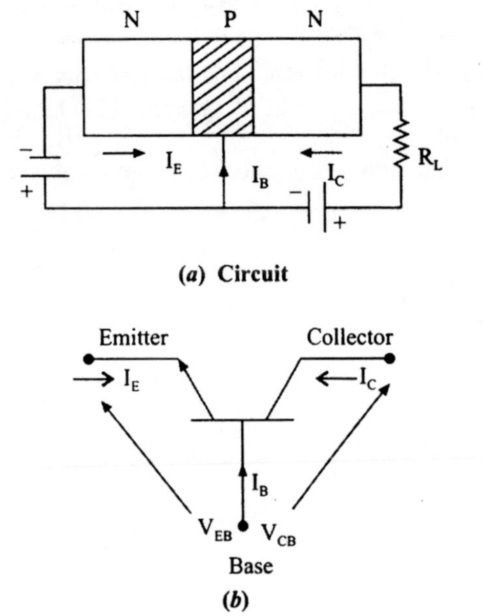

(*a*) Circuit

(*b*) Schematic represention (common base).
Fig. 2.7. N-P-N junction transistor.

Figures 2.6 and 2.7 show common base connection. However CE connection is more convenient. In CE connection, the emitter is common to the input and output circuits. CE connection is shown in Fig. 2.8 for N-P-N transistor.

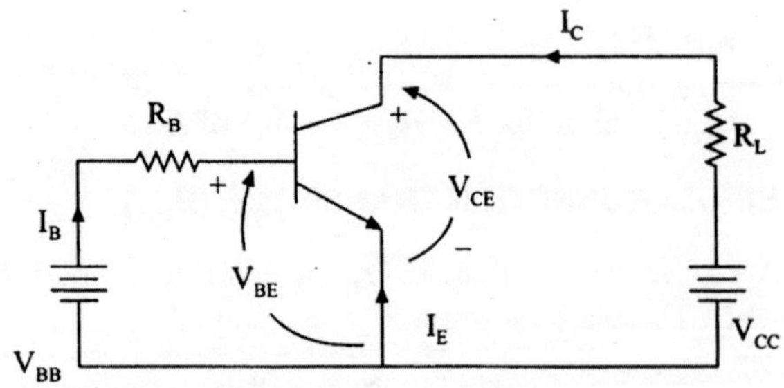

Fig. 2.8. CE connection of N-P-N transistor.

When used as an amplifier a transistor operates in active region. In this operation emitter junction is forward based and collector junction is reverse biased. I_B, I_C and I_E are related as

$$I_B = -(I_C + I_E) \qquad ...(2.22)$$

If reverse saturation current I_{C0} in neglected.

$$I_C = -\alpha I_E \qquad ...(2.23)$$

From Equations (2.22) and (2.23)

$$I_C = \frac{\alpha I_B}{1-\alpha} \qquad ...(2.24)$$

Transistor α is about 0.98. Therefore I_C is very large as compared to I_B. Depending on the bias the modes of operation of a BJT are given in Table 2.2.

TABLE 2.2. Modes of operation in BJT

Mode	***Junction bias***	
	Emitter-Base	***Base Collector***
Active (forward)	Forward	Reverse
Cut off	Reverse	Reverse
Saturation	Forward	Forward
Reverse active	Reverse	Forward

Figure 2.9 shows Ebers-Moll model for P-N-P transistor

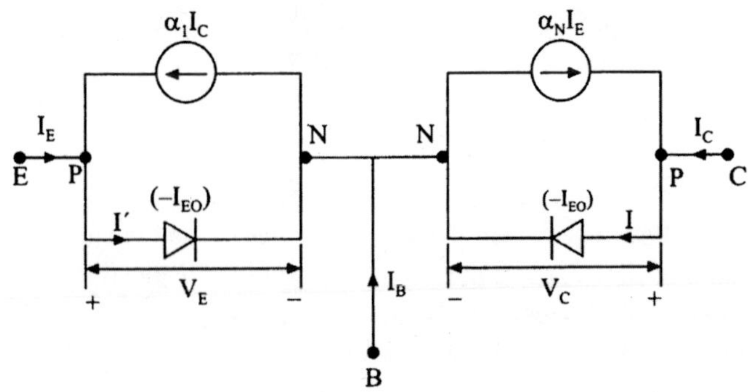

Fig 2.9. Ebers-Moll model for P-N-P transistor.

This model has two ideal diodes placed back to back with reverse saturation currents $-I_{Eo}$ and $-I_{Co}$. Two dependent current controlled sources shunt the ideal diodes. From Fig. 2.9.

$$I_C = -\alpha_N I_E + I = -\alpha_N I_E - I_{Co}\,(e^{V_C/V_T} - 1) \qquad ...(2.25)$$

$$I_E = -\alpha_I I_C - I_{Eo}\,(e^{V_E/V_T} - 1) \qquad ...(2.26)$$

2.4.1. *h* Parameter Equivalent Circuit

The h parameters of a transistor are the easiest to measure. Therefore a transistor is represented by h parameter equivalent circuit. The subscripts used are

$i = 11$ = input, $0 = 22$ = output

$f = 21$ = forward transfer, $r = 12$ = reverse transfer

Another subscript (b or e or c) in added to denote common base or common emitter or common collector configuration. Thus

$h_{ie} = h_{11e}$ = input impedance in CE connection.

$h_{fb} = h_{21b}$ = short-circuit forward current gain in common base connection.

Figure 2.10 shows CE amplifier connection and equivalent circuit. The terminal equations for this circuit are

$$V_{be} = h_{ie}\, i_b + h_{re}\, v_{ee} \qquad ...(2.27)$$

$$i_c = h_{fe}\, i_b + h_{oe}\, v_{ce} \qquad ...(2.28)$$

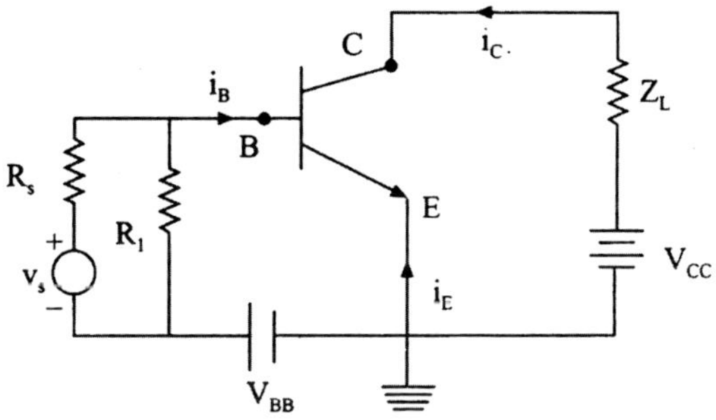

(a) **Circuit**

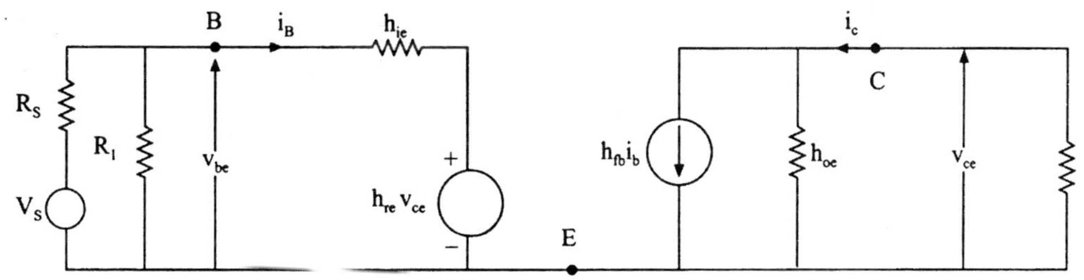

(b)

Fig. 2.10. CE Amplifier *(a)* Circuit *(b)* Equivalent circuit

Figure 2.11 shows CB circuit and equivalent circuit. The terminal equation are

$$v_{eb} = h_{ib}\, i_e + h_{rb}\, v_{cb} \quad ...(2.29)$$

$$i_c = h_{fb}\, i_e + h_{ob}\, v_{cb} \quad ...(2.30)$$

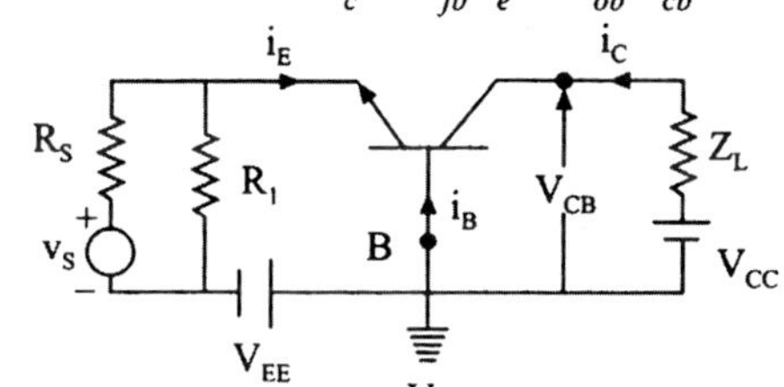

(a)

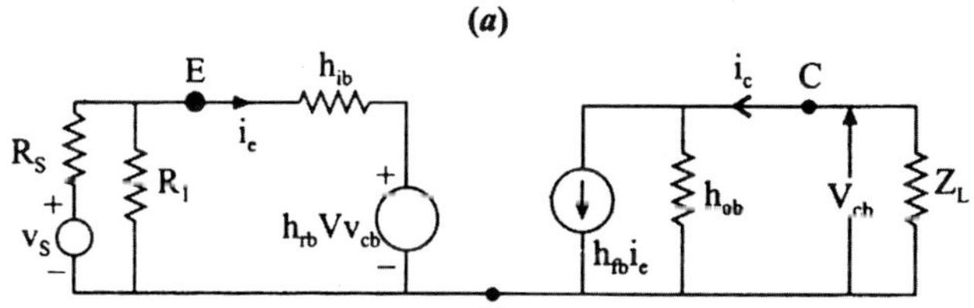

(b)

Fig. 2.11. CB amplifier *(a)* circuit, *(b)* Equivalent circuit

Figure 2.12 shows the circuit and *h* parameter equivalent circuit for common collector configuration. The terminal equation are

$$v_{bc} = h_{ic}\, i_b + h_{rc}\, v_{ec} \quad ...(2.31)$$

$$i_e = h_{fc}\, i_b + h_{oc}\, v_{ec} \quad ...(2.32)$$

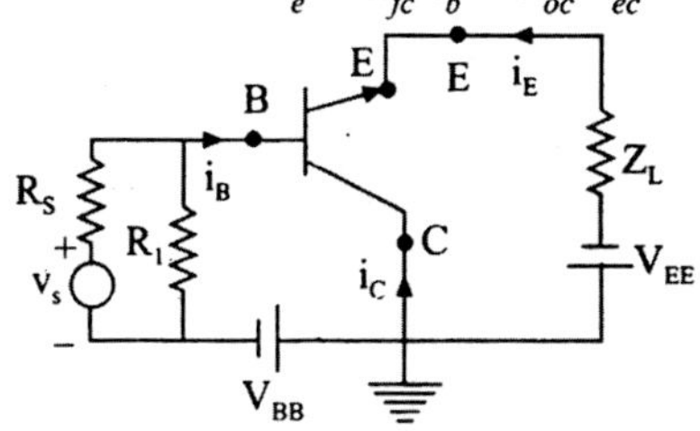

(a)

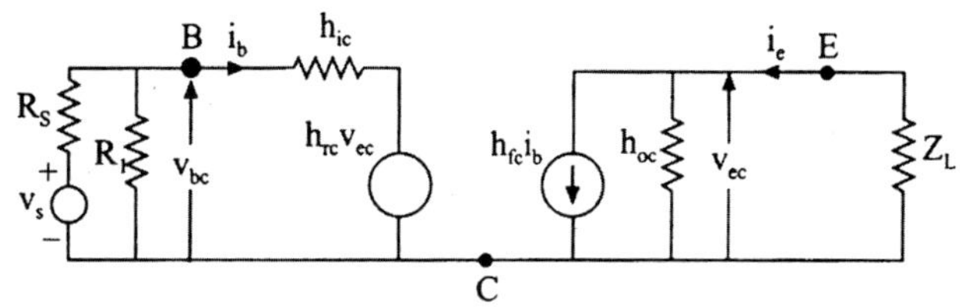

(b)

Fig. 2.12. CC amplifier circuit
***(a)* circuit, *(b)* equivalent circuit**

TABLE 2.3. Comparison of CE, CB, CC circuits

	CE	*CB*	*CC*
Current gain	good	less than 1	high
Voltage gain	good	high	less than 1
Power gain	high	good	good
Input Impedance	good	low	high
Output Impedance	good	high	low
Phase Shift	180° for both current and voltage	nil	nil
Application	Power Amplification	Radio frequency Isolation	Impedance matching

2.5. FIELD EFFECT TRANSISTOR (FET)

FET is a voltage controlled semiconductor device. Its main features are

1. It is a unipolar device because its operation depends on the flow of majority carriers.
2. Its input resistance is very high (about a few mega ohms)

3. It has thermal stability and is less noisy
4. It is well suited for controlled switching between a conducting and non-conducing states. It is very useful in digital circuits.

The main disadvantage of FET is relatively small gain bandwidth product as compared to that in BJT.

FET can be junction field effect transistor (JFET) or Metal oxide semiconductor FET (MOSFET). MOSFET is also called insulated gate FET.

2.5.1. JFET

Figures 2.13 and 2.14 show symbols and circuits of *n* channel JFET and *p* channel JFET.

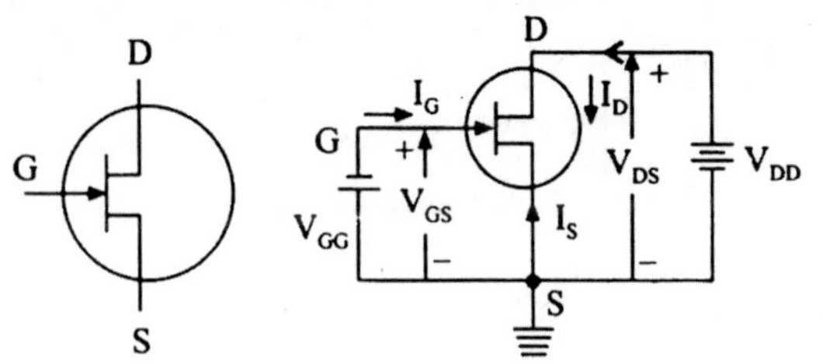

(*a*) (*b*)

Fig. 2.13. *n* channel JFET (*a*) symbol, (*b*) circuit

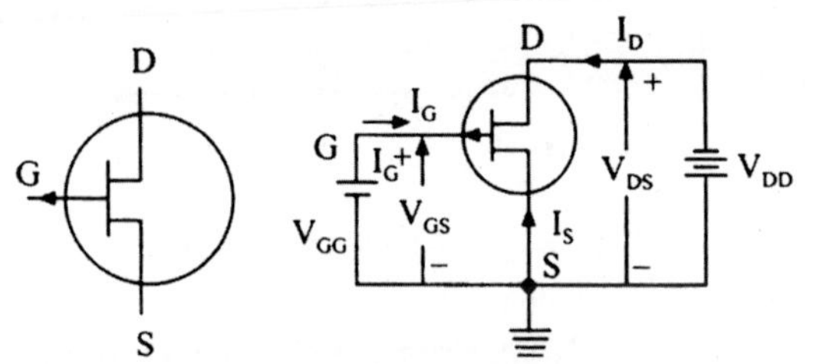

(*a*) (*b*)

Fig. 2.14. *p* channel JFET (*a*) Symbol, (*b*) Circuit

The saturation current I_{DS} can be written as

$$I_{DS} = I_{DSS}\left[1 - \frac{V_{GS}}{V_{GS(off)}}\right]^2 \qquad ...(2.33)$$

where $V_{GS(off)} = -V_p$

JFET equivalent current is shown in Fig. 2.15.

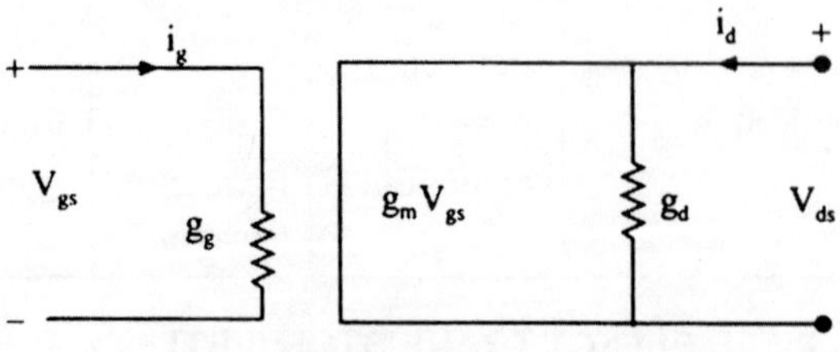

Fig 2.15. JFET equivalent circuit

$$g_m = g_{mo}\left(1 - \frac{V_{GS}}{-V_p}\right) \qquad ...(2.34)$$

where $g_{mo} = \dfrac{2I_{DSS}}{V_p}$.

2.5.2. MOSFET

MOSFET has extremely high input impedance. It can be depletion type or enhancement type. Figure 2.16 shows symbols for depletion type MOSFETs.

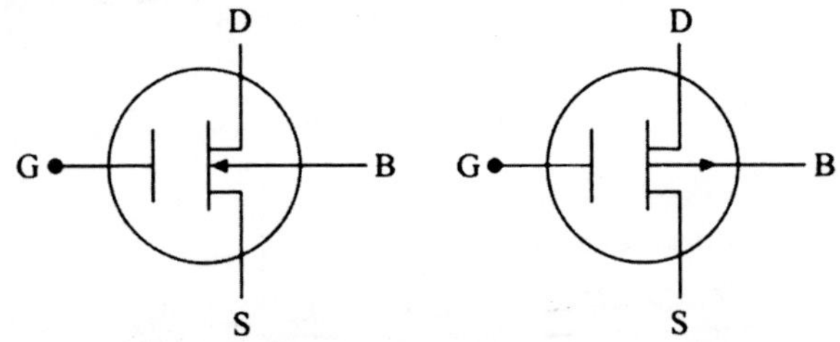

(*a*) *n* channel (*b*) *p* channel

Fig. 2.16. Symbol for Depletion type MOSFET

Figure 2.17 shows symbols for enhancement type MOSFET.

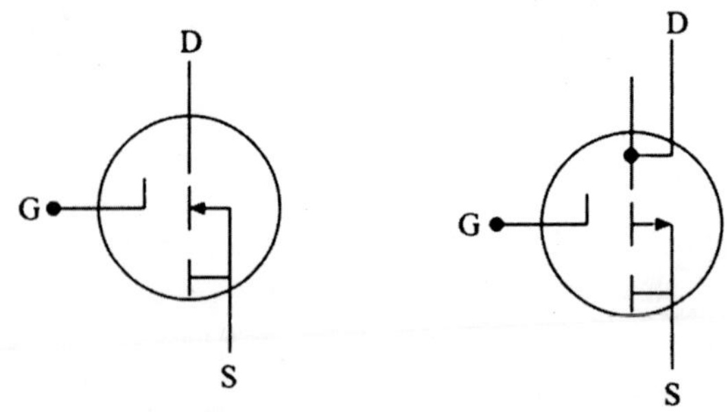

(*a*) *n* channel (*b*) *p* channel

Fig 2.17. Symbol for enhancement type MOSFET

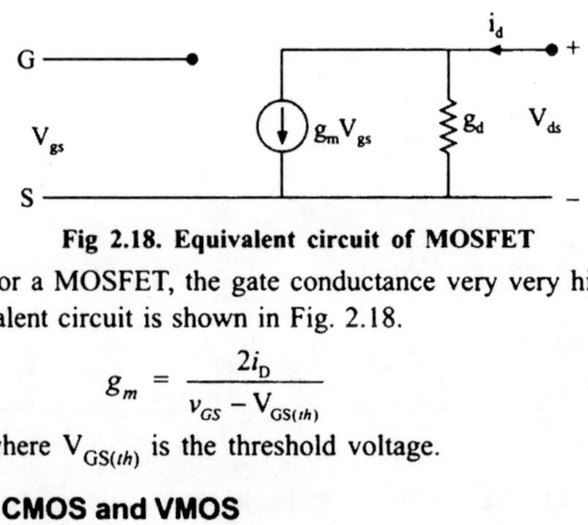

Fig 2.18. Equivalent circuit of MOSFET

For a MOSFET, the gate conductance very very high. Its equivalent circuit is shown in Fig. 2.18.

$$g_m = \frac{2i_D}{v_{GS} - V_{GS(th)}} \qquad ...(2.35)$$

where $V_{GS(th)}$ is the threshold voltage.

2.5.3 CMOS and VMOS

A CMOS (complementary MOSEFET) circuit has *p* channel and *n* channel devices connected as shown in Fig. 2.19. When V_{in} is low, device 1 is on but device 2 is off giving high V_0. When V_{in} is high, device 1 is off but 2 is on giving low V_0. The phase of V_0 is always opposite to that of V_{in}. Thus it is an inverter. It has very low power consumption and is widely used in wrist watches and pocket calculators.

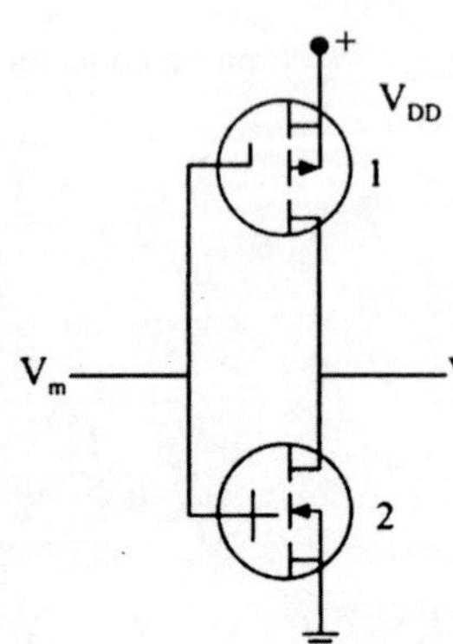

Fig 2.19. CMOS

A VMOS (vertical MOSFET) has a conducting channel in the shape of a V groove. This device can handle large currents and is used in circuits requiring high load power, *i.e.*, audio amplifiers, RF amplifiers etc.

2.6. ZENER DIODE

It is a special diode built to operate in breakdown region. Its *v-i* characteristic and symbol are shown in Fig. 2.20. It is used in voltage requlator circuits. It is also called avalanche diode.

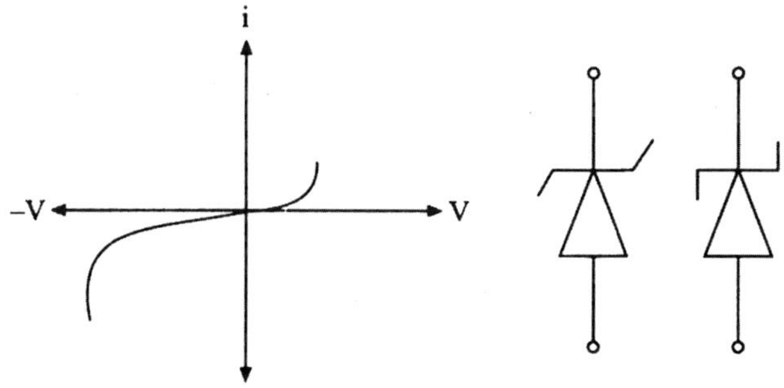

(a) v-i characteristics ***(b) symbols***

Fig. 2.20. Zener diode.

2.7. VARACTOR DIODE

When a diode is reverse biased, the incremental capacitance is defined as

$$C_T = \left|\frac{dQ}{dv}\right| \quad ...(2.36)$$

The voltage varible capacitance C_T of a reverse biased diode is used in tuning of LC circuits. The diode made for this application is known as varactor diode. Its symbol and equivalent circuit are shown in Fig. 2.21. R_S is ohmic resistance of diode. The reverse biased resistance R_r is some time neglected becuse it is very high

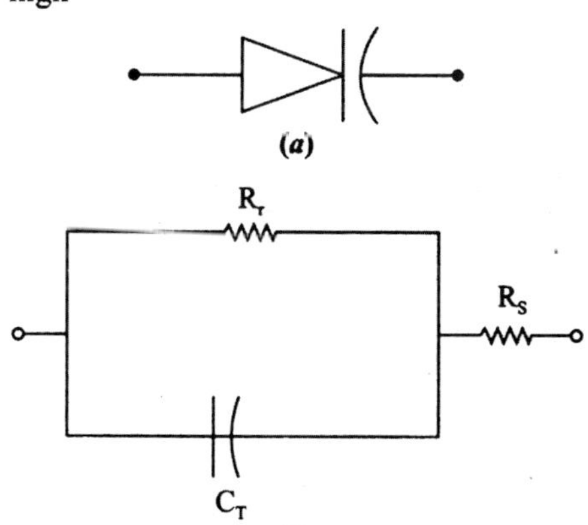

(a) ***(b)***

Fig. 2.21. Varactor diode (*a*) symbol, (*b*) equivalent circuit.

2.8. TUNNEL DIODE

It is a diode formed by highly doped *p* and *n* materials. The concerntration of impurities is very large (say I part in 10^3 as compared to 1 part in 10^8 in diodes). Moreover the width of junction barrier is very small. Figure 2.22 shows its *v-i* characteristics and symbol. It is used in pulse and digitial circuits. Its characteristics resemble letter 'N'.

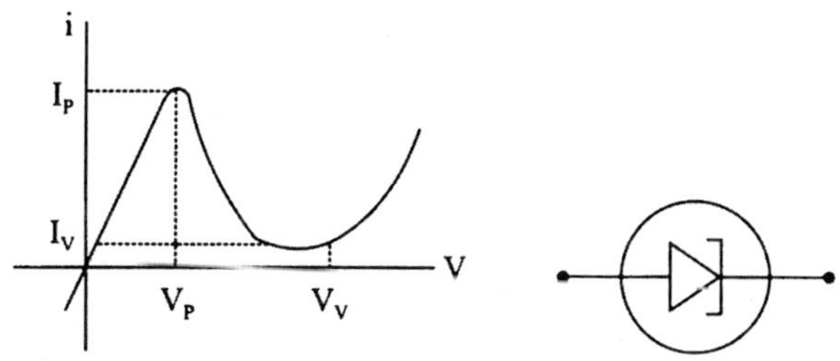

(a) ***(b)***

Fig. 2.22. Tunnel diode (*a*) *v-i* characteristics, (*b*) symbol.

2.9. PIN DIODE

It has P and N regions with an intrinsic (*i*) region between them (Fig. 2.23). The effective width of depletion layer increases by the width of *i* layer. Its capacitance is lower than that of varactor diode (for the same reverse bias). Its impedance varies with the applied bias. It is used at microwave frequencies.

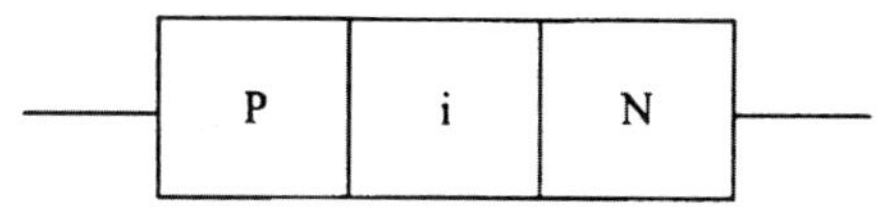

Fig. 2.23. PIN Diode.

2.10. PHOTO DIODE AND SOLAR CELL

It is reverse biased P-N Junction. The junction is exposed to radiant energy source through a window in the capsule. Thus it responds to photon absorption. It can be operated in third or fourth quadrant. When operated in third quadrant *v* and *i* are both negative. Thus the external source delivers power to the device. However in fouth quadrant, *v* is positive but *i* is negative. Thus the device delivers power to the external circuit. This operation in fourth quadrant refers to its operation as a solar call to convert solar energy to electrical energy.

Photo diodes are used in light detection system, light operated switches etc. Solar cells are used in conversion of solar energy to electrical energy. Figure 2.24 shows the symbol and *v-i* characteristic.

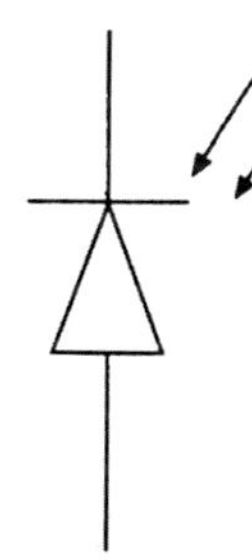

Fig. 2.24 (*a*) symbol

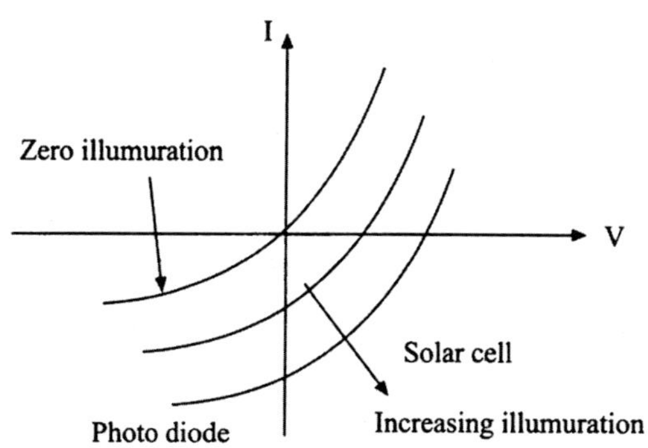

(*b*) *v-i* characteristic

Fig. 2.24. Photo diode and solar cell.

2.11. LIGHT EMITTING DIODE (LED)

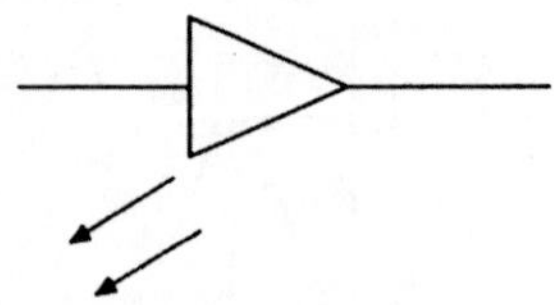

Fig. 2.25. Symbol of LED.

In an LED an electric current causes injection of minorty carries into regions of crystal where they can recombine with majority carriers resulting in emission of light. The light output varies as (curent)^n where n varies from 1.2 to 1.5. Some materials used in LED are gallium arsenide, ZnS and InSb. It is used for display in digital watches, calculators etc. This effect is known as electro luminiscence. Figure 2.25 shows the symbol of LED.

2.12. SCHOTTKY DIODE

It has an aluminium silicom junction. The silicon is generally n type. Its reverse recovery time is nearly zero and it can switch off faster than p-n diode. Figure 2.26 shows its symbol and v-i characteristic.

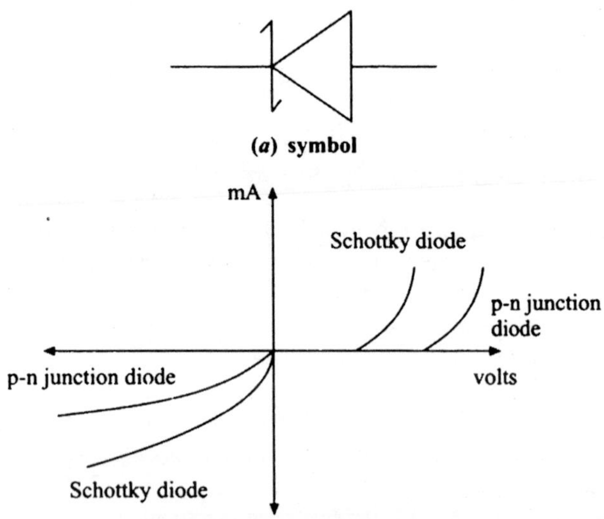

(*a*) symbol

(*b*) v-i characteristic

Fig. 2.26. Schottky diode.

2.13. LINEAR INTEGRATED CIRCUIT

It is a complete amplifier circuit on a single chip. The common IC operational amplifier is composed of a number of transistor stages on a single chip and is basically a voltage controlled voltage source. The terminals of an op-amp are shown in Fig. 2.27.

Terminals 2 and 3 are input terminals. The – sign at terminal 2 indicates that it is inverting input terminal. The + sign at terminal 3 indicates that is non-inverting terminal. + V_{cc} and $-V_{cc}$ are terminals for connections to positive and negative terminal of power supply. V_{out} is the output terminal. All amps have the following features :

1. Very high open loop gain
2. Very high input impedance
3. Very low output impedance
4. Broad bandwidth.
5. High stability
6. Low drift due to changes in temperture.

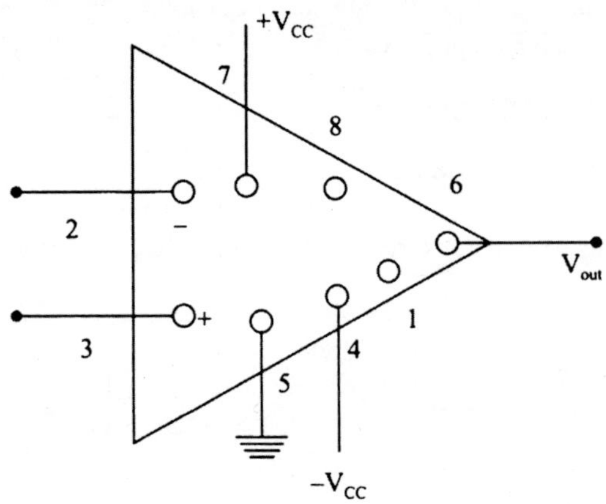

Fig. 2.27. Operational amplifier.

Common mode rejection ratio (CMRR) is defined as

$$\text{CMRR} = \frac{\text{Voltage gain for differential inputs}}{\text{Voltage gain for common mode inputs}} \quad ...(2.37)$$

A typical op-amp has a very high CMRR (around 10000 or so) CMRR can also be expressed in decibels. The decibel value is denoted by CMRR′.

$$\text{CMRR}' = 20 \log (\text{CMRR}) \quad ...(2.38)$$

At a frequency $f = f_{unity}$, voltage gain $A_v = 1$

op-amp is mostly used with feedback.

2.14. OPTO-ELECTRONICS

2.14.1. Photo Conductive, Photo emissive and Photo Voltaic Devices

A photo conductive device is one whose resistance changes due to change in illumination level on its surface. It has high resistance in darkness and resistance decreases with increase in illumination. Light dependent resistance (LDR) is one such device. It is made of semiconductor materials.

A photo voltaic device converts solar radiation into electrical energy. A photo volaic cell is commonly known as solar cell. The voltage and current rating of solar cell is very small. A large number of these cells are connected in series and parallel to form a solar array.

2.14.2. Photo Detector

It also converts light rays into electrical energy and is used in optical fibre communication system and as infrared sensor, photo diode, photo conductor, and avalanche photo diode are some types of photo detectors.

2.14.3. Laser Diode

Laser means light amplification by stimulated emission of radiation. It is monochromatic coherent light source having high optical power level, very small dimensions (the laser beam is as thin as hair) and very fast response (less than one nano second). A laser beam can travel long distance without divergence. Types of lasers are crystal lasers, gas lasers, liquid lasers and semiconductor lasers.

2.15. VACUUM TUBES

Space change limited current in vacuum diode = $ke_b^{1.5}$

Current equation for vacuum triode

$$i_b = k(e_b + \mu e_c)^{1.5}$$

μ, r_p and g_m in a triode are related as $\mu = r_p\, g_m$

Vacumm diode has two elements called cathode and plate sealed in evaluated glass tube.

Vacuum triode has three elements called cathode, grid and plate. Grid is a fine mesh structure between cathode and anode and near the cathode. Grid is usually kept at negative potential and grid voltage controls plate current. The inter electrode capacitance in a grid affects its operation in amplifier circuits.

Tetrode has four elements. The fourth element called screen grid eleminates interelectrode capacitance but leads to secondary emission. Pentode has a fifth element called suppressor grid which is kept at zero potential. Suppressor grid eliminates secondary emission.

OBJECTIVE TYPE QUESTIONS

A. Choose the most appropriate alternative

1. The resistivity of intrinsic semiconductor material is about
   (*a*) 10^5 ohm-m (*b*) 10^3 ohm-m
   (*c*) 100 ohm-m (*d*) 1 ohm-m
2. When a large number of atoms are brought together to form a crystal
   (*a*) the energy levels of inner shell electrons are affected appreciably by the presence of other neighbouring atoms
   (*b*) the energy levels of outer shell electrons are affected appreciably by the presence of other neighbouring atoms
   (*c*) the energy lends of both inner and outer shell electrons are affected appreciably by the presence of other neighbouring atoms
   (*d*) none of above
3. In which material do conduction and valence bands overlap
   (*a*) insulators
   (*b*) conductors
   (*c*) both conductor and semiconductor
   (*d*) semiconductors
4. At 0 K the forbidden energy gap in intrinstic semiconductor is about
   (*a*) 10 eV (*b*) 6 eV
   (*c*) 1 eV (*d*) 0.2 eV
5. In intrinsic semiconductor the increase in conductivity per degree increase in temperature is about
   (*a*) 2% (*b*) 6%
   (*c*) 15% (*d*) 25%
6. In intrinsic semiconductor, the fermi level
   (*a*) lies at the centre of forbidden energy gap
   (*b*) is near the conduction band
   (*c*) is near the valence band
   (*d*) may be anywhere in the forbidden energy gap
7. The ratio of impurity atoms to intrinsic semiconductor atoms in an extrinsic semiconductor is about
   (*a*) 1 : 10 (*b*) 1 : 1000
   (*c*) 1 : 100000 (*d*) 1 : 10^8
8. The addition of *n* type impurity to intrinsic material creates allowable energy levels
   (*a*) slightly below conduction band
   (*b*) slightly above conduction band
   (*c*) slightly below valence band
   (*d*) slightly above valence band
9. The addition of *p* type impurity to intrinsic material creates allowable energy levels
   (*a*) slightly below conduction band
   (*b*) slightly above conduction band
   (*c*) slightly below valence band
   (*d*) slightly above valence band
10. In a *p* type material the fermi level is 0.3 eV above valence band. The concentration of acceptor atoms is increased. The new position of fermi level is likely to be
    (*a*) 0.5 eV above valence band
    (*b*) 0.28 eV above valence band
    (*c*) 0.1 eV above valence band
    (*d*) below the valence band
11. The number of valence electrons in a silicon atom is
    (*a*) 4 (*b*) 2
    (*c*) 1 (*d*) 0
12. The number of protons in a silicon atom is
    (*a*) 32 (*b*) 28
    (*c*) 14 (*d*) 4
13. The most commonly used semiconductor material is
    (*a*) silicon (*b*) germanium
    (*c*) mixture of silicon and germanium
    (*d*) none of above
14. The presence of some holes in an intrinsic semiconductor at room temperature is due to
    (*a*) valence electrons (*b*) doping
    (*c*) free electrons (*d*) thermal energy
15. The merging of a hole and an electron is called
    (*a*) recombination (*b*) covalent bonding
    (*c*) thermal union (*d*) none of the above
16. At room temperature a semiconductor material is
    (*a*) perfect insulator (*b*) conductor
    (*c*) slightly conducting (*d*) any of above
17. The amount of time between the creation and disapperance of a hole in an intrisic semiconductor material is called

(*a*) life cycle (*b*) recombination time
(*c*) life time (*d*) half life

18. The types of carriers in a semiconductor are
(*a*) 1 (*b*) 2
(*c*) 3 (*d*) 4

19. When a voltage is applied to a semiconductor crystal the free electrons will flow
(*a*) towards positive terminal
(*b*) towards negative terminal
(*c*) either towards positive terminal or negative terminal
(*d*) towards positive terminal for 1 μs and towards negative terminal for next 1 μs

20. As temperature increases the number of tree electrons and holes in an intrinsic semiconductor
(*a*) increases
(*b*) decreases
(*c*) remains the same
(*d*) may increase of decrease

21. Holes act like
(*a*) positive charges (*b*) neutral atoms
(*c*) negative charges (*d*) crystals

22. The number of valence electrons in a donor atom is
(*a*) 2 (*b*) 3
(*c*) 4 (*d*) 5

23. The number of free electons in acceptor atom is
(*a*) 1 (*b*) 2
(*c*) 3 (*d*) 4

24. To produce *p* type semiconductor we add
(*a*) acceptor atoms (*b*) donor atoms
(*c*) pentavalent impurity (*d*) trivalent impurity

25. How many free electrons does a *p* type semiconductor has?
(*a*) only those produced by thermal energy
(*b*) only those produced by doping
(*c*) those produced by doping as well as thermal energy
(*d*) any of above

26. In an *n* type semiconductor
(*a*) number free electrons and holes is equal
(*b*) number of free elecrons is much greater than the number of holes
(*c*) number of free electrons may be equal or less than the number of holes
(*d*) number of holes is much greater than the number of free electrons

27. If E is energy level of electrons and E_F is fermi level, and T = 0 and E >E_F, then
(*a*) the probability of finding occupied quantum state of energy higher than E_F is zero
(*b*) all quantum states with energies greater than E_F are occupied
(*c*) some quantum states with energies greater than E_F are occupied
(*d*) majority of quantum states with energies greater than E_F are occupied

28. If E is energy level of electron and E_F is fermi level, then
(*a*) all quantum states with E less than E_F will be occupied at T = 0
(*b*) all quantum states with E less than E_F will be empty at T = 0
(*c*) some quantum stataes with E less than E_F will be occupied at T = 0
(*d*) any of above

29. For most metals, fermi level E_F is less than
(*a*) 0.1 eV (*b*) 2 eV
(*c*) 5 eV (*d*) 10 eV

30. The density of states (*i.e.*, number of states per eV per m^3) in the conduction band for energy level E is proportional to
(*a*) $\sqrt{E}$ (*b*) E
(*c*) $E^{1.5}$ (*d*) E^2

31. Atomic number of germanium is
(*a*) 24 (*b*) 28
(*c*) 32 (*d*) 36

32. The current density J, free election mobility μ_n, hole mobility μ_p magnitude of free electron and hole concentration n_i and electric field E and charge an electron *e*, in intrinsic semiconductor are related as
(*a*) $J = (\mu_n + \mu_p)\, en_i E$ (*b*) $J = \frac{(\mu_n + \mu_p)\, en_i}{E}$
(*c*) $J = \frac{1}{e}(\mu_n + \mu_p)\, n_i E$ (*d*) $J = (\mu_n + \mu_p)\frac{eE}{n_i}$

33. In an intrinsic semiconductor, the intrinsic charge concentration at any absolute temperature T is proportional to
(*a*) T (*b*) T^2
(*c*) T^3 (*d*) T^4

34. Forbidden energy gap in germanium at 0 K is about
(*a*) 10 eV (*b*) 5 eV
(*c*) 2 eV (*d*) 0.72 eV

35. At absolute temperature
(*a*) the forbidden energy gap in germanium is higher than that in silicon
(*b*) the forbidden energy gap in germanium and silicon are equal
(*c*) the forbidden energy gap in silicon is higher than that in germanium
(*d*) none of above

36. As temperature increases
(*a*) the forbidden energy gaps in silicon and germanium increase
(*b*) the forbidden energy gaps in silicon and germanium decrease
(*c*) the forbidden energy gap in silicon decreases while that in germanium decreases
(*d*) the forbidden energy gap in silicon increases while that in germanium decreases

37. If μ_n is mobility of free electron and μ_p is mobility of holes then for silicon at 300 K
(*a*) $\mu_n = \mu_p$ (*b*) $\mu_n > \mu_p$
(*c*) $\mu_n < \mu_p$
(*d*) μ_n may be more or less than v_p

38. The intrinsic resistivity of silicon at 300 K is about
(*a*) 1 Ω cm (*b*) 400 Ω cm
(*c*) 10000 Ω cm (*d*) 230000 Ω cm

39. In intrinsic semiconductor at 300 K, the magnitude of free electron concentration in silicon is about
(*a*) 15×10^{14} per cm^3 (*b*) 5×10^{12} per cm^3
(*c*) 1.45×10^{10} per cm^3 (*d*) 1.45×10^{6} per cm^3

40. In *n* type semiconductor, the free electron concentration
(*a*) is nearly equal to density of donor atoms
(*b*) is much greater than density of donor atoms
(*c*) is much less than density of donor atoms
(*d*) may be equal or more or less than density of donor atoms

41. In *p* type semiconductor, the hole concentration
(*a*) is nearly equal to density of acceptor atoms
(*b*) is much greater than density of acceptor atoms
(*c*) is much less than density of acceptor atoms
(*d*) may be equal or more or less than density of acceptor atoms

42. Diffusion constants D_p, D_n, mobility μ_p, μ_n and absolute temperature T are related as
(*a*) $\frac{D_p}{\mu_p} = \frac{D_n}{\mu_n} = \frac{1}{11600}$ (*b*) $\frac{D_p}{\mu_p} = \frac{D_n}{\mu_n} = \frac{11600}{T}$
(*c*) $\frac{D_p}{\mu_p} = \frac{\mu_n}{D_n} = \frac{T}{11600}$ (*d*) $\frac{D_n}{\mu_n} = \frac{\mu_p}{D_p} = \frac{T}{11600}$

43. The range of life time of carriers (electrons and holes) is
(*a*) 1 μs to 100 μs
(*b*) 1 nano sec to 1μs
(*c*) 1 nano sec to hundreds of μs
(*d*) none of the above

44. In a specimen of *n* type semiconductor, the initial concentration of holes and electrons is p_{no} and n_{no}. When the specimen is subjected to radiation, the hole and electron concentration increase to p_n and n_n. Then
(*a*) $p_n - p_{no} = n_n - n_{no}$ (*b*) $p_n + p_{no} = n_n + n_{no}$
(*c*) $p_n + n_n = p_{no} + n_{no}$ (*d*) $p_n - n_{no} = n_n - p_{no}$

45. Hall effect can be used
(*a*) to find type of semiconductor (whether *p* or *n*)
(*b*) to find carrier concentration
(*c*) to measure conductivity
(*d*) all of the above

46. Conductivity σ, mobility μ and Hall coefficient K_H are related as
(*a*) $\mu = \sigma K_H$ (*b*) $\sigma = \mu K_H$
(*c*) $K_H = \mu\sigma$ (*d*) $K_H = (\sigma\mu)^{1.1}$

47. Hall coefficient K_H and charge density ρ are related as
(*a*) $K_H = \frac{1}{\rho}$ (*b*) $K_H = \rho$
(*c*) $K_H = \left(\frac{1}{\rho}\right)^{1.2}$ (*d*) $K_H = (\rho)^{1.2}$

48. In an *n* type semiconductor the fermi level is 0.35 eV below the conduction band. The concentration of donor atoms is increased to three times. The new position of fermi level will be
(*a*) 0.35 eV below conduction band
(*b*) about 0.32 eV below conduction band
(*c*) about 0.32 eV above conduction band
(*d*) about 0.1 eV below conduction band

49. An electron rises through a voltage of 100 V. The energy acquired by it will be
(*a*) 100 eV (*b*) 100 joules
(*c*) $(100)^{1.2}$ eV (*d*) $(100)^{1.2}$ joules

50. Which statement is false as regards holes?
(*a*) Holes exist in conductors as well as semiconductors
(*b*) Holes constitute positive charges
(*c*) Holes exist only in semiconductors
(*d*) Holes and electrons recombine

51. In all metals
(*a*) conductivity decreases with increase in temperature
(*b*) current flow is by electrons as well as by holes
(*c*) resistivity decreases with increase in temperature
(*d*) The gap between valence and conduction bands is small

52. Which impurity will give *p* type semiconductor when added to intrinsic semiconductor?
(*a*) Phosphorous (*b*) Boron
(*c*) Arsenic (*d*) Antimony

53. Which of the following has highest conductivity?
(*a*) Silver (*b*) Aluminium
(*c*) Tungsten (*d*) Platinum

54. Which of the following is not a semiconductor?
(*a*) Silicon (*b*) Calcium-arsenide
(*c*) Germanium (*d*) Zinc sulphide

55. The concentration of minority carriers in a semiconductor depends mainly on
(*a*) the extent of doping (*b*) temperature
(*c*) the applied bias (*d*) none of above

56. The product *k*T (where *k* is Boltzmann's constant and T is absolute temperature) at room temperature is about
(*a*) 0.051 eV (*b*) 0.026 eV
(*c*) 0.01 eV (*d*) 0.001 eV

57. The probability giving distribution of electrons over a range of allowed energy levels is known as
(*a*) Maxwell's distribution
(*b*) Fermi-Dirac distribution
(*c*) Richardson-Dushman distribution
(*d*) none of above

58. The current due to thermionic emission is proportional to
(*a*) T (*b*) T^2
(*c*) T^3 (*d*) T^4

59. The thermionic emission current is given by
(*a*) Fermi-Dirac distribution
(*b*) Maxwell's equation
(*c*) Richardson-Dushman equation
(*d*) Langmuir Child Law

60. Photo electric emission can occur only if
(*a*) wavelength of incident radiation is equal to threshold value
(*b*) wavelength of incident radiation is less than threshold value
(*c*) frequency of incident radiation is less than threshold frequency
(*d*) none of above

61. In photo electric emission the maximum kinetic energy of emitted election is proportional to
(*a*) $\sqrt{f}$ (*b*) f
(*c*) f^2 (*d*) f^3

62. The amount of photoelectric emission current depends on
(*a*) frequecy of incident radiation
(*b*) intensity of incident radiation
(*c*) both frequency and intensity of incident radiation
(*d*) none of above

63. Photoelectric emitters in photo tubes are generally made of
(*a*) alkali metals
(*b*) metals
(*c*) semiconductors
(*d*) metal and semiconductors

64. As per Einstein's equation, the velocity of emitted electron in photoelectric emission is given by the equation
(*a*) $\frac{1}{2}mv^2 = hf - U_w$ (*b*) $\frac{1}{2}mv^2 \le hf - U_w$
(*c*) $\frac{1}{2}mv^2 = hf + U_w$ (*d*) $\frac{1}{2}mv^2 \le hf + U_w$

65. In photoelectric emission, the threshold frequency f_0, work function U_w and Planck's constant h are related as
(*a*) $f_0 = \frac{U_w}{h}$ (*b*) $U_w = \frac{f_0}{h}$
(*c*) $f_0 = (U_w)(h)^2$ (*d*) $h = (U_w)(f_0)$

66. It E is electric field intensity, the current density due to field emission is proportional to
(*a*) E (*b*) E^2
(*c*) $E^{2.5}$ (*d*) $E^{3.2}$

67. In commercial electron tubes the current produced by the cathode at 1000 K is about
(*a*) 0.01 A per cm² of cathode surface
(*b*) 0.1 A per cm² of cathode surface
(*c*) 1 A per cm² of cathode surface
(*d*) 5 A per cm² of cathode surface

68. The temperature of cathode is increased from 2500 K to 2600 K. The increase in thermionic emission current is about
(*a*) 0.1% (*b*) 4%
(*c*) 50% (*d*) 150%

69. In energy band diagram of *p* type semiconductor the acceptor energy level is
(*a*) in valence band
(*b*) in conduction band
(*c*) slightly above valence band
(*d*) slightly below conduction band

70. In energy band diagram of *n* type semiconductor, the donor energy level is
(*a*) in valence band
(*b*) in conduction band
(*c*) slightly above valence band
(*d*) slightly below conduction band

71. Fermi level is the amount of energy which
(*a*) a hole can have at room temperature
(*b*) an electron can have at room temperature
(*c*) must be given to an electron to move to conduction band
(*d*) none of above

72. Match List I with List II using the given codes

List I	List II
A. Silicon	1. Donor Impurity
B. Antimony	2. Acceptor Impurity
C. Gallium	3. Most Commonly used semiconductor material
D. Germanium	4. Atomic number 32

Codes:

	A	B	C	D
(*a*)	3	1	2	4
(*b*)	2	3	1	4
(*c*)	1	2	3	4
(*d*)	4	3	2	1

73. Match List I with List II using the given codes

List I	List II
A. Silicon	1. is trivalent
B. Arsenic	2. rarely used as semiconductor
C. Indium	3. is pentavalent
D. Germanium	4. Intrinsic semiconductor

Codes:

	A	B	C	D
(*a*)	1	2	3	4
(*b*)	4	3	2	1
(*c*)	4	3	1	2
(*d*)	2	1	3	4

74. Match List I with List II using the given codes

List I	List II
A. *p* type semiconductor	1. Silicon
B. *n* type semiconductor	2. holes are majority carriers

C. Intrinsic semiconductor	3. electrons are majority carriers
D. Tungsten	4. is a metal

Codes:

	A	B	C	D
(*a*)	2	3	1	4
(*b*)	1	2	3	4
(*c*)	4	3	2	1
(*d*)	3	1	2	4

75. Match List I with List II using the given codes

List I	List II
A. Copper	1. Produces discrete energy level just above valence band
B. Rubber	2. Produces discrete energy level just below conduction band
C. Antimony	3. Large forbidden gap
D. Boron	4. Valence and conduction bands overlap

Codes:

	A	B	C	D
(*a*)	1	2	4	3
(*b*)	4	1	2	3
(*c*)	4	3	1	2
(*d*)	2	1	3	4

76. When *a p-n* function is forward biased
(*a*) the current flow is due to electrons only
(*b*) the majority carriers in both *p* and *n* materials are driven toward the junction
(*c*) the majority carriers in both *p* and *n* materials are away from the junction
(*d*) both (*a*) and (*c*)

77. When a reverse bias is applied to a *p-n* junction, the width of depletion layer
(*a*) decreases
(*b*) increases
(*c*) remains the same
(*d*) may increase or decrease

78. When diodes are connected in series to increase voltage rating the peak inverse voltage per junction
(*a*) should not exceed half the breakdown voltage
(*b*) should not exceed the breakdown voltage
(*c*) should not exceed one-third the breakdown voltage
(*d*) may be equal or less than breakdown voltage

79. The number of *p-n* junctions in a semiconductor diode are
(*a*) 0 (*b*) 1
(*c*) 2 (*d*) 1 or 2

80. At room temperature the barrier potential in a silicon diode is
(*a*) 0.1 V (*b*) 0.3 V
(*c*) 0.7 V (*d*) 1 V

81. Surface leakage current is a part of
(*a*) reverse current (*b*) forward current
(*c*) reverse breakdown (*d*) forward breakdown

82. The voltage at which avalanche occurs is known as
(*a*) cut in voltage (*b*) barrier voltage
(*c*) breakdown voltage (*d*) depletion voltage

83. As the reverse voltage is increased, the depletion layer
(*a*) becomes narrow (*b*) widens
(*c*) remains the same (*d*) reduce to zero

84. A reverse voltage of 18 V is applied to a semiconductor diode. The voltage across the depletion layer is
(*a*) 0 (*b*) 0.7 V
(*c*) about 10 V (*d*) 18 V

85. A semiconductor diode is biased in forward direction and carrying current I. The current due to holes in *p* material is
(*a*) I (*b*) 0
(*c*) less than I (*d*) 0.5 I

86. When a diode is not conducting its bias is
(*a*) forward (*b*) zero
(*c*) reverse (*d*) zero or reverse

87. When a diode is forward biased, the diode current is
(*a*) high (*b*) low
(*c*) zero (*d*) low or zero

88. The cut in voltage of a diode is nearly equal to
(*a*) applied forward voltage
(*b*) applied reverse voltage
(*c*) barrier potential
(*d*) none of above

89. The reverse saturation current in a semiconductor diode consists of
(*a*) avalanche current
(*b*) zener current
(*c*) minority carrier current
(*d*) minority carrier current and surface leakage current

90. A silicon diode is forward biased and total applied voltage is 5 V. The voltage across *p-n* junction is
(*a*) 5 V (*b*) Slightly less than 5 V
(*c*) 0.7 V (*d*) 0

91. A potential of 7 V is applied to a silicon diode. A resistance of 1 k ohm is also in series with the diode. The current is
(*a*) 7 mA (*b*) 6.3 mA
(*c*) 0.7 mA (*d*) 0

92. A voltage of 9 V is applied in forward direction to a semiconductor diode in series with a load resistance of 1000 Ω. The voltage across the load resistance is zero. It indicates that
(*a*) diode is short circuited
(*b*) diode is open circuited
(*c*) resistor is open circuited
(*d*) diode is either o.c. or s.c.

93. The breakdown voltage in a zener diode
(*a*) is almost constant

(*b*) is very small
(*c*) may destroy the diode
(*d*) decreases with increase in current

94. The voltage across a zener diode
(*a*) is constant in forward direction
(*b*) is constant is reverse direction
(*c*) is constant in both forward and reverse directions
(*d*) none of the above

95. Which of the following is basically a voltage controlled capacitance?
(*a*) zener diode (*b*) diode
(*c*) varactor diode (*d*) LED

96. In a varactor diode the increase in width of depletion layer results in
(*a*) decrease in capacitance
(*b*) increase of capacitance
(*c*) no change in capacitance
(*d*) either (*a*) or (*b*)

97. A varactor diode is
(*a*) reverse biased (*b*) forward biased
(*c*) biased to breakdown (*d*) unbiased

98. When the light falling on a photo diode increases, the reverse minority carrier current
(*a*) increases (*b*) increases or decreases
(*c*) decreases (*d*) remains the same

99. A photo diode is
(*a*) forward biased
(*b*) reverse biased
(*c*) either forward or reverse biased
(*d*) unbiased

100. To ensure that a zener diode does not get destroyed
(*a*) the applied voltage should not exceed breakdown volatge
(*b*) the current should not exceed rated current
(*c*) the current should be less than magnitude of barrier potential
(*d*) both (*a*) and (*b*)

101. The number of doped regions is PIN diode is
(*a*) 1 (*b*) 2
(*c*) 3 (*d*) 1 or 2

102. In a zener diode
(*a*) the forward current is very high
(*b*) sharp breakdown occurs at a certain reverse voltage
(*c*) the ratio v/i can be negative
(*d*) there are two *p-n* junctions

103. What is the effect of cut in voltage on the wave form of output as compared to input in a semiconductor diode?
(*a*) the duration of output wave form is less than 180°
(*b*) output voltage is less than input voltage
(*c*) output voltage is more than input voltage
(*d*) both (*a*) and (*b*)

104. When a high voltage reference is required it is advantageous to use two or more zener diodes in series to allow
(*a*) higher voltage
(*b*) higher dissipation
(*c*) lower temperature coefficient
(*d*) all the above

105. If the reverse voltage across a *p-n* junction is increased three times, the junction capacitance
(*a*) will decrease
(*b*) will increase
(*c*) will decrease by an approximate factor of about 2
(*d*) will increase by an approximate factor of about 2

106. The depletion layer in a reverse biased *p-n* junction is due to the presence of
(*a*) electrons
(*b*) holes
(*c*) both electrons and holes
(*d*) immobile ions

107. The cut in voltage
(*a*) is the same for silicon and germanium diodes
(*b*) is a forward voltage
(*c*) is a reverse voltage
(*d*) is a forward voltage below which the current is very small

108. When a *p-n* junction is reverse based
(*a*) holes and electrons move away from the junction
(*b*) holes and electrons move towards the junction
(*c*) holes more towards junction and elections move away from junction
(*d*) holes move away from junction and electrons move towards junction

109. The saturation current in a semiconductor diode
(*a*) depends on forward voltage
(*b*) depends on reverse voltage
(*c*) is due to thermally generated minority carriers
(*d*) none of above

110. When a *p-n* junction is forward biased
(*a*) the width of depletion layer increases
(*b*) the width of depletion layer decreases
(*c*) the majority carriers move away form the junction
(*d*) the current is very small

111. In a reverse biased *p-n* junction, the reverse bias is 4 V. The junction capacitance is about
(*a*) 0.1 F (*b*) 4 μF
(*c*) 10 nF (*d*) 20 pF

112. The dynamic resistance of a forward biased *p-n* diode
(*a*) varies inversely with current
(*b*) varies directly with current
(*c*) is constant
(*d*) is either constant or varies directly with current

113. A *p-n* junction diode has
(*a*) low forward and high reverse resistance
(*b*) a non-linear *v-i* characteristics
(*c*) zero forward current till the forward voltage reaches cut in value
(*d*) all the above

114. Typical value of reverse current in a semiconductor diode is
(*a*) 1 A (*b*) 0.1 A
(*c*) 10 mA (*d*) 1 mA

115. An increase in junction temperature of a semiconductor diode
(*a*) causes a small increase in reverse saturation current
(*b*) causes large increase in reverse saturation current
(*c*) does not affect reverse saturation current
(*d*) may cause an increase or decrease in reverse saturation current depending on rating of diode

116. An extrinsic semiconductor sample has 6 billion silicon atoms and 3 million pentavalent impurity atoms. The number of electrons and holes is
(*a*) 3 million each
(*b*) 6 billion each
(*c*) 3 million free electrons and very small number of holes
(*d*) 3 million holes and very small number of free electrons

117. An intrinsic silicon sample has 2 million free electrons. The number of holes in the sample is
(*a*) 2 million (*b*) almost zero
(*c*) more than 2 million (*d*) less than 2 million

118. An intrinsic silicon sample has 1 million free electrons at row temperature. As the temperature is increased
(*a*) the number of free electrons increases
(*b*) the number of free electrons increases but the number of holes decreases
(*c*) the number of free electrons and holes increase by the same amount
(*d*) the number of free electrons and hole increase but not by the same amount

119. The factor n in the equation for calculating current for a silicon diode is
(*a*) 1
(*b*) 2
(*c*) 2.5
(*d*) 2 for low levels of current and 1 for high levels of current

120. At any point on the v-i characteristics of a semiconductor diode, the slope of i-v characteristics is called
(*a*) resistance of diode
(*b*) conductance of diode
(*c*) incremental resistance of diode
(*d*) incremental conductance of diode

121. In the equation $i = I_0\ (e^{V/\eta V_T} - 1)$, V_T =
(*a*) $\dfrac{11600}{T}$ (*b*) $\dfrac{T}{11600}$
(*c*) $T \times 11600$ (*d*) $\dfrac{11600}{T^2}$

122. A diode is operating in forward region and the forward voltage and current are $v = 3 + 0.3 \sin \omega t$ volts and $i = 5 + 0.2 \sin \omega t$ mA. The average power dissipated is
(*a*) 15 mW (*b*) about 15 mW
(*c*) 1.5 mW (*d*) about 1.5 mW

123. Which of these has highly doped p and n region?
(*a*) PIN diode (*b*) Tunnel diode
(*c*) Schottky diode (*d*) Photo diode

124. Which of these has 3 layers?
(*a*) PIN diode (*b*) Zener diode
(*c*) Schottky diode (*d*) Photo diode

125. Which of these has a layer of intrinsic semiconductor?
(*a*) Zener diode (*b*) PIN diode
(*c*) Photo diode (*d*) Schottky diode

126. Which of these has degenerate p and n materials?
(*a*) Zener diode (*b*) PIN diode
(*c*) Tunnel diode (*d*) Photo diode

127. In which of the following is the width of junction barrier very small?
(*a*) Tunnel diode (*b*) Photo diode
(*c*) PIN diode (*d*) Schottky diode

128. In degenerate p type semiconductor material, the fermi level
(*a*) is in the valence band
(*b*) is in conduction band
(*c*) is at the centre in between valence and conduction bands
(*d*) is very near conduction band

129. In a degenerate n type semiconductor material, the fermi level
(*a*) is in the valence band
(*b*) is in conduction band
(*c*) is at the centre in between valence and conduction bands
(*d*) is very near conduction band

130. Which of these has semiconductor-metal junction?
(*a*) Pin diode (*b*) Photo diode
(*c*) Tunnel diode (*d*) Schottky diode

131. As compared to an ordinary semiconductor diode, a Schottky diode
(*a*) has lower cut in voltage
(*b*) higher cut in voltage
(*c*) lower reverse saturation current
(*d*) both (*b*) and (*c*)

132. As compared to an ordinary semiconductor diode, a Schottky diode
(*a*) has higher reverse saturation current
(*b*) has higher reverse saturation current and higher cut in voltage
(*c*) has higher reverse saturation current and lower cut in voltage
(*d*) has lower reverse saturation current and lower cut in voltage

133. In which of these is reverse recovery time nearly zero
(*a*) diode (*b*) tunnel diode
(*c*) Schottky diode (*d*) PIN diode

134. In which of the following, does a negative resistance region exist in the v-i characteristics?

(*a*) PIN diode (*b*) Schottky diode
(*c*) Tunnel diode (*d*) Zener diode

135. Which of these has peak and valley points in *v-i* curve?
(*a*) Tunnel diode (*b*) Zener diode
(*c*) PIN diode (*d*) Schottky diode

136. Which of these is used in seven segment display?
(*a*) PIN diode (*b*) LED
(*c*) Photo diode (*d*) Tunnel diode

137. The light output of LED varies as (current)^n. The value of n is about
(*a*) 0.5 (*b*) 1
(*c*) 1.3 (*d*) 2.1

138. Which of these is also called 'hot carrier diode"?
(*a*) PIN diode (*b*) LED
(*c*) Photo diode (*d*) Schottky diode

139. Match List I with List II using the given codes

List I	List II
A. Zener diode	1. used in seven segment display
B. Tunnel diode	2. used in voltage regulator circuits
C. LED	3. used is solar energy utilization
D. Photovoltaic cell	4. used in digital circuits

Codes:

	A	B	C	D
(*a*)	1	3	2	4
(*b*)	2	3	1	4
(*c*)	2	4	1	3
(*d*)	3	2	4	1

140. Match the diodes in List I with symbol in List II using the given codes

List I	List II
A. Zener diode	1. 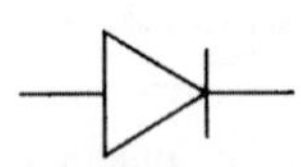
B. Tunnel diode	2. 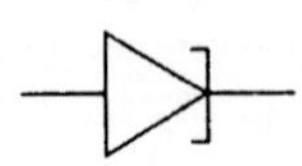
C. Schottky diode	3. 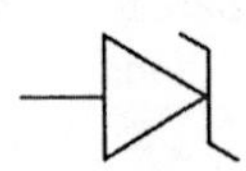
D. Diode	4. 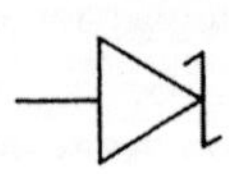

Codes:

	A	B	C	D
(*a*)	1	3	2	4
(*b*)	2	3	1	4
(*c*)	4	3	2	1
(*d*)	3	2	4	1

141. Match List I with List II using the given codes

List I	List II
A. Tunnel diode	1. is also known as hot carrier diode
B. PIN diode	2. used with reverse bias
C. Zener diode	3. has an intrinsic layer
D. Schottky diode	4. *v-i* curve has shape of N

Codes:

	A	B	C	D
(*a*)	2	3	1	4
(*b*)	4	3	2	1
(*c*)	3	2	4	1
(*d*)	1	2	3	4

142. A transistor has two *p-n* junctions. The batteries should be connected such that
(*a*) both junctions are forward biased
(*b*) both junctions are reverse biased
(*c*) one junction is forward biased and the other is reverse biased
(*d*) none of above

143. The value of α in a transistor
(*a*) is always equal 1
(*b*) is less than 1 but more than 0.9
(*c*) is about 0.4
(*d*) is about 0.1

144. When a *p-n-p* transistor is properly biased to operate in active region the holes from emitter
(*a*) diffuse through base into collector region
(*b*) recombine with electrons in base
(*c*) recombine with electrons in emitter
(*d*) none of above

145. A Varactor diode has
(*a*) a fixed capacitance
(*b*) a fixed inductance
(*c*) a voltage variable capacitance
(*d*) a current variable inductance

146. A zener diode is used in
(*a*) voltage regular circuit
(*b*) amplifier circuits
(*c*) both voltage regulator and amplifier circuit
(*d*) none of above

147. A varactor diode is used for
(*a*) tuning
(*b*) rectification
(*c*) amplification
(*d*) rectification and amplification

148. The number of doped regions in a bipolar transistor is
(*a*) 1 (*b*) 2
(*c*) 3 (*d*) 4

149. In an *n-p-n* transistor the majority carriers in the base are
(*a*) electrons
(*b*) holes

(*c*) both holes and electrons
(*d*) either holes or electrons.

150. In a bipolar transistor the barrier potential
(*a*) 0
(*b*) a total of 0.7 V
(*c*) 0.7 V across each depletion layer
(*d*) 0.35 V

151. In a bipolar transistor, the emitter base junction has
(*a*) forward bias (*b*) reverse bias
(*c*) zero bias (*d*) zero of reverse bias

152. In a bipolar transistor, the base collector junction has
(*a*) forward bias (*b*) reverse bias
(*c*) zero bias (*d*) zero or reverse bias

153. The current gain of a transistor is the ratio of
(*a*) emitter current to base current
(*b*) emitter current to collector current
(*c*) collector current to base current
(*d*) collector current to emitter current

154. In a bipolar transistor which current is largest
(*a*) collector current
(*b*) base current
(*c*) emitter current
(*d*) base current or emitter current

155. In a bipolar transistor which current is smallest
(*a*) collector current
(*b*) base current
(*c*) emitter current
(*d*) any of the three currents

156. In a bipolar transistor
(*a*) the base is thick and highly doped
(*b*) the base is thin and highly doped
(*c*) the base is thick and lightly doped
(*d*) the base is thin and lightly doped

157. The power dissipation in a transistor is the product of
(*a*) emitter current and emitter to base voltage
(*b*) emitter current and collector to emitter voltage
(*c*) collector current and collector to emitter voltage
(*d*) none of above

158. In a bipolar transistor
(*a*) recombination in base regions of both *n-p-n* and *p-n-p* transistor is minimum.
(*b*) recombination in base regions of both *n-p-n* and *p-n-p* transistors is high
(*c*) recombination in base region of *n-p-n* transistor is minimum but that in *p-n-p* transistor is high
(*d*) recombination in base region of *p-n-p* transistor is minimum but that in *n-p-n* transistor is high

159. For bipolar transistor

(*a*) $\beta_{dc} = \dfrac{\alpha_{dc}}{1-\alpha_{dc}}$ (*b*) $\beta_{dc} = \dfrac{\alpha_{dc}}{1+\alpha_{dc}}$

(*c*) $\beta_{dc} = \dfrac{1-\alpha_{dc}}{\alpha_{dc}}$ (*d*) $\beta_{dc} = \dfrac{1+\alpha_{dc}}{\alpha_{dc}}$

160. If α_{ac} for a transistor is 0.98, β_{dc} equal to
(*a*) 51 (*b*) 49
(*c*) 47 (*d*) 45

161. Figure 2.28 shows characteristics curves for bipolar transistor. These curves are
(*a*) output characteristics of *n-p-n* transistor (common base)
(*b*) output characteristics of *p-n-p* transistor (common base)
(*c*) output characteristics of *n-p-n* transistor (common emitter)
(*d*) output characteristics of *p-n-p* transistor (common emitter)

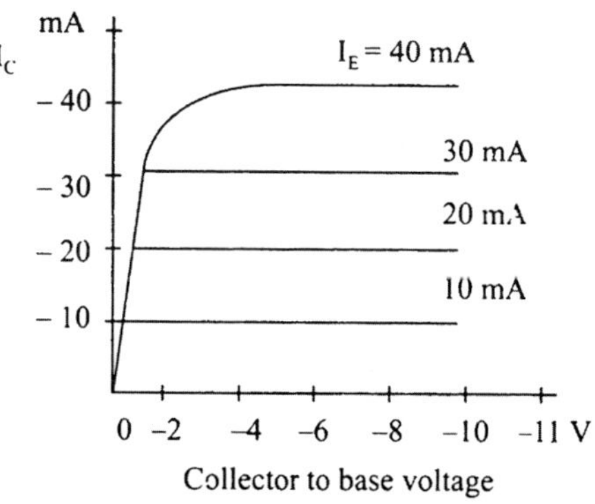

Fig. 2.28.

162. In *p-n-p* transistor the current I_E has two components viz. I_{EP} due to injection of holes from *p*-region to *n*-region and I_{En} due to injection of electrons from *n*-region to *p*-region. Then
(*a*) I_{Ep} and I_{En} are almost equal
(*b*) $I_{Ep} >> I_{En}$
(*c*) $I_{En} >> I_{Ep}$
(*d*) either (*a*) or (*c*)

163. In a bipolar transistor, emitter efficiency is about
(*a*) 0.99 (*b*) 0.9
(*c*) 0.8 (*d*) 0.7

164. In bipolar transistor de current gain is

(*a*) $\dfrac{I_C}{I_E}$ (*b*) $\dfrac{I_C}{I_B}$

(*c*) $\dfrac{I_E}{I_B}$ (*d*) $\dfrac{I_E}{I_C}$

165. The output characteristics of a bipolar transistor has three distinct regions. They are known as
(*a*) saturation region, active region and breakdown region
(*b*) inactive region, active region and breakdown region
(*c*) inactive region, saturation region and active region
(*d*) inactive region, saturation region and breakdown region

166. In which region of a CE bipolar transistor is collector current almost constant
(*a*) saturation region
(*b*) active region
(*c*) breakdown region
(*d*) both saturation and active regions

167. In a CE bipolar transistor operating in active region, collector current is independent of
(*a*) V_{CE} (*b*) I_B
(*c*) both V_{CE} and I_B (*d*) none of above

168. In the saturation region of CE output characteristics of *n-p-n* transistor, V_{CE} is about
(*a*) 0.5 V (*b*) 15 V
(*c*) – 0.5 V (*d*) –15 V

169. CE saturation resistance of *n-p-n* transistor is
(*a*) $\frac{V_{CB}}{I_E}$ (*b*) $\frac{V_{CE}}{I_C}$
(*c*) $\frac{V_{CE}}{I_E}$ (*d*) $\frac{V_{CE}}{I_B}$

170. In forward active region of operation of BJT
(*a*) both junctions are forward biased
(*b*) both junctions are reverse biased
(*c*) emitter base junction is forward biased and base collector junction is reverse biased
(*d*) emitter base junction is reverse biased and base collector junction is reverse biased

171. In which mode of BJT operation are both junctions reverse biased
(*a*) active (*b*) saturation
(*c*) cut off (*d*) reverse active

172. In which mode of BJT operation are both junctions forward biased
(*a*) active (*b*) saturation
(*c*) cut off (*d*) reverse active

173. Between which regions does BJT act like switch
(*a*) cut off and saturation
(*b*) cut off and forward active
(*c*) forward active and reverse active
(*d*) saturation and active

174. In which condition does BJT behave like a closed switch
(*a*) cut off (*b*) saturation
(*c*) forward active (*d*) reverse active

175. In what condition does BJT act like an open switch
(*a*) cut off (*b*) saturation
(*c*) active (*d*) both (*b*) and (*c*)

176. The derating factor for a BJT transistor is about
(*a*) 0.5 mW/°C (*b*) 2.5 mW/°C
(*c*) 10 mW/°C (*d*) 25 mW/°C

177. For BJT transistor, the maximum power dissipation is specified as 350 mW if ambient temperature is 25°C. If ambient temperature is 60°C the maximum power dissipation should be limited to about
(*a*) 100 mW (*b*) 250 mW
(*c*) 450 mW (*d*) 600 mW

178. The turn off time of a bipolar transistor is about
(*a*) 0.5 ns (*b*) 10 ns
(*c*) 70 ns (*d*) 150 ns

179. In CE connection, the leakage current of a transistor is about
(*a*) 10×10^{-9} A (*b*) 5×10^{-6} A
(*c*) 200×10^{-6} A (*d*) 5×10^{-3} A

180. In a bipolar junction transistor the base region is made very thin so that
(*a*) recombination in base region is minimum
(*b*) electric field gradient in base is high
(*c*) base can be easily fabricated
(*d*) base can be easily biased

181. In a *p-n-p* transistor the main current carriers and the mechanism of flow respectively are
(*a*) electrons, drift (*b*) holes, drift
(*c*) holes, diffusion (*d*) electrons, diffusion

182. In a bipolar junction transistor $\alpha_{dc} = 0.98$, $I_{CO} = 2$ μA and $I_B = 15$ μA. The collector current I_C is
(*a*) 635 μA (*b*) 735 μA
(*c*) 835 μA (*d*) 935 μA

183. When a *p-n-p* transistor is operation in active region, the current in the *n* region is due to
(*a*) only holes (*b*) only electrons
(*c*) mainly holes (*d*) mainly electrons

184. In common base configuration, the input characteristics of bipolar junction transistor are drawn between
(*a*) V_{EB} and I_E (*b*) V_{CB} and I_C
(*c*) I_C and V_{CE} (*d*) I_E and V_{CE}

185. In common base connection, the output characteristics of a bipolar junction transistor is drawn between
(*a*) I_C and V_{CB} (*b*) I_C and V_{CE}
(*c*) I_E and V_{CB} (*d*) I_E and V_{CE}

186. In CE configuration, the output characteristics of a bipolar junction transistor is drawn between
(*a*) I_C and V_{CB} (*b*) I_E and V_{CB}
(*c*) I_C and V_{CE} (*d*) I_E and V_{CE}

187. In an *n-p-n* transistor biased for operation in forward active region
(*a*) emitter is positive with respect to base
(*b*) collector is positive with respect to base
(*c*) base is positive with respect to emitter and collector is positive with respect to base
(*d*) none of above

188. In a *p-n-p* transistor operating in forward active mode
(*a*) base is positive with respect to emitter and collector
(*b*) base is negative with respect to emitter and collector
(*c*) emitter is positive with respect to base and base is positive with respect to collector
(*d*) emitter is negative with respect to base and base is positive with respect to collector

189. Figure 2.29 shows the terminals of a transistor in plastic package TO 18. Then

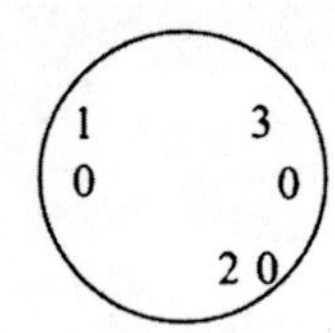

Fig. 2.29.

(*a*) terminals 1, 2, 3 are emitter, collector, base respectively
(*b*) terminals 1, 2, 3 are emitter, base, collector respectively
(*c*) terminals 1, 2, 3 are base, emitter, collector, respectively
(*d*) terminals 1, 2, 3, are collector, emitter, base respectively

190. In a transistor operating in forward active mode
(*a*) depletion layer between emitter and base is thin and that between base and collector is also thin
(*b*) both depletion regions are thick
(*c*) depletion layer between emitter and base is thin and that between base and collector is thick
(*d*) depletion layer between emitter and base is thick and that between base and collector is thin

191. It is required to trace the output characteristics of a CE bipolar transistor on a CRO screen. The proper method is
(*a*) apply the voltage drop across collector resistance to Y input, disconnect sweep generator and apply V_{CE} to X input
(*b*) apply voltage drop across collector resistance to Y input
(*c*) apply V_{CE} to X input
(*d*) apply V_{CE} to Y input, disconnect the sweep generator and apply voltage drop across collector resistance to X input

192. In the schematic representation of bipolar junction transistor the direction of arrow shows the direction of flow of
(*a*) holes
(*b*) electrons
(*c*) holes in *p-n-p* and holes in *n-p-n*
(*d*) electrons in *p-n-p* and holes in *n-p-n*

193. The relation between thermionic emission current and temperature is known as
(*a*) Richardson-Dushman equation
(*b*) Langmuir Child law
(*c*) Ohm's law
(*d*) Boltzmann's law

194. The relation between plate current and plate voltage of a vacuum diode is called
(*a*) Richardson-Dushman equation
(*b*) Langmuir Child law
(*c*) Ohm's law
(*d*) Boltzmann's law

195. In the vacuum diode equation $i_b = ke_b^{1.5}$, the current is
(*a*) temperature limited current
(*b*) space charge limited current
(*c*) any of above
(*d*) none of above

196. In a triode the potential of grid (with respect to cathode) is usually
(*a*) zero (*b*) negative
(*c*) positive (*d*) zero or positive

197. The potential of suppressor grid (with respect to cathode) is usually
(*a*) zero (*b*) negative
(*c*) positive (*d*) zero or positive

198. Secondary emission occurs in
(*a*) diode (*b*) triode
(*c*) tetrode (*d*) pentode

199. In a triode
(*a*) grid is nearer to cathode than anode
(*b*) grid is nearer to anode than cathode
(*c*) grid is equidistant from anode and cathode
(*d*) any of above

200. Which is correct for a vacuum triode?
(*a*) $\mu = r_p g_m$ (*b*) $r_p = \mu g_m$
(*c*) $g_m = \mu r_p$ (*d*) $\mu = \sqrt{r_p g_m}$

201. If i_b is plate current, e_b is plate voltage and e_c is grid voltage the *v-i* curve of a vacuum triode is
$i_b = 0.003\,(e_b + ke_c)^n$
Typical values of *k* and *n* are
(*a*) 0.5 and 1 (*b*) – 2 and 1
(*c*) 30 and 1 (*d*) 30 and 1.5

202. When the gate terminal of MOSFET is positive, it is said to operate in
(*a*) depletion mode (*b*) conduction mode
(*c*) enhancement mode (*d*) none of above

203. The channel of JFET consists of
(*a*) *p* type material only (*b*) *n* type material only
(*c*) conduction material
(*d*) either *p* or *n* type material

204. In *n* type MOSFET, the substrate
(*a*) is *p* type (*b*) is *n* type
(*c*) is metallic (*d*) may be *p* or *n* type

205. Which of the following exhibits negative resistance in a portion of its characteristics?
(*a*) Tunnel diode (*b*) Zener diode
(*c*) JFET
(*d*) MOSFET and Tunnel diode

206. As compared to an ordinary *p-n* diode, the extent of impurity atoms in a tunnel diode
(*a*) is more
(*b*) is less
(*c*) may be more or less
(*d*) is almost the same

207. Which of the following is used for generating time varying wave forms?
(*a*) MOSFET (*b*) PIN diode
(*c*) Tunnel diode (*d*) UJT

208. A JFET
(*a*) is current controlled device
(*b*) has low input resistance
(*c*) has high gate current
(*d*) is a voltage controlled device

209. In a JFET the width of channel is controlled by
(*a*) gate voltage (*b*) drain current
(*c*) source current (*d*) all the above

210. In *n* channel JFET, the gate voltage is made more negative
(*a*) the channel width will increase

(*b*) the channel width would decrease
(*c*) the channel width and drain current will decrease
(*d*) the channel width will decrease and drain current will increase

211. In a JFET
(*a*) the source gate junction is forward biased and gate drain junction is reverse biased
(*b*) both the junctions are forward biased
(*c*) both the junctions are reverse biased
(*d*) the source gate junction is reverse biased and gate drain junction is forward biased

212. The carriers is *n* channel JFET are
(*a*) free electrons and holes
(*b*) holes
(*c*) free electrons of holes
(*d*) free electrons

213. Compared to bipolar junction transistor, a JFET has
(*a*) lower input impedance
(*b*) high input impedance and high voltage gain
(*c*) higher voltage gain
(*d*) high input impedance and low voltage gain

214. In a JFET
(*a*) drain current is very nearly equal to source current
(*b*) drain current is much less than source current
(*c*) drain current may be equal to or less than source current
(*d*) drain current may be even more than source amount

215. The drain characteristics of JFET are drawn between
(*a*) V_{GS} and V_{DS} for different values of drain current
(*b*) drain current and V_{GS} for different values of V_{DS}
(*c*) drain current and V_{DS} for different values of V_{GS}
(*d*) drain current and V_{GS} for one value of V_{DS}

216. In a JFET avalanche breakdown occurs when $V_{DS} = 22V$ and $V_{GS} = 0$. If $V_{GS} = -1$ V, the avalanche breakdown will occur at
(*a*) $V_{DS} = 22$ V
(*b*) V_{DS} more than 22 V
(*c*) V_{DS} equal to or more than 22 V
(*d*) V_{DS} less than 22 V

217. The normal operation of JFET is
(*a*) in constant voltage region
(*b*) constant current region
(*c*) both constant voltage and constant current regions
(*d*) either constant voltage or constant current region

218. In a JFET, the drain current is maximum when
(*a*) $V_{GS} = 0$
(*b*) V_{GS} is not zero but is slightly negative
(*c*) V_{GS} is negative
(*d*) V_{GS} is negative and equal to V_{DS}

219. Which of the following is known as insulated gate FET?
(*a*) JFET
(*b*) MOSFET
(*c*) both JFET and MOSFET
(*d*) none of above

220. The units for transconductance are
(*a*) ohms (*b*) amperes
(*c*) volts (*d*) siemens

221. Trans conductance indicates the
(*a*) control of output voltage by input voltage
(*b*) control of output current by input voltage
(*c*) control of input voltage by output current
(*d*) control of input resistance by input voltage

222. In a JFET
(*a*) gate source cut of voltage is equal to V_{DS}
(*b*) gate source cut off voltage is equal to pinch off voltage
(*c*) gate source cut off voltage is twice the pinch off voltage
(*d*) gate source cut off voltage is equal to negative of pinch of voltage

223. *n*-channel FETs are better as compared to *p*-channel FET because
(*a*) they are more efficient
(*b*) they have high switching time
(*c*) they have higher input impedance
(*d*) mobility of electrons is more than that of holes

224. In an insulated gate FET, the polarity of inversion layer is the same as that of
(*a*) minority carriers in source
(*b*) majority carriers in source
(*c*) charge on gate electrode
(*d*) minority carriers in drain

225. The behaviour of a JFET is similar to that of
(*a*) NPN transistor (*b*) PNP transistor
(*c*) SCR (*d*) Vacuum triode

226. If the gate of JFET is reverse biased, the width of depletion region
(*a*) becomes zero
(*b*) is uniform
(*c*) is more near the source
(*d*) is more near the drain

227. The word enhancement mode is associated with
(*a*) tunnel diode (*b*) MOSFET
(*c*) JFET (*d*) photo diode

228. As compared to bipolar junction transistor, a FET
(*a*) is less noisy
(*b*) has better thermal stability
(*c*) has higher input resistance
(*d*) all the above

229. Which of the following are voltage controlled devices?
(*a*) Vacuum triode (*b*) FET
(*c*) SCR (*d*) both (*a*) and (*b*)

230. In an *n*-channel JFET, the gate is
(*a*) *n* type (*b*) *p* type
(*c*) either *n* or *p* (*d*) partly *n* and partly *p*

231. The depletion layer around *p-n* junction in JFET consists of
(*a*) holes (*b*) electrons
(*c*) immobile charges (*d*) none of above

232. In an *n* channel JFET
(*a*) I_D, I_S and I_G are considered positive when flowing into the transistor
(*b*) I_D and I_S are considered positive when flowing into transistor and I_G is considered positive when flowing out of it
(*c*) I_D, I_S, I_G are considered positive when flowing out of transistor
(*d*) I_S and I_G are considered positive when flowing into transistor and I_D is considered positive when flowing out of it

233. Transconductance characteristics of JFET depict the relation between
(*a*) I_D and V_{GS} for different value of V_{DS}
(*b*) I_D and V_{DS} for different values of V_{GS}
(*c*) V_{GS} and V_{DS} for different values of I_D
(*d*) V_{DS} and I_D for different values of V_{GS}

234. In an *n*-channel JFET, $V_{GS} = V_{GS(off)}$. Then
(*a*) I_D is zero
(*b*) I_D may be zero or positive
(*c*) I_D is positive
(*d*) I_D may be zero or negative

235. The drain characteristics of JFET in operating region, are
(*a*) inclined upwards
(*b*) almost flat
(*c*) inclined downwards
(*d*) inclined upwards or downwards

236. In a JFET V_{DS} exceeds the rated value. Then it operates in
(*a*) active region
(*b*) ohmic region
(*c*) cut off region
(*d*) either cut off or active region

237. A JFET operates in ohmic region when
(*a*) $V_{GS} = 0$
(*b*) V_{GS} is less than pinch off voltage
(*c*) V_{GS} is positive
(*d*) $V_{GS} = V_{DS}$

238. A JFET behaves as a constant current source when
(*a*) $V_{GS} = 0$
(*b*) V_{GS} is less than pinch off voltage
(*c*) $V_{GS} = V_{DS}$
(*d*) V_{GS} is more than pinch off voltage

239. In a JFET, the drain current is maximum when
(*a*) $V_{GS} = V_{DS}$
(*b*) V_{GS} = pinch off voltage
(*c*) $V_{GS} = 0$
(*d*) V_{GS} is negative

240. In an *n*-channel JFET
(*a*) I_D and V_{DS} are positive but V_{GS} is negative
(*b*) I_D and V_{GS} are positive but V_{DS} is negative
(*c*) V_{DS} and V_{GS} are positive but I_D is negative
(*d*) I_D, V_{DS} and V_{GS} are all positive

241. Which of the following can be operated with positive as well as negative gate voltage?
(*a*) JFET
(*b*) both JFET and MOSFET
(*c*) MOSFET
(*d*) neither JFET nor MOSFET

242. An *n*-channel depletion type MOSFET has
(*a*) lightly doped *p* substrate and highly doped *n* source and drain
(*b*) highly doped *p* substrate and highly doped *n* source and drain
(*c*) highly doped *p* substrate and lightly doped *n* source and drain
(*d*) lightly doped *n* substrate and highly doped *n* source and drain

243. Match List I with List II using the given codes

List I	List II
A NPN transistor	1. 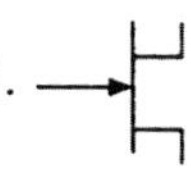
B PNP transistor	2. 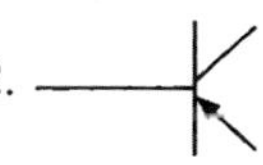
C *n*-channel JFET	3. 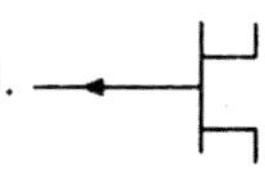
D *p*-channel JFET	4. 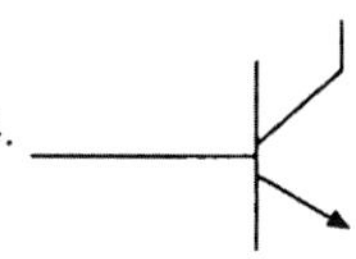

Codes:

	A	B	C	D
(*a*)	1	2	3	4
(*b*)	3	4	2	1
(*c*)	4	2	1	3
(*d*)	1	2	4	3

244. Match List I with List II using the given codes

List I	List II
A *n*-Channel JFET	1. 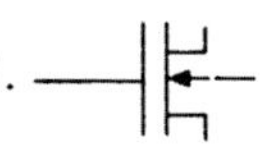
B *p*-Channel JFET	2. 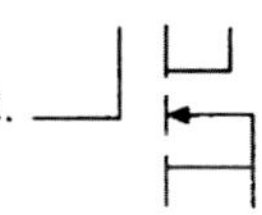
C Depletion MOSFET	3. 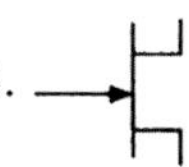
D Enhancement MOSFET	4. 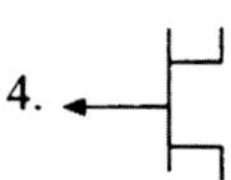

Codes:

	A	B	C	D
(*a*)	4	3	2	1
(*b*)	1	3	4	2
(*c*)	3	4	1	2
(*d*)	2	4	3	1

245. An enhancement mode MOSFET is off when the gate voltage is
(*a*) Zero
(*b*) negative
(*c*) less than threshold value
(*d*) none of above

246. An enhancement mode MOSFET is on when the gate voltage is
(*a*) zero (*b*) positive
(*c*) high
(*d*) more than threshold value

247. Which of the following devices has a silicon dioxide layer?
(*a*) NPN transistor (*b*) Tunnel diode
(*c*) JFET (*d*) MOSFET

248. In which *n* type device does *p* substrate extend upto silicon dioxide layer
(*a*) JFET
(*b*) Depletion type MOSFET
(*c*) Enhancement type MOSFET
(*d*) both (*b*) and (*c*)

249. Which of the following devices has substrate
(*a*) JFET
(*b*) Depletion type MOSFET
(*c*) Enhancement type MOSFET
(*d*) both (*b*) and (*c*)

250. Match List I with List II using the given codes

List I | **List II**

A *n*-channel enhancement MOSFET
1. 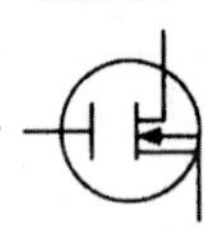

B *p*-channel enhancement MOSFET
2. 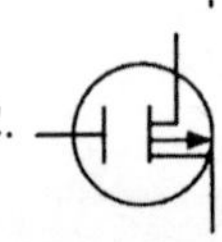

C *n*-channel depletion MOSFET
3. 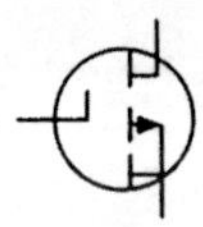

D *p*-channel depletion MOSFET
4. 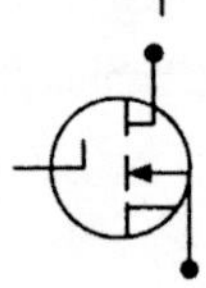

Codes:

	A	**B**	**C**	**D**
(*a*)	1	4	3	2
(*b*)	2	3	4	1
(*c*)	4	3	1	2
(*d*)	1	3	2	4

251. What is the function of silicon dioxide layer is MOSFETs
(*a*) to provide high input resistance
(*b*) to increase current carriers
(*c*) to provide high output resistance
(*d*) both (*a*) and (*c*)

252. In an *n*-channel MOSFET, the substrate is connected
(*a*) to negative terminal of battery
(*b*) to positive terminal of battery
(*c*) to either positive or negative terminal of battery
(*d*) none of above

253. Which quantity controls the effectiveness of LED in emitting light
(*a*) applied voltage (*b*) current
(*c*) extent of doping (*d*) temperature

254. In a solar cell, the photo voltaic voltage is the voltage at which the resultant current is
(*a*) positive (*b*) zero
(*c*) negative (*d*) rated current

255. For radiating ultraviolet rays, LEDs use
(*a*) zinc sulphide (*b*) gallium asenide
(*c*) gallium phosphide (*d*) none of above

256. In the circuit of Fig. 2.30 the function of resistor R and diode D are

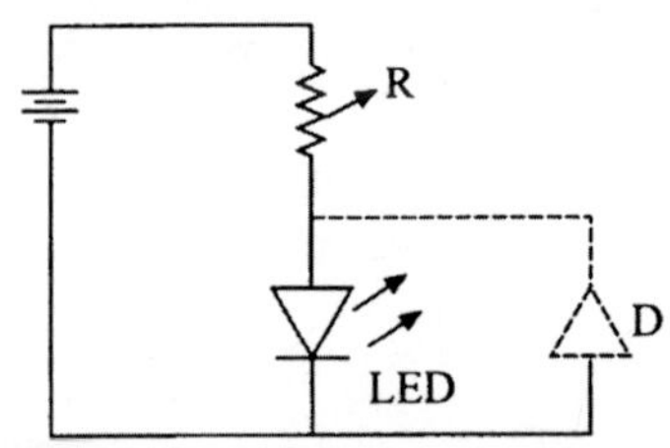

Fig. 2.30.

(*a*) to limit the current and to protect LED against over voltage
(*b*) to limit the voltage and to protect LED against over current
(*c*) to limit the current and protect LED against reverse breakdown voltage
(*d*) none of above

257. When the *i*-*v* curve of a photo diode passes through origin the illumination is
(*a*) maximum (*b*) minimum
(*c*) zero (*d*) equal to rated value

258. Photo conductive devices uses
(*a*) metallic conductors
(*b*) good quality insulators
(*c*) semiconductors
(*d*) either (*a*) or (*c*)

259. In which device does the extent of light controls the conduction
(*a*) photovoltaic cell (*b*) Photo electric relay
(*c*) LED
(*d*) photo sensitive device

260. Light dependent resistor is
(*a*) photo resistive device (*b*) photo voltaic device
(*c*) photo emissive device (*d*) either (*a*) or (*c*)

261. In a photo emissive device the emission efficiency is increased by
(*a*) coating the cathode by an active material
(*b*) coating the cathode by an insulating material

(*c*) heating the cathode
(*d*) cooling the anode

262. Light dependent resistors are
(*a*) highly doped semiconductor
(*b*) intrinsic semiconductor
(*c*) lightly doped semiconductor
(*d*) either (*a*) or (*c*)

263. An LED is
(*a*) an ohmic device (*b*) a display device
(*c*) a voltage regulated device
(*d*) all the above

264. In LED the radiation is in
(*a*) visible region (*b*) infrared region
(*c*) both (*a*) and (*b*) (*d*) neither (*a*) nor (*b*)

265. The emission of light is LED due to
(*a*) emission of holes
(*b*) emission of electrons
(*c*) generation of electromagnetic radiations
(*d*) conversion of heat energy into illumination

266. In a photo transistor the photo current is generated at
(*a*) emitter base junction (*b*) collector base junction
(*c*) collector (*d*) either (*a*) or (*b*)

267. The range of visible light is
(*a*) 300 to 2000 Å (*b*) 2000 to 4000 Å
(*c*) 4000 to 7700 Å (*d*) more than 10000 Å

268. An LED has a rating of 2 V and 10 mA. It is used along with 6 V battery. The range of series resistance is
(*a*) 0 to 200 Ω (*b*) 200 Ω to 400 Ω
(*c*) 200 Ω and above (*d*) 400 Ω and above

269. The sensitivity of human eyes is maximum at
(*a*) white portion of spectrum
(*b*) green portion of spectrum
(*c*) red portion of spectrum
(*d*) violet portion of spectrum

270. In a photo diode the current is due to
(*a*) majority carriers
(*b*) minority carriers
(*c*) both majority and minority carriers
(*d*) either (*a*) or (*b*)

271. AE 139 is a
(*a*) tunnel diode
(*b*) germanium power transistor
(*c*) photo conductive cell
(*d*) silicon diode

272. *n* type semiconductors
(*a*) are negatively charged
(*b*) are produced when indium is added as impurity to germanium
(*c*) are produced when phosphorous is added as impurity to silicon
(*d*) none of above

273. Consider the following statement associated with bipolar junction transistor and JFET
1. The former has higher input impedance than the later
2. The former has higher frequency stability than the later
3. The later has lower noise figure than the former
4. The later has higher power rating than the former

Of these statements
(*a*) 1 and 2 are correct (*b*) 2 and 3 are correct
(*c*) 3 and 4 are correct (*d*) 1 and 4 are correct

274. The collector to emitter cut of current (I_{CEO}) of a transistor is related to collector to base cut off current (I_{CBO}) as
(*a*) $I_{CEO} = I_{CBO}$ (*b*) $I_{CEO} = \alpha I_{CBO}$
(*c*) $I_{CEO} = \dfrac{I_{CBO}}{1-\alpha}$ (*d*) $I_{CEO} = \dfrac{I_{CBO}}{1+\alpha}$

275. The *i-v* characteristics of tunnel diode is best represented as

(*a*) 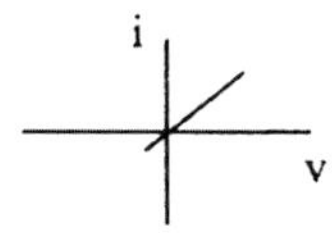

(*b*) 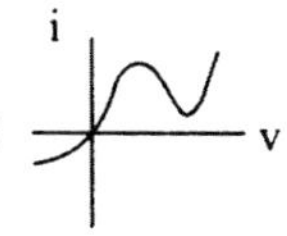

(*c*) 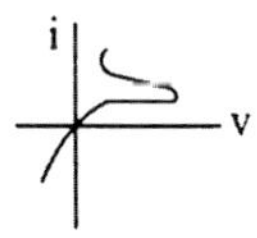

(*d*) 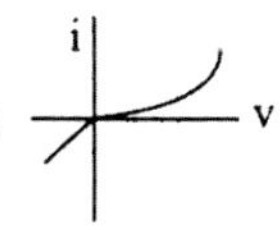

276. When a BJT is operated under saturated condition
(*a*) both junctions are forward biased
(*b*) both junctions are reverse biased
(*c*) CB junctions is forward biased and EB junction is reverse biased
(*d*) EB junction is forward biased and CB junction is reverse biased

277. In an ideal junction transistor the impurity concentration in emitter (E), base (B) and collector (C) is such that
(*a*) E > B > C (*b*) B > C > E
(*c*) C > E > B (*d*) C = E = B

278. Of the various compacitances associated with a junction transistor the gain band width product is affected to maximum extent by
(*a*) base collector parasitic capacitor
(*b*) base collector space charge layer capacitance
(*c*) base emitter space charge layer capacitance
(*d*) base emitter diffusion capacitance

279. An *n*-channel JFET has I_{DS} whose value is
(*a*) maximum for $V_{GS} = 0$ and minimum for V_{GS} negative and large
(*b*) minimum for $V_{GS} = 0$ and maximum for V_{GS} negative and large
(*c*) maximum for $V_{GS} = 0$ and minimum for V_{GS} positive and large
(*d*) minimum for $V_{GS} = 0$ and maximum for V_{GS} positive and large

280. Which of the following characteristics of a silicon *p-n* junction diode make it suitable for use as ideal diode
1. It has low saturation current
2. It has high value of cut in voltage
3. It can withstand large reverse voltage

4. When compared with germanium diode, silicon diode shows a lower degree of temperature dependence under reverse conditions

Select the answer using the code given below

(*a*) 1 and 2 (*b*) 1, 2, 3, 4

(*c*) 2, 3, 4 (*d*) 1, 3

281. At room temperature the current in an intrinsic semi-conductor is due to

(*a*) holes (*b*) electrons

(*c*) ions (*d*) holes and electrons

282. Which of the following is true as regards photo emission?

(*a*) velocity of emitted electron is dependent on light intensity

(*b*) rate of photo emission is inversely proportional to light intensity

(*c*) maximum velocity of electron increases with decreasing wave length

(*d*) both holes and electrons are produced

283. Measurement of hall coefficient enables the determination of

(*a*) mobility of charge carriers

(*b*) type of conductivity and concentration of charge carriers

(*c*) temperature coefficient and thermal conductivity

(*d*) fermi level and forbidden energy gap .

284. The output *v-i* characteristic of enhancement type MOSFET has

(*a*) only an ohmic region

(*b*) only a saturation region

(*c*) an ohmic region at low voltage value and a saturation region at high voltage

(*d*) a saturation region at low voltage value and an ohmic region at high voltage

285. Match List I with List II using the given codes

List I (Characteristics)	List II (Device)
A Voltage controlled device	1. BJT
B Current controlled device	2. UJT
C Conductivity modulation device	3. FET
D Negative conductance device	4. Impatt diode

Codes :

	A	B	C	D
(*a*)	2	3	1	4
(*b*)	2	3	4	1
(*c*)	3	1	2	4
(*d*)	3	1	4	2

286. The kinetic energy of free electrons in a metal is (where *k* is de Broglie wave number of the electrons)

(*a*) $\propto \frac{1}{k}$ (*b*) $\propto \frac{1}{k^2}$

(*c*) $\propto k$ (*d*) $\propto k^2$

287. For BJT under saturation condition

(*a*) $I_C = \beta I_B$

(*b*) $I_C > \beta I_B$

(*c*) I_C is independent of all other parameters

(*d*) $I_C < \beta I_B$

288. A Schottky diode clamp is used along with switching BJT for

(*a*) reducing the power dissipation

(*b*) reducing the switching time

(*c*) increasing the value of β

(*d*) reducing the base current

289. For an *n*-channel JFET for V_{DS} constant and if V_{GS} is made more negative, pinch off would occur at

(*a*) higher value of drain current

(*b*) saturation value of drain current

(*c*) zero drain current

(*d*) gate current equal to drain current

290. Consider the following statements:

The function of oxide layer in an IC device is to

1. mask against diffusion or ion implant
2. insulate the surface electrically
3. increase the melting point of silicon
4. produce a chemically stable protective layer

Of these statements

(*a*) 1, 2, 3 are correct (*b*) 1, 3, 4 are correct

(*c*) 2, 3, 4 are correct (*d*) 1, 2, 4 are correct

291. Figure 2.31 shows small signal common base transistor circuit. The current source I and resistor R on the output side are

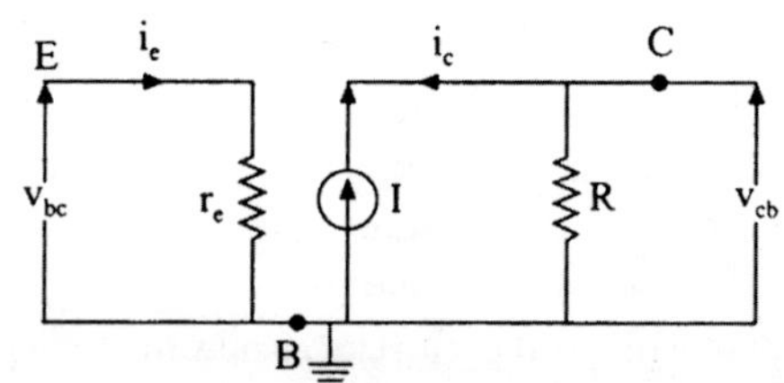

Fig. 2.31.

(*a*) $\frac{\alpha}{1-\alpha} i_e$ and r_e (*b*) $\frac{\alpha}{1-\alpha} i_e$ and $(1-\alpha) r_e$

(*c*) $(1-\alpha) i_e$ and $(1-\alpha) r_e$ (*d*) αi_e and r_e

292. For the transistor equivalent circuit of Fig. 2.32 match List I with List II using the given codes

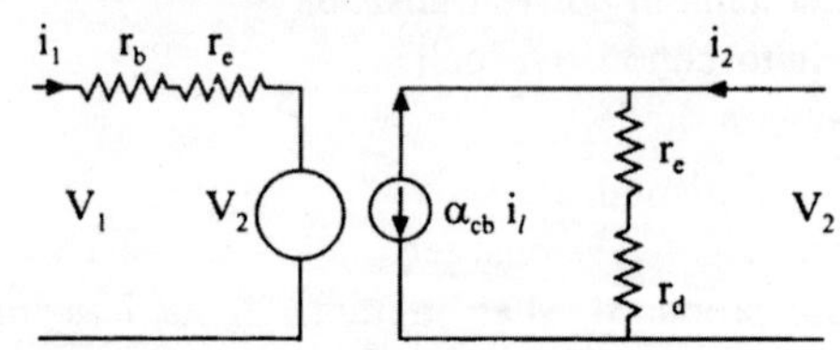

Fig. 2.32.

List I (Parameters)	List II (Values)
A. h_{22}	1. $r_b + r_e$
B. h_{11}	2. α_{cb}
C. h_{21}	3. $\frac{1}{r_e + r_d}$

Codes:

	A	B	C
(*a*)	3	1	2
(*b*)	1	3	2
(*c*)	2	3	1
(*d*)	3	2	1

293. For a photo conductor with equal electron and hole mobilities and perfect ohmic contacts at the ends, an increase in illumination results in
(*a*) a change in O.C. voltage
(*b*) a change in S.C. current
(*c*) a reduction in resistance
(*d*) an increase in resistance

294. Consider the following statements about conditions that make a metal semiconductor contact rectifying
1. N type semiconductor with work function ϕ_S more than work function ϕ_M of metal
2. N type semiconductor with work function ϕ_S less than work function Q_M of metal
3. P type semiconductor with work function ϕ_S more than work function Q_M of metal
4. P type semiconductor with work function ϕ_S less than work function Q_M of metal

Of these statements
(*a*) 1 and 3 are correct (*b*) 2 and 3 are correct
(*c*) 1 and 4 are correct (*d*) 2 and 4 are correct

295. Gold is often diffused into silicon P-N junction devices to
(*a*) increase recombination rate
(*b*) reduce recombination rate
(*c*) make silicon a direct gap semi conductor
(*d*) make silicon semi metal

296. A sample of N type semiconductor has electron density of $6.25 \times 10^8/cm^3$ at 300 K. If the intrinsic concentration of carriers in thisl sample is $2.5 \times 10^{13}/cm^3$ at this temperature, the hole density works out to be
(*a*) $10^6/cm^3$ (*b*) $10^8/cm^3$
(*c*) $10^{10}/cm^3$ (*d*) $10^{12}/cm^3$

297. The current in a *p-n* junction diode with V volts applied *p* to region relative to *n* region (where I_0 is reverse saturation current, *m* is ideality factor, *k* is Boltzmann constant, T is absolute temp and *q* is charge on electron) is
(*a*) $I_0 (e^{qV/mkT} - 1)$ (*b*) $I_0 e^{-qV/mkT}$
(*c*) $I_0 (1 - e^{qV/mkT})$ (*d*) $I_0 (e^{qV/mkT} - 1)$

298. The small signal input impedance of a transistor whose output is shorted for the measuring signal is
(*a*) $h_{11} = \frac{v_i}{i_1} / v_2 = 0$ (*b*) $h_{12} = \frac{v_i}{i_2} / v_1 = 0$
(*c*) $h_{21} = \frac{i_2}{i_1} / v_2 = 0$ (*d*) $h_{22} = \frac{i_2}{v_2} / i_1 = 0$

299. If both emitter base and collector base junctions of a BJT are forward biased, the transistor is in
(*a*) active region (*b*) saturated region
(*c*) cut off region (*d*) inverse mode

300. If the drift velocity of holes under a field gradient of 100 V/m is 5 m/s, their mobility (in SI units) is
(*a*) 0.05 (*b*) 0.5
(*c*) 50 (*d*) 500

301. The *i-v* characteristics of a FET is shown in Fig. 2.33. In which region is the device biased for small signal amplification

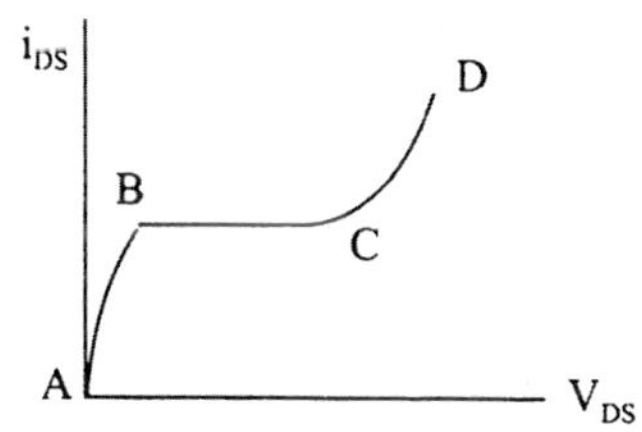

Fig. 2.33.

(*a*) AB (*b*) BC
(*c*) CD (*d*) BD

302. Typical values of *h* parameters at about 1 mA collector current for small signal audio amplifier in CE configuration are

	h_{ie}	h_{re}	h_{fe}	h_{oe}
(*a*)	100 Ω	10^{-1}	50	1 m mho
(*b*)	5 kΩ	10^{-4}	200	20 m mho
(*c*)	5 kΩ	0	20	3 m mho
(*d*)	100 kΩ	10^{-2}	100	10 m mho

303. For a junction FET in the pinch off region as the drain voltage is increased, the drain current
(*a*) becomes zero (*b*) abruptly decreases
(*c*) abruptly increases (*d*) remains constant

304. Some commonly used power devices are given in List I and the approximate maximum frequencies to which they can be used for switching large current is given in List II match List I and List II using the given codes

List I (Power device)	List II (Max. frequency)
A. Power Bipolar	1. MHz
B. Power MOSFET	2. 100 kHz
C. SCR	3. 10 kHz
D. Insulated gate bipolar transistor	4. 1 kHz

Codes:

	A	B	C	D
(*a*)	2	1	3	4
(*b*)	3	4	2	1
(*c*)	3	1	4	2
(*d*)	4	1	3	2

305. The conductivity of an intrinsic semiconductor is
(*a*) generally less than that of a doped semiconductor
(*b*) given by $\sigma_1 = en_i (\mu_n - \mu_p)$
(*c*) given by $\sigma_1 = en_i (\mu_p + \mu_n)$
(*d*) given by $\sigma_1 = n_i (\mu_n + \mu_p)$

306. Hall effect is observed in a specimen when it is carrying current and is placed in a magnetic field. The resultant electric field inside the specimen is
(*a*) normal to both current and magnetic field

(*b*) in the direction of current
(*c*) anti-parallel to magnetic field
(*d*) in arbitrary direction

307. In p type semiconductors the conduction due to holes ($= \sigma_p$) is (where e = charge on hole, μ_p is hole mobility and p is hole concentration)

(*a*) $\frac{pe}{\mu_p}$ (*b*) $\frac{\mu_p}{pe}$
(*c*) $pe\mu_p$ (*d*) $\frac{1}{pe\mu_p}$

308. The main reason why Ni – Zn ferrites are used for audio and TV transformers is
(*a*) they have large saturation magnetisation
(*b*) they are easy to fabricate by rolling
(*c*) they are free from pores and voids
(*d*) they have high elect-resistivity

309. If the energy gap of a semiconductor is 1.1 eV, then it would be
(*a*) opaque to visible light
(*b*) transparent to visible light
(*c*) transparent to ultraviolet radiation
(*d*) transparent to infrared radiation

310. The on voltage and forward break over voltage of an SCR depend on
(*a*) gate current alone
(*b*) band gap of semiconductor alone
(*c*) gate current and semiconductor band gap respectively
(*d*) semiconductor band gap and gate current respectively

311. An incremental model of a solid state device is one which represents the
(*a*) ac property of the device at desired operating point
(*b*) dc property of the device at all operating points
(*c*) complete ac and dc behaviour at all operating points
(*d*) ac property of the device at all operating points

312. The threshold voltage of a MOSFET can be lowered by
1. using thinner gate oxide
2. reducing the substrate concentration
3. increasing the substrate concentration
Of the above statement
(*a*) 3 alone is correct (*b*) 1 and 2 are correct
(*c*) 1 and 3 are correct (*d*) 2 alone is correct

313. Which of the following statements regarding two transistor model of *p-n-p-n* device is correct?
(*a*) It explains only the turn on portion of the device characteristic
(*b*) It explains only the turn off portion of the device characteristic
(*c*) It explains only the negative region portion of the device characteristic
(*d*) It explains all the regions of the device characteristics

314. In a piezo electric crystal, applications of a mechanical stress would produce
(*a*) plastic deformation of the crystal
(*b*) magnetic dipoles in the crystal
(*c*) electrical polarization in the crystal
(*d*) shift in the Fermi level

315. Which of the following elements act as donor impurities?
1. Gold 2. Phosphorous 3. Boron
4. Antimony 5. Arsenic 6. Indium
Select the answer using the following codes
(*a*) 1, 2 and 3 (*b*) 1, 2, 4 and 6
(*c*) 3, 4, 5 and 6 (*d*) 2, 4 and 5

316. Consider the following statements
1. Acceptor level lies close to the valence band
2. Donor level lies close to the valence band
3. n type semiconductor behaves as a conductor at 0 K
4. p type semiconductor behaves as an insulator at 0 K
Of these statements
(*a*) 2 and 3 are correct (*b*) 1 and 3 are correct
(*c*) 1 and 4 are correct (*d*) 3 and 4 are correct

317. Match List I (devices) with List II (characteristic) and select the answer using given codes

List I	**List II**
A BJT	1. Voltage controlled negative resistance
B MOSFET	2. High current gain
C Tunnel diode	3. Voltage regulation
D Zener diode	4. High input impedance

Codes:

	A	B	C	D
(*a*)	1	4	2	3
(*b*)	2	4	1	3
(*c*)	2	3	1	4
(*d*)	1	3	2	4

318. Which one of the following pairs of semiconductors and current carriers is correctly matched?
(*a*) intrinsic : number of electrons = number of holes
(*b*) p type : number of electrons > number of holes
(*c*) n type : number of electrons < number of holes
(*d*) Bulk : Neither electrons nor holes

319. In a junction transistor the recombination of electron and holes occurs in **[IES 2010]**
(*a*) base region only (*b*) emitter region only
(*c*) collector region only (*d*) all the three regions

320. Consider the following statements as regards FET
(*a*) Its operation depends on flow of majority carriers only
(*b*) It has high input resistance
(*c*) It is suitable for high frequency
(*d*) Its operation depends on flow of both majority and minority carrier
Which of the above are correct?
(*a*) 1, 2, 3, 4 (*b*) 1 and 2 only
(*c*) 2 and 3 only (*d*) 3 and 4 only
[IES 2010]

321. When donor atoms are added to intrinsic semiconductor, it

(*a*) increases the energy band gap of semiconductor
(*b*) decreases the energy band gap of semiconductor
(*c*) introduces a new narrow band gap near conduction band
(*d*) introduces a new discrete energy level below the conduction band **[IES 2010]**

322. Consider the following statements
1. Acceptor energy level is formed very close to conduction band
2. Effective mass of free electron is same as that of hole
3. Magnitude of charge of free electron is same as that of hole
4. Addition of donor impurities adds holes to semiconductor

Which of above are correct?
(*a*) 1, 3 (*b*) 2, 3
(*c*) 2, 4 (*d*) 3, 4 **[IES 2009]**

323. As per Hall effect if a specimen carrying current I is placed in a transverse magnetic field B, then an electric field E is induced in the specimen in the direction
(*a*) parallel to I
(*b*) perpendicular to B and parallel to I
(*c*) parallel to I and B
(*d*) perpendicular to both B and I **[IES 2009]**

324. JFET can be used as voltage variable resistor
(*a*) at pinch off condition
(*b*) beyond pinch off voltage
(*c*) well below pinch off voltage
(*d*) for any value of V_{DS} **[IES 2009]**

325. The symbol in Fig. 2.34 is for

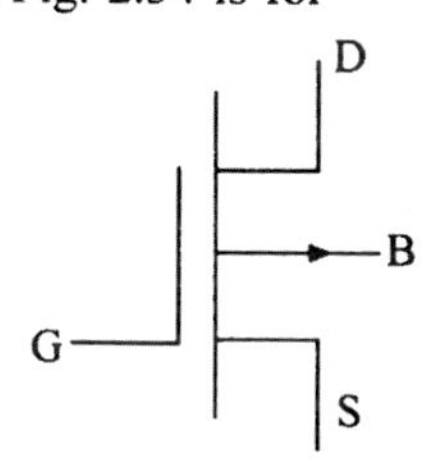

Fig. 2.34.

(*a*) *p*-channel depletion MOSFET
(*b*) *p*-channel enhancement MOSFET
(*c*) complementary MOSFET
(*d*) *p*-channel JFET **[IES 2009]**

326. The maximum power dissipation capacity of a transistor is 50 mW. If V_{CE} is 10 V, what is safe collector current?
(*a*) 5 mA (*b*) 2.5 mA
(*c*) 10 mA (*d*) 25 mA **[IES 2009]**

327. Which of the following is used in microprocessors
(*a*) JFET (*b*) BJT
(*c*) MOSFET (*d*) CMOS **[IES 2009]**

328. Which of the following is NOT LED material
(*a*) Ga As (*b*) GaP
(*c*) SiC (*d*) SiO_2 **[IES 2009]**

329. The minimum energy of a photon required for intrinsic excitation is equal to
(*a*) energy of bottom of conduction band
(*b*) energy of top of valence band
(*c*) forbidden gap energy
(*d*) Fermi energy **[IES 2009]**

330. Assuming that electron mobility in intrinsic silicon is 1500 cm^2/Vs at room temperature (300 K) and corresponding volt equivalent of temperature V_T is 25.9 mV what is the approximate value of diffusion constant?
(*a*) 40 cm^2/s (*b*) 4 cm^2/s
(*c*) 400 cm^2/s (*d*) 4000 cm^2/s **[IES 2008]**

331. In a MOSFET the transfer characteristics can be used to determine which of the following parameters
(*a*) Threshold voltage and output
(*b*) Transconductance and output resistance
(*c*) Threshold voltage and transconductance
(*d*) Transconductance and channel length modulation parameter **[IES 2008]**

332. Assuming an operating temperature of 300 K and corresponding V_T = 26 mV, what is the change in semiconductor silicon diode forward voltage V_D to produce a 10 : 1 change in dioide current I_B while operating in forward bias region. (< 25 mA)
(*a*) 60 mV (*b*) 120 m V
(*c*) 180 mV (*d*) 240 mV **[IES 2008]**

333. When a junction diode is used as a half wave rectifier with purely resistive load and sinusoidal input voltage, what is the value of conduction angle (where ϕ is ignition angle corresponding to cut in voltage).
(*a*) π (*b*) $\pi - \phi$
(*c*) $\pi - 2\phi$ (*d*) slightly more than π **[IES 2008]**

334. What is the most noticable effect of a small increase in temperature in CE connected BJT
(*a*) increase in I_{CBO}
(*b*) increase in output resistance
(*c*) decrease in forward current gain
(*d*) increase in forward current gain **[IES 2008]**

335. The data sheet of a JFET indicates that I_{DSS} = 15 mA and V_{GS} = – 5 V. What is I_D for V_{GS} = – 2 V?
(*a*) 58.8 mA (*b*) 29.4 mA
(*c*) 9.6 mA (*d*) 5.4 mA **[IES 2008]**

336. Consider the following statements
n type silicon can be made
1. by adding impurity of phosphorous
2. by adding impurity of arsenic
3. by adding impurity of boron
4. by adding impurity of aluminium

Which of the above statements are correct?
(*a*) 1, 3 (*b*) 3, 4
(*c*) 1, 2 (*d*) 1, 2, 3, 4 **[IES 2008]**

337. Consider the following statements for *p-n* junction diode
1. It is an active component
2. Depletion layer width decreases with forward biasing
3. In reverse biasing, the saturation current increases with increase in temperature

Which of the above statements are correct?

(*a*) 1, 2, 3 (*b*) 1, 2
(*c*) 2, 3 (*d*) 1, 3 **[IES 2008]**

338. Consider the following statements pertaining to tunnel diode
1. Impurity concentration is high
2. Carrier velocities are low
3. It has current controlled *v-i* characteristics

Which of the above statements are correct?
(*a*) 1 only (*b*) 2, 3
(*c*) 1, 3 (*d*) 1, 2 **[IES 2008]**

339. For a semiconductor the conductivity is a function of product of number of charge carriers and their mobilities. If the temperature of a slab of intrinsic semiconductor is increased, how does conductivity change
(*a*) decreases (*b*) increases
(*c*) remains unaffected
(*d*) increases or decreases depending on rise in temperature **[IES 2007]**

340. An *n* type semiconductor is illuminated by a steady flux of photons with energy greater than band gap energy. The change in conductivity $\Delta\sigma$ obeys the relation
(*a*) $\Delta\sigma = 0$ (*b*) $\Delta\sigma = e(\mu_n + \mu_p)\,\Delta n$
(*c*) $\Delta\sigma = e(\mu_n \Delta n - \mu_p \Delta p)$ (*d*) $\Delta\sigma = e\mu_n \Delta n$
[IES 2007]

341. Consider the following statements
Electrical conductivity of a metal has negative temperature coefficient because
1. Electron concentration increases with temperature
2. Electron mobility decreases with temperature
3. Electron lattice scattering increases with temperature

Which of the above are correct?
(*a*) 1, 2, 3 (*b*) 1, 2
(*c*) 2, 3 (*d*) 1, 3 **[IES 2006]**

342. Consider the following statements
FETs when compared to BJTs have
1. high input impedance
2. current flow due to majority carrier only
3. low input impedance
4. current flow due to minority carrier only

Which of the above are correct?
(*a*) 1, 4 (*b*) 2, 3
(*c*) 3, 4 (*d*) 1, 2 **[IES 2006]**

343. Which of the following can be determined using Hall crystal?
1. Concentration of holes in *p* type semiconductor
2. Concentration of electrons in *n* type semiconductor
3. Temperature of any setup with any semiconductor
4. Diffusion constant and life time of minority carriers

Which of the above are correct?
(*a*) Only 1, 2 (*b*) Only 1, 2, 4
(*c*) Only 3, 4 (*d*) Only 3, 4 **[IES 2006]**

344. What are the different methods followed to take *p-n-p-n* device from conducting to non-conducting state
1. Reducing anode current to less than holding current
2. Reducing gate current to zero
3. Reducing gate voltage to zero
4. Reducing anode voltage below the holding value

Which of the above are correct?
(*a*) 1, 2 (*b*) 2, 3
(*c*) 1, 4 (*d*) 3, 4 **[IES 2006]**

345. Compared to *p-n* junction with $N_A = N_D = 10^{14}/cm^3$ which one of the following statements is true for a *p-n* junction with $N_A = N_D = 10^{20}/cm^3$
(*a*) Reverse breakdown voltage is lower and depletion capacitance is lower
(*b*) Reverse breakdown voltage is higher and depletion capacitance is lower
(*c*) Reverse breakdown voltage is lower and depletion capacitance is higher
(*d*) Reverse breakdown voltage is higher and depletion capacitance is higher **[Gate 2010]**

346. In Fig. 2.35 op-amp is ideal. BJT has $V_{BE} = 0.6$ V and $\beta = 150$. Determine whether feedback is positive or negative and find V.

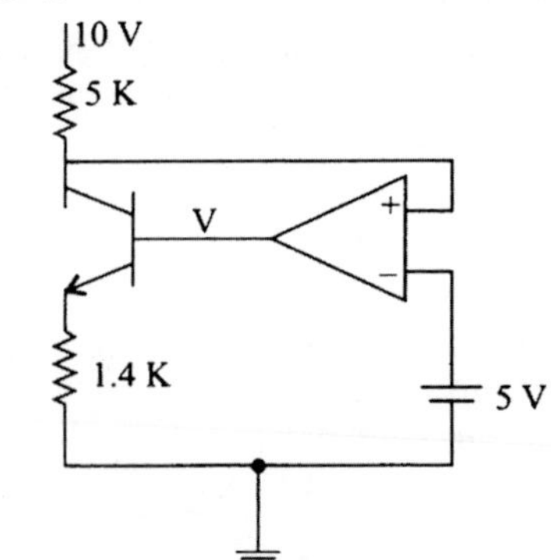

Fig. 2.35.

(*a*) Positive feedback, V = 10 V
(*b*) Positive feedback, V = 0 V
(*c*) Negative feedback, V = 5 V
(*d*) Negative feedback, V = 2 V **[Gate 2009]**

347. The ratio of mobility to diffusion coefficient in a semiconductor has the units of
(*a*) V^{-1} (*b*) cm V^{-1}
(*c*) V cm^{-1} (*d*) V.s **[Gate 2009]**

348. Silicon is doped with boron to a concentration of 4×10^{17} atoms/cm^3. Assume intrinsic carrier concentration to be $1.5 \times 10^{10}/cm^3$ and kT/*q* to be 25 mV at 300 K. Compared to undoped silicon, Femi level in doped silicon
(*a*) goes down by 0.13 eV (*b*) goes up by 0.13 eV
(*c*) goes down by 0.427 eV (*d*) goes up by 0.427 eV
[Gate 2008]

349. The electron and hole concentration in an intrinsic semiconductor are n_i per cm^3 at 300 K. If acceptor impurities are introduced with a concentration of N_A per cm^3 (where $N_A >> n_i$) the electron concentration per cm^3 at 300 K will be
(*a*) n_i (*b*) $n_i + N_A$
(*c*) $N_A - n_i$ (*d*) $\frac{n_i^2}{N_A}$ **[Gate 2007]**

350. In a *p-n* junction diode under reverse bias, the magnitude of electric field is maximum at

(*a*) edge of depletion region on the *p*-side
(*b*) edge of depletion region on the *n*-side
(*c*) *p-n* junction
(*d*) centre of depletion region on *n*-side **[Gate 2007]**

351. Group 1 lists four types of *p-n* junction diodes. Match each device in Group 1 with one of the options in Group 2 to indicate the bias condition of that device in normal mode of operation

	Group 1		Group 2
P	Zener diode	1.	Forward bias
Q	Solar cell	2.	Reverse bias
R	Laser diode		
S	Avalanche Photodiode		

(*a*) P-1, Q-2, R-1, S-2 (*b*) P-2 Q-1, R-1, S-2
(*c*) P-2 Q-2 R-2 S-1 (*d*) P-2 Q-1 R-2 S-2 **[Gate 2007]**

352. A *p-n* junction has a built in potential of 0.8 V. The depletion layer width at a reverse bias of 1.2 V is 2 μm. If reverse bias is 7.2 V, the depletion layer width will be
(*a*) 4 μm (*b*) 4.9 μm
(*c*) 8 μm (*d*) 12 μm **[Gate 2007]**

353. The majority carriers in an *n* type semiconductor have an average drift-velocity *v* in a direction perpendicular to a uniform magnetic field B. The electric field E due to Hall effect acts in the direction.
(*a*) $v \times B$ (*b*) $B \times v$
(*c*) along *v* (*d*) opposite to *v* **[Gate 2006]**

354. Find the correct match between group 1 and group 2

	Group 1		Group 2
E	Varactor diode	1.	Voltage reference
F	PIN diode	2.	High frequency switch
G	Zener diode	3.	Tuned circuits
H	Schottky diode	4.	Current controlled attenuator

(*a*) E-4 F-2 G-1 H-3 (*b*) E-2 F-4 G-1 H-3
(*c*) E-3 F-4 G-1 H-2 (*d*) E-1 F-3 G-2 H-4 **[Gate 2006]**

355. A heavily doped *n* type semiconductor has following data
Hole-electron mobility ratio 0.4
Doping concentration 4.2×10^8 atoms/cm^3
Intrinsic concentration 1.5×10^4 atoms/cm^3
The ratio of conductance of *n* type semiconductor to that of intrinsic semiconductor of the same material at the same temperature will be
(*a*) 0.00005 (*b*) 2000
(*c*) 10000 (*d*) 20000 **[Gate 2006]**

356. The values of voltage across a tunnel diode corresponding to peak and valley currents are V_p and V_v respectively. The range of tunnel diode voltage for which the slope of *i-v* would be negative is
(*a*) $v < 0$ (*b*) $0 \le v \le v_p$
(*c*) $v_p \le v \le v_v$ (*d*) $v \ge v_v$ **[Gate 2006]**

357. The concentration of minority carriers in an extrinsic semiconductor under equilibrium is
(*a*) directly proportional to doping concentration
(*b*) inversely proportional to doping concentration
(*c*) directly proportional to intrinsic concentration
(*d*) inversely proportional to intrinsic concentration **[Gate 2006]**

358. Under low level injection assumption, the injected minority carrier current for extrinsic semiconductor
(*a*) diffusion current (*b*) drift current
(*c*) recombination current (*d*) induced current

B. The following questions have two statements viz Assertion A and Reason R. Examine these statements and decide if A and R are individually true and if R is a correct explanation of A. Select the answer using the following codes
(*a*) Both A and R are true and R is correct explanation of A
(*b*) Both A and R are true but R is NOT a correct explanation of A
(*c*) A is true but R is false
(*d*) A is false but R is true.

359. A. When a photo conductive device is exposed to light, its bulk resistance increases.
R. When exposed to light, electron hole pairs are generated in the photo conductive device.

360. A. *p-n* junction is used as rectifier.
R. *p-n* junction has low resistance in forward direction and high resistance in reverse direction.

361. A. The conductivity of *p* type semiconductor is higher than that of intrinsic semiconductor.
R. The addition of donor impurity creates additional energy levels below conductor band.

362. A. Intrinsic semiconductor is an insulator at 0 K .
R. Fermi level in intrinsic semiconductor is in the center of forbidden energy band.

363. A. The resistivity of intrinsic semiconductor decreases with increases in temperature.
R. The forbidden gap decreases with increase in temperature.

364. A. In *p* type semiconductor conduction is mainly due to holes.
R. In *p* type material the holes are majority carriers.

365. A. Hall effect is used to find the type of semiconductor.
R. When a specimen of semiconductor carrying current I lies in a magnetic field the force on electrons and holes is in opposite directions.

366. A. Germanium is more commonly used than silicon.
R. Forbidden gap in germanium is less than that in silicon.

367. A. Silicon is less sensitive to changes in temperature than germanium.
R. It is more difficult to produce minority carriers in silicon than in germanium.

368. A. A p-n junction has high resistance in reverse direction.
R. When a reverse bias is applied to p-n junction, the width of depletion layer increases.

369. A. A decrease in temperature increases the reverse saturation current in p-n diode.
R. When a diode is reverse biased surface leakage current flows.

370. A. The reverse saturation current in a semiconductor diode is 4n A at 20°C and 32n A at 50°C.
R. The reverse saturation current in a semiconductor diode doubles for every 10°C rise in temperature.

371. A. When a high reverse voltage is applied to a p-n junction the diode breaks down.
R. High reverse voltage causes avalanche effect.

372. A. When forward biased a p-n junction has low resistance.
R. The ratio $\frac{dv}{di}$ is called dynamic resistance.

373. A. When reverse voltage across a p-n junction is increased, the junction capacitance decreases.
R. Capacitance of any layer is inversely proportional to thickness.

374. A. A high junction temperature may destroy a diode.
R. As temperature increases the reverse saturation current increases.

375. A. In a BJT, the base region is very thick.
R. In p-n-p transistor most of holes given off by emitter diffuse through the base.

376. A. In n-p-n transistor conduction is mainly due to electrons.
R. In n type materials electrons are majority carriers.

377. A. In a BJT base current is very small.
R. In a BJT recombination in base region is high.

378. A. In a BJT, α_{dc} is about 0.98.
R. In a BJT, recombination in base region is minimum.

379. A. In p-n-p transistor collector current is termed negative.
R. In a p-n-p transistor holes are majority carriers.

380. A. In an n-p-n transistor as the electrons enter the collector region, they are accelerated towards the collector terminal.
R. Emitter base junction, in BJT, is forward biased.

381. A. In CE connection of n-p-n transistor, V_{CE} is positive.
R. In BJT, the base collector junction is reverse biased.

382. A. In active region of CE output characteristics of BJT, collector current is nearly constant.
R. Base current in CE connection is very small.

383. A. In design of circuit using BJT, a derating factor is used.
R. As the ambient temperature increases, heat dissipation becomes slower.

384. A. BJT can be used as a switch.
R. In forward active mode emitter base junction is forward biased and base collector junction is reverse biased.

385. A. FET has characteristics very similar to that of pentode.
R. Both FET and pentode are voltage controlled devices.

386. A. JFET can be used as a current source.
R. In beyond pinch off region the current in JFET is nearly constant.

387. A. JFET is a voltage controlled device.
B. The drain current can be controlled by controlling V_{GS}.

388. A. The hybrid π model of a transistor can be reduced to h parameter model and vice versa.
R. Hybrid π and h parameter models are interrelated as both of them describe the same device.

389. A. When light falls at junction of p-n photo diode, its p side becomes positive and n side becomes negative.
R. When a photo diode is short-circuited, the current in the external circuit flows from p side to n side.

390. A. In Hall effect the O.C. transverse voltage developed by a current carrying semiconductor with a steady magnetic field perpendicular to the current direction has opposite signs for n and p semiconductors.
R. The magnetic field pushes both holes and electrons in the same direction.

391. A. Two transistors one n-p-n and the other p-n-p are idential in all respects (doping, construction, shape, size). The n-p-n transistor will have better frequency response.
R. The electron mobility is higher than hole mobility.

392. A. Power transistors are more commonly of silicon n-p-n type.
R. The fabrication of silicon n-p-n transistors is easy.

393. A. When a photo conductive device is exposed to light, it bulk resistance increases.
R. When exposed to light, electron hole pairs are generated in the photo conductive device.

394. A. At room temperature the fermi level in p type semiconductor lies nearer to the valence band and in n type semiconductor it lies nearer to the conduction band.
R. At room temperature, the p type semiconductor is rich in holes and n type semiconductor is rich in electrons.

395. A. Oxide coated cathodes are very commonly used.
R. Work function of oxide coated cathode is 1 eV whereas it is 4.5 eV for pure tungsten.

396. A. The frequency of light used for photo electric emission is high.
R. As per Einstein's equation

$$0.5\ mv^2 \le hf - U_w$$

397. A. Field emission is substantially independent of temperature.
R. When a high electric field is created at metal surface field emission may occur.

398. A. The amount of photo electric emission depends on the intensity of incident light.
R. Photo electric emission can occur only if frequency of incident light is less than threshold frequency.

399. A. Alkali metals are used as emitters in photo tubes.
R. Alkali metals have low work functions.

400. A. In intrinsic semiconductors, the charge concentration increases with temperature.
R. At higher temperatures, the forbidden energy gap in semiconductors is lower.

401. A. The conductivity of an *n* type semiconductor increases both with increase in temperature and increase in density of donor atoms.
R. Diffusion of carriers occurs in semiconductors.

402. A. Silicon is preferred over germanium in manufacture of semiconductor devices.
R. Forbidden gap in silicon is more than that in germanium.

403. A. When *p-n* junction is forward biased, steady current flows.
R. When a *p-n* junction is forward biased, free electrons cross the junction from *n* to *p* and holes from *p* to *n*.

404. A. In reverse biased *p-n* junction, the reverse saturation current is nearly constant if the reverse voltage is less than critical value.
R. The total reverse current is sum of reverse saturation current and surface leakage current.

405. A. When used as rectifier the reverse breakdown voltage should not be exceeded.
R. A high inverse voltage can destroy a *p-n* junction.

406. A. The reverse current in a *p-n* junction is nearly constant.
R. The reverse breakdown voltage of a *p-n* diode depends on the extent of doping.

407. A. The forward resistance of a *p-n* diode is not constant.
R. The *v-i* characteristics of *p-n* diode is non-linear.

408. A. The forward dynamic resistance of *p-n* diode varies inversely with current.
R. The forward dynamic resistance of *p-n* diode varies with the operating voltage.

409. A. The behaviour of FET is in many respects similar to that of a pentode.
R. FETs and Vacuum triode are voltage controlled devices.

410. A. FET is a unipolar device.
R. BJT is bipolar device.

411. A. A JFET behaves as a resistor when $V_{GS} < V_P$.
R. When $V_{GS} < V_P$, the drain current in a JFET is almost constant.

412. A. When V_{DS} is more than rated value the drain current in a JEFT is very high.
R. When V_{DS} is more than rated value, avalanche breakdown occurs.

413. A. A VMOS can handle much larger current than other field effect transistors.
R. In a VMOS the conduction channel is very narrow.

414. A. When a zener diode breaks down, the voltage across it is constant.
R. The upper limit of zener current is determined by power handling capacity.

415. A. Tunnel diode is used in many pulse and digital circuits.
R. The *i-v* curve of a Tunnel diode resembles letter 'N'.

416. A. The capacitance of a reverse biased PIN diode is lower than that of reverse biased *p-n* diode.
R. A PIN diode has an intrinsic layer between *p* and *n* regions.

417. A. In a Schottky diode the reverse recovery time is almost zero.
R. A Schottky diode has aluminium silicon junction.

418. A. In an intrinsic semiconductor electron mobility in conduction band is different from hole mobility in valence band.
R. In an intrinsic semiconductor electrons and holes are created solely by thermal excitation across the energy gap. **[IES 2010]**

419. A. At very high temperatures both *p* and *n* type semiconductors behave as intrinsic semiconductor.
R. In *n* type semiconductors the majority carriers are electron and in *p* type semiconductor the majority carriers are holes whereas in intrinsic semiconductor number of holes and electrons are equal. **[IES 2010]**

420. A. In intrinsic semiconductor the concentration of electrons and holes increases with increase in temperature.
R. Law of mass action holds good in case of semiconductors. **[IES 2010]**

421. A. Silicon is mainly used for making ICs and not Ge.
R. In silicon SiO_2 layer which acts as eliminator can be formed for isolation. Corresponding oxide layer cannot be formed in Ge. **[IES 2010]**

422. A. Short-circuit current gain of BJT, in common base configuration increases with increase in reverse bias collector to base voltage.
R. With increase in reverse bias collector to base voltage, effective base width decreases. **[IES 2010]**

423. A. BJT is basically a current amplifier.
R. The most simplified model of a BJT has a current dependent source in the output circuit whose magnitude directly depends on input current. **[IES 2010]**

424. A. If we have two *p-n-p* and *n-p-n* transistors of identical construction, the *n-p-n* transistor will have better frequency response characteristics as compared to *p-n-p*.
R. The diffusion constant of electrons is higher than that of holes. **[IES 2010]**

425. A. An unbiased *p-n* junction develops a built in potential at the junction with *n* side positive and *p* side negative.

R. The *p-n* function behaves as a battery and supplies current to a resistance connected across its terminals. **[IES 2009]**

426. A. Cutin voltage for Germanian diode is higher than that for silicon diode.

R. Germanian diode has higher reverse saturation current than silicon diode. **[IES 2009]**

427. A. For same drain current rating *n*-channel MOSFET occupies more area than *p*-channel MOSFET.

R. Electron mobility is much higher than hole mobility. **[IES 2009]**

428. A. A heavily doped semiconductor can exhibit positive temperature coefficient of resistance.

R. Carrier mobility decreases with increase in temperature. **[IES 2007]**

429. A. Resistors and capacitors fabricated using IC technology have poor tolerances with respect to their absolute values.

R. As all the components of IC are fabricated simultaneously their ratio of tolerance is low. **[IES 2007]**

430. A. An *n* type semiconductor behaves as a intrinsic semiconductor at very high temperatures.

R. Breaking of covalent bands becomes a very significant phenomenon at high temperature. **[IES 2006]**

431. A. Electron mobility in metals decreases with increasing temperature.

R. In metals electron concentration is high. **[IES 2006]**

C. State whether the following are true or false.

432. In a conductor the conduction and valence bands overlap.

433. A metal loses electrons at room temperature.

434. If E (*i.e.*, available energy state) = E_F (*i.e.*, fermi level), then probability that state E will be occupied is 0.5 for any temperature T.

435. Fermi level is the maximum energy that an electron can possess at 0 K.

436. 1 eV = 1.602×10^{-19} joules.

437. Work function is the maximum energy required by the fastest electron at 0 K to escape from the metal surface.

438. Work function of oxide coated cathode is much lower than that of tungsten cathode.

439. The current due to thermionic emission is given by Richardson-Dushman equation.

440. The energy to cause thermionic emission is supplied by heating the cathode.

441. When electronic emission occurs due to bombardment of high velocity electrons on a metal surface, it is called secondary emission.

442. Secondary emission is always detrimental.

443. Electrons can be emitted from a metal surface due to high electric field.

444. The energy of one quantum of light equal to *hf.*

445. Electrons within a metal have energy levels from zero to fermi lever E_F.

446. Photoelectric emission can occur only if the frequency of light is more than threshold frequency.

447. The amount of photo electric emission current depends on the frequency of incident light.

448. Photoelectric effect occurs only in semiconductors and not in metals.

449. In a vacuum triode $\mu = r_p g_m$.

450. The conductivity of germanium increases by about 6 percent per degree increase in temperature.

451. In intrinsic semiconductor magnitude of free electron and hole concentrations are equal.

452. As temperature increases the forbidden gap in silicon increases.

453. Mobility of electrons and holes are equal.

454. The drift velocity of electrons and holes is proportional to electric field strength.

455. Recombination produces new electron-hole pairs.

456. Holes and electrons move in opposite directions.

457. At room temperature kT = 0.03 eV.

458. Fermi level in intrinsic semiconductor is at the centre of forbidden energy band.

459. Donor energy level is *n* type semiconductor is very near valence band.

460. Indium, aluminium, arsenic are all *p* type impurities.

461. The addition of impurity in extrinsic semiconductor is about 1 part in 10^8 parts.

462. In *p* type semiconductor holes are majority carriers.

463. Pentavalent impurity creates *n* type semiconductor.

464. The rate of change of excess carrier density is proportional to carrier density.

465. The mean life time of carrier may range from 10^{-9} seconds to hundreds of μ-seconds.

466. Hall effect can be used to find the type of semiconductor.

467. Hall coefficient is reciprocal of charge density.

468. Mobility is directly proportional to Hall coefficient.

469. When a *p-n* junction in forward biased, the current remains zero till the applied voltage overcomes the barrier potential.

470. The depletion layer consists of immobile ions.

471. When *p-n* junction is reverse biased the holes in *p* material move towards the junction and electrons is *n* material move away from the junction.

472. When a *p-n* junction is forward biased, The width of depletion layer decreases.

473. Junction temperature is always the same as room temperature.

474. An increase in temperature increases the width of depletion layer.

475. The reverse breakdown voltage of a diode depends on the extent of doping.

476. The reverse saturation current depends on the reverse bias.

477. The reverse saturation current of a diode does not depend on temperature.

478. In a forward biased *p-n* junction current enters *p* material as hole current and leaves *n* material as electron current of the same magnitude.

479. When avalanche breakdown occurs covalent bonds are not affected.

480. The *i-v* characteristics of a diode may be linear or non-linear.

481. An increases of reverse voltage decreases the junction capacitance.

482. Varactor diode is forward biased when used.

483. Dynamic resistance of diode is $\frac{dv}{di}$.

484. In *p*-channel JFET, V_{GS} is positive.

D. Fill in the blanks with appropriate words

485. An *n* type silicon sample having electron mobility μ_n = twice the hole mobility μ_p is subjected to a steady illumination such that electron concentration doubles from its initial thermal equilibrium value. As a result the conductivity of the sample increases by a factor of ....

486. The forward dynamic resistance of a junction diode varies .... as the forward current.

487. The transit time of current carriers through the channel of a FET decides its..........characteristics.

488. For the 2N 338 transistor the manufacturer specifies P_{max} = 100 mW at 25°C ambient temperature and maximum junction temperature of 125°C. Its thermal resistance is ....

489. An *n* channel has I_{DSS} = 1 mA and V_P = – 5V. The maximum g_m is ....

490. A semiconductor sample has a charge density of 1.5×10^{-4} C/cm^3. The hall coefficient is ....

491. A silicon diode has a reverse saturation current of 4 nA at 20°C. The reverse saturation current at 50°C is ....

492. A silicon diode has a reverse saturation current of 4 nA when reverse voltage is 2 V. If reverse voltage is 4 V, the reverse saturation current is ....

493. When used as rectifier the reverse breakdown voltage should not be ....

494. In *p-n* diode the reverse break down occurs because a large number of .... carriers appear in the depletion layer.

495. At 25°C the barrier potential for a silicon diode is ....

496. If *n* is the number of free electrons per m^3 of metal, fermi level is proportional to..........

497. Work function is usually expressed in ....

498. Electron volt is a unit of ....

499. The maximum energy which an electron may possess at 0 K is called..........

500. Work-function of pure tungsten is about ....

501. The emission current in photo electric emission depends on ....

502. As compared to germanium diode silicon diode has higher ....voltage and higher .... voltage rating.

503. The conductivity of a *p* type semiconductor is nearly equal to ....

504. In forward direction, the resistance of a *p-n* diode is ....

505. A pure silicon sample has 2 million free electrons. The number of holes in the sample is ....

506. In diode IN4001, the typical values of forward voltage and forward current are 0.93 V and 1 A. The bulk resistance is ....

507. An extrinsic semiconductor has 6 billion silicon atoms and 3 million pentavalent impurity atoms. The number of free electrons is about..........

508. In a *p-n-p* transistor the majority carriers are ....

509. In a BJT, the emitter base junction is ....biased.

510. In a BJT, the collector base junction is ....biased.

511. In schematic representation of *p-n-p* transistor, the arrow head points.......... the base.

512. In active region of BJT (CE connection), the collector current is ....

513. In a BJT, α_{dc} is about ....

514. If $\alpha_{dc} = 0.99$, β_{dc} = ...

515. In saturation region of C_E transistor, V_{CE} is ....

516. In breakdown region of CE transistor, the collector current is ....

517. When used as amplifier a BJT is used in .... region.

518. When both junctions of a BJT are forward biased, it is in .... mode.

519. FET is a .... device.

520. The terminals of FET are known as ....

521. When $V_{GS} < V_p$, a JFET operates in .... region.

522. MOSFETs have a very thin .... layer.

523. In a VMOS the channel has the shape of ....

524. FET is ....noisy than BJT.

525. A zener diode is used in .... circuit.

526. A Schottky diode has .... junction.

527. A .... diode has negligible reverse recovery time.

528. A .... diode has its *i-v* curve in the shape of N.

529. Seven segment display generally uses ....

ANSWERS WITH HINTS AND EXPLANATION

1. (*d*)
2. (*b*) Inner shell electrons are tightly bound to the nucleus. Their energy levels cannot be affected by presence of other atoms.
3. (*b*) This is the reason of high conductivity of conductors.
4. (*c*) 5. (*b*) 6. (*a*) 7. (*d*)
8. (*a*) 9. (*d*)
10. (*b*) Addition of acceptor atom brings fermi level closer to valence band.
11. (*a*) 12. (*c*)
13. (*a*) Germanium is rarely used.
14. (*d*) 15. (*a*)
16. (*c*) At 0 K a semiconductor is perfect insulator. At room temperature it is slightly conducting.
17. (*c*)
18. (*b*) Holes and electrons.
19. (*a*) Since electrons are negatively charged they will flow towards positive terminal.
20. (*a*) Increase of temperature increases conductivity of semiconductors.
21. (*a*)
22. (*d*) Therefore it can donate one electron.
23. (*c*) 24. (*d*) 25. (*a*)
26. (*b*) Therefore in *n* type semiconductor electrons are majority carriers.
27. (*a*) 28. (*a*) 29. (*d*) 30. (*a*)
31. (*c*) 32. (*a*) 33. (*c*) 34. (*d*)
35. (*c*)
36. (*b*) Therefore conductivity increases.
37. (*b*) 38. (*d*) 39. (*c*) 40. (*a*)
41. (*a*) 42. (*a*) 43. (*c*) 44. (*a*)
45. (*d*) 46. (*a*) 47. (*a*)
48. (*b*) Therefore conductivity increases.
49. (*a*) When 1 electron rises through 1 V, energy = 1 eV.
50. (*a*) Holes do not exist in conductors.
51. (*a*) In all metals conductivity decreases (and resistance increases) with increase in temperature.
52. (*b*)
53. (*a*) Silver has the highest conductivity (and the lowest resistivity) in all metals.
54. (*b*)
55. (*b*) Minority carriers are generated due to thermal excitation.
56. (*b*) 57. (*b*) 58. (*b*) 59. (*c*)
60. (*b*) The frequency of incident radiation has to be more than threshold value. Wavelength is inversely proportional to frequency. Therefore wavelength of incident radiation must be less than threshold value.
61. (*b*)
62. (*b*) Only the intensity of incident radiation governs the amount of photoelectric emission.
63. (*a*)
64. (*b*) Some electrons have less energy than Fermi level E_1 and, therefore, more energy than U_w to escape. Therefore this equation has inequality sign.
65. (*a*) 66. (*b*) 67. (*b*) 68. (*d*)
69. (*c*) 70. (*d*) 71. (*c*)
72. (*a*) Silicon is the most commonly used semiconductor material. Antimony is pentavalent and hence donor impurity. Gallium is trivalent and hence acceptor impurity. Atomic number of Germanium is 32.
73. (*c*) Both silicon and germanium are semiconductor materials but Germanium is rarely used. Arsenic is pentavalent (donor) impurity and indium is trivalent (acceptor) impurity.
74. (*a*) In *p* type semiconductor holes are majority carriers, and in *n* type semiconductor electrons are majority carriers, silicon is intrinsic semiconductor and tungsten is a metal.
75. (*c*) Copper is a conductor and, therefore, valence and conduction bonds overlap. Rubber is insulator and has large forbidden gap, antimony is pentavalent and produces discrete energy level just above valence band, boron is trivalent and produces discrete energy level just below conduction band.
76. (*b*)
77. (*b*) Therefore the resistance is very high.
78. (*c*)
79. (*b*) Semiconductor diode has one *p-n* junction, BJT has two and SCR has three *p-n* junctions.
80. (*c*) This 0.7 V is also called cut in voltage. When forward voltage is 0.7 or less there is no current.
81. (*a*) 82. (*c*) 83. (*b*)
84. (*d*) Almost whole of reverse voltage is across depletion layer.
85. (*c*) Small current is due to minority carriers also. Therefore, hole current is less than I.
86. (*d*) Diode conducts only when forward biased. There is no conduction when bias is zero or reverse.
87. (*a*) 88. (*c*) 89. (*d*) 90. (*c*)
91. (*b*) $I = \frac{7-0.7}{1000}$.
92. (*b*) Since there is no voltage across load resistor, the current is zero and diode is open-circuited. If diode is short-circuited or resistor is open-circuited almost 9 V will appear across resistor.

93. (*a*) Therefore it is used in voltage regulator circuits.

94. (*b*) Zener diode is always reverse biased.

95. (*c*) Therefore varactor is used for tuning in TV.

96. (*a*) Capacitance is inversely proportional to distance between plates.

97. (*a*) **98.** (*a*) **99.** (*b*) **100.** (*b*)

101. (*b*) A pin diode has p and n doped regions separated by intrinsic layer.

102. (*b*) When reverse voltage equals breakdown value it start conducting and voltage does not increase further.

103. (*d*) Since the duration of output waveform is less than 180°, output voltage is less than input voltage.

104. (*d*)

105. (*c*) Increase of reverse voltage widens the depletion layer and junction capacitance decreases. However the decrease in capacitance is not proportional to increase in voltage.

106. (*d*) **107.** (*d*)

108. (*a*) Holes and electrons move away from junction and therefore resistance increases to a high value.

109. (*c*)

110. (*b*) Forward voltage decreases the width of depletion layer leading to low resistance.

111. (*d*) **112.** (*a*)

113. (*d*) A *p-n* function has all these features.

114. (*d*)

115. (*b*) Reverse saturation current doubles for every 10°C rise in junction temperature.

116. (*c*) When pentavalent impurity is added, the number of free electrons is very large as compared to number of holes.

117. (*a*) In intrinsic semiconductor the number of free electrons is equal to number of holes.

118. (*c*) **119.** (*d*) **120.** (*d*) **121.** (*b*)

122. (*b*) The contribution of sine terms to power dissipation in zero.

123. (*b*) Tunnel diode has heavily doped *p* and *n* regions leading to very thin depletion layer.

124. (*a*) *p* layer, *i* layer and *n* layer.

125. (*b*) The intrinsic layer is in between *p* and *n* layers.

126. (*c*) Tunnel diode has heavily doped *p* and *n* layers called degenerate *p* and *n* materials.

127. (*a*) Schottky diode has very small depletion layer.

128. (*a*) This is due to heavy doping.

129. (*b*) This is due to heavy doping.

130. (*d*)

131. (*a*) Cut in voltage in Schottky diode is about 0.3 as compared to 0.7 in ordinary semiconductor diode.

132. (*c*) This is due to high electron concentration in metals.

133. (*c*) In Schottky diode there is no charge storage and hence almost zero reverse recovery time.

134. (*c*) **135.** (*a*) **136.** (*b*) **137.** (*c*)

138. (*d*)

139. (*c*) Zener is used in voltage regulator circuits, tunnel diode is used in digital circuits since its reverse recovery time is nearly zero. LED is used in seven segment display and photo voltaic cell is used in converting solar energy to electrical energy.

140. (*d*)

141. (*b*) Schottky diode is also called hot carrier diode. PIN diode has intrinsic layer, zener is always used with reverse bias and tunnel diode has *v-i* curve of the shape of N.

142. (*c*) Emitter-base junction is forward biased and base collector junction in reverse biased.

143. (*b*) α is about 0.98.

144. (*a*) This is the reason that collector current is nearly equal to emitter current.

145. (*c*) The applied bias governs the width of depletion layer. Therefore capacitance varies with bias.

146. (*a*) Zener is used only in voltage regulator circuits.

147. (*a*) Its voltage dependent capacitance is used for tuning.

148. (*c*) *p, n, p* or *n, p, n*.

149. (*a*) Emitter is *n* type and emits electrons which difffuse through the base.

150. (*c*) Since there are two *p-n* junctions, there are two depletion layers and 0.7 V across each layer.

151. (*a*) **152.** (*b*) **153.** (*c*)

154. (*c*) Emitter current is larger, collector current is slightly less than emitter current and base current is very small.

155. (*b*) **156.** (*d*)

157. (*c*) Maximum power dissipation occurs at collector junction.

158. (*a*) Base is very thin and therefore recombination is minimum in both *p-n-p* and *n-p-n* transistors.

159. (*a*)

160. (*b*) $\beta_{\alpha c} = \dfrac{\alpha_{\alpha_c}}{1-\alpha_{\alpha_c}} = \dfrac{0.98}{0.02} = 49.$

161. (*b*)

162. (*b*) Emitter is *p*-type in *p-n-p* transistor. Therefore holes are majority carriers.

163. (*a*) **164.** (*b*) **165.** (*a*)

166. (*b*) It is used as amplifier when it operates in this region.

167. (*a*) **168.** (*a*) **169.** (*b*) **170.** (*c*)

171. (*c*) **172.** (*b*)

173. (*a*) In cut off mode a BJT is an open switch and in saturation mode it is a closed switch.

174. (*b*) **175.** (*a*)

176. (*b*) For every 1°C rise in ambient temperature the power dissipation must be reduced by 2.5 mW so that transistor is safe.

177. (*b*) 350 – 2.5 (60 – 25) ≅ 250 mW.

178. (*c*) **179.** (*c*)

180. (*a*) Since recombinaiton in base region is minium, $I_C \cong I_E$

181. (*c*)

182. (*b*) $I_C = + \alpha_{dc} \left(\frac{I_B}{1-\alpha_{dc}} \right) + I_{co} \simeq 735$ mA.

183. (*c*) Sicne emitter is *p*-type, emitter current is mainly due to holes. This current diffuses through the base.

184. (*a*) **185.** (*a*) **186.** (*c*)

187. (*c*) In forward active mode emitter base junction in forward biased and basecollector junction is reverse biased.

188. (*c*) **189.** (*b*) **190.** (*c*) **191.** (*a*)

192. (*a*) **193.** (*a*) **194.** (*b*) **195.** (*b*)

196. (*b*) **197.** (*a*) **198.** (*c*) **199.** (*a*)

200. (*a*)

201. (*d*) *k* has to be much higher than 1 so that grid is effective in controlling plate current *n* is not 1 because it is a non-linear device.

202. (*c*)

203. (*d*) The channel in JFET may be *n* type or *p* type. If channel is *n* type, gate is *p* type and if channel is *p* type gate is *n* type.

204. (*a*) **205.** (*a*)

206. (*a*) Tunnel diode has heavily doped *p* and *n* regions.

207. (*d*) Its output is used to trigger SCR.

208. (*d*) **209.** (*a*) **210.** (*c*) **211.** (*c*)

212. (*d*) In *n* type semiconductors carriers are electrons.

213. (*d*) JFET is voltage controlled device. Therefore its input impedance is high. But voltage gain is lower than in BJT.

214. (*a*) Since gate current is very small. I_D and I_S are almost equal.

215. (*c*) **216.** (*d*)

217. (*b*) In major portion of drain characteristics I_D is constant.

218. (*a*) **219.** (*b*)

220. (*d*) Its units are the same as the units of conductance.

221. (*b*) **222.** (*d*) **223.** (*d*) **224.** (*b*)

225. (*d*) Both JFET and vaccum triode are voltage controlled devices.

226. (*d*)

227. (*b*) MOSFET may be depletion mode or enhancement mode.

228. (*d*) **229.** (*d*)

230. (*b*) Since channel is *n* type gate must be *p* type.

231. (*c*) Depletion layer always has immobile charges.

232. (*a*) All currents are assumed positive when flowing into JFET.

233. (*a*) **234.** (*a*)

235. (*b*) The drain current is almost constant. Therefore characteristics is flat.

236. (*c*) **237.** (*b*) **238.** (*d*) **239.** (*c*)

240. (*a*) **241.** (*c*) **242.** (*a*)

243. (*c*) In all symbols current comes out of *p* layer.

244. (*c*) **245.** (*c*) **246.** (*d*)

247. (*d*) The SiO_2 layer provides very high input impedance.

248. (*c*)

249. (*d*) All MOSFETs have substrate.

250. (*c*) **251.** (*a*) **252.** (*a*) **253.** (*d*)

254. (*b*) **255.** (*a*)

256. (*c*) Resistance limits current and diode is reverse connected and therefore protects LED against reverse breakdown.

257. (*c*) *i* is plotted on *y*-axis and *v* on *x*-axis. When passing through origin current can be zero only.

258. (*c*)

259. (*d*) The resistance of photo sensitive device depends on the light.

260. (*a*) Resistance of LDR depends on light.

261. (*a*) **262.** (*c*) **263.** (*b*) **264.** (*d*)

265. (*c*) **266.** (*b*) **267.** (*c*)

268. (*d*) $R = \frac{6-2}{10 \times 10^{-3}} = 400\ \Omega$. R must be at least 400 Ω so that current in LED does not exceed 10 mA.

269. (*b*) **270.** (*b*) **271.** (*b*)

272. (*c*) n type semi conductor is produced when pentavalent impurity is added.

273. (*b*) **274.** (*c*)

275. (*b*) Tunnel diode characteristics resembles alphabet N.

276. (*a*) **277.** (*a*) **278.** (*d*) **279.** (*a*)

280. (*b*)

281. (*d*) Intrinsic material has equal number of holes and electrons.

282. (*c*) As wavelenghth decreases, frequency increases and maximum velocity of electron increases.

283. (*b*) Hall effect can be used to find type and concentration of charge carriers.

284. (*c*)

285. (*d*) FET is voltage controlled device (gate voltage is controlling quantity) BJT is current controlled device (base current is the controlling quantity) UJT has negative resistance region in its characteristics.

286. (*b*) **287.** (*d*)

288. (*b*) Schottky diode has very low switching time.

289. (*c*)

290. (*d*) Oxide layer cannot have any effect on melting point of silicon. Moreover before melting silicon breaks down.

291. (*d*) $i_c = \alpha\, i_e$.

292. (*a*)

293. (*c*) Increase in illumination reduces resistance of a photo conductor.

294. (*a*) **295.** (*b*) **296.** (*b*) **297.** (*d*)

298. (*a*) **299.** (*b*)

300. (*a*) Mobility $= \dfrac{\text{drift velocity}}{\text{field gradient}}$.

301. (*b*) Small signal amplifier operation is in constant region of characteristics.

302. (*b*) **303.** (*d*) **304.** (*c*) **305.** (*c*)

306. (*a*) **307.** (*c*) **308.** (*a*) **309.** (*a*)

310. (*d*)

311. (*a*) Incremental model is used for ac response at one operating point.

312. (*c*) Increasing the substrate concentration lowers threshold voltage. Gate oxide layer has no effect an threshold voltage.

313. (*d*)

314. (*c*) In piezo electric materials mechanical stress produces electric polarization.

315. (*d*) Only antimony, arsenic and phosphorous are pentavalent.

316. (*c*) **317.** (*b*) **318.** (*a*)

319. (*a*) Recombination occurs in only base region.

320. (*b*) FET is unipolar device and operation depends on majority carriers. Its input resistance is very high.

321. (*d*) A new discrete energy level below conduction band is added.

322. (*b*) Statements 1 and 4 are wrong. Acceptor energy level is close to valence band. Donor impurity donates electrons.

323. (*d*) Field is perpendicular to both B and I. Hall effect is used to find whether the semiconductor is *p* or *n* type.

324. (*c*) At well below pinch off voltage drain to source resistance depends on gate voltage V_{GS}. Therefore JFET can be used as voltage controlled resistor.

325. (*a*) See Fig. 2.16 (*b*).

326. (*a*) Safe current $= \dfrac{50\times10^{-3}}{10} = 5$ mA.

327. (*d*) CMOS has high speed and uses lesser power.

328. (*d*)

329. (*c*) Photon has to jump from valence band to conduction band. It requires forbidden gap energy.

330. (*a*) $\dfrac{D_n}{\mu_n} = V_T$ or $D_n = 25.9 \times 10^{-3} \times 1500 \simeq 40$ cm²/s.

331. (*c*) Transfer characteristic is also called trans-conductance curve. It is plotted between I_D and V_{GS}. The intercept of this curve on *x*-axis gives threshold voltage. Transconductance $g_m = \dfrac{\partial I_D}{\partial V_{GS}}$.

332. (*b*) Diode current for silicon is given by

$$I = I_0\,[e^{V/2V_T} - 1] \simeq I_0[e^{V/2V_T}]$$

The voltage ΔV required to increase diode current 10 times is

$$10 = e^{\Delta V/2V_T}$$

or

$$\Delta V = 2V_T \ln 10 = 2 \times 26 \ln(10)\text{ mV} = 120\text{ mV}.$$

333. (*c*) Diode can conduct only when input voltage is more than cut in voltage. So the conduction angle is π − 2ϕ as seen in Fig. 2.36.

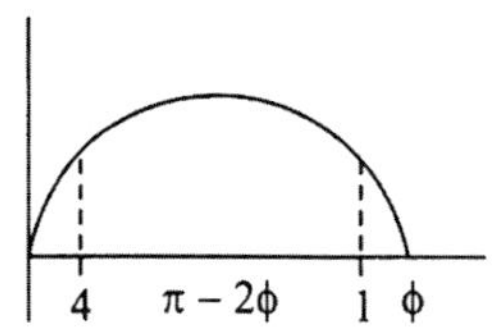

Fig. 2.36.

334. (*a*) I_{CBO} doubles for every 10°C increase in temperature.

335. (*d*)

$$I_D = I_{DSS}\left[1 - \frac{V_{GS}}{V_{GS(H)}}\right]^2$$

$$I_D = 15\left[1 - \left(\frac{-2}{-5}\right)\right]^2 = 5.4\text{ mA}.$$

336. (*c*) To make *n* type silicon pentavalent impurity is added to intrinsic silicon, phosphorous and arsenic are pentavalent. Boron and aluminium are trivalent.

337. (*a*) All the three statements are correct. See section 2.3.

338. (*a*) In tunnel diode impurity concentration is high. It has voltage controlled *v-i* curve. Carrier velocity is high. See section. 2.8.

339. (*b*) As temperature increases, number of charge carriers increases and conductivity increases.

340. (*b*) The number of new hole electron pairs is equal to number of photons, *i.e.*, $\Delta p = \Delta n$.

341. (*c*) In metals all electrons are free eletrons and change of temperature cannot change electron concentration. Hence statement 1 is wrong.

342. (*d*) FETs have current due to majority carrier only. They have high input impedance (See section 2.5).

343. (*a*) Half effect is used to find type of charge carrier and concentration.

344. (*c*) When *p-n-p-n* device is conducting, gate loses control.

345. (*c*) As doping level is increased, reverse breakdown voltage decreases. Depletion width also decreases.

Capacitance $= \frac{\in A}{d}$. Therefore capacitance increases.

346. (*d*) Current through 5 k resistor $= \frac{10-5}{5\times10^3} = 1$ mA

Voltage drop across 1.4 k resistor
$= 1.4 \times 10^3 \times 1 \times 10^{-3} = 1.4$ V

V = voltage drop across 1.4 k resistor + 0.6 = 2 V.

347. (*a*) $\frac{D}{\mu} = V_T$ or $\frac{\mu}{D} = \frac{1}{V_T}$. Hence unit is V^{-1}.

348. (*c*) Boron is pentavalent and hence fermi level goes down.

$$E_F - E_V = kT \ln \frac{N_V}{N_A} = 25 \times 10^{-3} \ln \frac{4\times10^{17}}{1.5\times10^{10}}$$

$$= 0.427 \text{ eV}$$

349. (*d*) $N_D + p = N_A + n$

$N_D = 0$ and $p >> n$

Since $np = n_i^2$, we get $n = \frac{n_i^2}{p} = \frac{n_i^2}{N_A}$.

350. (*c*) Electric field is maximum at junction.

351. (*b*) Zener and photodiode are reverse biased in normal mode. Other two are forward biased.

352. (*a*) Junction potential = built in potential + reverse bias and depletion width ∝ (junction potential)$^{0.5}$

$$z = k\,(0.8 + 1.2)^{0.5}$$

$$x = k\,(0.8 + 7.2)^{0.5}$$

or $x = 4$ μm.

353. (*b*) Hall effect

Electric force + magnetic force = 0

$$qE + qv \times B = 0$$

or $E = -v \times B = B \times v$.

354. (*c*) Varactor diode is used for tuning. Schottkey diode is a high freuqency switch, zener is used as a voltage reference.

355. (*d*) For *n* type $\sigma_n = nq\mu_n$

For intrinsic $\sigma_i = n_i\, q(\mu_n + \mu_p)$.

$$\frac{\sigma_n}{\sigma_i} = \frac{n\mu_n}{n_i(\mu_n + \mu_p)} = \frac{4.2\times10^8 \mu_n}{1.5\times10^4 \times \mu_n \left(1 + \frac{\mu_p}{\mu_n}\right)}$$

$$= \frac{4.2\times10^8}{1.5\times10^4 \times 1.4} = 20000.$$

356. (*c*) See Fig. 2.22 (*a*).

357. (*b*) $np = n_i^2$ where n_i is constant.

For *n* type semiconductor *p* is minority carrier concentration $p = \frac{n_i^2}{n}$ or $p \propto \frac{1}{n}$.

358. (*a*) It is a diffusion current.

359. (*d*) A is wrong because exposure to light decreases bulk resistance.

360. (*a*)

361. (*b*) A refers to *p* type semiconductor while R refers to n type semiconductor. Both A and R are correct but independent.

362. (*a*) **363.** (*a*) **364.** (*a*) **365.** (*a*)

366. (*d*) A is wrong because Germanium is rarely used.

367. (*a*)

368. (*a*) The increase in reverse resistance is due to widening of depletion layer.

369. (*d*) A is wrong because decrease in junction temperature decreases reverse saturation current.

370. (*a*) At 20°C, 4 *n*A, at 30°C, 8 *n*A, at 40°C, 16 *n*A, at 50°C, 32 *n*A.

371. (*a*) Avalanche breakdown occurs at high reverse voltage.

372. (*b*) Both A and R are correct but independent.

373. (*a*) Increase in reverse voltage increases the width of depletion layer and decrease of capacitance of layer.

374. (*a*) Both A and R are correct. The diode is destroyed due to high reverse current.

375. (*d*) Base region is thin.

376. (*a*)

377. (*c*) Recombination in base region is very low.

378. (*a*)

379. (*b*) Both A nad R are correct but independent.

380. (*b*) Both A nad R are correct but independent.

381. (*a*) **382.** (*a*)

383. (*a*) Increase of ambient temperature lowers the heat dissipation capacity. Therefore derated operation of BJT is necessary lest BJT should be destroyed.

384. (*b*) Both A and R are correct but independent.

385. (*a*) **386.** (*a*) **387.** (*a*) **388.** (*a*)

389. (*a*)

390. (*c*) Holes and electrons move is opposite directions. Therefore R is wrong.

391. (*a*) Therefore mostly *n-p-n* transistors are used.

392. (*a*)

393. (*d*) When exposed to light bulk resistance decreases. Therefore A is wrong.

394. (*a*)

395. (*a*) Lower work function leads to higher emission current. Therefore oxide coated cathodes are commonly used.

396. (*a*)

397. (*b*) Both A and R are correct but independent.

398. (*c*) Photo electric emission occurs only if frequency is higher than threshold frequency thus R is wrong.
399. (*a*) Low work function permits easy emission.
400. (*a*)
401. (*b*) Both A and R are correct but independent.
402. (*a*) Wider forbidden gap in silicon makes it less sensitive to temperature than germanium.
403. (*a*)
404. (*b*) Both A and R are correct but independent.
405. (*a*) PIV is an important parameter in rectifier operation.
406. (*b*)
407. (*a*) If *v-i* characteristis is non-linear, the ratio *v*/*i* is not constant.
408. (*b*) 409. (*a*)
410. (*b*) Both are correct but independent.
411. (*c*)
412. (*a*) Avalanche breakdown causes high current.
413. (*c*) R is wrong.
414. (*b*) Both A and R are correct but independent.
415. (*a*) 416. (*a*) 417. (*a*)
418. (*b*) Both are correct but independent.
419. (*b*) Both are correct but independent.
420. (*b*) Conductivity of both silicon and germanium increases with increase in temperature. Therefore, number of electrons and holes increases with increase in temperature.
421. (*a*) Enhancement type MOSFETs are commonly used in digital ICs. They have SiO_2 layer.
422. (*a*) 423. (*a*) See Fig. 2.11.
424. (*a*)
425. (*c*) See Fig. 2.3. The *p-n* junction cannot supply current because no external source of energy is available.
426. (*d*) Cut in voltage for silicon 0.7 V, cut in voltage for Germanium 0.3 V. Reverse saturation current in Germanium diode is more than that in silicon diode.
427. (*d*) P channel MOSFET occupies more area than N channel MOSFET. Electron mobility is more than hole mobility.
428. (*a*) Heavily doped semiconductor can exhibit metallic properties (and positive temperature coefficient of resistance). The resistance increases because carrier mobility decreases with increase in temperature.
429. (*a*)
430. (*a*) At high temperatures breaking of covalent bonds causes n_i to be much higher than N_D.
431. (*b*) As temperature increases electron mobility decreases due to increased vibrations of electrons in metals.
432. (T) This is the reason for high conductivity of metals.
433. (F) 434. (T) 435. (T)
436. (T) One electron has a charge 1.602×10^{-19} C, when it falls through 1 V, energy is 1 eV = 1.602×10^{-19} J.
437. (F) Work function is the minimum energy required by the fastest electron at 0 K to escape from the metal surface.
438. (T) Therefore emission current from oxide coated cathode is more.
439. (T) 440. (T) 441. (T)
442. (F) Sometimes it can be useful also.
443. (T) This is called field emission.
444. (T) 445. (T) 446. (T)
447. (F) It depends on intensity of incident light.
448. (F) 449. (T) 450. (T) 451. (T)
452. (F) With increase in temperature, the forbidden gap decreases.
453. (F) Mobility of electrons is more than that of holes.
454. (T)
455. (F) Due to recombination the number of electron-hole pairs is reduced.
456. (T) 457. (T) 458. (T)
459. (F) It is near conduction band.
460. (F) 461. (T) 462. (T) 463. (T)
464. (T) 465. (T) 466. (T) 467. (T)
468. (T) 469. (T) 470. (T)
471. (F) Both holes and electrons move away from junction.
472. (T)
473. (F) When the device is being used, junction temperature is higher than room temperature.
474. (F) With increase in temperature width of depletion layer decreases.
475. (T) 476. (F)
477. (F) It increases with increase of temperature.
478. (T)
479. (F) Covalent bonds are broken.
480. (F) It is always non-linear. 481. (T)
482. (F) It is used with reverse bias.
483. (T) 484. (T) 485. Two 486. inversely
487. switching 488. 1000° C/W
489. 0.4×10^{-3} siemens 490. 6666.67
491. 32 *n*A 492. 4 *n*A
493. exceeded 494. minority
495. 0.7V 496. $n^{2/3}$ 497. eV 498. energy
499. fermi energy 500. 4.5 eV
501. intensity 502. cut in, peak inverse
503. $p \mu_p e$ 504. small
505. 2 million 506. 0.23 Ω
507. 3 million 508. holes
509. forward 510. reverse
511. towards 512. constant
513. 0.98 514. 99 515. very small
516. very high 517. active
518. saturation 519. unipolar
520. source, gate, drain 521. ohmic
522. oxide 523. V 524. less
525. voltage regulator 526. aluminium silicon
527. Schottky 528. Tunnel
529. LEDs

CHAPTER

3

SIGNALS AND SYSTEMS

3.1. LINEAR SYSTEMS

A linear system is one in which the relationship between input and output is linear. All linear systems possess the property of superposition, *i.e.*, if input u_1 gives output y_1 and input u_2 gives output y_2, then input $(u_1 + u_2)$ will give output $y_1 + y_2$. Superposition implies homogeneity, *i.e.*, if input u gives output y, then input ku gives output ky where k is a rational number. In general a system is linear if a transformation T satisfies the relation.

$$\text{T}(\alpha u_1 + \beta u_2) = \alpha \text{T}(u_1) + \beta \text{T}(u_2) \quad ...(3.1)$$

where α and β are constant.

3.2. CONTINUOUS AND DISCRETE TIME SYSTEMS

A continuous time system is one in which the input $u(t)$ and output $y(t)$ are functions of a continuous real variable t which is usually a time variable. In a discrete time system an input sequence $u(k)$ is transformed into output sequence $y(k)$.

RLC circuits are continuous time systems. On the other hand, digital circuits are discrete time systems. The primitive elements in a discrete time system are adders, multipliers, storage elements (flip flops) etc.

3.3. Z TRANSFORM

Let x denote a finite sequence. Z transform X(z) of the complex variable z is defined by forming the polynomial.

$$X(z) = x_l z^{-1} + x_{l+1} z^{-(1+1)} + ... + x_m z^{-m}$$

$$= \sum_{k=l}^{m} x_k z^{-k} \quad ...(3.2)$$

In Eq. (3.2) the beginning and end indices l and m are integers including $-\infty$ and $+\infty$.

Z transform possesses the properties of linearity, shifting and convolution. Z transform can be inverted to give the original sequence $\{x_k\}$ by the methods of direct division, partial fractions.

3.4. LAPLACE TRANSFORM

$$\pounds f(t) = F(s) = \int_0^{\infty} f(t)\, e^{-st}\, dt$$

$$\pounds^{-1} F(s) = f(t)$$

$$\pounds[a f_1(t) + b f_2(t)] = aF_1(s) + bF_2(s)$$

$$\pounds\left[\frac{d}{dt} f(t)\right] = sF(s) - f(0^-)$$

$$\pounds\left[\frac{d^2}{dt^2} f(t)\right] = s^2F(s) - sf(0^{-1}) - \frac{df}{dt}(0^-)$$

$$\pounds\left[\int f(t)\, dt\right] = \frac{F(s)}{s} + \frac{f^{-1}(0)}{s}$$

where $f^{-1}(0) = \int_{-\infty}^{0} f(t)\, dt$

$$\pounds[f(t - T)] = e^{-sT} F(s)$$

Initial value theorem $\underset{t \to 0}{\text{Lt}}\ f(t) = \underset{s \to \infty}{\text{Lt}}\ [sF(s)]$

Final value theorem $\underset{t \to \infty}{\text{Lt}}\ f(t) = \underset{s \to 0}{\text{Lt}}\ [sF(s)]$

Convolution Integral

$$f(t) = \pounds^{-1}[F_1(s)\, F_2(s)]$$

$$= \int_0^t f_1(\tau)\, f_2(t - \tau)\, d\tau$$

$$= \int_0^t f_2(\tau)\, f_1(t - \tau)\, d\tau$$

where τ is dummy variable for t.

Laplace transforms of some common functions are

	$f(t)$	$F(s)$
1.	$u(t)$	$\frac{1}{s}$
2.	t	$\frac{1}{s^2}$
3.	$\frac{t^{n-1}}{(n-1)!}$, n integer	$\frac{1}{s^n}$
4.	e^{at}	$\frac{1}{s-a}$

5.	te^{at}	$\dfrac{1}{(s-a)^2}$
6.	$\dfrac{1}{(n-1)!}t^{n-1}e^{at}$	$\dfrac{1}{(s-a)^n}$
7.	$\dfrac{1}{a-b}(e^{at}-e^{bt})$	$\dfrac{1}{(s-a)(s-b)}$
8.	$1-e^{+at}$	$\dfrac{-a}{s(s-a)}$
9.	$\dfrac{1}{\omega}\sin\omega t$	$\dfrac{1}{s^2+\omega^2}$
10.	$\cos\omega t$	$\dfrac{s}{s^2+\omega^2}$
11.	$1-\cos\omega t$	$\dfrac{\omega^2}{s(s^2+\omega^2)}$
12.	$\sin(\omega t+\theta)$	$\dfrac{s\sin\theta+\omega\cos\theta}{s^2+\omega^2}$
13.	$\cos(\omega t+\theta)$	$\dfrac{s\cos\theta-\omega\sin\theta}{s^2+\omega^2}$
14.	$e^{-at}\sin\omega t$	$\dfrac{\omega}{(s+a)^2+\omega^2}$
15.	$e^{-at}\cos\omega t$	$\dfrac{s+a}{(s+a)^2+\omega^2}$
16.	$\sinh\alpha t$	$\dfrac{\alpha}{s^2-\alpha^2}$
17.	$\cosh\alpha t$	$\dfrac{s}{s^2-a^2}$

3.5. STATE VARIABLE ANALYSIS

In this approach an nth order system is described by n simultaneous first order equations.

In general the order system of the state of a linear time invariant system can be described by the following n first order differential equations with constant coefficients in terms of n state variables $x_1, x_2, \ldots x_n$.

$$\begin{bmatrix} \dot{x}_1 \\ \dot{x}_2 \\ \vdots \\ \dot{x}_n \end{bmatrix} = \begin{bmatrix} a_{11} & a_{12} & \cdots & a_{1n} \\ a_{21} & a_{22} & \cdots & a_{2n} \\ \cdot & \cdot & \cdot & \cdot \\ a_{n1} & a_{n2} & \cdots & a_{nn} \end{bmatrix} \begin{bmatrix} x_1 \\ x_2 \\ \cdot \\ x_n \end{bmatrix} + \begin{bmatrix} b_{11} & \cdots & b_{1m} \\ b_{21} & \cdots & b_{2m} \\ \cdot & \cdot & \cdot \\ b_{n1} & \cdots & b_{nm} \end{bmatrix} \begin{bmatrix} u_1 \\ u_2 \\ \cdot \\ u_m \end{bmatrix} \quad \text{...(3.3)}$$

where $\dot{x}_1 = \dfrac{dx_1}{dt},\ \dot{x}_2 = \dfrac{dx_2}{dt} \ldots$

or $$\dot{x} = AX + Bu \quad \text{...(3.4)}$$

The state variables in a network are inductor currents and capacitor voltages.

Taking Laplace transform of Eq. (3.4), we get

$$X(s) = (sI - A)^{-1}x(0^-) + (sI - A)^{-1}Bu(s) \quad \text{....(3.5)}$$

where X(s) is the Laplace transform of unforced response. $x(0^-)$ is the initial condition vector. Taking inverse transform of Eq. (3.5), we get

$$x(t) = \phi(t)\,x(0^-) + \pounds^{-1}[\phi(s)\,Bu(s)] \quad \text{...(3.6)}$$

where $\phi(t) = e^{At}$ is the state transition matrix.

3.6. FOURIER SERIES, FOURIER TRANSFORM

A periodic function $f(t)$ with time period T can be expressed as

$$\begin{aligned} f(t) &= a_0 + a_1\cos\omega_1 t + a_2\cos 2\omega_1 t \\ &\quad + a_3\cos 3\omega_1 t + \ldots b_1\sin\omega_1 t \\ &\quad + b_2\sin 2\omega_1 t + b_3\sin 3\omega_1 t + \ldots \end{aligned} \quad \text{...(3.7)}$$

where $\omega_1 = \dfrac{2\pi}{T}$ and T is the time period of the function $f(t)$

$$a_0 = \frac{1}{T}\int_0^T f(t)\,dt \quad \text{...(3.8)}$$

$$a_n = \frac{2}{T}\int_0^T f(t)\cos n\,\omega_1 t\,dt \quad \text{...(3.9)}$$

$$b_n = \frac{2}{T}\int_0^T f(t)\sin n\omega_1 t\,dt \quad \text{...(3.10)}$$

A function has half wave symmetry if $f(t) = -f\left(t + \dfrac{T}{2}\right)$

A function contains only sine terms if $f(-t) = -f(t)$

A function contains only consine terms if $f(-t) = f(t)$

Exponential form of Fourier series

$$f(t) = \sum_{n=-\infty}^{\infty} F_n\, e^{jn\omega_1 t} \quad \text{...(3.11)}$$

where $$F_n = \frac{1}{T}\int_{-T/2}^{T/2} f(t)\,e^{-jn\omega_1 t}\,dt \quad \text{...(3.12)}$$

$$\text{Fourier transform } F(j\omega) = \int_{-\infty}^{\infty} f(t)\,e^{j\omega t}\,dt \quad \text{...(3.13)}$$

$f(t)$ and F($j\omega$) constitute the Fourier transform pair. The properties of Laplace transform apply to the properties of Fourier transform with $s = j\omega$.

3.7. PROBABILITY AND RANDOM SIGNALS

Probability of an event A = P(A)

$$= \frac{\text{Number of possible favourable outcomes}}{\text{Total number of possible equally likely outcomes}} \quad \text{...(3.14)}$$

Mutually exclusive events: Occurrence of one outcome precludes the occurrence of the other. If A and B are mutually exclusive

$$P(AB) = 0$$
$$P(A+B) = P(A)+P(B)$$

In general if A and B are two events

$$P(A+B) = P(A)+P(B)-P(AB)$$

Bay's theorem $P(A_j/B_k) = \frac{P(A_j)}{P(B_k)} P\left(\frac{B_k}{A_j}\right)$

Cummulative distribution function is the probability that the outcome of an experiment will be one of the outcomes for which $X(\lambda) \le x$ where x is any given number.

Cummulative distribution function $F_X(x) \equiv P(X < x)$

Moreover $0 \le F(x) \le 1$

$$F(-\infty) = 0, F(\infty) = 1$$
$$F(x_1) \le F(x_2) \quad \text{if} \quad x_1 < x_2$$

Probability density function $f_x(x) = \frac{d}{dx} F_X(x)$

$$f(x) \ge 0 \text{ for all } x$$
$$\int_{-\infty}^{\infty} f(x)\,dx = 1$$
$$F(x) = \int_{-\infty}^{x} f(x)\,dx$$

$P(x \le X \le x + dx) = f(x)\,dx$

$$P(a < X < b) = \int_a^b f(x)\,dx$$

Mean of random variable X = expected value of X = E(X) or $\overline{X}$

$$\overline{X} = E(X) = \sum_i x_i\, P(x_i) = m$$

Variance $\sigma^2 = E[(x-m)^2] = \int_{-\infty}^{\infty} (x-m)^2 f(x)\,dx$

σ is called standard deviation.

If average value $m = 0$, $\sigma^2 = E(x^2)$

Mean and variance of sum of random varaibles.

If $Z = X + Y$

$$m_z = m_x + m_y$$

and $\sigma_z^2 = \sigma_x^2 + \sigma_y^2$

Tchebycheff's Inequality $P(|X| \ge \epsilon) = \frac{\sigma^2}{\epsilon^2}$

Binomial Distribution

If an experiment is repeated n times, p is the probability of success in one trial and $q = 1 - p$, the probability of r successes in n trials is

$$f(x) = P(X = r) = {}^nC_r (p)^r (q)^{n-r}$$

For a binomial distribution, mean = np

variance = npq

Poisson's Distribution

If X is a random variable which can assume values 0, 1, 2, ... probability of function X can be expressed by Poisson's distribution

$$f(x) = P(X = r) = \frac{\lambda^r e^{-\lambda}}{r!}$$

where λ is a positive constant.

For Poisson's distribution mean = λ = variance

If n is large and p is small (close to zero). Binomial distribution can be approximated by Poisson's distribution.

Normal or Gaussian Distribution

$$f(x) = \frac{1}{\sigma\sqrt{2\pi}} e^{-(x-m)^2/(2\sigma^2)}$$

where m is mean value and σ^2 is variance associated with $f(x)$.

Probability of error between limits a and b is

$$P_{ab} = \frac{h}{\sqrt{\pi}} \int_a^b e^{-h^2x^2 dx}$$

The constant h is the measure of precision

The error function of u (written as $erf\,u$) is defined as

$$erf\ u = \frac{2}{\sqrt{\pi}} \int_0^u e^{-u^2}\,du$$

Complementary error function

$$erfc\ u = 1 - erf\ u = \frac{2}{\sqrt{\pi}} \int_u^{\infty} e^{-u^2}\,du$$

Uniform Distribution $f(x) = \frac{1}{b-a}$ for $a \le x \le b$

$= 0$ otherwise

Rayleigh Distribution

$$f(r) = \frac{r}{a^2} e^{-r^2/2a^2} \quad \text{for} \quad 0 \le r \le \infty$$
$$= 0 \quad \text{for} \quad r < 0$$

The covariance μ of two random variables X and Y is

$$\mu = E\{(X - x_m)(Y - y_m)\}$$

Correlation

Correlation between waveforms $v_1(t)$ and $v_2(t)$ is written as $R_{12}(\tau)$ and is defined as

$$R_{12}(\tau) = \lim_{T\to\infty} \frac{1}{T} \int_{-T/2}^{T/2} v_1(t)\, v_2(t+\tau)\,dt$$

If $v_1(t)$ and $v_2(t)$ are periodic with same fundamental period T_0, then correlation (also called average cross correlation) is

$$R_{12}(\tau) = \frac{1}{T} \int_{-T_0/2}^{T_0/2} v_1(t)\, v_2(t+\tau)\,dt$$

Auto Correlation

The corelation of a function with itself is called autocorrelation, written as R(τ).

Thus if $v_1(t) = v_2(t) = v(t)$

$$R_{12}(\tau) = \lim_{T\to\infty} \frac{1}{T} \int_{-T/2}^{T/2} v(t)\, v(t+\tau)\, d\tau$$

Properties of Auto Correlation

1. Autocorrelation for $\tau = 0$ is the average power P of signal.
2. Autocorrelation function exhibits conjugate symmetry *i.e.*, $R(\tau) = R^*(\tau)$
3. The maximum value of autocorrelation function R(τ) occurs at origin, *i.e.*,
   $R(o) \geq R(\tau)$ for all τ
4. Autocorrleation R(τ) of periodic waveform is also periodic with the same period.
5. Power spectral density function S(ω) and autocorrelation function of a power signal are Fourier transform pair.

Central Limit Theorem

The probability density of a sum of N independent random variables tends to approach a gaussian density as the number N increases.

Parseval's Theorem

If Fourier transform of a single sample pulse $P_1(t)$ is $P_1(f)$, then the normalised energy of the pulse is

$$E_1 = \int_{-\infty}^{\infty} P_1(f) P_1^*(f) df$$

$$= \int_{-\infty}^{\infty} |P_1(f)|^2 df$$

The energy in the range *df* at frequency *f* is

$$dE_1 = |P_1(f)|^2 df$$

Power spectral density of NRZ (non-return to zero) signal

$$G(f) = V_b^2 T_b \left[\frac{\sin \pi f T_b}{\pi f T_b}\right]^2$$

Power spectral density of biphase signal

$$G(f) = V_b^2 T_b \left[\frac{\sin^2 \pi f (T_b/2)}{\pi f\, T_b/2}\right]^2$$

Method of Least Squares

If $s_1, s_2, s_3, \ldots s_n$ are the *n* observed values of a quantity, S, *i.e.*, sum of squares of their deviations from a number *q* is

$$S = (s_1 - q)^2 + (s_2 - q)^2 + \ldots + (s_n - q)^2$$

S is minimum when *q* is equal to the arithmetic mean of the observed values. This is called the method of least squares.

OBJECTIVE TYPE QUESTIONS

A. Choose the most appropriate alternative

1. The signal defined by the equations $u(t-a) = 0$ for $t < a$ and $u(t-a) = 1$ for $t \geq a$ is
(*a*) a unit step function
(*b*) a shifted unit step function originating at $t = a$
(*c*) a pulse function
(*d*) None of the above

2. The signal defined by the equations $f(t) = 0$ for $t < 0$, $f(t) = E$ for $0 \leq t \leq a$ and $f(t) = 0$ for $t > a$ is
(*a*) a step function
(*b*) a pulse function
(*c*) a shifted step function originating at $t = a$
(*d*) none of the above

3. The Laplace transform of a unit step function is
(*a*) 1 (*b*) $1/s$
(*c*) s (*d*) $1/s^2$

4. A pulse function having magnitude E and duration from t = 0 to $t = a$ can be represented as
(*a*) sum of two pulse functions
(*b*) difference of two pulse functions
(*c*) difference of two step functions each of magnitude E and one originating at $t = 0$ and the other at $t = a$
(*d*) none of the above.

5. Laplace transform of a pulse function of magnitude E and duration from $t = 0$ to $t = a$ is
(*a*) E/*s* (*b*) $\frac{E}{s} e^{-as}$
(*c*) $E\left(\frac{1}{s} - \frac{1}{s} e^{-as}\right)$ (*d*) $\frac{E}{s^2}(1 - e^{as})$

6. The function $f(t) = E\,[u(t-a) - u(t-b)]$ is
(*a*) a pulse function originating at $t = a$
(*b*) a step function originating at $t = 0$
(*c*) an exponetential function originating at $t = a$
(*d*) a pulse function originating at $t = a$ and duration $(b - a)$

7. For the function $u(t-a) = 0$ for $t < a$ and $u(t-a) = 1$ for $t \geq a$, the Laplace transform is
(*a*) $\frac{a}{s}$ (*b*) $\frac{e^{-at}}{s}$
(*c*) $\frac{e^{-a}}{s}$ (*d*) se^{-a}

8. For the function $f(t) - e^{-at}$, the Laplace transform is

(a) $\frac{a}{s}$ (b) $\frac{1}{as}$

(c) $\frac{1}{s-a}$ (d) $\frac{1}{s+a}$

9. An impulse train is
(a) a number of pulses
(b) a number of pulses spaced from each other
(c) a number of pulses all originating together
(d) none of the above

10. The integral of a unit impulse is
(a) a unit step function
(b) a ramp function of slope 1
(c) a pulse function of magnitude 1
(d) none of the above

11. The integral of a unit step function is
(a) a unit impulse function
(b) a unit pulse function
(c) a ramp function of slope 1
(d) none of the above

12. The Laplace transform of impulse $\delta(t)$ is
(a) 1 (b) $1/s$
(c) s (d) $1/s^2$

13. The function $\delta(t - b)$ is
(a) an impulse function
(b) a step function originating at $t = b$
(c) an impulse function orginating at $t = b$
(d) none of the above

14. The area of an impulse is
(a) 1 (b) 0
(c) ∞ (d) any finite value

15. The function $\delta'(t - b)$ is
(a) a step function originating at $t = b$
(b) an impulse function originating at $t = b$
(c) a unit doublet orginating at $t = b$
(d) none of the above

16. The Laplace transform of $f(t)$ is

(a) $\int_0^\infty f(t)dt$ (b) $\int_0^\infty f(t)e^{-st}\,dt$

(c) $\int_{-\infty}^\infty f(t)e^{-st}\,dt$ (d) none of the above

17. If $f_1(t)$ and $f_2(t)$ are two functions of time and a and b are constants, then
(a) $£[af_1(t) + bf_2(t)] = aF_1(s) + bF_2(s)$
(b) $£[af_1(t) + bf_2(t)] = aF_1(s) - bF_2(s)$
(c) $£[af_1(t) + bf_2(t)] = [aF_1(s)/bF_2(s)]$
(d) None of the above

18. If function $f(t)$ has an initial value $f(0^-)$ at $t = 0^-$, the Laplace transform of $\frac{d}{dt}[f(t)]$ is
(a) $sF(s) - f(0^-)$ (b) $sF(s) + f(0^-)$
(c) $s^2F(s) - f(0^-)$ (d) $s^2F(s) + f(0^-)$

19. As per time displacement theorem in Laplace transformation, displacement in the time domain by T becomes
(a) division by s in the s domain
(b) division by e^{-sT} in the s domain
(c) multiplication by s in the s domain
(d) multiplication by e^{-sT} in the s domain

20. The initial value theorem is
(a) $\underset{t\to 0}{\text{Lt}}\, f(t) = \underset{s\to 0}{\text{Lt}}\,[sF(s)]$
(b) $\underset{t\to 0}{\text{Lt}}\, f(t) = \underset{s\to \infty}{\text{Lt}}\,[F(s)/s]$
(c) $\underset{t\to 0}{\text{Lt}}\, f(t) = \underset{s\to \infty}{\text{Lt}}\,[sF(s)]$
(d) $\underset{t\to 0}{\text{Lt}}\, f(t) = \underset{s\to 0}{\text{Lt}}\,[F(s)/s]$

21. The final value theorem is
(a) $\underset{t\to \infty}{\text{Lt}}\, f(t) = \underset{s\to 0}{\text{Lt}}\,[sF(s)]$
(b) $\underset{t\to \infty}{\text{Lt}}\, f(t) = \underset{s\to \infty}{\text{Lt}}\,[sF(s)]$
(c) $\underset{t\to \infty}{\text{Lt}}\, f(t) = \underset{s\to 0}{\text{Lt}}\,[F(s)/s]$
(d) $\underset{t\to \infty}{\text{Lt}}\, f(t) = \underset{s\to \infty}{\text{Lt}}\,[F(s)/s]$

22. In Laplace transform, multiplication by e^{-at} in time domain becomes
(a) translation by a in s domain
(b) translation by $(-a)$ in s domain
(c) multiplication by e^{-as} in s domain
(d) none of the above

23. The Laplace transform of (t^{n-1}) where n is integer is
(a) $1/s^n$ (b) $(n-1)!/s^n$
(c) $\frac{1}{(n-1)!s^n}$ (d) $n!/s^n$

24. The inverse Laplace transform of $1/(s - a)^2$ is
(a) e^{at} (b) $t\,e^{at}$
(c) t^2e^{at} (d) e^{at}/t

25. The inverse Laplace transform of $\frac{-a}{s(s-a)}$ is
(a) e^{at} (b) $-e^{at}$
(c) $1 - e^{at}$ (d) $-1 + e^{at}$

26. The Laplace transform of $\sin \omega t$ is
(a) $\frac{1}{s^2 + \omega^2}$ (b) $\frac{\omega}{s^2 + \omega^2}$
(c) $\frac{1}{\omega(s^2 + \omega^2)}$ (d) none of the above

27. The Laplace transform of cos ωt is .

(a) $\frac{s}{(s^2+\omega^2)}$ (b) $\frac{1}{s^2+\omega^2}$

(c) $\frac{1}{s(s^2+\omega^2)}$ (d) $\frac{s^2}{s^2+\omega^2}$

28. The Laplace transform of sinh αt is

(a) $\frac{\alpha}{s^2-\alpha^2}$ (b) $\frac{\alpha}{s^2+\alpha^2}$

(c) $\frac{1}{\alpha(s^2-\alpha^2)}$ (c) $\frac{1}{\alpha(s^2+\alpha^2)}$

29. The inverse Laplace transform of $\frac{1}{(s+1)^3}$ is

(a) t^2e^{-t} (b) $\frac{t^2}{2}e^{-t}$

(c) te^{-t} (c) $\frac{t}{2}e^{-t}$

30. If $I(s) = \frac{5(s+250)}{s(s+100)}$, initial value of $i(t)$ is

(a) 5 A (b) 12.5 A

(c) 0.05 A (d) 1250 A

31. If $I(s) = \frac{5(s+250)}{s(s+100)}$, the final value of $i(t)$ is

(a) 5 A (b) 12.5 A

(c) 0.05 A (d) 1250 A

32. If $f(t) = A\,\delta(t-a)$, F(s) is

(a) $A\,e^{-as}$ (b) $A\,e^{as}$

(c) $A\,a\,e^{-as}$ (d) $Aa\,e^{as}$

33. The wavefrom shown in Fig. 3.1 can be written as

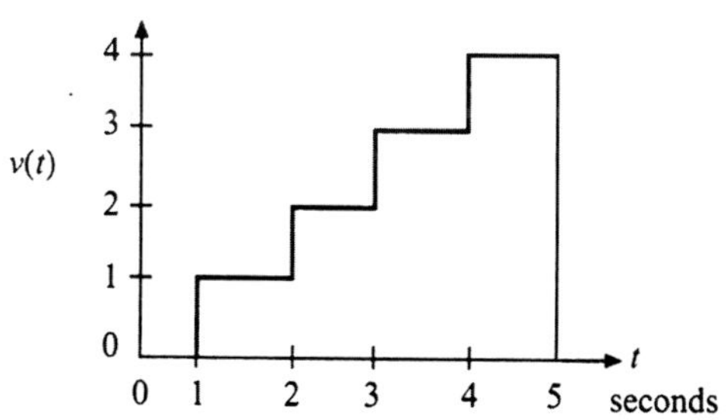

Fig. 3.1.

(a) $v(t) = u(t-1) + u(t-2) + u(t-3) + u(t-4)$

(b) $v(t) = u(t-1) + u(t-2) + u(t-3) + u(t-4) - u(t-5)$

(c) $v(t) = u(t-1) + u(t-2) + u(t-3) + u(t-4) - 4u(t-5)$

(d) $v(t) = u(t-1) + u(t-2) + u(t-3) + u(t-4) + 4u(t-5)$

34. The function shown in Fig. 3.2 can be written as

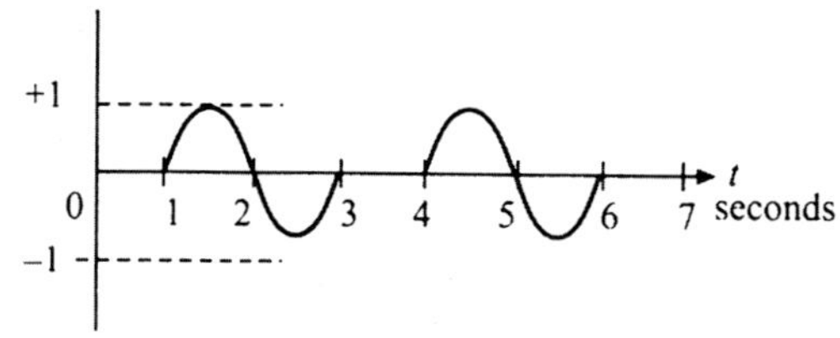

Fig. 3.2.

(a) sum of two sinusoidal functions

(b) sum of two sinusoidal functions one originating at $t = 0$ and the other at $t = 1$

(c) difference of two sinusoidal functions one originating at $t = 0$ and the other at $t = 1$

(d) none of the above

35. The function $A\,e^{st}$ where $s = \sigma + j\omega$ represents

(a) A phasor whose magnitude changes with time

(b) A phasor whose frequency is ω

(c) A phasor rotating in counter clockwise direction at angular frequency ω and whose magnitude increases with t

(d) A phasor rotating in clockwise direction at angular frequency ω and whose magnitude increases with t

36. State space analysis is applicable to

(a) time invariant systems

(b) time varying systems

(c) both time invariant and time varying systems

(d) none of the above

37. In RLC circuits the state variables generally selected are

(a) votlages across capacitors

(b) currents through resistances and voltages across capacitors

(c) currents through resistances and capacitances

(d) currents through inductances and voltages across capacitances

38. The tree selected for the formation of state equations contains

(a) all voltage sources

(b) all capacitors

(c) all inductors and current sources

(d) all voltage sources and maximum number of capacitors

39. For formation of state equations, the inductors and current sources

(a) should be in tree

(b) should be in cotree

(c) may be in tree or cotree

(d) should be in both tree and cotree

40. The state equations are in the form

(a) $\dot{X} = AX + Bu$ (b) $X = A\dot{X} + Bu$

(c) $\ddot{X} = AX + Bx$ (d) either (a) or (b)

41. For formation of state equations, resistances of the circuit

(a) should be in tree

(*b*) should be in cotree
(*c*) may be in tree or cotree
(*d*) none of the above

42. The solution of state equations using Laplace transform is
(*a*) $x(t) = \phi(t)\, x(0) + £^{-1}\,[\phi(s)\, Bu(s)]$
(*b*) $x(t) = \phi(s)\, x(0) + £^{-1}\,[\phi(s)\, Bu(s)]$
(*c*) $x(t) = e^{At}x(0) + \int_0^t e^{A(t-\tau)}Bu(\tau)\, d\tau$
(*d*) (*a*) and (*c*)

43. $(sI - A)^{-1}$ is equal to
(*a*) *adj*(*s*I – A)/*det* (*s*I – A)
(*b*) *del* (*s*I – A)/*adj*(*s*I – A)
(*c*) [*adj*(sI – A)][*det*(sI – A)]
(*d*) none of the above

44. A voltage wave $v = 10 + 20 \sin \omega t + 7.5 \sin 3\omega t$ is applied to a series combination of two elements. The current is $i = 5 \sin (\omega t + 20°) + 1.5 \sin (3\omega t + 10°)$. The elements are
(*a*) R and C (*b*) R and L
(*c*) L and C (*d*) both inductances

45. An ac circuit has an impedance of $(3 + j6)$ ohm for fundamental. The impedance for fifth harmonic is
(*a*) $(15 + j30)\,\Omega$ (*b*) $(15 + j6)\,\Omega$
(*c*) $(3 + j30)\,\Omega$ (*d*) $(3 + j1.2)\,\Omega$

46. An RLC series circuit has a variable inductance. The value of L for resonance conditions at fundamental frequency is 0.18 H. For resonance conditions at third harmonic frequency the value of inductance is
(*a*) 0.54 H (*b*) 0.06 H
(*c*) 0.02 H (*d*) 1.62 H

47. A voltage wave having 5% fifth harmonic content is applied to a series R-C circuit. The percentage fifth harmonic content in the current wave will be
(*a*) 5%
(*b*) more than 5%
(*c*) less than 5%
(*d*) equal or more than 5%

48. A voltage wave containing 10% third harmonic is applied to a series R-L circuit. The percentage third harmonic content in the current wave will be
(*a*) more than 10%
(*b*) 10% or less than 10%
(*c*) less than 10%
(*d*) 10% or more or less

49. A complex wave is $5 + 5 \sin \omega t$. Its rms value is
(*a*) 7.07 V (*b*) 5 V
(*c*) 10 V (*d*) 6.12 V

50. In an ac circuit the fundamental component of current wave lags the corresponding voltage wave by 20°. The third harmonic component of current wave lags the corresponding voltage by an angle.
(*a*) less than 20° (*b*) more than 20°
(*c*) equal to 20°
(*d*) equal to or more than 20°

51. An ac circuit has an impedance of $(2 - j9)\,\Omega$ for third harmonic. The impedance for fundamental is
(*a*) $6 - j27\,\Omega$ (*b*) $\frac{2}{3} - j3\,\Omega$
(*c*) $2 - j27\,\Omega$ (*d*) $2 - j3\,\Omega$

52. An ac network has a power factor of 0.8 leading if the applied wave is of fundamental frequency. If the applied wave contains third and fifth harmonics, the overall power factor will be
(*a*) 0.8 leading
(*b*) less than 0.8 leading
(*c*) more than 0.8 leading
(*d*) 0.8 leading or more

53. An ac network has a power factor of 0.8 lagging for fundamental frequency. If the applied voltage contains thrid and fifth harmonics, the overall power factor will be
(*a*) more than 0.8 lagging
(*b*) 0.8 lagging
(*c*) less than 0.8 lagging
(*d*) 0.8 lagging or less

54. A voltage wave is $v = 50 \sin \omega t$. Its average value calculated over full one cycle is
(*a*) zero (*b*) 35.36 V
(*c*) 31.85 V (*d*) none of the above

55. A voltage wave is $i = 100 \sin (\omega t)$. Its average value calculated over one half cycle is
(*a*) zero (*b*) 70.72 V
(*c*) 63.70 V (*d*) none of the above

56. An ac sinusoidal wave has an rms value of 10 V. The peak-to-peak value is
(*a*) 10 V (*b*) 0 V
(*c*) 14.14 V (*d*) 28.28 V

57. The peak factor is
(*a*) rms value/peak value
(*b*) rms value/average value
(*c*) peak value/rms value
(*d*) peak-to-peak value/rms value

58. For an ac sinusoidal wave, the rms value is 10 A. For the same wave delayed by 60° in each half cycle, the rms value is likely to be
(*a*) 10 A (*b*) 7.07 A
(*c*) 6.35 A (*d*) 3.5 A

59. In a complex wave, the negative half of the wave is a reproduction of the positive half wave. Then
(*a*) the wave contains only fundamental and third harmonic
(*b*) the wave does not contain even harmonics
(*c*) the wave does not contain odd harmonics
(*d*) the wave does not contain triplen harmonics

60. For the wave $i = I_0 + I_{1m} \sin \omega + I_{3m} \sin 3\omega t$, the rms value is

(a) $I = (I_0^2 + I_{1m}^2 + I_{3m}^2)^{0.5}$

(b) $I = \left(I_0^2 + \frac{I_{1m}^2}{2} + \frac{I_{3m}^2}{2} \right)^{0.5}$

(c) $I = 0.707 (I_0^2 + I_{1m}^2 + I_{3m}^2)^{0.5}$

(d) none of the above

61. If $f = 0$

(a) $X_L = 0$ and $X_C = 0$ (b) $X_L = \infty, X_C = \infty$

(c) $X_L = 0$ and $X_C = \infty$ (d) $X_L = \infty$ and $X_C = 0$

62. A wave $f(t)$ has half wave symmetry and time period equal to T. Then

(a) $f(t) = f\left(t \pm \frac{T}{2}\right)$ (b) $f(t) = -f\left(t + \frac{T}{2}\right)$

(c) $f(t) = f\left(t - \frac{T}{2}\right)$ (d) $f(t) = -f\left(t \pm \frac{T}{2}\right)$

63. A function will have only sine terms if

(a) $f(t) = -f(t)$ (b) $f(-t) = f(t)$

(c) $f(-t) = -f(t)$ (d) none of the above

64. If a function has only cosine terms, it must satisfy the equation

(a) $f(t) = -f(t)$ (b) $f(-t) = f(t)$

(c) $f(-t) = -f(t)$ (d) none of the above

65. The exponential form of Fourier series is

(a) $f(t) = F_0 + \sum_{n=1}^{\infty} [F_n e^{jn\omega_1 t} + F_{-n} e^{-jn\omega_1 t}]$

(b) $f(t) = \sum_{n=-\infty}^{\infty} F_n e^{jn\omega_1 t}$

(c) $f(t) = \sum_{n=1}^{\infty} F_n e^{jn\omega_1 t}$

(d) (a) and (b)

66. If $f(t)$ is an even function, the coefficients F_n in the exponential form of Fourier series

(a) are real

(b) are imaginary

(c) are complex

(d) may be real or imaginary

67. If $f(t)$ is an odd function, the coefficients F_n in the exponential form of Fourier series

(a) are real

(b) are imaginary

(c) are complex

(d) may be real or imaginary

68. $\cos (n\omega_1 t) =$

(a) $0.5 (e^{jn\omega_1 t} + e^{-jn\omega_1 t})$

(b) $0.5 (e^{jn\omega_1 t} + e^{-jn\omega_1 t})$

(c) $\frac{0.5}{j} (e^{jn\omega_1 t} + e^{-jn\omega_1 t})$ (d) $\frac{0.5}{j} (e^{jn\omega_1 t} - e^{-jn\omega_1 t})$

69. The coefficients F_n in the exponential form of Fourier series are

(a) independent of frequency

(b) functions of frequency

(c) even function of frequency

(d) odd functions of frequency

70. In plotting line spectrum

(a) a graph of complex number F_n versus frequency is plotted

(b) a graph of $|F_n|$ versus frequency is plotted

(c) a graph of the angle of F_n versus frequency is plotted

(d) two graphs *viz* $|F_n|$ versus frequency and the phase angle of F_n versus frequency are plotted

71. If $f(t)$ is in volts, then $F(j\omega)$ is in

(a) volts (b) volt seconds

(c) volts/sec (d) volt-sec^2

72. $F[c_1 f_1(t) + c_2 f_2(t)] =$

(a) $c_1 F_1(j\omega) + c_2 F_2(j\omega)$

(b) $c_1 F_1(j\omega) - c_2 F_2(j\omega)$

(c) $c_1 F_1(j\omega) + c_2 F_2(j\omega) + c_1 c_2$

(d) $c_1 F_1(j\omega) + c_2 F_2(j\omega) + c_1/c_2$

73. If $f(t)$ is an odd function, $F(j\omega) =$

(a) $2\int_0^{\infty} f(t) \cos \omega t \, dt$ (b) $\int_0^{\infty} f(t) \cos \omega t \, dt$

(c) $2j\int_0^{\infty} f(t) \sin (\omega t) \, dt$ (d) $j\int_0^{\infty} f(t) \sin (\omega t) \, dt$

74. If $f(t)$ is an even function, $F(j\omega) =$

(a) $2\int_0^{\infty} f(t) \cos (\omega t) \, dt$ (b) $\int_0^{\infty} f(t) \cos (\omega t) \, dt$

(c) $2j\int_0^{\infty} f(t) \sin (\omega t) \, dt$ (d) $-2j\int_0^{\infty} f(t) \sin (\omega t) \, dt$

75. If $\mathfrak{F} f(t) = F(j\omega)$, $\mathfrak{F} f(t - a) =$

(a) $F(j\omega) e^{-j\omega a}$ (b) $F(j\omega) e^{j\omega a}$

(c) $\frac{1}{a} F(j\omega) e^{-j\omega a}$ (d) $aF(j\omega) e^{j\omega a}$

76. If $\mathfrak{F} f(t) = F(j\omega)$, then $\mathfrak{F}\left[\frac{d^n}{dt^n} f(t)\right] =$

(a) $(j\omega)^n F(j\omega)$ (b) $F(j\omega)/(j\omega)^n$

(c) $(j\omega)^{n-1} F(j\omega)$ (d) $F(j\omega)/(j\omega)^{n-1}$

77. Fourier transform of the function $f(t) = 1$ is

(a) $2\pi \delta(\omega)$ (b) $\pi \delta(\omega)$

(c) $\delta(\omega)/2\pi$ (d) $\delta(\omega)/\pi$

78. If $F(j\omega)$ is the Fourier transform of $f(t)$ then $f(t) =$

(*a*) $\frac{1}{2\pi}\int_{-\infty}^{\infty} F(j\omega)\, e^{j\omega t}\, dt$ (*b*) $\frac{1}{2\pi}\int_{-\infty}^{\infty} F(j\omega)\, e^{-j\omega t}\, dt$

(*c*) $\frac{1}{2\pi}\int_{0}^{\infty} F(j\omega)\, e^{j\omega t}\, dt$ (*d*) $\frac{1}{2\pi}\int_{0}^{\infty} F(j\omega)\, e^{-j\omega t}\, dt$

79. If $F(j\omega)$ is the Fourier transform of $f(t)$, then

(*a*) $\mathfrak{F}\, f(-t) = F(j\omega)$ (*b*) $\mathfrak{F}\, f(-t) = F^*(j\omega)$

(*c*) $\mathfrak{F}\, f(-t) = F(-j\omega)$ (*d*) $\mathfrak{F}\, f(-t) = F^*(-j\omega)$

80. A voltage $v = 100 \sin \omega t + 10 \sin 5\,\omega t$ is applied to a pure capacitor having capacitance of 1 μF. If $\omega = 314$ rad/sec, the current through the capacitor is

(*a*) $0.0314 \cos 314t + 0.0157 \cos 1570t$

(*b*) $0.0314 \sin 314t + 0.0157 \sin 1570t$

(*c*) $0.0314 \cos 314t + 0.0314 \cos 1570t$

(*d*) $0.0157 \cos 314t + 0.0157 \cos 1570t$

81. A voltage $v = 5 + 50 \sin \omega t + 5 \sin 5\,\omega t$ is applied to a pure capacitor of capacitance 1 μF. If $f = 314$ rad/sec, current is

(*a*) $1 + 0.0157 \cos 314\,t + 0.00785 \cos 1570t$

(*b*) $0.0157 \cos 314t + 0.00785 \cos 1570t$

(*c*) $0.0157 \sin 314\,t + 0.00785 \sin 1570t$

(*d*) $0.0157 \sin (314t + 45°) + 0.00785 \sin (1570t + 45°)$

82. If $i = 10 + 12 \cos 500t - 3.5 \sin 500t$, then in amplitude – phase form $i =$

(*a*) $10 + 12.5 \sin (500t + 106.20°)$

(*b*) $5 + 12.5 \sin (500t + 106.20°)$

(*c*) $10 + 11 \sin (500t + 106.20°)$

(*d*) $10 + 12.5 \sin (500t + 10°)$

83. If a function $f(t)$ is Laplace transformable, then

(*a*) $\int_{0}^{\infty} |f(t)|\, e^{-\sigma_1 t}\, dt < \infty$ for a real positive σ_1

(*b*) $\int_{-\infty}^{\infty} |f(t)|\, e^{-\sigma_1 t}\, dt < 0$ for a real positive σ_1

(*c*) $\int_{0}^{\infty} |f(t)|\, e^{-\sigma_1 t}\, dt < 0$ for real positive σ_1

(*d*) $\int_{-\infty}^{\infty} |f(t)|\, e^{-\sigma_1 t}\, dt < \infty$ for a real positive σ_1

84. If Laplace transform of $f(t)$ is $F(s)$, then $\pounds\left[\frac{d}{dt} f(t)\right] =$

(*a*) $sF(s) - f(0^-)$ (*b*) $\frac{1}{s} F(s) - f(0^-)$

(*c*) $sF(s) + f(0^-)$ (*d*) $\frac{1}{s} F(s) + f(0^-)$

85. If $i(t)$ is a time varying current, $\int_{-\infty}^{0^-} i(t)\, dt$ is

(*a*) initial current (*b*) initial charge

(*c*) initial flux linakges (*d*) initial voltage

86. If $v(t)$ is a time varying voltage, $\int_{-\infty}^{0^-} v(t)\, dt$ is

(*a*) initial current (*b*) initial voltage

(*c*) initial charge (*d*) initial flux linkages

87. If Laplace transform of $f(t)$ is $F(s)$, then $\pounds\,[tf(t)] =$

(*a*) $\frac{dF(s)}{ds}$ (*b*) $\frac{-dF(s)}{ds}$

(*c*) $(s)\frac{dF(s)}{ds}$ (*d*) $-s\frac{dF(s)}{ds}$

88. If Laplace transform of $f(t)$ is $F(s)$, then $\pounds\, f(t-a)\, u(t-a) = 0$

(*a*) $e^{as} F(s)$ (*b*) $e^{-as} F(s)$

(*c*) $-e^{as} F(s)$ (*d*) $-e^{-as} F(s)$

89. The derivative of unit step function is

(*a*) unit impulse (*b*) ramp with slope 1

(*c*) impulse (*d*) either (*a*) or (*b*)

90. A unit impulse voltage is applied to an inductance at $t = 0$. The current at $t = 0$ will be

(*a*) 0 (*b*) L amperes

(*c*) ∞ (*d*) $\frac{1}{L}$ amperes

91. The integral of $k\,\delta(t)$ is

(*a*) a step function of magnitude k

(*b*) a step function of magnitude $1/k$

(*c*) a ramp of slope k

(*d*) a ramp of slope $1/k$

92. The integral of $k\,u(t)$ is

(*a*) a ramp of slope k

(*b*) a ramp of slope $1/k$

(*c*) $k\,\delta(t)$ (*d*) $\frac{1}{k}\delta(t)$

93. The function in Fig. 3.3 can be written as

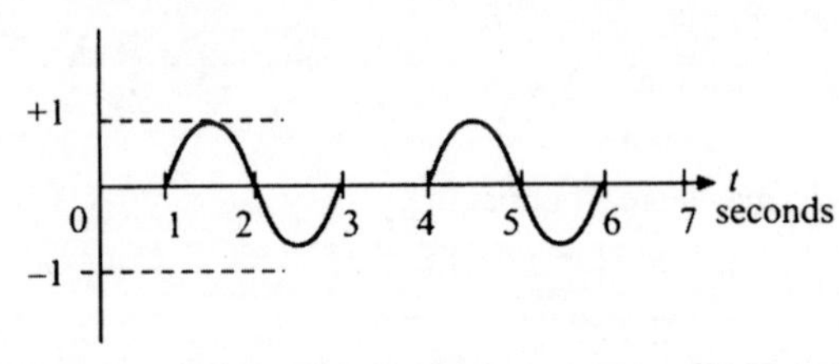

Fig. 3.3.

(*a*) $u(t-1) \sin \pi(t-1) - u(t-3) \sin \pi(t-3) + u(t-4) \sin \pi(t-4) - u(t-6) \sin \pi(t-6)$

(*b*) $u(t-1) \sin \pi(t-1) - u(t-3) \sin \pi(t-3) + u(t-4) \sin \pi(t-4)$

(*c*) $u(t-1) \sin \pi(t-1) + u(t-4) \sin \pi(t-4)$

(*d*) $u(t-1) \sin 2\pi(t-1) + u(t-4) \sin 2\pi(t-4)$

94. A gate function which occurs at $t = t_0$ and lasts for duration T can be written as

(a) $u(t - t_0 - T)$
(b) $u(t-t_0) + u(t - t_0 - T)$
(c) $u(t - t_0) - u(t - t_0 - T)$
(d) $u(t - t_0) - u(t - t_0 + T)$

95. Figure 3.4 shows a square wave $i(t)$. Then $I(s) =$

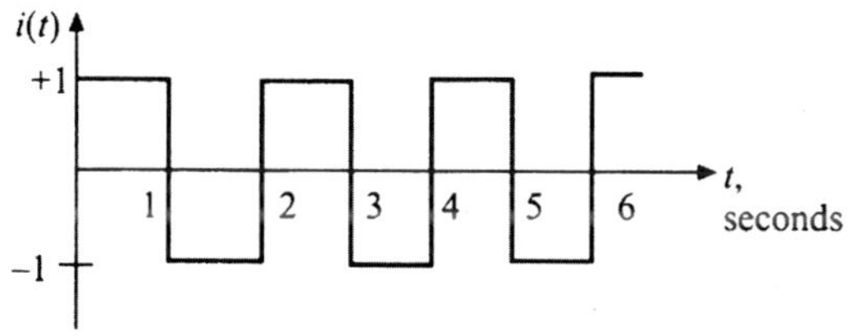

Fig. 3.4.

(a) $\frac{1}{s}\left[\frac{1-e^{-s}}{1+e^{-s}}\right]$
(b) $\frac{1}{s}\left[\frac{1+e^{-s}}{1-e^{-s}}\right]$
(c) $\frac{1}{s^2}\left[\frac{1-e^{-s}}{1+e^{-s}}\right]$
(d) $\frac{1}{s^2}\left[\frac{1+e^{-s}}{1-e^{-s}}\right]$

96. If a number of even functions are added, the resultant sum is

(a) even function
(b) odd function
(c) either even or odd function
(d) mixture of odd and even functions

97. If a number of odd functions are added, the resultant sum is

(a) even function
(b) odd function
(c) either even or odd function
(d) mixture of odd and even functions

98. If a function $f(t)$ is an odd function, its fourier series

(a) contains only sine terms
(b) contains only cosine terms
(c) constains a constant and sine terms only
(d) constains a constant and cosine terms only

99. Principle of superposition is applicable to

(a) linear systems only
(b) non-linear systems only
(c) both linear and non-linear systems
(d) linear systems and some non-linear systems

100. If u is input, y is output and a, b are constants, the system $y = au + b$

(a) is linear
(b) is non-linear
(c) may be linear or non-linear depending on values of a and b
(d) may be linear or non-linear depending on value of u

101. Figure 3.5 shows a discrete time system consisting of a unit delay system, a multiplier and a summer, such that $y(k) = x(k - 1) + 0.5\,x(k)$. This system

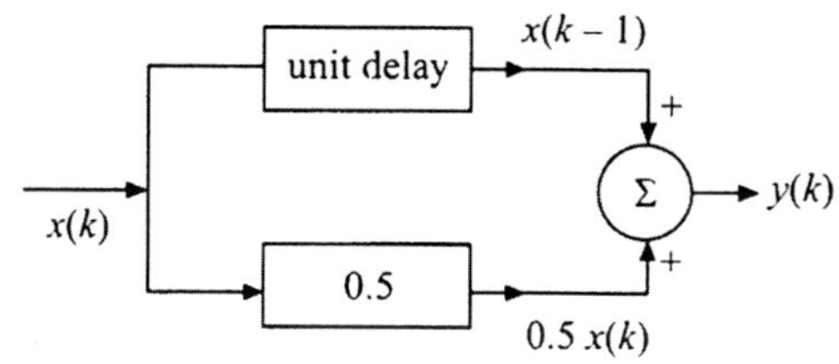

Fig. 3.5.

(a) is linear
(b) is non-linear
(c) may be linear or non-linear depending on value of x
(d) impossible to judge

102. Figure 3.6 shows a series, R-C circuit fed by a current source $i(t)$. There is an initial voltage v_0 across the capacitor. The system

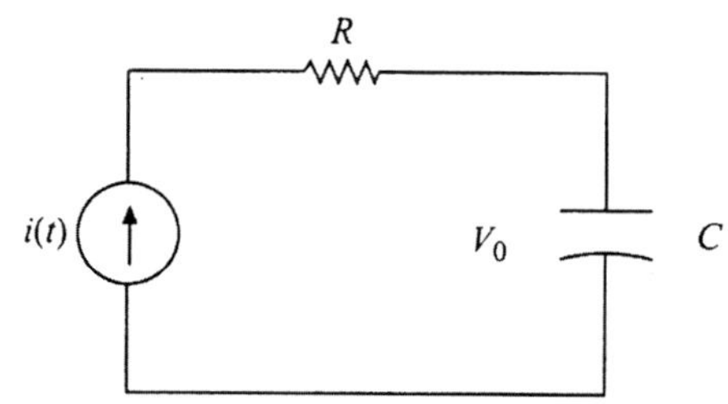

Fig. 3.6.

(a) is linear
(b) is non-linear
(c) may be linear or non-linear
(d) is linear only if v_0 is zero

103. Out of the three transforms viz Z-transform Laplace transform and Fourier transform

(a) all the three are used in continuous time domain
(b) all the three are used in both continuous time domain and discrete time domain
(c) Z transform is used in continuous time domain while Laplace and Fourier transforms are used in discrete time domain
(d) Z transform is used is discrete time domain while Laplace and Fourier transforms are used in continuous time domain

104. If $x_k = 0$ for $k < 0$ and $x_k = 2^k$ for $k \geq 0$, X(z) *i.e.*, Z transform of sequence x is

(a) $\frac{z}{z-2}$
(b) $\frac{1}{z-2}$
(c) $\frac{1}{(z-2)^2}$
(d) $\frac{1}{(z-2)^{-1}}$

105. If $x_k = 2^k$ for $k < 0$ and $x_k = 0$ for $k > 0$, Z transform of the sequence x is

(a) $\frac{z}{z-2}$
(b) $\frac{1}{z-2}$
(c) $\frac{2}{(z-2)}$
(d) $\frac{2}{2-z}$

106. If $x_k = \left(\frac{-1}{2}\right)^k$ for $k \geq 0$ and $x_k = 0$ for $k < 0$, Z trans-form of the sequence x is

(a) $\frac{z}{z+0.5}$ (b) $\frac{1}{z+0.5}$

(c) $\frac{2}{z+0.5}$ (d) $\frac{1}{2z+0.5}$

107. Z transform of $[a(x_k) + b(y_k)] =$

(a) $aX(z) - bY(z)$ (b) $aX(z) + bY(z)$

(c) $aX(z) + bY(z) + a/b$ (d) $aX(z) + bY(z) - a/b$

108. Z transform of $[(X_{k \pm k_0})] =$

(a) $z^{\pm k_0} X(z)$ (b) $z^{\mp k_0} X(z)$

(c) $\frac{X(z)}{z^{k_0}}$ (d) none of the above

109. If X(z) is the Z transform of sequence x_k, the Z transform of sequence (kx_k) is

(a) $z\frac{d}{dz}X(z)$ (b) $-z\frac{d}{dz}X(z)$

(c) $\frac{d}{dz}X(z)$ (d) $-\frac{d}{dz}X(z)$

110. If $X(z) = (1 - az^{-1})$, and $|\alpha| < |z|$, the initial value x_0 is

(a) 1 (b) 0

(c) 2 (d) ∞

111. If $X(z) = 2az^{-1}/(1 - az^{-1})^3$ and $|a| < |z|$ then the inital value x_0 is

(a) 0 (b) 1

(c) ∞ (d) 2

112. If $X(z) = (1 - a)z^{-1}/[(1 - z^{-1})(1 - az^{-1})]$ and $1 < |z|$, the final value of x_k is

(a) 0 (b) 1

(c) ∞ (d) 2

113. If $H(z) = \frac{1 + z^{-1}}{1 - \frac{5}{6}z^{-1} + \frac{1}{6}z^{-2}}$ the poles of $H(z)$ are at

(a) $z = \frac{1}{2}$ and $z = \frac{1}{3}$ (b) $z = 1$ and $z = 3$

(c) $z = \frac{-1}{2}$ and $z = \frac{-1}{3}$ (d) $z = -1$ and $z = -3$

114. The function $(\sin x)/x$

(a) has a period 2π, decays with increasing x and has zeros at $n\pi$, $n = \pm 1, \pm 2$

(b) has a period π

(c) has a period $\pi/2$

(d) has a period 2π, decays with increasing x, is an even function and has zeros at $n\pi$, $n = \pm 1, \pm 2$ ...

115. The function sinc x is equal to

(a) $\frac{\sin x}{x}$ (b) $x \sin x$

(c) $x^2 \sin x$ (d) $\frac{x}{\sin x}$

116. The signum function written as $[sgn(t)]$ is defined as

(a) $sgn(t) = -1$ for $t < 0$, $= 0$ for $t = 0$ and $= 1$ for $t > 0$

(b) $sgn(t) = 1$ for $t < 0$, $= 0$ for $t = 0$ and $= -1$ for $t > 0$

(c) $sgn(t) = 0$ for $t < 0$, $= 1$ for $t = 0$ and $= 0$ for $t > 0$

(d) $sgn(t) = 0$ for $t < 0$, $= 1$ for $t \geq 0$

117. If $f(t) = sgn(t)$, $F(j\omega) =$

(a) 1 (b) $\frac{1}{j\omega}$

(c) 2 (d) $\frac{2}{j\omega}$

118. For the discrete time system of Fig. 3.7

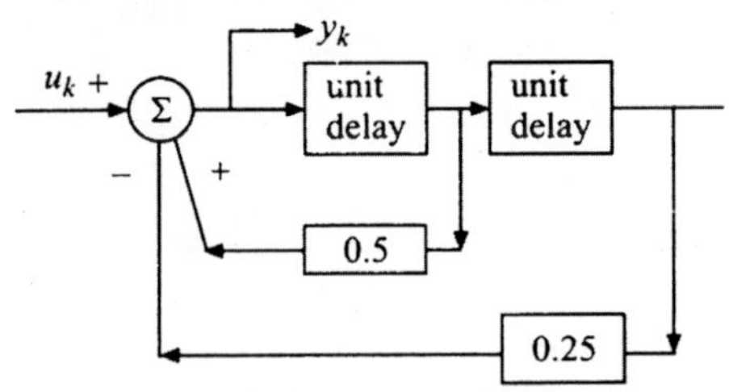

Fig. 3.7.

(a) $y_k - 0.5y_{k-1} - 0.25\,y_{k-2} = u_k$

(b) $y_k + 0.5y_{k-1} - 0.25y_{k-2} = u_k$

(c) $y_k - 0.5y_{k-1} + 0.25\,y_{k-2} = u_k$

(d) $y_k + 0.5y_{k-1} + 0.25\,y_{k-2} = u_k$

119. For the discrete time system of Fig. 3.8

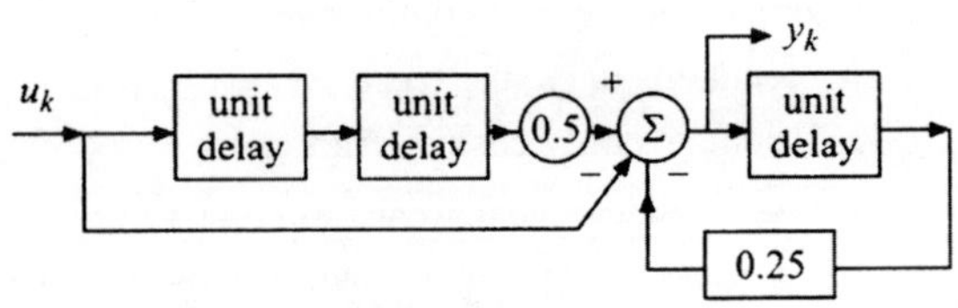

Fig. 3.8.

(a) $y_k = 0.5\,u_{k-2} - u_k - 0.25\,y_{k-1}$

(b) $y_k = 0.5\,u_{k-1} - u_k - 0.25y_{k-1}$

(c) $y_k = 0.5\,u_k - u_{k-1} - 0.25\,y_{k-1}$

(d) $y_k = 0.5\,u_{k-1} - u_{k-1} - 0.25\,y_{k-1}$

120. The energy associated with a function $f(t)$ is $E = \int_{-\infty}^{\infty} f^2(t)\,dt$. In terms of Fourier transform, E =

(a) $\frac{1}{2\pi}\int_{-\infty}^{\infty} |F(j\omega)|^2\,d\omega$ (b) $\frac{1}{\pi}\int_{-\infty}^{\infty} |F(j\omega)|^2\,d\omega$

(c) $\frac{1}{2\pi}\int_{-\infty}^{\infty} |F(j\omega)|\,d\omega$ (d) $\frac{1}{\pi}\int_{-\infty}^{\infty} |F(j\omega)|\,d\omega$

121. If p is the probability of success and q is the probability of failure, then the probability that there are r successes in n trials is

(a) ${}^{n}c_r p^r q^{n-r}$ (b) ${}^{n}c_r p^{n-r} q^r$

(c) ${}^{n}p_r p^r q^{n-r}$ (d) ${}^{n}c_r p^n q^{n-r}$

122. As per normal distribution the probability of an error between limits a and b is

(a) $P_{ab} = \frac{h}{\sqrt{\pi}} \int_a^b e^{-h^2x^2}\, dx$ (b) $P_{ab} = \frac{h}{\pi} \int_a^b e^{-h^2x^2}\, dx$

(c) $P_{ab} = \frac{h}{\sqrt{\pi}} \int_a^b e^{-hx}\, dx$ (d) $P_{ab} = \frac{h}{\pi} \int_a^b e^{-hx}\, dx$

123. If n is the number of observations and ρ is the residue, the standard deviation σ =

(a) $\frac{\Sigma\rho^2}{n}$ (b) $\left[\frac{\Sigma\rho^2}{n}\right]^{0.5}$

(c) $\frac{1}{n}\left[\Sigma\rho^2\right]^{0.5}$ (d) $\left[\frac{\Sigma\sigma^4}{n}\right]^{0.25}$

124. The data of speed of train V and resistance to motion R is

V	10	20	30	40	50
R	8	10	15	21	30

The law $R = a + bV^2$ is of the form

(a) $6.7 + 0.0092\, V^2$ (b) $3 + 0.001\, V^2$

(c) $3.2 + 0.0012\, V^2$ (d) $5.2 + 0.006\, V^2$

125. The data about p the pull required to lift a weight w by a pulley block is

w	50	70	100	120
p	12	15	21	25

The linear law $p = a + bw$ is

(a) $3.2 + 0.171\, w$ (b) $2.28 + 0.1879\, w$

(c) $1.2 + 0.25\, w$ (d) $0.6 + 0.3\, w$

126. Double integration of a unit step function would lead to

(a) an impulse (b) a parabola

(c) a ramp (d) a doublet

127. The effective value of the waveform given in Fig. 3.9 is

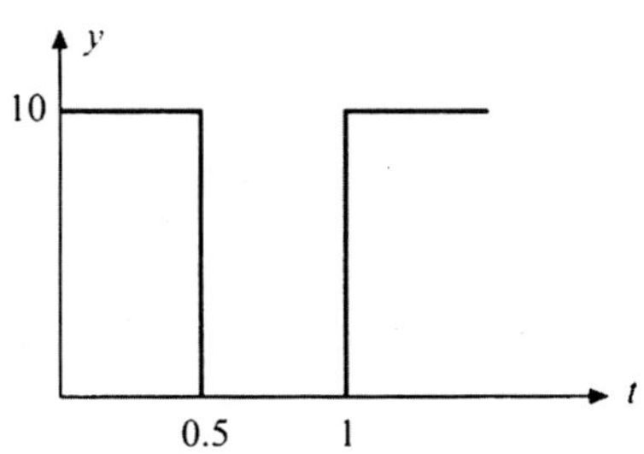

Fig. 3.9.

(a) 5 (b) 2.5

(c) $\sqrt{2.5}$ (d) $\sqrt{5.0}$

128. If $f(t) = -f(-t)$ and $f(t)$ satisfies Drichlet conditions, then $f(t)$ can be expanded in Fourier series containing

(a) only sine terms

(b) only cosine terms

(c) cosine terms and constant

(d) sine terms and constant

129. Fourier transform $F(j\omega)$ of an arbitrary signal has the property

(a) $F(j\omega) = F(-j\omega)$ (b) $F(j\omega) = -F(-j\omega)$

(c) $F(j\omega) = F^*(-j\omega)$ (d) $F(j\omega) = -F^*(-j\omega)$

130. The inverse Fourier transform of the function $F(\omega) = \frac{2}{j\omega}$ is

(a) $\sin \omega t$ (b) $\cos \omega t$

(c) $sgn(t)$ (d) $u(t)$

131. The following table gives some time functions and the Laplace transforms

	$f(t)$	F(s)
1.	$\delta(t)$	s
2.	$u(t)$	$1/s$
3.	$t\, u(t)$	$2/s^2$
4.	$t^2\, u(t)$	$2/s^3$

Of these correctly matched pairs are

(a) 2 and 4 (b) 1 and 4

(c) 3 and 4 (d) 1 and 2

132. The final value of $£^{-1} \frac{2s+1}{s^4 + 8s^3 + 16s^2 + s}$ is

(a) ∞ (b) 2

(c) 1 (d) zero

133. The two inputs to an analogue multipler are $x(t)$ and $y(t)$ with Fourier transforms $X(f)$ and $Y(f)$ respectively. The output $Z(t)$ will have a transform $Z(f)$ given by

(a) $X(f) \cdot Y(f)$ (b) $X(f) + Y(f)$

(c) $X(f)/Y(f)$ (d) $\int_0^t Y(\lambda)\, X(t-\lambda)\, d\lambda$

134. For the signal in Fig. 3.10 (a) the Fourier transform is $\frac{2 \sin \omega T_1}{\omega}$. Then the Fourier transform of the signal in Fig. 3.10 (b) is

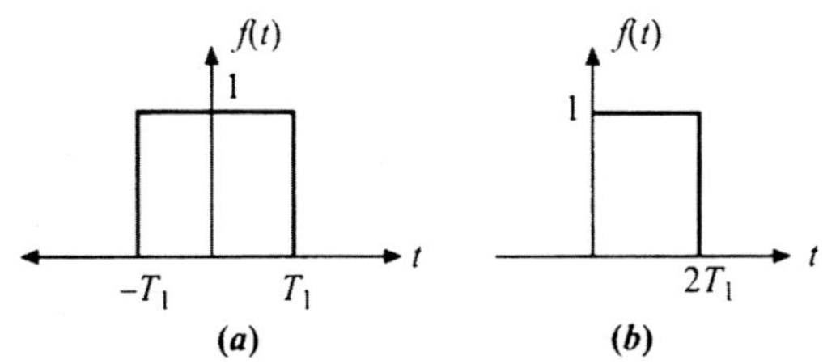

Fig. 3.10.

(a) $2\frac{\sin \omega T_1}{\omega} e^{j\omega T_1}$ (b) $2\frac{\sin \omega T_1}{\omega} e^{-j\omega T_1}$

(c) $\frac{\sin \omega T_1}{\omega} e^{-j\omega T_1}$ (d) $\frac{\sin \omega T_1}{\omega} e^{-j\omega(T_1-2)}$

135. In Laplace transforms $s = \sigma + j\omega$. Which of the following represents true nature of σ. Select your answer using the given codes

1. σ has a damping effect
2. σ is responsible for convergence of integral

$$\int_0^{\infty} f(t)\, e^{-st}\, dt$$

3. σ has a value less than zero

Codes:

(*a*) 1, 2 and 3 (*b*) 1 and 2
(*c*) 2 and 3 (*d*) 1 and 3

136. Which of the following is the correct Laplace transform of the signal shown in Fig. 3.11

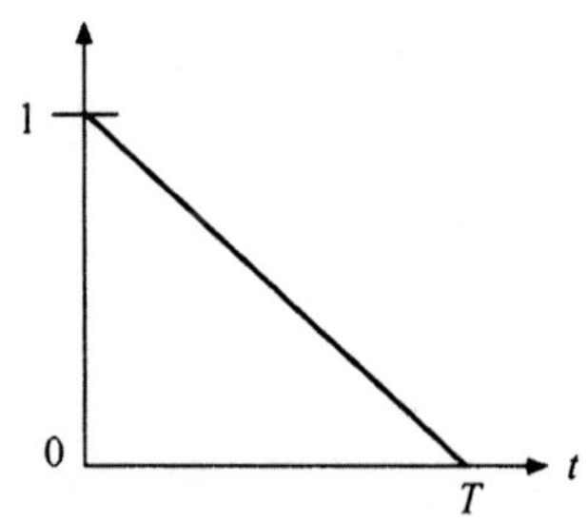

Fig. 3.11.

(*a*) $\frac{1}{Ts^2}[1 - e^{-Ts}(1 + Ts)]$ (*b*) $\frac{1}{Ts^2}[e^{-Ts} - 1 + Ts]$

(*c*) $\frac{1}{Ts^2}[e^{-Ts} + 1 - Ts]$ (*d*) $\frac{1}{Ts^2}[1 - e^{-Ts} + Ts]$

137. Which of the following represents a stable system

1. Impulse response decreases exponentially
2. Area within the impulse response in finite
3. Eigen values of the system are positive and real
4. Roots of the characteristic equation of the system are real and positive

Select the answer using the following codes

Codes:

(*a*) 1 and 4 (*b*) 1 and 3
(*c*) 2, 3 and 4 (*d*) 1, 2

138. Match List I and List II and select the correct answer using the given codes

List I **List II**

A. 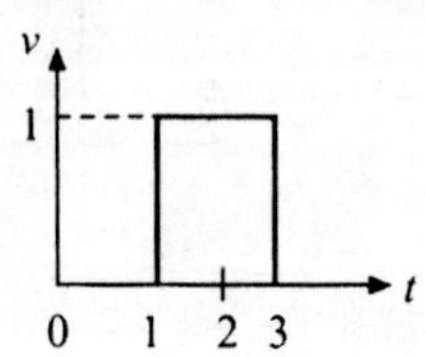

1. $v(t) = u(t + 1)$

B. 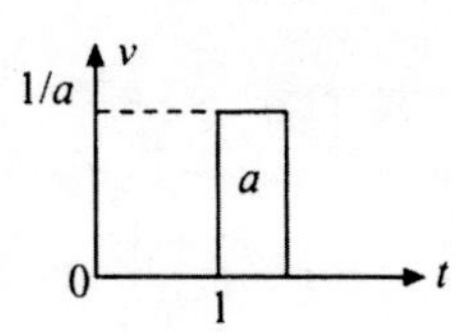

2. $v(t) = u(t) - 2u(t-1) + 2u(t-2) - 2u(t-3)...$

C. 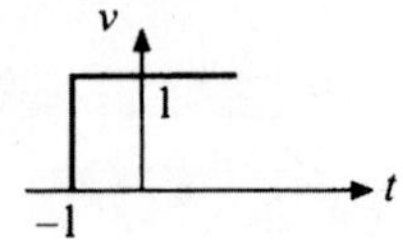

3. $v(t) = u(t-1) - u(t-3)$

D. 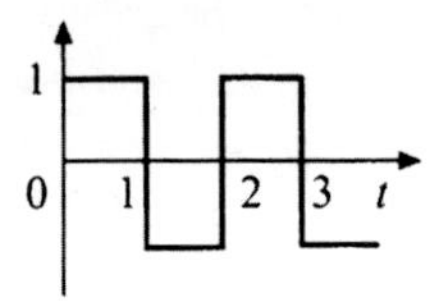

4. $\underset{a \to 0}{\text{Lt}}\ v(t) = \delta(t - 1)$

Fig. 3.12.

Codes:

	A	B	C	D
(*a*)	3	4	2	1
(*b*)	3	4	1	2
(*c*)	4	3	2	1
(*d*)	4	3	1	2

139. Match List I (source of signal) with List II (type of signal) and select the correct answer using the given codes

List I	**List II**
A. Output of signal generator	1. Modulated
B. Error signal from a synchro	2. Digital
C. Output of a JK flip-flop	3. Analog .
D. Signal received by radar	4. Stochastic

Codes:

	A	B	C	D
(*a*)	1	3	2	4
(*b*)	3	1	2	4
(*c*)	3	1	4	2
(*d*)	1	3	4	2

140. The amplitude of the first odd harmonic of the square wave shown in Fig. 3.13 is

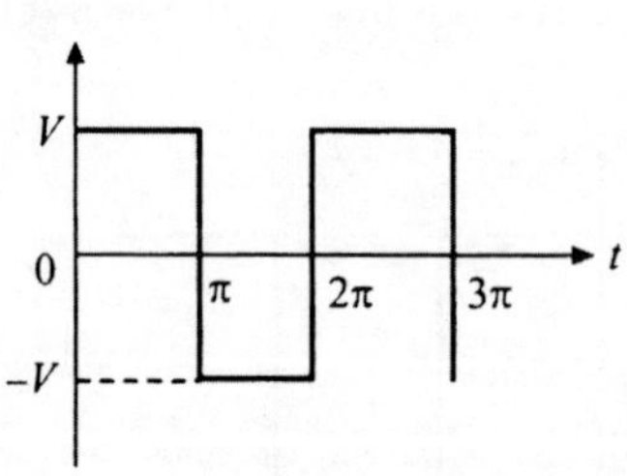

Fig. 3.13.

(*a*) $\frac{4V}{\pi}$ (*b*) $\frac{2V}{3\pi}$

(*c*) $\frac{V}{\pi}$ (*d*) 0

141. Figure 3.14 shows a periodic triangular wave. The Fourier series will have

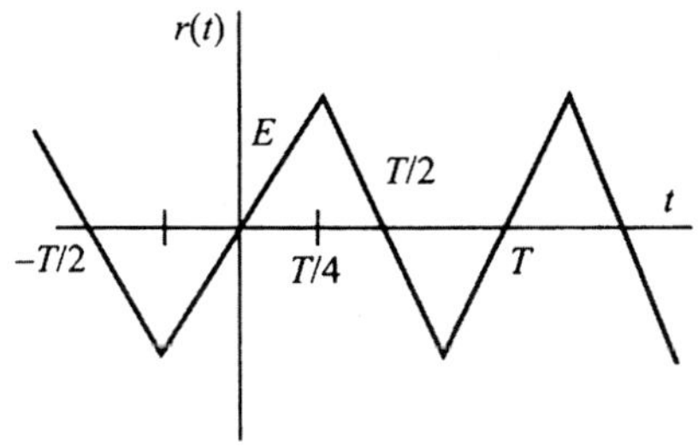

Fig. 3.14.

(a) all cosine terms (b) all sine terms
(c) odd cosine terms (d) odd sine terms

142. Fourier transform of the unit step function (*i.e.*, $u(t) = 1$ for $t \geq 0$ and $u(t) = 0$ for $t < 0$) is

(a) $\pi\,\delta(\omega)$

(b) $\dfrac{1}{j\omega}$

(c) $\dfrac{1}{j\omega} + \pi\delta(\omega)$

(d) $\dfrac{1}{j\omega} + 2\pi\delta(\omega)$

143. The Fourier transform of $f(t) = \cos \omega_0 t$ is

(a) $\pi[\delta\omega_0 + \delta(-\omega_0)]$

(b) $\pi[\delta(\omega - \omega_0) + \delta(\omega + \omega_0)]$

(c) $\pi[\delta\omega_0 - \delta(-\omega_0)]$

(d) $\pi[\delta(\omega - \omega_0) - \delta(\omega + \omega_0)]$

144. If $f(t)$ and $F(j\omega)$ form a transform pair, then as per symmetry in Fourier transforms

(a) $F(jt) \leftrightarrow 2\pi f(-\omega)$ (b) $F(jt) \leftrightarrow \pi f(\omega)$

(c) $F(jt) \leftrightarrow \dfrac{\pi}{2} f(\omega)$ (d) $F(jt) = \dfrac{\pi}{2} f(-\omega)$

145. If $x(t)$ and its first derivative are Laplace transformable and Laplace transform of $x(t)$ is $X(s)$, then $\underset{t \to 0}{\text{Lt}}\ x(t)$ is

(a) $\underset{s \to \infty}{\text{Lt}}\ sX(s)$ (b) $\underset{s \to 0}{\text{Lt}}\ sX(s)$

(c) $\underset{s \to 0}{\text{Lt}}\ \dfrac{X(s)}{s}$ (d) $\underset{s \to 0}{\text{Lt}}\ \dfrac{X(s)}{s}$

146. If $\delta(t)$ denotes a unit impulse, Laplace transform of $\dfrac{d^2}{dt^2}[\delta(t)]$ is

(a) 1 (b) s^2
(c) s (d) s^{-2}

147. The unit step response of a system is $(1 - e^{at})\,u(t)$. Then its impulse response is

(a) $e^{-\alpha t}u(t)$ (b) $\alpha e^{-\alpha t}u(t)$

(c) $\dfrac{1}{\alpha}e^{-\alpha t}u(t)$ (d) $-\alpha e^{-\alpha t}u(t)$

148. Given that $h(t) = 10e^{-10t}u(t)$ and $e(t) = \sin 10\,t\,u(t)$. The Laplace transform of $f(t) = \int_0^{\tau} h(t - \tau)\,e(\tau)\,d\tau$ is

(a) $\dfrac{10}{(s+10)(s^2+100)}$ (b) $\dfrac{10(s+10)}{s^2+100}$

(c) $\dfrac{100}{(s+100)(s^2+100)}$ (d) $\dfrac{1}{(s+10)(s^2+100)}$

149. Match the following using the given codes

List-I (f_k)	**List-II [F(z)]**
A. 1	1. $\left(-z\dfrac{d}{dz}\right)^n (1 - z^{-1})^{-1}$
B. k^n	2. $(1 - \alpha z^{-1})^{-1}$
C. α^k	3. $-\ln(1 - z^{-1})^{-1}$
D. $\dfrac{1}{k}$	4. $(1 - z^{-1})^{-1}$

Codes:

	A	B	C	D
(a)	1	4	2	3
(b)	2	3	1	4
(c)	4	1	2	3
(d)	1	4	3	2

150. If $I(s) = \dfrac{5(s+250)}{s(s+100)}$, final value of $i(t)$ is

(a) 0 (b) 2.5
(c) 12.5 (d) ∞

151. If $I(s) = \dfrac{5(s+250)}{s(s+100)}$, initial value of $i(t)$ is

(a) 0 (b) 5 A
(c) 10 A (d) −5 A

152. Match List I and List II using the given codes

List I	**List II**
A. t	1. $1/s$
B. e^{at}	2. $\dfrac{s}{s^2 + \omega^2}$
C. $u(t)$	3. $\dfrac{1}{s - a}$
D. $\cos \omega t$	4. $\dfrac{1}{s^2}$

Codes:

	A	B	C	D
(a)	1	3	2	4
(b)	1	4	2	3
(c)	2	4	1	3
(d)	4	3	1	2

153. Match List I and List II using the given codes

List I	List II
$f(t)$	F(s)
A. $\cos \omega t$	1. $s/(s^2 - \omega^2)$
B. $\sin \omega t$	2. $\omega/(s^2 - \omega^2)$
C. $\cosh \omega t$	3. $\omega/(s^2 + \omega^2)$
D. $\sinh \omega t$	4. $s/(s^2 + \omega^2)$

Codes:

	A	B	C	D
(a)	1	3	2	4
(b)	4	2	1	3
(c)	4	3	1	2
(d)	3	2	1	4

154. Fourier transform of $e^{-a|t|}$ is

(a) $\dfrac{a}{a^2 + \omega^2}$ (b) $\dfrac{2a}{a^2 + \omega^2}$

(c) $\dfrac{\omega}{a^2 + \omega^2}$ (d) $\dfrac{2\omega}{a^2 + \omega^2}$

155. If $v(t) = 0$ for $t < 0$ and $v(t) = e^{-\alpha t}$ for $t \geq 0$, $V(j\omega) =$

(a) $\dfrac{1}{\alpha - j\omega}$ (b) $\dfrac{1}{\alpha + j\omega}$

(c) $\dfrac{\alpha}{\alpha - j\omega}$ (d) $\dfrac{\alpha}{\alpha + j\omega}$

156. If $f(t) = 1$, $F(j\omega)$

(a) 2π (b) π

(c) $2\pi\delta(\omega)$ (d) $\pi\delta(\omega)$

157. Consider the following sets of values of E, R and C for the circuit in Fig. 3.15

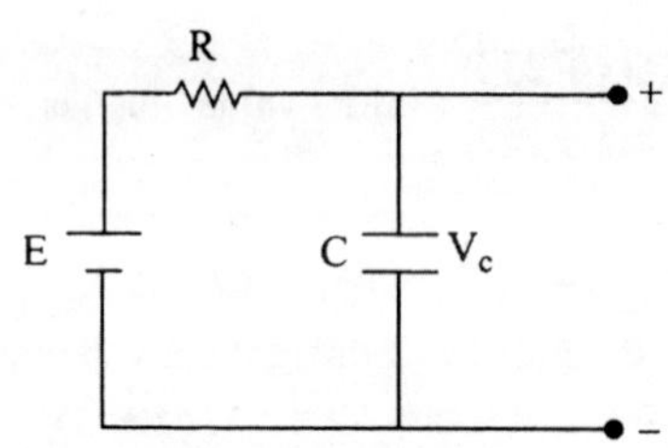

Fig. 3.15.

1. 2 V, 1 Ω.1.25 F 2. 1.6 V, 0.8 Ω, 1 F
3. 1.6 V, 1 Ω, 0.8 F 4. 2 V, 1.25 Ω, 1 F

Which of these sets of values will ensure that the state equation $\dfrac{dV_c}{dt} = 1.25\, V_c + 2$ is valid

(a) 1 and 4 (b) 1 and 2

(c) 3 and 4 (d) 2 and 3

158. The period of the function $\cos \dfrac{\pi}{4}(t - 1)$ is

(a) $\dfrac{1}{8}s$ (b) 8s

(c) 4s (d) $\dfrac{1}{4}s$

159. Match List I (Fourier transform) with List II (Functions of time) and select the correct answer using the given codes

List I	List II
A. $\dfrac{\sin k\omega}{\omega}$	1. A constant
B. $e^{-j\omega}$	2. Exponential function
C. $1/(j\omega + 2)^2$	3. t multiplied exponential function
D. $k\,\delta(\omega)$	4. rectangular pulse
	5. impulse function

Codes:

	A	B	C	D
(a)	4	5	3	1
(b)	4	5	3	2
(c)	3	4	2	1
(d)	3	4	2	5

160. Laplace transform of $\sin(\omega t + \alpha)$ is

(a) $\dfrac{\alpha}{s^2 + \alpha^2} \exp\left(\dfrac{s}{\alpha\omega}\right)$ (b) $\dfrac{\omega}{s^2 + \omega^2} \exp\left(\dfrac{s}{\alpha\omega}\right)$

(c) $\dfrac{\omega}{s^2 + \alpha^2} \exp\left(\dfrac{s}{\alpha\omega}\right)$ (d) $\dfrac{\omega}{s^2 - \alpha^2} \exp\left(\dfrac{s}{\alpha\omega}\right)$

161. The function $f(t)$ shown in Fig. 3.16 will have its Laplace transform as

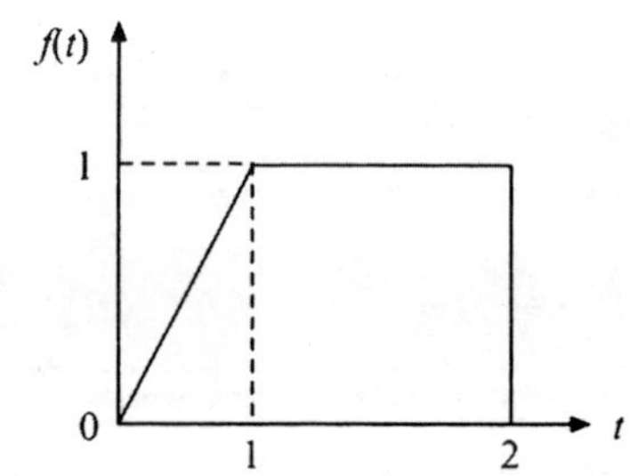

Fig. 3.16.

(a) $\dfrac{1}{s^2} - \dfrac{1}{s} e^{-s} - \dfrac{1}{s^2} e^{-2s}$ (b) $\dfrac{1}{s^2}(1 - e^{-s} - e^{-2s})$

(c) $\dfrac{1}{s}(1 - e^{-s} - e^{-2s})$ (d) $\dfrac{1}{s^2}(1 - e^{-s} - se^{-2s})$

162. Inverse Laplace transform of $\dfrac{2s + 5}{s^2 + 5s + 6}$ is

(a) $2\exp(-2.5\,t)\cosh(0.5t)$ (b) $\exp(-2t) + \exp(-3t)$

(c) $2\exp(-2.5t)\sinh(0.5t)$ (d) $2\exp(-2.5t)\cos 0.5\,t$

163. The Fourier series representation of a periodic current is $(2 + 6\sqrt{2}\cos\omega t + \sqrt{48}\sin 2\omega t)$ A. The effective value is

(a) $(2 + 6 + \sqrt{24})$ (b) 8 A

(c) 6 A (d) 2 A

164. Given that $\pounds[f(t)] = \dfrac{s+2}{s^2+1}$ and

$\pounds[g(t)] = \dfrac{s^2+1}{(s+3)(s+2)}$ and

$h(t) = \int_0^t f(\tau)\, g(t-\tau)\, d\tau$, then $\pounds[h(t)]$ is

(a) $\dfrac{s^2+1}{s+3}$ (b) $\dfrac{1}{s+3}$

(c) $\dfrac{s^2+1}{(s+3)(s+2)} + \dfrac{s+2}{s^2+1}$ (d) None of the above

165. If $[A] = \begin{bmatrix} 3 & 2 & 0 \\ -5 & 6 & 7 \\ 2 & 8 & 9 \end{bmatrix}$ and $[B] = \begin{bmatrix} 5 & 7 & -2 \\ 4 & 8 & 3 \\ 6 & 5 & 1 \end{bmatrix}$, then $[A+B]$ is

(a) $\begin{bmatrix} 8 & 9 & -2 \\ -1 & 14 & 10 \\ 8 & 13 & 10 \end{bmatrix}$ (b) $\begin{bmatrix} 7 & 8 & -2 \\ -1 & 14 & 10 \\ 2 & 6 & 5 \end{bmatrix}$

(c) $\begin{bmatrix} 6 & 8 & -2 \\ 0 & 2 & 4 \\ 6 & 8 & 9 \end{bmatrix}$ (d) $\begin{bmatrix} 5 & 6 & 7 \\ 9 & 8 & 8 \\ 0 & 2 & 5 \end{bmatrix}$

166. If $[A] = \begin{bmatrix} 3 & 0 & 0 \\ 5 & 8 & 4 \\ 6 & 3 & 1 \end{bmatrix}$, det [A] =

(a) –12 (b) 12

(c) 20 (d) –20

167. The z transform of $f_k = a^k$ is

(a) $\dfrac{1}{1-az^{-1}}$ (b) $\dfrac{a}{1-az^{-1}}$

(c) $\dfrac{1}{1-az}$ (d) $\dfrac{a}{1-az}$

168. If $\pounds[f(t) = F(s)$, then $\pounds[f(t-T)] =$

(a) $e^{st}F(s)$ (b) $e^{-st}F(s)$

(c) $\dfrac{F(s)}{1+e^{st}}$ (d) $\dfrac{F(s)}{1-e^{-st}}$

169. If $F[j\omega]$ is Fourier transform of $f(t)$, then Fourier transform of $f(-t) =$

(a) $-F(j\omega)$ (b) $\dfrac{1}{F(j\omega)}$

(c) $F^*(j\omega)$ (d) $\dfrac{1}{F^*(j\omega)}$

170. The Eigen values of matrix $\begin{bmatrix} 0 & 1 \\ 1 & 0 \end{bmatrix}$ are

(a) 1, 1 (b) –1, –1

(c) $j, -j$ (d) 1, –1

171. If $\pounds f(t) = \dfrac{\omega}{s^2+\omega^2}$, the value of $\underset{t\to\infty}{\text{Lt}}\, f(t)$

(a) cannot be determined (b) 0

(c) 1 (d) ∞

172. Laplace transform of $e^{\alpha t}\cos(\alpha t)$ is

(a) $\dfrac{s-\alpha}{(s-\alpha)^2+\alpha^2}$ (b) $\dfrac{s+\alpha}{(s+\alpha)^2+\alpha^2}$

(c) $\dfrac{1}{(s-\alpha)^2}$ (d) none of the above

173. A probability density function is given by $p(x) = Ke^{-x^2/2}$ for $-\infty < x < \infty$. The value of K should be

(a) $\dfrac{1}{\sqrt{2\pi}}$ (b) $\sqrt{\dfrac{2}{\pi}}$

(c) $\dfrac{1}{2\sqrt{\pi}}$ (d) $\dfrac{1}{\pi\sqrt{2}}$

174. The trignometric Fourier series of an even function of time does not have

(a) dc term (b) cosine term

(c) sine term (d) odd harmonic terms

175. If $\pounds[f(t)] = \dfrac{2(s+1)}{s^2+2s+5}$ then, $f(0^+)$ and $f(\infty)$ are given by

(a) 0 and 2 (b) 2, 0

(c) 0, 1 (d) $\dfrac{2}{5}, 0$

176. For a second order system, damping ratio ζ is such that $0 < \zeta < 1$. Then the roots of characterstic equation are

(a) real but not equal (b) real and equal

(c) complex conjugate (d) imaginary

177. Final value theorem is used to find

(a) steady state value of system output

(b) initial value of output

(c) transient behaviour of output

(d) none of the above

178. e^{At} can be expanded as

(a) $\sum_{k=0}^{\infty} \dfrac{A_k t^k}{(k+1)!}$ (b) $\sum_{k=0}^{\infty} \dfrac{A_k t^k}{k!}$

(c) $\sum_{k=0}^{\infty} \dfrac{A_k t^{k-1}}{(k+1)!}$ (d) $\sum_{k=0}^{\infty} \dfrac{A_k t^{k+1}}{k!}$

179. The current in a circuit with 10 Ω resistance is $i = 3 + 4\sin(100f + 45°) + 4\sin(300t + 60°)$ A The rms current and power dissipated are

(a) $\sqrt{41}$ A and 410 W (b) $\sqrt{35}$ A and 350 W

(c) 5 A and 250 W (d) 11 A and 1210 W

180. The Laplace transform of unit ramp function starting at $t = a$ is

(a) $\dfrac{1}{(s+a)^2}$ (b) $\dfrac{e^{-as}}{(s+a)^2}$

(c) $\frac{e^{-as}}{s^2}$ (d) $\frac{a}{s^2}$

181. The Fourier series of an odd periodic function contains
(a) odd harmonics only (b) even harmonics only
(c) cosine terms only (d) sine terms only

182. An excitation is applied to a system at $t = T$ and the response is zero for $-\infty < t < T$. This system is
(a) non-causal (b) stable
(c) causal (d) unstable

183. In a network $V(s) = Z(s)\, I(s)$ where $V(s)$, $Z(s)$ and $I(s)$ are Laplace transforms of $v(t)$, $z(t)$ and $i(t)$. The voltage $v(t)$ is
(a) $v(t) = z(t)\, i(t)$
(b) $v(t) = \int_0^t i(\tau)\, z(t-\tau)\, d\tau$
(c) $v(t) = \int_0^t i(\tau)\, z(t+\tau)\, d\tau$
(d) $v(t) = z(t) + i(t)$

184. The Laplace transform of a voltage across a capacitor is $V_c(s) = \frac{1}{s^2+1}$. If capacitor is 0.5 F, the current through capacitor at $t = 0^+$ is
(a) 0 A (b) 2 A
(c) 0.5 A (d) 1 A

185. If $F(s) = \frac{10}{s^2+4s+4}$, $f(t) =$
(a) $10te^{-2t}$ (b) $10t^2e^{-2t}$
(c) $10e^{-2t}$ (d) $5t^2 e^{-2t}$

186. If $f(t) = \delta(t-a)$, $F(s) =$
(a) 0 (b) 1
(c) e^{-as} (d) e^{as}

187. If $f_1(t) \leftrightarrow F_1(j\omega)$ and $f_2(t) \leftrightarrow F_2(j\omega)$ then $[a_1 f_1(t) + a_2 f_2(t)] \leftrightarrow$
(a) $a_1 F_1(j\omega) + a_2 F_2(j\omega)$ (b) $a_1 a_2 F_1(j\omega) F_2(j\omega)$
(c) $a_1 F_1(j\omega) - a_2 F_2(j\omega)$
(d) $a_1 F_1{*}(j\omega) + a_2 F_2{*}(j\omega)$

188. If $f(t) \leftrightarrow F_1(j\omega)$ and $f_2(t) \leftrightarrow F_2(j\omega)$, then $f_1(t) f_2(t) \leftrightarrow$
(a) $F_1(j\omega) F_2(j\omega)$ (b) $F_1(j\omega) * F_2(j\omega)$
(c) $\frac{1}{2\pi} F_1(j\omega) * F_2(j\omega)$ (d) none of the above

189. If $f(t) \leftrightarrow F(j\omega)$, $f(-t) \leftrightarrow$
(a) $-F(j\omega)$ (b) $F(-j\omega)$
(c) $F(j\omega)$ (d) $0.5\,F(j\omega)$

190. If $f(t) \leftrightarrow F(j\omega)$, $\frac{d^n}{dt^n} f(t) \leftrightarrow$
(a) $F(j\omega)$ (b) $[F(j\omega)]^n$
(c) $(j\omega)^n F(j\omega)$ (d) $\frac{F(j\omega)}{(j\omega)^n}$

191. If $F(s)$ is the Laplace transform of $f(t)$, then Laplace transform of $\frac{df(t)}{dt} =$
(a) $\frac{F(s)}{s}$ (b) $sF(s)$
(c) $s^2F(s)$ (d) $\frac{F(s)}{s^2}$

192. If $F(s) = \frac{s+3}{(s+1)[(s+1)^2+3]}$, the coefficient of term e^{-t} in $f(t)$ will be
(a) 1 (b) 0
(c) 0.5 (d) 2/3

193. For the system in Fig. 3.17

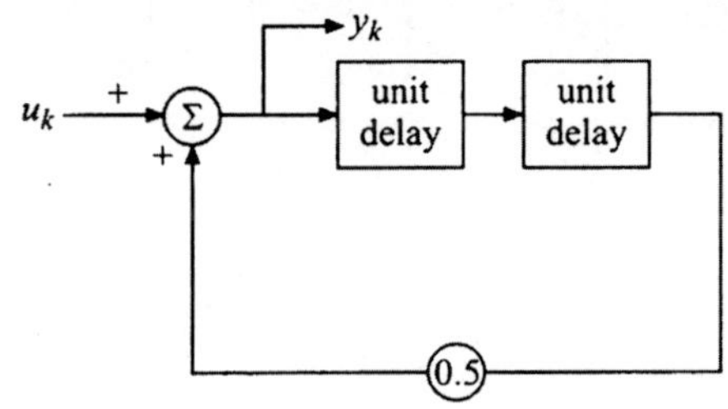

Fig. 3.17.

(a) $y_k = u_k + 0.5 y_{k-2}$ (b) $y_k = u_k + 0.5 u_{k-2}$
(c) $y_k = u_k - 0.5 y_{k-2}$ (d) $y_k = u_k - 0.5 u_{k-2}$

194. For the differential equation $(D^3 - D^2 + D - 1)[y(t)] = 0$ the roots of auxiliary equation are
(a) $-j, j, 1$ (b) $-j, j, 2$
(c) $1, 2, j$ (d) $1, 2, -j$

195. Which of the following is not true for impulse function $\delta(t)$
(a) $\delta(t) = 0$ for $t \neq 0$
(b) $\delta(t)$ is undefined for $t = 0$
(c) $\int_{-\infty}^{\infty} \delta(t)\, dt = 1$ (d) Laplace transform is $\frac{1}{s}$

196. If $f(k) \leftrightarrow F(z)$, then $k^n f_k \leftrightarrow$
(a) $[F(z)]^n$ (b) $\left(-z\frac{d}{dz}\right)^n F(z)$
(c) $[F(z)]^{1/n}$ (d) $\left(z\frac{d}{dz}\right)^n F(z)$

197. If F_n represents Fourier series coefficient of $f(t)$, then Fourier series coefficient of $f(t+\tau) =$
(a) $F_n e^{-j\omega_0\tau}$ (b) $F_n e^{j\omega_0\tau}$
(c) $F_n e^{jn\omega_0\tau}$ (d) $F_n e^{-jn\omega_0\tau}$

198. The energy E associated with time function $f(t)$ is
(a) $E = \int_0^{\infty} f^2(t)\, dt$ (b) $E = \int_{-\infty}^{\infty} f^2(t)\, dt$
(c) $E = 0.5\int_0^{\infty} f^2(t)\, dt$ (d) $E = 0.5\int_{-\infty}^{\infty} f^2(t)\, dt$

199. If $f(t) \leftrightarrow F(j\omega)$, and $f(t)$ is real, then $F(-j\omega) =$

(*a*) $F(j\omega)$ (*b*) $F^*(j\omega)$
(*c*) $F^*(-j\omega)$ (*d*) none of the above

200. If $f(t) \leftrightarrow F(j\omega)$, and $f(t)$ is real, then $\int_{-\infty}^{\infty} f^2(t)\, dt =$

(*a*) $\int_{-\infty}^{\infty} |F(j\omega)|^2\, d\omega$ (*b*) $\frac{1}{2\pi}\int_{-\infty}^{\infty} |F(j\omega)|^2\, d\omega$

(*c*) $\int_{0}^{\infty} |F(j\omega)|^2\, d\omega$ (*d*) $\frac{1}{2\pi}\int_{0}^{\infty} |F(j\omega)|^2\, d\omega$

201. In terms of signum function $sgn(t)$, unit step function $u(t) =$

(*a*) $1 + sgn(t)$ (*b*) $1 - sgn(t)$
(*c*) $0.5 + 0.5\, sgn(t)$ (*d*) $0.5 - 0.5\, sgn(t)$

202. The function shown in Fig. 3.18 is

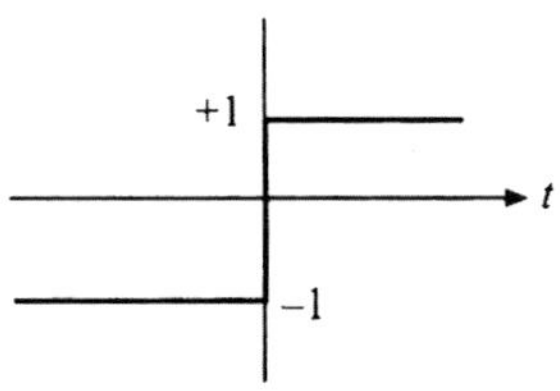

Fig. 3.18.

(*a*) impulse function (*b*) signum function
(*c*) unit step function (*d*) none of the above

203. For the signum function $sgn(t)$, $F(j\omega) =$

(*a*) $\frac{1}{j\omega}$ (*b*) $\frac{2}{j\omega}$
(*c*) $j\omega$ (*d*) $2j\omega$

204. If $F(s) = \frac{2s+3}{(s+1)(s+2)}$, the terms in $f(t)$ will have

(*a*) e^{-t} and e^{-2t} (*b*) e^{t} and e^{2t}
(*c*) te^{t} and te^{-2t} (*d*) none of the above

205. If $F(s) = \frac{1}{(s+a)^4}$, $f(t) =$

(*a*) $\frac{te^{-at}}{6}$ (*b*) $\frac{t^2 e^{-at}}{6}$
(*c*) $\frac{t^3 e^{-at}}{6}$ (*d*) $\frac{t^4 e^{-at}}{6}$

206. If S_n is the power associated with each frequency component and $S(f)$ is the sum of all these powers, normalised power spectral density is

(*a*) $\frac{dS(f)}{df}$ (*b*) $\frac{d[S(f)]^2}{df}$
(*c*) $\frac{d[\sqrt{S(f)}]}{df}$ (*d*) none of the above

207. A signal $m(t)$ is multiplied by a sinusoidal waveform of frequency f_c such that $v(t) = m(t)\cos 2\pi f_c t$
If Fourier transform of $m(t)$ is $M(f)$, Fourier transform of $v(t)$ is

(*a*) $0.5\, M(f+f_c)$ (*b*) $0.5M(f-f_c)$
(*c*) $0.5\, M(f+f_c) + 0.5\, M(f-f_c)$
(*d*) $0.5\, M(f+f_c) - 0.5\, M(f-f_c)$

208. The term 'energy spectral density' is associated with

(*a*) periodic waveform
(*b*) non-periodic waveform
(*c*) both periodic and non-periodic waveform
(*d*) none of the above

209. If average correlation between $v_1(t)$ and $v_2(t)$ is $R_{12}(\tau)$ and average correlation between $v_2(t)$ and $v_1(t)$ is $R_{21}(\tau)$ then

(*a*) $R_{12}(\tau) = R_{21}(\tau)$ (*b*) $R_{21}(\tau) = R_{12}(-\tau)$
(*c*) $R_{12}(\tau) = -R_{21}(\tau)$ (*d*) none of the above

210. Auto correlation $R(\tau)$ of a function $v(t)$ is defined as

(*a*) $R(\tau) = \lim_{T\to\infty} \frac{1}{T}\int_{-T/2}^{T/2} v(t)\, v(t+\tau)\, dt$

(*b*) $R(\tau) = \lim_{T\to\infty} \frac{0.5}{T}\int_{-T/2}^{T/2} v(t)\, v(t+\tau)\, dt$

(*c*) $R(\tau) = \frac{1}{T}\int_{-T/2}^{T/2} v(t)\, v(t+\tau)\, dt$

(*d*) none of the above

211. Auto correlation for $\tau = 0$ is equal to

(*a*) average power S of waveform
(*b*) square of average power S of waveform
(*c*) half of average power S of waveform
(*d*) square root of average power S of waveform

212. If $R(\tau)$ is autocorrelation of a waveform $v(t)$ and $R(0)$ is autocorrelation for $\tau = 0$, then

(*a*) $R(0) > R(\tau)$ (*b*) $R(0) \geq R(\tau)$
(*c*) $R(0) < R(\tau)$ (*d*) $R(0) \leq R(\tau)$

213. A function having frequency f is to be sampled. The sampling time T should be

(*a*) $T = \frac{1}{2f}$ (*b*) $T > \frac{1}{2f}$
(*c*) $T < \frac{1}{2f}$ (*d*) $T \geq \frac{1}{2f}$

214. Autocorrelation function

(*a*) is an even function of τ
(*b*) is an odd function of τ
(*c*) may be an even or odd function of τ
(*d*) is both an odd and even function of τ

215. For a periodic waveform, the power spectral density and correlation function form

(*a*) a Laplace transform pair (*b*) a Fourier transform pair
(*c*) either (*a*) or (*b*) (*d*) neither (*a*) nor (*b*)

216. An experiment is repeated N times. One event A occurs N_A times. Then

(a) $P(A) = \frac{N_A}{N}$ (b) $P(A) = \lim_{N \to \infty} \frac{N_A}{N}$

(c) $P(A) = \lim_{N \to 0} \frac{N_A}{N}$ (d) $P(A) = \lim_{N \to \infty} \frac{N}{N_A}$

217. A dice is tossed. The probability that either 1 or 2 will appear is

(a) $\frac{1}{2}$ (b) $\frac{1}{3}$

(c) $\frac{1}{6}$ (d) $\frac{1}{8}$

218. Consider the following as regards cumulative distribution function F(x)

1. $0 \le F(x) \le 1$
2. $F(-\infty) = 0$
3. $F(\infty) = 1$
4. $F(x_1) \le F(x_2)$ If $x_1 < x_2$

Out of above which are correct

(a) 1 and 2 only (b) 1, 2 and 3 only

(c) 1, 2, 3 and 4 (d) 2 and 4

219. Consider the following as regards probability distribution function f(x)

(1) $f(x) \ge 0$ for all x (2) $\int_{-\infty}^{\infty} f(x)\, dx = 1$

(3) cumulative distribution function

$$F(x) = \int_{-\infty}^{\infty} f(x)\, dx$$

Which of the above are correct

(a) 1 only (b) 1 and 2 only

(c) 1, 2 and 3 (d) 1 and 3 only

220. X and Y are two random variables and Z = X + Y. Let m_z, m_x, m_y represent mean of Z, X and Y. Then

(a) $m_z = m_x + m_y$ (b) $m_z \le m_x + m_y$

(c) $m_z < m_x + m_y$ (d) $m_z > m_x + m_y$

221. X and Y are two random variables and Z = X + Y. Let σ_x^2, σ_y^2 and σ_z^2 be variance of X, Y and Z. Then

(a) $\sigma_z^2 = \sigma_x^2 + \sigma_y^2$ (b) $\sigma_z^2 \le \sigma_x^2 + \sigma_y^2$

(c) $\sigma_z^2 < \sigma_x^2 + \sigma_y^2$ (d) $\sigma_z^2 > \sigma_x^2 + \sigma_y^2$

222. A voltage V(t) is a gaussian ergodic random process with a mean of zero and a variance of 4 volt^2. If it is measured by a dc meter. The reading will be

(a) 0 (b) 4

(c) 2 (d) $\sqrt{2}$

223. A voltage V(t) which is a gaussian ergodic random process with a mean of zero and a variance of 4 volt^2 is measured by a true rms meter. The reading will be

(a) 0 (b) 4

(c) 2 (d) 16

224. A voltage v(t) which is a gaussian ergodic random process with a mean of zero and a variance of 4 volt^2 is measured by a meter which first squares and then reads its dc component. The reading will be

(a) 0 (b) 4

(c) 16 (d) 2

225. The auto correlation of a sampling function is a

(a) triangular function (b) gate function

(c) signum function (d) none of the above

226. A signal is $x + f(t)$ where x is constant and $f(t)$ is a power signal with zero mean value. The power of the signal is

(a) x^2 (b) $x^2 + f^2(t)$

(c) more than $x^2 + f^2(t)$ (d) less than $x^2 + f^2(t)$

227. A box has 4 white and 3 red balls. Two balls are taken out in succession. What is the probability that both are white

(a) $\frac{4}{7}$ (b) $\frac{1}{2}$

(c) $\frac{2}{7}$ (d) $\frac{1}{7}$

228. Consider the following statements

1. If ensemble and time averages of a random process are identical, the process is ergodic
2. If ensemble and time average of a random process are not identical, the process is ergodic
3. An ergodic process is stationary
4. A stationary process is necessarily ergodic

Which of the above statements are correct?

(a) 1 only (b) 1 and 3 only

(c) 1, 3 and 4 only (d) 1 and 4 only

229. A stationary process has

(a) all statistical properties independent of time

(b) all statistical properties dependent of time

(c) zero variance

(d) ensemble average equal to time average

230. If the response of LTI continuous time system to unit step input is $\left(\frac{1}{2} - \frac{1}{2}e^{-2t}\right)$, then impulse response is

(a) $\frac{1}{2} - \frac{1}{2}e^{-2t}$ (b) e^{-2t}

(c) $1 - e^{-2t}$ (d) constant

[IES 2010]

231. Consider the following statements regarding the use of Laplace transform and Fourier transform in circuit analysis

1. Both make solution of circuit problems easy and simple
2. Both are applicable for study of circuit behaviour for $t = -\infty$ to ∞

3. Both convert differential equation to algebraic equation
4. Both can be used for transient and steady state analysis

Which of the above are correct?

(a) all (b) 2, 3, 4
(c) 1, 2, 4 (d) 1, 3, 4 **[IES 2010]**

232. The output of a linear system for step input is $t^2 e^{-2t}$. The transfer function is

(a) $\frac{s}{(s+2)^2}$ (b) $\frac{2s}{(s+2)^3}$
(c) $\frac{s}{s^2(s+2)}$ (d) $\frac{1}{(s+2)^3}$ **[IES 2010]**

233. z and Laplace transform are related by

(a) $s = \ln z$ (b) $s = \frac{\ln z}{T}$
(c) $s = z$ (d) $s = \frac{T}{\ln z}$ **[IES 2010]**

234. Convolution of two sequences $X_1(x)$ and $X_2(x)$ is written as

(a) $X_1(z) * X_2(z)$ (b) $X_1(z) X_2(z)$
(c) $X_1(z) + X_2(z)$ (d) $X_1(z) / X_2(z)$ **[IES 2010]**

235. Consider the following statements :
Fourier series of any periodic function $x(t)$ can be obtained if

1. $\int_0^1 |x(t)|\, dt < \infty$
2. Finite number of discontinuous exist within finite interval t

Which of the above are correct?

(a) 1 only (b) 2 only
(c) Both 1 and 2 (d) Neither 1 nor 2 **[IES 2010]**

236. If $x(\omega) = \delta(\omega - \omega_0)$, then $x(t) =$

(a) $e^{-j\omega_0 t}$ (b) δt
(c) $\frac{1}{2\pi} e^{j\omega_0 t}$ (d) 1 **[IES 2010]**

237. For distortionless transmission through LTI system, phase of $H(\omega)$ is

(a) constant
(b) one
(c) zero
(d) linealy dependent on ω **[IES 2010]**

238. Number of state varibles of discrete time system described by

$$y(n) = \frac{3}{4} y(n-1) + \frac{1}{8} y(n-2) = x(n) \text{ is}$$

(a) 2 (b) 3
(c) 4 (d) 1 **[IES 2010]**

239. Which one of the following is a Drichlet condition?

(a) $\int_{t_1}^{\infty} |x(t)| = \infty$
(b) Signal $x(t)$ must have finite number of maxima and minima is the expansion interval
(c) $x(t)$ can have an infinite number of discontinuties in the expansion interval
(d) $x^2(t)$ must be absolutely summable **[IES 2010]**

240. A signal $x_1(t)$ and $x_2(t)$ and the real and imaginary parts of a complex valued signal $x(t)$. What form of waveform is $x(t)$

(a) real symmetric (b) complex symmetric
(c) asymmetric
(d) conjugate symmetric **[IES 2009]**

241. A function of one or more variables which conveys information on the nature of physical phenomanon is called

(a) noise (b) interference
(c) system (d) signal **[IES 2009]**

242. What is the period of the sinusoidal signal $x(n) = 5\cos(0.2\pi x)$

(a) 10 (b) 5
(c) 1 (d) 0 **[IES 2009]**

243. When $y(t) \xrightarrow{FT} Y(j\omega)$, $x(t) \xrightarrow{FT} X(j\omega)$, $h(t) \xrightarrow{FT} H(j\omega)$ what is $Y(j\omega)$

(a) $\frac{X(j\omega)}{H(j\omega)}$ (b) $X(j\omega) H(j\omega)$
(c) $X(j\omega) + H(j\omega)$ (d) $X(j\omega) - H(j\omega)$ **[IES 2009]**

244. If the v-i characteristic of circuit is $v(t) = t\, i(t) + 2$, the circuit is of which type

(a) linear and time invariant
(b) linear and time variant
(c) Non-linear and time invariant
(d) Non-linear and time variant **[IES 2009]**

245. Which of the following is a periodic function?

(a) $\sin(10\pi t) + \sin(20\pi t)$
(b) $\sin(10t) + \sin(20\pi t)$
(c) $\sin(10\pi t) + \sin(20t)$
(d) $\sin(10t) + \sin(20t)$ **[IES 2008]**

246. What is the average power of periodic non-sinusoidal voltages and currents?

(a) Average power of fundamental component alone
(b) Sum of average powers of harmonics excluding fundamental
(c) Sum of average powers of sinusoidal components including fundamental
(d) Sum of rms power of sinusoidal components including fundamental **[IES 2008]**

247. Which one is a correct relation?

(a) $F(at) \leftrightarrow aF(\omega/a)$ (b) $F(at) \leftrightarrow aF(a\omega)$
(c) $F(t/a) \leftrightarrow aF(\omega/a)$
(d) $F(at) \leftrightarrow (1/a) F(\omega/a)$ **[IES 2008]**

248. Fourier transform of a function is equal to its two sided laplace transform evaluated
(*a*) on the real axis of *s* plane
(*b*) on a line parallel to the real axis of *s* plane
(*c*) on the imaginary axis of *s* plane
(*d*) on a line parallel to imaginary axis of *s* plane
[IES 2008]

249. A random variable X is defined by double exponential distribution $p_x(x) = ae^{-b|x|}$ for $-\infty < x < \infty$ where *a* and *b* are positive constants. What is the relation between *a* and *b* so that $p_x(x)$ is a probability density function
(*a*) $a = \frac{b}{2}$ (*b*) $b = \frac{a}{2}$
(*c*) $a = b$ (*d*) $a = \frac{1}{b}$ **[IES 2008]**

250. What is the spectral density of white noise?
(*a*) a constant (*b*) $\delta(\omega)$
(*c*) $[\delta(\omega)]^2$ (*d*) a step function in ω
[IES 2008]

251. The convolution of $f(t)$ with itself is given to be $\int_0^t F(\tau)\, d\tau$.
Then what is $f(t)$?
(*a*) unit ramp function (*b*) 1
(*c*) unit step function (*d*) unit impluse function
[IES 2008]

252. Which of the following is impulse response of the system if step response is given by
$$C(t) = 0.5\,(1 - e^{-2t})\,u(t)$$
(*a*) $e^{-2t}\,u(t)$ (*b*) $0.5\,\delta(t) + e^{-2t}\,u(t)$
(*c*) $0.5\,\delta(t) - 0.5\,e^{-2t}\,u(t)$ (*b*) $0.5\,e^{-2t}\,u(t)$
[IES 2008]

253. A voltage having Laplace transform $\frac{4s^2 + 3s + 2}{7s^2 + \cos + 5}$ is applied to a 2H inductor having zero initial currents. What is the current at $t = \infty$?
(*a*) Zero (*b*) $\frac{1}{5}$ A
(*c*) $\frac{2}{7}$ A (*d*) $\frac{2}{5}$ A **[IES 2008]**

254. What is the Laplace transform of $\cos \omega_0 t$?
(*a*) $\frac{\omega_0}{s^2 + \omega_0^2}$ (*b*) $\frac{s}{s^2 + \omega_0^2}$
(*c*) $\frac{\omega_0}{(s + \omega_0)^2}$ (*d*) $\frac{s}{(s + \omega_0)^2}$ **[IES 2008]**

255. If $f(t)$ is an even function, what is its Fourier transform $F(j\omega)$
(*a*) $\int_0^\infty f(t) \cos (2\omega t)\, dt$ (*b*) $2\int_0^\infty f(t) \cos (\omega t)\, dt$
(*c*) $2\int_0^\infty f(t) \sin (\omega t)\, dt$ (*d*) $2\int_0^\infty f(t) \sin (2\omega t)\, dt$
[IES 2008]

256. Which one of the following systems described by the following input-output relations is non-linear
(*a*) $y(n) = n\,x(2n)$ (*b*) $y(n) = x(n^2)$
(*c*) $y(n) = n^2\,x(n)$ (*d*) $y(n) = x^2(n)$ **[IES 2008]**

257. The relation between input $x(t)$ and output $y(t)$ of a continuous time signal is given by
$$\frac{dy(t)}{dt} + 3y\,(t) = x(t)$$
What is the forced response if $x(t)$ = K(a constant)
(*a*) K (*b*) $\frac{K}{3}$
(*c*) 3K (*d*) 0 **[IES 2007]**

258. What is the Laplace transform of a delayed unit impulse function $\delta(t - 1)$?
(*a*) 1 (*b*) zero
(*c*) e^{-s} (*d*) *s* **[IES 2007]**

259. What is the output for $t \to \infty$ for a system that has a transfer function $G(s) = \frac{2}{s^2 - s - 2}$ when subjected to step input
(*a*) –1 (*b*) 1
(*c*) 2 (*d*) unbounded
[IES 2007]

260. The response of a linear time-invarient system to a unit step input is $s(t) = (1 - e^{-t/RC})\,u(t)$ where $u(t)$ is the unit step input. What is the impulse response?
(*a*) $e^{-t/RC}$ (*b*) $e^{-t/RC}\,u(t)$
(*c*) $\frac{1}{RC} e^{-t/RC}\,u(t)$ (*d*) $\delta(t)$ **[IES 2007]**

261. If a linear time invariant system is excited by a true random signal like white noise, the ouput will have which of the following properties?
(*a*) output will be white noise
(*b*) output will be periodic
(*c*) output will not be random
(*d*) output will be correclated or coloured noise
[IES 2007]

262. The output of a linear system for any input can be computed in which one of the followng ways
(*a*) only by summation of impulse responses byconvolution integral
(*b*) only by summation of step responses by convolution integral
(*c*) neither (*a*) nor (*b*)
(*d*) either (*a*) or (*b*) **[IES 2007]**

263. Which of the following is correct? Energy of a power signal is
(*a*) finite (*b*) zero
(*c*) infinite (*d*) between 1 and 2
[IES 2007]

264. Which of the following rule determines the mapping of *s* plane to *z* plane?

(a) Right half of s plane maps into outside of unit circle in z plane
(b) Left half of s plane maps into inside of unit circle
(c) Imaginary axis in s plane maps into the circumference of the unit circle
(d) All of the above **[IES 2007]**

265. Which of the following is the correct statement? The system characterised by equation $y(t) = ax(t) + b$ is
(a) linear for any value of b (b) linear if $b > 0$
(c) linear if $b < 0$ (d) non-linear **[IES 2006]**

266. What is the Laplace transform of $x(t) = (-e^{-2t}\, u(t))\,(t\, u(t))$
(a) $\frac{-1}{s^2(s+2)}$ (b) $\frac{-1}{s^2(s-2)}$
(c) $\frac{1}{s^2(s-2)}$ (d) $\frac{-1}{s(s-2)}$ **[IES 2006]**

267. For a random variable x having PDF as shown in Fig. 3.19 what are the values of mean and variance?

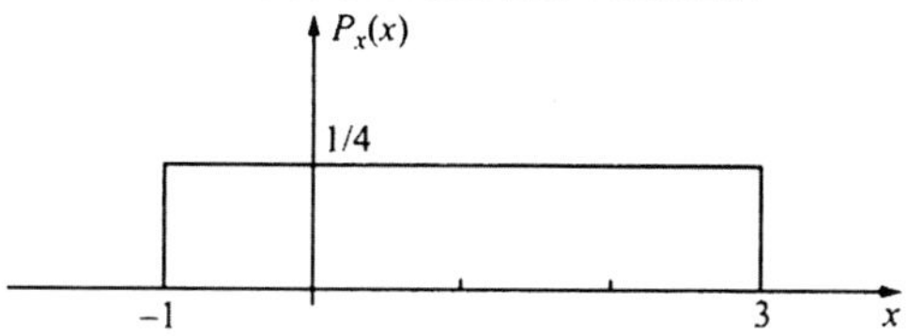

Fig. 3.19.

(a) $\frac{1}{2}$ and $\frac{2}{3}$ (b) 1 and $\frac{4}{3}$
(c) 1 and $\frac{2}{3}$ (d) 2 and $\frac{4}{3}$
[IES 2006]

268. For the waveform in Fig. 3.20 the Laplace transform is $\frac{1}{s^2}[1 + Ae^{-s} + Be^{-4s} + Ce^{-6s} + De^{-8s}]$. What is the value of D.

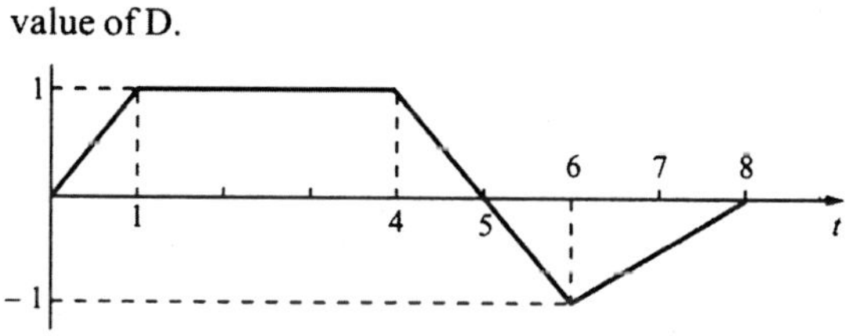

Fig. 3.20.

(a) –0.5 (b) –1.5
(c) 0.5 (d) 2.0 **[IES 2006]**

269. Which of the following is correct? The continuous time system $y(t) = x(t^2)$ is
(a) casual, linear and time varying
(b) casual non-linear and time varying
(c) non-casual, non-linear and time invariant
(d) non-casual and time variant **[IES 2006]**

270. Two discrete time systems with inpulse responses $h_1(n) = \delta(n-1)$ and $h_2(n) = \delta(n-2)$ are connected in cascade. The overal impulse response is
(a) $\delta(n-1) + \delta(n-2)$ (b) $\delta(n-4)$
(c) $\delta(n-3)$ (d) $\delta(n-1)\,\delta(n-2)$
[Gate 2010]

271. A continuous time LTI system is described by

$$\frac{d^2y(t)}{dt^2} + 4\frac{dy(t)}{dt} + 3y(t) = \frac{2d\,x(t)}{dt} + 4x(t)$$

Assuming zero initial condition, the response $y(t)$ for input $x(t) = e^{-2t}\,u(t)$ will be
(a) $(e^{t} - e^{-3t})\,u(t)$ (b) $(e^{-t} - e^{-3t})\,u(t)$
(c) $(e^{-t} + e^{3t})\,u(t)$ (d) $(e^{t} + e^{3t})\,u(t)$
[Gate 2010]

272. The Fourier series of a real periodic function has only
P cosine terms if it is even
Q sine terms if it is even
R cosine terms if it is odd
S sine terms if it is odd
Which of the above are correct?
(a) P and S (b) P and R
(c) Q and S (d) Q and R
[Gate 2009]

273. A function is given by $f(t) = \sin^2 t + \cos 2t$. Which of the following is true?
(a) It has frequency components 0 and $\frac{1}{2\pi}$ Hz
(b) It has frequency components 0 and $\frac{1}{\pi}$ Hz
(c) It has frequency components $\frac{1}{2\pi}$ and $\frac{1}{\pi}$ Hz
(d) It has frequency components 0, $\frac{1}{2\pi}$ and $\frac{1}{\pi}$ Hz
[Gate 2009]

274. Given that F(s) is one sided Laplace transform of $f(t)$. The Laplace transform of $\int_0^1 f(\tau)\,d\tau$ is
(a) $sF(s) - f(0)$ (b) $F(s)/s$
(b) $\int_0^s F(\tau)\,d\tau$ (d) $\frac{1}{s}[F(s) - f(0)]$
[Gate 2009]

275. The input and output of a continuous time system are $x(t)$ and $y(t)$. Which of the following correspands to causal system?
(a) $y(t) = x(t-2) + x(t+4)$ (b) $y(t) = (t-4)\,x(t+5)$
(c) $y(t) = (t+4)\,x(t-1)$ (d) $y(t) = (t+5)\,x(t+5)$
[Gate 2008]

276. The impulse response of LTI continuous time system is $h(t) = e^{\alpha t}\,u(t) + e^{\beta t}\,u(-t)$ where α and β are constants. This system is stable if
(a) α is positive and β is positive
(b) α is negative and β is negative
(c) α is positive and β is negative
(d) α is negative and β is positive **[Gate 2008]**

277. $p_x(x) = Me^{-2|x|} + Ne^{-3(x)}$ is the probability density function for a real random variable X on the entire X-axis. M and N are both positive real numbers. M and N are given by

(*a*) $M + \frac{2}{3}N = 1$ (*b*) $2M + \frac{N}{3} = 1$

(*c*) $M + N = 1$ (*d*) $M + N = 3$ **[Gate 2008]**

278. A signal $x(t)$ is described by

$x(t) = 1 \quad \text{for} -1 \le t \le 1$

$= 0$ otherwise

Two frequencies at which Fourier transform becomes zero are

(*a*) $\pi, 2\pi$ (*b*) $0.5\pi, 1.5\pi$

(*c*) $0, \pi$ (*d*) $2\pi, 2.5\pi$ **[Gate 2008]**

279. The impulse response $h(t)$ for a LTI continuous time system is $h(t) = e^{-2t}\, u(t)$. The frequency response $H(\omega)$ of this system in terms of angular frequency ω is given by $H(\omega) =$

(*a*) $\frac{1}{1 + j2\omega}$ (*b*) $\frac{\sin \omega}{\omega}$

(*c*) $\frac{1}{2 + j\omega}$ (*d*) $\frac{j\omega}{2 + j\omega}$ **[Gate 2008]**

B. The following questions have two statements viz. Assertion A and Reason R. Examine these statements and answer as per following codes

(a) Both A and R are true and R is correct explanation of A

(b) Both A and R are true but R is not correct explanation of A

(c) A is true but B is false

(d) A is false but R is true

280. A. For the function shown in Fig. 3.21.

Laplace transform is $e^{-ax}\left(\frac{1}{s}\right)$

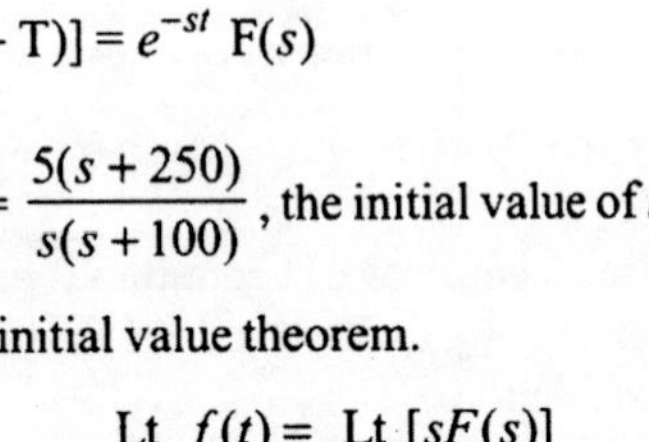

Fig. 3.21.

R. $\int [f(t - T)] = e^{-st}\, F(s)$

281. A. If $I(s) = \frac{5(s + 250)}{s(s + 100)}$, the initial value of $i(t)$ is 5A.

R. As per initial value theorem.

$$\underset{t \to 0}{\text{Lt}}\ f(t) = \underset{s \to \infty}{\text{Lt}}\ [sF(s)]$$

282. A. If $I(s) = \frac{s - 1}{s^2 + 3s + 2}$, the final value of $i(t)$ is 2 A.

R. As per final value theorem

$$\underset{t \to 0}{\text{Lt}}\ f(t) = \underset{s \to \infty}{\text{Lt}}\ [sF(s)].$$

283. A. Impulse function is an important function in network analysis.

R. Convolution integral enables us to find the response of a network to an arbitrary input in terms of the impulse response.

284. A. Transient analysis uses Laplace transformation.

R. Laplace transform method is suitable for solution of integro differential equations.

285. A. The modified ramp function of Fig. 3.22 (*a*) can be represented as sum of two ramp functions of Fig. 3.22 (*b*)

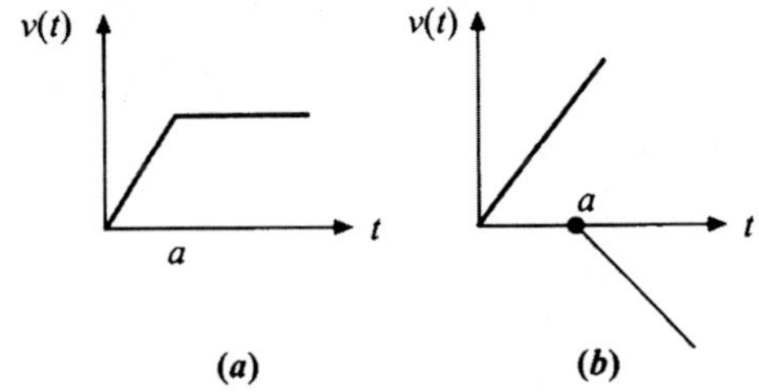

Fig. 3.22.

R. If $f(t) = t$, $F(s) = 1$

286. A. If $f(t) = e^{at}t$, $F(s) = \frac{1}{s - a}$

R. Displacement in time domain by T becomes multiplication by e^{-st} in s domain

287. A. $\pounds[e^{-at} f(t)] = F(s + a)$

R. In use of Laplace transform method, initial conditions may be neglected.

288. A. $\pounds[af_1(t) + bf_2(t)\} = aF_t(s) - bF_2(s)$

R. Initial value theorem enables us to find the value of $f(t)$ at $t = 0$ directly from $F(s)$.

289. A. $\delta(t - b)$ is an impulse occurring at $t = b$.

R. Integral of unit impulse gives unit step function.

290. A. The function $\delta'(t - b)$ is equal to 0 for $t \ne b$.

R. A number of impulses spaced from one another form an impulse train.

291. A. Integral of a unit step function is a ramp of slope unity.

R. If $f(t) = \sin \omega t$, $F(s) = \frac{1}{s^2 + \omega^2}$.

292. A. $\int_{-\infty}^{\infty} \delta(1 - b)dt = 1$ for $t \ge b$

$= 0$ for $t < b$

R. An impulse has a very high magnitude but very short duration.

293. A. Transient periods are of short duration but can result in dangerously high voltages and currents.

R. Circuit equations in transient analysis are integro differential equations.

294. A. A sweep circuit operates by virtue of recurring transient phenomenon

R. Transisents are always undesirable

295. A. If $f(t) = 1 - \cos \omega t$, $F(s) = \dfrac{\omega^2}{s(s^2 + \omega^2)}$

R. If $f(t) = \cos \omega t$, $F(s) = \dfrac{s}{s^2 + \omega^2}$

296. A. Laplace transform can be used to evaluate integrals.

R. Laplace transform can be used to solve differential equations.

297. A. A non-sinusoidal wave can be expressed in terms of sine waves of different frequencies which are multiples of the frequency of fundamental.

R. If negative half of a complex wave is a reproduction of the positive half, the even harmonics are absent.

298. A. The rms value of $v = 1 + \sin \omega t$ is $\sqrt{1.5}$.

R. If $i = I_0 + I_{1m} \sin \omega_1 t + I_{3m} \sin 3\omega_1 t$, then

$$I = \left(I_0^2 + \frac{I_{1m}^2}{2} + \frac{I_{3m}^3}{2} \right)^{0.5}$$

299. A. The wave shown in Fig. 3.23 does not contain the dc component and even harmonics.

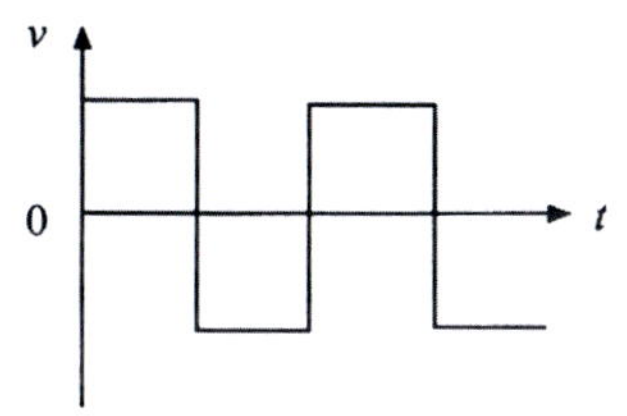

Fig. 3.23.

R. If $f(-t) = f(t)$ the wave has only cosine terms.

300. A. If a wave $v = A_1 + A_2 \sin \omega t$ is applied to a pure capacitor the current wave is

$$i = A_3 \sin(\omega t + 90°).$$

R. A capacitor presents infinite impedance to dc.

301. A. The Fourier series of the wave shown in Fig. 3.24 does not contain even harmonics.

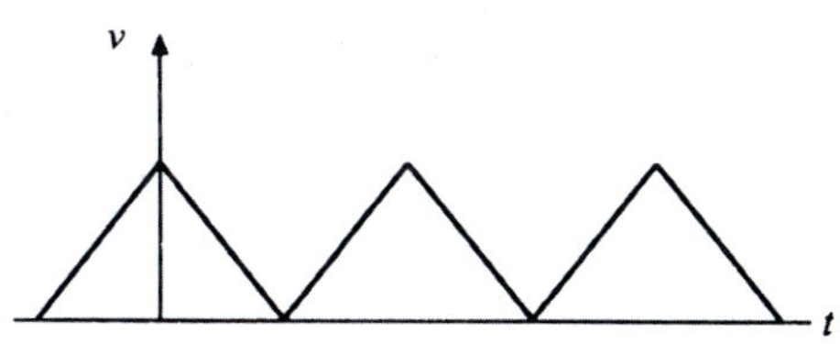

Fig. 3.24.

R. If $f(t) = -f(t \pm T/2)$ the function is said to have half wave symmetry.

302. A. Fourier series can also be written in exponential form.

R. sin ($n\ \omega t$) and cos ($n\ \omega t$) can be expressed as sum or difference of exponentials.

303. A. The exponential form of Fourier series is

$$f(t) = F_0 + \sum_{n=1}^{\infty} (F_n e^{jn\omega_1 t} + F_{-n} e^{-jn\omega_1 t}).$$

R. If $f(t)$ is an even function, the coefficients F_n are real.

304. A. When a function $f(t)$ is represented as exponential Fourier series, the set of complex coefficients F is called frequency spectrum of $f(t)$.

R. Frequency spectrum is also called line spectrum

305. A. For the determinant $A = \begin{vmatrix} a_{11} & a_{12} & a_{13} \\ a_{21} & a_{22} & a_{23} \\ a_{31} & a_{32} & a_{33} \end{vmatrix}$

the minor for a_{11} is

$$M = \begin{vmatrix} a_{22} & a_{23} \\ a_{32} & a_{33} \end{vmatrix}$$

R. The minor of any element of a determinant a_{jk} is the determinant which remains when the row and column corresponding to a_{jk} are deleted.

306. A. In complex frequency $s = \sigma + j\omega$, the terms σ and ω are neper frequency and radian frequency.

R. If $\omega = 0$, the graph of Ke^{st} will be decaying exponential if $\sigma < 0$.

307. A. The function ke^{st} with both σ and ω positive and $\sigma > 0$ depicts a sinusoid whose amplitude increases with time.

R. If $\sigma = 0$, ke^{st} becomes a sinusoid

308. A. In order that $f(t)$ is Laplace transformable, it is necessary that $\int_0^{\infty} | f(t) | e^{-\sigma_1 t} dt < 0$ for real positive σ_1.

R. If $f(t)$ is known we can find F(s) and vice versa.

309. A. Heaviside partial expansion gives a simple procedure to find inverse Laplace transform of the terms having a complex conjugate pair of roots.

R. If $I(s) = P(s)/Q(s)$ and all roots of $Q(s) = 0$ are simple, $i(t)$ will have terms with exponentials having real exponents only.

310. A. The scaling theorem in Laplace transform relates the scale changes in s domain to the consequent changes in t domain

R. If $I(s) = \dfrac{1}{s(s+1)}$, $i(t) = 1 - e^{-t}$.

311. A. The unit impulse function is also known as Dirac delta function.

R. If a voltage $\delta(t - a)$ is applied to an inductor L, the current is $\dfrac{1}{L} u(t - a)$.

312. A. The response of a network to an impulse is determined only by the parameters of the network.

R. If $f(t)$ is a unit parabolic function, its Laplace transform is $1/s^3$.

313. A. A half cycle of sine wave can be expressed as sum of sine wave and another sine wave shifted by T/2 where T is the time period.

R. The function $u(t - t_0) - u(t - t_0 - T)$ is a gate function occurring at $t = t_0$ and of duration T.

314. A. If $I(s) = \frac{2s+10}{s(s+2)}$, the final value of $i(t) = 10$.

R. If $I(s) = \frac{2s+10}{s(s+2)}$, the initial value of $i(t) = 2$.

315. A. The conditions under which it is possible to write Fourier series of a periodic function are called Drichlet conditions.

R. If $f(t) = -f(-t)$, it is referred to as odd symmetry.

316. A. If an event A can happen in m ways and another event B can happen in n ways, both can happen is mn ways.

R. Two events are mutually exclusive if happening of either precludes the occurrence of other.

317. A. The number $\in$ such that the probability that an error is between $-\in$ and $+\in$ is 1 is called probable error of single observation.

R. The standard deviation is the rms value of the deviation from the arithmetic mean.

318. A. Standard deviation is the abcissa of a point of inflection on the probability curve.

R. The equation for probability curve is $y = \frac{h}{\sqrt{\pi}} e^{-h^2x^2}$.

319. A. In the curve $y = \frac{h}{\sqrt{\pi}} e^{-h^2x^2}$, a high value of h indicate a high peak and rapid fall in the curve.

R. If the number of ways an event may result can be analysed into a successes and b failures, each equally likely to occur, the probability of success in a single trial is

$$\frac{a}{(a+b)}.$$

C. State whether the following are True or False

320. A unit step function is used to represent the closing a switch in a constant voltage system.

321. A pulse function can be represented as difference of two equal step functions.

322. For exponential function e^{-at}, the Laplace transform $1/(s-a)$.

323. $\delta(t-b)$ is a pulse function.

324. The integral of a unit impulse is unit step function.

325. The integral of unit step function is a ramp of slope unity.

326. The function $\delta'(t-b)$ is a unit doublet.

327. A number of impulses spaced from one another form an impulse train.

328. The triangular wave of Fig. 3.25 can be written as

$$v(t) = u(t) - tu(t) + (t-1)\,u(t-1)$$

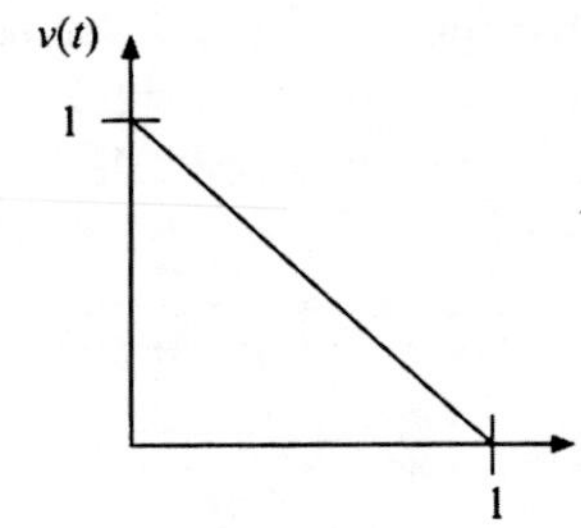

Fig. 3.25.

329. The real part of complex frequency s corresponds to oscillations.

330. The function $A\,e^{(\sigma + j\omega)t}$ represents a rotating phasor having a magnitude increasing with time.

331. Short-circuit is the dual of open circuit.

332. The n state variables can be considered as n components of a state vector.

333. $\dot{x} = AX + Bu$ is a state equation.

334. In the state equation $\dot{x} = AX + Bu$, A is a $n \times m$ matrix.

335. State variables which describe a system are a unique set.

336. $(sI - A)^{-1} = adj\,(sI - A)/det\,(sI - A)$.

337. The eigen values of $n \times n$ matrix A are the roots of the characteristic equation

$$1\,\lambda I - A1 = 0$$

338. The minor M_{ij} of an $n \times n$ matrix is the determinant of $(n-1) \times (n-1)$ matrix formed by deleting the ith row and jth column of $n \times n$ matrix.

339. For matrix A, $A^{-1}A = I$.

340. State variable formulation is very suitable for computer solution.

341. For a wave $v = V_{1m} \sin(\omega t + \theta) - V_{3m} \sin(3\omega t + \theta_3)$, the rms value is

$$(0.5\,V_{1m}^2 + 0.5\,V_{3m}^2)^{0.5}$$

342. If a complex voltage wave is applied to a pure inductance, the resulting current wave is much smoother than the voltage wave.

343. When a complex voltage wave is applied to, capacitor the resulting current wave is more distorted than the voltage wave.

344. An ac wave with superimposed dc is symmetrical about x-axis

345. If $f(-t) = f(t)$, the function $f(t)$ has only cosine terms.

346. Half wave symmetry means $f(t) = -f(t \pm T/2)$.

347. If $f(t)$ is an even function, then in the form $f(t) = \sum_{n=-\infty}^{\infty} F_n e^{jn\omega_1 t}$ coefficients F_n are real.

348. In the representation $f(t) = \sum_{n=-\infty}^{\infty} F_n e^{jn\omega_1 t}$ the set of complex coefficients F_n is the frequency spectrum of $f(t)$.

349. To plot frequency spectrum we draw two graphs *viz* $|F_n|$ versus frequency and ϕ_n versus frequency.

350. In the periodic train of rectangular pulses (Fig. 3.26), $F_0 = (V_0/T)\,d$.

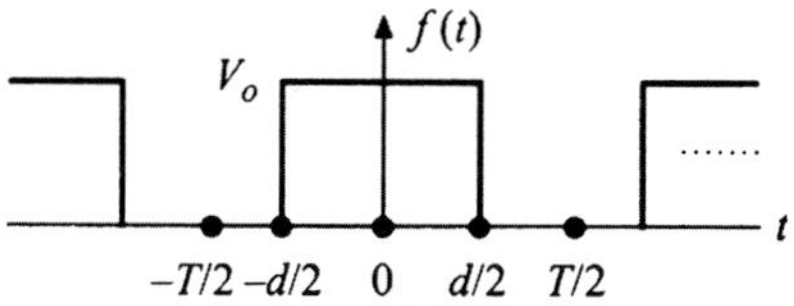

Fig. 3.26.

351. In Fig. 3.26 $f(t)$ is an even function.

352. In Fig. 3.26, the phase angle of F_n is either 0 or π.

353. In Fig. 3.26 the ratio T/d is the duty factor.

354. The units of $F(j\omega)$ are volt-seconds.

355. If $f(t)$ is an even function

$$F(j\omega) = 2\int_0^\infty f(t)\cos\omega t\,dt$$

356. If $f(t)$ is an odd function

$$F(j\omega) = 2\int_0^\infty f(t)\sin\omega t\,dt$$

357. $\mathfrak{F}[f(t-a)] = F(j\omega)e^{-j\omega a}$

358. $\mathfrak{F}\left[\frac{d^n}{dt^n} f(t)\right] = (j\omega)^n F(j\omega)$

359. For the single rectangular pulse of Fig. 3.27

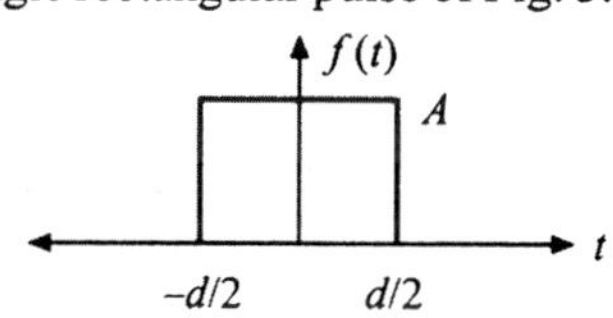

Fig. 3.27.

$$F(j\omega) = [Ad\sin(\omega d/2)]/(\omega d/2)$$

360. If $f(t) = 1$, $F(j\omega) = 2\pi\,\delta(\omega)$.

361. If $v(t) = 0$ for $t < 0$ and $e^{-\alpha t}$ for $t \geq 0$

$$V(j\omega) = 1/(\alpha + j\omega).$$

362. If $x_k = 0$ for $k < 0$ and $= 2^k$ for $k \geq 0$

$$X(z) = z/(z-2).$$

363. If $x_k = 2k$ for $k \leq 0$ and $= 0$ for $k > 0$

$$X(z) = 2/(2-z).$$

364. Z transform is a non-linear operation.

365. If X(z) is the z transform of x_k, then the z transform of kx_k is

$$-z\frac{d}{dz}X(z).$$

366. If $X(z) = (1 - \alpha z^{-1})^{-1}$ and $|\alpha| < |z|$, the initial value $x_0 = 0$.

367. If x_k is one sided to the right

$$x_0 = \operatorname*{Lt}_{x\to\infty} X(z)$$

368. If $x_k = \alpha^k$ and $k \geq 0$

$$X(z) = (1 - \alpha z^{-1})^{-1} \text{ with } |\alpha| < |z|.$$

369. If $x_k = \alpha^k$ and $k < 0$,

$$X(z) = -(1 - \alpha z^{-1}) \text{ with } |z| < |\alpha|.$$

370. If $x_k = \frac{1}{k}$ and $k > 0$,

$$X(z) = -\text{In}\,(1 - z^{-1}) \text{ with } 1 < |z|.$$

371. If two mutually exclusive events A and B can happen in m and n ways, then either A or B can happen in $(m + n)$ ways.

372. If the number of ways an event may result is analysed into m successes and n failures, each equally likely to occur, the probability of success in a single trial is $m/(m + n)$.

373. The standard deviation of a set of observations is the rms value of the deviation from the arithmetic mean.

D. Fill in the Blanks with suitable words

374. If [A] is a square matrix and $[A]^{-1}$ is its inverse, then AA^{-1} = ...

375. The inverse A^{-1} of a square matrix A is...........

376. The determinant of a square matrix is the product of its.............

377. If A and B are two matrices, $[AB]^T$ =

378. An identity matrices is a.......... matrix in which each diagonal element is 1.

379. The function $\delta'(t-b)$ is known as ....

380. The Laplace transform of $\delta'(t-b)$ is ....

381. The integral of unit impulse gives ....

382. The integral of unit step function gives a ....

383. For $t < 0$, the magnitude of impulse $\delta(t)$ is ....

384. If $f(t) = \sin\omega t$, $F(s)$ = ...

385. If $f(t) = e^{-at}$, $F(s)$ = ...

386. If $f(t) = 0$ for $t \leq 0$ and $f(t) = t^2$ for $t > 0$, $F(s)$ = ...

387. The parameter s ....

388. As per initial value theorem $\operatorname*{Lt}_{t\to 0} f(t)$ = ...

389. If $I(s) = \frac{1}{(s+3)^2}$, $i(t)$ = ...

390. If $i(t) = \cosh\alpha t$, $I(s)$ = ...

391. The standard form of state equation is X = ...

392. The state variables which describe a system are ....

393. $[sI - A]^{-1}$ = ...

394. The rms value of $v = 10\cos t - 2\cos 3t$ is ....

395. The rms value of $i = 20 + 10\sin t$ is ....

396. If $f(t) + f(-t) = 0$, Fourier series will have only ....terms

397. If $f(t)$ has half wave symmetry, Fourier series does not have ....

398. For the wave shown in Fig. 3.28 a_0 = ...

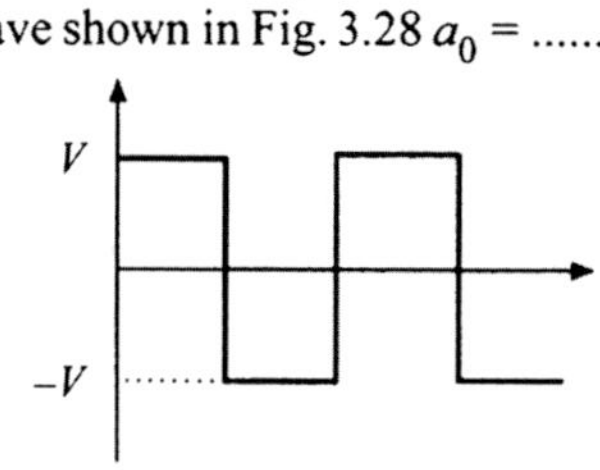

Fig. 3.28.

399. For the wave shown in Fig. 3.28, a_n = ...

400. The set of Fourier coefficients F_n is called ....

401. $F(j\omega)$ = ...

402. $\pounds[af(t)\}$ = ...

403. $\pounds[f(t-a)]$ = ...

404. If $f(t) = 1$, Fourier transform is ....

405. If $V_0(j\omega) = \frac{8}{1+j\omega}$, $V_0(t)$ = ...

406. For exponentially decaying pulse $e^{-\alpha t}$, $V(j\omega)$ = ...

407. Z transform of 1 is ....

408. If n is the number of deviations and ρ is residual error, σ = ...

409. The probability curve is given by y = ...

410. In terms of parameter h, standard deviation σ = ...

411. If σ is small, the probability curve will have ....

412. If σ is standard deviation, probable error of single observation is ....

413. In normal distribution the percentage of observations which fall within σ of arithmetic mean is ....

ANSWERS WITH HINTS AND EXPLANATIONS

1. (*b*) $u(t)$ is a unit step function, $u(t-a)$ is a unit step function shifted in time by a.

2. (*b*) It is a pulse lasting for $t = a$.

3. (*b*) See Table in section 3.4.

4. (*c*) $v(t) = E\,u(t) - E\,u(t-a)$.

5. (*c*) $\pounds\, v(t) = \pounds\,[E\,u(t) - E\,u(t-a)] = \frac{E}{s} - \frac{E}{s}e^{-as}$

6. (*d*) **7.** (*b*) See section 3.4.

8. (*d*) See Table in section 3.4.

9. (*b*) **10.** (*a*)

11. (*c*) $\int 1\, dt = t$. **12.** (*a*) See section 3.4.

13. (*c*) $\delta(t)$ is an impulse at $t = 0$, $\delta(t-b)$ is an impulse originating at $t = b$.

14. (*a*) **15.** (*c*)

16. (*b*) See section 3.4. **17.** (*a*) See section 3.4.

18. (*a*) See section 3.4.

19. (*d*) $\pounds f(t-1) = e^{-st}F(s)$.

20. (*c*) See section 3.4. **21.** (*a*) See section 3.4.

22. (*a*) $\pounds\, e^{-at}f(t) = F(s+a)$.

23. (*b*) See table in section 3.4.

24. (*b*) See table in section 3.4.

25. (*c*) $\frac{-a}{s(s-a)} = \frac{1}{s} - \frac{1}{s-a}$.

26. (*b*) See table in section 3.4.

27. (*a*) See table in section 3.4.

28. (*a*) See table in section 3.4.

29. (*b*) If $F(s) = \frac{1}{(s-a)^n}$, $f(t) = \frac{1}{(n-1)!}t^{n-1}e^{at}$. Substitute $a = -1$ and $n = 3$.

30. (*a*) $i(0) = \underset{s\to 0}{\text{Lt}}\,[s\,F(s)]$. Therefore,

$$i(0) = \frac{5\left(1+\frac{250}{s}\right)}{1+\frac{100}{s}} = 5.$$

31. (*b*) $i(\infty) = \underset{s\to 0}{\text{Lt}}\,[s\,F(s)]$. Therefore,

$$i(\infty) = \frac{5(250)}{100} = 12.5.$$

32. (*a*) It is an impulse having area A and originating at $t = a$.

33. (*c*) **34.** (*d*) **35.** (*c*) **36.** (*c*)

37. (*d*) **38.** (*d*) **39.** (*b*)

40. (*a*) See section 3.5.

41. (*c*) **42.** (*d*) **43.** (*a*)

44. (*a*) DC component is absent in current and current is leading the voltage. Hence R and C.

45. (*c*) R is independent of ω and $X_1 \propto \omega$.

46. (*c*) $\omega_r \propto \frac{1}{\sqrt{L}}$ or $L \propto \frac{1}{\omega_r^2}$.

47. (*b*) $I_C = V(j\omega C)$.

48. (*c*) $I_1 = \frac{V}{j\omega L}$.

49. (*d*) RMS value $= \left[5^2 + \frac{5^2}{2}\right]^{0.5}$.

50. (*b*) For third harmonic $\theta = \tan^{-1}\frac{3\omega L}{R}$.

51. (*c*) $X_c = \frac{-j}{\omega C}$.

52. (*c*) As frequency increases, X_c decreases and power factor improves.

53. (*c*) As frequency increases, X_L increases and power factor becomes worse.

54. (*a*) Average of sinusoid over full cycle is zero.

55. (*c*) $I_{av} = \frac{2\times 100}{\pi}$

56. (*d*) Peak-to-peak value = $(2\sqrt{2})$ (rms value).

57. (*c*) **58.** (*a*) **59.** (*b*) **60.** (*b*)

61. (*c*) $X_L = j\omega L$ and $X_c = \frac{1}{j\omega C}$.

62. (*d*) **63.** (*c*) **64.** (*b*) $\cos(-\theta) = \cos\theta$.

65. (*d*) Both *a* and *b* are exactly the same except the way of writing.

66. (*a*) **67.** (*b*) **68.** (*a*) **69.** (*b*)

70. (*d*)

71. (*b*) Since we integrate with respect to time to find $F(j\omega)$, we get volt-secs.

72. (*a*) **73.** (*c*) **74.** (*a*) **75.** (*a*)

76. (*a*) **77.** (*a*) **78.** (*a*) **79.** (*b*)

80. (*a*) For fundamental $X_C = \dfrac{1}{314 \times 1 \times 10^{-6}}$ and for 5th harmone $X_C = \dfrac{1}{5 \times 314 \times 1 \times 10^{-6}}$ ohms.

81. (*b*) $X_C = \dfrac{1}{\omega C}$ and $I = V\omega C$

82. (*a*) Use trignometric theorem.

83. (*a*)

84. (*a*) See section 3.4.

85. (*b*) Integral of current is charge.

86. (*d*) Since $e = \dfrac{N d\phi}{dt}$, Integraly *e* is flux linkages.

87. (*b*) **88.** (*b*) See section 3.4.

89. (*a*) **90.** (*d*) $i = \dfrac{1}{L}\int e\,dt$.

91. (*a*) **92.** (*a*) **93.** (*a*) **94.** (*c*)

95. (*a*) **96.** (*a*) **97.** (*b*) **98.** (*a*)

99. (*a*) Superposition is valid if response is linearly related to cause.

100. (*b*)

101. (*a*) All the elements are linear.

102. (*d*) Due to presence of v_0 the response will not be proportional to cause. Hence v_0 must be zero if it is a linear system.

103. (*d*) **104.** (*a*) **105.** (*d*) **106.** (*a*)

107. (*b*) Z transform possesses the property of linearity.

108. (*a*) **109.** (*b*) **110.** (*a*) **111.** (*a*)

112. (*b*)

113. (*a*) $H(z) = \dfrac{1 + \dfrac{1}{z}}{1 - \dfrac{5}{6z} + \dfrac{1}{6z^2}}$

$$= \frac{z(z+1)}{6z^2 - 5z + 1} = \frac{z(z+1)}{\left(z - \frac{1}{2}\right)\left(z - \frac{1}{3}\right)}.$$

Hence poles are at $z = \dfrac{1}{2}$ and $z = \dfrac{1}{3}$.

114. (*d*) **115.** (*a*) **116.** (*a*) **117.** (*d*)

118. (*c*) $u_k + 0.5y_{k-1} - 0.25\,y_{k-2} = y_k$

or $y_k - 0.5y_{k-1} + 0.25y_{k-2} = u_k$.

119. (*a*) u_k has two unit delays and a multiplier of 0.5, y_k has a unit delay and a multiplier of 0.25.

120. (*a*) **121.** (*a*) See section 3.7.

122. (*a*) See section 3.7. **123.** (*b*)

124. (*a*) Use method of least squares.

125. (*b*)

126. (*b*) $\int 1\,dt = t$ and $\int t\,dt = \dfrac{t^2}{2}$.

127. (*d*) Effective value $= \sqrt{(10)^2\,(0.5)}$.

128. (*a*) Since $f(-t) = -f(t)$ only sine terms will be present because $\sin(-x) = -\sin x$.

129. (*b*) **130.** (*c*)

131. (*a*) See Table in section 3.4.

132. (*c*) Final value $= \underset{s \to 0}{\text{Lt}}\,[sf(s)]$

$$= \underset{s \to 0}{\text{Lt}}\,\frac{2s+1}{s^3 + 8s^2 + 16s + 1} = 1.$$

133. (*d*) **134.** (*b*) **135.** (*b*)

136. (*b*) It is a sum of three functions, *i.e.*, a unit step function originating at $t = 0$, a ramp with slope $\dfrac{-1}{T}$ originating at $t = 0$ and another ramp with slope $\dfrac{1}{T}$ and originating at $t = T$.

Thus $\mathcal{L} f(t) = \dfrac{1}{s} - \dfrac{1}{Ts^2} + \dfrac{1}{Ts^2}e^{-Ts}$

$$= \frac{1}{Ts^2}[e^{-Ts} - 1 + Ts].$$

137. (*d*) **138.** (*b*)

139. (*b*) Output of signal generator is analog and that of JK flip-flop is digital.

140. (*a*) Fourier coefficient

$$b_n = \frac{1}{\pi}\int_0^{\pi} V \sin n\omega_1\, d(\omega_1 t) + \frac{1}{\pi}\int_{\pi}^{2\pi} (-v) \sin n\omega_1 t\, d(\omega_1 t)$$

$$= \frac{2V}{n\pi}(1 - \cos n\pi).$$

For $n = 1$, $b_n = \dfrac{4V}{\pi}$.

141. (*b*) **142.** (*c*) **143.** (*b*) **144.** (*a*)

145. (*a*) It is initial value theorem.

146. (*b*) **147.** (*b*) **148.** (*c*) **149.** (*c*)

150. (*c*) $\underset{s \to 0}{\text{Lt}}\; sI(s) = \dfrac{5(250)}{100} = 12.5$.

151. (*b*) $\underset{s \to \infty}{\text{Lt}}\; sI(s) = \dfrac{5\left(1 + \dfrac{250}{s}\right)}{\left(1 + \dfrac{100}{s}\right)} = 5$ A.

152. (*d*) $£(t) = \frac{1}{s^2}$.

153. (*c*) **154.** (*b*) **155.** (*b*) **156.** (*c*)

157. (*d*)

158. (*b*) $f = \frac{1}{8}$ Hz and T = 8*s*.

159. (*a*) **160.** (*b*)

161. (*d*) The function is sum of three functions, *i.e.*, a ramp of slope 1 originating $t = 0$, a ramp of slope -1 originating at $t = 1$ and a negative unit step function originating at $t = 2$, *i.e.*,

$$f(t) = tu(t) - tu(t-1) - u(-2)$$

$$F(s) = \frac{1}{s^2} - \frac{1}{s^2}e^{-s} - \frac{1}{s}e^{-2s}$$

162. (*b*) $F(s) = \frac{2s+5}{(s+3)(s+2)} = \frac{1}{s+2} + \frac{1}{s+3}$

$f(t) = e^{-2t} + e^{-3t}$.

163. (*b*) $I = (2^2 + 6^2 + 24)^{0.5} = \sqrt{64} = 8$.

164. (*b*)

165. (*a*) Add elements individually.

166. (*a*) del. $|A| = 3(8-12) = -12$.

167. (*a*) **168.** (*b*) See section 3.1.

169. (*c*)

170. (*d*) Eigen values of matrix A are roots of characteristic equation. $[\lambda I - A] = 0$ where λ is a scalar.

Therefore $\begin{bmatrix} \lambda & 0 \\ 0 & \lambda \end{bmatrix} - \begin{bmatrix} 0 & 1 \\ 1 & 0 \end{bmatrix} = \begin{bmatrix} \lambda & -1 \\ -1 & \lambda \end{bmatrix}$ or

$\lambda^2 - 1 = 0$ or $(\lambda + 1)(\lambda - 1) = 0$. The roots are -1 and $+1$.

171. (*a*)

172. (*a*) See section 3.4.

173. (*a*) K can be found by integrating and equating the area under the curve equal to 1.

174. (*c*) An even function cannot have sine terms because for an even function $f(-x) = f(x)$.

175. (*b*) Use initial and final value theorems.

176. (*c*) **177.** (*a*) **178.** (*b*)

179. (*c*) $I = \sqrt{3^2 + \frac{4^2}{2} + \frac{4^2}{2}} = 5$, $P = I^2R = 250$ W.

180. (*c*) See section 3.4.

181. (*d*) For an odd function $f(-x) = -f(x)$. Hence only sine terms.

182. (*c*) **183.** (*b*)

184. (*c*) $v(t) = \cos t$ and $i = C\frac{dv}{dt} = 0.5 \cos t$.

At $t = 0^+, i = 0.5$ A

185. (*a*) $F(s) = \frac{10}{(s+2)^2}$

186. (*c*) It is an impulse originating at $t = a$..

187. (*a*) **188.** (*c*) **189.** (*b*) **190.** (*c*)

191. (*b*) See section 3.4.

192. (*d*) Coefficient of $e^{-t} = \left|\frac{s+3}{(s+1)^2+3}\right|_{s=-1} = \frac{2}{3}$.

193. (*a*) y_k has two unit delays and then a multiplier of 0.5 before being fedback.

194. (*a*)

195. (*d*) The Laplace transform of $\delta(t)$ is 1.

196. (*d*) **197.** (*c*) **198.** (*b*) **199.** (*b*)

200. (*b*) **201.** (*c*) **202.** (*b*) **203.** (*b*)

204. (*a*) $(s + 1)$ term will give e^{-t} and $(s + 2)$ term will give e^{-2t}.

205. (*c*) See Table in section 3.4.

206. (*a*)

207. (*c*) It is a modulation process. The resultant has $(f + f_c)$ and $(f - f_c)$ terms.

208. (*b*) **209.** (*b*) **210.** (*a*) **211.** (*a*)

212. (*b*)

213. (*c*) Sampling frequency must be more than $2f$. Therefore

$$T < \frac{1}{2f}.$$

214. (*a*) **215.** (*b*) **216.** (*b*) **217.** (*c*)

218. (*c*) **219.** (*c*) **220.** (*a*) **221.** (*a*)

222. (*a*) DC meter reads mean value.

223. (*c*) rms value = $\sqrt{4}$

224. (*b*) **225.** (*b*) **226.** (*b*)

227. (*c*) The probability that first ball is white is $\frac{4}{7}$ and the probability that second ball is white is $\frac{3}{6}$. The probability that both are white is $\frac{4}{7} \times \frac{3}{6} = \frac{2}{7}$.

228. (*b*) **229.** (*a*)

230. (*b*) Impulse is differential of unit step function. Hence impluse response

$$= \frac{d}{dt}\left(\frac{1}{2} - \frac{1}{2}e^{-2t}\right) = e^{-2t}.$$

231. (*c*)

232. (*b*) $x(t)$ Input = $u(t)$

$y(t)$ = Output = $t^2 e^{-2t}$

$X(s) = \frac{1}{s}$, $Y(s) = \frac{2}{(s+2)^3}$

Transfer fucntion $\frac{Y(s)}{X(s)} = \frac{\left[\frac{2}{(s+2)^3}\right]}{1/s} = \frac{2s}{(s+2)^3}$.

233. (*b*) $Z = e^{sT}$ or $\ln Z = sT$ or $s = \dfrac{\ln z}{T}$.

234. (*b*) Convolution in time domain means multiplication in frequency domain.

235. (*b*) Fourier series can be written if a function satisfies. Drichlet condition. These are :

The function

1. has a finite number of discontinuities
2. possesses a finite number of maxima and minima
3. be absolutely convergent, *i.e.*,

$$\int_0^T |X(t)|\, dt < \infty$$

236. (*c*) $F[1] = 2\pi\,\delta(\omega)$

$$F[e^{j\omega_0 t}] = 2\pi\,\delta(\omega - \omega_0)$$

or $$F\left[\frac{1}{2\pi} e^{j\omega_0 t}\right] = \delta(\omega - \omega_0)$$

237. (*d*) Phase should be linearly dependent on ω.

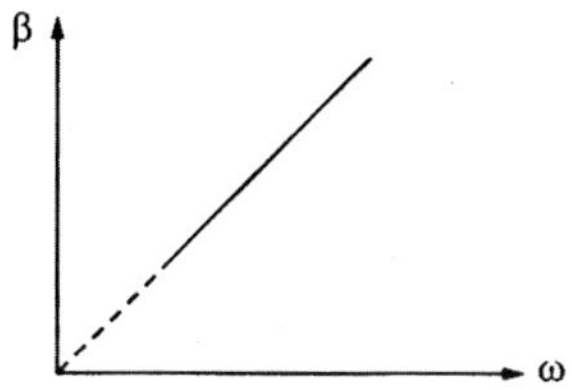

238. (*a*) Order of difference equations = 2 = number of state variables

239. (*b*) See Answer to Q. 236.

240. (*d*) Complex signal is conjugate symmetric.

241. (*d*) Only signal contains information.

242. (*a*) $\omega = \dfrac{2\pi}{N}$ or $N = \dfrac{2\pi}{\omega} = \dfrac{2\pi}{0.2\pi} = 10.$

243. (*b*) In time domain $y(t) = x(t)\,h(t)$

In frequency domain $Y(j\omega) = X(j\omega)\,H(j\omega)$

244. (*d*) The factor $t\,i\,(t)$ makes it time variant. The factor 2 makes it non-linear.

245. (*a*) A function is periodic if the ratio of time periods of its parts is a rational number.

$$T_1 = \frac{2\pi}{10\pi} = \frac{1}{5};\quad T_2 = \frac{2\pi}{20\pi} = \frac{1}{10};$$

$$\frac{T_1}{T_2} = 2.$$

246. (*c*) Total power is the sum of average power of each component including fundamental.

247. (*d*)

248. (*c*) The properties of Laplace transform apply to properties of Fourier transform with $s = j\omega$.

249. (*a*) p_x is a probability density function if

$$\int_{-\infty}^{\infty} p_x(x)\,dx = 1$$

or $$\int_{-\infty}^{\infty} ae^{-b|X|}\,dx = 1$$

or $$2\int_0^{\infty} ae^{-b|X|}\,dx = 1$$

or $$2a\left[\frac{e^{-bx}}{-b}\right]_0^{\infty} = 1$$

or $$\frac{2a}{b} = 1 \text{ or } a = \frac{b}{2}.$$

250. (*a*) Constant

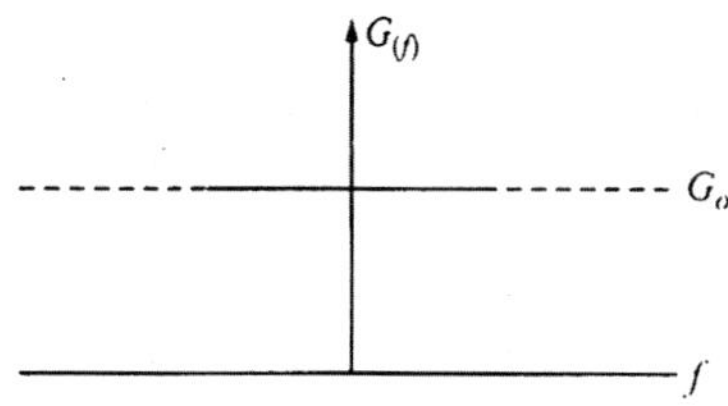

251. (*c*) Convolution of $f(t)$ is $\int_{-\infty}^{\infty} f(\tau)\,f(t-\tau)\,d\tau$

If $f(t) = u(t)$

$$\text{Convolution of } f(t) = \int_{-\infty}^{\infty} u(\tau)\,u(t-\tau)\,d\tau$$

$$= \int_0^t u(\tau)\,d\tau.$$

252. (*a*) Impulse response $= \dfrac{d}{dt}$ (step response)

$$= \frac{d}{dt}[0.5\,(1 - e^{-2t})\,u(t)]$$

$$= 0.5\,[0 - (-2)\,e^{-2t}]\,u(t)$$

$$= e^{-2t}\,u(t).$$

253. (*b*) $I(s) = \dfrac{V(s)}{2s} = \dfrac{4s^2 + 3s + 2}{2s\,[7s^2 + 8s + 5]}$

$$i(\infty) = \underset{s\to 0}{\text{Lt}}\ sI(s)$$

$$= \underset{s\to 0}{\text{Lt}}\ \frac{4s^2 + 3s + 2}{14s^2 + 16s + 10} = \frac{2}{10} = \frac{1}{5}.$$

254. (*b*) See Table in section 3.4.

255. (*b*) Even function × Even function = Even function

Even function × Odd function = Odd function

$$F(j\omega) = \int_{-\infty}^{\infty} f(t)\,e^{-j\omega t}\,dt$$

$$= \int_{-\infty}^{\infty} f(t)[\cos \omega t - j\sin \omega t]\, dt$$

$$F(j\omega) = \int_{-\infty}^{\infty} f(t) \cos \omega t\, dt$$

$$= 2\int_{0}^{\infty} f(t) \cos \omega t\, dt.$$

256. (*d*) Since output is square of input.

257. (*b*) $\frac{dy(t)}{dt} + 3y(t) = 0 + K$ or $y(t) = K/3$.

258. (*c*) See section 3.4.

259. (*d*) $Y(s) = \frac{2}{s(s^2 - s - 2)}$

$$\underset{t\to\infty}{\text{Lt}}\ y(t) = \underset{s\to 0}{\text{Lt}}\ sY(s)$$

$$= \underset{s\to 0}{\text{Lt}} \frac{2}{s^2 - s - 2} = \text{infinite.}$$

260. (*c*) Impulse response is $\frac{d}{dt}$ (unit step response)

$$\text{Impulse response} = \frac{d}{dt}(1 - e^{-t/RC})\, u(t)$$

$$= \left(0 - \left(\frac{1}{RC}\right)e^{-t/RC}\right)u(t) = \frac{1}{RC} e^{-t/RC}\, u(t).$$

261. (*a*) System is linear time invariant. Hence output is same as input.

262. (*a*) Convolution integral enables us to find response to any arbitrary input in terms of impulse response.

263. (*c*) Energy is time integral of power. Since time is not defined here, energy is infinite.

264. (*d*)

265. (*d*) The factor *b* makes it non-linear.

266. (*b*) $£ - e^{+2t}\, u(t) = \frac{-1}{s-2}$

$$£\ tu\,(t) = \frac{1}{s^2}$$

$$£\ x_1(t)\, x_2(t) = X_1(x)\, X_2(s)$$

$$\text{Hence } £[-e^{+2t}\, u(t)]\,[t\, u(t)] = \frac{-1}{s^2(s-2)}.$$

267. (*b*) Mean value $= \int_{-\infty}^{\infty} x\, p_x(x)\, dx = \frac{1}{4}\int_{-1}^{3} x \cdot dx$

$$= \frac{1}{4}\left|\frac{x^2}{2}\right|_{-1}^{3} = 1$$

$$\text{Variance} = \int_{-\infty}^{\infty} (x-1)^2\, p_x(x)\, dx$$

$$= \frac{1}{4}\int_{-1}^{3} (x-1)^2\, dx = \frac{4}{3}$$

268. (*a*) $f(t) = r(t) - r(t-1) - r(t-4) + 1.5\, r\,(t-6) - 0.5\, r\,(t-8)$

$$F(s) = \frac{1}{s^2}[1 - e^{-s} - e^{-4s} + 1.5e^{-6s} - 0.5e^{-8s}]$$

Hence $D = -0.5$.

269. (*d*) For system to be casual impulse response must be zero for $t < 0$, *i.e.*, $h(t) = 0$ for $t < 0$.

In this case impulse response is found to be non-zero for $t < 0$. Evidently system is non-causal. As seen system is time variant also as output is dependent on time.

270. (*c*) Z transform of $h_1(n) = H_1(z) = z^{-1}$

Z transform of $h_2(n) = H_2(z) = z^{-2}$

Overall response in Z domain $= z^{-1}\, z^{-2} = z^{-3}$

Overall response in discrete time domain $= \delta(n-3)$.

271. (*b*) Taking Laplace transform of given equation

$$s^2\, Y(s) + 4s\, Y(s) + 3\, Y(s) = 2s\, X(s) + 4\, X(s)$$

$$\text{or} \quad \frac{Y(s)}{X(s)} = \frac{2s+4}{s^2+4s+3} = \frac{2(s+2)}{(s+3)(s+1)}$$

$$x(t) = e^{-2t}\, u(t)$$

$$X(s) = \frac{1}{s+2}$$

$$Y(s) = \frac{2(s+2)}{(s+3)(s+1)(s+2)}$$

$$= \frac{2}{(s+1)(s+3)}$$

$$= \frac{1}{s+1} - \frac{1}{s+3}$$

Taking inverse transform

$$y(t) = (e^{-t} - e^{-3t})\, u(t).$$

272. (*a*)

273. (*b*) $f(t) = 0.5\,(1 - \cos 2t) + \cos 2t = 0.5 + 0.5 \cos 2t$

frequency of constant form = 0

frequency of second term $= \frac{2}{2\pi} = \frac{1}{\pi}$

274. (*b*) Laplace of $\int_0^t f(\tau)\, d\tau = \frac{F(s)}{s}$ neglecting initial condition.

275. (*c*) A system is causal if output at any time depends only on values of input at present time and in past.

276. (*d*) The graph's

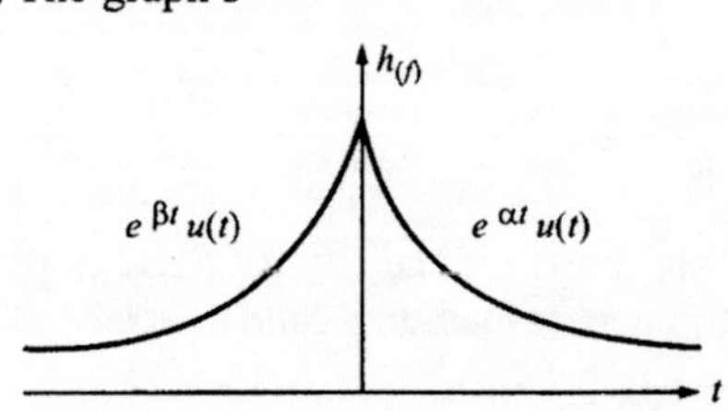

Evidently α should be –ve and β should be positive only then $\int_{-\infty}^{\infty} h(t)\, dt$ is less than infinite

277. (*a*) $\int_{-\infty}^{\infty} (Me^{-2|x|} + Ne^{-3|x|})\,dx = 1$

or $\int_{0}^{\infty} (Me^{-2|x|} + Ne^{-3|x|})\,dx = \frac{1}{2}$

or $\frac{M}{2} + \frac{N}{3} = \frac{1}{2}$

or $M + \frac{2N}{3} = 1.$

278. (*a*)

$$X(\omega) = \int_{-\infty}^{\infty} x(t)\,e^{-j\omega t}\,dt = \int_{-1}^{1} e^{-j\omega t}\,dt$$

$$= \left|\frac{e^{-j\omega}}{-j\omega}\right|_{-1}^{1} = \frac{e^{-j\omega}}{-j\omega} - \frac{e^{j\omega}}{-j\omega}$$

$$= \frac{1}{j\omega}(e^{j\omega} - e^{-j\omega})$$

$$X(\omega) = 0 \text{ if } e^{j\omega} - e^{-j\omega} = 0$$

or $e^{j\omega} - \frac{1}{e^{j\omega}} = 0$ or $e^{2j\omega} - 1 = 0$

or $e^{2j\omega} = \pm 1$ or $\omega = \pi, 2\pi$

279. (*c*)

$$H(\omega) = \int_{-\infty}^{\infty} h(t)\,e^{-j\omega t}\,dt$$

$$= \int_{0}^{\infty} e^{-2t}\,e^{-j\omega t}\,dt = \int_{0}^{\infty} e^{-(2+j\omega)t}\,dt$$

$$= \frac{1}{2+j\omega}\left| e^{-(2+j\omega)t}\right|_{0}^{\infty} = \frac{1}{2+j\omega}$$

280. (*a*) A follows from the theorem in R.

281. (*a*) Initial value theorem when applied to I (*s*) gives 5 A.

282. (*d*) Final value = $\underset{s\to 0}{\text{Lt}}\ \frac{s(s-1)}{s^2+3s+2} = 0.$

283. (*a*) **284.** (*a*)

285. (*c*) If $f(t) = t$, $F(s) = \frac{1}{s^2}$. Hence R is wrong.

286. (*b*) Both are correct but independent.

287. (*b*) Initial conditions are taken into account.

288. (*d*) £ $[af_1(t) + bf_2(t)] = aF_1(s) + bF_2(s)$. Hence A is wrong.

289. (*b*) Both are correct but independent.

290. (*b*)

291. (*c*) If $f(t) = \sin \omega t$, $F(s) = \frac{\omega}{s^2+\omega^2}$. Hence A is wrong.

292. (*b*) **293.** (*b*)

294. (*c*) Many circuits operate due to transient phenomena. Hence R is wrong.

295. (*a*) $F(s) = \frac{1}{s} - \frac{s}{s^2+\omega^2} = \frac{\omega^2}{s(s^2+\omega^2)}$

296. (*b*)

297. (*b*) Both are correct but independent.

298. (*a*) $\sqrt{1+\frac{1}{2}} = \sqrt{1.5}.$

299. (*b*)

300. (*a*) Since capacitor presents infinite impedance to dc, voltage component A_1, does not cause any current.

301. (*b*) **302.** (*a*)

303. (*b*) Both are correct but independent.

304. (*b*) **305.** (*a*) **306.** (*b*)

307. (*b*) Both are correct but independent.

308. (*b*) **309.** (*b*) **310.** (*b*)

311. (*b*) For an inductor L, the current is $\frac{1}{L}\int v\,dt$ and integral of $\delta(t-a)$ is $u(t-a)$.

312. (*b*) **313.** (*b*)

314. (*d*) Final value = 5. Hence A is wrong, initial value is 2. Hence R is correct.

315. (*b*) **316.** (*b*) **317.** (*b*) **318.** (*b*)

319. (*b*) Both are correct but independent.

320. (T) **321.** (T)

322. (F) £ $e^{-at} = \frac{1}{s+a}$.

323. (F) It is an impulse function occurring at $t = b$.

324. (T) **325.** (T) **326.** (T)

327. (T) **328.** (T)

329. (F) The real part of *s* corresponds to change in amplitude.

330. (T) **331.** (T) **332.** (T) **333.** (T)

334. (F) A is $n \times n$ matrix.

335. (F) It is not a unique set.

336. (T) **337.** (T) **338.** (T)

339. (T) **340.** (T) **341.** (T)

342. (T) **343.** (T)

344. (F) It is not symmetrical about *x*-axis.

345. (T) **346.** (T) **347.** (T)

348. (T) **349.** (T) **350.** (T)

351. (T) **352.** (T)

353. (F) Duty factor is $\frac{d}{t}$.

354. (T) **355.** (T) **356.** (F)

357. (T) **358.** (T) **359.** (T)

360. (T) **361.** (T) **362.** (T)

363. (T) **364.** (F) Z transform is a linear operation.

365. (T) **366.** (F) **367.** (T)

368. (T) **369.** (T) **370.** (T)

371. (T) **372.** (T) **373.** (T)

374. I **375.** $\frac{\text{adj A}}{\det \text{A}}$

376. eigen values

377. $B^T A^T$ **378.** diagonal

379. unit doublet

380. se^{-bs}

381. unit step function

382. ramp of slope unity

383. zero **384.** $\frac{\omega}{s^2 + \omega^2}$

385. $\frac{1}{s + a}$ **386.** $\frac{2}{s^3}$ **387.** $\sigma + j\omega$

388. $\underset{s \to \infty}{\text{Lt}} [sF(s)]$

389. te^{-3t}

390. $\frac{s}{s^2 - \alpha^2}$

391. $AX + Bu$

392. not unique

393. $\frac{\text{adj}(sI - A)}{\det(sI - A)}$ **394.** $\sqrt{52}$

395. $\sqrt{450}$ **396.** sine

397. even harmonics **398.** zero

399. zero

400. frequency spectrum of $f(t)$

401. $\int_{-\infty}^{\infty} f(t)\, e^{-j\omega t}\, dt$ **402.** $aF(j\omega)$

403. $F(j\omega)\, e^{-j\omega t}$

404. $2\pi\, \delta(\omega)$

405. $8e^{-t}$ **406.** $\frac{1}{\alpha + j\omega}$

407. $(1 - z^{-1})^{-1}$

408. $\sqrt{\frac{\Sigma \rho^2}{n}}$ **409.** $\frac{h}{\sqrt{\pi}} e^{-h^2 x^2}$

410. $\frac{1}{\sqrt{2h}}$ **411.** high peak

412. $0.6745\,\sigma$ **413.** 68.3%.

CHAPTER 4

NETWORK THEORY

4.1. BASIC CONCEPTS AND UNITS

Force F (newtons) = m (kg) × a (m/s^2)

Work W (joules) = F (newtons) × d (metres)

Torque T (newton metres) = F (newtons) × r (metres)

Kinetic energy (joules) = 0.5 m (kg) × v^2 (m/s)^2

Potential energy (joules) = 9.81 × m (kg) × h (metres)

Power (watts) = 2π × n (rev/sec) × T (newton-metres)

Current I (ampere) = dQ/dt

Charge Q (coulmb) = I (ampere) × t (seconds)

Heat required to raise the temperature of m (kg) by θ (kelvin)

= m (kg) × C (joules/kg/K) × θ

Electrical power P (watts) = $I^2R = VI = \dfrac{V^2}{R}$

Electrical energy (joules or watt-secs) = V I t

4.2. PARAMETERS OF ELECTRIC CIRCUITS

Ohm's law $I = \dfrac{V}{R}$ or $V = IR$ or $R = \dfrac{V}{I}$

Resistance of conductor $R = \dfrac{\rho l}{a}$

Resistances in series $R = R_1 + R_2 + R_3 + \ldots$

Resistances in parallel $\dfrac{1}{R} = \dfrac{1}{R_1} + \dfrac{1}{R_2} + \dfrac{1}{R_3} + \ldots$

$$R_t = R_0 (1 + \alpha_0 t)$$

$$\alpha_t = \frac{\alpha_0}{1 + \alpha_0 t}$$

Conductance = 1/Resistance

Induced emf $v = L\dfrac{di}{dt}$

Energy stored in inductance (joules) = 0.5 Li^2

$$L = N\frac{d\phi}{di} = N\frac{\phi}{i}$$

MMF = NI amperes

Magnetic field strength $H = \dfrac{NI}{l}$ amperes/metre

Reluctance $R = \dfrac{l}{\mu a}$

Permeability of free space $\mu_0 = 4\pi \times 10^{-7}$ H/m

$$\mu = \mu_0 \mu_r$$

Inductance $L = \dfrac{N^2 \mu a}{l}$

Inductances in series $L_s = L_1 + L_2 + L_3 + \ldots$

Inductances in parallel $= \dfrac{1}{L_p} = \dfrac{1}{L_1} + \dfrac{1}{L_2} + \dfrac{1}{L_3} + \ldots$

Mutual inductance $M = k\sqrt{L_1 L_2}$

Coupled coils in series aiding connection

Total inductance $L_a = L_1 + L_2 + 2M$

Coupled coils in series opposing connection

Total inductance $L_b = L_1 + L_2 - 2M$

Q (coulomb) = C (farads) × V (volts)

Capacitance of parallel plate capacitor

$$= \frac{\in \text{(farads/m)} \times A(m^2)}{d\text{(metre)}}$$

Energy stored in capacitor = 0.5 CV^2

Capacitances in series $\dfrac{1}{C_s} = \dfrac{1}{C_1} + \dfrac{1}{C_2} + \dfrac{1}{C_3} + \ldots$

Capacitances in parallel $C_p = C_1 + C_2 + C_3 + \ldots$

Capacitance of multiplate capacitor with n plates

$$= \frac{\in (n-1) A}{d}$$

Capacitance of isolated sphere of radius $r = 4\pi \in_0 r$

Capacitance of cylindrical capacitor $= \dfrac{2\pi \in l}{\ln (R/r)}$

Insulation resistance of cable $= \dfrac{\rho}{2\pi l} \ln (R/r)$

4.3. ELECTROLYSIS, STORAGE CELLS

m, *i.e.*, mass of element liberated from electrolyte = z (grams/coulomb) × I (amperes) × t (seconds)

Terminal voltage of a cell $V = E - Ir$

$$\text{Ampere hour efficiency} = \frac{\text{Ampere hour on discharge}}{\text{Ampere hour on charge}}$$

$$\text{Energy efficiency of cell} = \frac{\text{Watt hour on discharge}}{\text{Watt hour on charge}}$$

$$= \text{Ampere hour efficency} \times \frac{\text{Terminal voltage on discharge}}{\text{Terminal voltage on charge}}$$

4.4. DC CIRCUITS

$$\text{KCL} = \sum_{j=1}^{j=p} I_j = 0 \text{ at every node}$$

$$\text{KVL} = \sum_{j=1}^{j=r} V_j = 0 \text{ for a closed loop}$$

Thevenin's theorem: Load current $= \dfrac{V_{TH}}{R_{TH} + R_L}$

Norton's theorem: Load current $= I_N \dfrac{R_N}{R_N + R_L}$

Superposition Theorem: In a circuit having more than one voltage and current source, the total current or voltage in any branch is the algebraic sum of currents or voltages in that branch produced by each source acting independently.

Maximum power transfer condition $R = r$

Millman's Theorem: A number of current sources in parallel may be replaced by a single current source whose current strength is equal to algebraic sum of individual source currents and source resistance is equal to parallel combination of individual source resistances.

Substitution Theorem: In a network any branch may be substituted by another branch without disturbing the currents and voltages in other branches of the network provided the new branch has the same internal voltage when carrying the same current.

Compensation Theorem: in a linear network, if a resistance R carrying a current I is changed to $(R + \Delta R)$, then the change in current in any branch of the network can be found by assuming that an emf $(-I\,\Delta R)$ has been introduced in the changed branch.

Reciprocity Theorem: If a source of emf E located at one point in a network composed of linear bilateral circuit elements produces a current I at a selected point in the network, the same source of emf E acting at the second point will produce the same current at the first point.

Star–Delta Conversion

$$R_{ab} = \frac{R_a R_b + R_b R_c + R_c R_a}{R_c}$$

$$R_{bc} = \frac{R_a R_b + R_b R_c + R_c R_a}{R_a}$$

$$R_{ca} = \frac{R_a R_b + R_b R_c + R_c R_a}{R_b}$$

Delta–Star Conversion

$$R_a = \frac{R_{ab} R_{ca}}{R_{ab} + R_{bc} + R_{ca}}$$

$$R_b = \frac{R_{bc} R_{ab}}{R_{ab} + R_{bc} + R_{ca}}$$

$$R_c = \frac{R_{ca} R_{bc}}{R_{ab} + R_{bc} + R_{ca}}$$

4.5. SINGLE PHASE AC CIRCUITS

Frequency $f = \dfrac{1}{T}$ (Hz)

T = Time period (seconds)

$\omega = 2\pi f$ (rad/sec)

$$\text{Average value} = F_{av} = \frac{1}{t_2 - t_1}\int_{t_1}^{t_2} f(t)\,dt$$

$$I = \text{RMS current} = \left[\frac{1}{T}\int_0^T i^2\,dt\right]^{0.5}$$

For sinusoidal wave

Average value = 0.637 × Peak value

RMS value = 0.707 × Peak value

$$\text{Form factor} = \frac{\text{RMS value}}{\text{Average value}} = 1.11$$

Peak-to-peak value = 2 × Peak value

$$\text{Peak factor} = \frac{\text{Peak value}}{\text{RMS value}} = 1.414$$

Operator $j = 1\angle 90°$

$j^2 = 1\angle 180° = -1$

$j^3 = 1\angle 270°$

$j^4 = 1$

$a + jb = A\angle\theta = A\cos\theta + jA\sin\theta$

where $A = (a^2 + b^2)^{0.5}$ and $\theta = \tan^{-1}(b/a)$

$(A\angle\alpha)(B\angle\beta) = AB\angle(\alpha+\beta)$

$$\frac{A\angle\alpha}{B\angle\beta} = \frac{A}{B}\angle(\alpha - \beta)$$

$$\frac{1}{a + jb} = \frac{a - jb}{a^2 + b^2}$$

$[A\angle\alpha]^n = A^n\angle n\alpha$

$[A\angle\alpha]^{1/n} = A^{1/n}\angle(\alpha/n)$

$\ln[A\angle\alpha] = \ln A + j\alpha$

For pure resistance R, $I = \dfrac{V}{R}$, phase angle = 0

Active power = VI

For pure inductance L,

Inductive reactance $X_L = 2\pi fL$ ohms

$$I = \frac{V}{X_L}$$

Phase angle = 90°, current lagging

Active Power = 0

For pure capacitance C

Capacitive reactance $X_C = \frac{1}{2\pi fC}$ ohms

$$I = \frac{V}{X_C}$$

Phase angle = 90°, current leading

Active Power = 0

For R-L series circuit

Impedance $Z = R + jX_L = Z \angle \theta$ ohms

Admittance $Y = \frac{1}{Z} = G + jB = Y \angle -\theta$ siemens

Conductance $G = \frac{R}{R^2 + X_L^2}$ siemens

Susceptance $B = \frac{-X_L}{R^2 + X_L^2}$ siemens

Phase angle $\theta = \tan^{-1}(X_L/R)$, current lagging

$P = VI\cos\theta$

$Q = VI\sin\theta$

For R-C series circuit

Impedance $Z = R - jX_C = Z \angle -\theta$ Ohms

Admittance $Y = \frac{1}{Z} = G + jB = Y \angle \theta$ siemens

Conductance $G = \frac{R}{R^2 + X_C^2}$ siemens

Susceptance $B = \frac{X_C}{R^2 + X_C^2}$ siemens

Phase angle $\theta = \tan^{-1}\frac{X_C}{R}$, current leading

$P = VI\cos\theta$

$Q = VI\sin\theta$

For R-L-C series circuit

X = reactance $= X_L - X_C = 2\pi f L - \frac{1}{2\pi f C}$ ohms

Z = Impedance $= R + j(X_L - X_C) = R + jX = Z \angle \theta$ ohms

Y = Admittance $= \frac{1}{Z} = G + jB = Y \angle -\theta$ siemens

Conductance $G = \frac{R}{R^2 + X^2}$ siemens

Susceptance $B = \frac{-X}{a^2 + X^2}$ siemens

Phase angle $\theta = \tan^{-1}(X/R)$

Current lagging if X is positive

Current leading if X is negative

Power factor $= \cos\theta = (R/Z)$

$\sin\theta = (X/Z)$

S = Apparent power = VI volt-ampere

P = Active power = $VI\cos\theta$ watts

Q = Reactive power = $VI\sin\theta$ vars

Q is positive if $X_L > X_C$

Q is negative if $X_L < X_C$

For a circuit having R and L is parallel

Admittance $Y = \frac{1}{R} - \frac{j}{X_L} = G + jB = Y \angle -\theta$ siemens

Phase angle $\theta = \tan^{-1}\left(\frac{R}{X_L}\right)$, current lagging

Impedance $Z = \frac{RX_L^2}{R^2 + X_L^2} - j\frac{R^2X_L}{R^2 + X_L^2}$

Current $I = \frac{V}{Z} = VY$

For a circuit having R and C in parallel

Admittance $Y = \frac{1}{R} + \frac{j}{X_C} = G + jB = Y \angle \theta$ siemens

Phase angle $\theta = \tan^{-1}(R/X_C)$, current leading

Impedance $Z = \frac{RX_C^2}{R^2 + X_C^2} - j\frac{R^2X_C}{R^2 + X_C^2}$ ohms

Current $I = \frac{V}{Z} = VY$

For a circuit having R, L, C in parallel

Admittance $Y = \frac{1}{R} + j\left[\frac{1}{X_C} - \frac{1}{X_L}\right]$

$= Y \angle \theta = G + jB$ siemens

$I = VY$

Phase angle $\theta = \tan^{-1}\frac{\left[\frac{1}{X_C} - \frac{1}{X_L}\right]}{1/R}$

Current lagging if $X_L < X_C$

Current leading if $X_L > X_C$

For impedances $Z_1 = R_1 + jX_1$ and $Z_2 = R_2 + jX_2$ in series

Total impedance $Z = (R_1 + R_2) + j(X_1 + X_2)$

$= |Z| \angle \theta$

$|Z| = [(R_1 + R_2)^2 + (X_1 + X_2)^2]^{0.5}$

$\theta = \tan^{-1}[(X_1 + X_2)/(R_1 + R_2)]$

Current lagging if θ is positive

Current leading if θ is negative

For resonance in series circuit $f_r = \dfrac{1}{2\pi\sqrt{LC}}$

Quality factor $Q = \dfrac{\omega_r L}{R} = \dfrac{1}{R}\sqrt{\dfrac{L}{C}}$

Bandwidth $\Delta f = f_2 - f_1 = \dfrac{R}{2\pi L} = \dfrac{f_r}{Q}$

$$I = \frac{V}{R}, \text{ Phase angle} = 0$$

$$Q = \frac{2\pi\,(\text{total stored energy})}{\text{energy dissipated per cycle}}$$

Fractional frequency deviation $\delta = \dfrac{\omega - \omega_0}{\omega_0}$

For resonance in two branch parallel circuit (one branch having R and L in series and the other branch having C)

$$\omega_0 = \left[\frac{1}{LC} - \frac{R^2}{L^2}\right]^{0.5}$$

$$Z = \frac{L}{RC}$$

4.6. AC NETWORK ANALYSIS

All laws and theorems of dc circuits are applicable to ac networks. However, it is necessary to use phasor addition, subtraction, multiplication, division etc.

Tellegen's theorem: Sum of products of instantaneous branch voltages and currents in a network is zero, *i.e.*,

$\sum_{k=1}^{b} v_k i_k = 0$, where $v_1, v_2, \ldots v_b$ are branch voltages and $i_1, i_2, \ldots i_b$ are branch currents.

4.7. NON-SINUSOIDAL WAVES

$$\text{RMS Value } V = \left[V^2 + \frac{V_{1m}^2}{2} + \frac{V_{3m}^2}{2} + \ldots\right]^{0.5}$$

$$f(t) = a_0 + a_1 \cos\omega_1 t + a_2 \cos\omega_2 t + a_3 \cos\omega_3 t + \ldots + b_1 \sin\omega_1 t + b_2 \sin\omega_2 t + b_3 \sin\omega_3 t + \ldots$$

$$a_0 = \frac{1}{T}\int_0^T f(t)\,dt$$

$$a_n = \frac{2}{T}\int_0^T f(t)\cos n\omega_1 t\,dt$$

$$b_n = \frac{2}{T}\int_0^T f(t)\sin n\omega_1 t\,dt$$

A function has half wave symmetry if

$$f(t) = -f\left(t \pm \frac{T}{2}\right)$$

A function contains only sine terms if

$$f(-t) = -f(t)$$

A function contains only cosine terms if

$$f(-t) = f(t)$$

Exponential form of Fourier series

$$f(t) = \sum_{n=-\infty}^{\infty} F_n e^{jn\omega_1 t}$$

where $F_n = \dfrac{1}{T}\displaystyle\int_{-T/2}^{T/2} f(t)\,e^{-jn\omega_1 t}\,dt$

Fourier transform $F(j\omega) = \displaystyle\int_{-\infty}^{\infty} f(t)\,e^{-j\omega t}\,dt$

$f(t)$ and $F(j\omega)$ constitute the Fourier transform pair. The properties of Laplace Transform apply to the properties of Fourier transform with $s = j\omega$.

4.8. POLYPHASE NETWORKS

Three phase star connection $I_L = I_p$ and $V_L = \sqrt{3}V_p$

Three phase delta connection $I_L = \sqrt{3}I_p$ and $V_L = V_p$

For both star and delta connections $P = \sqrt{3}\,V_L I_L \cos\theta$

$$Q = \sqrt{3}\,V_L\,I_L \sin\theta$$

Neutral shift $V_{on} = \dfrac{V_{an}Y_a + V_{bn}Y_b + V_{cn}Y_c}{Y_a + Y_b + Y_c}$

Two wattmeter, method $P = W_1 + W_2$

$$\tan\theta = \frac{\sqrt{3}\,(W_2 - W_1)}{(W_1 + W_2)}$$

4.9. NETWORK TRANSIENTS

Rise of current in RL series circuit $i(t) = \dfrac{E}{R}(1 - \epsilon^{-tR/L})$

Time constant $T = \dfrac{L}{R}$ (seconds)

$$v_R = E\,(1 - \epsilon^{-tR/L})$$

$$v_L = E\,\epsilon^{-tR/L}$$

Decay of current in RL series circuit

$$i(t) = \frac{E}{R}\epsilon^{-tR/L}$$

$$v_R = E\epsilon^{-tR/L}$$

$$v_L = -E\epsilon^{-tR/L}$$

Charging a capactior $i(t) = \dfrac{E}{R}\epsilon^{-t/RC}$

Time constant $T = RC$ (seconds)

Discharging a capacitor

$$i(t) = -\frac{E}{R}\in^{-t/RC}$$

Step response of R-L-C series circuit

Damping ratio $\varsigma = \frac{R}{R_{Cr}} = \frac{R}{2\sqrt{L/C}}$

Critical resiscance $R_{Cr} = 2\sqrt{L/C}$

Natural frequency $\omega_n = \frac{1}{\sqrt{LC}}$

If $\left(\frac{R}{2L}\right)^2 > \frac{1}{LC}$, response in overdamped

If $\left(\frac{R}{2L}\right)^2 = \frac{1}{LC}$, response is critically damped

If $\left(\frac{R}{2L}\right)^2 < \frac{1}{LC}$, response is under damped

$$\omega_d = \omega_n\sqrt{1-\varsigma^2}$$

$$i(t) = \frac{E - q_0/C}{\omega_d L}\in^{-\varsigma\omega_n t}\sin(\omega_d t)$$

Response of R-L series circuit to sinusoidal excitation

If $v = V_m \sin(\omega t + \phi)$

$$i(t) = \frac{V_m}{(R^2+\omega^2L^2)^{0.5}}\sin(\omega t+\phi+-\theta)$$

$$-\frac{V_m\sin(\phi-\theta)}{(R^2+\omega^2L^2)^{0.5}}\in^{-Rt/L}$$

Response of RC series circuit to sinusoidal excitation

$$v = V_m\sin(\omega t+\phi)$$

$$i(t) = \frac{V_m}{\left(R^2+\frac{1}{\omega^2C^2}\right)^{0.5}}[\sin(\omega t+\phi+\theta)]$$

$$-\frac{V_m}{\left(R^2+\frac{1}{\omega^2C^2}\right)^{0.5}}\sin(\phi+\theta)\in^{-t/RC}$$

Response of inductance L to unit impulse $v = \delta(t-b)$

$$i_L = \frac{1}{L}u(t-b)$$

Response of capacitor to unit impulse $i = \delta(t-a)$

$$v_C = \frac{1}{C}u(t-a)$$

Response of series R-L circuit to impulse

$$v = A\,\delta(t)$$

$$i(t) = \frac{A}{L}\in^{-Rt/L}u(t)$$

A complicated waveform can be broken up into two or more simple waveforms

s-domain circuits

$$£\,[v(t) = Ri(t)] \text{ or } V(s) = RI(s)$$

$$£\left[v(t) = L\frac{di(t)}{dt}\right]$$

or $V(s) = sLI(s) - Li(0^-)$

$$£\left[v(t) = \frac{1}{C}\int_{-\infty}^{t} i(t)\,dt\right]$$

or $$V(s) = \frac{1}{C_s}I(s) + \frac{v(0^-)}{s}$$

Dual Quantities

Quantity	***Dual Quantity***	***Quantity***	***Dual Quantity***
Current	voltage	Node	Mesh
Current source	voltage source	KVL	KCL
Resistance	conductance	Tree branch	Link
Inductance	capacitance	charge	flux
Impedance	admittance	branches in series	branches in parallel
Short-circuit	open circuit		

4.10. TWO PORT NETWORKS

Open circuit Impedance Parameters

$$\begin{bmatrix}V_1\\V_2\end{bmatrix} = \begin{bmatrix}z_{11} & z_{12}\\ z_{21} & z_{22}\end{bmatrix}\begin{bmatrix}I_1\\I_2\end{bmatrix}$$

Short-circuit Admittance Parameters

$$\begin{bmatrix}I_1\\I_2\end{bmatrix} = \begin{bmatrix}y_{11} & y_{12}\\ y_{21} & y_{22}\end{bmatrix}\begin{bmatrix}V_1\\V_2\end{bmatrix}$$

Hybrid (h) Parameters

$$\begin{bmatrix}V_1\\I_2\end{bmatrix} = \begin{bmatrix}h_{11} & h_{12}\\ h_{21} & h_{22}\end{bmatrix}\begin{bmatrix}I_1\\V_2\end{bmatrix}$$

Inverse Hybrid (g) Parameters

$$\begin{bmatrix}I_1\\V_2\end{bmatrix} = \begin{bmatrix}g_{11} & g_{12}\\ g_{21} & g_{22}\end{bmatrix}\begin{bmatrix}V_1\\I_2\end{bmatrix}$$

Transmission Parameters

$$\begin{bmatrix}V_1\\I_1\end{bmatrix} = \begin{bmatrix}A & B\\ C & D\end{bmatrix}\begin{bmatrix}V_2\\-I_2\end{bmatrix}$$

Inverse Transmission Parameters

$$\begin{bmatrix}V_2\\I_2\end{bmatrix} = \begin{bmatrix}A' & B'\\ C' & D'\end{bmatrix}\begin{bmatrix}V_1\\-I_1\end{bmatrix}$$

Parameter Relations for Reciprocal and Symmetrical Networks

Parameter	***Condition for reciprocity***	***Condition for electrical symmetry***
z	$z_{12} = z_{21}$	$z_{11} = z_{22}$
y	$y_{12} = y_{21}$	$y_{11} = y_{22}$
h	$h_{12} = -h_{21}$	$h_{11}h_{22} - h_{12}h_{21} = 1$
g	$g_{12} = -g_{21}$	$g_{11}g_{22} - g_{12}g_{21} = 1$
A, B, C, D	AD − BD = 1	A = D
A′, B′, C′, D′	A′ D′ − B′ C′ = 1	A′ = D′

Image Parameters: If the driving point impedance at port 1 with impedance Z_{12} connected across port 2 is Z_{11} and driving point impedance at port 2 with impedance Z_{11} connected across port 1 is Z_{12}, then Z_{11} and Z_{12} are image impedances of network.

$$Z_{11} = [(AB)/(CD)]^{0.5}$$

$$Z_{12} = [(BD)/(AC)]^{0.5}$$

Image transfer parameter

$$\theta = \log_e \left[\frac{V_1 I_1}{-V_2 I_2}\right]^{0.5}$$

$$= \cosh^{-1}\sqrt{AD} = \sinh^{-1}\sqrt{BC}$$

$$= \tanh^{-1}\sqrt{BC/AD}$$

$$Z_{11} = \sqrt{(Z_{oc1})(Z_{sc1})}$$

$$Z_{12} = \sqrt{(Z_{oc2})(Z_{sc2})}$$

$$\theta = \tanh^{-1}\sqrt{\frac{z_{sc1}}{z_{oc1}}} = \tanh^{-1}\sqrt{\frac{z_{sc2}}{z_{oc2}}}$$

Symmetrical network

$$Z_{11} = Z_{12} = Z_0$$

For a symmetrical network

$$\frac{V_1}{V_2} = \frac{I_1}{-I_2} = e^{\gamma}$$

$$\gamma = a + j\beta$$

ABCD parameters of a symmetrical network are

$$A = D = \cosh\gamma$$

$$B = Z_0 \sinh\gamma$$

$$C = \frac{1}{Z_0}\sinh\gamma$$

4.11. FILTER NETWORKS

$$\text{Attenuation} = \frac{1}{2}\log_e \frac{P_1}{P_2} \text{ nepers}$$

$$= \frac{1}{2}\log_e \left[\frac{I_1}{I_2}\right]^2 \text{ nepers}$$

$$\log_e e^{\alpha} = \alpha \text{ nepers}$$

Attenuation in decibels = 8.686 α

In pass band $-1 < \frac{Z_1}{4Z_2} < 0$

For constant k filters $Z_1 Z_2 = k = R_0^2$

Low Pass Filter

$$f_c = \frac{1}{\pi\sqrt{LC}} \qquad R_0 = \sqrt{L/C}$$

$$L = \frac{R_0}{\pi f_c} \qquad C = \frac{1}{\pi f_c R_0}$$

High Pass Filter

$$f_c = \frac{1}{4\pi\sqrt{LC}} \qquad R_0 = \sqrt{L/C}$$

$$L = \frac{R_0}{4\pi f_c} \qquad C = \frac{1}{4\pi f_c R_0}$$

Band Pass Filter

$$f_{c2} - f_{c1} = \frac{1}{\pi\sqrt{L_1 C_2}} \qquad \sqrt{\frac{L_2}{C_1}} = \sqrt{\frac{L_1}{C_2}} = R_0$$

$$L_1 = \frac{R_0}{\pi(f_{c2} - f_{c1})} \qquad C_1 = \frac{f_{c2} - f_{c1}}{4\pi R_0 f_{c2} f_{c1}}$$

$$L_2 = \frac{R_0(f_{c2} - f_{c1})}{4\pi f_{c2} f_{c1}} \qquad C_2 = \frac{1}{\pi R_0 (f_{c2} - f_{c1})}$$

Band Elimination Filter

$$f_{c2} - f_{c1} = \frac{1}{4\pi\sqrt{L_2 C_1}} \qquad \sqrt{\frac{L_1}{C_2}} = \sqrt{\frac{L_2}{C_1}} = R_0$$

$$L_1 = \frac{R_0 (f_{c2} - f_{c1})}{\pi(f_{c2})(f_{c1})} \qquad C_1 = \frac{1}{4\pi R_0 f_{c2} f_{c1}}$$

$$L_2 = \frac{R_0}{4\pi (f_{c2} - f_{c1})} \qquad C_1 = \frac{f_{c2} - f_{c1}}{\pi R_0 (f_{c2})(f_{c1})}$$

***m*-derived Low Pass Filter**

$$f_{\infty} = \frac{1}{\pi\sqrt{(1-m^2)LC}} = \frac{f_c}{\sqrt{1-m^2}}$$

***m*-derived High Pass Filter**

$$f_{\infty} = \frac{\sqrt{1-m^2}}{4\pi\sqrt{LC}} = \left(\sqrt{1-m^2}\right) f_c$$

4.12. NETWORK FUNCTIONS, POLES AND ZEROS

Network functions of two port network

Driving point impedance $Z(s) = \frac{V(s)}{I(s)}$

Driving point admittance $Y(s) = \frac{1}{Z(s)} = \frac{I(s)}{V(s)}$

Network functions of two port network

Driving point Impedance at port 1 = $Z_{11}(s) = \frac{V_1(s)}{I_1(s)}$

Driving point impedance at port 2 = $Z_{22}(s) = \frac{V_2(s)}{I_2(s)}$

Transfer function $G_{12}(s) = \frac{V_2(s)}{V_1(s)}$

Transfer function $Z_{12}(s) = \frac{V_2(s)}{I_1(s)}$

Transfer function $Y_{12}(s) = \frac{-I_2(s)}{V_1(s)}$

Transfer function $\alpha_{12}(s) = \frac{-I_2(s)}{I_1(s)}$

Restrictions on Pole-zero locations for Driving Point Functions

1. The coefficients in the polynomials $p(s)$ and $q(s)$ of the network function $N(s) = \frac{p(s)}{q(s)}$ must be real and positive.
2. Poles and zeros must be conjugate if complex or imaginary.
3. The real part of all poles and zeros must be either negative or zero.
4. If the real part of a pole or zero is zero, that pole or zero must be simple.
5. The polynomials $p(s)$ and $q(s)$ should not have missing terms between the highest and lowest degree unless all odd or even terms are missing.
6. The degrees of $p(s)$ and $q(s)$ may differ by either zero or one only.
7. The terms of the lowest degree in $p(s)$ and $q(s)$ may differ in degree by one at the most.

Restrictions on Pole-zero locations for Transfer Functions

1. The coefficients of the different terms in the polynomials $p(s)$ and $q(s)$ must be real and those for $q(s)$ must be positive.
2. If the poles and zeros are complex or imaginary they must be conjugate.
3. The real part of pole must be negative or zero. If the real part is zero, then that pole must be simple.
4. The polynomial $q(s)$ should not have any missing term between those of highest and lowest degrees unless all the even or odd terms are missing.
5. The polynomial $p(s)$ may have missing terms between those of highest and lowest degrees.
6. The degree of $p(s)$ may be as small as zero, independent of the degree of $q(s)$.
7. For Z_{12} and Y_{12}, the maximum degree of $p(s)$ is the degree of $q(s)$ plus one.
8. For $G_{12}(s)$ and $\alpha_{12}(s)$ the maximum degree of $p(s)$ is the degree of $q(s)$.

4.13. SYMMETRICAL COMPONENTS

$$\begin{aligned}
\text{Operator}\quad a &= 1\angle 120° = -0.5 + j0.866\\
\text{Operator}\quad a^2 &= 1\angle 240° = -0.5 - j0.866\\
\text{Operator}\quad a^3 &= 1\angle 360° = 1\\
\text{Operator}\quad a^4 &= 1\angle 480° = a\\
1 + a &= 1\angle 60° = 0.5 + j0.866 = -a^2\\
1 - a &= \sqrt{3}\ \angle -30° = 1.5 - j0.866\\
1 + a^2 &= 1\angle -60° = 0.5 - j0.866 = -a\\
1 - a^2 &= \sqrt{3}\ \angle 30° = 1.5 + j0.866\\
1 + a + a^2 &= 0
\end{aligned}$$

$$\begin{bmatrix} V_a \\ V_b \\ V_c \end{bmatrix} = \begin{bmatrix} 1 & 1 & 1 \\ 1 & a^2 & a \\ 1 & a & a^2 \end{bmatrix} \begin{bmatrix} Va_0 \\ Va_1 \\ Va_2 \end{bmatrix}$$

$$\begin{bmatrix} V_{a0} \\ V_{a1} \\ V_{a2} \end{bmatrix} = \frac{1}{3}\begin{bmatrix} 1 & 1 & 1 \\ 1 & a & a^2 \\ 1 & a^2 & a \end{bmatrix} \begin{bmatrix} Va \\ Vb \\ Vc \end{bmatrix}$$

$$S = 3\,(V_{a0}\, I_{a0}^{*} + V_{a1}\, I_{a1}^{*} + V_{a2}^{*}\, I_{a2}^{*})$$

4.14 NETWORK SYNTHESIS

Theorem 1: H(s) is a positive real function if and only if

(i) H(s) is real when s is real
(ii) $\text{Re}[H(s)] \geq 0$ when $\text{Re}(s) \geq 0$

Theorem 2: A practical and useful set of necessary and sufficient conditions for H(s) to be positive real is

(i) H(s) has no poles and zeros in the right half s plane, *i.e.*, its poles and zeros can only be of the form $-\sigma \pm j\omega$ where σ and ω are real and positive.
(ii) The poles of $H(s)$ on imaginary axis are simple with real and positive residues.
(iii) $\text{Re}[H(j\omega)] \geq 0$ for $0 \leq \omega \leq \infty$

Hurwitz Polynomial: A Hurwitz polynomial possesses the following properties:

(i) All the coefficients are non-negative.
(ii) Both its even and odd parts have roots on the $j\omega$ axis.
(iii) The continued fraction expansion of the ratio of odd to even or even to odd parts of a Hurwitz polynomial yields all positive quotient terms.

Properties of LC immittance H(s) (*i.e.* Reactance function)

(i) All poles and zeros are simple and lie on $j\omega$ axis only
(ii) It has a pole or zero at origin.
(iii) It has a pole or zero at infinity.
(iv) Its poles and zeros alternate on the $j\omega$ axis.
(v) It is always a quotient of even to odd or odd to even polynomials. If the odd polynomial is numerator, then H(s) has a pole at $s = 0$.
(vi) $\text{Re}[H(j\omega)] = 0$ for $\omega \geq 0$
(vii) The residues at all poles are real and positive.
(viii) The slope of $dH(j\omega)/d\omega$ is always positive.

Properties of RC Impedance or RL Admittance Functions

(i) All poles are simple and lie on negative real axis.
(ii) The poles and zeros are interlaced.
(iii) The lowest critical frequency, (*i.e.*, the one nearest the origin) is a pole.
(iv) The highest critical frequency is a zero.
(v) The residues at all poles must be real and positive.
(vi) The slope $dZ/d\sigma$ is negative.
(vii) $Z(\infty) \leq Z(0)$.

Properties of RL Impedance or RC Admittance Functions

(*i*) The properties of RL admittances are the same as the properties of RC Impedances. The properties of RC admittances are the same as the properties of RL impedances.

(*ii*) If a given driving point impedance function is positive real, has all its poles and zeros on the negative real axis in the *s* plane with poles and zeros interlaced and the smallest and highest critical frequencies are of opposite kinds (*i.e.*, one pole and the other zero), then Z(*s*) can be realized as an RL network or RC network but not both.

(*iii*) If the smallest critical frequency of Z(*s*) is a pole Z(*s*) can be realised as an RC network. If the smallest critical frequency is a zero, it can be realized as an RL network.

(*iv*) An RL impedance function can be realized as RC admittance function. An RL admittance function can be realized as an RC impedance function.

Brune's Synthesis of RLC Networks

The synthesis of minimum function was first suggested by 0 Brune. A positive real function is a minimum function if

(*i*) It has no poles or zeros on the imaginary axis of *s* plane.

(*ii*) It has real, finite and positive values at $s = 0$ and $s = \infty$

(*iii*) Its real part vanishes for atleast one finite frequency, *i.e.*, $\text{Re}[Z(j\omega_0)] = 0$ for some $w_0 \geq 0$.

This synthesis procedure requires the use of mutually coupled inductors.

Bott Duffin Synthesis of RLC Networks

This synthesis procedure does not require mutually coupled inductors. It uses Richard's transformation. R Richard showed that if Z(*s*) is a positive real function, then R(*s*) is also positive real where

$$R(s) = \frac{kZ(s) - sZ(k)}{kZ(k) - sZ(s)}$$

and *k* has only positive real values.

OBJECTIVE TYPE QUESTIONS

A. Choose the most appropriate alternative

1. The drift velocity of electrons is
   (*a*) very small as compared to speed of light
   (*b*) equal to speed of light
   (*c*) almost equal to speed of light
   (*d*) half the speed of light
2. If absolute potential of *A* is –8 V and that of *B* is –16 V, V_{AB} is equal to
   (*a*) 24 V (*b*) 8 V
   (*c*) –8 V (*d*) –24 V
3. One ampere means the flow of
   (*a*) One coulomb each minute
   (*b*) One electron per second
   (*c*) One coulomb per second
   (*d*) One coulomb each hour
4. In metals the flow of current is due to
   (*a*) electrons (*b*) ions
   (*c*) holes (*d*) electrons and ions
5. In gases the flow of current is due to
   (*a*) positive ions only (*b*) electrons only
   (*c*) electrons, positive ions and negative ions
   (*d*) electrons and positive ions
6. One coulomb charge is equal to the charge on
   (*a*) 6.24×10^{18} electrons (*b*) 6.24×10^{24} electrons
   (*c*) 6.24×10^{18} atoms (*d*) none of the above
7. A resistor is rated at 0.75 watt, 1 kΩ. Its current rating is
   (*a*) 1 mA (*b*) 27.386 mA
   (*c*) 1.268 mA (*d*) 5.393 mA
8. One commercial unit of energy is equal to
   (*a*) 1 watt-sec (*b*) 1 kilowatt-sec
   (*c*) 1 kilowatt-hour (*d*) One watt-hour
9. If J, I and a denote current density, current and area of cross-section of a conductor, then
   (*a*) $J = Ia$ (*b*) $J = I/a^2$
   (*c*) $J = I/a$ (*d*) $3 = Ia^2$
10. If Δs is an element of area and ΔI is the current through this element, current density J is defined as
    (*a*) $J = \Delta I/\Delta s$ (*b*) $J = \lim_{\Delta s \to 0} \Delta I/\Delta s$
    (*c*) $J = \lim_{\Delta s \to \infty} \Delta I/\Delta s$ (*d*) none of the above
11. A current is flowing through a conductor with non-uniform area of cross-section. Then
    (*a*) current will be different at different cross-sections
    (*b*) current will be the same at all the cross-sections
    (*c*) current will be different but current density will be same at all the cross-sections
    (*d*) current will be the same but current density will be different at different cross-sections
12. When a current flows through a conductor,
    (*a*) an electric field exists at all points within the conductor
    (*b*) electric field exists only at all points on the surface of the conductor
    (*c*) electric field exists only at ends of the conductor
    (*d*) any of above depending on magnitude of current
13. Ohm's law is valid for
    (*a*) dc voltage and current
    (*b*) ac voltage and current
    (*c*) all voltages and currents
    (*d*) dc and ac voltages and currents
14. For a non-linear resistance
    (*a*) *v-i* plot is not a straight line
    (*b*) resistance increases as current increases

(c) resistance decreases as current increases
(d) current decreases as voltage is increased

15. The correct relation between energy and charge is
(a) Energy = voltage/charge
(b) Charge = energy × voltage
(c) Energy = voltage (charge)$^{0.5}$
(d) Energy = voltage × charge

16. The maximum current and voltage ratings for a 0.5 W, 10 kΩ resistor are
(a) 7.07 mA and 70.7 V
(b) 70.7 mA and 70.7 V
(c) 0.707 mA and 70.7 V
(d) 7.07 mA and 70.7 × 10^3 V

17. The correct relation between joules and calories is
(a) calories = joules/4.156
(b) joules = calories/4.156
(c) calories = joules × 10^7/4.156
(d) joules = calories × 10^7/4.156

18. In the circuit of Fig. 4.1, the source supplies, 2.25 mA. The value of R must be

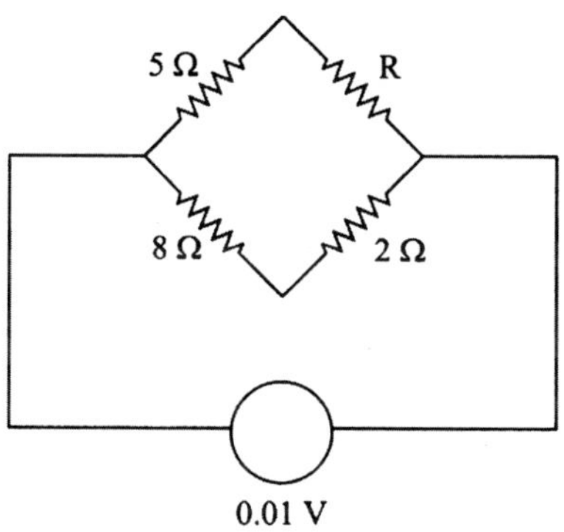

Fig. 4.1.

(a) 5 Ω (b) 3 Ω
(c) 1.5 Ω (d) 0.5 Ω

19. The electrical energy required to heat a bucket of water to a certain temperature is 2 kWh. If heat losses, are 25%, the energy input is
(a) 2.67 kWh (b) 3 kWh
(c) 2.5 kWh (d) 3.5 kWh

20. The resistance of human body is about
(a) 5000 Ω (b) 2000 Ω
(c) 1000 Ω (d) 200 Ω

21. As the diameter of a wire is doubled, the resistance becomes
(a) twice (b) one-half
(c) one-fourth (d) four times

22. Two resistances of 2 Ω each are connected in series. A shunt having zero resistance is connected across the combination. The overall resistance will be
(a) 4 Ω (b) 1 Ω
(c) 0.5 Ω (d) zero

23. A specimen of a material has a resistivity of 1.78 × 10^8 ohm-m. The material is
(a) semiconductor
(b) conductor
(c) semiconductor or conductor
(d) insulator

24. As temperature increases, the temperature coefficient of a metal
(a) decreases (b) increases
(c) remains unchanged (d) any of above

25. The resistance of a 230 V, 100 W lamp is
(a) 529 Ω (b) 2300 Ω
(c) 5290 Ω (d) 23 Ω

26. Which of the following is the poorest conductor of electricity
(a) copper (b) aluminium
(c) manganin (d) nichrome

27. If the diameter of a wire is doubled, the current carrying capacity of the wire becomes
(a) half (b) twice
(c) four times (d) one-fourth

28. There resistances of 15 Ω, 10 Ω and 3 Ω are connected in parallel. The overall resistance is
(a) less than 3 Ω (b) between 15 and 10 Ω
(c) between 10 and 3 Ω (d) more that 3 Ω

29. Nichrome is used in
(a) heater coils (b) underground cables
(c) overhead wires (d) lamp filaments

30. The colour code of a carbon resistance is yellow-violet-orange silver. The value of resistance is
(a) 57000 Ω ± 5% (b) 47000 Ω ± 10%
(c) 47000 Ω + 5% (d) 57000 Ω ± 5%

31. Lead becomes a superconductor at
(a) 9.1 K (b) 7 K
(c) 5 K (d) 3 K

32. The *i*-*v* relation for a varistor is
(a) $i = kv^n$ (b) $v = ki^n$
(c) $i = kv^3$ (d) $v = ki^3$

33. Figure 4.2 shows the *i*-*v* characteristics of a material. The material is likely to be

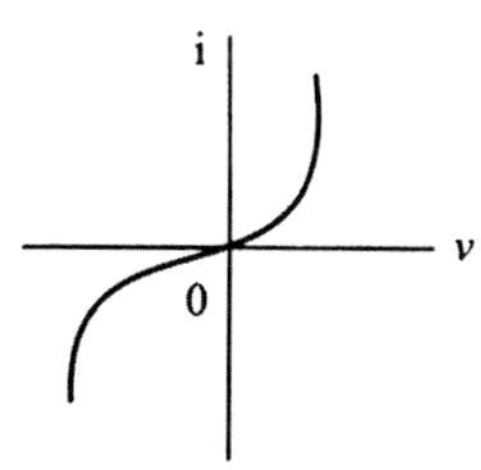

Fig. 4.2.

(a) copper (b) carbon
(c) zinc oxide (d) rubber

34. A conductor has a cylindrical shape with a diameter *d*. Its resistance is 2 Ω. It is recast into a cylindrical conductor of diameter 0.3*d*. The resistance of new conductor is likely to be

(*a*) less than 2 Ω (*b*) less than 0.6 Ω
(*c*) between 2 and 100 Ω (*d*) more than 100 Ω

35. In Fig. 4.3, the value of R should be

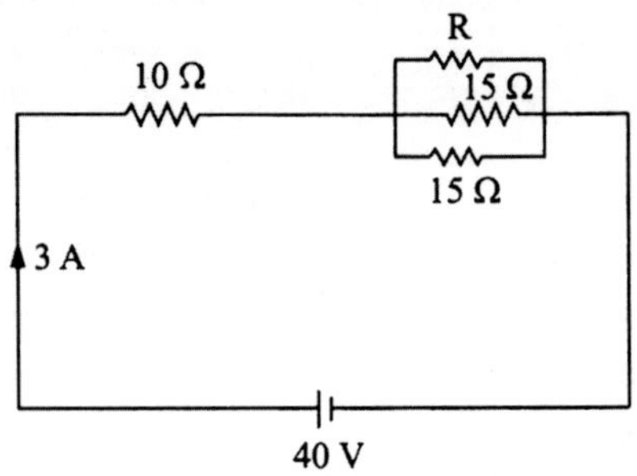

Fig. 4.3.

(*a*) 12 Ω (*b*) 6 Ω
(*c*) 3 Ω (*d*) 1.5 Ω

36. A thermistor is used for
(*a*) over voltage protection (*b*) temperature alarm circuit
(*c*) automatic light control (*d*) none of the above

37. A capacitor stores 0.15 C at 5V. Its capacitance is
(*a*) 0.75 F (*b*) 0.75 μF
(*c*) 0.03 F (*d*) 0.03 μF

38. For making a capacitor, the dielectric should have
(*a*) high $\in_r$
(*b*) low $\in_r$
(*c*) $\in_r = 1$
(*d*) $\in_r$ neither too high nor too low

39. If a sheet of a mica is inserted between the plates of an air capacitor, the capacitance
(*a*) will increase
(*b*) will remain the same
(*c*) will decrease
(*d*) may increase or decrease

40. A 0.5 μF capacitor is connected across a 10 V battery. After a long time, the circuit current and voltage across capacitor will be
(*a*) 0.5 A and 0 V (*b*) 0 A and 10 V
(*c*) 20 A and 5 V (*d*) 0.05 A and 5 V

41. A capacitor can be represented by a capacitance C in parallel with a resistance R. For a good capacitor, the resistance R should be
(*a*) very high (*b*) high
(*c*) very low (*d*) low

42. Cells are connected in series to increase
(*a*) current-rating
(*b*) both voltage and current rating
(*c*) the life of cell
(*d*) voltage rating

43. The storage battery used in power stations and substations is generally
(*a*) nickle cadmium battery
(*b*) lead acid battery
(*c*) zinc carbon battery
(*d*) zinc carbon or lead acid battery

44. When a lead acid battery is being charged, the specific gravity of the electrolyte
(*a*) decreases
(*b*) increases
(*c*) either increases or decreases
(*d*) remains the same

45. The output voltage of a charger is
(*a*) less than battery voltage
(*b*) same as battery voltage
(*c*) more than battery voltage
(*d*) same or more than battery voltage

46. The capacity of a battery is expressed in terms of
(*a*) current (*b*) voltage
(*c*) Ah (*d*) current and voltage

47. When a lead acid battery is fully charged the colours of its positive and negative plates are
(*a*) grey and dark brown (*b*) dark grey and brown
(*c*) grey and brown (*d*) dark brown and grey

48. In a lead acid battery
(*a*) the numbers of positive and negative plates are equal
(*b*) the number of positive plates may be equal to or more than the number of negative plates
(*c*) the number of positive plates is more than the number of negative plates
(*d*) the number of negative plates is more than the number of positive plates

49. The active elements on the positive and negative plates of lead acid battery are
(*a*) lead peroxide and lead
(*b*) lead and lead peroxide
(*c*) lead peroxide and lead sulphate
(*d*) lead and lead

50. In Fig. 4.4 the open circuit emf is

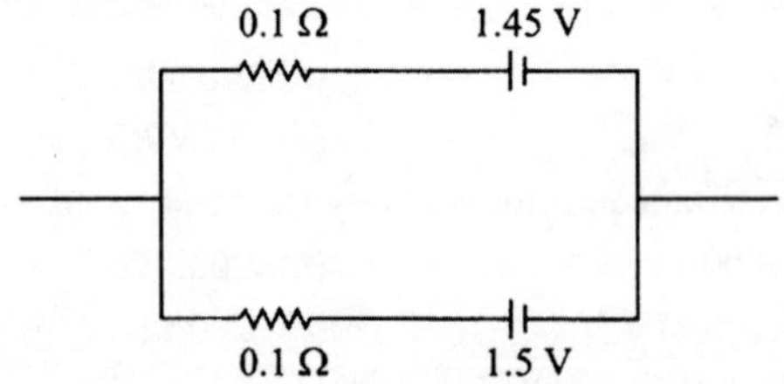

Fig. 4.4.

(*a*) 1.5 V (*b*) 1.45 V
(*c*) 1.475 V (*d*) 1.485 V

51. A parallel plate capacitor with air as dielectric has a capacitance C. If the dielectric has a dielectric constant $\in_r$ and the remaining dimensions are the same, the capacitance will be
(*a*) $C/\in_r$ (*b*) $\in_r C$
(*c*) $\in_r^2 C$ (*d*) $\in_r^3 C$

52. A capacitor having capacitance C is raised to voltage V. The energy stored will be
(*a*) CV^2 (*b*) V^2/C
(*c*) $0.5V^2C$ (*d*) $0.5C^2V$

53. Two capacitors of 20 μF each are in series. The total capacitance is
(*a*) 5 μF (*b*) 10 μF
(*c*) 40 μF (*d*) 400 μF

54. A parallel plate capacitor has plate area A and distance between the plates is *d*. It has two dielectrics each of thickness *d* and area 0.5 A. The dielectric constants are $\in_1$ and $\in_2$. The total capacitance is given by the equation

(*a*) $C = \frac{0.5 \in_0 A}{d}(\in_1 + \in_2)$

(*b*) $C = \frac{\in_0 A}{d}(\in_1 + \in_2)$

(*c*) $\frac{1}{C} = \frac{d}{2\in_0 A}\left(\frac{1}{\in_1} + \frac{1}{\in_2}\right)$

(*d*) $\frac{1}{C} = \frac{2d}{\in_0 A}\left(\frac{1}{\in_1} + \frac{1}{\in_2}\right)$

55. Two capacitors each of capacitance C and breakdown voltage V are joined in series. The capacitance and breakdown voltage of the combination is
(*a*) 0.5 C and 2 V (*b*) 0.5 C and 0.5 V
(*c*) C and V (*d*) 2 C and 2 V

56. The distance between the plates of a parallel plate capacitor is *d*. The dielectric is in two parts, each of equal thickness but dielectric constants $\in_1$ and $\in_2$ respectively. The total capacitance will be proportional to

(*a*) $\frac{\in_1 \in_2}{\in_1 + \in_2}$ (*b*) $\in_1 - \in_2$

(*c*) $\frac{\in_1}{\in_2}$ (*d*) $\in_1 \in_2$

57. A copper coil has a resistance of 40 ohm at 20°C. At 100°C the resistance is likely to be
(*a*) 44 Ω (*b*) 54 Ω
(*c*) 84 Ω (*d*) 104 Ω

58. A wire has a resistance R ohms. If another wire of the same material and same weight has double the diameter (as compared to the first wire), the resistance of the second wire will be
(*a*) 0.5 R (*b*) 0.25 R
(*c*) 0.125 R (*d*) R/16

59. A wire has a diameter *d* and resistance R. If the diameter is made 3*d*, the length and material remaining the same, the new value of resistance will be
(*a*) R/3 (*b*) 3 R
(*c*) R/9 (*d*) 9 R

60. A cable has an insulation resistance of 10^{10} ohms. If the length is doubled, the insulation resistance will be
(*a*) 2×10^{10} Ω (*b*) 0.5×10^{10} Ω
(*c*) 10^{10} Ω (*d*) 4×10^{10} Ω

61. The inductance of an air cored coil is 2 H. The number of turns is made 3 times, all other quantities remaining the same. The new value of inductance will be

(*a*) $\frac{2}{9}$ H (*b*) $\frac{2}{3}$ H
(*c*) 6 H (*d*) 18 H

62. The current rating of a cable depends on
(*a*) length of cable
(*b*) diameter of cable
(*c*) both length and diameter of cable
(*d*) none of the above

63. An air cored coil has an inductance of 0.3 H.
If another coil has an iron core of $\mu_r = 500$, the remaining quantities being the same, the inductance of the second coil will be
(*a*) 0.3 H (*b*) 0.3×500 H
(*c*) $\frac{0.3}{500}$ H (*d*) none of the above

64. Two capacitors of 1 μF and 2μF capacitance are connected in parallel across a 30 V dc battery. After the capacitors have bean charged, the charges across the two capacitors will be
(*a*) 30 μC each
(*b*) 60 μC each
(*c*) 30 μC and 60 μC respectively
(*d*) 60 μC and 30 μC respectively

65. Two capacitors of 1 μF and 2 μF are connected in series across a 30 V dc battery. After the capacitors have been charged, the charges across the two capacitors will be
(*a*) 10 μC each (*b*) 20 μC each
(*c*) 10 μC and 20 μC (*d*) 20 μC and 10 μC

66. Two capacitors of 16 μF each are connected in series. Another capacitor of 16 μF is connected in parallel with this combination. The total capacitance will be
(*a*) 48 μF (*b*) 32 μF
(*c*) 24 μF (*d*) $\frac{32}{3}$ μF

67. A square current wave of magnitude 1 A is applied to an element for 5 seconds. The voltage across the element rises linearly from 0 to final value linearly. The element is
(*a*) resistance (*b*) inductance
(*c*) capacitance (*d*) R and C in series

68. Two capacitors of 2 μF and 4 μF capacitance are connected in series across a 30 V dc battery. After the capacitors have been charged, the voltages across them will be
(*a*) 15 V each
(*b*) 10 V and 20V respectively
(*c*) 20 V and 10 V respectively
(*d*) 30 V each

69. An ideal current source has
(*a*) infinite source resistance
(*b*) zero source resistance
(*c*) large value of source resistance
(*d*) finite value of source resistance

70. In a practical voltage source, the terminal voltage
(*a*) cannot be less than source voltage
(*b*) cannot be higher than source voltage
(*c*) is always less than source voltage
(*d*) is always equal to source voltage

71. Kirchhoff s laws are applicable to
(*a*) dc only
(*b*) ac sinusoidal wave only
(*c*) dc and ac sinusoidal waves
(*d*) all wave shapes

72. In a practical current source, the load current
(*a*) is always equal to source current
(*b*) is always less than source current
(*c*) is always equal to or less than source current
(*d*) cannot be higher than source current

73. The current through a 10 Ω resistance in a dc circuit having two emf sources is 3 A. If this resistance is changed to 20 Ω, the new value of current will be
(*a*) less than 3A
(*b*) equal to or less than 3A
(*c*) equal to or less or more than 3A
(*d*) 1.5 A

74. When determining thevenin's resistance of a circuit
(*a*) all sources must be open-circuited
(*b*) all sources must be short-circuited
(*c*) all voltage sources must be open-circuited and all current sources must be short-circuited
(*d*) all sources must be replaced by their internal resistances

75. A source is delivering maximum power to a resistance through a network. The ratio of power delivered to the source power.
(*a*) is always 0.5 (*b*) may be 0.5 or less
(*c*) may be 0.5 or less or more (*d*) may be 0.5 or more

76. The units for electrochemical equivalent are
(*a*) grams (*b*) grams per coulomb
(*c*) gram-coulomb (*d*) grams per ampere

77. A circuit is replaced by its Thevenin's equivalent to find current through a certain branch. If V_{TH} = 10 V and R_{TH} = 20 Ω, the current through the branch
(*a*) will always be 0.5 A
(*b*) will always be less than 0.5 A
(*c*) will always be equal to or less than 0.5 A
(*d*) may be 0.5 A or more or less

78. The electrochemical equivalent of a material
(*a*) depends on its valency
(*b*) depends on its atomic weight
(*c*) depends on its valency and atomic weight
(*d*) depends on the degree of hardness of the material

79. In an electrochemical process the mass of material deposited at an electrode is proportional to
(*a*) time for which current is passed
(*b*) magnitude of current
(*c*) electric charge through the electrolyte
(*d*) none of the above

80. To determine the polarity of voltage drop across a resistor, it is sufficient to know
(*a*) direction of current through the resistor
(*b*) direction and magnitude of current through the resistor
(*c*) direction and magnitude of current and value of resistance
(*d*) magnitude of current through the resistance

81. Three resistance of 15 Ω each are connected in delta. The resistances of equivalent star will have a value of
(*a*) 15 Ω (*b*) 5 Ω
(*c*) 5/3 Ω (*d*) 45 Ω

82. If $V = 100 \angle 16°$, the $\sqrt{(V)}$ is
(*a*) $10 \angle 32°$ (*b*) $10 \angle 8°$
(*c*) $10 \angle 16°$ (*d*) $10 \angle 4°$

83. A heater is rated at 230 V, 5 A ac. Then
(*a*) both the current and voltage are peak values
(*b*) both the current and voltage are rms values
(*c*) the current is peak value and voltage is rms value
(*d*) the current is rms value and voltage is peak value

84. Two voltages are $v_1 = 100 \sin(\omega t + 15°)$ and $v_2 = 60 \cos \omega t$, then
(*a*) v_1 is leading v_2 by 15° (*b*) v_1 is leading v_2 by 75°
(*c*) v_2 is leading v_1 by 75° (*d*) v_2 is laading v_1 by 15°

85. A capacitor is needed for an ac circuit of 230 V, 50 Hz. The peak voltage rating of the capacitor should be
(*a*) 230 V (*b*) 0.5 × 230 V
(*c*) $\sqrt{2}$ × 230 V (*d*) $230/\sqrt{2}$ V

86. If $V = 20 + j20$, then $\log_e A$ is
(*a*) $3.342 + j45$ (*b*) $28.28 \angle 45°$
(*c*) $400 \angle 45°$ (*d*) $3.342 + j0.785$

87. A sinusoidal voltage has peak-to-peak value of 100 V. The rms value is
(*a*) 50 V (*b*) 70.7 V
(*c*) 35.35 V (*d*) 141.41 V

88. If $A = 7 \angle 15$, A^2 is
(*a*) $49 \angle 225°$ (*b*) $49 \angle 30°$
(*c*) $49 \angle 15°$ (*d*) $49 \angle 7.5°$

89. In a resistive ac circuit the current.......... the voltage by............
(*a*) lags, 90° (*b*) lags, 45°
(*c*) lags, 0° (*d*) leads 90°

90. In a purely inductive circuit the current.......... the voltage by............

(*a*) lags, 0° (*b*) leads, 90°

(*c*) lags, 90° (*d*) lags, 45°

91. In a purely capactive circuit, the current .... the voltage by ....

(*a*) lags, 0° (*b*) leads, 90°

(*c*) lags, 90° (*d*) lags, 45°

92. In an RL circuit, the angle θ by which the current lags the voltage is

(*a*) $0 > \theta > -90°$ (*b*) 90°

(*c*) 0° (*d*) $0 < \theta < 90°$

93. In an R-C circuit, the impedance is 40 Ω at a frequency of 100 Hz. At 200 Hz the impedance should be

(*a*) 40 Ω (*b*) 80 Ω

(*c*) 20 Ω

(*d*) more than 20 Ω but less than 40 Ω

94. In an R-L circuit, the impedance is 30 Ω at a frequency of 100 Hz. At 200 Hz, the impedance should be

(*a*) 30 Ω or more

(*b*) 60 Ω or more

(*c*) 15 Ω or less

(*d*) more than 30 Ω but less than 60 Ω

95. In an ac series RLC circuit the maximum phase difference between any two voltages can be

(*a*) 90° (*b*) 45°

(*c*) –90° (*d*) 180°

96. An RLC series circuit is excited by 200 V, 100 Hz supply. The current is 20 A. If $R = 10\ \Omega$, $X_L = 50\ \Omega$, X_C is

(*a*) 10 Ω (*b*) 14.14 Ω

(*c*) 7.07 Ω (*d*) 50 Ω

97. An RLC series circuit has $R = 30\ \Omega$, $X_L = 30\ \Omega$ and $X_C = 30\ \Omega$. Its impedance is

(*a*) $30\angle 0°$ (*b*) $30\angle 45°$

(*c*) $42.42\angle 45°$ (*d*) $90\angle 45°$

98. In an ac circuit, the maximum and minimum values of power factor can be

(*a*) 2 and 0 (*b*) 1 and 0

(*c*) 0 and –1 (*d*) 1 and –1

99. A choke coil has a resistance to reactance ratio of 0.2. The power factor of the choke coil is

(*a*) less than 0.2 lagging (*b*) less than 0.2 leading

(*c*) about 1 (*d*) about 0

100. Which statement is true for an ac circuit

(*a*) active power is always less than apparent power

(*b*) active power cannot be more than apparent power

(*c*) active power is always more than reactive power

(*d*) active power cannot be more than reactive power

101. In an RLC series circuit having finite values of R, L, C, the phase difference between applied voltage and current is

(*a*) always 0° (*b*) between 0 and 180°

(*c*) between 0 and 90° (*d*) always 90°

102. Two passive elements are connected in series and a dc voltage is applied to the circuit. The time variation of the current is shown in Fig. 4.5. The two elements are

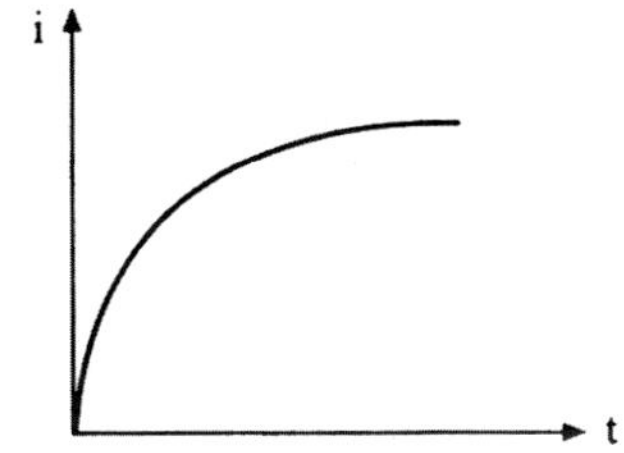

Fig. 4.5.

(*a*) one resistance and the other inductance

(*b*) both resistances

(*c*) one inductance and the other capacitance

(*d*) one resistance and the other capacitance

103. The resistance of the circuit shown is Fig. 4.6 is

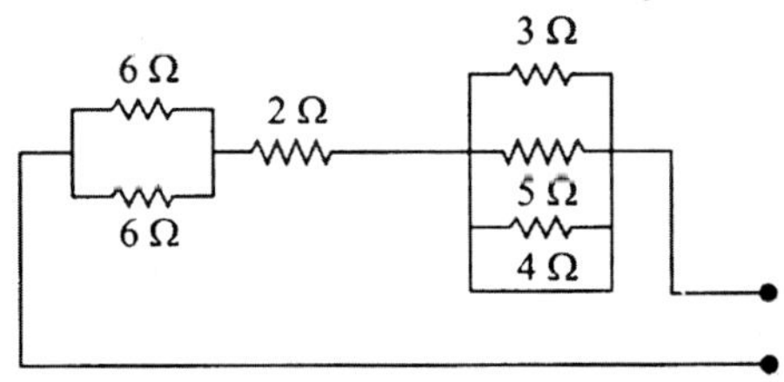

Fig. 4.6.

(*a*) 5 Ω (*b*) more than 6 Ω

(*c*) more than 4 Ω (*d*) between 6 Ω and 7 Ω

104. The current waveshape shown in Fig. 4.7 (*a*) is applied to a circuit element. The voltage across the element is shown in Fig. 4.7 (*b*). The element is

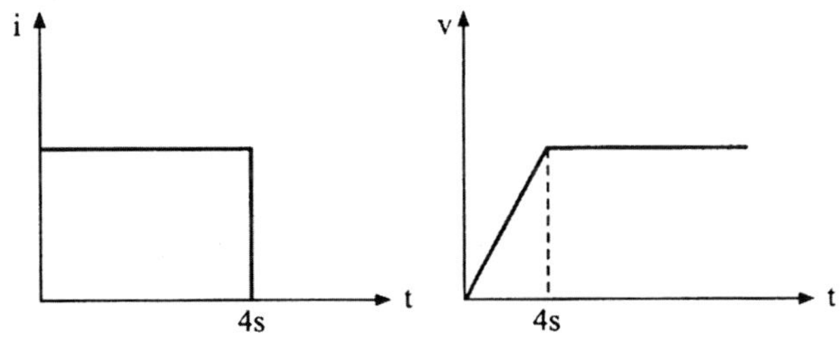

Fig. 4.7.

(*a*) R (*b*) L

(*c*) C (*d*) semiconductor

105. In Fig. 4.8, the charge on 0.8 μF capacitor is

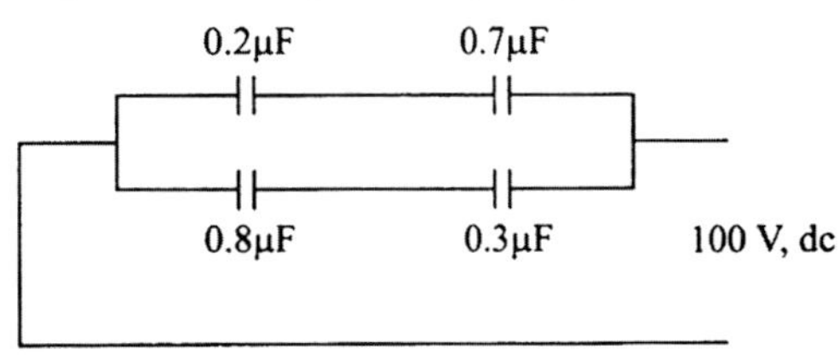

Fig. 4.8.

(*a*) 100 μC (*b*) 50 μC

(*c*) 40 μC (*d*) 20 μC

106. The temperature coefficient of copper is 0.00393 at 20°C. At 50°C the temperature coefficient is

(*a*) 0.0042 (*b*) 0.0035

(*c*) 0.0025 (*d*) 0.0015

107. A bulb rated at 60 W, 120 V is used for 30 minutes. The charge associated with this operation is
(*a*) 3600 C (*b*) 900 C
(*c*) 7200 C (*d*) 60 C

108. A 60 W, 230 V bulb is used for 45 minutes. The electrical energy consumed in commercial units is
(*a*) 2700 units (*b*) 0.45 units
(*c*) 4.5 units (*d*) 0.045 units

109. A thermal power plant has an efficiency of 0.4. If 0.6 kg of coal is burnt to generate 1 kWh, the calorific value of coal is
(*a*) about 3500 kcal/kg (*b*) about 2500 kcal/kg
(*c*) about 1500 kcal/kg (*d*) about 7500 kcal/kg

110. The resistance of the aluminium conductor ($\rho = 2.83 \times 10^{-8}$ Ω-m) in Fig. 4.9 is

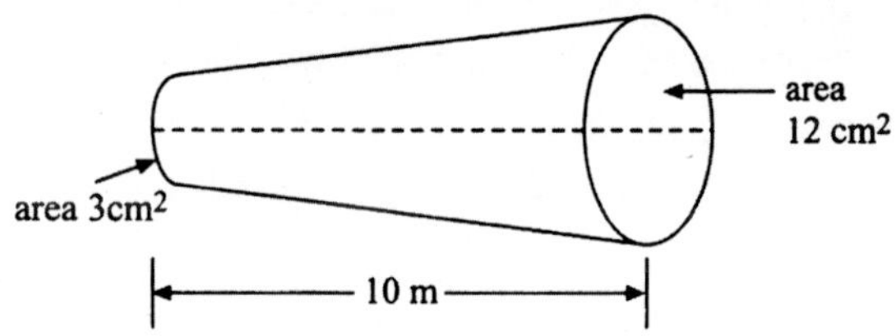

Fig. 4.9.

(*a*) 0.001 Ω (*b*) less than 0.001 Ω
(*c*) more than 0.001 Ω (*d*) more than 0.002 Ω

111. The frame of electric motor is connected to three plates having earthing resistance of 10 Ω, 20 Ω and 30 Ω respectively. The materials of the three plates are copper, aluminium and iron respectively. The percentage of energy dissipated by 10 Ω earthing plate
(*a*) is more than 50% of total energy
(*b*) is less than 50% of total energy
(*c*) depends on the materials of the three plates
(*d*) may be more or less than 50% of total energy

112. A sawtooth current wave is passed through a resistance
(*a*) the wave shapes of voltages across resistance and power will also be sawtooth
(*b*) the waveshape of voltage across resistance will be sawtooth
(*c*) the waveshape of power will be sawtooth
(*d*) neither the voltage waveshape nor power waveshape will be sawtooth

113. The voltage of single phase supply to residential customers in India is
(*a*) 400 V (*b*) 230 V
(*c*) 2.2 kV (*d*) 11 kV

114. The distributors for residential areas are
(*a*) single phase (*b*) three phase three wire
(*c*) three phase 4 wire (*d*) (*a*) or (*b*)

115. The minimum permissible size of aluminium cable for lighting circuit is
(*a*) 1.2 sq. mm (*b*) 1.5 sq. mm
(*c*) 2.2 sq. mm (*d*) 3.5 sq. mm

116. The purpose of earthing electric appliances is
(*a*) to provide safety against shock
(*b*) to ensure that appliance works satisfactorily
(*c*) to ensure that appliance gets full voltage
(*d*) all the above

117. The total lumens emitted by a source of 1 candela are
(*a*) 0.4 π (*b*) 4 π
(*c*) 2 π (*d*) 8 π

118. The function a choke in a fluorescent lamp is
(*a*) at the time of starting only
(*b*) during the working time but not at starting time
(*c*) both at the time of starting and during working
(*d*) none of the above

119. Sometimes the rotating devices in a room having fluorescent lamps appear to be stationany. This is due to
(*a*) defect in the lamp (*b*) low voltage
(*c*) high voltage (*d*) stroboscopic effect

120. In a tungsten filament lamp
(*a*) the current at the instant of switching is less than operating current
(*b*) the current at the instant of switching is more than operating current
(*c*) the current at the instant of switching is equal to operating current
(*d*) the current at the instant of switching may be less than or equal to operating current

121. When a source is delivering maximum power to a load through a network, the ratio of power delivered to the source power
(*a*) may be 50% or less than 50%
(*b*) is 50%
(*c*) may be 50% or more than 50%
(*d*) may be 50% or more or less than 50%

122. A resistance of 5 ohms is connected in one branch of a network. The current in this branch is 2.2 A. If the resistance is increased to 7.5 Ω, the remaining circuit being the same, the current in this branch will be
(*a*) 2.2 A
(*b*) less than 2.2 A
(*c*) 2.2 A or less than 2.2 A
(*d*) 2.2 A or less or more than 2.2 A

123. In a network
(*a*) the number of tree branches is always less than the number of links
(*b*) the number of tree branches is always more than the number of links
(*c*) the number of tree branches and links are always equal
(*d*) the number of tree branches has no relation with the number of links

124. A dependent source
(*a*) may be a current source or voltage source
(*b*) is neither a current source nor a voltage source
(*c*) is always a voltage source
(*d*) is always a current source

125. In the network shown in Fig. 4.10, the voltage drops across R_2 and R_3 are 10 V and 16 V with polarities as shown. The current in R_1

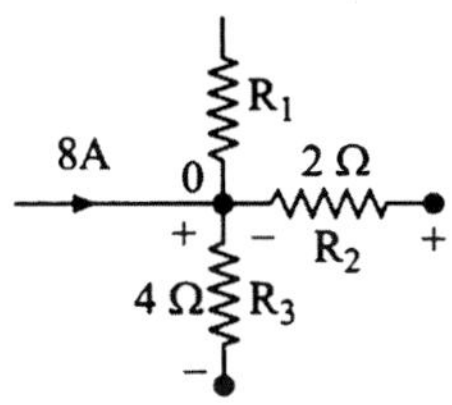

Fig. 4.10.

(*a*) is 9 A and directed towards node 0
(*b*) is 9 A and directed away from node 0
(*c*) is 9 A but direction of current cannot be found
(*d*) is impossible to be found

126. In the network of Fig. 4.11, the voltage drops across R_1 R_2, R_3 are 6 V, 8 V and 12 V with polarities as shown. The value of R_3 is

(*a*) 12 Ω (*b*) 8 Ω
(*c*) 5 Ω (*d*) 3 Ω

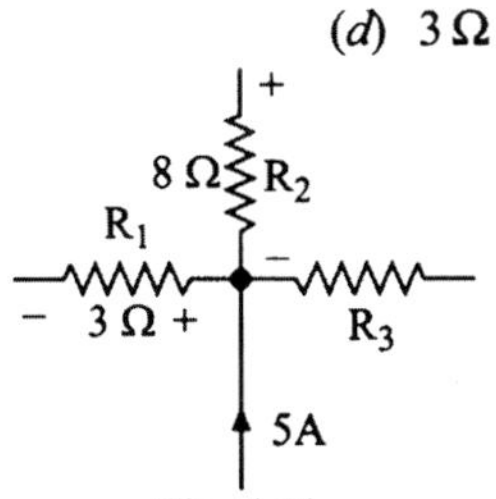

Fig. 4.11.

127. In Fig. 4.12 the voltage drop across the 10 Ω resistance is 10 V. The resistance R

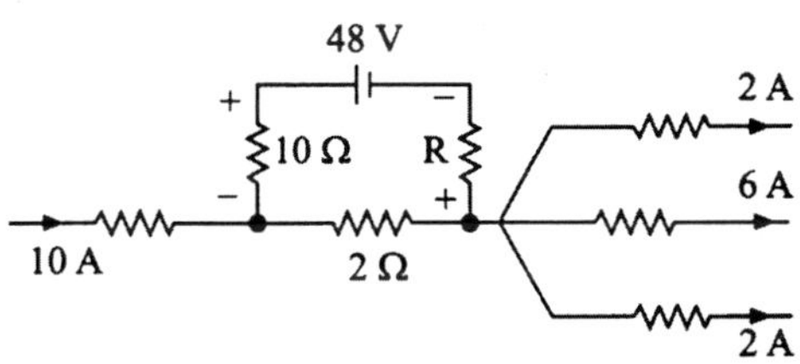

Fig. 4.12.

(*a*) is 16 Ω (*b*) is 8 Ω
(*c*) is 4 Ω (*d*) cannot be found

128. To find the current in a 20 Ω resistance connected is a circuit, Norton's theorem is used. I_N = 7.5 A. The current through 20 Ω resistance

(*a*) is 7.5 A (*b*) is less than 7.5 A
(*c*) may be 7.5 A or less
(*d*) may be 7.5 A or more or less

129. Maximum power transfer theorem finds application in

(*a*) power circuits (*b*) distribution circuits
(*c*) Communication circuits
(*d*) both power and communication circuits

130. In the circuit of Fig. 4.13, Thevenin's resistance is

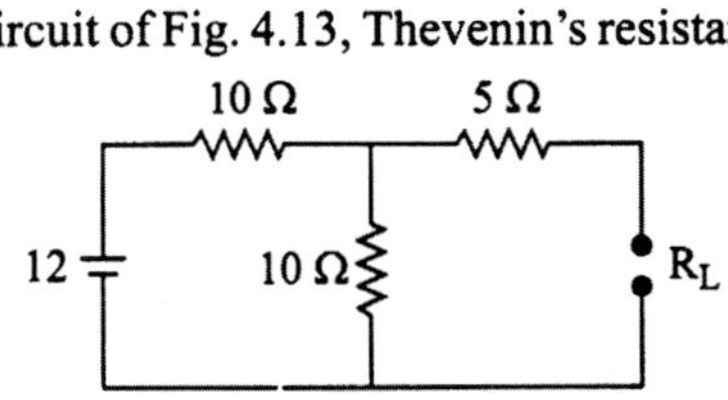

Fig. 4.13.

(*a*) 25 Ω (*b*) 10 Ω
(*c*) 13.33 Ω (*d*) 6 Ω

131. To find current in a resistance connected in a network, Thevenin's theorem is used. V_{TH} = 20 V and R_{TH} = 5 Ω. The current through the resistance

(*a*) is 4 A
(*b*) is 4 A or less
(*c*) is less than 4 A
(*d*) may be 4 A or less or more than 4 A

132. The input resistance of the network in Fig. 4.14 is

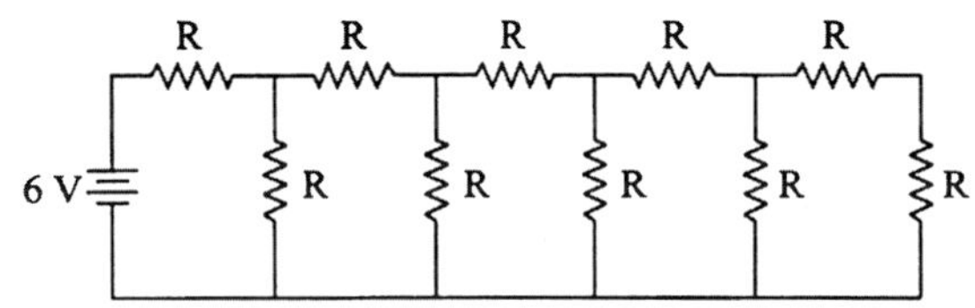

Fig. 4.14.

(*a*) 1.5 R (*b*) $\frac{13}{8}$ R
(*c*) $\frac{34}{21}$ R (*d*) $\frac{89}{55}$ R

133. A current of 1 A in the coil of an air cored electromagnet causes a flux density of 0.2 T. If the current is 2 A, B =

(*a*) 0.2 T (*b*) 0.4 T
(*c*) 0.4 T or less (*d*) 0.4 T or more

134. A current of 1 A in the coil of an iron cored electromagnet causes B = 0.5 T. If current is 2A, B =

(*a*) 0.5 T (*b*) 1 T
(*c*) 1 T or more (*d*) 1 T or less

135. Permeability is analogous to

(*a*) conductivity (*b*) resistivity
(*c*) retentivity (*d*) coercivity

136. A material for permanent magnet should have

(*a*) high retentivity (*b*) high permeability
(*c*) low hysteresis loss (*d*) high hysteresis loss

137. The relative permeability of an iron specimen is 400 when B = 0.2 T. If B = 0.4 T, μ_r will be

(*a*) 400 (*b*) less than 400
(*c*) more than 400 (*d*) more or less than 400

138. In an air cored coil a current of 0.7 A causes a flux of 5×10^{-6} Wb. If the core is made of cast iron and the coil is similar, the same current can set up a flux of about

(*a*) 5×10^{-6} Wb (*b*) 5×10^{-8} Wb
(*c*) 10^{-2} Wb (*d*) 5×10^{-4} Wb

139. Tesla is a unit of

(*a*) ϕ (*b*) B
(*c*) H (*d*) mmf

140. The unit of permeability is

(*a*) T (*b*) Wb
(*c*) H/M (*d*) H

141. In a specimen of cast iron, a field strength of 400 AT/m causes a flux density of 0.6 T. In a specimen of silicon steel, the same value of H would cause B to be

(*a*) 0.6 T (*b*) more than 0.6 T
(*c*) less than 0.6 T (*d*) 0.6 T or less than 0.6 T

142. As B is increased from zero to about 2 T, μ_r, of a ferromagnetic material

(*a*) increases
(*b*) decreases
(*c*) first increases then decreases
(*d*) remains the same

143. The self-inductances of two coils are 27 H and 3 H. If the winding of coils is such that 50% of flux of one links the other, the mutual inductance is

(*a*) 30 H (*b*) 24 H
(*c*) 9 H (*d*) 4.5 H

144. Two coils X and Y have self-inductances of 5 mH and 10 mH and mutual inductance of 3 mH. If the current in coil X changes at a steady rate of 100 A/s, the emf induced in coil Y is

(*a*) 0.3 V (*b*) 0.5 V
(*c*) 1 V (*d*) 1.5 V

145. Three inductors each 30 mH are connected in delta. The value of inductance of each arm of equivalent star is

(*a*) 10 mH (*b*) 15 mH
(*c*) 30 mH (*d*) 90 mH

146. Two coils are so wound that 60% of the flux of one links the other. The coefficient of coupling is

(*a*) 1.67 (*b*) 0.6
(*c*) 0.36 (*d*) 0.216

147. The total inductance of two coils is 60 mH in series aiding connection and 40 mH in series opposing connection. The mutual inductance between coils is

(*a*) 20 mH (*b*) 10 mH
(*c*) 5 mH (*d*) 2.5 mH

148. If the iron core of a coil is removed so that it becomes an air cored coil, its inductance

(*a*) will increase
(*b*) will decrease
(*c*) will remain the same
(*d*) may increase or decrease

149. Two coils are connected in parallel. Their fluxes are aiding. If the connections of one of the coils are reversed, the total inductance of the arrangement

(*a*) will decrease
(*b*) may increase or decrease
(*c*) will remain the same
(*d*) will increase

150. Three identical coils each having self-inductance L are connected in series so that the fluxes of all the three are in the same direction. If mutual inductance between each pair of coils is M, the total inductance of the arrangement is

(*a*) 3 L + 9 M (*b*) 1.5 L + 3 M
(*c*) 3 L + 3 M (*d*) 3 L + 6 M

151. A coil is wound with a wire having a diameter *d*. Its inductance is 10 mH. If it is rewound on the same core with a wire having diameter 0.5 *d* and total weight of the wire is the same as before, the inductance will be

(*a*) 5 mH (*b*) 10 mH
(*c*) 160 mH (*d*) 40 mH

152. Two coupled coils with self-inductances of 4 mH and 16 mH are connected in series. The maximum possible value of inductance can be

(*a*) 36 mH (*b*) 20 mH
(*c*) 28 mH (*d*) 64 mH

153. The Laplace transform of a function is $\frac{E}{s}e^{-as}$.

The function is

(*a*) E sin *at* (*b*) Ee^{-at}
(*c*) E*u*(*t* – *a*) (*d*) E cos *at*

154. If $f(t) = r(t-a)$, F(*s*) =

(*a*) e^{-as/s^2} (*b*) $a/(s+a)$
(*c*) $1/(s+a)$ (*d*) $e^{-as/s}$

155. A step voltage E is applied to an R-L series circuit. At *t* = 0, the current in the circuit is

(*a*) E/R (*b*) E/L
(*c*) infinity (*d*) zero

156. In an RC series circuit excited by a dc voltage E, the current at *t* = 0 is

(*a*) E/R (*b*) E/C
(*c*) zero (*d*) infinity

157. A step voltage E is applied to a series R-L circuit. The rate of change of current is maximum at *t* =

(*a*) zero (*b*) infinity
(*c*) L/R (*d*) R/L

158. In a series RLC circuit, the value of resistance for critical damping is

(*a*) L/C (*b*) $(L/C)^{0.5}$
(*c*) $1/\sqrt{LC}$ (*d*) $2/\sqrt{L/C}$

159. In an R-L-C series circuit $R = 2/\sqrt{L/C}$. If R is doubled

(*a*) The response will become underdamped
(*b*) The response will become overdamped
(*c*) The response will be critically damped
(*d*) The response may be under damped or over damped

160. An RLC series circuit is exited by a step voltage E. The circuit current at *t* = 0 is

(*a*) E/R (*b*) $E/\sqrt{LC}$
(*c*) zero (*d*) infinity

161. Elements R, L, C are connected in parallel and excited by a dc current I. If i_R, i_L and i_C are the currents through R, L, C respectively, then at *t* = 0

(*a*) $i_R > i_L > i_C$ (*b*) $i_L > i_R > i_C$
(*c*) $i_C = 0$ (*d*) $i_C = I$

162. A series RL circuit is excited by a voltage $v(t) = Ve^{-\alpha t}$ such that $\alpha = R/L$. Then the circuit current is

(*a*) $\frac{V}{R} te^{-\alpha t}$ (*b*) $\frac{V}{L} te^{-\alpha t}$

(*c*) $\frac{V}{R} e^{-\alpha t}$ (*d*) $\frac{V}{L} te^{\alpha t}$

163. A series RL circuit is excited by a voltage $v(t) = V_m \sin(\omega t + \phi)$. The transient current will be zero if

(*a*) $\phi = 0$ (*b*) $\omega = 0$

(*c*) $\phi = \tan^{-1}(\omega L/R)$ (*d*) $\phi - \tan^{-1}(\omega L/R) = \pi/2$

164. A unit impulse voltage $\delta(t - b)$ is applied to an inductor with inductance L. The current through L is

(*a*) $\frac{1}{L}\delta(t - b)$ (*b*) $\frac{1}{L}u(t - b)$

(*c*) $\sqrt{\delta(t - b)}$ (*d*) $\frac{t}{L}(t - b)$

165. A unit impulse current $\delta(t - b)$ is applied to a capacitance C. The voltage across capacitance is

(*a*) $\frac{1}{C}\delta(t - b)$ (*b*) $C\delta(t - b)$

(*c*) $Cu(t - b)$ (*d*) $\frac{1}{C}u(t - b)$

166. An RLC series circuit is excited by a step voltage E. The initial current through L is $i(0^-)$. In the *s* domain circuit the element L will be replaced by

(*a*) an impedance *s*L

(*b*) an impedance *s*L in series with current source $i(0^-)$

(*c*) an impedance *s*L in series with a source $\frac{1}{L}i(0^-)$

(*d*) an impedance *s*L in series with a voltage source L $i(0^-)$

167. The dual of quantity charge is

(*a*) current (*b*) voltage

(*c*) flux (*d*) $d\phi/dt$

168. The dual of a node is

(*a*) mesh (*b*) principal node

(*c*) tree branch (*d*) link

169. In Fig. 4.15 the switch is closed at $t = 0$. At $t = 0^+$. The value of current supplied by battery is

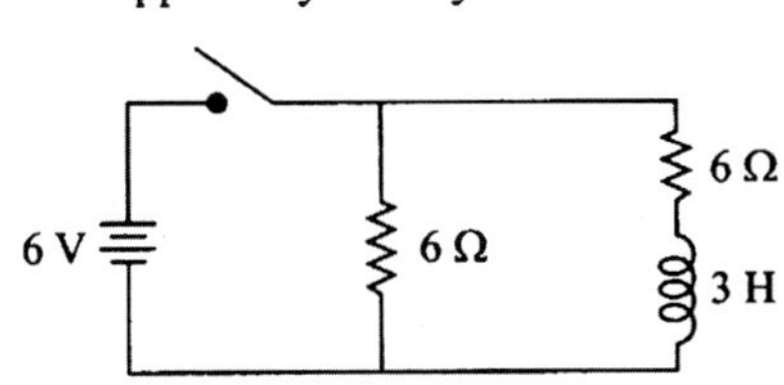

Fig. 4.15.

(*a*) 1A (*b*) 2 A

(*c*) zero (*d*) infinity

170. In Fig. 4.15 the battery current after a long time after closing of switch is

(*a*) 1 A (*b*) 2 A

(*c*) 0 (*d*) impossible to find

171. In Fig. 4.15 the current supplied by battery at $t = 0.5s$ is

(*a*) 1.1 A (*b*) 1.9 A

(*c*) 0 (*d*) 1.632 A

172. The function $\delta'(t - b)$ is a

(*a*) unit impulse (*b*) unit step

(*c*) unit doublet (*d*) none of above

173. In the circuit of Fig. 4.16 the switch is open for a long time. At $t = 0$ the switch is closed. At $t = 0^+$ the current supplied by battery is

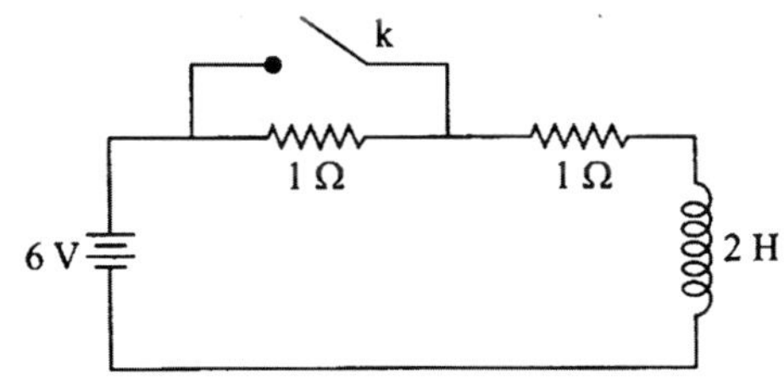

Fig. 4.16.

(*a*) 6 A (*b*) 3 A

(*c*) 3.6 A (*d*) 1.5 A

174. In the circuit of Fig. 4.17 the switch is closed at $t = 0$. At $t = 0^+$ the current through inductance is

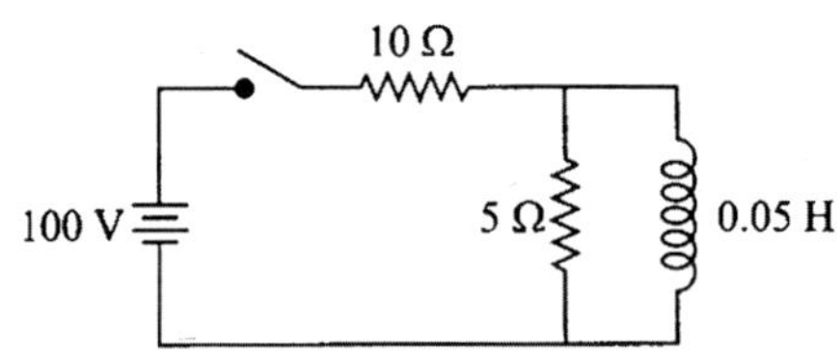

Fig. 4.17.

(*a*) 0 (*b*) 10 A

(*c*) 6.67 A (*d*) 5.1 A

175. In the circuit of Fig. 4.17 the current through 5 Ω resistance at $t = 0^+$ is

(*a*) 0 A (*b*) 10 A

(*c*) 6.67 A (*d*) 5.1 A

176. In the circuit of Fig. 4.17 the current through L at $t = \infty$ is

(*a*) 0 A (*b*) 10 A

(*c*) 6.67 A (*d*) 5.1 A

177. In the circuit of Fig. 4.17 the current through 5 Ω resistance at $t = \infty$ is

(*a*) 0 A (*b*) 10 A

(*c*) 6.61 A (*d*) 5.1 A

178. In the circuit of Fig. 4.18, the switch is closed at $t = 0$. At $t = 0^+$ the current through C is

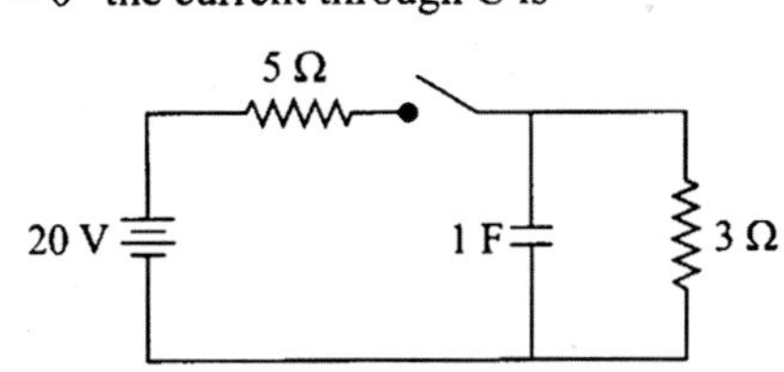

Fig. 4.18.

(*a*) 4 A (*b*) 2.5 A

(*c*) 3.1 A (*d*) 0

179. In the circuit of Fig. 4.18 the current through C at $f = \infty$ is

(a) 4 A (b) 2.5 A
(c) 3.1 A (d) 0

180. In the circuit of Fig. 4.18 the current through 3 Ω resistance at $t = 0^+$ is

(a) 4 A (b) 2.5 A
(c) 3.1 A (d) 0

181. In the circuit of Fig. 4.18 the current through 3 Ω resistance at $t = \infty$ is

(a) 4 A (b) 2.5 A
(c) 3.1 A (d) 0 A

182. In the circuit of Fig. 4.18 the voltage across C at $t = 0^+$ is

(a) 0V (b) 20V
(c) very large (d) 7.5V

183. In the circuit of Fig. 4.18 the voltage across capacitor at $t = \infty$ is

(a) 0V (b) 20V
(c) very large (d) 7.5V

184. In the circuit of Fig. 4.18 the current through 5 Ω resistance at $t = 0^+$ is

(a) 4 A (b) 2.5 A
(c) 3.1 A (d) 0 A

185. In the circuit of Fig. 4.18 the current through 5 Ω resistance at $t = \infty$ is

(a) 4 A (b) 2.5 A
(c) 3.1 A (d) 0 A

186. An RLC series circuit is excited by a step voltage E. The units of roots of characteristic equation are

(a) seconds (b) 1/seconds
(c) Tesla (d) henry

187. In the circuit of Fig. 4.19, the switch is closed at $t = 0$. At $t = 0^+$ the current supplied by battery is

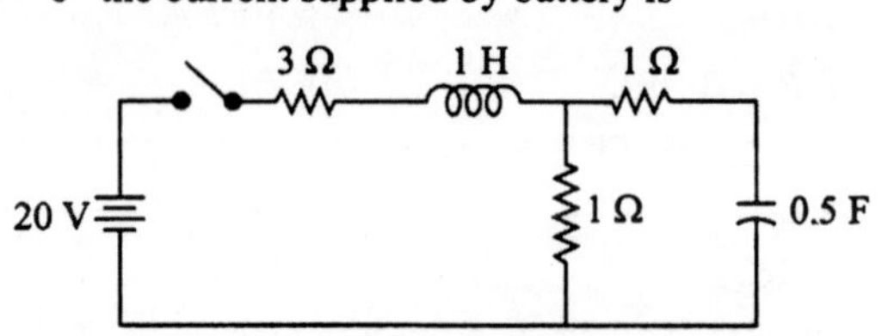

Fig. 4.19.

(a) 0 A (b) 5 A
(c) less than 5 A (d) infinity

188. In the circuit of Fig. 4.19, the current supplied by battery at $t = \infty$ is

(a) 0 A (b) 5 A
(c) less than 5 A (d) very large

189. For representing two sinusoidal voltages on a phasor diagram

(a) both voltages must be of the same frequency
(b) both voltages must be sinusoidal but may be of different frequencies
(c) one of the voltages must be of 50 Hz
(d) the frequency of one must be integral of that of other

190. The peak factor is the ratio of

(a) peak and rms values
(b) peak and average values
(c) rms and average values
(d) none of the above

191. If two voltages are $v_1 = 100 \sin(\omega t + 30°)$ and $v_2 = \cos \omega t$ then

(a) v_1 is leading v_2 by 30° (b) v_2 is leading v_1 by 30°
(c) v_2 is leading v_1 by 60° (d) v_1 is leading v_2 by 60°

192. A sinusoidal voltage is written as

$$v = I_m (R^2 + X_C^2)^{0.5} \sin 314\, t.$$

Its peak value and frequency are

(a) I_m and 50
(b) $I_m (R^2 + X_C^2)^{0.5}$ and 50
(c) $0.707\, I_m (R^2 + X_C^2)^{0.5}$ and 50
(d) $\sqrt{2}\, I_m (R^2 + X_C^2)^{0.5}$ and 50

193. If the peak value of output of a half wave rectifier is 100 V, the rms value is

(a) 100 V (b) 50V
(c) 70.7V (d) 35.35V

194. If $A = 10 + j\,10$, $\log_e A$ is

(a) $14.14 \angle 45°$ (b) $200 \angle 45°$
(c) $2.65 + j45$ (d) $2.65 + j0.785$

195. If $A = 16 \angle 64°$, $(A)^{0.5}$ is

(a) $4 \angle 64°$ (b) $4 \angle 8°$
(c) $4 \angle 128°$ (d) $4 \angle 32°$

196. If $A = 4 \angle 7°$, A^2 is

(a) $16 \angle 14°$ (b) $16 \angle 49°$
(c) $16 \angle 7°$ (d) $16 \angle 3.5°$

197. An RL series circuit has an impedance of 20 Ω when frequency is 25 Hz. At $f = 50$ Hz, the impedance will be

(a) 40 Ω
(b) 20 Ω
(c) less than 40 Ω but more than 20 Ω
(d) less than 20 Ω but more than 10 Ω

198. An RLC series circuit has R = 8 Ω, $X_L = 8$ Ω and $X_C - 8$ Ω, its impedance is

(a) 8 + j16 Ω (b) 8 + j8 Ω
(c) 8 Ω (d) 24 Ω

199. The impedance of an RC series circuit is 12 Ω at $f = 50$ Hz. At $f = 200$ Hz, the impedance will be

(a) more than 12 Ω
(b) less than 3 Ω
(c) more than 3 Ω but less than 12 Ω
(d) more than 12 Ω but less then 24 Ω

200. An RC series circuit has R = 20 Ω and X_C = 20 Ω. Then Z =

(a) 40 Ω (b) 28.28 Ω
(c) 20 Ω (d) 1 Ω

201. In an RC series circuit R = 100 Ω and X_C = 10 Ω. In this circuit

(*a*) the current and voltage are in phase
(*b*) the current leads the voltage by about 6°
(*c*) the current leads the voltage by about 84°
(*d*) the current lags the voltage by about 6°

202. The maximum phase difference between the applied voltage and current in an RLC series current with finite values of R, L, C and *f* can be

(*a*) less than 90° (*b*) less than 45°
(*c*) 180° (*d*) 30°

203. A circuit has two branches in parallel and each branch has only one element. The applied voltage and source current are

$v = 200 \sin(200t + 30°)$
and $i = 50 \sin(200t + 50°)$.

If one branch has capacitor, the other branch contains

(*a*) resistance
(*b*) inductance
(*c*) capacitance
(*d*) resistance or capacitance

204. A network has two branches in parallel. One branch contains impedance Z_1 and the other branch has impedance Z_2. If it is fed from an ac voltage V of frequency *f*, the current through Z_1 depends, on

(*a*) V, Z_1, Z_2 (*b*) V, Z_1
(*c*) Z_1, Z_2 (*d*) V, Z_1, f

205. Two sinusoidal voltage sources

$v_1 = 50 \sin(100t)$
and $v_2 = 50 \sin(100t + \pi)$

are connected in parallel and feed an inductance X_L = 2 Ω. The current through X_L is

(*a*) 25A (*b*) about 17.5A
(*c*) zero (*d*) 50 A

206. An ac circuit has two branches in parallel. The current in one branch is 5A. Then the current of source

(*a*) must be less than 5A
(*b*) must be more than 5A
(*c*) may be equal or more or less than 5A
(*d*) 5A

207. Two impedances $10 - j10$ and $10 + j10$ are connected in parallel. Their combined impedance is

(*a*) $5 + j5$ (*b*) $5 - j5$
(*c*) $10 + j0$ (*d*) $20 - j10$

208. An ac circuit has two branches in parallel. The impedances of the two branches are equal. The p.f. of one branch is 0.8 lagging and that of other is 0.6. lagging. The overall p.f. is about

(*a*) zero (*b*) 0.5 lagging
(*c*) 1 (*d*) 0.7 lagging

209. An ac circuit has two branches in parallel. The impedances of the two branches are equal. Their power factors are 0.8 lagging and 0.8 leading. The overall p.f. is

(*a*) 1 (*b*) zero
(*c*) about 0.7 lagging (*d*) about 0.7 leading

210. A circuit has two branches 1 and 2 in parallel. If $|Q_1|$ = 8 kvar and $|Q_2|$ = 6 kvar, the magnitude of total kvar drawn from supply will be

(*a*) 14 kvar (*b*) 2 kvar
(*c*) zero (*d*) 14 or 2 kvar

211. In the circuit of Fig. 4.20

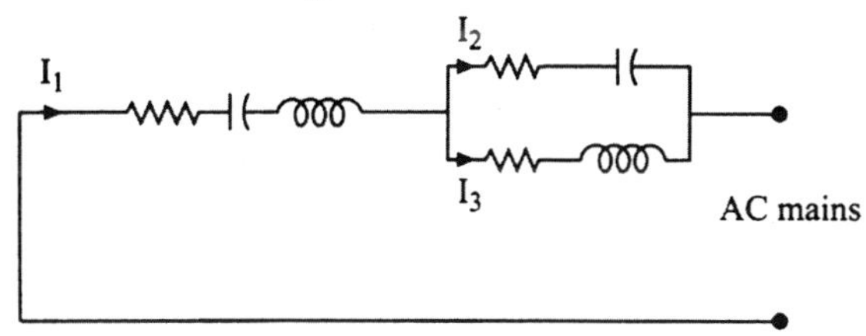

Fig. 4.20.

(*a*) I_2 will always lag I_3
(*b*) I_2 may be in phase with or lag I_3
(*c*) I_3 will always lag I_2
(*d*) I_2 and I_3 will be in phase

212. A resistance and an inductance are connected in parallel and fed from 50 Hz ac mains. Each branch takes a current of 5 A. The current supplied by source is

(*a*) 10 A (*b*) 7.07A
(*c*) 5 A (*d*) zero

213. An inductance having X_L = 5 Ω and a capacitance having X_C = 5 Ω are connected in parallel across 100 V, 50 Hz supply. The current drawn from source is

(*a*) zero (*b*) 20 A
(*c*) 40 A (*d*) 28.28 A

214. A series RLC circuit is at resonance at 200 Hz. If capacitance is increased to four times, the circuit will be in resonance at

(*a*) 200 Hz (*b*) 400 Hz
(*c*) 100 Hz (*d*) 800 Hz

215. A series RLC circuit has a resonant frequency of 1000 Hz. The maximum voltage across C is, likely to occur at a frequency of about

(*a*) 1000 Hz (*b*) 2000 Hz
(*c*) 1025 Hz (*d*) 975 Hz

216. A series RLC circuit has a resonant frequency of 2000 Hz. The maximum voltage across L is likely to occur at a frequency of about

(*a*) 2000 Hz (*b*) 4000 Hz
(*c*) 2025 Hz (*d*) 1975 Hz

217. An RLC series circuit is fed from 100 V ac supply. Inductance is 1H and Q = 7.5. At resonance, the voltage across inductance is

(*a*) 750 V (*b*) 100 V
(*c*) about 13.5 V (*d*) 1500 V

218. A two branch parallel circuit has a 20 Ω resistance and 1H inductance in one branch and a 100 μF capacitor in the second branch.

At resonance, the input impedance of the circuit is

(a) 500 Ω (b) 50 Ω
(c) 20 Ω (d) 5 Ω

219. A two branch parallel circuit has a 10 Ω resistance and 0.5 H inductance in one branch and a 100 μF capacitor in the second branch. It is fed from 100 V ac supply. At resonance, the source current is

(a) 10 A (b) 1 A
(c) 0.2 A (d) 0.05 A

220. In a series RLC circuit the dB level of currents at half power frequencies as compared to current at resonance is

(a) +3 (b) –3
(c) +0.5 (d) –0.5

221. An RLC series circuit has R = 7.07 Ω, L = 0.707 H and C = 7.07 μF. At half power frequencies the circuit impedance is likely to be

(a) 7.07 Ω (b) 10 Ω
(c) 14.14 Ω (d) 20 Ω

222. An RLC series circuit has Q = 100 and ω_0 = 20 rad/sec. The bandwidth is

(a) 0.2 rad/sec (b) 2 rad/sec
(c) 20 rad/sec (d) 2000 rad/sec

223. The Q factor of a series RLC circuit is

(a) $\frac{1}{R}\sqrt{\frac{C}{L}}$ (b) $R\sqrt{\frac{C}{L}}$
(c) $R\sqrt{\frac{L}{C}}$ (d) $\frac{1}{R}\sqrt{\frac{L}{C}}$

224. In a series RLC circuit, the current at resonance is I_0. The total energy stored in the circuit at resonance is

(a) LI_0^2 (b) $2LI_0^2$
(c) $\sqrt{2}\,LI_0^2$ (d) $0.5LI^2$

225. In a series RLC circuit, the resonant frequency is ω_0. The fractional frequency deviation is defined as

(a) $\frac{\omega_0 - \omega}{\omega}$ (b) $\frac{\omega_0 - \omega}{\omega_0}$
(c) $\frac{\omega - \omega_0}{\omega_0}$ (d) $\frac{\omega - \omega_0}{\omega}$

226. Which statement is incorrect for parallel resonance

(a) p.f. is 1 (b) Y is minimum
(c) B is zero (d) Z is minimum

227. The internal impedance of a source is

$3 + j7\,\Omega.$

For maximum power transfer, load impedance should be

(a) $3 + j7\,\Omega$ (b) $3 - j7\,\Omega$
(c) $7 + j3\,\Omega$ (d) $7 - j3\,\Omega$

228. The internal impedance of a source is

$3 + j4\,\Omega$

It is desired to supply maximum power to a resistive load. The load resistance should be

(a) 3 Ω (b) 4 Ω
(c) 7 Ω (d) 5 Ω

229. Kirchhoff's laws are applicable to

(a) dc circuits
(b) circuits with sinusoidal excitation only
(c) dc circuits and circuits with sinusoidal excitation
(d) circuits with any excitation

230. Kirchhoff's laws are applicable for

(a) steady state conditions only
(b) transient conditions only
(c) both steady state and transient condition
(d) both steady state and transient condition but with some constraints

231. An ac network has two sources of the same frequency. The currents in one branch when the sources act independently are 10 ∠ 30°A and 10 ∠ 60°A. When both sources act, the current in the branch should be

(a) 19.32 ∠45° (b) 15 ∠45°
(c) 20 ∠45° (d) zero

232. In an ac network the sum of currents entering a node is 10 ∠ 45°. The sum of currents leaving the node is

(a) 10 ∠45°A (b) 10 ∠225°A
(c) 10 ∠–45°A (d) 10 ∠–225°A

233. Reciprocity theorem is applicable to

(a) circuits with one source
(b) circuits with only one independent source
(c) circuits with any number of sources
(d) circuits with any number of independent sources

234. Tellegen's theorem is applicable to

(a) circuits with passive elements only
(b) circuits with linear elements only
(c) circuits with time invariant elements only
(d) circuits with active or passive, linear or non-linear and time invarient or time varying elements

235. Three impedances $(3 + j4)\,\Omega$ each are connected in star. The impedances of equivalent delta circuit will be

(a) $3 - j4\,\Omega$ each (b) $9 + j12\,\Omega$ each
(c) $9 - j\,12\,\Omega$ each (d) $3 + j4, 3 - j4$ and $4 - j3\,\Omega$

236. Thevenin's impedance of an ac network is in general

(a) a resistance (b) R and L in series
(c) R and C in series
(d) either R and L or R and C in series

237. In an ac network. Thevenin's impedance and Norton's impedance are

(a) always the same (b) generally the same
(c) sometimes the same (d) always different

238. A network has two sources of different frequencies. The only method of analysis which can be used to find current and voltages of different branches is the use of

(a) Kirchhoff's laws (b) superposition theorem
(c) Thevenin's theorem (d) Tellegen's theorem

239. A circuit is fed by two sources of the same frequency and values $100\angle 0°$ V and $20\angle 30°$V. The current in a particular branch is $5\angle 19°$ A. If the sources were $200\angle 0°$ and $40\angle 30°$V, the current in the branch should be

(a) $5\angle 19°$A (b) $10\angle 19°$A
(c) $10\angle 38°$A (d) $10\angle 9.5°$A

240. A circuit is fed by two sources of different frequencies and magnitudes $100\angle 0°$ V and $50\angle 0°$. The current in a particular branch is $10\angle 20°$ A. If the magnitudes of the sources are changed to $200\angle 0°$ V and $100\angle 0°$ V, the current in that branch should be

(a) $20\angle 20°$A (b) $10\angle 20°$A
(c) $14.14\angle 20°$A (d) $10\angle -20°$A

241. The network shown in Fig. 4.21 is

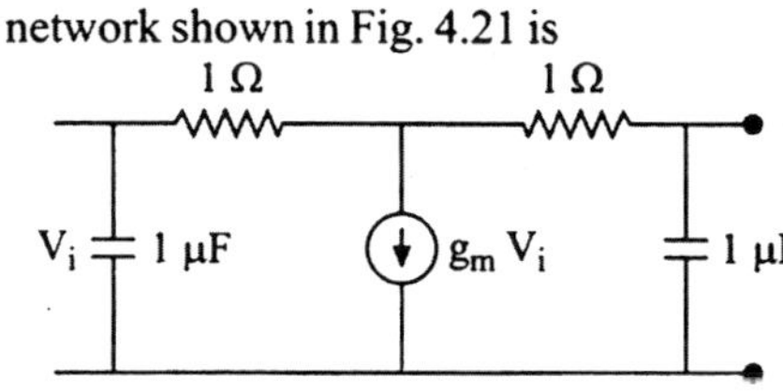

Fig. 4.21.

(a) reciprocal
(b) reciprocal and symmetrical
(c) non-reciprocal (d) unsymmetrical

242. Norton's equivalent circuit

(a) is same as Thevenin's circuit
(b) has no relation with Thevenin's circuit
(c) is dual of Thevenin's circuit
(d) is reciprocal of Thevenin's circuit

243. Thevenin's equivalent circuit

(a) can be drawn only in time domain
(b) can be drawn only in frequency domain
(c) can be drawn both in time and frequency domain
(d) can be drawn in frequency domain with some restrictions

244. Thevenin's theorem is used to find current in a branch of a network. It is now desired to find current in another branch of the network

(a) the same Thevenin's circuit can be used
(b) the same Thevenin's circuit with some modification can be used
(c) it is necessary to find new Thevenin's circuit
(d) either (b) or (c)

245. In the network of Fig. 4.22 it is desired that maximum power be supplied to Z_2. The values of Z_1 and Z_2 are fixed. However the transformation ratio 'a' of transformer can be changed. The value of 'a' for maximum power transfer is

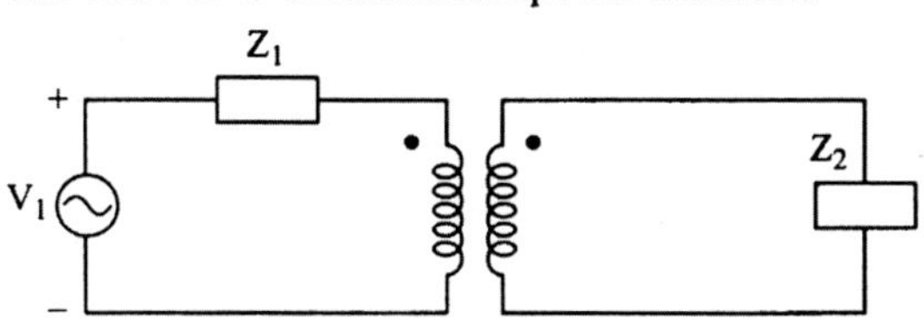

Fig. 4.22.

(a) $|Z_2|/|Z_1|$ (b) $[|Z_2|/|Z_2|]^{0.5}$
(c) $[|Z_2|/|Z_1|]^2$ (d) $[|Z_2|/|Z_1|]^{0.75}$

246. For the network of Fig. 4.23, z_{11} =

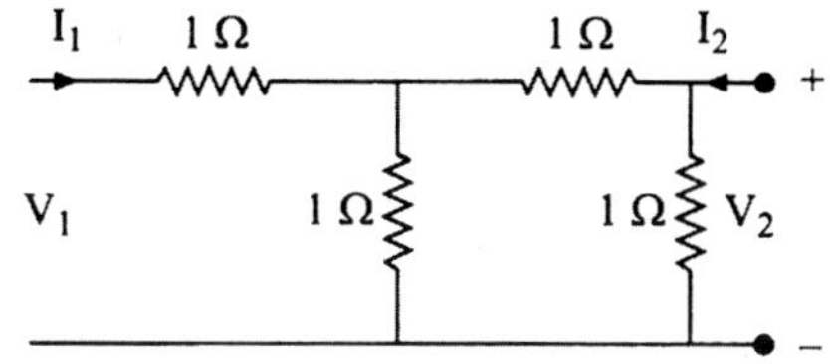

Fig. 4.23.

(a) 5/3 Ω (b) 3/2 Ω
(c) 2 Ω (d) 2/3 Ω

247. The impedance parameter z_{21} is

(a) transfer impedance at port 1 with port 2 open-circuited.
(b) transfer impedance at port 1 with port 2 short-circuited
(c) transfer impedance at port 2 with port 1 open-circuited
(d) transfer impedance at port 2 with port 1 short-circuited

248. For a symmetrical network

(a) $h_{11} = h_{22}$ (c) $h_{12} = h_{22}$
(c) $h_{11}h_{22} - h_{12}h_{21} = 0$ (d) $h_{11}h_{22} - h_{12}h_{21} = 1$

249. In terms of ABCD parameters, the image parameter Z_{11} is equal to

(a) $\sqrt{AC/BD}$ (b) $\sqrt{BD/AC}$
(c) $\sqrt{AB/CD}$ (d) $\sqrt{AD/BC}$

250. For the network of Fig. 4.24, [z] matrix is

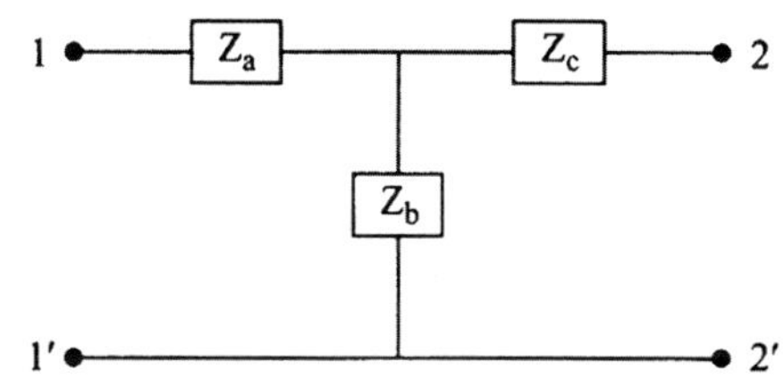

Fig. 4.24.

(a) $\begin{bmatrix} Z_a + Z_c & Z_b \\ Z_b & Z_b + Z_c \end{bmatrix}$ (b) $\begin{bmatrix} Z_a + \dfrac{Z_b Z_c}{Z_b + Z_c} & Z_b \\ Z_b & Z_a \end{bmatrix}$

(c) $\begin{bmatrix} Z_a + Z_b & Z_b \\ Z_b & Z_b + Z_c \end{bmatrix}$ (d) $\begin{bmatrix} Z_a + Z_b & -Z_b \\ Z_b & Z_b + Z_c \end{bmatrix}$

251. Which of the following is correct?

(a) $y_{11} = z_{11}/|z|$ (b) $y_{11} = z_{22}/|z|$
(c) $z_{11} = y_{11}/|y|$ (d) $z_{22} = y_{22}/|y|$

252. For a symmetrical network

(a) $z_{11} = z_{22}$ (b) $z_{12} = z_{21}$
(c) $z_{11} = z_{22}$ and $z_{12} = z_{21}$ (d) $z_{11}z_{12} - z_{12}^2 = 0$

253. If a two port network is represented by an equivalent π network the admittance of series arm in terms of ABCD parameters is equal to

(a) C (b) 1/C
(c) (A – 1)/B (d) 1/B

254. If a two port network is represented by an equivalent T network, the impedance of shunt arm in terms of ABCD parameters is

(a) 1/C (b) (A – 1)/C
(c) 1/B (d) (D – 1)/C

255. The parameter A of a two port is equal to

(a) V_2/V_1 with $I_1 = 0$
(b) V_1/I_2 with $V_2 = 0$
(c) I_1/I_2 with $V_2 = 0$
(d) V_1/V_2 with $I_2 = 0$

256. For the two port of Fig. 4.25, z_{11} is

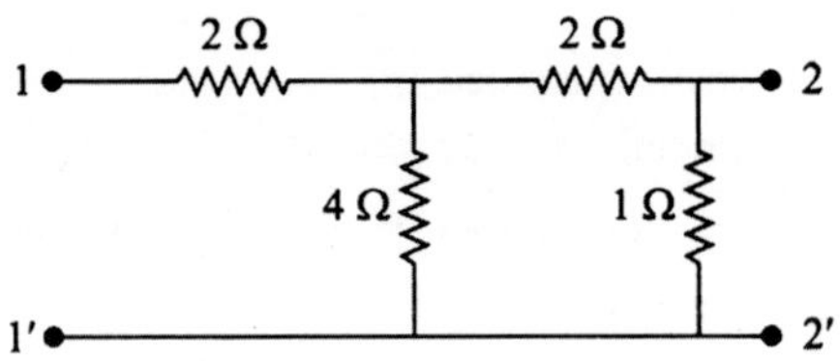

Fig. 4.25.

(a) 26/7 Ω (b) 10/3 Ω
(c) 6 Ω (d) 2 Ω

257. For the two port of Fig. 4.25, z_{22} is

(a) 26/7 Ω (b) 6/7 Ω
(c) 2/3 Ω (d) 1 Ω

258. Which of the statement is incorrect

(a) *h* and *g* matrices are inverse of each other
(b) *z* and *y* matrices are inverse of each other
(c) for reciprocal networks AD – BC = 1
(d) ABCD parameters are used in transistor circuit

259. In open circuit impedance parameters, the independent variables are

(a) I_1 and I_2 (b) V_1 and V_2
(c) V_1 and I_1 (d) V_2 and I_2

260. In *h* parameter representation, the independent variables are

(a) I_1 and V_2 (b) I_1 and V_1
(c) V_1 and V_2 (d) I_1 and I_2

261. In terms of *z* parameters, the condition for the network to be passive is

(a) $z_{11} = z_{22}$ (b) $z_{12} = z_{21}$
(c) $z_{12} = 0$ (d) $z_{22} = 0$

262. When analysing two port networks in cascade, it is more covenient to use

(a) *z* parameters (b) *h* parameters
(c) *T* parameters (d) *y* parameters

263. When analysing two port networks in parallel it is more convenient to use

(a) *z* parameters (b) *h* parameters
(c) T parameters (d) *y* parameters

264. A constant *k* high pass π section has a characteristic impedance of 300 Ω at $f = \infty$. At $f = f_c$, the characteristic impedance will be

(a) zero (b) ∞
(c) 300 Ω (d) more than 300 Ω

265. A constant *k* high pass filter has $f_c = 3000$ Hz. At $f = 1000$ Hz, the phase shift is

(a) zero (b) –π
(c) 2π (d) more than π

266. A constant *k* band pass filter has a pass band from 1000 to 4000 Hz. The resonant frequency of series and shunt arms is

(a) 3000 Hz (b) 2500 Hz
(c) 2000 Hz (d) 1500 Hz

267. A constant *k* low pass T section filter has a characteristic impedance of 600 Ω at zero frequency. At $f = f_c$, the characteristic impedance is

(a) 600 Ω (b) zero
(c) infinite (d) more than 600 ohms

268. A constant k low pass filter has $f_c = 1000$ Hz. At $f = 500$ Hz, the phase shift is

(a) π (b) zero
(c) more than π (d) less than π

269. A constant *k* low pass filter has $f_c = 500$ Hz. At $f = 1000$ Hz, the attenuation is

(a) zero
(b) infinity
(c) may be zero or infinity
(d) more than zero but less than infinity

270. An *m* derived low pass filter has $f_c = 1000$ Hz and $m = 0.6$. This filter will have infinite attenuation at

(a) $f = 1250$ Hz
(b) all frequencies above 1000 Hz
(c) $f = 1562.5$ Hz
(d) $f = 1666.67$ Hz

271. A constant *k* band stop filter has cutoff frequencies of 1000 and 4000 Hz. The attenuation will be maximum when frequency is

(a) 4000 Hz (b) 3000 Hz
(c) 2000 Hz (d) 1000 Hz

272. An *m* derived and constant *k* filter section connected in cascade give

(a) proper impedance matching
(b) proper phase shift characteristics
(c) proper attenuation characteristics
(d) all the above features

273. An *m* derived low pass filter has $f_c = 1000$ Hz, $f_\infty = 1250$ Hz and $m = 0.6$. If *m* is increased, then

(a) both f_c and f_∞ will increase
(b) both f_c and f_∞ will decrease
(c) f_c will remain constant but f_∞ will decrease
(d) f_c will remain constant but f_∞ will increase

274. If $F(s) = \dfrac{s+a}{(s+a)^2 + \omega_0^2}$, then $f(t) =$

(*a*) $\cos \omega_0 t$ (*b*) $\sin \omega_0 t$

(*c*) $e^{at} \cos \omega_0 t$ (*d*) $e^{-at} \cos \omega_0 t$

275. The transfer function $G_{12}(s)$ for a two port is

(*a*) V_2/V_1 (*b*) $V_2(s)/V_1(s)$

(*c*) $V_2(s)/I_1(s)$ (*d*) $I_2(s)/I_1(s)$

276. For the ladder network of Fig. 4.26, open circuit driving point impedance at port 1 =

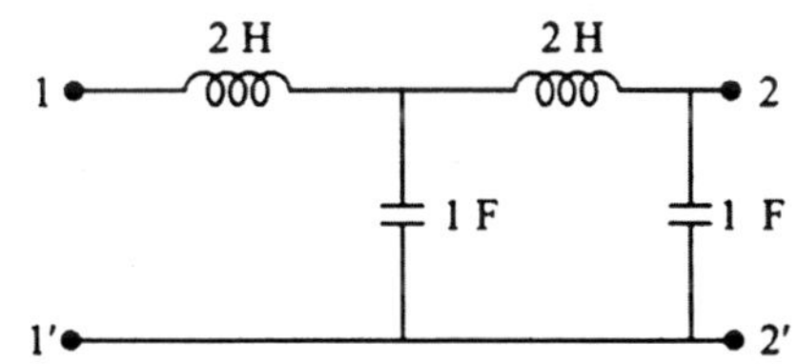

Fig. 4.26.

(*a*) $2s + \cfrac{1}{s + \cfrac{1}{2s + \cfrac{1}{s}}}$ (*b*) $2s + \cfrac{1}{2s + \cfrac{1}{s + \cfrac{1}{2s}}}$

(*c*) $s + \cfrac{1}{2s + \cfrac{1}{s + \cfrac{1}{2s}}}$ (*d*) $2s + \cfrac{1}{s + \cfrac{1}{s + \cfrac{1}{2s}}}$

277. For stable network, the real part of poles and zeros of driving point function must be

(*a*) negative (*b*) negative or zero

(*c*) zero (*d*) positive

278. The poles with greater displacement from the real axis correspond to

(*a*) higher frequencies of oscillation

(*b*) stable response

(*c*) lower frequencies of oscillation

(*d*) unbounded output

279. The response to sinusodial excitation can be found by

(*a*) $s = j\omega$ in the network function

(*b*) setting $s - j\omega$ in the network function

(*c*) taking inverse transform of network function

(*d*) none of the above

280. For an RLC series circuit, Z (*s*) will be of the form

(*a*) $\dfrac{5(s^2+5)}{s^2+s+1}$ (*b*) $\dfrac{5s}{s^2+5}$

(*c*) $\dfrac{3(s^2+s+1)}{s^2+1}$ (*d*) $\dfrac{s^2+10s+100}{s}$

281. The phase difference $(j\omega - p_t)$ means

(*a*) phasor directed from p_1 to $j\omega$

(*b*) phasor directed from origin to p_1

(*c*) phasor directed from origin to $j\omega$ point

(*d*) phasor directed from $j\omega$ to p_1

282. An RLC series circuit has $Y(s) = \dfrac{2s}{(s-p_1)(s-p_1^*)}$. If the poles p_1 and p_1^* are close to $j\omega$ axis, then

(*a*) Q is high (*b*) Q is low

(*c*) ω_0 is high (*d*) bandwidth is large

283. For the two port network of Fig 4.27, the voltage transfer ratio in *s* domain is

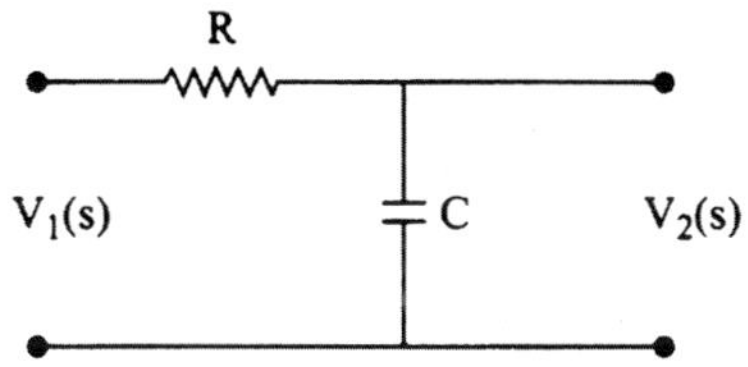

Fig. 4.27.

(*a*) $\dfrac{s}{1+RCs}$ (*b*) $\dfrac{1}{1+RCs}$

(*c*) $\dfrac{s^2}{1+RCs}$ (*d*) $\dfrac{1}{1+RCs^2}$

284. The poles of a reactance function

(*a*) lie on $j\omega$ axis only

(*b*) lie on negative real axis only

(*c*) may lie anywhere on *s* plane

(*d*) lie on $j\omega$ axis or on negative real axis

285. The function $Z(s) = \dfrac{s(s^2+a)}{(s^2+1)(s^2+16)}$

(*a*) is a reactance function

(*b*) is an RC function

(*c*) is an RL function

(*d*) may be reactance function or RC function

286. The function $(s+2)(s+4)/[(s+1)(s+3)]$

(*a*) is a reactance function

(*b*) is an RC function

(*c*) may be a reactance function or an RC function

(*d*) may be an RC function or an RL function

287. An RC impedance function has a constant value at $s = 0$. The last element in the first cauer synthesis

(*a*) will be a capacitor

(*b*) may be R or C

(*c*) will be R

(*d*) will depend on the behaviour of function at $s = \infty$

288. An RL impedance function is to be synthesised in second Foster form. The proper function to be expanded is

(*a*) Z(*s*) (*b*) *s*Z(*s*)

(*c*) 1/Z(*s*) (*d*) Z(*s*)/*s*

289. If a function Z(*s*) has a pole at infinity, then

(*a*) the degrees of numerator and denominator are equal

(*b*) the degree of numerator is one higher than that of denominator

(*c*) the degree of denominator is one higher than that of numerator

(*d*) the difference between the degrees of numerator and denominator is 0 or 1

290. A reactance function
(*a*) has a pole at origin
(*b*) has a zero at origin
(*c*) may have a pole or zero at origin
(*d*) has a pole or zero at origin and a pole or zero at infinity.

291. The poles of an RC function
(*a*) are simple and lie on negative real axis
(*b*) are simple and lie on $j\omega$ axis
(*c*) must be complex conjugate
(*d*) may be anywhere on s plane

292. An RC impedance function has a pole at $s = 0$. The first element in the first Foster form of synthesis
(*a*) is R (*b*) may be R or C
(*c*) is C (*d*) is R and C in parallel

293. In a minimum function
(*a*) the degree of numerator is higher than that of denominator
(*b*) the degree of numerator is lower than that of denominator
(*c*) the degrees of numerator and denominator are equal
(*d*) the denominator is a constant

294. In an LC immittance H(s)
(*a*) the slope of $dH(j\omega)/d\omega$ is always negative
(*b*) the slope of $dH(j\omega)/d\omega$ is always positive
(*c*) the slope of $dH(j\omega)/d\omega$ may be positive or negative
(*d*) the slope of $dH(j\omega)/d\omega$ is constant

295. In the synthesis of RLC network
(*a*) the number of elements in Bott-Duffin synthesis and Brune's synthesis are always equal
(*b*) the number of elements in Bott-Duffin synthesis is more than that in Brune's synthesis
(*c*) the number of elements in Bott-Duffin synthesis is less than that in Brune's synthesis
(*d*) the number of elements in Bott-Duffin synthesis may be more or less than that in Brune's synthesis

296. Which synthesis procedure requires the use of mutual inductance
(*a*) Foster's synthesis (*b*) Cauer synthesis
(*c*) Bott-Duffin synthesis (*d*) Brune's synthesis

297. Figure 4.28 shows an RC circuit. The magnitude plot V_2/V_1

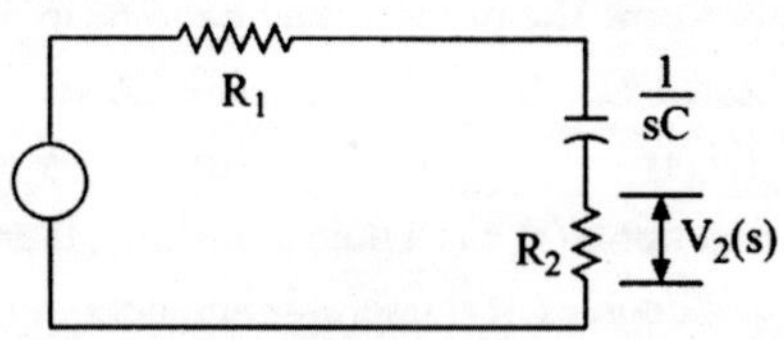

Fig. 4.28.

(*a*) will rise exponentially to $R_2/(R_1 + R_2)$ as $\omega \to \infty$
(*b*) will rise linearly to $R_2/(R_1 + R_2)$ as $\omega \to \infty$
(*c*) will decrease to zero as $\omega \to \infty$
(*d*) will be 1 at $\omega = 0$ and then decrease as $\omega \to \infty$

298. In Fig. 4.28 the phase plot
(*a*) will be 90° at $\omega = 0$ and 0° at $\omega = \infty$
(*b*) will be 0° at $\omega = 0$ and 90° at $\omega = \infty$
(*c*) will be 45° at $\omega = 0$ and 0° at $\omega = \infty$
(*d*) will be 45° at $\omega = 0$ and 90° at $\omega = \infty$

299. In a Hurwitz polynomial
(*a*) all coefficients are non-negative
(*b*) both the odd and even parts have roots on $j\omega$ axis
(*c*) the continued fraction expansion of the ratio of odd to even or even to odd parts of a Hurwitz polynomial yields all positive quotient terms
(*d*) all the above

300. The function H(s) is positive real if
(*a*) H(s) has no poles or zeros in right half plane
(*b*) The poles of H(s) on the imaginany axis are simple with real and positive residues
(*c*) $\text{Re}[H(j\omega)] \geq 0$ for $0 \leq \omega \leq \infty$
(*d*) all the above

301. A dc network has an open circuit voltage of 5 V. When its terminals are short circuited a current of 5 A flows through the short circuit. If the short circuit is removed and a resistance of 1.5 Ω is connected across its terminals, the current will be
(*a*) 1 A (*b*) 2 A
(*c*) 2.5 A (*d*) 5 A

302. Figure 4.29 shows a parallel resonant circuit having resonance frequency of 15 MHZ. At this frequency

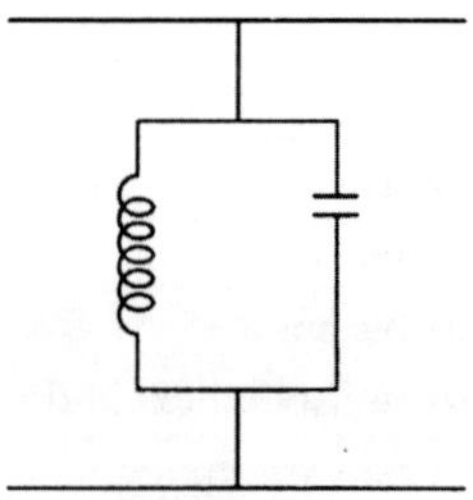

Fig. 4.29.

(*a*) its impedance is minimum
(*b*) its impedance is maximum
(*c*) its impedance is infinite
(*d*) its impedance is zero

303. Figure 4.30 shows a pole zero plot of I(s). The likely current response in time domain is

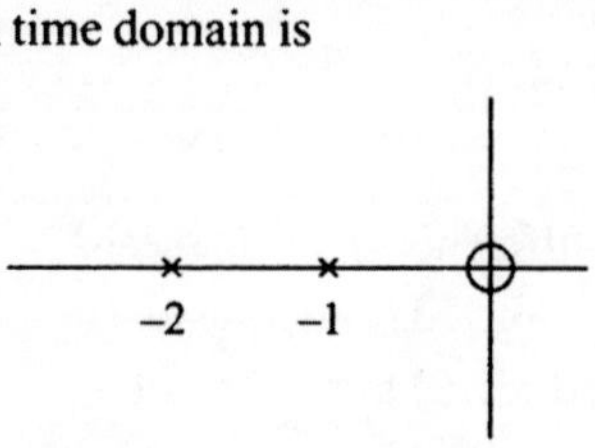

Fig. 4.30.

(*a*) $i(t) = e^{-t} - e^{-3t}$ (*b*) $i(t) = -e^{-t} + 2e^{-2t}$
(*c*) $i(t) = e^{-t} - 2e^{-2t}$ (*d*) $i(t) = e^{-t} - e^{-4t}$

304. In Fig. 4.31 the ratio V(*s*)/I(*s*) is

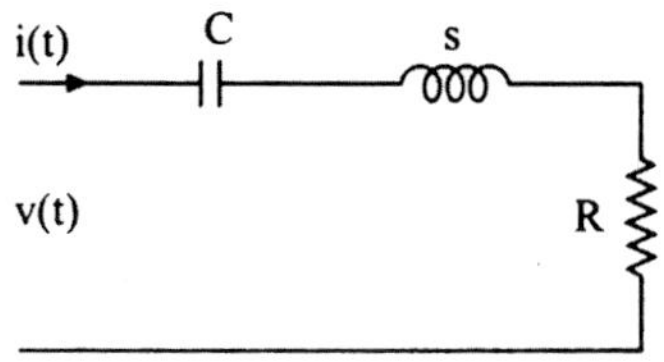

Fig. 4.31.

(*a*) R (*b*) R + *s*L

(*c*) $R + sC + \frac{1}{sL}$ (*d*) $R + sL + \frac{1}{sC}$

305. A series RLC circuitt has Q = 5. A voltage of 10 V rms is applied to it. At resonance, the voltage across C is

(*a*) 10 V rms (*b*) 10 V peak

(*c*) 50 V rms (*d*) 50 V peak

306. The circuit in Fig. 4.32 is switched on at *t* = 0. At any time *t*, *i*(*t*) =

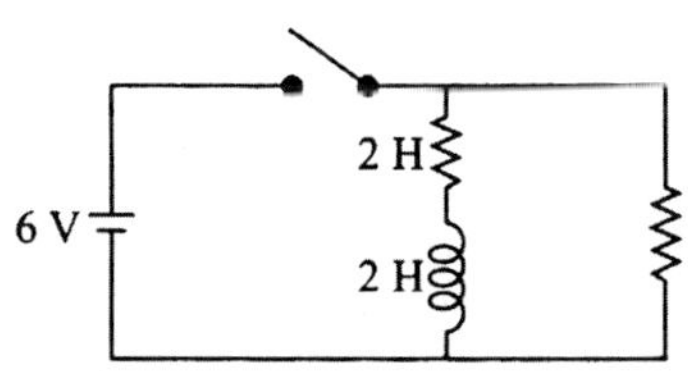

Fig. 4.32.

(*a*) $3 + 3e^{-t}$ (*b*) $6 - 3e^{-t}$

(*c*) $3 - 3e^{-t}$ (*d*) $6 - 6e^{-t}$

307. In Fig. 4.33 the battery has remained switched on for a long time. At *t* = 0 the switch is closed. The current *i*(*t*) is likely to be

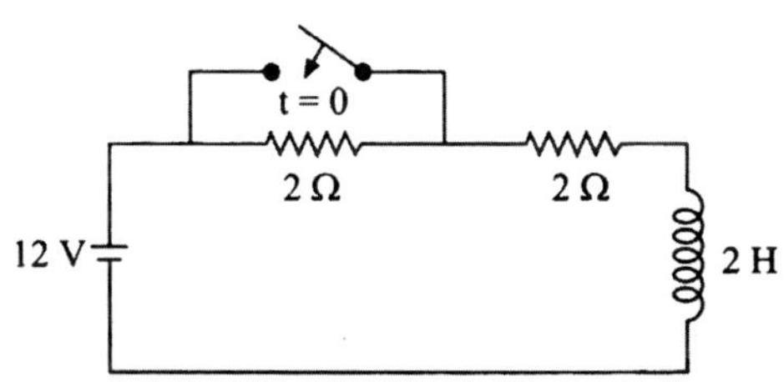

Fig. 4.33.

(*a*) 6 A (*b*) $6 - 3e^{-t}$

(*c*) $6 + 3e^{-t}$ (*d*) $3 - 6e^{-t}$

308. Two 2 port circuits are cascaded. To determine the response, it is more convenient to use

(*a*) O.C. impedance parameters

(*b*) S.C. admittance parameters

(*c*) ABCD parameters

(*d*) *h* parameters

309. Two two port circuits are connected in parallel. To determine the response, it is more convenient to use

(*a*) O.C. impedance parameters

(*b*) S.C. admittance parameters

(*c*) ABCD parameters

(*d*) *h* parameters

310. In Fig. 4.34 the current I

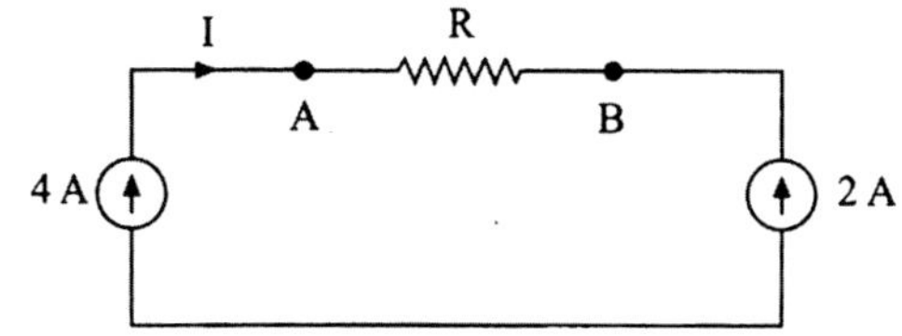

Fig. 4.34.

(*a*) is 2 A from A to B

(*b*) is 2 A from B to A

(*c*) depends on the value of R

(*d*) none of the above

311. In Fig. 4.35, the current I

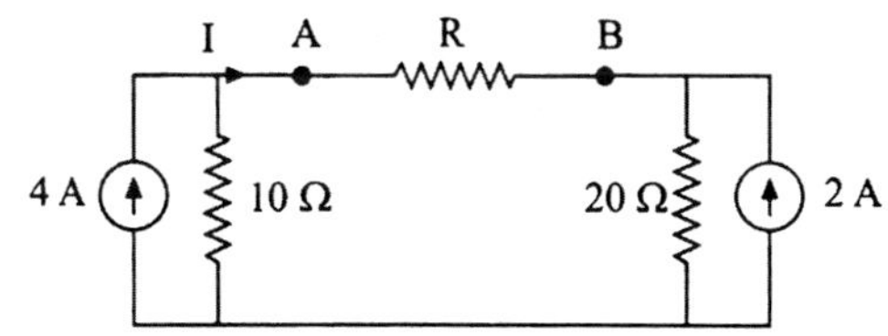

Fig. 4.35.

(*a*) is 2 A from A to B

(*b*) is 2 A from B to A

(*c*) depends on the value of R

(*d*) none of the above

312. In the voltage equation

$v = 500\, e^{-5t} \sin(2\pi\, 50t)$, the frequency is

(*a*) 2π × 50 Hz (*b*) 50 Hz

(*c*) $\frac{2\pi \times 50}{5}$ Hz (*d*) (2π × 50) 5 Hz

313. In the voltage equation *v* = 500 sin (1000*t*) V, the rms value of voltage is

(*a*) 500 V (*b*) $500\sqrt{2}$ V

(*c*) $\frac{500}{\sqrt{2}}$ (*d*) 1.1 × 500 V

314. Three capacitors each 9 μF are connected in delta. The value of each capacitor in equivalent star connection is

(*a*) 3 μF (*b*) 6 μF

(*c*) 27 μF (*d*) 81 μF

315. In Fig. 4.36, the short-circuit current through terminal AB will be

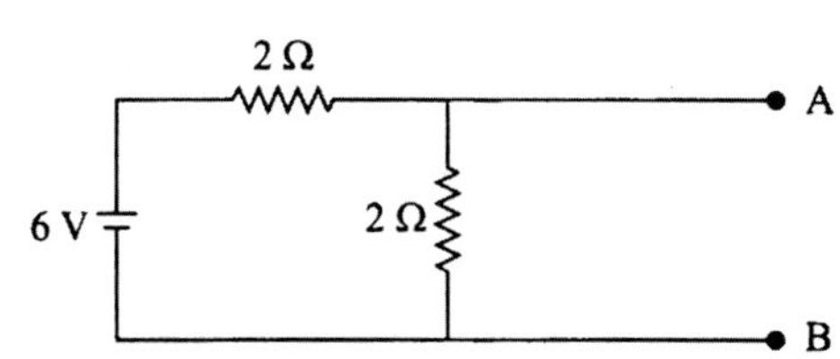

Fig. 4.36.

(*a*) 3 A (*b*) 1.5 A

(*c*) 2 A (*d*) 1 A

316. In Fig. 4.37 the short circuit current through terminal AB will be

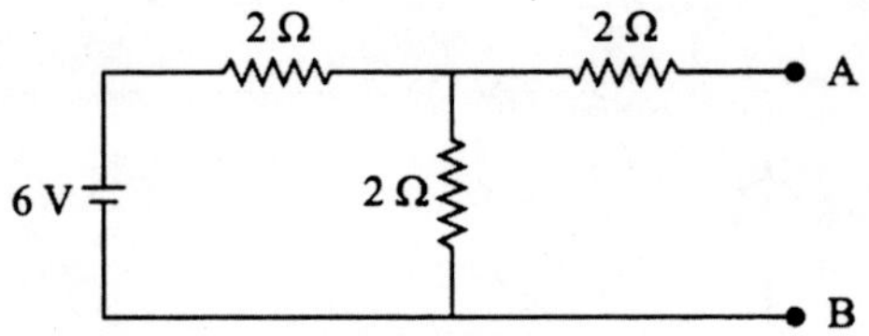

Fig. 4.37.

(*a*) 1.5 A (*b*) 3 A
(*c*) 2 A (*d*) 1 A

317. In Fig. 4.38 the short-circuit current through terminal AB will be

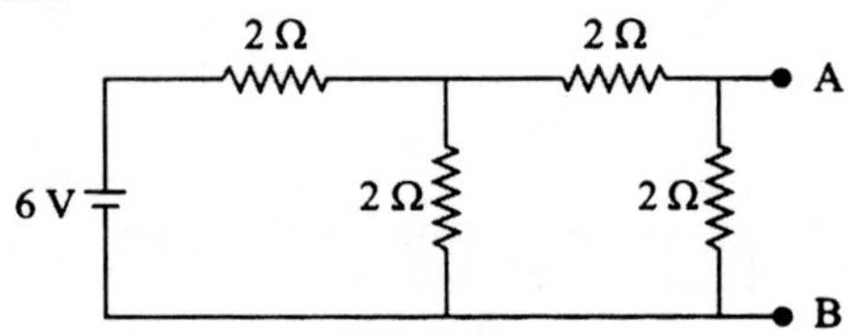

Fig. 4.38.

(*a*) 1.5 A (*b*) 3 A
(*c*) 2 A (*d*) 1 A

318. A series circuit has R = 5 Ω and C = 10 μF. It is switched on to a 12 V dc battery at $t = 0$. The current in the circuit will be maximum

(*a*) at $t = RC$ (*b*) at $t = \infty$
(*c*) at $t = 0^+$ (*d*) none of the above

319. A series circuit has R = 5 Ω and C = 10 μF. AT $t = 0$ it is switched on to a 10 V dc battery. The maximum current in the circuit is

(*a*) 2 A (*b*) 1 A
(*c*) 10^6 A (*d*) 10 μA

320. A series R, L, C circuit has R = 2 Ω, L = 1 H, C = 1 μF. It is connected to an ac variable frequency source of peak voltage 10 V. At resonance, the power supplied by the source will be

(*a*) 50 W (*b*) 25 W
(*c*) 12.5 W (*d*) 6.25 W

321. Network transients are due to

(*a*) circuit resistance
(*b*) circuit voltage
(*c*) changes in energy stored in L and C
(*d*) circuit admittance

322. In a series R-L circuit when excited by a 1 volt battery the current is $1 - e^{-t}$. If it is excited by a source 2 sin t, the current will be

(*a*) $\sqrt{2} \sin (t - 45°) + e^{-t}$
(*b*) $\sqrt{2} \sin (t + 45°) + e^{-t}$
(*c*) $\sin (t - 45°) + e^{-t}$
(*d*) $\sin (t + 45°)\, e^{-t}$

323. In an R-L circuit excited by a battery of voltage E, the current and voltage across resistor are $1 - e^{-t}$ and $2(1 - e^{-t})$ respectively. If there were an initial current of 2 A in the inductance, the voltage across resistance would have been

(*a*) $1 - e^{-t}$ (*b*) $2 - 2e^{-t}$
(*c*) $2 + 2e^{-t}$ (*d*) $4 - 4e^{-t}$

324. An R-L circuit has an initial current of 5 A. In the *s*-domain equivalent circuit, the voltage source to take into account the initial current has a strength of

(*a*) 5 L volts (*b*) 5/L volts
(*c*) 5 L^2 volts (*d*) 5/L^2 volts

325. In an R-C circuit, there is an initial voltage of 3 V across the capacitor. In *s* domain equivalent circuit, the voltage source to take into account the initial voltage has a strength of

(*a*) $\frac{3}{s}$ (*b*) $3s$
(*c*) $\frac{3}{Cs}$ (*d*) $3Cs$

326. A series circuit having R = 3 Ω and L = 2H is excited by a 10 V battery. At $t = 0^+$, di/dt is

(*a*) 5A/s (*b*) $\frac{10}{3}$ A/s
(*c*) 20 A/s (*d*) 15 A/s

327. In an R-C series circuit excited by a voltage E, the charge across capacitor at $t = 0^+$ is

(*a*) 0 (*b*) CE
(*c*) $CE(1 - e^{-t/RC})$ (*d*) none of the above

328. An R-C series circuit is charged by a voltage E. After a long time the battery is removed and a short circuit is placed across R-C. Then

(*a*) the polarity of voltage across R is same during discharging as during charging
(*b*) the polarity of voltage across R during discharging is opposite to that during charging
(*c*) the polarity of voltage across R depends on time constant RC
(*d*) none of the above

329. In a one port network $Z(s) = \frac{s+2}{s+3}$. It is excited by a ułit step current. The initial and final values of voltage across the network impedance are respectively

(*a*) 1 and $\frac{2}{3}$ (*b*) $\frac{2}{3}$ and 1
(*c*) 0 and $\frac{2}{3}$ (*d*) 1 and 0

330. A series RL circuit has $Z(s) = \frac{1}{s+4}$. If a voltage 4 sin 4t is applied, the steady state current will be

(*a*) $4 \sin (4t - 45°)$ (*b*) $\frac{1}{\sqrt{2}} \sin (4t - 45°)$
(*c*) $\frac{1}{\sqrt{2}} \sin 4t$ (*d*) $\frac{1}{2} \sin (4t - 45°)$

331. A series RC circuit has R = 5 Ω and C = 10 μF. The current in the circuit is 5 sin 20000t. The applied voltage is

(*a*) $25\sqrt{2} \sin (20000t + 45°)$
(*b*) $25\sqrt{2} \sin (20000t - 45°)$

(c) $25\sqrt{2}\sin 20000t$

(d) $25\sqrt{2}\sin(20000t - 90°)$

332. In Fig. 4.39, the voltage $V_1 = 4\angle 0$ V. The source voltage V_2 is

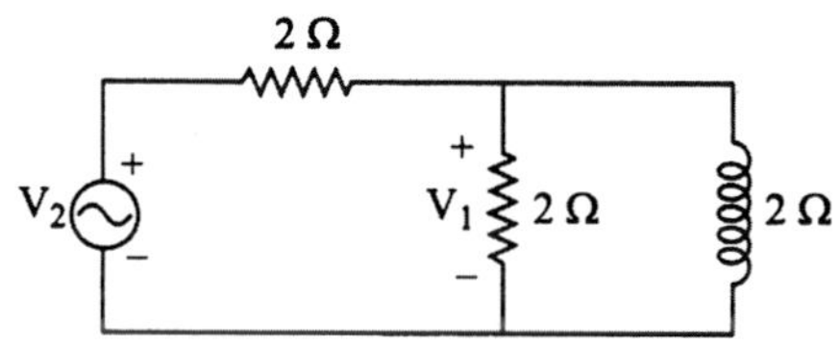

Fig. 4.39.

(a) $8 - j8$ volts (b) $4 - j4$ volts
(c) $8 - j4$ volts (d) $8 - j16$ volts

333. The circuit in Fig. 4.40 is switched on at $t = 0$. After a long time the current through 5 Ω resistance is

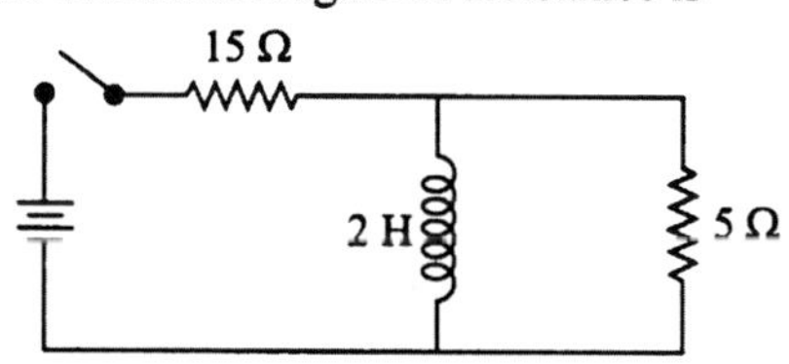

Fig. 4.40.

(a) 1 A (b) 4 A
(c) 10 A (d) 0

334. In Fig. 4.40 the current supplied by battery immediately after switching on of the circuit is

(a) 1 A (b) 4 A
(c) 10 A (d) 0

335. A series RLC circuit is supplied from a variable frequency ac source. The resonant frequency is ω_0. Voltage across C is maximum at a frequency

(a) equal to ω_0 (b) slightly less than ω_0
(c) slightly more than ω_0 (d) none of the above

336. A series RLC circuit is supplied from variable frequency source. The resonant frequency is ω_0. Voltage across L is maximum at a frequency

(a) equal to ω_0
(b) slightly less than ω_0
(c) slightly more than ω_0
(d) none of the above

337. The switch in Fig. 4.41 is initially closed. At $t = 0$ the switch is opened. At $t = 0^+$, the current through inductance is

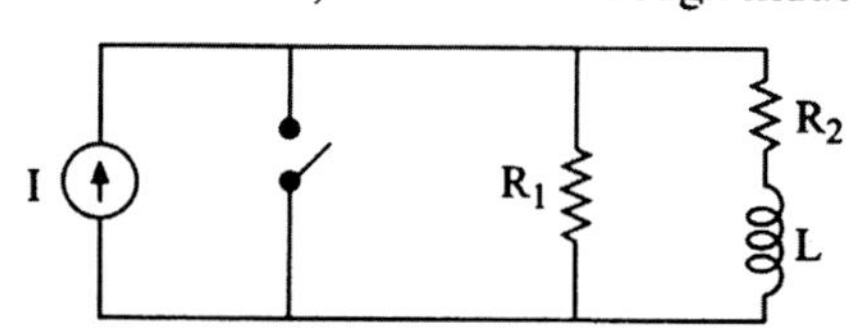

Fig. 4.41.

(a) $\dfrac{IR_1}{R_1 + R_2}$ (b) I

(c) 0 (d) $\dfrac{IR_2}{R_1 + R_2}$

338. The switch in Fig. 4.42 is initially closed. At $t = 0$ switch is opened. At $t = 0^+$, the current through R is

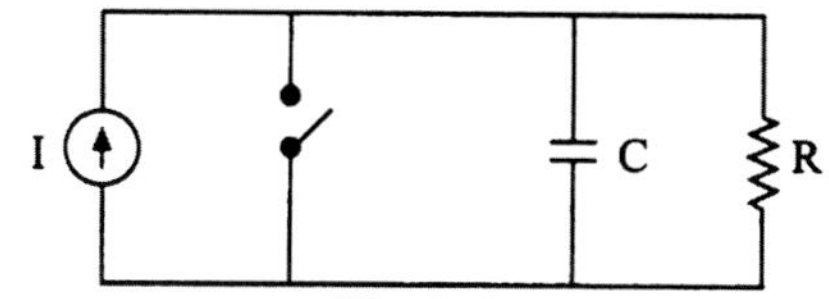

Fig. 4.42.

(a) I (b) 0

(c) $Ie^{-t/RC}$ (d) $\dfrac{IR}{R + C}$

339. In Fig. 4.42, the current through R, after the steady state condition is reached, is

(a) I (b) 0

(c) $\dfrac{IR}{R + C}$ (d) infinite

340. In an underdamped RLC series circuit, the natural frequency is ω_n. The frequency of damped oscillations is

(a) ω_n (b) less than ω_n
(c) more than ω_n (d) $2\omega_n$

341. An RLC series circuit is underdamped. To make it overdamped, the value of R

(a) has to be increased
(b) has to be decreased
(c) has to be increased to infinity
(d) has to be reduced to zero

342. Two RLC series circuit has same damping ratio. Circuit 1 has higher natural frequency as compared to circuit 2. Then

(a) transients in circuit 2 will decay faster
(b) transients in circuit 1 will decay faster
(c) transients in both circuits will decay at the same rate
(d) transients in any one may decay faster, than in the other

343. In Fig. 4.43, the switch has been in closed position for a long time. At $t = 0$ the switch is opened. At $t = 0^+$, the current i_c is

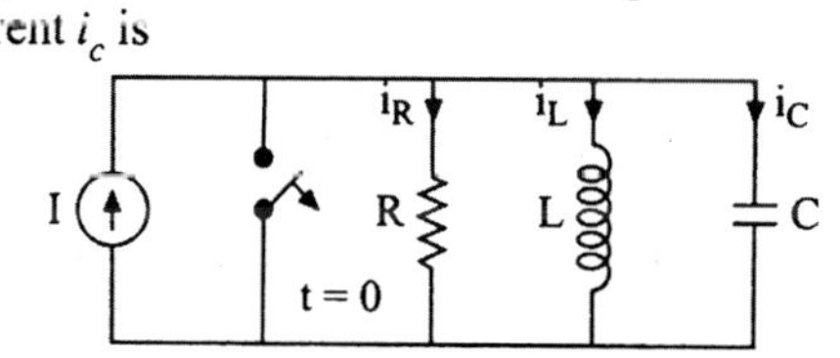

Fig. 4.43.

(a) I (b) 0

(c) infinite (d) $\dfrac{IR}{LC}$

344. In Fig. 4.43, i_L at $t = 0^+$ is

(a) 0 (b) I

(c) infinite (d) $\dfrac{IL}{RC}$

345. In Fig. 4.43 which is correct at any time t

(a) $I = i_R + i_L$ (b) $I = i_R + i_L + i_C$
(c) $I = i_R + i_L - i_C$ (d) $I = i_R - i_L - i_C$

346. In Fig. 4.43, the current i_L at $t = \infty$ is

(*a*) I (*b*) 0

(*c*) $\frac{IR}{L}$ (*d*) $\frac{IL}{RC}$

347. A 10 ohm resistance and a 10 ohm capacitor are connected in series across a 100 V ac supply. The rms voltage across resistance is

(*a*) $100\sqrt{2}$ V (*b*) 100V

(*c*) $\frac{100}{\sqrt{2}}$ V (*d*) none of the above

348. In Fig. 4.44, the switch has been in closed position for a long time. At $t = 0$, the switch is opened. At $t = 0^+$, the current through R_1 is

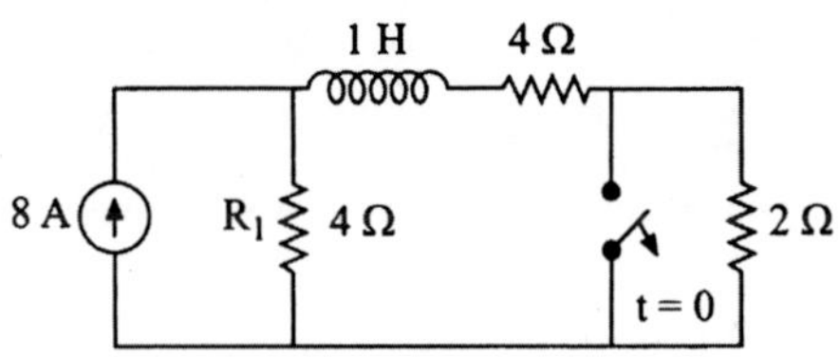

Fig. 4.44.

(*a*) 8 A (*b*) 4 A

(*c*) 6 A (*d*) 2 A

349. Twelve resistances each 2 ohm are connected to form a cube, one resistance being in each arm of cube. The resistance between any two diagonally opposite corners is

(*a*) 1 Ω (*b*) 2 Ω

(*c*) 4 Ω (*d*) 8 Ω

350. An R-L series circuit has R = 5 Ω and L = .01 H. It is fed by ac 50 Hz supply. The rms current in the circuit is 5A. The active power supplied by the source is

(*a*) 25 W (*b*) 50 W

(*c*) 125 W (*d*) 250 W

351. Figure 4.45 shows a dc circuit fed by a current source. With respect to terminals AB. Thevenin's voltage and Thevenin's resistance are

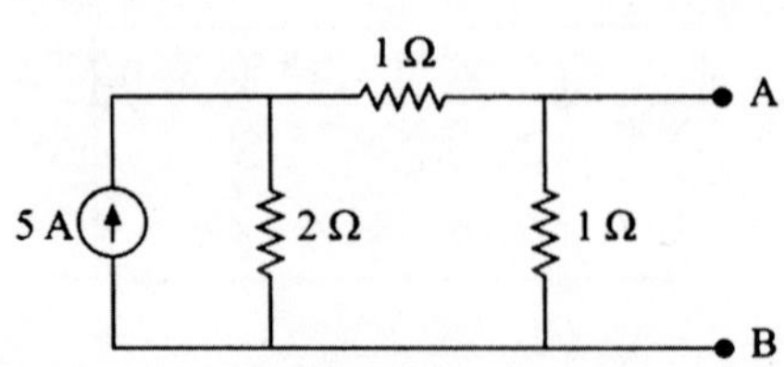

Fig. 4.45.

(*a*) $V_{TH} = 5$ V $R_{TH} = 0.75\ \Omega$

(*b*) $V_{TH} = 2.5$ V $R_{TH} = 0.75\ \Omega$

(*c*) $V_{TH} = 2.5$ V $R_{TH} = 1\ \Omega$

(*d*) $V_{TH} = 5$ V $R_{TH} = 1\ \Omega$

352. An ac voltage 20 sin 1000*t* feeds an R-C series circuit. R = 8 Ω and active power is 4 W. The power factor is

(*a*) 1 (*b*) 0.8

(*c*) 0.6 (*d*) 0.4

353. For a half wave rectified voltage, the form factor and peak factor respectively are

(*a*) 1.11 and $\sqrt{2}$ (*b*) 1.57 and 2

(*c*) 1.11 and 1.11 (*d*) 1.57 and $\sqrt{2}$

354. For a full rectified sine wave, the form factor and peak factor respectively are

(*a*) 1.57 and $\sqrt{2}$ (*b*) 1.57 and 2

(*c*) 1.11 and $\sqrt{2}$ (*d*) 1.11 and 2

355. A current wave is $i = 100\,e^{-60t}$. The initial and final values of current are

(*a*) 100 and 50 (*b*) 100 and 0

(*c*) 0 and 100 (*d*) 100 and 100

356. For the wave shown in Fig. 4.46, the average value is

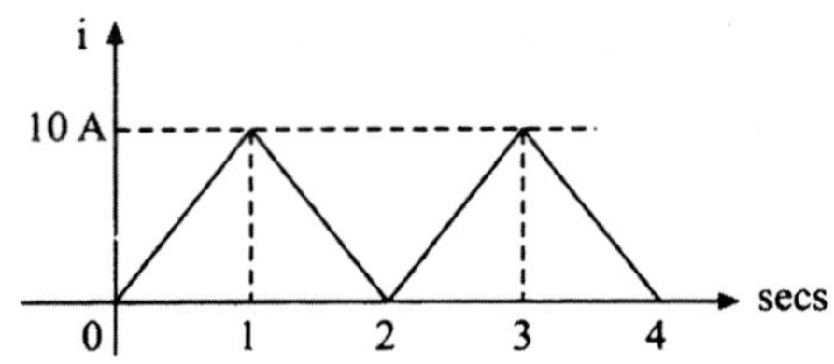

Fig. 4.46.

(*a*) 7.07 A (*b*) 5 A

(*c*) 2.5 A (*d*) 0.5 A

357. For the wave shown in Fig. 4.47 the average value is

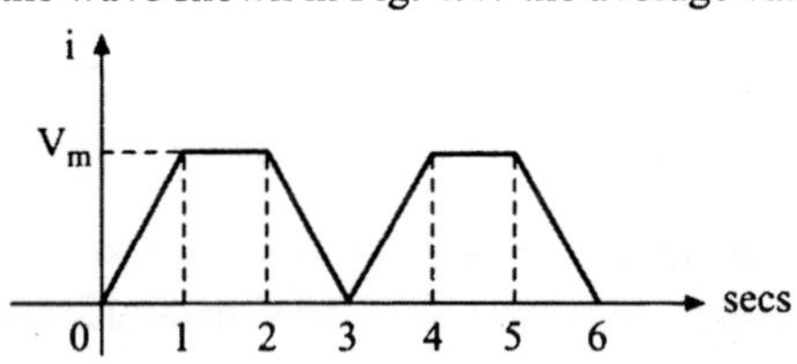

Fig. 4.47.

(*a*) $0.5\,V_m$ (*b*) $\frac{2}{3}\,V_m$

(*c*) $0.75\,V_m$ (*d*) $0.91V_m$

358. For the current wave $i = 10 + 10 \sin \theta$, the rms value is

(*a*) 10 A (*b*) 12.247 A

(*c*) 20 A (*d*) 14.14 A

359. Two voltages 50 V each have a phase difference of 45°. The rms value of resultant sum of the two voltages is

(*a*) 70.7 V (*b*) 80.4 V

(*c*) 92.4 V (*d*) 100 V

360. For the wave shown in Fig. 4.48, the average and rms values are

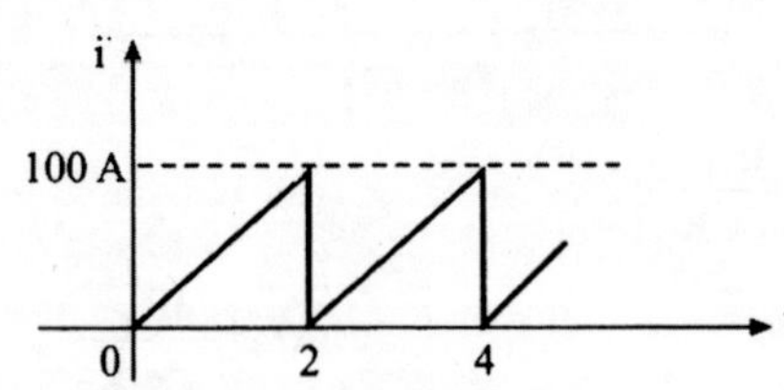

Fig. 4.48.

(*a*) 50 A and 57.73 A respectively
(*b*) 57.73 A and 50 A respectively
(*c*) 25 A and 57.73 A respectively
(*d*) 57.73 A and 70.7A respectively

361. For the wave shown in Fig. 4.49, the average value is

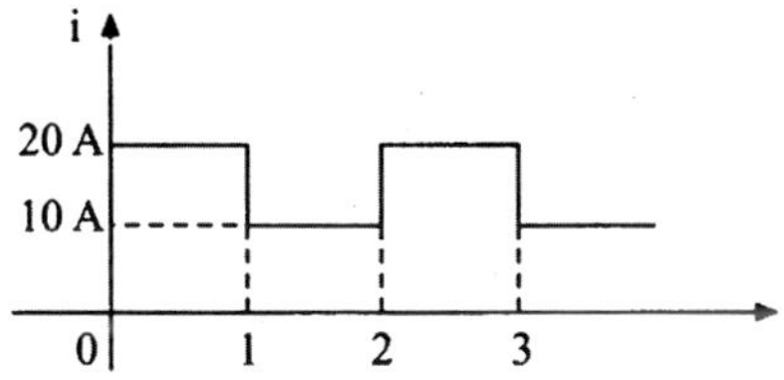

Fig. 4.49.

(*a*) 14.14 A (*b*) 15 A
(*c*) 17.17 A (*d*) 19.19 A

362. Two current sources each of rms value 10 A have frequencies 40 Hz and 50 Hz respectively. They are connected in series, the expression for resultant wave is
(*a*) $i = 10 \sin 40t + 10 \sin 50t$
(*b*) $i = 14.14 \sin 40t + 14.14 \sin 50t$
(*c*) $i = 14.14 \sin 80\pi t + 14.14 \sin \pi t$
(*d*) $i = 20 \sin 80\pi t + 20 \sin 100\pi t$

363. Two currents have the same peak value. Wave 1 is sinusoidal and wave 2 is rectangular. The ratio of their heating effects are
(*a*) 1 : 1 (*b*) $1 : \sqrt{2}$
(*c*) 1 : 2 (*d*) 1 : 1.5

364. A 50 Hz current has an amplitude of 10 A. The rate of change of current at $t = 0.005$ after $i = 0$ and is increasing is
(*a*) 2221.44 A/s (*b*) 0
(*c*) –2221.44 A/s (*d*) –3141.6 A/s

365. If A = 3 + *j*1 and B = 4 + *j*3, A + B =
(*a*) $7\angle 45°$ (*b*) $10\angle 45°$
(*c*) $8.062\angle 29.7°$ (*d*) $8.062\angle 73°$

366. If A = 3 + *j*1 and B = 4 + *j*3, A – B =
(*a*) $8.062\angle 29.7°$ (*b*) $10\angle 45°$
(*c*) $2.236\angle 243.43°$ (*d*) $6.72\angle 243.43°$

367. If A = 3 + *j*1, A^4 =
(*a*) $100\angle 18.43°$ (*b*) $100\angle 73.72°$
(*c*) $10\angle 18.43°$ (*d*) $10\angle 73.72°$

368. A series resonant circuit has parameters 10 Ω, 1 H and 0.1 F. It is excited by a voltage $v = 100 \sin \omega t$. At resonant frequency, the power dissipated in the circuit is
(*a*) 1000 W (*b*) 707 W
(*c*) 500 W (*d*) 414.14 W

369. A series resonant circuit is fed by a voltage having rms value V. At resonance, the voltage across inductance V_L and voltage across capacitance V_C are related as
(*a*) $V_L > V_C$ (*b*) $V_L < V_C$
(*c*) $V_L = V_C$ (*d*) $V_L = 0.5V_C$

370. A voltage $v = 100 \sin (100\pi t + 45°)$ V is applied to an impedance 10 + *j*10 ohm. The current is
(*a*) $10\angle 45°$A (*b*) $7.07\angle 45°$A
(*c*) $5\angle 45°$A (*d*) $5\angle 0$A

371. If P = 250 W and *pf* = 0.707 lagging, Q =
(*a*) 250 vars (*b*) –250 vars
(*c*) 176.7 vars (*d*) – 176.75 vars

372. A series R-L circuit has R = 1 Ω and L = 1 H. It is excited by a 10 V battery at $t = 0$. The steady state condition is reached at
(*a*) $t = 1s$ (*b*) $t = 0.1s$
(*c*) $t = 0s$ (*d*) $t = \infty$

373. The current wave shown in Fig. 4.50 is applied by a 0.1 F capacitor at $t = 0$. At $t = 0^+$, the voltage across capacitor is

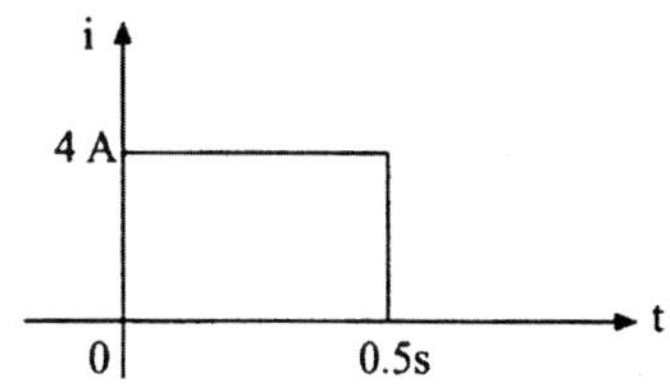

Fig. 4.50.

(*a*) 0 (*b*) 10 V
(*c*) 4 V (*d*) 20 V

374. The current wave shown in Fig. 4.50 is applied at $t = 0$ to a 0.1F capacitor. At $t = 1s$, the voltage across capacitor is
(*a*) 0V (*b*) 10V
(*c*) 20V (*d*) 40V

375. In Fig 4.51, the value of R should be

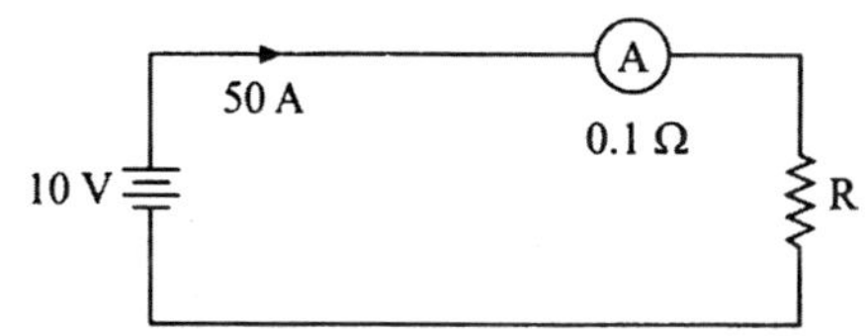

Fig. 4.51.

(*a*) 0.2 Ω (*b*) 0.1 Ω
(*c*) 0.05 Ω (*d*) 0.01 Ω

376. In Fig. 4.52, the value of R should be

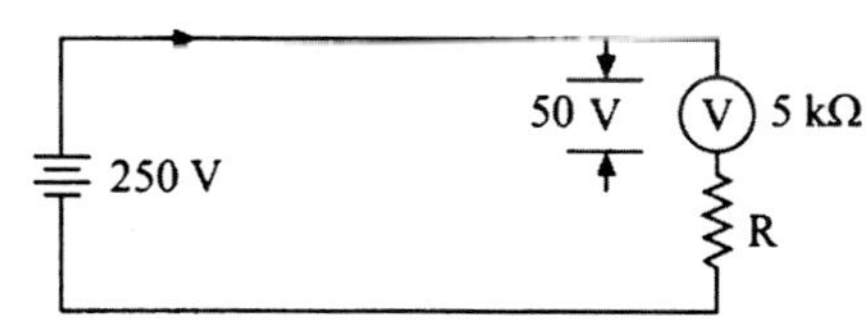

Fig. 4.52.

(*a*) 5 kΩ (*b*) 50 kΩ
(*c*) 20 kΩ (*d*) 10 kΩ

377. The frame of an electric motor is earthed through three earthing plates resistances 30 Ω, 20 Ω and 10 Ω respectively. During fault the energy dissipated by three plates are W_1, W_2 and W_3. Then
(*a*) $W_1 > W_2 > W_3$ (*b*) $W_1 < W_2 < W_3$
(*c*) $W_1 = W_2 = W_3$ (*d*) $W_1 + W_3 < W_2$

378. The power dissipated in 4 Ω resistance in Fig. 4.53 is

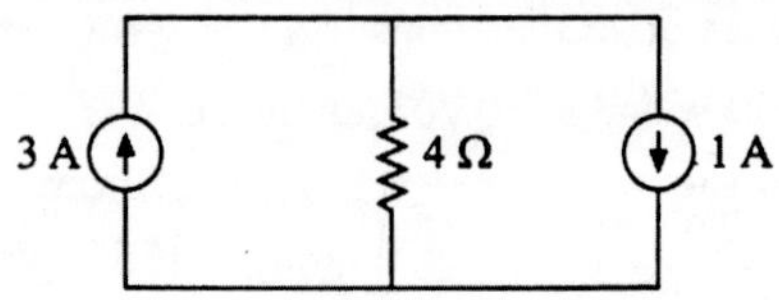

Fig. 4.53.

(*a*) 64 W (*b*) 32 W
(*c*) 16 W (*d*) 8 W

379. In Fig. 4.53, the power associated with 3 A source is
(*a*) 36 W (*b*) 24 W
(*c*) 16 W (*d*) 8 W

380. In Fig. 4.53, the power dissipated in 4 Ω resistance is P_1 and power associated with 3A source is P_2. Then
(*a*) $P_1 = P_2$ (*b*) $P_1 > P_2$
(*c*) $P_1 < P_2$
(*d*) P_1 may be equal or less than P_2

381. The current wave of Fig. 4.54, is passed through a 3H inductor. The voltage across inductance during the period 0 to 2 seconds

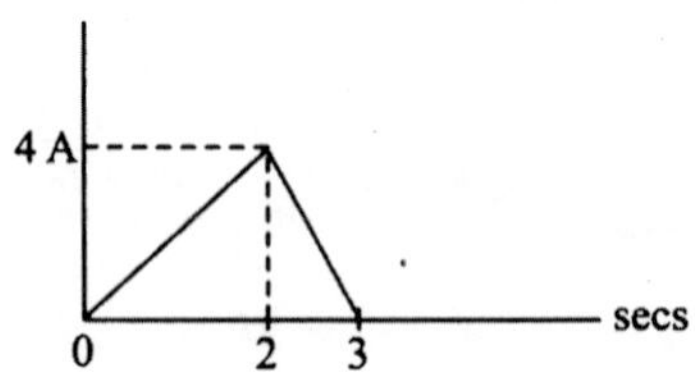

Fig. 4.54.

(*a*) is 3 V (*b*) is 6 V
(*c*) increases from 0 at $t = 0$ to 6 V at $t = 2$ seconds
(*d*) increases from 0 at $t = 0$ to 6 V at $t = 2$ seconds and then decreases

382. The current wave in Fig. 4.54, is passed through 3H inductor. The voltage across inductance during the period $2 < t < 3$ seconds is
(*a*) 12V (*b*) –12V
(*c*) 6V (*d*) –6V

383. The current in an RL series circuit is $1 - e^{-t}$ amperes. Then di/dt is
(*a*) $1 - e^{-t}$ (*b*) $-e^{-t}$
(*c*) e^{t} (*d*) ee^{-2t}

384. In an RL series circuit the inductance is 1 H and current at an instant is 0.6 A. The energy stored in magnetic field is
(*a*) 0.36 J (*b*) 0.18 J
(*c*) 0.09 J (*d*) 0.03 J

385. In a resistor the current is i and the voltage is v. The rate of dissipation of heat at any time t is
(*a*) vi (*b*) vit
(*c*) vt (*d*) vit^2

386. The resistance of a bulb at the time of full brightness is 1200 Ω. Then the resistance at room temperature is about
(*a*) 1200 Ω (*b*) 120 Ω
(*c*) 600 Ω (*d*) 10 Ω

387. In a network
(*a*) Number of links = Number of branches – Number of nodes + 1
(*b*) Number of links = Number of branches – Number of nodes
(*c*) Number of links = Number of branches – Number of nodes – 1
(*d*) Number of links = Number of branches – Number of nodes – 2

388. A tie set schedule shows
(*a*) incidence of branches to nodes
(*b*) incidence of branches to loops
(*c*) incidence of nodes to loops
(*d*) none of the above

389. A cut set schedule shows
(*a*) incidence of branches to loops
(*b*) incidence of branches to nodes
(*c*) incidence of nodes to loops
(*d*) none of the above

390. The strength of a controlled voltage source depends on
(*a*) value of current elsewhere in the circuit
(*b*) value of voltage elsewhere in the circuit
(*c*) value of current or voltage elsewhere in the circuit
(*d*) none of the above

391. In the set of equations

$$\begin{bmatrix} a_{11} & a_{12} & a_{13} \\ a_{21} & a_{22} & a_{23} \\ a_{31} & a_{32} & a_{33} \end{bmatrix} \begin{bmatrix} x_1 \\ x_2 \\ x_3 \end{bmatrix} = \begin{bmatrix} k_1 \\ k_2 \\ k_3 \end{bmatrix}$$

$x_1 =$

(*a*) $\dfrac{\begin{bmatrix} k_1 & a_{12} & a_{13} \\ k_2 & a_{22} & a_{23} \\ k_3 & a_{32} & a_{33} \end{bmatrix}}{\begin{bmatrix} a_{11} & a_{12} & a_{13} \\ a_{21} & a_{22} & a_{23} \\ a_{31} & a_{32} & a_{33} \end{bmatrix}}$ (*b*) $\dfrac{\begin{bmatrix} a_{11} & k_1 & a_{13} \\ a_{21} & k_2 & a_{23} \\ a_{31} & k_3 & a_{33} \end{bmatrix}}{\begin{bmatrix} a_{11} & a_{12} & a_{13} \\ a_{21} & a_{22} & a_{23} \\ a_{31} & a_{32} & a_{33} \end{bmatrix}}$

(*c*) $\dfrac{\begin{bmatrix} a_{11} & a_{12} & a_{13} \\ a_{21} & a_{22} & a_{23} \\ a_{31} & a_{32} & a_{33} \end{bmatrix}}{\begin{bmatrix} k_1 & a_{12} & a_{13} \\ k_2 & a_{22} & a_{23} \\ k_3 & a_{32} & a_{33} \end{bmatrix}}$ (*d*) $\dfrac{\begin{bmatrix} a_{11} & a_{12} & k_1 \\ a_{21} & a_{22} & k_2 \\ a_{31} & a_{32} & k_3 \end{bmatrix}}{\begin{bmatrix} a_{11} & a_{12} & a_{13} \\ a_{21} & a_{22} & a_{23} \\ a_{31} & a_{32} & a_{33} \end{bmatrix}}$

392. In the circuit of Fig. 4.55, the current through battery E_2 is

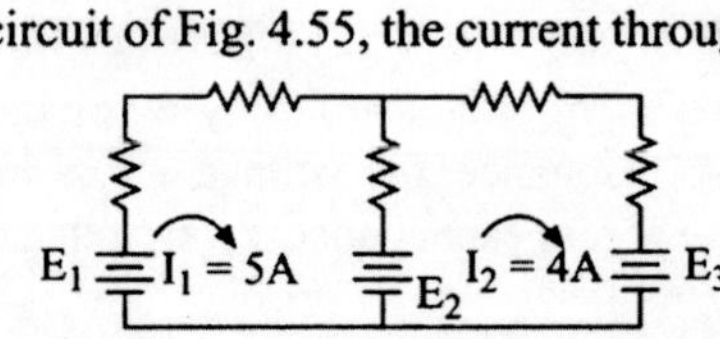

Fig. 4.55.

(*a*) 1 A discharging (*b*) 1 A charging
(*c*) 9 A discharging (*d*) 9 A charging

393. In the Fig. 4.55, the current through battery E_3 is
(*a*) 4 A charging
(*b*) 4 A discharging
(*c*) 4 A charging or discharging depending on value of E_3
(*d*) 4 A charging or discharging depending on values of E, E_2, E_3

394. A dc network has 3 independent nodes and 4 loops. In nodal analysis, the number of equations is
(*a*) 4 (*b*) 3
(*c*) 2 (*d*) 1

395. The current source in Fig. 4.56, can be replaced by

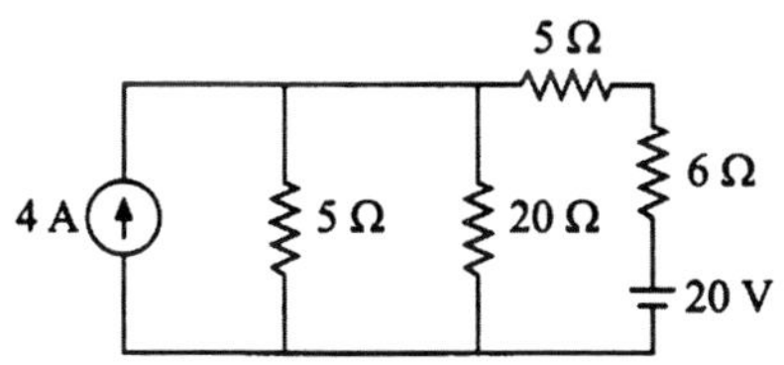

Fig. 4.56.

(*a*) a voltage source of 20 V in series with 5 Ω resistance
(*b*) a voltage source of 16 V in series with 4 Ω resistance
(*c*) a voltage source of 20 V in series with 4 Ω resistance
(*d*) a voltage source of 16 V in series with 5 Ω resistance

396. In Fig. 4.57, the current through 10 V battery is

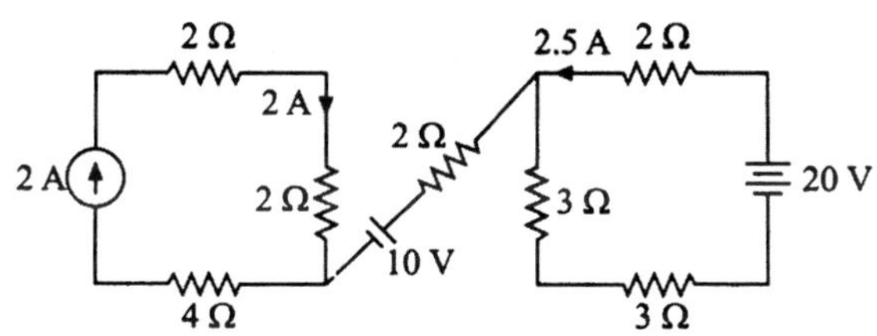

Fig. 4.57.

(*a*) 0 (*b*) 0.5 A charging
(*c*) 0.5A discharging (*d*) none of the above

397. In the analogy between electric and magnetic circuits, flux is analogous to
(*a*) current
(*b*) voltage
(*c*) rate of change of current
(*d*) rate of change of voltage

398. In the analogy between electric and magnetic circuits permeability is analogus to
(*a*) resistivity (*b*) conductivity
(*c*) electric field intensity (*d*) current density

399. Hysteresis loss is proportional to
(*a*) f and $B_m^{1.6}$ (*b*) f and B_m
(*c*) f^2 and B_m (*d*) f^2 and B_m^2

400. Eddy current loss is proportional to
(*a*) f and B_m^2 (*b*) f^2 and B_m^2
(*c*) f^2 and B_m (*d*) f and B_m

401. In Fig. 4.58, the total inductance of the circuit is

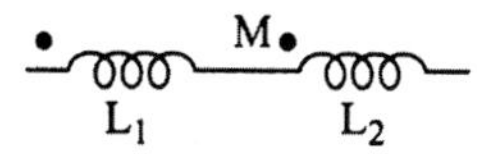

Fig. 4.58.

(*a*) $L_1 + L_2$ (*b*) $L_1 + L_2 + M$
(*c*) $L_1 + L_2 + 2M$ (*d*) $L_1 - L_2 - M$

402. In Fig. 4.59 the total inductance of the circuit is

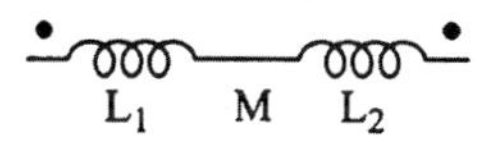

Fig. 4.59.

(*a*) $L_1 + L_2$ (*b*) $L_1 + L_2 - M$
(*c*) $L_1 + L_2 - 2M$ (*d*) $L_1 + L_2 - M^2$

403. Two coupled coils lie in parallel planes such that 50% of magnetic flux produced by one links the other. The coefficient of coupling is
(*a*) 0.5 (*b*) 0.25
(*c*) 0.125 (*d*) $\sqrt{0.5}$

404. Two coupled coils A and B are connected in series. The total inductances in series aiding and series opposing connections are L_A and L_B. The mutual inductance M is
(*a*) $\dfrac{L_A + L_B}{4}$ (*b*) $\dfrac{L_A - L_B}{4}$
(*c*) $\dfrac{L_A + L_B}{2}$ (*d*) $\dfrac{L_A - L_B}{2}$

405. For the three coupled coils shown in Fig. 4.60, KVL equation is

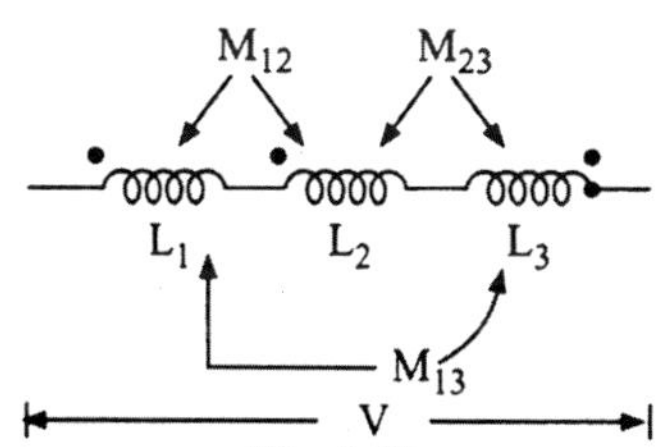

Fig. 4.60.

(*a*) $v = (L_1 + L_2 + L_3)\dfrac{di}{dt}$
(*b*) $v = (L_1 - L_2 - L_3 - M_{13})\dfrac{di}{dt}$
(*c*) $v = (L_1 + L_2 + L_3 + 2M_{12} - 2M_{23} - 2M_{13})\dfrac{di}{dt}$
(*d*) $v = (L_1 + L_2 + L_3 - 2M_{12} + 2M_{23} + 2M_{13})\dfrac{di}{dt}$

406. Two similar coupled coils are connected in series. The total inductance in series aiding connection is 100 mH. If current is 3A, the energy stored in magnetic field is
(*a*) 0.9 J (*b*) 0.45 J
(*c*) 0.225 J (*d*) 1.8 J

407. Two coupled coils have self inductances L_1 and L_2. The coefficient of coupling is k. The mutual inductance is
(*a*) $kL_1 L_2$ (*b*) $k^2\sqrt{L_1 L_2}$
(*c*) $k\sqrt{L_1 L_2}$ (*d*) $\sqrt{kL_1 L_2}$

408. Two coils have their axes perpendicular to each other. The coefficient of coupling k is

(*a*) 1

(*b*) 0.5

(*c*) more than 0 but less than 1

(*d*) 0

409. Two similar coils have self inductance of 1 mH each. Coefficient of coupling is 0.5. The mutual inductance M is

(*a*) 0.25 mH (*b*) 0.5 mH

(*c*) 0.707 mH (*d*) 1 mH

410. KVL equation for the circuit of Fig. 4.61 is

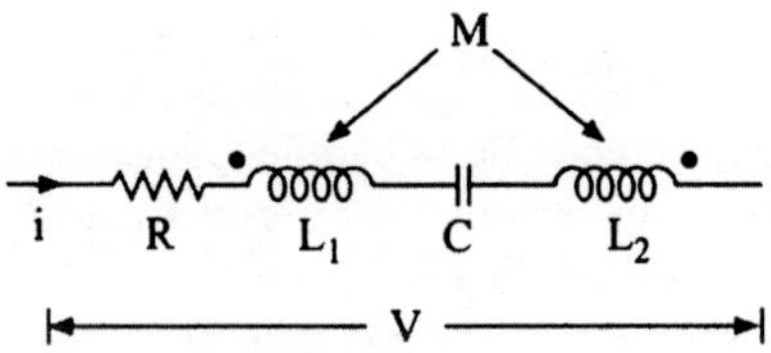

Fig. 4.61.

(*a*) $v = Ri + L_1 \frac{di}{dt} + L_2 \frac{di}{dt} - M\frac{di}{dt} + \frac{1}{C}\int idt$

(*b*) $v = Ri - \frac{1}{C}\int idt + (L_1 + L_2 - 2M)\frac{di}{dt}$

(*c*) $v = Ri + \frac{1}{C}\int idt + (L_1 + L_2 - 2M)\frac{di}{dt}$

(*d*) $v = Ri + \frac{1}{C}\int idt + (L_1 + L_2 + 2M)\frac{di}{dt}$

411. For the circuit of Fig. 4.62

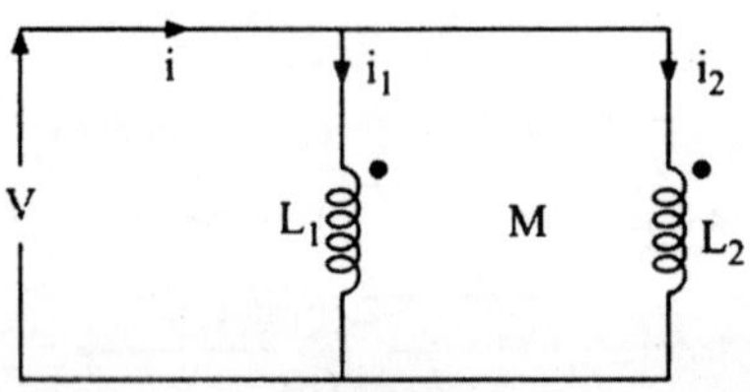

Fig. 4.62.

(*a*) $v = L_1 \frac{di_1}{dt} - M\frac{di}{dt}$ (*b*) $v = L_1 \frac{di_1}{dt} - M\frac{di_2}{dt}$

(*c*) $v = L_1 \frac{di_1}{dt} + M\frac{di}{dt}$ (*d*) $v = L_1 \frac{di_1}{dt} + M\frac{di_2}{dt}$

412. In Fig. 4.63

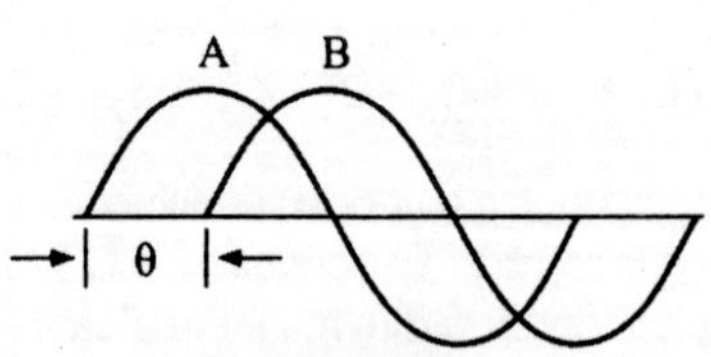

Fig. 4.63.

(*a*) Wave A is lagging wave B by $\angle\theta$

(*b*) Wave A is leading wave B by $\angle\theta$

(*c*) Wave A is leading wave B by $\angle\theta$ and the two waves have same rms values

(*d*) Wave A is lagging wave B by $\angle\theta$ and the two waves have different rms values

413. Two sinusoidal waves 1 and 2 have peak values V_{m1} and V_{m2}. Wave 1 is leading wave 2 by $\angle\theta$. Which of the following is correct

(*a*) $v_1 = V_{m1} \sin\theta$ and $v_2 = V_{m2}\cos\theta$

(*b*) $v_1 = V_{m1} \sin\omega t$ and $v_2 = V_{m2}\sin(\omega t - \theta)$

(*c*) $v_1 = V_{m1} \sin(\omega t + \theta/_2)$ and $v_1 - v_{m2} \sin\omega t$

(*d*) none of the above

414. If the time is counted from the instant when an ac wave is at its peak value, we get a

(*a*) sine wave (*b*) cosine wave

(*c*) distorted sine wave (*d*) none of the above

415. In drawing phasor diagram

(*a*) rms values are used

(*b*) peak values are used

(*c*) either rms or peak values are used

(*d*) either rms or peak or average values are used

416. $j^6 =$

(*a*) j (*b*) j^2

(*c*) j^3 (*d*) $-j$

417. Two phasors having rms values V_1 and V_2 are added to give a phasor having rms value V_3. V_3 is maximum if phase angle between V_1 and V_2 is

(*a*) 90° (*b*) 180°

(*c*) 270° (*d*) 360°

418. For addition and subtraction operations, phasors

(*a*) must be represented in polar form

(*b*) must be represented in rectangular form

(*c*) may be represented in any form

(*d*) none of the above

419. If $B = 4 + j3$, $(B)^{1/3} =$

(*a*) $2\angle 30°$ (*b*) $2\angle 36.87°$

(*c*) $1.71\angle 12.3°$ (*d*) $1.71\angle 30°$

420. If $A = 3 + j1$, $A^4 =$

(*a*) $3.16\angle 18.4°$ (*b*) $100\angle 73.72°$

(*c*) $100\angle 18.4°$ (*d*) $3.16\angle 73.72°$

421. For a rectified sinusoidal wave clipped at 0.707 its peak value (Fig. 4.64), the average value is

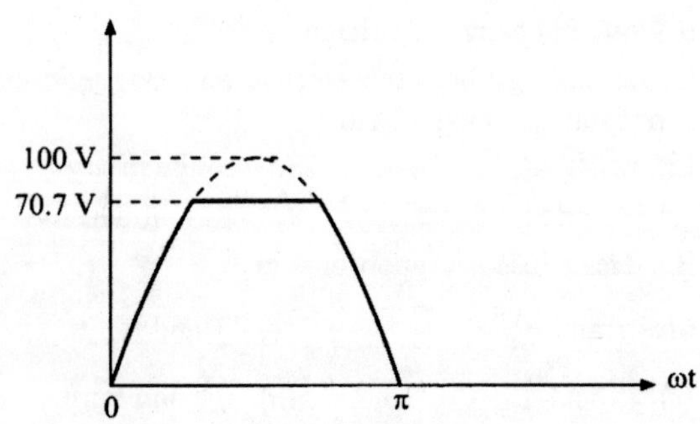

Fig. 4.64.

(*a*) 70.7 V (*b*) 63.7 V

(*c*) 54 V (*d*) 15 V

422. The rms value of wave in Fig. 4.65 is

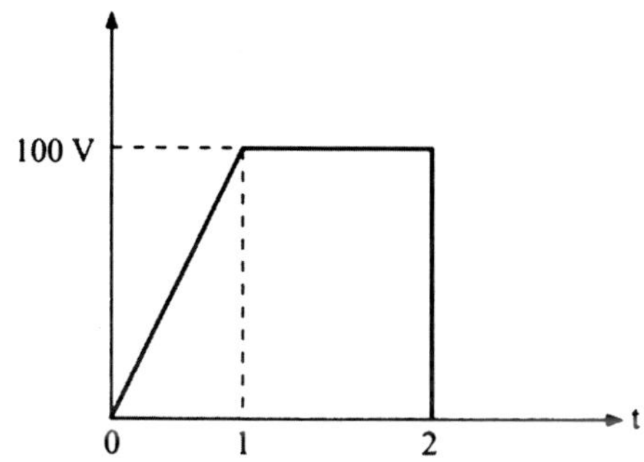

Fig. 4.65.

(*a*) about 95 V (*b*) about 80 V
(*c*) about 50 V (*d*) about 25 V

423. The average value of wave in Fig. 4.65 is
(*a*) 90 V (*b*) 75 V
(*c*) 60 V (*d*) 50 V

424. Two voltages are $50 \angle 0$ V and $75 \angle -60°$ V. The sum of these voltages is
(*a*) $109 \angle -60°$ V (*b*) $109 \angle -25°$ V
(*c*) $109 \angle -36.6°$ V (*d*) $100 \angle -50.1°$ V

425. Wave A = 100 sin ωt and Wave B = 100 cos ωt. Then
(*a*) rms values of the two waves are equal
(*b*) rms value of A is more than that of B
(*c*) rms value of A is less than that of B
(*d*) rms values of the two may or may not be equal

426. The capacitance of a parallel plate capacitor varies as per the equation $C(t) = C_0(1 - \cos \omega t)$. It is connected to a battery of V volts. The equation for current is
(*a*) $\omega C_0 V(1 - \cos\omega t)$ (*b*) $\omega C_0 V(1 - \sin\omega t)$
(*c*) $\omega C_0 V(1 - \sin 2\omega t)$ (*d*) $\omega C_0 V \sin \omega t$

427. Which of the following is correct?
(*a*) voltage across a capacitance can change instantaneously
(*b*) voltage across a capacitance cannot change instantaneously
(*c*) charge across a capacitance is always constant
(*d*) charge across a capacitance can change instantaneously

428. The flux linkage ψ, self-inductance L and current i are related as
(*a*) $\psi = L/i$ (*b*) $\psi = Li$
(*c*) $\psi = Li^2$ (*d*) $\psi = \frac{1}{2} Li^2$

429. The current voltage relation for an inductance L is
(*a*) $v = L \int_{-\infty}^{t} i dt$ (*b*) $i = L \int_{-\infty}^{t} v dt$
(*c*) $i = \frac{1}{L} \int_{-\infty}^{t} v dt$ (*d*) $v = \frac{1}{L} \int_{-\infty}^{t} i dt$

430. Flux linkages ψ and voltage v are related as
(*a*) $v = \int_{-\infty}^{t} \psi dt$ (*b*) $\psi = \int_{-\infty}^{t} v dt$
(*c*) $\psi = \frac{dv}{dt}$ (*d*) $\psi = \frac{d^2 v}{dt^2}$

431. Current denisly J, conductivity σ and electric field E in a conductor are related as
(*a*) $J = \sigma E$ (*b*) $\sigma = JE$
(*c*) $J = \frac{E}{\sigma}$ (*d*) $J = \sigma E^2$

432. In a conductor, voltage v, conductance G and charge q are related as
(*a*) $v = \frac{1}{G} \frac{dq}{dt}$ (*b*) $v = G \frac{dq}{dt}$
(*c*) $v = Gq$ (*d*) $q = vG$

433. The energy dissipated W in a resistance having voltage v and current i is given by
(*a*) $W = \int_{-\infty}^{t} vi\, dt$ (*b*) $W = vit^2$
(*c*) $W = \int_{-\infty}^{t} \frac{v}{i} dt$ (*d*) $W = \int_{-\infty}^{t} vi^2\, dt$

434. Pico-farad is equal to
(*a*) 10^{-6} farad (*b*) 10^{-12} farad
(*c*) 10^{-3} farad (*d*) 10^{-9} farad

435. Giga watt is equal to
(*a*) 10^{12} W (*b*) 10^{9} W
(*c*) 10^{6} W (*d*) 10^{3} W

436. Terra-watt is equal to
(*a*) 10^{3} W (*b*) 10^{6} W
(*c*) 10^{9} W (*d*) 10^{12} W

437. Henry is equivalent to
(*a*) Volts/ampere (*b*) Weber/volt
(*c*) Weber/ampere (*d*) Weber/ampere^2

438. Volt is equivalent to
(*a*) joule/coulomb (*b*) joule/watt
(*c*) joule/ampere (*d*) joule/volt

439. Watt is equivalent to
(*a*) joule/ampere (*b*) joule/second
(*c*) joule/volt (*d*) joule/coulomb

440. The equivalent unit of joule is
(*a*) newton (*b*) newton-metre
(*c*) newton-metre^2 (*d*) none of the above

441. The model of a system
(*a*) is unique
(*b*) depends on desired degree of accuracy
(*c*) depends on the type of analysis
(*d*) both (*b*) and (*c*)

442. A capacitor whose capacitance

$$C_t = C_0(1 - \cos \omega t)$$

is fed by a sinusoidal voltage $v = V_0 \sin \omega t$. The equation for current i is

(*a*) $\omega C_0 V_0 \cos \omega t$

(*b*) $\omega C_0 V_0 \sin \omega t$

(*c*) $\omega C_0 V_0 (\cos \omega t - \cos^2 \omega t)$

(*d*) none of the above

443. A 1 μF capacitor is connected to 12 V battery. The energy stored in the capacitor is

(*a*) 12×10^{-6} J (*b*) 24×10^{-6} J

(*c*) 48×10^{-6} J (*d*) 72×10^{-6} J

444. A current increasing at the rate of 1 A/s is passed through a 1 H inductor. At $t = 0$, $i = 0$. The flux linkages after 1 sec are

(*a*) 0.5 Wb-turn (*b*) 1 Wb turn

(*c*) 2 Wb turn (*d*) 4 Wb turn

445. A current increasing at the rate of 1 A/s is passed through 1 F capacitor. At $t = 0$, $i = 0$. The charge stored after 1 second is

(*a*) 0.5 C (*b*) 1 C

(*c*) 2 C (*d*) 4 C

446. Fig. 4.66 shows an R-L circuit. The current is $i = 1 - e^{-t}$. The rate of change of current at $t = 1$ second is

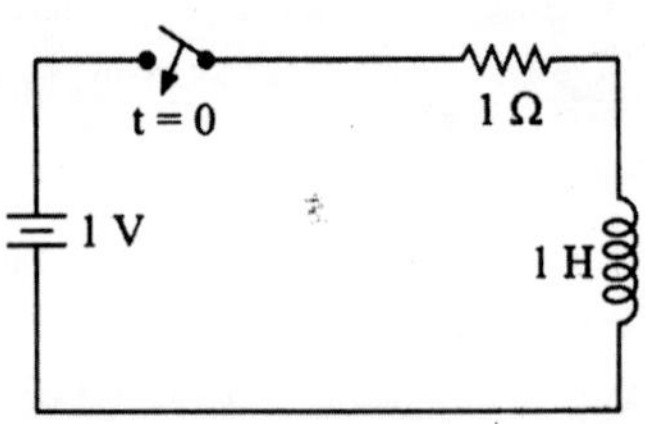

Fig. 4.66.

(*a*) 1 A/s (*b*) 0.63 A/s

(*c*) 0.37 A/s (*d*) 0

447. In Fig. 4.67, the capacitor is charged to 1V. At $t = 0$ the switch is closed so that $i = e^{-t}$. When $i = 0.37$A, the voltage across capacitor is

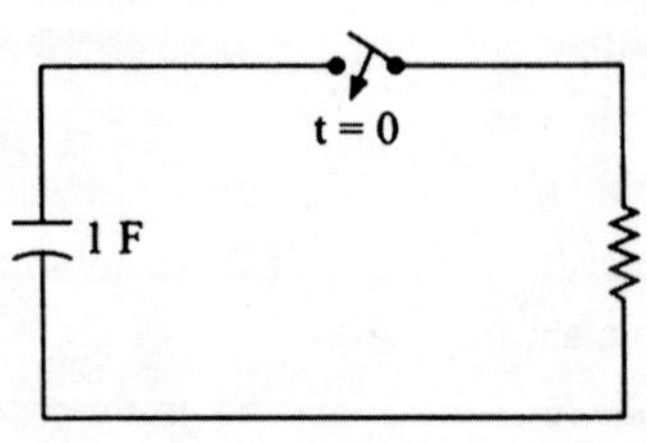

Fig. 4.67.

(*a*) 1V (*b*) 0.63 V

(*c*) 0.37 V (*d*) 0.185 V

448. In a coupled coil the primary has 100 turns and secondary has 200 turns. The primary produces, a flux $\phi = e^{-t}$. The coefficient of coupling is 1. The voltage induced in secondary is

(*a*) $200\, e^{-t}$ V (*b*) $200\, e^{-t}$ V

(*c*) 200 V (*d*) none of the above

449. For node 1 in Fig. 4.68, KCL equation is

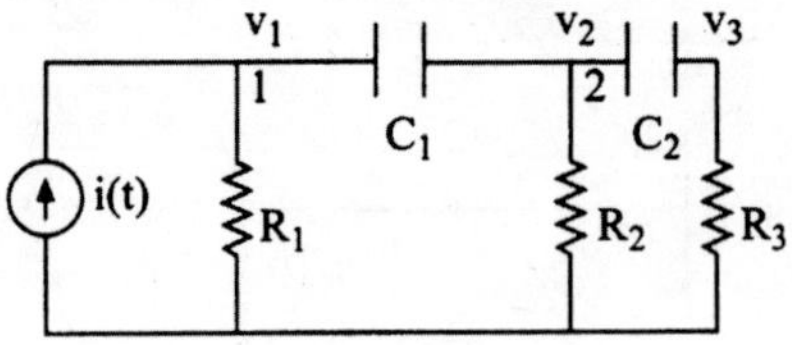

Fig. 4.68.

(*a*) $i(t) = \dfrac{v_1}{R_1} + C_1 \dfrac{d}{dt}(v_2 - v_1)$

(*b*) $i(t) = -\dfrac{v_1}{R_1} + C_1 \dfrac{d}{dt}(v_1 - v_2)$

(*c*) $i(t) = \dfrac{v_1}{R_1} + C_1 \dfrac{d}{dt}(v_1 - v_2)$

(*d*) none of the above

450. For node 2 in Fig. 4.68, KCL equation is

(*a*) $0 = \dfrac{v_2}{R_2} + C_1 \dfrac{d}{dt}(v_1 - v_2)$

(*b*) $0 = \dfrac{v_2}{R_2} + C_1 \dfrac{d}{dt}(v_2 - v_1) + C_2 \dfrac{d}{dt}(v_2 - v_3)$

(*c*) $0 = -\dfrac{v_2}{R_2} + C_1 \dfrac{d}{dt}(v_2 - v_1) + C_2 \dfrac{d}{dt}(v_2 - v_3)$

(*d*) $0 = -\dfrac{v_2}{R_2} - C_1 \dfrac{d}{dt}(v_2 - v_1) + C_2 \dfrac{d}{dt}(v_2 - v_3)$

451. For the network of Fig. 4.69 KVL for first loop is

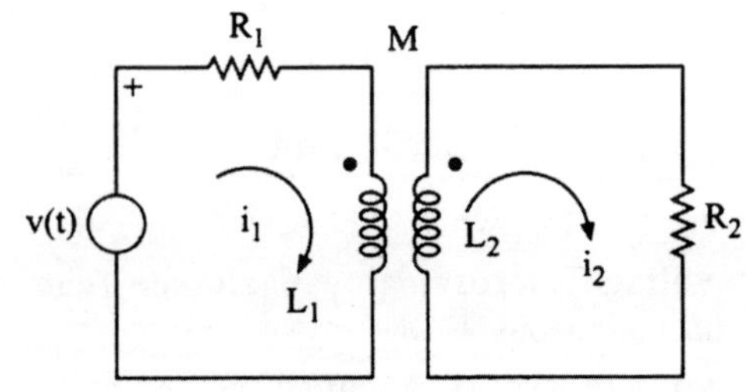

Fig. 4.69.

(*a*) $v(t) = R_1 i_1 + L_1 \dfrac{di_1}{dt} + M \dfrac{di_2}{dt}$

(*b*) $v(t) = R_1 i_1 - L_1 \dfrac{di_1}{dt} - M \dfrac{di_2}{dt}$

(*c*) $v(t) = R_1 i_1 + L_1 \dfrac{di_1}{dt} - M \dfrac{di_2}{dt}$

(*d*) $v(t) = R_1 i_1 - L_1 \dfrac{di_1}{dt} + M \dfrac{di_2}{dt}$

452. KVL for second loop in Fig. 4.69 is

(*a*) $0 = R_2 i_2 - L_2 \dfrac{di_2}{dt} - M \dfrac{di_1}{dt}$

(*b*) $0 = R_2 i_2 + L_2 \dfrac{di_2}{dt} - M \dfrac{di_1}{dt}$

(*c*) $0 = R_2 i_2 - L_2 \frac{di_2}{dt} + M\frac{di_1}{d}$

(*d*) $0 = R_2 i_2 + L_2 \frac{di_2}{dt} + M\frac{di_1}{dt}$

453. Two capacitors of 1 F each are connected in parallel. Another IF capacitor is connected in series with the parallel combination. The total capacitance is

(*a*) 3 F (*b*) 2 F

(*c*) $\frac{2}{3}$ F (*d*) $\frac{1}{3}$ F

454. In Fig. 4.70 the switch has been in position 1 for a long time. At $t = 0$, the switch is moved to position 2. The equation for transient current is

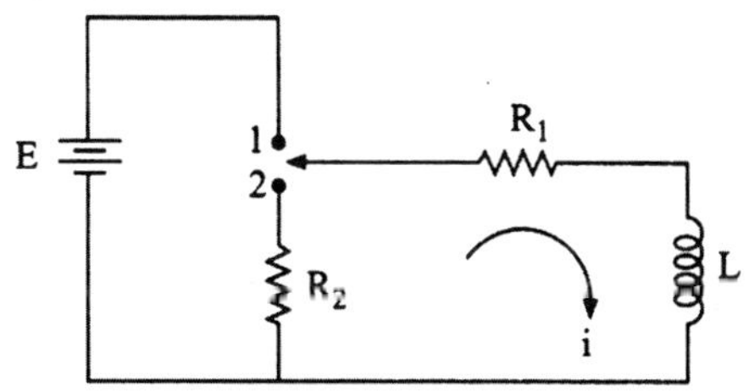

Fig. 4.70.

(*a*) $i = \frac{E}{R_2} e^{-R_1 t/L}$ (*b*) $i = \frac{E}{R_1} e^{-R_2 t/L}$

(*c*) $i = \frac{E}{R_1} e^{-(R_1 + R_2)t/L}$ (*d*) $i = \frac{E}{R_1 + R_2} e^{-(R_1 + R_2)t/L}$

455. In a series RLC circuit the phase angle θ of impedance $Z \angle \theta$ is positive. The circuit will behave as

(*a*) Resistive circuit
(*b*) RC circuit
(*c*) RL circuit
(*d*) either RL or RC circuit

456. A variable resistance R and capacitive reactance X_c are connected in series. As R is varied, the locus of Z is

(*a*) a straight line parallel to Y-axis in 4th quadrant
(*b*) a straight line parallel to X-axis is 4th quadrant
(*c*) a straight line parallel to Y-axis in second quadrant
(*d*) a straight line parallel to Y-axis in first quadrant

457. A variable resistance R and capacitive reactance X_c are fed by an ac voltage. The current locus is

(*a*) a semi-circle in 4th quadrant
(*b*) a semi-circle in first quadrant
(*c*) a straight line in 4th quadrant
(*d*) a straight line in first quadrant

458. For a series circuit of variable R and X_L, the impedance locus is

(*a*) a straight line parallel to X-axis in first quadrant
(*b*) a straight line parallel to X-axis in fourth quadrant
(*c*) a straight line parallel to Y-axis in second quadrant
(*d*) a straight line parallel to Y-axis in first quadrant

459. For a series circuit of R and variable X_L, the admittance locus is

(*a*) a semi-circle in first quadrant
(*b*) a semi-circle in fourth quadrant
(*c*) a straight line in first quadrant
(*d*) a straight line in fourth quadrant

460. In the circuit of Fig. 4.71,

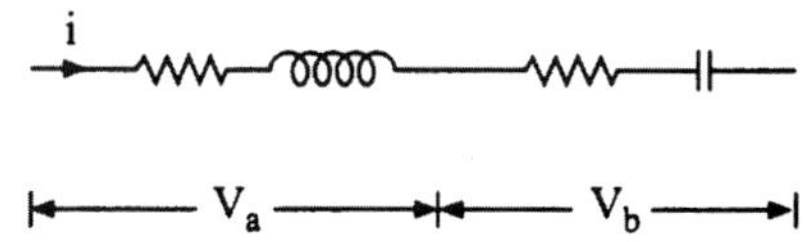

Fig. 4.71.

(*a*) V_a would always lag V_b
(*b*) V_a would always lead V_b
(*c*) V_a and V_b can be in phase
(*d*) phase angle between V_a and V_b would always be 90°

461. In the circuit of Fig 4.72.

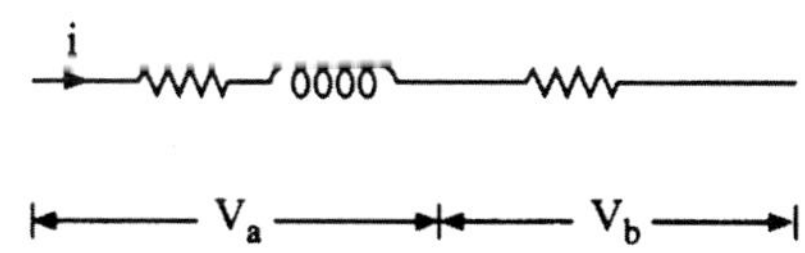

Fig. 4.72.

(*a*) V_a would always lag V_b
(*b*) V_a would always lead V_b
(*c*) V_a and V_b would always be in phase
(*d*) V_a may lag or lead V_b

462. A current 10 sin (200*t*) A flows through a 5 mH inductance. The expression for voltage is

(*a*) 10 sin (200*t* + 90°) (*b*) 10 sin (200*t* – 90°)
(*c*) 10 sin (1000*t* + 90°) (*d*) 10 sin (1000*t* – 90°)

463. The three sides of an admittance triangle denote

(*a*) G, B, Y (*b*) R, X, Z
(*c*) G, B, Z (*d*) R, X, Y

464. In a series L-C circuit fed by ac voltage, voltage drops across L and C are 40 V and 30 V respectively. The supply voltage is

(*a*) 70 V (*b*) 10 V
(*c*) 50 V (*d*) 1.333 V

465. In a series ac circuit a voltage of 20 V at 25 Hz causes a current of 0.20 A, while the same voltage at 75 Hz causes a current of 0.12 A. The circuit consists of

(*a*) R and C in parallel (*b*) R and C in series
(*c*) R and L in series (*d*) either (*a*) or (*b*)

466. The equation $\sum_{k=1}^{b} v_k i_k = 0$ where $v_1, v_2 \ldots v_b$ are the instantaneous branch voltages and $i_2 \ldots i_b$ are the instantaneous branch currents pertains to

(*a*) compensation theorem
(*b*) bisectional theorem
(*c*) Tellegen's theorem
(*d*) Reciprocity theorem

467. In Fig. 4.73, the current I_4 is

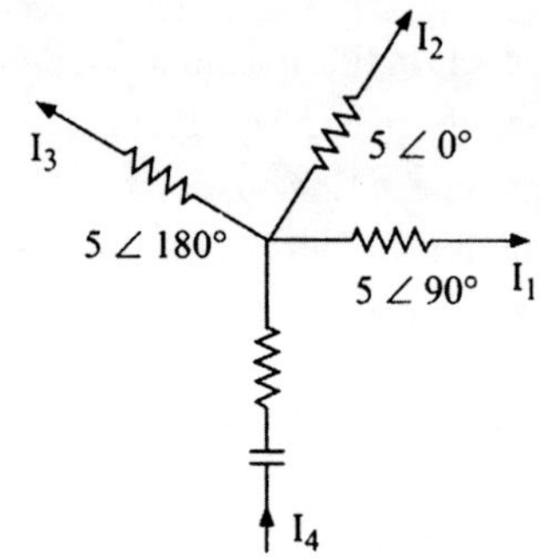

Fig. 4.73.

(*a*) 10∠0 (*b*) 5∠90°
(*c*) 5∠270° (*d*) 10∠90°

468. In Fig. 4.74, the magnitude of V

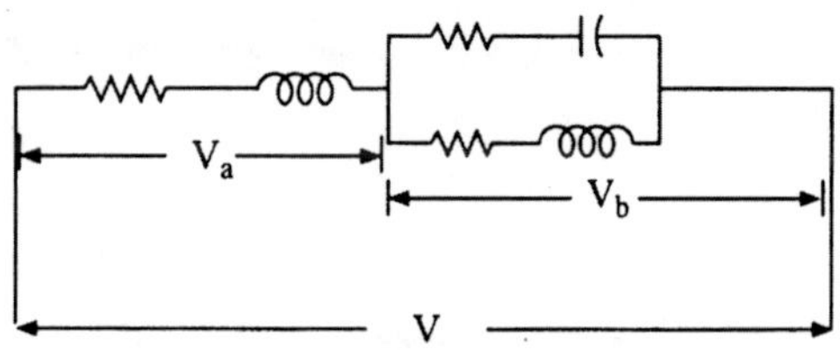

Fig. 4.74.

(*a*) must be more than that of V_a
(*b*) must be more than that of V_b
(*c*) must be more than that of either V_a or V_b
(*d*) can be equal to or less than or more than that of V_a or V_b

469. A two branch parallel tuned circuit has 100 Ω resistance and 0. 1 H inductance in one branch and 10^{-6} F capacitance in the other. At resonance frequency, the dynamic resistance is

(*a*) 10 Ω (*b*) 100 Ω
(*c*) 1000 Ω (*d*) 10000 Ω

470. A two branch parallel tuned circuit has a coil of resistance 100 Ω and inductance 0.005 H in one branch and a 10 μF capacitor in the second branch. At resonance frequency the dynamic resistance is

(*a*) 1 Ω (*b*) 10 Ω
(*c*) 100 Ω (*d*) more than 100 Ω

471. A two branch parallel tuned circuit has a coil of resistance 100 Ω and inductance 0.005 H in one branch and a 10 *n*F capacitor in the second branch. The resonance frequency is

(*a*) 100 rad/sec
(*b*) 1000 rad/sec
(*c*) 10000 rad/sec
(*d*) more than 10000 rad/sec

472. The resonant frequency of a series resonant circuit is

(*a*) arithmetic mean of half power frequencies
(*b*) independent of inductance of the circuit
(*c*) independent of capacitance of the circuit
(*d*) geometric mean of half power frequencies

473. In a series resonant circuit Q is more than 10. Then the lower half power frequency ω_1 and resonant frequency ω_0 are related as

(*a*) $\omega_1 = \omega_0$ + bandwidth
(*b*) $\omega_1 = \omega_0$ + 0.5 (bandwidth)
(*c*) $\omega_1 = \omega_0$ – 0.5 (bandwidth)
(*d*) $\omega_1 = \omega_0$ + bandwidth

474. A two branch tuned circuit has a coil of resistance R and inductance L in one branch and capacitance C in the second branch. If R is increased, the dynamic resistance

(*a*) increases (*b*) decreases
(*c*) remains constant
(*d*) may increase or decrease

475. A two branch parallel tuned circuit has a coil of resistance R and inductance L in one branch and capacitance C in the second branch. If L is increased, the dynamic resistance

(*a*) decreases (*b*) increases
(*c*) remains constant
(*d*) may increase or decrease

476. A two branch parallel tuned circuit has a capacitance C in one of the two branches. As C is decreased, the dynamic resistance

(*a*) increases (*b*) decreases
(*c*) remains constant
(*d*) may increase or decrease

477. A series resonant circuit has R = 10 Ω, L = 1 mH and C = 1 μF. If R is increased to 20 Ω, the resonant frequency

(*a*) increases
(*b*) decreases
(*c*) remains constant
(*d*) may increase or decrease

478. A series resonant circuit has inductance L. If L is increased, the resonant frequency

(*a*) increases
(*b*) decreases
(*c*) remains constant
(*d*) may increase or decrease

479. A coil of resistance R and inductance L is connected in series with 10 μF capacitor. The resonant frequency is 1000 Hz. Another 10 μF capacitor is connected in parallel with the above capacitor. The new resonance frequency will be

(*a*) 1000 Hz
(*b*) more than 1000 Hz
(*c*) less than 1000 Hz
(*d*) may be more or less than 1000 Hz

480. A coil of resistance R and inductance L is connected in series with 5 μF capacitor. The source frequency is equal to resonance frequency and circuit current is 0.5 A. Another 5 μF capacitor is connected in parallel with the above capacitor. The circuit current will be

(*a*) 0.5 A (*b*) more than 0.5 A
(*c*) less than 0.5 A (*d*) 0.5 A or less

481. A two branch parallel resonant circuit has a coil of resistance R and inductance L in one branch and a capacitor C in the other branch. The source voltage is V. At resonance, the circulating current is

(*a*) $\frac{V}{R}$ (*b*) $\frac{VRC}{L}$

(*c*) $V\sqrt{\frac{L}{C}}$ (*d*) $V\sqrt{\frac{C}{L}}$

482. A coil of resistance R and inductance L has a capacitor C in parallel with it. The source frequency is adjusted to a value equal to resonant frequency. Another capacitor C is added in parallel with the above circuit. The source current will

(*a*) increase (*b*) decrease

(*c*) remain the same

(*d*) may increase or decrease

483. A resistance R, inductance L and capacitance C are in series. The source frequency is adjusted to be equal to resonant frequency. The lower half power frequency is ω_1. Another resistance R is added in series with the circuit. The new lower half power frequency will be

(*a*) ω_1 (*b*) less than ω_1

(*c*) more than ω_1

(*d*) either ω_1 or less than ω_1

484. A wave trap is a

(*a*) series tuned circuit

(*b*) a parallel tuned circuit

(*c*) either a series tuned circuit or parallel tuned circuit

(*d*) none of the above

485. A two branch parallel tuned circuit has a coil of resistance R and inductance L in one branch and capacitance C in the second branch. The bandwidth is 200 radians/sec. A load of resistance R_L is connected in parallel. The new bandwidth will be

(*a*) 200 rad/sec (*b*) more than 200 rad/sec

(*c*) less than 200 rad/sec (*d*) 200 rad/sec or less

486. A series resonant circuit has R = 2 Ω, L = 0.1 H and C = 10 μF. At resonance applied voltage = 10 ∠ 0 V. Voltage across inductance is likely to be

(*a*) $10\angle 0$ V (*b*) $10\angle 45°$ V

(*c*) $100\angle 90°$ V (*d*) $100\angle -90°$ V

487. If $\pounds[f(t)] = F(s)$, $\pounds\, f(t - T)$

(*a*) $e^{sT}F(s)$ (*b*) $e^{-sT}F(s)$

(*c*) $F(s) - e^{-sT}$ (*d*) $F(s) + e^{-sT}$

488. A dc machine has 6 poles all connected in series. Each pole has 9 H inductance and field current is 2A. Total inductance and total energy stored are

(*a*) 9 H and 18 J (*b*) 18 H and 36 J

(*c*) 54 H and 108 J (*d*) 108 H and 216 J

489. The energy stored in a coil is 108 J. It is opened suddenly in 10 ms. The power dissipated instantaneously across the blades of switch is

(*a*) 108 W (*b*) 1080 W

(*c*) 10800 W (*d*) 108000 W

490. Two similar mutually coupled coils have a total inductance of 900 mH and coefficient of coupling 0.5. The self-inductance of each coil is

(*a*) 300 mH (*b*) 150 mH

(*c*) 75 mH (*d*) 50 mH

491. Two coils having self-inductances of 10 mH and 40 mH are mutually coupled. The maximum possible mutual inductance is

(*a*) 40 mH (*b*) 20 mH

(*c*) 10 mH (*d*) 5 mH

492. If $f(t) = \delta(t)$, $F(s) =$

(*a*) $\frac{1}{s^2}$ (*b*) $\frac{1}{s}$

(*c*) 1 (*d*) s

493. The integral of a unit step function is

(*a*) an impulse function

(*b*) a ramp function

(*c*) a ramp function of slope 1

(*d*) none of the above

494. In $f(t) = \delta(t - T)$, $F(s) =$

(*a*) 1 (*b*) $\frac{1}{s}$

(*c*) e^{-Ts} (*d*) $\frac{e^{-Ts}}{s}$

495. If $I(s) = \frac{s+1}{(s+4)(s+3)}$, $i(t) =$

(*a*) $3e^{-4t} - e^{-5t}$ (*b*) $3e^{-4t} - 2e^{-3t}$

(*c*) $2e^{-4t} - 5e^{-2t}$ (*d*) $3e^{-4t} - e^{-t}$

496. If $I(s) = \frac{1}{s(s+4)^2}$, $i(t) =$

(*a*) $\frac{1}{16} - \frac{e^{-4t}}{16}$ (*b*) $\frac{1}{16} - \frac{e^{-4t}}{16} - \frac{te^{-4t}}{4}$

(*c*) $\frac{1}{16}t - \frac{e^{-4t}}{16}$ (*d*) $-\frac{e^{-4t}}{16} - \frac{te^{-4t}}{4}$

497. If $V(s) = \frac{1}{s^2 + 4s + 8}$, $v(t) =$

(*a*) $\frac{\sin 2t}{2}$ (*b*) $\frac{e^{-2t}\sin 2t}{2}$

(*c*) $0.5e^{-2t}$ (*d*) $0.5e^{2t}$

498. A second order system will have

(*a*) all resistances

(*b*) resistances and inductances

(*c*) resistances, inductances and capacitances

(*d*) either (*b*) or (*c*)

499. The response shown in Fig. 4.75, is

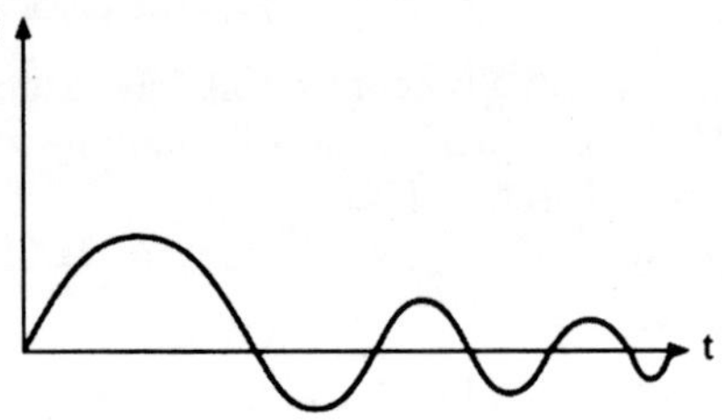

Fig. 4.75.

(*a*) over damped (*b*) critically damped
(*c*) under damped (*d*) either (*a*) or (*c*)

500. Damping ratio =
(*a*) Actual resistance/critical resistance
(*b*) Critical resistance/actual resistance
(*c*) Actual resistance × critical resistance
(*d*) none of the above

501. In an RLC underamped circuit

(*a*) $\frac{R}{2L} = \xi\omega_n$ (*b*) $\frac{R}{L} = \xi\omega_n$

(*c*) $\frac{R}{4L} = \xi\omega_n$ (*d*) $\frac{R}{3L} = \xi\omega_n$

502. The dc voltage applied to an R-L series circuit is suddenly changed from V_1 to V_2. The expression for transient current is

(*a*) $\frac{V_1}{R} + \left(\frac{V_2}{R} - \frac{V_1}{R}\right)e^{-Rt/L}$ (*b*) $\frac{V_2}{R} + \frac{V_1}{R}e^{-Rt/L}$

(*c*) $\frac{V_2}{R} + \left(\frac{V_1}{R} - \frac{V_2}{R}\right)e^{-Rt/L}$ (*d*) either (*a*) or (*c*)

503. A coil has a resistance of 5 Ω and inductance of 0.1 mH. The applied voltage is suddenly increased from 10 V to 20 V. After steady state conditions have reached, the current is
(*a*) 2 A
(*b*) more than 2 A but less than 4 A
(*c*) 4 A
(*d*) more than 4 A

504. The Laplace transforms of voltage function shown in Fig. 4.76 is

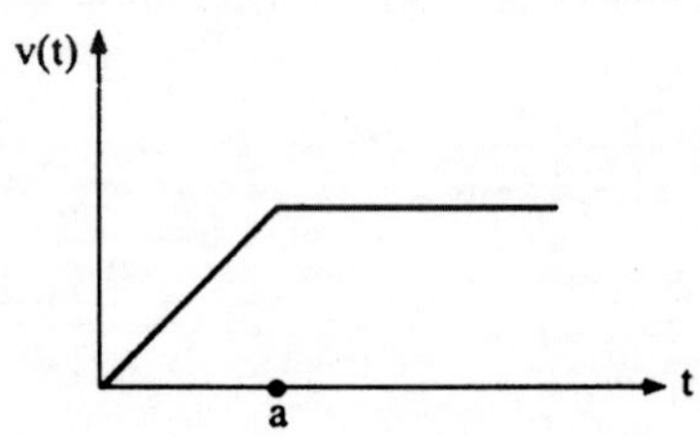

Fig. 4.76.

(*a*) $\frac{1}{s^2}e^{-as}$ (*b*) $\frac{1}{s^2}e^{as}$

(*c*) $\frac{1}{s^2} - \frac{1}{s^2}e^{-ax}$ (*d*) $\frac{1}{s^2} + \frac{1}{s^2}e^{-as}$

505. Which of the following is correct for a driving point function?
(*a*) The real parts of all poles must be negative
(*b*) The real parts of all poles and zeros must be negative
(*c*) The real parts of all poles and zeros must be negative or zero
(*d*) The real parts of all zeros must be negative

506. If a driving point function $N(s) = p(s)/q(s)$, then
(*a*) the degrees of $p(s)$ and $q(s)$ may differ by one only
(*b*) the degrees of $p(s)$ and $q(s)$ may differ by zero or one only
(*c*) the degrees of $p(s)$ and $q(s)$ be equal
(*d*) the degrees of $p(s)$ and $q(s)$ must not be equal

507. If a driving point function $N(s) = p(s)/q(s)$, then
(*a*) the terms of lowest degree in $p(s)$ and $q(s)$ may differ in degree by zero only
(*b*) the terms of lowest degree in $p(s)$ and $q(s)$ may differ in degree by one at the most
(*c*) the terms of lowest degree in $p(s)$ and $q(s)$ may differ in degree by two at the most
(*d*) the terms of lowest degree in $p(s)$ and $q(s)$ may differ in degree by three at the most

508. Which is correct for a driving point function?
(*a*) If the real part of a pole is zero, that pole must be simple
(*b*) If the real part of a pole or zero is zero, that pole or zero must be simple
(*c*) If the real part of a zero is zero, that zero must be simple
(*d*) If the real part of a pole is zero, that pole must not be simple

509. If the ratio $p(s)/q(s)$ represents a transfer function, then
(*a*) the degree of $p(s)$ may be as small as zero, independent of degree of $q(s)$
(*b*) the degrees of $p(s)$ and $q(s)$ must be equal
(*c*) the degrees of $p(s)$ and $q(s)$ must not be equal
(*d*) the degrees $p(s)$ and $q(s)$ must differ by at least two

510. Which is correct for a transfer function?
(*a*) The poles and zero must be complex
(*b*) If the poles and zeros are complex or imaginary, they must be conjugate
(*c*) If the poles and zeros are complex, they must be simple
(*d*) If a zero is imaginary, it must be simple

511. For the time domain response shown in Fig. 4.77, the pole

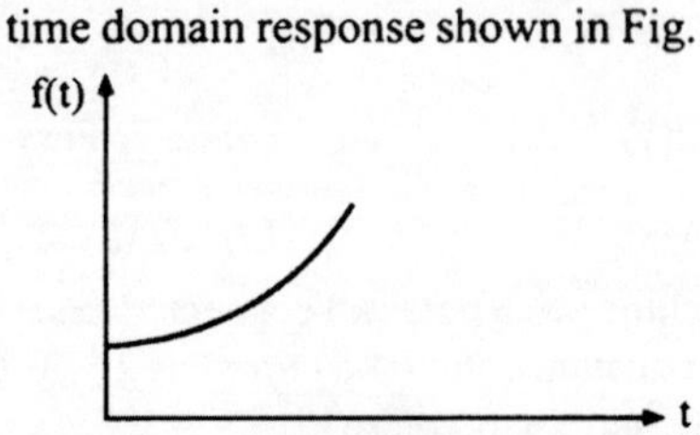

Fig. 4.77.

(*a*) lies in left half plane
(*b*) lies on imaginary axis
(*c*) lies along the positive real axis
(*d*) may be complex or imaginary

512. For the time domain response shown in Fig. 4.78. the pole

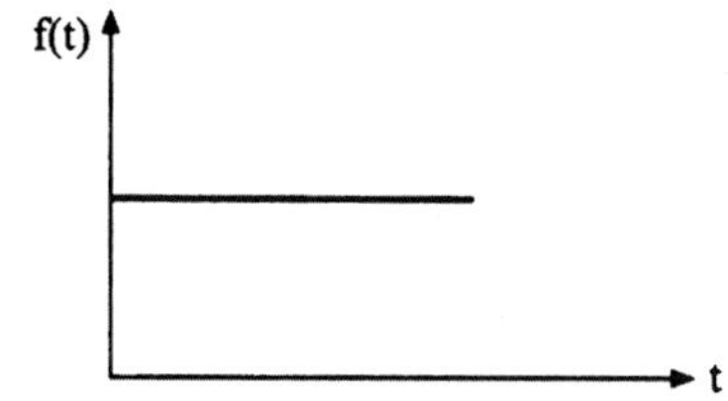

Fig. 4.78.

(*a*) lies in left half plane
(*b*) lies in right half plane
(*c*) lies at origin
(*d*) may be complex or imaginary

513. For the time domain response shown in Fig. 4.79. the pole

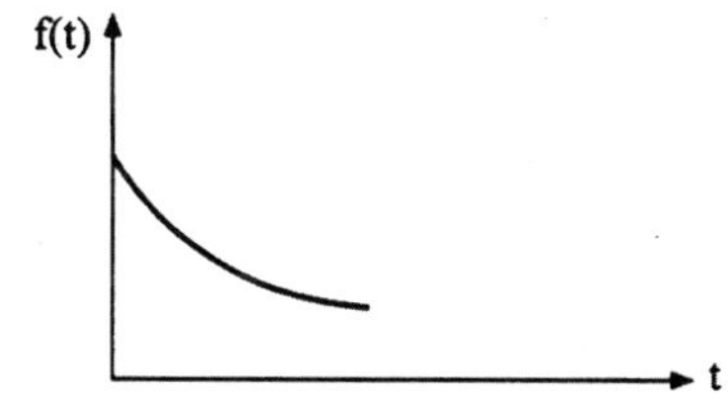

Fig. 4.79.

(*a*) lies on negative real axis
(*b*) lies on positive real axis
(*c*) lies on imaginary axis
(*d*) is complex or imaginary

514. In Fig. 4.80, $i = I_m \sin \omega t$. If $v = 0$, $\omega =$

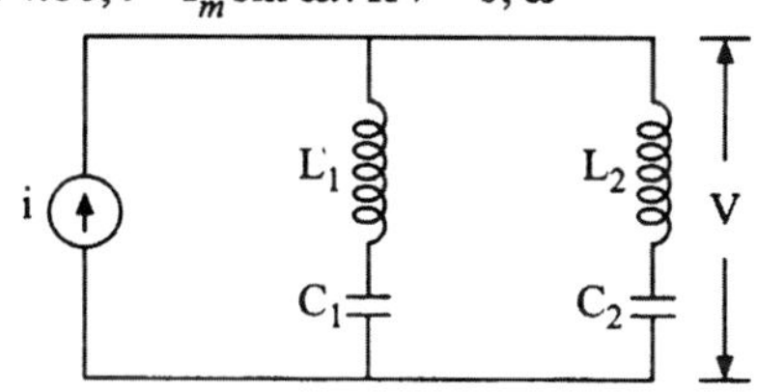

Fig. 4.80.

(*a*) $\dfrac{1}{\sqrt{L_1 C_1}}$ and $\dfrac{1}{\sqrt{L_2 C_2}}$

(*b*) $\sqrt{\dfrac{L_1}{L_2 C_1}}$ and $\sqrt{\dfrac{L_2}{L_1 C_1}}$

(*c*) $\dfrac{1}{\sqrt{L_1 C_1}}$

(*d*) $\dfrac{1}{\sqrt{L_2 C_2}}$

515. A coil has resistance R and inductance L. At $\omega = \infty$, the phase angle between voltage and current is

(*a*) 0 (*b*) 180°
(*c*) 45° (*d*) 90°

516. The polar plot of $P(j\omega) = \omega^2 \angle 45°$ as ω varies from $-\infty$ and ∞ is

(*a*) a straight line parallel to *x*-axis
(*b*) a straight line parallel to *y*-axis
(*c*) a line passing through origin and inclined at 45°
(*d*) a line in first quadrant passing through origin and inclined at 45° to positive *x*-axis

517. The zeros of a minimum phase function lie on

(*a*) imaginary axis
(*b*) left half plane or origin
(*c*) left half plane or imaginary axis
(*d*) right half plane or origin

518. The sum of positive real functions

(*a*) is a positive real function
(*b*) may be a positive real function or not
(*c*) cannot be a positive real function
(*d*) either (*b*) or (*c*)

519. If H(*s*) is positive real, 1/H(s)

(*a*) is positive real
(*b*) is not positive real
(*c*) may or may not be positive real
(*d*) must possess some other properties also if it is to be positive real

520. For H(*s*) to be positive real, the condition Re $[H(j\omega)] \geq 0$ for $0 \leq \omega \leq \infty$ is

(*a*) a necessary condition
(*b*) a sufficient condition
(*c*) a necessary and sufficient condition
(*d*) is not a necessary condition

521. The poles and zeros of a positive real function

(*a*) may lie any where in *s* plane
(*b*) may have positive or negative real part
(*c*) must occur in conjugate pair
(*d*) must have either negative or zero real part

522. In a Hurwitz polynomial

(*a*) all coefficients are non-negative
(*b*) the coefficients may be positive or negative
(*c*) all coefficients are negative
(*d*) all coefficients are non-negative and all zeros should be either in left half plane or simple on the imaginary axis

523. The function Z(*s*) has a pole at infinity if

(*a*) the degrees of numerator and denominator of Z(*s*) are equal
(*b*) the numerator of Z(*s*) is one degree higher than denominator
(*c*) the denominator of Z(*s*) is one degree higher than numerator
(*d*) the degrees of numerator and denominator are either equal or differ by one

524. When a function Z(*s*) having a pole at infinity is realized, we get

(*a*) a capacitance in series
(*b*) a resistance in series
(*c*) an inductance in series
(*d*) a resistance or inductance in series

525. A function Z(*s*) has a zero at infinity if
(*a*) the degrees of numerator and denominator are equal
(*b*) the numerator of Z(*s*) is one degree higher than denominator
(*c*) the denominator of Z(*s*) is one degree higher than numerator
(*d*) the degree of numerator and denominator are either equal or differ by one

526. If Z(*s*) has a pole at origin, the realization leads to
(*a*) a resistance in series
(*b*) an inductance in series
(*c*) a capacitance in series
(*d*) a resistance or capacitance in series

527. In a reactance function
(*a*) all poles are simple
(*b*) all poles and zeros are simple
(*c*) all poles and zeros are simple and lie on *j*ω axis only
(*d*) all poles and zeros lie on negative real axis

528. The realization of a reactance function gives
(*a*) a network having inductances only
(*b*) a network having capacitances only
(*c*) a network having resistances only
(*d*) a network having inductances and capacitances only

529. The realization of the term Ks yields
(*a*) a resistance of value K ohms
(*b*) a capacitance of value K farads
(*c*) an inductance of value K henry
(*d*) either (*b*) or (*c*)

530. For a reactance function H(*s*)
(*a*) Re[H(*j*ω)] is positive for $\omega \geq 0$
(*b*) Re[H(*j*ω)] is zero for $\omega \geq 0$
(*c*) Re[H(*j*ω)] is positive or zero for $\omega \geq 0$
(*d*) Re[H(*j*ω)] is zero or negative for $\omega \geq 0$

531. For a reactance function H(*s*), the slope $dH(j\omega)/d\omega$ is
(*a*) 0 (*b*) always positive
(*c*) always negative (*d*) either 0 or positive

532. Figure 4.81 shows the variation of Z(ω) for a positive real function. The function is

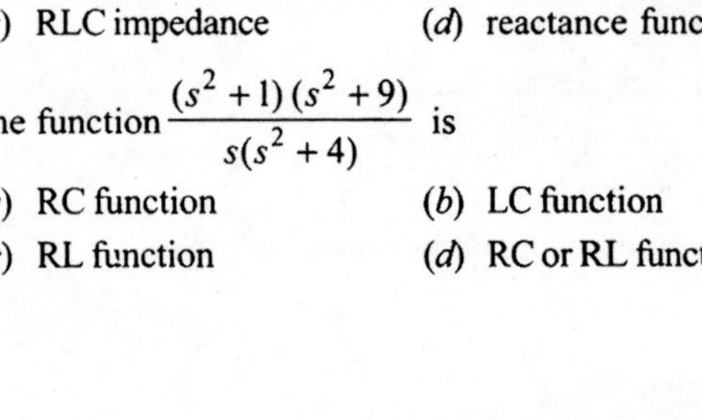

Fig. 4.81.

(*a*) RC impedance (*b*) RL impedance C
(*c*) RLC impedance (*d*) reactance function

533. The function $\frac{(s^2+1)(s^2+9)}{s(s^2+4)}$ is
(*a*) RC function (*b*) LC function
(*c*) RL function (*d*) RC or RL function

534. A reactance function is always a quotient of
(*a*) even to odd polynomials
(*b*) odd to even polynomials
(*c*) even to odd or odd to even polynomials
(*d*) even to even polynomials

535. The realization of a term $\frac{K_0}{s}$ in the impedance function will give
(*a*) an inductance of value K_0 in series
(*b*) an inductance of value $1/K_0$ in series
(*c*) a capacitance of value K_0 in series
(*d*) a capacitance of value $1/K_0$ in series

536. An impedance function has term $\frac{2K_2 s}{s^2+\omega_2^2}$.

The realization would give
(*a*) an inductance of value $2K_2/\omega_2^2$
(*b*) a capacitance of value $1/2K_2$
(*c*) an inductance $2K_2/\omega_2^2$ and capacitance $\frac{1}{2K_2}$ in parallel
(*d*) none of the above

537. If an admitance Y(*s*) has a term K*s*, the realization would give
(*a*) an inductance of value K
(*b*) a resistance of value K
(*c*) a capacitance of value K
(*d*) either (*a*) or (*c*)

538. The function $\frac{s^2+1}{s(s^2+2)}$ is an
(*a*) LC function (*b*) RC function
(*c*) RL function (*d*) LC or RC function

539. Which of the following is true for RC impedance?
(*a*) The lowest critical frequency is a pole
(*b*) The lowest critical frequency is zero
(*c*) The lowest critical frequency may be pole or zero
(*d*) The lowest critical frequency is a pole but cannot be at $s = 0$

540. In an RC impedance function
(*a*) the highest critical frequency is a pole
(*b*) the highest critical frequency is a zero
(*c*) the highest critical frequency is a pole or zero
(*d*) the highest critical frequency is a pole at $s = 0$

541. If $Z(s) = \frac{s^2+1}{s}$, the network has
(*a*) 1 H inductor and 1 F capacitor in parallel
(*b*) 1 H inductor and 1 F capacitor in series
(*c*) 1 H inductor and 1 Ω resistor in series
(*d*) 1 H inductor and 1 Ω resistor in parallel

542. If $Y(s) = \frac{s^2+1}{s}$, the network has

(a) 1 H inductor and 1 F capacitor in parallel
(b) 1 H inductor and 1 F capacitor in series
(c) 1 H inductor and 1 Ω resistance is series
(d) 1 H inductor and 1 Ω resistance in parallel

543. For a network having 1 Ω resistance and 1 F capacitor in series, Z(s) =

(a) $\frac{s+1}{s}$ (b) $\frac{s}{s+1}$
(c) $1+s$ (d) $\frac{1}{s+1}$

544. If Y(s) = 1 + s, the network has

(a) 1 Ω resistor and 1 F capacitor in series
(b) 1 Ω resistor and 1 F capacitor in parallel
(c) 1 Ω resistor and 1 H inductor in series
(d) 1 Ω resistor and 1 H inductor in parallel

545. For an RC impedance function

(a) all poles lie on $j\omega$ axis
(b) all poles are simple
(c) all poles are simple and lie on negative real axis
(d) all poles are simple and lie on $j\omega$ axis

546. For an RC impedance function

(a) the highest critical frequency is a zero
(b) the highest critical frequency is a pole
(c) the highest critical frequency may be a zero or pole
(d) none of the above

547. For an R-C impedance function, the residues at all poles are

(a) real (b) imaginary
(c) real and positive (d) positive

548. For an RC impedance function Z,

(a) $Z(\infty)=Z(0)$ (b) $Z(\infty)<Z(0)$
(c) $Z(\infty)\le Z(0)$ (d) $Z(\infty)>Z(0)$

549. For an RC admittance function

(a) all poles and zeros are simple
(b) all poles and zeros are simple and lie on negative real axis
(c) all poles and zeros lie on imaginary axis
(d) all poles and zeros lie on positive real axis

550. For an RC admittance function

(a) the lowest critical frequency is a zero
(b) the highest critical frequency is a pole
(c) the lowest critical frequency is zero and the highest critical frequency is a pole
(d) the lowest critical frequency is a pole

551. For an RC admittance function Y, the slope $dY/d\sigma$

(a) is positive (b) is negative
(c) may be positive or negative
(d) 0

552. For an RC admittance function Y,

(a) $Y(\infty)<Y(0)$ (b) $Y(0)<Y(\infty)$
(c) $Y(\infty)$ may be lesser or greater than Y(0)
(d) $Y(\infty)=Y(0)$

553. If $Z_{RC}(s)$ has a pole at $s = 0$, the first element in first Foster realization is

(a) a capacitor (b) a resistor
(c) R-C in series
(d) either a capacitor or resistor

554. If $Z_{RC}(s)$ is constant at $s = 0$, the first element in second Foster realization is

(a) R (b) C
(c) R and C is series (d) R and C in parallel

555. If $Z_{RC}(s)$ is constant at $s=\infty$, the last element in first Foster realization is

(a) R (b) C
(c) R and C in series (d) R and C in parallel

556. If $Z_{RC}(s) = 0$ at $s = \infty$ the last element in second Foster realization will be

(a) R (b) C
(c) R and C in series (d) R and C in parallel

557. If $Z_{RC}(s)=0$ at $s=\infty$, the first element in First Cauer form is

(a) R (b) C
(c) R and C in series (d) R and C in parallel

558. If $Z_{RC}(s)$ = constant at $s = 0$, the last element is First Cauer realization is

(a) R (b) C
(c) R and C in series (d) R and C in parallel

559. An RL impedance function can also be realized as

(a) RC impedance function (b) RC admittance function
(c) LC impedance function (d) LC admittance function

560. An RL admittance function can also be realized as

(a) RC admittance function (b) RC impedance function
(c) LC admittance function (d) LC impedance function

561. A minimum function has

(a) poles and zeros on imaginary axis of s plane
(b) no poles or zeros on imaginary axis of s plane
(c) only poles but no zeros on imaginary axis of s plane
(d) only zeros but no poles on imaginary axis of s plane

562. In a minimum function

(a) the degrees of numerator and denominator are equal
(b) the degree of numerator and denominator are unequal
(c) the degree of numerator is one more than degree of denominator
(d) the degree of numerator is one less than degree of denominator

563. The synthesis of minimum function was suggested by

(a) O Brune (b) R. Richards
(c) Bott and Duffin (d) none of the above

564. As compared to the number of network elements in Brune synthesis, the number of network elements in Bott-Duffin realization is

(a) more (b) less
(c) equal (d) any of the above

565. If $Z(s)$ is positive real, then $R(s) = \dfrac{kZ(s) - sZ(k)}{kZ(k) - sZ(s)}$ where k has only positive real values is also positive real. This hypothesis was first given by

(a) O' Brune (b) R. Richards
(c) Bott and Duffin (d) P.S. Laplace

566. If operator, 'a' = $1 \angle 120°$ then $(1 + a) =$

(a) $-a$ (b) $-a^2$
(c) $\sqrt{2}$ (d) -1

567. If operator, 'a' = $1 \angle 120°$ then $(1 + a - a^2) =$

(a) $-a$ (b) $-a^2$
(c) 1 (d) 0

568. If operator, 'a' = $1 \angle 120°$ then $(1 - a) =$

(a) $\sqrt{3}$ (b) $\sqrt{3} \angle -30°$
(c) $\sqrt{3} \angle +30°$ (d) $\sqrt{3} \angle 60°$

569. If operator, 'a' = $1 \angle 120°$ then $(1 + a^2) =$

(a) $\sqrt{3}$ (b) $\sqrt{3} \angle 30°$
(c) $\sqrt{3} \angle -30°$ (d) $\sqrt{3} \angle 60°$

570. If operator, 'a' = $1 \angle 120°$ then $(1 + a^2) =$

(a) $1 \angle -30°$ (b) $1 \angle 30°$
(c) $1 \angle -60°$ (d) $1 \angle 60°$

571. If V_{a1}, V_{a2}, V_{a0} are the symmetrical components of V_a, then

(a) $V_b = V_{a1} + V_{a2} + V_{a0}$
(b) $V_b = a^2V_{a1} + aV_{a2} + V_{a0}$
(c) $V_b = aV_{a1} + a^2V_{a2} + V_{a0}$
(d) $V_b = aV_{a0} + a^2V_{a1} + V_{a2}$

572. If V_{a1}, V_{a2}, V_{a0} are the symmetrical components of V_a, then

(a) $V_c = V_{a1} + V_{a2} + V_{a0}$
(b) $V_c = a^2V_{a1} + aV_{a2} + V_{a0}$
(c) $V_c = aV_{a1} + a^2V_{a2} + V_{a0}$
(d) $V_c = aV_{a0} + a^2V_{a1} + V_{a2}$

573. If $V_a = 10 \angle 0, V_b = 10 \angle -120°$ and $V_c = 10 \angle -240°$, then V_{a2} and V_{a0} are

(a) 0 and 0 respectively
(b) 0 and 10 respectively
(c) 0 and 0 (d) 10 and 10

574. If I_a, I_b, I_c are line currents and I_{a0} is the zero sequence component of I_a, then

(a) $I_{a0} = I_a$ (b) $I_{a0} = I_a + I_b + I_c$
(c) $I_{a0} = \frac{1}{3}(I_a + I_b + I_c)$
(d) $I_{a0} = \frac{1}{3}(I_a + aI_b + a^2I_c)$

575. If I_a, I_b and I_c are line currents and I_{a1} is the positive sequence component of I_a, then

(a) $I_{a1} = I_a + I_b + I_c$ (b) $I_{a1} = \frac{1}{3}(I_a + I_b + I_c)$
(c) $I_{a1} = \frac{1}{3}(I_a + aI_b + a^2I_c)$
(d) $I_{a1} = \frac{1}{3}(I_a + a^2I_b + aI_c)$

576. If I_a, I_b and I_c are line currents and I_{a2} is the negative sequence component of I_a, then

(a) $I_{a2} = I_a + I_b + I_c$ (b) $I_{a2} = \frac{1}{3}(I_a + I_b + I_c)$
(c) $I_{a2} = \frac{1}{3}(I_a + a^2I_b + I_c)$
(d) $I_{a2} = \frac{1}{3}(I_a + aI_b + a^2I_c)$

577. If V_{a0}, V_{a1}, V_{a2} are sequence components of V_a and I_{a0}, I_{a1}, I_{a2} are sequence components of I_a, then total 3 phase power P is given by

(a) $P = V_{a0}I_{a0} + V_{a1}I_{a1} + V_{a2}I_{a2}$
(b) $P = V_{a0}I_{a0}{}^* + V_{a1}I_{a1}{}^* + V_{a2}I_{a2}{}^*$
(c) $P = 3(V_{a0}I_{a0} + V_{a1}I_{a1} + V_{a2}I_{a2})$
(d) $P = 3(V_{a0}I_{a0}{}^* + V_{a1}I_{a1}{}^* + V_{a2}I_{a2}{}^*)$

578. In a 3-phase system, the zero sequence component of line to line voltages is zero

(a) if line-to-line voltages are balanced
(b) if line-to-line voltages are unbalanced
(c) if line-to-line voltages are balanced or unbalanced
(d) if degree of unbalance in line-to-line voltage is small

579. If sequence currents $I_{a1} = 2 \angle -90°$, $I_{a2} = 0.2 \angle 90°$ and $I_{a0} = 1.8 \angle 90°$, $I_a =$

(a) 0 (b) 1
(c) 4 (d) $\sqrt{8}$

580. If $V_{a1} = V_{b2} = 100$ V and $V_{a0} = 0$, then V_a, V_b, V_c are V_c

(a) 200, – 100 and – 100 V respectively
(b) 100, 100, 300 V respectively
(c) 200, 100, 100 V respectively
(d) 100, – 100, 100 V respectively

581. In the circuit of Fig. 4.82, the power consumed in resistance R is measured when one source is acting at one time. These values are 18 W, 50 W and 98 W. When all sources act together the maximum and minimum power can be

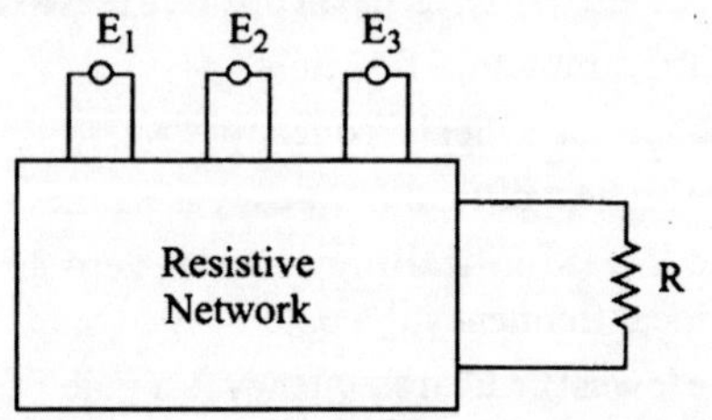

Fig. 4.82.

(a) 98 W and 18 W (b) 166 W and 18 W
(c) 450 W and 2 W (d) 166 W and 2 W

582. In a two terminal network the o.c. voltage measured at the given terminals is 100 V and s.c. current at the same terminals

5 A. If a load of 80 Ω resistance is connected at the terminals, load current is

(*a*) 1A (*b*) 1.25 A
(*c*) 6A (*d*) 6.25A

583. Consider the following statements

1. Tellegen's theorem is applicable to any lumped network
2. Reciprocity theorem is applicable to linear bilaeral network
3. Thevenin theorem is applicable to two terminal linear active network
4. Norton's theorem is applicable to linear active network

Which of the above statements are correct?

(*a*) 1, 2 and 3 (*b*) 1, 2, 3 and 4
(*c*) 1, 2 and 4 (*d*) 3 and 4

584. Which of the following theorems can be conveniently used to find power consumed in 10 Ω resistor in Fig. 4.83

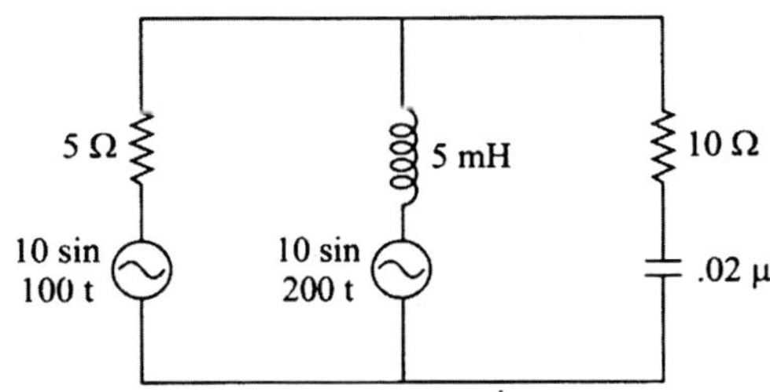

Fig. 4.83.

(*a*) Thevenin theorem
(*b*) Maximum power transfer theorem
(*c*) Millman's theorem
(*d*) Superposition theorem

585. A network has 10 nodes and 17 branches. The number of different node pairs are

(*a*) 1 (*b*) 9
(*c*) 10 (*d*) 45

586. In the circuit shown in Fig. 4.84 the current I of sinusoidal source is

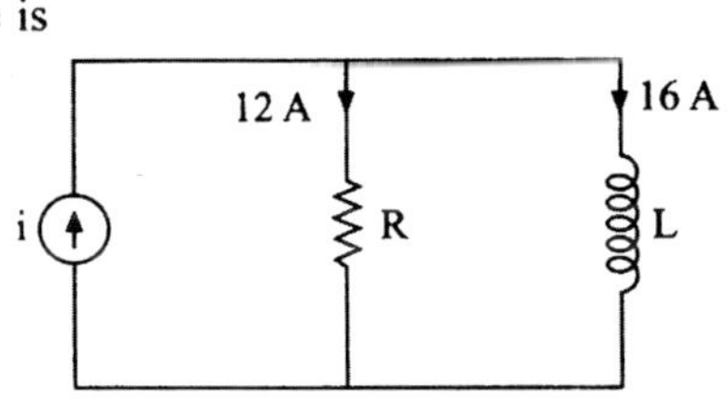

Fig. 4.84.

(*a*) 25 A (*b*) 4 A
(*c*) 20 A
(*d*) Not determinable from given data

587. A 3-phase balanced supply feads 3-phase unbalanced load. Power supplied to the load can be measured by using

1. 2 wattmeters 2. one wattmeter
3. 3 wattmeters

Which of the above statements is correct?

(*a*) 1 and 2 (*b*) 1 and 3
(*c*) 2 and 3 (*d*) 3 alone

588. In Fig. 4.85, power in 6 Ω resistor is zero then V =

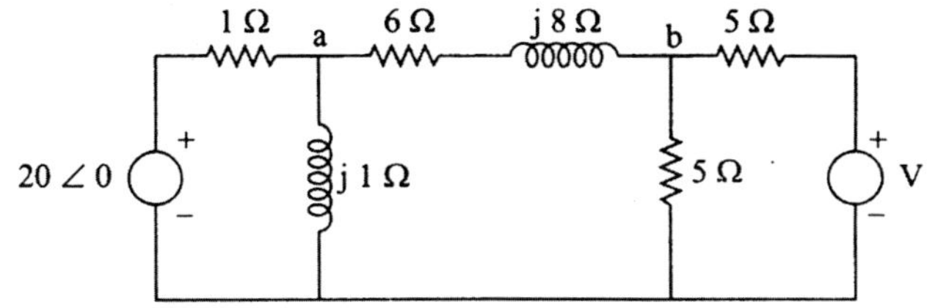

Fig. 4.85.

(*a*) $20\sqrt{2}\angle 45°$ (*b*) $20\angle 30°$
(*c*) $20\angle 45°$ (*d*) $20\sqrt{2}\angle 30°$

589. A capacitor used for power factor correction is single phase circuit decreases

(*a*) power factor
(*b*) line current
(*c*) both line current and p.f.
(*d*) the line current and increases p.f.

590. For 2 port reciprocal network, o.c. output voltage divided by input current is equal to

(*a*) B (*b*) z_{12}
(*c*) $1/y_{12}$ (*d*) h_{12}

591. In Fig. 4.86, the transmission parameters are A = C = 1, B = 2 and D = 3. Then Z_{in} =

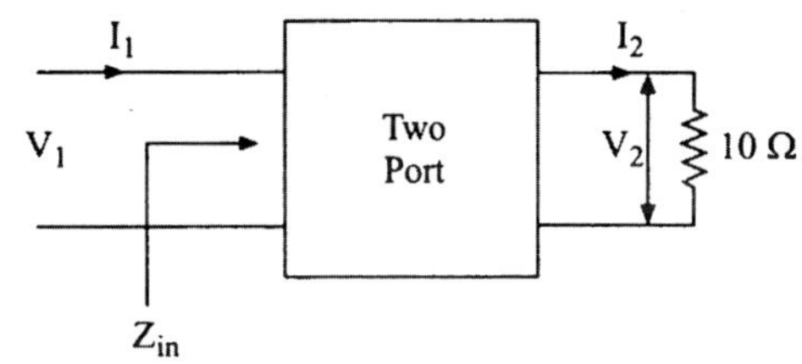

Fig. 4.86.

(*a*) 12/13 Ω (*b*) 13/12 Ω
(*c*) 3 Ω (*d*) 4 Ω

592. An R-L series circuit is excited by a voltage $v(t)$. The circuit was initially relaxed. Which of the following sources of $v(t)$ would cause maximum current at $t = 0^+$

(*a*) unit step (*b*) unit impulse
(*c*) unit ramp (*d*) unit step plus unit ramp

593. In Fig. 4.87 the switch *S* is initially open and steady state conditions are reached. At $t = 0$ switch is closed. The initial current through 2C capacitor is

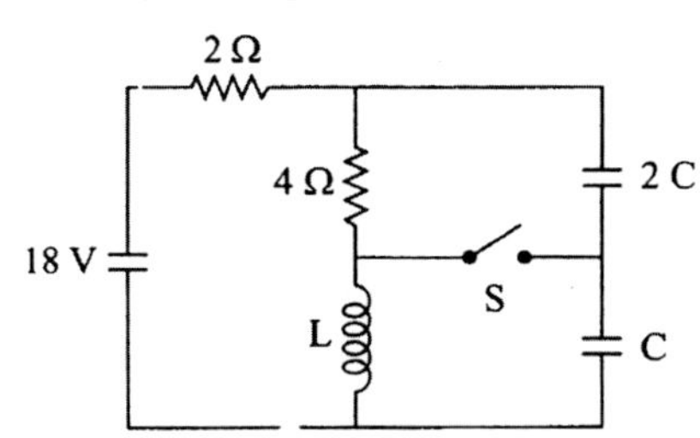

Fig. 4.87.

(*a*) zero (*b*) 1 A
(*c*) 2 A (*d*) 3 A

594. In Fig. 4.88, the switch is closed at $t = 0$. When current through inductor is 6 A, the rate of change of current through resistor is 6 A/s. The value of L is

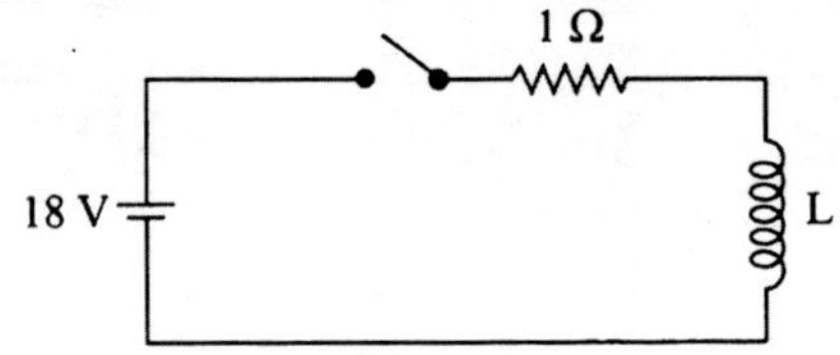

Fig. 4.88.

(*a*) 1 H (*b*) 2 H
(*c*) 3 H (*d*) 4 H

595. Consider the following:
If a network has impedance $(1 - j)$ at a specific frequency, the circuit would consist of series
1. R and C 2. R and L
3. R, L and C
Which of these statements is correct?
(*a*) 1 and 2 (*b*) 1 and 3
(*c*) 1, 2 and 3 (*d*) 2 and 3

596. In Fig. 4.89, $V_1 = 40$ V when R= 10 Ω, if R = 0, V_2 will be

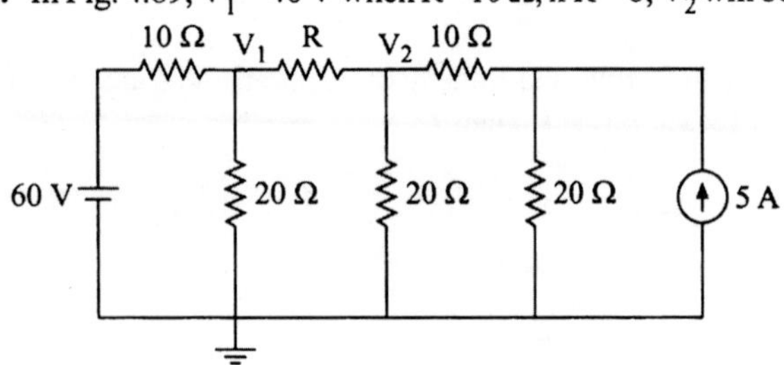

Fig. 4.89.

(*a*) 40V (*b*) 30V
(*c*) 20V (*d*) 10V

597. A network has only independent current sources and resistances. If values of all resistors are doubled, the values of node voltages
(*a*) will become half
(*b*) will remain unchanged
(*c*) will become double
(*d*) cannot be determined unless circuit configuration and values of resistors are known

598. In the transformer shown in Fig. 4.90, the inductance measured across terminal 1 and 2 was 4 H with terminals 3 and 4 open. It was 3 H when terminals 3 and 4 were short circuited. The coefficient of coupling is

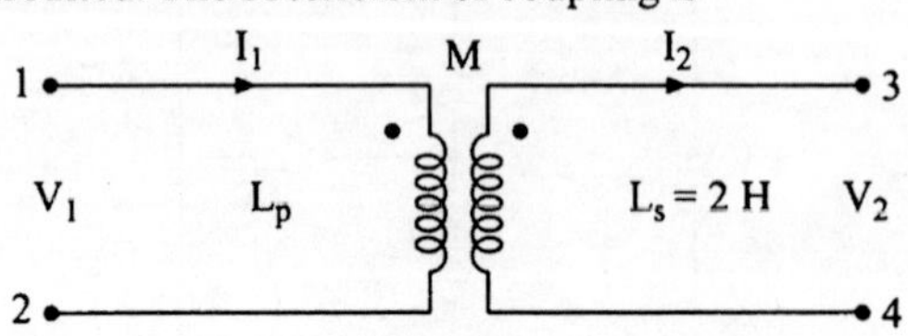

Fig. 4.90.

(*a*) 1 (*b*) 0.707
(*c*) 0.5
(*d*) indeterminate due to insufficient data

599. The circuit in Fig. 4.91 will act as ideal current source with respect to terminals A and B when frequency is

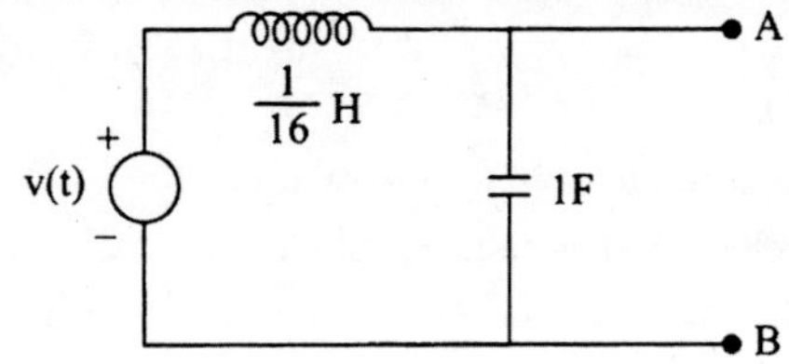

Fig. 4.91.

(*a*) 0 (*b*) 1 rad/sec
(*c*) 4 rad/sec (*d*) 16 rad/sec

600. In Fig. 4.92, $Z(s) = \dfrac{3(s^2 + 6s + 8)}{s^2 + 4s + 3}$. The values of C and R are

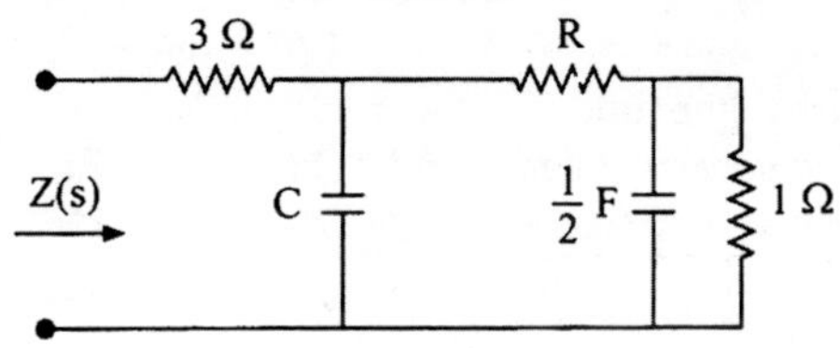

Fig. 4.92.

(*a*) $\frac{1}{6}$ F and 4 Ω (*b*) $\frac{2}{9}$ F and 4 Ω
(*c*) $\frac{2}{3}$ F and $\frac{1}{2}$ Ω (*d*) $\frac{1}{2}$ F and 1 Ω

601. The circuit in Fig. 4.93 (*a*) is replaced by that in Fig. 4.93 (*b*). If I remains same, R_0 will be

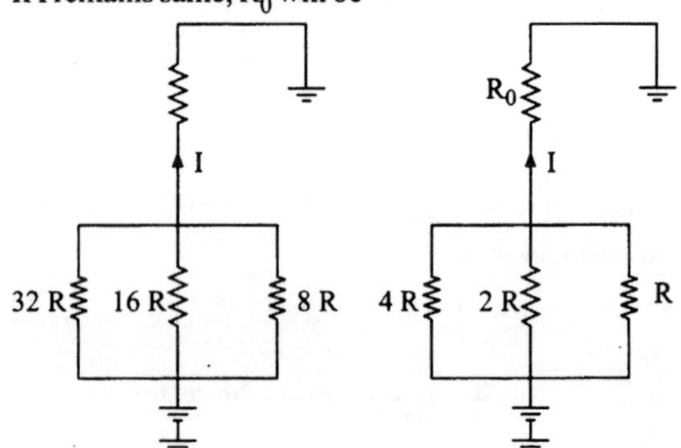

Fig. 4.93.

(*a*) zero (*b*) R
(*c*) 2R (*d*) 4R

602. If a resistance R of 1Ω is connected between A and B in Fig. 4.94, current flowing through R is

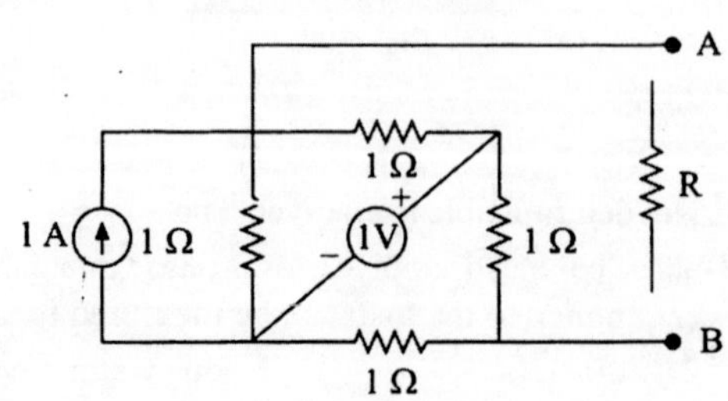

Fig. 4.94.

(*a*) 1A (*b*) 0.5A
(*c*) 0.25 A (*d*) 0.125 A

603. If the π and T circuits in Fig. 4.95 are equivalent, then R_1, R_2, R_3 respectively are

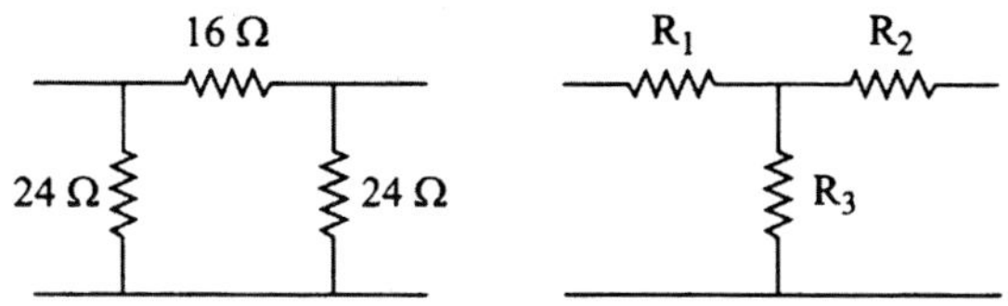

Fig. 4.95.

(*a*) $9\Omega, 6\Omega, 6\Omega$ (*b*) $6\Omega, 6\Omega, 9\Omega$

(*c*) $9\Omega, 6\Omega, 9\Omega$ (*d*) $6\Omega, 9\Omega, 6\Omega$

604. In Fig. 4.96, power dissipated in 30 Ω resistance will be maximum when value of R =

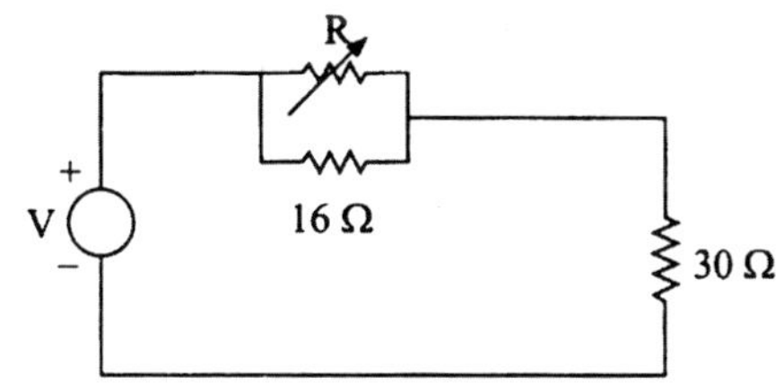

Fig. 4.96.

(*a*) 30 Ω (*b*) 16 Ω

(*c*) 9 Ω (*d*) zero

605. In Fig. 4.97, I = 2 A if R = 20 Ω. When R = 10 Ω, current I is

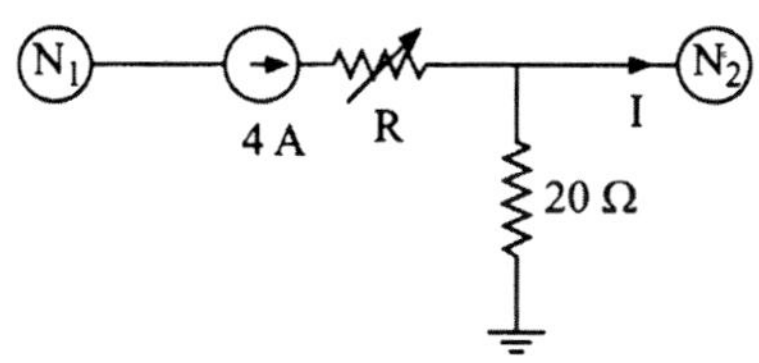

Fig. 4.97.

(*a*) 1A (*b*) 2A

(*c*) 2.5 A (*d*) 3A

606. The average value of periodic function $v(t)$ in Fig. 4.98 is

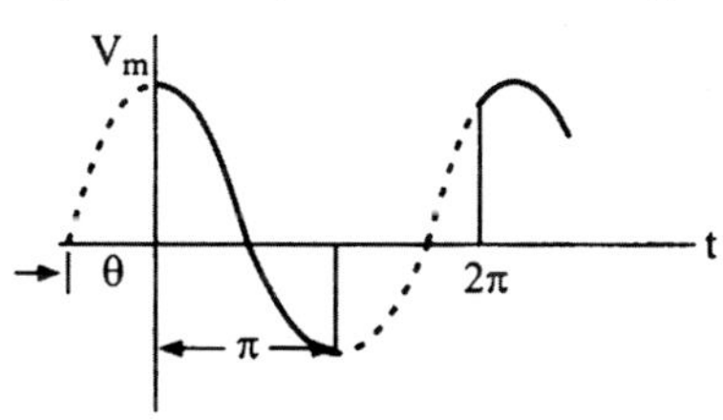

Fig. 4.98.

(*a*) $\frac{V_m}{\pi}\cos\theta$ (*b*) $\frac{V_m}{\pi}\sin\theta$

(*c*) $\frac{2V_m}{\pi}\cos\theta$ (*d*) $\frac{V_m}{\pi}$

607. Consider the following statements regarding driving point admittance function

$$Y(s) = \frac{s^2 + 2.5s + 1}{s^2 + 4s + 3}$$

1. It is an admittance of RL network
2. Poles and zeros alternate on the negative real axis of plane
3. The lowest critical frequency is a pole
4. Y(0) = 1/3

Which of these statements are correct?

(*a*) 1, 2 and 3 (*b*) 2 and 4

(*c*) 1 and 3 (*d*) 1, 2, 3 and 4

608. The RLC circuit for driving point admittance function $\dfrac{\frac{1}{RLs}}{\frac{1}{R}+\frac{1}{Ls}} + Cs$ is

(*a*)

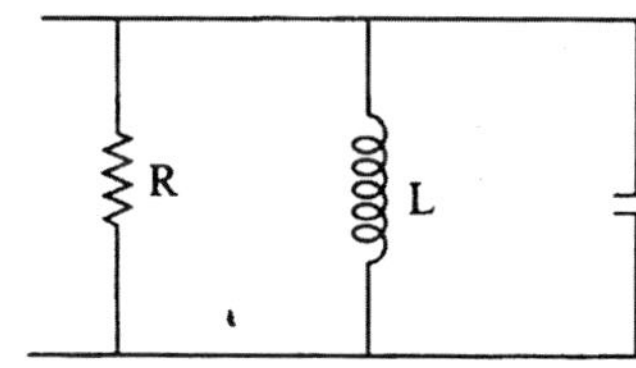

(*b*)

(*c*)

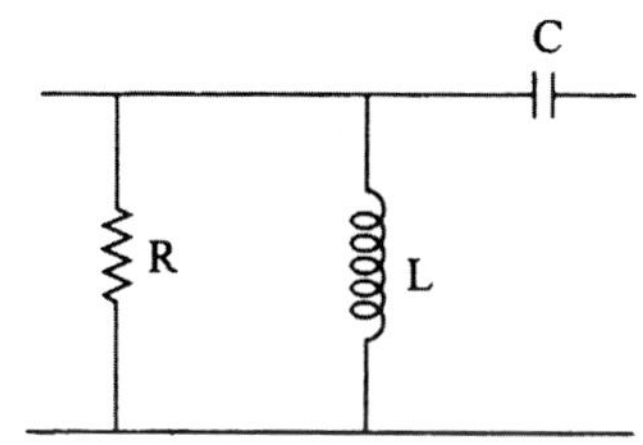

(*d*)

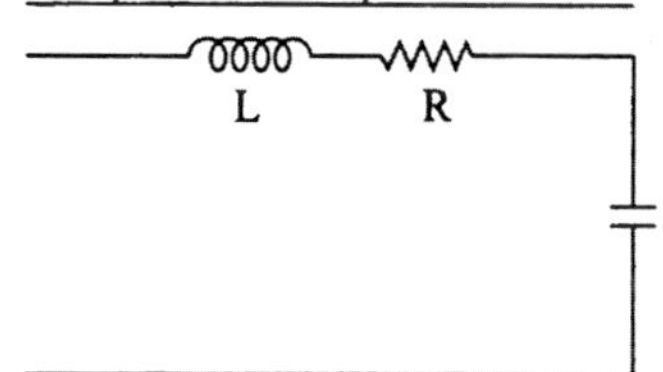

Fig. 4.99.

609. Match List I with List II for the driving point synthesis and select the correct answer using the given codes

List I (Form)	List II (Network)
A. Cauer I	1. L in series arm and C in shunt arm of ladder
B. Cauer II	2. C in series arm and L in shunt arm of ladder
C. Foster I	3. Series combination of L and C in parallel
D. Foster II	4. Parallel combination of L and C in series

Codes:

	A	B	C	D
(*a*)	1	2	3	4
(*b*)	1	2	4	3
(*c*)	2	1	4	3
(*d*)	2	1	3	4

610. Poles and zeros of a driving point function of a network are simple and alternate on $j\omega$ axis. The network consists of

(*a*) R and C (*b*) L and C

(*c*) R and L (*d*) R, L and C

611. Figure 4.100, shows a dc circuit. The Thevenin's equivalent circuit at terminals $a - b$ is

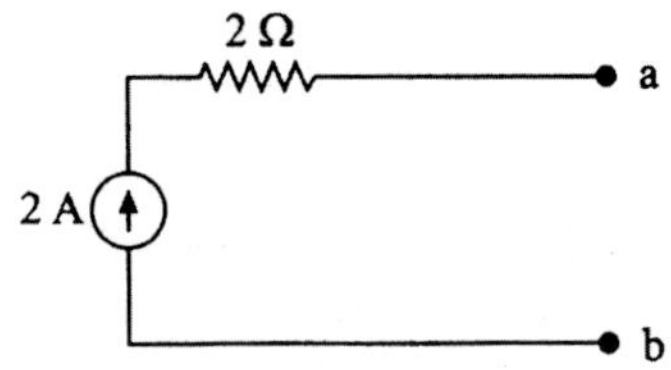

Fig. 4.100.

(*a*)

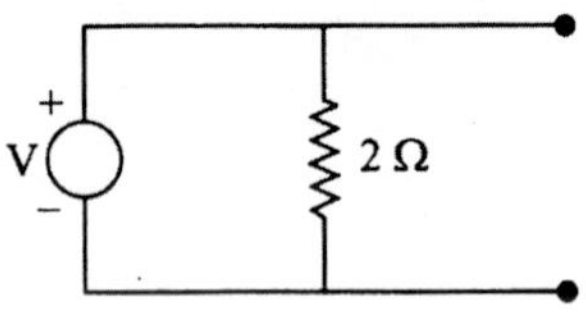

(*b*)

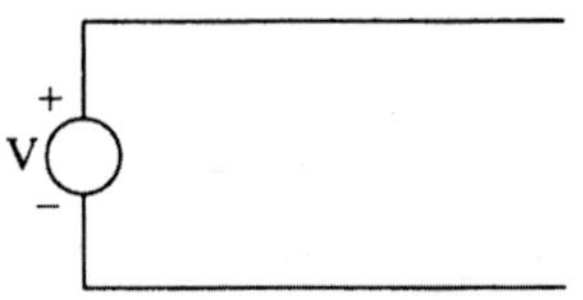

(*c*)

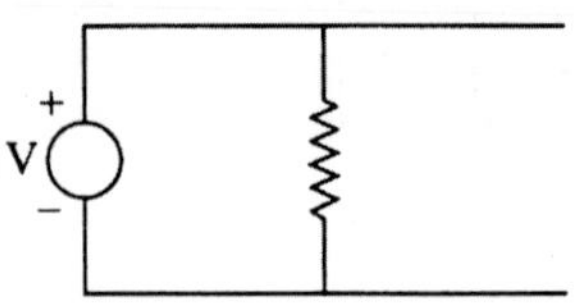

(*d*) not feasible

612. A battery charger can drive a current of 5 A with 1 Ω resistance connected at its output. If it can charge 2 V battery at 7 A rate, its Thevenin's equivalent circuit will be

(*a*) 7.5 in series with 0.5 Ω resistance

(*b*) 1.25 V in series with 1.5 Ω resistance

(*c*) 7.5 V in parallel with 0.5 Ω resistance

(*d*) 12.5 V in parallel with 1.5 Ω resistance

613. For the network in Fig. 4.101, the correct loop equation for loop 3 is

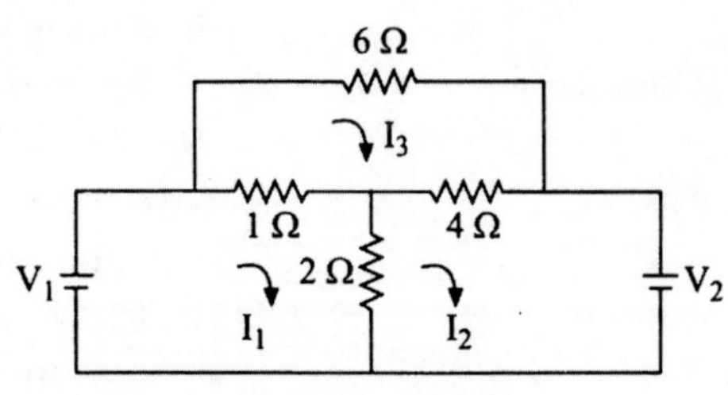

Fig. 4.101.

(*a*) $-I_1 + 4I_2 + 11I_3 = 0$

(*b*) $I_1 + 4I_2 + 11I_3 = 0$

(*c*) $-I_1 - 4I_2 + 11I_3 = 0$

(*d*) $I_1 - 4I_2 + 6I_3 = 0$

614. For graph of Fig. 4.102 one set of fundamental cutsets would be

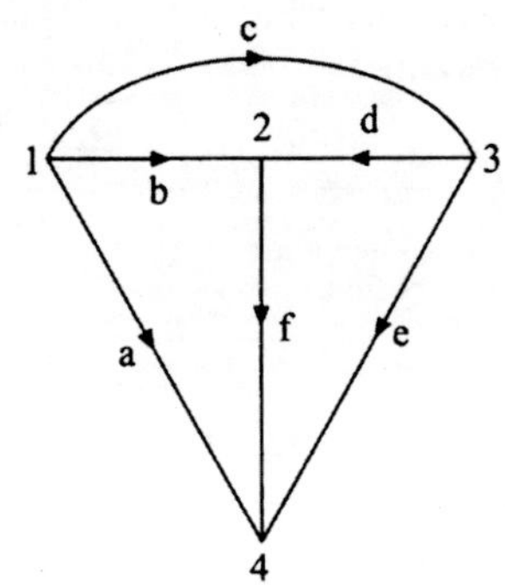

Fig. 4.102.

(*a*) *abc, cde, afe* (*b*) *afde, cde, abde*

(*c*) *cbfe, afe, bdf* (*d*) *cbd, abde, cde*

615. A particular current is made up of two components 10 A dc and sinusoidal current of peak value 14.14 A. The average of resultant current is

(*a*) zero (*b*) 24.14 A

(*c*) 10 A (*d*) 14.14 A

616. For the two port in Fig. 4.103, y_{11} ,= 8 mho, $y_{12} = y_{21} = -6$ mho and $y_{22} = 6$ mho. The values of Y_A, Y_B, Y_C respectively are

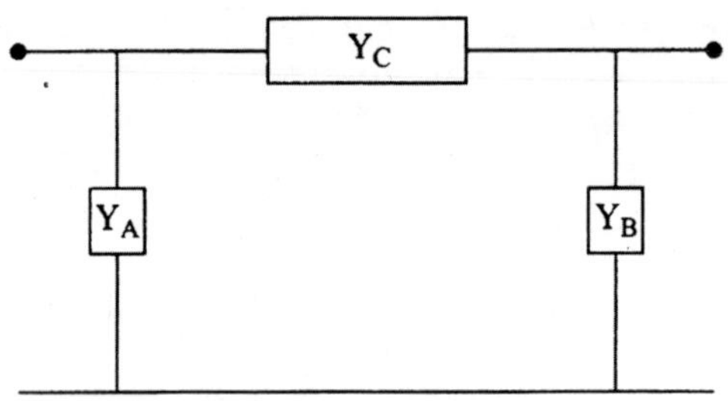

Fig. 4.103.

(*a*) 2, 6, –6 mho (*c*) 2, 0, 6 mho

(*b*) 2, 6, 0 mho (*d*) 2, 6, 8 mho

617. The transfer function $\dfrac{s}{s+a}$ is for

(*a*) low pass filter (*b*) notch filter

(*c*) high pass filter (*d*) band pass filter

618. The model of a transistor in CE connection is shown in Fig. 4.104, Match List I with List II and select the correct answer using the given codes

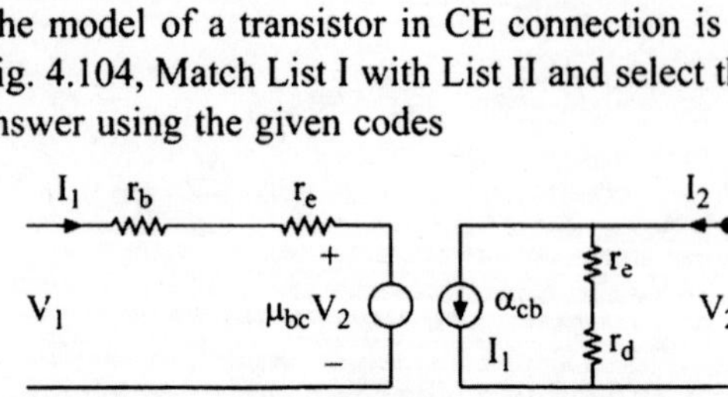

Fig. 4.104.

List I	List II
A. h_{22}	1. $r_b + r_e$
B. h_{11}	2. α_{cb}
C. h_{21}	3. $\dfrac{1}{r_e + r_d}$

Codes:

	A	B	C
(a)	3	1	2
(b)	1	3	2
(c)	2	3	1
(d)	3	2	1

619. In Fig. 4.105, $i(t)$ is a unit step current. The steady state value of $v(t)$ is

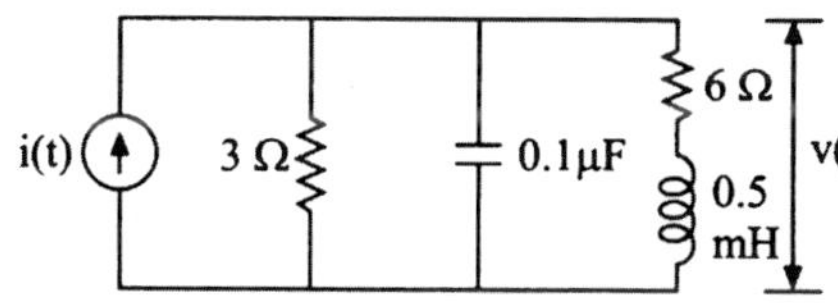

Fig. 4.105.

(a) 2V (b) 3V
(c) 6V (d) 9V

620. The pole locations of three networks and their impulse responses are given in Fig. 4.106. Match List I with List II and select the correct answer using the given codes

List I **List II**

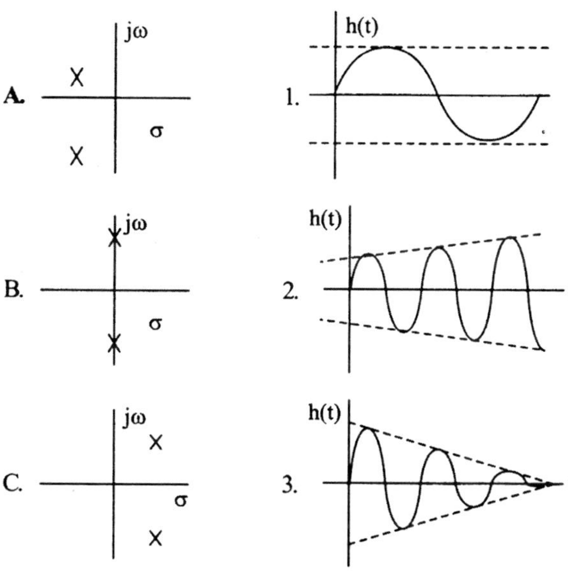

Fig. 4.106.

Codes:

	A	B	C
(a)	1	2	3
(b)	2	1	3
(c)	3	1	2
(d)	3	2	1

621. Match List I with List II and select the correct answer using the given codes

List I **List II**

A. 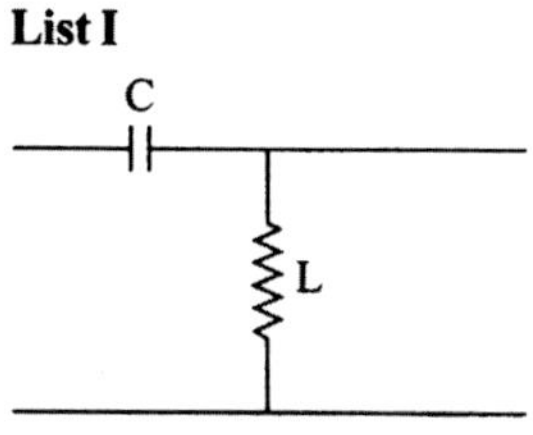

1. Band reject (notch) filter

B. 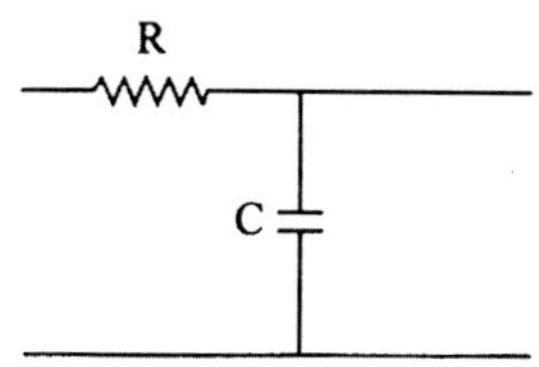

2. Band pass filter

C. 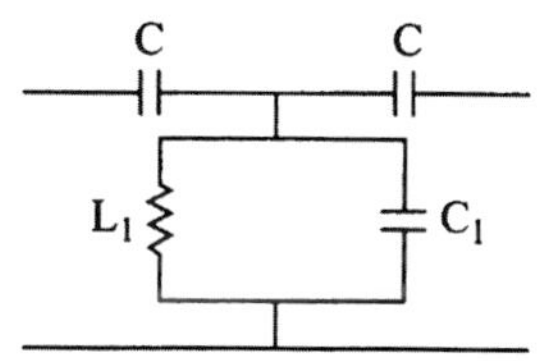

3. High pass filter

D. 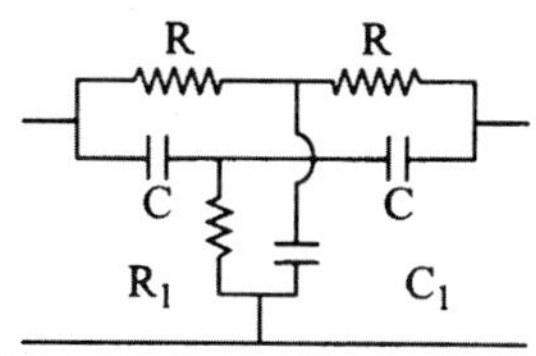

4. Low pass filter

Codes:

	A	B	C	D
(a)	3	4	2	1
(b)	3	4	1	2
(c)	4	3	1	2
(d)	4	3	2	1

622. Consider the following from the point of view of possible realization as driving point impedances using passive elements

1. $\dfrac{1}{s\,(s+5)}$ 2. $\dfrac{s+3}{s^2\,(s+5)}$

3. $\dfrac{s^2+3}{s(s^2+5)}$ 4. $\dfrac{s+5}{s(s+3)}$

Among these

(a) 1, 2, 4 are realisable (b) 1, 2, 3 are realisable
(c) 3, 4 are realisable (d) none is realisable

623. An RC driving point impedance function has zeros at $s = -2$ and $x = -5$. The admissible poles are

(a) $s = 0$ and $s = -6$ (b) $s = -1$ and $s = -3$
(c) $s = 0$ and $s = -1$ (d) $s = -3$ and $s = -4$

624. Two networks are cascaded through an ideal buffer. If t_{d1} and t_{d2} are delay times of network, the overall delay time is

(a) $\sqrt{t_{d_1} + t_{d_2}}$ (b) $\sqrt{t_{d1}^2 + t_{d2}^2}$

(c) $t_{d_1} + t_{d_2}$ (d) $\dfrac{t_{d_1} + t_{d_2}}{2}$

625. A periodic voltage $v(t) = 1 + 4 \sin \omega t + 2 \cos \omega t$ is applied across a 1 ohm resistance. The power dissipated is

(a) 1 W (b) 11 W
(c) 21 W (d) 24.5 W

626. If Fourier transform of $f(t)$ is $F(j\omega)$, then which pairs of time functions and Fourier transforms are matched

1. $f(t+2)$ $e^{j2\omega}F(j\omega)$
2. $f(-0.5t)$ $2F(-2j\omega)$

3. $\int_{\infty}^{t} f(t)\, dt$ $\qquad$ $F(j\omega)\left[\frac{1}{j\omega} + \pi\delta\omega\right]$

Select the answer using the following codes

(*a*) 1 and 2 (*b*) 1 and 3
(*c*) 2 and 3 (*d*) 1, 2 and 3

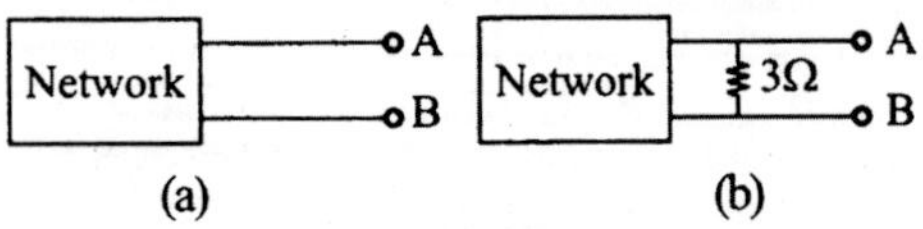

Fig. 4.107.

627. The Thevenin's equivalent of network in Fig. 4.107 (*a*) is a 10 V source is series with 2 Ω resistance. If a 3 Ω resistance is connected across AB as shown in Fig. 4.107 (*b*) the Thevenin's equivalent is

(*a*) 10V in series with 1.2 Ω resistance
(*b*) 6 V in series with 1.2 Ω resistance
(*c*) 10 V in series with 5 Ω resistance
(*d*) 6 V in series with 5 Ω resistance

628. A certain network has two ideal voltage sources and a large number of ideal resistances. The power consumed in one of the resistances is 4 W when either of two sources is active and the other source is short circuited. When both sources are active, the power consumed in the same resistor would be

(*a*) zero or 16 W (*b*) 4 W or 8 W
(*c*) zero or 8 W (*d*) 8 W or 16 W

629. In Fig. 4.108 the effective resistance faced by voltage source is

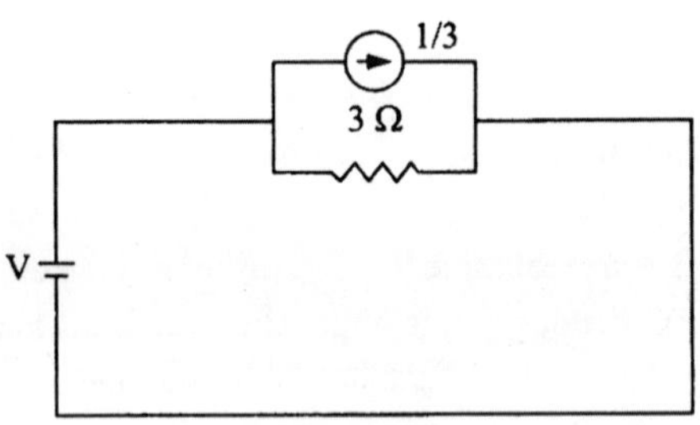

Fig. 4.108.

(*a*) 1 Ω (*b*) 2 Ω
(*c*) 3 Ω (*d*) 3.3 Ω

630. The graph of a network has 6 branches with 3 tree branches. The minimum number of equations required for solution of network is

(*a*) 2 (*b*) 3
(*c*) 4 (*d*) 5

631. The circuit in Fig. 4.109, is to be scaled to an impedance level of 5 kΩ and resonant frequency of 5×10^6 rad/sec. The possible combination of R, L, C is

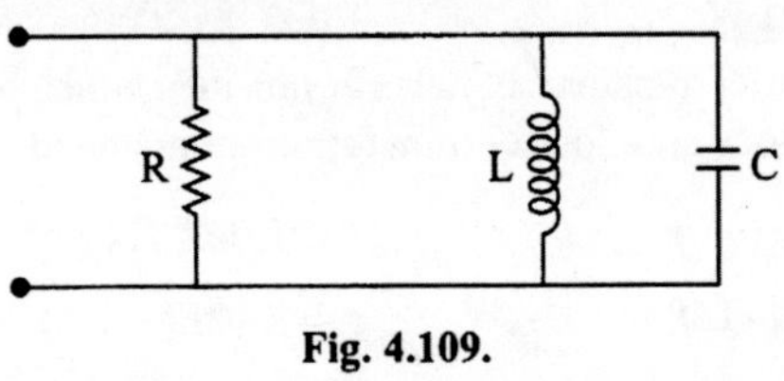

Fig. 4.109.

(*a*) 2.5 Ω, 0.2 mH, 200 pF (*b*) 5 Ω, 0.2 mH, 200 pF
(*c*) 5 kΩ, 0.2 mH, 200 μF (*d*) 5 kΩ, 0.1 mH, 0.4 μF

632. A parallel RLC network has R = 4 Ω, L = 4H, C = 0.25 F. Then at resonance Q =

(*a*) 1 (*b*) 10
(*c*) 20 (*d*) 40

633. An RLC series circuit is excited by an ac voltage $v = 1 \sin t$. If L = 10H, C = 0.1 F peak value of voltage across R is

(*a*) 0.707 V
(*b*) 1 V
(*c*) 1.414 V
(*d*) indeterminate as value of R is not given

634. A magnetic circuit has an iron length of 20 cm and air gap length 0.1 cm. If $\mu_r = 200$ then

(*a*) mmf for iron and air gap are equal
(*b*) mmf for iron is much less than that for air gap
(*c*) mmf for iron is much more than that for air gap
(*d*) either (*a*) or (*c*)

635. An RL series circuit is initially relaxed. A step voltage E is applied. If τ is time constant, voltage across R and L will be equal at *t* =

(*a*) $\tau \ln 2$ (*b*) $\tau \ln \frac{1}{2}$
(*c*) $\frac{1}{2} \ln 2$ (*d*) $\frac{1}{2} \ln \frac{1}{2}$

636. The pole zero plot of a filter is shown in Fig. 4.110. It is

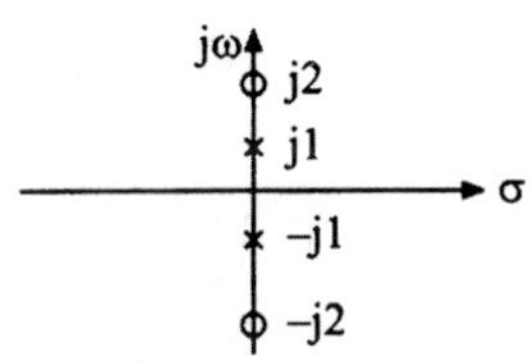

Fig. 4.110.

(*a*) low pass filter (*b*) high pass filter
(*c*) band pass filter (*d*) all pass filter

637. A resistance coil has self-inductance L, resistance R and capacitance C. The impedance across the coil is

(*a*) $\frac{R}{(R + sL)\, sC + 1}$ (*b*) $\frac{R + sL}{sC\,(R + sL) + 1}$

(*c*) $\frac{R}{sL\,(R + sC)}$ (*d*) $\frac{sC}{(R + sL) + sC + 1}$

638. In Fig. 4.111, the total power consumed is

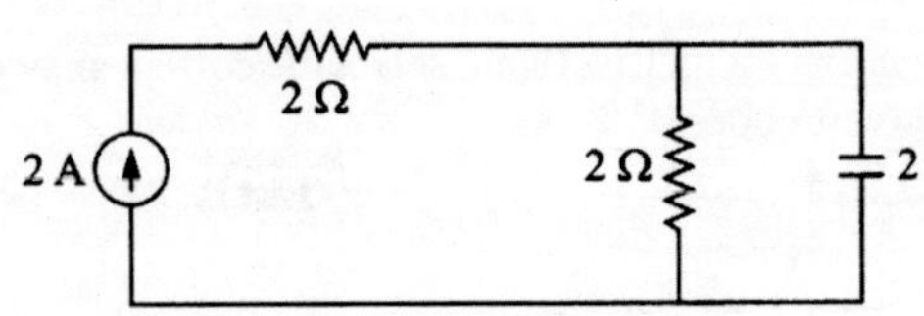

Fig. 4.111.

(*a*) 10 W (*b*) 12 W
(*c*) 16 W (*d*) 20 W

639. In Fig. 4.112, I = 2 A, then V =

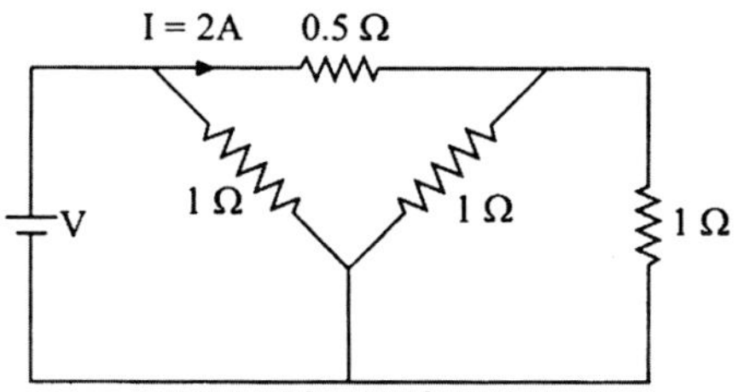

Fig. 4.112.

(*a*) 5V (*b*) 3V
(*c*) 2V (*d*) 1V

640. In Fig. 4.113, the effective resistance across AB is

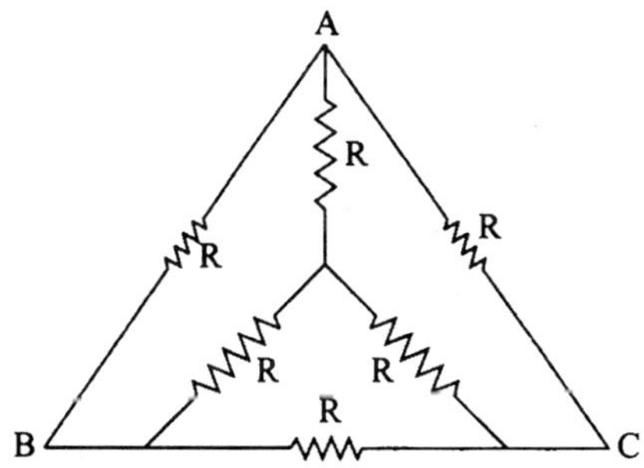

Fig. 4.113.

(*a*) R (*b*) R – 1
(*c*) $\frac{R}{2}$ (*d*) $\frac{6}{11}R$

641. In active filters inductances are avoided because they
(*a*) are always associated with some resistance
(*b*) are bulky and unsuitable for miniaturization
(*c*) are non-linear
(*d*) saturate quickly

642. The magnitude response of a normalised Butterworth low pass filter is
(*a*) linear starting with value of unity at zero frequency and 0.707 at cut off frequency
(*b*) non-linear all through but with values of unity at zero frequency and 0.707 at cut off frequency
(*c*) linear upto cutoff frequency and non-linear thereafter
(*d*) non linear upto cutoff frequency and linear thereafter

643. Which of the following pairs are correctly matched?
1. Brune's realization: realisation with ideal transformer
2. Cauer realisation: ladder realisation
3. Bott Duffin realisation: realisation with non-ideal transformer

Select the answer using the following codes
(*a*) 1, 2 and 3 (*b*) 2 and 3
(*c*) 1 and 3 (*d*) 1 and 2

644. The polynomial
$P(s) = (s - 1)(s^2 + 1)(s + 2)(s + 3)$ is
(*a*) Hurwitz but not strictly Hurwitz
(*b*) not Hurwitz
(*c*) strictly Hurwitz (*d*) anti Hurwitz

645. A delta connection contains three impedances of 60 Ω each. The impedances of equivalent star connection will be
(*a*) 15 Ω (*b*) 20 Ω
(*c*) 30 Ω (*d*) 40 Ω

646. Substitution theorem applies to
(*a*) linear network (*b*) non-linear network
(*c*) linear time invariant network
(*d*) any network

647. In Fig. 4.114, S is closed at $t = 0$. At $t = t_1$, i = 1A and di/dt = 1 A/s. Then the value of L is

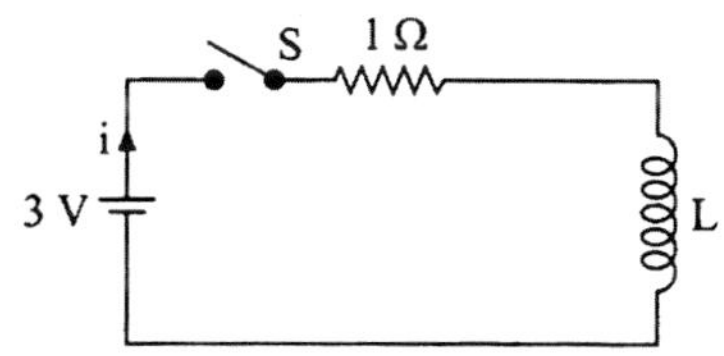

Fig. 4.114.

(*a*) 1 H (*b*) 2 H
(*c*) 3 H (*d*) 4 H

648. In Fig. 4.115, the switch is closed at $t = 0$. At $t = 0^+$, $di/dt = 4$ A/s. Then V =

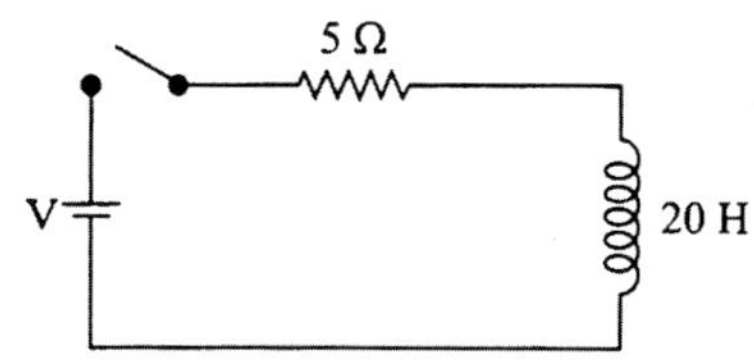

Fig. 4.115.

(*a*) 10 V (*b*) 20 V
(*c*) 40 V (*d*) 80 V

649. In Fig. 4.116, the tree branches are *bdf*. The fundamental loops include

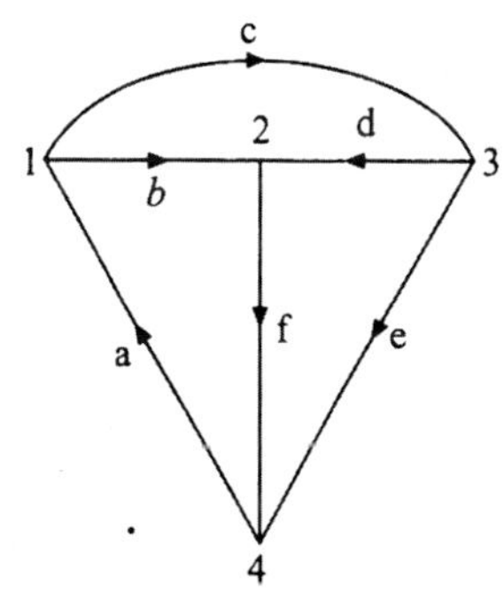

Fig. 4.116.

(*a*) *abc, def, bdea* (*b*) *cea, bdea, abc*
(*c*) *cdb, def, bfa* (*d*) *abde, def, cdb*

650. For Fig. 4.117, the incidence matrix A is

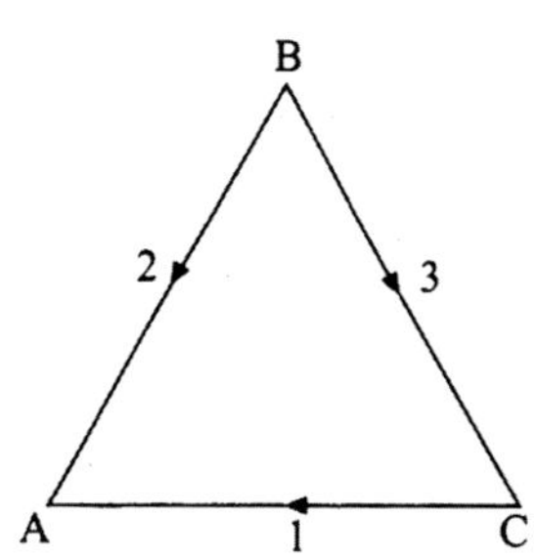

Fig. 4.117.

(a) $\begin{bmatrix} -1 & -1 & 0 \\ 0 & 1 & 1 \\ -1 & 0 & -1 \end{bmatrix}$ (b) $\begin{bmatrix} 1 & 0 & -1 \\ 1 & 1 & 0 \\ 0 & -1 & 1 \end{bmatrix}$

(c) $\begin{bmatrix} -1 & -1 & 0 \\ 0 & 1 & 1 \\ 1 & 0 & -1 \end{bmatrix}$ (d) $\begin{bmatrix} 1 & 0 & 1 \\ -1 & 1 & 0 \\ 0 & -1 & 1 \end{bmatrix}$

651. Consider the following units

1. sec^{-1}
2. rad^2/sec^2
3. sec
4. Ω

The units of $\frac{R}{L}$, $\frac{1}{LC}$, RC and $\sqrt{\frac{L}{C}}$ are respectively

(a) 1, 2, 4, 3 (b) 3, 2, 1, 4
(c) 2, 4, 1, 3 (d) 1, 2, 3, 4

652. RMS value of $i(t) = 10\,[1 + \sin(-t)]$ is

(a) 10 (b) $10/\sqrt{2}$
(c) $\sqrt{150}$ (d) $10/\sqrt{2}$

653. A system has transfer function $\frac{10}{(s+1)(s+2)}$.
The dc gain is

(a) 1 (b) 2
(c) 5 (d) 10

654. In Fig. 4.118, the ratio V_{24}/V_{13} is

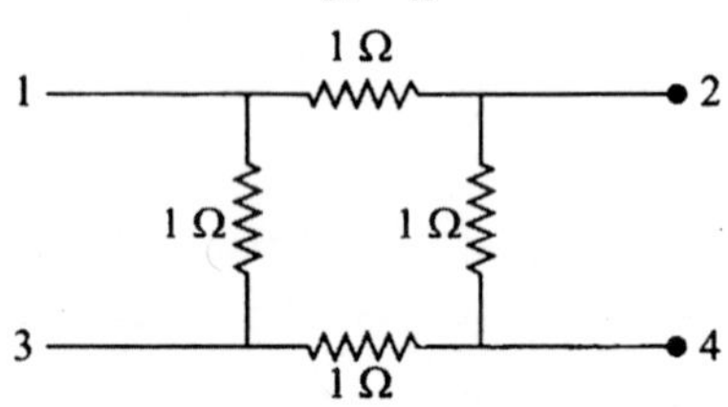

Fig. 4.118.

(a) $\frac{1}{3}$ (b) $\frac{2}{3}$
(c) $\frac{3}{4}$ (d) $\frac{4}{3}$

655. In a linear bilateral passive element two port, which of the following holds for z parameters

(a) $z_{11} = z_{22}$ (b) $z_{12} z_{22} = z_{11} z_{22}$
(c) $z_{11} z_{12} = z_{22} z_{21}$ (d) $z_{12} = z_{21}$

656. In Fig. 4.119, capacitor is used to develop pulses of short duration and large duty cycle across R_2 when switch is closed. Which of the following combinations is best suited?

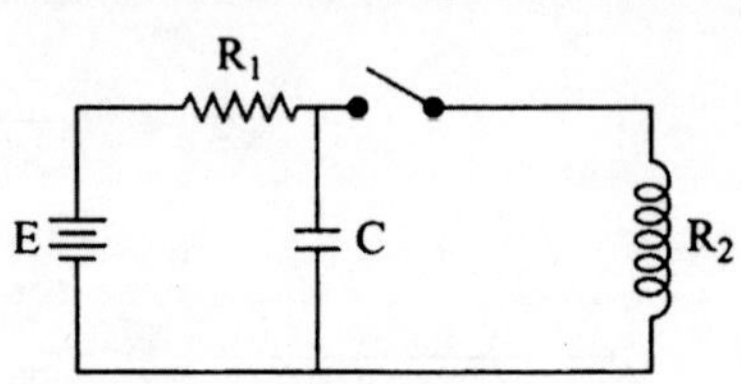

Fig. 4.119.

(a) R_1 and R_2 large but C small
(b) R_1 and R_2 small but C large
(c) R_1 and C large but R_2 small
(d) R_2 and C large but R_1 small

657. In Fig. 4.120, the switch is closed at $t = 0$. After some time $i = 6$ A and rate of change of current was 4A/s. The value of inductor is

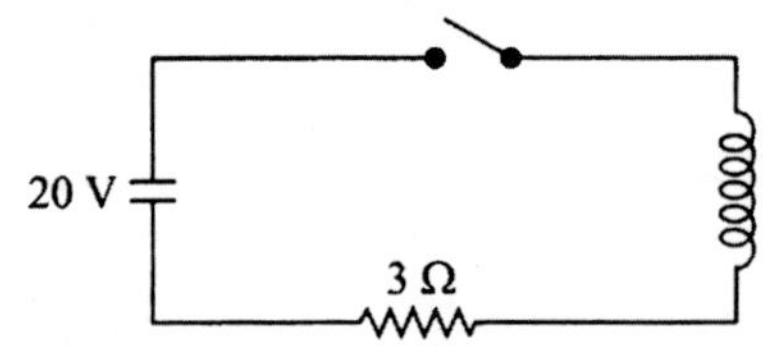

Fig. 4.120.

(a) 1 H (b) 1.5 H
(c) indeterminate (d) 0.5 H

658. In Fig. 4.121, $i(0^-)$ and $v(0^-)$ are zero. The switch is closed at $t = 0$. $i(0^+)$ and $i(\infty)$ are

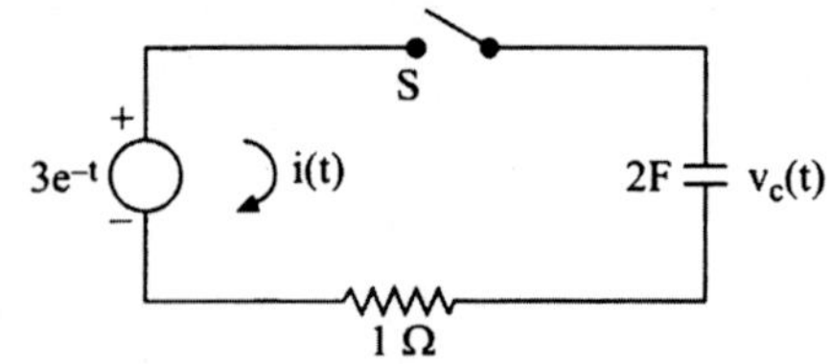

Fig. 4.121.

(a) 3 A and 1.5 A (b) 1.5 A and zero
(c) 3 A and zero (d) 1.5 A and 3 A

659. Match List I with List II and select the correct answer using the given codes. (Capacitor is charged initially)

List I **List II**

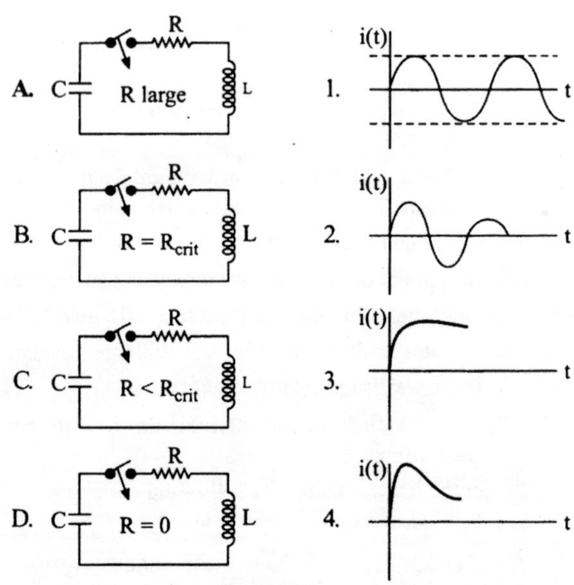

Fig. 4.122.

Codes:

	A	B	C	D
(a)	1	2	3	4
(b)	3	4	2	1
(c)	4	3	2	1
(d)	3	4	1	2

660. In Fig. 4.123, the switch is initially open. At $t = 0$ switch is closed. $v_c(0^+)$ and $i_L(0^+)$ are

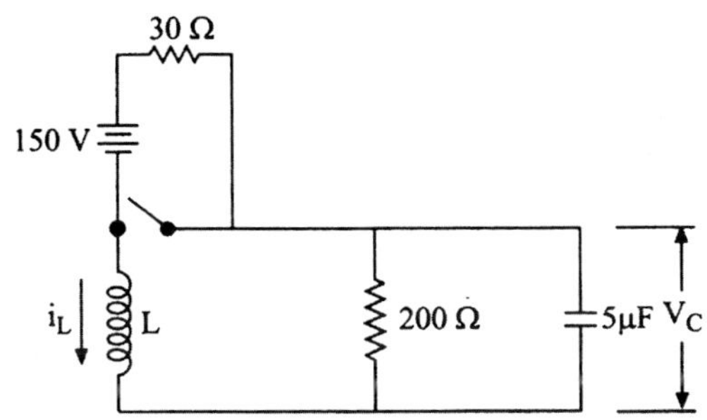

Fig. 4.123.

(*a*) 60V and –0.3A (*b*) 150 V and zero
(*c*) zero and 0.3A (*d*) 90 V and –0.3A

661. In Fig. 4.124, the time constant for the capacitor charging is

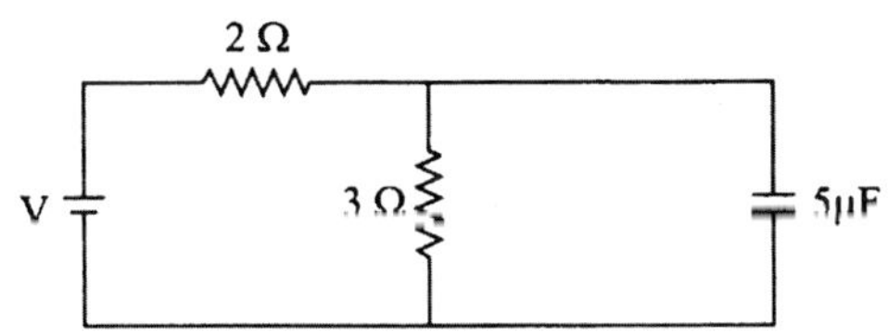

Fig. 4.124.

(*a*) 6 μs (*b*) 10 μs
(*c*) 15 μs (*d*) 25 μs

662. In Fig. 4.125, E = 1V (rms value). The average power is 250 mW. Then phase angle between E and I is

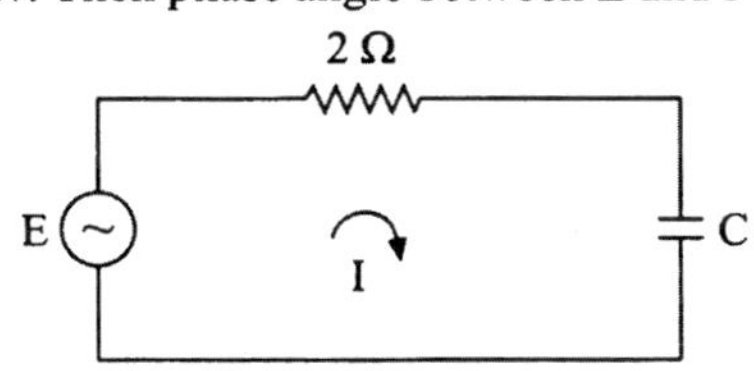

Fig. 4.125.

(*a*) 90° (*b*) 60°
(*c*) 45° (*d*) 30°

663. A 3-phase star connected load has one capacitor C in R phase and two equal resistances in Y and B phases. The phase sequence is RBY. Then

(*a*) $V_{BO} > V_{YO}$ (*b*) $V_{BO} < V_{YO}$
(*c*) $V_{BO} = V_{YO}$ (*d*) $V_{BO} = V_{BY}$

664. In Fig. 4.126, the voltmeter is ideal. The transformer has two identical windings with perfect coupling. The reading of voltmeter is

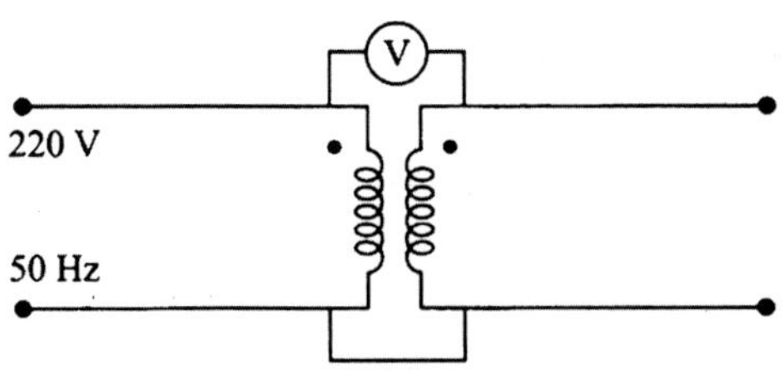

Fig. 4.126.

(*a*) 440 V (*b*) 220 V
(*c*) 110 V (*d*) zero

665. In Fig. 4.127, the current through 2 Ω resistance is

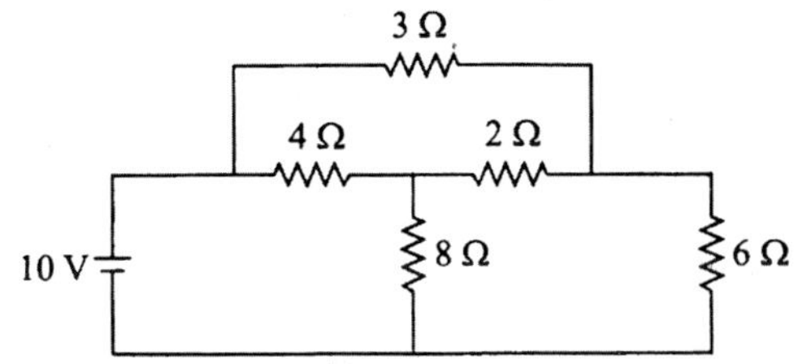

Fig. 4.127.

(*a*) zero (*b*) –2A
(*c*) 2A (*d*) 1A

666. Laplace transform method enables us to find the response in

(*a*) steady state
(*b*) transient state
(*c*) both steady and transient state
(*d*) transient state provided forcing functions do not exist

667. Ideally an attenuator pad should not change

(*a*) voltage level
(*b*) impedance level
(*c*) power level
(*d*) voltage, power or impedance level

668. The filter in Fig. 4.128, uses ideal op-amp. It is

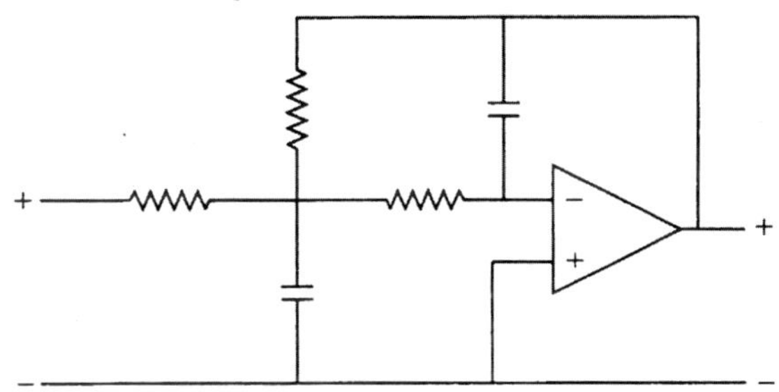

Fig. 4.128.

(*a*) low pass filter (*b*) high pass filter
(*c*) band pass filter (*d*) band stop filter

669. The design of wave filter is based on characteristic impedance which is

(*a*) pure resistance
(*b*) pure reactance
(*c*) pure resistance in pass band and pure reactance in attenuation band
(*d*) pure reactance in pass band and pure resistance in attenuation band

670. Match List I with List II and select the correct answer using the given codes.

List I	**List II**
A. Poles and zeros of driving point LC network	1. lie on real axis
B. Canonic LC network has	2. a zero
C. The number of Canonic networks for the given driving point function	3. minimum number of elements

D.	The first critical frequency nearest the origin for RL driving point impedance	4.	four
		5.	minimum number of elements
		6.	alternate
		7.	either a pole a zero
		8.	three

Codes:

	A	B	C	D
(*a*)	1	5	8	7
(*b*)	6	5	3	2
(*c*)	6	5	4	2
(*d*)	1	3	4	7

671. The driving point impedance of a network is $Z(s) = \dfrac{s^2 + 2s + 6}{s(s+3)}$. The number of energy storing elements in the network is

(*a*) 1 (*b*) 2
(*c*) 3 (*d*) 4

672. Which one of the following is positive real function?

(*a*) $\dfrac{s(s^2+4)}{(s^2+1)(s^2+6)}$ (*b*) $\dfrac{s(s^2-4)}{(s^2+1)(s^2+6)}$

(*c*) $\dfrac{s^3+3s^2+2s+1}{4s}$ (*d*) $\dfrac{s^2(s^4+3s^2+1)}{(s+1)(s+2)(s+3)}$

673. In Fig. 4.129 V_{AB}=

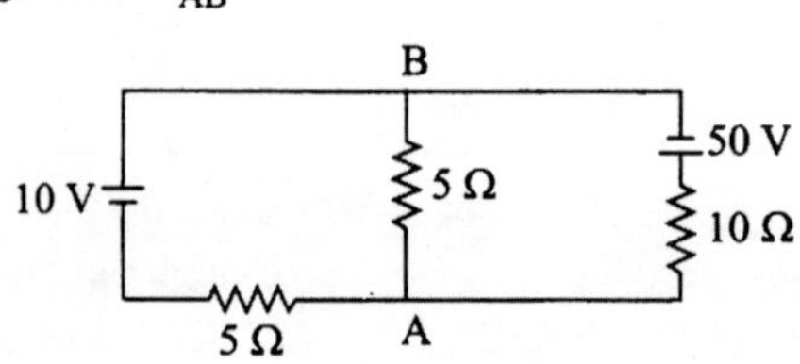

Fig. 4.129.

(*a*) 6V (*b*) 10V
(*c*) 25V (*d*) 40V

674. In Fig. 4.130, which value of Z_L will cause maximum power to be transferred to the load

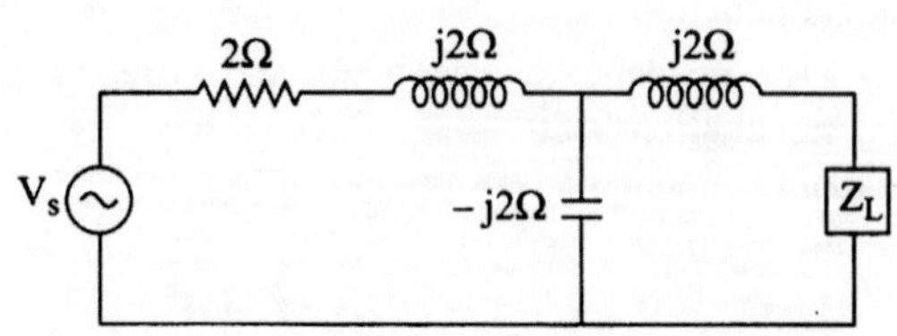

Fig. 4.130.

(*a*) $2+j2\,\Omega$ (*b*) $2-j2\,\Omega$
(*c*) $-j2\,\Omega$ (*d*) $2\,\Omega$

675. For circuit in Fig. 4.131, which combination of V_{TH} and R_{TH} represents the circuit as seen from *ab*

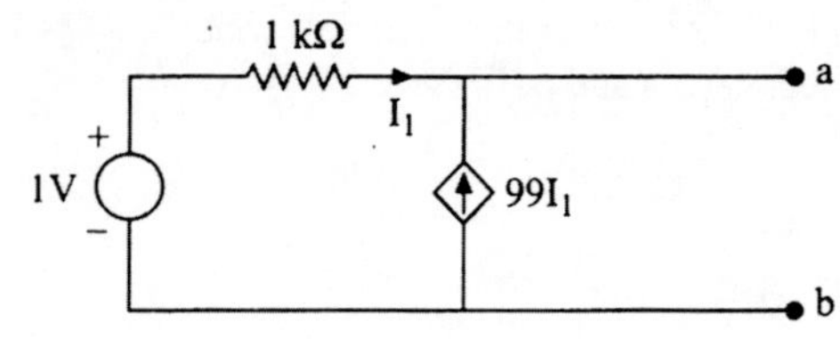

Fig. 4.131.

(*a*) 1V, 10 Ω (*b*) 1V, 1000 Ω
(*c*) 1 mV, 1000 Ω (*d*) 1 mV, 10 Ω

676. Match List I and II for Fig. 4.132 and select the correct answer using the given codes

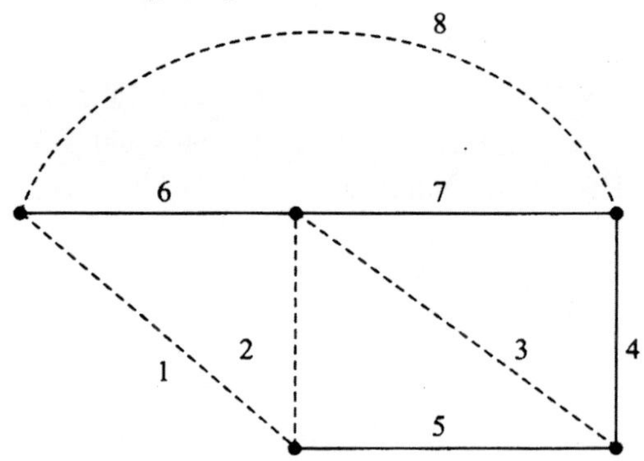

Fig. 4.132.

List I **Branches**	**List II**
A. 1, 2, 3, 4	1. Twigs
B. 4, 5, 6, 7	2. links
C. 1, 2, 3, 8	3. Fundamental cutset
D. 1, 4, 5, 6, 7	4. Fundamental loop

Codes:

	A	B	C	D
(*a*)	3	1	2	4
(*b*)	2	3	1	4
(*c*)	3	2	4	1
(*d*)	1	4	3	2

677. Two impedances are connected in series. Three voltmeters are connected, one across each impedance and one across combination. The readings of all the three are equal. The phase angle between voltages across the two impedances is

(*a*) 30° (*b*) 60°
(*c*) 90° (*d*) 120°

678. For a given voltage four heating coils will give maximum heat when connected

(*a*) in parallel
(*b*) in series
(*c*) with two parallel pairs in series
(*d*) with one pair in parallel and the other two in series

679. A black box has one of the elements out of R, L, C. When a 220 V ac supply is connected the current supplied by source is I. When a 0.1F capacitor is connected in parallel with the box, the current is 2 I. The element is

(*a*) resistance (*b*) inductance

(c) 0.1F capacitance

(d) not identifiable on the basis of given data

680. A two port network is symmetrical if

(a) $z_{11} z_{12} - z_{12} z_{21} = 1$ (b) $AD - BC = 1$

(c) $h_{11} h_{22} - h_{12} h_{21} = 1$ (d) $y_{11} y_{22} - y_{12} y_{21} = 1$

681. Which pair is correctly matched?

(a) symmetrical two port $AD - BC = 1$

(b) Reciprocal two port $z_{11} = z_{12}$

(c) Inverse hybrid parameter A, B, C, D

(d) hybrid parameter $(V_1 I_2) = f(I_1, V_2)$

682. An ideal transformer has a turn ratio of 2 : 1. Taking *hv* side as port 1 and *lv* side as port 2, transmission parameters of transformer are

(a) $\begin{bmatrix} 0 & -2 \\ 0.5 & 0 \end{bmatrix}$ (b) $\begin{bmatrix} -2 & 0 \\ 0 & 0.5 \end{bmatrix}$

(c) $\begin{bmatrix} 2 & 0 \\ 0 & 0.5 \end{bmatrix}$ (d) $\begin{bmatrix} 0.5 & 0 \\ 0 & 2 \end{bmatrix}$

683. A two port network has $V_1 = 24I_1, + 8I_2$ and $V_2 = 8I_1 + 32I_2$. Which of the following networks is represented by above equation

(a)
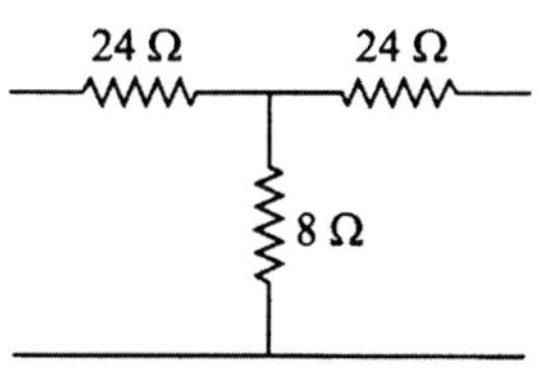

(b)
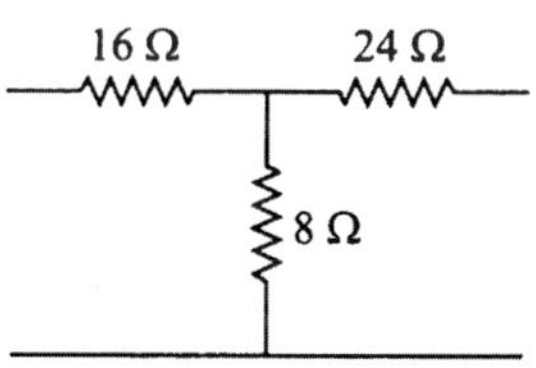

(c)
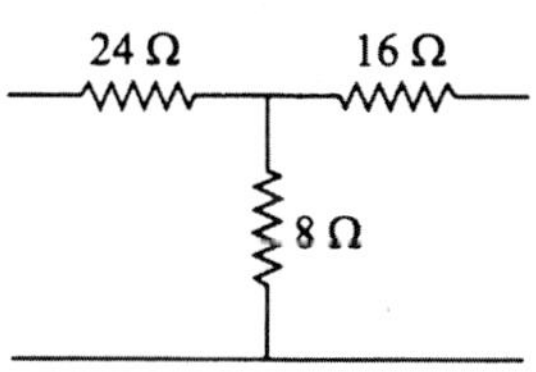

(d)
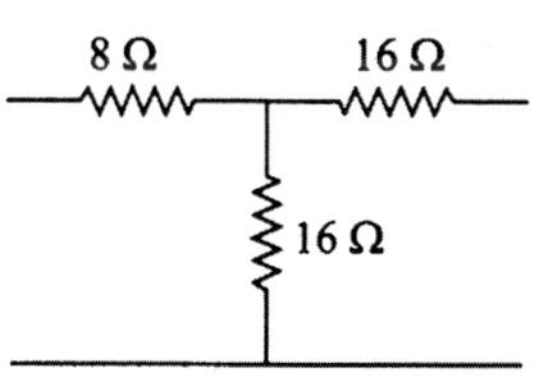

684. The circuit in Fig. 4.133 has reached steady values with switch open. If switch is closed at $t = 0$ the voltage across 2 μF capacitance at $t = 0^+$ is

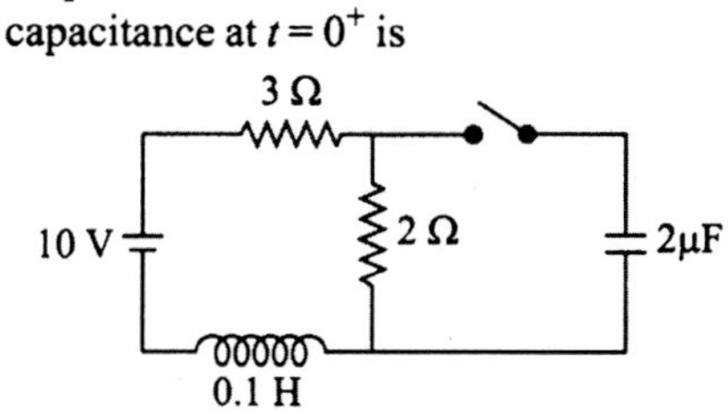

Fig. 4.133.

(a) zero (b) 4V

(c) 6V (d) 10V

685. The response of a network is shown in Fig 4.134. Which of following represents location of poles

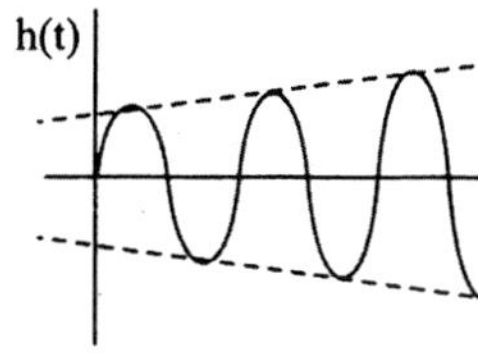

Fig. 4.134.

(a)
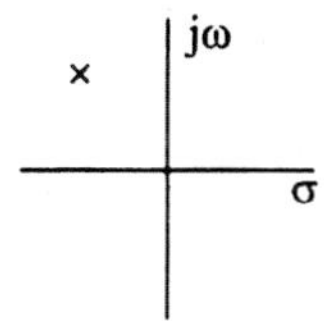

(b)
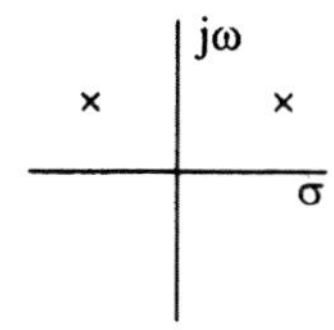

(c)
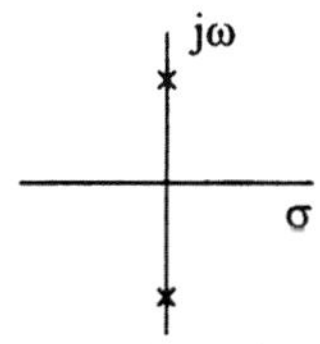

(d)
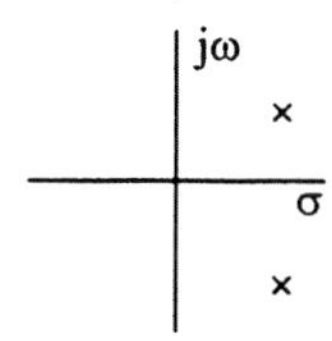

686. A step voltage is applied to RLC series circuit having R = 2 Ω, L = 1H and C = 1F. The transient current response will be

(a) Overdamped

(b) Critically damped

(c) Under damped

(d) any of above depending on magnitude of voltage

687. When a current source I is suddenly connected across a two terminal relaxed RC circuit at time $t = 0$, the voltage across the current source is shown in Fig. 4.135. The RC circuit is

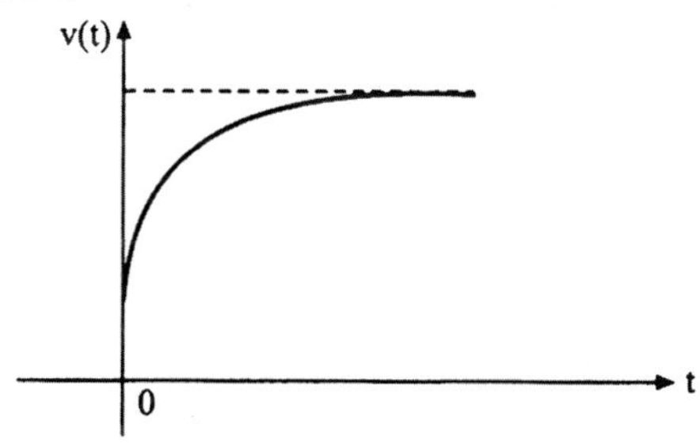

Fig. 4.135.

(a) a series combination of R and C

(b) a parallel combination of R and C

(c) a series combination of R and parallel combination of R and C

(d) a pure capacitor

688. A resistance R is connected to a voltage source V_s having internal resistance R_s. A voltmeter of resistance R_m is used to measure the voltage across R. The voltmeter will read

(a) $\dfrac{V_s R_s R_m}{R_s R_m + R_s R + R R_m}$

(b) $\dfrac{V_s R}{R + R_s}$

(*c*) $\dfrac{V_s R_m}{R + R_s + R_m}$

(*d*) $\dfrac{V_s R R_m}{R_s R + R_s R_m + R R_m}$

689. When two coupled coils of equal self-inductance are connected in series in one way the net inductance is 12 mH and when they are connected in the other way the net inductance is 4 mH. The maximum value of inductance when they are connected in parallel is

(*a*) 2mH (*b*) 3mH
(*c*) 4mH (*d*) 6mH

690. Which of the following polynomials is not Harwitz?

(*a*) $P_1 = s^8 + 2s^6 + 4s^4$
(*b*) $P_2 = s^6 - 3s^4 + 2s^2 + 1$
(*c*) $P_3 = s^4 + 3s^3 + 3s^2 + 2s + 1$
(*d*) $P_4 = s^7 + 2s^6 + 2s^4 + 4s^3 + 8s^2 + 8s + 4$

691. For a voltage transfer function H(*s*), realizable by RLC network, the following statements are made.

1. H(*s*) cannot have a pole at $s = 0$
2. H(*s*) cannot have pole at $s = \pm j4$
3. H(*s*) cannot have a pole at $s = \infty$
4. H(*s*) cannot have pole at $s = +2$

Of these statements

(*a*) 3 and 4 are correct (*b*) 2 and 4 are correct
(*c*) 1 and 2 are correct (*d*) 1 and 3 are correct

692. The first critical frequency nearest the origin of the complex frequency plane for an RL driving point impedance will be

(*a*) a zero in left half plane
(*b*) a zero in right half plane
(*c*) a pole in left half plane
(*d*) a pole or zero in left half plane

693. If $F_1(s)$ and $F_2(s)$ are two positive real functions, then the function which is always positive real is

(*a*) $F_1(s) F_2(s)$ (*b*) $F_1(s) - F_2(s)$
(*c*) $\dfrac{F_1(s) F_2(s)}{F_1(s) + F_2(s)}$ (*d*) $F_1(s)/F_2(s)$

694. Match List I with List II and give the correct answer using the given codes

List I	List II
A. Reactance theorem	1. Foster
B. Driving point impedance	2. Bartlett
C. Continued fraction expansion	3. Cauer
D. Bisection Theorem	4. Positive real function

Codes:

	A	B	C	D
(*a*)	1	4	3	2
(*b*)	1	2	3	4
(*c*)	2	3	1	4
(*d*)	3	4	2	1

695. Double integration of unit step function leads to

(*a*) an impulse (*b*) a parabola
(*c*) a ramp (*d*) a doublet

696. In Fig 4.136 the rms value is

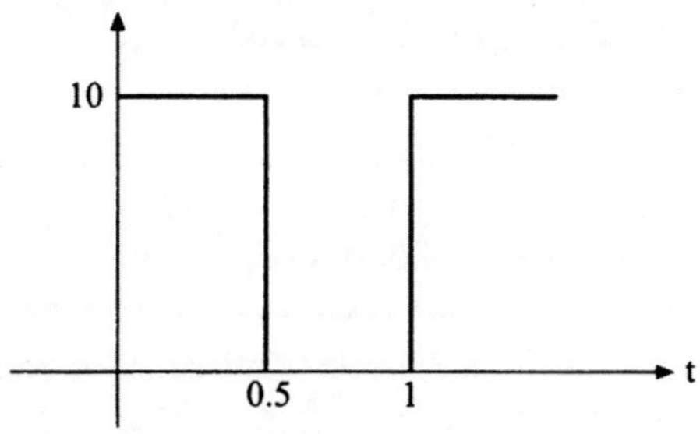

Fig. 4.136.

(*a*) 5 (*b*) 2.5
(*c*) $\sqrt{2.5}$ (*d*) $\sqrt{50}$

697. If $f(t) = -f(-t)$ and $f(t)$ satisfies Drichlet conditions, then $f(t)$ can be expanded in Fourier series having

(*a*) sine terms only (*b*) cosine terms only
(*c*) cosine and constant term (*d*) sine and constant term

698. Fourier transform $F(j\omega)$ of a signal $f(t)$ has the property

(*a*) $F(j\omega) = F(-j\omega)$ (*b*) $F(j\omega) = -F(-j\omega)$
(*c*) $F(j\omega) = F^*(-j\omega)$ (*d*) $F(j\omega) = -F^*(-j\omega)$

699. The inverse Fourier transform of $F(\omega) = \dfrac{1}{j\omega} = \pi\delta(\omega)$ is

(*a*) sin ωt (*b*) cos ωt
(*c*) *sgn*(*t*) (*d*) *u*(*t*)

700. The following table gives some Laplace transforms and the functions

f(*t*)	F(*s*)
1. $\delta(t)$	s
2. $u(t)$	$1/s$
3. $tu(t)$	$2/s^2$
4. $t^2 u(t)$	$2s^3$

Of these the correctly matched pairs are

(*a*) 2 and 4 (*b*) 1 and 4
(*c*) 3 and 4 (*d*) 1 and 2

701. The final value of $£^{-1} \dfrac{2s+1}{s^4 + 8s^3 + 16s^2 + s}$ is

(*a*) ∞ (*b*) 2
(*c*) 1 (*d*) zero

702. For a transmission line open-circuit and short-circuit impedances are 20 Ω and 5 Ω. Then characteristic impedance is

(*a*) 100 Ω (*b*) 50 Ω
(*c*) 25 Ω (*d*) 10 Ω

703. The equivalent resistance between terminals *x* and *y* in Fig 4.137 is

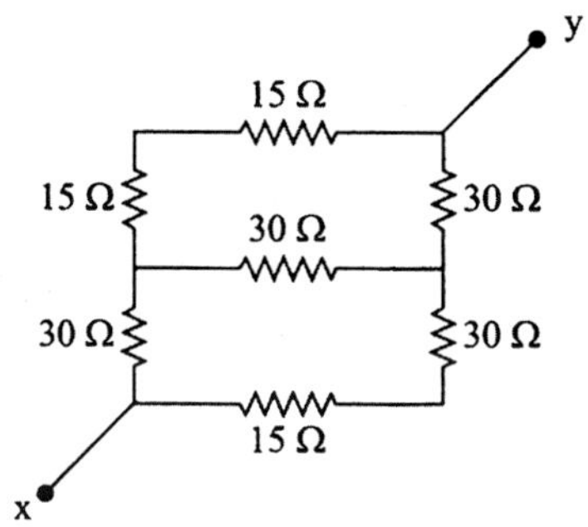

Fig. 4.137.

(*a*) 150 Ω (*b*) 45 Ω
(*c*) 55 Ω (*d*) 30 Ω

704. In Fig. 4.138 the initial current through L_1 and initial voltage across C are zero and switch is open. At $t = 0$ the switch is closed. i_{L1} and V_c are found at $t = 0^+$ and $t = \infty$. Which of the following sets is correct.

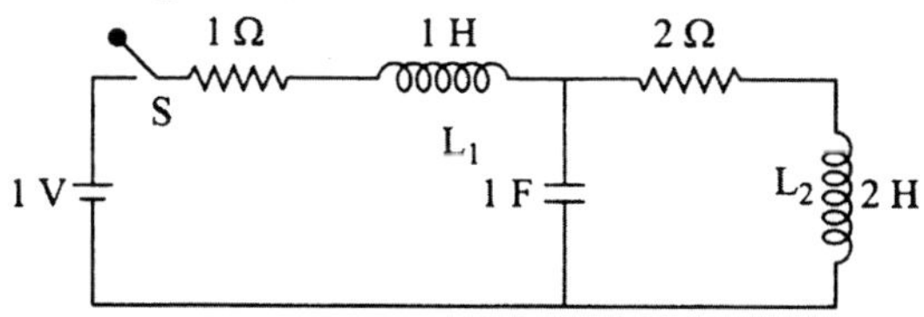

Fig. 4.138.

	$i_{L1}(0^+)$	$i_{L1}(\infty)$	$v_c(0^+)$	$V_c(\infty)$
(*a*)	1/3A	1/3A	2/3V	2/3V
(*b*)	0	1/3A	0	lV
(*c*)	1/3 A	0	2/3V	0
(*d*)	0	1/3 A	0	2/3 V

705. The dual of network in Fig. 4.139 is

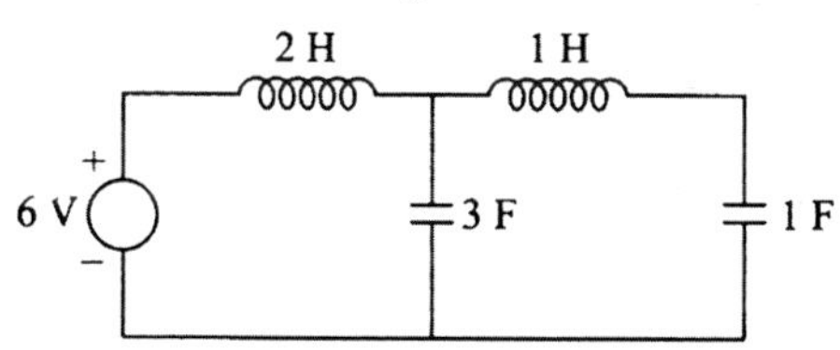

Fig. 4.139.

(*a*)
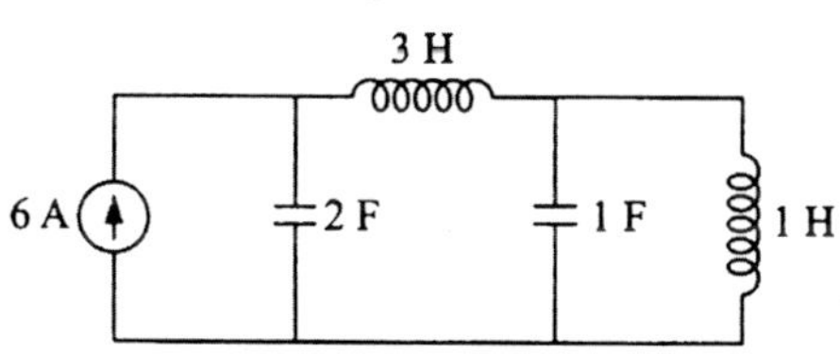

(*b*)
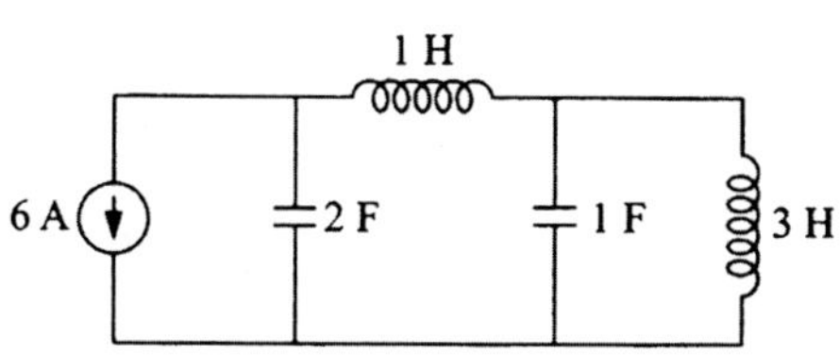

(*c*)
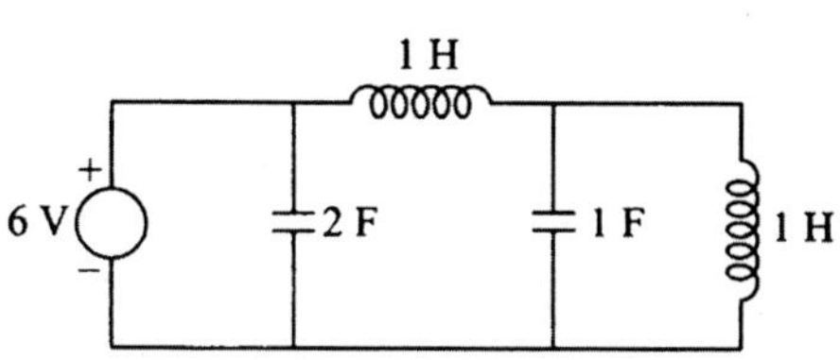

(*d*)
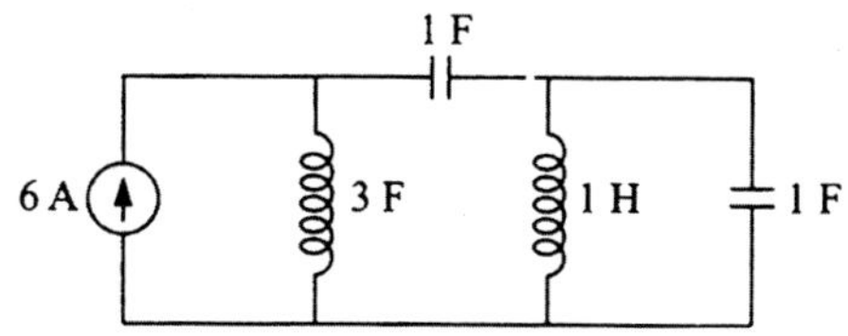

706. In the graph of Fig. 4.140 the total number of trees is

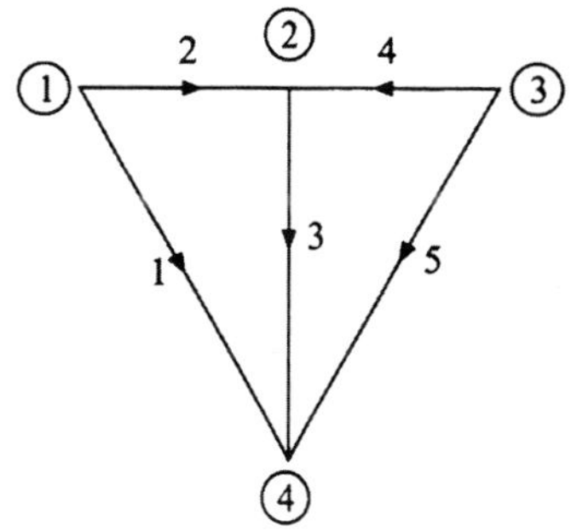

Fig. 4.140.

(*a*) 4 (*b*) 5
(*c*) 6 (*d*) 8

707. In Fig. 4.140 the number of tree branches and links respectively are

(*a*) 3 and 2 (*b*) 2 and 3
(*c*) 4 and 1 (*d*) 1 and 4

708. For the star network of Fig. 4.141, the equivalent delta network is

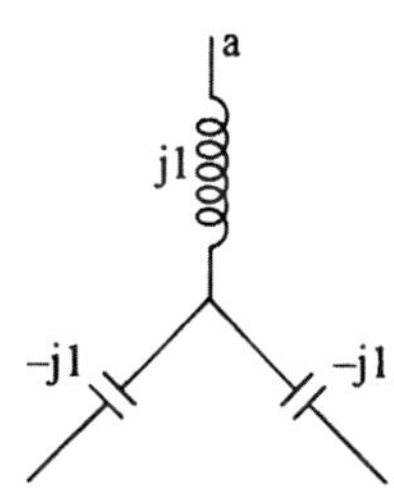

Fig. 4.141.

(*a*)
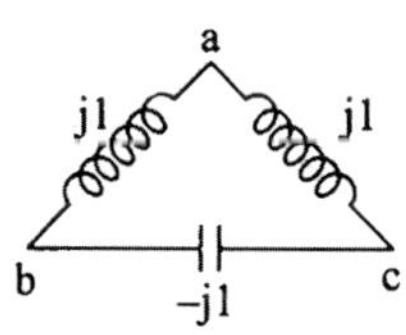

(*b*)
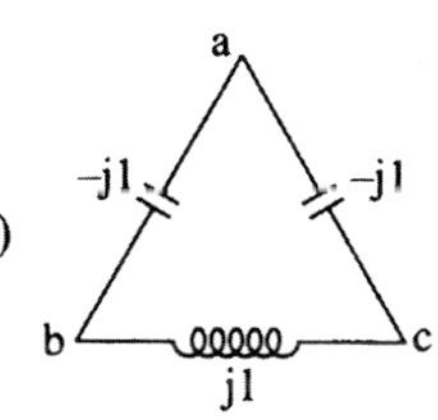

(*c*)
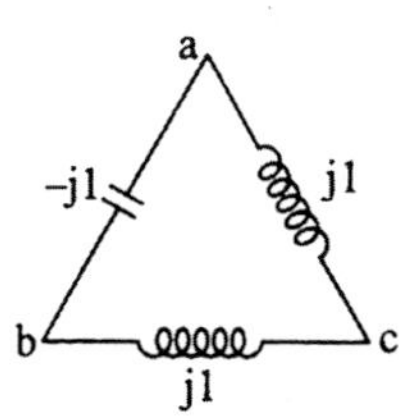

(*d*)
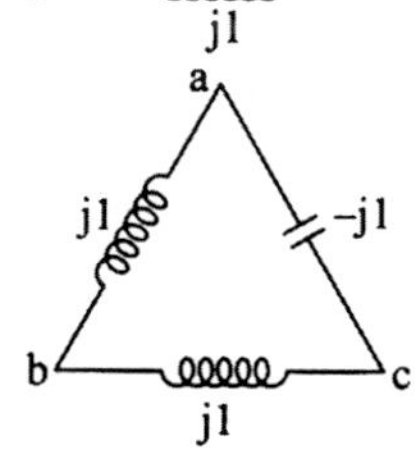

709. The average value of delayed full wave rectified sine wave (Fig. 4.142) is

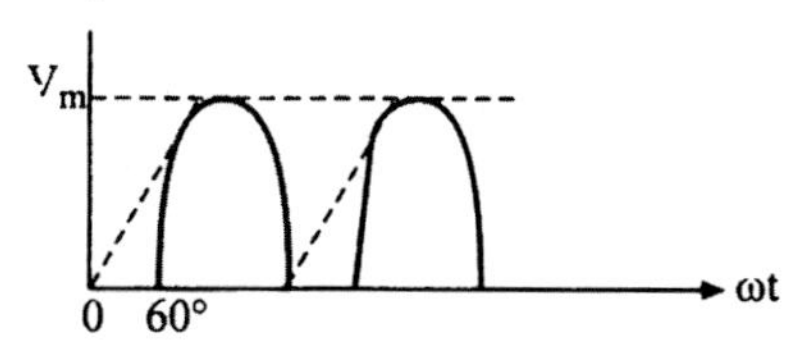

Fig. 4.142.

(*a*) $\frac{V_m}{\pi}$ (*b*) $\frac{2V_m}{\pi}$

(*c*) $\frac{3V_m}{2\pi}$ (*d*) $\frac{V_m}{2\pi}$

710. The relation AD – BC = 1 where A, B, C, D are transmission parameters of a network is valid for

(*a*) every network

(*b*) passive but not reciprocal network

(*c*) passive and reciprocal network

(*d*) both active and passive networks

711. A T network is inserted between source and R as shown in Fig 4.143. The resistance seen by the source remains the same with or without T network when R is

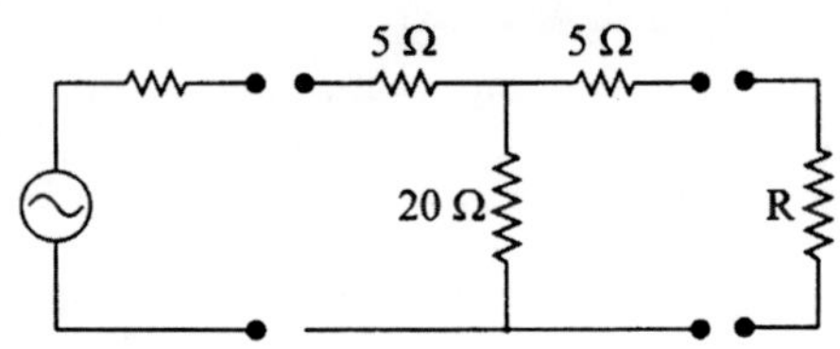

Fig. 4.143.

(*a*) 5 Ω (*b*) 10 Ω

(*c*) 15 Ω (*d*) 20 Ω

712. The *y* parameters of a four terminal network are $\begin{bmatrix} 4 & 2 \\ 1 & 1 \end{bmatrix}$. A one ohm resistance is connected as shown in Fig. 4.144. The new *y* parameters are

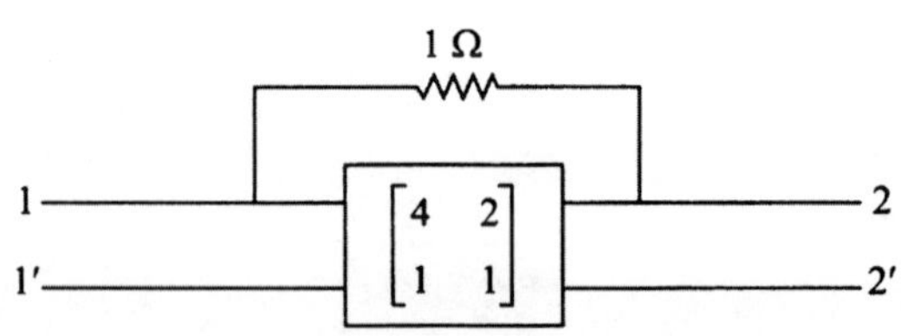

Fig. 4.144.

(*a*) $\begin{bmatrix} 5 & 1 \\ 0 & 2 \end{bmatrix}$ (*b*) $\begin{bmatrix} 4 & 3 \\ 2 & 2 \end{bmatrix}$

(*c*) $\begin{bmatrix} 3 & 2 \\ 1 & 1 \end{bmatrix}$ (*d*) $\begin{bmatrix} 4 & 2 \\ 1 & 1 \end{bmatrix}$

713. The time constant of the circuit in Fig 4.145 is

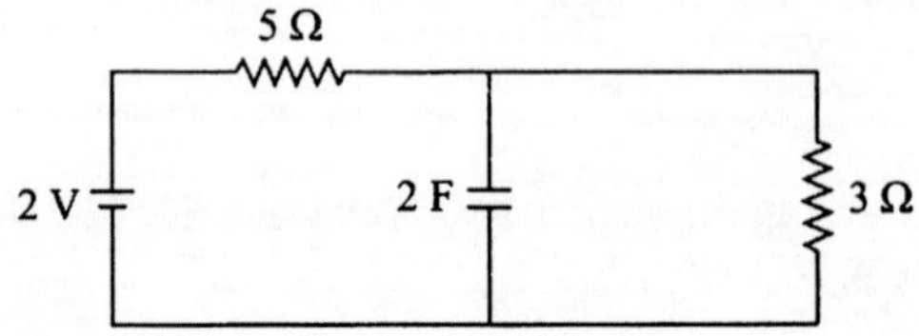

Fig. 4.145.

(*a*) 6 sec (*b*) 10 sec

(*c*) $\frac{15}{4}$ sec (*d*) 16 sec

714. The time constant of the circuit in Fig. 4.146 is

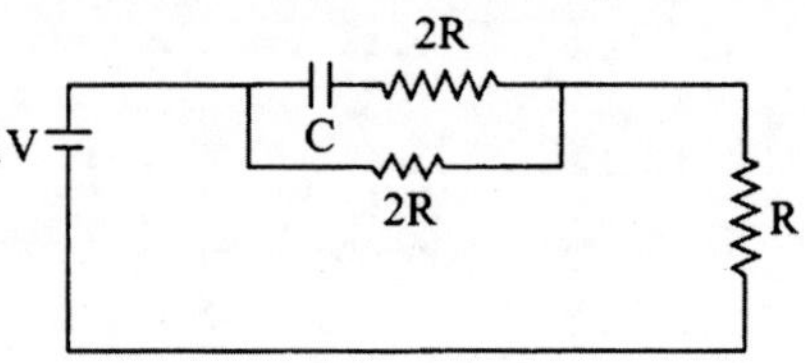

Fig. 4.146.

(*a*) RC (*b*) 2 RC

(*c*) 3 RC (*d*) 5 RC

715. Speecial circuits using neon tubes, contact relays and capacitors are used in many domestic appliances. The role of capacitor in Fig. 4.147 is to

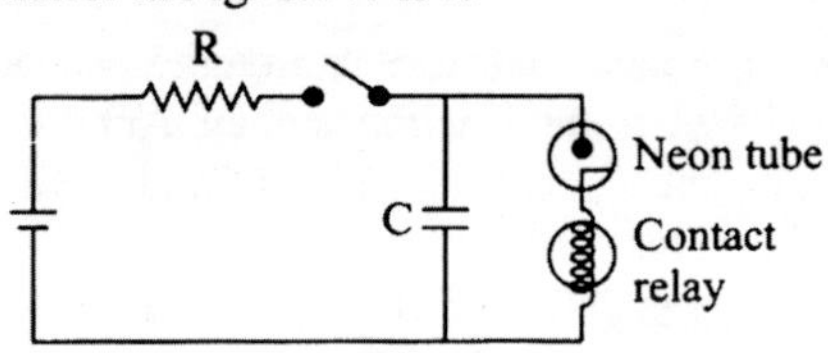

Fig. 4.147.

(*a*) reduce power loss

(*b*) function as resistor when relay is on

(*c*) delay the operation of the contact relay after switching on

(*d*) open circuit the relay terminals

716. Match List I with List II and select correct answer using the given code

List I **List II**

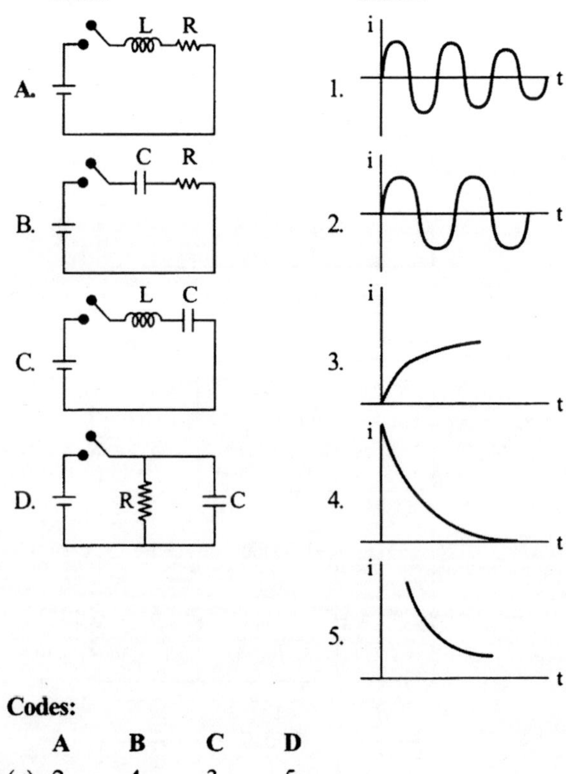

Codes:

	A	B	C	D
(*a*)	2	4	3	5
(*b*)	3	4	2	5
(*c*)	5	2	3	1
(*d*)	3	1	2	4

717. An RLC series circuit has R = 1Ω, L = 1 H, C = 1F. The damping ratio is

(*a*) more than 1 (*b*) 1

(*c*) 0.5 (*d*) zero

718. The transfer function V_2/V_1 for Fig. 4.148 has

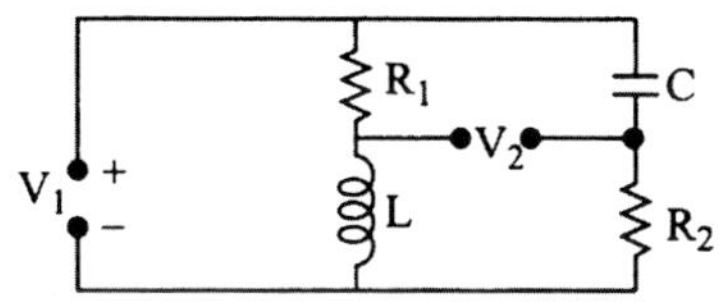

Fig. 4.148.

(*a*) a zero at origin (*b*) a zero at infinity

(*c*) no zero (*d*) a zero at $j1$

719. Consider the following statements

1. Current through inductor cannot change abruptly
2. Voltage across capacitor cannot change abruptly
3. Initial value of $f(t)$ is $\underset{s \to 0}{\text{Lt}}\ sF(s)$
4. Final value of $f(t)$ is $\underset{s \to \infty}{\text{Lt}}\ sF(s)$

Of the above statements

(*a*) 3 and 4 are correct (*b*) 1 and 4 are correct

(*c*) 1 and 2 are correct (*d*) 2 and 3 are correct

720. The transfer function $H(s) = \dfrac{s^2 - 5s + 100}{s^2 + 5s + 100}$ represents

(*a*) a high pass filter (*b*) band elimination filter

(*c*) a resonator (*d*) an all pass filter

721. In Fig. 4.149, V reads 45 V. Then I =

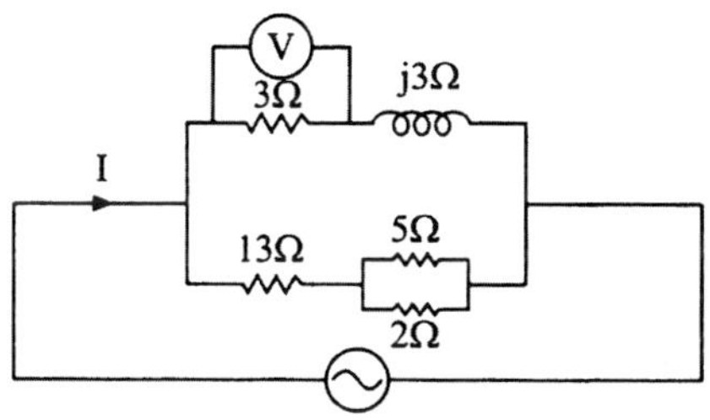

Fig. 4.149.

(*a*) 10.1A (*b*) 13.3 A

(*c*) 16.1 A (*d*) 19.4 A

722. In Fig 4.150, I =

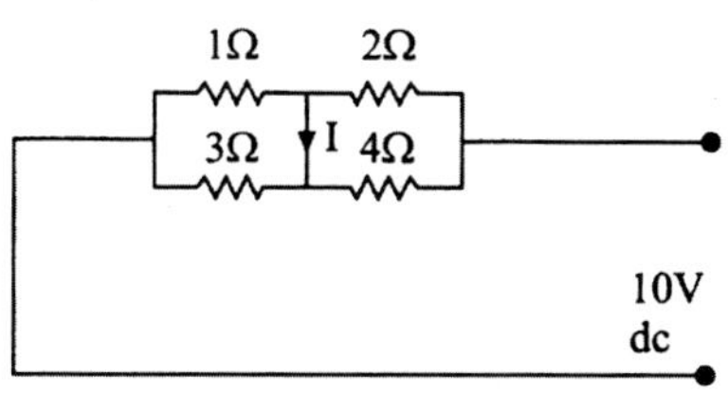

Fig. 4.150.

(*a*) $\dfrac{-2}{5}$ A (*b*) $\dfrac{24}{5}$ A

(*c*) $\dfrac{18}{5}$ A (*d*) $\dfrac{2}{5}$ A

723. The circuit in Fig. 4.151 is

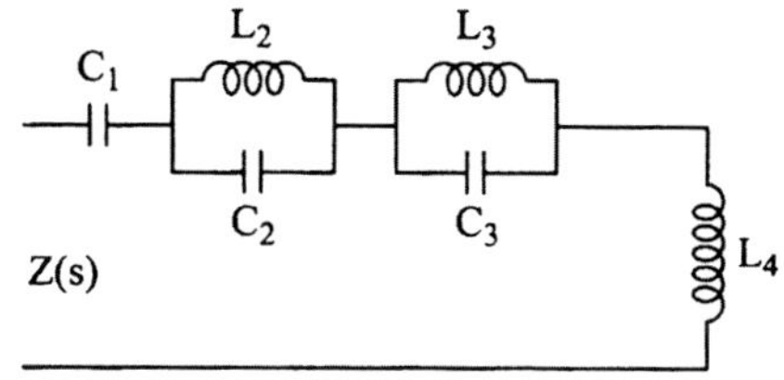

Fig. 4.151.

(*a*) Cauer I form (*b*) Foster I form

(*c*) Cauer II form (*d*) Foster II form

724. A two terminal black box contains one of the elements R, L, C. The black box is connected to 220V ac supply. The current is I. When a capacitance of 0.1 F is connected in series between the source and box, the current is 2I. The element is

(*a*) resistance

(*b*) inductance

(*c*) capacitance of 0.5 F

(*d*) not identifiable from the given data

725. For a transfer function $H(s) = P(s)/Q(s)$, where $P(s)$ and $Q(s)$ are polynomials in s

(*a*) the degrees of $P(s)$ is always greater than degree of $Q(s)$

(*b*) the degrees of $P(s)$ and $Q(s)$ are equal

(*c*) the degree of $P(s)$ is independent of degree of $Q(s)$

(*d*) the maximum difference in degrees of $P(s)$ and $Q(s)$ is 1

726. For the mechanical system of Fig. 4.152, the correct differential equation is (where x = displacement in m, M = mass in kg, B = damping constant, N-sec/m K is spring constant N/m and F = force, newton).

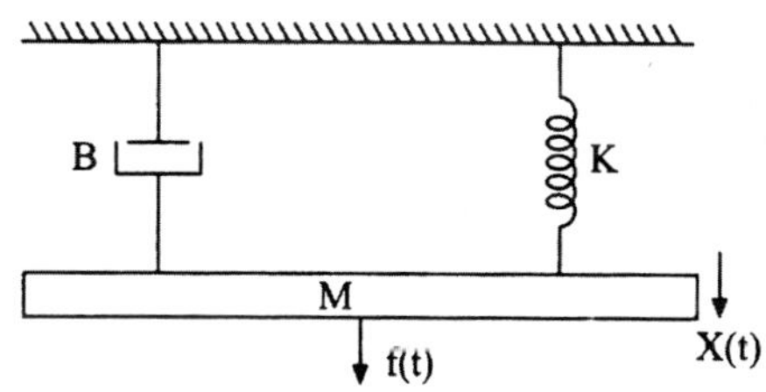

Fig. 4.152.

(*a*) $f(t) - M\dfrac{d^2x(t)}{dt^2} - B\dfrac{dx(t)}{dt} - Kx(t) = 0$

(*b*) $f(t) - M\dfrac{dx(t)}{dt} - Bx(t) - K\int x(t)\, dt = 0$

(*c*) $f(t) + M\dfrac{dx(t)}{dt} + Bx(t) + K\int x(t)\, dt = 0$

(*d*) $f(t) - M\dfrac{d^2x(t)}{dt^2} - \dfrac{1}{B}\dfrac{dx(t)}{dt} - \dfrac{1}{K}x(t) = 0$

727. In any network of linear impedances the current flowing at any point is equal to algebraic sum of currents caused to flow at that point by each of the emfs taken separately with all other emfs reduced to zero. The statement represents

(*a*) KCL (*b*) Norton's theorem

(*c*) Thevenin's theorem (*d*) Superposition theorem

728. Which of the following theorems is a manifestation of law of conservation of energy

(*a*) Tellegen's theorem (*b*) Reciprocity theorem

(*c*) Thevenin's theorem (*d*) Norton's theorem

729. In Fig. 4.153, the current through R is 1 A when $V_A = 0$ and $V_B = 15$ V. If both V_A and V_B are increased by 15 V each, current through R will be

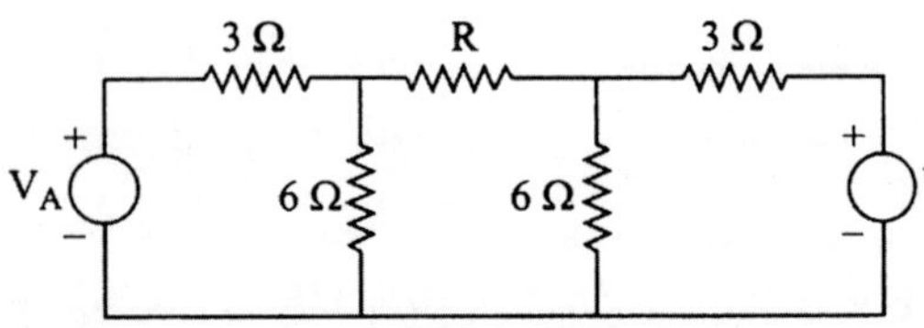

Fig. 4.153.

(*a*) 1A (*b*) 0.5 A

(*c*) 3 A (*d*) 0.33 A

730. If the network in Fig. 4.154 (*a*), (*b*) are equivalent at terminals A-B, then voltage V and resistance R are

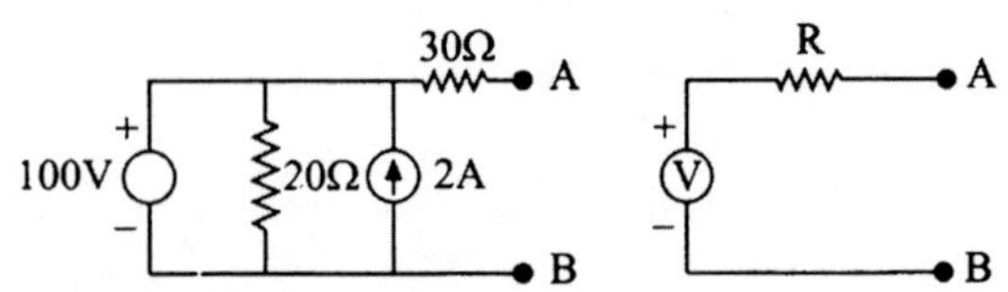

Fig. 4.154.

(*a*) 100 V and 12 Ω (*b*) 60 V and 12 Ω

(*c*) 100 V and 30 Ω (*d*) 60 V and 30 Ω

731. Match List I with List II and give correct answer using the given codes

List I	**List II**
(Theorems)	**(Property of Network)**
A. Reciprocity	1. Impedance matching
B. Tellegen's	2. Bilateral
C. Superposition	3. $\sum V_{jk}\, I_{jk} = 0$
D. Maximum power transfer	4. Linear
	5. Non-linear

Codes:

	A	B	C	D
(*a*)	1	2	3	4
(*b*)	1	2	3	5
(*c*)	2	3	4	1
(*d*)	2	3	5	1

732. In Fig. 4.155, the steady state current through 1 H inductance is

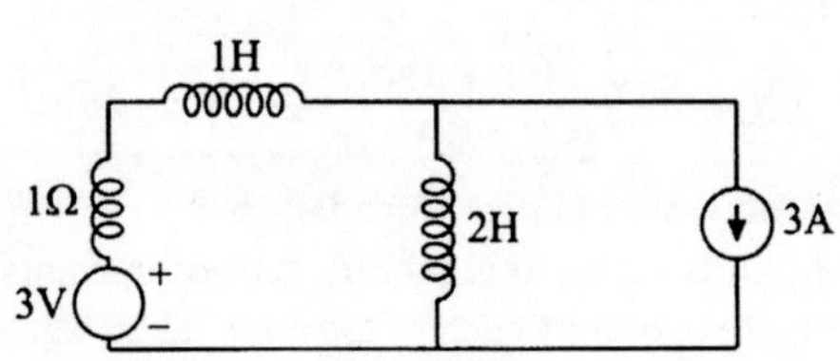

Fig. 4.155.

(*a*) zero (*b*) 3 A

(*c*) 5 A (*d*) 6 A

733. For the circuit in Fig. 4.156, fed by a unit step voltage, $v_c =$

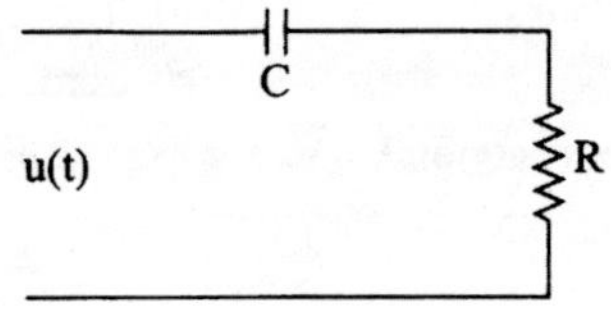

Fig. 4.156.

(*a*) $1 - e^{-t/RC}$ (*b*) $e^{-t/RC}$

(*c*) $e^{t/RC}$ (*d*) 1

734. In Fig. 4.157, capacitor is uncharged. The switch in closed at $t = 0$. At $t = 0^+$, di/dt is

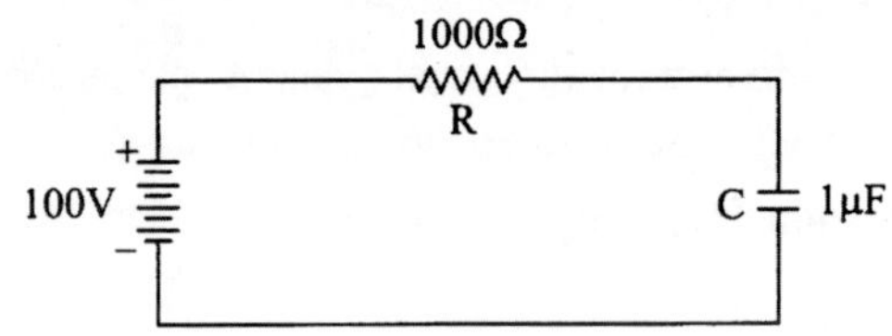

Fig 4.157.

(*a*) 100 A/s (*b*) −100 A/s

(*c*) 1000 A/s (*d*) −1000 A/s

735. For the circuit of Fig. 4.158 the time constant is

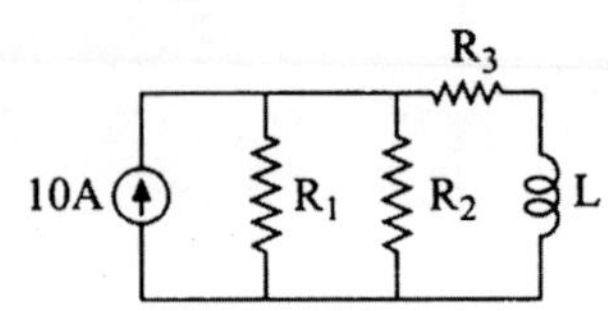

Fig. 4.158.

(*a*) $\dfrac{L}{R_3 + \dfrac{R_1 R_2}{R_1 + R_2}}$ (*b*) $\dfrac{L}{R_1 + R_2 + R_3}$

(*c*) $\dfrac{L}{\dfrac{1}{R_1} + \dfrac{1}{R_2} + \dfrac{1}{R_3}}$ (*d*) $\dfrac{L/R_1 R_2}{R_1 + R_2 + R_3}$

736. A series RLC circuit is overdamped when

(*a*) $\dfrac{R^2}{4L^2} > \dfrac{1}{LC}$ (*b*) $\dfrac{R^2}{4L^2} = \dfrac{1}{LC}$

(*c*) $\dfrac{R^2}{4L^2} < \dfrac{1}{LC}$ (*d*) *R* is infinite

737. The circuit in Fig. 4.159 is in steady state with switch closed. At $t = 0$ switch is opened. The current *i* is

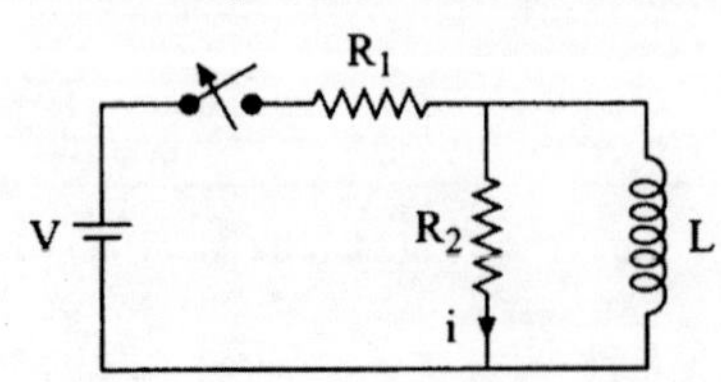

Fig. 4.159.

(*a*) $\dfrac{V}{R_2} \in^{-R_1 t/L}$ (*b*) $\dfrac{V}{R_1} \in^{-R_2 t/L}$

(*c*) $\dfrac{VR_2}{R_1+R_2}\epsilon^{-R_1t/L}$ (*d*) $\dfrac{V}{R_1+R_2}\epsilon^{-R_2t/L}$

738. For the network in Fig. 4.160, which of the following expressions gives the value of net inductance between terminals A and D

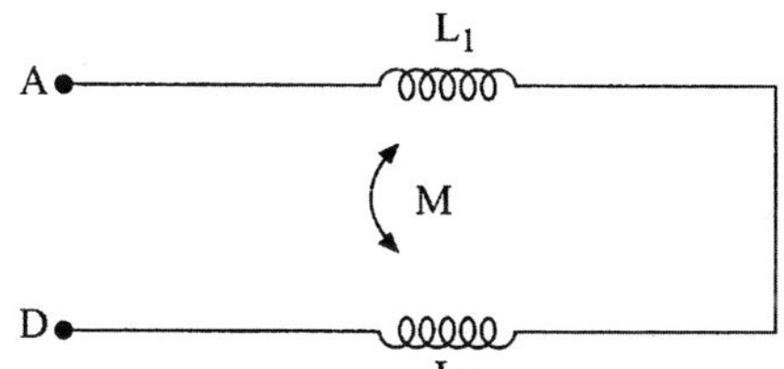

Fig. 4.160.

1. $L_1 \pm L_2$ 2. $L_1+L_2 \pm M$
3. L_1+L_2-2M 4. L_1+L_2+2M

Select the answer using the following codes

(*a*) 1 only (*b*) 1 and 2
(*c*) 3 and 4 (*d*) 2 and 4

739. The voltage transfer function of two port circuits connected in cascade can be easily found from

(*a*) product of individual A B C D matrices
(*b*) product of voltage transfer functions of two networks
(*c*) sum of *z* matrics
(*d*) sum of *h* matrics

740. A choke coil of resistance R and inductance L is shunted by capacitor C. The dynamic resistance is

(*a*) R/LC (*b*) C/RL
(*c*) L/RC (*d*) 1/RLC

741. In Fig. 4.161 the effective capacitance between A and B is

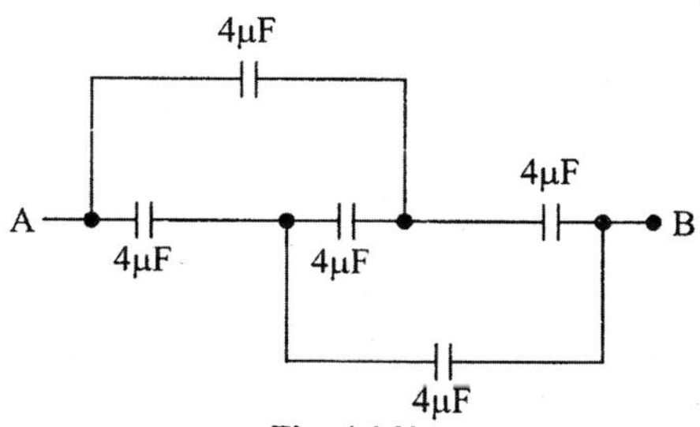

Fig. 4.161.

(*a*) 2 μF (*b*) 4 μF
(*c*) 8 μF (*d*) 10 μF

742. In a series resonant circuit, the resonant frequency is

(*a*) geometric mean of half power frequencies
(*b*) arithmetic mean of half power frequencies
(*c*) difference of half power frequencies
(*d*) none of the above

743. In low pass filter of Fig. 4.162 $R_0 = 1\Omega$. and cut off frequency is 1 rad/sec. The values of L and C are

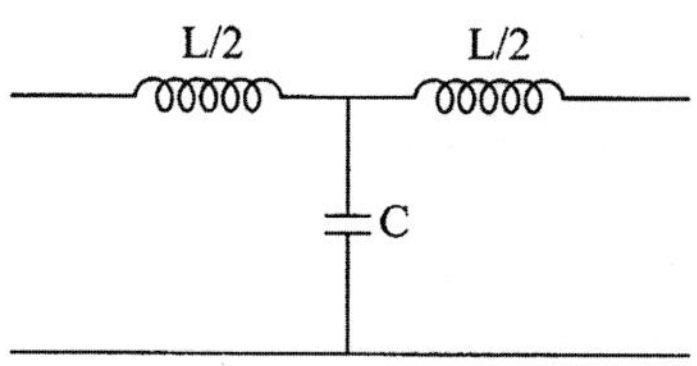

Fig. 4.162.

(*a*) 1 H and 2 F (*b*) 1H and 1 F
(*c*) 2 H and 1 F (*d*) 2 H and 2 F

744. The pass band of a constant *k* filter with Z_1 and Z_2 as series and shunt arm impedances is given by

(*a*) $-1 < \dfrac{Z_1}{4Z_2} < 0$ (*b*) $-2 < \dfrac{Z_1}{4Z_2} < -1$

(*c*) $-1 < \dfrac{Z_1}{4Z_2} < 1$ (*d*) $0 < \dfrac{Z_1}{4Z_2} < 1$

745. For an RC driving point impedance function, the poles and zeros

(*a*) should alternate on real axis
(*b*) should alternate only on negative real axis
(*c*) should alternate on imaginary axis
(*d*) none of the above

746. Current having waveform in Fig. 4.163 flows through 10 Ω resistance. Average power is

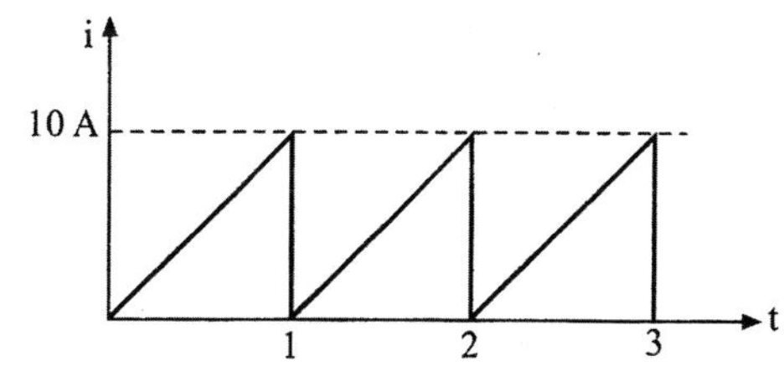

Fig. 4.163.

(*a*) 1000 W (*b*) 500 W
(*c*) 333.3 W (*d*) 250 W

747. In Fig. 4.164 the current *I* is

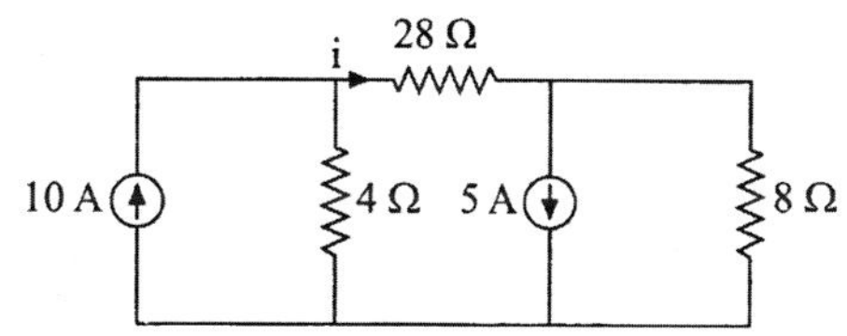

Fig. 4.164.

(*a*) 1 A (*b*) 2 A
(*c*) 4 A (*d*) 8 A

748. In a series RLC circuit excited by $e = E_m \sin \omega t$ it is seen that $LC < 1/\omega^2$. Then

(*a*) I lags E
(*b*) I leads E
(*c*) I and E are in phase
(*d*) voltage across L and C are equal

749. If $e = 110 \sin(\omega t + \pi/3)$ and $i = 5 \sin(\omega t - \pi/3)$, the impedance of the circuit is

(*a*) 22 Ω (*b*) 16 Ω
(*c*) 30.8 Ω (*d*) 11 Ω

750. In Fig. 4.165, a voltmeter of internal resistance 1200 Ω is connected across 600 Ω resistance and it reads 5V. The value of R is

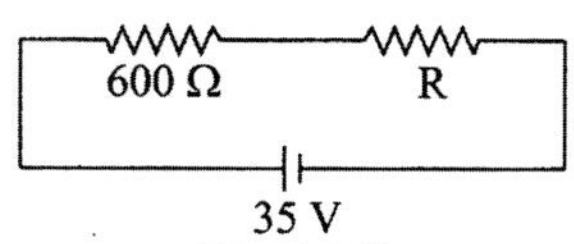

Fig. 4.165.

(*a*) 1200 Ω (*b*) 2400 Ω
(*c*) 3600 Ω (*d*) 7200 Ω

751. A first order circuit is excited by a dc source. The current through any element can be written as (i_f and i_i are the final and initial values of current respectively)

(*a*) $i_i - (i_f - i_i)\, e^{-t/T}$ (*b*) $i_f - (i_f - i_i)\, e^{-t/T}$
(*c*) $i_f - (i_i - i_f)\, e^{-t/T}$ (*d*) $i_i - (i_i - i_f)\, e^{-t/T}$

[IES 2010]

752. In the circuit in Fig. 4.166 the switch has been in closed position for a long time. At $t = 0$ the switch is closed. The current $i_s\,(0^+)$ through the switch is

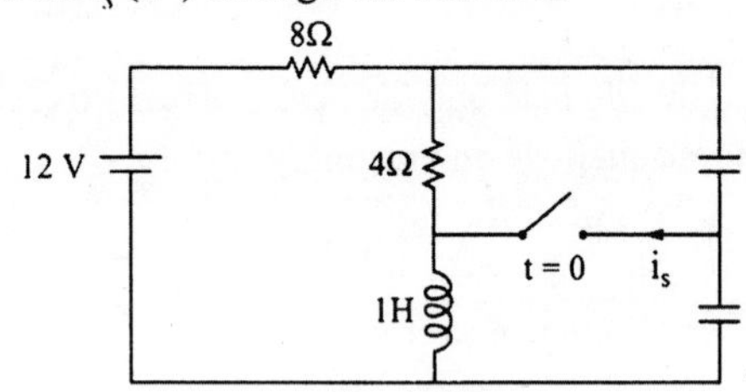

Fig. 4.166.

(*a*) 1 A (*b*) $\frac{2}{3}$ A
(*c*) $\frac{1}{3}$ A (*d*) 0 A **[IES 2010]**

753. The value of current in 80 Ω resistor in Fig. 4.167 is

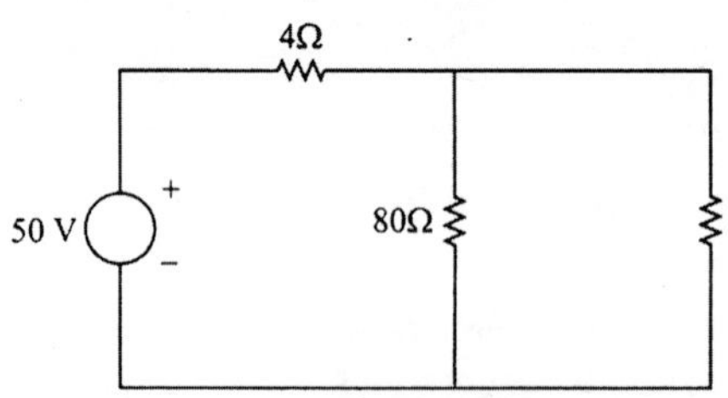

Fig. 4.167.

(*a*) 0.5 A (*b*) 2 A
(*c*) 5 A (*d*) 20 A **[IES 2010]**

754. A voltage source of 240 V having an internal impedance of $(3 - j4)$ ohm is supplying power to a complex load impedance. The maximum power transferred to the load is

(*a*) 2.4 kW (*b*) 3.6 kW
(*c*) 4.8 kW (*d*) 6 kW **[IES 2010]**

755. The power dissipated in 1 ohm resistor in Fig. 4.168 is 1 W due to 5 V source acting alone and 576 W due to 30 A source acting alone. The total power absorbed by 1 ohm resistor due to both sources is

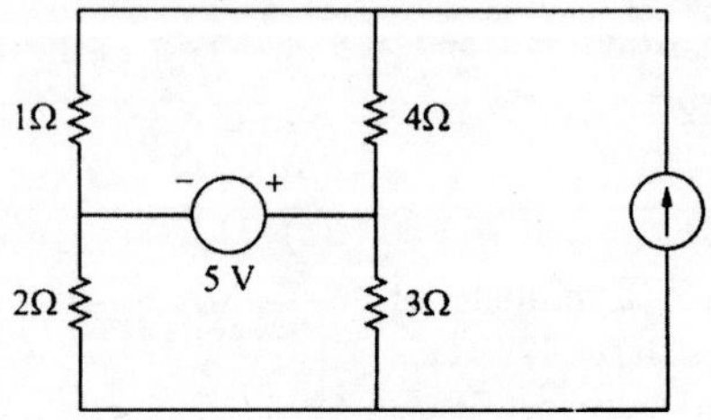

Fig. 4.168.

(*a*) 577 W (*b*) 575 W
(*c*) 625 W (*d*) 579 W **[IES 2010]**

756. The driving point impedance of a network is $Z(s) = \dfrac{2s+1}{s(s+1)}$. The foster realization is

(*a*)
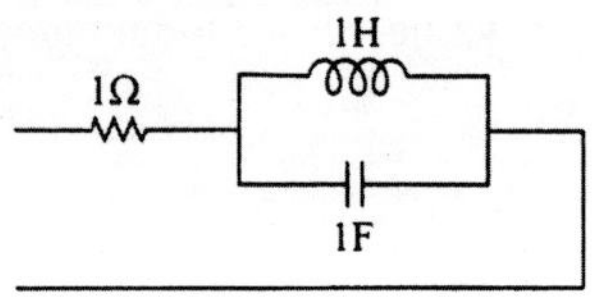

(*b*)
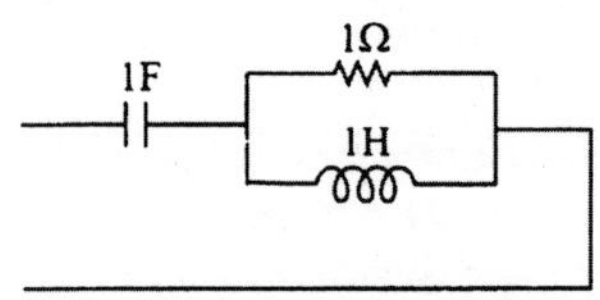

(*c*)
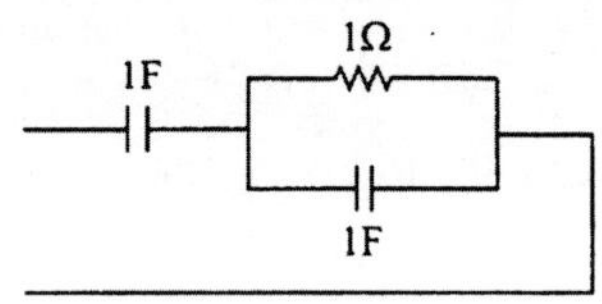

(*d*)
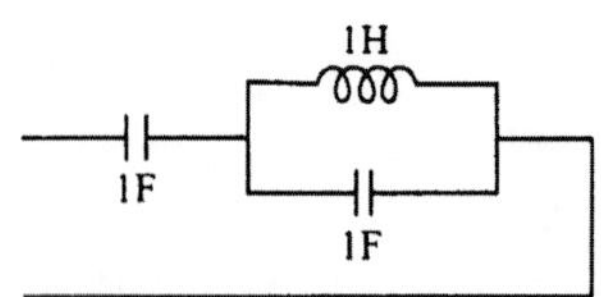

[IES 2010]

757. A two port network has ABCD parameters as $\begin{bmatrix} 7 & 8 \\ 3 & 4 \end{bmatrix}$. If two such networks are cascaded, the overall ABCD parameters are

(*a*) $\begin{bmatrix} 14 & 16 \\ 6 & 8 \end{bmatrix}$ (*b*) $\begin{bmatrix} 73 & 88 \\ 33 & 40 \end{bmatrix}$

(*c*) $\begin{bmatrix} 1 & 1 \\ 1 & 1 \end{bmatrix}$ (*d*) $\begin{bmatrix} 40 & 64 \\ 9 & 16 \end{bmatrix}$ **[IES 2010]**

758. Consider the following statements

1. Power factor will be unity
2. Current in the circuit will be maximum
3. Current in the circuit will be minimum

Which of these statements are correct for RLC parallel circuit in resonance?

(*a*) 1, 2, 3 (*b*) 1 and 2 only
(*c*) 2 and 3 only (*d*) 1 and 3 only

[IES 2010]

759. For a series RLC network the characteristic equation is

$$s^2 + \frac{R}{L}s + \frac{1}{LC} = 0$$

If $\dfrac{R}{2L} = \alpha$ and $\dfrac{1}{\sqrt{LC}} = \beta$. Then under the condition $\beta^2 > \alpha^2$, the system will be

(*a*) critically damped (*b*) under damped
(*c*) undamped (*d*) over damped

[IES 2009]

760. An RC series circuit with $R = 3\ \Omega$ and $X_C = 4\ \Omega$ at 50 Hz is supplied with voltage $50 + 141.4 \sin 314t$. RMS current in the circuit is approximately

(*a*) 5 A (*b*) 10 A
(*c*) 20 A (*d*) 22.36 A **[IES 2009]**

761. An RLC series circuit is supplied by a voltage source. Then its dual network will be described by the differential equation

(*a*) $v(t) = Ri(t) + C\frac{di(t)}{dt} + \frac{1}{C}\int i(t)\,dt$

(*b*) $v(t) = \frac{1}{G}i(t) + C\frac{di(t)}{dt} + \frac{1}{L}\int i(t)\,dt$

(*c*) $i(t) = Gv(t) + C\frac{dv(t)}{dt} + \frac{1}{L}\int v(t)\,dt$

(*d*) $v(t) = Ri(t) + L\frac{di(t)}{dt} + C\int i(t)\,dt$ **[IES 2009]**

762. In a circuit with 12 circuit elements and five nodes the minimum number of mesh equations is
(*a*) 24 (*b*) 12
(*c*) 10 (*d*) 8 **[IES 2009]**

763. If the connection of two two-port network is such that transmission matrix of overall network is product of individual matrices, what type of connection is
(*a*) series (*b*) cascade
(*c*) parallel (*d*) none of these **[IES 2009]**

764. Consider the following network functions

$$N(s) = \frac{s(s+2)}{(s+4)(s+1+j1)(s+1-j1)}$$

To make N(*s*) a rational network function, it is essential to include
(*a*) zero at origin (*b*) zero at infinity
(*c*) pole at origin (*b*) pole at infinity **[IES 2009]**

765. In the circuit shown in Fig. 4.169 the switch is open for a long time. At $t = 0$ switch is closed. What is the current through the switch?

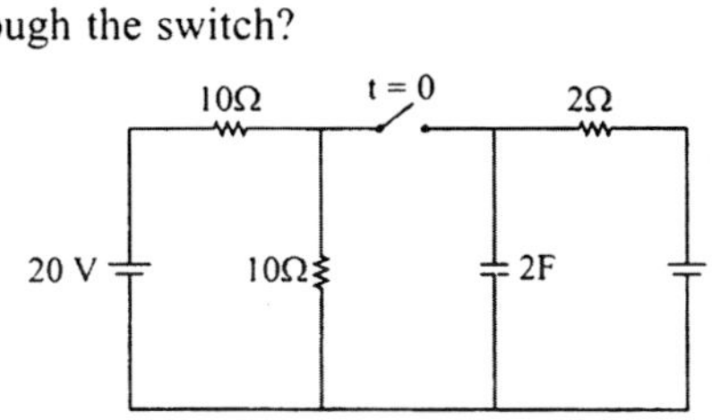

Fig. 4.169.

(*a*) zero (*b*) 1 A
(*c*) 2 A (*d*) 5 A **[IES 2008]**

766. Which of the following represents linear mathematical model for the physical system in Fig. 4.170?

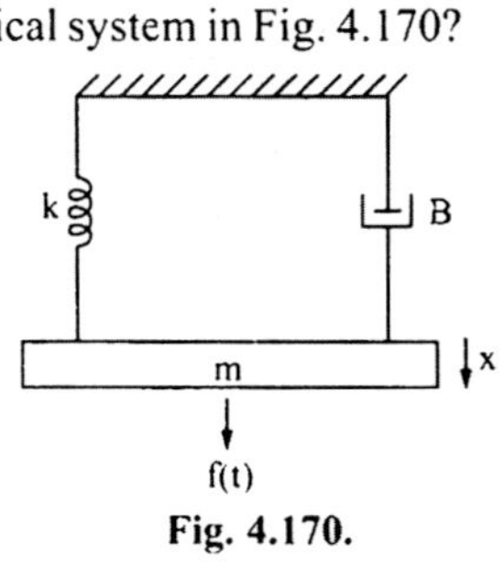

Fig. 4.170.

(*a*) $m\frac{d^2x}{dt^2} + B\frac{dx}{dt} + kx = f(t)$

(*b*) $m\frac{d^2x}{dt^2} + B\frac{dx}{dt} + kx = 0$

(*c*) $m\frac{d^2x}{dt^2} + B\frac{dx}{dt} + kx + f(t) = 0$

(*d*) $m\frac{d^2x}{dt^2} + B\frac{dx}{dt} - kx - f(t) = 0$ **[IES 2008]**

767. A two port network has $z_{11} = \frac{13}{35}$, $z_{12} = z_{21} = \frac{2}{35}$ and $z_{22} = \frac{3}{35}$. Then y_{11} and y_{12} are
(*a*) 3, –2 (*b*) 3, 2
(*c*) 13, –2 (*d*) 13, 2 **[IES 2008]**

768. For the circuit in Fig. 4.171 what is natural frequency

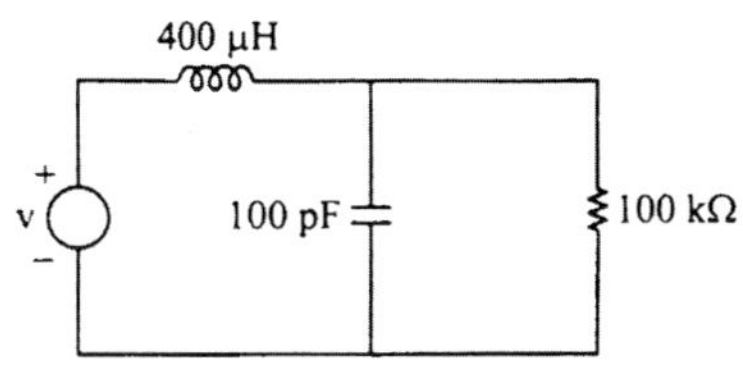

Fig. 4.171.

(*a*) 1 M rad/sec (*b*) 2 M rad/sec
(*c*) 3 M rad/sec (*d*) 5 M rad/sec **[IES 2008]**

769. Match List I and List II and select correct answer using the given code

List I	**List II**
A. Norton's theorem	1. Effects of independent sources in a linear circuit are additive
B. Superposition theorem	2. Law of non-accumulation of charge
C. Thevenin's theorem	3. Current source with shunt resistor
D. KCL	4. Voltage source with series resistor

Codes:

	A	B	C	D
(*a*)	2	4	1	3
(*b*)	3	1	4	2
(*c*)	2	1	4	3
(*d*)	3	4	1	2

[IES 2008]

770. Figure 4.172 shows a graph. Which of the following is not a tree of this graph

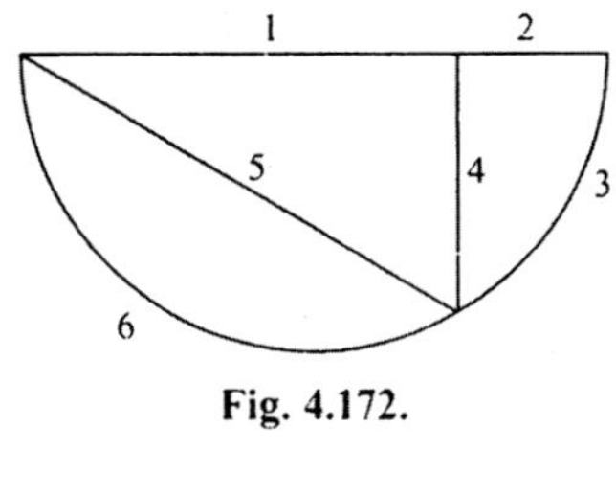

Fig. 4.172.

(a) 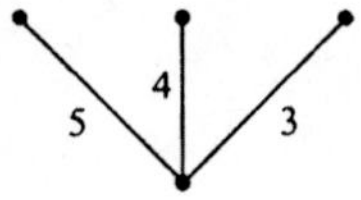

(b) 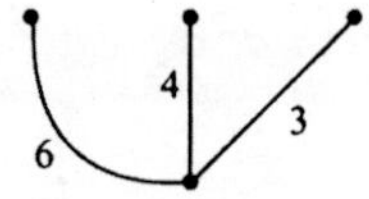

(c) 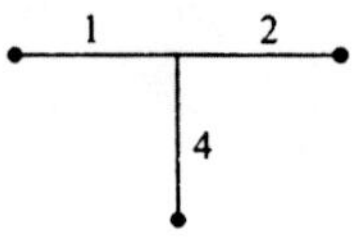

(d) 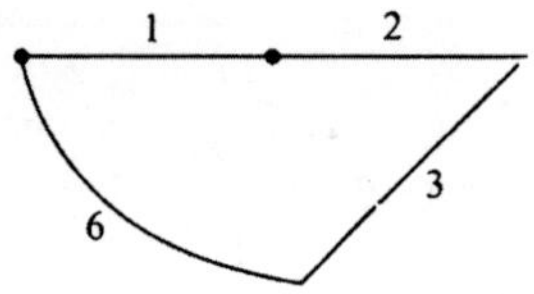

[IES 2008]

771. For the network is Fig. 4.173 which of the following is not a tree

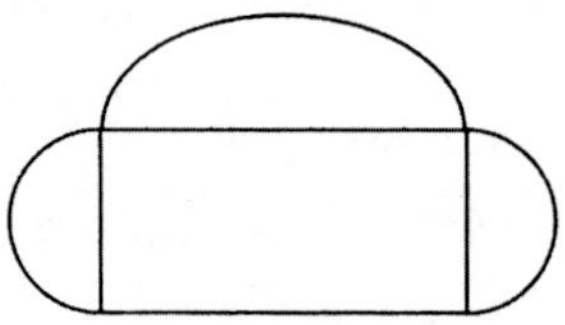

Fig. 4.173.

(a) 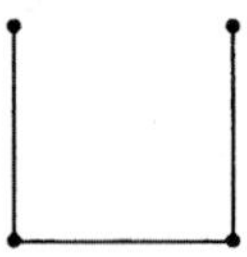
(b) 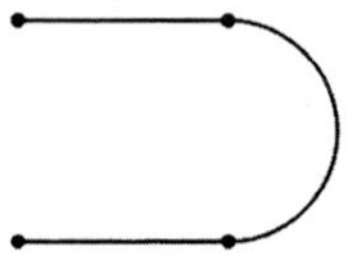

(c) 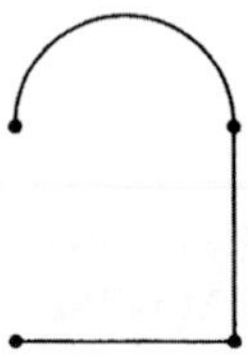
(d) 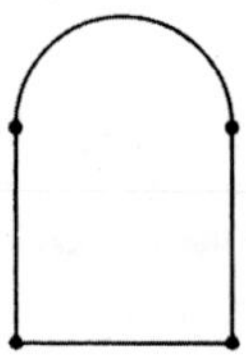

[IES 2006]

772. For the network is Fig. 4.174 what is o.c. impedance $z_{11}(s)$

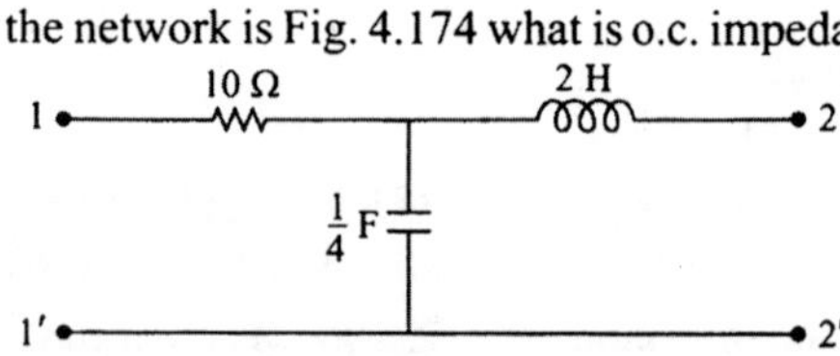

Fig. 4.174.

(a) $10+2s$ (b) $10-\frac{4}{s}$

(c) $10+\frac{4}{s}$ (d) $10-2s$ **[IES 2006]**

773. What is the total number of trees for the graph in Fig. 4.175.

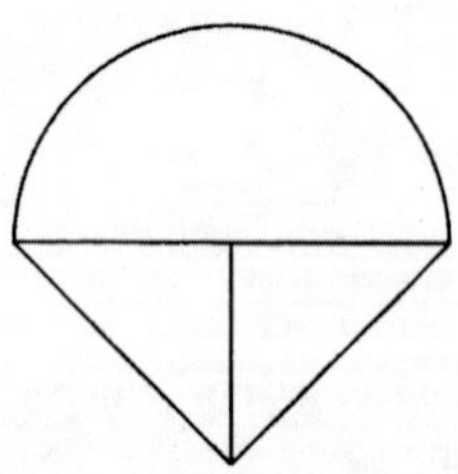

Fig. 4.175.

(a) 4 (b) 8

(c) 12 (d) 16 **[IES 2006]**

774. Consider the following

Superposition theorem is applicable to linear networks in determining

1. Current response 2. Voltage response
3. Power response

Which of the above statements are correct?

(a) 1 and 2 (b) all

(c) 1 and 3 (d) 2 and 3 **[IES 2005]**

775. An RLC series circuit has R = 20Ω and voltage across inductor is twice that across the capacitor. Current lags voltage by 45°. Then inductive reactance is

(a) 10 Ω (b) 20 Ω

(c) 40 Ω (d) 60 Ω **[IES 2005]**

776. If the scattering matrix of a two port is

$$[s]=\begin{bmatrix}0.2\angle 0 & 0.9\angle 90^\circ\\ 0.9\angle 90^\circ & 0.1\angle 90^\circ\end{bmatrix}$$

Then the network is

(a) lossless and reciprocal

(b) lossless but not reciprocal

(c) not lossless but reciprocal

(d) neither lossless nor reciprocal **[Gate 2010]**

777. In the circuit of Fig. 4.176 the switch is open for a long time and is closed at $t = 0$. The current $i(t)$ at $t = 0^+$ is

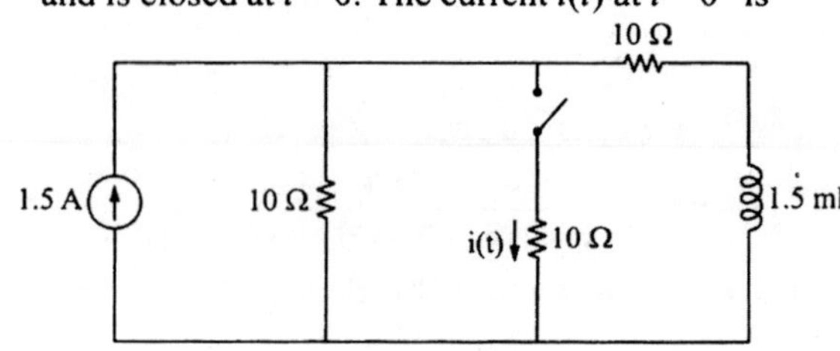

Fig. 4.176.

(a) $i(t)=0.5-0.125\,e^{-1000t}$ (b) $i(t)=1.5-0.125\,e^{-1000t}$

(c) $i(t)=0.5-0.5\,e^{-1000t}$ (d) $i(t)=0.375e^{-1000t}$

[Gate 2010]

778. The current i in Fig. 4.177 is

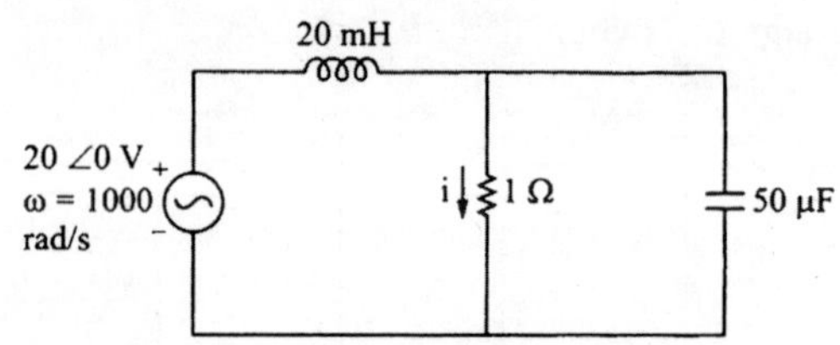

Fig. 4.177.

(a) $-j\,1$ A (b) $j\,1$ A

(c) 0 A (d) 20 A **[Gate 2010]**

779. In Fig. 4.178, the power supplied by voltage source is

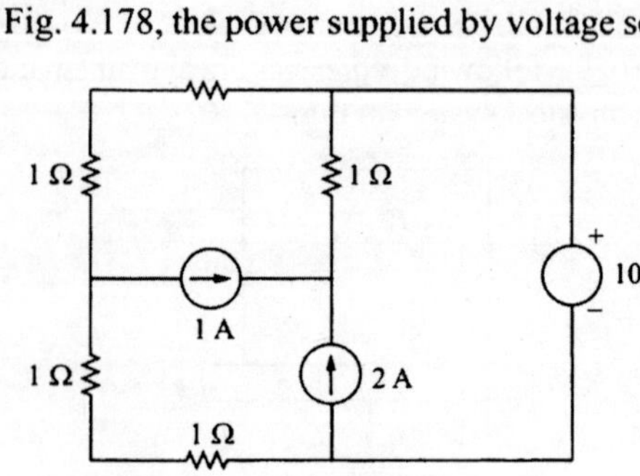

Fig. 4.178.

(a) 0 W (b) 5 W

(c) 10 W (d) 100 W **[Gate 2010]**

780. In the interconnection of ideal sources shown in Fig. 4.179 it is known that 60 V source is absorbing power.

Which of the following can be the value of I

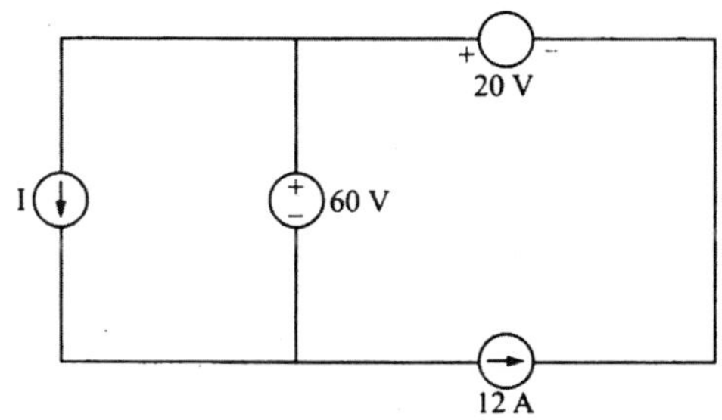

Fig. 4.179.

(a) 10A (b) 13A

(c) 15A (d) 18A **[Gate 2009]**

781. If the transfer function in Fig. 4.180 is $\frac{V_0(s)}{V_1(s)} = \frac{1}{2 + sCR}$

The value of load resistance is

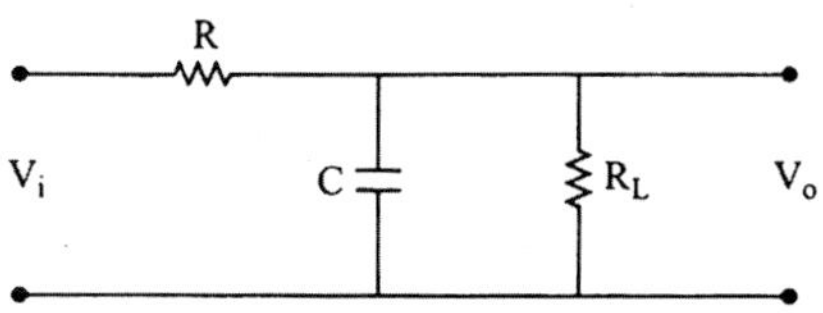

Fig. 4.180.

(a) $\frac{R}{4}$ (b) $\frac{R}{2}$

(c) R (d) 2R **[Gate 2009]**

782. An ac source with rms voltage 20 V and internal impedance $(1 + 2j)$ feeds a load $Z_L = 7 + j$ in Fig. 4.181.

The reactive power absorbed by load is

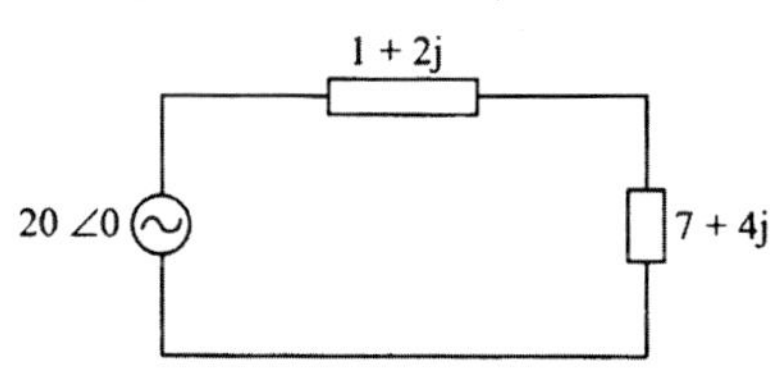

Fig. 4.181.

(a) 8 Var (b) 16 Var

(c) 24 Var (d) 32 Var **[Gate 2009]**

783. In Fig. 4.182 the switch was in position *a* for a long time. At $t = 0$ switch is moved to position *b*. The current $i(t)$ is

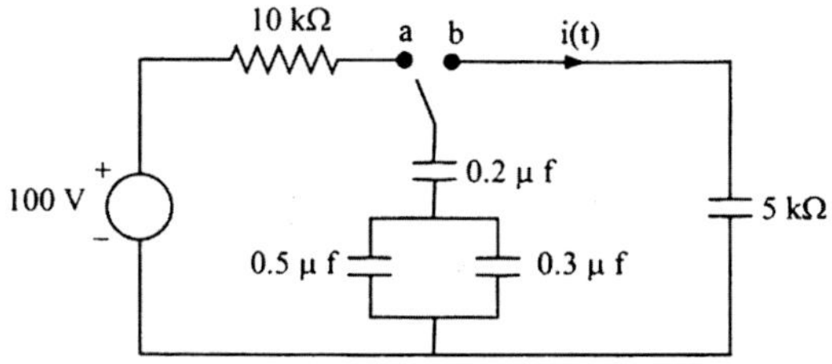

Fig. 4.182.

(a) $0.2\, e^{-125t}\, u(t)$ mA (b) $20\, e^{-1250t}\, u(t)$ mA

(c) $0.2\, e^{-1000t}\, u(t)$ mA (d) $20\, e^{-1000t}\, u(t)$ mA

[Gate 2009]

784. Thevenin impedance between nodes P and Q in Fig. 4.183 is

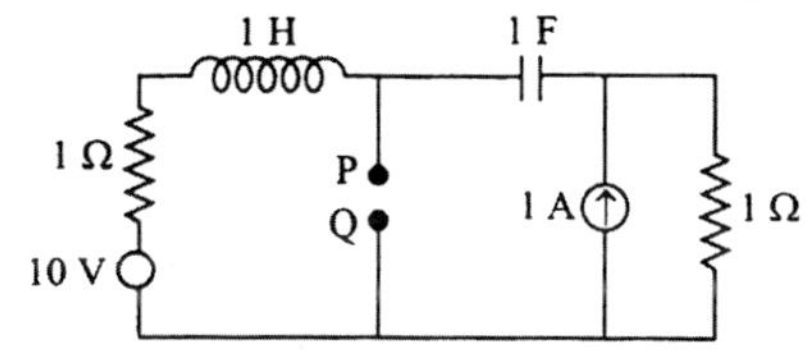

Fig. 4.183.

(a) 1 (b) $1 + s + \frac{1}{s}$

(c) $2 + s + \frac{1}{s}$ (d) $\frac{s^2 + s + 1}{s^2 + 2s + 1}$

[Gate 2008]

785. The driving point impedance of the network in Fig. 4.184 is given by

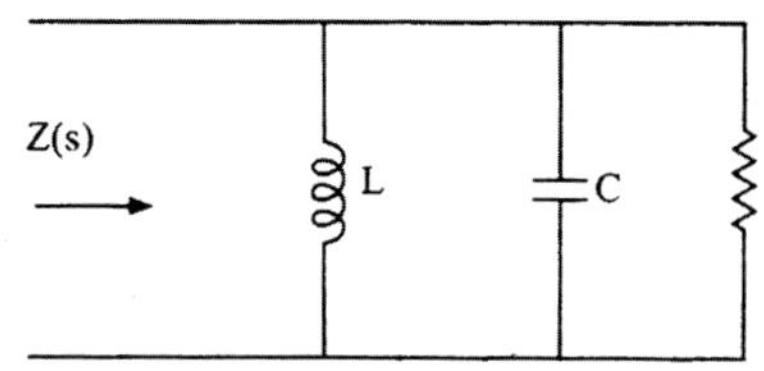

Fig. 4.184.

$Z(s) = \frac{0.2s}{s^2 + 0.1s + 2}$. The component values are

(a) L = 5 H, R = 0.5 Ω, C = 0.1 F

(b) L = 0.1 H, R = 0.5 Ω, C = 5 F

(c) L = 5 H, R = 2 Ω, C = 0.1 F

(d) L = 0.1 H, R = 2 Ω, C = 5 F **[Gate 2008]**

786. The series RLC circuit in Fig. 4.185 is excited by impulse

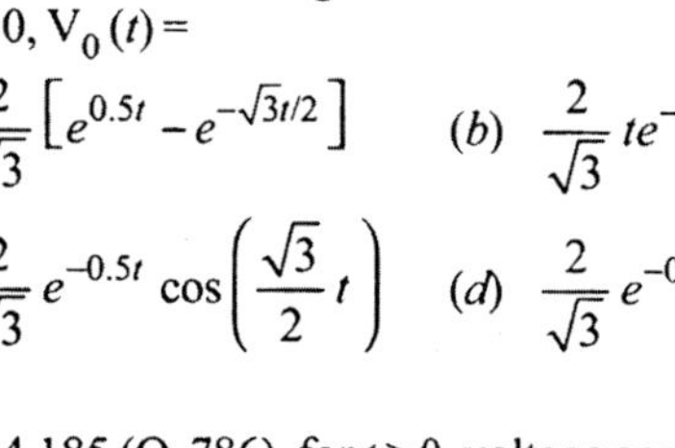

Fig. 4.185.

For $t > 0$, $V_0(t) =$

(a) $\frac{2}{\sqrt{3}}\left[e^{0.5t} - e^{-\sqrt{3}t/2}\right]$ (b) $\frac{2}{\sqrt{3}}\, te^{-0.5t}$

(c) $\frac{2}{\sqrt{3}}\, e^{-0.5t} \cos\left(\frac{\sqrt{3}}{2} t\right)$ (d) $\frac{2}{\sqrt{3}}\, e^{-0.5t} \sin\left(\frac{\sqrt{3}t}{2}\right)$

[Gate 2008]

787. In Fig. 4.185 (Q. 786), for $t > 0$, voltage across R is

(a) $\frac{1}{\sqrt{3}}\left[e^{-\sqrt{3}t/2} - e^{-0.5t}\right]$ (b) $\frac{2}{\sqrt{3}}\, e^{-0.5t} \sin\left(\frac{\sqrt{3}}{2} t\right)$

(c) $e^{-0.5t}\left[\cos\frac{\sqrt{3}}{2}t - \frac{1}{\sqrt{3}}\sin\frac{\sqrt{3}}{2}t\right]$

(*d*) $\frac{2}{\sqrt{3}} e^{-0.5t} \cos\left(\frac{\sqrt{3}}{2}t\right)$ **[Gate 2008]**

788. In a two port network the following measurements were made

S_1 open S_2 closed $A_1 = 0, A_2 = 1A, V_1 = 4.5\text{ V}, V_2 = 1.5\text{ V}$

S_1 closed S_2 open $A_1 = 4A, V_1 = 6V, V_2 = 6V, A_2 = 0A$

z parameter matrix is

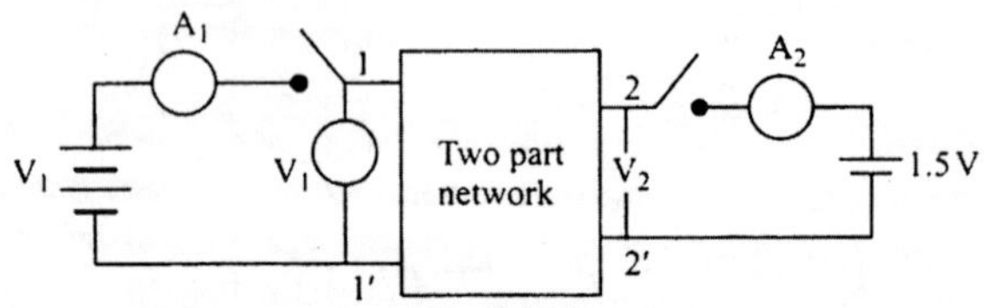

Fig. 4.186.

(*a*) $\begin{bmatrix} 1.5 & 1.5 \\ 4.5 & 1.5 \end{bmatrix}$ (*b*) $\begin{bmatrix} 1.5 & 4.5 \\ 1.5 & 4.5 \end{bmatrix}$

(*c*) $\begin{bmatrix} 1.5 & 4.5 \\ 1.5 & 1.5 \end{bmatrix}$ (*d*) $\begin{bmatrix} 4.5 & 1.5 \\ 1.5 & 4.5 \end{bmatrix}$ **[Gate 2008]**

789. For Fig. 4.186 (Q. 788) *h* parameter matrix is

(*a*) $\begin{bmatrix} -3 & 3 \\ -1 & 0.67 \end{bmatrix}$ (*b*) $\begin{bmatrix} -3 & -1 \\ 3 & 0.67 \end{bmatrix}$

(*c*) $\begin{bmatrix} 3 & 3 \\ 1 & 0.67 \end{bmatrix}$ (*d*) $\begin{bmatrix} 3 & 1 \\ -3 & -0.67 \end{bmatrix}$

[Gate 2008]

790. Two series resonant filters are shown in Fig. 4.187. Let 3 dB bandwidth of fiiter 1 be B_1 and that of filter 2 be B_2. The value of B_1/B_2 is

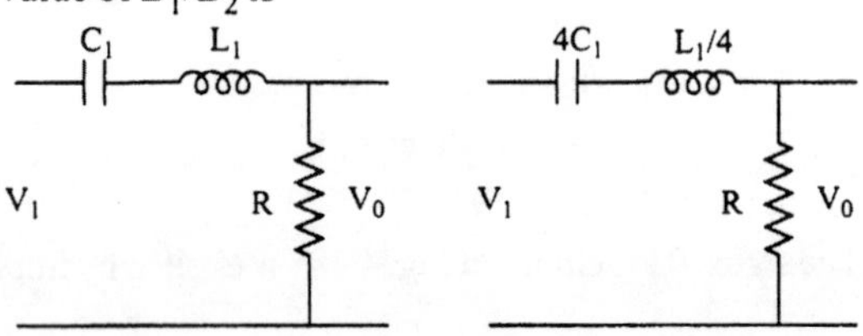

(*i*) (*ii*)

Fig. 4.187.

(*a*) 4 (*b*) 1

(*c*) $\frac{1}{2}$ (*d*) $\frac{1}{4}$ **[Gate 2007]**

791. For the circuit in Fig. 4.188, Thevenin's voltage and resistance looking into X-Y are

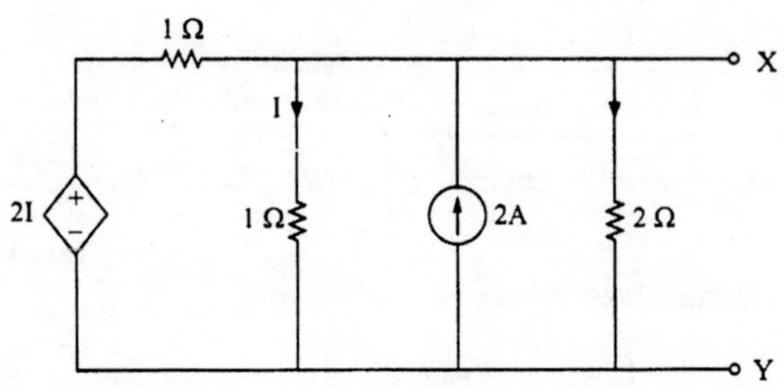

Fig. 4.188.

(*a*) 4/3 V, 2 Ω (*b*) 4 V, 2/3 Ω

(*c*) 4/3 V, 2/3 Ω (*d*) 4 V, 2 Ω **[Gate 2007]**

792. In Fig. 4.189 $V_C = 0$ at $t = 0$. For $t > 0$, the current i_c is

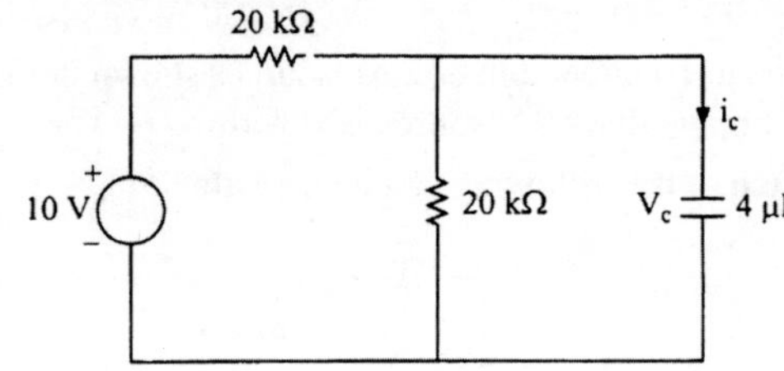

Fig. 4.189.

(*a*) $0.5\, e^{-25t}$ mA (*b*) $0.25\, e^{-25t}$ mA

(*c*) $0.5\, e^{-12.5t}$ mA (*d*) $0.25\, e^{-12.5t}$ mA

[Gate 2007]

793. In Fig. 4.190, Voltage $V_{AB} =$

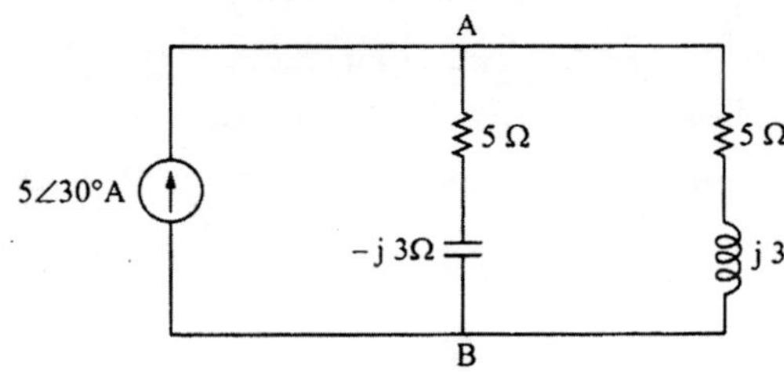

Fig. 4.190.

(*a*) 0 (*b*) $5 \angle 30°$

(*c*) $12.5 \angle 30°$ (*d*) $17 \angle 30°$ **[Gate 2007]**

794. The circuit in Fig. 4.191 is

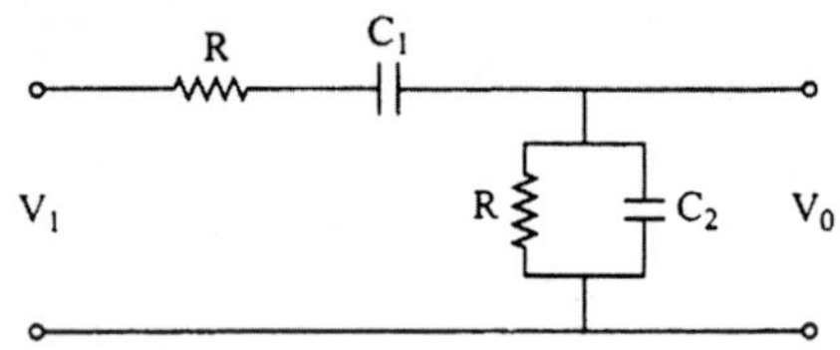

Fig. 4.191.

(*a*) low pass filter (*b*) high pass filter

(*c*) band pass filter (*d*) band reject filter

[Gate 2007]

795. A two port network is represented by ABCD parameters as

$$\begin{bmatrix} V_1 \\ I_1 \end{bmatrix} = \begin{bmatrix} A & B \\ C & D \end{bmatrix} \begin{bmatrix} V_2 \\ -I_2 \end{bmatrix}$$

It is terminated by load impedance R_L The input impedance as seen at part 1 is

(*a*) $\frac{A + BR_L}{C + DR_L}$ (*b*) $\frac{AR_L + C}{BR_L + D}$

(*c*) $\frac{DR_L + A}{BR_L + C}$ (*d*) $\frac{B + AR_L}{D + CR_L}$

[Gate 2006]

796. In the two port network shown in Fig. 4.192, z_{12} and z_{21} are

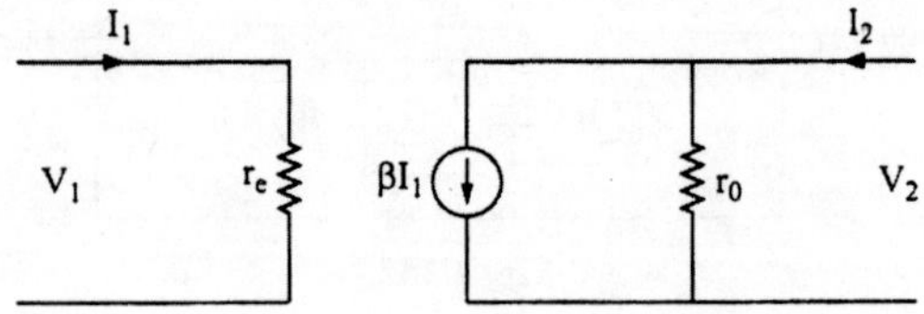

Fig. 4.192.

(*a*) r_e and βr_0 (*b*) 0 and $-\beta r_0$

(*c*) 0 and βr_0 (*d*) r_e and $-\beta r_0$

[Gate 2006]

797. A 2 mH inductor with some initial current is represented as in Fig. 4.193. The value of initial current is

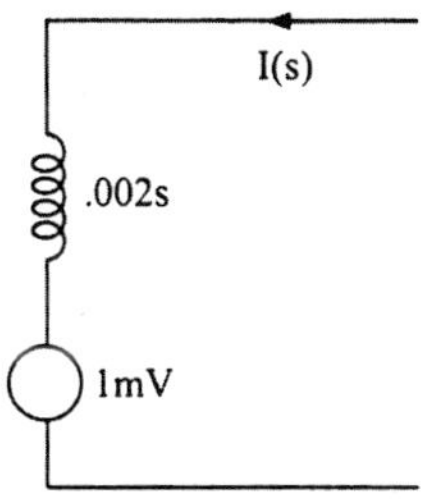

Fig. 4.193.

(*a*) 0.5 A (*b*) 2 A

(*c*) 1 A (*d*) 0 A **[Gate 2006]**

798. In Fig. 4.194 all capacitors are initially uncharged. If $v_1(t) = 10\,u(t)$, $v_0(t)$ is

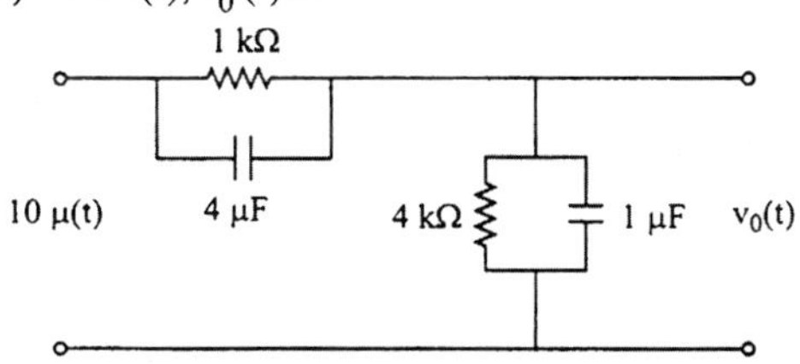

Fig. 4.194.

(*a*) $8e^{-t/0.004}$ V (*b*) $8(1 - e^{-t/.004})$ V

(*c*) $8\,u(t)$ V (*d*) 8V **[Gate 2006]**

799. A negative resistance R_{neg} is connected to a passive network *N* is Fig. 4.195. For $Z_2(s)$ to be positive real

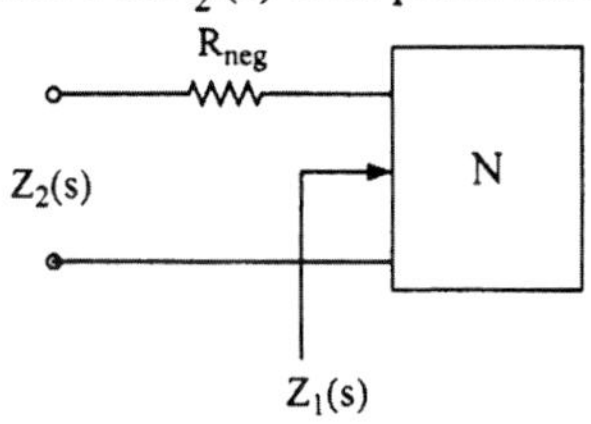

Fig. 4.195.

(*a*) $|R_{neg}| \le \text{Re } Z_1(j\omega)$ (*b*) $|R_{neg}| \le Z_1(j\omega)$

(*c*) $|R_{neg}| \le \text{Im } Z_1(j\omega)$ (*d*) $|R_{neg}| \le \angle Z_1(j\omega)$

for all values of ω. **[Gate 2006]**

800. If $R_1 = R_2 = R_4 = R$ and $R_3 = 1.1R$, in the bridge circuit in Fig. 4.196, the reading of ideal voltmeter connected between *a* and *b* is

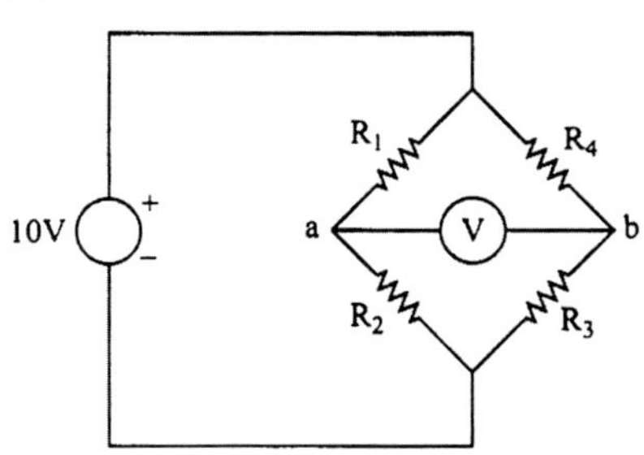

Fig. 4.196.

(*a*) 0.238 V (*b*) 0.138 V

(*c*) –0.238 V (*d*) 1 V **[Gate 2005]**

B. The following questions have two statements, *i.e.*, Assertion (A) and Reasons (R) Examine the statements and give answer as per following codes

(*a*) Both A and R are true and R is correct explanation of A

(*b*) Both A and R are true and R is not the correct explanation of A

(*c*) A is true but R is false

(*d*) A is false but R is true

801. **A.** For networks in Fig. 4.197 (*a*) and (*b*) sum of products of branch voltages and branch currents at any time is zero.

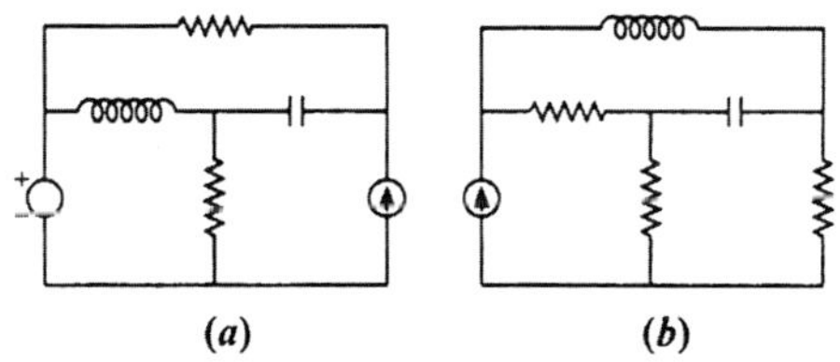

Fig. 4.197.

R. The networks in Fig. 4.197 (*a*) and (*b*) are not the same structurally.

802. **A.** When a square periodic wave is applied to an RC circuit, the voltage across capacitor is observed to be a triangular periodic wave.

R. The RC circuit works as integrator and its time constant is much larger than time period of input wave.

803. **A.** $Z(s) = \dfrac{k(s^2+1)(s^2+5)}{(s^2+2)(s^2+10)}$ is not an LC function.

R. The highest powers of numerator and denominator do not differ by 1.

804. **A.** Laplace transform of $f(t) = e^{-at}\sin \omega t$ is

$$\frac{\omega}{(s+a)^2 + \omega^2}.$$

R. If Laplace transform of $f(t)$ is F(*s*), then Laplace transform of $e^{-at} f(t)$ is F(*s* + *a*).

805. **A.** Two terminal black boxes of R and C can be identified by plotting their static V – I characteristics.

R. The V – I characteristic of resistance is a straight line of slope R passing through origin.

806. **A.** In high Q circuits poles of Y(*s*) lie close to ω axis in complex frequency plane.

R. Q is inversely proportional to damping factor.

807. **A.** In Fig. 4.198 the coefficients

$$a_n = \frac{4}{\pi(1-4n^2)}$$

$n = 1, 2, \ldots$ and $b_n = 0$

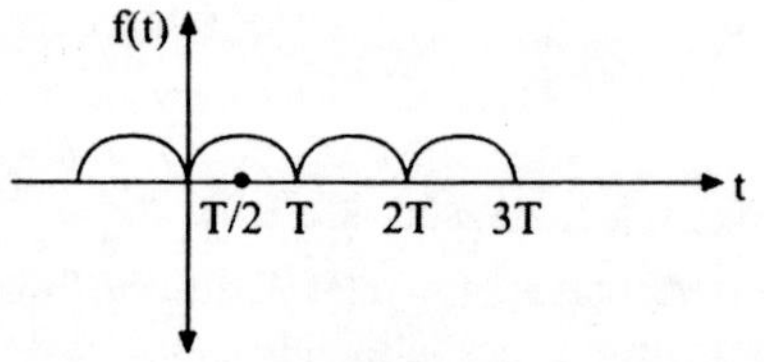

Fig. 4.198.

R. The periodic function is symmetrical about t = T and t = T/2.

808. A. The impedance of a series resonant circuit is minimum at resonance.

R. Resonance condition implies unity power factor condition.

809. A. For a physically realisable driving point function, the degree of numerator and denominator should be equal.

R. The sum of positive real functions is real.

810. The characteristic impedance of a linear passive two port network is $-j\ 1\ \Omega$. The network is terminated by 1 Ω resistance. Then

A. The network cannot transmit power to load.

R. Since the characteristic impedance is pure capacitance, it cannot consume power.

811. A. The response of a network to unit impulse can be obtained directly from the network function.

R. The response of a network for any input can be obtained from impulse response.

812. A. A series RLC circuit resonates when excited by variable frequency source.

R. Resonant frequency is geometric mean of half power frequencies.

813. A. A graph is planar if it has a dual.

R. The dual of a graph can be found by window dot method.

814. A. Superposition theorem can be used to find the output of a full wave rectifier excited by sinusoidal signal sources of different frequencies connected in series.

R. Superposition theorem is valid for all linear systems.

815. A. If $Z_1(s)$ and $Z_2(s)$ are positive real then $Z_1(s) + Z_2(s)$ as well as $1/Z_1(s)$ and $1/Z_2(s)$ are positive real.

R. The poles of a positive real function are real or occur in conjugate pairs.

816. A. $Z(s) = \dfrac{s+4}{s^2+2s+5}$ is positive real.

R. For a positive real function $\text{Re}[H(j\omega)] \ge 0$ for $0 \le \omega \le \infty$.

817. A. The function $Z(s) = \dfrac{1}{s^2+1}$ is not positive real.

R. The degrees of numerator and denominator differ by 2.

818. A. The polynomial $s^3 + 6s^2 + 12 + 8$ is Hurwitz.

R. In a Hurwitz polynomial all coefficients are non-negative.

819. A. In a series resonant circuit current is maximum at resonance.

R. The inductive and capactive reactances are equal.

820. A. A parallel resonant circuit is also called and resonant circuit.

R. In a parallel resonant circuit impedance is maximum at resonance.

821. A. A band elimination filter attenuates signals whose frequencies lie between the cut off frequencies.

R. A band elimination circuit is very different from band pass circuit.

822. A. When fed by a dc voltage a capacitor behaves as short circuit at $t = 0$ and open circuit at $t = \infty$.

R. The impedance of a capacitor is $\dfrac{1}{sC}$.

823. A. When excited by a unit step voltage, an inductor behaves as open circuit at $t = 0$.

R. Inductance has the property of inertia.

824. A. In KVL equations involving mutually coupled circuits the sign of M terms can be positive or negative.

R. Dot convention helps in determining sign of M terms in KVL equations.

825. A. Thevenin's theorem helps us to find current in a branch of a network for different values of impedances of that branch.

R. Thevenin's equivalent circuit replaces a network by a voltage source is series with impedance.

826. A. Thevenin's theorem and Norton's theorem are dual of each other.

R. Voltage source can be converted into current source and vice versa.

827. A. Millman's theorem helps in replacing a number of current sources in parallel by a single current source.

R. Maximum power transfer theorem is applicable only for dc circuits.

828. A. In drawing an electric circuit analogous to a mechanical circuit, spring is replaced by capacitance.

R. The behaviour of a spring and capacitor are similar.

829. A. Energy stored in inductance is $\dfrac{1}{2}$ LI^2 and that in capacitance is $\dfrac{1}{2}$ CV^2.

R. Inductance and capacitance are dual of each other.

830. A. Power is the rate at which energy is supplied.

R. Energy = VI t.

831. A. A tree of graph contains all the nodes of the graph.

R. The number of links is always less than the number of tree branches.

832. A. The number of basic loops is equal to number of links.

R. The graph theory helps in choosing independent variables, in circuit analysis.

833. A. A graph can have many trees.

R. The number of tree branches is equal to number of nodes.

834. A. Tie set schedule is used in nodal analysis.

R. Loop currents are fictitous quantities.

835. **A.** A branch with no current through it can be replaced by an open circuit or short-circuit.
R. Reciprocity theorem is applicable only if the circuit has one source.

836. **A.** The inductance of an iron cored solenoid is not constant.
R. BH curve of an iron specimen is non-linear.

837. **A.** Most of magnetic circuits have an air gap.
R. An air gap has high reluctance.

838. **A.** An impulse has very high magnitude but very small duration.
R. A pulse function is always rectangular.

839. **A.** The Laplace transform of a ramp is $1/s^2$.
R. The integral of a unit step function gives an impulse function.

840. **A.** In an R-L circuit having high inductance, the rise of current is very slow.
R. The time constant of R-L circuit is L/R.

841. **A.** In a parallel circuit with three branches having R, L and C respectively and fed by a step current I, the current through inductance is always zero.
R. The time constant of R-L circuit is L/R.

842. **A.** A complicated waveform can be replaced by sum or difference of two or more waveforms.
R. The method of superposition is applicable only to linear systems.

843. **A.** The s domain impedance of a series R-L circuit is $R + sL$ irrespective of initial conditions.
R. Complex frequency $\sigma = s + j\omega$.

844. **A.** Charge is dual of flux.
R. Short-circuit is dual of open circuit.

845. **A.** The dual of a circuit can be drawn by window-dot method.
R. The principle of duality is of great help in circuit analysis

846. **A.** KCL is dual of KVL.
R. Node is dual of mesh.

847. **A.** Peak-to-peak value of a sinusoidal wave is 2.828 times the rms value.
R. Peak factor of a sinusoidal wave is 1.414.

848. **A.** Operator j has a value $1\angle -90°$ or $1\angle 90°$.
R. Operator 'a' has a value $1\angle 120°$.

849. **A.** If $A = 0.5\angle 20°$, $A^2 = 0.25\angle 40°$.
R. When drawing a phasor diagram, sinusoids may have the same or different frequencies.

850. **A.** In an R-L circuit having low resistance, current lags the voltage by nearly 90°.
R. Power factor of a series RL circuit is R/Z.

851. **A.** In series R-L-C circuit, the current may lead or lag the voltage depending on the value of R.
R. Phase angle of an R-L series circuit is

$$\tan^{-1}(\omega L/R).$$

852. **A.** In a circuit having high inductance, reactive power may be more than apparent power.
R. Reactive power = VI sin θ.

853. **A.** In a parallel circuit having two branches, each branch current may be more than source current.
R. Calculations in ac circuits use rules of complex algebra.

854. **A.** Q factor of a series resonant circuit is

$$\frac{1}{R}\sqrt{\frac{L}{C}}.$$

R. High Q means better selectivity.

855. **A.** In state space analysis, capacitors are taken to be in tree.
R. State variable analysis is very suitable for computer solution.

856. **A.** Ideal current sources and ideal voltage sources do not exist in reality.
R. All sources have internal impedances. **[IES 2009]**

857. **A.** KCL states that sum of currents entering any node is equal to sum of currents leaving that node.
R. KCL is based on law of conservation of charge. **[IES 2008]**

858. **A.** In a linear network superposition theorem can be used to calculate power dissipation in a resistive branch of the circuit. **[IES 2008]**
R. Superposition theorem is valid for currents and voltages and is therefore valid for power dissipation.

859. **A.** Power factor is defined as the ratio of apparent power to average power in an ac circuit. **[IES 2008]**
R. Magnitude of power factor is always less than unity.

860. **A.** Sinusoidal response in an ac circuit can be represented in rectangular form as $v = V\cos\theta + jV\sin\theta$ exactly in a way similar to vectors which we come across for force analysis in mechanics.
R. Phasors are nothing but vectors. **[IES 2008]**

861. **A.** A high pass R-C circuit has a lower cutoff frequency but does not have upper cutoff frequency.
R. Transfer function of high pass R-C circuit is

$$|H| = \frac{1}{\left[1+\left(\frac{f_1}{f}\right)^2\right]^{0.5}},\ \theta = \text{arc tan}\,(f_1/f)$$

where $f_1 = \dfrac{1}{2\pi RC}$ **[IES 2008]**

862. **A.** If a network contains a diode, Thevenin's theorem cannot be applied.
R. Thevenin's theorem is applicable to linear and bilateral networks. **[IES 2005]**

863. **A.** KVL states that in a closed path of network, the algebraic sum of all voltages in a single direction is zero.
R. Law of conservation of charge is the basis for this law. **[IES 2003]**

864. A. Superposition theorem is used to calculate current in a particular branch of a linear network by considering the effect of each independent source taken one at a time.

R. In a linear network the behaviour of circuit does not vary depending on the source. **[IES 2003]**

865. A. Maximum power transfer from a source with complex internal impedance will occur if source impedance is same as load impedance.

R. The efficiency of maximum power transfer cannot exceed 50%. **[IES 2003]**

866. A. The current in a series RLC circuit driven by sinusoidal voltage source may lead, lag or in phase with voltage.

R. Series resonance does not imply unity p.f. condition. **[IES 2003]**

867. A. A two terminal network representing a given driving point reactance function is said to be canonic or a fundamental network.

R. A driving point reactance function is totally specified by location of poles and zeros. **[IES 2003]**

868. A. When a sinusoidal voltage is switched on to a circuit consisting of RLC elements, the response consists of two components expressed by free response and forced response.

R. The response at $t = 0^+$ is called free response and the one at $t = \infty$ is called forced response as it consists of values of R, L, C and applied voltage and frequency. **[IES 2002]**

869. A. A series $R_1 - L$ and series $R_2 - C$ circuits are connected in parallel. Parallel resonance will occur at all frequencies if $R_1^2 = R_2^2 = L/C$.

R. An inductor must be operated below self-resonant frequency. **[IES 2002]**

870. A. A circuit having reactance is said to be at resonance if voltage and current are in phase.

R. At resonance power factor is zero. **[IES 2001]**

C. State if the following are True or False

871. $I = dq/dt$.

872. A*h* is the unit of quantity of electricity.

873. Energy = Voltage × Charge.

874. Magnetic effect of current is used in generators but not in motors.

875. Threshold of perception occurs at a current of about 1 mA.

876. A non-linear element has zero resistance.

877. If $V_A = -10$ V and $V_B = -15$ V, $V_{AB} = 5$ V.

878. In liquids charge carriers are electrons only.

879. If current is less than 20 mA and time of contact is very very small, the effect of current on human body is likely to be temporary.

880. The current density J at a point and current I are related as

$$J = \underset{\Delta s \to 0}{\text{Lt}} (\Delta I/\Delta s).$$

881. Force = Electric intensity × Current.

882. When a current flows through a coil of insulated wire, the coil shows the properties of a magnet.

883. Drift velocity of electrons is generally less than 1 mm/s.

884. Heat in calories = $I^2R\,t/4.186$.

885. In solids the current flow is due to electrons only.

886. Thermistor has a positive temperature coefficient.

887. The units of resistivity are ohm-metre.

888. In all insulators the bonding is covalent.

889. Bus bars are always meant for very heavy currents.

890. In all pure metals, temperature coefficient of resistance is positive.

891. In ionic bonding, the valence electron is donated by one atom and accepted by the other atom.

892. As the number SWG increases, area of cross-section decreases.

893. Colour code of carbon resistors has only two bands.

894. In a varistor the *i-v* graph is linear.

895. Carbon resistances are used in very high values.

896. Resistivity of semiconductors is around, 100 Ω–m.

897. As temperature increases, the temperature coefficient of metals increases.

898. At very low temperatures, semiconducting materials behave as insulators.

899. In carbon resistor colour code, the fourth band indicates tolerance.

900. At a termperature of 7 K, lead is a superconductor.

901. Zinc oxide is a non-linear resistance.

902. As per BIS, solid insulators are classified into 6 categories.

903. *v-i* relation for a capacitor is $v = \frac{1}{C}\int_a^t i\,dt$.

904. Insulation resistance of a cable is directly proportional to its length.

905. If all other parameters are the same, a capacitor with mica as dielectric has higher capacitance than air capacitor.

906. Dielectric constant of air is about 5.

907. Transformer oil is an insulator.

908. Discharge breakdown in insulators is due to voids.

909. The capacitance of a cylindrical capacitor is directly proportional to its length.

910. $\epsilon_0 = 8.85 \times 10^{-12}$ F/m.

911. A dielectric does not have leakage current.

912. Class C solid insulators should not be used above 100°C.

913. Paper capacitors are used in power supplies for filtering.

914. The current flowing through a pure capacitor is called displacement current.

915. $C = (2)\,(\text{energy})/V^2$.

916. In solid dielectrics the rate of deterioration doubles for every 0–12°C rise in temperature.

917. A dielectric should have poor heat conductivity.

918. Permittivity has the units of F/m.

919. Dielectric materials have high specific resistance.

920. Stray capacitance is important at low frequencies.

921. When cells are connected in series, emfs of all cells must be equal.

922. When cells having different emfs are connected in parallel, a circulating current flows.

923. The electrolyte in a dry cell is completely dry.

924. The zinc can of a dry cell serves as positive terminal.

925. The electrolyte in nickle cadmium cell is dilute potassium hydroxide.

926. A nickle cadmium cell is an alkaline cell.

927. In a lead acid cell, the positive plate has spongy lead.

928. When a lead acid cell is in fully charged condition the positive and negative plates are similar.

929. When a lead acid cell discharges, the specific gravity of electrolyte increases.

930. A zinc carbon cell is a primary cell.

931. When a lead acid cell is in fully charged condition the specific gravity of electrolyte is about 1.22.

932. In an ideal current source, the source resistance is zero.

933. An ideal voltage source cannot be converted into an equivalent current source.

934. In using KCL it is necessary that current leaving a node is written as negative.

935. KCL is a restatement of principle of conservation of energy.

936. When tracing through a resistance we always get a voltage drop.

937. When sources of different frequencies feed a network the only method of solution is superposition theorem.

938. When calculating Norton's resistance, all current sources are short-circuited.

939. Maximum power transfer theorem is applicable to circuits with one source only.

940. In using mesh current method it is necessary that at least one mesh current links every element.

941. RMS value of a square wave is the same as peak value.

942. Wave A starts its positive half cycle at $\omega t = 30°$ and wave B starts its positive half cycle at $\omega t = 75°$ then wave B is leading wave A by 45°.

943. For a sine wave the average value is calculated over one-half the cycle.

944. Peak-to-peak value of a sine wave is equal to twice the peak value.

945. The sum of two sine waves of different frequencies is a sine wave.

946. The terms RMS and average values apply only to sine waves.

947. For a delayed half wave rectified wave, the average and rms values are lower than those for half wave.

948. As per Euler's identity

$$e^{j\omega t} = \cos \omega t + j \sin \omega t$$

949. In drawing phasor diagrams, RMS values are used.

950. RMS value is always higher than average value.

951. When ac current is zero, no electrons are in motion.

952. In a circuit with capacitance connected to ac source, the rate of change of capacitor voltage is constant.

953. In a purely resistive circuit the instantaneous power varies from zero to maximum twice in each cycle.

954. The number of electrons in motion at the instant of positive peak of a symmetrical ac wave is the same as that at the instant of negative peak.

955. In a purely inductive ac circuit, the supply voltage and voltage induced across inductance are equal and opposite at all instants.

956. In a purely inductive ac circuit the power curve is a sinusoid of the same frequency as the voltage.

957. For an ac circuit having capacitance C, instantaneous current i and instantaneous voltage e are related as $i = C\frac{de}{dt}$.

958. In a series RC circuit $Z = (R^2 + \omega^2 C^2)^{0.5}$.

959. In a series R, L, C circuit the impedance triangle is the same when $X_L < X_C$ and $X_L > X_C$.

960. The units for apparent power are watts.

961. In a series circuit with X_L constant and R variable the current locus lies in the third quadrant.

962. In a series RLC circuit, p.f. is leading if $X_C < X_L$.

963. In a series circuit with fixed R and variable X_L, the current locus lies in the fourth quadrant.

964. A series RC circuit has fixed R and C and variable f. The current locus will lie in the first quadrant.

965. A series circuit has fixed resistance and variable reactance. The impedance locus will be a straight line parallel to Y-axis.

966. A series RC circuit has fixed X_C and variable R. The current locus is in the first quadrant.

967. When R and L are connected in parallel, $R > |Z|$.

968. When R and C are connected in parallel, phase angle of Y is positive.

969. All network theorems of dc are applicable to ac also with different quantities taken as phasors.

970. In ac circuits maximum power is delivered to a load if load impedance is conjugate of source impedance.

971. When an ac circuit has sources of different frequencies, the only method of analysis is superposition theorem.

972. In a series RLC circuit p-f is lagging when $f <$ resonance frequency.

973. If R, L, C are in parallel, the currents through L and C are equal in magnitude at resonance.

974. In a parallel resonant circuit, the circuit current at resonance is maximum.

975. For a series RLC circuit bandwidth is

$$\omega_2 - \omega_1 = R/2L.$$

976. When a series RL circuit is switched on to a dc voltage, the rate of change of current is maximum at $t = 0$.

977. During discharging a capacitor through a resistance, voltages across R and C are equal in magnitude.

978. When a series RL circuit having power factor angle θ is switched on to an ac voltage $V_m \sin(\omega t + \phi)$, the transient term is zero if $\phi = \theta$.

979. For any given two port network, any parameter set can be determined from measurements made at the two ports.

980. In admittance parameter set, the dependent variables are the currents at the two ports.

981. h parameters are used in transistor analysis.

982. When two port networks are connected in cascade, it is more convenient to use hybrid parameters.

983. In any two port network, input impedance is independent of load impedance.

984. In attenuation band, β is an odd multiple of π.

985. For a high pass filter, $\beta = \pi$ in attenuation band.

986. In a passive band pass filter, the resonant frequencies of series and shunt arms are equal.

987. For a high pass filter attenuation = 0 at $f = 0$.

988. One limitation of constant k filters is that Z_0 is not constant in the pass band.

989. For ideal performance, a filter should be terminated in characteristic impedance.

990. A band elimination constant k filter has series resonant circuit in series arms.

991. When a constant voltage source and constant current source are in series, the current source is ineffective.

992. R, L and C are in parallel. The circuit is switched on to a constant current source I at $t = 0$. At $t = \infty$ the current through inductance is I.

993. A series RL circuit is switched on to an exponential source $Ve^{-\alpha t}$ at $t = 0$. The currant $i(t) = \frac{V}{L} e^{-\alpha t}$.

994. A series RL circuit is switched on to an impulse. A $\delta(t)$ at $t = 0$. Then

$$i(t) = \frac{A}{L} e^{-Rt/L} u(t).$$

995. For an m-derived high pass filter

$$f_\infty = [\sqrt{1 - m^2}\,] f_c.$$

996. For an m-derived filter, attenuation is infinite when $f = f_\infty$.

997. In a composite filter terminating half sections are used for impedance matching.

998. In a driving point function poles and zeros must be conjugate if complex or imaginary.

999. In a driving point function, the degrees of numerator and denominator polynomials in s may differ by either zero or one only.

1000. For a transfer function the coefficients of the numberator and denominator polynomials in s must be real.

1001. Complex poles give rise to oscillatory terms in the response.

1002. When a circuit is excited by ac sinusoidal source, the admittance is obtained by letting $s = j\omega$ in $Y(s)$.

1003. If the transfer function has zeros only on $j\omega$ axis or in left half plane, it is a minimum phase function.

1004. A network is causal if impulse response is zero for $t < 0$.

1005. Paley Wiener criterion states that a necessary and sufficient condition for an amplitude function $|H(j\omega)|$ to be causal is that

$$\int_{-\infty}^{\infty} \frac{|\log H(j\omega)|}{1 + \omega^2} d\omega < \infty.$$

1006. The low pass filter characteristic shown in Fig. 4.199 is realizable.

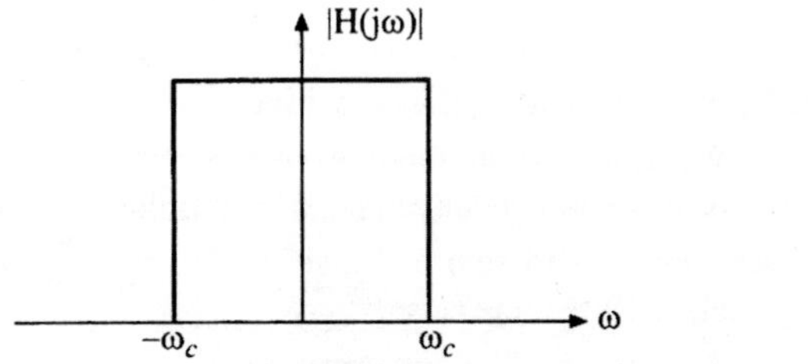

Fig. 4.199.

1007. The Gaussion filter characteristic Fig. 4.200 is realizable.

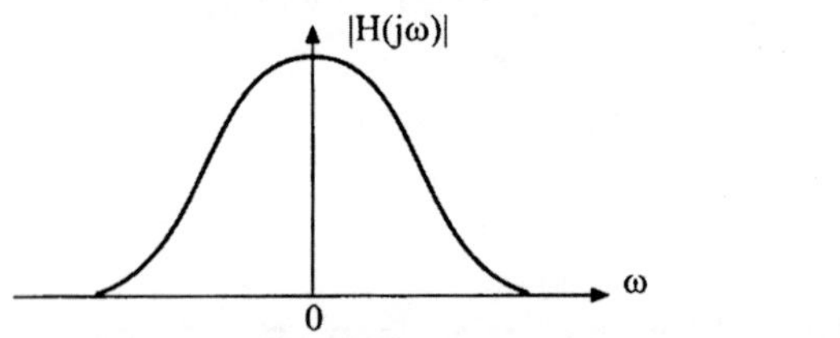

Fig. 4.200.

1008. The sum of positive real functions is positive real.

1009. The poles and zeros of a positive real functions are real or occur in conjugate pair.

1010. If the numerator of $Z(s)$ is one degree higher than denominator, $Z(s)$ has a pole at infinity.

D. Fill in the blanks with suitable words.

1011. If $a = 1 \angle 120°$, $1 + a + a^2 = ...$

1012. If $a = 1 \angle 120°$, $1 - a^2 = ...$

1013. Transpose of product of two matrices is equal to............... of the matrices in ....order.

1014. In three-phase three wire systems, zero sequence current is.............

1015. In a 3-phase system the zero sequence component of line voltages is.............

1016. In classical analogue, torque is analogous to ....

1017. In classical analogue, potential energy is analogous to............... energy.

1018. In mobility analogue velocity is analogous to ....

1019. In mobility analogue, mass is analogous to ....

1020. In classical analogue, displacement is analogous to...............

1021. Bott-Duffin synthesis uses...... transformation.

1022. In a minimum function, the degree of numerator is.......... the degree of denominator.

1023. Synthesis of minimum function is also called .... synthesis.

1024. The properties of RL impedance are the same as properties of RC ....

1025. The first Foster realisation of an RC driving point impedance will have .... as first element if $Z_{RC}(s)$ at $s = 0$ is a pole.

1026. An RC impedance function is constant at $s = 0$. The first element in second Foster realisation is a...............

1027. The poles and zeros of an RC impedance function............... on............... real axis.

1028. An LC driving point impedance function has a............... or............... at origin.

1029. In a reactance function all the poles and zeros are............... and lie on............... axis.

1030. In a reactance function the residues at all the poles are............... and ....

1031. In a Hurwitz polynomial, all the coefficients are...............

1032. The sum of positive real functions is...............

1033. In a positive real function the poles and zeros are............... or occur in ....

1034. In a driving point function the degrees of numerator and denominator polynomial may differ by either............... or............... only.

1035. In a driving point function, the real part of all poles and zeros must be either........... or..........

1036. If $Z_{RC}(s)$ has a zero $s = \infty$ the last element in second cauer form is a...............

1037. The four forms of realization (*i.e.*, first and second Foster, first and second cauer) are.........

1038. In $Z_{RC}(s)$, the slope $dZ/d\sigma$ is...............

1039. A function whose circuit has only inductances and capacitances is known as ....function.

1040. For an LC immittance function H(*s*), the slope $dH(j\omega)/d\omega$ is always...............

1041. For a transfer function the degree of numerator polynomial is .... of the degree of denominator polynomial.

1042. In *m* derived filters, *m* is always such that.......

1043. A constant *k* band elimination filter consists of............ circuits in series arms and............... circuit is shunt arms.

1044. In attenuation band of constant *k* filter, phase constant β =...............

1045. If γ is propagation constant, the parameter A in transmission parameter set is equal to.........

1046. If γ is propagation constant, the parameter C in transmission parameter set is equal to........

1047. In terms of transmission parameters of a two port network, the open circuit impedance a two port 1 is...............

1048. In terms of transmission parameters of a two port network, the image parameter Z_{11} at port 1 =............

1049. For a reciprocal two port network the *z* parameter condition is...............

1050. For a reciprocal two port network the ABCD parameter condition is...............

1051. In terms of the readings W_1 and W_2 in two wattmeter method of power measurement in 3-phase circuits, power factor =...............

1052. In two wattmeter method of power measurement in 3-phase resistive load, the readings W_1 and W_2 are ....

1053. In two wattmeter method of power measurement in 3-phase circuits, one of the wattmeter reads zero when power factor = ...

1054. When an unbalanced star connected load is fed by three-phase 3-wire system, the phenomenon of............... occurs.

1055. Fourier transform of 1 is ....

1056. For a rectangular pulse of magnitude A and width *d* extending from $\frac{-d}{2}$ to $\frac{d}{2}$, Fourier transform = ...

1057. If $F(j\omega)$ is Fourier transform $f(t)$ the Fourier transform of $f(-t)$ = ...

1058. In exponential form of Fourier series $f(t)$ = ...

1059. If $f(-t) = -f(t)$, Fourier series of $f(t)$ will have only ....terms.

1060. For a wave $i = I_0 + I_{1m} \sin \omega t$, rms value = ...

1061. A complex wave has 20% third harmonic. If it is applied to an inductance L, the percentage harmonic content in current wave will be...............

1062. The standard form of state equation is...............

1063. Tellegen's theorem in equation form is ....

1064. Norton's theorem replaces a network by a........

1065. In a three branch parallel circuit having R in one arm, L in second and C in third, at resonance, the currents in L and C are ....

ANSWERS WITH HINTS AND EXPLANATIONS

1. (*a*) Drift velocity of electrons is very small.
2. (*b*) $V_{AB} = V_A - V_B$ or $V_{AB} = -8 - (-16) = 8$ V.
3. (*c*) As per definition one ampere means 1 coulomb per second.
4. (*a*) 5. (*c*)
6. (*a*) As per definition $1C = 6.24 \times 10^{18}$ electrons.
7. (*b*) $0.75 = I^2 \times 1 \times 1000$ or $I = 27.386$ mA.
8. (*c*) Consumers are charged on the basis of kWh of energy used.
9. (*c*) Current density = current/area of cross-section.
10. (*b*)
11. (*d*) Current density = current/area of cross-section. Same current throughout the conductor. Since area of cross-section is non-uniform, current density will be different at different cross-sections.
12. (*a*)
13. (*c*) KCL and KVL are valid for voltages and currents of all waveshapes during transiant as well as steady state conductions.
14. (*a*) In a non-linears resistance *v-i* graph is a curve.

15. (*d*) energy = V × I × *t* since I × *t* = Q, energy = Voltage × charge.

16. (*a*) $P = \frac{V^2}{R}$ or $V = \sqrt{PR} = \sqrt{0.5 \times 100}$

$= 70.7 \text{ V and } I = \frac{V}{R} = \frac{70.7}{10000 \text{ A}}$.

17. (*a*) Calories and joules are related by J, *i.e.*, mechanical equivalent of heat.

18. (*b*) The current through 8 and 2 ohm resistances is $\frac{0.01}{10} = 0.001$ A or 1 mA. Therefore current through (5 + R) ohm is 1.25 m A. $(5+R) = \frac{0.01}{1.25 \times 10^{-3}}$ or R = 3 ohm.

19. (*a*) Since 25% of input is lost, 0.75 Input = 2 or input $= \frac{2}{0.75} = 2.67$ kWh.

20. (*c*)

21. (*c*) Since diameter is doubled, area of cross-section becomes four times and resistance is inversely proportional to area of cross-section.

22. (*d*) $\frac{1}{R} = \frac{1}{2} + \frac{1}{2} + \frac{1}{0} = \infty$. Hence R = 0.

23. (*d*) Insulators have high resistivity.

24. (*a*) If α_t is temperature coefficient at *t*°C and α_0 is temperature coefficient at 0°C. $\alpha_t = \frac{\alpha_0}{1 + \alpha_0 t}$.

25. (*a*) $P = \frac{V^2}{R}$ or $R = \frac{V^2}{R}$ or $R = \frac{230 \times 230}{100} = 529$ ohm.

26. (*d*) Nichrome has highest resistivity.

27. (*c*) Since diameter is doubled, area of cross-section becomes four times current carrying capacity is proportional to area of cross-section.

28. (*a*) For resistances is parallel, overall resistance is less than the smallest resistance.

29. (*a*) 30. (*b*) 31. (*b*) 32. (*a*)

33. (*c*)

34. (*d*) Since new diameter is 0.3 *d*, the new area of cross-section $\alpha\, (0.3d)^2$. The volume remains the same.

Therefore length increase by a factor $\frac{1}{(0.3)^2}$.

Hence $\frac{\text{New resistance}}{2}$

$$= \frac{\rho \frac{l}{(0.3)^2}\left[\frac{1}{\frac{\pi}{4}(0.3d)^2}\right]}{\rho\, l \left[\frac{1}{\frac{\pi}{4} d^2}\right]} = 123.5\,.$$

35. (*b*) The resistance of parallel combination $= \frac{40}{3} - 10$ $= 3.33\ \Omega$; $\frac{1}{3.33} = \frac{1}{12} + \frac{1}{15} + \frac{1}{15}$ or R = 6 Ω

36. (*b*) The resistance of a thermistor decreases with increases in temperature. It is used to monitor hot spot temperature of electric machines.

37. (*c*) Q = CV or 0.15 = C(5) or C = 0.03F.

38. (*a*) A high ϵ_r will lead to high value of capacitance.

39. (*a*) $C = \frac{\epsilon_o \epsilon_r A}{t}$. For mica ϵ_r is more than that of air.

40. (*b*) When the capacitor is fully charged, *i* = 0 and voltage across capacitor is equal to battery voltage.

41. (*a*) Very high R will mean low dielectric loss.

42. (*d*) When cells are in series, current remains the same but voltage increases.

43. (*b*)

44. (*b*) Specific gravity of electrolyte is highest when battery is fully charged and is lowest when discharged.

45. (*c*) The charger has to supply current to battery for charging. This is possible only if voltage of charger is more than that of battery.

46. (*c*) 47. (*d*) 48. (*d*) 49. (*a*)

50. (*c*) Since emfs of two cells are not equal, a circulating current flows. Circulating current $= \frac{1.5 - 1.45}{0.2} =$ 0.25 A

OC cmf = 1.5 – (0.25) (0.1) = 1.475V.

51. (*b*) Capacitance $\propto \epsilon_r$.

52. (*c*) Energy stored = 0.5 CV^2.

53. (*b*) $C = \frac{20 \times 20}{40} = 10\ \mu$F.

54. (*a*) The combination is equal to two capacitors in parallel.

$$C = \frac{\epsilon_o \epsilon_1 (0.5\text{ A})}{d} + \frac{\epsilon_o \epsilon_2 (0.5\text{A})}{d}.$$

55. (*a*) Since the capacitors are in series, the total capacitance becomes 50%. Voltage across each is V.

56. (*a*) The combination is equal to two capacitors in series.

$$C = \frac{\left[\epsilon_o \epsilon_1 \left(\frac{A}{0.5d}\right)\right]\left[\epsilon_o \epsilon_2 \left(\frac{A}{0.5d}\right)\right]}{\epsilon_o \epsilon_1 \frac{A}{0.5d} + \epsilon_o \epsilon_2 \frac{A}{0.5d}}$$

Hence $C \propto \frac{\epsilon_1 \epsilon_2}{\epsilon_1 + \epsilon_2}$.

57. (*b*) Temperature coefficient of copper = 0.00393

$R_{100} = 40\,(1 + 0.00393 \times 80) = 54\ \Omega$.

58. (*d*) Since diameter is double, area of cross-section is four times and length is one-fourth.

$$R_2 = \frac{\frac{\rho l}{4}}{4a} = \frac{\rho l}{16a} = \frac{R}{16}.$$

59. (*c*) Since diameter is $3d$, area of cross-section is increased 9 times. Therefore resistance is $\frac{R}{9}$.

60. (*b*) Insulation resistance of cable is inversely proportional to length.

61. (*d*) Inductance $\propto N^2$.

62. (*b*) Current rating depends only on area of cross-section and hence diameter.

63. (*b*) Inductance $\alpha\ \mu_r$.

64. (*c*) $Q_1 = 1 \times 10^{-6} \times 30 = 30\ \mu C, Q_2 = 2 \times 10^{-6} \times 30 = 60\ \mu C$.

65. (*b*) Since capacitors are in series, charge will be same.

Total capacitance $= \frac{1 \times 2}{3} = \frac{2}{3}\ \mu F$ charge

$- \frac{2}{3} \times 30 = 20\ \mu C.$

66. (*c*) $C = \frac{16 \times 16}{32} + 16 = 24\ \mu F.$

67. (*c*) Charge across capacitance increases linearly. Since capacitance is constant, voltage increases linearly.

68. (*c*) Charge will be the same. Since $Q = CV$ voltage will be inversely proportional to Q. Hence 20 V and 10 V.

69. (*a*)

70. (*b*) A practical voltage source has some source resistance. Because of voltage drop across source resistance, terminal voltage cannot be higher than source voltage. If current is zero, terminal voltage and source voltage are equal.

71. (*d*)

72. (*b*) In a current source the source resistance is in parallel with the source. Some current flows through this source resistance.

73. (*a*) The current will be less than 3 A. It may not be 1.5 A because other resistance in the circuit also affect the value of current.

74. (*d*) **75.** (*b*) **76.** (*b*)

77. (*b*) $I = \frac{V_{TH}}{R_{TH} + R} = \frac{10}{20 + R}$. Hence less than 0.5 A.

78. (*c*) **79.** (*c*)

80. (*a*) Polarity of voltage depends only on direction of current and nothing else.

81. (*b*) $R = \frac{15 \times 15}{45} = 5\ \Omega.$

82. (*b*) $\sqrt{V} = \sqrt{100}\ \underline{|16/2} = 10 \angle 8°.$

83. (*b*) In heaters, the ratings are specified in terms of rms values.

84. (*c*) $v_1 = 100 \sin(\omega t + 15°)$ and $v_2 = 60 \cos \omega t = 60 \sin(\omega t + 90°)$. Hence v_2 is leading v_1 by $(90 - 15) = 75°$.

85. (*c*) Peak voltage rating $= \sqrt{2}$ (rms voltage rating).

86. (*d*) $20 \times j20 = 28.28 \angle 45° \log_e V$

$= \log_e (28.28) + j \frac{45° \times \pi}{180} = 3.342 + j.785.$

87. (*c*) Peak value = 50 V, rms value $= \frac{50}{\sqrt{2}} = 35.35$ V.

88. (*b*) $V^2 = 7 \times 7\ \underline{|2 \times 15} = 49 \angle 30°$.

89. (*c*) In a resistive circuit the current is in phase with voltage, *i.e.*, lags by 0°.

90. (*c*) In capacitive circuit current lags the voltage by 90°.

91. (*b*) In capacitive circuit current leads the voltage by 90°.

92. (*d*) In RL circuit the current lags the voltage by an angle which is greater than 0 but less than 90°.

93. (*d*) $Z = \sqrt{R^2 + \frac{1}{\omega^2 c^2}}$. Since frequency is doubled, $\frac{1}{\omega^2 c^2}$ becomes one-fourth but R^2 remains the same.

94. (*d*) $Z = (R^2 + \omega^2 L^2)^{0.5}$.

95. (*d*) Voltage across inductance leads the current by 90° and voltage across capacitance lags the current by 90°.

96. (*d*) $Z = \frac{200}{20} = 10\ \Omega = 10 + j(50 - X_C)$.

Hence $X_C = 50\ \Omega$.

97. (*a*) $Z = 30 + j(30 - 30) = 30 \angle \Omega$.

98. (*b*) Power factor is maximum and equal to 1 for purely resistive load. Power factor is minimum and equal to zero for purely reactive load.

99. (*a*) $\frac{R}{X} = 0.2$ or $\tan\theta = \frac{X}{R} = 5$, Therefore $\cos\theta = 0.196$ lagging.

100. (*b*) Active and apparent power are equal for resistive circuit. Otherwise active power is less than apparent power.

101. (*d*)

102. (*a*) In an RL circuit excited from dc source the current rises exponentially with a time constant $\frac{L}{R}$.

103. (*a*) The parallel combination of 3, 5 and 4 ohm is short-circuted. Therefore their net resistance is zero.

104. (*c*) When a current pulse is applied to a capacitor the voltage rises linearly and becomes constant at the end of pulse.

105. (*c*) Each of parallel combination of capacitors is equal to 1 μF. Therefore voltage across each capacitor is 50 V. $Q = 0.8 \times 50 = 40\ \mu C$.

106. (*b*) The temperature coefficient decreases slightly with increase in temperature.

The equation is $\alpha_1 = \dfrac{\alpha_0}{1+\alpha_0 t}$.

107. (*b*) $Q = It = \dfrac{60}{120} \times 30 \times 60 = 900C$.

108. (*d*) $kWh = \dfrac{60 \times 45}{1000 \times 60} = .045$.

109. (*a*) $1\ kWh = 36 \times 10^5\ J = \dfrac{36 \times 10^5}{4.186}$ cal. Therefore calorific

value $= \dfrac{36 \times 10^5}{4.186 \times 0.4 \times 0.6} = 3500$ k cal.

110. (*b*)

111. (*a*) The parallel combination of 30 Ω and 20 Ω is 12 Ω. Since 12 Ω and 10 Ω are in parallel, the 10 Ω plate draws more than 50% current and dissipates more than 50% energy.

112. (*b*) $V = IR$ and $P = I^2 R$. Therefore waveshape of V is the same as that of I but waveshape of P is different.

113. (*b*) **114.** (*c*) **115.** (*b*)

116. (*a*) Earthing prevents electric shock because the fault current flows through the earth conductor and not through body.

117. (*b*)

118. (*c*) At the time of starting the choke provides a high voltage. During working there is voltage drop across choke so that lamp gets proper voltage.

119. (*d*)

120. (*b*) Resistance at the time of switching on is minimum and increases due to increase of temperature.

121. (*a*) The network can be replaced by Thevenin's equivalent circuit. The load gets maximum power from the network but Thevenin's theorem does not gurantee that this will be half of maximum power.

122. (*b*) **123.** (*d*) **124.** (*a*)

125. (*b*) Current through $R_2 = 5$ A towards 0. Current through R_3 is 4 A away from node. Total incoming current = 13 A. Current going away from 0 and flowing through $R_1 = 9$ A.

126. (*d*) From the given data the current through R_3 is 4 A.

127. (*a*) Current supplied by battery $= \dfrac{10}{10} = 1$ A. Therefore current through 2 Ω resistance is 11 A. Voltage across $R = 48 - 10 - 11 \times 2 = 16$ V.

128. (*b*) Norton's equivalent is as shown in Fig. 4.201. Some current will flow through R_N. Therefore current through 20 Ω resistance is less than 7.5 A.

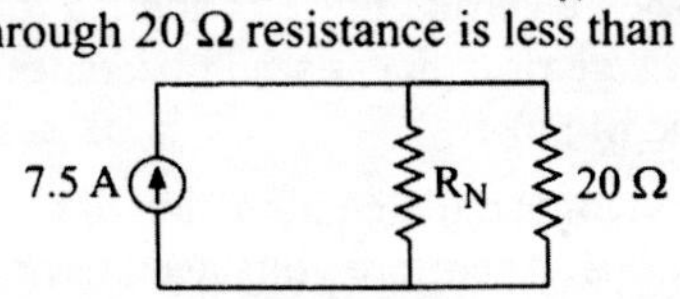

Fig. 4.201.

129. (*c*) In power and distribution circuits efficiency has to be very high.

130. (*b*) $R_{TH} = \dfrac{10 \times 10}{20} + 5 = 10$ ohm.

131. (*c*) $I = \dfrac{20}{5 + R_{TH}}$.

132. (*d*)

133. (*b*) In air cored coil flux density increases linearly with current because there is no saturation.

134. (*d*) If saturation starts B will be less than 1T.

135. (*a*) $R = \dfrac{\rho l}{a}$ and Reluctance $= \dfrac{l}{\mu a}$.

136. (*a*) Only high retentivity will ensure permanent magnetism.

137. (*d*)

138. (*d*) The permeability of cast iron is about 100.

139. (*b*) **140.** (*c*)

141. (*b*) Since μ_r for silicon steel is more than that of cast iron, flux density will be more than 0.6 T.

142. (*c*)

143. (*d*) $M = 0.5\sqrt{(27 \times 3)}$.

144. (*a*) $e = M\dfrac{di_1}{dt} = \dfrac{3}{1000} \times 100 = 0.3$ V.

145. (*a*) $\dfrac{30 \times 30}{30 + 30 + 30}$ mH.

146. (*b*)

147. (*c*) $M = \dfrac{60 - 40}{4}$ mH.

148. (*b*) μ_r for air is 1 while μ_r for iron is more than 1.

149. (*a*) Mutual inductance term will become negative.

150. (*d*) Total inductance $= L + L + L + M + M + M + M + M + M = 3L + 6M$.

151. (*c*) Since diameter is halved, area of corss-section is one-fourth. The number of turns becomes four times. Inductance $\propto N^2$.

152. (*a*) Inductance $= 4 + 16 + 2\sqrt{4 \times 16} = 36$ mH.

153. (*c*) **154.** (*a*)

155. (*d*) Inductance does not allow the current to increase instantaneously.

156. (*a*) Since the capacitor does not have any charge initially, $i = \dfrac{E}{R}$.

157. (*a*) $i = \dfrac{E}{R}(1 - e^{-Rt/L})$ or $\dfrac{di}{dt} = \dfrac{E}{L}e^{-Rt/L}$.

158. (*d*)

159. (*b*) Increase of resistance increases damping.

160. (*c*).

161. (*d*) Capacitor behaves as a short-circuit at $t = 0$.

162. (*b*) **163.** (*c*) **164.** (*b*) **165.** (*d*)

166. (*d*) **167.** (*c*) **168.** (*a*)

169. (*a*) At $t = 0$, current in R-L branch is zero.

170. (*b*) At $t = \infty$ the circuit has effectively two 6 Ω resistances in parallel.

171. (*d*) Time constant of R-L branch $= \frac{L}{R} = 0.5$ seconds.

Current in 6 Ω resistance is 1 A and current in R-L branch is 63.2% of final current of 1 A.

172. (*c*)

173. (*b*) Inductance does not allow the current to change instantaneously.

174. (*a*) Current through an inductance cannot change instantaneously.

175. (*c*) Current through inductance is zero. Current through 5 ohm resistance $= \frac{100}{15}$ A.

176. (*b*) At $t = \infty$ inductance behaves as short-circuit.

177. (*a*) The 5 ohm resistance is short-circuited by inductance.

178. (*a*) Capacitor behaves as a short-circuit at $t = 0$.

179. (*d*) At $t = \infty$ capacitor behaves as open circuit.

180. (*d*) The 3 Ω resistance is short-circuited by capacitor at $t = 0$.

181. (*b*) Since capacitor is open-circuted at $t = \infty$, current 20/8 A.

182. (*a*) Voltage across a capacitor cannot change instantaneously.

183. (*d*) Voltage across capacitor = voltage across 3 Ω resistance $= \frac{20}{5+3} \times (3) = 7.5$ V.

184. (*a*) Since 3 ohm resistance is short-circuited by capacitor, I = 20/5 A.

185. (*b*) **186.** (*b*)

187. (*a*) Current through an inductance cannot change instantaneously.

188. (*b*) At $t = \infty$, inductance is short-circuited and capacitor is open-circuited.

189. (*a*) **190.** (*a*)

191. (*c*) $v_2 = \sin(\omega t + 90°)$.

192. (*b*) **193.** (*b*)

194. (*d*) A = 14.14 ∠ 45°. $\log_e A = \log_e 14.14 \times j\frac{45 \times \pi}{180}$

$= 2.65 + j0.785$.

195. (*d*) $\sqrt{A} = \sqrt{16}\,\underline{|64°/2}$.

196. (*a*) $A^2 = 4^2\underline{|2 \times 7°}$.

197. (*c*) $Z = \sqrt{R^2 \times X_L^2}$. When frequency is doubled, X_L becomes twice but R remains the same.

198. (*c*) $Z = 8 + j(8 - 8)$.

199. (*c*) $Z = \sqrt{R^2 + X_C^2}$. When f is made four times, X_C becomes one-fourth but R remains the same.

200. (*b*) $Z = \sqrt{20^2 + 20^2} = 28.28$ ohm.

201. (*b*) In RC circuit the current leads the voltage, $\theta = \tan^{-1}\frac{10}{100} \simeq 6°$.

202. (*a*)

203. (*a*) The current is leading the voltage. Therefore it is an RC circuit.

204. (*d*) In parallel branches, the current through any branch depends only on voltage, impedance and frequency only.

205. (*c*) Since voltages have a phase difference of 180°, current is zero.

206. (*c*) 5 A current may be (say) 3 +j4A. The current in the other branch is $x + jy$. The total current is $(3 + x) + (4 \pm y)$. This may be 5 A or more or less.

207. (*c*) $Z = \frac{(10 - j10)(10 + j10)}{10 - j10 + 10 + j10} = 10 + j0$.

208. (*d*) **209.** (*a*)

210. (*d*) Both Q_1 and Q_2 may be positive or negative.

211. (*c*) I_3 is inductive current and must lag I_2 which is capacitive current.

212. (*b*) The current is $5 - j5 = \sqrt{5^2 + 5^2} = 7.07$ A.

213. (*a*) The total current is $\frac{V}{j5} - \frac{V}{j5} = 0$.

214. (*c*) $f_r = \frac{1}{2\pi\sqrt{LC}}$. If C becomes four times, f is reduced to half.

215. (*d*) Maximum voltage across capacitance occurs at a frequency slightly less than resonant frequency.

216. (*c*) The maximum voltage across inductance occurs at a frequency slightly above the resonant frequency.

217. (*a*) Voltage across inductance at resonance = QV = 7.5 × 100 V.

218. (*a*) At parallel resonance circuit impedance $= \frac{L}{RC}$.

219. (*c*) At parallel resonance source current $= \frac{V}{(L/RC)}$.

220. (*b*) At half power frequencies, current is –3 dB.

221. (*b*) At half power frequency the circuit impedance is $\sqrt{2}$ times resistance.

222. (*a*) Band width $= \frac{\omega_0}{Q}$.

223. (*d*)

224. (*a*) Total energy $= \frac{1}{2}LI_0^2 + \frac{1}{2}LI_0^2 = LI_0^2$.

225. (*c*)

226. (*d*) At parallel resonance Z is maximum and Y is minimum.

227. (*b*) Load impedance should be conjugate of source impedance.

228. (*d*) $R_1 = \sqrt{3^2 + 4^2}$.

229. (*d*) **230.** (*c*) **231.** (*a*)

232. (*a*) As per KCL sum of currents entering a node is equal to sum of currents leaving a node.

233. (*b*) **234.** (*d*)

235. (*b*) Delta impedances

$$= \frac{3(3+j4)(3+j4)}{3+j4} = 9 + j12 \text{ ohm.}$$

236. (*d*) Thevenin's impedance may be inductive or capacitive.

237. (*a*) **238.** (*b*)

239. (*b*) Since strengths of both sources are doubled, the current would become twice.

240. (*a*) **241.** (*c*) **242.** (*c*) **243.** (*c*)

244. (*c*) Thevenin's equivalent circuit is with respect to particular terminals only.

245. (*b*)

246. (*a*) $z_{11} = 1 + \frac{1 \times 2}{3} = \frac{5}{3}\,\Omega$.

247. (*a*) **248.** (*d*) **249.** (*c*) **250.** (*c*)

251. (*b*) **252.** (*c*) **253.** (*d*) **254.** (*a*)

255. (*d*)

256. (*a*) $z_{11} = 2 + \frac{4 \times 3}{7} = \frac{26}{7}$.

257. (*b*) $z_{22} = \frac{1 \times 6}{7}$.

258. (*d*) Transistor circuits are analysed using *h* parameters.

259. (*a*) **260.** (*a*) **261.** (*b*) **262.** (*c*)

263. (*d*)

264. (*b*) For constant *k* high pass π section $Z_{0\pi}$

$$= \frac{R_0}{\sqrt{1 - \left(\frac{f_c}{f}\right)^2}}. \text{ At } f = f_c, Z_{0\pi} = \text{infinite.}$$

265. (*b*)

266. (*c*) Resonant frequency of arms in constant *k* band pass filter $= \sqrt{(f_{c1})(f_{c2})}$.

267. (*b*) $Z_{0T} = R_0\,(1 - f^2/f_c^2)^{0.5}$. At $f = f_0$, $Z_{0T} = 0$.

268. (*d*) **269.** (*d*)

270. (*a*) For *m* derived low pass filter $f_c = (\sqrt{1-m^2})\, f_\infty$.

271. (*c*) **272.** (*b*) **273.** (*d*) **274.** (*d*)

275. (*b*)

276. (*a*) *s* domain impedance of each inductance = 2*s* and that of each capacitor is $\frac{1}{s}$.

277. (*b*)

278. (*a*) As distance from real axis increases, the frequency of oscillation increases.

279. (*a*)

280. (*d*) $Z(s) = R + sL + \frac{1}{sC} = \frac{s^2LC + RCs + 1}{sC}$ which is similar to $\frac{s^2 + 10s + 100}{s}$.

281. (*a*) **282.** (*d*)

283. (*b*) $\frac{V_2(s)}{V_1(s)} = \frac{\frac{1}{sC}}{R + \frac{1}{sC}} = \frac{1}{1 + RCs}$.

284. (*a*) A reactance function means that the network is composed of only L and C. Since there is no resistance, poles lie on *j*ω axis only.

285. (*a*) **286.** (*d*) **287.** (*c*) **288.** (*d*)

289. (*b*) **290.** (*d*) **291.** (*a*) **292.** (*b*)

293. (*c*) **294.** (*b*) **295.** (*b*) **296.** (*d*)

297. (*a*) As $\omega \to \infty$ $\frac{1}{sC} \to 0$. Therefore $\frac{V_2(s)}{V_1(s)} \to \frac{R_2}{R_1 + R_2}$.

298. (*a*) **299.** (*d*) **300.** (*d*)

301. (*b*) $R_{TH} = \frac{5}{5} = 1\,\Omega$. $I = \frac{5}{1 + 1.5} = 2$ A.

302. (*b*)

303. (*b*) $I(s) = \frac{s}{(s+1)(s+2)} = \frac{-1}{s+1} + \frac{2}{s+2}$. Therefore $i(t) = -e^{-t} + 2e^{-2t}$.

304. (*d*)

305. (*c*) $V_C = QV = 50$ V rms.

306. (*b*) $i(t) = \frac{6}{2} + \frac{6}{2}(1 - e^{t}) = 6 - 3e^{-t}$.

307. (*b*) $i(t) = \frac{12}{12}\left(1 - \frac{2}{4}e^{-t}\right) = 6 - 3e^{-t}$.

308. (*c*) ABCD matrix of two-four terminal network is the product of individual ABCD matrices.

309. (*b*) **310.** (*a*) **311.** (*c*) **312.** (*b*)

313. (*c*) **314.** (*c*)

315. (*a*) No current will flow through 2 Ω resistance in parallel with short-circuit.

316. (*d*) Battery current $= \frac{6}{2 + \frac{2 \times 2}{4}} = 2$ A,

$$I_{AB} = \frac{2\times 2}{4} = 1\text{ A}.$$

317. (*d*)

318. (*c*) In RC circuit excited by a dc battery of voltage E, $i(t) = \frac{E}{R}e^{-t/RC}$ and is maximum at $t = 0^+$.

319. (*a*) Maximum current $= \frac{E}{R} = 2$ A.

320. (*b*) $I = \frac{10/\sqrt{2}}{2}$ and $P = I^2R = \left(\frac{10}{2\sqrt{2}}\right)^2 (2) = 25$ W.

321. (*c*) **322.** (*a*)

323. (*c*) From the given data R = 2 Ω, L = 2 H and E = 2 V. Because of initial current of 2 A, $i(t) = 1 - e^{-t} + 2e^{-t} = 1 + e^{-t}$. Therefore $V_R = 2(1 + e^{-t})$.

324. (*a*) **325.** (*a*)

326. (*a*) $i = \frac{10}{3}\left(1 - e^{\frac{-3t}{2}}\right) = \frac{10}{3}(1 - e^{-1.5t})$.

$$\frac{di}{dt} = 5e^{-1.5}t.\ \text{At } t = 0^+\ \frac{di}{dt} = 5\text{A/s}$$

327. (*a*) Voltage across a capacitor cannot change instantaneously.

328. (*b*) Direction of current during discharging is opposite to that during charging.

329. (*a*) $V(s) = [I(s)][Z(s)] = \frac{1}{s}\left(\frac{s+2}{s+3}\right)$.

Initial value $= \underset{s\to\infty}{\text{Lt}}\,[sF(s)] = 1$ V

Final value $= \underset{s\to 0}{\text{Lt}}\,[sF(s)] = \frac{2}{3}$ V.

330. (*b*) R = 4 Ω, L = 1 H, ω = 4 rad/sec, $X_l = 4\ \Omega$

$$I = \frac{4/\sqrt{2}}{(R^2 + X_L^2)^{0.5}} = \frac{4/\sqrt{2}}{\sqrt{32}} = \frac{1}{2}\text{ A}.$$

Peak current $= \sqrt{2}\,I = \frac{1}{\sqrt{2}}$ A, $\phi = \tan^{-1}\frac{X_L}{R} = 45°$

$$i = \frac{1}{\sqrt{2}}\sin(4t - 45°).$$

331. (*b*) R = 5 Ω, $X_C = \frac{1}{\omega C} = \frac{10^6}{20000\times 10} = 5\ \Omega$.

$\theta = \tan^{-1}\frac{X_C}{R} = 45°$, current leads voltage.

$$Z = \sqrt{R^2 + X_C^2} = \sqrt{50}$$

$$V = 5\sqrt{50}\,(\sin 20000\,t - 45°)$$

$$= 25\sqrt{2}\sin(20000t - 45°)$$

332. (*c*) $I_R = 2\angle OA$, $I_L = 2\angle -90° = -j2$A.

Total current $= 2 - j2$.

$V_2 = 2(2 - j2) + 4\angle 0 = 8 - j4$ volts.

333. (*d*) At $t = \infty$ inductance behaves as short-circuit.

334. (*a*) At $t = 0$, inductance behaves as O.C.

335. (*b*) **336.** (*c*)

337. (*c*) At $t = 0^+$ inductance behaves as O.C.

338. (*b*) At $t = 0^+$ capacitor behaves as short-circuit.

339. (*a*) At $t = \infty$ capacitor behaves as O.C.

340. (*b*) In underdamped circuit the frequency of oscillations is less than resonant frequency.

341. (*a*) For overdamping R is more.

342. (*b*)

343. (*a*) Capacitor behaves as short-circuit at $t = 0^+$.

344. (*a*) Inductance behaves as O.C. at $t = 0^+$.

345. (*b*) As per KCL sum of currents entering a node is equal to sum of currents leaving the node.

346. (*a*) At $t = \infty$, inductance behaves as short-circuit.

347. (*c*) $I = \frac{100}{\sqrt{10^2 + 10^2}} = \frac{10}{\sqrt{2}}$ A and

$$V_R = \frac{10}{\sqrt{2}}\times 10 = \frac{100}{\sqrt{2}}\text{ V}.$$

348. (*b*) **349.** (*b*)

350. (*c*) $P = I^2R = 5^2(5) = 125$ W.

351. (*b*) $V_{TH} = \frac{5\times 2}{4}\times 1 = 2.5$ V, $R_{TH} = \frac{1\times 3}{4} = 0.75\ \Omega$.

352. (*d*) $\frac{V_R^2}{8} = 4$ or $V_R^2 = 32$

or $V_R = 4\sqrt{2} = V\cos\theta = \frac{20}{\sqrt{2}}\cos\theta$

or $\cos\theta = \frac{(4\sqrt{2})(\sqrt{2})}{20} = 0.4$.

353. (*b*) **354.** (*c*)

355. (*b*) $i(0) = 100e^{-0} = 100$, $i(\infty) = 100e^{-\infty} = 0$.

356. (*b*) Since area of triangle is half of the rectangle enclosing it, average value = 0.5 × peak value.

357. (*b*) **358.** (*b*)

359. (*c*) RMS value $= \sqrt{(50 + 50\times 0.707)^2 + (50\times 0.707)^2}$.

360. (*a*) Area of triangle is half of rectangle. So average value = 0.5 × 100 = 50 A.

$$(\text{RMS value})^2 = \frac{1}{2}\int_0^2 (50t)^2\,dt = 57.73\text{ A}.$$

361. (*b*) Average value = 10 + 0.5 × 10 = 15 A.

362. (*c*) Peak value = 14.14 A; $\omega_1 = 2\pi\times 40$, $\omega_2 = 2\pi\times 50$.

363. (*c*) If peak value is I_m, heating effect of sine wave $\propto \frac{I_m^2}{2}$ and heating effect of rectangular wave $\propto I_m^2$.

364. (*b*) $i = 10 \sin 314.16t, \frac{di}{dt} = 10 \times 314.16 \cos \omega t$. At $t = 0.005, i = 314.16 \cos(314.16 \times .005) = 0$.

365. (*c*) $A + B = 7 + j4 = 8.062 \angle 29.7°$.

366. (*c*) $A - B = -1 - j2 = 2.236 \angle 243.43°$.

367. (*b*) $A = 3 + j1 = 3.162 \angle 18.43°, A^4 = 100 \angle 73.72°$.

368. (*c*) $V = \frac{100}{\sqrt{2}}$ and $P = \frac{V^2}{R} = \frac{100 \times 100}{2 \times 10} = 500$ W.

369. (*c*) At resonance $X_L = X_C$. Therefore $I X_L = I X_C$ or $V_L = V_C$.

370. (*d*) $Z = \sqrt{200} \angle 45° \,\Omega,\ V = \frac{100}{\sqrt{2}} \angle 45°$

$$I = \frac{V}{Z} = \frac{100 \angle 45°}{(\sqrt{2})(\sqrt{200}) \angle 45°} = 5 \angle 0\text{A}.$$

371. (*a*) Apparent power $= \frac{200}{0.707}$ sin $\theta = 0.707$, $Q = 250$ vars and lagging vars are taken as positive.

372. (*d*) $i = 10\,(1 - e^{-t})$. It reaches steadily state value at $t = \infty$.

373. (*a*) Voltage across a capacitor cannot change instantaneously.

374. (*c*) $Q = 4 \times 0.5 = 2$ C and $V = \frac{Q}{C} = \frac{2}{0.1} = 20$ V.

375. (*b*) $\frac{10}{0.1 + R} = 50$ or $R = 0.1\,\Omega$.

376. (*c*) $I = \frac{50}{5 \times 1000} = 0.01 \text{ A} = \frac{250}{5000 + R}$

or $R = 20000\,\Omega$.

377. (*b*) Energy dissipated by the plate with lowest resistance is maximum.

378. (*c*) Current through 4 Ω resistance = 2 A.

379. (*b*) Voltage across 4 Ω resistance = source voltage = 8 V. Power = 8 × 3 = 24 W.

380. (*c*)

381. (*b*) $\frac{di}{dt} = 2$ A/s, $v_L = L\frac{di}{dt} = 3 \times 2 = 6$ V.

382. (*b*) $\frac{di}{dt} = -4$A/s, $v_L = L\frac{di}{dt} = -12$ V.

383. (*c*) $\frac{di}{dt} = 0 - (-e^{-t}) = e^{-t}$.

384. (*b*) Energy $= 0.5LI^2 = 0.5 \times 1 \times 0.6^2 = 0.18$ J.

385. (*a*) Rate of heat = rate of energy = power = *vi*.

386. (*b*) **387.** (*a*) **388.** (*b*) **389.** (*b*)

390. (*c*) A controlled voltage source may be a voltage source or current source and its strength may depend on another voltage or current.

391. (*a*)

392. (*b*) 1 A current is flowing into the battery. Therefore it is charging current.

393. (*a*) 4 A current is flowing into the battery. Therefore it is charging current.

394. (*c*) The number of nodal equations is one less than the number of independent nodes.

395. (*b*) The parallel combination of 20 Ω and 5 Ω is 4 Ω. Therefore strength of current source is 4 × 4 = 16 A and source resistance is 4 ohms.

396. (*a*) Since there is no return path, the current is zero.

397. (*a*) **398.** (*b*) **399.** (*a*) **400.** (*b*)

401. (*c*) Inductances are in series aiding connection. M terms will become twice because of mutual inductance between L_1 and L_2 and that between L_2 and L_1.

402. (*c*) **403.** (*a*) **404.** (*b*)

405. (*c*) M_{12} is positive while M_{23} and M_{13} are negative because of dots shown in Fig. 4.60.

406. (*c*) Energy $= 0.5\,LI^2$.

407. (*c*)

408. (*d*) When axes are perpendicular there is no mutual coupling.

409. (*b*) $M = k\sqrt{L_1 L_2}$.

410. (*c*) The dots indicate that M term is KVL equation will be negative.

411. (*d*)

412. (*b*) Wave A starts its cycle θ° earlier than wave B.

413. (*b*) A leading B means B lagging A by ∠ θ.

414. (*b*) **415.** (*a*)

416. (*b*) $j^6 = j^4 \cdot j^2 = j^2$ because $j^4 = 1$.

417. (*d*) Sum of phasors is maximum if they are in phase and 360° = 0.

418. (*b*) Addition and subtraction is not possible in polar form.

419. (*c*) $4 + j3 = 5 \angle 36.87°$

$$(B)^{1/3} = (5)^{1/3} \underline{|36.87/3} = 1.71 \underline{|12.3°}$$

420. (*b*) $3 + j1 = \sqrt{10} \angle 18.43°$

$$A^4 = (\sqrt{10})^4 \underline{|4 \times 18.43°} = 100 \underline{|73.72°}.$$

421. (*c*) $$V_{av} = \frac{1}{\pi}\left[\int_0^{\pi/4} V_m \sin \omega t\, d(\omega t) + \int_{\pi/4}^{3\pi/4} 0.707 V_m d(\omega t) + \int_{3\pi/4}^{\pi} V_m \sin d(\omega t)\right] = 0.54\, V_m.$$

422. (*b*) RMS value

$$= \left[\frac{1}{2}\left\{\int_0^1 (100t)^2\, dt + \int_1^2 100\, dt\right\}\right]^{0.5} \simeq 80\text{V}.$$

423. (*b*) Average value

$$= \frac{1}{2}\left[\left\{\int_0^1 100\,t\,dt + \int_1^2 100\,dt\right\}\right] = 75\text{ V}.$$

424. (*c*) $50\angle 0 = 50 + j0,\ 75\angle -60° = 37.5 - j64.95$

Sum $= (50 + 37.5) - j64.95 = 87.5 - j64.95$

$= 109\angle -36.6°$.

425. (*a*) Since peak values are equal, their rms values are equal irrespective of phase difference.

426. (*d*) **427.** (*b*) **428.** (*b*)

429. (*c*) Since $v = L\frac{di}{dt},\ i = \int_{-\infty}^{i} v\,dt.$

430. (*b*) $v = \frac{d\psi}{dt}$ or $\psi = \int_{t-\infty}^{t} v\,dt.$

431. (*a*)

432. (*a*) $v = Ri - \frac{1}{G}\frac{dq}{dt}$

433. (*a*) **434.** (*b*) **435.** (*b*) **436.** (*d*)

437. (*c*) Inductance is flux per ampere.

438. (*a*) Volt $= \frac{\text{joule}}{\text{coulomb}} = \frac{vit}{it} = v.$

439. (*b*)

440. (*b*) Both joule and Newton-metre are units of energy.

441. (*d*)

442. (*c*) $i = C\frac{dv}{dt}$ or $i = C_0(1 - \cos\omega t)\,\omega V_0\cos\omega t$

$= C_0\omega t = C_{0\omega}V_0(\cos\omega t - \cos^2\omega t).$

443. (*d*) Energy $\frac{1}{2}CV^2 = \frac{1}{2}\times 1\times 10^{-6}\times 144 = 72\times 10^{-6}$ J.

444. (*b*)

445. (*a*) $q = \int i\,dt.$

446. (*c*) $\frac{di}{dt} = e^{-t}$. At $t = 1$ second, $\frac{di}{dt} = e^{-t} = \frac{1}{e} = 0.37$ A/s.

447. (*c*) During discharge of capacitor $v_C = v_R = 0.37 \times 1$

$= 0.37$ V.

448. (*a*) $e = N\frac{d\theta}{dt}.$

449. (*c*) $i(t)$ is the incoming current. The currents leaving node 1 are $\frac{v_1}{R_1}$ and $C_1\frac{d}{dt}(v_1 - v_2).$

450. (*b*)

451. (*c*) M term is negative because i_1 is entering the dotted terminal and i_2 is leaving the dotted terminal.

452. (*b*) emf induced in loop 2 due to i_1 is $-M\frac{di_1}{dt}.$

453. (*c*) $\frac{(1+1)(1)}{3} = \frac{2}{3}$ F.

454. (*c*) When switch is in position 2, time constant is $\frac{L}{(R_1 + R_2)}.$

455. (*c*) In RL circuits phase angle of Z is positive.

456. (*b*) **457.** (*b*) **458.** (*a*) **459.** (*b*)

460. (*b*) Voltage across inductance leads the current and voltage across capacitance lags the current.

461. (*b*) Voltage across inductance leads the current and voltage across resistance is in phase with the current.

462. (*a*) $Z = jX_L = j\omega L = \frac{200\times 5}{1000} = = 1\angle 90°$ and $V = IX_L$.

463. (*a*) Admittance triangle depicts admittance Y and its components G and B.

464. (*b*) $|V| = IX_L - IX_C$.

465. (*c*) As f increases impedance increases and current decreases in RL circuit.

466. (*c*)

467. (*b*) $I_4 = I_1 + I_2 + I_3$ or $I_4 = 5\angle 90° + 5\angle 0 + 5\angle 180°$

$= 5\angle 90°.$

468. (*d*) V depends on magnitudes and phase angles of V_a and V_b.

469. (*c*) Dynamic resistance $= \frac{L}{RC} = \frac{0.1}{100\times 10^{-6}}\ 1000\ \Omega.$

470. (*d*) Dynamic resistance

$$= \frac{L}{RC} = \frac{0.005}{100\times 10\times 10^{-9}}\ 5000\ \Omega.$$

471. (*d*) $\omega_0 = \left[\frac{1}{LC} - \frac{R^2}{L^2}\right]^{0.5}$

$$= \left[\frac{1}{.005\times 10^{-9}} - \frac{100^2}{0.005^2}\right]^{0.5}$$

$= 44.67\times 10^4$ rad/sec.

472. (*d*) **473.** (*c*)

474. (*b*) Dynamic resistance $= \frac{L}{RC}.$

475. (*b*) Dynamic resistance $= \frac{L}{RC}$ and increases as L is increased.

476. (*a*) Dynamic resistance $= \frac{L}{RC}$ and increases as C is decreased.

477. (*c*) f_r depends only on L and C.

478. (*b*) $\omega_r = \frac{1}{\sqrt{LC}}$. As L increases ω_r decreases.

479. (*c*) Total capacitance increases and therefore ω_r decreases.

480. (*c*) Since C increases, the circuit will not be in resonance. Therefore current decreases (in series resonant circuit current is maximum at resonance.).

481. (*d*)

482. (*a*) The circuit will not be in resonance. In parallel resonant circuit the current is minimum at resonance. Therefore current will increase.

483. (*b*) As R increases, Q decreases, bandwidth $= \frac{\omega_0}{Q}$.

Therefore bandwidth increases. Hence lower half power frequency will be less than ω_1.

484. (*b*) It has R and L in one branch and a variable capacitance in second branch.

485. (*b*) Loading increases bandwidth of parallel tuned circuit.

486. (*c*) Voltage across inductance = QV and leads the applied voltage.

487. (*b*)

488. (*c*) Total L = 6 × 9 = 54 H. Energy $= \frac{1}{2}\ 54 \times 2^2 = 108$ J.

489. (*c*) $P = \frac{\text{Energy}}{\text{Time}} = \frac{108}{10 \times 10^{-3}} = 10800$ W.

490. (*a*) $M = k\sqrt{L_1 L_2}$, $L_1 = L_2 = L$, $M = kL$.

Total inductance $= L_1 + L_2 + 2M$

$= 2L + 2kL = 900$ mH

or $L(2 \times 2 \times 0.5) = 900$ or $L = 300$ mH.

491. (*b*) Maximum mutual inductance $= \sqrt{L_1\ L_2} = 20$mH.

492. (*c*) **493.** (*c*) **494.** (*c*)

495. (*b*) $I(s) = \frac{s+1}{(s+4)(s+3)}$

$= \frac{3}{s+4} - \frac{2}{s+3}$; $i(t) = 3e^{-4t} - 2e^{-3t}$.

496. (*b*) $I(s) = \frac{1}{s(s+4)^2} = \frac{1}{16s} - \frac{1}{16(s+4)} - \frac{1}{4(s+4)^2}$.

$i(t) = \frac{1}{16} - \frac{e^{-4t}}{16} - \frac{te^{-4t}}{4}$.

497. (*b*) **498.** (*c*) **499.** (*c*) **500.** (*a*)

501. (*a*) $\zeta = \frac{R}{2\sqrt{L/C}}$ and $\omega_n = \frac{1}{\sqrt{LC}}$.

Therefore, $\frac{R}{2L} = \xi\omega_n$

502. (*c*) The final steady state current is $\frac{V_2}{R}$ and transient term must depend on both V_1 and V_2.

503. (*c*) The final steady state current in RL circuit depends only on final voltage (*i.e.*, 20 V) and resistance.

504. (*c*) The wave can be written as $t\,u(t) - (t - a)\,u(t - a)$.

505. (*c*) It is a necessary condition for response to be stable.

506. (*b*) **507.** (*b*) **508.** (*b*) **509.** (*a*)

510. (*b*)

511. (*c*) The response is unstable. Therefore pole is on positive real axis.

512. (*c*) **513.** (*a*) **514.** (*a*)

515. (*d*) When $\omega = \infty$, $X_L = \omega L = \infty$.

Therefore $\theta = \tan^{-1}\frac{\omega L}{R} = 90°$.

516. (*d*) **517.** (*c*)

518. (*a*) It is a property of positive real functions.

519. (*a*)

520. (*a*) H(*s*) must possess two more properties if it is positive real.

521. (*d*) Only then it is possible to synthesize by R, L and C elements.

522. (*d*)

523. (*b*) If numerator is one degree higher than denominator *z*(*s*) tends to infinity as s tends to infinity.

524. (*c*) If we divide numerator by denominator we get term *s*L. Therefore, the network has inductance in series.

525. (*b*)

526. (*c*) In such a function the denominator is one degree higher than the numerator. Therefore, we get a term $\frac{1}{sC}$. Hence, the network has capacitance in series.

527. (*c*) Only then the network has only L and C. If poles and zeros are not on $j\omega$ axis, resistance will be there.

528. (*d*) Both X_L and X_C are reactances.

529. (*c*) Impedance of inductance K is *s*K.

530. (*b*) **531.** (*b*) **532.** (*d*) **533.** (*b*)

534. (*c*)

535. (*d*) Impedance of capacitance $\frac{1}{K_0}$ is $\frac{K_0}{s}$.

536. (*c*)

537. (*c*) Admittance of a capacitance K is K*s*.

538. (*a*) **539.** (*a*) **540.** (*b*)

541. (*b*) $\frac{s^2+1}{s} = s + \frac{1}{s}$ which means 1H inductance and 1F capacitor is series.

542. (*a*) $Y(s) = s + \frac{1}{s}$ which means 1H inductance and 1F capacitor in parallel.

543. (*a*) $Z(s) = 1 + \frac{1}{s} = \frac{s+1}{s}$.

544. (*b*) **545.** (*c*) **546.** (*a*) **547.** (*c*)

548. (*c*) **549.** (*b*) **550.** (*c*) **551.** (*a*)

552. (*a*) **553.** (*a*) **554.** (*a*) **555.** (*a*)

556. (*b*) **557.** (*b*) **558.** (*a*)

559. (*b*) RL impedance function and RC admittance have similar forms.

560. (*b*) RL admittance function and RC impedance function have similar forms.

561. (*b*) Poles and zeros on imaginary axis lead to reactance function. A minimum function cannot have poles and zeros on imaginary axis.

562. (*a*) If degrees of numerator and denominator were unequal, a pole or zero can be removed. Such a function cannot be a minimum function. Therefore degrees of numerator and denominator must be equal.

563. (*a*) The procedure is called Brune's synthesis.

564. (*a*) To avoid mutually coupled inductors, more elements are needed.

565. (*b*)

566. (*b*) Since $1 + a + a^2 = 0$, $1 + a = -a^2$.

567. (*d*)

568. (*b*) $1 - a = 1 - 1\angle 120° = 1 + 0.5 - j0.866 = 1.5 - j\,0.866$
$= \sqrt{3}\angle -30°$.

569. (*b*) $1 - a^2 = 1 + 0.5 + j\,0.866 = \sqrt{3}\angle +30°$.

570. (*c*) $1 + a^2 = -a = 1\angle -60°$.

571. (*b*) **572.** (*c*)

573. (*c*) Since it is a balanced system of voltages, Va_2 and Va_0 must be zero.

574. (*c*) 575. (*c*) **576.** (*c*)

577. (*d*) There are a total of 9 components (three for each phase). Moreover, in general, $P = VI^*$.

578. (*c*) Zero sequence component of line-to-line voltage is always zero irrespective of degree of unbalance.

579. (*a*) $I_a = I_{a1} + I_{a2} + I_{a0}$
$= 2\angle -90 + 0.2\angle 90 + 1.8\angle 90°$
$= 2\angle -90° + 2\angle 90° = 0$.

580. (*a*)

581. (*c*) Let $R = 1\Omega$. Then $I_1 = 3\sqrt{2}$ A, $I_2 = 5\sqrt{2}$ A and $I_3 = 7\sqrt{2}$A.
$(I_1 + I_2 + I_3)^2 R = (15/\sqrt{2})^2\ R = 450$ W
$(I_1 + I_2 - I_3)^2\ R = (1/\sqrt{2}) = 2$ W.

582. (*a*) $R_{TH} = \dfrac{100}{5} = 20$ ohm. $I = \dfrac{100}{80 + 20} = 1$A.

583. (*b*)

584. (*d*) Since the frequencies of the two sources are different only superposition theorem can be used.

585. (*b*) One node is reference node. Therefore the system has 9 node pairs.

586. (*c*) Currents in resistance and inductance are out of phase by 90°. Hence $I = \sqrt{12^2 + 16^2} = 20$ A.

587. (*b*) A minimum of 2 wattmeters is required to measure 3 phase power. Of course power can be measured by putting one wattmeter in each phase.

588. (*a*) Power in 6 Ω resistor can be zero only if current in 6 Ω resistor is zero. Therefore, voltages of nodes *a* and *b* are equal $V_a = \dfrac{20\angle 45°}{\sqrt{2}} = V_b$. Since equal resistances of 5 Ω are on two sides of node *b*,
$V = \dfrac{2(20\angle 45°)}{\sqrt{2}} = 20\sqrt{2}\angle 45°$.

589. (*d*) Capacitor is used to improve (increase) p.f. Therefore, line current is decreased for the same active power.

590. (*b*)

591. (*a*) $Z_{input} = \dfrac{AZ_L + B}{CZ_L + D} = \dfrac{10 + 2}{10 + 3} = \dfrac{12}{13}$.

592. (*b*) An impulse has very high rate of rise. Therefore it causes maximum current.

593. (*b*)

594. (*b*) $i = 18\,(1 - e^{t/L}) = 6;\ \dfrac{di}{dt} = 18e^{-t/l} = 6$.
Therefore, L = 2 H.

595. (*b*) RC circuit has impedance of the form $(1 - j)$. Moreover, R, L, C circuit can also have impedance of the form $1 - j$ if $X_C > X_L$.

596. (*c*) **597.** (*c*) **598.** (*c*)

599. (*c*) When ω = 4 rad/sec ωL and $\dfrac{1}{\omega C}$ are equal.

600. (*a*)

601. (*d*) Net resistance of circuit in Fig. (*a*) is 4.571 R and that of circuit in Fig. (*b*) is 0.571 R. Hence, R_0 must be 4R.

602. (*c*)

603. (*b*) Find R_1, R_2, R_3, by delta star conversion.

604. (*d*) When R = 0, circuit current $= \dfrac{V}{30}$ and power dissipated $= \dfrac{V^2}{30}$ watts, which is the maximum possible value.

605. (*b*)

606. (*a*) Average value

$$= \frac{1}{2\pi}\int_{\theta}^{\pi+\theta} V_m \sin \omega t\ d(\omega t) = \frac{V_m}{\pi}\cos\theta.$$

607. (*b*) **608.** (*b*) **609.** (*a*)

610. (*b*) In network having only L and C, poles and zeroes of driving point function are simple and alternate on *j*ω axis.

611. (*d*) A current source with series resistance is not feasible.

612. (*a*) **613.** (*c*) **614.** (*a*)

615. (*c*) The average of sinusoid over 1 cycle is zero.

616. (*c*) **617.** (*c*) **618.** (*a*)

619. (*a*) In steady state inductance is short-circuited and capacitance is open-circuited. Current through 6 Ω resistance = $\frac{1 \times 3}{9}$ and V(*t*) = $\frac{3}{9} \times 6 = 2$ V.

620. (*c*) Poles on $j\omega$ axis lead to sustained oscillations. Poles with positive real part give unbound response and poles with negative real part give bound response.

621. (*c*)

622. (*c*) In realisable functions degrees of numerator and denominator can differ by a maximum of 1.

623. (*b*) In RC functions poles and zeros are interlaced.

624. (*c*) In cascaded networks delay times are added.

625. (*b*) Power = $\frac{1^2}{1} + \frac{\left(\frac{4}{\sqrt{2}}\right)^2}{1} + \frac{\left(\frac{2}{\sqrt{2}}\right)^2}{1} = 11$ W.

626. (*a*)

627. (*b*) $V_{TH} = \frac{10 \times 3}{5} = 6$ V and $R_{TH} = \frac{3 \times 2}{5} = 1.2\,\Omega$.

628. (*a*) If currents of both sources add, total current would be twice and power is four time, *i.e.*, 16 W. If currents subtract, net current and power are zero.

629. (*b*)

630. (*b*) Three tree branches mean 3 loops and hence 3 equations.

631. (*b*) If L = 0.2 mH and C = 200 pF, $\omega_0 = 5 \times 10^6$ rad/sec and if R is 5 kΩ, impedance level is 5 kΩ.

632. (*a*) $\frac{1}{\sqrt{LC}} = 1$ rad/sec. For parallel resonant circuit $Q = \frac{\omega_0 C}{G} = 1$.

633. (*b*) It is a series resonant circuit and peak voltage across R is equal to peak value of source voltage.

634. (*a*) MMF for air = $\frac{B}{4\pi \times 10^{-7}} \times 0.1$

MMF for iron = $\frac{B \times 20}{200(4\pi \times 10^{-7})} = \frac{B \times (0.1)}{4\pi \times 10^{-7}}$.

635. (*a*) $v_u = E(1 - e^{-t/\tau})$ and $v_1 = E e^{-t/\tau}$ or $1 - e^{-t/\tau} = e^{-t/\tau}$. This happens at $t = \tau \ln 2$.

636. (*d*)

637. (*b*) The coil has R and L in series and C in parallel.

638. (*a*) Use superposition theorem.

639. (*c*)

640. (*c*) Convert the inner star into delta.

641. (*b*) Active filters consist of electronic components and have very small size.

642. (*a*)

643. (*d*) Bott Duffin realisation does not use transformer.

644. (*b*) **645.** (*b*) **646.** (*c*)

647. (*b*) $i = 18(1 - e^{-t/L})$ and $\frac{di}{dt} = 18\, e^{-t/L} = 6$. Hence, L = 2 H.

648. (*d*) At $t = 0^+$, $\frac{di}{dt} = 4 = \frac{V}{L} = \frac{V}{20}$.

649. (*c*) **650.** (*c*).

651. (*d*) $\frac{L}{R}$ has units of seconds. Therefore, $\frac{R}{L}$ has units of $\sec^{-1}$ $\frac{1}{\sqrt{LC}}$ has units of rad/sec. Therefore, $\frac{1}{LC}$ has units of rad^2/sec^2, RC is a time constant and has units of sec. $\sqrt{\frac{L}{C}}$ has units of ohms.

652. (*c*) RMS value = $\sqrt{10^2 + \left(\frac{10}{\sqrt{2}}\right)^2} = \sqrt{150}$.

653. (*c*) DC gain, *i.e.*, gain at $s = 0$ equals $\frac{10}{2} = 5$.

654. (*a*) **655.** (*d*)

656. (*c*) Large values of R_i and C mean large time of charging C and hence large duty cycle. Small value of R_2 ensures that duration of pulses is small.

657. (*d*) Steady state current = $\frac{20}{3}$ A.

Therefore, $i = \frac{20}{3}(1 - e^{-3t/L}) = 6$ A

and $\frac{di}{dt} = \frac{20}{3}\left(\frac{3}{L}\right)e^{-3t/L} = 4$ A/s.

From these equations L = 0.5 H.

658. (*c*) At $t = 0$ applied voltage = 3 V and current must be 3 A. At $t = \infty$ source voltage is zero and current must be zero.

659. (*b*) **660.** (*a*)

661. (*a*) Effective resistance is $\frac{2 \times 3}{5} = 1.2\,\Omega$.

Hence, RC = 6 μs.

662. (*c*) $EI \cos\theta = 0.25$ or $I \cos\theta = 0.25$

$Z \cos\theta = 2$ or $\frac{E}{I}\cos\theta = 2$ or $\frac{1}{I}\cos\theta = 2$.

From above equations $\cos\theta = 0.707$ and $\theta = 45°$.

663. (*c*)

664. (*d*) The secondary voltage is equal to primary voltage and the two voltages oppose each other.

665. (*a*) **666.** (*c*) **667.** (*b*)

668. (*a*) Capacitor in shunt gives low pass filter.

669. (*c*) **670.** (*c*)

671. (*b*) Two energy storing elements give s^2 terms.

672. (*a*)

673. (*a*) Use superposition theorem.

674. (*d*) As seen from Z_L, the circuit impedance

$= j2 + \frac{(2 + j2)(-j2)}{2 + j2 - j2} = 2\,\Omega$.

675. (*a*) **676.** (*a*) **677.** (*d*)

678. (*a*) When in parallel each will get full voltage and hence maximum heat.

679. (*b*) **680.** (*c*) **681.** (*d*)

682. (*c*) For the given data $V_1 = 2V_2$ and $I_1 = 0.5\ I_2$.

683. (*b*)

684. (*a*) At $t = 0^-$, $V_c = 0$ and voltage across a capacitor cannot change instantaneously.

685. (*d*) Since response is unbound, poles are in right half plane.

686. (*b*) $R_{cr} = 2\sqrt{\frac{L}{C}} = 2\ \Omega$. Since $R = R_{cr}$, critically damped response.

687. (*c*) Initially capacitor behaves as short-circuit and finally as open circuit.

688. (*d*) Effective resistance of R and R_m is $\frac{R\,R_m}{R + R_m}$.

The reading is

$$\frac{V_s}{R_s + \frac{R}{R + R_m}}\left[\frac{RR_m}{R + R_m}\right] = \frac{V_s\,RR_m}{R_sR + R_sR_m + RR_m}$$

689. (*b*) $M = \frac{12 - 4}{4} = 2$ H; $L_1 = L_2 = L = 4$ H. When in parallel effective inductance $= \frac{L^2 - M^2}{2L - 2M} = \frac{16 - 4}{4} = 3$ H.

690. (*b*) One coefficient is negative. Hence not Hurwitz.

691. (*a*) **692.** (*a*) **693.** (*c*) **694.** (*a*)

695. (*b*) Integral of unit step function is ramp and integral of ramp is doublet.

696. (*d*) RMS value $= \sqrt{10^2 \times 0.5} = \sqrt{50}$.

697. (*a*) Since $f(-t) = -f(t)$ only sine terms exist.

698. (*b*) **699.** (*c*) **700.** (*a*)

701. (*c*) Use final value theorem.

702. (*d*) $Z_0 = \sqrt{Z_{oc}\,Z_{sc}}$.

703. (*d*) Use star-delta conversion.

704. (*d*)

705. (*a*) L in series gives C in parallel of the same value and vice versa.

706. (*d*) **707.** (*a*) **708.** (*a*)

709. (*c*) $V_{av} = \frac{2}{2\pi}\left[\int_{\pi/3}^{\pi} V_m \sin\omega t\ d(\omega t)\right]$

$$= \frac{V_m}{\pi}\left(1 + \cos\frac{\pi}{3}\right) = \frac{3V_m}{2\pi}.$$

710. (*c*)

711. (*c*) If R = 15 Ω resistance, seen by the source

$$= \frac{20 \times 20}{40} + 5 = 15\ \Omega.$$

712. (*a*)

713. (*c*) Effective resistance

$$= \frac{5 \times 3}{5 + 3} = \frac{15}{8}\ \Omega;\ RC = \frac{15}{8} \times 2 = \frac{15}{4}\ \text{sec}.$$

714. (*b*) **715.** (*c*) **716.** (*b*)

717. (*c*) $R_{cr} = 2\sqrt{\frac{L}{C}} = 2\ \Omega$, $\xi = \frac{1}{2} = 0.5$.

718. (*a*) **719.** (*c*) **720.** (*d*) **721.** (*d*)

722. (*d*) **723.** (*b*) **724.** (*b*) **725.** (*c*)

726. (*a*) **727.** (*d*) **728.** (*a*) **729.** (*a*)

730. (*c*) **731.** (*c*)

732. (*b*) In steady state inductance behaves as short-circuit.

733. (*a*) At $t = 0$, $v_c = 0$ and at $t = \infty$, $v_c = 1$. This condition is satisfied only by $1 - e^{-t/RC}$.

734. (*b*) $RC = 10^{-3}$, $i = \frac{100}{1000}e^{-t/10^{-3}} = 0.1\,e^{-1000t}$.

Therefore, $\frac{di}{dt}$ at $t = 0^+$ equals -100 A/s.

735. (*a*) **736.** (*a*)

737. (*b*) The initial current through inductance is $\frac{V}{R_1}$.

738. (*c*) **739.** (*a*) **740.** (*c*) **741.** (*b*)

742. (*a*) **743.** (*d*) **744.** (*a*) **745.** (*b*)

746. (*c*) RMS current $= I = 1\int_0^1 (10\,t)^2\ dt = 5.773$ A,

$$P = I^2R = 333.3\ \text{W}.$$

747. (*b*)

748. (*b*) At frequencies less than resonant frequency current leads voltage.

749. (*a*) $|Z| = \frac{V}{I} = \frac{\text{Peak value of V}}{\text{Peak value of I}}$.

750. (*b*) $\frac{1200 \times 600}{1800} = 400\ \Omega$,

$$R = \frac{35 - 5}{5} \times 400 = 2400\ \text{ohm}.$$

751. (*b*) In an R-L series circuit excited by a dc voltage V, the current is $\frac{V}{R}(1 - e^{-t/T}) = \frac{V}{R} - \frac{V}{R}e^{-t/T}$. In this circuit the final current is $\frac{V}{R}$ and initial current is 0. Thus, the current can be written as $i_f - (i_f - i_i)\,e^{-t/T}$.

752. (*c*) At $t = 0^-$, the circuit can be redraw as in Fig. 4.202 (*a*). At $t = 0^-$, inductance behaves as short-circuit and capacitors as open circuits. Therefore, current supplied by battery is $12/(8+4) = 1$ A. Voltages across capacitors are 8/3 V and 4/3 V.

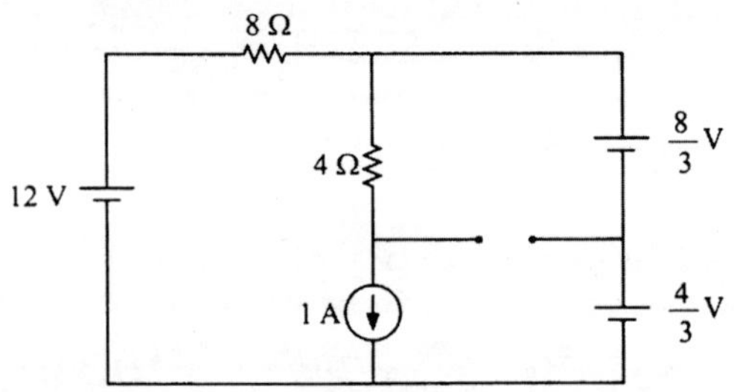

Fig. 4.202 (*a*)

At $t = 0^+$ the circuit can be redrawn as in Fig. 4.202 (*b*).

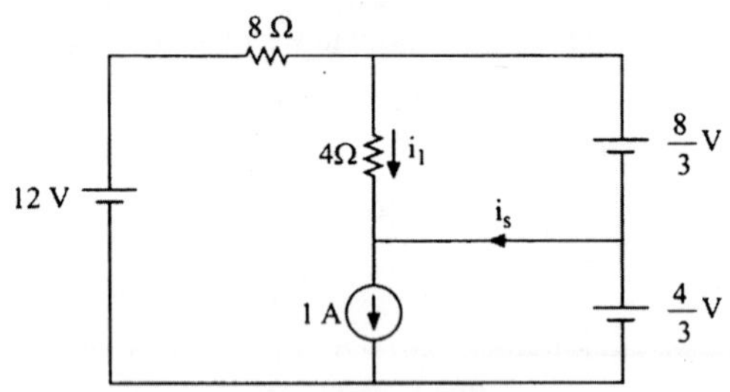

Fig. 4.202 (*b*)

$$i_i = \frac{8}{3}\Big/4 = \frac{2}{3}\text{ A}$$

$$i_i + i_s = 1\text{ A} \quad \text{or} \quad i_s = 1 - \frac{2}{3} = \frac{1}{3}\text{ A}$$

753. (*a*) Current supplied by source $= \dfrac{50}{4 + \dfrac{80 \times 20}{80 \times 20}} = 2.5$ A

Current through 80 Ω resistor $= \dfrac{2.5 \times 20}{80 + 20} = 0.5$ A.

754. (*c*) Load impedance for maximum power transfer should be complex conjugate of source impedance.

Hence $Z_L = 3 + j4\ \Omega$

$$I = \frac{240}{3 - j4 + 3 + j4} = 40\text{ A}$$

$$P = 40^2(3) = 4800\text{ W} = 4.8\text{ kW}.$$

755. (*c*) Due to 5 V source, current in 1 ohm resistor

$$= \sqrt{\frac{1}{1}} = 1\text{A}.$$

Due to 30 A source, current in 1 ohm resistor

$$= \sqrt{\frac{576}{1}} = 24\text{ A}.$$

Total current in 1 ohm resistor = 24 + 1 = 25 A

$P = i^2 R = 25^2(1) = 625$ W.

756. (*c*) Z(*s*) has poles at $s = 0$ and $s = 1$ and zero as $s = -0.5$. It cannot be LC or RLC network. Hence, (*c*).

757. (*b*) $ABCD = \begin{bmatrix} 7 & 8 \\ 3 & 4 \end{bmatrix}\begin{bmatrix} 7 & 8 \\ 3 & 4 \end{bmatrix} = \begin{bmatrix} 73 & 88 \\ 33 & 40 \end{bmatrix}$.

758. (*d*) In RLC parallel network resonance leads to unity power factor and minimum current.

759. (*b*) $\beta^2 > \alpha^2$

or $\dfrac{1}{LC} > \dfrac{R^2}{4L^2}$ or $\dfrac{R^2LC}{4L^2} < 1$ or $\dfrac{R}{2\sqrt{\dfrac{L}{C}}} < 1$

Then $\delta = \dfrac{R}{2\sqrt{\dfrac{L}{C}}} < 1$

Hence, underdamped.

760. (*d*) $|Z| = |3 - j4| = 5\ \Omega$

$$I = \sqrt{\left(\frac{50}{3}\right)^2 + \left(\frac{141.4}{\sqrt{2}\,(5)}\right)^2} = 26\text{ A}$$

Hence (*d*) is the nearest correct value.

761. (*c*) Dual of $v(t)$ is $i(t)$. Therefore, $i(t)$ should appear on LHS.

762. (*d*) $l = b - n + 1 = 12 - 5 + 1 = 8$.

763. (*b*) In cascade connection ABCD matrix of overall network is product of individual ABCD matrices.

764. (*b*) In a rational network, number of zeros and poles is equal. Hence, add zero at infinity.

765. (*a*) At $t = 0^-$ the circuit is as below

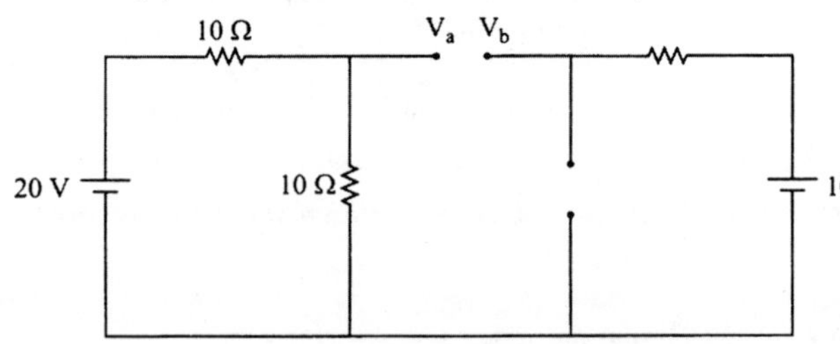

Fig. 4.203.

Capacitor behaves as open circuit at $t = 0^-$

$$V_a = \frac{20 \times 10}{20} = 10\text{ V}; \quad V_b = 10\text{ V}$$

Hence, $i = 0$.

766. (*a*) At balance $f(t) = m\dfrac{d^2x}{dt^2} + B\dfrac{dx}{dt} + kx$.

767. (*a*) $|z| = z_{11}z_{22} - z_{12}z_{21} = \dfrac{1}{35}$

$$y_{11} = \frac{z_{22}}{|z|} = \frac{3 \times 35}{35} = 3$$

$$y_{12} = -\frac{z_{12}}{|z|} = -\frac{2}{35} \times 35 = -2$$

768. (*d*) $\omega_n = \dfrac{1}{\sqrt{LC}} = \left[\dfrac{1}{400 \times 10^{-6} \times 10 \times 10^{-12}}\right]^{0.5} = \dfrac{10^9}{200}$

$= 5 \times 10^6$ rad/sec.

769. (*b*) Norton theorem replaces a network by a current source and Therenin's theorem replaces a network by a voltage source.

770. (*d*) Tree does not have a loop.

771. (*d*) Tree does not have a loop.

772. (*c*) $z_{11}(s) = 10 + \dfrac{1}{\frac{1}{4}s} = 10 + \dfrac{4}{s}$

773. (*d*) Number of trees $= n^{n-2} = 4^{4-2} = 16$

where n is the number of nodes.

774. (*a*) Superposition theorem is not valid for calculating power response directly.

775. (*c*) $IX_L = 2IX_C$ or $X_L = 2X_C$

$$\tan\theta = \frac{X_L - X_C}{R}$$

or $\tan 45° = \dfrac{X_L - X_C}{R}$

or $\dfrac{X_L - X_C}{R} = 1$

or $X_L - X_C = R$

or $2X_C - X_C = R$ or $X_C = 20\,\Omega$

$X_L = 2X_C = 40\,\Omega$

776. (*c*) For reciprocal network $s_{12} = s_{21}$

For lossless reciprocal networks

$|s_{11}| = (s_{22})$

and $|s_{11}|^2 + |s_{12}|^2 = 1$

777. (*a*) Immediately after closing of switch current

$i(t) = i(0^+) = 0.75/2 = 0.375$ A

$$i(\alpha) = \frac{1.5}{3} = 0.5 \text{ A}$$

Looking backward $R = 10 + \dfrac{10}{2} = 15\,\Omega$

$$i(t) = i(\alpha) - [i(\alpha) - i(0^+)]\, e^{-Rt/L}$$

$$= 0.5 - [0.5 - 0.375]\, e^{-15t/15\times10^{-3}}$$

$$= 0.5 - 0.125\, e^{-1000t}$$

778. (*a*) $j\omega L = j1000 \times 20 \times 10^{-3} = +j20$

$$X_C = \frac{1}{1000 \times 50 \times 10^{-6}} = -j20$$

$$Z = j20 + \frac{1(-j20)}{1 - j20}$$

$$= \frac{j20 + 400 - j20}{1 - j20} = \frac{400}{1 - j20}\,\Omega$$

$$i = \frac{20\angle 0}{\frac{400}{1 - j20}}\left[\frac{-j20}{1 - j20}\right]$$

$$= \frac{20\,(1 - j20)\,(-j20)}{400\,(1 - j20)}$$

$= -j1.$

779. (*a*) Applying KVL to outer loop

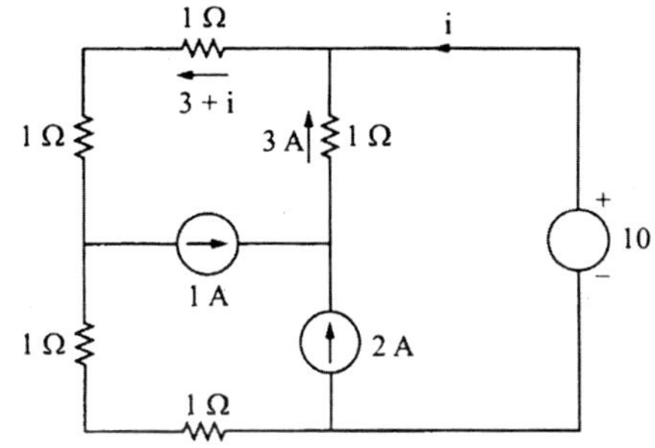

Fig. 4.204.

$$(3 + i)1 + (3 + i)\,1 + (2 + i)\,i + (2 + i) = 10$$

or $10 + 4i = 10$ or $i = 0$

Hence, power = 0.

780. (*a*) Since 60 V source is absorbing power

I + current through 60 V source = 12

or I < 12 A.

781. (*c*) $Z(s) = \dfrac{R_L \times \frac{1}{sC}}{R_L + \frac{1}{sC}} = \dfrac{R_L}{sC}\,\dfrac{sC}{1 + R_L\,sC} = \dfrac{R_L}{1 + R_L\,sC}$

$$\frac{V_o(s)}{V_i(s)} = \frac{Z(s)}{R + Z(s)}$$

$$= \left[\frac{R_L}{1 + R_L sC}\right]\left[\frac{1 + R_L sC}{R + R_L + R_L RsC}\right]$$

or $\dfrac{V_o(s)}{V_i(s)} = \dfrac{R_L}{R + R_L + R_L\,RsC} = \dfrac{1}{2 + sCR}$

Therefore $R = R_L$.

782. (*c*) $I = \dfrac{20}{1 + 2j + 7 + 4j} = \dfrac{20}{8 + 6j} = 2\angle -36.87°$

Reactive power = VI sin θ = 20 × 2 sin (–36.87°)

= 24 Var

783. (*b*) Equivalent capacitance $= \dfrac{0.2 \times 0.8}{1} = 0.16\ \mu F$

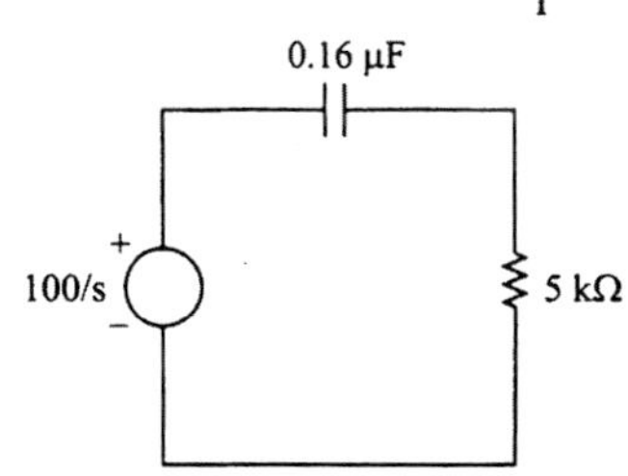

Fig. 4.205.

initial voltage across capacitor = 100 V

$RC = 0.16 \times 10^{-6} \times 5 \times 10^3 = 0.8 \times 10^{-3}$

$$i(t) = \frac{100}{5000}\, e^{-t/RC}$$

$$= \frac{1}{50}\, e^{-t/0.8\times10^{-3}} \text{ A}$$

$$= 20\, e^{-1250t}\, u(t) \text{ mA}$$

784. (*a*)

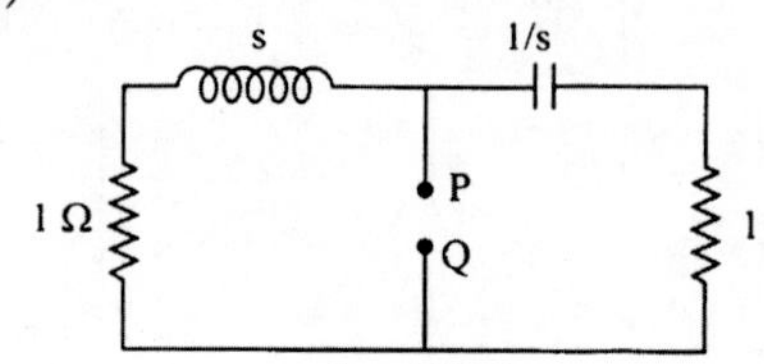

Fig. 4.206.

$$Z_{TH} = (1+s) \,||\, \left(1+\frac{1}{s}\right) = \frac{(1+s)\left(1+\frac{1}{s}\right)}{1+s+1+\frac{1}{s}} = 1$$

785. (*d*) $Y(s) = \dfrac{s^2+0.1s+2}{0.2s} = 5s + 0.5 + \dfrac{10}{s}$

Comparing with $Cs + \dfrac{1}{R} + \dfrac{1}{Ls}$

$C = 5\text{ F}, R = 2\,\Omega, L = 0.1\text{ H}.$

786. (*d*)

$$V_0(s) = \frac{\frac{1}{s}}{s+1+\frac{1}{s}} = \frac{1}{s^2+s+1}$$

$$= \frac{1}{(s+0.5)^2 + \left(\frac{\sqrt{3}}{2}\right)^2}$$

$$v_0(t) = \frac{2}{\sqrt{3}} e^{-0.5t} \sin\left(\frac{\sqrt{3}t}{2}\right).$$

787. (*c*) $V_0(s) = \dfrac{1}{s+1+\frac{1}{s}} = \dfrac{s}{s^2+s+1}$

Taking inverse laplace transform

$$v_0(t) = e^{-0.5t}\left[\cos\frac{\sqrt{3}t}{2} - \frac{1}{\sqrt{3}} \sin\frac{\sqrt{3}t}{2}\right].$$

788. (*c*) $V_1 = z_{11} I_1 + z_{12} I_2$

$V_2 = z_{21} I_1 + z_{22} I_2$

$$z_{11} = \left|\frac{V_1}{I_1}\right|_{I_2=0} = \frac{6}{4} = 1.5$$

$$z_{12} = \left|\frac{V_1}{I_1}\right|_{I_1=0} = \frac{4.5}{1} = 4.5$$

$$z_{21} = \left|\frac{V_2}{I_1}\right|_{I_2=0} = 1.5$$

$$z_{22} = \left|\frac{V_2}{I_2}\right|_{I_1=0} = 1.5$$

Hence, $[z] = \begin{bmatrix} 1.5 & 4.5 \\ 1.5 & 1.5 \end{bmatrix}.$

789. (*a*) $V_1 = h_{11} I_1 + h_{12} V_2$

$I_2 = h_{21} I_1 + h_{22} V_2$

$$h_{12} = \left.\frac{V_1}{V_2}\right|_{I_1=0} = \frac{4.5}{1.5} = 3$$

$$h_{22} = \left|\frac{I_2}{V_2}\right|_{I_1=0} = \frac{1}{1.5} = 0.67$$

$$h_{11} = \frac{|z|}{z_{22}}$$

$$= \frac{z_{11}z_{22} - z_{12}z_{21}}{z_{22}} = \frac{1.5\times1.5 - 4.5\times1.5}{1.5}$$

$$= -\frac{4.5}{1.5} = -3$$

$$h_{21} = \frac{-z_{21}}{z_{22}} = \frac{-1.5}{1.5} = -1.$$

790. (*d*) 3 dB bandwidth $= \dfrac{R}{L}$

Then $B_1 = \dfrac{R}{L_1}$

$$B_2 = \frac{R}{L_1/4} = \frac{4R}{L_1}$$

$$\frac{B_1}{B_2} = \frac{R}{L_1}\times\frac{L_1}{4R} = \frac{1}{4}.$$

791. (*d*) Applying KCL at node X

$$2 = \frac{V_{xy}}{2} + \frac{V_{xy}}{1} + \frac{V_{xy} - 2I}{1}$$

where $I = \dfrac{V_{xy}}{1}$

or $2 = \dfrac{V_{xy}}{2} + \dfrac{V_{xy}}{1} + \dfrac{V_{xy}}{1} - \dfrac{2V_{xy}}{1}$

or $V_{xy} = V_{TH} = 4\text{ V}$

If we short-circuit xy, $I_{sC} = 2$ A

$$R_{TH} = \frac{V_{TH}}{2} = 2\,\Omega.$$

792. (*a*) At $t = 0^+$, capacitor behaves as short-circuit

$$i_c(0^+) = \frac{10}{20\times10^3} = 0.5\text{ mA}$$

At $t = \infty$, $i_c = 0$

Time constant

$$T = \frac{20000\times20000}{40000} = 4\times10^{-6} = 40\times10^{-3}s$$

$$i_c(t) = i_c(\infty) - [i_c(\infty) - i_c(0^+)]\,e^{-t/T}$$

$$= 0 - (0 - 0.5)\,e^{-t/40\times10^{-3}}$$

$$= +0.5e^{-25t}\text{ mA}$$

793. (*d*) $Z = \dfrac{(5-j3)(5+j3)}{5-j3+5+j3} = \dfrac{25+j15-j15s+9}{10} = 3.4$

$V_{AB} = (5\angle30)(3.4) = 17\angle30°$

794. (*c*) At $\omega = 0$ C_1 and C_2 are O.C.

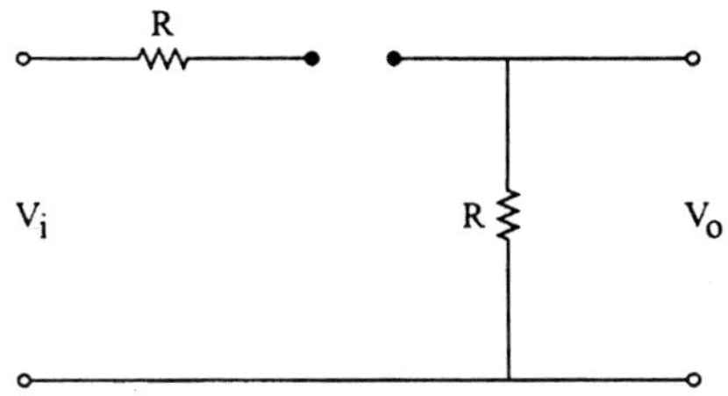

Fig. 4.207.

and $V_0 = 0$

At $\omega = \infty$, C_1 and C_2 are short-circuited.

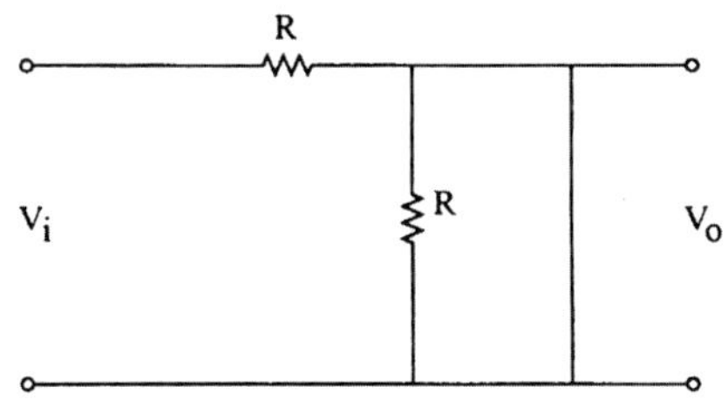

Fig. 4.208.

and $V_0 = 0$

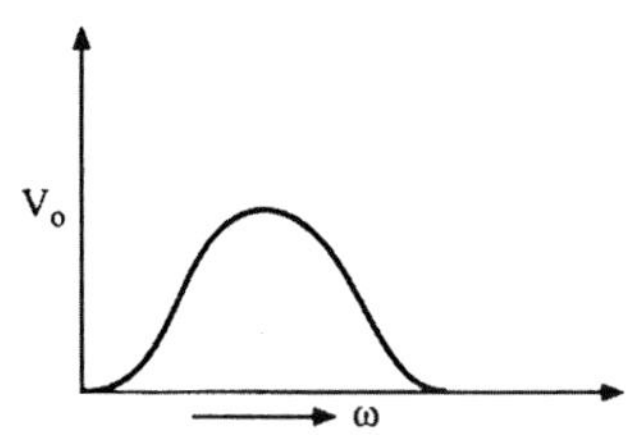

Fig. 4.209.

Therefore, band pass filter because response is as shown.

795. (*d*) $V_1 = AV_2 - BI_2$

$I_1 = CV_2 - D\,I_2$

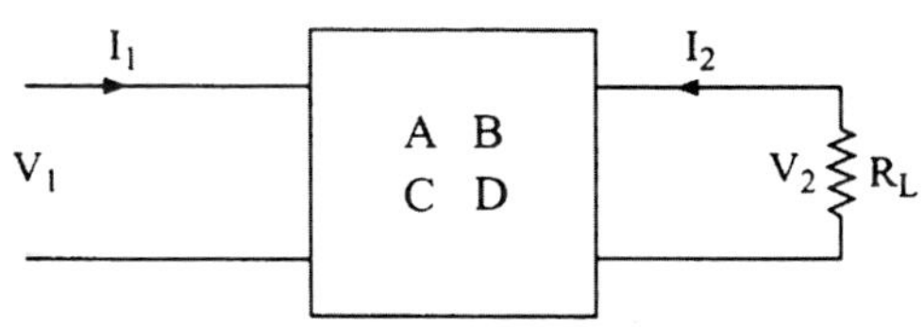

Fig. 4.210.

$$V_2 = -I_2 R_L$$

Then $V_1 = -A\,I_2 R_L - BI_2$

$$Z_1 = -C\,I_2 R_L - DI_2$$

$$Z_{11} = \frac{V_1}{I_1} = \frac{AR_L + B}{CR_L + D}$$

796. (*b*) $V_1 = z_{11} I_1 + z_{12} I_2$

$V_2 = z_{21} I_1 + z_{22} I_2$

$$z_{12} = \left.\frac{V_1}{I_2}\right|_{I_1=0}$$

when $I_1 = 0$, $V_1 = 0$. Hence, $z_{12} = 0$

$$z_{21} = \left.\frac{V_1}{I_1}\right|_{I_2=0}$$

when $I_2 = 0$, $V_2 = -\beta I_1 r_0$. Hence, $Z_{21} = -\beta r_0$.

797. (*a*) $v = \dfrac{Ldi}{dt}$

$$V(s) = sL\,I(s) - LI(0)$$

$$LI(0) = 1 \times 10^{-3}\,V$$

$$I(0) = \frac{1\times10^{-3}}{2\times10^{-3}} = 0.5\,A$$

798. (*b*) Finally in steady state, all capacitors behave as O.C.

$$v_c(\infty) = \frac{4}{1+1}\times 10 = 8\,V$$

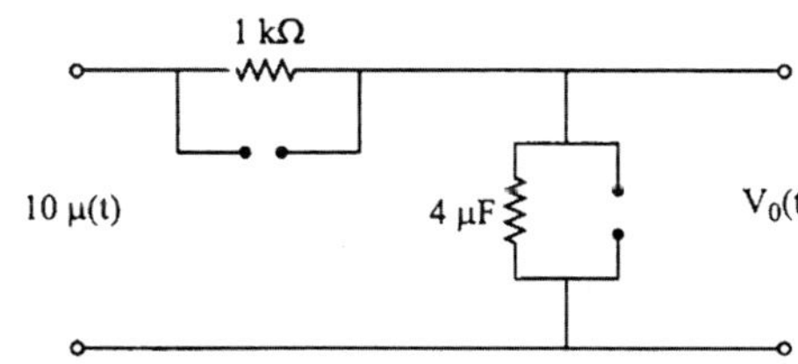

Fig. 4.211.

Initially all capacitors behave as s.c. therefore, $sv_c(0) = 0$

$$C_{eq} = 4 + 1 = 5\,\mu F = 5 \times 10^{-6}\,F$$

$$R_{eq} = \frac{1\times 4}{1+4} = \frac{4}{5}\,k\Omega = \frac{4000}{5}\,\Omega$$

$$T = RC$$

$$= \frac{4000}{5}\times 5\times 10^{-6}$$

$$= 4\times 10^{-3}\text{ secs} = 0.004s$$

$$v_c(t) = v_c(\infty) - [v_c(\infty) - v_c(0)]e^{-t/T}$$

$$= 8 - (8-0)\,e^{-t/0.004}$$

$$= 8\,[1 - e^{-t/0.004}]\,V.$$

799. (*a*) If $Z_2(s)$ is positive real, total resistance should be positive. Thus real power of $Z_1(s)$ should be more than R_{neg} or $R_{neg} \leq \text{Re}\,Z_1(j\omega)$.

800. (*c*) Since $R_1 = R_2 = R$, $V_a = 5\,V$

$$V_b = \frac{R_4}{R_4 + R_3}(10)$$

$$= \frac{1.1\,R}{1.1\,R + R}(10) = \frac{1.1}{2.1}(10) = 5.238\,V$$

$$V_a - V_b = -0.238\,V$$

801. (*c*) A is statement of Tellegen's theorem while R is wrong.

802. (*a*) **803.** (*b*) **804.** (*a*)

805. (*d*) A is wrong because static V-I characteristics do not identify the elements.

806. (*a*) **807.** (*a*)

808. (*a*) Since *p f* is unity, impedance is minimum.

809. (*d*) The degrees can differ by a maximum of 1.

810. (*d*)

811. (*a*) Convolution intepral can be used to find the response for any arbitrary input interms of impulse response.

812. (*b*) Both are correct but independent of each other.

813. (*b*)

814. (*d*) Sinusoidal sources of different frequencies cannot be used for a rectifier circuit.

815. (*b*)

816. (*d*) A is not positive real.

817. (*a*) If degrees of numerator and denominator differ by more than 1, the function is not positive real.

818. (*a*) **819.** (*a*) **820.** (*a*) **821.** (*b*)

822. (*a*)

823. (*a*) Because of inertia the current cannot change instantaneously. Therefore, it behaves as O.C. at $t = 0$.

824. (*b*) **825.** (*a*)

826. (*a*) Thevenin's circuit uses a voltage source and Norton's circuit uses a current source.

827. (*c*) Maximum power transfer theorem is applicable to ac circuits also.

828. (*a*)

829. (*a*) In principle of duality L is dual of C and I is dual of V.

830. (*a*) $\frac{d\text{ Energy}}{dt} = \text{VI} = \text{Power.}$

831. (*c*) The number of links has no relation to number of tree branches.

832. (*b*) Both are correct but independent.

833. (*c*)

834. (*d*) Tie set schedule is used in loop analysis.

835. (*b*)

836. (*a*) Since B-H curve is non-linear, the inductance of iron cored solenoid is not constant.

837. (*b*) **838.** (*c*)

839. (*c*) Integral of unit impulse gives unit step function.

840. (*a*) A high value of time constant results in slow rise of current.

841. (*d*) The current through inductance is zero at $t = 0$ and equal to I at $t = \infty$.

842. (*a*)

843. (*d*) The *s* domain equivalent depends on initial conditions.

844. (*b*) Both are correct but independent.

845. (*b*) **846.** (*a*) **847.** (*a*)

848. (*d*) $j = 1\angle 90°$

849. (*c*) In a phasor diagram all phasors must be of the same frequency.

850. (*a*)

851. (*d*) The current may lead or lag the voltage depending on values of L and C.

852. (*d*) Reactive power cannot be more than apparent power.

853. (*b*) **854.** (*b*) **855.** (*b*) **856.** (*a*)

857. (*a*) **858.** (*d*)

859. (*d*) Power factor = $\frac{\text{Average power}}{\text{Apparent power}}$ but magnitude of power factor is always less than unity.

860. (*a*) **861.** (*c*) **862.** (*a*)

863. (*c*) Law of conservation of charge is basis of KCL not KVL.

864. (*a*)

865. (*d*) Load impedance must be conjugate of source impedance.

866. (*c*) At series resonance power factor is unity.

867. (*b*)

868. (*c*) The response at $t = 0^+$ is sum of free response and forced response.

869. (*c*) $\omega_0 = \frac{1}{\sqrt{LC}}\left[\frac{R_1^2 - \frac{L}{C}}{R_2^2 - \frac{L}{C}}\right]^{0.5}$

If $R_1^2 = R_2^2 = L/C$, resonance occurs as all frequencies.

870. (*c*) At resonance voltage and current are in phase and power factor is unity.

871. (T) **872.** (T)

873. (T) Energy = voltage × current × time = voltage × charge.

874. (F) Magnetic effect of current is used in both generators and motors.

875. (T) Threshold of perception means current which causes small sensation in body.

876. (F) A non-linear element has a resistance which depends on the magnitude of current.

877. (T) $V_{AB} = V_A - V_B = 10 - (-15) = 5$ V.

878. (F) **879.** (T) **880.** (T) **881.** (F)

882. (T) **883.** (T)

884. (T) Calories and joules are related by mechanical equivalent of heat which is 4.186.

885. (T)

886. (F) Temperature coefficient of thermistor is negative.

887. (T) **888.** (F)

889. (F) Bus bars are used for heavy as well as light currents.

890. (T) **891.** (T) **892.** (T)

893. (F) It has four bands.

894. (F) The *v-i* graph, in varistor, is non-linear.

895. (T)

896. (F) Resistivity of semiconductors is about 1 ohm-m.

897. (F) As temperature increases the temperature coefficient of metal decreases.

898. (T) **899.** (T) **900.** (T)

901. (T) It is used in lightning arresters due to its non-linear resistance property.

902. (F) The number of categories is seven.

903. (T) $v = \frac{q}{C} = \frac{1}{C}\int_0^t i\,dt$

904. (F) Insulation resistance is inversely proportional to length.

905. (T) $C \propto \epsilon$.

906. (F) ϵ_r for air is 1.

907. (T) **908.** (T) **909.** (T) **910.** (T)

911. (F) All dielectrics have some leakage current.

912. (F) For class C insulations temperature limit is 180°C.

913. (T) **914.** (T)

915. (T) energy $= \frac{1}{2}CV^2$ or $C = \frac{2\text{ energy}}{V^2}$.

916. (T) **917.** (F) **918.** (T) **919.** (T)

920. (F) At high frequencies the reactances of stray capacitance is small. Therefore, stray capacitance assumes importance at high frequencies.

921. (F)

922. (T) To avoid circulating current cells in parallel should be identical.

923. (F) The electrolyte is moist.

924. (F) Zinc can serves as negative terminal.

925. (T) **926.** (T)

927. (F) The positive plate has lead peroxide.

928. (F) They are similar in discharged condition.

929. (F) During discharging specific gravity decreases.

930. (T) **931.** (T)

933. (F) In ideal current source, the source resistance is infinite.

933. (T)

934. (F) It may be written as positive or negative as per convenience.

935. (F) KCL is restatement of principle of conservation of charge.

936. (F) It depends on direction of current.

937. (T)

938. (F) All current sources are open-circuited.

939. (F) **940.** (T) **941.** (T)

942. (F) B is lagging A by 45°.

943. (T) The average value over full cycle is zero.

944. (T) **945.** (F)

946. (F) These terms are applicable to all waveshapes.

947. (T) **948.** (T) **949.** (T) **950.** (T)

951. (T) **952.** (F) **953.** (T)

954. (T) Only directions are opposite.

955. (T) **956.** (F) **957.** (T)

958. (F) $Z = \sqrt{R^2 + \frac{1}{\omega^2 C^2}}$.

959. (F) The quadrants are different.

960. (F) Apparent power has units of volt-amperes.

961. (F) It lies in fourth quadrant.

962. (F) If $X_c < X_L$ power factor is lagging.

963. (T) **964.** (T) **965.** (T) **966.** (T)

967. (T)

968. (T) I = YV. Since I leads V, phase angle of Y must be positive.

969. (T) **970.** (T) **971.** (T)

972. (F) When $f < f_r$, $X_C > X_L$ and p.f. is leading.

973. (T)

974. (F) In parallel resonance current is minimum.

975. (F) $\omega_2 - \omega_1 = \frac{R}{L}$.

976. (T) **977.** (T) **978.** (T) **979.** (T)

980. (T) **981.** (T)

982. (F) It is more convenient to use transmission parameters.

983. (F) **984.** (T)

985. (F) $\beta = -\pi$.

986. (T)

987. (F) Attenuation is infinite at $f = 0$ for high pass filter.

988. (T) **989.** (T)

990. (F) It has parallel resonant circuit in series arms.

991. (F) The voltage source is ineffective.

992. (T) At $t = \infty$ inductance behaves as short-circuit.

993. (F) **994.** (T) **995.** (T) **996.** (T)

997. (T) **998.** (T) **999.** (T) **1000.** (T)

1001. (T) **1002.** (T) **1003.** (T) **1004.** (T)

1005. (T)

1006. (F) It is not realizable.

1007. (F) It is realizable because $|\log|H(j\omega)| = \omega^2$.

1008. (T) **1009.** (T) **1010.** (T) **1011.** Zero

1012. $\sqrt{3}\angle 30°$

1013. Product, transposes, reverse.

1014. zero **1015.** zero

1016. voltage **1017.** electrostatic

1018. voltage **1019.** capacitance

1020. charge **1021.** Richard's

1022. equal to
1023. Brune
1024. admittance
1025. Capacitor
1026. resistor
1027. alternate, negative
1028. pole, zero
1029. simple, imaginary
1030. real, positive
1031. non-negative
1032. positive real
1033. real, conjugate pair
1034. zero, one
1035. negative, zero
1036. capacitor
1037. canonic
1038. negative
1039. reactance
1040. positive
1041. independent
1042. $0 < m < 1$
1043. parallel tuned, series tuned
1044. π
1045. $\cosh \gamma$
1046. $\sinh \gamma / Z_0$
1047. A/C
1048. $(AB/CD)^{0.5}$
1049. $z_{12} = z_{21}$
1050. $AD - BC = 1$
1051. $\cos\left[\tan^{-1}\frac{\sqrt{3}(W_2 - W_1)}{(W_1 + W_2)}\right]$
1052. equal
1053. 0.5
1054. neutral shift
1055. $2\pi\delta(\omega)$
1056. $Ad\frac{\sin(\omega d/2)}{(\omega d/2)}$
1057. $F^*(j\omega)$
1058. $\sum_{n=-\infty}^{\infty} F_n e^{jn\omega_j t}$
1059. sine
1060. $\left[I_0^2 + \frac{I_{1m}^{\ 2}}{2}\right]^{0.5}$
1061. 20/3%
1062. $\dot{X} = AX + Bu$
1063. $\sum_{k=1}^{b} v_k i_k = 0$
1064. Current source
1065. equal

CHAPTER

5

ELECTROMAGNETIC THEORY

5.1. VECTOR ANALYSIS

Rectangular coordinates

$$A = \hat{x}A_x + \hat{y}A_y + \hat{z}A_z$$

$$\hat{x} = \text{unit vector in } x \text{ direction} = \frac{A_x}{|A_x|}$$

$$\hat{y} = \text{unit vector in } y \text{ direction} = \frac{A_y}{|A_y|}$$

$$\hat{z} = \text{unit vector in } z \text{ direction} = \frac{A_z}{|A_z|}$$

$$|A| = \sqrt{A_x^2 + A_y^2 + A_z^2}$$

Scalar or Dot Product $A.B = |A||B|\cos\theta$

If A and B are at right angle $A.B = 0$

Line integral of force F along a path from P_1 to P_2 is

$$W = \int_{P_1(\text{start})}^{P_2(\text{end})} F.dl$$

Surface integral $\psi = \iint_{\text{Area A}} B.\hat{n}\, ds$

where $\hat{n}$ is unit normal to surface area ds

volume intgeral $M = \iiint \rho\, dv$

Vector or Cross Product $A \times B = \hat{n} AB \sin\theta$

where $\hat{n}$ is unit vector normal to plane containing A and B

Conversion between rectangular and polar coordinates

$$x = r\cos\theta$$

$$y = r\sin\phi$$

$$r = (x^2 + y^2)^{0.5} \text{ and } \phi = \tan^{-1}\left(\frac{y}{x}\right).$$

Conversion from rectangular to cylindrical coordinates

$$A_r = A_x \frac{x}{\sqrt{x^2 + y^2}} + A_y \frac{y}{\sqrt{x^2 + y^2}}$$

$$A_\phi = -A_x \frac{y}{\sqrt{x^2 + y^2}} + A_y \frac{x}{\sqrt{x^2 + y^2}}$$

Conversion from Cylindrical to Rectangular Coordinates

$$A_x = A_r \cos\phi - A_\phi \sin\phi$$

$$A_\phi = A_r \sin\phi + A_\phi \cos\phi$$

$$A_z = A_z$$

Conversion from Rectangular to Spherical Coordinates

$$A_r = A_x \frac{x}{\sqrt{x^2 + y^2 + z^2}} + A_y \frac{y}{\sqrt{x^2 + y^2 + z^2}} + A_z \frac{z}{\sqrt{x^2 + y^2 + z^2}}$$

$$A_\phi = A_x \frac{xz}{\sqrt{(x^2 + y^2)}\sqrt{(x^2 + y^2 + z^2)}}$$

$$+ A_y \frac{yz}{\sqrt{(x^2 + y^2)}\sqrt{(x^2 + y^2 + z^2)}} - A_z \frac{\sqrt{x^2 + y^2}}{\sqrt{(x^2 + y^2 + z^2)}}$$

$$A_\phi = -A_x \frac{y}{\sqrt{x^2 + y^2}} + A_y \frac{x}{\sqrt{x^2 + y^2}}$$

Conversion from Spherical to Rectangular Coordinates

$$A_x = A_r \sin\theta\cos\phi + A_\theta \cos\theta\cos\phi - A_\phi \sin\phi$$

$$A_y = A_r \sin\theta\sin\phi + A_\theta \cos\theta\sin\phi + A_\phi \cos\phi$$

$$A_z = A_r \cos\theta - A_\theta \sin\theta$$

5.2. ELECTROSTATICS

Coulomb's Law

$$F = \hat{r}\frac{Q_1 Q_2}{4\pi\in r^2}$$

where $\hat{r}$ is a unit vector pointing in the direction of line joining the charges Q_1 and Q_2

Electric field intensity E due to a charge

$$Q_1 = \frac{\hat{r}Q_1}{4\pi\in r^2}$$

Superposition of Electric Fields: The total electric field at a point, due to a number of charges, is the vector sum of component fields at that point.

Electric potential due to a point charge

$$Q = \frac{Q}{4\pi\in r}$$

Relation between Electric field intensity E and potential V

$$E = -\nabla V, \textit{ i.e.}, -\text{gradient V}$$

$$V = -\int E \cdot dl$$

Line integral of a static electric field around any closed path is zero, *i.e.*, $\oint E \cdot dl = 0$. Such a field is called conservative field or lamellar field. Potential difference between any two points of a conservative field is independent of path.

Electric field intensity and Potential due to distributed charges

Line charge $E = \int_c \frac{\rho_l \, dl}{4\pi \in r^2} \hat{r}$ and $V = \int_c \frac{\rho_l \, dl}{4\pi \in r}$

Surface charge $E = \iint_s \frac{\rho_s \, ds}{4\pi \in r^2} \hat{r}$ and $V = \iint_s \frac{\rho_s \, ds}{4\pi \in r}$

Volume charge $E = \iiint_v \frac{\rho_v \, dv}{4\pi \in r^2} \hat{r}$ and $V = \iiint_v \frac{\rho_v \, dv}{4\pi \in r}$

If n point charges, line charge, surface charge and volume charge are simultaneously present

$$V = \frac{1}{4\pi \in} \left[\sum_1^n \frac{Q_n}{r_n} + \int_c \frac{\rho_l \, dl}{r} + \iint_s \frac{\rho_s \, ds}{r} + \iiint_v \frac{\rho_v \, dv}{r} \right]$$

Gauss's Law

Electric flux through any surface equals the surface integral of flux density over the surface, *i.e.*,

$$\psi = \iint D \cdot ds$$

Motion of particle with charge e and mass m in electric field having potential difference V

$$v = \sqrt{\frac{2\,eV}{m}}$$

Boundary Relations for Static Electric Fields

Field Component	***Boundary relation***	***Condition***
Tangential	$E_{t1} = E_{t2}$	Any two media
Tangential	$E_{t1} = 0$	Medium 1 is a dielectric, medium 2 is a conductor
Normal	$D_{n1} - D_{n2} = \rho_s$	Any two media with charge at boundary
Normal	$D_{n1} = D_{n2}$	Any two media with no charge at boundary
Normal	$D_{nl} = \rho_s$	Medium 1 is dielectric Medium 2 is a conductor with surface charge

5.3. MAXWELL'S DIVERGENCE EQUATION FROM GAUSS'S LAW

$$\nabla \cdot D = \rho$$

In a region free from charge $\rho = 0$ and

$$\nabla \cdot D = 0$$

Divergence Theorem

$$\oint_s \nabla \cdot ds = \oint_v \nabla \cdot D \, dv$$

Poisson's equation $\nabla^2 V = -\frac{\rho}{\in}$

If $\rho = 0$, Poisson's equation reduces to Laplace equation, *i.e.*,

$$\nabla^2 V = 0$$

Electric current in terms of charge density ρ, area A and drift velcoity v_d

$$I = v_d \rho A$$

Current density $J = \frac{I}{A} = v_d \rho$

If current density is non-uniform

$$J = \underset{\Delta s \to 0}{\text{Lt}} \frac{\Delta I}{\Delta s}$$

Ohm's law at a point $J = \sigma E$

Continuity relation between current density J and charge density ρ at a point

$$\nabla \cdot J = \frac{\partial \rho}{\partial t}$$

For steady currents $\frac{\partial \rho}{\partial t} = 0$ and $\nabla \cdot J = 0$

5.4. ELECTROMAGNETISM

Force F on a wire with current I in magnetic field B

$$F = (I \times B)L$$

Biot Savart Law $dB = \frac{\mu}{4\pi} \frac{Idl \sin \theta}{r^2}$

Force between two parallel linear conductors

$$F = \frac{\mu_0 I_1 I_2}{2\pi R} L$$

Magnetic flux over any surface $\psi_m = \iint_s B \cdot ds$

Magnetic flux over closed surface

(Gauss's law in integral form)

$$\iint_x B \cdot ds = 0$$

Gauss's law in differential form $\nabla \cdot B = 0$

Torque on a loop $T = m \times B = IAB \sin \theta$

where m is magnetic moment

Ampere's Law $\oint Hdl = I = \iint_s J \cdot ds$

i.e., Line integral of H around a single closed path is equal to current enclosed.

$$\text{curl } H = J$$

or $\nabla \times H = J$

Magnetic vector potential due to a current distribution

$$A = \frac{\mu}{4\pi} \iiint \frac{J}{r} dv \quad B = \nabla \times A$$

Boundary Relations for Magnetic Fields

Field Component	*Boundary relation*	*Condition*
Normal	$B_{n1} = B_{n2}$	Any two media
Normal	$\mu_{r1} H_{n1} = \mu_{r2} H_{n2}$	Any two media
Tangential	$\mu_{t1} - H_{t2} = K$	Any two media with thin current sheet at boundary
Tangential	$H_{t1} = H_{t2}$	Any two media with no current sheet at boundary
Tangential	$H_{t1} = K_2$	$H_{t2} = 0$, also medium 2 has a thin current sheet at boundary, H_{t1} and K_2 are normal to each other

Energy in a magnet $W_m = \int_0^B H dB$

Reluctance of magnetic circuit $= \dfrac{NI}{\iint B \cdot ds} = \dfrac{MMF}{Flux}$

Permeance $= \dfrac{1}{\text{Reluctance}}$

Energy stored in a gap $W_m = \dfrac{B^2 Ag}{2\mu_0}$

where A = gap area and g is width of gap

Comparison of equations involving polarization P and magnetisation M

Description of Equation	*For electric fields*	*For magnetic fields*
Dipole moment	$P = \dfrac{QL}{v}$	$M = \dfrac{Q_m L}{v}$
Flux density	$D = \epsilon_0 E + P$	$B = \mu_0 (H + M)$
Permittivity and Permeability	$\epsilon = \epsilon_0 + \dfrac{P}{E}$	$\mu = \mu_0 \left(1 + \dfrac{M}{H}\right)$
Relative permittivity and relative permeability	$\epsilon_r = 1 + \dfrac{P}{\epsilon_0 E}$	$\mu_r = 1 + \dfrac{M}{H}$
Susceptability	$\psi = \epsilon_r - 1$	$\psi_m = \mu_r - 1$
Relation to polarization charge density and to pole	$\nabla \cdot P = -\rho_p$	$\nabla \cdot M = -\rho_m$
density and equivalent current density	$\nabla \cdot P = \rho$	$\nabla \times M = J'$
Relation to surface charges and poles and equivalent current sheet	$P = \rho_{sp}$ $D = \rho_s$	$M = \rho_m$ $M = K'$
Poisson's equations	$\nabla^2 V = -\dfrac{\rho}{\epsilon}$	$\nabla^2 u = \nabla \cdot M = -\nabla \cdot H$
Scalar and vector potentials	$V = \dfrac{1}{4\pi\epsilon_0} \iiint \dfrac{\rho - \nabla . P}{r} dv$	$A = \dfrac{\mu_0}{4\pi} \iiint \dfrac{J + \nabla \times M}{r} dv$

Laplace Equation in Rectangular Coordinate

$$\frac{\partial^2 V}{\partial x^2} + \frac{\partial^2 V}{\partial y^2} + \frac{\partial^2 V}{\partial z^2} = 0$$

Laplace Equation in Cylindrical Cordinates (r, ϕ, z)

$$\frac{\partial^2 V}{\partial r^2} + \frac{1}{r}\frac{\partial V}{\partial r} + \frac{1}{r^2}\frac{\partial^2 V}{\partial \phi^2} + \frac{\partial^2 V}{\partial z^2} = 0$$

Laplace Equation in Spherical Coordinates (r, θ, ϕ)

$$\frac{\partial^2 V}{\partial r^2} + \frac{1}{r^2}\frac{\partial^2 V}{\partial \theta^2} + \frac{1}{r^2 \sin^2\theta}\frac{\partial^2 V}{\partial \theta^2} + \frac{2}{r}\frac{\partial V}{\partial r} + \frac{\cot\theta}{r^2}\frac{\partial V}{\partial \theta} = 0$$

Faraday's Law

General case of induction

Total induced emf $e = \oint_l (v \times B) \cdot dl - \iint \frac{\partial B}{\partial t} \cdot ds$

Motional — Time change

Motional emf $e = \oint_l (v \times B) \cdot dl$

Transformer emf $e = -\iint \frac{\partial B}{\partial t} \cdot ds$

Stokes Theorem $\oint_l E \cdot dl = \iint (\nabla \times E) \cdot ds$

5.5. MAXWELL'S EQUATIONS IN INTEGRAL FORM

Case/Quantity	*From Ampere mmf, Amperes*	*From Faraday emf, volts*	*From Gauss*	
			Electric flux, coulombs	*Magnetic flux webers*
General	$F = \oint H \cdot dl = \int_s \left(J + \frac{\partial D}{\partial t}\right) \cdot ds = I_{total}$	$emf = \oint E \cdot dl = -\int_s \frac{\partial B}{\partial t} \cdot ds$	$\psi = \oint_s D \cdot ds = \int_v \rho dv$	$\psi_m = \oint_s B \cdot ds = 0$
Free space	$F = \oint H \cdot dl = \int_s \frac{\partial D}{\partial t} \cdot ds = I_{disp}$	$emf = \oint E \cdot dl = -\int_s \frac{\partial B}{\partial t} \cdot ds$	$\psi = \oint_s D \cdot ds = 0$	$\psi_m = \oint_s B \cdot ds = 0$
Harmonic variation	$F = \oint H \cdot dl = (\sigma + j\omega\epsilon) \int_s E \cdot ds = I_{total}$	$emf = \oint E \cdot dl - j\omega\mu \int_s H.ds$	$\psi = \oint_s D \cdot ds = \int_v \rho dv$	$\psi_m = \oint_s B \cdot ds = 0$
Steady	$F = \oint H \cdot dl = \int_s J \cdot ds = I_{cond}$	$emf = \oint E \cdot dl = 0$	$\psi = \oint_s D \cdot ds = \int_v \rho dv$	$\psi_m = \oint_s B \cdot ds = 0$
Static	$U = \oint H \cdot dl = 0$	$emf = \oint E \cdot dl = 0$	$\psi = \oint_s D \cdot ds = \int_v \rho dv$	$\psi_m = \oint_s B \cdot ds = 0$

5.6. MAXWELL'S EQUATIONS IN DIFFERENTIAL FORM

Case/Quantity	*From Ampere Electric current*	*From Faraday Electric potential*	*From Gauss*	
			Electric flux,	*Magnetic flux*
General	$\nabla \times H = J + \frac{\partial D}{\partial t}$	$\nabla \times E = -\frac{\partial B}{\partial t}$	$\nabla \cdot D = \rho$	$\nabla \cdot B = 0$
Free space	$\nabla \times H = \frac{\partial D}{\partial t}$	$\nabla \times E = -\frac{\partial B}{\partial t}$	$\nabla \cdot D = 0$	$\nabla \cdot B = 0$
Harmonic variation	$\nabla \cdot H = (\sigma + j\omega \epsilon) E$	$\nabla \times E = -j\omega \mu H$	$\nabla \cdot D = \rho$	$\nabla \cdot B = 0$
Steady	$\nabla \times H = J$	$\nabla \times E = 0$	$\nabla \cdot D = \rho$	$\nabla \cdot B = 0$
Static	$\nabla \times H = 0$	$\nabla \times E = 0$	$\nabla \cdot D = \rho$	$\nabla \cdot B = 0$

5.7. TRANSMISSION LINES

Classification

1. TEM mode (transverse electromagnetic mode) E and H are entirely transverse. All two conductor lines including coaxial lines belong to this category. Power flows along and between conductors.
2. Higher mode type: E and H or both have components in the direction of transmission. Hollow single conductor wave guides, dielectric rods and optical fibres belong to this category. Power flows in space inside hollow conductor wave guide or in or close to dielectric rod or fibre.
3. TEM space wave. Between antennas of a radio link Power radiates through space.

Wave Equation in E_y (linearly polarized wave, *i.e.*, E in *y* direction and travelling in *x* direction) for loss less medium

$$\frac{\partial^2 E_y}{\partial t^2} = \frac{1}{\mu\epsilon} \frac{\partial^2 E_y}{\partial x^2}$$

Wave Equation in H_z for loss less medium

$$\frac{\partial^2 H_z}{\partial t^2} = \frac{1}{\mu\epsilon} \frac{\partial^2 H_z}{\partial x^2}$$

Wave velocity $v = \frac{1}{\sqrt{\mu\epsilon}}$

Travelling wave $E_y = \sin(\beta x \pm \omega t)$

In the above equation, + sign is associated with a wave to the left (*i.e,* in the negative direction of *x* or backward wave) and – sign is associated with a wave to the right (*i.e.*, in the positive direction of *x* or forward wave)

Velocity $v = \frac{dx}{dt} = \frac{\omega}{\beta} = \frac{2\pi f}{2\pi/\lambda} = \lambda f$

Wave equation for Lossy medium (*i.e.*, σ finite)

$$\frac{\partial^2 E_y}{\partial x^2} = (j\omega\mu\sigma - \omega^2 \mu \epsilon) E_y$$

Transmission line Wave equations in V and I

$$\frac{\partial^2 V}{\partial x^2} - ZYV = 0$$

$$\frac{\partial^2 I}{\partial x^2} - ZYI = 0$$

Propagation constant $\gamma = \sqrt{ZY} = \alpha + j\beta$

Attenuation constant $\alpha = \text{Re}\sqrt{ZY}$ Neper/m

Phase constant $\beta = \text{Im}\sqrt{ZY}$ rad/m

Characteristic Impedance $Z_0 = \frac{V}{I} = \sqrt{\frac{Z}{Y}}$ ohms

$$= \sqrt{\frac{j\omega\mu}{\sigma + j\omega\epsilon}} \text{ ohms} \quad \text{(Conducting medium)}$$

Intrinsic impedance of medium

$$Z_0 = \frac{E}{H} = \sqrt{\frac{\mu}{\epsilon}} \text{ ohms}$$

Intrinsic impedance of free space

$$Z_0 = \frac{E}{H} = \sqrt{\frac{\mu_0}{\epsilon_0}} = 120\pi = 376.731\ \Omega$$

Relative Phase Velocity

$$p = \frac{\text{Phase velocity}}{\text{Velocity of light}} = \frac{\sqrt{\mu_0 \epsilon_0}}{\sqrt{\mu\epsilon}} = \frac{1}{\sqrt{\mu_r \epsilon_r}}$$

Index of reflection $\eta = \frac{1}{p} = \frac{C}{v} = \sqrt{\mu_r \epsilon_r}$

Wave stationary in space $E_y = 2E_0 \cos \omega t \sin \beta x$

The total electric field E_y consists of incident Wave E_0 and reflected wave E_1

For a pure standing wave $|E_1| = |E_0|$

Voltage standing wave ratio (VSWR) $= \frac{E_0 + E_1}{E_0 - E_1}$

Depth in which the wave attenuates to 36.8% of its initial value is called depth of penetration δ.

δ, *i.e.*, depth of penetration $= \frac{1}{\sqrt{f\pi\mu\sigma}} = \sqrt{\frac{2}{\omega\mu\sigma}}$

Instaneous Poynting Vector $S = E \times H$

S is surface power density

Average Poynting Vector $= \frac{1}{2}\text{Re}\, E \times H^*$

Total energy density W is the sum of electric and magnetic energies

$$W = W_e + W_m = \frac{1}{2}\epsilon E^2 + \frac{1}{2}\mu H^2 \text{ J/m}^2$$

Polarization parameters

$$\cos 2\gamma = \cos 2\epsilon \cos 2\tau$$

$$\tan\delta = \frac{\tan 2\epsilon}{\sin 2\tau}$$

$$\tan 2\tau = \tan 2\gamma \cos\delta$$

$$\sin 2\epsilon = \sin 2\gamma \sin\delta$$

Snell's law $\sin\theta_t = \frac{\eta_1}{\eta_2}\sin\theta_i$

For lossless medium $\eta = \sqrt{\mu_r \epsilon_r}$

Snell's law for loss less medium

$$\sin\theta_t = \sqrt{\frac{\mu_1 \epsilon_1}{\mu_2 \epsilon_2}}\sin\theta_i$$

5.8. WAVE GUIDES

All transmission lines which can convey electromagnetic waves only in higher order modes are called wave guides, *e.g.*, hollow single conductor, dielectric rods and optical fibres. The wave is guided either in the space inside hollow conductor or in or close to dielectric rod or fibre.

At low frequencies electromagnetic energy cannot be transmitted by such a hollow single conductor. However, at high frequencies this is possible. This transition or low frequency cut off occurs when wave length is of the same order as diameter of pipe. The critical wave length is called cut off wave length. In case two conducting sheets having a spacing b cut off wave length.

$$\lambda_{OC} = 2b/n$$

$n = 1, 2, 3...$ Each value of n corresponds to a particular higher order mode

scalar wave equation in H_x for hollow rectangular wave guide

$$\frac{\partial^2 H_x}{\partial y^2} + \frac{\partial^2 H_x}{\partial z^2} + k^2 H_x = 0$$

Solution of above equation is

$$H_x(y, z, x, t) = H_0 \cos\frac{n\pi y}{y_1}\cos\frac{m\pi z}{z_1}e^{-\gamma x}$$

where m and n are same or different integers. If $m = 1, n = 0$ we have TE_{10} mode. If $m = 2, n = 0$, we have TE_{20} mode. If $m = n = 1$, we have TE_{11} mode

TE mode requirement $E_x = 0$

Transverse wave impedance $Z_{yz} = \frac{\omega\mu}{\beta} = \frac{Z_d}{\sqrt{1 - (\lambda_0/\lambda_{oc})^2}}$

At low freqnencies ω is small, γ is real, guide opaque and wave does not propagate.

At intermediate frequencies, ω is intermediate, γ = 0 and transition condition occurs (cut off).

At high frequencies ω is high, γ is imaginary and wave propagation occurs.

Cut off frequency f_c is

$$f_c = \frac{1}{2\sqrt{\mu \in}} \sqrt{\left(\frac{n}{y_1}\right)^2 + \left(\frac{m}{z_1}\right)^2} \text{ Hz}$$

Cut off wave length λ_c is

$$\lambda_c = \frac{2}{\sqrt{(n/y_1)^2 + (m/z_1)^2}}, \text{ m}$$

where y_1 and z_1 are dimensions of hollow rectangular wave guide.

For Hollow cylindrical wave guide

$$\text{Cut off frequency } f_c = \frac{1}{2\pi\sqrt{\mu \in}} \frac{k'_{nr}}{r_0}, \text{ Hz}$$

$$\text{Cut off wave length } \lambda_c = \frac{2\pi r_0}{k'_{nr}}, \text{ m}$$

Attenuation when frequency is less than cut off frequency

$$\alpha = \frac{2\pi}{\lambda_0} \sqrt{\left(\frac{\lambda_0}{\lambda_c}\right)^2 - 1} \text{ Nepers/m}$$

Attenuation at frequencies higher than cut off

$$\alpha = \frac{\text{Re } Z_c \int |H_{t1}|^2 \, dl}{2 \text{ Re } Z_{yz} \iint |H_{t2}|^2 \, ds} \text{ Nepers/m}$$

5.9. CAVITY RESONATORS

A resonator is an energy storage device. Thus it is equivalent to a resonant circuit element. A cavity resonator is a completely enclosed rectangular box. Its inductance and capacitance are very low so that the resonant frequency is very high (corresponding to very short wave length).

For TE_{lmn} wave (where l refers to half cycle variations of field in x direction, m in z direction and n in y direction) and $n = 0$.

$$\text{Resonant wavelength } \lambda = \frac{2}{\sqrt{(l/x_1)^2 + (m/z_1)^2}}$$

where x_1 and z_1 are dimensions of resonator.

For TE_{110} mode in a square box (*i.e.*, $x_1 = z_1$)

$$\text{Resonant wavelength } \lambda = \frac{2}{\sqrt{2/x_1^2}} = 1.41 x_1, \text{ metres}$$

$$\text{Also for } TE_{110} \text{ mode, } Q = \frac{2}{\delta}\left[\frac{\text{Volume of cavity}}{\text{Interior surface area of cavity}}\right]$$

where $\delta = 2 \text{ Re } (Z_c/\omega\mu_0) = 1/e$ = depth of penetration

For copper $\delta = 6.6 \times 10^2/\sqrt{f}$

If $x_1 = z_1 = 100$ mm and $y_1 = 50$ mm

Resonant wavelength = 141 mm and Q = 17500

A microwave oven is a cavity resonator with food as absorber.

5.10. ANTENNAS

Basic Radiation Equation $\dot{I}L = Q\dot{v}$ Am/s

where $\dot{I}$ = rate of change of current, A/s

L = length of current element, m

Q = charge, C

$\dot{v}$ = acceleration of charge, m/s^2

The radiation is perpendicular to acceleration of charge and the radiated power is proportional to square of $\dot{I}L$ or $Q\dot{v}$

Antenna converts electromagnetic photons into circuit current or vice versa. Thus it is a transducer which converts photon energy into electron energy or vice versa.

$$\text{Main beam efficiency} = \frac{\Omega_M}{\Omega_m}$$

where Ω_M is Beam'component area due to main lobe and Ω_m is Beam component area due to minor lobes

$$\text{D, } i.e., \text{ Directivity} = \frac{4\pi}{\Omega_A}$$

where $\Omega_A = \Omega_M + \Omega_m$

Gain = kD

where k is efficiency factor of antenna ($0 \le k \le 1$).

If antenna is 100% efficient, $k = 1$ and Gain = D.

For short dipole antenna, Magnetic field H at any point P is

$$H = \frac{P}{\mu_0} \nabla \times A \text{ A/m}$$

where A is vector potential at point P, Wb/m

$$\text{Intrinsic impedance of free space} = \sqrt{\frac{\mu_0}{\epsilon_0}} = 376.7 \text{ Ohm}$$

OBJECTIVE TYPE QUESTIONS

A. Choose the most appropriate alternative

1. Farad is the same as
(a) coulomb^2/N (b) coulomb^2/N-m
(c) coulomb^2/m (d) coulomb/m

2. Permittivity has the units
(a) farads/metre (b) coulomb/m
(c) farad/m^2 (d) coulomb/m^2

3. $\epsilon_0 =$
(a) 8.85×10^{-12} Fm^{-1} (b) 8.85×10^{-12} Cm^{-1}
(c) 8.85×10^{-10} Fm^{-1} (d) 8.85×10^{-10} Cm^{-1}

4. Coulomb's law
(a) is a vector equation
(b) is a scalar equation
(c) may be a vector or scalar equation
(d) is a phasor equation

5. Force between two point charges Q_1 and Q_2 having a distance r between them ($\hat{r}$ is a unit vector pointing in the direction of line joining the two charges) =
(a) $\dfrac{\hat{r} Q_1 Q_2}{4\pi\epsilon r^2}$ (b) $\dfrac{-\hat{r} Q_1 Q_2}{4\pi\epsilon r^2}$
(c) $\dfrac{\hat{r} Q_1 Q_2}{4\pi r^2}$ (d) $\dfrac{-\hat{r} Q_1 Q_2}{4\pi r^2}$

6. Coulomb's force on a charge Q_1 due to a charge Q_2
(a) is directed from Q_2 to Q_1
(b) is directed from Q_1 to Q_2
(c) may be directed from Q_2 to Q_1 or from Q_1 to Q_2
(d) is proportional to Q_1/Q_2

7. The units of relative permittivity are
(a) Fm^{-1} (b) Fm^{-2}
(c) Cm^{-1} (d) dimensionless

8. Coulomb's force has components
(a) in x direction only (b) in y direction only
(c) in z direction only (d) in x, y, z directions

9. If two point charges each 2C are separated by 1 mm distance, the coulomb's force will be
(a) 36×10^{15} newtons (b) 10^{12} newtons
(c) 36×10^{6} newtons (d) 36×10^{-6} newtons

10. If two point charges each 1C are separated by 1 mm distance, the coulomb's force
(a) will be a very small force
(b) will be a small force
(c) will be sufficient to lift a 20 storey building
(d) will be sufficient to lift all the buildings in India simultaneously

11. The units of electric field intensity are
(a) NC^{-1} (b) Fm^{-1}
(c) Nm^{-1} (d) FC^{-1}

12. Electric field intensity at a point distant r metres from an electric charge Q
(a) is proportional to Q and r
(b) is proportional to Q and $1/r$
(c) is proportional to Q and $1/r^2$
(d) is proportional to Q and $1/r^3$

13. Which of the following is correct?
(a) Field of a function f is the region over which f exists and may vary with position. Function f is a scalar
(b) Field of a function f is the region over which f exists and may vary with time. Function f is a vector
(c) Field of a function f is the region over which f exists and may vary with time or position or both. Function f may be a scalar or vector
(d) Field of a function f is the region over which f exists and may vary with time or position or both. Function f may be a scalar or vector or both

14. The total electric field at a point, due to a number of charges, is
(a) scalar sum of individual component fields at that point
(b) vector sum of individual component fields at that point
(c) scalar or vector sum of individual component fields at that point
(d) scalar or vector sum or product of individual component fields at that point

15. A long wire has a charge density 1×10^{-9} cm^{-1}. The electric field at a point at radial distance 2 m will be about
(a) 6 Vm^{-1} (b) 60 Vm^{-1}
(c) 600 Vm^{-1} (d) 0.6 Vm^{-1}

16. An electric dipole consists of a charge q and another charge $-q$ separated by a distance a. The electric field intensity at a point P at perpendicular distance r from the centre O of the line joining the two charges
(a) is directed along line OP towards O
(b) is directed along line OP towards P
(c) is directed along a line parallel to the line joining $+q$ and $-q$ and directed in the direction of $+q$
(d) is directed along a line parallel to the line joining $+q$ and $-q$ and directed in the direction of $-q$

17. The unit newton metres per coulomb is the same as
(a) volts (b) amperes
(c) volts/metre (d) watts

18. The electric potential at a point P situated at distance r from a charge Q is
(a) directly proportional to r
(b) directly proportional to Q
(c) directly proportional to Q and inversely proportional to r
(d) is inversely proportional to Q and directly propotional to r

19. Total electric potential, at a point, due to a number of charges
(a) is the phasor sum of individual potentials at that point
(b) is the algebraic sum of individual potentials at that point
(c) is either phasor sum or algebraic sum of individual potentials at that point
(d) none of the above

20. A charge is uniformly distributed over a spherical volume and the volume charge density is ρ. The electric potential at a point P situated at distance r from the centre of sphere is

(a) $\frac{1}{4\pi\epsilon_0}\iiint \frac{\rho}{r}\,dv$ (b) $\frac{1}{\epsilon_0}\iiint \frac{\rho}{r}\,dv$

(c) $\frac{1}{2\pi\epsilon_0}\iiint \frac{\rho}{r}\,dv$ (d) $\frac{1}{2\pi}\iiint \frac{\rho}{r}\,dv$

21. Which of the following equations is correct
(a) E = grad V (b) E = –grad V
(c) V = grad E (d) V = – grad E
where E is electric field intensity and V is potential.

22. Capacitance of an isolated sphere of radius r_1 situated in air is equal to
(a) $\epsilon_0 r_1$ (b) $2\pi\epsilon_0 r_1$
(c) $4\pi\epsilon_0 r_1$ (d) r_1

23. The electric field around a negative charge is directed
(a) towards the charge
(b) away from the charge
(c) towards the charge or away from the charge depending on the magnitude of the charge
(d) none of the above

24. The units of electric displacement is
(a) Cm^{-2} (b) Cm^{-1}
(c) Fm^{-1} (d) C

25. The relation between electric flux density D and electric field intensity E is
(a) $D = \epsilon E$ (b) $E = \epsilon D$
(c) $\epsilon = DE$ (d) $D = \epsilon E^2$

26. Electric field lines and equipotential lines
(a) are always parallel
(b) are always orthogonal
(c) may be parallel or orthogonal
(d) may be inclined at any angle

27. An isolated charged sphere (radius r) has a surface charge density σ C/m^2. Its potential will be
(a) $\frac{\sigma r}{\epsilon_0}$ (b) $\frac{\sigma r^2}{\epsilon_0}$
(c) $\frac{\epsilon_0 r^2}{\sigma}$ (d) $\frac{\sigma}{\epsilon_0 r}$

28. The force between two charges is 120 N. If the distance between the charges is doubled the force will be
(a) 120 N (b) 60 N
(c) 30 N (d) 15 N

29. The ability of charged bodies to exert force on one another is attributed to the existence of
(a) atoms (b) protons
(c) electrons (d) electric field

30. A field line and equipotential surface are
(a) always parallel (b) always at 90°
(c) inclined at $\angle\theta$ (d) any of the above

31. The charge on an isolated conductor resides
(a) at the surface of the conductor
(b) inside the conductor
(c) partly at the surface and partly inside the conductor
(d) with uniform distribution over the volume of the conductor

32. If a charged particle having charge e is released in an electric field E, the force on the particle is equal to
(a) eE (b) E/e
(c) E^2/e (d) eE^2

33. A charged particle having mass m and charge e is moved through a potential difference V. If the initial velocity of particle is zero, the final velocity is
(a) $(2eV/m)^{0.5}$ (b) $(eV/m)^{0.5}$
(c) $(eV/2m)^{0.5}$ (d) $V(e/m)^{0.5}$

34. An electron (charge 1.6×10^{-19} C and $m = 0.91 \times 10^{-30}$ kg) falls through a potential difference of 1 V. If initial velocity is zero, the final velocity of electron is about
(a) 1 m/s (b) 1 km/s
(c) 100 km/s (d) 600 km/s

35. Potential and electric field at a point distant r from a dipole vary as
(a) $1/r$ and $1/r^2$ respectively
(b) r and $\sqrt{r}$ respectively
(c) $1/r^2$ and $1/r^3$ respectively
(d) r^2 and r^3 respectively

36. A dielectric is placed in an electric field E. If D is flux density in dielectric and P is polarization, then
(a) $D = \epsilon_0 E + P$ (b) $D = \epsilon_0 E - P$
(c) $D = \epsilon_0 (E + P)$ (d) $D = \epsilon_0 (E - P)$

37. The relation between absolute permittivity ϵ_0, permittivity of dielectric ϵ, polarization P and electric field E is
(a) $\epsilon = \epsilon_0 + P/E$ (b) $\epsilon = \epsilon_0 - P/E$
(c) $\epsilon = \epsilon_0 (1 + P/E)$ (d) $\epsilon = \epsilon_0 (1 - P/E)$

38. In isotropic media with polarization P and electric field E
(a) P and E are in the same direction and ϵ is scalar
(b) P and E may or may not be in the same direction but ϵ is a scalar
(c) P and E are not in the same direction and ϵ is not a scalar
(d) P and E are in the same direction but ϵ is not a scalar

39. At a boundary between two different media
(a) electric field is continuous
(b) electric field may change abruptly in magnitude
(c) electric field may change abruptly in direction
(d) electric field may change abruptly in both direction and magnitude

40. At the boundary of two dielectrics
(a) tangential components of electric field on both sides of boundary are the same
(b) tangential components of electric field are zero
(c) tangential components of electric field on the two sides of boundary may be finite or zero
(d) tangential componenents of electric field on both sides of boundary may be the same or different

41. At a boundary of a dielectric and a conductor
(a) normal component of flux density is zero
(b) normal componenet of flux density is continuous
(c) normal component of flux density is equal to surface charge density on the conductor
(d) normal component of flux density may be zero or continuous

42. At the boundary between two dielectrics, a change in the direction of field line occurs. If ϵ_1 and ϵ_2 are the relative permittivities of the two dielectrics and α_1 and α_2 are the angles which the field line makes with normal to the boundary then
(*a*) $(\tan\alpha_1)(\epsilon_2) = (\tan\alpha_2)(\epsilon_1)$
(*b*) $(\tan\alpha_1)(\epsilon_1) = (\tan\alpha_2)(\epsilon_2)$
(*c*) $(\sin\alpha_1)(\epsilon_1) = (\sin\alpha_2)(\epsilon_2)$
(*d*) $(\sin\alpha_1)(\epsilon_2) = (\sin\alpha_2)(\epsilon_1)$

43. Let α_1 and α_2 denote the angles of field lines with respect to normal to the boundary of two dielectrics with relative permittivities ϵ_1 and ϵ_2. If $\epsilon_1 = 1$ and $\epsilon_2 = 4$, then
(*a*) $\alpha_1 > \alpha_2$ (*b*) $\alpha_1 = \alpha_2$
(*c*) $\alpha_1 < \alpha_2$
(*d*) α_1 may be more or less than α_2

44. Static electric field line at a dielectric-conductor boundary is
(*a*) parallel to conductor surface
(*b*) perpendicular to conductor surface
(*c*) inclined to conductor surface
(*d*) either perpendicular or inclined to conductor surface

45. A parallel plate capacitor has plates of area A. The dielectric between the plates is in two parts with thickness d_1 and d_2 and relative permittivities ϵ_1 and ϵ_2. The total capacitance is

(*a*) $\dfrac{\epsilon_1 \epsilon_0 A}{d_1} + \dfrac{\epsilon_1 \epsilon_0 A}{d_2}$ (*b*) $\dfrac{\epsilon_1 \epsilon_0 A}{d_1}\left(1 + \dfrac{d_2}{d_1}\right)$

(*c*) $\dfrac{\left(\dfrac{\epsilon_1 \epsilon_0 A}{d_1}\right)\left(\dfrac{\epsilon_2 \epsilon_0 A}{d_2}\right)}{\dfrac{\epsilon_1 \epsilon_0 A}{d_1} + \dfrac{\epsilon_2 \epsilon_0 A}{d_2}}$ (*d*) $\dfrac{\dfrac{\epsilon_1 \epsilon_0 A}{d_1} + \dfrac{\epsilon_2 \epsilon_0 A}{d_2}}{\left(\dfrac{\epsilon_1 \epsilon_0 A}{d_1}\right)\left(\dfrac{\epsilon_2 \epsilon_0 A}{d_2}\right)}$

46. The units of dielectric strengths are the same as that of
(*a*) electric field intensity (*b*) voltage
(*c*) electric flux density (*d*) electric charge

47. Two spherical drops of water each having a charge Q and surface potential V are combined to form a bigger drop. The surface potential of the bigger drop
(*a*) will be V
(*b*) will be 2 V
(*c*) may be less than or more than 2 V
(*d*) will be more than V but less than 2 V

48. A parallel plate capacitor is charged so that the electric field in the dielectric is E. If the permittivity of dielectric is ϵ, the energy density *i.e.*, energy per unit volume is
(*a*) $0.5 \epsilon E$ (*b*) $0.5 \epsilon E^2$
(*c*) $0.5 \epsilon^2 E$ (*d*) $0.5 E^2/\epsilon$

49. A parallel plate capacitor has air as dielectric. The electric field in the dielectric is E. If air is replaced by a dielectric of permittivity ϵ, the dielectric gets polarized to polarization P. The additional energy stored in dielectric is
(*a*) 0.5 P/E (*b*) 0.5 PE
(*c*) PE (*d*) $0.5 P^2/E$

50. Taking the earth's radius as 6370 km and electrosphere height as 25 km, the capacitance of earth electrosphere combination is about
(*a*) 1 F (*b*) 0.2 F
(*c*) 0.01 F (*d*) 1 μF

51. Which of the following is not correct as regards properties of an accurate electric field map in a single medium of uniform permittivity?
(*a*) The capacitance of any field cell is the same
(*b*) The capacitance per unit depth of field cell is equal to permittivity of medium
(*c*) The energy stored in any cell of the same kind is the same
(*d*) The average field intensity E in any cell of the same kind is directly proportional to cell width

52. Divergence of D
(*a*) is a scalar point function
(*b*) is a vector point function
(*c*) is a phasor
(*d*) may be a scalar or vector point function

53. Air flows with uniform velocity *v* through a tube open at both ends. Then
(*a*) $\nabla \cdot v = 0$ (*b*) $\nabla \cdot v =$ constant
(*c*) $\nabla \cdot v = 0$ or constant (*d*) $\nabla \cdot v$ is proportional to *v*

54. In a region free from charge
(*a*) $\nabla \cdot D = 0$ (*b*) $\nabla \cdot D =$ constant
(*c*) $\nabla \cdot D = 0$ or constant (*d*) $\nabla \cdot D = \infty$

55. Divergence theorem as applied to D is
(*a*) $\oint_s D.ds = \oint_v \nabla \cdot D\, dv$ (*b*) $\oint_s D.ds = \oint \nabla \cdot D\, dl$
(*c*) $\oint_s D.ds = \oint_v \nabla \cdot D\, dv$ (*d*) either (*a*) or (*b*)

56. Poisson's equation is
(*a*) $\nabla^2 V = -\rho/\epsilon$ (*b*) $\nabla^2 V = \rho/\epsilon$
(*c*) $\nabla^2 V = -\epsilon/\rho$ (*d*) $\nabla^2 V = \epsilon/\rho$

57. Laplace's equation is
(*a*) $\nabla^2 V = 0$ (*b*) $\nabla^2 V = \rho/\epsilon$
(*c*) $\nabla^2 V = \infty$ (*d*) $\nabla^2 V = -\rho/\epsilon$

58. Six capacitors of different capacitances $C_1, C_2, C_3, C_4, C_5, C_6$ are connected in series. $C_1 > C_2 > C_3 > C_4 > C_5 > C_6$. The total cpacitance is almost equal to
(*a*) C_1 (*b*) C_3
(*c*) C_4 (*d*) C_6

59. A composite capacitor has a total thickness of 10 mm. The dielectric is in two parts with thicknesses 5 mm each but relative permittivity 6 and 2. If flux densities are D_1 and D_2 then
(*a*) $D_1 = D_2$ (*b*) $D_1 > D_2$
(*c*) $D_1 < D_2$ (*d*) impossible to judge

60. A composite capacitor has two dielectrics. The upper one has thickness 20 mm and $\epsilon_r = 6$ and the second (lower) one has thickness 10 mm and $\epsilon_r = 12$. If a voltage of 100 V is applied, the voltages across upper and lower dielectrics are
(*a*) 50 V and 50 V (*b*) 80 V and 20 V
(*c*) 40 V and 60 V (*d*) 20 V and 80 V

61. In question 60, the electric field intensities in the two dielectrics, if 200 V is applied across the capacitor, are
(*a*) 6 kV/m each (*b*) 8 kV/m and 2 kV/m
(*c*) 8 kV/m and 4 kV/m (*d*) 2 kV/m and 8 kV/m

62. One capacitor has dielectric of $\in_r = 6$ and electric field intensity is 50 V/m. Another capacitor has $\in_r = 100$ and E = 10 V/m. The ratio of energies stored in the two capacitors is
(*a*) 1 (*b*) 1.5
(*c*) 2 (*d*) 3

63. A particle having charge *e* is released in an electric field E. If there are no collisions (restraints) the velocity of particle
(*a*) will be constant
(*b*) will keep on increasing
(*c*) will keep on decreasing
(*d*) may be constant or decrease

64. A particle having charge *e* is released in a homogenous gaseous medium having constant electric field E. The velocity of particle
(*d*) will be constant and proportional to E
(*b*) will be constant and proportional to E^2
(*c*) will keep on increasing
(*d*) will keep on decreasing

65. If a conducting medium of cross-section Am^2 has many free to move charged particles of volume density ρ and drift velocity *v*, the current I =
(*d*) $v\rho A$ (*b*) $\rho A/v$
(*c*) $v^2\rho A$ (*d*) $v\rho^2 A$

66. Current density at a surface Δ*s* normal to current direction is
(*a*) $\underset{\Delta s \to 0}{\text{Lt}} \left(\frac{\Delta I}{\Delta s}\right)$ (*b*) $\underset{\Delta s \to \infty}{\text{Lt}} \left(\frac{\Delta I}{\Delta s}\right)$
(*c*) $\underset{\Delta s \to 0}{\text{Lt}} \left(\frac{\Delta s}{\Delta I}\right)$ (*d*) $\underset{\Delta s \to \infty}{\text{Lt}} \left(\frac{\Delta s}{\Delta I}\right)$

67. In a conductor, the current density at a point
(*a*) is a scalar point function
(*b*) is a vector point function
(*c*) may be scalar or vector point function
(*d*) always tends to infinity

68. A cylindrical conductor has a length *d* and resistance R. If electric field is E, the current I =
(*a*) E*d*/R (*b*) E/*d*R
(*c*) $E^2 d/R$ (*d*) $\sqrt{E}\,(d/R)$

69. If *J* is current density, σ is conductivity and E is electric field intensity, ohm's law at a point is
(*a*) $J = \sigma E$ (*b*) $J = E/\sigma$
(*c*) $J = \sigma E^2$ (*d*) $J = \sigma/\sqrt{E}$

70. If σ, ρ and μ denote conductivity, density and mobility of electrons in a metal, then
(*a*) $\sigma = \rho\mu$ (*b*) $\sigma = \rho\sqrt{\mu}$
(*c*) $\sigma = \rho/\mu$ (*d*) $\sigma = \mu/\rho$

71. In terms of current density J, KCL is
(*a*) $\oint_s J \cdot ds = 0$ (*b*) $\oint_s J \cdot ds = \text{constant}$
(*c*) $\oint_s J \cdot ds = \text{Infinity}$
(*d*) $\oint_s J \cdot ds = 0$ or constant

72. If *J* is current density and ρ is charge density at a point, the equation of conservation of charge is
(*a*) $\nabla \cdot J = \rho$ (*b*) $\nabla \cdot J = \frac{\partial \rho}{dt}$
(*c*) $\nabla \cdot J = \frac{-\partial \rho}{dt}$ (*d*) $\nabla \cdot J = -\frac{(\partial \rho)^2}{dt}$

73. A cylindrical conductor has an inner conductor and an outer conductor with insulator in between. The currents in the inner and outer conductors are in opposite directions. If E_t and E_n denote the fields tangential and normal to the boundary, then in the conductor
(*a*) $E_n = 0$ (*b*) $E_t = 0$
(*c*) both E_n and E_t are finite (*d*) either E_t or E_n is zero

74. In a coaxial cylindrical conductor, the space between inner and outer conductors has a dielectric. The currents in inner and outer conductors are in opposite directions. In the dielectric
(*a*) field is tangential only
(*b*) field is normal to the boundary
(*c*) field is partly tangential and partly normal to the boundary
(*d*) field is either tangential or normal to the boundary

75. Two conductors A and B are similar in dimensions and are carrying equal currents. Conductivity of A is twice that of B. If E_A and E_B are electric field intensities in conductors A and B
(*a*) $E_A = E_B$ (*b*) $E_A = 2E_B$
(*c*) $E_A = 0.5E_B$
(*d*) E_A may be equal or less than E_B

76. As regards current mapping, which statement is wrong
(*a*) current lines and equipotentials intersect at right angles
(*b*) the total current through any cross-section of a continuous current tube is constant
(*c*) current tubes are continuous
(*d*) In a uniform current distribution potential varies inversely with distance

77. To find the conductance of a conductor with non-uniform current distribution, the conductor is divided into a number of conductor cells. If G_0 is the conductance of each cell, N is the number of cells in parallel and *n* is the number of cells in series, the conductance G of the conductor is given by
(*a*) $G = (N/n)G_0$ (*b*) $G = (n/N)G_0$
(*c*) $G = (N/nG_0)$ (*d*) $G = (G_0/nN)$

78. The current tubes and conductivity σ in conducting media are analogous to
(*a*) flux tubes and permittivity ∈ in insulating media
(*b*) flux tubes and relative permittivity $\in_r$ in insulating media
(*c*) electric charge and relative permittivity $\in_r$ in insulating media
(*d*) none of the above

79. Atmospheric conductivity near earth during fair weather is about
(*a*) 10^{-14} mho m^{-1} (*b*) 10^{-4} mho m^{-1}
(*c*) 10^{-8} mho m^{-1} (*d*) impossible to find

80. It is known that direction of current changes in flowing from one medium to other. If α_1 and α_2 are angles of direction of current with respect to normal to the boundary of two conducting media having conductivities σ_1 and σ_2, then
(*a*) $(\sigma_2)(\tan\alpha_1) = (\sigma_1)(\tan\alpha_2)$
(*b*) $(\sigma_1)(\tan\alpha_1) = (\sigma_2)(\tan\alpha_2)$
(*c*) $(\sigma_1)(\cos\alpha_1) = (\sigma_2)(\cos\alpha_2)$
(*d*) $(\sigma_1)(\cos\alpha_2) = (\sigma_2)(\cos\alpha_1)$

81. If J_{nl}, J_{n2} denote components of current density normal to boundary and E_{tl}, E_{t2} denote component of field tangential to boundary of two conducting media 1 and 2 respectively, then
(*a*) $J_{nl} = J_{n2}$ and $E_{tl} = E_{t2}$
(*b*) $E_{tl} = E_{t2}$ and J_{nl}, J_{n2} are not equal
(*c*) $J_{nl} = J_{n2}$ but E_{tl}, E_{t2} are not equal
(*d*) $J_{nl} \neq J_{n2}$ and $E_{t2} \neq E_{t2}$

82. A lightning stroke has a current of 100 kA and a duration of 200 ms. If potential is 100 kV, the energy involved is
(*a*) 2000 MJ (*b*) 20 MJ
(*c*) 0.2 MJ (*d*) 0.01 MJ

83. Conductance is analogous to
(*a*) flux (*b*) current
(*c*) reluctance (*d*) permeance

84. An air gap is usually inserted in magnetic circuits to
(*a*) increase the flux (*b*) decrease the flux
(*c*) increase mmf (*d*) prevent saturation

85. The relative permeability of ferromagnetic material is
(*a*) 1 (*b*) about 1
(*c*) less than 1 (*d*) more than 1000

86. If the area of hysteresis loop of a material is large, the hysteresis loss in the material will be
(*a*) zero (*b*) low
(*c*) zero or low (*d*) high

87. Silicon steel is used in electric machines because it has
(*a*) low hysteresis loss (*b*) low retentivity
(*c*) low coercivity (*d*) all the above

88. Hard steel is suitable for making permanent magnets because
(*a*) it has good retentivity
(*b*) it is mechanically strong
(*c*) it has big hysteresis loop
(*d*) all the above

89. The aim of shielding an instrument is
(*a*) to prevent damage due to moisture
(*b*) to provide mechanical protection
(*c*) to increase the range of instrument
(*d*) to reduce the effect of stray magnetic fields

90. In a stationary conductor, an emf can be induced by
(*a*) a changing magnetic field
(*b*) electrons
(*c*) steady magnetic field
(*d*) all the above

91. If a copper wire is moved in a magnetic field
(*a*) an emf is induced in the wire
(*b*) a current flows in the wire
(*c*) an emf is induced and current flows in the wire
(*d*) the wire is magnetised

92. A collapsing field around a coil
(*a*) tends to oppose the decay of current in the coil
(*b*) helps the decay of current in the coil
(*c*) has no effect on decay of current in the coil
(*d*) any of the above

93. Voltage is a form of
(*a*) potential energy (*b*) kinetic energy
(*c*) either potential or kinetic energy
(*d*) none of the above

94. The direction of induced emf is given by
(*a*) Fleming's right hand rule
(*b*) Cork screw rule
(*c*) Fleming's right hand rule
(*d*) KVL

95. The magnitude of emf induced in a wire does not depend on
(*a*) length of wire (*b*) diameter of wire
(*c*) speed of wire (*d*) flux density of field

96. The vector force on a wire carrying current I and lying in a magnetic field B is (if L is length of wire)
(*a*) $F = (I \times B) L$ (*b*) $F = BIL$
(*c*) $F = (L \times B) I$ (*d*) $F = (I \times L) B$

97. In vector rotation, the field ΔB at a point P due to current element of length Δl and current I such that R is vector distance from P to ΔL is
(*a*) $dB = \dfrac{\mu}{4\pi}\dfrac{IR \times \Delta L}{|R|^3}$ (*b*) $dB = \dfrac{\mu}{4\pi}\dfrac{I \times \Delta L}{|R|^3}$
(*c*) $dB = \dfrac{\mu}{4\pi}\dfrac{R \times \Delta L}{I|R|^3}$ (*d*) $dB = \dfrac{\mu}{8\pi}\dfrac{IR \times \Delta L}{|R|^3}$

98. Magnetic field at a point P at perpendicular distance *r* from a long wire carrying current I
(*a*) $\dfrac{\mu I}{2\pi r}$ (*b*) $\dfrac{\mu I}{4\pi r}$
(*c*) $\dfrac{\mu I}{8\pi r}$ (*d*) $\dfrac{\mu I r}{4\pi}$

99. A loop of radius *r* is carrying current I. The field at a distance *z* along the axis of the loop is
(*a*) $B = \dfrac{\mu I r^2}{2(r^2 + z^2)^{1.5}}$ (*b*) $B = \dfrac{\mu I r^2}{2(r^2 + z^2)^{0.5}}$
(*c*) $B = \dfrac{\mu I r^2}{(r^2 + z^2)^{1.5}}$ (*d*) $B = \dfrac{\mu I r}{2(r^2 + z^2)^{1.5}}$

100. Two parallel wires are carrying currents in the opposite directions. Then
(*a*) the magnetic field between the wires is stronger than the magnetic field outside
(*b*) the magnetic field between the wires is the same as that outside
(*c*) the magnetic field outside is stronger than that between the wires
(*d*) the magnetic field between the wires is zero

101. Magnetic flux over a surface having flux density B is

(a) $\phi = \iint B \cdot ds$ (b) $\phi = \int B \cdot ds$

(c) $\phi = \iint B\, ds$ (d) $\phi = \iiint B \cdot ds$

102. Gauss's law in integral form is

(a) $\oint_s B \cdot ds = 0$ (b) $\oint B \cdot dl = 0$

(c) $\oint_s B\, ds = 0$ (d) $\oint B dl = 0$

103. If m is magnetic moment and B is flux density, the torque T on a loop of side L is

(a) $m \times B$ (b) mB

(c) $(m \times B)L$ (d) $(m \times B)/L$

104. Gauss's law in differential form is

(a) $\nabla \cdot B = 0$ (b) $\nabla B = 0$

(c) $\nabla \times B = 0$ (d) none of the above

105. The flux densities at the centre and end of a solenoid having N turns and carrying current I and length l of solenoid much larger than radius, r, are

(a) $\frac{\mu NI}{l}$ and $\frac{0.5\mu NI}{l}$

(b) $\frac{\mu NI}{l}$ and $\frac{\mu NI}{l}$

(c) $\frac{0.5\mu NI}{l}$ and $\frac{\mu NI}{l}$

(d) $\frac{0.5\mu NI}{l}$ and $\frac{0.25\mu NI}{l}$

106. Inductance of a long solenoid having length l, area of cross-section a and turns N is

(a) $\mu N^2 a/l$ (b) $\mu N^2 l/a$

(c) $\mu Na/l$ (d) $N^2 a/\mu l$

107. A toroid has a radius R and has N turns. If the radius of the coil is r and permeability is μ, the inductance is

(a) $\mu N^2 r^2/2R^2$ (b) $\mu N^2 r/2R$

(c) $\mu N^2 r^2/R$ (d) $\mu N^2 r^2/2R$

108. A coaxial line has a current I. The outer conductor is the return conductor. The flux density at a radius r from the centre (inside the line) is

(a) $\mu I/\pi r$ (b) $\mu I/2\pi r$

(c) $\mu I/4\pi r$ (d) $\mu Ir/2\pi$

109. A coaxial line has an inner conductor of outside radus r and an outer conductor of inside radius R. If relative permeabilty of medium inside the coaxial line is μ_r, the inductance per unit length is

(a) $\mu \log_{10}(R/r)$ (b) $0.46\, \mu_r \log_{10}(R/r)$

(c) $0.92\, \mu \log_{10}(R/r)$ (d) $0.46\, \mu_r \ln(R/r)$

110. A two wire line has r as the radius of each wire and R as the distance between the centres of wires. If μ_r is the relative permeability of medium, the inductance per unit length is

(a) $0.92\, \mu_r \log_e(R/r)$ (b) $0.92\, \mu \log_{10}(R/r)$

(c) $0.92\, \mu_r \log_{10}(R/r)$ (d) $0.92\, \mu \log_e(R/r)$

111. Ampere's law is

(a) $\oint H \cdot dl = I$ (b) $\oint H \cdot ds = I$

(c) $\oint H dl = I$ (d) none of the above

112. A long wire is carrying current I. The value of H at a radius of 1 m from wire is 1 A/m. Then I is

(a) 4π amperes (b) π amperes

(c) 2π amperes (d) $\frac{\pi}{2}$ amperes

113. A solid wire of radius R carries a current I. The value of H at any point distance r (and $r < R$) is

(a) $H = \frac{Ir^2}{2\pi R^2}$ (b) $H = \frac{Ir}{2\pi R}$

(c) $H = Ir/2\pi R^2$ (d) $H = Ir/\pi R^2$

114. A solid wire carries a current I. At any point distance r from the centre and inside the wire, the magnetic field H

(a) is constant

(b) increases linearly as r increases

(c) decreases linearly as r increases

(d) may increase or decrease depending on value of I

115. A solid wire carries a current I. At a point outside the wire and at distance r from the centre, the magnetic field is proportional to

(a) r (b) $1/r$

(c) r^2 (d) $1/r^2$

116. If H is magnetic field intensity, I is current and J is current density

(a) $\nabla \cdot H = J$ (b) $\nabla \times H = J$

(c) $\nabla \times H = I$ (d) none of the above

117. A solid wire of radius r is carrying current I. If current density is J, then outside the wire

(a) $\nabla \times H = I$ (b) $\nabla \times H = J$

(c) $\nabla \times H = 0$ (d) $V \times H = 0$ or J

118. In terms of B, H and I, the inductance of an inductor carrying current I is

(a) $L = \iiint \frac{B \cdot H dv}{2I^2}$ (b) $L = \iiint \frac{B \cdot H dv}{I^2}$

(c) $L = \iiint \frac{B \cdot H dv}{I}$ (d) $L = \iiint \frac{B \cdot H dv}{4I^2}$

119. In terms of current density J and conductivity σ, the resistance of a conductor carrying current I is

(a) $R = \iiint \frac{(J^2/\sigma)\, dv}{\pi I^2}$ (b) $R = \iiint \frac{(J^2/\sigma)\, dv}{I^2}$

(c) $R = \iiint (J^2\sigma)\, dv/I^2$ (d) none of the above

120. In terms of E and D, the capacitance C is

(a) $C = \frac{\iiint E \cdot D dv}{V}$ (b) $C = \frac{\iiint ED dv}{V^2}$

(c) $C = \frac{\iiint E \cdot D dv}{V^2}$ (d) $C = \frac{\iiint ED dv}{V}$

121. If a particle having charge e moves in a field B with velocity v, the Lorentz force F =

(a) $e(v \times B)$ (b) $e(vB)$

(c) $e(v/B)$ (d) $e(v \cdot B)$

122. A magnetic field is at right angle to the motion of a charged particle having charge e and velocity v and mass m. If flux density is B
(a) the motion of particle is in straight line
(b) the motion of particle is in a circular path of radius mv/eB
(c) the motion of particle is in ellipsoidal path
(d) any of above depending on values of v and B

123. A single turn loop having magnetic moment m is in a field B. The torque on the loop is
(a) $m \times B$ (b) mB
(c) m/B (d) B/m

124. Vacuum is an example of
(a) diamagnetic medium (b) non-magnetic medium
(c) paramagnetic medium (d) none of the above

125. Aluminium is
(a) non-magnetic (b) diamagnetic
(c) paramagnetic (d) ferri magnetic

126. Ferri magnetic materials
(a) are less magnetic than ferromagnetic materials
(b) are more magnetic than ferromagnetic materials
(c) have almost same magnetic properties as ferromagnetic materials
(d) none of the above

127. Super paramagnetic materials are used in
(a) some special machines
(b) audio and video tapes
(c) instruments
(d) all the above devices

128. When a field is present, the moments of adjacent atoms of anti-ferromagnetic materials
(a) align in opposite directions
(b) do not align
(c) align in the same direction
(d) none of the above

129. The electrical conductivity of ferrites is
(a) high (b) low
(c) infinite (d) neither low nor high

130. Above curie temperature, a ferromagnetic material behaves as a
(a) non-magnetic material (b) diamagnetic material
(c) para magnetic material (d) ferrimagnetic material

131. If m is magnetic dipole moment in volume v, magnetisation M =
(a) $\frac{m}{v}$ (b) mv
(c) $\frac{v}{m}$ (d) mv^2

132. The relation between flux density B, magnetisation M and magnetic field H is
(a) $B = \mu_0(H + M)$ (b) $B = \mu_0 H + M$
(c) $B = \mu_0 M + H$ (d) none of the above

133. Magnetic energy density =
(a) $0.5\,\mu H$ (b) $0.5\,\mu H^2$
(c) $0.5\ mH^3$ (d) $0.5\,\mu\sqrt{H}$

134. In terms of magnetisation M, flux denstiy B and field strength H, magnetic energy density =
(a) $0.5\,(\mu_0 H^2 + MB)$ (b) $0.5\,(\mu_0 H^2 - MB)$
(c) $0.5\,(\mu_0 H + MB)$ (d) $0.5\,(\mu_0 H - MB)$

135. When a magnetic field goes from a medium of relative permittivity μ_{r1} to another medium of relative permittivity μ_{r2} a change in direction occurs. If α_1 and α_2 are angles of the directions in the two media with respect to normal to the boundary then
(a) $(\tan \alpha_1)\,\mu_{r1} = (\tan \alpha_2)\,\mu_{r2}$
(b) $(\tan \alpha_2)\,\mu_{r1} = (\tan \alpha_1)\,\mu_{r2}$
(c) $(\sin \alpha_1)\,\mu_{r1} = (\sin \alpha_2)\,\mu_{r2}$
(d) $(\sin \alpha_1)\,\mu_{r2} = (\sin \alpha_2)\,\mu_{r1}$

136. Figure 5.1 shows the B-H curve for a ferromagnetic material. Maximum value of μ occurs at

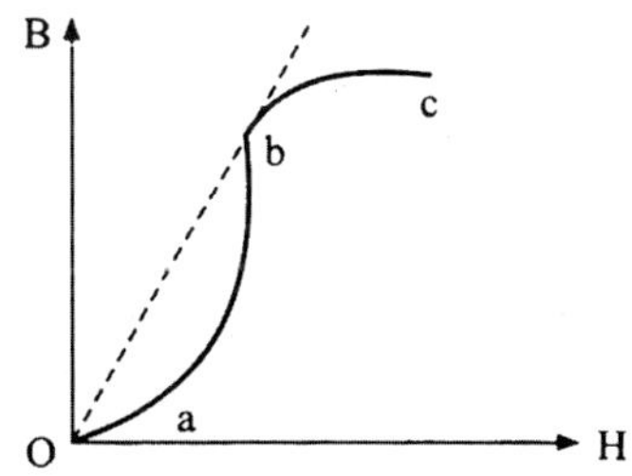

Fig. 5.1.

(a) point a (b) point b
(c) point c (d) point O

137. In Fig. 5.1 the minimum value of μ is at point
(a) a (b) b
(c) c (d) O

138. The magnetic energy per unit volume of a specimen brought to flux denstiy B from zero =
(a) $\int_0^B H dB\ Jm^{-3}$ (b) $0.5 \int_0^B H dB\ Jm^{-3}$
(c) $2 \int_0^B H dB\ Jm^{-3}$ (d) $\int_0^H H dH\ Jm^{-3}$

139. If the initial magnetisation curve is retraced and no hysteresis is present, the area of hysteresis loop is
(a) zero (b) very small
(c) very large (d) infinite

140. The dimensions of product BH are that of
(a) energy (b) energy density
(c) power (d) power density

141. Which of the these is not a permanent magnetic material?
(a) Chrome steel (b) Silicon steel
(c) Alnico (d) Platinum cobalt

142. In magnetisation, the word deperming means
(a) magnetisation
(b) demagnetisation
(c) removing the permanent magnetisation so that residual flux density is reduced to zero under the condition of zero H
(d) Increasing the permanent magnetisation to the maximum possible limit

143. The unit for reluctance is
(*d*) H (*b*) H^{-1}
(*c*) A (*d*) ampere turns

144. If l is the distance between two points, a is the cross-sectional area of magnetic circuit and μ is the permeability, the reluctance between these points =
(*d*) $l/\mu a$ (*b*) $\mu l/a$
(*c*) $\frac{\mu a}{l}$ (*d*) $\frac{1}{\mu} al$

145. Two magnetic blocks having reluctance, $\mathfrak{R}_1$ and $\mathfrak{R}_2$ are connected in series. The total reluctance is
(*a*) $\mathfrak{R}_1 + \mathfrak{R}_2$
(*b*) $\mathfrak{R}_1 \mathfrak{R}_2/(\mathfrak{R}_1 + \mathfrak{R}_2)$
(*c*) more than $\mathfrak{R}_1$, but less than $\mathfrak{R}_1 + \mathfrak{R}_2$
(*d*) none of the above

146. The units of permeance are
(*a*) H (*b*) H^{-1}
(*c*) amper turns (*d*) weber

147. Which of the following is not true for magnetic field mapping?
(*d*) Magnetic flux tubes are continuous
(*b*) A magnetic flux tube is parallel to the field
(*c*) The field and magnetic potential lines intersect at right angles
(*d*) At the boundary between air and iron, the field lines on the air side of the boundary are parallel to the boundary surface

148. A toroid has N turns carrying current I. The radius of toroid is r and permeability of material is μ. The flux density B =
(*a*) $\mu NI/2\pi r$ (*b*) $NI/2\pi\mu r$
(*c*) $NI/\pi\mu r$ (*d*) $NI/\mu r$

149. A magnetic circuit has an air gap of width g and area A. If flux density is B, the total energy stored in the gap is
(*a*) B^2Ag/μ_0 (*b*) $B\,Ag/2\mu_0$
(*c*) $B^2Ag/2\mu_0$ (*d*) $B^2Ag/2A\mu_0$

150. Laplace equation in rectangular coordinates is
(*a*) $\frac{\partial^2 V}{\partial x^2} + \frac{\partial^2 V}{\partial y^2} + \frac{\partial^2 V}{\partial z^2} = \text{constant}$
(*b*) $\frac{\partial^2 V}{\partial x^2} + \frac{\partial^2 V}{\partial y^2} + \frac{\partial^2 V}{\partial z^2} = \text{infinite}$
(*c*) $\frac{\partial^2 V}{\partial x^2} + \frac{\partial^2 V}{\partial y^2} + \frac{\partial^2 V}{\partial z^2} = 0$
(*d*) $\frac{\partial^2 V}{\partial x^2} + \frac{\partial^2 V}{\partial y^2} + \frac{\partial^2 V}{\partial z^2} = 0 \text{ or constant}$

151. A parallel plate capacitor has its plates vertical so that the Laplace equation reduces to $\frac{\partial^2 V}{\partial x^2} = 0$ (there is no variation of potential in x and y directions). Then the solution of this equation is (if potential between the plates is V_1. and separation is d)
(*a*) $V = \frac{V_1}{d} x$ (*b*) $V = \frac{V_1}{d^2} x^2$
(*c*) $V = \frac{V_1}{\sqrt{d}} \sqrt{x}$ (*d*) $V = \frac{V_1 x^3}{d^3}$

152. Lapalce equation in cylindrical coordinates is
(*a*) $\frac{\partial^2 V}{\partial r^2} + \frac{1}{r}\frac{\partial V}{\partial r} + \frac{1}{r^2}\frac{\partial^2 V}{\partial \phi^2} + \frac{\partial^2 V}{\partial z^2} = 0$
(*b*) $\frac{\partial^2 V}{\partial r^2} + \frac{\partial V}{\partial r} + \frac{1}{r^2}\frac{\partial^2 V}{\partial \phi^2} + \frac{\partial^2 V}{\partial z^2} = 0$
(*c*) $\frac{\partial^2 V}{\partial r^2} + \frac{1}{r}\frac{\partial V}{\partial r} + \frac{1}{r^3}\frac{\partial^2 V}{\partial \phi^2} + \frac{\partial^2 V}{\partial z^2} = 0$
(*d*) $\frac{\partial^2 V}{\partial r^2} + \frac{1}{r}\frac{\partial V}{\partial r} + \frac{1}{r^2}\frac{\partial^2 V}{\partial \phi^2} + \frac{1}{z^2}\frac{\partial^2 V}{\partial z^2} = 0$

153. In rectangular coordinates, Poisson's equation is
(*a*) $\frac{\partial^2 V}{\partial x^2} + \frac{\partial^2 V}{\partial y^2} + \frac{\partial^2 V}{\partial z^2} = 0$
(*b*) $\frac{\partial^2 V}{\partial x^2} + \frac{\partial^2 V}{\partial y^2} + \frac{\partial^2 V}{\partial z^2} = -\frac{\rho}{\epsilon}$
(*c*) $\frac{\partial^2 V}{\partial x^2} + \frac{\partial^2 V}{\partial y^2} + \frac{\partial^2 V}{\partial z^2} = \frac{\rho}{\epsilon}$
(*d*) none of the above

154. When a wire moves with velocity v in a magnetic field B, the emf induced in the wire is
(*a*) $\oint_l (v \times B) \cdot dl$ (*b*) $\oint_l (v \cdot B) \cdot dl$
(*c*) $\oint_l vB\, dl$ (*d*) none of the above

155. The total emf induced due to motion as well as transformer action is given by
(*a*) $\oint_L (v \times B) \cdot dl - \oint_s \frac{\partial B}{\partial t} \cdot ds$
(*b*) $\oint_s \frac{\partial B}{\partial t} \cdot ds - \oint_L (v \times B) \cdot dl$
(*c*) $\oint_L (v \cdot B) \cdot dl - \oint_s \frac{\partial B}{\partial t} \cdot ds$
(*d*) none of the above

156. If a rectangular loop of length l and radius r is rotating at angular velocity ω in a field B, the motional emf is
(*a*) $2\,\omega r l\, B \sin \omega t$ (*b*) $\omega r l\, B \sin \omega t$
(*c*) $\omega r^2 l\, B \sin \omega t$ (*d*) $2\omega^2 r l\, B \sin \omega t$

157. Stokes theorem is
(*a*) $\oint_l E \cdot dl = \int_s (\nabla \times E).\, ds$
(*b*) $\oint_l E \cdot dl = \int_s (\nabla \cdot E).\, ds$
(*c*) $\oint_l E \times dl = \int_s (\nabla \times E).\, ds$
(*d*) $\oint_l E \times dl = \int_s (\nabla \cdot E) ds$

158. Maxwell's equation from Faraday in differential form is

(a) $\nabla \times E = \frac{-\partial B}{\partial t}$ (b) $\nabla \cdot E = \frac{-\partial B}{\partial t}$

(c) $\nabla E = -\frac{\partial B}{\partial t}$ (d) $\nabla \times E^2 = \frac{-\partial B}{\partial t}$

159. The current through a pure capacitor is

(a) displacement current
(b) conduction current
(c) partly displacement current and partly conduction current
(d) either conduction current or displacement current

160. The current through a capacitor having a leaky dielectric is

(a) only displacement current
(b) only conduction current
(c) partly displacement current and partly conduction current
(d) either dispalcement current or conduction current

161. The displacement current is equal to

(a) $\in \frac{dD}{dt}$ (b) $\in \frac{dE}{dt}$

(c) $\frac{1}{\in}\frac{dE}{dt}$ (d) $\frac{1}{\in}\frac{dD}{dt}$

162. Maxwell's equation from Ampere in integral form is

(a) $\oint_l H \cdot dl = \oint_s \left(J + \frac{\partial D}{\partial t}\right) ds$

(b) $\oint_l H \times dl = \oint_s \left(J + \frac{\partial D}{\partial t}\right) ds$

(c) $\oint_l H \cdot dl = \oint_s \left(J - \frac{\partial D}{\partial t}\right) ds$

(d) none of the above

163. Maxwell's equation from ampere in differential form is

(a) $\nabla \cdot E = \sigma E + \in \frac{\partial E}{\partial t}$ (b) $\nabla \times E = \sigma E + \in \frac{\partial E}{\partial t}$

(c) $\nabla \times E = \sigma E + \in \frac{\partial B}{\partial t}$ (d) $\nabla \times E = \frac{1}{\sigma} E + \in \frac{\partial E}{\partial t}$

164. Maxwell's equation from Faraday in integral form is

(a) $\oint_l E.dl = -\int_s \frac{\partial B}{\partial t} \cdot ds$ (b) $\oint_l E \times dl = -\int_s \frac{\partial B}{\partial t} \cdot ds$

(c) $\oint_l E \cdot dl = -\int_s \frac{\partial B}{\partial t} \times ds$ (d) none of the above

165. Taking the partial derivative of a phasor having angular frequency ω is equivalent to

(a) multiplying the phasor by $j\omega$
(b) multiplying the phasor by ω
(c) dividing the phasor by $j\omega$
(d) dividing the phasor by ω

166. The phasor form of Maxwell's equation from Ampere is

(a) $\nabla \cdot H = (\sigma + j\omega\in)E$ (b) $\nabla \times H = (\sigma + j\omega\in)E$

(c) $\nabla^2 H = (\sigma + j\omega\in)E$ (d) $\nabla \cdot H = (\sigma - j\omega\in)E$

167. The phenomenon of dielectric hysteresis

(a) is important in dc only
(b) is important in ac only
(c) is important in dc as well as ac
(d) none of the above

168. If the dielectric constant ∈ is expressed as a complex quantity $\in' - j\in''$, the total capacitor current is

(a) $J_{total} = (\sigma + \omega\in'')E + j\omega\in' E$
(b) $J_{total} = (\sigma + \omega\in')E + j\omega\in'' E$
(c) $J_{total} = (\sigma - \omega\in'')E + j\omega\in' E$
(d) $J_{total} = (\sigma + \omega\in')E + j\omega\in'' E$

169. If conductivity is σ and dielectric constant (as a complex quantity) is $\in' - j\in''$, the loss angle δ is given by the equation

(a) $\tan\delta = (\sigma + \omega\in'')/\omega\in'$ (b) $\tan\delta = (\sigma + \omega\in')/\omega\in''$
(c) $\tan\delta = (\sigma - \omega\in'')/\omega\in'$ (d) $\tan\delta = (\sigma - \omega\in')/\omega\in$

170. If electric field intensity in a capacitor is E conductivity is σ and dielectric constant tan $\in = \in' - j\in''$, the power dissipated per unit volume at frequency ω is

(a) $(\sigma + \omega\in'')E$ (b) $(\sigma + \omega\in'')E^2$
(c) $(\sigma + \omega\in)E$ (d) $(\sigma + \omega\in')E^2$

171. A non-conducting medium has ∈ = 4 and loss tangent = 0.001. The power dissipated per unit volume at 9 MHz and with E = 1 kVm^{-1} is likely to be about

(a) 200 W/m^3 (b) 20 W/m^3
(c) 2 W/m^3 (d) 0.2 W/m^3

172. A parallel plate capacitor has a lossy dielectric with tan δ = 0.004. The area of plates is 350 mm^2 and distance between plates is 12 mm, $\in_r$ = 14. At 300 V (rms) and 3 MHz, the power loss is about

(a) 100 W (b) 8.6 W
(c) 2.2 W (d) 1.2 W

173. In transverse electromagnetic mode transmission lines

(a) E is entirely transverse
(b) H is entirely transverse
(c) both E and H are entirely transverse
(d) none of the above

174. If a plane wave is travelling in + x direction, the electric field E_y is only in y direction and magnetic field H_z is only in z direction. Then

(a) $\frac{\partial E_y}{\partial x} = -\mu \frac{\partial H_z}{\partial t}$ (b) $\frac{\partial E_y}{\partial x} = +\mu \frac{\partial H_z}{\partial t}$

(c) $\frac{\partial E_y}{\partial x} = -\frac{\partial H_z}{\partial t}$ (d) $\in \frac{\partial E_y}{\partial x} = -\mu \frac{\partial H_z}{\partial t}$

175. For a plane wave travelling in + x direction so that E has only component E_y in y direction, the wave equation in E_y is

(a) $\frac{\partial^2 E_y}{\partial t^2} = \frac{\mu \partial^2 E_y}{\partial x^2}$ (b) $\frac{\partial^2 E_y}{\partial t^2} = \frac{1}{\mu\in}\frac{\partial^2 E_y}{\partial x^2}$

(c) $\frac{\partial^2 E_y}{\partial t^2} = \mu\in \frac{\partial^2 E_y}{\partial x^2}$ (d) $\mu\in \frac{\partial^2 E_y}{\partial t^2} = \frac{-\partial^2 E_y}{\partial x^2}$

176. For a plane wave travelling in + x direction, the magnetic field H has a component H_z in z direction only. The wave equation in H_z is

(a) $\frac{\partial^2 H_z}{\partial t^2} = \frac{-1}{\mu\in}\frac{\partial^2 H_z}{\partial x^2}$ (b) $\frac{\partial^2 H_z}{\partial t^2} = \frac{1}{\mu\in}\frac{\partial^2 H_z}{\partial z^2}$

(c) $\frac{\partial^2 H_z}{\partial t^2} = \frac{1}{\mu \in} \frac{\partial^2 H_z}{\partial x^2}$ (d) $\frac{\partial^2 H_z}{\partial t^2} = \frac{\mu \partial^2 H_z}{\partial x^2}$

177. Wave velocity v in a medium having dielectric constant $\in$ and permeability μ is

(a) $v = \mu \in$ (b) $v = (\mu \in)^{0.5}$

(c) $v = \left(\frac{1}{\mu \in}\right)^{0.5}$ (d) $v = (\mu / \in)^{0.5}$

178. In free space the velocity of electromagnetic wave is about

(a) 3000 m/μs (b) 300 m/μs

(c) 30 m/μs (d) 3 m/μs

179. For wave travelling in negative x direction

(a) $\frac{dx}{dt} = v$ (b) $\frac{dx}{dt} = -v$

(c) $\frac{d^2 x}{dt^2} = -v$ (d) $\frac{d^2 x}{dt^2} = v$

180. Which of the following waves does not travel in positive x direction?

(a) $E_y = \sin(\omega t - \beta x)$ (b) $E_y = \cos(\omega t - \beta x)$

(c) $E_y = \cos(\omega t + \beta x)$ (d) $E_y = \cos(\beta x - \omega t)$

181. Wave equation for a linearly polarised wave (E in y direction) travelling in x direction is

(a) $\frac{\partial^2 E_y}{\partial x^2} = (j\omega\mu\sigma - \omega^2 \mu \in) E_y$

(b) $\frac{\partial^2 E_y}{\partial x^2} = (j\omega\mu\sigma + \omega^2 \mu \in) E_y$

(c) $\frac{\partial^2 E_y}{\partial x^2} = j(\omega\mu\sigma + \omega^2 \mu \in) E_y$

(d) $\frac{\partial^2 E_y}{\partial x^2} = (j\omega\mu\sigma + \omega\mu \in) E_y$

182. Transmission line wave equations from circuit point of view are

(a) $\frac{d^2 V}{dx^2} - ZYV = 0$ and $\frac{d^2 I}{dx^2} - ZYI = 0$

(b) $\frac{d^2 V}{dx^2} + ZYV = 0$ and $\frac{d^2 I}{dx^2} + ZYI = 0$

(c) $\frac{d^2 V}{dx^2} - ZYV = 0$ and $\frac{d^2 I}{dx^2} + ZYI = 0$

(d) $\frac{d^2 V}{dx^2} + ZYV = 0$ and $\frac{d^2 I}{dx^2} - ZYI = 0$

183. Propagation constant γ =

(a) ZY (b) $\sqrt{ZY}$

(c) $(ZY)^{1.5}$ (d) $(ZY)^{-0.5}$

184. Characteristic impedance Z_0 =

(a) $\sqrt{ZY}$ (b) ZY

(c) $\sqrt{Z/Y}$ (d) $\sqrt{Y/Z}$

185. Intrinsic impedance of medium Z_0 =

(a) $\sqrt{\mu \in}$ (b) $\sqrt{\mu / \in}$

(c) $\sqrt{\in / \mu}$ (d) $\in \mu$

186. Intrinsic impedance of free space is about

(a) 377 Ω (b) 37.7 Ω

(c) 3.77 Ω (d) 0.377 Ω

187. Intrinsic impedance Z_0 of a conducting medium =

(a) $\frac{j\omega\mu}{\sigma + j\omega\in}$ (b) $\sqrt{\frac{j\omega\mu}{\sigma + j\omega\in}}$

(c) $\frac{\sigma + j\omega\in}{j\omega\mu}$ (d) $\sqrt{\frac{\sigma + j\omega\in}{j\omega\mu}}$

188. Characteristic impedance Z_0 for a loss less line =

(a) $\sqrt{\frac{L}{C}}$ (b) $\sqrt{\frac{C}{L}}$

(c) $\sqrt{LC}$ (d) $\frac{1}{\sqrt{LC}}$

189. In a loss less line with ρ_v as the reflection coefficient for the voltage, voltage standing wave ratio (VSWR) =

(a) $\frac{1 + |\rho_v|}{1 - |\rho_v|}$ (b) $\frac{1 - |\rho_v|}{1 + |\rho_v|}$

(c) $\left(\frac{1 + \rho_v}{1 - \rho_v}\right)^2$ (d) $\sqrt{\frac{1 + \rho_v}{1 - \rho_v}}$

190. In a loss less line having ρ_v as the reflection coefficient for voltage and τ_v as transmission coefficient for voltage

(a) $\tau_v = 1 + \rho_v$ (b) $\tau_v = 1 - \rho_v$

(c) $\tau_v = 1 + \rho_v^{0.5}$ (d) $\tau_v = \sqrt{1 + \rho_v}$

191. For a loss less line short-circuited at load end, the transmission coefficients for voltage and current are

(a) 0 and 2 respectively (b) 2 and 0 respectively

(c) 1 each (d) none of the above

192. For a loss less line open-circuited at far end, the transmission coefficients for voltage and current are

(a) 0 and 2 respectively (b) 2 and 0 respectively

(c) 1 each (d) none of the above

193. If μ_r and $\in_r$ are relative permeability and relative permittivity respectively, the relative phase velocity and index of reflection are

(a) $\frac{1}{\sqrt{\mu_r \in_r}}$ and $\sqrt{\mu_r \in_r}$ respectively

(b) $\sqrt{\mu_r \in_r}$ and $\frac{1}{\sqrt{\mu_r \in_r}}$ respectively

(c) $\mu_r \in_r$ and $\frac{1}{\mu_r \in_r}$ respectively

(d) $\frac{1}{\mu_r \in_r}$ and $\mu_r \in_r$ respectively

194. If E_0 is the amplitude of incident wave and E_1 is the amplitude of reflected wave, then for a pure standing wave

(a) $|E_1| = |E_0|$

(b) $|E_1| > |E_0|$

(c) $|E_1| < |E_0|$

(d) $|E_1|$ may be equal to or less than $|E_0|$

195. Poynting vector (instantaneous value) S =
(*a*) E × H (*b*) E.H
(*c*) $\sqrt{E \times H}$ (*d*) $\sqrt{E.H}$

196. Units of Poynting vector are
(*a*) W (*b*) Wm^{-1}
(*c*) Wm^{-2} (*d*) Wm^{-3}

197. Power P and Poynting vector S are related as
(*a*) $P = \oint_l S.dl$ (*b*) $P = \oint_s S.ds$
(*c*) $P = \oint_v S.dv$ (*d*) none of the above

198. If a coaxial line has a perfectly conducting solid tube as the outer conductor, then
(*a*) both electric and magnetic fields are confined internally
(*b*) electric field is confined internally
(*c*) magnetic field is confined internally
(*d*) neither electric nor magnetic field is confined internally

199. In a wave disgnated as TE_{mn}
(*a*) *m* and *n* are the number of half cycle variations in the larger and smaller transverse dimensions of wave guide respectively
(*b*) *m* and *n* are the number of half cycle variations in the smaller and larger transverse dimensions of the wave guide respectively
(*c*) *m* and *n* are the numbers of half cycle variations in any transverse dimension of the wave guide
(*d*) none of the above

200. If θ_i is angle of incidence and θ_t is angle of refraction, as a wave goes from one loss less medium to another, Snell's law is
(*a*) $\sin\theta_t = \sqrt{\frac{\mu_1 \epsilon_1}{\mu_2 \epsilon_2}} \sin\theta_i$ (*b*) $\sin\theta_t = \sqrt{\frac{\mu_2 \epsilon_2}{\mu_1 \epsilon_1}} \sin\theta_i$
(*c*) $\tan\theta_t = \sqrt{\frac{\mu_1 \epsilon_1}{\mu_2 \epsilon_2}} \tan\theta_i$ (*d*) $\tan\theta_t = \sqrt{\frac{\mu_2 \epsilon_2}{\mu_1 \epsilon_1}} \tan\theta_i$

201. For TE_{10} mode in a hollow rectangular wave guide
(*d*) E_x, E_z, H_y are not zero (*b*) E_y, H_x, H_z are not zero
(*c*) E_x, H_x, H_z are not zero (*d*) E_x, E_z, H_z are not zero

202. For TE_{11} mode in a hollow rectangular wave guide
(*a*) only $E_x = 0$ (*b*) only $E_z - 0$
(*c*) only $H_x = 0$ (*d*) only $H_z = 0$

203. Wave impedance is
(*a*) ratio of electric field component to magnetic field component at the same point of wave
(*b*) ratio of magnetic field component to electric field component at the same point of wave
(*c*) product of electric and magnetic field components at the same point of wave
(*d*) square root of the ratio of electric field component to magnetic field component at the same point of wave

204. Retina of a human eye contains a bundle of about
(*a*) 1000 optical fibres
(*b*) million optical fibres
(*c*) 10 million optical fibres
(*d*) 100 million optical fibres

205. For a cavity resonator Q =
(*a*) 2π (total energy stored/decrease in energy in 1 cycle)
(*b*) total energy stored/decrease in energy in 1 cycle
(*c*) 2π (decrease in energy in 1 cycle/total energy stored)
(*d*) π (total energy stored/decrease in energy in 1 cycle)

206. Which of the following is correct?
(*a*) Electromagnetic compatibility chamber is a resonant cavity
(*b*) Microwave oven is a resonant cavity
(*c*) Both electromagnetic compatibility chamber and microwave oven are resonant cavities
(*d*) None of the above

207. If I denotes time changing current, L is the length of current element, Q is charge and *v* is the acceleration of charge, the basic radiation equation for antennas is
(*a*) $\dot{I}L = Q\dot{v}$ (*b*) $\dot{I}v = QL$
(*c*) $\dot{I}/L = Q\dot{v}$ (*d*) $\dot{I}/L = Q\dot{v}$

208. Main beam efficiency of an antenna is
(*a*) $\frac{\text{Main lobe beam area solid angle}}{\text{Minor lobe beam area solid angle}}$
(*b*) $\frac{\text{Minor lobe beam area solid angle}}{\text{Main lobe beam area solid angle}}$
(*c*) $\frac{\text{Beam area solid angle}}{\text{Main lobe beam area solid angle}}$
(*d*) $\frac{\text{Main lobe beam area solid angle}}{\text{Beam area solid angle}}$

209. If Ω_a denotes beam area solid angle, directivity of an antenna is
(*a*) $4\pi/\Omega_a$ (*b*) π/Ω_a
(*c*) $\Omega_a/4\pi$ (*d*) $4\pi/\Omega_a$

210. A short dipole antenna means
(*a*) one whose electrical length is short as compared to wavelength
(*b*) one which is physically short
(*c*) one which is physically very very short
(*d*) any of the above

211. Which of the following statements regarding electric flux are true?
1. It begins on positive charges and terminates on negative charges
2. Flux is in the same direction as electric field intensity
3. Flux is proportional to electric field intensity
4. In SI systems the total flux emnating from a charge Q coulombs is Q coulombs

(*a*) 1 only (*b*) 1 and 2 only
(*c*) 1, 2, 3 only (*d*) 1, 2, 3 and 4

212. For a loss less transmission line
1. series resistance is zero
2. shunt conductance is zero
3. series resistance is infinite
4. shunt conductance is infinite

Which are correct?
(*a*) 1 and 2 (*b*) 2 and 3
(*c*) 2 and 4 (*d*) 1 and 4

213. A $\lambda/4$ transformer is used for
(*a*) light-loads
(*b*) high frequency loads
(*c*) connecting high impedance line to low impedance loads
(*d*) all the above

214. The dimensions of ϵ_0 are
(*a*) $M^{-1}L^{-3}T^{3}A^{4}$ (*b*) $M^{-1}L^{-3}T^{4}A^{2}$
(*c*) $M^{-2}L^{-3}T^{4}A^{4}$ (*d*) none of the above

215. The quantity $\dfrac{1}{\sqrt{\epsilon_0 \mu_0}}$ is SI units has the
(*a*) value 330 m/s (*b*) value 1.73×10^4 m/s
(*c*) dimensions LT^{-1} (*d*) none of the above

216. Joule/coulomb is a unit of
(*a*) electric field potential (*b*) potential
(*c*) charge (*d*) none of the above

217. An infinite number of concentric rings carry a charge Q each alternatively positive and negative as shown in Fig. 5.2. The radii of rings are 1, 2, 4, 8... metres in geometric progression. The potential at the centre is

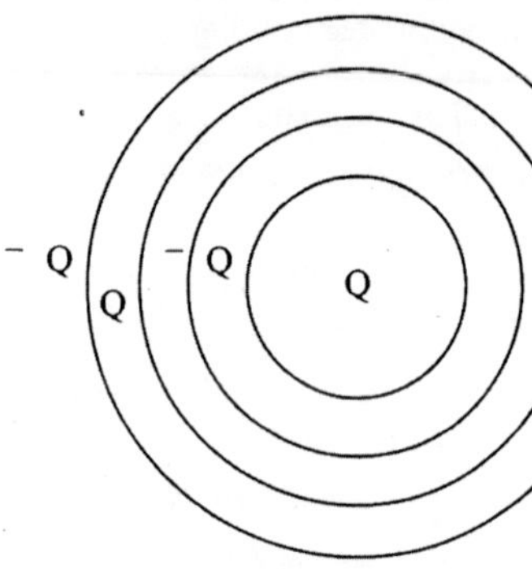

Fig. 5.2.

(*a*) zero (*b*) $\dfrac{Q}{12\pi\epsilon_0}$
(*c*) $\dfrac{Q}{8\pi\epsilon_0}$ (*d*) $\dfrac{Q}{6\pi\epsilon_0}$

218. Which of the following is not electromagnetic in nature
1. α rays 2. X-rays
3. γ-rays 4. cathode rays
(*a*) 1 and 2 only (*b*) 2 and 3 only
(*c*) 3 and 4 only (*d*) 1 and 4 only

219. Two concentric rings 1 and 2 carry equal and uniform charge densities. They are revolving at same angular speed ω about centre O. The ratio of flux densities due to these two rings at centre O is
(*a*) 1 : 1 (*b*) 1 : 2
(*c*) 1 : 4 (*d*) 2 : 1

220. Two electrons moving parallel to each other with the same velocity will
(*a*) attract each other if they move in the same direction
(*b*) repel each other more strongly when moving in the same direction than when moving in opposite direction
(*c*) repel each other less strongly when moving in the same direction than when moving in opposite direction
(*d*) none of the above

221. In a rectangular cavity resonator having width *a*, height *b* and length *d* such that $a > b > d$, the dominent mode is
(*a*) TE_{101} (*b*) TM_{101}
(*c*) TE_{110} (*d*) TM_{110}

222. Two linearly polarised plane waves A and B travelling in free space in positive *x* direction meet at a point. The wave A is polarised in *y* direction and wave B is polarised in *z* direction. The resulting wave is linearly polarised. Therefore, A and B
(*a*) have equal amplitude and are in time phase
(*b*) have equal amplitudes but need not be in time phase
(*c*) may have different amplitudes but must be in time phase
(*d*) none of the above

223. The transmission loss for a 3 GHz microwave system is 130 dB over a certain distance. If frequency is doubled the transmission loss will be
(*a*) 136 dB (*b*) 133 dB
(*c*) 127 dB (*d*) 124 dB

224. Which quantity is analogus to permeability
(*a*) current density (*b*) conductivity
(*c*) resistivity (*d*) consistency

225. Which of the following quantities have same units in electromagnetic and electrostatic systems?
1. Current 2. Electric energy
3. Electric power
Select the answer as per following codes
(*a*) 1, 2 and 3 (*b*) 1 and 2
(*c*) 1 and 3 (*d*) 2 and 3

226. A two port network has a loss of 6 dB. If P_1 and P_2 are input and output powers then
(*a*) $\dfrac{P_2}{P_1} = 4$ (*b*) $\dfrac{P_2}{P_1} = 2$
(*c*) $\dfrac{P_2}{P_1} = 0.707$ (*d*) $\dfrac{P_2}{P_1} = 0.5$

227. The type of time variation $\left(\dfrac{\partial}{\partial t}\right)$ implied with reference to field quantities in Maxwell's equation is
(*a*) sinusoidal
(*b*) non-sinsoidal
(*c*) both sinusoidal and non-sinusoidal
(*d*) exponential

228. Lorentz's force law for a point charge *q* is
(*a*) $\vec{F} = q(\vec{E} + v.\vec{B})$ (*b*) $\vec{F} = q(\vec{E} + v \times \vec{B})$
(*c*) $\vec{F} = q(\vec{B} + v \times \vec{E})$ (*d*) $\vec{F} = q(\vec{B} + v \cdot \vec{E})$

229. The equation $\nabla \times E = \dfrac{\partial B}{\partial t}$ is a generalisation of
(*a*) Ampere law (*b*) Gauss law
(*c*) Faraday'slaw (*d*) Biot Savart law

230. The skin depth δ due to radio frequency current is related to ω, μ, and σ as (*a*) $\delta = \frac{2}{\omega\mu\sigma}$ (*b*) $\delta = \frac{2}{\sqrt{\omega\mu\sigma}}$

(*c*) $\delta = \sqrt{\frac{2}{\omega\mu\sigma}}$ (*d*) $\delta = \frac{\sqrt{2}}{\omega\mu\sigma}$

231. For an open wire line having L and C as inductance and capacitance per unit length, excited by a source of frequency *f*, the correct equation is

(*a*) $\frac{dV(x)}{dx} = -j\omega LI(x)$ (*b*) $\frac{dV(x)}{dx} = j\omega LI(x)$

(*c*) $\frac{dV(x)}{dx} = j\omega CI(x)$ (*d*) $\frac{dV(x)}{dx} = -j\omega CI(x)$

232. Consider the following statements for a line having finite values of R, L, G, C

1. by proper choice of parameters, the line can be made loss less even if R and G are finite
2. the line, in general, will have amplitude and phase distortion

Of the above statements

(*a*) 1 alone is true (*b*) 2 alone is true
(*c*) both 1 and 2 are true (*d*) both 1 and 2 are false

233. For a lossy transmission line short-circuited at receiving end, input impedance is

(*a*) $Z_0 \coth \gamma l$ (*b*) $Z_0 \cot \gamma l$
(*c*) $Z_0 \tanh \gamma l$ (*d*) $Z_0 \tan \gamma l$

234. A short grounded vertical antenna has length $L = 0.05\,\lambda$ at a frequency *f*. Its radiation resistance is R ohms. At a frequency 2*f*, the radiation resistance will be

(*a*) R/2 ohms (*b*) R ohms
(*c*) 2R ohms (*d*) 4R ohms

235. The mouth diameters of two parabolic reflectors are in the ratio 2 : 1. Their power gains with respect to a resonant half wave dipole are in the ratio

(*a*) 2 : 1 (*b*) 1 : 4
(*c*) 4 : 1 (*d*) 1 : 2

236. The dominant mode in a wave guide is characterised by

(*a*) longest cut off wavelength
(*b*) shortest cut off wavelength
(*c*) infinite attenuation
(*d*) zero attenuation

237. The energy spent in moving a charge of 10 C from point *a* to point *b* is 50 J. The potential difference between *a* and *b* is

(*a*) 2 V (*b*) 5V
(*c*) 10 V (*d*) 100V

238. Point P is at a large distance from the centre O of a short dipole formed by two point charges lying in horizantal plane. If θ is the angle between OP and dipole axis, then component of electric field at P is proportional to

(*a*) cos θ (*b*) sin θ
(*c*) tan θ (*d*) none cf the above

239. Two concentric spherical conducting shells are held at different potentials. Their centre coincides with the origin of spherical coordinate system. Which one of the following statements is true for E field in annular gap

(*a*) is purely radial and independent of (*r*, θ, ϕ)
(*b*) is purely radial, indenpendent of (θ, ϕ) and varies as 1/*r*
(*c*) the field lines depend on (θ, ϕ) but not on *r*
(*d*) none of the above

240. A conductor having surface charge density σ is embelled in a dielectric having permittivity ∈. The field in dielectric is E. The pressure *p* on the conductor surface is equal to electric charge density in medium. Then *p* is

(*a*) $\sigma^2/2$ (*b*) $\frac{\sigma}{2\epsilon}$

(*c*) $\frac{\sigma}{4\pi\epsilon}$ (*d*) $\frac{\sigma^2}{2\epsilon}$

241. A generator of 50 ohm internal impedance and operating at 1 GHz feeds a 75 Ω load through a coaxial line of $Z_0 = 50\ \Omega$. The VSWR on the line is

(*a*) 0.5 (*b*) 1.5
(*c*) 2.5 (*d*) 1.75

242. A loss less line having $Z_0 = 50$ ohm is terminated in (*a*) 25 Ω (*b*) 100 Ω loads. As regards voltage standing wave patterns in the two cases which statement correct

(*a*) the two patterns will be identical and cannot be distinguished
(*b*) VSWR will be higher in case (*a*)
(*c*) VSWR will be higher in case (*b*)
(*d*) the two patterns will be identical except for a relative spatial shift of quarter wave length in the two cases

243. A 50 ohm loss less line is terminated in 100 Ω load. It is excited by 30 MHz source of internal resistance 50 Ω. For maximum power transfer the length of line is

(*a*) 5 m (*b*) 1.25 m
(*c*) 2.5 m (*d*) 10 m

244. A quarter wave transformer made of air filled coaxial cable matches two lines of 50 Ω and 72 Ω characteristic impedance. The inner conductor of line is 10 mm in diameter, the diameter of outer conductor is

(*a*) 16 mm (*b*) 20 mm
(*c*) 27 mm (*d*) 32 mm

245. The input impedance of a short-circuited line of length less than quarter wave length is

(*a*) purely resistive (*b*) purely inductive
(*c*) purely capacative (*d*) complex

246. A pair of lossless lines is terminated by resistance = $2R_0$ where R_0 is characteristic impedance. It is excited by a source of voltage E and source resistance $2R_0$. The voltage build up at middle of line is given by (t_0 is the time taken by wave to go from one end to the other end)

(*a*)

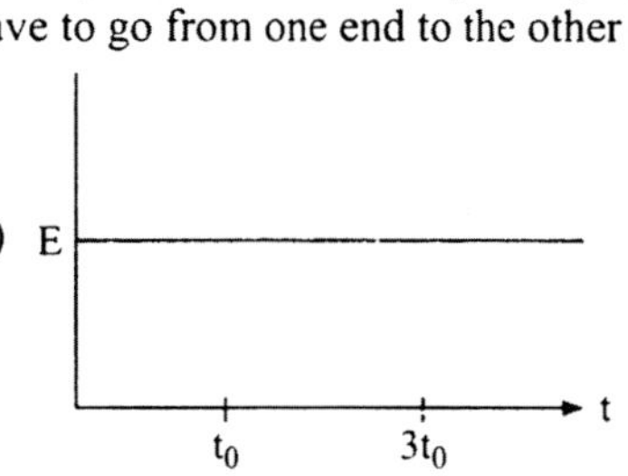

(b)

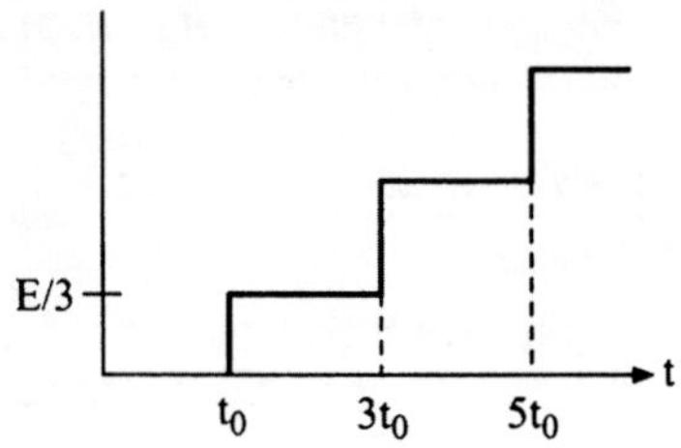

(c)

(d)

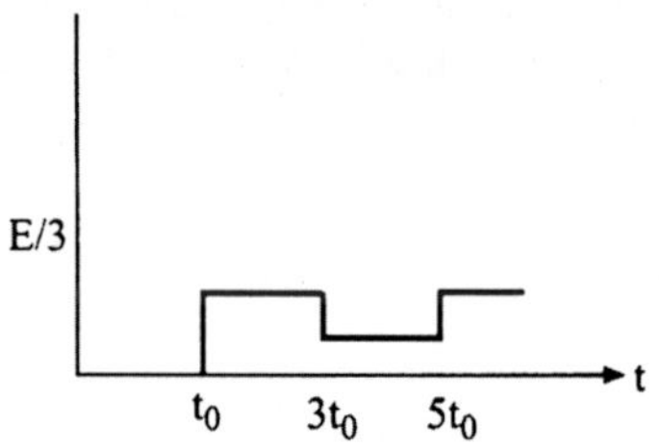

Fig. 5.3.

247. The antenna used commonly for telecast in UHF band is
(a) dipole antenna (b) turnstile antenna
(c) Yagi antenna (d) rhombic antenna

248. A short dipole and half wave dipole have radiation resistances of 0.4 Ω and 72 Ω respetively. If the former requires 10 A rms current for output power P, the current required by the half wave dipole for same power is
(a) 0.25 A (b) 0.5 A
(c) 0.75 A (d) 1.25 A

249. The electric field of an electromagnetic wave at a point in free space is in $+y$ direction and magnetic field is in $-x$ direction. The direction of power flow is in
(a) $+x$ direction (b) $+y$ direction
(c) $+z$ direction (d) $-z$ direction

250. The o.c. and s.c. impedances of a line are 20 Ω and 5 Ω respectively. Then Z_0 of the line =
(a) 100 Ω (b) 50 Ω
(c) 25 Ω (d) 10 Ω

251. A long straight wire carries a current I_1 and a rectangular loop carries current I_2 as shown in Fig. 5.4. The net force on loop will be in

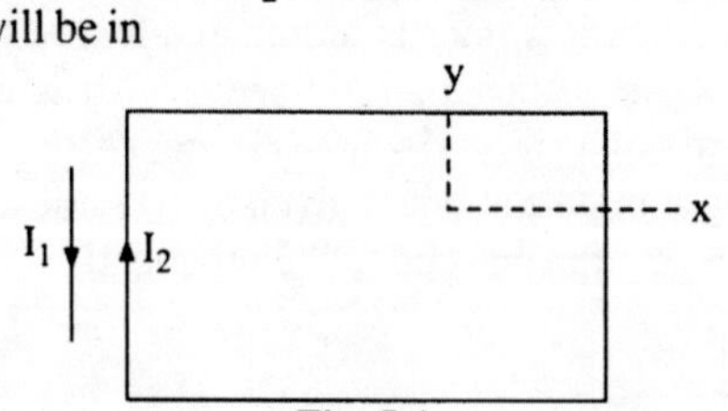

Fig. 5.4.

(a) $+x$ direction (b) $-x$ direction
(c) $+y$ direction (d) $-y$ direction

252. A line has inductance 0.294 μ H/m and capacitance 60 pF/m. Its characteristic impedance is
(a) 49 Ω (b) 60 Ω
(c) 70 W (d) 140 W

253. When a dielectric is placed in an electric field, E, then D, E and P (polarization) are related as
(a) $\vec{D} = \epsilon_0 \vec{E} \times \vec{P}$ (b) $\vec{D} = \epsilon_0 \vec{E} - \vec{P}$
(c) $\vec{D} = \epsilon_0 (\vec{E} + \vec{P})$ (d) $\vec{D} = \epsilon_0 \vec{E} + \vec{P}$

254. The reflection coefficient of a line in terms of load impedance Z_L and characteristic impedance Z_0 is
(a) $\dfrac{Z_L + Z_0}{Z_0 - Z_L}$ (b) $\dfrac{Z_0 Z_L}{Z_0 - Z_L}$
(c) $\dfrac{Z_L - Z_0}{Z_L + Z_0}$ (d) $\dfrac{Z_L - Z_0}{Z_0 Z_L}$

255. For a TEM wave, the propagation constant of line in terms of its parameters R, L, G, C is
(a) $(R + j\omega L)(G + j\omega C)$
(b) $\{(R + j\omega L)(G + j\omega C)\}^{0.5}$
(c) $(R + j\omega L)(G - j\omega C)$
(d) $(R + j\omega L)(G - j\omega C)^{0.5}$

256. The electric field at equipotential surface is
(a) always parallel to surface
(b) zero
(c) unity
(d) always perpendicular to surface

257. For a wave equation in a conducting medium, match List I and List II and select the answer using the given codes

List I (Parameter)	List II (Expression)
A. Propagation constant	1. $(2/\omega\mu\sigma)^{0.5}$
B. Skin depth	2. $\left(\dfrac{\omega\mu_0}{2\sigma}\right)^{0.5}$
C. Sheet resistance	3. $(j\omega\mu_0\sigma)^{0.5}$
	4. $(\omega\mu_0\sigma/2)^{0.5}$

Codes:

	A	B	C
(a)	3	2	1
(b)	3	1	2
(c)	2	3	4
(d)	2	4	3

258. An electromagnetic wave is obliquely incident at the surface of a dielectric medium 2 (constants μ_0, ϵ_2) from dielectric 1 (constants μ_0, ϵ_1). θ_i incidence angle and θ_c is critical angle. For total internal reflection
(a) $\epsilon_1 > \epsilon_2$ and $\theta_i < \theta_c$ (b) $\epsilon_1 < \epsilon_2$ and $\theta_i > \theta_c$
(c) $\epsilon_1 < \epsilon_2$ and $\theta_i = \theta_c$ (d) $\epsilon_1 > \epsilon_2$ and $\theta_i > \theta_c$

259. A circular coil of 100 turns and radius 0.1 m carries a current 0.1 A. Its magnetic dipole moment is
(a) 0.0001 π (b) 0.001 π
(c) 0.01 π (d) 0.1 π

260. The absorption of em waves in atmosphere is due to
(a) nitrogen and oxygen
(b) nitrogen and hydrogen
(c) oxygen and water vapour
(d) nitrogen and water vapour

261. Match List I with List II and given correct answer using the given codes

List I (Antennas)	**List II (Application)**
A. Cassegrain antenna	1. Large bandwidth
B. Yagi antenna	2. Direction finding
C. Parabolic reflector antenna	3. Radar
D. Loop antenna	4. Directional transmission Codes

Codes:

	A	B	C	D
(*a*)	1	4	2	3
(*b*)	3	1	4	2
(*c*)	2	4	3	1
(*d*)	3	2	4	1

262. What are the directions of E and H in TEM mode of transmission?
(*a*) Both E and H are entirely transverse to direction of propagation
(*b*) E is entirely transverse but H has a component in the direction of propagation
(*c*) H is entirely transverse but E has a component in the direction of propagation
(*d*) Both E and H have components in the direction of propagation

263. In an ideal transmission line with matched Load, the VSWR and reflection coefficient are respectively
(*a*) 1 and 1 (*b*) 0 and 1
(*c*) infinity (*d*) 1 and 0

264. An antenna of input resistance 73 ohms is connected to 50 ohm line. Neglecting losses, the efficiency is
(*a*) 0.81 (*b*) 0.19
(*c*) 0.97 (*d*) 0.27

265. Which of the following antennas are frequency independent
1. folded dipole 2. half wave dipole
3. parabolic reflector 4. helical antenna
Select your answer as per following codes
(*a*) 2 and 4 (*b*) 3 and 4
(*c*) 1, 3 and 4 (*d*) 1, 2 and 3

266. The critical angle for an electromagnetic wave passing from quartz ($\in_r = 4$) to air ($\in_r = 1$) is
(*a*) 15° (*b*) 30°
(*c*) 45° (*d*) 90°

267. Consider the following features
1. easier to use
2. lower power losses
3. higher opening frequencies possible
The advantages of wave guides over coaxial lines include
(*a*) 1 and 2 (*b*) 1 and 3
(*c*) 2 and 3 (*d*) 1, 2 and 3

268. In a medium $\in_r = 4$ and $\mu_r = 1$. The velocity of electromagnetic waves in this medium is
(*a*) 0.75×10^8 m/s (*b*) 1.5×10^8 m/s
(*c*) 3×10^8 m/s (*d*) 6×10^8 m/s

269. Relative permittivities of polysterene and air are 2.7 and 1 respectively. A wave is incident at an angle 30° and goes from air to polystererne. The angle of transmission is
(*a*) 10.2° (*b*) 17.7°
(*c*) 35.4° (*d*) 55.2°

270. A wave incident at an angle of 30° goes from polystrene into air. For polystrene $\in_r = 2.7$ and $\mu_r = 1$. For air $\mu_r = 1$. The angle of transmission is
(*a*) 10.2° (*b*) 17.7°
(*c*) 35.4° (*d*) 52.2°

271. A plane wave is incident from water onto water-air interface at 45°. For water $\in_r = 81$ and $\mu_r = 1$. The critical angle is about
(*a*) 6.4° (*b*) 12.8°
(*c*) 25° (*d*) 50°

272. A wave is incident from a more dense medium into less dense medium. Incident angle and critical angle are θ_i and θ_{ic}. For total internal reflection
(*a*) $\theta_i < \theta_{ic}$ (*b*) $\theta_i < \theta_{ic}$
(*c*) $\theta_i \geq \theta_{ic}$ (*d*) $\theta_i = 0$

273. If angle of incidence is 60° and critical frequency is 60 MHz, MUF is
(*a*) 45.9 MHz (*b*) 111.7 MHz
(*c*) 120 MHz (*d*) 150 MHz

274. Which of the following is incorrect for a microwave link?
(*a*) It uses line of sight transmission
(*b*) It uses reflector type antennas
(*c*) It can operate from 1 MHz onwards
(*d*) It uses repeaters to increase the range

275. An electric charge of 100 C is placed in a sphere of 10 m radius. The electric displacement density (C/m^2) is
(*a*) 0.0833 (*b*) 0.833
(*c*) 1.666 (*d*) 10

276. An isolated conducting sphere has a radius $\frac{1}{4\pi\in_0}$ metres. If medium is air, the capacitance is
(*a*) 0 (*b*) 1 F
(*c*) 4π F (*d*) 5 F

277. Index of refraction η is
(*a*) $\sqrt{\mu_r \in_r}$ (*b*) $\sqrt{\mu \in}$
(*c*) $\sqrt{\frac{\mu_r}{\in_r}}$ (*d*) $\sqrt{\frac{\mu}{\in}}$

278. Paraffin has $\in_r = 2.1$ and $\mu_r = 1$. Index of refraction for paraffin is
(*a*) 1 (*b*) 1.45
(*c*) 2.1 (*d*) 3.2

279. Phase velocity of a wave in an unbounded medium of paraffin ($\in_r = 2.1$ and $\mu_r = 1$) is
(*a*) 207 m/μs (*b*) 300 m/μs
(*c*) 100 m/μs (*d*) zero

280. For a medium $\mu_r = 1$ and $\in_r = 81$. The wave velocity in this medium will be
(*a*) 300 m/μs (*b*) 100 m/μs
(*c*) 33.3 m/μs (*d*) 3.33 m/μs

281. For a medium having $\mu_r = 1$ and $\in_r = 81$, the index of refraction is
(*a*) 81 (*b*) 9
(*c*) 3 (*d*) 1.732

282. In a dispersive medium
(*a*) the phase velocity is constant
(*b*) the phase velocity is zero
(*c*) the phase velocity is a function of frequency
(*d*) none of the above

283. In non-dispersive medium
(*a*) phase velocity = group velocity
(*b*) phase velocity > group velocity
(*c*) phase velocity < group velocity
(*d*) phase velocity is either less or more than group velocity

284. If σ is condcutivity, ω is angular frequency and $\in$ is permittivity, then for a dielectric
(*a*) $\frac{\sigma}{\omega\in} < 1$ (*b*) $\frac{\sigma}{\omega\in} < 0.1$
(*c*) $\frac{\sigma}{\omega\in} < 0.01$ (*d*) none of the above

285. If σ is conductivity, ω is frequency and $\in$ is permittivity, the quantity $\sigma/\omega \in$
(*a*) is dimensionless (*b*) is always imaginary
(*c*) has unit of Siemens
(*d*) has dimensions $M^{-1}L^{-2}T^{-3}$

286. If σ is conductivity, ω is frequency and $\in$ is permittivity, the ratio $\frac{\sigma}{\omega\in}$ for a conductor is
(*a*) less than 1 (*b*) less than 10
(*c*) more than 1 (*d*) more than 100

287. The standing wave envolpe for a reflection coefficient of 1 is like
(*a*) a sine wave (*b*) a rectified sine wave
(*c*) a displaced sine wave (*d*) none of the above

288. The values of reflection coefficient and VSWR, for an open-circuited line, respectively are
(*a*) 1 and 0 (*b*) 1 and 1
(*c*) 1 and infinity (*d*) 0 and 0

289. In a dielectric
(*a*) the conduction current and displacement current are almost equal
(*b*) the displacement current is much greater than conduction current
(*c*) the conduction current is zero
(*d*) the conduction current is much greater than displacement current

290. A 75 Ω transmission line is to be terminated in two resistive loads R_1 and R_2 such that standing wave patterns in the two cases are the same. The values of R_1 and R_2 are
(*a*) 250 and 200 Ω (*b*) 225 and 25 Ω
(*c*) 100 and 150 Ω (*d*) 50 and 125 Ω

291. For a transmission line the input impedances in short-circuit and open-circuit terminations are 100 Ω and 64 Ω respectively. The characteristic impedance is
(*a*) 80 Ω (*b*) 164 Ω
(*c*) 36 Ω (*d*) 64 Ω

292. An ideal quarter wave line has characteristic impedance of 50 Ω and load impedance of 100 Ω. The input impedance is
(*a*) 25 Ω (*b*) 50 Ω
(*c*) 75 Ω (*d*) 100 Ω

293. A current element of a very small length and carrying a current $I_0 I^{j\omega t}$ radiates a total average power P_r into space. P_r is proportional to
(*d*) I_0 (*b*) I_0^2
(*c*) I_0^3 (*d*) $\sqrt{I_0}$

294. A point P_1 has coordinates (2, 0, 2) and point P_2 has coordinates (0, 2, 1). The vector R_{12} is
(*d*) $-2\hat{x} + 2\hat{y} - 1\hat{z}$ (*b*) $2\hat{x} - 2\hat{y} + 1\hat{z}$
(*c*) $-2\hat{x} - 2\hat{y} - 1\hat{z}$ (*d*) $2\hat{x} - 2\hat{y} - 1\hat{z}$

295. The distance between two points having coordinates (2, 0, 2) and (0, 2, 1) is
(*a*) 1.5 (*b*) 3
(*c*) 3.6 (*d*) 4

296. If $\hat{x}$ denotes unit vector in x direction $\hat{x} \cdot \hat{x}$ =
(*d*) 1 (*b*) 0
(*c*) 2 (*d*) 0.1

297. Vector $\vec{A} = \hat{x}A_x + \hat{y}A_y + \hat{z}A_z$ and vector $\vec{B} = \hat{x}B_x + \hat{y}B_y + \hat{z}B_z$. Then $\vec{A} \cdot \vec{B}$ =
(*a*) $A_xB_x + A_yB_y + A_zB_z$ (*b*) $A_xB_y + B_xA_y + A_zB_z$
(*c*) $A_xB_x - A_yB_y - A_zB_z$ (*d*) $-A_xB_x - A_yB_y - A_zB_z$

298. A force field F is in x direction and is increasing linearly with distance x. Work done by force in moving an object from $x = 1$ to $x = 2$ is
(*a*) 1.5 (*b*) 3
(*c*) 3.5 (*d*) 4.5

299. A radial force field decreases with distance as $F = \hat{r}\, r^{-2}$. The work done in moving from a point $r = \sqrt{2}$ to $2\sqrt{2}$ by direct radial path is
(*d*) $\frac{1}{2}$ (*b*) $\frac{1}{\sqrt{2}}$
(*c*) $\frac{1}{2\sqrt{2}}$ (*d*) $\frac{1}{4}$

300. A force field $F = \hat{r}\, r^{-2}$. Two paths exist from points A (1, 1) to B (2, 2). One is direct radial path and other is from (1, 1) to (2, 1) and then from (2, 1) to (2, 2). The work done in following first path is W_1 and that in following second path is W_2 then
(*a*) $W_1 = W_2$ (*b*) $W_2 > W_1$
(*c*) $W_2 < W_1$
(*d*) W_2 may be equal or greater than W_1

301. In cylindrical coordinates system (r, ϕ, z), an elemental length dL is given by
(*a*) $dr^2 + r^2d\phi^2 + dz^2$ (*b*) $(dr^2 + r^2d\phi^2 + dz^2)^{0.5}$
(*c*) $dr^2 + d\phi^2 + dz^2$ (*d*) $(dr^2 + d\phi^2 + dz^2)^{0.5}$

302. In rectangular coordinate system, elemental length dl and elemental volume dv are respectively
(*a*) $(dx^2 + dy^2 + dz^2)^{0.5}$ and $dx\, dy\, dz$
(*b*) $dx^2 + dy^2 + dz^2$ and $(dx\, dy\, dz)^{1/3}$
(*c*) $(dx^2 + dy^2 + dz^2)^{0.5}$ and $(dx\, dy\, dz)^{1/3}$
(*d*) none of the above

303. In cylindrical cordinate system $(r\ \phi, z)$, elemental volume dv is given by
(a) $dr\ d\phi\ dz$ (b) $dr\ r\ d\phi\ dz$
(c) $(dr\ d\phi\ dz)^{1/3}$ (d) $(dr\ r\ d\phi\ dz)^{1/3}$

304. In spherical coordinates system (r, θ, ϕ) an element length dl is given by
(a) $dr^2 + r^2\ d\theta^2 + r^2 \sin^2\theta\ d\phi^2$
(b) $dr + d\theta^2 + d\phi^2$
(c) $(dr^2 + d\theta^2 + d\phi^2)^{0.5}$
(d) $(dr^2 + r^2 d\theta^2 + r^2 \sin^2\theta d\phi^2)^{0.5}$

305. In spherical coordinates (r, θ, ϕ), an elemental volume dv is given by
(a) $dr\ r\ d\theta\ r \sin\theta\ d\phi$ (b) $dr\ d\theta\ d\phi$
(c) $dr\ r\ d\theta\ r \cos\theta d\phi$ (d) $(dr\ d\theta\ d\phi)^{1/3}$

306. A vector r has angles α, β, γ between r and the axis x, y, z. The direction cosines, are
(a) $\cos\alpha, \cos\beta, \cos\gamma$ (b) $\sin\alpha, \sin\beta, \sin\gamma$
(c) $\tan\alpha, \tan\beta, \tan\gamma$ (d) $\sec\alpha, \sec\beta, \sec\gamma$

307. If $\hat{r}$ is a unit vector and rectangular unit vectors are $\hat{x}, \hat{y}, \hat{z}$ and α is the angle between r and x-axis, then dot product $\hat{x} \cdot \hat{r}$ is
(a) $\cos\alpha$ (b) $\sin\alpha$
(c) 0 (d) 1

308. Two + 1C charges are separated by 1 mm distance. The coulomb force is
(a) $\simeq 9.2 \times 10^{11}$ metric tons (b) $\simeq 0$
(c) $\simeq$ 1 metre ton (d) $\simeq$ 1 N

309. A negative point charge 1×10^{-8} C is situated at origin. If medium is air, electric field intensity at a point 3 m from origin is
(a) 10 V/m in negative r direction
(b) 10 V/m in positive r direction
(c) 1 V/m in negative r direction
(d) 1 V/m in positive r direction

310. A line charge 1×10^{-8} C/m is uniformly distributed along a line of length 4 m. To calculate electric field intensity at a point situated at distance r along a perpendicular from line charge and $r >> 4$, the line charge is replaced by a point charge q situated at centre of line charge. The magnitude of q should be
(a) 4×10^{-8} C (b) 2×10^{-8} C
(c) 1×10^{-8} C (d) 0.5×10^{-8} C

311. The electric field around a positive charge is
(a) inwards
(b) outwards
(c) inwards or outwards depending on magnitude of charge
(d) generally outwards

312. Newtons/Coulomb is the unit of
(a) electric field intensity (b) electric potential
(c) electric flux density (d) electric current

313. Consider the following statements
1. E and V lines are always perpendicular to each other
2. E and V lines are perpendicular to each other except at singular points
3. In a uniform field V varies linearly with distance

Which of the above statements are correct?
(a) 1 only (b) 2 and 3
(c) 1 and 3 (d) 1, 2 and 3

314. For a linear charge distribution ρ_l the potential at any point P at distance r is
(a) $\frac{1}{\epsilon_0}\int \frac{\rho_l}{r} dl$ (b) $\frac{1}{4\pi\epsilon_0}\int \frac{\rho_l}{r} dl$
(c) $\frac{1}{4\pi\epsilon_0}\int (\rho_l)\, r dl$ (d) $\frac{\epsilon_0}{4\pi}\int \frac{\rho_l}{r} dl$

315. A well collinated electron beam constitutes a
(a) uniform line charge
(b) non-uniform charge distribution
(c) uniform surface charge
(d) uniform volume charge

316. If potential decreases by 2V/m in x direction and 1 V/m in y direction, the electric field E is
(a) $2.24 \angle 26.6°$ V/m (b) $-2.24 \angle 26.6°$ V/m
(c) $2.24 \angle 63.4°$ V/m (d) $-2.24 \angle 63.4°$ V/m

317. A positive charge Q is uniformly distributed over a spherical shell of radius r_1. The medium is air. The electric field for $r < r_1$ is
(a) $\frac{Q}{4\pi\epsilon_0 r}$ (b) $\frac{Q}{4\pi\epsilon_0 r^2}$
(c) zero (d) infinite

318. A positive charge Q is uniformly distributed over a spherical shell of radius r_1. The medium is air. The electric potential for $r < r_1$
(a) is constant (b) varies inversely with r
(c) varies directly with r (d) zero

319. The capacitance of a single isolated sphere of radius r in free space is
(a) $4\pi\epsilon_0 r$ (b) $2\pi\epsilon_0 r$
(c) $\frac{r}{4\pi\epsilon_0}$ (d) $\frac{4\pi\epsilon_0}{r}$

320. The capacitance of earth-electrosphere combination is about
(a) 2.2 F (b) 1.1 F
(c) 0.18 F (d) 0.001 F

321. Surface charge denstiy on the surface of earth is about
(a) 3.2×10^{-9} C/m^2 (b) 3.2×10^{-12} C/m^2
(c) 3.2×10^{-6} C/m^2 (d) 3.2×10^{-3} C/m^2

322. The electrostatic energy stored in earth-electrosphere combination is about
(a) 7×10^6 J (b) 7×10^9 J
(c) 7×10^{12} J (d) 7×10^{15} J

323. If m_0 is the rest mass of a particle, c is velocity of light and m is mass at velocity v, then
(a) $m = \frac{m_0}{1 + (v/c)^2}$ (b) $m = \frac{m_0}{\sqrt{1 + (c/v)^2}}$
(c) $m = \frac{m_0}{\sqrt{1 - (v/c)^2}}$ (d) $m = \frac{m_0}{1 - (c/v)^2}$

324. In Fig. 5.5, the potential at point P ($r >> l$) is

(*a*) $\dfrac{Ql \cos\theta}{4\pi \in_0 r^2}$

(*b*) $\dfrac{Q \cos\theta}{4\pi \in_0 r^2}$

(*c*) $\dfrac{Ql}{4\pi \in_0 r^2}$

(*d*) $\dfrac{Ql \cos\theta}{4\pi \in_0 r}$

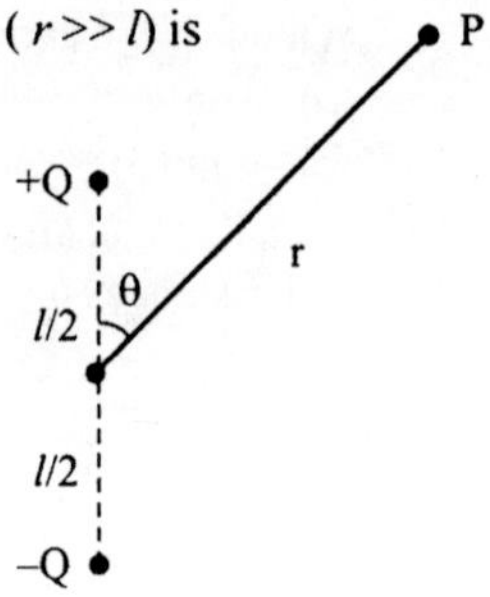

Fig. 5.5.

325. In Fig. 5.5, the total electric field at point P($r >> l$) varies as

(*a*) $\dfrac{1}{r}$ (*d*) $\dfrac{1}{r^2}$

(*c*) $\dfrac{1}{r^3}$ (*d*) $\dfrac{1}{r^4}$

326. In Fig. 5.6, which of the following equation is correct

(*a*) $\dfrac{\tan\alpha_2}{\tan\alpha_1} = \dfrac{\in_1}{\in_2}$

(*b*) $\dfrac{\tan\alpha_1}{\tan\alpha_2} = \dfrac{\in_1}{\in_2}$

(*c*) $\dfrac{\sin\alpha_1}{\sin\alpha_2} = \dfrac{\in_1}{\in_2}$

(*d*) $\dfrac{\sin\alpha_2}{\sin\alpha_1} = \dfrac{\in_1}{\in_2}$

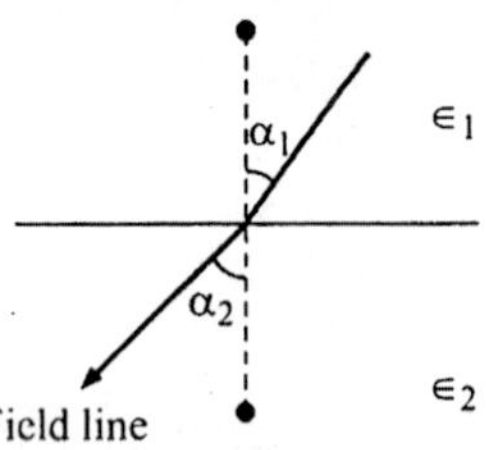

Fig. 5.6.

327. In Fig. 5.6, medium 1 is air ($\in_1 = \in_0$) and medium 2 has $\in_{r2} = 4$. If $\alpha_1 = 30°$, α_2 is

(*a*) 15° (*d*) 30°

(*c*) about 42° (*d*) about 67°

328. The energy stored in a 12 V, 100 Ah storage battery is about

(*a*) 4×10^6 J (*b*) 4×10^3 J

(*c*) 40 J (*d*) 4 J

329. A coaxial transmission line has inner conductor radius *a* and outer radius *b*. If ρ_l is the charge per unit length on inner condcutor and $\in$ is the permittivity, the potential difference between conductors is

(*a*) $\dfrac{\rho_l}{4\pi\in} \ln \dfrac{b}{a}$ (*b*) $\dfrac{\rho_l}{2\pi\in} \ln \dfrac{b}{a}$

(*c*) $\dfrac{\rho_l}{\pi\in} \ln \dfrac{b}{a}$ (*c*) none of the above

330. A coaxial transmission line has *a* and *b* as inner and outer radii and $\in$ as permittivity of dielectric. The capacitance per unit length is

(*a*) $\dfrac{2\pi\in}{\ln (b/a)}$ (*b*) $\dfrac{4\pi\in}{\ln (b/a)}$

(*c*) $\dfrac{\ln (b/a)}{2\pi\in}$ (*d*) $\dfrac{\ln (b/a)}{4\pi\in}$

331. Consider the following statements about electric flux tubes

1. Flux tubes from a point charge Q contain a flux $\dfrac{\Delta\Omega}{4\pi}$ Q where $\Delta\Omega$ is the solid angle subtended by the tube
2. Eletric flux is constant over any cross-saction of the tube
3. A flux tube originates on a positive charge and ends on an equal negative charge

Of the above statements, the correct statements are

(*a*) 1 and 2 only

(*b*) 1 and 3 only

(*c*) 2 and 3 only

(*d*) 1, 2 and 3

332. Consider the following properties of a field cell in a single medium of uniform permittivity

1. The capacitance per unit depth of field cell is equal to permittivity of medium
2. The average flux density D in any cell of the same kind is directly proportional to width of cell
3. The energy stored in any cell of the same kind is the same

Which of the above statements are correct?

(*a*) 1 only (*b*) 1 and 2 only

(*c*) 1 and 3 only (*d*) 1, 2 and 3

333. The dimensions of resistance are

(*a*) $\dfrac{ML^2}{I^2T^3}$ (*b*) $\dfrac{ML}{I^2T^3}$

(*c*) $\dfrac{M^2L}{I^2T^3}$ (*d*) $\dfrac{M^2L}{I^2T^2}$

334. The conductivity of human blood is about

(*a*) 0.7 siemens/m (*b*) 25 siemens/m

(*c*) 100 siemens/m (*d*) 0.007 siemens/s

335. If 10 V (rms) 50 Hz voltage is applied to a 100 m long copper wire and the current is 5 A the electron excursion distance back and forth is about

(*a*) 10^{-6} m (*b*) 10^{-8} m

(*c*) 10^{-3} m (*d*) 0.1 m

336. The equation of conservation of charge at a point (where J is current density and ρ is charge density) is

(*a*) $\nabla \cdot J = \dfrac{\partial\rho}{\partial t}$ (*b*) $\nabla \cdot J = \dfrac{-\partial\rho}{\partial t}$

(*c*) $\nabla J = \dfrac{\partial\rho}{\partial t}$ (*d*) $\nabla J = \dfrac{-\partial\rho}{\partial t}$

337. Consider the following statements regarding current mapping

1. Current lines and equipotentials intersect at right angles
2. In a uniform current distribution, the potential varies linearly with distance
3. The total current through any cross-section of a continuous current tube is a constant

Which of the above statements are correct?

(*a*) 1 and 2 only

(*b*) 1 and 3 only

(*c*) 1, 2 and 3

(*d*) 2 and 3 only

338. Figure 5.7. shows a block of a conducting material. Out of the three directions of current flow viz. left to right, top to bottom and front to back, which direction has lowest resistance

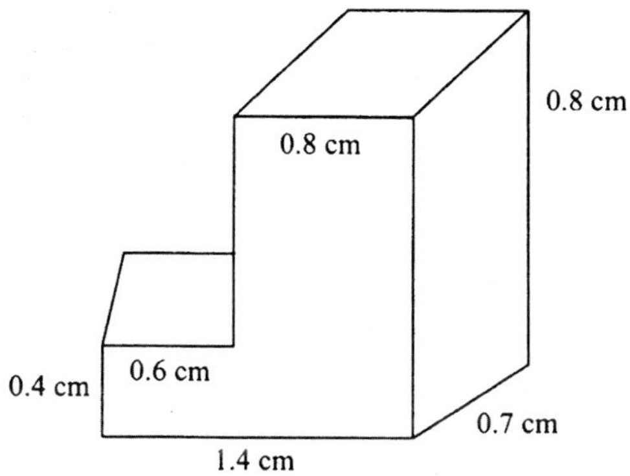

Fig. 5.7.

(*a*) left to right (*b*) top to bottom
(*c*) front to back (*d*) either (*a*) or (*b*)

339. In which direction of current flow in Fig. 5.7 is the resistance maximum
(*a*) left to right (*b*) top to bottom
(*c*) front to back (*d*) either (*a*) or (*b*)

340. A lightning stroke has a current of 100 kA and duration 400 ms. If potential is 90 MV, the energy involved is
(*a*) 3.6×10^6 J (*b*) 3.6×10^9 J
(*c*) 3.6×10^{12} J (*d*) 3.6×10^{15} J

341. A circular wire loop of radius R carries a current I. The magnetic flux density at centre of wire is
(*a*) $\frac{\mu I}{4\pi R}$ (*b*) $\frac{\mu I}{2\pi R}$
(*c*) $\frac{\mu I}{4R}$ (*d*) $\frac{\mu I}{2R}$

342. The magnetic field H at a radius of 1 m from a long wire is 1 A/m. The current in the wire is
(*a*) 2 π amperes (*b*) π amperes
(*c*) 4 π amperes (*d*) 0.5 π amperes

343. A solid cylindrical conductor of radius R carries a current I. The magnetic field inside the conductor at a distance *r* from the centre
(*a*) $\frac{Ir}{2\pi R^2}$ (*b*) $\frac{Ir}{2\pi R}$
(*c*) $\frac{I}{2\pi R^2}$ (*d*) $\frac{Ir^2}{2\pi R}$

344. The curl of the gradient of a scalar function is equal to
(*a*) 1 (*b*) 0
(*c*) infinity (*d*) 2π

345. μ_r for Mumetal is about
(*a*) 100 (*b*) 10000
(*c*) 100000 (*d*) 10000000

346. Which of the following relations is correct for Fig. 5.8.

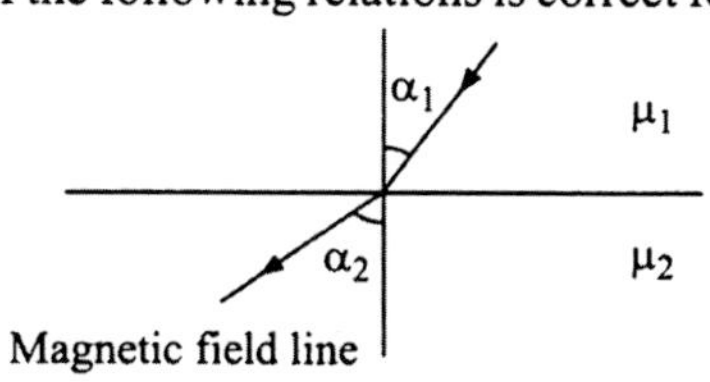

Fig. 5.8.

(*a*) $\frac{\tan\alpha_1}{\tan\alpha_2} = \frac{\mu_1}{\mu_2}$ (*b*) $\frac{\sin\alpha_1}{\sin\alpha_2} = \frac{\mu_1}{\mu_2}$
(*c*) $\frac{\cos\alpha_1}{\cos\alpha_2} = \frac{\mu_1}{\mu_2}$ (*d*) $\frac{\tan\alpha_2}{\tan\alpha_1} = \frac{\mu_1}{\mu_2}$

347. A magnetic field line is nearly tangent to iron air boundary. If field line goes from iron to air, the direction of field line in air will be
(*a*) tangent to surface
(*b*) normal to surface
(*c*) at an angle 45° to surface
(*d*) at about 20° to surface

348. The saturation region in initial magnetisation curve of a ferromagnetic material is also called
(*a*) easy magnetisation
(*b*) hard magnetisation
(*c*) relative magnetisation
(*d*) maximum magnetisation

349. The magnetic energy per unit volume of a specimen of ferromagnetic material brought to saturation from originally unmagnetised condition is euqal to
(*a*) H.B (*b*) $\int_0^B H dB$
(*c*) 0.5 H.B (*d*) $0.5 \int_0^B H\, dB$

350. Energy per unit volume has the dimensions
(*a*) $\frac{M}{LT^2}$ (*b*) $\frac{M^2}{LT^2}$
(*c*) $\frac{M}{L^2T^2}$ (*d*) $\frac{M^2}{L^2T^2}$

351. The units of coercivity are
(*a*) A (*b*) A/m
(*c*) A/m^2 (*d*) Wb/m^2

352. The units of retentivity are
(*a*) A (*b*) A/m
(*c*) A/m^2 (*d*) Wb/m^2

353. If μ is permeability, A is area and *l* is length of magnetic circuit, permeance =
(*a*) $\frac{l}{\mu A}$ (*b*) $\frac{\mu A}{l}$
(*c*) $\mu A l$ (*d*) $\frac{\mu l}{A}$

354. The units for permeance are
(*a*) Henry (*b*) Ampere
(*c*) $\frac{1}{\text{Henry}}$ (*d*) $\frac{1}{\text{Ampere}}$

355. Consider the following statements about magnetic field mapping
1. Magnetic flux tubes are continuous
2. In a uniform field, the potential varies linearly with distance
3. The boundary between air and a ferromagnetic material may be regarded as an equipotential with respect to air side of boundary but not, in general, with respect to ferromagnetic material side

Which of the above statements are correct?
(a) 1 only (b) 1 and 2
(c) 1, 2 and 3 (d) 1 and 3

356. An air capacitor is immersed in a large tank filled with a liquid of uniform conductivity σ and the dc resistance between the plates is measured. If this resistance is R, the capacitance of air capacitor is given by
(a) $\frac{\sigma R}{\epsilon_0}$ (b) $\frac{\epsilon_0}{\sigma R}$
(c) $\frac{R}{\epsilon_0 \sigma}$ (d) $\frac{\epsilon_0 \sigma}{R}$

357. A magnetic circuit is 50 cm long and has an air gap of 2 mm. If μ_r = 1000 and ampere turns required for iron portion of circuit are 100, the ampere turns for air gap are about
(a) 4000 (b) 400
(c) 40 (d) 4

358. The magnetic gap force is proportional to
(a) B (b) B^2
(c) $B^{2.5}$ (d) B^3

359. If a magnetic monopole is found to exist, the ratio of the weight of this monopole to that of a proton would be about
(a) 10^{12} (b) 10^6
(c) 10^3 (d) 10^{-3}

360. A conductor is placed vertically with respect to ground and carries a current I upwards. The image conductor (below ground) will have a current
(a) I in downward direction
(b) I in upward direction
(c) I/2 is downward direction
(d) I/2 in upward direction

361. Stoke's theorem is
(a) $\oint_l E \cdot dl = \int_s (\nabla \times E) \cdot ds$
(b) $\oint_l E \cdot dl = \int_v (\nabla \times E)\, dv$
(c) $\oint_l E \cdot dl = \int_l (\nabla \times E)\, dl$
(d) $\oint_l E \cdot dl = \int_l (\nabla E)\, dl$

362. Which of the following is correct?
(a) the displacement current density and conduction current density are in phase
(b) the displacement current density leads the conduction current density by about 45°
(c) the displacement current density leads the conduction current density by 90°
(d) the displacement current density lags the conduction current density in 90°

363. In a good insulator the dc conduction current is negligible. In such an insulator
(a) the ac current due to dielectric hysteresis is also negligible
(b) the ac current due to dielectric hysteresis may be appreciable
(c) the ac current due to dielectric hysteresis would be in phase with displaement current
(d) the ac current due to dielectric hysteresis would lag displacement current by 45°

364. An air filled coaxial line has the outer radius equal to twice the inner radius. Its characteristic impedance is about
(a) 400 Ω (b) 40 Ω
(c) 4 Ω (d) 0.4 Ω

365. An air filled coaxial line has a characteristic impedance of 40 ohms. A coaxial line of the same dimensions but filled with a dielectric of ϵ_r = 4 will have a characteristic impedance of
(a) 40 Ω (b) 20 Ω
(c) 10 Ω (d) 2.5 Ω

366. For a space paper the resistance is
(a) 376.7 ohm per square cm
(b) 376.7 ohm per square metre
(c) 376.7 ohm per square
(d) none of the above

367. If σ is conductivity and h is thickness of space paper in metres, then
(a) $\sigma = \frac{h}{376.7}$ (b) $\sigma = \frac{1}{376.7h}$
(c) $\sigma = \frac{376.7}{h}$ (d) $\sigma = \frac{1}{\sqrt{376.7h}}$

368. The intrinsic impedance of Ferrite-Barium titanate is
(a) equal to intrinsic impedance of free space
(b) half the intrinsic impedance of free space
(c) double the intrinsic impedance of free space
(d) an integer fraction of the intrinsic inpedance of free space

369. In a Ferrite-titanate medium
(a) both matching and attenuation are independent of frequency
(b) matching is independent of frequency but attenuation depends on frequency
(c) matching dependes on frequency but attenuation is independent of frequency
(d) both matching and attenuation are dependent on frequency

370. A power divider is required when
(a) a wave guide feeds two antennas
(b) a wave guide feeds three antennas
(c) a wave guide feeds four antennas
(d) a wave guide feeds two or more antennas

371. Consider the following statements relating to cavity resonator
1. The cavity resonator does not possess as many modes as the corresponding wave guide
2. The resonant frequencies of cavities are closely spaced
3. The resonant frequency of a cavity resonator can be changed by altering its dimensions

Which of the above statements are correct?
(a) 2 and 3 (b) 1, 2 only
(c) 3 only (d) none of these

[IES 2010]

372. Given a range of frequencies which of the following systems is best for transmission line load matching
(*a*) single stub
(*b*) double stub
(*c*) single stub with adjustable position
(*d*) quarter wave transformer **[IES 2010]**

373. In cylindrical wave guide the attenuation will be minimum at a frequency which is $\sqrt{3}$ times cutoff frequency for the following modes of operations.
1. TE_{10} 2. TM_{11}
3. TM_{10} 4. TE_{11}
Which of the above are correct?
(*a*) 1, 2, 3, 4 (*b*) 2, 3 only
(*c*) 1, 2 only (4) 3, 4 only **[IES 2010]**

374. Consider the following statements
1. (Electric or magnetic) field must have two orthogonal linear components
2. The two components must have the same magnitude
3. The two components must have time phase difference of odd multiples of 90°
Which of above are necessary condition for a time harmonic wave to be circularly polarized at a given point in space?
(*a*) 1, 2 only (*b*) 2, 3 only
(*c*) 1, 2, 3 (*d*) 1, 3 only **[IES 2010]**

375. The mode with lowest cutoff frequency for an E-M wave propagating between two perfectly conducting parallel plates of infinite extent is
(*a*) TE_{10} (*b*) TM_{10}
(*c*) TM_{01} (*d*) TEM **[IES 2010]**

376. Consider the following statements regarding skin depth in a conductor
1. In increases as frequency increases
2. It is inversely proportional to square root of μ and σ
3. It is inversely proportional to square root of frequency
4. It is directly proportional to square root of μ and σ
Which of the above statements are correct?
(*a*) 1, 2 only (*b*) 3 and 4 only
(*c*) 2, 3 only (*d*) 1, 2, 3, 4 **[IES 2010]**

377. In a three element Yagi antena
(*a*) All the three elements are of equal length
(*b*) The driven and director are equal in length but the reflector is longer than them
(*c*) Reflector is longer than driven element which in turn is longer than director
(*d*) Director is longer than driven element which in turn is longer them reflector **[IES 2010]**

378. A short current element has length $l = 0.03\ \lambda$ where λ is the wavelength. The radiation resistance for uniform current distribution is
(*a*) $0.072\ \pi^2\ \Omega$ (*b*) $80\pi^2\ \Omega$
(*c*) 72 Ω (*d*) 80 Ω **[IES 2010]**

379. A line of characteristic impedance 50 ohms is terminated at one end by $+j50$ ohms. VSWR is
(*a*) 1 (*b*) ∝
(*c*) 0 (*d*) j **[IES 2010]**

380. There are two conducting plates of sizes 1 × 1 m and 2 × 2 m ratio of capacitance of second to that of first is
(*a*) 4 (*b*) 2
(*c*) 0.5 (*d*) 0.25 **[IES 2010]**

381. The intrinsic impedance of copper at 3 GHz [with $\mu = 4\pi \times 10^{-7}$ H/m, $\in = 10^{-9}/36\pi$ and $\sigma = 5.8 \times 10^7$ mho/m] is
(*a*) $0.02\ e^{\pi/4}$ ohm (*b*) $0.02\ e^{\pi/2}$ ohm
(*c*) $0.2\ e^{\pi/2}$ ohm (*d*) $0.2\ e^{\pi/4}$ ohm **[IES 2010]**

382. The flux and potential functions due to a line charge and due to two concentric circular conductors are of the following form
(*a*) Concentric circular equipotential lines and straight radial flux lines
(*b*) Concentric circular flux lines and straight equipotential lines
(*c*) Equipotential due to line charge are concentric cylinders and equipotential due to two conductors are straight lines
(*d*) Equipotentials due to line charge are straight flat surfaces and those due to two conductors are concentric cylinders **[IES 2010]**

383. An electric charge Q coulomb is located as origin. Electric potential V and electric field intensity E at any point x, y, z are
(*a*) E and V are both scalars
(*b*) E and V are both vectors
(*c*) E is a scalar and V is a vector
(*d*) E is a vector and V is a scalar **[IES 2010]**

384. Consider the following statements regarding a transmission line
(*a*) Its attenuation is constant and is independent of frequency
(*b*) Its attenuation varies linearly with frequency
(*c*) Its phase shift varies linearly with frequency
(*d*) Its phase shift is constant and is independent of frequency
Which of the above are correct for distortionless line?
(*a*) 1, 2, 3, 4 (*b*) 2 and 3 only
(*c*) 1 and 3 only (*d*) 3 and 4 only **[IES 2010]**

385. Consider the following
In a parallel plate capacitor charge is constant while the dielectric material is replaced by a different dielectric consider.
1. stored energy
2. electric field intensity
3. capacitance
Which of these change?
(*a*) 1 only (*b*) 1 and 2 only
(*c*) 2 and 3 only (*d*) 1, 2, 3 **[IES 2010]**

386. The magnetic flux density B and vector magnetic potential A are related as
(*a*) $B = \nabla \times A$ (*b*) $A = \nabla \times B$
(*c*) $B = \nabla \cdot A$ (*d*) $A = \nabla \cdot B$ **[IES 2010]**

387. Method of images is applicable to
(*a*) Electrosratic fields only
(*b*) Electrodynamic fields only

(c) neither (a) nor (b)
(d) both (a) and (b) **[IES 2009]**

388. Who developed the concept of time varying electric field producing a magnetic field?
(a) Gauss (b) Faraday
(c) Hertz (d) Maxwell **[IES 2009]**

389. A single turn loop is situated in air with a uniform magnetic field normal to its plane. The area of loop is $5m^2$ and rate of change of flux density is 2 $Wb/m^2/s$. What is the emf appearing at the terminals of loop
(a) –5 V (b) –10 V
(c) –0.4 V (d) 0V **[IES 2009]**

390. Which of the following equations results from the circuital form of Ampere's load?
(a) $\nabla \times E = -\frac{\partial B}{\partial F}$ (b) $\nabla \cdot B = 0$
(c) $\nabla \cdot D = \rho$ (d) $\nabla \times H = J + \frac{\partial D}{\partial E}$
[IES 2009]

391. In which direction is the plane wave travelling $\bar{E} = 50 \sin (10^8 t + 2z)\, \bar{a}_r$ V/m where $\bar{a}_r$ is the unit vector in y direction) travelling
(a) along y direction (b) along $-y$ direction
(c) along z direction (d) along $-z$ direction
[IES 2009]

392. For plane wave propagating in free space on two conductor line what is the relation between phase velocity V_p group velocity V_g and speed of light C
(a) $V_p > C > V_g$ (b) $V_p < C < V_g$
(c) $V_p = C = V_g$ (d) $V_p < V_g < C$ **[IES 2009]**

393. Consider the following statements
For a 10 m long common power line connecting a switch to a light bulb
(1) It is a distributed circuit
(2) Time delay for propagation through it is negligible
(3) It is in the form of a shielded coaxial cable of circular cross-section
(4) As the intensity of lamp varies, the input impedance of this line also changes
Which of above are correct?
(a) 1 only (b) 1 and 2
(c) 2 and 3 (d) 2 and 4 **[IES 2009]**

394. It is required to match a 200 ohm load to a 450 ohm transmission line. To reduce SWR to 1 what must be the characteristic impedance of the quarter wave transformer if it is directly connected to load
(a) 90 k ohm (b) 300 ohm
(c) $\frac{9}{4}$ ohm (d) $\frac{3}{2}$ ohm **[IES 2009]**

395. The load end of a quarter wave transformer gets open circuited. What will be the input impedance of transformer
(a) Zero (b) Infinite
(c) Finite and positive (d) Finite and negative
[IES 2009]

396. A transmission line has a characteristic impedance of 50 ohm and a resistance of 0.1 ohm. If the line is distortionless the attenuation constant (Np/m) is
(a) 500 (b) 5
(c) 0.014 (d) 0.002 **[Gate 2010]**

397. The electric field component of a time harmonic plane EM wave travelling in a non-magnetic lossless dielectric medium has an amplitude of 1 V/m. If $\epsilon_r = 4$, the magnitude of time average power density vector (in W/m^2) is
(a) $\frac{1}{30\pi}$ (b) $\frac{1}{60\pi}$
(c) $\frac{1}{120\pi}$ (d) $\frac{1}{240\pi}$ **[Gate 2010]**

398. If a vector field $\vec{V}$ is related to another vector field $\vec{A}$ through $\vec{V} = \nabla \times \vec{A}$ which of the following is true
(a) $\oint_c \vec{V} \cdot \vec{dl} = \iint_s \vec{A} \cdot \vec{ds}$
(b) $\oint_c \vec{A} \cdot \vec{dl} = \iint_s \vec{V} \cdot \vec{ds}$
(c) $\oint_c \nabla \times \vec{V} \cdot \vec{dl} = \iint_s \nabla \times \vec{A} \cdot \vec{ds}$
(d) $\oint_c \nabla \times \vec{A} \cdot \vec{dl} = \iint_s \vec{V} \cdot \vec{ds}$ **[Gate 2009]**

399. A transmission line branches into two lines each of length $\lambda/4$ as shown in Fig. 5.9. Each branch line is terminated in 50 ohm impedance lines are lossless and have Z_0 as shown, find Z_1 as seen by source
(a) 200 Ω (b) 100 Ω
(c) 50 Ω (d) 25 Ω **[Gate 2009]**

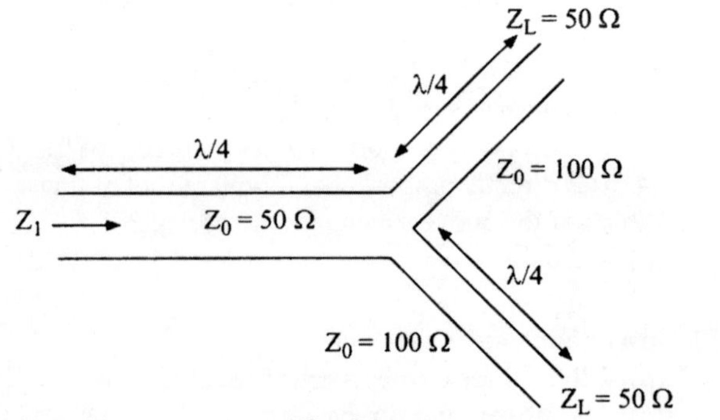

Fig. 5.9.

400. A rectangular waveguide has dimension a = 4 cm and b = 3 cm. It is required to be operated in TE_{11} mode. The minimum operating frequency is
(a) 6.25 GHz (b) 6 GHz
(b) 5 GHz (d) 3.75 GHz **[Gate 2008]**

401. One end of a lossless line having Z_0 = 75 ohm and l = 1 cm is shorted. At 3 GHz the input impedance at other end is
(a) 0 (b) resistive
(c) capacitive (d) Inductive **[Gate 2008]**

402. A uniform plane wave in free space is normally incident on an infinitely thick dielectric slab [ϵ_r = 9]. The reflection coefficient is
(a) 0 (b) 0.3
(c) 0.5 (d) 0.8 **[Gate 2008]**

403. At 20 GHz the gain of a parabolic dish antenna of 1 m dia and 70% efficiency is

(*a*) 15 dB (*b*) 25 dB
(*c*) 35 dB (*d*) 45 dB **[Gate 2008]**

404. For static electric and magnetic fields in an inhomogenous source free medium which of the following represent two maxwell's equations

(*a*) $\nabla \cdot E = 0$, $\nabla \times B = 0$ (*b*) $\nabla \cdot E = 0$, $\nabla \cdot B = 0$
(*c*) $\nabla \times E = 0$, $\nabla \times B = 0$ (*d*) $\nabla \times E = 0$, $\nabla \cdot B = 0$ **[Gate 2008]**

405. An air filled rectangular waveguide is 3 cm × 2 cm Wave impedance in TE_{20} mode at $f = 30$ GHz is

(*a*) 308 ohm (*b*) 355 ohm
(*c*) 400 ohm (*d*) 461 ohm

Assume that intrinsic impedance of free space is 377 ohm. **[Gate 2007]**

406. The $\vec{H}$ field (in A/m) of a plane wave propagating in free space is given by

$$\vec{H} = \hat{x}\frac{5\sqrt{3}}{\eta_0}\cos(\omega t - \beta z) + \hat{y}\frac{5}{\eta_0}\sin\left(\omega t - \beta z + \frac{\pi}{2}\right)$$

The time average power flow density in watts is

(*a*) $\frac{\eta_0}{100}$ (*b*) $\frac{100}{\eta_0}$
(*c*) $80\eta_0^2$ (*d*) $\frac{50}{\eta_0}$ **[Gate 2007]**

407. A load of 50 ohm is connected in shunt in a 2 wire line of $Z_0 = 50$ ohm as shown in Fig. 5.10.

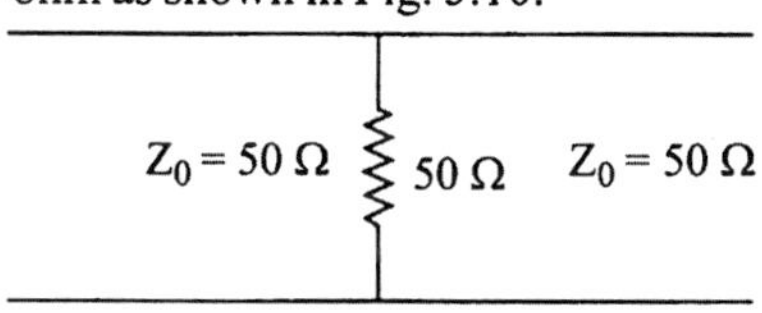

Fig. 5.10.

The two port scattering parameter matrix (S matrix) of shunt element is

(*a*) $\begin{bmatrix} -\frac{1}{2} & \frac{1}{2} \\ \frac{1}{2} & -\frac{1}{2} \end{bmatrix}$ (*b*) $\begin{bmatrix} 0 & 1 \\ 1 & 0 \end{bmatrix}$

(*c*) $\begin{bmatrix} -\frac{1}{3} & \frac{2}{3} \\ \frac{2}{3} & -\frac{1}{3} \end{bmatrix}$ (*d*) $\begin{bmatrix} \frac{1}{4} & -\frac{3}{4} \\ -\frac{3}{4} & \frac{1}{4} \end{bmatrix}$ **[Gate 2007]**

408. The parallel branches of a 2 wire transmission line (Fig. 5.11) are terminated in 100 ohm and 200 ohm resistance as shown. $Z_0 = 50$ ohm and each section has a length of $\lambda/4$. The voltage reflection coefficient at input is

(*a*) $-j\frac{7}{5}$ (*b*) $-\frac{5}{7}$
(*c*) $j\frac{5}{7}$ (*d*) $\frac{5}{7}$ **[Gate 2007]**

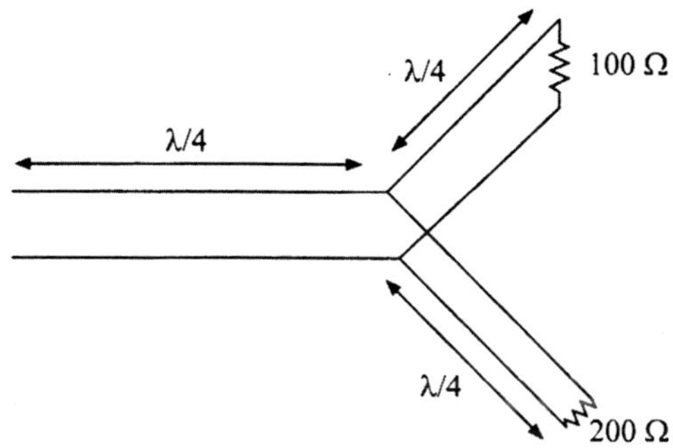

Fig. 5.11.

409. A right circularly polarized (RCP) plane wave is incident at an angle of 60° to the normal on an air-dielectric surface as shown in Fig. 5.12. If the reflected wave is linearly polarized, the relative dielectric constant is

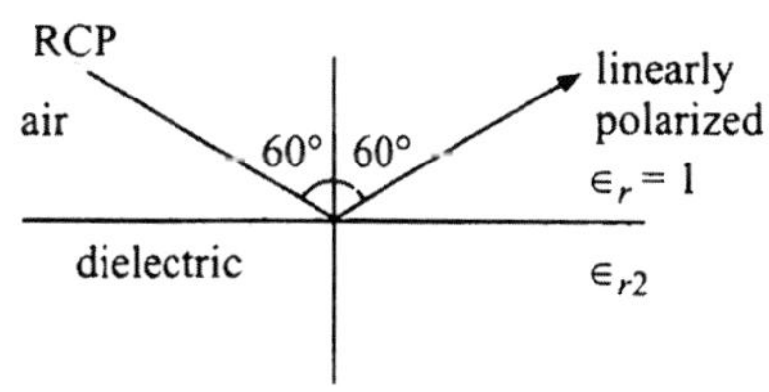

Fig. 5.12.

(*a*) $\sqrt{2}$ (*b*) $\sqrt{3}$
(*c*) 2 (*d*) 3 **[Gate 2007]**

410. A medium of $\in_r = 2$ forms an interface with free space. A point source of e.m. energy is located in the medium at 1 m from interface. Due to total internal reflection the transmitted beam has a circular cross-section over the surface. The area of beam cross-section at the interface is given by

(*a*) $2\pi\, m^2$ (*b*) $\pi^2\, m^2$
(*c*) $\frac{\pi}{2} m^2$ (*d*) $\pi\, m^2$ **[Gate 2006]**

411. When a plane wave travelling in free space is incident normally on a medium having $\in_r = 4$, the fraction of power transmitted into medium is given by

(*a*) $\frac{8}{9}$ (*b*) $\frac{1}{2}$
(*c*) $\frac{1}{3}$ (*d*) $\frac{5}{6}$ **[Gate 2006]**

412. A rectangular wave guide having TE_{10} as dominant mode has cutoff frequency of 18 GHz for TE_{30} mude. The inner broad wall dimension of wave guide is

(*a*) $\frac{5}{3}$ cm (*b*) 5 cm
(*c*) $\frac{5}{2}$ cm (*d*) 10 cm **[Gate 2006]**

413. A mast antenna consisting of a 50 m long vertical conductor operates over a perfectly conducting ground plane. It is base fed at a frequency of 600k Hz. Radiation resistance is

(*a*) $\frac{2\pi^2}{5}$ (*b*) $\frac{\pi^2}{5}$

(*c*) $\frac{4\pi^2}{5}$ (*d*) $20\pi^2$ **[Gate 2006]**

414. A 30 V battery with zero internal resistance is connected to a coaxial line of $Z_0 = 50$ ohms at $t = 0$ sec and terminated in a resistance load. It takes 400 μs for an em wave to travel from source to load and vice versa. At $t = 400$ μs the voltage at far end is 40 V. Load resistance is

(*a*) 25 ohms (*b*) 50 ohms
(*c*) 75 ohms (*d*) 100 ohms **[Gate 2007]**

415. A transmission line is feeding 1 watt power to a horn antenna having a gain of 10 dB. The antenna is matched to transmission line. The total power radiated

(*a*) 10 W (*b*) 1 W
(*c*) 0.1 W (*d*) 0.01 W **[Gate 2007]**

416. Multiple number of antennas are arranged in arrays in order to enhance which property

(*a*) both bandwidth and directivity
(*b*) Only directivity
(*c*) Only bandwidth
(*d*) neither bandwidth nor directivity **[IES 2009]**

417. Consider the following statements regarding antenna

1. It is a transducer
2. Its performance is essentially frequency sensitive
3. It is a reciprocal device
4. With increase in effective aperture area, width of radiated beam increases

Which of above are correct?

(*a*) 1, 2, 3 (*b*) 1, 2, 4
(*c*) 2 and 3 (*d*) 1 and 4 **[IES 2009]**

418. What is the minimum value of VSWR which may exist on a transmission line

(*a*) less than zero (*b*) zero
(*c*) one (*d*) 10 **[IES 2008]**

419. As the aperture area of an antenna increases, its gain

(*a*) increases (*b*) decreases
(*c*) remains steady (*d*) behaves unpredictably

420. A lossless line having characteristic impedance Z_0 and length $l < \lambda/4$ is terminated by an open circuit. What is input impedance

(*a*) $j Z_0 \tan \beta l$ (*b*) $j Z_0 \cot \beta l$
(*c*) $-j Z_0 \tan \beta l$ (*d*) $-j Z_0 \cot \beta l$ **[IES 2008]**

B. The following questions have two statements viz Assertion (A) and Reason (R). Examine them carefully and answer as per the following codes

(*a*) Both A and R are true and R is correct explanation of A
(*b*) Both A and R are true but R is not correct explanation of A
(*c*) A is true but R is false
(*d*) A is false but R is true.

421. A. The line integral of $\vec{B}$ around a closed contour, in a region containing steady current, is μ_0 times the total net current, flowing through the contour of integration.

R. The curl of $\vec{B}$ is equal to μ_0 times current density vector $\vec{J}$.

422. A. Conductors do not permit propagation of waves more than a short distance into the conductor at microwave frequencies.

R. The relaxation time constant for conductors is much smaller than the period of centimetric EM waves.

423. A. If conductivitiy is not a function of frequency, then the skin depth in a good conductor decreases as $f^{0.5}$ while in a good dielectric it is independent of frequency.

R. In a good dielectric conduction current is very small as compared to displacement current but in a good conductor the reverse is true.

424. A. A single conductor wave guide having a simply connected contour and homogenously filled with a non-magnetic dielectric cannot support TEM wave.

R. If a TEM wave is assumed to exist within a single conductor wave guide having a simply connected contour and homogenously filled with a non-magnetic dielectric it would violate Maxwell's equation.

425. A. In a loss less transmission line the voltage and current distribution along the line is always constant.

R. The voltage and current distributions in an open line are such that at a distance λ/4 from the load end, the line looks like a series resonant circuit.

426. A. As per Maxwell's theorem $\nabla \cdot B$ is zero but $\nabla \cdot D$ is not necessarily zero.

R. An electric charge can be isolated but a magnetic pole can not be isolated.

427. A. Helical antennas can be used as feeds for large parabolic reflectors to obtain circular polarisation.

R. Parabolic reflectors reverse the sense of polarisation of the wave during reflection.

428. A. The total emf induced in a circuit is equal to the time rate of change of total magnetic flux linking the circuit.

R. The induced current in a loop is always so directed as to produce a flux opposing the change in the flux density.

429. A. The net charge within a conductor is always zero.

R. A conductor has a large number of free electrons.

430. A. The solution of Poisson's equation is the same as the solution of Laplace equation.

R. Laplace equation is a special case of Poisson's equation for source free regions.

431. A. In cylindrical coordinates a point P is specified by r, ϕ, z where ϕ is measured from x-axis.

R. In spherical coordinates a point P specified by r, θ, ϕ where r is measured from origin, θ is measured from z-axis and ϕ is measured from x-axis.

432. A. In terms of spherical coordinate system the rectangular coordinate distances are $x = r \sin\theta \cos\phi$, $y = r \sin\theta$ $\sin\theta$ and $z = r\cos\theta$.

R. The dot products of the unit vector $\hat{r}$ with the rectangular unit vectors $\hat{x}$, $\hat{y}$, $\hat{z}$ are equal to direction cosines as given by

$$\hat{x} \cdot \hat{r} = \sin\theta \cos\phi$$

$\hat{y} \cdot \hat{r} = \sin\theta \sin\phi$

$\hat{z} \cdot \hat{r} = \cos\theta$

433. A. Coulomb's force is proportional to the magnitude of the charges.
R. Coulomb's force is inversely proportional to square of separation distance r.

434. A. Coulomb's force $F \propto \frac{Q_1 Q_2}{r^2}$.
R. Newton's force $F \propto -\frac{M_1 M_2}{r^2}$.

435. A. Coulomb's force holds solids together.
R. One coulomb is a very small amount of charge.

436. A. Electric field intensity at a point distant r from a charge
$Q_1 = \frac{Q_1}{4\pi\epsilon_0 r^2}$.
R. Coulomb's force $\propto \frac{Q_1 Q_2}{r^2}$.

437. A. Total resultant field at a point is the vector sum of individual component fields at that point.
R. Superposition theorem is applicable to all systems where cause and effect are linearly related.

438. A. When we move perpendicular to electric field intensity we travel along equipotential line.
R. Equipotential and field lines are orthogonal.

439. A. Potential difference is independent of the path taken.
R. Absolute potential at a point distant r from a charge
$Q = \frac{Q}{4\pi\epsilon_0 r}$.

440. A. Line integral of static electric field around a closed path is zero.
R. Potential rise is always opposite to the direction of E.

441. A. The surface of a conductor is an equipotential.
R. E meets the conducting surface at right angles.

442. A. Total electric potential at a point is the vector sum of individual potentials at that point.
R. In a uniform field V varies linearly with distance.

443. A. $\nabla V = -E$.
R. Line integral of electric field between any two points gives the potential difference between the two points.

444. A. Electric flux $\psi = \iint D \cdot ds$
R. $D = \epsilon_0 E$.

445. A. No charge exists on the inside of a hollow conductor.
R. Gauss's law states that total electric dispalcement flux over a sphere is equal to charge enclosed by sphere.

446. A. An electromagnetic compatibility chamber is a large completely enclosed metal room to provide shielding against outside sources of radio frequency interference.
R. Inside a metal cage the field is zero.

447. A. Capacitance of a single isolated sphere of radius $r_1 = 4\pi\epsilon_0 r_1$.
R. Capacitance of concentric spheres
$= \frac{4\pi\epsilon_0 r_1 r_2}{r_2 - r_1}$.

448. A. When a dielectric is placed in an electric field, the molecules of dielectric start behaving like dipoles.
R. Polarization $P = \frac{\text{Net dipole moment}}{\text{volume}}$.

449. A. $D = \epsilon_0 E + P$.
R. $\epsilon = \epsilon_0 + \frac{P}{E}$.

450. A. The tangential components of electric field are the same on both sides of a boundary between two dielectrics.
R. The normal component of flux density is continuous across a charge free boundary between two dielectrics.

451. A. If medium 1 is a dielectric and medium 2 is a conductor, then at the boundary $E_{t1} = 0$.
R. If medium 1 is a dielectric and medium 2 is a conductor $D_{n1} = \rho_s$.

452. A. Force due to infinitesimal current carrying element dl carrying current I is
$dF = IB\, dl \sin\theta$.
R. Biot savart law $dB = \frac{\mu}{4\pi} \frac{I\, dl \sin\theta}{r^2}$.

453. A. Magnetic flux $\psi_m = \iint B \cdot ds$.
R. $\oint_s B \cdot ds = 0$.

454. A. Magnetic moment $T = m \times B$.
R. Flux density at the centre of a solenoid is twice that at the end.

455. A. Inductance of a long solenoid $= \frac{\mu N^2 A}{l}$.
R. Inductance of a toroid $= \frac{\mu N^2 r^2}{2R}$.

456. A. $\oint H \cdot dl = I$.
R. $B = \mu H$.

457. A. $\nabla \times H = J$.
R. $B = \nabla \times A$.

458. A. When a paramagnetic material is brought near the pole of a strong bar magnet, it is attracted to it.
R. When a diamagnetic specimen is brought near the pole of a magnet, it is attracted to it.

459. A. Ferrites are used as cores in high frequency applications
R. A ferrite is a ferrimagnetic material with low electrical conductivity.

460. A. The normal compoenent of flux density B is continuous across the boundary between two media.
R. Tangential components of H are continuous across the boundary between two media provided the boundary has no current sheet.

461. A. Capacitance between two parallel plates of area A and separation d is $\frac{\in A}{d}$ for large A/d ratio.

R. Fringing electric field can be neglected for large A/d ratio. **[IES 2010]**

462. A. In solving boundary value problems, the method of image is used.

R. By this technique conducting surface can be removed from solution domain. **[IES 2010]**

463. A. Velocity of light in any medium is slower than that in vacuum.

R. Dielectric constant of vacuum in unity and is lesser than that of any other medium. **[IES 2010]**

464. A. The impedance of a matched load is equal to Z_0 of line.

R. A matched termination absorbs the entire power incident on it. **[IES 2010]**

465. A. Capacitance of a solid conducting spherical body of radius a is $4\pi\epsilon_0 a$ in free space.

R. $\nabla \times H = j\omega\in E + J$. **[IES 2009]**

466. A. The expression $E = -\nabla V$ is not valid for time varying field.

R. The curl of a gradient is identically zero. **[IES 2009]**

467. A. A z-directed reactangular wave guide with cross-section 3 cm × 1 cm will support propagation at 4 GHz.

R. $k_z^2 + \left(\frac{m\pi}{3}\right)^2 + \left(\frac{n\pi}{1}\right)^2 = \left(\frac{2\pi}{\lambda}\right)^2$. **[IES 2009]**

468. A. As the length of resonant antenna is increased the number of lobes increases and the direction of major lobes is closer and closer to the direction of dipole.

R. As length increases the current distribution along wire become more and more uniform. **[IES 2009]**

469. A. Time varying electric field produces magnetic field.

R. Time varying magnetic field produces electric field. **[IES 2006]**

470. A. Potential everywhere on a conducting surface of infinite extent is zero.

R. Displacement density on a conducting surface is normal to the surface. **[IES 2006]**

C. State whether the following are True or False

471. The force between point charges is proportional to product of magnitudes of charges.

472. Benjamin Franklin was ihe first person to use the names positive and negative charges to designate the two different kinds of charge.

473. Coulomb's law has the same form as Newton's law of gravitation.

474. Electric field intensity at a point due to a charge at a distance r from the point is inveresely proportional to r.

475. Electric field around a positive charge is outward.

476. When a number of charges are present in a region, the electric field at a point is the algebraic sum of individual fields due to the various charges at that point.

477. $\rho_v = \underset{\Delta Q \to 0}{\text{Lt}} \Delta Q / \Delta v$.

478. Potential is a scalar quantity.

479. The potential due to a point charge at a distance r from the charge is inversely proportional to r.

480. Potential difference is single valued.

481. For a static field E, $\oint E \cdot dl = 0$.

482. The surface of a conductor is an equipotential.

483. In a uniform field, V varies linearly with square of distance.

484. grad V = – E.

485. Electric flux through any surface equals the integral of flux density over the surface.

486. For an isolated sphere of radius r, $C = 4\pi \in_0 r$.

487. The energy acquired by an electron in falling through a potential difference of 1 volt is 1 eV.

488. A medium is linear with respect to an electrostatic field if $D \propto E$.

489. In an isotropic material, the properties are independent of the direction.

490. Electric field at a point distant r from a dipole varies as $1/r^2$.

491. When a dielectric is placed in an electric field, the atoms or molecules of dielectric behave as dipoles.

492. If P is polarization, E is electric field, $\in$ is permittivity and $\in_0$ is absolutely permittivity, $\in = \in_0 - P/E$.

493. The tangential components of electric field on both sides of a boundary between two dielectrics are the same.

494. Tangential electric field at a dielectric conductor boundary is infinite.

495. Normal component of the flux density at a 'dielectric conductor boundary is zero.

496. Dielectric strength of air is about 50 kV/cm.

497. Dielectric strength of different dielectric materials is less than that of air.

498. If a breakdown of an air capacitor occurs, the capacitor is damaged permanently.

499. As compared to a capacitor, the energy density of a lead acid battery is low.

500. A tube of flux originates on a positive charge and ends on an equal negative charge.

501. Capacitance per unit depth of a field cell capacitor is equal to permittivity.

502. Laplace equation is $\nabla^2 V = 0$.

503. Current density at a point $J = \underset{\Delta s \to 0}{\text{Lt}} \frac{\Delta I}{\Delta s}$.

504. Ohm's law at a point is $J = \sigma E$.

505. KCL $\oint J \cdot ds = 0$.

506. For steady currents $\nabla \cdot J = 0$.

507. At a conductor to conductor boundary, component of current density normal to boundary is the same in the two conductors.

508. At a conductor to conductor boundary, component of field tangent to the boundary is the same in the two conductors.

509. The relation of magnetic field direction to the current direction is given by right hand rule.

510. Inductance of a long solenoid is proportional to number of turns in the solenoid.

511. Amperes's law $\oint H.dl = I$.

512. $\nabla \times H = J$.

513. A wire is carrying a current I. Then outside the wire $\nabla \times H = I/a$.

514. The magnetic field inside a solenoid is the same as near the ends.

515. If an inductor is carrying a current I, its inductance

$$L = \left(\iiint B.H\, dv\right) / I^2.$$

516. If a capacitor is charged to votlage V, its capacitance C is

$$C = \left(\iiint E.D\, dv\right) / V^2.$$

517. For a particle with charge e moving with velocity v in a field of flux density B, Lorentz force F is

$$F = e(v \times B)$$

518. When a charged particle enters at right angles to a field B, the particle is deflected into circular path.

519. If a particle with charge e is moving with velocity v in a combined electric field E and magnetic field B, Lorentz force F is

$$F = e[E + (v + B)].$$

520. A vector function whose divegence is zero every whre is called solenoidal.

521. div (grad f) $= \nabla^2 f$.

522. If F is a vector function $\nabla \cdot (\nabla \times F) = 0$.

523. If a vector field has no curl, the field is lamellar.

524. When a magnetic field B is applied to an atom with moment B, the torque T is

$$T = m \cdot B$$

525. In ferrimagnetic materials the magnetic moments of adjacent atoms are equal and opposite.

526. Ferrites are ferrimagnetic materials with low electrical conductivity.

527. Ferrimagnetic materials are more magnetic than ferromagnetic materials.

528. Aluminium is a diamagnetic material.

529. Bismuth is a paramagnetic material.

530. In anti-ferromagnetic materials, the magnetic moments of adjacent atoms are unequal.

531. If H is magnetic field intensity and M is magnetisation

$$\nabla \cdot H = -\nabla \cdot M.$$

532. The units of ∇. H are Ampere per metre.

533. In a single medium the magnetic field is continuous.

534. Normal component of magnetic flux density B is continuous over the boundary between two media.

535. In the case of any two media with no current sheet at boundary, tangential components of H are continuous across the boundary.

536. In an iron crystal there are six directions of easy magnetisation.

537. The saturation region of B–H curve is the region of hard magnetisation.

538. The product BH has the dimensions of energy.

539. As H is increased, the relative permeability of a ferromagnetic material increases.

540. Permeance has the same units as inductance.

541. When two magnetic paths are in parallel, the total permeance is sum of individual permeances.

542. If a function V is a solution of Laplace equation $\nabla^2 V = 0$ and boundary conditions are also satisfied, it is a unique solution.

543. When both transformer emf and motional emf are present, the total induced emf is

$$\text{Induced emf} = \oint (v \times B).dl + \int_s \frac{\partial B}{\partial t}.ds.$$

544. Stokes theorem is 'line integral of a vector function around a closed contour C is equal to integral of the normal component of curl of that function over any surface having contour C as its boundary edge.

545. In empty space the magnetic field is due to conduction current.

546. The differential form of Maxwell's equation from Ampere is

$$\nabla \times H = J + \delta D/\delta t$$

547. Displacement current density leads the conduction current density by 90°.

548. If δ is loss angle of dielectric and θ is $p.f$ angle, $\delta = \theta$.

549. Maxwell's equation in free space $\oint_s B \cdot ds = 0$.

550. Maxwell's equation from Faraday, in phasor form is

$$\nabla \times E = j\omega\mu H.$$

551. Maxwell's equation, from Ampere, in phasor form is

$$\nabla \times H = (\sigma + j\omega \epsilon) E.$$

552. In a TEM wave, E and H are transverse to the direction of propagation.

553. Wave equation in E_y is

$$\frac{\partial^2 E_y}{\partial t^2} = \frac{1}{\mu \epsilon} \frac{\partial^2 E_y}{\partial x^2}.$$

554. The wave $E_y = \sin(\beta x - \omega t)$ travels in the negative x direction.

555. Wave equation in E_y for a plane wave in conducting medium is

$$\frac{\partial^2 E_y}{\partial x^2} = (j\omega\mu\sigma - \omega^2 \mu \epsilon) E_y.$$

556. Transmission line wave equation in V is

$$\frac{\partial^2 V}{\partial x^2} - YZV = 0.$$

557. Transmission line characteristic impedance $= \sqrt{\frac{Y}{Z}}$.

558. If Z_L and Z_0 are load impedance and characteristic impedance for a line, the reflection coefficient for voltage =

$$\frac{Z_0 - Z_L}{Z_0 + Z_L}.$$

559. If Z_L and Z_0 are load impedance and characteristic impedance, the transmission coefficient for voltage

$$= 2Z_L/(Z_L + Z_0).$$

560. If ρ_v is reflection coefficient for voltage, voltage standing wave ratio is

$$VSWR = \frac{1+|\rho_v|}{1-|\rho_v|}.$$

D. Fill in the Blanks with suitable words

561. Coulomb's law is of the same form as Newton's law of ....

562. Any field for which the line integral around a closed path is zero is called........ field.

563. In spherical coordinates a point P is specified by ....

564. The first scientist to accurately measure the force between electric charges was ....

565. A wave in which E and H are transverse to the direction of propagation is called .... wave.

566. For a non-conducting medium Maxwell's equation (in differential form) from Ampere's law is ....

567. If μ, is permeability and $\in$ is permittivity of a medium, the wave velocity v = ...

568. VSWR stands for ....

569. If a transmission line is terminated in a resistance equal to characterisitic impedance of line, there is no ....

570. The action of a $\lambda/4$ transformer is like that of ....

571. A medium in which wave velocity is a function of frequency is called .... medium.

572. At a frequency of about 1 kHz the ground behaves as ....

573. The process of isolating a transmission line from its surroundings is called ....

574. A wave travelling in z direction and having electric field at all times in y direction is called ..... wave.

575. A wave travelling in z direction and having electric field with x and y component is said to be ....

576. For parallel polarization, the incident angle which makes Fresnel reflection coefficient zero is called ....

577. Transmission lines which can convey electromagnetic wave only in higher order modes are called ....

578. Iris can act as a ....

579. A resonator is an energy .... device.

580. An EMC chamber is lined with absorber to simulate ....

581. When beam area Ω_A is separated into Ω_M (due to main lobe) and Ω_m (due to minor lobes), the ratio Ω_M/Ω_m is called..........

582. The smaller the beam area Ω_A, greater is ......

583. $\in$ is called.............

584. Electric field intensity is a......... quantity.

585. Electric field intensity is negative of......

586. Volt is the same as............

587. The tangent to a line of force at any point gives the direction of.......... at that point.

588. The equipotential lines due to a concentrated charge are........

589. No work is required in moving a charge along .... surface.

590. Electric flux is measured in terms of number of.............

591. The charge on a conductor resides on its ....

592. $\iint D.ds =$ ....

593. Dielectric materials have high ....

594. The current flowing through a capacitor is called ....

595. The units for permittivity are ....

596. The direction of magnetic field due to a current carrying conductor is given by ....

597. Lines of magnetic flux .... each other.

598. The field pattern of a solenoid is similar to that of a ....

599. Reluctance is analogous to ....

600. $1/\mu$ is the reluctance of......... of magnetic material.

ANSWERS WITH HINTS AND EXPLANATIONS

1. (*b*) Capacitance is charge/voltage. The unit of charge is coulomb. The dimension of voltage is W/A or N-m/ As or N-m/ coulomb. Therefore, farad is the same is coulomb^2/N-m.

2. (*a*) **3.** (*a*)

4. (*a*) Coulomb's law defines force between charges and force is a vector.

5. (*a*) This is Coulomb's law.

6. (*a*)

7. (*d*) It is only a numeric having no units.

8. (*d*) Force can exist in all the three directions.

9. (*a*) $F = \frac{2 \times 2}{4\pi (8.85 \times 10^{-12})(10^{-3})^2} N = 36 \times 10^{+15}$ N

10. (*d*) It is a very very high force.

11. (*a*) Electric field intensity is force per unit charge or NC^{-1}.

12. (*c*) Electric field intensity $= \frac{Q}{4\pi \epsilon_0 r^2}$.

13. (*c*)

14. (*b*) Since electric field has a direction we have to take vector sum.

15. (*a*) Field near a line charge

$$= \frac{\rho}{2\pi \epsilon_0 r} = \frac{1\times 10^{-9}}{2\pi \times 8.85\times 10^{-12}\times 2} \simeq 6 \text{ V/m}.$$

16. (*d*) Electric field is directed towards negative charge.

17. (*a*) $\text{Volt} = \frac{\text{Watts}}{\text{Ampere}} = \frac{\text{N - m/s}}{\text{Ampere}} = \frac{\text{N - m}}{\text{Coulomb}}.$

18. (*c*) $\text{Potential} = \frac{Q}{4\pi \epsilon_0 r}.$

19. (*b*) Potential is a sealer quantity, therefore, we have to take algebraic sum.

20. (*a*)

21. (*b*) E is in a direction opposite to grad V.

22. (*c*) $E = \frac{Q}{4\pi\epsilon_0 x^2}, \quad V = -\int_{\infty}^{r_1} E\cdot dx = \frac{Q}{4\pi\epsilon r_1},$

$$C = \frac{Q}{V} = 4\pi\epsilon_0 r_1.$$

23. (*a*) E is always directed away from positive charge and towards negative charge.

24. (*a*)

25. (*a*) $D = \epsilon E$. This is a standard equation in electrostatics relating D and E.

26. (*b*) **27.** (*a*)

28. (*c*) Force between charges $\propto \frac{1}{(\text{distance})^2}.$

29. (*c*)

30. (*b*) Equipotential surface has to be perpendicular to field line because potential along an equipotential surface is constant.

31. (*a*) Charge on an isolated conductor resides on its surface. This follows from Gauss' law.

32. (*a*) Force is charge × field intensity.

33. (*a*)

34. (*d*) $\text{Final velocity} = \left(\frac{2\,eV}{m}\right)^{0.5}.$

35. (*c*) For a point charge potential and field vary as $\frac{1}{r}$ and $\frac{1}{r^2}$. But for a dipole potential and field vary as $\frac{1}{r^2}$ and $\frac{1}{r^3}$.

36. (*a*) **37.** (*a*) **38.** (*a*) **39.** (*d*)

40. (*a*) Tangential electric field is continuous across a boundary of two dielectrics.

41. (*c*)

42. (*a*) $\frac{E_1}{D_1}\tan\alpha_1 = \frac{E_2}{D_2}\tan\alpha_2$

or $\frac{\tan\alpha_1}{\epsilon_1} = \frac{\tan\alpha_2}{\epsilon_2}$ or $(\epsilon_1)\tan\alpha_2 = (\epsilon_2)\tan\alpha_1.$

43. (*c*) $\frac{\tan\alpha_1}{1} = \frac{\tan\alpha_2}{4}$ or $4\tan\alpha_1 = \tan\alpha_2$ or $\alpha_1 < \alpha_2.$

44. (*b*)

45. (*c*) The combination is equivalent to two capacitors in series. Their capacitances are

$$C_1 = \frac{\epsilon_1\epsilon_0 A}{d_1} \text{ and } C_2 = \frac{\epsilon_2\epsilon_0 A}{d_2}$$

The total capacitance is $\frac{C_1C_2}{C_1 + C_2}.$

46. (*a*) Both dielectric strength and electric field intensity have the units of V/m.

47. (*d*)

48. (*b*) $\text{Energy} = \frac{1}{2}CV^2 = \frac{1}{2}\frac{\epsilon A}{d}V^2 = \frac{1}{2}\frac{\epsilon Ad}{d^2}V^2.$

$$\text{Energy density} = \frac{\text{Energy}}{\text{Volume}} = \frac{1}{2}\epsilon\frac{V^2}{d^2} = \frac{1}{2}\epsilon E^2.$$

49. (*b*) Increase in ϵ is $\frac{P}{E}$. Since energy $= \frac{1}{2}CV^2$, and increase is C is proportional to increase in ϵ.

$$\text{Increase in energy} = \frac{1}{2}\frac{P}{E}V^2 = \frac{1}{2}\frac{P}{E}\cdot\frac{E^2}{d^2} = \frac{1}{2}\frac{PE}{d^2}.$$

50. (*b*) **51.** (*d*) **52.** (*a*)

53. (*a*) When air flows with uniform velocity through a tube open at both ends, divergence of velocity is zero.

54. (*a*) In general $\nabla D = \rho$. If $\rho = 0$, $\nabla D = 0$.

55. (*a*) See Section 5.3.

56. (*a*) See Section 5.3.

57. (*a*) See Section 5.3.

58. (*d*) If C is total capacitance

$$\frac{1}{C} = \frac{1}{C_1} + \frac{1}{C_2} + \frac{1}{C_3} + \frac{1}{C_4} + \frac{1}{C_5} + \frac{1}{C_6}.$$

59. (*a*) Electric flux densities have to be the same in the two dielectrics.

60. (*b*)

$$C_1 \propto \frac{6}{20} \text{ or } C_1 \propto 0.3 \text{ and } C_2 \propto \frac{12}{10} \text{ or } C_2 \propto 1.2.$$

Since Q is the same, voltage is inversely proportional to C. Hence voltage division is 80 V and 20 V.

61. (*c*) $V_1 = 160\text{ V}, V_2 = 40\text{ V}, E_1 = \frac{160}{20\times 10^{-3}} = 8\text{ kV/m},$

$$E_2 = \frac{40}{10\times 10^{-3}} = 4\text{ kV/m}.$$

62. (*b*) $\text{Energy} = \frac{1}{2}CV^2 = \frac{1}{2}\frac{\epsilon_0\epsilon_r A}{d}V^2$

$$= \frac{1}{2}\epsilon_0\epsilon_r Ad\left(\frac{V}{d}\right)^2 = \frac{1}{2}\epsilon_0\epsilon_r AdE^2.$$

$$\frac{\text{Energy stored in first capacitor}}{\text{Energy stored in second capacitor}} = \frac{6\times 50^2}{100\times 10^2} = 1.5.$$

63. (*b*) The particle will receive an acceleration equal to $\frac{eE}{m}$ and, therefore, its velocity will keep on increasing.

64. (*a*) Because of collisions the velocity will be restricted to a constant value equal to mobility × electric field intensity.

65. (*a*) $I = v\rho A$ or $\left(\frac{m}{s}\frac{C}{m^3}m^2 = \frac{C}{s} = \text{ampere}\right)$

66. (*a*) 67. (*b*)

68. (*a*) Voltage = (electric field intensity) *d* and $I = \frac{Ed}{R}$.

69. (*a*) $J = \sigma E$ is Ohm's law at a point.

70. (*a*)

71. (*a*) Integral of current density J over a closed surface must be zero. This is KCL.

72. (*c*) This is continuity relation between current density J and charge density ρ at a point (See Section 5.3).

73. (*a*) Field normal to conductor is zero in this case.

74. (*b*)

75. (*c*) Since conductivity of A is twice that of B, resistance of A is half that of B. Since current is same, potential difference in A is half of that in B. Therefore, electric field intensity in A is also half of that in B.

76. (*d*)

77. (*a*) Conductances in parallel are added directly. For conductances in series, reciprocals of conductances are added.

78. (*a*) 79. (*a*)

80. (*a*) At boundary of conducting media

$$J_{n1} = J_{n2} \text{ and } E_{t1} = E_{t2} \text{ or } \frac{J_{t1}}{\sigma_1} = \frac{J_{t2}}{\sigma_2}.$$

$$\text{Therefore } \frac{J_{r1}}{\sigma_1 J_{n1}} = \frac{J_{t2}}{\sigma_2 J_{r2}} \text{ or } \frac{\tan\alpha_1}{\tan\alpha_2} = \frac{\sigma_1}{\sigma_2}.$$

81. (*a*)

82. (*a*) Energy = $VIt = V \times Q = 100 \times 10^3 \times 100 \times 10^3 \times 200 \times 10^{-3} = 2000$ MJ.

83. (*d*)

84. (*d*) MMF required for air gap is more. Therefore, saturation is prevented.

85. (*d*)

86. (*d*) Hysteresis loss is proportional to area of hysteresis loop.

87. (*a*) Low hysteresis loss is necessary in all electric machines.

88. (*a*) A material will retain magnetism if retentivity is good.

89. (*d*) Shielding prevents stray magnetic fields.

90. (*a*) It is called Transformer emf.

91. (*a*) Motional emf will be induced. Current can not flow because the circuit is not closed.

92. (*a*) This is as per Lenz's law.

93. (*a*) 94. (*a*)

95. (*b*) emf = BIL sin θ.

96. (*a*) This is Lorentz force equation in vector notation.

97. (*a*) This is derived from Biot Savart law.

98. (*a*) 99. (*a*) 100. (*a*)

101. (*a*) Flux is surface integral of B.

102. (*a*) 103. (*a*) 104. (*a*) 105. (*a*)

106. (*a*) 107. (*d*) 108. (*b*) 109. (*b*)

110. (*c*) 111. (*a*)

112. (*c*) $H = \frac{I}{2\pi R}$. If H = 1 A/m and R = 1, I = 2π.

113. (*c*)

114. (*b*) As we move from the centre towards periphery, the magnetic flux increases.

115. (*b*)

116. (*b*) See Section 5.4.

117. (*c*) 118. (*b*) 119. (*b*) 120. (*c*)

121. (*a*)

122. (*b*) Force on particle = *ev*B. This is balanced by centrifugal force $\frac{mv^2}{R}$. Therefore, $evB = \frac{mv^2}{R}$ or $R = mv/eB$.

123. (*a*)

124. (*b*) Vacuum is non-magnetic.

125. (*c*) 126. (*a*) 127. (*b*)

128. (*a*) In ferromagnetic materials the moments align in the same direction. In anti-ferromagnetic materials the moments align in opposite directions.

129. (*b*) Ferrites must have low conductivity to reduce eddy current loss.

130. (*a*) Magnetism vanishes above the curie temperature.

131. (*a*)

132. (*a*) See Eqn.(1.31).

133. (*b*) 134. (*a*)

135. (*b*) For a plane boundary between two media with permeability μ_1 and μ_2, we have $B_{n1} = B_{n2}$ and $H_{t1} = H_{t2}$ which give $\frac{\tan\alpha_1}{\tan\alpha_2} = \frac{\mu_1}{\mu_2} = \frac{\mu_{r1}}{\mu_{r2}}$.

136. (*b*) 137. (*d*)

138. (*a*) The magnetic energy per unit volume of a specimen brought to flux density B from original un-magnetised condition (B = 0) is given by integral of initial magnetisation curve, *i.e.*, $\int_0^B H\,dB$.

139. (*a*) When there is no hysteresis loss, there is no hysteresis loop.

140. (*b*) The product BH is energy density per unit volume.

141. (*b*) Silicon steel is used to make electromagnets for electric machines.

142. (*c*)

143. (*b*) Reluctance = $\frac{l}{\mu a}$ and inductance of solenoid = $\frac{N^2\mu a}{l}$. Therefore, units for reluctance are H^{-1}.

144. (*a*) Reluctance is directly proportional to length and inversely proportional to μ and area of cross-section.

145. (*a*) Reluctances are added as resistances.

146. (*a*) Permeance is reciprocal of reluctance.

147. (*d*) 148. (*a*)

149. (*c*) Energy denstiy = $0.5\frac{B^2}{\mu_0}$. Total energy = energy density × volume of gap = $\frac{0.5B^2Ag}{\mu_0}$.

150. (*c*) See Page 200.
151. (*a*) Voltage at any point is equal to potential gradient × distance.
152. (*a*) See Page 200.
153. (*b*) See Section 5.3.
154. (*a*) When only motional emf exists emf

$$= \oint (Lv \times B) \cdot dl.$$

155. (*a*) This is Faraday's law and gives total emf (due to motion as well as due to changing flux).
156. (*a*) emf= $-\int_s \frac{\partial B}{\partial t} \cdot ds = 2\omega r l \sin \omega t$.
157. (*a*) See Page 200.
158. (*a*) See Table 5.6.
159. (*a*) Conduction current flows through a conductor only. However current does flow through capacitor. Maxwell gave it the name of displacement current.
160. (*c*) A leaky capacitor is represented by a capacitor C in parallel with resistor R, (Fig. 1.5). Current through C is displacement current and current through R is conduction current.
161. (*b*)
162. (*a*) See Table 5.5.
163. (*b*)
164. (*a*) See Table 5.5.
165. (*a*) Integral of sin ωt cos ωt = sin (ωt + 90°).
166. (*b*) See Table 5.6.
167. (*b*) The phenomenon of dielectric hysteresis in dielectrics is similar to that of magnetic hysteresis is ferromagnetic materials. It results in energy loss when dielectric is excited by a.c. supply.
168. (*a*) **169.** (*a*) **170.** (*b*)
171. (*c*) $P = \omega C V^2 \tan \delta$

$$= 2\pi \times 9 \times 10^6 \times \frac{8.85 \times 10^{-12} \times 1}{1} \times 10^6 \times 0.001 \simeq 2 \text{ W/m}^3.$$

172. (*b*) Find C and use the equation $P = \omega C V^2 \tan \delta$.
173. (*c*) See Section 5.7.
174. (*a*)
175. (*b*) See Section 5.7.
176. (*c*) See Section 5.7.
177. (*c*) **178.** (*b*)
179. (*b*) When a wave travels in negative *x* direction, *x* decreases as *t* increases.
180. (*c*) For wave travelling in +*x* direction $\frac{dx}{dt}$ must be positive.
181. (*a*) This is wave equation in conducting medium.
182. (*a*) **183.** (*b*) **184.** (*c*)
185. (*b*) See Section 5.7.
186. (*a*) See Section 5.7.
187. (*b*) See Section 5.7.
188. (*a*) **189.** (*a*) **190.** (*a*)
191. (*a*) At short-circuit voltage is zero and current is reflected back.
192. (*b*) At open circuit voltage is twice the incident wave and current is zero.
193. (*a*) See Section 5.7.
194. (*a*) **195.** (*a*)
196. (*c*) Poynting vector gives power per square metre.
197. (*b*) Since Poynting vector gives power per square metre, we have to take surface integral.
198. (*a*) **199.** (*a*)
200. (*a*) See Section 5.7.
201. (*b*) For TE mode $E_x = 0$.
202. (*a*) For TE mode $E_x = 0$.
203. (*a*)
204. (*d*) Retina has 100 million optical fibres which act as polyrod antennas and photon detection devices.
205. (*a*) As per definition for a cavity resonator Q

$$= 2\pi \left[\frac{\text{total energy stored}}{\text{decrease in energy in one cycle}}\right].$$

206. (*c*) Both are resonant cavities.
207. (*a*)
208. (*a*) Total antenna beam area can be separated to component area due to main lobe and component area due to minor lobes. Main beam efficiency is the ratio of these two area (*i.e.*, solid angles).
209. (*a*) Total solid angle is 4π.
210. (*a*) The word short is used to compare the length of dipole with wavelength.
211. (*d*) All are correct.
212. (*a*) Series resistance has to be zero and shunt conductance has to be zero for zero losses.
213. (*c*) It is used for impedance matching.
214. (*b*)
215. (*c*) It has the dimension of velocity, *i.e.*, metres/sec, or LT^{-1}.
216. (*a*) Joule/coulomb = $\frac{\text{V.A.S.}}{\text{A.S.}}$ = voltage. Thus it is the unit of electric field potential. Only the word potential does not signify the correct word.
217. (*d*) **218.** (*d*) **219.** (*a*) **220.** (*a*)
221. (*a*) **222.** (*c*) **223.** (*b*) **224.** (*b*)
225. (*d*)
226. (*a*) $\frac{P_2}{P_1} = 4$, Loss in dB = 10 log 4 = 6.
227. (*c*) Max-well's equations are valid for sinusoidal as well as non-sinusoidal time variation.
228. (*b*)
229. (*c*) See Table 5.6
230. (*c*) See Section 5.7.
231. (*a*) $\frac{dV(x)}{dx}$ is voltage drop and $-j\omega LI(x)$ is also voltage drop.
232. (*b*) If R and G are finite line cannot be loss less. Hence only 2 is correct.
233. (*c*) **234.** (*d*) **235.** (*c*) **236.** (*a*)
237. (*b*) Potential difference = $\frac{\text{Energy}}{\text{Charge}} = \frac{V.I.t}{I.t}$.

238. (*a*)

239. (*b*) Field is inversely proportional to *r*.

240. (*d*)

241. (*b*) $\rho_v = \dfrac{75-50}{75+50} = 0.2$, VSWR $= \dfrac{1+0.2}{1-0.2} = 1.5$.

242. (*d*) Since $|\rho_v|$ is the same, VSWR pattern is the same. However positions of voltage maxima and voltage minima are not the same in the two cases.

243. (*b*) In this case $\lambda = 10$ m. Input impedance of $\dfrac{\lambda}{8}$ line terminated in resistive load is equal to characteristic resistance. Hence length should be $\dfrac{\lambda}{8} = 1.25$ m.

244. (*c*) Required $Z_0 = \sqrt{50 \times 72} = 60$ ohm. Then $60 = \dfrac{60}{1} \ln \dfrac{R}{10}$ or $R = 27$ mm.

245. (*b*) Input impedance of a short-circuited line of length $\dfrac{\lambda}{4}$ is purely inductive.

246. (*b*) Reflection coefficient is $\dfrac{1}{3}$.

247. (*c*) **248.** (*c*) **249.** (*d*)

250. (*d*) $Z_C = \sqrt{(Z_{0c})(Z_{sc})}$.

251. (*a*) Since currents are in opposite direction, force is repulsive.

252. (*c*) $Z_0 = \sqrt{\dfrac{L}{C}} = \left[\dfrac{0.294 \times 10^{-6}}{60 \times 10^{-12}}\right]^{0.5} = 70$ ohm.

253. (*d*) 254. (*c*)

255. (*b*) $\gamma = \sqrt{ZY}$.

256. (*d*) There is no potential change along a equipotential surface. Therefore, electric field has to be perpendicular to it.

257. (*b*) **258.** (*d*)

259. (*d*) Magnetic dipole moment of coil = current × N × area of coil.

260. (*c*) **261.** (*b*) **262.** (*a*)

263. (*d*) When Z_0 and load impedance are exactly identical, there is no reflection, *i.e.*, $\rho_v = 0$ and VSWR = 1.

264. (*a*) **265.** (*c*)

266. (*b*) Critical angle $= \sin^{-1}\sqrt{\dfrac{\epsilon_2}{\epsilon_1}} = \sin^{-1}\sqrt{\dfrac{1}{4}} = 30°$.

267. (*c*) Wave guides have lower power losses and can be used at very high frequencies but are not easy to use.

268. (*b*) $v = \dfrac{1}{\sqrt{\mu \epsilon}}$

$$= \left[\frac{1}{4\pi \times 10^{-7} \times 1 \times 8.85 \times 10^{-12} \times 4}\right]^{0.5} = 1.5 \times 10^8 \text{ m/s.}$$

269. (*b*) $\sin\theta_r = \sqrt{\dfrac{\mu_1 \epsilon_1}{\mu_2 \epsilon_2}} \sin\theta_i$

$$= \sqrt{\frac{1}{2.7}}\,(0.5) = 0.304 \text{ or } \theta_t = 17.7°.$$

270. (*d*) $\sin\theta_t = \sqrt{2.7}\,(0.5) = 0.822$ or $\theta_t = 52.2°$.

271. (*a*) Critical angle $= \sin^{-1}\sqrt{\dfrac{1}{81}} = 6.4°$.

272. (*c*) **273.** (*c*) **274.** (*c*)

275. (*a*) $\dfrac{100}{4\pi(10)^2} \simeq 0.0833$.

276. (*b*) $C = 4\pi\epsilon_0 r = 4\pi\epsilon_0\left(\dfrac{1}{4\pi\epsilon_0}\right) = 1\text{F}$.

277. (*a*) See Section 5.7.

278. (*b*) $n = \sqrt{2.1 \times 1} = 1.45$.

279. (*a*) Velocity $= \dfrac{1}{\sqrt{\mu\epsilon}}$.

280. (*c*)

281. (*b*) $\eta = \sqrt{81 \times 1} = 9$.

282. (*c*) 283. (*a*)

284. (*c*) If $\dfrac{\sigma}{\omega\epsilon} > 100$, the material is classified as conductor.

If $\dfrac{1}{100} < \dfrac{\sigma}{\omega\epsilon} < 100$, it is quasi conductor and if $\dfrac{\sigma}{\omega\epsilon} < \dfrac{1}{100}$ it is a dielectric.

285. (*a*) It is dimensionless.

286. (*d*)

287. (*b*) VSWR = infinite.

288. (*c*)

289. (*c*) In an ideal dielectric conduction current is zero.

290. (*b*) For both R_1 and R_2, VSWR = 3.

291. (*a*) $Z_0 = \sqrt{100 \times 64}$.

292. (*a*) Input impedance $= \dfrac{Z_0^2}{Z_L} = \dfrac{50 \times 50}{100} = 25$ ohm.

293. (*b*) **294.** (*a*)

295. (*b*) $[(2-0)^2 + (0-2)^2 + (2-1)^2]^{0.5} = 3$.

296. (*a*) **297.** (*a*)

298. (*a*) $F.dl = x dx$, Work $= \displaystyle\int_1^2 F.dl = \int_1^2 x.dx = 1.5$.

299. (*c*) $F.dl = r^2 dr$, Work $= \int F.dl$

$$= \int_{\sqrt{2}}^{2\sqrt{2}} \frac{dr}{r^2} = \left| -\frac{1}{r} \right|_{\sqrt{2}}^{2\sqrt{2}} = \frac{1}{2\sqrt{2}}.$$

300. (*a*) Work done does not depend on the path because line integral depends only on the end points.

301. (*b*) **302.** (*a*) **303.** (*b*) **304.** (*a*)

305. (*a*)
306. (*a*) Direction cosines are cos α, cos β and cos γ.
307. (*a*) Dot product of unit vector with rectangular coordinate vectors are equal to direction cosines.
308. (*a*) Coulomb force

$$= \frac{1}{8.85\times10^{-12}\times10^{-6}} \text{ N} \simeq 9.2\times10^{11} \text{ metric tons.}$$

309. (*a*) $E = \frac{10^{-8}}{4\pi\times8.85\times10^{-12}\times9} \simeq 10$ V/m. Since charge is negative, E is inwards, *i.e.*, negative *r* direction.
310. (*a*) If distance is large, line charge can be replaced by a point charge $=1\times10^{-8}\times4=4\times10^{-8}$.
311. (*b*) **312.** (*a*) **313.** (*b*) **314.** (*b*)
315. (*a*)
316. (*a*) $\sqrt{2^2+1^2}\angle\tan^{-1}0.5$.
317. (*c*) The electric field inside the sphere is zero.
318. (*a*)
319. (*a*) $E = \frac{Q}{4\pi\epsilon_0 x^2}$, $V = -\int_{\infty}^{r} E.dx = \frac{Q}{4\pi\epsilon_0 r}$,

Capacitance $= \frac{Q}{V} = 4\pi\epsilon_0 r$.

320. (*c*) **321.** (*a*) **322.** (*c*) **323.** (*c*)
324. (*a*) **325.** (*c*) **326.** (*b*)
327. (*d*) $\tan\alpha_2 = 4\tan\alpha_1$.
328. (*a*) Energy $= V.It = 12\times100\times3600 \simeq 4\times10^6$ J.
329. (*b*) **330.** (*d*)
331. (*d*) All the statements are correct.
332. (*c*) **333.** (*a*) **334.** (*a*) **335.** (*a*)
336. (*b*) See Section 5.3.
337. (*c*) All the statements are correct.
338. (*b*) Resistance $\propto \frac{l}{a}$. Find average length and average area of cross-section for all the three directions.
339. (*a*)
340. (*c*) Energy $= VIt = 90\times10^6\times100\times10^3\times400\times10^{-3}$
$= 3.6\times10^{12}$ J.
341. (*d*) **342.** (*a*) **343.** (*a*) **344.** (*b*)
345. (*c*) **346.** (*a*)
347. (*b*) If μ_r is taken as 5000, $\tan\alpha_2 = 5000\tan\alpha_1$.
348. (*b*) **349.** (*b*) **350.** (*a*)
351. (*b*) The units for coercivity are the same as for H, *i.e.*, A/m.
352. (*d*) The units for retentivity are the same as for flux density, *i.e.*, W/m^2.
353. (*b*) Permeance is reciprocal of reluctance.
354. (*a*) Units for premeance $= \frac{\frac{H}{m}\times m^2}{m} = H$.
355. (*c*) **356.** (*b*)
357. (*b*) Since flux is same MMF $\propto$ reluctance. Reluctance of iron $= \frac{0.5}{\mu_0\times1000\times a}$.

Reluctance of air gap $= \frac{2\times10^{-3}}{\mu_0 a}$,

MMF for air gap

$$= \frac{2\times10^{-3}}{\mu_0 a}\left(\frac{1000\mu_0 a}{0.5}\right)(100) = 400.$$

358. (*b*) Magnetic gap force $= \frac{0.5\,B^2A}{\mu_0}$.
359. (*a*) **360.** (*a*)
361. (*a*) See Section 5.4.
362. (*c*) Capacitor current leads conduction current by 90°.
363. (*b*)
364. (*b*) $Z_0 = \frac{50}{\sqrt{\epsilon_r}}\ln\frac{R}{r} = \frac{60}{1}\ln 2 \simeq 40\ \Omega$.
365. (*b*) $Z_0 \propto \frac{1}{\sqrt{\epsilon_r}}$.
366. (*c*) **367.** (*b*) **368.** (*a*) **369.** (*b*)
370. (*d*)
371. (*c*) Resonant frequency

$$\propto \left[\left(\frac{l\pi}{a}\right)^2 = \left(\frac{m\pi}{b}\right)^2 + \left(\frac{n\pi}{c}\right)^2\right]^{0.5}$$

It is seen that resonant frequency can be changed by changing dimensions of cavity. Moreover cavity resonator has as many modes as corresponding waveguide and resonant frequencies of cavities are not closely spaced.
372. (*b*) Single stub and quarter wave transformer are suitable only for one frequency. Double stub is suitable for a range of frequencies.
373. (*d*) In uniform cylinderical waveguide TE_{10} and TM_{11} modes have high attenuation.
374. (*c*) At $z = 0$
$E_x = E_1 \sin\omega t$
$E_y = E_2 \sin(\omega t + \delta)$
where δ is time phase difference between E_x and E_y.
For circular polarization $E_1 = E_2 = E_0$ and $\delta = \pm 90°$.
Hence all the three statements are correct.
375. (*d*) Only TEM wave will propagate between two perfectly conducting parallel plates of infinite extent.
376. (*c*) As given in section 5.7

Skin depth $\delta = \frac{1}{\sqrt{f\pi\mu\sigma}}$

Hence only 2 and 3 are correct.
377. (*c*) Reflector longer than driven element longer than director.
378. (*a*) Radiation resistance $= 80\pi^2\left(\frac{l}{\lambda}\right)^2$
$= 80\pi^2(0.03)^2 = 0.072\pi^2$.
379. (*b*) Reflection coefficient $= \frac{Z_L - Z_0}{Z_L + Z_0} = \frac{j50-50}{j50+50}$

Magnitude of reflection coefficient = 1

$$\text{VSWR} = \frac{1+1}{1-1} = \infty$$

380. (*a*) Capacitance $= \frac{\in A}{d}$. Area of second set of plates is 4 times that of first.

381. (*a*) $\omega\in = \frac{2\pi\times3\times10^9\times10^{-9}}{36\pi} = \frac{1}{6}$

Intrinsic impedance $\eta = \sqrt{\frac{j\omega\mu}{\sigma + j\omega\in}}$

Since $\omega\in << \sigma$, $\eta = \sqrt{\frac{j\omega\mu}{\sigma}} = \sqrt{\frac{\omega\mu}{\sigma}}\, e^{j\pi/4}$

$$= \left[\frac{2\pi\times3\times10^9\times4\pi\times10^{-7}}{5.8\times10^7}\right]^{0.5} e^{j\pi/4}$$

$$= 0.02\, e^{j\pi/4}.$$

382. (*c*) If follows from laws of electrostatics.

383. (*d*) These follow from laws of electrostatics.

384. (*c*) To avoid frequency distortion, attenuation must be constant and independent of frequency. To avoid phase distortion the phase constant must vary linearly with frequency.

385. (*d*) $C = \frac{\in A}{d}$ So capacitance changes

$Q = CV = \text{constant}$

or $V \propto \frac{1}{C}$ and Electric field intensity $= \frac{V}{d}$

Energy $= \frac{1}{2}CV^2$

Thus all the three change.

386. (*a*) $\nabla\cdot B = 0$ and $\nabla\cdot(\nabla\times A) = 0$

Hence $B = \nabla\times A$.

387. (*a*) Applicable only to electrostatic field.

388. (*b*)

389. (*b*) $e = -N\frac{d\phi}{dt} = (-1)(5\times2) = -10$ V.

390. (*d*) See Table in section 5.6.

391. (*d*) It is in $-z$ direction.

392. (*c*) For plane wave in free space $V_p = C = V_g$. These three are different in wave quides.

393. (*c*) Time is negligible. Statement 3 is also true. Length is small so it is lumped circuit.

394. (*b*) $Z_0^2 = 200\times450$ or $Z_0 = 300$ ohm.

395. (*a*) $Z_0^2 = $ (Input impedance) (infinity)

or input impedance $= \frac{Z_0^2}{\text{Infinity}} = 0.$

396. (*d*) In distortionless line $\frac{L}{C} = \frac{R}{G}$

$$Z_0 = \sqrt{\frac{L}{C}} = \sqrt{\frac{R}{G}}$$

$$\alpha = \sqrt{RG} = \sqrt{R}\cdot\frac{\sqrt{R}}{Z_0}$$

$$= \frac{R}{Z_0} = \frac{0.1}{50} = 0.002 \text{ Np/m}.$$

397. (*c*) $\eta = \sqrt{\frac{\mu}{\in}} = \sqrt{\frac{\mu_0}{4\in_0}} = \left[\frac{4\pi\times10^{-7}}{8.85\times10^{-12}}\right]^{0.5} = 60\pi$

$$P_{av} = \frac{1^2}{2\eta} = \frac{1}{120\pi}.$$

398. (*b*) This is Stobes theorem

$$\oint_c \vec{A}\cdot\vec{dl} = \iint_s (\nabla\times\vec{A})\cdot\vec{ds} = \iint_s \vec{V}\cdot\vec{ds}$$

399. (*d*) For each branch line input impednace

$$= \frac{Z_0^2}{Z_L} = \frac{100^2}{50} = 200\ \Omega.$$

Load impedance as seen by in main line = 200 || 200 = 100 ohm

$$Z_1 = \frac{Z_0^2}{Z_L} = \frac{50^2}{100} = 25\ \Omega.$$

400. (*a*) Minimum operating frequency $= \frac{c}{2}\sqrt{\left(\frac{1}{a}\right)^2 + \left(\frac{1}{b}\right)^2}$

$$= \frac{3\times10^8}{2}\left[\left(\frac{1}{0.04}\right)^2 + \left(\frac{1}{0.03}\right)^2\right]^{0.5}$$

$$= 6.25 \text{ GHz}.$$

401. (*d*) $\beta = \frac{2\pi}{\lambda}$ and $\lambda = \frac{c}{f} = \frac{3\times10^8}{3\times10^9} = \frac{1}{10} = 0.1$

$$\beta = \frac{2\pi}{0.1} \text{ and } \beta l = \frac{2\pi}{0.1}\left(\frac{1}{100}\right) = \frac{2\pi}{10} = \frac{\pi}{5}$$

Input impedance $= Z_0\left[\frac{Z_L + j\,Z_0 \tan\beta l}{Z_0 + j\,Z_L \tan\beta l}\right]$

$Z_L = 0$, $Z_0 = 75$ ohm

Input impedance $= 75\left[\frac{0 + j75\tan\frac{\pi}{5}}{75}\right] = j\,55\ \Omega$

Hence it is inductive.

402. (*c*) $\in_r$ for free space = 1

Reflection coefficient $= \left|\frac{1-\sqrt{9}}{1+\sqrt{9}}\right| = \left|\frac{1-3}{1+3}\right| = 0.5$

403. (*d*) Gain $= \eta\pi^2\left[\frac{D}{\lambda}\right]^2 = 0.7\pi^2\left[\frac{1\times20\times10^9}{3\times10^8}\right]^2 = 3.705$

Gain in $dB = 10\log_{10}(30705) = 45\ dB$.

404. (*b*) ρ, *i.e.*, charge density is zero in source free medium. See Table in section 5.6.

405. (*c*) For TE_{20} mode $\lambda = a = 3\times10^{-2}$ m

$$f_c = \frac{C}{\lambda_c} = \frac{3\times10^8}{3\times10^{-2}} = 10^{10}\text{ Hz}$$

$$\eta = \frac{\text{Intrinsic impedance of free space}}{\left[1-\left(\frac{f_c}{f}\right)^2\right]^{0.5}}$$

$$= \frac{377}{\left[1-\left(\frac{10^{10}}{3\times10^{10}}\right)^2\right]^{0.5}} = 400\text{ ohm.}$$

406. (*d*) For free space

$$E = \eta_0 H \text{ and } \rho = \frac{E^2}{2\eta_0}$$

$$\text{Hence } \rho = \frac{\eta_0^2 H^2}{2\eta_0} = \frac{\eta_0 H^2}{2}$$

$$\text{For given wave, } H = \frac{1}{\eta_0}\left[(5\sqrt{3})^2 + 5^2\right]^{0.5} = \frac{10}{\eta_0}$$

$$\text{Hence } \rho = \frac{\eta_0}{2}\left(\frac{10}{\eta_0}\right)^2 = \frac{50}{\eta_0}\text{ watts.}$$

407. (*b*) In S matrix diagonal elements represent reflection coefficients. In this case no power is reflected on either port hence $S_{11} = S_{22} = 0$.
The off diagonal elements represent transmission coefficients. Since there is perfect matching $S_{12} = S_{21} = 1$.

408. (*d*) For line terminated in 100 ohm resistance

$$Z_{i1} = \frac{50^2}{100} = 25\text{ ohm}$$

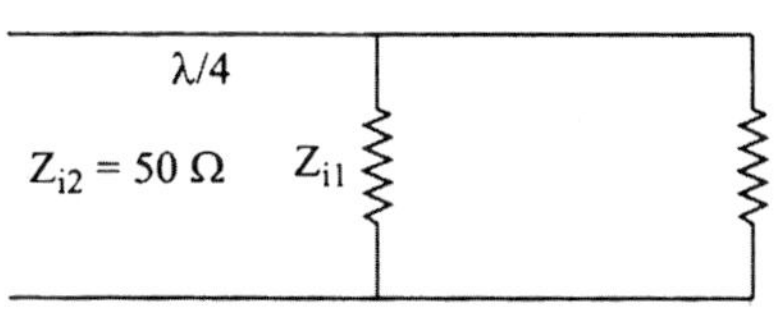

Fig. 5.13.

For line terminated in 200 ohm resistance

$$Z_{i2} = \frac{50^2}{200} = 12.5\text{ ohm}$$

Parallel combination of Z_{i1} and Z_{i2}

$$= \frac{25\times12.5}{25\times12.5} = \frac{25}{3}\text{ ohm}$$

$$\text{For main line } Z_i = \frac{Z_0^2}{25/3} = \frac{50^2}{25/3} = 300\text{ ohm}$$

Voltage reflection coefficient at input

$$= \frac{300-50}{300+50} = \frac{5}{7}.$$

409. (*d*)

$$\tan\theta = \sqrt{\frac{\epsilon_{r2}}{\epsilon_{r1}}} \text{ or } \tan 60° = \sqrt{\frac{\epsilon_{r2}}{1}}$$

$$\text{or } \sqrt{3} = \sqrt{\epsilon_{r2}} \text{ or } \epsilon_{r2} = 3.$$

410. (*d*) $\epsilon_r = 2$

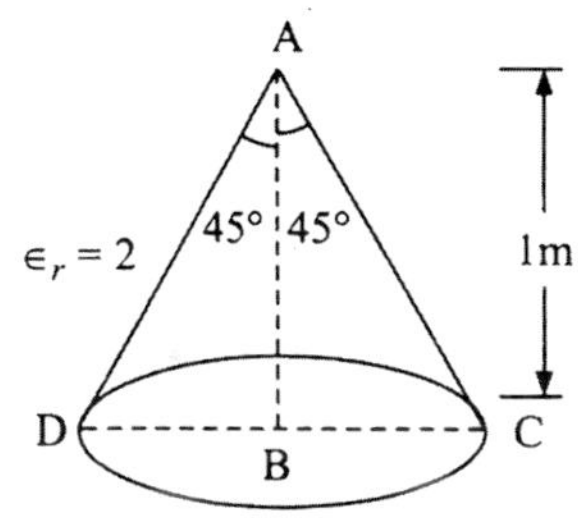

Fig. 5.14.

$$\sin\theta = \frac{1}{\sqrt{\epsilon_r}} = \frac{1}{\sqrt{2}}$$

$$0 = 45°$$

$$BD = AB = 1\text{ m}$$

$$\text{area} = \pi r^2 = \pi(BD)^2 = \pi\, m^2.$$

411. (*a*) Transmission coefficient

$$= \frac{n_2 - n_1}{n_2 + n_1} = \frac{\sqrt{\frac{\mu_0}{\epsilon_0\epsilon_r}} - \sqrt{\frac{\mu_0}{\epsilon_0}}}{\sqrt{\frac{\mu_0}{\epsilon_0\epsilon_r}} + \sqrt{\frac{\mu_0}{\epsilon_0}}}$$

$$= \frac{\frac{1}{2}-1}{\frac{1}{2}+1} = -\frac{1}{3}$$

$$T = 1-\left(-\frac{1}{3}\right)^2 = \frac{8}{9}$$

$$P_t = TP_1 \text{ or} = \frac{P_t}{P_1} = \frac{8}{9}.$$

412. (*c*)

$$\lambda_c = \frac{3\times10^8}{18\times10^9} = \frac{1}{60} = \frac{2a}{3}$$

$$a = \frac{1}{40}\text{ m} = 2.5\text{ cm}.$$

413. (*a*)

$$\lambda = \frac{3\times10^8}{600\times10^3} = 500\text{ m}$$

Radiation resistance

$$= 40\pi^2\left(\frac{dl}{\lambda}\right)^2 = 40\pi^2\left(\frac{50}{500}\right)^2 = \frac{2}{5}\pi^2.$$

414. (*d*) Reflected wave = 40 – 30 = 10 V

$$\text{Reflection coefficient} = \frac{10}{30} = \frac{1}{3} = \frac{Z_L - 50}{Z_L + 50}$$

or $Z_L = 100$ ohms.

415. (*b*) Antenna radiates the power it receives.

416. (*b*) It is only for directivity.

417. (*a*) Antenna is a transducer, is frequency sensitive. It converts electrical energy into EM energy and vice versa. Thus it is reciprocal.

418. (*c*) Range of VSWR is 1 to infinity because VSWR = E_{max}/E_{min}.

419. (*a*) Gain of antenna $= \dfrac{4\pi A}{\lambda^2}$

where A is aperture area.

420. (*d*) Input impedance $= Z_0 \left[\dfrac{Z_L + jZ_0 \tan \beta l}{Z_0 + j Z_L \tan \beta l}\right]$

Since $Z_L = \infty$

Input impedance $= Z_0 \left[\dfrac{1 + j\dfrac{Z_0}{Z_L} \tan \beta l}{\dfrac{Z_0}{Z_L} + j \tan \beta l}\right]$

$= -j Z_0 \cot \beta l.$

421. (*b*) Both are correct but independent.

422. (*b*)

423. (*b*) Both are true but independent.

424. (*d*) **425.** (*b*)

426. (*a*) In general $\nabla \cdot D = \rho$.

427. (*c*) R is not correct.

428. (*b*) A is Faraday's law and B is Lenz's law. Both are independent.

429. (*b*) Both are correct but independent.

430. (*d*) If $\rho = 0$. Poisson's equation reduces to Laplace equation but solution of Poisson's equation is not the same as that of Laplace equation.

431. (*b*) Both are true but independent (See Section 5.1).

432. (*a*)

433. (*b*) Coulomb's force $\propto \dfrac{Q_1 Q_2}{r^2}$.

434. (*b*)

435. (*c*) One Coulomb is very big charge.

436. (*a*) Field intensity is Coulomb's force per unit charge.

437. (*a*) **438.** (*a*)

439. (*b*) Both are true but independent.

440. (*b*) **441.** (*a*)

442. (*d*) A is wrong because electric potential is a scalar.

443. (*a*) **444.** (*b*)

445. (*a*) Charge resides on the surface of sphere.

446. (*a*) Since the field inside the metal cage is zero, shielding is provided.

447. (*a*) **448.** (*b*)

449. (*a*) $\in = \in_0 + \dfrac{P}{E}$ or $\in E = \in_0 E + P$ or $D = \in_0 E + P$.

450. (*b*) **451.** (*b*)

452. (*b*) Both are true but independent.

453. (*b*) **454.** (*b*)

455. (*b*) Both are true but independent.

456. (*b*) **457.** (*b*)

458. (*c*) Diamagnetic specimen is repelled by magnet.

459. (*a*) Since conductivity is low, eddy currents are avoided. So ferrite can be used at high frequencies.

460. (*b*) Both are correct but independent.

461. (*a*) Fringing is high if $\dfrac{A}{d}$ is small.

462. (*a*) Method of images is a very well known technique to solve many problems like effect of earth on capacitance of line etc.

463. (*a*) In any medium velocity of hight $= \dfrac{c}{\sqrt{\in_r}}$

For all medium $\in_r > 1$. For vaccum $\in_r = 1$. Hence velocity of light in any medium is lower than that in vaccum.

464. (*a*) When $Z_0 = Z_L$ reflection is zero.

465. (*b*) Both are correct but independent.

466. (*b*) Both are correct but independent.

467. (*d*) A is wrong.

468. (*c*) As length increases current distribution becomes non-uniform.

469. (*b*) Both are correct but indepedent.

470. (*b*) A conducting surface is an equipotential surface. Since potential at infinite point is zero, potential of infinite conducting surface is zero. Both statements are correct but independent.

471. (T) This is as per Coulomb's law.

472. (T) **473.** (T)

474. (F) $E \propto \dfrac{1}{r^2}$.

475. (T)

476. (F) Electric field is a vector quantity, so we have to take vector sum.

477. (T) **478.** (T) **479.** (T) **480.** (T)

481. (T) **482.** (T)

483. (F) It varies linearly with distance.

484. (T) $E = -\text{grad } V$ or $\text{grad } V = -E$.

485. (T) **486.** (T)

487. (T) This is the definition of electron volt.

488. (T)

489. (T) It is the property of isotropic material.

490. (T) **491.** (T)

492. (F) $\in + \in_0 + \dfrac{P}{E}$.

493. (T) **494.** (F)

495. (F) Normal component of flux density at dielectric conductor boundary is equal to surface charge density on the conductor.

496. (F) It is about 30 kV/cm.

497. (F) $\in_r$ for different materials is more than 1.

498. (F)

499. (F) Energy density in a battery may be higher.

500. (T) **501.** (T) **502.** (T) **503.** (T)

504. (T) **505.** (T) **506.** (T) **507.** (T)

508. (F) **509.** (T)

510. (F) Inductance $\propto N^2$.

511. (T) **512.** (T) **513.** (F)

514. (F) Magnetic field at the ends is half of that at centre.

515. (T) **516.** (T) **517.** (T) **518.** (T)

519. (T) **520.** (T) **521.** (T) **522.** (T)

523. (T) **524.** (T)

525. (F) There is anti-parallel orientation of unequal moments (See Table 1.8).
526. (T)
527. (F) Ferrimagnetic materials are less magnetic than ferromagnetic materials.
528. (F) Aluminium is para magnetic.
529. (F) Bismuth is diamagnetic.
530. (F) Anti-ferromagnetic materials have anti-parallel orientation of equal moments.
531. (T) **532.** (F) **533.** (T) **534.** (T)
535. (T) **536.** (T) **537.** (T)
538. (F) It has the dimension of energy density.
539. (F) μ_r first increases, then decreases.
540. (T) **541.** (T) **542.** (T)
543. (F) See Table 5.5.
544. (T) **545.** (F) **546.** (T) **547.** (T)
548. (F) $\delta = 90 - \theta$.
549. (T)
550. (T) See Section 5.6.
551. (T) **552.** (T) **553.** (T)
554. (F) If travels in positive x direction.
555. (T) **556.** (T)
557. (F) $Z_0 = \sqrt{\frac{Z}{Y}}$.
558. (F) Reflection coefficient $= \frac{Z_L - Z_0}{Z_L + Z_0}$.

559. (T) **560.** (T)
561. gravitation
562. conservative or lamellar
563. r, θ, ϕ
564. coulomb
565. TEM **566.** $\nabla \times H = \frac{\partial D}{\partial t}$
567. $\frac{1}{\sqrt{\mu \epsilon}}$
568. voltage standing wave ratio
569. reflected wave **570.** resonator
571. dispersive **572.** conductor
573. shielding **574.** linearly polarised
575. elliptically polarised **576.** Brewster angle
577. waveguides **578.** matching device
579. storage **580.** free space
581. main beam efficiency **582.** directivity
583. permittivity **584.** vector
585. potential gradient **586.** joules/coulomb
587. electric field intensity
588. circular **589.** equipotential
590. lines of force **591.** surface
592. Q **593.** resistivity
594. displacement current **595.** F/m
596. right hand rule **597.** repel
598. bar magnet **599.** resistance
600. cubic metre

CHAPTER 6

ELECTRICAL AND ELECTRONIC INSTRUMENTS

6.1. ERROR ANALYSIS

6.1.1. Error and Correction

The actual or true value of the quantity being measured is called the true value. The difference between the measured (or observed) value and true value is the error. The difference between the true value and measured value is the correction.

6.1.2. Significant Figures

The various measurements are expressed as decimal numbers. The number of significant figures gives information regarding magnitude and precision.

6.1.3. Types of Errors

Errors can be classified as Gross errors, Systematic errors and Random errors.

Gross errors are caused by human mistakes in reading the instrument and recording the value. Systematic errors can be further classified as instrument errors, environmental errors and observational errors.

The instrument errors may be due to inherent shortcoming of instruments, (*e.g.*, due to construction, calibration or operation of instrument). Loadings of instruments may also result in errors. Changes in temperature, humidity, pressure etc. may cause environmental errors. Observational errors include error in reading due to parallax, cramped scale etc.

Random errors (also called residual errors) are those errors which are not included in the above errors. These errors may occur due to a variety of factors.

6.1.4. Errors in Combination of Quantities

Let $y = u \pm v$

and $\pm du \pm dv$ be errors in u and v respectively. Then error dy is given by

$$\frac{dy}{y} = \pm\left(\frac{u}{y}\frac{du}{u} + \frac{u}{y}\frac{dv}{v}\right) \quad ...(6.1)$$

If $y = uv$

$$\frac{dy}{y} = \pm\left(\frac{du}{u} + \frac{dv}{v}\right) \quad ...(6.2)$$

Equation (6.2) is also valid if $y = \frac{u}{v}$

If $y = u^n$

then $\frac{dy}{y} = \pm n\frac{du}{u}$

If $y = u^n v^m$

then $\frac{dy}{y} = \pm\left(n\frac{du}{u} + m\frac{dv}{v}\right)$

6.1.5. Statistical Analysis

If n reading have values $x_1, x_2, x_3, ... x_n$, the arithmetic mean $\bar{x}$ is

$$\bar{x} = \frac{x_1 + x_2 + ... + x_n}{n} = \frac{\Sigma x}{n}$$

The deviations from mean are

$d_1 = x_1 - \bar{x}$, $d_2 = x_2 - \bar{x}$, $d_3 = x_3 - \bar{x}$ etc.

$d_1, d_2, d_3......$ may be positive or negative

$$\text{Average deviation} = \frac{|d_1| + |d_2| + ... + |d_n|}{n} = \frac{\Sigma |d|}{n}$$

The standard deviation σ is

$$\sigma = \left[\frac{d_1^2 + d_2^2 + ... + d_n^2}{n}\right]^{0.5} = \left[\frac{\Sigma d_1^2}{n}\right]^{0.5}$$

Variance = mean square deviation = σ^2

Probable error = $\pm$ 0.6745 σ

6.2. UNITS AND DIMENSIONS

6.2.1. SI Units

SI system of units consists of seven base units. These are given in Table 6.1.

Table 6.1. SI units

Quantity	*SI Units*	*Symbol*
Length	metre	m
Mass	kilogram	kg
Time	second	s
Electric current	ampere	A
Themodynamic temperature	kelvin	k
Luminous intensity	candela	cd
Amount of substance	mole	mol

Two supplementary units, *i.e.*, radian (rad.) and steradian (*sr*) for measuring plane angle and solid angle, respectively, form part of SI system.

6.2.2. Derived Units

Some derived width have been named after famous scientists, *e.g.*, Newton (derived unit of force), Joule (derived unit of energy), Watt (derived unit of power), Coulomb (derived unit of electric charge), Ohm (derived unit of resistance), Volt (derived unit of potential difference and emf), Siemen (derived unit of conductance, and admittance), Hertz (derived unit of frequency) etc.

6.2.3. Multiples of units

The various multipliers used in SI system are given in Table 6.2.

Table 6.2. Multipliers and prefixes in SI System

Prefix	*Factor*	*Symbol*
atto	10^{-18}	a
femto	10^{-15}	f
pico	10^{-12}	p
nano	10^{-9}	n
micro	10^{-6}	μ
milli	10^{-3}	m
kilo	10^{3}	k
mega	10^{6}	M
giga	10^{9}	G
tera	10^{12}	T

6.2.4. Dimensions of Quantities

Different units can be represented dimensionally in terms of length L, mass M, mass M, time T and current, I, *e.g.*,

$$\text{Velocity} = \frac{\text{Length}}{\text{Time}} = \frac{L}{T} = LT^{-1}$$

$$\text{Acceleration} = \frac{\text{Velocity}}{\text{Time}} = \frac{LT^{-1}}{T} = LT^{-2}$$

$$\text{Force} = \text{mass} \times \text{acceleration} = MLT^{-2} \text{ etc.}$$

6.3. ABSOLUTE AND SECONDARY INSTRUMENTS

Absolute instruments are used only in standard laboratories. They give the value of quantity being measured in terms of deflection of instrument and the constant of instrument *e.g.*, Rayleigh current balance measures current in terms of force between coils. Tangent galvanometer measures current in terms of tangent of angle of deflection, radius and number of turns of galvanometer coil and horizontal component of earth's magnetic field. Secondary instruments have a calibrated scale and give the value of quantity being measured directly.

6.4. CLASSIFICATION OF SECONDARY INSTRUMENTS

6.4.1. Classification according to Effect of Current

The different effects of electric current used for operation of instruments are : Magnetic effect (ammeters, voltmeters, wattmeters), Electromagnetic Induction effect (AC ammeters, voltmeters, wattmeters, energy meters), Heating effect (ammeters, voltmeters), chemical effect (dc ampere hour meter), electrostatic effect (voltmeters)

6.4.2. Classification According to Mounting

Switch board (or panel) instruments are mounted on switch boards and panels. They are mostly installed in vertical position and are used in power plants, industries. Portable instruments are designed for use in laboratories and are kept in horizontal position.

6.4.3. Analog and Digital Instrument

In analog instrument the pointer moves on calibrated scale and gives the value of the quantity being measured. A digital instrument has a display of numbers indicating the value.

6.4.4. Classification as per Accuracy Class

An per BIS 1248 (1968), eight accuracy classes are 0.05, 0.1, 0.2, 0.5, 1, 1.5, 2.5 and 5. These figures indicate the percentage error, *e.g.*, instrument of accuracy class 0.5 can have an error upto ± 0.5% of full scale deflection.

6.4.5. Classification as Indicating, Integrating and Recording

An indicating instrument indicates the value of quantity being measured, *e.g.*, ammeter, voltmeter, wattmeter. An integrating instrument measures the total quantity of electricity or electrical energy supplied during a certain time. Energy meter is an integating instrument. A recording instrument gives a continuous record of the variations of quantity by means of a path traced out by a pen (attached to the moving system) on a sheet of paper.

6.5. DEFLECTING, CONTROLLING AND DAMPING MECHANISM

The deflecting mechanism produces the deflecting torque which causes movement of pointer. The deflecting torque can be produced by magnetic, induction, heating, chemical and electrostatic effect.

The function of controlling mechanism is to limit the deflection of the moving system so that the deflection is proportional to the quantity being measured. Controlling torque is mostly provided by a spring and sometimes by gravity.

The function of damping mechanism is to bring the pointer to final position quickly. Damping can be by air friction, fluid friction and eddy currents.

6.6. PERMANENT MAGNET MOVING COIL INSTRUMENT

This instrument has a permanent magnet which provides the magnetic field. A small light weight coil, air cored or mounted on aluminium former, is pivoted in the field of magnet.

$$\text{Deflection} = \frac{k}{C} I$$

where k and C are constants and I is the current being measured.

These instruments have advantages of linear scale, low power consumption, good sensitivity, good damping and high torque/weight ratio. However, they can be used only for dc and are more expensive than moving iron instruments.

6.7. GALVANOMETERS

6.7.1. D'Arsonval Galvanometer

It is also a permanent magnet moving coil instrument and is used to indicate null position in dc potentiometers and dc bridges.

$$\text{Deflection } \theta = \frac{NBIA}{C} \text{ radians}$$

where N is the number of turns in coil, B is flux density Wb/m^2, I is current through coil, A is area of coil and C is spring constant.

$$\text{Static sensitivity} = \frac{\theta}{I} = \frac{NBA}{C} \text{ rad/A}$$

The dynamic response of D'Arsonval Galvanometer is given by the equation

$$J\frac{d^2\theta}{dt^2} + D\frac{d\theta}{dt} + C\theta = GI$$

where J is moment of inertia of moving system, kg-m^2, D is damping constant N-m/rad. s^{-1}, C is spring constant N-m/rad and G is displacement constant (= NBA) Newton-metre/A.

The actual response depends on the nature of roots of characteristic equation.

If $\frac{D^2}{4J^2} < \frac{C}{J}$ or $D^2 < 4$ CJ, response is underdamped

If $\frac{D^2}{4J^2} = \frac{C}{J}$ or $D^2 = 4$ CJ, response is critically damped

If $\frac{D^2}{4J^2} > \frac{C}{J}$ or $D^2 > 4$ CJ, response is over damped

If R_g is resistance of galvanometer, the external resistance R_f required for critical damping is

$$R_f = \frac{G^2}{2(CJ)^{0.5}} - R_g$$

6.7.2. Vibration Galvanometer

It is also a permanent magnet moving coil instrument used to indicate null position in ac potentiometers and ac bridges operating at 50 or 60 Hz. Its response is very sharp and it can detect even very small ac currents. Many times it is used along with a mirror and a scale.

6.7.3. Ballistic Galvanometer

Its construction is very similar to D'Arsonval galvanometer. It is an intergrating instrument used to measure quantity of electricity, *i.e.*, charge. When magnetic flux linking the coil changes, an emf is induced in the coil, this emf being proportional to the charge.

6.8. MOVING IRON INSTRUMENT

It is very commonly used as ac and dc ammeter and voltmeter. It has one fixed coil and two iron vanes, one fixed and one movable. When current passes through the coil, the two vanes are magnetised in the same direction and repel each other.

Deflecting torque = kI^2

Its advantages include robust construction, low cost and suitability for both ac and dc. However, it has non-linear scale and its operation is affected by stray magnetic field, change of frequency, change of temperature and change of inductance of coil. The effect of change in inductance L is written as

$$\theta = 0.5\frac{I^2}{C}\frac{dL}{d\theta}$$

6.9. RECTIFIER INSTRUMENT

It consists of a bridge rectifier (copper oxide or silicon) and a permanent magnet moving coil instrument (Fig. 6.1)

For a full wave rectifier

$$I_{dc} = \frac{2I_m}{\pi} = \frac{2\sqrt{2}I}{\pi} = 0.9\ I$$

Therefore ac sensitivity is 0.9 times dc sensitivity.

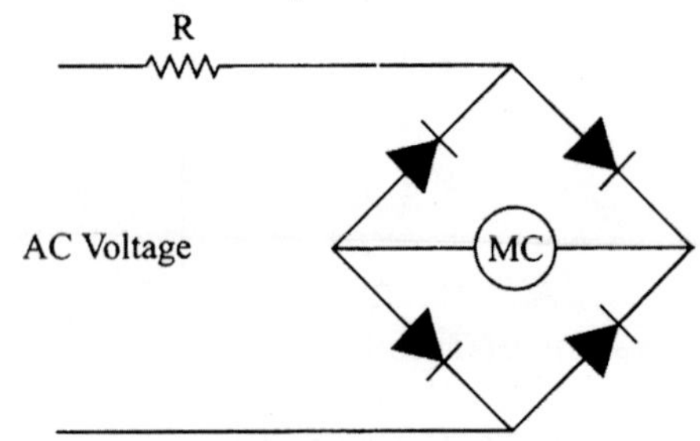

Fig. 6.1. Rectifier Instrument.

6.10. THERMOCOUPLE INSTRUMENT

It uses seeback effect, *i.e.*, when two wires of different materials are joined to form two junctions and these junctions are kept at different temperatures, a current flows in the circuit. They can be used even for very high frequencies upto 50 MHz. The commonly used junctions are copper-constantan and iron-constantan. Their overload capacity is very small.

6.11. ELECTROSTATIC VOLTMETER

It is used for measurement of high voltage. It is based on the principle that there is a force of attraction between two or more charged bodies. Quadrant electrometer, multicellular electrostatic voltmeter and attracted disc electrometer are some constructions based on this principle. The deflection is proportional to square of applied voltage. They are suitable for both ac and dc.

6.12. EXTENSION OF RANGE OF AMMETERS AND VOLTMETERS

6.12.1. Extension of Range of Ammeter

A low resistance shunt is connected in parallel with ammeter as shown in Fig. 6.2(*a*). Figure 6.2(*b*) shows the equivalent circuit. Figure 6.2 (*c*) shows a multi range ammeter using shunt resistance R_{s1}, R_{s2}, R_{s3}

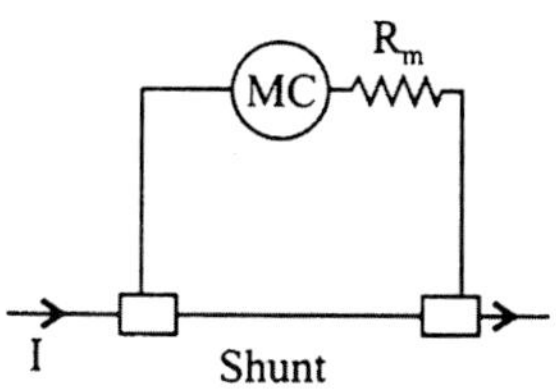

(a) Connection

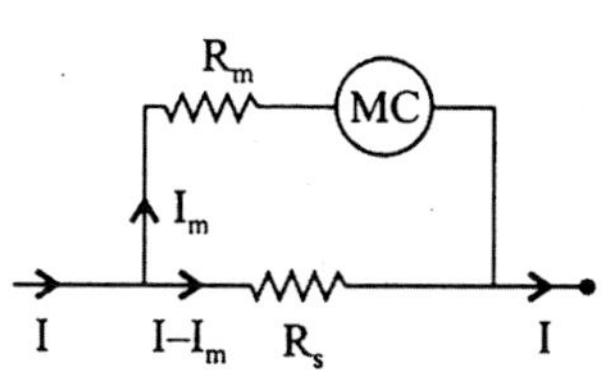

(b) Equivalent circuit

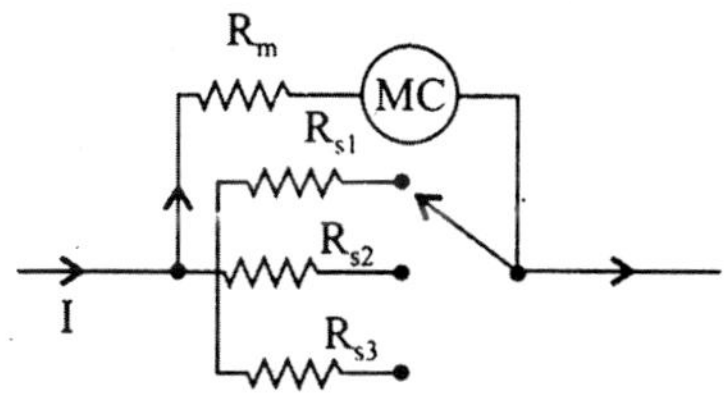

(c) Multi range ammeter

Fig. 6.2. Extension of range of ammeter.

$$R_s = \frac{I_m R_m}{I - I_m}$$

where R_s is the resistance of shunt, I_m, is the ammeter current for full scale deflection, I is the new current range and R_m is the resistance of moving coil.

6.12.2. Extension of Range of Voltmeter

As shown in Fig. 6.3 (*a*) the range of a voltmeter can be extended by connecting a high resistance R in series with the meter. R_m is the resistance of the coil of the meter.

$$\text{Meter current} = \frac{V}{R + R_m}$$

Figure 6.3 (*b*), shows a multi range voltmeter.

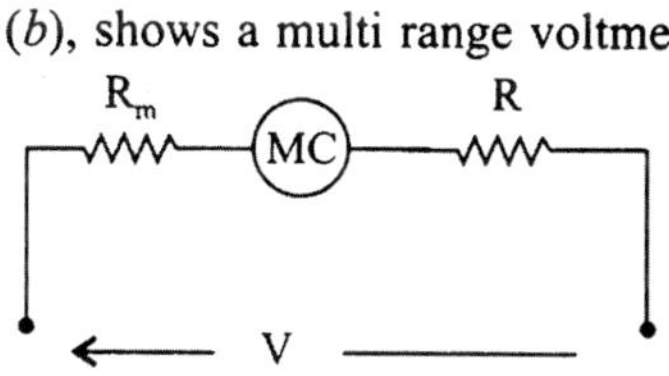

(a) Single range

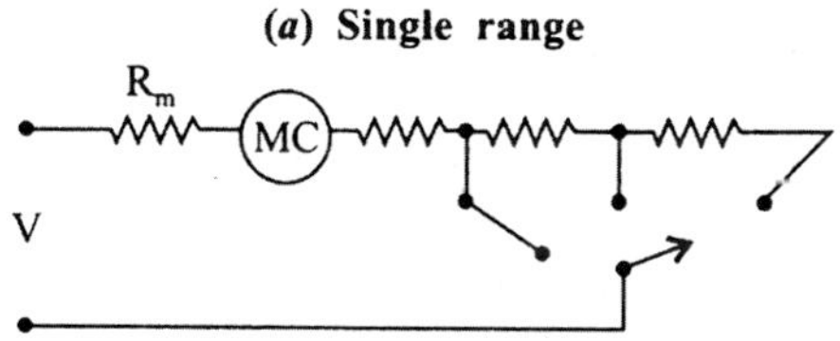

(b) Multi range

Fig. 6.3. Extension of voltmeter range.

6.13. DYNAMOMETER WATTMETER

Figure 6.4 shows the construction. It has two fixed coils connected in series and one moving coil. The fixed coils carry the current and moving coil is excited by the circuit voltage. All coils are air cored to avoid hysteresis and eddy current errors.

$$\text{Deflection } \theta = \frac{I_1 I_2 \cos\varphi}{C} \frac{dM}{d\theta}$$

where I_1 and I_2 are the currents through the fixed and moving coils, cos ϕ is power factor, C is spring constant, M is mutual inductance between coils, Since I_2 is proportional to circuit voltage, the reading is proportional to VI cos ϕ.

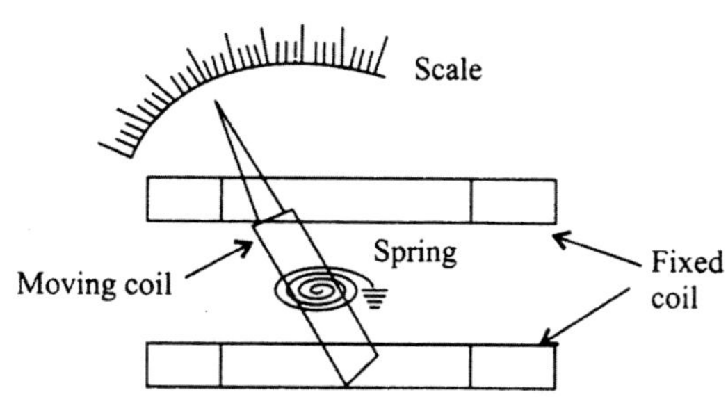

Fig. 6.4. Dynamometer wattmeter.

A dynamometer wattmeter can be used for both ac and dc, has good accuracy and has linear scale.

6.14. INDUCTION WATTMETER

Figure 6.5 (*a*) shows the construction and Fig. 6.5 (*b*) shows the phasor diagram.

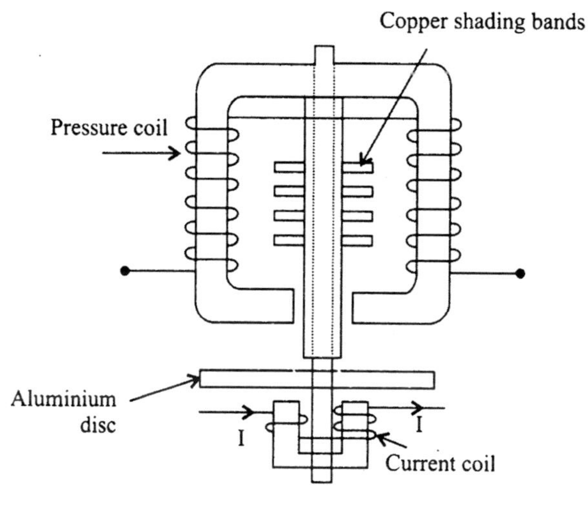

(*a*)

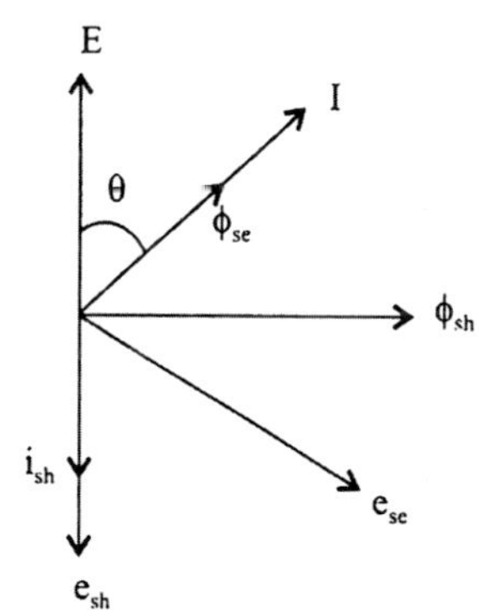

(*b*)

Fig. 6.5. Induction wattmeter (a) construction (b) phasor diagram.

The fluxes of the two magnets induce eddy currents in the disc. The interaction of eddy currents and fluxes creates the deflecting torque.

$$\text{Deflecting torque} \propto \phi_{sh} i_{sc} - \phi_{se} i_{sh}$$
$$\propto \text{VI} \cos\theta$$

It is suitable for ac only.

6.15. WATTMETER CONNECTIONS FOR HIGH AND LOW IMPEDANCE LOADS

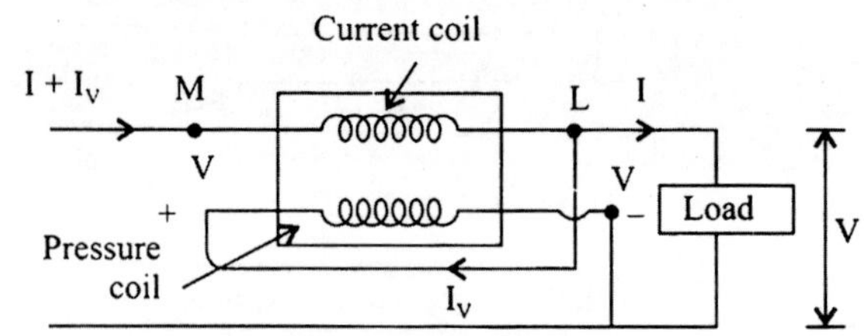

(*a*) **Low impedance loads.**

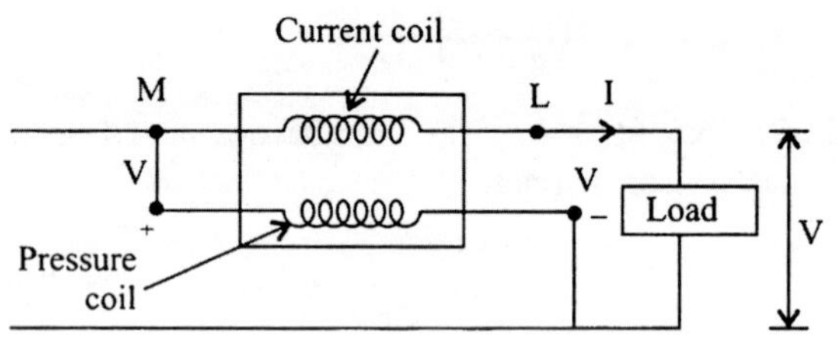

(*b*) **High impedance loads.**

Fig. 6.6. Wattmeter connection.

In the connection in Fig. 6.6 (*a*),

$$\text{error} = \frac{V^2}{\text{Resistance of pressure coil}}$$

In the connection of Fig. 6.6 (*b*) error = I^2 (Resistance of current coil)

6.16. AC SINGLE PHASE ENERGY METER

Its construction is similar to that of induction wattmeter. However, it has only one pressure coil on the central limb of upper electromagnet. A brake magnet provides braking torque.

Deflection Torque $\propto$ VI cos θ

Braking torque $\propto$ N, *i.e.*, speed of disc

Thus N $\propto$ VI cos θ

or $\int N\,dt \propto (VI \cos\theta)\,dt$

Therefore, number of revolutions $\propto$ energy consumed. Friction compensation and creep compensation are also provided.

6.17. OHM-METER

It consists of a battery, a moving coil instrument (MC) and a variable resistance R (Fig. 6.7). When terminals AB are open circuited, current in zero and R_x is infinite. When terminal AB are short-circuited the current is maximum (=E/R) and R_x is zero. The scale is calibrated to indicate the unknown resistance directly.

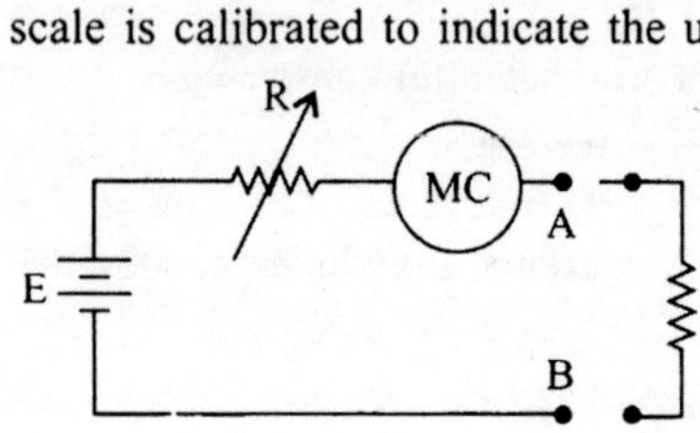

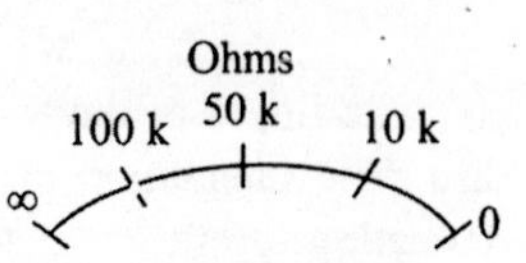

(*a*) **Circuit** (*b*) **Scale**

Fig. 6.7. Ohm meter.

6.18. WHEAT STONE BRIDGE

It is very suitable for measurement of medium resistance. (1 ohm to 10^5 ohm).

Figure 6.8 shown the circuit

At balance $I_1P = I_2R_x$

$I_1Q = I_2S$

or $\dfrac{P}{Q} = \dfrac{R_x}{S}$

or $R_x = \dfrac{P}{Q}S$

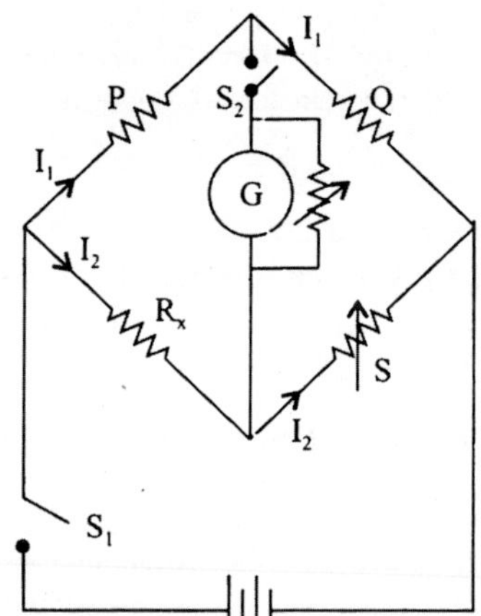

Fig. 6.8. Wheat stone bridge

It is necessary that battery key S_1 is closed first followed by galvanometer key S_2 to avoid sudden galvanometer deflection due to self-induced emf when R_x has appreciable self-inductance.

6.19. KELVIN'S DOUBLE BRIDGE

When measuring low resistance (*e.g.*, resistance of shunt, ammature winding etc.) it is necessary to eliminate resistance of contacts and leads. Low resistances have four terminals (two for current and two for voltage). Kelvin's double bridge is for measurement of low resistance (less than 1 ohm). Figure 6.9 shows the circuit.

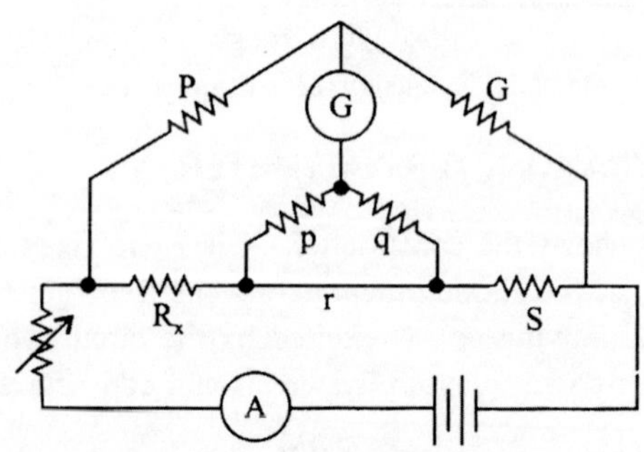

Fig. 6.9. Kelvin's double bridge

$$R_x = \frac{P}{Q}(S) + \frac{qr}{p+q+r}\left[\frac{P}{Q} - \frac{p}{q}\right]$$

Since resistance r is negligible and $\frac{P}{Q} = \frac{p}{q}$

$$R_x = \frac{P}{Q} S$$

6.20. MEASUREMENT OF HIGH RESISTANCE (MORE THAN 100 k OHM) LOSS OF CHARGE METHOD

Capacitor C is first charged from battery E (by closing S_1) and then discharged through R (by opening S_1 and closing S_2).

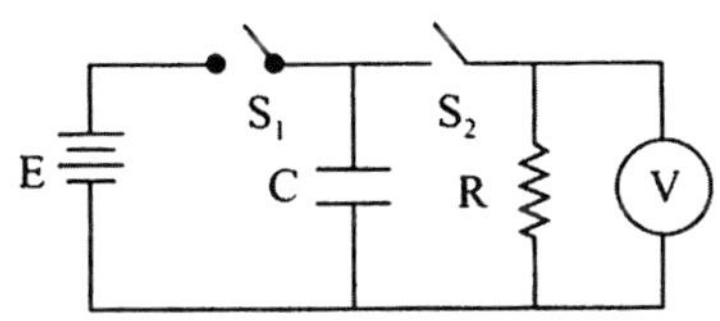

Fig. 6.10. Loss of charge method

During discharging

$$V = iR = Ee^{-t/RC}$$

or
$$R = \frac{t}{C \ln (E/v)}$$

It is necessary that resistance of voltmeter is very high as compared to the value of R.

6.21. MEGGAR

It is very suitable for measurement of very high resistance and is commonly used to measure insulation resistance of a cable. It has a small hand driven permanent magnet gernerator which produces about 500 V. The principle is the same as that of ohm-meter. The scale is calibrated to indicate the resistance directly.

6.22. DC POTENTIOMETER

6.22.1. Principle

It has a slide wire MN, battery E, variable resistance R, standard cell SC and a galvanometer connected as shown. If l_1 and l_2 are lengths of slide wire when standard cell and unknown emf are connected respectively.

$$\frac{\text{EMF of standard cell}}{\text{Unknown emf}} = \frac{l_1}{l_2}$$

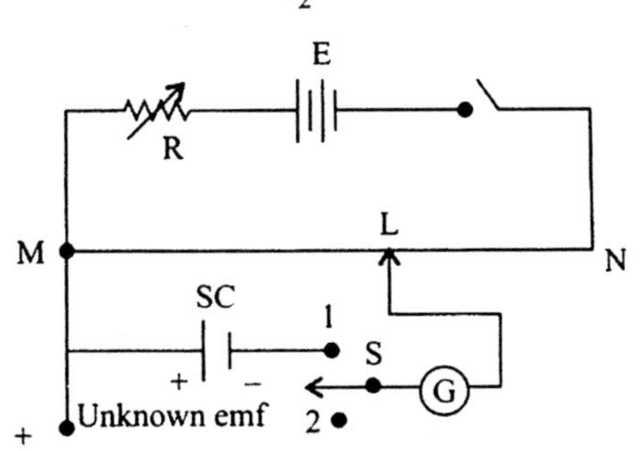

Fig. 6.11. DC Potentiometer

A crompton potentiometer is a commercial form of dc potentiometer. It has 14 coils and a slide wire in the form of a circle. A standard cell of 1.01859 V is used for calibration.

6.22.2. Calibration of Voltmeter

It is necessary to use a volt ratio box because potentiometer can measure upto 1.5 V only. Figure 6.12 (*a*) shows volt ratio box and Fig. 6.12 (*b*) shows the circuit for calibration of voltmeter.

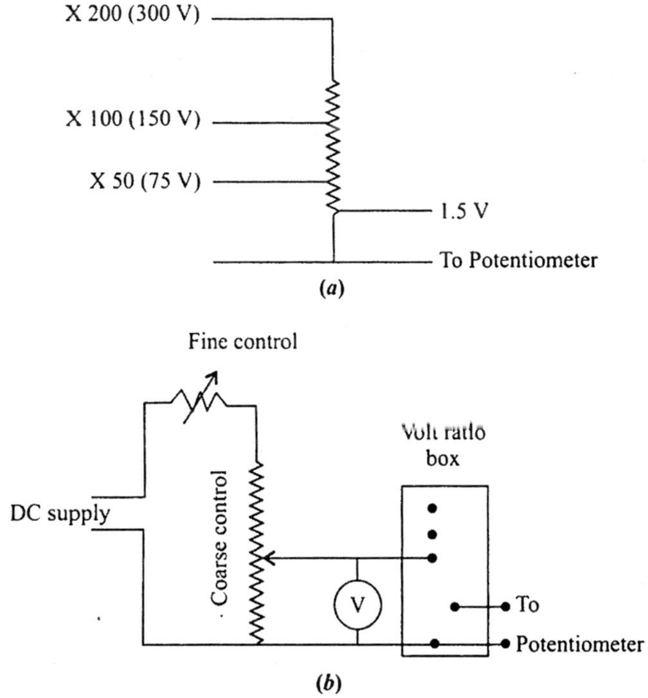

Fig. 6.12.

The reading of voltmeter is the observed value. The reading of dc potentiometer multiplied by the ratio of volt ratio box gives the accurate value.

6.22.3 Calibration of Ammeter

Figure 6.13 shows the circuit. The ammeter is connected in series with a standard resistance S, a variable resistance and battery. The reading of potentiometer divided by resistance S gives the accurate value. The reading of ammeter is the observed value.

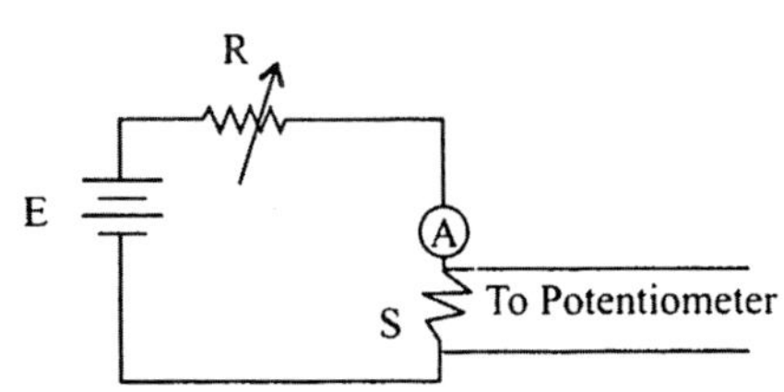

Fig. 6.13.

6.22.4 Measurement of Resistance

Figure 6.14 shows the circuit R_x is unknown resistance S is standard resistance.

$$\frac{R_x}{S} = \frac{\text{EMF across } R_x}{\text{EMF across } S}$$

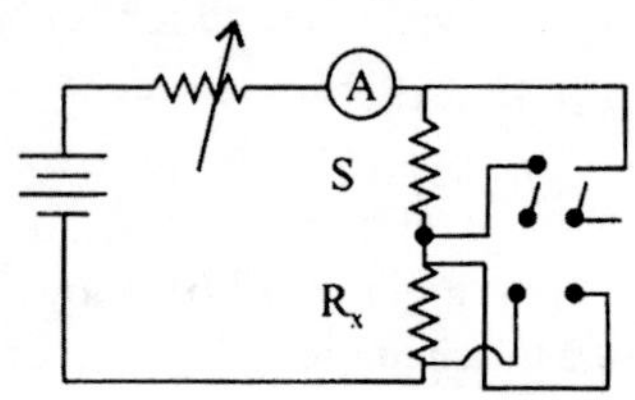

Fig. 6.14.

6.23. ELECTRONIC INSTRUMENTS

6.23.1. Average Reading Voltmeter

Figure 6.15 shows the circuit. Diode provides half wave rectification. The average value of ac voltage is applied across R, is amplified by dc amplifier and read by moving coil instrument (MC).

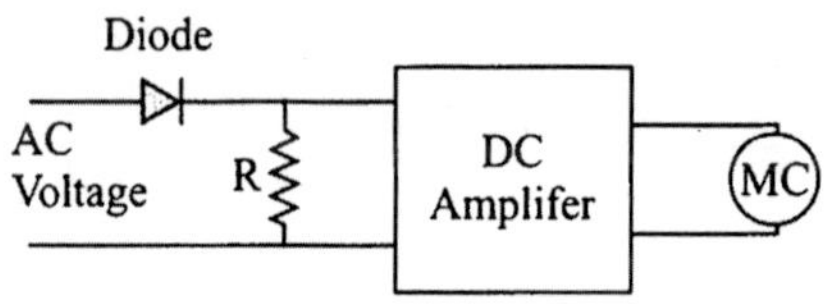

Fig. 6.15. Average reading electronic voltmeter.

6.23.2. Peak Reading Voltmeter

Figure 6.16 shows the circuit. Diode provides half wave rectification. The capacitor gets charged to peak value of ac voltage. After amplification it is applied to moving coil instrument (MC).

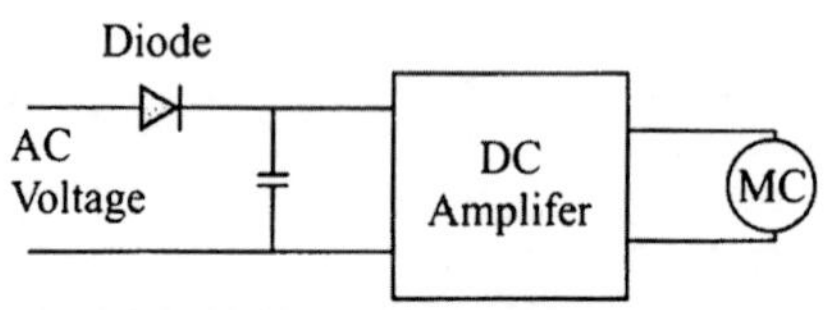

Fig. 6.16. Peak readig voltmeter.

6.23.3. RMS Electronic Voltmeter

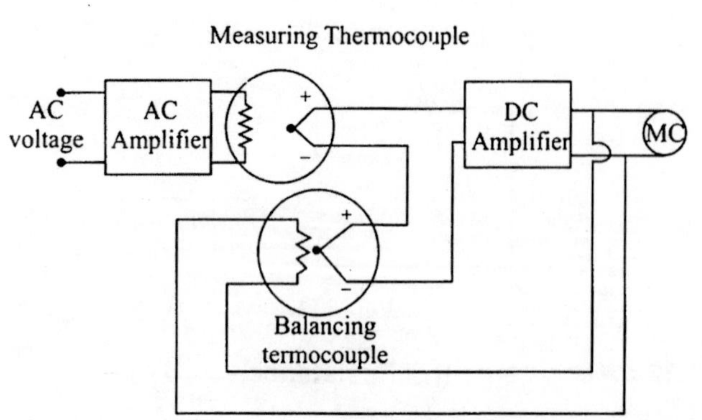

Fig. 6.17. RMS Electronic Voltmeter.

Figure 6.17 shows the circuit. The two thermocouples form a bridge circuit connected to input of dc amplifier. When ac input voltage is applied; the balance is upset and difference voltage is fed to heating element of balancing thermocouple. The bridge balance is re-established when feedback current delivers sufficient heat to balancing thermocouple so that voltage of two thermocouples are the same. The dc current output is proportional to rms value of input voltage.

6.24. CATHODE RAY TUBE

6.24.1. Construction

Figure 6.18 shows the construction.

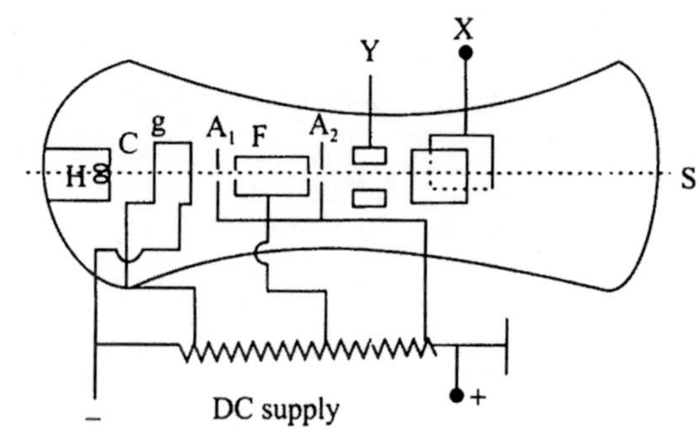

Fig. 6.18. Cathode Ray Tube.

The main parts are electron gun, deflection plates and screen.

The electron gun produces a focussed beam of high velocity electrons. It has a heater (H), cathode (C), control grid (*g*) accelerating anodes (A_1, A_2) and focussing electrode.

Electrons are emitted by indirectly heated cathode. The control grid is kept at negative potential and its function is to control the number of electrons and thus adjust intensity of display. A_1 and A_2 accelerate the electrons. A_1 and A_2 are connected together and maintained at high potential with respect to cathode. The focussing electrode F serves to focus the electrons so as to form a sharp and narrow beam.

X and Y deflecting plates serve to deflect the electron beam in horizontal and vertical planes respectively. Fluorescent screen S is for display. Its inner surface is coated with a special material phosphor. This material emits light when stimulated by radiation.

6.24.2. Electrostatic Deflection

$$\text{Deflection } D = \frac{LVl_d}{2dV_A}$$

where L is distance from centre of vertical deflecting plate to screen

l_d is length of vertical deflecting plates.

V is voltage applied to deflecting plates.

V_A is voltage between accelerating anode and cathode

d is distance between vertical deflecting plates.

$$\text{Deflection Sensitivity } \frac{D}{V} = \frac{Ll_d}{2dV_A} \text{ metres/volt}$$

6.25. CATHODE RAY OSCILLOSCOPE

6.25.1. Construction

Figure 6.19 shows the block diagram of CRO.

The input signal is amplified by vertical amplifier and fed to *Y* deflecting plates. The time base generator produces a saw tooth voltage so as to drive the electron beam at a steady speed across the screen and when the beam reaches the right end of screen it is brought back to starting position on left end of screen. Trigger circuit produces a trigger pulse to start the time base generator. The horizontal amplifier amplifies the input to horizontal deflecting plates.

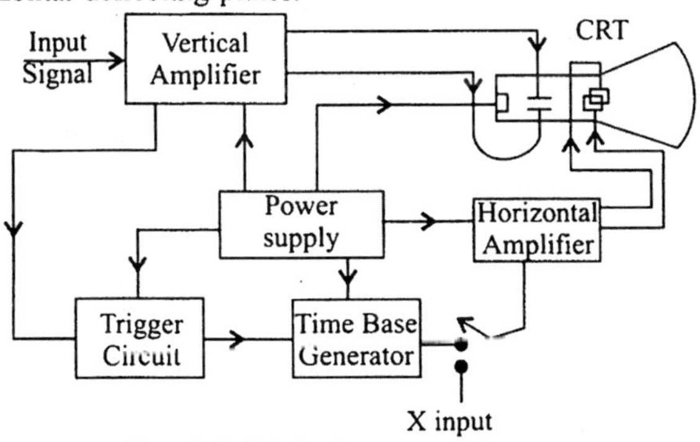

Fig. 6.19. Cathode ray oscilloscope.

6.25.2. Voltage Measurement

The unknown voltage is applied to Y plates. X plates are excited by time base generator. A trace is obtained on the screen. Using the markings on the screen and gain of vertical amplifier, the peak-to-peak value (and hence rms value) of input signal can be measured.

6.25.3. Measurement of Phase

One of the signals is fed to Y plates and the other to the X plates. (The time base generator is switched off). If θ is phase angle.

$$\sin\theta = Y_1/Y_2$$

where Y_1 and Y_2 are vertical intercepts of the display.

6.25.4. Measurement of Frequency

Frequencies of two signals can be compared by feeding them to X and Y deflecting plates (time base generator is switched off). The display is called Lissajous Figures. The shape of display depends on the ratios of the two frequencies.

6.26. DIGITAL STORAGE OSCILLOSCOPE

A storage cathode tube has many disavantages. (1) It can preserve the stored waveform only for a finite amount of time. (2) The image will be stored only when power supply is on. (3) The stored trace is not fine. (4) The writing rate of storage tube is rather slow. (5) Storage tube is expensive. (6) Only one image can be stored.

In a digital storage oscillosope the waveform is digitized, stored in a digital memory and then retrieved for display as and when required. The digitizing process is done by an A/D converter.

6.27. FUNCTION GENERATOR

It is a very versatile instrument which can give sine, triangular, square and sawtooth waves over a wide frequency range from a fraction of hertz to several hundred kilohertz. Its main parts are frequency control network, integrator (using op-amp), voltage comparator, multivibrator, output amplifier etc.

6.28. WAVE ANALYZER

This instrument is designed to measure relative amplitudes of single frequency components in a complex wave. It acts like a frequency selective voltmeter which is tuned to one frequency and rejects all frequencies. It has a filter section with a very narrow pass band that can be tuned to the frequency required to be measured. Its main components are : input attenuator, driver amplifier, high Q active filter, output amplifier etc.

6.29. DATA ACQUISITION SYSTEM

An analog data acquistion system consists of transducer (to convert physical quantity to electrical signal), signal conditioner (to amplify and modify the signal), visual display device (for display), graphical recording instrument (to keep a permanent record) and magnetic tape instrumentation.

6.30. AC BRIDGES

6.30.1. Maxwell's Bridge

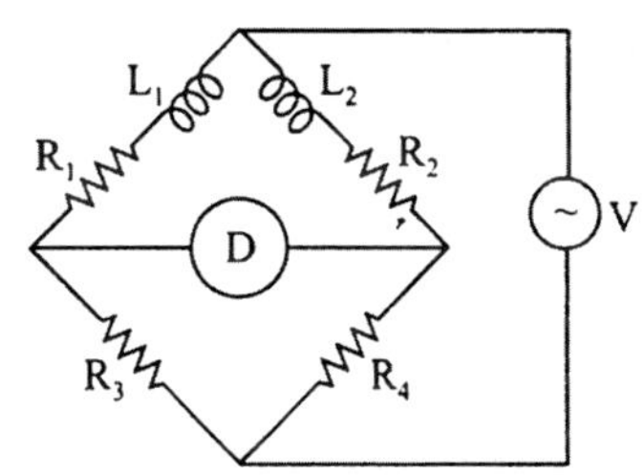

Fig. 6.20. Maxwell's Bridge.

This bridge is used to measure an inductance. The unknown inductance is compared with a known self-inductance, Figure 6.20 shows the circuit. At balance

$$\frac{R_1}{R_2} = \frac{L_1}{L_2} = \frac{R_3}{R_4}$$

6.30.2. Maxwell Wien Bridge

This bridge is used to measure inductance by comparing it with a standard capacitance. Figure 6.21 shows the circuit.

At balance

$$L = CR_2R_3$$

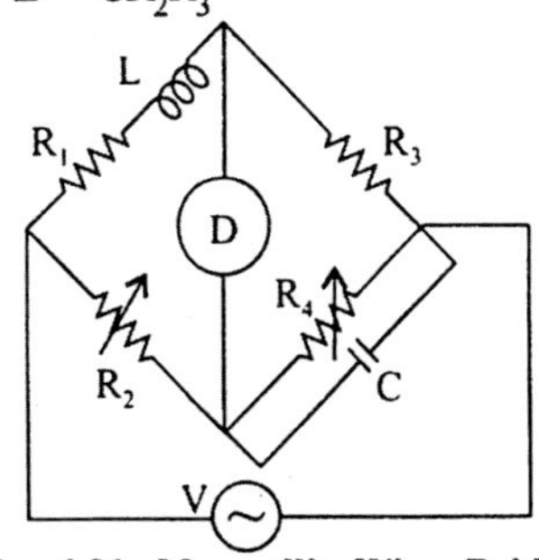

Fig. 6.21. Maxwell's Wien Bridge.

6.30.3. Hay's Bridge

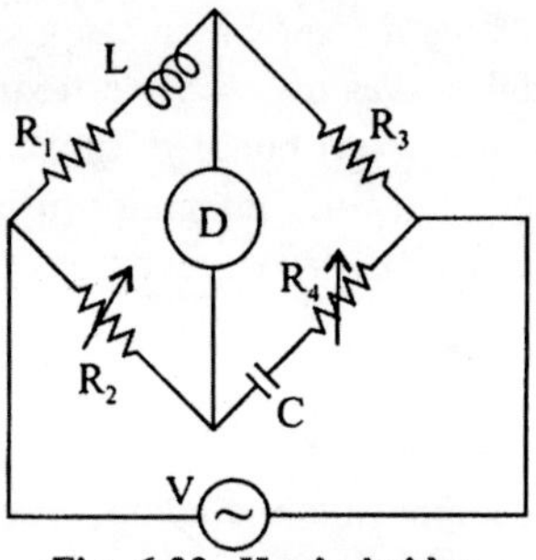

Fig. 6.22. Hay's bridge.

This is a modification of Maxwell-Wien Bridge and is particularly suitable for measuring an inductance having a comparatively low resistance (*i.e.*, large time constant L/R). Figure 6.22 shows the circuit

$$L = (R_2 R_3 C)/(1+\omega^2 R_4^2 C^2)$$

$$L_1 = (R_2 R_3 R_4 \omega^2 C^2)/(1+\omega^2 R_4^2 C^2)$$

6.30.4. Owen's Bridge

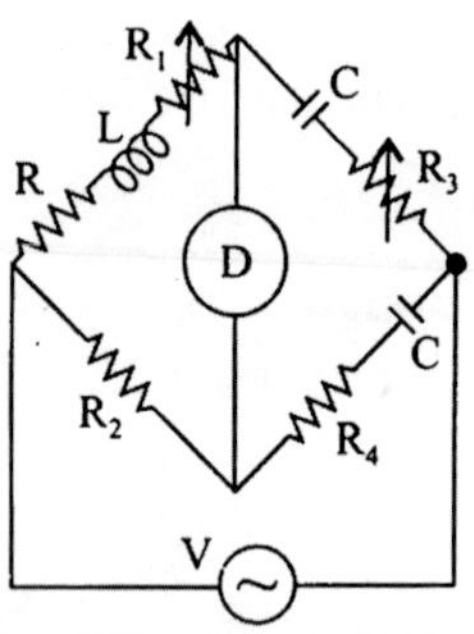

Fig. 6.23. Owen's Bridge.

$$R = R_2(C_s/C) - R_1$$

$$L = R_2 R_3 C_s$$

6.30.5. Anderson Bridge

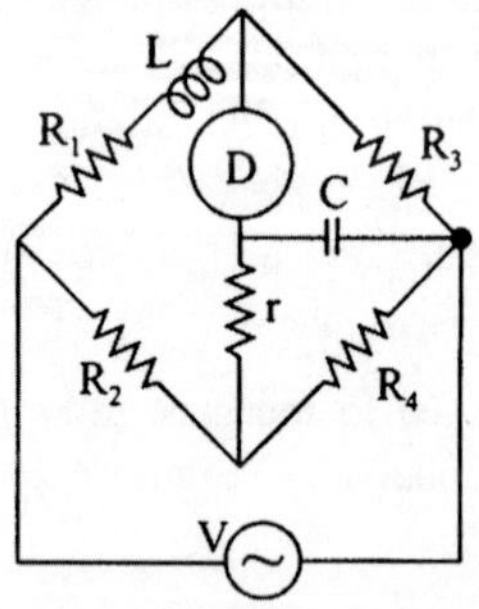

Fig. 6.24. Anderson Bridge.

$$R_1 = R_2 R_3 / R_4$$

$$L = \frac{CR_3[r(R_2+R_4)+R_2R_4]}{R_4}$$

6.30.6. De Sauty's Bridge

This bridge is very simple but can be used to measure only pure capacitance (*i.e.*, capacitors having air or mica as dielectric). The balance is obtained by varying R_1 or R_2

$$C_1 = \frac{R_2 C_2}{R_1}$$

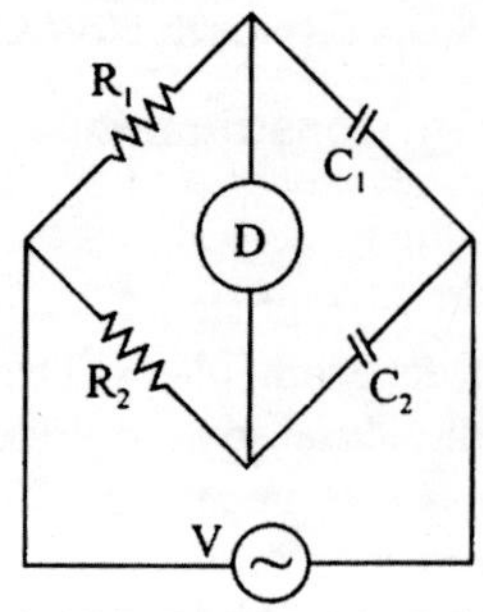

Fig. 6.25. De Sauth's Bridge.

6.30.7. Modified De Sauty's Bridge

It is used for imperfect capacitors, r_1 and r_2 represent small resistances representing loss components of capacitors C_1 and C_2 as shown in Fig. 6.26. The balance is obtained by varying R_1, R_2, R_3, R_4

$$\frac{R_1}{R_2} = \frac{R_3 + r_1}{R_4 + r_2} = \frac{C_2}{C_1}$$

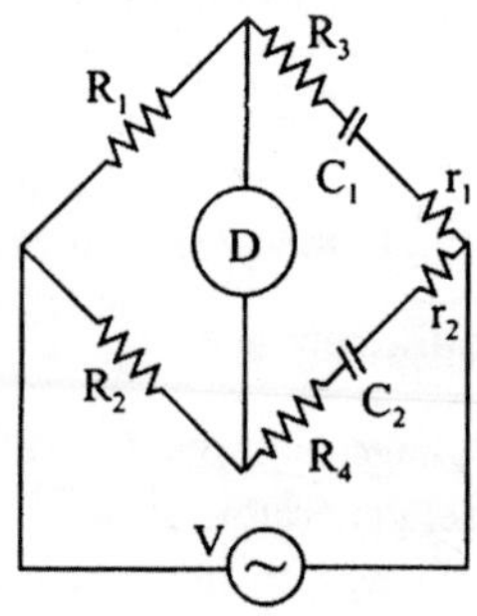

Fig. 6.26. Modified De Sauty's bridge.

6.30.8. Universal RLC Bridge

It is capable of measuring R, L and C by using the following bridge combinations

Measurement of R	Wheat stone bridge
Measurement of L	Maxwell-Wien bridge
Measurement of C	Modified De Sauty's bridge

6.30.9. Wien Bridge

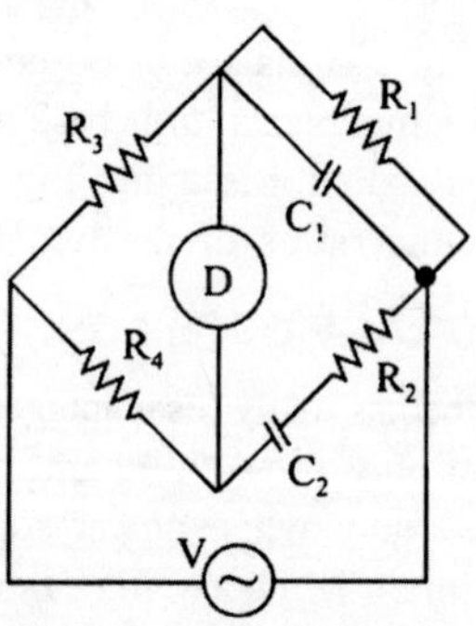

Fig. 6.27. Wien Bridge.

Figure 6.27 shows the circuit. The balance is obtained by varying R_2, R_3, R_4

$$C_1 = \frac{R_4 C}{R_2(1+\omega^2/R_2^2 C_2^2)} \qquad R_1 = \frac{R_3(1+\omega^2 R_2^2 C_2^2)}{\omega^2 R_2 R_4 C_2^2}$$

6.30.10. Schering Bridge

It is very widely used for measurements of small capacitance and dielectric loss. Its use at high voltage is more common because measurements of small capacitance can be done. A step up transformer is used to give high voltage supply. Earthed screens are used to avoid errors due to intercapacitance between high and low voltage arms of bridge. VG is vibration galvanometer.

$$C_1 = \frac{C_2 R_4}{R_3} \quad \text{and} \quad \tan\delta = \frac{1}{\omega C_1 R_1}$$

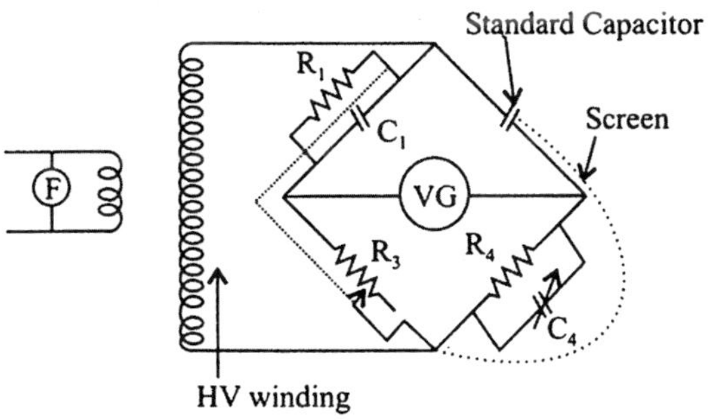

Fig. 6.28. Schering Bridge.

6.30.11. Heaviside Bridge for Measurement of Mutual Inductance

Figure 6.29 shows the circuit. L_2 is self-inductance of secondary of coil whose mutual inductance M is required to be measured. L_1 is self-inductance

$$M = (L_1 R_4 - L_2 L_3)/(R_3 + R_4)$$

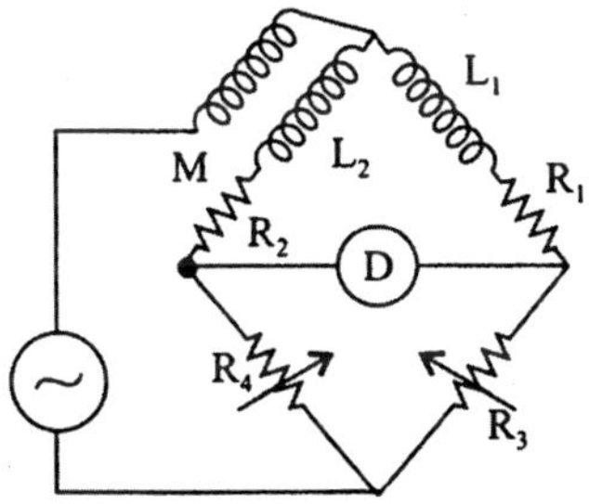

Fig. 6.29. Heaviside Bridge.

6.30.12. Heaviside Campbell Equal Ratio Bridge

The secondary of mutual inductance is in two parts the primary reacting with both of them. L_1 is self-inductance to be measured

$$R_1 = R_2 + r$$
$$L_1 = 2(M_1 + M_2)$$

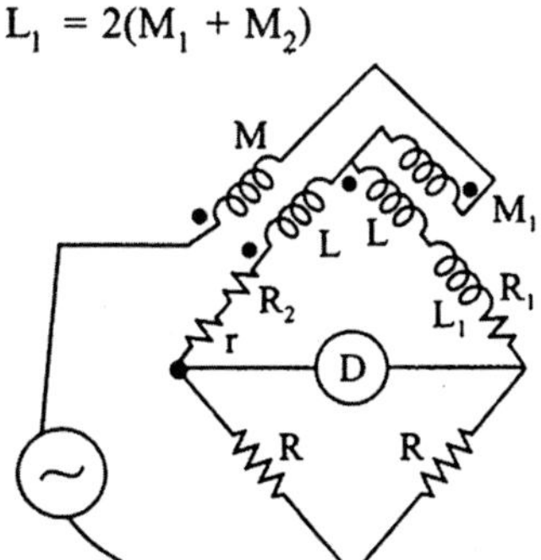

Fig. 6.30. Heaviside campbell equal ratio bridge.

6.30.13. Wagner's Earthing Device

This device is used to eliminate error due to electrostatic coupling Z_1, Z_2, Z_3, Z_4 and arms of bridge. X and Y are two variable impedances of Wagner's earthing device. Operation in as under.

Switch S connected to *b* and null point obtained by varying Z_1 and Z_2

Switch S connected to *c* and null point obtained by varying X and Y.

Switch S connected to *b* and null point obtained by varying Z_1 and Z_2.

Switch S connected *c* and null point obtained again.

The above procedure is repeated till null point is obtained with switch S in position *b* or *c*.

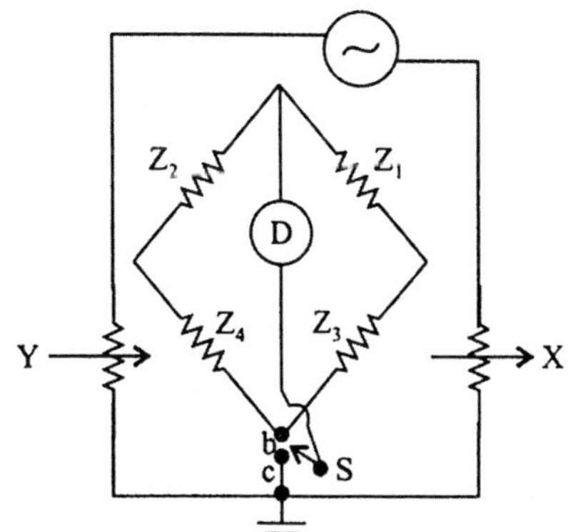

Fig. 6.31. Wagner's Earthing Device.

6.31. TRANSDUCERS

Many times it is convenient to measure a non-electrical quantity (*e.g.*, temperature, pressure, displacement etc.) by converting it to an electrical quantity (emf or current or resistance). A transducer converts an input signal into an output signal of a different form.

6.31.1. Metallic Resistance Thermometer

It consists of a platinum coil wound on a mica or ceramic former. For temperatures about 0°C, Callender's equation is

$$t = \left(\frac{R_t - R_0}{R_{100} - R_0}\right)100 + \delta\left(\frac{t}{100} - 1\right)\left(\frac{t}{100}\right)$$

For temperatures below 0°C, Callender-Van Dusan equation is used and

$$t = \left(\frac{R_t - R_0}{R_{100} - R_0}\right)100 + \delta\left(\frac{t}{100} - 1\right)\left(\frac{t}{100}\right) + \beta\left(\frac{t}{100} - 1\right)\left(\frac{t}{100}\right)^3$$

6.31.2. Thermistor

It is metal oxide and is a semiconductor. It has a negative temperature coefficient

$$R_T = R_0 \in^{\beta\left(\frac{1}{T} - \frac{1}{T_0}\right)}$$

It is very commonly used to measure temperature of inner parts of electric machines.

6.31.3. Thermocouple Transducer

It works on See back effect. Some pairs used as thermocouples are Copper constantan (–250°C to 400°C), Iron-Constantan (– 200 to 850°C), Chromel-Aluminium (– 200 to 1100°C), Platinum rhodium-platinum (0 to 1400°C).

6.31.4. Radiation Thermometer

E_0, *i.e.*, Energy absorbed by detector is

$$E_0 = \sigma(T^4 - T_d^4)$$

6.31.5. Potentiometric Position (Displacement) Transducer

It consists of a potentiometer fed by a dc voltage. The slider of the potentiometer is mechanically coupled to the output of the system whose displacement is required to be measured. The output voltage is a measure of the displacement.

6.31.6. Capacitance Displacement Transducer

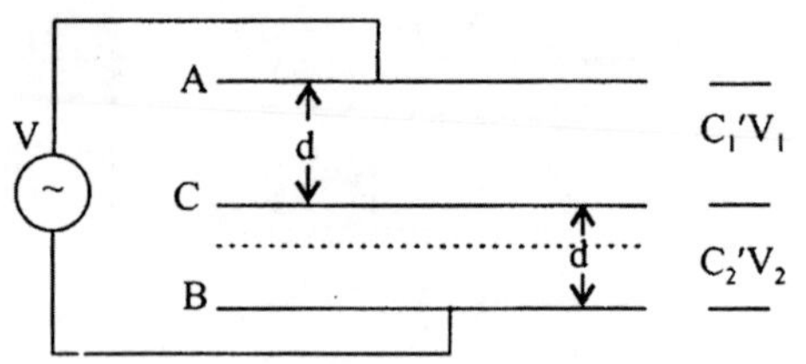

Fig. 6.32.

Figure 6.32 shows the configuration. Initially the movable plate C is at a distance *d* from each of the fixed plates A and B. After plate C moves by a distance *x*

$$\Delta V = V_1 - V_2 = \frac{Vx}{d}$$

6.31.7. Capacitance Thickness Transducer

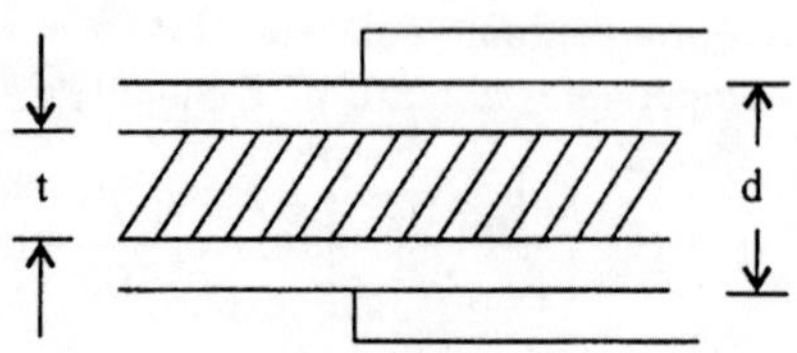

Fig. 6.33.

It is used to measure thickness *t* of the dielectric sheet passing through the air grap between two plates (Fig. 6.33).

$$\frac{\Delta C}{C} = \frac{\Delta t}{t}(N)\left[\frac{1}{1 - N\frac{\Delta t}{t}}\right]$$

6.31.8. Linear Variable Differential Transformer (LVDT)

It is based on the principle of mutual inductance and is used to measure displacement. Figure 6.34 shows the configuration. It has a primary winding P and two exactly similar secondary windings S_1 and S_2. Primary winding is excited by ac voltage. When position of core is symmetrical (with respect to S_1, S_2) $E_0 = 0$. The sign and magnitude of E_0 is a measure of the displacement of core.

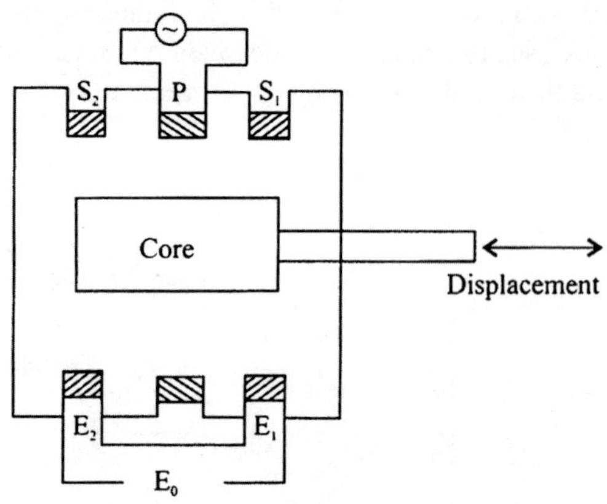

Fig. 6.34

6.31.9. Strain Gauge

When a metallic foil is stressed its resistance changes

$$\frac{dR/R}{dl/l} = 1 + 2\mu + \frac{d\rho/\rho}{dl/l} = G$$

G is called gauge factor.

Strain gauge is very commonly used to measure strain in structures. A metallic strain gauge has a gauge factor of about 2. Semiconductor strain gauge has a very high gauge factor (about 100).

6.31.10. Piezo electric Transducer

These transducers use the phenomenon of piezo electricity to measure force, strain, torque and aceleration.

6.31.11. Hall Effect Transducer

Hall voltage $E_H = (k_H BI)/d$

Thus E_H is a measure of flux density B. It is used to for mapping of magnetic field around conductors carrying current.

6.31.12. Tachometer (Tachogenerator)

A dc tachogenerator is a small dc generator used to measure angular speed of a shaft.

Output voltage E is proportional to speed of shaft.

An ac tachogenerator is a two phase induction motor with squirrel cage rotor.

6.32. RF VOLTAGE MEASUREMENT

A radio frequency voltage is measured by rectifying the voltage, amplifying the dc output voltage and measuring the dc output voltage of amplifier. Since RF signals extend into gigahertz region it is not possible to amplify the signal directly. Therefore, it must be rectified before amplification. Ordinary diodes are not suitable for rectification because they have excessive capacitance and excessive reverse recovery time. Therefore, either Schottky diodes or point contact diodes are used.

OBJECTIVE TYPE QUESTIONS

A. Choose the most appropriate alternative

1. If σ standard deviation, probable error
   (*a*) σ (*b*) $\pm 1.19\sigma$
   (*c*) $\pm 0.6745\sigma$ (*d*) $\pm 0.4125\sigma$
2. If *n* measurements of a quantity are made, the arithmetic mean is
   (*a*) $\frac{\Sigma x}{n}$ (*b*) $\frac{\sqrt{\Sigma x^2}}{n}$
   (*c*) $\frac{(\sqrt{\Sigma x})^2}{n}$ (*d*) none of the above
3. If *n* measurements of a quantity *x* are made and $\bar{x}$ is the arithmetic mean, the average deviation is
   (*a*) $\frac{\sum_{i=1}^{n}(x_i - \bar{x})}{n}$ (*b*) $\frac{\sum_{i=1}^{n}(\bar{x} - x_i)}{n}$
   (*c*) $\frac{\sum_{i=1}^{n}\sqrt{x_i^2 - \bar{x}^2}}{n}$ (*d*) $\frac{\sum_{i=1}^{n}|x_i - \bar{x}|}{n}$
4. If $d_1, d_2, \ldots\ldots d_n$ are *n* individual deviations, the standard deviation is
   (*a*) $\sqrt{\frac{d_1^2 + d_2^2 + \ldots.. d_n^2}{n}}$ (*b*) $\sqrt{\frac{d_1^2 + d_2^2 + \ldots.. d_n^2}{n-1}}$
   (*c*) $\sqrt{\frac{d_1^2 + d_2^2 + \ldots.. d_n^2}{n-2}}$ (*d*) none of the above
5. If σ is standard deviation, variance is
   (*a*) $\sqrt{\sigma - 1}$ (*b*) $\sigma^{1.5}$
   (*c*) σ^2 (*d*) σ^3
6. A 0–150 voltmeter has a guaranteed accuracy of 1% of full scale reading. If actual reading is 83 V, the percentage error is
   (*a*) 0.83% (*b*) 1%
   (*c*) 1.81% (*d*) none of the above
7. The current passing through a resistance of 100 ± 0.2% Ω is 2 ± 0.5% A. The limiting error in computation of power will be
   (*a*) ± 0.7% (*b*) ± 0.9%
   (*c*) ± 1.2% (*d*) ± 0.3%
8. Two resistances R_1 = 18.7 Ω and R_2 = 3.624 Ω are in series. The total resistance to the appropriate number of significant figures is
   (*a*) 22.324 Ω (*b*) 22.32 Ω
   (*c*) 22.3 Ω (*d*) 22 Ω
9. A current of 3.18 A is flowing through resistance of 35.68 Ω. The voltage drop across resistance to the appropriate number of significant figures is
   (*a*) 113.4624 V (*b*) 113.462 V
   (*c*) 113.46 V (*d*) 113 V
10. A current of 3.25 A is flowing through resistance of 34.98 Ω. The voltage drop across resistance to the appropriate number of significant figures is
   (*a*) 113.685 V (*b*) 113.68 V
   (*c*) 113.6 V (*d*) 114 V
11. Two numbers are 131 ± 5 and 629 ± 3. The sum of the number with range of doubt is
   (*a*) 760 ± 8 (*b*) 760 ± 2
   (*c*) 760 + 8 (*d*) 760 – 8
12. If A = 629 ± 8 and B = 131 ± 3, A – B with range of doubt is
   (*a*) 498 ± 8 (*b*) 498 ± 3
   (*c*) 498 ± 11 (*d*) 498 ± 13
13. If A = 131 ± 4 and B = 106 ± 4, the percentage range of doubt in A – B is
   (*a*) 4% (*b*) 8%
   (*c*) 16% (*d*) 32%
14. A digital voltmeter has a read out range from 0 to 999 counts. If the full scale reading is 9.999 V, the resolution is
   (*a*) 1 V (*b*) 0.01 V
   (*c*) 1 mV (*d*) 1 μV
15. Four capacitors are in para in parallel. Their values are 36.3 μF, 3.85 μF, 34.002 μF and 850 *n*F with an uncertainty of digit in the last place. The total capacitance in significant figures is
   (*a*) 7.50 μF (*b*) 7.5 ± 0.1 μF
   (*c*) 75.1 ± 0.1 μF (*d*) 75.002 ± 0.1 μF
16. What would be the reading of voltmeter in Fig. 6.35. The voltmeter has 0–1 V scale and is a 20000 Ω/V meter.

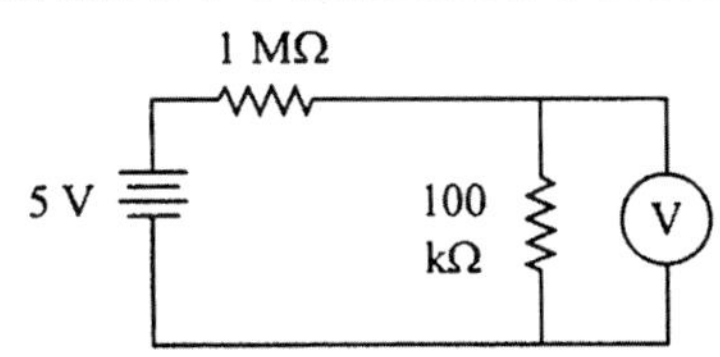

Fig. 6.35.

   (*a*) 0.4545 V (*b*) 0.082 V
   (*c*) 0.82 V (*d*) 4.54 V
17. Two resistances are : R_1 = 36 Ω ± 1.8 Ω and R_2 = 75 Ω ± 3.75 Ω. The sum $R_1 + R_2$ along with limiting error is
   (*a*) 111 Ω ± 1.8 Ω (*b*) 111 Ω ± 3.75 Ω
   (*c*) 111 Ω ± 5.55 Ω (*d*) 111 Ω ± 1.95 Ω
18. A generated voltage is equally dependent on three resistances R_1, R_2, R_3 and its value is $V = (R_1R_2)/R_3$, If tolerance of each resistor is 0.1%, the maximum error in voltage is
   (*a*) 0.3% (*b*) ± 0.3%
   (*c*) 0.1% (*d*) ± 0.1%
19. Two resistances are R_1 = 36 ± 5% and R_2 = 75 ± 5%. If they are connected in parallel, the limiting error in percent will be
   (*a*) 5% (*b*) 10%
   (*c*) 15% (*d*) 20%

20. The power input and output in a device are 3650 W and 3385 W with each reading in doubt by ± 10 W. The percentage uncertainty in losses is
(*a*) ± 7.55% (*b*) ± 3.77%
(*c*) ± 11.32% (*d*) ± 15.09%

21. As per BIS, the number of accuracy classes of instruments is
(*a*) 5 (*b*) 6
(*c*) 7 (*d*) 8

22. If eddy current damping is to be used, the disc of instrument should be
(*a*) non-conducting and non-magnetic
(*b*) conducting and magnetic
(*c*) conducting and non-magnetic
(*d*) non-conducting and magnetic

23. The household energy meter is
(*a*) indicating instrument
(*b*) recording instrument
(*c*) integrating instrument
(*d*) none of the above

24. An instrument of accuracy class 1 can have an error equal to
(*a*) + 1% of reading
(*b*) ± 1% of reading
(*c*) ± 1% of full scale deflection
(*d*) + 1% of full scale deflection

25. A moving iron instrument gives correct reading when used at
(*a*) low frequencies
(*b*) high frequencies
(*c*) only one frequency
(*d*) all frequencies up to a certain value

26. A moving coil instrument has a resistance of 0.6 Ω and full scale deflection at 0.1 A. To convert it into an ammeter of 0-15 A range, the resistance of shunt should be
(*a*) 0.6 Ω (*b*) 0.06 Ω
(*c*) 0.1 Ω (*d*) 0.004 Ω

27. A moving coil meter has a resistance of 50 ohms and a full scale deflection current of 20 m A. To convert it into a voltmeter of 0-100 V range, the series resistance should be
(*a*) 5000 Ω (*b*) 5050 Ω
(*c*) 4950 Ω (*d*) 505 Ω

28. The switch board instruments
(*a*) should be mounted in vertical position
(*b*) should be mounted in horizontal position
(*c*) should be mounted at about 45°
(*d*) should be mounted at about 80°

29. The chemical effect of current is used in
(*a*) de ampere hour meter (*b*) de ammeter
(*c*) ac energy meter (*d*) all of the above

30. In portable instruments, the controlling torque is provided by
(*a*) spring (*b*) gravity
(*c*) eddy current (*d*) either (*a*) or (*c*)

31. If most of instruments, damping is provided by
(*a*) fluid friction (*b*) spring
(*c*) eddy current (*d*) either (*a*) or (*c*)

32. An ammeter is
(*a*) secondary instrument (*b*) absolute instrument
(*c*) recording instrument (*d*) integrating instrument

33. The pointer of an indicating instrument should be
(*a*) very light
(*b*) very heavy
(*c*) neither heavy nor light
(*d*) heavy

34. A permanent magnet moving coil meter can be used for
(*a*) ac only (*b*) dc only
(*c*) both ac and dc (*d*) low frequency ac and dc

35. The function of shunt in an ammeter is to
(*a*) by pass the current
(*b*) increase sensitivity of meter
(*c*) increase resistance of meter
(*d*) all the above

36. A meter has a resistance 1 Ω and full scale deflection current of 1 mA. To convert into an ammeter of 1 A range, the shunt resistance should be
(*a*) 1000 Ω (*b*) 1 Ω
(*c*) 0.01 Ω (*d*) 0.001 Ω

37. The multiplier resistance and meter coil in a voltmeter are in
(*a*) series (*b*) parallel
(*c*) series- parallel (*d*) either (*b*) or (*c*)

38. A meter with a resistance of 100 Ω and a full scale deflection current of 1 mA is to be converted into voltmeter of 0-5 V range. The multiplier resistance should be
(*a*) 490 Ω (*b*) 5000 Ω
(*c*) 4900 Ω (*d*) 5100 Ω

39. A moving iron instrument can be used for
(*a*) dc only
(*b*) ac only
(*c*) both dc and ac
(*d*) dc and high frequency ac

40. The scale of hot wire ammeter is
(*a*) linear (*b*) non-linear
(*c*) cramped (*d*) absolutely linear

41. For measuring current at a high frequency we should use
(*a*) moving iron instrument
(*b*) electrostatic instrument
(*c*) thermocouple instrument
(*d*) moving coil instrument

42. An electrostatic voltmeter is suitable for
(*a*) low voltage at high frequency
(*b*) high voltage
(*c*) low voltage at low frequency
(*d*) all ac and dc voltages

43. The accuracy of a multimeter is about
(*a*) ± 0.5% (*b*) ± 3%
(*c*) ± 0.01% (*d*) ± 10%

44. The most commonly used moving iron instruments are
(*a*) repulsion type (*b*) attraction type
(*c*) a combination of attraction and repulsion type
(*d*) none of the above

45. A moving coil instrument has a resistance of 0.5 Ω and a scale deflection of 0.1 A. To convert it into an ammeter of 0-10 A the shunt resistance should be
(*a*) 0.004 Ω (*b*) 0.005 Ω
(*c*) 0.05 Ω (*d*) 0.1 Ω

46. A meter coil has a resistance of 100 Ω and a full scale deflection current of 1 mA. To convert it into an ammeter of 0-200 V range, the series resistance should be
(*a*) 20 kΩ (*b*) 19.9 kΩ
(*c*) 1.9 kΩ (*d*) 20.1 kΩ

47. Consider the following features of an instrument
1. suitable for both ac and dc
2. free from hysteresis and eddy current errors
3. No waveform and frequency errors
4. it is essentially a voltmeter

This instrument is
(*a*) moving iron (*b*) moving coil
(*c*) electrostatic (*d*) dynamometer

48. The resistance of a dc galvanometer is 20 Ω. A current of 10 mA causes full scale deflection. To convert into 0-300 V voltameter, we have to use a series resistance of
(*a*) 30000 Ω (*b*) 29980 Ω
(*c*) 6000 Ω (*d*) 15 Ω

49. The coil of a moving iron instrument has a resistance of 500 Ω and an inductance of H. It reads 250 V when a 250 V dc is applied. If series resistance is 2000 Ω, its reading when fed by 250 V, 50 Hz ac will be
(*a*) 260 V (*b*) 252 V
(*c*) 250 V (*d*) 248 V

50. A sinusoidal ac voltage of rms value 100 V is applied to a rectifying device which offers 10 Ω resistance in forward direction and infinite resistance in reverse direction. The reading on hot wire ammeter will be
(*a*) 10 A (*b*) 5 A
(*c*) 3.185 A (*d*) 1.414 A

51. A sinusoidal ac voltage of amplitude 100 V is applies to a rectifying device which offers 10 Ω resistance in forward direction and infinite resistance in backward direction. A moving coil ammeter is also connected in the circuit. The reading of ammeter will be
(*a*) 10 A (*b*) 5 A
(*c*) 3.185 A (*d*) 1.414 A

52. In air friction damping
(*a*) the air pressure opposes the motion of piston when it is moving inwards
(*b*) the air pressure opposes the motion of piston when it is moving outwards
(*c*) the air pressure opposes the motion of piston both when it is moving inwards as well as outwards
(*d*) the air pressure may oppose or aid the motion of pistion depending on the design

53. Panel type instruments are used in
(*a*) power stations (*b*) automobiles
(*c*) industries (*d*) all the above places

54. Heating effect of current is used in
(*a*) ammeters
(*b*) voltmeters
(*c*) both ammeters and voltmeters
(*d*) wattmeters

55. Magnetic effect of current is used in
(*a*) ammeters
(*b*) voltmeters
(*c*) wattmeters
(*d*) ammeters, voltmeters, wattmeters and energy meters

56. Consider the following instruments
1. Tangent galvanometer
2. Rayleigh current balance
3. Household energy meter
4. Hot wire ammeter

Out of above absolute instruments are
(*a*) 1, 2 and 3 (*b*) 1, 2 and 4
(*c*) 1 and 2 (*d*) 2, 3 and 4

57. Match List I and List II and give correct answer as per the given codes

List I	List II
A. Tangent galvanometer	1. Can be used for dc only
B. Moving coil instrument	2. Intergrating instrument
C. Household energy meter	3. Used for voltage measurement only
D. Electrotatic instrument	4. Absolute instrument

Codes:

	A	B	C	D
(*a*)	1	2	3	4
(*b*)	4	1	2	3
(*c*)	1	2	4	3
(*d*)	4	1	3	2

58. Which of the following instruments has non-linear scale?
1. Moving coil instrument
2. Moving iron instrument
3. Electrostatic instrument
4. Rectifier instrument

Select the answer as per following codes
(*a*) 2 only (*b*) 2 and 3 only
(*c*) 2, 3 and 1 (*d*) 1, 2, 3 and 4

59. Which of the following instruments can be used for dc as well ac upto a few MHz
1. Moving iron instrument
2. Thermocouple instrument
3. Induction instrument
4. Moving coil instrument

Select the correct answer as per following codes
(*a*) 2 only (*b*) 2 and 3 only
(*c*) 1, 2 and 3 only (*d*) 2 and 3 only

60. Two ammeters A and B are connected in parallel in a circuit carrying 100 A. Their resistance are 0.8 mΩ and 1 mΩ. Their reading will be
(*a*) 55.56 and 44.44 A respectively

(b) 44.44 A and 55.56 A respectively
(c) 60 A and 40 A respectively
(d) 40 A and 60 A respectively

61. If a shunt of 200 Ω resistance is used with a galvanometer of 1000 Ω resistance, the multiplying power is
(a) 6 (b) 5
(c) 1/5 (d) 1/6

62. An ammeter has a resistance of 50 ohms and a shunt of 0.01 Ω. The deflection is θ. If the shunt resistance is increased to 0.02 Ω, the deflection will be
(a) θ (b) 2 θ
(c) 0.5 θ (d) 0.25 θ

63. A D'Arsonval galvanometer is a
(a) moving coil instrument
(b) moving iron instrument
(c) dynamometer instrument
(d) induction instrument

64. The coil of a D'Arsonval galvanometer has N turns. If flux density is B, current is I and area of coil is A, the deflecting torque is
(a) NBI/A (b) NBIA
(c) NIA/B (d) N^2IAB

65. If C is spring constant and θ is deflection, controlling torque is
(a) C θ (b) C/θ
(c) θ/C (d) $C^2\theta$

66. If J is moment of media, D is damping constant, C is spring constant, G is displacement constant and I is current through the coil of D'Arsonval galvanometer, the equation of motion is
(a) $J\frac{d^2\theta}{dt^2} - D\frac{d\theta}{dt} + C\theta = GI$
(b) $J\frac{d^2\theta}{dt^2} + D\frac{d\theta}{dt} + C\theta = GI$
(c) $J\frac{d^2\theta}{dt^2} + D\frac{d\theta}{dt} + GI = C\theta$
(d) $J\frac{d^2\theta}{dt^2} + D\frac{d\theta}{dt} - C\theta = GI$

67. The response of D'Arsonval galvanometer is over damped if
(a) $D^2 > 4CJ$ (b) $D^2 > 2CJ$
(c) $D^2 < 4CJ$ (d) $D^2 > 2CJ$

68. The damping ration, in D'Arsonval galvanometer, is equal to
(a) $\frac{D}{\sqrt{CJ}}$ (b) $\frac{D}{2\sqrt{CJ}}$
(c) $\frac{\sqrt{D}}{2CJ}$ (d) $\frac{D}{\sqrt{2CJ}}$

69. For critical damping, the resistance in D'Arsonval galvanometer circuit should be equal to
(a) $\frac{G^2}{2\sqrt{CJ}}$ (b) $\frac{G}{2\sqrt{CJ}}$
(c) $\frac{G^2}{2CJ}$ (d) $\frac{\sqrt{G}}{2CJ}$

70. A vibration galvanometer is a
(a) moving coil instrument
(b) moving iron instrument
(c) dynamometer instrument
(d) induction instrument

71. Consider the following galvanometers
1. D'Arsonval 2. Ballistir 3. Vibration
Which of these are permanent magnet moving coil instrument. Answer as per the following code
(a) 1 and 2 only (b) 1, 2 and 3
(c) 2 and 3 only (d) 1 and 3 only

72. In a ballistic galvanometer, a charge of 100 μC gives a first swing of 25°. The charge required to charge required to cause a first swing of 50° is
(a) 200 μC (b) 400 μC
(c) 50 μC (d) 25 μC

73. The deflection of a D'Arsonval galvanometer is measured through a mirror and scale. The mirrors is 1 m away from the light source. The mirror is mounted on the moving system. If mirror rotates through 0.2 rad the displacement on the scale is
(a) 0.2 m (b) 0.4 m
(c) 0.1 m (d) 0.05 m

74. A moving iron instrument has
(a) two iron vanes, one fixed and one movable
(b) two movable iron vanes
(c) two aluminium vanes, one fixed and one movable
(d) two vanes, one fixed made of aluminium and the other movable made of iron

75. The method of frequency compensation in moving iron voltmeter is
(a) to use a capacitor in series with coil
(b) to connect a capacitor in parallel with coil
(c) to connect a capacitor in parallel with series resistance
(d) none of the above

76. A moving iron voltmeter reads 250 V when connected to 250 V dc. If connected to 250 V 50 Hz, the voltmeter reading is likely to be
(a) 250 V (b) 252 V
(c) 248 V (d) 210 V

77. A rectifier voltmeter using bridge rectifier and PMMC meter is calibrated to read rms value of sine wave. A triangular wave is applied to it and it reads 6.82 V. The rms value of triangular wave is
(a) 6.82 V (b) about 7.1 V
(c) about 6.5 V (d) about 10 V

78. The resistance in the circuit of the moving coil of a dynamometer wattmeter should be
(a) very low (b) low
(c) high (d) almost zero

79. A dynamometer wattmeter can be used for
(a) dc only (b) ac only

(c) dc and 50 Hz ac
(d) dc and high frequency

80. An induction wattmeter can be used for
(a) dc only
(b) 50 Hz ac only
(c) both ac and dc
(d) high frequency ac only

81. The pressure coil of a wattmeter should be connected to the supply side of the current coil when
(a) load impedance in high
(b) load impedance is low
(c) voltage is high
(d) voltage is low

82. In low power factor wattmeter, the pressure coil is connected
(a) to the supply side of current coil
(b) to the load side of current coil
(c) in either (a) or (b)
(d) across the mains

83. In low power factor wattmeter, the compensating coil is connected
(a) in series with current coil
(b) in parallel with current coil
(c) in series with pressure coil
(d) in parallel with pressure coil

84. In 3-phase power measurement by two wattmeter method, both wattmeters had equal readings. The power factor of load is
(a) 1 (b) 0.5 lagging
(c) 0.5 leading (d) 0

85. In 3-phase power measurement by two wattmeter method. The reading of one wattmeter is zero. The power factor of load is
(a) 1 (b) 0.5
(c) 0 (d) 0.8

86. The adjustment of position of shading bands. In an energy meter, is done to provide
(a) friction compensation (b) creep compensation
(c) braking torque (d) none of the above

87. A polyphase wattmeter basically consists of two wattmeters. The moving system of two wattmeters
(a) are common (b) are separate
(c) are generally common (d) are generally separate

88. Which instrument needs creep compensation?
(a) Dynamometer wattmeter
(b) Induction wattmeter
(c) Energy meter (d) Both (b) and (c)

89. Which instrument is based on induction principle?
(a) Dynamometer wattmeter
(b) Induction wattmeter
(c) Energy meter (d) Both (b) and (c)

90. The electromagnets of an induction wattmeter
(a) are laminated (b) are not laminated
(c) may be laminated (d) must not be laminated

91. The dynamometer principle can be used for
(a) ammeter only (b) voltmeter only
(c) wattmeter only
(d) ammeter, voltmeter and wattmeter

92. If φ_{sh} is flux of shunt magnet, ϕ_{se} is flux of series magnet, i_{sh} is eddy current due to ϕ_{sh} and i_{se} is eddy current due to ϕ_{se}, the torque in an induction wattmeter is proportional to
(a) $\phi_{se}\, i_{se} + \phi_{se}\, i_{sh}$ (b) $\phi_{sh}\, i_{se} + \phi_{se}\, i_{sh}$
(c) $\phi_{sh}\, i_{sh} + \phi_{se}\, i_{se}$ (d) $\phi_{sh}\, i_{sh} + \phi_{se}\, i_{se}$

93. If M is mutual inductance between fixed and moving coils and θ is deflection of dynamometer instrument, then
(a) $\theta \propto \dfrac{dM}{d\theta}$ (b) $\theta \propto M$
(c) $\theta \propto \left(\dfrac{dM}{d\theta}\right)^2$ (d) $\theta \propto M^2$

94. The scale of a dynamometer is linear when used as
(a) ammeter (b) voltmeter
(c) wattmeter (d) (a) and (c)

95. Consider the following statements as regards induction and dynamometers wattmeters
1. Dynamometers wattmeters can be used for dc only
2. Induction wattmeter has higher torque and longer length of scale than dynamometer wattmeter
3. Induction wattmeter is less accurate, has more power consumption and more weight than dynamometer wattmeter

Which of the above statements are correct?
(a) 1 only (b) 1and 2 only
(c) 2 and 3 only (d) 1 and 3 only

96. Two wattmeter method for 3-phase power measurement can be used only for
(a) balanced loads
(b) unbalanced loads
(c) both balanced and unbalanced loads
(d) star connected loads

97. In two wattmeter method of 3-phase power measurement, when does one wattmeter read negative
(a) when power factor is less than 0.5 lagging
(b) when power factor is less than 0.5 leading
(c) when power factor is less than 0.5
(d) when power factor is unity

98. In Fig. 6.36, the reading of wattmeter will be proportional to

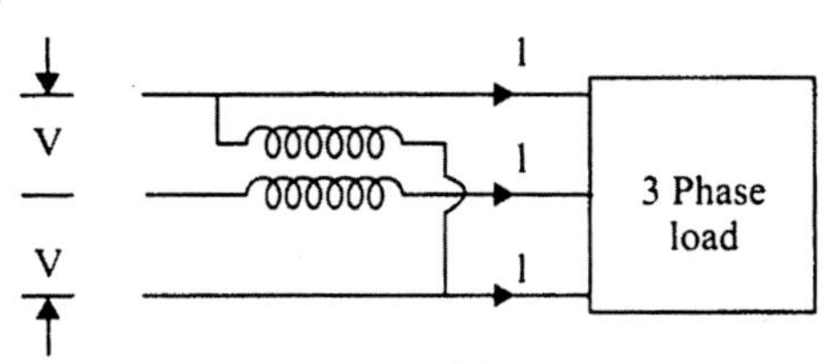

Fig. 6.36.

(a) V I cos θ (b) $\sqrt{3}$ V I cos θ
(c) $\sqrt{3}$ V I sin θ (d) V I sin θ

99. In an ac single phase energy meter, the potential coil is put on
(*a*) both side limbs of upper magnet
(*b*) central limb of upper megnet
(*c*) one side limb of upper magnet
(*d*) either (*a*) or (*b*)

100. During testing of an energy meter, the energy recorded by meter was 0.08 kWh and correct energy was found to be 0.09 kWh. The error and correction respectively are
(*a*) – 0.01 kWh and 0.01 kWh
(*b*) 0.01 kWh and – 0.01 kWh
(*c*) – 0.01 kWh and ± 0.01 kWh
(*d*) ± 0.01 kWh and 0.01 kWh

101. In household energy meters the flux of series magnet is in phase with current but the flux of shunt magnet lags the voltage by 88°. The reading of energy meters
(*a*) will be correct
(*b*) will have a negative error
(*c*) will have a positive error
(*d*) may have positive or negative error

102. An ohmmeter is a
(*a*) moving coil instrument
(*b*) moving iron instrument
(*c*) dynamometer instrument
(*d*) induction instrument

103. A resistance is measured by voltmeter-ammeter method. The voltmeter is 0-250 V, ± 1% accuracy and ammeter is 0-5 A, ± 1% accuracy. The readings of voltmeter and ammeter are 100 V and 2 A respectively. The error in the measured resistance can be
(*a*) ± 1% (*b*) ± 2%
(*c*) ± 5% (*d*) ± 10%

104. When a capacitor is connected to the terminals of an ohm-meter, the pointer indicated a low resistance initially and finally come to infinity position. This shows that capacitor is
(*a*) short-circuited (*b*) all right
(*c*) faulty (*d*) open circuited

105. In measuring a resistance by voltmeter-ammeter method, the voltmeter can be connected either across supply or across the resistance. If the resistance is low, the voltmeters should be connected
(*a*) across the supply
(*b*) across the resistance
(*c*) in either of two position
(*d*) preferably across the supply

106. For measuring a very high resistance we should use
(*a*) Kelvin's double bridge
(*b*) Wheatstone bridge
(*c*) Megger
(*d*) either (*a*) or (*c*)

107. The effect of lead resistance is important in measurement of
(*a*) resistance of shunt
(*b*) resistance of shunt winding
(*c*) resistance of ammeter
(*d*) both (*a*) and (*c*)

108. The electrical power to a megger is provided by
(*a*) battery
(*b*) inverter
(*c*) permanent magnet dc generator
(*d*) ac generator

109. In a megger the controlling torque is provided by
(*a*) spring (*b*) gravity
(*c*) coil (*d*) eddy current

110. The operating voltage of a megger is about
(*a*) 100 V (*b*) 12 V
(*c*) 6 V (*d*) 1 V

111. The loss of charge method can be used to measure
(*a*) resistance of shunt
(*b*) insulation resistance of cable
(*c*) resistance of shunt winding
(*d*) both (*b*) and (*c*)

112. To measure a resistance of about 8 ohms we should use
(*a*) Kelvin's double bridge (*b*) Wheatstone bridge
(*c*) Megger (*d*) either (*a*) or (*c*)

113. A high resistance was measured by loss of charge method. The voltmeter resistance was 10^{10} ohms. The calculations gave the value of unknown resistance as 0.8×10^9 ohms. The true value of resistance is
(*a*) $0.8 \times 10^9\ \Omega$ (*b*) less than $0.8 \times 10^9\ \Omega$
(*c*) more than $0.8 \times 10^9\ \Omega$ (*d*) either (*b*) or (*c*)

114. Which of the following devices should be used for accurate measurement of a low dc voltage?
(*a*) small range moving coil voltmeter
(*b*) dc potentiometer
(*c*) small range thermo couple voltmeter
(*d*) hot wire instrument

115. It is required to measure the true open circuit emf of a battery. The best device is
(*a*) dc voltmeter
(*b*) ammeter and known resistance
(*c*) dc potentiometer
(*d*) either (*a*) or (*c*)

116. A voltage of about 200 V dc can be measured
(*a*) directly by dc potentiometer
(*b*) a dc potentiometer in conjunction with a volt rations box
(*c*) a dc potentiometer in conjunction with a known resistance
(*d*) both (*b*) and (*c*)

117. A direct current can be measured
(*a*) directly by dc potentiometer
(*b*) by dc potentiometer in conjunction with a standard resistance
(*c*) by dc potentiometer in conjunction with volt rations box
(*d*) none of the above

118. To measure a resistance by a dc potentiometer
(*a*) it is necessary to standardise the potentiometer
(*b*) it is not necessary to standardise the potentiometer
(*c*) it is necessary to use a voltration box
(*d*) none of the above

119. A phase shifting transformer is used in conjunction with
(*a*) DC slide wire potentiometer
(*b*) Crompton potentiometer
(*c*) AC coordinate potentiometer
(*d*) Drysdale potentiometer

120. Basically a potentiometer is a device for
(*a*) comparing two voltage
(*b*) measuring a voltage
(*c*) comparing two currents
(*d*) measuring a current

121. In order to achieve accuracy, the slide wire of a potentiometer should be
(*a*) as long as possible
(*b*) as short as possible
(*c*) neither too long nor too short
(*d*) very thin

122. To measure an unknown ac voltage by an ac potentiometer it is desirable that supply for the potentiometer is taken
(*a*) from a battery
(*b*) from a source which is not the same as unknown voltage
(*c*) from the same source as the unknown voltage
(*d*) either (*a*) or (*b*)

123. The stator of phase shifting transformer for use in conjunction with an ac potentiometer has a
(*a*) single phase winding
(*b*) two phase winding
(*c*) 3 phase winding
(*d*) either (*a*) or (*c*)

124. In an ac coordinate potentiometer, the currents in in-phase and quadrature potentiometer are adjusted to be
(*a*) out of phase by 90° (*b*) in phase
(*c*) out of phase by 45° (*d*) out of phase 180°

125. Crompton potentiometer has
(*a*) 14 coils and a slide wire all having the same resistance
(*b*) 14 coils and a slide wire but the resistance of slide wire is half that of each coil
(*c*) 14 coils each of different resistance
(*d*) 14 coils out of which 7 coils have resistance which is twice that of remaining 7 coils

126. The volt ratio box is basically a
(*a*) potential divider
(*b*) current transformer
(*c*) a high resistance voltmeter
(*d*) either (*a*) or (*c*)

127. A standard cell of 1.0185 V is used with a slide wire potentiometer. The balance was obtained at 60 cm. When an unknown emf was connected, the balance was obtained at 82 cm. The magnitude of unknown emf is
(*a*) 1.39 V (*b*) 0.745 V
(*c*) 13.9 V (*d*) 7.45 V

128. A universal RLC bridge uses
(*a*) Maxwell bridge for measurement of inductance and De-Sauty's bridge for measurement of capacitance
(*b*) Maxwell's bridge for measurement of inductance and modified Desauty bridge for measurement of capacitance
(*c*) Maxwell Wien bridge for measurement of inductance and modified De-Sauty bridge for measurement of capacitance
(*d*) Maxwell Wien bridge for measurement of inductance and Wien bridge for measurement of capacitance

129. For measurements of high voltage capacitors the suitable bridge is
(*a*) Wien bridge
(*b*) modified De-Sauty bridge
(*c*) Schering bridge
(*d*) either (*a*) or (*b*)

130. In an Anderson bridge the unknown inductance is measured in terms of
(*a*) known inductance and resistance
(*b*) known inductance
(*c*) known capacitance and resistance
(*d*) known capacitance

131. Wagner earthing device is used to eliminate error due to
(*a*) electromagnetic coupling
(*b*) electrostatic coupling
(*c*) both electrostatic and electromagnetic coupling
(*d*) either (*a*) or (*b*)

132. For measuring mutual inductance we can use
(*a*) Anderson bridge (*b*) Maxwell's bridge
(*c*) Heaviside bridge (*d*) either (*a*) or (*b*)

133. For measuring inductance with high Q, we should use
(*a*) Maxwell's bridge (*b*) Maxwell Wien bridge
(*c*) Hay's bridge (*d*) Either (*a*) or (*b*)

134. If the current in a capacitor leads the voltage by 80°, the loss angle of capacitor is
(*a*) 80° (*b*) 10°
(*c*) 100° (*d*) 170°

135. In Schering bridge, the potential of detector above earth potential is about
(*a*) a few volts (*b*) 10 kV
(*c*) 1000 V (*d*) either (*a*) or (*b*)

136. To avoid the effect of stray magnetic fields in ac bridges we can use
(*a*) magnetic screening
(*b*) Wagner earthing device
(*c*) wave filter (*d*) transformer

137. If an inductance L is connected in one arm of bridge and resistance R_1, R_2, R_3 in the other three arms
(*a*) the bridge cannot be balanced
(*b*) the bridge can be balanced
(*c*) the bridge is balanced for some specified value of frequency
(*d*) the bridge is balanced for some specific values of R_1, R_2, R_3

138. A power factor meter has
(*a*) one current and one pressure circuit
(*b*) one current circuit and two pressure circuits
(*c*) two current circuits and two pressure circuits
(*d*) two current circuits and one pressure circuit

139. The two pressure coils of a single phase power factor meter have
(*a*) the same dimension and the same numbers of turns
(*b*) the same dimension but different number of turns
(*c*) different dimension and different numbers of turns
(*d*) different dimension but the same number of turns

140. In a single-phase power factor meter the phase difference between currents in the two pressure coils is
(*a*) exactly 0° (*b*) approximate 0°
(*c*) exactly 90° (*d*) approximate 90°

141. In a dynamometer 3 phase power factor meter, the planes of the two moving coils are at
(*a*) 60° (*b*) 0°
(*c*) 120° (*d*) 90°

142. In a vibrating reed frequency meter the natural frequency of two adjacent reeds have a difference of
(*a*) 0.25 Hz (*b*) 0.5 Hz
(*c*) 1.25 Hz (*d*) 1 Hz

143. For use at 50 Hz, the range of scale of vibrating reed frequency meter is
(*a*) 25–75 Hz (*b*) 40–60 Hz
(*c*) 47–53 Hz (*d*) 42–58 Hz

144. In Weston frequency meter, the magnetic axes of the two fixed coils are
(*a*) parallel (*b*) perpendicular
(*c*) inclined at 60° (*d*) inclined at 15°

145. A weston frequency meter is
(*a*) moving coil instrument
(*b*) moving iron instrument
(*c*) dynamometer instrument
(*d*) induction instrument

146. In a Weston synchronoscope the deflecting torque is zero when
(*a*) currents in fixed and moving coils are in phase
(*b*) currents in fixed and moving coils are out of phase by 90°
(*c*) currents in fixed and moving coils are out of phase by 60°
(*d*) none of the above

147. A Weston synchronoscope is a
(*a*) moving coil instrument
(*b*) moving iron instrument
(*c*) dynamometer instrument
(*d*) induction instrument

148. In weston synchronoscope the fixed coils are connected across
(*a*) bus bars (*b*) incoming alternator
(*c*) lamp (*d*) either (*a*) or (*b*)

149. In a Weston synchronoscope the moving coil is connected across
(*a*) bus bars (*b*) incoming alternator
(*c*) lamp (*d*) fixed coils

150. The power factor of single phase load can be calculated if instruments available are
(*a*) one voltmeter, one ammeter
(*b*) two ammeters
(*c*) two voltmeters
(*d*) one voltmeter, one ammeter and one wattmeter

151. In a CRO, the time base signal is applied to
(*a*) X plates
(*b*) Y plates
(*c*) either X or Y plates
(*d*) alternately X and Y plates

152. The numbers of accelerating anodes in a CRO is
(*a*) 1 (*b*) 2
(*c*) 3 (*d*) 4

153. If E_b is the potential of accelerating anode, the energy of electrons as they emerge from election gun is proportional to
(*a*) E_b (*b*) $\sqrt{E_b}$
(*c*) $E_b^{1.5}$ (*d*) E_b^2

154. In a CRO the Y plates are oriented
(*a*) horizontally
(*b*) vertically
(*c*) at an angle of 20° of horizontal
(*d*) at an angle of 70° of horizontal

155. If E_y is the voltage applied to Y plates, the vertical deflection of beam is proportional to
(*a*) E_y (*b*) $\sqrt{E_y}$
(*c*) $E_y^{1.5}$ (*d*) E_y^2

156. The wave applied to X plates, for obtaining trace of voltage wave, is
(*a*) sawtooth wave (*b*) sinusoidal wave
(*c*) rectangular wave (*d*) (*a*) or (*c*)

157. If two time varying signals of equal magnitude and zero phase displacement are applied to X and Y plates respectively; of CRO, the display will be
(*a*) a straight line (*b*) a circle
(*c*) an ellipse (*d*) a square

158. If two time varying signals of equal magnitudes and 90° phase displacement are applied to X and Y plates of CRO, the display will be
(*a*) a straight line (*b*) a circle
(*c*) an ellipse (*d*) a square

159. In electronic voltmeters measuring device is a
(*a*) moving coil instrument (*b*) moving iron instrument
(*c*) dynamometer instrument
(*d*) none of the above

160. A voltmeter using thermocouples measures
(*a*) rms value (*b*) peak value
(*c*) average value (*d*) peak to peak value

161. When measuring phase angle between two waves using a CRO, the time base generator is connected to
(*a*) X plate (*b*) Y plates
(*c*) either X or Y plates (*d*) neither X or Y plates

162. The accuracy of measurement using CRO is about
(*a*) ± 0.5% (*b*) ± 0.05%
(*c*) ± 1% (*d*) ± 3%

163. The material used in RTD is
(*a*) copper (*b*) aluminium
(*c*) silver (*d*) platinum

164. The response of a thermistor over the whole of its temperatures range is
(*a*) linear (*b*) non-linear
(*c*) almost linear (*d*) (*a*) or (*c*)

165. If temperature increases by about 50°, the resistance of thermistor changes to
(*a*) half of its initial value (*b*) twice its initial value
(*c*) eight times its initial value
(*d*) one-eighth of its initial value

166. The resistance of resistance strain gauge is about
(*a*) 5 Ω (*b*) 50 Ω
(*c*) 120 Ω (*d*) 0.5 Ω

167. The gauge factor of a semi conductor stain gauge is about
(*a*) 2 (*b*) 10
(*c*) 100 (*d*) 1000

168. LVDT is a
(*a*) pressure transducer (*b*) displacement transducer
(*c*) velocity transducer (*d*) acceleration transducer

169. In a thermocouple temperature transducer, the output voltage is usually measured by
(*a*) high resistance ammeter
(*b*) high resistance voltmeter
(*c*) high resistance milli voltmeter
(*d*) none of the above

170. A capacitance transducer can be used to measure
(*a*) thickness of sheet (*b*) displacement
(*c*) level of fluids (*d*) all of the above

171. A land line telemetry system is suitable for distance upto
(*a*) about 1 km only (*b*) about 10 km only
(*c*) about 100 km only (*d*) any of above

172. The speed transducer using a drag cup rotor has the main advantage of
(*a*) very high accuracy (*b*) low energy requirement
(*c*) cheap construction (*d*) all the above

173. In a strain measuring device using a strain gauge, the output quantity is
(*a*) voltage (*b*) resistance
(*c*) impedance (*d*) either (*a*) or (*b*)

174. An E transformer is a
(*a*) displacement transducer
(*b*) velocity transducer
(*c*) acceleration transducer
(*d*) pressure transducer

175. The range of RTD using platinum is about
(*a*) – 100°C to 700°C (*b*) – 200°C to 1000°C
(*c*) 100°C to 1000°C (*d*) about 500°C

176. For measurement of surface temperature of solids the element of RTD is usually in the shape of
(*a*) bar (*b*) coil
(*c*) flat grid (*d*) either (*a*) or (*b*)

177. Which of the following combination is not used in thermocouple transducer measurement of temperature?
(*a*) iron-constantan (*b*) chromel-aluminium
(*c*) copper-constantan (*d*) iron-silver

178. A radiation thermometer is suitable for temperatures
(*a*) below 500°C (*b*) above 100°C
(*c*) above 500°C (*d*) above 1400°C

179. In a platinum RTD, the change in resistance for a temperature rise of 50°C is about
(*a*) 2% (*b*) 10%
(*c*) 20% (*d*) 40%

180. In a semiconductor strain gauge, the ohmic electrical contacts are generally made of
(*a*) copper wire (*b*) gold wire
(*c*) aluminium wire (*d*) nickel wire

181. Which of the following is not a piezo electric material?
(*a*) Rochell salt (*b*) quartz
(*c*) lithium sulphate (*d*) sodium bicarbonate

182. In a CRO which of the following is not a part of electron gun
(*a*) cathode (*b*) grid
(*c*) accelerating anode (*d*) X – Y plates

183. Magnetic deflection is inversely proportional to
(*a*) voltage (*b*) $(\text{voltage})^{0.5}$
(*c*) $(\text{voltage})^{1.5}$ (*d*) $(\text{voltage})^2$

184. Magnetic deflection sensitivity is defined as
(*a*) deflection/flux density
(*b*) $(\text{deflection/flux density})^{0.5}$
(*c*) $(\text{Deflection})^{0.5}$/ flux density
(*d*) none of above

185. In a CRO, the frequency of output of time base generator
(*a*) is only one constant value
(*b*) is adjustable
(*c*) varies automatically (*d*) (*a*) or (*c*)

186. The dimensions of force are
(*a*) MLT^{-2} (*b*) $ML^{-1}T^{-2}$
(*c*) MLT^{-1} (*d*) ML^2T^{-2}

187. The dimensions of EMF are
(*a*) $ML^2I^{-1}T^{-3}$ (*b*) $MLI^{-1}T^{-3}$
(*c*) $ML^2I^{-2}T^{-3}$ (*d*) $ML^3I^{-1}T^{-3}$

188. Magnetic flux density has the dimensions
(*a*) $MI^{-1}T^{-2}$ (*b*) $M^{-1}I^{-2}T^{-2}$
(*c*) $MI^{-3}T^{-2}$ (*d*) $MI^{-1}T^{-3}$

189. Magnetic field intensity has the dimensions
(*a*) IL (*b*) I^2L
(*c*) IL^{-1} (*d*) IL^{-2}

190. Current density has the dimensions
(*a*) IL^{-1} (*b*) IL^{-2}
(*c*) IL^{-3} (*d*) I^2L^{-2}

191. Figure 6.37 shows a Wheatsone bridge with RTD in one arm. If resistance of RTD = 1020 Ω, the voltage V_o is

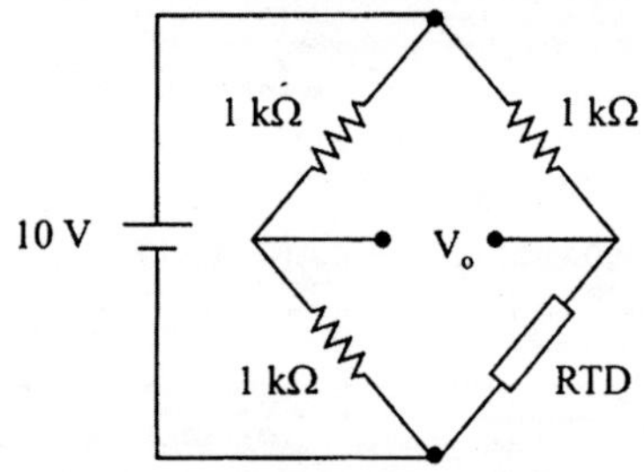

Fig. 6.37.

(*a*) 49 mV (*b*) 51 mV
(*c*) 48 mV (*d*) 45 mV

192. The advantage of semiconductor strain gauge over resistance strain gauge is
(*a*) it is more sensitive (*b*) it is more linear
(*c*) it is less temperature dependent
(*d*) it is cheaper

193. In a resistance strain gauge, G = 2, stress = 1050 kg/cm^2, Young's modules = 2.1 × 10^6 kg/cm^2, R = 1000 Ω. The value of Δ R will be
(*a*) 2 Ω (*b*) 3 Ω
(*c*) 4 Ω (*d*) 1 Ω

194. Schering bridge is used to measure
(*a*) capacitance and dielectric loss
(*b*) inductance
(*c*) resistance (*d*) mutual inductance

195. LVDT is used to measure
(*a*) displacement (*b*) pressure
(*c*) temperature (*d*) any of above

196. Which of the following is active transducer
(*a*) photo electric (*b*) photo voltaic
(*c*) photo conductive (*d*) photo emissive

197. Match list I with list II and select the answer, using the given codes

List I	**List II**
A. Resistance	1. $M^{-1}L^{-2}T^4I^2$
B. Inductance	2. $ML^2T^{-2}I^{-2}$
C. Capacitance	3. $ML^2T^{-3}I^2$

Codes:

	A	B	C
(*a*)	1	2	3
(*b*)	2	1	3
(*c*)	2	3	1
(*d*)	3	2	1

198. Which of the following methods can be used to measure power in a single phase circuit without using wattmeter?
1. One voltmeter, one ammeter
2. Two voltmeters, two ammeter
3. Three voltmeters
4. Three ammeters

(*a*) 1 and 2 only (*b*) 2, and 4 only
(*c*) 1 and 3 only (*d*) 3 and 4 only

199. Match List I with List II and select the answer using the codes

List I	**List II**
A. Digital counter	1. Measurement of harmonics
B. Schering Bridge	2. Measurement of frequency
C. Meggar	3. Measurement of loss angle of dielectric
D. Spectrum analyser	4. Measurement, of insulation resistance

Codes:

	A	B	C	D
(*a*)	1	2	3	4
(*b*)	4	3	2	1
(*c*)	2	3	4	1
(*d*)	4	1	3	2

200. Which of the following is true about an impulse voltage?
(*a*) it is a unidirectional voltage
(*b*) in chopped impulse flashover does not occur
(*c*) time taken to rise is equal to time taken to fall
(*d*) rms value is always less than 50% of average value value

201. A digital voltmeter can count from 0 to 9999. If full scale reading is 9.999 V, the resolution of full scale reading is
(*a*) 0.001 (*b*) 1000
(*c*) 2 digit (*d*) 1 mV

202. The input impedance of CRO is about
(*a*) zero (*b*) 10 Ω
(*c*) 100 Ω (*d*) one mega ohm

203. In bridge circuit of Fig. 6.38, equivalent resistance between A and B is

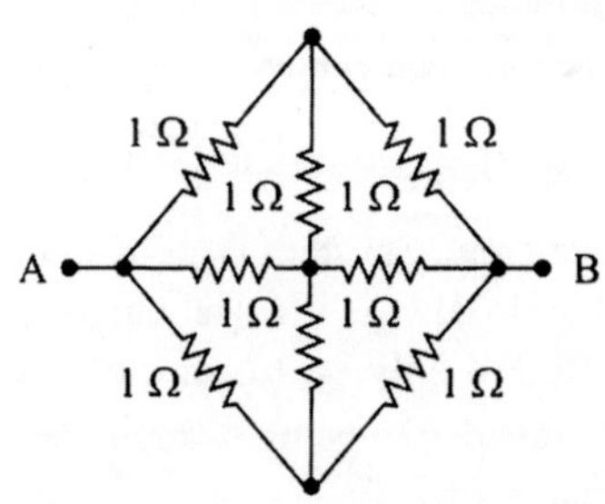

Fig. 6.38.

(*a*) 1 Ω (*b*) 0.5 Ω
(*c*) 2 Ω (*d*) 1.5 Ω

204. Two meters X and Y require 40 mA and 50 mA for full scale deflection. Then
(*a*) X is more sensitive
(*b*) Y is more sensitive
(*c*) both are equally sensitive
(*d*) not possible to determine from given data

205. Match List I with List II and select the answer using the codes

List I	List II
A. Electro-encephalo graph	1. Diagnostic tool for heart aliment
B. Electro-cardiograph	2. Diagnostic tool for brain aliment
C. Sphygmo-manometer	3. used for measuring BP
D. Stethoscope	4. used to hear pulse/heart beat

Codes:

	A	B	C	D
(*a*)	2	1	3	4
(*b*)	3	1	2	4
(*c*)	2	1	4	3
(*d*)	1	2	3	4

206. It is required to study transients during switching process. Which CRO should be preferred

(*a*) ordinary CRO (*b*) dual beam CRO
(*c*) dual trace CRO (*d*) storage oscilloscope

207. Match List I with List II and select the answer, using the given coils

List I	List II
A. Resistance	1. $M^{-1}L^{-2}T^{2}I^{2}$
B. Inductance	2. $ML^{2}T^{-2}I^{-2}$
C. Capacitance	3. $M^{-1}L^{-2}T^{4}I^{2}$
D. Reluctance	4. $ML^{2}T^{-3}I^{-2}$

Codes:

	A	B	C	D
(*a*)	1	2	3	4
(*b*)	2	1	3	4
(*c*)	4	2	3	1
(*d*)	2	4	3	1

208. Which of the following closely represents the resistance-temperature characteristics of thermistor

(*a*) $R_T = a\, R_0\, e^{b/t}$ (*b*) $R_T = a\, R_o\, e^{-b/t}$
(*c*) $R_T = a\, R_0 e^{-b/t}$ (*d*) $R_T = a\, R_0\, e^{\,b/t}$

209. Which of the following statements is correct for two wattmeters method of power measurement in 3-phase circuit

(*a*) Power can be measured by this method only for star connected loads
(*b*) When two wattmeters have equal readings power factor is 0.5
(*c*) When power factor is 1, one of the wattmeters readings is zero
(*d*) When the two wattmeter read equal and opposite, p.f is zero

210. A temperature sensitive transducer is subjected to sudden temperature change. It takes 10 seconds for transducer to reach equilibrium condition (five times time constant). The time taken to reach half of temperature difference will be

(*a*) 1.38 s (*b*) 5 s
(*c*) 8.62 s (*d*) 9.8 s

211. The precision of an instrument indicates its ability to reproduce a certain reading with a given

(*a*) drift (*b*) resolution
(*c*) shift (*d*) consistency

212. In ac bridge measurements. 'Wagner ground' means

(*a*) a special RC connection to eliminate stray magnetic effects
(*b*) a special RC connection to eliminate stray capacitance effects
(*c*) an unwanted and unintended ground connection
(*d*) a large metal plate buried in ground connected to one corner of bridge

213. Which of the following quantities have same dimension in *emu* and *esu*?

1. current
2. electrical power
3. electrical energy

Select the answer as per following codes:

(*a*) 1 only (*b*) 1 and 2
(*c*) 2 and 3 (*d*) 1 and 3

214. In a moving iron voltmeter the coil has a resistance of 20 Ω and an inductance of 0.3 H. The swamping resistance in 2000 Ω. To ensure correct reading with dc as well as 50 Hz ac, the value of capacitor to be connected in parallel with swamping resistance should be

(*a*) 0.0125 μF (*b*) 0.0311 μF
(*c*) 0.0521 μF (*d*) 0.0622 μF

215. Figure 6.39 shows an RC potentiometer to measure ac voltage. It is desired that V_0/V_i should be independent of frequency. The value of C should be

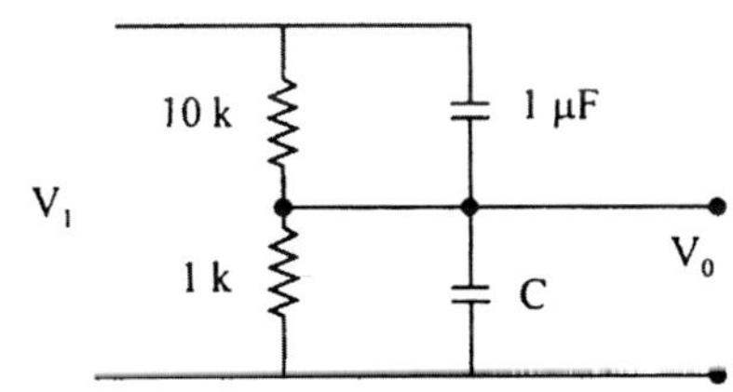

Fig. 6.39.

(*a*) 10 μF (*b*) 11 μF
(*c*) 0.1 μF (*d*) 0.09 μF

216. The circuit of Fig. 6.40 is used to measure the value R_x. Ammeter resistance R_a = 0.1 Ω voltmeter resistance R_v = 5000 Ω. If R_{x1} and R_{x2} are the measured values of R_x with switch in position 1 and 2, then mean of R_{x1} and R_{x2} is

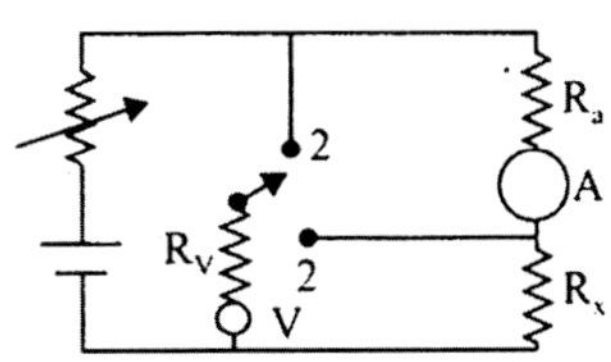

Fig. 6.40.

(*a*) 9.98 Ω (*b*) 10 Ω
(*c*) 10.04 Ω (*d*) 10.1 Ω

217. If reference sound pressure P_0 is 2×10^{-5} N/m^2, a sound pressure of 90 dB is equal to
(*a*) 0.632 N/m^2 (*b*) 0.707 N/m^2
(*c*) 0.835 N/m^2 (*d*) 0.925 N/m^2

218. With a variable reluctance tachometer having 60 rotor teeth the count reads 3600 counts per minute. The speed in RPM is
(*a*) 36 (*b*) 60
(*c*) 360 (*d*) 3600

219. A capacitance transducer has two plates each of area 5 mm^2. The coil gap is 2 mm. Displacement sensitivity due to change in gap length is
(*a*) 11.1 pF/cm (*b*) 44.2 pF/cm
(*c*) 52.3 pF/cm (*d*) 66.3 pF/cm

220. In an experiment with a thermocouple with one junction at 0°C, the following data was obtained at three value of θ°C

θ°C	20°C	100°C	300°C
e (mill volts)	0.194	0.85	1.65

Assuming parabolic relation, the neutral temperature for this thermocouple is
(*a*) 300 °C (*b*) 333 °C
(*c*) 366 °C (*d*) 399 °C

221. Consider the following statements
1. Thermistor is more sensitive that plantinum resistance thermometer
2. The resistance of thermistor is solely a function of its absolute temperature whether the source of heat is external or internal
3. Thermistor has linear temperature-resistance relationship
4. Thermistor has linear negative temperature coefficient

Of the statements, the correct statements are
(*a*) 1 and 2 (*b*) 3 and 4
(*c*) 1, 2, 3, 4 (*d*) 1, 2 and 4

222. Shaft encoder is used to measure
(*a*) angular position (*b*) linear position
(*c*) linear velocity (*d*) linear acceleration

223. A Hall effect transducer has $K_H = -1 \times 10^{-8}$. If B = 1 Wb/m^2, I = 3 A and bismuth slab is 2 mm wide, the Hall voltage is
(*a*) -7.5×10^{-6} V (*b*) -15×10^{-6} V
(*c*) -20×10^{-4} V (*d*) -22.5×10^{-4} V

224. The advantage of LVDT over inductive bridge type transducer is that
(*a*) it gives high output for small changes in core position
(*b*) it can be used over a temperature range of –265°C to 600°C
(*c*) it can measure displacement of moving object
(*d*) the error due to stray magnetic field is less

225. The resistance of a 125 Ω strain gauge changes by 1 ohm for 4000 micro strain. The gauge factor is
(*a*) 1.5 (*b*) 2
(*c*) 2.5 (*d*) 3

226. The output of a Piezo electric crystal has
(*a*) low amplitude and low impedance
(*b*) high amplitude and low impedance
(*c*) low amplitude and high impedance
(*d*) high amplitude and high impedance

227. PCM employing 4 bit code is used to send data having frequency range from 0 to 2 kHz. The minimum required bandwidth of carrier channel is
(*a*) 2 kHz (*b*) 4 kHz
(*c*) 8 kHz (*d*) 16 kHz

228. Consider the following statements regarding electromagnetic flowmeter
1. AC excitation is used to avoid polarisation of fluid
2. The meter calibration is changed as viscosity of fluid changes
3. Stainless steel pipes can be used for measuring the flow of liquid of high conductivity

Of the above statements
(*a*) 1, 2, 3 are correct (*b*) 1 and 2 are correct
(*c*) 2 and 3 are correct (*d*) 1 and 3 are correct

229. The percentage limiting error in an instrument reading 8.3 V and having guaranteed accuracy of 1% of the full scale reading and range of 0–150 V is
(*a*) 1.81% (*b*) 0.181%
(*c*) 18.1% (*d*) 0.0018%

230. A meter having 0–1000 μA range and 1000 Ω resistance is required to be changed to 0–100 mA meter. The value of shunt resistance is about
(*a*) 1 Ω (*b*) 10 Ω
(*c*) 99 Ω (*d*) 100 Ω

231. A moving coil meter has a shunt of 0.02 Ω resistance. If coil resistance is 1000 Ω and a potential difference of 500 mV is required across it to cause full scale deflection, the current through shunt is
(*a*) 2.5 A (*b*) 25 A
(*c*) 0.25 A (*d*) 0.025 A

232. In the Wheatstone bridge in Fig. 6.41, each resistance is increased by 0.05%. The value of v_0 will be

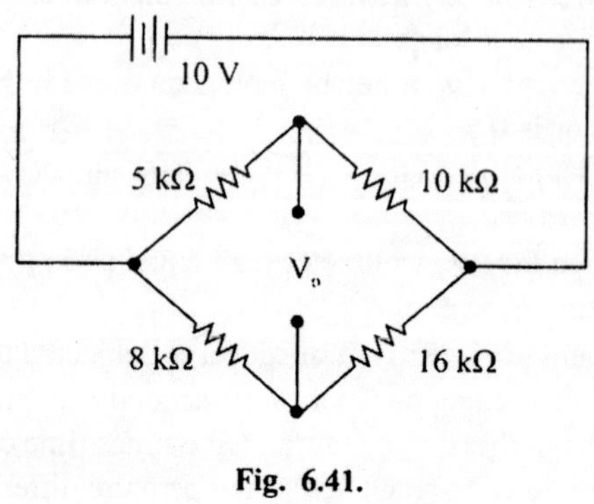

Fig. 6.41.

(*a*) 50 mV (*b*) 5 mV
(*c*) 0.1 mV (*d*) zero

233. In Fig. 6.42 the output V_0 is

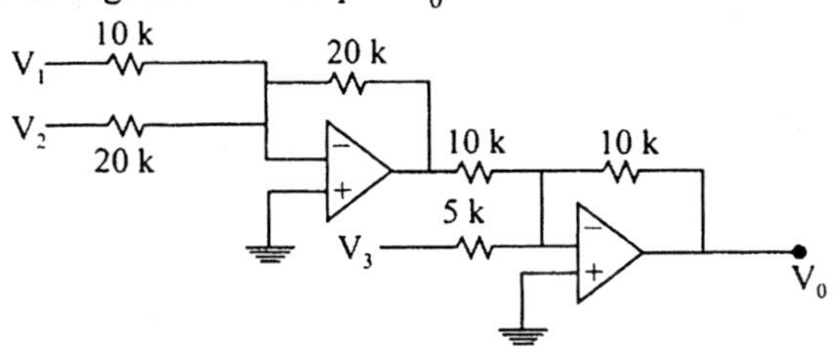

Fig. 6.42.

(*a*) $2V_1 + V_2 - 2V_3$ (*b*) $2(V_1 + V_2) - V_3$
(*c*) $V_1 + V_2 + V_3$ (*d*) $V_1 + 2V_2 - V_3$

234. Consider the following statements regarding electro dynamometer type indicating instrument
It can be used for measuring
1. Direct current 2. Alternating current
3. Direct voltage 4. Alternating voltage
Of these statements
(*a*) 1 and 2 are correct (*b*) 3 and 4 are correct
(*c*) 1 and 3 are correct (*d*) 1, 2, 3, 4 are correct

235. A full wave rectifier type meter is used to measure the wave shape shown in Fig. 6.43. The voltmeter is calibrated to read rms value of sine wave. The reading of voltmeter will be

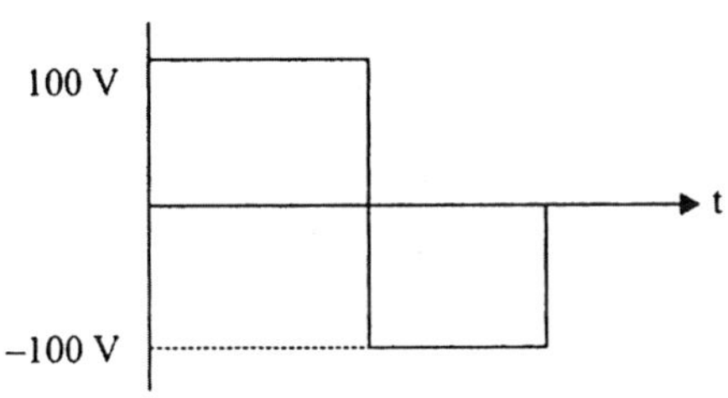

Fig. 6.43.

(*a*) $\frac{50}{\sqrt{2}}$ I (*b*) $\frac{100}{\sqrt{2}}$ V
(*c*) 100 V (*d*) 111 V

236. The current through a resistance R is a shown in Fig. 6.44. The computed value of power is

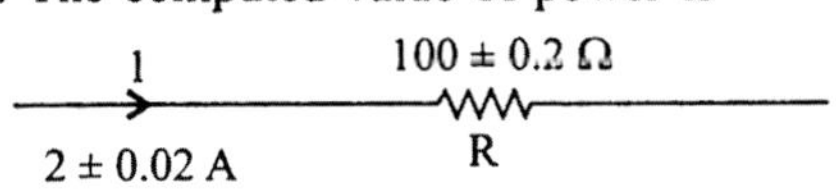

Fig. 6.44.

(*a*) 400 ± 0.42 W (*b*) 400 ± 4.6 W
(*c*) 400 ± 8.85 W (*d*) 400 ± 10.65 W

237. The most serious problems in ac bridge measurement is
(*a*) eddy currents (*b*) leakage current
(*c*) residual magnetism (*d*) stray fields

238. A peak responding ac voltmeter using a half wave rectifier has been calibrated to read 1 V for 1 V rms sine wave input. If input is 1 V dc the reading will be
(*a*) 0.707 V (*b*) 0.707 or 0 V
(*c*) 1.414 V (*d*) 1.414 or 0 V

239. The Lissajous pattern observed on screen of CRO is a straight line inclined at 45° to *x* axis. If X-plate input is 2 sin ω*t*, the Y-plate input is
(*a*) 2 sin ω*t* (*b*) 2 sin (ω*t* + 45°)
(*c*) 2 sin (ω*t* – 45°) (*d*) $2\sqrt{2}$ sine (ω*t* + 45°)

240. An oscilloscope has an input impedance consisting of 1 MΩ and 20 pF is parallel. A high impedance probe connected to input of this oscilloscope has a 10 MΩ series resistance. This 10 MΩ series resistance
(*a*) may not be shunted
(*b*) should be shunted by 2 pF capacitor
(*c*) should be shunted by 20 pF capacitor
(*d*) should be shunted by 200 pF capacitor

241. In a digital measuring device the input electrical signal is in the frequency range of dc to *f* Hz. It must be sampled at a rate of
(*a*) *f* times/sec (*b*) 0.5 *f* times/sec
(*c*) 0.25 *f* times/sec (*d*) 2 *f* times/sec

242. If the secondary burden of a CT 15 VA and secondary current is 5 A, the load impedance of CT is
(*a*) 0.6 Ω (*b*) 5 Ω
(*c*) 6 Ω (*d*) 10 Ω

243. Digital instruments are preferred to analogue instruments due to the following qualities
A. narrow band width
B. better accuracy
C. better resolution
D. cost
The correct arrangement of these qualities in order of their decreasing importance are
(*a*) A, B, C, D (*b*) B, C, D, A
(*c*) B, C, A, D (*d*) C, B, A, D

244. Without a spectrum analyser it is not possible to determine
(*a*) carrier frequency
(*b*) antenna pattern
(*c*) pulse width
(*d*) spurius signal strength and its location

245. Desauty bridge is very widely used deu is
(*a*) simplicity
(*b*) perfect balance for imperfect capacitor
(*c*) perfect balance for air capacitors
(*d*) maximum sensitivity

246. In Fig. 6.45, Z_1 = 200 ∠ 60° Ω, Z_2 = 400 ∠– 90°Ω, Z_3 = 300 ∠ 0°, Then Z_4 for bridge to be balanced is

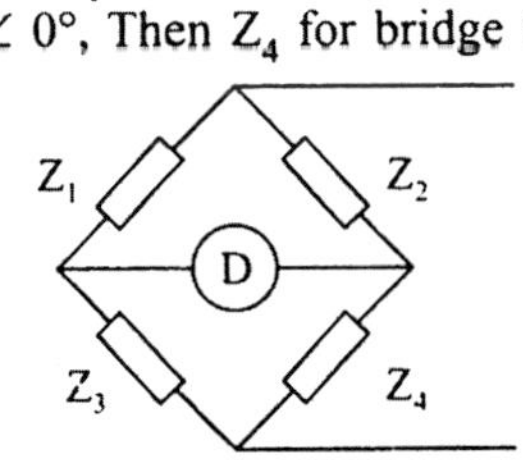

Fig. 6.45.

(*a*) 150 ∠ 30° Ω (*b*) 400 ∠ – 90° Ω
(*c*) 300 ∠ 90° Ω (*d*) 150 ∠ – 150° Ω

247. A single element gauge bridge has two fixed resistances R_1 and R_2 of 120 Ω each and a variable resistance which gives full deflection at 120 Ω for zero strain and 120.6 Ω for strain *e*. If gauge factor is 2, strain *e* is
(*a*) 0.0025 (*b*) 0.0035
(*c*) 0.005 (*d*) 0.0075

248. Capacitive transducer is superior to inductive type for measurement of capacitance because of
(*a*) absence of non-linearity
(*b*) high frequency response
(*c*) small size
(*d*) high accuracy

249. Measurement of non-electrical quantity involves the following sub-systems
1. Amplifier block 2. Display block
3. Instrumentation block 4. Transducer block
The correct sequence of these block is
(*a*) 4, 1, 3, 2 (*b*) 4, 3, 1, 2
(*c*) 1, 3, 2, 4 (*d*) 1, 2, 3, 4

250. Four voltmeters have following specifications
(*a*) 100 V, 2 mA (*b*) 100 V, 100 Ω/V
(*c*) 100 V, 1 mA (*d*) 100 V, 0.1 mA
If accuracy is main consideration which is the best.

251. The phenomenon of creep occurs in
(*a*) energy meters (*b*) voltmeters
(*c*) ammeters (*d*) wattmeters

252. Potentiometer method of dc voltage measurement is more accurate than direct measurement using a voltmeter because
(*a*) it loads the circuit to maximum extent
(*b*) it loads the circuit moderately
(*c*) it does not load the circuit at all
(*d*) it uses centre zero galvanometer instead of voltmeters

253. Which of the following instruments is used almost exclusively to measure radio frequency currents?
(*a*) moving coil (*b*) rectifier
(*c*) moving iron (*d*) thermo couple

254. Which one of the following is true?
(*a*) alternating voltage and current contain odd and even harmonics in general
(*b*) the inductance in a circuit reduces the effect of harmonics and makes current wave more sinusoidal than voltage wave
(*c*) alternating voltage and current wave forms, in general, do not contain odd harmonics
(*d*) inductance of a circuit does not alter wave shape at all

255. Insertion of a dielectric material between the plates of an air capacitor
(*a*) increases the capacitance
(*b*) decreases capacitance
(*c*) has no effect on capacitance
(*d*) increases breakdown voltage

256. Match List I with List II and select the correct answer using the given codes

List I (Instrument)	**List II (Measured quantity)**
A. Hot wire	1. only voltage
B. PMMC	2. Phase
C. Electrostatic	3. True RMS value
D. Lock in Amplifier	4. dc value

Codes:

	A	B	C	D
(*a*)	3	2	1	4
(*b*)	1	4	2	3
(*c*)	3	4	1	2
(*d*)	2	3	4	1

257. Which one of the following effects is used in measuring power?
(*a*) Seeback (*b*) Ferranti
(*c*) Induction (*d*) Hall

258. An average responding voltmeter using full wave rectifier has been calibrated to read 1 V for 1 V rms sinusoidal input. For 1 V peak-to-peak square wave input, the meter reading will be
(*a*) $\frac{\pi}{4\sqrt{2}}$ V (*b*) zero V
(*c*) $\frac{\pi}{2\sqrt{2}}$ V (*d*) $\pi\sqrt{2}$ V

259. The sensitivity of electromagnetic digital flow meter can be increased by increasing
(*a*) circumferential width of rotor teeth
(*b*) radial depth of rotor teeth
(*c*) thickness of rotor teeth
(*d*) number of teeth

260. A digital frequency counter can be converted to DVM by addition of a suitable
(*a*) voltage controlled oscillator
(*b*) D/A converter
(*c*) power amplifier
(*d*) operational amplifier

261. Match List I and List II and give the answer using the codes

List I (Instrument)	**List II (Parameters)**
A. Moving coil galvanometer	1. High inertia, soft spring control
B. Ballistic galvanometer	2. No control, high damping
C. Vibration galvanometer	3. Soft spring control, critical critical damping
D. Moving coil flux meter	4. Low inertia, tuned spring control

Codes:

	A	**B**	**C**	**D**
(*a*)	4	2	3	1
(*b*)	1	3	2	4
(*c*)	2	1	4	3
(*d*)	3	1	4	2

262. Consider the following statements regarding phase sensitive detector
1. The detector reads zero when phases of input signal and reference signal are different
2. The detector acts as a linear rectifier

3. The detector can be used as vector voltmeter

of the above statements which are correct

(*a*) 1, 2, 3 (*b*) 1 and 2
(*c*) 2 and 3 (*d*) 1 and 3

263. Consider the following statements regarding resistance strain gauges

1. The gauge factor is in the range of 2 to 5
2. The resistance changes due to change in specific resistance with strain
3. The resistance increases with increase in temperature

Of the above statements, which are correct?

(*a*) 1, 2, 3 (*b*) 1 and 2
(*c*) 2 and 3 (*d*) 1 and 3

264. Thermistors are made of

(*a*) pure metals (*b*) pure insulators
(*c*) sintered mixture of metallic oxides
(*d*) pure semiconductors

265. Acoustic strain gauges

(*a*) have good linearity of measurement
(*b*) do not have long term stability
(*c*) have a large range of measurement
(*d*) have an accuracy range of 0.1–1% depending on environment

266. Match List I and List II and answer using the given codes

List I (Transducer)	List II (Quantity measured)
A. LVDT	1. Pressure
B. Bourden gauge	2. Temperature
C. Strain gauge	3. Displacement
D. Thermistor	4. Stress

Codes:

	A	B	C	D
(*a*)	4	3	2	1
(*b*)	3	2	1	4
(*c*)	4	1	3	2
(*d*)	3	1	4	2

267. A piezo electric force transducer has a charge sensitivity of 20 pC/N. It is connected to a charge amplifier and overall gain of transducer and amplifier is 50 mV/N. The gain of amplifier is

(*a*) 1 mV/pC (*b*) 1.5 mV/pC
(*c*) 2.5 mV/pC (*d*) 4 mV/pC

268. The dead zone in a pyrometer is 0.125% of span. The calibration is 400°C to 1000°C. What temperature change might occur before it is detected

(*a*) 0.125°C (*b*) 0.75°C
(*c*) 0.4°C (*d*) 1°C

269. To measure moisture in wood, the most suitable method is

(*a*) electric conduction
(*b*) electrical-capacitive
(*c*) absorption of radiation
(*d*) equilibrium-moisture vs humidity

270. In dc tacho generation used for measurement of speed of shaft, frequent calibration is necessary because

(*a*) contacts wear off
(*b*) strength of permanent magnet decreases with age
(*c*) armature current produces beating effect
(*d*) there is back emf

271. A permanent magnet moving coil measures

(*a*) true rms value (*b*) dc value
(*c*) peak value (*d*) rms plus dc value

272. Consider the following statements about a moving iron instrument

1. It is an unpolarised instrument
2. It has permanent magnet
3. It has a fixed coil
4. It has two iron vanes

Of these statements

(*a*) 1, 2, 3 are correct (*b*) 1, 2, 4 are correct
(*c*) 1, 3 and 4 correct (*d*) 2, 3, 4 are correct

273. The coil of moving coil meter has 100 turns, is 40 mm long and 30 mm wide. The control torque in is 240×10^{-6} N-m on full scale. If B = 1 Wb/m^2, range of meter is

(*a*) 1 mA (*b*) 2 mA
(*c*) 3 mA (*d*) 4 mA

274. Four terminal approach is used in measuring low resistance because it

(*a*) eliminates thermo electric energy
(*b*) eliminates stray capacitance
(*c*) eliminates stray inductance
(*d*) eliminates the effect of leads and contacts

275. A 1 mA meter is to be converted to 1 A ammeter. The meter resistance is 100 Ω. The value of shunt resistance is

(*a*) 0.001 Ω (*b*) 0.1001 Ω
(*c*) 100 Ω (*d*) 1000 Ω

276. A 0–200 V voltmeter has an accuracy of 0.75% of full scale reading. If voltage measured is 100 V, The error is

(*a*) 3% (*b*) 2%
(*c*) 1.5% (*d*) 0.75%

277. Consider the following statements about mutual inductances:

1. The value should remain constant under all conditions of use
2. The mutual inductance should not vary with temperature
3. The inductance should be independent of excitation
4. They have two coils wound on a bobbin of hard magnetic material, coils being separated by a flange

Out of the above, correct statements are

(*a*) 2, 3, 4 (*b*) 1, 2, 3
(*c*) 1, 3, 4 (*d*) 1, 2, 4

278. Two resistance R_1 and R_2 are connected in series $R_1 = 528 \pm 5\ \Omega$ and $R_2 = 325 \pm 3\Omega$. The total resistance will be

(a) 853 ± 2 Ω (b) 853 ± 5 Ω
(c) 853 ± 3 Ω (d) 853 ± 8 Ω

279. Consider the following statements.
In a differential amplifier using JFET, the differential output offset is due to
1. mismatch between FET parameters
2. difference in values of resistances used in circuit
3. variation is supply voltage

Of the statements, the correct statements are
(a) 1 and 2 (b) 1, 2, 3
(c) 2, 3 (d) 1, 3

280. An average responding rectifier type electronic voltmeter has its scale calibrated is terms of rms value of sine wave. If 100 V peak value square wave is fed to this voltmeter, the reading will be
(a) 111 V (b) 100 V
(c) 90.9 V (d) 70.7 V

281. Wagner earthing device eliminates
(a) all stray capacitance in the circuit
(b) stray capacitances between components
(c) stray capacitance between detector terminal and ground
(d) mutual coupling between complying

282. Kelvin's double bridge is used to measure low resistances because
(a) it has high sensitivity
(b) there is no thermoelectric emf
(c) resistance variation due to temperature is accounted for
(d) effect of contact and lead resistances is eliminated

283. In Fig. 6.46 the values of R_x and C_x are

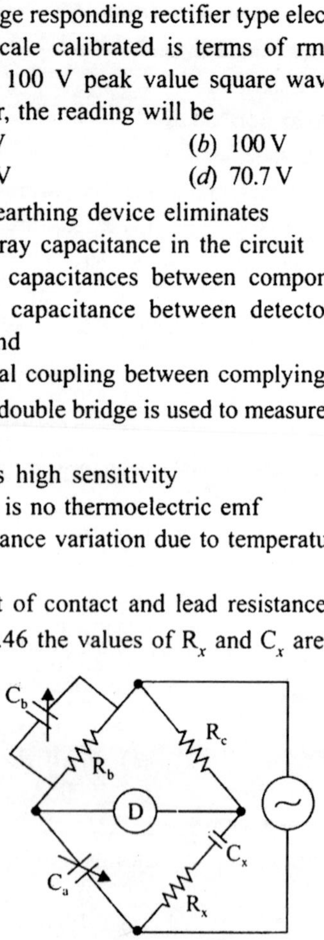

Fig. 6.46.

(a) $R_x = \frac{C_b}{C_a} R_c$ and $C_x = \frac{R_b}{R_a} C_a$

(b) $R_x = \frac{C_a}{C_b} R_c$ and $C_x = \frac{R_c}{R_b} C_a$

(c) $R_x = \frac{R_b}{R_c} C_a$ and $C_x = \frac{C_h}{C_a} R_c$

(d) $R_x = \frac{R_c}{R_b} C_a$ and $C_x = \frac{C_a}{R_b} R_c$

284. A wire strain gauge has a gauge factor of 2, resistance of 125 Ω and length 1 m. If length of wire changes by 0.005 m, change in resistance will be
(a) 0.25 Ω (b) 0.5 Ω
(c) 1.25 Ω (d) 2.5 Ω

285. Which of the following is not transducer in true sense?
(a) thermocouple (b) piezo electric pick up
(c) photo electric cell (d) LCD

286. Match List I with List II and answer using given codes

List I Transducer	List II Operation
A. Bourden tube	1. Fluid flow to resistance change
B. Hot wire ammeter	2. Velocity to pressure
C. Hydrometer	3. Pressure to displacement
D. Pitot tube	4. Specific gravity to displacement

Codes:

	A	B	C	D
(a)	1	3	2	4
(b)	3	1	2	4
(c)	3	1	4	2
(d)	4	2	3	1

287. Pirani gauge is used to measure
(a) very low pressure
(b) high pressure
(c) pressure in the range of 1 atmosphere
(d) fluid flow

288. In a thermocouple two metal junctions between metals M_1 and M_2 are kept at temperature T_1 and T_2. The thermocouple emf is produced because
(a) M_1, M_2 are similar and T_1 and T_2 are equal
(b) M_1, M_2 are similar and T_1 and T_2 are unequal
(c) M_1, M_2 are dissimilar but T_1 , T_2 are equal
(d) M_1, M_2 are dissimilar and T_1, T_2 are unequal

289. Minute cracks in metals and minute faults in welded joints can be detected using
(a) IR rays (b) UV rays
(c) α rays (d) X-ray/γ-rays

290. An acclerometer has a seismic mass of 0.05 kg and spring constant of 3×10^3 N/m. If maximum displacement of mass is ± 1 mm, the maximum accleration which can be measured by
(a) 30 m/s^2 (b) 60 m/s^2
(c) 90 m/s^2 (d) 120 m/s^2

291. The flow rate of electrically conducting liquid without any suspended particle cannot be measured by
(a) turbine flow meter
(b) electromagnetic flow meter
(c) ultra sonic flow meter
(d) thermistor based heat flow meter

292. The channel required for FM telemetry is
(a) same as required for AM telemetry
(b) smaller than that required by AM telemetry
(c) 100 times that required for AM telemetry
(d) 10 times that required for AM telemetry

293. Which of the following draws minimum power from source
(a) PMMC (b) electronic voltmeter
(c) electrostatic voltmeter (d) moving iron voltmeter

294. A moving coil meter has 10 resistance and requires 40 mA for full deflection. The shunt resistance required to convert it into 0–2 A ammeter is
(*a*) 0.1021 Ω (*b*) 0.2041 Ω
(*c*) 0.256 Ω (*d*) 0.4210 Ω

295. The use of thermocouple meters for ac measurement leads to a scale which is
(*a*) linear (*b*) square law
(*c*) logarithmic (*d*) exponential

296. A dynamometer wattmeter is connected in ac circuit. The measured power is
(*a*) rms power (*b*) average power
(*c*) peak power (*d*) instantaneous power

297. The dimensions of permittivity are
(*a*) $L^{-3} M^{-3} T^2 Q^2$ (*b*) $L^{-1} M^{-1} T^2 Q^2$
(*c*) $L^2 MT^{-1} Q^{-1}$ (*d*) $L^{-2} M^{-1} T^2 Q^2$

298. The D'Arsonval movement can be converted into audio frequency ac ammeter by adding a
(*a*) thermocouple (*b*) transducer
(*c*) chopper (*d*) rectifier

299. The standard deviation σ in terms of *n* observations having deviations from mean d_m can be written as
(*a*) $\sigma = \frac{\Sigma d_m}{n}$ (*b*) $\sigma = \sqrt{\frac{\Sigma d_m^2}{n}}$
(*c*) $\sqrt{\frac{\Sigma d_m^2}{n-1}}$ (*d*) $0.6745\sqrt{\frac{\Sigma d_m^2}{n-1}}$

300. The time base of a high frequency CRO whose screen diameter is 10 cm is set at 10 μ sec/cm. The lowest frequency which can be fully displayed on the screen is
(*a*) more than 100 kHz (*b*) equal to 100 kHz
(*c*) equal to 10 kHz (*d*) less than 10 kHz

301. In Fig. 6.47, the position of voltmeter and ammeter are exchanged. It may result in damage to

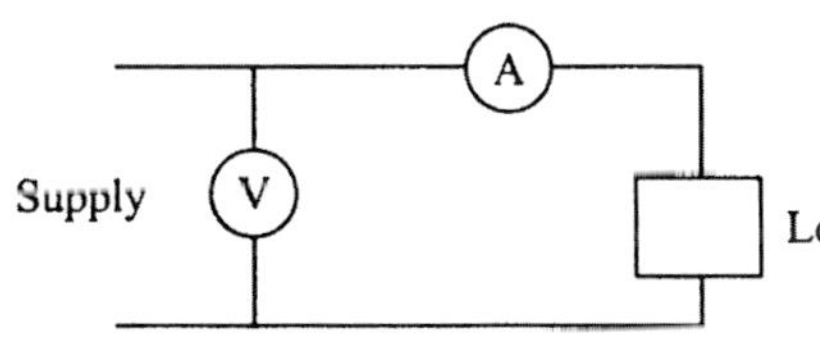

Fig. 6.47.

(*a*) both the instruments (*b*) ammeter
(*c*) voltmeter (*d*) neither of two

302. It is necessary to measure the amplitudes of two signals as also the phase difference between them. This can be done by using a
(*a*) phase meter (*b*) wave form meter
(*c*) vector voltmeter (*d*) digital voltmeter

303. An amplifier has to amplify pulses of 1 μ-s duration. The minimum bandwidth of amplifier must be
(*a*) 10 kHz (*b*) 10 MHz
(*c*) 1 kHz (*d*) 1 MHz

304. A diode peak reading VTVM and thermocouple meter are connected across output of an amplitude modulated circuit. In the absence of modulation both meters read 10 V. When sinusoidal AM is applied, VTVM reads 15 V. The reading of thermocouple meter will be
(*a*) 10 V (*b*) 12.5 V
(*c*) 14.1 V (*d*) 15 V

305. The dynamic characteristics of capacitive transducer are similar to those of
(*a*) L P filter (*b*) H P filter
(*c*) notch filter (*d*) band stop filter

306. Which of the following is used as detector in audio frequency ac bridges
(*a*) ac voltmeter (*b*) CRO
(*c*) head phones (*d*) vibration galvanometer

307. Which is the displacement transducer with excellent sensitivity, linearity and resolution
(*a*) incremental encoder (*b*) absolute encoder
(*c*) LVDT (*d*) strain gauge

308. A 24 mm long conductor has a resistance of 128 Ω. The change in resistance is 13.3 Ω and change in length is 1.6 mm under tension. The gauge factor is
(*a*) 1.2 (*b*) 1.6
(*c*) 2.1 (*d*) 2.6

309. The temperature coefficient of resistance for a thermistor is
(*a*) low and negative (*b*) low and positive
(*c*) high and negative (*d*) high and positive

310. A variable reluctance type tachometer has 150 teeth on rotor. The counter records 13500 pulse/sec. The rotational speed is
(*a*) 4800 RPM (*b*) 5400 RPM
(*c*) 6000 RPM (*d*) 7200 RPM

311. A linear displacement digital transducer generally uses
(*a*) straight binary code (*b*) BCD
(*c*) gray code (*d*) hexadecimal code

312. For measuring temperatures below 800°C with high accuracy, the instrument is
(*a*) optical pyrometer
(*b*) thermistor based
(*c*) Ga As junction diode thermometer
(*d*) platinum resistance thermometer

313. If low pressure of 10^{-6} mm of Hg is required to be measured, we should use
(*a*) compound pressure gauge
(*b*) thermocouple vacuum gauge
(*c*) Pirani gauge
(*d*) ionization type vacuum gauge

314. The sensitivity of a voltmeter using 0–5 m A meter movement is
(*a*) 50 Ω/V (*b*) 100 Ω/V
(*c*) 200 Ω/V (*d*) 500 Ω/V

315. Two sinusoidal signals of equal amplitude and frequency are applied to X and Y plate of CRO respectively. The observed Lissajous pattern is a straight line. The phase shift between signals is

(*a*) zero (*b*) 90°
(*c*) either zero or 180° (*d*) either 90° or 270°

316. A dynamometer wattmeter is connected in ac circuit. The reading will be
(*a*) V–A product (*b*) average power
(*c*) peak power (*d*) instantaneous power

317. DC voltage of the order of a few mV can be accurately of measured by
(*a*) PMMC voltmeters
(*b*) moving iron voltmeter
(*c*) null balancing potentiometer
(*d*) electrostatic voltmeter

318. The accuracy of an instrument is determined by
(*a*) closeness of indicated value to the actual value
(*b*) repeatability of measured value
(*c*) speed with which the reading approaches final value
(*d*) least value which can be measured

319. Two resistators $R_1 = 36\ \Omega \pm 5\%$ and $R_2 = 75\ \Omega \pm 5\%$ are connected in series. The total resistance is
(*a*) $111 \pm 0\ \Omega$ (*b*) $111 \pm 2.778\ \Omega$
(*c*) $111 \pm 5.55\ \Omega$ (*d*) $111 \pm 7.23\ \Omega$

320. In Fig. 6.48, C_0 is distributed capacitance and C is tuning capactitor. If $C = C_1$ for fundamental frequency and $C = C_2$ for second harmonic, then C_0 =

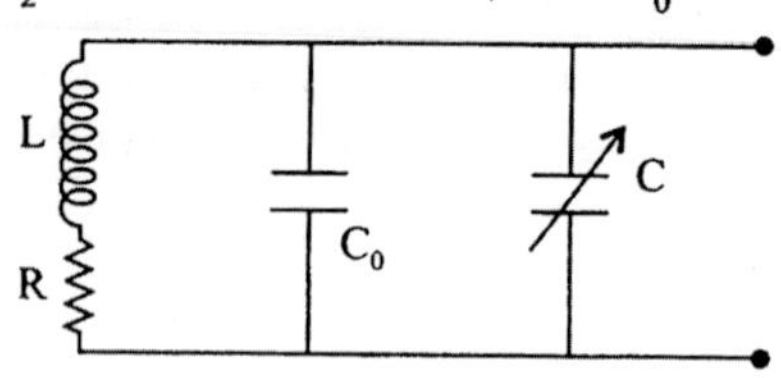

Fig. 6.48.

(*a*) $\frac{C_1 - 2C_2}{3}$ (*b*) $\frac{C_1 - 4C_2}{3}$

(*c*) $\frac{C_1 - C_2}{3}$ (*d*) $C_1 - 2C_2$

321. In Fig. 6.49, $Z_a = 100 \angle 50°$, $Z_b = 300 \angle -90°$ and $Z_c = 200 \angle 0°$. For balanced condition Z_α =

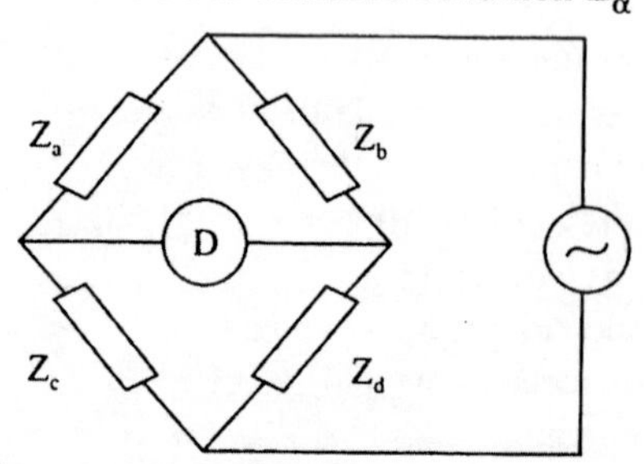

Fig. 6.49.

(*a*) $600 \angle -40°$ (*b*) $600 \angle 140°$
(*c*) $600 \angle -140°$ (*d*) $150 \angle 140°$

322. An LVDT is used to measure displacement. The LVDT feeds a voltmeter of 0–5 V rang through a 250 gain amplifier. For a displacement 0.5 mm the output of LVDT is 2 mV. The sensitivity of instrument is
(*a*) 0.1 V/mm (*b*) 0.5 V/mm
(*c*) 1 V/mm (*d*) 5 V/mm

323. It is required to measure temperature in the range of 1300–1520°C. The most suitable thermocouple is
(*a*) chromel constantan (*b*) iron constantan
(*c*) chromel alumel (*d*) platinum - rhodium

324. It is required that a strain gauge has gauge factor of 100. The proper strain gauge is
(*a*) constantan strain gauge
(*b*) Nichrome strain gauge
(*c*) semiconductor strain gauge
(*d*) alloy strain gauge

325. It is required to measure angular position. Out of the transducers
1. Circular potentiometer 2. LVDT
3. E pick up 4. Synchro
The proper devices are
(*a*) 1, 2, 3, 4 (*b*) 2 and 3
(*c*) 1, 2, 4 (*d*) 1 and 4

326. Bridge circuits using RTD in temperature measurement usually employ 3 lead system so as to obtain
(*a*) higher sensitivity
(*b*) better impedance matching
(*c*) compensation for signal wire resistance
(*d*) reduction in power consumption

327. It is required to measure pressure using LVDT. The additional device needed is
(*a*) strain gauge (*b*) pitot tube
(*c*) bourden tube (*d*) rotameter

328. A resistance wire strain gauge has gauge factor 2. If stress is 200 giga newton/m^2, the percentage change in gauge resistance due to stress is
(*a*) 0.1 (*b*) 0.2
(*c*) 0.4 (*d*) 0.5

329. A variable reluctance tachometer has 180 teeth on rotor. The speed of shaft on which it is mounted in 1200 rpm. The frequency of output pulses is
(*a*) 1800 per second (*b*) 3600 per second
(*c*) 4800 per second (*d*) 5600 per second

330. The coding system generally used in digital telemetry is
(*a*) PPM (pulse and position modulation)
(*b*) PAM (pulse and amplitude modulation)
(*c*) PCM (Pulse code modulation)
(*d*) PDM (pulse duration modulation)

331. Consider the following operation is respect of wheatstone bridge (K_b, denotes battery key, K_g denotes galvanometer key)
1. open K_b 2. close K_g
3. close K_b 2. open K_g
The correct sequence of operations is
(*a*) 1, 2, 3, 4 (*b*) 3, 1, 2, 4
(*c*) 4, 3, 2, 1 (*d*) 3, 2, 4, 1

332. Loading effect is primarily caused by instruments having
(*a*) high resistance (*b*) high sensitivity
(*c*) low sensivity (*d*) high range

333. Match List I and List II and select the answer using the given codes

List I	**List II**
A. Former	1. Produces deflecting torque
B. Coil	2. Provides base for coil
C. Core	3. Makes the magnetic field radial
D. Spring	4. Provides controlling torque

Codes:

	A	B	C	D
(*a*)	1	2	3	4
(*b*)	2	1	3	4
(*c*)	1	2	4	3
(*d*)	2	1	4	3

334. Measuring of unknown voltage with a dc potentiometer loses its advantage of open circuit measurement when
(*a*) primary circuit battery is changed
(*b*) standardisation is done to compensate for drift
(*c*) voltage is larger than range of potentiometer
(*d*) range reduction by factor of 10 is employed

335. Figure 6.50 shows the equivalent circuit of resistor. The resistor is non-inductive when

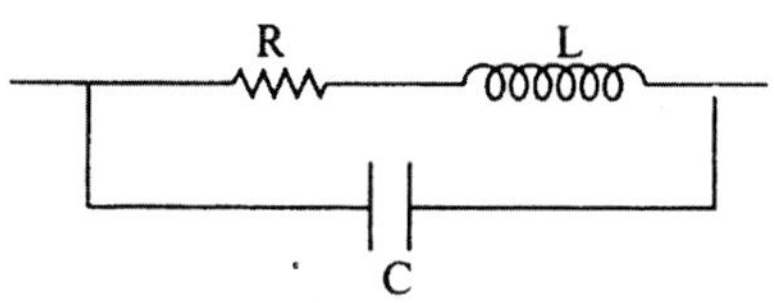

Fig. 6.50.

(*a*) R = L/C (*b*) LR = $\sqrt{L/C}$
(*c*) L = CR^2 (*d*) C = LR^2

336. The difference between measured value and true value is called
(*a*) gross error (*b*) relative error
(*c*) probable error (*d*) absolute error

337. Harmonic distortion analyser
(*a*) measures amplitude of each harmonic
(*b*) measures rms value of fundamental frequency component
(*c*) measures rms value of all harmonics except fundamental frequency
(*d*) displays rms value of each harmonic on screen of CRO

338. If the secondary winding of CT is opened when primary is carrying current
(*a*) transformer will burn immediately
(*b*) there will be weak flux density in core
(*c*) there will be very high induced voltage in secondary
(*d*) there will be high current in secondary

339. A dual trace CRO has
(*a*) one electron gun
(*b*) two electron guns
(*c*) one electron gun with one two pole switch
(*d*) two electron guns with one two pole switch

340. A $3\frac{1}{2}$ digit voltmeter having a resolution of 100 mV can be used to measure a maximum of
(*a*) 100 V (*b*) 200 V
(*c*) 1000 V (*d*) 5000 V

341. A coil resonates at 1 MHz with a resonating capacitance of 72 pF. At 500 pF, resonance is obtained with a resonating capacitor of 360 pF. The self-capacitance of coil is
(*a*) 12 pF (*b*) 24 pF
(*c*) 30 pF (*d*) 72 pF

342. Match List I with List II give correct answer using the given code

List I	**List II**
A. Anderson Bridge	1. Low resistance
B. Kelvin Bridge	2. Medium resistance
C. Schering Bridge	3. Inductance
D. Wheatstone Bridge	4. Capacitance

Codes:

	A	B	C	D
(*a*)	4	2	3	1
(*b*)	3	2	4	1
(*c*)	3	1	4	2
(*d*)	4	1	3	2

343. In the balanced bridge shown in Fig. 6.51 X should be

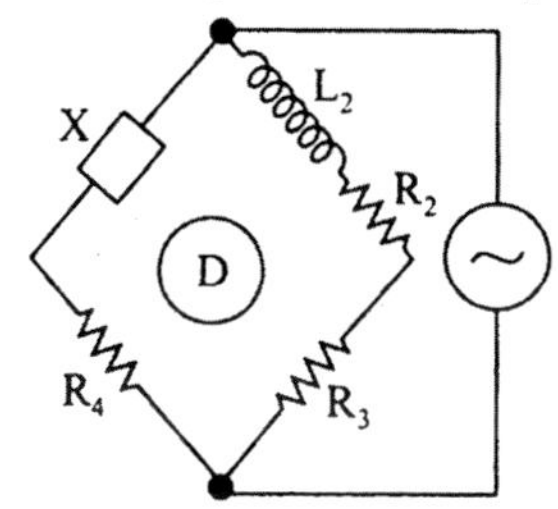

Fig. 6.51.

(*a*) a self-inductance having resistance
(*b*) a capacitance
(*c*) a non-inductive resistance
(*d*) an inductance and capacitance in parallel

344. A 300 V voltmeter has an accuracy of ± 2% of full scale deflection. When the reading is 222 V, the actual voltage
(*a*) lies between 217.56 and 226.44 V
(*b*) lies between 217.4 and 226.6 V
(*c*) lies between 216 and 228 V
(*d*) is exact 222 V

345. The ac bridge in Fig. 6.52 is balance. $Z_1 = 100 \angle 30°$, $Z_2 = 150 \angle 0°$, $Z_3 = 250 \angle -40°$.
Then Z_4 =

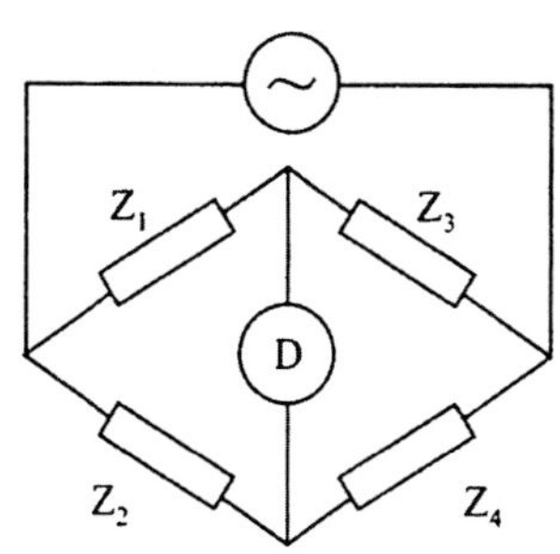

Fig. 6.52.

(*a*) 375 ∠ 70° Ω (*b*) 375 ∠ – 70° Ω
(*c*) 150 ∠ 0 Ω (*d*) 150 ∠ 20° Ω

346. The device having highest photo sensitivity is
(*a*) photo conductive cell (*b*) photo voltaic cell
(*c*) photo diode (*d*) photo transistor

347. Load cell uses
(*a*) piezo electric crystal (*b*) capacitor
(*c*) mutual inductance (*d*) strain gauge

348. A data signal having frequency components from dc to 500 Hz is to be sent through pulse code modulation using an 8 digit code. The minimum carrier channel bandwidth is
(*a*) 1 kHz (*b*) 2 kHz
(*c*) 4 kHz (*d*) 8 kHz

349. Match List I and List II and select correct answer using the given codes

List I	**List II**
A. PMMC	1. Square law type scale
B. Moving iron	2. Very good response
C. Thermo-couple	3. Linear Scale
D. Electrostatic	4. Voltmeter

Codes:

	A	B	C	D
(*a*)	4	1	2	3
(*b*)	3	2	1	4
(*c*)	4	2	1	3
(*d*)	3	1	2	4

350. An ammeter of 0-25 A range has a guaranteed accuracy of 1% of full scale reading. The current measured is 5 A. The limiting error is
(*a*) 2% (*b*) 2.5%
(*c*) 4% (*d*) 5%

351. The Y plates of a CRO are excited by a voltage 2 sin 100*t* and the X plates are not connected. The display would be
(*a*) a vertical line (*b*) a horizontal line
(*c*) sine wave (*d*) slanting line

352. In Fig. 6.53 the bridge is balanced when Z_4 is

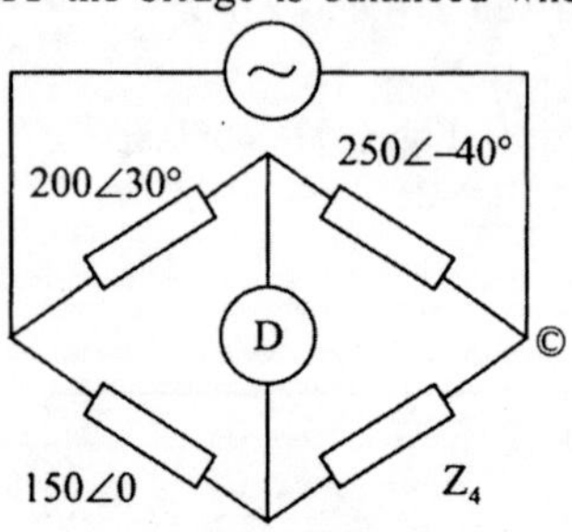

Fig. 6.53.

(*a*) 120 ∠ 70° (*b*) 187.5 ∠– 10°
(*c*) 187.5 ∠ – 70° (*d*) 333.3 ∠ – 70°

353. Radiation pyrometer is used to measure temperature in the range of
(*a*) – 200 to 500°C (*b*) 0 to 500°C
(*c*) 500 to 1200°C (*d*) 1200 to 2500°C

354. Magnetic flux can be measured by
(*a*) capacitive pick up (*b*) inductive pick up
(*c*) resistive pick up (*d*) Hall effect pick up

355. Two voltmeters are each 0–300 V range. Their internal resistances are 10 kΩ and 20 kΩ. They are connected is series and a voltage of 400 V is applies. Their readings respectively will be
(*a*) 266.6 and 133.3 V (*b*) 133.3 and 266.6 V
(*c*) 200 V each (*d*) 100 and 300 V

356. The increase in intensity of trace in a CRO causes a change in
(*a*) beam current (*b*) frequency of time base
(*c*) gain of Y amplifier (*d*) accelerating voltage

357. Which of the following does not use the principle of change in inductance
(*a*) LVDT
(*b*) inductance displacement transducer
(*c*) strain gauge
(*d*) reluctance pick up

358. Which of the following is not correct?
(*a*) voltmeter should have a very high resistance
(*b*) an ammeter should have a very low resistance
(*c*) a shunt should have a very low resistance
(*d*) an electronic voltmeter draws appreciable current from source

359. Both X and Y plates of CRO are excited by ac signals. The display is an ellipse which makes one rotation in 5 seconds. The means that
(*a*) the frequency of one input is 5 times that of other
(*b*) the difference between the frequencies of the two signals in one-fifth of hertz
(*c*) the difference between the frequencies of the two signals is 5 Hz
(*d*) either (*a*) or (*c*)

360. A sinusoidal signal is measured by CRO. The scale is set at 4 mV/cm. If vertical distance between positive and negative peaks is 6 cm, the peak value of ac signal is
(*a*) 6 mV (*b*) 12 mV
(*c*) 24 mV (*d*) 3 mV

361. Which of the following effects is used in measurement of magnetic flux
(*a*) Hall effect (*b*) See back effect
(*c*) piezo electric effect (*d*) photo electric effect

362. Two ammeters A and B both 0–10 A have internal resistance of 1 Ω and 0.5 Ω respectively. They are connected in parallel. If total current is 15 A, then
(*a*) $I_A = I_B = 7.5$ A (*b*) $I_A = 5$ A, $I_B = 10$ A
(*c*) $I_A = 10$ A, $I_B = 5$ A (*d*) $I_A = 9$ A, $I_B = 6$ A

363. Which of the following is not correct in respect of electrostatic instrument
(*a*) It is essentially a voltmeter
(*b*) it is suitable for only high voltages
(*c*) it has a uniform scale
(*d*) it can be used for both ac and dc

364. The current through the coil of an electro magnet is doubled. The magnetic field around the coil
(*a*) is halved (*b*) is doubled
(*c*) becomes four times (*d*) remains the same

365. A digital voltmeter has a $4\frac{1}{2}$ digit display. The 1 V range can measure voltage up to
(*a*) 0.9999 V (*b*) 1.9999 V
(*c*) 0.199 V (*d*) 9.999 V

366. Which of these instruments has the highest power consumption?
(*a*) dynamometer (*b*) electrostatic
(*c*) moving coil (*d*) hot wire

367. The deflection sensitivity of a CRO is generally expressed in
(*a*) meter/V (*b*) cm/V
(*c*) V/meter (*d*) V/cm

368. The time base generator in CRO gives an output which is
(*a*) low frequency sinusoidal
(*b*) high frequency sinusoidal
(*c*) low frequency rectangular
(*d*) high frequency sawtooth

369. A Q meter uses the principle of
(*a*) variation of self-inductance
(*b*) variation of mutual inductance
(*c*) series resonance
(*d*) none of above

370. RTD is generally suitable for temperature up to
(*a*) about 800°C (*b*) about 1500°C
(*c*) about 200°C (*d*) about 50°C

371. A single phase energy meter has the rating 1200 resolution/kWh. If a 500 W electric gadget is used for 4 hours, the energy meter will make
(*a*) 1200 revolutions (*b*) 1800 revolutions
(*c*) 2100 revolutions (*d*) 2400 revolutions

372. Absolute instruments are used in
(*a*) laboratories (*b*) industries
(*c*) power houses (*d*) standard laboratories

373. In a Ballistic galvanometer
(*a*) inertia is small and damping is also small
(*b*) inertia is large and damping is critical
(*c*) inertia is large and damping is small
(*d*) both inertia and damping are large

374. An electrodynamometer, instrument has the advantage of
(*a*) freedom from hysteresis and eddy currents because air cored coils are used
(*b*) freedom from hysteresis and eddy currents because cores of coils have special ferromagnetic material
(*c*) high torque/weight ratio
(*d*) low cost

375. Which of the following voltmeters would you use for measuring voltage acress 20 kΩ resistance?
(*a*) voltmeter having a resistance of 5 kΩ
(*b*) voltmeter having a sensitivity of 1 kΩ/V
(*c*) voltmeter having sensitivity of 10 kΩ/V
(*d*) any of the above

376. A CT is used to feed current coils of wattmeter, energy meter, power factor meter and an ammeter. Then
(*a*) all the coils are in series
(*b*) all coils are in parallel
(*c*) the coils are series-parallel
(*d*) either (*a*) or (*b*)

377. In which of the following instruments is the deflecting torque proportion to $\frac{dL}{dQ}$ (where L is inductance of coil and Q is deflection)
(*a*) PMMC (*b*) electrostatic
(*c*) moving iron (*d*) both (*a*) and (*b*)

378. Gauge factor of a strain gauge is
(*a*) unit change in resistance per unit change in length
(*b*) change is resistance per unit change in length
(*c*) unit change in length per unit change in resistance
(*d*) none of the above

379. To measure dielectric loss you would use
(*a*) Anderson bridge (*b*) Kelvin bridge
(*c*) Schering bridge (*d*) Maxwell's bridge

380. The 10 MHz in the specification of CRO means that
(*a*) the sweep frequency is 10 MHz
(*b*) the frequency of input signal should not be more than 10 MHz
(*c*) the V-amplifier has been designed for 10 MHz
(*d*) none of the above

381. To avoid creep in an energy meter
(*a*) two shading bands are put on shunt magnet
(*b*) two shading bands are put on series magnet
(*c*) two holes are cut in the disc on opposite sides of spindle
(*d*) the potential coil is made of very thin wire

382. A 0–200 V voltmeter has a sensitivity of 1 kΩ per volt. The resistance of voltmeter is
(*a*) 1 kΩ (*b*) 200 kΩ
(*c*) 100 kΩ (*d*) 50 kΩ

383. Two ammeters having resistances of 0.5 Ω and 0.25 Ω and having full reading of 25 A each are connected in parallel. To ensure that neither of them goes beyond scale, the total current should not be more than
(*a*) 50 A (*b*) 37.5 A
(*c*) 25 A (*d*) 40 A

384. Two voltmeters 0–250 V each have resistances of 5 kΩ and 10 kΩ respectively and are connected in series. To ensure that neither of them goes beyond scale the total voltage which can be applied should be less than
(*a*) 500 V (*b*) 400 V
(*c*) 375 V (*d*) 250 V

385. Burden of an instrument transformer is expressed in
(*a*) A (*b*) V – A
(*c*) Ω (*d*) watt

386. Which these instrument does not have a control spring?
(*a*) moving iron ammeter
(*b*) PMMC ammeter
(*c*) electrostatic voltmeters
(*d*) power factor meter

387. Which of these instruments is dynamometers type?
(*a*) wattmeter (*b*) power factor meter
(*c*) frequency meter (*d*) both (*a*) and (*b*)

388. In a current transformers
(*a*) the number of turns in primary and secondary windings are nearly equal
(*b*) the number of primary turns is very small as compared to number of secondary turns
(*c*) the number of secondary turns is very small as compared to the number of primary turns
(*d*) either (*a*) or (*c*)

389. Which of the following are needed both for protection and metering?
(*a*) wattmeters (*b*) instrument transformers
(*c*) energy meters (*d*) power factor meters

390. The sensitivity of a potentiometer can be increased by
(*a*) increasing the length of wire
(*b*) decreasing the length of wire
(*c*) increasing the current in the wire
(*d*) decreasing the current in the wire

391. A meter has a square law scale. For 2 A current the deflection in 90°. For a deflection of 45°, the current is
(*a*) 1 A (*b*) 1.5 A
(*c*) 1.414 A (*d*) 1.61 A

392. In a hot wire instrument, the deflection is proportional to
(*a*) peak value of ac current
(*b*) average value of ac current
(*c*) rms value of ac current
(*d*) either (*a*) or (*b*)

393. Which the following can be used for measuring capacitance?
(*a*) De-Sauty bridge (*b*) Maxwell's bridge
(*c*) Anderson bridge (*d*) Hay's bridge

394. Which of these has a magnetic brake?
(*a*) thermocouple ammeter
(*b*) energy meter
(*c*) dynamometer wattmeter
(*d*) frequency meter

395. For a current transformer the standard secondary rating is
(*a*) 1 A or 5 A (*b*) 2.5 A
(*c*) 10 A (*d*) 5 A or 10 A

396. The angle δ of for a mica capacitor is about
(*a*) 25° (*b*) 15°
(*c*) 5° (*d*) 1°

397. Three-phase four wire induction energy meter is used to measure
(*a*) energy in 3-phase balanced star load
(*b*) energy in 3-phase balanced delta load
(*c*) energy in 3-phase unbalanced load
(*d*) either (*a*) or (*b*)

398. Which of these has poor overload capacity
(*a*) PMMC instrument
(*b*) hot wire instrument
(*c*) dynamometer instrument
(*d*) both (*a*) and (*c*)

399. Induction wattmeter can be used in
(*a*) ac circuit only
(*b*) dc circuit only
(*c*) both ac and dc circuit
(*d*) ac 3-phase only

400. To increase Q factor of coil, the wire should be
(*a*) long (*b*) thin
(*c*) thick (*d*) long and thin

401. In measurements using CRO, lissajous patterns are used to
(*a*) measure magnitude of peak voltage
(*b*) measure frequency
(*c*) measure impedance
(*d*) measure current

402. The torque in induction wattmeters is due to
(*a*) capacitive current (*b*) electrostatic effect
(*c*) hall effect (*d*) eddy currents

403. When deflecting torque and braking torque in an energy meter are zero
(*a*) the disc rotates at steady speed
(*b*) the disc is stationary
(*c*) the disc accelerates
(*d*) the disc retards

404. Which of the following measurements using a dc potentiometer do not require standardisation of potentiometer
(*a*) measurement of unknown voltage
(*b*) measurement of resistance
(*c*) measurement of unknown current
(*d*) calibration of voltmeter

405. The current in a circuit is measured using a 150 : 1 CT. If the ammeter reads 0.6 A, the circuit current is
(*a*) 250 A (*b*) 90 A
(*c*) 156 A (*d*) 144 A

406. The instrument used to check insulation of household wiring is
(*a*) megger (*b*) ohm-meter
(*c*) multimeter (*d*) high resistance voltmeter

407. In measurements using strain gauge, the dummy strain gauge is used to
(*a*) compensate for resistance of strain gauge
(*b*) compensate for change in temperature
(*c*) compensate for error due to change in voltage of battery
(*d*) compensate for error in voltmeter

408. Two wattmeter method of 3-phase power measurement can be used to find
(*a*) total active power

(*b*) total active power and power factor
(*c*) total active power and reactive power
(*d*) total active power, power factor, apparent power, and reactive power

409. Which bridge can be used to measure frequency?
(*a*) Maxwell (*b*) Kelvin
(*c*) Schering (*d*) Wien

410. Use of high permeability core in moving iron instrument helps in
(*a*) reducing size of meter
(*b*) increasing sensitivity
(*c*) reducing temperature error
(*d*) all the above

411. Which of the following transformers has low primary current?
(*a*) current transformer
(*b*) potential transformer
(*c*) distribution transformer
(*d*) both (*a*) and (*b*)

412. The term 'Poisson's ratio' is connected with
(*a*) strain gauge (*b*) LVDT
(*c*) Bourden tube (*d*) capacitance transducer

413. It is required to measure a low magnitude high frequency current. The proper instrument is
(*a*) moving iron (*b*) dynamometer
(*c*) thermocouple (*d*) induction

414. A resistance strain gauge is usually
(*a*) wire grid type (*b*) foil type
(*c*) any of these (*d*) none of these

415. In which of the transformer is the secondary nearly short-circuited under normal operating conditions
(*a*) CT
(*b*) PT
(*c*) distribution transformer
(*d*) power transformer

416. A resistance strain gauge has a resistance of about
(*a*) 100 Ω (*b*) 25 Ω
(*c*) 10 Ω (*d*) 1 Ω

417. In semiconductor strain gauge
(*a*) resistance increases with strain
(*b*) resistance decreases with strain
(*c*) resistance increases with strain in *p*-type and decreases with strain in *n*-type
(*d*) resistance increases with strain in *n*-type and decreases with strain in *p*-type

418. Piezo electric effect can be used to measure
(*a*) force (*b*) strain
(*c*) acceleration (*d*) all the above

419. A transducer must have
(*a*) good dynamic response
(*b*) good sensitivity
(*c*) good resolution
(*d*) all the above

420. The signal to noise ratio in telemetry should be better than
(*a*) 5 dB (*b*) 10 dB
(*c*) 25 dB (*d*) 60 dB

421. Consider the following statements regarding thermistor
1. It has high sensitivity
2. It has a linear relationship with temperature
3. It is a resistive device
4. It can be used as a time delay device

Which of the above statements are correct?
(*a*) 1, 2, 3 and 4 (*b*) 2, 3 and 4
(*c*) 1, 3, 4 (*d*) 1, 2 and 3 **[IES 2010]**

422. A strain gauge of resistance 120 ohm is mounted along the axial direction of an axially loaded specimen of steel (E = 200 GPa). The percentage change in length of rod due to loading is 3% and corresponding change in resistivity is 0.3%. For Poisson's ratio of 0.3 the gauge factor is
(*a*) 1.3 (*b*) 1.5
(*c*) 1.7 (*d*) 2.0 **[IES 2010]**

423. The incorrect statement is
(*a*) Thermistor has high sensitivity
(*b*) Thermocouple does not require an external electrical source for its operation
(*c*) Platinum has linear R-T relationship
(*d*) Thermistor does not require an external electrical source for its operation **[IES 2010]**

424. Match list I (transducers) and list II (type of transducer) and give correct answer using the codes.

List I	**List II**
A. LVDT	1. Resistive
B. Strain gauge	2. Inductive
C. Dielectric gauge	3. Capaciture
D. Thermocouple	4. Self-generating

Code :

	A	B	C	D
(*a*)	2	3	1	4
(*b*)	4	3	1	2
(*c*)	2	1	3	4
(*d*)	4	1	3	2

[IES 2010]

425. Two milliammeters with full scale currents 41 mA and 10 mA are in parallel. They read 0.5 mA and 2.5 nA. Their internal resistances are in the ratio
(*a*) 1 : 10 (*b*) 10 : 1
(*c*) 1 : 5 (*d*) 5 : 1 **[IES 2010]**

426. The value of resistance as measured by Wheatstone bridge is 10 K using 10 V source. The source resistance is measured by same bridge using 15 V source. The value of resistance is
(*a*) 15 K (*b*) 15.5 K
(*c*) 16.6 K (*d*) 10 K **[IES 2010]**

427. Wheatstone bridge method of measuring resistance is ideally suited for measurement of resistance value in the range

(a) 0.001 to 1 ohm (b) 0.1 to 100 ohm
(c) 100 to 10 k ohm (d) 100 k ohm to 10 M ohm **[IES 2009]**

428. What is the major cause of creep in energy meter?
(a) Over compensation for friction
(b) Mechanical vibrations
(c) Excessive voltage across potential coil
(d) Strong magnetic field **[IES 2009]**

429. Which of the following meters cannot measure ac quantities?
(a) Themocouple (b) Hot wire
(c) PMMC (d) Electrodynamometer **[IES 2009]**

430. Which of the following meters does require an external source?
(a) PMMC (b) Hot wire ammeter
(c) Electronic voltmeter
(d) Electrodynamic wattmeter **[IES 2009]**

431. Which of the following is not true for thermistors?
(a) They have negative temperature coefficient of resistance
(b) Their sensing element is made of sintered ceramic which are oxides of metals in the form of beads
(c) Their variation of resistance with temperature is linear
(d) The resistance at ambient temperature may range from 100 ohm to 100 kohm **[IES 2009]**

432. Which of the following is used at the first stage of an electronic voltmeter?
(a) BJT (b) SCR
(c) FET (d) UJT **[IES 2009]**

433. What is the advantage of an electronic voltmeter over non-electronic voltmeter
(a) Low power consumption
(b) Low input impedance
(c) Ability to measure wide range of voltages and resistances
(d) Large portability **[IES 2009]**

434. An average response rectifier type electronic voltmeter has 10 V dc applied to it. What is its reading?
(a) 7.1 V (b) 10 V
(c) 11.1 V (d) 22.2 V **[IES 2009]**

435. Which oscillator is used for generation of high frequencies?
(a) R–C phase shift oscillator
(b) Wien bridge Oscillator
(c) L–C Oscillator
(d) Blocking Oscillator **[IES 2009]**

436. Which transducer does require a high input impedance pre-amlifier for proper measurement?
(a) Thermocouple (b) Piezo electric
(c) Thermistor (d) LVDT **[IES 2009]**

437. In a digital voltmeter the oscillator frequency is 400 kHz. The ramp voltage falls from 8 V to 0 V in 20 ms. What is the number of pulses counted by counter?
(a) 800 (b) 2000
(c) 4000 (d) 8000 **[IES 2009]**

438. Which of the following transducers is most suitable for monitoring continuous variations in vary fine thickness of material?
(a) Diaphragm (b) Capacitor
(c) LVDT (d) Piezo electric **[IES 2009]**

439. In LVDT there are two secondary coils connected for a single output. Which of the following is correct?
(a) The coils are in series and in phase opposition
(b) The coils are parallel and in phase opposition
(c) The coils are in series and in same phase condition
(d) The coils are in parallel and in same phase condition **[IES 2008]**

440. A resistance strain of gauge factor 2 is used as a transducer element. Neglecting piezo resistive effects, what is the value of Poisson's ratio?
(a) 0.5 (b) 1
(c) 1.6 (d) 2 **[IES 2008]**

441. Match List I (transducers) with List II (temperature range) and select correct answer using the codes.

List I	List II
A. Copper-Constantan	1. –200°C to 850°C
B. (Platinum-Rohdium)-Platinum	2. –200°C to 350°C
C. (Rohdium-Indium)-Indium	3. 0 to 1000°C
D. Iron-Constantan	4. 1200 to 2000°C

Code :

	A	B	C	D
(a)	1	4	3	2
(b)	2	3	4	1
(c)	1	3	4	2
(d)	2	4	3	1

[IES 2008]

442. A coil is tuned to resonance at 500 kHz with a resonating capacitor of 36 pF. At 250 kHz resonance obtained with a capacitor of 160 pF. What is the self-capacitance of coil?
(a) 2.66 pF (b) 5.33 pF
(c) 8 pF (d) 10.66 pF **[IES 2008]**

443. The accuracy of a 0-10 mA meter is ±2%. What is the accuracy when taking a reading of 5mA?
(a) ± 0.5% (b) ± 10%
(c) ± 2% (d) ± 4% **[IES 2008]**

444. Match List I (term) with List II (statement) and select correct answer using the give codes.

List I
A. Relative error
B. Precision
C. Calibration
D. Resolution

List II
1. The ability of device to give identical output when repeat measurements are made with same input

2. The ratio of difference between measured value and true value to the true value of measurand
3. The smallest increment in measurand that can be detected with certainty by the instrument
4. The process of making adjustments on the scale so that instrument readings conform to an accepted standard.

Code :

	A	B	C	D
(*a*)	2	3	4	1
(*b*)	4	1	2	3
(*c*)	4	3	2	1
(*d*)	2	1	4	3

[IES 2008]

445. Two capacitances $C_1 = 150 \pm 2.4$ μF and $C_2 = 120 \pm 1.5$ μF are connected in parallel. What is the limiting error of resultant capacitance?
(*a*) 0.9 μF (*b*) 1.95 μF
(*c*) 3.9 μF (*d*) 4.8 μF **[IES 2008]**

446. The expression for mean torque T of an electrodynamic wattmeter can be expressed as $T \propto M^a V^b Z^c$ where M is mutual inductance between fixed and moving coils, V is applied voltage and Z is impedance of coil. What are the values of *a*, *b* and *c*?
(*a*) $a = 1, b = 2, c = -2$ (*b*) $a = 1, b = 1, c = -1$
(*c*) $a = 1, b = -2, c = 2$ (*d*) $a = 2, b = 2, c = -1$
[IES 2008]

447. If R, L, C are the parameters of a resistor, what is the condition for resistor to be non-inductive?
(*a*) $L = CR$ (*b*) $L = CR^2$
(*c*) $C = LR^2$ (*d*) $C = R/L$ **[IES 2008]**

448. Which is a passive transducer?
(*a*) Piezo electric (*b*) Thermocouple
(*c*) Photovoltaic cell (*d*) LVDT **[IES 2008]**

449. A 0-250 V voltmeter has a guaranteed accuracy of 1% of full scale reading. The voltage measured is 100 V. The limiting error is
(*a*) 1.25% (*b*) 25%
(*c*) 0.25% (*d*) 2.5% **[IES 2008]**

450. Which of the following materials is not used as piezo-electric transducer?
(*a*) Rochelle salt (*b*) Lithium sulphate
(*c*) Barium titarate (*d*) Tungsten oxide
[IES 2008]

451. Which of the following modulation techniques is the most efficient for pulse telemetry?
(*a*) PAM (*b*) PCM
(*c*) PDM (*d*) PPM **[IES 2008]**

452. Illumination is measured using which are of following.
(*a*) Millivoltmeter (*b*) Stroboscope
(*c*) Luxmeter (*d*) pH meter **[IES 2008]**

453. A capacitive transducer is made of two concentric cylinders. The outer diameter of inner cylinder is 4 mm and dielectric is air. The inner diameter of outer cylinder is 4.1 mm. When 150 V is applied between the cylinders, the electric field intensity is
(*a*) 1500 V/mm (*b*) 3000 V/mm
(*c*) 4500 V/mm (*d*) 6000 V/mm **[IES 2008]**

454. A single even parity cheek bit in telemetry data transmission system has which capability?
(*a*) Only detection of single error
(*b*) Both detection and correction of single error
(*c*) Detection of an even number of errors only
(*d*) Detection of any number of errors **[IES 2008]**

455. Hay's bridge is suitable for measuring inductance of which one of the following inductors
(*a*) having Q less then 10
(*b*) having Q greater than 10
(*c*) of any value of Q
(*d*) having range of reactance very large **[IES 2007]**

456. Match List I (bridge) and List II (Quantity) and select correct answer using the codes.

	List I		**List II**
A.	Wheatsone Bridge	1.	Capacitance
B.	Wien Bridge	2.	Very low resistance
C.	Kelvin Double Bridge	3.	Resistance
D.	Schering Bridge	4.	Frequency

Code :

	A	B	C	D
(*a*)	3	4	2	1
(*b*)	2	1	3	4
(*c*)	3	1	2	4
(*d*)	2	4	3	1

[IES 2007]

457. The current through a 20 ohm resister is measured by hot wire ammeter. If $i = 2 + 4 \sin 314t$, what is the reading?
(*a*) 2 A (*b*) 3.46 A
(*c*) 4 A (*d*) 2.83 A **[IES 2007]**

458. The AC bridge in Fig. 6.54 is supplied with 10 kHz source. What is the value of C_x?

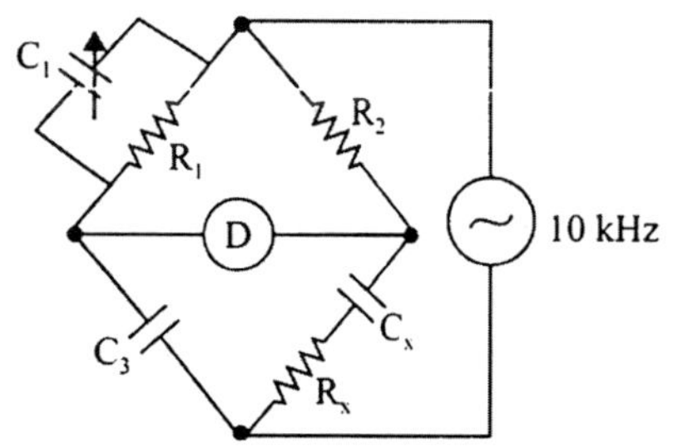

Fig. 6.54.

(*a*) $C_3 \dfrac{R_1}{R_2}$ (*b*) $C_3 \dfrac{R_2}{R_1}$

(*c*) $C_3 \dfrac{R_x}{R_1}$ (*d*) $C_3 \dfrac{R_x}{R_x}$ **[IES 2008]**

459. A circle is formed on the screen of CRO when two time-varying signals of same frequency and same magnitude

are applied to x and y plates of CRO. What is the phase difference?
(a) 0 (b) 90°
(c) 180° (d) 45° [IES 2007]

460. It is desired to make accurate measurement of voltage using CRO. Which of the following terms should be taken into consideration in measurement.
1. Electrostatic deflection type CRT
2. Magnetic deflection type CRT
3. Small deflection for measurement at centre of screen
4. Large deflection to cover entire screen

Select the correct answer using the following codes
(a) 1 and 4 (b) 2 and 3
(c) 2 and 4 (d) 1 and 3 [IES 2007]

461. Which of the following is not true of digital instruments?
(a) Loading of circuit under measurement is less
(b) Accuracy is better
(c) Free from observation error
(d) Can present the reading in overall contest of range [IES 2007]

462. Which of the following is true for a potentiometer transducer?
(a) It is a zero order displacement transducer
(b) It is a first order displacement transducer
(c) It is a zero order temperature transducer
(d) It is a second order displacement transducer [IES 2007]

463. Which one is correct? The generated emf of a tachogenerator is
(a) directly proportional to speed
(b) inversely proportional to speed
(c) proportional to square root of speed
(d) proportional to square of speed [IES 2007]

464. The best use of a resistance thermometer is in which range
(a) between 600°C to 1100°C
(b) between 1100°C to 1400°C
(c) between 1400°C to 1800°C
(d) above 1800°C [IES 2007]

465. A strain gauge with a gauge factor of 2 and resistance 120 ohm undergoes a strain of 10^{-5}. What is the change in resistance?
(a) 240 ohm (b) 24 ohm
(c) 2.4×10^{-3} ohm (d) 2.4×10^{-2} ohm [IES 2007]

466. An ac LVDT is given 6.3 V input and produces 5.2 V for range of 0.5 inch. When the core is –0.25 inch from centre what is the output?
(a) –2 V (b) +2 V
(c) –2.6 V (d) +2.6 V [IES 2007]

467. Which of the following has minimum loading effect on the quantity under measurement?
(a) PMMC (b) CRO
(c) Hot wire (d) Electrodynamometer [IES 2007]

468. A meter having a sensitivity of 2 kΩ/V is used for measurement of voltage across a circuit having output resistance of 1 kΩ and open circuit voltage of 8 V. What is the reading of meter on its 10 V scale?
(a) 5.72 V (b) 6.51 V
(c) 7.62 V (d) 7.91 V [IES 2006]

469. Consider the following statements in respect of two types of wattmeter connections in Fig. 6.55.

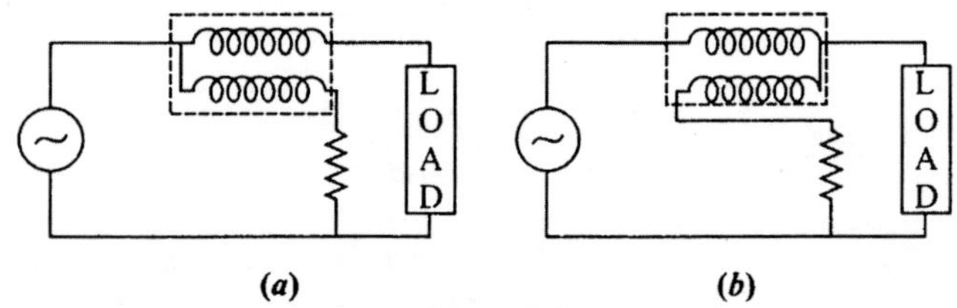

Fig. 6.55.

(a) Figure 6.55 (a) is used when applied voltage is high and current is low
(b) Figure 6.55 (b) is used when applied voltage is low and current is high
(c) If order of connections is statements 1 and 2 are in opposite order, gross error will increase
(d) If order of connections in statements 1 and 2 are in opposite order, gross error will be correct

Which of the above statements are correct?
(a) 1 and 2 only (b) 1 and 3 only
(c) 1, 2 and 3 only (d) 1, 2 and 4 only [IES 2006]

470. Which of the following is most suitable for measurement of linear displacement?
(a) Strain gauge (b) LVDT
(c) Piezo electrice (d) Microphone [IES 2006]

B. The following questions have an Assertion (A) and Reasoning (R). Read them carefully and select the answer using the following codes
(a) Both A and R are true and R is correct explanation of A
(b) Both A and R are true but R is not correct explanation of A
(c) A is true R is false
(d) A is false R is true

471. A. Absolute instruments are used in standard laboratories.
R. Absolute instruments do not require calibration.

472. A. Secondary instruments require calibration at frequent intervals.
R. Secondary instruments have many errors.

473. A. Many instruments use eddy current damping.
R. Eddy currents set up a magnetic flux which opposes the original movement which generates them.

474. A. If the reading of an instrument is low, the error is large.
R. The error of an instrument is certain percentage of its full scale deflection.

475. A. The scale a PMMC ammeter is linear.
R. The deflecting torque of PMMC is proportional to current in the coil.

476. A. Shunt of an ammeter has a low resistance.
R. Shunt may be connected in series or in parallel with ammeter.

477. A. Moving iron instrument can be used on ac or dc.
R. In a moving iron instrument deflecting torque is proportional to square of current.

478. A. Thermocouple instruments can be used for ac only.
R. Thermocouple instrument has a non-linear scale.

479. A. Electrodynamic instrument is not used as ammeter or voltmeter.
R. Electrodynamic instrument is more costly than PMMC and moving iron instrument.

480. A. When a wattmeter is used for low impedance loads, the pressure coil is connected across the load.
R. A wattmeter measures apparent power.

481. A. A low power factor wattmeter has special constructional features to ensure accurate measurement.
R. Extension of wattmeter range in dc circuit can be easily done by using current and potential transformer.

482. A. Energy meters use induction principle.
R. Induction wattmeter is less accurate than a dynamometer wattmeter.

483. A. A 3-phase wattmeter consists of two wattmeters mounted in one case.
R. Power measurement in a 3-phase circuit can be conveniently done by two wattmeter method.

484. A. Meggar is used to check insulation resistance of cables.
R. Insulation resistance of cables is very high.

485. A. Double Kelvin's bridge is used to measure low resistance.
R. In low resistance, the contact and lead resistance are very important.

486. A. In loss of charge method, it is preferable to use an electronic voltmeter.
R. The resistance of electronic voltmeter is very high.

487. A. A dc potentiometer measures the true open circuit emf of a source.
R. At balance point, a dc potentiometer does not draw any current from the source.

488. A. If the resistance of shunt of an ammeter is increased, the range of ammeter is reduced.
R. If the series resistance of a voltmeter is increased, the range of voltmeter is increased.

489. A. For best results the system of a D'Arsonval galvanometer should be slightly under damped.
R. Degree of damping affects the time taken by the instrument to come to final position.

490. A. Damping in a vibration galvanometer is small.
R. Vibration galvanometer is a PMMC instrument.

491. A. A ballistic galvanometer is used to measure charge.
R. In a flux meter the damping is large.

492. A. CRT is the most important component of a CRO.
R. When an electron beam strikes the screen of CRT, a spot of light is produced.

493. A. In a CRT, grid is kept at negative potential with respect to cathode.
R. The function of grid in an electronic tube is to control the number of electrons.

494. A. The trigger circuit of a CRO is excited by the same signal as Y plates.
R. The time base generator develops a saw tooth voltage.

495. A. CRO can be used to study waveshape accurately.
R. Input impedance of a CRO is very high.

496. A. When frequencies of signals fed to X and Y plates of CRO are multiples of each other, Lissajous pattern is displayed on screen.
R. In a CRO, the vertical deflection is proportional to voltage applied to Y plates.

497. A. An RTD uses platinum coil.
R. Platinum is a very stable material.

498. A. A thermocouple transducer is based on See back effect.
R. In a thermocouple transducer used for temperature measurement, the cold junction is usually kept in ice bath.

499. A. A radiation pyrometer is very suitable for high temperature measurement.
R. When temperature is high, a physical contact between hot body and thermometer is not desirable.

500. A. Capacitance effect can be used to measure thickness of sheets, displacement and pressure.
R. Capacitance between two plates depends on the area of plates, distance between plates and permittivity of medium.

501. A. The sign and magnitude of displacement can be measured by an LVDT.
R. LVDT is based on the principle of mutual inductance.

502. A. The speed of a shaft can be measured by using a dc tachogenerator.
R. A two-phase servo meter can also be used as a speed transducer.

503. A. Hall effect is observed in all metals but is very appreciable in semiconductors.
R. Hall effect transducer is very suitable for measurement of magnetic flux density.

504. A. A transducer needs an output measuring or display device for measuring a non-electrical quantity.
R. In some transducers the output voltage is too small and needs amplification before it can be measured.

505. A. A photovoltaic cell is greatly affected by temperature.
R. Both germanium and silicon can be used for photovoltaic cells.

506. A. Strain measurement using strain gauge invariably requires a dummy strain gauge.

R. The resistance of strain gauge depends on temperature.

507. A. A semiconductor strain gauge is more accurate than a resistance strain gauge.
R. The gauge factor of a semiconductor strain gauge is about 100 where as the gauge factor of a resistance strain gauge is only about 2.

508. A. Thermistors are commonly used to measure hot spot temperatures in electric machines.
R. Thermistor has the advantages of high temperature coefficient, small size and high speed of response.

509. A. A hot wire ammeter has a cramped scale.
R. The heating is proportional to square of current.

510. A. In hot wire instrument the material of hot wire is generally plantinum iridium.
R. A hot wire ammeter is suitable for ac only.

511. A. An electrostatic voltmeter is free from hysteresis and eddy current errors.
R. An electrostatic voltmeter has a number of fixed and moving vanes in parallel.

512. A. In two wattmeter method of 3-phase power measurement, one of the wattmeters would read negative if power factor of load in less then 0.5.
R. In two wattmeter method of 3-phase power measurement, both the wattmeters would show equal readings if power factor of load is unity.

513. A. The gall tinsley potentiometer is a coordinate potentiometer and has two exactly similar potentiometers.
R. An ac potentiometer may measure the unknown voltage in polar form or rectangular coordinate form.

514. A. Maxwell bridge is for measuring inductance of medium magnitude.
R. In Maxwell's bridge the unknown inductance is measured by comparing it with a capacitance.

515. A. Hay's bridge in particularly suitable for measuring inductance having a comparatively low resistance.
R. In Hay's bridge the value of frequency of supply must be known.

516. A. De sauty's bridge is suitable only for pure capacitor.
R. Capacitors are mostly perfect.

517. A. A universal RLC bridge has a battery as also a high frequency ac source.
R. RLC bridge uses a Wien bridge arrangement for measuring capacitance.

518. A. Schering bridge is a low voltage bridge.
R. Schering bridge is very commonly used for measurement of capacitance.

519. A. A power factor meter operates on dynamometer principle.
R. A power factor meter does not have a control spring.

520. A. Wien bridge can be used for measuring frequency.
R. A frequency meter for 50 Hz system has a frequency range of 40 to 60 Hz.

521. A. A thermocouple is an active transducer.
R. It is activated by the temperature gradient. **[IES 2010]**

522. A. VTVM gives more accurate reading as compared to accuracy of PMMC used in it while measuring voltage across high resistance.
R. The electronic circuits used in VTVM make it more accurate by increasing its sensitivity. **[IES 2008]**

523. A. If a piece of metal is made to have a temperature gradient between its ends an emf exists between these ends.
R. Electrons at the hot end move towards cold end. **[IES 2007]**

524. A. Instrument probes are used in conjunction with measuring instruments to improve frequency response.
R. A probe introduces attenuation.

525. A. Electrodynamometer type instruments have higher power consumption.
R. The current under measurement has to provide the field flux. **[IES 2006]**

526. A. CRO is capable of measurement at high frequencies.
R. It uses high gain amplifier for its operation. **[IES 2005]**

527. A. A ballistic galvanometer does not produce a steady state deflection and its pointer oscillates with decreasing amplitude.
R. The ballistic galvanometer is made to have critical damping. **[IES 2004]**

528. A. An electronic voltmeter reads voltage accoss a high resistance used in an electronic circuit more accurately than a PMMC voltmeter though electronic voltmeter also uses PMMC instrument as indicating device.
R. Accuracy of electronic voltmeter is less than that of PMMC instrument. **[IES 2004]**

529. A. VTVM is preferred to voltmeter for measuring voltage.
R. VTVM has inbuilt amplifier and gives accurate results. **[IES 2003]**

530. A. Operational amplifiers are commonly used in instrumentation.
R. Operational amplifiers do not load the circuit due to their high input impedance. **[IES 2003]**

C. State whether the following are True or False

531. Rayleigh current balance is an absolute instrument.
532. Induction wattmeter is an absolute instrument.
533. DC ampere hour meter is an indicating instrument.
534. Panel instruments are generally mounted vertically.
535. As per BIS there are 6 accuracy classes of instruments.
536. Watthour meter is an integrating instrument.
537. In indicating instruments the controlling torque is provided by a spring.
538. The scale of PMMC instrument is linear.
539. A moving iron instrument can be used for dc as well as ac.
540. The moving iron instruments are generally repulsion type.

541. Moving iron instrument can be used for ac only.
542. The scale of moving iron instrument is linear.
543. The scale of hot wire instrument is linear.
544. A thermocouple instrument can be used over a very wide frequency range.
545. A multimeter can be used as ohmmeter also.
546. Electrostatic voltmeter is suitable only for low voltages.
547. A hot wire instrument is not affected by stray magnetic fields.
548. The scale of thermocouple instrument is linear.
549. In a dynamometer instrument, the moving coil is pressure coil.
550. The scale of dynamometer wattmeter is non-linear.
551. A dynamometer wattmeter is suitable for both dc and ac.
552. In an induction wattmeter, the shading bands are made of non-conducting material.
553. Eddy currents are essential for the deflecting torque of an induction wattmeter.
554. An induction wattmeter can be used for ac only.
555. If the load current is high, the pressure coil of wattmeter should be connected directly across the load.
556. In a low power factor wattmeter, compensation for pressure coil current is provided.
557. In a low power factor wattmeter, the resistance of pressure coil is more than in ordinary wattmeter.
558. Two wattmeter method is suitable only for balanced loads.
559. A single phase energy meter operates on induction principle.
560. The braking torque in single phase energy meter is proportional to speed of disc.
561. In measuring resistance using voltmeter and ammeter, the resistance of voltmeter and ammeter also affect the accuracy of measurement.
562. The accuracy of measurement by ohmmeter is very good.
563. A megger has a hand driven dc generator.
564. The total insulation resistance and surface leakage resistance and surface leakage resistance.
565. In a megger the moving coil meter has a permanent magnet.
566. Wheat stone bridge is suitable for low resistance.
567. The resistance of series field winding is a low resistance.
568. Loss of charge method is suitable for medium resistance.
569. The resistance of voltmeter affects the accuracy of measurement in loss of charge method.
570. To measure an emf by dc potentiometer it is necessary to standardise the potentiometer with the help of a standard cell.
571. A volt ratio box should have low resistance.
572. While calibrating an ammeter by a dc potentiometer it is necessary to use a standard resistance.
573. The emf of a Weston cell is 1.286 V.
574. In a phase shifting transformer the currents in the two stator winding are in phase.
575. The coils of Drysdale polar potentiometer are wound non-inductively.
576. An ac bridge uses galvanometer as detector.
577. A Maxwell's bridge has the unknown inductance in one arm and resistance in the other three arms.
578. In Maxwell's bridge the unknown inductances is measured by comparing it with a standard capacitor.
579. Hay's bridge in suitable for inductances having low L/R ration.
580. Schering bridge operates at a high voltage of about 10 kV.

D. Fill in the blanks with suitable words

581. An instrument of accuracy class 1.5 can have an error of ± 1.5% of ....
582. Thread suspension is also known as ....
583. Instruments using fluid friction damping have to be used in .... position.
584. Damping disc in instruments is generally made of ....
585. Most instruments use .... damping.
586. When reading an indicating instrument, error due to .... must be avoided.
587. The resistance of shunt is very ....
588. The scale of moving iron instrument is ....
589. The moving iron construction generally used is .... type.
590. For high voltage measurement we use .... voltmeter.
591. A hot wire instrument is not affected by ....
592. In PMMC the damping torque is due to ....
593. The scale. of .... instrument is linear.
594. A rectifier instrument is combination of .... and .... ammeter.
595. The main feature of a thermocouple instrument is that it can be used for ....
596. An energy meter operates on .... principle.
597. A dynamometer wattmeter has .... coils.
598. In induction wattmeters .... help in ensuring that flux of voltage magnet is lagging the applied voltage by 90°.
599. An induction wattmeter has .... coil.
600. In low power factor wattmeter the compensating coil is connected in .... with the .... coil.
601. In two wattmeter method of 3-phase power measurement the voltage across each pressure coil is .... voltage.
602. A polyphase wattmeter has a laminated iron sheet between the two component wattmeters to provide ....
603. One wattmeter method of power measurement in a 3-phase circuit is only for .... load.
604. Insulation resistance of wires and cables is a .... resistance.
605. In megger a guard terminal is provided so that .... resistance is eliminated in measurement.
606. In wheatstone bridge measurement, ... key should be closed before .... key.
607. In wheatstone bridge measurement resistance of galvanometer should be kept .... in the beginning.

608. In loss of charge method, a high value of unknown resistance R means that fall in voltage would be ....

609. To calibrate a voltmeter using a potentiometer it is necessary to use a ....

610. Drysdals potentiometer, measures an ac voltage in .... from.

611. In Gall Tinsley potentiometer, the currents in the two potentiometers are in ....

612. The stator of a phase shifting transformer has a .... field.

613. In drysdale potentiometer, the potentiometer is fed from .... of phase shifting transformer.

614. When measuring an unknown resistance by a dc potentiometer, the unknown resistance and standard resistance are connected in ....

615. A 50 Hz ac bridge uses a .... detector.

616. To balance an ac bridge, the voltage drops must be equal in .... as well as in ....

617. Schering bridge is very commonly used to measure properties of ....

618. The use of .... eliminates error due to stray magnetic field.

619. Heaviside Campbell equal ratio bridge uses two ....

620. The universal RLC bridge uses .... bridge configuration for measurement of capacitance.

ANSWERS WITH HINTS AND EXPLANATIONS

1. (*c*) **2.** (*a*)

3. (*d*) To find average deviation we have to take summation of magnitudes of individual magnitudes.

4. (*a*) **5.** (*c*)

6. (*c*) Error $= \dfrac{150\times 1}{100} = 1.5$ V. % error $= \dfrac{1.5}{83}\times 100 = 1.81\%$.

7. (*c*) Limiting error $= 2\times 0.5 + 0.2 = \pm 1.2\%$ (we have to take 2×0.5 because power $\propto I^2$).

8. (*c*) Since second and decimal places of R_1 are not known, we can take only first decimal place of R_2.

9. (*d*) **10.** (*d*)

11. (*a*) The numbers can be $131+5$, $131-5$, $629+3$ and $629-3$. Therefore the sum can be 760 ± 8.

12. (*c*)

13. (*d*) $(A-B) = 25\pm 8$. Therefore % range of doubt

$$= \frac{8}{25}\times 100 = 32\%.$$

14. (*c*) Resolution is the smallest change in the quantity being measured which will cause an observable change in output. In this case it is 1 mV.

15. (*b*)

16. (*b*) Effective resistance of circuit

$$= 1\ \text{M}\Omega + \frac{100\times 20}{120}\ \text{k}\Omega = 1.016667\ \text{M}\Omega.$$

$$V = \frac{5}{1.016667\times 10^6}\times 16.6667\times 10^3 = 0.082\ \text{V}.$$

17. (*c*) $36+75=111$, $1.8+3.75=5.55$. Therefore $R_1+R_2 = 111\pm 5.55$ ohm.

18. (*b*) The tolerances are additive.

19. (*c*)

20. (*a*) Losses $= 3650-3358 = 265$ W uncertainty $= \pm 20$ W.

$$\text{Percentage uncertainty} = \frac{20}{265}\times 100 = \pm 7.55\%.$$

21. (*d*) The 8 classes are given in section 6.4.4.

22. (*c*) Disc must be conducting otherwise eddy currents cannot flow. It must be non-magnet *ic*, so that it does not affect the measurement.

23. (*c*) It integrates energy over a period of time.

24. (*c*) See Section 6.4.4.

25. (*c*) If frequency changes reactance of coil changes and reading is affected.

26. (*d*) Resistance of shunt $= \dfrac{0.6\times 0.1}{15-0.1} = 0.004\ \Omega$.

27. (*c*) $20\times 10^{-3} = \dfrac{100}{R+50} = 4950$ ohm.

28. (*a*) **29.** (*a*)

30. (*a*)

31. (*c*) Eddy current damping is very common.

32. (*a*) **33.** (*a*) **34.** (*b*)

35. (*a*) It bypasses a major portion of circuit current so that only the current of meter can flow through the meter.

36. (*d*) Resistance of shunt $= \dfrac{1\times 10^{-3}}{1-1\times 10^{-3}} = 10^{-3}\Omega = 0.001$ ohm.

37. (*a*)

38. (*c*) $R = \dfrac{5}{1\times 10^{-3}} - 100 = 4900$ ohm.

39. (*c*) Its deflecting torque $\propto I^2$. So it can be used for both d.c. and a.c.

40. (*b*) Since deflection torque $\propto I^2$, scale is non-linear.

41. (*c*) **42.** (*b*) **43.** (*b*)

44. (*a*) Repulsion type have better accuracy.

45. (*b*) Resistance of shunt $= \dfrac{0.5\times 0.1}{10-0.1} = 0.005$ ohm.

46. (*b*) $R = \dfrac{200}{10\times 10^{-3}} - 100 = 19900$ ohm.

47. (*c*) Only electrostatic instrument has all these features.

48. (*b*) $R = \dfrac{300}{10\times 10^{-3}} - 20 = 29980$ ohm.

49. (*d*) Meter current for d.c. $= \dfrac{250}{2000+500} = 0.1$ A.

Meter current for a.c. $= \dfrac{250}{2500+j314.16}$

$= 0.09922$ A. Reading $= \frac{0.09922}{0.1} \times 250 = 248$ V.

50. (*b*) Hot wire ammeter indicates rms value $= \frac{100}{10} \times 0.5 = 5$ A.

51. (*b*) Moving coil instrument indicates average value $= \frac{100}{\pi(10)} = 3.185$ A.

52. (*c*) **53.** (*d*)

54. (*c*) **55.** (*d*)

56. (*c*) Only tangent galvanometer and Rayleigh Current balance are absolute instruments. Energy meter and ammeter are secondary instruments.

57. (*b*)

58. (*b*) Moving coil and Rectifier Instruments have linear scale. Moving iron and electrostatic have non-linear scale.

59. (*a*) Thermocouple instrument is based on heating effect and can be used for very high frequencies also.

60. (*a*) Reading of A $= 100 \times \frac{1}{1+0.8} = 55.56$ A. Reading of B $= 100 \times \frac{0.8}{1.8} = 44.44$ A.

61. (*a*) If current through galvanometer is I. current through shunt $= \frac{I \times 1000}{200} = 5$ I. Total current 6 I. Hence multiplying factor is 6.

62. (*b*) Since shunt resistance is doubled, current through ammeter is doubled (current division for resistances is parallel). Therefore reading is 2θ.

63. (*a*)

64. (*b*) See section 6.7.1.

65. (*a*)

66. (*b*) See section 6.7.1.

67. (*a*) See section 6.7.1.

68. (*b*) **69.** (*a*) **70.** (*a*) **71.** (*b*)

72. (*a*) Swing is proportional to charge.

73. (*b*) Diplacement $= (1)(2\theta) = 0.4$ m.

74. (*a*) **75.** (*c*)

76. (*c*) The inductance of coil will allow slightly less ac current. Hence reading slightly less.

77. (*b*) Average value $= \frac{6.82}{1.11} = 6.144$ V. Peak value of triangular ware is twice average value $= 2 \times 6.144 = 12.288$ V. RMS value of triangular ware is peak value divided by $\sqrt{3}$. Therefore reading

$$= \frac{12.288}{\sqrt{3}} \simeq 7.1 \text{ V}.$$

78. (*c*) Moving coil is voltage coil and must have high resistance.

79. (*c*)

80. (*b*) Induction principle is valid for ac only. Induction instruments are calibrated for 50 Hz ac.

81. (*a*) See Fig. 6.6.

82. (*b*) Low p.f. means comparatively high current. Hence connection of Fig. 6.6 (*a*).

83. (*c*)

84. (*a*) Readings of two wattmeters are same when $\theta = 0$ or p.f. = 1.

85. (*b*) When p.f. = 0.5, $\theta = 60°$ and one of the wattmeters reads zero.

86. (*a*)

87. (*a*) The moving system has to be common since reading depends on the action of both wattmeters.

88. (*c*)

89. (*d*) Energy meter is also based on induction principle.

90. (*a*)

91. (*d*) Dynamometer principle can be used for ammeter, voltmeter and wattmeter. However it is generally used for wattmeters only.

92. (*b*) **93.** (*a*) **94.** (*c*)

95. (*c*) **96.** (*c*)

97. (*c*) When p.f. is less than 0.5 (leading or lagging) one of the wattmeters reads negative.

98. (*c*) **99.** (*b*) **100.** (*a*)

101. (*b*) Since angle is less than 90°, reading will be less. Therefore error is negative and correction is positive.

102. (*a*)

103. (*c*) Error is voltmeter can be ± 2.5 V which for 100 V reading is ± 2.5%. Error in ammeter can be ± 0.05 A which for 2 A reading is ± 2.5%. Hence total error can be ± 5%.

104. (*b*) Capacitor behaves as short-circuit initially and open circuit finally when fed by dc supply.

105. (*b*) Since resistance is low current is high and voltage drop across ammeter is high. Therefore, voltmeter should be connected across resistance.

106. (*c*)

107. (*d*) Lead resistance is important when measuring low resistance. Resistance of shunt and ammeter is low.

108. (*c*) **109.** (*c*) **110.** (*a*)

111. (*b*) It is for measuring very high resistance.

112. (*b*)

113. (*c*) Let unknown resistance be R. Then $\frac{R \times 10^{10}}{R + 10^{10}} = 0.8 \times 10^9$. Evidently R is more than 0.8×10^9 ohm.

114. (*b*) **115.** (*c*) **116.** (*b*) **117.** (*b*)

118. (*b*) **119.** (*d*) **120.** (*a*) **121.** (*a*)

122. (*c*) **123.** (*b*) **124.** (*a*) **125.** (*a*)

126. (*a*) DC potentiometer can measure voltages upto about 1.5 V. Volt ratio box is used to reduce the unknown voltage so that it can be measured by dc potentiometer.

127. (*a*) Unknown emf $= \frac{82}{60} \times 1.0185$.

128. (*c*) **129.** (*c*) **130.** (*c*)

131. (*b*) See section 6.30.13.

132. (*c*) **133.** (*c*)
134. (*b*) Loss angle $\delta = 90 - 80 = 10°$.
135. (*a*) **136.** (*a*)
137. (*a*) To balance the bridge both magnitudes and phase angles of voltages have to balanced.
138. (*b*) **139.** (*a*) **140.** (*c*) **141.** (*c*)
142. (*b*)
143. (*c*) Generally the frequency does not fall below 47 Hz and does not exceed 53 Hz.
144. (*b*) **145.** (*b*) **146.** (*b*) **147.** (*c*)
148. (*b*) **149.** (*a*)
150. (*d*) Since $\cos\theta = \frac{W}{VI}$, we need to measure active power, voltage and current.
151. (*a*) **152.** (*b*) **153.** (*b*) **154.** (*a*)
155. (*a*) See section 6.24.2.
156. (*a*) It is called time base signal.
157. (*a*) **158.** (*b*)
159. (*a*) The a.c. voltage is rectified and measured by PMMC instrument.
160. (*a*) Since the thermocouple is based on heating effect, rms value is measured.
161. (*d*) Time base signal is not connected. The two waves are fed to X and Y plates respectively.
162. (*d*) **163.** (*d*)
164. (*b*) It is a non-linear device. The change of resistance with temperature is non-linear. See section 6.31.2.
165. (*d*) **166.** (*c*) **167.** (*c*)
168. (*b*) It is used to measure displacement.
169. (*c*) The output voltage is very small. So we need a milli voltmeter.
170. (*d*) In all these, change of capacitance occurs.
171. (*a*)
172. (*b*) A drag cup rotor has low inertia. Therefore it requires little amount of energy.
173. (*a*) The measurement is done using an unbalanced bridge.
174. (*a*) **175.** (*a*)
176. (*c*) The flat grid is attached to the surface of solid.
177. (*d*)
178. (*d*) It is suitable for only very high temperatures.
179. (*c*) **180.** (*b*) **181.** (*d*)
182. (*d*) Electron gun consists of cathode, grid and accelerating anode.
183. (*b*) Deflection $\propto \frac{1}{(V_A)^{0.5}}$; where V_A is the potential difference between anode and cathode.
184. (*a*)
185. (*b*) Time base signal frequency is adjustable so that voltages of different frequencies can be measured.
186. (*a*) $F = ma = MLT^{-2}$.
187. (*a*) EMF = workdone/charge.
188. (*a*) Flux density = flux/area
189. (*c*) Magnetic field intensity = MMF/length = IL^{-1}.
190. (*b*) Current density = current/area IL^{-2}.
191. (*a*) $10\left(\frac{1020}{1020+1000}\right) - 5$
192. (*a*)
193. (*d*) $\frac{dR}{R}$ = G (strain) = G (Stress/Young's modulus) or

$$dR = \frac{2\times1050\times1000}{2.1\times10^6} = 1 \text{ ohm.}$$

194. (*a*) **195.** (*a*) **196.** (*d*) **197.** (*d*)
198. (*d*) Single phase power can be measured using either 3 voltmeters or 3 ammeters.
199. (*c*)
200. (*a*) Impulse is unidirectional (either positive or negative).
201. (*d*)
202. (*d*) CRO has very high input impedance.
203. (*a*) **204.** (*d*) **205.** (*a*)
206. (*d*) We can store the trace.
207. (*c*)
208. (*a*) See section 6.31.2.
209. (*d*) When p.f. is zero, active power is zero and the two readings are equal and opposite.
210. (*a*) The initial change is fast.
211. (*d*) **212.** (*b*) **213.** (*c*)
214. (*c*) $C = 0.41L/R^2$ or $C = (0.41\times 0.3)/(2000)^2 = 0.0311 \times 10^{-6}$ F.
215. (*a*) Since resistances are in the ratio of 10 : 1, capacitances should be in the ratio of 1 : 10 because $X_C \propto \frac{1}{C}$.
216. (*c*)
217. (*a*) $90 = 20\log\frac{P}{2\times10^{-5}}$.
218. (*b*) If speed is 60, the counter will count $60\times60 = 3600$ counts/per minute.
219. (*a*) **220.** (*b*)
221. (*d*) Thermistor is a non-linear device.
222. (*a*)
223. (*b*) Hall voltage

$$= \frac{K_H BI}{d} = \frac{-1\times10^{-8}\times1\times3}{2\times10^{-3}} = -15\times10^{-6}\text{ V.}$$

224. (*a*)
225. (*b*) Gauge factor $G = \frac{dR}{R}\left[\frac{1}{\text{strain}}\right]$

$$\text{or } G = \frac{1}{125\times4000\times10^{-6}} = 2.$$

226. (*c*)
227. (*b*) Bandwidth required is twice the required frequency range.
228. (*d*)
229. (*c*) Error = 1% of 150 = 1.5 V, % error = $\frac{1.5}{8.3}\times100 = 18.1\%$
230. (*b*) Shunt resistance

$$= \frac{1000\times10^{-6}\times1000}{100\times10^{-3} - 1000\times10^{-6}} = 10 \text{ ohm,}$$

231. (*b*) $I = \dfrac{500\times10^{-3}}{0.02} = 25$ A.

232. (*d*)

233. (*a*) Output of first op-amp is $-2V_1 - V_2$. Therefore,

$$V_0 = -(-2V_1 - V_2)\frac{10}{10} - V_3 - \left(\frac{10}{5}\right) = 2V_1 + V_2 - 2V_3.$$

234. (*d*)

235. (*d*) Peak value of sine wave = 100 × 1.11 = 111 V.

236. (*c*) % error in current is 2% and % error in resistance is 0.2%. Since $P = I^2R$, % error in power = 2 × 2% + 0.2% = 4.2%.

237. (*d*) **238.** (*b*)

239. (*a*) When signals to X and Y plates are in phase, the display is a straight line inclined at 45°.

240. (*b*)

241. (*d*) Sampling frequency must be twice the input frequency.

242. (*a*) $V = \dfrac{15}{5} = 3V, R = \dfrac{3}{5} = 0.6\,\Omega.$

243. (*c*) **244.** (*a*) **245.** (*a*)

246. (*d*) $Z_4 = \dfrac{Z_2Z_3}{Z_1} = \dfrac{(400\angle-90)(300\angle0)}{200\angle60}$

or $Z_4 = 600\angle-150°$ ohm.

247. (*a*) $\dfrac{dR}{R} = G\,(\text{strain})$ or strain $= \dfrac{dR}{2R} = \dfrac{0.6}{2\times120} = 0.0025.$

248. (*b*) **249.** (*a*)

250. (*d*) Since the current drawn is minimum, the accuracy is maximum.

251. (*d*) See section 6.16.

252. (*c*) Since the circuit is not loaded, the accurate value of dc voltage is measured.

253. (*d*) **254.** (*b*)

255. (*a*) $C = \dfrac{\epsilon_0\epsilon_r A}{d}$

256. (*c*)

257. (*c*) Induction wattmeter.

258. (*a*) **259.** (*d*)

260. (*b*) D/A converter changes digital output.

261. (*d*) **262.** (*c*)

263. (*d*) The change in resistance is due to change in length and area of cross-section.

264. (*c*) **265.** (*d*) **266.** (*d*)

267. (*c*) $\dfrac{50}{20} = 2.5$ mV/pC.

268. (*b*) Span = 600°, Dead zone $= \dfrac{0.125}{100}\times600 = 0.75$°C.

269. (*a*) **270.** (*b*) **271.** (*b*)

272. (*c*) It does not have a permanent magnet.

273. (*b*) NBAI = Controlling torque or

$$\frac{100\times1\times40\times30}{10^6}\,I = 240\times10^{-6} \text{ or } I = 2 \text{ mA}.$$

274. (*d*)

275. (*b*) Shunt resistance $= \dfrac{100^{-3}\times100}{1-10^{-3}} = 0.1001$ ohm.

276. (*c*) Error $= \dfrac{0.75\times200}{100} = 1.5$ V, % error $= \dfrac{15}{100}\times100 =$ 1.5%

277. (*b*)

278. (*d*) Both errors can be positive or negative simultaneously. Hence 853 ± 8 Ω.

279. (*a*) Variation in supply voltage affects both amplifiers equally.

280. (*a*) **281.** (*c*)

282. (*d*) Effect of lead and contact resistances is very important in measurement of low resistances.

283. (*a*)

284. (*c*) $\dfrac{dR}{R} = G\dfrac{dl}{l}$ or $dR = (125)(2)\left(\dfrac{0.005}{1}\right) = 1.25$ ohm.

285. (*d*) **286.** (*b*) **287.** (*a*)

288. (*d*) The two materials in thermocouple are different and the two junctions are kept at different temperatures.

289. (*d*)

290. (*b*) $0.05\times a = 3\times10^3\times10^{-3}$ or $a = 60$ m/s^2.

291. (*d*) **292.** (*c*) **293.** (*b*)

294. (*b*) Shunt resistance $= \dfrac{10\times40\times10^{-3}}{2-40\times10^{-3}} = 0.2041$ ohm.

295. (*b*) Thermocouple meter is based on heating effect which varies as I^2.

296. (*b*) All wattmeters measure average power over a cycle.

297. (*a*)

298. (*d*) Rectifier converts ac to dc which is measured by D'Arsonval meter.

299. (*b*) See section 6.1.5.

300. (*c*)

301. (*b*) Very high current will flow through ammeter because ammeter resistance is very low.

302. (*b*) A wave form meter is suitable to study amplitudes, phase angles etc. of different wave forms.

303. (*d*) Time period of 1 μs corresponds to frequency of MHz.

304. (*a*) **305.** (*b*) **306.** (*c*) **307.** (*c*)

308. (*b*) $\dfrac{dR}{R} = G\dfrac{dl}{l}$ or $\dfrac{13.3}{128} = G\left(\dfrac{1.6}{24}\right)$ or $G \approx 1.6$.

309. (*c*)

310. (*b*) RPM $= \dfrac{13500}{150}\times60 = 5400.$

311. (*c*) **312.** (*d*) **313.** (*c*)

314. (*c*) $\dfrac{1}{5\times10^{-3}} = 200$ ohm/V.

315. (*c*)

316. (*b*) Wattmeter measures average power over a cycle.

317. (*c*) **318.** (*a*)

319. (*c*) Error in $R_1 = \dfrac{36\times5}{100} = 1.8\,\Omega$, error in $R_2 = \dfrac{75\times5}{100}$ $= 3.75\,\Omega$.

Total maximum error $\pm(1.8+3.75) = \pm5.55\,\Omega$.

320. (*b*)

321. (*c*) $Z_d = \dfrac{Z_b Z_c}{Z_a} = \dfrac{(300\angle -90)(200\angle 0)}{100\angle 50} = 600\angle -140°.$

322. (*c*) Final output $= 2 \times 10^{-3} \times 250 = 0.5$ V.

Sensitivity $= \dfrac{0.5\text{ V}}{0.5\text{ mm}} = 1$ V/mm.

323. (*d*)

324. (*c*) Only semiconductor strain gauge can have a gauge factor of 100. For all other strain gauges, gauge factor is about 2 only.

325. (*d*) LVDT and E pickup are suitable for linear displacement.

326. (*c*) **327.** (*c*)

328. (*a*) $\dfrac{dR}{R} = \dfrac{2(100\times10^6)}{200\times10^9}$ or $\dfrac{dR}{R} = 1 \times 10^{-3}$ or 0.1%.

329. (*b*) $f = \dfrac{1200}{60} \times 180 = 3600$/sec.

330. (*c*)

331. (*d*) Battery key must be closed before galvanometer key. Moreover galvanometer key must be opened before battery key. This is to avoid damage to galvanometer.

332. (*c*) **333.** (*b*)

334. (*d*) When volt ratio box is used to reduce range, the advantage of open-circuit measurement is lost.

335. (*c*) **336.** (*d*) **337.** (*c*)

338. (*c*) Since the primary of C T always remains excited, a high voltage is induced in secondary if the secondary is opened.

339. (*c*) **340.** (*a*) **341.** (*b*) **342.** (*c*)

343. (*a*) Since two branches have resistances, one branch has R_2 and L_2, X must also have R and L to obtain balance.

344. (*c*) Error = ± 6V. Therefore, the actual value would be between 222 ± 6 V.

345. (*b*) $Z_4 = \dfrac{Z_2 Z_3}{Z_1} = \dfrac{(150\angle 0)(250\angle -40)}{200\angle 30}$

or $Z_4 = 375\angle -70°$ ohm.

346. (*a*) **347.** (*d*) **348.** (*c*) **349.** (*d*)

350. (*d*) Error $= \dfrac{25\times1}{100} = 0.25$ A, % error $= \dfrac{0.25}{5} \times 100 = 5\%$.

351. (*a*)

352. (*c*) $Z_4 = \dfrac{(150\angle 0)(250\angle -40)}{200\angle 30°} = 187.5\angle -70°$ ohm.

353. (*d*) **354.** (*d*)

355. (*b*) The readings will be proportional to their internal resistances, *i.e.*, $\dfrac{10}{30} \times 400$ and $\dfrac{20}{30} \times 400$.

356. (*a*)

357. (*c*) In strain gauge change in resistance occurs.

358. (*d*) An electronic voltmeter draws neglible current.

359. (*b*)

360. (*b*) Peak value $= \dfrac{4\times6}{2} = 12$ mV, Peak to peak value will be 24 mV.

361. (*a*)

362. (*b*) Currents divide in inverse proportion of resistances.

$I_A = \dfrac{15\times0.5}{1.5}$ and $I_B = \dfrac{15\times1}{1.5}$.

363. (*c*) The scale of electrostatic voltmeter is not uniform.

364. (*b*) $B \propto H$.

365. (*c*) **366.** (*d*) **367.** (*a*) **368.** (*d*)

369. (*c*) **370.** (*a*)

371. (*d*) $0.5 \times 4 = 2$ kWh $= 2 \times 1200$ revolutions.

372. (*d*) **373.** (*c*) **374.** (*a*)

375. (*c*) Resistance of voltmeter should be very high as compared to 20 kΩ.

376. (*a*) **377.** (*c*) **378.** (*a*) **379.** (*c*)

380. (*a*)

381. (*c*) When a hole comes under a pole of upper electromagnet the disc becomes stationary. Thus creep is avoided.

382. (*b*) 200 × 1 kΩ.

383. (*b*) $I\left(\dfrac{0.5}{0.5+0.25}\right) = 25$ or $I = 37.5$ A.

384. (*c*) $V\left(\dfrac{10}{10+5}\right) = 250$ or $V = 375$ V.

385. (*b*) **386.** (*d*) **387.** (*d*) **388.** (*b*)

389. (*b*) Instrument transformers are used to measure high currents and voltages. They also feed the relays for protection.

390. (*a*) **391.** (*c*) Current $= \left(\dfrac{90}{45}\right)^{0.5}$

392. (*c*) **393.** (*a*)

394. (*b*) Braking torque, in an energy meter, is provided by a magnet.

395. (*a*)

396. (*d*) Mica capacitor has very low loss and loss angle δ is very very small.

397. (*c*)

398. (*b*) The wire is very delicate and has little over load capacity.

399. (*a*) Induction effect occurs with ac only.

400. (*c*) Q is high if resistance is low. Therefore wire should be thick.

401. (*b*)

402. (*d*) Eddy currents are induced in the disc.

403. (*a*)

404. (*b*) Unknown resistance is compared with a standard resistance. Therefore standardistation of potentiometer is not necessary.

405. (*b*) 150 × 0.6 = 90 A.

406. (*a*) **407.** (*b*) **408.** (*d*) **409.** (*d*)

410. (*b*) **411.** (*b*) **412.** (*a*) **413.** (*c*)

414. (*c*)

415. (*a*) CT is operated with secondary short-circuited otherwise a high emf would be induced across secondary.

416. (*a*) **417.** (*c*) **418.** (*d*) **419.** (*d*)

420. (*d*)

421. (*c*) Thermistor is a non-linear device. The change of resistance with temperature is non-linear as seen in section 6.31.2.

422. (*c*) As given in section 6.31.9

$$G = 1 + 2\mu + \frac{\frac{dp}{p}}{\frac{dl}{l}} = 1 + 2\times0.3 + \frac{0.003}{0.03} = 1.7$$

423. (*d*) Thermistor does require electrical source in many application.

424. (*c*) LVDT is inductive (operates on mutual inductance) strain gange is resistive. Dielectric gauge is capacitive. Thermocouple does not require electrical source. (See section 6.31)

425. (*b*) Let R_1 and R_2 be internal resitances. When in parallel, voltage is same across each at full deflection

$$\frac{V}{1\times10^{-3}} = R_1 \quad \text{and} \quad \frac{V}{10\times10^{-3}} = R_2$$

$\frac{R_1}{R_2} = 10$. Hence ratio is 10 : 1.

426. (*d*) The voltage of source does not affect the measurement.

427. (*c*) Wheatstone bridge is suitable for medium resistances (1 ohm to 100 k ohm). See section 6.18.

428. (*a*) Creep means slow movement of energy meter disc even when current in current coil is zero. Its cause is over compensation for friction.

429. (*c*) PMMC is only for dc (See section 6.6)

430. (*c*) See Section 6.23. Amplifier requires external source.

431. (*c*) The variation of resistance with temperature is non-linear (See section 6.31.2).

432. (*c*) FET has very high input impedance. Therefore it draws very little current from source. This is an essential feature of an electronic voltmeter.

433. (*a*) Electronic voltmeter has high input impedance and, therefore, draws very little current and consumes very less power.

434. (*c*) It measures rms value

$$V_{rms} = \frac{V_m}{\sqrt{2}} \qquad V_{dc} = \frac{2V_m}{\pi}$$

$$\therefore \quad V_{rms} = \frac{\pi V_{dc}}{2\sqrt{2}} = \frac{\pi(10)}{2\sqrt{2}} = 11.1 \text{ V}$$

435. (*c*) At high frequencies only L-C oscillaters are used. R-C. Phase shift and Wien bridge oscillator are for audio frequencies.

436. (*b*) Since output is small, a high input preamplifier is needed.

437. (*d*) $T = \frac{1}{400\times10^3}$

Number of pulses

$$= \frac{20\times10^{-3}}{T} = 20\times10^{-3}\times400\times10^3 = 8000$$

438. (*b*) Capacitance transducer is very sensitive and can measure very small thickness.

439. (*a*) See Fig. 6.34. The secondaries are in series and inphase opposition.

440. (*a*) Using the Equation given in section 6.31.9 and neglecting $\frac{d\rho}{\rho}$ we have

$G = 1 + 2\mu = 2$ or $\mu = 0.5$

441. (*b*)

442. (*b*) $f_r = \frac{1}{2\pi\sqrt{L(C + C_s)}}$

where C_s is self-capacitance of coil

$$500 \times 10^6 = \frac{1}{2\pi\sqrt{L(36\times10^{-12} + C_s)}}$$

$$250 \times 10^6 = \frac{1}{2\pi\sqrt{L(160\times10^{-12} + C_s)}}$$

From above equation

$$4 = \frac{160\times10^{-12} + C_S}{36\times10^{-12} + C_S} \quad \text{or } C_s = 5.33 \text{ pF}$$

443. (*d*) $\text{Error} = 10\times\frac{2}{100} \pm 0.2\text{mA}$

$$\text{Accuracy} = \pm\frac{0.2}{5}\times100 = \pm 4\%$$

444. (*d*)

445. (*c*) The capacitances in parallel are added. Limiting error is sum of errors, *i.e*,. $\pm 2.4 \pm 1.5 = \pm 3.9\ \mu$F.

446. (*b*) $T = I_1 I_2 \frac{dM}{d\theta}$ or $T \propto \frac{V^2}{Z^2}M$

or $T = MV^2 Z^{-2}$ Hence $a = 1$, $b = 2$, $c = -2$

447. (*b*)

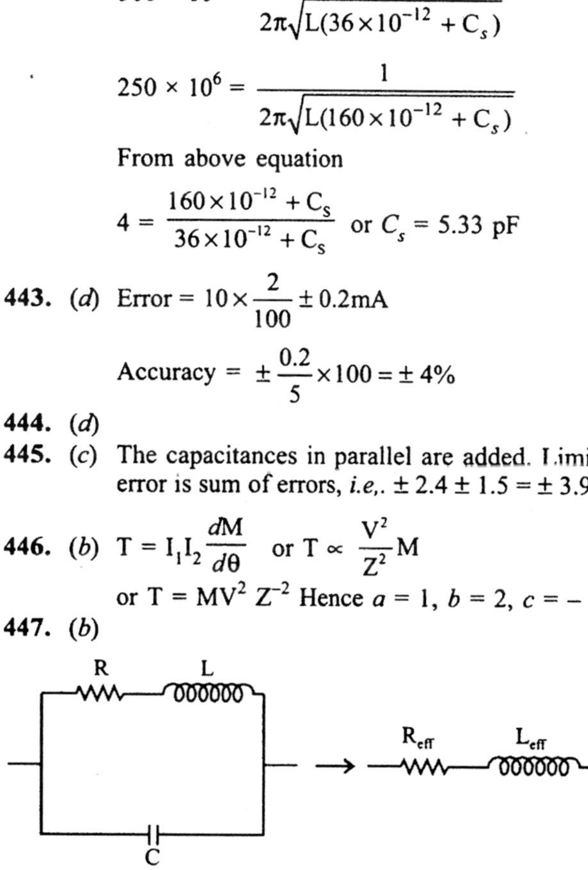

$$Z = \frac{\left(\frac{1}{j\omega c}\right)(R + j\omega L)}{R + j\omega L + \frac{1}{j\omega c}}$$

$$= \frac{R + j\omega(L - \omega^2 L^2C - CR^2)}{1 + \omega^2C^2R^2 - 2\omega^2LC + \omega^4L^2C^2}$$

L and C are generally very small. Hence $\omega^2LC << 1$ and $\omega^4L^2C^2 << 1$

Hence $z = \dfrac{R + j\omega(L - CR^2)}{1 + \omega^2C(CR^2 - 2L)}$

$L_{eff} = \dfrac{L - CR^2}{1 + \omega^2C(CR^2 - 2L)}$

Resistor will be non-inductive if $L - CR^2 = 0$ or $L = CR^2$.

448. (*d*) Only LVDT is passive. In all others internal voltage is generated. LVDT requires external supply.

449. (*d*) Error $= 250 \times \frac{1}{100} = 2.5\text{ V}$

$\%$ error $= \frac{2.5}{100} \times 100 = 2.5\%$

450. (*d*) Tungsten oxide

451. (*d*) PCM

452. (*c*) Luxmeter

453. $E = \frac{V}{d} = \frac{150}{4.1 - 4} = 1500\text{ V/mm}$

454. (*a*)

455. (*b*) Hay bridge is suitable for high $\frac{L}{R}$ and hence high Q.

456. (*a*) Wheatstone bridge is for resistance, Kelvin double bridge is for very low resistance, Schering bridge is for capacitance and Wien bridge is for frequency.

457. (*b*) Hot wire ammeter reads rms value

$I = \left[2^2 + \frac{4^2}{2}\right] = 3.45\text{ A}$

458. (*a*) At balance

$$R_2\left[\frac{1}{j\omega C_3}\right]\left[R_x + \frac{1}{j\omega C_x}\right]\left[\frac{R_1}{1 + j\omega R_1C_1}\right]$$

or $j\omega R_2C_x(1 + j\omega R_1C_1) = j\omega R_1C_3(1 + j\omega R_xC_x)$

Separating real and imaginary parts

$\omega R_2C_x = \omega R_1C_3$ or $C_x = \frac{R_1 C_3}{R_2}$.

459. (*b*) Circle is formed when $\theta = 90°$ or $270°$.

460. (*a*) Electrostatic deflection type CRT with large deflection.

461. (*d*) A digital instrument cannot present the reading in overall context of range.

462. (*a*)

463. (*a*) Emf $\propto \frac{P\phi ZN}{60\text{ A}}$

A tachogenerator has permanent magnets. Hence flux is constant. Hence Emf $\propto$ speed.

464. (*a*)

465. (*c*) $G = \frac{dR/R}{dl/l}$ or $dR = G\frac{dl}{l} \cdot R$

or $dR = 2 \times 10^{-5} \times 120 = 2.4 \times 10^{-3}$ ohm

466. (*c*) Output $= \frac{0.25}{0.5} \times 5.2 = -2.6\text{ V}$

467. (*b*) CRO has minimum loading effect due to its very high input impedance.

468. (*c*) Resistance of meter = 10 × 2 = 20 k ohm

Reading $= \dfrac{\text{Resistance of meter} \times 8}{\text{Resistance of meter + Output resistance}}$

$= \frac{8 \times 20}{20 + 1} = 7.62\text{ V}$.

469. (*c*) Connection (*a*) is used for low load current because voltage drop current coil will be low connection (*b*) is used for high load current because voltage drop across current coil is high.

470. (*b*) LVDT is a very suitable and widely used tranducer to measure linear displacement.

471. (*b*) Both are correct but independent.

472. (*a*) **473.** (*a*) **474.** (*a*)

475. (*a*) Since the deflecting torque is proportional to current, the scale is linear.

476. (*c*) Shunt is always connected in parallel with ammeter.

477. (*a*)

478. (*d*) Thermocouple instrument can be used for both ac and dc.

479. (*a*)

480. (*c*) Wattmeter measures active power.

481. (*c*) CT and PT cannot be used in dc.

482. (*b*) Both are correct but independent.

483. (*a*) **484.** (*a*) **485.** (*a*)

486. (*a*) If resistance of voltmeter is not high, accuracy is affected.

487. (*a*)

488. (*b*) Both are correct but independent.

489. (*b*) **490.** (*b*) **491.** (*b*)

492. (*b*) Both are correct but independent.

493. (*a*) **494.** (*b*) **495.** (*a*)

496. (*b*) **497.** (*a*) **498.** (*b*)

499. (*a*) **500.** (*a*)

501. (*b*) Both are correct but independent.

502. (*b*) **503.** (*b*) **504.** (*b*)

505. (*b*)

506. (*a*) The dummy strain gauge compensates the error due to change in temperature because temperature affects both the active and dummy strain gauges equally.

507. (*a*) Higher gauge factor results in better accuracy.
508. (*a*)
509. (*a*) Since heating ∝ I^2, scale is cramped.
510. (*a*) Hot wire instruments are suitable for both ac and dc.
511. (*b*)
512. (*b*) Both are correct but independent.
513. (*b*)
514. (*c*) Maxwell bridge does not use a capacitor.
515. (*b*)
516. (*c*) Capacitors are mostly imperfect.
517. (*c*) Modified De Sauty bridge is used to measure capacitance.
518. (*d*) Schering bridge is a high voltage bridge.
519. (*b*)
520. (*c*) Frequency range is 47 to 53 Hz.
521. (*a*) **522.** (*a*)
523. (*b*)
524. (*c*) Probe does not introduce attenuation. It does not alter or load the test circuit.
525. (*a*)
526. (*b*) Both are individually correct.
527. (*c*) R is wrong.
528. (*c*). R is wrong. Electronic voltmeter is more accurate.
529. (*a*)
530. (*a*)
531. (T) **532.** (F)
533. (F) It is an integrating instrument.
534. (T)
535. (F) There are 8 classes of accuray as per BIS.
536. (T) **537.** (T)
538. (T) **539.** (T)
540. (T)
541. (F) Moving iron instruments can be used for both dc and ac.
542. (F) Since reading is proportional to I^2, the scale is non-linear.
543. (F) Since reading is proportinal to I^2, the scale is non-linear.
544. (T) **545.** (T)
546. (F) It is suitable for high voltage.
547. (T) **548.** (F)
549. (T)
550. (F) It is linear.
551. (T)
552. (F) The shading bands are of conducting material.
553. (T) **554.** (T)
555. (T) **556.** (T)
557. (T)
558. (F) It is suitable for both balanced and unbalanced loads.
559. (T) **560.** (T)
561. (T) **562.** (F) It is very low.
563. (T) **564.** (T)
565. (T)
566. (F) It is suitable for medium resistances.
567. (T)
568. (F) It should for high resistances.
569. (T) **570.** (T)
571. (F) It should have high resistance.
572. (T)
573. (F) It is 1.01859.
574. (F) **575.** (T)
576. (F)
577. (F) See section 6.30.1.
578. (F) See section 6.30.1.
579. (F) It is suitable for high $\frac{L}{R}$ ratio inductances.
580. (T)
581. Full scale deflection **582.** taut band
583. vertical
584. aluminium **585.** air friction
586. parallax **587.** low
588. non-linear **589.** repulsion
590. electrostatic **591.** stray magnetic fields
592. eddy currents **593.** moving coil
594. bridge rectifier **595.** high frequencies
596. induction **597.** air cored
598. shading bands **599.** uniform
600. series, pressure **601.** line
602. magnetic screening **603.** balanced
604. high **605.** surface leakage
606. battery, galvanometer
607. low **608.** slow
609. volt ratio box **610.** polar
611. quadrature **612.** rotating
613. rotor **614.** series
615. vibration galvanometer
616. magnitude, phase **617.** insulators
618. magnetic screen **619.** mutual
620. modified De-Sauty

CHAPTER 7

ANALOG ELECTRONIC CIRCUITS

7.1. DIODE CIRCUITS

7.1.1. Half Wave Rectifier

Figure 7.1 (*a*) shows the circuit

$$i = \frac{E_m \sin \omega t}{R_f + R_L} \text{ for } 0 \le \omega t \le \pi$$

$$i = 0 \text{ for } \pi \le \omega t \le 2\pi$$

where R_f is forward resistance of diode

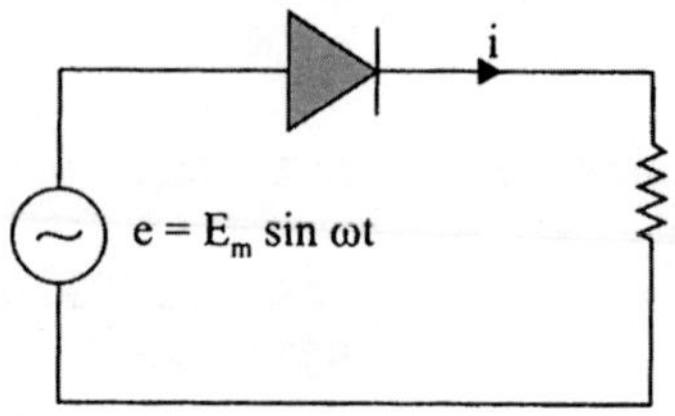

(*a*) Circuit

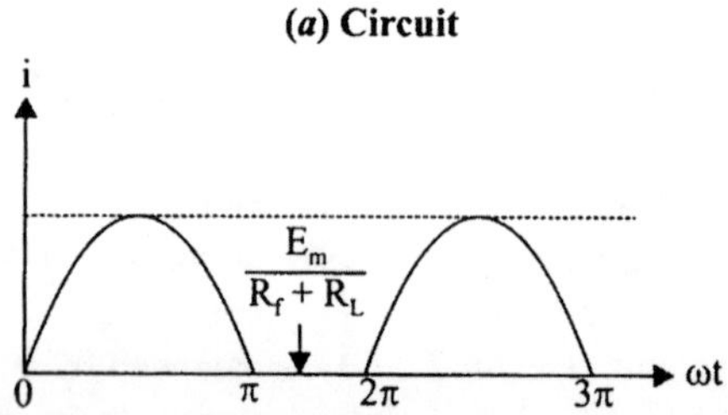

(*b*) Current wave form

Fig. 7.1. Half wave rectifier.

Average current $I_{dc} = \frac{1}{2\pi}\int_0^{2\pi} i d(\omega t) = \frac{E_m}{\pi(R_f + R_L)}$

RMS current $I = \left[\frac{1}{2\pi}\int_0^{2\pi} i^2 d(\omega t)\right]^{0.5} = \frac{0.5E_m}{R_f + R_L}$

Average output voltage $V_{dc} = I_{dc} R_L = \frac{E_m R_L}{\pi(R_f + R_L)}$

RMS output voltage $V = IR_L = \frac{0.5 E_m R_L}{R_f + R_L}$

P_{dc}, *i.e.*, dc power delivered to the load

$$= I_{dc}^2 R_L = \frac{E_m^2 R_L}{\pi^2 (R_f + R_L)^2}$$

P_{in}, *i.e.*, total power input to the circuit

$$= I^2(R_f + R_L) = \frac{E_m^2}{4(R_f + R_L)}$$

Rectifier efficiency $= \frac{P_{dc}}{P_{in}} = \left(\frac{2}{\pi}\right)^2 \frac{R_L}{R_f + R_L} = \frac{0.406}{1 + R_f/R_L}$

If $R_L >> R_f$, efficiency $\simeq 0.406$

$$\gamma = \text{Ripple factor} = \left[\frac{I^2}{I_{dc}^2} - 1\right]^2 = 1.21$$

Peak inverse voltage (PIV) $= E_m$

Output voltage at no load $V_{dco} = E_m/\pi$

$$V_{dc} = V_{dco} - I_{dc} R_f$$

7.1.2. Full Wave Rectifier (Using Centre Tapped Transformer)

Figure 7.2. shows the circuit

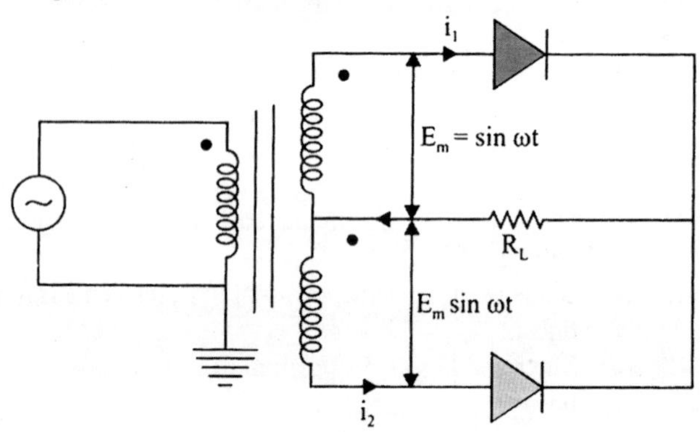

(*a*) Circuit

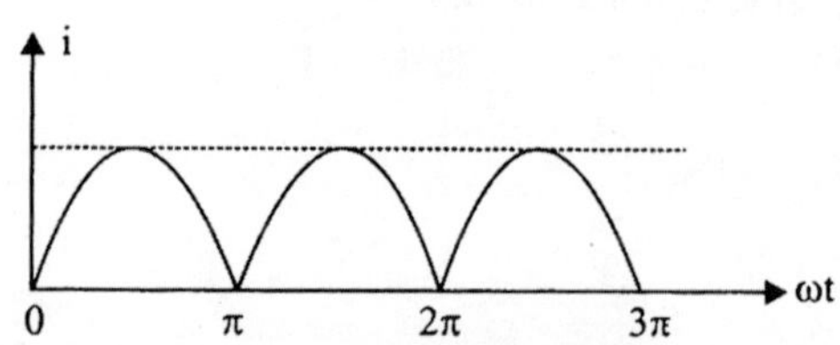

(*b*) Waveshape of current

Fig. 7.2. Full wave rectifier.

$$i_1 = \frac{E_m \sin \omega t}{R_f + R_L} \text{ for } 0 \le \omega t \le \pi$$

$$i_1 = 0 \text{ for } \pi \le \omega t \le 2\pi$$

$$i_2 = 0 \text{ for } 0 \le \omega t \le \pi$$

$$i_2 = \frac{E_m \sin \omega t}{R_f + R_L} \text{ for } \pi \le \omega t \le 2\pi$$

Average current $I_{dc} = \dfrac{2E_m}{\pi(R_f + R_L)}$

RMS current $I = \dfrac{E_m}{\sqrt{2}(R_f + R_L)}$

Average voltage $V_{dc} = I_{dc}\,R_L = \dfrac{2E_m R_L}{\pi(R_f + R_L)}$

RMS voltage $V = IR_L = \dfrac{E_m R_L}{\sqrt{2}(R_f + R_L)}$

DC power delivered to load $P_{dc} = I_{dc}^2 R_L = \dfrac{4E_m^2 R_L}{\pi^2 (R_f + R_L)^2}$

Power input $P_{in} = I^2(R_f + R_L) = \dfrac{E_m^2}{2(R_f + R_L)}$

Rectifier Efficiency $= \dfrac{P_{dc}}{P_{in}} = \dfrac{0.812}{1 + R_f/R_L}$

Ripple factor $\gamma = \left[\dfrac{I^2}{I_{dc}^2} - 1\right]^{0.5} = 0.483$

Output voltage at no load $V_{dco} = \dfrac{2E_m}{\pi}$

$$V_{dc} = V_{dco} - I_{dc}\,R_f = \frac{2E_m}{\pi} - I_{dc}R_f$$

Peak inverse voltage (PIV) $= 2E_m$

7.1.3. Bridge Rectifier

Figure 7.3 (*a*), shows the circuit. It is also a full wave rectifier.

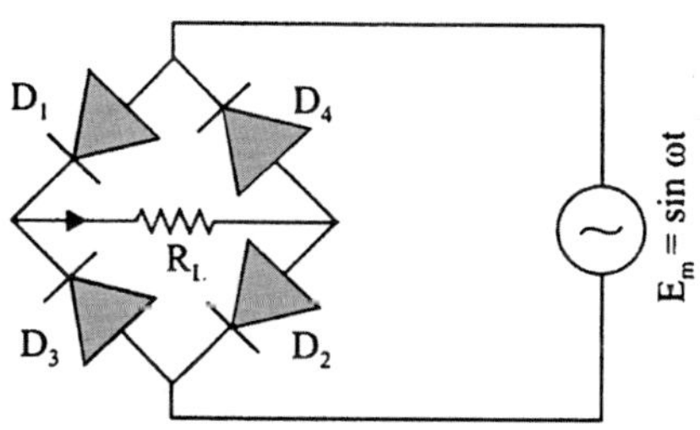

(*a*) Circuit

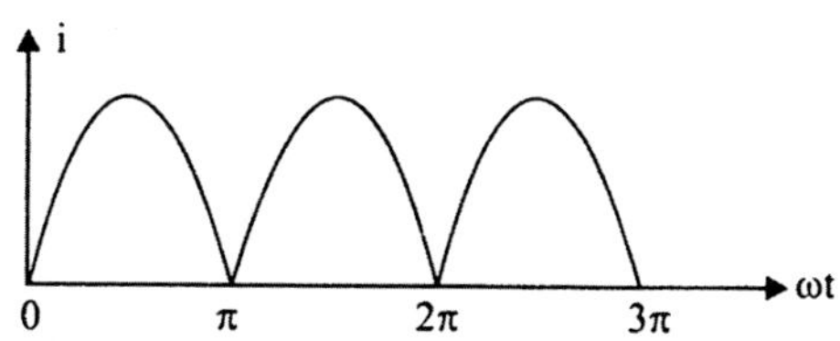

(*b*) Wave shape of current

Fig. 7.3. Bridge rectifier.

$$i = \frac{E_m \sin \omega t}{2R_f + R_L}$$

$$I_{dc} = \frac{1}{2\pi}\int_0^{2\pi} \frac{E_m \sin \omega t}{2R_f + R_L} = \frac{2E_m}{\pi(2R_f + R_L)}$$

$$I = \left[\frac{1}{2\pi}\int_0^{2\pi} i^2 d(\omega t)\right] = \frac{E_m}{\sqrt{2}(2R_f + R_L)}$$

$$V_{dc} = I_{dc}\,R_L = \frac{2E_m R_L}{\pi(2R_f + R_L)}$$

$$V = IR_L = \frac{E_m R_L}{\sqrt{2}(2R_f + R_L)}$$

$$P_{dc} = I_{dc}^2 R_L = \frac{4E_m^2}{\pi^2(2R_f + R_L)^2}$$

$$P_{in} = I_2(2R_f + R_L) = \frac{E_m^2}{2(2R_f + R_L)}$$

Efficiency $\dfrac{P_{dc}}{P_{in}} = \dfrac{0.812}{1 + 2R_f/R_L}$

$$\gamma = \left[\frac{I^2}{I_{dc}^2} - 1\right]^{0.5} = 0.483$$

$$V_{dco} = \frac{2E_m}{\pi}$$

$$V_{dc} = V_{dco} - I_{dc}(2R_f)$$

Peak inverse voltage $= E_m$

7.1.4. Three Phase Rectifier

It is used when power requirement is high. The output voltage waveshape is more smooth than that of single phase rectifier. The lowest ac frequency present in the output is 300 Hz. The ripple is 4.2%. Figure 7.4 shows the circuit.

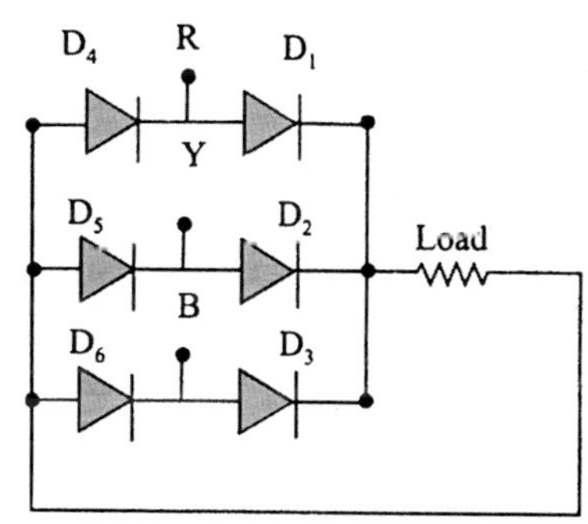

(*a*) Circuit

Max. value of line to line voltage

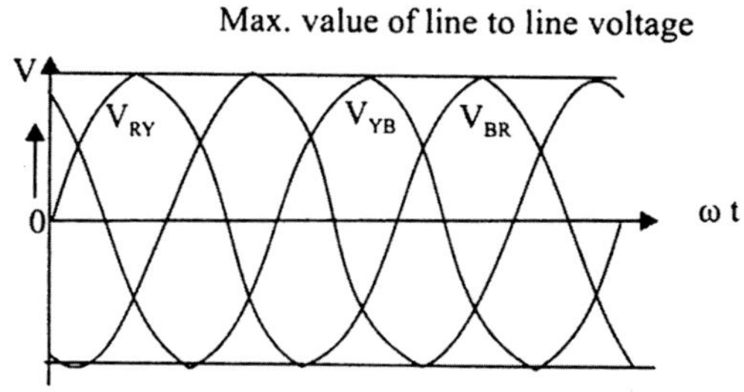

(*b*) Wave shape of input voltage

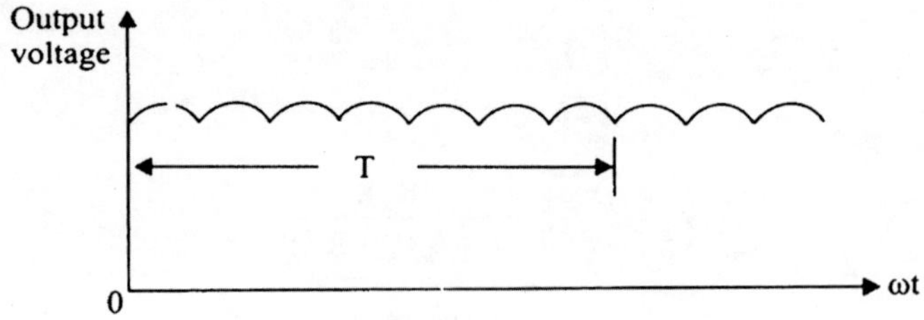

(*c*) **Wave shape of output voltage**

Fig. 7.4. Three-phase full wave rectifier circuit

7.1.5. Rectifier Filters

Figure 7.5 to 7.8 show the configurations.

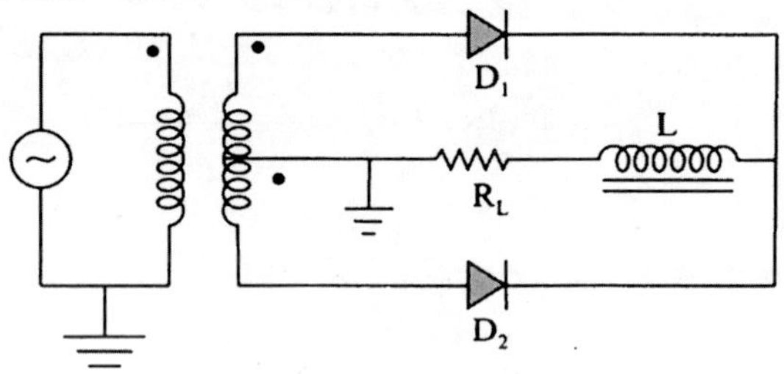

Fig. 7.5. Inductance filter

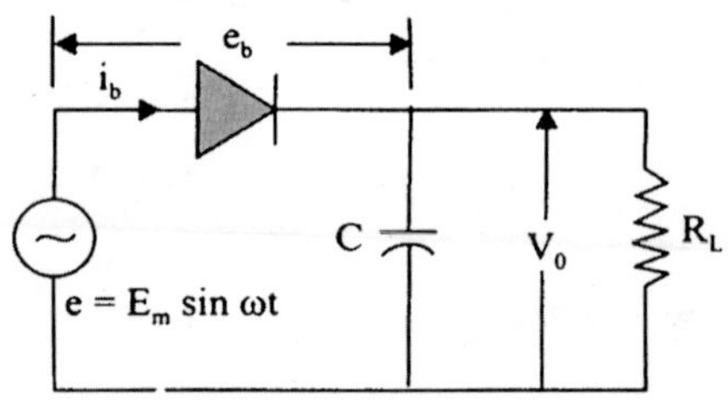

(*a*) **Circuit**

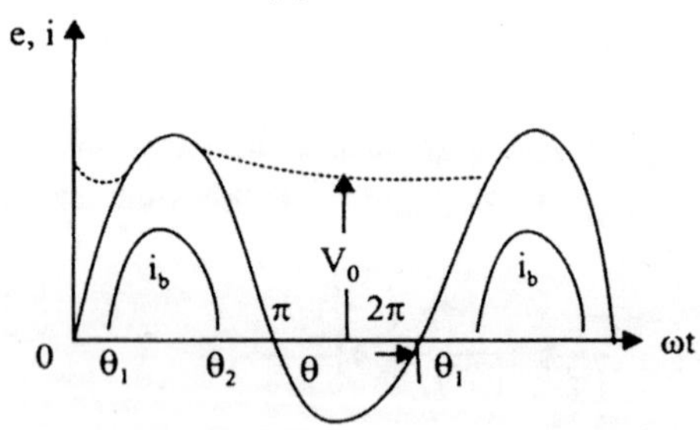

(*b*) **Current and voltage wave-forms**

Fig. 7.6. Half-wave rectifier with capacitance filter

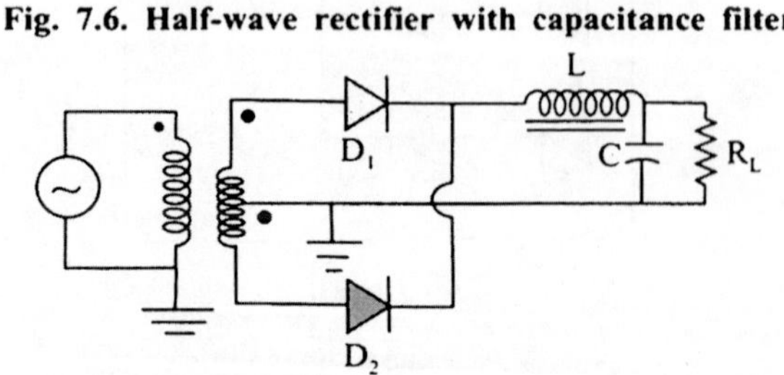

Fig. 7.7. L-section filter

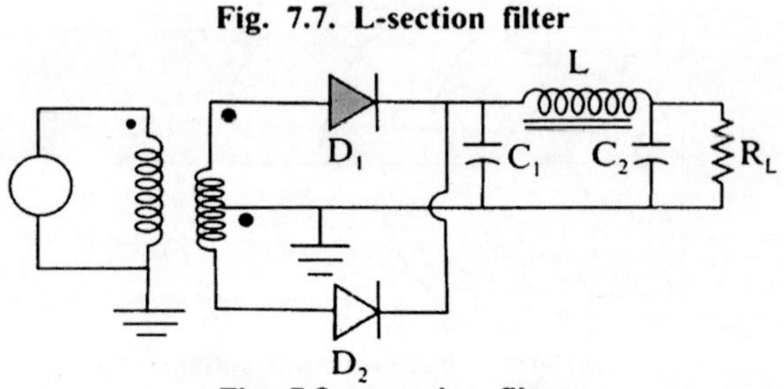

Fig. 7.8 π section filter

7.1.6. Voltage Doubler

Figures 7.9 and 7.10 show the circuits.

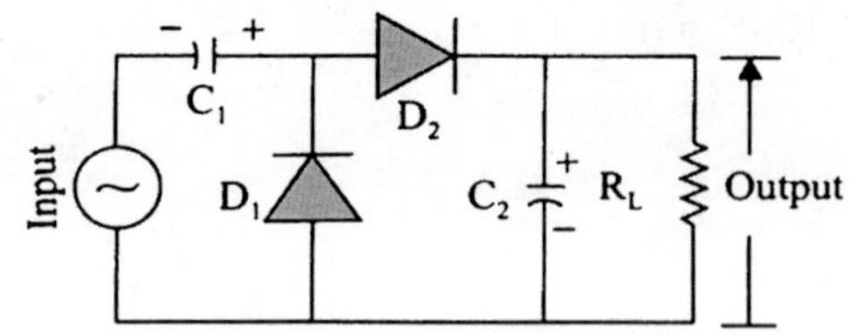

Fig. 7.9. Half-wave voltage doubler circuit

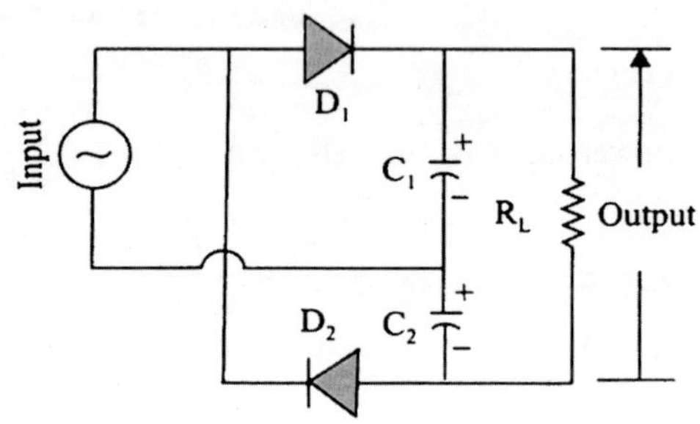

Fig. 7.10. Full-wave voltage doubler circuit

7.1.7. Voltage Multiplier

Figures 7.11, 7.12 and 7.13 show the circuits.

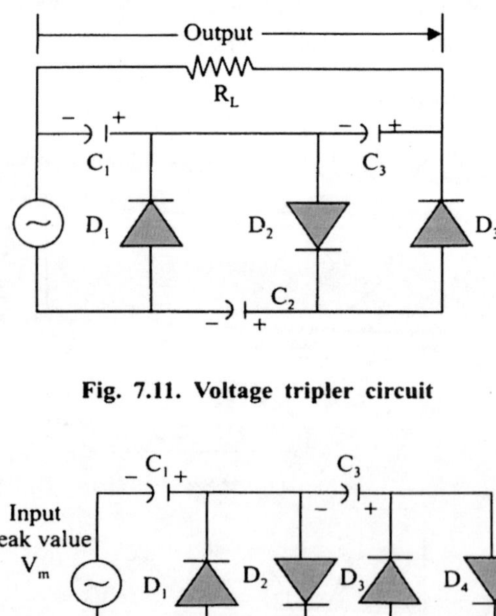

Fig. 7.11. Voltage tripler circuit

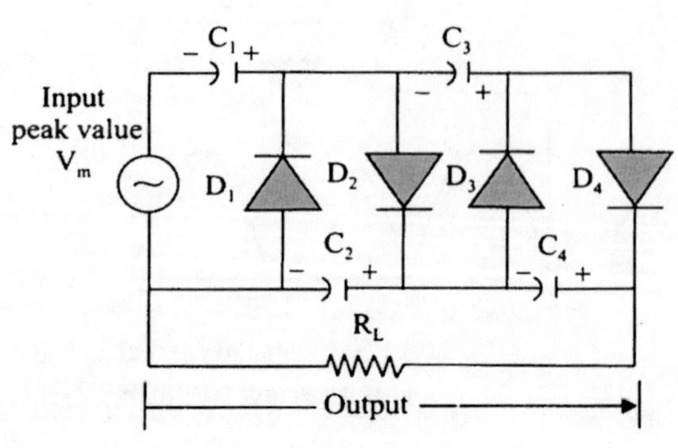

Fig. 7.12 Voltage quadrupler circuit

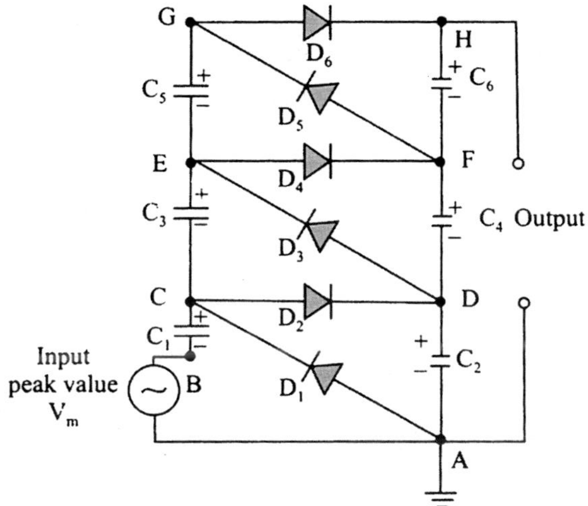

Fig. 7.13. Voltage multiplier circuit

7.1.8. LIMITER CIRCUITS (CLIPPER CIRCUITS)

Figure 7.14 to 7.19, show the circuits. A limiter removes the signal voltage below or above a certain specified level.

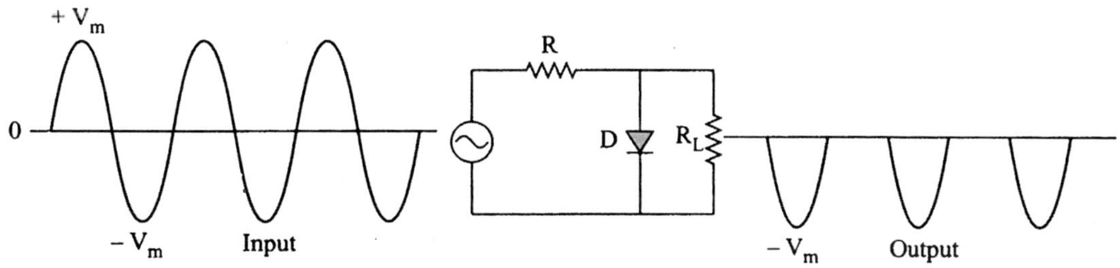

Fig. 7.14. Positive limiter circuit

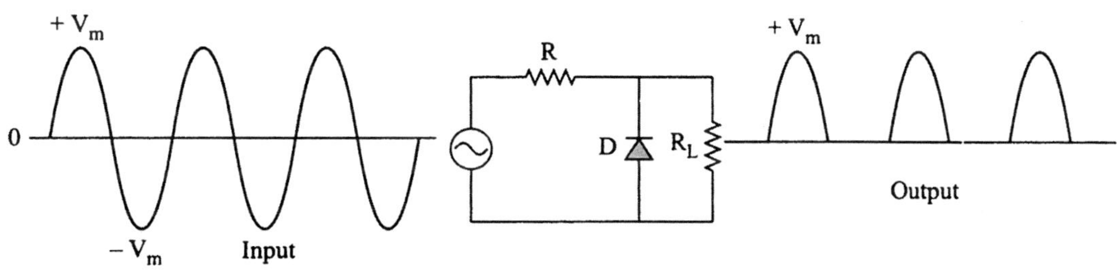

Fig. 7.15. Negative limiter circuit

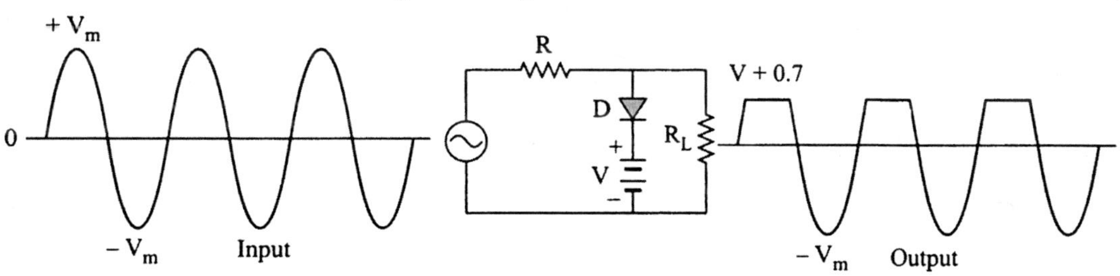

Fig. 7.16. Biased positive limiter circuit

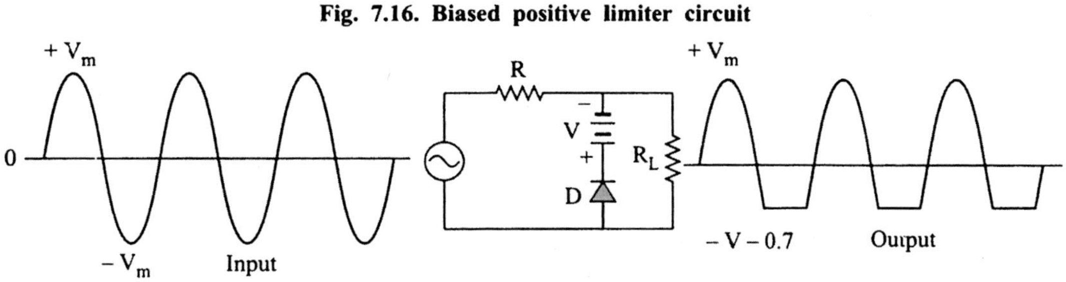

Fig. 7.17. Biased negative limiter circuit

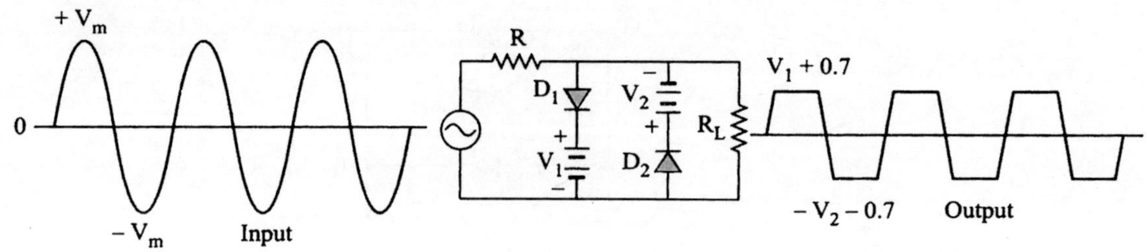

Fig. 7.18. Combination limiter circuit

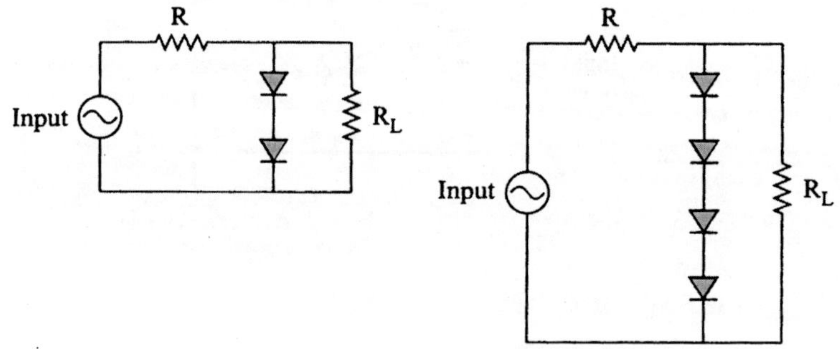

Fig. 7.19. Biased positive limiter circuit using diodes

7.1.9. Clamper Circuits

A clamper is used to add a dc voltage to a signal. Figures 7.20 and 7.21, show the circuits

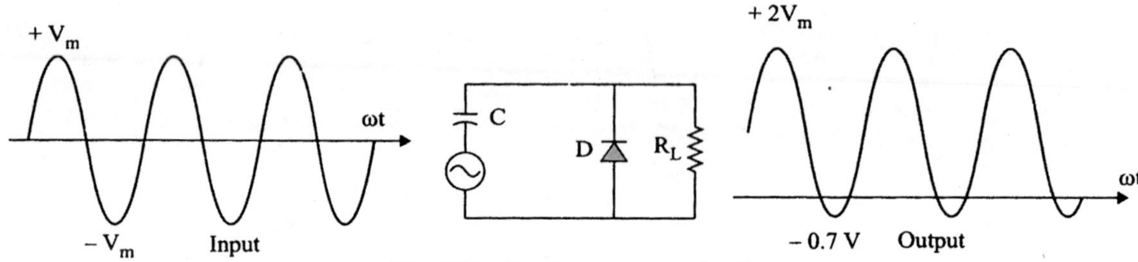

Fig. 7.20. Positive clamper circuit

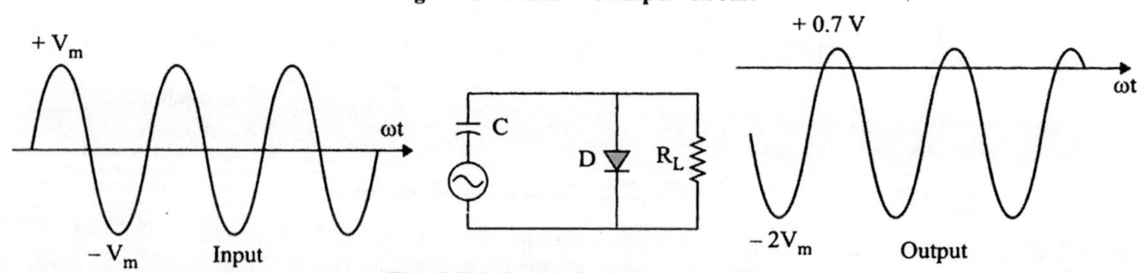

Fig. 7.21. Negative clamper circuit

7.2. BJT AMPLIFIER CIRCUITS

7.2.1. Voltage and Current Symbols

Figure 7.22 shows a CE transistor amplifier circuit. The terminology of voltages and currents is shown in Table 7.1.

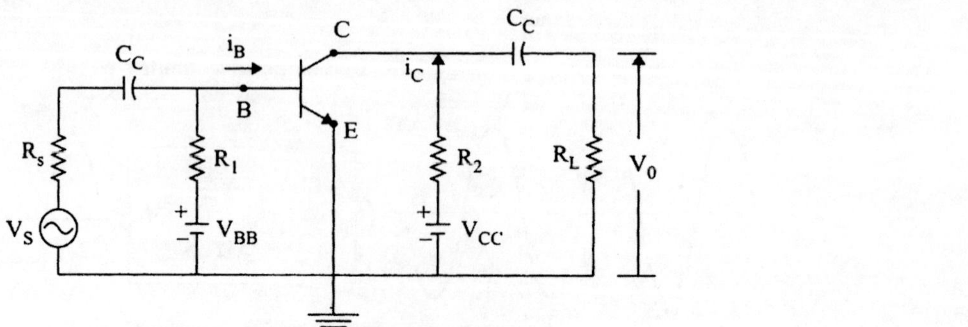

Fig. 7.22. Transistor amplifier circuit

TABLE 7.1. Transistor Voltage and Current Symbols

	Total instantaneous value	***Average value (or quiescent value)***	***AC or time-varying component***	
			Instantaneous value	***RMS value***
Collector voltage	v_{CE}	V_{CE}	v_{ce}	V_{ce}
Base voltage	v_{BE}	V_{BE}	v_{be}	V_{be}
Base current	i_B	I_B	i_b	I_b
Collector current	i_C	I_C	i_c	I_c
Emitter current	i_E	I_E	i_e	I_e
	These are denoted by lower case letters with capital subscripts	These are denoted by capital letters with capital subscripts	These are denoted by lower case letters with lower case subscripts	These are denoted by capital letters with lower case subscripts

The total instantaneous value is the sum of the average (or quiescent) value and time-varying component, *e.g.*, $i_B = I_B + i_b$.

7.2.2. Graphical Analysis

It is done by drawing dc and ac load lines. Figure 7.23 shows the graphical analysis.

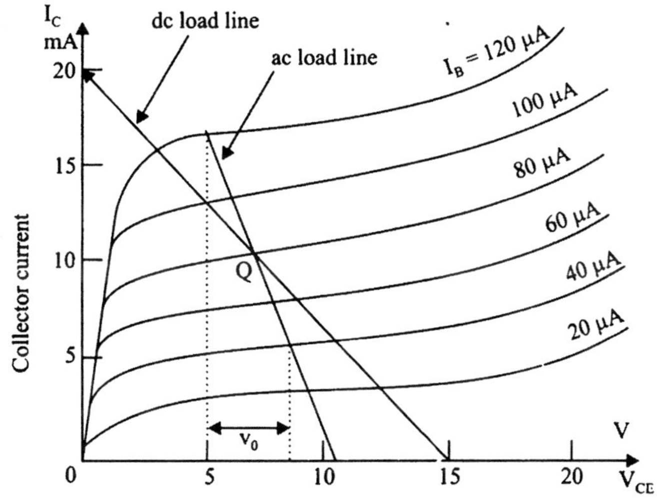

Fig. 7.23. Graphical analysis of transistor amplifier circuit

7.2.3. Hybrid Parameter Equivalent Circuit

Figure 7.24 shows CE amplifier and its *h* parameter equivalent circuit.

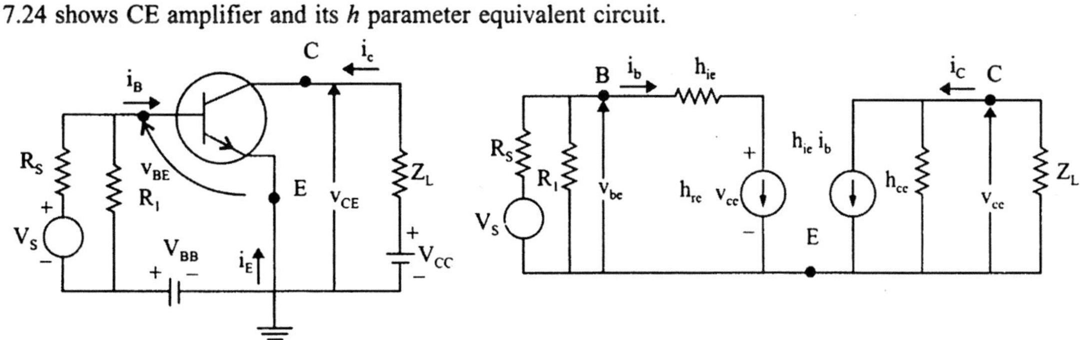

(*a*) **Circuit** (*b*) **Equivalent circuit**

Fig. 7.24. CE amplifier

$$v_{be} = h_{ie}\, i_b + h_{re}\, v_{ce}$$
$$i_c = h_{fe}\, i_b + h_{oe}\, v_{ce}$$

Figure 7.25 shows CB amplifier and its equivalent circuit.

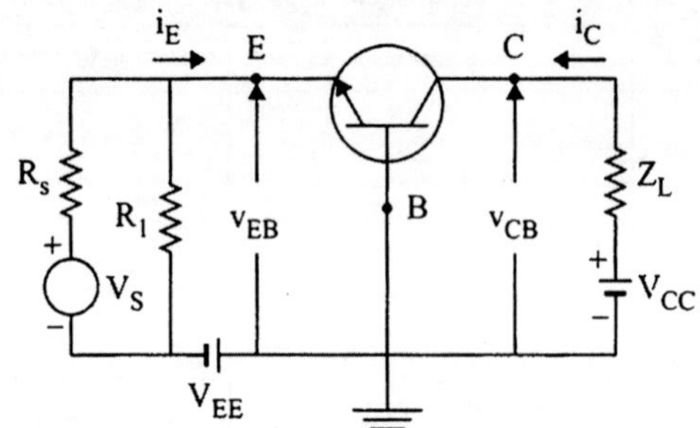

(*a*) **Circuit**

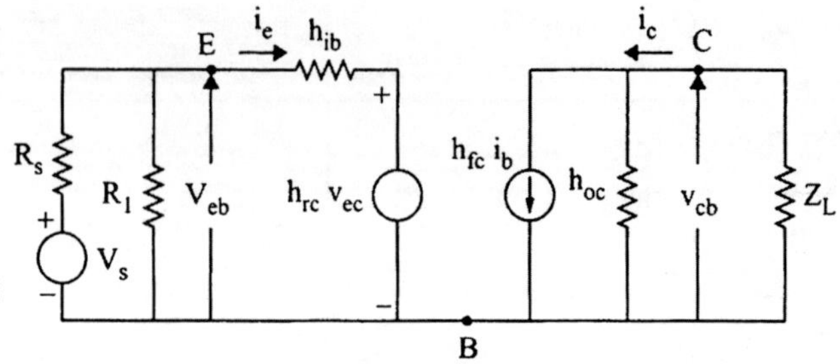

(*b*) **Equivalent circuit**

Fig. 7.25. CB amplifier

$$v_{eb} = h_{ib}\, i_e + h_{rb}\, v_{cb}$$
$$i_c = h_{fb}\, i_e + h_{ob}\, v_{cb}$$

Figure 7.26 shows CC amplifier and its equivalent circuit.

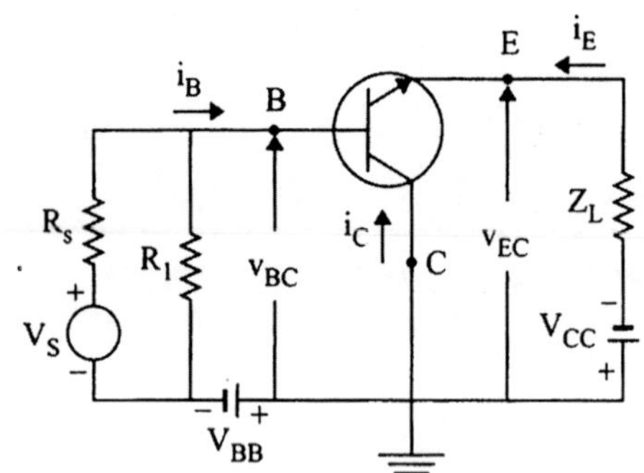

(*a*) **Circuit**

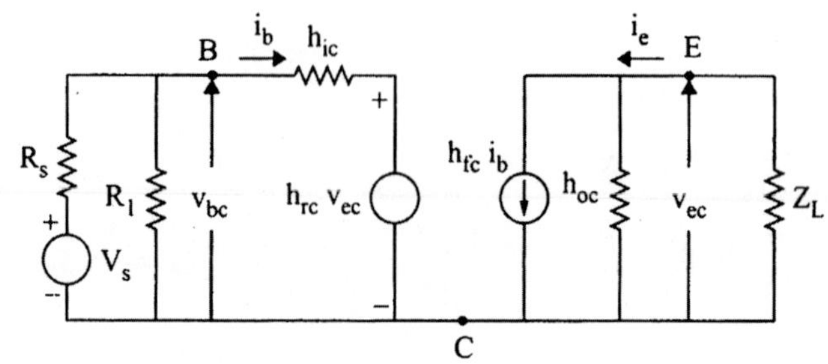

(*b*) **Equivalent circuit**

Fig. 7.26. CC amplifier

$$v_{bc} = h_{ic}\, i_b + h_{rc}\, v_{ec}$$
$$i_e = h_{fc}\, i_b + h_{oc}\, v_{ec}$$

7.2.4. Transistor Linear Circuit Analysis

It is more convenient to write the equations of Section 7.2.3 in terms of generalised notation *i, v,* etc. Table 7.2 gives the correlation of *i, v,* etc. with circuit configuration of Figures 7.24 to 7.26.

TABLE 7.2. Correlation of Generalized Two-port Notation to Transistor Amplifier Configurations

Generalized notation	*Units*	*CE*	*CB*	*CC*
i_1	A	i_b	i_e	i_b
v_1	V	v_{be}	v_{eb}	v_{bc}
i_2	A	i_c	i_c	i_e
v_2	V	v_{ce}	v_{cb}	v_{ec}
h_{11}	ohms	h_{ie}	h_{ib}	h_{ic}
h_{12}	—	h_{re}	h_{rb}	h_{rc}
h_{21}	—	h_{fe}	h_{fb}	h_{fc}
h_{22}	siemens	h_{oe}	h_{ob}	h_{oc}

The conversion formulae of *h* parameters are given in Table 7.3.

TABLE 7.3. Conversion Formulae for *h* Parameters

	CE	*CC*	*CB*	*CC*
h_{11}	$h_{ie} = \frac{h_{ib}}{1+h_{fb}}$	$h_{ic} = \frac{h_{ib}}{1+h_{fb}}$	$h_{ib} = \frac{h_{ic}}{1+h_{fe}}$	$h_{ic} = h_{ie}$
h_{12}	$h_{re} = \frac{\Delta h_b - h_{rb}}{1+h_{fb}}$	$h_{rc} = 1$	$h_{rb} = \frac{\Delta h_e - h_{re}}{1+h_{fe}}$	$h_{rc} = 1 - h_{re}$
h_{21}	$h_{fe} = \frac{-h_{fb}}{1+h_{fb}}$	$h_{fc} = \frac{-1}{1+h_{fb}}$	$h_{fb} = \frac{-h_{fe}}{1+h_{fe}}$	$h_{fc} = -1 - h_{fe}$
h_{22}	$h_{oe} = \frac{h_{ab}}{1+h_{fb}}$	$h_{oc} = \frac{h_{ob}}{1+h_{fb}}$	$h_{oc} = \frac{h_{oe}}{1+h_{fe}}$	$h_{oc} = h_{oe}$
where	$\Delta h_b = h_{ib}\, h_{ab} - h_{fb}\, h_{rb}$ $\Delta h_e = h_{ie}\, h_{oe} - h_{fe}\, h_{re}$			

$$v_1 = h_{11} i_1 + h_{12} v_2$$
$$i_2 = h_{21} i_1 + h_{22} v_2$$
$$v_2 = Z_L i_2$$

Current gain $$A_i = \frac{i_2}{i_1} = \frac{-h_{21}}{1+h_{22}Z_L}$$

Voltage gain $$A_v = \frac{v_2}{v_1} = \frac{-h_{21}Z_L}{h_{11}+Z_L(h_{11}h_{22}-h_{12}h_{21})}$$

Power gain $$A_p = \frac{-v_2 i_2}{v_1 i_1} = A_v A_i$$

Input impedance $$Z_i = \frac{v_i}{i_1} = \frac{Z_L(h_{11}h_{22}-h_{12}h_{21})+h_{11}}{h_{22}Z_L+1}$$

Output impedance $$Z_0 = \left.\frac{v_2}{i_2}\right|_{v_s=0} = \frac{R'_s + h_{11}}{R'_s h_{22} + h_{11}h_{22} - h_{12}h_{21}}$$

where $$R'_s = \frac{R_s R_1}{R_s + R_1}$$

Table 7.4 gives comparison of CE, CB and CC configurations.

TABLE 7.4. Comparison of CE, CB and CC amplifier configurations

	CE	*CB*	*CC*
A_i	good	less than 1	high
A_v	good	high	less than 1
A_p	high	good	good
Z_i	good	low	high
Z_p	good	high	low
Phase shift	180° for both current and voltage	nil	nil
Application	Power amplification	Radio frequency isolation	Impedance matching

7.2.5. SELF BIAS CIRCUIT (VOLTAGE DIVIDER BIAS)

Figure 7.27 shows self bias circuit and its equivalent circuit.

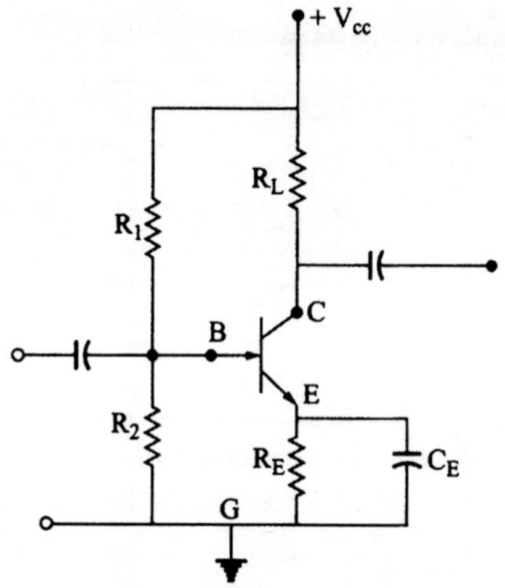

(a)

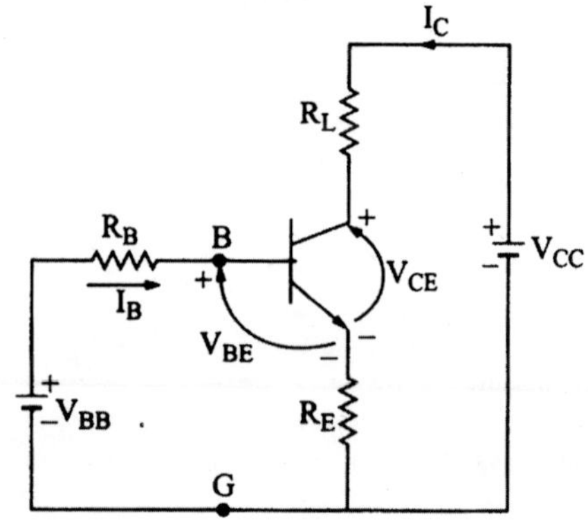

(b)

Fig. 7.27 (a) Self-bias circuit (b) Equivalent circuit

$$V_{BB} = \frac{R_2}{R_1 + R_2} V_{CC}$$

$$R_B = \frac{R_1 R_2}{R_1 + R_2}$$

$$I_B \simeq \frac{V_{BB} - V_{BE}}{R_B + \beta R_E} \simeq \frac{V_{BB} - V_{BE}}{\beta R_E}$$

$$\text{Stability factor } S = (1+\beta)\left[\frac{1 + R_B/R_E}{1 + \beta + R_B/R_E}\right]$$

7.2.6. r_e Transistor Model

AC resistance of diode Junction $r_e = \frac{0.025}{I_E}$

Figure 7.28 shows CB amplifier and its r_e model equivalent circuit.

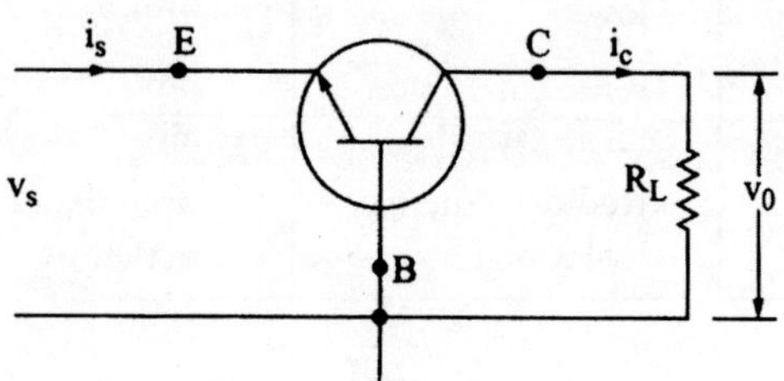

(a) Circuit

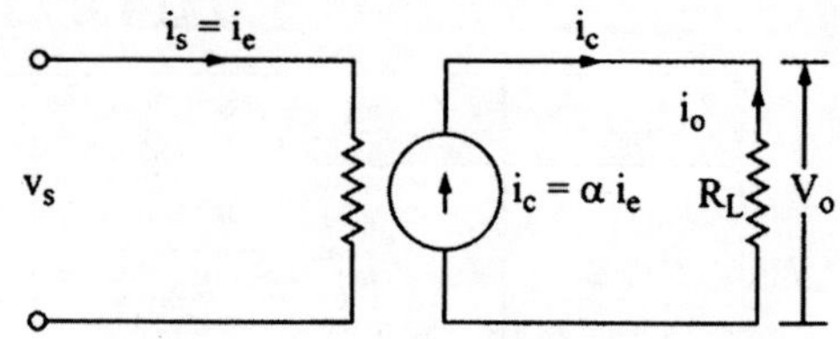

(b) Equivalent circuit based on r_e model

Fig. 7.28. CB amplifier circuit.

Voltage gain $A_v = \frac{\alpha R_L}{r_e}$

Current gain $A_i = \alpha$

Power gain $A_p = A_v A_i$

Figure 7.29 shows CE amplifier and its r_e model equivalent circuit.

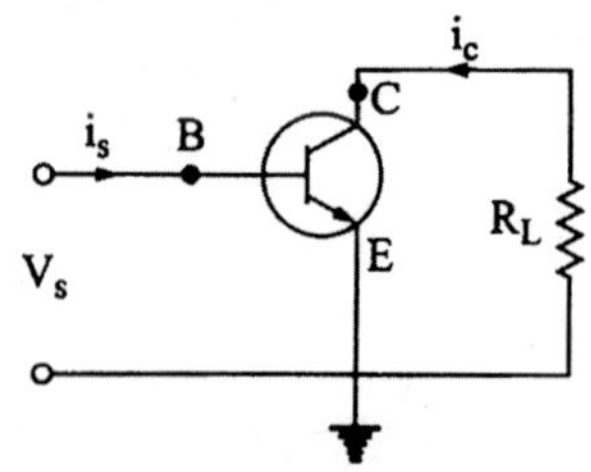

(a) Circuit

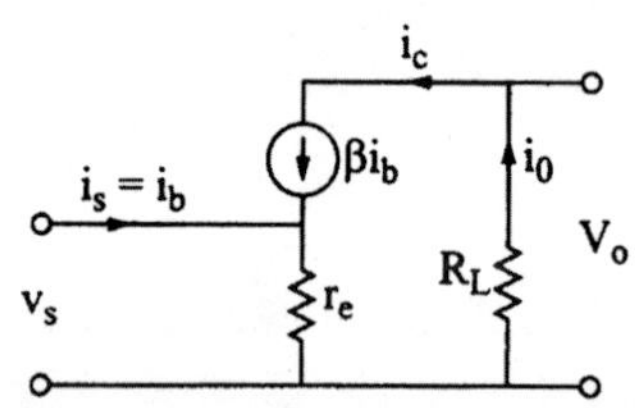

(b) Equivalent circuit based on r_e model

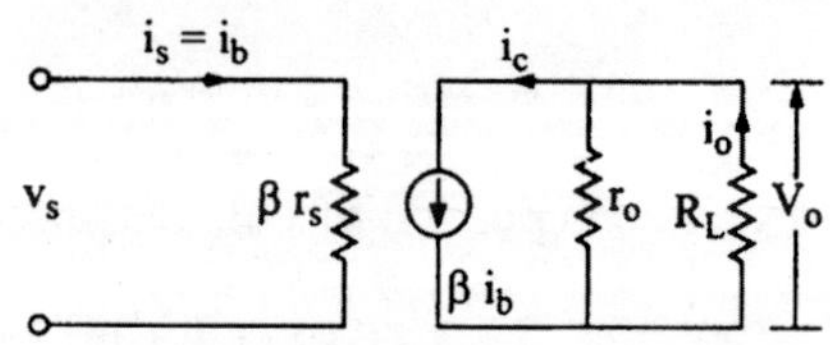

(c) Modified equivalent circuit

Fig. 7.29. CE amplifier

$$A_v = \frac{-r_0 R_L}{(r_0 + R_L) r_e}$$

$$A_i = -\beta\left(\frac{r_0}{r_0 + R_L}\right)$$

$$A_p = A_v A_i$$

Input impedance $Z_i = \beta r_e$

Output impedance $Z_0 = r_0$

Figure 7.30 shows CC amplifier circuit and r_e model equivalent circuit.

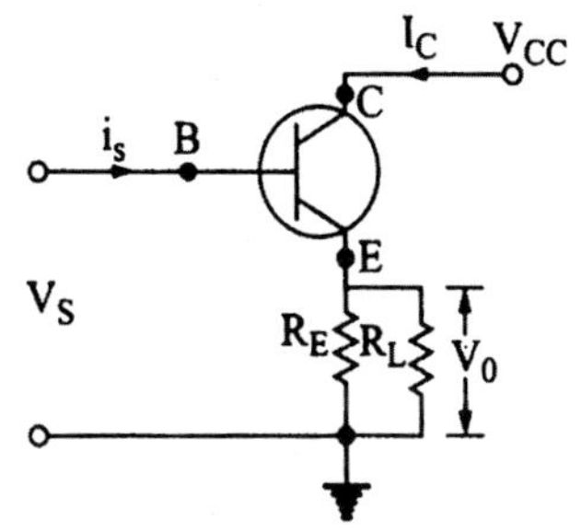

(*a*) Circuit

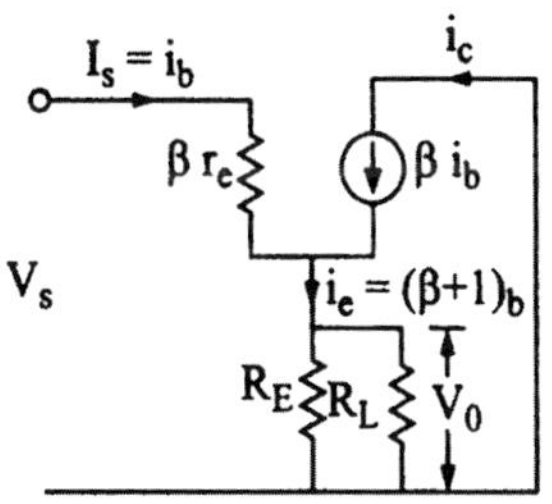

(*b*) Equivalent-circuit based on r_e model

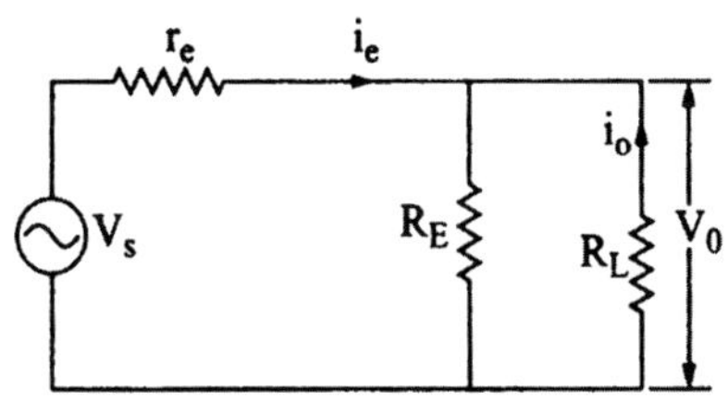

(*c*) Equivalent circuit as seen from output

Fig. 7.30. CC amplifier.

$$A_v = \frac{R_e'}{r_e + R_e'}$$

where $$R_e' = \frac{R_E R_L}{R_E + R_L}$$

$$A_i = \frac{\beta R_E}{R_E + R_L}$$

Input impedance $Z_i = \beta(r_e + R_e')$

Output impedance $$Z_0 = \frac{r_e R_E}{r_e + R_E}$$

7.2.7. Transformer Coupled Amplifier

Figure 7.31 shows transformer coupling for input and output

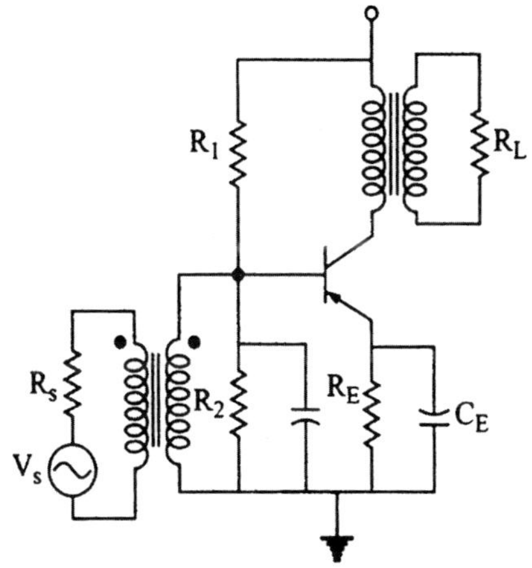

Fig. 7.31. Transform coupling.

$$R_L' = n^2 R_L$$

where $$n = \frac{N_1}{N_2}$$

7.2.8. RC Coupled Amplifier

Figure 7.32 shows CE RC coupled amplifier stages. Figure 7.33 shows the equivalent circuit

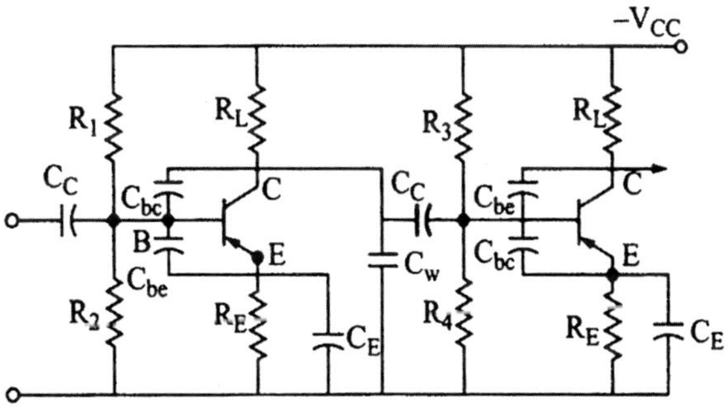

Fig. 7.32. RC coupled amplifier.

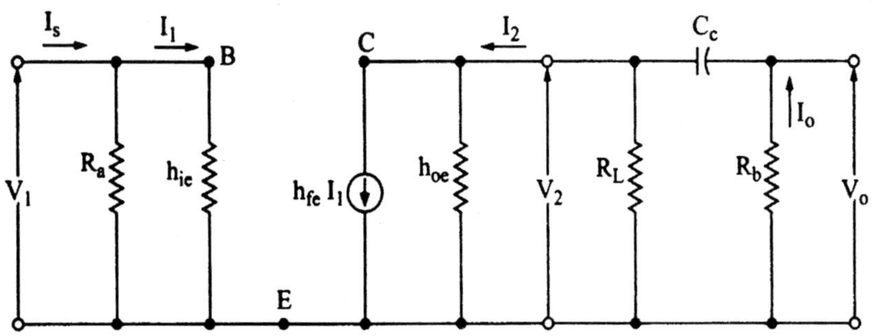

(*a*) Low frequency

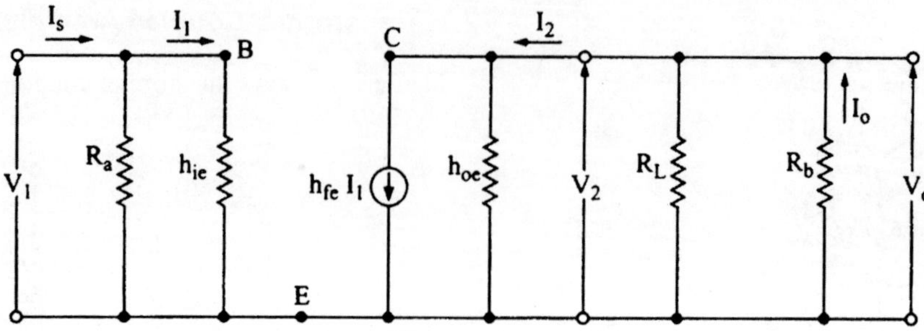

(b) Mid-band frequency

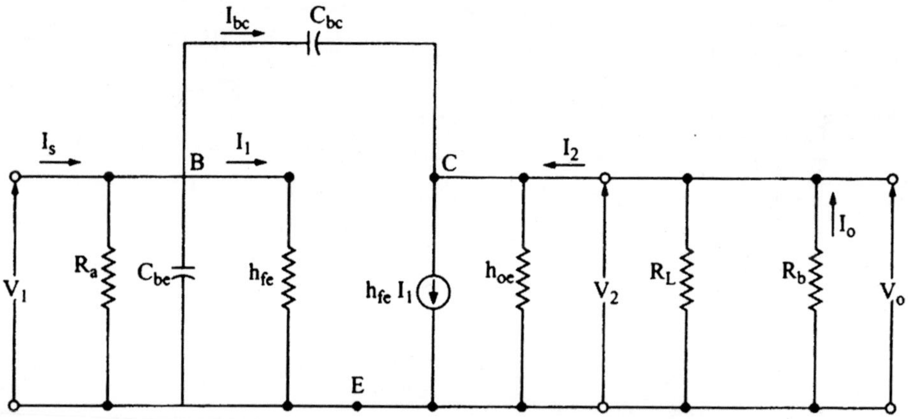

(c) High frequency

Fig. 7.33. CE amplifier equivalent circuit.

(a) Midband response

$$A_{im} = \left(\frac{R_a}{R_a + h_{ie}}\right)\left(\frac{-h_{fe}}{1 + h_{oe}R'_L}\right)\left(\frac{R_L}{R_L + R_b}\right)$$

$$A_{vm} = \frac{-h_{fe}R'_L}{R'_L h_{ie}h_{oe} + h_{ie}}$$

where $R_L' = \dfrac{R_L R_b}{R_L + R_b}$

(b) Low frequency response

$$A_{il} = \left(\frac{R_a}{R_a + h_{ie}}\right)\left(\frac{-h_{fe}}{1 + h_{oe}R_L}\right)\left(\frac{R_L}{R_L + R_b + 1/j\omega C_c}\right)$$

where $Z_L = \dfrac{R_L[R_b + (1/j\omega C_c]}{R_L + R_b + 1/j\omega C_c}$

also $A_{il} = \dfrac{A_{im}}{1 - j(\omega_1/\omega)}$

where $\omega_1 = \dfrac{1 + h_{oe}R_L}{C_c(R_L + R_b + h_{oe}R_L R_b)}$

$$A_{vl} = \frac{A_{vm}}{1 - j(\omega_1/\omega)}$$

(c) High frequency response $A_{ih} = \dfrac{A_{im}}{1 + j\omega/\omega_2}$

$$A_{vh} = A_{vm}$$

where $\omega_2 = \dfrac{R_a + h_{ie}}{C_{in}R_a h_{ie}}$

The frequency response is shown in Fig. 7.34.

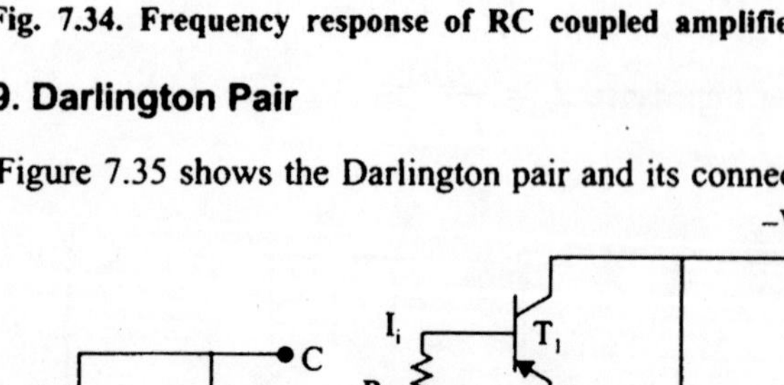

Fig. 7.34. Frequency response of RC coupled amplifier.

7.2.9. Darlington Pair

Figure 7.35 shows the Darlington pair and its connections

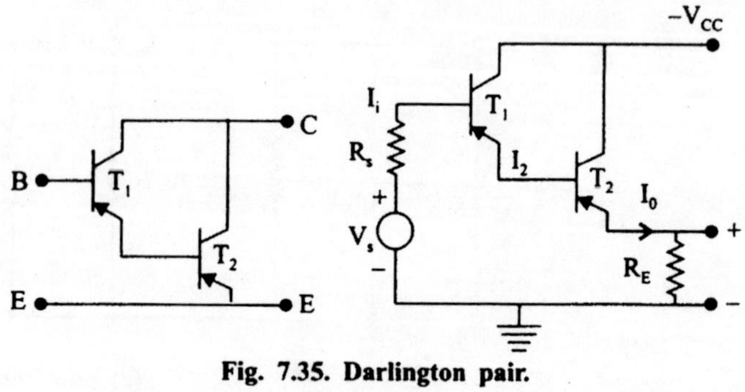

Fig. 7.35. Darlington pair.

$$A_i = (A_{i1})(A_{i2}) = \frac{(1+h_{fe})^2}{(1+h_{oe}h_{fe}R_E)}$$

A_v = voltage gain of T_2

$$\text{Input Impedance} = \frac{h_{fe}R_E(h_{ie}h_{oe}+h_{fe})+h_{ie}}{1+h_{oe}h_{fe}R_E}$$

$$\text{Output Impedance} = \frac{R_s+h_{ie}}{(1+h_{fe})^2}+\frac{h_{ie}}{1+h_{fe}}$$

7.3. POWER AMPLIFIER

Class A operation: Collector current exists for 360° of input signal cycle.

Class B operation: Collector current exists for 180° of input signal cycle.

Class C operation: Collector current exists for less than 180° of input signal cycle.

Maximum Efficiency of class A amplifier = 25%

Maximum Efficiency of class B and C amplifier

$$= \frac{\theta - \sin\theta}{4\sin\frac{\theta}{2} - 2\theta\cos\frac{\theta}{2}}$$

If $\theta = 180°$ (class B operation), maximum efficiency = 78.5%

Push pull amplifier: It has high efficiency and low distortion. Figure 7.36 shows the circuit

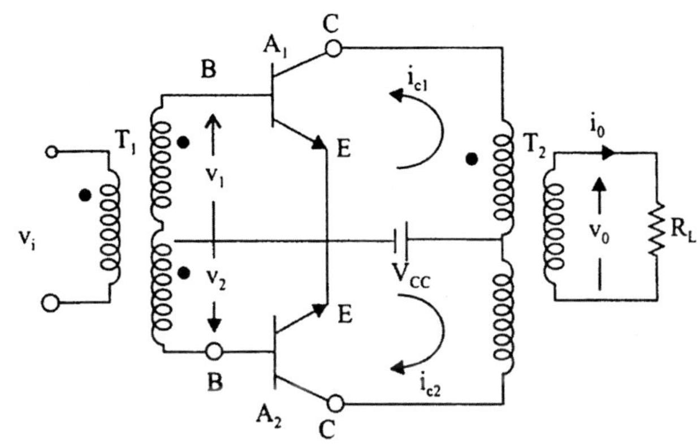

Fig. 7.36. Push-pull amplifier circuit.

7.4. FET AMPLIFIER

7.4.1. Self Bias Circuit

Figure 7.37 shows the self bias circuit for FET.

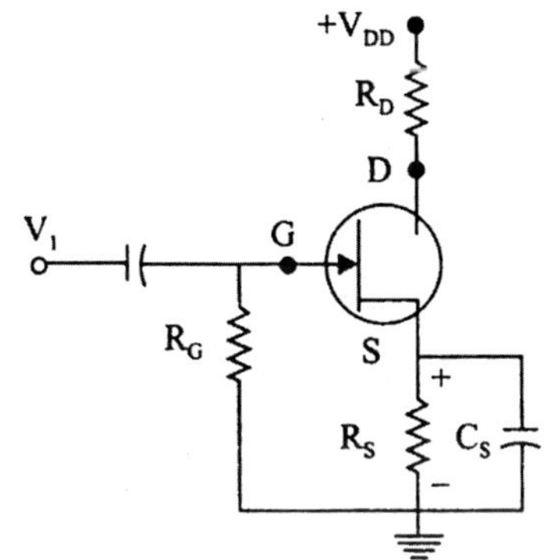

Fig. 7.37. Biasing circuit for FET.

7.4.2. FET Voltage and Current Symbols

Table 7.5 shows FET voltage and current symbols.

TABLE 7.5. FET Voltage and Current Symbols

	Total instantaneous value	*Average value (or quiescent value)*	*AC or time-varying component*	
			Instantaneous value	*RMS value*
Drain voltage Drain current Gate voltage Gate current	v_{DS} i_D v_{GS} i_G These are denoted by lower-case letters with capital subscripts	V_{DS} I_D V_{GS} I_G These are denoted by capital letters with capital subscripts	v_{ds} i_d v_{gs} i_g These are denoted by lower-case letters with lower-case subscripts	V_{ds} I_d V_{gs} I_g These are denoted by capital letters with lower-case subscripts
Total instantaneous value is sum of quiescent value and time varying component, *e.g.*, $i_D = I_D + i_d$.				

7.4.3. JFET Equivalent Circuit

Figure 7.38 shows JFET equivalent circuit.

$$g_{mo} = \frac{2I_{DSS}}{V_P}; \; g_m = g_{mo}\left(1-\frac{v_{GS}}{-V_P}\right)$$

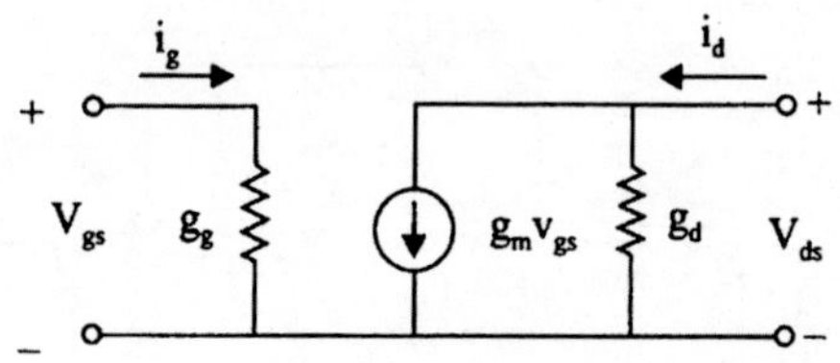

Fig. 7.38. JEFT equivalent circuit.

7.4.4. MOSFET Equivalent Circuit

Figure 7.39 shows MOSFET equivalent circuit.

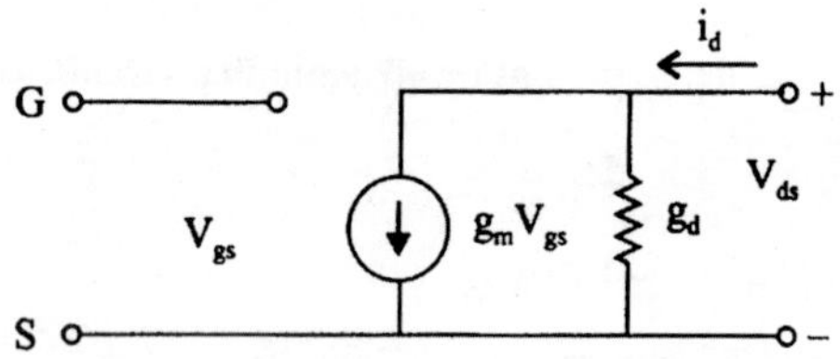

Fig. 7.39. MOSFET equivalent circuit.

$$g_m = \frac{2i_D}{v_{GS} - V_{GS(th)}}$$

Typical values of FET parameters are given is Table 7.6.

TABLE 7.6. Typical values of FET parameters

	JFET	*MOSFET*
g_m	0.1×10^{-3} to 0.01 siemens	0.1×10^{-3} to 0.02 siemens
r_d	0.1 to 1 M-ohm	1 to 50 k-ohm
g_g	$< 10^{-8}$ siemens	$< 10^{-10}$ siemens

7.4.5. Common Source JFET Amplifier

Figure 7.40 shows the circuit and equivalent circuit

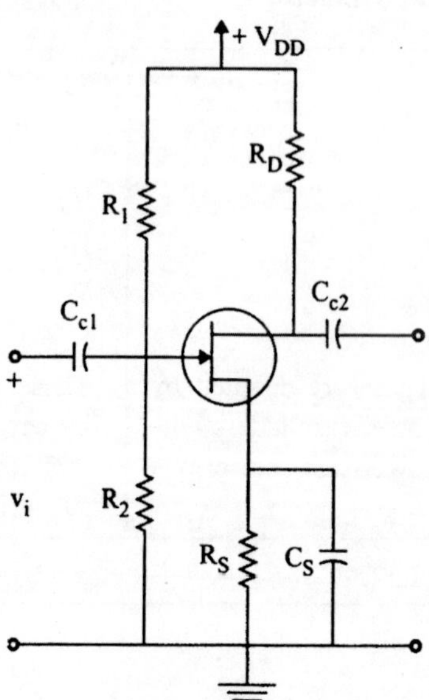

***(a)* Circuit**

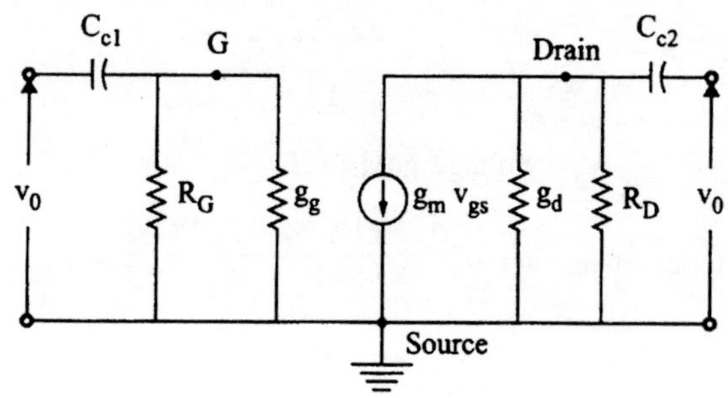

***(b)* Equivalent circuit**

Fig. 7.40. Common source JFET amplifier.

$$A_v = \frac{-\mu R_D}{R_D + r_d} \cong \frac{-g_m}{G_d} = -g_m R_d$$

7.4.6. Common Source MOSFET Amplifier

Figure 7.41 shows the circuit and the equivalent circuit

$$A_v = \frac{-\mu R_D}{R_D + r_d}$$

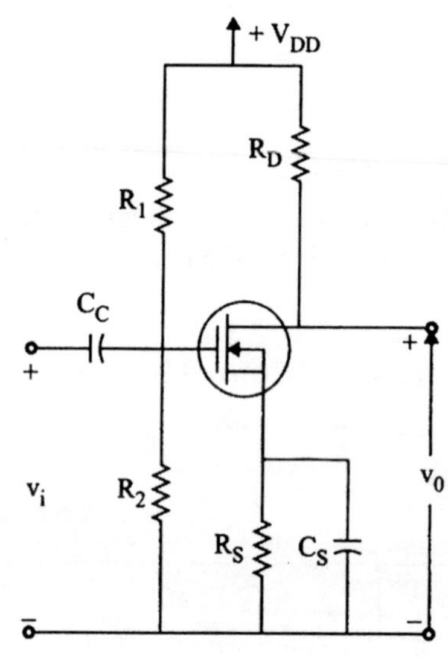

***(a)* Circuit**

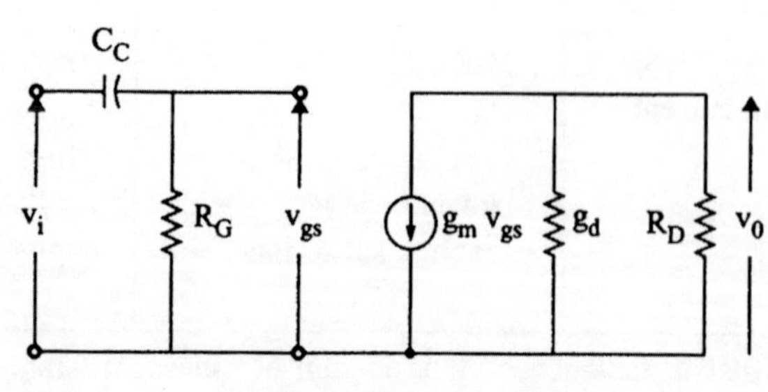

***(b)* Equivalent circuit**

Fig. 7.41. Common source MOSFET amplifier.

7.5. FEEDBACK AMPLIFIERS

Figure 7.42 shows four types of feedback.

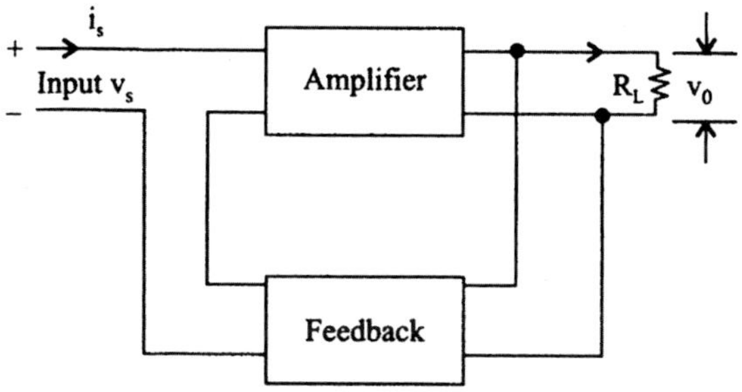

(*a*) **Series-parallel (SP)**

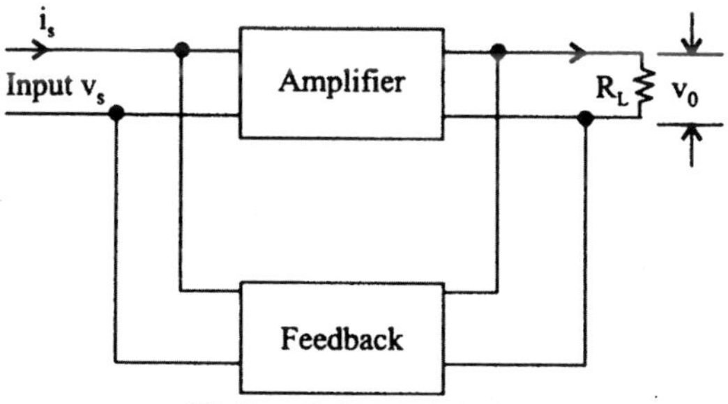

(*b*) **Parallel-parallel (PP)**

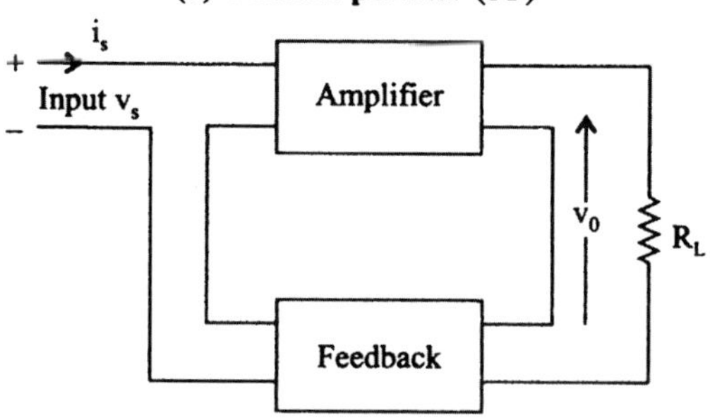

(*c*) **Series-series (SS)**

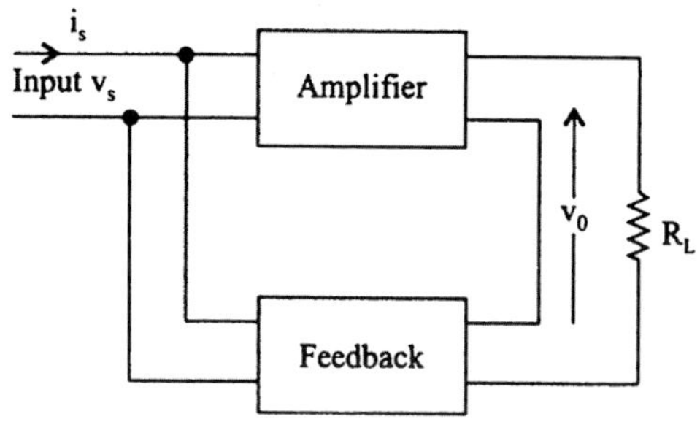

(*d*) **Parallel-series (PS)**

Fig. 7.42. Types of Feedback

In Fig. 7.42 (*a*), *i.e.*, series-parallel connection the input to feedback is in parallel with the load. The feedback signal and system input are in series. In Fig. 7.42 (*b*) called parallel-parallel connection the feedback is in parallel with inputs and outputs. In Fig. 7.42 (*c*) called series-series connection the feedback is in series with input and output. In Fig. 7.42 (*d*) called parallel-series connection the output of feedback is in parallel with system input. Table 7.5 shows a comparison of these four feedback connections.

The connection of Fig. 7.42 (*a*), is the most common of feedback connection used widely is amplifier circuits.

TABLE 7.7. Comparison of Feedback Circuits

	SP	PP	SS	PS
Stable ratio obtained	$\frac{v_o}{v_s}$	$\frac{v_o}{i_s}$	$\frac{i_o}{v_s}$	$\frac{i_o}{i_s}$
Effect on input impedance	increase	decrease	increase	decrease
Effect on output impedance	decrease	decrease	increase	increase
Distortion	decrease	decrease	decrease	decrease
Bandwidth	increase	increase	increase	increase
Application	voltage amplifier	current to voltage converter	voltage to current converter	current amplifier

For Fig. 7.42 (*a*), also called series voltage feedback

$$\frac{v_o}{v_s} = A'_v = \frac{A_v}{1+\beta A_y}$$

Negative feedback stabilizes gain

$$\frac{dA'_v}{A'_v} = \left[\frac{1}{1+A_v\beta}\right]\frac{dA_v}{A_v}$$

The fractional change in A_v is densensitized by a factor $(1+ A_v\beta)$. Negative feedback decreases lower cut off frequency by $(1+ A_v\beta)$ and increases upper cutoff frequency by $(1+ A_v\beta)$. Thus bandwidth is increased. Input impedance is increased, distortion is decreased and output impedance is decreased.

7.6. OSCILLATORS

7.6.1. Barkhausen Criterion

$1 + A_v\beta = 0$	or	$A_v\beta = -1$
i.e., $\lvert A_v\beta\rvert = 1$	and	Phase angle of $A_v\beta = 0$ or a multiple of 2π

7.6.2. Phase Shift Oscillator

Figure 7.43 shows phase shift oscillators using BJT, FET and op-amp respectively. These are used for frequencies of few hertz to several hundred hertz.

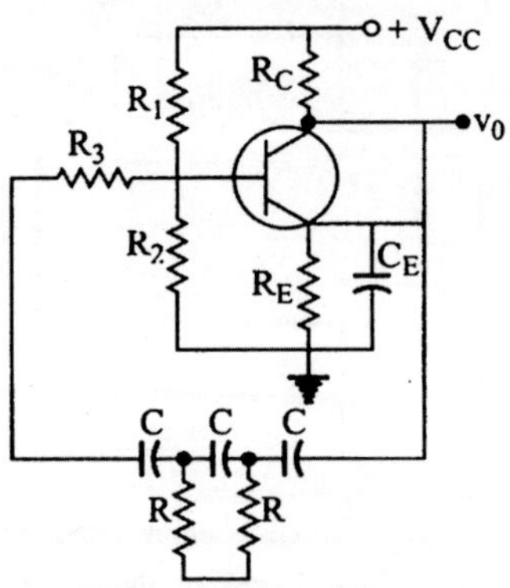

(a) RC phase shift oscillator using BJT

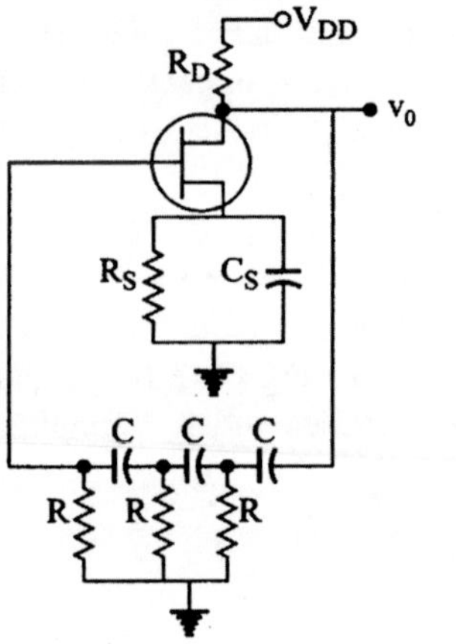

(b) Phase shift oscillator using FET

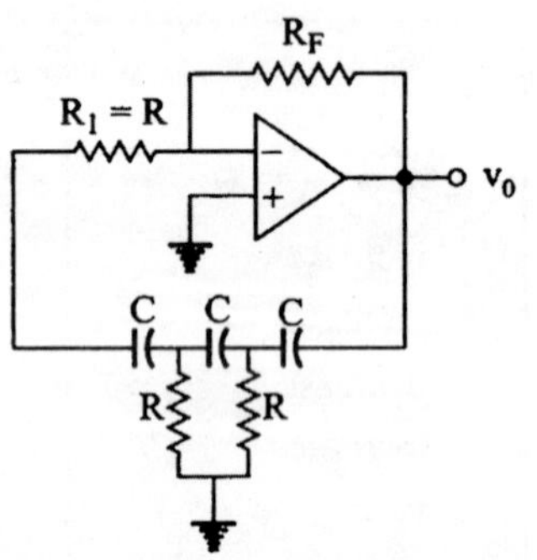

(c) Phase shift oscillator using operational amplifier

Fig. 7.43.

Frequency of oscillation $f = \dfrac{1}{2\pi RC\sqrt{6}}$

Voltage gain $|A_v|$ must be more than 29.

7.6.3. LC Oscillators

These oscillators using parallel tuned LC circuit for feedback. They are used for frequency range of 1 Hz to 500 MHz or so.

Hartley oscillators (Fig. 7.44).

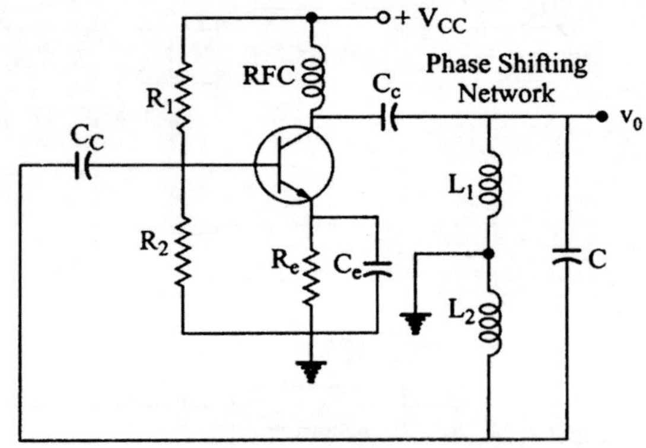

(a) Hartley oscillator using BJT

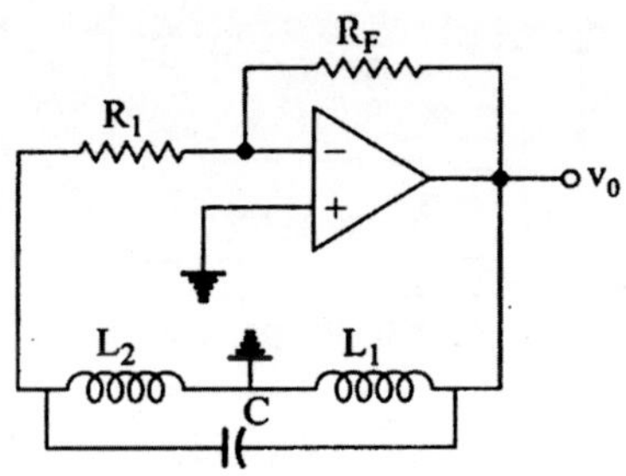

(b) Hartley oscillator using operational amplifier

Fig. 7.44.

$$f = \frac{1}{2\pi\sqrt{(L_1 + L_2)\,C}}$$

Colpitt's oscillator (Fig. 7.45 and Fig. 7.46)

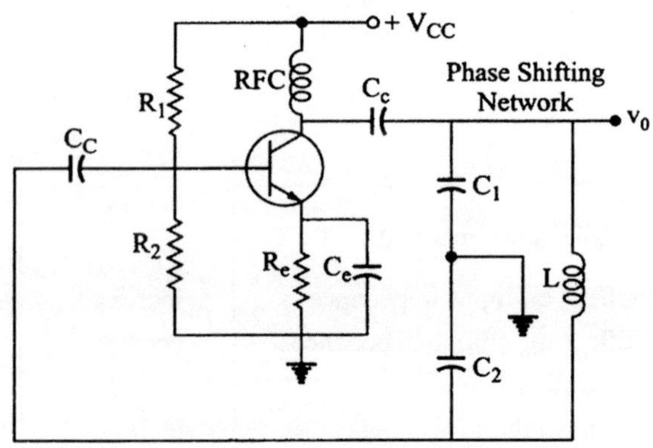

Fig. 7.45 Colpitt's oscillator using BJT

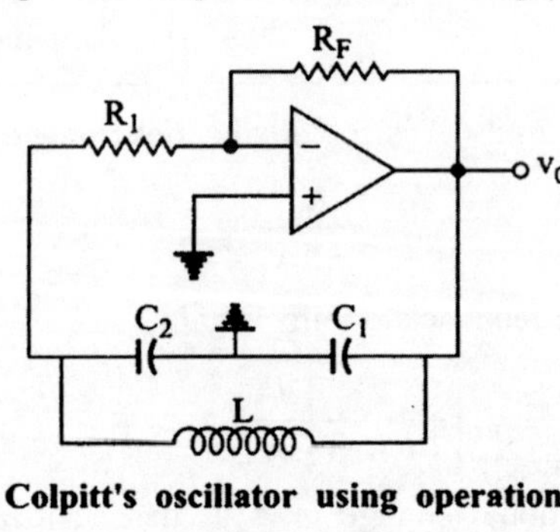

Fig. 7.46. Colpitt's oscillator using operational amplifier.

$$f = \frac{1}{2\pi\left[L\dfrac{C_1C_2}{C_1+C_2}\right]^{0.5}}$$

Clapp's oscillator [Fig. 7.47 (*a*) and (*b*)]

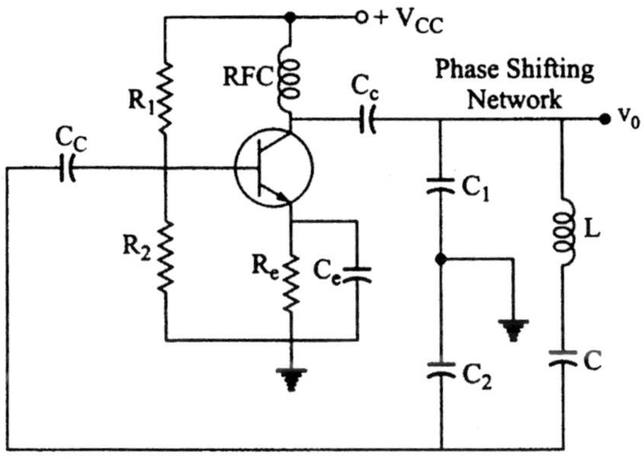

(*a*) **Clapp's oscillator using BJT.**

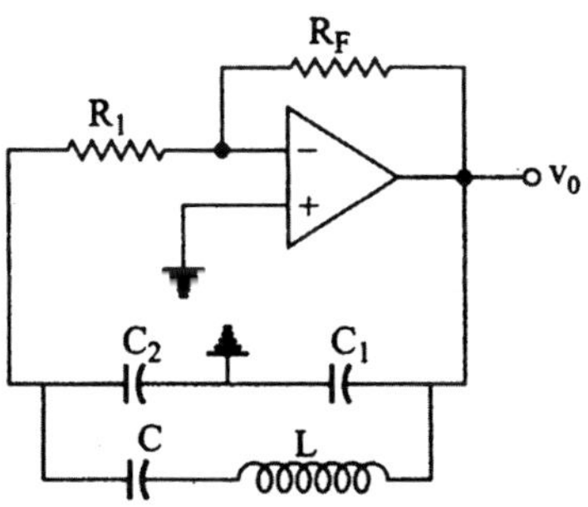

(*b*) **Clapp's oscillator using operational amplifier**

Fig. 7.47.

$$f = \frac{1}{2\pi\sqrt{LC}}$$

7.6.4. Crystal Oscillator

Crystal has piezoelectric property. When used in oscillator circuit a crystal provides very good frequency stability.

Figure 7.48 shows equivalent circuit of crystal. C_h is the capacitance of the crystal holder (or electrodes). It can be connected in place of inductance L is Fig. 7.45 and Fig. 7.46. The frequency of series resonance of the crystal is the oscillator frequency.

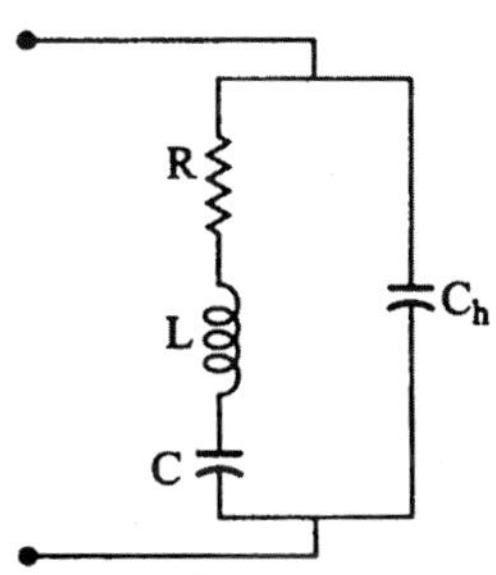

Fig. 7.48 Equivalent circuit of crystal

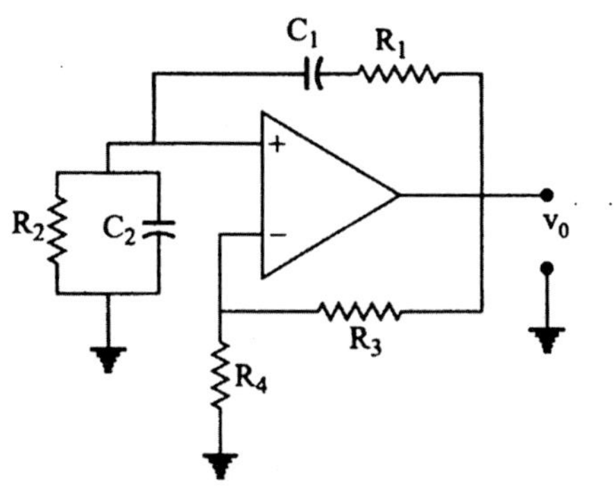

Fig. 7.49. Wienbridge oscillator

7.6.5. Wien Bridge Oscillator

It is used for audio freqeuncies (20 Hz to 20 kHz). Figure 7.49 shows the circuit.

Generally $R_1 = R_2 = R$ and $C_1 = C_2 = C$

The bridge is balanced when $R_3 = 2R_4$

$$f = \frac{1}{2\pi RC}$$

7.7. OPERATIONAL AMPLIFIER

7.7.1. Virtual Ground

An op-amp is used as a feedback amplifier. Figure 7.50 shows the circuit and equivalent circuit.

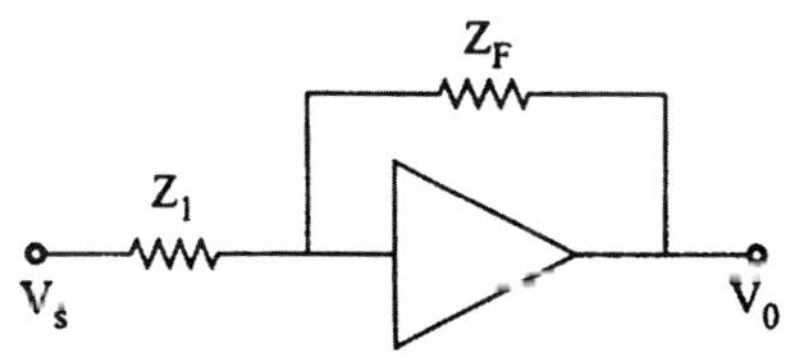

(*a*) **Circuit**

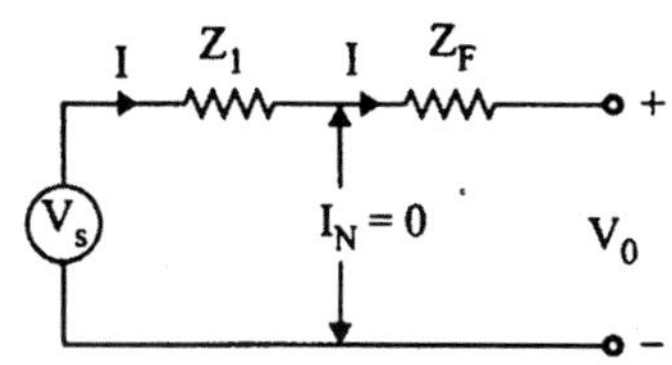

(*b*) **Equivalent circuit**

Fig. 7.50. Feedback operational amplifier.

Because of very high input impedance the current input to op-amp is nearly zero. The source current I flows entirely through Z_F. Thus there is a virtual ground or virtual short circuit at the input as shown by double headed arrow in Fig. 7.50 (*b*).

7.7.2. Inverting Amplifier (Fig. 7.51)

Closed loop gain $A'_v = -\frac{R_F}{R_1}$

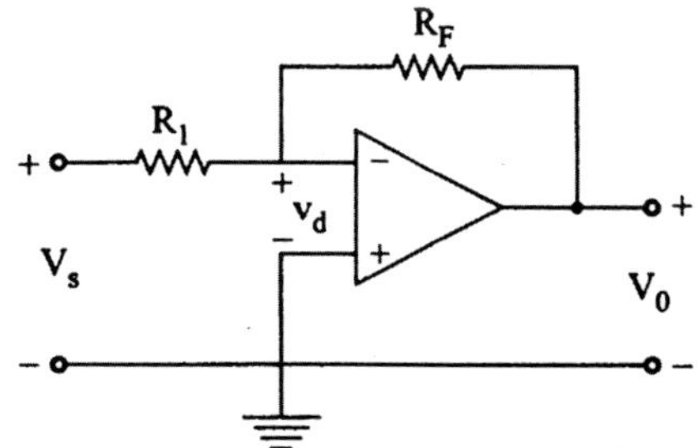

Fig. 7.51. Inverting amplifier.

Closed loop output impedance $Z'_0 = \frac{Z_0}{1 + A_v\beta}$

where Z_0 is open loop output impedance of op-amp, A_v is open loop voltage gain of op-amp and

$$\beta = R_1/(R_F + R_1)$$

Closed loop input impedance = R_1

Closed loop upper cut off frequency $f_2' = (1 + A_v\beta)f_2$

where f_2 is open loop upper cut off frequency of op-amp.

7.7.3. Sing Changer

If $R_F = R_1$, then $v_0/v_s = -1$, *i.e.*, sign of input is changed without change in magnitude.

7.7.4. Phase Shifter

If impedance Z_F is used in place of R_F in Fig. 7.51, any change in phase (± 180°) can be obtained).

7.7.5. Non-inverting Amplifier (Fig. 7.52)

Feedback factor $\beta = \dfrac{R_1}{R_1 + R_2}$

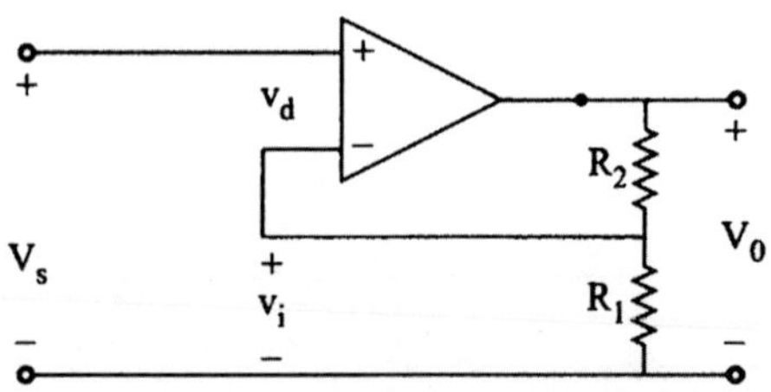

Fig. 7.52. Non-inverting amplifier.

$$A_v' = \frac{R_1 + R_2}{R_1}$$

Closed loop input impedance $Z_i' = (1 + A_v\beta)Z_i$

where Z_i is the open loop input impedance

closed loop output impedance $Z_0' = \dfrac{Z_0}{1 + A_v\beta}$

closed loop upper cut frequency $f_2' = (1 + A_v\beta)f_2$

7.7.6. AC Amplifier (Fig. 7.53)

Critical frequency $f = \dfrac{1}{2\pi RC}$

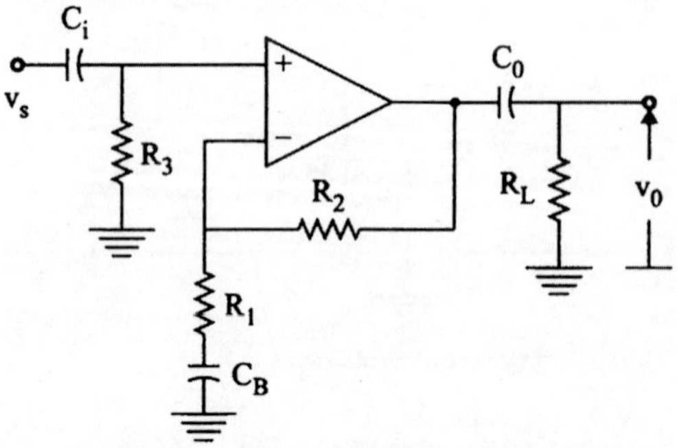

Fig. 7.53. AC amplifier using op-amp.

The circuit has 3 critical frequencies (one corresponding to C_i and R_3, second corresponding to R_L and C_0 and third corresponding to R_1 and C_B). The highest of these frequencies is the lower cut off frequency.

7.7.7. Source Follower (Fig. 7.54)

$$v_0 = v_s$$

It is also called voltage follower, unity gain amplifier, buffer amplifier or isolation amplifier.

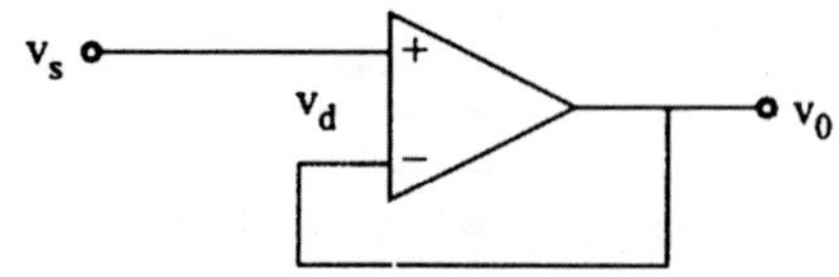

Fig. 7.54. Source follower.

7.7.8. Summing Amplifier (Fig. 7.55)

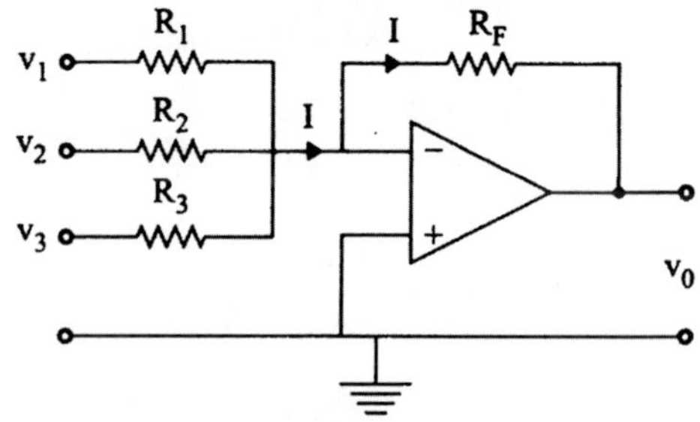

Fig. 7.55. Summing amplifier.

$$v_0 = -R_F\left[\frac{v_1}{R_1} + \frac{v_2}{R_2} + \frac{v_3}{R_3}\right]$$

If $R_1 = R_2 = R_3 = R_F$

$$v_0 = -(v_1 + v_2 + v_3)$$

7.7.9. Difference Amplifier (Fig. 7.56)

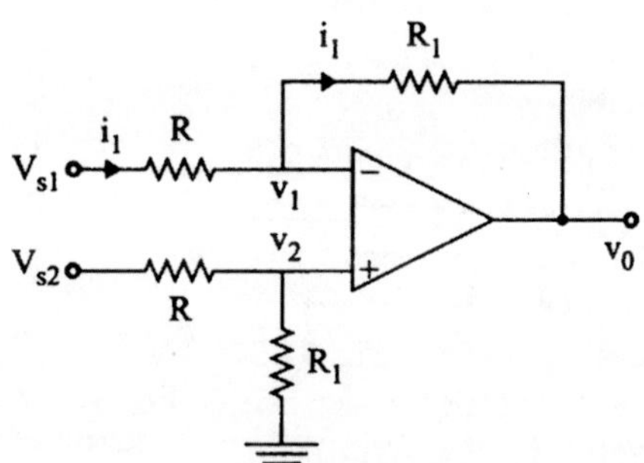

Fig. 7.56. Difference amplifier.

$$v_0 = \frac{R_1}{R}(v_{s2} - v_{s1}) = A_v'(v_{s2} - v_{s1})$$

$$A_v' = \frac{v_0}{v_{s2} - v_{s1}}$$

$$v_{com} = 0.5(v_{s1} + v_{s2})$$

$$\text{CMMR} = \frac{A_v'}{\text{Common mode gain}}$$

7.7.10. Integrating Amplifier (Fig. 7.57)

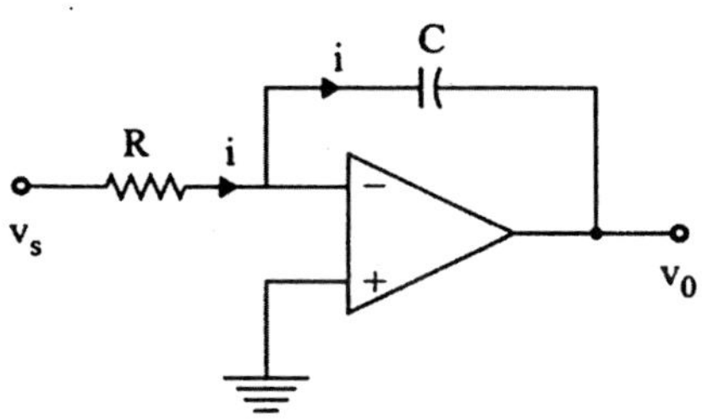

(*a*) Circuit

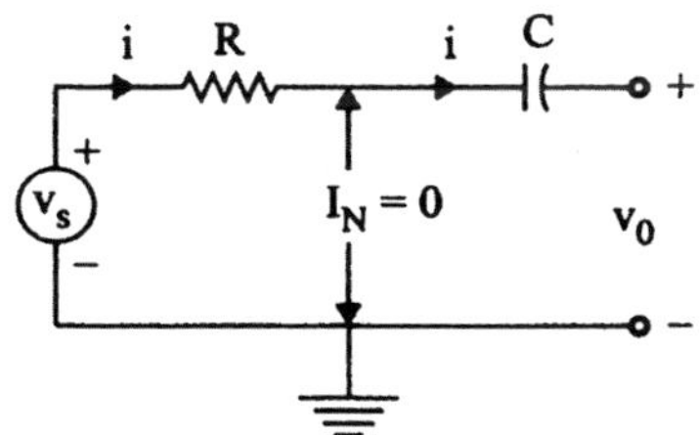

(*b*) Equivalent circuit

Fig. 7.57. Integrating amplifier.

$$v_0 = \frac{1}{RC}\int v_s dt$$

7.7.11. Differentiating Amplifier (Fig. 7.58)

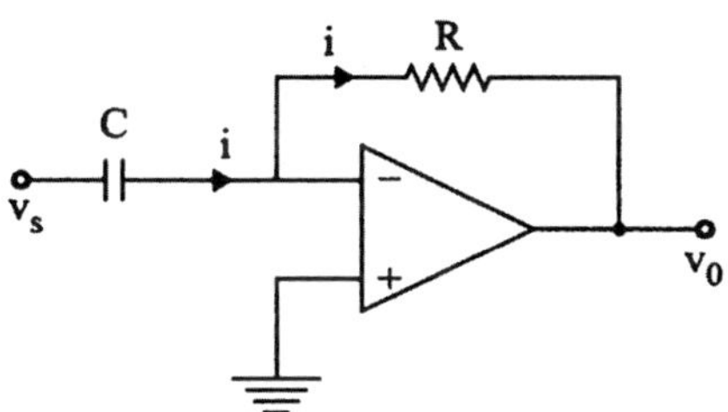

(*a*) Circuit

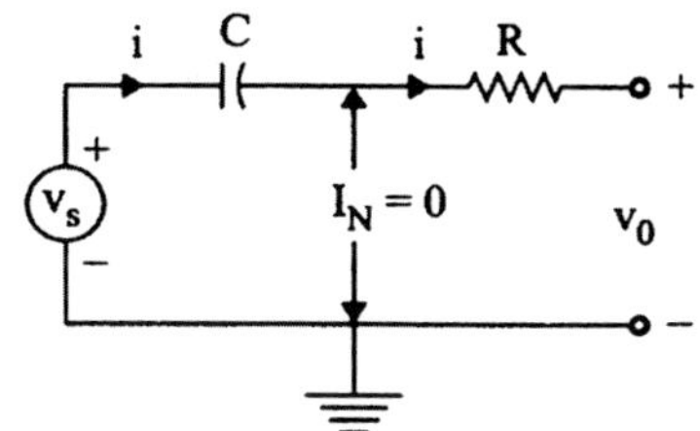

(*b*) Equivalent circuit

Fig. 7.58. Differentiating amplifier.

$$v_0 = -RC\frac{dv_s}{dt}$$

7.7.12. Logarithmic Amplifier (Fig. 7.59)

$$v_0 = V_T \ln (v_i)$$

where $$V_T = \frac{\text{Absolute temperature}}{11600}$$

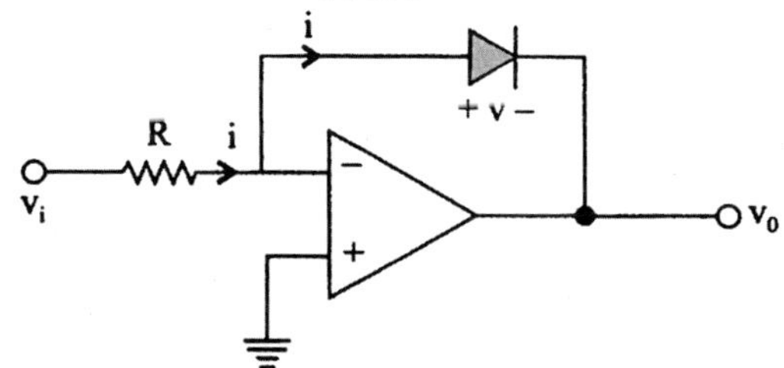

Fig. 7.59. Logarithmic amplifier.

7.7.13. Exponential Amplifier (Fig. 7.60)

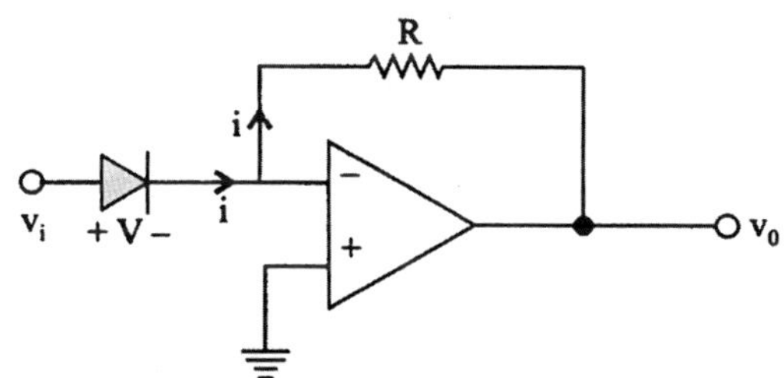

Fig. 7.60. Exponential amplifier.

$$v_0 = -RI_0 \ln^{-1}\left(\frac{v_i}{V_T}\right)$$

7.7.14. Active Positive Limiter (Fig. 7.61)

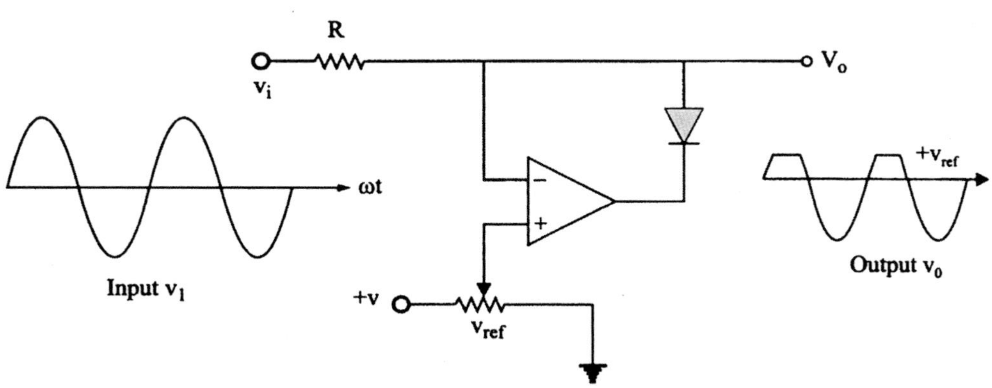

Fig. 7.61. Active positive limiter circuit.

7.7.15. Active Positive Clamper (Fig. 7.62)

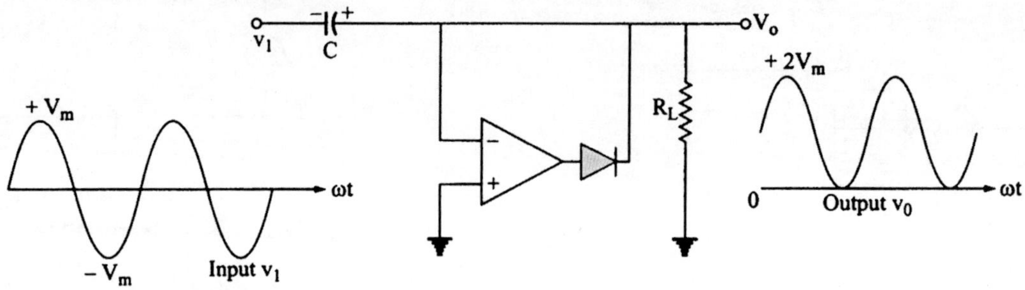

Fig. 7.62. Active positive clamper circuit.

7.7.16. Active Peak Detector (Fig. 7.63)

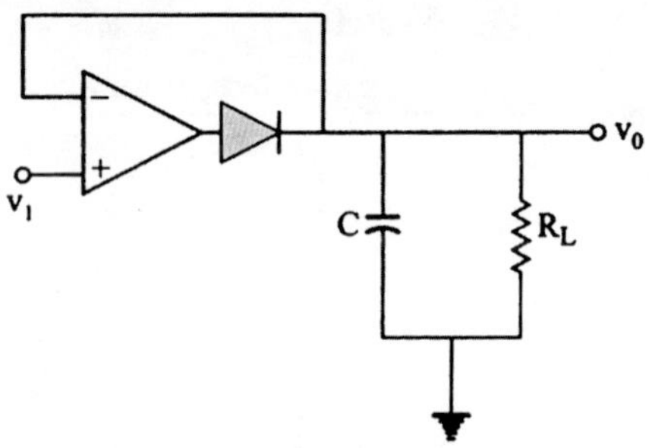

Fig. 7.63. Active peak detector circuit.

7.7.17. Comparator (Fig. 7.64)

In this application op-amp is used in open loop configuration.

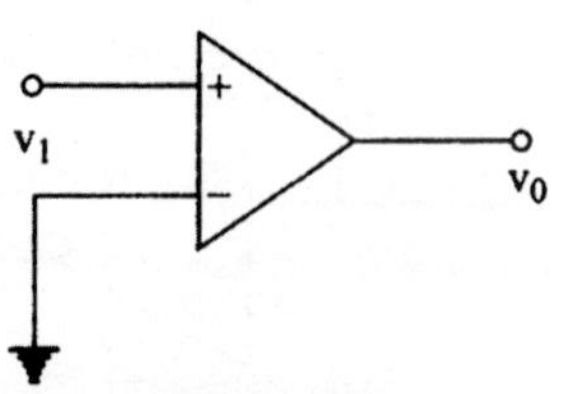

(a) Circuit

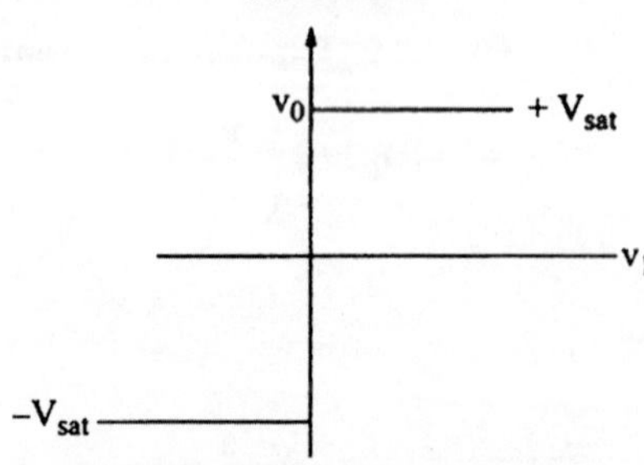

(b) Waveshape of output voltage

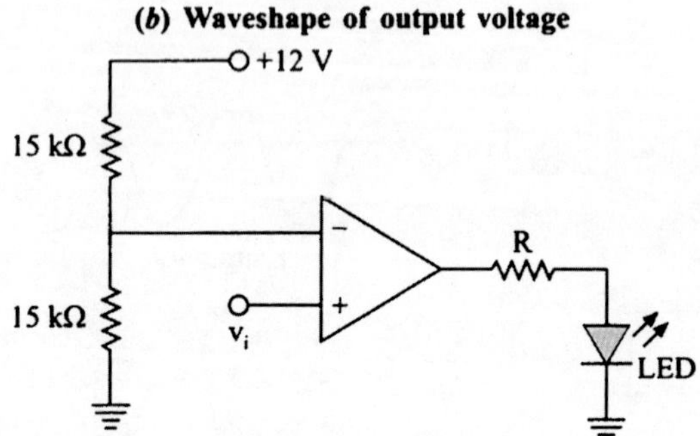

(c) LED on when $v_i > +6$ V

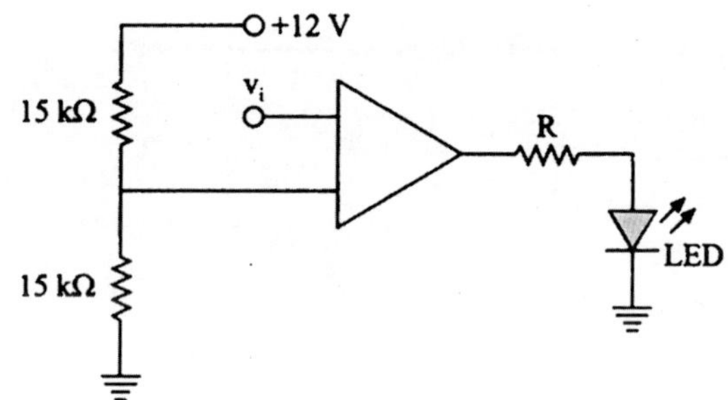

(d) LED on when $v_i < +6$ V

Fig. 7.64. Comparator.

7.7.18. One Pole Active Low Pass Filter (Fig. 7.65)

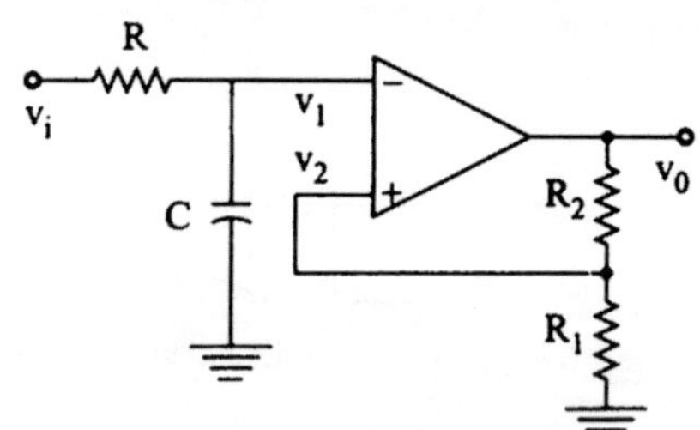

Fig. 7.65. One pole active low pass filter.

$$\text{Cut of frequency } f_c = \frac{1}{2\pi RC}$$

$$\frac{v_0}{v_i} = \frac{1 + R_2/R_1}{\left[1 + (f/f_c)^2\right]^{0.5}}$$

when $f < f_c$, $\dfrac{v_0}{v_i} = 1 + \dfrac{R_2}{R_1} = A'_v$

when $f = f_c$, $\dfrac{v_0}{v_i} = 0.707\, A'_v$

when $f > f_c$, $\dfrac{v_0}{v_i} << A'_v$

Above f_c gain decreases 20 dB per decade increase in frequency.

7.7.19. Tow Pole Active Low Pass Filter (Fig. 7.66)

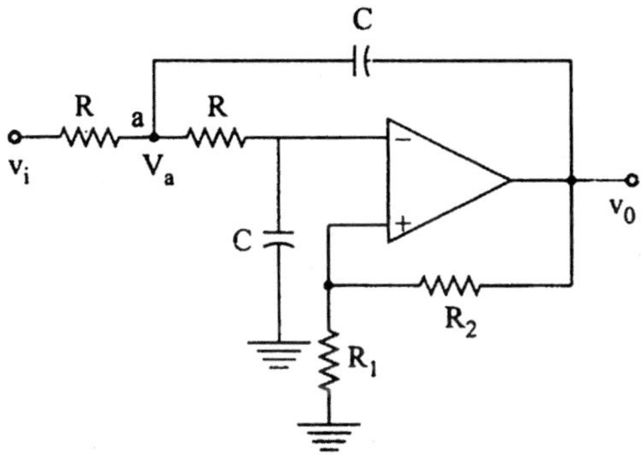

Fig. 7.66. Two pole active low pass filter.

$$\frac{V_0(s)}{V_i(s)} = \frac{(A'_v)\omega_0^2}{s^2 + \frac{\omega_0 s}{Q} + \omega_0^2}$$

where $\omega_0^2 = \frac{1}{R^2C^2}$ and $Q = \frac{1}{3 - A'_v}$

For Butterworth response $R_2 = 0.586\ R$

Above f_c, gain decreases 40 dB per decade increase in frequency.

7.7.20. One Pole Active High Pass Filter (Fig. 7.67)

Cut of frequency $f_c = \frac{1}{2\pi RC}$

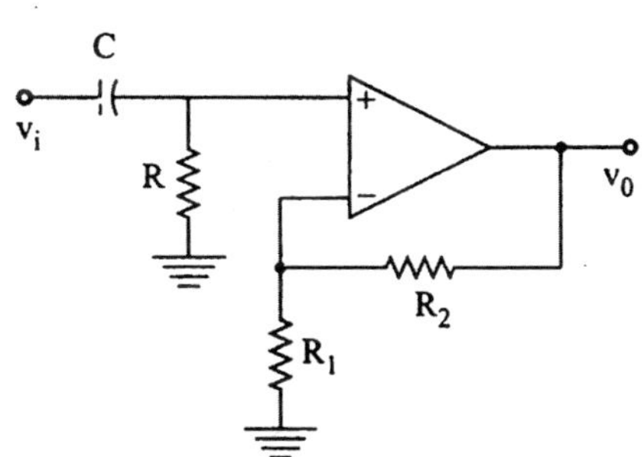

Fig. 7.67. One pole active high pass filter.

$$\frac{v_0}{v_i} = \left(1 + \frac{R_2}{R_1}\right)\left[\frac{jf/f_c}{1 + jf/f_c}\right]$$

when $f > f_c$, $\frac{v_0}{v_i} = 1 + \frac{R_2}{R_1}$

when $f = f_c$, gain decreases by 3 dB

when $f < f_c$, gain decreases by 20 dB per decade decrease in frequency.

7.7.21. Two Pole Active High Pass Filter (Fig. 7.68)

$$\frac{V_0(s)}{V_i(s)} = \frac{A'_v s^2}{s^2 + \frac{\omega_0 s}{Q} + \omega_0^2}$$

$$A'_v = 1 + \frac{R_2}{R_1}, \omega_0^2 = \frac{1}{R^2C^2} \text{ and } Q = \frac{1}{3 - A'_v}$$

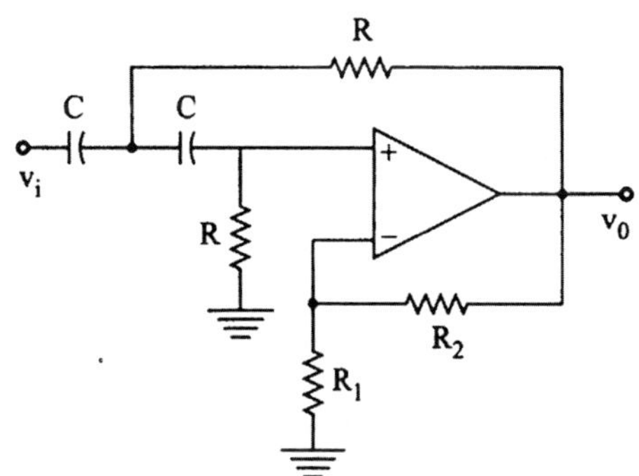

Fig. 7.68. Two pole active high pass filter

For Butterworth response $R_2 = 0.586\ R_1$

when $f > f_c$, gain $= 1 + \frac{R_2}{R_1}$

when $f = f_c$, gain decreases by 3 dB

when $f < f_c$, gain decreases 40 dB per decade decrease in frequency.

7.7.22. Parameters of Op-amp

Table 7.6 shows parameters of some op-amp.

TABLE 7.6. Parameters of op-amps at 25°C

Number	*Max. input bias current nA*	*Max. input offset current nA*	*Short circuit output current nA*	*Max. input offset voltage mV*	f_{unity} *MHz*	*SR V/μs*	*Open loop gain dB*	*Minimum CMRR' dB*
LF351	0.2	0.1	10	10	4	13	88	70
LF353	0.2	0.1	10	10	4	13	88	70
LF356	0.2	0.05	20	5	5	12	94	85
LF411A	200	100	20	0.5	4	15	88	80
LM12	300	100	10000	7	0.7	9	94	75
LM301A	250	50	10	7.5	1	0.5	108	707
LM307	250	50	10	7.5	1	0.5	108	70
LM308	7	1	5	7.5	0.3	0.15	108	80

Number	Max. input bias current nA	Max. input offset current nA	Short circuit output current nA	Max. input offset voltage mV	f_{unity} MHz	SR V/μs	Open loop gain dB	Minimum CMRR' dB
LM675	2000	500	3000	10	5.5	8	90	70
LM741C	500	200	25	6	1	0.5	100	70
LM1458	500	200	20	6	1	0.5	104	70
OP64E	300	100	20	0.75	200	200	100	110
TL072	0.2	0.05	10	10	3	13	88	70
TL082	0.2	0.01	10	3	3	13	94	80

7.8. POWER SUPPLIES

7.8.1 Shunt Regulator (Fig. 7.69)

$$V_0 = \frac{R_1+R_2}{R_1} V_z$$

$$I_s = \frac{V_I - V_0}{R_s}$$

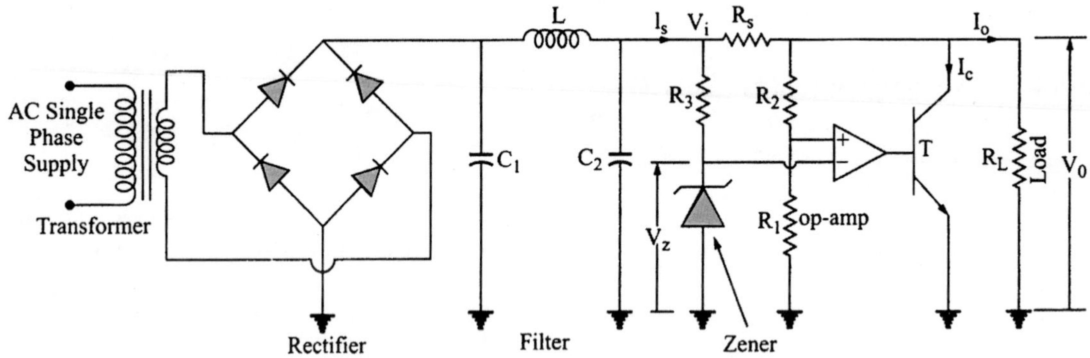

Fig. 7.69. Shunt Regulator.

$$I_0 = \frac{V_0}{R_L}$$

$$I_C = I_s - I_0$$

7.8.2. Series Regulator (Fig. 7.70)

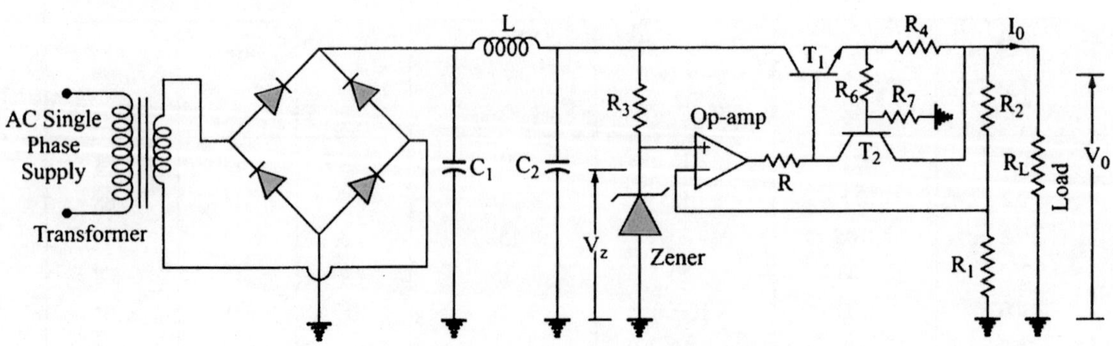

Fig. 7.70. Series Regulator.

$$V_0 = \frac{R_1+R_2}{R} V_z$$

7.8.3. Regulator

IC regulator LM 78 XX series regulators are available for output voltage XX = 05, 06, 08, 10, 12, 15, 18, 24 V. The series LM79 XX given negative output voltages (XX = –05, –06, –08, –10, –12, –15, –18, –24 V). By combining LM78 XX and LM79 XX we can have dual voltage supply (Fig. 7.71).

The diodes are to ensure that dual regulators turn on under all operating conditions. The capacitors C_i (typical value 0.22 μF) are to prevent oscillations which may be caused due to inductance of leads. Capacitors C_0 improve frequency response (typical value 0.1 μF).

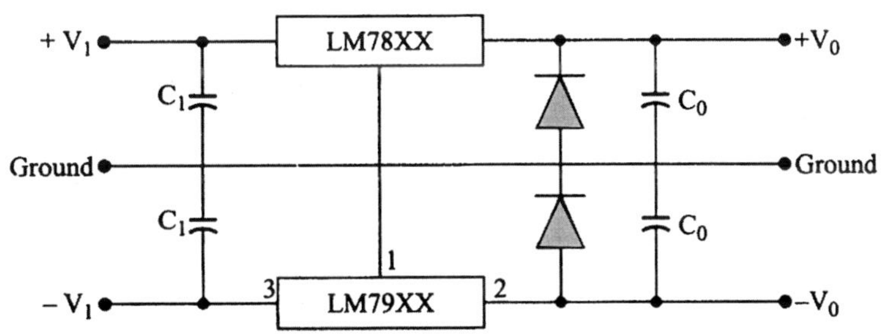

Fig. 7.71. Regulator for Dual output voltage.

OBJECTIVE TYPE QUESTIONS

A. Select the most appropriate alternative

1. The load impedance Z_L of a CE amplifier has R and L in series. The phase difference between output and input will be
   (*a*) 180°
   (*b*) 0
   (*c*) more the 90° but less than 180°
   (*d*) more than 180° but less than 270°
2. The maximum theoretical efficiency for class A amplifier is
   (*a*) 12.5% (*b*) 25%
   (*c*) 50% (*d*) 75%
3. If it is desired to have low output impedance in an amplifier circuit then we should use
   (*a*) common collector circuit
   (*b*) common base circuit
   (*c*) common emitter circuit
   (*d*) either (*b*) or (*c*)
4. In the graphical analysis of an amplifier circuit, the slope of dc load line depends on
   (*a*) load resistance only
   (*b*) collector supply voltage only
   (*c*) both load resistance and collector supply voltage
   (*d*) none of the above
5. Which amplifier has the highest power gain
   (*a*) common emitter (*b*) common collector
   (*c*) common base (*d*) both CC and CE
6. Which amplifier has a voltage gain of less than 1
   (*a*) Common emitter (*b*) Common collector
   (*c*) Common base (*d*) both CB and CC
7. In the analysis of a common emitter amplifier we may neglect
   (*a*) h_{re} (*b*) h_{oe}
   (*c*) h_{fe} (*d*) h_{ie}
8. In class *C* operation of an amplifier circuit, the collector current exists for
   (*a*) 360° of input wave
   (*b*) 180° of input wave
   (*c*) more than 180° of input wave
   (*d*) less than 180° of input wave
9. For class B operation the theoretical maximum efficiency is
   (*a*) 100% (*b*) 78.5%
   (*c*) 50% (*d*) 27.5%
10. A push pull amplifier is a
    (*a*) power amplifier
    (*b*) voltage amplifier
    (*c*) current amplifier
    (*d*) both power and current amplifier
11. If the mid band gain of an amplifier is 40 dB, the gain at half power frequency is
    (*a*) 37 dB (*b*) 30 dB
    (*c*) 20 dB (*d*) 13 dB
12. The purpose of impedance matching an amplifier circuit is to achieve
    (*a*) high efficiency
    (*b*) reduced distortion
    (*c*) maximum power transfer
    (*d*) none of die above
13. A transistor amplifier is operating in class A mode. If a transformer is connected for impedance matching, the efficiency will
    (*a*) increase
    (*b*) decrease

(c) not be affected
(d) may increase or decrease

14. At mid band frequencies the coupling capacitor of an RC coupled CE amplifier may be considered as
(a) open circuit
(b) short circuit
(c) a reactance equal to load resistance
(d) none of the above

15. As compared to a BJT amplifier circuit, a JFET amplifier circuit
(a) has low input impedance
(b) has high input impedance
(c) has very low input impedance
(d) may have either low or high input impedance

16. Which of the following amplifier circuits has highest input impedance
(a) circuit using BJT
(b) circuit using JFET
(c) circuit using MOSFET
(d) either (a) or (b)

17. A transistor has two p-n junctions. The batteries should be connected such that
(a) both junctions are forward biased
(b) both junctions are reverse biased
(c) one junction is forward biased and the other is reverse biased
(d) either (a) or (b)

18. The theretical maximum efficiency of a half wave diode rectifier is
(a) 40.6% (b) 50%
(c) 81.2% (d) slightly less than 100%

19. As compared to full wave rectifier, a half wave rectifier
(a) has higher ripple factor
(b) has lower ripple factor
(c) may have higher or lower ripple factor
(d) has same ripple factor

20. A full wave rectifier using centre tapped transformer and a bridge rectifier use similar diodes and have equal no load output voltage. Under equal load conditions
(a) Output voltage of bridge rectifier will be less than that of other
(b) Output voltage of bridge rectifier will be more than that of other
(c) Output voltage of both will be exactly equal
(d) Output voltage of any one may be more than that of other

21. A rectifier circuit has load resistance R. It uses a capacitance filter having capacitance C. For low ripple in the output
(a) RC should be 1
(b) RC should be low
(c) RC should be high
(d) RC should be neither low nor high

22. The open loop gain of an amplifier is 200. If negative feedback with $\beta = 0.2$ is used, the closed loop gain will be
(a) 200 (b) 40.12
(c) 4.878 (d) 2.2

23. The open loop gain of an amplifier is 50 but is likely to decrease by 20% due to various factors. If negative feedback with $\beta = 0.1$ is used the change in gain will be about
(a) 20% (b) 2%
(c) 0.5% (d) 0.05%

24. An RC coupled amplifier has an open loop gain of 200 and a lower cutoff frequency of 50 Hz. If negative feedback with $\beta = 0.1$ is used the lower cut off frequency will be
(a) about 50 Hz (b) about 5 Hz
(c) about 2.38 Hz (d) about 70.5 Hz

25. A Hartley oscillator uses
(a) a tapped inductor (b) a tapped capacitor
(c) both (a) and (b) (d) neither (a) and (b)

26. A Colpitt's oscillator uses
(a) a tapped inductor
(b) an inductor and two capacitors
(c) both (c) and (b)
(d) either (a) or (b)

27. A clapp's oscillator is
(a) a modified Hartley oscillator
(b) a modified Colpitt's oscillator
(c) a modified RC oscillator
(d) none of the above

28. An RC oscillator uses
(a) one RC combination
(b) two RC combinations.
(c) at least three RC combinations
(d) either (a) or (b)

29. A Wien bridge oscillator uses Wien bridge for
(a) tuning and phase shift
(b) tuning
(c) phase shift
(d) either tuning or phase shift

30. A Wien bridge is suitable for
(a) audio frequencies only
(b) radio frequencies only
(c) both audio and radio frequencies
(d) very very low frequencies only

31. The equivalent circuit of an op-amp has
(a) a virtual ground at input
(b) a virtual ground at output
(c) virtual ground at input as well as at output
(d) none of the above

32. CMRR of a difference amplifier is equal to
(a) $(A'_v)(A_{com})$ (b) A'_v / A_{com}
(c) A_{com} / A'_v (d) $(A_v)(A_{com})^2$

33. An op-amp has
(a) low input and output impedance

(*b*) low input impedance and high output impedance
(*c*) low output impedance and high input impedance
(*d*) high output impedance and high input impedance

34. An op-amp is fed from a signal source through resistance R_1. If it is to work as sign changer, the feedback resistance R_f should be
(*a*) equal to R_1 (*b*) twice R_1
(*c*) half R_1 (*d*) either (*b*) or (*c*)

35. For an op-amp source follower circuit
(*a*) input and output voltages have same magnitude
(*b*) output and input voltages are in phase
(*c*) output and input voltage are in phase but have different magnitudes
(*d*) output and input voltages have same magnitude and are in phase

36. The input to an op-amp integrating amplifier is a constant voltage. The output will be
(*a*) a constant voltage (*b*) a sawtooth wave
(*c*) a ramp (*d*) ac voltage

37. The element in the feedback circuit of a logarithmic amplifier using op-amp is
(*a*) a resistor (*b*) a capacitor
(*c*) an inductor (*d*) a diode

38. A potential of 7 V is applied to a silicon diode. A resistance of 1 kΩ is also connected in series. If the diode is forward biased, the current in the circuit is
(*a*) 7 mA (*b*) 6.3 mA
(*c*) 0.7 mA (*d*) 0

39. A forward voltage of 9 V is applied to a diode in series with a 1 kΩ load resistor. The voltage across load resistor is zero. It indicates that
(*a*) diode is short circuited
(*b*) diode is open circuited
(*c*) resistor is open circuited
(*d*) diode is either short circuited or open circuited

40. In a half wave diode rectifier circuit the current flows in the load circuit for
(*a*) 90° in each cycle (*b*) 180° in each cycle
(*c*) 270° in each cycle (*d*) 360° in each cycle

41. In a full wave diode rectifier circuit each diode carries current for
(*a*) 90° in each cycle (*b*) 180° in each cycle
(*c*) 270° in each cycle (*d*) 360° in each cycle

42. If input frequency is 50 Hz, the frequency of output wave in a half wave diode rectifier circuit is
(*a*) 25 Hz (*b*) 50 Hz
(*c*) 100 Hz (*d*) 200 Hz

43. If input frequency is 50 Hz, the frequency of output wave in a full wave diode rectifier circuit is
(*a*) 25 Hz (*b*) 50 Hz
(*c*) 100 Hz (*d*) 200 Hz

44. The dc load current in a bridge rectifier circuit is 10 mA. The dc load current through each diode is
(*a*) 40 mA (*b*) 20 mA
(*c*) 10 mA (*d*) 5 mA

45. In a bridge rectifier circuit the rms value of input ac voltage is 10 V. The PIV across each diode is
(*a*) 7.07 V (*b*) 14.14 V
(*c*) 10 V (*d*) 28.28 V

46. A bridge rectifier circuit has a dc load current of 10 mA and a filter capacitance of 1000 μF The peak to peak ripple voltage is
(*a*) 0.01 V (*b*) 0.05 V
(*c*) 0.1 V (*d*) 0.5 V

47. In a rectifier circuit the filter capacitance is increased. Then the ripple will
(*a*) increase (*b*) decrease
(*c*) remain the same (*d*) either (*a*) or (*b*)

48. In an amplifier circuit the emitter is at ac ground potential in a
(*a*) CB circuit (*b*) CC circuit
(*c*) CE circuit (*d*) both CD and CE circuit

49. In an amplifier the voltage gain is the ratio of
(*a*) output voltage and emitter resistance
(*b*) output voltage and input current
(*c*) output voltage and input voltage
(*d*) none of the above

50. In a self bias circuit for CE amplifier, the base voltage is
(*a*) equal to supply voltage
(*b*) more than supply voltage
(*c*) equal to or more than supply voltage
(*d*) less than supply voltage

51. The self bias provides
(*a*) stable Q point (*b*) large voltage gain
(*c*) high input impedance (*d*) high base current

52. In a self bias circuit for CE amplifier, an increase in emitter resistance R_E results in
(*a*) increase in emitter voltage
(*b*) decrease in emitter voltage
(*c*) increase in emitter current
(*d*) decrease in emitter current

53. In a self bias circuit for CE amplifier, the emitter resistance is made three times the original value. The collector current will
(*a*) remain the same (*b*) become three times
(*c*) will decrease (*d*) will become one-third

54. In a self bias circuit for CE amplifier the collector supply voltage is made twice. The collector current will
(*a*) decrease (*b*) increase
(*c*) remain the same (*d*) increase or decrease

55. For dc the current through coupling capacitor in CE amplifier circuit is
(*a*) very high (*b*) high
(*c*) low (*d*) zero

56. For high frequencies a capacitor is like
(*a*) an open circuit (*b*) a short circuit
(*c*) either open circuit or short circuit
(*d*) none of the above

57. A bypass capacitor produces
(*a*) dc ground (*b*) ac ground
(*c*) both dc and ac ground (*d*) none of above

58. In deriving ac equivalent circuit for an amplifier circuit we short circuit
(*a*) all resistors (*b*) all transistors
(*c*) all inductors (*d*) all capacitors

59. When the ac base voltage in a CE amplifier circuit is too high, the ac emitter current is
(*a*) zero (*b*) constant
(*c*) alternating (*d*) distorted

60. A CE amplifier has a high input signal. In the wave shape of emitter current
(*a*) positive and negative half cycles will be symmetrical
(*b*) positive half cycle will be elongated and negative half cycle will be compressed
(*c*) positive half cycle will be compressed and negative half cycle will be elongated
(*d*) either of the half cycles may be elongated or compressed

61. It is desired to reduce distortion in a CE amplifier circuit. We should
(*a*) reduce emitter current
(*b*) reduce collector current
(*c*) reduce ac base voltage
(*d*) reduce base emitter voltage

62. The ac collector current is nearly equal to
(*a*) ac base current (*b*) ac signal current
(*c*) ac supply current (*d*) ac emitter current

63. In a CE amplifier, the output voltage is equal to product of
(*a*) ac collector current and ac collector resistance
(*b*) ac base current and ac collector resistance
(*c*) ac emitter current and ac emitter resistance
(*d*) ac collector current and source resistance

64. In a CE amplifier the input impedance is equal to the ratio of
(*a*) ac base voltage to ac base current
(*b*) ac base voltage to ac emitter current
(*c*) ac emitter voltage to ac collector current
(*d*) ac collector voltage to ac collector current

65. In a CE amplifier circuit, the output voltage
(*a*) is very small (*b*) is always constant
(*c*) depends on β (*d*) none of the above

66. In CE amplifier circuit, voltage gain is directly proportional to
(*a*) β
(*b*) collector supply voltage
(*c*) base resistance (*d*) none of above

67. The ac input voltage of an amplifier
(*a*) is always equal to ac signal voltage
(*b*) may be equal to or more than ac signal voltage
(*c*) is always less than ac signal voltage
(*d*) may be equal or more or less than ac signal voltage

68. In a two stage CE amplifier circuit, the ac collector resistance of the first stage depends on
(*a*) load resistance
(*b*) input impedance of first stage
(*c*) input impedance of second stage
(*d*) all the above

69. In a CE amplifier circuit the emitter by pass capacitor is removed. The ac output voltage
(*a*) increases (*b*) remains the same
(*c*) increases or decreases (*d*) decreases

70. In a CE amplifier the collector resistance is short circuited. The ac output voltage will
(*a*) increase (*b*) decrease
(*c*) increase or decrease (*d*) remain the same

71. The purpose of connecting a coupling capacitor in the output circuit of an amplifier is
(*a*) to increase voltage gain
(*b*) to increase collector current
(*c*) to block dc collector voltage
(*d*) all of the above

72. In a CE amplifier circuit the ac voltage between emitter and ground
(*a*) is very high
(*b*) may be high or low
(*c*) is zero
(*d*) depends on circuit configuration

73. In a CE amplifier circuit the dc voltage between emitter and ground
(*a*) is always zero
(*b*) cannot be zero
(*c*) may or may not be zero
(*d*) is negative

74. In a CE amplifier circuit the load resistance is open circuited. The output voltage will
(*a*) increase (*b*) decrease
(*c*) become zero (*d*) increase or decrease

75. In a CE amplifier the output coupling capacitor is short circuited. The dc collector voltage
(*a*) will increase
(*b*) may increase or decrease
(*c*) will remain the same (*d*) will decrease

76. In a CE amplifier the dc load line is the same as ac load line when
(*a*) ac collector resistance and dc collector resistance are equal
(*b*) ac collector resistance and dc emitter resistance are equal
(*c*) ac collector resistance is very high
(*d*) ac collector resistance is one half of dc collector resistance

77. In a CE amplifier, the collector resistance is 6 kΩ and load resistance is 3 kΩ. The ac load resistance is
(*a*) 9 kΩ (*b*) 6 kΩ
(*c*) 3 kΩ (*d*) 2 kΩ

78. In a CE amplifier the ac signal is increased. Which of the following will not change
(*a*) ac base voltage
(*b*) ac collector voltage
(*c*) ac emitter voltage
(*d*) ac collector voltage and ac emitter voltage

79. In a CE amplifier
(*a*) both ac and dc load lines have the same slope
(*b*) the ac load line has more slope than dc load line
(*c*) the ac load line has less slope man dc load line
(*d*) either of the two load lines may have more slope than the other

80. In a CE amplifier circuit
(*a*) Q point must lie on both dc and ac load lines
(*b*) Q point lies on dc load line
(*c*) Q point lies on ac load line
(*d*) Q point lies may or may not lie on both dc and ac load lies

81. For maximum unclippled output from a CE amplifier, the Q point should be
(*a*) at the centre of dc load line
(*b*) at the centre of ac load line
(*c*) at the centre of collector current
(*d*) none of the above

82. An increase in ambient temperature means that maximum power rating of transistor
(*a*) will increase
(*b*) will decrease
(*c*) may increase or decrease
(*d*) may increase or remain the same

83. Heat sink results in
(*a*) slower dissipation of heat to atmosphere
(*b*) faster dissipation of heat to atmosphere
(*c*) lower ambient temperature
(*d*) lower transistor power

84. Which configuration is suitable for impedance matching
(*a*) CB (*b*) CE
(*c*) CC (*d*) CB and CE

85. Which configuration has high input impedance and low output impedance
(*a*) CB (*b*) CE
(*c*) CC (*d*) CB and CE

86. In which configuration is ac collector voltage zero
(*a*) CE (*b*) CB
(*c*) CC (*b*) none of the above

87. In which configuration does output ac voltage appear between emitter and ground
(*a*) CE (*b*) CB
(*c*) CC (*d*) both CE and CB

88. If a CE amplifier drives a low load resistance directly the result will be
(*a*) overloading of amplifier
(*b*) underloading of amplifier
(*c*) overloading or underloading of amplifier depending on circuit
(*d*) none of the above

89. In a CE amplifier it is desired that the signal should be able to swing equally in both directions. Then the Q point should be
(*a*) at the centre of ac load line
(*b*) at the centre of dc load line
(*c*) at the centre of collector voltage
(*d*) at the centre of collector current

90. An amplifier has a large ac input signal. The clipping occurs on both the peaks. The output voltage will be nearly a
(*a*) sine wave (*b*) square wave
(*c*) tringular wave (*d*) (*a*) or (*c*)

91. A Darlington pair consist of
(*a*) two CB amplifiers
(*b*) two CC amplifiers
(*c*) two cascaded CC amplifiers
(*d*) two cascaded CE amplifiers

92. In a push pull circuit
(*a*) each transistor conducts for 180°
(*b*) each transistor conducts for more than 180° but less than 360°
(*c*) each transistor conducts for less than 180°
(*d*) the period of conduction of each transistor depends circuit configuration

93. A Darlington pair has $\beta = 300$. If $R_E = 1$ kΩ and $R_L = 100$ kΩ, the input impedance at the base is approxmaitely
(*a*) 300 k ohm (*b*) 3 M ohm
(*c*) 100 k ohm (*d*) 30 k ohm

94. The main application of enhancement mode MOSFET is in
(*a*) amplifier circuits (*b*) oscillator circuits
(*c*) switching circuits (*d*) all of the above

95. A buffer amplifier should have
(*a*) high output impedance and high input impedance
(*b*) high output impedance and low input impedance
(*c*) low output impedance and low input impedance
(*d*) low output impedance and high input impedance

96. The main advantage of CMOS circuit is
(*a*) high gain
(*b*) high output impedance
(*c*) low power consumption
(*d*) high gain and high output impedance

97. The phase angle of a lag circuit is
(*a*) between 0 and –90° (*b*) between 0 and +90°
(*c*) 90° (*d*) more than 90°

98. The phase angle of lead circuit is
(*a*) between 0 and –90° (*b*) between 0 and +90°
(*c*) –90° (*d*) more than 90°

99. An oscillator requires an amplifier
(*a*) with negative feedback

(*b*) with positive feedback
(*c*) with either positive or negative feedback
(*d*) none of the above

100. A Wien bridge oscillator uses
(*a*) negative feedback
(*b*) positive feedback
(*c*) both negative and positive feedback
(*d*) none of the above

101. The input voltage for starting oscillations in an oscillator is caused by
(*a*) negative feedback
(*b*) positive feedback
(*c*) noise voltage in resistors
(*d*) a special generator

102. When the circuit is switched on, the loop gain of a Wien bridge oscillator is
(*a*) 1 (*b*) more than 1
(*c*) less than 1 (*d*) about 0.5

103. In an oscillator the total phase shift around the loop must be
(*a*) 180° (*b*) 90°
(*c*) 270° (*d*) 0°

104. The most widely used LC oscillator is
(*a*) Hartley oscillator (*b*) Crystal oscillator
(*c*) Colpitt's oscillator (*d*) Clapp's oscillator

105. Quartz crystal has
(*a*) very small Q (*b*) small Q
(*c*) very high Q (*d*) Q less than 1

106. In a crystal
(*a*) the frequencies of series and parallel resonance are equal
(*b*) the frequency of series resonance is more than frequency of parallel resonance
(*c*) the frequency of parallel resonance is more than frequency of series resonance
(*d*) any of the above

107. The stray capacitance between the last stage and the first stage of an oscillator may cause
(*a*) unwanted low frequency oscillations
(*b*) unwanted high frequency oscillations
(*c*) unwanted oscillations at low and high frequencies
(*d*) large feedback

108. Which of the following values represents the open loop gain of typical op-amp?
(*a*) 100 (*b*) 10,000
(*c*) 100,000 (*d*) 10,00,000

109. Which of the following best represents the input impedance of an actual op-amp?
(*a*) 10 ohm (*b*) 10 k ohm
(*c*) 1 M ohm (*d*) 100 M ohm

110. Which of the following best represents the output impedance of an actual op-amp?
(*a*) 10 ohm (*b*) 100 ohm
(*c*) 10 k ohm (*d*) 100 k ohm

111. Which of the following best represents the band width of an actual op-amp?
(*a*) 100 Hz to 1 MHz (*b*) 0 Hz to 100 k Hz
(*c*) 0 Hz to 10 MHz (*d*) 0 Hz to infinity

112. Ac signals are given to both inverting and non-inverting terminals of an op-amp. When will the output be maximum?
(*a*) when the phase difference between the inputs is 0°
(*b*) when the phase difference between the inputs is 90°
(*c*) when the phase difference between the inputs is 180°
(*d*) when the phase difference between the inputs is 0 or 180°

113. Equal amplitude ac signals are given as inputs to both inverting and non-inverting terminals of an op-amp. The output will be zero when the phase difference between inputs is
(*a*) 0° (*b*) 90°
(*c*) 180° (*d*) 270°

114. In an op-amp a common mode signal is applied to
(*a*) inverting terminal
(*b*) non-inverting terminal
(*c*) both terminals
(*d*) one or both the terminals

115. A negative feedback can be of
(*a*) 2 types (*b*) 3 types
(*c*) 4 types (*d*) only 1 type

116. The purpose of using negative feedback in op-amps is to
(*a*) stabilize over all gain
(*b*) increase output
(*c*) stabilize over all gain and increase output
(*d*) increase output and decrease input

117. An inverting voltage feedback is a
(*a*) voltage amplifier
(*b*) current amplifier
(*c*) current to voltage converter
(*d*) voltage to current converter

118. A non-inverting voltage feedback amplifier is a
(*a*) voltage amplifier
(*b*) current amplifier
(*c*) current to voltage converter
(*d*) voltage to current converter

119. In a linear op-amp circuit
(*a*) the product of gain and band width is constant
(*b*) input can be dc only
(*c*) input can be ac only (*d*) saturation may occur

120. A virtual ground is a ground for
(*a*) voltage
(*b*) current
(*c*) both voltage and current
(*d*) voltage but not for current

121. In an inverting amplifier using op-amp, the current input to the inverting terminal
(*a*) is equal to the current through feedback resistance

(*b*) is less than the current through feedback resistance
(*c*) is more than the current through feedback resistance
(*d*) may be equal to, less or more than current through feedback resistance

122. An op-amp integrating circuit uses
(*a*) an inductor
(*b*) a capacitor
(*c*) both inductor and capacitor
(*d*) none of the above

123. In op-amp integrating amplifier the input is fed to
(*a*) inverting input
(*b*) non-inverting input
(*c*) both inverting and non-inverting input
(*d*) either inverting or non-inverting input

124. The input to integrating op-amp is a rectangular wave. The output will be
(*a*) a sine wave (*b*) a rectangular wave
(*c*) a triangular wave (*d*) none of the above

125. In an op-amp differentiator
(*a*) the amplitude of output is proportional to rate of change of input
(*b*) the amplitude of output is proportional to input
(*c*) output occurs when input is finite and constant
(*d*) polaity of input and output is the same

126. Match List I and List II and give the correct answer using the given codes

List I (Amplifier quantity)	List II (Symbol)
A. Total instantaneous value of emitter current	1. I_E
B. Quiescent value of emitter current	2. i_E
C. Instantaneous value of ac component of emitter current	3. I_e
D. RMS value of ac component of emitter current	4. i_e

Codes:

	A	B	C	D
(*a*)	1	2	4	3
(*b*)	2	1	4	3
(*c*)	2	1	3	4
(*d*)	2	3	4	1

127. In a common emitter circuit all voltages are measured with respect to
(*a*) base (*b*) emitter
(*c*) collector (*d*) base or collector

128. In a BJT CE amplifier circuit the value of coupling capacitor should be such that
(*a*) its reactance is small for the lowest frequency in the input and output
(*b*) its reactance is high for the lowest frequency in the input and output
(*c*) its reactance is zero for the lowest frequency in the input and output
(*d*) its reactance is infinite for the lowest frequency in the input and output

129. In the CE circuit shown in Fig. 7.72 the slope of ac load line will be

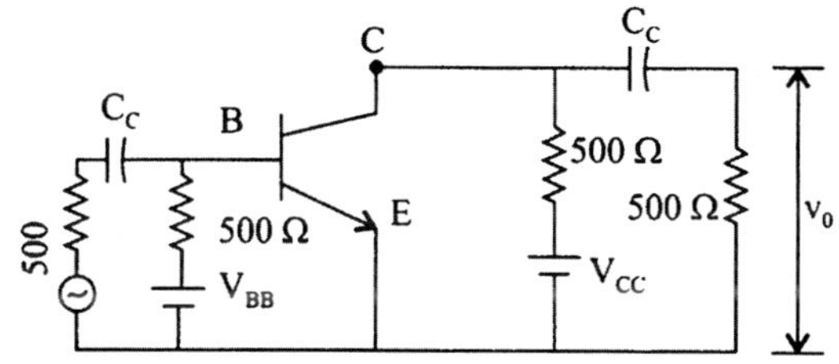

Fig. 7.72.

(*a*) $-\frac{1}{500}$ (*b*) $-\frac{1}{250}$
(*c*) $-\frac{1}{750}$ (*d*) $-\frac{1}{1000}$

130. In the CE circuit of Fig. 7.72 the slope of dc load line is
(*a*) $-\frac{1}{500}$ (*b*) $-\frac{1}{250}$
(*c*) $-\frac{1}{2000}$ (*d*) $-\frac{1}{1000}$

131. Which of the following of parameters of CE circuit has the unit of ohms?
(*a*) h_{ie} (*b*) h_{re}
(*c*) h_{fe} (*d*) h_{oe}

132. Which of the following *h* parameters is an admittance?
(*a*) h_{ie} (*b*) h_{re}
(*c*) h_{fe} (*d*) h_{oe}

133. Match List I and List II and give the correct answer using the given codes

List I (*h* parameter of BJT amplifier)	List II (Units)
A. h_{ie}	1. siemens
B. h_{oe}	2. no units
C. h_{fe}	3. ohms

Codes:

	A	B	C
(*a*)	1	2	3
(*b*)	3	1	2
(*c*)	2	1	3
(*d*)	3	2	1

134. Match List I and List II and give correct answer using the given codes

List I (*h* parameter in CE configuration)	List II (Typical values)
A. h_{ie}	1. 49
B. h_{re}	2. 25×10^{-6} siemens
C. h_{fe}	3. 2.4×10^{-4}
D. h_{oe}	4. 1000 ohms

Codes:

	A	B	C	D
(*a*)	1	2	4	3
(*b*)	4	3	2	1
(*c*)	3	4	1	2
(*d*)	4	3	1	2

135. Match List I and List II and give correct answer using the given codes

List I (BJT amplifier)	**List II** (Characteristics)
A. CE	1. good voltage gain, good current gain
B. CC	2. voltage gain less than I, high current gain
C. CB	3. current gain less than 1, voltage gain high

Codes:

	A	B	C
(*a*)	2	1	3
(*b*)	1	2	3
(*c*)	3	1	2
(*d*)	3	2	1

136. Fig. 7.73 shows the self bias circuit for CE amplifier and its equivalent circuit. V_{BB} and R_B respectively are

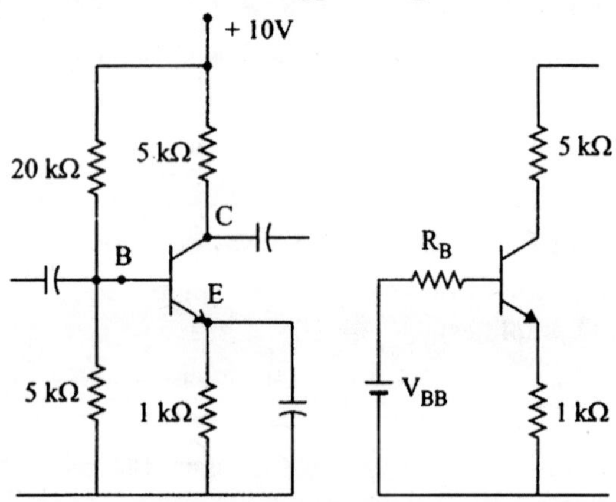

Fig. 7.73.

(*a*) 10 V and 4 k ohm (*b*) 5 V and 4 k ohm
(*c*) 2 V and 4 k ohm (*d*) 2 V and 10 k ohm

137. In the circuit of Fig. 7.73 (*a*) $\beta = 50$ and $V_{BE} = 0.5$ V. The quiescent value of base current I_B is

(*a*) about 28 μA (*b*) about 28 mA
(*c*) about 0.28 μA (*d*) about 2.8 A

138. In the circuit of Fig. 7.73 $\beta = 100$ and quiescent value of base current is 20 μA. The quiescent value of collector current is

(*a*) 0.2 μA (*b*) 2 μA
(*c*) 2 mA (*d*) 20 mA

139. In the r_e model of a BJT amplifier, the ac resistance of diode at room temperature is about (I_E is quiescent emitter current)

(*a*) $\frac{0.025}{I_E}$ (*b*) $\frac{0.25}{I_E}$
(*c*) $\frac{25}{I_E}$ (*d*) $\frac{250}{I_E}$

140. If ac resistance of diode is r_e, then in CE, BJT amplifier the input impedance is

(*a*) r_e (*b*) $\beta\, r_e$
(*c*) $\frac{r_e}{\beta}$ (*d*) $(\beta r_e)^2$

141. As per miller theorem the circuit of Fig. 7.74 (*a*) can be replaced by that in Fig. 7.74 (*b*) If $V_2/V_1 = K$, the Y_1 and Y_2 are equal to

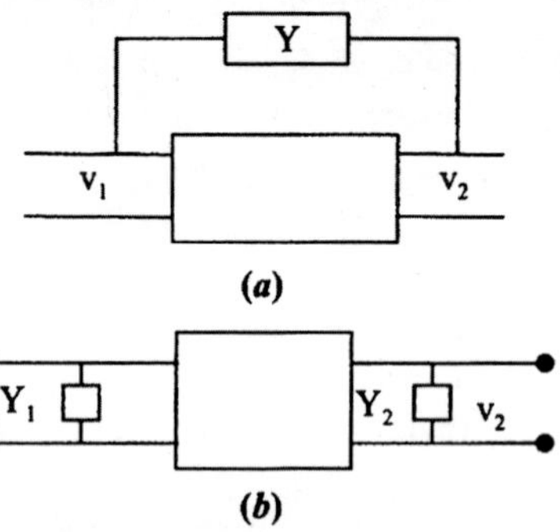

Fig. 7.74.

(*a*) $Y(1 - K)$ each

(*b*) $Y\left(1-\frac{1}{K}\right)$each

(*c*) $Y_1 = Y(1-K)$ and $Y_2 = Y\left(1-\frac{1}{K}\right)$

(*d*) $Y_1 = Y\left(1-\frac{1}{K}\right)$ and $Y_2 = (1-K)$

142. Match List I and List II and select the correct answer using the given codes

List I (FET quantity)	**List II** (Symbol)
A. Total instantaneous value of drain voltage	1. v_{ds}
B. Quiescent value of drain voltage	2. V_{ds}
C. Instantaneous value of ac component of drain voltage	3. V_{DS}
D. RMS value of ac component of drain voltage	4. v_{DS}

Codes:

	A	B	C	D
(*a*)	2	1	4	3
(*b*)	4	3	1	2
(*c*)	1	2	4	3
(*d*)	3	4	1	2

143. In Fig. 7.75, V_G =

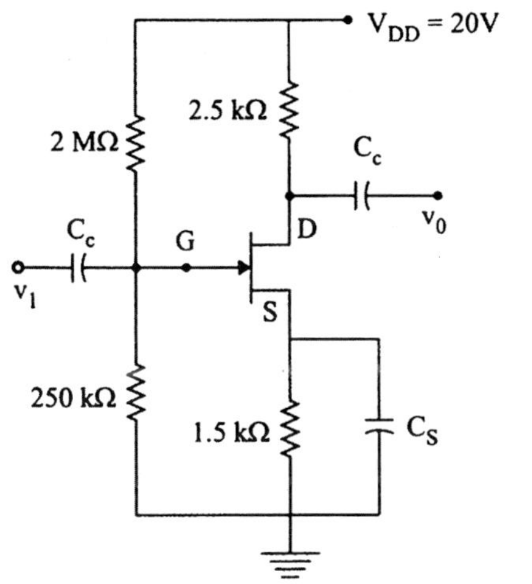

Fig. 7.75.

(*a*) 2.22 V (*b*) 22.2 V
(*c*) 0.22 V (*d*) 0.022 V

144. In Fig. 7.75, I_D = 4 mA. Then V_D =
(*a*) 20 V (*b*) 10 V
(*c*) 30 V (*d*) 1 V

145. In Fig. 7.75, I_D = 4 mA. Then V_S =
(*a*) 20 V (*b*) 16 V
(*c*) 6 V (*d*) 0.6 V

146. In Fig. 7.75, I_D = 4 mA, Then V_{DS} =
(*a*) 20 V (*b*) 16 V
(*c*) 4 V (*d*) 0.4 V

147. Fig, 7.76 shows a self bias circuit for FET amplifier, I_D = 4m A. Then I_S =

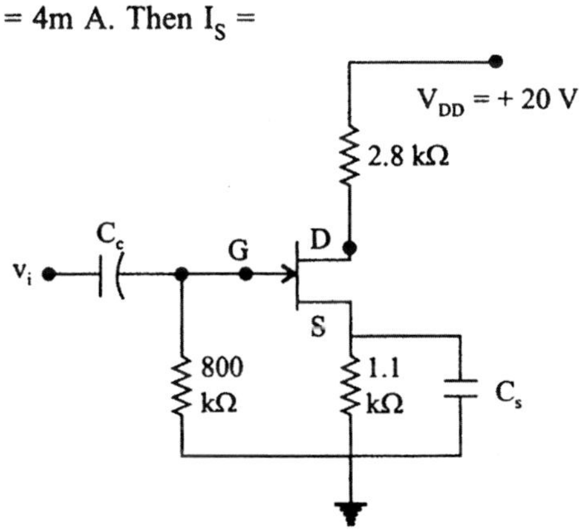

Fig.7.76.

(*a*) 4 mA (*b*) 2 mA
(*c*) 0.4 mA (*d*) 0.2 mA

148. In Fig. 7.76 I_D = 4 mA, Then V_S =
(*a*) 20 V (*b*) 4.4 V
(*c*) 0.4 V (*d*) 0.44 V

149. In Fig. 7.76 I_D = 4 mA, Then V_{GS} =
(*a*) 4.4 V (*b*) –4.4 V
(*c*) 0.44 V (*d*) –0.44 V

150. In Fig. 7.76 the potential of gate terminal is
(*a*) 4 V (*b*) –4 V
(*c*) 0 (*d*) 20 V

151. In which amplifier circuit is gate conductance negligible
(*a*) JFET amplifier
(*b*) MOSFET amplifier
(*c*) both JFET and MOSFET amplifier
(*d*) none of die above

152. Match List I and List II and select the correct answer using the codes

List I (FET parameters)	List II (Typical values)
A. g_m for JFET	1. 25 kΩ
B. r_d for JFET	2. 0.01 siemen
C. r_d for MOSFET	3. 0.5 MΩ
D. g_g for JFET	4. 10^{-8} siemen

Codes:

	A	B	C	D
(*a*)	1	2	3	4
(*b*)	2	3	1	4
(*c*)	4	2	3	1
(*d*)	4	3	2	1

153. Match List I and List II and select the correct answer using the given codes

List I (Double Rectifier)	List II (Number of Diodes)
A. Full wave Bridge rectifier	1. 1
B. Half wave rectifier	2. 2
C. Full wave rectifier using centre tapped transformer	3. 4

Codes:

	A	B	C
(*a*)	1	2	3
(*b*)	3	1	2
(*c*)	2	1	3
(*d*)	3	1	2

154. Match List I and List II and select the correct answer using the given codes

List I (Diode rectifier)	List II (Expression for output voltage)
A. Bridge	1. $V_{dc} = \frac{2E_m}{\pi} - I_{dc}(2R_f)$
B. Half wave	2. $V_{dc} = \frac{2E_m}{\pi} - I_{dc}R_f$
C. Full wave using centre tapped transformer	3. $V_{dc} = \frac{E_m}{\pi} - I_{dc}R_f$

Codes:

	A	B	C
(*a*)	1	2	3
(*b*)	3	2	1
(*c*)	1	3	2
(*d*)	3	1	2

155. Match List I and List II and select the correct answer using the given codes (R_f is forward resistance of diode and R_L is load resistance)

List I (Diode rectifier)	**List II (Efficiency if $\frac{R_f}{R_L} = 0.25$)**
A. Half wave	1. 0.54
B. Bridge	2. 0.65
C. Full wave using centre tapped transformer	3. 0.325

Codes:

	A	B	C
(*a*)	1	2	3
(*b*)	2	1	3
(*c*)	3	2	1
(*d*)	3	1	2

156. In the circuit of Fig. 7.77 $v_1 = 8$ V and $v_2 = 0$. Which diode will conduct (Assume ideal diodes)

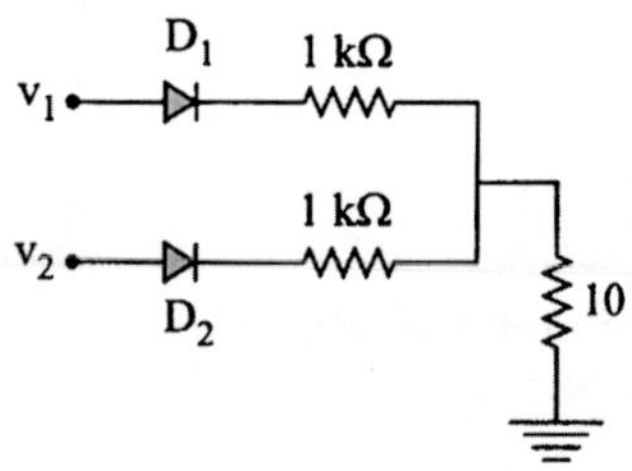

Fig. 7.77.

(*a*) D_1 only (*b*) D_2 only
(*c*) both D_1 and D_2 (*d*) neither D_1 or D_2

157. In Fig. 7.77 $v_1 = 8$ V and $V_2 = 4$ V. Which diode will conduct

(*a*) D_2 only (*b*) D_1 only
(*c*) both D_1 and D_2 (*d*) neither D_1 nor D_2

158. In Fig. 7.77 $v_1 = 8$ V and $v_2 = 8$ V. Which diode will conduct

(*a*) D_1 only (*b*) D_2 only
(*c*) both D_1 and D_2 (*d*) neither D_1 nor D_2

159. In the circuit of Fig. 7.78 both diodes are ideal. If $v_1 = 10$ V and $v_2 = 10$ V which diode will conduct

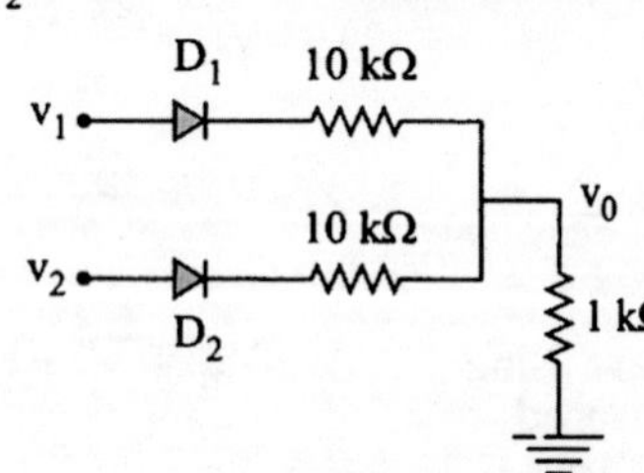

Fig. 7.78.

(*a*) D_1 only (*b*) D_2 only
(*c*) both D_1 and D_2 (*d*) neither D_1 nor D_2

160. In the circuit of Fig. 7.78, $v_1 = v_2 = 10$ V Then

(*a*) $I_1 = I_2$ (*b*) $I_1 > I_2$
(*c*) $I_2 > I_1$ (*d*) $I_1 = 0$

161. Match List I and List II and select the correct answer using the given codes

List I **List II**

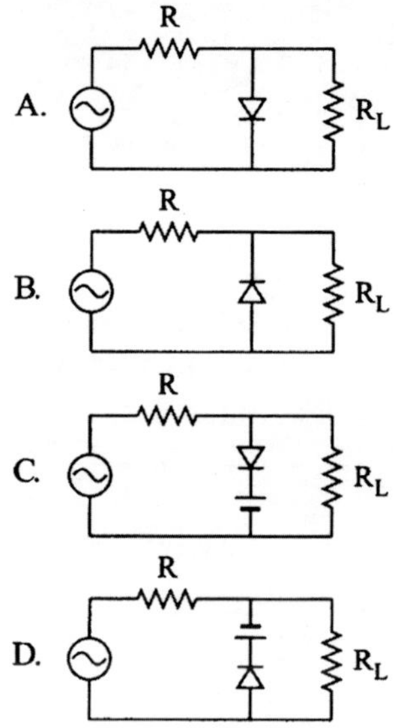

1. Negative limiter
2. Positive Limiter
3. Biased negative limiter
4. Biased positive limiter

Fig 7.79.

Codes:

	A	B	C	D
(*a*)	1	2	3	4
(*b*)	2	1	4	3
(*c*)	2	1	3	4
(*d*)	1	2	4	3

162. Match List I and List II and select the correct answer using the codes

List I (Diode circuit) **List II (Name of the circuit)**

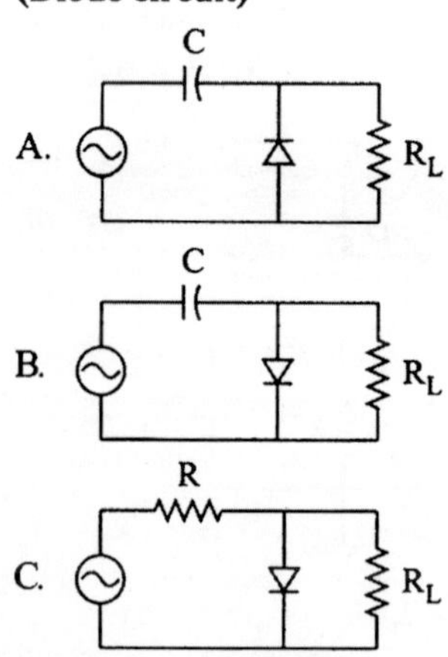

1. Positive limiter
2. Positive clamper
3. Negative clamper

Fig. 7.80.

Codes:

	A	B	C
(*a*)	2	3	1
(*b*)	1	2	3
(*c*)	2	1	3
(*d*)	3	1	2

163. Which op-amp circuit uses a resistance in series with input and capacitor in feedback path

(*a*) differenting amplifier (*b*) integrating amplifier
(*c*) logarithmic amplifier (*d*) exponential amplifier

164. Which op-amp circuit uses a capacitor in series with input and resistance in feedback path?
(*a*) differentiating amplifier
(*b*) integrating amplifier
(*c*) logarithmic amplifier
(*d*) none of the above

165. Which op-amp circuit uses a diode as one the circuit components
(*a*) integrating amplifier (*b*) differentiating amplifier
(*c*) logarithmic amplifier (*d*) summing amplifier

166. In which of the following circuits is op-amp used in open loop configuration
(*a*) comparator (*b*) summing amplifier
(*c*) integrating amplifier (*d*) logarithmic amplifier

167. In the circuit of Fig. 7.81, v_0 =

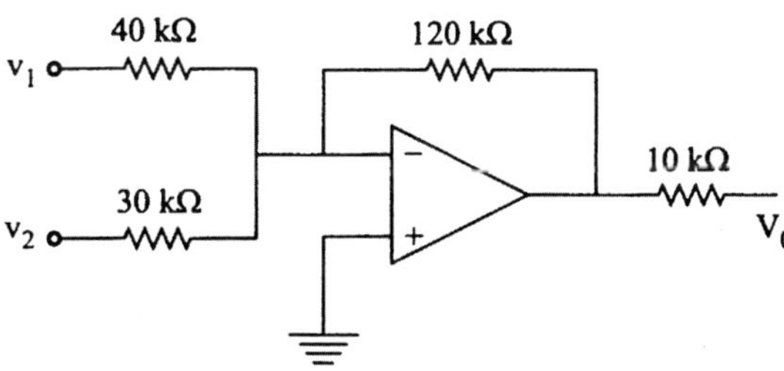

Fig. 7.81.

(*a*) $3\ v_1 + 4\ v_2$ (*b*) $-3\ v_1 - 4\ v_2$
(*c*) $4\ v_1 + 3\ v_2$ (*d*) $-4\ v_1 - 3\ v_2$

168. A difference amplifier using op-amp has closed loop gain = 50. If input is 2 V to each of inverting and non-inverting terminals, output is 5 m V. Then CMRR =
(*a*) 20000 (*b*) 2000
(*c*) 200 (*d*) 20

169. An op-amp has zero gain for common mode inputs. Then CMRR =
(*a*) zero (*b*) infinite
(*c*) high (*d*) low

170. The slew rate of an ideal op-amp, is
(*a*) very very slow (*b*) slow
(*c*) fast (*d*) infinitely fast

171. An ideal op-amp requires
(*a*) zero input offset voltage
(*b*) low input offset voltage
(*c*) high input offset voltage
(*d*) infinite input offset voltage

172. The open loop gain of an ideal op-amp is
(*a*) infinite (*b*) very high
(*c*) neither high nor low (*d*) low

173. The output impedance of an ideal op-amp is
(*a*) zero (*b*) low
(*c*) high (*d*) infinite

174. Which of the following op-amp circuits has zero resistance in the feedback path?
(*a*) inverting amplifier (*b*) summing amplifier
(*c*) source follower (*d*) integrating amplifier

175. In the circuit Fig. 7.82 LED will be on when v_1 is

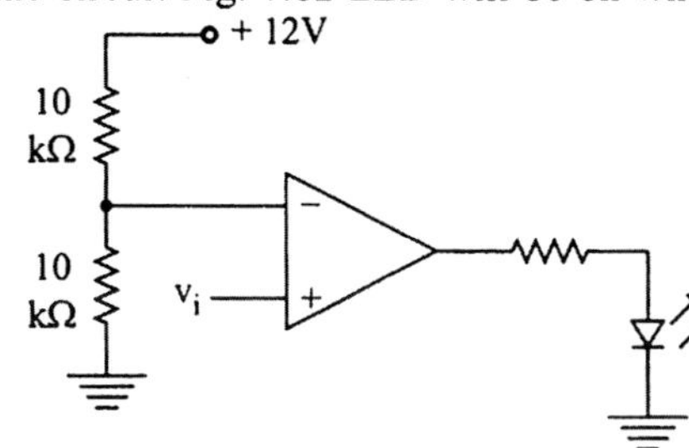

Fig. 7.82.

(*a*) less than 6 V (*b*) more than 6 V
(*c*) less than 12 V (*d*) more than 12 V

176. In a one pole active low pass filter, the cutoff frequency is f_c. Then
(*a*) gain decreases at the rate of 20 dB per decade increase in frequency above f_c
(*b*) gain decreases at the rate of 20 dB per decade decrease in frequency below f_c
(*c*) gain decreases at the rate of 40 dB per decade increase in frequency above f_c
(*d*) gain decreases at the rate 40 dB per decade decrease in frequency below f_c

177. Which of the following uses a capacitor in the feedback path?
(*a*) active low pass filter (*b*) active high pass filter
(*c*) both (*a*) or (*b*) (*d*) neither (*a*) nor (*b*)

178. Which of the following oscillator circuits uses three capacitors in the phase shifting network?
(*a*) Hartley (*b*) Clapp
(*c*) Colpatt (*d*) both (*b*) and (*c*)

179. Which of the following oscillators is suitable for frequencies in the range of mega hertz?
(*a*) RC phase shin (*b*) Wien bridge
(*c*) Hartley (*d*) Both (*a*) and (*c*)

180. An ideal power supply has
(*a*) zero internal resistance
(*b*) very high internal resistance
(*c*) high output resistance (*d*) both (*b*) and (*c*)

181. An ideal op-amp has
(*a*) infinite input and output resistance
(*b*) low input and output resistance
(*c*) low input and high output resistance
(*d*) infinite input resistance and zero output resistance

182. Two transistors have same value of α but different gain (G) bandwidth (B) product. The transistors have similar geometries. However one of them is germanium and the other is silicon. The transistor with lower G B product
(*a*) is germanium
(*b*) is silicon
(*c*) may be either of two
(*d*) cannot be identified on the basis of above data

183. A Darlinghon pair is used for
(*a*) low distortion (*b*) high frequency range
(*c*) high power gain (*d*) high current gain

184. Figure 7.83 shows small signal model of CB amplifier circuit. Then I and R are respectively equal to

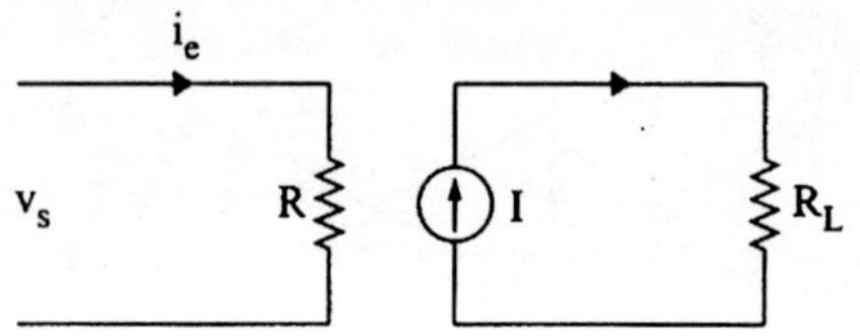

Fig. 7.83.

(a) $\alpha\, i_e$ and r_e (b) $\frac{\alpha}{1-\alpha}\, i_e$ and $r_e(\alpha - 1)$

(c) $\alpha\, i_e$ and $r_e(\alpha - 1)$ (d) $(\alpha - 1)\, i_e$ and r_e

185. A two stage amplifier has an upper cutoff frequency of 2 MHz and a lower cutoff frequency of 30 Hz. If both stages are similar, their individual upper and lower cutoff frequencies respectively are

(a) 4 MHz and 60 MHz (b) 3 MHz and 20 Hz

(c) 3 MHz and 60 Hz (d) 4 MHz and 20 Hz

186. In Fig. 7.84 the collector to emitter breakdown voltage of the transistor (BV_{CEO}) should be

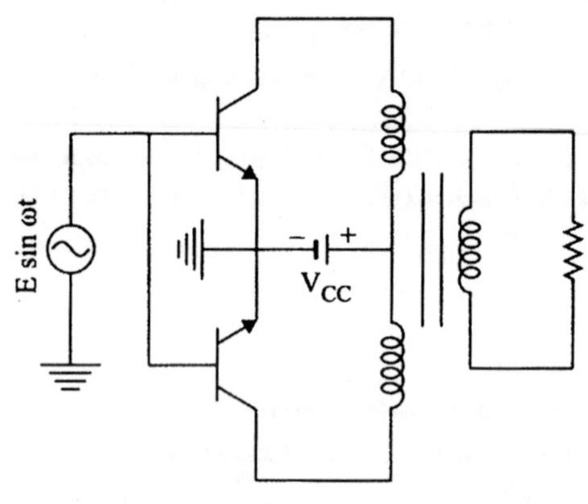

Fig. 7.84.

(a) more than V_{CC} (b) more than 0.5 V_{CC}

(c) more than V_{CC} + E (d) more than 2 V_{CC}

187. In the amplifier circuit Fig. 7.85 the quiescent point voltage and current are

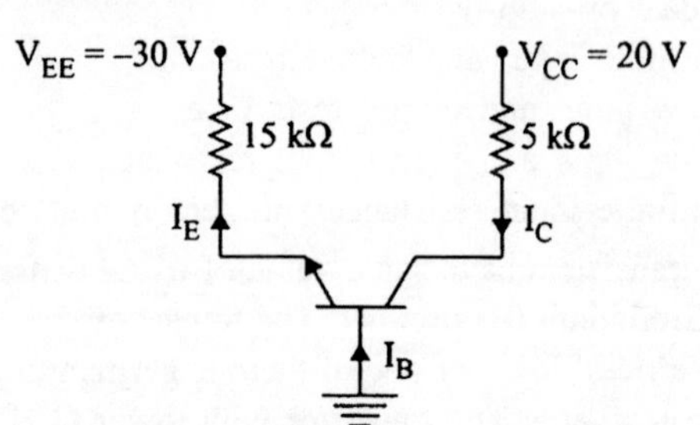

Fig. 7.85.

(a) 12 V, 5 mA (b) 12 V, 2 mA

(c) 10 V, 2 mA (d) 10 V, 5 mA

188. The input impedance of op-amp circuit of Fig. 7.86 is

(a) 120 k ohm (b) 110 k ohm

(c) infinity (d) 10 k ohm

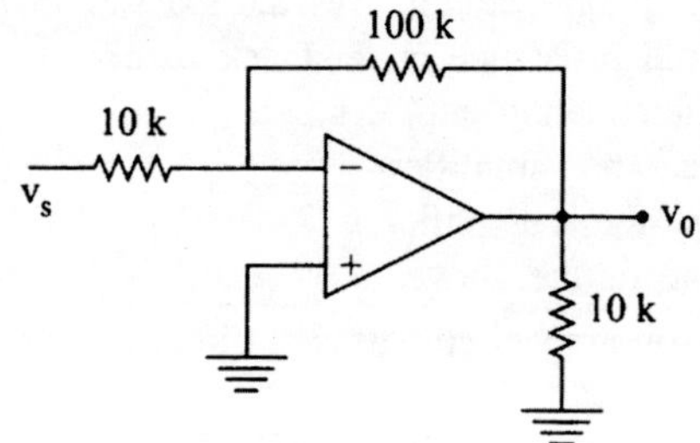

Fig. 7.86.

189. The circuit in Fig. 7.87 is

(a) positive clamper

(b) negative clamper

(c) positive peak clipper

(d) negative peak clipper

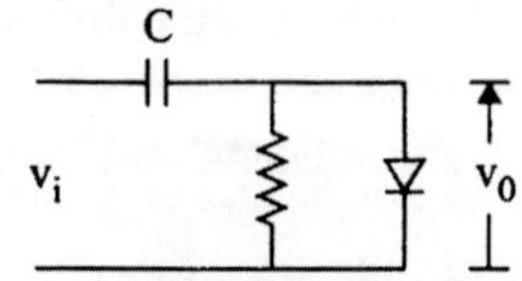

Fig. 7.87.

190. Examine the two lists

List I Output wave form	List II Circuit and input signal
1. Triangular wave	Integrating circuit and square wave
2. Impulse wave	Differentiating circuit and step wave
3. Sawtooth wave	Differentiating circuit and triangular wave

Which of the above pairs are correctly matched?

(a) 2 and 3 (b) 1 and 2

(c) 1 only (d) 1 and 3

191. Figure 7.88 shows a zener circuit and characteristic of the zener used. If R_2 is the lowest load resistance, then for proper voltage regulation R_1 must be

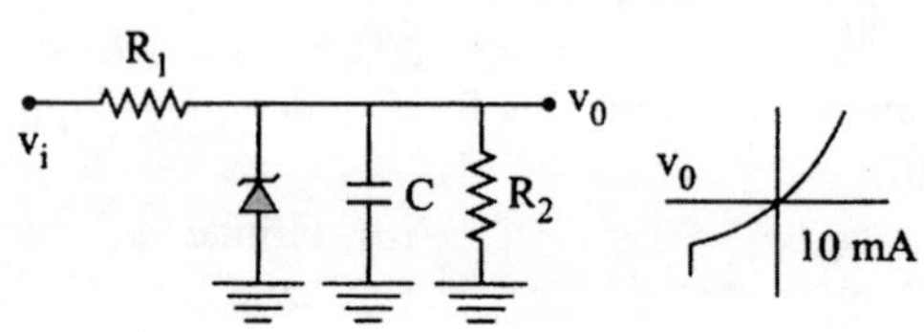

Fig. 7.88.

(a) more than $\frac{V_0 + .01R_2}{(V_i - V_0)R_2}$

(b) more than $\frac{(V_i - V_0)R_2}{V_0 + 0.01R_2}$

(c) less than $\frac{(V_i - V_0)R_2}{V_0 + 0.01R_2}$

(d) less than $\frac{V_0 + .01R_2}{(V_i - V_0)R_2}$

192. Study the circuit of Fig. 7.89 and examine the following statements

1. It is an active low pass filter
2. it is a second order filter
3. the change in gain is 40 dB/decade

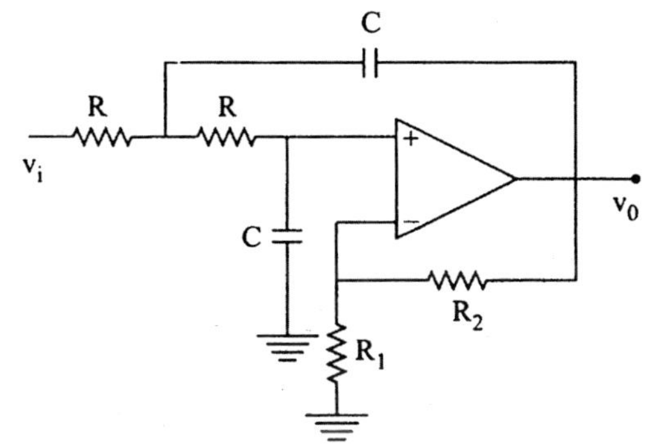

Fig. 7.89.

Which of the above statements are correct?

(*a*) 1, 2 and 3 (*b*) 1 and 2
(*c*) 1 and 3 (*d*) 2 and 3

193. As compared to a full wave diode rectifier circuit using centre tapped transformer, the bridge diode rectifier circuit has the main advantage of

(*a*) higher current carrying capacity
(*b*) higher efficiency
(*c*) lower ripple factor
(*d*) lower PIV

194. In the circuit of Fig. 7.90, PIV can be up to

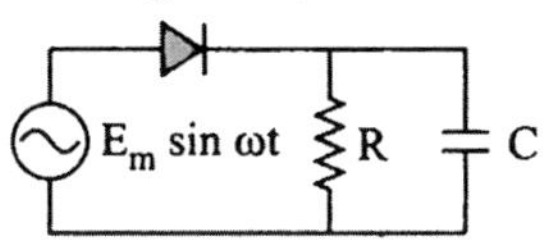

Fig. 7.90.

(*a*) E_m (*b*) $2\,E_m$
(*c*) $0.5\,E_m$ (*d*) $\sqrt{2}\,E_m$

195. The circuit in Fig. 7.91 is

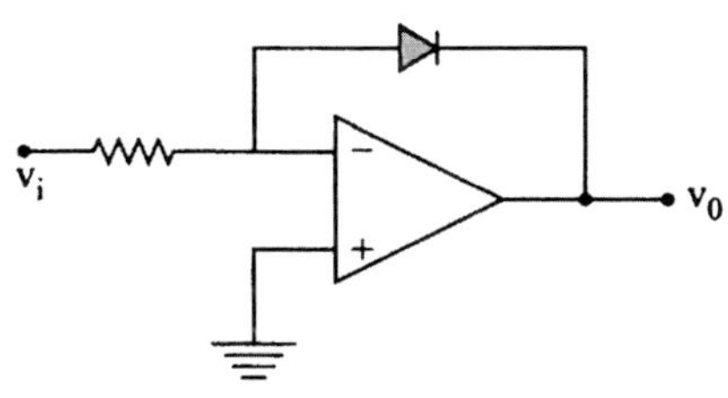

Fig. 7.91.

(*a*) rectifier
(*b*) voltage to frequency converter
(*c*) frequency to voltage converter
(*d*) logarithmic amplifier

196. Match List I and List II and select the correct answer using the given codes

List I (Circuit)	**List II (Characteristics)**
A. Darlington pair	1. uses pnp and npn transistors, used as power amplifier
B. CB amplifier	2. Voltage gain nearly 1 and high input impedance
C. Cascade amplifier	3. low input impedance, used for high frequency isolation

Codes:

	A	B	C
(*a*)	1	2	3
(*b*)	2	3	1
(*c*)	2	1	3
(*d*)	3	1	2

197. Examine the circuit in Fig. 7.92 and match lists I and II Select correct answer using the given codes

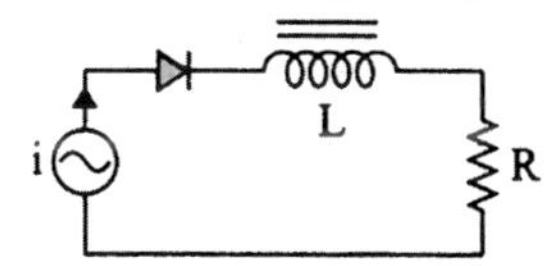

Fig. 7.92.

List I	**List II** (Value of $\tan^{-1}\frac{\omega L}{R}$)
A. 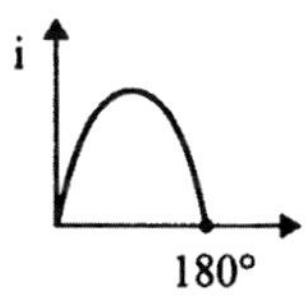	1. 90°
B. 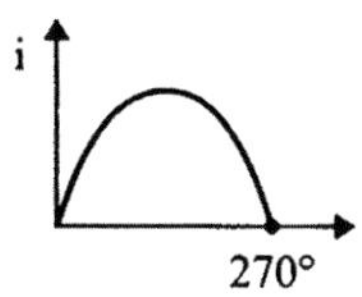	2. 0°
C. 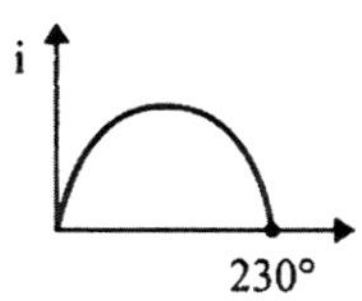	3. 50°

Codes:

	A	B	C
(*a*)	1	2	3
(*b*)	2	1	3
(*c*)	3	1	2
(*d*)	3	2	1

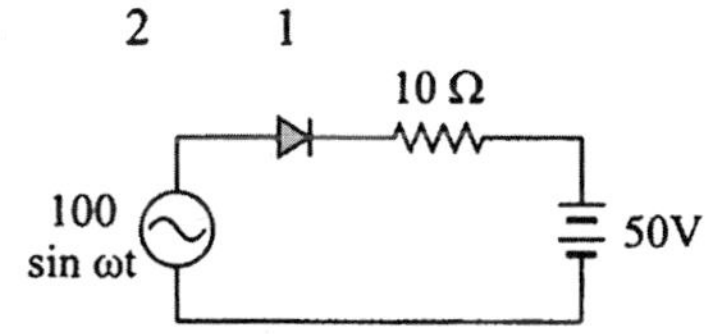

Fig. 7.93.

198. In the circuit of Fig. 7.93 diode will conduct

(*a*) from 0 to 180° in the positive half cycle
(*b*) from 30° to 180° in the positive half cycle
(*c*) from 50° to 160° the positive half cycle
(*d*) from 30° to 150° in positive half cycle

199. In the circuit of Fig. 7.94 the diode

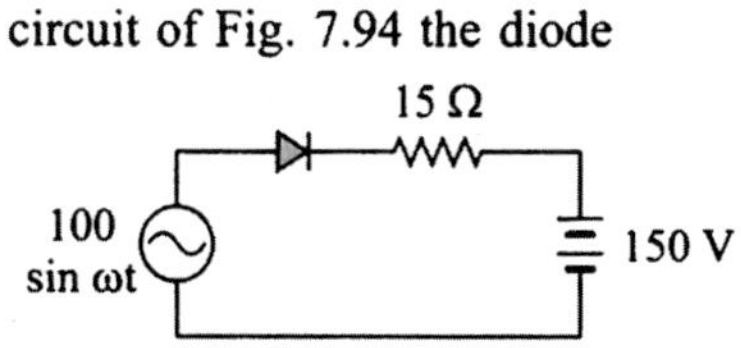

Fig. 7.94.

(*a*) will conduct in the whole cycle
(*b*) will conduct in the positive half cycle

(*c*) will not conduct
(*d*) will conduct from 30° to 150° in the positive half cycle

200. Examine the circuit of Fig. 7.95 and the following statements

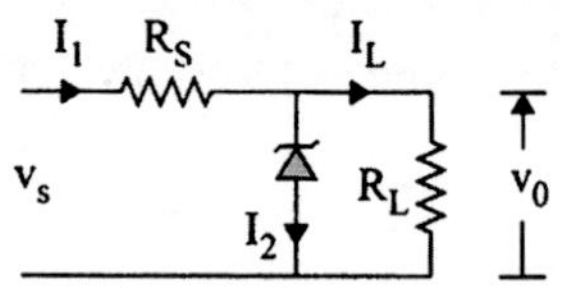

Fig. 7.95.

1. V_s is more than the zener breakdown voltage
2. $I_L = I_1 - I_2$
3. R_s is less than zener nominal resistance

Which of the above statements are correct
(*a*) 1, 2, 3 (*b*) 1, 2
(*c*) 1 only (*d*) 2 and 3

201. The ideal characteristics of a stabilizer is
(*a*) constant output voltage with low internal resistance
(*b*) constant output current with low internal resistance
(*c*) constant output voltage with high internal resistance
(*d*) constant internal resistance and constant output current

202. The circuit in Fig 7.96 is

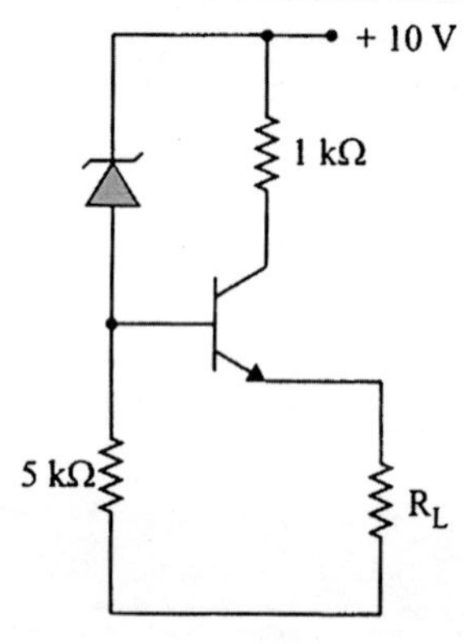

Fig. 7.96.

(*a*) CE amplifier
(*b*) CB amplifier
(*c*) Constant voltage source
(*d*) Constant current source

203. If the op-amp in Fig. 7.97 has a unity gain frequency of 1 MHz, the cutoff frequency is

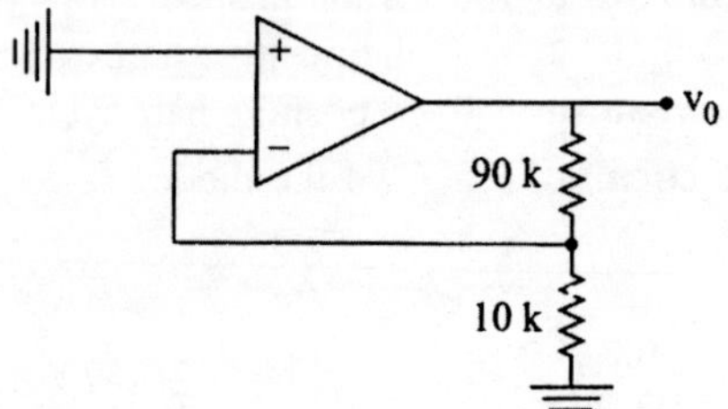

Fig. 7.97.

(*a*) 100 kHz (*b*) 1 MHz
(*c*) 10 MHz (*d*) 90 MHz

204. The minimum value of h_{fe} of a transistor used in RC phase shift oscillator is
(*a*) 54.4 (*b*) 45.4
(*c*) 44.5 (*d*) 29

205. When $Z_1 << Z$, the circuit of Fig. 7.98 works as

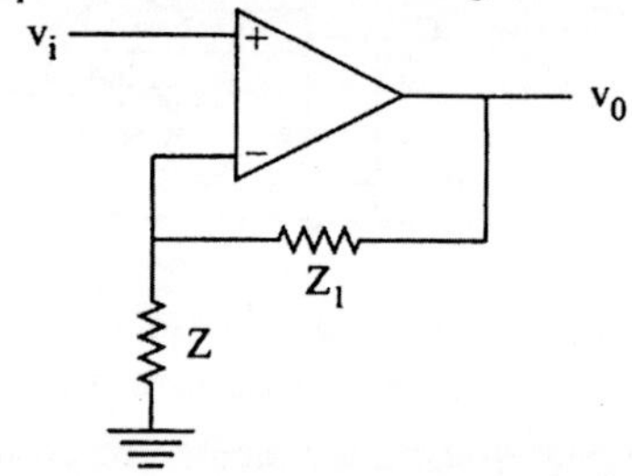

Fig. 7.98.

(*a*) inverting amplifier (*b*) Schmitt trigger
(*c*) non-inverting amplifier (*d*) source follower

206. For the circuit in Fig. 7.99 the output wave shape is

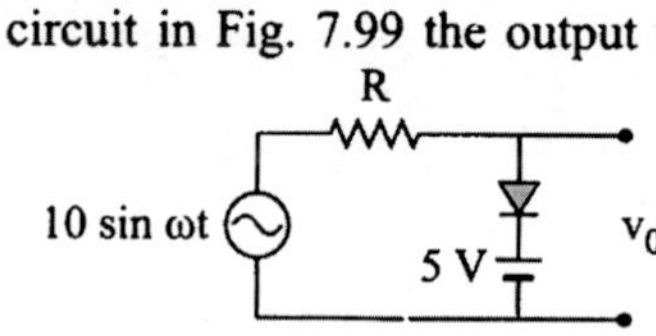

Fig. 7.99.

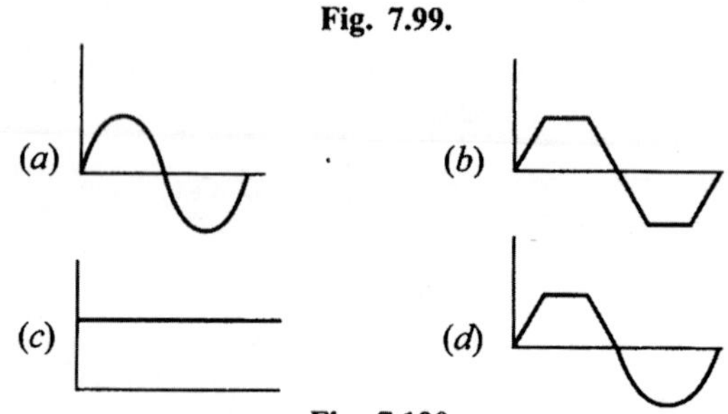

Fig. 7.100.

207. In Fig. 7.101, V_0 =

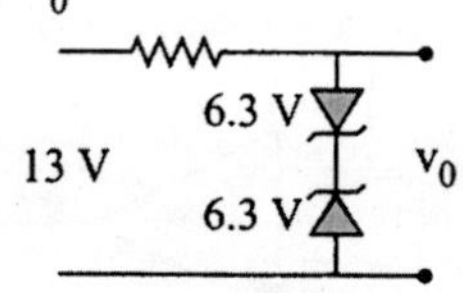

Fig. 7.101.

(*a*) zero (*b*) 5.7 V
(*c*) 6.9 V (*d*) 12.6 V

208. It is desired to reduce distortion in an amplifier from 8% to 2% by using 5% negative feedback. The open and closed loop gains respectively are
(*a*) 60 and 15 (*b*) 15 and 60
(*c*) 6 and 1.5 (*d*) 1.5 and 6

209. In the amplifier circuit of Fig. 7.102 h_{fe} = 100 and h_{ie} = 1000 Ω. The voltage gain of amplifier is about

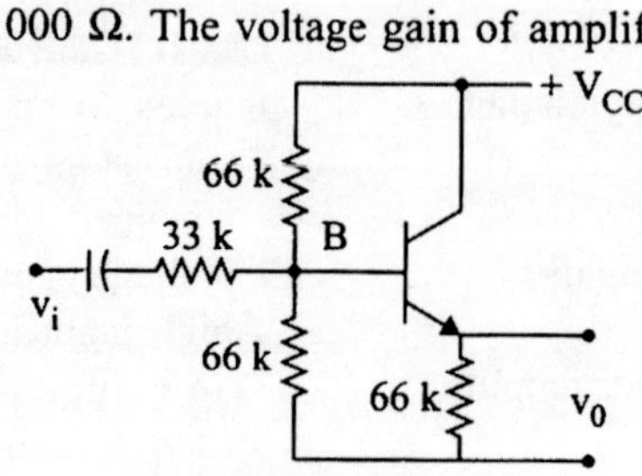

Fig. 7.102.

(*a*) 0.33 (*b*) 0.5
(*c*) 0.66 (*d*) 1

210. Figure 7.103 shows a portion of linear *v-i* characteristics of diode. If applied voltage is +6 V, the current will be

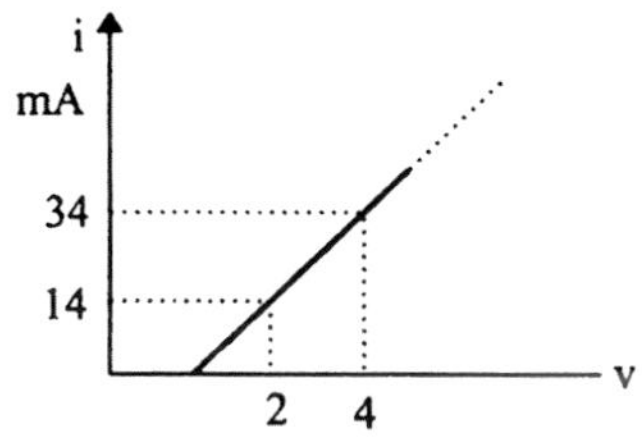

Fig. 7.103.

(*a*) 42 mA (*b*) 48 mA
(*c*) 51 mA (*d*) 54 mA

211. Match List I and List II and select the correct answer using the given codes

List I (Oscillator)	**List II (Features)**
A. Wien Bridge	1. RF oscillator, uses two inductors and one capacitor
B. Colpitt's	2. LC oscillator for radio frequencies. uses three capacitances and one inductance
C. Hartley	3. RC oscillator for audio frequencies
D. Clapp	4. RF oscillator, uses two capacitances and one inductance

Codes:

	A	B	C	D
(*a*)	2	1	4	3
(*b*)	2	4	1	3
(*c*)	3	4	1	2
(*d*)	3	1	4	2

212. Figure 7.104 is a 24 V_r stabilized power supply. The zener is 24 V, 600 mW. The minimum zener current is 10 mA. Proper values R and maximum load current are

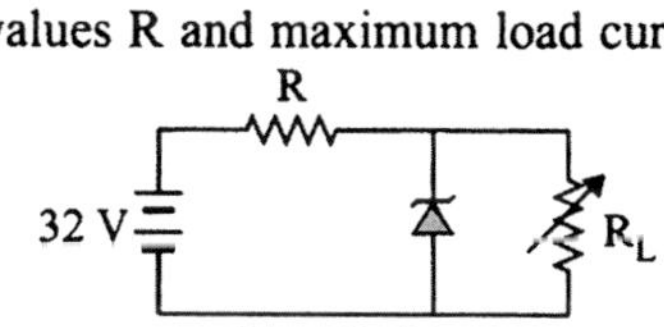

Fig. 7.104.

(*a*) 320 Ω, 10 mA (*b*) 400 Ω, 15 mA
(*c*) 400 Ω, 10 mA (*d*) 320 Ω, 15 mA

213. Consider the following statements : A clamper circuit
1. adds or subtracts a dc voltage to a wave form
2. does not change the wavefrom
3. amplifies the waveform

Which are correct?

(*a*) 1, 2 (*b*) 1, 3
(*c*) 1, 2, 3 (*d*) 2, 3

214. Match List I and List II and select the correct answer using the codes

List I (Amplifier coupling)	**List II (Feature)**
A. RC	1. High voltage gain and impedance matching
B. Inductive	2. Ability to amplify dc and low frequency signals
C. Transformer	3. Minimum non-linear distortion
D. Direct	4. Low collector supply voltage can be used

Codes:

	A	B	C	D
(*a*)	4	1	3	2
(*b*)	3	4	1	2
(*c*)	1	2	3	4
(*d*)	4	3	2	1

215. In Fig. 7.105 the voltage drop across diode is 0.7 V. Then the voltage across diode during negative half cycle is

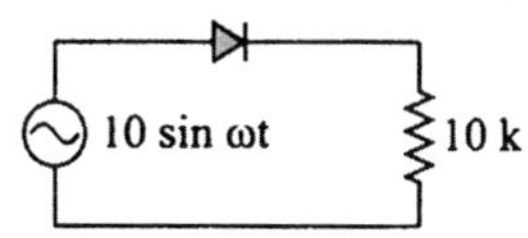

Fig. 7.105.

(*a*) 10 V (*b*) 9.3 V
(*c*) zero (*d*) 10.7 V

216. In Fig. 7.106 The cut in voltage of diode is 0.7 V and its bulk resistance is 20 Ω. The peak value of circuit current of during positive and negative half cycles are

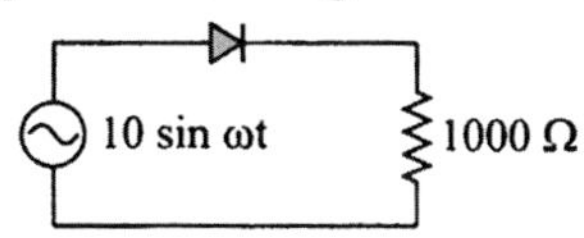

Fig. 7.106.

(*a*) 10 mA and zero (*b*) 9.12 mA and zero
(*c*) 9.3 mA and zero (*d*) zero and 9.12 mA

217. Figure 7.107 uses 10 V zener diode. The minimum and maximum current through series resistance are

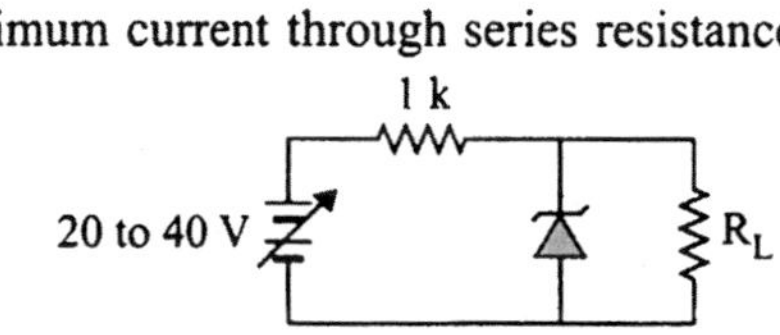

Fig. 7.107.

(*a*) 10 mA and 30 mA (*b*) 20 mA and 40 mA
(*c*) 0 and 30 mA (*d*) 0 and 40 mA

218. The knee voltages is silicon and germanium diodes respectively are

(*a*) 0.3 V and 0.7 V (*b*) 0.7 V and 0.3 V
(*c*) 0.7 V each (*d*) 0.3 V each

219. At the peak of negative half cycle in Fig. 7.108

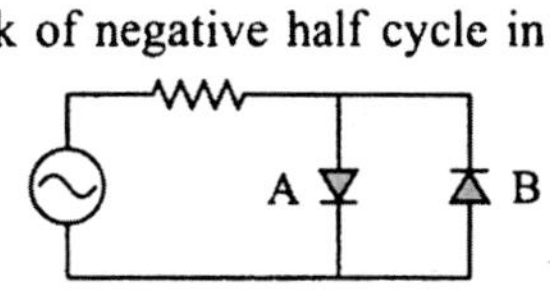

Fig. 7.108.

(*a*) diode A is forward biased and diode B is reverse biased
(*b*) both diodes are reverse biased
(*c*) both diodes are forward biased
(*d*) diode A is reverse biased and diode B is forward biased

220. An amplifier has input impedance of 4 kΩ and output impedance of 80 kΩ. It is used in negative feedback circuit with 10% feedback. If open loop gain is 90, the closed loop input and output impedances are
(*a*) 40 kΩ and 8 kΩ respectively
(*b*) 4 kΩ and 80 kΩ respectively
(*c*) 0.4 kΩ and 800 kΩ respectively
(*d*) 8 kΩ and 160 kΩ respectively

221. In Fig. 7.109, $V_{EB} = 0.6$ V, $\beta = 99$. Then V_C and I_C are

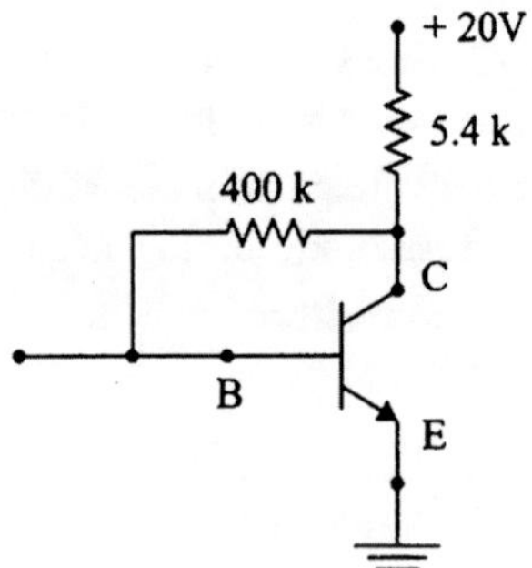

Fig. 7.109.

(*a*) 9.3 V and 1.98 m A respectively
(*b*) 4.6 V and 1.98 m A respectively
(*c*) 9.3 V and 0.02 m A respectively
(*d*) 4.6 V and 0.02 m A respectively

222. In the op-amp circuit shown in Fig. 7.110 the current through load resistance R_L is

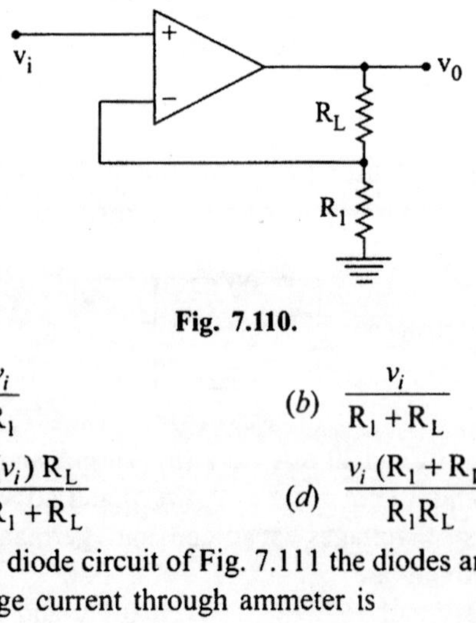

Fig. 7.110.

(*a*) $\frac{v_i}{R_1}$ (*b*) $\frac{v_i}{R_1 + R_L}$
(*c*) $\frac{(v_i) R_L}{R_1 + R_L}$ (*d*) $\frac{v_i (R_1 + R_L)}{R_1 R_L}$

223. In the diode circuit of Fig. 7.111 the diodes are ideal. The average current through ammeter is

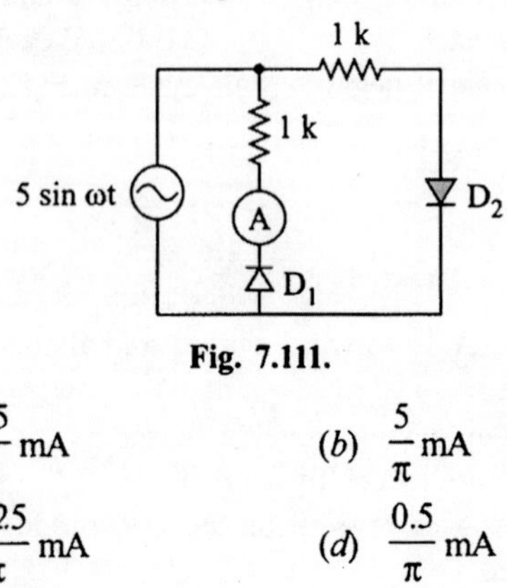

Fig. 7.111.

(*a*) $\frac{2.5}{\pi}$ mA (*b*) $\frac{5}{\pi}$ mA
(*c*) $\frac{0.25}{\pi}$ mA (*d*) $\frac{0.5}{\pi}$ mA

224. In the diode circuit shown in Fig. 7.111 the diodes are ideal. The impedance seen by source is
(*a*) 1 k ohm (*b*) 2 k ohm
(*c*) 0.5 k ohm (*d*) infinite

225. In the op-amp circuit of Fig. 7.112, V_0 =
(*a*) V_i (*b*) $3 V_i$
(*c*) $5 V_i$ (*d*) $0.5 V_i$

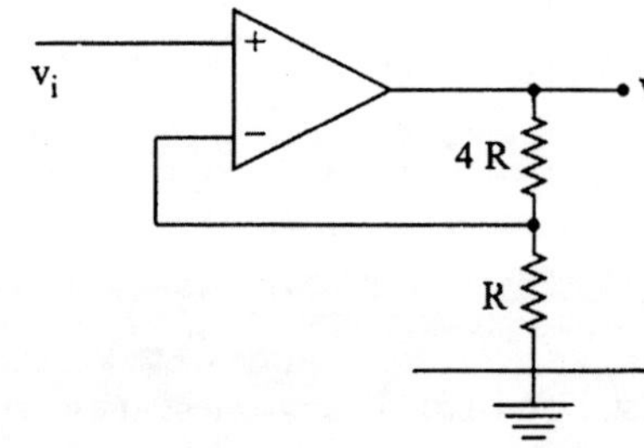

Fig. 7.112.

226. In the op-amp circuit of Fig. 7. 113, V_0 =

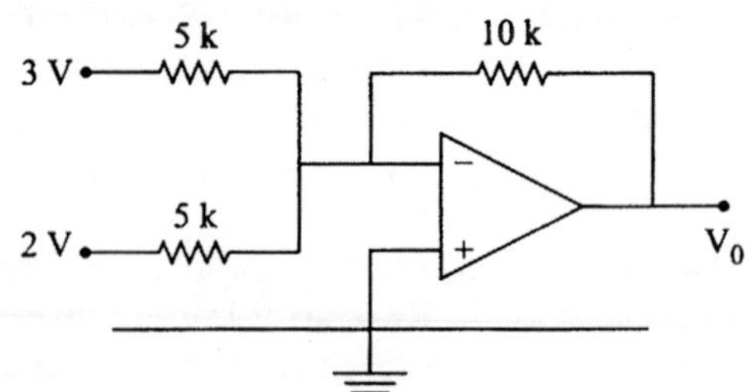

Fig. 7.113.

(*a*) 10 V (*b*) –10 V
(*c*) 5 V (*d*) –5 V

227. In the op-amp circuit of Fig. 7.114, V_0 =

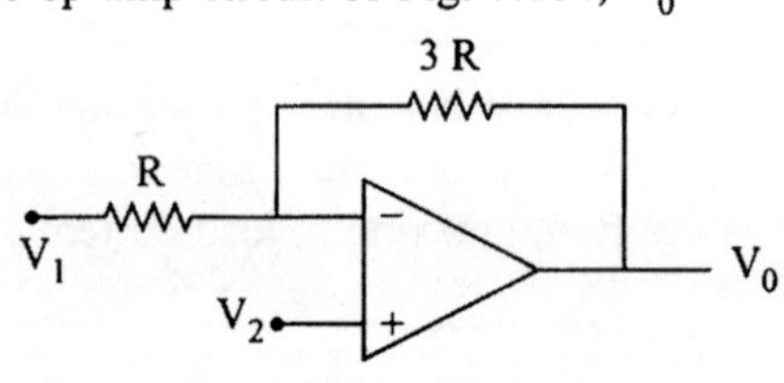

Fig. 7.114.

(*a*) $3 V_1 - 4 V_2$ (*b*) $-3 V_1 + 4 V_2$
(*c*) $3(V_1 - V_2)$ (*d*) $-3(V_1 - V_2)$

228. A half wave diode circuit using ideal diode has an input voltage 20 sin ω*t* volts. Then average and rms values of output voltage are
(*a*) $\frac{10}{\pi}$ V and 10 V (*b*) $\frac{20}{\pi}$ V and 10 V
(*c*) $\frac{10}{\pi}$ V and 5 V (*d*) $\frac{20}{\pi}$ V and 5 V

229. A bridge diode circuit using ideal diodes has an input voltage of 20 sin ω*t* volts. The averages and rms values of output voltage are
(*a*) $\frac{40}{\pi}$ V and $\frac{20}{\sqrt{2}}$ V (*b*) $\frac{20}{\pi}$ and $\frac{20}{\sqrt{2}}$ V
(*c*) $\frac{40}{\pi}$ V and 10 V (*d*) $\frac{20}{\pi}$ V and 10 V

230. A 120 V, 50 Hz source feeds a half wave rectifier circuit through a 4 :1 step down transformer. The average output voltage is
(*a*) 30 V (*b*) 13.5 V
(*c*) 10 V (*d*) 7.07 V

231. In a full wave rectifier circuit using centre tapped transformer the peak voltage across half the secondary winding is 30 V. Then PIV is
(*a*) 30 V (*b*) 60 V
(*c*) 15 V (*d*) 10 V

232. In a full wave rectifier circuit using centre tapped transformer, the peak voltage across half the secondary winding is 40 V. If diodes are ideal, the average output voltage is
(*a*) 40 V (*b*) 25.5 V
(*c*) 20 V (*d*) 2.55 V

233. A full wave rectifier circuit using centre tapped transformer, the input frequency a 50 Hz. The frequency of output is
(*a*) 100 Hz (*b*) 50 Hz
(*c*) 25 Hz (*d*) none of the above

234. A full wave rectifier circuit using centre tapped transformer and a bridge rectifier are fed at 100 V, 50 Hz. The frequencies of outputs in these two rectifiers respectively are
(*a*) 100 Hz each
(*b*) 50 Hz each
(*c*) 100 Hz and 50 Hz respectively
(*d*) 50 Hz and 100 Hz respectively

235. A half wave diode rectifier and full wave diode bridge rectifier both have an input frequency of 50 Hz. The frequencies of outputs respectively are
(*a*) 100 Hz and 50 Hz (*b*) 50 Hz and 100 Hz
(*c*) 100 Hz each (*d*) 50 Hz each

236. A half wave diode rectifier uses a diode having forward resistance of 50 ohms. The load resistance is also 50 ohms. Then the voltage regulation is
(*a*) 25% (*b*) 50%
(*c*) 100% (*d*) 200%

237. A full wave diode rectifier using centre tapped transformer has diodes with forward resistance of 50 ohms each. The load resistance is also 50 ohms. The voltage regulation is
(*a*) 25% (*b*) 50%
(*c*) 100% (*d*) 200%

238. A full wave bridge diode rectifier uses diodes having forward resistance of 50 ohms each. The load resistance is also 50 ohms. The voltage regulations is
(*a*) 25% (*b*) 50%
(*c*) 100% (*d*) 200%

239. Which of (he following components is essential for a voltage multiplier circuit
(*a*) resistor (*b*) inductor
(*c*) capacitor (*d*) both (*a*) and (*c*)

240. A voltage doubler circuit uses
(*a*) two diodes
(*b*) two capacitors
(*c*) two diodes and two capacitors
(*d*) none of the above

241. In a half wave voltage doubler circuit the input frequency is 50 Hz. The ripple frequency is
(*a*) 25 Hz (*b*) 50 Hz
(*c*) 100 Hz (*d*) 150 Hz

242. In a full wave voltage doubler circuit the input frequency is 50 Hz. The ripple frequency is
(*a*) 25 Hz (*b*) 50 Hz
(*c*) 100 Hz (*d*) 200 Hz

243. A voltage multiplier circuit using diodes and capacitors is suitable for
(*a*) high voltage high current device
(*b*) high voltage low current device
(*c*) low voltage low current device
(*d*) low voltage high current device

244. A half wave diode rectifier has a capacitance input filter. If input voltage is $V_m \sin \omega t$, PIV is
(*a*) V_m (*b*) $2 V_m$
(*c*) $3 V_m$ (*d*) $4 V_m$

245. A bridge rectifier circuit has input of 50 Hz frequency. The load resistance is R_L and filter capacitance is C. For good output wave shape, the time constant R_LC should be at least equal to
(*a*) 10 ms (*b*) 20 ms
(*c*) 50 ms (*d*) 100 ms

246. In a diode rectifier with a capacitance input filter, a surge current is likely to flow when the circuit is switched on. The cause of this surge current is
(*a*) the non-linear behaviour of diodes
(*b*) the capacitor acts as a short circuit at the instant of switching
(*c*) the inductance of leads
(*d*) both (*a*) and (*b*)

247. In Fig. 7.115, the function of resistor R is

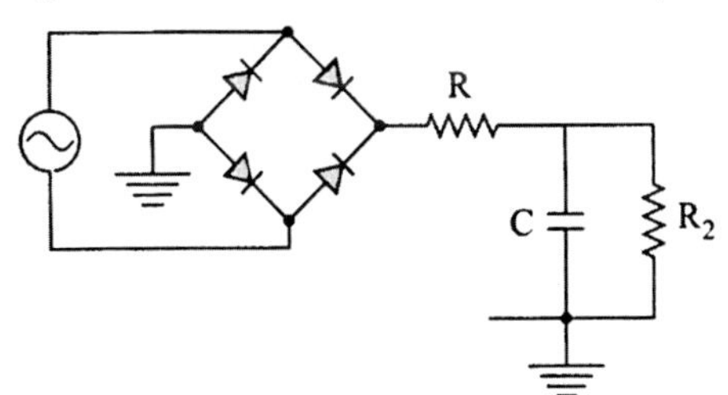

Fig. 7.115.

(*a*) to reduce ripples in the output
(*b*) to improve voltage regulation
(*c*) to limit the surge current
(*d*) either (*a*) or (*c*)

248. It has been found that in a rectifier circuit with RC filter one RC section reduces ripple by 15%. If two RC sections are used in cascade the reduction in ripple would be

(*a*) 15% (*b*) 30%
(*c*) 150% (*d*) 225%

249. In rectifiers the use of RC filters is recommended when
(*a*) load current is high
(*b*) load current is small
(*c*) load current is neither high nor small
(*d*) either (*a*) or (*c*)

250 A voltage doubler circuit is fed by a voltage $V_m \sin \omega t$. The output voltage will be nearly $2 V_m$ only if
(*a*) load resistance is small
(*b*) load resistance is large
(*c*) load resistance is neither small nor large
(*d*) either (*d*) or (*c*)

251. A voltage triplet circuit uses
(*a*) 2 diodes and 2 capacitors
(*b*) 3 diodes and 3 capacitors
(*c*) 2 diodes and 3 capacitors
(*d*) 3 diodes and 2 capacitors

252. A voltage tripler circuit and voltage quadrupler circuit use identical components. Then
(*a*) voltage regulation in both circuits will be same
(*b*) voltage regulation in both may or may not be same
(*c*) voltage regulation in quadrupler circuit will be better than in tripler circuit
(*d*) voltage regulation in quadrupler circuit will be poorer than in tripler circuit

253. In a commercially available good power supply the voltage regulation is about
(*a*) 1% (*b*) 5%
(*c*) 10% (*d*) 20%

254. In Fig. 7.116, the minimum and maximum load currents are

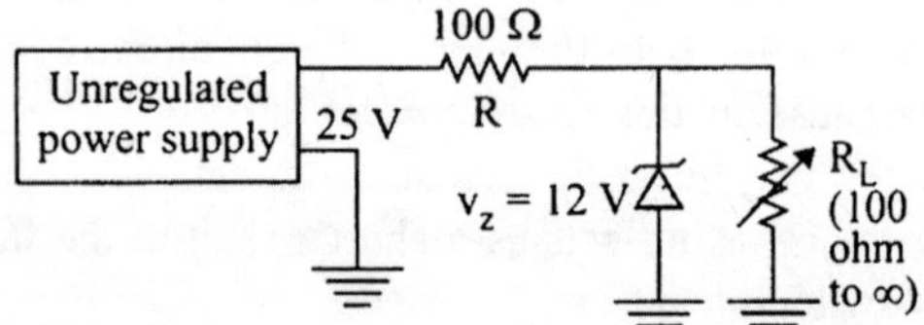

Fig. 7.116.

(*a*) 0 and 60 m A (*b*) 0 and 120 m A
(*c*) 10 m A and 60 m A (*d*) 10 m A and 120 m A

255. In Fig. 7.116 the current through resistor R
(*a*) is always equal to 130 m A
(*b*) may vary between 130 m A and 10 m A
(*c*) may vary between 65 m A and 10 m A
(*d*) is always equal to 65 m A

256. In Fig. 7.116 the zener current
(*a*) is always 130 m A
(*b*) may vary between 10 m A and 130 m A
(*c*) may vary between 65 m A and 130 m A
(*d*) is always equal to 65 m A

257. In Fig. 7.117 as the load resistance is changed
(*a*) the zener current remain constant
(*b*) the load current remains constant
(*c*) the output voltage remains constant
(*d*) the output voltage and current through R remain constant

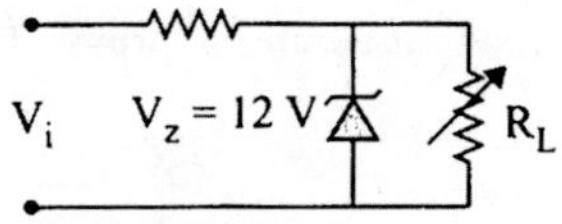

Fig. 7.117.

258. In Fig. 7.117 the zener has a resistance of 5 ohms. As the load resistance is varied, the output voltage
(*a*) will remain constant at 12 V
(*b*) will vary between 12.05 and 12.65 V
(*c*) will vary from 10 to 20 V
(*d*) will remain constant at 12.05 V

259. In Fig. 7.118 V may vary from 15 to 20 V and load current may vary from 20 m A to 100 m A. Suitable maximum value of R is

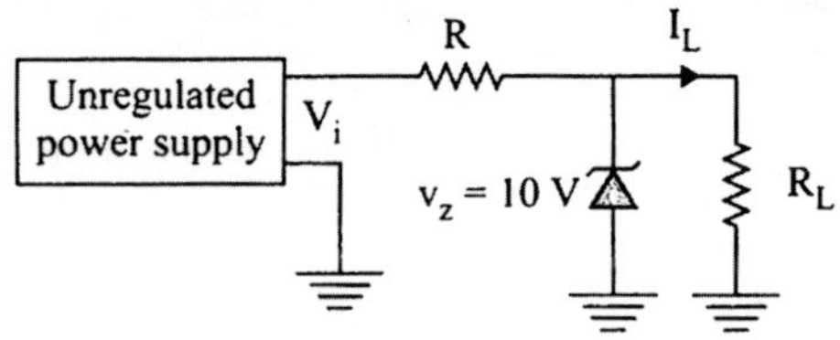

Fig. 7.118.

(*a*) 250 Ω (*b*) 500 Ω
(*c*) 100 Ω (*d*) 200 Ω

260. The circuit of Fig. 7.119 is

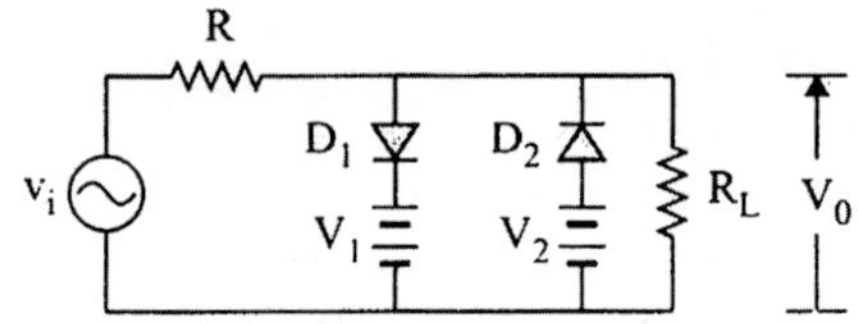

Fig, 7.119.

(*a*) positive clipper (*b*) negative clipper
(*c*) combination clipper (*b*) either (*a*) or (*b*)

261. In Fig. 7.119, D_1 turns on when
(*a*) v_i is more than V_1
(*b*) v_i is less than V_1
(*c*) v_i is between V_1 and V_2
(*d*) none of the above

262. In Fig. 7.119, D_2 turns on when
(*a*) v_i is more positive than V_1
(*b*) v_i. is less positive than V_1
(*c*) v_i is more negative than V_2
(*d*) v_i is less negative than V_2

263. A clamper
(*a*) adds a dc component to the input signal
(*b*) removes signal voltages above or below a specified value
(*c*) both (*a*) and (*b*)
(*d*) either (*a*) or (*b*)

264. A clipper
(*a*) adds a dc component to the input signal
(*b*) removes signal voltages above or below a specified value
(*c*) both (*a*) and (*b*)
(*d*) either (*a*) or (*b*)

265. In Fig. 7.119, v_i is a sinusoidal wave. Which of the following best describes the output v_0

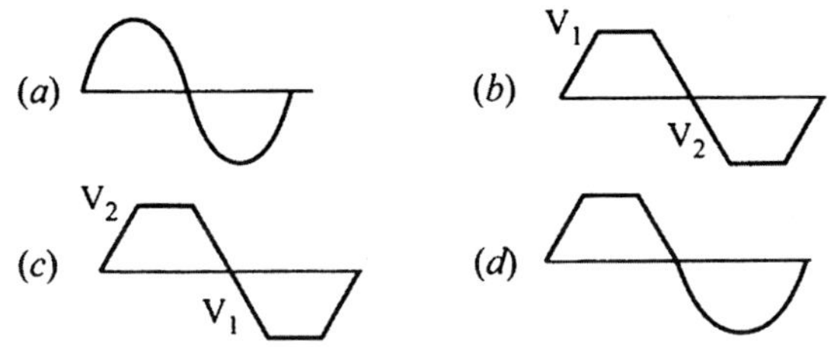

266. In Fig. 7.119 it is desired that output wave is symmetrical. Then
(*a*) both diodes should be similar and $|V_1| = |V_2|$
(*b*) both diodes should be similar
(*c*) $|V_1| = |V_2|$
(*d*) either (*a*) or (*b*)

267. To prevent a dc return between source and load, it is necessary to use
(*a*) resistor between source and toad
(*b*) inductor between source and load
(*c*) capacitor between source and load
(*d*) either (*a*) or (*b*)

268. Which of the following couplings may cause unwanted clamping
(*a*) resistive (*b*) inductive
(*c*) capacitive (*d*) both (*a*) or (*b*)

269. In Fig 7.120 the trace on the CRO will be

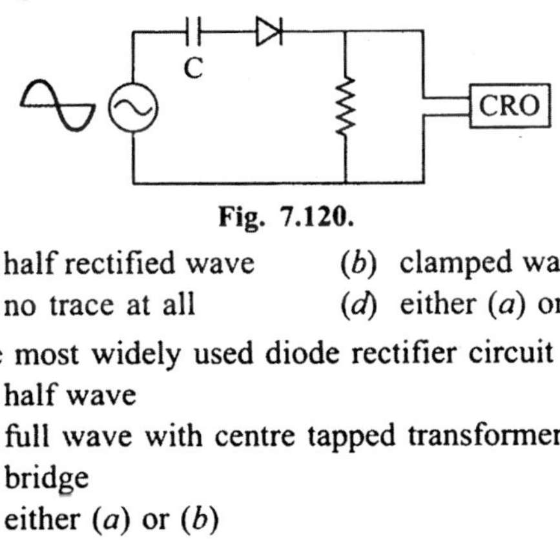

Fig. 7.120.

(*a*) half rectified wave (*b*) clamped wave
(*c*) no trace at all (*d*) either (*a*) or (*b*)

270. The most widely used diode rectifier circuit is
(*a*) half wave
(*b*) full wave with centre tapped transformer
(*c*) bridge
(*d*) either (*a*) or (*b*)

271. Which component is allowed to pass through it by a choke filter
(*a*) dc (*b*) ac
(*c*) both dc and ac (*d*) only high frequency ac

272. To protect the diodes in a rectifier and capacitor input filter circuit it is necessary to use
(*a*) surge resistor (*b*) surge inductor
(*c*) surge capacitor (*d*) both (*a*) or (*b*)

273. In a voltage regulated power supply the zener operates in the breakdown region when (V_{in} is input voltage and V_z is zener breakdown voltage)
(*a*) $V_{in} < V_z$ (*b*) $V_{in} = V_z$
(*c*) $V_{in} > V_z$ (*d*) both (*a*) or (*b*)

274. A capacitor input filter relies on
(*a*) average detection (*b*) peak detection
(*c*) both (*a*) and (*b*) (*d*) neither (*a*) nor (*b*)

275. In Fig. 7.121 what is the base current if $V_{BE} = 0.7$ V

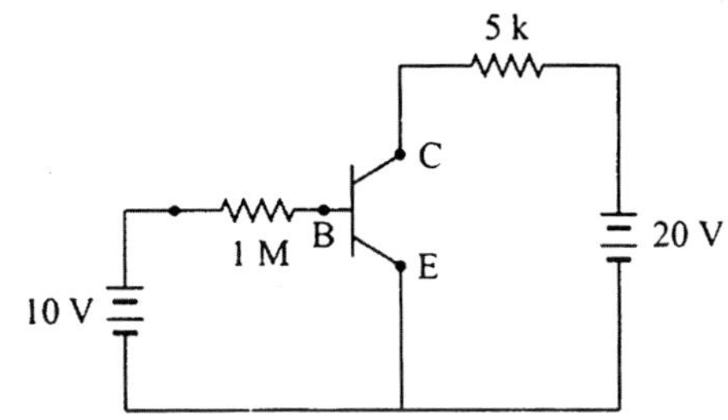

Fig. 7.121.

(*a*) 10 mA (*b*) 9.3 mA
(*c*) 10 μA (*d*) 9.3 μA

276. For a base current of 10 μA, what is the value of collector current in Fig. 7.121 if $\beta_{dc} = 110$
(*a*) 10 μA (*b*) 100 μA
(*c*) 1 mA (*d*) 10 mA

277. In Fig. 7.121 base current is 10 μA and $\beta_{dc} = 100$. Then $V_{CE} =$
(*a*) 5 V (*b*) 10 V
(*c*) 15 V (*d*) 20 V

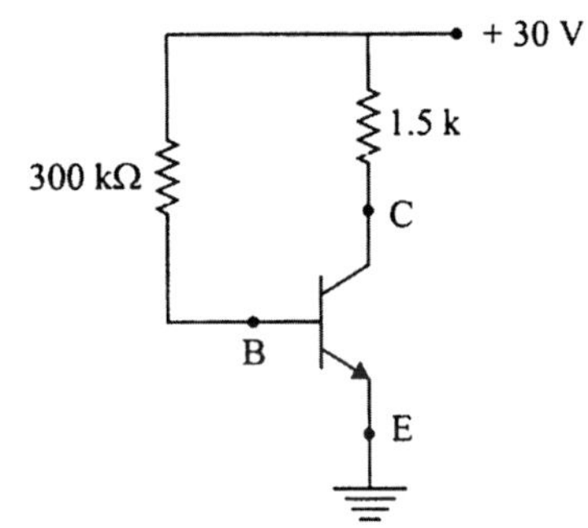

Fig. 7.122.

278. In Fig. 7.122 the dc load line will meet the *y* and *x* axis respectively at
(*a*) 20 mA and 30 V (*b*) 15 mA and 20 V
(*c*) 25 mA and 40 V (*d*) 10 mA and 20 V

279. In Fig. 7.122 $V_{BE} = 0.7$ V. The base current is
(*a*) above 97 μA (*b*) about 97 mA
(*c*) about 75 mA (*d*) about 75 mA

280. In Fig. 7.122 what is value of I_C if $\beta_{dc} = 100$. Neglect V_{BE}
(*a*) 5 mA (*b*) 5 μA
(*c*) 10 mA (*d*) 10 μA

281. In Fig. 7.122 the coordinates of Q point on the load line are (Neglect V_{BE})
(*a*) $I_C = 10$ mA, $V_{CE} = 15$ V
(*b*) $I_C = 5$ mA, $V_{CE} = 22.5$ V
(*c*) $I_C = 5$ mA, $V_{CE} = 15$ V
(*d*) $I_C = 10$ mA, $V_{CE} = 15$ V

282. In Fig. 7.122 V_{CE} (sat.) = 0.1 V. Then I_C (sat) =
(*a*) 10 mA (*b*) 20 mA
(*c*) 30 mA (*d*) 40 mA

283. A transistor in CE mode is in saturation region. Then the collector is almost at

(*a*) ground potential
(*b*) a potential equal to V_{CC}
(*c*) a potential equal to 0.5 V_{CC}
(*d*) a potential equal to 2 V_{CC}

284. In a transistor in CE mode, V_{CC} = +30 V. If the transistor is in cut off region, V_{CE} =
(*a*) +30 V (*b*) +20 V
(*c*) 10 V (*d*) zero

285. In a transistor in CE mode the collector current with the transistor in cutoff region is
(*a*) zero
(*b*) high
(*c*) equal to saturated value
(*d*) none of the above

286. Two CE stages are coupled through a capacitor. To calculate the quiescent base current of the two transistors, the capacitor is treated as
(*a*) open circuit (*b*) short circuit
(*c*) neither (*a*) nor (*b*) (*d*) either (*a*) or (*b*)

287. Two CE stages, 1 and 2 are coupled through a capacitor. V_{CC}, is the same for both. Base resistances R_{B1} and R_{B2} are such that $R_{B1} > R_{B2}$. Then
(*a*) base currents for both stages are equal
(*b*) base current of stage 1 is more than that of stage 2
(*c*) base current of stage 2 is more than that of stage 1
(*d*) either (*b*) or (*c*)

288. In Fig. 7.123, transistor β_{dc} = 100 and LED voltage when it is conducting is 2 V. Then the base current which saturates the transistor is

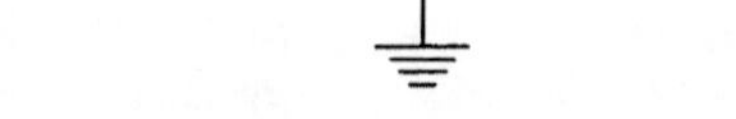

Fig. 7.123.

(*a*) 18 mA (*b*) 1.8 mA
(*c*) 0.18 mA (*d*) 0.018 mA

289. In Fig. 7.123 V_{BE} = 0.7 V. If base current required to saturate the transistor is 0.1 mA, V_i =
(*a*) 5 V (*b*) 5.7 V
(*c*) 10 V (*d*) 10.7 V

290. In Fig. 7.124 the saturation collector current is

(*a*) $\frac{V_{CC}}{R_C}$ (*b*) $\frac{V_{CC}}{R_E}$

(*c*) $\frac{V_{CC}}{R_C + R_E}$ (*d*) none of the above

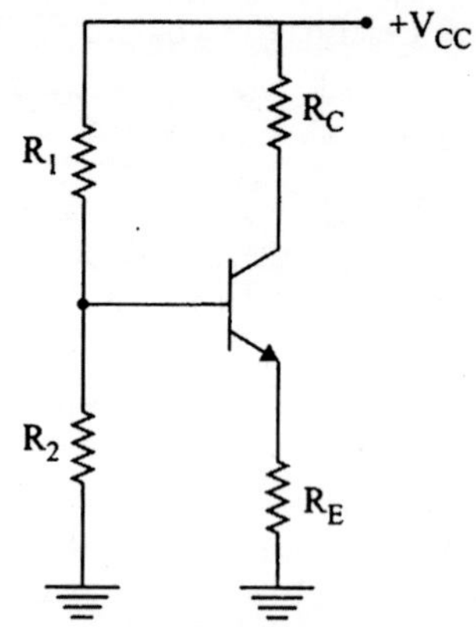

Fig. 7.124.

291. In Fig. 7.124 if the transistor is cutoff, the collector voltage is equal to
(*a*) V_{CC} (*b*) 0.5 V_{CC}
(*c*) $V_{CC} \frac{R_1}{R_1 + R_2}$ (*d*) zero

292. In Fig. 7.124, V_{CC} = + 30 V, R_1 = 200 kΩ and R_2 = 100 kΩ. If V_{BE} = 0.7 V, the voltage a cross R_E =
(*a*) + 20 V (*b*) + 19.3 V
(*c*) + 10 V (*d*) + 9.3 V

293. In Fig. 7.124, voltage across R_2 = +10 V. If V_{BE} = 0.7 V and R_E 10 kΩ, current through R_E is
(*a*) 1 mA (*b*) 0.93 mA
(*c*) 2 mA (*d*) 1.86 mA

294. In Fig, 7.124 V_{CC}, = + 30 V, I_C corresponding to Q point is 1 mA. If R_C = 5 kΩ and R_E = 5 kΩ, V_{CE} corresponding to Q point is
(*a*) 30 V (*b*) 20 V
(*c*) 25 V (*d*) 5 V

295. In a transistor the minimum and maximum values of β_{dc}, are 100 and 121. The proper value of β_{dc}. to be used to locate Q point is
(*a*) 100 (*b*) 121
(*c*) 110.5 (*d*) 110

296. In a transistor β_{dc} is likely to vary between 100 and 400. The proper value of β_{dc}. to be used to locate Q point is
(*a*) 100 (*b*) 400
(*c*) 250 (*d*) 200

297. In a BJT circuit a pnp transistor is replaced by npn transistor. To analyse the new circuit
(*a*) all calculations done earlier have to be repeated
(*b*) replace all calculated voltages by reverse values
(*c*) replace all calculated currents by reverse values
(*d*) replace all calculated voltages and currents by reverse values

298. In the graphical analysis of CE amplifier circuit, the upper end of load line is called
(*a*) saturation point (*b*) cutoff point
(*c*) Q point (*d*) none of the above

299. In the graphical analysis of CE amplifier circuit, the intersection of base current and dc load line is called
(*a*) saturation point (*b*) cutoff point
(*c*) Q point (*d*) operating point

300. The most commonly used bias in BJT amplifier circuits is
(*a*) voltage divider bias (*b*) emitter bias
(*c*) collector bias (*d*) collector feedback bias

301. In a BJT amplifier circuit the size of coupling capacitor depends on
(*a*) lowest frequency
(*b*) highest frequency
(*c*) average of lowest and highest frequencies
(*d*) none of the above

302. In Fig. 7.125 we need an ac ground. The proper value of C is

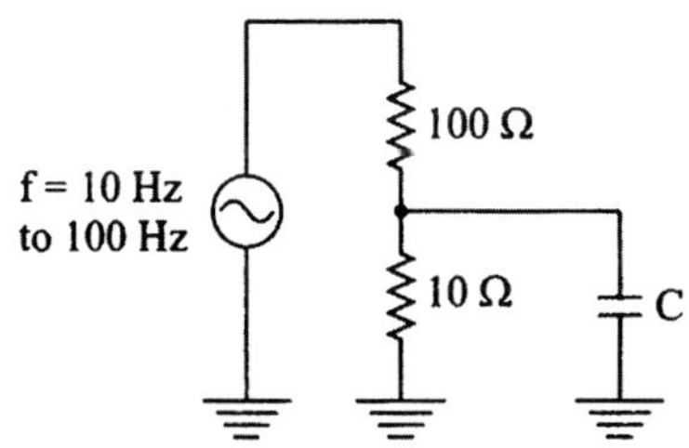

Fig. 7.125.

(*a*) 1000 μF (*b*) 100 μF
(*c*) 10 μF (*d*) 1 μF

303. To draw dc equivalent circuit for a CE amplifier circuit we should
(*a*) reduce all ac sources to zero
(*b*) open all capacitors
(*c*) reduce all ac sources to zero and open all capacitors
(*d*) reduce all ac sources to zero and short all capacitors

304. To draw ac equivalent circuit for a CE amplifier we should
(*a*) reduce all dc sources to zero
(*b*) short all capacitors
(*c*) reduce all dc sources to zero and short all capacitors
(*d*) reduce all dc sources to zero and open all capacitors

305. The *h* parameter equivalent to ac resistance r_e is
(*a*) h_{ib} (*b*) h_{fe}
(*c*) h_{ob} (*d*) h_{rb}

306. The coupling capacitor in amplifier circuits
(*a*) affects dc biasing
(*b*) does not affect dc biasing
(*c*) affects dc biasing to some extent
(*d*) both (*a*) or (*c*)

307. The value of parameter r_e used in r_e transistor model
(*a*) increases with increase in temperature
(*b*) decreases with increase in temperature
(*c*) is not affected by increase in temperature
(*d*) either (*b*) or (*c*)

308. In the circuit of Fig. 7.126, the value of r_e is (neglecting v_{be})

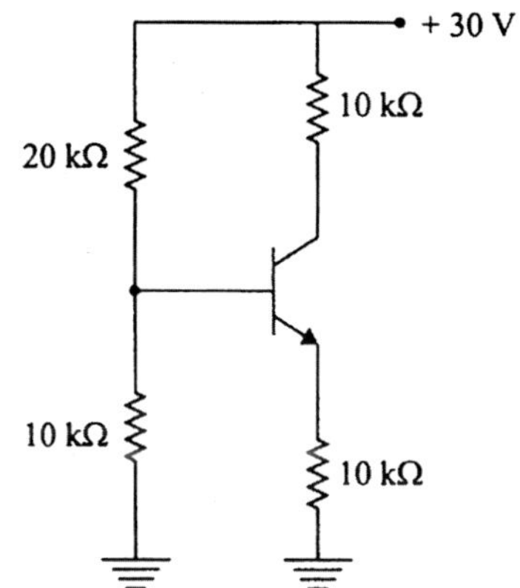

Fig. 7.126.

(*a*) 100 Ω (*b*) 75 Ω
(*c*) 50 Ω (*d*) 25 Ω

309. In the CE equivalent circuit of Fig. 7.127, the voltage gain is

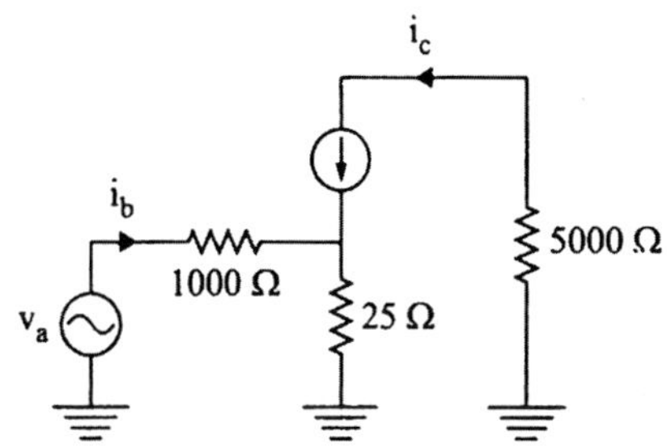

Fig. 7.127.

(*a*) 200 (*b*) 40
(*c*) 5 (*d*) 1000

310. The r_e model equivalent circuit shown in Fig. 7.128 is for

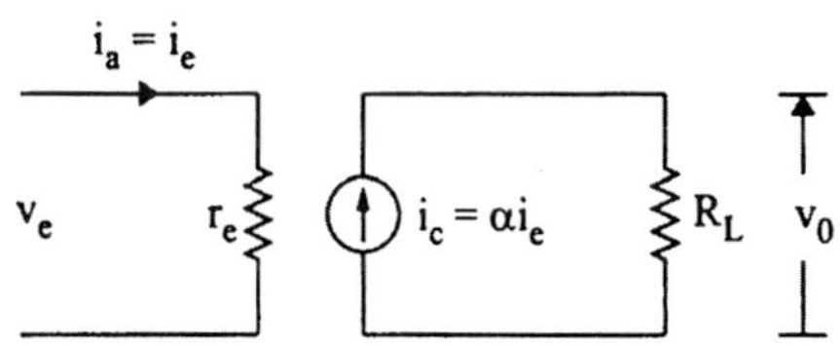

Fig. 7.128.

(*a*) CE amplifier (*b*) CC amplifier
(*c*) CB amplifier (*d*) none of the above

311. In the equivalent circuit of Fig. 7.128 voltage gain is
(*a*) α (*b*) $\frac{R_L}{r_e}$
(*c*) $\frac{\alpha R_L}{r_e}$ (*d*) $\frac{r_e}{R_L}$

312. In Fig. 7.128, current gain is
(*a*) α (*b*) r_e
(*c*) $\frac{R_L}{r_e}$ (*d*) $\frac{r_e}{R_L}$

313. The equivalent circuit shown in Fig. 7.129 is for
(*a*) CE (*b*) CB
(*c*) CC (*d*) None of the above

314. In Fig. 7.129 the voltage gain is

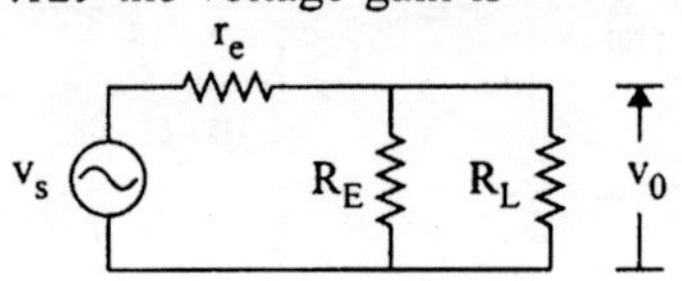

Fig. 7.129.

(*a*) $\frac{r_e R_L}{r_e + R_E}$

(*b*) r_e

(*c*) $\left[\frac{R_E R_L}{R_E + R_L}\right] \Big/ \left[r_e + \frac{R_E R_L}{R_E + R_L}\right]$

(*d*) $\frac{R_L}{r_e}$

315. In Fig. 1.129 output impedance of amplifier is

(*a*) R_E (*b*) $\frac{R_E R_L}{R_E + R_L}$

(*c*) $\frac{r_e R_L}{r_e + R_E}$ (*d*) $r_e + R_E$

316. A CB amplifier has $\alpha = 0.98$ and $R_L = 600\ \Omega$. If $I_E = 3.5$ mA, the current gain is

(*a*) 1 (*b*) 0.98

(*c*) 0.98×3.5 (*d*) $\frac{3.5}{0.98}$

317. A CB amplifier has $I_E = 3.5$ mA, $\alpha = 0.98$ and $R_L = 600\ \Omega$. The input impedance

(*a*) 0.98 Ω (*b*) 0.025 Ω

(*c*) $\frac{0.025}{3.5 \times 10^{-5}} \Omega$ (*d*) 600 Ω

318. A CB amplifier has $r_e = 6\ \Omega$, $R_L = 600\ \Omega$ and $\alpha = 0.98$. The voltage gain is

(*a*) 100 (*b*) 600×0.98

(*c*) 98 (*d*) 6

319. A C E amplifier has $r_e = 6\ \Omega$, $\beta = 100$ and $\alpha = 0.98$. The input impedance is

(*a*) 6 Ω (*b*) 100 Ω

(*c*) 600 Ω (*d*) 98 Ω

320. In a CC circuit the current gain is

(*a*) $\frac{\beta R_E}{R_E + R_L}$ (*b*) $\frac{R_E + R_L}{R_E}$

(*c*) $\frac{R_E + R_L}{R_L}$ (*d*) β

321. To by pass a 100 ohm emitter resistor at 10 Hz the proper size of capacitor is

(*a*) 10 μF (*b*) 100 μF

(*c*) 1000 μF (*d*) 1 μF

322. The disadvantage of direct coupled amplifiers is

(*a*) drift (*b*) large size of transistor

(*c*) low voltage gain (*d*) both (*b*) and (*c*)

323. Typical values of voltage gain for CE, CB and CC amplifiers are

(*a*) 100, 100 and 1 respectively

(*b*) 100, 1 and 100 respectively

(*c*) 100, 1 and 1 respectively

(*d*) 1, 100 and 100 respectively

324. Typical values of current gains for CE, CB and CC amplifiers are

(*a*) 120, 120 and 1 respectively

(*b*) 120, 1 and 120 respectively

(*c*) 1, 120 and 120 respectively

(*d*) 1, 1 and 120 respectively

325. In all base driven amplifiers

(*a*) ac collector voltage is 180° out of phase with ac base voltage

(*b*) ac emitter voltage is 180° out of phase with ac base voltage

(*c*) ac collector voltage is in phase with ac base voltage

(*d*) none of the above

326. In a CE amplifier the ac emitter resistance

(*a*) is zero (*b*) is high

(*c*) is very high (*d*) may have any value

327. In a base driven amplifier the voltage gain is the ratio of

(*a*) ac collector voltage to ac base voltage

(*b*) ac emitter voltage to ac base voltage

(*c*) ac collector voltage to ac emitter voltage

(*d*) ac base voltage to ac emitter voltage

328. The output voltage waveform of a CE amplifier is fed to a dc coupled CRO. The trace on the screen will be

(*a*) dc output voltage

(*b*) ac output voltage

(*c*) sum of dc and ac output voltage

(*d*) either (*a*) or (*b*)

329. The disadvantage of transformer coupling in amplifier is

(*a*) cost of transformer at audio frequencies

(*b*) size of transformer at audio frequencies

(*c*) size of transformer at radio frequencies

(*d*) cost and size of transformer at audio frequencies

330. Transformer coupling is mostly used in

(*a*) RF amplifier

(*b*) audio amplifier

(*c*) both audio and RF amplifier

(*d*) none of the above

331. Which circuit is called swamped amplifier

(*a*) CC (*b*) CE

(*c*) CB (*d*) both (*a*) and (A)

332. Which circuit is called emitter follower

(*a*) CC (*b*) CE

(*c*) CB (*d*) both (*a*) and (*b*)

333. The main advantage of CC amplifier is

(*a*) High input impedance and low output impedance

(*b*) Low input impedance and high output impedance

(*c*) High voltage gain

(*d*) High current gain

334. Which connection is rarely used

(*a*) CE (*b*) CB

(*c*) CC (*d*) both (*a*) and (*b*)

335. In a CE amplifier the Q point is very close to cutoff point on the load line. This results in
(*a*) saturation clipping
(*b*) cutoff clipping
(*c*) both saturation and cutoff clipping
(*d*) either saturation clipping or cutoff clipping

336. In a CE amplifier the Q point is very close to saturation point on the dc load line. This results in
(*a*) positive clipping of collector current
(*b*) negative clipping of collector current
(*c*) both positive and negative clipping of collector current
(*d*) either positive or negative clipping of collector current

337. In a CE amplifier the ac cutoff voltage is 9 V and slope of ac load line is –0.5 mA/V. The ac saturation current is
(*a*) 9 mA (*b*) 4.5 mA
(*c*) 3 mA (*d*) 18 mA

338. To have a centred Q point in CE amplifier
(*a*) ac resistance of collector circuit
$$= \frac{\text{Quiescent collector voltage}}{\text{Quiescent collector current}}$$
(*b*) ac resistance of emitter circuit
$$= \frac{\text{Quiescent collector voltage}}{\text{Quiescent collector current}}$$
(*c*) ac resistance of collector and emitter circuits
$$= \frac{\text{Quiescent collector voltage}}{\text{Quiescent collector current}}$$
(*d*) ac resistance of collector and emitter circuits
$$= \frac{\text{Quiescent collector current}}{\text{Quiescent collector voltage}}$$

339. Output of a phase splitter is
(*a*) a pair of sine waves of unequal amplitude
(*b*) a pair of sine waves of equal amplitudes
(*c*) a pair of sine waves of equal amplitudes and opposite phase
(*d*) a pair of sine waves of unequal amplitudes and opposite phase

340. In a CE class A amplifier the quiescent current and voltage arc I_{CQ} and V_{CEQ}. The maximum power output can be
(*a*) $(V_{CEQ})(I_{CQ})$ (*b*) $\frac{(V_{CEQ})(I_{CQ})}{2}$
(*c*) $\frac{(V_{CEQ})(I_{CQ})}{4}$ (*d*) $2(V_{CEQ})(I_{CQ})$

341. If $V_{CC} = 15$ V, $V_{CBQ} = 5$ V and $I_{CQ} = 50$ m A, the efficiency in class A operation is
(*a*) 100% (*b*) 50%
(*c*) 25% (*d*) 16.7%

342. The thermal resistance of a BJT is about
(*a*) 100°C/W (*b*) 200°C/W
(*c*) 20°C/W (*d*) 2°C/W

343. If the thermal resistance of a transistor is lower it means
(*a*) higher junction temperature
(*b*) lower junction temperature
(*c*) higher or lower junction temperature depending on circuit conditions
(*d*) higher or lower junction temperature depending on transistor

344. A power transistor dissipates 5 W. If ambient temperature is 30° and case to air thermal resistance is 10°C/W, the case temperature is
(*a*) 80°C (*b*) 20°C
(*c*) 32°C (*d*) 305°C

345. A transistor has a maximum power dissipation of 350 m W at an ambient temperature of 25°C. If derating factor is 2 m W/°C, the maximum power dissipation for 40°C ambient temperature is
(*a*) 300 mW (*b*) 330 mW
(*c*) 350 mW (*d*) 380 mW

346. A transistor has a power rating of 8 W for a case temperature of 25°C. If derating factor is 30 m W/°C the power rating for 55°C case temperature is
(*a*) 8 W (*b*) 7.5 W
(*c*) 7.1 W (*d*) 6.8 W

347. The quiescent collector current I_C and collector to emitter voltage V_{CE} in a CE connection are the values when
(*a*) ac signal in zero (*b*) ac signal is low
(*c*) ac signal is negative (*d*) either (*a*) or (*b*)

348. The horizontal intercept of dc load line is the same as ideal
(*a*) cutoff point (*b*) saturation point
(*c*) operating point (*d*) quasi saturation point

349. When the ac input signal is very large the instantaneous operating point in CE amplifier moves all the way between
(*a*) ac saturation and ac cutoff points
(*b*) ac operating and ac cutoff points
(*c*) ac saturation and ac operating points
(*d*) none of the above

350. Most of the transistor data sheets specify the power rating for an ambient temperature of
(*a*) 10°C (*b*) 25°C
(*c*) 50°C (*d*) 75°C

351. The derating factor, for ambient temperature higher than specified, for a transistor is about
(*a*) 2.5 m W/°C (*b*) 0.5 m W/°C
(*c*) 10 m W/°C (*d*) 20 m W/°C

352. In a power amplifier the collector current flows for 270° of the input cycle. The operation is
(*a*) class A (*b*) class B
(*c*) class AB (*d*) class C

353. The circuit of Fig. 7.130 is called
(*a*) current mirror (*b*) voltage mirror
(*c*) impedance inversion (*d*) phase spilling

354. In Fig. 7.130 $I_C \approx$
(*a*) I_1 (*b*) I_B
(*c*) $0.5\ I_1$ (*d*) $0.5\ I_B$

355. In Fig. 7.130 V_{CC} = 20 V and R = 10 kΩ. If V_{BE} = 0.7 V, I_1 =

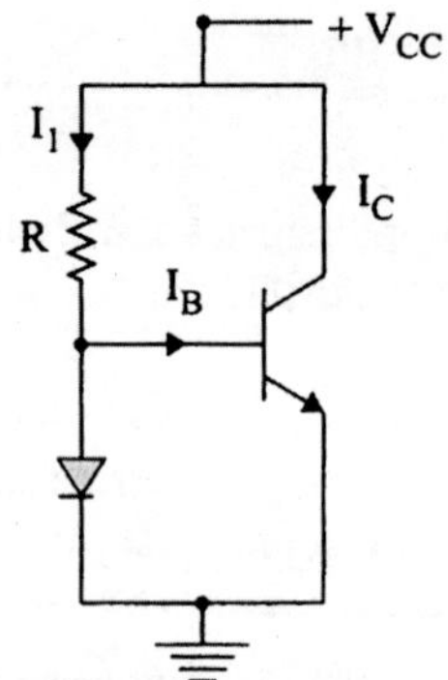

Fig. 7.130.

(*a*) 1.93 mA (*b*) 19.3 mA
(*c*) 0.193 mA (*d*) 193 mA

356. For a push pull circuit the most favoured biasing method is
(*a*) self bias (*b*) emitter bias
(*c*) diode bias (*d*) collector bias

357. The circuit of Fig. 7.131 is called

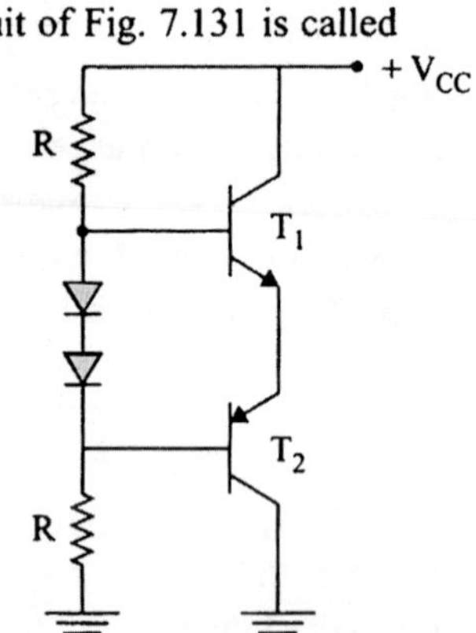

Fig. 7.131.

(*a*) diode bais (*b*) current mirror
(*c*) complementary mirror (*d*) none of the above

358. In Fig. 7.131
(*a*) the collector currents of T_1 and T_2 are equal
(*b*) the collector current of T_1 is more than that of T_2
(*c*) the collector current of T_1 is double that of T_2
(*d*) the collector currents of T_2 is more than that of T_1

359. In Fig. 7.131 V_{CEQ} for each transistor is
(*a*) V_{CC} (*b*) 0.5 V_{CC}
(*c*) 0.75 V_{CC} (*d*) 0.25 V_{CC}

360. In a push pull circuit the two transistors are
(*a*) both pnp
(*b*) both npn
(*c*) one pnp and other npn
(*d*) either (*a*) or (*b*)

361. In a current mirror
(*a*) the *i–v* curve of compensating diode is similar to collector curve of transistor
(*b*) the *i–v* curve of compensating diode is similar to transconductance curve of transistor
(*c*) either (*a*) or (*b*)
(*d*) neither (*a*) nor (*b*)

362. Class C amplifier is suitable for
(*a*) a narrow frequency band
(*b*) a wide frequency band
(*c*) audio frequencies
(*d*) either (*b*) or (*c*)

363. In Class C operation the collector current looks like
(*a*) full sine wave (*b*) half sine wave
(*c*) narrow pulses (*d*) nearly half sine wave

364. Which power amplifier can deliver maximum load power
(*a*) Class A (*b*) Class AB
(*c*) Class B (*d*) Class C

365. When narrow current pulses drive a high Q resonant circuit the voltage across the circuit looks like
(*a*) narrow pulses (*b*) half sine wave
(*c*) almost sin wave (*d*) either (*a*) or (*b*)

366. In a class C power amplifier the input signal has a frequency of 250 kHz. If the collector current pulses are 0.1 μs wide, the duty cycle of current wave form is
(*a*) 50% (*b*) 25%
(*c*) 2.5% (*d*) 0.25%

367. Figure 7.132 shows current pulses of a class C power amplifier. The fundamental frequency of these pulses is

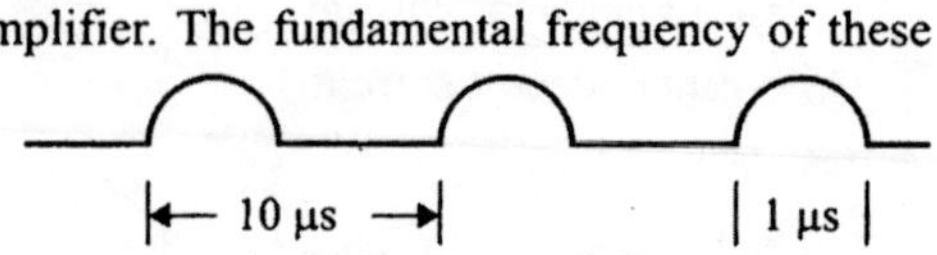

Fig. 7.132.

(*a*) 100 kHz (*b*) 1000 kHz
(*c*) 10 kHz (*d*) 1 kHz

368. A class C amplifier uses
(*a*) high Q tuned circuit (*b*) low Q tuned circuit
(*c*) untuned circuit (*d*) either (*b*) or (*c*)

369. In a class C power amplifier
(*a*) resonant frequency of tuned circuit is equal to fundamental frequency of pulsed wave form
(*b*) resonant frequency of tuned circuit is half the fundamental frequency of pulsed waveform
(*c*) resonant frequency of tuned circuit has no relation to the fundamental frequency of pulsed wave form
(*d*) either (*b*) or (*c*)

370. In a class C power amplifier the input frequency of ac signal is 1 MHz. If tank circuit has C = 1000 pF, the value of L =
(*a*) 25 μH (*b*) 10 μH
(*c*) 100 μH (*d*) 200 μH

371. In a tuned class C power amplifier the quiescent value of collector to emitter voltage is equal to (V_{CC} denotes collector battery voltage)
(*a*) V_{CC} (*b*) 0.5 V_{CC}
(*c*) 0.25 V_{CC} (*d*) either (*b*) or (*c*)

372. Class A and Class B power amplifiers predominate at
(*a*) audio frequencies (*b*) radio frequencies
(*c*) all frequencies
(*d*) audio and radio frequency

373. The input to negative clamper has a peak value of V. Then in the output the positive and negative peaks are nearly
(*a*) 0 and $-2V_p$ respectively
(*b*) V_p and $-V_p$ respectively
(*c*) 2 V_p and 0 respectively
(*d*) none of the above

374. Class C operation finds application in
(*a*) RF power amplifiers (*b*) frequency multipliers
(*c*) both (*a*) and (*b*) (*d*) neither (*a*) nor (*b*)

375. If the output current wave shape of class C circuit has a period of 1 μs and a pulse width of 0.006 μs, the duty cycle is
(*a*) 6% (*b*) 0.6%
(*c*) 0.006% (*d*) 0.06%

376. In a class C operation
(*a*) V_{CC} is much larger than $V_{CE(sat.)}$
(*b*) V_{CC} is much smaller than $V_{CE(sat.)}$
(*c*) V_{CC} and $v_{CE(sat.)}$ are nearly equal
(*d*) none of the above

377. In a class C operation V_{CC} 30 V and $V_{CE(sat.)}$ = 2 V. The efficiency is
(*a*) 93.75% (*b*) 88%
(*c*) 6.66% (*d*) 3.33%

378. In a class C operation V_{CC} = 40 V, R_L = 50 Ω. The maximum load power can be
(*a*) 64 W (*b*) 32 W
(*c*) 16 W (*d*) 8 W

379. In a class C operation V_{CC} = 40 V and $V_{CC(sat.)}$ = 0.4 V. The efficiency is
(*a*) 99% (*b*) 49.5%
(*c*) 1% (*d*) 0.5%

380. A BJT amplifier has a power gain of 0.5 from source to base. The power gain in dB is
(*a*) –0.3 dB (*b*) –3 dB
(*c*) –6 dB (*d*) –10 dB

381. In a BJT amplifier the power gain from base to collector is 4000. The power gain in dB is
(*a*) 3.6 (*b*) 36
(*c*) 7.2 (*d*) 72

382. A buffer amplifier should have
(*a*) high input impedance and low output impedance
(*b*) high input impedance and high output impedance
(*c*) low input impedance and high output impedance
(*d*) low input impedance and low output impedance

383. Which of the following can be used as a buffer amplifier
(*a*) CE (*b*) CC
(*c*) CB (*d*) both (*a*) and (*c*)

384. FET can be used in
(*a*) buffer amplifier circuits
(*b*) cascade amplifier circuits
(*c*) digital circuits
(*d*) all the above circuits

385. A dc amplifier can operate
(*a*) all the way to zero frequency
(*b*) only at low frequencies
(*c*) only at high frequencies
(*d*) both at low and high frequencies

386. A low noise device is most important
(*a*) at front end of equipment
(*b*) in the last stage of equipment
(*c*) in the middle stages of equipment
(*d*) both (*b*) and (*c*)

387. The gain of an FET amplifier can be changed by changing
(*a*) g_m (*b*) g_g
(*c*) R_d (*d*) none of the above

388. For the circuit of Fig. 7.133 the input and output Miller resistances are

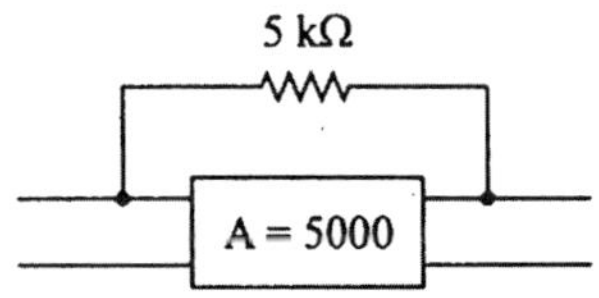

Fig. 7.133.

(*a*) 5 kΩ each
(*b*) 1 ohm and 5 kΩ respectively
(*c*) 5 kΩ and 1 ohm respectively
(*d*) 1 ohm each

389. In Fig. 7.134 the input and output Miller capacitances are

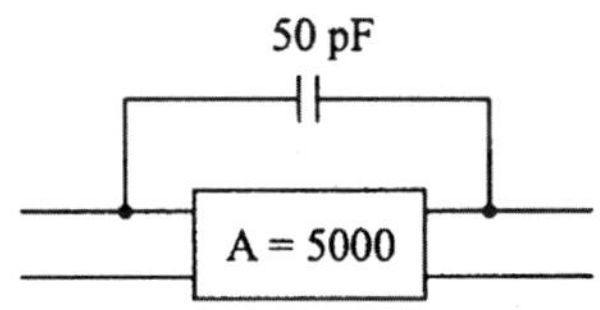

Fig. 7.134.

(*a*) 0.25 μF each
(*b*) 500 pF each
(*c*) 0.25 μF and 500 pF respectively
(*d*) 500 pF and 0.25 μF

390. The critical frequency of the circuit of Fig. 7.135 is

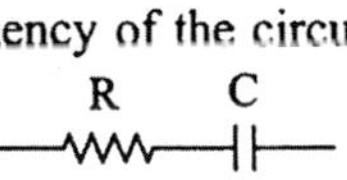

Fig. 7.135.

(*a*) $f = \frac{1}{2\pi C}$ (*b*) $f = \frac{1}{2\pi RC}$
(*c*) $f = \frac{R}{2\pi C}$ (*d*) $f = \frac{C}{2\pi R}$

391. At high frequencies the output circuit of a transistor amplifier reduces to
(*a*) a lead network (*b*) a lag network
(*c*) either (*a*) or (*b*) (*d*) neither (*a*) nor (*b*)

392. For the circuit of Fig. 7.136 the critical frequency is

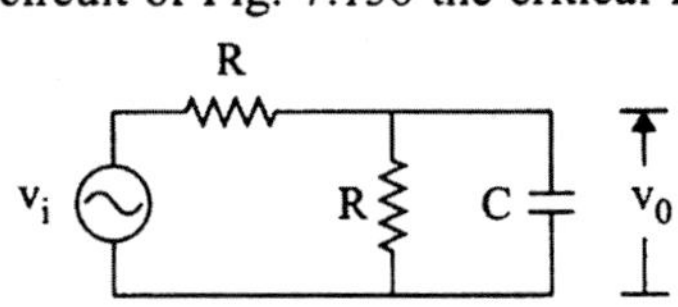

Fig. 7.136.

(*a*) $\frac{1}{2\pi RC}$ (*b*) $\frac{1}{\pi RC}$

(*c*) $\frac{2}{\pi RC}$ (*d*) $\frac{4}{\pi RC}$

393. Which of the following are difficult to be fabricated in a monolithic IC
(*a*) inductors (*b*) large capacitors
(*c*) small capacitors (*d*) both (*a*) and (*b*)

394. The current through common emitter resistor of a differential amplifier is called
(*a*) differential current (*b*) emitter currant
(*c*) ground current (*d*) tail current

395. If r_e is ac resistance of diode, the Thevenin input impedance of a differential amplifier is
(*a*) r_e (*b*) βr_e
(*c*) $2 \beta r_e$ (*d*) $0.5 \beta r_e$

396. By connecting external resistances to a differential amplifier we can adjust
(*a*) voltage gain
(*b*) band width
(*c*) both voltage gain and bandwidth
(*d*) none of the above

397. For input and later stages. BIFET op-amp use
(*a*) JFETs and MOSFETs respectively
(*b*) MOSFETs and JFETs respectively
(*c*) JFETs and BJTs respectively
(*d*) BJTs and JFETs respectively

398. An op-amp uses
(*a*) RC coupling (*b*) direct coupling
(*c*) transformer coupling (*d*) (*b*) or (*c*)

399. Generally op-amp is used in
(*a*) open loop (*b*) closed loop
(*c*) cascaded loop (*d*) either (*a*) or (*c*)

400. A sine wave with slew rate distortion looks like
(*a*) step wave (*b*) square wave
(*c*) triangular wave (*d*) cosine wave

401. Which of the following are linear ICs
(*a*) RF and IF amplifies (*b*) wide band amplifier
(*c*) audio amplifiers (*d*) all the above

402. In Fig. 7.137 the dc emitter current of each transistor is about

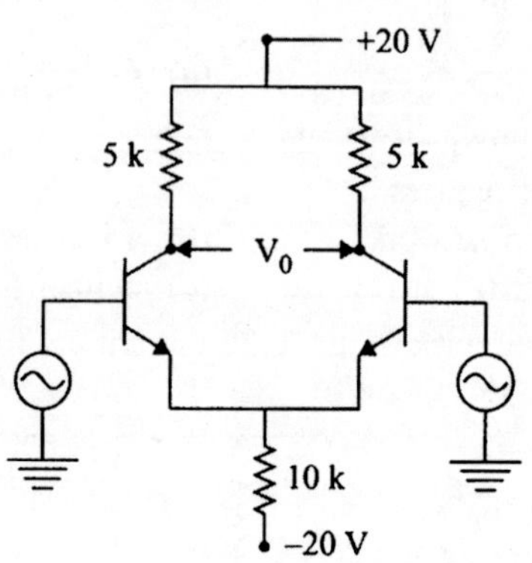

Fig. 7.137.

(*a*) 4 mA (*b*) 2 mA
(*c*) 3 mA (*d*) 1 mA

403. In Fig. 7.137, V_0 =
(*a*) 20 V (*b*) 10 V
(*c*) 5 V (*d*) 0

404. Positive feedback is mainly used in
(*a*) oscillators (*b*) thyristors
(*c*) voltage regulators (*d*) both (*a*) and (*b*)

405. In an RC phase shift oscillator, the total phase shift of the three RC lead networks is
(*a*) 360° (*b*) 180°
(*c*) 90° (*d*) 0

406. Wien bridge uses
(*a*) lead network (*b*) lag network
(*c*) lead-lag network (*d*) both (*a*) and (*b*)

407. High frequency oscillations can be caused by
(*a*) electric coupling
(*b*) magnetic coupling
(*c*) ground loop
(*d*) all of the above

408. In a voltage regulator circuit using zener diode the load voltage is nearly constant only if
(*a*) zener current is very small
(*b*) zener current is very large
(*c*) zener current is neither small nor large
(*d*) any of the above

409. If the changes in zener current in a voltage regulator circuit are too large, it can be corrected by using a
(*a*) emitter follower (*b*) CE amplifier
(*c*) capacitor (*d*) either (*a*) or (*b*)

410. In an op-amp comparator, the positive or negative output is
(*a*) saturated (*b*) amplified
(*c*) unsaturated (*d*) any of the above

411. In a transistor I_E = 10 mA, I_{CO} = 0.5 μA and α = 0.995. Then I_{CEO} =
(*a*) 100 μA (*b*) 25 μA
(*c*) 10.1 mA (*d*) 10.5 mA

412. The approximate value of input impedance of CE amplifier with emitter resistance R_E is given by
(*a*) $h_{ie} + A_i R_E$ (*b*) $h_{ie} + (1 + h_{fe}) R_E$
(*c*) h_{ie} (*d*) $(1 + h_{fe})R_E$

413. In an amplifier with a gain of –1000 and feedback factor β = –0.1, the change in gain is 20% due to temperature. The change in gain for feedback amplifier will be
(*a*) 10% (*b*) 5%
(*c*) 0.2% (*d*) 0.01%

414. In Fig. 7.138 v_i = 10 m V dc maximum. The maximum possible dc output offset voltage is

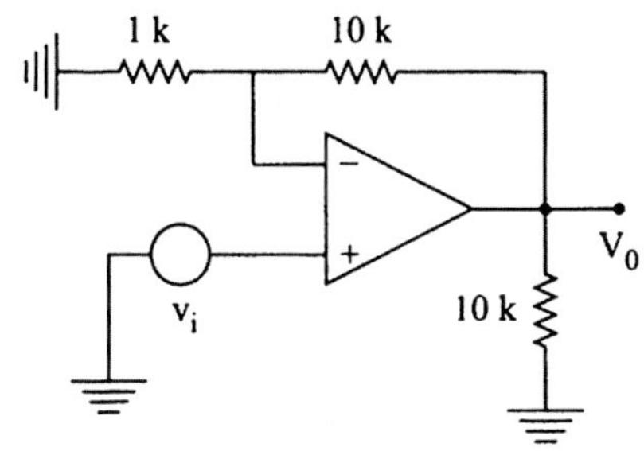

Fig. 7.138.

(a) 60 mV dc (b) 110 m V dc
(c) 130 m V dc (d) 150 m V dc

415. Match List I and List II and select the correct answer using the given codes

List I	List II
A. RC coupled amplifier	1. Very low drift
B. Tuned amplifier	2. Flat frequency response from zero frequency onwards
C. Chopper stabilized amplifier	3. Flat frequency response with upper and lower cutoff frequency
D. DC amplifier	4. Peak in gain frequency response

Codes:

	A	B	C	D
(a)	4	3	1	2
(b)	3	4	2	1
(c)	3	4	1	2
(d)	4	3	2	1

416. For the amplifier circuit of Fig. 7.139, the *h* parameters of transistor are $h_{ib} = 25\ \Omega$, $h_{fb} = 0.999$, $h_{ob} = 10 - 6\ \Omega$. The voltage gain is

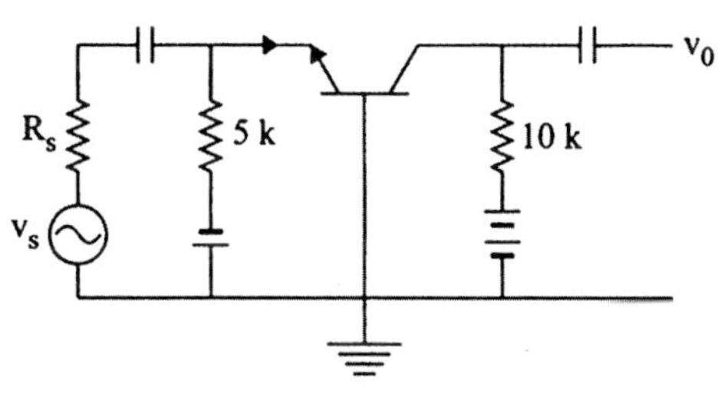

Fig. 7.139.

(a) 0.999 (b) 1.98
(c) 20 (d) 400

417. The circuit of Fig. 7.140 is

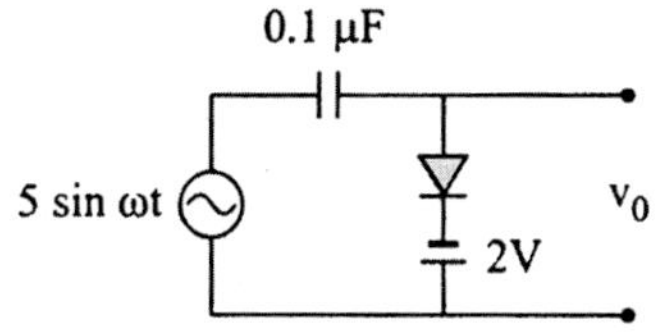

Fig. 7.140.

(a) clipper, sine wave clipped at –2 V
(b) clamper
(c) clamper, sine wave clamped at zero V
(d) clipper, sine wave clipped at +2 V

418. In the circuit of Fig. 7.141, I_c is about

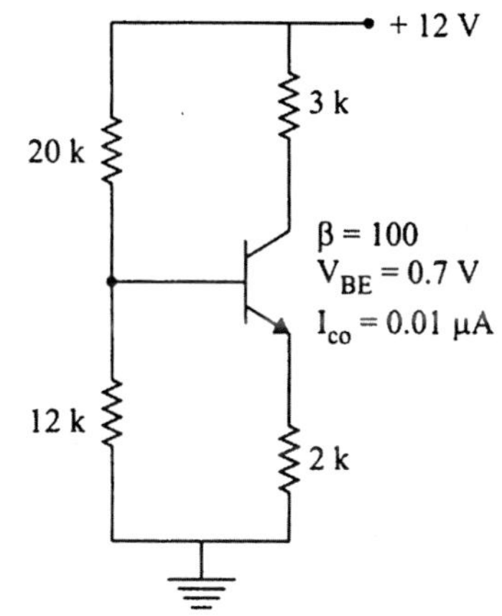

Fig. 7.141.

(a) 0.375 mA (b) 0.185 mA
(c) 1.85 mA (d) 5.6 mA

419. The ripple factor of a power supply is

(a) $\dfrac{P_{dc}}{P_{ac}}$ (b) $\left[\left(\dfrac{I_{rms}}{I_{dc}}\right)^2 - 1\right]^{0.5}$

(c) $\left[\left(\dfrac{I_{dc}}{I_{rms}}\right)^2 - 1\right]^{0.5}$ (d) $\dfrac{I_{dc}}{I_{rms}}$

420. A class B amplifier
1. is biased just at cutoff
2. has a high theoretical efficiency of 78.5% because its quiscent current is low
3. is biased at mid point of load line

Of the above statements
(a) 2 and 3 are correct (b) 1 alone is correct
(c) 1 and 2 are correct (d) 2 alone is correct

421. The main advantage of a crystal oscillator is
(a) It can oscillate at any frequency
(b) It gives a high output voltage
(c) Its frequency of oscillation is almost constant
(d) It operates at a low dc voltage

422. In Fig. 7.142 which diode will conduct and what will be the value of V_0

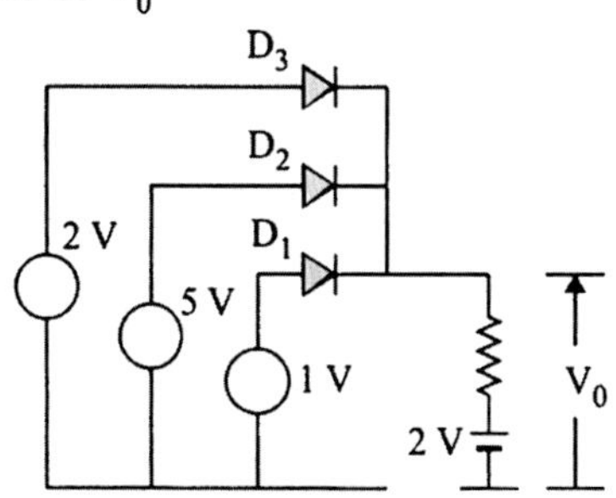

Fig. 7.142.

(a) D_3, 1 V (b) D_1, 2 V
(c) D_2, 5 V (d) D_1, 5 V

423. In a biased JFET (Fig. 7.143) the shape of the channel is as shown because

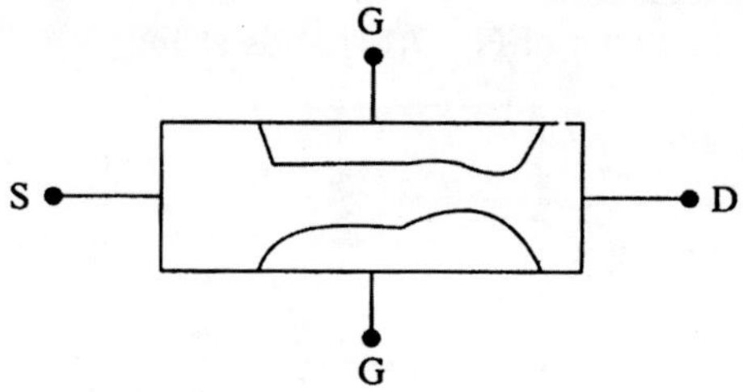

Fig. 7.143.

(*a*) it is the property of material used
(*b*) the drain end is more reverse biased than source end
(*c*) drain end is more forward biased than source end
(*d*) impurity profile changes with distance

424. In a single stage differential amplifier, the output offset voltage is mainly dependent on the mismatch of
(*a*) V_{BE}, I_B and β (*b*) V_{BE} and I_B
(*c*) I_B and β (*d*) V_{BE} and β

425. In Fig. 7.144, V_C is approximately equal to

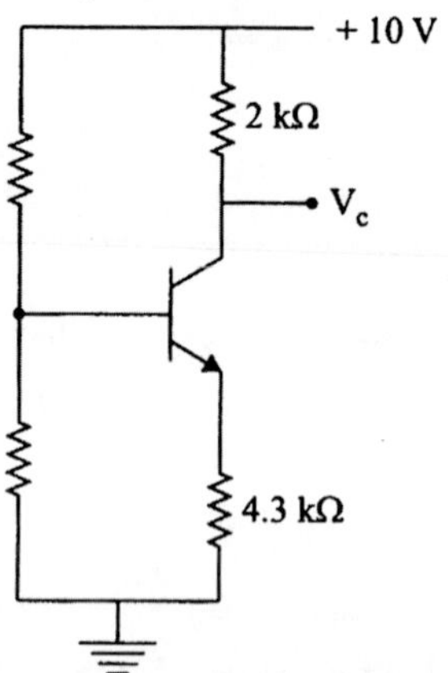

Fig. 7.144.

(*a*) 2 V (*b*) 4.6 V
(*c*) 8 V (*d*) 8.6 V

426. A second order active band pass filter can be obtained by cascading LP second order filter having cutoff frequency f_{OH} with a second order HP filter having cutoff frequency f_{OL} provided
(*a*) $f_{OL} < f_{OH}$ (*b*) $f_{OL} > f_{OH}$
(*c*) $f_{OH} = f_{OL}$ (*d*) $f_{OH} \leq f_{OL}$

427. A Hartley oscillator is used for
(*a*) very low frequencies
(*b*) radio frequencies
(*c*) micro wave frequencies
(*d*) audio frequencies

428. The output V_0 in Fig. 7.145 is
(*a*) –100 V (*b*) –100 m V
(*c*) 10 V (*d*) 10 m V

429. A class C amplifier has input frequency of 200 kHz. If width of collector pulses is 0.1 us, duty cycle is
(*a*) 1% (*b*) 2%
(*c*) 10% (*d*) 20%

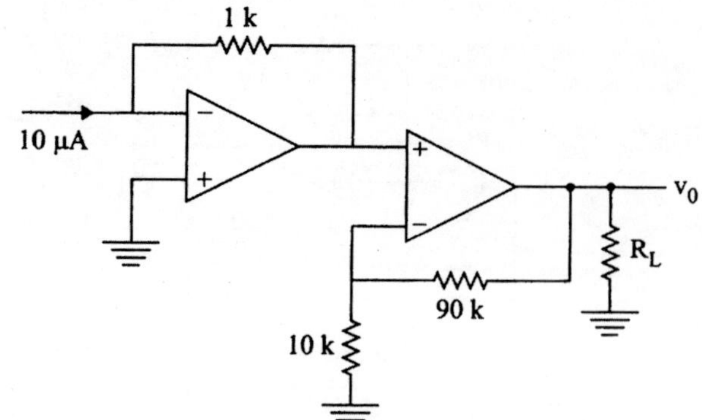

Fig. 7.145.

430. In Fig. 7.146 gain band width product is

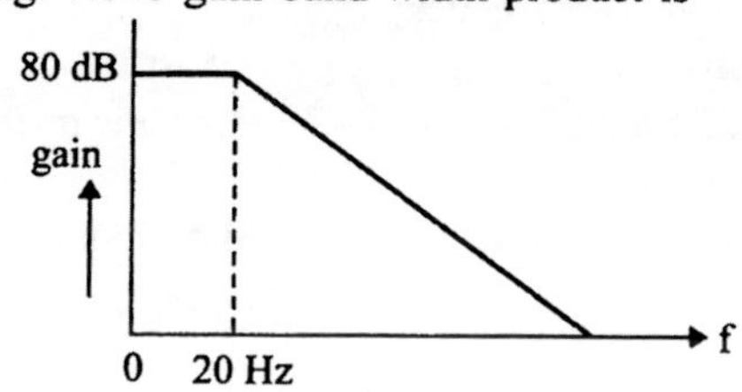

Fig. 7.146.

(*a*) 200 Hz (*b*) 200 MHz
(*c*) 200 kHz (*d*) 2 MHz

431. In Fig. 7.147 v_0 =

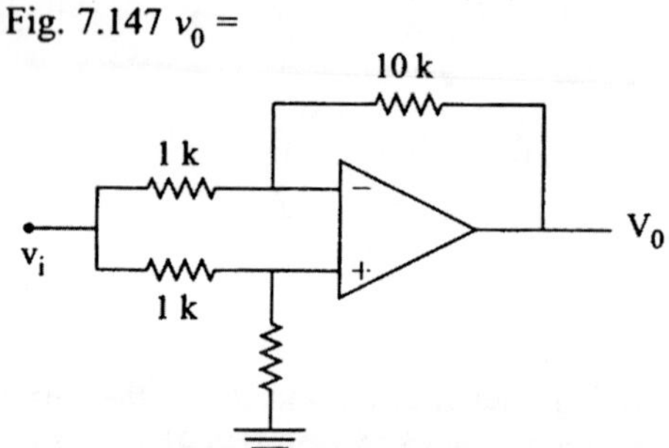

Fig. 7.147.

(*a*) 11 v_i (*b*) 10 v_i
(*c*) v_i (*d*) 0

432. For CE amplifier of 7.148, the slope of ac load line is

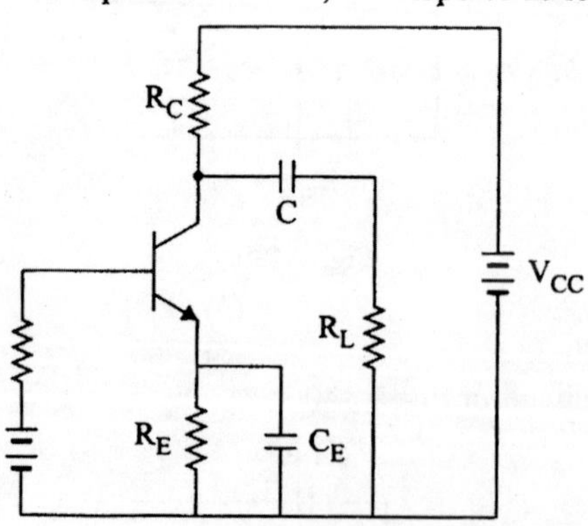

Fig. 7.148.

(*a*) $-\left(\frac{1}{R_L} + \frac{1}{R_C}\right)$ (*b*) $-\frac{1}{R_L + R_C}$

(*c*) $-\frac{1}{R_L}$ (*d*) $-\frac{1}{R_C}$

433. In Fig. 7.149 the approximate voltages of

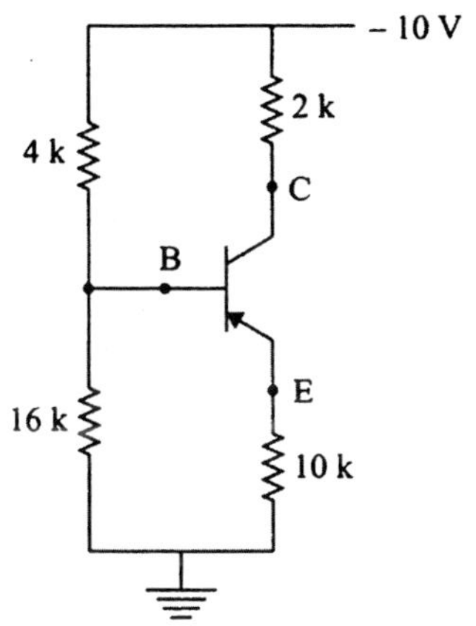

Fig. 7.149.

(*a*) base and emitter are – 8 V and – 7.3 V
(*b*) base and collector are – 8 V and – 5 V
(*c*) collector and emitter are –5 V and –7.3 V
(*d*) base, emitter and collector are –8, –7.3 and –5 V

434. Consider the following statements :
The lower cutoff frequencies of an RC coupled amplifier depend on
1. input and output coupling capacitor
2. emitter by pass capacitors
3. junction capacitances

Which of the above are correct?
(*a*) 1 only (*b*) 2 only
(*c*) 1 and 2 (*d*) 2 and 3

435. Which of the following power amplifiers has highest efficiency?
(*a*) class A (*b*) class AB
(*c*) class B (*d*) class C

436. The voltage gain an amplifier without and with feedback are 100 and 20. The negative feedback is
(*a*) 4% (*b*) 5%
(*c*) 20% (*d*) 80%

437. If the Q of a single stage single tuned amplifier is doubled, the band width will
(*a*) remain the same (*b*) become half
(*c*) become double (*d*) become four times

438. In every practical oscillator the loop gain is slightly more than unity and amplitude of oscillations is limited by
(*a*) magnitude of loop gain
(*b*) onset of non linearity
(*c*) magnitude of gain
(*d*) β

439. In Fig. 7.150 the source resistance of current source I_{EE} is infinite. Then common mode gain is
(*a*) infinite (*b*) zero
(*c*) indeterminate (*d*) $\frac{V_{i1}+V_{i2}}{V_{EE}}$

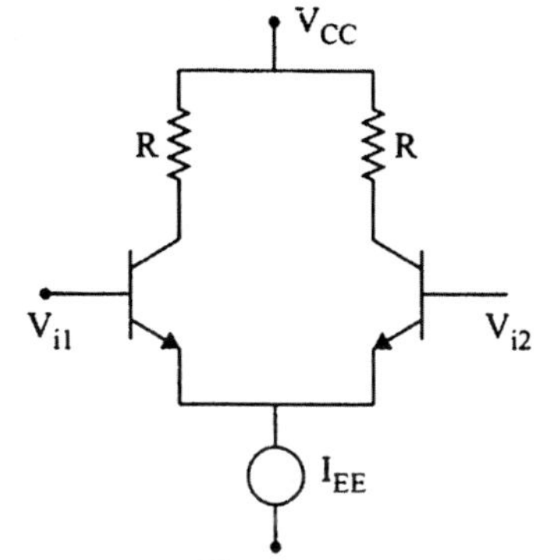

Fig. 7.150.

440. In Fig. 7.151, V_0 =
(*a*) – 1 V (*b*) 2 V
(*c*) +1 V (*d*) +15 V

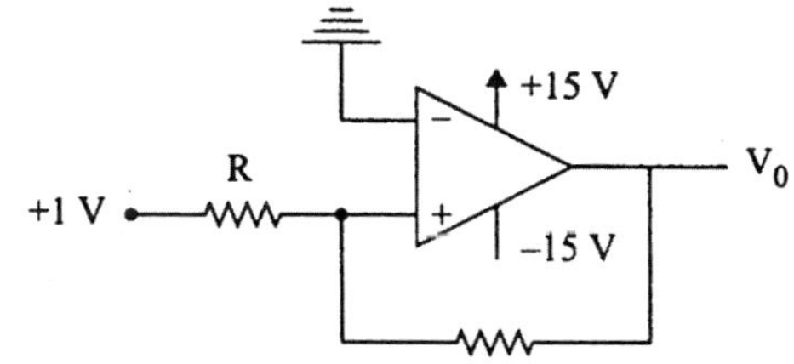

Fig. 7.151.

441. The emitter resistor R_E in the emitter of CE amplifier stabilizes the dc operating point against variation in
(*a*) temperature only
(*b*) β of transistor
(*c*) both temperature and β
(*d*) none of the above

442. The circuit of Fig. 7.152 is

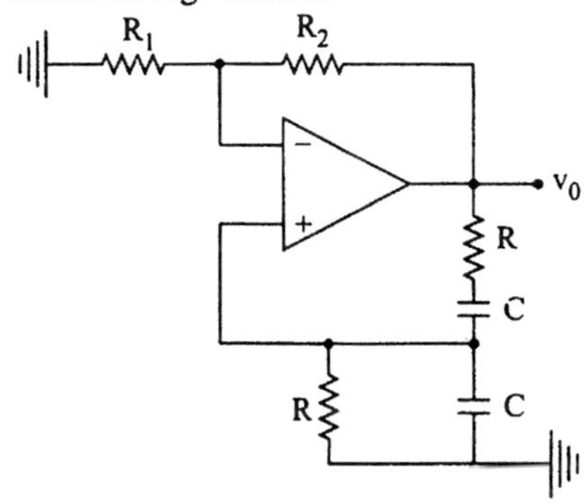

Fig. 7.152.

(*a*) Precision integrator
(*b*) Hartley oscillator
(*c*) Butlerworth high pass filter
(*d*) Wien bridge oscillator

443. Assume that op-amp in Fig. 7.153 is ideal. If input v_i is triangular the output v_0 will be

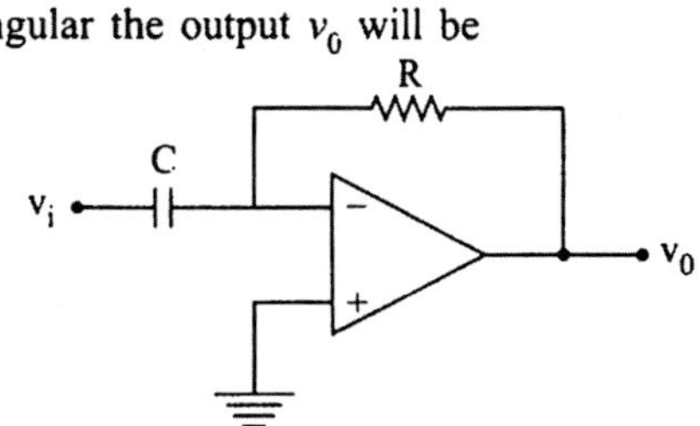

Fig. 7.153.

(*a*) square wave (*b*) sine wave
(*c*) triangular wave (*d*) parabolic wave

444. Negative feedback in amplifier
(*a*) reduces gain
(*b*) increases frequency and phase distortions
(*c*) reduces bandwidth
(*d*) increases noise

445. The early effect in BJT is caused by
(*a*) fast turn on
(*b*) fast turn off
(*c*) large collector base reverse bias
(*d*) large emitter base forward bias

446. Cross-over distortion behaviour is a characteristic of
(*a*) class A output stage
(*b*) class B output stage
(*c*) class AB output stage
(*d*) common base output stage

447. A dc power supply has no load voltage of 30 V and full load voltage of 25 V at full load current of 1 A. The output resistance and voltage regulation respectively are
(*a*) 5 Ω and 25% (*b*) 25 Ω and 20%
(*c*) 5 Ω and 16.7% (*d*) 25 Ω and 16.7%

448. An amplifier has open loop gain of 100, input impedance 1 k Ω and output impedance 100 Ω. If negative feedback with $\beta = 0.99$ is used, the new input and output impedances are
(*a*) 10 Ω and 1 Ω (*b*) 10 Ω and 10 k Ω
(*c*) 100 k Ω and 1 Ω (*d*) 100 k Ω and 10 Ω

449. The static characteristic of an adequately forward biased *p-n* junction is a straight line if the plot is of
(*a*) log I vs log V (*b*) log I vs V
(*c*) I vs log V (*d*) I vs V

450. Two indentical FETs each having parameters g_m and r_d are connected is parallel. The composite FET has parameters
(*a*) 0.5 g_m and 2 r_d (*b*) 0.5 g_m and 0.5 r_d
(*c*) 2 g_m and 2 r_d (*d*) 2 g_m and 0.5 r_d

451. In a differential amplifier, CMRR can be improved by using an increased
(*a*) emitter resistance (*b*) collector resistances
(*c*) power supply voltages (*d*) source resistance

452. In a series regulated power supply circuit the voltage gain of pass transistor satisfies the condition
(*a*) $A_v \to \infty$ (*b*) $1 << A_v << \infty$
(*c*) $A_v \approx 1$ (*d*) $A_v << 1$

453. For full wave rectification four diode bridge rectifier is claimed to have following advantages :
1. Less expensive transformer
2. Small size transformer
3. Suitable for higher voltage applications
Which of these are correct?
(*a*) 1 and 2 only (*b*) 1 and 3 only
(*c*) 2 and 3 only (*d*) 1, 2, 3

454. In Fig. 7.154 the diode is ideal (zero cut in voltage, zero forward resistance). Then the current through diode is

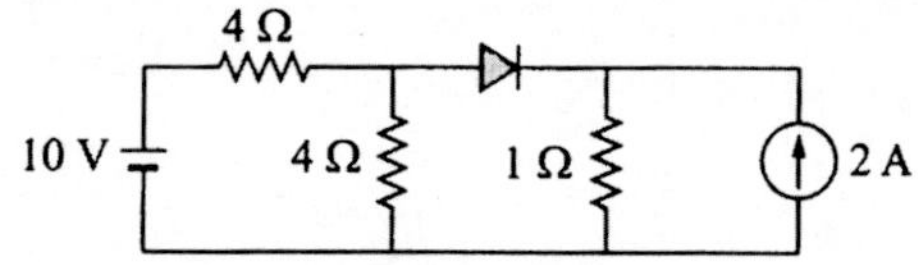

Fig. 7.154.

(*a*) 0 (*b*) 4 A
(*c*) 1 A (*d*) none of the above

455. A BJT is said to be operating in saturation region if
(*a*) both junctions are reverse biased
(*b*) both junctions are forward biased
(*c*) base emitter junction is forward biased and base collector junction is reverse biased
(*d*) base collector junction is forward biased and base emitter junction is reverse biased

456. A class A transformer coupled power amplifier is to deliver 10 W output. The power rating of transistor should not be less than
(*a*) 5 W (*b*) 10 W
(*c*) 20 W (*d*) 40 W

457. For transistor 2 N 338 the manufacturer specifies P_{max} = 100 m W at 25°C free air temperature and maximum junction temperature of 125°C. Its thermal resistance is
(*a*) 10°C/W (*b*) 100°C/W
(*c*) 1000°C/W (*d*) 5000°C/W

458. In Fig. 7.155, secondary winding has 40 turns. For maximum power transfer to 2 ohm resistance the number of turns in primary is

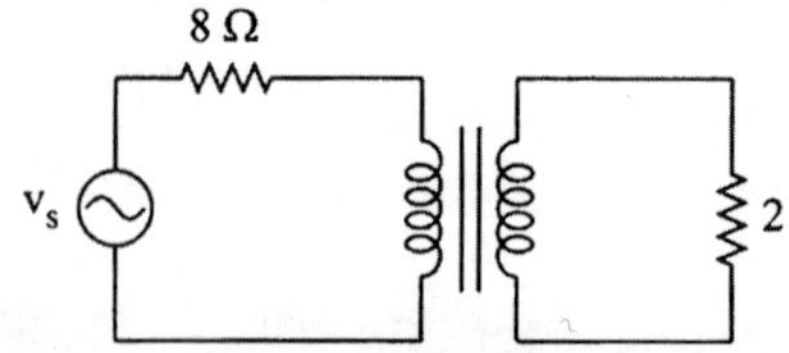

Fig. 7.155.

(*a*) 20 (*b*) 40
(*c*) 80 (*d*) 160

459. For the amplifier in Fig. 7.156 β = 800. The mid band voltage gain V_0/V_i is
(*a*) 0 (*b*) < 1
(*c*) ≈ 1 (*d*) 800

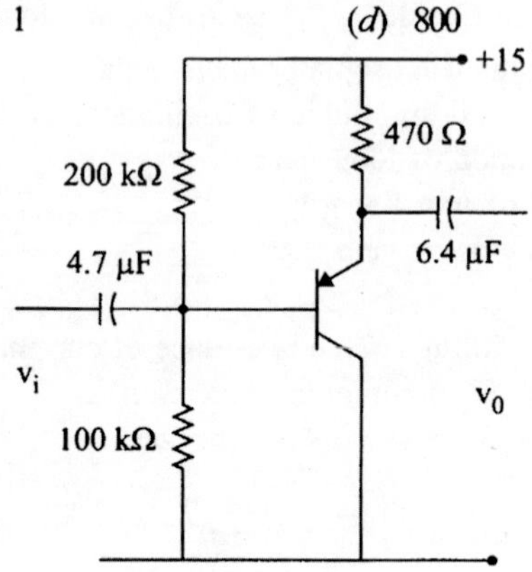

Fig. 7.156.

460. In a transistor push pull amplifier
(*a*) there is no dc present in the output
(*b*) there are no even harmonics in the output
(*c*) both (*a*) and (*b*)
(*d*) neither (*a*) nor (*b*)

461. Consider the following statements
When compared to a bridge rectifier, a centre tapped full wave rectifier
1. Has larger transformer utilisation factor
2. Can be used for floating output terminals *i.e.* no input terminal is grounded
3. Needs two diodes instead of four
4. Needs diodes of lower PIV rating

Which of the above are correct?
(*a*) 1 and 2 only (*b*) 1, 2 and 3
(*c*) 3 only (*d*) 3 and 4 only **[IES 2010]**

462. Transistor work's in three regions
1. cutoff 2. active
3. saturation

When used as a switch in digital logic gates regions in which it works are
(*a*) 1 and 2 (*b*) 2 and 3
(*c*) 1 and 3 (*d*) 1, 2, 3 **[IES 2010]**

463. A rectifier (without filter) with fundamental ripple frequency equal to twice the main frequency has ripple factor 0.482 and efficiency 81.2%. The rectifier is
1. bridge 2. full wave (non bridge)
3. Half wave **[IES 2010]**

Which of these are correct
(*a*) 2 and 3 (*b*) 2 only
(*c*) 1 and 2 (*d*) 1, 2, 3

464. The output of op-amp in Fig. 7.157 is

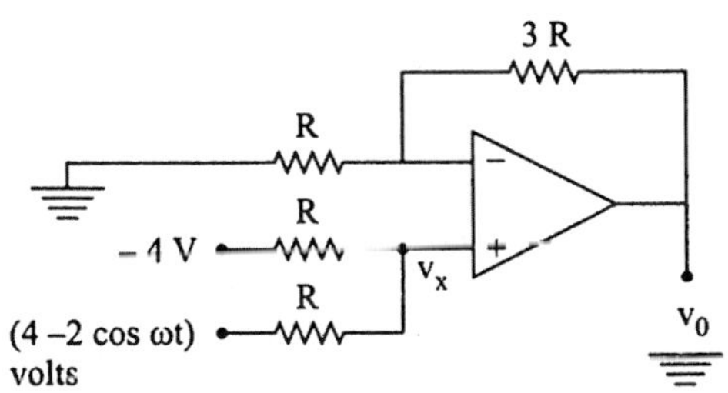

Fig. 7.157.

(*a*) – 0.75 V (*b*) – cos ω*t* volts
(*c*) – 8 cos ω*t* volts (*d*) – 4cos ω*t* volts

465. Consider the circuit is Fig. 7.158. For $v_i = v_m \sin \omega t$ the output voltage for $R_L \to \infty$ will be

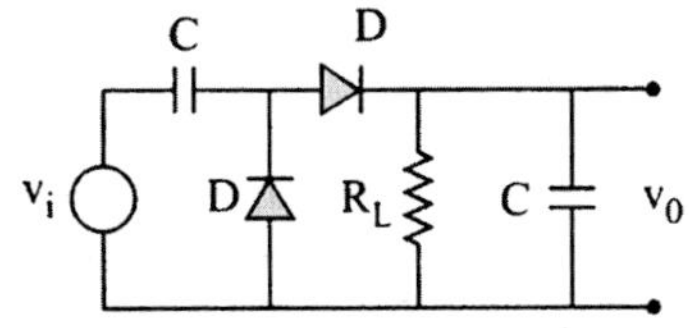

Fig. 7.158.

(*a*) zero (*b*) V_m
(*c*) $2V_m$ (*d*) $-V_m$ **[IES 2010]**

466. The circuit shown in Fig. 7.159 is

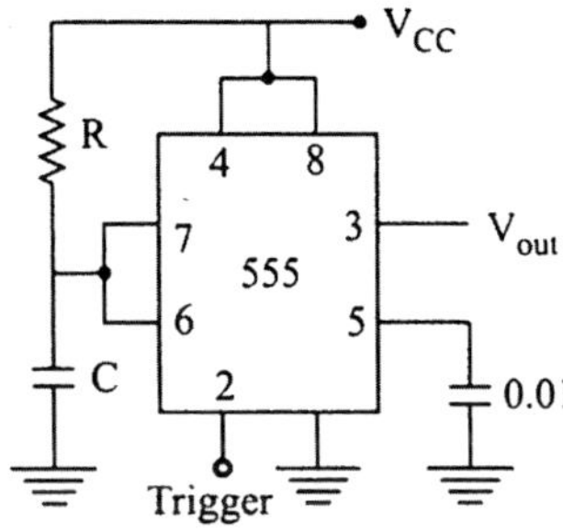

Fig. 7.159.

(*a*) Bistable multivibrator
(*b*) Monostable multivibrator
(*c*) Free running multivibrator
(*d*) Ramp generator **[IES 2010]**

467. The correct waveform for V_0 for the natural in Fig. 7.160 as

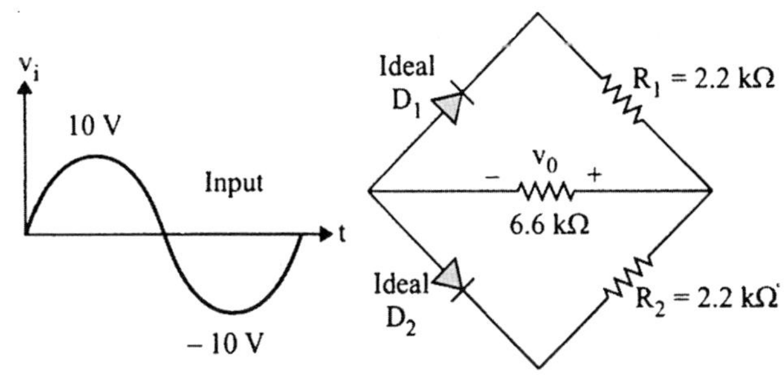

Fig 7.160.

(*a*)
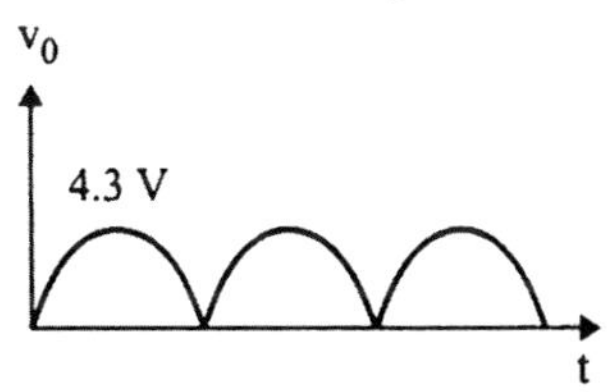

(*b*)
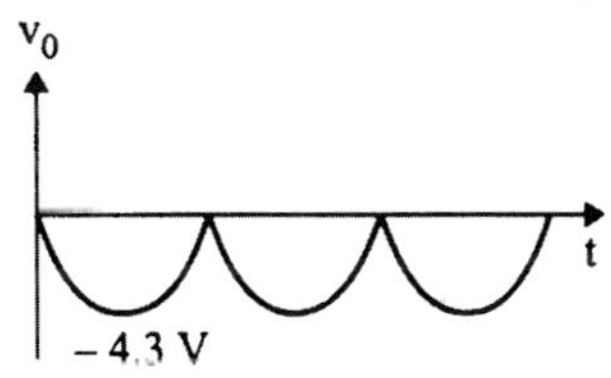

(*c*)
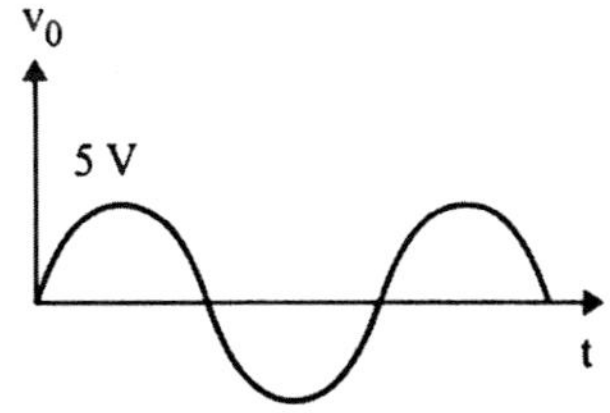

(*d*)
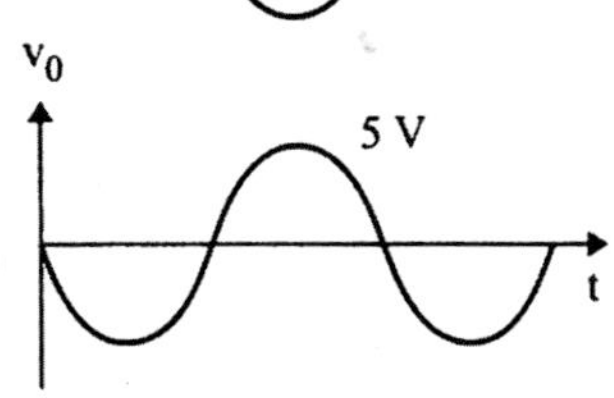

[IES 2010]

468. Consider the following
1. Oscillator 2. Emitter follower
3. Cascaded amplifier 4. Power amplifier

Which of the above use feedback amplifier?
(*a*) 1 and 2 (*b*) 1 and 3
(*c*) 2 and 4 (*d*) 3 and 4 **[IES 2009]**

469. Why are npn transistors preferred over pnp transistors?
(*a*) Leakage current in npn transistors is less than in pnp transistor
(*b*) Mobility of majority carrier in npn transistor is more than mobility of majority carrier in pnp transistor
(*c*) Bias voltage in npn transistor is less than that in pnp transistor
(*d*) Bias voltage in npn is greater than in pnp transistor

470. To draw ac equivalent circuit of a transistor
1. dc sources are shorted
2. ac sources are shorted
3. dc sources are opened
4. ac sources are connected to dc sources

Which of the above are correct?
(*a*) 2 and 4 (*b*) 1 and 2
(*c*) 1 only (*d*) 3 and 4 **[IES 2009]**

471. Which of the following is not correct with regard to power amplifier?
(*a*) The collector current is large
(*b*) They are used as the front end of multistage amplifier
(*c*) They are used near the end of multistage amplifier
(*d*) They have a power rating > 0.5 W **[IES 2009]**

472. Consider the following statements regarding class B power amplifier (complementary symmetry type)
1. Efficiency is higher than that of class A
2. Power output is low
3. Crossover distortion is present
4. Standby power dissipation is absent

Which of the above are correct?
(*a*) 1, 2, 3 (*b*) 1, 2, 4
(*c*) 1, 3, 4 (*d*) 2, 3, 4 **[IES 2009]**

473. Consider the following
1. Distortion 2. Gain
3. Bias stabilisation 4. Sensitivity
5. Frequency response

Which of the above should be kept in view while designing a good power amplifier?
(*a*) 1, 2, 3 (*b*) 1, 3, 5
(*c*) 2, 3, 4 (*d*) 4, 5 **[IES 2009]**

474. Consider the following
1. Coupling capacitor
2. Emitter by pass capacitor
3. Emitter to base diffusion capacitance
4. Stray capacitances

Which of the above components in RC coupled amplifier control lower cutoff frequency?
(*a*) 1, 2 (*b*) 2, 3
(*c*) 3, 4 (*d*) 1, 4

475. Which of the following is a regulated power supply
(*a*) IC 555 (*b*) IC 844
(*c*) IC 3080 (*d*) IC 723 **[IES 2009]**

476. Consider the following regarding op-amp
1. All types of negative feedback reduce non-linear distortion
2. All types of negative feedback reduce output feedback voltage
3. Non-inverting (current and voltage) feedback increases input impedance
4. Inverting (current and voltage) feedback decreases input impedance

Which of the above are correct?
(*a*) 1 only (*b*) 2, 3
(*c*) 2, 4 (*d*) 1, 2, 3, 4

477. Consider the following statements
The bias stability of an emitter bias amplifier circuit is improved by
1. decreasing the value of R_B
2. increasing the value of R_E
3. decreasing the value of R_E
4. increasing the value of R_B
5. increasing the value of R_C

Which the above are correct?
(*a*) 1 and 2 (*b*) 2 and 3
(*c*) 3 and 4 (*d*) 4 and 5 **[IES 2009]**

478. Which of the following is true for CE amplifier if the emitter resistor is made equal to zero
1. gain will increase
2. stability will increase
3. gain will decrease
4. stability will decrease

Select the correct answer from the following codes
(*a*) 1 and 2 (*b*) 2 and 3
(*c*) 3 and 4 (*d*) 1 and 4 **[IES 2009]**

479. Which circuit is used to convert sine wave to square wave
(*a*) astable multivibrator
(*b*) monostable multivibrator
(*c*) bistable multivibrator
(*d*) Schmilt trigger **[IES 2009]**

480. Which transistor model is most preferred to analyse transistor circuit both at mid band and high frequencies
(*a*) *h* parameter model (*b*) *y* parameter model
(*c*) *s* parameter model (*d*) hybrid *h* model

481. Which of the following describe the correct properties of an emitter follower circuit **[IES 2009]**
1. It is a voltage series feedback circuit
2. It is a current series feedback circuit
3. Its gain is less than unity
4. Its output impedance is very low

Select the correct answer using the following code
(*a*) 1, 3 and 4 (*b*) 2, 3 and 4
(*c*) 2 and 3 (*d*) 2 and 4

482. Which of the following types of negative feedback increases input resistance and decreases output resistance of an amplifier

(*a*) current series feedback
(*b*) voltage series feedback
(*c*) current shunt feedback
(*d*) current series feedback **[IES 2009]**

483. Which of the following oscillator is well suited for generation of audio frequency waves of wide range
(*a*) RC phase shift (*b*) Wien bridge
(*c*) Colpitt's (*d*) Hartley

484. Consider the following statements about a good power supply
1. AC ripple should be high
2. Voltage stability factor should be low
3. Temperature stability factor should be low

Which of the above statements are correct?
(*a*) 1, 2, 3 (*b*) 2 only
(*c*) 3 only (*d*) 2 and 3 only

485. Which of the following are non-linear applications of op-amp
1. Current to voltage converter
2. Comparator 3. Peak defect.
4. Limiter

Select your answer from the following codes
(*a*) 1, 2, 3 (*b*) 2, 3, 4
(*c*) 1, 3, 4 (*d*) 1, 2, 4 **[IES 2009]**

486. Which of the following does not show non-linear *v-i* characteristics
(*a*) Schottky diode
(*b*) turned diode
(*c*) Thermistor at a fixed temperature
(*d*) *p-n* junction **[IES 2009]**

487. Which of the following conditions must be satisfied for a transistor to remain under saturation
1. collector to base junction should be forward biased
2. collector base junction should be reverse biased
3. Emitter to base junction should be reverse biased
4. Emitter to base junction should be forward biased

Select your answer using the following code
(*a*) 1 and 2 (*b*) 1 and 3
(*c*) 2 and 3 (*d*) 1 and 4 **[IES 2009]**

488. Consider the following statement
Bias stabilisation in BJT circuit is very important because it
1. provides high voltage gain
2. ensures large bandwidth
3. keeps the operating point unchanged even when temperature changes

Which of the above are correct?
(*a*) 1 and 2 (*b*) 2 and 3
(*c*) 3 only (*d*) 1 and 3 **[IES 2008]**

489. In an oscillator circuit which of the following limits the amplitude of oscillations
(*a*) onset of non-linearity
(*b*) power supply voltage
(*c*) oscillation frequency
(*d*) temperature of active device **[IES 2008]**

490. Consider the following statements
Tuned amplifiers
1. are wide band amplifiers
2. are used in radio transmitters and receivers
3. performance is determined by Q of the circuit

Which of the above statements are correct?
(*a*) 1 and 2 (*b*) 2 and 3
(*c*) 1 and 3 (*d*) 1, 2, 3

491. A feedback amplifier has an open loop gain of –100. If 4% of output is feedback in a degenerative loop what is the closed loop gain
(*a*) –33.3 (*b*) –25
(*c*) –20 (*d*) +25 **[IES 2008]**

492. An amplifier has gain A = 100 ∠180°, upper cut off frequency 100 kHz, lower cutoff frequency 1 kHz. A negative feedback of β = 0.1 is added. Which of the following is not correct?
(*a*) Gain $= \frac{100}{11}$
(*b*) Lower cutoff frequency $\frac{1000}{11}$ Hz
(*c*) Upper cut off frequency 1.1 MHz
(*d*) dB of feedback $20 \log_{10} 11$ **[IES 2008]**

493. What is the efficiency of class B push pull amplifier if V_m is the peak load voltage and V_{cc} is collector supply voltage
(*a*) $\eta = \frac{\pi}{4}\frac{V_m}{V_{cc}} \times 100\%$ (*b*) $\eta = \frac{\pi}{2}\frac{V_m}{V_{cc}} \times 100\%$
(*c*) $\eta = \frac{\pi}{2}\frac{V_{cc}}{V_m} \times 100\%$ (*d*) $\eta = \frac{V_{cc} - V_m}{V_m} \times 100\%$
[IES 2008]

494. Which components are needed to construct a dc power supply to supply 6 V dc to a taperecorder from 230 V ac
1. Step down transform
2. Diodes
3. Resisters and capacitor
4. Three pin voltage stabilizer

Select your answer using the following codes
(*a*) 1, 2, 3 (*b*) 1, 4
(*c*) 3, 4 (*d*) 1, 2, 3, 4

495. The circuit in Fig. 7.161 is

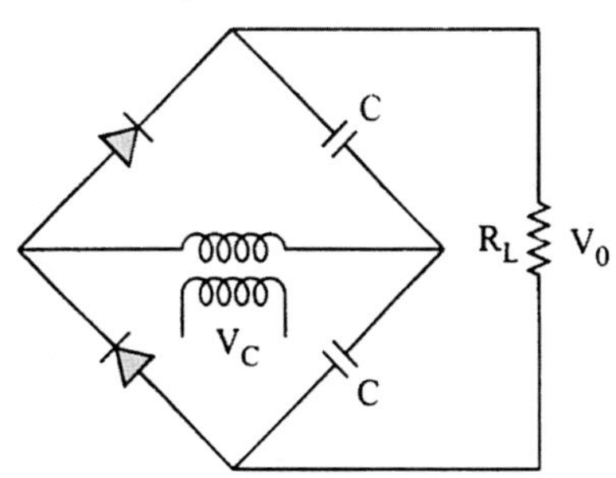

Fig. 7.161.

(*a*) bride rectifier (*b*) voltage doubler
(*c*) rectifier with filter (*d*) comparator

496. Which of the following components control the high frequency response of RC coupled amplifier

1. Parasitic capacitances of transistor
2. Coupling capacitance
3. Stray capacitances
4. Wiring capacitance

Select your answer using the following codes

(*a*) 1, 2 (*b*) 2, 3
(*c*) 3, 4 (*d*) 1, 3, 4 **[IES 2008]**

497. The circuit is Fig. 7.162 is

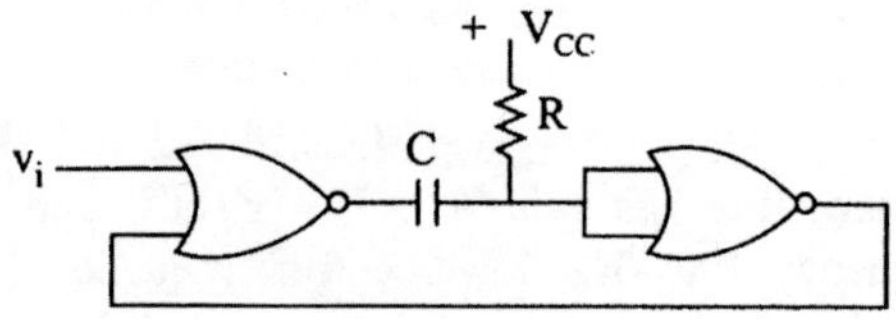

Fig. 7.162.

(*a*) Bistable multivibrator
(*b*) Schmilt trigger
(*c*) Monostable multivibrator
(*d*) astable multivibrator

498. The transfer characteristics in Fig. 7.163 is for which of the following

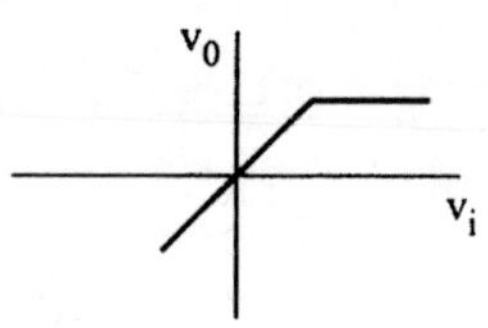

Fig. 7.163.

(*a*) Peak clipper
(*b*) Bottom clipper
(*c*) Clamper
(*d*) Two level clipper **[IES 2008]**

499. Match List I (circuit) with List II (property) and select the answer using the codes

List-I	List-II
A. Preamplifier	1. Non-linear circuit
B. Power amplifier	2. Lumped, linear, passive, bilateral, finite circuit
C. Rectifier circuit	3. Large signal amplifier
D. Purely resistive circuit	4. Small signal amplifier

Codes :

	A	B	C	D
(*a*)	4	2	1	3
(*b*)	1	3	4	2
(*c*)	4	3	1	2
(*d*)	1	2	4	3

500. In a class A amplifier $V_{CE(max)} = 15$ V, $V_{CE(min)} = 1$ V. The conversion efficiency for a series fed load is

(*a*) 25% (*b*) 23.33%
(*c*) 12.5% (*d*) 11.67%

501. In the silicon BJT circuit in Fig. 7.163 assume that emitter area of transistor Q_1 is half of transistor Q_2
The value of current I_O is approximately

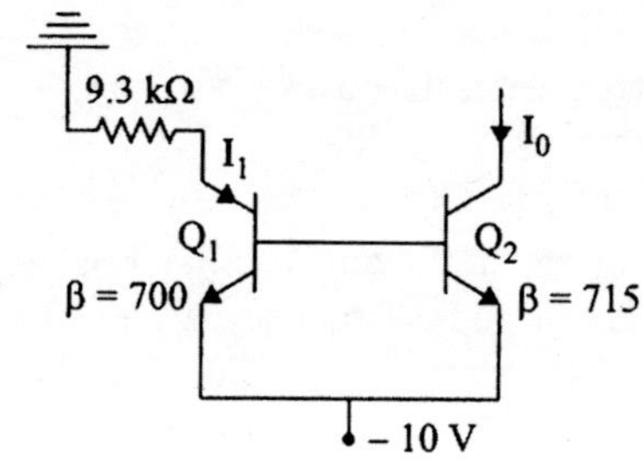

Fig. 7.163.

(*a*) 0.5 mA (*b*) 2 mA
(*c*) 9.3 mA (*d*) 15 mA **[Gate 2010]**

502. Assume the op-amp Fig. 7.164 is ideal.

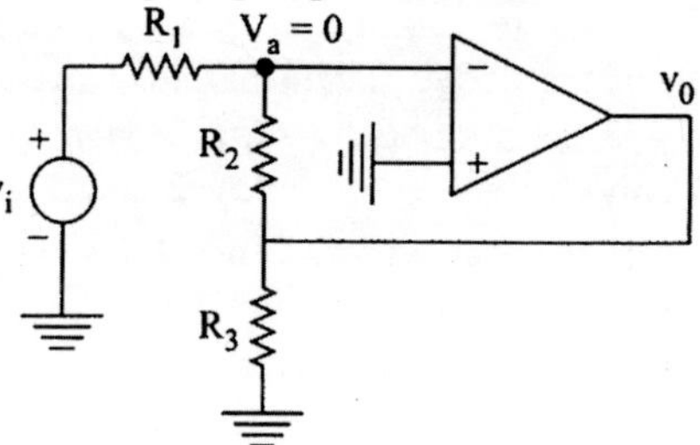

Fig. 7.164.

The voltage gain is

(*a*) $\frac{R_2}{R_1}$ (*b*) $\frac{R_3}{R_1}$

(*c*) $-\left[\frac{R_2 \| R_3}{R_1}\right]$ (*d*) $-\frac{R_2 + R_3}{R_1}$

[Gate 2010]

503. The transfer characteristics for the precision rectifier circuit in Fig. 7.165 is (Assume ideal op-amp and practical diode)

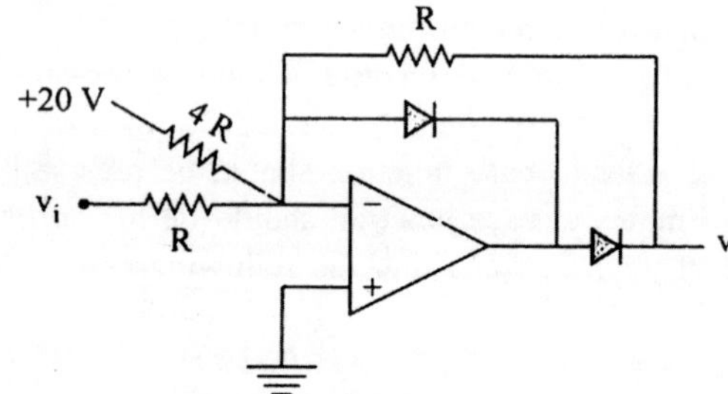

Fig. 7.165.

(*a*)

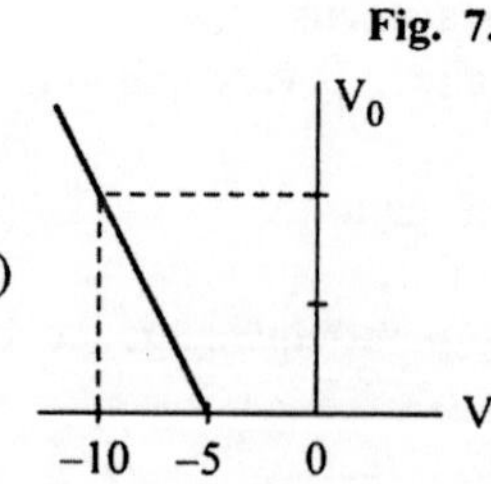

(*b*)

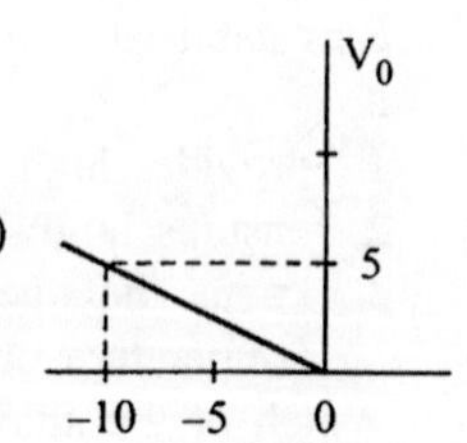

(*c*)

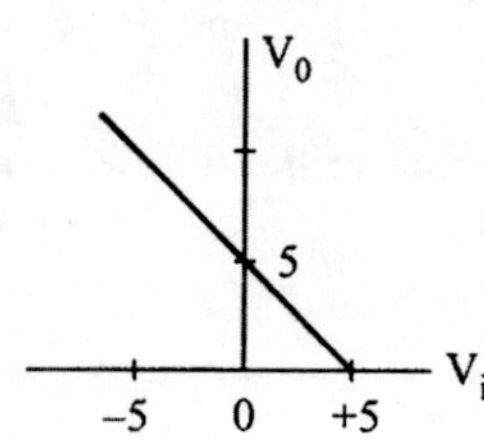

(*d*)

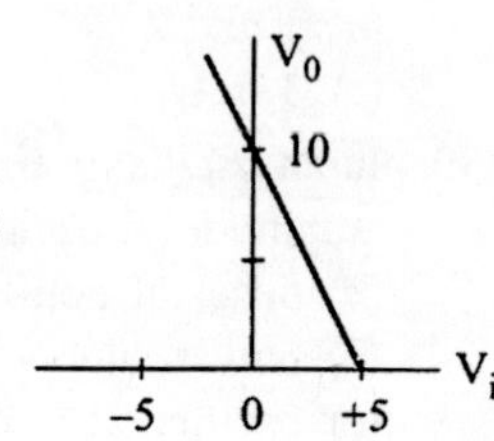

[Gate 2010]

504. Figure 7.166 shows a CE amplifier circuit. The circuit parameters are

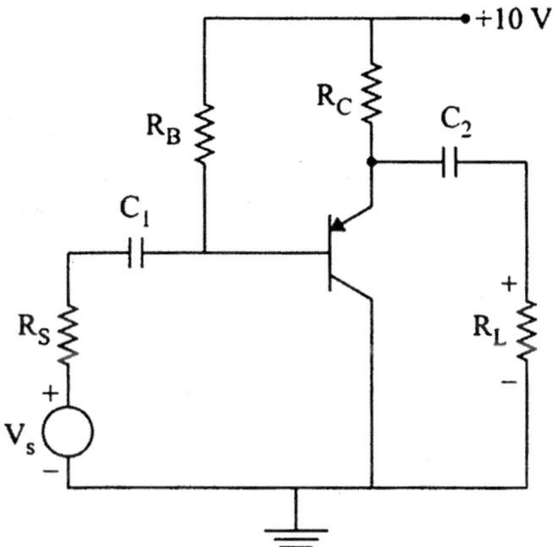

Fig. 7.166.

$\beta = 10, g_m = 0.3861$ A/V
$r_o = \infty, R_s = 1$ kΩ
$R_B = 93$ kΩ, $R_c = 250$ Ω
$R_L = 1$ kΩ, $C_1 = \infty$
$C_2 = 4.7$ μF

The resistance as seen by sources

(*a*) 250 Ω (*b*) 1258 Ω
(*c*) 95 kΩ (*d*) ∞

505. In the circuit of problem 504 the lower cutoff frequency due to C_2 is

(*a*) 33.9 Hz (*b*) 27.1 Hz
(*c*) 13.6 Hz (*d*) 16.9 Hz **[IES 2010]**

506. In Fig. 7.167 the diode is ideal. The voltage V is given by

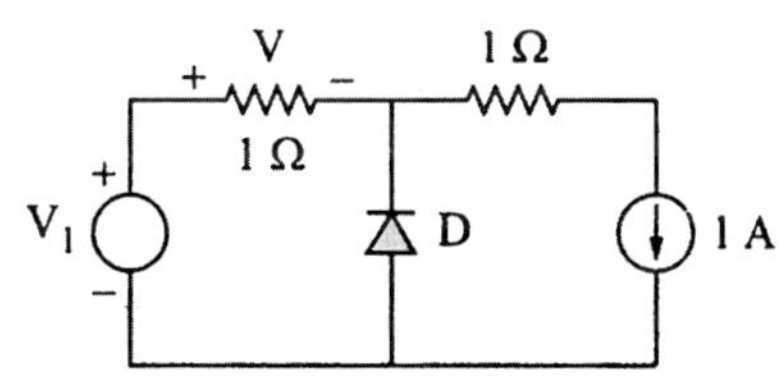

Fig. 7.167.

(*a*) min (V_i, 1) (*b*) max (V_i, 1)
(*c*) min ($-V_i$, 1) (*d*) max ($-V_i$, 1) **[IES 2009]**

507. Figure 7.168 shows a circuit using ideal op-amp. *i-v* characteristic for diode is $i = I_0\left[e^{V/V_T} - 1\right]$ where V_T = 25 mV, I_0 = 1 μA and V is voltage across diode (taken positive for forward bias)

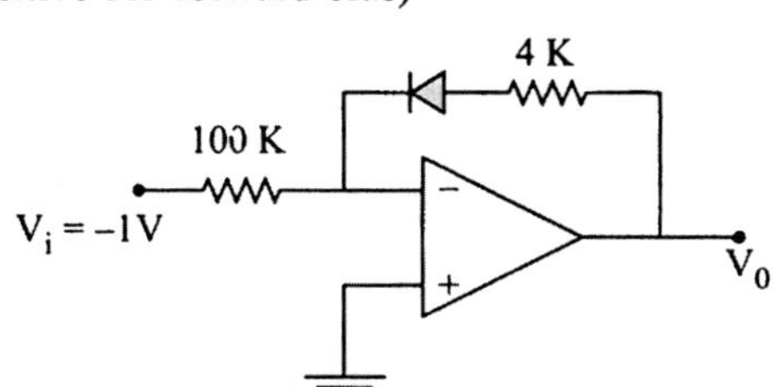

Fig. 7.168.

The output voltage V_0 is

(*a*) 0 (*b*) 0.1 V
(*c*) 0.4 V (*d*) 1.1. V **[Gate 2008]**

508. The circuit in Fig. 7.169 is

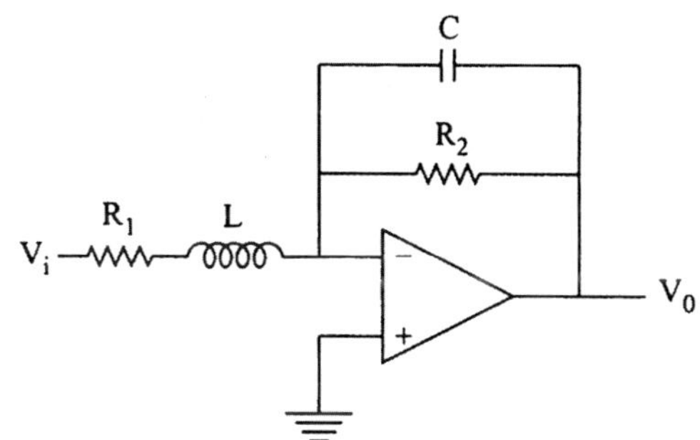

Fig. 7.169.

(*a*) low pass filter
(*b*) high pass filter
(*c*) band pass filter
(*d*) band stop filter **[Gate 2008]**

509. Consider the Schmilt trigger circuit of Fig. 7.170.

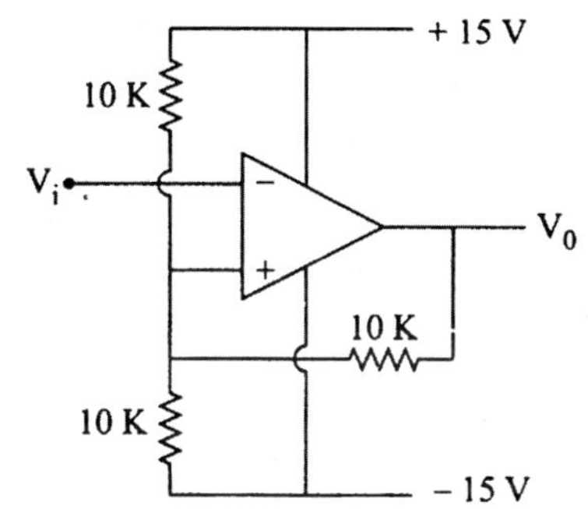

Fig. 7.170.

A triangular wave which goes form −12 to 12 V is applied to inverting input. Assume that V_0 swings from +15 V to −15 V. The voltage at non-inverting terminal switches between

(*a*) −12 and +12V (*b*) −7.5 and +7.5 V
(*c*) −5 and +5 V (*d*) 0 and 5 V **[Gate 2008]**

510. In op-amp circuit of Fig. 7.171, V_0 is **[Gate 2007]**

(*a*) −2 V (*b*) −1 V
(*c*) −0.5 V (*d*) 0.5 V

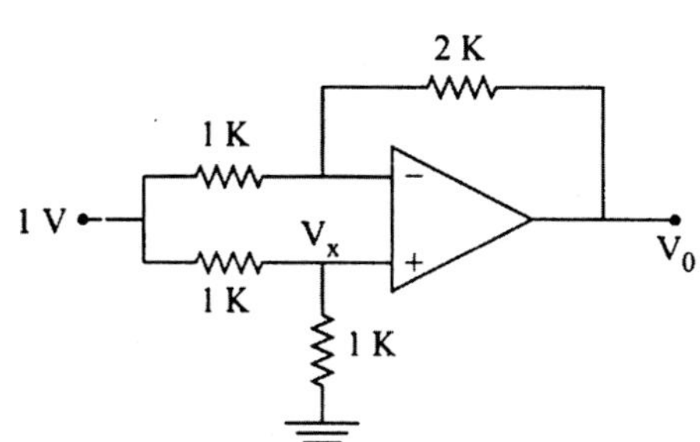

Fig. 7.171.

B. The following questions have two statements A (assertion) and R (reason). Read them and give correct answer as per following code

(a) Both A and R are correct and R is correct explanation for A

(b) Both A and R are correct but R is not correct explanation for A

(c) A is correct R is wrong

(d) A is wrong R is correct

511. A. A good op-amp should have very large band width.
R. An ideal op-amp should have infinite input resistance and zero output resistance.

512. A. In a full wave rectifier output the lowest ac component is at twice the input frequency.
R. In a full wave rectifier the output waveform consists of half sinusoids.

513. A. A dc stabilizer has constant output voltage irrespective of variations in input voltage and load current.
R. A dc series regulator uses a zener diode across the load to stabilize the output voltage.

514. A. A class C amplifier is a tuned amplifier which needs a tank circuit as the load in the collector circuit of the transistor for its proper operation.
R. In class C operation collector current flows for less than 180° of the ac cycle and thus flows in pulses.

515. A. In the circuit shown Fig. 7.172 the diode is used to protect the active devices. Transistor T goes into saturation when a positive pulse is applied at base and non-conducting otherwise.
R. When the current in an inductor is suddenly stopped, a high voltage is induced across it.

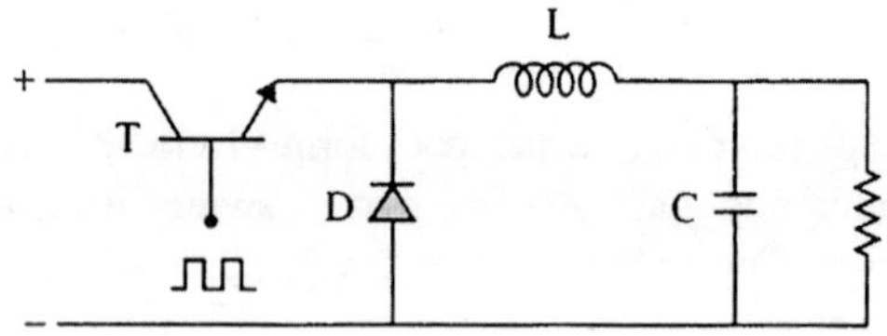

Fig. 7.172.

516. A. A Darlington amplifier has very low output impedance
R. The circuit has a low resistance R_E between emitter and ground.

517. A. The circuit in Fig. 7.173 produces repetitive narrow pulses when input is fed with sine or triangular waveform having peak value more than + V.
R. The high gain op-amp produces voltages at two levels. If input is more than $+V_1$ the output is +15 V otherwise the output is −15V.

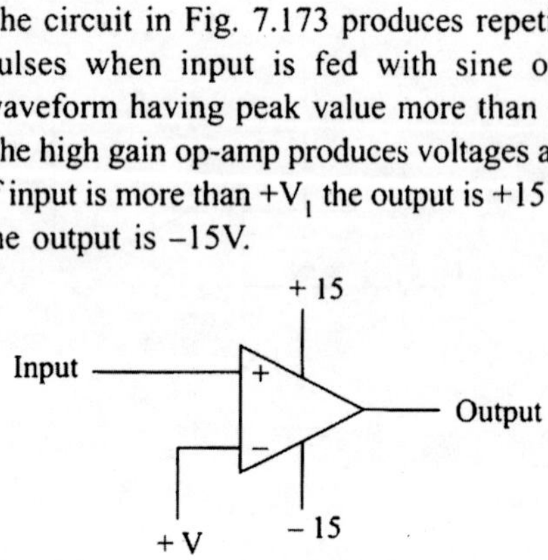

Fig. 7.173.

518. A. A large negative feedback is deliberately introduced in an amplifier to make its gain independent of variations in parameters of active device and other components.
R. A large negative feedback results in a high value of return difference compared to unity which makes the feedback gain inversely proportional to feedback factor.

519. A. For large signal variations an amplifier circuit has to be analysed graphically.
R. The output characteristics of a transistor is non-linear.

520. A. Manufacturer's quote *h* parameters of transistor.
R. *h* parameters of a transistor are easiest to measure.

521. A. CE amplifier is the most widely used BJT amplifier.
R. CE amplifier has zero phase difference between input and output.

522. A. A self bias circuit is the most commonly used bias circuit for bipolar junction transistors.
R. A self bias circuit limits the increase in collector or emitter current due to change in junction temperature.

523. A. An ac amplifier uses coupling capacitors at input and output.
R. The capacitive reactance varies inversely with frequency. It is infinite for dc.

524. A. A common collector amplifier is used for matching a high impedance source to a low impedance load.
R. A common collector amplifier has a near unity voltage gain.

525. A. Direct coupled amplifiers are rarely used.
R. Direct coupled amplifiers suffer from the disadvantage of drift.

526. A. An op-amp amplifier can amplify dc signal also.
R. An op-amp amplifier is a direct coupled amplifier.

527. A. Transformer coupling has the disadvantage of the weight of transformer and saturation of core.
R. Transformer coupling is very suitable for impedance matching.

528. A. In an RC coupled amplifier the junction capacitances can be neglected at low frequencies but not at high frequencies.
R. Capacitive reactance is inversely proportional to frequency.

529. A. A tuned amplifier uses a parallel tuned circuit.
R. A tuned amplifier is used to amplify a narrow band of frequencies.

530. A. In class A amplifier the distortion is minimum.
R. In class A amplifier collector current exists for 360° of input cycle.

531. A. Negative feedback reduces the gain of an amplifier.
R. Negative feedback is very commonly used in amplifier circuits.

532. A. Negative feedback reduces the band width of an amplifier.
R. Negative feedback stabilizes the gain of an amplifier.

533. A. In CE amplifier the emitter is at ground potential for ac signals.
R. A CE amplifier has near unity voltage gain.

534. A. In CC amplifier the collector is at ac ground.
R. CC amplifier is also called emitter follower.

535. A. In Darlington pair, the overall β is product of individual βs.
R. The voltage gain of a base driven amplifier is equal to ac collector voltage divided by ac base voltage.

536. A. The horizontal intercept of dc load line is the same as ideal cutoff point.
R. When a transistor is in cutoff region, collector current is zero.

537. A. The ac load line is different from dc load line.
R. The ac load resistance seen by collector and emitter is different from dc load resistance.

538. A. If we overdrive an amplifier the output signal may get clipped.
R. To dissipate heat to atmosphere a heat sink is sometimes attached to the case of transistor.

539. A. In a push pull circuit each transistor stays in the active region for one half cycle.
R. A push pull amplifier has low distortion.

540. A. The maximum efficiency of a class B push pull amplifier is 78.5%.
R. A current mirror uses a compensating diode in parallel with the emitter diode.

541. A. In class C amplifier the efficiency is more than that for other classes of amplifiers.
R. In class C amplifier we can get high efficiency over a wide band.

542. A. Class A and Class B amplifier predominate in audio applications.
R. Class C amplifier is used at radio frequencies.

543. A. A class C amplifier can also be used as frequency multiplier.
R. When narrow current pulses drive a resonant circuit the output can be a near perfect sine wave.

544. A. The duly cycle of a class C amplifier affects the power dissipation of transistor.
R. At higher harmonics the efficiency of a class C amplifier becomes poor.

545. A. In a circuit using FET, gate is approximately at dc ground.
R. Self bias stabilizes the quiescent operating point against changes in JFET parameters.

546. A. In a common source amplifier with source terminal at ac ground the voltage gain is about $g_m r_d$.
R. A common source amplifier is a source follower circuit.

547. A. The gain of an FET amplifier can be changed by changing g_m.
R. Most receivers use automatic gain control to counteract changes in volume caused by fading.

548. A. If power gain doubles the bel power gain increases by 3 dB.
R. The bel power gain of cascaded stages is equal to sum of each bel power gain.

549. A. As per Miller theorem a capacitance C from input to output appears as a much larger capacitance across input terminals.
R. In a lag network the output voltage appears across capacitor.

550. A. In monolithic IC the components are part of one chip
R. An op-amp is a high gain dc amplifier.

551. A. In a differential amplifier the current through common emitter resistor is called tail current.
R. CMRR of an op-amp indicates how well is a differential signal amplified.

552. A. An op-amp has high voltage gain, high input impedance and low output impedance.
R. Negative feedback increases output impedance.

553. A. A resistor is a passive load and a transistor is an active load.
R. Op-amp is mostly used in feedback mode.

554. A. In an op-amp the voltage gain and band width can be adjusted as per requirement.
R. Large value capacitor can also be fabricated on a chip.

555. A. A sine wave with slew rate distortion looks rectangular.
R. The maximum rate of output voltage change is called slew rate.

556. A. Linear ICs include audio amplifier, wideband amplifier, RF amplifier and IF amplifier.
R. The error voltage in a feedback circuit is very small.

557. A. In an op-amp the gain band width product equals f_{inity}.
R. The change in open loop gain of an amplifier has almost no effect on closed loop gain.

558. A. The main use of positive feedback is in oscillator circuits.
R. In an oscillator the loop gain Aβ is kept slightly more than unity initially.

559. A. Electric coupling, magnetic coupling and ground loop can cause unwanted high frequency oscillations.
R. Hartley oscillator uses a tapped inductor.

560. A. The main use of Miller integrator is to generate ramp.
R. An inverting amplifier can handle more man one input at the same time.

561. A. The small signal analysis of a transistor is done to obtain current gain, voltage gain and conversion efficiency of amplifier. **[IES 2009]**
R. The small signal analysis of a transistor amplifier uses small signal parameters of transistor.

562. A. A rectifier with inductor filter is more efficient for high load current. **[IES 2009]**
R. In rectifier with inductor filter we can use a large choke to reduce ripple. Large choke has higher dc resistance which will reduce dc output voltage for high load current.

563. A. It is not desirable to drive a transistor into hard saturation in high speed switching currents.

R. It may not be possible to bring transistor from hard saturation to cutoff state. **[IES 2009]**

564. **A.** Complementary transistors are used for class B push pull amplifier.

R. We do not need transformer if we use complementary transistor in class B push pull amplifier. **[IES 2008]**

565. **A.** A high value of resistor is connected across the diode in a clamper circuit for better performance.

R. If this resistor is not used, the circuit will respond to sudden decease in the magnitude of input impulse train but the response to sudden increase in amplitude will cause improper response. **[IES 2008]**

566. **A.** Many power devices use the case itself as one of the principal current carrying terminals.

R. Such a connection maximises the thermal resistance between the working semiconductor and case.

567. **A.** The three electrodes source, drain and gate for FET correspond to Emitter, base collector BJT. Input signal is applied to gate and gate voltage controls the electric field in channel.

R. In FET the gate and channel act as the plates of a capacitor. A change of one polaity at gate causes an equal and opposite change at channel. Therefore conductivity of channel can be increased or decreased by gate voltage. **[IES 2007]**

568. **A.** The self bias technique used for JFET to fix operating point cannot be used for MOSFET.

R. The voltage drop across R_s is such that it reverse biases the gate. **[IES 2005]**

569. **A.** It is possible to use the technique of additive amplification to get wide band amplifier.

R. In additive amplifier the overall gain is the sum of gain of individual amplifier. **[IES 2005]**

570. **A.** In Darlington connection two transistors are connected in cascade in CE configuration.

R. Darlington connection aims at making current gain high almost equal to product of individual current gain. **[IES 2005]**

C. State whether the following are true (T) or false (F)

571. Efficiency of a half wave diode rectifier is less man 40%.

572. In a capacitor filter the time constant R_LC should be small.

573. The ripple factor of a half wave rectifier is more than 2.

574. The ripple factor of a full wave rectifier is about 0.5.

575. In a rectifier only a portion of input power is converted into dc power. The rest is ac power.

576. The PIV in a bridge rectifier is less than that in full wave rectifier with centre tapped transformer.

577. The efficiency of a full wave rectifier using centre tapped transformer is twice that in full wave bridge rectifier.

578. The ripple factor of a full wave rectifier is about 0.5.

579. As the ratio R_f/R_L increases the efficiency of a rectifier increases.

580. A CE amplifier is the most common BJT amplifier.

581. In CE amplifier the base current is very high.

582. In an amplifier the stray capacitances assume impedance at low frequencies.

583. In graphical analysis of BJT amplifier the slope of ac line is more than that of dc load line.

584. The units of transistor h parameters h_{11} and h_{22} are the same.

585. DC amplifiers have a tendency to be unstable.

586. Saturation of iron core causes distortion.

587. Maximum efficiency of class B power amplifier is 50%.

588. The h parameter equivalent circuit of BJT has one voltage source and one current source.

589. The h parameter h_{11} is the same in CE and CC circuit.

590. In calculating input impedance, the load impedance is considered to be connected at the output.

591. Efficiency of class A power amplifier is very poor.

592. When load is coupled to class A amplifier through transformer, efficiency decreases.

593. In calculating output impedance of an amplifier the source is replaced by an open circuit.

594. In CE circuit the parameter h_{12} is very small.

595. CC circuit is suitable for impedance matching.

596. The parameter h_{11} for CB circuit is higher, than that for CE circuit.

597. FET amplifier is more difficult to analyse than BJT amplifier.

598. Tuned amplifiers are used at radio frequencies.

599. A typical value of g_m for a FET is about 25 μS.

600. In feedback amplifier the closed loop gain is nearly independent of open loop gain and depends only on the feedback factor.

601. Feedback factor may be less or more than 1.

602. With negative feedback the closed loop gain is always less than open loop gain.

603. The changes in ambient temperature affect the open loop gain of amplifier.

604. When negative feedback is used the fractional changes in closed loop gain are desensitized by the factor $(1 + A_v\beta)$.

605. Barkhausen cirterion is $1 + A_v\beta = 0$.

606. For Q >> 1, the output frequency of LC oscillator is equal to resonant frequency.

607. Negative feedback reduces noise originating at the amplifier input.

608. Negative feedback lowers the lower cutoff frequency and increase the upper cutoff frequency of amplifier.

609. A ripple in collector supply voltage of the output stage of amplifier causes noise.

610. The parameter β of a BJT change by 300%. If self bias is used in the amplifier using this BJT, the quiescent value of I_C and V_{CE} remain almost the same.

611. A limiter circuit can be built using a diode and resistance.

612. In a biased limiter the clipping level can be adjusted by adjusting the voltage of battery.

613. In a combination limiter with sinusoidal input voltage the output wave can be nearly square.

614. A diode can have a very high forward resistance.

615. A differential amplifier consists of two CB amplifiers.
616. f_{unity} of an op-amp is about 1 kHz.
617. CMRR is the ratio of differential voltage gain to common mode gain.
618. CMRR of an op-amp is about 0.1.
619. The input impedance of a non-inverting amplifier is very low.
620. An ideal op-amp has zero slew rate.
621. The gain of a source follower is unity.
622. An integrating amplifier has resistance in feedback path.
623. A logarithmic amplifier has diode in the feedback path.
624. An ideal op-amp can amplify dc signals also.
625. The open loop gain of an actual op-amp is about 100.
626. A differentiating amplifier has capacitor in the input path.
627. If f is more than f_{unity} the gain of op-amp becomes less than 1.
628. An exponential amplifier has diode in feedback path.

D. Fill in the Blanks with appropriate words.

629. In a source follower the input and output are................. .
630. Feedback in used to..................the ....
631. In a sign changer the phase difference between input and output is................. .
632. ....across a virtual ground is ....
633. From application point of view an IC may be .... or ....
634. Current through a virtual ground is................. .
635. In an amplifier using negative feedback, the closed loop gain is almost independent of ....
636. Two forms of IC technology are.................. and ....
637. A dc regulated supply uses a ....
638. An inductance presents ....to high frequency ac components.
639. The voltage across a zener diode remains ....
640. Ripple factor is a measure of lack of ....of output wave form.
641. In a capacitor filter used with half wave rectifier, the voltage across a diode (during negative half cycle) decays.................. .
642. If the gain of an amplifier is real, constant and independent of frequency, the distortion is.................. .
643. Voltage gain is less than 1 in..............circuit.
644. A MOSFET amplifier has very high...................... .
645. In class C operation the efficiency is very ....
646. Interelectrode capacitances are important at ....
647. At low frequencies the..................assumes importance.
648. CB circuit can be used for.................. .
649. Non-linear distortion occurs due to.................. nature of electronic devices.
650. A tuned amplifier is used at.................. .
651. In push pull circuit each transistor conducts for..................
652. Positive feed back leads to.................. .
653. Colpitt's oscillator is very suitable for ....
654. In a feedback oscillator the phase angle of $A_v\beta$ must be..................or a multiple of ....
655. If β = 0.04, the closed loop gain will be about ....irrespective of the value of open loop gain.
656. Negative feedback causes a..............in gain.
657. Oscillators using crystal have good................. .
658. The voltage gain of a base driven amplifier is equal to ac collector voltage divided by ac..................voltage.
659. The voltage gain of an emitter follower is less than................. .
660. In a CE amplifier ac emitter resistance to ground is.................. .
661. When a collector is at ac ground, the circuit is called.................. .
662. In a CE amplifier the phase of ac emitter voltage is the same as phase of ac................ voltage.
663. We can multiply individual βs to get the overall β of a................. .
664. The main problem of dc amplifier is ....
665. The quiescent collector current and voltage are I_C and V_{CE} when there is no................. .
666. If an amplifier is overdriven, the output signal is ....
667. From dc equivalent circuit of an amplifier we can find................. .
668. If the input signal is very large the instantaneous operating point may move all the way to .... and ....
669. If ambient temperature is more than 25°C we have to................the electronic devices.
670. The maximum efficiency of a transformer less class A amplifier is................ .
671. To get maximum unclipped signal the Q point should be at the................of ac load line.
672. A low thermal resistance means it is easy for ....to flow from transistor junction to surrounding air.
673. In a current mirror the current voltage curve of the compensating diode is similar to ....curve.
674. Diode bias is used in linear................ .
675. A class B push pull amplifier has less non-linear distortion because................cancel out.
676. In a negative clamper the positive peak and negative peak in the output are approximately at................ and ....
677. Class A and class B amplifiers are mainly used at ....
678. In FET self bias circuit the drain current flows through................resistor to produce the bias voltage.
679. In FET circuit, source follower is also called ....circuit.
680. The input resistance of a source follower is very................

ANSWERS WITH HINTS AND EXPLANATION

1. (*d*) It is 180° for purely resistive load and between 180° and 270° for R-L load.
2. (*b*)
3. (*a*) Therefore CC configuration is used for impedance matching.
4. (*a*) Slope is $\frac{-1}{R_L}$.
5. (*a*) 6. (*b*) 7. (*a*)
8. (*d*) In class C operation output is in the form of short duration pulses.
9. (*b*) 10. (*a*)
11. (*a*) Half power frequency is –3 dB frequency
12. (*c*) 13. (*a*) 14. (*b*) 15. (*b*)
16. (*c*)
17. (*c*) Emitter base junction is forward biased and base-collector junction is reverse biased.
18. (*a*) 19. (*a*)
20. (*a*) Since two diodes are in series in bridge rectifier, voltage drop is more.
21. (*c*)
22. (*c*) Closed loop gain = $\frac{200}{1+0.2\times 200}$.
23. (*c*) The stabilisation of gain is one of the advantages of closed loop system.
24. (*c*) New lower cut off frequency = $\frac{50}{1+200\times 0.1}$.
25. (*a*) 26. (*b*) 27. (*b*)
28. (*c*) One R C combination can give a phase shift of less then 90°. Therefore 3 R C combinations are required for 180° phase shift.
29. (*a*) 30. (*a*) 31. (*a*) 32. (*b*)
33. (*c*) High input impedance ensures minimum loading of source. Low output impedance is required for impedance matching at output.
34. (*a*) 35. (*d*)
36. (*c*) Integral of a constant is *t*.
37. (*d*)
38. (*b*) $i = \frac{7-0.7}{1000}$ A = 6.3 mA.
39. (b) Since load voltage is zoo, the current is zero and diode is open-circuited. If load was open-circuited, load voltage would be about 9 V.
40. (*b*) Current flows during positive half cycle.
41. (*b*) 42. (*b*)
43. (*c*) There are two output waves for one input wave.
44. (*d*) Since both diodes contribute equal load current, each diode current is 5 mA.
45. (*b*) PIV is $\sqrt{2}\times 10 = 14.14$ V.
46. (*c*) Peak tc peak ripple voltage

$$= \frac{I_{dc}}{f_c} = \frac{10\times 10^{-3}}{100\times 1000\times 10^{-6}} = 0.1 \text{ V}.$$

47. (*b*) 48. (*c*)
49. (*c*) Voltage gain is always the ratio of output and input voltages.
50. (*d*)
51. (*a*) Stable Q point is the advantage of self bias circuit.
52. (*d*) Emitter voltage is the same. Increase of emitter resistance decreases emitter current.
53. (*d*) The emitter current becomes one-third. Since collector current in BJT is almost the same as emitter current, collector current also becomes one-third.
54. (*b*)
55. (*d*) dc remains confined within BJT Only ac signal comes out.
56. (*b*) $X_C = \frac{1}{\omega C}$. If ω is very high, X_C is nearly zero.
57. (*b*) 58. (*d*)
59. (*d*) A high signal causes distortion because *v*-*i* characteristics of BJT is non-linear.
60. (*b*) 61. (*d*) 62. (*d*)
63. (*a*) Since output is taken from collector terminal, output voltage is equal to collector current multiplied by collector resistance.
64. (*a*) Input is applied to base with emitter grounded. The input impedance is the ratio of ac base voltage to ac base current.
65. (*c*) 66. (*a*) 67. (*d*)
68. (*c*) Since the first amplifier is feeding the second stage, the input impedance of second stage is the ac collector resistance of first stage.
69. (*d*)
70. (*b*) Since the output is taken across collector resistance, short-circuiting collector resistance will decrease output voltage.
71. (*c*) Capacitor acts as oc for dc.
72. (*c*) Common emitter means emitter at ground potential.
73. (*b*) Since dc is flowing through emitter resistance, dc voltage cannot be zero.
74. (*a*) 75. (*d*)
76. (*c*) When ac collector resistance is very high, the net resistance is the same as dc collector resistance.
77. (*d*) $\frac{6\times 3}{6+3} = 2$ kΩ.

78. (*c*)

79. (*b*) Since ac collector resistance is less than dc collector relistance, slope of ac load line is less than that of dc load line.

80. (*a*) **81.** (*b*)

82. (*b*) Increase of ambient temperature means that heat dissipation is slower

83. (*b*) Surface area of heat dissipating surface increases. Therefore heat dissipation is faster

84. (*c*) CC has high input impelance and low output impedance. Hence suitable for impedance matching.

85. (*c*) **86.** (*c*) 87. (*c*)

88. (*a*) Low load resistance means high current and hence over loading.

89. (*a*)

90. (*b*) When a sinusoidal voltage is clipped on both sides it resembles a square wave.

91. (*c*)

92. (*a*) Since each transistor conducts for 180°. Therefore high efficiency but low distortion.

93. (*b*)

94. (*c*) Because of high switching speed MOSFET is very suitable for digital circuit.

95. (*d*) So that it can couple high impedance source to low impedance load and maximum power can be transferred.

96. (*c*) Low power consumption is a big advantage in digital circuits.

97. (*a*) **98.** (*b*)

99. (*b*) Positive feedback is necessary for sustained oscillations.

100. (*c*)

101. (*c*) The amplifier in the oscillator amplifies the noise voltages. However phase shift around the closed loop is zero at only one frequency. Therefore only this frequency appears at output.

102. (*b*) The initial loop gain must be more than 1.

103. (*d*) **104.** (*a*) **105.** (*c*) **106.** (*c*)

107. (*b*) **108.** (*c*) **109.** (*c*) **110.** (*b*)

111. (*b*)

112. (*c*) For non-inverting terminal input, output is in phase with input. For inverting terminal input, output is 180° out of phase with input. If inputs have 180° phase difference, outputs will be in phase and additive.

113. (*a*) **114.** (*c*)

115. (*c*) Series voltage, series current, shunt voltage and shunt current.

116. (*a*) **117.** (*c*) **118.** (*a*) **119.** (*a*)

120. (*d*) No current can flow through virtual ground. However voltage is zero.

121. (*a*)

122. (*b*) Capacitor in feedback path provides integration.

123. (*a*)

124. (*c*) Integral of constant is *t*.

125. (*a*) **126.** (*b*) **127.** (*b*) **128.** (*a*)

129. (*b*) $\frac{500 \times 500}{500+500} = 250\ \Omega$, slope is $\frac{-1}{250}$.

130. (*a*) **131.** (*a*) **132.** (*d*)

133. (*b*) h_{ie} is expressed in ohms, h_{oe} is an admittance and hence expressed in Siemens, h_{fe} has no units.

134. (*d*) **135.** (*b*)

136. (*c*) $V_{BB} = \frac{5}{20+5}(10) = 2$ V, $R_B = \frac{20 \times 5}{20+5} = 4$ kΩ.

137. (*a*) $I_B = \frac{2-0.5}{4000+50\times1000} \simeq 28\ \mu A$

138. (*c*) $I_C = \beta I_B = 2$ mA.

139. (*a*) **140.** (*b*) **141.** (*c*) **142.** (*b*)

143. (*a*) $V_G = 20 \times \frac{250\times10^3}{2\times10^6+250\times10^3} = 2.22$ V.

144. (*b*) $V_D = 20 - 4 \times 10^{-3} \times 2.5 \times 10^3 = 10$ V.

145. (*c*) $V_S = 4 \times 10^3 \times 1.5 \times 10^{-3} = 6$ V.

146. (*c*) $V_{DS} = V_D - V_S = 10 - 6 = 4$ V.

147. (*a*) Since gate current is negligible $I_S = I_D = 4$ mA.

148. (*b*) $V_S = I_S R_S = 4 \times 10^{-3} \times 1.1 \times 10^3 = 4.4$ V.

149. (*b*) $V_{GS} = -V_S = -4.4$ V.

150. (*c*) **151.** (*b*) **152.** (*b*)

153. (*b*) See Figs. 7.1, 7.2 and 7.3.

154. (*c*) See Eqs. in sections 7.1.1, 7.1.2 and 7.1.3.

155. (*d*) See Eqs. in sections 7.1.1, 7.1.2 and 7.1.3.

156. (*a*) Since only v_1 is positive and more than 0.7 V only D_1 will conduct.

157. (*b*) D1 will conduct and the output voltage will be about 7 V. Therefore D_2 will be reverse biased and will not conduct.

158. (*c*) Both D_1 and D_2 are forward biased.

159. (*c*) Both D_1 and D_2 are forward biased.

160. (*c*) Current in diode D_2 will be more because resistance in series with D_2 is less.

161. (*b*) See section 7.1.8.

162. (*a*) See section 7.1.8 and 7.1.9.

163. (*b*) See Fig. 7.57.

164. (*a*) See Fig. 7.58.

165. (*c*) See Fig. 7.59.

166. (*a*) See Fig. 7.64.

167. (*b*) See Fig. 7.55.

168. (*a*) CMRR $= \frac{2}{5\times10^{-3}} \times 50.$

169. (*b*) Since common mode signal is reduced to zero, CMRR is infinite.

170. (*d*) Slew rate signifies the fastest change which can occur in output voltage. In an ideal op-amp the output can change instantaneously.

171. (*a*) An ideal op-amp has identical transistors and hence does not require input offset voltage.

172. (*a*) An ideal op-amp has infinite open loop gain.

173. (*a*) An ideal op-amp has zero output impedance and can deliver any amount of load current without any voltage drop.

174. (*c*) See Fig. 7.54.

175. (*b*) Voltage input to inverting input is $12 \times \frac{10}{20} = 6$ V. Therefore v_i should be more than 6 V.

176. (*a*) See section 7.7.18.

177. (*a*) See Fig. 7.66.

178. (*b*) See Fig. 7.47.

179. (*c*) Only LC oscillators are suitable for MHz range.

180. (*a*) An ideal power supply should have zero voltage drop.

181. (*d*) **182.** (*b*) **183.** (*d*) **184.** (*a*)

185. (*b*) **186.** (*b*) **187.** (*c*)

188. (*d*) Due to the presence of virtual ground at input, the resistance in the series path of input of inverting amplifier is input impedance.

189. (*a*) **190.** (*b*) **191.** (*c*)

192. (*a*) The circuit is a second order active low pass filter for which gain is 40 dB/decade.

193. (*d*) For bridge rectifier, peak inverse voltage is equal to peak value of ac voltage while for the circuit using centre tapped transformer peak inverse voltage is $2E_m$.

194. (*b*) Peak inverse voltage can be up to $2E_m$ since capacitor can store charge equal to E_m.

195. (*d*) See Fig. 7.59.

196. (*b*) **197.** (*b*)

198. (*d*) Diode will conduct when instantaneous value of ac voltage is more than 50 V.

199. (*c*) Diode will always be reverse biased.

200. (*b*)

201. (*a*) A stabilizer must give constant output voltage. This is possible only if internal resistance is low.

202. (*c*) **203.** (*a*) **204.** (*c*)

205. (*d*) If $Z_1 << Z$, it resembles Fig. 7.54.

206. (*d*) It is a biased limiter circuit. During positive half clipping is at 5 V due to battery.

207. (*c*) Output voltage is $\simeq 6.3 + 0.7$ V.

208. (*a*) Closed loop gain = $\frac{60}{1+60\times 0.05} = 15$.

209. (*d*) Voltage gain

$$\simeq \frac{h_{fe}\,R_E}{h_{ie}+h_{fe}R_E} = \frac{100\times 66\times 1000}{1000+66000\times 100} \simeq 1.$$

210. (*d*) $i = 14 + \frac{6-2}{4-2} \times 20 = 54$ mA.

211. (*c*) See section 7.6.

212. (*d*) Max. zener current = $\frac{600}{24} = 25$ mA, $R = \frac{32-24}{25\times 10^{-3}}$ = 320 ohm. Maximum load current = 25 – 10 = 15 mA.

213. (*a*) Clamper does not amplify the waveform. Hence 1 and 2 are correct.

214. (*b*) See section 7.2.

215. (*a*) PIV= 10 V.

216. (*b*) During positive half cycle current

$I = \frac{10-0.7-I\times 20}{1000}$ or $I \simeq 9.12$ mA. During negative half cycle diode is reverse biased and current is zero.

217. (*a*) Minimum current = $\frac{20-10}{1000}$ A =10 mA, Maximum current = $\frac{40-10}{1000}$ A or 30 mA.

218. (*b*) **219.** (*d*)

220. (*a*) Closed loop input impedance = 4 (1 + 0.1 × 90) = 40 kΩ, closed loop output impedance = $\frac{80}{1+0.1\times 90}$ = 8 kΩ.

221. (*a*) $I_C \simeq \frac{20-0.6}{5.4+\frac{400}{99}} \simeq 1.98$ mA,

$V_C = 20 - 1.98 \times 10^{-3} - 5.4 \times 10^3 \simeq 9.3$ V.

222. (*a*)

223. (*b*) $I_{dc} = \frac{5}{\pi\times 1000}\ \text{A} = \frac{5}{\pi}$ mA.

224. (*a*) Only one diode conducts at one time.

225. (*c*) $v_0 = v_i \frac{(4R+R)}{R} = 5\,v_i$.

226. (*b*) $v_0 = -10\left(\frac{3}{5}+\frac{2}{5}\right)$.

227. (*b*)

228. (*a*) $V_{dc} = \frac{20}{\pi}$ and V = 0.5 × 20 = 10 V

229. (*a*) $V_{dc} = \frac{2\times 20}{\pi}$, $V = \frac{20}{\sqrt{2}}$.

230. (*b*) $V_{dc} = \frac{120}{4}\left(\frac{\sqrt{2}}{\pi}\right) = 13.5$ V.

231. (*b*) In full wave circuit using centre tapped transformer PIV = 2 V_m.

232. (*b*) $V_{dc} = \frac{2 \times 40}{\pi} = 25.5$ V.

233. (*a*) There are two output waves in one cycle of input wave.

234. (*a*) Output wave in each case has two cycles for one cycle of input wave.

235. (*b*)

236. (*c*) No load output voltage $= \frac{V_m}{\pi}$, output voltage at full load $= \frac{V_m(50)}{\pi(50+50)}$.

237. (*c*) No load output voltage $= \frac{2V_m}{\pi}$, output voltage at full load $= \frac{2V_m}{\pi}\left(\frac{50}{50+50}\right)$.

238. (*d*) No load output voltage $= \frac{2V_m}{\pi}$, output voltage at full load $= \frac{2V_m}{\pi}\left(\frac{50}{2\times 50+50}\right)$.

239. (*c*)

240. (*c*) See Figs. 7.9 and 7.10.

241. (*b*) **242.** (*c*)

243. (*b*) It is a high voltage device. However current has to be low since diode rated currents are low.

244. (*b*) Because capacitor charge is retained.

245. (*d*) Time constant $R_L C$ must be about 5 × 20 ms.

246. (*b*) **247.** (*c*)

248. (*d*) One filter reduces ripple to 0.15 of initial value. The second filter reduces ripple to 0.15 × 0,15 of initial value.

249. (*b*)

250. (*b*) If load resistance is large; current would be low and voltage drop would be low.

251. (*b*) See Fig. 7.11

252. (*d*) Since voltage quadrupler circuit uses more components, voltage drop will be more and voltage regulation poorer.

253. (*a*) A good power supply does not allow output voltage to vary by more than 1%.

254. (*b*) When $R_L = \infty$, $I_L = 0$, When $R_L = 100\ \Omega$, $I_L = \frac{12}{100}$ A or 120 mA.

255. (*a*) $I_R = \frac{25-12}{100}$ A.

256. (*b*) When $I_L = 0$, Zener current = 130 mA, when $I_L = 120$ mA, zener current = 10 mA.

257. (*d*) Output voltage = 12 V, $I_R = \frac{V_i - 12}{R}$.

258. (*b*) $V_0 = 12 + I_z \times 5$, Max $I_z = 130$ mA, minimum $I_z = 10$ mA.

259. (*b*)

260. (*c*) See Fig. 7.18.

261. (*a*) D_1 is forward biased when $v_i > V_1$.

262. (*c*) **263.** (*a*) **264.** (*b*) **265.** (*b*)

266. (*a*)

267. (*c*) Capacitor offers infinite impedance to dc.

268. (*c*) **269.** (*c*)

270. (*c*) Bridge rectifier has the advantage of lower PIV. Moreover it does not require centre tapped transformer.

271. (*a*) X_L is zero for dc.

272. (*a*) Resistor reduces surge current.

273. (*c*) Zener breakdown when input voltage is more than zener breakdown voltage.

274. (*b*)

275. (*d*) $I_B = \frac{10-0.7}{10^6} = 9.3\ \mu A$.

276. (*c*) $I_C = 10 \times 100\ \mu A = 1$ mA.

277. (*c*) $I_C = 1$ mA and $V_{CE} = 20 - 5 \times 10^3 \times 1 \times 10^{-3} = 15$ V.

278. (*a*)

279. (*a*) $I_B \simeq \frac{30-0.7}{300\times 10^3}$ A $\simeq 97\ \mu A$.

280. (*c*) Since $I_B \simeq 100\ \mu A$, $I_C = \beta I_B \simeq 10$ mA.

281. (*a*) **282.** (*b*) **283.** (*a*)

284. (*a*) When transistor is in cutoff region $V_{CE} \simeq V_{CC}$.

285. (*a*) When transistor is cutoff, I_C is zero.

286. (*a*) Capacitor has infinite impedance for dc.

287. (*c*) Since $R_{B2} < R_{B1}$, $I_{B2} > I_{B1}$.

288. (*c*) $\frac{20-2}{100\times 1000}$ A.

289. (*b*)

290. (*c*) Since $I_E \simeq I_C$, $I_C = \frac{V_{CC}}{R_C + R_E}$.

291. (*a*) Since transistor is cutoff, $I_C = 0$ and $V_C = V_{CC}$.

292. (*d*) Voltage across $R_E = \frac{30\times 100}{200+100} - 0.7$.

293. (*b*) Current through $R_E = \frac{10-0.7}{10\times 10^3} = 0.93$ mA.

294. (*b*)

295. (*a*) Lowest value of β should be used to locale Q point.

296. (*a*)

297. (*d*) All voltages and currents have reverse polarity.

298. (*a*) Current is maximum at saturation point.

299. (*c*)

300. (*a*) Voltage divider bias stabilizes I_C.

301. (*a*)

302. (*a*) X_C should be less than 10% of 10 kΩ. There $X_C < 1000\ \Omega$.

303. (*c*) **304.** (*c*) **305.** (*a*)

306. (*b*) Capacitor is open-circuit for dc.

307. (*a*) $r_e = \frac{\eta T}{11600\ I}$.

308. (*d*) $I_E = \frac{10}{10 \times 10^3} = 1$ mA, $r_e = \frac{0.025}{1 \times 10^{-3}} = 25\Omega$.

309. (*a*) Voltage gain $\simeq \frac{5000}{25} = 200$.

310. (*c*) **311.** (*c*) **312.** (*a*) **313.** (*c*)

314. (*c*) **315.** (*c*)

316. (*b*) Current gain of CB amplifier is α.

317. (*c*) $r_e = \frac{0.025}{3.5 \times 10^{-3}}$. Input impedance of CB amplifier equals r_e.

318. (*c*) A_v of CB amplifier $\simeq \frac{\alpha R_L}{r_e} = \frac{0.98 \times 600}{6} = 98$.

319. (*c*) Input impedance of CE amplifier is $\beta r_e = 100 \times 6 = 600\ \Omega$.

320. (*a*)

321. (*c*) $X_C < 0.1 \times 100$ or $X_C < 10\ \Omega$ or $10 = \frac{1}{2\pi(10)\ C}$ or $C \simeq 1000\ \mu F$.

322. (*a*) The drift is amplified by different stages.

323. (*a*) **324.** (*b*) **325.** (*a*)

326. (*a*) Emitter is at ground potential for ac signals.

327. (*a*)

328. (*c*) Due to direct coupling between output and CRO both dc and ac output voltage will appear on screan of CRO

329. (*d*) **330.** (*c*) **331.** (*a*) **332.** (*a*)

333. (*a*) Therefore it is suitable for impedance matching.

334. (*b*) **335.** (*b*) **336.** (*a*)

337. (*b*) AC saturation current is cutoff voltage × slope of ac load line.

338. (*c*)

339. (*c*) A phase splitter gives TWO voltages which have equal magnitudes but are out of phase by 180°.

340. (*b*) **341.** (*d*) **342.** (*b*) **343.** (*b*)

344. (*a*) $\frac{T - 30}{5} = 10$ or T = 80°C.

345. (*c*) Derating improves reliability but does not affect maximum power dissipation.

346. (*c*) Power rating = $8 - (55 - 25)(30 \times 10^{-3}) = 7.1$ W.

347. (*a*) Quiescent-operation means operation when ac signal is not applied.

348. (*a*) **349.** (*a*) **350.** (*b*) **351.** (*a*)

352. (*c*) In class A operation I_C exists for 360°. In class B operation I_C exists for 180°. In between there is the class AB.

353. (*a*) **354.** (*a*)

355. (*a*) $I_1 = \frac{20 - 0.7}{10 \times 1000}$ A = 19.3 mA.

356. (*c*) **357.** (*c*) **358.** (*a*) **359.** (*b*)

360. (*c*) **361.** (*b*) **362.** (*a*)

363. (*c*) Since collector current exists for less than 180°, it looks like narrow pulses.

364. (*d*) Since efficiency of class C operation is high, maximum load power is delivered.

365. (*c*)

366. (*c*) $f = \frac{1}{0.1 \times 10^{-6}} = 10^7$ Hz,

$$\text{Duty cycle} = \frac{250 \times 10^3}{10^{-6}} \times 100 = 2.5\%.$$

367. (*a*) **368.** (*a*) **369.** (*a*)

370. (*a*) $10^{12} = \frac{1}{2\pi\sqrt{LC}}$ or L= 25 μH.

371. (*a*) **372.** (*a*) **373.** (*a*) **374.** (*c*)

375. (*b*) Duty cycle

$$= \frac{\text{Pulse width} \times 100}{\text{Time period}} = \frac{0.006 \times 100}{1} = 0.6\%.$$

376. (*a*) **377.** (*a*) **378.** (*c*) **379.** (*a*)

380. (*b*)

381. (*b*) Gain in dB = 10 log (4000).

382. (*a*) To provide impedance matching input impedance should be high and output impedance should below.

383. (*b*) Because of high input impedance and low output impedance, CC amplifier is suitable as buffer amplifier.

384. (*a*) **385.** (*a*) **386.** (*a*)

387. (*a*) Gain is proportional to g_m.

388. (*b*) **389.** (*c*)

390. (*b*) Critical frequency of RC circuit = $\frac{1}{2\pi RC}$.

391. (*b*)

392. (*b*) The two resistances are in parallel. Critical

$$\text{frequency} = \frac{1}{2\pi(0.5R)\ C} = \frac{1}{\pi\ RC}.$$

393. (*d*) **394.** (*d*) **395.** (*c*) **396.** (*c*)

397. (*c*) **398.** (*b*) **399.** (*b*) **400.** (*c*)

401. (*d*)

402. (*d*) Total $I_E = \dfrac{V_{EE} - 0.7}{R_E + \dfrac{R_B}{2\beta_{dc}}} \simeq \dfrac{V_{EE}}{R_E}$ = 2 mA. It divides equally between the two transistors.

403. (*d*) Both collector terminals are at the same potential.

404. (*d*) In oscillators, the oscillations are generated due to positive feedback. In thyristors anode current increases due to positive feedback.

405. (*b*) Each RC network gives 60° phase shift.

406. (*c*) **407.** (*d*) **408.** (*a*) **409.** (*a*)

410. (*a*) **411.** (*a*) **412.** (*c*)

413. (*c*) $A_{vf} = \dfrac{-1000}{1+(-1000)(-.1)}$ = 9.9. When gain changes,

$A_{vf} = \dfrac{-800}{1+(-800)(-0.1)}$ = 9.877. Percentage change

in $A_{vf} = \dfrac{9.9-9.877}{9.9} \simeq 0.2\%$.

414. (*b*) Maximum dc output offset voltage $= (10)\left(\dfrac{10+1}{1}\right)$

=110 mV dc.

415. (*c*)

416. (*d*) It is a common base circuit.

$A_v \simeq \dfrac{-(-0.999)\times 10\times 10^3}{25} \simeq 400.$

417. (*b*) See section 7.1.9.

418. (*c*) **419.** (*b*) **420.** (*d*)

421. (*c*) Crystal oscillator has the advantage of frequency stability.

422. (*c*) Since battery connected to anode of D_2 is of 5 V it is forward biased. D_1 and D_3 are reverse biased. Output voltage is 5 V.

423. (*b*) **424.** (*b*) **425.** (*c*)

426. (*a*) If f_{OL} is less than f_{OH}, a pass band will exist from f_{OL} to f_{OH}.

427. (*b*) LC oscillators are used for radio frequencies.

428. (*b*) Input to non-inverting op-amp is $-10 \times 10^{-6} \times 10^3$ = −10 mV. Therefore output $= (-10)\left(\dfrac{90+10}{10}\right)$ = −100 mV.

429. (*b*) $T = \dfrac{1}{200\times 10^3}$ s = 5 μs.

Duly cycle $= \dfrac{0.1\times 100}{5} = 2\%$.

430. (*c*)

431. (*c*) $v_0 = \dfrac{-10}{1}v_i + \dfrac{10+1}{1}v_i = v_i$.

432. (*a*) For ac, R_L and R_C appear in parallel. Hence $\dfrac{1}{R} = \dfrac{1}{R_L} + \dfrac{1}{R_C}$ and slope is $\dfrac{-1}{R}$.

433. (*a*) Base voltage $= \dfrac{-10\times 16}{16+4}$ = − 8 V. Emitter voltage = −8 + 0.7 = −7.3 V.

434. (*a*)

435. (*d*) Class C has more efficiency than all other classes.

436. (*a*) $20 = \dfrac{100}{1+100(\beta)}$ or (β) = 0.04 or 4%.

437. (*b*) $Q = \dfrac{f_r}{\text{bandwidth}}$. If Q is doubles, bandwidth becomes half.

438. (*b*) **439.** (*a*)

440. (*b*) $v_0 = 1\times\dfrac{R+R}{R} = 2$ V.

441. (*c*) **442.** (*d*)

443. (*a*) It is a differentiating amplifier.

444. (*a*) **445.** (*c*) **446.** (*b*)

447. (*b*) Output resistance $= \dfrac{\text{full load output voltave}}{\text{output current}} = 25$,

Voltage regulation $= \dfrac{30-25}{25} = 20\%$.

448. (*c*) Input impedance with feedback = (1 kΩ) (1 + 100 + .99) = 100 kΩ. Output impedance with feedback

$= \dfrac{100}{1+100\times 0.99} \simeq 1\Omega.$

449. (*b*) **450.** (*d*) **451.** (*a*) **452.** (*c*)

453. (*d*) **454.** (*c*)

455. (*b*) For saturation both junctions have to be forward biased.

456. (*c*) Efficiency is 50%. Therefore input is 20 W.

457. (*c*) Thermal resistance $= \dfrac{125-25}{100\times 10^{-3}}$ = 1000°C/W.

458. (*c*) $8 \times \left(\dfrac{40}{N_1}\right)^2 = 2$ or $N_1 = 80$.

459. (*c*) **460.** (*c*)

461. (*c*) Only 3 is correct. 1 is wrong, 4 is wrong.

462. (*d*) In ECL logic gate, transistor works in active and cutoff regions. In other digital gates it works in saturation and cutoff regions. Hence all the three regions are correct. Hence (*d*).

463. (*c*) These are the characteristics of a full wave rectifier. Hence only 1 and 2 are correct.

464. (*d*) $V_0 = \left(1+\dfrac{3R}{R}\right)v_x = 4v_x$

v_x due to −4 V input $= \dfrac{R}{2R}(-4) = -2$ V

v_x due to (4 − 2 cos ωt) input

$= \dfrac{R}{2R}(4 - 2\cos\omega t) = 2 - \cos\omega t$ volts

Total $v_x = -2 + 2 - \cos \omega t = -\cos \omega t$

$v_0 = 4\, v_x = -4 \cos \omega t$

465. (*c*) It is half wave voltage doubler circuit (Fig. 7.9) $v_0 = 2\, V_m$.

466. (*b*) It is a monostable multivibrator having $t = T = 1.1\, R_C$.

467. (*a*) The circuit works as full wave rectifier.

The path of current is

Positive half cycle R_1, 6.6 kΩ, D_2

Negative half cycle R_2, 6.6 kΩ, D_1

In each half cycle 6.6 kΩ and 2.2 kΩ resistance are in parallel $\dfrac{6.6 \times 2.2}{6.6 + 2.2} = 1.65$ kΩ

$$V_m = \frac{10 \times 1.65}{2.2 + 1.65} = 4.3 \text{ V.}$$

468. (*a*) Oscillator works on feedback principle.

Emitter follower has voltage series feed back.

469. (*b*) In npn, electrons are majority carriers while in pnp holes are majority carriers. Mobility of electrons is about 3 time that of holes.

470. (*c*) DC sources are shorted (See Fig. 7.24)

471. (*b*) Power amplifier in last stage not as front end.

472. (*c*) Class B power amplifier has higher efficiency than class A. Since only one transistor is on at one time, standly power dissipation is absent. Due to cut in voltage crossover distortion in present.

473. (*a*) 1, 2 and 3 are most important.

474. (*a*) Only coupling capacitor and emitter bypass capacitor control the lower cutoff frequency. Stray capacitances affect higher frequencies only.

475. (*d*) IC 723 is a regulated power supply.

476. (*d*) all are correct.

477. (*a*) Bias stability $= (1+\beta)\left(\dfrac{1 + R_B/R_E}{1 + \beta + R_B/R_E}\right)$

If $\beta >> \dfrac{R_B}{R_E}$

Bias stability $\simeq 1 + \dfrac{R_B}{R_E}$

Thus decreasing R_B and increasing R_E improves bias stability because bias stability $\simeq 1$.

478. (*d*) Emitter resistor gives negative feedback which decreases gain and increases stability. Reducing emitter resistance to zero increases gain and decreases stability.

479. (*d*) Schmilt trigger converts the wave to square wave.

480. (*d*)

481. (*c*) See Table 7.4.

482. (*b*) Voltage series feedback increases input impedance and decreases output impedance (See Table 7.7).

483. (*b*) See section 7.6.5.

484. (*d*) AC ripple should be as low as possible in a good power supply.

485. (*b*) Current to voltage converter is a linear application.

486. (*c*) At fixed temperature thermistor has linear *v-i* curse.

487. (*d*) Both functions should be forward based (See Table 2.2).

488. (*c*) Bias stabilisation is only to keep operating point fixed.

489. (*a*) Onset of non-linearity limits amplitude of oscillations.

490. (*b*) Tuned amplifier is a narrow band amplifier.

491. (*c*) Closed loop gain

$$= \frac{A_v}{1 + A_v\beta} = \frac{-100}{1 + (-100)\left(-\dfrac{4}{100}\right)} = -20.$$

492. (*d*) Gain $= \dfrac{-100}{1 + (-100)\left(-\dfrac{1}{100}\right)} = \dfrac{-100}{11} = \left|\dfrac{100}{11}\right|$

Lower cutoff frequency $= \dfrac{1000}{11}$ Hz

Upper cutoff frequency = 100 × 11 kHz = 1.1 MHz

dB of feedback $= \log_{10} B = 20 \log_{10}(0.1)$

Hence (*d*) is not correct.

493. (*a*) Output $= \dfrac{I_m V_m}{2}$, dc power input $= \dfrac{2 I_m V_{CC}}{\pi}$

$$\eta = \left(\frac{I_m V_m}{2} \Big/ \frac{2 I_m V_{CC}}{\pi}\right) \times 100 = \frac{\pi}{4} \frac{V_m}{V_{CC}} \times 100\%$$

494. (*d*) All are needed. Stepdown transformer to step down the voltage, diodes convert ac to dc, resistor and capacitor for filtering and stabilizer to stabilize the voltage.

495. (*b*) It is a voltage doubler. See Fig. 7.10.

496. (*d*) At high frequency, the reactance of coupling capacitor is very small and the coupling capacitor acts as a short circuit.

497. (*c*) It is a monostable multivibrator. It is also called one-shot.

498. (*a*) It is seen that positive peak has been applied.

499. (*c*) Preamplifier is a small signal amplifier. Power amplifier is a large signal amplifier. Purely resistive circuit is lumped, linear, bilateral.

500. (*b*) $\eta = \dfrac{25(15-1)}{15} = 23.33\%$

The figure 25 indicates the theoretical maximum efficiency of class A amplifier.

501. (*b*) It is a current mirror circuit $9.3 \times 10^3\, I_1 + 0.7 = 0 - (-10) = 10$ or $I_1 \simeq 1$ mA

Since emitter area of Q_1 is half that of Q_2 $I_0 = 2I_1 = 2$ mA.

502. (*c*) Since there is a virtual short circuit at input of op-amp $v_a = 0$. Then R_2 and R_3 are effectively in parallel

Therefore $\frac{v_i - 0}{R_1} = \frac{0 - V_0}{R_2 \| R_3}$ or $\frac{V_0}{V_i} = \frac{R_2 \| R_3}{R_1}$ = voltage gain.

503. (*b*) When $V_i = -10$ V

$$\frac{V_0 - 0}{R} = \frac{0-20}{4R} + \frac{(-10)}{R} \text{ or } V_0 = -5 + 10 = 5 \text{ V}$$

When $V_i = -5$ V

$$\frac{V_0 - 0}{R} = \frac{0-20}{4R} + \frac{0-(-5)}{R} \text{ or } V_0 = -5 + 5 = 0 \text{ V}$$

When $V_i > -5$ both diodes conduct and $V_0 = 0$.

504. (*b*) The equivalent circuit is as under

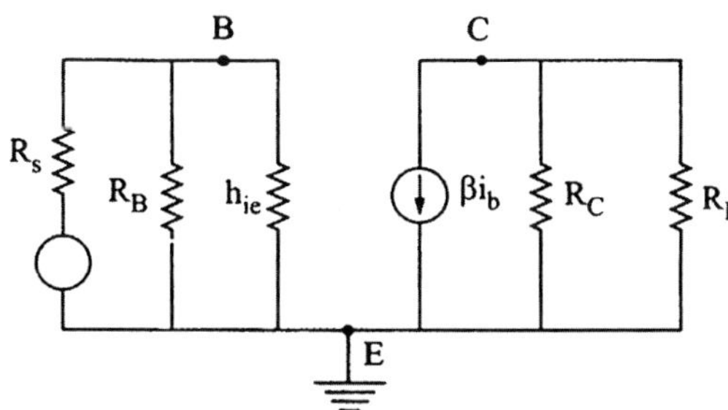

$$h_{ie} = \frac{\beta}{g_m} = \frac{100}{0.3861} = 259 \text{ ohm}$$

Resistance as seen by source = $R_s + (R_B \| h_{ie})$

$$= 1000 + \frac{93000 \times 259}{93000 + 259} = 1258 \text{ ohm}$$

505. (*b*) Lower cutoff frequency due to C_2

$$= \frac{1}{2\pi(R_C + R_C)C_2}$$

$$= \frac{1}{2\pi(250+1000)(4.7 \times 10^{-6})} = 27.1 \text{ Hz}$$

506. (*a*) When diode is conducting $V = V_i$

When diode is not conducting current is 1 A and V = 1 volt.

507. (*b*) $i = I_0(e^{V/V_T} - 1) = \frac{0-(-1)}{100 \times 10^3} = 10^{-5}$

or $10^{-6}\left[e^{\frac{V}{25 \times 10^{-3}}} - 1\right] = 10^{-5}$ or V = 0.06 V

$$\frac{V_0 - V}{4 \times 10^3} = 10^{-5} \text{ or } V_0 = 4 \times 10^3 \times 10^{-5} + 0.06$$

or $V_0 = 0.04 + 0.06 = 0.1$ V

508. (*a*) Since there is virtual shot circuit at input of op-amp

$$\frac{V_i}{R_1 + sL} = \frac{-V_0}{\left(R_2 \times \frac{2}{Cs}\right) \Big/ R_2 + \frac{1}{Cs}} = \frac{-V_0}{R_2/1 + R_2C_2s}$$

or $$\frac{V_0}{V_i} = \frac{-R_2}{(R_1 + sL)(1 + R_2C_2s)}$$

This expression for $\frac{V_0}{V_i}$ is the transfer function of low pass filter.

509. (*c*) Let V be the voltage at non-inverting terminal Applying KCL at non-inverting terminal

$$\frac{15 - V}{10 \times 10^3} + \frac{V_0 - V}{10 \times 10^3} = \frac{V - (-15)}{10 \times 10^{-3}}$$

or $15 - V + V_0 - V = V + 15$ or $V = \frac{V_0}{3}$

Since V_0 swin s between +15 and –15, V swings between +5 and –5 V.

510. (*c*) Using voltage division rule, $V_x = 0.5$ V

$$V_0 = 1\left(\frac{-2k}{1k}\right) + 0.5\left(1 + \frac{2k}{1k}\right) = -2 + 1.5 = -0.5 \text{ V}$$

511. (*b*) Both are correct but independent.

512. (*a*) Since output consists of half sinusoids, output frequency is twice the input frequency.

513. (*b*) Both are correct but independent.

514. (*a*) **515.** (*a*)

516. (*c*) R is not correct.

517. (*a*) It is a comparator circuit.

518. (*a*)

519. (*a*) Since the characteristics is non-linear, graphical analysis is needed.

520. (*a*)

521. (*c*) There is 180° phase difference between input and output.

522. (*a*) **523.** (*a*)

524. (*b*) Both are correct but independent.

525. (*a*) **526.** (*a*)

527. (*b*) Both are correct but independent.

528. (*a*) At high frequencies X_C is very small.

529. (*a*) **530.** (*a*)

531. (*b*) Both are correct but independent.

532. (*d*) A is wrong because negative feedback increases the bandwidth.

533. (*c*) CE amplifier has good voltage gain.

534. (*a*)

535. (*b*) Both are correct but independent.

536. (*a*) **537.** (*a*)

538. (*b*) Both are correct but independent.

539. (*a*)

540. (*b*) Both are correct but independent.

541. (*c*) R is not correct because bandwidth of class C amplifier is narrow.

542. (*b*) Both ace correct but independent.

543. (*b*)

544. (*b*) Both are correct but independent.

545. (*b*)
546. (*c*) R is wrong.
547. (*b*) **548.** (*b*) **549.** (*b*)
550. (*b*) Both are correct but independent.
551. (*c*)
552. (*c*) Negative feedback reduces output impedance.
553. (*b*) Both are correct but independent.
554. (*c*) Large value capacitor cannot be fabricated on a chip.
555. (*d*) A is wrong.
556. (*b*) Both are correct but independent.
557. (*b*) **558.** (*b*) **559.** (*b*)
560. (*b*) Both are correct but independent.
561. (*d*) Small signal analysis is not used to obtain conversion efficiency.
562. (*a*)
563. (*c*) A transistor can be brought from hard saturation to cutoff state but it takes more time. Therefore it is not possible in fast switching circuits.
564. (*b*) Both are correct but independent.
565. (*a*)
566. (*c*) R is not correct.
567. (*a*)
568. (*d*) Self bias circuit can be used.
569. (*d*) Additive technique cannot give wide band amplifier.
570. (*d*) Dashington pair used common collector connection.
571. (T)
572. (F) Time constant should be large.
573. (F) Ripple factor of half wave rectifier is 1.21.
574. (T) **575.** (T) **576.** (T)
577. (F) Efficiency of full wave rectifier with centre tapped transformer is slightly higher than that of bridge rectifier.
578. (T)
579. (F) As $\frac{R_f}{R_L}$ increases efficiency decreases.
580. (T)
581. (F) Base current is small.
582. (F) Stray capacitances assume importance at high frequencies because $X_C = \frac{1}{\omega C}$.
583. (T)
584. (F) The units of h_{11} are ohms and units of h_{22} Siemens.
585. (T) Due to drift. **586.** (T)
587. (F) Maximum efficiency for class B-operation can be 78.5%.
588. (T) **589.** (T) **590.** (T) **591.** (T)
592. (F) Transformer coupling increases efficiency.
593. (F) It is replaced by short circuit.
594. (T) **595.** (T)
596. (F) It is lower.
597. (F) FET amplifier is easier to analyse since gate current is negligible.
598. (T)
599. (F) g_m is about 0.1×10^{-3} Siemens.
600. (T)
601. (F) Feedback factor is much less than 1.
602. (T) **603.** (T) **604.** (T) **605.** (T)
606. (T)
607. (F) It has no effect on noise originating at amplifier input.
608. (T) **609.** (T) **610.** (T) **611.** (T)
612. (T) **613.** (T)
614. (F) Forward resistance of diode is low.
615. (F) It consists of two CE amplifiers.
616. (F) It is about I MHz.
617. (T) **618.** (F) **619.** (F)
620. (F) An ideal op-amp has infinitely fast slew rate.
621. (T)
622. (F) It has capacitance in feedback path.
623. (T) **624.** (T)
625. (F) It is very high.
626. (T) **627.** (T)
628. (F) It has resistance in feedback path.
629. inphase **630.** stabilise, gain
631. 180°
632. voltage, zero **633.** analog, digital
634. zero **635.** open loop gain
636. monolithic, hybrid **637.** zener diode
638. high impedance **639.** constant
640. smoothness **641.** exponentially
642. zero **643.** common collector
644. input impedance **645.** high
646. high frequencies **647.** coupling capacitance
648. radio frequency isolation
649. non-linear **650.** radio frequencies
651. 180° **652.** instability
653. high frequencies **654.** zero, 360°
655. 25 **656.** reduction
657. gain stability **658.** base
659. one **660.** zero
661. common collector **662.** base
663. darlington pair **664.** drift
665. input signal **666.** clipped
667. quiescent values **668.** saturation, cutoff
669. derate **670.** 25%
671. centre **672.** heat
673. transconductance **674.** integrated circuits
675. even harmonics **676.** $0, -2V_m$
677. audio frequencies **678.** source
679. common drain **680.** high

CHAPTER 8

POWER ELECTRONICS

8.1. THYRISTOR

Thyristor (also called silicon controlled rectifier or SCR) consists of 4 alternate *p* and *n* layers (*i.e.*, *p-n-p-n*) forming three *p-n* junctions. The anode terminal is outside the *p* layer and cathode is outside the *n* layer (Fig. 8.1*a*). A contact welded to inner *p* layer (*i.e.*, p_2) forms the gate. Its symbol is as shown in Fig. 8.1. (*b*).

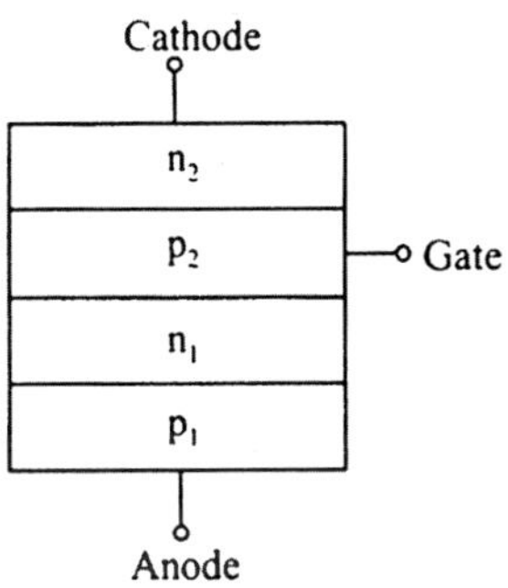

(*a*) Configuration

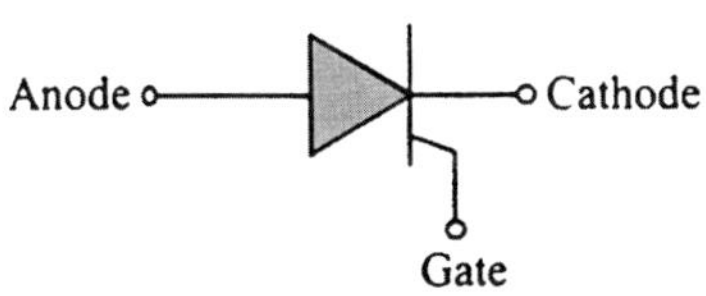

(*b*) Symbol

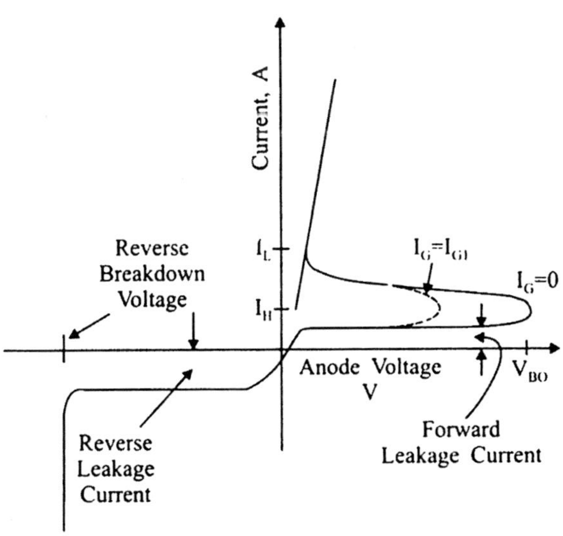

(*c*) v-i characteristics

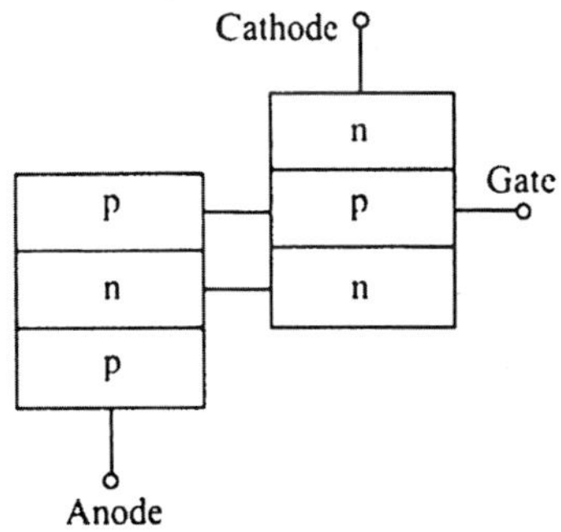

(*d*) Two transistor model

Fig. 8.1. Thyristor

The *v-i* charatcteristic of a thyristor is shown in Fig. 8.1 (*c*). The beheviour of the Thyristor can be explained by two transistor model (Fig. 8.1 *d*).

Thyristor can be turned on by applying a positive gate signal. Other triggering methods are $\frac{dv}{dt}$ triggering, high temperature triggering and light triggering. When a thyristor is triggered the gate loses control. It can be turned off by decreasing the current to less than holding current. The turn on time of Thyristors is less than about 3 μ-s and turn off time is between 10-300 μ-s.

Thyristors are available upto about 10 kV and 1200 A rating. For high voltage and high current circuits series and parallel connection of thyristors are used. When connected in series, the voltage across the thyristors are not equal. When connected in parallel, the currents through them are non-equal. Therefore, thyristors in series need static and dynamic equalising circuits. Derated operation improves reliability. A snubber circuit is used to limit $\frac{dv}{dt}$ across the thyristor.

It is necessary to turn on and turn off a thyristor at proper instant. The turning on is done by applying a positive gate signal. The turn off methods are called commutation methods. Commutation can be natural or forced. When an SCR is connected to an ac source, the current goes through its natural current zero at the end of every half cycle and a reverse voltage automatically appears across it. This is called natural commutation. In dc circuits there is no natural current zero and forced commutation is necessary. Forced commutation methods are Resonant commutation (Class A commutation), Self commutation (Class B commutation), Auxiliary commutation

(Class C commutation), Complementary commutation (Class D commutation) and external pulse commutation (Class E commutation). Natural commutation is also called line commutation or class F commutation.

8.2. THYRISTOR FAMILY

(*a*) DIAC. A DIAC is a two electrode device and can conduct in either direction. Its construction, symbol and *v-i* characteristics are as shown in Fig. 8.2. It is used in triggering TRIACS.

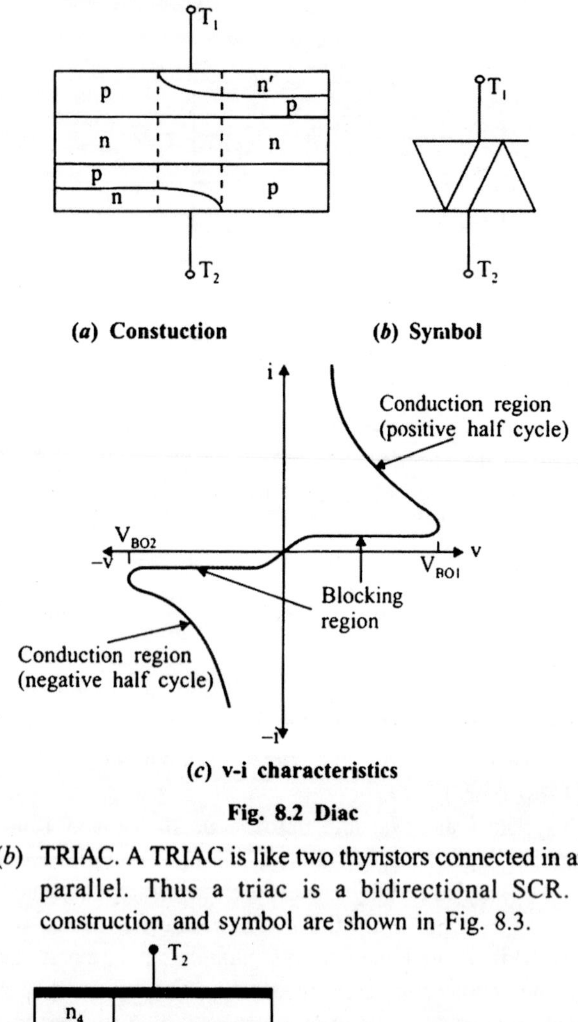

(*a*) Constuction (*b*) Symbol

(*c*) v-i characteristics

Fig. 8.2 Diac

(*b*) TRIAC. A TRIAC is like two thyristors connected in anti-parallel. Thus a triac is a bidirectional SCR. Its construction and symbol are shown in Fig. 8.3.

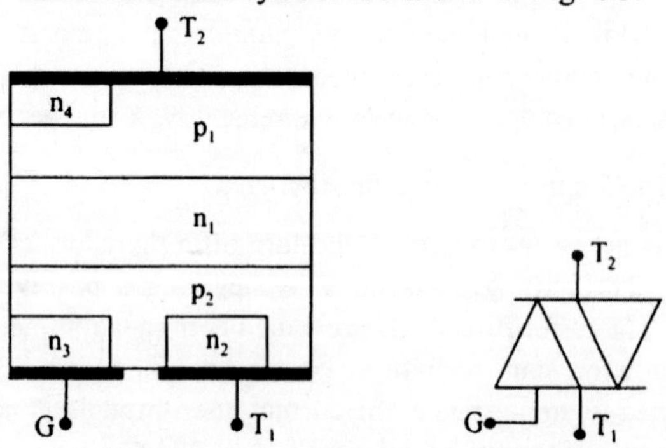

(*a*) Construction (*b*) Symbol

Fig. 8.3 Triac

(*c*) Amplifying Gate Thyristor (Fig. 8.4). It has a main thyristor Th_m and auxiliary thyristor (Th_a). The auxiliary thyristor Th_a is turned on by gate signal. The amplified output of Th_a acts as a gate signal for main Thyristor Th_m

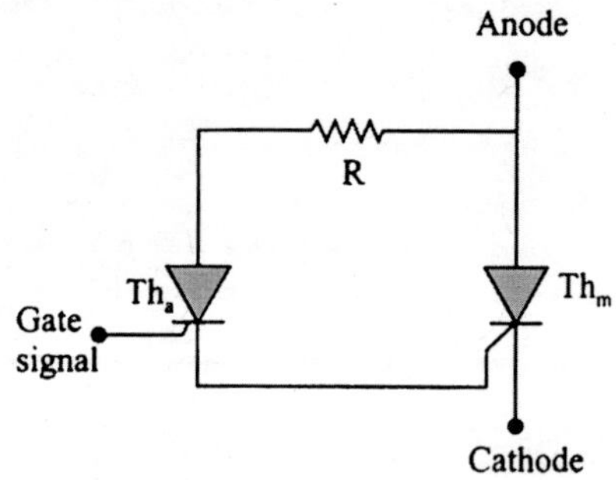

Fig. 8.4. Amplifying gate thyristor

(*d*) Gate turn off Thyristor (GTO). It is a special thyristor which can be turned on by a positive gate signal and turned off by negative gate signal. Figure 8.5 shows two symbols for GTO.

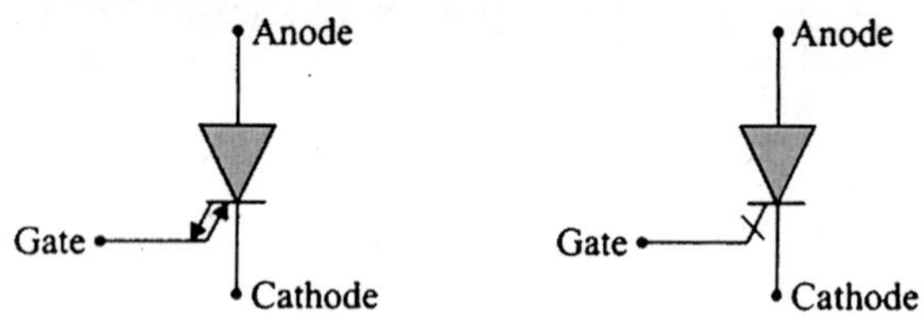

Fig. 8.5. Symbols for gate turn off thyristor

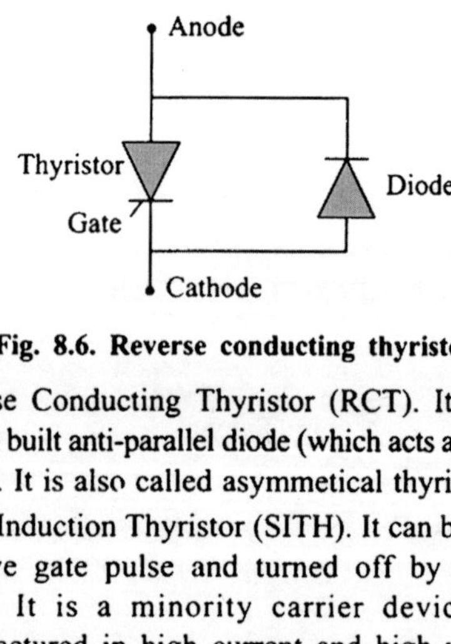

Fig. 8.6. Reverse conducting thyristor

(*e*) Reverse Conducting Thyristor (RCT). It is a thyristor with in built anti-parallel diode (which acts as free wheeling diode). It is also called asymmetical thyristor (Fig. 8.6).

(*f*) Static Induction Thyristor (SITH). It can be turned on by positive gate pulse and turned off by negative gate pulse. It is a minority carrier device. It can be manufactured in high current and high voltage ratings and its on state voltage drops is low (Fig. 8.7)

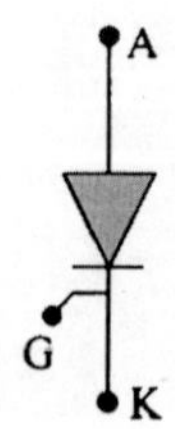

Fig. 8.7. Symbol for SITH

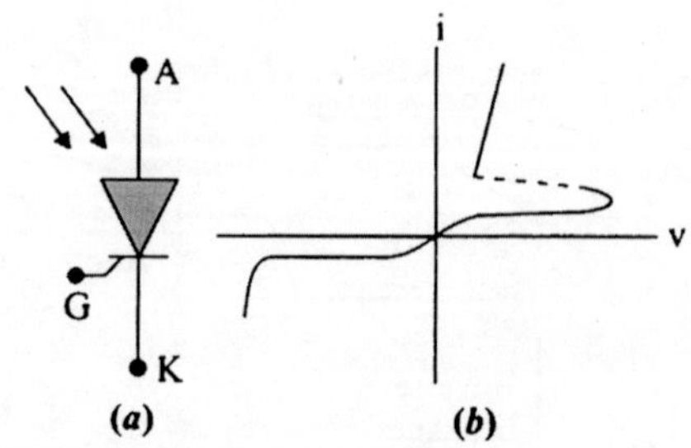

(*a*) (*b*)

Fig. 8.8. (*a*) Symbol (*b*) *i-v characteristics of LASCR*

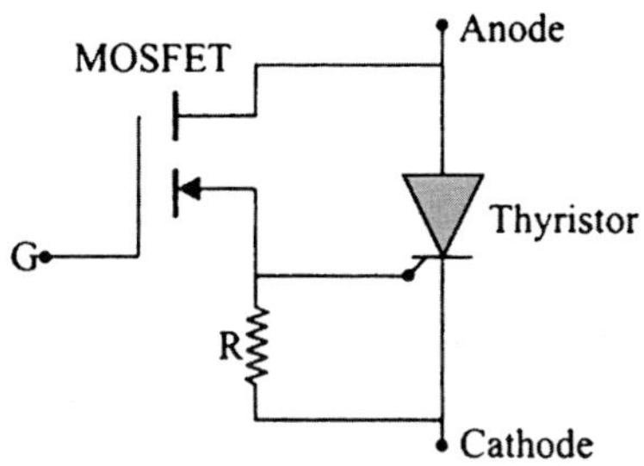

Fig. 8.9. FET controlled thyristor

(*g*) Light Activated SCR (LASCR). Its symbol and *v-i* charactersitics are shown in Fig. 8.8. It can be turned on by direct radiation of light on the silicon wafer. Light radiation creates electron hole pairs which cause triggering.

(*h*) FET controlled Thyristor (FET-CHT). Figure 8.9 shows its symbol. It is an SCR and MOSFET in one single unit and connected in parallel. When a voltage of about 3V is applied to gate of MOSFET, a triggering pulse is generated internally and SCR turns on.

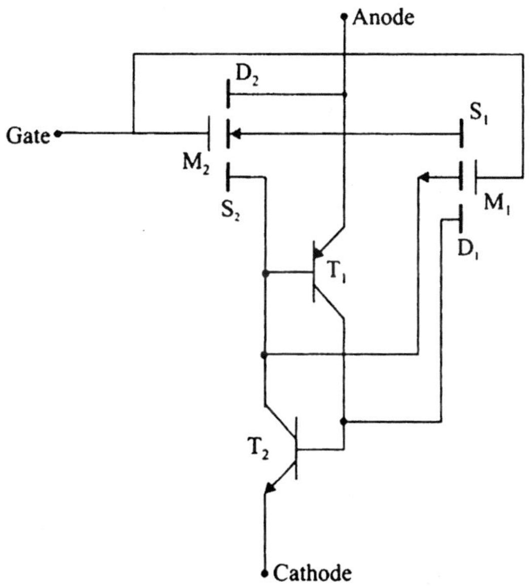

(*a*) **equivalent circuit**

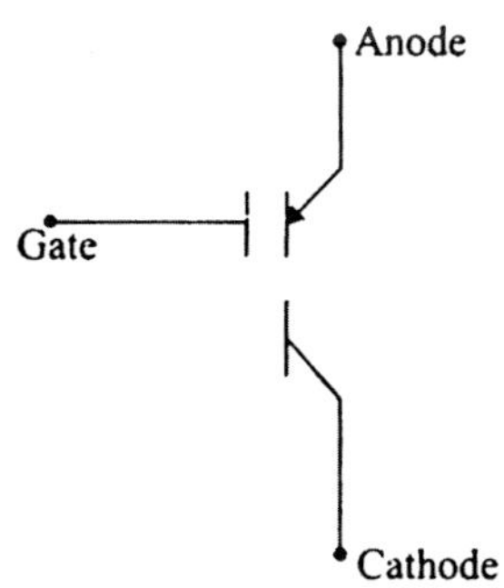

(*b*) **symbol**

Fig. 8.10. MOS controlled thyristor

(*i*) MOS Controlled Thyristor (MCT). It combines the features of four layered thyristor and MOS gate structure. The four layered structure is NPNP (and not PNPN as in a thyristor). It can be represented in two transistor analogy by NPN transistor T_2 and PNP transistor T_1 (Fig. 8.10). The MOS gate structure has a *p* channel MOSFET M_1 and *n* channel MOSFET M_2. Since the 4 layer structure is NPNP, the refrence terminal is anode for all gate signals.

(*j*) Programmable Unijunction Transistor (PUT). It is 4 layer 3 terminal device but the gate is connected to N terminal as shown in Fig. 8.11. Figure 8.11(*c*) shows its *v-i* characteristics. When operated in negative resistance region it can be used as a Relaxation oscillator.

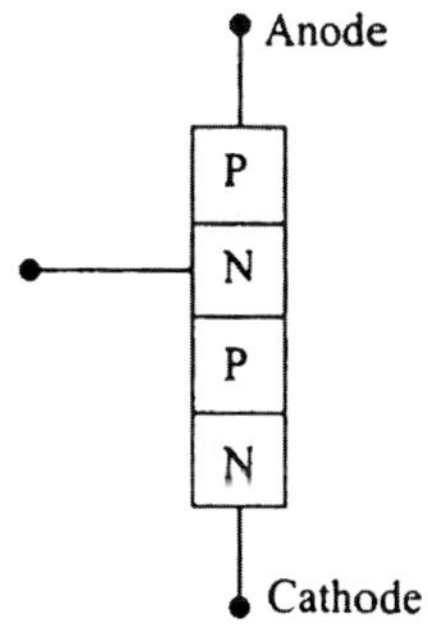

(*a*) **Schematic diagram**

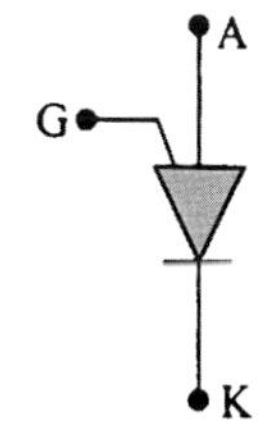

(*b*) **Symbol**

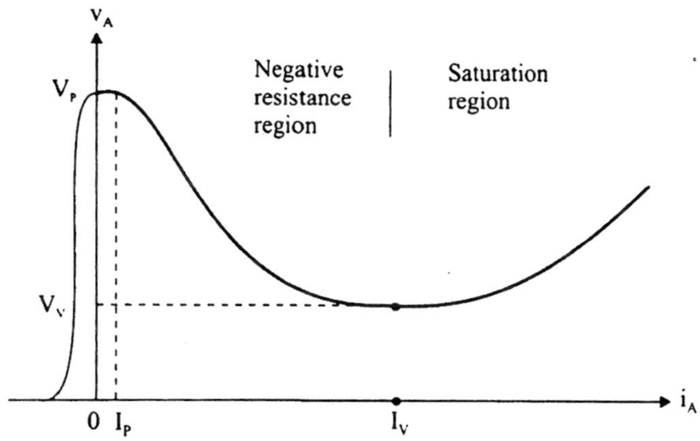

(*c*) ***v-i* characteristics of programmable unijuction transistor**

Fig. 8.11. Programmable unijunction transistor

(*k*) Silicon Unilateral Switch SVS (Fig. 8.12). It is a programmable unijuction transistor and an inbuilt avalanche diode. It switches at a fixed voltage determined by its avalanche diode.

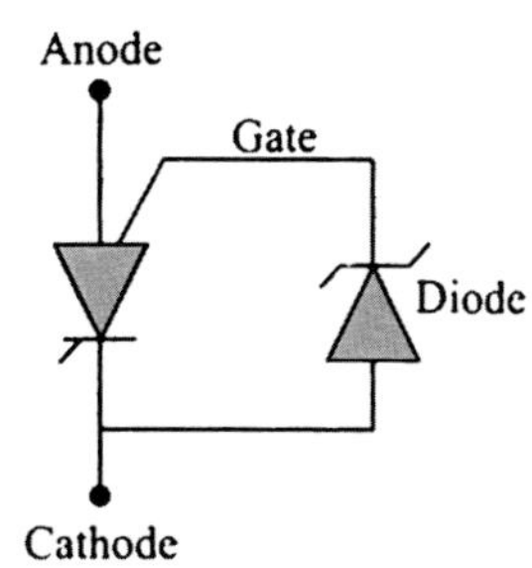

(*a*) **Equivalent circuit**

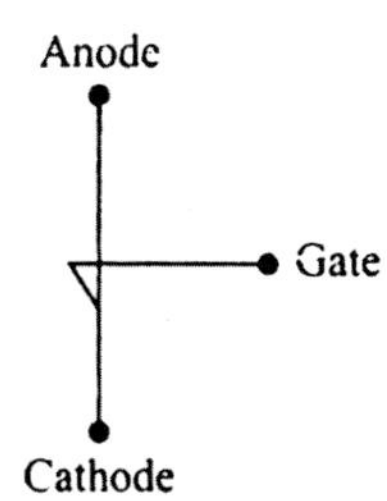

(*b*) **Symbol**

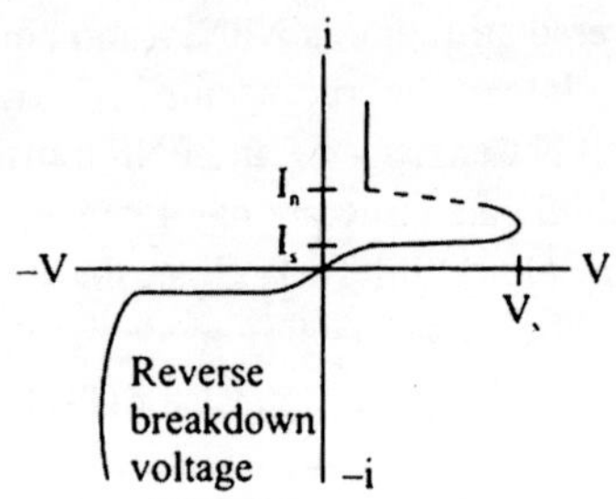

(*c*) **v-i characteristics**

Fig. 8.12. SUS

(*l*) Silicon Bilateral Switch SBS (Fig. 8.13). It consists of two silicon unilateral switches connected in anti-parallel. It can conduct in both directions. SUS and SBS find applications in high speed switching circuits, digital circuits, dc power supply circuits, pulse generators etc.

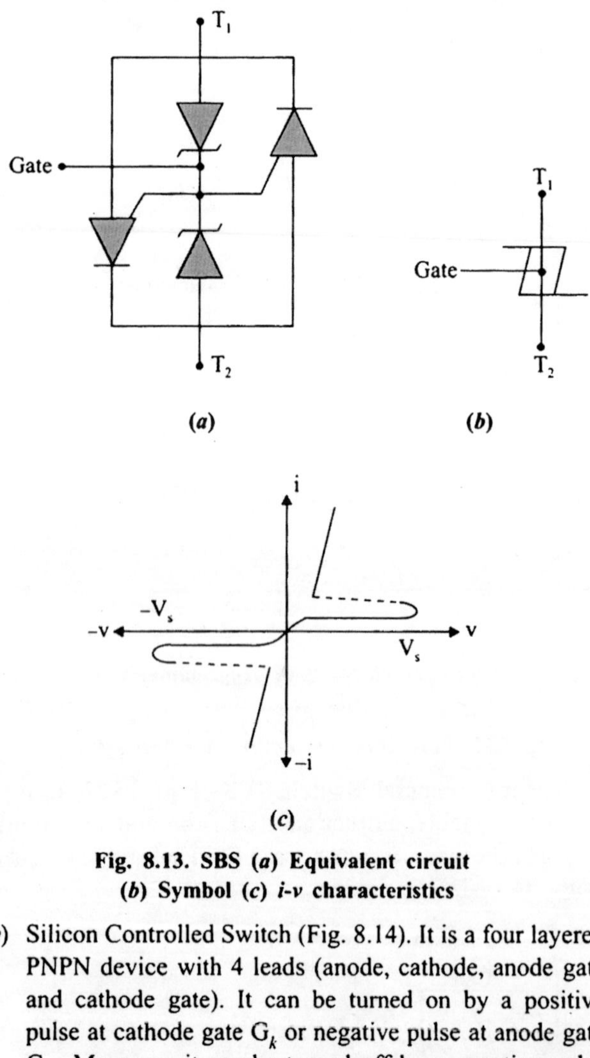

Fig. 8.13. SBS (*a*) Equivalent circuit (*b*) Symbol (*c*) *i-v* characteristics

(*m*) Silicon Controlled Switch (Fig. 8.14). It is a four layered PNPN device with 4 leads (anode, cathode, anode gate and cathode gate). It can be turned on by a positive pulse at cathode gate G_k or negative pulse at anode gate G_A. Moreover it can be turned off by a negative pulse at cathode gate G_k or positive pulse at anode gate G_A. It has a low turn off time (about 1 to 10 μ). It is used in digital circuits.

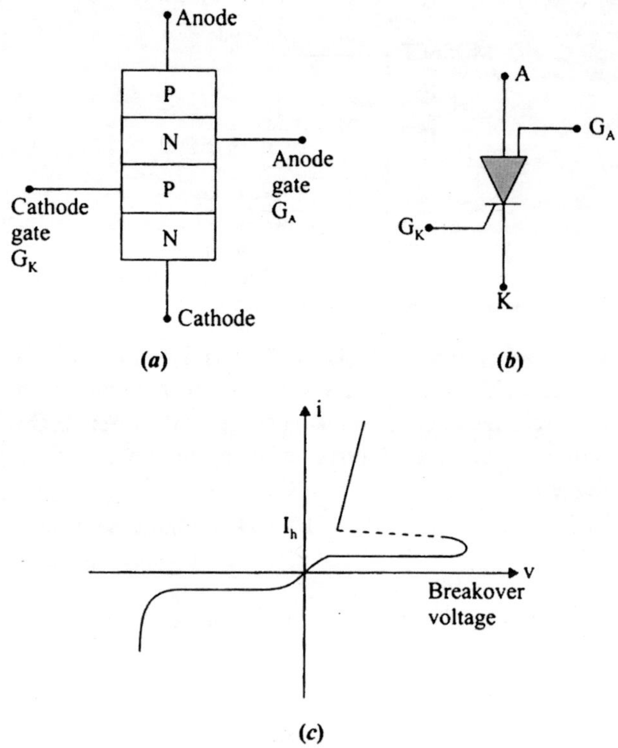

Fig. 8.14. Silicon controlled switch (*a*) configuration (*b*) symbol (*c*) *i-v* characteristics

(*n*) Shockley Diode (Fig. 8.15). It is a four layered PNPN device without a gate terminal. Its characteristics is similar to that of SCR with zero gate current. Figure 8.14. (*d*) shows one application of Shockley diode. It is acting as a trigger switch for Thyristor.

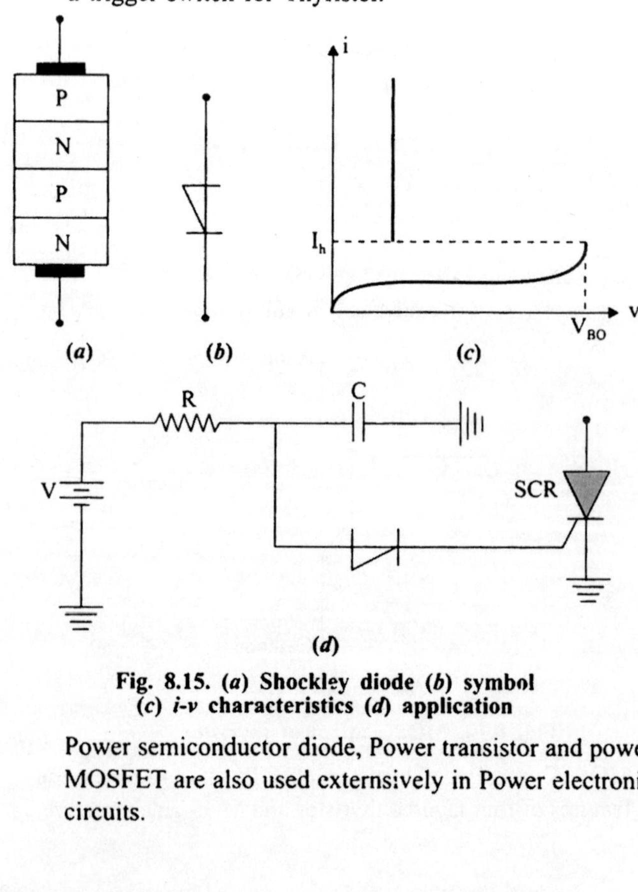

Fig. 8.15. (*a*) Shockley diode (*b*) symbol (*c*) *i-v* characteristics (*d*) application

Power semiconductor diode, Power transistor and power MOSFET are also used externsively in Power electronic circuits.

8.3. CONTROLLED RECTIFIERS

8.3.1. Performance Indices of Rectifiers

$$V_{dc} = \frac{1}{2\pi}\int_{\alpha}^{2\pi} v\, d(\omega t) \quad ...(8.1)$$

$$I_{dc} = V_{dc}/R \quad ...(8.2)$$

$$P_{dc} = V_{dc} I_{dc} \quad ...(8.3)$$

$$V_{rms} = \left[\frac{1}{2\pi}\int_{\alpha}^{2\pi} v^2 d(\omega t)\right]^{0.5} \quad ...(8.4)$$

$$I_{rms} = V_{rms}/R \quad ...(8.5)$$

$$P_{ac} = \text{output ac power} = V_{rms} I_{rms} \quad ...(8.6)$$

$$\text{Rectification efficiency} = \frac{P_{dc}}{P_{ac}} \quad ...(8.7)$$

RMS value of ac component of output voltage

$$V_{ac} = \left(V_{rms}^2 - V_{dc}^2\right)^{0.5} \quad ....(8.8)$$

$$\text{Form factor FF} = \frac{V_{rms}}{V_{dc}} \quad ...(8.9)$$

$$\text{Ripple factor RF} = \frac{V_{ac}}{V_{dc}} \quad ...(8.10a)$$

$$= \left[\left(\frac{V_{rms}}{V_{dc}}\right)^2 - 1\right]^{0.5} = (FF^2 - 1)^{0.5} \quad ...(8.10b)$$

$$\text{TUF, i.e., Transformer Utilization Factor} = \frac{P_{dc}}{V_s I_s} \quad ...(8.11)$$

$$\text{Displacement factor} = \cos\theta \quad ...(8.12)$$

$$\text{Harmonic factor} = \left[\left(\frac{I_s}{I_{s1}}\right)^2 - 1\right]^{0.5} \quad ...(8.13)$$

$$\text{Input power factor} = \frac{I_{s1}}{I_s}\cos\theta \quad ...(8.14)$$

$$\text{Crest factor} = \frac{I_s(\text{peak})}{I_s} \quad ...(8.15)$$

In the Eqs. (8.1) to (8.15), v is the instantaneours ac voltage at input of rectifier. V_{dc} is the average output voltage, R is load resistance, I_{dc} is average load current, P_{dc} is output dc power, V_{rms} is rms output voltage, I_{rms} is rms load current, α is the firing angle of thyristor, P_{ac} is output ac power, V_s and I_s are rms voltage and current at secondary of transformer feeding the rectifier circuit, θ is the phase angle between fundamental components of input voltage and current, I_{s1} is the fundamental component of input current.

An ideal rectifier should have 100% efficiency, $V_{ac} = 0$, RF = 0, TUF = 1, PF = 1 and DF = 1.

8.3.2 Single Phase Half Wave Converter (Resistive Load)

Figure 8.16 shows the circuit. Thyristor can be triggered at any angle α in positive half cycle. Conduction takes place from α to π in each positive half cycle. If

$$v = V_m \sin \omega t,$$

$$V_{dc} = \frac{V_m}{2\pi}(1 + \cos\alpha) \quad ...(8.16)$$

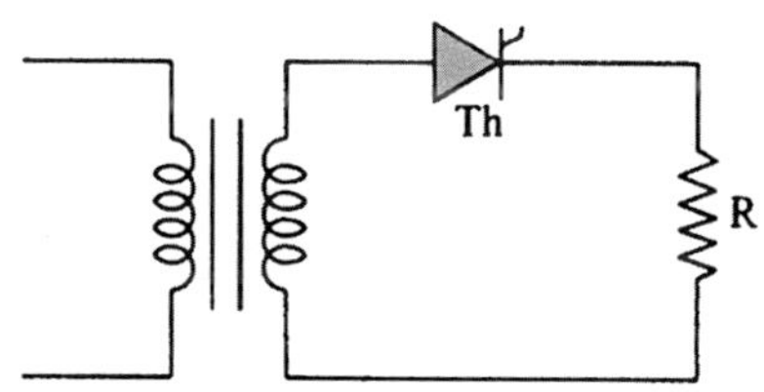

Fig. 8.16 Single phase half wave converter (Resistive load)

$$V_{rms} = V_m\left[\frac{\pi - \alpha}{4\pi} + \frac{\sin 2\alpha}{8\pi}\right]^{0.5} \quad ...(8.17)$$

$$\text{Input} \quad VA = \frac{V_m}{\sqrt{2}}\frac{V_{rms}}{R} = \frac{V_m^2}{\sqrt{2}R}\left[\frac{\pi - \alpha}{4\pi} + \frac{\sin 2\alpha}{8\pi}\right]^{0.5} \quad ...(8.18)$$

$$\text{Input power factor} \quad \frac{V_{rms}}{V_m/\sqrt{2}} = \sqrt{2}\left[\frac{\pi - \alpha}{4\pi} + \frac{\sin 2\alpha}{8\pi}\right]^{0.5} \quad ...(8.19)$$

$$\text{Peak inverse voltage} = V_m \quad ...(8.20)$$

8.3.3. Single Phase Half Wave Converter with RL Load and Free Wheeling Diode

Figure 8.17 shows the circuit. The free wheeling diode helps in improving the waveshape of load current and power factor.

$$V_{dc} = \frac{V_m}{2\pi}(1 + \cos\alpha) \quad ...(8.21)$$

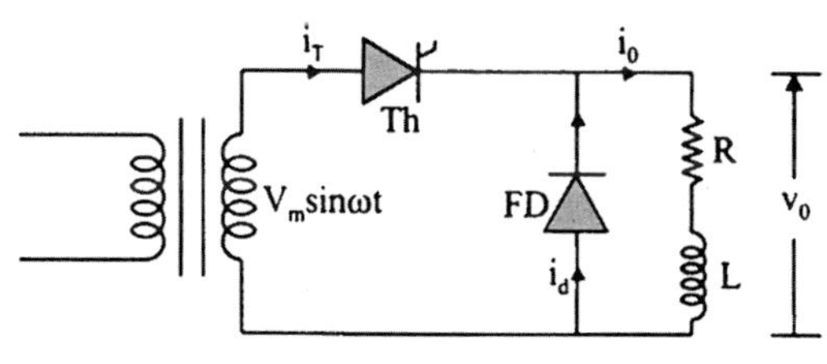

Fig. 8.17. Single phase half wave converter with R-L load and free wheeling diode

8.3.4. Single Phase Full Wave Converter (M-2 Connection)

Figure 8.18 shows the circuit for resistance load

$$V_{dc} = \frac{V_m}{\pi}(1 + \cos\alpha) \quad ...(8.22)$$

$$V_{rms} = V_m\left[\frac{\pi - \alpha}{2\pi} + \frac{\sin 2\alpha}{4\pi}\right]^{0.5} \quad ...(8.23)$$

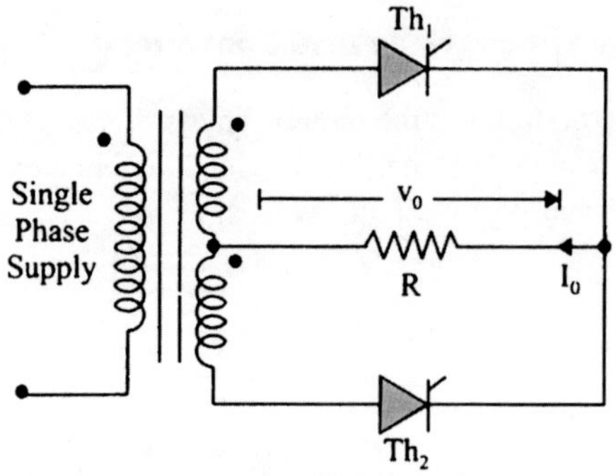

(*a*) **Circuit**

Fig. 8.18. Single phase full wave converter with centre tapped transformer and resistive load

Figure 8.19 shows this circuit for R-L load

$$V_{dc} = \frac{2V_m}{\pi}\cos\alpha \quad ...(8.24)$$

$$V_{rms} = \frac{V_m}{\sqrt{2}} \quad ...(8.25)$$

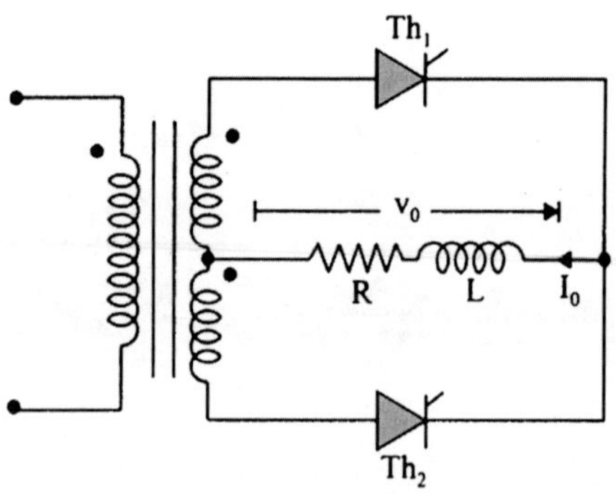

(*a*) **Circuit**

Fig. 8.19. Single phase full wave converter with centre tapped transformer and R-L load

8.3.5. Single Phase Fully Controlled Bridge Converter (B-2 Connection)

Figure 8.20 shows the circuit for resistive load.

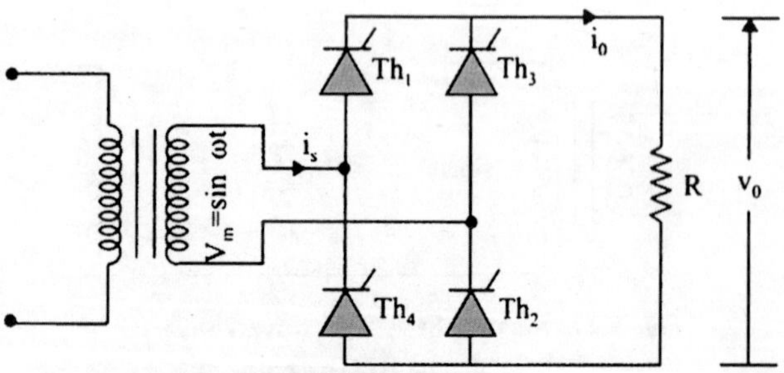

(*a*) **Circuit**

Fig. 8.20. Single phase fully controlled bridge converter supplying resistive load

Equations for V_{dc} and V_{rms} are the same as Eqs. (8.22) and (8.23) respectively.

Figure 8.21 (*a*) shows the connection for RLE (load). RLE load means load consisting of resistance R, inductance L and motor (back emf E). It can operate in quadrants 1 and 4 as shown in Fig. 8.21 (*b*).

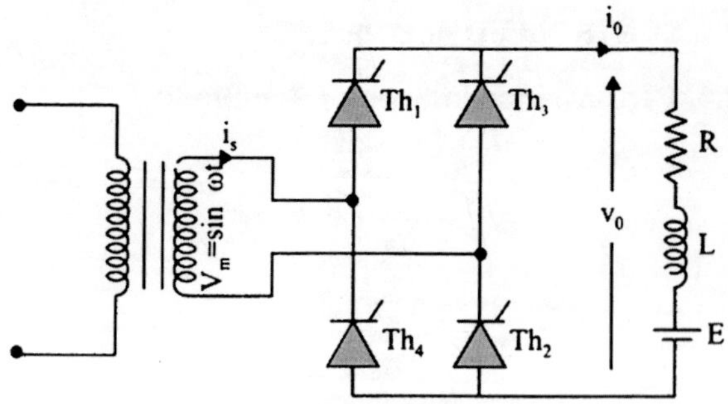

(*a*) **Circuit**

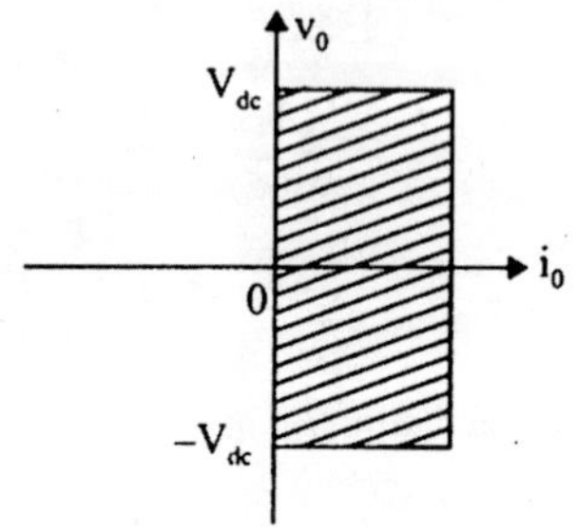

(*b*) **Quadrants of operation**

Fig. 8.21. Single phase fully contolled bridge converter with RLE load

$$V_{dc} = \frac{2V_m}{\pi}\cos\alpha \quad ...(8.26)$$

$$V_{rms} = \frac{V_m}{\sqrt{2}} \quad ...(8.27)$$

8.3.6. Single Phase Semi Converter

A single phase fully controlled bridge converter uses 4 thyristors. A semi converter has two thyristors and two diodes. Thus it is cheaper than full converter. Figure 8.22 shows its circuit for RLE load. It operates only in first quadrant

$$V_{dc} = \frac{V_m}{\pi}(1+\cos\alpha) \quad ...(8.28)$$

$$V_{rms} = \frac{V_m}{\sqrt{2}}\left[\frac{1}{\pi}\left(\pi-\alpha+\frac{\sin 2\alpha}{2}\right)\right]^{0.5} \quad ...(8.29)$$

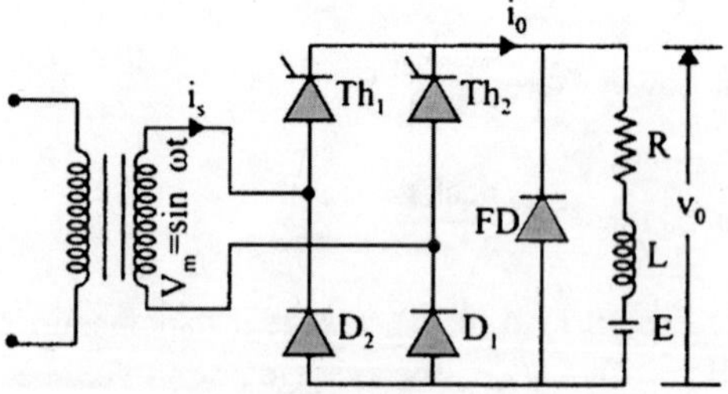

(*a*) **Circuit**

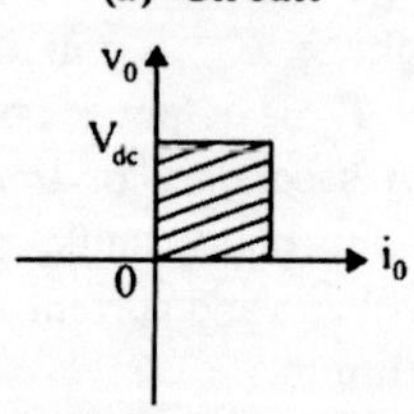

(*b*) **Quadrants of operation**

Fig. 8.22. Single phase semiconverter with RLE load

8.3.7. Single Phase Dual Converter

A dual converter can operate in all the four quadrants. It consists of two fully controlled bridge converters connected in anti-parallel as shown in Fig. 8.23.

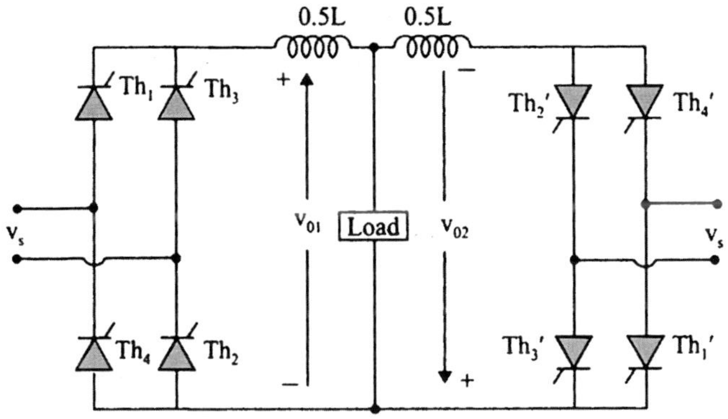

(*a*) Circuit

Fig. 8.23. Single phase dual converter

One converter operates as rectifier and the second converter operates as inverter. If α_1 and α_2 are the firing angles of the two,

$$V_{dc1} = \frac{2V_m}{\pi}\cos\alpha_1 \qquad ...(8.30)$$

$$V_{dc2} = \frac{2V_m}{\pi}\cos\alpha_2 \qquad ...(8.31)$$

Since one converter is rectifying and the other is inverting

$$V_{dc1} = -V_{dc2} \qquad ...(8.32)$$

or $\alpha_1 + \alpha_2 = \pi$

The single phase dual converter can be operated in non-circulating current mode or circulating current mode. In non-circulating current mode only one converter operates at one. In circulating current mode both converters operate simultaneously. The instantaneous ciculating current i_r is given by

$$i_r = \frac{2V_m}{\omega L}(\cos\omega t - \cos\alpha_1) \qquad ...(8.33)$$

where L is the inductance connected in the circuit to limit the circulating current. The circulating current helps in maintaining continuous conduction of both converters irrespective of load.

8.3.8. Three Phase Fully Controlled Bridge Converter

This converter uses 6 thyristors (Fig. 8.24). Three thyristors Th_1, Th_3, Th_5 form positive group and the remaining three thyristors (*i.e.*, Th_2, Th_4, Th_6) form the negative group. The firing frequency is six times the line frequency. There are 6 pulses of output in one cycle of input wave. Thus it is six pulse converter.

$$V_{dc} = \frac{3\sqrt{3}}{\pi}V_m\cos\alpha \qquad ...(8.34)$$

$$V_{rms} = \sqrt{3}\,V_m\left(0.5 + \frac{3\sqrt{3}}{4\pi}\cos 2\alpha\right)^{0.5} \qquad ...(8.35)$$

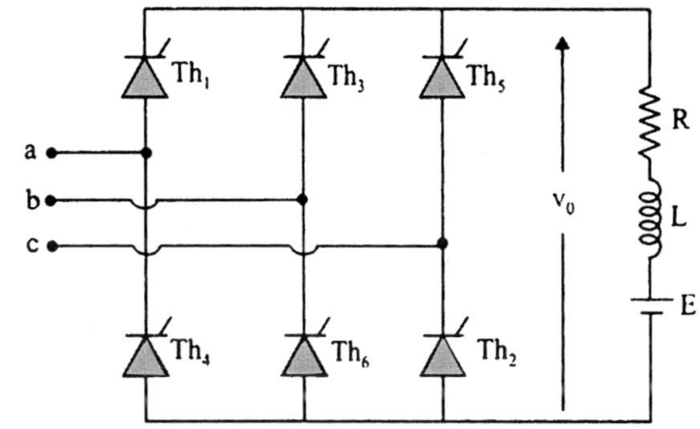

Fig. 8.24. Three phase fully controlled bridge converter

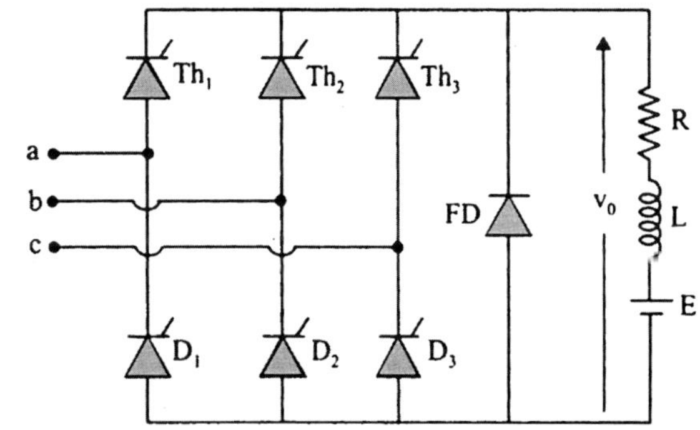

Fig. 8.25 Three phase semi converter

8.3.9. Three Phase Semi Converter

This converter uses 3 thyristors and 3 diodes for bridge as shown in Fig. 8.25. Free wheeling diode FD helps in improving the wave shape of load current. When $\alpha \le \frac{\pi}{3}$, the output voltage wave is continuous

$$V_{dc} = \frac{3\sqrt{3}\,V_m}{2\pi}(1+\cos\alpha) \qquad ...(8.36)$$

$$V_{rms} = \sqrt{3}\,V_m\left[\frac{3}{4\pi}\left(\frac{2\pi}{3} + \sqrt{3}\cos^2\alpha\right)\right]^{0.5} \qquad ...(8.37)$$

When $\alpha > \frac{\pi}{3}$, the output voltage wave is discontinuous

$$V_{dc} = \frac{3\sqrt{3}\,V_m}{2\pi}(1+\cos\alpha) \qquad ...(8.38)$$

$$V_{rms} = \sqrt{3}\,V_m\left[\frac{3}{4\pi}\left(\pi - \alpha + \frac{\sin 2\alpha}{2}\right)\right]^{0.5} \qquad ...(8.39)$$

8.3.10. Effect of Source Inductance on Converter Operation

The source inductance does not allow instantaneous commutation of current. The current in the outgoing thyristor can not change from full value to zero instantaneously. Similarly the current of the incoming thyristor cannot increase from zero to full value instantaneously. Thus there is a period of overlap μ. This angle μ is called overlap angle.

When source inductance is taken into account, the output voltage V_0 of single phase fully controlled bridge converter is

$$V_0 = \frac{2V_m}{\pi}\cos\alpha - \frac{\omega L_s}{\pi} I_0 \quad ...(8.40)$$

In 3-phase converter, the average voltage reduction due to overlap $= \frac{3\omega L_s}{\pi} I_0$.

8.4. INVERTERS

An inverter is a dc-ac converter. It may be single phase or 3-phase depending on whether the output is single phase or 3-phase. An inverter which feeds an ac system is line commutated. The ac line voltage is used for commutation. However an inverter feeding an isolated ac load requires forced commutation.

Inverters are also classified as series inverter, parallel inverter and bridge inverter.

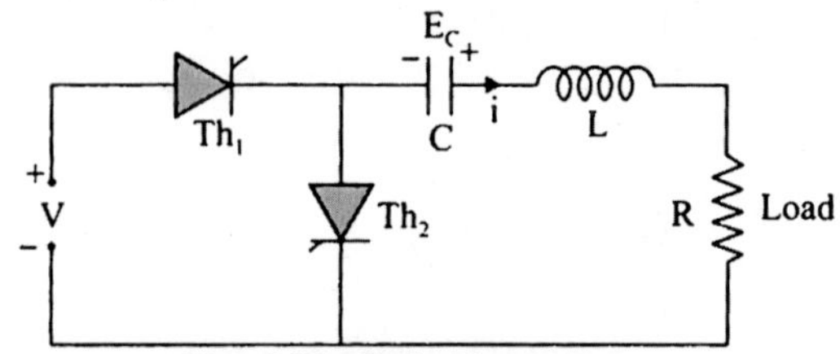

Fig. 8.26. Series inverter.

8.4.1. Series Inverter

Figure 8.26 shows the basic series inverter circuit. The commutating elements L and C are such that R, L and C form an under damped circuit. The output frequency of is given by

$$f = \left[\frac{1}{2\left[\frac{T}{2} + T_{off}\right]}\right] \text{Hz} \quad ...(8.41)$$

where $\frac{T}{2}$ in the time period of oscillations and T_{off} is the time gap between turn off of one thyristor and turn on of second thyri;stor. Moreover

$$T = \frac{2\pi}{\left(\frac{1}{LC} - \frac{R^2}{4L}\right)^{0.5}} \quad ...(8.24)$$

Modified series inverter uses two inductors and two capacitors to avoid intermittent operation of battery.

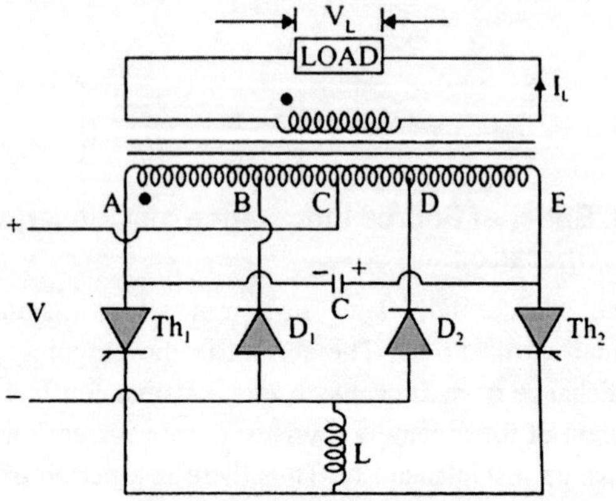

Fig. 8.27. Parallel inverter

8.4.2. Parallel Inverter

Figure 8.27 shown the circuit. The commutating capacitor is effectively in parallel with the load. This circuit uses two thyristors Th_1, Th_2 and two feedback diodes D_1, D_2. The function of diodes is to feedback the reactive power to the battery.

8.4.3. Bridge Inverter

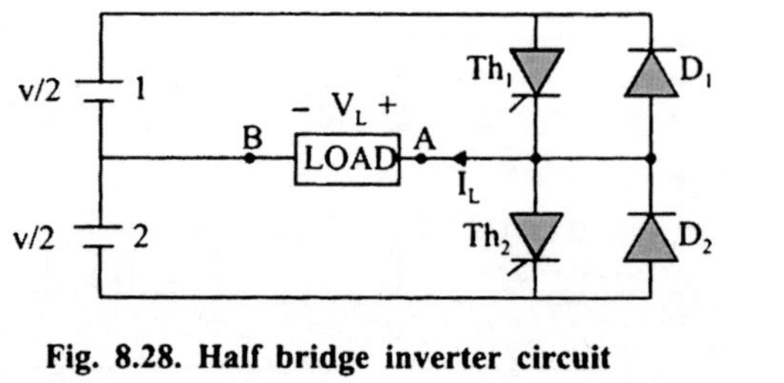

Fig. 8.28. Half bridge inverter circuit

A bridge inverter uses thyristors in bridge configuration. A half bridge inverter uses to thyristors and two diodes as shown in Fig. 8.28. Two batteries form a 3 wire dc supply. Each thyristor is gated at a frequency $f = \frac{1}{T}$. The output frequency $\omega = \frac{2\pi}{T}$ rad/sec. The gate pulses of Th_1, Th_2 are out of phase by 180°.

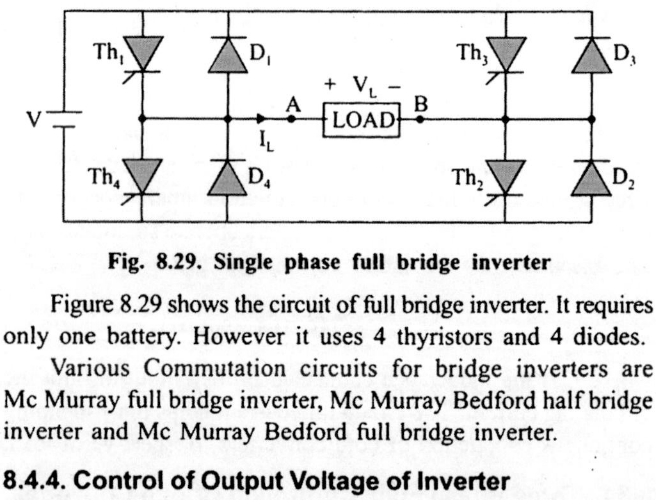

Fig. 8.29. Single phase full bridge inverter

Figure 8.29 shows the circuit of full bridge inverter. It requires only one battery. However it uses 4 thyristors and 4 diodes.

Various Commutation circuits for bridge inverters are Mc Murray full bridge inverter, Mc Murray Bedford half bridge inverter and Mc Murray Bedford full bridge inverter.

8.4.4. Control of Output Voltage of Inverter

The most commonly used method is pulse width modulation. This is classified as single pulse width modulation, multiple pulse width modulation and sinusoidal pulse width modulation.

8.5. CHOPPERS

A chopper is a dc-dc converter. It converts a fixed dc voltage into a variable dc voltage.

8.5.1. Principle

Figure 8.30 shows the principle of operation of chopper. The thyristor conduct for period T_{on} and is off for period T_{off} For a resistive load the average load voltage is

$$V_{av} = V\frac{T_{on}}{T_{on}+T_{off}} = \alpha V \qquad \text{....(8.43)}$$

RMS outpur Voltage $V_L = \sqrt{\alpha}\, V$ ...(8.44)

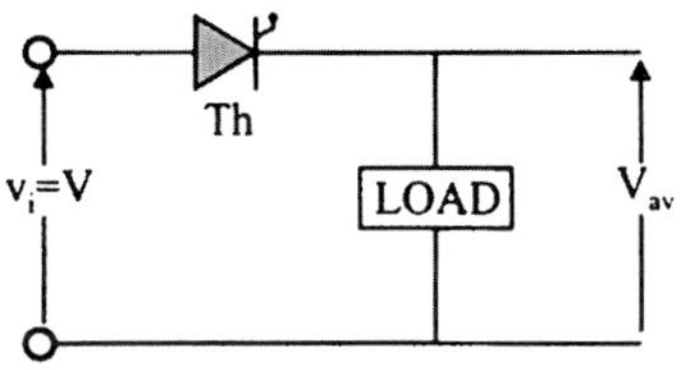

(*a*) **Circuit**

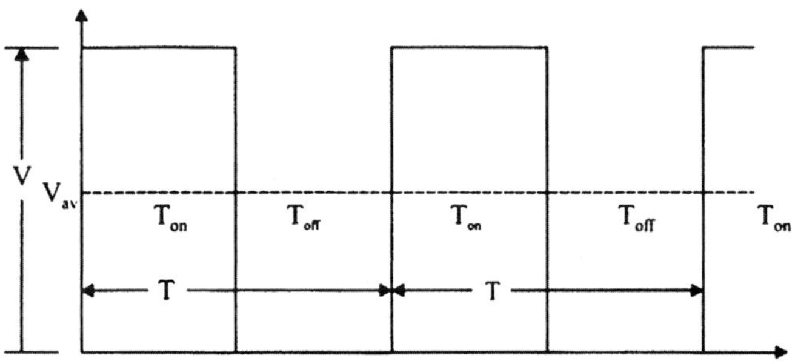

(*b*) **On and off periods**

Fig. 8.30. Chopper

$$\text{Power output} \quad P_0 = \frac{V_L^2}{R} = \frac{\alpha V^2}{R} \qquad \text{...(8.45)}$$

$$\text{where} \qquad \alpha = \frac{T_{on}}{T_{on}+T_{off}} = \frac{T_{on}}{T} = \text{Duty cycle}$$

$$f = \frac{1}{T} = \text{Chopper frequency}$$

8.5.2. Control Technique

Two methods are used to control the duty cycle α. They are called constant frequency system and variable frequency system.

In constant frequency system, the frequency f (and therefore T) is kept constant. T_{on} is varied to change the cycle α. This method is also called time ratio control (TRC) or pulse width modulation (PWM).

In variable frequency system the chopper frequency f (and therefore T) is varied but T_{on} is kept constant. This method is also called frequency modulation.

8.5.3. Step Down Chopper with RLE Load

The circuit is Fig. 8.30 is called step down chopper because the output voltage is always less than the input voltage. When the load consists of resistance R, inductance L and back emf E, the current cannot decrease or increase instantaneously. Instead the current rises or decays exponentially between maximum value I_{max} and minimum value I_{min}. The maximum variation of current is given by

$$(I_{max} - I_{min})_{max} = \frac{V}{R}\tanh\frac{RT}{4L} \qquad \text{...(8.46}a\text{)}$$

$$= \frac{V}{R}\tanh\frac{R}{4fL} \qquad \text{...(8.4}b\text{)}$$

If $4fL >> R$, $\tanh\frac{R}{4fL} = \frac{R}{4fL}$ and

$$(I_{max} - I_{min})_{max} = \frac{R}{4fL} \qquad \text{...(8.47)}$$

If the ratio $\frac{R}{L}$ is high (*i.e.*, time constant $\frac{L}{R}$ is low) or back emf E is high or duty cycle is low the current may become discontinuous. This condition should be avoided.

8.5.4. Classification

Choppers are classified as class A, B, C, D, E. This classification is as per the quadrants in which the choppers operate and is shown in Fig. 8.31.

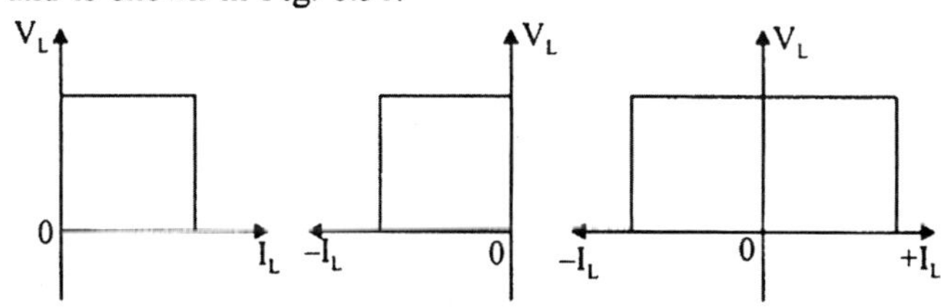

(*a*) **Class A** (*b*) **Class B** (*c*) **Class C**

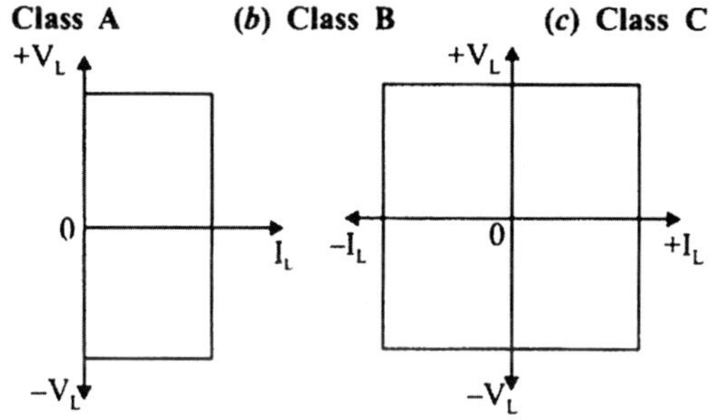

(*d*) **Class D** (*e*) **Class E**

Fig. 8.31. Quadrants of operation of choppers

8.5.5. Communtation Methods

Since the chopper is a dc-dc converter, forced commutation is necessary. The commutation methods are auxiliary commutation and load commutation. Jones chopper is an auxiliary commutated chopper. In this circuit a tapped auto transformer is used to complete the commutation process with high reliability. Morgan chopper also uses auxiliary commutation. In this circuit a saturable reactor is used to ensure reliable commutation.

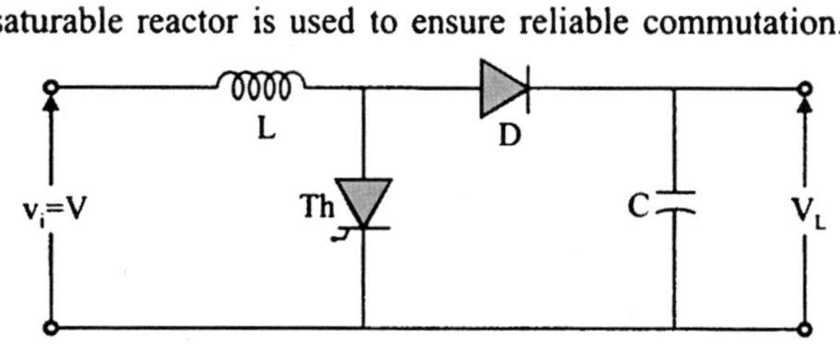

Fig. 8.32 Step up chopper

8.5.6. Step Up Chopper

Figure 8.32 shows a step up chopper. Thyristor Th is on for period T_{on} and off for period T_{off}

$$V_L = V\frac{T_{on}+T_{off}}{T_{off}} \qquad \text{...(8.48)}$$

8.6. AC REGULATOR

An ac regulator is an ac-ac converter. It converts constant ac voltage into a variable ac voltage. AC regulators use line commutation.

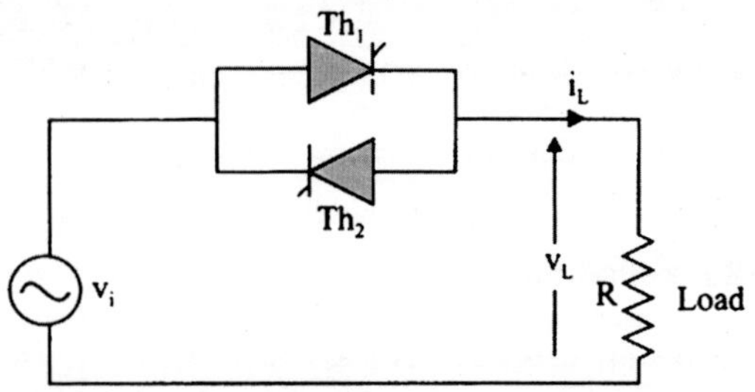

Fig. 8.33. AC regulator with resistance load

Figure 8.33 shows the single phase ac regulator circuit with resistance load. Th_1 conducts during positive half cycles and Th_2 conducts during negative half cycles. Thyristors are switched on for n cycles and switched off for m cycles. RMS output voltage V_L is

$$V_L = V\left(\frac{n}{n+m}\right)^{0.5} = V\sqrt{\alpha} \qquad ...(8.49)$$

The above medhod of control is called integral cycle (because n and m are integers). This method is used for loads having high mechanical inertia and high thermal constant.

Another control method is called phase control. In this method each thyristor is swiched on at $\angle\alpha$. By variation of α, the output voltage is varied.

8.7. CYCLO CONVERTER

Cyclo converter is a frequency changer. It changes ac voltage at a frequency f into an ac variable voltage at a frequency diferent from f.

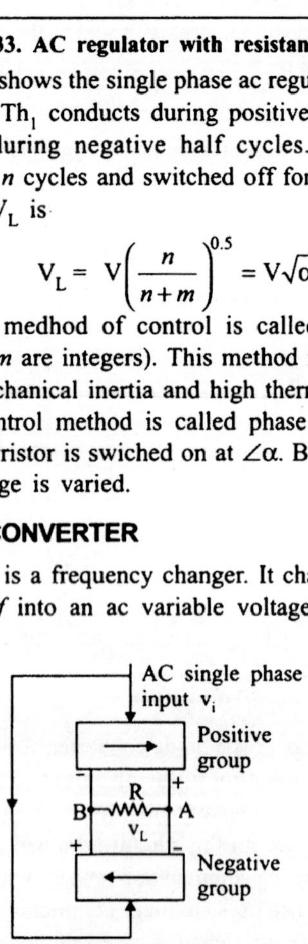

(a)

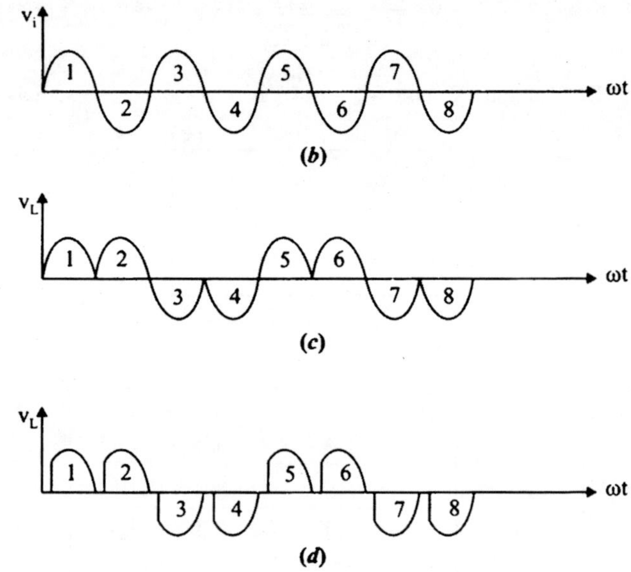

Fig. 8.34. Single phase cyclo converter (a) diagram, (b) input (c) output $f_0 = f/2$ and $\alpha = 0$, (d) output $f_0 = f/2$ and delayed firing

Figure 8.34 illustrates the principle of cyclo conveter. It consists of two groups of rectifiers one forming the positve group and the other forming the negative group. The positive group allows all positive half waves to pass through it and also converts negative half wave into positive half wave. The firing is such that half waves 1, 2, 5 and 6 pass through it. Its output will be half waves 1, 2, 5, 6 as shown in Fig. 8.34 (c). The negative group allows all negative half waves coming to it and also converts positive half wave into negative half waves. Its output is half waves 3, 4, 7, 8. Thus we get the putput shown in Fig. 8.34 (b). If we want variable output voltage, firing angle α can be varied over the range $0 \le \alpha \le \pi$ to give the waveshape shown in Fig. 8.34 (d). By proper firing of thyristors, we can get any frequency which is submultiple of input frequency Line commutation is used in the above conversion. Some other circuits of cyclo converters are single phase bridge configuration, three phase to single phase-phase bridge circuit and three-phase to three-phase bridge circuit.

OBJECTIVE TYPE QUESTIONS

A. Choose the most appropriate alternative

1. The rated current of a thyristor is 40 A. The hodling current will be about:
(a) 20 A (b) 1 A
(c) 100 mA (d) 10 mA

2. The main utility of SCR is due to
(a) gate triggering
(b) high temperature triggering
(c) $\frac{dv}{dt}$ triggering
(d) high break down voltage

3. The two-transistor model of a thyristor
(a) one n-p-n and other p-n-p
(b) both p-n-p
(c) both n-p-n
(d) one n-p-n and other UJT

4. The gate current required to turn on an SCR is
(a) few amperes
(b) a few mA
(c) almost equal to anode current
(d) about 50% of anode current

5. Natural commutation can be used in:
(a) DC circuits only
(b) AC circuits only
(c) both AC and DC circuits
(d) none of the above

6. Thyristors connected in series need:
(a) static equalizing circuit
(b) dynamic equalizing circuit
(c) both static and dynamic equalizing circuits
(d) none of the above

7. A triac can be turned on
(a) by applying a positive signal to gate
(b) by applying a negative signal to gate
(c) by applying either positive or negative signal to gate
(d) none of the above

8. A relaxation oscillator uses
(a) tunnel diode
(b) UJT
(c) both tunnel diode and UJT
(d) PIN diode

9. The number of leads in an SCR are:
(a) 2 (b) 3
(c) 4 (d) 5

10. When a thyristor is conducting, the voltage drop across it
(a) is absolutely constant
(b) decreases with increase in load current
(c) increases slightly with increase in load current
(d) any of the above

11. Current derating may be necessary when thyristors are operating in
(a) series (b) parallel
(c) series or parallel (d) none of the above

12. When thyristors are connected in series and parallel, it may be necessary to have
(a) current derating
(b) voltage derating
(c) both current and voltage derating
(d) none of the above

13. An RC snubber circuit is used to protect a thyristor against
(a) false triggering (b) failure to turn on
(c) switching transients (d) failure to commutate

14. A triac is equivalent to
(a) two thyristors in series
(b) two thyristors in parallel
(c) one thyristor and one diode
(d) one thyristor and one transistor

15. A sillicon controlled switch has:
(a) 2 leads (b) 3 leads
(c) 4 leads (d) 5 leads

16. The method of triggering SCS is
(a) application of positive pulse at cathode gate or negative pulse at anode gate
(b) application of pulse at anode gate
(c) application of pulse at cathode gate
(d) application of low voltage at anode

17. The minimum gate current whch can turn on SCR is called
(a) trigger current (b) holding current
(c) junction current (d) breakover current

18. To turn off an SCR, it is necessary to reduce its current to less than
(a) trigger current (b) holding current
(c) breakover current (d) none of the above

19. The gate lead, in a thyristor is welded to:
(a) p layer to which anode is connected
(b) n layer nearest to the anode terminal
(c) p layer nearest to the cathode terminal
(d) outside n layer

20. When a thyristor is conducting, the voltage drop across it is about
(a) 1 V (b) 10 V
(c) 100 V (d) 0.1 V

21. A thyristor is reverse biased. A positive gate pulse is applied. The thyristor
(a) will be turned on
(b) may or may not turn on
(c) will not rurn on
(d) will turn on after some time

22. In a thyristor, the forward breakover voltage
(a) is constant
(b) may be constant or may depend on gate current
(c) decreases as gate current is increased
(d) increases as gate current is increased

23. The number of leads in DIAC are
(a) 2 (b) 3
(c) 4 (d) 6

24. When a UJT is used as a relaxation oscillator the current should be
(a) less than I_p
(b) more than I_v
(c) more than I_p but less then I_v
(d) negative

25. The factor η for a UJT is about
(a) 0.1 (b) 0.2
(c) 0.6 (d) 0.9

26. The number of doped regions in a SCR is
(a) 2 (b) 3
(c) 4 (d) 5

27. In an ordinary thyristor
(a) turn on and turn off times are equal
(b) turn on time is more than turn off time
(c) turn off time is more than turn on time
(d) turn on time may be equal, less than or more than turn off time

28. A thyristor circuit is feeding an R-L load. The turn on time can be reduced by
(a) decreasing R
(b) decreasing L
(c) increasing L
(d) decreasing R and L together

29. In an SCR, the anode current is controlled by
(*a*) gate current only
(*b*) external circuit only
(*c*) both gate current and external circuit
(*d*) none of the above

30. The number of gates in a SCR is
(*a*) 1 (*b*) 3
(*c*) 4 (*d*) 6

31. A LASCR has
(*a*) 4 semiconductor layers and 3 junctions
(*b*) 3 semiconductor layer and 2 junctions
(*c*) 2 semiconductor layers and 2 junctions
(*d*) any of these depending on the rating

32. The average load current supplied by a thyristor depends on
(*a*) firing angle
(*b*) firing frequency
(*c*) magnitude of gate current
(*d*) all the above

33. A voltage V_{BB} is applied across the terminals of a UJT. The emitter voltage at peak point is
(*a*) ηV_{BB} (*b*) $(\eta + 1)V_{BB}$
(*c*) $\eta V_{BB} + V_D$ (*d*) V_D

34. When an SCR is in forward blocking mode, the applied voltage
(*a*) appears across only one junction
(*b*) is equally shared by two junctions
(*c*) is equally shared by three junctions
(*d*) may appear across one or more junctions

35. The condition for underdamped oscillations in a series RLC circuit is
(*a*) $R < \frac{L}{C}$ (*b*) $R^2 < \frac{4L}{C}$
(*c*) $R^2 < \frac{2L}{C}$ (*d*) $R < \sqrt{\frac{L}{C}}$

36. When an SCR is reverse biased
(*a*) two junctions are reverse biased and one junction is forward biased
(*b*) all the three junctions are reverse biased
(*c*) one juction is reverse biased and two junctions are forward biased
(*d*) any of above depending on the magnitude of reverse bias

37. In a controlled rectifier a free wheeling diode is necessary if the load is
(*a*) inductive (*b*) resistive
(*c*) capacitive (*d*) any of above

38. The number of thyristors in a single phase fully controlled bridge rectifier is
(*a*) 16 (*b*) 8
(*c*) 4 (*d*) 2

39. In a single phase semiconverter the number of thyristors is
(*a*) 16 (*b*) 8
(*c*) 4 (*d*) 2

40. A single phase full converter can operate in
(*a*) 4 quadrants (*b*) 3 quadrants
(*c*) 2 quadrants (*d*) 1 quadrant

41. If 4 quadrant operation is required we need
(*a*) dual converter (*b*) full converter
(*c*) semiconverter (*d*) any of above

42. In a single phase full wave controlled rectifier using centre tap transformer, the voltage across each half of secondary is $V_m \sin \omega t$. The peak inverse voltage is
(*a*) $2 V_m$ (*b*) V_m
(*c*) $0.5 V_m$ (*d*) $0.25 V_m$

43. A sigle phase fully controlled bridge converter is fed by a voltage $V_m \sin \omega t$. The average and rms values of output voltage are
(*a*) $\frac{2V_m}{\pi}\cos\alpha$ and $\frac{V_m}{\sqrt{2}}$ (*b*) $\frac{V_m}{\pi}\cos\alpha$ and $\frac{V_m}{\sqrt{2}}$
(*c*) $\frac{2V_m}{\pi}\cos\alpha$ and V_m (*d*) $\frac{V_m}{\pi}\cos\alpha$ and $\sqrt{2}\, V_m$

44. In a single phase full converter fed by a source having inductance, the number of thyristors the number of thyristors conducting during overlap is
(*a*) 1 (*b*) 2
(*c*) 3 (*d*) 4

45. A single phase fully controlled bridge rectifier is fed by a source having inductance. The output voltage during overlap period is
(*a*) V_m
(*b*) $0.5 V_m$
(*c*) zero
(*d*) $0.5 V_m$ – voltage drop across inductance

46. The effect of source inductance in a 3-phase fully controlled bridge converter is to
(*a*) decrease the output voltage
(*b*) increase the output voltage
(*c*) decrease the output voltage and increase the output current
(*d*) increase the output voltage and decrease the output current

47. A single phase semiconverter has a free wheeling diode. If the firing angle is α and the load is purely resistive, the periods of conduction and free wheeling respectively are
(*a*) $\pi - \alpha$ and 0 (*b*) $\pi - \alpha$ and α
(*c*) $\pi + \alpha$ and 0 (*d*) $\pi + \alpha$ and α

48. HVDC transmission systems generally use
(*a*) 12 pulse converters
(*b*) 6 plulse converters
(*c*) 3 plulse converters
(*d*) either 12 or 6 plulse converters

49. In a single phase converter using a centre tapped transformer and with a highly inductive load, the firing angles is α. Each thyristor conducts for a period
(*a*) π (*b*) α
(*c*) $\pi - \alpha$ (*d*) $\frac{\pi}{2}$

50. In rectifier circuits, the rms value of ac component of output voltage is equal to
(*a*) V_{rms} (*b*) $0.5\,V_{rms}$
(*c*) $(V_{rms} - V_{dc})^{0.5}$ (*d*) $(V_{rms}^2 - V_{dc}^2)^{0.5}$

51. In rectifier circuits, ripple factor RF and form factor FF are related as
(*a*) $RF = (FF^2 - 1)^{0.5}$ (*b*) $RF = FF$
(*c*) $RF = FF - 1$ (*d*) $RF = (FF - 1)^{0.5}$

52. A single phase fully controlled bridge converter is feeding an RLE load. Input voltage is $V_m \sin \omega t$ and firing angle is α. If average current is I_0, then
(*a*) $\frac{2V_m}{\pi}\cos\alpha = RI_0 + E$ (*b*) $\frac{V_m}{\pi}\cos\alpha = RE_0 + E$
(*c*) $\frac{2V_m}{\pi}\cos\alpha = E_0 - RI_0$ (*d*) $\frac{V_m}{\pi}\cos\alpha = E - RI_0$

53. A single phase semiconverter is supplying RLE laod and has a free wheeling diode across the load. The firing angle is α. The free wheeling diode conducts for
(*a*) $\alpha°$ in each half cycle
(*b*) $(\pi - \alpha)°$ in each cycle
(*c*) $\alpha°$ in one full cycle
(*d*) $(\pi - \alpha)°$ in each half cycle

54. In a single phase dual converter, α_1 and α_2 are firing angles of the two converters. Then
(*a*) $\alpha_1 - \alpha_2 = \frac{\pi}{2}$ (*b*) $\alpha_1 + \alpha_2 = \pi$
(*c*) $\alpha_1 - \alpha_2 = \pi$ (*d*) $\alpha_1 + \alpha_2 = \frac{\pi}{2}$

55. A single phase converter is fed by a source having an inductance L. If load current is I_C, the decrease in output voltage due to overlap is
(*a*) $\frac{0.5\omega L}{\pi} I_0$ (*b*) $\frac{\omega L}{4\pi} I_0$
(*c*) $\frac{\omega L}{\pi} I_0$ (*d*) $\frac{\omega L}{3\pi} I_0$

56. An *n* pulse rectifier is fed by a source having an inductance L. If load current is I_0, the voltage drop due to overlap is
(*a*) $\frac{n\omega L}{2\pi} I_0$ (*b*) $\frac{n\omega L}{\pi} I_0$
(*c*) $\frac{n^2\omega L}{\pi} I_0$ (*d*) $\frac{n\omega L}{3\pi} I_0$

57. The commutation method, in an inverter, is
(*a*) line commutation (*b*) forced commutation
(*c*) either (*a*) or (*b*) (*d*) none of the above

58. In a series inverter supplying a load resistance R, the commutating elements L and C should be such that
(*a*) $R^2 = \frac{4L}{C}$ (*b*) $R^2 < \frac{4L}{C}$
(*c*) $R^2 > \frac{4L}{C}$ (*d*) $R^2 < \frac{2L}{C}$

59. In a series inverter
(*a*) the commutating elements carry full load current
(*b*) the commutating elements carry up to 50% of load current
(*c*) the commutating elements do not carry load current
(*d*) any of the above

60. In a modified series inverter
(*a*) the maximum output frequency is less than frequency of damped oscillations
(*b*) the output frequency can be even higher than the frequency of damped oscillations
(*c*) the output frequency cannot exceed the frequency of damped oscillations
(*d*) any of the above

61. The advantage of modified series inverter with two inductors and two capacitors is
(*a*) the intermittent operation of battery is avoided
(*b*) efficency of circuit is more
(*c*) the output frequency is low
(*d*) all the above

62. In a modified series inverter with two inductors and two capacitors, the load current
(*a*) has 2 parallel paths
(*b*) has only one parallel path
(*c*) can have three parallel paths
(*d*) has four parallel paths

63. In a three phase series inverter, the firing frequency is
(*a*) three times output frequency
(*b*) six times output frequency
(*c*) twice the output frequency
(*d*) equal to output frequency

64. In an *n* stage high frequency series inverter having a time period T, each stage is active for a time interval of
(*a*) T (*b*) $\frac{T}{n}$
(*c*) $\frac{2T}{n}$ (*d*) $\frac{T}{\sqrt{n}}$

65. In a parallel inverter
(*a*) each thyristor is turned on twice during each cycle
(*b*) each thyristor is turned on once during each cycle
(*c*) each thyristor is turned on either once or twice in each load cycle
(*d*) each thyristor may be turned on up to four times in each load cycle

66. In a bridge inverter
(*a*) an output transformer is essential
(*b*) an output transformer is not essential
(*c*) an output transformer cannot be connected
(*d*) an output transformer, if used, can improve the efficiency of the circuit

67. A half bridge inverter
(*a*) needs a 3 wire dc supply
(*b*) may use a two wire or 3 wire dc supply
(*c*) has higher output voltage if fed from 3 wire dc supply
(*d*) has higher efficiency if fed from 3 wire dc supply

68. A single phase half bridge inverter required to feed R-L loads meeds
(*a*) 2 thyristors
(*b*) 4 thyristors
(*c*) 2 thyristors and 2 diodes
(*d*) 4 thyristors and 4 diodes

69. A single phase full bridge inverter for R-L loads needs
(*a*) 4 thyristors
(*b*) 4 thyristors and 4 diodes
(*c*) 4 thyristors and 2 diodes
(*d*) 8 thyristors

70. The load circuit of a single phase full bridge inverter
(*a*) must be underdamped
(*b*) must be overdamped
(*c*) may be underdamped or overdamped
(*d*) may be any type of load

71. McMurrary full bridge inverter uses
(*a*) auxiliary commutation
(*b*) complementary commutation
(*c*) natural commutation
(*d*) any of the above

72. McMurray Bedford full bridge inverter uses
(*a*) natural commutation
(*b*) auxiliary commutation
(*c*) complementary commutation
(*d*) any of the above

73. In a 3-phase bridge inverter, the gating signals for the three phases have a phase difference of
(*a*) 120° (*b*) 60°
(*c*) 240° (*d*) any of above

74. All other parameters remaining the same, a 3-phase bridge inverter with 180° mode of operation
(*a*) has the same peak thyristor current as in 120° mode of operation
(*b*) has higher peak thyristor current than in 120° mode of operation
(*c*) has lower peak thyristor current than in 120° mode of operation
(*d*) any of the above

75. In a bridge inverter with single pulse width modulation
(*a*) the output frequency is equal to frequency of reference signal
(*b*) the output frequency is equal to frequency of control signal
(*c*) either (*a*) or (*b*)
(*d*) neither (*a*) nor (*b*)

76. If A_r and A_c are the amplitudes of reference and control signals, the modulation index M is
(*a*) $M = \frac{A_r}{A_c}$ (*b*) $M = \frac{A_c}{A_r}$
(*c*) $M = \sqrt{\frac{A_r}{A_c}}$ (*d*) $M = \sqrt{\frac{A_c}{A_r}}$

77. A chopper
(*a*) converts constant voltage dc into ac and then into variable voltage dc
(*b*) converts constant voltage dc into variable voltage dc directly
(*c*) converts ac of one frequency into ac of another frequency
(*d*) converts ac to dc

78. The duty cycle of a step down chopper is
(*a*) $\frac{T_{on}}{T_{on}+T_{off}}$ (*b*) $\frac{T_{off}}{T_{on}+T_{off}}$
(*c*) $\frac{T_{on}+T_{off}}{T_{on}}$ (*d*) $\frac{T_{on}+T_{off}}{T_{off}}$

79. If a step down chopper is fed by a dc voltage V and T_{on} and T_{off} are on and off periods, then $V\frac{T_{on}}{T_{on}+T_{off}}$ is
(*a*) average output voltage
(*b*) rms output voltage
(*c*) average or rms output voltage depending on load
(*d*) neither of the above

80. A step down dc chopper has an input voltage V. If duty cycle is α and the load is a resistance R, the rms value of output voltage is
(*a*) αV (*b*) $\sqrt{\alpha}\, V$
(*c*) $\alpha^2 V$ (*d*) $\frac{\alpha V}{R}$

81. In pulse width modulation of chopper
(*a*) T is kept constant and T_{on} is varied
(*b*) T_{on} is kept constant T is varied
(*c*) both T and T_{on} are varied
(*d*) either T or T_{on} is varied

82. The efficiency of a chopper circuit is about
(*a*) 80% or more (*b*) around 50%
(*c*) around 20% (*d*) around 5%

83. When a dc chopper feeds an RLE load, the load current, during steady state operation
(*a*) remains constant
(*b*) varies between maximum and minimum values
(*c*) may remain constant or vary
(*d*) is constant if R is constant

84. A dc chopper feeds an RLE load and has a duty cycle of 1. The load current will be such that
(*a*) $I_{max} = I_{max} = \frac{V-E}{R}$ (*b*) $I_{max} > I_{max}$
(*c*) $I_{min} > 0$ (*d*) $I_{max} < 0$

85. A dc chopper is feeding an RLE load. The condition for maximum ripple is
(*a*) $\alpha = 1$ (*b*) $\alpha = 0.8$
(*c*) $\alpha = 0.5$ (*d*) $\alpha < 0.5$

86. A dc chopper is feeding an RLE load. The maximum steady state ripple is
(*a*) $\frac{V}{4fL}$ (*b*) $\frac{V}{2fL}$
(*c*) $\frac{\sqrt{V}}{4fL}$ (*d*) $\frac{V}{4f\sqrt{L}}$

87. A dc chopper is feeding an RLE load. The current can become discontinous if

(a) $\frac{R}{L}$ is high (b) α is low
(c) E is high (d) any of the above

88. Class A and class B chopper
(a) operate in first quadrant
(b) operate in second quadrant
(c) class A chopper operates in first quadrant while class B operates in second quadrant
(d) may operate in either first or second quadrant

89. A class D chopper
(a) can operate in first quadrant only
(b) can operate in second quadrant only
(c) can operate in first or fourth quadrant
(d) can operate in all the quadrants

90. A class E chopper
(a) can operate in second quadrant only
(b) can operate in first or third quadrant
(c) can operate in all the four quadrants
(d) can operate in second or third quadrant

91. In auxiliary commutated chopper fed by voltage V and having L and C commutating elements, the peak value of capacitor current is

(a) $\frac{VC}{L}$ (b) $\frac{VC}{\sqrt{L}}$
(c) $V\sqrt{\frac{C}{L}}$ (d) $\sqrt{\frac{VC}{L}}$

92. In auxiliary commutated chopper, the peak value of thyristor current is
(a) equal to peal value of load current
(b) equal to peak value of capacitor current
(c) sum of peak value of load current and capacitor current
(d) neither of above

93. In auxiliary commutated chopper, the charging and discharging time of capacitor
(a) are constant
(b) depend on load current
(c) may be constant or may depend on load current
(d) neither of above

94. In load commutated chopper, the commutating element
(a) is capacitor
(b) is capacitor and inductor
(c) is capacitor, inductor and auxiliary thyristor
(d) any of above depending on load

95. A load commutated chopper circuit has
(a) one thyristor (b) two thyristors
(c) four thyristors (d) six thyristors

96. In Jone chopper, the commutating element
(a) is a capacitor
(b) is a capacitor and an inductor
(c) is a capacitor and an auto transformer
(d) none of the above

97. Which chopper circuit uses saturable reactor?
(a) auxiliary commutated (b) Jones chopper
(c) Morgan chopper (d) load commutated

98. A step up chopper can give an output voltage
(a) higher than input voltage
(b) lower than input voltage
(c) both higher and lower input voltage
(d) all the above

99. In integral cycle control of ac regulators, the period of integral cycle control, *i.e.*, $(m + n)\,T$ should be
(a) less than mechanical or thermal time constant of the system
(b) more than the mechanical or thermal time constant of the system
(c) may be more or less than the mechanical or thermal time constant
(d) should be very high

100. In integral cycle control of ac regulator, the input power factor is equal to
(a) duty cycle α (b) α^2
(c) $\sqrt{\alpha}$ (d) $(\alpha)^{1/3}$

101. In single phase half wave regulator, the average current over one full cycle
(a) is always positive
(b) may be positive or negative
(c) may be negative
(d) is always negative

102. In a single phase full wave regulator feeding a resistive load, the average current over one full cycle
(a) is always zero
(b) cannot be zero
(c) may be sometimes zero
(d) is always positive

103. In a single phase half wave ac regulator feeding a resistive load, the rms load current
(a) is always positive
(b) may be positive or negative
(c) is always negative
(d) can be even zero

104. In a single phase full wave ac regulator, the firing angles in the two half cycles
(a) are always equal
(b) are sometimes equal
(c) are never equal
(d) may be equal or unequal

105. In a single phase full wave ac regulator using two thyristors only, the gate circuits of the two thyristors
(a) must be isolated
(b) may or may not be isolated
(c) should be preferably isolated
(d) should not be isolated

106. In a single phase foil wave ac regulator feeding R-L load, the gate pulses

(a) are always of short duration
(b) may be of short duration
(c) should be rectangular with width $(\pi - \alpha)$
(d) cannot be rectangular with width $(\pi - \alpha)$

107. In a single phase full wave regulator, the rms current in each thyristors is I_r. The rms load current is equal to
(a) $2I_r$ (b) $\sqrt{2}I_r$
(c) I_r (d) $\frac{I_r}{\sqrt{2}}$

108. In a single phase full wave regulator feeding an R-L load, the power factor angle of load is θ. The range of control of firing angle α is
(a) $0 \le \alpha \le \pi$ (b) $\theta \le \alpha \le \pi$
(c) $0 \le \alpha \le (\pi - \theta)$ (d) $\theta \le \alpha \le (\pi + \theta)$

109. A single phase full wave regulator feeds a resistive load. The firing angle of thyristors is 90°. The input power factor is
(a) 1 (b) 0.707 lagging
(c) 0.5 lagging (d) 0.2 lagging

110. A single phase full wave regulator feeds R-L load. The best gating signal is
(a) short duration pulses (b) long duration pulses
(c) pulse train (d) either (a) or (b)

111. In a 3-phase full wave regulator feeding a star connected resistance load, firing angle is 75°. Then the number of thyristors conducting at one time is
(a) 1 (b) 2
(c) 3 (d) 4

112. In a 3-phase full wave regulator feeding a star connected resistance load the input voltage is 400 V line-to-line. The firing angle is 160°. The line-to-line output voltage would be
(a) 400 V (b) about 100 V
(c) about 20 V (d) zero

113. In a 3-phase full wave regulator feeding resistance load connected in star, the possible range of firing angle α is
(a) $0 \le \alpha \le 180°$ (b) $0 \le \alpha \le 150°$
(c) $0 \le \alpha \le 120°$ (d) $0 \le \alpha \le 90°$

114. A cyclo converter for single phase operation has
(a) one positive group and one negative group converters
(b) two positive group and two negative group converters
(c) three positive group and three negative group converters
(d) none of above

115. A cyclo converter is
(a) ac-dc converter (b) dc-ac converter
(c) dc-dc converter (d) ac-ac converter

116. A cyclo converter can be
(a) step down (b) step up
(c) step down or step up (d) neither of above

117. A cyclo converter uses
(a) natural commutation
(b) forced commutation
(c) either natural or forced commutation
(d) both natural and forced commutation together

118. In a single phase cyclo converter with centre tapped transformer
(a) all thyristors carry currents for equal duration
(b) the durations of currents in all thyristors may or may not be equal
(c) the durations of currents in all thyristors cannot be equal
(d) the duration of currents in all thyristors must be equal for satisfactory operation

119. A setp down cyclo converter has an input frequency f. If it uses natural commutation, the output frequency
(a) must be a sub-multiple of input frequency
(b) can be any frequency
(c) must be an even sub-multiple of input frequency
(d) must be an odd sub-multiple of input frequency

120. The power factor at the input terminals of a cyclo converter is generally
(a) low and leading (b) low and lagging
(c) high and leading (d) high and lagging

121. A single phase cyclo converter with centre tapped input transformer requires
(a) 2 thyristors (b) 4 thyristors
(c) 8 thyristors (d) either 2 or 4 thyristors

122. In a three-phase to single phase full wave bridge cyclo converter, the number of thyristors is
(a) 36 (b) 18
(c) 12 (d) 8

123. A single phase cyclo converter feeds a resistance heating load. The heating power is provided by
(a) only the fundamental component of output wave
(b) only higher harmonics in the output wave
(c) both fundamental and higher harmonics in the output wave
(d) either (a) or (b)

124. A bridge type single phase cyclo converter changes the frequency f to $f/3$. Then one half wave of output contains
(a) 3 full waves of input (b) 3 half waves of input
(c) 6 full waves of input (d) 6 half waves of input

125. A single phase step up cycloconverter changes 50 Hz to 100 Hz. Then one half wave of input will give rise to
(a) one half wave of output
(b) one full wave of output
(c) two full waves of output
(d) either (b) or (c)

126. In a cyclo converter it is possible to vary
(a) only output frequency
(b) only output voltage
(c) both output voltage and frequency but not simultaneously
(d) both output voltage and output frequency simultaneously

127. A dc circuit breaker must use
(*a*) natural commutation
(*b*) forced commutation
(*c*) both natural and forced commutation
(*d*) either natural or forced commutation

128. A NAND gate has 3 inputs and one output. The number of thyristors required for this gate are
(*a*) 1 (*b*) 2
(*c*) 3 (*d*) 3 or 4

129. In controlled heating circuits, thyristors act as
(*a*) rectifier (*b*) inverter
(*c*) ac regulator (*d*) chopper

130. An electric heater is controlled by thyristors through on-off control. If $\alpha = 0.4$, the heating is
(*a*) 16% maximum (*b*) 40% of maximum
(*c*) 60% of maximum (*d*) 84% of maximum

131. A single phase electric heater uses phase controlled thyristors. The firing angle is 90°. The degree of heating is
(*a*) 25% of maximum (*b*) 50% of maximum
(*c*) 75% of maximum (*d*) none of the above

132. In dielectric heating, the thyristor circuits consist of
(*a*) rectifier-chopper combination
(*b*) controlled rectifier
(*c*) ac regulator
(*d*) rectifier-inverter combination

133. In brushless excitation system of modern alternators, the main exciter is
(*a*) dc generator
(*b*) ac generator
(*c*) either dc or ac generator
(*d*) induction generator

134. A static excitation system uses
(*a*) transformer and thyristors
(*b*) transformer and ac exciter
(*c*) dc exciter and ac exciter
(*d*) transformer and dc exciter

135. Present-day HVDC converters are all
(*a*) 3 pulse converters
(*b*) 6 pulse converters
(*c*) 12 pulse converters
(*d*) either 6 pulse or 12 pulse converters

136. Thyristors can be used for control of
(*a*) dc separately excited motor
(*b*) dc shunt motor
(*c*) dc series motor
(*d*) all the above

137. A dc separately excited motor has constant field current. The armature is fed from a single phase supply through a full converter. When $\alpha = 0$, speed is 500 rpm. If $\alpha = 45°$, the speed is likely to be
(*a*) about 350 rpm (*b*) about 250 rpm
(*c*) about 175 rpm (*d*) about 125 rpm

138. A dc separately excited motor has a constant field current. The armature is fed from a single phase supply through a semi-converter. When $\alpha = 0$, speed is 600 rpm. If $\alpha = 90°$, the speed is likely to be
(*a*) about 300 rpm (*b*) about 150 rpm
(*c*) about 75 rpm (*d*) about zero

139. A dc series motor is coupled to a load whose torque is proportional to speed. It is fed by a single phase ac supply through a full converter. When speed is 500 rpm, firing angle is 45°. It is desired to raise the speed. It will be necessary to
(*a*) decrease the firing angle
(*b*) increase the firing angle
(*c*) either decrease or increase the firing angle depending on the required speed
(*d*) decrease the load torque

140. A 3-phase induction motor is fed by a 3-phase ac regulator to change the stator voltage. The variation in speed will be more if
(*a*) load torque is constant
(*b*) load torque increases with increase in speed
(*c*) load torque decreases with increase in speed
(*d*) either (*a*) or (*b*)

141. A 3-phase induction motor is controlled by voltage and frequency control such that the ratio V/f is constant. At 50 Hz, the starting torque is 100 N-m. At 25 Hz, the starting torque
(*a*) will be less than 100 N-m
(*b*) will be more than 100 N-m
(*c*) may be more or less than 100 N-m depending on motor parameters
(*d*) very small

142. A 3-phase induction motor is controlled by voltage and frequency control such that the ratio V/f is constant. At 50 Hz, the speed is 1440 rpm. At 25 Hz, the speed is likely to be
(*a*) about 1350 rpm (*b*) about 720 rpm
(*c*) about 360 rpm (*d*) either (*b*) or (*c*)

143. The total number of leads in SCR, DIAC and TRIAC respectively are
(*a*) 3, 2, 3 (*b*) 2, 3, 3
(*c*) 3, 3, 2 (*d*) 3, 2, 4

144. The leads in a TRIAC are denoted by
(*a*) anode cathode, gate (*b*) T_1, T_2, gate
(*c*) T_1, T_2, anode (*d*) T_1, T_2, T_3

145. The total number of leads in reverse conducting Thyristor is
(*a*) 2 (*b*) 3
(*c*) 1 (*d*) 4

146. The total number of leads in MOS controlled thyristor is
(*a*) 2 (*b*) 3
(*c*) 4 (*d*) 2 or 3

147. A static induction thyristor
(*a*) is a majority carrier device
(*b*) is a minority carrier device
(*c*) is either a majority carrier or minority carrier device
(*d*) has a total number of 2 leads

148. The total number of leads in SUS, SBS and SCS respectively are

(a) 3, 3, 4 (b) 3, 3, 3
(c) 2, 3, 3 (d) 3, 3, 5

149. An SUS is

(a) exactly similar to SCR but with higher rating
(b) similar to UJT
(c) PUT and avalanche diode connected in anti-parallel
(d) SCR and avalanche diode connected in parallel

150. Which of these devices can conduct in both directions

(a) SCR (b) SUS
(c) SBS (d) UJT

151. The number of layers in a shockley diode are

(a) 2 (b) 3
(c) 4 (d) 2 or 3

152. A shockley diode can be used

(a) in place of SCR
(b) as a trigger switch for SCR
(c) in place of UJT
(d) both in place of SCR and as a trigger switch for SCR

153. The number of leads in a power semi-conductor diode are

(a) 2 (b) 3
(c) 4 (d) 2 or 3

154. If we connect two silicon unilateral switches in anti-parallel, we get

(a) SCR (b) SBS
(c) SCS (d) UJT

155. The symbol in Fig. 8.35 is for

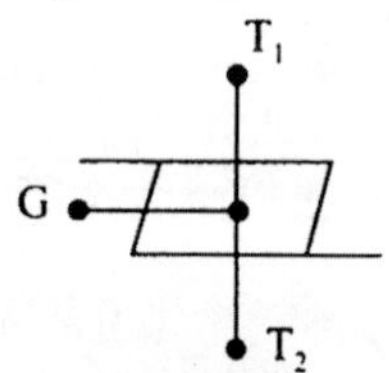

Fig. 8.35.

(a) SUS (b) SBS
(c) SCS (d) shockley diode

156. The symbol in Fig. 8.36 is for

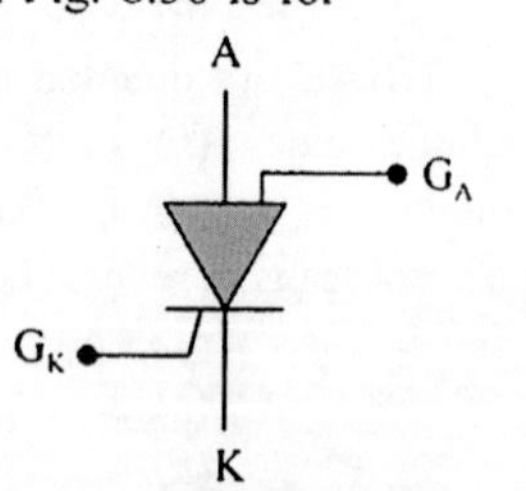

Fig. 8.36.

(a) SUS (b) SBS
(c) SCS (d) none of the above

157. The symbol in Fig. 8.37 is for

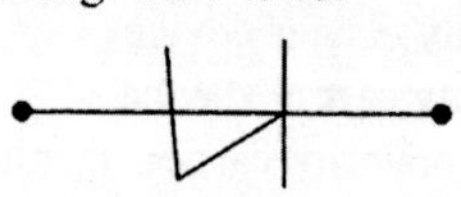

Fig. 8.37.

(a) SUS (b) SBS
(c) Shockley diode (d) UJT

158. The symbol in Fig. 8.38 is for

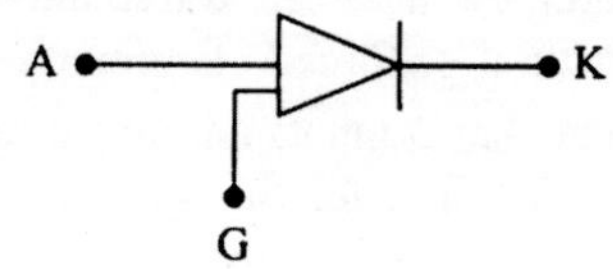

Fig. 8.38.

(a) UJT (b) PUT
(c) SCS (d) SBS

159. The symbol in Fig. 8.39 is for

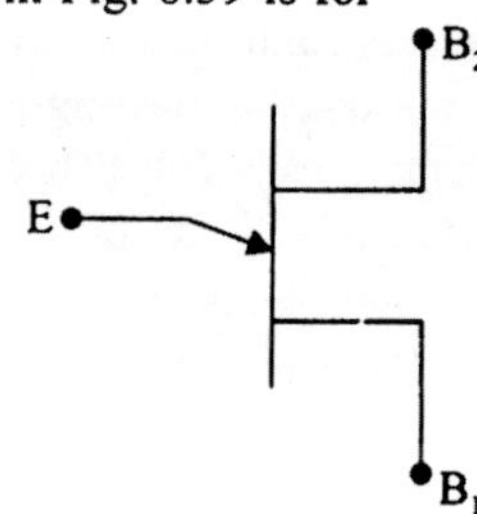

Fig. 8.39.

(a) UJT (b) PUT
(c) Shockley diode (d) TRIAC

160. Which of the following devices exhibit negative resistance characteristics

(a) UJT (b) PUT
(c) both UJT and PUT (d) UJT, TUT and SBS

161. A UJT exhibits negative resistance characteristics when

(a) current is less than valley current
(b) current is less than peak current
(c) current is more than peak current but less than valley current
(d) none of the above

162. Which of these devices is used in relaxation oscillator?

(a) UJT (b) SCR
(c) DIAC (d) TRIAC

163. The characteristics of which device has peak point and valley point

(a) SCR (b) PUT
(c) SCS (d) Schockley diode

164. Lead mounting for SCR is used for

(a) rating less than mA
(b) rating about 1A of less
(c) rating of 20 A or less than 20 A
(d) rating 100 A or less

165. The stud in a stud mounted SCR acts as

(a) anode (b) cathode
(c) gate (d) anode or cathode

166. In a stud mounted SCR, the stud is

(a) in thermal contact with heat sink
(b) insulated from heat sink
(c) in electrical and thermal contact with heat sink
(d) either in electrical or thermal contact with heat sink

167. In disc mounting of SCR

(a) SCR is clamped on one side of disc shaped heat sink

(*b*) SCR is clamped between two disc shaped heat sinks
(*c*) the SCR has low current rating
(*d*) none of the above

168. Which of these devices have four layers?
(*a*) UJT (*b*) SCR
(*c*) TRIAC (*d*) SCR and TRIAC

169. In thermal ohm's law, temperature and power loss are analogous to
(*a*) V and I respectively of ohm's law
(*b*) I and V respectively of ohm's law
(*c*) R and V respectively of ohm's law
(*d*) I and R respectively of ohm's law

170. In a power diode the reverse recovery time is the time from the instant the forward current is zero to the instant when reverse recovery current has decayed to
(*a*) 50% of peak reverse current
(*b*) 25% of peak reverse current
(*c*) 10% of peak reverse current
(*d*) zero value

171. The cut in voltages for germanium and silicon diodes respectively as
(*a*) 0.3 V and 0.7 V (*b*) 0.7 V and 0.3 V
(*c*) 0.6 V and 0.9 V (*d*) 0.5 V and 0.7 V

172. When a power diode is conducting, the voltage drop across it is about
(*a*) 1 V (*b*) 5 V
(*c*) 20 V (*d*) 50 V

173. The terminals of a power BJT are called
(*a*) emitter, base, collector
(*b*) emitter, base, drain
(*c*) source, base, drain
(*d*) source, base, collector

174. The terminals of a power MOSFET are called
(*a*) emitter, base, collector (*b*) source, gate, drain
(*c*) source, base, drain (*d*) emitter, gate, drain

175. Which statement is correct
(*a*) BJT and MOSFET are current controlled devices
(*b*) BJT is voltage controlled and MOSFET is current controlled device
(*c*) BJT and MOSFET are voltage controlled devices
(*d*) BJT is current controlled and MOSFET is voltage controlled device

176. Which of the following is correct for a BJT
(*a*) $\beta = \frac{\alpha}{1-\alpha}$ (*b*) $\alpha = \frac{\beta}{1-\beta}$
(*c*) $\beta = \frac{\alpha}{1+\alpha}$ (*d*) $\alpha = \frac{\beta}{1+\beta}$

177. For a BJT $\alpha = 0.98$, then $\beta =$
(*a*) 0.02 (*b*) 0.5
(*c*) 4.9 (*d*) 49

178. As compared to BJT, MOSFET has
(*a*) lower switching losses but higher conduction losses
(*b*) higher switching losses and higher conduction losses
(*c*) higher switching losses and lower conduction losses
(*d*) lower switching losses and lower conduction losses

179. The turn on characteristics of a power transistor can be assumed to be linear, as shown in Fig. 8.40. During turn on time T, voltage reduces linearly from V to 0 and current rises linearly from 0 to I. The energy loss during time T is

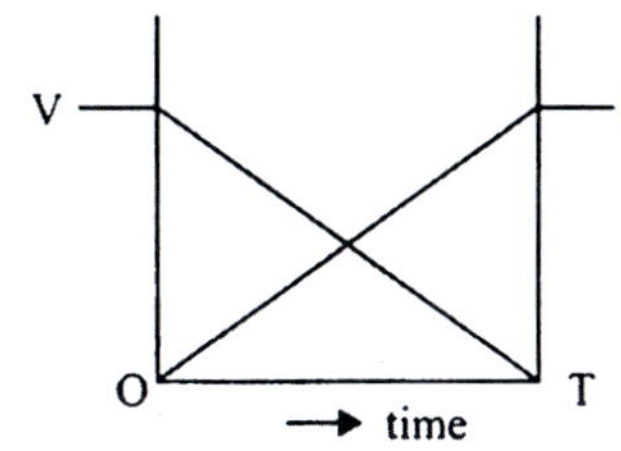

Fig. 8.40.

(*a*) VIT (*b*) $\frac{VIT}{2}$
(*c*) $\frac{VIT}{3}$ (*d*) $\frac{VIT}{6}$

180. Which of the following devices has metal-silicon junction
(*a*) general purpose power diode
(*b*) Schottky diode
(*c*) SCR
(*d*) MOSFET

181. As compared to a *p-n* junction diode
(*a*) a Schottky diode has lower cut in voltage and lower reverse saturation current
(*b*) a Schottky diode has higher cut in voltage and higher reverse saturation current
(*c*) a Schottky diode has lower cut in voltage and higher reverse saturation current
(*d*) a Schottky diode has higher cut in voltage and lower saturation current

182. The symble in Fig. 8.41 is for

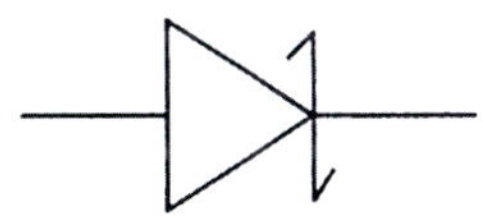
Fig. 8.41.

(*a*) DIAC (*b*) P-N junction diode
(*c*) Schottky diode (*d*) none of the above

183. In an SCR, the anode current consists of
(*a*) only holes
(*b*) only electrons
(*c*) both electrons and holes
(*d*) either electrons or holes

184. A thyristor has its gate current zero. To start conduction the forward voltage must exceed
(*a*) about 0.5 V
(*b*) forward breakover voltage
(*c*) on state voltage
(*d*) peak working off state forward voltage

185. A surge voltage
(*a*) has very high magnitude and very long duration
(*b*) has very high magnitude and very small duration
(*c*) has very high magnitude very small duration and can be positive or negative
(*d*) can be only positive

186. Forward $\frac{dv}{dt}$ rating of an SCR is
(*a*) the maximum rate of rise of anode voltage which will trigger the SCR
(*b*) the maximum rate of rise of anode voltage which will not trigger the SCR if gate, signal is applied
(*c*) the maximum rate of rise of anode voltage which will not trigger the SCR if gate signal is not applied
(*d*) either (*b*) *or* (*c*)

187. Surge current rating of an SCR is
(*a*) the maximum repetitive surge current which the SCR can withstand
(*b*) the maximum non-repetitive surge current which the SCR can withstand
(*c*) the maximum repetitive or non-repetitive surge current which the SCR can withstand
(*d*) the maximum surge current which the SCR can withstand if both surge current and steady current exist together

188. During forward blocking state of SCR, the voltage and current respectively are
(*a*) high and high (*b*) low and low
(*c*) high and low (*d*) low and high

189. Between which points does UJT exhibit negative resistance characteristics
(*a*) between peak and valley points
(*b*) before peak point
(*c*) after peak point
(*d*) after valley point

190. UJT can be used for
(*a*) firing an SCR
(*b*) turning off an SCR
(*c*) both firing and turning off of SCR
(*d*) none of the above

191. when UJT is used for thyristor gale signal, the output of UJT is
(*a*) sine wave (*b*) square wave
(*c*) saw tooth wave (*d*) either (*a*) or (*c*)

192. A thyristor has maximum allowable junction temperature of 110°C and the ambient temperature is 40°C. If the maximum internal power dissipation is 50 W, the thermal resistance is
(*a*) 2°C/W (*b*) 1.4°C/W
(*c*) 0.7W/°C (*d*) 2.2°C/W

193. A thyristor has internal power dissipation of 40 W and is operated at an ambient temperature of 20°C. If thermal resistance is 1.6°C/W, the junction temperature is
(*a*) 114°C (*b*) 164°C
(*c*) 94°C (*d*) 84°C

194. A thyristor has a maximum allowable junction temperature of 120°C and the ambient temperature is 40°C. If thermal resistance is 1.6°C/W. the maximum allowable internal power dissipation is
(*a*) 20 W (*b*) 50 W
(*c*) 92 W (*d*) 128 W

195. If t_{on} and t_{off} are the turn on and turn off times of a thyristor, then
(*a*) t_{on} and t_{off} are almost equal
(*b*) $t_{on} >> t_{off}$
(*c*) $t_{off} >> t_{on}$
(*d*) t_{on} is about twice t_{off}

196. Typical values of $\frac{di}{dt}$ (A / μsec) and $\frac{dv}{dt}$ (v/μ sec) for an SCR are in the range of
(*a*) 1 to 10 (*b*) 25 to 500
(*c*) 0.1 to 1 (*d*) 1000 to 2000

197. In a thyristor the gate current is increased, then
(*a*) anode current will increase
(*b*) anode current will decrease
(*c*) anode current will remain constant
(*d*) anode current may increase or decrease

198. An SCR has a turn an time of 4 μs. The gate pulse should have a duration of about
(*a*) 2 μs (*b*) 5 μs
(*c*) 20 μs (*d*) 50 μs

199. From which instant is the turn off time of a thyristor defined
(*a*) when anode voltage becomes zero
(*b*) when anode current becomes zero
(*c*) when both anode voltage and current become zero
(*d*) none of the above

200. A thyristor has a turn on time of 6 μs. If the anode circuit is inductive, the turn on time will be
(*a*) 6 μs (*b*) less than 6 μs
(*c*) more than 6 μs (*d*) either 6 μs or less

201. An SCR is triggered at 80° in the positive half cycle only. The rms anode current is 25 A. If the firing angle is changed to 160°, the rms current is likely to be
(*a*) 25 A (*b*) 12.5 A
(*c*) less than 25 A (*d*) less than 12.5 A

202. An SCR is triggered at 40° in th positive half cycle only. The average anode current is 50 A. If the firing angle is changed to 80°, the average anode current is likely to be
(*a*) 50 A (*b*) 25 A
(*c*) less than 50 A but more than 25 A
(*d*) less than 25 A

203. A thyristor needs protection against
(*a*) high $\frac{dv}{dt}$
(*b*) high $\frac{di}{dt}$
(*c*) both high $\frac{dv}{dt}$ and high $\frac{di}{dt}$
(*d*) either high $\frac{dv}{dt}$ or high $\frac{di}{dt}$

204. The snubber circuit consists of
(*a*) a capacitor connected in series with thyristor
(*b*) a capacitor connected in parallel with thyristor

(*c*) a resistor and capacitor in series connected across thyristor
(*d*) none of the above

205. In a thyrite arrester $i = kv^{\alpha}$. The value of α is
(*a*) about 0.5 (*b*) about 2
(*c*) about 5 (*d*) about 20

206. The characteristics of a non-linear resistance is $i = kv^4$. If i becomes 100 times, v becomes
(*a*) about 100 times (*b*) about 10 times
(*c*) about 3 times (*d*) about twice

207. Current chopping leads to
(*a*) over voltage
(*b*) over current
(*c*) both over voltage and over current
(*d*) either over voltage or over current

208. A 750 A circuit uses thyristor of current rating 175 A. If derating is 25%, the number of thyristors in parallel is
(*a*) 3 (*b*) 4
(*c*) 6 (*d*) 10

209. A 3 kV circuit uses SCR of 800 V rating. If derating is 25%, the number of SCRs in series is
(*a*) 4 (*b*) 5
(*c*) 6 (*d*) 8

210. In which of the following commutation methods is the overall circuit under damped
(*a*) Class A (*b*) Class B
(*c*) Class C (*d*) both Class A and Class B

211. The difference between Resonant commutation and self-commutation is
(*a*) both these methods are based on entirely different principles
(*b*) in resonant commutation, the commutating elements carry load current but in self-commutation, the commutating elements do not carry load current
(*c*) resonant commutation uses L and C but self-commutation uses only capacitor
(*d*) none of the above

212. Which of these commutation methods is used in AC-DC converters?
(*a*) Class A (*b*) Class C
(*c*) Class D (*d*) Class F

213. Which of these commutation methods uses an external pulse for commutation?
(*a*) Class A (*b*) Class C
(*c*) Class E (*d*) Class F

214. Which of these commutation methods uses two loads?
(*a*) Class A (*b*) Class B
(*c*) Class C (*d*) Class D

215. Which of these commutation methods uses an auxiliary SCR?
(*a*) Class A (*b*) Class B
(*c*) Class C (*d*) Class D

216. When thyristors are connected in parallel, the current distribution may become non-uniform due to
(*a*) inductive effect of current carrying conductors
(*b*) capacitive effect of current carrying conductors
(*c*) both inductive and capacitive effects
(*d*) none of the above

217. The specified maximum junction temperature of thyristors is about
(*a*) 50°C (*b*) 80°C
(*c*) 120°C (*d*) 200°C

218. The static equalisation circuit for thyristors connected in series is
(*a*) one resistor in series with each thyristor
(*b*) one resistor in parallel with each thyristor
(*c*) one resistor and one capacitor in parallel with each thyristor
(*d*) any of the above

219. Two thyristors of the same rating and specifications
(*a*) will have equal on state voltage drops
(*b*) will have equal forward blocking voltage for the same current
(*c*) will have different on state voltage drops
(*d*) may have equal or different on state voltage drops and equal or different forward blocking voltages for the same current

220. Two thyristors of same rating and same specifications
(*a*) will have equal turn on and turn off periods
(*b*) will have equal turn on but unequal turn off periods
(*c*) may have equal or unequal turn on and turn off periods
(*d*) will have unequal turn on and turn off periods

221. The dynamic equalising circuit consists of a series combination of capacitor C and resistor R_C across each thyristor. This resistance R_C. (along with parallel connected diode)
(*a*) limits charging current
(*b*) limits discharging current
(*c*) limits both charging and discharging currents
(*d*) none of the above

222. The value of capacitor C for dynamic equalising circuit of series connected thyristors is determined by
(*a*) forward characteristics of thyristors
(*b*) reverse recovery characterstics of thyristors
(*c*) both forward and reverse recovery characteristics of thyristors
(*d*) none of the above

223. The circuit in Fig. 8.42 is for

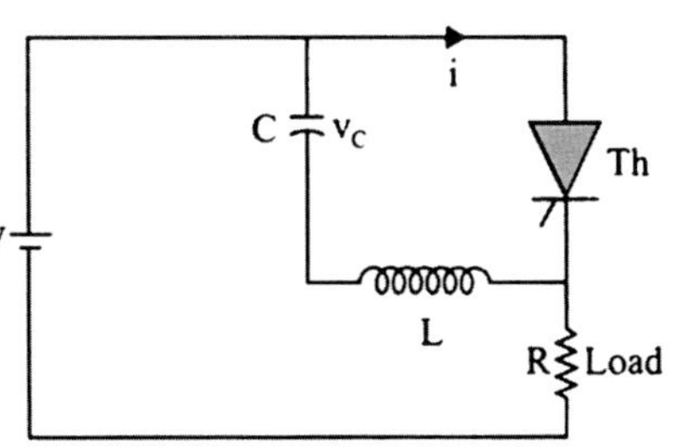

Fig. 8.42.

(*a*) resonant commutation (*b*) self-commutation
(*c*) load commutation (*d*) auxiliary commutation

224. The circuit in Fig. 8.43 is for

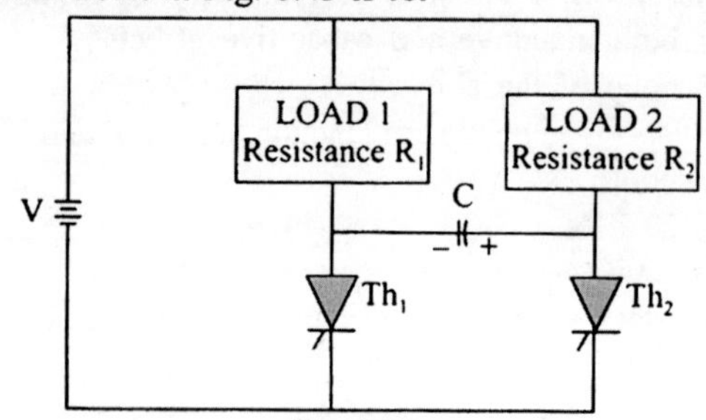

Fig. 8.43.

(*a*) resonant commutation (*b*) line commutation
(*c*) auxiliary commutation
(*d*) complementary commutation

225. Which of these triggering methods for series connected SCRs use LASCR in the gate circuit
(*a*) resistance triggering (*b*) simultaneous triggering
(*c*) sequential triggering (*d*) optical triggering

226. Ripple factor RF and form factor FF are related as
(*a*) $RF = FF^2 - 1$ (*b*) $RF = (FF^2 - 1)^{0.5}$
(*c*) $RF = \sqrt{FF - 1}$ (*d*) $RF = (FF - 1)^{0.5}$

227. A rectifier is fed by a transformer. The rms voltage and current at secondary of transformer are V_s amd I_s. Output dc power is $V_{dc}\,I_{dc}$ and output ac power is $(V_{rms})\,(I_{rms})$. Transformer utilisation factor TUF is
(*a*) $TUF = \dfrac{V_{dc}I_{dc}}{V_sI_s}$ (*b*) $TUF = \dfrac{(V_{rms})(I_{rms})}{V_sI_s}$
(*c*) $TUF = \dfrac{P_{dc}I_{dc}}{(V_{rms})(I_{rms})}$ (*d*) $\dfrac{(V_{rms})(I_{rms})}{P_{dc}I_{dc}}$

228. An ideal rectifier should have
(*a*) 100% efficiency, zero ripple factor, zero harmonic factor and unity displacement factor
(*b*) 100% efficiency, zero ripple factor, zero harmonic factor and zero displacement factor
(*c*) 100% efficiency, zero ripple factor, unity harmonic factor and unity displacement factor
(*d*) 100% efficiency, unity ripple factor, zero harmonic factor and unity displacement factor

229. A single phase half wave converter is feeding a resistor load. The input voltage is $v = V_m \sin \omega t$. If firing angle is zero, the output dc voltage V_{dc} and output rms voltage V_{rms} respectively are
(*a*) $\dfrac{V_m}{\pi}$ and $\dfrac{V_m}{\sqrt{2}}$ (*b*) $\dfrac{V_m}{\pi}$ and $\dfrac{V_m}{2}$
(*c*) $\dfrac{V_m}{2\pi}$ and $\dfrac{V_m}{\sqrt{2}}$ (*d*) $\dfrac{V_m}{2\pi}$ and $\dfrac{V_m}{2}$

230. A single phase half wave converter is feeding a resistive load. Just before triggering, the voltage across thyristor is
(*a*) zero (*b*) about 1 V
(*c*) same as input voltage (*d*) twice the input voltage

231. A single phase half wave rectifier is feeding a resistive load. Input voltage $v = V_m \sin \omega t$. The output dc voltage is V_{dc} and output rms voltage is V_{rms}. If firing angle is 180°, V_{dc} and V_{rms} respectively are
(*a*) 0 and 0 (*b*) $\dfrac{V_m}{\pi}$ and $\dfrac{V_m}{\sqrt{2}}$
(*c*) 0 and $-V_m$ (*d*) $\dfrac{-V_m}{\pi}$ and $\dfrac{-V_m}{\sqrt{2}}$

232. A single phase half wave converter is used to charge a battery of 50 V. The thyristor is continuously fired by dc signal. Input voltage is $v = V_m \sin \omega t$. The thyristor will conduct
(*a*) from 0 to 180° (*b*) when $v > 50$ V
(*c*) from 0 to 90° (*d*) When $v < 50$ V

233. A single phase half wave rectifier is feeding a resistive load. When $\alpha = 0$, dc output voltage is 127 V. When $\alpha = 30°$, the dc output voltage will be about
(*a*) 127 V (*b*) 120 V
(*c*) 110 V (*d*) 100 V

234. A single phase half wave rectifier is feeding a resistive load. As firing angle is increased
(*a*) dc output voltage and rms output voltage decrease in the same proportion
(*b*) dc output voltage and rms output voltage decrease
(*c*) dc output voltage and rms output voltage increase
(*d*) dc output voltage arid rms output voltage increase in the same proportion

235. A two winding transformer is feeding a single phase half wave rectifier circuit. The load is purely resistive. The rms value of transformer secondary current is I_s and rms value of load current in I_{rms}. Then
(*a*) $I_s = I_{rms}$ irrespective of the value of firing angle
(*b*) $I_s = 0.5\ I_{rms}$ irrespective of the value of firing angle
(*c*) $I_s = I_{rms}$ if firing angle is less than 30°
(*d*) $I_s = I_{rms}$ if firing angle is more than 30°

236. A single phase half wave converter is feeding a resistive load. Output dc power is P_{dc} and total power delivered to load is P_{ac}, then
(*a*) $P_{dc} = P_{ac}$ for all values of firing angles
(*b*) $P_{dc} = P_{ac}$ if firing angle is less than 180°
(*c*) $P_{dc} < P_{ac}$ for all values of firing angles
(*d*) $P_{dc} < P_{ac}$ if firing angle is less than 180°

237. A single phase half wave rectifier circuit has a free wheeling diode. The free wheeling diode will conduct only if
(*a*) load is purely resistive
(*b*) load is purely inductive
(*c*) load is combination of R and L
(*d*) load is purely inductive or combination of R and L

238. A single phase half wave controlled rectifier circuit has a free wheeling diode. The load is a combination of R and L. The firing angle is α. The period of conduction of SCR and free wheeling diode respectively are
(*a*) $\alpha < \omega t < \pi$ and $\pi < \omega t < (2\pi + \alpha)$
(*b*) $\alpha < \omega t < \pi$ and $\pi < \omega t < 2\pi$
(*c*) $\alpha < \omega t < (\pi + \alpha)$ and $(\pi + \alpha) < \omega t < 2\pi$
(*d*) $\alpha < \omega t < \pi$ and $\pi < \omega t < (2\pi - \alpha)$

239. A single phase half wave controlled rectifier circuit is feeding on RL load. The circuit has a free wheeling diode. The input voltage is $v = V_m \sin \omega t$. If i_0 is the load current during the conduction of free wheeling diode, the KVL equation during the condition of free wheeling diode is

(a) $V_m \sin \omega t = Ri_0 + \frac{di_0}{dt}$ (b) $0 = Ri_0 + L\frac{di_0}{dt}$

(c) $0 = Ri_0 - L\frac{di_0}{dt}$ (d) $V_m \sin \omega t = Ri_0 - L\frac{di_0}{dt}$

240. A single phase half wave controlled rectifier circuit has an R-L load. A free wheeling diode is also in the circuit. When free wheeling diode is conducting the SCR
(a) is forward biased
(b) is reverse biased
(c) may be forward biased or reverse biased
(d) forward biased initially but reverse biased afterwards

241. If V_{dc1} is the dc output voltage in half wave controlled rectifier circuit feeding resistive load and V_{dc2} is the dc output voltage in full wave controlled rectifier circuit (M – 2 connection) feeding resistive load, then
(a) $V_{dc1} = 0.5\ V_{dc2}$ for all values of firing angle
(b) $V_{dc1} = V_{dc2}$ for all values of firing angle
(c) $V_{dc1} \le 0.5\ V_{dc2}$ for all values of firing angle
(d) $V_{dc1} \ge 0.5\ V_{dc2}$ for all values of firing angle

242. It is possible to use a free wheeling diode in
(a) single phase half wave controlled rectifier as well as in single phase full wave controlled ractifier (M – 2 connection)
(b) single phase half wave controlled rectifier but not in single phase full wave controlled ractifier (M – 2 connection)
(c) single phase full wave controlled rectifier (M – 2 connection) but not in single phase half wave controlled rectifier
(d) all controlled rectifier circuits

243. In a single phase full wave converter (M – 2 connection), the rms transformer voltage is 300 V across each half of secondary. The PIV for each thyrisior is

(a) 300 V (b) $\frac{300}{\sqrt{2}}$ V

(c) $600\sqrt{2}$ V (d) $\frac{300}{2}$ V

244. In a single phase full bridge converter (B – 2 connection), the input voltage is $v = V_m \sin \omega t$. The output dc voltage for resistive load is

(a) $\frac{1}{\pi}\int_0^{\pi} V_m \sin \omega t\, d(\omega t)$ (b) $\frac{1}{2\pi}\int_{\alpha}^{\pi} V_m \sin \omega t\, d(\omega t)$

(c) $\frac{1}{\pi}\int_{\alpha}^{\pi} V_m \sin \omega t\, d(\omega t)$ (d) $\frac{1}{\pi}\int_{\alpha}^{\pi+\alpha} V_m \sin \omega t\, d(\omega t)$

245. In a single phase full wave converter (M – 2 connection) feeding on R–L load, the input voltage is $v = V_m \sin \omega t$. The expression for dc output voltage is

(a) $\frac{1}{\pi}\int_0^{\pi} V_m \sin \omega t\, d(\omega t)$ (b) $\frac{1}{\pi}\int_{\alpha}^{\pi} V_m \sin \omega t\, d(\omega t)$

(c) $\frac{1}{\pi}\int_{\alpha}^{\pi+\alpha} V_m \sin \omega t\, d(\omega t)$ (d) $\frac{1}{\pi}\int_{\alpha}^{\pi-\alpha} V_m \sin \omega t\, d(\omega t)$

246. In a single phase full wave converter (M – 2 connection) feeding on R-L load the input voltage is $v = V_m \sin \omega t$. The expression for rms output voltage is

(a) $\left[\frac{1}{\pi}\int_{\alpha}^{\pi} V_m^2 \sin^2 \omega t\, d(\omega t)\right]^{0.5}$

(b) $\left[\frac{1}{\pi}\int_{\alpha}^{\pi+\alpha} V_m^2 \sin^2 \omega t\, d(\omega t)\right]^{0.5}$

(c) $\left[\frac{1}{\pi}\int_{\alpha}^{\pi-\alpha} V_m^2 \sin^2 \omega t\, d(\omega t)\right]^{0.5}$

(d) $\left[\frac{1}{2\pi}\int_{\alpha}^{\pi+\alpha} V_m^2 \sin^2 \omega t\, d(\omega t)\right]^{0.5}$

247. In a single phase full wave converter (M – 2 connection) feeding a highly inductive load, the firing angle for each thyristor is α° in the respective half cycle. The period of conduction of each thyristor is
(a) π – α (b) π
(c) π + α (d) ω – 2α

248. In Fig 8.44, the load is highly inductive. The firing angle is α° for each of the thyristors. The periods of conduction of Th_1 and Th_2 respectively are

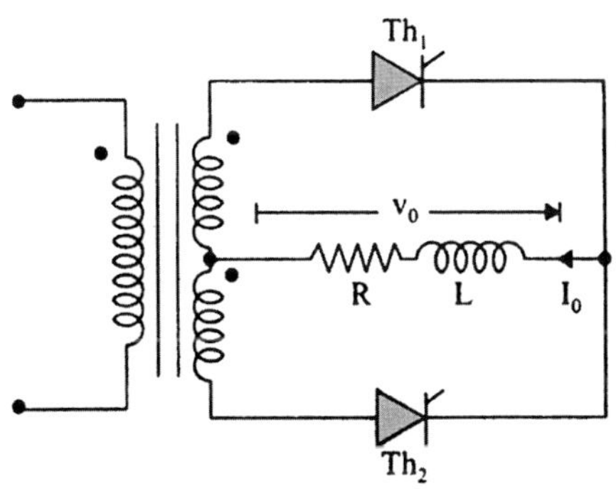

Fig. 8.44.

(a) 0 to π and π to 2π
(b) α to π and π + α to 2π
(c) α to π + α and π + α to 2π + α
(d) α to π – α and π + α to 2π – α

249. In Fig. 8.44, the average value of load current is 15 A. The load is highly inductive. The peak value of thyristor current is

(a) 15 A (b) $15\sqrt{2}$A

(c) $\frac{15}{\sqrt{2}}$A (d) 30 A

250. In Fig. 8.44 the average load current is 15 A. The rms value of transformer secondary current is

(*a*) 15 A (*b*) 10.61 A
(*c*) 7.5 A (*d*) 14.14 A

251. In Fig. 8.44 the rms voltage across each half of secondary is 200 V. The peak inverse voltage across each of thyristor is
(*a*) 200 V (*b*) $\sqrt{2}(200)$ V
(*c*) $\sqrt{2}(400)$ V (*d*) 800 V

252. In a circuit using a full wave converter (M – 2 connection) the PIV of each thyristor is 400 V, For the same output voltage and fully controlled bridge converter, PIV will be
(*a*) 400 V (*b*) 800 V
(*c*) 200 V (*d*) $200\sqrt{2}$ V

253. A fully controlled bridge converter can operate in quadrants
(*a*) 1 (*b*) 1 and 2
(*c*) 1 and 3 (*b*) 1 and 4

254. A fully controlled bridge converter feeds a highly inductive load such that load current can be assumed to be constant. The wave shape of input current is
(*a*) dc (*b*) sinusoidal
(*c*) rectangular (*d*) none of the above

255. A fully controlled bridge converter is feeding an RLE load. The load current is I_0. The input voltage is $V_m \sin \omega t$. The KVL equation to find firing angle α is
(*a*) $\frac{2V_m}{\pi} \cos \alpha = E + I_o R$
(*b*) $\frac{V_m}{\pi} \cos \alpha = E + I_o R$
(*c*) $\frac{2V_m}{\pi} \cos \alpha = E - I_o R$
(*d*) $\frac{V_m}{\pi} \cos \alpha = E - I_o R$

256. A fully controlled bridge converter is feeding an RLE load. For some circuit conditions firing angle is 47°. If connections of battery are reversed, the firing angle is likely to be
(*a*) 47° (*b*) less than 47°
(*c*) more than 47°
(*d*) either more or less than 47°

257. In a full wave converter (M – 2 connection) feeding an RLE load, the peak thyristor current is 15 A. For fully controlled bridge circuit and same load current, the peak thyristor current will be
(*a*) 30 A (*b*) $15\sqrt{2}$ A
(*c*) 15 A (*d*) $\frac{15}{\sqrt{2}}$ A

258. In a semiconverter feeding a resistive load, the firing angle is α. Each thyristor conducts for
(*a*) π radians (*b*) (π – α) radians
(*c*) (π + α) radians (*d*) (π – 0.5α) radians

259. In a semiconverter feeding a resistive load
(*a*) the waveshapes of output voltage and output current are exactly similar
(*b*) the waveshapes of output voltage and output current cannot be similar
(*c*) the waveshapes of output voltage and output current are similar only for low values of firing angle
(*d*) the waveshapes of output voltage and output current may be similar or dissimilar

260. In a semiconveter feeding a resistive load,
(*a*) the waveshapes of input voltage and input current are always similar
(*b*) the waveshaps of input voltage and input current can never be similar
(*c*) the waveshapes of input voltage and input current are similar only if firing angle is zero
(*d*) the waveshapes of input voltage and input current are similar only if firing angle is high

261. In a fully controlled bridge converter feeding a resistive load the waveshapes of output voltage and output current
(*a*) are always similar
(*b*) can never be similar
(*c*) may be similar or dissimilar
(*d*) are similar only if α = 0

262. For a single phase half wave converter feeding a resistive load, the firing angle α. The waveshape in Fig. 8.45 is for

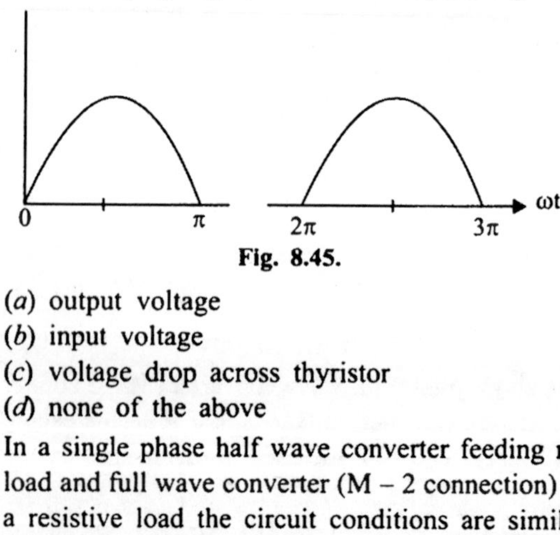

Fig. 8.45.

(*a*) output voltage
(*b*) input voltage
(*c*) voltage drop across thyristor
(*d*) none of the above

263. In a single phase half wave converter feeding resistive load and full wave converter (M – 2 connection) feeding a resistive load the circuit conditions are similar. The firing angle is α. Then the waveshapes of voltages across thyristors
(*a*) are similar
(*b*) are not similar
(*c*) may be similar or dissimilar
(*d*) are similar only for some values of π

264. A semiconverter feeding an RLE load operates in
(*a*) first quadrant only
(*b*) first and second quadrants
(*c*) first and fourth quadrants
(*d*) first and third quadrants

265. In a single phase semiconverter feeding a highly inductive load, the load current is
(*a*) sinusoidal (*b*) nearly constant
(*c*) absolutely constant (*d*) rectangular

266. A single phase semiconverter is feeding a highly inductive load. The circuit has a free wheeling diode across the load and firing angle is α. During one cycle of input voltage, the total duration of conduction of free wheeling diode is
(*a*) α (*b*) 2α
(*c*) 4α (*d*) 0.5α

267. A single phase semiconverter is feeding a highly inductive load and has free wheeling diode across the load. The waveshapes of output voltage and output current
(*a*) are similar
(*b*) are not similar
(*c*) may be similar or dissimilar
(*d*) are similar only if firing angle is zero

268. A single phase semiconverter is feeding a highly inductive load and has a free wheeling diode across the load. The waveshape of input current is
(*a*) sinusoidal
(*b*) rectangular
(*c*) sinusoidal or rectangular
(*d*) triangular

269. A single phase dual converter consists of
(*a*) two fully controlled converters connected in parallel
(*b*) two fully controlled converters connected in anti-parallel
(*c*) two fully controlled semiconverters connected in anti-parallel
(*d*) two fully controlled semiconverters connected in parallel

270. When a single phase dual converter is operating in circulating current mode
(*a*) both converters are working as rectifiers simultaneously
(*b*) one converter is always working as rectifier and the other as inverter
(*c*) only one converter is working at one time
(*d*) both converters are either working as rectifiers or as inverters simultaneously

271. As regards single phase dual converter
(*a*) the change in operation from one quadrant to the other is faster in non-circulating current mode
(*b*) the change in operation from one quadrant to the other is faster in circulating current mode
(*c*) the time response to change in operation from one quadrant to other is the same in both the modes
(*d*) the time response to change in operation from one quadrant to the other may be faster in any of the two modes

272. A single phase series converter is used for
(*a*) high voltage output
(*b*) high current output
(*c*) high voltage and high current output
(*d*) improving the circuit efficiency

273. Two semiconverters are connected in series to form a single phase series converter. The input $v = V_m \sin \omega t$ and α_1, α_2 are the firing angles of the two converters. If $\alpha_1 = \alpha_2 = 0$, the dc output voltage is
(*a*) $\frac{V_m}{\pi}$ (*b*) $\frac{2V_m}{\pi}$
(*c*) $\frac{4V_m}{\pi}$ (*d*) $\frac{8V_m}{\pi}$

274. Two single phase semiconverters are connected in series to form a series converter. The input is $v = V_m \sin \omega t$ and α_1, α_2 are the firing angles. If $\alpha_1 = 0$, the dc output voltage is
(*a*) $\frac{V_m}{\pi}(1+\cos\alpha_2)$ (*b*) $\frac{V_m}{\pi}(2+\cos\alpha_2)$
(*c*) $\frac{V_m}{\pi}(3+\cos\alpha_2)$ (*d*) $\frac{V_m}{\pi}(4+\cos\alpha_2)$

275. In a 3-phase fully controlled converter the firing frequency is
(*a*) 3 times the line frequency
(*b*) 6 times the line frequency
(*c*) 9 times the line frequency
(*d*) 12 times the line frequency

276. A 3-phase fully controlled converter is a
(*a*) 3 pulse converter (*b*) 6 pulse converter
(*c*) 2 pulse converter (*d*) 12 pulse converter

277. A 3-phase fully controlled bridge converter is fed by a 3-phase system having phase voltage $v = V_m \sin \omega t$. The firing angle is α. The dc output voltage is
(*a*) $V_{dc} = \frac{3\sqrt{3}}{\pi} V_m \cos\alpha$ (*b*) $V_{dc} = \frac{3V_m}{\pi} \cos\alpha$
(*c*) $V_{dc} = \frac{\sqrt{3}V_m}{\pi} \cos\alpha$ (*d*) $V_{dc} = \frac{3\sqrt{3}}{2\pi} V_m \cos\alpha$

278. In a 3-phase full converter the average load current is 150 A. The average thyristor current is
(*a*) 150 A (*b*) 75 A
(*c*) 50 A (*d*) 25 A

279. In a 3-phase full converter feeding a highly inductive load, the average load current is 150 A. The peak current through thyristor is
(*a*) 150 A (*b*) 75 A
(*c*) 50 A (*d*) 300 A

280. In a 3-phase full converter, the firing angle is less than 60°. The instantaneous output voltage
(*a*) will have positive part only
(*b*) will have both positive and negative parts
(*c*) may have both positive and negative parts
(*d*) will have negative part if load is inductive

281. In a 3-phase full converter, the instantanous output voltage can have negative part if
(*a*) firing angle is less than 60°
(*b*) firing angle is less than 90°
(*c*) firing angle is more than 60°
(*d*) firing angle is more than 90°

282. In a 3-phase semiconverter, the firing frequency is
(*a*) equal to line frequency
(*b*) 3 times the line frequency
(*c*) 6 times the line frequency
(*d*) 12 times the line frequency

283. In a 3-phase semiconverter, the free wheeling diode comes into operation only if firing angle
(*a*) is zero (*b*) is 60°
(*c*) is more than 60° (*d*) is more than 90°

284. In a 3-phase semi-converter, the dc output voltage for a certain firing angle is 50% of the maximum possible value. The firing angle is
(*a*) 30° (*b*) 60°
(*c*) 90° (*d*) 120°

285. In a 3-phase semiconverter, the output voltage becomes discontinuous for
(*a*) firing angle > 30° (*b*) firing angle > 60°
(*c*) firing angle > 90° (*d*) firing angle > 120°

286. A 3-phase semiconverter when supplying a certain load has a rectification efficiency of 0.6 and transformer utilisation factor of 0.35. When a 3-phase full converter is supplying the same load
(*a*) rectification efficiency will be more than 0.6 and TUF will be more than 0.35
(*b*) rectification efficiency will be more than 0.6 but TUF will be less than 0.35
(*c*) rectification efficiency will be less than 0.6 and TUF will be less than 0.35
(*d*) rectification efficiency will be less than 0.6 but TUF will be more than 0.35

287. The phenomenon of overlap in converter operation is due to
(*a*) source resistance
(*b*) source inductance
(*c*) both source resistance and source inductance
(*d*) high value of firing angle

288. Which converter is used in HVDC transmission?
(*a*) 3 pulse (*b*) 6 pulse
(*c*) 12 pulse (*d*) 24 pulse

289. For a 12 pulse operation, the two 6 pulse units are fed by
(*a*) Y – Y and Y – Y transformers
(*b*) Δ – Δ and Δ – Δ transformers
(*c*) Y – Y and Y – Δ transformers
(*d*) Y – Δ and Y – Δ transformers

290. A 3-phase dual converter consists of
(*a*) two 3-phase semiconverters connected in parallel
(*b*) two 3-phase semiconverters connected in anti-parallel
(*c*) two 3-phase full converters connected in anti-parallel
(*d*) two 3-phase full converters connected in parallel

291. In a dual converter operated in circulating current mode, the circulating current flows
(*a*) when load current is zero
(*b*) when load current is non-zero
(*c*) even when load current is zero
(*d*) none of the above

292. Inverters find applications in
(*a*) HVDC transmission
(*b*) UPS
(*c*) variable speed ac drives
(*d*) all the above

293. In a series inverter
(*a*) the commutating elements L and C are in parallel
(*b*) the commutating elements L and C are in series with load
(*c*) the commutating L and C may be in series or parallel with the load
(*d*) none of the above

294. If 0.5 T is the time period of oscillations and T_{off} is the time between turn off of one thyristor and turn on of second thyristor, the frequency of output f of a series inverter is
(*a*) $f = \dfrac{1}{2(0.5T + T_{off})}$ Hz (*b*) $f = \dfrac{1}{0.5T + T_{off}}$ Hz
(*c*) $f = \dfrac{2}{0.5T + T_{off}}$ Hz (*d*) $f = \dfrac{1}{T + T_{off}}$ Hz

295. If the frequency is less than the damped frequency in a series inverter
(*a*) the distortion of output voltage waveform is low
(*b*) the distortion of output voltage waveform is high
(*c*) the distortion of output voltage may be high or low depending on the load impedance
(*d*) none of the above

296. In the basic series inverter circuit, the source supplies load
(*a*) only in positive half cycle
(*b*) both in positive and negative half cycles
(*c*) only in part of positive half cycle
(*d*) in positive half cycle and part of negative half cycle

297. In the basic series inverter circuit, the source current has
(*a*) low harmonic content
(*b*) high harmonic content
(*c*) low or high harmonic content depending on firing angle
(*d*) zero harmonic content

298. A 3-phase series inverter has
(*a*) 12 thyristors (*b*) 6 thyristors
(*c*) 3 thyristors (*d*) 3 or 6 thyristors

299. A high frequency series inverter consists of
(*a*) a number of series inverters in series
(*b*) a number of series inverters in parallel
(*c*) a number of parallel inverters in parallel
(*d*) a number of parallel inverters in series

300. In a parallel inverter
(*a*) the commutating capacitor C is across the primary of transformer and the load is across the secondary

(*b*) the commutating capacitor and load are connected across the secondary of transformer
(*c*) the commutating capacitor C and load are connected in series and the combination is across the secondary of transformer
(*d*) none of the above

301. In a parallel inverter the turn off time t_q of thyristor and values of commutating elements L and C are related as
(*a*) $t_q = \pi\sqrt{LC}$ (*b*) $t_q = \frac{\pi}{2}\sqrt{LC}$
(*c*) $t_q = \frac{\pi}{3}\sqrt{LC}$ (*d*) $t_q = \frac{\pi}{4}\sqrt{LC}$

302. In a half bridge inverter, the free wheeling diodes are needed
(*a*) even when load is resistive
(*b*) only when load is inductive
(*c*) only when load is inductive or capacitive
(*d*) for all types of loads

303. In a half bridge inverter the gate pulses of the two thyristors have a phase displacement of
(*a*) 60° (*b*) 90°
(*c*) 120° (*d*) 180°

304. In a half bridge inverter each thyristor is gated at a frequency f. The output frequency is
(*a*) $2f$ (*b*) f
(*c*) $0.5f$ (*d*) $0.25f$

305. The waveshape of output voltage of half bridge inverter is
(*a*) sinusoidal (*b*) square
(*c*) triangular (*d*) (*a*) or (*b*)

306. In a full bridge circuit supplying an RLC under damped circuit
(*a*) the waveshapes of output voltage and load current are square
(*b*) the waveshapes of output voltage and load current are sinusoidal
(*c*) the waveshapes of output voltage is square but the waveshape of load current is nearly sinusoidal
(*d*) none of the above

307. In a full bridge inverter, the waveshape of load current depends on
(*a*) relative values of load parameters R, L and C
(*b*) relative values of load parameters R and L
(*c*) relative values of load parameters L and C
(*d*) gating ferquency

308. In a full bridge single phase inverter feeding RLC load
(*a*) one thyristor and one diode conduct simultaneously
(*b*) two thyristors conduct simultaneously
(*c*) two thyristors or two diodes conduct simultaneously
(*d*) four thyristors conduct simultaneously

309. In a half bridge inverter circuit fed by a battery of voltage V, the rms value of fundamental component of output voltage is
(*a*) $\sqrt{2}V$ (*b*) $\frac{\sqrt{2}V}{\pi}$
(*c*) $\frac{2V}{\pi}$ (*d*) $\frac{V}{\sqrt{2}\pi}$

310. In a full bridge inverter fed by a battery of voltage V, the rms value of fundamental component of output voltage is
(*a*) $\frac{\sqrt{2}\,V}{\pi}$ (*b*) $\frac{2\sqrt{2}\,V}{\pi}$
(*c*) $\frac{2\,V}{\pi}$ (*d*) $\frac{4\,V}{\pi}$

311. A full bridge inverter is fed by a battery of 30 V. The rms value of output voltage is
(*a*) 15 V (*b*) $15\sqrt{2}$ V
(*c*) 30 V (*d*) $30\sqrt{2}$ V

312. McMurray full bridge inverter uses
(*a*) 4 thyristors and 2 diodes
(*b*) 4 thyristors and 4 diodes
(*c*) 8 thyristors and 4 diodes
(*d*) 8 thyristors and 8 diodes

313. Modified McMurray full bridge inverter uses
(*a*) 4 thyristors and 2 diodes
(*b*) 4 thyristors and 4 diodes
(*c*) 8 thyristors and 4 diodes
(*d*) 8 thyristors and 8 diodes

314. Which inverter circuit uses coupled inductors
(*a*) Modified McMurray full bridge inverter
(*b*) McMurray Bedford half bridge inverter
(*c*) Modified McMurray Bedford half bridge inverter
(*d*) both (*b*) and (*c*)

315. The two modes of operation of 3-phase bridge inverters are known as
(*a*) 60° mode and 120° mode
(*b*) 90° mode and 120° mode
(*c*) 120° mode and 180° mode
(*d*) 120° mode and 240° mode

316. In a 3-phase bridge inverter with 180° mode of operation the numbers of thyristors conducting at one time are
(*a*) 1 (*b*) 2
(*c*) 3 (*d*) 4

317. In a 3-phase bridge inverter with 120° mode of operation the number of thyristors conducting at one time are
(*a*) 1 (*b*) 2
(*c*) 3 (*d*) 4

318. A 3-phase bridge inverter is fed by 400 V battery. The load is star connected and has a resistance of 10 Ω per phase. The peak value of load current is
(*a*) 40 A (*b*) 20 A
(*c*) 10 A (*d*) 5 A

319. A 3-phase bridge inverter is fed by 400 V battery. The load is star connected and has a resistance of 10 Ω per phase. The peak value of thyristor current is

(a) $\frac{400}{10}$ A (b) $\frac{400}{15}$ A

(c) $\frac{400}{20}$ A (d) $\frac{400}{30}$ A

320. In a certain inverter circuit each thyristor carries 26.67 A for 60°, 13.33 A for 120° and 0 A for 180° of the cycle. The rms current of thyristor is
(a) 20 A (b) 13.33 A
(c) 6.67 A (d) 3.33 A

321. In a step down chopper using time ratio control, the input voltage is V and duty cycle is α. The load is purely resistive having resistance R. The average and rms load currents are

(a) $\frac{\alpha V}{R}$ and $\frac{\alpha^2 V}{R}$ (b) $\frac{\alpha V}{R}$ and $\frac{\sqrt{\alpha}\,V}{R}$

(c) $\frac{\sqrt{\alpha}\,V}{R}$ and $\frac{\alpha V}{R}$ (d) $\frac{\alpha V}{R}$ and $\frac{(\alpha)^{1.5} V}{R}$

322. A separately excited dc motor is run from a dc source through a chopper. At 600 rpm and rated torque, the duty cycle of chopper is 0.8. At 600 rpm and half the rated torque, the duty cycle is likely to be
(a) 0.9 (b) 0.8
(c) 0.75 (d) 0.6

323. Figure 8.46 shows a chopper feeding RLE load. The free wheeling diode conducts when

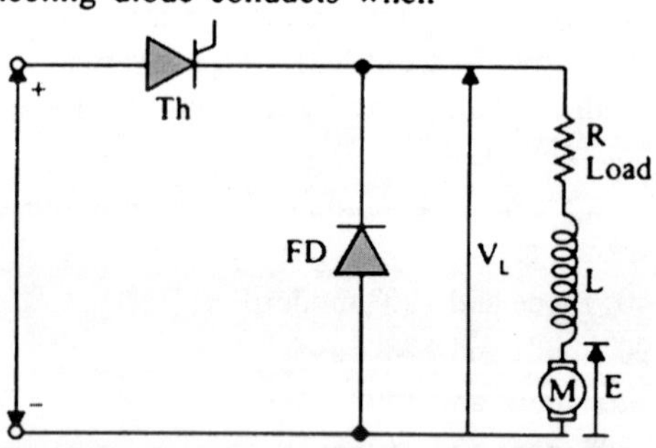

Fig. 8.46. Step down chopper with RLE load.

(a) thyristor is on
(b) thyristor is off
(c) both when thyristor is on and thyristor is off
(d) partly when thyristor is off and partly when thyristor is on

324. In the circuit of Fig. 8.46, the waveshape of toad current when thyristor is conducting is
(a) constant
(b) increasing exponentially
(c) decreasing exponentially
(d) either (b) or (c)

325. In the circuit of Fig. 8.46, the load current when thyristor is off
(a) is constant
(b) increases exponentially
(c) decreases exponentially
(d) zero

326. In a step down chopper using pulse width modulation, $T_{on} = 3 \times 10^{-3}$ and $T_{off} = 1 \times 10^{-3}$ s. The chopping frequency is
(a) 333.33 (b) 250
(c) 500 (d) 1000

327. Which chopper circuit operates in all the 4 quadrants ?
(a) A (d) C
(c) D (d) E

328. Which chopper circuit operates in quadrants 1 and 4
(a) A (b) B
(c) D (d) E

329. A class A chopper has an input voltage of 200 V. T_{on} = 1.5 ms and T_{off} = 4.5 ms. The duty cycle, average output voltage and ripple factor respectively are
(a) 0.5, 100 V and 1 (b) 0.25, 50 V and $\sqrt{3}$
(c) 0.25, 50 V and 1 (d) 0.5, 100 V and $\sqrt{3}$

330. A capacitor C farads is charged to a voltage V. It discharges through a thyristor, a diode and an inductance L. The peak current in the circuit is

(a) VCL (b) $V\sqrt{\frac{C}{L}}$

(c) $V\sqrt{\frac{L}{C}}$ (d) $\frac{V}{\sqrt{LC}}$

331. In Fig. 8.47, thyristor Th is in off state. When thyristor is turned on, the peak thyristor current can be

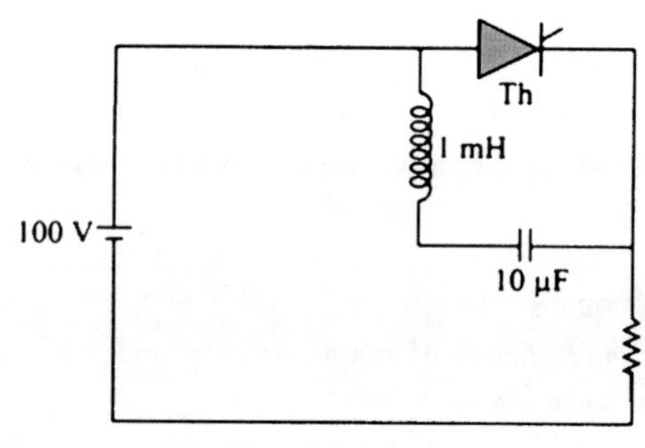

Fig. 8.47.

(a) 5 A (b) 10 A
(c) 15 A (d) 20 A

332. A dc chopper V = 500 V feeds an electric traction dc series motor. The chopping frequency is 400 Hz. The series inductance required to limit the maximum current swing to 10 A is
(a) 20 mH (b) about 31 mH
(c) about 51 mH (d) about 92 mH

333. In a multiphase chopper, the different choppers operate in
(a) series and all simultaneously
(b) parallel and one at a time
(c) parallel and simultaneously
(d) either (b) or (c)

334. Multiphase choppers are used in
(a) low power circuits
(b) high power circuit

(c) both low and high power circuits
(d) medium power circuits

335. A single phase half wave regulator uses
(a) 2 diodes and 2 thyristors
(b) 2 thyristors
(c) 1 diode and 1 thyristor
(d) 2 diodes

336. If peak value of input voltage is V_m and firing angle is α, the average output voltage for single phase half wave and single phase full wave regulators are
(a) $\frac{V_m}{2\pi}(\cos\alpha - 1)$ and $\frac{V_m}{2\pi}(\cos\alpha + 1)$
(b) $\frac{V_m}{\pi}(\cos\alpha - 1)$ and $\frac{V_m}{2\pi}(\cos\alpha + 1)$
(c) $\frac{V_m}{\pi}(\cos\alpha - 1)$ and $\frac{V_m}{\pi}(\cos\alpha + 1)$
(d) $\frac{V_m}{2\pi}(\cos\alpha + 1)$ and $\frac{V_m}{2\pi}(\cos\alpha - 1)$

337. In a single phase full wave regulator, the firing angles in the positive and negative half cycles are generally
(a) equal
(b) different
(c) equal or different
(d) different but sometimes equal

338. In a single phase full wave ac regulator the rms value of load current is 20 A. Load is purely resistive. Then
(a) average thyristor current will be less than 20 A but rms thyristor current will be more than 20 A
(b) both the average and rms thyristor currents will be less than 20 A
(c) both the average and rms thyrsitor currents will be more than 20 A
(d) rms thyristor current will be 20 A and average thyristor current will be 10 A

339. A 3-phase ac regulator uses
(a) 3 thyristors (b) 6 thyristors
(c) 9 thyristors (d) 12 thyristors

340. Which of the following devices has two gates?
(a) SCR (b) SCS
(c) SUS (d) SBS

341. A string of series connected thyristors needs equalising circuit to
(a) ensure proper firing
(b) ensure proper current division
(c) ensure proper voltage division
(d) none of the above

342. A single phase full converter gives maximum and minimum output voltage at firing angles of
(a) 0 and 180° respectively
(b) 180° and 0° respectively
(c) 0 and 90° respectively
(d) 90° and 0° respectively

343. To protect an SCR against high *di/dt* we connect
(a) a capacitance in series with SCR
(b) a capacitance in parallel with SCR
(c) an inductance in parallel with SCR
(d) an inductance in series with SCR

344. Match List I with List II and select the correct answer by using the given codes

List I — **List II**

A. BJT — 1. 

B. SCR — 2. 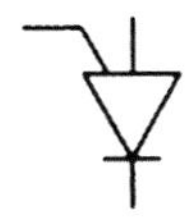

C. UJT — 3. 

D. PUT — 4. 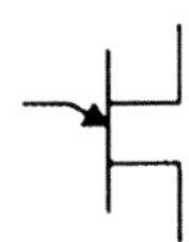

Codes:

	A	B	C	D
(a)	1	2	3	4
(b)	2	3	1	4
(c)	1	3	4	2
(d)	4	2	1	3

345. Match List I with List II and select the correct answer by using the given codes

List I — **List II**

A. DIAC — 1. 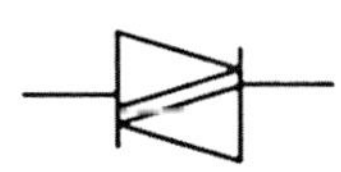

B. TRIAC — 2. 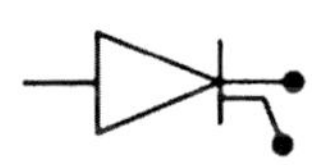

C. SIT — 3. 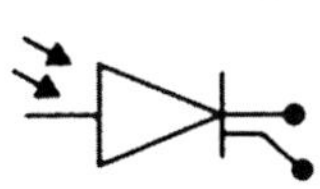

D. LASCR — 4. 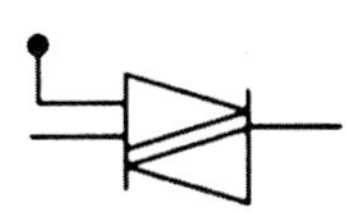

Codes:

	A	B	C	D
(a)	4	2	3	1
(b)	1	2	3	4
(c)	1	4	2	3
(d)	3	2	1	4

346. Match List I with List II and select the correct answer by using the given codes

List I	List II
A. SUS	1. 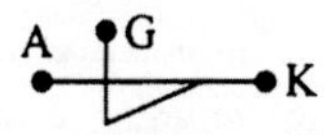
B. SBS	2. 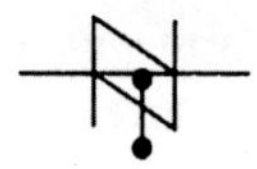
C. SCS	3. 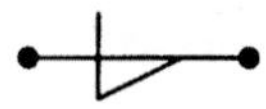
D. Schottky diode	4. 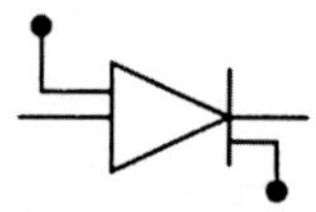

Codes:

	A	B	C	D
(a)	1	2	3	4
(b)	1	3	4	2
(c)	4	2	3	1
(d)	1	2	3	4

347. Match List I with List II and select the correct answer by using the given codes

List I	List II
A. GTO	1. Widely used in digital computers
B. UJT	2. Used for free wheeling
C. Power Diode	3. Can be turned off by negative gate signal
D. MOSFET	4. Used in generating triggering pulses

Codes:

	A	B	C	D
(a)	1	2	3	4
(b)	4	2	3	1
(c)	3	4	2	1
(d)	2	1	3	4

348. Match List I with List II and select the correct answer by using the given codes

List I

A. Line commutation
B. Resonant commutation
C. Complementary commutation
D. Self-commutation

List II

1. 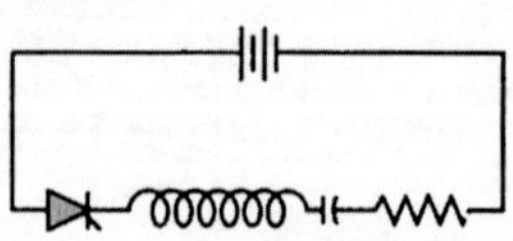

2. 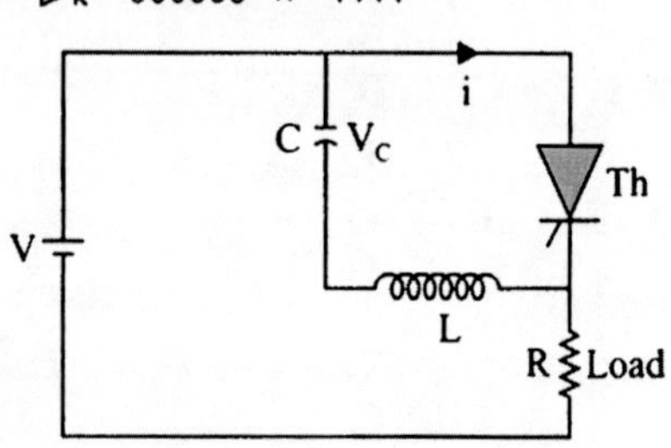

3. 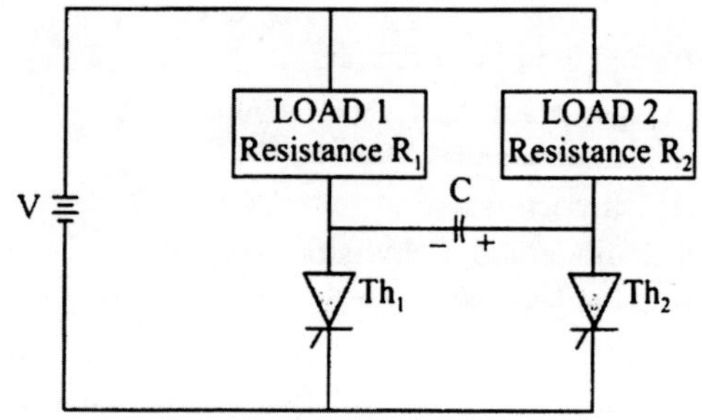

4. 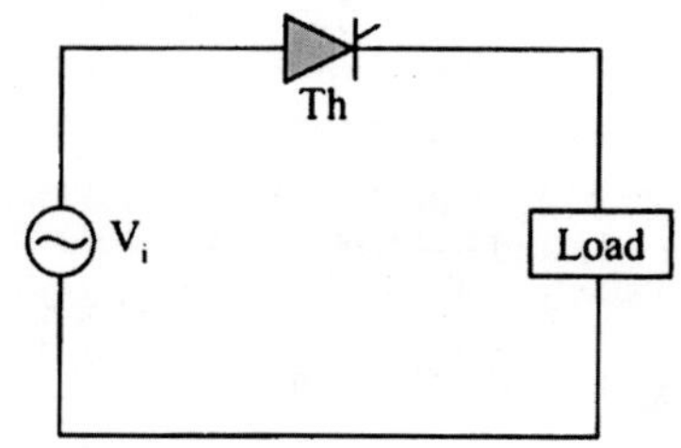

Codes:

	A	B	C	D
(a)	1	4	2	3
(b)	1	2	3	4
(c)	4	1	3	2
(d)	2	3	4	1

349. Match List I with List II and select the correct answer by using the given codes

List I (Circuits)	List II (Output dc voltage)
A. Single phase half wave converter	1. $\frac{V_m}{\pi}(\cos\alpha + 1)$
B. Fully controlled bridge converter (1 phase) with RLE load	2. $\frac{2V_m}{\pi}\cos\alpha$
C. 3-phase full converter	3. $\frac{3\sqrt{3}}{\pi}V_m\cos\alpha$
D. Single phase full wave converter (M-2 a connection)	4. $\frac{V_m}{2\pi}(1+\cos\alpha)$

Codes:

	A	B	C	D
(a)	1	2	3	4
(b)	4	3	1	2
(c)	4	2	3	1
(d)	1	2	4	3

350. Match List I with List II and select the correct answer by using the given codes

List I

A. Step up chopper
B. Class E chopper
C. Class D chopper
D. Class B chopper

List II

1. 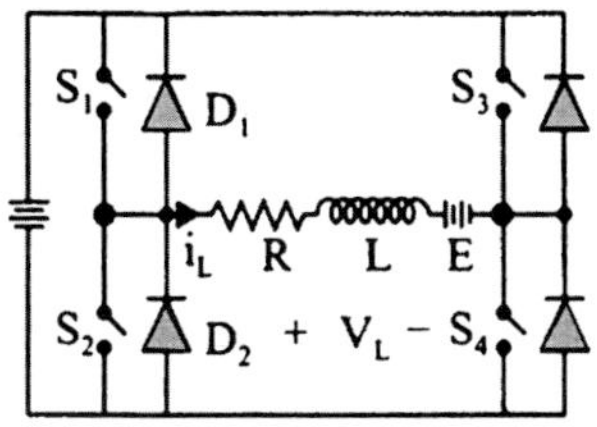

2. 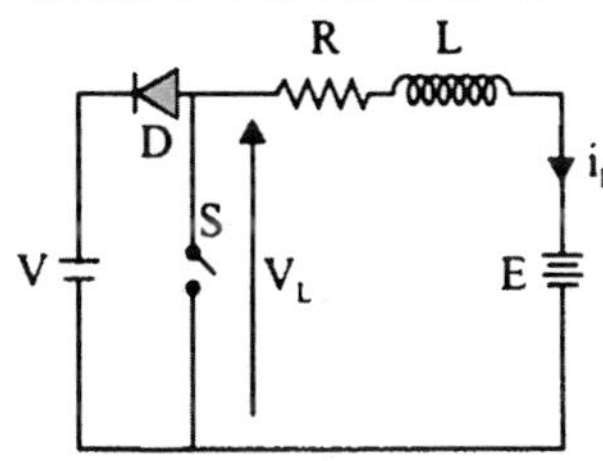

3. 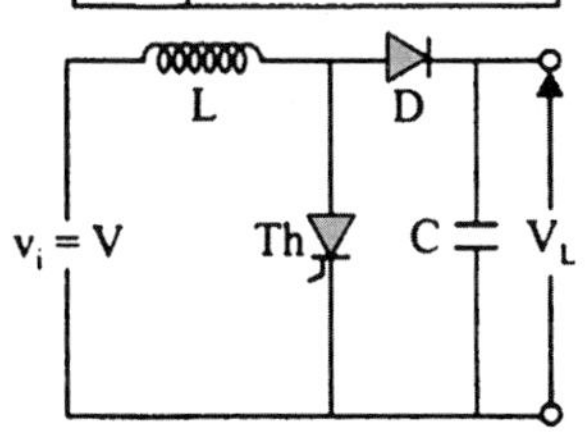

4. 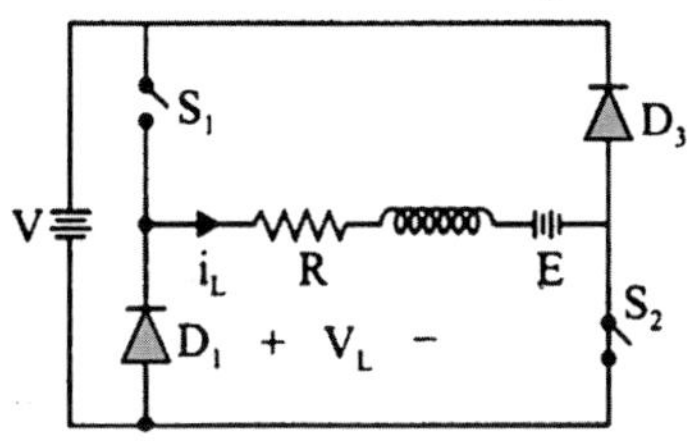

Codes:

	A	B	C	D
(*a*)	1	3	4	2
(*b*)	3	1	4	2
(*c*)	1	2	3	4
(*d*)	4	2	1	3

B. These questions have two statements, Asseretion (A) and Reason (R). Read them carefully and give correct answer as per the following codes:

(a) Both A and R are correct and R is correct explanation of A

(b) Both a and R correct but R is not correct explanation of A

(c) A is correct but R is wrong

(d) A is wrong but R is correct

351. A. A high *dv*/*dt* may cause triggering of a thyristor.
R. A high *dv*/*dt* results in charging current in the thyristor.

352. A. Gate triggering is the most commonly used method for triggering of an SCR.
R. Even a small gate current is sufficient to turn on an SCR.

353. A. Light triggering is very suitable for HVDC transmission.
R. Light triggering has the advantage of complete electrical isolation of gate circuit.

354. A. The gate current of an SCR is always in the form of pulses.
R. The gate losses, in pulse triggering, are low.

355. A. The turn on time of SCR is about 1 to 4 μs.
R. The turn off time of SCR is about 10 to 300 μs.

356. A. When the anode current of an SCR starts increasing during turn on process, the anode cathode voltage drop starts decreasing.
R. During turn on the reverse biased junction in SCR breaks down.

357. A. The surge current which an SCR can withstand is much higher than rms on state current.
R. The duration of surge current is very small.

358. A. When an SCR is conducting, the voltage drop across it is about 1 V.
R. When an SCR is conducting the junction temperature is about 200° C.

359. A. A DIAC has four layers but only two terminals.
R. A DIAC can conduct in both directions.

360. A. A TRIAC is a bidirectional SCR.
R. A TRIAC is a four layer 3 terminal device and can conduct in both directions.

361. A. Typical value of thermal resistance from case to sink of an SCR is about 0.3° C/W.
R. Heak sinks of thyristors are generally made of aluminium.

362. A. Snubber circuit has an RC series circuit in parallel with SCR.
R. Voltage across a capacitor cannot change instantanously.

363. A. Inverters and choppers use fast switching thyristors.
R. Fast switching SCR has low turn off time.

364. A. SUS is a PUT and avalanche diode connected between gate and cathode.
R. SBS has two SUS inbuilt together and connected in anti-parallel.

365. A. A GTO can be turned on and turned off by gate pulses.
R. A GTO has higher latching and holding current than SCR.

366. A. A GTO circuit is more compact than SCR circuit.
R. A GTO can be turned off by negative gate pulse.

367. A. A transistor requires a continous base signal for conduction but a thyristor requires a gate pulse.
R. Transistors find widespread application in power electronic circuits.

368. A. A transistor has lower voltage drop during conduction as compared to thyristor.
R. Transistors arc manufactured in very high voltage and current ratings.

369. A. Power transistor has higher *dV*/*dt* rating than SCR.
R. Power transistor has lower switching time than SCR.

370. A. An ideal rectifier should have unity harmonic factor.
R. An ideal rectifier should have unity displacement factor.

371. A. In the circuit of Fig. 8.48 thyristor will conduct for 180° during positive half cycle if it is continuously fired.
R. A thyristor can conduct only when it is forward biased.

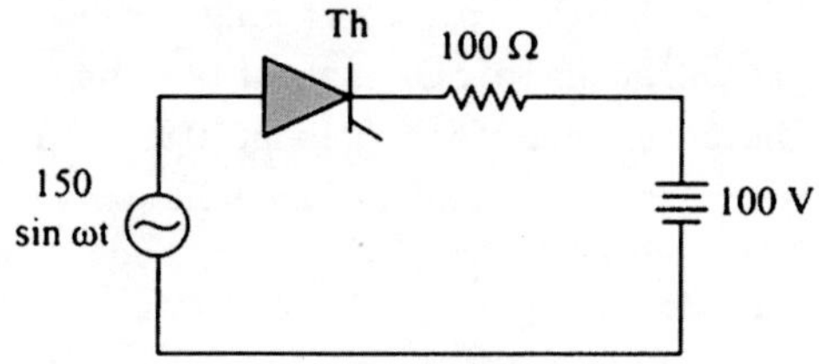

Fig. 8.48.

372. A. A half wave controlled rectifier has poorer efficiency and higher ripple factor than a full wave controlled rectifier.
R. The use of a free wheeling diode in half wave controlled rectifier circuit improves the waveform of load current and circuit power factor.

373. A. A fully controlled bridge converter can operate in first and fourth quadrant.
R. A semiconverter is cheaper than a full converter.

374. A. A dual converter may be circulating current type or non circulating current type.
R. In circulating current dual converter only one converter can be operated at one time.

375. A. An inverter does not require forced commutation.
R. A series inverter is a forced commutation inverter.

376. A. McMurray Bredford half bridge inverter uses complementary commutation.
R. 3-phase inverter can have two modes of operation.

377. A. The output voltage of an inverter can be controlled by PWM technique.
R. In a current source inverter the input current is constant.

378. A. If the duty cycle of chopper is low the current would not become discontinuous.
R. If the time constant L/R of the load in chopper is low the current may become discontinuous.

379. A. A chopper always requires forced commutation.
R. A Morgan chopper uses a saturable reactor.

380. A. A class C chopper can operate in first and second quadrants.
R. A class B chopper operates only in second quadrant.

381. A. An ac regulator uses natural commutation.
R. An ac regulator can be used for static on load tap changing of transformer.

382. A. A cyclo converter may be line commutated or forced commutated.
R. A stepup cyclo converter requires forced commutation.

383. A. Thyristors can be used in controlled heating, excitation systems of alternators, speed control of motors and HVDC.
R. A static var system using thyristors is very commonly used in high voltage ac systems.

C. State whether the following are True (T) or False (F)

384. A crowbar circuit is generally needed for protection under fault conditions.
385. Induction heating systems require low frequency inverters.
386. Thyristors are not suitable for logic circuits.
387. Thyristors are suitable for dc circuit breakers but not for ac circuit breakers.
388. The amount of dielectric heating is inversely proportional to frequency.
389. For an illumination control circuit it is advisable to use a matched Triac-diac pair.
390. HVDC systems are preferable for only long distances.
391. The output frequency of a cyclo converter can be changed by changing the firing angle.
392. A single phase bridge type cyclo converter uses eight thyristors.
393. A cycolo converter cannot give a frequency higher than input frequency.
394. A cyclo converter is a group of controlled rectifiers.
395. The output wave of a cyclo converter consists of a number of segments derived from the input wave.
396. In integral cycle control the period of on-off control should be more than mechanical time constant of load.
397. In a single phase full wave ac regulator the firing angles during both half cycles are equal.
398. In a single phase full wave regulator feeding R-L load, short gate pulses are not suitable.
399. On load tap changer using thyristors can have a maximum of two taps.
400. In a single phase full wave regulator feeding R-L load, the minimum firing angle can be zero.
401. In a multiphase chopper all choppers operate together.
402. When the core of a saturable reactor is saturated, the inductance is high.
403. A class B chopper operates in third quadrant.
404. A chopper cannot be line commutated.
405. A chopper is a high efficiency circuit.
406. In sinusoidal PWM for inverters the width of different pulses are not the same.
407. In 180° mode of opcration of a 3-phase bridge inverter, two thyristors conduct at one time.
408. All inverters use forced commutation.
409. Half bridge and full bridge inverters of one type use different commutation methods.
410. A current source inverter has simple commutation circuit.
411. A semiconverter operates in two quadrants.
412. In a dual converter both converters work as rectifiers.
413. The use of free wheeling diode in controlled rectifier improves the waveshape of load current.
414. A reactor is used in a dual converter operating in non-circulating current mode.
415. As the firing of a thyristor is delayed, the output voltage of a controlled rectifier decreases.
416. An SBS can conduct is one direction only.
417. Thyristors connected in series need static and dynamic stabilizing circuits.
418. A DIAC has two terminals, anode and cathode.
419. In a thyristor the gate lead is connected to the n layer near the p layer to which anode is connected.

420. The application of gate signal decreases the break over voltage of on SCR.

D. Fill in the blanks with suitable words

421. An SCR has .... layers and .... terminals.
422. As the magnitude of gate current in an SCR is increased, the forward break over voltage ....
423. The two methods of commutation of thyristors are.................and.................. .
424. The working of thyristor can be explained by .... model.
425. When a thyristor is forward biased the number of p-n junctions with forward bias and reverse bias are .... and .... respectively.
426. Gate current.................. turn off a thyristor.
427. Reverse bias.................the width of.................
428. When a thyristor is in reverse blocking mode, ...is positive with respect to............
429. When voltage across a reverse bias layer is high, ... break down occurs.
430. Light traggering is used in the device called.................
431. In................triggering the gate losses are low.
432. The turn on time of an SCR is about I to 4...................
433. To turn off an SCR its anode-cathode current must be reduced to a value less than..................
434. When an SCR is in forward blocking anode a small current called..................current flows.
435. In a thyristor there is an internal........ .... feedback.
436. A.................. voltage has high magnitude and small duration.
437. The ratio temperature difference per watt of power dissipation is called...................
438. A DIAC has...................terminals and can conduct in...................
439. A TRIAC is a .... SCR.
440. When a potential V_{BB} is applied to a UJT, the potential of emitter is μV_{BB}. The quantity μ is called ....
441. A UJT exhibits .... characteristics.
442. An oscillator using UJT is called ....
443. The main application of .... is in triggering circuit of ....
444. In resonant commutation, the RLC circuit must be ....
445. Auxiliary commutation is also called .... commutation.
446. The circuit used for *dv*/*dt* protection of SCR is called .... circuit.
447. For equalisation of potentials across SCRs connected in series during turn on and turn off, a .... equalising circuit is used.
448. String efficiency is always .... than 1.
449. A GTO can be turned off by a .... gate current.
450. A reverse conducting thyristor has a .... connected in .... with SCR.
451. In a PUT the parameters .... and .... can be controlled.
452. A silicon controlled switch has.................. layers and..................terminals.
453. During conduction the voltage drop across an SCR is..................than that in a transistor.
454. Turn on time of tranistor is..................than that of SCR.
455. In a single phase half wave controlled rectifier the output voltage is maximum when firing angle is...................
456. If the input voltage to a single phase half wave controlled rectifier is $V_n \sin \omega t$, the peak inverse voltage is...................
457. When a half wave controlled rectifier feeds an RL load, the use of .... improves the waveshapes of load current and power factor.
458. In a controlled full wave rectifier using centre tapped transformer, the voltage from midpoint to each terminal is $V_m \sin \omega t$. Then peak inverse voltage is...................
459. A single phase semiconverter uses.................. thyristors and..................diodes.
460. A fully controlled bridge converter operates in .... and .... quadrants.
461. A dual converter can operate in.................. quadrants.
462. A dual converter is generally operated in..................mode.
463. Inverters can be broadly classified in two categories..................and...................
464. A half bridge inverter requires...................dc supply.
465. The two methods of chopper control are called .... and ....
466. A low duty cycle in a chopper may result in...................
467. A class E chopper can operate in.................. quadrants.
468. A single phase full wave ac regulator uses two .... connected in ....
469. A cyclo converter converts ac at 50 Hz to...................ac.
470. A step drown cyclo converter uses .... commutation.

ANSWERS WITH HINTS AND EXPLANATIONS

1. (*d*) The holding current is much less than 1% of rated current.
2. (*a*) 3. (*a*) See Fig. 8.1.
4. (*b*) Only a very small gate current is required to trigger SCR into conduction.
5. (*b*) 6. (*c*) 7. (*c*) 8. (*b*)
9. (*b*) Anode, cathode and gate.
10. (*c*)
11. (*b*) Since current does not divide equally, current derating is necessary.
12. (*c*) Thyristors in series do not share the voltage equally and require voltage derating. Thyristors in parallel need current derating.

13. (*c*) Especially high $\frac{dv}{dt}$ during switching.
14. (*b*)
15. (*c*) Anode, cathode, anode gate and cathode gate.
16. (*a*) **17.** (*a*) **18.** (*b*)
19. (*b*) See Fig. 8.1. **20.** (*a*)
21. (*c*) Thyristor can turn on only when it is forward biased.
22. (*c*) See Fig. 8.1.
23. (*a*) T_1 and T_2 as shown in Fig. 8.2.
24. (*c*) **25.** (*c*) **26.** (*c*) *p–n–p–n*.
27. (*c*) **28.** (*b*)
29. (*b*) Once SCR has been triggered the gate has no control. Only external resistance controls the anode current.
30. (*a*) **31.** (*a*)
32. (*a*) The firing angle controls the period of conduction and hence average load current.
33. (*c*) **34.** (*a*) **35.** (*b*) **36.** (*a*)
37. (*a*) **38.** (*c*) See Fig. 8.20.
39. (*d*) See Fig. 8.22. **40.** (*c*) See Fig. 8.22.
41. (*a*) See Section 8.3.7. **42.** (*a*)
43. (*a*) See Section 8.3.5. **44.** (*d*)
45. (*c*) During the period of overlap all the four thyristors conduct and load is effectively short-circuited. Output voltage is zero.
46. (*a*)
47. (*a*) Free wheeling diode conducts only when load is inductive. If load is purely resistive free wheeling diode does not come into operation.
48. (*a*) To reduce ripple, 12 pulse converters are used in HVDC.
49. (*a*) **50.** (*d*) **51.** (*a*) **52.** (*a*)
53. (*a*)
54. (*b*) See Eq. (8.32). **55.** (*c*) See Eq. (8.40).
56. (*a*) **57.** (*c*)
58. (*b*) Load should be under damped.
59. (*a*) **60.** (*b*) **61.** (*a*) **62.** (*a*)
63. (*b*) **64.** (*b*) **65.** (*a*) **66.** (*b*)
67. (*a*) See Fig. 8.28. **68.** (*c*) See Fig. 8.28.
69. (*b*) See Fig. 8.29.
70. (*d*) **71.** (*a*) **72.** (*c*) **73.** (*a*)
74. (*b*) **75.** (*a*) **76.** (*a*)
77. (*b*) See Section 8.5. **78.** (*a*) See Section 8.5.
79. (*a*)
80. (*b*) See Eq. (8.44). **81.** (*a*) **82.** (*a*)
83. (*b*) See Section 8.4.3.
84. (*a*) **85.** (*c*)
86. (*a*) See Eq. (8.47).
87. (*d*) See Section 8.4.3.
88. (*c*) See Fig. 8.31. **89.** (*c*)
90. (*c*) See Fig. 8.31.
91. (*c*) **92.** (*c*) **93.** (*b*) **94.** (*a*)
95. (*c*)
96. (*c*) See Section 8.4.5.
97. (*c*)
98. (*a*) See Eq. (8.48). **99.** (*a*) **100.** (*c*)
101. (*d*) The input current during positive half cycle is less than that during negative half cycle.
102. (*a*) **103.** (*a*) **104.** (*a*) **105.** (*a*)
106. (*c*) **107.** (*b*) **108.** (*b*) **109.** (*b*)
110. (*c*) To ensure proper triggering pulse train is necessary because load is R-L,
111. (*b*)
112. (*d*) In a three-phase full wave regulator, output voltage is zero if $\alpha \geq 150°$.
113. (*b*) Since output voltage is zero if $\alpha \geq 150°$, the possible range of firing angles is $0 \leq \alpha \leq 150°$.
114. (*a*)
115. (*d*) It changes ac at frequency *f* into ac at a frequency different from *f*.
116. (*c*) The output frequency can be more or less than input frequency.
117. (*c*) **118.** (*b*) **119.** (*a*) **120.** (*b*)
121. (*b*) **122.** (*c*) **123.** (*c*) **124.** (*b*)
125. (*b*) Since output frequency is 2*f*, one half wave of input must equal one full wave of output.
126. (*d*)
127. (*b*) Natural commutation is possible only in ac circuits.
128. (*c*)
129. (*c*) The output voltage is controlled ac.
130. (*b*) $V_L = V\sqrt{\alpha}$. Therefore $V_L^2 = V^2 (\alpha)$. Heating is proportional to V_L^2 and hence proportional to α.
131. (*b*) Conduction occurs only during 50% of wave.
132. (*d*) Input ac is rectified and then converted to high frequency ac.
133. (*b*) **134.** (*a*)
135. (*c*) To reduce ripple 12 pulse converters are used.
136. (*d*)
137. (*d*) Average output voltage $\propto \cos \alpha$ and therefore reduced to 0.707 of the value for $\alpha = 0$. Speed is nearly proportional to voltage.
138. (*a*) Average output voltage is reduced to 50% if $\alpha = 90°$. Hence speed is also reduced to 50%.
139. (*a*)
140. (*b*) If load torque is constant, the change in speed is very small.
141. (*b*)
142. (*b*) In this case synchronous speed at 50 Hz is 1500 rpm. Therefore synchronous speed at 25 Hz is 750 rpm. Since slip is about 3-5%, speed is about 720 rpm.
143. (*a*) See Section 8.2.
144. (*b*) See Fig. 8.3. **145.** (*b*) See Fig. 8.6.
146. (*b*) See Fig. 8.10. **147.** (*b*) See Section 8.2.
148. (*a*) See Section 8.2. **149.** (*c*) See Section 8.2.
150. (*c*) **151.** (*c*) See Section 8.2. **152.** (*b*)
153. (*a*) A diode has 2 leads irrespective of its rating.
154. (*b*)
155. (*b*) See Fig. 8.13. **156.** (*c*) See Fig. 8.14.
157. (*c*) See Fig. 8.15. **158.** (*b*) See Fig. 8.11.
159. (*a*) **160.** (*c*) **161.** (*c*) **162.** (*a*)

163. (*b*) See Fig. 8.11.
164. (*b*) **165.** (*a*) **166.** (*c*) **167.** (*b*)
168. (*b*) See Fig. 8.1 and 8.3.
169. (*a*) **170.** (*b*) **171.** (*a*) **172.** (*a*)
173. (*a*) In a BJT terminals are emitter, base and collector irrespective of rating.
174. (*b*) In a MOSFET terminals are source, gate, drain irrespective of rating.
175. (*d*) **176.** (*a*)
177. (*d*) $\beta = \frac{\alpha}{1-\alpha} = \frac{0.98}{1-0.98} = 49$.
178. (*a*) **179.** (*d*) **180.** (*b*) **181.** (*c*)
182. (*c*)
183. (*c*) Both electrons and holes contribute to current.
184. (*b*) With zero gate current, conduction can be started only if forward voltage exceeds breakover voltage.
185. (*c*) **186.** (*c*) **187.** (*b*)
188. (*c*) Voltage is high but current is nearly zero.
189. (*a*)
190. (*a*) UJT relaxation oscillator is used to provide gate pulses to SCR for triggering.
191. (*c*)
192. (*b*) Thermal resistance $= \frac{100-40}{50}$
193. (*d*) $1.6 = \frac{T-20}{40}$ or T = 84°C.
194. (*b*) $1.6 = \frac{120-40}{P}$ or P = 50 W.
195. (*c*) Turn on time is less than 3 μ-s while turn off time is between 10-300 μ-s.
196. (*b*)
197. (*c*) Gate current does not affect the magnitude of anode current.
198. (*b*) Duration of gate pulse should be slightly more than turn time of SCR.
199. (*b*)
200. (*c*) Inductive load increases the turn on time.
201. (*d*)
202. (*c*) $I_{dc} = \frac{V_m}{2\pi}(1+\cos\alpha)$. Therefore

$$\frac{I_{dc} \text{ for } 80° \text{ firing angle}}{I_{dc} \text{ for } 40° \text{ firing angle}} = \frac{1+\cos 80°}{1+\cos 40°} = \frac{1.174}{1.766}.$$

203. (*c*) Both high $\frac{dv}{dt}$ and high $\frac{di}{dt}$ can destroy SCR.
204. (*c*) **205.** (*c*)
206. (*c*) $v \propto (i)^{1/4} \propto (100)^{1/4} \simeq 3$.
207. (*a*) $e = L\frac{di}{dt}$. In current chopping $\frac{di}{dt}$ is high. Hence *e* is high.
208. (*c*) $0.25 = 1 - \frac{750}{n_p \times 175}$ or $n_p = 6$.
209. (*b*) $0.25 = 1 - \frac{3000}{n_s \times 800}$.
210. (*d*) **211.** (*a*)
212. (*d*) Class F is natural commutation AC-DC converters use natural (or line) commutation.
213. (*c*) **214.** (*d*)
215. (*c*) Class C commutation uses an auxiliary SCR and is also called auxiliary commutation.
216. (*a*) **217.** (*c*) **218.** (*b*) **219.** (*d*)
220. (*c*) Turn on and turn off times of similar SCRs may not be same.
221. (*b*) Discharge current flows through R_c.
222. (*b*) Value of C is proportional to maximum difference between reverse charges of SCRs in series.
223. (*b*) **224.** (*d*)
225. (*d*) Optical triggering uses LASCR.
226. (*b*) **227.** (*a*) **228.** (*a*)
229. (*b*) See Section 8.3.2. (*d*) and put α = 0.
230. (*c*) Before triggering the whole of applied voltage is across SCR and load current is zero.
231. (*a*) Put α = 180° in Eqs. (8.16) and (8.17). Otherwise also α = 180° means no conduction. Hence zero output voltage.
232. (*b*) Thyristor can conduct only when it is forward biased. This occurs only when $v > 50$ V.
233. (*b*) Using Eq. (8.16)

$$\frac{V_{dc} \text{ for } 30° \text{ firing anlge}}{V_{dc} \text{ for } 0° \text{ firing anlge}} = \frac{1+\cos 30°}{1+\cos 0} = 0.933.$$

Hence V_{dc} for 30° firing angle = 0.933 × 127 ≃ 120 V.
234. (*b*) As seen from Eqs. (8.16) and (8.17) average output voltage and rms output voltage do not decrease in the same proportion.
235. (*a*) See Eq. (8.17). **236.** (*d*) **237.** (*d*)
238. (*a*) SCR conducts from α to π and free wheeling diode from π to (2π + α).
239. (*b*) Load is short-circuited by free wheeling diode.
240. (*b*) Free wheeling diode is forward biased only when SCR is reverse biased.
241. (*a*) See Eqs. (8.16) and (8.22).
242. (*d*)
243. (*c*) PIV is twice the peak voltage across each half of secondary.
244. (*c*) SCRs conduct from α to π in each half cycle.
245. (*c*) Due to inductive load the period of conduction is α to π + α in each half cycle.
246. (*b*)
247. (*b*) Since conduction is from α to π + α, period of conduction is π in each half cycle.
248. (*c*)
249. (*a*) Because of highly inductive load, output current is nearly constant.
250. (*b*) Rms secondary current $= \frac{15}{\sqrt{2}}$.
251. (*c*) PIV = $\sqrt{2}$(2 × 200).
252. (*c*) In bridge connection PIV is half of that in M-2 connection.

253. (*d*) **254.** (*c*) **255.** (*a*) **256.** (*c*)
257. (*c*) In both the connections the output voltage and peak thyristor current are the some. Both are full wave rectifiers.
258. (*b*) The conduction is from α to π.
259. (*a*) **260.** (*c*)
261. (*a*) For resistive load output current is always proportional to output voltage.
262. (*d*) **263.** (*a*)
264. (*a*) Semiconverter operates only in first quadrant.
265. (b)
266. (*b*) Free wheeling diode coducts for $\alpha°$ in each half cycle.
267. (*b*) When free wheeling diode conducts output current exists but output voltage is zero.
268. (*b*) **269.** (*b*) See Fig. 8.23. **270.** (*b*)
271. (*b*)
272. (*a*) Series connection increases the voltage.
273. (*c*) Output voltage is $\frac{2V_m}{\pi}$ for each.
274. (*c*)
275. (*b*) It is a 6 pulse converter.
276. (*b*)
277. (*a*) See Eq. (8.34).
278. (*c*) Each thyristor conducts for one-third of one cycle.
279. (*a*) **280.** (*a*) **281.** (*c*)
282. (*b*) Since the circuit has 3 thyristors, firing frequency is three times the line frequency.
283. (*c*)
284. (*c*) $V_{dc} \propto (1 + \cos \alpha)$. Maximum possible output voltage occurs when $\alpha = 0$.
285. (*b*) **286.** (*a*)
287. (*b*) Source inductance delays turn off and hence causes overlap.
288. (*c*)
289. (*c*) 30° phase difference is necessary.
290. (*c*) **291.** (*c*) **292.** (*d*)
293. (*b*) See Fig. 8.26. **294.** (*a*) See Eq. (8.41).
295. (*b*) **296.** (*a*) **297.** (*b*) **298.** (*b*)
299. (*b*) A high frequency series inverter has many series inverters in parallel. Only one inverter operates at one time and they operate in fixed sequence.
300. (*a*) See Fig. 8.27.
301. (*c*) **302.** (*c*) **303.** (*d*) **304.** (*b*)
305 (*b*) **306.** (*c*) **307.** (*a*) **308.** (*c*)
309. (*b*) **310.** (*b*) **311.** (*c*) **312.** (*c*)
313. (*d*) **314.** (*d*)
315. (*c*) In 180° mode each thyristor conducts for 180° and in 120° mode each thyristor conducts for 120°.
316. (*c*) **317.** (*b*)
318. (*b*) Two resistances are in series at any time.
319. (*b*) **320.** (*b*)
321. (*b*) See Eqs. (8.43) and (8.44).
322. (*c*)
323. (*b*) FD is forward biased only when thyristor is off.
324. (*b*) **325.** (*c*)
326. (*b*) $f = \frac{1}{T_{on} + T_{off}}$ **327.** (*d*) See Fig. 8.31.
328. (*c*) See Fig. 8.31. **329.** (*b*) **330.** (*b*)
331. (*c*)
332. (*b*) $(I_{max} - I_{min}) = \frac{V}{4fL}$ or $10 = \frac{500}{4 \times 400 \times L}$
or $L = 31 \times 10^{-3}$ H.
333. (*b*) Only one chopper operates at one time.
334. (*b*)
335. (*c*) Regulation is possible only in positive half cycle.
336. (*a*)
337. (*a*) Firing angles are kept equal to get symmetrical output voltage.
338. (*b*) Each thyristor conducts for 50% of time.
339. (*b*) Two thyristos for each phase.
340. (*b*) See Fig. 8.14.
341. (*c*)
342. (*a*) Output voltage is maximum if $\alpha = 0$ and is minimum if $\alpha = 180°$.
343. (*d*) Current through inductance cannot change instantaneously.
344. (*c*) See Section 8.2. **345.** (*c*) See Section 8.2.
346. (*d*) See Section 8.2.
347. (*c*) **348.** (*c*) **349.** (*c*) See Section 8.3.
350. (*b*)
351. (*a*) The charging current may cause triggering.
352. (*a*) **353.** (*a*)
354. (*a*) Gate current in the form of pulses is commonly used because gate losses are low.
355. (*b*) Both are correct but independent
356. (*a*)
357. (*a*) Since duration of surge current is small, the associated energy is small.
358. (*c*) Junction temperature is about 120°C. Hence R is wrong.
359. (*b*) Both are correct but independent.
360. (*a*) Since a triac can conduct in both directions it is bi-directional SCR.
361. (*b*) Both are correct but independent.
362. (*a*) Since voltage across capacitor cannot change instantaneously it protects SCR against high $\frac{dv}{dt}$.
363. (*a*) Low turn off time enables fast switching.
364. (*b*)
365. (*b*) Both are correct but independent.
366. (*a*) Since commutation circuit is not required, GTO circuit is compact.
367. (*c*) Transistors are used but not very widely.
368. (*c*)
369. (*d*) Power transistor does not have high $\frac{dv}{dt}$ rating.
370. (*d*) Ideal rectifier should have zero harmonic factor and unity displacement factor.
371. (*d*) In Fig. 8.48 SCR will conduct only when $v > 100$ V.
372. (*b*) Both are correct but independent.

373. (*b*) Both are correct but independent.
374. (*c*) In circulating current dual converter both converters operate together.
375. (*d*) Some inverters require forced commutation.
376. (*b*) Both are correct but independent.
377. (*b*)
378. (*c*) If duty cycle is low, current may become discontinuous.
379. (*b*) Both are correct but independent.
380. (*b*) Both are correct but independent.
381. (*b*) **382.** (*a*) **383.** (*b*) **384.** (T)
385. (F) It uses high frequency inverter.
386. (F) Thyristors can be used for logic circuits.
387. (F) They are suitable for both dc and ac circuit breakers.
388. (F) It is proportional to frequency.
389. (T) **390.** (T) **391.** (F) **392.** (T)
393. (F) Step up cycloconverter can give a frequency higher than input frequency.
394. (T) **395.** (T)
396. (F) It should be less than mechanical time constant.
397. (T) **398.** (T)
399. (F) Any number of taps are possible.
400. (F) Range of control of α is $\theta \leq \alpha \geq \pi$.
401. (F) They operate in a sequence.
402. (F) Since core is saturated, change of flux is small and inductance is low.
403. (F) It operates in second quadrant as seen in Fig. 8.31.
404. (T) **405.** (T) **406.** (T)
407. (F) In 180° mode three thyristors conduct simultaneously.
408. (F) Line commutated inverters are also used especially in HVDC.
409. (F) **410.** (T)
411. (F) It operates only in one quadrant.
412. (F) One operates as rectifier and the other as inverter.
413. (T)
414. (F) Reactor is used only for circulating current mode.
415. (T)
416. (F) It is a bilateral switch.
417. (T)
418. (F) They are designated as T_1 and T_2 because dia can conduct in both directions.
419. (F) See Fig. 8.1.
420. (T)
421. four, three **422.** decreases
423. natural commutation and forced commutation
424. two transistors
425. two, one **426.** cannot
427. increases, depletion layer
428. cathode, anode **429.** avalanche
430. LASCR
431. pulse **432.** microseconds
433. holding current
434. reverse saturation **435.** positive
436. surge
437. thermal resistance
438. two, both directions **439.** bidirectional
440. intrinsic stand off ratio
441. negative resistance
442. relaxation oscillator
443. DIAC, TRIAC
444. underdamped **445.** Voltage
446. snubber **447.** dynamic
448. less
449. negative
450. diode, anti-parallel **451.** V_p, η
452. four, four **453.** more
454. less
455. zero **456.** V_m
457. free wheeling diode
458. $2V_m$ **459.** two, two
460. first, fourth
461. four
462. circulating current
463. line commutated, force commutated
464. three wire
465. pulse width modulation, frequency modulation
466. discontinuous operation
467. four
468. thyristors, anti-parallel
469. variable frequency **470.** line

CHAPTER 9

DIGITAL ELECTRONICS

9.1. BINARY NUMBER SYSTEM

The binary number system has only two digits 0 and 1. Therefore its base is 2. The abbreviation of binary digit is bit. The binary number 1011 has 4 bits, 100111 has 6 bits and 11001111 has 8 bits. A string of 8 bits is called a byte. A byte is the basic unit of data in computers. In computers, the data is stored and processed in strings of 8 bits or some multiples (16, 24, 32 etc.). Table 9.1 shows the 16 combinations of a 4 bit binary word.

Table 9.1. Binary, Decimal, Hexadecimal and Octal Equivalence

Binary				*Decimal*	*Hexadecimal*	*Octal*
0	0	0	0	0	0	0
0	0	0	1	1	1	1
0	0	1	0	2	2	2
0	0	1	1	3	3	3
0	1	0	0	4	4	4
0	1	0	1	5	5	5
0	1	1	0	6	6	6
0	1	1	1	7	7	7
1	0	0	0	8	8	10
1	0	0	1	9	9	11
1	0	1	0	10	A	12
1	0	1	1	11	B	13
1	1	0	0	12	C	14
1	1	0	1	13	D	15
1	1	1	0	14	E	16
1	1	1	1	15	F	17

9.1.1. Binary to Decimal Conversion

1. Write the binary number.
2. Write weights $2^0, 2^1, 2^2$, under the binary digits starting from the bit on right hand side.
3. Cross out weights under zeros.
4. Add the remaining weights.

9.1.2. Decimal to Binary Conversion

The method called double dabble involves successive division by 2 and recording the remainder (the remainder is 0 or 1). The division is stopped when we get a quotient of 0 with a remainder of 1. The remainders when read upwards give the equivalent binary number.

9.1.3. Binary Arithmetic

(*a*) Binary Addition

$0+0 = 0$

$0+1 = 1+0 = 1$

$1+1 = 10$ *i.e.*, 1 + 1 equals 0 with a carry of 1 to next higher column

$1+1+1 = 11$ *i.e.*, 1 + 1 + 1 equals 1 with a carry of 1 to next higher column.

(*b*) Binary Subtraction

$0-0 = 0$

$1-0 = 1$

$1-1 = 0$

$10-1 = 1$

(*c*) Binary Multiplication

$0 \times 0 = 0$

$0 \times 1 = 0$

$1 \times 0 = 0$

$1 \times 1 = 1$

Binary multiplication involves forming partial products, shifting successive partial products left by one place and then adding all the partial products.

(*d*) Binary Division. The method is the same long division procedure as in decimal system.

9.1.4. Signed Binary Numbers

The most significant bit is the sign bit (0 for positive and 1 for negative). The numbers may be 4 bit, 8 bit, 16 bit etc. In every case the leading (MSB) bit represents sign and the remaining bits represent magnitude.

9.1.5. 1's Complement and 2's Complement

The 1's complement of a binary number is obtained by complementing each bit (0 for 1 and 1 for 0). It is written as $\bar{A}$.

The 2's complement of a binary number is obtained by adding 1 to be 1's complement. It is written as A′. If we lake 2's complement twice, we get original number (*i.e.*, A″ = A).

9.1.6. 2's Complement Addition and Subtraction

When A and B are to be added, the B bits are not inverted so that, we get

$$S = A + B$$

When B is to be subtracted from A, the computer hardware forms 2's complement of B and then adds it to A thus

$$S = A + B' = A + (-B) = A - B$$

9.1.7. Binary Fractions

To convert binary to decimal, the weights of binary bits after decimal are written as

$$0.1011 = 1 \times 2^{-1} + 0 \times 2^{-2} + 1 \times 2^{-3} + 1 \times 2^{-4}$$

$$= 1 \times \frac{1}{2} + 0 \times \frac{1}{4} + 1 \times \frac{1}{8} + 1 \times \frac{1}{16}$$

$$= 0.5 + 0 + 0.125 + 0.0625 = 0.6875 \text{ (decimal)}$$

To convert decimal to binary successive multiplication by 2 is used. The integer parts when read downwards give the fraction part of binary.

9.1.8. Double Precision Numbers

In 16 bit representation the numbers from +32767 to −32766 can be stored in each register. To store numbers greater than these numbers double precision system is used. In this system two storage locations are used to represent each number. The format is

First word	S	Higher Order Bits
Second word	O	Low order bits

S is sign bit and O is zero. Thus numbers of 31 bit lengths can be stored in 16 bit registers. For still bigger numbers triple precision can be used.

9.1.9. Floating Point Numbers

The numbers are expressed by using a mantissa and an exponent. The mantissa has 10 bit length and exponent has 6 bit length. The left most bit of mantissa is sign bit. The binary point is to the right of sign bit. The 6 bit exponent has a base of 2. The exponent can represent numbers from 0 to 63. To express negative exponents the decimal number 32 (equivalent to binary number 100000) is added to exponent. This is called excess - 32 notation. Thus the number

Mantissa	Exponent
0111001101	100111

is equivalent to

Mantissa +0. 111001101

Exponent 100 111

Subtracting 100000 from 100111 we get 000 111. Thus the above number is

$$0.11100\,1101_2 \times 2^7 = 1110011.01_2 = 115.25_{10}$$

The advantage of floating point representation is that very large and very small numbers can be stored.

9.1.10. Hexadecimal Numbers and Octal Numbers

The base of hexadecimal numbers is 16. It uses 16 digits, *i.e.*, 0, 1, 2, 3, 4, 5, 6, 7, 8, 9, A, B, C, D, E, F. Table 9.1 gives decimal-binary-hexadecimal- octal equivalence.

9.1.10.1. Hexadecimal to Binary Conversion

Convert each hexadecimal digit to its 4 bit binary equivalent using Table 9.1

7	B	F	Hexadecimal number
↓	↓	↓	
0111	1011	1111	Binary numbers

9.1.10.2. Binary to Hexadecimal Conversion

Convert each 4 bit binary into equivalent hexadecimal using Table 9.1

1111	1101	1000	0110	Binary number
↓	↓	↓	↓	
F	D	8	6	Hexadecimal number

9.1.10.3. Hexadecimal to Decimal Conversion

The weights of different bits of hexadecimal number are 16^0, 16^1, 16^2 etc. starting with the bit on extreme right. Sum of all digits multiplied by the their weights is the equivalent decimal numbers

E	7	F	6	Hexadecimal
16^3	16^2	16^1	16^0	Weights

$$E7F6 = E \times 16^3 + 7 \times 16^2 + F \times 16^1 + 6 \times 16^0$$

$$= 14 \times 4096 + 7 \times 256 + 15 \times 16 + 6 \times 1 = 59382_{10}$$

9.1.10.4. Decimal to Hexadecimal Conversion

The method involves successive division by 16 and writing hexadecimal equivalents of the remainders. Then read the remainders upwards, *e.g.*,

16	5390	
16	336 remainder	14 : E
16	21 remainder	0 : 0
16	1 remainder	5 : 5
	0 remainder	1 : 1

$$5390_{10} = 150E_{16}$$

9.1.10.5. Octal Numbers

The base is 8 and the digits used are 0, 1, 2, 3, 4, 5, 6, 7. Table 9.1 gives binary-decimal-hexadecimal-octal equivalent. It was used in early mini computers.

9.1.11. Binary Coded Decimal (BCD)

BCD is used in pocket calculators, electronic counters, digital voltmeters digital clocks etc. In BCD 8421, each decimal number is replaced by its equivalent 4 bit code. Table 9.2 gives BCD 8421.

Table 9.2. BCD 8421 Code

Decimal	0,	1,	2,	3,	4,
8421 BCD	0000,	0001,	0010,	0011,	0100,
Decimal	5,	6,	7,	8,	9
8421 BCD	0101,	0110,	0111,	1000,	1001

The combinations 1010, 1011, 1100, 1101, 1110 and 1111 are not used in BCD 8421. To represent any decimal number in BCD, replace each decimal digit by 4 bit BCD code using Table 9.2, *e.g.*,

1	1	5	decimal
0001	0001	0101	BCD

9.1.12. Gray Code

It is an unweighted code. The bit positions do not have any specific weights assigned to them. The important characteristic of this code is that only a single bit change occurs when going from one code to next. It is used in special applications, *e.g.*, shaft encoders. Table 9.3 gives Gray code.

Table 9.3. Gray Code

Decimal	0,	1,	2,	3,	4,
Binary	0000,	0001,	0010,	0011,	0100,
Gray	0000,	0001,	0011,	0010,	0110,
Decimal	5,	6,	7,	8,	9
Binary	0101,	0110,	0111,	1000,	1001
Gray	0111,	0101,	0100,	1100,	1101

9.1.13. Excess-3 Code

Excess-3 is a digital code obtained by adding 3 to each decimal digit and then converting the result to four bit binary. Table 9.4 gives Excess-3 code

Table 9.4. Excess-3 Code

Decimal	0	1	2	3	4
Excess-3	0011,	0100,	0101,	0110,	0111,
Decimal	5	6	7	8	9
Excess-3	1000,	1001,	1010,	1011,	1100

The combinations 0000, 0001,0010, 1101, 1110 and 1111 are invalid in this code.

To convert decimal to excess-3 code, add 3 to each decimal digit and then convert into binary, *e.g.*,

4	6	decimal
+3	+3	
7	9	
↓	↓	
0111	1001	excess-3

9.1.14. ASCII Code (American Standard Code for Information Interchange).

It is a 7 bit alphanumeric code used extensively for printers and terminals of computer systems.

9.1.15. EBCDIC Code (Extended Binary Coded Decimal Interchange Code)

It is an 8 bit alphanumeric code used in most of large computers.

9.2. LOGIC GATES

Figure 9.1 gives the summary of digital gates.

Logic Function	Truth Table Inputs A	B	Output Y	Boolean Expression	Symbol
OR	0 0 1 1	0 1 0 1	0 1 1 1	$Y = A + B$	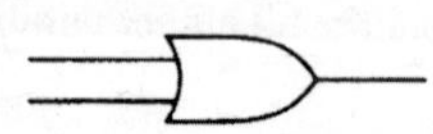
AND	0 0 1 1	0 1 0 1	0 0 0 1	$Y = A . B$	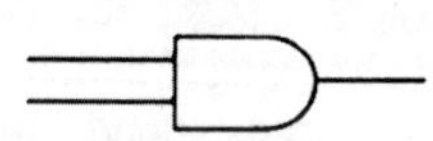
NOT (Inverter)	0 1		1 0	$Y = \overline{A}$	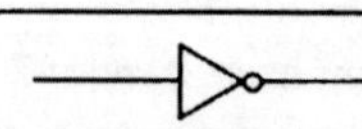
NOR	0 0 1 1	0 1 0 1	1 0 0 0	$Y = \overline{A + B}$	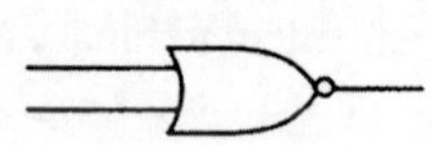
NAND	0 0 1 1	0 1 0 1	1 1 1 0	$Y = \overline{AB}$	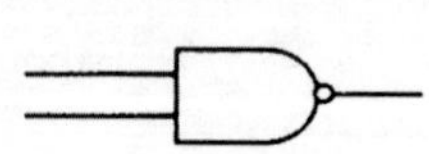
XOR	0 0 1 1	0 1 0 1	0 1 1 0	$Y = A \oplus B$	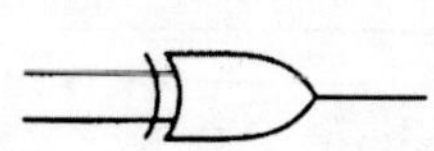
XNOR	0 0 1 1	0 1 0 1	1 0 0 1	$Y = A \odot B$ or $Y = \overline{A \oplus B}$	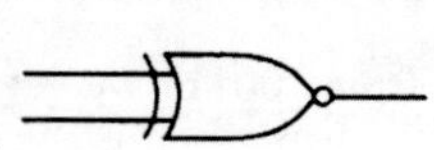

Fig. 9.1.

9.3. BOOLEAN ALGEBRA

9.3.1. Commutative Law

$$A + B = B + A$$
$$A . B = B . A$$

Associative Law

$$A + (B + C) = (A + B) + C$$
$$A . (B . C) = (A . B) . C$$

Distributive Law

$$A + (B . C) = (A + B) . (A + C)$$
$$A . (B + C) = A . B + A . C.$$

Table 9.5 gives Boolean Laws.

Table 9.5. Boolean Laws

1.	$A + A$	=	A
2.	$A + 1$	=	1
3.	$A + 0$	=	A
4.	$A + \bar{A}$	=	1
5.	$A . A$	=	A
6.	$A . 1$	=	A
7.	$A . 0$	=	0
8.	$A . \bar{A}$	=	0
9.	$\bar{\bar{A}}$	=	A
10.	$A + A . B$	=	A
11.	$A (A + B)$	=	A
12.	$A (\bar{A} + B)$	=	$A . B$
13.	$A + \bar{A} . B$	=	$A + B$
14.	$\bar{A} + A . B$	=	$\bar{A} + B$
15.	$\bar{A} + A . \bar{B}$	=	$\bar{A} + \bar{B}$

9.3.2. DeMorgan's Theorem

First theorem $\overline{A + B} = \bar{A} . \bar{B}$

Second theorem $\overline{AB} = \bar{A} + \bar{B}$

9.4. LOGIC SIMPLIFICATION

A logic function can be written in Sum of Products (SOP) or Product of Sums (POS form).

For two variables A, B the fundamental products are $\bar{A}\bar{B}$, $\bar{A}B$, $A\bar{B}$ and AB. For three variables A, B, C, the fundamental products are $\bar{A}\bar{B}\bar{C}$, $\bar{A}\bar{B}C$, $\bar{A}B\bar{C}$, $\bar{A}BC$, $A\bar{B}\bar{C}$, $A\bar{B}C$, $AB\bar{C}$, and ABC.

A typical example of SOP form is

$$Y = \bar{A}B\bar{C} + A\bar{B}\bar{C} + AB\bar{C} + ABC$$

Each term is called minterm.

To express SOP in minterms substitute 0 for a letter with bar and 1 for a letter without bar. Express the resultant binary number into equivalent decimal and write it as a subscript of *m, e.g.*

Minterm designation for $A\bar{B}CD$ is m_{11} as seen below

A	$\bar{B}$	C	D		
1	0	1	1	=	11

Following the above method

$$Y = \sum m(1,3,5,7) = \bar{A}\bar{B}C + \bar{A}BC + A\bar{B}C + ABC$$

Product of Sum form is written using max terms. In this notation unbarred letters represent 0s and barred letters represent 1s, *e.g.*,

Decimal 0 means binary 000 and term is $A + B + C$

Decimal 4 means binary 100 and term is $\bar{A} + B + C$

$$Y = \prod M(0,1,3,4)$$
$$= (A + B + C)(A + B + \bar{C})(A + \bar{B} + \bar{C})(\bar{A} + B + C)$$

A hybrid function has both SOP and POS terms.

A logic equation can be simplified by using Boolean laws and Demorgan's theorem. Another method is to use Karnaugh's map.

9.5. KARNAUGH MAP

9.5.1. Karnaugh Map for Sum of Product Expression

A K map for 2 variables has 4 cells, one for 3 variables has 8 cells and that for 4 variables has 16 cells. Figure 9.2 shows 3 variable K maps and grouping of terms.

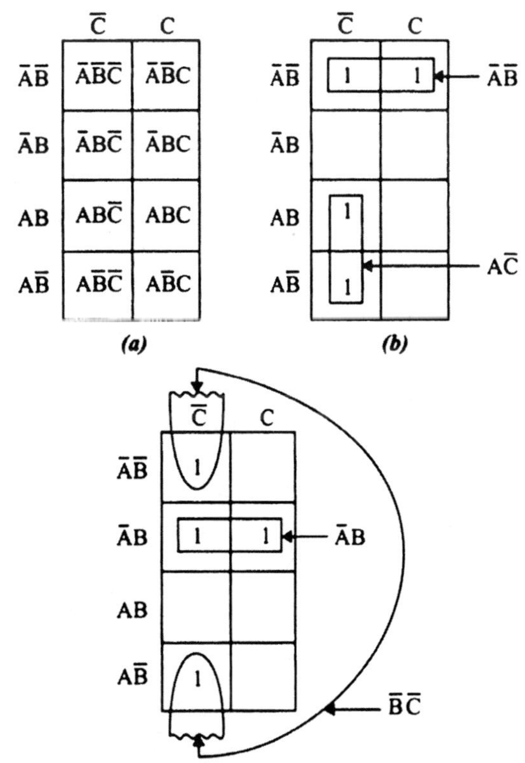

(c)

Fig. 9.2. Three variable K map (*a*) ordering of terms (*b*) K map for Y = $A\bar{B}\bar{C} + \bar{A}\bar{B}\bar{C} + \bar{A}\bar{B}C + AB\bar{C}$
(*c*) K map for Y = $\bar{A}\bar{B}\bar{C} + \bar{A}B\bar{C} + \bar{A}BC + A\bar{B}\bar{C}$

9.5.2. Karnaugh Map for Product of Sum Expression

Figure 9.3 shows K map for the Boolean expression.

$Y = \Pi M (0, 1, 3, 5, 6, 7, 10, 14, 15)$ and the grouping of terms.

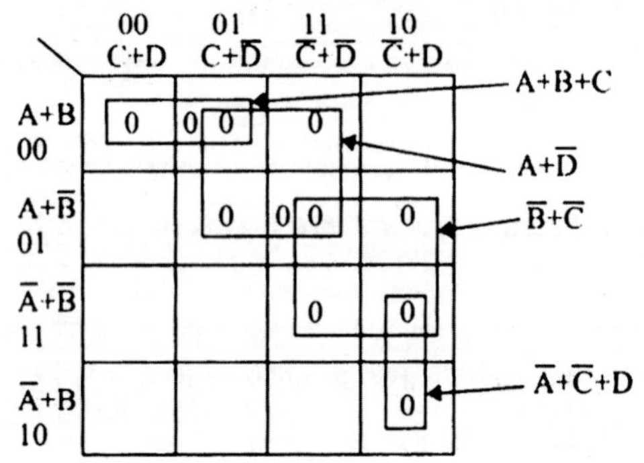

Fig. 9.3.

9.6. LOGIC FAMILIES

9.6.1. Classification

Digital circuits are classified as small scale integration (SSI), medium scale integration (MSI), Large scale integration (LSI), Very large scale integration (VLSI) and ultra large scale integration (ULSI). Table 9.6 gives this classification.

Table 9.6. Levels of integration of digital ICs

Category	*Numbers of equivalent basic gates on a single chip*	*Total number of components on single chip*
SSI	Less than 1 2	Less than 100
MSI	12-99	100-999
LSI	100-999	1000-9999
VLSI	1000-9999	10,000 to 99,999
ULSI	10,000 and more	100,000 and more

9.6.2. Resistor Transistor Logic (RTL)

It uses resistors and transistors. It was one of the earliest technologies and is not used nowadays. Figure 9.4 shows this logic.

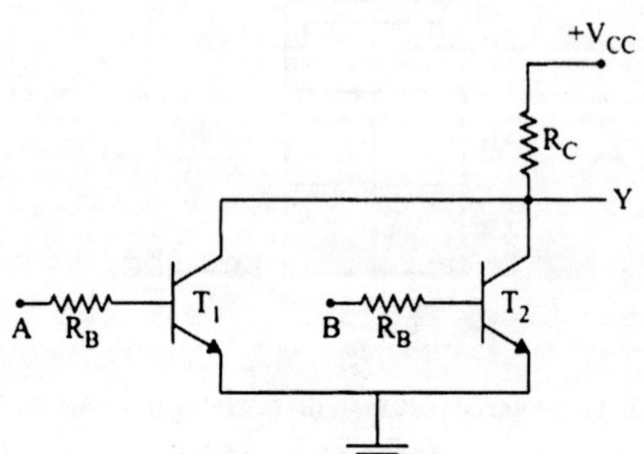

(a) **circuit**

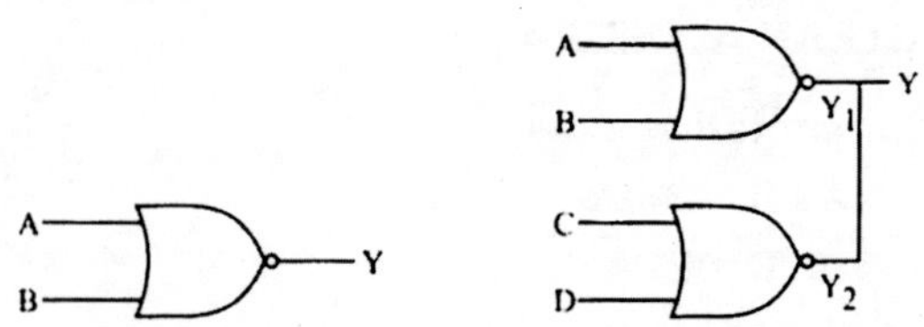

***(b)* symbol** ***(c)* wired OR connection**

Fig. 9.4. Resistor transistor logic.

For Fig. 9.4 (*c*)

$$Y = \overline{(A+B)}\,\overline{(C+D)} = \overline{A+B+C+D}$$

9.6.3. Direct Coupled Logic (DCL)

It is same as RTL (Fig. 9.4) except that base resistances R_B are not used in this case. Noise margin of this circuit is very poor leading to false operation.

9.6.4. Integrated Injection Logic (I^2L)

Figure 9.5 shows I^2L logic. It is a bipolar technology capable of very high component densities. It has low power requirement and good switching speed.

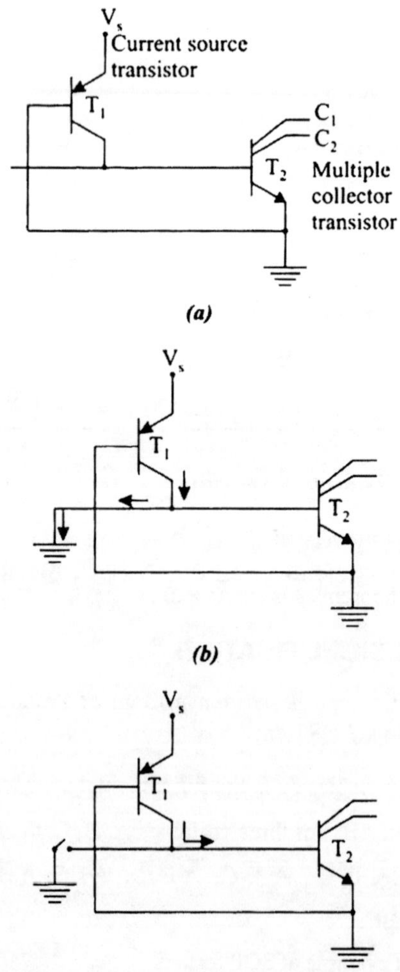

Fig. 9.5. I^2L logic, (*a*) gate configuration, (*b*) closed switch simulates a Low input to base of T_2, (*c*) open switch simulates a High input to base of T_2.

9.6.5. Diode Transistor Logic (DTL)

Figure 9.6 shows 2 input NAND gates using DTL. This logic uses transistor and diodes.

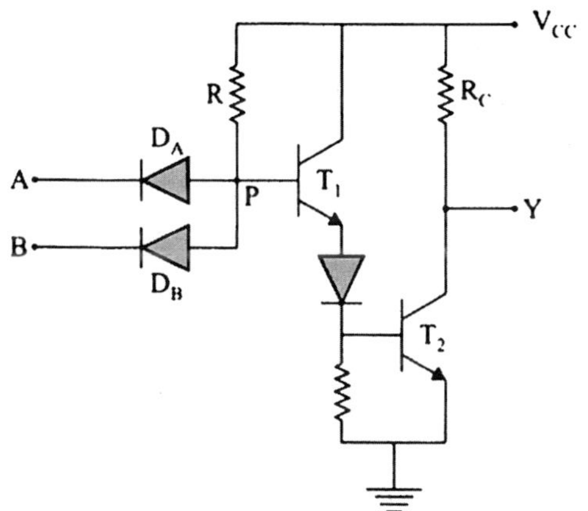

(a) **circuit**

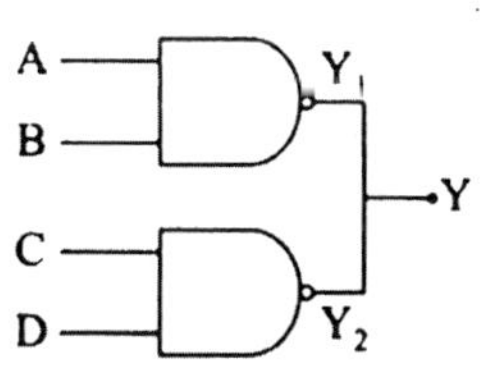

(b) **wired logic**

Fig. 9.6. Diode transistor logic

Figure 9.6 (*b*) shows two DTL NAND gates with outputs connected together. Thus

$$Y = Y_1 Y_2 = (\overline{AB})(\overline{CD}) = \overline{AB+CD}$$

9.6.6. Transistor Transistor Logic (TTL)

Figure 9.7 shows TTL NAND gate with a totem pole output. The totem pole output means that transistor T_4 sits atop T_3 so as to give low output impedance.

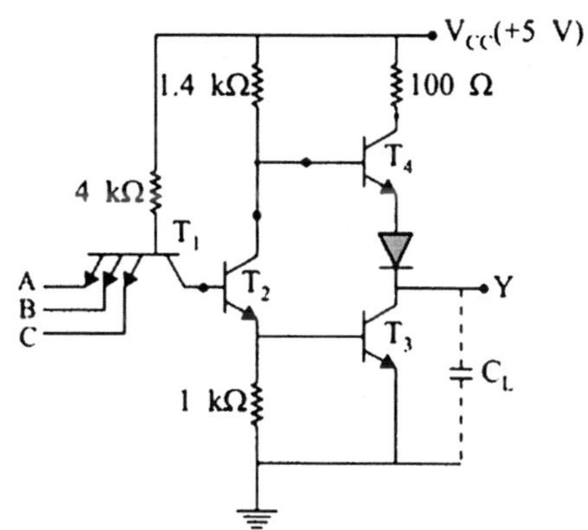

Fig. 9.7. TTL NAND gate with totem pole output

TTL family includes standard TTL, high speed TTL, low power TTL, Schottky TTL, low power Schottky TTL, advanced low power TTL and advanced Schottky TTL. TTL logic has low power dissipation, good noise immunity and reliable operation. However its propagation delay is high.

9.6.7. Emitter Coupled Logic (ECL)

ECL has low propagation delay but consumes more power. Figure 9.8 shows ECL 4 input OR-NOR gate

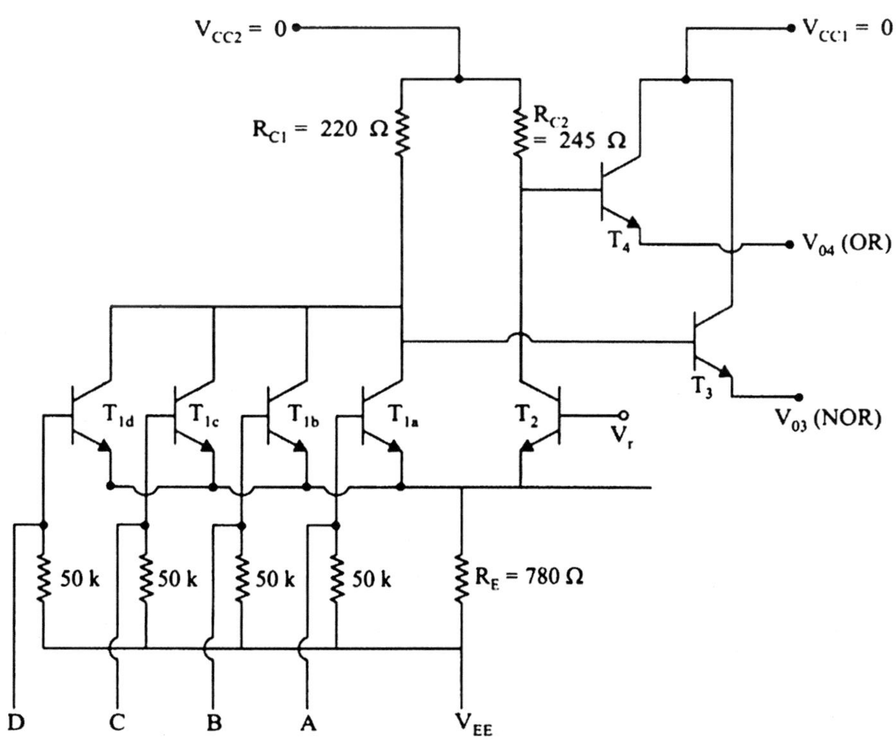

Fig. 9.8. ECL 4 input OR-NOR gate.

9.6.8. CMOS Logic

Logic gate using MOS devices have slow speed but very small size and simple geometry. Therefore tens of thousands of MOSFETs can be packed in a fraction of square cm area giving VLSI circuit. Figure 9.9 shows 2 input CMOS NAND gate.

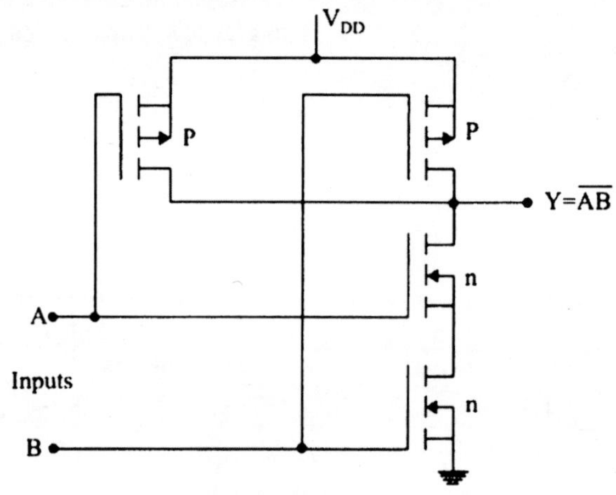

Fig. 9.9. Two input CMOS NAND gate.

Table 9.7 gives a comparison of TTL, ECL and CMOS circuits.

Table 9.7.

	TTL		*ECL*		*CMOS*
	Series 54AS/74AS	*Series 54ALS/74ALS*	*Series 10000H*	*Series III*	*Series 74HC*
Typical value of output voltage					
High. V	3.5	3.4	−0.9	−0.9	V_{CC} = −0.1
Low. V	0.5	0.4	−1.75	−1.75	0.1
Noise margin					
Low. V	1.15	1.15	0.27	0.2	2.3
High. V	1.95	1.95	0.23	0.2	2.7
Fan out	33	20	90	70	More than 50
Supply voltage, V	5	5	−5.2	5.2	6
Power dissipation per gate	8 mW	1.2 mW	25 mW	60 mW	2.5 μW
Propagation delay time, ns	1.7	4	1.1	0.7	8

9.7. ARITHMETIC CIRCUITS

9.7.1. Half Adder

Figure 9.10 shows the circuit and symbol of half adder. It consists of one AND and one XOR gate. Table 9.8 shows its truth table.

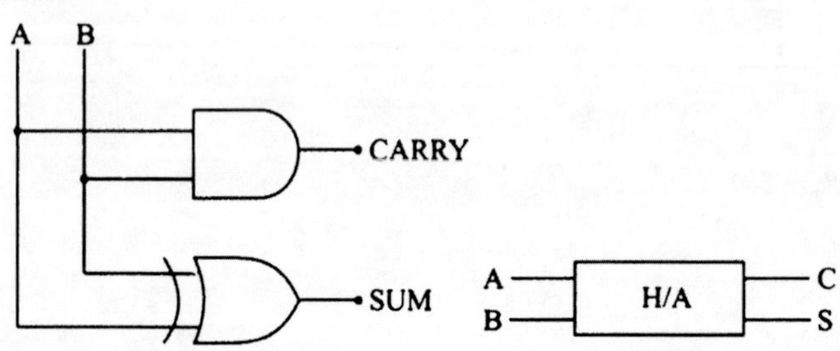

(*a*) **logic circuit** (*b*) **symbol**

Fig. 9.10. Half adder.

Table 9.8. Truth table for half adder

A	*B*	*S*	*C*
0	0	0	0
0	1	1	0
1	0	1	0
1	1	0	1

Logic equations for SUM and CARRY are

$$S = \bar{A}B + A\bar{B}$$

$$C = AB$$

A half adder can add only two one bit numbers. It cannot accept a carry from a previous addition.

9.7.2. Full Adder

A full adder has two half adders and an OR gate. Figure 9.11 shows a full adder. Table 9.9 shows its truth table.

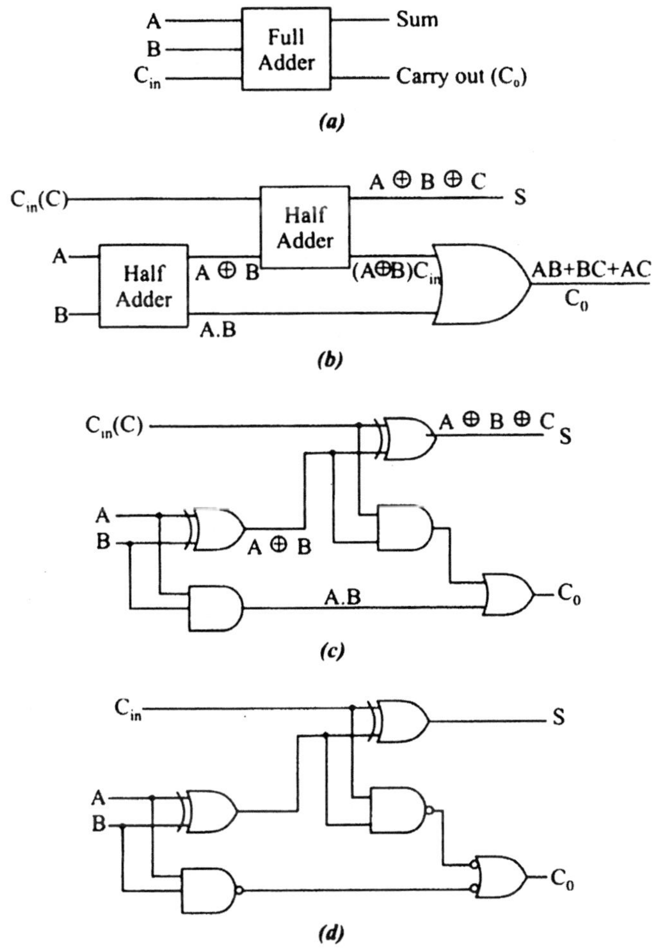

Fig. 9.11. Full adder (*a*) block diagram and symbol, (*b*) as combination of two half adders (*c*) logic diagram, (*d*) logic diagram using XOR and NAND gates.

Table 9.9. Truth table for adding A, B and a carry C_{in}

Inputs			*Outputs*	
C_{in}	*A*	*B*	*Sum*	*Carry*
0	0	0	0	0
0	0	1	1	0
0	1	0	1	0
0	1	1	0	1
1	0	0	1	0
1	0	1	0	1
1	1	0	0	1
1	1	1	1	1
Carry + A + B			Sum	Carry out

9.7.3. Four Bit Binary Adder

Figure 9.12 shows the circuit of 4 bit binary adder. It has one half adder and 3 full adders. It can add two four bit binary numbers.

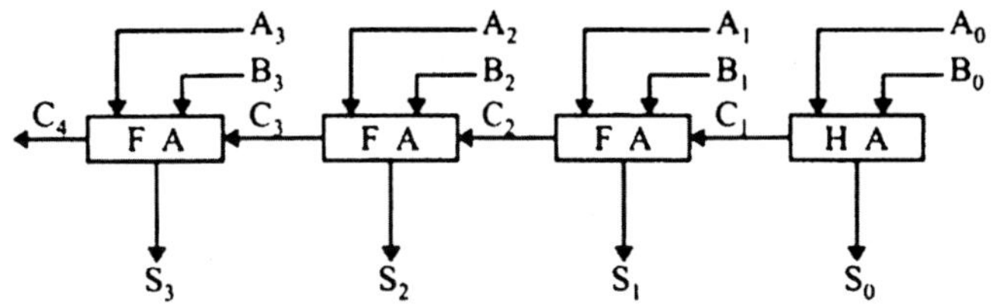

Fig. 9.12. Binary adder circuit.

9.7.4. Half Subtracter

Figure 9.13 shows the symbol and circuit of half subtractor. It uses one AND gate, one NOT gate and one XOR gate.

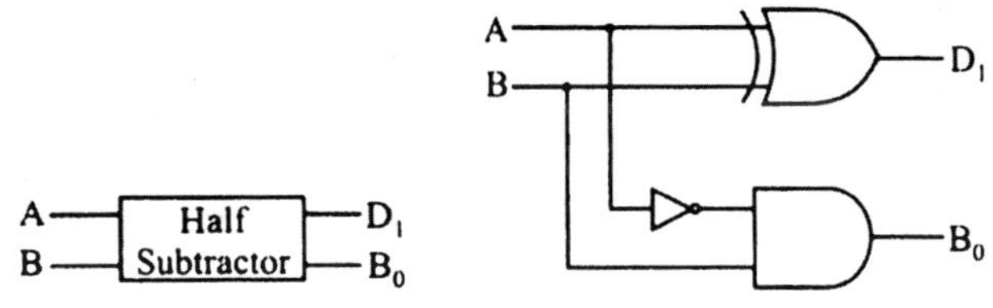

(*a*) block diagram **(*b*) logic circuit**

Fig. 9.13. Half subtractor.

9.7.5. Full Subtractor

Figure 9.14 shows the symbol and circuit of full subtractor. Table 9.10 shows its Truth table

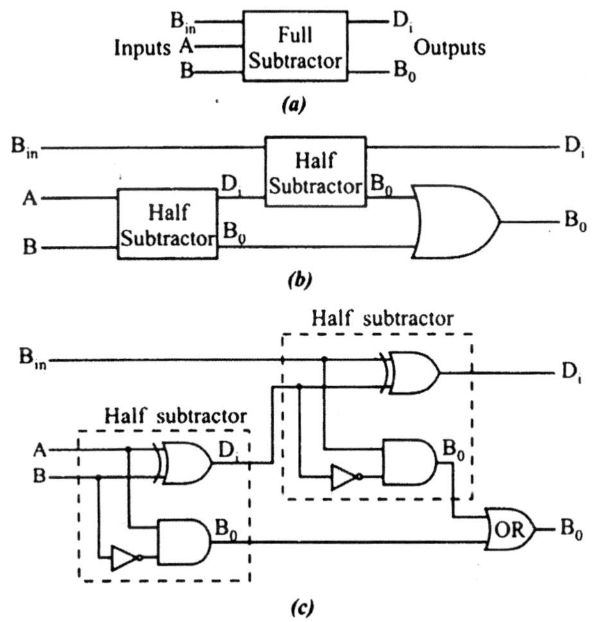

Fig. 9.14. Full subtracter (*a*) block diagram, (*b*) as combination of two half Subtractors and OR gale, (*c*) logic diagram.

Table 9.10. Truth table for full subtractor

Inputs			*Outputs*	
A	*H*	B_{in}	D_i	B_0
0	0	0	0	0
0	0	1	1	1
0	1	0	1	1
0	1	1	0	1
1	0	0	1	0
1	0	1	0	0
1	1	0	0	0
1	1	1	1	1

9.7.6. 2's Complement Adder/Subtracter

Figure 9.15 shows 2's complement, adder/subtractor. It is widely used in computers.

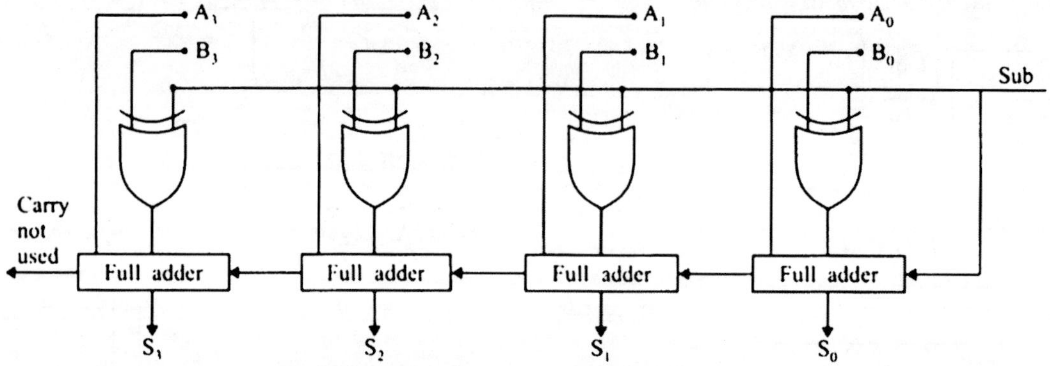

Fig. 9.15. 2's complement adder/subtractor.

9.8. ENCODERS, DECODERS AND DISPLAY DEVICES

The display devices used in digital electronics are LED display, LCD display and vacuum fluorescent display. Figure 9.16 shows 7 segment display.

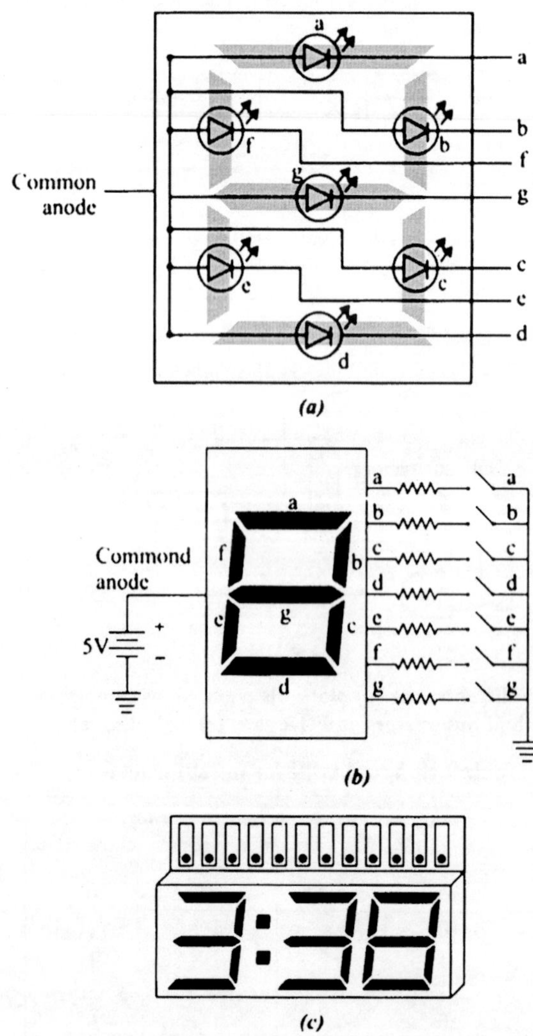

Fig. 9.16. Seven segment display *(a)* Indicator, *(b)* Schematic diagram, *(c)* Multidigit display.

9.8.1. Decimal to Binary Encoder

An encoder converts alphanumeric characters to binary codes. Figure 9.17 shows decimal to binary encoder for 10 inputs. When push button 4 is pressed OR gate Y_2 has high input and output is

$$Y_3 Y_2 Y_1 Y_0 = 0100$$

When push button 7 is pressed OR gates Y_2, Y_1, Y_0 have high inputs and output is

$$Y_3 Y_2 Y_1 Y_0 = 0111$$

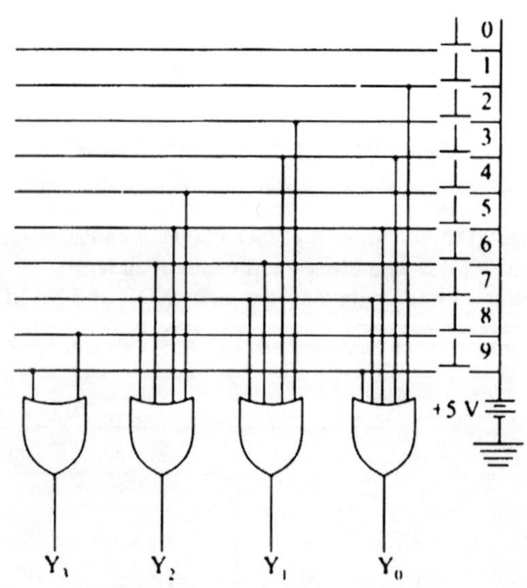

Fig. 9.17. Decimal to binary encoder for 10 inputs (*i.e.*, 0-9).

In a priority encoder priority is given to the highest order decimal digit in the inputs and other inputs are ignored, *e.g.*, if decimals 8 and 4 are pressed together, the priority encoder will convert decimal 8 into binary and ignore decimal 4.

9.8.2. Decoder

A decoder converts binary words into alphanumeric characters. Fig. 9.18 shows 4 line to 16 line decoder. TTL MSI 74154 is a 4 line to 16 line digital IC decoder.

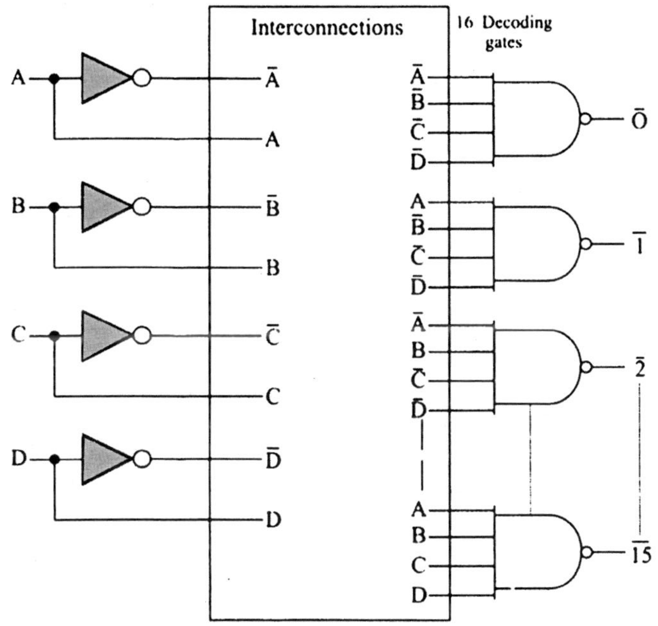

Fig. 9.18. Logic for a 4-line-to-16-line decoder.

9.9. MULTIPLEXERS AND DEMULTIPLEXERS

9.9.1. Multiplexer

A multiplexer is a device having many inputs and one output. It is also called data selector. By applying control signals any one of the inputs can be steered to the output. Figure 9.19 shows the logic circuit for 4 input MUX. As seen from Table 9.11 the select input S_1, S_0 determine the selected input.

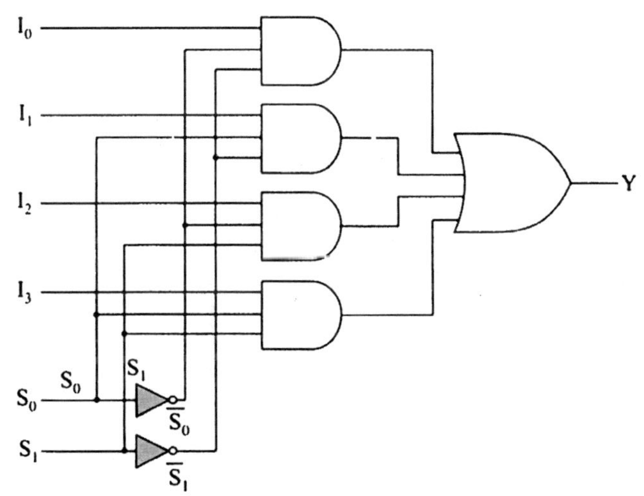

Fig. 9.19. Logic circuit for 4 input Mux.

Table 9.11. Data Selection for 4 :1 Mux.

S_1	S_0	Input selected
0	0	I_0
0	1	I_1
1	0	I_2
1	1	I_3

A multiplexer can also be used as a logic function generator. Figure 9.20 shows an 8 : 1 Mux used to generate a logic function

$$Y(A, B, C) = \bar{C}\bar{B}A + \bar{C}BA + C\bar{B}A + CB\bar{A}$$

A single multiplexer can handle a maximum of 16 inputs. When the number of inputs is more than 16, a multiplexer tree is used. Figure 9.21 shows a 32 : 1 multiplexer tree using two 16 : 1 and one 2 : 1 multiplexer.

Fig. 9.20.

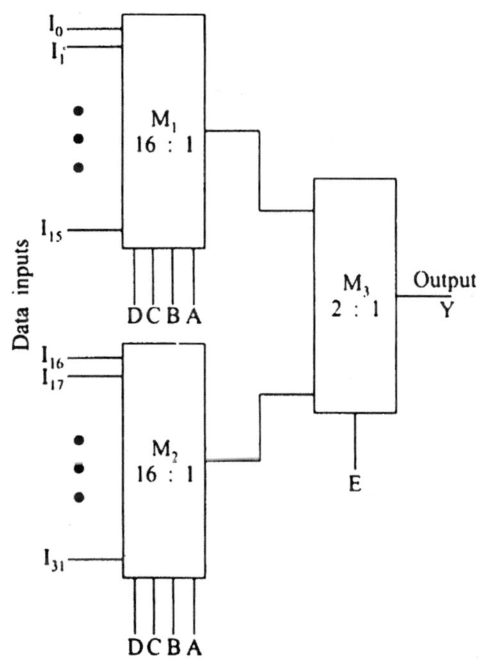

Fig. 9.21.

9.9.2. Demultiplexer

A demultiplexer takes in data from one line and directs it to any of its N outputs depending on the status of select inputs. Figure 9.22 shows the logic circuit of 1 : 4 demultiplexer.

When the number of outputs is more than 16, a demultiplexer tree is required.

Multiplexer and Demultiplexers are available as IC packages.

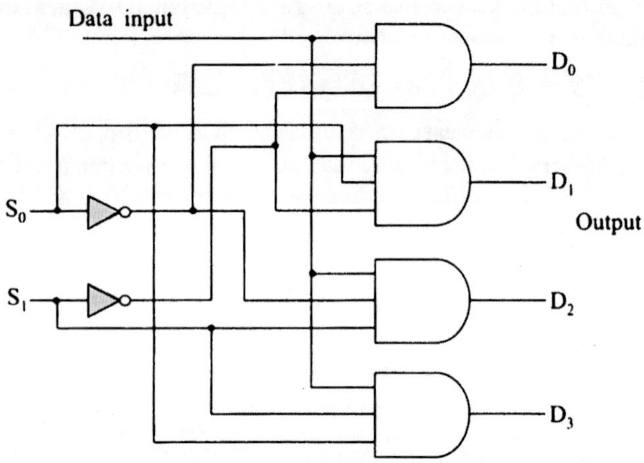

Fig. 9.22. Logic circuit of 1:4 demultiplexer.

9.10. LATCHES AND FLIP-FLOPS

Flip-flop is a basic building block for a sequential logic circuit. Flip-flops are combined to form counters, shift registers and various memory devices. Some of these are shown in the following diagrams.

Table 9.12. NOR latch

R	*S*	*Q*	*Comment*
0	0	NC	No change
0	1	1	Set
1	0	0	Reset
1	1		Race

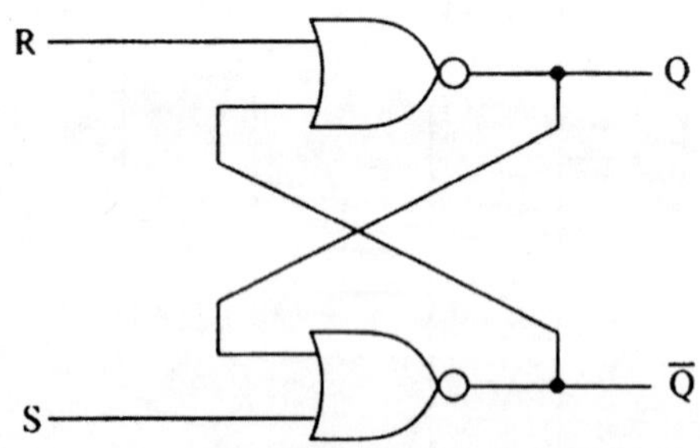

Fig. 9.23. RS latch using NOR gates.

Table 9.13. NAND latch

Inputs		*Output*		*Comments*
S	*R*	*Q*	$\bar{Q}$	
0	0			Race
0	1	1	0	Set
1	0	0	1	Reset
1	1	NC	NC	No change

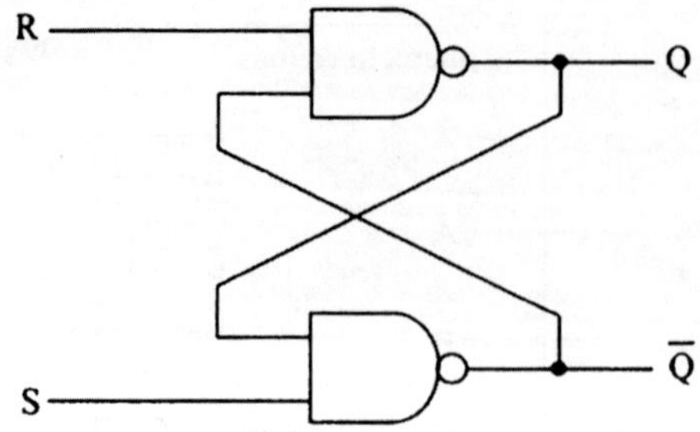

Fig. 9.24. RS latch using NAND gates.

Table 9.14. Clocked RS latch

Inputs			*Output*		*Comments*
CLK	*S*	*R*	*Q*	$\bar{Q}$	
1	0	0	NC		No change
1	0	1	0	1	Reset or cleared to 0
1	1	0	1	0	Set to 1
1	1	1			Race Not allowed condition

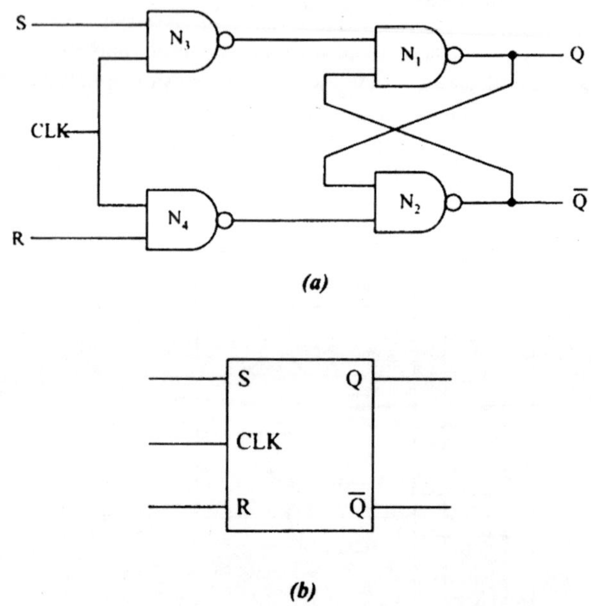

Fig. 9.25. Clocked RS flip-flop using NAND gates.

Table 9.15. Truth table of clocked D latch

Input		*Output*	
CLK	*D*	*Q*	$\bar{Q}$
0	×	NC	
1	0	0	1
1	1	1	0
× indicates does not matter			

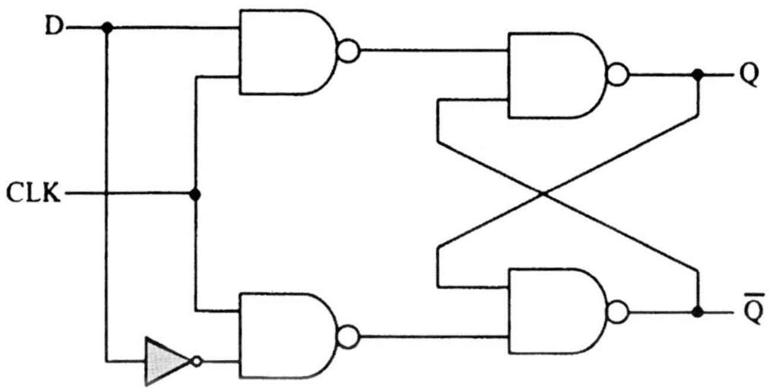

Fig. 9.26. Clocked D flip-flop.

Table 9.16. Truth table for an edge triggered D flip-flop

CLK	*D*	*Q*
0	×	No change
1	×	No change
↓	×	No change
↑	0	0
↑	1	1

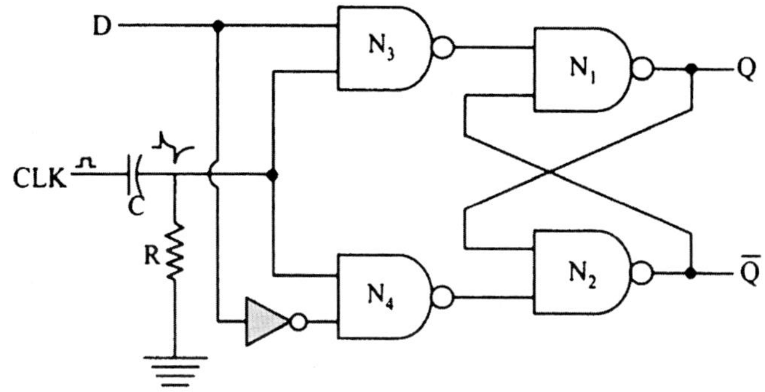

Fig. 9.27. Edge triggered D flip-flop.

Table 9.17. Edge triggered D flip-flop with preset and clear.

Preset	*Clear*	*CLK*	*D*	*Q*
0	0	×	×	Race
0	1	×	×	1
1	0	×	×	0
1	1	0	×	No change
1	1	1	×	No change
1	1	↓	×	No change
1	1	↑	0	0
1	1	↑	1	1

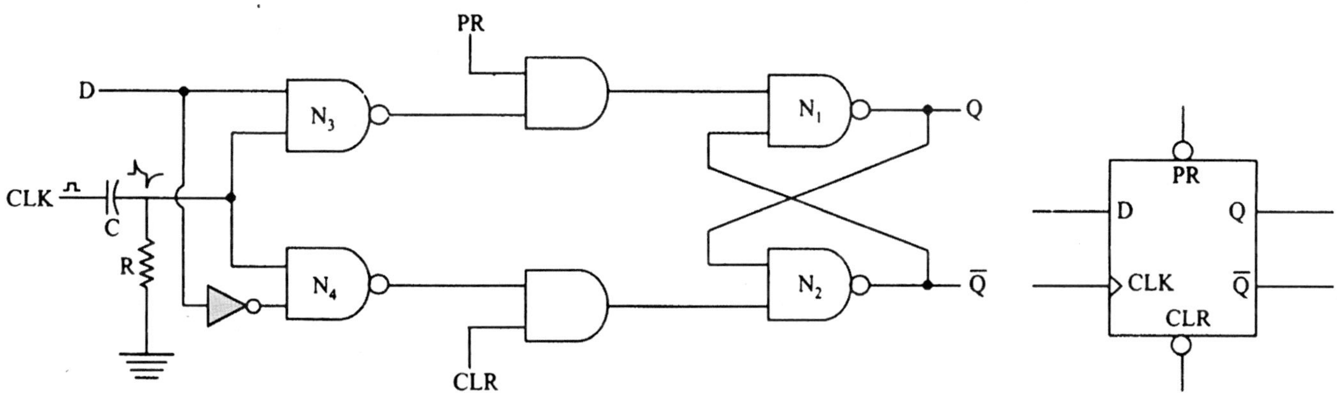

(a) *(b)*

Fig. 9.28. Edge triggered D flip-flop with preset and clear.

Table 9.18. Truth table of JK master slave flip-flop.

Inputs					*Output*
PR	*CLR*	*CLK*	*J*	*K*	*Q*
0	0	×	×	×	Race condition
0	1	×	×	×	1
1	0	×	×	×	0
1	1	×	0	0	No change
1	1	⎍	0	1	0
1	1	⎍	1	0	1
1	1	⎍	1	1	Toggle

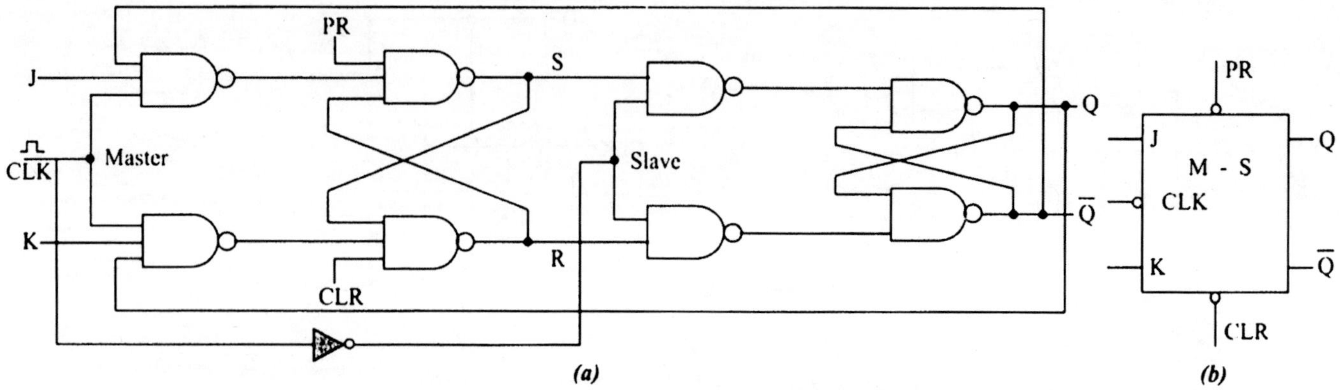

Fig. 9.29. JK Master slave flip-flop (*a*) circuit, (*b*) symbol.

Table 9.19. Truth table for Positive edge triggered JK flip-flop.

CLK	*J*	*K*	*Q*
0	×	×	No change
1	×	×	No change
↓	×	×	No change
×	0	0	No change
↑	0	1	0 (reset)
↑	· 1	0	1 (set)
↑	1	1	Toggle

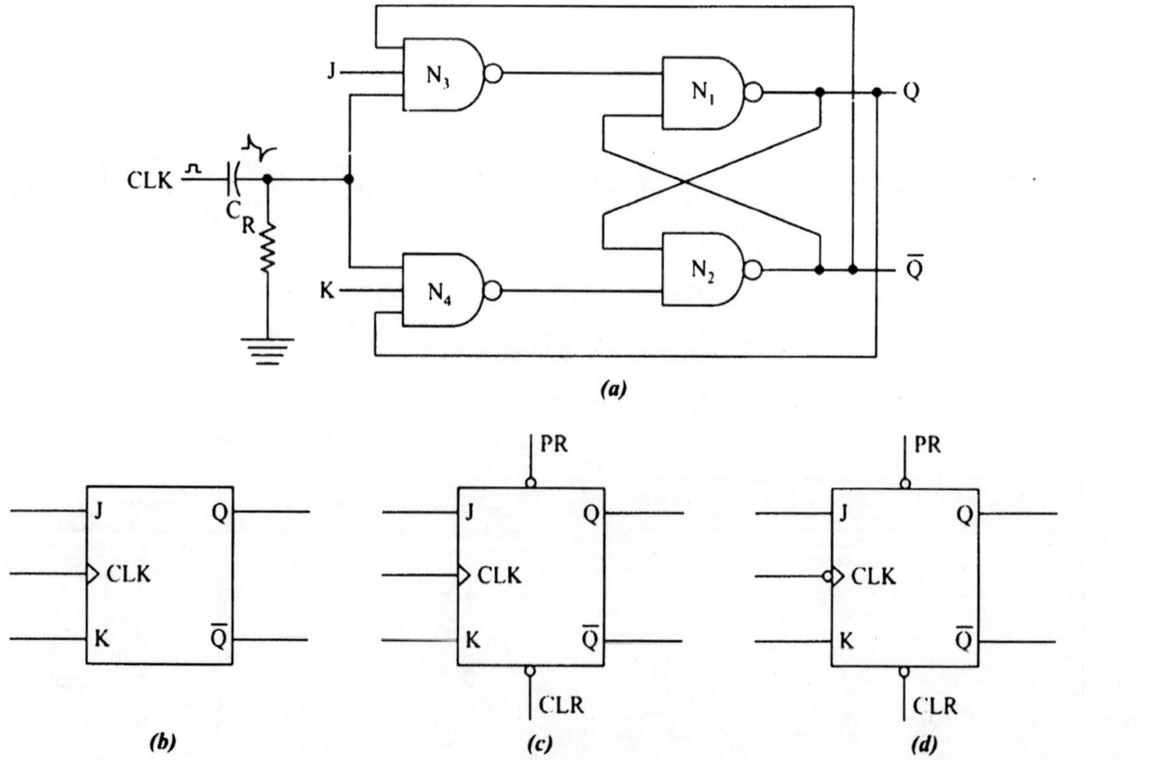

Fig. 9.30. Edge trigerred JK flip-flop (*a*) circuit, (*b*) symbol for positive edge triggered JK flip-flop, (*c*) symbol for positive edge triggered JK flip-flop preset and clear, (*d*) symbol for negative edge triggered JK flip-flop with preset and clear.

Table 9.20 Truth table of negative edge triggered T flip-flop.

Inputs		*Output*
CLK	***T***	***Q***
0	×	No change
1	×	No change
↑	×	No change
×	0	No change
↓	0	No change
↓	1	Toggles

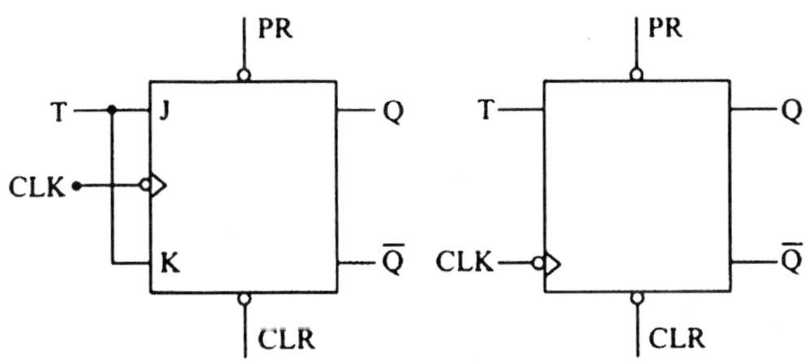

Fig. 9.31. (*a*) **Conversion of JK flip-flop to T flip-flop,** (*b*) **symbol for negative edge triggered T flip-flop with preset and clear.**

9.11. COUNTERS

A group of memory elements is called a register. A counter is a register capable of counting the number of clock pulses which have arrived at the clock input. Some commonly used counters are shown below.

Table 9.21. 4 Bit ripple counter

Count	Q_3	Q_2	Q_1	Q_0
0	0	0	0	0
1	0	0	0	1
2	0	0	1	0
3	0	0	1	1
4	0	1	0	0
5	0	1	0	1
6	0	1	1	0
7	0	1	1	1
8	1	0	0	0
9	1	0	0	1
10	1	0	1	0
11	1	0	1	1
12	1	1	0	0
13	1	1	0	1
14	1	1	1	0
15	1	1	1	1

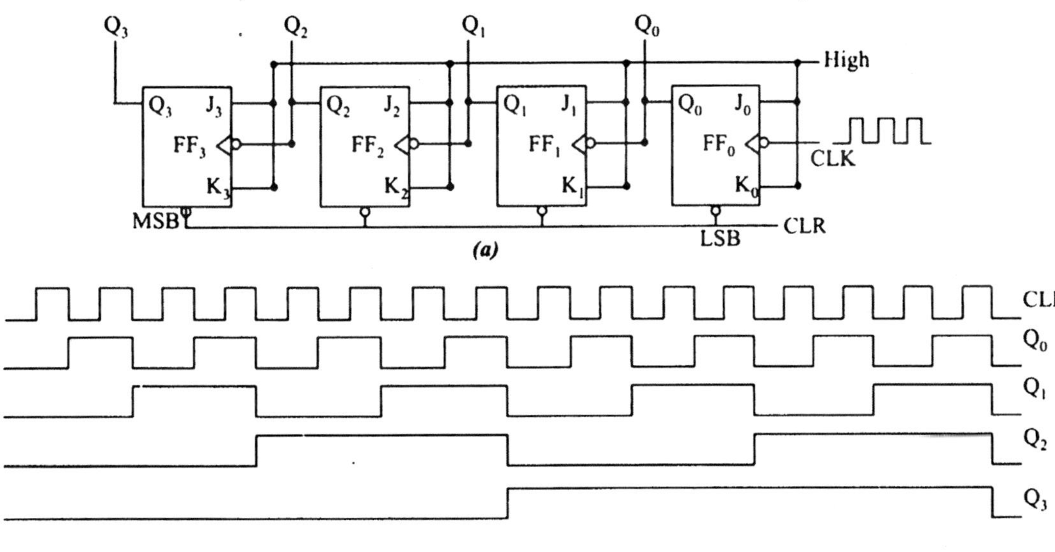

Fig. 9.32. Ripple counter (*a*) **circuit,** (*b*) **timing diagram.**

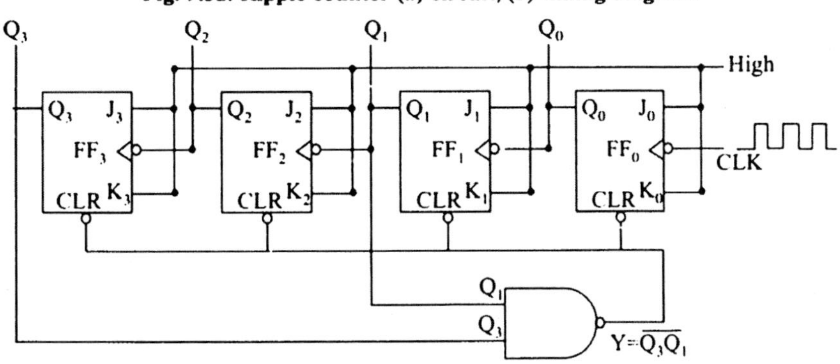

(*a*)

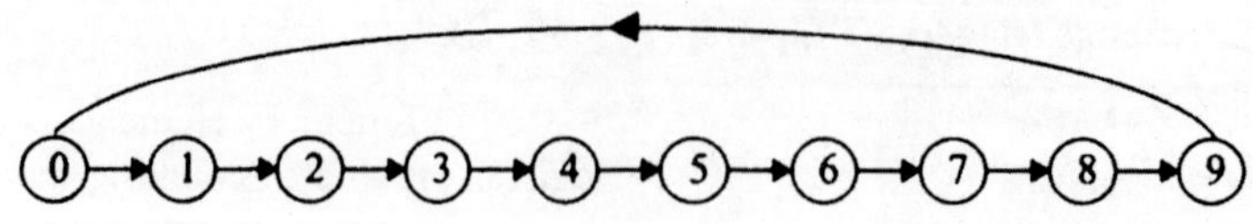

(b)

Fig. 9.33. Decade counter (*a*) circuit, (*b*) state diagram.

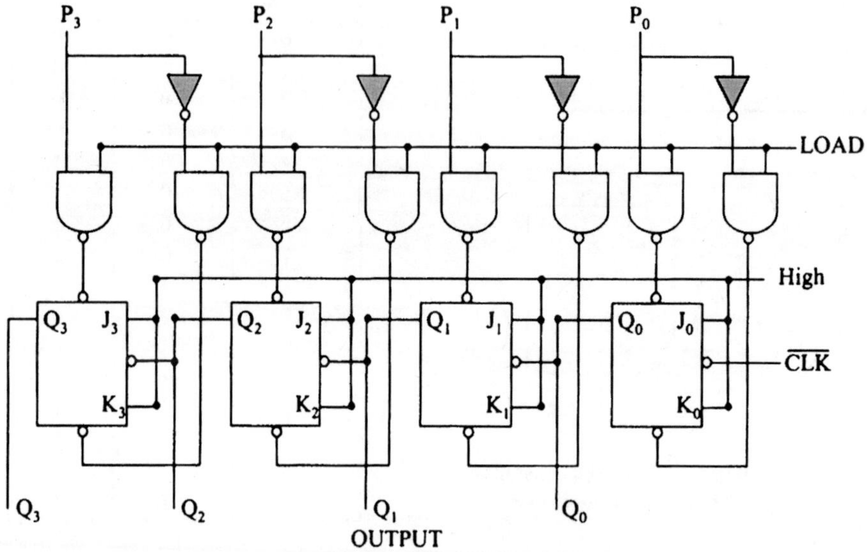

Fig. 9.34. Presettable counter.

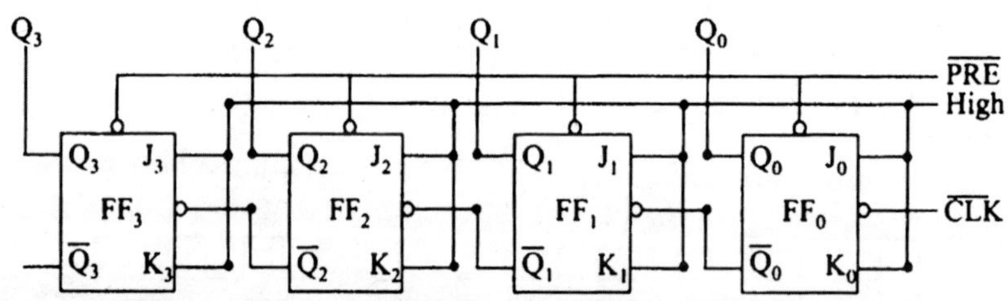

Fig. 9.35. Down counter.

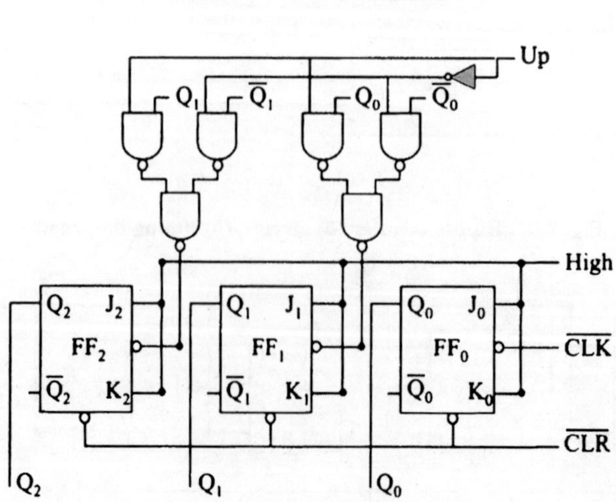

Fig. 9.36. Up-down counter.

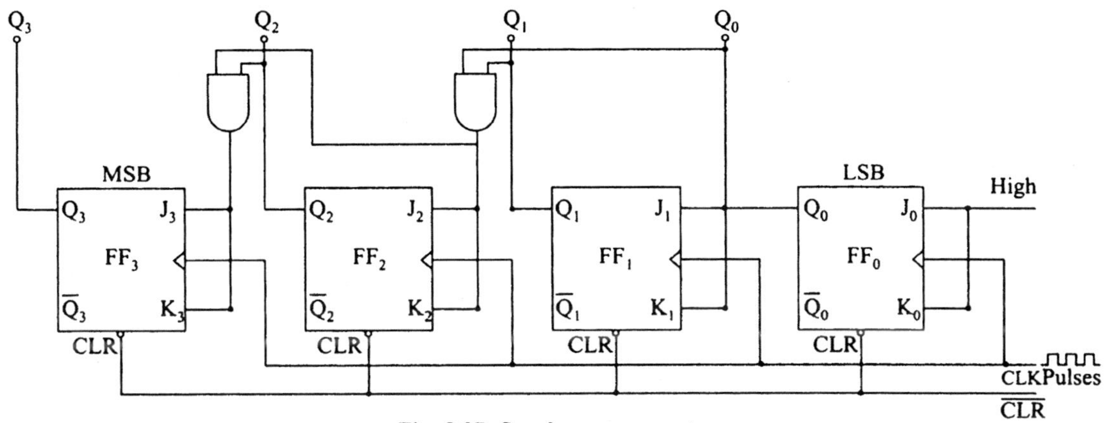

Fig. 9.37. Synchronous counter.

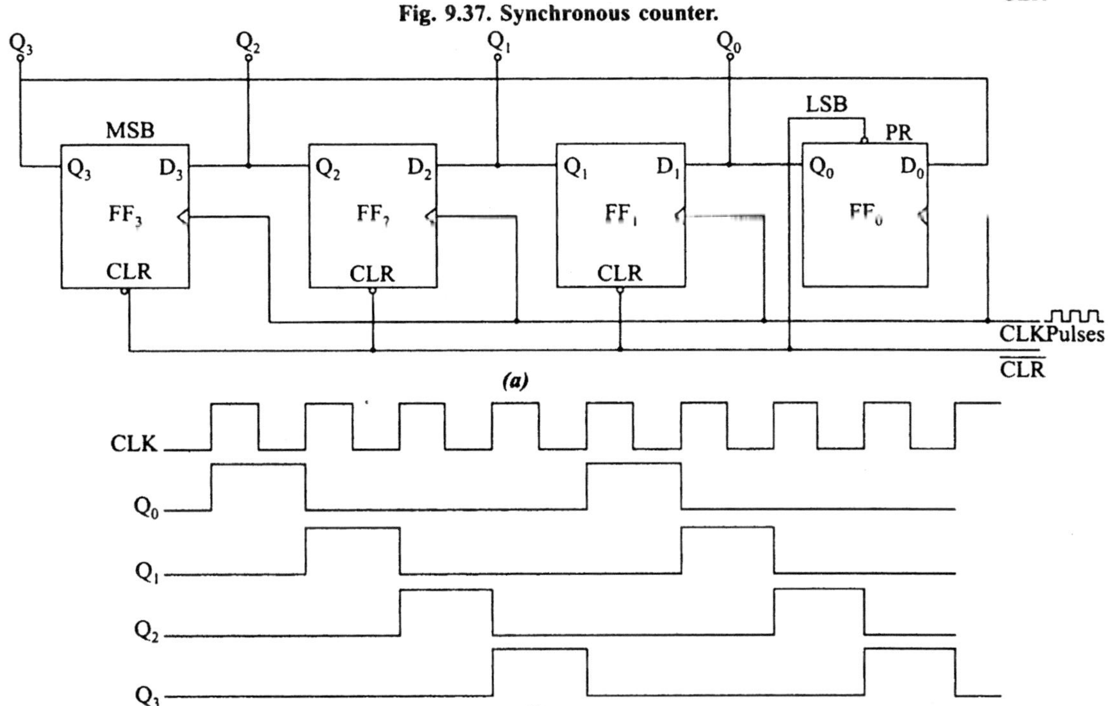

(a)

(b)

Fig. 9.38. Ring counter (*a*) circuit, (*b*) waveform/timing diagram.

Table 9.22. Counting Sequence of 4 bit Synchronous Counter

Number of clock pulses	*Counting sequence*				*Equivalent decimal*
	Q_3	Q_2	Q_1	Q_0	
0	0	0	0	0	0
1	0	0	0	1	1
2	0	0	1	0	2
3	0	0	1	1	3
4	0	1	0	0	4
5	0	1	0	1	5
6	0	1	1	0	6
7	0	1	1	1	7
8	1	0	0	0	8
9	1	0	0	1	9
10	1	0	1	0	10
11	1	0	1	1	11
12	1	1	0	0	12
13	1	1	0	1	13
14	1	1	1	0	14
15	1	1	1	1	15

9.12. SHIFT REGISTERS

A register is a device capable of storing a bit. The data can be serial or parallel. A register can convert a data from serial to parallel and vice versa. Shifting the bits to left or right is an important aspect of arithmetic operations. The circuits of some shift registers are as under

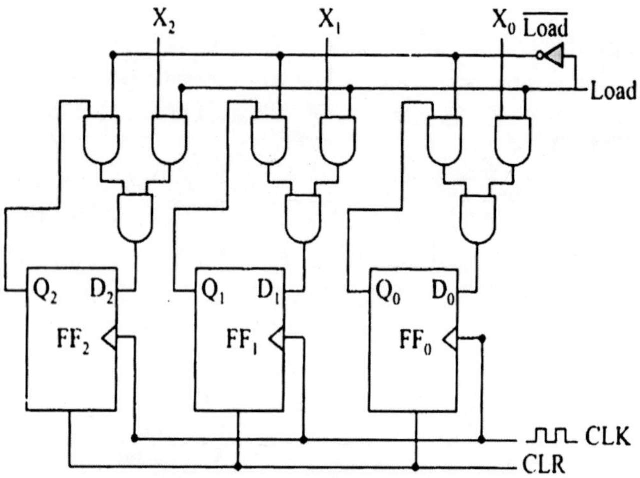

Fig. 9.39. 3 bit controlled buffer register.

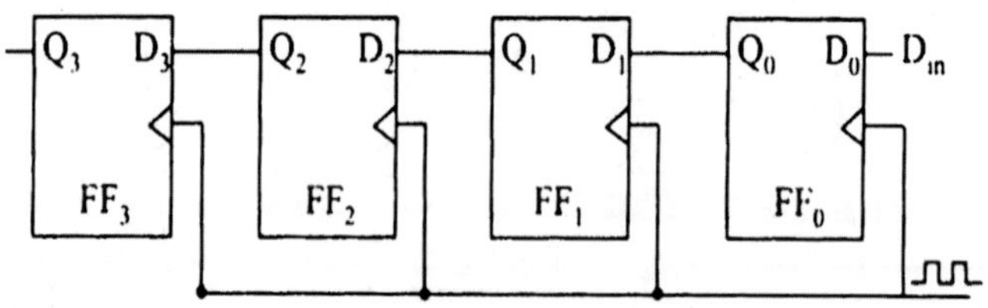

Fig. 9.40. Shift left register.

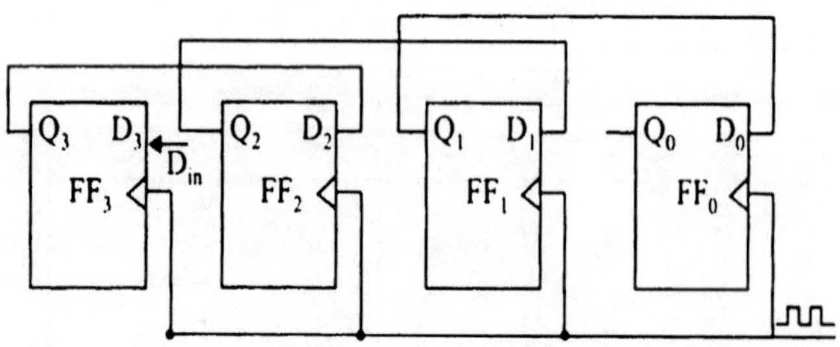

Fig. 9.41. Shift right register.

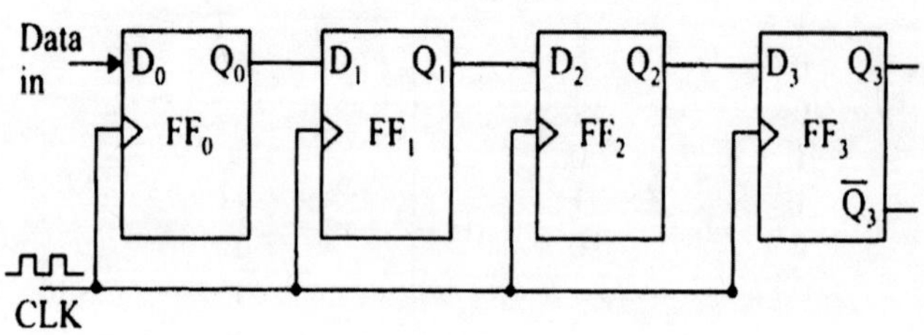

Fig. 9.42. 4 bit serial in-serial out shift register.

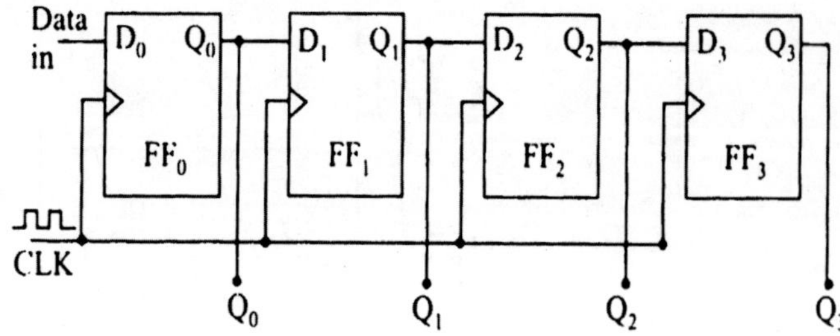

Fig. 9.43. 3-bit serial in-parallel out shift register.

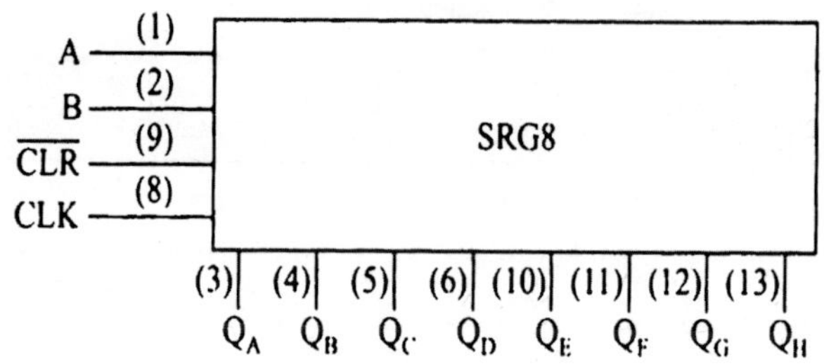

Fig. 9.44. 8 bit serial in-parallel out shift register IC 74164.

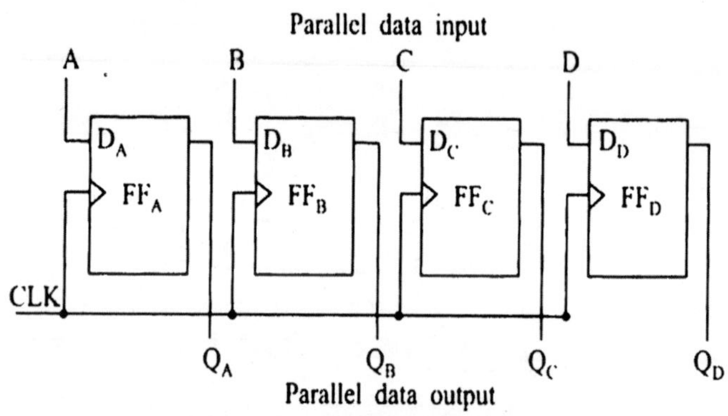

(a)

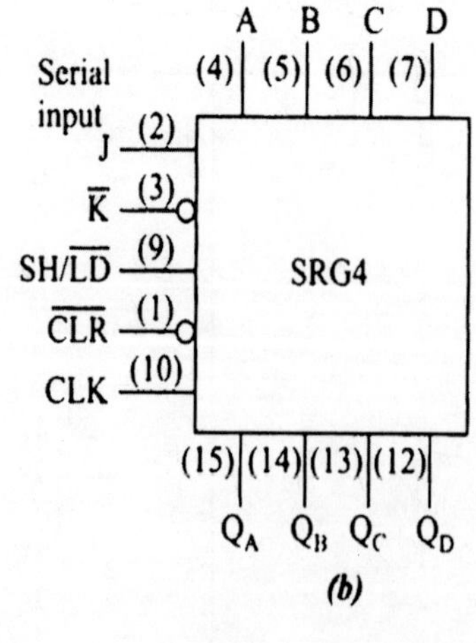

(b)

Fig. 9.45. (*a*) Parallel in-parallel out shift register, (*b*) logic symbol of IC 74195 4 bit shift register.

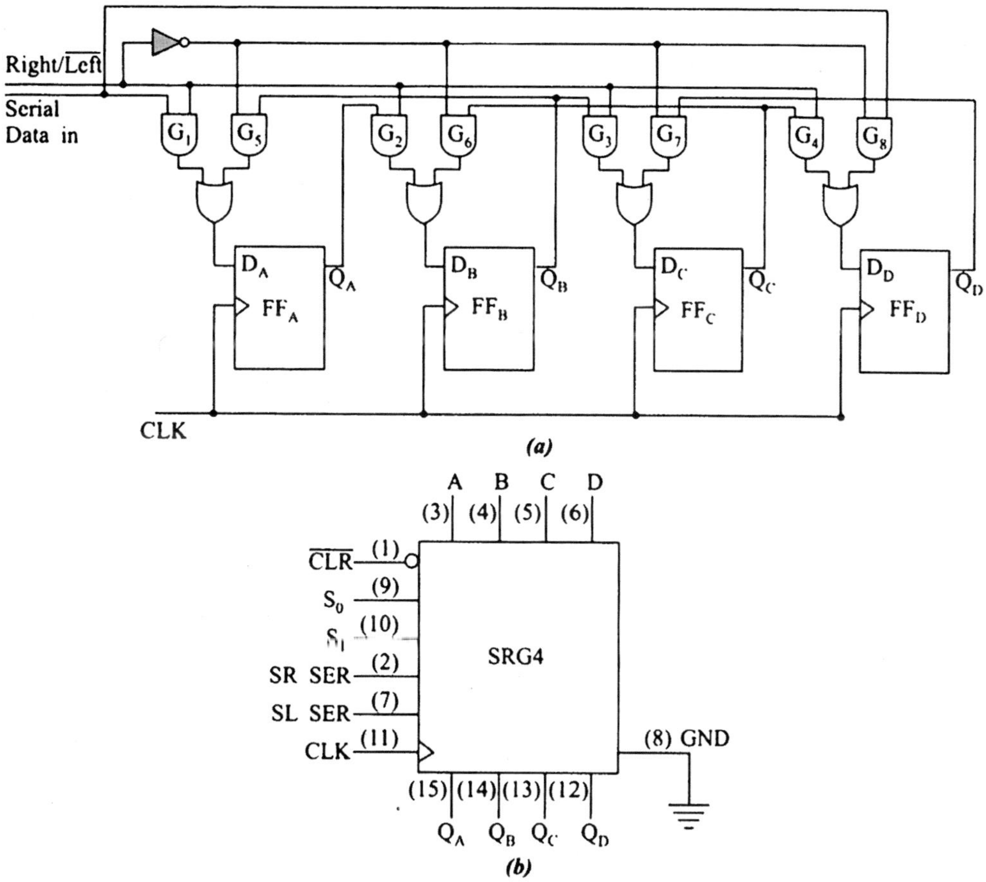

Fig. 9.46. *(a)* **4 bit universal shift register,** *(b)* **symbol of IC 74194 4 bit universal shift register.**

9.13. DIGITAL MEMORY

Memory size M × N where M is the number of locations (or addresses) and N is the number of bits per location.

ROM : Read only memory. It is a permanent or semi-permanent memory.

RAM : Random access memory: Both read and write operations are possible.

Mask ROM : It is permanently programmed during manufacture and the contents cannot be changed.

PROM : Programmable read only memory. In this memory the user can store data using an instrument called PROM programmer.

EPROM : Erasable PROM. It can be reprogrammed by first erasing the existing program. It can be either UV EPROM (ultra violet erasable PROM. The erasure is achieved by using ultra violet light) or EEPROM (electrically erasable PROM. The erasure and programming is done by electrical pulses)

Dynamic ROM Cell : The data is stored in a capacitor. It is shown in Fig. 9.47.

Programmed Logic Array (PLA) : It is also a programmed storage device but internal structure is different from ROM. It is also called field programmable array (FPLA). Figure 9.48 shows a simple example of PLA.

Programmable Logic Device (PLD) : An IC which can be used to implement logic functions.

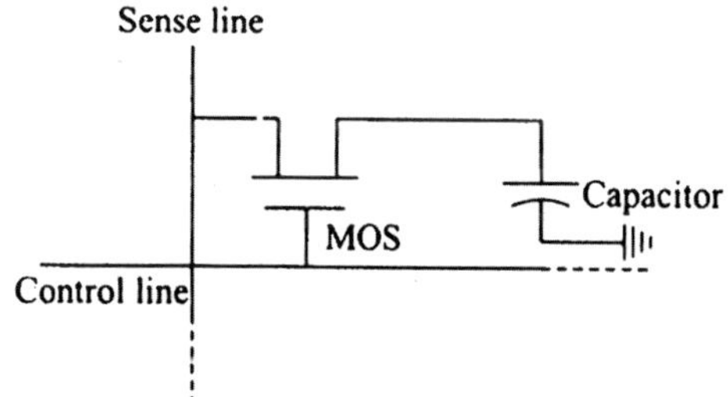

Fig. 9.47. Dynamic RAM cell.

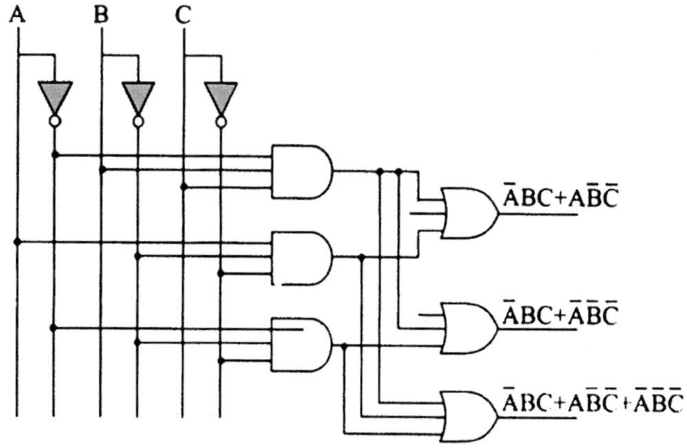

Fig. 9.48. A simple example of PLA.

Programmable Array Logic (PAL). It is a special class of PLD. The inputs to AND gates are programmable and inputs to OR gates are hard wired.

9.14. D/A CONVERTER

9.14.1. Weighted Resister Network

Figure 9.49 shows the circuit.

$$V_0 = \frac{8V_A + 4V_B + 2V_C + V_D}{15}$$

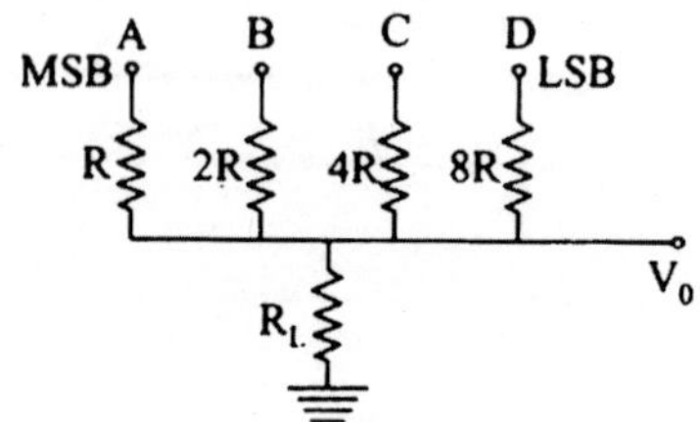

Fig. 9.49. Weighted resistor network for D/A conversion.

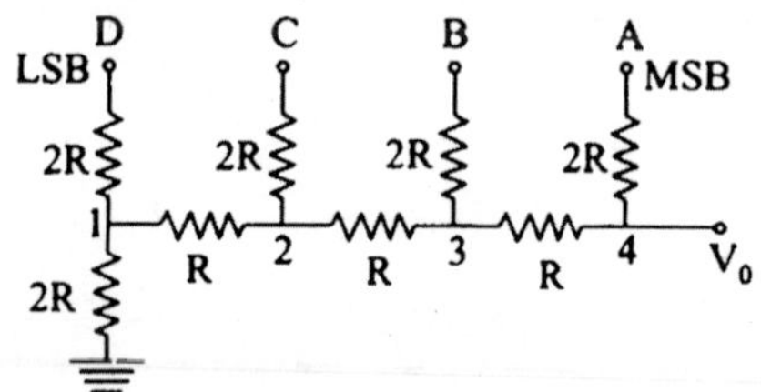

Fig. 9.50. Binary ladder network for D/A conversion.

9.14.2. Binary Ladder Network

Figure 9.50 shows the circuit.

From each of the nodes 1, 2, 3, 4 the resistance to source and ground is 2R each.

9.15. A/D CONVERTER

9.15.1. Successive Approximation A/D Converter

Figure 9.51 *(a)* shows the block diagram and Fig. 9.51 *(b)* shows the output waveform. The main components are op-amp comparator, control, logic, successive approximation register and D/A converter.

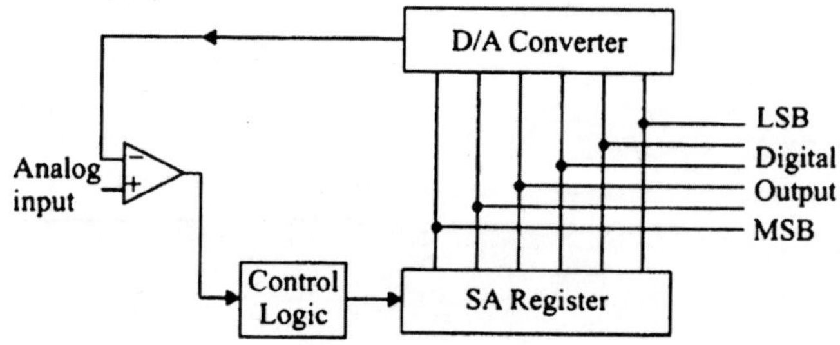

***(a)* Block diagram**

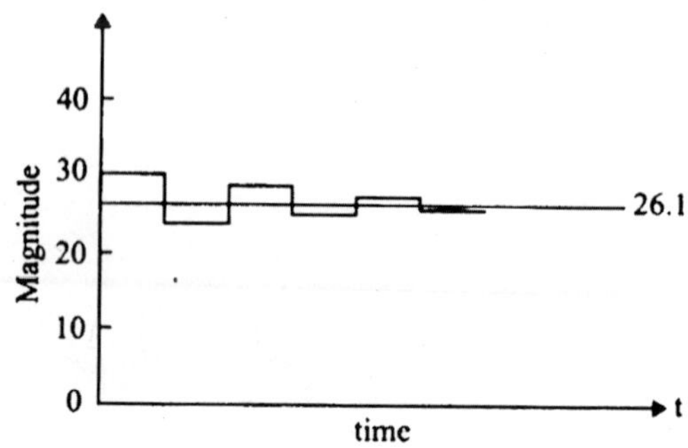

***(b)* Waveforms**

Fig. 9.51. Successive approximation A/D converter.

OBJECTIVE TYPE QUESTIONS

A. Select the most appropriate alternative.

1. The number of digits in octal system is
   (*a*) 8 (*b*) 7
   (*c*) 8 or 7 (*d*) 10
2. The number of digits in hexadecimal system is
   (*a*) 15 (*b*) 16
   (*c*) 10 (*c*) 8
3. The number of bits in a nibble is
   (a) 8 (*b*) 4
   (*c*) 2 (*d*) 16
4. The digit F in hexadecimal system is equivalent to number............. in decimal system
   (*a*) 16 (*b*) 15
   (*c*) 17 (*d*) 8
5. Which of the following binary numbers is equivalent to decimal 10
   (a) 1000 (*b*) 1100
   (*c*) 1010 (*d*) 1001
6. The number FF in hexadecimal system is equivalent to number............. in decimal system
   (*a*) 256 (*b*) 255
   (*c*) 240 (*d*) 239
7. ICs are
   (a) analog (*b*) digital
   (*c*) both analog and digital (*d*) mostly analog
8. The rate of change of digital signals between High and Low level is
   (*a*) very fast (*b*) fast
   (*c*) slow (*d*) very slow
9. Digital circuits mostly use
   (*a*) Diodes
   (*b*) Bipolar transistors
   (*c*) Diodes and bipolar transistors
   (*d*) Bipolar transistors and FETs
10. Two voltages are –5 V and –10 V. In positive logic
    (*a*) –5 V is 1 and –10 V is 0
    (*b*) –10 V is 1 and –5 V is 0
    (*c*) –5 V is 1 in some circuits and 0 in others
    (*d*) –10 V is 1 in some circuits and 0 in others
11. Logic pulser
    (*a*) generates short duration pulses

(*b*) generates long duration pulses
(*c*) generates sinusoidal voltage
(*d*) generates ideal pulses

12. The number of LED display indicators in logic probe are
(*a*) 1 (*b*) 2
(*c*) 1 or 2 (*d*) 4

13. The output impedance of a logic pulser is
(*a*) low (*b*) high
(*c*) may be low or high (*d*) cannot say

14. A current tracer responds to
(*a*) steady current only
(*b*) pulsating current only
(*c*) both pulsating and steady current
(*d*) none of the above

15. Logic analyser is
(*a*) a multi channel oscilloscope
(*b*) similar to logic pulser
(*c*) similar to current tracer
(*d*) none of the above

16. Which of these are two state devices?
(a) lamp (*b*) punched card
(*c*) magnetic tape (*d*) all the above

17. In the decimal number 27, the digital 2 represents
(a) 2 (*b*) 20
(*c*) 0.2 (*d*) 200

18. In hexadecimal number system, A is equal to decimal number
(a) 10 (*b*) 11
(*c*) 17 (*d*) 18

19. Hexadecimal number F is equal to octal number
(a) 15 (*b*) 16
(*c*) 17 (*d*) 18

20. Hexadecimal number E is equal to binary number
(*a*) 1110 (*b*) 1101
(*c*) 1001 (*d*) 1111

21. Binary number 1101 is equal to octal number
(*a*) 17 (*b*) 16
(*c*) 15 (*d*) 14

22. Octal number 12 is equal to decimal number
(*a*) 8 (*b*) 9
(*c*) 10 (*d*) 11

23. 100101_2 is equal to decimal number
(*a*) 47 (*b*) 37
(*c*) 27 (*d*) 17

24. Decimal number 10 is equal to binary number
(*a*) 1110 (*b*) 1010
(*c*) 1001 (*d*) 1000

25. Binary number 110011011001 is equal to decimal number
(*a*) 3289 (*b*) 2289
(*c*) 1289 (*d*) 289

26. The binary addition 1 + 1 =
(*a*) 11 (*b*) 10
(*c*) 111 (*d*) 100

27. 1111 + 11111 =
(a) 101111 (*b*) 101110
(*c*) 111111 (*d*) 011111

28. Binary multiplication $1 \times 0 =$
(*a*) 1 (*b*) 0
(*c*) 10 (*d*) 11

29. $11001_2 - 10001_2 =$
(*a*) 10000 (*b*) 01000
(*c*) 00100 (*d*) 0001

30. $1011_2 \times 101_2 = \ldots\ldots_{10}$
(*a*) 55 (*b*) 45
(*c*) 35 (*d*) 25

31. $11101_2 \times 10001_2 = \ldots\ldots_2$
(*a*) 111101101 (*b*) 111101100
(*c*) 11110 (*d*) 1100110

32. –8 is equal to signed binary number
(*a*) 10001000 (*b*) 00001000
(*c*) 10000000 (*d*) 11000000

33. 1's complement of 11100110 is
(*a*) 00011001 (*b*) 10000001
(*c*) 00011010 (*d*) 00000000

34. 2's complement of binary number 0101 is
(*a*) 1011 (*b*) 1111
(*c*) 1101 (*d*) 1110

35. 2's complement number 00011111 $= \ldots\ldots_{10}$
(*a*) +31 (*b*) –31
(*c*) +11 (*d*) – 11

36. 2's complement of – 24 =
(*a*) 11101000 (*b*) 01001000
(*c*) 01111111 (*d*) 00111111

37. $0.1011_2 = \ldots\ldots_{10}$
(*a*) 0.6875_{10} (*b*) 0.6800_{10}
(*c*) 0.0100_{10} (*d*) 0.5000_{10}

38. $39.12_{10} = \ldots\ldots_2$
(*a*) 100111.0001111 (*b*) 100011.0001111
(*c*) 100001.000111 (*d*) 001111.0001111

39. $110.11_2 \times 110_2 = \ldots\ldots_{10}$
(*a*) 40.5 (*b*) 30.5
(*c*) 20.5 (*d*) 10.5

40. In floating point representation, the number of bits in mantissa is
(*a*) 10 (*b*) 9
(*c*) 8 (*d*) 7

41. In floating point representation the number of bits in exponent is
(*a*) 8 (*b*) 7
(*c*) 6 (*d*) 5

42. In floating point representation, mantissa is expressed in
(*a*) excess 3 notation (*b*) gray code
(*c*) excess 32 notation
(*d*) 2's complement

43. In floating point number 0110100000010101, the exponent is equal $\ldots\ldots_{10}$
(*a*) 11 (*b*) – 11
(*c*) 10 (*d*) – 10

44. In floating point representation, the accuracy is
(*a*) 10 bit (*b*) 9 bit
(*c*) 8 bit (*d*) 6 bit

45. $7BF_{16}$ = ...$_2$
(*a*) 0111 1011 1110 (*b*) 0111 1011 1111
(*c*) 0111 1011 0111 (*d*) 0111 1011 0011

46. $E7F6_{16}$ = ...$_{10}$
(*a*) 600000_{10} (*b*) 59382_{10}
(*c*) 9382_{10} (*d*) 382_{10}

47. 268_{10} = ...$_{16}$
(a) $10A_{16}$ (*b*) $10B_{16}$
(*c*) $10C_{16}$ (*d*) $10D_{16}$

48. $AECFI_{16} + 15ACD_{16}$ = ...$_{16}$
(*a*) $C47BB_{16}$ (*b*) $C47BE_{16}$
(*c*) $A234F_{16}$ (*d*) $A1111_{16}$

49. $1274_{16} - 3A7_{16}$ = ...$_{16}$
(*a*) ACA_{16} (*b*) ACB_{16}
(*c*) ECD_{16} (*d*) AAA_{16}

50. $A_{16} \times 2_{16}$ = ...$_{16}$
(*a*) 16_{16} (*b*) 15_{16}
(*c*) 14_{16} (*d*) 13_{16}

51. 71_8 = ...$_2$
(*a*) 111000_2 (*b*) 111001_2
(*c*) 100001_2 (*d*) 110001_2

52. 1001101.1011_2 = ...$_8$
(*a*) 115.22_8 (*b*) 115.54_8
(*c*) 215.54_8 (*d*) 100.01_8

53. FF_{16} when converted to 8421 BCD =
(*a*) 0000 0101 0101 (*b*) 0010 0101 0101
(*c*) 1111 0101 0101 (*d*) 1000 0101 0101

54. Decimal number 9 in gray code =
(*a*) 1100 (*b*) 1101
(*c*) 110 (*d*) 1111

55. 11011 in gray code = ...$_2$
(*a*) 10010_2 (*b*) 11111_2
(*c*) 11100_2 (*d*) 10001_2

56. Decimal 8 in excess-3 code =
(*a*) 1000 (*b*) 1001
(*c*) 1011 (*d*) 1111

57. Decimal number 46 in excess 3 code =
(*a*) 1000 1001 (*b*) 0111 1001
(*c*) 0111 1111 (*d*) 1000 1111

58. In ASCII, letter B is coded as
(*a*) 1000001 (*b*) 1000010
(*c*) 1000100 (*d*) 1001000

59. The number of bits in ASCII is
(*a*) 12 (*b*) 10
(*c*) 9 (*d*) 7

60. The number of bits in EBCDIC is
(*a*) 12 (*b*) 10
(*c*) 8 (*d*) 6

61. The parity bit is
(*a*) always 1 (*b*) always 0
(*c*) 1 or 0 (*d*) none of the above

62. In 2 out of 5 code, decimal number 8 is
(*a*) 11000 (*b*) 10100
(*c*) 1100 (*d*) 1010

63. If the number of information bits is *m*, the number of parity bits *p* in the Hamming code is given by the equation
(*a*) $2^p = m + p + 1$ (*b*) $2^p \geq m + p + 1$
(*c*) $2^p = m + p$ (*d*) $2^p \geq m + p$

64. If number of information bits is 11, the number of parity bits in Hamming code is
(*a*) 5 (*b*) 4
(*c*) 3 (*d*) 2

65. If number of information bits is 4, the parity bits in Hamming code are located at bit positions .... from the LSB.
(*a*) 1, 2, 5 (*b*) 1, 2, 4
(*c*) 1, 2, 3 (*d*) 1, 2

66. The seven bit Hamming code as received is 0010001. If even parity has been used, the correct code is
(*a*) 0010001 (*b*) 1110001
(*c*) 0110001 (*d*) 0011001

67. For a 4096 × 8 EPROM, the number of address lines is
(*a*) 14 (*b*) 12
(*c*) 10 (*d*) 8

68. 23.6_{10} = ...$_2$
(*a*) 11111.10011 (*b*) 10111.10011
(*c*) 00111.101 (*d*) 10111.1

69. BCD number 01100111 = ...$_{10}$
(*a*) 66 (*b*) 67
(*c*) 68 (*d*) 69

70. The total number of input words for 4 input OR gate is
(*a*) 20 (*b*) 16
(*c*) 12 (*d*) 8

71. In a 4 input OR gate, the total number of High outputs for the 16 input states are
(*a*) 16 (*b*) 15
(*c*) 14 (*d*) 13

72. In a 4 input AND gate, the total number of High outputs for 16 input states are
(*a*) 16 (*b*) 8
(*c*) 4 (*d*) 1

73. A buffer is
(*a*) always non-inverting (*b*) always inverting
(*c*) inverting or non-inverting
(*d*) none of the above

74. An AND gate has two inputs A and B and one inhibit input S. Output is 1 if
(*a*) A = 1, B = 1, S = 1 (*b*) A = 1, B = 1, S = 0
(*c*) A = 1, B = 0, S = 1 (*d*) A = 1, B = 0, S = 0

75. An AND gate has two inputs A and B and one inhibit input S. Out of total 8 input states, output is 1 in
(*a*) 1 state (*b*) 2 states
(*c*) 3 states (*d*) 4 states

76. In a 3 input NOR gate, the number of states in which output is 1 equals
(*a*) 1 (*b*) 2
(*c*) 3 (*d*) 4

77. Which of these are universal gates?
(*a*) only NOR (*b*) only NAND
(*c*) both NOR and NAND (*d*) NOR, NAND, OR

78. In a 3 input NAND gate, the number of states in which output is 1 equals
(*a*) 8 (*b*) 7

(*c*) 6 (*d*) 5

79. A XOR gate has inputs A and B and output Y. Then the output equation is

(*a*) $Y = AB$ (*b*) $Y = AB + \bar{A}B$

(*c*) $Y = \bar{A}B + A\bar{B}$ (*d*) $Y = A\bar{B} + \bar{A}\bar{B}$

80. For the circuit of Fig. 9.52, the output equation is

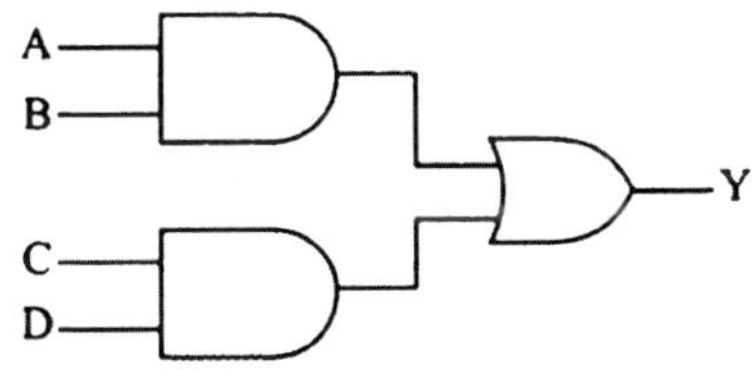

Fig. 9.52.

(*a*) $Y = ABCD$ (*b*) $Y = AB + CD$

(*c*) $Y = A + BCD$ (*d*) $Y = AB + \overline{CD}$

81. A 14 pin NOT gate IC has..................NOT gates

(*a*) 8 (*b*) 6

(*c*) 5 (*d*) 4

82. A 14 pin AND gate IC has..................AND gates

(*a*) 8 (*b*) 6

(*c*) 4 (*d*) 2

83. Symbol in Fig. 9.53 is IEEE symbol for

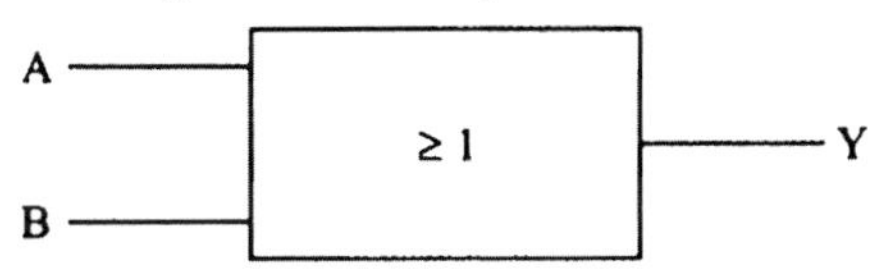

Fig. 9.53.

(*a*) AND (*b*) OR

(*c*) NAND (*d*) NOR

84. The first contribution to logic was made by

(*a*) George Boole (*b*) Copernicus

(*c*) Aristotle (*d*) Shannon

85. As per Boolean Algebra, inputs can be interchanged in

(*a*) OR gates (*b*) AND gates

(*c*) both OR and AND gates (*d*) none of the above

86. Boolean algebra obeys

(*a*) commutative law (*b*) associative law

(*c*) distributive law

(*d*) commutative, associative and distributive law

87. $A + (B.C) =$

(*a*) $A.B + C$ (*b*) $A.B + A.C$

(*c*) A (*d*) $(A + B).(A + C)$

88. $A.0 =$

(*a*) 1 (*b*) A

(*c*) 0 (*d*) A or 1

89. $A + A.B =$

(*a*) B (*b*) A.B

(*c*) A (*d*) A or B

90. Demorgan's first theorem is =

(*a*) $A.\bar{A} = 0$ (*b*) $\bar{\bar{A}} = A$

(*c*) $\overline{A+B} = \bar{A}.\bar{B}$ (*d*) $\overline{AB} = \bar{A} + \bar{B}$

91. Demorgan's second theorem is

(*a*) $A.\bar{A} = 0$ (*b*) $\bar{\bar{A}} = A$

(*c*) $\overline{A+B} = \bar{A}.\bar{B}$ (*d*) $\overline{AB} = \bar{A} + \bar{B}$

92. Which of the following is true

(*a*) SOP is a two level logic

(*b*) POS is a two level logic

(*c*) both SOP and POS are two level logic

(*d*) Hybrid function is two level logic

93. The problem of logic race occurs in

(*a*) SOP functions (*b*) POS functions

(*c*) Hybrid function (*d*) SOP and POS functions

94. In which function is each term known as min term

(*a*) SOP (*b*) POS

(*c*) Hybrid (*d*) both SOP and POS

95. In which function is each term known as max term

(*a*) SOP (*b*) POS

(*c*) Hybrid (*d*) SOP and Hybrid

96. In the expression A + BC, the total number of min terms will be

(*a*) 2 (*b*) 3

(*c*) 4 (*d*) 5

97. The min term designation for ABCD is

(*a*) m_8 (*b*) m_{10}

(*c*) m_{14} (*d*) m_{15}

98. For the min term designation $Y = \Sigma m(1, 3, 5, 7)$ the complete expression is

(*a*) $Y = \bar{A}\bar{B}C + A\bar{B}C$

(*b*) $Y = \bar{A}\bar{B}C + A\bar{B}C + ABC + \bar{A}BC$

(*c*) $Y = \bar{A}\bar{B}C + \bar{A}\bar{B}C + \bar{A}BC + A\bar{B}C$

(*d*) $Y = \bar{A}\bar{B}\bar{C} + ABC + \bar{A}\bar{B}C + A\bar{B}C$

99. The function $Y = AC + BD + EF$ is

(*a*) POS (*b*) SOP

(*c*) Hybrid (*d*) none of the above

100. The expression $Y = \Pi M(0, 1, 3, 4)$ is

(*a*) POS (*b*) SOP

(*c*) Hybrid (*d*) none of the above

101. $AB + A\bar{B} =$

(*a*) B (*b*) A

(*c*) 1 (*d*) 0

102. In a four variable Karnaugh map eight adjacent cells give a

(*a*) two variable term (*b*) single variable term

(*c*) three variable term (*d*) four variable term

103. A Karnaugh map with 4 variables has

(*a*) 2 cells (*b*) 4 cells

(*c*) 8 cells (*d*) 16 cells

104. In a Karnaugh map for an expression having 'don't care terms' the don't cares can be treated as

(*a*) 0 (*b*) 1

(*c*) 1 or 0 (*d*) none of the above

105. The term VLSI generally refers to a digital IC having

(*a*) more than 1000 gates

(*b*) more than 100 gates

(*c*) more than 1000 but less than 9999 gates

(*d*) more than 100 but less than 999 gates

106. Typical size of digital IC is about
(*a*) 1″ × 1″ (*b*) 2″ × 2″
(*c*) 0.1″ × 0.1″ (*d*) 0.001″ × 0.001″

107. A digital clock uses.................. chip
(*a*) SSI (*b*) LSI
(*c*) VLSI (*d*) MSI

108. Digital technologies being used nowadays are
(*a*) DTL and EMOS
(*b*) TTL, ECL, CMOS and RTL
(*c*) TTL, ECL and CMOS
(*d*) TTL, ECL. CMOS and DTL

109. A TTL circuit with totem pole output has
(*a*) high output impedance
(*b*) low output impedance
(*c*) very high output impedance
(*d*) any of above

110. TTL uses
(*a*) multi emitter transistors
(*b*) multi collector transistors
(*c*) multi base transistors
(*d*) multi emitter or multi collector transistors

111. Advanced low power Schottky is a part of
(*a*) ECL family (*b*) CMOS family
(*c*) TTL family (*d*) none of the above

112. For wired AND connection we should use
(*a*) TTL gates with active pull up
(*b*) TTL gates with open collector
(*c*) TTL gates without active pull up and with open collector
(*d*) any of the above

113. Time delay of a TTL standard family is about
(*a*) 180 ns (*b*) 50 ns
(*c*) 18 ns (*d*) 3 ns

114. As compared to TTL, ECL has
(*a*) lower power dissipation
(*b*) lower propagation delay
(*c*) higher propagation delay
(*d*) higher noise margin

115. As compared to TTL, CMOS logic has
(*a*) higher speed of operation
(*b*) higher power dissipation
(*c*) smaller physical size
(*d*) all the above

116. When a TTL is driving N CMOS gates it is necessary to ensure that
(*a*) V_{OL} (for TTL) ≤ V_{IL} (for CMOS) and V_{OH} (for TTL) ≥ V_{IH} (for CMOS)
(*b*) V_{OL} (for TTL) ≥ V_{IL} (for CMOS)
(*c*) V_{OH} (for TTL) ≤ V_{IH} (for CMOS)
(*d*) V_{OL} (for TTL) ≥ V_{IL} (for CMOS) and V_{OH} (for TTL) ≤ V_{IH} (for CMOS)

117. 74HCTOO series is
(*a*) NAND IC
(*b*) interface between TTL and CMOS
(*c*) inverting IC
(*d*) NOR IC

118. CD 4010 is a
(*a*) inverting buffer (*b*) non-inverting hex buffer
(*c*) NOR IC (*d*) NAND IC

119. Current requirement of a piezo buffer is about
(*a*) 100 mA (*b*) 20 mA
(*c*) 4 mA (*d*) 0.4 mA

120. TSL inverter has
(*a*) one input (*b*) two inputs
(*c*) one or two inputs (*d*) three inputs

121. Parallel adder is
(*a*) sequential circuit
(*b*) combinational circuit
(*c*) either sequential or combination circuit
(*d*) none of the above

122. The inputs to a 3 bit binary adder are 111_2 and 110_2. The output will be
(*a*) 101 (*b*) 1101
(*c*) 1111 (*d*) 1110

123. A half adder can be used only for adding
(*a*) 1*s* (*b*) 2*s*
(*c*) 4*s* (*d*) 8*s*

124. A 3 bit binary adder should use
(*a*) 3 full adders
(*b*) 2 full adders and 1 half adder
(*c*) 1 full adder and 2 half adders
(*d*) 3 half adders

125. When two 4 bit parallel adders are cascaded, we get
(*a*) 4 bit parallel adder (*b*) 8 bit parallel adder
(*c*) 16 bit parallel adder (*d*) none of the above

126. The widely used binary multiplication method is
(*a*) repeated addition (*b*) add and shift
(*c*) shift and add (*d*) any of above

127. When microprocessor processes both positive and negative numbers, the representation used is
(*a*) 1's complement (*b*) 2's complement
(*c*) signed binary (*d*) any of above

128. Decimal – 90 equals .... in 8 bit 2's complement
(*a*) 1000 1000 (*b*) 1010 0110
(*c*) 1100 1100 (*d*) 0101 0101

129. In 2's complement addition, the carry generated in the last stage is
(*a*) added to LSB
(*b*) neglected
(*c*) added lo bit next to MSB
(*d*) added to the bit next to LSB

130. The number of inputs and outputs in a full adder are
(*a*) 2 and 1 (*b*) 2 and 2
(*c*) 3 and 3 (*d*) 3 and 2

131. In a 7 segment display the segments *a, c, d, f, g* are lit. The decimal number displayed will be
(*a*) 9 (*b*) 5
(*c*) 4 (*d*) 2

132. In a 7 segment display, LEDs *b* and *c* lit up. The decimal number displayed is
(*a*) 9 (*b*) 7
(*c*) 3 (*d*) 1

133. BCD input 1000 is fed to a 7 segment display through a BCD to 7 segment decoder/driver. The segments which will lit up are
(*a*) *a, b, d* (*b*) *a, b, c*
(*c*) all (*d*) *a, b, g, c, d*

134. A device which converts BCD to seven segment is called
(*a*) encoder (*b*) decoder
(*c*) multiplexer (*d*) none of these

135. Which device uses nematic fluid?
(*a*) LED (*b*) LCD
(*c*) VF display (*d*) none of these

136. +12 V supply is given to plates *a, d, g,* and 0 V to other plates and grid of a VF display. The plates which will glow are
(*a*) *a, d, g* (*b*) *b, c, e, f*
(*c*) all of these (*d*) none of these

137. Which of these is the most recent display device?
(*a*) LED (*b*) LCD
(*c*) VF (*d*) (*a*) and (*c*)

138. VF glows with.................... colour when activated.
(*a*) Red (*b*) Orange
(*c*) Bluish green (*d*) none of these

139. Which display device resembles vacuum tube?
(*a*) LED (*b*) LCD
(*c*) VF (*d*) none of these

140. Which device changes parallel data to serial data?
(*a*) decoder (*b*) multiplexer
(*c*) demultiplexer (*d*) flip-flop

141. Which device changes serial data to parallel data?
(*a*) counter (*b*) multiplexer
(*c*) demultiplexer (*d*) flip-flop

142. A 1 of 4 multiplexer requires....................data select line
(*a*) 1 (*b*) 2
(*c*) 3 (*d*) 4

143. It is desired to route data from many registers to one register. The device needed is
(*a*) decoder (*b*) multiplexer
(*c*) demultiplexer (*d*) counter

144. Which device has one input and many outputs?
(*a*) multiplexer (*b*) demultiplexer
(*c*) counter (*d*) flip-flop

145. Two 16 : 1 and one 2 : 1 multiplexers can be connected to form a
(*a*) 16 : 1 multiplexer (*b*) 32 : 1 multiplexer
(*c*) 64 : 1 multiplexer (*d*) 8 : 1 multiplexer

146. A flip-flop is a
(*a*) combinational circuit (*b*) memory element
(*c*) arithmetic element
(*d*) memory or arithmetic element

147. In a D latch
(*a*) data bit D is fed to S input and $\bar{D}$ to R input
(*b*) data bit D is fed to R input and $\bar{D}$ to S input
(*c*) data bit D is fed to both R and S inputs
(*d*) data bit $\bar{D}$ is not fed to any input

148. In a D latch
(*a*) a high D sets the latch and low D resets it
(*b*) a low D sets the latch and high D resets it
(*c*) race can occur
(*d*) none of the above

149. In a positive edge triggered JK flip-flop
(*a*) High J and High K produce inactive state
(*b*) Low J and High K produce inactive state
(*c*) Low J and Low K produce inactive state
(*d*) High J and Low K produce inactive state

150. In a positive edge triggered D flip-flop
(*a*) D input is called direct set
(*b*) Preset is called direct reset
(*c*) Present and clear are called direct set and reset respectively
(*d*) D input overrides other inputs

151. In a JK master slave flip-flop
(*a*) master is clocked when clock is low
(*b*) slave is clocked when clock is high
(*c*) master is clocked when clock is high and slave is clocked when clock is low
(*d*) master is clocked when clock is low and slave is clocked when clock is high

152. In a positive edge triggered JK flip-flop, J = 1, K = 0 and clock pulse is rising. Q will
(*a*) be 0 (*b*) be 1
(*c*) show no change (*d*) toggle

153. For edge triggering in flip-flops, manufacturers use
(*a*) RC circuit (*b*) direct coupled design
(*c*) either RC circuit or direct coupled design
(*d*) none of the above

154. In a JK flip-flop toggle means
(*a*) set Q = 1 and $\bar{Q}$ = 0
(*b*) set Q = 0 and $\bar{Q}$ = 1
(*c*) change the output to the opposite state
(*d*) no change in output

155. A mod 4 counter will count
(*a*) from 0 to 4
(*b*) from 0 to 3
(*c*) from any number *n* to *n* + 4
(*d*) none of the above

156. A counter has N flip-flops. The total number of states is
(*a*) N (*b*) 2N
(*c*) 2^N (*d*) 4N

157. A counter has a modulus of 10. The number of flip-flops is
(*a*) 10 (*b*) 5
(*c*) 4 (*d*) 3

158. In a ripple counter,
(*a*) whenever a flip-flop sets to 1, the next higher FF toggles
(*b*) whenever a flip-flop sets to 0, the next higher FF remains unchanged
(*c*) whenever a flip-flop sets to 1, the next higher FF faces race condition
(*d*) whenever a flip-flop sets to 0, the next higher FF faces race condition

159. A counter has 4 flip-flops. It divides the input frequency by
(*a*) 4 (*b*) 2
(*c*) 8 (*d*) 16

160. A decade counter skips
(*a*) binary states 1000 to 1111
(*b*) binary states 0000 to 0011
(*c*) binary states 1010 to 1111
(*d*) binary states 1111 and higher

161. The number of flip-flops needed for a Mod 7 counter are
(*a*) 7 (*b*) 5
(*c*) 3 (*d*) 1

162. A presettable counter with 4 flip-flops can start counting from
(*a*) 0000
(*b*) 1000
(*c*) any number from 0000 to 1000
(*d*) any number from 0000 to 1111

163. In a programmable counter
(*a*) the starting number for counting can be programmed
(*b*) the modulus can be programmed
(*c*) both the starting number and modulus can be programmed
(*d*) either the starting number or modulus can be programmed

164. A 4 bit down counter can count from
(*a*) 0000 to 1111 (*b*) 1111 to 0000
(*c*) 000 to 111 (*d*) 111 to 000

165. A 3 bit up-down counter can count from
(*a*) 000 to 111
(*b*) 111 to 000
(*c*) 000 to 111 and also from 111 to 000
(*d*) none of the above

166. IC counters are
(*a*) synchronous only
(*b*) asynchronous only
(*c*) both synchronous and asynchronous
(*d*) none of the above

167. Shifting digits from left to right and vice versa is needed in
(*a*) storing numbers (*b*) arithmetic operations
(*c*) counting (*d*) storing and counting

168. The basic storage element in a digital system is
(*a*) flip-flop (*b*) counter
(*c*) multiplexer (*d*) encoder

169. The simplest register is
(*a*) buffer register (*b*) shift register
(*c*) controlled buffer register (*d*) bidirectional register

170. The basic shift register operations are
(*a*) serial in – serial out (*b*) serial in – parallel out
(*c*) parallel in – serial out (*d*) all the above

171. A universal shift register can shift
(*a*) from left to right (*b*) from right to left
(*c*) both from left to right and right to left
(*d*) none of the above

172. In a shift left register, shifting a bit by one bit means
(*a*) division by 2 (*b*) multiplication by 2
(*c*) subtraction of 2 (*d*) any of above

173. An 8 bit binary number is to be entered into an 8 bit serial shift register. The number of clock pulses required is
(*a*) 1 (*b*) 2
(*c*) 4 (*d*) 8

174. The number of flip-flops required to construct a register to store number 102_{10} in binary form is
(*a*) 2 (*b*) 4
(*c*) 8 (*d*) 10

175. An 8 bit data is to be entered into a parallel in register. The number of clock pulses required is
(*a*) 8 (*b*) 4
(*c*) 2 (*d*) 1

176. Which is correct with respect to computer memory?
(*a*) only a particular address can have a particular memory
(*b*) any address can have any memory
(*c*) to read a word from computer memory, it is not necessary to specify a location
(*d*) none of the above

177. Words having 8 bits are to be stored into computer memory. The number of lines required for writing into memory are
(*a*) 1 (*b*) 2
(*c*) 4 (*d*) 8

178. EPROM can be
(*a*) UVPROM
(*b*) EEPROM
(*c*) both UVPROM and EEPROM
(*d*) none of the above

179. EEPROM is also known as
(*a*) UVPROM (*b*) EAPROM
(*c*) both UVPROM and EAPROM
(*d*) none of the above

180. The access time of ROM using bipolar transistors is about
(*a*) 1 sec (*b*) 1 milli sec
(*c*) 1μ sec (*d*) 1 n sec

181. The density of dynamic RAM is
(*a*) the same as static RAM
(*b*) less than that of static RAM
(*c*) more than that of static RAM
(*d*) either equal or less than that of static RAM

182. Which memory requires periodic recharging?
(*a*) All ROMS (*b*) All RAMS
(*c*) Static RAM (*d*) Dynamic RAM

183. RAM can be expanded to
(*a*) increase word size
(*b*) increase word number
(*c*) increase word size or increase word number
(*d*) none of the above

184. Which memory is available in all technologies?
(*a*) PROM (*b*) EEPROM
(*c*) ROM (*d*) EPROM

185. Which memory does not require programming equipment
(*a*) RAM (*b*) EPROM
(*c*) EEPROM (*d*) UVPROM

186. In CCD
(*a*) a small charge is deposited for logical 1
(*b*) a small charge is deposited for both logical 1 and 0

(*c*) a small charge is deposited for logical 0 and large charge for logical 1
(*d*) none of the above

187. The internal structure of PLA is similar to
(*a*) RAM (*b*) ROM
(*c*) both RAM and ROM (*d*) neither RAM nor ROM

188. As access time is decreased, the cost of memory
(*a*) remains the same (*b*) increases
(*c*) decreases
(*d*) may increase or decrease

189. If an analog voltage is expressed in binary using 4 bits, each successive binary count would represent
(*a*) $\frac{1}{2}$ of total voltage (*b*) $\frac{1}{4}$ of total voltage
(*c*) $\frac{1}{15}$ of total voltage (*d*) $\frac{1}{16}$ of total voltage

190. If number of bits is N, the % resolution in analog to digital conversion is
(*a*) $\frac{100}{N}$ (*b*) $\frac{100}{2^N}$
(*c*) $\frac{100}{2^N - 1}$ (*d*) $\frac{100}{2^N + 1}$

191. In a 4 bit D/A converter, the offset error is the output voltage when input digital voltage is
(*a*) 1111 (*b*) 0000
(*c*) either 1111 or 0000 (*d*) none of the above

192. D/A converters are generally
(*a*) weighted resistor network
(*b*) binary ladder network
(*c*) either (*a*) or (*b*)
(*d*) neither (*a*) nor (*b*)

193. Quantization error occurs in
(*a*) D/A converter (*b*) A/D converter
(*c*) both D/A and A/D converter
(*d*) neither D/A nor A/D converter

194. Which is known as flash converter
(*a*) weighted resistor D/A converter
(*b*) parallel A/D converter
(*c*) stair step A/D converter
(*d*) up-down counter type A/D converter

195. Which converters uses integrating op-amp?
(*a*) parallel A/D converter
(*b*) single slope A/D converter
(*c*) dual slope A/D converter
(*d*) both (*b*) and (*c*)

196. ADC 0804 is a
(*a*) dual slope A/D converter
(*b*) single slope A/D converter
(*c*) successive approximation A/D converter
(*d*) either (*a*) or (*b*)

197. A binary ladder network D/A converter requires
(*a*) resistors of one value only
(*b*) resistors of many different values
(*c*) resistors of two different values
(*d*) none of the above

198. In an op-amp a virtual ground is a ground for
(*a*) voltage only (*b*) current only
(*c*) both voltage and current
(*d*) neither voltage nor current

199. The accuracy of A/D conversion is generally
(*a*) $\pm\frac{1}{2}$ LSB (*b*) $\pm$LSB
(*c*) $\pm\frac{5}{4}$ LSB (*d*) none of the above

200. The number of counter states which an 8 bit stair step A/D converter has to pass through before conversion takes place is equal to
(*a*) 1 (*b*) 8
(*c*) 255 (*d*) 256

201. A 4 input AND gate is equivalent to
(*a*) 4 switches in parallel
(*b*) 2 switches in series and 2 in parallel
(*c*) three switches in parallel and one in series
(*d*) 4 switches in series

202. A NOR gate is a combination of
(*a*) OR gate and AND gate
(*b*) AND gate and NOT gate
(*c*) OR gate and NOT gate
(*d*) two NOT gates

203. DeMorgan's first theorem shows the equivalence of
(*a*) OR gate and Exclusive OR gate
(*b*) NOR gate and Bubbled AND gate
(*c*) NOR gate and NAND gate
(*d*) NAND gate and NOT gate

204. If A = 0101, then A′ is
(*a*) 1010 (*b*) 1011
(*c*) 1001 (*d*) 0110

205. In 2's complement representation the number 11100101 represents the decimal number
(*a*) +37 (*b*) −31
(*c*) +27 (*d*) −27

206. Using 2's complement, the largest positive and negative number which can be stored with 8 bits are
(*a*) + 128 and − 127 (*b*) + 128 and − 128
(*c*) + 127 and − 128 (*d*) + 127 and − 127

207. For the binary number 11101000, the equivalent hexadecimal number is
(*a*) F9 (*b*) F8
(*c*) E9 (*d*) E8

208. Binary number 11001 is equivalent to decimal number
(*a*) 35 (*b*) 15
(*c*) 105 (*d*) 25

209. A 4 bit transistor register has output voltage of high-low-high-low. The binary number stored and its decimal equivalent are
(*a*) 0101 and 5 (*b*) 0101 and 10
(*c*) 1010 and 5 (*d*) 1010 and 10

210. An 8 bit transistor register has output voltage of low-high-low-high-low-high-low-high. The decimal number stored is
(*a*) 105 (*b*) 95
(*c*) 85 (*d*) 75

211. An OR gate has 4 inputs. One input is high and the other three are low. The output
(*a*) is low
(*b*) is high
(*c*) is alternately high and low
(*d*) may be high or low depending on relative magnitude of inputs

212. An AND gate has four inputs. One of the inputs is low and other inputs are high. The output
(*a*) is low (*b*) is high
(*c*) is alternately low and high
(*d*) may be high or low depending on relative magnitudes of inputs

213. SSI refers to ICs with
(*a*) less man 12 gates on the same chip
(*b*) less than 8 gates on the same chip
(*c*) less than 6 gates on the same chip
(*d*) less than 3 gates on the same chip

214. Medium scale integration refers to ICs with
(*a*) more than 12 but less than 30 gates on the same chip
(*b*) more than 50 gates on the same chip
(*c*) more than 20 but less than 100 gates on the same chip
(*d*) more than 12 but less than 100 gates on the same chip

215. Large scale integration refers to
(*a*) more than 50 gates on the same chip
(*b*) more than 80 gates on the same chip
(*c*) more than 100 gates on the same chip
(*d*) none of the above

216. It is desired to display the digit 7 using a seven segment display. The LEDs to be turned on are
(*a*) *a, b, c* (*b*) *b, c, d*
(*c*) *c, d, e* (*d*) *a, b, d*

217. If all the LEDs in a seven segment display are turned on, the number displayed is
(*a*) 7 (*b*) 3
(*c*) 0 (*d*) 8

218. If we need a low noise device, we should use
(*a*) BJT (*b*) FET
(*c*) thyristor (*d*) UJT

219. The main advantage of CMOS circuit is
(*a*) high gain
(*b*) high output impedance
(*c*) low power consumption
(*d*) high gain and high output impedance

220. If we need a device to shut off a few ampere current in a few nano seconds, the proper devices is
(*a*) VMOS (*b*) CMOS
(*c*) BJT (*d*) UJT

221. CMOS is extensively used in
(*a*) pocket calculators (*b*) digital wrist watches
(*c*) satellites (*d*) all of the above

222. ALU carries out
(*a*) arithmetic operations
(*b*) logic operations
(*c*) both arithmetic and logic operations
(*d*) none of the above

223. ALU processes
(*a*) decimal numbers (*b*) binary numbers
(*c*) hexadecimal numbers (*d*) all the above

224. A half adder adds
(*a*) 2 bits (*b*) 3 bits
(*c*) 4 bits (*d*) 2 or 3 bits

225. A full adder adds
(*a*) 2 bits (*b*) 3 bits
(*c*) 4 bits (*d*) any number of bits

226. The output of a half adder is
(*a*) SUM (*b*) CARRY
(*c*) SUM and CARRY (*d*) none of the above

227. The output of a full adder is
(*a*) SUM (*b*) CARRY
(*c*) SUM and CARRY (*d*) none of the above

228. A 3 bit binary adder can add
(*a*) 2 bits (*b*) 2 binary numbers
(*c*) 3 bits (*d*) 8 bits

229. A binary adder has
(*a*) one half adder
(*b*) one half adder and 2 full adders
(*c*) two half adders
(*d*) one half adder and many full adders

230. A 16 bit binary adder has
(*a*) 16 half adders (*b*) 16 full adders
(*c*) one half adder and 15 full adders
(*d*) 8 half adders and eight full adders

231. If A is 1011, A″ is
(*a*) 1011 (*b*) 0100
(*c*) 1100 (*d*) 1010

232. In 2's complement form, – 2 is
(*a*) 1011 (*b*) 1110
(*c*) 1100 (*d*) 1010

233. In an R-S latch, to set the output to high
(*a*) R is low and S is high (*b*) R and S are high
(*c*) R and S are low (*d*) R is high and S is low

234. In an R-S latch, race condition occurs when
(*a*) R is low and S is high (*b*) R and S are high
(*c*) R and S are low (*d*) R is high and S is low

235. R-S latch can be built with
(*a*) NOR gates (*b*) NAND gates
(*c*) NOR or NAND gates (*d*) none of the above

236. In a clocked latch, clock drives
(*a*) R NAND gate
(*b*) S NAND gate
(*c*) both R and S NAND gates
(*d*) either R or S NAND gates

237. In a clocked NAND latch, race condition occur when
(*a*) R and S are high and CLK is low
(*b*) R and CLK are high and S is low
(*c*) R, S, CLK are high
(*d*) R, S, CLK are low

238. In a D latch
(*a*) D drives R and $\bar{D}$ drives S
(*b*) D drives S and $\bar{D}$ drives R

(c) D drives both R and S
(d) $\bar{D}$ drives both R and S

239. Which of the following is suseptible to race condition?
(a) R-S latch (b) D latch
(c) both R-S and D latches (d) none of the above

240. In level clocking the output can change
(a) on rising edge of clock cycle
(b) on falling edge of clock cycle
(c) during entire half cycle of the clock
(d) none of the above

241. In a JK Master slave flip-flop
(a) both master and slave are positively clocked
(b) both master and slave are negatively clocked
(c) master is positively clocked and slave is negatively clocked
(d) master is negatively clocked and slave is positively clocked

242. A mode-10 counter can divide the clock frequency by a factor of
(a) 10 (b) 100
(c) 1000 (d) 10000

243. A three state switch has three outputs. These are
(a) low, low and high (b) low, high, high
(c) low, floating, low (d) low, high, floating

244. Which has the lowest propagation delay time?
(a) ECL (b) TTL
(c) CMOS (d) PMOS

245. Which has the highest power dissipation per gate?
(a) ECL (b) TTL
(c) CMOS (d) PMOS

246. The most commonly used logic family is
(a) ECL (b) TTL
(c) CMOS (d) PMOS

247. The power dissipated per gate
(a) is constant at all frequencies
(b) increases with frequency
(c) decreases with frequency
(d) may increase or decrease with frequency

248. Full adder circuit can be implemented by
(a) Multiplexers (b) Half adders
(c) AND or OR gates (d) decoders

249. In a 4 bit full adder how many half adders and OR gates are required?
(a) 8 and 4 (b) 7 and 4
(c) 7 and 3 (d) 8 and 3

250. Which type of gate is in Fig. 9.54?

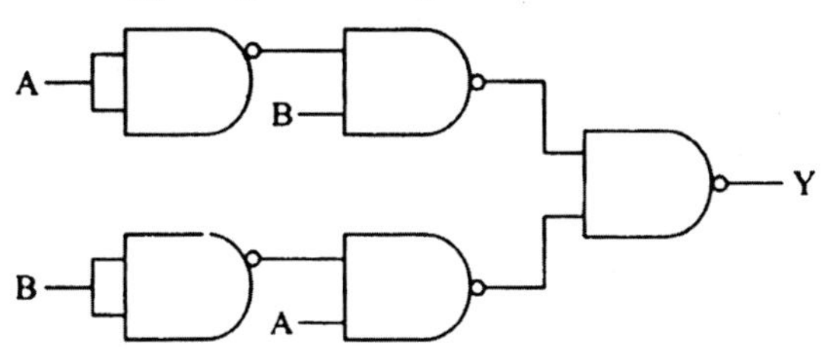

Fig. 9.54.

(a) NOR (b) OR
(c) NAND (d) EX-OR

251. In JK flip-flop toggle state exists if
(a) J = 0, K = 1 (b) J = 1, K = 0
(c) J = 0, K = 0 (d) J = 1, K = 1

252. Which of the following is error correcting code?
(a) EBCDIC (b) Gray
(c) Hamming (d) ASCII

253. In a JK flip-flop output $Q_n = 1$ and it does not change when clock pulse is applied. The possible combination of J_n and K_n is (X denotes don't care)
(a) X and 0 (b) X and 1
(c) 0 and X (d) 1 and X

254. The counter shown in Fig. 9.55 is

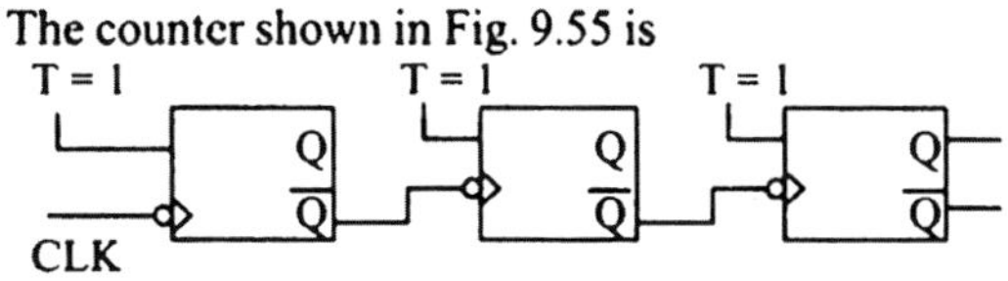

Fig. 9.55.

(a) Synchronous (b) Johnson
(c) Ring (d) None

255. The dual of A + [B + (AC)] + D is
(a) A + [(B (A + C))] + D (b) A [B + AC] D
(c) A + [B (A + C)] D (d) A [B (A + C)] D

256. The open wired circuit in Fig. 9.56 works as a

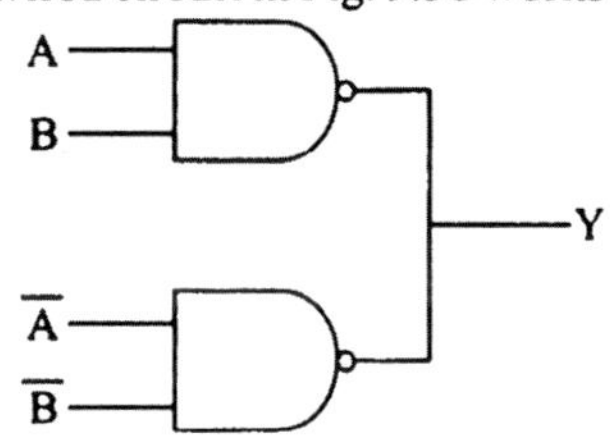

Fig. 9.56.

(a) Ex-NOR gate (b) AND gate
(c) XOR gate (d) NOR gate

257. The minimized version of logic circuit of Fig. 9.57 is

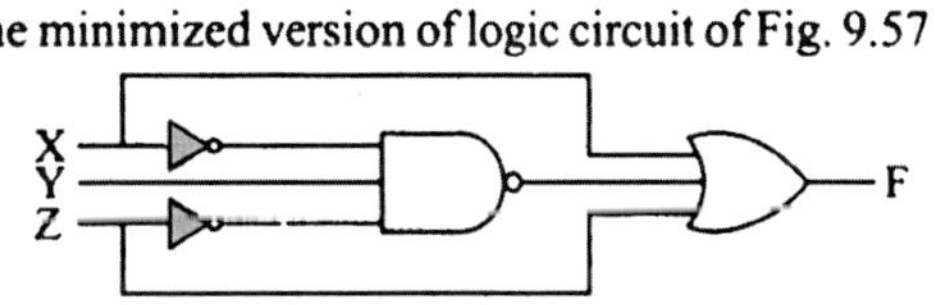

Fig. 9.57.

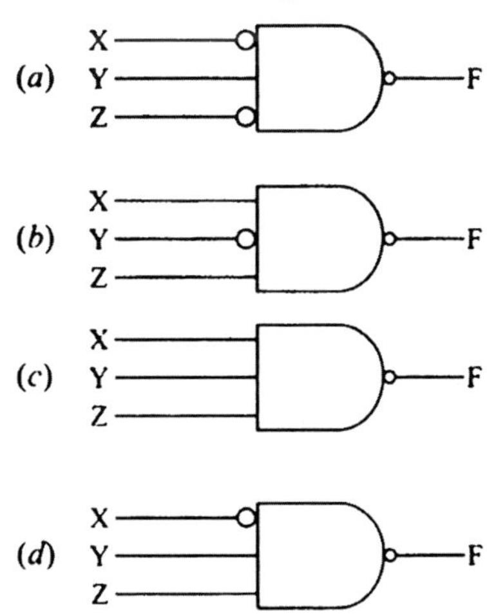

Fig. 9.58.

258. Match List I with List II and select the correct answer using the given codes

List I	List II
A. Multiplexer	1. Sequential memory
B. De-multiplexer	2. Converts decimal number to binary
C. Shift register	3. Data selector
D. Encoder	4. Routes single input to many outputs

Codes :

	A	B	C	D
(*a*)	3	4	1	2
(*b*)	4	3	1	2
(*c*)	3	4	2	1
(*d*)	1	2	3	4

259. ECL has high switching speed because the transistors are
(*a*) switching between cutoff and saturation regions
(*b*) switching between cutoff and active regions
(*c*) switching between active and saturation regions
(*d*) any of above

260. Which of the following is coincidence logic circuit?

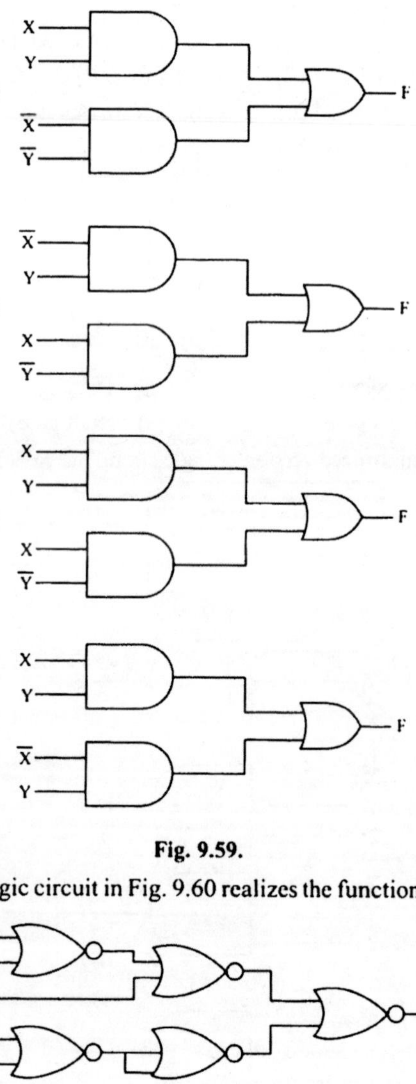

Fig. 9.59.

261. The logic circuit in Fig. 9.60 realizes the function

Fig. 9.60.

(*a*) $(A+B+C)(D\bar{E})$ (*b*) $(A+\overline{B+C})(\bar{D}E)$
(*c*) $(A+B+C)(\bar{D}\bar{E})$ (*d*) $(\overline{A+B}+C)(\bar{D}\bar{E})$

262. Two numbers in excess-3 code are added and the result is less than 9. To get equivalent binary
(*a*) 0011 is subtracted (*b*) 0011 is added
(*c*) 0110 is subtracted (*d*) 0110 is added

263. Which of the following is the truth table for half adder?

(*a*)

A	B	S	C
0	0	0	1
0	1	0	1
1	0	1	0
1	1	1	0

(*b*)

A	B	S	C
0	0	1	1
0	1	1	0
1	0	0	1
1	1	0	1

(*c*)

A	B	S	C
0	0	0	0
0	1	1	0
1	0	1	0
1	1	0	1

(*d*)

A	B	S	C
0	0	0	0
0	1	0	0
1	0	0	0
1	1	1	1

264. A is even or B is true. Negation of above statement is
(*a*) A is odd or B is false (*b*) A is odd and B is false
(*c*) A is even or B is false (*d*) A is even and B is false

265. The difference between sequential and combinational circuits is
(*a*) Combinational circuits store bits
(*b*) Combinational circuits have memory
(*c*) Sequential circuits store bits
(*d*) Sequential circuits have memory

266. Examine the following truth table. The product of sum form for this truth table is

X	Y	Z
0	0	1
0	1	0
1	0	1
1	1	0

(*a*) $(\bar{X}+\bar{Y})(X+Y)$ (*b*) $(X+\bar{Y})(\bar{X}+\bar{Y})$
(*c*) $(\bar{X}+Y)(\bar{X}+\bar{Y})$ (*d*) $(X+\bar{Y})(X+Y)$

267. A seven bit code (even parity) 1111101 may have error. The correct code is
(*a*) 1111101 (*b*) 1111111
(*c*) 0101011 (*d*) 1100011

268. A seven bit (even parity) code 0010001 may have error. The correct code is
(*a*) 0010001 (*b*) 1110001
(*c*) 0011001 (*d*) 1111110

269. Which of the following flip-flops cannot be converted to D flip-flop?
(*a*) JK (*b*) SR
(*c*) Master slave (*d*) none of the above

270. Which of the following is not a characteristic of a flip-flop?
(*a*) It is a bistable device
(*b*) It has two outputs
(*c*) Its two outputs are complement of each other
(*d*) It has one input terminal

271. The boolean expression for shaded area in Fig. 9.61 is

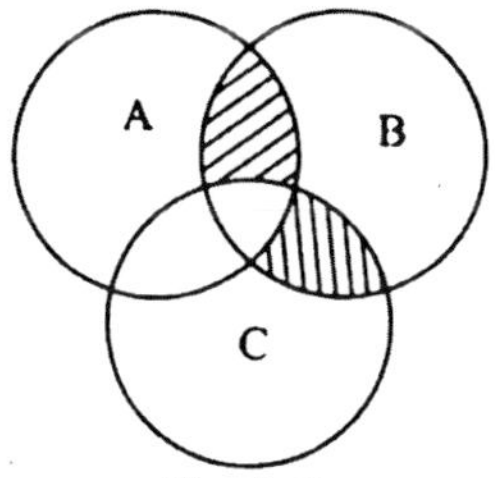

Fig. 9.61.

(*a*) AB + AC (*b*) $ABC + \overline{A}\overline{B}\overline{C}$
(*c*) $AB\overline{C} + \overline{A}BC$ (*d*) None of the above

272. For the given truth table, the correct boolean expression is

X	Y	Z	F
1	1	1	0
0	0	0	0
1	0	0	0

(*a*) $XYZ + \overline{X}YZ + X\overline{Y}\overline{Z}$
(*b*) $\overline{X}\overline{Y}\overline{Z} + XYZ + \overline{X}YZ$
(*c*) $\overline{X}\overline{Y}\overline{Z} + X\overline{Y}Z + XY\overline{Z}$
(*d*) $XYZ + X\overline{Y}Z + XY\overline{Z}$

273. In a gate output is Low if and only if all inputs are High. The gate is
(*a*) OR (*b*) AND
(*c*) NOR (*d*) NAND

274. Which of the following is the truth table for AND gate?

(*a*)

Input		Output
0	0	1
0	1	0
1	0	0
1	1	1

(*b*)

Input		Output
0	0	0
0	1	1
1	0	1
1	1	0

(*c*)

Input		Output
0	0	0
0	1	0
1	0	0
1	1	1

(*d*)

Input		Output
0	0	1
0	1	1
1	0	1
1	1	0

275. The inputs A and B of Fig. 9.62 are applied to a two input NOR gate. The output wave form is

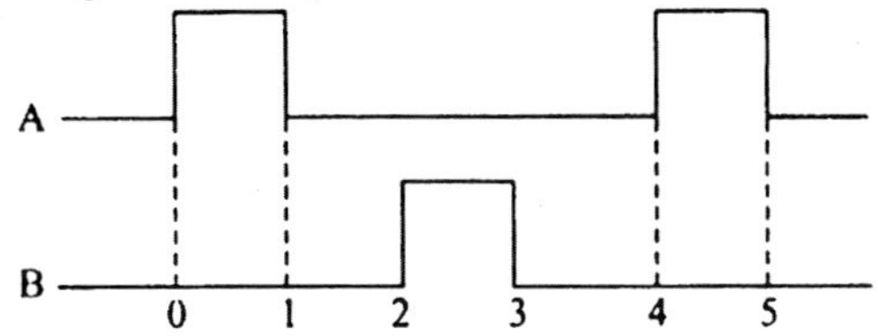

Fig. 9.62.

(*a*)

(*b*)
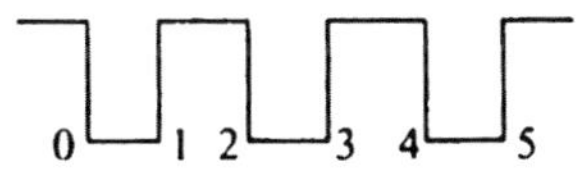

(*c*)
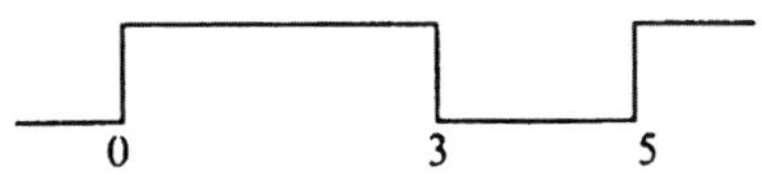

(*d*)
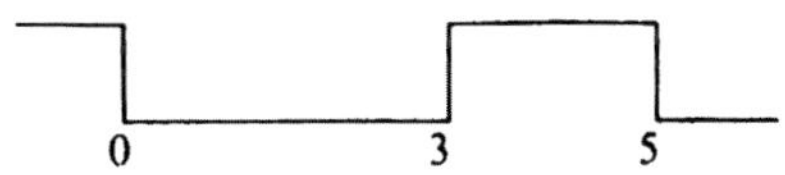

Fig. 9.63.

276. The inputs A, B, C of Fig. 9.64 are applied to a 3 input NOR gate. The output is

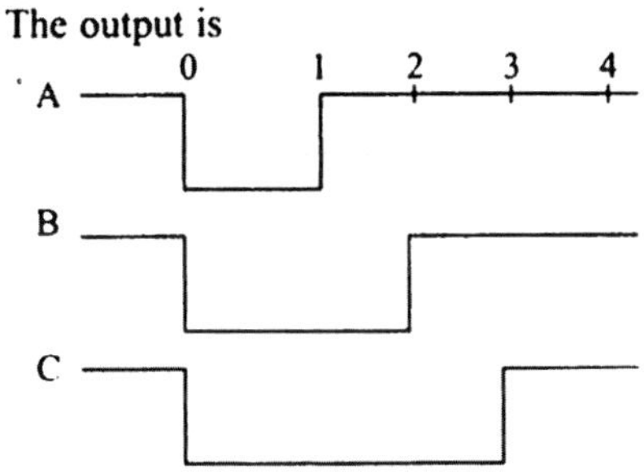

Fig. 9.64.

(*a*) HIGH from 0 to 4
(*b*) LOW from 0 to 4
(*c*) HIGH from 0 to 1 and LOW from 2 to 4
(*d*) LOW from 0 to 1 and HIGH from 2 to 4

277. Inputs A and B of Fig. 9.65 are applied to a NAND gate. The output is LOW.

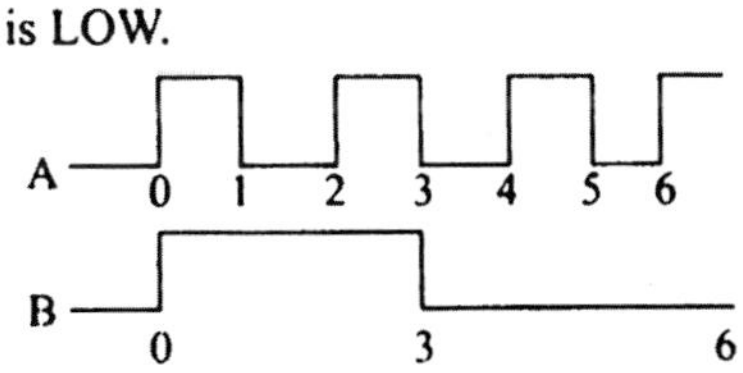

Fig. 9.65.

(*a*) from 0 to 6 (*b*) from 0 to 2
(*c*) from 0 to 1 and 2 to 3 (*d*) from 1 to 2 and 3 to 4

278. A (A + B) =
(*a*) A (*b*) B
(*c*) $\overline{A}$ (*d*) $\overline{B}$

279. $\overline{A} + A\overline{B}$
(*a*) A (*b*) $\overline{A}$
(*c*) $\overline{A} + \overline{B}$ (*d*) $A + \overline{B}$

280. $A(\bar{A} + B) =$

(*a*) A (*b*) $\bar{A}B$

(*c*) AB (*d*) $A\bar{B}$

281. The total number of fundamental products of three variables is

(*a*) 3 (*b*) 5

(*c*) 8 (*d*) 16

282. In a three input AND gate A = 1, B = 1, C = 0. The output Y =

(*a*) ABC (*b*) $\bar{A}BC$

(*c*) $AB\bar{C}$ (*d*) $A\bar{B}C$

283. The minterm designation for ABCD is

(*a*) m_{11} (*b*) m_{13}

(*c*) m_{15} (*d*) m_{16}

284. The complement of exclusive OR function is

(*a*) $AB + \bar{A}B$ (*b*) $AB + \bar{A}\bar{B}$

(*c*) $\bar{A}B + A$ (*d*) $AB + A\bar{B}$

285. Four inputs A, B, C, D are fed to a NOR gate. The output of NOR gate is fed to an inverter. The output of inverter is

(*a*) A + B + C + D (*b*) $\overline{A + B + C + D}$

(*c*) ABCD (*d*) $\overline{ABCD}$

286. In Fig. 9.66, A = B = 1 and C = D = 0. Then Y =

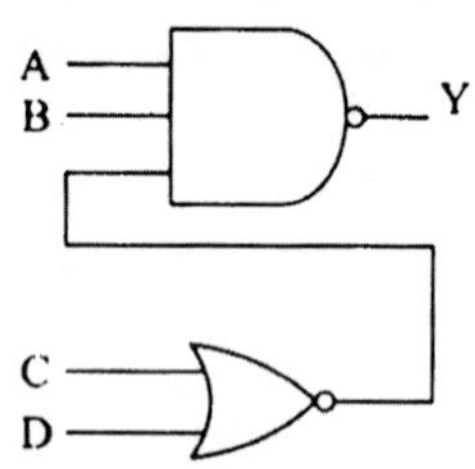

Fig. 9.66.

(*a*) 1 (*b*) 0

(*c*) either 1 or 0 (*d*) indeterminate

287. $\bar{A}\bar{B}\bar{C}\bar{D} + \bar{A}\bar{B}C\bar{D} =$

(*a*) ABC (*b*) $\bar{A}\bar{B}\bar{C}\bar{D}$

(*c*) $\bar{A}\bar{B}\bar{D}$ (*d*) ABCD

288. Max term designation for A + B + C is

(*a*) M_0 (*b*) M_1

(*c*) M_3 (*d*) M_4

289. $\overline{\overline{AB} + \overline{AC}} =$

(*a*) A + B + C (*b*) ABC

(*c*) $\bar{A}BC$ (*d*) $\overline{A + B + C}$

290. In a four variable K map eight adjacent cells give

(*a*) single variable term (*b*) two variable term

(*c*) three variable term (*d*) four variable term

291. In a three variable K map four adjacent cells give

(*a*) single variable term (*b*) two variable term

(*c*) three variable term (*d*) either (*a*) or (*b*)

292. The number of cells in a 4 variable K map is

(*a*) 16 (*b*) 8

(*c*) 4 (*d*) 32

293. K maps can be drawn

(*a*) only for 2 variables (*b*) only for 3 variables

(*c*) only for 4 variables

(*d*) for any number of variables

294. A NOR gate has 3 inputs A, B, C. For which combination of inputs is output HIGH

(*a*) A = B = C = 0 (*b*) A = B = C = 1

(*c*) A = B = 1 and C = 0 (*d*) A = C = 1 and B = 0

295. The circuit of Fig. 9.67 performs

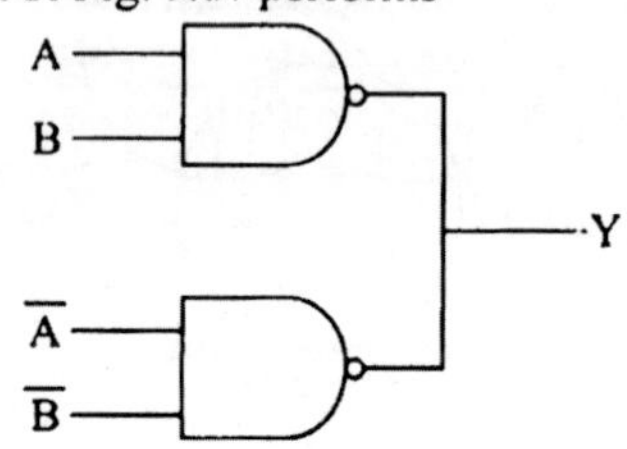

Fig. 9.67.

(*a*) OR function

(*b*) AND function

(*c*) exclusive OR function

(*d*) exclusive NOR function

296. The series 54 H / 74 H denotes

(*a*) Standard TTL

(*b*) High speed TTL

(*c*) Low Power TTL

(*d*) Low power Schottky TTL

297. Which of the following finds application in pocket calculators?

(*a*) TTL (*b*) CMOS

(*c*) ECL (*d*) both (*a*) and (*c*)

298. Which of the following is non-saturating?

(*a*) TTL (*b*) CMOS

(*c*) ECL (*d*) both (*a*) and (*b*)

299. The factors which govern fan out of CMOS gates are

(*a*) maximum permissible propagation delay

(*b*) input capacitance of each load

(*c*) both (*a*) and (*b*)

(*d*) neither (*a*) nor (*b*)

300. The advantages of high density IC are

(*a*) more circuits on chip

(*b*) high operating speed

(*c*) low power consumption

(*d*) both (*a*) and (*b*)

301. 74 H C series can sink upto 4 mA. The 74 L S series has $I_{IL(max)}$ = 0.4 m A. How many 74 L S inputs be driven by 74 H C output

(*a*) 100 (*b*) none

(*c*) 8 (*d*) 10

302. The function of interface circuit is

(*a*) to match output conditions of driver with input conditions of load device

(*b*) to match output conditions of load device with input conditions of driver

(c) both (a) and (b)
(d) neither (a) nor (b)

303. The circuit in Fig. 9.68 is

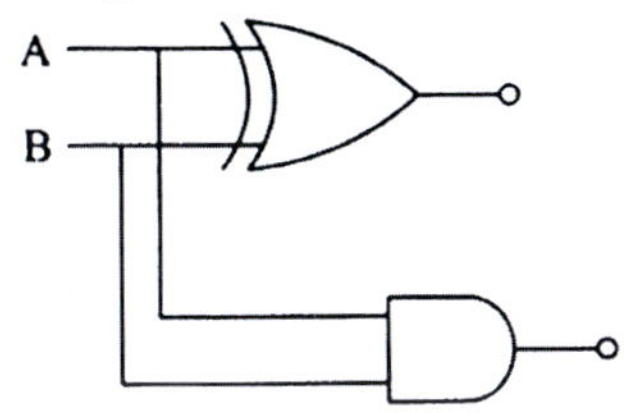

Fig. 9.68.

(a) half adder (b) full adder
(c) half subtractor (d) full subtractor

304. A full subtractor has a total of
(a) 2 inputs (b) 3 inputs
(c) 4 inputs (d) 5 inputs

305. 9's complement of 12_{10} is
(a) 21_{10} (b) 87_{10}
(c) 78_{10} (d) 76_{10}

306. 9's complement of 56_{10} is
(a) 43_{10} (b) 84_{10}
(c) 65_{10} (d) 53_{10}

307. 10's complement of 16_{10} is
(a) 83_{10} (b) 84_{10}
(c) 38_{10} (d) 48_{10}

308. BCD equivalent of -8_{10} is
(a) 00110 (b) 10110
(c) 11000 (d) 01000

309. The circuit of Fig. 9.69 is

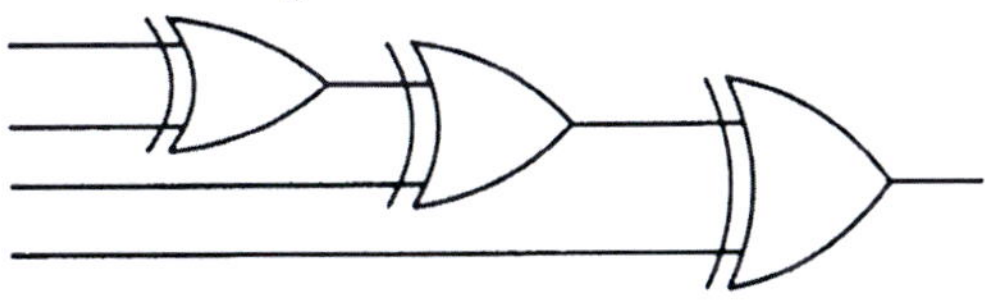

Fig. 9.69.

(a) full adder
(b) magnitude comparator
(c) parity detector
(d) none of the above

310. The number of inputs and outputs of a full adder are
(a) 3 and 2 respectively (b) 2 and 3 respectively
(c) 4 and 2 respectively (d) 2 and 4 respectively

311. A parity detector can detect
(a) single error only
(b) single error or odd number of errors
(c) two errors
(d) single or two errors

312. A five bit binary adder is used to add 01001 and 00111. The adders from LSB position are numbered as FA_0, FA_1 FA_2, FA_3 and FA_4. The outputs of FA_2 are
(a) SUM = 0 (b) SUM. = 0, CARRY = 1
(c) SUM = 1, CARRY = 0 (d) SUM = 1, CARRY = 1

313. The input to a parity detector is 1001. The output is
(a) 0 (b) 1
(c) 0 or 1 (d) indeterminate

314. One XOR gate can work as
(a) one bit magnitude comparator
(b) two bit magnitude comparator
(c) either (a) or (b)
(d) neither (a) nor (b)

315. In a 7 segment LED display, the minimum number of segments is activated when the input decimal number is
(a) 0 (b) 1
(c) 2 (d) 3

316. Which of them radiates emission?
(a) LED only (b) LCD only
(c) both LED and LCD (d) neither LED nor LCD

317. Which of them contains an organic fluid?
(a) LED (b) LCD
(c) both LED and LCD (d) neither LED nor LCD

318. In which of these does thermionic emission occur
(a) LED (b) LCD
(c) VF display (d) None of these

319. Which of them has 10 inputs and 4 outputs
(a) Decimal to BCD encoder (b) BCD to decimal decoder
(c) Octal to binary encoder (d) all encoders

320. A 4 line to 16 line decoder has
(a) 16 inputs and 4 outputs (b) 4 inputs and 16 outputs
(c) either (a) or (b) (d) neither (a) nor (b)

321. In a BCD to 7 segment decoder the minimum and maximum number of outputs active at any time is
(a) 2 and 7 (b) 3 and 7
(c) 1 and 6 (d) 3 and 6

322. The expression $Y(A, B, C) = \Sigma m(1, 3, 5, 6)$ is to be realized using a multiplexer. Then
(a) use 8 : 1 multiplexer and ground input lines 1, 3, 5, 6
(b) use 8 : 1 multiplexer and ground input lines 0, 2, 4, 7
(c) use 8 : 1 multiplexer and ground input lines 0, 1, 2, 3
(d) use 8 : 1 multiplexer and ground input lines 4, 5, 6, 7

323. The function $Y = A\bar{B}\bar{C} + AB\bar{C} + A\bar{B}C + \bar{A}BC$ is to be realized using discrete gates. The inputs available are A, B, C. We need a total of
(a) 8 gates (b) 6 gates
(c) 10 gates (d) 5 gates

324. The number of select lines in a 16 : 1 multiplexer are
(a) 4 (b) 3
(c) 2 (d) 1

325. Out of multiplexer and demultiplexer, which has select inputs
(a) multiplexer only
(b) demultiplexor only
(c) both multiplexer and demultiplexer
(d) none of these

326. Out of multiplexer and demultiplexer which can be used as a logic function generator
(a) multiplexer only (b) demultiplexer only
(c) both (a) and (b) (d) none of these

327. As applied to a flip-flop the word 'edge triggered' means
(a) flip-flop can change state when clock transition occurs
(b) flip-flop can change state when clock signal goes from LOW to HIGH only

(*c*) flip-flop can change state when clock signal goes from HIGH to LOW only
(*d*) none of the above

328. Out of S, R, J, K, Preset, Clear inputs to flip-flops, the synchronous inputs are
(*a*) S, R, J, K only (*b*) S, R, Preset, Clear only
(*c*) Preset, Clear only (*d*) S R only

329. The device 'one shot' has
(*a*) two stable states (*b*) one stable state
(*c*) either 1 or 2 stable states (*d*) no stable state

330. A J K flip-flop can be converted to D flip-flop by
(*a*) connecting both J and K terminals to D
(*b*) connecting J terminal to D and leaving K open
(*c*) connecting K terminal to D and leaving J open
(*d*) connecting J terminal to D and K terminal to D through an inverter

331. Out of latch and flip-flop, which has clock input
(*a*) latch only (*b*) flip-flop only
(*c*) both latch and flip-flop (*d*) none of these

332. An SR flip-flop can be built using
(*a*) NOR gate only
(*b*) NAND gate only
(*c*) either NOR or NAND gates
(*d*) neither NOR nor NAND gates

333. Out of S R and J K flip-flops, which is susceptible to race condition
(*a*) SR (*b*) JK
(*c*) both SR and JK (*d*) none of these

334. In a flip-flop which input determines the state to which output will transition
(*a*) control input
(*b*) clock input
(*c*) both control and clock inputs
(*d*) either control or clock input

335. To count 1000 bottles in a milk plant, the minimum number of flip-flops required is
(*a*) 12 (*b*) 10
(*c*) 8 (*d*) 6

336. A 4 bit ripple counter starts in 0000 state. When the counter reads 0010 the number of clock pulses which have occurred is
(*a*) 2 (*b*) 18
(*c*) 2 or 18 (*d*) 2 or 18 or 34

337. In a 4 bit ripple counter using flip-flops with t_{pd} = 40 ns, the maximum frequency can be
(*a*) 1.25 MHz (*b*) 3.25 MHz.
(*c*) 6.25 MHz (*d*) 12.5 MHz

338. Using the same flip-flops
(*a*) a synchronous flip-flop can operate at higher frequency than ripple counter
(*b*) a ripple counter can operate at higher frequency than sychronous counter
(*c*) both ripple and synchronous counter can operate at the same frequency
(*d*) cannot determine

339. For a Mod-64 synchronous counter the number of flip-flops and AND gates needed is
(*a*) 6 and 4 respectively (*b*) 4 and 6 respectively
(*c*) 7 and 5 respectively (*d*) 5 and 7 respectively

340. In which counter does the maximum frequency depend on the modulus
(*a*) synchronous (*b*) ripple
(*c*) both synchronous and ripple
(*d*) neither synchronous nor ripple

341. Three Mod-16 counters are connected in cascade. The maximum counting range is
(*a*) 16 (*b*) 256
(*c*) 4096 (*d*) none of the above

342. Figure 9.70 shows a negative logic AND gate. If positive logic is used this gate is equivalent to

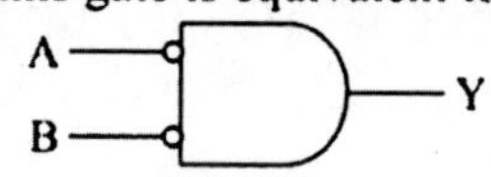

Fig. 9.70.

(*a*) AND gate (*b*) OR gate
(*c*) NOR gate (*d*) NAND gate

343. Figure 9.71 shows a logic circuit. The minimum Boolean expression for this circuit is

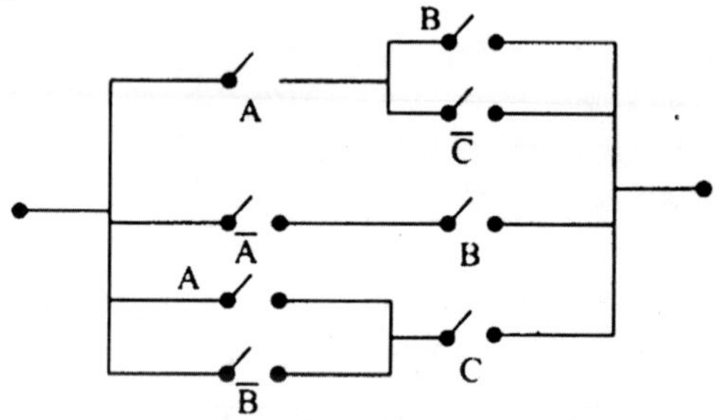

Fig. 9.71.

(*a*) A + B (*b*) A + B + C
(*c*) AB + C (*d*) AB + AC + BC

344. The number of switching functions of 3 variables are
(*a*) 3 (*b*) 8
(*c*) 16 (*d*) 32

345. When signed numbers are used in binary arithmetic, which of the following has unique representation for zero
(*a*) sign-magnitude (*b*) 1's complement
(*c*) 2's complement (*d*) 9's complement

346. A carry look ahead adder is frequently used for addition because
(*a*) it costs less (*b*) it is faster
(*c*) it is more accurate (*d*) is uses fewer gates

347. The greatest negative number which can be stored is 8 bit computer using 2's complement arithmetic is
(*a*) – 256 (*b*) – 128
(*c*) – 255 (*d*) – 127

348. A retriggerable one shot is one which
(*a*) can be triggered only once
(*b*) has two quasi stable states
(*c*) cannot be retriggered till the complete pulse appears at output
(*d*) can be retriggered while the output is being generated

349. Match List I and List II and give correct answer using the given codes

List I	**List II**
A interface	1. A measure of rate of data transmission
B A bit	2. A binary digit
C Baud speed	3. The common boundary between various sections and subsections
	4. A/D converter

Codes:

	A	B	C
(*a*)	3	2	1
(*b*)	3	1	2
(*c*)	2	4	1
(*d*)	4	2	3

350. If memory chip size is 256 × 1 bits, the number of chips required to make 1 k byte memory is
(*a*) 8 (*b*) 12
(*c*) 24 (*d*) 32

351. A 6 bit ladder D/A converter has input 101001. The output is (assume 0 = 0 V and I = 10 V)
(*a*) 4.23 (*b*) 6.41
(*c*) 5.52 (*d*) 9.23

352. An *n* bit ADC using V_r as reference has a resolution (in volts) of
(*a*) $\frac{V_r}{2^n}$ (*b*) $V_r(n)$
(*c*) $\frac{V_r}{2^n - 1}$ (*d*) $2V_r(n)$

353. Which one of the following is D/A conversion technique?
(*a*) successive approximation
(*b*) weighted resistor
(*c*) dual slope
(*d*) single slope

354. Match List I and List II and select the correct answer using the given codes

List I (Logic)	**List II (Characteristic)**
A RTL	1. High fan out
B CMOS	2. Highest speed of operation
C I^2L	3. High noise immunity
D ECL	4. Lowest product of power and delay

Codes:

	A	B	C	D
(*a*)	4	3	2	1
(*b*)	4	1	2	3
(*c*)	3	1	4	2
(*d*)	3	4	1	2

355. The figure of merit for a logic family is
(*a*) gain band width product
(*b*) (propagation delay time) (power dissipation)
(*c*) (fan out) (power dissipation)
(*d*) (noise margin) (power dissipation)

356. A full adder is to be implemented using half adders and OR gates. A 4 bit parallel adder without any initial carry requires
(*a*) 8 half adders and 4 OR gates
(*b*) 8 half adders and 3 OR gates
(*c*) 7 half adders and 3 OR gates
(*d*) 1 half adders and 4 OR gates

357. Which of the following is equivalent to AND – OR realization
(*a*) NAND–NOR (*b*) NOR–NOR
(*c*) NOR–NAND (*d*) NAND–NAND

358. A combination circuit is one in which the output depends on
(*a*) input combination at that lime
(*b*) input combination and previous output
(*c*) input combination and previous input
(*d*) present output and previous output

359. The open collector gates in Fig. 9.72 are tied together. The output Y =

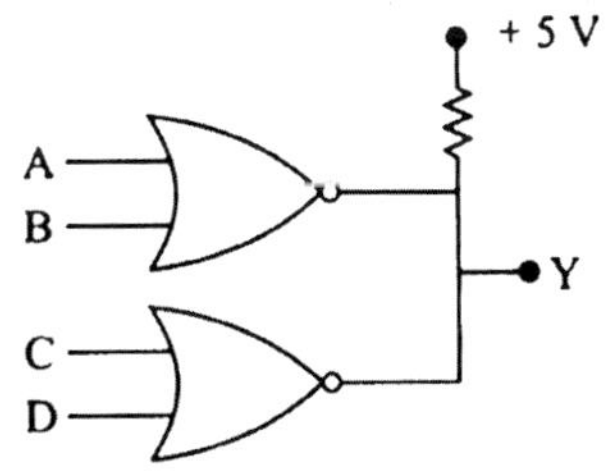

Fig. 9.72.

(*a*) $\overline{A+B}+\overline{C+D}$ (*b*) $\overline{A+B+C+D}$
(*c*) $(A+B)(C+D)$ (*d*) $AB+CD$

360. A divide by 78 counter can be obtained by
(*a*) 6 numbers of mod-13 counters
(*b*) 13 numbers of mod-6 counters
(*c*) one mod-13 counter followed by mode-6 counter
(*d*) 13 number of mod-13 counters

361. In a sequential circuit the output at any instant depends on
(*a*) present inputs only
(*b*) past inputs only
(*c*) past outputs only
(*d*) past output and present input

362. In a 4 bit weighted resistor D/A converter, resistor value corresponding to LSB is 32 k ohm. The resistor value corresponding to MSB is
(*a*) 32 k ohm (*b*) 16 k ohm
(*c*) 8 k ohm (*d*) 4 k ohm

363. A 10 bit D/A converter gives a maximum output of 10.23 V. The resolution is
(*a*) 10 mV (*b*) 20 mV
(*c*) 15 mV (*d*) 25 mV

364. The disadvantage of counter type ADC as compared to comparator ADC is that
(*a*) resolution is low (*b*) conversion time is more
(*c*) circuit is more complex (*d*) stability is low

365. The following switching functions are to be implemented using a decoder
$f_1 = \Sigma m(1, 2, 4, 8, 10, 14)$
$f_2 = \Sigma m(2, 5, 9, 15)$
$f_3 = \Sigma m(2, 4, 5, 6, 7)$

The minimum configuration of decoder is

(*a*) 2 to 4 line (*b*) 3 to 8 line
(*c*) 4 to 16 line (*d*) 5 to 32 line

366. Flash ADC is

(*a*) serial ADC
(*b*) parallel ADC
(*c*) series-parallel ADC
(*d*) successive approximation ADC

367. A 4 bit synchronous counter uses flip-flops with a delay time of 15 ns each. The time required for change of slate is

(*a*) 15 ns (*b*) 30 ns
(*c*) 45 ns (*d*) 60 ns

368. Which of the following is best suited for parity checking and parity generation?

(*a*) AND, OR, NOT gates
(*b*) XOR, Exclusive NOR gate
(*c*) NAND gates
(*d*) NOR gates

369. The inputs to logic gate are 0. The output is 1. The gate is

(*a*) NAND or XOR (*b*) OR or XOR
(*c*) AND or XOR (*d*) NOR or Ex-NOR

370. The circuit in Fig. 9.73 is

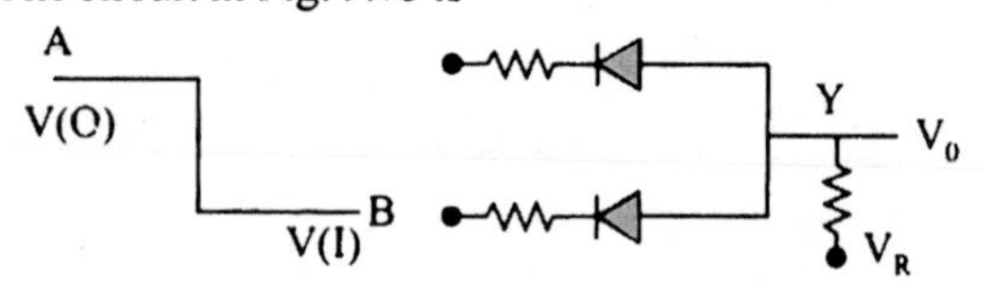

Fig. 9.73.

(*a*) positive logic OR gate (*b*) negative logic OR gate
(*c*) negative logic AND gate (*d*) positive logic AND gate

371. The circuit of Fig. 9.74 gives the output Y =

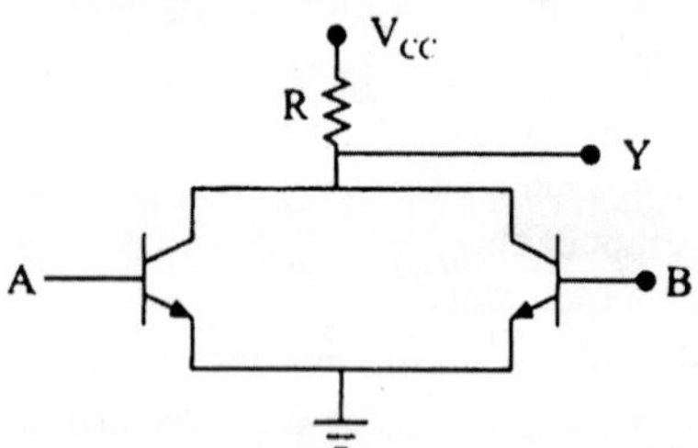

Fig. 9.74.

(*a*) AB (*b*) A+B
(*c*) $\bar{A}B$ (*d*) $\overline{A+B}$

372. For the K map in Fig. 9.75 the simplified Boolean expression is

AB \ CD	00	01	11	10
00	1	1		1
01	1	1		1
11			1	1
10				

Fig. 9.75.

(*a*) $\bar{A}\bar{C} + \bar{A}\bar{D} + ABC$ (*b*) $\bar{A}C + \bar{A}\bar{D} + ABC$
(*c*) $\bar{A}C + \bar{A}D + ACD$ (*d*) $\bar{A}\bar{C} + \bar{A}\bar{D} + AB\bar{C}$

373. For the K map of Fig. 9.76, the simplified Boolean expression is

AB \ CD	00	01	11	10
00				
01			1	1
11		1	1	1
10		1		

Fig. 9.76.

(*a*) $A\bar{C}D + BC$ (*b*) $A\bar{B}D + BC$
(*c*) $A\bar{C}D + AC$ (*d*) $A\bar{C}D + AC + BC$

374. For the K map of Fig. 9.77 the simplified Boolean expression is

AB \ CD	00	01	11	10
00				
01		1		
11		1	1	
10			1	

Fig. 9.77.

(*a*) $ACD + ABD + B\bar{C}D$ (*b*) $ACD + B\bar{C}D$
(*c*) $ACD + ABD$ (*d*) none of the above

375. For the K map of Fig. 9.78 the simplified Boolean expression is

AB \ C	0	1
00	1	1
01		
11	1	
10	1	

Fig. 9.78.

(*a*) $A\bar{C}+\bar{A}\bar{B}$
(*b*) $A\bar{B}\bar{C} + \bar{A}\bar{B}\bar{C} + \bar{A}\bar{B}C + AB\bar{C}$
(*c*) $A\bar{C} + \bar{A}\bar{B} + AC$
(*d*) $A\bar{C} + \bar{A}\bar{B} + AB$

376. For the K map of Fig. 9.79 the simplified Boolean expression is

A \ B	0	1
0	1	
1	1	

Fig. 9.79.

(*a*) $\bar{A}\bar{B} + A\bar{B}$ (*b*) $\bar{B}$

(*c*) $\bar{A}$ (*d*) $A\bar{B}$

377. For the logic circuit of Fig. 9.80 the simplified Boolean equation is

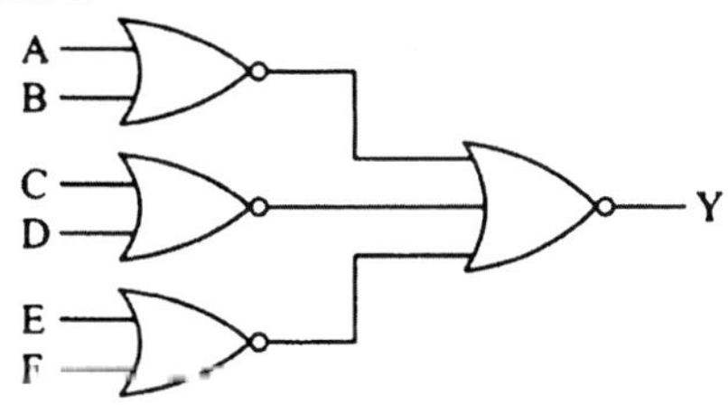

Fig. 9.80.

(*a*) $Y = (A+B)(C+D)(E+F)$

(*b*) $Y = A+B+C+D+E+F$

(*c*) ABCDEF

(*d*) ABC + DEF

378. For the logic circuit of Fig. 9.81 the simplified Boolean expression is

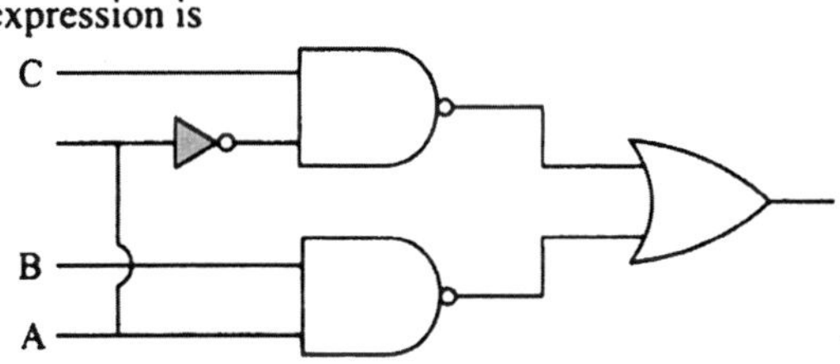

Fig. 9.81.

(*a*) $Y = AB + A\bar{C}$ (*b*) $Y = (AB)(A\bar{C})$

(*c*) $Y = ABC$ (*d*) $Y = A\bar{B} + \bar{A}C$

379. For the K map of Fig. 9.82 the simplified Boolean expression is

AB \ C	0	1
00	1	1
01	1	1
11		1
10		

Fig. 9.82.

(*a*) $\bar{A} + BC$ (*b*) $A + BC$

(*c*) $\bar{A} + B\bar{C}$ (*d*) $\bar{A} + B\bar{C}$

380. For the logic circuit of Fig. 9.83 the simplified Boolean expression is

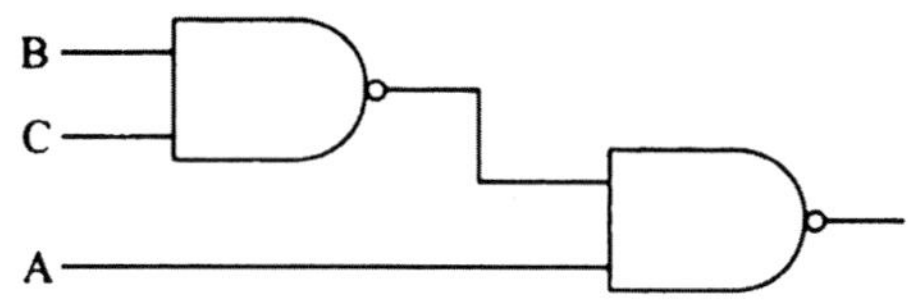

Fig. 9.83.

(*a*) $A + BC$ (*b*) $\bar{A} + BC$

(*c*) $\bar{A} + \bar{B}C$ (*d*) $A + B\bar{C}$

381. For the logic circuit of Fig. 9.84 the simplified Boolean expression is

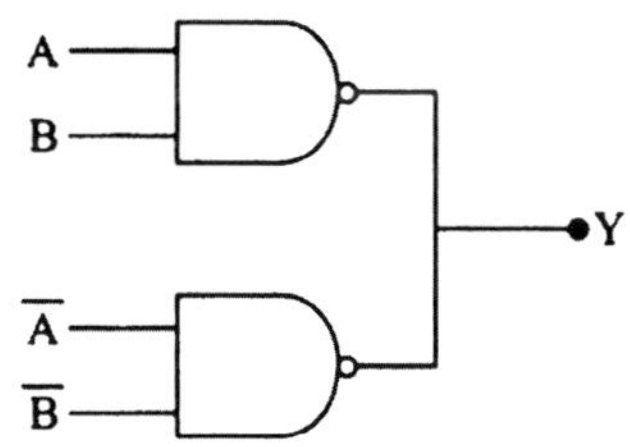

Fig. 9.84.

(*a*) $A \oplus B$ (*b*) $A + B$

(*c*) AB (*d*) $\bar{A} + B$

382. For K map of Fig. 9.85 the simplified Boolean expression is

AB \ C	0	1
00	1	1
01		1
11		
10	1	1

Fig. 9.85.

(*a*) $Y = \bar{A}B\bar{C} + A\bar{B}\bar{C} + \bar{A}\bar{B}C$

(*b*) $Y = AC + \bar{B}$

(*c*) $Y = \bar{A}\bar{B}\bar{C} + ABC$

(*d*) $Y = A\bar{B}C + AB\bar{C} + ABC$

383. The simplified logic expression for the circuit of Fig. 9.86 is

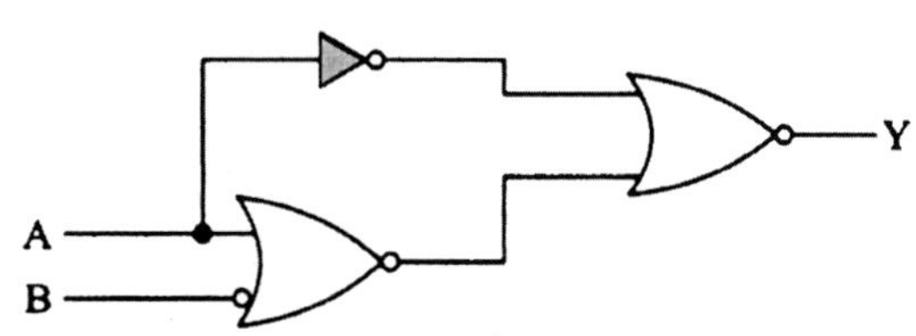

Fig. 9.86.

(*a*) $Y = A$ (*b*) $Y = AB$

(*c*) $Y = A + B$ (*d*) $Y = \bar{A} + B$

384. For the truth table of Fig. 9.87 Y =

(*a*) $A + B + C$ (*b*) $\bar{A} + BC$

(*c*) $\bar{A}$ (*d*) $\bar{B}$

A	B	C	Y
0	0	0	1
0	0	1	1
0	1	0	0
0	1	1	0
1	0	0	1
1	0	1	1
1	1	0	0
1	1	1	0

Fig. 9.87.

385. The Boolean expression for the circuit of Fig. 9.88 is

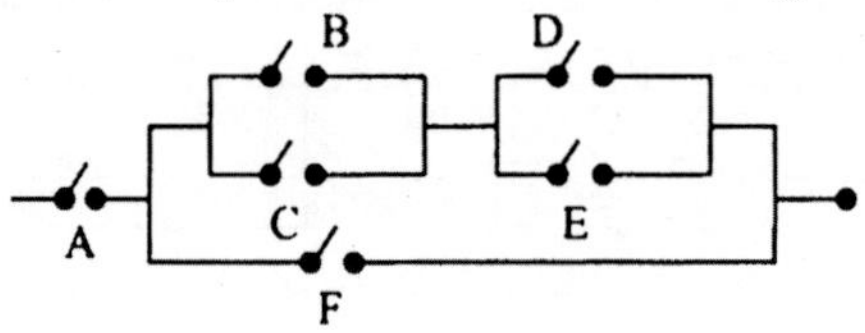

Fig. 9.88.

(*a*) A [F + (B + C) (D + E)]
(*b*) A [F + (B + C) (DE)]
(*c*) A + F + (B + C) (D + E)]
(*d*) A [F + (BC) (DE)]

386. The number of memory locations which 14 address bits can access is

(*a*) 1024 (*b*) 2048
(*c*) 4096 (*d*) 16384

387. The number of address lines in EPROM 4096 × 8 is

(*a*) 2 (*b*) 4
(*c*) 8 (*d*) 12

388. 2's complement of a given 3 or more bit number of non-zero magnitude is the same as the original number if all bits except

(*a*) MSB are zero (*b*) LSB are zero
(*c*) MSB are 1 (*d*) LSB are 1

389. Match List I and List II and select the answer using the given codes

List I (Memories)	**List II (Characteristic)**
A Static P L	1. erasable, programmable
B CCD	2. ultra high speed
C ECL	3. stores large volume of data
D GAL	4. Does not need refreshing
	5. Non-volatile

Codes:

	A	**B**	**C**	**D**
(*a*)	4	3	2	1
(*b*)	4	2	3	1
(*c*)	5	1	2	3
(*d*)	3	5	2	1

390. Which of the following characteristic are necessary for a sequential circuit

(*a*) it must have 6 gates
(*b*) it must have feedback
(*c*) its output should depend on past value

Which of the above statements are correct?

(*a*) 1, 2, 3 (*b*) 1, 2
(*c*) 2, 3 (*d*) 1, 3

391. Which of the circuits in Fig. 9.89 converts J K flip-flop to T flip-flop

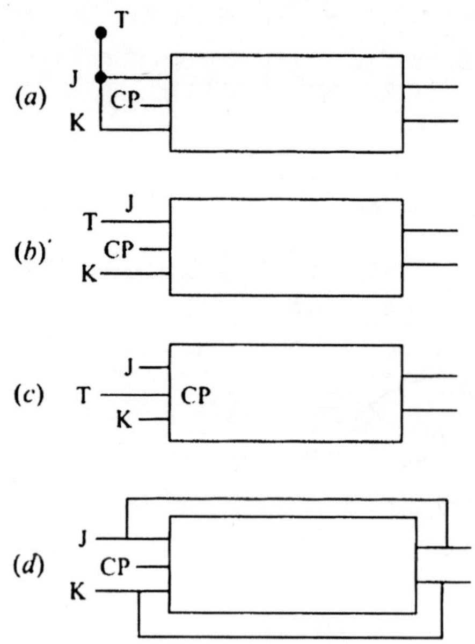

Fig. 9.89.

392. Consider the following statements Timer 555 can be used as

1. monostable multivibrator
2. bistable multivibrator
3. astable multivibrator

Which of the above statements are correct?

(*a*) 1 and 2 (*b*) 1 and 3
(*c*) 2 and 3 (*d*) 1, 2, 3

393. A 3 input NAND gate is to be used as inverter. Which of the following will give better results

(*a*) the two unused inputs are left open
(*b*) the two unused inputs are connected to 0
(*c*) the two unused inputs are connected to 1
(*d*) none of the above

394. Match List I and List II and select the answer using the given code

List I (Logic family)	**List II (Characteristic)**
A TTL	1. Maximum power consumption
B ECL	2. Highest packing density
C NMOS	3. Least power consumption
D CMOS	4. Saturated Logic

Codes:

	A	**B**	**C**	**D**
(*a*)	1	4	2	3
(*b*)	1	4	3	2
(*c*)	4	1	2	3
(*d*)	4	1	3	2

395. For the shaded area in Venn diagram of Fig. 9.90, the Boolean expression is

(*a*) $AB + BC + CA$ (*b*) $ABC + A\overline{B}C$
(*c*) $ABC + \overline{A}\overline{B}\overline{C}$ (*d*) $\overline{A}\overline{B}\overline{C} + AB$

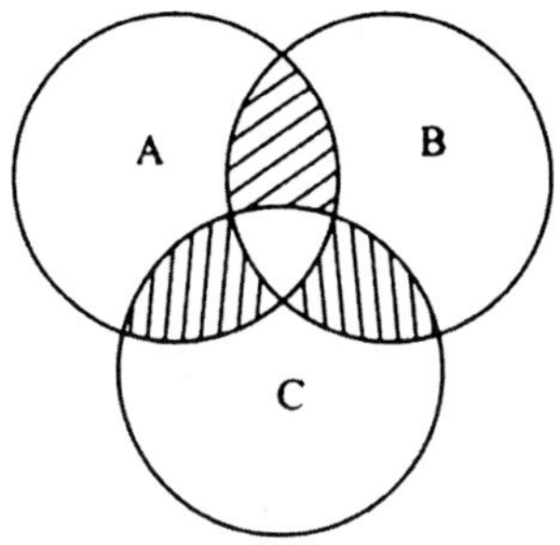

Fig. 9.90.

396. For the logic circuit of Fig. 9.91. the minimized expression is

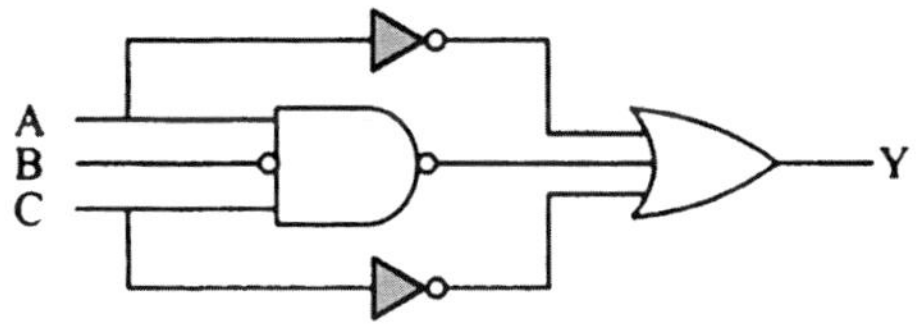

Fig. 9.91.

(a) $Y = \overline{A\bar{B}C}$ (b) $Y = A + B + C$
(c) $Y - A + B$ (d) $Y = A\bar{B}C$

397. The circuit of Fig. 9.92 realizes the function

(a) $Y = (\bar{A} + \bar{B})\,C + \overline{DE}$ (b) $Y = \bar{A} + \bar{B} + \bar{C} + \bar{D} + \bar{E}$
(c) $AB + C + DE$ (d) $AB + C\,(D + E)$

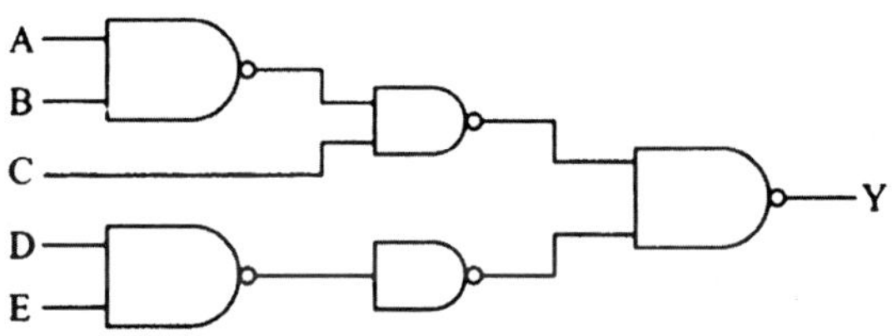

Fig. 9.92.

398. State transition table and state transition diagram form part of design steps for

(a) combination circuits (b) sequential circuits
(c) delay circuits (d) all the above

399. A 4 bit DAC gives an output of 4.5 V for input of 1001. If input is 0110, the output is

(a) 1.5 V (b) 2.0 V
(c) 3.0 V (d) 4.5 V

400. ADAC has a full scale output of 5 V. If accuracy is ± 0.2% the maximum error for an output of 1 V is

(a) 5 mV (b) 10 mV
(c) 2 mV (d) 20 mV

401. For the logic circuit of Fig. 9.93 Y =

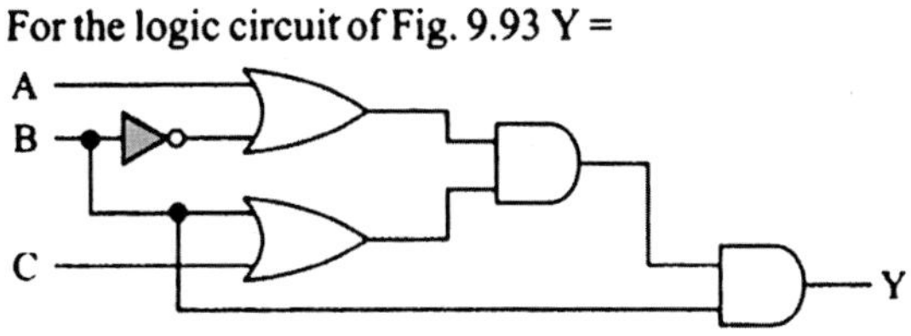

Fig. 9.93.

(a) $A + BC$ (b) BC
(c) AB (d) $AB + C$

402. For the K map in Fig. 9.94, the simplified expression is

(a) $\bar{x}_1\bar{x}_2\bar{x}_4 + x_2x_4 + x_1\bar{x}_3$
(b) $x_2x_4 + \bar{x}_2\bar{x}_4 + x_1\bar{x}_3$
(c) $\bar{x}_1\bar{x}_2\bar{x}_4 + \bar{x}_1x_2 + \bar{x}_3x_4 + x_2x_4$
(d) $\bar{x}_1\bar{x}_2 + \bar{x}_2\bar{x}_3 + x_3x_4$

x_3x_4 \ x_1x_2	00	01	11	10
00	1		d	d
01		1	d	
11		d	1	
10	1	d		d

Fig. 9.94.

403. In a digital system there are three inputs A B C. Output should be high when at least two inputs are high. Then output =

(a) $AB + BC + AC$
(b) $ABC + AD\bar{C} + \bar{A}BC + A\bar{B}C$
(c) $AB + BC + A\bar{C}$
(d) $A\bar{B} + B\bar{C} + \bar{A}C$

404. In digital circuits Schottky transistors are preferred over normal transistors because of their

(a) lower propagation delay
(b) lower power dissipation
(c) higher propagation delay
(d) higher power dissipation

405. A full adder can be made out of

(a) two half adders
(b) two half adders and an OR gate
(c) two half adders and a NOT gate
(d) three half adders

406. For the design of a sequential circuit having 9 states, minimum number of memory elements required is

(a) 3 (b) 4
(c) 5 (d) 9

407. A ripple counter has 4 bits and uses flip-flops with propagation delay lime of 25 ns. The maximum possible time for change of state will be

(a) 25 ns (b) 50 ns
(c) 75 ns (d) 100 ns

408. A 12 bit ADC is used to convert analog voltage of 0 to 10 V into digital. The resolution is

(a) 2.44 mV (b) 24.4 mV
(c) 1.2 V (d) none of the above

409. The hexadecimal number $(3E8)_{16}$ is equal to decimal number

(a) 1000 (b) 982
(c) 768 (d) 323

410. While obtaining minimal SOP expression

(a) all do not care terms are ignored
(b) all do not care terms are treated as 1
(c) all do not care terms are treated as 0
(d) only such do not care terms which aid minimization are treated as 1

411. The complement of Boolean expression $A\,B\,(\bar{B}C + A\,C)$ is

(*a*) $(\bar{A}+\bar{B})+(B+\bar{C})\,(\bar{A}+\bar{C})$

(*b*) $\overline{AB}+(B+\bar{C})\,(\bar{A}+\bar{C})$

(*c*) $(\bar{A}+\bar{B})+(B+\bar{C})\,(\bar{A}+\bar{C})$

(*d*) none of the above

412. The circuit of Fig. 9.95 is

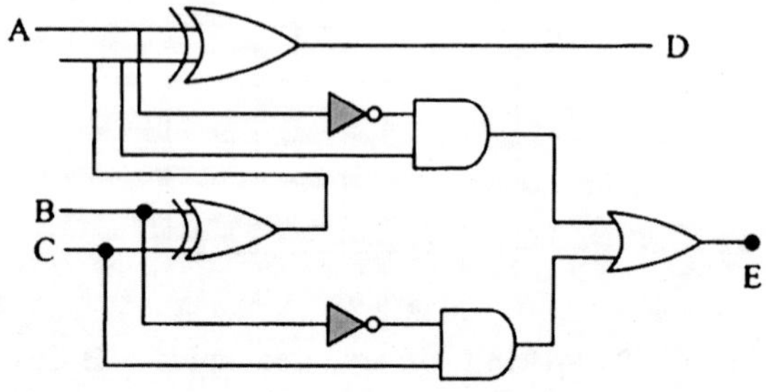

Fig. 9.95.

(*a*) full adder (*b*) full subtractor
(*c*) shift register (*d*) decade counter

413. Match List I and List II and give correct answer using the given codes

List I	List II
A 45	1. 10110100
B 90	2. 11010010
C 180	3. 01011010
D 210	4. 00101101
	5. 10101000

Codes:

	A	B	C	D
(*a*)	3	4	5	2
(*b*)	4	3	5	2
(*c*)	4	3	1	2
(*d*)	3	4	2	1

414. A transistor is operated as non-saturated switch to eliminate

(*a*) turn on time (*b*) turn off time
(*c*) storage time (*d*) delay time

415. In Fig. 9.96 $R_C = R_L = 1\ k\Omega$, then $V_0 =$

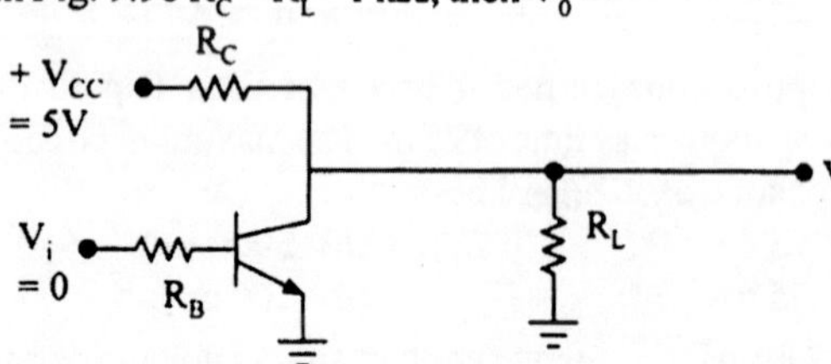

Fig. 9.96.

(*a*) 5V (*b*) 2.5V
(*c*) 1V (*d*) 0V

416. In Fig. 9.97, Y =

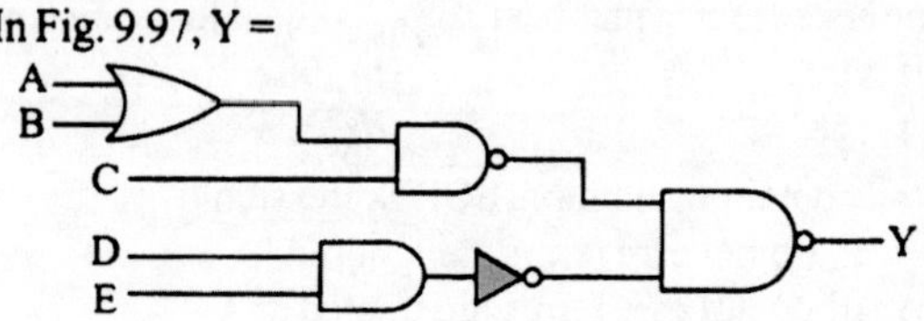

Fig. 9.97.

(*a*) (A + B) C + DE (*b*) AB + C (D + E)
(*c*) (A + B) C + D + E (*d*) none of the above

417. TTL circuit with active pull up is preferred because of its suitability for

(*a*) wired AND operation
(*b*) bus operated system
(*c*) wired logic operation
(*d*) reasonable dissipation and speed of operation

418. The resolution of an *n* bit DAC with a maximum input of 5 V is 5 mV. The value of *n* is

(*a*) 8 (*b*) 9
(*c*) 10 (*d*) 11

419. The inputs to a NAND gate are as shown in Fig. 9.98. The waveform of output is

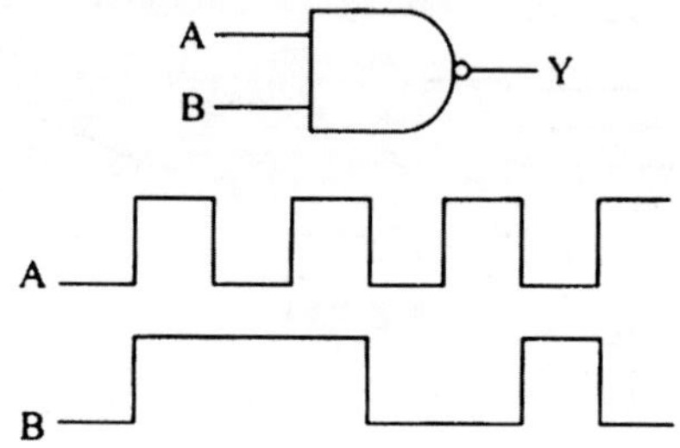

Fig. 9.98.

(*a*)
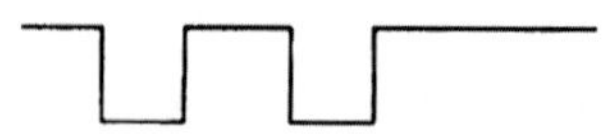

(*b*)
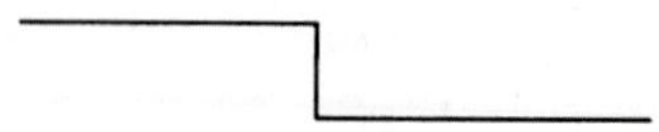

(*c*)
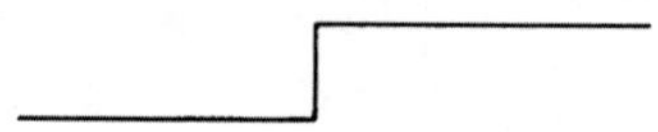

(*d*)
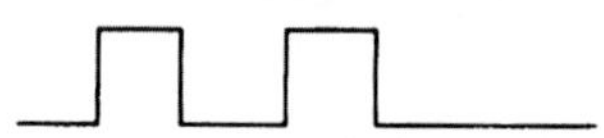

Fig. 9.99.

420. In the circuit of Fig. 9.100, Y =

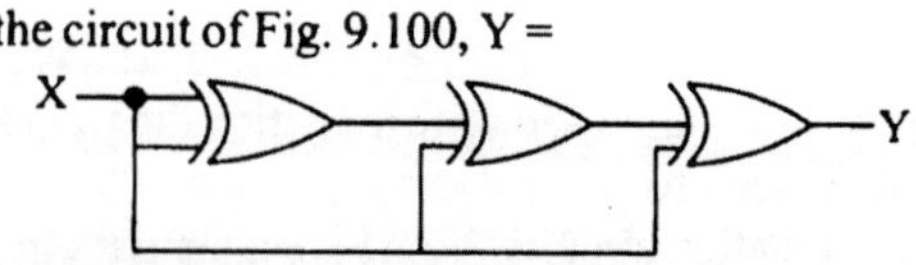

Fig. 9.100.

(*a*) 0 (*b*) 1
(*c*) X (*d*) $\bar{X}$

421. In the circuit of Fig. 9.101. $V_0 =$

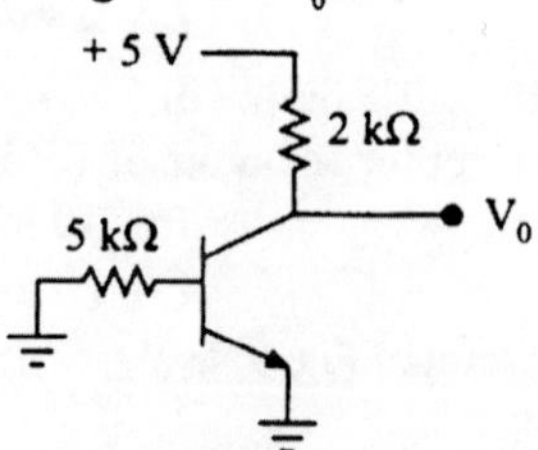

Fig. 9.101.

(*a*) 5V (*b*) 3.1V
(*c*) 2.5V (*d*) 0

422. Consider the following statements

1. ECL has least propagation delay
2. TTL has largest fan out
3. CMOS has highest noise margin
4. TTL has lowest power dissipation

Which of these are correct?

(a) 1 and 3 (b) 2 and 4
(c) 3 and 4 (d) 1 and 2

423. For a logic family
V_{OH} is the minimum output high level voltage
V_{OL} is the maximum output low voltage level
V_{IH} is the minimum acceptable input high level voltage
V_{IL} is the maximum acceptable input low level voltage
The correct relationship is
(a) $V_{IH} > V_{OH} > V_{IL} > V_{OL}$
(b) $V_{OH} > V_{IH} > V_{IL} > V_{OL}$
(c) $V_{IH} > V_{OH} > V_{OL} > V_{IL}$
(d) $V_{OH} > V_{IH} > V_{OL} > V_{IL}$

424. The initial state of a Mod-16 counter is 0110. Alter 37 clock pulses the state of counter will be
(a) 1011 (b) 0110
(c) 0101 (d) 0001

425. Symmetrical square wave of time period 100 μs can be obtained from square wave of time period 10 μs by using
(a) divide by 5 circuit
(b) divide by 2 circuit
(c) divide by 5 circuit followed by divide by 2 circuit
(d) BCD counter

426. For a particular type of memory the access time and cycle time are 200 ns each. The maximum rate at which data can be accessed is
(a) 2.5×10^6/s (b) 5×10^6/s
(c) 0.2×10^6/s (d) 10^6/s

427. Karnaugh map is used to
(a) minimise the number of flip-flops in a digital circuit
(b) minimise the number of gates only in a digital circuit
(c) minimise the number of gates and fan in a digital circuit
(d) design gates

428. The logic circuit of Fig. 9.102 is equivalent to

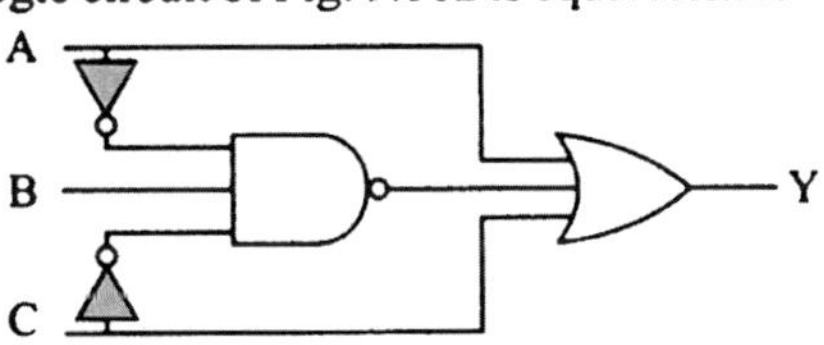

Fig. 9.102.

(a)
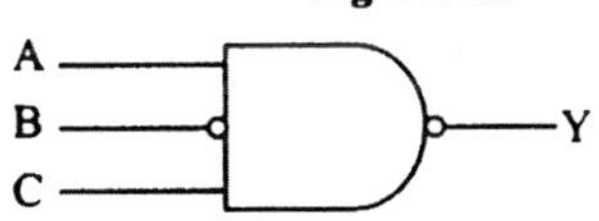

(b)
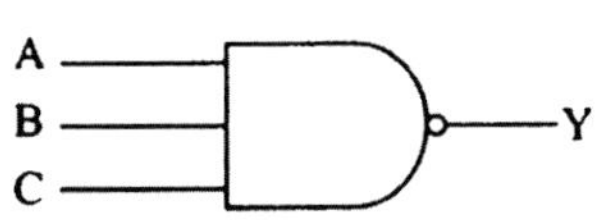

(c)
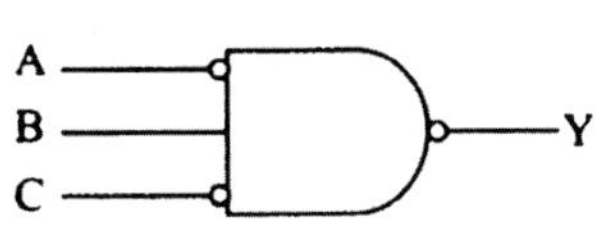

(d)
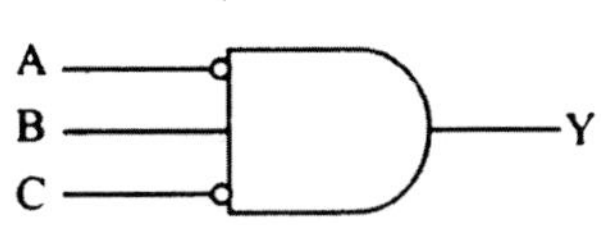

Fig. 9.103.

429. A the logic circuit of Fig. 9.104 realizes the function

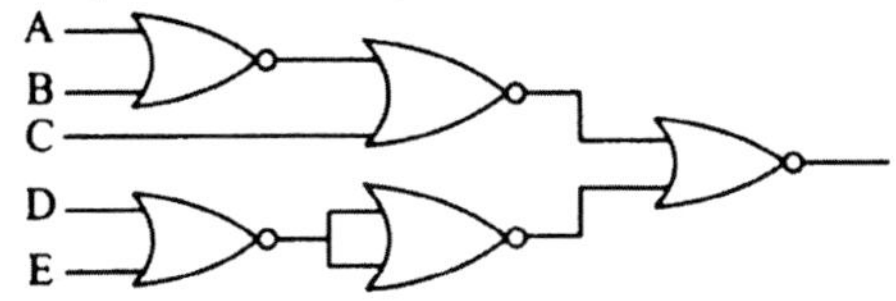

Fig. 9.104.

(a) $\overline{(A+B+C)}(\overline{D})(\overline{E})$ (b) $\overline{(A+B+C)}(D)(E)$
(c) $\overline{(A+B+C)}(\overline{D})(E)$ (d) $(A+B+\overline{C})(D)(E)$

430. Which of the following statements defines a full adder?
(a) It adds two binary digits and produces their SUM and CARRY
(b) It adds three inputs (two binary digits and a carry) and produces their SUM and CARRY
(c) It has two inputs and two outputs
(d) None of the above

431. If A = B = 1, the outputs P and Q in Fig. 9.105 are

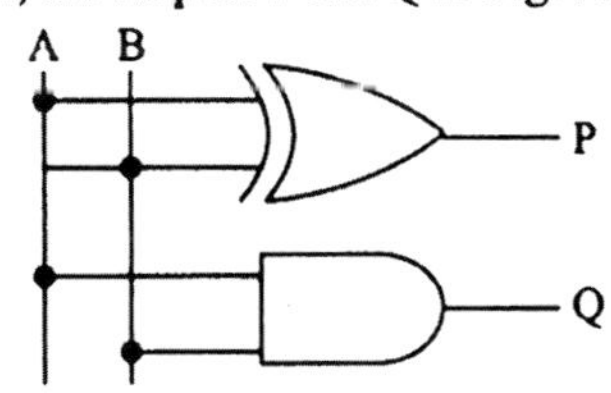

Fig. 9.105.

(a) P = Q = 0 (b) P = 0, Q = 1
(c) P = 1, Q = 0 (d) P = Q = 1

432. Which one of the following can be used as parallel to series converter?
(a) Decoder (b) Digital counter
(c) Multiplexer (d) Demultiplexer

433. Consider the following statements. A multiplxer
1. selects one of the several inputs and steers it to the output
2. routes the data from a single input to many outputs
3. converts parallel data into serial data
4. is a combinational circuit

Of the above statements which are correct?
(a) 1, 2, 4 (c) 2, 3, 4
(c) 1, 3, 4 (d) 1, 2, 3

434. Consider the following statements
1. Race around condition can occur in J K flip-flop when both inputs are 1
2. A flip-flop is used to store one bit of information
3. A transparent latch is a D type flip-flop
4. Master slave flip-flop is used to store two bits of information

Which of the above are correct
(a) 1, 2, 3 (b) 1, 3, 4
(c) 1, 2, 4 (d) 2, 3, 4

435. A ring counter with 5 flip-flops will have
(a) 5 states (b) 10 states
(c) 32 states (d) infinite states

436. A crystal is frequently used in digital circuits for timing purpose because of its
(a) low cost (b) high frequency stability

(c) simple circuitry
(d) ability to set frequency at desired value

437. Which one of the following is correct?
(a) RAM is a Volatile memory but ROM is non-volatile memory
(b) RAM is non-volatile memory but ROM is volatile memory
(c) Both ROM and RAM are volatile but in ROM data is not lost when power is switched off
(d) Both ROM and RAM are non-volatile but in RAM data is lost when power is switched off

438. In a 4 bit parallel in parallel out shift register A = 1, B = 1, C = 0, D = 1. The data output after 3 clock pulses is
(a) 1101 (b) 1001
(c) 0101 (d) 0001

439. In a shift register the data is loaded in one operation but shifted out one bit a time. The shift register is
(a) serial in – serial out (b) parallel in – serial out
(c) serial in – parallel out (d) parallel in – parallel out

440. The applications of shift registers are
1. Time delay
2. Ring counter
3. Serial to parallel data conversion
4. Parallel to serial data conversion

Which of the above are correct?
(a) 1, 2, 3 (b) 2, 3, 4
(c) 1, 2, 3, 4 (d) 1, 2, 4

441. Figure 9.106 shows a 3 bit shift register using TTL flip-flops. Initially all the flip-flops are set to 0 state. After 8 clock pulses

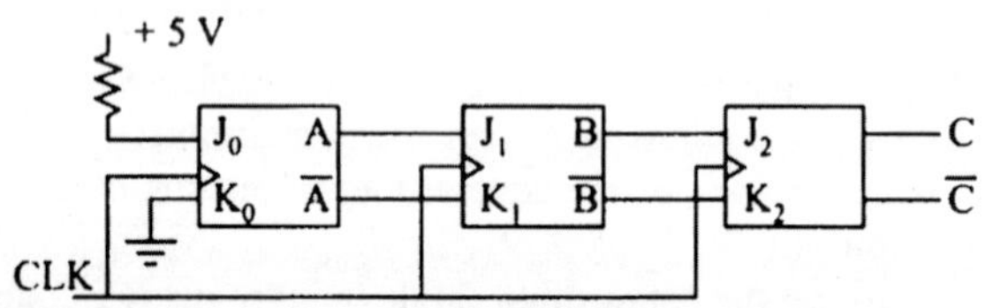

Fig. 9.106.

(a) A = 1, B = 1, C = 1 (b) A = 1, B = 1, C = 0
(c) A = 0, B = 1, C = 1 (d) A = 0, B = 0, C = 1

442. A 4 bit ripple counter uses flip-flops with propagation delay of 50 ns each. The maximum clock frequency which can be used is
(a) 5 MHz (b) 10 MHz
(c) 20 MHz (d) 25 MHz

443. Figure 9.107 shows a 4 bit serial in parallel out right shift register. The initial contents as shown are 0110. After 3 clock pulses the contents will be

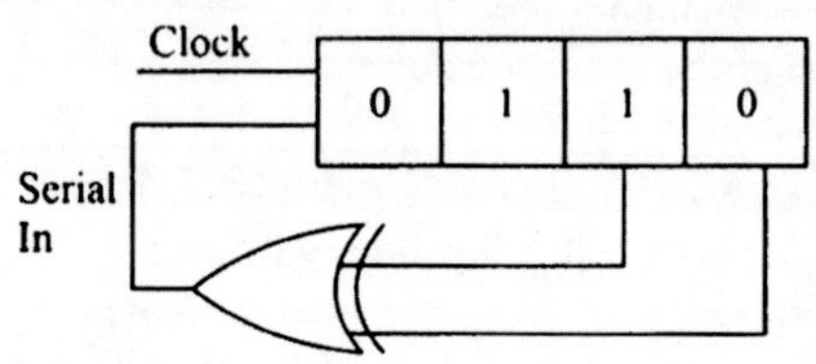

Fig. 9.107.

(a) 0000 (b) 0101
(c) 1010 (d) 1111

444. Read the following statements. Dual slope ADC provides
1. higher speed as compared to other ADC
2. very good accuracy
3. good rejection of power supply hum
4. better resolution as compared to other ADC for the same number of bits

Which of the above are correct?
(a) 1 and 2 (b) 2 and 3
(c) 1, 2, 3 (d) 1, 2, 3, 4

445. A pulse train with a 1 MHz frequency is counted using a 1024 modulus ripple counter using JK flip-flops. The maximum propagation delay for each flip-flop should be
(a) 1 μs (b) 0.5 μs
(c) 0.2 μs (d) 0.1 μs

446. Match the following for TTL and give correct answer using the given codes

A.	$V_{OH(min)}$	1. 2.4 V
B.	$V_{IH(min)}$	2. 1.5 V
C.	$V_{OL(max)}$	3. 0.4 V
		4. 2.0 V
		5. 0.8 V

Codes:

	A	B	C
(a)	1	3	4
(b)	1	4	3
(c)	5	4	3
(d)	1	5	2

447. Match the following and select correct answer using the given codes

List I	**List II**
A Flash converter ADC	1. requires conversion time of a few μs
B Dual slope ADC	2. requires DAC
C Successive approximation ADC	3. minimizes the effect of power supply interference
	4. requires complex hardware
	5. is a tracking ADC

Codes:

	A	B	C
(a)	2	3	4
(b)	4	3	2
(c)	5	4	3
(d)	1	2	5

448. Schottky clamping is resorted to in TTL gates
(a) to reduce propagation delay
(b) to increase noise margin
(c) to increase packing density
(d) to increase fan out

449. A pulse train can be delayed by a finite number of clock periods by using
(a) serial in–serial out shift register
(b) parallel in–serial out shift register
(c) serial in–parallel out shift register
(d) parallel in–parallel out shift register

450. A 12 bit ADC is operating with 1 μs clock period. Total conversion time is 14 μs. ADC is
(a) flash type (b) counting type

(c) integrating type
(d) successive approximation type

451. A dynamic RAM cell which holds 5 V has to be refreshed every 20 ms so that the stored voltage does not fall by more than 0.5 V. If the cell has a constant discharge current of 0.1 p A, the storage capacitance of cell is

(a) 4×10^{-6} F (b) 4×10^{-9} F
(c) 4×10^{-12} F (d) 4×10^{-15} F

452. A 10 bit ADC with a full scale output voltage of 10.24 V is to be designed to have ± LSB/2 accuracy. If ADC is calibrated at 25°C, the maximum net temperature coefficient of ADC should not exceed

(a) ±200 μV/°C (b) ±400 μV/°C
(c) ±600 μV/°C (d) ±800 μV/°C

453. A memory system of size 16 k bytes is to be designed using memory chips which have 12 address lines and 4 data lines each. The number of such chips required to design the memory system is

(a) 2 (b) 4
(c) 8 (d) 16

454. Each cell of a static RAM has

(a) 4 MOS transistors
(b) 4 MOS transistors and 2 capacitors
(c) 2 MOS transistors and two capacitors
(d) 1 MOS transistor and 1 capacitor

455. In standard TTL the totem pole stage refers to

(a) multiemitter input stage
(b) phase splitter
(c) output buffer
(d) open collector output stage

456. Inverter 74 ALS04 has following specifications $I_{OH\ max} = -0.4$ mA, $I_{OLmax} = 8$ mA, $I_{IH\ max} = 20$ μA, $I_{IL\ max} = -0.1$ mA. The fan out of this inverter is

(a) 10 (b) 20
(c) 60 (d) 100

457. For the gate in Fig. 9.108 the output will be

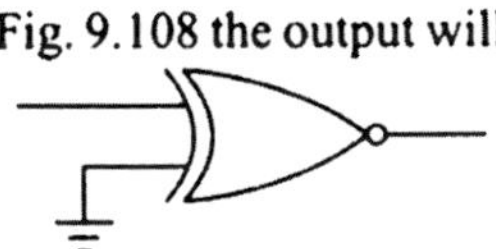

Fig. 9.108.

(a) 0 (b) 1
(c) A (d) $\bar{A}$

458. For the NMOS gate in Fig. 9.109, F =

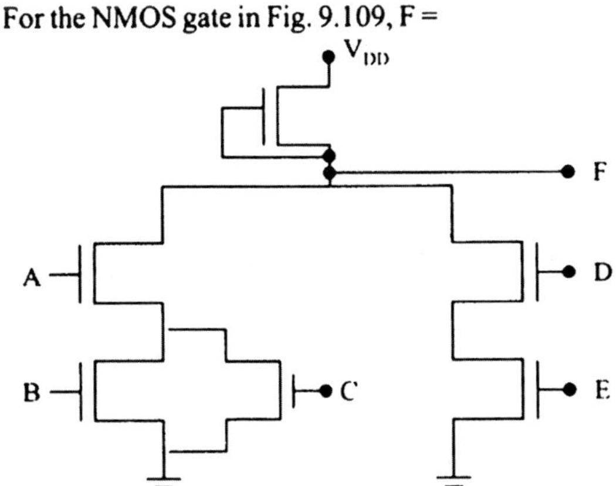

Fig. 9.109.

(a) $\overline{ABCDE}$ (b) $(AB + \bar{C})(\bar{D} + \bar{E})$
(c) $\overline{A(B + C) + DE}$ (d) $\overline{A + B}\,C + \bar{D}\,\bar{E}$

459. The boolean function A + BC is reduced form of

(a) AB + BC (b) $\bar{A}B + \bar{A}BC$
(c) (A + B)(A + C) (d) (A + C)B

460. The minimum number of 2 input NAND gates required to implement boolean function $A\bar{B}C$ if A, B, C are available is

(a) 2 (b) 3
(c) 5 (d) 6

461. The noise margin of a TTL gate is about

(a) 0.2 V (b) 0.4 V
(c) 0.6 V (d) 0.8 V

462. In Fig. 9.110 A = 1, B = 1. B is now changed to a sequence 101010...... The outputs X and Y will be

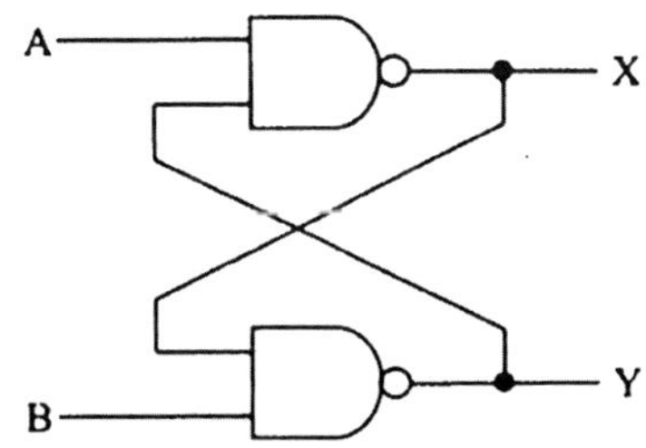

Fig. 9.110.

(a) fixed at 0 and 1 respectively
(b) X = 1010........ and Y = 0101 ...
(c) X = 1010......... and Y = 1010
(d) fixed at 1 and 0 respectively

463. The advantage of using dual slope ADC in digital voltmeter is that

(a) its conversion time is small
(b) its accuracy is high
(c) its output is in BCD
(d) it does not require a comparator

464. For $AB + \bar{A}C + BC = AB + \bar{A}C$ the dual form is

(a) $(A + B)(\bar{A} + C)(B + C) = (A + B)(\bar{A} + C)$
(b) $(\bar{A} + \bar{B})(A + \bar{C})(\bar{B} + \bar{C}) = (\bar{A} + \bar{B})(A + \bar{C})$
(c) $(A + B)(A + C)(B + C) = (\bar{A} + \bar{B})(\bar{A} + C)$
(d) none of the above

465. The modulus of counter in Fig. 9.111 is

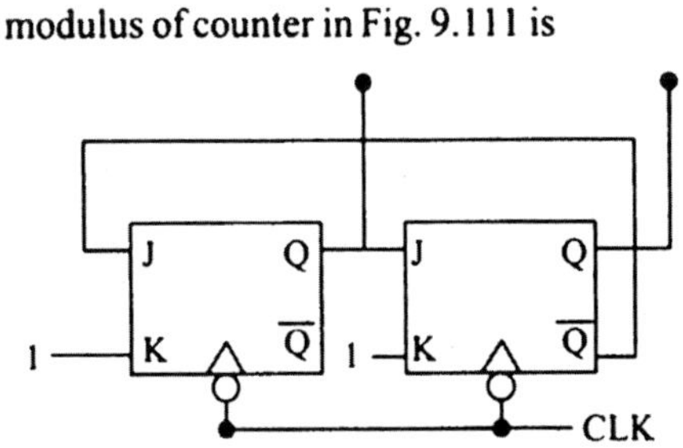

Fig, 9.111.

(a) 1 (b) 2
(c) 3 (d) 4

466. Figure 9.112 shows a K-map for a Boolean function. The number of essential prime implicants is

AB \ CD	00	01	11	10
00	1	1		1
01				1
11	1			
10	1			1

Fig. 9.112.

(*a*) 4 (*b*) 5
(*c*) 6 (*d*) 8

467. $Y = A + \bar{A}B$ is the same as
(*a*) $Y = AB$ (*b*) $Y = A + B$
(*c*) $Y = \bar{A} + \bar{B}$ (*d*) $Y = \bar{A} + B$

468. Commercial ECL gates use two ground lines and one negative supply to
(*a*) reduce power consumption
(*b*) increase fan out
(*c*) reduce loading effect
(*d*) eliminate the effect of power line glitches on the biasing circuit

469. In Fig. 9.113, Y =

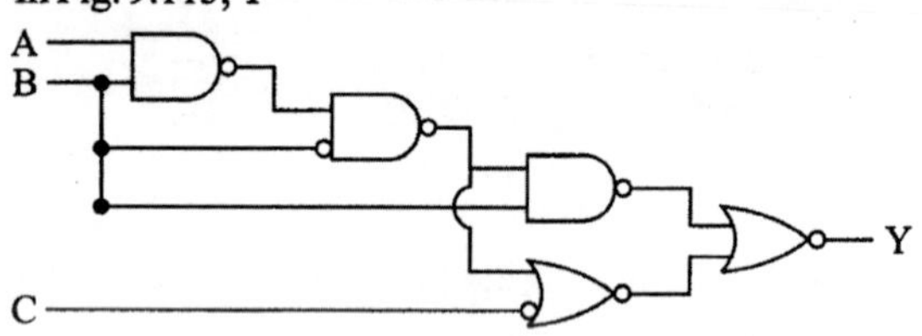

Fig. 9.113.

(*a*) A (*b*) B
(*c*) A+B+C (*d*) ABC

470. For the DAC in Fig. 9.114, V_0 =

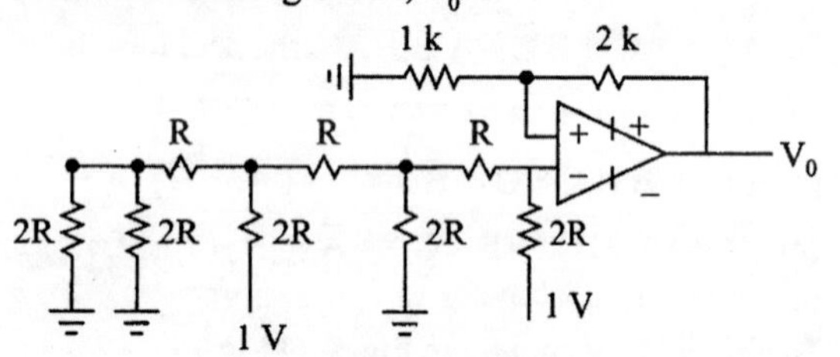

Fig. 9.114.

(*a*) 10V (*b*) 5V
(*c*) 4V (*d*) 8V

471. In a mod-12 counter the input clock frequency is 10 kHz. The output frequency is
(*a*) 0.833 kHz (*b*) 1.0 kHz
(*c*) 0.91 kHz (*d*) 0.77 kHz

472. In a NAND SR latch S = R = 1. Then
(*a*) $Q = 1, \bar{Q} = 0$
(*b*) $Q = 0, \bar{Q} = 1$
(*c*) both Q and $\bar{Q}$ will he same as before
(*d*) the latch will be set

473. In a NAND SR latch S = 0, R = 1, then
(*a*) Q = 0 (*b*) $Q = 1, \bar{Q} = 0$
(*c*) it is race condition (*d*) none of the above

474. For a NAND SR latch what is the normal resting state of inputs
(*a*) S = R = 1 (*b*) S = 0, R = 1
(*c*) S = 1, R = 0 (*d*) S = R = 0

475. For a NOR SR latch the normal resting state is
(*a*) S = R = 1 (*b*) S = 0, R = 1
(*c*) S = 0, R = 0 (*d*) S = 1, R = 0

476. The race condition in NOR SR latch occurs when
(*a*) S = R = 1 (*b*) S = 0, R = 1
(*c*) S = 0, R = 0 (*d*) S = 1, R = 0

477. In Fig. 9.115, the flip-flop is

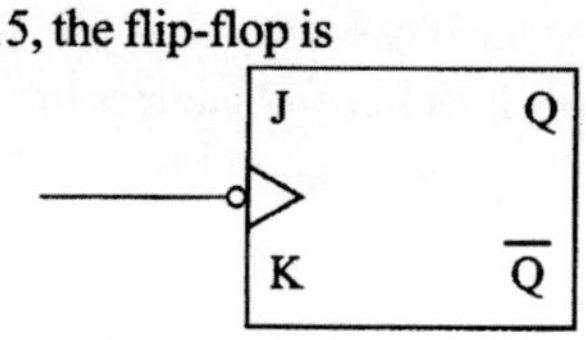

Fig. 9.115.

(*a*) negative edge triggered
(*b*) positive edge triggered
(*c*) level triggered
(*d*) either (*a*) or (*c*)

478. In flip-flops the duration of clock pulse is about
(*a*) 3 ns (*b*) 10 ns
(*c*) 25 ns (*d*) 50 ns

479. In a J K flip-flop the input J = K = 1 causes flip-flop to
(*a*) set (*b*) reset
(*c*) no change (*d*) toggle

480. To convert SR latch to D latch
(*a*) connect both S and R to D
(*b*) connect D to S directly and D to R through inverter
(*c*) connect D to R directly and D to S through inverter
(*d*) connect S to D and leave R open

481. To convert J K flip-flop to D flip-flop
(*a*) connect D to both J and K
(*b*) connect D to J directly and D to K through inverter
(*c*) connect D to K directly and D to J through inverter
(*d*) connect D to K and leave J open

482. A 4 hit ripple counter is in 0000 state. The clock pulses are applied and then removed. The counter reads 0011. The number of clock pulses which have occurred are
(*a*) 3 (*b*) 3 or 19
(*c*) 3 or 19 or 35 (*d*) none of the above

483. The counter in Fig. 9.116 is

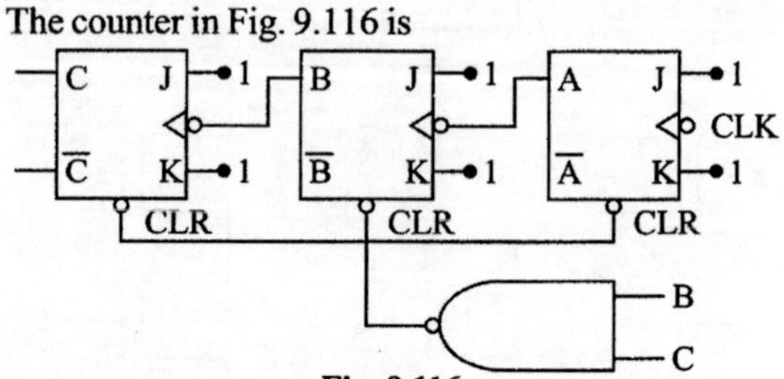

Fig. 9.116.

(*a*) Mod 3 (*b*) Mod 6
(*c*) Mod 8 (*d*) Mod 7

484. A 4 bit synchronous counter has flip-flops having propagation delay of 50 ns each and AND gates having propagation delay of 20 ns each. The maximum frequency of clock pulses can be

(*a*) 20 MHz (*b*) 50 MHz
(*c*) 14.3 MHz (*d*) 5 MHz

485. As the number of flip-flops are increased, the total propagation delay of

(*a*) ripple counter increases but that of synchronous counter remains the same
(*b*) both ripple and synchronous counters increase
(*c*) both ripple and synchronous counters remain the same
(*d*) ripple counter remains the same but that of synchronous counter increases

486. Read the following statements

1. The circuitry of ripple counter is more complex than that of synchronous counter
2. The maximum frequency of operation of ripple counter depends on the modulus of the counter
3. The maximum frequency of operation of synchronous counter does not depend on the modulus of the counter

Which of the above statements are correct?

(*a*) 1 only (*b*) 1, 2
(*c*) 1, 2, 3 (*d*) 2, 3

487. The number of logic devices required in a Mod 16 synchronous counter are

(*a*) 4 (*b*) 5
(*c*) 6 (*d*) 7

488. A JK flip-flop has t_{pd} = 12 ns. The largest modulus of a ripple counter using these flip-flops and operating at 10 MHz is

(*a*) 16 (*b*) 64
(*c*) 128 (*d*) 256

489. For a Mod 64 parallel counter we need

(*a*) 6 flip-flops
(*b*) 6 flip-flops and 2 AND gates
(*c*) 6 flip-flops and 4 AND gates
(*d*) none of the above

490. The maximum counting range of a four stage counter using IC 74193 is

(*a*) 0 to 1023 (*b*) 0 to 4093
(*c*) 0 to 65535 (*d*) 0 to 131071

491. The reason for glitches on the outputs of decoding gates of a synchronous counter is

(*a*) FFs changing states together
(*b*) FFs changing states one at a time
(*c*) AND gates not functioning properly
(*d*) none of the above

492. A 6 bit DAC uses binary weighted resistors. If MSB resistor is 20 k ohm, the value of LSB resistor is

(*a*) 20 k ohm (*b*) 80 k ohm
(*c*) 320 k ohm (*d*) 640 k ohm

493. An 8 bit DAC has a full scale output of 2 m A and full scale error of ± 0.5%. If input is 10000000 the range of outputs is

(*a*) 994 to 1014 μA (*b*) 990 to 1020 μA
(*c*) 800 to 1200 μA (*d*) none of the above

494. A voltage DAC is generally slower than current DAC

(*a*) because of more accuracy
(*b*) because of higher resolution
(*c*) because of response time of op-amp current to voltage converter
(*d*) none of the above

495. Some ADCs are

1. Parallel 2. Updown
3. Single slope 4. Dual slope
5. Successive approximation
6. Stair step

Out of above which ones do not use DAC

(*a*) 1 only (*b*) 1, 3, 4
(*c*) 1, 2, 3 (*d*) 1, 5, 6

496. Out of 5 M × 8, 1 M × 16, 2 M × 16 and 3 M × 8 memories, which memory can store more bits

(*a*) 5 M × 8 (*b*) 2 M × 16
(*c*) 3 M × 8 (*d*) 1 M × 16

497. Out of SAM and RAM

(*a*) access time in both does not depend on address location
(*b*) In SAM access time depends on address location but not in RAM
(*c*) In RAM access time depends on address location but not in SAM
(*d*) In both the access time depends on storage location

498. The process of entering data into ROM is called

(*a*) writing (*b*) burning
(*c*) decoding (*d*) registering

499. The main drawbacks of EEPROM are

(*a*) low density (*b*) high cost
(*c*) both low density and high cost
(*d*) none of the above

500. The advantages of flash memory over EEPROM are

(*a*) higher density
(*b*) lower cost
(*c*) both higher density and lower cost
(*d*) none of the above

501. Hexadecimal representation of 657_8 is

(*a*) 1AFH (*b*) D78H
(*c*) D71H (*c*) 32FH **[IES 2010]**

502. Boolean expression

$$\overline{B+C} + \overline{\bar{A} + \bar{B} + C} + \overline{A + \bar{B}} + \bar{C} + ABC$$

reduces to

(*a*) A (*b*) B
(*c*) C (*d*) A + B + C **[IES 2010]**

503. The complement of expression

$$Y = ABC + AB\bar{C} + \bar{A}\bar{B}C + \bar{A}BC$$

is

(*a*) $(A + \bar{B})(A + \bar{C})$ (*b*) $(\bar{A} + B)(A + C)$
(*c*) $(A + \bar{B})(\bar{A} + C)$ (*d*) $(\bar{A} + \bar{B})(A + \bar{C})$

[IES 2010]

504. The standard SOP expression for Boolean expression $Y = A\bar{B} + AC + BC$ is

(*a*) $\bar{A}BC + ABC + A\bar{B}C + A\bar{B}\bar{C}$
(*b*) $AB\bar{C} + \bar{A}BC + ABC$

(c) $A\overline{B}C + AB\overline{C} + ABC$

(d) $\overline{A}BC + AB\overline{C} + A\overline{B}C$ **[IES 2010]**

505. The figure of merit of a logic family is given by the product of

(a) Gain and bandwidth
(b) Propagation delay time and power dissipation
(c) Fan out and propagation delay time
(d) Noise margin and power dissipation

506. In NOR-NOR configuration the minimum number of NOR gates required to implement the switching function $X + X\overline{Y} + X\overline{Y}Z$ is

(a) 5 (b) 3
(c) 2 (d) 0 **[IES 2010]**

507. Match List I and List II and select the proper answer using the given code.

	List I		List II
A	HTL	1.	High fan out
B	CMOS	2.	Highest speed of operation
C	I^2L	3.	High noise immunity
D	ECL	4.	Lowest speed power product

Codes:

	A	B	C	D
(a)	3	4	1	2
(b)	2	1	4	3
(c)	3	1	4	2
(d)	2	4	1	3

[IES 2010]

508. A D/A converter has 5 V full scale output voltage and an accuracy of ± 0.2%. The maximum error for any output is

(a) 5 mV (b) 10 mV
(c) 20 mV (d) 1.0 mV **[IES 2010]**

509. The output X in Fig. 9.117 is

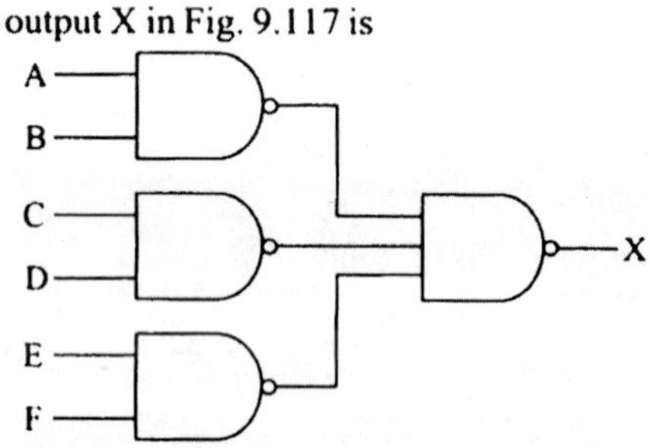

Fig. 9.117.

(a) $AB + CD + EF$ (b) $\overline{AB} + \overline{C}D + \overline{EF}$
(c) $(A + B)(C + D)(E + F)$
(d) $(\overline{A + B})(\overline{C + D})(\overline{E + F})$ **[IES 2010]**

510. The circuit in Fig. 9.118 is used to implement the function $Z = f(A, B) = \overline{A} + B$. The values of I and J are

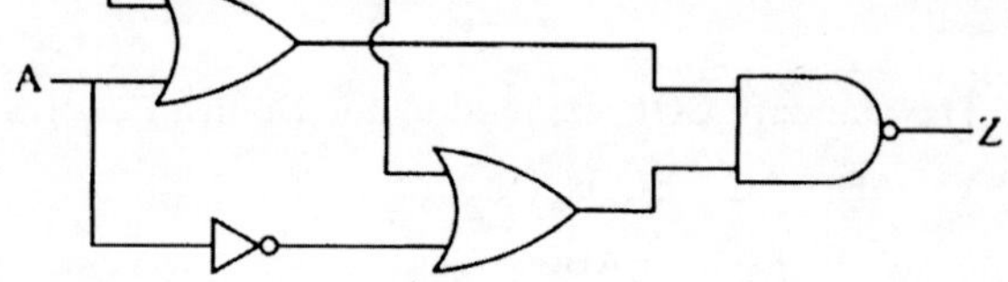

Fig. 9.118.

(a) I = 0 and J = B (b) I = 1 and J = B
(c) I = B and J = 1 (d) I = B and J = 0 **[IES 2010]**

511. Boolean expression for output of logic circuit in Fig. 9.119 is

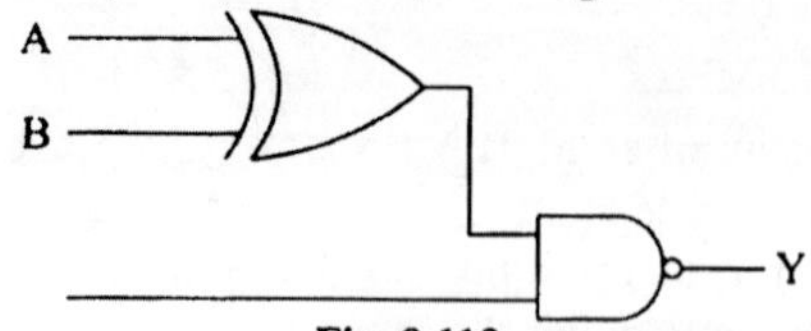

Fig. 9.119.

(a) $Y = \overline{AB} + AB + \overline{C}$ (b) $Y = \overline{A} + A\overline{B} + \overline{C}$
(c) $Y = A \oplus B + \overline{C}$ (d) $Y = AB + \overline{C}$ **[IES 2010]**

512. A ROM is used to store the table for multiplication of two 8 bit unsigned integers. The size of ROM required is

(a) 256 × 1 (b) 64 k × 8
(c) 4 k × 16 (d) 64 k × 16 **[IES 2010]**

513. In the circuit of Fig. 9.120, X is given by

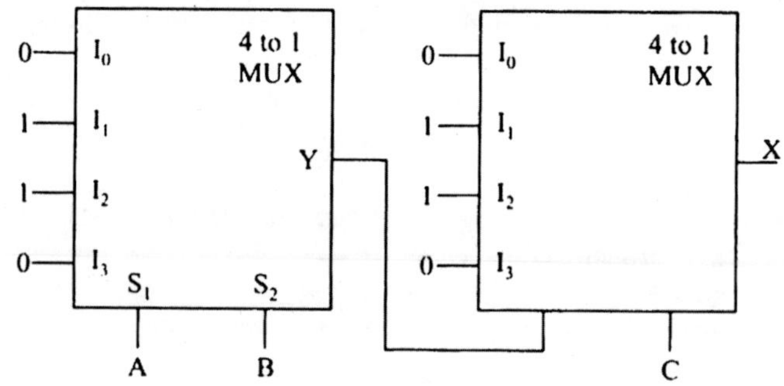

Fig. 9.120.

(a) $X = A\overline{B}\overline{C} + \overline{A}B\overline{C} + \overline{A}\overline{B}C + ABC$
(b) $X = \overline{A}BC + A\overline{B}C + AB\overline{C} + ABC$
(c) $X = AB + BC + AC$
(d) $X = \overline{AB} + \overline{BC} + \overline{AC}$ **[IES 2010]**

514. The correct statement is

(a) PROM contains a programmable AND array and a fixed OR array
(b) PLA contains a fixed AND array and a programmable OR array
(c) PROM contains a fixed AND array and programmable OR array
(d) PLA contains a programmable AND array and a fixed OR array

515. A 4 × 1 MUX is used to implement a 3 input Boolean function as shown in Fig. 9.121. The function is

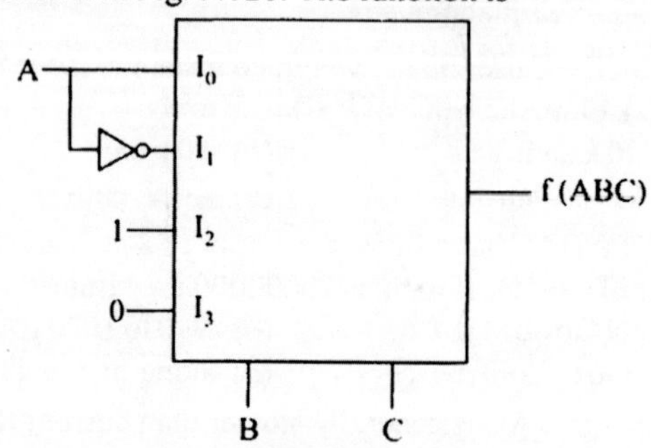

Fig. 9.121.

(a) $f(ABC) = \Sigma(1, 2, 4, 6)$ (b) $f(ABC) = \Sigma(1, 2, 6)$
(c) $f(ABC) = \Sigma(2, 4, 5, 6)$ (d) $f(ABC) = \Sigma(1, 5, 6)$ [IES 2010]

516. Processing of MOS ICs is less expensive than bipolar IC
(a) because they use cheaper components
(b) because they need no component isolation
(c) because they require much less diffusion steps
(d) because they have very high packing density [IES 2010]

517. A 10 bit A – D converter is used to digitize an analog signal in 0–5 V range. The max. peak-to-peak ripple voltage which can be allowed in DC supply voltage is
(a) 100 mV (b) 50 mV
(c) 25 mV (d) 5 mV [IES 2010]

518. Which of the following is correct?
(a) $A\bar{A} = 1$ (b) $A + AB = A + B$
(c) $A + \bar{A}B = A + B$ (d) $A(A + B) = B$ [IES 2010]

519. Which are the ultimate purposes of minimizing logic expression?
(a) to get a small size expression
(b) to reduce the number of variables in the expression
(c) to implement the function with minimum hardware
(d) to reduce the function to make if feasible for hardware implementation.
Select the answer from the following codes :
(a) 1 only (b) 2 and 3
(c) 3 only (d) 3 and 4 [IES 2009]

520. Which of the following factors are responsible to design IC logic gates to operate at a fixed supply voltage of 5 V.
1. Low heating of IC gates
2. Compatibility with other logic gates
3. Satisfactory and safe operation
4. Standardisation form
Select the answer using the following codes :
(a) 1 only (b) 2 only
(c) 2 and 3 (d) 3 and 4 [IES 2009]

521. Which of the following is not correct.
(a) Propagation delay is the time required for a gate to change its state
(b) Noise immunity is the amount of noise which can be applied to the input of a gate without causing the gate to change state
(c) Fan in of a gate is always equal to fan out of gate
(d) Operating speed is the maximum frequency at which digital data can be applied to the speed. [IES 2009]

522. Which are universal gate?
1. NAND
2. NOR
3. XOR
Select your answer from the following codes :
(a) 1 and 2 (b) 1 and 3
(c) 2 and 3 (d) 1, 2, 3 [IES 2009]

523. Which of the following outputs are available in TTL?
1. Open collector output
2. Totem pole output
3. Tristate output
Select your answer using the following gates :
(a) 1 only (b) 1, 2
(c) 2 and 3 (d) 1, 2, 3 [IES 2009]

524. Which logic family can be operated using a supply voltage from 3 V to 15 V?
(a) TTL (b) ECL
(c) PMOS (d) CMOS [IES 2009]

525. Which of the following come under the class of combinational logic circuits?
1. Full adder 2. Full subtractor
3. Half adder 4. JK flip-flop
5. Converter
Select your answer using the following codes :
(a) 1 only (b) 3 and 4
(c) 4 and 5 (d) 1, 2, 3

526. Consider a multiplexer with X and Y as data input and Z as control. Z = 0 selects input X and Z = 1 selects input Y. What are the connections required to implement $f = T + R$ without any other hardware?
(a) R to X, 1 to Y and T to Z
(b) T to X, R to Y and T to Z
(c) T to X, R to Y and 0 to Z
(d) R to X, 0 to Y and T to Z [IES 2009]

527. With which decoder is it possible to obtain many code conversions?
(a) 2 line to 4 line (b) 3 line to 8 line
(c) Not possible with decoder
(d) 4 line to 16 line [IES 2009]

528. Match List I (application of circuits) and List II (circuit name) and select the correct answer using the code.

	List I		List II
A	Divider	1.	Astable multivibrator
B	Clips input voltage at two predetermined levels	2.	Schmitt trigger
C	Square wave generator	3.	Bistable multivibrator
D	Narrow current pulse generator	4.	Blocking oscillator

Codes:

	A	B	C	D
(a)	4	2	1	3
(b)	3	2	1	4
(c)	4	1	2	3
(d)	3	1	2	4

529. Consider the following statements :
For master slave J-K flip-flop
1. Toggle frequency is the maximum clock frequency at which the flip-flop will toggle reliably
2. data input must process the clock triggering edge transition time by some minimum time
3. data input must remain fixed for a given time after the clock triggering edge transition time for reliable operation
4. propagation delay time is equal to rise time and fall time of data
Which statements are correct?
(a) 1, 2 and 3 (b) 1 and 2
(c) 2 and 3 (d) 3 and 4 [IES 2009]

530. Consider the following statements :

1. A flip-flop is used to store 1 bit of information
2. Race around condition occurs in JK flip-flop when both inputs are 1
3. Master slave configuration is used in flip-flops to store 2 bits of information
4. A transparent latch consists of D type flip-flop

Which of the above are correct?

(*a*) 1 only (*b*) 1, 3 and 4
(*c*) 1, 2 and 4 (*d*) 2 and 3 **[IES 2009]**

531. Which of the following flip-flops is used as a latch?

(*a*) JK (*b*) RS
(*c*) T (*d*) D **[IES 2009]**

532. Which of the following should be satisfied to call an astable multivibrator circuit using discrete components as a digital circuit?

1. A flip-flop is always a digital circuit.
2. Only when we assign 1 and 0 to the high and low level of output a flip-flop is called a digital circuit.
3. Only if the power supply voltage is maintained between + 5 V or – 5 V it is called a digital circuit.
4. Only if it is IC form following the technology of IC manufacture it is called a digital circuit.

Select the correct answer from the codes given below :

(*a*) 1 only (*b*) 2 and 3
(*c*) 2 only (*d*) 3 and 4

533. Consider the following statements regarding registers and latches

1. Registers are made of edge triggered FFs whereas latches are made from level triggered FFs.
2. Registers are temporary storage devices whereas latches are not.
3. A latch always uses cross coupled feedback connections
4. A register stores a binary used but latches do not

Which of the above are correct?

(*a*) 1 only (*b*) 1 and 3
(*c*) 2 and 3 (*d*) 3 and 4 **[IES 2009]**

534. Which of the following capabilities are available in universal shift register?

1. shift left
2. shift right
3. parallel load
4. serial add

Select your answer using the following code :

(*a*) 2 and 4 (*b*) 1, 2 and 3
(*c*) 1, 2 and 4 (*d*) 1, 3 and 4 **[IES 2009]**

535. Match List I (Type of ADC) and List II (Characteristics) and select the correct answer using the code.

	List I		**List II**
A	Flash counter	1.	Integrating type
B	Successive approximation	2.	Fastest converter
C	Counter ramp	3.	Maximum conversion time = N bits
D	Dual slope	4.	Uses a DAC in feedback path

Codes:

	A	B	C	D
(*a*)	2	3	4	1
(*b*)	1	3	4	2
(*c*)	2	4	3	1
(*d*)	1	4	3	2

[IES 2009]

536. Match the logic gates in column A with their equivalents in column B.

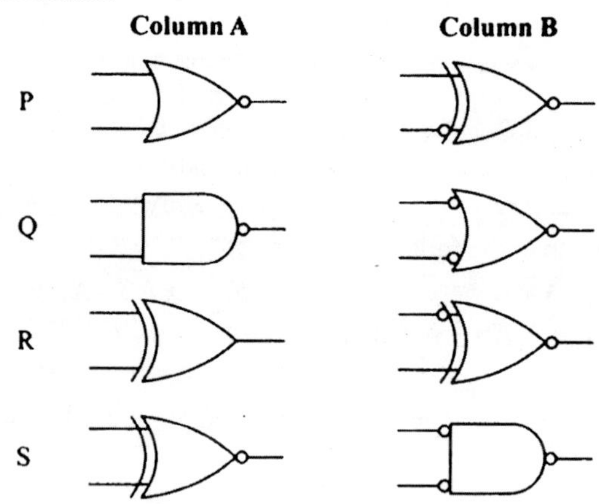

(*a*) P-2, Q-4, R-1, S-3 (*b*) P-4, Q-2, R-1, S-3
(*c*) P-2, Q-4, R-3, S-1 (*d*) P-4, Q-2, R-3, S-1 **[Gate 2010]**

537. For output F to be 1 input combination in Fig. 9.122 should be

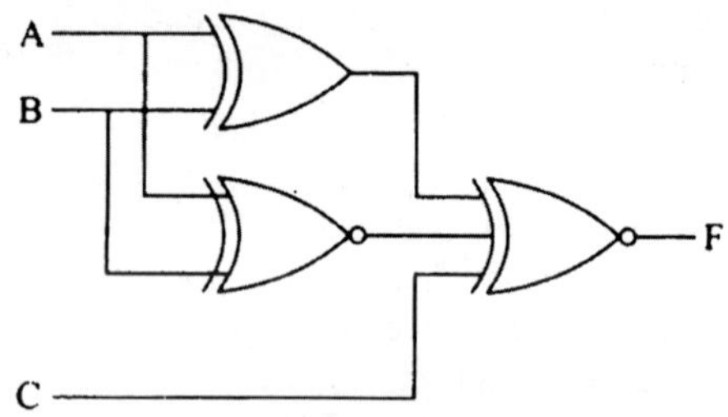

Fig. 9.122.

(*a*) A = 1, B = 1, C = 0 (*b*) A = 1, B = 0, C = 0
(*c*) A = 0, B = 1, C = 0 (*d*) A = 0, B = 0, C = 1 **[Gate 2010]**

538. Boolean function realized by logic circuit in Fig. 9.123 is

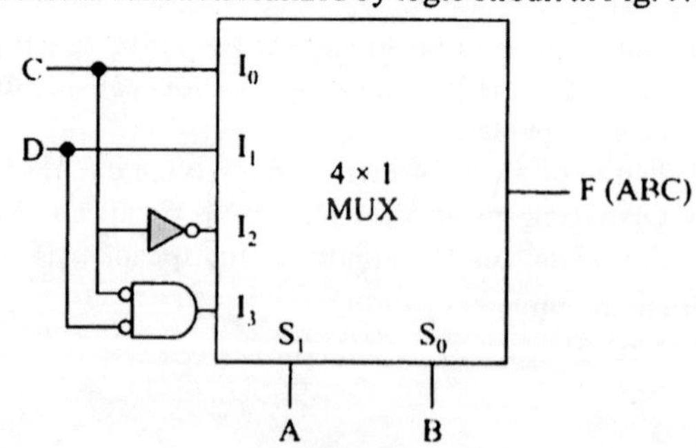

Fig. 9.123.

(*a*) $F = \Sigma m(0, 1, 3, 5, 9, 10, 14)$
(*b*) $F = \Sigma m(2, 3, 5, 7, 8, 12, 13)$
(*c*) $F = \Sigma m(1, 2, 4, 5, 11, 14, 15)$
(*d*) $F = \Sigma m(2, 3, 5, 7, 8, 9, 12)$

539. The full forms of TTL and CMOS with reference to logic families are

(a) Triple transistor logic and chip metal oxide semiconductor
(b) Tristate transistor logic and chip metal oxide semiconductor
(c) Transistor transistor logic and complementary metal oxide semiconductor
(d) Tristate transistor logic and complementary metal oxide silicon **[Gate 2009]**

540. If X = 1 in logic equation

$$[X + Z\{\bar{Y} + (\bar{Z} + X\bar{Y})\}]\{\bar{X} + \bar{Z}(X + Y)\} = 1$$

then

(a) $Y = Z$ (b) $Y = \bar{Z}$
(c) $Z = 1$ (d) $Z = 0$

541. What are the minimum number of 2 to 1 MUX required to generate a 2 input AND gate and a 2 input EX-OR gate
(a) 1 and 2 (b) 1 and 3
(c) 1 and 1 (d) 2 and 2 **[Gate 2009]**

542. Figure 9.124 gives NAND and NOR latches. Inputs P_1, P_2 for both the latches are first made 0, 1 and after a few seconds made 1, 1. Corresponding stable outputs Q_1, Q_2 are

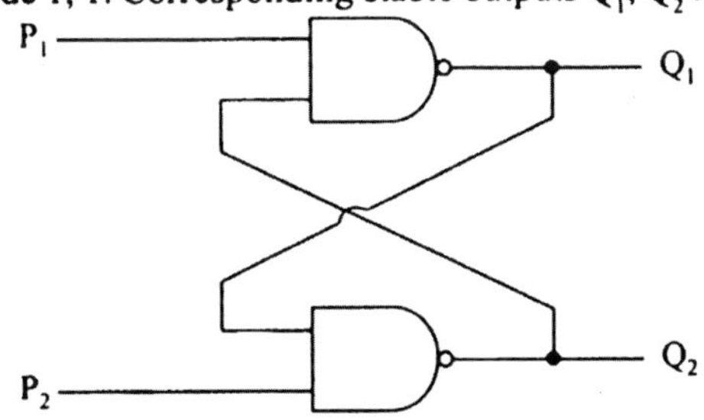

(a)

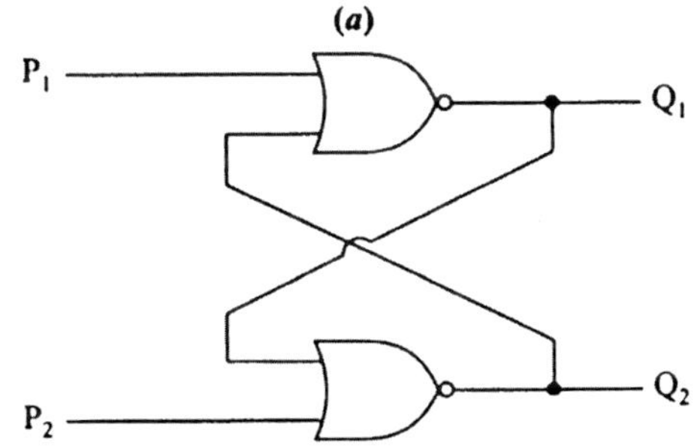

(b)

Fig. 9.124.

(a) NAND first 0, 1 then 0, 1; NOR first 1, 0 then 0, 0
(b) NAND first 1, 0 then 1, 0; NOR first 1, 0 then 1, 0
(c) NAND first 1, 0 then 1, 0; NOR first 1, 0 then 0, 0
(d) NAND first 1, 0 then 1, 1; NOR first 0, 1 then 0, 1 **[Gate 2009]**

543. What are the counting states (Q_1, Q_2) for the counter in Fig. 9.125.

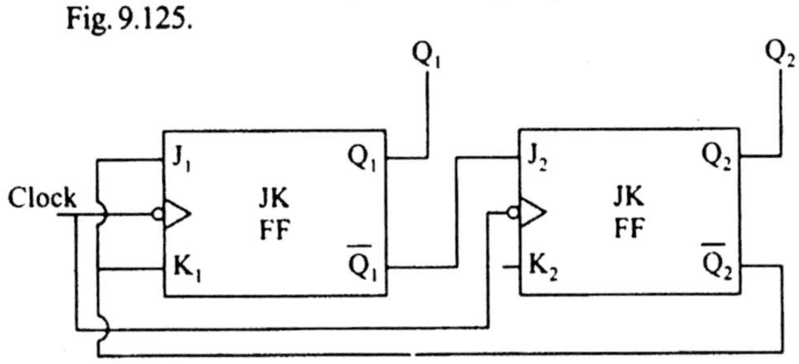

Fig. 9.125.

(a) 11, 10, 00, 11, 10 (b) 01, 10, 11, 00, 01
(c) 00, 11, 01, 10, 11 (d) 01, 10, 00, 01, 10 **[Gate 2005]**

544. Two numbers represented in signed 2's complement form are P = 11101101 and = 11100110. It Q is subtracted from P, the value obtained in signed 2's complement form is
(a) 100000111 (b) 100000111
(c) 11111001 (d) 111111001 **[Gate 2008]**

545.

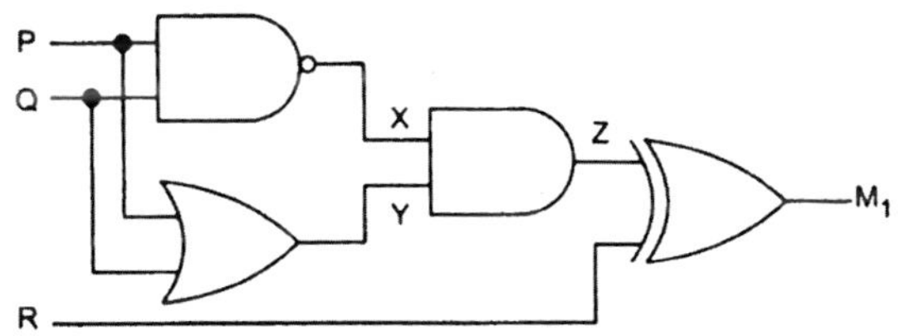

Fig. 9.126.

Which of the following represents relation between P, Q, R, M_1?
(a) M_1 = (P OR Q) XOR R
(b) M_1 = (P AND Q) XOR R
(c) M_1 = (P NOR Q) XOR R
(d) M_1 = (P XOR Q) XOR R **[Gate 2008]**

546.

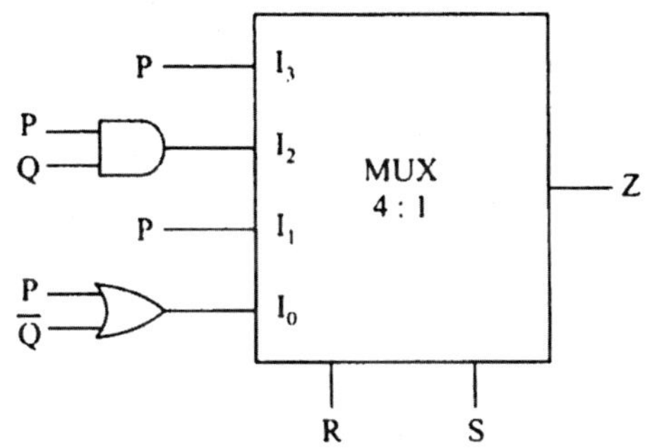

Fig. 9.127.

In Fig. 9.127, $I_0 - I_3$ are inputs to MUX, R, S are control bits. Then Z =

(a) $PQ + P\bar{Q}S + \bar{Q}\bar{R}\bar{S}$ (b) $P\bar{Q} + PQ\bar{R} + \bar{P}\bar{Q}\bar{S}$
(c) $P\bar{Q}\bar{R} + \bar{P}QR + PQRS$
(d) $PQ\bar{R} + PQR\bar{S} + P\bar{Q}\bar{R}S + \bar{Q}\bar{R}\bar{S}$ **[Gate 2008]**

547. Expression

$$Y = \bar{A}\bar{B}\bar{C}D + \bar{A}BC\bar{D} + A\bar{B}\bar{C}D + AB\bar{C}\bar{D}$$

can be minimize to **[Gate 2007]**

(a) $Y = \bar{A}\bar{B}\bar{C}D + \bar{A}BC + A\bar{C}D$
(b) $Y = \bar{A}\bar{B}\bar{C}D + BC\bar{D} + A\bar{B}\bar{C}D$
(c) $Y = \bar{A}BC\bar{D} + \bar{B}\bar{C}D + A\bar{B}\bar{C}D$
(d) $Y = \bar{A}BC\bar{D} + \bar{B}\bar{C}D + AB\bar{C}\bar{D}$

548. X = 01110 and Y = 11001 are two numbers in 2's complement form. Sum of X and Y in 2's complement form using 6 bits is
(a) 100111 (b) 001000
(c) 000111 (d) 101001 **[Gate 2007]**

549. Boolean function Y = AB × CD is to be realized using 2 input NAND gates only. The minimum number of gates required is

(*a*) 2 (*b*) 3
(*c*) 4 (*d*) 5 **[Gate 2007]**

550. In D-A converter of Fig. 9.128, V_R = 10 V and R = 10 kΩ current *i* =

(*a*) 31.25 μA (*b*) 62.5 μA
(*c*) 125 μA (*d*) 250 μA **[Gate 2007]**

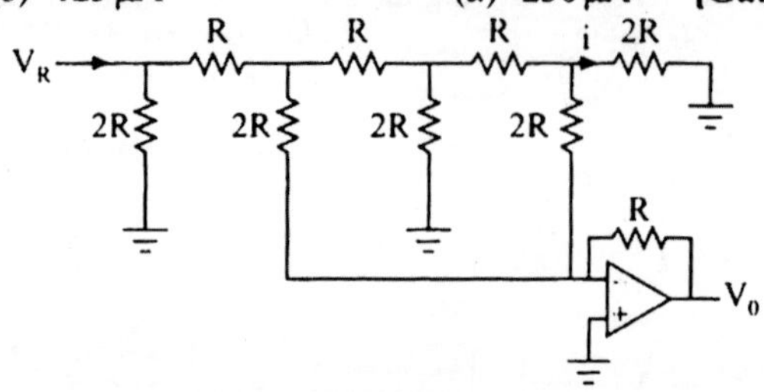

Fig. 9.128.

B. The following questions have two statements Assertion (A) and Reason (R). Examine them and answer as per the following code.

(*a*) Both A and R are correct and R is correct explanation of A
(*b*) Both A and R are correct but R is not correct explanation of A
(*c*) A is true, R is false
(*d*) A is false, R is true

551. **A.** Decoder is not a combinational circuit.
R. Every memory element of semiconductor form has a decoder.

552. **A.** XOR gate is not universal gate.
R. It is not possible to realize any Boolean function using XOR gates only.

553. **A.** 2's complement arithmetic is preferred in digital computers.
R. The hardware required to obtain 2's complement is simple.

554. **A.** ECL gate has very high speed of operation.
R. Transistors in ECL do not go into saturation region.

555. **A.** The carry look ahead adder is very fast.
R. The carry look ahead adder generates the carry and sum digits directly.

556. **A.** Tristate logic is used for bus oriented systems.
R. The outputs of a tristate logic are 0, 1 and indeterminant.

557. **A.** Even if TTL gates and CMOS gates used in a realization have 'the same power supply of + 5 V, suitable circuit is needed to interconnect them.
R. V_{OH}, V_{OL}, V_{IH} and V_{IL} of a TTL gate are respectively 2.4, 0.4, 2 and 0.8 V respectively. If supply voltage is + 5 V. V_{IL} and V_{IH} for CMOS gate (for the supply voltage of + 5 V) are 1.5 V and 3.5 V respectively.

558. **A.** A PROM can be used as a synchronous counter.
R. Each memory location in a PROM can be read synchronously.

559. **A.** Master slave JK flip-flop is commonly used in high speed synchronous circuitry.
R. Master slave JK flip-flop uses two JK flip-flops in cascade.

560. **A.** A demultiplexer cannot be used as a decoder.
R. A multiplexer selects one of many outputs whereas a decoder selects an output corresponding to coded input.

561. **A.** Schottky transistors are preferred over normal transistors in digital circuits.
R. A schottky transistor when used as a switch, switches between cutoff and active region.

562. **A.** The access time of memory is lowest in the case of DRAM.
R. DRAM uses refreshing cycle.

563. **A.** R-2R ladder type DAC has higher speed of operation than weighted resistor DAC.
R. R-2R ladder type DAC uses only two different values of resistors whereas weighted resistor network uses many different values of resistors.

564. **A.** The propagation delay in ECL is minimum.
R. Transistors used in ECL switch between active and cutoff regions.

565. **A.** A demultiplexer can be used as a decoder.
R. A demultiplexer can be built by using AND gates only.

566. **A.** The output of an 8 bit ADC is 80 H if input is 2.5 V.
R. ADC has an output range of 00 to FF H if input range is – 5 to + 5 V.

567. **A.** Synchronous counter has higher speed of operation than ripple counter.
R. Synchronous counter uses high speed flip-flops.

568. **A.** It is possible that a digital circuit gives the same output for different input voltages.
R. A digital circuit is also called logic circuit.

569. **A.** A binary number is a string of zeros and ones only.
R. The base in a binary system is 2.

570. **A.** The use of 2's complement has simplified the computer hardware for arithmetic operations.
R. 2's complement is obtained by adding 1 to 1's complement.

571. **A.** In computers the data is stored in hexadecimal form.
R. Hexadecimal representation is short as compared to binary representation.

572. **A.** BCD code is used in pocket calculators.
R. BCD code is slower than binary.

573. **A.** Gray code is used in shaft position encoding.
R. In gray code only a single bit change occurs when going from one code to another.

574. **A.** Hamming code is commonly used, for error correction.
R. In Hamming code the number of parity bits increases as the number of information bits increases.

575. **A.** Different symbols are used for different gates.
R. IEEE symbols are the same as traditional symbols for gates.

576. **A.** DeMorgan's first theorem replaces a NOR gate by bubbled AND gate.
R. Double complement of a variable equals the variable.

577. A. In hybrid digital circuits the problem of logic race can occur.
R. In two level logic there is no problem of logic race.

578. A. The output of a NOR gate is equal to the complement of the OR of input variables.
R. A XOR gate is a universal gate.

579. A. Boolean expressions can be easily simplified using Karnaugh map.
R. Karnaugh map can be drawn for minterms as well as max terms.

580. A. In totem pole output the output impedance is low.
R. TTL gate with active pull up should not be used in wired AND connection.

581. A. TTL uses multiple emitter transistor.
R. Multiple emitter transistors with about 60 emitters have been developed.

582. A. TTL is a very popular logic in SSI and MSI category.
R. In Schottky TTL the power dissipation is less than in ordinary TTL.

583. A. ECL is fast as compared to TTL.
R. ECL dissipates less power than TTL.

584. A. CMOS devices have very high speed.
R. CMOS devices have very small physical size and simple geometry.

585. A. CMOS devices have very low power consumption.
R. CMOS devices have high noise margin.

586. A. The TTL output acts as a current sink in low state.
R. The TTL input current is largest in low state.

587. A. The advantages of totem pole output are: fast switching and low power consumption.
R. IC packages available are DIP, surface mount and J lead surface mount.

588. A. Power drain of CMOS increases with operating frequency.
R. All unused CMOS inputs should be tied either to a fixed voltage level (o or V_{DD}) or to another input.

589. A. A high density IC has more diodes, transistors and resistors per unit surface area.
R. A high density IC can handle high voltages.

590. A. In a magnitude comparator IC the cascading inputs provide a means to cascade the 4 bit comparators.
R. A magnitude comparator has three outputs, $A = B$, $A > B$ and $A < B$.

591. A. In a BCD to 7 segment decoder the number of outputs active at one time is always the same.
R. A decoder can be used to interface BCD input to LED display.

592. A. An encoder converts key dips to binary code.
R. An encoder has more inputs than output.

593. A. A multiplexer can be used for data routing.
R. A multiplexer has one input and many outputs.

594. A. A 4 input variable logic circuit can be implemented using a 8 : 1 multiplexer.
R. When a multiplexer is used as a logic function generator, the logic design is simple.

595. A. An SR latch has the problem of Race condition
R. When designing a digital circuit Race condition should be avoided.

596. A. A presettable counter can be preset to any desired starting point.
R. The maximum frequency of a ripple counter depends on the modulus.

597. A. Div-64 counter is a Mod-64 counter and divides the input frequency by 64.
R. A Mod-64 counter can be obtained by cascading Mod-16 and Mod-4 counters.

598. A. In a parallel in–serial out shift register data is loaded one bit-at a time.
R. A serial in–serial out shift register can be used to introduce a time delay.

599. A. In a serial in-serial out shift register, access is available only to the left most or right most flip-flops.
R. If the output of a shift register is fedback to serial input it can be used as a ring counter.

600. A. The resolution of a DAC depends on the number of bits.
R. Low resolution leads to fine control.

601. A. TTL NAND gate in tristate output configuration can be used for a bus arrangement with more than one gate output connected to a common line.
R. The tristate configuration has a control input which can detech a logic level (0/1) from coming on to bus line. **[IES 2010]**

602. A. When all the inputs of a NAND gate are shorted to get one input one output gate it becomes an inverter.
R. When all inputs of a NAND gate are are logic 0 level, the output is at logic 1 level. **[IES 2009]**

603. A. D flip-flops are used as buffer register. **[IES 2008]**
R. D flip-flops are free from race around condition.

604. A. To obtain high switching speed in BJT based logic circuits transistors are operated in active region.
R. In active region a transistor works as a linear element. **[IES 2006]**

605. A. The switching speed of ECL is very high. **[IES 2003]**
R. The devices in ECL operate in active region.

606. A. When transistor switches are to be used in an application where speed is a premium, it is better to reduce storage time.
R. It is comparatively easy to reduce storage time than the rise time and fall time of a transistor. **[IES 2003]**

607. A. Asynchronous digital circuits are difficult to design.
R. External clock is used for synchronisation of asynchronous digital circuits. **[IES 2003]**

608. A. Master slave JK flip-flop is preferred to an edge triggered JK flip-flop in high speed circuits.
R. Master slave JK flip-flop is free from race around condition. **[IES 2002]**

609. A. A Look ahead carry adder is a fast adder.
R. A parallel carry adder generates sum digit directly from input digits. **[IES 2001]**

610. A. ECL gate has highest speed of operation as compared to other logic families.
R. ECL gate dissipates more power. **[IES 2001]**

C. State whether the following are True or False

611. If the inputs to a 2 input XOR gale are high and high, the output is high.
612. NAND and NOR gates are universal gates.
613. In a NOT gate the output is always the opposite of the input.
614. Both OR and AND gates can have only two inputs.
615. Each term in POS form is called max terms.
616. In two level logic, logic race does not occur.
617. Karnaugh map cannot be drawn when the number of variables is more than 4.
618. Logic hardware is available only in NAND and NOR.
619. TTL is used in electronic calculators.
620. A multiple emitter transistor has many emitters and collectors.
621. ECL is a saturating logic.
622. In TTL the transistors are driven into and out of saturation.
623. SSI is used in flip-flops.
624. Wired AND connection can be used in TTL with totem pole output.
625. If noise margin is high, there is lesser chance of false operation.
626. A 4 bit binary adder has 4 full adders.
627. Adder is a combinational logic circuit.
628. Binary multiplication can be done by repeated addition.
629. A full adder has two outputs SUM and CARRY.
630. If the inputs to a 3 bit binary adder are 111_2 and 111_2, the output will be 110_2.
631. A VF display operates on the principle of a vacuum diode.
632. TTL XOR gate is very suitable as LCD driver.
633. Zero suppression is not used in actual practice.
634. An encoder converts decimal numbers to binary and other codes.
635. A multi level decoder is more costly as compared to single level decoder.
636. A 16 : 1 multiplexer has 4 select input lines.
637. Available multiplexer IC package can have a maximum of 8 inputs.
638. Multiplexer can be used as logic function generator.
639. A decoder can be used as a demultiplexer.
640. An SR flip-flop can be built using NOR gates or NAND gates.
641. A clock signal coordinates the working of different flip-flops.
642. NAND gates can be used to produce edge triggering.
643. In a JK master slave flip-flop race condition does not occur.
644. Schmitt trigger is used for wave shaping.
645. Computers use RC circuit for edge triggering.
646. In a synchronous counter all flip-flops are clocked together.
647. A ripple counter has propagation delay.
648. A decade counter has 4 flip-flops and skips states 10 to 15.
649. If modulus is less than 2^N, some states of the counter are skipped by using NAND gates.
650. A 4 bit down counter starts counting from 1111 irrespective of modulus.
651. A universal shift register can shift left or right.
652. Shift left and shift right operations occur in a calculator also.
653. In parallel in–parallel out shift register all bits are loaded and read simultaneously.
654. A parallel in–parallel out shift register can be used to introduce delay in digital circuits.
655. Serial in–serial out shift register can be built using D flip-flops.
656. For an N bit ADC, the percentage resolution is $[1/2^N - 1)]100$.
657. Dual slope ADC uses an op-amp comparator.
658. A quantization error is an essential aspect of ADC.
659. A/D parallel converter is also called flash converter.

D. Fill in the Blanks with suitable words.

660. Excess-3 code for decimal number 14 = ...
661. In floating point system the exponent is written in .... notation.
662. If inverters are added to the inputs of AND gate the logic function produced is ....
663. An OR gate can be converted into NAND gate by adding .... to the inputs of the OR gate.
664. If both inputs to an exclusive OR gate are low, the output will be ....
665. An inverter is also called .... gate.
666. A NOR gate is basically a .... gate followed by .... gate.
667. A minterm boolean expression is also known as .... form.
668. In a Karnaugh map a quad eliminates .... variables.
669. A group of wires carrying digital signals is called ....
670. Max term expression directly yield .... logic circuit.
671. An output of +2 V from TTL IC is considered ....
672. LSI and VLSI devices use .... technology.
673. The most important advantage of CMOS is the ....
674. If a CMOS has +10 V supply a typical high output voltage is .... V.
675. In 2's complement representation MSB is the .... bit.
676. One OR gate can work as .... comparator.
677. A 2's complement adder can .... and ....
678. Calculators use .... number system.
679. The decimal number –6 in 4 bit 2's complement representation is...................... .

680. In BCD number system subtraction is done by .... method.
681. In LED the radiation is in .... region.
682. Logic circuit for decimal to BCD encoding has .... gates.
683. Power consumption of LCD is ....
684. RII and RBO on a decoder/driver are for ....
685. The three electrodes of a VF display are called ....
686. A multiplexer with 32 data input lines needs .... select lines.
687. A demultiplexer changes.......... data into .... data.
688. A .... can be used as demultiplexer.
689. In a multiplexer used as logic function generator, ... lines are used as input lines.
690. PROM stands for Programmable read only ....
691. A positive edge triggered flip-flop changes state on the .... transition of the clock pulses.
692. A group of memory elements is called ....
693. Latches and flip-flops are .... logic elements.
694. Clear signal is the same as.......... signal.
695. A .... is a suitable device for waveshaping.
696. A counter having N flip-flops has a modulus of ....
697. The clock signal drives all the flip-flops of a .... counter.
698. A counter is a .... logic circuit.
699. A Mod-10 counter is also called .... counter.
700. If a binary word 1101 is entered serially in a 4 bit serial in–parallel out shift register (which is initially clear) the Q output after 2 clock pulses is ....
701. Serial loading means entering one bit per ....
702. The number of clock pulses required to enter 8 bit data serially is ....
703. The two principle functions of a shift register are .... and ....
704. Copying information into computer memory is called .... into memory.
705. ROM is a .... memory.
706. ROM is a .... density memory device.
707. CD-ROM stands for ....
708. A DAC consists of a .... network and .... amplifier.
709. IC A/D converter is based on .... principle.
710. A typical value of conversion time of an ADC is about ....

ANSWERS WITH HINTS AND EXPLANATIONS

1. (*a*) The octal system has 8 digits 0 to 7.
2. (*b*) It has 16 digits 0 to 15.
3. (*b*) 4. (*b*)
5. (*c*) $1010 = 8 + 0 + 2 + 0 = 10$
6. (*b*) $FF = 15 \times 16 + 15 = 255$.
7. (*c*) Op-amp is an analog IC.
8. (*a*) 9. (*c*) 10. (*a*) 11. (*a*)
12. (*b*) 13. (*a*)
14. (*b*) The indicator of current tracer glows when its tip is held over a pulsating current path.
15. (*a*) It is a multi channel oscilloscope.
16. (*d*) Each has 2 states.
17. (*b*) $2 \times 10 = 20$.
18. (*a*)
19. (*c*) F = 15 in decimal notation = 17 in octal notation $(8 + 7 = 15)$.
20. (*a*) E = 14 in decimal or 1110 $(8 + 4 + 2 + 0 = 14)$ is binary.
21. (*c*) 1101 = 13 in decimal = 15 (*i.e.*, $8 + 5$) in octal.
22. (*c*) 12 in octal = $8 + 2$, *i.e.*, 10 in decimal.
23. (*b*) $32 + 4 + 1 = 37$ in decimal.
24. (*b*) $1010 = 8 + 2 = 10$ in decimal.
25. (*a*) $2048 + 1024 + 128 + 64 + 16 + 8 + 1 = 3289$ in decimal.
26. (*b*)
27. (*b*) See Section 9.13.
28. (*b*) See Section 9.13.
29. (*b*) See Section 9.13.
30. (*a*) 1011 = 11 in decimal and 101 = 5 in decimal, $11 \times 5 = 55$.
31. (*a*) See Section 9.13.
32. (*a*)
33. (*a*) Replace 1 by 0 and 0 by 1.
34. (*a*) 1's complement of 0101 is 1010 and 2's complement is $1010 + 1 = 1011$.
35. (*a*) 36. (*a*)
37. (*a*) $0.5 + 0 + 0.125 + 0.0625 = 0.6875$.
38. (*a*)
39. (*a*) 110.11 = 6.75 and 110 = 6 and $6.75 \times 6 = 40.5$ in decimal.
40. (*a*) 41. (*c*) 42. (*c*) 43. (*b*)
44. (*b*)
45. (*b*) $7BF_{16} = 7 \times 16^2 + 11 \times 16^1 + 15 \times 16^0 = 1983$ in decimal = 0111 1011 1111 in binary.
46. (*b*) E7 F6 in hexadecimal = $14 \times 16^3 + 7 \times 16^2 + 15 \times 16 + 6$ = 59382 in decimal.
47. (*c*) 10C in hexadecimal = $1 \times 16^2 + 12 = 268$ in decimal.
48. (*b*) Convert to decimal, add and change the result to hexadecimal.
49. (*c*) Convert to decimal, subtract and change the result to hexadecimal.
50. (*c*) $A_{16} = 10$, $2_{16} = 2$, $10 \times 2 = 20$ in decimal = 14 in hexadecimal.
51. (*b*) 71 in octal = $7 \times 8 + 1 = 57$ in decimal = 111001 in binary.
52. (*b*)
53. (*b*) FF in hexadecimal = $15 \times 16 + 15 = 255$ in decimal = 0010 0101 0101 in BCD.
54. (*b*) See Table 9.3.
55. (*a*) Change to decimal and then to binary.
56. (*c*) Decimal 8 in excess 3 code = decimal 11 = 1011 in binary.

57. (*b*) Decimal 46 in excess-3 code = 46 + 33 = 79 in decimal = 0111 1001 in 4 bit binary.

58. (*b*)

59. (*d*) ASCII is a 7 bit code.

60. (*c*) EBCDIC is an 8 bit code.

61. (*c*) **62.** (*b*) **63.** (*b*)

64. (*b*) $2^p \geq m + p + 1$. If $m = 11, p$ must be 4 to satisfy this equation.

65. (*b*) The number of parity bits is 3. These bits are located at $2^0, 2^1, 2^2$, *i.e.*, 1, 2, 4th bits starting from LSB.

66. (*d*)

67. (*b*) $4096 = 2^{12}$.

68. (*b*) 23 = 10111 and 0.6 = 10011. Hence 10111 . 10011.

69. (*b*) 0110 = 6 and 0111 = 7. Hence 67.

70. (*b*) $2^4 = 16$.

71. (*b*) OR gate gives high output when one or more inputs are high.

72. (*d*) In only one combination are all inputs high.

73. (*c*)

74. (*b*) All AND inputs must be 1 and inhibit 0 for output to be 1.

75. (*a*) Only one input, *i.e.*, A = 1, B = 1 and S = 0 gives output 1.

76. (*a*) Only one input, *i.e.*, *A* = 0, B = 0 and C = 0 gives 1 as output.

77. (*c*) Both NAND and NOR are called universal gates.

78. (*b*) Only one input, *i.e.*, A = 1, B = 1 and C = 1 gives low output.

79. (*c*) XOR gate recognises odd number of 1's.

80. (*b*) The inputs A and B are ANDed. Inputs B and C also ANDed. The outputs of two AND gates are ORed.

81. (*b*) 6 Input pins, 6 output pins, 1 supply pin and 1 ground pin. Hence 6 NOT gates.

82. (*c*) Each AND gates requires 3 pins, 2 inputs and one output, 1 supply pin and 1 ground pin. Hence 4 AND gates.

83. (*b*) **84.** (*c*) **85.** (*c*) **86.** (*d*)

87. (*d*) This is distributive law.

88. (*c*) 1.0 = 0 and 0.0 = 0.

89. (*c*) A.B = 1 or 0, A + A.B = 0 if A is zero and A + A.B = 1 if A = 1. Hence A + A.B = A.

90. (*c*)

91. (*d*) See Section 9.3.

92. (*c*) **93.** (*c*) **94.** (*a*) **95.** (*b*)

96. (*d*) The min terms are $\bar{A}BC + ABC + A\bar{B}\bar{C} + A\bar{B}C + AB\bar{C}$.

97. (*d*) ABCD = 1111 = m_{15}

98. (*d*) Decimal number 1 = binary number 001 = $\bar{A}\bar{B}C$. Decimal number 7 = binary number 111 = ABC, Decimal number 3 = binary number 011 = $\bar{A}BC$ Decimal number 5 = binary number 101 = $A\bar{B}C$. Hence result.

99. (*b*) It is sum of products.

100. (*a*) This is product of sums expression.

101. (*b*) $AB + A\bar{B} = A(B + \bar{B}) = A.1 = A$.

102. (*b*)

103. (*d*) $2^4 = 16$.

104. (*c*)

105. (*c*) See Table 9.6.

106. (*c*) **107.** (*b*) **108.** (*b*) **109.** (*b*)

110. (*a*) See Table 9.7.

111. (*c*) **112.** (*c*) **113.** (*c*)

114. (*b*) See Table 9.7.

115. (*c*) **116.** (*a*) **117.** (*b*) **118.** (*b*)

119. (*c*)

120. (*b*) Data input and control input.

121. (*b*) Adder is a combinational circuit.

122. (*b*) Rules for binary addition are given in Section 9.1.3.

123. (*a*) See Fig. 9.10.

124. (*b*) **125.** (*b*) **126.** (*b*)

127. (*b*) 2's complement representation requires simple electronic circuitry.

128. (*b*) + 90 in binary is 01011010. Its 2's complement is 10100110.

129. (*b*)

130. (*d*) See Fig. 9.11. **131.** (*b*) See Fig. 9.16.

132. (*d*) See Fig. 9.16.

133. (*c*) 1000 equals decimal 8. Therefore all segments will lit up.

134. (*b*) Decoder converts binary/BCD to alphanumeric.

135. (*b*) **136.** (*d*)

137. (*c*) VF display can operate at very low voltages, has low power consumption and very long life.

138. (*c*)

139. (*c*) It is similar to triode.

140. (*b*) **141.** (*c*)

142. (*b*) $2^2 = 4$. Hence 2 select lines.

143. (*b*)

144. (*b*) Demultiplexer takes data from one line and directs it to any of its N output depending on the status of its select lines.

145. (*b*)

146. (*b*) It is a memory element used in digital circuits.

147. (*a*) See Fig. 9.26.

148. (*a*) See Table 9.15. **149.** (*c*) See Table 9.19.

150. (*c*)

151. (*c*) See Table 9.18. **152.** (*b*) See Table 9.18.

153. (*b*)

154. (*c*) See Table 9.18.

155. (*b*) Mod-4 counter has 4 states, 0 to 3.

156. (*c*) One flip-flop means 2 states and N flip-flops means 2^N states.

157. (*c*) $2^3 = 8$ and $2^4 = 16$. Therefore 4 flip-flops are needed. Some states will be skipped to give a modulus of 10.

158. (*a*) In a ripple counter the effect ripples through the counter.

159. (*d*) $2^4 = 16$.

160. (*c*) A decade counter counts from 0 to 9. It has 4 flip-flops. The states skipped are 10 to 15 or 1010 to 1111.

161. (*c*) $2^3 = 8$. If 3 flip-flops are used the modulus can be a maximum of 8.

162. (*d*) Since 4 flip-flops are used it can count from 0000 to 1111. By pre setting it can start from any number.

163. (*b*)

164. (*b*) It can count from 15 to 0 or 1111 to 0000.

165. (*c*) Since it is an up-down counter it can count upwards or downwards.

166. (*c*) 167. (*b*)

168. (*a*) Storing can be done only in memory and flip-flop is a memory element.

169. (*a*) 170. (*d*)

171. (*c*) Both left to right and right to left operations are possible in universal shift register.

172. (*b*) In binary shift one bit left means multiplication by 2.

173. (*d*) In serial shift register one pulse is needed to store each bit.

174. (*d*)

175. (*d*) In a parallel in register only one pulse is needed to enter data.

176. (*b*) 177. (*d*)

178. (*c*) Both ultra violet and electric pulses are used for erasing.

179. (*b*) 180. (*c*)

181. (*c*) Dynamic RAM has higher packing density and requires lesser space.

182. (*d*) 183. (*c*) 184. (*a*) 185. (*a*)

186. (*c*) 187. (*d*) 188. (*b*)

189. (*c*) Since 1111 = decimal 15, each successive binary count represents $\frac{1}{15}$ of total voltage.

190. (*c*) 191. (*b*) 192. (*b*)

193. (*b*) Analog input can have any value but digital value can have only 2^N discrete levels (for N bits). Hence quantization error in A-D conversion.

194. (*b*) 195. (*c*) 196. (*c*) 197. (*c*)

198. (*a*) 199. (*a*)

200. (*d*) $2^8 = 256$.

201. (*d*) All the switches have to be closed so that the circuit can be made. In AND gate all the inputs have to be high for output to be high.

202. (*c*) OR and NOT = NOR.

203. (*b*) See Section 9.3.2.

204. (*b*) $A' = 1010 + 1 = 1011$.

205. (*d*) A = 11100101. Therefore $\bar{A} = 0001\ 1010$ and $A' = \bar{A} + 1 = 00011011 = 16 + 8 + 2 + 1 = 27$. Therefore $A = -27$.

206. (*c*) Largest positive number is 0111 111 = +127 and largest negative number is 1000 0000 = −128.

207. (*d*) 1101000 = 128 + 64 + 32 + 8 = 232 in decimal = E 8 in hexadecimal.

208. (*d*) 11001 = 16 + 8 + 1 = 25 in decimal.

209. (*d*) 1010 in binary and 8 + 2 = 10 in decimal.

210. (*c*) 01010101 in binary and 64 + 16 + 4 + 1 = 85 in decimal.

211. (*b*) In OR any input high means high output.

212. (*a*) In AND gate all inputs must be high to give high output.

213. (*a*) See Table 9.6. 214. (*d*) See Table 9.6.

215. (*c*) See Table 9.6. 216. (*a*) See Fig. 9.16.

217. (*d*) See Fig. 9.16. 218. (*b*)

219. (*c*) See Table 9.7.

220. (*a*) 221. (*d*) 222. (*c*) 223. (*b*)

224. (*c*) See Fig. 9.10. 225. (*b*) See Fig. 9.11.

226. (*c*) See Fig. 9.10. 227. (*c*) See Fig. 9.11.

228. (*c*) 229. (*d*) 230. (*c*)

231. (*a*) Taking 2's complement twice gives the original number.

232. (*b*) A = 1110, $\bar{A} = 0001$, $A' = 0001 + 1 = 0010 = 2$. Therefore $A = -2$.

233. (*a*)

234. (*b*) See Table 9.12.

235. (*c*) 236. (*c*) 237. (*c*) 238. (*b*)

239. (*a*) See Table 9.12 and Table 9.13.

240. (*c*)

241. (*c*) See Fig. 9.29.

242. (*a*) Decade counter is divide by 10 counter.

243. (*d*) Third state is floating.

244. (*a*) See Table 9.7. 245. (*a*) See Table 9.7.

246. (*b*) 247. (*b*) 248. (*a*) 249. (*c*)

250. (*d*)

251. (*d*) See Table 9.19.

252. (*c*) Hamming code is widely used for error correction.

253. (*a*) 254. (*d*)

255. (*d*) In taking dual OR is replaced by AND and vice versa.

256. (*c*)

257. (*a*) The boolean equation is $F = \overline{\overline{XY\bar{Z}}} + X + Z = \bar{\bar{X}} + \bar{Y} + \bar{\bar{Z}} + X + Z = X + X + \bar{Y} + Z + Z = X + \bar{Y} + Z$. The circuit in Fig. 9.58 (*a*) gives $F = \overline{\bar{X}Y\bar{Z}} = \bar{\bar{X}} + \bar{Y} + \bar{\bar{Z}} = X + \bar{Y} + Z$.

258. (*a*)

259. (*b*) High speed is obtained because saturation state is avoided.

260. (*a*) 261. (*d*) 262. (*c*)

263. (*c*) See Table 9.8.

264. (*b*) **265.** (*d*) **266.** (*d*) **267.** (*b*)

268. (*c*) **269.** (*d*)

270. (*d*) Flip-flop has more than 1 input.

271. (*c*) **272.** (*b*)

273. (*d*) See Fig. 9.1. **274.** (*c*) See Fig. 9.1.

275. (*b*)

276. (*c*) NOR gate gives high output if all inputs are Low. For all other combinations of inputs, output is Low.

277. (*c*) NAND gate gives Low output if all inputs are High. For other combinations of inputs, output is High.

278. (*a*) See Table 9.5. **279.** (*c*) See Table 9.5.

280. (*c*) See Table 9.5.

281. (*c*)

282. (*c*) Since C = 0 we get $\overline{C}$. Therefore $Y = A B \overline{C}$.

283. (*c*) ABCD = 1111 = m_{15}.

284. (*c*) See Fig. 9.1 Complement of XOR is X NOR.

285. (*a*) Output = $\overline{\overline{A+B+C+D}} = A+B+C+D$.

286. (*b*) The CD inputs when fed to NOR gate give output 1.

Therefore $Y = \overline{1.1.1} = 0$.

287. (*c*) $\overline{A}\overline{B}\overline{C}\overline{D} + \overline{A}\overline{B}C\overline{D} = \overline{A}\overline{B}\overline{D}(C+\overline{C}) = \overline{A}\overline{B}\overline{D}$.

288. (*a*) $A+B+C = 000 = M_0$.

289. (*b*) $Y = \overline{\overline{A}+\overline{B}+\overline{C}+\overline{C}} = \overline{\overline{A}}\,\overline{\overline{B}}\,\overline{\overline{C}} = ABC$.

290. (*a*) **291.** (*a*)

292. (*a*) $2^4 = 16$.

293. (*d*)

294. (*a*) NOR gate gives High output when all inputs are Low.

295. (*c*)

296. (*b*) It denotes high speed TTL.

297. (*b*) Because of low power consumption CMOS is used in pocket calculators.

298. (*c*) Since it is non-saturating, ECL has low propagation delay.

299. (*c*) **300.** (*d*)

301. (*d*) $\dfrac{4}{0.4} = 10$.

302. (*a*) Output of driver stage has to be matched with input of load.

303. (*a*)

304. (*b*) See Fig. 9.14.

305. (*b*) 99 – 12 = 87.

306. (*a*) 99 – 56 = 43.

307. (*b*)

308. (*c*) Decimal 8 = 1000 and – 8 = 11000.

309. (*c*)

310. (*a*) Inputs are carry from lower bits and two other bits. Outputs are SUM and CARRY.

311. (*b*) **312.** (*b*)

313. (*a*) Since the number of 1's is even, output is 0.

314. (*a*)

315. (*b*) Only 2 segments are lit as seen in Fig. 9.16.

316. (*a*) **317.** (*b*) **318.** (*c*) **319.** (*a*)

320. (*b*)

321. (*a*) Minimum number of outputs when input is decimal 1 and maximum number of outputs when input is decimal 8.

322. (*b*)

323. (*a*) Three NOT gates, four AND gates and one OR gate, *i.e.*, total of 8 gates.

324. (*a*) $2^4 = 16$.

325. (*c*) Both need select inputs as discussed in Section 9.9.

326. (*a*)

327. (*a*) Edge triggering means the instant when clock transition occurs.

328. (*a*) Preset and clear inputs are not applied in any fixed sequence.

329. (*b*) One shot means one stable state.

330. (*d*)

331. (*b*) This the main difference between latch and flip-flop. Only flip-flop has clock input.

332. (*c*) See Section 9.10.

333. (*a*) See Section 9.10.

334. (*a*)

335. (*b*) $2^9 = 512$ and $2^{10} = 1024$. Hence at least 10 flip-flops are required.

336. (*d*) At pulse 1 the state changes to 0001 and at pulse 2 it changes to 0010. Since it has a total of 16 states, the counter returns to 0000 after 16 and 32 pulses.

337. (*c*) Total time delay = 40 × 4 = 160 × 10^{-9} s. Hence

$$f = \frac{1}{160 \times 10^{-9}} = 6.25 \text{ MHz}.$$

338. (*a*) In a synchronous counter clock pulses are applied to all flip-flops simultaneously. Hence minimum time delay and high frequency.

339. (*a*) $2^6 = 64$. Hence 6 flip-flops. The number of AND gates is 6 – 2 = 4. See Fig. 9.37.

340. (*b*) Since propagation delays of all flip-flops are added in ripple counter, the maximum frequency depends on the number of flip-flops which depends on modulus.

341. (*c*) 16 × 16 × 16 = 4096.

342. (*c*) $Y = \overline{A}\,\overline{B} = \overline{A+B}$.

343. (*b*) Output $A(B+\overline{C}) + \overline{A}B + C(A+\overline{B}) = AB + A\overline{C} + \overline{A}B + AC + \overline{B}C = B + A + \overline{B}C = A + B + C$.

344. (*b*) $2^3 = 8$.

345. (*c*)

346. (*b*) In look ahead carry adder the carry is directly derived from the gates when original inputs are being added. Hence the addition is fast. This process requires more gates and is costly.

347. (*b*) The largest negative number is 1000 0000 = – 128.

348. (*d*) **349.** (*a*)

350. (*d*) $(1024 \times 8)/(256 \times 1) = 32$.

351. (*b*) Output $= \dfrac{101001}{111111} \times 10 = 6.41$ V.

352. (*c*)

353. (*b*) See Section 9.14.

354. (*c*) **355.** (*b*) **356.** (*c*) **357.** (*d*)

358. (*a*) Combination circuit does not have memory.

359. (*a*)

360. (*c*) Modulus 13 × modulus 6 = modulus 78.

361. (*d*)

362. (*d*) See Fig. 9.49.

363. (*a*) Resolution $= \dfrac{10.23}{2^{10}-1}$ V or 10 mV.

364. (*a*) **365.** (*c*) **366.** (*b*)

367. (*a*) In a synchronous counter clock input is applied to all flip-flops simultaneously. Hence total delay time is 15 ns.

368. (*b*)

369. (*d*) See Fig. 9.1.

370. (*b*) Since V (1) is lower state than V (0) it is a negative logic circuit. Since diodes are in parallel, it is an OR gate.

371. (*d*) When any input is High, the corresponding transistor conducts and output is Low. Hence it is a NOR gate.

372. (*a*) See Fig. 9.3. **373.** (*a*) See Fig. 9.3.

374. (*b*) **375.** (*a*) **376.** (*b*)

377. (*a*) $Y = \overline{\overline{A+B} + \overline{C+D} + \overline{E+F}}$

$= (A + B)(C + D)(E + F)$.

378. (*a*) **379.** (*a*)

380. (*b*) Output $= \overline{A\overline{BC}} = \overline{A} + BC$.

381. (*a*) **382.** (*b*) **383.** (*a*)

384. (*d*) $Y = \overline{A}\,\overline{B}\,\overline{C} + \overline{A}\,\overline{B}C + A\overline{B}\,\overline{C} + A\overline{B}C$

$= \overline{A}\,\overline{B}\,(\overline{C} + C) + A\overline{B}\,(\overline{C} + C)$

$= \overline{A}\,\overline{B} + A\overline{B} = \overline{B}\,(A + \overline{A}) = \overline{B}$.

385. (*a*) B and C in parallel give B + C. Similarly D and E in parallel give D + E. (B + C) in series with (D + E) give (B + C) (D + E). Since F is in parallel we get F + (B + C) (D + E). Finally A is in series. Therefore we get A [F + (B + C) (D + E)].

386. (*d*) $2^{14} = 16384$.

387. (*d*) $2^{12} = 4096$.

388. (*a*) **389.** (*c*)

390. (*c*) The number of gates may be any.

391. (*a*) **392.** (*b*)

393. (*c*) Unused inputs should be given high input.

394. (*c*) **395.** (*a*)

396. (*a*) $Y = \overline{A\overline{B}C} + \overline{A} + \overline{C}$

$= \overline{A} + B + \overline{C} + \overline{A} + \overline{C} = \overline{A} + \overline{C} + B = \overline{A\overline{B}C}$.

397. (*a*) $Y = \overline{\overline{\overline{ABC}\,DE}} = \overline{\overline{ABC}} + \overline{DE} = \overline{ABC} + \overline{DE}$

or $Y = (\overline{A} + \overline{B})\,C + \overline{DE}$.

398. (*b*)

399. (*c*) Output $= \dfrac{0110}{1001} \times 4.5 = \dfrac{6}{9} \times 4.5 = 3$ V.

400. (*b*) Maximum error $= \dfrac{5 \times 0.2}{100} \times 1000 = 10$ mV.

401. (*c*) **402.** (*b*) **403.** (*b*)

404. (*a*) Schottky transistors have low switching time and hence low propagation delay.

405. (*b*) See Fig. 9.11.

406. (*b*) $2^3 = 8$ and $2^4 = 16$. Hence minimum of 4 memory elements.

407. (*d*) In ripple counter all the delays are added.

408. (*a*) $\dfrac{10}{2^{12}-1} \times 1000 = 2.44$ mV.

409. (*a*) $3 \times 16^2 + 14 \times 16^1 + 8 = 1000$.

410. (*d*) **411.** (*c*)

412. (*b*) See Fig. 9.14.

413. (*b*) Convert to binary and compare.

414. (*b*)

415. (*b*) Transistor is off. $V_0 = \dfrac{R_L}{R_L + R_C} V_{CC} = \dfrac{1}{2} \times 5 = 2.5$ V.

416. (*a*) $Y = \overline{\left[\overline{(A+B)C}\right]DE}$

$= \overline{\overline{(A+B)C}} + \overline{\overline{DE}} = (A + B)C + DE$.

417. (*d*)

418. (*c*) $\left[\dfrac{5}{2^N - 1}\right] 1000 = 5$ or N = 10.

419. (*a*) NAND gate gives high output when both inputs are Low.

420. (*a*) $X\,\overline{X} = 0$.

421. (*a*) Transistor is off and circuit current is zero.

422. (*a*) See Table 9.7.

423. (*b*)

424. (*a*) After 37 clock pulses the state will be the same as after (37 – 32), *i.e.*, 5 clock pulses. 6 + 5 = 11. Hence 1011.

425. (*c*) Frequency has to be divided by 10.

426. (*a*) **427.** (*b*)

428. (*c*) $Y = \overline{\overline{A}\overline{B}\overline{C}} + A + C$

$= \overline{\overline{A}} + A + \overline{B} + \overline{\overline{C}} + C = A + \overline{B} + C.$

In Fig. 9.103 (*c*) $Y = \overline{\overline{A}B\overline{C}} = A + \overline{B} + C$.

429. (*a*) **430.** (*b*)

431. (*b*) It is a half adder 1 + 1 = 10. Therefore SUM = P = 0 and Carry = Q = 1.

432. (*c*) It converts parallel data into serial data.

433. (*c*) **434.** (*d*)

435. (*c*) $2^5 = 32$.

436. (*b*) **437.** (*a*) **438.** (*a*) **439.** (*b*)

440. (*c*) Shift registers are used for all these applications.

441. (*a*)

442. (*a*) Time delay = $50 \times 4 \times 10^{-9}$ s, $f = \dfrac{1}{50 \times 4 \times 10^{-9}} = 5$ MHz.

443. (*c*) Output of XOR gate is input to register.

444. (*b*)

445. (*d*) Total delay = $\dfrac{1}{10^{-6}}$ = 1 μs = 10 × delay of one flip-flop. Therefore delay for one flip-flop = 0.1 μs.

446. (*b*) **447.** (*b*) **448.** (*a*) **449.** (*a*)

450. (*c*)

451. (*d*) $Q = 20 \times 10^{-3} \times 0.1 \times 10^{-12} = 2 \times 10^{-15}$ C and

$C = \dfrac{Q}{v} = \dfrac{2 \times 10^{-15}}{0.5} = 4 \times 10^{-15}$ F.

452. (*a*) LSB = $\dfrac{10.24 \times 1000}{1024}$ = 10 mV, accuracy = 5 mV, temp. coeff. = $\dfrac{5 \times 1000}{25}$ = 200 μV/°C.

453. (*c*) $\dfrac{16 \times 1024 \times 8}{4096 \times 4} = 8$.

454. (*a*) **455.** (*c*)

456. (*b*) Fan out in low state = $\dfrac{8}{0.1}$ = 80. Fan out in high state $= \dfrac{0.4 \times 1000}{20}$ = 20. Hence fan out is smaller of two values, *i.e.*, 20.

457. (*d*) If A = 0, Y = 1 and A = 1, Y = 0. Therefore $Y = \overline{A}$.

458. (*c*) B + C are in parallel and A is in series with this parallel combination. Similarly D + E are in series. Then D, E are in parallel with A, B and C. $Y = \overline{A(B + C) + DE}$.

459. (*c*) **460.** (*c*) **461.** (*b*) **462.** (*a*)

463. (*b*)

464. (*a*) Change OR to AND and vice versa.

465. (*c*) Third clock pulse resets the counter to 00 state. Hence mod is 3.

466. (*a*)

467. (*b*) Verify by preparing truth table.

468. (*d*) **469.** (*b*) **470.** (*a*)

471. (*a*) mode 12 counter is divide by 12 counter. Output frequency = $\dfrac{10}{12}$ = 0.833 kHz.

472. (*c*) See Table 9.13. **473.** (*b*) See Table 9.13.

474. (*a*) See Table 9.13. **475.** (*c*) See Table 9.12.

476. (*a*) See Table 9.12.

477. (*c*) The small circle indicates negative and small triangle indicates edge triggering.

478. (*a*)

479. (*d*) See Table 9.19.

480. (*b*) **481.** (*b*)

482. (*c*) After 16 pulses the counter resets. 0011 is 3. Hence clock pulses can be 3 or 3 + 16 or 3 + 16 + 16.

483. (*b*) When counter is 110 the counter resets. Hence mod-6.

484. (*c*) Maximum delay = 50 + 20 = 70 × 10^{-9} s. Hence $f = \dfrac{1}{70 \times 10^{-9}}$ = 14.3 MHz.

485. (*a*) In ripple counter the clock pulses are applied to one flip-flop only. Hence as the number of flip-flops increases the delay increases. In synchronous counter clock pulses to all flip-flops are applied simultaneously.

486. (*d*) Circuitry of ripple counter is simpler than that of synchronous counter.

487. (*c*) 4 Flip-flops and 2 AND gates.

488. (*d*) Number of flip-flops = $\dfrac{1}{12 \times 10^{-9} \times 10 \times 10^{6}} = \dfrac{1000}{120}$ = 8.333 say 8. Modulus = 2^8 = 256.

489. (*c*) $2^6 = 64$. So we need 6 flip-flops and (6 – 2) = 4 AND gates.

490. (*c*) IC 74193 is a divide by 16 counter. Since 4 stages are used, counting range = 16^4 = 65536.

491. (*b*)

492. (*d*) Resistances are R, 2R, 4R, 8R, 16R and 32R. LSB resistance = 32R = 32 × 20 = 640 K ohm.

493. (*a*) 10000000 = 128, 11111111 = 255. If there is no error, output = $\dfrac{128}{255}$ × 2000 = 1004 μA. Maximum error = $\dfrac{2000 \times 0.5}{100}$ = ± 10 μA. Hence range of output 994 to 1014 μA.

494. (*c*) **495.** (*b*)

496. (*a*) Bits are 40M, 16M and 32M.

497. (*b*) **498.** (*b*) **499.** (*c*) **500.** (*c*)

501. (*a*) Octal number 657 is equivalent to decimal number

6	5	7	Octal number
8^2	8^1	8^0	Weights

$657_8 = 6 \times 8^2 + 5 \times 8^1 + 7 \times 8^0 = 6 \times 64 + 5 \times 8 + 7 = 431$

$431_{10} = 16 \underline{| \ 431 \ } = 1AF_H$
$\quad 16 \underline{| \ 26-15 \ }$
$\quad\quad 1-10$

502. (*b*) $\overline{A+\overline{B}+C}+\overline{\overline{A}+\overline{B}+C}+\overline{A+\overline{B}+\overline{C}}+ABC$

$= \overline{A}B\overline{C} + AB\overline{C} + \overline{A}BC + ABC$

$= \overline{A}B(\overline{C} + C) + AB(\overline{C} + C)$

$= \overline{A}B + AB = B(\overline{A} + A) = B.$

503. (*d*) $ABC + AB\overline{C} + \overline{A}BC + \overline{A}\overline{B}C$

$= AB(C + \overline{C}) + \overline{A}C(B + \overline{B}) = AB + \overline{A}C$

Complement of $AB + \overline{A}C$

$= \overline{AB + \overline{A}C} = (\overline{A} + \overline{B})(A + \overline{C}).$

504. (*a*) All the variables have to be in each term

$Y = A\overline{B} + AC + BC$

$= A\overline{B}(C + \overline{C}) + AC(B + \overline{B}) + BC(A + \overline{A})$

$= A\overline{B}C + A\overline{B}\overline{C} + ACB + AC\overline{B} + ABC + \overline{A}BC$

$= A\overline{B}C + \overline{A}BC + A\overline{B}\overline{C} + ABC$

505. (*b*) It is called speed-power product.

506. (*d*) $X + X\overline{Y} + X\overline{Y}Z = X(1 + \overline{Y} + \overline{Y}Z) = X$

We do not require any logic gate to implement this only a wire is sufficient.

507. (*c*) CMOS has high fan out, ECL has highest speed of operation I^2L has lowest speed power product.

508. (*b*) Max. error $= \dfrac{5 \times 0.2}{100} = 10$ mV.

509. (*a*) $X = \overline{(\overline{A}B)(\overline{C}\overline{D})(\overline{E}\overline{F})}$

$= \overline{\overline{AB}} + \overline{\overline{CD}} + \overline{\overline{EF}} = AB + CD + EF$

510. (*b*) $Z = (A + I)(\overline{A} + J)$

Let I = 1 and J = B, then $Z = (A + 1)(\overline{A} + B) = \overline{A} + B$.

511. (*a*) $Y = \overline{A \oplus B\,.\,C} = \overline{A \oplus B} + \overline{C} = A \odot B + \overline{C}$

or $Y = \overline{A}\overline{B} + AB + \overline{C}$.

512. (*b*) Size of ROM = M × N, where M = Number of locations, N = Number of bits at each location. N is given as 8.

513. (*a*)

A	B	Y
0	0	0
0	1	1
1	0	1
1	1	0

or $Y = \overline{A}B + A\overline{B} = A \oplus B$

Y	C	X
0	0	0
0	1	1
1	0	1
1	1	0

$X = \overline{Y}C + Y\overline{C} = Y \oplus C$

or $X = \overline{A \oplus B}\,C + (A \oplus B)\overline{C}$

$= (\overline{A}\overline{B} + AB)C + (\overline{A}B + A\overline{B})\overline{C}$

$= \overline{A}\overline{B}C + ABC + \overline{A}B\overline{C} + A\overline{B}\overline{C}$

$= A\overline{B}\overline{C} + \overline{A}B\overline{C} + \overline{A}\overline{B}C + ABC$

514. (*c*) PROM contains fixed AND array and programmable OR array.

515. (*a*) $f(ABC) = A\overline{B}\overline{C} + \overline{A}\overline{B}C + B\overline{C}$

A \ BC	$\overline{B}\overline{C}$ 00	$\overline{B}C$ 01	BC 11	$B\overline{C}$ 10
0		1		1
1	1			1

$f(ABC) = \Sigma\,(1, 2, 4, 6)$

516. (*d*) High packing density.

517. (*d*) Max. peak-to-peak ripple $= \dfrac{5}{2^n - 1} = \dfrac{5}{2^{10} - 1} \simeq 5$ mV.

518. (*c*) See Table 9.5.

519. (*d*) Hardware implementation is the aim.

520. (*c*) Compatibility and safe operation are most important.

521. (*c*) Fan out and fan in may not be same.

522. (*a*) NAND and NOR

523. (*d*) See Fig. 9.7.

524. (*d*) CMOS has this capability.

525. (*d*) Flip-flop and counter are sequential circuits.

526. (*a*) $f = X\overline{Z} + YZ$. Let X = R, Y = 1, Z = T, then

$f = R\overline{T} + 1\,.\,T = R\overline{T} + T.$

As seen in Table 9.5 $R\overline{T} + T = R + T$.

527. (*d*) **528.** (*d*)

529. (*a*) Propagation delay time is between 50% values.

530. (*c*) Only 1 bit of information can be stored.

531. (*d*) When it is not edge triggered D flip-flop can be used as a latch.

532. (*a*) Flip-flop is always a digital circuit.

533. (*b*) Both are permanent storage devices. Both store binary word. Thus statements 2 and 4 are wrong.

534. (*b*) Serial addition is not possible in universal shift register.

535. (*a*) Flash counter is fastest converter. Dual slope A/D converter uses op-amp as integrating amplifier.

536. (*d*) P is NOR gate and is equivalent to Invert AND gate (Number 4 in column B) as per DeMorgan's theorem. Q is NAND gate and is equivalent to invert OR gate (Number 2 in column B as per DeMorgan's theorem. R is XOR gate and is equivalent to 3 in column B.

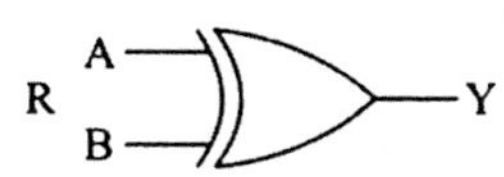

$Y = \overline{A}B + A\overline{B}$

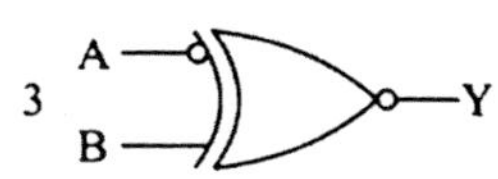

$Y = \overline{A}B + \overline{\overline{A}}\,\overline{B} = \overline{A}B + A\overline{B}$

537. (*a, b c*) All the three *a, b, c* are correct.

$$F = (A \oplus B) \odot (A \odot B) \odot C$$

$$= [(A \oplus B)(A \odot B) + (A \odot B)(A \oplus B)] \odot C$$

$$= [(A \oplus B)(A \odot B) \odot C = 0 \odot C = 0.C + 1.\bar{C} = \bar{C}$$

Hence C must be 0 so that $F = \bar{C} = 1$.

538. (*d*) $F(A, B, C, D) = \bar{A}\bar{B}C + \bar{A}BD + A\bar{B}\bar{C} + AB(\bar{C}\bar{D})$

$= \bar{A}\bar{B}C(D+\bar{D}) + \bar{A}BD(C+\bar{C}) + A\bar{B}\bar{C}(D+\bar{D}) + ABC\bar{D}$

Karnaugh map for this expression is as below:

AB \ CD	00	01	11	10
00			1	1
01		1	1	
11	1			
10	1	1		

$F = \Sigma m\,(2, 3, 5, 7, 8, 9, 12)$

539. (*c*)

540. (*d*) $[1 + Z\{\bar{Y} + (\bar{Z} + X\bar{Y})\}]\{0 + \bar{Z}(1+Y)\} = 0$

or $1[\bar{Z}(1+Y)] = 1$ or $\bar{Z} = 1$ or $Z = 0$.

541. (*a*) One 2 × 1 MUX can generate AND gate. Two 2 × 1 MUX are required for EX-OR gate.

542. (*c*) NAND gates inputs (0, 1), (1, 1); outputs (1, 0), (1, 0) NOR gates inputs (0, 1), (1, 1); Outputs (1, 0), (0, 0).

543. (*a*)

Clock	J_1	K_1	J_2	K_2	Q_1	Q_2
0	1	1	1	1	0	0
1	1	1	1	1	1	1
2	0	0	0	1	1	0
3						

Thus the counting sequence is 111000111000.

544. (*b*) Signed 2's complement of P = 11101101.

Hence P = 00010011

Signed 2's complement of Q = 11100110

P − Q = P + 2's complement of Q

P	00010011
+(−Q)	11100110
	11111001
2's complement	00000111

545. (*d*) $M_1 = [\overline{PQ}(P+Q)] \oplus R$

$= [(\bar{P}+\bar{Q})(P+Q)] \oplus R = (P \oplus Q) \oplus R$

546. (*a*) $Z = PRS + PQR\bar{S} + P\bar{R}S + (P+\bar{Q})\bar{R}\bar{S}$

Karnaugh map is as under:

PQ \ RS	00	01	11	10
00	1			
01				
11	1	1	1	1
10	1		1	1

PQ, $\bar{Q}\bar{R}\bar{S}$, $P\bar{Q}S$

Hence $Z = PQ + P\bar{Q}S + \bar{Q}\bar{R}\bar{S}$.

547. (*d*) $Y = \bar{A}BC\bar{D} + AB\bar{C}\bar{D} + \bar{B}\bar{C}D$

AB \ CD	00	01	11	10
00		1		
01				1
11	1			
10		1		

548. (*c*)

X = 0 1 1 1 0

Y = 1 1 0 0 1

Sum X + Y = 0 0 1 1 1

In 6 bit representation 000111.

549. (*c*) $Y = AB + CD = \overline{\overline{AB}} + \overline{\overline{CD}}$

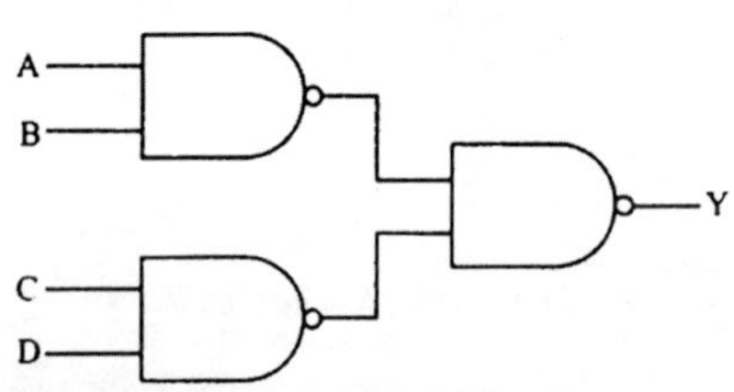

550. (*b*) Because of virtual ground at inverting terminal of op-amp, last 2 resistances of 2R each are in parallel and in series with R. Net resistance 2R. This 2R resistance is in parallel with next 2R and in series with R. After solving complete circuit it is found that net resistance of complete circuit is R. So

$I = \dfrac{10}{10 \times 10^3} = 1$ mA. The subdivision of 1 mA current is as shown below.

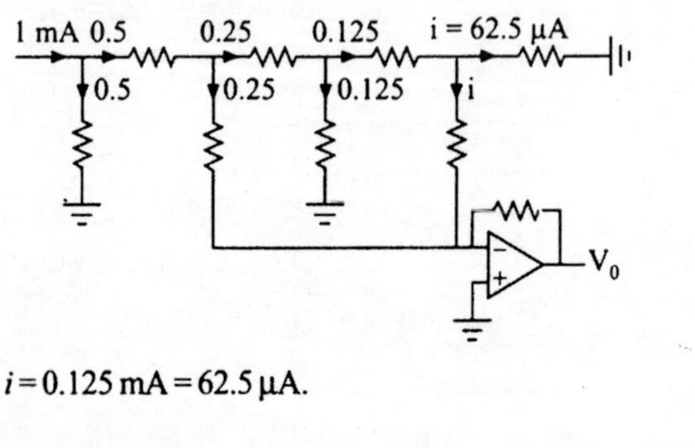

$i = 0.125$ mA $= 62.5$ µA.

551. (*b*) Both are correct but independent.

552. (*a*) A universal gate is one which can be used to realize any Boolean function using that type of gates only. Only NOR and NAND are universal gates.

553. (*c*)

554. (*a*) Since transistors in ECL do not enter saturation slate, switching is fast.

555. (*c*)

556. (*c*) Outputs of tristate logic are 0, 1 and high impedance state.

557. (*a*) Interfacing is necessary and interfacing depends on gate parameters like V_{OH}, V_{OL}, I_{IH}, I_{IL}.

558. (*d*) PROM is a memory. It is not a counter.

559. (*d*) M-S, J-K flip-flop is not used very commonly.

560. (*d*) Demux can be used as a decoder.

561. (*a*) Schottky transistors are fast switching because saturation is avoided.

562. (*d*) DRAM has lower speed than SRAM.

563. (*b*) Both are correct but independent.

564. (*a*) Since saturation is avoided switching speed is high.

565. (*c*) Demultiplexer requires NOT gates also in addition to AND gates.

566. (*c*)

567. (*c*) Higher speed is due to the reason that clock pulses in synchronous counter are applied simultaneously to all the flip-flops.

568. (*b*) Both are correct but independent.

569. (*a*)

570. (*b*) Both are correct but independent.

571. (*d*) Data is stored in binary form only.

572. (*b*) Both are correct but independent.

573. (*a*) **574.** (*b*)

575. (*c*) IEEE symbols are different than traditional symbols.

576. (*b*) **577.** (*b*)

578. (*c*) XOR gate is not a universal gate.

579. (*b*) Both are correct but independent.

580. (*b*) **581.** (*b*)

582. (*c*) In Schottky TTL storage time is reduced.

583. (*c*) ECL dissipates more power than TTL.

584. (*d*) CMOS has high packing density but low speed.

585. (*b*) Both are correct but independent.

586. (*b*) **587.** (*b*)

588. (*b*) Both are correct but independent.

589. (*c*) R is not correct.

590. (*b*)

591. (*d*) A is not correct.

592. (*b*)

593. (*d*) Multiplexer has many inputs but one output.

594. (*b*) Both are correct but independent.

595. (*b*) **596.** (*b*) **597.** (*b*)

598. (*d*) A is not correct.

599. (*b*) Both are correct but independent.

600. (*b*)

601. (*c*) A is correct but R is wrong.

602. (*b*) Both are correct but independent.

603. (*b*) Both are correct but independent.

604. (*b*) To obtain high switching speed a transistor is not allowed to go to saturation state. Otherwise, storage delay time would be high. Thus, transistor operates in active or cutoff state. In active region transistor can be treated as a linear element.

605. (*a*) When a transistor enters saturation state, the storage delay time is high. ECL operates only in active and cutoff states thus ensuring high speed.

606. (*a*) When a transistor is in saturation region, the time in that region is called storage time.

607. (*c*) External clock is not needed in asynchronous digital circuits.

608. (*b*) Both are correct but independent.

609. (*a*)

610. (*b*) Both are correct but independent.

611. F, Output is Low.

612. T **613.** T

614. F, Any number of inputs are possible.

615. T **616.** T

617. F, K map can be drawn for more than 4 variables also.

618. F, Other gates are also available.

619. F, Calculators use CMOS.

620. F, It has many emitters but one collector.

621. F, It is a non-saturating logic. Hence highest speed of operation.

622. T **623.** T

624. F, No it cannot be used.

625. T

626. F, No see Fig. 9.12.

627. T **628.** T

629. T

630. F, 111 + 111 = 1110.

631. F, It operates on the principle of vacuum triode.

632. F

633. F, Zero suppression is commonly used.

634. T

635. T, A multi level decoder may be cheaper than single level decoder.

636. T, $2^4 = 16$.

637. F, 16 : 1 IC multiplexer is also available.

638. T **639.** T

640. T, See Fig. 9.23 and Fig. 9.24.

641. T **642.** T **643.** T **644.** T

645. F, No, gates are used for edge triggering.

646. T **647.** T **648.** T **649.** T

650. F, The starting count depends on modulus. 4 bit decade down counter starts counting from 1010.

651. T **652.** T **653.** T

654. F, No time delay in this shift register since all bits are loaded and read simultaneously.

655. T **656.** T **657.** T **658.** T

659. T **660.** T

661. excess 32 **662.** NOR

663. inverters **664.** High

665. NOT **666.** OR, NOT

667. sum of products **668.** two

669. bus **670.** OR-AND

671. undefined **672.** MOS

673. low power consumption

674. +10 **675.** sign

676. one bit **677.** add, subtract

678. BCD **679.** 1010

680. 10's complement **681.** infra-red

682. OR **683.** very small

684. zero suppression **685.** plate, grid, cathode

686. 5 **687.** serial, parallel

688. decoder **689.** select

690. memory **691.** low to high

692. register **693.** bistable

694. reset **695.** schmitt trigger

696. 2^N **697.** synchronous

698. sequential **699.** decade

700. 0100 **701.** clock pulse

702. 8 **703.** data storage, shifting

704. writing **705.** non-volatile

706. high

707. compact disc read only memory

708. resistor, summing

709. successive approximation

710. 10 μs

CHAPTER 10

CONTROL SYSTEMS

10.1. OPEN AND CLOSED LOOP SYSTEMS

An open loop system is one in which the output has no effect on the control action. As the input or the parameters of the device change, the output changes. The output is neither measured nor fedback to the input.

A closed loop system (also feed back system) is one in which a fraction of the output is fed back to the controller so as to keep the output constant.

10.2. ANALOGOUS SYSTEMS

The differential equations in electrical, mechanical, thermal and many other systems are similar. Therefore these systems are analogous.

10.2.1. Mechanical Translational Systems

When a force f is applied to a mass M, the velocity u is given by

$$f = M\frac{du}{dt} \qquad ...(10.1)$$

Equation (10.2) is similar to v-i relation for inductance, *i.e.*,

$$v = L\frac{di}{dt} \qquad ...(10.2)$$

Thus mass in analogous to inductance, force to voltage and velocity to current. This is called force voltage or classical analogue. For compliance C_m (reciprocal of C_m is K, *i.e.*, the spring stiffness) the elongation x is given by

$$x = C_m f$$

$$\text{or} \qquad f = Kx \qquad ...(10.3)$$

If the two terminals of spring are moving with velocities u_2 and u_1 the relative velocity $u = u_2 - u_1$. Then

$$f = K\int(u_2 - u_1)dt = K\int u dt = \frac{1}{C_m}\int u dt \qquad ...(10.4)$$

Equation (10.4) is similar to v–i relation for capacitance C, *i.e.*,

$$v = \frac{1}{C}\int i dt \qquad ...(10.5)$$

Thus spring compliance is analogous to capacitance.

For mechanical resistance (friction)

$$f = Bu \qquad ...(10.6)$$

where B is viscous friction coefficient or mechanical resistance. Equation (10.6) is similar v-i equation for resistance R, *i.e.*,

$$v = Ri$$

Thus mechanical resistance is analogous to electrical resistance. Table 10.1 lists analogous quantities in force-voltage analogy.

Table 10.1 Analogous Quantities in Force-voltage Analogy

Mechanical quantity	*Electrical analogous quantity*
Force f	Voltage v
Velocity u	Current i
Displacement x	Charge q
Mass M	Inductance L
Mechanical resistance B	Resistance R
Compliance C_m	Capacitance C
Spring constant $K = \frac{1}{C_m}$	$\frac{1}{C}$
Kinetic energy $W_k = \frac{1}{2}Mu^2$	Magnetic energy $W_m = \frac{1}{2}Li^2$
Potential energy $W_p = \frac{1}{2C_m}x^2$	Electrostatic energy $= \frac{1}{2C}q^2$
Dissipated power $p = Bu^2$	Dissipated power $p = Ri^2$
Mechanical impedance Z_m	Impedance Z
Mechanical mobility Y_m	Admittance Y

Figure 10.1 shows a mechanical system, the mechanical circuit diagram and classical electrical analogue

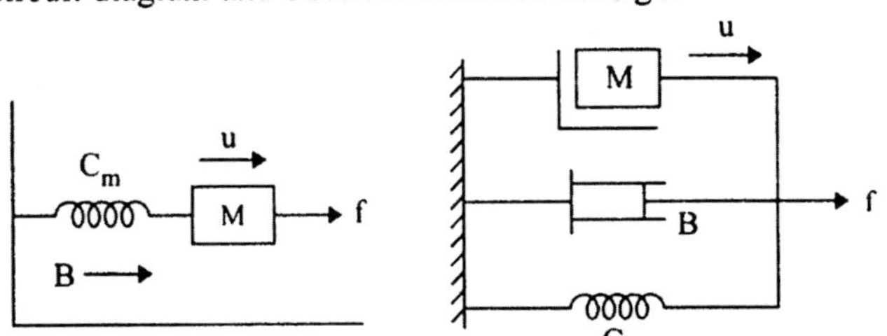

(*a*) Mechanical system **(*b*) Mechanical circuit diagram**

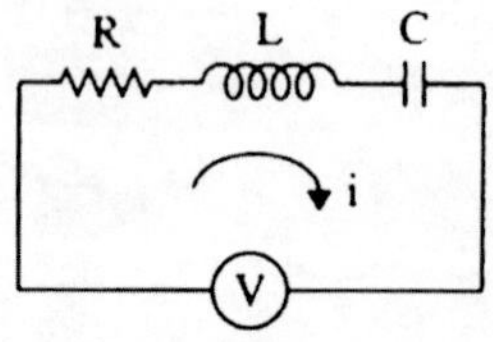

(c) Classical electrical analogue

Fig. 10.1

The equations for the systems in Fig. (10.1) (*a* and *c*) are

$$f = Bu + M\frac{du}{dt} + \frac{1}{C_m}\int u dt \qquad ...(10.7)$$

$$v = Ri + L\frac{di}{dt} + \frac{1}{C}\int i dt \qquad ...(10.8)$$

Consider the mechanical circuit diagram shown in Fig. 10.2 (*a*).

The equation relating force (f) and displacement (x) is

$$M\frac{d^2x}{dt^2} + B\frac{dx}{dt} + Kx = f \qquad ...(10.9)$$

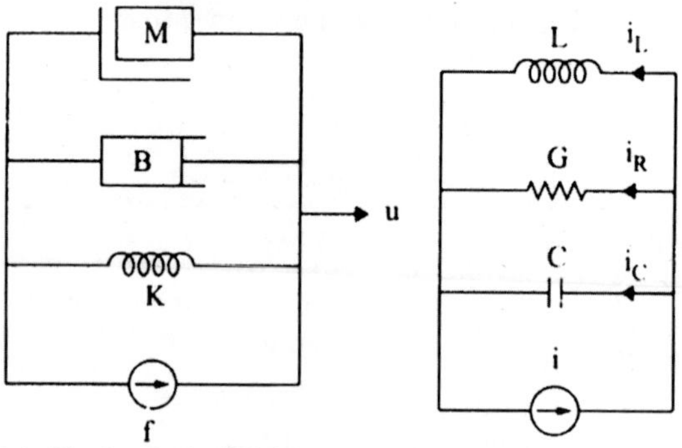

(a) Mechanical circuit **(b) Mobility analogue**

Fig. 10.2

For the parallel RLC circuit of Fig. 10.2 (*b*)]

$$i = i_C + i_R + i_L$$

or

$$i = C\frac{dv}{dt} + \frac{v}{R} + \frac{1}{L}\int v dt \qquad ...(10.10)$$

But

$$v = \frac{d\psi}{dt} \qquad ...(10.11)$$

where ψ is magnetic flux linkage.

Substituting Eq. (10.11) into Eq. (10.10), we get

$$C\frac{d^2\psi}{dt^2} + \frac{1}{R}\frac{d\psi}{dt} + \frac{\psi}{L} = i \qquad ...(10.12)$$

Equations (10.9) and (10.12) are similar. Thus f is analogous to current, velocity to voltage, mass to capacitance compliance to inductance and mechanical resistance to conductance (or 1/R). This is called mobility or force-current analogy. The force-voltage and force-current analogies are dual of each other. Table 10.2 lists analogous quantities in force-current analogy.

Table 10.2 Analogous Quantities in Force Current (Mobility) Analogue

Mechanical quantity	*Electrical quantity*
Force f	Current i
Velocity u	Voltage v
Displacement x	Magnetic flux linkage ψ
Mass M	Capacitance C
Mechanical resistance B	Conductance G
Compliance C_m	Inductance L
Kinetic energy $W_k = \frac{1}{2}Mu^2$	Electrostatic energy $-\frac{1}{2}Cv^2$
Potential energy $W_p = \frac{1}{2}C_m f^2$	Magnetic energy $= \frac{1}{2}Li^2$
Dissipated power $p = Bu^2$	Dissipated power $= Gv^2$
Mechanical mobility Y_m	Impedance Z
Mechanical impedance Z_m	Admittance Y

10.2.2. Mechanical Rotational Systems

Analogy between mechanical rotational system and electrical system can be as per torque-voltage or torque current basis. Table 10.3 lists the analogous quantities in both cases.

Table 10.3 Analogy between Mechanical Rotational Quantities and Electrical Quantities

Mechanical rotational quantity	*Analogous quantity*	
	Torque-voltage or Classical analogue	*Torque-current or mobility analogue*
Torque	Voltage v	Current i
Angular velocity $\frac{d\theta}{dt}(= \theta')$	Current i	Voltage v
Angular displacement x	Charge q	Magnetic flux ϕ
Moment of inertia J	Inductance L	Capacitance G
Rotational resistance B_r	Resistance R	Conductance G
Rotational compliance C_r	Capacitance C	Inductance L
Kinetic energy $W_k = \frac{1}{2}J\frac{d\theta}{dt}$	Magnetic energy $\frac{1}{2}Li^2$	Electrostatic energy $\frac{1}{2}Cv^2$
Potential energy $W_p = \frac{1}{2C_r}\theta^2$	Electrostatic energy $\frac{1}{2C}q^2$	Magnetic energy $\frac{1}{2L}\phi^2$
Rotational impedance Z_r	Impedance Z	Admittance Y
Rotational molarity Y_r	Admittance Y	Impedance Z
Dissipated power $p = B_r(\theta')^2$	Dissipated power $p = Ri^2$	Dissipated power $p = Gv^2$

10.2.3. Thermal, Hydraulic and Air Flow System Analogies

For Thermal System

$$\frac{dQ}{dt} = C_t \frac{d\theta}{dt} \quad \text{...(10.13)}$$

When Q is heat energy, θ is temperature and C_t is thermal capacitance.

Electrical equation analogous to Eq. (10.13) is

$$i = C\frac{dv}{dt} \quad \text{...(10.14)}$$

Another set of analogous equations is

$$\frac{dQ}{dt} = G_t(\theta_1 - \theta_2) \quad \text{...(10.15)}$$

$$i = G(v_1 - v_2) \quad \text{...(10.16)}$$

Thus analogy exists between electrical and thermal systems. Similar analogies also exist between electrical and hydraulic systems and between electrical and air flow (pneumatic) systems.

Table 10.4 lists these analogies.

Table 10.4 Analogy Between Thermal, Hydraulic, Air Flow and Electrical Systems

Quantities in Thermal systems	*Quantities in hydraulic systems*	*Quantities in air flow systems*	*Analogous electrical quantities*
Heat energy Q	Amount of liquid m^3	Amount of air, m^3	Charge, q
Rate of flow of heat Energy	Rate of liquid flow, m^3/s	Rate of air flow, m^3/s	Current, i
Temperature, 0°C	Head, m	Head, m	Voltage, v
Thermal resistance	Resistance	Resistance	Resistance
Thermal capacitance	Capacitance	Capacitance	Capacitance

10.3. TRANSFER FUNCTION

It is the ratio of Laplace transform of one quantity at the output of the Laplace transform of another quantity at the input. It can be the ratio of two voltages, two currents, a voltage and current or current and voltage. For the two-port network (Fig. 10.3) four transfer functions can be defined. These are listed in Table 10.5.

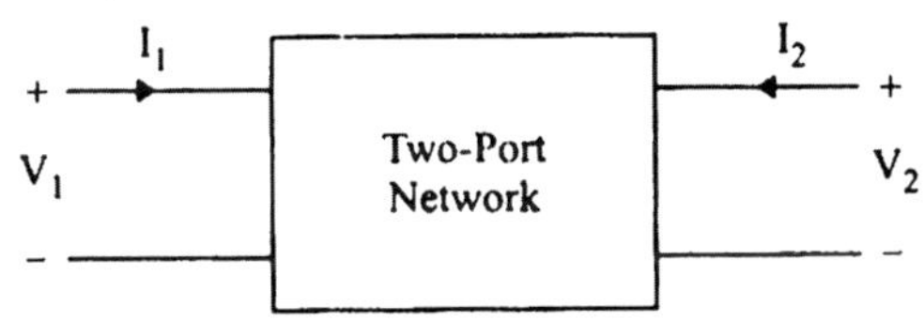

Fig. 10.3. Two port Network

Table 10.5 Transfer Functions of Two-Port Network

Transfer function	*Symbol*
$V_2(s)/V_1(s)$	$G_{12}(s)$
$V_2(s)/I_1(s)$	$Z_{12}(s)$
$-I_2(s)/V_1(s)$	$Y_{12}(s)$
$-I_2(s)/I_1(s)$	$\alpha_{12}(s)$

In defining a transfer function all initial conditions are assumed to be zero.

Equation (10.9) can be written as

$$F(s) = Ms^2X(s) + BsX(s) + KX(s)$$

The transfer function, *i.e.*, ratio of output (i.e. force) in s domain is

$$G(s) = \frac{X(s)}{F(s)} = \frac{1}{Ms^2 + Bs + K} \quad \text{...(10.17)}$$

For the series RLC circuit in Fig. 10.4

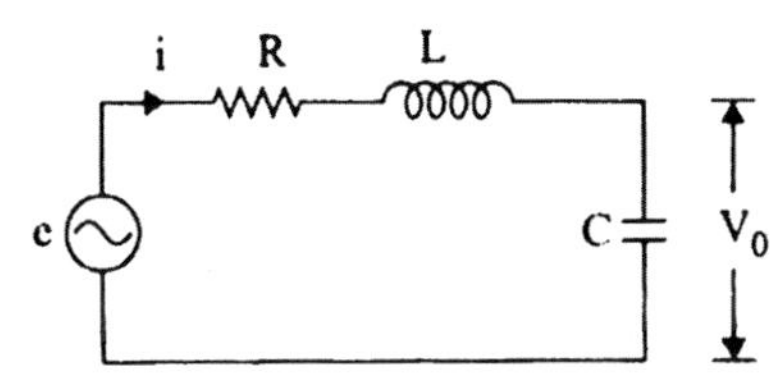

Fig. 10.4.

$$e = Ri + L\frac{di}{dt} + \frac{1}{C}\int i\,dt$$

or $$E(s) = RI(s) + sLI(s) + \frac{I(s)}{Cs}$$

and $$v_0 = \frac{1}{C}\int i\,dt$$

or $$V_0(s) = \frac{I(s)}{Cs}$$

$$\text{Transfer function } \frac{V_0(s)}{E(s)} = \frac{1}{LCs^2 + RCs + 1} \quad \text{...(10.18)}$$

The procedure for obtaining transfer function is

1. Write differential equation of the system.
2. Obtain Laplace transform of differential equation, assuming that initial conditions are zero.
3. Obtain the ratio of output to input.

A transfer function has the following features.

1. It is property of system and independent of the magnitude or type of input.
2. If transfer function is known, we can find the output for any given input.

3. The transfer function does not convey any information about the system itself. Many different systems may have the same transfer function.

10.4. POLES AND ZEROS

In general the transfer function of a system can be written as:

$$\text{Transfer function} \quad G(s) = \left.\frac{\mathcal{L}(\text{output})}{\mathcal{L}(\text{input})}\right|_{\text{zero initial conditions}}$$

$$\text{or } G(s) = \frac{p(s)}{q(s)} = \frac{b_0 s^m + b_1 s^{m-1} + \ldots + b_{m-1} s + b_m}{a_0 s^n + a_1 s^{n-1} + \ldots + a_{n-1} s + a_n} \quad \ldots(10.19)$$

where a and b are real and positive for network having passive elements and no controlled sources. The equation $p(s) = 0$ has m roots and the equations $q(s) = 0$ has n roots. Equation (10.19) can be written as

$$G(s) = \frac{b_0}{a_0} \frac{(s - z_1)(s - z_2) \ldots (s - z_m)}{(s - p_1)(s - p_2) \ldots (s - p_n)} \quad \ldots(10.21)$$

When $z_1, z_2, \ldots z_m, p_1, p_2, \ldots, p_n$ are complex frequencies and the ratio b_0/a_0 is the scale factor.

When variable s has values $z_1, z_2, \ldots, z_m$, the transfer function becomes zero. Therefore the complex frequencies $z_1, z_2, \ldots, z_m$ are called zeros of the transfer function. When variables s has values $p_1, p_2, \ldots, p_m$ the transfer function becomes infinite. Hence $p_1, p_2, \ldots, p_n$ are called poles of the function.

When a pole or zero is not repeated it is said to be simple or distinct. When r poles or zeros have the same value, the pole or zero is said to be of multiplicity r. The poles and zeros are together known as critical frequencies. In Eq. (10.19) the highest power of s is n. Thus it is an nth order system.

10.5. CLOSED LOOP TRANSFER FUNCTION

In Fig. 10.5 R(s) is input, C(s) is output, B(s) is feedback signal and E(s) is error signal.

$$\frac{C(s)}{R(s)} = \frac{G(s)}{1 + G(s)\,H(s)} \quad \ldots(10.22)$$

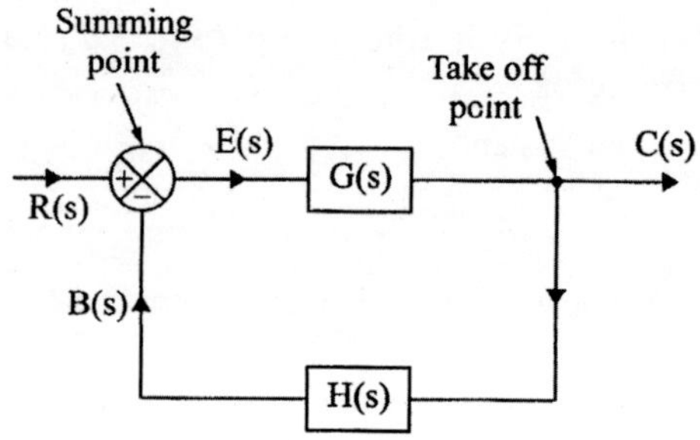

Fig. 10.5.

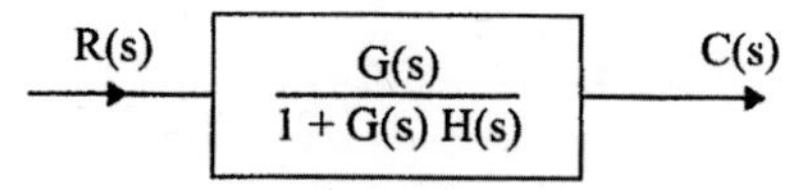

Fig. 10.6.

$C(s)/R(s)$ is called closed loop transfer function. The system of Fig. 10.5 can be reduced to that shown in Fig. 10.6.

For a multi input system, the response can be obtained by using superposition principle. For the two input system of Fig. 10.7

$$C(s) = \frac{G_2(s)}{1 + G_1(s)\,G_2(s)\,H(s)} [G_1(s)R(s) + U(s)] \quad \ldots(10.23)$$

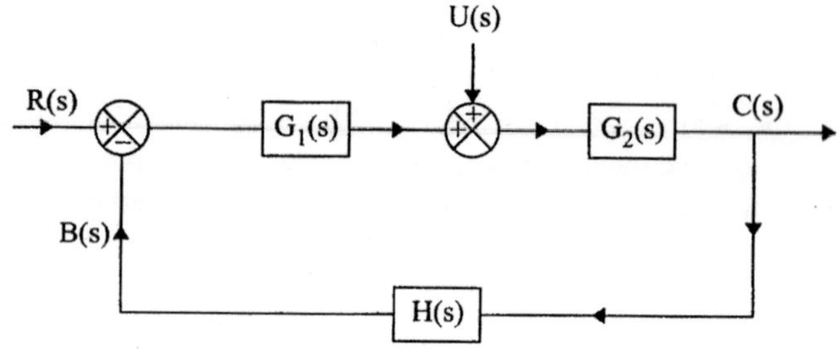

Fig. 10.7

10.6. BLOCK DIAGRAM REDUCTION

The rules for block diagram reduction are given in Table 10.6.

Table 10.6 Block Diagram Algebra

Rule	*Original block diagram*	*Equivalent block diagram*
1. Blocks in Cascade	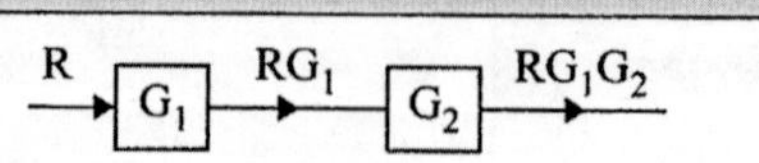	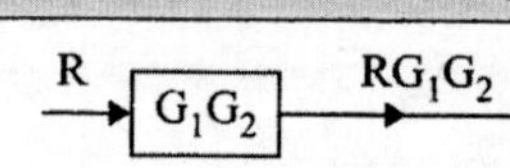
2. Moving a Summing point after a block	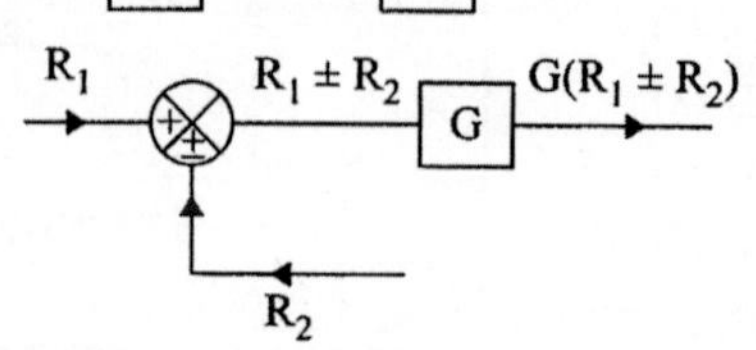	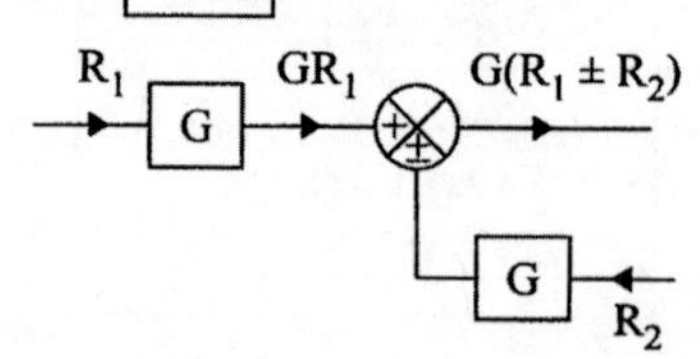
3. Moving a Summing point ahead of a block	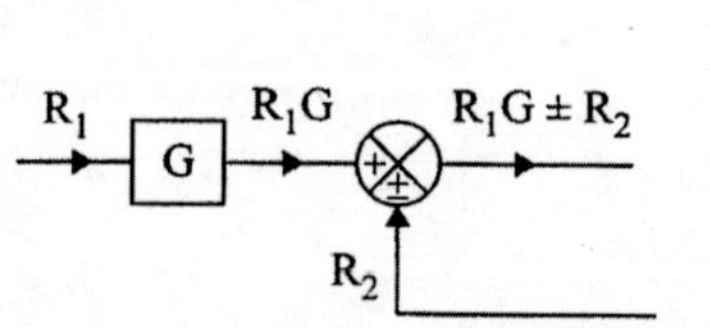	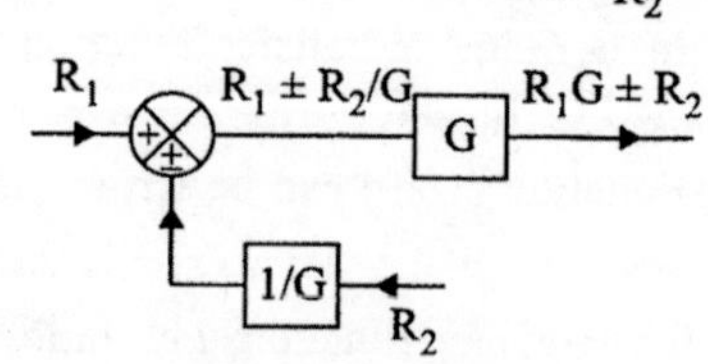
4. Moving take off point after a block	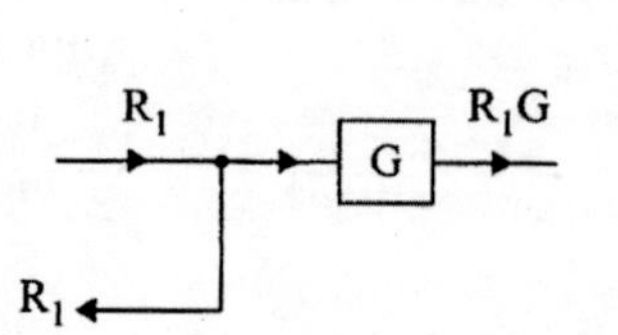	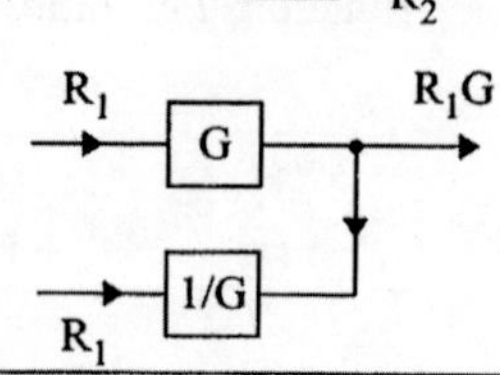

Rule	Original block diagram	Equivalent block diagram
5. Moving a take off point ahead of a block	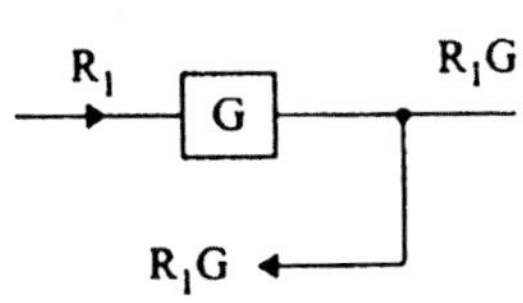	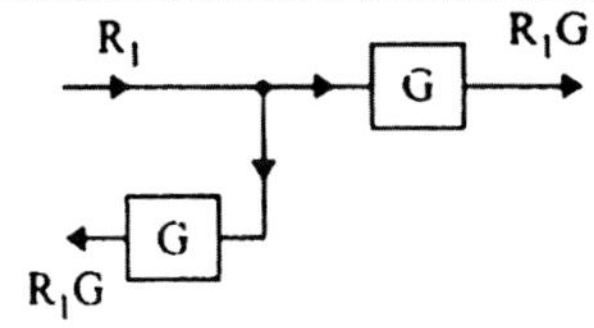
6. Eliminating a feedback loop	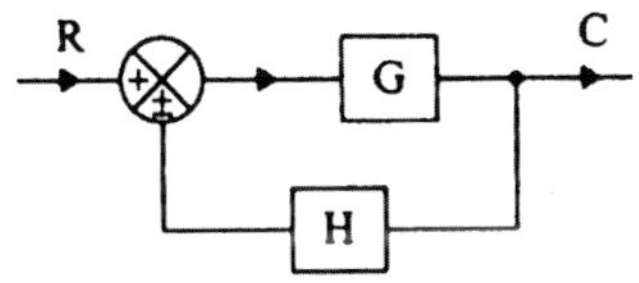	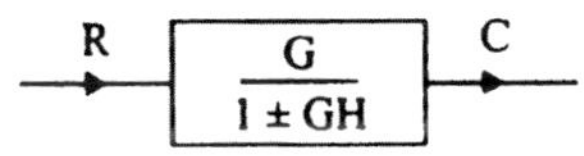

In simplifying a block diagram the important points to remember are : (1) The product of transfer functions in the forward direction must remain the same. (2) The product of transfer functions around the loop must remain the same.

10.7. SIGNAL FLOW GRAPH

10.7.1. Notations

Node : A system variable is equal to sum of all incoming signals at the nodes. The outgoing signals from the node do not affect the value of node variable. In Fig. 10.8 (*a*) $x_3 = a_{13}x_1 + a_{23}x_2$.

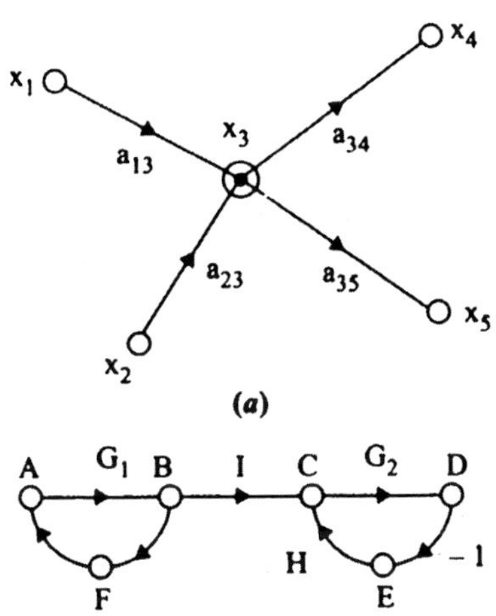

Fig. 10.8

The node may be a summing point or transmitting point. Node variable x_3 is a summing point. A node which has only outgoing branches is called input node or source. x_1 and x_2 are input nodes in Fig. 10.8. If a node has only incoming branches it is called output node or sink *e.g.* nodes x_4 and x_5 in Fig. 10.8 (*a*).

Branch : As a signal travels along a branch (from one node to other in the direction of arrow) it gets multiplied by the gain or transmittance of that branch.

Path : The connected branches in the direction of arrows such that no node is traversed more than once.

Loop : A path which originates and terminates at the same node. In Fig. 10.8 (*b*) CDEC is a loop.

Non-touching loops : The loops which do not have a common node. In Fig. 10.8 (*b*), CDEC and ABFA are non-touching loops.

Forwath path gain : It is the product of gains encountered in a path.

Loop gain : It is the product of branch gains encountered in traversing the loop. For loop CDEC, the loop gain $(G_2)(-1)(H)$ or $-G_2H$.

10.7.2. Drawing the Signal Flow Graph

Consider the system of following set of equations

$$\begin{aligned} x_2 &= a_{12}x_1 + a_{32}x_3 + a_{42}x_4 \\ x_3 &= a_{23}x_2 + a_{33}x_3 \\ x_4 &= a_{34}x_3 + a_{44}x_4 \end{aligned} \qquad ...(10.24)$$

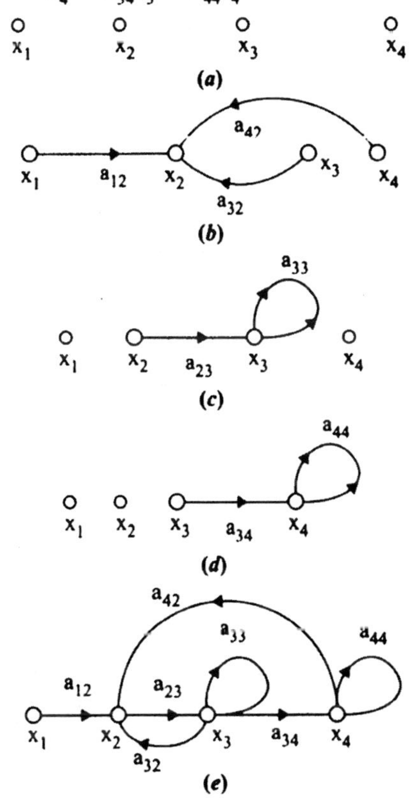

Fig. 10.9. Signal flow graph for Eq. (10.24).

In Eq. (10.24) the input variable is x_1 and output variable is x_4. The drawing of signal flow graph is shown in Fig. 10.9. The first step is to locate the nodes x_1, x_2, x_3, x_4 (Fig. 10.9 *a*). As indicated by the first equation in (10.24) x_2 is the sum of three signals $a_{12}x_1$, $a_{32}x_3$ and $a_{42}x_4$. This is shown in Fig. 10.9 (*b*). Fig. 10.9 (*c*) and Fig. 10.9 (*d*) show the signal flow graphs for second and third part of Eq. (10.24). The complete signal flow graph is shown in Fig. 10.9 (*e*).

10.8. MASON'S GAIN FORMULA

$$T = \frac{1}{\Delta}\sum_k P_k \Delta_k \qquad ...(10.25)$$

where T = overall system gain
P_k = path gain of kth forward path
Δ = determinant of graph
= 1 − (sum of loop gains of all individual loops)
+ (sum of gain products of all possible combinations of two non-touching loops)
− (sum of gain products of all possible combinations of three non-touching loops)
+ ....
Δ_k = the value of Δ for the part of graph not touching the kth forward path.

10.9. CONTROL SYSTEM COMPONENTS

10.9.1. D.C. Servomotor

DC servomotor is generally a permanent magnet dc motor. Its speed is almost directly proportional to applied voltage for a given load torque. Three types of constructions used for dc servo motors are slotted armature construction (armature surface has slots which house armature conductors), surface wound armature construction (the winding is placed on armature surface) and surface wound rotating armature with stationary core.

10.9.2. DC Tacho Generator

It is a small permanent magnet dc generator. Its emf is directly proportional to speed. It is very commonly used in speed feedback and position control systems to give a voltage signal proportional to speed.

10.9.3. Brushless dc Motor

It is a dc motor without brushes. The conversion of applied dc voltage to ac is obtained by solid state electronic devices. In a brushless dc motor armature is placed on stator and the permanent magnet on the rotor.

10.9.4. AC Servomotor

It is a two phase induction motor with special design features. Voltages equal in magnitude and having a phase difference of 90° are applied to the two stator windings. The phase relationship of the two stator voltages determines the direction of rotation. The motor has high resistance (and therefore low X/R ratio) so that torque speed characteristic is linear.

10.9.5. AC Tachometer

It has drag cup construction and is also called drag cup generator. The output voltage is proportional to speed.

10.9.6. Potentiometer

It is a transducer to convert mechanical displacement into a voltage signal. Potentiometers for both linear and angular displacements are manufactured. It may be wire wound, carbon composition and conductive plastic type.

10.9.7. Encoder

It converts linear or angular displacement into digital signal. It consists of a light source, rotary disc (with alternate opaque and transparent sectors) stationary mask and sensor

10.9.8. Synchro

Also called selsyn it converts angular position of shaft into an electric signal. It consists of 3 phase alternator (called synchro transmitter) and synchro control transformer.

10.9.9. Stepper Motor

It is a special synchronous motor. When an electrical pulse is applied from the control unit it rotates through an angle (called step).

10.10. TRANSIENT RESPONSE

10.10.1. First Order System (Fig. 10.10)

For a first order system the transfer function is

$$\frac{C(s)}{R(s)} = \frac{1}{Ts+1} \quad ...(10.26)$$

If the input is a step function, $R(s) = \frac{1}{s}$

$$C(s) = \frac{1}{s(Ts+1)}$$

$$C(t) = 1 - e^{-t/T} \quad ...(10.27)$$

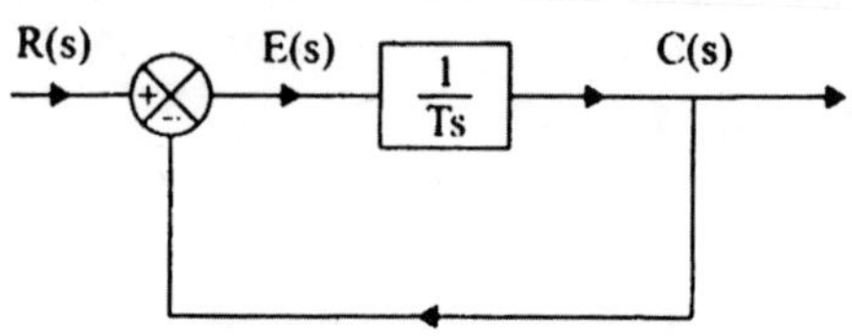

Fig. 10.10 First order system

Figure 10.11 shows the transient response. The output increases exponentially from zero to the final value 1 with a time constant T.

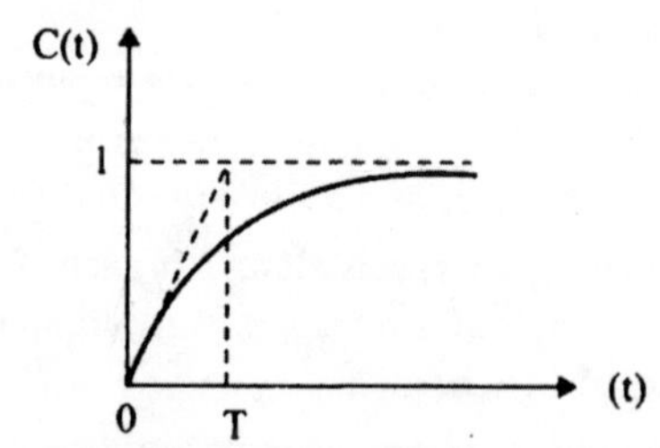

Fig. 10.11. Transient response of first order system for step input.

The slope of response at $t = 0$.

$$\left.\frac{dC}{dt}\right|_{t=0} = \left.\frac{1}{t}e^{-t/T}\right|_{t=0} = \frac{1}{T} \quad ...(10.28)$$

$$\text{error} = e(t) = r(t) - C(t) = e^{t/T}$$

$$\text{Steady state error} = \lim_{t\to\infty} e(t) = 0$$

For a ramp input $R(s) = \frac{1}{s^2}$

$$C(s) = \frac{1}{s^2 (Ts+1)} \quad ...(10.29)$$

$$C(t) = t - T(I - e^{-t/T})$$

$$e(t) = r(t) - c(t) = T(I - e^{-t/T}) \quad ...(10.30)$$

Steady state error $= \lim_{t\to\infty} e(t) = T$

Figure 10.12 shows the response and steady state error

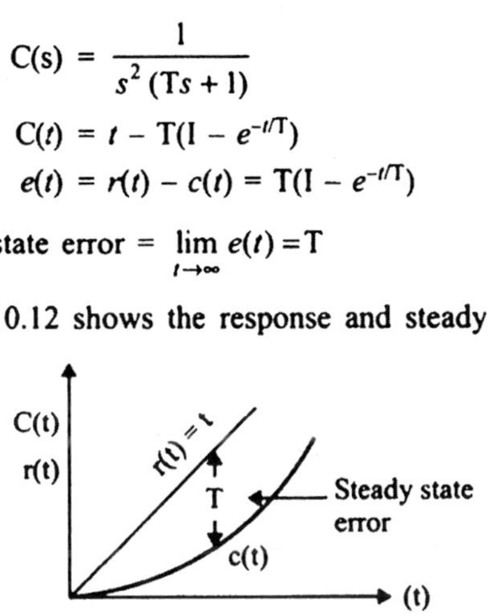

Fig. 10.12. Response of first order system to ramp input

10.10.2. Second Order System (Fig. 10.13)

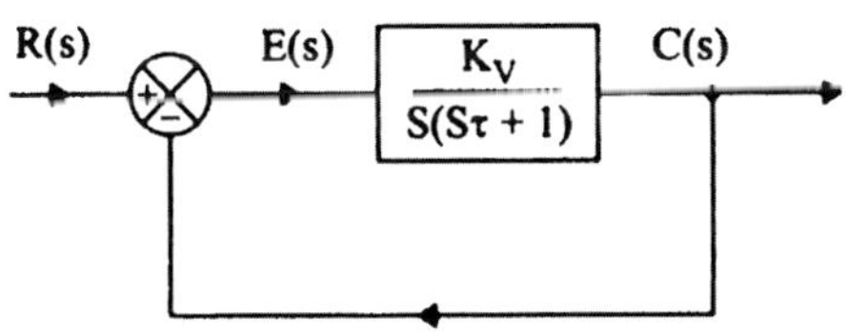

Fig. 10.13 Second order system

For a unit step input $R(s) = \frac{1}{s}$, the transfer function can be expressed as

$$\frac{C(s)}{R(s)} = \frac{\omega_n^2}{s^2 + 2\varsigma\omega_n s + \omega_n^2} = \frac{p(s)}{q(s)} \quad ...(10.31)$$

w here ς = damping ratio (or damping factor) $= \frac{1}{2\sqrt{(K_v \tau)}}$

ω_n = undamped natural frequency $= \sqrt{(K_v/\tau)}$

$$C(t) = 1 - \frac{e^{-\xi\omega_n t}}{\sqrt{1-\zeta^2}} \sin\left[\omega_n\sqrt{(1-\xi^2)}\,t + \tan^{-1}\frac{\sqrt{(1-\xi^2)}}{\xi}\right] \quad ...(10.32)$$

The response is shown in Fig. 10.14 for an under damped system ($\xi < 1$). Most of the systems are under damped to achieve high response speed and other performance measures.

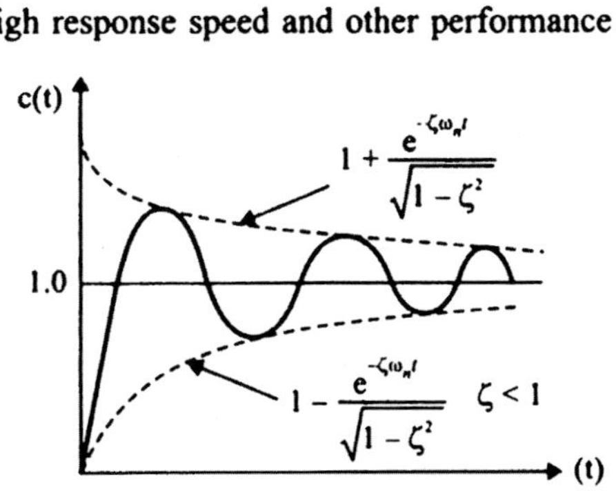

Fig. 10.14. Response of underdamped second order system to unit step input.

If $\xi = 1$ the response is

$$C(t) = I - e^{-\omega_n t} - \omega_n t e^{-\omega_n t} \quad ...(10.33)$$

The time response specifications are

1. **Delay Time t_d** : The time taken to reach 50% of final value in the first attempt.
2. **Rise Time t_r** : For underdamped system it is the time taken to rise from 0 to 100% of final value (for over-damped it is the time taken to rise take from 10% to 90% of final value).
3. **Peak Time (t_p)** : It is the time to reach the peak value.
4. **Peak Overshoot (M_p)** : It is the normalised difference between the peak value and the steady state value, *i.e.*,

   Peak percent overshoot $= \frac{C(t_p) - C(\infty)}{C(\infty)} \times 100$

5. **Setting Time t_s** : It is the time to reach and stay within a specified tolerance band (generally 2% or 5%) of final value.
6. **Steady State Error e_{ss}** : It is the difference (*i.e.*, error) between actual output and desired output as t tends to infinity.

$$e_{ss} = \lim_{t\to\infty} [r(t) - c(t)] \quad ...(10.34)$$

Figure 10.15 shows the various specifications

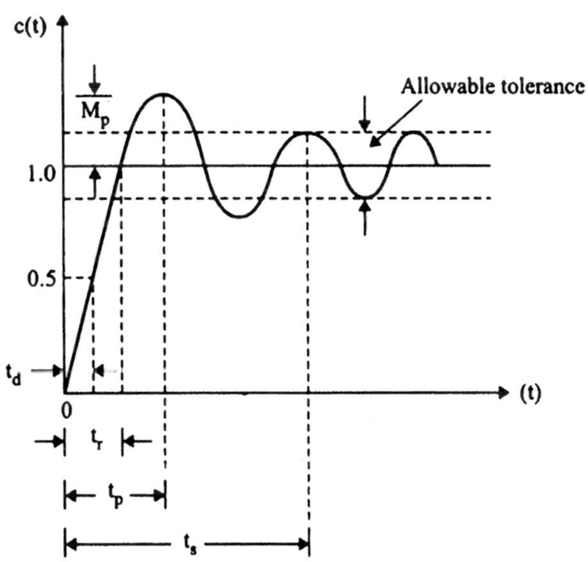

Fig. 10.15. Time response specifications of second order system.

The open loop transfer function of a unity feedback system can be written in following, two standard forms

$$G(s) = \frac{K(T_{z1}s+1)(T_{z2}s+1)\ldots}{s^n(T_{p1}s+1)(T_{p2}s+1)\ldots} \quad \text{i.e., time constant form} \quad ...(10.35\ a)$$

$$= \frac{K'(s+z_1)(s+z_2)\ldots}{s^n(s+p_1)(s+p_2)\ldots} \quad \text{i.e., pole zero form} \quad ...(10.35\ b)$$

In Eq. (10.35) if $n = 0$ it is a type 0 system
if $n = 1$ it is a type 1 system
if $n = 2$ it is a type 2 system

The steady state errors for the above systems and various inputs are as under.

Table 10.7. Steady State Errors

Input signal	***Steady state error***		
	Type 0 system	***Type 1 system***	***Type 2 system***
Unit step	$\frac{1}{1+K_p}$	0	0
Unit ramp	∞	$\frac{1}{K_v}$	0
Unit parabolic	∞	∞	$\frac{1}{K_a}$
In Table 10.7	$K_p = \lim_{s\to 0} G(s)$	$K_v = \lim_{s\to 0} sG(s)$	$K_a = \lim_{s\to 0} s^2G(s)$

10.11. STABILITY CONCEPT

(*a*) If all the roots of the characteristic equation have negative real parts, the system is stable.

(*b*) If any roots of characteristic equation has a positive real part or if there is a repeated root on the $j\omega$ axis the system is unstable.

(*c*) If condition (*a*) is satisfied except that one or more non-repeated roots lie on $j\omega$ axis, the system is limitedly stable.

Two other terms used in stability are :

Absolutely Stable : With respect to a system parameter if the system is stable for all values of this parameter.

Conditionally Stable : With respect to a system parameter it the system is stable for only some specified range of values of this parameter.

Figure 10.16 shows some locations of poles and time domain response.

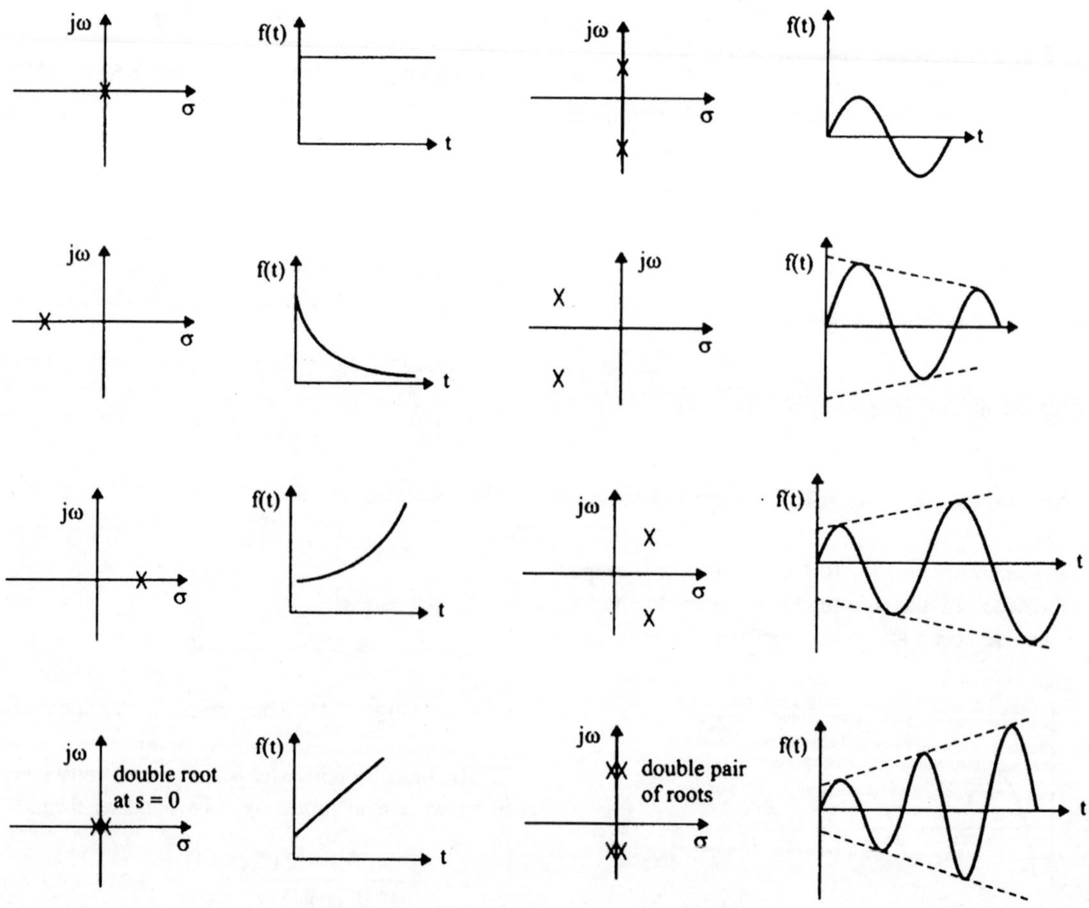

Fig. 10.16. Pole locations and time domain response.

Relative Stability : It is a measure of how fast the transients die out. The settling time is inversely proportional to the real part (negative) of the pole. As a pole moves away from the imaginary axis the stability improves. This is shown in Fig. 10.17 for some pole locations.

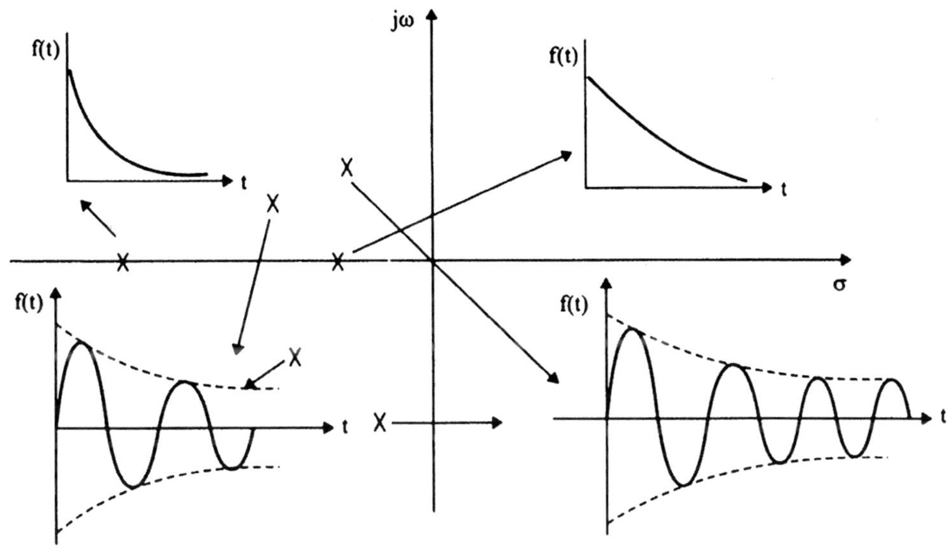

Fig. 10.17. Relative Stability

10.12. HURWITZ STABILITY CRITERION

For n the order system the characteristic equation is

$$q(s) = a_0 s^n + a_1 s^{n-1} + a_2 s^{n-2} + \dots + a_n = 0 \quad \text{...(10.36)}$$

For this system to the stable it is necessary and sufficient that n determinants formed from the coefficients $a_0, a_1, \dots, a_n$ of the characteristic equation be positive where these determinants are taken as the principal minors of the following determinant called Hurwitz determinant.

$$\begin{bmatrix} a_1 & a_0 & 0 & 0 & 0 & 0 & \cdot & \cdot & 0 & 0 \\ a_3 & a_2 & a_1 & a_0 & 0 & 0 & \cdot & \cdot & 0 & 0 \\ a_5 & a_4 & a_3 & a_2 & a_1 & a_0 & \cdot & \cdot & 0 & 0 \\ \cdot & \cdot & \cdot & \cdot & \cdot & \cdot & \cdot & \cdot & \cdot & \cdot \\ \cdot & \cdot & \cdot & \cdot & \cdot & \cdot & \cdot & \cdot & \cdot & \cdot \\ a_{2n-1} & a_{2n-2} & a_{2n-3} & \cdot & \cdot & \cdot & \cdot & \cdot & a_{n+1} & a_n \end{bmatrix} \quad \text{...(10.37)}$$

The coefficients with indices more than n or with negative indices are replaced by zeros.

In other words, necessary and sufficient conditions for system to be stable are

$$\Delta_1 = a_1 > 0$$

$$\Delta_2 = \begin{vmatrix} a_1 & a_0 \\ a_3 & a_2 \end{vmatrix} > 0$$

$$\Delta_3 = \begin{vmatrix} a_1 & a_2 & a_0 \\ a_3 & a_2 & a_1 \\ a_5 & a_4 & a_3 \end{vmatrix} > 0$$

$$\Delta_n = \text{Complete determinant (10.37)} > 0$$

If $\Delta_{n-1} = 0$ the system is limitedly stable

10.13. ROUTH STABILITY CRITERION

The coefficients of the characteristic equation (10.36) are arranged into an array called Routh array as below:

$$\begin{array}{c|ccccccc} s^n & a_0 & a_2 & a_4 & a_6 & \cdot & \cdot \\ s^{n-1} & a_1 & a_3 & a_5 & \cdot & \cdot & \cdot \\ s^{n-2} & b_1 & b_2 & b_3 & \cdot & \cdot & \cdot \\ s^{n-3} & c_1 & c_2 & c_3 & \cdot & \cdot & \cdot \\ s^{n-4} & d_1 & d_2 & d_3 & \cdot & \cdot & \cdot \\ \cdot & \cdot & \cdot & \cdot & \cdot & \cdot & \cdot \\ \cdot & \cdot & \cdot & \cdot & \cdot & \cdot & \cdot \\ \cdot & \cdot & \cdot & \cdot & \cdot & \cdot & \cdot \\ s^2 & e_1 & e_2 \\ s^1 & f_1 \\ s^0 & g_1 \end{array} \quad \text{...(10.38)}$$

The coefficient $b_1, b_2 \dots$ are found as

$$b_1 = \frac{a_1 a_2 - a_2 a_3}{a_1}$$

$$b_2 = \frac{a_1 a_4 - a_0 a_5}{a_1}$$

$$b_3 = \frac{a_1 a_6 - a_0 a_7}{a_1}$$

$$c_1 = \frac{b_1 a_3 - a_1 b_2}{b_1}$$

$$c_2 = \frac{b_1 a_5 - a_1 b_3}{b_1}$$

$$c_3 = \frac{b_1 a_7 - a_1 b_4}{b_1}$$

$$d_1 = \frac{c_1 b_2 - b_1 c_4}{c_1}$$

$$d_2 = \frac{c_1 b_3 - b_1 c_3}{c_1} \quad \text{and so on}$$

This process is continued till we get a zero as the last coefficient in the third row. In this process the missing terms are taken as zero. In the process all the elements of a row can be divided by a positive number during the process to reduce the computational work.

Routh stability criterion is

For a system to be stable it is necessary and sufficient that each term of the first column of Routh array (Eq. 10.38) be positive if $a_0 > 0$. If this condition is not satisfied the system is unstable and the number of sign changes of the terms of first column of Routh array corresponds to the number of roots of the characteristic equation in the right half s-plane.

10.14. ROOT LOCUS ANALYSIS

The characteristic equation of closed loop system is

$$1 + G(s)\,H(s) = 0$$

or $$G(s)\,H(s) = -1$$

This means $|G(s)\,H(s)| = 1$ ...(10.39 *a*)

and $\angle G(s)\,H(s) = \pm\, 180°(2k + 1)$ ...(10.39 *b*)

where $k = 0, 1, 2, ...$

The values of s which satisfy both the magnitude and angle criterion (Eq. 10.39 *a* and *b*) are the roots of the equation $1 + G(s)\,H(s) = 0$. A plot of the points on s plane satisfying the angle condition (Eq. 10.39 *b*) is called rootlocus. The roots of the characteristic equation corresponding to a given value of gain can be found from the magnitude criterion (Eq. 10.39 *a*).

10.15. COMPENSATION

Compensation means changing the system dynamics to satisfy the given specifications.

Figure 10.18 shows series and feedback compensation.

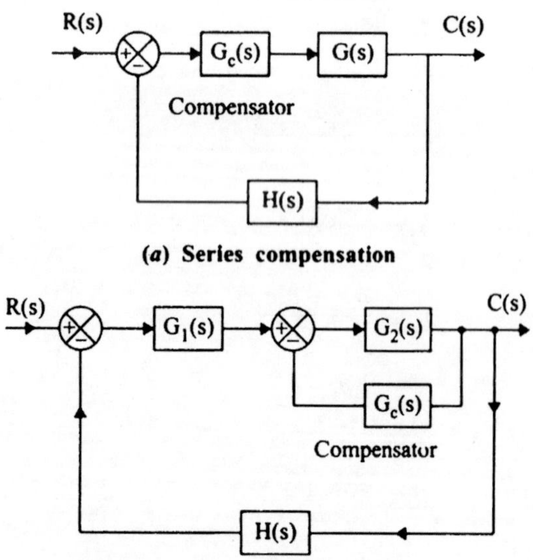

(*a*) **Series compensation**

(*b*) **Feedback compensation**

Fig. 10.18

10.15.1. Lead Compensator

Figure 10.19 (*a*) shows a lead compensator.

$$\frac{E_0(s)}{E_i(s)} = \frac{R_2}{R_1 + \dfrac{R_1/sC}{R_1 + \dfrac{1}{sC}}}$$

$$= \frac{s + 1/R_1C}{s + \dfrac{1}{\left(\dfrac{R_2}{R_1 + R_2}\right)R_1C}} = \alpha\left[\frac{\tau s + 1}{\alpha\tau s + 1}\right] \quad ...(10.40)$$

where $\tau = R_1C$ and $\alpha = \dfrac{R_2}{R_1 + R_2} < 1$

Figure 10.19 (*b*) the pole and zero of lead compensator

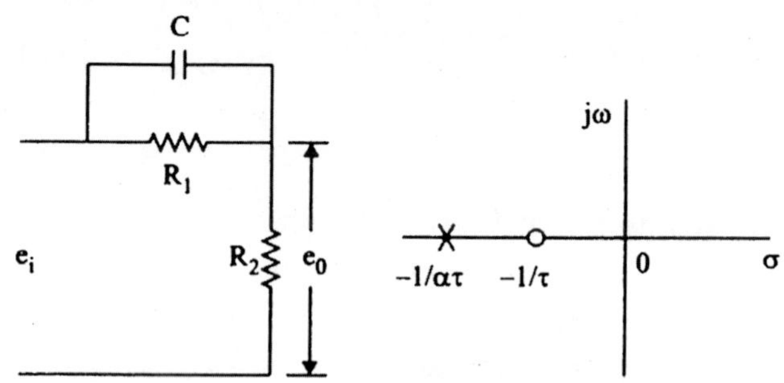

(*a*) circuit **(*b*) location of pole and zero**

Fig. 10.19. Lead compensator

In a lead compensator the output leads the input. A lead compensator speeds up the transient response and increases the stability margin. It increases the system error constant to a limited extent.

10.15.2. Lag Compensator

Figure 10.20 shows a lag compensator and its pole-zero plot.

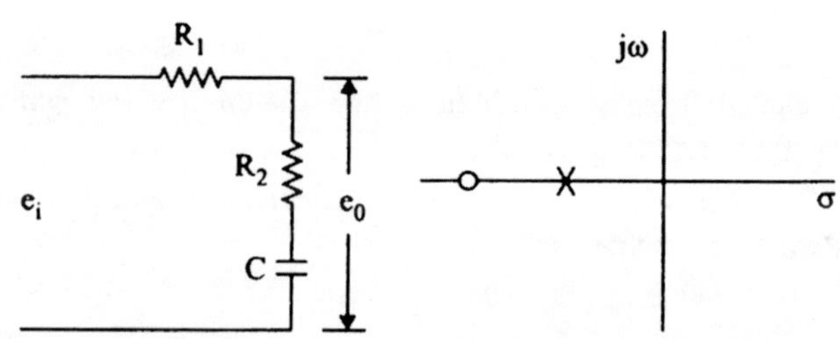

(*a*) circuit **(*b*) location of pole and zero**

Fig. 10.20. Lag compensator

$$\frac{E_0(s)}{E_i(s)} = \frac{R_2 + \dfrac{1}{sC}}{R_1 + R_2 + \dfrac{1}{sC}}$$

$$= \frac{1}{\dfrac{R_1 + R_2}{R_2}}\left[\frac{s + 1/R_2C}{s + \dfrac{1}{\left(\dfrac{R_1 + R_2}{R_2}\right)R_2C}}\right] = \frac{\tau s + 1}{\beta\tau s + 1} \quad ...(10.41)$$

When $\tau = R_2C$ and $\beta = \dfrac{R_1 + R_2}{R_2} > 1$.

The output always lags the input. A lag compensator improves the steady state behaviour of the system while preserving the transient response.

10.15.3. Lag Lead Compensator

It is a series connection of a lag compensator and a lead compensator. Figure 10.21 shows its circuit and pole-zero diagram.

The transfer function of lag-lead compensator is

$$G_c(s) = \frac{(s+1/\tau_1)(s+1/\tau_2)}{(s+1/\beta\tau_1)(s+\beta/\tau_2)}, \beta > 1 \qquad ...(10.42)$$

where $\tau_1 = R_1C_1$, $\tau_2 = R_2C_2$ and $\beta > 1$ such that

$$\frac{1}{R_1C_1} + \frac{1}{R_2C_2} + \frac{1}{R_2C_1} = \frac{1}{\beta\tau_1} + \frac{\beta}{\tau_2}$$

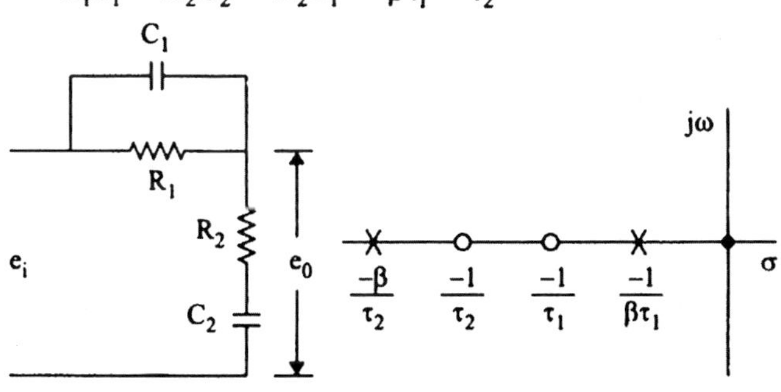

(*a*) circuit (*b*) location of pole and zero

Fig. 10.21. Lag-lead compensator

A lag-lead compensator provides lag compensation for some frequencies and lead compensation for some other frequencies. It is used when both transient and steady state response require improvement.

10.16 POLAR PLOTS

The sinusoidal transfer function $G(j\omega)$ is

$$G(j\omega) = |G(j\omega)| \angle G(j\omega) = M\angle\phi \qquad ...(10.43)$$

Thus $G(j\omega)$ can be written as a phasor of magnitude M and phase angle ϕ (measured positively in counterclock direction). As ω is varied from 0 to ∞ the values of M and ϕ change. The tip of phasor $G(j\omega)$ traces a locus in the complex plane. This locus is called polar plot. For the function $G(j\omega) = \frac{1}{1+j\omega T}$, the polar plot is shown in Fig. 10.22.

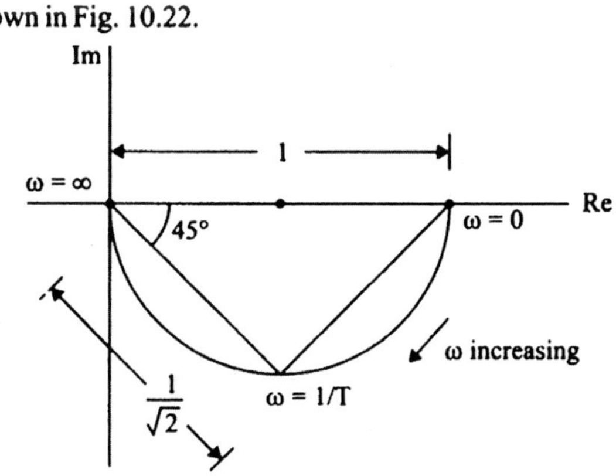

Fig. 10.22. Polar plot of $\frac{1}{1+j\omega T}$.

10.17. BODE PLOT

$$G(j\omega) = |G(j\omega)|e^{j\phi(\omega)} \qquad ...(10.44)$$

$$\ln G(j\omega) = \ln |G(j\omega)| + j\phi(\omega) \qquad ...(10.45)$$

$$\text{Also } \log G(j\omega) = \log |G(j\omega)| + 0.434 j\phi(\omega) \qquad ...(10.46)$$

where ln denotes natural logarithm and log denotes logarithm to base 10.

Two graphs are plotted viz. 20 log $|G(j\omega)|$ vs log ω and $\phi(\omega)$ vs log ω. A semilog graph paper is used. The use of logarithmic scale for frequency on x-axis enables us to draw the graphs for a very wide range of frequencies than possible with linear frequency axis. These plots are called Bode plots.

10.18. NYQUIST STABILITY CRITERION

$$\text{Let } G(s)H(s) = K\frac{(s+z_1)(s+z_2)...(s+z_m)}{(s+p_1)(s+p_2)...(s+p_n)}, m \le n \qquad ...(10.47)$$

$$\text{Then} \quad q(s) = 1 + G(s)H(s)$$

$$= 1 + K\frac{(s+z_1)(s+z_2)...(s+z_m)}{(s+p_1)(s+p_2)...(s+p_n)}$$

$$= \frac{(s+p_1)(s+p_2)...(s+p_n) + K(s+z_1)(s+z_2)...(s+z_m)}{(s+p_1)(s+p_2)...(s+p_n)}$$

$$\text{or} \quad q(s) = \frac{(s+z'_1)(s+z'_2)...(s+z'_n)}{(s+p_1)(s+p_2)...(s+p_n)} \qquad ...(10.48)$$

It is seem from Eq. (10.48) that zeros of $q(s)$ at $-z_1' - z_2' ... -z_n'$ are the roots of the characteristic equation and poles of $q(s)$ at $-p_1, -p_2 .... -p_n$ are the same as open loop poles.

A contour called Nyquist contour encircling the complete right half plane is drawn. Nyquist stability criterion is as under :

If the contour ΓGH of the open loop transfer function $G(s)H(s)$ corresponding to Nyquist contour in the s plane encircles the point $(-1 + j0)$ in the counterclockwise directions as many times as the number of right half s plane poles of $G(s)\,H(s)$, the closed loop system is stable. Moreover if the contour does not encircle $(-1 + j0)$ point the closed loop system is stable.

10.19. GAIN MARGIN AND PHASE MARGIN

10.19.1 Gain Margin

It is the factor by which the gain of the system can be increased to bring the system to the limit of stability. If the gain, at the frequency at which phase angle is 180°, is a then gain margin is $1/a$. Generally gain margin is expressed is decibels. Thus

$$\text{Gain Margin} = -20 \log a \, dB \qquad ...(10.49)$$

For stable system a is less than 1. Therefore log a is negative and gain margin is positive. The frequency at which phase angle becomes 180° is called phase cross-over frequency.

10.19.2 Phase Margin

The frequency at which $|G(j\omega)\,H(j\omega)| = 1$ is called gain cross-over frequency. The phase margin is the additional phase angle lag, at the gain cross over frequency, required to bring the system to the limit of stability. If phase angle, at gain cross-over frequency is $-180 + \phi$, then phase margin is $\phi°$.

A high value of gain and phase margin means a very stable but sluggish system. If gain margin is close to 1 or phase margin is zero, the system is highly oscillatory.

10.20 CONTROLLERS

10.20.1 Proportional Controller

In a proportional controller the output is proportional to the actuating error signal (Fig. 10.23), *i.e.*,

$$c(t) = K_p e(t) \quad ...(10.50\,a)$$

or

$$\frac{C(s)}{E(s)} = K_p \quad ...(10.50\,b)$$

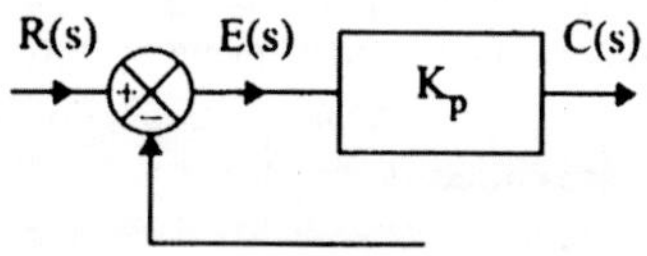

Fig. 10.23. Proportional controller

10.20.2 Integral Controller

In an integral controller the rate of change of output is proportional to actuating error signal (Fig. 10.24), *i.e.*,

$$\frac{dc(t)}{dt} = K_i e(t)$$

or

$$c(t) = K_i \int_0^t e(t)\, dt \quad ...(10.51\,a)$$

i.e.,

$$\frac{C(s)}{E(s)} = \frac{(K_i)}{s} \quad ...(10.51\,b)$$

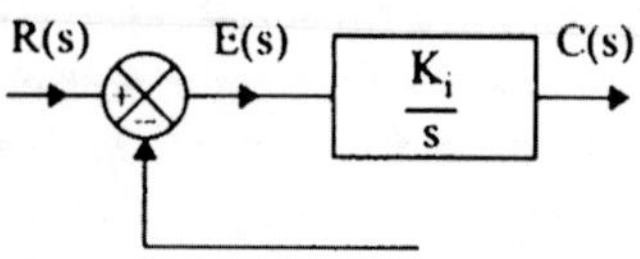

Fig. 10.24. Integral controller

10.20.3 PI Controller

In PI (proportional plus integral controller) the control action is defined as

$$c(t) = K_p\, e(t) + \frac{K_p}{T_i}\int_0^t e(t)\, dt \quad ...(10.52\,a)$$

i.e.,

$$\frac{C(s)}{E(s)} = K_p\left[1 + \frac{1}{T_i s}\right] \quad ...(10.52\,b)$$

This is shown in Fig. 10.25.

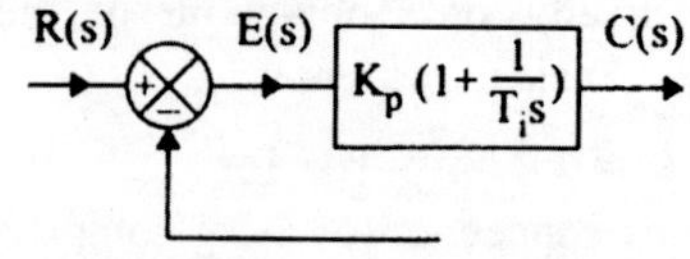

Fig. 10.25. PI Controller

10.20.4 Proportional Plus Derivative Controller (Fig. 10.26)

It is combination of proportional and derivative control. The control action can be written as

$$c(t) = K_p e(t) + K_p T_d \frac{de(t)}{dt} \quad ...(10.53\,a)$$

or

$$\frac{C(s)}{E(s)} = K_p(1 + T_d s) \quad ...(10.53\,b)$$

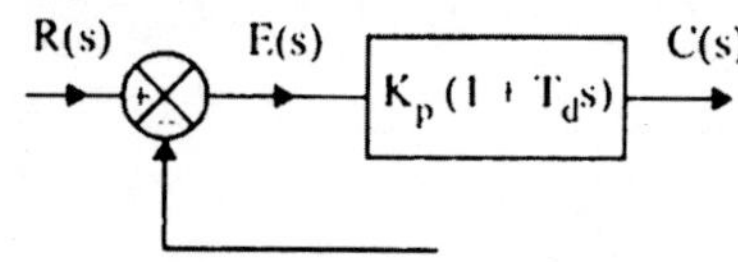

Fig. 10.26. Proportional derivative controller

10.20.5 PID Controller (Proportional Plus Integral Plus Derivative Controller)

It is a combination of proportional, integral and derivative control actions. The control action can be written as

$$c(t) = K_p\, e(t) + \frac{K_p}{T_i}\int_0^t e(t)dt + K_p T_d \frac{de(t)}{dt} \quad ...(10.54\,a)$$

or

$$\frac{C(s)}{E(s)} = K_p\left(1 + \frac{1}{T_i s} + T_d s\right) \quad ...(10.54\,b)$$

Figure 10.27 shows block diagram of PID controller.

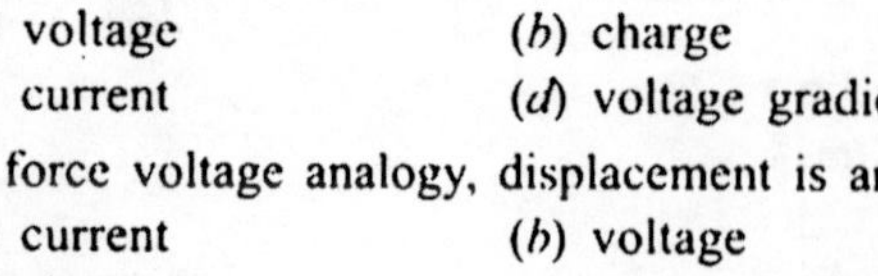

Fig. 10.27. PID controller

OBJECTIVE TYPE QUESTIONS

A. Choose the most appropriate alternative.

1. In force voltage analogy the quantity analogous to spring constant K is
   (*a*) R (*b*) C
   (*c*) L (*d*) 1/C
2. Mechanical impedance is the ratio of
   (*a*) rms force to rms velocity
   (*b*) rms force to rms displacement
   (*c*) rms velocity to rms displacement
   (*d*) none of the above
3. In Force current analogy, displacement x is analogous to
   (*a*) charge (*b*) magnetic flux linkage
   (*c*) electrostatic energy (*d*) voltage
4. Time rate of change in heat energy is analogous to
   (*a*) voltage (*b*) charge
   (*c*) current (*d*) voltage gradient
5. In force voltage analogy, displacement is analogous to
   (*a*) current (*b*) voltage
   (*c*) charge (*d*) none of the above
6. In classical analogue, torque is analogous to
   (*a*) current (*b*) voltage
   (*c*) charge (*d*) voltage gradient
7. In force current analogue, mechanical mobility is analogous to
   (*a*) impedance Z (*b*) admittance Y
   (*c*) reactance X (*d*) conductance G
8. Temperature is analogous to
   (*a*) voltage (*b*) flux
   (*c*) charge (*d*) none of above

9. For the equation $\frac{dQ}{dt} = C_t \frac{d\theta}{dt}$ where Q is heat energy, C_t is thermal capacitance and θ is temperature, the analogous electrical equation is

(a) $i = C\frac{dv}{dt}$ (b) $v = L\frac{di}{dt}$

(c) $e = N\frac{d\phi}{dt}$ (d) none of the above

10. For the equation $\frac{dQ}{dt} = G_t(\theta_1 - \theta_1)$ where Q is heat energy G_t is thermal conductance and θ is temperature, analogous electrical equation is

(a) $i = G(v_1 - v_2)$ (b) $Q = C\frac{dv}{dt}$

(c) $e = N\frac{d\phi}{dt}$ (d) $v = L\frac{di}{dt}$

11. For the function $G(s) = \frac{1}{1-e^{-s}}$ the poles are at

(a) $s = 0$ and $s = \infty$

(b) $s = j2\pi\ \pi$ where $n = 0, 1, 2$

(c) $s = \pm j2n\pi$ where $n = 0, 1, 2...$

(d) none of the above.

12. For the system of Fig. 10.28 the transfer function $\frac{E_0(s)}{E_i(s)} =$

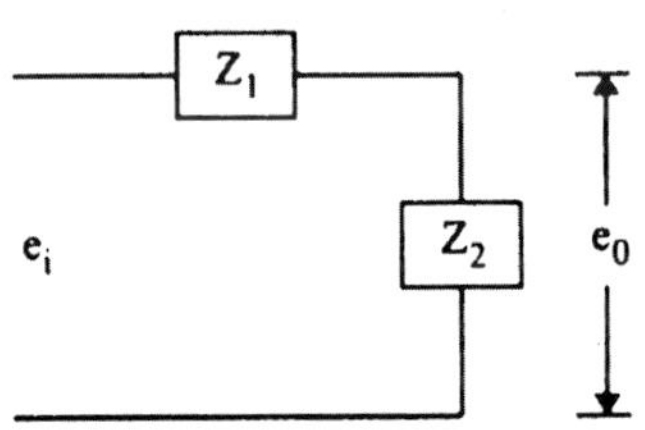

Fig. 10.28.

(a) $\frac{Z_2(s)}{Z_1(s)}$ (b) $\frac{Z_2(s)}{Z_1(s) + Z_2(s)}$

(c) $\frac{Z_1(s)}{Z_2(s)}$ (d) $\frac{Z_1(s)}{Z_1(s) + Z_2(s)}$

13. The resistance R of a liquid flow system is defined as

(a) $\frac{\text{change in flow rate}}{\text{change in level difference}}$

(b) $\frac{\text{change in level difference}}{\text{change in flow rate}}$

(c) change in head × change in flow rate

(d) none of the above

14. In liquid flow systems the capacitance of a water tank is defined as

(a) $\frac{\text{change in liquid stored}}{\text{change in head}}$

(b) $\frac{\text{change in head}}{\text{change in liquid stored}}$

(c) $\frac{\text{change in flow rate}}{\text{change in head}}$

(d) none of the above

15. The units of thermal capacitance are

(a) k cal/°C (b) °C/k cal

(c) °C sec/k cal (d) none of the above

16. If C is thermal capacitance, m is mass of liquid and c is specific heat, then

(a) $C = mc$ (b) $c = mC$

(c) $m = cC$ (d) $C = (mc)^2$

17. For a second order system delay time t_d is the time required to reach

(a) half the final value the very first time

(b) the final value the very first time

(c) 90% of the final value the very first time

(d) none of the above

18. For under damped second order systems the rise time is the time required for the response to rise from

(a) 0% to 100% of its final value

(b) 10% to 90% of its final value

(c) 5% to 95% of its final value

(d) either (a) or (b)

19. For the block diagram of Fig. 10.29, the transfer function C(s)/R(s) is

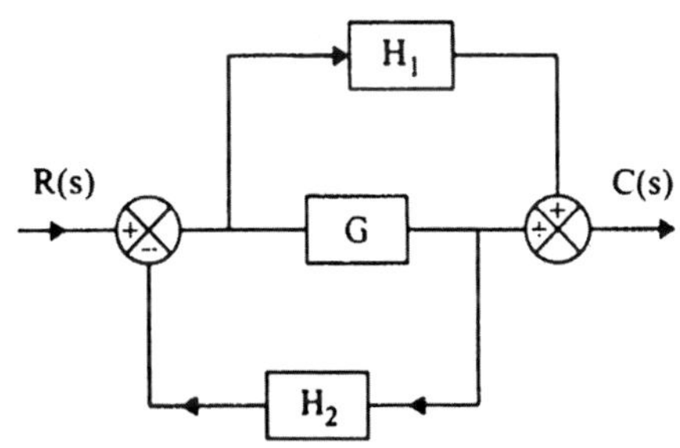

Fig. 10.29.

(a) $\frac{GH_1}{1 + GH_2}$ (b) $\frac{G + H_1}{1 + GH_2}$

(c) $\frac{G + H_2}{1 + H_2(G + H_2)}$ (d) $\frac{G}{1 + GH_2} + H_1$

20. For the system of Fig. 10.30. The transfer function C(s)/R(s) is

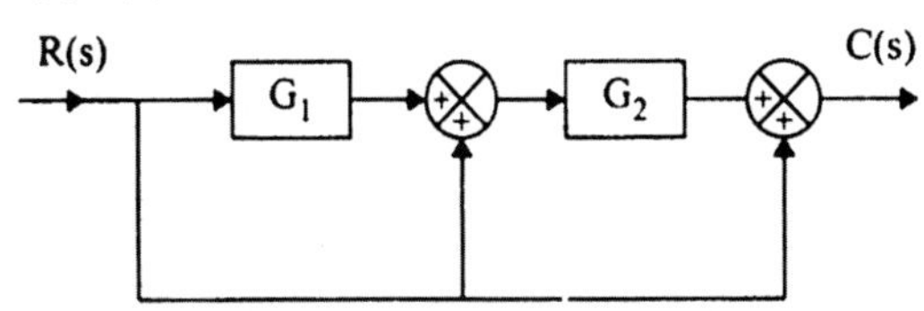

Fig. 10.30.

(a) $G_1 + G_2 + 1$ (b) $G_1 G_2 + 1$

(c) $G_1 G_2 + G_2 + 1$ (d) $G_1 G_2 + G_1 + 1$

21. For the system in Fig. 10.31, the transfer function C(s)/R(s) is

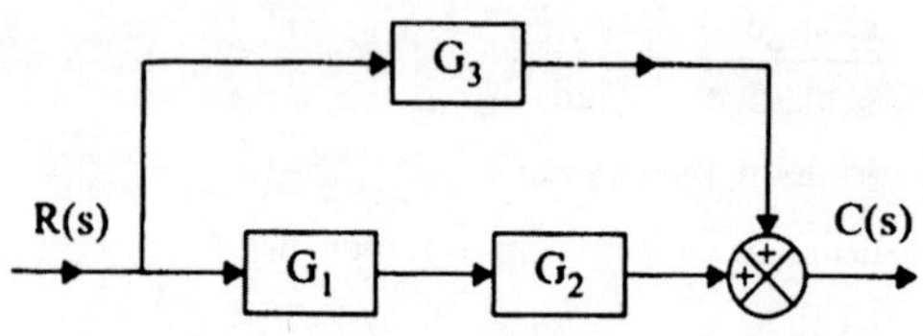

Fig. 10.31.

(a) $G_1 + G_2 + G_3$ (b) $G_1 G_2 + G_3$
(c) $G_1 G_2 G_3$ (d) $(G_1 + G_2) G_3$

22. For a first order system having transfer function $\frac{1}{1+sT}$, the unit step response is
(a) $1 - e^{-t/T}$ (b) $e^{-t/T}$
(c) $e^{-t/T} - 1$ (d) $1 + e^{-t/T}$

23. For a first order system having transfer function $\frac{1}{1+sT}$, the unit ramp response is
(a) $T(1 - e^{-t/T})$ (b) $t - T + Te^{-t/T}$
(c) $t + T + T e^{-t/T}$ (d) $t + 2Te^{-t/T}$

24. For a first order system having transfer function $\frac{1}{1+sT}$, the unit impulse response is
(a) $e^{-t/T}$ (b) $T e^{-t/T}$
(c) $\frac{1}{T}e^{-t/T}$ (d) $T^2 e^{-t/T}$

25. A proportional controller is basically
(a) an amplifier with adjustable gain
(b) an integrating amplifier
(c) an amplifier with infinite gain
(d) an amplifier with almost zero gain

26. In an integral controller
(a) the output is proportional to input
(b) the rate of change of output is proportional to input
(c) the output is proportional to rate of change of input
(d) none of the above

27. In a proportional plus integral control action
(a) proportional gain K_p is adjustable but integral time T_i is constant
(b) proportional gain K_p is constant but integral time T_i is adjustable
(c) both proportional gain K_p and integral time T_i are adjustable
(d) both proportional gain K_p and integral time T_i are constant

28. Which control action is also called rate control?
(a) proportional
(b) derivative
(c) integral
(d) proportional plus integral

29. Derivative control action is effective
(a) only during steady state periods
(b) only during transient state periods
(c) during both steady state and transient periods
(d) either during steady state or during transient state periods

30. Which control action can never be used alone?
(a) proportional (b) derivative
(c) integral (d) both (b) and (c)

31. For a type 1 system and unit step input, the steady state error is
(a) 0 (b) 1
(c) $\frac{1}{1+K_p}$ (d) ∞

32. For a type 0 system and unit ramp input, the steady state error is
(a) 0 (b) ∞
(c) 1 (d) $1/K_v$

33. For a type 1 system and unit parabolic input, the steady state error is
(a) 0 (b) ∞
(c) 1 (d) $1/K_v$

34. A system has its two poles on the negative real axis and one pair of poles lies on $j\omega$ axis. The system is
(a) stable (b) unstable
(c) limitedly stable (d) either (a) or (c)

35. A compensator has its pole and zero as shown in Fig. 10.32. This is a

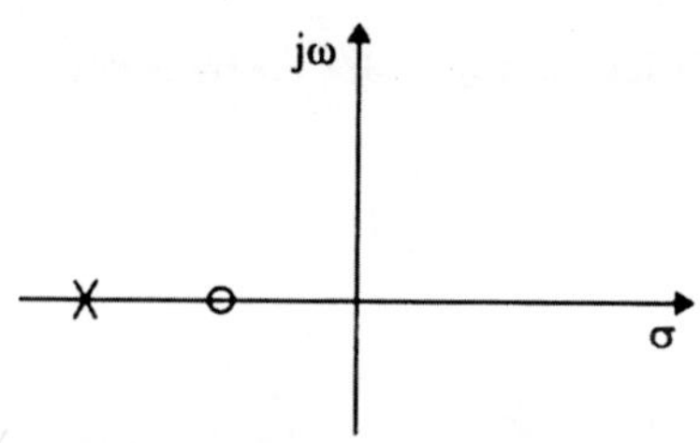

Fig. 10.32.

(a) lag compensator (b) lead compensator
(c) lead-lag compensator (d) series compensator

36. A lead compensator
(a) speeds up the transient response
(b) increases the stability margin
(c) increases the stability margin and speeds up the transient-response
(d) none of the above

37. The pole-zero configuration of Fig. 10.33 cam be for

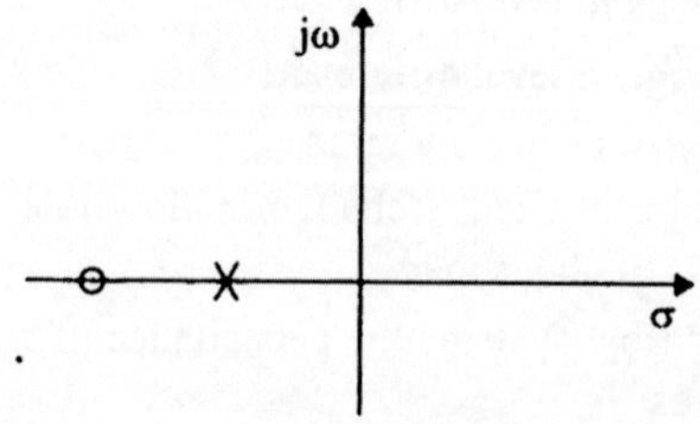

Fig. 10.33.

(a) lag compensator (b) lead compensator
(c) lead-lag compensator (d) series compensator

38. A lag compensator
(a) speeds up the transient response of the system

(*b*) improves steady state behaviour of the system
(*c*) improves steady state behaviour and speeds up the transient response of the system
(*d*) improves steady state behaviour and preserves the transient response of the system

39. The frequency at which phase angle is 180° is called
(*a*) phase cross-over frequency
(*b*) stability limit frequency
(*c*) frequency of limited stability
(*d*) gain margin frequency

40. A system has high gain and phase margins. The system is
(*a*) very stable
(*b*) sluggish
(*c*) very stable and sluggish
(*d*) oscillatory

41. A system is highly oscillatory if
(*a*) gain margin is high
(*b*) gain margin is close to 1
(*c*) gain margin is close to 1 or phase margin is zero
(*d*) gain margin is high and phase margin is 180°

42. A system has $\frac{Y(s)}{X(s)} = \frac{s+1}{s^2+5s+6}$. If $x(t) = t^2$, Y(s) is

(*a*) $\frac{2(s+1)}{s^2+5s+6}$ (*b*) $\frac{2s^2(s+1)}{s^2+5s+6}$

(*c*) $\frac{2(s+1)}{s^3(s^2+5s+6)}$ (*d*) $\frac{s+1}{2s^3(s^2+5s+6)}$

43. In the system in Fig. 10.34, $x(o) = 0$ and $\dot{x}(c) = 0$, At $t = 0$ the unit impulse $\delta(t)$ is applied. X(s) =

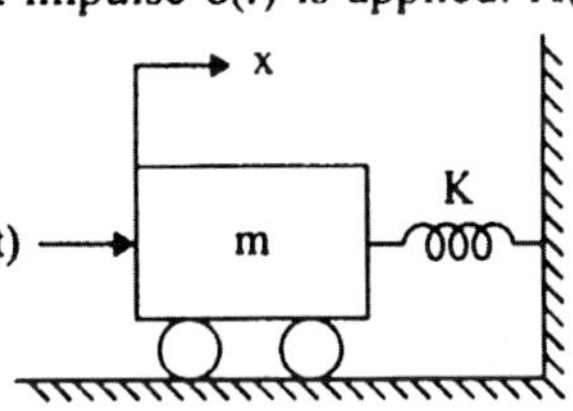

Fig. 10.34.

(*a*) $\frac{1}{ms+K}$ (*b*) $\frac{K}{ms}$

(*c*) $\frac{1}{ms^2+K}$ (*d*) $\frac{1}{ms^3+K}$

44. A motor is coupled to a load through gear ratio *n*. If T is the motor torque, J_m and J_L are moment of inertia of rotor and load, then torque to inertia ratio referred to motor shaft is

(*a*) $\frac{T}{J_m + J_L}$ (*b*) $\frac{T}{J_m + nJ_L}$

(*c*) $\frac{T}{J_m + n^2J_L}$ (*d*) $\frac{T}{n^2J_m + J_L}$

45. For Fig. 10.35, $\frac{C(s)}{R(s)}$ =

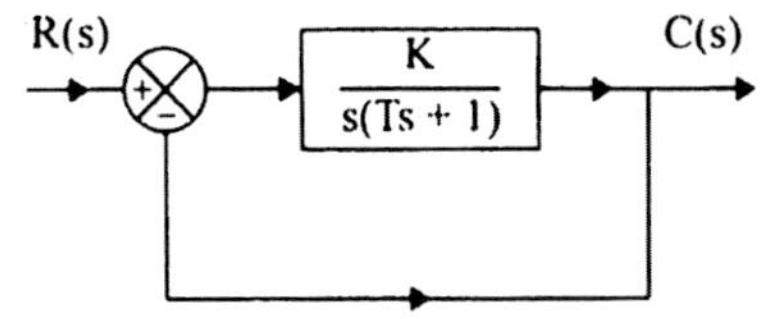

Fig. 10.35.

(*a*) $\frac{Ks}{s(Ts+1)}$ (*b*) $\frac{Ks}{Ts+1}$

(*c*) $\frac{K}{Ts^2+s+K}$ (*d*) $\frac{K}{Ts^3+s^2+Ks}$

46. For Fig. 10.36, C(*s*)/R(*s*) =

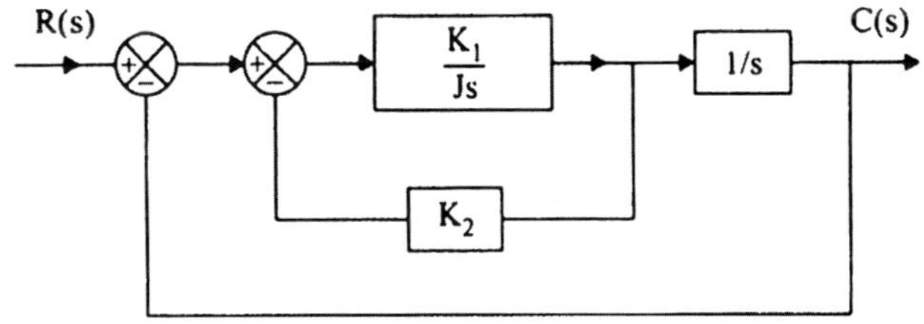

Fig. 10.36.

(*a*) $\frac{K_1}{Js^2+K_1K_2s+K_1}$ (*b*) $\frac{K_1}{Js^3+K_1K_2s+K_1}$

(*c*) $\frac{K_1K_2}{Js^2+K_1K_2+K_1}$ (*d*) none of the above

47. For Fig. 10.37, C(*s*)/R(*s*) =

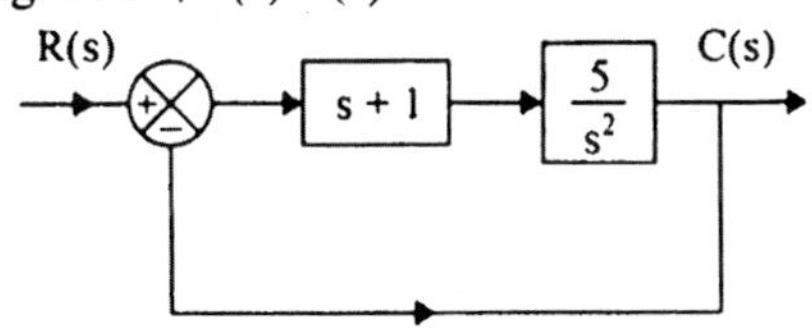

Fig. 10.37.

(*a*) $\frac{5s+5}{s^3+5s+5}$ (*b*) $\frac{5s+5}{s^3+5s^2+5}$

(*c*) $\frac{5s}{s^2+5s+5}$ (*d*) $\frac{5}{s^2+5s+5}$

48. For Fig. 10.38, the transfer function X(*s*)/P(*s*) =

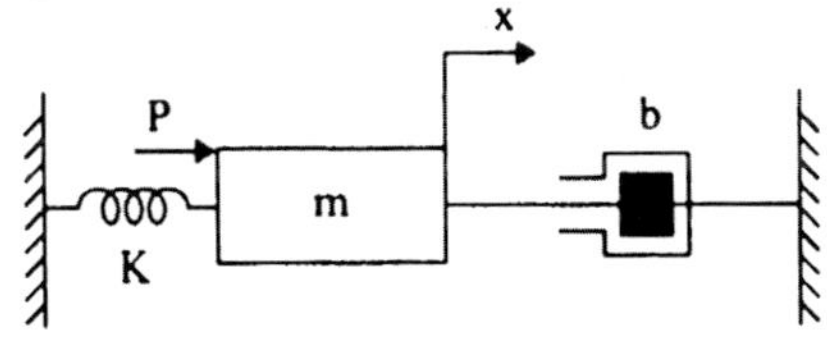

Fig. 10.38.

(*a*) $\frac{1}{ms+b+K}$ (*b*) $\frac{1}{ms^2+bs+K}$

(*c*) $\frac{K}{ms^2+bs+K}$ (*d*) $\frac{b}{ms^2+bs+K}$

49. In Fig. 10.38, P = 3 kg force. Then X(*s*) =

(*a*) $\frac{3}{ms+b+K}$ (*b*) $\frac{3}{ms^2+bs+K}$

(*c*) $\frac{3}{s(ms^2+bs+K)}$ (*d*) $\frac{3}{s^2(ms^2+bs+K)}$

50. In Fig. 10.38, $P = P_m \sin \omega t$. Then $X(s) =$

(*a*) $\dfrac{P_m \omega}{(s^2 + \omega^2)(ms^2 + bs + K)}$

(*b*) $\dfrac{P_m}{(s^2 + \omega^2)(ms^2 + bs + K)}$

(*c*) $\dfrac{P_m \omega^2}{(s^2 + \omega^2)(m^2 + bs + K)}$

(*d*) $\dfrac{P_m \omega}{s^2(ms^2 + bs + K)}$

51. A second order system had damping ratio ξ and natural frequency ω_n. The unit step response is

(*a*) $\dfrac{\omega_n}{s^2 + 2\xi\omega_n s + \omega_n^2}$ (*b*) $\dfrac{1}{s^2 + 2\xi\omega_n s + \omega_n^2}$

(*c*) $\dfrac{\omega_n^2}{s^2 + 2\xi\omega_n s + \omega_n^2}$ (*d*) $\dfrac{\omega_n^3}{s^2 + 2\xi\omega_n s + \omega_n^2}$

52. For Fig. 10.39, the equation of motion is

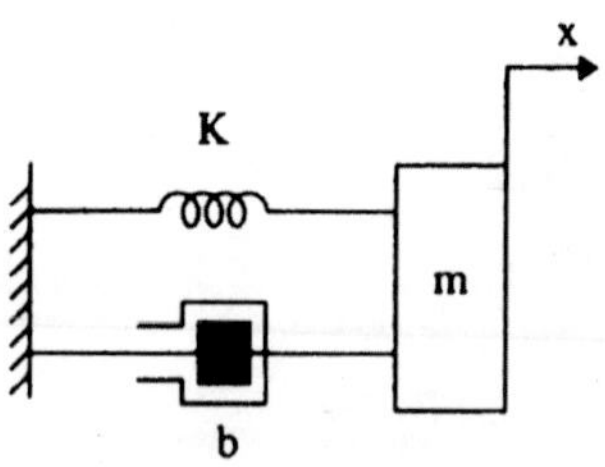

Fig. 10.39.

(*a*) $m\dfrac{d^2x}{dt^2} + b\dfrac{dx}{dt} + Kx = 0$ (*b*) $m\dfrac{dx}{dt} + bx + K = 0$

(*c*) $m\dfrac{d^2x}{dt^2} + (b + K)x = 0$ (*d*) none of the above

53. In a second order system $\omega_n = 10$ rad/sec and $\xi = 0.1$, then $\omega_d =$

(*a*) 10.1 rad/sec (*b*) 9.95 rad/sec
(*c*) 9 rad/sec (*d*) 1 rad/sec

54. For Fig. 10.40, $C(s)/R(s) =$

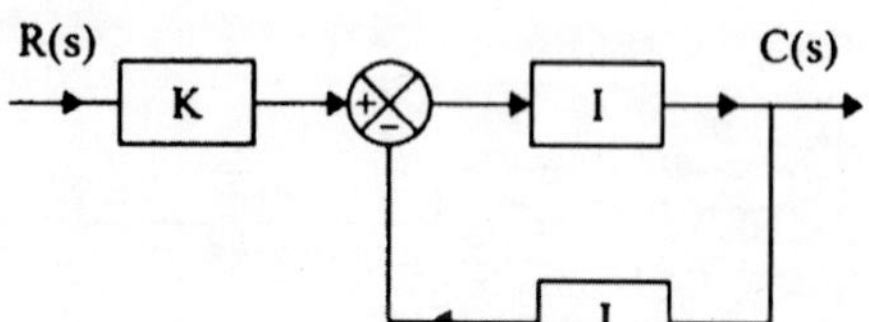

Fig. 10.40.

(*a*) K (*b*) 0.5 K
(*c*) 2 K (*d*) 0.25 K

55. A thermometer requires 1 minute to indicate 98% of its final response to a step input. If it is a first order system the time constant is

(*a*) 1 minute (*b*) 0.5 minute
(*c*) 0.25 minute (*d*) 0.1 minute

56. A unity feedback system has open loop transfer function $\dfrac{1}{(s+5)s}$. The closed loop transfer function is

(*a*) $\dfrac{1}{s^2 + 5s}$ (*b*) $\dfrac{1}{5s^2 + s}$

(*c*) $\dfrac{1}{s^2 + 5s + 1}$ (*d*) $\dfrac{s}{s^2 + 5s + 1}$

57. A unit feedback system has an open loop transfer function $\dfrac{2s+1}{s^2}$. The closed loop transfer function is

(*a*) $\dfrac{s^2}{2s + 1}$ (*b*) $\dfrac{2s + 1}{s^2 + 2s + 1}$

(*c*) $\dfrac{2s}{s^2 + 5s + 1}$ (*d*) $\dfrac{1}{s^2 + 5s + 1}$

58. For the system of Fig. 10.41 the closed loop transfer function is

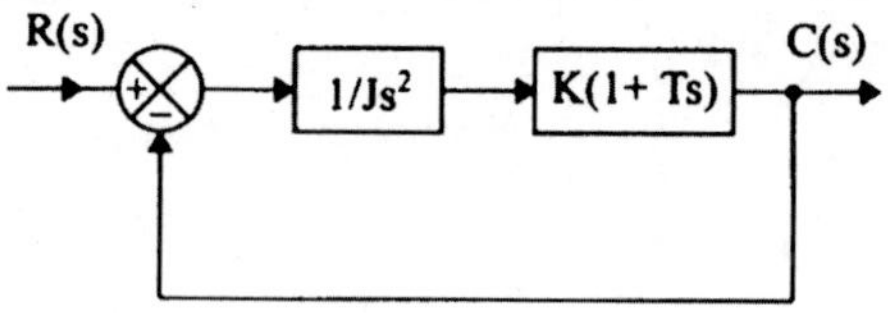

Fig. 10.41.

(*a*) $\dfrac{K(1 + T_s)}{Js^2}$ (*b*) $\dfrac{K}{Js^2(1 + T_s)}$

(*c*) $\dfrac{KTs^2 + K}{Js^2 + KTs + K}$ (*d*) $\dfrac{KTs + K}{Js^2 + KTs + K}$

59. For Fig. 10.42, $C(s)/R(s) =$

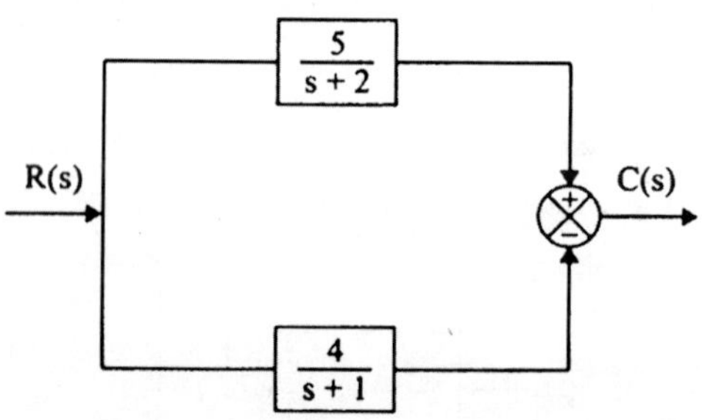

Fig. 10.42.

(*a*) $\dfrac{4(s+2)}{(s+2)(s+1)}$ (*b*) $\dfrac{s - 3}{(s+2)(s+1)}$

(*c*) $\dfrac{9s + 13}{(s+2)(s+1)}$ (*d*) $\dfrac{1}{(s+2)(s+1)}$

60. The system of Fig. 10.42

(*a*) has two poles and one zero in left half plane
(*b*) has two poles and one zero in right half plane
(*c*) has two poles in left half plane and a zero in right half plane
(*d*) none of the above

61. The block diagram of Fig. 10.43 is for

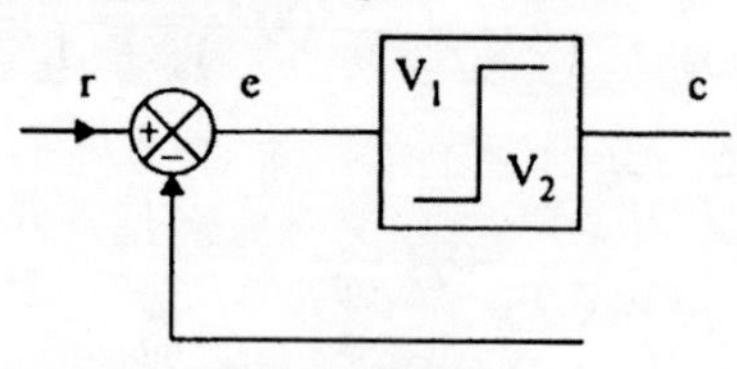

Fig. 10.43.

(a) On-off controller
(b) On-off controller with differential gap
(c) On-off controller with integral gap
(d) Three position controller

62. In some on-off controllers a differential gap is provided
(a) to make the manufacture process easier
(b) to prevent too often frequent operation of controller
(c) both (a) and (b)
(d) neither (a) nor (b)

63. In a simple on-off controller with differential gap, the magnitude of differential gap determines
(a) the level of accuracy (b) the life of controller
(c) both (a) and (b) (d) neither (a) and (b)

64. In response to a unit step input the controller output in Fig. 10.44 is for

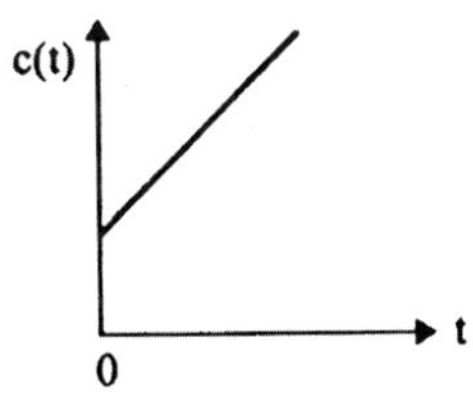

Fig. 10.44.

(a) proportional controller
(b) proportional plus integral controller
(c) integral controller
(d) none of the above

65. Match List I and List II and select correct answer using the given codes

List I (Controller) / **List II (Controller output for unit step input)**

A. Proportional — 1.

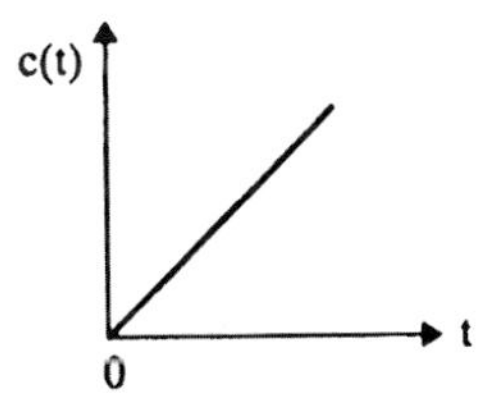

B. Proportinoal plus derivative — 2.

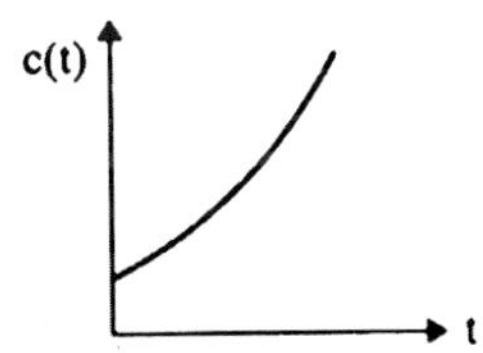

C. Proportinoal plus derivative plus integral — 3.

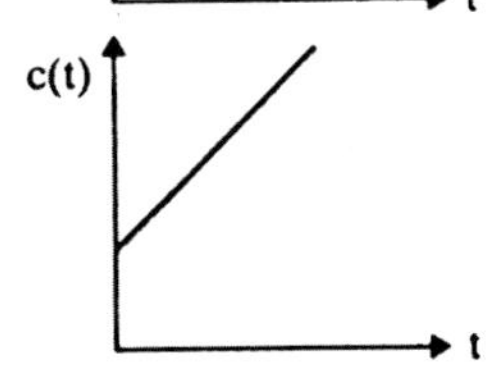

Codes:

	A	B	C
(a)	1	2	3
(b)	1	3	2
(c)	3	2	1
(d)	2	3	1

66. Integral control action
(a) removes offset
(b) may lead to oscillatory response
(c) both (a) and (b)
(d) neither (a) nor (b)

67. Whether a linear system is stable or unstable
(a) is a property of the system only
(b) depends on the input function only
(c) both (a) and (b)
(d) either (a) or (b)

68. Examine the following statements
1. Pneumatic systems are fire proof
2. Pneumatic systems are explosion proof
3. Hydraulic systems are explosion proof
4. Hydraulic systems are fire proof

Which the above statement are correct?
(a) 1 only (b) 1 and 2 only
(c) 1, 2, 3, 4 (d) 2, 3, 4 only

69. Examine the following statements
1. Air and gases are compressible
2. Air, gases and some liquids are compressible
3. Oil can function as hydraulic fluid as well as lubricator

Which of the above statements are correct?
(a) 3 only (b) 2 and 3 only
(c) 1 and 3 only (d) 2 only

70. Gas flow resistance is defined as
(a) $\frac{\text{Change in gas pressure difference}}{\text{Change in gas flow rate}}$
(b) $\frac{\text{Change in gas flow rate}}{\text{Change in gas pressure difference}}$
(c) (Change in gas pressure difference) (Change in gas flow rate)
(d) none of the above

71. The capacitance of a gas pressure vessel is defined as
(a) $\frac{\text{Change in gas stored}}{\text{Change in gas pressure}}$
(b) $\frac{\text{Change in gas pressure}}{\text{Change in gas stored}}$
(c) (Change in gas stored) (Change in gas pressure)
(d) none of the above

72. Examine the following statements
1. A small size hydraulic actuator can develop good torque
2. It is easy to maintain a hydraulic system leakproof
3. Both linear and rotary hydraulic actuators are available
4. Hydraulic power is more easily available than electric power

Which of the above statements are correct?
(a) all (b) 1 and 2 only
(c) 1 and 3 only (d) 2, 3 and 4 only

73. In op-amp circuit of Fig. 10.45, $E_o(s)/E_i(s)$ =
(a) $-\frac{R_2}{R_1}\left(\frac{1}{R_2Cs+1}\right)$ (b) $-\frac{R_2C}{R_1}$

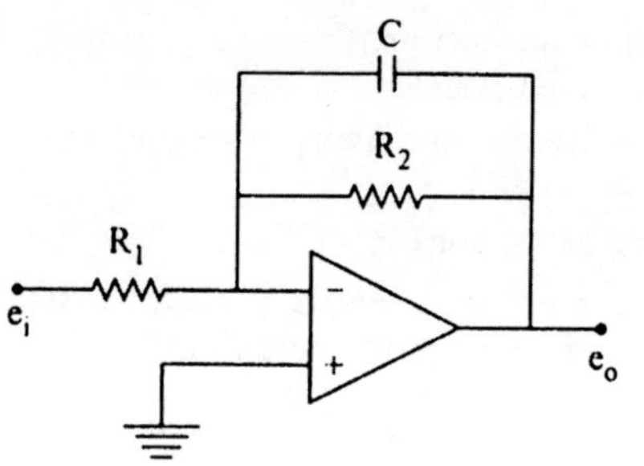

Fig. 10.45.

(c) $\dfrac{R_1}{R_2Cs+1}$ (d) $-\dfrac{R_1}{R_2}\left(\dfrac{1}{R_2Cs+1}\right)$

74. For op-amp circuit of Fig. 10.456, $E_o(s)/E_i(s)$ =

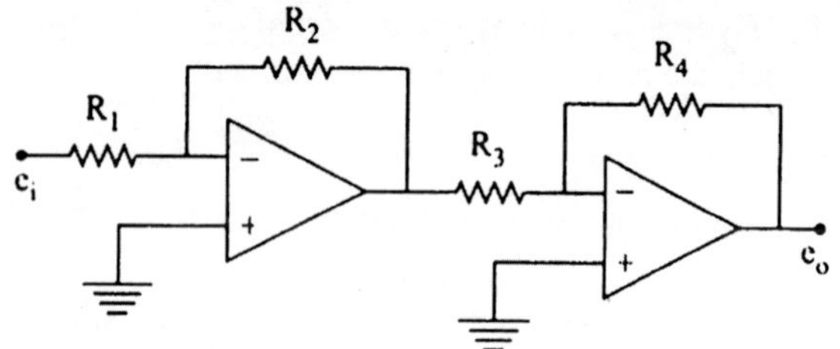

Fig. 10.46.

(a) $-\left(\dfrac{R_2}{R_1}\right)\left(\dfrac{R_3}{R_4}\right)$ (b) $\dfrac{R_2R_3}{R_1R_4}$

(c) $-\left(\dfrac{R_2}{R_1}\right)\left(\dfrac{R_4}{R_3}\right)$ (d) $\dfrac{R_2R_4}{R_1R_3}$

75. For the op-amp circuit of Fig. 10.47, $E_o(s)/E_i(s)$ =

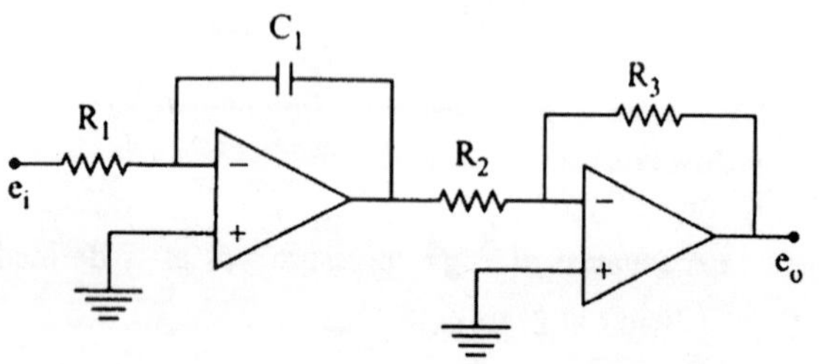

Fig. 10.47.

(a) $-\dfrac{R_3}{R_2}\left(\dfrac{1}{R_1C_1s}\right)$ (b) $\dfrac{R_3}{R_2}\left(\dfrac{1}{R_1C_1s}\right)$

(c) $\dfrac{R_3}{R_2}\left(\dfrac{C_1s}{R_1}\right)$ (d) $-\dfrac{R_3}{R_2}\dfrac{(C_1s)}{R_1}$

76. A system has transfer function $\dfrac{K_1K_2}{1+sT}$. If a sinusoidal input of frequency ω is applied, the steady state output is of the form

(a) $\dfrac{K_1K_2}{1+j\omega T}$ (b) $K_1\,K_2\,(1+j\omega T)$

(c) $K_1K_2/j\omega T$ (d) $\dfrac{K_1K_2}{1+j\omega^2 T}$

77. A network has $G(s) = \dfrac{s+1/T_1}{s+1/T_2}$. Then

(a) it is lead network

(b) it is lag network

(c) it is lead network if $T_1 > T_2$ and lag network $T_1 < T_2$

(d) it is lead network if $T_1 < T_2$ and and lag network if $T_1 > T_2$

78. For any test point s on the real axis, the sum of angular contributions of the complex conjugate poles is

(a) 90° (b) 180°

(c) 270° (d) 360°

79. The root locus branches

(a) start from open loop pole and terminate at zeros

(b) start from open loop zeros and terminate at poles

(c) may start from pole or zero and terminate at another pole or zero

(d) none of the above

80. The number of branches of root locus plot is equal to

(a) the number of roots of characteristic equation

(b) double the number of roots of characteristic equation

(c) the number of roots of characteristic equation minus one

(d) the number of roots of characteristic equation plus one

81. If zeros at infinity are included in the count, the number of zeros of G(s) H(s) is

(a) equal to number of poles

(b) one more than the number of poles

(c) one less than the number of poles

(d) none of the above

82. In root locus analysis the breakway and beak in points

(a) lie on the real axis

(b) either lie on the real axis or occur in complex conjugate pairs

(c) always occur in complex conjugate pairs

(d) none of the above

83. The system in Fig. 10.48.

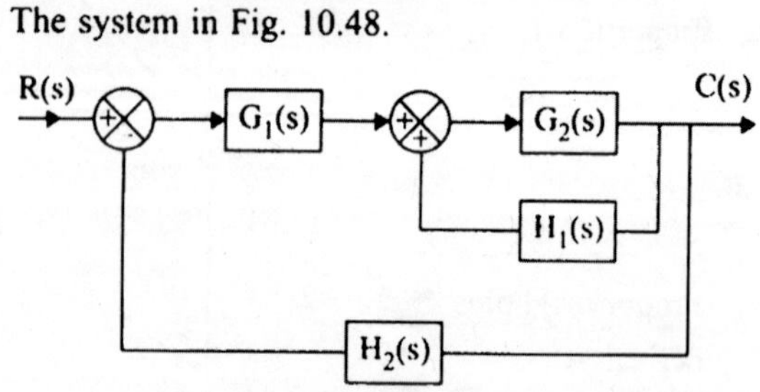

Fig. 10.48.

(a) will be unstable due to the presence of positive feedback in the inner loop

(b) may be stabilized by the negative feedback outer loop

(c) can never be stable

(d) will always be stable

84. In a minimum phase system

(a) all poles lies in the left half plane

(b) all zeroes lie in the left half plane

(c) all poles and zeroes lie in the left half plane

(d) all except one pole or zero lie in the left half plane

85. Transport lag usually exists in
(*a*) thermal systems
(*b*) hydraulic systems and thermal systems
(*c*) pneumatic systems
(*d*) thermal, hydraulic and pneumatic systems

86. For Fig. 10.49, $C(s)/R(s)$ =

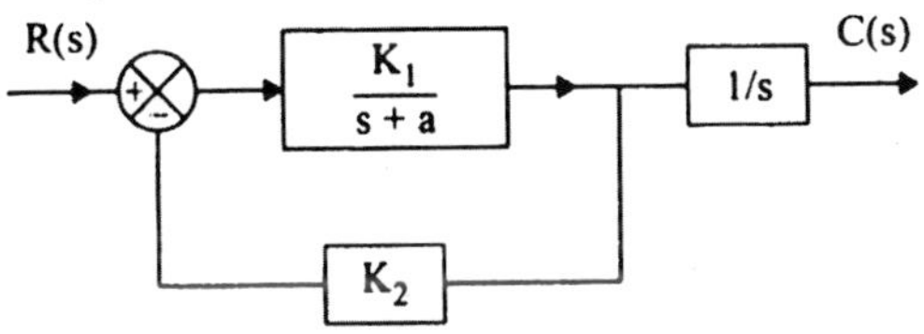

Fig. 10.49.

(*a*) $\frac{K_1}{s+a+K_1K_2}$ (*b*) $\frac{K_1}{s(s+a+K_1K_2)}$
(*c*) $\frac{K_2}{s(s+a+K_1K_2)}$ (*d*) $\frac{K_1K_2}{s(s+a+K_1K_2)}$

87. For the system in Fig. 10.50 characteristic equation is

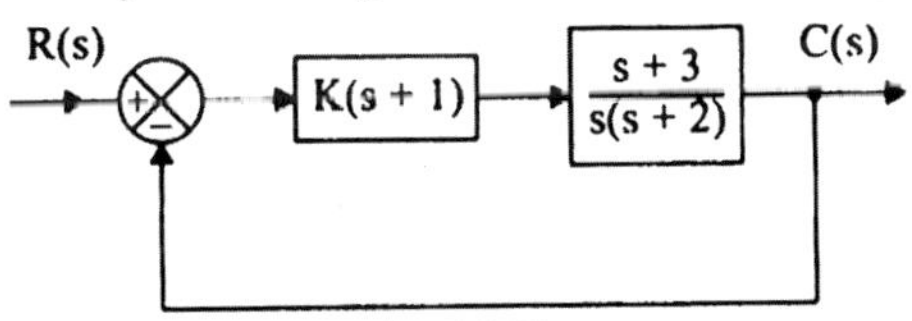

Fig. 10.50.

(*a*) $1+\frac{K(s+1)(s+3)}{s(s+2)}=0$

(*b*) $1+\frac{K(s+2)(s+3)}{s(s+1)}=0$

(*c*) $K(s + 1)(s + 3) = 0$
(*d*) $s(s + 2) = 0$

88. For the system of Fig. 10.50, the break away and break in points are determined from the condition $\frac{dK}{ds}=0$. Then K =

(*a*) $\frac{-s(s+1)}{(s+2)(s+3)}$ (*b*) $\frac{-s(s+2)}{(s+1)(s+3)}$

(*c*) $\frac{s(s+1)}{(s+2)(s+3)}$ (*d*) $\frac{s(s+2)}{(s+1)(s+3)}$

89. For the system in Fig. 10.51 the root locus

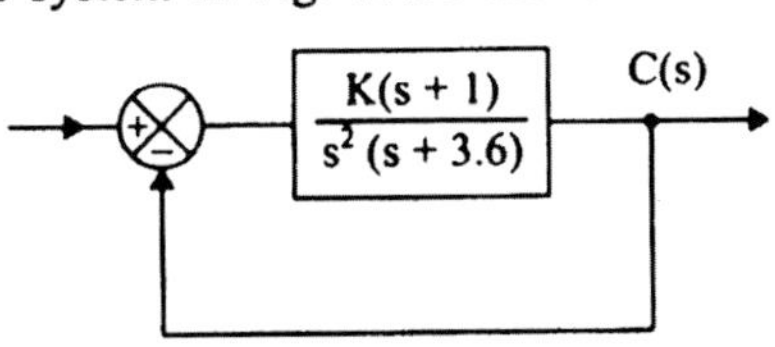

Fig. 10.51.

(*a*) is on real axis between points $s = -1$ and $s = -3.6$
(*b*) is on real axis between points $s = 0$ and $s = -1$
(*c*) is on real axis between points $s = -3.6$ and $s = \infty$
(*d*) is on real axis between points $s = 0$ and $s = -3.6$

90. The addition of a pole to the open loop transfer function pulls the root locus
(*a*) to the right
(*b*) to the left and making the system more stable
(*c*) towards right or left depending on the position of new pole
(*d*) none of these above

91. The addition of a zero to the open loop transfer function
(*a*) pulls the root locus to the left and makes the system more stable
(*b*) pulls the root locus to the right and makes the system less stable
(*c*) either (*a*) or (*b*)
(*d*) neither (*a*) nor (*b*)

92. For the system of Fig. 10.52, the closed loop poles are located at

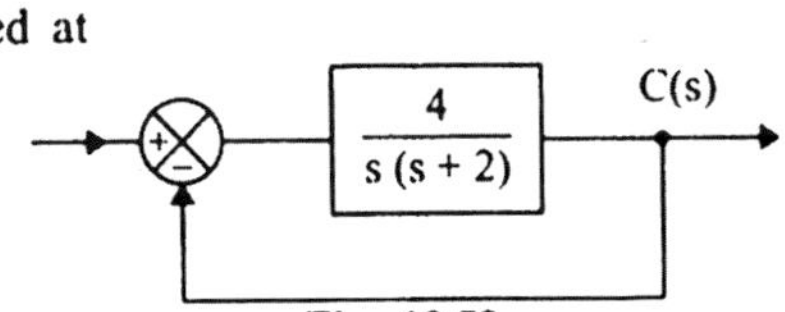

Fig. 10.52.

(*a*) $s = 0$ and $s = -2$ (*b*) $s = 0,\ s = -1 \pm j\sqrt{3}$
(*c*) $s = -1 \pm j\sqrt{3}$
(*d*) $s = -2$ and $s = -1 \pm j\sqrt{3}$

93. For the system of Fig. 10.52, the damping ratio of closed loop poles is
(*a*) 1.5 (*b*) 1
(*c*) 0.5 (*d*) 0.25

94. For the system of Fig. 10.52, the undamped natural frequency of closed loop poles is
(*a*) 2 rad/sec (*b*) 1 rad/sec
(*c*) 0.5 rad/sec (*d*) 0.25 rad/sec

95. The angular location of poles depends on
(*a*) undamped natural frequency
(*b*) damping ratio
(*c*) both (*a*) and (*b*)
(*d*) neither (*a*) nor (*b*)

96. The distance of poles from origin depends on
(*a*) undamped natural frequency
(*b*) damping ratio
(*c*) both (*a*) and (*b*)
(*d*) neither (*a*) nor (*b*)

97. If damping ratio is 0.5, the lines joining complex poles with origin are inclined to negative real axis at
(*a*) ±90° (*b*) ±60°
(*c*) ±45° (*d*) ±30°

98. If poles lie in first quadrant, damping ratio is
(*a*) 1 (*b*) more than 1
(*c*) less than 1 (*d*) zero

99. If the poles lie in fourth quadrant, the damping ratio is
(*a*) 1 (*b*) less than 1
(*c*) zero (*d*) more the 1

100. The damping ratio of a pair of complex poles when expressed in terms of angle θ measured form $j\omega$ axis is equal to

(*a*) $\sin\theta$ (*b*) $\cos\theta$
(*c*) $\tan\theta$ (*d*) $\cot\theta$

101. The transfer function $G_C = 5\left(\frac{s+2.9}{s+5.4}\right)$ can be for

(*a*) lead compensator (*b*) lag compensator
(*c*) either (*a*) or (*b*) (*d*) neither (*a*) nor (*b*)

102. The transfer function $G_C = 3\left(\frac{s+0.05}{s+0.005}\right)$ can be for

(*a*) lead compensator (*b*) lag compensator
(*c*) neither (*a*) nor (*b*) (*d*) either (*a*) or (*b*)

103. The transfer function $G_C = 5\left(\frac{s+2.39}{s+8.45}\right)\left(\frac{s+0.1}{s+0.03}\right)$ can be for

(*a*) lead compensator (*b*) lag compensator
(*c*) lead-lag compensator (*d*) either (*a*) or (*b*)

104. The compensator of Fig. 10.53 is a

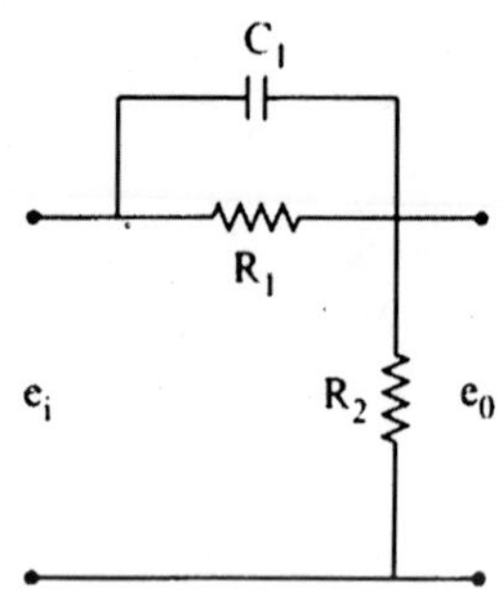

Fig. 10.53.

(*a*) lag compensator (*b*) lead compensator
(*c*) lag-lead compensator (*d*) none of the above

105. The compensator in Fig. 10.54 is a

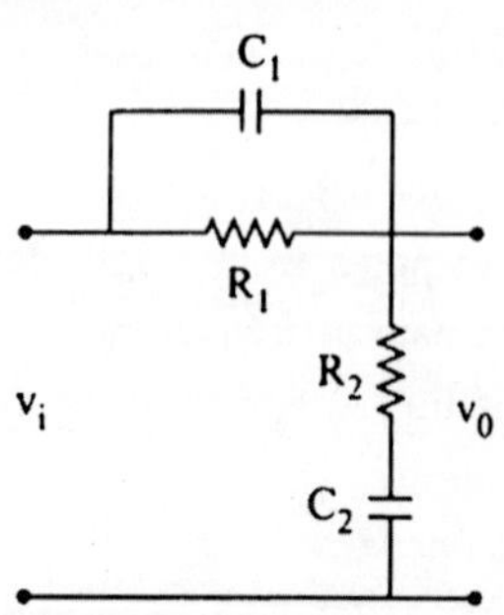

Fig. 10.54.

(*a*) lag compensator (*b*) lead compensator
(*c*) lag-lead compensator (*d*) none of the above

106. The system of Fig. 10.55

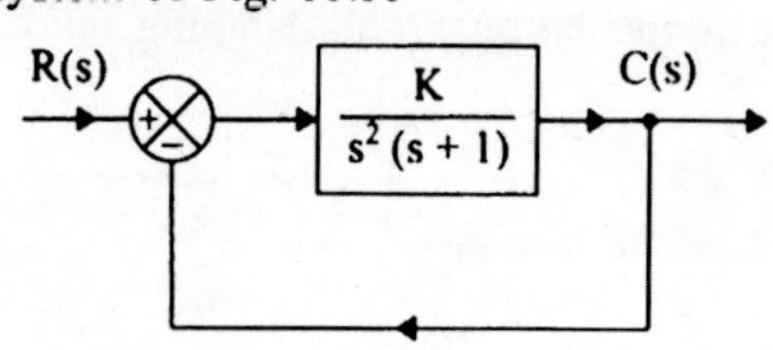

Fig. 10.55.

(*a*) is unstable for all values of K
(*b*) is stable for all values of K
(*c*) can be stable for some values of K
(*d*) data is insufficient to determine stability

107. For the system of Fig. 10.56, the closed loop poles are at

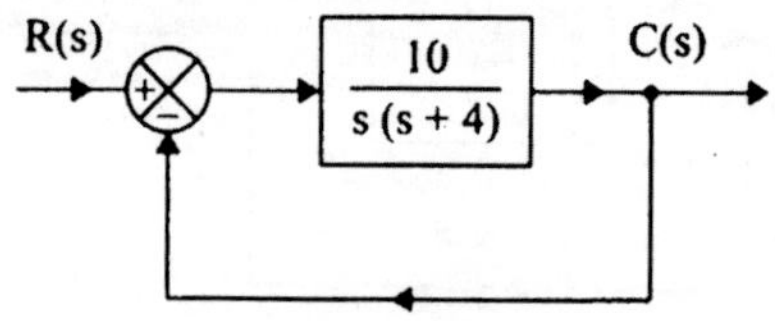

Fig. 10.56.

(*a*) 0 and –4 (*b*) $-2 \pm j\sqrt{6}$
(*c*) $2 \pm j\sqrt{6}$ (*d*) $-3 \pm j\sqrt{8}$

108. The log magnitude curve for a constant gain K is a

(*a*) horizontal straight line
(*b*) horizontal straight line of magnitude 20 log K decibels
(*c*) an inclined line having slope K
(*d*) an inclined line having slope –K

109. In Bode diagrams an octave is a frequency band from

(*a*) ω_1 to 10 ω_1 (*b*) ω_1 to 8 ω_1
(*c*) ω_1 to 4 ω_1 (*d*) ω_1 to 2 ω_1

110. In Bode diagram (log magnitude plot) the factor $\frac{1}{j\omega}$ in the transfer function gives a line having slope

(*a*) – 20 *dB* per octave (*b*) – 10 *dB* per octave
(*c*) – 6 *dB* per octave (*d*) – 2 *dB* per octave

111. In Bode diagram (log magnitude plot) the factor $(j\omega)^n$ in the transfer function gives a line having slope

(*a*) 20 *dB*/decade (*b*) 20 *dB*/decade
(*c*) $\frac{20}{n}$ *dB*/decade (*d*) – 20*n* *dB*/decade

112. In Bode diagram (log magnitude plot) of the factor $1/(1 + j\omega T)$ in the transfer function, the log magnitude for $\omega << \frac{1}{T}$ can be approximated by

(*a*) 0 *dB* (*b*) 20 log ω T *dB*
(*c*) – 20 log ωt *dB* (*d*) none of the above

113. At corner frequency the phase angle of factor $1/(1 + j\omega T)$ is

(*a*) + 90° (*b*) –90°
(*c*) +45° (*d*) –45°

114. In log magnitude Bode diagram the slope of high frequency asymptote of $(1 + j\omega T)$ is

(*a*) 20 *dB* per decade (*b*) 10 *dB* per decade
(*c*) 20 *dB* per octave (*d*) 10 *dB* per octave

115. For quadratic factor

$$\frac{1}{1+2\xi\left(j\frac{\omega}{\omega_n}\right)+\left(j\frac{\omega}{\omega_n}\right)^2}$$

In Bode log magnitude plot the high frequency asymptote is a straight line having die slope
(*a*) –20 *dB* per decade (*b*) +20 *dB* per decade
(*c*) –40 *dB* per decade (*d*) +40 *dB* per decade

116. For a quadratic factor in denominator of transfer function the phase angle at corner frequency (in Bode phase angle plot)
(*a*) is –90°
(*b*) is +90°
(*c*) depends on the damping factor
(*d*) none of the above

117. The phase angle curve of G(*j*ω) H(*j*ω) can be drawn
(*a*) by adding the phase angle curves of individual factors
(*b*) by subtracting the phase angle curves of individual factors
(*c*) by multiplying the phase angle curves of individual factors
(*d*) by dividing the phase angle curves of individual factors

118. The open loop transfer function of a system is $\frac{K}{s(Js+F)}$.
If the frequency at the intersection of –40 *dB*/decade line and 0 *dB* line is ω, then
(*a*) $\omega = \frac{K}{J}$ (*b*) $\omega^2 = \frac{K}{J}$
(*c*) $\omega^3 = \frac{K}{J}$ (*d*) $\omega^4 = \frac{K}{J}$

119. The polar plot of G(*j*ω) = 1/*j*ω is
(*a*) positive imaginary axis
(*b*) positive real axis
(*c*) negative imaginary axis
(*d*) negative real axis

120. If $G(j\omega) = \frac{1}{1+j\omega T}$, the polar plot, as the frequency is varied from 0 to infinity is
(*a*) a semicircle (*b*) a circle
(*c*) a straight line (*d*) none of the above

121. Consider the following statements about transport-lag
1. It is of non-minimum phase behaviour
2. It normally exits in thermal, hydraulic and pneumatic systems
3. It has an excessive phase lag with no attenuation at high frequency

Which of the above statements are correct?
(*a*) 1 only (*b*) 1 and 2
(*c*) all (*d*) 2 and 3

122. For the transport lag $G(j\omega) = e^{-j\omega T}$, the magnitude is always equal to
(*a*) 0 (*b*) 1
(*c*) 10 (*d*) 0.5

123. For the transport lag $G(j\omega) = e^{-j\omega T}$, the phase angle
(*a*) is constant
(*b*) varies linearly with frequency
(*c*) varies linearly with frequency for low frequencies only
(*d*) varies as per square law

124. For the transport lag $G(j\omega) = e^{-j\omega T}$, the polar plot is
(*a*) a semicircle (*b*) a circle
(*c*) a unit circle (*d*) none of the above

125. For the transfer function $G(j\omega) = \frac{e^{-j\omega L}}{1+j\omega T}$, the polar plot is a
(*a*) semicircle (*b*) circle
(*c*) spiral (*d*) none of the above

126. For type 2 system, the magnitude and phase angle of the term $(j\omega)^2$ in the denominator, at ω = 0, are respectively
(*a*) 0 and –90° (*b*) 0 and +90°
(*c*) infinity and –180° (*d*) infinity and +180°

127. The polar plot in Fig. 10.57 is for the term

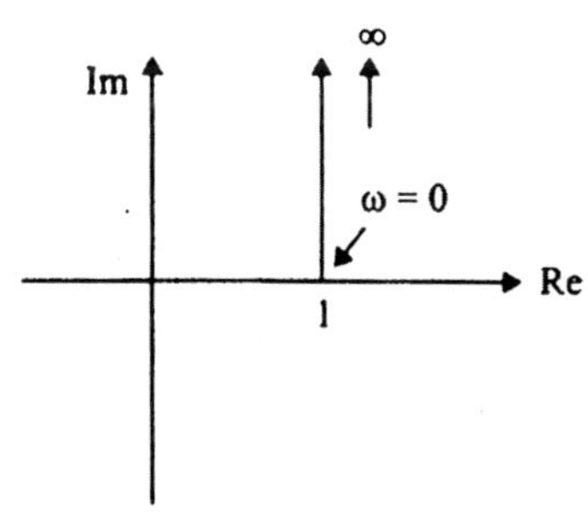

Fig. 10.57.

(*a*) *j*ωT (*b*) 1 + *j*ωT
(*c*) $\frac{1}{j\omega T}$ (*d*) $\frac{1}{1+j\omega T}$

128. The polar plot of Fig. 10.58 is for the term

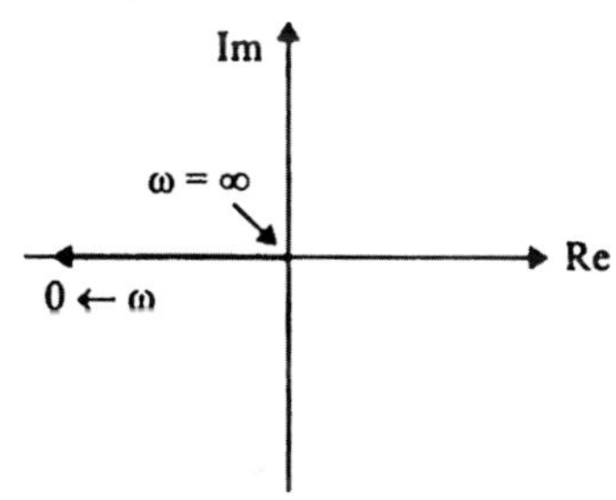

Fig. 10.58.

(*a*) $(j\omega)^2$ (*b*) $1 + (j\omega)^2$
(*c*) $\frac{1}{(j\omega)^2}$ (*d*) $\frac{1}{(1+j\omega)}$

129. If Z is the number of zeros of 1 + G(*s*) H(*s*) in the right half *s* plane,
N is die number of clockwise encirclements of (–1 + *jo*) point
P is the number of poles of G(*s*) H(*s*) in right half *s* plane
The Nyquist stability criterion can be expressed as
(*a*) Z = N + P (*b*) Z = P/N
(*c*) Z = N – P (*d*) Z = PN

130. If Nyquist plot has no encirclement of (–1 + *jo*) point, the system is stable if

(*a*) there are no poles of G(*s*) H(*s*) in right half plane
(*b*) there is one pole of G(*s*) H(*s*) in right half plane
(*c*) there are no zeros of G(*s*) H(*s*) in right half plane
(*d*) none of the above

131. A conditionally stable system is stable for the value of gain between two critical values. It is unstable if
(*a*) the gain is increased beyond upper critical value
(*b*) the gain is decreased below lower critical value
(*c*) both (*a*) and (*b*)
(*d*) none of the above

132. If $G(s) = \frac{1}{s^3 + s^2 + 1}$, there are
(*a*) two poles of G(*s*) in the left half plane
(*b*) two poles of G(*s*) in the right half plane
(*c*) three poles of G(*s*) in the right half plane
(*d*) four poles of G(*s*) in the right half plane

133. If phase angle of open loop transfer function becomes $-180°$ at frequency ω_1, then gain margin is equal to
(*a*) $|G(j\omega_1)|$ (*b*) $\frac{1}{|G(j\omega_1)|}$
(*c*) $1+|G(j\omega_1)|$ (*d*) $\frac{1}{1+|G(j\omega_1)|}$

134. The gain margin for a stable system
(*a*) has a positive decibel value
(*b*) has a negative decibel value
(*c*) has a large negative decibel value
(*d*) none of the above

135. A conditionally stable system has
(*a*) one phase cross over frequency
(*b*) two phase cross over frequencies
(*c*) two or more phase cross-over frequencies
(*d*) one or more phase cross-over frequencies

136. For a stable system having two or more gain cross-over frequencies the phase margin is measured
(*a*) at lowest gain cross-over frequency
(*b*) at highest gain cross-over frequency
(*c*) at either of the gain cross-over frequencies
(*d*) none of the above

137. The relative stability of a system is given by
(*a*) gain margin alone
(*b*) phase margin alone
(*c*) both phase and gain margins
(*d*) either (*a*) or (*b*)

138. For a stable system
(*a*) both phase and gain margin are negative
(*b*) both phase and gain margin are positive
(*c*) phase margin is positive but gain margin is negative
(*d*) gain margin is positive but phase margin is negative

139. If a system is to follow arbitrary inputs accurately the bandwidth should be
(*a*) large (*b*) small
(*c*) very small (*d*) neither small nor large

140. From the view point of noise, bandwidth should
(*a*) be large
(*b*) not be too large
(*c*) should be as large as possible
(*d*) should be infinite

141. As the bandwidth increases, the cost of components
(*a*) generally decreases
(*b*) generally increases
(*c*) may increase or decrease
(*d*) does not change

142. Consider the following systems

System I, $\frac{C(s)}{R(s)} = \frac{1}{s+1}$

System II, $\frac{C(s)}{R(s)} = \frac{1}{3s+1}$

Where C(*s*) is the output and R(*s*) is the input. Of these systems
(*a*) System I has larger bandwidth and faster speed of response than system II
(*b*) System II has larger bandwidth and faster speed of response than system I
(*c*) System I has larger bandwidth but system II faster speed of response
(*d*) System II has larger bandwidth but system I has faster speed of response

143. The constant magnitude loci on G(*s*) plane are
(*a*) a family of straight horizontal lines
(*b*) a family of straight inclined lines with positive slope
(*c*) a family of straight inclined lines with negative slope
(*d*) a family of circles

144. For any transfer function, the constant phase angle loci are
(*a*) a family of circles (*b*) a family of straight lines
(*c*) either (*a*) or (*b*) (*d*) neither (*a*) nor (*b*)

145. Nichols chart consists of
(*a*) Constant magnitude loci
(*b*) Constant phase angle loci
(*c*) Magnitude and phase angle loci in log-magnitude versus phase diagram
(*d*) none of the above

146. The log magnitude plot of Fig. 10.59 is for

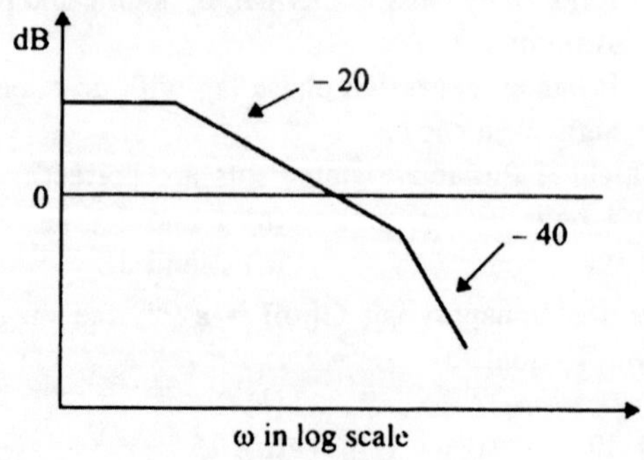

Fig. 10.59.

(*a*) type 0 system
(*b*) type 1 system

(c) type 2 system
(d) either type 0 or type 1 system

147. For the system in Fig. 10.60, the gain margin is
(a) 0 (b) low
(c) $+\infty$ (d) none of the above

148. For the system of Fig. 10.60, $|G(j\omega)| = 0$ dB at ω =

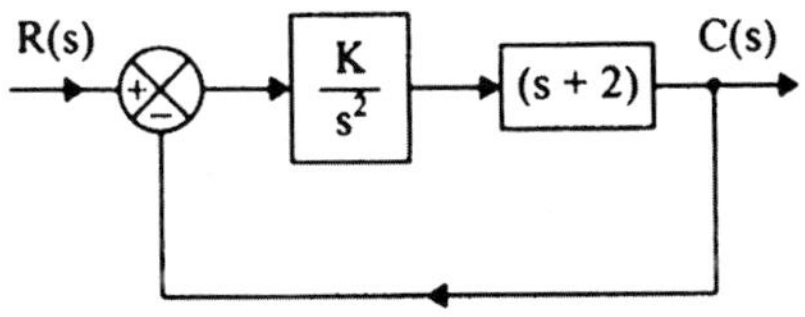

Fig. 10.60.

(a) 0 (b) 1 rad/sec
(c) ∞ (d) 2.38 rad/sec

149. A lead compensator is basically a
(a) high pass filter (b) low pass filter
(c) band stop filter (d) none of the above

150. A lag compensator provides attenuation in
(a) very low frequency range
(b) low frequency range
(c) high frequency range
(d) none of the above

151. The primary function of lag compensator is to provide sufficient
(a) gain margin
(b) phase margin
(c) both gain and phase margins
(d) either (a) or (b)

152. A lag compensator is essentially a
(a) low pass filter (b) high pass filter
(c) band pass filter (d) either (b) or (c)

153. PI-D controlled system has
(a) P and I actions in forward path and D action in feedback path
(b) P and I actions in feedback path and D action in forward path
(c) All the three (i.e. P, I and D) actions in forward path
(d) All the three (i.e. P, I and D) actions in feedback path

154. I-PD controlled system has
(a) I, P and D actions in forward path
(b) I, P and D actions in feedback path
(c) I action in forward path but P and D actions in feedback path
(d) I action in feedback path but P and D actions in forward path

155. Modeling error may occur due to
(a) neglecting the non-linear characteristics of plant
(b) inaccuracy of parameters
(c) change of plant characteristics with time
(d) any of above

156. If the characteristic equation is $(\lambda + 1)(\lambda + 2)(\lambda + 3) = 0$, the eigen values are
(a) –1, –2, –3 (b) 1, 2, 3
(c) 0, 1, 2 (d) 0, 2, 3

157. For an $n \times n$ matrix A, the eigen values are the roots of the characteristic equation
(a) $|I - A| = 0$ (b) $|\lambda I - A| = 0$
(c) $|\lambda A - I| = 0$ (d) $|A - I| = 0$

158. Figure 10.61 shows a root locus plot. Consider the following statements for this plot

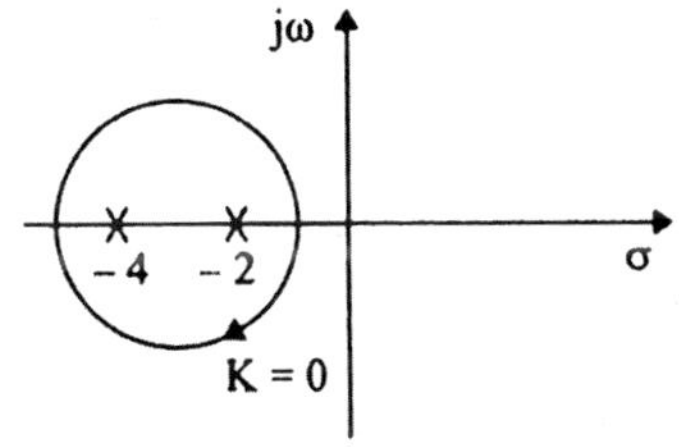

Fig. 10.61.

1. The open loop system is a second order system
2. The system is overdamped for K > 1
3. The system is stable for all values of K

Which of the above statements are correct?
(a) 1, 2, 3 (b) 1, 3
(c) 2, 3 (d) 1, 2

159. A system has 12 poles and 2 zeros. Its high frequency asymptote in its magnitude plot has a slope of
(a) –200 dB/decade (b) –240 dB/decade
(c) –280 dB/decade (d) –320 dB/decade

160. The effect of adding poles and zeros can be determined quickly by
(a) Nichols chart (b) Nyquist plot
(c) Bode plot (d) Root locus

161. Magnetic amplifier is used for
(a) voltage amplification (b) power amplification
(c) current amplification (d) frequency amplification

162. In Fig. 10.62, x_6 =

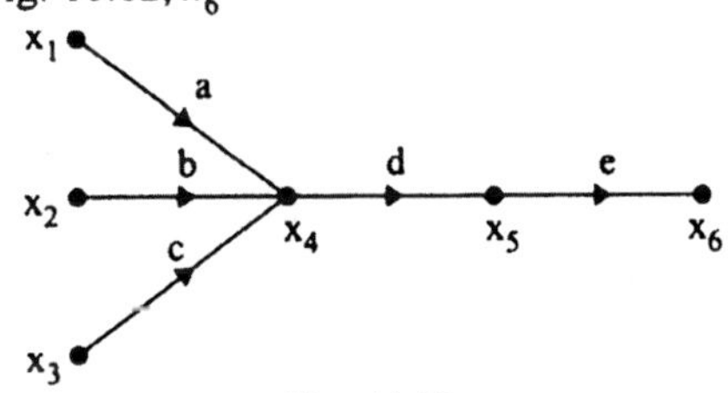

Fig. 10.62.

(a) $de(ax_1 + bx_2 + cx_3)$
(b) $(a + b + c)(x_1 + x_2 + x_3)(d + e)$
(c) $(ax_1 + bx_2 + cx_3)(d + e)$
(c) $a\,b\,c\,d\,e\,(x_1 + x_2 + x_3)$

163. For Fig. 10.63, the transfer function of signal flow graph is

(a) $\dfrac{T_{12}}{1 - T_{22}}$

(b) $\dfrac{T_{22}}{1 - T_{12}}$

(c) $\dfrac{T_{12}}{1 + T_{22}}$

(d) $\dfrac{T_{22}}{1 + T_{12}}$

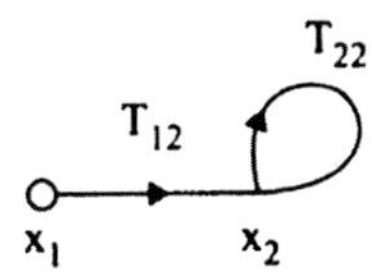

Fig. 10.63.

164. A second order system exhibits 100% overshoot. The damping ratio is

(*a*) 0 (*b*) 1
(*c*) < 1 (*d*) > 1

165. A stepper motor is

(*a*) a two-phase induction motor
(*b*) is a kind of rotating amplifier
(*c*) is an electromagnetic transducer used to convert an angular position of shaft into electrical signal
(*d*) is an electromechanical device which actuates a train of step angular movements in response to a train of input pulses on one to one basis

166. For the second order system $2\frac{d^2y}{dt}+4\frac{dy}{dt}+8y=8x$, the damping ratio is

(*a*) 0.1 (*b*) 0.25
(*c*) 0.333 (*d*) 0.5

167. For the circuit of Fig. 10.64, $\frac{v_0}{v_i}$ =

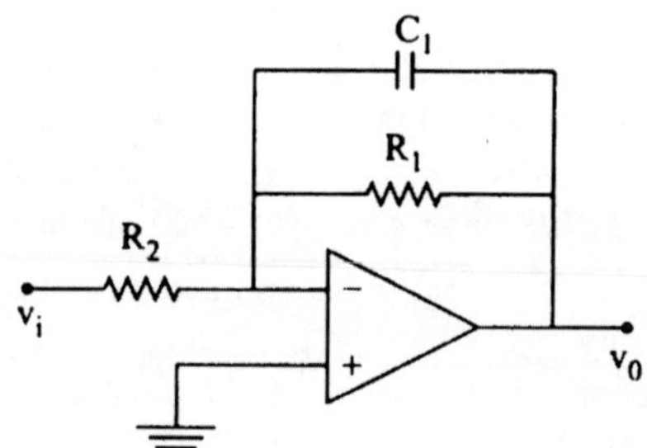

Fig. 10.64.

(*a*) $\frac{1}{R_2(1+j\omega C_1R_1)}$ (*b*) $\frac{-1}{R_2(1+j\omega C_1R_1)}$

(*c*) $\frac{-R_1/R_2}{1+j\omega R_1C_1}$ (*d*) $\frac{R_1/R_2}{1+j\omega R_1C_1}$

168. Match List I and List II and select the correct answer using the codes

List I	List II
A. Synchro	1. Amplifier
B. Amplidyne	2. Actuator
C. Servo	3. Compensator
D. RC network	4. Transducer

Codes:

	A	B	C	D
(*a*)	1	2	3	4
(*b*)	4	3	2	1
(*c*)	3	2	4	1
(*d*)	4	1	2	3

169. Match List I and List II and select the answer using the codes

List I (Location of characteristic roots)	List II (System characteristic)
A. (–1 + *j*), (–1 –*j*)	1. Marginally stable
B. (–2 + *j*), (–2 –*j*), (2 *j*) (–2*j*)	2. Unstable
C. –*j*, *j*, –1, 1	3. Stable

Codes:

	A	B	C
(*a*)	1	2	3
(*b*)	2	3	1
(*c*)	3	1	2
(*d*)	1	3	2

170. Match List I and List II and select the correct answer using the codes

List I (Roots in *s* plane) — **List II (Impulse response)**

A. Two imaginary roots — 1.

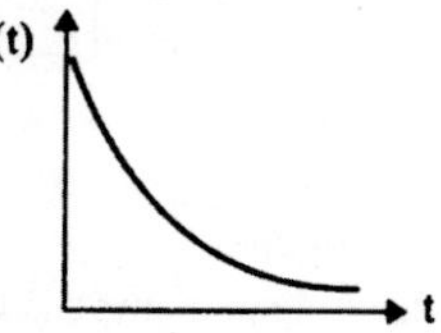

B. Two complex roots in right half plane — 2.

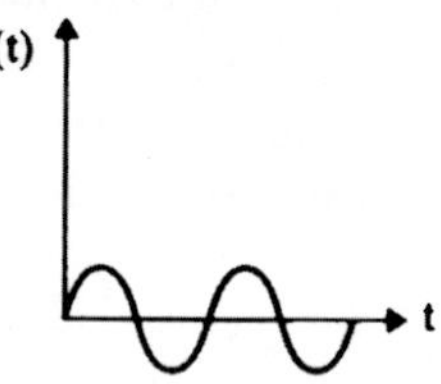

C. A single root on negative real axis — 3.

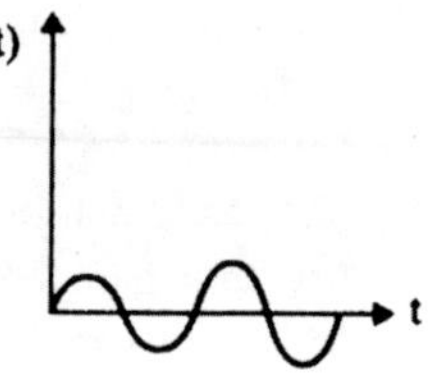

D. A single root at origin — 4.

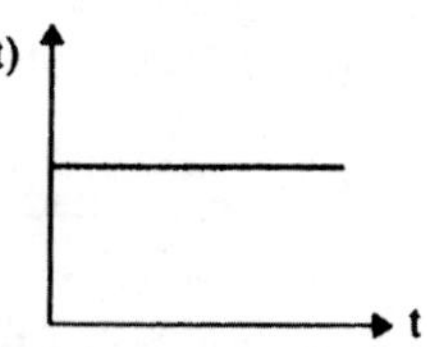

Codes:

	A	B	C	D
(*a*)	2	3	1	4
(*b*)	1	2	3	4
(*c*)	4	3	2	1
(*d*)	3	2	4	1

171. For type 1 system the velocity lag error is

(*a*) inversely proportional to bandwidth of system
(*b*) directly proportional to gain constant
(*c*) inversely proportional to gain constant
(*d*) independent of gain constant

172. The system in Fig. 10.65 is

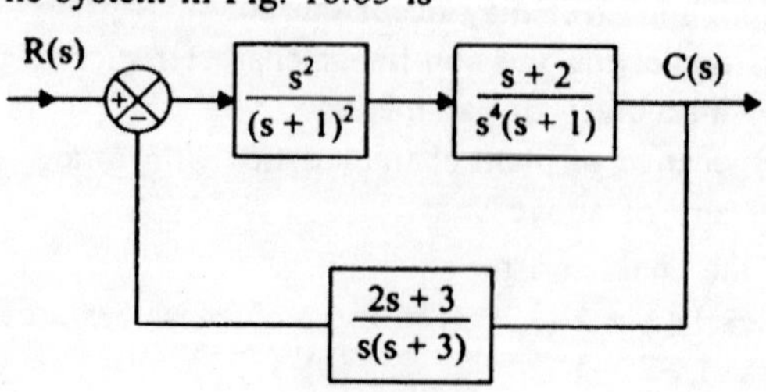

Fig. 10.65.

(*a*) type 0 system (*b*) type 1 system
(*c*) type 2 system (*d*) type 3 system

173. In the signal flow graph of Fig. 10.66 the number of forward paths and pairs of non-touching loops are respectively

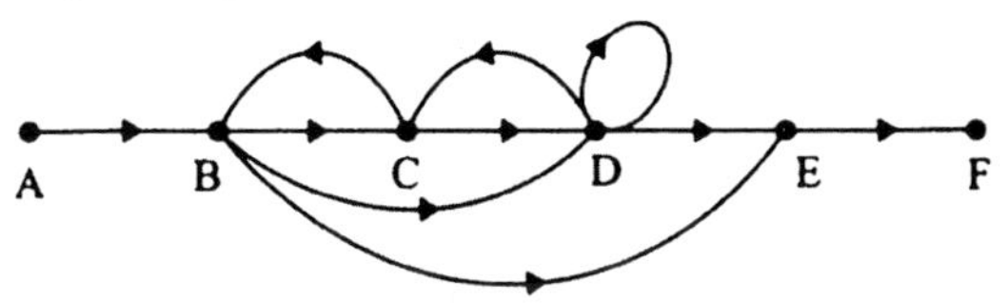

Fig. 10.66.

(*a*) 3 and 1 (*b*) 3 and 2
(*c*) 4 and 2 (*d*) 2 and 4

174. Figure 10.67 shows a pole zero diagram. The transfer function G(*j* 1) is

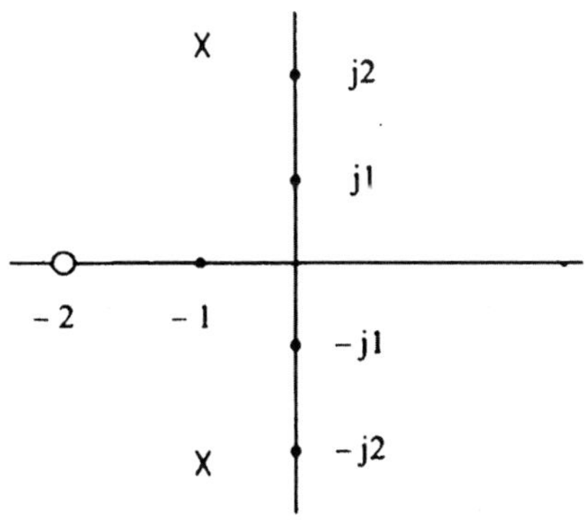

Fig. 10.67.

(*a*) 0.5 ∠ 0 (*b*) 2.7 ∠ –31°
(*c*) 2 ∠ 45° (*d*) 2 ∠ 67.4°

175. The open loop transfer function of a unity feedback system is $G(s) = \frac{1}{(s+2)^2}$. The poles of closed loop system are at

(*a*) –2, – 2 (*b*) – 2, –1
(*c*) – 2 ±*j*1 (*d*) – 2, 2

176. The a.c. motor used in servo applications is
(*a*) single-phase induction motor
(*b*) two-phase induction motor
(*c*) three-phase induction motor
(*d*) synchronous motor

177. In position control system the device used for providing rate-feedback voltage is
(*a*) potentiometer (*b*) synchro-transmitter
(*c*) synchro transformer (*d*) techo generator

178. A synchro transmitter receiver unit is a
(*a*) two-phase ac device (*b*) three-phase ac device
(*c*) dc device (*d*) one-phase ac device

179. Match List I and List II and select the correct answer using the given codes.

List I	**List II**
A. Servo motor	1. Error detector
B. Amplidyne	2. Transducer
C. Potentiometer	3. Actuator
D. Flapper valve	4. Power amplifier

Codes:

	A	**B**	**C**	**D**
(*a*)	3	1	2	4
(*b*)	4	3	2	1
(*c*)	3	4	1	2
(*d*)	3	4	2	1

180. The purpose of series quadrature windings in amplidyne is to
(*a*) neutralise the effect of armature reaction
(*b*) reduce commutation difficulties
(*c*) increase the gain
(*d*) increase response time

181. Figure 10.68 shows a control system. The maximum value of gain for which the system is stable is

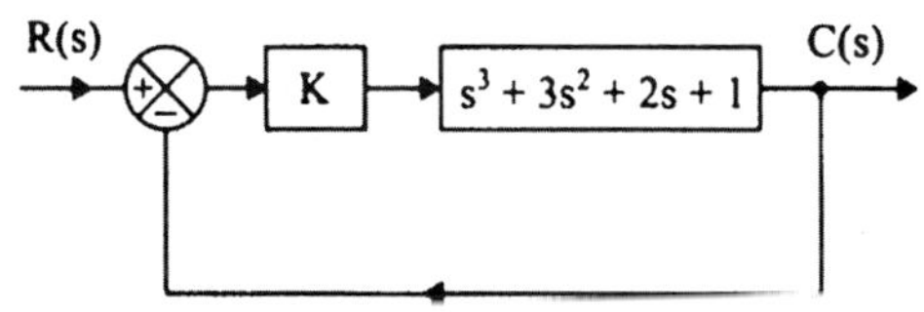

Fig. 10.68.

(*a*) $\sqrt{3}$ (*b*) 3
(*c*) 4 (*d*) 5

182. The system matrix of continuous time system described in state variable form is

$$A = \begin{bmatrix} x & 0 & 0 \\ 0 & y & -1 \\ 0 & 1 & -2 \end{bmatrix}$$

The systems is stable for all values of x and y satisfying

(*a*) $x < \frac{1}{2}, y < \frac{1}{2}$ (*b*) $x > \frac{1}{2}, y < 0$
(*c*) $x < 0, y < 2$ (*d*) $x < 0, y < \frac{1}{2}$

183. For the function $G(s)\,H(s) = \frac{1}{s(s+1)(s+0.5)}$ the phase cross-over frequency is
(*a*) 0.5 rad/sec (*b*) 0.707 rad/sec
(*c*) 1.732 rad/sec (*d*) 2 rad/sec

184. In Fig. 10.69 the input is $x(t) = A \sin \omega t$. The steady state output $y(t)$ =

Fig. 10.69.

(*a*) $A \sin(\omega t + \phi)$ where $\phi = \tan^{-1} |G(j\omega)|$
(*b*) $|G(j\omega)|\, A \sin[\omega t + \angle G(j\omega)]$
(*c*) $|G(j\omega)|\, A \sin[2\omega t + \angle G(j\omega)]$
(*d*) $AG(j\omega) \sin[\omega t + \angle G(j\omega)]$

185. An open loop system has a forward path transfer function $(42.25)/s(s + 6.5)$. The unit step response of the system starting from rest will have its maximum value at a time equal to
(*a*) 0 sec (*b*) 0.56 sec
(*c*) 5.6 sec (*d*) infinity

186. Match the system open loop transfer function in List I with steady error produced by ramp input in List II and select the answer using the given codes (Assume unity feedback system)

List I	List II
(*a*) $\dfrac{30}{s^2+6s+9}$	1. Zero
(*b*) $\dfrac{30}{s^2+6s}$	2. 0.2
(*c*) $\dfrac{30}{s^2+9s}$	3. 0.3
(*d*) $\dfrac{s+1}{s^2}$	4. infinity

Codes:

	A	B	C	D
(*a*)	1	2	3	4
(*b*)	1	3	2	4
(*c*)	4	3	2	1
(*d*)	4	2	3	1

187. Figure 10.70 shows pole-zero plot. If steady state gain is 2 the transfer function G(*s*) is

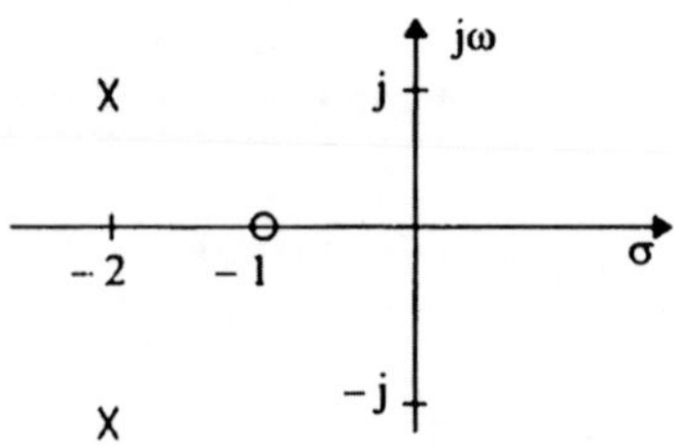

Fig. 10.70.

(*a*) $\dfrac{2(s+1)}{s^2+4s+5}$ (*b*) $\dfrac{5(s+1)}{s^2+4s+4}$

(*c*) $\dfrac{10(s+1)}{s^2+4s+3}$ (*d*) $\dfrac{10(s+1)}{(s+2)^2}$

188. $G(s) = \dfrac{1}{s^2+0.2s+1}$. It is required that response settles to within 2% of final value for a step input. The settling time is

(*a*) 20 sec (*b*) 40 sec

(*c*) 35 sec (*d*) 45 sec

189. In terms of ξ and ω_n the settling time of a second order systems for 2% criterion and step input is equal to

(*a*) $\dfrac{1}{\xi\omega_n}$ (*b*) $\dfrac{2}{\xi\omega_n}$

(*c*) $\dfrac{3}{\xi\omega_n}$ (*d*) $\dfrac{4}{\xi\omega_n}$

190. It terms of ξ and ω_n, the settling time (5% criterion) of a second order system for a step input, is equal to

(*a*) $\dfrac{1}{\xi\omega_n}$ (*b*) $\dfrac{2}{\xi\omega_n}$

(*c*) $\dfrac{3}{\xi\omega_n}$ (*d*) $\dfrac{4}{\xi\omega_n}$

191. The maximum shift which can be provided by a lead compensator with transfer function $G_c(s) = \dfrac{1+6s}{1+2s}$ is

(*a*) 15° (*b*) 30°

(*c*) 45° (*d*) 60°

192. A compensating network has $G_c(s) = \dfrac{1+\alpha Ts}{1+Ts}$. If it is a phase lag circuit the value of α should be

(*a*) 0 (*b*) between 0 and 1

(*c*) 1 (*d*) more than 1

193. A feedback amplifier has forward gains A_1, A_2 and feedback factor β as shown in signal flow graph of Fig. 10.71. Because of feedback as shown the signal noise ratio at output will

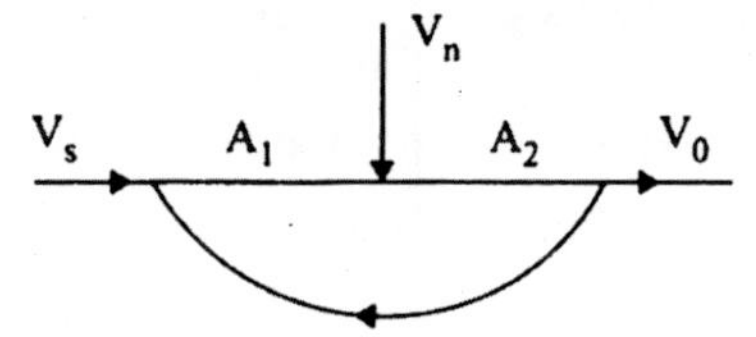

Fig. 10.71.

(*a*) increase (*b*) decrease

(*c*) be unaffected (*d*) depend of value of β

194. Which of the following is open loop system?

(*a*) wind screen wiper

(*b*) aqualing

(*c*) respiratory system of animal

(*d*) system for controlling anti-rocket missiles

195. If $G(s)\,H(s) = \dfrac{K(s+6)}{(s+3)(s+5)}$ the break away and break in points in the root locus are located at

(*a*) – 2 and – 1 (*b*) – 2.47 and – 3.77

(*c*) – 4.27 and – 7.73 (*d*) none of the above

196. If $G(s)\,H(s) = \dfrac{K(s+10)}{s(s+8)(s+16)(s+72)}$, the closed loop pole will be located at $s = -12$ when K =

(*a*) 4355 (*b*) 5760

(*c*) 9600 (*d*) 9862

197. An electro mechanical device which actuates a train of step angular movements in response to a train of input pulses on one to one basis is

(*a*) synchro control transformer

(*b*) LVDT

(*c*) stepper motor

(*d*) tacho generator

198. Which of the following is not correct for two phase servo motor?

(*a*) rotor diameter is small

(*b*) applied voltages are seldom balanced

(*c*) rotor resistance is low

(*d*) torque speed curve is linear

199. Which of the following is used to obtain output position in a position control system?
(*a*) strain gauge (*b*) load cell
(*c*) synchro (*d*) thermistor

200. Consider the following expressions which indicate the step or impulse response of an initially relaxed system
1. $[5 - 4e^{-2t}]u(t)$ 3. $\{e^{-2t} + 5\}\,u(t)$
2. $\delta(t) + 8e^{-2t}\,u(t)$ 4. $\delta(t) + 4e^{-2t}\,u(t)$
Those which correspond to the step and impulse response of the same system include
(*a*) 1 and 3 (*b*) 1 and 4
(*c*) 2 and 4 (*d*) 2 and 3

201. The correct sequence of events to improve system stability is
(*a*) insert derivative action, use negative feed back, reduce gain
(*b*) reduce gain, use negative feed back, insert derivative action
(*c*) reduce gain, insert derivative feed back, use negative feed back
(*d*) use negative feed back, reduce gain, insert derivative action

202. The transient response of a second order system is given by $C(t) = 1 - \frac{e^{-4t}}{\sqrt{1-\zeta^2}} \sin\left[\omega_n\left(\sqrt{1-\zeta^2} + \theta\right)\right]$.
For 5% criterion the settling time is
(*a*) 0.25 sec (*b*) 0.75 sec
(*c*) 1.25 sec (*d*) 4 sec

203. Consider the systems with following open loop transfer functions
1. $\frac{36}{s(s+3.6)}$
2. $\frac{100}{s(s+5)}$
3. $\frac{6.25}{s(s+4)}$
If unity feedback is used, the increasing order of time taken for unit step response to settle is
(*a*) 1, 2, 3 (*b*) 3, 1, 2
(*c*) 2, 3, 1 (*d*) 3, 2, 1

204. Match List I with List II and select the correct answer using the given codes

List I (Characteristic equation)	List II (Degree of damping)
A. $s^2 + 15s + 56.25$	1. Undamped
B. $s^2 + 5s + 6$	2. under damped
C. $s^2 + 20.25$	3. critically damped
D. $s^2 + 4.55s + 42.25$	4. over damped

Codes:

	A	B	C	D
(*a*)	1	2	3	4
(*b*)	2	3	1	4
(*c*)	3	4	1	2
(*d*)	1	2	4	3

205. For the system with $T(s) = \frac{4}{s^2 + 2s + 4}$, the maximum resonance peak is
(*a*) 4 (*b*) 4/3
(*c*) 2 (*d*) $2/\sqrt{3}$

206. For the compensator in Fig. 10.72 the maximum phase lead is

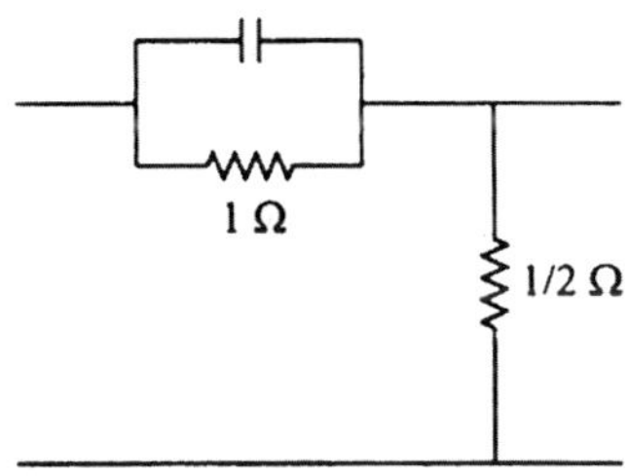

Fig. 10.72.

(*a*) $\sin^{-1}(1/3)$ (*b*) 30°
(*c*) 45° (*d*) 60°

207. For the control system is Fig. 10.73, the value of K for critical damping is

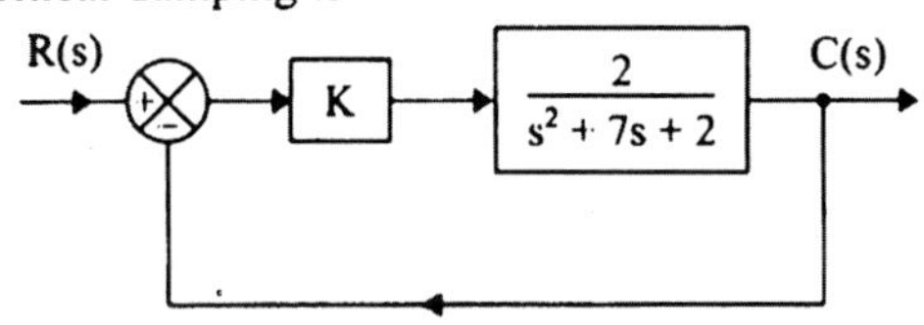

Fig. 10.73.

(*a*) 1 (*b*) 5.125
(*c*) 6.831 (*d*) 10

208. Far the mechanical system in Fig. 10.74 the transfer function $\frac{X(s)}{F(s)} =$

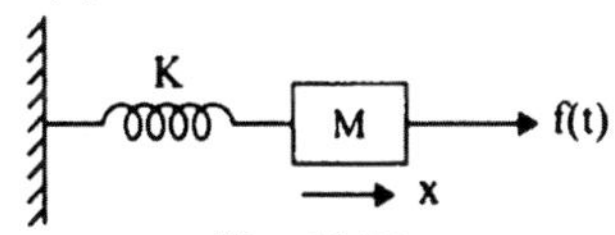

Fig. 10.74.

(*a*) $\frac{1}{Ms^2 + K}$ (*b*) $\frac{Ms^2 + 1}{K}$
(*c*) $\frac{K}{Ms^2 + 1}$ (*d*) $\frac{1}{Ks^2 + M}$

209. The state equation of a system is $X = \begin{vmatrix} 0 & 1 \\ -20 & -9 \end{vmatrix} X + \begin{bmatrix} 0 \\ 1 \end{bmatrix}$.
The poles of this system are located at
(*a*) –1, – 9 (*b*) –1, – 20
(*c*) –4, – 5 (*d*) –9, – 20

210. A unity feedback system has the following open loop frequency response

ω	2	3	4	5	6	8	10
$\lvert G(j\omega)\rvert$	7.5	4.8	3.15	2.25	1.70	1	0.64
$\angle G(j\omega)$	–118°	–130°	–140°	–150°	–157°	–170°	–180°

The gain margin and phase margin are
(*a*) 0 *d*B, – 170° (*b*) 3.86 *d*B, – 180°
(*c*) 0 *d*B, 10° (*d*) 3.86 *d*B, 10°

211. For the system in Fig. 10.75, $R(s) = \frac{1}{s}$. Then steady state error is

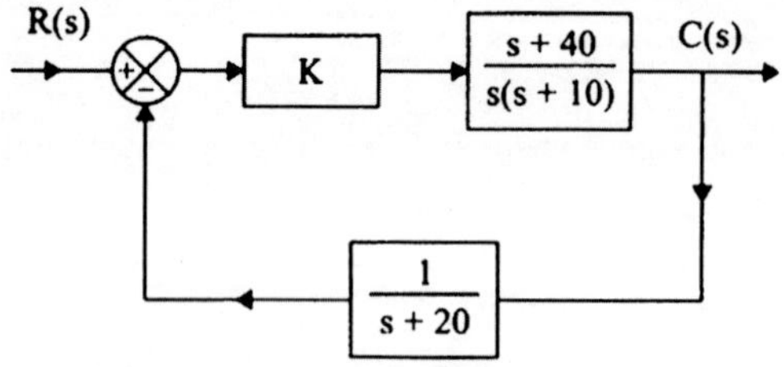

Fig. 10.75.

(*a*) $\frac{1}{1+K}$ (*b*) K

(*c*) Zero (*d*) 1

212. Figure 10.76 shows the magnitude plot for a system. The steady state error for unit step input is

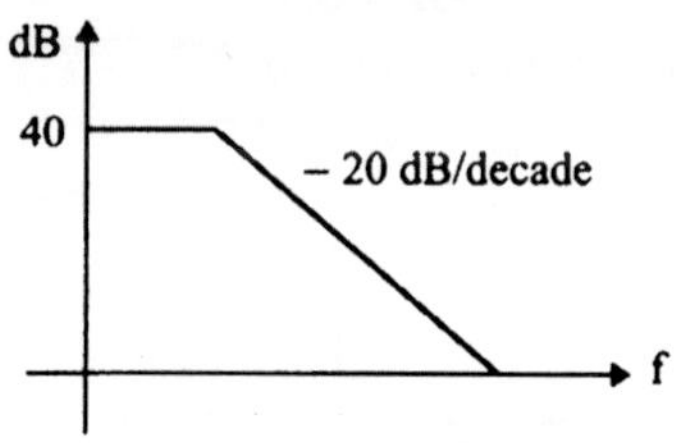

Fig. 10.76.

(*a*) 1/101 (*b*) 1/100

(*c*) 1/41 (*d*) 1/40

213. A unity feedback system has $G(s) = \frac{10}{s(s+1)(s+2)}$. If input $r(t) = 1 + 2t + \frac{3}{2}t^3$, the steady error is

(*a*) zero (*b*) 0.4

(*c*) 4 (*d*) infinity

214. Figure 10.77 shows a unity feedback system. The ratio of time constants for open loop and closed loop response are

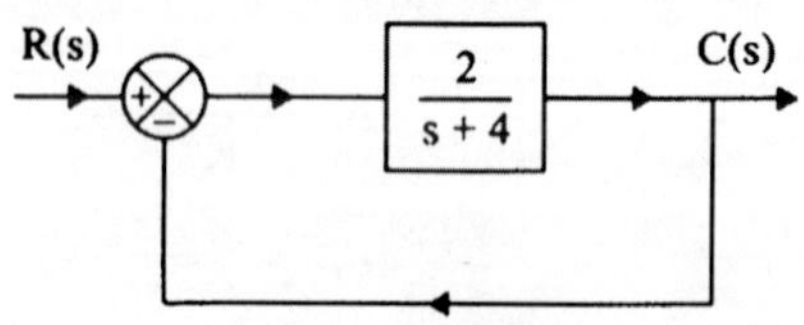

Fig. 10.77.

(*a*) 1 : 1 (*b*) 2 : 1

(*c*) 3 : 2 (*d*) 2 : 3

215. For the system in Fig. 10.78, G varies slightly. The ratio of open loop sensitivity to closed loop sensitivity is

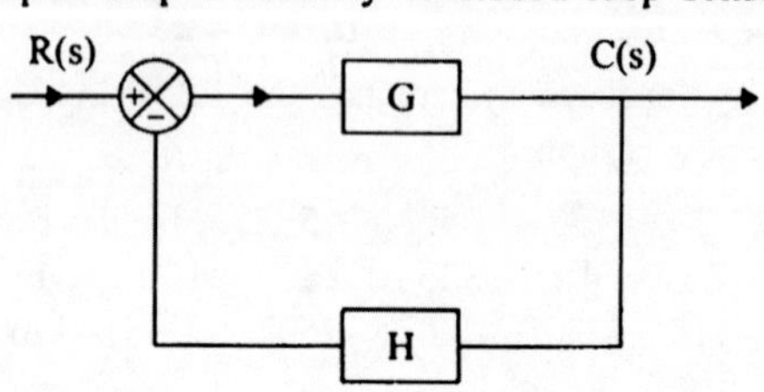

Fig. 10.78.

(*a*) 1 : (1 + GH) (*b*) $1 : (1 + GH)^{-1}$

(*c*) 1 : (1 – GH) (*d*) $1 : (1 - GH)^{-1}$

216. In case of synchro error detector the electrical zero position of control transformer is obtained when angular displacement between rotors is

(*a*) zero (*b*) 45°

(*c*) 90° (*d*) 180°

217. For the characteristic equation $s^5 + s^4 + 2s^3 + 2s^2 + 3s + 15 = 0$, the number of roots in left half s plane are

(*a*) 1 (*b*) 2

(*c*) 3 (*d*) 5

218. Match List I and List II and select the correct answer using the codes

List I	List II
A. Lag compensation	1. Exterior of unit circle $\lvert z \rvert = 1$
B. Lead compensation	2. PD control
C. Stability	3. Non-minimum phase
D. Instability	4. Left half s plane
	5. PI control

Codes:

	A	B	C	D
(*a*)	2	5	1	3
(*b*)	3	1	2	5
(*c*)	1	2	3	4
(*d*)	5	2	4	1

219. Match List I and List II and select the correct answer using the given codes

List I (Transfer function)	List II (Description)
A. $\frac{1-s}{1+s}$	1. Non-minimum phase
B. $\frac{1-s}{(1+s)(1+2s)(1+3s)}$	2. Minimum Phase
C. $\frac{1-3s}{(1+s)(1+2s)(1+4s)}$	3. All pass system

Codes:

	A	B	C
(*a*)	1	3	2
(*b*)	3	1	2
(*c*)	3	2	1
(*d*)	2	1	3

220. Consider the following statements about variable reluctance stepper motor used in control systems

1. The static torque acting on rotor is a function of angular misalignment between stator and rotor teeth
2. There are two positions of zero torque, *i.e.*, $\theta = 0$ and $\frac{180°}{T}$ where T is the number of rotor teeth
3. Both zero torque positions are stable
4. When stator is excited rotor is pulled into minimum reluctance position

Of these statements, the correct ones are

(*a*) 2, 3, 4 (*b*) 1, 2, 3

(*c*) 1, 2, 4 (*d*) 1, 3, 4

221. For the system in Fig. 10.79, the ratio C(s)/R(s) =

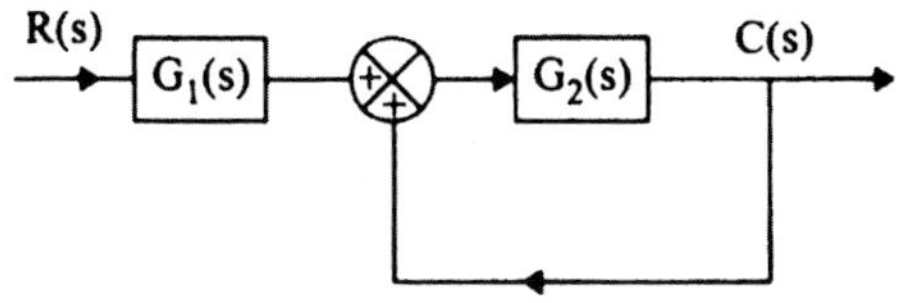

Fig. 10.79.

(a) $\dfrac{G_1(s)\,G_2(s)}{1-G_2(s)}$ (b) $\dfrac{G_1(s)}{1+G_1(s)\,G_2(s)}$

(c) $\dfrac{G_2(s)}{1+G_1(s)\,G_2(s)}$ (d) $\dfrac{1}{1+G_1(s)\,G_2(s)}$

222. The loop transfer function of a feed back system is G(s) H(s) = $\dfrac{K}{s(s+2)(s^2+2s+2)}$. The number of asymptotes of its root loci are

(a) 1 (b) 2

(c) 3 (d) 4

223. In a two phase ac servo motor rotor resistance is R and rotor reactance is X. The speed curve will be linear if

(a) $\dfrac{X}{R} << 1$ (b) $\dfrac{X}{R} >> 1$

(c) $\dfrac{X}{R} = 1$ (d) $X^2 = R$

224. The radius and centre of M circles are given by

(a) $\dfrac{M}{M^2-1}, \left(\dfrac{-M^2}{M^2-1}, 0\right)$ (b) $\dfrac{M^2}{M^2-1}, \left(\dfrac{-M}{M^2-1}, 0\right)$

(c) $\dfrac{M^2}{M^2-1}, \left(\dfrac{-M^2}{M-1}, 0\right)$ (d) $\dfrac{M^2}{M^2-1}, \left(\dfrac{M^2}{M^2-1}, 0\right)$

225. For the system in Fig. 10.80, $\dfrac{C(s)}{R(s)}$ =

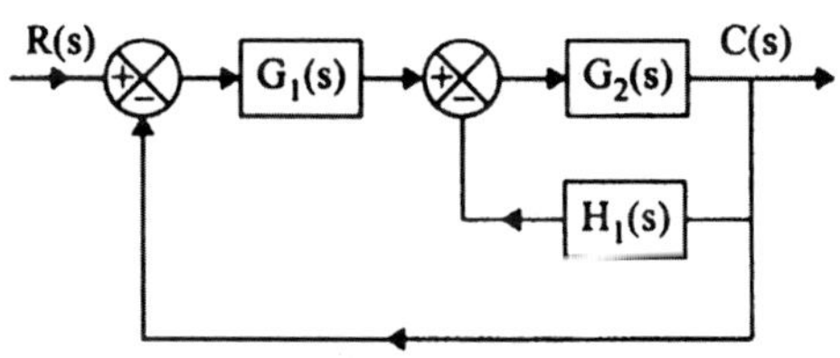

Fig. 10.80.

(a) $\dfrac{G_1(s)G_2(s)}{1+G_1(s)[G_2(s)+H_1(s)]}$

(b) $\dfrac{G_1(s)G_2(s)}{1+G_2(s)[G_1(s)+H_1(s)]}$

(c) $\dfrac{G_1(s)+G_2(s)}{1+G_1(s)G_2(s)+H_1(s)}$

(d) none of the above

226. Which of the following can act as error detecting devices

1. a pair of potentiometers
2. a pair of synchros
3. a differential transformer
4. a metadyne
5. a control transformer

Select the answer using the following codes

(a) 1, 2, 5 (b) 2, 3, 4, 5

(c) 1, 3, 4, 5 (d) 1, 2, 3, 4

227. The first column of a Routh array is

s^5	1
s^4	2
s^3	1.5
s^2	−1/3
s^1	10
s^0	2

How many roots of the corresponding characteristic equation are in left half s-plane

(a) 2 (b) 3

(c) 4 (d) 5

228. The overall transfer function for a unity feedback system is $\dfrac{4}{s^2+4s+4}$. Which of the following statements about this system are correct

1. Position error constant K_p is 4
2. System is type 1
3. Velocity error constant is finite

Select the answer as per following code

(a) 1, 2, 3 (b) 1, 2

(c) 2, 3 (d) 1, 3

229. A negative feedback system has G(s) = $\dfrac{1}{s+1}$ and H(s) = $\dfrac{K}{s(s+2)}$. The closed loop system is stable for

(a) K > 20 (b) 15 < K < 19

(c) 8 ≤ K ≤ 14 (d) K < 6

230. For a closed loop system

$$\frac{C(s)}{R(s)} = \frac{2600K(s+25)}{s^4+125s^3+5100s^2+65000s+65000K}$$

The imaginary axis intercepts of the root loci will be

(a) $\pm j\,22.8$ (b) $\pm j\,2.28$

(c) $\pm j\,1.14$ (d) $\pm j\,11.4$

231. For the system shown in Fig. 10.81 the time response to unit step input is

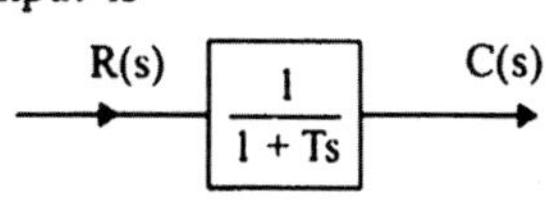

Fig. 10.81.

(a) $c(t) = \dfrac{1}{T}e^{-t/T}$ (b) $c(t) = T(1 - e^{-t/T})$

(c) $c(t) = 1 - e^{-t/T}$ (d) $c(t) = T\,e^{-t/T}$

232. A system has $\dfrac{C(s)}{R(s)} = \dfrac{4}{s^2+1.6s+4}$

For unit step input the settling time for 2% tolerance band is

(a) 1.6 (b) 2.5

(c) 4 (d) 5

233. Consider the following statements :
The gain margin and phase margin of an unstable system may respectively be
1. positive, positive
2. positive, negative
3. negative, positive
4. negative, negative

Of these statements
(*a*) 1, 4 are correct
(*b*) 1 and 2 are correct
(*c*) 1, 2, 3 are correct
(*d*) 2, 3, 4 are correct

234. A system with transfer function
$$\frac{600}{s(s+1)(s+15)(s+20)}$$
may be approximated by the system

(*a*) $\frac{2}{s(s+1)}$
(*b*) $\frac{40}{s(s+20)}$
(*c*) $\frac{600}{(s+15)(s+20)}$
(*d*) $\frac{40}{(s+1)(s+20)}$

235. Figure 10.82 shows a hybrid servo system. The components 1, 2, 3, 4 are respectively

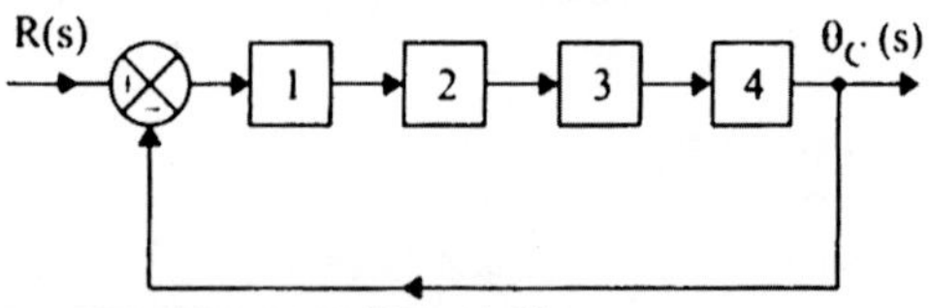

Fig. 10.82.

(*a*) amplifier, demodulator, dc servo motor and load
(*b*) demodulator, amplifier, dc servo motor and load
(*c*) demodulator, dc servo motor, amplifier and load
(*d*) amplifier, dc servo motor, demodulator and load

236. Match List I and List II and select the answer using the given code

List I (Controller)	List II (Application)
A. On-off	1. steam kettle with different settings
B. Proportional	2. Elevator
C. Cascade	3. Robot-positioning
D. Digital	4. Domestic Refrigerator

Codes:

	A	B	C	D
(*a*)	4	2	3	1
(*b*)	2	4	1	3
(*c*)	3	1	4	2
(*d*)	4	2	1	3

237. The signal flow graph of a system is shown in Fig. 10.83. The effect of disturbance T_D can be reduced by

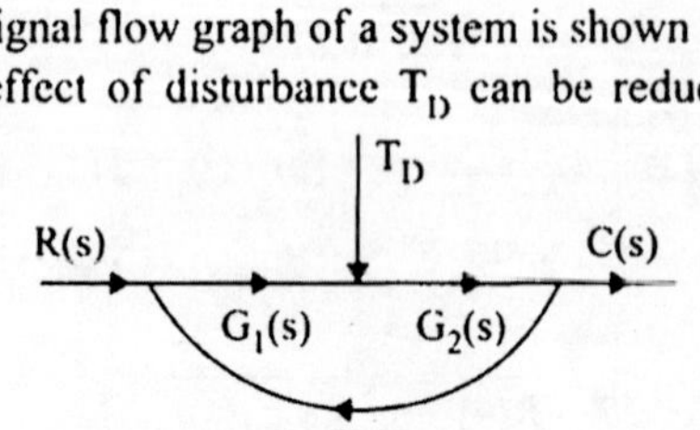

Fig. 10.83.

(*a*) increasing $G_2(s)$
(*b*) decreasing $G_2(s)$
(*c*) increasing $G_1(s)$
(*d*) decreasing $G_1(s)$

238. In Fig. 10.84, the transfer function $\frac{C(s)}{N(s)}$ =

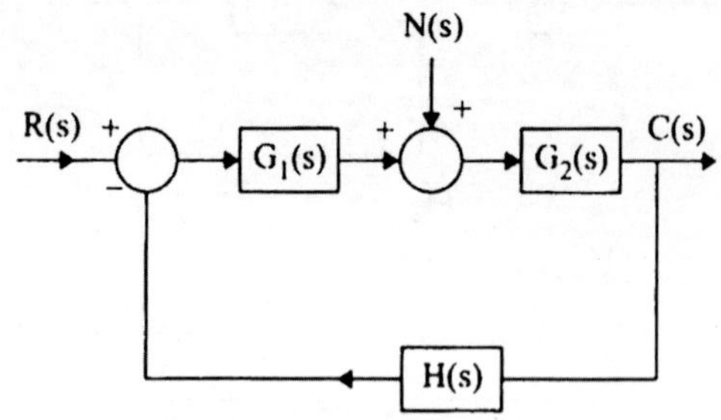

Fig. 10.84.

(*a*) $\frac{G_1(s)\,G_2(s)}{1+G_1(s)G_2(s)H(s)}$
(*b*) $\frac{G_1(s)}{1+G_1(s)H(s)}$
(*c*) $\frac{G_2(s)}{1+G_2(s)H(s)}$
(*d*) $\frac{G_2(s)}{1+G_1(s)G_2(s)H(s)}$

239. Which of the following can be used as a rotating amplifier in a control system
1. Amplidyne
2. Separately excited dc generator
3. Synchro
4. Self-excited dc generator

Select the answer from the following codes
(*a*) 3 and 4
(*b*) 1 and 2
(*c*) 1, 2 and 3
(*d*) all

240. While forming Routh's array, the situation of a row of zeros indicates that the system
(*a*) has symmetrically located roots
(*b*) is not sensitive to variations in gain
(*c*) is stable
(*d*) is unstable

241. For the second order system having following differential equation
$$J\frac{d^2\theta_0}{dt^2}+F\frac{d\theta_0}{dt}+K\theta_0 = K_2\theta_i$$
(When θ_0 and θ_i are output and input) the natural frequency is

(*a*) $\sqrt{\frac{K}{J}}$
(*b*) $\sqrt{\frac{J}{K}}$
(*c*) $\sqrt{KJ}$
(*d*) $\frac{1}{\sqrt{KJ}}$

242. If I $(s) = \frac{10s^2(s+7)}{(s+1)(s+8)(s+10)}$, initial and final values of $i(t)$ are
(*a*) zero and 1
(*b*) zero and 10
(*c*) 10 and zero
(*d*) 70 and 80

243. A system has two poles as shown in Fig. 10.85. The zeros are at infinity. If input is unit step, the steady state output is

(*a*) $\frac{1}{4}$
(*b*) $\frac{1}{2}$

(c) $\frac{1}{\sqrt{2}}$ (d) 1

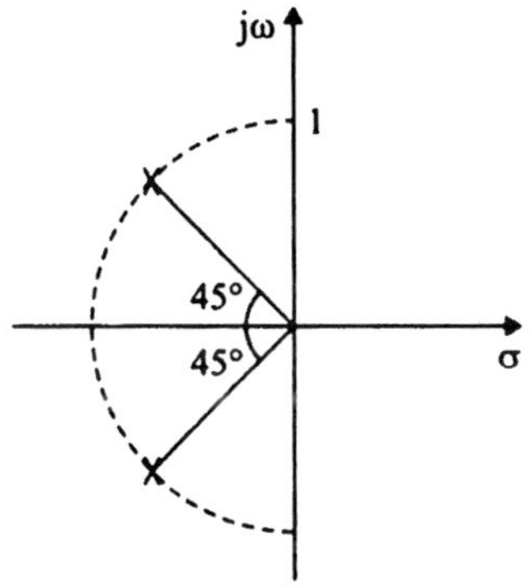

Fig. 10.85.

244. For the log magnitude Bode plot of Fig. 10.86, the transfer function is

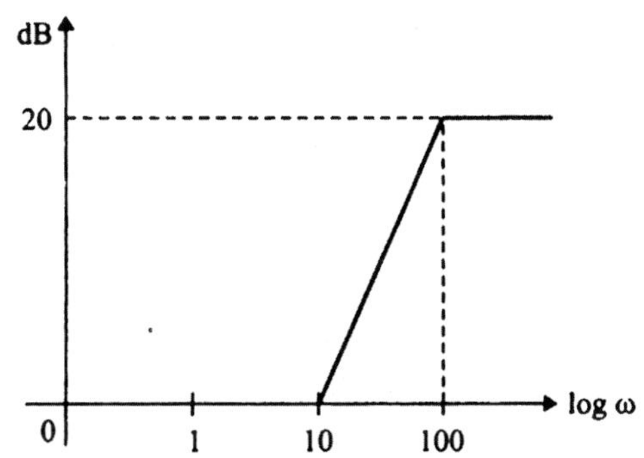

Fig. 10.86.

(a) $\frac{s-10}{s+100}$ (b) $\frac{s+10}{s-100}$

(c) $\frac{s-10}{s-100}$ (d) $\frac{s+10}{s+100}$

245. For the system of Fig. 10.87, the transfer function is

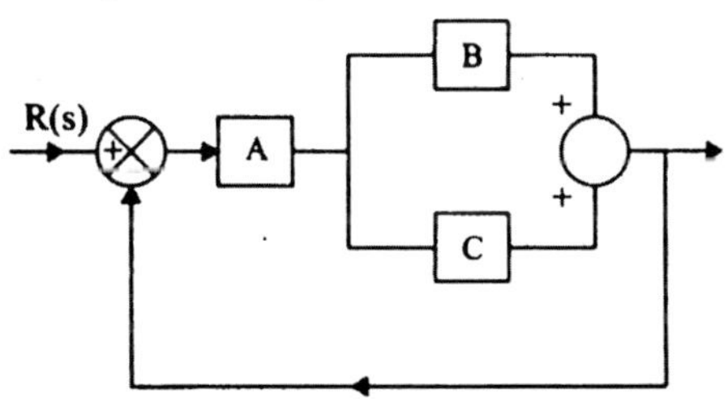

Fig. 10.87.

(a) $\frac{ABC}{1+ABC}$ (b) $\frac{A+B+C}{1+AB+AC}$

(c) $\frac{AB+AC}{ABC}$ (d) $\frac{AB+AC}{1+AB+AC}$

246. For the signal flow graph of Fig. 10.88 consider the following statements

1. It has three forward paths
2. It has three individual loops
3. It has two non-touching loops

Out of above the Correct statements are

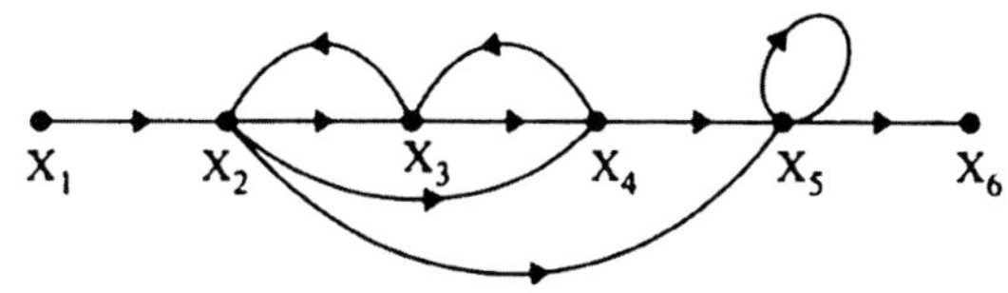

Fig. 10.88.

(a) 1, 2, 3 (b) 1, 2

(c) 2, 3 (d) 1, 3

247. If $G(s)\,H(s) = \frac{K}{s^2(s^2+2s+2)}$, the angle of departure of root locus at $s = -1 + j$ is

(a) zero (b) 90°

(c) −90° (d) −180°

248. In the field controlled motor the entire damping comes from

(a) armature resistance

(b) back emf

(c) motor friction and load

(d) field resistance

249. Which of the following rotors are used in two phase ac servomotor?

1. solid iron rotor
2. squirrel cage rotor
3. drag cup rotor

Of the above the correct ones are

(a) 1, 2, 3 (b) 1, 2

(c) 2, 3 (d) 1, 3

250. If $F(s) = \frac{s+a}{(s+a)^2+\omega_0^2}$, then $f(t) =$

(a) $\cos \omega_0 t$ (b) $\sin \omega_0 t$

(c) $e^{at} \cos \omega_0 t$ (d) $e^{-at} \cos \omega_0 t$

251. For stable network the real parts of poles and zeros of driving point functions must be

(a) negative (b) negative or zero

(c) zero (d) positive

252. The poles with greater displacement from the real axis correspond to

(a) higher frequencies of oscillations

(b) stable response

(c) lower frequencies of oscillations

(d) unbounded output

253. The response to sinusoidal excitation can be found by

(a) setting $s = j\omega$ in the network function

(b) setting $s = -j\omega$ the network function

(c) taking inverse transform of network function

(d) none of the above

254. For an RLC series circuit Z(s) is of the form

(a) $\frac{5(s^2+5)}{s^2+s+1}$ (b) $\frac{5s}{s^2+5}$

(c) $\frac{3(s^2+s+1)}{s^2+1}$ (d) $\frac{s^2+10s+100}{s}$

255. The phase difference $(j\omega - p_1)$ means

(a) phasor drawn from p_1 to $j\omega$

(*b*) phasor drawn from origin to p_1
(*c*) phasor directed from origin to $j\omega$ point
(*d*) phasor directed from $j\omega$ to p_1

256. Bode magnitude plot is drawn between
(*a*) magnitude of network function and ω
(*b*) *d*B magnitude and log ω
(*c*) *d*B magnitude and ω
(*d*) $\log_e$ (magnitude) and log ω

257. An RLC series circuit has

$$Y(s) = \frac{2s}{(s-p_1)(s-p_1^\bullet)}$$

If Poles p_1 and $p_1^\bullet$ are close to $j\omega$ axis, then
(*a*) resonant frequency is high
(*b*) Q is high
(*c*) band width is large
(*d*) Q is low

258. For the circuit of Fig. 10.89, $V_2(s)/V_1(s) =$

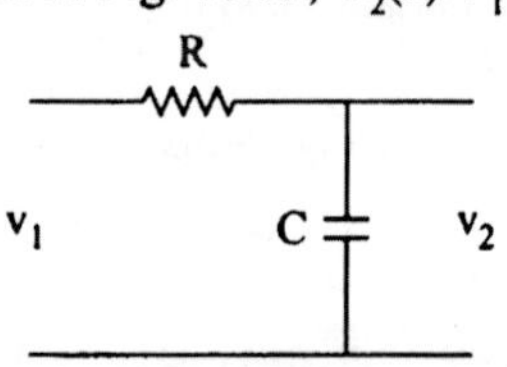

Fig. 10.89.

(*a*) $\frac{s}{1+RCs}$ (*b*) $\frac{1}{1+RCs}$
(*c*) $\frac{s^2}{1+RCs}$ (*d*) $\frac{1}{1+RCs^2}$

259. A reactance function
(*a*) has, a pole at origin
(*b*) has a zero at origin
(*c*) may have a pole or zero at origin
(*d*) has a pole or zero at origin and a pole or zero at infinity

260. The poles of an RC function
(*a*) are simple and lie on $j\omega$ axis
(*b*) are simple and lie on negative real axis
(*c*) may be complex conjugate
(*d*) may be anywhere is *s* plane

261. For a feedback system

$$\frac{C(s)}{R(s)} = \frac{G(s)}{1+G(s)H(s)}$$

Then the velocity error constant K_v is given by
(*a*) $K_v = \lim_{s\to o} G(s)H(s)$
(*b*) $K_v = \lim_{s\to o} \frac{sG(s)}{1+G(s)H(s)}$
(*c*) $K_v = \lim_{s\to o} sG(s)$
(*d*) $K_v = \lim_{s\to o} s[1+G(s)H(s)]$

262. Figure 10.90 shows a magnitude Bode plot. The transfer function for this plot is

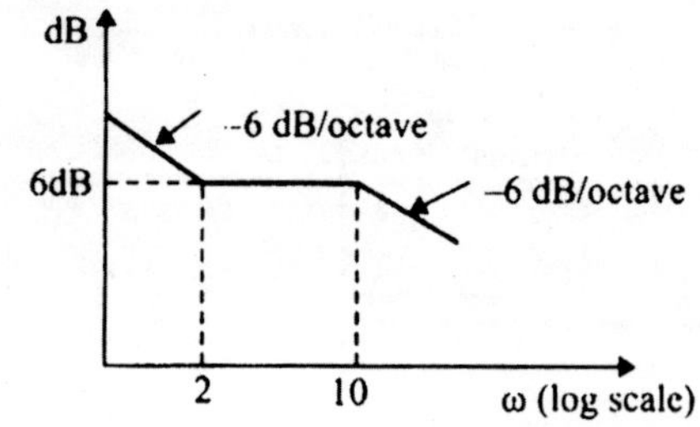

Fig. 10.90.

(*a*) $\frac{4(1+0.5s)}{s(1+0.1s)}$ (*b*) $\frac{4s(1+0.5s)}{(1+0.1s)}$
(*c*) $\frac{4(1+2s)}{s(1+10s)}$ (*d*) $\frac{4s(1+2s)}{(1+10s)}$

263. A phase lead network with $G_C(s) = \frac{s+1/T}{s+1/aT}$. where $a < 1$, gives maximum phase lead at a frequency equal to
(*a*) $\frac{1}{T}$ (*b*) $\frac{1}{aT}$
(*c*) $\frac{1}{T\sqrt{a}}$ (*d*) $\frac{1}{a\sqrt{T}}$

264. M circles are symmetrical with respect to
(*a*) real and imaginary axis
(*b*) M = 1 straight line and real axis
(*c*) M = 1 straight line and imaginary axis
(*d*) M = 1 straight line

265. For the feedback system with closed loop transfer function $\frac{C(s)}{R(s)} = \frac{\omega_n^2}{s^2+2\xi\omega_n s+\omega_n^2}$, the settling time for 2% tolerance is
(*a*) $\frac{2}{\xi\omega_n}$ (*b*) $\frac{\xi\omega_n}{2}$
(*c*) $\frac{4}{\xi\omega_n}$ (*d*) $4\,\xi\omega_n$

266. In a two-phase servomotor, if R is rotor resistance, X is rotor reactance, L is rotor length, D is rotor diameter then
(*a*) $\frac{X}{R}$ and $\frac{L}{D}$ are both small
(*b*) $\frac{X}{R}$ is large but $\frac{L}{D}$ is small
(*c*) $\frac{X}{R}$ is small but $\frac{L}{D}$ is large
(*d*) $\frac{X}{R}$ and $\frac{L}{D}$ are both large

267. Consider the following statements relating to synchros
1. Rotor of control transformer is either disc shaped or umbrella shaped
2. Rotor of transmitter has low magnetic reluctance
3. Transmitter and control transformer pair act as error detector

Out of above the correct statements are
(a) 1, 2, 3 (b) 1 and 2
(c) 2 and 3 (d) 1 and 3

268. First column elements of Routh's tabulation are 3, 5, $\frac{-3}{4}, \frac{1}{2}$, 2. If means that there
(a) is one root in left half s plane
(b) are two roots in left half s plane
(c) are two roots in right half s plane
(d) is one root in right half s plane

269. The response $\frac{d^2C(t)}{dt^2}+4\frac{dC(t)}{dt}+5C(t)=0$ is
(a) undamped (b) under damped
(c) critically damped (d) oscillatory

270. The M circle for M = 1 is the
(a) straight line at $x = -0.5$
(b) critical point (– 1, jo)
(c) circle with $r = 0.33$
(d) circle with $r = 0.67$

271. The radius of constant N circle for N = 1 is
(a) 2 (b) $\sqrt{2}$
(c) 1 (d) $\frac{1}{\sqrt{2}}$

272. The polar plot of a transfer function passes through (– 1, 0) point. The gain margin is
(a) zero (b) –1 dB
(c) 1 dB (d) infinity

273. The transfer function G(s) of a PID controller is
(a) $K\left[1+\frac{1}{T_i s}+T_d s\right]$ (b) $K[1 + T_i s + T_d s]$
(c) $K\left[1+\frac{1}{T_i s}+\frac{1}{T_d s}\right]$ (d) $K\left[1+T_i s+\frac{1}{T_d s}\right]$

274. Consider the following statements
1. The effect of feedback is to reduce system error
2. Feedback increases the system gain at one frequency but reduces the system gain as another frequency
3. Feedback can cause an originally stable system to become unstable

Which of the above statements are correct?
(a) 1, 2, 3 (b) 1, 2
(c) 2, 3 (d) 1, 3

275. Match List I and List II and give the answer using the codes

List I (Functional components)	List II (Devices)
A. Error detector	1. Three-phase FHP induction motor
B. Servo motor	2. Synchro transmitter and control transformer
C. Amplifier	3. Tacho generator
D. Feedback	4. Armature controlled FHP dc motor
	5. Amplidyne

Codes:

	A	B	C	D
(a)	2	4	1	5
(b)	4	2	5	3
(c)	2	4	5	3
(d)	1	2	3	5

276. Match List I and List II and select the correct answer using the codes

List I	List II
A. Bode	1. Asymptotic plots
B. Evans	2. Polar plots
C. Nyquist	3. Root locus technique
	4. M and N circles

Codes:

	A	B	C
(a)	1	4	2
(b)	2	3	4
(c)	3	1	4
(d)	1	3	2

277. Which of the following represents constant magnitude locus for M = 1?
(a) $x = -0.5$ (b) $x = 0$
(c) $x^2 + y^2 = 1$ (d) $(x + 1)^2 + y^2 = 1$

278. Which of the following features is not associated with Nichols chart?
(a) (0 dB, –180°) points on Nichols chart represent critical point (–1, 0)
(b) It is symmetric about –180°
(c) M loci are centered about (0 dB, –180°) point
(d) The frequency at intersection of $G(j\omega)$ locus and M = +3 dB locus gives band width of closed loop system

279. Which of the following are applications of position control systems?
(a) control of sheet metal thickness in hot rolling mill
(b) missile guidance of ships
(c) radar tracking system
(d) all the above

280. Coulomb friction is
(a) the force of sliding friction between dry surfaces
(b) the force of friction between moving surfaces separated by viscous fluid
(c) the force required to initiate motion between two contacting surfaces
(d) all the above

281. The force required to initiate motion between two contacting surfaces is called
(a) coulomb friction (b) viscous friction
(c) stiction (d) none of the above

282. Friction force acts in a direction
(a) opposite to that of velocity
(b) same as that of velocity
(c) at right angle to that of velocity
(d) partly opposite and partly at right angle to the direction of velocity

283. In a mechanical accelerometer, the steady state displacement is a measure of
(*a*) constant input acceleration
(*b*) constant input velocity
(*c*) either (*a*) or (*b*)
(*d*) neither (*a*) nor (*b*)

284. Generally a servo motor operates at
(*a*) high torque and high speed
(*b*) low torque and low speed
(*c*) low torque and high speed
(*d*) high torque and low speed

285. For the block diagram of Fig. 10.91 the equation describing system dynamics is

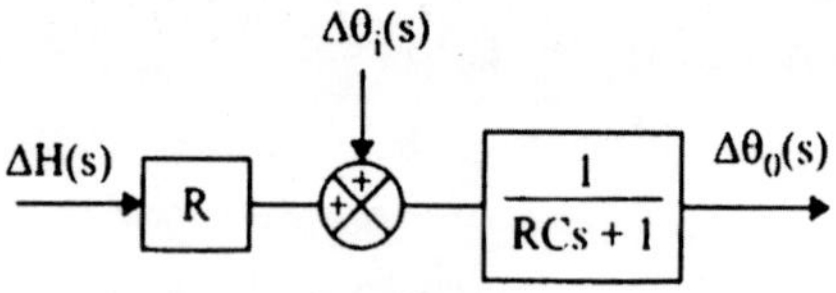

Fig. 10.91.

(*a*) $(1 + RCs)\,\Delta\theta_o(s) = \Delta\theta_i(s) + R\,\Delta H(s)$
(*b*) $(1 + RCs)\,[\Delta H(s)] = \Delta\theta_i(s) - \Delta\theta_o(s)$
(*c*) $(1 + RCs)\,\Delta\theta_o(s) = R\,\Delta H(s) - \Delta\theta_i(s)$
(*d*) $(1 + RCs)\,\Delta\theta_i(s) = H(s)[R] - \Delta\theta_o(s)$

286. For Fig. 10.92, time constant RC = τ. Then $\frac{V_o(s)}{V_i(s)} =$

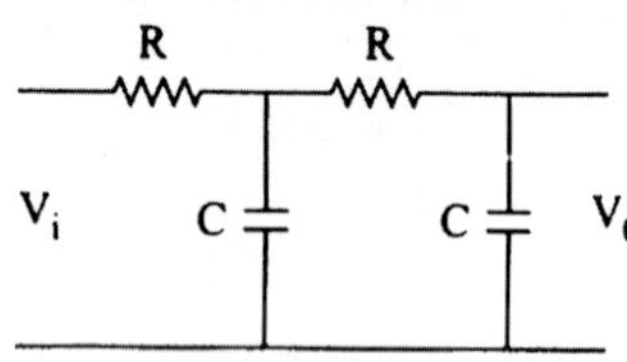

Fig. 10.92.

(*a*) $\tau^2 s^2 + 2\tau s + 1$ (*b*) $\frac{1}{\tau^2 s^2 + 2\tau s + 1}$
(*c*) $\tau^2 s^2 + 1$ (*d*) $\frac{1}{\tau^2 s^2 + 1}$

287. In an armature controlled dc motor let J be equivalent moment of inertia of motor and load, f_0 be equivalent viscous friction coefficient of motor and load and θ be angular displacement of motor shaft. Then torque equation is

(*a*) $J\frac{d^2\theta}{dt^2} + f_0\frac{d\theta}{dt}$ = Torque developed
(*b*) $J\frac{d^2\theta}{dt^2} - f_0\frac{d\theta}{dt}$ = Torque developed
(*c*) $J\frac{d^2\theta}{dt^2} + f_0(\theta)$ = Torque developed
(*d*) $J\frac{d^2\theta}{dt^2} - f_0(\theta)$ = Torque developed

288. In Fig. 10.93 C(*s*) =

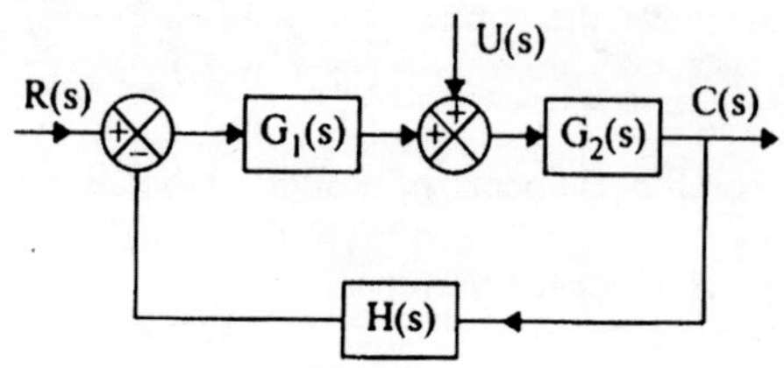

Fig. 10.93.

(*a*) $\frac{G_2(s)}{1+G_1(s)G_2(s)}[G_1(s)+U(s)]$
(*b*) $\frac{G_2(s)}{1+G_1(s)G_2(s)H(s)}[G_1(s)R(s)+U(s)]$
(*c*) $\frac{G_1(s)}{1+G_1(s)G_2(s)H(s)}[G_1(s)R(s)+U(s)]$
(*d*) $\frac{G_1(s)}{1+G_1(s)G_2(s)H(s)}[G_1(s)R(s)-U(s)]$

289. If block diagrams of Fig. 10.94 (*a*) and (*b*) are to be equivalent, then

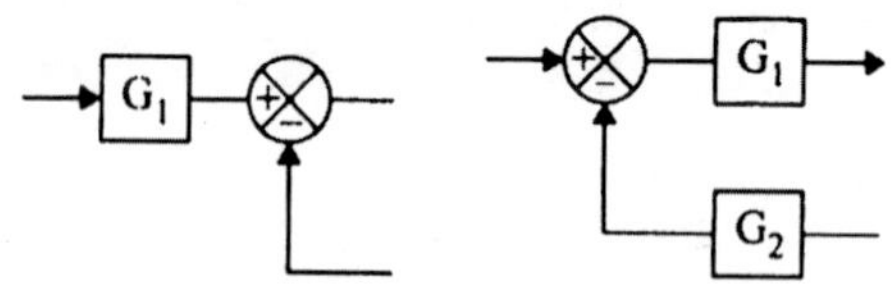

(*a*) (*b*)

Fig. 10.94.

(*a*) $G_2 = G_1$ (*b*) $G_2 = \frac{1}{G_1}$
(*c*) $G_2 = G_1^2$ (*d*) $G_2 = \frac{1}{G_1^2}$

290. If block diagrams of Fig. 10.95 (*a*) and (*b*) are to be equivalent, then

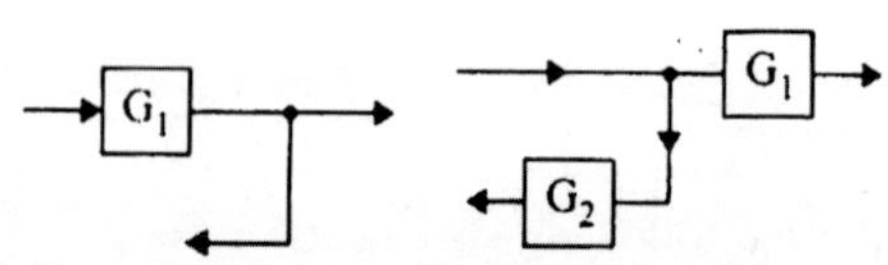

(*a*) (*b*)

Fig. 10.95.

(*a*) $G_2 = G_1$ (*b*) $G_2 = -G_1$
(*c*) $G_2 = \frac{1}{G_1}$ (*d*) $G_2 = \frac{-1}{G_1}$

291. In Fig. 10.96, $\frac{C(s)}{U(s)} =$

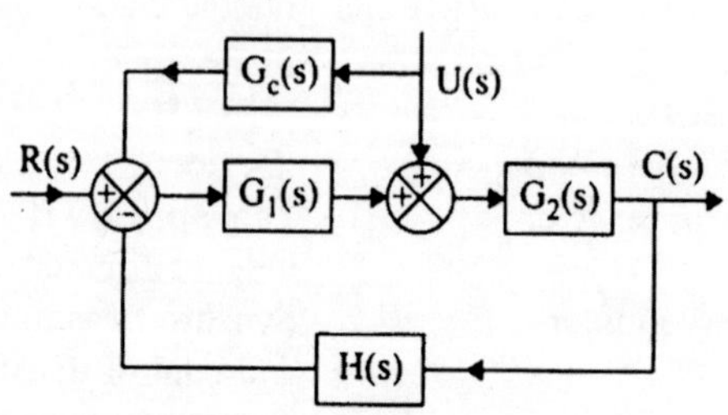

Fig. 10.96.

(*a*) $\frac{G_2(s)+G_1(s)G_2(s)G_c(s)}{1+G_1(s)G_2(s)H(s)}$

(b) $\dfrac{G_1(s)G_2(s)G_c(s)}{1+G_1(s)G_2(s)H(s)}$

(c) $\dfrac{G_1(s)G_2(s)+G_c(s)}{1+G_1(s)G_2(s)H(s)}$

(d) $\dfrac{G_1(s)G_c(s)+G_2(s)}{1+G_1(s)G_2(s)H(s)}$

292. In the signal flow graph of Fig. 10.97, the number of forward paths is

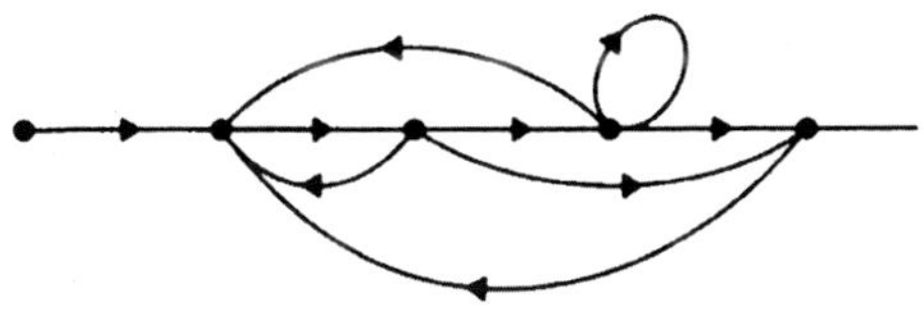

Fig. 10.97.

(a) 1 (b) 2
(c) 3 (d) 4

293. In Fig. 10.97, the number of individual loops is
(a) 2 (b) 4
(c) 5 (d) 6

294. In Fig. 10.97, the combinations of two non-touching loops is
(a) 1 (b) 2
(c) 3 (d) 4

295. In Fig. 10.97 the number of combination of three non-touching loops is
(a) 0 (b) 1
(c) 2 (d) 3

296. Which of the following is not a necessary advantage of negative feedback system?
(a) Feetback improves speed of response
(b) Feedback improves stability
(c) Feedback reduces effect of variations in system parameters
(d) Feedback reduces the effect of external disturbances

297. Consider the following statements regarding negative feedback
1. It increases system gain
2. It improves sensitivity
3. It improves system stability
Which of these are not correct?
(a) all (b) 1, 2
(c) 1, 3 (d) 1 only

298. For the unity feed back system of Fig. 10.98, the closed loop transfer function is

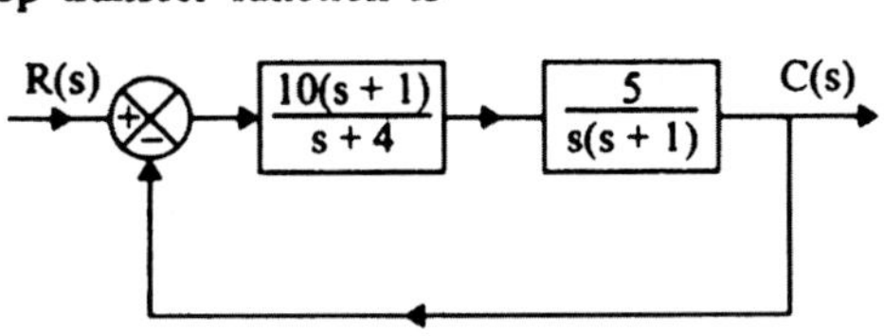

Fig. 10.98.

(a) $\dfrac{50}{s^2+4s+50}$ (b) $\dfrac{50}{s+50}$

(c) $\dfrac{50}{s(s+4)}$ (d) $\dfrac{50}{s^2+50}$

299. For Fig. 10.99, $\dfrac{C(s)}{R(s)} =$

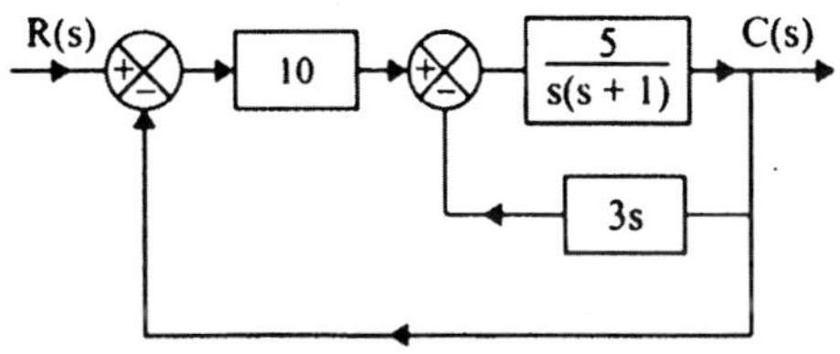

Fig. 10.99.

(a) $\dfrac{50}{s^2+50}$ (b) $\dfrac{50}{s^2+16s+50}$

(c) $\dfrac{50}{s^2+4s+50}$ (d) $\dfrac{50}{s^2+16s}$

300. In control systems the word controller includes
(a) error detector and feedback element
(b) feedback element and control elements
(c) error detector and control elements
(d) error detector, control elements and feedback elements

301. In control systems the output of sensor, usually, is
(a) analog electrical signal
(b) digital electrical signal
(c) mechanical signal
(d) analog or digital electrical signal

302. The most commonly used devices for differencing and amplification, in control systems, are
(a) BJT amplifiers (b) FET amplifiers
(c) operational amplifiers (d) either (a) or (b)

303. DC servo motors are generally
(a) permanent magnet dc motors
(b) shunt field motors
(c) separately excited dc motors
(d) any of above

304. Which of the following is correct?
(a) speed torque characteristics of permanent magnet dc servo motor is more flat than that of wound field dc servo motor
(b) speed torque characteristics of wound field dc servo motor is more flat than that of permanent magnet dc servo motor
(c) speed torque characteristics of wound field and permanent magnet dc servo motor are equally flat
(d) any of above

305. Which dc servo motor has minimum inertia?
(a) slotted armature construction
(b) surface wound armature
(c) surface wound with stationary rotor
(d) either (a) or (b)

306. Which dc servo motor requires strongest permanent magnet?
(a) slotted armature

(*b*) surface wound armature
(*c*) surface wound with stationary rotor
(*d*) either (*a*) or (*b*)

307. DC tachnogenerator can be used in
(*a*) speed control system
(*b*) position control system
(*c*) both (*a*) and (*b*)
(*d*) neither (*a*) nor (*b*)

308. The current rating of dc tachogenerator is usually
(*a*) small (*b*) high
(*c*) medium (*d*) either (*b*) or (*c*)

309. The permanent magnet field system of brushless dc servo motor is generally on
(*a*) stator
(*b*) rotor
(*c*) partly on stator and partly on rotor
(*d*) either (*a*) or (*b*)

310. In a two-phase servo motor the two stator windings have a phase difference of
(*a*) 30° (*b*) 45°
(*c*) 90° (*d*) 135°

311. In a two-phase servo motor the voltages applied to the two stator windings
(*a*) are equal in magnitude and are in phase
(*b*) are unequal in magnitude but are in phase
(*c*) are equal in magnitude but have a phase difference of 90°
(*d*) are unequal in magnitude and have a phase difference of 90°

312. In a two-phasor servo motor, the direction of rotation of the rotating field depends on
(*a*) magnitudes of voltages applied to the two-stator windings
(*b*) phase relation of voltages applied, to the two-stator windings
(*c*) both magnitude and phase relationship of voltages applied to the two-stator windings
(*d*) none of the above

313. The shape of torque-speed curve of a two-phase servo motor mostly depends on
(*a*) ratio of rotor reactance to rotor resistance
(*b*) ratio of stator reactance to rotor resistance
(*c*) ratio of rotor reactance to rotor resistance and also the ratio of stator reactance to stator resistance
(*d*) magnitude of stator voltages

314. An ac tachometer has a drag cup construction so that
(*a*) rotor inertia is low
(*b*) rotor body is conducting
(*c*) rotor is easy to fabricate
(*d*) both (*a*) and (*b*)

315. The number of stator field windings in an ac tachometer are
(*a*) 1 (*b*) 2
(*c*) 3 (*d*) 4

316. In Fig. 10.100, of potentiometer $V_o = V_i (R_o/R_t)$ only when

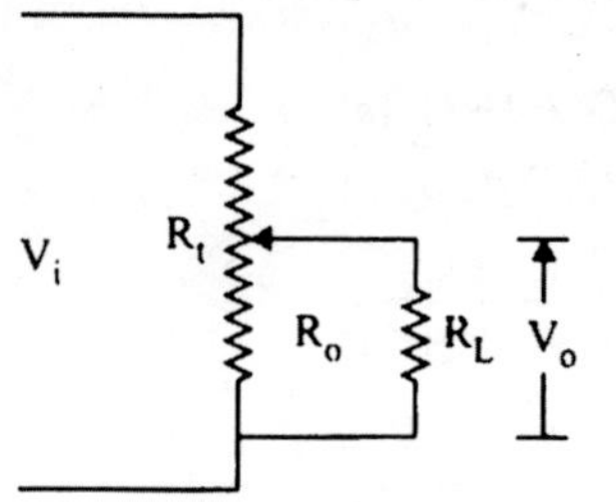

Fig. 10.100.

(*a*) $R_L = 0$ (*b*) $R_L = \infty$
(*c*) R_L is neither low nor high
(*d*) R_L is low but not zero

317. The resolution of a wire wound potentiometer is about
(*a*) 0.001% (*b*) 0.1%
(*c*) 1% (*d*) 5%

318. Power handling capacity of wire wound potentiometer used in control systems is about
(*a*) 0.1 W (*b*) 5 W
(*c*) 25 W (*d*) 100 W

319. Optical encoders most commonly used in control systems are
(*a*) absolute encoders (*b*) secondary encoders
(*c*) incremental encoders (*d*) none of the above

320. The sensor generally used in incremental optical encoder is
(*a*) LED (*b*) LCD
(*c*) photo diode (*d*) either (*a*) or (*b*)

321. In a synchro transmitter which coil acts as primary
(*a*) rotor
(*b*) stator
(*c*) partly rotor and partly stator
(*d*) tertiary

322. In a synchro transmitter
(*a*) input is angular position of rotor shaft and output is stator voltages
(*b*) input is stator voltages and output is angular position of shaft
(*c*) input is rotor voltage and output is angular position of rotor shaft
(*d*) input is angular position of rotor shaft and output is rotor voltage

323. In a synchro transmitter-control transformer, the voltage induced in rotor winding of control transformer is zero when the angular displacement between rotors of transmitter and transformer is
(*a*) 0° (*b*) 90°
(*c*) 180° (*d*) either (*a*) or (*c*)

324. A stepper motor is essentially
(*a*) a synchronous motor (*b*) induction motor
(*c*) dc motor (*d*) either (*b*) or (*c*)

325. Stepper motors find applications in
(*a*) X-Y plotters
(*b*) numerically controlled machining equipment

(*c*) printers
(*d*) all the above

326. In a pneumatic bellows
(*a*) The input surface, output surface and side walls are corrugated
(*b*) The input surface and output surface are flat but the side walls are corrugated
(*c*) The input surface and side walls are flat but the output surface is corrugated
(*d*) The input surface and output surface are corrugated but the side walls are flat

327. If ΔP is pressure difference, A is area of each flat surface, K is stiffness and Δx is displacement of movable surface from reference of bellows, then
(*a*) $\frac{\Delta X(s)}{\Delta P(s)} = \frac{A}{K}$ (*b*) $\frac{\Delta X(s)}{\Delta P(s)} = \frac{K}{A}$
(*c*) $\frac{\Delta X(s)}{\Delta P(s)} = AK$ (*d*) $\frac{\Delta X(s)}{\Delta P(s)} = \sqrt{AK}$.

328. Bellows converts
(*a*) pressure difference into displacement
(*b*) pressure difference into voltage
(*c*) displacement into pressure difference
(*d*) either (*a*) or (*c*)

329. The action of bellows is similar to that of.
(*a*) mass (*b*) viscous friction
(*c*) spring (*d*) none of the above

330. In Fig. 10.101, $R_2 >> R_1$. Then $\frac{V_2(s)}{V_1(s)} \approx$

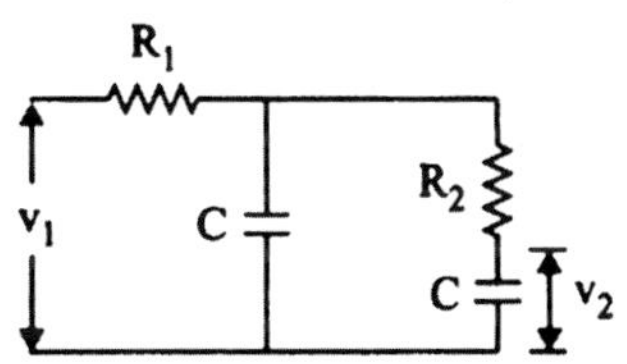

Fig. 10.101.

(*a*) $\frac{1}{1+R_2Cs}$ (*b*) $\frac{1}{1+R_1Cs}$
(*c*) $\frac{1}{1+R_2/Cs}$ (*d*) $\frac{1}{1+R_1/Cs}$

331. A first-order system has a time constant τ. For unit ramp input, the steady slate error is
(*a*) τ^2 (*b*) τ
(*c*) $\sqrt{\tau}$ (*d*) $1/\tau$

332. In a second order system with a unit step input, the speed of response is high if system is
(*a*) overdamped (*b*) undamped system
(*c*) critically damped (*d*) under damped

333. In a second order undamped system, the poles are on
(*a*) positive real axis (*b*) imaginary axis
(*c*) negative real axis (*d*) either (*a*) or (*c*)

334. For a second order system, the position of poles is shown in Fig. 10.102. This system is

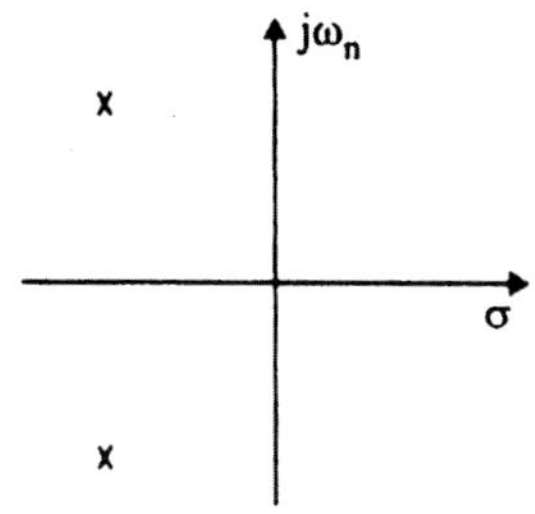

Fig. 10.102.

(*a*) undamped (*b*) overdamped
(*c*) under damped (*d*) critically damped

335. For a second order system with $0 < \xi < 1$, the angle which poles make with negative real axis is
(*a*) $\sin^{-1}\xi$ (*b*) $\cos^{-1}\xi$
(*c*) $\tan^{-1}\xi$ (*d*) $\cot^{-1}\xi$

336. For a second order system, the damping factor is varied from 0 to 1. The locus of poles is
(*a*) a straight line on imaginary axis
(*b*) a straight line on real axis
(*c*) a semi-circle
(*d*) a circle

337. For a second order system, ω_n is natural frequency, ω_d is frequency of damped oscillations and ξ is damping factor. Then
(*a*) $\omega_d = \omega_n \xi$ (*b*) $\omega_d = \omega_n (1-\xi)^2$
(*c*) $\omega_d = \omega_n (1-\xi)^{0.5}$ (*d*) $\omega_d = \omega_n (1-\xi^2)^{0.5}$

338. For a second order system, damping factor is 0.1. If ω_d is frequency of damped oscillations and ω_n is natural frequency, then
(*a*) $\omega_d \approx \omega_n$ (*b*) $\omega_d \approx \frac{1}{\omega_n}$
(*c*) $\omega_d \approx \sqrt{\omega_n}$ (*d*) $\omega_d \approx (\omega_n)^2$

339. In a second order system, peak overshoot is independent of
(*a*) damping factor
(*b*) natural frequency
(*c*) both damping factor and natural frequency
(*d*) neither damping factor nor natural frequency

340. In a second order system, the time constant τ of exponential envelopes depends
(*a*) only on damping factor
(*b*) only on natural frequency
(*c*) both on damping factor and natural frequency
(*d*) neither on damping factor nor on natural frequency

341. If a zero is added to a second order under damped system
(*a*) the peak overshoot is increased
(*b*) the peak overshoot is decreased
(*c*) the peak overshoot remains unchanged
(*d*) either (*a*) or (*b*)

342. For a second order system, the position of poles is in the shaded region shown in Fig. 10.103. The variation in damping factor is

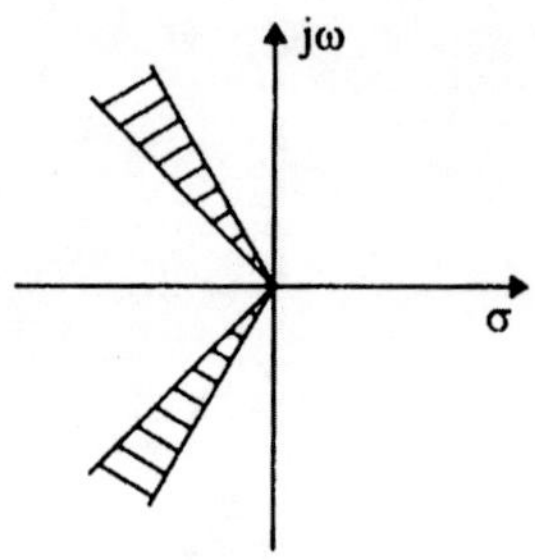

Fig. 10.103.

(*a*) from 0 to 1 (*b*) from about 0.3 to 0.7
(*c*) from 1 to 5 (*d*) from 0 to 3

343. In control systems, the magnitude of error voltage
(*a*) is very small
(*b*) is vary large
(*c*) is neither small nor large
(*d*) may be any value from 0 to about 50 V

344. If feedback factor is β the overall gain of a closed loop system is approximately equal to

(*a*) $\frac{1}{A_v\beta}$ (*b*) $\frac{1}{\beta}$

(*c*) β (*d*) $\sqrt{\beta}$

345. The effect of negative feedback on distortion and bandwidth are
(*a*) both are decreased
(*b*) both are increased
(*c*) distortion is reduced and bandwidth is increased
(*d*) distortion is increased and bandwidth is decreased

346. For very low frequencies, v_0/v_i in Fig. 10.104 equals

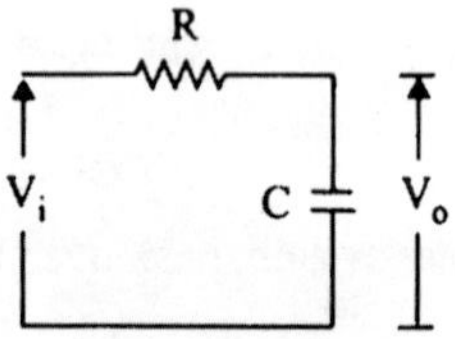

Fig. 10.104.

(*a*) 1 (*b*) 0
(*c*) ∞ (*d*) 0.5

347. In Fig. 10.104 the input frequency is such that R << X_C.. then
(*a*) v_0 lags v_i by 90°
(*b*) v_0 leads v_i by 90°
(*c*) v_0 and v_i are in phase
(*d*) either (*b*) or (*c*)

348. In Fig. 10.104, if R = X_C, voltage gain is
(*a*) 0 *d*B (*b*) 3 *d*B
(*c*) –3 *d*B (*d*) 20 *d*B

349. In Fig. 10.104, the input frequency is such that R = X_C. Then
(*a*) v_0 lags v_i by 45° (*b*) v_0 lags v_i by 90°
(*c*) v_0 leads v_i by 45° (*d*) v_0 leads v_i by 90°

350. In Fig. 10.104 the input frequency
f = 0.1 (1/2 π RC).
The voltage gain is
(*a*) 0 *d*B (*b*) 3 *d*B
(*c*) –3 *d*B (*d*) 20 *d*B

351. In Fig. 10.104 the input frequency
f = 10 (1/2 π RC).
The voltage gain is
(*a*) 0 *d*B (*b*) 3 *d*B
(*c*) –3 *d*B (*d*) –20 *d*B

352. Figure 10.105 shows a constant current source driving a parallel RC circuit. It is equivalent to

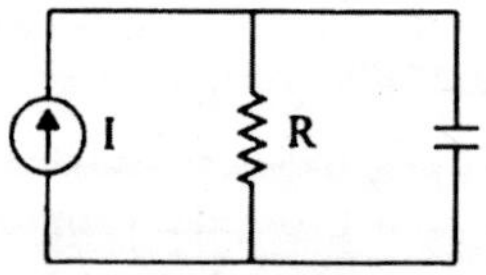

Fig. 10.105.

(*a*) a lag network (*b*) a lead network
(*c*) a lag-lead network (*d*) either (*a*) or (*b*)

353. The network of Fig. 10.106 is a

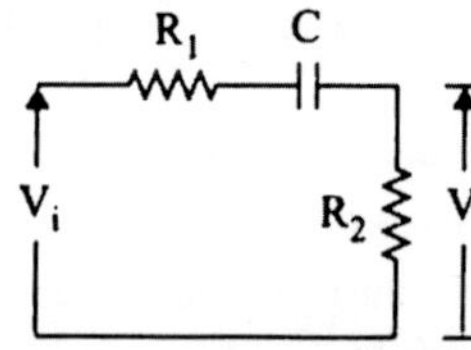

Fig. 10.106.

(*a*) lag network (*b*) lead network
(*c*) lag-lead network (*d*) any of above

354. In a lag network the input frequency is 100 f_c where f_c is critical frequency. The voltage gain is
(*a*) 0 *d*B (*b*) –20 *d*B
(*c*) –40 *d*B (*d*) –60 *d*B

355. When a unit step voltage drives a lag network the output
(*a*) remains constant at unit step value
(*b*) increases exponentially from zero to final value
(*c*) decreases exponentially from 1 to 0
(*d*) either (*b*) or (*c*) depending on values of R and C

356. Integral error compensation changes a second order system to
(*a*) first order system
(*b*) third order system
(*c*) fourth order system
(*d*) either (*a*) or (*b*) depending on the system

357. Consider following statements
1. Integral of absolute magnitude of error

$$= \int_0^T |e(t)|\,dt$$

2. Integral time-absolute error

$$= \int_0^T t\,|e(t)|\,dt$$

3. Integral time-square error = $\int_0^T t^2 |e(t)|\,dt$

Which of the above are correct?

(*a*) 1 and 3 (*b*) 1 only
(*c*) 1 and 2 (*d*) 2 and 3

358. For Fig. 10.107, $\frac{\theta_0(s)}{\theta_i(s)} =$

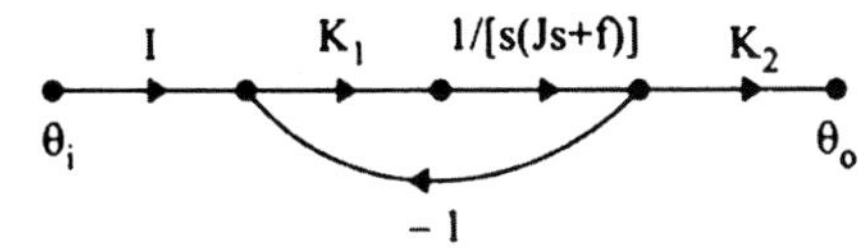

Fig. 10.107.

(*a*) $\frac{K_1K_2}{s(Js+f)}$ (*b*) $\frac{K_1K_2}{Js^2+sf+K_1}$

(*c*) $\frac{K_1K_2}{Js^2+sf+K_2}$ (*d*) $\frac{K_1K_2}{Js^2+sf+K_1K_2}$

359. For the system of Fig. 10.108, the characteristic equation is

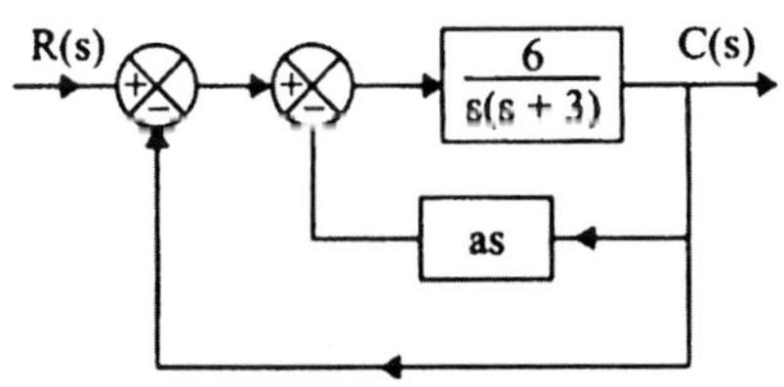

Fig. 10.108.

(*a*) $s^2 + s(3 + 6a) + 6 = 0$
(*b*) $s^2 + s(3+ 12a)+ 12 = 0$
(*c*) $s^2 + 3s = 0$ (*d*) $s^2 + 6s = 0$

360. If error voltage is $e(t)$, integral square error =

(*a*) $e^2(t)dt$ (*b*) $\int_0^\infty e^2(t)dt$

(*c*) $\left[\int_0^\infty e^2(t)dt\right]^{0.5}$ (*d*) none of the above

361. Robotic manipulator arms mostly use

(*a*) hydraulic actuator (*b*) pneumatic actuator
(*c*) electric motor actuator (*d*) none of the above

362. Consider the following statements :

1. If any root of characteristic equation has a positive real part the impulse response is unbounded and system is unstable.
2. If all the roots of a characteristic equation have negative real parts, the impulse response decays to zero.
3. If one or more non-repeated roots of characteristic equation are on $j\omega$ axis impulse response is bounded but the system is unstable.

Which of the above equations are correct?

(*a*) 1, 2 and 3 (*b*) 1 and 2 only
(*c*) 1 and 3 only (*d*) 2 and 3 only

363. Consider the following statements :

1. The positiveness of coefficients of characteristic equation is necessary as well as sufficient condition for stability of first and second order systems.
2. The positiveness of the coefficients of characteristic equation ensures negativeness of real roots but is not sufficient condition for stability of third and higher order systems.

Out of above statements

(*a*) only 1 is correct (*b*) both 1 and 2 are correct
(*c*) only 2 is correct (*d*) both are wrong

364. The entries in the first column of Routh array of a fourth order are 5, 2, – 0.1, 2, 1. The number of poles in the right half plane are

(*a*) 1 (*b*) 2
(*c*) 3 (*d*) 4

365. If any term in the first column of a Routh array becomes zero, then

(*a*) Routh criterion cannot be used to determine stability
(*b*) Routh criterion can be used by substituting a small positive number for zero and completing the array
(*c*) Routh criterion can be used by substituting a big positive number for zero and completing the array
(*d*) Routh criterion can be used by substituting a small negative number for zero and completing the array

366. Consider the following statements about root locus

1. The root locus is symmetrical about real axis
2. If a root locus branch moves along the real axis from an open loop pole to zero or to infinity, this root locus branch is called real root branch.
3. The breakaway points of the root locus are the solutions of $\frac{dK}{dS} = 0$

Which statements out of above are correct?

(*a*) 1 and 2 only (*b*) 1 and 3 only
(*c*) 1, 2 and 3 (*d*) 2 and 3 only

367. If G(*s*) H(*s*) = $\frac{K}{s(s+1)(s^2+4s+20)}$, the number of branches of root locus are

(*a*) 1 (*b*) 2
(*c*) 3 (*d*) 4

368. Consider the following statements about root locus

1. The intersection of root locus branches with the imaginary axis can be determined by the use of Routh criterion
2. Segments of real axis having an odd number of real axis open loop poles plus zeros to their right are not parts of root locus
3. The number of root locus branches terminating on infinity is always zero

Which of the above statements are correct?

(*a*) 1, 2 and 3 (*b*) 1 and 2 only
(*c*) 1 only (*d*) 2 and 3 only

369. The polar plot of Fig. 10.109 can be for G($j\omega$) =

(*a*) $\frac{1}{j\omega}$ (*b*) $\frac{1}{j\omega(1+j\omega T_1)}$

(c) $\frac{1}{j\omega^3(1+j\omega T_1)}$ (d) $\frac{1+j\omega T_1}{j\omega}$

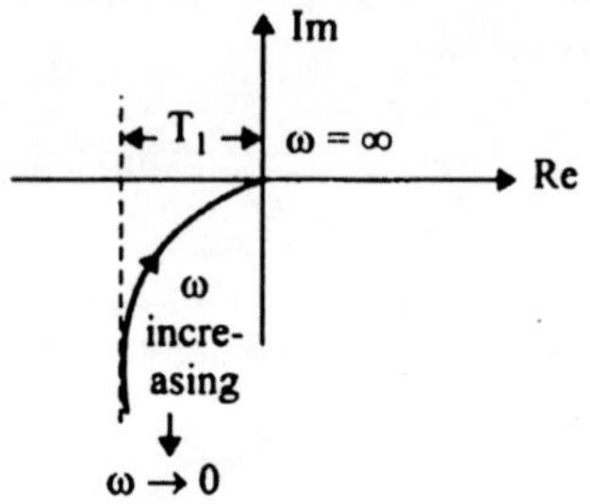

Fig. 10.109.

370. Consider the following statements about polar plots

1. Addition of a non-zero pole to a transfer function results in further rotation of polar plot through an angle of –90° as ω → ∞
2. Addition of a pole at the origin to a transfer function rotates the polar plot at zero and infinite frequencies by a further angle of -90°.

Of the above statements

(a) both are correct
(b) both are wrong
(c) 1 is correct and 2 is wrong
(d) 2 is correct and 1 is wrong

371. For the plot of Fig. 10.110, G (jω) =

(a) $\frac{1}{j\omega}$
(b) $\frac{1}{(j\omega)^2}$
(c) $\frac{1}{(j\omega)^2(1+j\omega T_1)}$
(d) $\frac{1}{(j\omega)^2(1+j\omega T_1)(1+j\omega T_2)}$

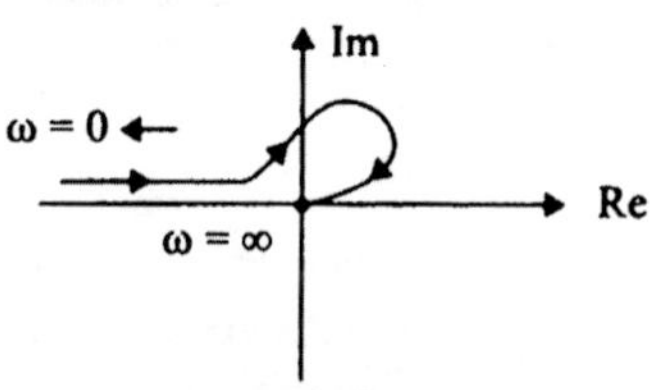

Fig. 10.110.

372. The polar plot of Fig. 10.111 is for G(jω) =

(a) $\frac{1}{j\omega(1+j\omega T_1)}$
(b) $\frac{j\omega}{(1+j\omega T_1)}$
(c) $\frac{1}{j\omega(1+j\omega T_1)(1+j\omega T_2)}$
(d) $\frac{j\omega(j\omega T_1)}{1+j\omega T_2}$

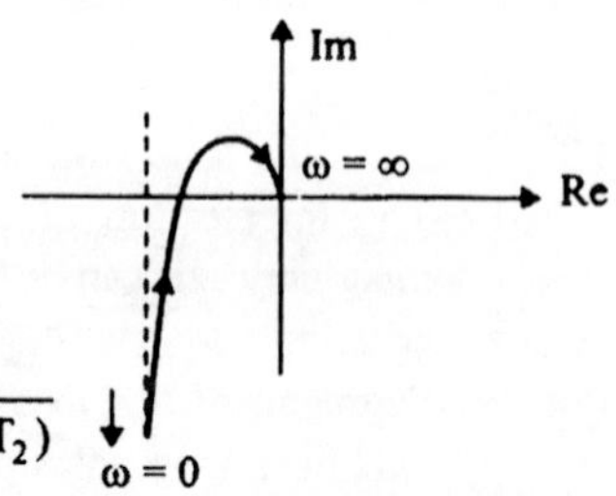

Fig. 10.111.

373. For very low frequencies, log magnitude for Fig. 10.112 is

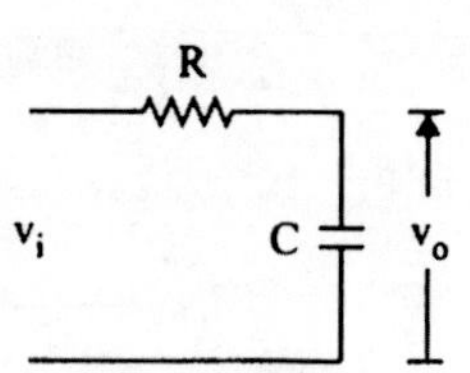

Fig. 10.112.

(a) 0 dB (b) –3 dB
(c) 3 dB (d) 10 dB

374. The log magnitude for the system of Fig. 10.112 is (where T = RC).

(a) 20 log (1 + ωT) (b) –20 log (1 + ωT)
(c) –10 log (1 + ω^2T^2) (d) –20 log (1 + ω^2T^2)

375. For a factor $\frac{1}{j\omega^2}$ in the transfer function, the log magnitude plot

(a) is a straight line with slope 0
(b) is a straight line with slope 20 dB/decade
(c) is a straight line with slope –20 dB/decade
(d) is a straight line with slope –40 dB/decade

376. For a factor $\frac{1}{(1+j2\xi\mu-u^2)}$, the log magnitude plot for $u >> 1$

(a) is a straight line with slope 20 dB/decade
(b) is a straight line with slope 0
(c) is a straight line with slope –20 dB/decade
(d) is a straight line with slope –40 dB/decade

377. The Bode plot of (1 + jωT) has

(a) slope of 20 dB/decade and phase angle + $\tan^{-1}(\omega T)$
(b) slope of –20 dB/decade and phase angle + $\tan^{-1}(\omega T)$
(c) slope of 20 dB/decade and phase angle – $\tan^{-1}(\omega T)$
(d) slope of +40 dB/decade and phase angle – $\tan^{-1}(\omega T)$

378. For the factor $\frac{1}{1+2j\omega}$ the corner frequency is

(a) 0.5 rad/sec (b) 1 rad/sec
(c) 2 rad/sec (d) 4 rad/sec

379. For the factor (1 + 0.5 jω) the corner frequency is

(a) 1 rad/sec (b) 0.5 rad/sec
(c) 2 rad/sec (d) 0.25 rad/sec

380. For $G(j\omega) = \frac{1-j\omega T}{1+j\omega T}$

(a) magnitude is 1 at all frequencies and phase angle varies from 0 to –180° as ω varies from 0 to ∞
(b) magnitude is T at all frequencies and phase angle 0
(c) magnitude is $(1 + \omega^2T^2)^{0.5}$ and phase angle varies from 0 to –180° as ω varies from 0 to ∞
(d) none of the above

381. The functions $\frac{1-j\omega T_3}{(1+j\omega T_1)(1+j\omega T_2)}$ and $\frac{1+j\omega T_3}{(1+j\omega T_1)(1+j\omega T_2)}$ have

(a) identical magnitude and phase curves
(b) identical magnitude curves but different phase curves
(c) different magnitude curves but identical phase curves
(d) different magnitude and phase curves

382. The pole zero plot of Fig. 10.113 is for

(a) non-minimum phase function
(b) minimum phase function
(c) all pass function (d) either (a) or (b)

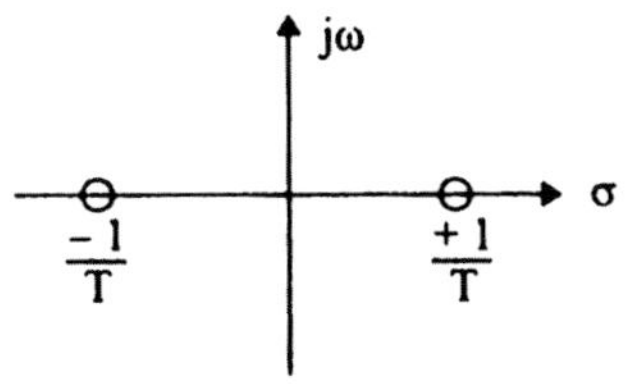

Fig. 10.113.

383. The pole zero plot of Fig. 10.114 is for

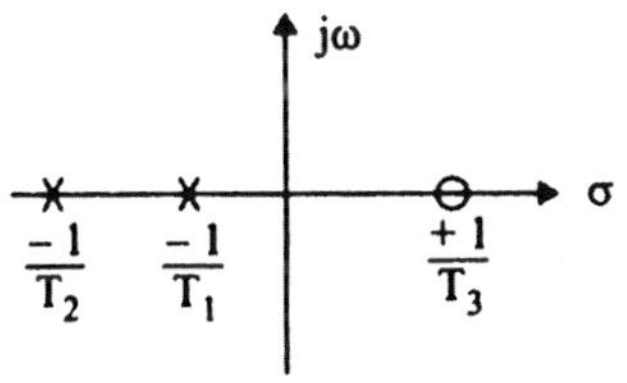

Fig. 10.114.

(*a*) non-minimum phase function
(*b*) minimum phase function
(*c*) all pass function
(*d*) either (*b*) or (*c*)

384. The pole zero plot of Fig. 10.115 is for

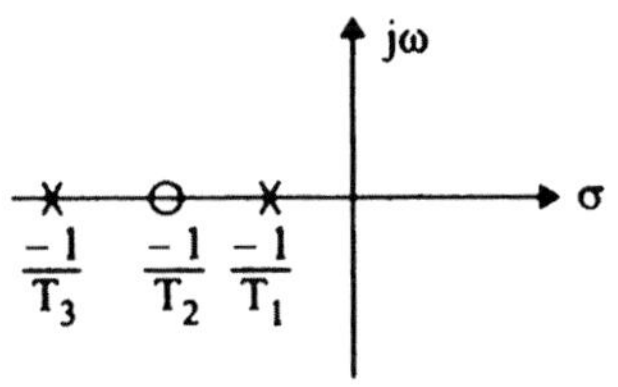

Fig. 10.115.

(*a*) non minimum phase function
(*b*) minimum phase function
(*c*) all pass function
(*d*) any of the above

385. The transfer function of transportation lag is
(*a*) $G(j\omega) = e^{j\omega T}$ (*b*) $G(j\omega) = e^{-j\omega T}$
(*c*) $G(j\omega) = (1 + j\omega T)$ (*d*) $G(j\omega) = (1 - j\omega T)$

386. The slope of log-magnitude asymptote changes by –40 *d*B/decade at a frequency ω_1. This means that
(*a*) a double pole is present
(*b*) a pair of complex conjugate poles is present
(*c*) either a double pole or a pair of complex conjugate poles is present
(*d*) a pole or zero at origin is present

387. The slope of log magnitude asymptote changes by – 40 *d*B/decade at a certain frequency and the error between asymptote and actual curve is –6 *d*B. This means that
(*a*) a double pole is present
(*b*) a pair of complex conjugate poles is present
(*c*) either a double pole or a pair of complex conjugate poles is present
(*d*) either a pole or zero at that frequency is present

388. The slope of log magnitude curve changes by –40 *d*B/decade at a certain frequency and the error between asymptote and actual curve is positive. This means that
(*a*) a double pole is present
(*b*) a pair of complex conjugate poles is present
(*c*) either (*a*) or (*b*)
(*d*) neither (*a*) nor (*b*)

389. The phase angle characteristic of Fig. 10.116 is for

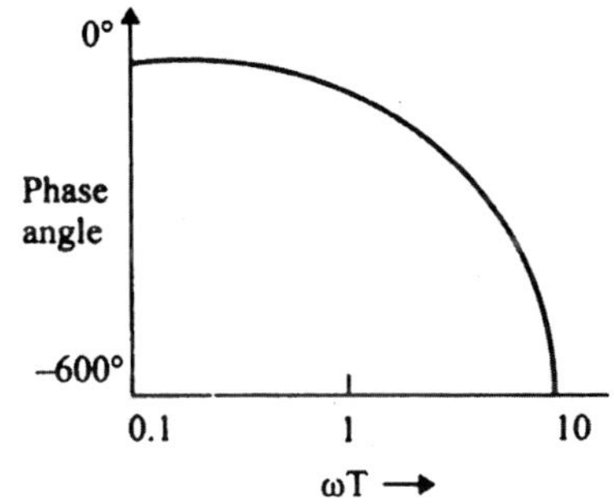

Fig. 10.116.

(*a*) all pass function
(*b*) transport lag
(*c*) minimum phase function
(*d*) none of the above

390. In Fig. 10.117, the curves 1, 2, 3 are respectively for

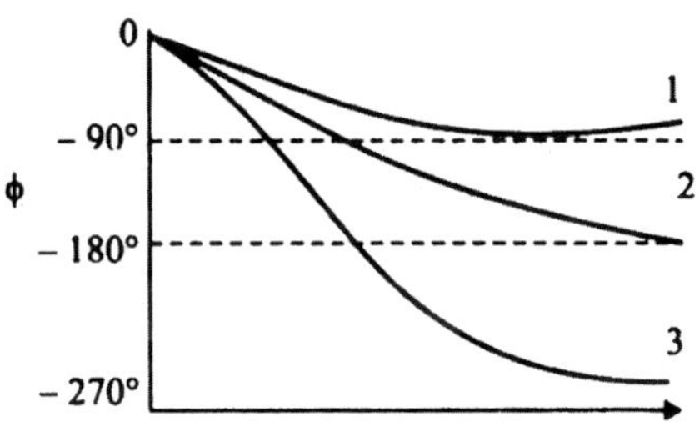

Fig. 10.117.

(*a*) all pass, minimum phase and non-minimum phase functions
(*b*) minimum phase, all pass and non-minimum phase functions
(*c*) non-minimum phase, all pass and minimum phase functions
(*d*) minimum phase, non-minimum phase and all pass functions

391. The polar plot of $G(s) = \dfrac{1}{(1+s)(1+2s)}$
(*a*) does not cross real axis
(*b*) crosses real axis
(*c*) crosses real axis at ω = 0.7 rad/sec
(*d*) crosses real axis at ω = 1.4 rad/sec

392. The polar plot of $G(s) = \dfrac{1}{s^2(1+s)(1+2s)}$
(*a*) does not cross real axis
(*b*) crosses real axis once
(*c*) crosses real axis twice
(*d*) crosses real axis 4 times

393. The Nyquist plot of Fig. 10.118 is for G(s) H(s) =

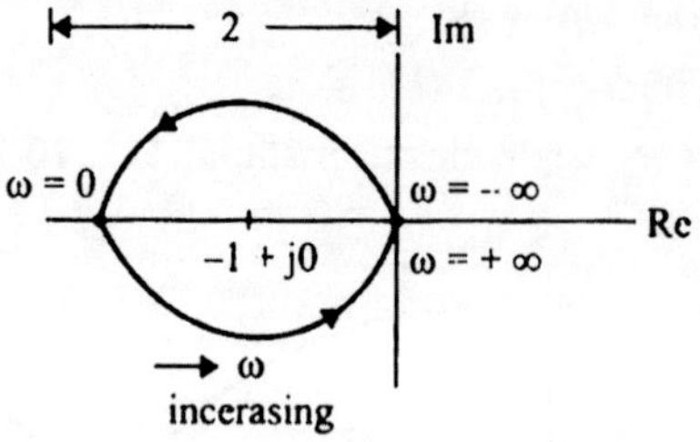

Fig. 10.118.

(a) $\frac{(s+3)}{(s+1)(s-1)}$ (b) $\frac{(s+3)}{(s+2)(s-1)}$

(c) $\frac{s+2}{(s+1)(s-1)}$ (d) $\frac{s+2}{(s+4)(s-1)}$

394. The system of Fig. 10.119 has

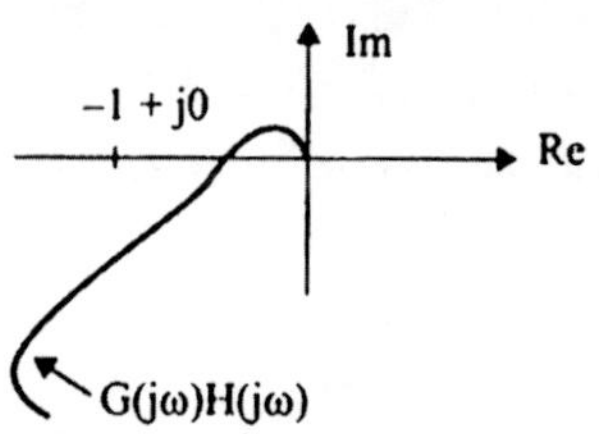

Fig. 10.119.

(a) good gain and phase margins
(b) good gain margin and poor phase margin
(c) poor gain and phase margins
(d) poor gain margin but good phase margin

395. The system of Fig. 10.120 has

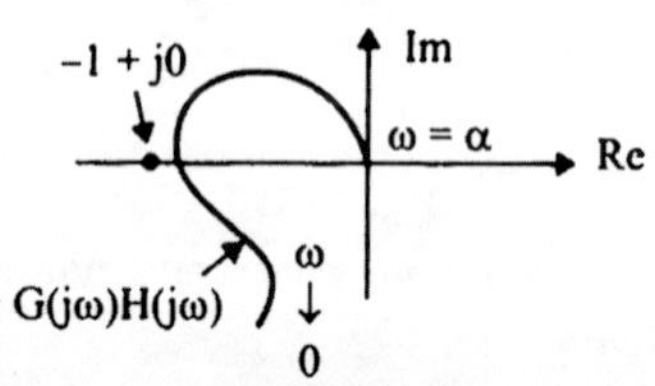

Fig. 10.120.

(a) good gain and phase margin
(b) poor gain and phase margin
(c) good gain margin but poor phase margin
(d) poor gain margin but good phase margin

396. Figure 10.121 shows three nyquist plots of G(s) H(s) = $e^{-sT}/[s(s+1)]$ for different values of T. The curves 1, 2 and 3 can be

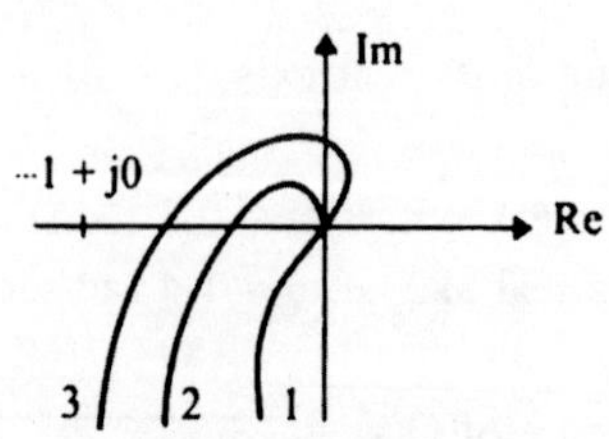

Fig. 10.121.

(a) for T = 0, 0.5 and 1 respectively
(b) for T = 1, 0.5 and 0 respectively
(c) for T= 0.5, 1 and 0 respectively
(d) for T = 0, 1 and 0.5 respectively

397. Figure 10.122 shows log magnitude versus ω plots of a system for different values of gain. For this system

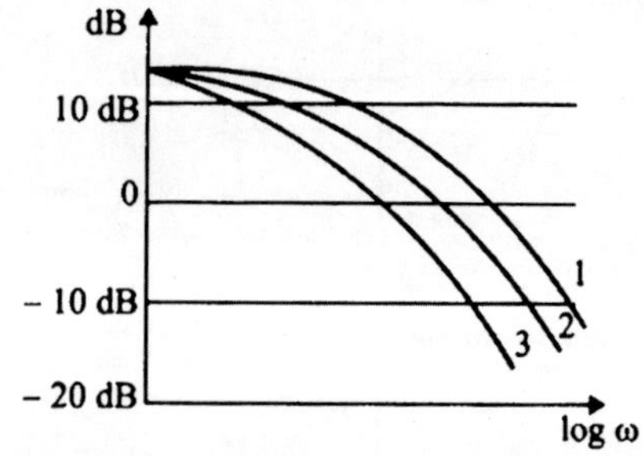

Fig. 10.122.

(a) curve 1 shows highest gain margin and curve 2 shows poorest gain margin
(b) gain margins shown by the three curves are equal
(c) curve 1 shows poorest gain margin and curve 3 shows highest gain margin
(d) curve 3 shows poorest gain margin while curve 1 shows highest gain margin

398. Figure 10.123 shows dB magnitude plots for type 0, type 1 and type 2 systems. Out of these

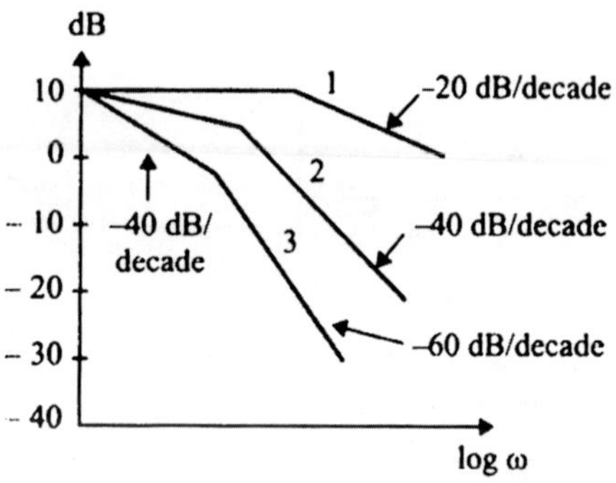

Fig. 10.123.

(a) curves 1, 2, 3 are for type 0, 1, 2 systems respectively
(b) curves 1, 2, 3 are for type 2, 1, 0 systems respectively
(c) curves 1, 2, 3 are for type 1, 0, 2 systems respectively
(d) curves 1, 2, 3 are for type 1, 2, 0 systems respectively

399. If G(s) H(s) = $\frac{1}{s^2+100}$, the closed loop poles are on

(a) negative real axis (b) jω axis
(c) right half plane (d) either (a) or (b)

400. The open loop frequency response of a unity feedback system is as under

ω	1	3	5	7	9	10
$\lvert G(j\omega)\rvert$	6	5.1	3	2	1.0	0.5
$\angle G(j\omega)$	–120°	–130°	–140°	–150°	–165°	–180°

The gain margin and phase margin respectively are

(a) 20 dB and 15° (b) 20 dB and 25°
(c) 6 dB and 15° (d) 6 dB and 25°

401. A linear time invariant system initially at rest when subjected to unit step input gives a response $y(t) = te^{-t}$ for $t > 0$. The transfer function of the system is

(a) $\frac{1}{(s+1)^2}$ (b) $\frac{1}{s(s+1)^2}$

(c) $\dfrac{s}{(s+1)^2}$ (d) $\dfrac{1}{s+1}$ **[IES 2010]**

402. Consider the function $F(s) = \dfrac{5}{s(s^2+3s+2)}$. The initial value of $f(t)$ is

(a) 5 (b) $\dfrac{5}{2}$

(c) $\dfrac{5}{3}$ (d) 0 **[IES 2010]**

403. In a unity feedback control system with $G(s) = \dfrac{4}{s^2+0.4s}$ when subjected to unit step input it is required that system response be settled within 2% tolerance band. The system setting time is

(a) 1 sec (b) 2 sec

(c) 10 sec (d) 20 sec **[IES 2010]**

404. In Fig. 10.124, the transfer function $d(s)$ to $y(s)$ is

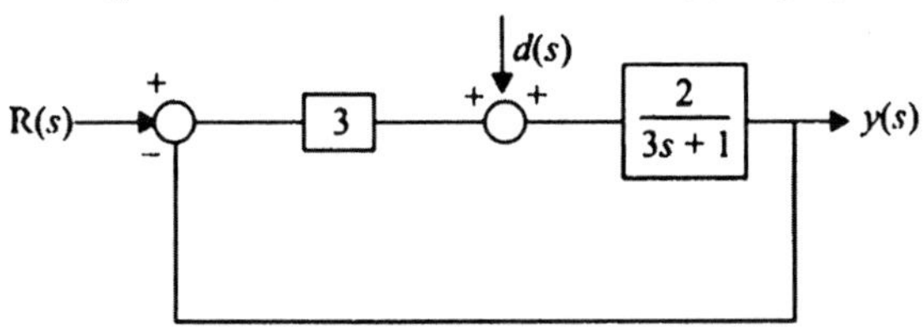

Fig. 10.124.

(a) $\dfrac{2}{3s+7}$ (b) $\dfrac{2}{3s+1}$

(c) $\dfrac{6}{3s+7}$ (d) $\dfrac{2}{3s+6}$ **[IES 2010]**

405. The characteristic equation of closed loop system in Fig. 10.125 is

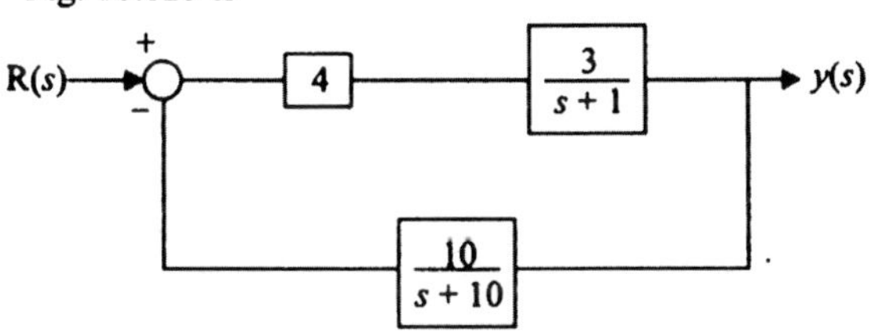

Fig. 10.125.

(a) $s^2 + 11s + 10 = 0$ (b) $s^2 + 11s + 130 = 0$

(c) $s^2 + 10s + 120 = 0$ (d) $s^2 + 10s + 12 = 0$

[IES 2010]

406. If the gain of a system is reduced to zero, the roots of the system in s-plane

(a) coincide with zeros (b) move away from zeros

(c) move away from poles (d) coincide with poles

[IES 2010]

407. In the Bode plot of a unity feedback control system the value of phase angle $G(j\omega)$ is $-90°$ at a gain cross-over frequency of the Bode plot. The phase margin of the system is

(a) $-180°$ (b) $180°$

(c) $-90°$ (d) $+90°$ **[IES 2010]**

408. The Nyquist plot of loop transfer function $G(s)\,H(s)$ of a closed loop system passes through $(-1, 0)$ point in $G(s)\,H(s)$ plane. The phase margin is

(a) 0 (b) 45°

(c) 90° (d) 180° **[IES 2010]**

409. For the Nichols plot in Fig. 10.126, the system is

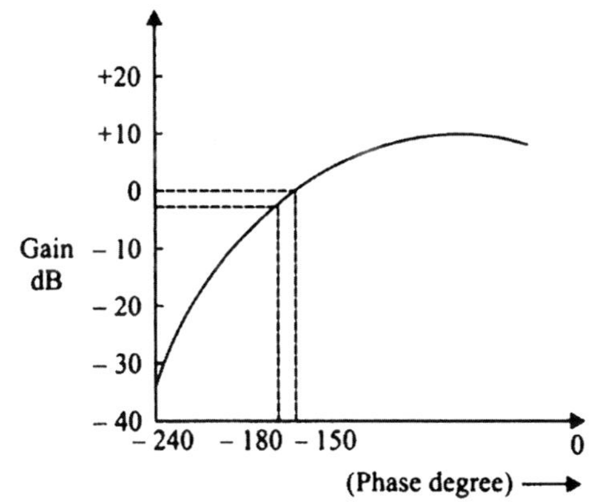

Fig. 10.126.

(a) unstable (b) stable

(c) overdampes (d) critically stable

[IES 2010]

410. The magnitude and phase of transfer function $G(s) = \dfrac{1}{s+1}$ at $\omega = 1$ are

(a) 0.707 and 45° (b) -3 dB and 0.78 rad

(c) 0.707 and $-45°$ (d) 3 dB and $-90°$

[IES 2010]

411. Which of the following is the response of a spring mass damper system with underdamping

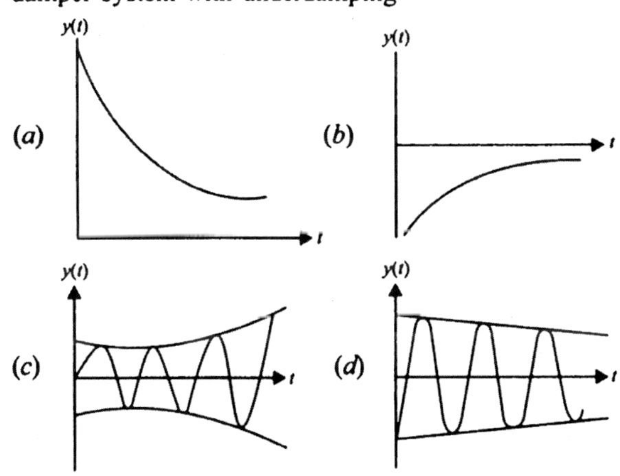

Fig. 10.127.

412. From the point of stability and response speed of a closed loop system, the approximate range of damping rate is

(a) 0 to 0.2 (b) 0.4 to 0.7

(c) 0.8 to 1 (d) 1.1 to 1.5 **[IES 2010]**

413. Transfer function $G(s) = \dfrac{1}{3s+1}$ has a corner frequency of

(a) 3 rad/sec (b) 0.33 rad/sec

(c) 1 rad/sec (d) 30 rad/sec

[IES 2010]

414. Which of the following represents pole-zero location for lead-compensator?

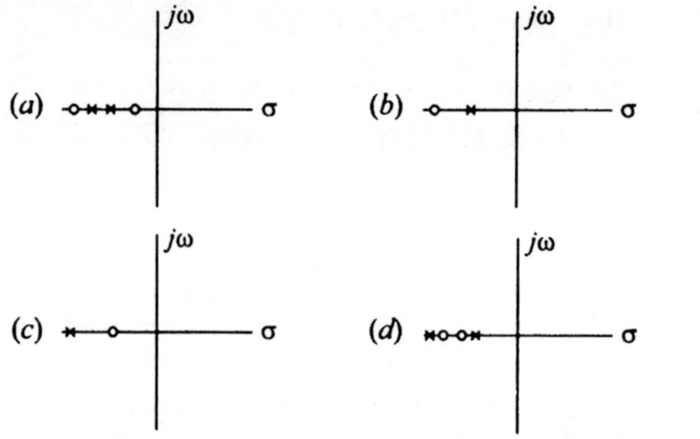

[IES 2010]

Fig. 10.128.

415. Transfer function for the state representation of continuous time LTI system

$\dot{q}(t) = A\, q(t) + bx(t)$

$y(t) = C\, q(t) + dx(t)$

is given by

(a) $C(sI - A)^{-1}\, b + d$ (b) $b(sI - A)^{-1}\, b + d$

(c) $a(sI - A)^{-1}\, b + d$ (d) $d(sI - A)^{-1}\, b + c$

[IES 2010]

416. Transfer function of a certain system is

$$\frac{Y(s)}{U(s)} = \frac{1}{s^4 + 5s^3 + 8s^2 + 6s + 3}$$

Which of the following is the A, B matrix pair of state variable representation

(a) $\begin{bmatrix} 0 & 1 & 0 & 0 \\ 0 & 0 & 1 & 0 \\ 0 & 0 & 0 & 1 \\ -3 & -6 & -8 & -5 \end{bmatrix}\begin{bmatrix} 0 \\ 0 \\ 0 \\ 1 \end{bmatrix}$

(b) $\begin{bmatrix} 0 & 1 & 0 & 0 \\ 0 & 0 & 1 & 0 \\ 0 & 0 & 0 & 1 \\ -5 & -8 & -6 & -3 \end{bmatrix}\begin{bmatrix} 0 \\ 0 \\ 0 \\ 1 \end{bmatrix}$

(c) $\begin{bmatrix} 1 & 0 & 0 & 0 \\ 0 & 1 & 0 & 0 \\ 0 & 0 & 1 & 0 \\ -3 & -6 & -8 & -5 \end{bmatrix}\begin{bmatrix} 1 \\ 0 \\ 0 \\ 0 \end{bmatrix}$

(d) $\begin{bmatrix} 0 & 1 & 0 & 0 \\ 0 & 0 & 1 & 0 \\ 0 & 0 & 0 & 1 \\ -3 & -6 & -8 & -5 \end{bmatrix}\begin{bmatrix} 1 \\ 0 \\ 0 \\ 0 \end{bmatrix}$ [IES 2009]

417. Consider the function $F(s) = \dfrac{\omega}{s^2 + \omega^2}$ where F(s) is the Laplace transform of $f(t)$. The steady state value of $f(t)$ is

(a) zero (b) one

(c) two (d) between – 1 and + 1

[IES 2009]

418. Transfer function of LTI system is $\dfrac{1}{s+1}$. What is the steady state value of unit impulse response

(a) zero (b) one

(c) two (d) infinite [IES 2009]

419. What are the requirements of a good control system?

(a) sensitive to parameter variation

(b) insensitive to input command

(c) neither sensitive to parameter variation nor sensitive to input command

(d) insensitive to parameter variation and sensitive to input commands [IES 2009]

420. How can bandwidth of a control system be increased?

(a) by the use of a phase lead network

(b) by the use of phase lag network

(c) by the use of both phase lead and phase lag networks

(d) by the use of cascaded amplifies in the system

[IES 2009]

421. Which of the following may lead to instability problem?

(a) large error (b) high selectivity

(c) high gain (d) noise [IES 2009]

422. A negative feedback system is supplied an input of 5 V. Forward gain is 1 and feedback gain is also 1. What is the output voltage?

(a) 1 V (b) 1.5 V

(c) 2 V (d) 2.5 V [IES 2009]

423. For what positive value of K does the polynomial

$s^4 + 8s^3 + 24s^2 + 32s + K$

have roots with zero real part

(a) 10 (b) 20

(c) 40 (d) 80 [IES 2009]

424. How many roots with positive real parts do the equation $s^3 + s^2 - s + 1 = 0$ have

(a) zero (b) one

(c) two (d) three [IES 2009]

425. The characteristic equation of a control system is $s^4 + 4s^3 + 4s^2 + 3s + K = 0$.

What is the value of K for which the system is marginally stable?

(a) $\dfrac{9}{16}$ (b) $\dfrac{19}{16}$

(c) $\dfrac{29}{16}$ (d) $\dfrac{39}{16}$ [IES 2009]

426. Transfer function of a controller is $K_p + K_d s + \dfrac{K_1}{s}$ where K_p, K_d and K_1 are constant. What type of controller is this?

(a) proportional

(b) proportional plus derivative

(c) proportional plus integral

(d) proportional plus integral plus derivative

[IES 2009]

427. Impulse response of linear time invariant system is $g(t) = e^{-t}$ for $t > 0$

(a) $\frac{1}{s}$ (b) $\frac{1}{s(s+1)}$

(c) $\frac{1}{s+1}$ (d) $\frac{s}{s+1}$ **[IES 2008]**

428. In a dc servomotor the back emf is equivalent to an 'electric friction' which tends to

(a) improve stability of motor

(b) slowly decrease stability of motor

(c) rapidly decreases stability

(d) have no effect on stability **[IES 2008]**

429. A system with gain margin close to unity or phase margin close to zero is

(a) relatively stable (b) oscillatory

(c) stable (d) highly stable **[IES 2008]**

430. Open loop transfer function of a unity feedback control system is $G(s) = Ke^{-Ts}$ where K and T are variables and greater than zero. The stability of closed loop system depends as

(a) K only (b) both K and T

(c) T only (d) neither K nor T **[IES 2008]**

431. The input to a controller is

(a) sensed signal

(b) error signal

(c) desired variable value

(d) signal of fixed amplitude not dependent on desired value **[IES 2008]**

432. The open loop transfer function of a unity feedback control system is

$$G(s) = \frac{K}{s(s+1)(s+5)}$$

What is the value of K for stable operation?

(a) $0 < K < 5$ only (b) $0 < K < 6$ only

(c) $1 < K < 5$ only (d) $0 < K < 30$ only **[IES 2008]**

433. Consider the unity feedback system having

$$G(s) = \frac{K}{(s^2 + 2s + 2)(s + 2)}$$

The system is marginally stable. What is the radian frequency of oscillation?

(a) $\sqrt{2}$ (b) $\sqrt{3}$

(c) $\sqrt{5}$ (d) $\sqrt{6}$ **[IES 2008]**

434. Characteristic equation of a control system is $s^5 + s^4 + 2s^3 + 2s^2 + 4s + 6 = 0$

What is the number of roots of the equation in right half of s-plane?

(a) zero (b) 1

(c) 2 (d) 3 **[IES 2008]**

435. A diaphragm type pressure sensor has two poles as shown in Fig. 10.129. Zeros are at infinity. What is the steady state deformation for a unit step input pressure?

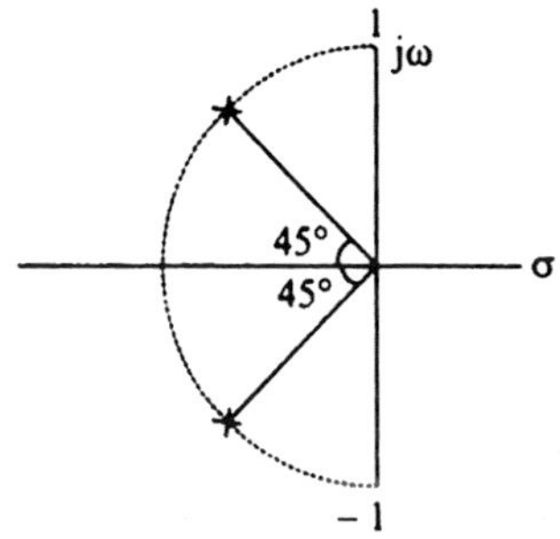

Fig. 10.129.

(a) 0.25 (b) 0.5

(c) 0.707 (d) 1 **[IES 2008]**

436. A second order system has $G(s) = \frac{16}{s^2 + 4s + 16}$. What is the time for first overshoot?

(a) $\frac{2\pi}{\sqrt{3}}$ secs (b) $\frac{\pi}{\sqrt{3}}$ secs

(c) $\frac{\pi}{2\sqrt{3}}$ secs (d) $\frac{\pi}{4\sqrt{3}}$ secs **[IES 2008]**

437. For the closed loop system in Fig. 10.130. What is the ratio of output frequencies

$$\frac{\omega\,(\text{for } K = 32)}{\omega\,(\text{for } K = 16)}$$

(a) 1.40 (b) 1.42

(c) 1.44 (d) 1.46 **[IES 2008]**

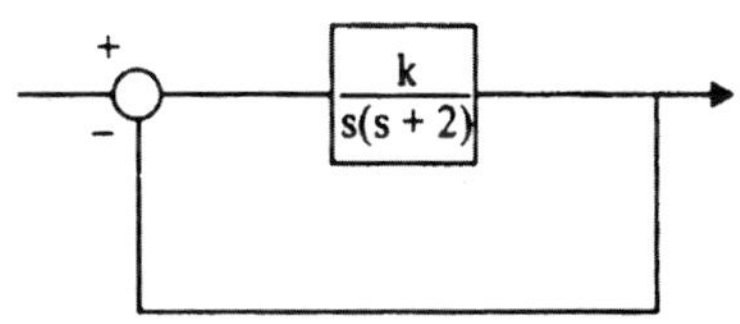

Fig. 10.130.

438. Match List I (shape of Nyquist plot) with List II (Gain margin) and select the correct answer using the codes.

	List I	List II
A.	The plot does not intersect negative real axis	1. < 0 dB
B.	The plot intersects negative real axis between 0 and (– 1, j0)	2. 0 dB
C.	The plot passes through the point (– 1, j0)	3. > 0 dB
D.	The plot encloses the point (– 1, j0)	4. ∞ dB

Codes:

	A	B	C	D
(a)	3	4	1	2
(b)	4	3	2	1
(c)	3	4	2	1
(d)	4	3	1	2

[IES 2008]

439. Figure 10.131 shows gain phase plot of a linear control system.
What are the gain and phase margin?
(*a*) GM > 0 *d*B, PM > 0 deg.
(*b*) GM > 0 *d*B, PM < 0 deg.
(*c*) GM < 0 *d*B, PM > 0 deg.
(*d*) GM < 0 *d*B, PM < 0 deg. **[IES 2008]**

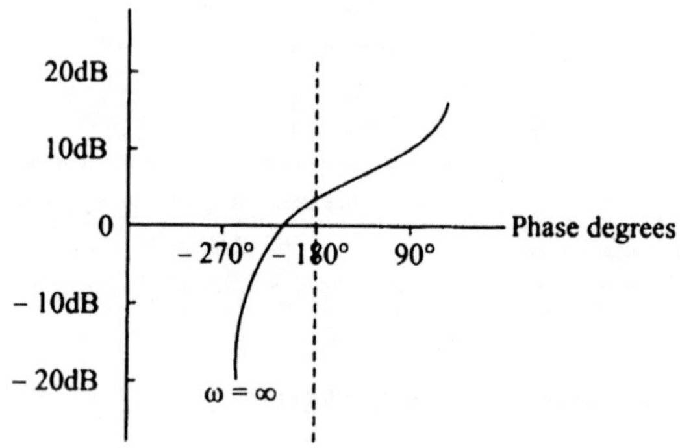

Fig. 10.131.

440. What is the gain margin of a system when the magnitude of the polar plot at phase cross-over is a
(*a*) $\frac{1}{a}$ (*b*) $-a$
(*c*) zero (*d*) a **[IES 2008]**

441. What is the damping ratio of a second order system when value of resonant peak is unity?
(*a*) $\sqrt{2}$ (*b*) unity
(*c*) $\frac{1}{\sqrt{2}}$ (*d*) zero **[IES 2008]**

442. The closed loop transfer function is $\frac{K}{s(s+1)(s+5)+K}$
What is the frequency of oscillation for marginally stable system?
(*a*) $\sqrt{5}$ rad/sec (*b*) $\sqrt{6}$ rad/sec
(*c*) 5 rad/sec (*d*) 6 rad/sec **[IES 2008]**

443. The closed loop transfer function of a control system has the following poles and zeros:

Poles	**Zeros**
$p_1 = -0.5$	$z_1 = -6$
$p_2 = -1.0$	$z_2 = -8$
$p_3 = -5$	
$p_4 = -10$	

The closed loop response can be approximated by considering only
(*a*) p_1 and p_2 (*b*) p_3 and p_4
(*c*) p_3 and z_1 (*d*) p_4 and z_2 **[IES 2008]**

444. Which of the following is correct for root loci?
(*a*) Root loci are symmetrical with respect to real axis
(*b*) Root loci are asymmetrical with respect to real axis
(*c*) Root locus lies in a section of real axis to the left of which an even number of poles and zeros of open loop system are present
(*d*) Root locus lies in a section of real axis to the right of which an even number of poles and zeros of open loop system are present **[IES 2008]**

445. Which of the following is correct?
The root locus is the path of the roots of the characteristic equation traced out in *s*-plane
(*a*) as the input of the system is changed
(*b*) as the output of the system is changed
(*c*) as the system parameter is changed
(*d*) as the sensitivity is changed **[IES 2008]**

446. Match List I and List II and give the answer using the code:

List I	**List II**
A. Servo motor	1. Error detector
B. Amplidyne	2. Transducer
C. Potentiometer	3. Actuator
D. Flapper valve	4. Power amplifier

Codes:

	A	**B**	**C**	**D**
(*a*)	2	4	1	3
(*b*)	3	1	4	2
(*c*)	2	1	4	3
(*d*)	3	4	1	2

[IES 2007]

447. Consider the following for PI compensator
1. It is equivalent to adding a zero at origin
2. It reduces overshoot
3. It improves order of the system by 1
4. It improves steady state error
Which of the above are correct?
(*a*) 1, 2, 3, 4 (*b*) 1, 2, 3 only
(*c*) 2, 3, 4 only (*d*) 1 and 4 only **[IES 2007]**

448. A controller transfer function is $C(s) = (2s + 1)/(0.2s + 1)$
What is its nature and parameter?
(*a*) Lag controller $\alpha = 10$ (*b*) Lag controller $\alpha = 2$
(*c*) Lead controller $\beta = 0.1$ (*d*) Lead controller $\beta = 0.2$ **[IES 2007]**

449. Which one of the following statements is correct for gain margin and phase margin of two closed loop systems having loop functions $G(s)\,H(s)$ and $e^{-sT}G(s)\,H(s)$?
(*a*) Both gain margin and phase margin of two systems are identical
(*b*) Both gain margin and phase margin of $G(s)\,H(s)$ will be more
(*c*) Gain margin of two systems are same but phase margin of $G(s)\,H(s)$ will be more
(*d*) Phase margin of two systems are same but gain margin of $G(s)\,H(s)$ will be less **[IES 2007]**

450. Consider the following statements as regard relative stability of a system
1. Phase margin is related to effective damping of system
2. Gain margin gives better estimate of damping ratio than phase margin
3. When expressed in *d*B, gain margin is negative for stable system

Which of the above are correct?

(a) 1 and 2 only (b) 1 and 3 only

(c) 2 and 3 only (d) 1, 2, 3 **[IES 2007]**

451. For system in Fig. 10.132 transfer function $\frac{Y(s)}{R(s)}$ is

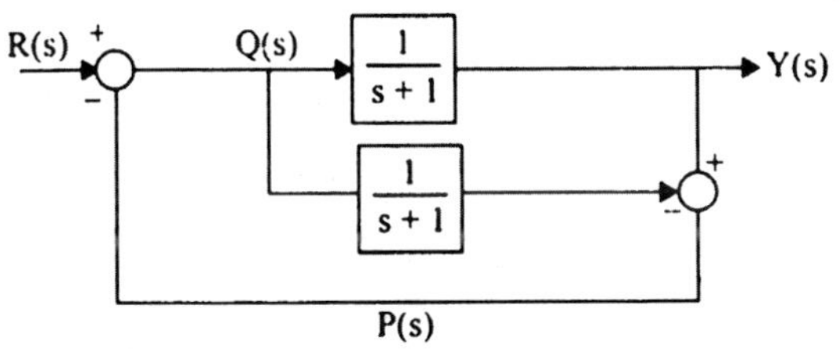

Fig. 10.132.

(a) 0 (b) $\frac{1}{s+1}$

(c) $\frac{2}{s+1}$ (d) $\frac{2}{s+3}$ **[Gate 2010]**

452. A system with transfer function $\frac{Y(s)}{X(s)} = \frac{s}{s+p}$ has an output $y(t) = \cos\left(2t - \frac{\pi}{3}\right)$ for input $x(t) = p\cos\left(2t - \frac{\pi}{2}\right)$

Then the system parameter p is

(a) $\sqrt{3}$ (b) $\frac{2}{\sqrt{3}}$

(c) 1 (d) $\frac{\sqrt{3}}{2}$ **[Gate 2010]**

453. For the asymptotic Bode plot in Fig. 10.133, the transfer function can be

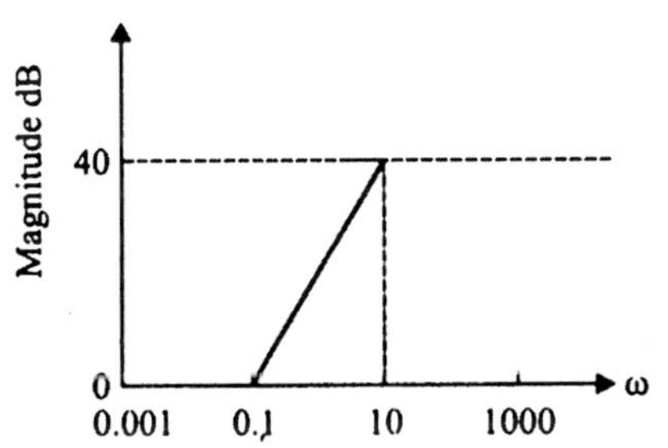

Fig. 10.133.

(a) $\frac{10s+1}{0.1s+1}$ (b) $\frac{100s+1}{0.1s+1}$

(c) $\frac{100s}{10s+1}$ (d) $\frac{0.1s+1}{10s+1}$ **[Gate 2010]**

454. A unity feedback closed loop system has $G(s) = \frac{1}{s^2+2s+2}$ and a controller $G_c(t)$ in feed forward path. For unit step input the transfer function of controller which gives minimum steady state error is

(a) $G_c(s) = \frac{s+1}{s+2}$ (b) $G_c(s) = \frac{s+2}{s+1}$

(c) $G_c(s) = \frac{(s+1)(s+4)}{(s+2)(s+3)}$ (d) $G_c(s) = 1 + \frac{2}{s} + 3s$

[Gate 2010]

455. For the signal flow graph of Fig. 10.134

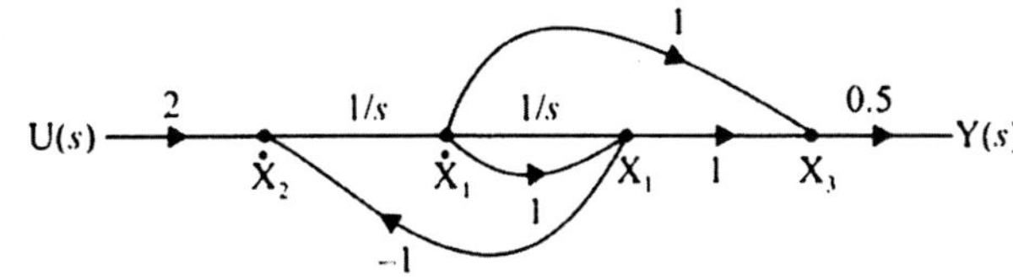

Fig. 10.134.

the state variable representation can be

(a) $\dot{x} = \begin{bmatrix} 1 & 1 \\ -1 & 0 \end{bmatrix} x + \begin{bmatrix} 0 \\ 2 \end{bmatrix} u$

$y = [0 \;\; 0.5]x$

(b) $\dot{x} = \begin{bmatrix} -1 & 1 \\ -1 & 0 \end{bmatrix} x + \begin{bmatrix} 0 \\ 2 \end{bmatrix} u$

$y = [0 \;\; 0.5]x$

(c) $\dot{x} = \begin{bmatrix} 1 & 1 \\ -1 & 0 \end{bmatrix} x + \begin{bmatrix} 0 \\ 2 \end{bmatrix} u$

$y = [0.5 \;\; 0.5]x$

(d) $\dot{x} = \begin{bmatrix} -1 & 1 \\ -1 & 0 \end{bmatrix} x + \begin{bmatrix} 0 \\ 2 \end{bmatrix} u$

$y = [0.5 \;\; 0.5]x$ **[Gate 2010]**

456. For system in Fig. 10.134, transfer function is

(a) $\frac{s+1}{s^2+1}$ (b) $\frac{s-1}{s^2+1}$

(c) $\frac{s+1}{s^2+s+1}$ (d) $\frac{s-1}{s^2+s+1}$ **[Gate 2010]**

457. The unit step response of an underdamped second order system has a steady state value – 2. Which of the following has these properties?

(a) $\frac{-2.24}{s^2+2.59s+1.12}$ (b) $\frac{-3.82}{s^2+1.91s+1.91}$

(c) $\frac{-2.24}{s^2-2.59s+1.12}$ (d) $\frac{-3.82}{s^2-1.91s+1.91}$

[Gate 2009]

458. A linear time invariant system has a rational transfer function with simple poles at $s = -2$ and -4 and zero at $s = -1$. A unit step is applied and in steady state output = 1. The impulse response is

(a) $[e^{-2t} + e^{-4t}]\, u(t)$

(b) $[-4e^{-2t} + 12e^{-4t} - e^{-t}]\, u(t)$

(c) $[-4e^{-2t} + 12e^{-4t}]\, u(t)$

(d) $[-0.5\, e^{-2t} + 1.5\, e^{-4t}]\, u(t)$ **[Gate 2008]**

459. Group I lists four transfer functions. Group II gives same possible responses. Match the two groups

Group I : $P = \frac{25}{s^2+25}$, $Q = \frac{36}{s^2+20s+36}$

$R = \frac{36}{s^2+12s+36}$, $S = \frac{36}{s^2+7s+49}$

Group II :

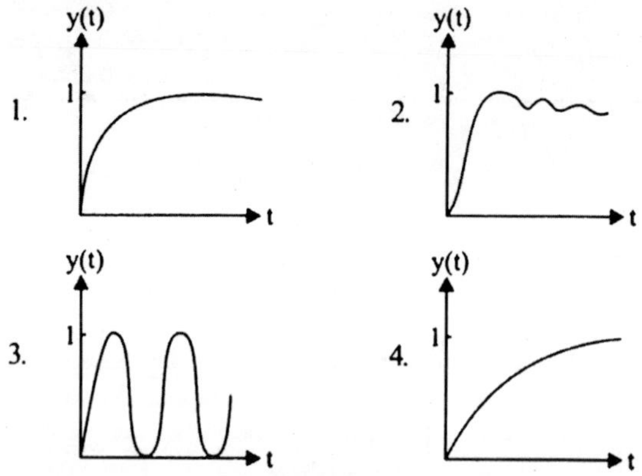

Fig. 10.135.

(*a*) P – 3, Q – 1, R – 4, S – 2
(*b*) P – 3, Q – 2, R – 4, S – 1
(*c*) P – 2, Q – 1, R – 4, S – 3
(*d*) P – 3, Q – 4, R – 1, S – 2 **[Gate 2008]**

460. A system has $G(s) = \dfrac{s+8}{s^2 + \alpha s - 4}$ where α is a parameter.

Consider the standard negative unity feedback in Fig. 10.136.

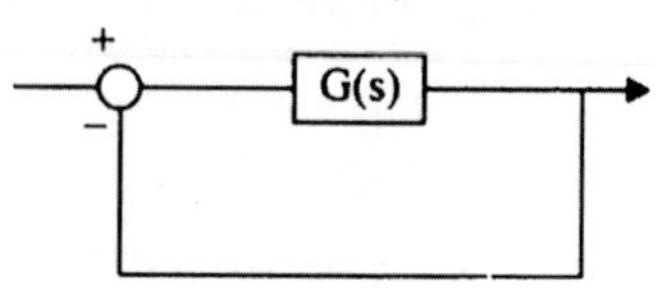

Fig. 10.136.

Which of following is true?

(*a*) The closed loop system is never stable for any value of α.
(*b*) For same positive values of α, system is stable but not for all positive values of α.
(*c*) For all positive values of α, system is stable.
(*d*) The system is stable for all values of α, positive and negative.

B. The following questions have two statements A (assertion) and R (reason). Read them and give correct answer as per following codes

(*a*) Both A and R are correct and R is correct explanation for A
(*b*) Both A and R are correct but R is not correct explanation for A
(*c*) A is correct R is wrong
(*d*) R is correct but A is wrong

461. A. When the forward gain of a system is increased, the transient response becomes oscillatory.
R. Increasing the forward gain reduces steady state error.

462. A. When the phase margin of a control system is improved using a lead network, noise transmission increases.
R. Phase lead compensation increases bandwidth.

463. A. When the performance of a stable second order system is improved by a PID controller, the system may become unstable.
R. PID controller increases the order of the system to 3.

464. A. The root locus of a control system is symmetrical about real axis.
R. For all polynomials with real coefficients, complex poles always occur in conjugate pairs.

465. A. When a PID controller is added to a system, the peak overshoot to step input cannot be determined easily.
R. Use of PID controller introduces a zero in the forward path.

466. A. The time response, usually, is slower with lag compensation.
R. The natural undamped frequency and bandwidth are increased.

467. A. If system output decayed exponentially from an initial value of 1000 V to 368 V in 10 minutes.
A. The time constant of the system is 10 minutes.
R. Time constant is the time taken by the system response to reach 63.2% of its final value when excited by step input.

468. A. The part of root locus on the real axis is not dependent on the poles and zeros which are not on the real axis.
R. Poles and zeros which are not on the real axis occur is conjugate pairs.

469. A. The transient performance of a control system is generally analysed by using unit step function as reference input.
R. Impulse is the most common input found in practice.

470. A. The time response of the system with $G(s) = \dfrac{36}{s^2 + 36s + 36}$ will have any overshoot for unit step input.
R. A critically damped system does not have any overshoot to a unit step input.

471. A. Nyquist stability criterion uses the number of encirclements around $(-1 + jo)$ point in GH plane to determine stability.
R. The $(-1 + jo)$ point in the s-plane lies in the Nyquist path.

472. A. A synchro control transformer has nearly constant impedance across its rotor terminals.
R. Rotor of a synchro control transformer is cylindrical so as to give uniform flux distribution.

473. A. If the Laplace transform is $G(s) = \dfrac{4}{s^2 + 2s + 4}$, the Laplace transform of impulse response is given by $C(s) = \dfrac{4}{s^2 + 2s + 4}$.
R. The Laplace transform for an impulse function is 1.

474. A. For a unity feedback system, with $G(s) = \frac{K}{s(s+2)}$, the settling time of step response is constant for all values of $K \geq 1$.

R. The real part of the roots for all values of $K \geq 1$ are negative.

475. A. Potentiometers cannot be used as error detectors in position control systems.

R. The resolution of a potentiometer places an upper limit on its accuracy.

476. A. If minimum phase system has a gain margin of 8 *dB* and phase margin of 21°, the system is stable.

R. For a minimum phase system both GM and PM must be positive for the system to be stable.

477. A. A large negative feedback is deliberately introduced in an amplifier to make its gain independent of variation in parameters of active device.

R. A large negative value of feedback results in a high value of return difference compared to unity which makes the feedback gain inversely proportional to the feedback factor.

478. A. The phase angle plot in Bode diagram is not affected by change in system gain.

R. The variation in gain has no effect on phase margin of the system.

479. A. The closeness of $G(j\omega)$ locus to $(-1 + jo)$ point can be used as a measure of margin of stability.

R. When the $G(j\omega)$ locus comes close to encircling $(-1 + jo)$ point the system becomes oscillatory.

480. A. Feedback control systems offer more accurate control than open loop systems.

R. The feedback path establishes a link for comparison of input and output and subsequent error correction.

481. A. An on off controller gives rise to oscillation of the output between two limits.

R. Location of a pair of poles on $j\omega$ axis gives rise to self-sustained oscillations in the output.

482. A. For an underdamped system with damping ratio ξ, the maximum overshoot is $e^{-\xi\pi/(1-\xi^2)^{0.5}}$.

R. Peak time of a second order under damped system = π/ω_d where ω_d is frequency of damped oscillations.

483. A. Temperature control of the passenger compartment of car uses both feed forward and feedback controls.

R. Feed forward control gives corrective action before the disturbance affects the output and feedback control applies corrective action after output changes.

484. A. Industrial robots improve productivity.

R. Robots can handle monotonous and complex jobs without error in operation.

485. A. In an automobile the amount of fuel admitted to the engine is adjusted as per the difference between desired and actual speed.

R. The control signal is the amount of fuel to be admitted to the engine.

486. A. When performance specifications are given in term of transient response characteristics, root locus method is a suitable method.

R. When performance specifications in terms of state variables are given, modern control methods can be used.

487. A. Points in the s-plane at which function $G(s)$ is not analytic are called singular points.

R. Singular points at which the function $G(s)$ or its derivative approach infinity are called poles.

488. A. The function $G(s) = \frac{K(s+2)}{(s+5)^2}$ has a double pole at $s = -5$.

R. If the function $[G(s)] (s + p)^n$ for $n = 1, 2, 3$ ... has a finite non-zero value at $s = -p$, then $s = -p$ is called a pole of order n.

489. A. If points at infinity are included, $G(s)$ has same number of poles and zeros.

R. The function $\frac{K(s+2)}{(s+5)^2}$ has a zero at $s = -2$.

490. A. Inverse Laplace transform can be found from $F(s)$ by inversion integral.

R. The time function $f(t)$ is assumed to be zero for $t < 0$.

491. A. $F(s)$ is valid throughout s-plane except at poles of $F(s)$.

R. $\mathcal{L}[f_1(t) + f_2(t)] = \mathcal{L}[f_1(t)] + \mathcal{L}[f_2(t)]$.

492. A. Laplace transform of impulse function is equal to area under the impulse.

R. An impulse function whose area is unity is called unit impulse function.

493. A. The concept of impulse function is useful in differentiating discontinuous functions.

R. $\delta(t - t_0) = \frac{d}{dt}(t - t_0)$.

494. A. If $f(t) = 1 - e^{-t}$, $\lim_{t \to \infty} f(t) = 1$

R. $\mathcal{L}[f(t)] = \frac{1}{s(s+1)}$.

495. A. Integration in time domain is converted into division in s domain

R. $f(0^+) = \lim_{s \to \infty} sf(s)$.

496. A. Most of actual systems are non-linear.

R. Non-linearity may be due to saturation, dead zone, square law etc.

497. A. In Fig. 10.137, open loop transfer function = $\frac{B(s)}{E(s)}$ = $G(s) H(s)$.

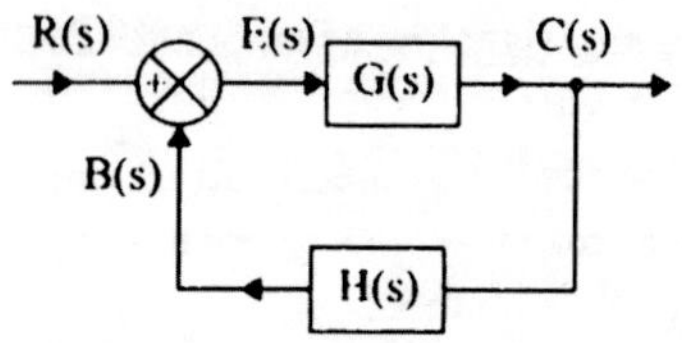

Fig. 10.137.

R. The ratio $\frac{C(s)}{E(s)} = G(s)$ is called feed forward transfer function.

498. A. In simplifying block diagram the product of transfer functions in feed forward direction must remain the same.

R. In simplifying block diagram the product of transfer functions around a closed loop must remain the same.

499. A. Positive feedback is not used in any circuit.

R. Positive feedback loads to instability.

500. A. In liquid system the ratio of change in liquid level to change in flow rate is called resistance of liquid flow.

R. In liquid level systems the ratio of change in liquid stored in a tank to the change in head is called capacitance of tank.

501. A. Derivative control action cannot be used alone.

R. Derivative control action is effective only during transient periods.

502. A. Derivative control action is also called rate control and has an anticipatory character.

R. Derivative control amplifies noise signals.

503. A. Integral control action is sometimes called reset control.

R. Reset rate is expressed as number of repeats per unit time.

504. A. If integral control action is included in the controller the offset can be eliminated.

R. Integral control action may lead to oscillatory response.

505. A. The steady state response of a stable, linear, time invariant system, to sinusoidal input depends on initial conditions.

R. Frequency response, in steady state, is obtained by replacing s in the transfer function by $j\omega$.

506. A. In control system velocity error is not an error in velocity but is an error in position due to a ramp input.

R. For a type 0 system velocity error is zero.

507. A. In root locus the breakaway and break in points either lie on real axis or occur in complex conjugate pairs.

R. All root locus asymptotes intersect on real axis.

508. A. A slight change in pole-zero configuration can cause only minor changes in root locus configuration.

R. Asymptotes show the behaviour of root loci for $|s| >> 1$.

509. A. Increasing the system gain improves steady state behaviour but may result in instability.

R. A compensator is inserted into a system to ensure specified performance.

510. A. For a given system only one of the static error constants is finite and significant.

R. If finite static error constant is large, the loop gain becomes smaller as ω approaches zero.

511. A. Integral wind up effect in controller causes excessive overshort.

R. Presence of saturation in controller and actuator deteriorate the PID control. **[IES 2010]**

512. A. Steady state error can be reduced by increasing integral gain.

R. Overshoot can be reduced by increasing derivative gain. **[IES 2010]**

513. A. The system having characteristic equation $4s^2 + 6s + 1 = 0$ gives rise to underdamped oscillations for step input.

R. The underdamped natural frequency of oscillation is 0.5 rad/s. **[IES 2008]**

514. A. The closed loop stability can be determined from the poles of an open loop system and polar plot of frequency response.

R. Unstable system has right half poles. **[IES 2006]**

515. A. In the error detector configuration using a synchro transmitter and system control transformer, the later is connected to error amplifier. **[IES 2004]**

R. Synchro control transformer has almost a uniform reluctance path between the rotor and stator.

516. A. The error detector in a position control system using synchro pairs employs synchro transmitter for reference signal and synchro control transformer for the feedback signal. **[IES 2004]**

R. Synchro control transformer rotor has a uniform magnetic reluctance.

517. A. The number of branches of root locus terminating on infinity is equal to number of open loop poles minus the number of zeros.

R. Segment of real axis having an odd number of real axis open loop poles plus zeros to their right are parts of root locus. **[IES 2004]**

518. A. In a shunt regulator, the control element is connected in shunt with the load to achieve constant output voltage.

R. The impedance of the control element varies to keep the total current flowing through the load and control element constant. **[IES 2003]**

519. A. A linear negative feedback control system is invariably stable if its open loop configuration is stable.

R. The negative feedback reduces the overall gain of the system. **[IES 2003]**

520. A. The bandwidth of a control system indicates the noise filtering characteristic of the system.

R. Bandwidth is a measure of ability of a control system to reproduce the input signal.

C. Indicate whether the following statements are True (T) or False (F)

521. Nyquist plot provides information about absolute stability as well as relative stability.

522. Transfer functions of even complicated components can be found by frequency response tests.

523. Nyquist stability criterion uses open loop frequency response characteristics.

524. In Bode plots, frequency is plotted on log scale.

525. The log magnitude curve for a constant gain K is a horizontal straight line at a magnitude 20 log K *d*B.

526. The log magnitude curve of factor $j\omega$ is a straight line having a slope of 20 *d*B/decade.

527. The phase angle of $\frac{1}{j\omega}$ is constant and equal to $-90°$ for all values of ω.

528. At corner frequency the phase angle of factor $\frac{1}{(1+j\omega T)}$ is $-45°$.

529. For low frequencies such that $\omega << \frac{1}{T}$, the log magnitude of factor $\frac{1}{1+j\omega T}$ is 0 *d*B.

530. In Bode plots the frequency at which two asymptotes meet is called corner frequency.

531. In Bode method of stability analysis, the error curve is not symmetric about corner frequency.

532. The transfer function $\frac{1}{(1+j\omega T)}$ has the characteristics of high pass filter.

533. A first order element yields almost exact duplication only for constant or slowly varying phenomena.

534. When time constant T in any factor is varied, the shapes of log magnitude and phase angle curves remain the same.

535. In Bode analysis the correction at corner frequency is 3 *d*B.

536. For overdamped systems a quadratic factor in transfer function can be replaced by two first order factors with real poles.

537. For a quadratic factor in transfer function the phase angle is not a function of damping factor.

538. The phase angle curve of $G(j\omega)\ H(j\omega)$ is drawn by adding the phase angles of individual factors.

539. One advantage of using polar plot is that it depicts the frequency response characteristics over the entire frequency.

540. The polar plot of $G(j\omega) = \frac{1}{j\omega}$ is the negative imaginary axis.

541. For $G(j\omega) = \frac{1}{1+j\omega T}$, as ω approaches infinity, the magnitude approaches zero and phase angle approaches $-90°$

542. The magnitude of transport lag factor is always zero.

543. The polar plot of transport lag is a unit circle.

544. At low frequencies the factor $e^{-j\omega T}$ and $\frac{1}{1+j\omega T}$ behave similarly.

545. For type 0 system the starting point (*i.e.*, $\omega = 0$) of polar plot is finite and is on positive real axis.

546. The complicated shapes in the polar plots are only due to time constants in the numerator of transfer functions.

547. Nyquist stability criterion is based on a theorem from the theory of complex variables.

548. In using Nyquist stability criterion it is not necessary to determine closed loop poles.

549. If the Nyquist plot does not encircle the $(-1 + jo)$ point the system is always stable.

550. If the Nyquist plot encircles the $(-1 + jo)$ point clockwise once or more number of times, the system is unstable.

551. If $G(s)\ H(s) = K\ (T_1\ s + 1)/[s^2\ (T_2\ s + 1)]$ the system stability depends on relative magnitudes of T_1 and T_2.

552. Gain margin and phase margin can be determined from Bode plots as well as Nyquist plot.

553. The gain margin of first and second order systems is infinite.

554. For an unstable system the gain margin is indicative of how much the gain should be decreased to make the system stable.

555. For a stable system having two or more gain cross-over frequencies the phase margin is measured at the highest cross-over frequency.

556. The phase margin and damping ratio have no relation.

557. Frequency response approach is a simple approach to control system design.

558. The frequency at which magnitude of closed loop frequency response is 3 *d*B below its zero frequency value is called critical frequency.

559. The slope of log magnitude curve near the cut off frequency is called cut off rate.

560. Bandwidth is proportional to speed of response.

561. The constant M loci on $G(s)$ plane are a family of circles.

562. M circles are symmetrical with respect to real axis.

563. N circles for $\alpha = \alpha_1$ and $\alpha = \alpha_1 \pm 180°n$ (where n = 1, 2, 3....) are the same.

564. For $0 < M < 1$, the centres of M circle lie to the right of origin.

565. Nichols chart consists of M and N loci in log magnitude versus phase diagram.

566. Nichol's chart is symmetric about –180° axis.

567. Accurate amplitude measurements can be made more easily than accurate phase shift measurements.

568. Lag compensation permits a high gain at low frequencies.

569. Lag compensation increases bandwidth of the system.

570. Polar plot of a lag-lead compensator is a circle.

D. Fill in the Blanks with suitable words.

571. Feedback....................the sensitivity of the system to changes in parameters.

572. Feedback.................... the speed of response.

573. The process by which the body temperature is regulated is essentially a.................... process.

574. Feedback results in....................of system gain.

575. Feedback....................the bandwidth of the system.

576. The component most commonly used in control systems for differencing and amplification is.................... .

577. The output of a sensor is mostly.................. .

578. Three types of actuators used in control systems are...................., ... and ....

579. DC servo motors mostly use .... magnet field system.

580. A large air gap in electric servo motor requires very strong ....

581. A servo motor used in control systems should have minimum ....

582. A dc techogenerator produces an output voltage proportional to ....

583. The.................... rating of a dc tachogenerator is very small.

584. A brushless dc motor uses .... devices.

585. An ac servo motor is generally a .... phase induction motor.

586. In an ac servo motor the ratio of rotor reactance to rotor resistance is ....

587. The torque-speed curve of an ac servo motor is almost ....

588. The two windings in an ac servo motor are called .... winding and .... winding.

589. The voltages applied to the two stator windings of an ac servomotor are out of phase by.................... .

590. In an ac servo motor the direction of rotation depends on .... of ....

591. The voltages applied to the stator windings of an ac servo motor are ....

592. The torque speed curve of an ac servo motor is linear in .... speed region.

593. An ac tachometer uses .... rotor construction.

594. An ac techometer has .... stator field coils.

595. A potentiometer is used in control systems to convert .... into a ....

596. In a potentiometer the output voltage varies linearly with displacement only if the load resistance is.................... .

597. In a potentiometer the smallest incremental change is called ....

598. The power rating of a potentiometer used in control systems is about ....

599. An optical encoder converts displacement into a .... or .... signal.

600. A synchro converts .... of .... into an electrical voltage.

601. The rotor of synchro transmitter is .... shaped.

602. The stator of a synchro transmitter has .... phase winding.

603. The stator winding of a synchro control transformer has .... impedance.

604. A stepper motor is a special type of .... motor.

605. In integral error compensation the output depends on .... input signal.

606. Integral error compensation changes a second order system to .... order system.

607. If $e(t)$ is error, then ISE = ...

608. If $e(t)$ is error then $\int_0^T |e(t)|\,dt$ is called ....

609. In a stable system, bounded input results into ....

610. For a system having a single pole on negative real axis, the impulse response ....

611. If any root of characteristic equation has a positive real part the system is ....

612. If the characteristic equation has repeated roots on $j\omega$ axis the system is ....

613. The impulse response shown in Fig. 10.138 indicates that characteristic equation of the system has .... at origins.

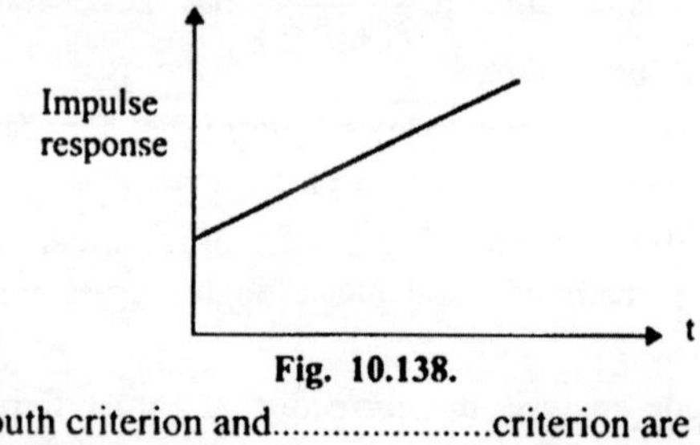

Fig. 10.138.

614. Routh criterion and....................criterion are equivalent.

615. Routh criterion does not give information about ....

616. Routh criterion does not distinguish between ....

617. Root locus is symmetrical about .... axis.

618. The number of root locus branches terminating at .... is equal to number of open loop poles minus zeros.

619. The number of root locus branches is equal to the number of ....

620. In root locus technique the asymptotes cross the real axis at a point called ....

ANSWERS WITH HINTS AND EXPLANATION

1. (*d*) See Table 10.1.
2. (*a*) In force voltage analogy force is analogous to voltage and velocity to current. Just as $\frac{v}{i}$ = Z, force/velocity = mechanical impedance.
3. (*b*) See Table 10.2.
4. (*c*) See Table 10.4.
5. (*c*) See Table 10.1.
6. (*b*) See Table 10.3.
7. (*a*) See Table 10.2.
8. (*a*) See Table 10.4.
9. (*a*)
10. (*a*) Temperatures θ_1 and θ_2 are analogous to voltage v_1 and v_1.
11. (*c*)
12. (*b*) $I(s) = \frac{E_i(s)}{Z_1(s)+Z_2(s)}$, $E_0(s) = [I(s)][Z_2(s)]$

    or $\frac{E_0(s)}{E_i(s)} = \frac{Z_2(s)}{Z_1(s)+Z_2(s)}$.
13. (*b*) Electrical resistance R = $\frac{v}{i}$, level difference or head is analogous to v and change in flow rate is analogous to current.
14. (*a*) Electrical C = $\frac{dQ}{dv}$ and dQ is analogous to change in amount of liquid and dv is analogous to change in head.
15. (*a*) Heat energy (kcal) is analogous to electric charge and temperature is analogous to voltage.

    $$\text{Thermal capacitance} = \frac{\text{Heat energy}}{\text{Temperature}} = k \text{ cal/°C}.$$
16. (*a*)
17. (*a*) See Section 10.10.2.
18. (*a*) See Section 10.10.2.
19. (*b*) Forward transfer function is G + H_1.
20. (*c*) Transfer function = $(G_1 + 1)\, G_2 + 1$

    $= G_1 G_2 + G_2 + 1$.
21. (*b*) G_1 and G_2 are cascaded.
22. (*a*) Since the input is unit step, the response is

    $$\mathcal{L}^{-1}\left[\frac{1}{s(1+sT)}\right] = 1 - e^{-t/T}.$$
23. (*b*) The response is $\mathcal{L}^{-1}\left[\frac{1}{s^2(1+sT)}\right] = t - T + Te^{-t/T}$.
24. (*c*) The response is

    $$\mathcal{L}^{-1}\left[\frac{1}{s+sT}\right] = \mathcal{L}^{-1}\left(\frac{1/T}{s+\frac{1}{T}}\right) = \frac{1}{T}e^{-t/T}.$$
25. (*a*)
26. (*b*) Output is proportional to integral of input or rate of change of output is proportional to input.
27. (*c*) 28. (*b*)
29. (*b*) Derivative of constant is zero. Hence derivative control action is not effective is steady state.
30. (*b*)
31. (*a*) See Table 10.7.
32. (*b*) See Table 10.7.
33. (*b*) See Table 10.7.
34. (*c*) Poles on j x-axis lead to oscillations.
35. (*b*) See Section 10.15.1.
36. (*c*) See Section 10.15.1.
37. (*a*) See Section 10.15.2.
38. (*d*) See Section 10.15.2
39. (*a*) 40. (*c*) 41. (*c*)
42. (*c*) $\mathcal{L}(t^2) = \frac{2}{s^2}$.
43. (*c*) From Eq. (10.7) $f = M\frac{du}{dt} + k\int u\,dt$

    or $f = \left(Ms + \frac{k}{s}\right)u$. Therefore displacement

    $$X(s) = \frac{1}{Ms^2 + k}.$$
44. (*c*)
45. (*c*) $$\frac{C(s)}{R(s)} = \frac{\frac{K}{s(Ts+1)}}{1+\frac{K}{s(Ts+1)}} = \frac{K}{Ts^2+K+1}.$$
46. (*a*) $$\frac{C(s)}{R(s)} = \frac{\frac{K_1}{s(Js+K_1K_2)}}{1+\frac{K_1}{s(Js+K_1K_2)}} = \frac{K}{Js^2+K_1K_2s+K_1}.$$
47. (*a*) $$\frac{C(s)}{R(s)} = \frac{\frac{5(s+1)}{s^2}}{1+\frac{5(s+1)}{s^2}} = \frac{5s+5}{s^2+5s+5}.$$
48. (*b*)
49. (*c*) If P = 3, $\mathcal{L}$(P) = $\frac{3}{s}$. Hence $X(s) = \frac{3}{s(ms^2+bs+K)}$.
50. (*a*) If P = P_m sin ω; $\mathcal{L}$ (P) = $\frac{\omega P_m}{s^2+\omega^2}$.
51. (*c*) Unit step response = $\frac{\omega_n^2}{s^2+2\zeta\omega_n s+\omega_n^2}$.
52. (*a*)

53. (*b*) $\omega_d = \omega_n\sqrt{1-\xi^2} = 10(1-0.01)^{0.5} = 9.95$ rad/s.

54. (*b*) $\frac{C(s)}{R(s)} = K\frac{1}{1+1} = 0.5$ K.

55. (*c*) 0.98 = 1 − $e^{1/T}$ or T = 0.25 minute.

56. (*c*) $\dfrac{\frac{1}{(5s^2+5s)}}{1+\frac{1}{5s^2+5s}} = \frac{1}{5s^2+5s+1}$.

57. (*b*) $\dfrac{\frac{2s+1}{s^2}}{1+\frac{2s+1}{s^2}} = \frac{2s+1}{s^2+2s+1}$.

58. (*d*) $\frac{C(s)}{R(s)} = \dfrac{K(1+Ts)\frac{1}{Js^2}}{1+K(1+Ts)\frac{1}{Js^2}}$

or $\frac{C(s)}{R(s)} = \frac{KTs+K}{Js^2+KTs+K}$.

59. (*b*) $\frac{C(s)}{R(s)} = \frac{5}{s+2} - \frac{4}{s+1} = \frac{s-3}{(s+2)(s+1)}$.

60. (*c*) Since $\frac{C(s)}{R(s)} = \frac{s-3}{(s+2)(s+1)}$ there are two poles at $s = -2$ and $s = -1$ and one zero at $s = +3$.

61. (*a*) **62.** (*b*) **63.** (*c*)

64. (*b*) The intercept on *y*-axis is due to proportional action and inclined line is due to integral action.

65. (*b*) **66.** (*c*)

67. (*a*) Stability is a property of the system only.

68. (*b*) **69.** (*c*)

70. (*a*) See Table 10.4.

71. (*a*) See T able 10.4.

72. (*c*)

73. (*a*) $Z_F(s) = \dfrac{R_2\left(\frac{1}{Cs}\right)}{R_2+\frac{1}{Cs}} = \frac{R_2}{1+R_2Cs}\ \frac{E_0(s)}{E_i(S)}$

$= -\frac{Z_F(s)}{R_1} = \frac{-R_2}{R_1}\left(\frac{1}{1+R_2Cs}\right)$.

74. (*d*) $\frac{E_0(s)}{E_i(s)} = \left(\frac{-R_2}{R_1}\right)\left(\frac{-R_4}{R_3}\right) = \frac{R_2R_4}{R_1R_3}$.

75. (*b*) $\frac{E_0(s)}{E_i(s)} = \left(\dfrac{-\frac{1}{Cs}}{R_1}\right)\left(\frac{-R_3}{R_2}\right) = \frac{R_3}{R_2}\left(\frac{1}{R_1Cs}\right)$.

76. (*a*) For sinusoidal input $j\omega$ replaces s in the transfer function.

77. (*c*) **78.** (*d*) **79.** (*a*) **80.** (*a*)

81. (*a*) **82.** (*b*)

83. (*b*) The relative extent of negative and positive feedbacks determines stability.

84. (*c*) For stability poles must be in left half plane. Zeros in left half plane lead to minimum phase shift.

85. (*d*)

86. (*b*) $\frac{C(s)}{R(s)} = \left(\dfrac{\frac{K_1}{s+a}}{1+\frac{K_1K_2}{s+a}}\right)\left(\frac{1}{s}\right)$

or $\frac{C(s)}{R(s)} = \frac{K_1}{s(K_1K_2+s+a)}$.

87. (*a*) Transfer function = $\dfrac{K(s+1)\frac{(s+3)}{s(s+2)}}{1+\frac{K(s+1)(s+3)}{s(s+2)}}$. Hence characteristic equation is $1+\frac{K(s+1)(s+3)}{s(s+2)} = 0$.

88. (*b*) **89.** (*a*) **90.** (*a*) **91.** (*a*)

92. (*c*) Transfer function = $\dfrac{\frac{4}{s(s+2)}}{1+\frac{4}{s(s+2)}} = \frac{4}{s^2+2s+4}$. The roots of equation $s^2 + 2s + 4 = 0$ are at $-1 \pm j\sqrt{3}$.

93. (*c*) The standard form of characteristic equation is $s^2 + \frac{R}{L}s + \frac{1}{LC} = 0$. Therefore $\frac{R}{L} = 2$ and $\frac{1}{LC} = 4$.

Also $\zeta = \dfrac{R}{2\sqrt{\frac{L}{C}}} = 0.5$.

94. (*a*) $\omega_n = \frac{1}{\sqrt{LC}} = \sqrt{4} = 2$ rad/sec.

95. (*b*) **96.** (*a*) **97.** (*b*) **98.** (*c*)

99. (*b*) **100.** (*a*)

101. (*a*) For a lead compensator $G_C(s) = \dfrac{s+\frac{1}{\tau}}{s+\frac{1}{\alpha\tau}}$ and $\alpha < 1$. Therefore $\frac{1}{\alpha\tau}$ is more than $\frac{1}{\tau}$.

102. (*b*) For a lag compensator $G_C(s) = \dfrac{s+\frac{1}{\tau}}{s+\frac{1}{\beta\tau}}$ and $\beta > 1$.

Therefore $\frac{1}{\beta\tau}$ is less than $\frac{1}{\tau}$.

103. (*c*) For a lead-lag compensator

$$G_C(s) = \left[\frac{s+\frac{1}{\tau_1}}{s+\frac{1}{\beta\tau_1}}\right]\left[\frac{s+\frac{1}{\tau_2}}{s+\frac{1}{\alpha\tau_2}}\right], \beta > 1 \text{ and } \alpha < 1.$$

104. (*b*) Obtain transfer function.

105. (*c*) Obtain transfer function.

106. (*a*)

107. (*b*) $\frac{C(s)}{R(s)} = \frac{\frac{10}{s(s+4)}}{1+\frac{10}{s(s+4)}} = \frac{10}{s^2+4s+10}$.

For the characteristic equation $s^2 + 4s + 10 = 0$, the roots are $-2 \pm j\sqrt{6}$.

108. (*b*) The factor K in the numerator of G ($j\omega$) gives is straight line at 20 $\log_{10}$ K in Bode plot.

109. (*d*) An octave is a frequency band from ω_1 to $2\omega_1$.

110. (*c*) For $\frac{1}{j\omega}$ term the slope is –20 dB/decade or – 6 dB/octave because –20 log 2 = – 6 dB.

111. (*b*) For $(j\omega)^n$ term the slope in (20 *n*) dB/decade.

112. (*a*) **113.** (*d*) **114.** (*a*)

115. (*c*) For quadratic factor in denominator, slope at high frequencies is – 40 dB/decade.

116. (*a*)

117. (*a*) When terms are multiplied, phase angles are added.

118. (*b*) **119.** (*c*)

120. (*a*) At $\omega = 0$, G ($j\omega$) = $1 \angle 0$, At $\omega = \infty$, $G(j\omega) = 0 \angle -90°$.

121. (*c*)

122. (*b*) Magnitude is 1.

123. (*b*) **124.** (*c*) **125.** (*c*)

126. (*c*) If $\omega = 0$, $\frac{1}{(j\omega)^2}$ has a magnitude of infinity and phase angle –180°.

127. (*b*) For the term $1 + j\omega t$, magnitude is 1 at $\omega = 0$ and infinite at $\omega = \infty$.

128. (*c*) For the term $\frac{1}{(j\omega)^2}$, magnitude is 0 at $\omega = \infty$ and infinite at $\omega = 0$. Moreover phase angle is –180°.

129. (*a*) **130.** (*a*) **131.** (*c*) **132.** (*b*)

133. (*b*) If gain is *a*, gain margin is $\frac{1}{a}$. In this case gain is $|G(j\omega_1)|$.

134. (*a*) For a stable system gain margin is positive. Therefore decibel value is also positive.

135. (*c*) **136.** (*b*) **137.** (*c*) **138.** (*b*)

139. (*a*) Arbitrary inputs require large bandwidth.

140. (*b*) If noise is to be eliminated, bandwidth cannot be large.

141. (*b*) **142.** (*a*) **143.** (*d*) **144.** (*a*)

145. (*c*) **146.** (*a*) **147.** (*c*) **148.** (*d*)

149. (*a*) See transfer function in Eq. (10.40).

150. (*c*)

151. (*b*) It changes phase.

152. (*a*) See transfer function in Eq. 10.41.

153. (*a*) **154.** (*c*)

155. (*d*) Any of these may cause modelling error.

156. (*a*) **157.** (*b*) **158.** (*b*)

159. (*a*) – (12 – 2) 20 or – 200 dB/decade.

160. (*c*) **161.** (*b*)

162. (*a*) The input x_1 is multiplied by *a*, x_2 by *b* and x_3 by *c*. The sum total is multiplied by *d* and *e*.

163. (*a*) **164.** (*a*)

165. (*d*) When input pulses are applied, the motor moves in steps.

166. (*d*) The standard form is $s^2 + 2\zeta\omega_n s + \omega_n^2 = 0$.

167. (*c*) $Z_F(j\omega) = \frac{R_1\left(\frac{1}{j\omega C_1}\right)}{R_1 + \frac{1}{j\omega C_1}} = \frac{R_1}{1+j\omega R_1 C_1}$

and $\frac{v_0}{v_i} = \frac{-Z_F(j\omega)}{R_2}$.

168. (*c*) RC network is a compensator. Hence D matches with 3.

169. (*c*) Pole in right half plane means unstable system.

170. (*a*) See Fig. 10.16.

171. (*c*) **172.** (*d*) **173.** (*a*)

174. (*a*) Join j_1 point with the zero and poles. Find magnitude and phase angles of all these lines.

175. (*c*) $\frac{1}{1+\frac{1}{(s+2)^2}} = \frac{(s+2)^2}{s^2+4s+5}$ for $s^2 + 4s + 5 = 0$, the roots are $-2 \pm j1$.

176. (*b*)

177. (*d*) Tachogenerator provides a voltage whose magnitude is proportional to speed.

178. (*d*) **179.** (*d*) **180.** (*a*) **181.** (*b*)

182. (*c*) **183.** (*b*)

184. (*b*) The magnitudes A and G($j\omega$) are multiplied and phase angles are added.

185. (*d*) Unit step response is

$$\frac{42.25}{s^2(s+6.5)} = \frac{-1}{s} + \frac{6.5}{s^2} + \frac{1}{s+6.5}.$$

Taking inverse transform, the time domain response is $-1 + 6.5t + e^{-6.5t}$. It reaches maximum value at t = infinity.

186. (*d*)

187. (*a*) The zero at –1 gives the term $(s + 1)$ in the numerator. The poles at $(-2 + j1)$ and $(-2 - j1)$ give $(s^2 + 4s + 5)$ term in the denominator. Gain is 2.

188. (*b*) **189.** (*d*) **190.** (*c*)

191. (*b*) The maximum phase lead is given by $\sin \phi_m = \frac{1-\alpha}{1+\alpha}$. In this case $\alpha = \frac{1}{3}$. Hence $\sin \phi_m = \frac{1-\frac{1}{3}}{1+\frac{1}{3}}$ $= 0.5$ or $\phi_m = 30°$.

192. (*b*) See Section 10.15.2.

193. (*c*) **194.** (*a*) **195.** (*c*) **196.** (*b*)

197. (*c*) See Section 10.9.

198. (*c*) In this motor ratio $\frac{X}{R}$ is small or R is high.

199. (*b*) **200.** (*a*) **201.** (*d*)

202. (*b*) For 5% criterion settling time $= 3T = \frac{3}{\zeta\omega_n}$. Since $\zeta\omega_n = 4$, $T = \frac{1}{4}$ and $3T = 0.75$ secs.

203. (*b*) $T = \frac{1}{\zeta\omega_n}$. For the three systems the values of $\zeta\omega_n$ are 1.8, 2.5 and 2 respectively. Hence values of T are 0.55, 0.4 and 0.5 respectively.

204. (*c*) For system A, $\zeta_{,} = 1$. Hence critically damped. For system B, $\zeta = 1.02$ Hence overdamped. For system C, $\zeta = 0$. Hence undamped and for system D, $\zeta = 0.35$. Hence under damped.

205. (*d*)

206. (*b*) $\frac{E_0(s)}{E_i(s)} = \frac{0.5(s+3)}{s+1}$. Hence $\alpha = \frac{1}{3}$. Also

$$\sin \phi_m = \frac{1-\frac{1}{3}}{1+\frac{1}{3}} = 0.5 \text{ and } \phi_m = 30°.$$

207. (*b*) $\frac{C(s)}{R(s)} = \frac{2K}{s^2+7s+2+2K}$ since $\zeta = 1$ $\omega_n = 3.5$.

Hence $2 + 2K = 3.5^2 = 12.25$ or $K = 5.125$.

208. (*a*) See Section 10.21.

209. (*c*) **210.** (*d*) **211.** (*c*) **212.** (*c*)

213. (*d*) See Table 10.7.

214. (*d*) For open loop time constant = 0.25. For closed loop $\frac{C(s)}{R(s)} = \frac{\frac{2}{x+4}}{1+\frac{2}{x+4}} = \frac{2}{s+6}$ and time constant $\frac{1}{6}$. Hence ratio of time constants is 3 : 2.

215. (*b*) **216.** (*c*) **217.** (*c*) **218.** (*d*)

219. (*b*) **220.** (*c*)

221. (*a*) Since positive feedback is used,

$$\frac{C(s)}{R(s)} = [G_1,(s)]\left[\frac{G_2(s)}{1-G_2(s)}\right].$$

222. (*d*)

223. (*a*) In a two-phase servo motor $\frac{X}{R} \angle < 1$ so that torque speed curve is almost linear.

224. (*a*)

225. (*b*) For inner loop, we have transfer function

$$= \frac{G_2(s)}{1+G_2(s)H_1(s)}$$

Then $$\frac{C(s)}{R(s)} = \frac{\frac{G_1(s)G_2(s)}{1+G_2(s)H_1(s)}}{1+\frac{G_1(s)G_2(s)}{1+G_2(s)H_1(s)}}$$

$$= \frac{G_1(s)G_2(s)}{1+G_1(s)G_2(s)+G_2(s)H_1(s)}.$$

226. (*a*)

227. (*b*) Total number of roots is 5. Since two sign changes in first column, two roots are in right half plane. Hence 3 roots in left half plane.

228. (*d*)

229. (*d*) Find characteristic equation.

230. (*a*) **231.** (*c*)

232. (*d*) $\zeta\omega_n = 0.8$ and $T = \frac{1}{0.8} = 1.25$ s. For 2% criterion, settling time = 4T = 5 seconds.

233. (*d*) For stable system both must be positive.

234. (*a*) Coefficient $(s + 15)$ gives the term e^{-15t} and the coefficient $(s + 20)$ gives the term e^{20t}. These can be neglected as compared to terms corresponding to s and $(s + 1)$.

235. (*a*) **236.** (*d*)

237. (*c*) If $G_1(s)$ is increased, input is amplified more while disturbance remains the same.

238. (*d*) Use signal flow graph.

239. (*b*) **240.** (*a*)

241. (*a*) Compare with standard form.

242. (*c*) Use initial and final value theorems.

243. (*d*) **244.** (*d*)

245. (*d*) Transfer function $= \frac{A(B+C)}{1+A(B+C)}$.

246. (*d*) **247.** (*d*) **248.** (*c*)

249. (*c*) Solid iron rotor would give rise to eddy currents.

250. (*d*) See Table of Laplace transforms.

251. (*b*) **252.** (*a*) **253.** (*a*)

254. (*d*) $Z(s) = R + sL + \frac{1}{Cs} = \frac{s^2LC+RCs+1}{Cs}$.

255. (*a*)

256. (*b*) Frequency is plotted on log scale and magnitude is taken in dB.

257. (*c*)

258. (*b*) $\frac{V_2(s)}{V_1(s)} = \frac{1/Cs}{R + \frac{1}{Cs}} = \frac{1}{1 + RCs}$.

259. (*d*) **260.** (*a*) **261.** (*a*) **262.** (*c*)

263. (*c*) **264.** (*b*)

265. (*c*) $T = \frac{1}{\zeta\omega_n}$ and settling time for 2% tolerance is 4T.

266. (*c*) **267.** (*c*)

268. (*c*) Since there are two sign changes in the first column, hence two roots in right half plane.

269. (*b*) $\zeta = \frac{4}{2\sqrt{5}}$ and is less than 1.

270. (*a*) **271.** (*d*)

272. (*a*) When polar plot passes through (–1, 0) the system is on limit of stability and gain margin is zero.

273. (*a*) See Eq. (10.54).

274. (*d*) **275.** (*c*)

276. (*d*) Bode plots are asymptotes and Nyquist criterion uses polar plot.

277. (*a*) **278.** (*d*) **279.** (*d*) **280.** (*a*)

281. (*c*) **282.** (*a*) **283.** (*a*) **284.** (*c*)

285. (*a*) Total input is $[\Delta H(s)]R + [\Delta Q_i(s)]$. This input gets multiplied by $\frac{1}{RCs + 1}$ to give $\Delta Q_O(s)$.

286. (*b*) $\frac{V_0(s)}{V_i(s)} = \frac{1}{(1 + \tau s)^2} = \frac{1}{\tau^2 s^2 + 2\tau s + 1}$.

287. (*a*) See Section 10.2.

288. (*b*) Use principle of superposition. Initially assume R(*s*) = 0 and find C(*s*). Then assume U(*s*) = 0 and find C(*s*). Add the two responses.

289. (*b*) Use Table 10.6.

290. (*a*) Use Table 10.6.

291. (*a*) Use block diagram algebra.

292. (*b*) **293.** (*c*) **294.** (*b*) **295.** (*a*)

296. (*b*) Feedback may or may not improve stability. Moreover feedback may also cause instability.

297. (*d*)

298. (*a*) $$\frac{C(s)}{R(s)} = \frac{\left[\frac{10(s+1)}{s+4}\right]\left[\frac{5}{s(s+1)}\right]}{1 + \left[\frac{10(s+1)}{s+4}\right]\left[\frac{5}{s(s+1)}\right]} = \frac{50}{s^2 + 4s + 50}.$$

299. (*b*) For the inner loop, transfer function

$$= \frac{\left[\frac{5}{s(s+1)}\right]}{1 + \left[\frac{15s}{s(s+1)}\right]} = \frac{5}{s^2 + 16s}.$$

Then $$\frac{C(s)}{R(s)} = \frac{\frac{5 \times 10}{s^2 + 16s}}{1 + \frac{50}{s^2 + 16s}} = \frac{50}{s^2 + 16s + 50}.$$

300. (*c*) **301.** (*d*)

302. (*c*) Op-amps are commonly used due to their several advantages.

303. (*a*) **304.** (*a*) **305.** (*c*) **306.** (*c*)

307. (*c*)

308. (*c*) Control systems require small current rating of components.

309. (*b*)

310. (*c*) See Section 10.9.4.

311. (*d*)

312. (*b*) See Section 10.9.4.

313. (*a*) **314.** (*d*) **315.** (*b*)

316. (*b*) If R_L is not infinite, the voltage ratio of potentiometer depends on the value of R_L. This is called loading effect.

317. (*b*) **318.** (*b*) **319.** (*c*) **320.** (*c*)

321. (*a*) **322.** (*a*) **323.** (*b*)

324. (*a*) See Section 10.9.9.

325. (*d*) Stepper motors are used in all these applications.

326. (*b*) **327.** (*a*)

328. (*a*) To measure pressure, it is convenient to convert pressure into displacement and then use a displacement transducer.

329. (*c*) Spring and bellows convert force into displacement.

330. (*a*) $\frac{V_2(s)}{V_1(s)} = \left[\frac{1}{1 + R_1Cs}\right]\left[\frac{1}{1 + R_2Cs}\right]$. Since $R_2 >> R_1$, the term $1 + R_1Cs$ is small as compared to $1 + R_2Cs$. Hence the factor $\frac{1}{1 + R_1Cs}$ can be neglected.

331. (*b*) See Section 10.10.1.

332. (*d*) An underdamped system with damping factor slightly less than 1 gives high speed of response.

333. (*b*) If poles are on imaginary axis the response is undamped and oscillatory.

334. (*c*) **335.** (*b*) **336.** (*c*) **337.** (*d*)

338. (*a*) $\omega_d = \omega_n(1 - 0.1)^2 = \omega_n$.

339. (*b*)

340. (*c*) $t = \frac{1}{\zeta\omega_n}$.

341. (*a*)

342. (*b*) Variation in damping factor is small.

343. (*a*) The error voltage is always very small. A high value of error voltage may cause instability.

344. (*b*) Closed loop gain = $\frac{A_v}{1+A_v\beta}$. Since 1 << A_v β, the gain is $\frac{1}{\beta}$.

345. (*c*) Negative feedback increases bandwidth and reduces distortion.

346. (*a*) For low frequencies X_c is very high as compared to R. Hence $\frac{V_o}{V_i} \simeq 1$.

347. (*a*) Voltage across capacitor lags by 90°.

348. (*c*) If R = X_C, V_o = 0.5 V_i. Hence gain is –3 dB.

349. (*a*) R = X_C, voltage drops across R and X_C are equal and at 90° with respect by each other. V_o lags V_i by 45°.

350. (*a*) X_C = 10 R. Hence $V_o \simeq V_i$.

351. (*d*) X_C = 0.1 R. Hence $V_o = 0.1\ V_i$ or $\frac{V_o}{V_i} = 0.1$ or –20 dB.

352. (*a*)

353. (*b*) V_0 will lead V_i. Hence lead network.

354. (*c*)

355. (*b*) At *t* = 0 the capacitor behaves as a short-circuit and output voltage is 0. At *t* = ∞ capacitor behaves as an open-circuit and $V_0 = V_i$.

356. (*b*) **357.** (*c*)

358. (*b*) $$\frac{\theta_0(s)}{\theta_i(s)} = (K_2)\left[\frac{\frac{K_1}{s(Js+f)}}{1+\frac{K_1}{s(Js+f)}}\right] = \frac{K_1 K_2}{Js^2+fs+K_1}$$

359. (*a*) Transfer function of inner loop

$$= \frac{\frac{6}{s(s+3)}}{1+\frac{6as}{s(s+3)}} = \frac{6}{s^2+3s+6as}.$$

Then $$\frac{C(s)}{R(s)} = \frac{\frac{6}{s^2+3s+6as}}{1+\frac{6}{s^2+3s+6as}}$$

or $$\frac{C(s)}{R(s)} = \frac{6}{s^2+3s+6as+6}.$$

The characteristic equation is $s^2 + 3s + 6as + 6 = 0$.

360. (*b*) Error is squared and integrated.

361. (*c*) **362.** (*a*) **363.** (*b*)

364. (*b*) Two sign changes and hence two poles in right half plane.

365. (*b*) **366.** (*c*) **367.** (*d*) **368.** (*b*)

369. (*b*) **370.** (*a*) **371.** (*d*) **372.** (*c*)

373. (*a*) If *f* is very low, X_C is very high and $V_0 \simeq V_i$.

374. (*c*)

375. (*d*) *j*ω term in denominator gives a straight line with a slope of –20 dB/decade and $\frac{1}{(j\omega)^2}$ gives a straight line with a slope of –40 dB/decade.

376. (*d*) **377.** (*a*) **378.** (*a*) **379.** (*c*)

380. (*a*) Magnitude = $\frac{\sqrt{1+\omega^2T^2}}{\sqrt{1+\omega^2T^2}} = 1$. When ω ≃ 0, phase angle is 0. For high values of a phase angle = –90° – 90° = –180°.

381. (*b*) Magnitude are the same, *i.e.*, $\frac{(1+\omega^2T_3^2)^{0.5}}{(1+\omega^2T_1^2)(1+\omega^2T_2^2)^{0.5}}$. However phase angles are different because the phase angle of $1 - j\omega T_3$ is negative but that of $1 + j\omega T_3$ is positive.

382. (*c*) **383.** (*a*) **384.** (*b*) **385.** (*b*)

386. (*c*) A double pole or pair of complex poles result in – 40 dB/decade slope in log magnitude graph.

387. (*a*) **388.** (*b*) **389.** (*b*) **390.** (*b*)

391. (*a*) It does not become real at any frequency.

392. (*a*) **393.** (*c*) **394.** (*b*)

395. (*a*) Since the plot is very near (–1 + *jo*) point, gain margin is poor.

396. (*a*) **397.** (*c*) **398.** (*a*)

399. (*b*) Poles are at *s* = ± *j*10.

400. (*c*)

401. (*c*) Input *x*(*t*) = *u*(*t*)

$$X(s) = \frac{1}{s}$$

output $y(t) = te^{-t}$

$$Y(s) = \frac{1}{(s+1)^2}$$

$$\text{Transfer function} = \frac{Y(s)}{X(s)} = \frac{s}{(s+1)^2}$$

402. (*d*) $$f(t)_{t\to 0} = \text{Lt}_{s\to\infty}\ sf(s)$$

$$= \text{Lt}_{s\to\infty} \frac{5}{s^2+3s+2}$$

$$= \text{Lt}_{s\to\infty} \frac{5/s^2}{1+\frac{3}{s}+\frac{2}{s^2}} = 0$$

403. (*d*) H(*s*) = 1

$$G(s) = \frac{4}{s^2+0.4s}$$

1 + G(*s*) H(*s*) = 0

or $$1 + \frac{4}{s^2+0.4s} = 0$$

or $s^2 + 0.4s + 4 = 0$

The standard equation is $s^2 + 2 + 2\zeta w_n s + \omega_m^2 = 0$

Thus $\Rightarrow 2\zeta\omega_n = 0.4$

or $\zeta\omega_n = 0.2$

Settling time with 2% of tolerance band

$$= \frac{4}{\zeta\omega_n} = \frac{4}{0.2} = 20 \text{ secs.}$$

404. (*a*) The system can be redrawn as under

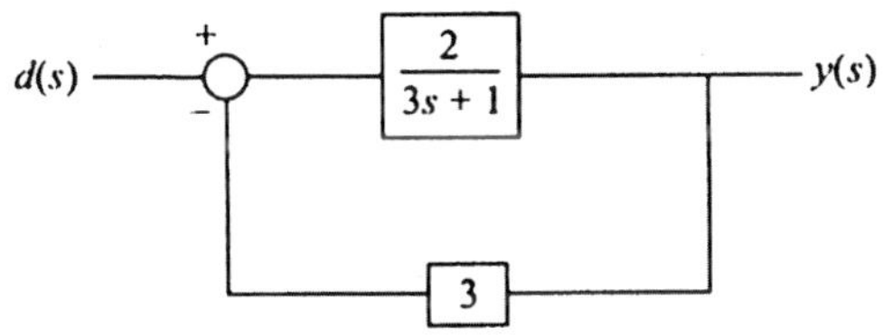

$$\frac{y(s)}{d(s)} = \frac{\frac{2}{3s+1}}{\frac{1+(3)\,2}{3s+1}} = \frac{2}{3s+7}$$

405. (*b*) $G(s) = \frac{(4)(3)}{s+1} = \frac{12}{s+1}$

$$H(s) = \frac{10}{s+10}$$

$1 + G(s)\,H(s) = 0$

or $1 + \left(\frac{12}{s+1}\right)\left(\frac{10}{s+10}\right) = 0$

or $1 + \frac{120}{s^2 + 11s + 10} = 0$

or $s^2 + 11s + 130 = 0$

406. (*d*)

407. (*d*) Phase angle of $G(j\omega) = -90°$

Phase margin $= 180° - 90° = +90°$

408. (*a*) Phase angle of $(-1, 0)$ point $= -180°$

Phase margin $= 180° - 180° = 0°$

409. (*b*) Gain margin, *i.e.*, gain when phase angle is $-180°$ is positive.

Phase margin $= 180° - 150° = 30°$

Hence stable.

410. (*c*) $G(j\omega) = \frac{1}{1 + jw}$

$$|G(j\omega)| = \frac{1}{\sqrt{1+\omega^2}}$$

At $\omega = 1$, $|G(j\omega)| = \frac{1}{\sqrt{2}} = 0.707$

$\angle G(j\omega) = -\tan^{-1}\omega = -\tan^{-1} 1 = -45°$

411. (*d*) In an underdamped system $y(t)$ decreases exponentially. The rate of decay depends on degree of damping.

412. (*b*) Damping ratio 0.4 to 0.7 is considered the best.

413. (*b*) $G(j\omega) = \frac{1}{3j\omega + 1}$

Corner frequency $= \frac{1}{3} = 0.33$ rad/sec

414. (*c*) See Fig. 10.19

415. (*a*) Taking Laplace transform

$$sq(s) = A\,q(s) + b\,Y(s)$$
$$y(s) = C\,q(s) + d\,X(s)$$

or $q(s)\,[sI - A] = b\,X(s)$

or $q(s) = [sI - A]^{-1}\,[b\,X(s)]$

$$y(s) = C[sI - A]^{-1}\,[b\,X(s)] + d\,X(s)$$
$$y(s) = \{C[sI - A]^{-1}\,b + d\}\,X(s)$$

or $\frac{y(s)}{X(s)} = C[sI - A]^{-1}\,b + d$

416. (*a*) $(s^4 + 5s^3 + 8s^2 + 6s + 3)\,Y(s) = U(s)$

Let $X_1 + Y$, $X_2 = \dot{X}_1$

$$X_3 = \dot{X}_2,\ X_4 = \dot{X}_3$$

Thus the given equation is

$$\dot{X}_4 = U(s) - 3X_1 - 6X_2 - 8X_3 - 5X_4$$

In matrix form

$$\begin{bmatrix}\dot{X}_1\\ \dot{X}_2\\ \dot{X}_3\\ \dot{X}_4\end{bmatrix} = \begin{bmatrix}0 & 1 & 0 & 0\\ 0 & 0 & 1 & 0\\ 0 & 0 & 0 & 1\\ -3 & -6 & -8 & -5\end{bmatrix}\begin{bmatrix}X_1\\ X_2\\ X_3\\ X_4\end{bmatrix} + \begin{bmatrix}0\\0\\0\\1\end{bmatrix} u(s)$$

417. (*d*) $f(t) = \sin t$

The value of $\sin t$ lies between -1 and $+1$.

418. (*a*) Steady state value = final value $= \lim_{s \to 0} \frac{s}{s+1} = 0.$

419. (*d*) A good control system must be insensitive to parameter variation but must be sensitive to input commands.

420. (*a*) Addition of phase lead network increases bandwidth.

421. (*c*) High gain (in feedback path) can cause instability.

422. (*d*) $G(s) = 1$, $H(s) = 1$.

$$\text{Output} = 5\left[\frac{G(s)}{1 + G(s) + H(s)}\right] = \frac{5 \times 1}{1 + 1 \times 1}$$

$= 2.5$ V.

423. (*d*) Routh array

s^4	1	24	K
s^3	8	32	
s^2	20	K	
s^1	$\frac{80-K}{20}$	0	
s^0	K		

For roots with zero real part

$$\frac{80 - K}{20} = 0$$

or $K = 80$.

424. (*c*) Routh arr·y

s^3	1	-1
s^2	1	1
s^1	-2	0
s^0	1	

Sign changes twice in first column. Hence two roots have positive real parts.

425. (*d*) Routh array

s^4	1	4	K
s^3	4	3	
s^2	$\frac{13}{4}$	K	
s^1	$\dfrac{\frac{39}{4} - 4K}{\frac{13}{4}}$	0	
s^0	K		

System is marginally stable if $\frac{39}{4} - 4K = 0$ or

$$K = \frac{39}{16}.$$

426. (*d*) It has 3 items, one proportional (*i.e.*, K_p), one derivative (*i.e.*, $K_d s$) and one integral (*i.e.*, $\frac{K_1}{s}$). Hence it is proportional plus integral plus derivative (*i.e.*, PID) controller.

427. (*c*) Transfer function = Laplace transform of impulse response

$$= \mathcal{L}[e^{-t}] = \frac{1}{s+1}$$

428. (*a*) Friction opposes motion and always improves stability.

429. (*b*) Gain margin close to 1 or phase margin close to zero indicates oscillatory system.

430. (*c*) Stability decreases as T increases.

431. (*b*) Controller operates on error signal.

432. (*d*) $$\frac{G(s)}{1 + G(s)H(s)} = \frac{\frac{K}{s(s+1)(s+2)}}{1 + \frac{K}{s(s+1)(s+5)}}$$

or $$\frac{G(s)}{1 + G(s)H(s)} = \frac{K}{s(s+1)(s+5) + K}$$

$$= \frac{K}{s^3 + 6s^2 + 5s + K}$$

Characteristic equation $s^3 + 6s^2 + 5s + K = 0$

s^3	1	5
s^2	6	K
s^1	$\frac{30 - K}{6}$	
s^0	K	

A system is stable if there is no change in sign of the elements in first column.

Hence $K < 30$

or $0 < K < 30$

433. (*d*) $$\frac{G(s)}{1 + G(s)H(s)} = \frac{\frac{K}{(s^2 + 2s + 2)(s+2)}}{1 + \frac{K}{(s^2 + 2s + 2)(s+2)}}$$

$$= \frac{K}{s^3 + 4s^2 + 6s + 4 + K}$$

Characteristic equation

$$s^3 + 4s^2 + 6s + 4 + K = 0$$

Routh array

s^3	1	6
s^2	4	$4 + K$
s^1	$\frac{24 - 4 - K}{6}$	0
s^0	$4 + K$	

A system is marginally stable if all elements in the row of s are zero

i.e., $24 - 4 - K = 0$

or $K = 20$

$4s^2 + 4 + K = 20$

or $4s^2 + 4 + 20 = 0$

or $4s^2 + 24 = 0$

or $4(j\omega)^2 + 24 = 0$

or $\omega^2 = 6$

or $\omega = \sqrt{6}$

434. (*c*) $s^5 + s^4 + 2s^3 + 2s^0 + 4s + 6 = 0$

Let $s = \frac{1}{z}$

Then the equation becames

$$6z^5 + 4z^4 + 2z^3 + 2z^2 + z + 1 = 0$$

Routh array

z^5	6	2	1
z^4	4	2	1
z^3	$\frac{8 - 12}{4} = -1$	$\frac{4 - 6}{4} = -0.5$	1
z^2	0(or $\in$)	1	
z	1		

Routh array

s^3	1	5
s^2	6	K
s^1	$\frac{30-K}{6}$	
s^0	K	

For marginally stable system $\frac{30-K}{6} = 0$

or $K = 30$

For frequency of sustained oscillation

$$6s^2 + K = 0$$

or $6s^2 + 30 = 0$

or $s^2 + 5 = 0$

$(j\omega)^2 + 5 = 0$

or $-\omega^2 + 5 = 0$

or $\omega = \sqrt{5}$

443. (*a*) p_1 and p_2 determine the approximate response became $\left|\frac{p_3}{p_2}\right| \geq 4$.

444. (*a*) Root loci are symmetrical about real axis.

445. (*c*) Root locus is the path when parameter is varied from 0 to ∞.

446. (*d*) Servo motor is actuator. Amplidyne is power amplifier. Potentiometer is error detector.

447. (*c*) $\frac{C(s)}{E(s)} = K_p + \frac{K_i}{s} = \frac{sK_p + K_i}{s}$

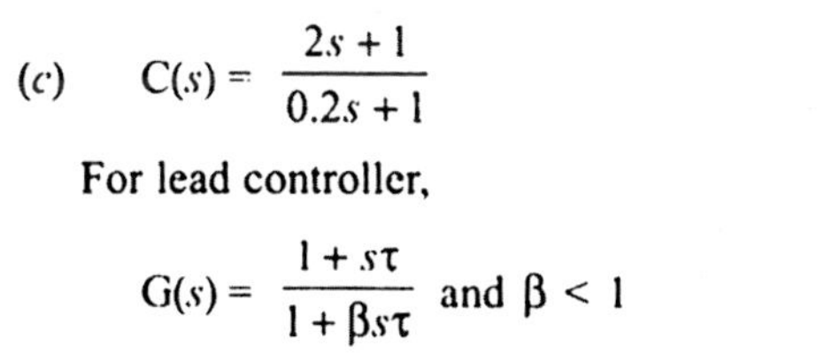

It is seen that one open loop pole at origin and a zero at negative real axis have been added.

448. (*c*) $C(s) = \frac{2s+1}{0.2s+1}$

For lead controller,

$$G(s) = \frac{1+s\tau}{1+\beta s\tau} \text{ and } \beta < 1$$

Therefore $\tau = 2$ and $\beta\tau = 0.2$

or $\beta = \frac{0.2}{2} = 0.1$

449. (*c*) $|e^{-st}| = 1$. Hence gain margin is same for both. However e^{-st} factor causes each point of G(*s*) H(*s*) to rotate by ωt radian in clockwise direction. As *t* increases phase margin reduces.

450. (*a*) Phase margin $= 180° - \tan^{-1}\frac{\sqrt{1-2\zeta^2}}{\zeta}$

Thus phase margin is related to damping.

For stable system, gain margin (dB) is positive.

451. (*b*) $p(s) = \left[\frac{1}{s+1} - \frac{1}{s+1}\right]Q(s) = 0$

Hence $Y(s) = \frac{Q(s)}{s+1} = \frac{R(s)}{s+1}$

Hence $\frac{Y(s)}{R(s)} = \frac{1}{s+1}$

452. (*b*) Phase difference between input and output

$$= \frac{-\pi}{3} - \left(-\frac{\pi}{2}\right) = 30°$$

$\omega = 2$ rad/sec

$$\phi = 90° - \tan^{-1}\frac{\omega}{p} = 30°$$

$$\tan^{-1}\frac{2}{p} = 60°$$

or $\frac{2}{p} = \tan^{-1} 60° = \sqrt{3}$

or $p = \frac{2}{\sqrt{3}}$

453. (*a*) $G(s)\,H(s) = \frac{K\left[1+\frac{s}{0.1}\right]}{\left[1+\frac{s}{10}\right]}$

20 log K = 0

or K = 1

Hence $\frac{10s+1}{0.1s+1}$.

454. (*d*) $r(t) = u(t)$

$$R(s) = \frac{1}{s}$$

Steady state error $= \underset{s\to 0}{\text{Lt}} \frac{s\,R(s)}{1+G(s)\,G_c(s)}$

$$= \underset{s\to 0}{\text{Lt}} \frac{s\frac{1}{s}}{1+G(s)\,G_c(s)}$$

$$= \underset{s\to 0}{\text{Lt}} \frac{1}{1+G(s)\,G_c(s)}$$

For $G_c(s) = 1 + \frac{2}{s} + 3s$, steady state error is zero.

For all others it is positive.

It is seen that sign in the first column changes twice. Hence two roots are in right half of s-plane.

435. (*b*) $G(s) = \dfrac{1}{(s + 0.707 + j0.707)(s + 0.707 - j0.707)}$

Input $= u(t)$

$\mathcal{L}$ (Input) $= \dfrac{1}{s}$

$$\text{Steady state error} = \underset{s \to 0}{\text{Lt}} \frac{s\left(\frac{1}{s}\right)}{1 + G(s)}$$

$$= \underset{s \to 0}{\text{Lt}} \frac{1}{1 + (s + 0.707 + j0.707)(s + 0.707 - j.707)}$$

$$= \frac{1}{1 + (0.707 + j0.707)(0.707 - j0.707)}$$

$$= \frac{1}{1 + 0.5 + 0.5} = \frac{1}{2} = 0.5$$

436. (*c*) The standard form is $\dfrac{\omega_n^2}{s^2 + j\omega_n s + \omega_n^2}$

Comparing the standard form with given equation

$\omega_n^2 = 16$

or $\omega_n = 4$ rad/sec

$2\zeta\omega_n = 4$

or $z = 0.5$

$$\text{Time for first overshoot} = \frac{\pi}{\omega_n\sqrt{1 - \zeta^2}}$$

$$= \frac{\pi}{4\sqrt{\frac{3}{4}}} = \frac{\pi}{2\sqrt{3}} \text{ secs.}$$

437. (*c*) $\dfrac{\dfrac{K}{s(s+2)}}{1 + G(s)H(s)} = \dfrac{\dfrac{K}{s(s+2)}}{1 + \dfrac{K}{s(s+2)}}$

$$= \frac{K}{s^2 + 2s + K}$$

Characteristic equation is $s^2 + 2s + K = 0$

When $K = 32$,

$\omega_n^2 = 32$

or $\omega_n = 4\sqrt{2}$

$2\zeta\omega_n = 2$

or $\zeta = \dfrac{1}{4\sqrt{2}} = \dfrac{1}{\sqrt{32}}$

$$\omega_d = \omega_n\sqrt{1 - \zeta^2}$$

$$= 4\sqrt{2}\left[1 - \frac{1}{32}\right]^{0.5}$$

$$= 4\sqrt{2}\sqrt{\frac{31}{32}}$$

When $K = 16$,

$\omega_n^2 = 16$

or $\omega_n = 4$

$2\zeta\omega_n = 2$

or $\zeta = \dfrac{1}{4} = 0.25$

$$\omega_d = 4\sqrt{1 - \frac{1}{16}}$$

$$= 4\sqrt{\frac{15}{16}}$$

$$\frac{\omega_d \text{ for } K = 32}{\omega_d \text{ for } K = 16} = \frac{\left[4\sqrt{2}\sqrt{\frac{31}{32}}\right]}{4\sqrt{\frac{15}{16}}}$$

$$= \frac{4\sqrt{2}\sqrt{31}\sqrt{16}}{(\sqrt{32})(4)\sqrt{15}}$$

$$= \frac{\sqrt{31}}{\sqrt{15}} = 1.44$$

438. (*b*) Gain margin $= 20 \log\left(\dfrac{1}{a}\right)$

When $a < 1$, Gain margin > 0 dB

When $a = 1$, Gain margin $= 0$

When $a > 1$ Gain margin < 0 dB

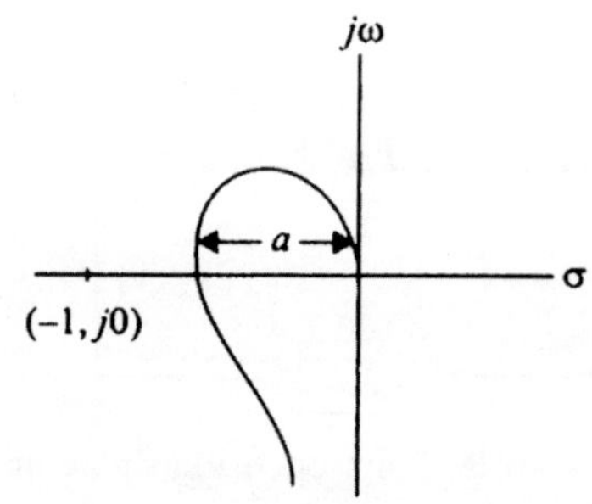

439. (*d*) From the Fig. 10.132 it is seen that gain margin is less than 0 and phase margin is also less than 0.

440. (*a*) If a = gain at phase crossover

Gain margin $= \dfrac{1}{a}$

441. (*d*) Magnitude of resonant peak $= e^{-\zeta\pi/(1-\zeta^2)} = 1$

or $\dfrac{-\zeta\pi}{\sqrt{1 - \zeta^2}} = 0$

or $\zeta = 0$

442. (*a*) $s(s + 1)(s + 5) + K = 0$

or $s^3 + 6s^2 + 5s + K = 0$

455. (*d*)
$$\dot{X}_2 = -X_1 + 2U$$
$$\dot{X}_1 = -X_1 + X_2$$
$$y = 0.5X_3 = 0.5(X_1 + X_2) = 0.5X_1 + 0.5X_2$$
$$\text{Hence } \dot{X} = \begin{bmatrix} -1 & 1 \\ -1 & 0 \end{bmatrix} X + \begin{bmatrix} 0 \\ 2 \end{bmatrix} u$$
$$y = [0.5 \;\; 0.5]\, X$$

456. (*c*) Gain of forward path
$$P_1 = 2\left(\frac{1}{s}\right)\left(\frac{1}{s}\right)(0.5) = \frac{1}{s^2}$$
$$P_2 = 2\left(\frac{1}{s}\right)(1)(0.5) = \frac{1}{s}$$
$$\Delta_1 = 1$$
$$\Delta_2 = 1$$
$$\Delta = 1 - \left[-\frac{1}{s} - \frac{1}{s^2}\right] = 1 + \frac{1}{s} + \frac{1}{s^2}$$
$$\text{Transfer function} = \frac{P_1\Delta_1 + P_2\Delta_2}{\Delta} = \frac{\frac{1}{s^2} + \frac{1}{s}}{1 + \frac{1}{s} + \frac{1}{s^2}} = \frac{s+1}{s^2 + s + 1}$$

457. (*b*) Steady state value = – 2
$$2\zeta\omega_n = 1.91$$
$$\omega_n^2 = 1.91$$
$$\text{or} \quad \omega_n = 1.4$$
$$\zeta = \frac{1.91}{2 \times 1.4} \text{ is less than 1.}$$

Hence underdamped.

458. (*c*)
$$H(s) = \frac{K(s+1)}{(s+2)(s+4)}$$
$$\text{and } R(s) = \frac{1}{s}$$
$$\text{Output} = \frac{K(s+1)}{s(s+2)(s+4)}$$
$$\underset{s \to 0}{\text{Lt}} \frac{s\,K(s+1)}{s(s+2)(s+4)} = 1 \text{ or } \frac{K}{8} = 1 \text{ or } K = 8$$
$$H(s) = \frac{8(s+1)}{(s+2)(s+4)} = \frac{-4}{s+2} + \frac{12}{s+4}$$
$$h(t) = (-4e^{-2t} + 12e^{-4t})\,u(t)$$

459. (*d*) The standard form is $\dfrac{\omega_n^2}{s^2 + 2\zeta\omega_n s + \omega_n^2}$

$$P = \frac{25}{s^2 + 25}$$

and $\zeta = 0$

Hence undamped.

$$Q = \frac{36}{s^2 + 20s + 36}$$

and $\zeta = \dfrac{20}{2 \times 6} = 1.67$

Hence overdamped.

$$R = \frac{36}{s^2 + 12s + 36}$$

and $\zeta = \dfrac{12}{2 \times 6} = 1$

Hence critically damped.

$$S = \frac{36}{s^2 + 7s + 49}$$

and $\zeta = \dfrac{7}{2 \times 7} < 1$

Hence underdamped.

460. (*c*)
$$G(s)\,H(s) = \frac{\frac{s+8}{s^3 + \alpha s - 4}}{1 + \frac{s+8}{s^2 + \alpha s - 4}} = \frac{s+8}{s^3 + \alpha s - 4 + s + 8}$$

Characteristic equation $s^2 + (\alpha + 1)s + 4 = 0$

Closed loop system is stable if $\alpha > -1$. Hence the system is stable for all positive values of α.

461. (*a*)

462. (*a*) More bandwidth means more noise.

463. (*a*) **464.** (*a*) **465.** (*a*) **466.** (*d*)

467. (*b*) **468.** (*a*)

469. (*a*) Both are correct but independent.

470. (*c*) System in A is overdamped.

471. (*c*) R is not correct.

472. (*a*)

473. (*a*) Since $\mathcal{L}[\delta(t)] = 1$, $C(s) = \dfrac{4}{s^2 + 2s + 4}$.

474. (*a*)

475. (*d*) Potentiometers are used as error detectors.

476. (*a*)

477. (*a*) Gain with feedbacks $= \dfrac{A_v}{1 + A_v\beta} \approx \dfrac{1}{\beta}$.

478. (*c*) R is not correct.
479. (*a*) **480.** (*a*)
481. (*b*) Both are correct but independent.
482. (*b*) **483.** (*a*) **484.** (*a*) **485.** (*b*)
486. (*b*) Both are correct but independent.
487. (*b*) **486.** (*a*)
489. (*b*) Both are correct but independent.
490. (*b*) **491.** (*b*) **492.** (*b*) **493.** (*a*)
494. (*b*) Both are correct but independent.
495. (*b*) **496.** (*b*) **497.** (*b*)
498. (*b*) Both are correct but independent.
499. (*d*) Positive feedback is used in oscillator circuits.
500. (*b*) See Table 10.4.
501. (*a*) **502.** (*b*) **503.** (*b*) **504.** (*b*)
505. (*d*) Steady state response does not depend on initial conditions.
506. (*b*) See Table 10.7.
507. (*b*)
508. (*d*) Even a slight change in pole-zero position may affect root locus appreciably.
509. (*b*) **510.** (*c*)
511. (*b*) Both are correct but independent.
512. (*b*) Integral controller improves steady state performance. Derivative controller improves response. Thus both correct but independent.
513. (*d*) $s^2 + 1.5s + 0.25 = 0$. Thus $\omega_n = 0.5$ and $2\zeta\omega_n = 1.5$ or $\zeta = (1.5)/(2 \times 0.5) = 1.5$. Hence overdamped system.
514. (*b*)
515. (*c*)
516. (*d*)
517. (*b*) Both are correct but independent.
518. (*c*) A is correct R is wrong.
519. (*b*)
520. (*d*)
521. T **522.** T **523.** T **524.** T
525. T **526.** T **527.** T **528.** T
529. T **530.** T
531. F, Error curve is symmetric.
532. F, When ω is high, the output is low.
533. T **534.** T **535.** T **536.** T
537. F, Damping ratio affects phase angle.
538. T **539.** T **540.** T **541.** T
542. (F) It is not always zero.
543. T **544.** T **545.** T **546.** T
547. T **548.** T
549. F, It is not necessary.
550. T **551.** T **552.** T **553.** T
554. T **555.** T
556. F, They are related.
557. T
558. F, It is not critical frequency.
559. T
560. T **561.** T **562.** T **563.** T
564. T **565.** T **566.** T **567.** T
568. T
569. F, Lag compensation improves steady state behaviour.
570. T
571. reduces **572.** improves
573. feedback **574.** loss
575. increases **576.** op-amp
577. electrical
578. electrical, hydraulic, penumatic
579. permanent **580.** permanent magnets
581. inertia **582.** speed
583. current **584.** solid state
585. two **586.** small
587. linear **588.** reference, control
589. 90° **590.** polarity, control signal
591. unbalance **592.** low
593. drag cup **594.** two
595. displacement, voltage
596. infinite **597.** resolution
598. 5 W **599.** digital, pulse
600. angular position, shaft
601. dumb-bell **602.** three
603. high **604.** synchronous
605. integral **606.** third
607. $\int_0^\infty e^2(t)dt$
608. integral of absolute magnitude of error
609. bounded output **610.** decreases exponentially
611. unstable **612.** unstable
613. double root **614.** Hurwitz
615. values of roots **616.** real and complex roots
617. real **618.** infinity
619. open loop poles **620.** centroid

MICROPROCESSORS AND COMPUTERS

11

CHAPTER

11.1. COMPUTER CLASSIFICATION

Computers can be classified as main frame computers, mini computer, micro computers and super computers.

Main Frame Computer

A main frame computer has a large storage capacity and can have a large number of terminals. It requires special dust-free air conditioned environment. These computers were very popular in 1960s and 1970s (before the advent of personal computers). They are still being used at many places for data storage and analysis. DEC 2050 is a main frame computer.

Mini Computer

It is very small and much cheaper as compared to main frame computer. It does require dust-free air conditioned environment but the requirements are not as stringent as for main frame computers. Due to the advancement of technology a mini computer can also store huge data. Most of the main frame computers have been replaced by mini computers. PDP-8, PDP-11 are mini computers.

Micro Computer

It is a computer which uses a microprocessor as its CPU. A typical microcomputer has three types of chips: microprocessor (one chip), memory (one or more chips), input-output (one or more chips). In small memory micro computers all these three are fabricated on a single chip, *e.g.*, 8048 micro computer has an 8 bit CPU, 1088 bytes of memory and 27 input output lines on a single chip.

Some examples of micro computers are 8080 (introduced by intel corporation USA in 1973), 8085, Z 80, M 6800, Intel 8086, Intel 8088, Pentium 3, Pentium 4 etc. Earlier version of micro computers used 8 bit microprocessor but the recent versions use 32 bit microprocessor. A micro computer is also called personal computers and is available as table top, laptop and palm top varieties.

Super Computer

It is very powerful and fast computer having enormous data storage and data processing capability. Super computers are used in weather forecasting, monitoring of rockets and satellites etc.

11.2. MICRO-COMPUTER ORGANISATION

Figure 11.1 shows a simple representation of a microcomputer. The meaning of different blocks is as under :

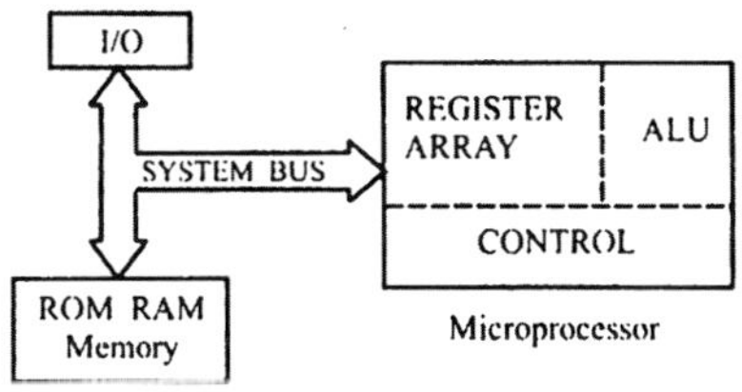

Fig. 11.1. Microcomputer system.

Arithmetic Logic Unit (ALU) : All the computing operations viz., addition, subtraction, exponentiation etc. as well as logic operations AND, OR, XOR etc. are done in ALU. The results of these operations are stored in registers or memory.

Register-Array : This part of the microprocessor contains the registers. These registers store data and partial results temporarily during the execution of the program. Some of these registers can be examined by the user through proper instructions.

Control Unit : The necessary timing and control units are provided by this unit. It also controls the flow of data between the micro-computer, memory and peripherals*.

Memory : This unit stores the program, data and results before and after the execution of the program. Two classes of memory are used, *i.e.*, ROM and RAM.

Read only memory (ROM) is permanently stored and the computer can read from this information as and when required. In general this information cannot be altered. Random access memory (RAM) is also known as *read-write memory.* The data in this memory can be read by computer as and when required. Moreover, the contents of this memory can be altered by the user.

Input-Output : The input-output devices are for communication with the user. The data, program and other information is fed to the computer through the input which is always a keyboard. The results are obtained through output. Various output devices are video monitor, printer, D/A converter, X-Y plotter etc.

System Bus : The communication channel between the microprocessor and peripherals is known as system bus. The double headed arrows denote the two way communication. All peripherals share the same bus but the microprocessor communicates with only one peripheral at one time as per the timing provided by the control unit. Bus is just a group of wires. A system may also have many buses.

* The input-output devices are known as peripherals.

11.3. TYPES OF BUSES IN 8085 MICRO-COMPUTER

A micro-computer is also known as microprocessing unit (MPU). Some of the tasks performed by it are:

1. Reads data (and instructions) from memory
2. Writes data (and instructions) into memory.
3. Accepts data (and instructions) from input devices.
4. Sends data, results to output devices.

Thus, microprocessor has to constantly communicate with memory and peripheral devices. For this communication the following steps are necessary :

1. Identification of peripheral or the memory location (along with its address).
2. Transfer of data.
3. Provision of timing and synchronisation signals.

8085 has three sets of communication lines for the above purpose. These are known as address bus, data bus and control bus.

11.3.1. Address Bus

It is a group of 16 lines denoted as A_0 to A_{15}. This bus in unidirectional, *i.e.*, bits flow in only one direction from MPU to memory and peripherals. The address bus is for identification of memory locations and peripherals.

Each memory location or peripheral is identified by a binary number (called address). The 8085 has 16 address lines and can address $2^{16} = 655536$ (commonly known as 64 K) memory locations.

11.3.2. Data Bus

8085 data bus is a group of 8 lines (denoted as AD_0 to AD_7). This bus is bidirectional, *i.e.*, data can flow in both directions between MPU and memory/peripherals. The transfer of data is done by the data bus. The 8 lines enable MPU to manipulate 8 bit data. This 8 bit data can range from 00000000 to 11111111 (*i.e., a* total of 2^8 or 256 numbers). On the other hand, 16 bit MPU (*e.g.*, intel 8086, Motorola 68000) have 16 data lines and 32 bit MPU (*e.g.*, intel 80386/486 have 32 data lines).

11.3.3. Control Bus

The control lines which carry synchronisation and timing signals are known as control bus.

It is important to note that address and data bus are group of wires but control bus consists of individual lines which provide a pulse to indicate the MPU operation.

To understand the operation let it be required to read an instruction from a certain memory location. Microprocessor places the 16 bit address on the address-bus. The address on the address bus is decoded by an external logic circuit and the memory location is identified. Microprocessor sends a pulse called $\overline{MEMR}$ (memory read) as the control signal through the control bus. This pulse activates the memory chip and the contents of that particular memory location (8 bit data) are placed on the data bus and transferred to microprocessor. This is shown in Fig. 11.2.

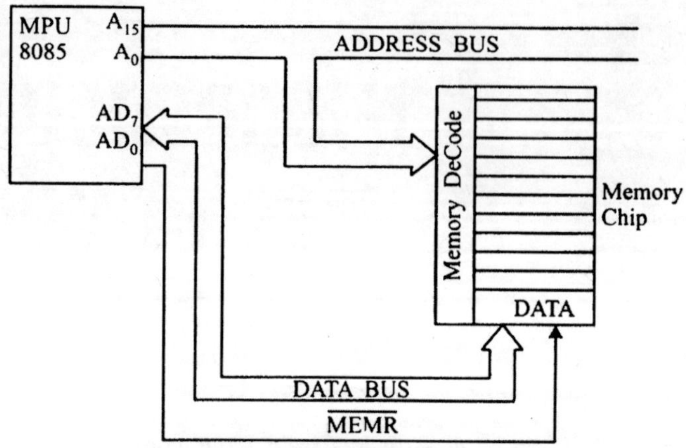

Fig. 11.2. Reading from memory.

11.4. MICROPROCESSOR 8085 ARCHITECTURE

The internal structure of a microprocessor is known as 'Architecture'. It is based on logic design which depends on the required operation. Figure 11.3 shows the architecture of 8085. The details of different blocks are as under.

11.4.1. Accumulator

It is an 8 bit register which stores intermediate answers during a computer operation. It is also called *register A*. The bidirectional arrow between the data bus and the accumulator means that the accumulator can send or receive data. The two state output of the accumulator drives the ALU.

11.4.2. Temporary Register

The 8 bit temporary register stores the operands of the arithmetic logic operations, *e.g.*, during an operation ADD B, the contents of B register are copied in the temporary register during one T state and added during another T state.

11.4.3. ALU and Flags

All the arithmetic and logic operations are done by ALU. The input to ALU are the contents of the accumulator and the temporary register. After the operations, the results are stored in the accumulator. The 5 flags (flip-flops) are set or reset according to the results of the operation. The flags indicate data conditions in the accumulator. The five flags are S, Z, AC, P and CY.

(*a*) **S-Sign Flag:** The sign flag is set when the contents of accumulator become negative during the execution of an instruction. This means that after an arithmetic or logic operation if bit D_7 of the result is 1, the sign flag is set, *i.e.*, the number will be viewed as a negative number. If bit D_7 is 0, the number will be viewed as positive. In chapter 9, we have discussed that to represent negative numbers in binary systems, the most significant bit is the sign bit followed by magnitude bits. In 8 bit representation the bits are D_0 to D_7 and the most significant bit is D_7. Thus the contents of D_7 indicate the sign (0 for the positive and 1 for negative).

(*b*) **Z-Zero Flag :** If an ALU operation results in 0, the zero flag is set. If the result is not 0, the flag is reset.

(c) **AC-Auxiliary Carry Flag :** This flag is used only internally for BCD* operations. When a carry is generated by digit D_3 and passed on to D_4, this flag is set.

* BCD stands for binary coded decimal.

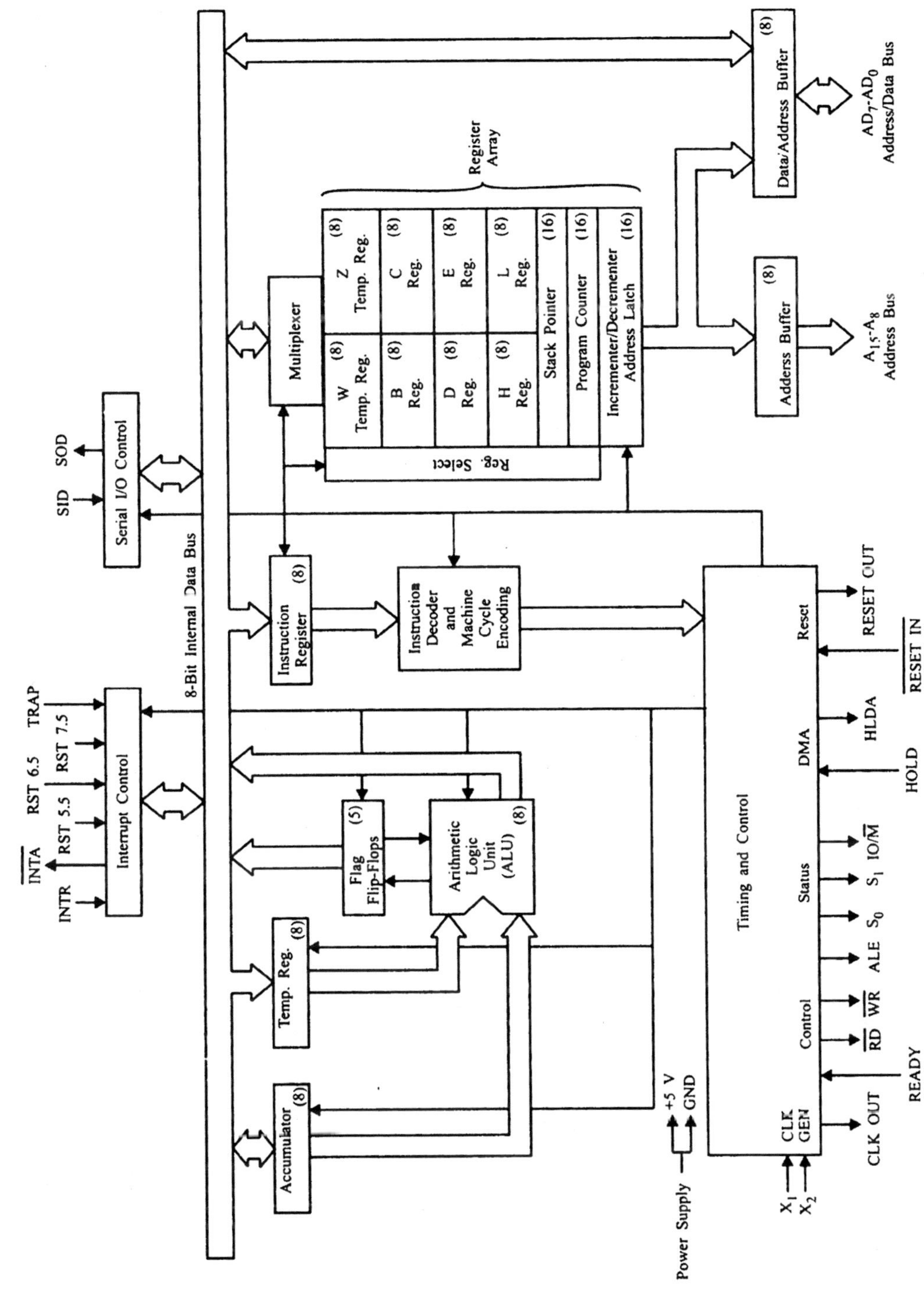

Fig. 11.3. Block diagram of MPU 8085.

(*d*) **P-Parity Flag :** During the execution of some instructions (*e.g.*, ADD), the result of ALU is checked for parity. If the result has an even number of 1s, the P flag is set and if an odd number of 1s, the P flag is reset, *e.g.*, the binary number 00000011 has even parity and this result would cause P flag to be set. (It is important to note that the number 00000011 denotes 3 in decimal notation and is an odd number but in binary notation its parity is even because it has two 1's in it).

(*e*) **CY-Carry Flag :** When an arithmetic operation results in Carry*, the carry-flag is set otherwise it is reset. This flag also serves as a borrow flag for subtraction.

The flag register is also known as F register. It has 8 bits but only 5 are used for the above 5 flags. The bit positions reserved for these flags in F register are as under :

* For the meaning of carry in binary notation see chapter 9.

D_7	D_6	D_5	D_4	D_3	D_2	D_1	D_0
S	Z		AC		P		CY

If the contents of F register are 01010001, it means that S = 0, Z = 1, AC = 1, P = 0 and CY = 1.

11.4.4. Instruction Register and Decoder

When an instruction is called from the memory, the *op* code of the instruction is stored in the instruction register. This *op* code then drives the instruction decoder and the machine cycle encoder.

Instruction cycle is the time required to complete the execution of an instruction. The execution of an instruction means fetching the instruction from the memory and executing the instruction. The fetch cycle has a constant duration but the execution cycle is of variable duration depending on the length of the instruction to be executed. Therefore, the total time of instruction cycle is variable. In the MPU, different instruction cycles with alternate fetch and execute cycles occur in continuation till the complete program is executed.

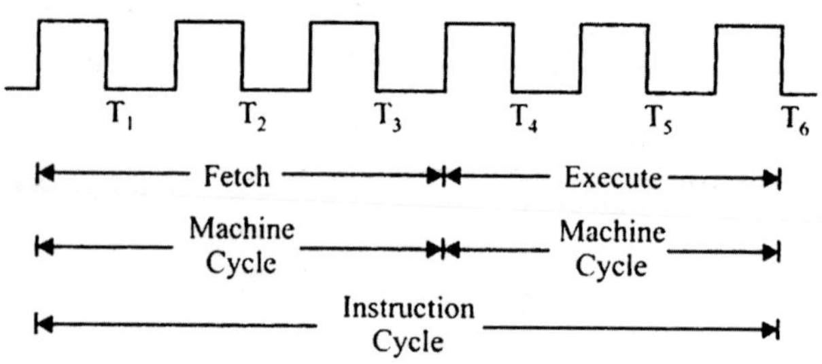

Fig. 11.4. Machine and instruction cycles.

The time required to accomplish a specific task is known as *machine cycle.* The fetch cycle has one machine cycle but the execute cycle may have one or more machine cycles depending on the length of instruction. Figure 11.4 shows two machine cycles. Each machine cycle consists of several clock cycles T_1, T_2 T_3 etc. These clock cycles are square wave pulses. In Fig. 11.4 the fetch cycle has 3 clock cycles T_1, T_2, T_3 and the execute cycle has another 3 clock cycles T_4, T_5, T_6. One T state is precisely equal to one clock period. If the frequency of clock is 3 MHz, the duration of one T state equals $\frac{1}{3 \times 10^6}$ seconds or 0.333 μs.

11.4.5. Register Array

Microprocessor 8085 has six registers B, C, D, E, H, L and two additional temporary registers W, Z. All these are 8 bit registers. In addition it has stack pointer, program counter, incrementer/decrementer address latch, 16 bit capacity each. Each of the registers B, C, D, E, H, L can hold 8 bit data. However, B-C, D-E, H-L may be combined to form register pairs to 16 bit data. When so combined they are known as extended registers. To load the register pair, letter X is added in the instruction, *e.g.*,

$$\text{LXI B, 90FFH}$$

means that register B and C are loaded with the upper and lower bytes to get

$$B = 90\text{H}$$

$$C = \text{FFH}$$

Thus we can write $\quad BC = 90\text{FFH}$

The pairing is always, B-C, D-E and H-L.

A stack is a portion of memory set aside for saving return addresses. The instructions that read and write into stack are called stack instructions. The stack pointer holds the address of the desired memory location. The contents of the stack pointer indicate which memory location is to be accessed, *e.g.*,

$$SP = 20\text{FFH}$$

means that stack pointer is indicating the memory location $M_{20\,FFH}$. Depending on the stack instruction, a byte is, then, read from or written into this memory location.

The program counter keeps track of the instruction being executed. Each time a new instruction is fetched from memory, the program counter is incremented to produce the next higher count. The program is stored at the beginning of the memory. The program counter is a 16 bit register which sends to the memory the address of the next instruction to be fetched and executed.

The incrementer/decrementer can add 1 or subtract 1 from the contents of the stack pointer or program counter.

11.4.6. Address Buffer and Address-Data Buffer

These are buffer registers each 8 bit capacity. A buffer register stores a digital word. The contents of the stack pointer or program counter can be loaded into the address buffer and the address-data buffer. The outputs from these buffer registers then, drive the external address bus and address data bus. The 8 bit internal data bus is also connected to these buffer registers through bi-directional connection. Thus, these buffer registers can send data to or receive data from the 8 bit internal data bus.

11.4.7. Serial I/O Control

Sometimes input-output devices work with serial data and not parallel data. In such cases the serial input data has to be converted to 8 bit parallel data before the micro-computer can use it. Moreover, the 8 bit output has to be converted to serial form before a serial output device can use it. The serial data enters at SID (serial input data) and serial output leaves at SOD (serial output data). Two instructions RIM and SIM are used in this regard. Each time a new bit reaches SID, we can execute RIM instruction to convert a serial input data stream into 8 bit parallel word. Similarly, for output we use SIM instruction to convert the data received from computer into serial output.

11.4.8. Interrupt Control

Suppose that someone is reading a book when the phone rings. He interrupts reading, answers the phone call and then resumes reading. Interrupt control is analogous to this situation. The input/output device may send an interrupt signal to the interrupt control unit to indicate that data is ready. The computer will temporarily stop what it is doing, receive the data and resume what it was doing.

When interrupt occurs, microprocessor completes the instruction it is currently executing. Then it branches to a subroutine which services the peripheral device. Upon completion of service subroutine it returns to the main program it was executing. INTR, RST 5.5, RST 6.5, RST 7.5 and TRAP are inputs for the interrupt signal. $\overline{\text{INTA}}$ is an output which means interrupt acknowledge. If more than one interrupt is active at the same time, MPU 8085 takes them in order of following priorities.

Signal	Priority
TRAP	1
RST 7.5	2
RST 6.5	3
RST 5.5	4
INTR	5

Thus TRAP has the highest priority, followed by RST 7.5 and so on. The signals TRAP, RST 7.5, RST 6.5, RST 5.5. are hardware restarts. MPU 8085 has 8 software restarts RST 0 to RST7. These are known as *software restarts* because these are instructions. When any of the hardware restart pins is active, the internal circuit of 8085 produces a hardware CALL to a predetermined vector location. The vector locations for hardware restarts are exactly half way between the software restart locations. In view of this the hardware restarts are numbered 5.5, 6.5 and 7.5.

11.4.9. Timing and Control Unit

This unit generates the control signals and synchronises all the operations with its clock. 8085 uses an oscillator to generate two phase clock signals CLK and $\overline{\text{CLK}}$ which synchronise all registers. The controller-sequencer also produces control signals which are needed for internal and external control. Before each computer run, a $\overline{\text{CLR}}$ signal is sent to program counter and CLR to instruction register. This is done to reset the program counter to 0000 and wipe out the last instruction in the instruction register. A clock signal CLK is sent to all buffer registers. A $\overline{\text{CLK}}$ signal also goes to program counter.

$\overline{\text{RD}}$ is the read control signal and $\overline{\text{WR}}$ is the write control signal. $\overline{\text{RD}}$ is used for both reading from memory as well as from input device. $\overline{\text{WR}}$ is used for both writing into memory as well into output device. Using IO/$\overline{\text{M}}$ (which stands for input-output/memory) two signals are generated from $\overline{\text{RD}}$ and $\overline{\text{WR}}$ each.

ALE stands for 'address latch enable'. To work properly MPU needs to be connected to one or more memory chips. Each memory chip has its own address latch. This latch stores the incoming address from the address bus and the address data bus. ALE signal comes out of MPU and goes to memory chip. The falling edge at ALE signal loads the address on the address bus and address data bus into the address latch of the memory chips.

S_0 and S_1 are output status signals. These two signals and IO/$\overline{\text{M}}$ signal indicate whether fetch, memory write, memory read or some other operation is being carried out.

Whenever data is fed into or taken out of microprocessor, accumulator is involved. The IN instruction causes accumulator to receive input data. The OUT instruction transfers data from the accumulator to output device. In both the cases going through accumulator slows down the data transfer.

DMA (direct memory access) allows direct access to memory to speed up data transfer. The control of buses is turned over to DMA controller which is a chip meant for high speed data transfer. HOLD and HLDA signals are used in DMA operations. When DMA controller is ready to take over the control, it sends a high HOLD signal to MPU which then floats its address, data and control buses. MPU also sends a high HLDA (hold acknowledge) signal to DMA controller. The DMA controller transfers data at high speed and then returns control back to MPU by sending back a low HOLD signal. This is illustrated in Fig. 11.5.

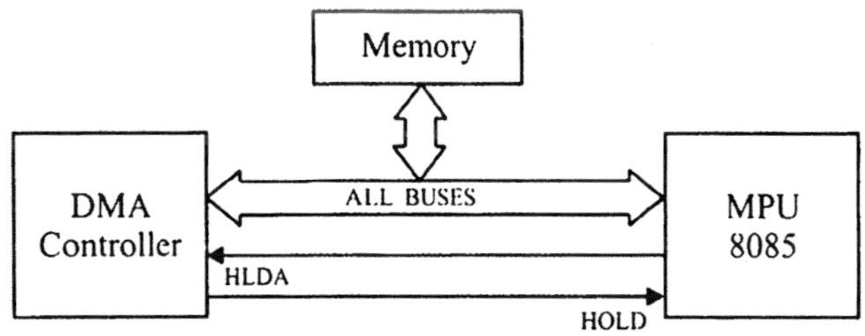

Fig. 11.5. Direct memory access.

$\overline{\text{RESET IN}}$ signal may come from the user reset button or some other source. When $\overline{\text{RESET IN}}$ is low, CPU will reset the program counter, instruction register and other circuits. The CPU remains in reset till $\overline{\text{RESET IN}}$ signal goes high. Then the processing of data begins.

RESET OUT signal when high, indicates that CPU is being reset (*i.e.*, program pounter, instruction register etc. are being reset to zero). This RESET OUT signal goes to peripheral chips. When RESET OUT goes low, processing begins.

11.4.10. Multiplexing Address/Data Buses

It is seen from Fig. 11.3 that signal lines $AD_7 - AD_0$ are bi-directional. These lines are multiplexed, *i.e.*, they are used for address bits during some T states and for data bits during other T states. Therefore, these lines are designated as Address/Data bus. The reason for multiplexing is that when 8085 was developed, the number of pins was to be limited to 40. Therefore, part of address bus was multiplexed with data bus.

11.5. PIN DIAGRAM OF MICROPROCESSOR 8085

Figure 11.6 shows the pin diagram of 8085 A (generally known as 8085). The functions of these pins are as under:

Pins 1 and 2 : A microprocessor needs an oscillator to generate signals which synchronise all operations. 8085 has an on-chip oscillator except for crystal (or RC or LC network). The crystal (or RC or LC network) is connected to pins 1 and 2. For frequency stability a crystal is preferred. The frequency is internally divided by 2 (because there are two clock signals CLK and $\overline{\text{CLK}}$). To operate a microprocessor at 3 MHz, the frequency of the crystal should be 6 MHz.

Pin 3 : The RESET OUT signal from pin 3 goes to the peripheral chips. When high, this signal indicates that program counter, instruction register etc. are being reset to zero. When the system is started, it is reset. When RESET OUT goes low, the processing starts.

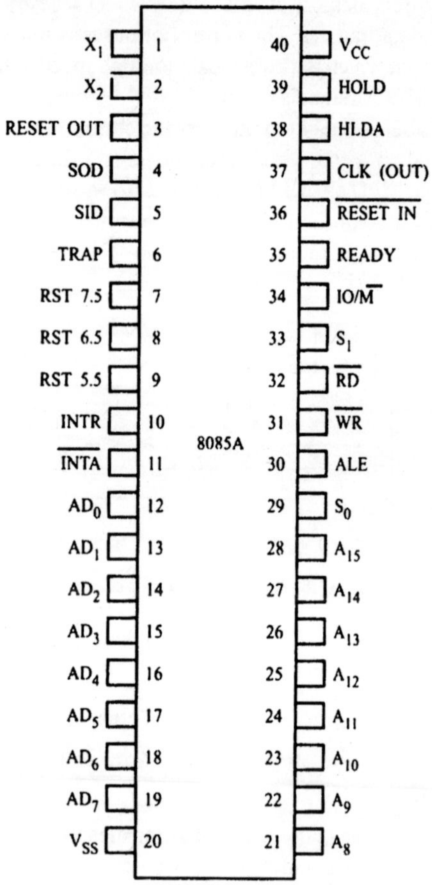

8085 Pinout

Fig. 11.6. Pin diagram of 8085.

Pin 4 : SOD. The serial output data leaves at this pin as discussed in section 11.4.7.

Pin 5 : SID. The serial input data enters at this pin as discussed in section 11.4.7.

Pins 6-11 : These pins are for interrupt control as discussed in section 11.4.8.

Pins 12 to 19 : These pins carry the lower 8 address bits or 8 data bits. These lines are multiplexed, *i.e.*, they are used for address bits during other *T* states. Because of this reason $AD_7 - AD_0$ are labelled address-data bus in Fig. 11.3.

Pin 20 : Ground connection.

Pins 21-28 : These are the rest of address bus.

Pin 29 : S_0. It takes out status signal.

Pin 30 : ALE or address latch enable as discussed in section 11.4.9.

Pins 31, 32, 34 : These three pins are labelled as $\overline{WR}$, $\overline{RD}$ and $IO/\overline{M}$ and work together. They stand for write, read, input-output/memory respectively. They are connected to memory and input-output chips. A low $IO/\overline{M}$ indicates memory operation and a high $IO/\overline{M}$ indicates that I-O instruction is being carried out, *i.e.*, low $IO/\overline{M}$ enables memory chips and high $IO/\overline{M}$ enables I-O chips.

$\overline{WR}$ And $\overline{RD}$ signals are active low. Thus low $\overline{WR}$ means write operation and low $\overline{RD}$ means read operation. Thus, we have the following combination :

Memory read : $IO/\overline{M}$ low, $\overline{WR}$ high, $\overline{RD}$ low

Memory write : $IO/\overline{M}$ low, $\overline{WR}$ low, $\overline{RD}$ high

Input read : $IO/\overline{M}$ high, $\overline{WR}$ high, $\overline{RD}$ low

Output write : $IO/\overline{M}$ high, $\overline{WR}$ low, $\overline{RD}$ high

Pin 33 : Status signal S_1 indicates status (along with S_0, $IO/\overline{M}$) as discussed in section 11.4.9.

Pin 35 : READY. The peripheral devices are rather slow and cannot keep pace with 8085. The 8085 addresses a peripheral device for receiving or sending data. If the device is not ready it sends a low READY signal to 8085. The MPU generates a number of wait states. When device is ready, it sends High READY signal to MPU so that MPU can complete data transfer.

Pin 36 : $\overline{\text{RESET IN}}$. When $\overline{\text{RESET IN}}$ is low, the CPU will reset the instruction register, program counter etc. and also send $\overline{\text{RESET OUT}}$ signal to pin 3. CPU remains in reset till $\overline{\text{RESET IN}}$ signal goes high. Then data transfer begins.

Pin 37 : CLK signal comes out of this pin and can be fed to peripheral devices to synchronise their working.

Pins 38 and 39 : These are for direct memory address as discussed in section 11.4.9.

Pin 40 : This pin is for connection of +5 V supply. The tolerance on the supply voltage is ± 5%. The power dissipation is about 1.5 W.

11.6. OSCILLATOR

The frequency limits specified for 8085 A are 500 kHz $\le f \le$ 3.125 MHz. Thus, it operates properly only when clock frequency is within these limits.

The crystal (or LC or RC circuit) is connected to pins 1 and 2. The oscillator output drives a flip-flop which produces two clock signals CLK and $\overline{CLK}$. These clock signal drive the internal circuits. This signal also comes out of pin 37 for connection to peripheral devices.

Since the oscillator is producing 2 clock signals, the oscillator frequency should be double the frequency of clock, *i.e.*, for 3 MHz the oscillator frequency should be 6 MHz. A crystal is the best choice because it gives good frequency stability. Figure 11.7 shows the crystal connections. Schmitt's trigger produces a perfectly rectangular output and is used in 8085.

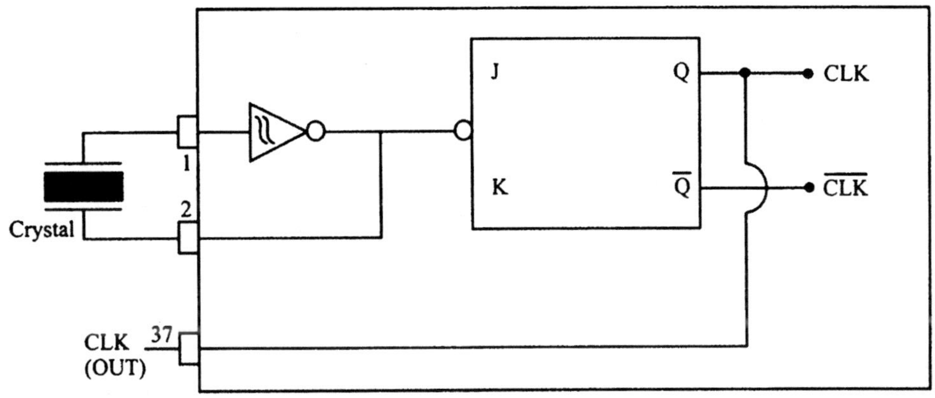

Fig. 11.7. Crystal driven Schmitt's trigger for 8085.

11.7. MICROPROCESSOR 8086

An important feature of a microprocessor is the speed of operation. It is expressed in million instructions per second'(MIPS). In case of microprocessors which provide instructions for floating point arithmetic, the speed is expressed in million floating point operations performed per second (MFLOPS).

In 8 bit microprocessor speed of operations, directly addressable memory, data handling capacity etc., are very limited, The 16 bit microporcessor have higher speed of operation, directly addressable memory, data handling capacity, etc. 8086 is a 16 bit 40 pin microprocessor.

11.7.1. Memory Organisation

8086 has 20 lines on which data and address are multiplexed as shown in Fig. 11.8. The lines AD_0 to AD_{15} are used for the 16 least significant bits of address. The remaining 4 most significant bits are sent over the lines $A_{16} - A_{19}$. The line $AD_0 - AD_{15}$ are also used for 16 bit data word. The lines $AD_0 - AD_7$ are used for least significant byte of the data word while lines $AD_8 - AD_{15}$ are used for the most significant byte of the data word. If only one byte of data is transferred either lines $AD_0 - AD_7$ or $AD_8 - AD_{15}$ are used depending on whether the byte corresponds to even or odd memory address respectively. Since we have 20 address lines, the largest address can be $2^{20} - 1$ or 1048575. Any two consecutive bytes form one 16 bit word. The address of the least significant byte is the address of the word. This is illustrated in Fig. 11.9. The word at address 04 has two bytes located at 04 and 05. The contents of this word are $OB4OE_{16}$. The least significant 8 bits viz. OE are at lower address, *i.e.*, 04 while the most significant 8 bit viz., OB4 are at address 05. A word can also begin at odd address, *e.g.*, in Fig. 11.9 the word at address 07 consists of two bytes located at address 07 and 08. The contents of this word are $OA5OF_{16}$ (OF at address 07 and OA5 at address 08).

Most instructions of microprocessor 8086 can operate both on bytes (8 bits) and words (16 bits), *e.g.*, we can move contents at address 04 to AL register (lower half of accumulator). Moreover, the contents of the complete word starting at location 04 (and located partly at 05) can be shifted to 16 bit accumulator (register AX). Since lines $AD_0 - AD_{15}$ are used for data as well as address, it is more convenient to refer them as $D_0 - D_{15}$ when used for data and $A_0 - A_{15}$ when used for address. The complete memory can be thought of as two banks of 512 k bytes each. $D_0 - D_7$ lines are used for lower bank and $D_8 - D_{15}$ for the higher bank. The lines A_0 and $\overline{BHE}$ are used to determine one of the three possible references to the memory as shown in Table 11.1. When A_0 is Low, higher bank is accessed.

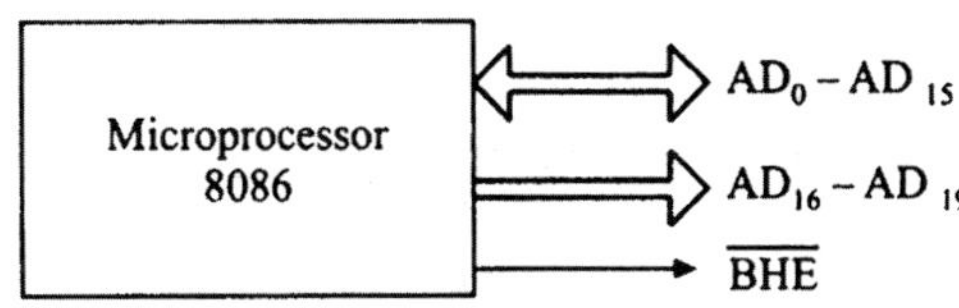

Fig. 11.8. Memory addresing signals in 8086.

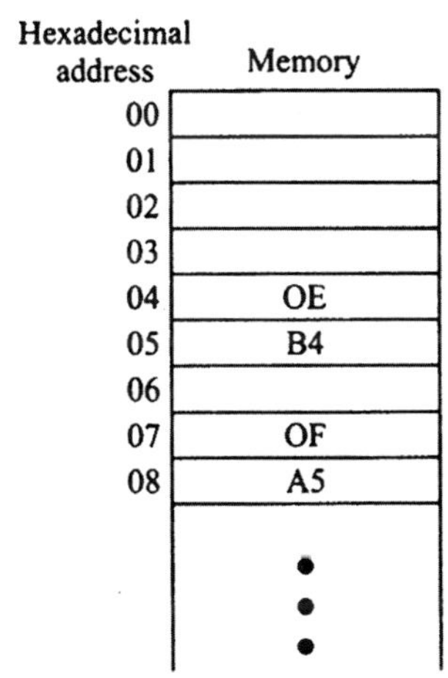

Fig. 11.9. Two words in 8086 memory.

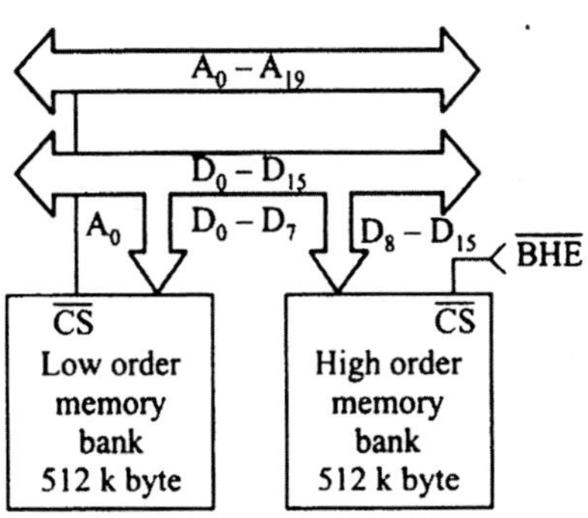

Fig. 11.10. Read/write process for word/byte in 8086.

Table 11.1 Memory selection in microprocessor 8086

$\overline{BHE}$	A_0	*Read/Write*
0	0	One 16 bit word
0	1	One byte from/to odd address
1	0	One byte from/to even address
1	1	NIL

11.7.2. Structure of 8086 Registers

Microprocessor 8086 has thirteen registers (each 16 bit capacity) and 9 flags. The registers can be grouped as

(*a*) General registers

(*b*) Pointer and Index registers

(c) Segment registers

(*d*) Instruction pointer and flags.

These are shown in Fig. 11.11.

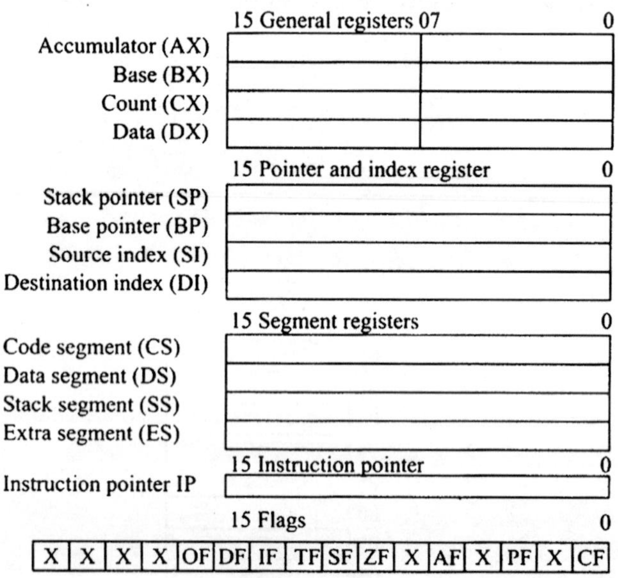

Fig. 11.11. 8086 registers.

(*a*) **General Registers :** The four general registers are accumulator (AX), Base (BX), Count (CX) and data (DX). They are used as operands in arithmetic and logic operations. Each of these registers can be used as a 16 bit register or two 8 bit registers.

(*b*) **Pointer and Index Registers :** These registers are stack pointer (SP), base pointer (BP), source index (SI) and destination index (DI). Each of these is 16 bit and serves to hold offsets of operands residing in the memory. The stack pointer has the same function as in microprocessor 8085. All push and pop operations derive their addresses from SP. It points to the current top of the stack. A stack refers to an area of contiguous locations in memory. There can be many such stack segments but only one of them is active at any time. SP points to this active stack.

Each stack has many data areas. BP register is used to hold the offset of the base of a data area in the current stack.

Many string* operations are possible in 8086. Strings can be moved from one area to another. SI points to the source string to be moved and DI points to the corresponding destination.

(*c*) **Segment Registers :** 8086 memory is divided into many segments. The four types of segments are code segments (CS), Data segments (DS), Stack segment (SS) and Extra segment (ES).

Each segment can be up to 64 k bytes long. A segment can begin at any memory address divisible by 16. The starting address of the currently active segments is held in segment registers, CS, DS and ES hold start of code, data stack and extra segments. Data area of the program, variables and constants may reside in data segment. Any stack used resides in stack segment. Program instructions reside in code segment. Some strings operated on by special instructions reside in extra segment.

(*d*) **Instruction Pointer and Flags :** Instruction pointer (IP) is a 16 bit register and points to the next instruction to be executed within the current code segment. As the program is executed, it is automatically updated.

The nine flags can be decomposed into two groups, *i.e.*, status flags and control flags. The status flags are carry (CF), Parity (PF), Auxiliary carry (AF), Zero (ZF), Sign (SF) and Overflow (OF). They are set (or reset) after an arithmetic or logical instruction executed. The control flags are Direction (DF), Interrupt enable (IF) and Trace (TF). These flags control the execution program. DF setting determines whether the string bytes are to be processed from low to high address or high to low address. It enables or disables the external interrupts. TF flag (when ON) forces 8086 to generate an interrupt after execution of each instruction.

Flags are shown in Fig. 11.12.

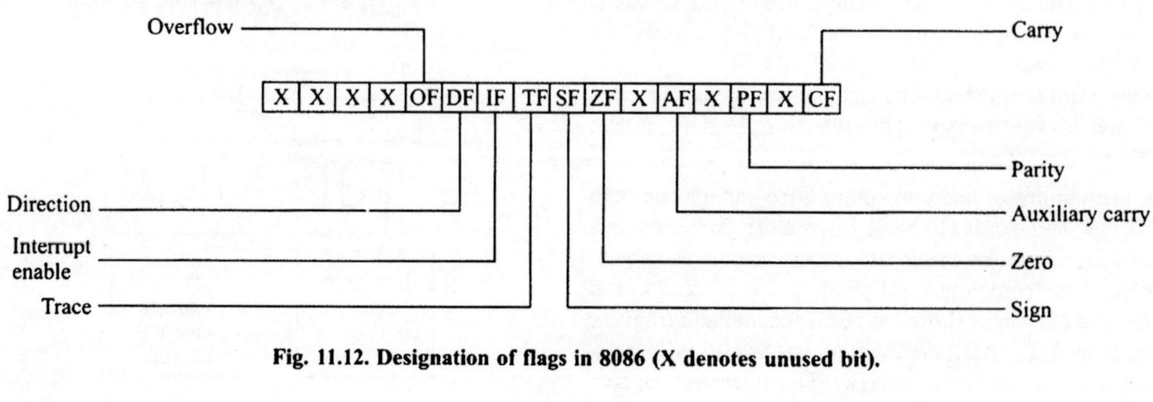

Fig. 11.12. Designation of flags in 8086 (X denotes unused bit).

* A string is a sequence of bytes or words.

11.7.3. Bus Structure of 8086

Figure 11.13 [(*a*) and (*b*)] depicts the pins in 8086. It is seen that the functions of pins depend on whether MN/$\overline{MX}$ is Low or High. Figure 11.13 (*a*) shows the conditions when MN/$\overline{MX}$ is high (connected to V_{CC}). This mode is called Minimum mode. Figure 11.13 (*b*) shows the conditions when MN/$\overline{MX}$ is grounded. This mode is called Maximum mode. The user can choose any of these two modes depending on the complexity of the system. The maximum mode is used to utilize the full power of 8086. For medium complexity systems minimum mode is used.

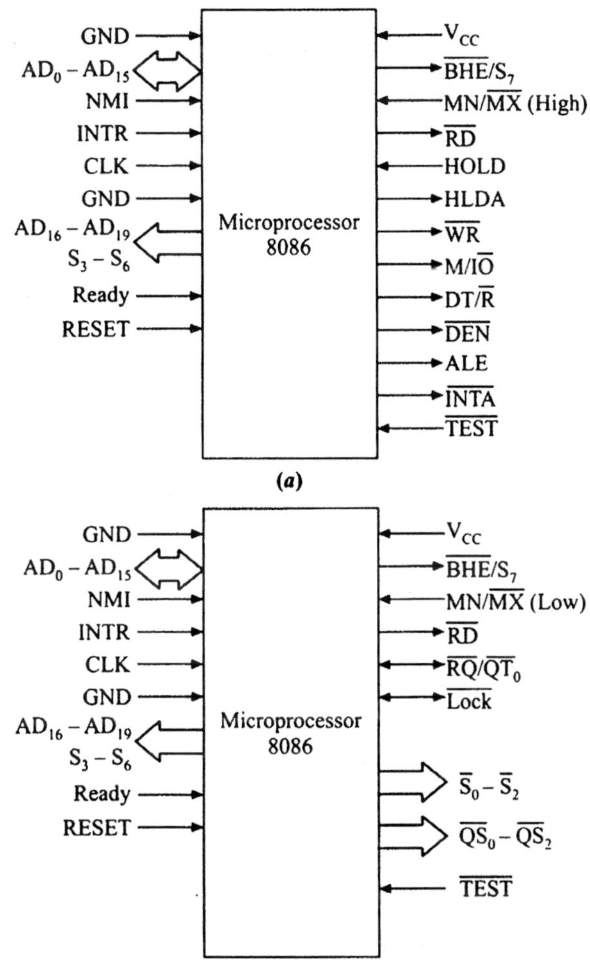

Fig. 11.13. Pins/signals in μ P 8086 (*a*) minimum mode, (*b*) maximum mode.

11.7.4. Bus Interface and Execution Units

The architecture of 8086 can be divided into two distinct parts, viz., Bus interface unit (BIU) and Execution Unit (EU). These two units interact with each other to accomplish the various tasks. BIU transfers instructions and data bytes between the processor and memory of I/O devices. The EU does the job of decoding and execution. BIU contains the 4 segment register, *i.e.*, code segment (CS), data segment (DS), stack segment (SS) and extra segment (ES). Each of these registers is 16 bit. These registers are loaded with the most significant 16 bits of the starting addresses of the memory segments which the microprocessor is handling at any particular time. BIU inserts zero for the lowest four bits of 20 bit starting address for a segment, BIU fetches successive instruction byte from the memory and inserts these into a 6 byte instruction queue. EU removes these instructions from this queue, decodes and executes them. This queue acts as First In First Out (FIFO) queue for the EU. BIU fetches instruction bytes independent of EU. Whenever the bus is idle and there is space for at least 2 bytes in the instruction queue, BIU fetches the 2 bytes pointed at by the IP from the current code segment. During execution of instruction, whenever, EU requires the next instruction byte, it gets it directly from the instruction queue. Thus the process of instruction fetch and execution are overlapped. This overlapping saves execution time. This overlap architectural feature of 8086 is called *instruction pipe lining.*

11.8. MICROPROCESSOR 8088

8088 is a 16 bit microprocessor with 8 bit data bus. Thus it needs to multiplex only 8 data bits with the 8 least significant bits address bus.

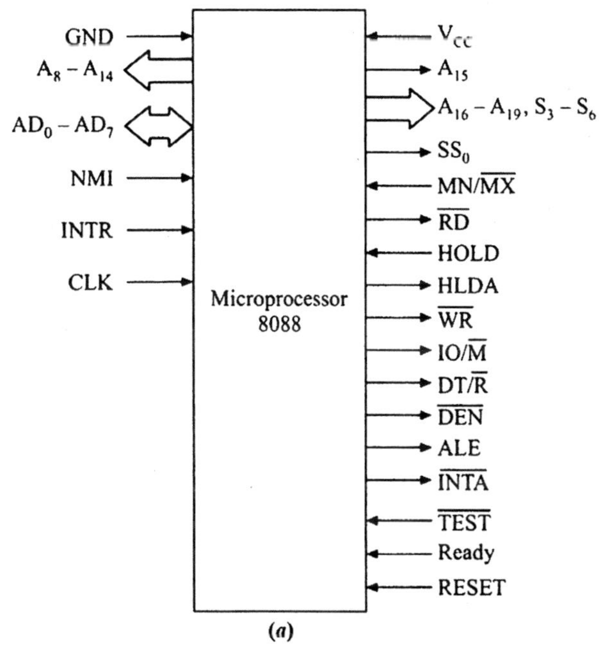

Fig. 11.14. Pins/signals in μP 8088 : (*a*) minimum mode.

The register set of 8088 is identical to that of 8086.

The memory of 8088 is in the sequence of bytes . The concept of two memory banks of 8086 is not applicable in this case. Evidently all addresses sent out by 8088 are in the form of byte addresses. To address a word, two consecutive memory read or write cycles, are performed and each cycle accesses one byte.

Figure 11.14 [(*a*) and (*b*)] shows the pins/signals in 8088 for the Minimum and Maximum modes. Since 8088 has an 8 bit data bus, only 8 bits are multiplexed with 8 least significant address bits. Since 8088 sees only one memory bank, the signal $\overline{BHE}$ is not required in 8088. Therefore, $\overline{BHE}$ pin (of 8086) is replaced by SS_0 pin in 8088. This SS_0 pin along with IO/$\overline{M}$ and DT/$\overline{R}$ are used to determine the current bus cycle. It is seen that M/$\overline{IO}$ line of 8086 is replaced by IO/$\overline{M}$ line in 8088. This is similar to that in

8085 because 8085 has also IO/M line for distinguishing between memory and I/O memory address bits.

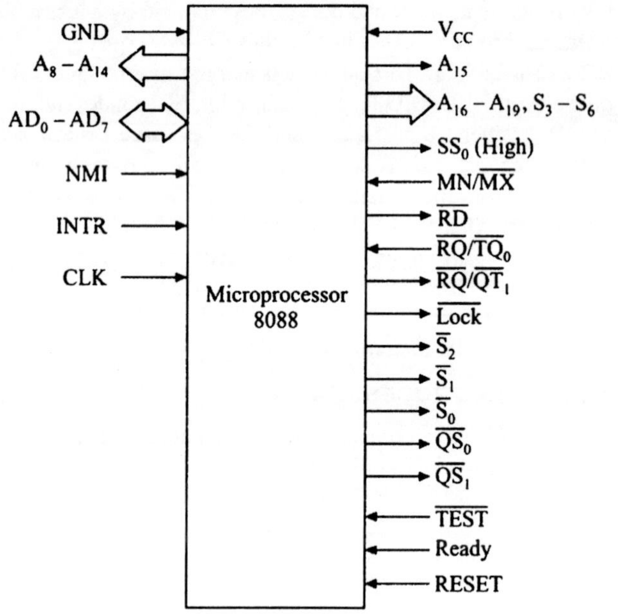

(*b*)

Fig. 11.14. Pins/signals in μP 8088 : (*b*) maximum mode.

11.9. MICROPROCESSOR PROGRAMMING

A microcomputer means a small size and rather cheap computer. To enable it to do computations it must do the following :

1. Receive instructions and data and store them
2. Carry out computations
3. Store the results of computations
4. Display the results of computations (and data if required)
5. Control all the operations.

The instructions and data are fed to the computer through an input device. The commonly used input devices are keyboard and mouse. The instructions, data and results are stored in the memory. As discussed in chapter 9 present-day computers use semiconductor memories. Hard disks and diskettes are used as external memory devices. The results are displayed by an output devices. Printers, LEDs*, CRT** displays etc. are the output devices. The computations and control functions are performed by the microprocessor.

The various inputs which cause the computer to do the specified task are called instructions. A complete set of instruction for a specific task is called instruction set. These series of instructions is called a computer programme***. A programme may just have one instruction. However most programs have many instructions.

A computer programme can be written in a number of languages. However a computer can understand a programme written in bits**** (*i.e.*, string of 0s and 1s). Such a programme (written in 0s and 1s) is called machine language programme. To write a programme in machine language is a very difficult task and many errors may also creep in rendering the program useless.

A programme can be very conveniently written in mnemonic operation codes and symbols and data can be written in decimal or hexadecimal notation. Such a language is called assembly language. A special device called assembler translates the assembly language programme into machine language. Development of a programme involves the following basic steps.

1. Problem definition : In this step the task to be performed is specified precisely. The input and output devices, system constraints (*e.g.*, execution time, accuracy level required etc.) are also specified.

2. Flow chart : The second stage in programme development is to write a flow chart. A flow chart depicts the way the computations will proceed to accomplish the given task. Standard symbols shown in Fig. 11.15 are used for the flow chart.

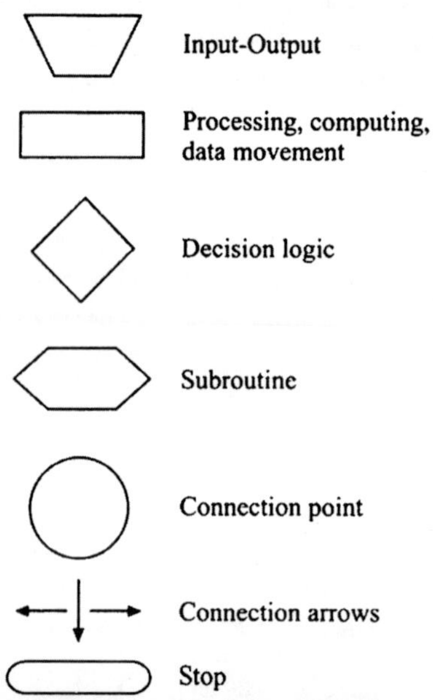

Fig. 11.15. Symbols for flow charts.

Many programmers consider a flow chart as unnecessary and develop an algorithm for the problem.

3. Algorithm : An algorithm is a precise methodology for the required computations. It contains all the necessary equations and other information. In fact a flow chart can also be considered an algorithm.

4. Codification, Translation and Debugging : Figure 11.16 shows a flow chart depicting the programming process. A programme is useless unless it has beet tested using some standard data and the results are found to be satisfactory. For this purpose the program has to be written in a proper language and then translated into machine language (by assembler). The data is fed in and the results are examined. The programme is good only if the results are satisfactory. If not the programme is modified. This part of the process is called debugging*. Once the program is

* Light emitting diode.

** Cathode ray tube.

*** A computer programme along with users instructions etc. is called software.

**** Bit means binary digit.

found to be ok it can be documented. Documentation involves writing user's instructions, comments etc.

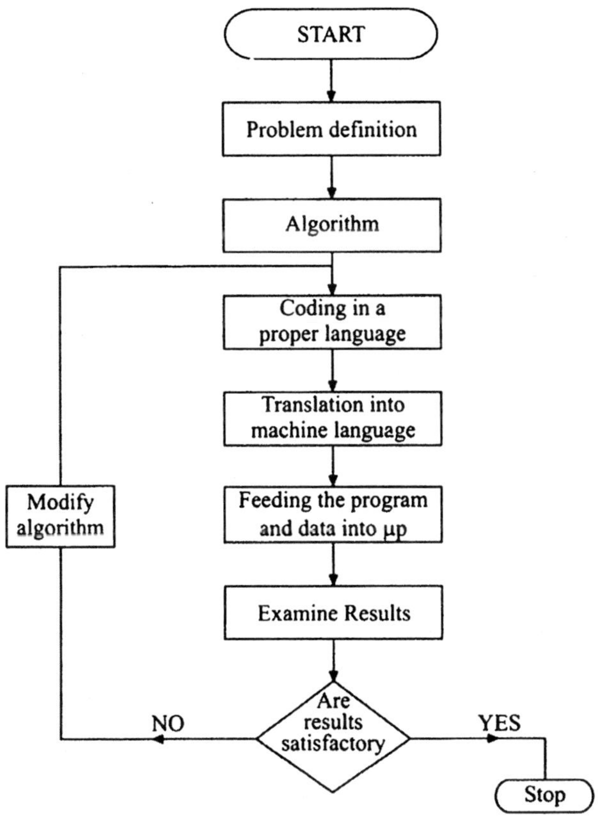

Fig. 11.16. Programming methodology.

11.10. MICROPROCESSOR INSTRUCTIONS

11.10.1. Instruction Format

A computer program consists of a number of instructions. Each instruction must contain the following information :

1. The Operation Code (op code) : This part of the instruction specifies the particular operation to be performed.

2. Source of Data : The operation may involve one operand or two operands. One operand operation is called monodic and two operands operation is called dyadic. The instruction must contain the source address from which the operands are to be called. Instructions like Multiply and Add involve two operands whereas an instruction like Complement or Shift involves only one operand.

3. Destination of Result : The address of result must be specified. Irrespective of the type of instruction and operands, the result is always one.

4. Address of Next Instruction : After an instruction is executed, the microprocessor has to execute the next instruction. This part of an instruction specifies the address of next instruction.

Figure 11.17 shows the format of a complete instruction.

Operation Code	Address of Operand 1	Address of Operand 2	Address of result	Address of next instruction

Fig. 11.17. Instruction format.

	Op code	(*a*)

First byte	Op code	(*b*)
Second byte	Operand/address	

First byte	Op code	(*c*)
Second byte	Operand/address	
Third byte	Operand/address	

Fig. 11.18. Instruction length (*a*) one byte, (*b*) two bytes, (*c*) three bytes.

11.10.2. Instruction Length

The length of an instruction may be one byte or two bytes or three bytes. Figure 11.18 shows these three lengths. A single byte instruction consists only operation code (op code). A two byte instruction has operation code in the first byte and operand/address in the second byte. These two bytes are always at consecutive addresses. A three byte instruction has operation code in the first byte and operands/addresses in the second and third bytes.

An instruction which contains the op code, source of data, destination of result and address of next instruction may be too long. An op code may have 4 bits. Each address may have 12 bits. Thus the complete instruction may be 40 bit long. Such an instrction can not be handled by an 8 bit or 16 bit microprocessor. Therefore, it is always necessary to reduce the length of an instruction. The methods to reduce instruction lengths are :

1. **Using program counter :** The program counter holds the address of instruction. The CPU increments the program counter after executing the instruction and fetches the next instruction from the next higher address in the computer memory.
2. **Implicit Source and Destination Address :** The source and destination addresses are made implicit (and not explicit). An implicit address may be register or memory location addressed through register.
3. **Common Destination and Source Addresses :** In this case the result replaces one of the operands (say stored in accumulator).
4. Limiting addresses to the registers or the contents of registers.

11.10.3. Addressing Methods

Each instruction specifies the required operation. Before the instruction can be executed the address of data has to be specified. The methods of specifying the address of data are called Addressing methods (or addressing modes). The addressing methods used in 8085 are :

* Debugging means elimination of errors.

1. Direct addressing
2. Register addressing
3. Register indirect addressing
4. Immediate addressing
5. Implicit addressing.

(*a*) **Direct Addressing [Fig. 11.19 (*a*)] :** In this method the address of the operand is specified in the instruction. When this method is used the instruction is 3 byte long [Fig. 11.18 (*c*)]. However IN or OUT instructions are only two bytes long. Some examples of direct addressing are LDA FFA, *i.e.*, load accumulator direct. The contents of memory location whose address is specified as FFA in second and third bytes are moved to accumulator (also known as register A). Another example is STA FFC, *i.e.*, store accumulator direct. The contents of accumulator are transferred to the memory location whose address is specified as FFC (second and third bytes of instruction). Both the above instructions can also be written in machine language (*i.e.*, binary code). This method is also called absolute addressing.

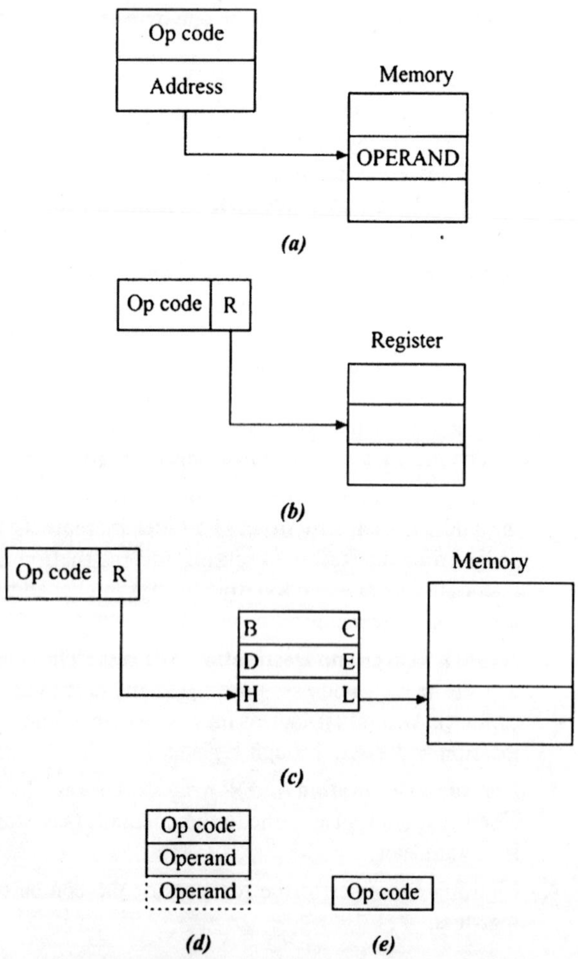

Fig. 11.19. Addressing methods (*a*) direct addressing, (*b*) register addressing, (*c*) register indirect addressing, (*d*) immediate addressing, (*e*) implicit addressing.

(*b*) **Register Addressing [Fig. 11.19 (*b*)]:** In this addressing method the operands are in general purpose registers*. Therefore it is sufficient to specify the register in which the operand is stored. MOV and ADD instructions in register addressing are written as

MOV	A, D	Move contents of register D to register A (*i.e.*, accumulator)
ADD	B	Add contents of register B to accumulator
MOV	C, A	Move contents of register A (*i.e.*, accumulator) to register C.

(*c*) **Register Indirect Addressing [Fig. 11.19 (*c*)] :** In this addressing mode the address of the operand (and not the operand) is in the specified register. In contrast the register addressing mode contained the operand in the specified register. In microprocessor 8085, the register pair which hold the address of operand is always H-L. A typical instruction using register indirect addressing is

MOV	A, M	Move contents of memory location, whose address is in registers H-L, to accumulator.

It is evident that this addressing mode uses letter M to indicate that contents of memory location whose address in register H-L is the operand.

(*d*) **Immediate Addressing [Fig. 11.19 (*d*)] :** In this addressing mode, a part of the instruction (second or third byte) contains the operand. Since the operand is not to be fetched from any memory location but is contained in the instruction itself this addressing mode is very fast. An example of immediate addressing is

ADD	5	Add 5 to the contents of accumulator
LDA	7	Load accumulator with number 7.

(*e*) **Implicit addressing [Fig. 11.19 (*e*)] :** This addressing mode is for instructions in which there is only one operand. It is assumed that operand is in register A (*i.e.*, accumulator). An example is

CMA Complement the contents of accumulator

This method is also known as Inherent addressing. The addressing modes are shown in Fig. 11.19.

11.11. GROUPING OF INSTRUCTIONS

The total group of instructions of a microprocessor is called instruction set. Within a set, the instructions can be grouped as

(*a*) Data transfer group
(*b*) Arithmetic group
(*c*) Logic group
(*d*) Branch group
(*e*) Stack, I/O and Machine Control group

(*a*) **Data Transfer Group :** The instructions in this group cause movement of data from one register to another, from register to memory and vice versa. When this instruction is carried out the source register continues to hold the data. Thus it is basically a copying instruction. The instructions in this group are Move, Load, Store and Exchange.

(*b*) **Arithmetic Group :** As the name suggest the instructions in this group are for arithmetic operations, *i.e.*, addition, subtraction, Incrementing. Decrementing etc.

* IN 8085, the general purpose registers are B, C, D, E, H, L.

(*c*) **Logic Group :** Many logic operations are necessary in a computer programme, AND, OR, XOR, Compare, Rotate, Complement are the logic operations required. This group of instructions contains these logic instructions.

(*d*) **Branch Group :** Some times it is necessary to change the sequence of operations in a computer programme. This group is to change the sequence. The branch instructions may be conditional or unconditional. The unconditional instructions do the specified operation without any condition. On the other hand, conditional instructions do the specified operation if some condition is satisfied. Jump, Call, Return, Restart are these instructions.

(*e*) **Stack, I/O, Machine Control Group :** This group contains instructions for maintaining the stack, reading the input data, writing the output etc.

11.12. PERSONAL COMPUTER (PC)

A typical PC has CPU box, keyboard and mouse, monitor, floppy disc drive and hard disc drive.

CPU Box : It is the heart and brain of computer. It contains all electronic circuits, power supply, floppy disc drive, hard disk drive, etc

Keyboard : It is very similar to the keyboard of a typewriter, but has some additional keys. The number of keys may be 84, 101 or 104.

Mouse : It is an input device and is used to move the cursor, to select options and to execute commands. It has two or three buttons.

Monitor (Video Display Unit) : It is the display unit of the computer and is like ordinary television. It may be black and white or coloured. When you type a character it is displayed at the cursor position.

Floppy Disc Drive : It is a memory storage device and stores information on the floppy disc (diskette). Most personal computers use 5.25" mini floppy disc drive or 3.5" micro floppy disc drive or both.

Hard Disc : It is also a storage device and is permanently fixed in the computer. Its storage capacity is 10 MB or higher.

In addition to above, PC may have some other devices also. These are Printer, Compact Disc (CD) drive, Zip drive and Scanner.

Printer : It is an output device used to print documents.

Some types of printers are dot matrix, inkjet, laser. Each printer has its own style of loading paper, ribbon, ink cartridge etc.

CD Drive : Compact disc drive is also a storage device. In most of CD drives the stored information cannot be changed. These are called CD-ROM drives.

Zip Drive : It is a modified floppy drive and can store about 100 MB of data.

Scanner: It is an input device and can recognise texts or graphics on the input document. Its applications include : as bar code reader to identify a product, as magnetic ink character to sort bank cheques and as optical reader to evaluate objective type answer sheets.

11.13. COMPUTER SOFTWARE

A computer programme is a finite set of instructions to solve a problem. It can be a system programme or application programme. The system programme directs the internal operation of computer. An application programme is for solving specific user oriented problems, *e.g.*, preparation of pay roll, preparation of water bill, preparation of electoral rolls etc. The total group of programmes which a computer needs to function is called software.

The main component of system programme is operating system. It is group of instructions which controls all operations (flow of programme and data, control of input/output, managing storage etc). Computer and assembler are also part of system programme. DOS, Windows, UNIX are some operating systems.

11.14. COMPUTER LANGUAGES

Every computer has its own language called machine language. The instructions of machine language are in binary number system. Writing a programme in machine language is very cumbersome. In assembly language symbols are used in place of sequences of 1's and 0's of binary. The operations are specified in symbolic codes called mnemonics. Before execution the assembly language programme is converted into machine language by assembler.

To enable easy writing of programmes many higher level languages have been developed. These languages are user-friendly and are easy to understand. Fortran, Cobol, Pascal, C, Java etc. are all high level languages. A programme written in high level language is called source programme. A compiler converts the source programme into equivalent machine language programme called object programme.

11.15. BASIC

Basic (Beginners all purpose symbolic instruction code) is a very simple and easy to understand language used by beginners.

11.15.1. Basic Definitions

Alphanumeric Characters	
Digit	0 1 2 3 4 5 6 7 8 9
Letters	A To Z
Special characters	$ * ( ) , . + / ' > < " ↑ ; ƀ = #
Statement numbers	One to five digits. Zero should not be used.
String Constant	Any string of alphanumeric or special characters included in quotation mark, *e.g.*, 'DIGITAL'.
Numeric Constant	A string of digits signed or unsigned (unsigned numbers are taken as positive) with or without decimal point. It may be followed by an exponent part, *e.g.*, 210, –156, 3.12 E 3.
Numeric Variable	It may be a letter, *e.g.*, P, letter followed by digit, *e.g.*, B5 or a string variable which is an alphabet followed by dollar sign, *e.g.*, B$.

Operators

Arithmetic Operators : Addition + , Subtraction –, Multiplication * , Division /, Exponentiation ** or ↑

Logic Operators : AND, OR, NOT

Relational Operators : =, <, >

Expression

A Constant, *e.g.*, – 5

A variable, *e.g.*, F, X(1, K)

A string of constants, variables and functions each separated by operators

e.g., A + B + M * N – P ** 2

1.25 * A ** B

The priority of various operations is

Operation	***Order***
Expression within parenthesis	I
Evaluation of function	II
Exponentiation	III
Multiplication and Division	IV
Addition and Subtraction	V

Library Functions

Trigonometric functions :
SIN (X), COS (X), TAN (X), ATN (X), COT (X).

Exponential functions :
EXP (X), *i.e.*, e^x, LOG 10 (X), LOG (X), SQR (X)

Arithmetic functions :
SGN (X), ABS (X), *i.e.*, determines the absolute value of X, INT (X), *i.e.*, calculates the integer part of X.

Basic Statements

READ, DATA, INPUT, LET, PRINT, END, STOP, GO TO, IF

System Commands :

RUN, SAVE, LIST, HELP, NEW, OLD, SCRATCH, BYE.

11.16. PASCAL

Character set : Letters A to Z and *a* to *z*
Digits 0 to 9
Arithmetic operators + – * / div mod
Logic operators AND or NOT.
Relational operators =, <, >,#,<=,> =
Parentheses ()
Index brackets []
Comment {}
Pointer ↑

Constants : Integer constants 217, –1016
Real Constants –12.56, 101.23
(Floating point constants)
0.67E7 = (0.67×10^7)

Identifier : In pascal the word identifier is used for names given to variables functions, constants, arrays, procedures etc.

Library functions : EXP (*i.e.*, e^x), LN (*i.e.*, natural log) SIN (*i.e.*, sine) COS (*i.e.*, cosine), ARCTAN (*i.e.*, $\tan^{-1}$) SQR (*i.e.*, square), SQRT (*i.e.*, square root), ROUND (*i.e.*, *x* rounded), TRUNC (*i.e.*, *x* truncated), ABS (*i.e.*, | *x* |)

11.17. FORTRAN 77

The most recent ANSI Fortran standard is FORTRAN 77.

Characters : Letters A to Z
Digits 0 to 9
$ ' ( ) * ƀ + – / , . : =

Constant : Integer constant also called fixed point constant is a constant without a decimal point, *e.g.*, 26, – 71.
Real constant also called floating point constant is a constant with a decimal point, *e.g.*, 19.3, – 6.27. Exponential form is used for very small or very large real constants, *e.g.*, 6.2 E – 5 (= 6.2×10^{-5}).

Character type constant, *e.g.*, EIL

Logical type constant .TRUE, and .FALSE.

Variables upto 6 alphanumeric characters

Arithmetic operations + –*/** (*i.e.*, exponentiation) and negation

Logical operation .AND., .OR., .NOT.

Trigonometric functions SIN, COS, TAN, A SIN, A COS, A TAN, SINH, COSH, TANH.

Other functions ABS, MAXO, MINO, SQRT, EXP, LOG, LOG 10.

11.19. LANGUAGE C

11.19.1. C Character Set

A character denotes any digit, alphabet or special symbol. The valid characters in C are :

Alphabets	Capital letters	A To Z
	Lower case letter	a to z.
Digits		0 to 9
Special symbols		@ # + – ( ) * < > / ? [ ] % etc.

11.19.2. C Constants

C constants can be further subdivided into primary and secondary constants.

Primary Constants : Integer constant, real constant and character constant.

Secondary Constants : Array, pointer, union, structure etc.

Integer Constant : An integer constant has one or more digits but no decimal point. It may be positive or negative. Commas and blanks are not permitted. For 16 bit computers the range of integer constants is –32768 to +32767.

Real Constant

A real constant has one or more digits with a decimal point. Commas and blanks are not allowed. It can be positive or negative. It can also be written in exponential form, *e.g.*, 427.23, 3.821, – 0.3 *e* – 3 are all real constants. The range of real constants in exponential form is – 3.4 *e* 38 to 3.4 *e* 38.

Character Constant

A character constant is a single alphabet or single digit or single special character enclosed within single inverted commas.

11.19.3. C Variables

A C variable is a combination of alphabets, digits or

underscores with alphabet as first character and up to 8 character long, *e.g.*, *cop_s*, *m_nth*, *m* 32_*o* etc. It should not have a blank or comma or special symbol except underscore (*i.e.*, _).

11.19.4. C Keywords

These are the words whose meaning has already been explained to C compiler. They are also called reserved words and the total number of keywords is 32. Some key words are auto, break, go to, union, while, for, if, int., switch.

11.19.5. C Instructions

C instructions are classified into 4 categories, viz., type declaration, input/output, arithmetic and control.

Type Declaration Instruction : It is used to declare the type of variables used in the programme. It is written generally in the beginning of the programme, *e.g.*, int, age, int *ml*, float *sr*, char a.

Arithmetic Instruction : It has a variable name on left hand side of = sign and variable name and constants on the right hand side of = sign. The variable names and constants on right hand side are connected by arithmetic operators +, –, *, /.

Arithmetic instructions may be integer mode (*i.e.*, all operands are either variables or constants), real mode (all operands are either real variables or constants) or mixed mode (some operands are real and some are integers).

The hierarchy of operations is first *, /, second + , – , third =

Input/Output Instruction : A number of library functions are available for input/output in C. These can be classified into control I/O functions, Disk I/O functions and Port I/O functions.

Control Instruction : These instructions enable us to specify the sequence in which the execution of instructions will be done. Control instructions can be classified into sequence control instruction, decision control instruction, loop control instruction and case control instruction.

Decision Control Instruction : In C language the decision control instruction uses if statement or if else statement or conditional operators.

The if statement causes execution when condition following if is true.

The if else statement causes execution of one group of instructions when condition following if is true and another group of instructions if the condition following if is false.

In C three logical operators && (*i.e.*, AND), || (*i.e.*, OR), ! (*i.e.*, NOT) are also used.

11.19.6. Loop Control

Loop control allows us to repeat a set instructions. C allows 3 methods of loop control, *i.e.*, using a 'for' statement using a 'while' statement and using a 'do-while' statement.

The using of 'for' loop is the most popular loop control in C. It allows us to specify three values viz. initial value of loop counter, increment in the value of loop counter and testing the loop counter to see if its value has reached the final value, *i.e.*, the desired number of repetitions has been achieved.

The 'while' statement is used when the repetitions are to be done for a fixed number of times, *e.g.*, to prepare the electricity bill of 150 houses.

The 'do-while' statement tests the condition after executing the statement within the loop. (While statement tests the condition before executing the statement within the loop.)

C also allows use of GO TO statement, functions, arrays and has many other features.

11.20. JAVA

11.20.1. Java Variables

Java variables can be instance variables, class variables and local variables. Instance variables are used to define an object's attributes. Class variables define attributes of an entire class of objects.

Local variables are used inside method definitions and can be used only when the method or block is being executed by Java interpreter.

A variable name can start with a letter, dollar sign or underscore character. A variable name cannot start with a number.

Java variables can be formed by joining many words together.

11.20.2. Data Types in Java

Data types in Java can be integer types or floating point number type.

Integer data type : These are further classified into four types viz. byte, short, int and long.

Floating point type : These are the numbers with decimal. These can be written as 256. 321 or 1.5 E 12. Floating type range is 1.4 E – 45 to 3.4 E + 38. Double type with range 4.9 E – 324 to 1.7 E + 308 is also possible.

11.20.3. Literal

Any number, text or other information which directly represents a value is called literal in Java. Numbers, characters, strings are all literals. Literal can be integer literal or float literal or boolean literal. Boolean literals are true and false. Character literals are expressed by a single character with quotation mark (*e.g.*, '*a*', '3' etc.). Java supports thousands of characters (in addition to 128 characters of ASCII). String literals are also used in Java. In Java strings can be combined, modified etc. Strings literals consist of a series of characters inside a double quotation mark.

11.20.4. Arithmetic Operators

Arithmetic operators in Java are addition, (+), subtraction (–), multiplication (*), division (/) and modulus (%).

The subtraction operator can also be used to negate a single operand and is equivalent to multiplying the operand by –1. Modulus operator produces the remainder of a division operation (*e.g.*, 29% 4 results in answer 1).

11.20.5. Assignment Operators in Java

Expression	Meaning
$x += y$	$x = x + y$
$x -= y$	$x = x - y$
$x *= y$	$x = x * y$
$x /= y$	$x = x/y$

11.20.6. Incrimenting and Decrementing

In Java increment operator is + + and decrement operator is – –. These operators can be placed before or after a variable name. When written before a variable name it is called prefix operator and when written after a variable name it is called post fix operator. In simple expressions prefix and post fix incrementing and decrementing do not change the result. However when these operations are part of larger expressions the results are different, *e.g.*, if

$x = 42$ and $y = x$ ++, then $y = 42$.

However if $x = 42$ and $y =$ ++ x, then $y = 43$.

11.20.7. Operator Precedence

In Java the operator precedence is as under : Increment and decrement operations, arithmetic operations, comparisons, logical operations and assignment operation.

11.21. MS OFFICE

11.21.1. MS WORD

MS WORD is used to prepare, edit and print documents (letter, memorandum, balance sheet, research paper etc.). It can add text, change look of document, add graphics, create tables, collaborate (*i.e.*, share) document with e-mail, transform documents into web pages, position text, graphics, tables anywhere on a page. It has spell check, bold face headings, shading, margin setting, mail merge, page formatting, font size, font style, colour and specific effects, indenting, character style, paragraph style, word count, character count, line count, paragraph count and many other facilities.

11.21.2. MS EXCEL

Excel helps in making decisions based on numbers. An electronic worksheet or spreadsheet is used, to perform calculations, store information in computer memory and display the information in required format. Excel is an integrated electronic spreadsheet program developed by Microsoft Corporation (USA). Some features of Excel are : date and time related functions, data base management, manipulation of character data, changing look of data, using formulae, adding graphics, using web, analyse data, use external data, auditing spread sheet, trace errors, view all cells used in a formula etc.

11.21.3. MS PowerPoint

PowerPoint is a presentation software which can help in 'slide based' presentations. It can help in creating black and white transparencies and colour transparencies for overhead projector, slides for slide projector, video slide show on computer screen, printed handouts, detailed speaker's notes, adjusting size of text, adding graphic bullets, choosing slide layout, zoom (*i.e.*, viewing slides from various angles and in various sizes) and many other facilities.

11.21.4. MS Access

It is a data base management system (DBMS) software. The facilities available in MS Access include creation of tables to store data, designing forms, creating queries, keeping track of data, designing data base, making tables, determining fields and field size, compressing data base, jumping from one data base to another etc.

11.21.5. MS Front Page

It is a software to create websites, nested subwebs (*i.e.*, a website nested inside another website), reserve files, creating content, publishing content, supporting HTML format, supporting multilingual features, organise web pages, importing content from websites etc.

11.21.6. MS Photo Draw

It is a software to draw and use pictures in different formats, clip art, shapes, texts, images from other graphic programs, line tools (straight, curved or freehand drawings), shadow effect, 3D effect etc.

11.21.7. MS Publisher

It is a software for DTP (Desk Top Publishing). It is used to design and print reports, books, catalogues, manuals etc. on a personal computer. The printing is generally done on laser printer to get best results. It has spell check, grammar check, zoom, scroll, save on web page, create table, rotate text, font size, colour schemes etc.

11.21.8. MS Out Look

It is a software to take control of information, search information, create appointments, events, vacations, trips, view and manage one day program, organise meetings, manage calender, manage e-mail, communicate through e-mail, share information etc.

11.22. MS DOS

DOS (Disc Operating System) is the first program which has to be loaded into PC memory before the PC can be used for any application. DOS controls all the operations in the PC, *e.g.*, creating files, changing files, displaying list of files, creating directory, changing directory, deleting files, formatting floppy disc. There are two types of DOS commands, *i.e.*, internal and external. The internal commands are automatically loaded into computer memory when DOS is loaded. These internal commands are used for creating, deleting, changing, renaming files etc. The external commands are used for relatively complex operations, *e.g.*, copying complete floppy disc, formatting the floppy disc etc.

11.23. WINDOWS

Windows is an operating system (developed by Microsoft Corporation, USA) for controlling the operation of computer and user interface. Windos 95 is a full-fledged operating system. Windows 2000 is an improved verision and has many additional features. It supports several multi media devices like sound, music, video, recording, media player etc. Windows enable us to use world wide web, e-mail etc.

11.24. COMPUTER VIRUS

Computer virus are types of software which spread, reproduce and damage the working of computer. They mostly spread through internet. If a floppy containing a virus is inserted into a computer, the virus may affect the hard disc on the computer. The virus may hide itself in the boot sector of floppy or hard disc. A virus infected computer may fail to boot or stop working or take long time to load or execute. Many times strange messages (*e.g.*, Happy birthday, I love you etc.) appear on the monitor. These could be due to computer virues.

Computer viruses are very large in number may be some thousands and are getting added daily. They can be classified into 8 major categories: boot sector, file, multipartite, trojan horse, file over write, polymorphic, stealth and macro. Many real time anti-virus programmes have been developed to scan the floppy discs/hard discs and remove viruses.

OBJECTIVE TYPE QUESTIONS

A. Select the most appropriate alternative

1. For an accurate clock period in 8085 it is preferable to use
   (a) crystal
   (b) crystal or LC resonant circuit
   (c) crystal or RC circuit
   (d) LC or RC circuit
2. Which of the following produces very wide drifts in clock frequency?
   (a) crystal
   (b) RC circuit
   (c) LC resonant circuit
   (d) crystal and RC circuit
3. When a peripheral device is not ready for data transfer it will send to the microprocessor a
   (a) low READY bit
   (b) high READY bit
   (c) low or high READY bit depending on whether input or output is activated
   (d) none of the above
4. DMA stands for
   (a) direct memory address
   (b) distributed memory address
   (c) direct memory accumulator
   (d) decimal machine address
5. In 8085 the pins for +5 V input and ground are
   (a) 20 and 40 respectively
   (b) 40 and 20 respectively
   (c) 1 and 2 respectively
   (d) 2 and 1 respectively
6. The five flags in 8085 are designated as
   (a) Z, CY, S, P and AC (b) D, Z, S, P, AC
   (c) Z, C, S, P, AC (d) Z, CY, S, D, AC
7. In 8085, the pairing of registers B, C, D, E, H, L is
   (a) B-C, D-E, H-L (b) B-D, C-E, H-L
   (c) B-C, D-L, H-E (d) B-H, D-E, C-L
8. In 8085, the pins for SID and SOD are
   (a) 4 and 5 respectively (b) 5 and 4 respectively
   (c) 3 and 4 respectively (d) 4 and 3 respectively
9. RESET OUT signal when high indicates that
   (a) program counter etc. are being set
   (b) program counter etc. are being reset
   (c) either (a) or (b)
   (d) none of the above
10. ALE stands for
    (a) address latch enable
    (b) accumulator latch enter
    (c) address latch enter
    (d) accumulator latch enable
11. In 8085
    (a) the upper 8 address bits appear on address bus and lower 8 bits on address data bus
    (b) the lower 8 address bits appear on address bus and the upper 8 address bits appear on address data bus
    (c) either upper or lower 8 address bits may appear at address bus
    (d) either upper or lower 8 address bits may appear at address data bus
12. In 8085
    (a) P flag is set when the result has even parity
    (b) P flag is set when the result has odd parity
    (c) P flag is reset when the result has odd parity
    (d) P flag is reset when the result has even parity
13. In 8085
    (a) all operations are on 8 bit data
    (b) majority of operations are on 8 bit data
    (c) all operations are on 8 bit or 4 bit data
    (d) either (a) or (c)
14. If 8085 is required to add two 32 bit numbers the numbers of addition sequences is
    (a) 1 (b) 2
    (c) 4 (d) 8
15. In 8085 the time required to add two 32 bit numbers is (if clock frequency is 5 MHz)
    (a) about 1 ns (b) about 1 μs
    (c) about 30 μs (d) about 100 μs
16. Consider the following tasks in a micro computer
    1. Receiving data and instructions
    2. Performing arithmetic computation
    3. Performing logical computations
    4. Storing data and instructions

    Which of the above are performed by ALU?
    (a) 2 only (b) 2 and 3 only
    (c) 1, 2 and 3 only (d) all of the above
17. In a micro computer the control bus
    (a) transmits and receives control signals between processor and other devices
    (b) transmits and receives control signals and also sends and receives data
    (c) transmits and receives control signals, sends and receives data and transmits address of the device
    (d) none of the above
18. Binary storage of decimal number 44 in 8 bits is
    (a) 00101100 (b) 10101100
    (c) 10101101 (d) 00101101
19. If word length is 16 bits the decimal equivalent of largest number which can be stored is
    (a) 32767 (b) 32766
    (c) 16383 (d) 16384
20. In one's complement 8 bit representation 11111111 represents
    (a) +0 (b) −0
    (c) +1 (d) −1
21. Which of the following has unique representation of 0?
    (a) sign magnitude (b) 1's complement
    (c) 2's complement (d) both (b) and (c)
22. In Intel 8085, the integer range per word length
    (a) – 128 to 127 (b) – 127 to 128

(c) −64 and 63 (d) −63 and 64

23. A real number consists of
(a) integer part
(b) integer part and fraction part
(c) integer part, fraction part along with positive or negative sign
(d) none of the above

24. Which of the following is not a real number?
(a) 16.23 (b) −16.23
(c) 40 (d) 40

25. The decimal equivalent of binary 1101.11 is
(a) 9.75 (b) 13.75
(c) 11.50 (d) 11.75

26. Binary number 1101.101 is equivalent to decimal number
(a) 13.5 (b) 13.75
(c) 13.625 (d) 13.875

27. The normalised form of 1.81×10^{-11} is
(a) 18.1×10^{-12} (b) 18.1×10^{-13}
(c) 0.181×10^{-10} (d) 0.0181×10^{9}

28. The normalised form of 0.0152×10^{18} is
(a) 0.152×10^{17} (b) 1.52×10^{16}
(c) 15.2×10^{15} (d) 152×10^{14}

29. In floating point representation decimal number 13.5 is
(a) 1101.1 (b) 1.1011×2^3
(c) 0.11011×2^4 (d) 11.011×2^2

30. The sign of a floating point number depends on
(a) only the sign of mantissa
(b) only the sign of exponent
(c) the sign of both mantissa and exponent
(d) either (a) or (b)

31. For 8 bit representation the minimum positive and minimum negative numbers are
(a) 2^{-128} each (b) 2^{-128} and 2^{-127}
(c) 2^{-127} and 2^{-128} (d) 2^{-127} each

32. A 37 bit mantissa has an accuracy of
(a) 6 decimal places (b) 8 decimal places
(c) 10 decimal places (d) 11 decimal places

33. 6 bytes means
(a) 6 bits (b) 24 bits
(c) 48 bits (d) 96 bits

34. If the absolute value of a number is less than decimal 1, the sign of exponent in floating point representation is
(a) zero (b) negative
(c) zero or negative (d) positive

35. If the sign bit of mantissa is 0 and the exponent is increased from a positive to a more positive number the result is
(a) a larger floating point number
(b) a smaller floating point number
(c) either (a) or (b) depending on the actual number
(d) a negative floating point number

36. If the mantissa has a sign bit of 0 and the exponent is changed from a negative number to a more negative number the result is
(a) a positive number and closer to zero
(b) a positive number and away from zero
(c) a negative number
(d) a negative number and closer to zero

37. Which of the following is true?
(a) During addition if the carry into and carry out of the sign bit do not match then an overflow has occurred.
(b) During addition if the carry into and carry out of the sign bit do not match then an overflow or underflow has occurred.
(c) During addition if the carry into and carry out of the sign bit match then an overflow or underflow has occurred.
(d) During addition if the carry into and carry out of the sign bit match then an overflow has occurred.

38. Consider the following statements for 8085
1. The most significant 8 bits of address are transmitted on the 8 lines on which data is transmitted.
2. Eight pins are dedicated to transmit the most significant 8 bits of memory address.
3. The least significant 8 bits of address are transmitted on the lines on which data is transmitted.
4. The data and least significant 8 bits are transmitted at different points in time.

Which of the above statements are correct?
(a) all (b) 1, 2 and 3 only
(c) 2, 3 and 4 only (d) 1, 3 and 4 only

39. In memory mapped I/O means that
(a) an input or output device is treated by the microprocessor as one memory location
(b) input or output devices are treated as distinct
(c) an address on the address bus may refer either to a memory location or to an I/O device
(d) none of the above

40. The general purpose register code for accumulator in 8085 is
(a) 111 (b) 000
(c) 001 (d) 1111

41. In 8085 stack pointer is
(a) 4 bit register (b) 8 bit register
(c) 16 bit register (d) 32 bit register

42. In a 3 byte instruction of 8085, the first, second and third byte respectively indicate
(a) operation code, low order byte of data/ address and high order byte of data/ address
(b) low order byte of data/address, high order byte of data/ address and operation code
(c) high order byte of data/address, low order byte of data/ address and operation code
(d) high order byte of data/address, operation code and low order byte of data/address

43. In direct addressing mode of 8085
(a) all instructions are 3 bytes long
(b) all instructions are 3 bytes long except IN instruction which is 2 bytes long
(c) all instructions are 3 bytes long except IN and OUT instructions which are 2 bytes long
(d) all instructions are 2 bytes long

44. The 8085 instruction LDA FFA is an example of
(a) Direct addressing

(*b*) Register addressing
(*c*) Immediate addressing
(*d*) Register indirect addressing

45. In 8085 the term 'absolute addressing' means
(*a*) direct addressing
(*b*) immediate addressing
(*c*) register addressing
(*d*) register indirect addressing

46. In 8085 the instruction MOV A, B is an example of
(*a*) direct addressing
(*b*) immediate addressing
(*c*) register addressing
(*d*) register indirect addressing

47. Match List I (Addressing modes) and List II (Instructions) and give correct answer using the codes

	List I		List II
A	Direct addressing	1.	MOV A, M
B	Register addressing	2.	MOV C, A
C	Register indirect addressing	3.	LDA 7
D	Immediate addressing	4.	STA FFC

Codes :

	A	B	C	D
(*a*)	1	2	3	4
(*b*)	4	2	1	3
(*c*)	2	3	4	1
(*d*)	3	2	1	4

48. In 8085, the instruction CMA is an example of
(*a*) implict addressing (*b*) direct addressing
(*c*) register addressing (*d*) immediate addressing

49. In 8085, which addressing mode is also called inherent addressing
(*a*) direct (*b*) register
(*c*) implicit (*d*) immediate

50. In 8085 the instruction MOV A, M is an example of
(*a*) direct addressing
(*b*) immediate addressing
(*c*) register addressing
(*d*) register indirect addressing

51. An instruction using immediate addressing mode in 8085 is
(*a*) 2 bytes long (*b*) 3 bytes long
(*c*) 2 or 3 bytes long (*d*) 1 or 2 or 3 bytes long

52. Read the following statements about time and space requirements of microprocessors
1. Space requirement is also known as memory requirement
2. Space requirement is expressed in terms of bytes
3. Time requirement is expressed in terms of states
4. Time requirement is expressed in terms of μ-s

Which of the above statements are correct?
(*a*) 1, 2 and 3 only (*b*) 1, 2 and 4 only
(*c*) 2 and 3 only (*d*) 1 and 4 only

53. A microprocessor can understand instruction written in
(*a*) machine language only
(*b*) mnemonics operation codes only
(*c*) high language only
(*d*) both machine language and mnemonics operation codes

54. Consider the following mnemonics
1. MOV 2. ADD 3. LXI
Which of the above are valid for 8085?
(*a*) 1 and 2 only (*b*) 2 only
(*c*) 1, 2 and 3 (*d*) 1 and 3 only

55. The instruction LDA 20 in 8085 means
(*a*) contents of memory location 20 are fetched into the accumulator
(*b*) the value 20 is fetched into accumulator
(*c*) either (*a*) or (*b*)
(*d*) neither (*a*) nor (*b*)

56. The instruction MVI A, 20 in 8085 means
(*a*) contents of memory location 20 are brought into the accumulator
(*b*) the value 20 is brought into accumulator
(*c*) either (*a*) or (*b*)
(*d*) neither (*a*) nor (*b*)

57. Consider the following statements about 8085
1. The instruction LDA uses direct addressing mode.
2. The instruction MVI uses direct addressing mode.
3. The instruction MVI uses immediate addressing mode.
4. The instruction LDA uses immediate addressing mode.

Which of the above are correct?
(*a*) 1 and 3 only (*b*) 3 and 4 only
(*c*) 1 and 2 only (*d*) 2 only

58. Consider the directive T : DS105. This directive
(*a*) directs the assembler to reserve a sequence of 105 bytes in the memory
(*b*) directs the assembler to reserve a sequence of 105 bits in the memory
(*c*) directs the assembler to reserve a sequence of consecutive 105 bytes in the memory
(*d*) directs the assembler to reserve a sequence of consecutive 105 bits in the memory

59. DS directive in 8085
(*a*) forces the assembler to reserve one byte of memory
(*b*) forces the assembler to reserve a specified number of bytes in the memory
(*c*) forces the assembler to reserve a specified number of consecutive bytes in the memory
(*d*) none of the above

60. Which of the following directives in 8085 causes two bytes to be reserved in the memory?
(*a*) DS (*b*) DB
(*c*) DW (*d*) EQU

61. Which of the following is not an assembler directive in 8085?
(*a*) DS (*b*) DB
(*c*) DQ (*d*) EQU

62. Which of the following directives in 8085 causes one byte to be reserved in the memory
(*a*) DS (*b*) DB
(*c*) DW (*d*) EQU

63. Which of the following assembler directives in 8085 does not cause any memory location to be used up?

(a) DS (b) DB
(c) DW (d) EQU

64. Which of the following is not treated as hexadecimal constant by assembler in 8085?
(a) 45 H (b) 6 AFH
(c) 234 (d) 64 H

65. Consider the following statements about constants in 8085
1. the assembler treats 64 H as hexadecimal constant
2. the assembler treats 154 as binary constant
3. the assembler treats 540 as octal constant
4. the assembler treats 138 as decimal constant

Which of the above are correct?
(a) 1, 2 and 3 only (b) 1, 3 and 4 only
(c) 1 and 3 only (d) 2 and 4 only

66. Which of the following is not treated as hexadecimal constant by assembler in 8085?
(a) 64 H (b) 5 AFH
(c) AFH (d) OCFH

67. Which of the following statements is written in plain English in 8085?
(a) Data definition (b) Assignment
(c) Conditional (d) Repetitive

68. The initial contents of ACC and CY in 8085 are

CY	ACC
1	10000001

After instruction RAL is executed once the contents of CY and ACC respectively will be
(a) 1 and 00000011 (b) 0 and 00000011
(c) 1 and 00000001 (d) 0 and 00000001

69. Let the initial contents of ACC and CY in 8085 be

CY	ACC
1	10000001

After the instruction RAL is executed twice the contents of CY and ACC respectively will be
(a) 1 and 00000011 (b) 0 and 00000111
(c) 1 and 00000111 (d) 0 and 00000011

70. Read the following statements about 8085
1. RAL instruction rotates ACC contents left by 1 bit
2. When RAL instruction is executed once the binary number gets doubled
3. STC means set CY to 1
4. STC means set CY to 0

Which of the above statements are correct?
(a) 1 and 2 only (b) 1, 2 and 3 only
(c) 1, 2 and 4 only (d) 1 and 4 only

71. To multiply a number by 8 in 8085 we have to use RAL instruction
(a) once (b) twice
(c) thrice (d) four times

72. When we use RRC instruction once in 8085, the number is
(a) multiplied by 2 (b) divided by 2
(c) multiplied by 4 (d) divided by 4

73. If initial value of PROD is zero and a program adds X to PROD, Y times we get the result
(a) X+Y (b) XY
(c) X/Y (d) X−Y

74. In 8085 the data is stored in the stack on
(a) first in first out basis (b) last in first out basis
(c) first in last out basis (d) last in last out basis

75. Consider the following operation in a computer
1. multiplication of scalar
2. inversion of matrix
3. printing a line of 120 characters on a printer

Which of the above is a subroutine?
(a) 1 and 2 only (b) 1 and 3 only
(c) 1, 2 and 3 (d) 2 only

76. In 8085 which instructions are useful for writing and using subroutines
(a) CALL (b) RET
(c) CALL and RET (d) none of the above

77. Which of the following is true?
(a) Intel 8080 was the ancestor of Intel 8085
(b) Intel 8080 was the ancestor of Intel Z 80
(c) Intel 8080 was the ancestor of both Intel 8085 and Z 80
(d) Intel 8085 was the ancestor of Intel 8080

78. Consider the following statements
1. In 8085 address and data busses are multiplexed.
2. In Z 80 address and data busses are multiplexed.
3. In Z 80 address and data busses are not multiplexed.
4. Z 80 has 16 bit address bus and 8 bit data bus.

Which of the above statements are correct?
(a) 1 and 2 only (b) 1, 3 and 4 only
(c) 1, 2 and 4 only (d) 2 and 4 only

79. The memory hierarchy in microprocessor based system is
(a) register, cache, primary, mass, off line back up
(b) cache, register, primary, mass, off line back up
(c) cache, register, mass, primary, off line back up
(d) register, cache, mass, primary, off line back up

80. In super computers the extra level of primary memory is generally called
(a) solid state device (b) single stage device
(c) digital state device (d) surface stage device

81. Programs like compilers, assemblers etc. are stored in a microcomputer in
(a) registers (b) cache memory
(c) primary memory (d) mass storage

82. Which of the following computer memories is fastest?
(a) cache (b) primary
(c) mass storage (d) off line back up

83. The size of cache memory in most microcomputers is about
(a) a few bytes (b) a few kilobytes
(c) a few mega bytes (d) a few giga bytes

84. A memory chip has 65536 cells in it. Then
(a) Time required to access cell 1 is the same as that to access cell 65536
(b) Time required to access cell 1 is less than that required to access cell 65536
(c) Time required to access cell 1 is more than that required to access cell 65536
(d) either (a) or (b)

85. While selecting a memory chip the feature which should be examined is

(*a*) capacity (*b*) ac characteristics
(*c*) dc characteristic (*d*) all the above

86. Which characteristic of a memory chip is called ac characteristics?
(*a*) power consumption (*b*) bus loading
(*c*) timing characteristic (*d*) all the above

87. Which characteristics of a memory chip is called dc characteristics?
(*a*) organisation
(*b*) timing characteristics
(*c*) physical dimensions
(*d*) power consumption and bus loading

88. In the words Kb or Mb used describing memory capacity, the letter b denotes
(*a*) bits
(*b*) bytes
(*c*) bits or bytes depending on type of memory
(*d*) bubbles

89. The organisation of Intel memory chip 127256 is
(*a*) 32 K × 8 (*b*) 16 K × 8
(*c*) 32 K × 4 (*d*) 16 K × 4

90. In a RAM chip with a total of 8096 words, the word addresses range from
(*a*) 1 to 3096 (*b*) 0 to 8095
(*c*) 1 to 8095 (*d*) 0 to 8096

91. The word size accessible at a time from the memory sub system depends on
(*a*) width of data bus (*b*) width of address bus
(*c*) width of control bus (*d*) both (*b*) and (*c*)

92. Microprocessor 8088 has
(*a*) 16 bit registers and 16 bit data bus
(*b*) 8 bit registers and 8 bit data bus
(*c*) 8 bit registers and 16 bit data bus
(*d*) 16 bit registers and 8 bit data bus

93. Microprocessor 386SX is a 16/32 bit microprocessor. It has
(*a*) 16 bit registers and 32 bit data bus
(*b*) 16 bit registers and 16 bit data bus
(*c*) 32 bit registers and 32 bit data bus
(*d*) 32 bit register and 16 bit data bus

94. In static RAM 6264 the number of address inputs are
(*a*) 10 (*b*) 12
(*c*) 13 (*d*) 15

95. As compared to operating current, the current drawn by a memory chip in stand by mode is
(*a*) about the same (*b*) about one-eighth
(*c*) about 1/64 (*d*) about 1/500

96. Most of the memory chips in static RAMS
(*a*) do not require any supply
(*b*) need 2 V supply
(*c*) need 5 V supply
(*d*) need 12 V supply

97. For most static RAMS the maximum access time is about
(*a*) 1 ns (*b*) 10 ns
(*c*) 100 ns (*d*) 1 μs

98. For most static RAMS the write pulse width should be at least
(*a*) 10 ns (*b*) 60 ns
(*c*) 300 ns (*d*) 1 μs

99. The read and write cycle times of cache memory are in the range of
(*a*) 1 – 3 ns (*b*) 25 ns – 45 ns
(*c*) 200 ns – 300 ns (*d*) 1 μs – 3 μs

100. 'Burst refresh' in DRAM is also called
(*a*) concentrated refresh (*b*) distributed refresh
(*c*) hidden refresh (*d*) none of the above

101. Consider the following statements about DRAM
1. Page mode read operation is faster than $\overline{RAS}$ read
2. $\overline{RAS}$ input remains active during column address strobe
3. The row and column addresses are stored into the internal buffers using $\overline{RAS}$ and $\overline{CAS}$ inputs respectively

Which of the above statements are correct?
(*a*) 1 and 3 only (*b*) all
(*c*) 1 and 2 only (*d*) 2 and 3 only

102. A nibble corresponds to
(*a*) 2 successive bits of data (*b*) 4 successive bits of data
(*c*) 8 successive bits of data
(*d*) 16 successive bits of data

103. The power requirements of a DRAM in active and standby modes is about
(*a*) 350 mW and 5 mW respectively
(*b*) 350 mW each
(*c*) 5 mW each
(*d*) 350 mW and 100 mW respectively

104. Which memory has read operation, byte erase, byte write and chip erase?
(*a*) RAM (*b*) UVEPROM
(*c*) EEPROM (*d*) Both (*b*) and (*c*)

105. Which of the following is also called Non-volatile RAM?
(*a*) Shadow RAM (*b*) UVEPROM
(*c*) EEPROM (*d*) Both (*b*) and (*c*)

106. A memory system has a 9 chips. If MTBF of one clip is 900000 hours, MTBF of complete memory system is
(*a*) 900 000 hours (*b*) 100 000 hours
(*c*) 81 00 000 hours (*d*) 1800 000 hours

107. A failure rate of 0.01 percent in 1000 hours in a memory corresponds to an MTBF of
(*a*) 1000 hours (*b*) 1000 00 hours
(*c*) 10 hours (*d*) 100 hours

108. A microprocessor is generally
(*a*) single chip SSI (*b*) single chip MSI
(*c*) single chip LSI (*d*) any of the above

109. IC (instruction cycle), FC (fetch cycle) and EC (executive cycle) are related as
(*a*) IC = FC – EC (*b*) IC = FC + EC
(*c*) IC = FC + 2EC (*d*) EC = IC + FC

110. In a microprocessor the clock signal
(*a*) is always generated internally
(*b*) is always supplied externally
(*c*) may be generated internally or supplied externally
(*d*) is mostly supplied externally

111. In a microprocessor
(*a*) all instructions need only one machine cycle
(*b*) some instructions need only one machine cycle while some other instructions need more than one machine cycle
(*c*) all instructions need only fraction of one machine cycle
(*d*) some instructions need less than one machine cycle

112. In a microprocessor
(*a*) one machine cycle is equal to one clock cycle
(*b*) one clock cycle consists of several machine cycles
(*c*) one machine cycle consists of several clock cycles
(*d*) one machine cycle is always less than one clock cycle

113. The total number of pins in 8085 are
(*a*) 20 (*b*) 30
(*c*) 40 (*d*) 50

114. Consider the instructions
1. MOV M, B
2. OUT 4
3. LDA X

Which of the above require memory write cycle?
(*a*) 1, 2, 3 (*b*) 1 and 2 only
(*c*) 1 and 3 only (*d*) 2 and 3 only

115. Consider the instructions
1. MOV M, B
2. OUT4
3. LDA X

Which of the above uses IOW cycle?
(*a*) 1 only (*b*) 1 and 2 only
(*c*) 2 only (*d*) 3 only

116. The number of interrupt lines in 8085 is
(*a*) 2 (*b*) 3
(*c*) 4 (*d*) 5

117. In 8085, high level on INTR line can be recognised only if
(*a*) TRAP and RST lines are high
(*b*) TRAP and RST lines are not high
(*c*) TRAP line is high but not RST line
(*d*) RST line is high but not TRAP line

118. In 8085
(*a*) RST 6.5 and RST 5.5 are high level sensitive interrupts
(*b*) RST 6.5 and RST 5.5 are low level sensitive interrupts
(*c*) RST 5.5 is low level sensitive interrupt but RST 5.5 is high level sensitive interrupt
(*d*) RST 5.5 is low level sensitive interrupt but RST 6.5 is high level sensitive interrupt

119. Which of the following is not an interrupt line in 8085?
(*a*) TRAP (*b*) RST 5.5
(*c*) RST 7.5 (*d*) RST 9.5

120. The number of categories in which registers are classified in μP are
(*a*) 2 (*b*) 3
(*c*) 4 (*d*) 5

121. Incompatibility between memory and I/O device may be due to
(*a*) timing according to which data transfer is to take place
(*b*) electrical characteristics of the two devices
(*c*) format of data transfer (*d*) all the above

122. The timing difference between a slow memory and fast processor can be resolved if
(*a*) processor is capable of waiting
(*b*) external buffer is used
(*c*) either (*a*) or (*b*) (*d*) neither (*a*) nor (*b*)

123. In 8085 the address space size is
(*a*) 64 k bits (*b*) 64 k bytes
(*c*) 32 k bits (*d*) 16 k bytes

124. In a typical micro computer data transfer takes place between
(*a*) μP and memory (*b*) μP and I/O devices
(*c*) memory and I/O devices (*d*) all the above

125. Which data transfer scheme is used when large block of data is to be transferred in a computer?
(*a*) DMA
(*b*) programmed data transfer
(*c*) either (*a*) or (*b*)
(*d*) neither (*a*) nor (*b*)

126. Which of the following is not a classification of programmed data transfer?
(*a*) synchronous transfer
(*b*) hand shaking
(*c*) interrupt driven transfer
(*d*) direct transfer

127. Which of the following is also called hand shaking programmed data transfer?
(*a*) synchronous transfer
(*b*) asynchronous transfer
(*c*) interrupt driven transfer
(*d*) both (*a*) and (*c*)

128. Which of the following is not a general purpose peripheral?
(*a*) I/O port
(*b*) programmable interrupt controller
(*c*) programmable CRT controller
(*d*) programmable interval timer

129. Which of the following is not a special purpose peripheral?
(*a*) programmable DMA controller
(*b*) programmable Floppy disc controller
(*c*) programmable Hard disc controller
(*d*) programmable CRT controller

130. 8255 A is a
(*a*) programmable peripheral interface
(*b*) I/O device
(*c*) memory chip (*d*) none of the above

131. 8255 A has three 8 bit ports A, B, C for peripheral data transfer. For programming 8255 A, these ports are grouped as
(*a*) Port A with Port B
(*b*) Port A with 4 most significant ports of C, port B with 4 least significant ports of C
(*c*) Port A with Port C
(*d*) none of the above

132. The operating modes of 8255 A are called
(*a*) mode 0 and mode 1
(*b*) mode 0, mode 1 and mode 2
(*c*) mode 0 and mode 2
(*d*) mode 0, mode 2 and mode 3

133. The chip 8259 is a
(*a*) programmable interrupt controller
(*b*) programmable peripheral interface
(*c*) I/O device
(*d*) memory chip

134. The chip 8257 is a
(*a*) programmable interrupt controller
(*b*) programmable DMA controller
(*c*) programmable peripheral interface
(*d*) none of the above

135. The chip 8254 is a 40 pin programmable interval timer. The number of its independent 16 bit counters is
(*a*) 2 (*b*) 3
(*c*) 4 (*d*) 5

136. The use of microprocessor control in automobiles results in
(*a*) improved fuel economy
(*b*) improved car control
(*c*) advanced information system for driver
(*d*) all the above

137. Which of these are 16 bit microprocessors?
(*a*) 80286 (*b*) MC 68000
(*c*) Z 8000 (*d*) all the above

138. The number of pins in microprocessor 8086 is
(*a*) 30 (*b*) 40
(*c*) 50 (*d*) 60

139. In 8086 the number of lines on which data and address is multiplexed is
(*a*) 8 (*b*) 16
(*c*) 20 (*d*) 32

140. In 8086 the number of bytes which can be addressed directly is about
(*a*) 1000 (*b*) 10000
(*c*) 100000 (*d*) one million

141. In 8086 the word address is
(*a*) the address of the least significant byte
(*b*) the address of most significant byte
(*c*) either (*a*) or (*b*) depending on the word
(*d*) neither (*a*) nor (*b*)

142. In 8086 a word 0A 50 FH is stored in 02 and 03 memory locations. Then
(*a*) 0FH is stored at location 02 and 0 A 5 at location 03 and address of word is 02
(*b*) 0FH is stored at location 03 and 0 A 5 at location 02 and address of word in 02
(*c*) 0FH is stored at location 02 and 0 A 5 at location 03 and address of word is 03
(*d*) 0FH is stored at location 03 and 0 A 5 at location 02 and address of word is 03

143. Match List I and List II for signal encoding for memory select in 8086 and give correct answer using the given codes

	List I		List II
	$\overline{BHE}$	A0	What is read/written
A.	0	0	1. 1 byte from/to odd address
B.	0	1	2. 1 byte from/to even address
C.	1	0	3. 1 16 bit word
D.	1	1	4. None

Codes:

	A	B	C	D
(*a*)	1	2	3	4
(*b*)	3	1	2	4
(*c*)	2	3	4	1
(*d*)	4	1	2	3

144. The number of registers and flags in 8086 are
(*a*) 13 and 5 respectively
(*b*) 9 and 5 respectively
(*c*) 13 and 9 respectively
(*d*) 9 and 9 respectively

145. In 8086 the instruction ADD AL, CH means
(*a*) the two 8 bit values in registers AL and CH are added and result placed in register AL
(*b*) the two 8 bit values in registers AL and CH are added and result placed in register CH
(*c*) the two 16 bit values in registers AL and CH are added and result placed in register AL
(*d*) the two 16 bit values in registers AL and CH are added and result placed in register CH

146. The memory segment registers in 8086 are denoted by
(*a*) AS, BS, CS, DS (*b*) BS, CS, SS, ES
(*c*) CS, DS, SS, ES (*d*) DS, ES, FS, SS

147. In 8086
(*a*) the registers SI and DI are used for string operation
(*b*) the registers CS and DS are used for string operation
(*c*) the registers SS and ES are used for string operation
(*d*) the registers SI and SP are used for string operation

148. In 8086 the flag which enables or disables external interrupts is
(*a*) IF (*b*) DF
(*c*) TF (*d*) CF

149. In microprocessor 8088
(*a*) 8 data bits are multiplexed with 8 least significant bits of address bus
(*b*) 8 data bits are multiplexed with 8 most significant bits of address bus
(*c*) 16 data bits are multiplexed with 16 bits of address bus
(*d*) none of the above

150. The register set and instruction set of 8088 is similar to that of
(*a*) 8085
(*b*) 8086
(*c*) partly 8085 and partly 8086
(*d*) none of the above

151. The number of primitive string operations in 8086 is
(*a*) 3 (*b*) 4
(*c*) 5 (*d*) 6

152. The signals in 8086 are classified into
(*a*) minimum mode and maximum mode
(*b*) decrement mode and increment mode
(*c*) low mode and high mode
(*d*) none of the above

153. The signal in 8086 are in minimum mode when
(*a*) MN/$\overline{MX}$ pin is tied to V_{CC}
(*b*) MN/$\overline{MX}$ pin is grounded

(c) MN/$\overline{MX}$ pin is left open
(d) none of the above

154. The signals in 8086 are in maximum mode when
(a) MN/$\overline{MX}$ pin is tied to V_{CC}
(b) MN/$\overline{MX}$ pin is grounded
(c) MN/$\overline{MX}$ pin is left open
(d) none of the above

155. Internally the two distinct units of 8086 are called
(a) bus interface unit and execution unit
(b) command unit and follow unit
(c) master unit and slave unit
(d) memory unit and operation unit

156. Match List I and List II and give the answer using the given codes

List I (µP)	List II (Number of bits in register)
A. 8085	1. 32 bit
B. 8086	2. 16 bit
C. 68000	3. 8/16 bit

Codes:

	A	B	C
(a)	1	2	3
(b)	2	3	1
(c)	3	2	1
(d)	3	1	2

157. Read the following statements as regards register pairs in microprocessor 8085
1. B represents B, C pair with B as high order register and C as low order register.
2. D represents D, E pair with D as high order register and E as low order register.
3. H represents H, L pair with H as high order register and L as low order register.

Which of the above statements are correct?
(a) all (b) none
(c) 1 and 2 only (d) 2 and 3 only

158. Match List I and List II as regards register pairs in microprocessor 8085 and give answer using the codes

List I (Bit Pattern)	List II (Register Pairs)
A. 00	1. SP
B. 01	2. B-C
C. 10	3. D-E
D. 11	4. H-L

Codes:

	A	B	C	D
(a)	1	2	3	4
(b)	2	3	4	1
(c)	2	3	1	4
(d)	4	3	2	1

159. Match List I and List II and give the answer using the codes

List I (Functional headings of instruction set of 8085)	List II (Functions performed)
A Branch group	1. Compares, rotates etc.
B Logic group	2. Moves, loads, stores etc.
C Data transfer group	3. Initiates conditional or unconditional jumps, calls etc.

Codes :

	A	B	C
(a)	1	2	3
(b)	2	3	1
(c)	3	1	2
(d)	1	3	2

160. Which of the following instructions in 8085 does not belong to data transfer group?
(a) LDA addr (b) STA addr
(c) XCHG (d) SUB *r*

161. The instruction MOV r_1, r_2 in 8085
(a) causes contents of registers r_2 to be moved to r_1 and uses register addressing
(b) causes contents of registers r_1 to be moved to r_2 and uses register addressing
(c) causes contents of registers r_1 to be moved to r_2 and uses register immediate addressing
(d) causes contents of registers r_2 to be moved to r_1 and uses register immediate addressing

162. The instruction MOVM, *r* in 8085
(a) causes content of register *r* to be moved to memory location whose address is in register H, L and uses register indirect addressing mode
(b) causes the content of memory location whose address is in register H, L to move to register *r* and uses register indirect addressing mode
(c) causes the content of register *r* to be moved to memory location whose address is in register H, L and uses immediate addressing mode
(d) causes the content of memory location whose address is in register H, L to move to register *r* and uses immediate addressing mode

163. In 8085 which of the following registers cannot be paired
(a) A (b) B
(c) C (d) both (b) and (c)

164. In 8085 which of the following is an 8 bit register
(a) A (b) SP
(c) PC (d) both (a) and (b)

165. Which of the following instructions in 8085 does not belong to data transfer group?
(a) L X I *rp*, data 16 (b) MOV M, *r*
(c) MVI M, data (d) ADC M

166. Which of the following 8085 instructions does not belong to arithmetic group?
(a) SHL D addr (b) SUI data
(c) ADC M (d) INR *r*

167. Which of the following 8085 instructions do not effect CY flag?
(a) SBB M (b) SBB *r*
(c) DCR M (d) ADC M

168. In which of the following 8085 instructions no condition flags are affected
(a) INX *rp* (b) INR *r*
(c) SBI data (d) INR M

169. In which of the following 8085 instructions is the 8 bit number in accumulator adjusted to form two four bit BCD numbers?
(*a*) DAD *rp* (*b*) DAA
(*c*) DCX *rp* (*d*) ADC M

170. In which of the following 8085 instructions are all condition flags except CY affected?
(*a*) INR *r* (*b*) SBB M
(*c*) ACI data (*d*) ADC *r*

171. In which group of 8085 instructions condition flags are not affected?
(*a*) data transfer group (*b*) arithmetic group
(*c*) logical group (*d*) both (*a*) and (*c*)

172. In which of the following logical group instructions of 8085 is the addressing mode 'immediate'?
(*a*) ANA *r* (*b*) ANA M
(*c*) ANI data (*d*) XRA *r*

173. In which of the following logical group instructions of 8085 do the contents of accumulator remain unchanged?
(*a*) ORA *r* (*b*) ORA M
(*c*) ORI data (*d*) CMP *r*

174. In which of the following instructions of 8085 are the contents of accumulator complemented?
(*a*) STC (*b*) CMC
(*c*) CMA (*d*) RAL

175. In which of the following 8085 instructions is the CY flag complemented?
(*a*) CMA (*b*) CMC
(*c*) STC (*d*) RRC

176. Which of the following instruction modes does not exist in microprocessor 8085?
(*a*) inherent (*b*) immediate
(*c*) absolute (*d*) indirect

177. Which of the following 8 bit microprocessors has maximum addressing modes?
(*a*) 8085 (*b*) Z 80
(*c*) 6800 (*d*) 6809

178. Which of the following methods does not cause any reduction in instruction length?
(*a*) using programme counter
(*b*) common source and destination address
(*c*) implicit source and destination address
(*d*) machine language programming

179. The addressing mode depicted in Fig. 11.20 is

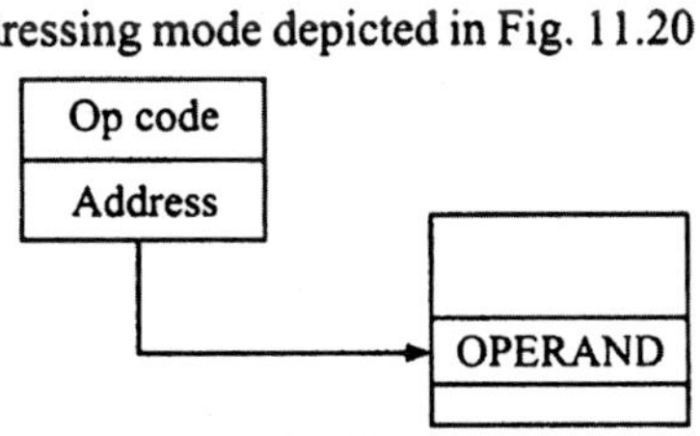

Fig. 11.20.

(*a*) direct (*b*) register
(*c*) register indirect (*d*) implicit

180. Which addressing mode is suitable only for these instructions in which there is only one operand?
(*a*) implicit (*b*) register
(*c*) direct (*d*) immediate

181. 'Rotate instruction' in microprocessor belongs to
(*a*) data transfer group (*b*) arithmetic group
(*c*) logic group (*d*) branch group

182. Which group of instructions is used to change sequence of operations in a computer program?
(*a*) data transfer group (*b*) branch group
(*c*) arithmetic group (*d*) logic group

183. The 8085 instruction 'PUSH PSW' belongs to the group
(*a*) data transfer (*b*) arithmetic
(*c*) logic (*d*) none of the above

184. Consider the following program for 8085
MVIA, OOH
MVIA, 53 H
CMA
The contents of accumulator at the end of this program will be
(*a*) OACH (*b*) 53 H
(*c*) OADH (*d*) 54 H

185. Consider the following program for 8085
MVIA, OOH
LXIH, OEAO FG
LDAZ
ADI 0005H
What does this program do?
(*a*) adds decimal number 5 to contents of memory
(*b*) subtracts decimal number 5 from contents of memory
(*c*) subtracts decimal number 15 from contents of memory
(*d*) adds decimal number 15 to contents of memory

186. Consider the following program for 8085
MOV A, D
RAL
MOVD, A
If initial contents of register D is decimal number 20, the final content of register D is
(*a*) decimal 20 (*b*) decimal 40
(*c*) decimal 10 (*d*) none of the above

187. Initially the number decimal 8 is stored. If instruction RAL is executed twice, the final number stored will be
(*a*) decimals 8 (*b*) decimal 16
(*c*) decimal 32 (*d*) decimal 2

188. Which of the following buses does not exist in 8085?
(*a*) data bus (*b*) address bus
(*c*) control bus (*d*) exchange bus

189. The operations which occur most frequently in a computer are
(*a*) fetch (*b*) execute
(*c*) both (*a*) and (*b*) (*d*) none of the above

190. Which of the following instructions have only one operand?
(*a*) multiply (*b*) add
(*c*) shift (*d*) subtract

191. The addressing mode depicted in Fig. 11.21 is

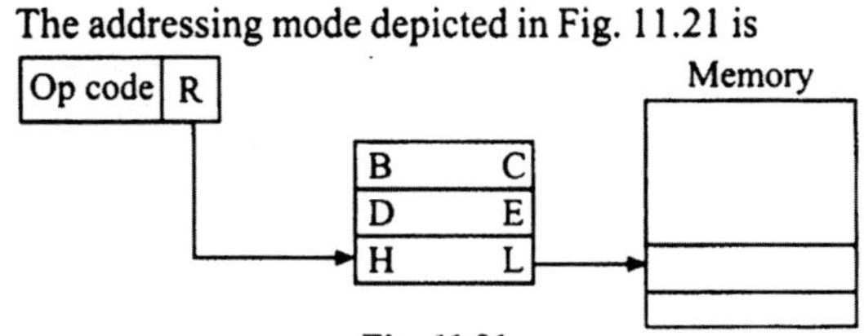

Fig. 11.21.

(*a*) direct (*b*) register
(*c*) register indirect (*d*) immediate

192. What does the instruction 'PCHL' in 8085 do?
(*a*) load the program counter with contents of register pair HL
(*b*) Double the contents of register pair H L
(*c*) shift the contents of register pair H L by one bit to the right
(*d*) shift the contents of register pair H L by one bit to left.

193. Which 8085 instruction causes exchange of contents of stack pointer with the contents of H L register pair
(*a*) XTHL (*b*) PCHL
(*c*) LHLD (*d*) MVIA

194. ALU IC 74181 can accept two words each having
(*a*) 2 bits (*b*) 4 bits
(*c*) 8 bits (*d*) 16 bits

195. The number of select lines in ALU IC 74181 is
(*a*) 2 (*b*) 4
(*c*) 8 (*d*) 16

196. In ALU IC 74181 the addition and subtraction is in
(*a*) 2's complement (*b*) 1's complement
(*c*) binary (*d*) none of the above

197. When two ALU IC 74181 are cascaded
(*a*) the carry out of first stage is connected to carry in of second stage
(*b*) the carry in of first stage connected to carry in of second stage
(*c*) the carry in of first stage is connected to carry out of second stage
(*d*) the carry out of first stage is connected to carry out of second stage

198. The number of logic or arithmetic operations which ALU IC 74181 can carry out is
(*a*) 4 (*b*) 8
(*c*) 16 (*d*) 32

199. A microprocessor uses 3 MHz oscillator. The duration of one T state is
(*a*) 1 μs (*b*) 0.666 μs
(*c*) 0.333 μs (*d*) 3 μs

200. The contents of F register in 8085 are 01010001. This means
(*a*) S = 0, Z = 1, AC = 1, P = 0 and CY = 1
(*b*) S = 1, Z = 1, AC = 1, P = 0 and CY = 1
(*c*) S = 1, Z = 0, AC = 1, P = 0 and CY = 1
(*d*) S = 1, Z = 0, AC = 0, P = 0 and CY = I

201. Which of the following is not an integer constant?
(*a*) 0 (*b*) −8
(*c*) +1074 (*d*) −8.1

202. Which of the following is a valid integer constant?
(*a*) 17.0 (*b*) −18.58
(*c*) 10, 114 (*d*) −810

203. Which of the following is not a valid real constant?
(*a*) −73.1 (*b*) 10, 251.3
(*c*) 10.23 (*d*) −5.7381

204. Which of the following is a valid real constant?
(*a*) −81.56 (*b*) 73
(*c*) 10, 254.1 (*d*) 42

205. The standard Fortran exponential form of decimal number 0.0000437 is
(*a*) 4.37 E−5 (*b*) 4.37 E−4
(*c*) 43.7 E−5 (*d*) 43.7 E−4

206. The standard Fortran exponential form of decimal number −0.000000023 is
(*a*) 2.3 E−8 (*b*) −2.3 E−8
(*c*) 2.3 E−7 (*d*) −2.3 E−7

207. The length of character constant P. *b* *b* .E. *b* C. is
(*a*) 9 (*b*) 6
(*c*) 10 (*d*) 7

208. Which of the following is a valid variable in Fortran 77?
(*a*) 3 A (*b*) ALPHA–3
(*c*) ALPHA 3 (*d*) BALPHA C

209. In Fortran 77 the total number of characters in a variable can be a maximum of
(*a*) 6 (*b*) 7
(*c*) 8 (*d*) 9

210. Which of the following is not a valid variable in Fortran 77?
(*a*) ERG (*b*) VS
(*c*) VS 3 A (*d*) 3 VSA

211. The statement IMPLICIT INTEGER (A – F) in Fortran identifies
(*a*) all variable names which begin with A, B, C, D, E or F as being of integer type
(*b*) all character names which begin with A, B, C, D, E, or F as being of integer type
(*c*) all variable names which begin with A and have 6 characters
(*d*) none of the above

212. Which one of the following is valid integer variable in Fortran 77?
(*a*) BRG (*b*) IBRG
(*c*) VS (*d*) PEC

213. Which of the following is not valid integer variable in Fortran 77?
(*a*) MBR (b) LOTS
(*c*) JAMMU (*d*) BRG

214. Which of the following is a valid real variable in Fortran 77?
(*a*) KPR (*b*) LINE
(*c*) JOB (*d*) RATE

215. Which of the following is not a valid real variable in Fortran 77?
(*a*) ANUT (*b*) MONEY
(*c*) AMONEY (*d*) BS

216. The correct form of expression $B^3 - 2AC$ in Fortran 77 is
(*a*) B ** 3 − 2 * A * C (*b*) B * 3 − 2 * A AC
(*c*) B ** 3 − 2 ** A * C (*d*) B *** 3 − 2 * A * C

217. Which of the following is a valid integer constant?
(*a*) 127 (*b*) 127.0
(*c*) 127. (*d*) 125 + 3

218. Which of the following is a valid real constant?
(*a*) 1281 (*b*) 1281.0
(*c*) K M (*d*) 11821

219. Which of the following is a valid integer variable in Fortran 77?

(a) K.M (b) KM – 1
(c) DELHI (d) $318

220. Which of the following is a valid real variable in Fortran 77?
(a) IT (b) GURGAON
(c) 6JUNE (d) SUNDAY

221. Which of the following arithmetic operations requires only one operand in Fortran 77?
(a) Addition (b) Multiplication
(c) Negation (d) Substraction

222. Which of the following is a valid expression in Fortran 77?
(a) B * – 3 (b) B * (– 3)
(c) B ** – 3 (d) B * + 2

223. In the expression 3 * * 2 ** 4 ** 2 in Fortran 77 which exponentiation is performed first
(a) 4 * * 2
(b) 2** 4
(c) 3 ** 2
(d) all exponentiations are performed together

224. In Fortran 77 the correct way to write mathematical expression $\frac{a}{bc} - d^2$ is
(a) A/(B * C) – D ** 2
(b) A/B – C – D**2
(c) A/BC – D**2
(d) none of the above

225. Which of the following is not a proper Fortran expression?
(a) B + A/C – D (b) (A + C)/(B + D)
(c) B + *A/C (d) A ** (B + 2)

226. In Fortran 77 FLOAT (– 28) gives the value
(a) 28 (b) – 28
(c) – 28.0 (d) none of the above

227. In Fortran 77 IFIX (– 73.2) gives the value
(a) 73 (b) – 73
(c) – 7 (d) – 73.2

228. Fortran expression for $\sqrt{3b^2 - 5a^2}$ is
(a) SQRT (3 * B ** 2 – 5 * A ** 2)
(b) SQRT 3 *B ** 2 – 5 * A ** 2
(c) SQRT(3 * b ** 2 – 5 * a ** 2)
(d) SQRT (3 * B ** 2) – SQRT (5 *A **2)

229. Which of the following is not an integer constant?
(a) 2.3 (b) 156231
(c) 10 (d) – 256

230. The proper exponential form of – 0.0032 in Fortran 77 is
(a) – 3.2E – 3 (b) – 0.0032
(c) – 0.00032 E1 (d) – 0.032 E – 1

231. Which of the following is a valid real constant in Fortran 77?
(a) 1.2 E 231 (b) 3E – 2
(c) – 3.2 E 2.3 (d) 1.2 E 23

232. Which of the following is a valid integer variable in Fortran 77?
(a) ALPHA (b) NEXT
(c) BANGLORE (d) ABC

233. Which of the following is a valid Fortran assignment statement?
(a) 10 = N (b) X + 2.1 = 5.6
(c) A = B (d) A = B = 1

234. If PRIN = 200.0 and BINT = 0.0525 the result of expression INCOME = PRIN * (1 + BINT) will be
(a) 210 (b) 210.5
(c) 210.5000 (d) 210

235. If INT = 50 and LOOT = 30, the result of assignment statement SAL = INT + LOOT is
(a) 80 (b) 80.
(c) 8.0 (d) .80

236. If F = 95.0 what will be the result of following Fortran 77 program?
REAL F, C
READ *, F
C = (5./9.)*(F – 32.)
(a) 35.0 (b) 35
(c) 25.0 (d) 25

237. If KMT = 100, LITRES = 5.0 , PCOST = 28.0 and OPCOST = 70.0 what will he the result of the following Fortran 77 program?
INTEGER KMT
REAL LITRES, PCOST, OPCOST, PKCOST READ *, KMT, LITRES, PCOST, OPCOST, PKCOST
KMPL = KMT/LITRES
PKCOST = (PCOST * LITRES + OPCOST)/ KMT
(a) 2 (b) 2.10
(c) 21 (d) 21.0

238. If RADIUS = 3.0 the result of the following Fortran 77 program will be
REAL RADIUS, CAREA, SAREA, RATIO, PI
PI = 3.1416
READ*, RADIUS
CAREA = PI * RADIUS * RADIUS
SAREA = 2 * RADIUS * RADIUS
RATIO = CAREA/SAREA
(a) 1.5708 (b) 1
(c) 28.2744 (d) 18.0

239. If N = 196, the result of the following program will be
INTEGER N, DIGIT 1, DIGIT 2, DIGIT 3, SUM
READ *, N
DIGIT 1 = N – (N/10)* 10
N = N/10
DIGIT 2 = N – (N/10)* 10
N = N/10
DIGIT 3 = N – (N/10)* 10
SUM = DIGIT 1 + DIGIT 2 + DIGIT 3
(a) 196 (b) 16
(c) 16.0 (d) 196.0

240. If J = 12 and X = 24.4 the result of the following Fortran 77 program will be
READ *, J, X
K = J**2
Z = 3 * X
Print *, J, K, Z
(a) 12, 144, 73.2 (b) 12.0, 144.0. 73.2
(c) 12.0, 144.0, 73 (d) 12. 144. 73

241. The forms of IF statements in Fortran 77 are called
(a) logical IF and block IF
(b) logical IF, block IF and arithmetic IF
(c) logic IF, block IF, arithmetic IF and negate IF
(d) logical IF and arithmetic IF

242. Consider the following logical IF statement in Fortran 77
IF (SALT. EQ. PEPPER) GO TO 11
GO TO 13
The above statement using arithmetic IF statement would be
(a) IF (SALT – PEPPER) 13, 11, 13
(b) IF (SALT – PEPPER) 11, 11, 13
(c) IF (SALT – PEPPER) 11, 13, 13
(d) IF (SALT – PEPPER) 13, 11, 11

243. Consider the following logical IF statement in Fortran 77
IF (SALT. GE. PEPPER) GO TO 11
GO TO 13
The above statement using arithmetic IF statement would be
(a) IF (SALT – PEPPER) 11, 11, 13
(b) IF (SALT – PEPPER) 13, 11, 13
(c) IF (SALT – PEPPER) 13, 11, 11
(d) IF (SALT – PEPPER) 11, 13, 13

244 Consider the following logical IF statement in Fortran 77
IF (SALT. LE. PEPPER) 11
GO TO 13
The above statement using arithmetic IF statement would be
(a) IF (SALT – PEPPER) 11, 11, 13
(b) IF (SALT – PEPPER) 13, 11, 11
(c) IF (SALT – PEPPER) 13, 13, 11
(d) IF (SALT – PEPPER) 11, 13, 13

245. Let JCOKE = 3 and LPEPSI = 5
consider the statement
IF (3 * JCOKE. EQ. LPESPSI)
JCOKE = JCOKE + 2
JCOKE = J COKE + 5
What will be the final value of JCOKE after the above statement is executed?
(a) 8 (b) 7
(c) 15 (d) 10

246. Let JCOKE = 20 and LPEPSI = 15. Consider the statement
IF (2 * JCOKE. LE. LPEPSI)
JCOKE = JCOKE + 5
JCOKE = JCOKE + 3
The value of JCOKE after the execution of above statement will be
(a) 25 (b) 23
(c) 15 (d) 17

247. Let JCOKE = 11 and LPEPSI= 19. Consider the statement
IF (JCOKE – LPEPSI) 5, 5, 7
5 JCOKE = JCOKE + 2
7 JCOKE = JCOKE + 5
The value of JCOKE after the execution of above statement will be
(a) 13 (b) 16
(c) 9 (d) 14

248. Let JCOKE = 98 and LPEPSI = 42. Consider the statement
IF (JCOKE – 3 * LPEPSI) 5, 6, 7
5 JCOKE = JCOKE + 5
6 JCOKE = JCOKE + 8
7 JCOKE = JCOKE + 11
The value of JCOKE after the execution of above statement will be
(a) 103 (b) 106
(c) 109 (d) none of the above

249. JCOKE = 3
JCOKE = JCOKE + 1
GO TO (5, 8, 9, 11, 15, 16 18, 20) JCOKE
After the execution of above statement, the control is transferred to statement number
(a) 8 (b) 11
(c) 16 (d) 20

250. LPEPSI = 2
2EPSI = LPEPSI * LPEPSI + 1
GOT TO (3, 5, 9, 11, 13, 17, 21, 23) LPEPSI
After the execution of above statement the control is transferred to statement number
(a) 3 (b) 9
(c) 13 (d) 23

251. Which of the following DO statements is invalid in Fortran 77?
(a) DO 16 I = 1, 15 (b) DO 16 I = 1, 10, 1
(c) DO 16 I = – 5, 5, 2 (d) none of the above

252. Consider the following DO statement in Fortran 77
DO 25 J= 1, 7
The number of DO loop executions in the above statement is
(a) 7 (b) 6
(c) 1 (d) 8

253. Consider the following DO statement in Fortran 77
DO 52 N = 1, 10, 3
The number of DO loop executions in the above statement is
(a) 1 (b) 10
(c) 3 (d) 4

254. Consider the following DO statement in Fortran 77
DO 12 A = 1.0, 5.5, 0.5
The number of DO loop executions in the above statement is
(a) 1 (b) 11
(c) 10 (d) 4

255. Consider the following DO statement in Fortran 77
DO 23 X= 10.0, 2.0
The number of DO loop executions in the above statement is
(a) 10 (b) 2
(c) 5 (d) zero

256. A computer program is used to read N and print the sum $1^2 + 2^2 + 3^2 + ... + N^2$. If N = 10, the print out will show the number
(a) 100 (b) 10
(c) 385 (d) 285

257. A computer program is used to read M and print the sum 2 + 4 + 6 + 8 + ... + 2M. If M = 9, the printout will show the number

(*a*) 9 (*b*) 18
(*c*) 90 (*d*) 100

258. A computer program is used to read N and compute N !. IF N = 6, the result will be

(*a*) 6 (*b*) 12
(*c*) 720 (*d*) 1040

259. Which of the following DO loop instructions is invalid in Fortran 77

(*a*)
```
DO 20 J = .....
:
DO 10 K =.....
:
10 CONTINUE
:
20 CONTINUE
```
(*b*)
```
DO 20 J =......
:
DO 10 K =.......
:
20 CONTINUE
:
10 CONTINUE
```
(*c*)
```
DO 30 I =......
:
30 CONTINUE
```
(*d*)
```
DO 40 N =......
:
40 CONTINUE
```

260. Read the following Fortran 77 program

```
MIX = 2
DO 13 LIX = 3, 8, 2
MIX = MIX + LIX
IF (MIX. GT.6) GO To 25
13 CONTINUE
25 MIX = 3 * MIX
```
The final result will be

(*a*) 10 (*b*) 20
(*c*) 30 (*d*) 15

261. Which of the following array names in Fortran 77 is invalid?

(*a*) A (6) (*b*) A(R)
(*c*) A (I) (*d*) A(N)

262. Which of the following array names is invalid in Fortran 77?

(*a*) B (2, 4) (*b*) B (M, 3)
(*c*) B (M, R) (*d*) B (3, 6)

263. Consider the following statement in Fortran 77
READ*, (X(I), I = 0, 4)
Which of the following is correct?

(*a*) 3 values would be read and stored in X (0), X (1), X (2)
(*b*) 4 values would be read and stored in X (0), X (1), X (2), X (3)
(*c*) 5 values would be read and stored in X (0), X (1), X (2), X (3), X (4)
(*d*) 5 values would be read and stored in X (1), X (2), X (3), X (4), X (5)

264. Consider the following statement in Fortran 77
READ*, [X (I), Y (I)], I = 0, 3, 2]
Which of the following is correct?

(*a*) 2 values could be read and stored as X (0), Y (0)
(*b*) 4 values could be read and stored as X (0), Y (0), X (2), Y (2)
(*c*) 6 values could be read and stored as X (0), Y (0) X (2) Y (2), X (3), Y (3)
(*d*) none of the above

265. Which of the following READ statements is correct in Fortran 77?

(*a*) READ *, HONEY (I), I = 1, 5
(*b*) READ *, (HONEY (I), I = 1, 5)
(*c*) READ * HONEY I, I = 1, 5
(*d*) READ HONEY I, I = 1, 5

266. Which of the following is correct in Fortran 77?

(*a*) PARAMETER PMOS = 9.8, NMOS = X
(*b*) PARAMETER (PMOS = 9.8, NMOS = X)
(*c*) PARAMETER PMOS = 9.8 NMOS = X
(*d*) PARAMETER PMOS, NMOS = 9.8, X

267. Read the following statements

1. Magnetic tape is a sequential device
2. Magnetic disc is a direct access device
3. Magnetic tape and disc are both sequential devices
4. Magnetic tape and disc are both direct access devices

Which the above statements are correct?

(*a*) 1 and 2 only (*b*) 1 and 3 only
(*c*) 2 and 4 only (*d*) 1, 2 and 3 only

268. Which of the following statements are executable in Fortran 77?

(*a*) arithmetic IF (*b*) assign
(*c*) call (*d*) all the above

269. Which of the following statements is non-executable in Fortran 77?

(*a*) DO (*b*) ASSIGN
(*c*) END (*d*) DIMENSION

270. Which of the following statements in Fortran 77 is non-executable?

(*a*) GOTO (*b*) RETURN
(*c*) STOP (*d*) PARAMETER

271. Consider the following symbols

1. letters A to Z and *a* to *z*
2. digits 0 to 9
3. logical operators AND, OR, NOT
4. Parenthesis ( )

Which of the above symbols are available in Pascal?

(*a*) 1 and 2 only (*b*) 1,2 and 3 only
(*c*) 1, 2 and 4 only (*d*) all

272. Consider the following strings

1. A
2. Z2
3. X3Y
4. P12R
5. BX

Which of the above strings are valid in Pascal?

(*a*) all (*b*) 1, 2, 3
(*c*) 1, 2, 4, 5 (*d*) 1, 2, 3, 5

273. Which of the following identifiers is invalid in Pascal?

(*a*) BETA (*b*) MAX 40
(*c*) 2ND (*d*) A MAX

274. Which of the following identifiers is valid in Pascal

(*a*) THETA (*b*) NUM.
(*c*) 1 ST (*d*) $BETA

275. Consider the following symbols for arithmetic operators for integers

1. +
2. div
3. mod
4. /

Which of the above are used in Pascal?

(*a*) 1, 2 and 4 only (*b*) all
(*c*) 1, 2, 3 only (*d*) 1 and 4 only

276. Which of the following integer expressions is correct in Pascal?
(*a*) – A + B + M (*b*) A div B* mod C
(*c*) A – 4.0 (*d*) A/B + D

277. Which of the following integer expressions is incorrect in Pascal?
(*a*) L div M + P (*b*) A * * B
(*c*) A div B – C (*d*) A mod B

278. Which of the following real expressions is incorrect in Pascal
(*a*) Y – X (*b*) (A + B)(C + D)
(*c*) X/Y – Z (*d*) P * Z – 5.0

279. Which of the following real expressions is correct in Pascal
(*a*) – Z + P * X (*b*) (X + Y)(A + B)
(*c*) Z ** M (*d*) ZP * M

280. In the integer expression – A * B div C + D which operation is done first in Pascal
(*a*) – (*b*) *
(*c*) div (*d*) +

281. Consider the real expression in Pascal language
((X * Z) + Y/(M * P + K) + X)/Y+Y)
What would be evaluated first?
(*a*) XZ and MP + K (*b*) XZ + Y and MP + K
(*c*) (MP + K + X)/Y (*d*) Y/(MP + K)

282. The expression $A^2 + B^2 - 3\,AB$ when written is Pascal should be written as
(*a*) A * A + B * B – 3 A * B
(*b*) A * A + B * B – 3.0 A * B
(*c*) A * A + B * B – 3.0*A * B
(*d*) A * A + B * A – 3. *A * B

283. The correct way of writing the expression $\frac{NV^2}{2} + \frac{GH}{3D}$ in Pascal is
(*a*) N * V * V/2 + G * H/3D
(*b*) N * V * V/2.0 + G * H/(3.0 * D)
(*c*) N * V * V/2. + G * H/3.0 * D
(*d*) N * V * V/2.0 + G * H/3.0 * D

284. When one operand is real and one is integer in Pascal, which of the following operations are not allowed
(*a*) + and – (*b*) *
(*c*) / (*d*) div and mod

285. Consider the following library functions
1. EXP (*i.e.*, e^x)
2. ARCTAN (*i.e.*, $\tan^{-1} x$)
3. SQRT (*i.e.*, $\sqrt{x}$)
4. ABS (*i.e.*, $|x|$)
Which of the above are allowed in Pascal
(*a*) all (*b*) 1 and 3 only
(*c*) 1, 3, 4 only (*d*) 1, 2 and 3 only

286. It is desired to evaluate the expression
$F = B\cos Y + C\cos^2 Y + A\cos^3 Y$
The Pascal equivalent to ensure minimum computer time should be written as
(*a*) F = B * COS (Y) + C * SQR (COS (Y) + A * COS (Y) * COS (Y) * COS (Y)
(*b*) F = B * COS (Y) + C * SQR [COS (Y)] + A * COS (Y) * SQR [COS (Y)]
(*c*) F = B * COS (Y) + C * COS(Y) * COS (Y) + A * COS (Y) * COS (Y) * COS (Y)
(*d*) X = COS (Y)
Z = SQR (X)
F = B * X + C * Z + A * X * Z

287. For the expression $\frac{\beta}{\sqrt{\beta^2 + \omega^2}} \cos(\omega t + \theta)$ the correct Pascal equivalent is
(*a*) BETA* COS (OMEGA * T + THETA)/ SQRT (SQR (BETA) + SQR (OMEGA))
(*b*) BETA COS (OMEGA * T + THETA) / SQRT [SQR (BETA) + SQR (OMEGA)]
(*c*) BETA COS (OMEGA T + THETA) / SQRT [SQR (BETA) + SQR (OMEGA)]
(*d*) none of the above

288. Match List I and List II and give correct answer using the codes

List I (Library function in Pascal)	List II (Type of argument)
A ROUND X	1. X in radians
B ABS X	2. X real
C SQR X	3. X real or integer
D SIN X	

Codes:

	A	B	C	D
(*a*)	2	3	3	1
(*b*)	2	3	1	3
(*c*)	3	2	2	1
(*d*)	2	3	2	1

289. If X = 2.5 and; Y = 2.5, which of the following expressions would give the result as 5/7
(*a*) X + 2.5/Y + 3.0 (*b*) (X + 2.5)/Y + 4.5
(*c*) (X + 2.5)/(Y + 4.5) (*d*) (X + 2.5)/Y + 3.5

290. Consider the following program in Pascal
X = 2.57
X = (X + 0.06) * 10
I = TRUNC (X)
X = I
X = X/10.0
The final value of X would be
(*a*) 2 (*b*) 2.6
(*c*) 2.63 (*d*) 26.3

291. In Pascal if I = 45, I div 10 * 10 =
(*a*) 45 (*b*) 40
(*c*) 0.45 (*d*) 0.4

292. In Pascal if I = 55, 10 * I div 10 =
(*a*) 55 (*b*) 50
(*c*) 5.5 (*d*) 5

293. If X = – 285.56 and the Pascal write statement used is WRITE (X : 8). Then the print out will appear as
(*a*) – 285.560 (*b*) – 2.85 E + 02
(*c*) – 2.85560 E + 02 (*d*) – 2.8 E + 02

294. The relational operators allowed in Pascal are
(*a*) =, >, <, > =, < = and < > (*b*) =, >, < and < >
(*c*) =, >, <, > = and < = (*d*) >, <, > = and < =

295. Which of the foilowing is not a valid logical expression in Pascal?

(*a*) A>=B (*b*) A<=B
(*c*) A=<B (*d*) A<>B

296. Which of the following is a valid logical expression in Pascal?

(*a*) 2.5:=A (*b*) (A+B)<<(C+D)
(*c*) A=<B (*d*) A>=B

297. Which of the following conditional statements is not valid in Pascal?

(*a*) if A>=B then B:=C+D+X
(*b*) if(N=45) THEN X:=X+2 ELSE X:=0
(*c*) if A<=B then A:=C*D
(*d*) if(A>=B) X:Y else X:=Z

298. In Pascal the repeat and while loops will terminate

(*a*) only when the logical expression controlling the loop becomes true for both repeat and while loops
(*b*) only when the logical expression controlling the loop becomes false for both repeat and while loops
(*c*) only when the logical expression controlling the loop becomes true for repeat loop and false for while loop
(*d*) only when the logical expression controlling the loop becomes false for repeat loop and true for while loop

299. Consider the series

$$\text{SUM} = x - \frac{x^3}{3!} + \frac{x^5}{5!} - \frac{x^7}{7!} + \ldots + (-1)^{n-1}\frac{x^{2n-1}}{(2n-1)!}$$

Which of the following is true for above series?

(*a*) $(i+1)^{\text{th}}$ term $=(-1)\dfrac{x}{(2i+1)(2i)} i^{\text{th}}$ term

(*b*) i^{th} term $=(-1)\dfrac{x}{(2i+1)(2i)}(i+1)^{\text{th}}$ term

(*c*) $(i+1)^{\text{th}}$ term $=\dfrac{(-1)\,x}{(2i+1)(2i)} i^{\text{th}}$ term

(*d*) $(i+1)^{\text{th}}$ term $=\dfrac{(-1)\,x^3}{(2i+1)(2i)} i^{\text{th}}$ term

300. In setting up 'for' loops in Pascal, the initial and final values of index may be

(*a*) integer constants
(*b*) Integer constants or integer variable names
(*c*) integer constants or integer variable names or integer expressions
(*d*) integer constants or real constants

301. Which one is true for writing array variable in Pascal?

(*a*) The general form of an array variable is an integer or a real variable name followed by subscripts enclosed within square brackets. The subscripts are separated by commas.
(*b*) The general form of an array variable must be an integer name followed by subscripts enclosed within square brackets. The subscripts are separated by commas.
(*c*) The subscripts in array variable must be separated by dots.
(*d*) The subscripts must be enclosed within parenthesis.

302. In Pascal an array declaration provides information to the Pascal processor to enable it to

(*a*) identify the variable as an array name
(*b*) reserve memory locations to store all the components of array
(*c*) check the values computed for subscripts expressions during program execution with the specified lower and upper limits of subscripts
(*d*) all the above

303. Read the following rules for scanning Boolean expressions in Pascal

1. The expressions are scanned from left to right
2. The operations are scanned in the order NOT, AND, OR
3. All operators (*i.e.*, NOT, AND and OR) are scanned together
4. In one scanning all the operators of only one type are scanned

Which of the above are correct?

(*a*) all (*b*) 1, 2 and 3 only
(*c*) 1, 2 and 4 only (*d*) 2, 3 and 4 only

304. Read the following statements about sets

1. If A and B are sets then A + B denotes union of sets A and B
2. The union of A and B is that set whose elements are in either A or B or both
3. If A and B are sets, then difference of sets A and B is denoted by A – B
4. A – B in the set obtained by removing elements of A from B

Which of the above are true?

(*a*) all (*b*) 1, 2 and 3 only
(*c*) 2 and 3 only (*d*) 1 and 4 only

305. Match List I and List II and give correct answer using the given code

List I (Set operation)	**List II** (Set operator)
A. Set equality	1. –
B. Set inclusion	2. =
C. Set intersection	3. <=
D. Set difference	4. *

Codes :

	A	B	C	D
(*a*)	2	3	4	1
(*b*)	3	2	4	1
(*c*)	1	2	3	4
(*d*)	4	2	3	1

306. Consider the following statements about In-fix and Post-fix notations in Pascal :

1. In In-fix notation operators appear between the operands on which they operate
2. In Post-fix notation operators appear after the operand
3. The In-fix expression A + B – C + D will appear as AB + C – D + in post-fix notation

Which of the above are correct?

(*a*) all (*b*) 1 and 2 only
(*c*) 1 and 3 only (*d*) 2 and 3 only

307. Read the following statements about files in Pascal :

1. A file is a data structure which consists of a sequence of components of the same type
2. The number of components in a file is variable
3. The components in a file can be accessed only sequentially starting from the beginning of the file.

Which of the above are correct?

(*a*) all (*b*) 1 and 2 only
(*c*) 2 and 3 only (*d*) 1 and 3 only

308. Consider the following statements about an array in Pascal

1. An array has a fixed predetermined number of components
2. Any component of an array can be accessed using an index
3. An array variable name refers to a group of quantities by a single name
4. When array variables are used an appropriate var declaration is required

Which of the above are correct?

(*a*) all (*b*) 1, 2, 3 only
(*c*) 2, 3, 4 only (*d*) 1, 2 and 4 only

309. Read the following statements about files in Pascal

1. A file can grow or shrink dynamically
2. A file which consists of only characters is called text file
3. Text files may be read and written using standard Pascal READ and WRITE procedures
4. Every file must have a beginning but may or may not have an end

Which of the above are correct

(*a*) all (*b*) 1, 2, 3 only
(*c*) 2, 3, 4 only (*d*) 1, 2 and 4 only

310. Consider the following statements about data structures in Pascal

1. Arrays and files are types of data structures
2. List structure is not a data structure in Pascal
3. Each element of a binary tree is called a node of the tree

Which of the above statements are correct?

(*a*) all (*b*) 1 and 2 only
(*c*) 1 and 3 only (*d*) 2 and 3 only

311. For a 16 bit computer the range of integer constants in C is

(a) -32768 to $+32767$ (*b*) -32767 to $+32768$
(*c*) -65536 to $+65535$ (*d*) -65535 to $+65536$

312. Consider the following rules about integer constant in C

1. An integer constant must have at least one digit
2. An integer constant must not have a decimal point
3. An integer constant may be positive or negative
4. The default sign is positive
5. A maximum of one comma or blank is allowed

Which of the above are correct?

(*a*) all (*b*) 1, 2, 3, 4 only
(*c*) 1, 2, 3, 5 only (*d*) 2, 3, 4, 5 only

313. Which of the following is not a valid integer constant in C (for 16 bit)?

(*a*) 321 (*b*) -162
(*c*) -65321 (*d*) $+1$

314. Which of the following is not a valid real constant in C?

(*a*) 426 (*b*) 321.0
(*c*) -621.231 (*d*) $+201.2314$

315. In C the range of real constants expressed in exponential form is

(*a*) $-2.3\,e\,37$ to $+2.3\,e\,37$ (*b*) $-3.4\,e\,38$ to $+3.4\,e\,38$
(*c*) $-3.4\,e\,37$ to $+3.4\,e\,37$ (*d*) $-2.3\,e\,38$ to $+2.3\,e\,38$

316. Consider the following statements about expressing real constants in exponential form in C

1. The mantissa and exponent are separated either by *e* or E
2. The mantissa and exponent may have positive or negative sign
3. Default sign of mantissa is positive
4. Default sign of exponent is negative

Which of the above statements are wrong?

(*a*) 1 only (*b*) 1 and 4 only
(*c*) 1, 3 and 4 only (*d*) 4 only

317. Which of the following is not a valid real constant in C?

(*a*) $1.2\,e\,2.0$ (*b*) $-2.3\,e-2$
(*c*) $-2.3\,e\,2$ (*d*) $-2.3\,\mathrm{E}\,2$

318. The maximum number of alphabets/digits or special symbols for character constants in C is

(*a*) 1 (*b*) 2
(*c*) 4 (*d*) 8

319. Which of the following is nota valid character constant in C?

(*a*) 'A' (*b*) '2'
(*c*) '=' (*d*) 2

320. The maximum number of characters for variable names in C is

(*a*) 4 (*b*) 8
(*c*) 12 (*d*) 16

321. Which of the following is not a valid variable name in C?

(*a*) 1*a* (*b*) *a* 12
(*c*) *ab* 123 (*d*) *abc* 123

322. Consider the following rules for constructing variable names in C :

1. It can have alphabets, digits or underscores but no other special character
2. The maximum number of characters is 8
3. The first character may be an alphabet or digit but not underscore

Which of the above are correct?

(*a*) all (*b*) 1 and 2 only
(*c*) 1 and 3 only (*d*) 2 and 3 only

323. Which of the following is not a valid variable name in C?

(*a*) *p_ce*1 (*b*) 1 *p_ce*
(*c*) *p*1*_ce* (*d*) *p_ced*

324. Which of the following is not a valid variable name in C?

(*a*) *p o, t* (*b*) *p o t*
(*c*) *po_t* (*d*) *p_ot*

325. Which of the following can not be used as a variable name in C?

(*a*) else (*b*) coal
(*c*) ram (*d*) vendy

326. Which of the following is not a keyword in C?
(*a*) pop (*b*) for
(*c*) if (*d*) union

327. The number of keywords in C is
(*a*) 4 (*b*) 8
(*c*) 16 (*d*) 32

328. In C the keywords are also called
(*a*) special words (*b*) reserved words
(*c*) class words (*d*) character words

329. Which of the following characters is allowed in variable names in C?
(*a*) comma (*b*) dot
(*c*) blank (*d*) underscore

330. Which of the following is not a key word in C?
(*a*) int (*b*) float
(*c*) return (*d*) capital

331. Which of the following is type declaration statement in C?
(*a*) int bar (*b*) $s = s + 1$
(*c*) king = horse +1 (*d*) prin = prin * prin

332. An arithmetic statement in C must be
(a) integer mode arithmetic statement
(*b*) real mode arithmetic statement
(*c*) mixed mode arithmetic statement
(*d*) any of the above

333. Which of the following is not a valid arithmetic statement in C?
(*a*) $x = x + 1$ (*b*) $x = y * z$
(*c*) $x * y = z$ (*d*) $z = y/x$

334. Consider the following C program
int ram
float alpha, gamma
ram = 6400
alpha = 0.562
gamma = alpha * ram + 1
Now consider the following statements about this program
1. ram is an integer variable
2. alpha, gamma are real variables
3. *, + are arithmetic operators
4. 0.562 may be an integer constant or real constant
Which of the above are correct?
(*a*) all (*b*) 1, 2, 3 only
(*c*) 2, 3, 4 only (*d*) 1, 2, 4 only

335. Which of the following operations will give the result 2 in C?
(*a*) 5/2 (*b*) 4.0/2
(*c*) 4.0/2.0 (*d*) 4/2.0

336. Consider the following in C :
1. An arithmetic operation between integer and an integer gives integer as the result
2. An arithmetic operation between a real and a real constant gives real constant as the result
3. An arithmetic operation between an integer constant and a real constant is not valid
Which of the above are correct?
(*a*) all (*b*) 1 and 2 only
(*c*) 2 and 3 only (*d*) 1 and 3 only

337. Which of the following is not correct in C?
(*a*) 6/4 = 1 (*b*) 2/5 = 0
(*c*) 2.0/5 = 0 (*d*) 2.0/5.0 = 0.4

338. Consider the expression $b = acd\ (xz)$. The correct C statement for above expression is
(*a*) $acd\,(x * z)$ (*b*) $a * c * d * (x\,z)$
(*c*) $a * c * d\,(x * z)$ (*d*) none of the above

339. Consider the following in C
1. Arithmetic operations can be performed on ints
2. Arithmetic operations can be performed on floats
3. Arithmetic operation can be performed on chars
Which of the above are correct?
(*a*) all (*b*) 1 and 2 only
(*c*) 1 and 3 only (*d*) 2 and 3 only

340. Consider the following in C
int k ;
float a ;
$k = 4.2$
$a = 300$
The values stored in computer memory for k and a are respectively
(*a*) 4.2 and 300
(*b*) 4 and 300 respectively
(*c*) 4.2 and 300.00 respectively
(*d*) 4 and 300.00 respectively

341. Consider the following program in C
int k;
float b ;
$k = 3$
$b = 10.1$
$k = k/b$
The final result would be k =
(*a*) 0.297 (*b*) 0.2
(*c*) 0 (*d*) 3

342. Consider the following program in C
int k ;
float a, b, c;
$k = a * b * c/100 + 2.5/2 - 0.3 * 9$
if $a - 5.0$, $b = 2.0$ and $c = 3.0$ the final result k =
(*a*) 1 (*b*) -1
(*c*) $+1.15$ (*d*) -1.15

343. In the expression $a = 5 * 2/3.2 + 4/8 - 2 + 5/9$ which operation will have the first priority in C
(*a*) 5 * 2 (*b*) 2/3.2
(*c*) 5 * 2/3.2 (*d*) 4/8

344. In the expression $b = 2.1/1.2 * 3.1 + 3/7.1 * 6$ which operation would have the first priority in C
(*a*) 2.1/1.2 (*b*) 1.2 * 3.1
(*c*) 2.1/1.2 * 3.1 (*d*) 3/7.1

345. Assuming that k is an integer what would the result of following expression in C
$k = 2 * 3/5 + 4/4 + 8 - 2 - 4/8$
(*a*) 8 (*b*) 7
(*c*) 7.5 (*d*) zero

346. Match the following Lists for C and give answer using the codes

	List I (Operator)	List II (Description)
A.	&&	1. Assignment
B.	||	2. Logical AND
C.	=	3. Logical OR

Codes:

	A	B	C
(a)	1	2	3
(b)	2	3	1
(c)	3	2	1
(d)	1	3	2

347. Match List I and List II for C and give the answer using the codes

	List I (Operator)	List II (Description)
A.	!=	1. Left Shift
B.	==	2. Right Shift
C.	<<	3. Equal to
D.	>>	4. Not equal to

Codes :

	A	B	C	D
(a)	1	2	3	4
(b)	2	3	4	1
(c)	4	1	3	2
(d)	4	3	1	2

348. Match List I and List II for C and give answer using the codes

	List I (Operator)	List II (Description)
A.	&&	1. Bitwise inclusive OR
B.	&	2. Bitwise AND
C.	||	3. Logical AND
D.	|	4. Logical OR

Codes:

	A	B	C	D
(a)	1	2	3	4
(b)	3	2	4	1
(c)	2	3	1	4
(d)	1	4	2	3

349. Match List I and List II in C and give answer using codes

	List I (Operator)	List II (Description)
A.	?:	1. Modulus
B.	[]	2. Array expression
C.	%	3. Conditional

Codes:

	A	B	C
(a)	3	2	1
(b)	1	2	3
(c)	1	3	2
(d)	2	3	1

350. What does the operator ( ) mean in C?

(a) function expression (b) array expression
(c) structure operator (d) increment/decrement

351. Read the following C program float *bs, gs, ha* ;

```
if(bs < 5000)
{
  ha = bs * 20/1000;
}
else
{
  ha = 500;
}
  gs = bs + ha ;
```

If *bs* =7000, the output will be *gs* =

(a) 7000.0 (b) 8400.0
(c) 7500.0 (d) 8000.0

352. Read the following C program

```
int bs, da, gs ;
if(bs< 10000)
{
da = bs * 10/100;
}
else
{
  da= 1500;
}
  gs = bs + da;
```

For *bs* = 9000 and 14000 respectively, the result of *gs* will be

(a) 9900 and 15500 respectively
(b) 10500 and 15400 respectively
(c) 9000 and 14000 respectively
(d) 9900 and 15400 respectively

353. Consider the following C program

```
int m1, m2, m3
pr = (m1 + m2 + m3)/3;
if(pr >= 60)
  print f(' first');
else
{
  if(pr >= 50)
  print f('second');
else
{
  if(pr>= 40)
  print f( 'third');
  else
  {
    print f('fail');
```

Assuming $m_1 = 20$, $m_2 = 80$ and $m_3 = 30$, the computer will print the result as

(a) first (b) second
(c) third (d) fail

354. Consider the following program in C int m_1, m_2, m_3

```
pr = (m1 + m2 + m3)/3;
if(pr>= 60)
  print f( 'first');
if((pr >= 50) && (pr < 60))
  print f( 'second');
if((pr >= 40) && (pr < 50))
  print f('third');
if(pr < 40)
  print f('fail');
```

Assuming $m_1 = 60$, $m_2 = 70$ and $m_3 = 30$, the result will be printed as

(*a*) first (*b*) second
(*c*) third (*d*) fail

355. The statement ! ($b < 5$) in C means

(*a*) $b = 5$ (*b*) $b > 5$
(*c*) $b >= 5$ (*d*) none of the above

356. Consider the following part of a C program

float *a, b*
$b = (a > 4\ ?\ 2 : 9)$;

Assuming that *a* has values of 3 and 6, the values stored respectively will be

(*a*) 2 and 9 (*b*) 2.0 and 9.0
(*c*) 2.0 and 2.0 (*d*) 9.0 and 2.0

357. In which computer language is 'break' statement used?

(*a*) Fortran 77 (*b*) Pascal
(*c*) C (*d*) none of the above

358. The loops 'for', 'while', 'do-while' are used in

(*a*) Fortran 77 (*b*) Pascal
(*c*) C (*d*) none of the above

359. In a C expression using assignment operators, relational operators and arithmetic operators, the hierarchy of operations (in the absence of parenthesis) is

(*a*) assignment, relational, arithmetic
(*b*) relational, assignment, arithmetic
(*c*) arithmetic, assignment, relational
(*d*) arithmetic, relational, assignment

360. The correct hierarchy of operations for the operators &&, = =, =, * is

(*a*) &&, = =, =, * (*b*) *, = =, &&, =
(*c*) *, &&, = =, = (*d*) =, &&, = =, *

361. Match List I and List II and give correct answer using the given codes

	List I **(Operator in C)**		**List II** **(Priority of Operation)**
A.	!	1.	1
B.	<>	2.	2
C.	*	3.	3
D.	=	4.	4

Codes :

	A	B	C	D
(*a*)	1	2	3	4
(*b*)	1	3	2	4
(*c*)	4	3	2	1
(*d*)	3	2	1	4

362. Consider the following statements about 'Functions' in C

1. A function can be called any number of times
2. A function gets called when the function name is followed by a semicolon
3. The order in which functions are defined in a program and the order in which they are called must be the same

Which of the above statements are correct?

(*a*) all (*b*) 1 and 2 only
(*c*) 1 and 3 only (*d*) 2 and 3 only

363. Consider the following about 'return' statement in C :

1. On executing the return statement, it immediately transfers the control back to the calling program
2. There is no restriction on the number of 'return' statements present in a function
3. The 'return' statement must be present at the end of called function

Which of the above are correct?

(*a*) all (*b*) 1 and 2 only
(*c*) 1 and 3 only (*d*) 2 and 3 only

364. The number of storage classes in C is

(*a*) 2 (*b*) 3
(*c*) 4 (*d*) 6

365. Which language has a preprocessor?

(*a*) Fortran 77 (*b*) Pascal
(*c*) C (*d*) none of above

366. Match List I and List II and give the answer using the codes

	List I **(Processor in C)**		**List II** **(Output)**
A.	Preprocessor	1.	Source code file
B.	Compiler	2.	Executable code in machine language
C.	Linker	3.	Relocatable object code

Codes :

	A	B	C
(*a*)	1	2	3
(*b*)	2	3	1
(*c*)	3	2	1
(*d*)	1	3	2

367. Consider the following about 'arrays' in C

1. An array is a collection of similar elements
2. An array cannot have 10 numbers 5 of which are ints and 5 are floats
3. An array can have ints and chars in the same array

Which of the above are correct?

(*a*) all (*b*) 1 and 2 only
(*c*) 1 and 3 only (*c*) 2 and 3 only

368. Which of the following is not a library string function in C?

(*a*) strlen (*b*) strcpy
(*c*) strdup (*d*) strbig

369. Consider the following features of 'structures' in C

1. The values of a structure variable can be assigned to another structure variable of the same type using the assignment operator
2. One structure can be nested within another structure
3. A variable structure can also be passed to a function

Which of the above are correct?

(*a*) all (*b*) 1 and 2 only
(*c*) 1 and 3 only (*d*) 2 and 3 only

370. Match List I and List II and give answer using the codes

	List I **(Bits operator in C)**		**List II** **(Meaning)**
A.	~	1.	Bitwise XOR
B.	&	2.	Bitwise AND
C.	|	3.	Bitwise OR
D.	^	4.	One's complement

Codes:

	A	B	C	D
(a)	1	2	3	4
(*b*)	4	3	2	1
(*c*)	2	3	1	4
(*d*)	4	2	3	1

371. Read the following statements about Java

1. It is well suited to design software which works in conjunction with internet
2. It is cross platform language
3. Java is close to programming language C, C ++ etc.
4. It is an object oriented programming language

Which of the above are correct?

(*a*) all (*b*) 1, 2 and 3 only
(*c*) 1, 2 and 4 only (*d*) 2, 3, 4 only

372. Java variables are called

(*a*) instance variables, class variables and global variables
(*b*) instance variables, class variables and local variables
(*c*) instance variables, class variables and sector variables
(*d*) global variables, class variables and local variables

373. Consider the following statements about Java variables

1. Java variable must start with a letter or underscore or dollor sign
2. Java variable cannot start with a digit
3. After the first character the variable can have any combination of letters or numbers
4. Capitalization of letters must be consistent

Which of the above are correct?

(*a*) all (*b*) 1, 2 and 3 only
(*c*) 2, 3 and 4 only (*d*) 1, 2 and 4 only

374. The four types of integers in Java are called

(*a*) byte, short, int and long
(*b*) bit, byte, short and long
(*c*) nibble, byte, short and long
(*d*) nibble, byte, int and long

375. The range of floating point numbers in Java is

(*a*) 1.4 E−45 to 3.4 E+38
(*b*) 3.4 E−45 to 1.4 E+38
(*c*) 1.4 E−38 to 3.4 E+45
(*d*) 3.4 E−38 to 1.4 E+45

376. The number of operators to accomplish basic arithmetic in Java is

(*a*) 4 (*b*) 5
(*c*) 6 (*d*) 7

377. In Java 31% 9 gives an answer

(*a*) 3 (*b*) 4
(*c*) 3/4 (*d*) 4/3

378. Match List I and List II and give answer using the codes

List I (Symbol in Java)	**List II** (Meaning)
A. &	1. Logical AND
B. ^	2. AND
11	
C. 11	3. XOR
D. &&	4. Logical OR

Codes:

	A	B	C	D
(*a*)	1	2	3	4
(*b*)	2	3	1	4
(*c*)	4	2	3	1
(*d*)	2	3	4	1

379. The symbol ! in Java means

(*a*) logical NOT (*b*) logical OR
(*c*) AND (*d*) logical XOR

380. Java's syntax is based on

(*a*) C (*b*) C++
(*c*) Fortran 77 (*d*) Pascal

381. Consider the following expressions in Java

int *a*, *b*, *c*,
a = 40;
b = *a*++;
c = ++*a*;

Now the values of *a*, *b*, *c* are

(*a*) 42, 40, 42 respectively (*b*) 42, 42, 42 respectively
(*c*) 40, 42, 42 respectively (*d*) 40, 40, 42 respectively

382. Match List I and List II and give the answer using the codes

List I (Symbols of Java comparison operator)	**List II** (Meaning)
A. ==	1. Greater than or equal to
B. !=	2. Equal
C. >=	3. Not equal

Codes:

	A	B	C
(*a*)	1	2	3
(*b*)	3	2	1
(*c*)	1	3	2
(*d*)	2	3	1

383. The operator precedence in Java is in the order

(*a*) arithmetic operation, increment and decrement operation, comparison, logical operation and assignment operation
(*b*) increment and decrement operation, arithmetic operation, comparison, logical operation and assignment operation
(*c*) comparison, arithmetic operation, increment and decrement operation, logical operation and assignment operation
(*d*) arithmetic operation, comparison operation, increment and decrement operation, logical operation and assignment operation

384. In Java it is possible to have

1. An array of integers
2. An array of string objects
3. An array of arrays
4. An array on integer and strings

Which of the above are correct?

(*a*) all (*b*) 1, 2 and 3 only
(*c*) 1 and 2 only (*d*) 1, 3 and 4 only

385. Consider the following in object oriented programming

1. An object is a self-contained element of a computer program which represents a related group of interests

2. Each object has a specific role in a program
3. All objects can work with other objects in defined ways
4. Objects are also called instances

Which of the above are correct?

(a) all (b) 1, 2 and 3 only
(c) 1 and 2 only (d) 1, 2 and 4 only

386. Which is not a valid Java variable?

(a) lot (b) c 1t
(c) s $ 2 (d) 2 s $

387. Which is a valid Java variable?

(a) 1 a b (b) 56 a
(c) a 56 (d) 5 a 6

388. Consider the following :

1. The first letter of a Java variable is lower case
2. Each successive word in Java variable begins with a capital letter
3. All other letters are lower case
4. A Java variable can have a digit as first character

Which of the above are correct?

(a) all (b) 1, 2, 3 only
(c) 1, 2, 4 only (d) 2, 3, 4 only

389. In Java which operator can be used even if there is a single operand

(a) + (b) –
(c) * (d) %

390. Assume that $x = 5$ and $y = 2$. Then in Java operation $x += y$, the result is

(a) $x = 7$ (b) $x = 2$
(c) $x = 3$ (d) $x = 2$

391. Assume that $x = 7$, the Java operation $x = x -$ means that final value of x is

(a) 7 (b) –7
(c) 6 (d) –6

392. Which of the following alphanumeric characters is not allowed in Basic

(a) capital letters A to Z
(b) lower case letters a to z
(c) digits 0 to 9
(d) special characters +, –, *, / etc.

393. Which of the following statement number is not valid in Basic?

(a) 12 (b) 521
(c) 662512 (d) 99999

394. A string constant in Basic is

(a) a string of capital alphabets A to Z, included in quotation mark
(b) a string of capital alphabets A to Z and digits 0 to 9 included in quotation mark
(c) a string of capital alphabets A to Z, digits 0 to 9 and special characters
(d) a string of capital alphabets A to Z, digits 0 to 9 and special characters included in quotation mark

395. A string variable in Basic is

(a) an alphabet followed by dollar sign
(b) a digit followed by dollar sign
(c) either (a) or (b)
(d) neither (a) nor (b)

396. Which of these are allowed in Basic?

(a) arithmetic operators +, –, *, / etc.
(b) logical operators AND, OR, NOT
(c) relational operators =, <, > etc.
(d) all the above

397. Which of the following is invalid in Basic?

(a) A ** B – C (b) 5.2** 4.3
(c) 5.2** (–4.3) (d) (–5.2) ** (4.3)

398. Which of the following is invalid in Basic?

(a) A ** B + – C (b) A ** B – C
(c) A * B – C (d) A * B – C ** D

399. Match List I with List II and give answer using the codes

	List I (Operation in Basic)		List II (Priority)
A.	Exponentiation	1.	1
B.	Multiplication and division	2.	2
C.	Addition and subtraction	3.	3
D.	Expression within parenthesis	4.	4

Codes :

	A	B	C	D
(a)	1	2	3	4
(b)	2	3	1	4
(c)	4	3	2	1
(d)	2	3	4	1

400. Consider the following program in Basic :

```
5    READ R
10   D=2*R
20   A=3.1416*D*D/4
30   PRINT A
```

Assuming that R = 3.0, the output A =

(a) 2.827 (b) 28.27
(c) 282.7 (d) 2827

401. Consider the following program in Basic :

```
10   READ X, Y, Z
15   DATA 20, 30, 10
20   S=X+Y+Z
30   A=S/3
35   PRINT S, A, X, Y, Z
40   END
```

The output will be

(a) 60, 20, 20, 30, 10 (b) 10, 30, 20, 20, 60
(c) 60, 20, 30, 10, 20 (d) 60, 30, 10, 20, 20

402. Consider the following ON GO TO statement in Basic

35 ON A GO TO 15, 25, 35

If the value of A is 2.6, the control will transfer to

(a) line 15
(b) line 25
(c) line 35
(d) any of the lines 15 or 25, 35

403. Consider the following ON GO TO statement in Basic

45 ON C GO TO 10, 20, 30

If the values of C are 1, 2, 3, 2.6 the control will be transferred to statement number

(a) 10, 20, 30, 20 respectively
(b) 10, 20, 20, 30 respectively
(c) 20, 30, 20, 10 respectively
(d) 10, 30, 20, 20 respectively

404. Consider the following statement in Basic

```
15    INPUT X, Y, Z
20    IF Y > X THEN 25
21    IF Z > Y THEN 30
25    PRINT Y
30    PRINT Z, X
35    END
```

If X, Y, Z are 150, 200, 250 then the output will be
(a) 200, 250, 150 respectively
(b) 150, 200, 250 respectively
(c) 250, 200, 150 respectively
(d) 200, 150, 250 respectively

405. Consider the statement

```
20    X = A + B/C + D * Y
```

Which of the following formulae does the above statement represent?

(a) $X = \frac{A+B}{C+DY}$ (b) $X = A + \frac{B}{C} + DY$

(c) $X = A + \frac{B}{C+DY}$ (d) none of the above

406. Virus is a
(a) computer hardware problem
(b) an operating system
(c) a computer program
(d) a computer language

407. A computer window has a bar at its top to bear the name of window. This bar is called
(a) title bar (b) status bar
(c) menu bar (d) control bar

408. The command ^C in MS WORD
(a) copies the text in clip board
(b) copies the text in secondary storage device
(c) copies the text in file
(d) none of the above

409. MICR stands for
(a) magnetic ink chart receipt
(b) magnetic ink character recognition
(c) magnetic ink chart recognition
(d) magnetic ink character receipt

410. 'WRITE' is a
(a) word processor (b) line editor
(c) text editor (d) column editor

411. Match List I and List II and give correct answer using the codes

	List I		List II
A.	CLP	1.	Measure of μP clock speed
B.	SHELL	2.	Double sided double density
C.	DSDD	3.	Allows you to specify file as command processor
D.	Megahertz	4.	Microcode stored in ROM

Codes :

	A	B	C	D
(a)	4	3	2	1
(b)	3	4	2	1
(c)	1	2	3	4
(d)	2	3	4	1

412. Match List I and List II and give answer using the codes

	List I		List II
A	Word wrap	1.	Interleaved execution of two or more programs by the same computer
B	Client	2.	automatically moves a word to the next line if it does not fit in the current line
C	Multitasking	3.	an application receiving data
D	Landscape	4.	orientation of worksheet

Codes:

	A	B	C	D
(a)	1	2	3	4
(b)	2	3	4	1
(c)	4	3	2	1
(d)	2	3	1	4

413. Zip floppy has a capacity of
(a) 50 MB (b) 500 MB
(c) 25 MB (d) 100 MB or 250 MB

414. For Disk partitioning the DOS command is
(a) F DISK (b) H DISK
(c) A DISK (d) D DISK

415. Which DOS command improves memory usage?
(a) MEM MAKER (b) MOVE
(c) MEM IM (d) none of the above

416. Match List I (DOS Command) and List II (Meaning) and give answer using codes

	List I		List II
A.	CLS	1.	Displays DOS version being used
B.	MSAV	2.	Partitions hard disk
C.	F DISK	3.	Anti-virus program
D.	VER	4.	Cleans screen

Codes :

	A	B	C	D
(a)	1	2	3	4
(b)	4	3	2	1
(c)	2	3	4	1
(d)	2	1	4	3

417. Match List I and List II and give answer using the codes

	List I (DOS commands)		List II (Meaning)
A.	DBLSPACE	1.	Deletes a directory and its sub-directories
B.	MEM	2.	Displays the amount of used and free memory
C.	SYS	3.	Puts a copy of DOS on disk
D.	DELTREE	4.	Compresses files to increase the storage capacity of disc

Codes:

	A	B	C	D
(*a*)	4	3	2	1
(*b*)	2	3	4	1
(*c*)	4	2	3	1
(*d*)	1	2	3	4

418. The port of PC to which mouse is connected is called
(*a*) serial port (*b*) parallel port
(*c*) input port (*d*) scanner port

419. Which is not a software?
(*a*) DOS (*b*) windows
(*c*) MS WORD (*d*) hard disk

420. The storage capacity of DSDD disk and DSHD disk are
(*a*) 1.2 MB each
(*b*) 360 KB each
(*c*) 1.2 MB and 360 KB respectively
(*d*) 360 KB and 1.2 MB respectively

421. Which of the following is not an internal DOS command?
(*a*) VER (*b*) COPY
(*c*) MEM (*d*) DATE

422. Match List I and List II and give answer using the codes

List I (DOS Command)	**List II (Meaning)**
A. ATTRIB	1. Displays all directories and subdirectories
B. SCAN DISK	2. Moves files to another directory
C. MOVE	3. Checks the status of disc
D. TREE	4. Hides a file

Codes :

	A	B	C	D
(*a*)	1	2	3	4
(*b*)	3	4	2	1
(*c*)	2	3	4	1
(*d*)	4	3	2	1

423. Match List I and List II and give answer using the codes

List I	**List II**
A. Title bar	1. Window menu
B. Control menu box	2. First row of window
C. Start-button	3. Box in title bar
D. Control	4. Button on mouse

Codes :

	A	B	C	D
(*a*)	3	2	4	1
(*b*)	2	3	4	1
(*c*)	3	2	1	4
(*d*)	2	3	1	4

424. Match List I and List II and give answer using the codes

List I	**List II**
A. True type	1. Type of calculator in window
B. FIND	2. Window command
C. WORD PAD	3. Name of a Font
D. STANDARD	4. Has many features of a simple word processor

Codes:

	A	B	C	D
(*a*)	1	2	3	4
(*b*)	4	3	2	1
(*c*)	3	2	1	4
(*d*)	3	2	4	1

425. Match List I and List II and give answer using the code

List I (Command in WORD)	**List II (Meaning)**
A. END	1. The bar displays the current position of cursor
B. HOME	2. Moves cursor to the beginning of current line
C. Status	3. Moves cursor to end of document

Codes:

	A	B	C
(*a*)	1	2	3
(*b*)	3	1	2
(*c*)	2	3	1
(*d*)	3	2	1

426. The two orientations of worksheet page are called
(*a*) portrait and landscape
(*b*) auto text and auto correct
(*c*) font and footer
(*d*) header and footer

427. Read the following statements about Excel
1. It supports many arithmetic functions like SUM, ABS, EXP, INT, MOD etc.
2. It supports many trigonometric functions
3. It supports IRR function to calculate interest rate
4. It supports PV function to calculate the present value of future receipts

Which of the above are correct?
(*a*) all (*b*) 1, 2 and 3 only
(*c*) 1 and 2 only (*d*) 1 only

428. Which of the following is not a data base function
(*a*) DAVERAGE (*b*) MAX
(*c*) DCOUNT (*d*) DSTDEV

429. The function to calculate the maximum value in the specified field of a data base is
(*a*) DM (*b*) DMAX
(*c*) MAX (*d*) DBMAX

430. The programming language in which Excel records the macro code is
(*a*) basic (*b*) visual basic
(*c*) video basic (*d*) basic 2

431. Which of the following is not a DTP program?
(*a*) page maker (*b*) coral draw
(*c*) coral ventura (*d*) coral key

432. Which of the following is the software for financial accounting?
(*a*) COBOL (*b*) MS WORD
(*c*) Page maker (*d*) dB III plus

433. Which of the following is a computer language?
(*a*) MS WORD (*b*) Page maker
(*c*) EXCEL (*d*) LISP

434. During processing the instructions, data, intermediate results and final results in a computer are held in
(*a*) CPU (*b*) ALU
(*c*) primary memory (*d*) secondary memory

435. Which of the following can be accessed only sequentially?
(*a*) floppy disk (*b*) hard disk
(*c*) magnetic tape (*d*) ROM

436. Which has volatile memory?
(*a*) magnetic tape (*b*) RAM
(*c*) diskette (*d*) hard disk

437. Match List I and List II and select answer using the codes

List I	List II
A. DMA Controller	1. Interface between computer and video display adapter
B. CRT controller	2. Used to read data directly from disk into computer memory
C. Interrupt controller	3. Supervises the operation of control signal

Codes :

	A	B	C
(*a*)	1	2	3
(*b*)	2	3	1
(*c*)	3	2	1
(*d*)	2	1	3

438. Match List I and List II and select correct answer using codes

List I	List II
A. Firm ware	1. Converts program in high level language to machine language
B. Compiler	2. Converts assembly language program into machine language
C. Assembler	3. Software embedded in hardware
D. Virtual addressing	4. Grouping information between various memories

Codes:

	A	B	C	D
(*a*)	1	2	3	4
(*b*)	2	3	4	1
(*c*)	3	4	1	2
(*d*)	3	1	2	4

439. Which of the following is operating system?
(*a*) MS Word (*b*) Ventura
(*c*) Java (*d*) Windows

440. A modem is connected to PC at
(*a*) serial port (*b*) parallel port
(*c*) SCSI port (*d*) input port

441. In a display specified as 600 × 400 the number of pixels across the display screen is
(*a*) 600 (*b*) 400
(*c*) 240000 (*d*) either (*a*) or (*b*)

442. If the resolution of a printer is 1200 dpi, the number of dots per square inch is
(*a*) 1200 (*b*) 800
(*c*) 1200 × 1200 (*d*) either (*a*) or (*b*)

443. A colour monitor using 16 bits at a resolution of 600 × 400 pixels needs a memory of
(*a*) 240000 B (*b*) 9600 B
(*c*) 6400 B (*d*) 480000 B

444. Which of the following is not an application software?
(*a*) MS Word (*b*) DBMS
(*c*) Worksheet (*d*) Lisp

445. If A = 20 and B = – 7 in PASCAL, A mod B =
(*a*) 6 (*b*) 2
(*c*) – 1 (*d*) 3

446. In a multi processor configuration two coprocessors are connected to host 8086 processor. The instruction sets of the two coprocessors
(*a*) must be same
(*b*) may overlap
(*c*) must be disjoint
(*d*) must be the same as that of host

447. An advantage of memory interfacing is that
(*a*) a larger memory is obtained
(*b*) effective speed of memory is increased
(*c*) cost of memory is reduced
(*d*) a non-volatile memory is obtained

448. Consider the following statements
1. A 33 MHz 486 has higher MIPS than 33 MHz 386
2. A 33 MHz 486 has higher MFLOPS than 33 MHz 386

Of these statements
(*a*) both are true (*b*) both are false
(*c*) 1 is true but 2 is false (*d*) 1 is false and 2 is true

449. A PC has typically
(*a*) 5 to 10 KB of main memory
(*b*) 10 to 100 KB of main memory
(*c*) 100 to 256 KB of main memory
(*d*) 256 K B to 1 MB of main memory

450. Expression + + i is equivalent to
(*a*) $i = i + 2$ (*b*) $i = i + 1$
(*c*) $i = i + i$ (*d*) $i = i + i + i$

451. Each instruction in assembly language program has the following fields
1. Label field 2. Mnemonic field
3. Operand field 4. Comment field

The correct sequence of these fields is
(*a*) 1, 2, 3, 4 (*b*) 1, 2, 4, 3
(*c*) 2, 1, 3, 4 (*d*) 2, 1, 4, 3

452. Consider the following statements
1. A total of about 1 million bytes can be directly accessed by 8086 microprocessor
2. The 8086 has thirteen 16 bit registers
3. The 8086 has eight flags
4. Compared to 8086, the 80286 provides a higher degree of memory protection

Of the above statements
(*a*) 1, 2, 3, 4 are correct (*b*) 1, 3, 4 are correct
(*c*) 1, 2, 4 are correct (*d*) 1, 2, 3 are correct

453. The interface chip for 8086 and 16 bit ADC is
(*a*) 8259 (*b*) 8255
(*c*) 8253 (*d*) 8251

454. For the Infix expression A + B – C + D the Post-fix notation is

(*a*) AB+C–D+ (*b*) +AB–C+D
(*c*) AB–C+D (*d*) none of the above

455. For the in-fix expression (A/B) * (C * F + (A – D) * E) the corresponding Post-fix notation is

(*a*) A/B *CF AD – E * + (*b*) AB/CF * AD – E * + *
(*c*) AB/C*FAD – E * + * (*d*) none of the above

456. Consider the following statements

1. An assembler is a program having assembly language program as input and machine language program as output
2. A macro assembler is an assembler with additional macro facilities
3. A cross assembler for a μP X is an assembler which executes on source micro computer with a different CPU than X but generates code for X
4. Assemblers generally do not provide macro facilities

Of the above statement

(*a*) all are correct (*b*) only 1, 2, 3 are correct
(*c*) only 1, 2, 4 are correct (*d*) only 1, 3, 4 are correct

457. To realize the full power of 8086 the pin of 8086 connected to ground is

(*a*) $\overline{DEN}$ (*b*) ALE
(*c*) INTR (*d*) MN/$\overline{MX}$

458. Cache memory size is about

(*a*) a few KB (*b*) about 500 KB
(*c*) about 1 MB (*d*) about 50 MB

459. In EPROM integrated dose is defined as

(*a*) UV light intensity × exposure time in seconds
(*b*) UV light intensity / exposure time in seconds
(*c*) exposure time in seconds / UV light intensity
(*d*) none of the above

460. In 8085 microprocessor with memory mapped I/O which of the following is true

(*a*) I/O devices have 16 bit addresses
(*b*) I/O devices are accessed during IN and OUT instructions
(*c*) There can be a maximum of 256 input and 256 output devices
(*d*) Logic operations cannot be performed

461. The following program is run on 8085

Memory address in hex	**Instruction**
2000	LXI SP, 100
2003	PUSH H
2004	PUSH D
2005	CALL 2050
2008	POP H
2009	HLT

After the completion of program, stack pointer and program counter will contain

(*a*) OFFC and 2050 respectively
(*b*) OFFC and 2003 respectively
(*c*) OFFDE and 2003 respectively
(*d*) OFFDE and 2050 respectively

462. Each cell of static RAM contains

(*a*) 4 MOS transistors
(*b*) 4 MOS transistors and 1 capacitor
(*c*) 2 MOS transistors
(*d*) 4 MOS transistors and 2 capacitors

463. An 8 bit wide 4 word sequential memory has

(*a*) Eight 4 bit shift registers
(*b*) Eight 8 bit shift registers
(*c*) Four 8 bit shift registers
(*d*) Four 4 bit shift registers

464. A 256 × 4 EPROM has

(*a*) 8 address pins and 4 data pins
(*b*) 8 address pins and 8 data pins
(*c*) 4 address pins and 8 data pins
(*d*) 4 address pins and 4 data pins

465. A μP contains 2 MHz crystal. The time taken to complete instructions with 1000 cycles is

(*a*) 2 ms (*b*) 1 ms
(*c*) 2 ns (*d*) 1 ns

466. An I/O processor controls the flow of information between

(*a*) cache memory and I/O devices
(*b*) main memory and I/O devices
(*c*) two I/O devices
(*d*) cache and main memory

467. An instruction used to set carry flag in a computer is classified as

(*a*) data transfer (*b*) arithmetic
(*c*) logical (*d*) program control

468. If CS = $\overline{A}_{15}$ A_{14} A_{13} is used as chip select logic of a 4 K RAM in 8085 system, its memory range is

(*a*) 3000 H – 3FFF H
(*b*) 7000 H – 7FFF H
(*c*) 5000 H – 5FFF H and 6000 H – 6FFF H
(*d*) 6000 H – 6FFF H and 7000 H – 7FFF H

469. Consider the following statements :

1. EEPROM has the problem of high write time
2. EEPROM has the problem of limited number of write cycles
3. Shadow RAM has the problem of high write time

Which of the above statements are correct?

(*a*) all (*b*) 1 and 2 only
(*c*) 1 and 3 only (*d*) 2 and 3 only

470. Intel 2004 is a

(*a*) ROM (*b*) volatile RAM
(*c*) NV RAM (*d*) none of the above

471. EEPROM permits

(*a*) read operation only
(*b*) read and byte erase operations
(*c*) read, byte erase and byte write operations
(*d*) read, byte erase, byte write and chip erase operations

472. A blank EPROM has

(*a*) all bits set to logical 0
(*b*) all bits set to logical 1
(*c*) half the total number of bits set to 0 and remaining half to logical 1
(*d*) either (*a*) or (*b*)

473. Read the following statements about address space in microprocessors
1. An address space is a set of all possible addresses which can be generated by a microprocessor
2. Each address in the address space allows a designer to provide at least one memory or I/O location in the system
3. Two types of address spaces are memory and I/O address space
4. Some microprocessors have only one type of address space

Which of the above are correct?
(*a*) all (*b*) 1, 2, 3 only
(*c*) 1, 2, 4 only (*d*) 1, 3, 4 only

474. Read the following statements
1. A μP is generally interfaced to several memory chips and I/O devices
2. Address space provided by μP is partitioned into sub sets of addresses
3. Address map and address partition refer to different terms in address space

Which of the above are correct?
(*a*) all (*b*) 1 and 2 only
(*c*) 2 and 3 only (*d*) 1, 3 only

475. For the floating point numbers 1.81×10^{-29} and 0.0152×10^{18} the normalised forms are
(*a*) 0.181×10^{-28} and 0.152×10^{17} respectively
(*b*) 18.1×10^{-30} and 0.152×10^{17} respectively
(*c*) 0.181×10^{-28} and 1.52×10^{16} respectively
(*d*) none of the above

476. For decimal number 13.5, the floating point representation in 4 byte excess 128 code is
(*a*) 01101100 00000000 00000000 10000100
(*b*) 00000000 01101100 00000000 10000100
(*c*) 00000000 01101100 10000100 00000000
(*d*) 01101100 00000000 10000100 00000000

477. Consider the following statements
1. The sign of a floating point number depends on the sign of mantissa only
2. If the absolute value of a number is less than 1, the sign of exponent is zero or positive
3. With sign bit of mantissa as 0, increasing the exponent from a positive to a more positive value gives a larger floating point number

Which of the above are correct?
(*a*) all (*b*) 1 and 2 only
(*c*) 1 and 3 only (*d*) 2 and 3 only

478. Which of the following is invalid integer variable name in Fortran?
(*a*) IAEG (*b*) I124
(*c*) AILG (*d*) MIAC

479. Which of the following real variable names is not valid in Fortran?
(*a*) BETA (*b*) ALPHA
(*c*) A (*d*) AB * 2

480. If it is desired to tabulate number of students in a class in Fortran language, a proper variable name is
(*a*) ST (*b*) STUD
(*c*) NSTUD (*d*) STD

481. Which of the following expressions is invalid in Fortran
(*a*) IA + IB (*b*) IB * IC/ID
(*c*) IA * B (*d*) − K * I + J

482. Consider the following rules in Fortran
1. A signed or unsigned real variable name or a real constant is a real expression
2. Two real expressions connected by an arithmetic operator is a real expression
3. A real variable or constant exponentiated by an integer constant or integer variable is not valid

Which of the above are correct?
(*a*) all (*b*) 1 and 2 only
(*c*) 1 and 3 only (*d*) 2 and 3 only

483. The expression A ** B/C + D is equivalent to
(*a*) AB/C + D (*b*) $A^B/C + D$
(*c*) $A^B/(C + D)$ (*d*) none of the above

484. If I = J = K = 13, then in Fortran
I = 10 − (I/10) * 10 + (J/5) * 4 + K ** (− 2) is equal to
(*a*) 8 (*b*) 1
(*c*) 6 (*d*) 5

485. Consider the following program

```
2   READ A, B, C, P, Q, R
4   LET Q = Q − (P * B/A)
6   LET R = R − (P * C/A)
8   IF Q = 0 THEN 10
    LET Y = R/Q
    LET X = (C − B * Y)/A
    PRINT X, Y
    PRINT NO SOLUTION
    STOP
```

If A, B, C, P, Q, R are 2, 3, 4, 5, 6, 7 respectively the print output will be
(*a*) 1, 2 (*b*) − 1, 2
(*c*) 2, 1 (*d*) 2, − 1

486. The number 0.004854 in normalised floating point mode is represented as
(*a*) | 4 | 8 | 5 | 4 | 0 | 2 | (*b*) | 0 | 4 | 8 | 5 | 0 | 1 |
(*c*) | 4 | 8 | 5 | 4 | 0 | 1 | (*d*) | 0 | 4 | 8 | 5 | 0 | 2 |

487. When .4546 E 5 and .5433 E 7 are to be added in normalized floating point mode
(*a*) none of the numbers is changed to an other form
(*b*) .4546 E 5 is changed .004546 E 7 and .5433 E 7 is not changed
(*c*) .5433 E 7 is changed to 54.33 E 5 and .4546 E 5 is not changed
(*d*) both the numbers are changed and their exponents are made equal to 6

488. When .9432 E − 4 is subtracted from .5452 E − 3 in normalized floating point mode
(*a*) none of the numbers is changed
(*b*) .9432 E − 4 is changed to .09432 E − 3 and .5452 E − 3 is not changed
(*c*) .5452 E − 3 is changed to 5.452 E − 4 but .9432 E − 4 is not changed

(*d*) both the numbers are changed and their exponents are made equal to – 5

489. If $x = 0.1396$ radians, the values of $1 - \cos x$ and $2 \sin^2 (x/2)$ using floating point arithmetic with a 4 digit mantissa are respectively
(*a*) .1000E – 1 each
(*b*) .9727 E – 2 each
(*c*) . 1000 E – 1 and .9727 E – 2 respectively
(*d*) .9727 E – 2 and .1000 E – 1 respectively

490. If quadratic equation $x^2 - 1000x + 25 = 0$ is solved using floating point arithmetic with 4 digit mantissa, the roots are
(*a*) .1000 E4 and .1000 E 4 respectively
(*b*) .0000 E4 and .0000 E4 respectively
(*c*) .1000 E4 and .0000 E4 respectively
(*d*) none of the above

491. When performing numerical calculations in floating point arithmetic
1. Error may occur due to finite representation of numbers
2. Error may occur due to rounding
3. Error may occur due to finite representation of inherently infinite process

Which of the above are correct?
(*a*) all (*b*) 1 and 2 only
(*c*) 1 and 3 only (*d*) 2 and 3 only

492. Read the following statement :
1. Multi-processing operation permits highly efficient parallel processing at several levels
2. Multi-processing involves use of several processing units jointly on a common program and severally on different programs
3. In multi-processing the assignment of facilities to tasks, loading of programs and monitoring of I/O activity is taken care of by operating system

Which of the above are correct?
(*a*) all (*b*) 1 and 2 only
(*c*) 1 and 3 only (*d*) 2 and 3 only

493. For converting a decimal number into octal number, the decimal number is expressed as a series of the type
(*a*) $\sum_{a=0}^{n} a_i 8^i$ with each a less than 8
(*b*) $\sum_{a=0}^{n} a_i 8$ with each a less than 8
(*c*) $\sum_{a=0}^{n} \frac{a_i}{8^i}$ with each a less than 8
(*d*) none of the above

494. To evaluate X = (A + B + 1 2) * (A + B – C) the sequence of operation is
(*a*) + + + – * = (*b*) * + + + – =
(*c*) + + – * + = (*d*) – + + + * =

495. The language PL/1 has
(a) one character set having 60 characters
(*b*) one character set having 48 characters
(*c*) one character set having 60 characters and another character set having 48 character
(*d*) none of the above

496. Simpson's rule for numerical integration is
(*a*) $S = \frac{h}{3}(f_1 + 4f_2 + 2f_3 + 4f_4 + 2f_5 + 4f_6 + ... + f_{n+1})$
(*b*) $S = \frac{h}{3}(f_1 + 4f_2 + 4f_3 + 2f_4 + 2f_5 + 4f_5 + ... + f_{n+1})$
(*c*) $S = \frac{h}{3}(f_1 + 3f_2 + 4f_3 + 3f_5 + 4f_5 + ... + f_{n+1})$
(*d*) $S = \frac{h}{3}(f_1 + 4f_2 + 3f_3 + 4f_4 + 3f_5 + ... + f_{n+1})$

497. Which one is not a valid subscripted variable?
(*a*) KET (2 * J, 3 * 1 – 4) (*b*) MET (J + 2, K – 4)
(*c*) A (2 I) (*d*) VEL (I)

498. Which of the following is not a legal subscripit?
(*a*) J + 2 (*b*) JA – 4
(*c*) – 3 (*d*) 3 * IA – 2

499. Which is not a valid dimension statement?
(*a*) DIMENSION KAT (10, 20)
(*b*) DIMENSION BAT (30, 20)
(*c*) DIMENSION KAT (K, 10)
(*d*) DIMENSION BAT (10, 5)

500. In the dimension statement DIMENSION A (50, 50), K (100, 100) the total number of locations reserved will be
(*a*) 2500 (*b*) 100000
(*c*) 125000 (*d*) 250000

501. The default parameter passing mechanism in C programming language is
(*a*) call by reference (*b*) call as random
(*c*) call by value (*d*) call by value result
[IES 2010]

502. The data type defined by user is
(*a*) built in data type (*b*) abstract data type
(*c*) homogenous data type (*d*) real data type
[IES 2010]

503. Insertion of a record is circularly linked list involves the modification of
(*a*) 4 pointers (*b*) 3 pointers
(*c*) 2 pointers (*d*) 1 pointer **[IES 2010]**

504. Which of the following data structures is used in a compiler to manage information about variables and their attributes?
(*a*) abstract syntax tree (*b*) linked list
(*c*) Parse table (*d*) symbol table
[IES 2010]

505. The number of edges in a regular graph of three D and N vertices is equal to
(*a*) ND (*b*) ND/2
(*c*) N + D (*d*) N^D **[IES 2010]**

506. A single bus structure is primarily found in
(*a*) main frames
(*b*) mini and micro computers
(*c*) super computers
(*d*) high performance machines **[IES 2010]**

507. The ability to halt the CPU temporarily and use this interval of time to send information to buses is called
(*a*) cycle stealing (*b*) interrupt

(c) poling (d) direct memory address [IES 2010]

508. Consider the foilowing statements :
1. Compilers
2. Design
3. Evaluation
4. Instruction set architecture

Which of above are included in the present definition of computer architecture to design a full computer system?
(a) 1, 2, 3 (b) 1, 3, 4
(c) 2, 3, 4 (d) 1, 2, 3, 4 [IES 2010]

509. Given that main memory access time is 1200 ns and cache access time is 100 ns, the average memory access time is not to exceed 120 ns. The hit ratio for cache memory must be at least
(a) 90% (b) 98%
(c) 80% (d) 75% [IES 2010]

510. An I/O processor controls the flow of information between
(a) cache memory and I/O devices
(b) main memory and I/O devices
(c) two I/O devices
(d) cache memory and main memory [IES 2010]

511. On receiving an interrupt from an I/O device, CPU
(a) halts for predetermined time
(b) branches off to the interrupt service routine after completion of current instruction
(c) branches off to interrupt service routine immediately
(d) hands over the control of address bus and data bus to the interrupting device [IES 2010]

512. In microprocessor-based system, DMA facility is required to increase the speed of data transfer between the
(a) microprocessor and I/O devices
(b) microprocessor and memory
(c) memory and I/O devices
(d) memory and reliability system [IES 2010]

513. Consider the following instruction of 8085 microprocessor
1. Move B, C 2. STA address
3. ORI cycle

The correct sequence in the decreasing order of their respective memory space requirement is
(a) 3, 2, 1 (b) 1, 3, 2
(c) 1, 2, 3 (d) 2, 3, 1 [IES 2010]

514. If the call instruction in 8085 in the main program is conditional then RETURN instruction in the subroutine can be
(a) conditional
(b) conditional or unconditional
(c) determined by DdA instruction
(d) unconditional [IES 2010]

515. When compared to RS232C serial part the USB (Universal serial bus)
(a) supports a lower range of peripherals
(b) supports a faster transfer rate
(c) does not support 'hot plug ability'
(d) controller in PC cannot detect the presence or absence of USB devices [IES 2010]

516. The 8085 has two registers called primary data pointers. These are registers
(a) B and C (b) D and E
(c) H and L (d) C and D [IES 2010]

517. READY signal is 8085 is useful when CPU communicates with
(a) slow peripheral device (b) fast peripheral device
(c) DMA controller chip (d) PPI chip [IES 2010]

518. The interface chip used for data transmission between 8086 and a 16 bit ADC is
(a) 8251 (b) 8253
(c) 8255 (d) 8259 [IES 2010]

519. Microprocessor-based system can perform many different functions because
(a) its operation is controlled by software
(b) it is a digital system
(c) it uses a RAM
(d) it can be controlled by input and output device [IES 2010]

520. Which instruction copies byte of data into the accumulator from the memory address given in the instruction
(a) LDA address (b) LDAXB
(c) LHLD address (d) STA address [IES 2010]

521. The output lines of microprocessors and memories are usually tristated because
(a) more than one device can transmit information over the data bus by enabling only one device at a time
(b) more than one device can transmit information over the data bus at the same time
(c) the data lines can be multiplexed for both input and output
(d) it increases the speed of data transmission over the data bus [IES 2010]

522. The purpose of a start bit in RS232 serial communication protocol is
(a) to synchronise receiver for receiving every byte
(b) to synchronise receiver for receiving a sequence of bytes
(c) as a parity bit
(d) to synchronise receiver for receiving last byte [IES 2010]

523. Consider the following :
1. Wind tunnel simulation
2. Real time video viewing

A computer is used for
(a) neither 1 nor 2 (b) both 1 and 2
(c) 1 only (d) 2 only [IES 2010]

524. Consider the following statements:
1. Strictly speaking C supports 1 dimensional array only
2. An array element may be an array by itself
3. Array elements need not occupy contiguous memory location

Which of the above are correct?
(a) 1 only (b) 2 only
(c) 1 and 2 (d) 2 and 3 [IES 2009]

525. What can be the maximum dimension of array in a C language program
(*a*) 3 (*b*) 4
(*c*) 5 (*d*) it is compiler dependent **[IES 2009]**

526. With reference to C language, which of the following statements are correct. Select your answer from the given codes.
1. An identifier may start with an underscore
2. An identifier may end with an underscore
3. IF is a valid identifier
4. The number of significant characters in an identifier is implementation dependent

(*a*) 1 and 2 only (*b*) 2 and 3 only
(*c*) 1, 2, 3, 4 (*d*) 3 and 4 only **[IES 2009]**

527. Consider the following statements :
1. Internal sorting is used if number of items to be sorted is very large
2. External sorting is used if number of items to be sorted is very large
3. External sorting needs auxiliary storage
4. Internal sorting needs auxiliary storage

Which of the above statements are correct?
(*a*) 1 and 2 (*b*) 2 and 3
(*c*) 3 and 4 (*d*) 2 and 4 **[IES 2009]**

528. Which of the following are correct statements?
1. Bus is a group of wires carrying information
2. Bus is needed to achieve reasonable speed of operation
3. Bus can carry data or address
4. A bus can be shared by more than one devices

Select the correct answer from the following codes:
(*a*) 1 only (*b*) 1 and 2 only
(*c*) 2, 3, 4 only (*d*) 1, 2, 3, 4 **[IES 2009]**

529. Which of following are included in architecture of computer?
1. Addressing modes, CPU
2. Instruction set, data formats
3. Secondary memory, operating system

Select the correct answer using the following codes:
(*a*) 1, 2 (*b*) 1, 3
(*c*) 2, 3 (*d*) 3 only **[IES 2009]**

530. Match List I (Type of memory) with List II (Used as) and select the correct answer using the given codes.

List I	List II
A. DRAM	1. Cache memory
B. SRAM	2. Main memory
C. Parallel access registers	3. BIOS memory
D. ROM	4. CPU registers

Codes :

	A	B	C	D
(*a*)	1	4	2	3
(*b*)	3	4	2	1
(*c*)	1	2	4	3
(*d*)	3	2	4	1

[IES 2009]

531. Which of the following are memory performance parameters?
1. Access time and latency
2. Block size and Block access time
3. Cycle time and bandwidth

Select the correct answer using the following codes:
(*a*) 1 only (*b*) 1, 2 only
(*c*) 2, 3 only (*d*) 1, 2, 3 **[IES 2009]**

532. What is the address space of 8086 CPU?
(*a*) one megabyte (*b*) 256 kilobytes
(*c*) 1 K megabyte (*d*) 64 kilobites **[IES 2009]**

533. Which of the following counters can be used to divide the clock frequency of a microprocessor by 5?
(*a*) 3 bit counter (*b*) 5 bit counter
(*c*) mode 3 counter (*d*) mode 5 counter **[IES 2009]**

534. Which of the following is used as the interface chip for data transmission between 8086 and 16 bit ADC?
(*a*) 8259 (*b*) 8255
(*c*) 8253 (*d*) 8251 **[IES 2009]**

535. What is the purpose of start bit in RS232 serial communication protocol?
(*a*) To synchronise receiver for receiving every byte
(*b*) To synchronise receiver for receiving a sequence of byte
(*c*) Acts as a parity bit
(*d*) To synchronise receiver for receiving the last byte **[IES 2009]**

536. The resolution of DAC depends on
(*a*) number of bits (*b*) monotonocity
(*c*) reference (*d*) value of resistance **[IES 2009]**

537. What is the purpose of DMA facility in microprocessor-based system?
(*a*) To increase speed of data transfer between μP and I/O devices
(*b*) To increase speed of data transfer between μP and memory
(*c*) To increase speed of data transfer between memory and I/O devices
(*d*) To improve reliability of system **[IES 2009]**

538. Personal computer cannot be used for
(*a*) game playing (*b*) weather forecasting
(*c*) office automation (*d*) home computing **[IES 2009]**

539. Parallel computers are divided into which of following architecture configurations?
1. Array processor 2. Data processors
3. Multiprocessor system 4. Pipeline computer

Give the answer using the following code :
(*a*) 1, 2, 3 only (*b*) 2, 3, 4 only
(*c*) 1, 3, 4 only (*d*) 1, 2, 4 only **[IES 2008]**

540. A microprogram is
(*a*) any source program run on microcomputers
(*b*) any set of instructions for primitive operation in a system
(*c*) a general name for macros used in assembly language programming
(*d*) a program of very small size **[IES 2008]**

541. A magnetic drum 8 inches diameter has 100 tracks and storage capacity of 200 bits/inch. What is the storage capacity?

(*a*) 8402 bits (*b*) 202400 bits
(*c*) 502400 bits (*d*) 1004800 bits **[IES 2008]**

542. If 8085 adds 87 H and 79 H then
(*a*) both CARRY and ZERO flags will be set to 0
(*b*) CARRY flag set to 0 and ZERO flag to 1
(*c*) CARRY flag set to 1 and ZERO flag to 0
(*d*) Both CARRY and ZERO flags set to 1 **[IES 2008]**

543. In 8085 which of the following machine cycles are not used in CALL instruction?
1. Instruction fetch 2. I/O
3. Memory read 4. Memory write
Select your answer using these codes:
(*a*) 2 only (*b*) 1 and 4
(*c*) 2, 3, 4 (*d*) None of these **[IES 2008]**

544. In 8085, if clock frequency is 5 MHz time required to execute an instruction of 18 T states is
(*a*) 3 μs (*b*) 3.6 μs
(*c*) 4 μs (*d*) 6 μs **[IES 2008]**

545. Which of the following interrupts is both level and edge sensitive?
(*a*) RST7.5 (*b*) RST5.5
(*c*) TRAP (*d*) INTR **[IES 2008]**

546. A group of personal computers are configured to work together for speeding up the execution of a single program in
(*a*) simulated evaluation (*b*) cluster computing
(*c*) network computing (*d*) client server computing **[IES 2008]**

547. A memory system of size 32 k bytes is required to be designed using memory chips which have 12 address lines and 4 data lines. What is the number of chips required?
(*a*) 4 (*b*) 8
(*c*) 16 (*d*) 32 **[IES 2008]**

548. Match List I and List II and select the correct answer using the codes.

List I	**List II**
A. Immediate addressing	1. LDA 30 FF
B. Implied addressing	2. MOV A, B
C. Register addressing	3. LX1H, 2050
D. Direct addressing	4. RRC

Codes:

	A	B	C	D
(*a*)	3	4	2	1
(*b*)	2	1	3	4
(*c*)	3	1	2	4
(*d*)	2	4	3	1

[IES 2008]

549. What does the minimising of the memory requirement of a program mean?
(*a*) maximizing the number of macro calls
(*b*) minimizing the number of macro calls
(*c*) maximizing depth of recursion
(*d*) both (*a*) and (*c*) **[IES 2007]**

550. In a micro computer why are wait states used?
(*a*) To make processor wait during DMA operation
(*b*) To make processor wait during interrupt processing
(*c*) To make processor wait during power shutdown
(*d*) To interface slow peripherals to processor **[IES 2007]**

551. In which unit is the performance of cache memory measured?
(*a*) Hz (*b*) bits/s
(*c*) cache constant (*d*) hit ratio **[IES 2007]**

552. Three memory chips are of size 1 kB, 2 kB, 4 kB. Their address bus is 10 bits. What are their data bus sizes?
(*a*) 8 bits, 16 bits, 24 bits respectively
(*b*) 8 bits, 16 bits, 32 bits respectively
(*c*) 8 bits, 16 bits, 64 bits respectively
(*d*) 8 bits, 16 bits, 128 bits respectively **[IES 2007]**

553. A memory system of 16 kB is required using chips having 12 address lines and 4 data lines. What is the number of chips?
(*a*) 2 (*b*) 4
(*c*) 8 (*d*) 16 **[IES 2007]**

554. Which of the following correctly defines temporal locality?
(*a*) Adjacent instructions for current instruction may be needed soon
(*b*) Current instruction being fetched may be needed again soon
(*c*) Instructions temporarily residing in memory
(*d*) none of above **[IES 2005]**

555. Consider the following statements :
1. Cache memory is low cost and fast
2. Cache memory fast but costly
3. Performance of cache memory is measured by hit ratio
Which of the above are correct?
(*a*) 1 and 2 (*b*) 2 and 3
(*c*) 3 and 1 (*d*) 1, 2, 3 **[IES 2005]**

556. Consider the following statements :
SIM instruction outputs the contents of accummulator to define
1. interrupt mask bit
2. interrupt pending bit
3. serial input data line
4. serial output data line
Which of above statements are correct?
(*a*) 1 and 2 (*b*) 2 and 3
(*c*) 3 and 4 (*d*) 1 and 4 **[IES 2005]**

557. What are the number of machine cycles *n* and the type of machine cycles carried out for PUSH B?
(*a*) $n = 2$, fetch and memory write
(*b*) $n = 3$, fetch and 2 memory write
(*c*) $n = 3$ fetch, memory write and read
(*d*) $n = 3$ fetch and 2 memory read **[IES 2005]**

558. In a 500 × 500 matrix, 95% of elements are zero and these elements are randomly distributed. Which is the appropriate data structure to store it efficiently.
(*a*) array (*b*) tree
(*c*) list (*d*) stack **[IES 2005]**

559. Which block replacement algorithm is generally not used in cache operation?

(*a*) LIFO (*b*) FIFO

(*c*) LRU (*d*) Random **[IES 2006]**

560. If PUSH and POP operations on a stack takes 1 unit time how much time would it take to delete an element at n^{th} position from bottom?

(*a*) 1 (*b*) n

(*c*) n^2

(*d*) cannot be determined from given data **[IES 2006]**

561. Self-referential structures of C programming language are very useful in applications which involve

(*a*) graphs (*b*) lists

(*c*) queues (*d*) stacks **[IES 2006]**

562. Which one is correct for ASCII code of all upper case English letters compared to all lower case English letters

(*a*) are all larger

(*b*) are all smaller

(*c*) are all equal

(*d*) some are larger, some smaller **[IES 2006]**

563. In 8085 assembly language program given below, the contents of accumulator after execution of program is

```
300    MVI A,45
3002   MOV B,A
3003   STC
3004   CMC
3005   RAR
3006   XRA B
```

(*a*) 00H (*b*) 45H

(*c*) 67H (*d*) E7H **[Gate 2010]**

564. In a microprocessor the service routine for a certain interrupt starts from a fixed location of memory which cannot be externally set but the interrupt can be delayed or rejected. Such an interrupt is

(*a*) non-maskable and non-vectored

(*b*) maskable and non-vectored

(*c*) non-maskable and vectored

(*d*) maskable and vectored **[Gate 2009]**

565. An 8085 executes the following instructions

```
2710   LXI H, 30 AOH
2713   DAD H
2714   PCHL
```

All addresses and constants are in H. Let PC be the contents of program counter and HL be the contents of HL register pair after executing PCHL. Then which of the following is correct?

(*a*) PC = 2715 H, HL = 30 A0H

(*b*) PC = 30 A0H, HL = 2715 H

(*c*) PC = 6140 H, HL = 6140 H

(*d*) PC = 6140 H, HL = 2715 H **[Gate 2008]**

566. Figure 11.22 shows astable multivibrator using IC 555. The circuit is oscillating steadily. V_C can vary between

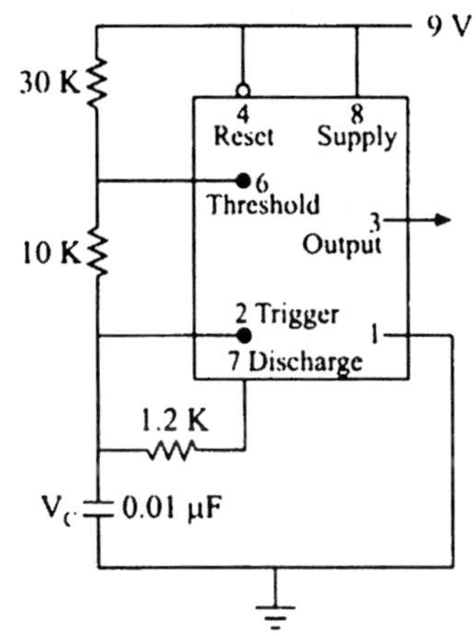

Fig. 11.22.

(*a*) 3 to 5 V (*b*) 3 to 6 V

(*c*) 3.6 to 6 V (*d*) 3.6 to 5 V **[Gate 2008]**

567. An 8085 program is as under

```
Line 1   MVI A, B5 H
2        MVI B, 0E H
3        XRI 69 H
4        ADD B
5        ANI 98 H
6        CPI 9 FH
7        STA 3010 H
8        HLT
```

After execution of ADD instruction contents of accumulator are

(*a*) C3H (*b*) EAH

(*c*) DCH (*d*) 69H **[IES 2007]**

568. In Q 567 status of CY and Z after execution of line 7 of the program will be

(*a*) CY = 0, Z = 0 (*b*) CY = 0, Z = 1

(*c*) CY = 1, Z = 0 (*d*) CY = 1, Z = 1 **[IES 2007]**

569. The following program starts at location 0100H

```
LXI SP, 00FF
LXI H, 0107
MVI A, 20 H
SUB M
```

The contents of accumulator when program counter reaches 0109 H is

(*a*) 20 H (*b*) 02 H

(*c*) 00 H (*d*) FF H **[IES 2005]**

570. In continuation of Q. 569, the following program exists from 0109 H onwards

```
ORI 40 H
ADD M
```

What will be contents of accumulator after the last instruction?

(*a*) 40 H (*b*) 20 H

(*c*) 60 H (*d*) 42 H **[IES 2005]**

B. The following questions have two statements. Assertion (A) and Reason (R). Read them and answer as per following code.
(a) Both A and R are correct and R is correct explanation of A
(b) Both A and R are correct but R is not correct explanation of A
(c) A is correct R is wrong
(d) A is wrong R is correct

571. A. The data bus and address bus of 8085 are multiplexed.
R. Multiplexing reduces the number of pins.

572. A. A dynamic RAM chip has much smaller cell than a static RAM chip.
R. In dynamic RAM one bit of information is stored as the charge on a capacitor.

573. A. Dynamic RAM needs refreshing.
R. Capacitors tend to loose charge.

574. A. Static RAMS are used where speed of operation is of primary concern.
R. Dynamic RAM has lower cost per bit than static RAM.

575. A. A RAM with access time of the order of hundreds of nano seconds is suitable for a control memory.
R. The time taken to execute an operation in a microcomputer is critically dependent on access time of control memory.

576. A. PC-XT is a personal computer.
R. PC-XT uses a floppy drive and has a medium-sized hard disc.

577. A. Microprocessors 8086 and 8088 are very similar in organisation and have identical instruction sets.
R. If clock speed is same the execution time of 8088 and 8086 are exactly equal.

578. A. 8086 is a 16 bit processor and has 40 pins.
R. 8085 is an 8 bit processor and has 40 pins.

579. A. Structures are useful in DBMS.
R. Structures have a very large number of applications besides DBMS.

580. A. The 'do-while' statement is used less frequently than the 'while' statement.
R. For most applications it is more natural to test for continuation of loop at the beginning than at the end of loop.

581. A. Most PCs use static RAM for main memory.
R. Static RAM is faster than dynamic RAM.

582. A. A PROM is a user programmable ROM.
R. After programming PROM behaves like ROM.

583. A. IC 74181 is an ALU in TTL 7400 series.
R. IC 74181 can accept two 4 bit words and can perform any 1 of possible 16 arithmetic or logic operations.

584. A. To add 8 bit words, two IC 74181 can be cascaded.
R. In IC 74181 a sum greater than 15 leads to high carry out.

585. A. Microprocessor 8085 has 16 address bus lines and 8 data bus lines.
R. Address bus in 8085 is unidirectional.

586. A. In 8085, the zero flag is set when ALU operation results in 0. If the result is not 0, this flag is reset.
R. In 8085 the S flag is set when the contents of accumulator become negative during execution of an instruction.

587. A. In 8085, P flag is for parity checking.
R. If the result of ALU has even parity, P flag is set and if the result has odd parity, P flag is reset.

588. A. Microprocessor 8085 has an 8 bit register called accumulator which stores intermediate answers during a computer operation.
R. Accumulator can send or receive data.

589. A. Microprocessor 8085 can address 65536 memory locations.
R. Microprocessor 8085 has 16 address lines.

590. A. An instruction cycle consists of fetch and execute cycles.
R. Fetch cycle has one machine cycle but execute cycle may have one or more machine cycles depending on length of instruction.

591. A. In 8085 SP and PC are 8 bit each.
R. 8085 is an 8 bit microprocessor.

592. A. Microprocessor 8085 has six 8 bit registers B, C, D, E, H L.
R. B-C, D-E, H-L may be combined to form register pairs to hold 16 bits data.

593. A. In microprocessor 8085 instruction LXIB, 90 FF H means register B and C are loaded with upper and lower bytes to get B = 90 H and C = FFH.
R. In 8085 the stack pointer indicates which memory location is to accessed.

594. A. If contents of F register in 8085 are 01010001, it means S = 0, Z = 1, AC = 1, P = 0 and CY = 1.
R. When an instruction is called from memory, the op code of instruction is stored in instruction register.

595. A. Microprocessor 8085 has facility of interrupt control.
R. TRAP, RST 7.5, RST 6.5, RST 5.5 and INTR are interrupt signals in 8085.

596. A. Microprocessor 8085 has on-chip oscillator with inbuilt crystal.
R. For frequency stability crystal oscillator is preferred.

597. A. In microprocessor 8085 DMA allows direct access to memory to speed up data transfer.
R. HOLD and HLDA signals are used for DMA operations.

598. A. Each memory chip has its own address latch.
R. ALE signal comes out of microprocessor 8085 and goes to memory chip.

599. A. In 8085 when RESET OUT goes low processing begins.
R. CPU remains reset till $\overline{\text{RESET IN}}$ signal goes high.

600. A. In 8085 $\overline{WR}$ and $\overline{RD}$ signals are active high.

R. LOW $\overline{WR}$ means write operation and low $\overline{RD}$ means read operation.

601. A. A two byte instruction of 8085 has an operation code in first byte and operand/address in the second byte.

R. If an instruction is too long it is possible to reduce its length by a number of methods.

602. A. If source and destination addresses are made implicit the length of instruction is reduced.

R. 8085 has only 3 addressing modes.

603. A. In direct addressing the address of operand is specified in the instruction.

R. Direct addressing is also called absolute addressing.

604. A. In register indirect addressing mode of 8085, the address of operand is in the specified register.

R. An example of register indirect addressing mode is MOV, A, M.

605. A. Logical group of instructions contains instructions for logic operations.

R. Data transfer group instruction is basically a copying instruction.

606. A. Interfacing means synchronisation of digital information transmission between input output devices and computer.

R. A computer with higher bit word size executes programs at a faster rate.

607. A. Magnetic tape can be accessed only sequentially.

R. Magnetic tape consists of magnetised spots or tracks.

608. A. Mnemonics are used in assembly level program.

R. Computer cannot understand mnemonics.

609. A. A higher level language is less efficient than machine language but more efficient than assembly language.

R. It is easier to learn and use higher level language than machine language.

610. A. Negative values of incremental operator in DO loop are allowed in Fortran 77 but not in earlier versions of Fortran.

R. Fortran 77 has better array facilities than earlier versions of Fortran.

611. A. A good algorithm should be fast, require minimal computer memory and time.

R. Many times an algorithm is prepared in the form of a flow chart.

612. A. The variable name 5KING is not valid in Fortran 77.

R. In Fortran 77 the first character of a variable name must be an alphabet.

613. A. In Fortran 77 the statement 'IMPLICIT INTEGER (A-E) identifies all variable name which begin with A, B, C, D, or E as being of integer type.

R. IMPICIT statement in Fortran 77 is an improvement over earlier versions of Fortran.

614. A. C has 32 keywords like auto, if, do, float etc.

R. Keywords cannot be used as variable names in C.

615. A. In C a character constant is either a single alphabet or digit or special symbol enclosed within single inverted commas.

R. In C the maximum length of a character constant can be 2 characters.

616. A. In C a variable name can have up to 8 characters and special symbols can also be part of variable name.

R. In C a type declaration at the beginning of program indicates the type of variable name used in the program.

617. A. In C arithmetic operations are permissible on ints, floats and chars.

R. In C every arithmetic operator must be written explicitly.

618. A. In C it is possible for functions to call themselves.

R. C makes use of 'Pointers'.

619. A. C allows the use of logical operators, AND, OR, NOT.

R. C has logical operators as well as bit wise operators like AND, OR etc.

620. A. C has a large number of library functions classified as arithmetic functions, data conversion functions, character classified functions, string manipulation functions, searching and sorting functions etc.

R. 8086 uses a segment : offset addressing scheme.

621. A. The main difference between a microprocessor and a microcontroller is that former does not have on chip main memory but the later has.

R. A microprocessor does not need memory to run program. **[IES 2010]**

622. A. A logic analyser offers a delayed sweep.

R. Because the logic analyser sweep is really a clock signal. **[IES 2010]**

623. A. When you turn on PC a process called POST (Power on self test) begins with an electrical signal.

R. The electrical signal restores left over data from the chips internal memory register. **[IES 2010]**

624. A. DMA is faster than either interrupt initiated or polling based I/O for very larger data transfer.

R. DMA takes control of system buses and needs no processor intervention during data transfer. **[IES 2007]**

625. A. Neumann machines are called control flow computers.

R. Instructions are executed sequentially as controlled by program counter. **[IES 2007]**

626. A. Data flow computers exploit maximum parallelism.

R. Data flow computers require no program counter. **[IES 2007]**

627. A. In a microprogrammed CPU each machine instruction is executed by a real time interpreter.

R. Real time interpreter helps to achieve high degree parallelism in microprogrammed control. **[IES 2007]**

628. A. The development of a microprocessor based product requires the design of program and hardware.

R. The design effort for an electronic product follows the same basic steps used in the development of software. [IES 2006]

629. A. Processor level design is very much heuristic process.
R. At this level of abstraction, components are very complex. [IES 2006]

630. A. Binary search function can be easily implemented using recursion.
R. Recursion is based on number of elements in array to be searched. [IES 2006]

C. Indicate whether the following statements are True (T) or False (F)

631. A microprocessor needs memory chips and input-output chip to form a micro-computer.
632. System bus is the communication channel between microprocessor and peripherals.
633. Contents of RAM cannot be altered.
634. 8085 data bus has 8 lines and is bidirectional.
635. 8 bit data can have 256 numbers ranging from 00000000 to 1111111 1.
636. Accumulator is an 8 bit register and is also known as register B.
637. The accumulator drives the ALU.
638. Temporary register in 8085 is a 16 bit register.
639. The input to ALU are the contents of accumulator and temporary register.
640. 8085 has 6 sign flags.
641. The S (sign) flag is set when the contents of accumulator become negative during an operation.
642. Z flag is reset if an ALU operation results in 0.
643. P flag is for parity.
644. Parity means the same as positive or negative number.
645. In 8085, F register has 8 bits.
646. Instruction cycle is the time required to complete the execution of an instruction.
647. Machine cycle is always the same as instruction cycle.
648. A machine cycle has one clock cycle.
649. One T state is precisely equal to one clock period.
650. The combination of registers in 8085 is B-C, D-E, H-L only.
651. The stack pointer holds the address of the desired memory location.
652. In 8085 eight address and data buses are multiplexed.
653. DMA stands for direct memory access.
654. 8085 needs +10 V supply.
655. In computers the secondary memory is faster than primary memory.
656. In computers the size of secondary memory is usually much larger than that of primary memory.
657. When the mains supply is switched off the contents of primary computer memory are lost.
658. In a computer the data transfer between hard disk and CPU is nearly the same as that between diskette and CPU.
659. Fortran 77, C and Pascal are high level languages.
660. It is difficult to write a program in a machine language.
661. It is possible to copy a file in the same directory.
662. DOS has a command named 'DIRECTORY'.
663. X COPY is an internal DOS command.
664. Windows can run several tasks simultaneously.
665. The first row of the window is called 'Title bar'.
666. The box on the external left of the title bar of menu is called control menu box.
667. In windows the middle button of mouse is not used.
668. One can format a disc using window explorer.
669. Plug and play is a hardware specification.
670. Windows automatically starts CD player when one inserts audio CD in CD-ROM drive.
671. Internet is a world wide network of computers where most of the information is freely available.
672. An e-mail message can be sent to many recipients.
673. Windows have a user-friendly program called 'Front Page Express' for creating web pages.
674. In a computer MPU is the centre of all operations and control.
675. Diskette can be used for primary storage as well as secondary storage in a computer.
676. RAM is a volatile memory device.
677. Computers 'Basic input output system' is generally stored in RAM chips.
678. In a computer the instructions, data, intermediate and final results during processing are held in ALU.
679. Mouse is a very suitable input device for computers.
680. Page maker is an application software for DTP.
681. Almost all high level languages have provision for logical operations.
682. Both ROM and RAM can be accessed randomly.
683. E-mail can be used to send text, graphics, and video.
684. A PC can be upgraded to a multi media PC by adding CD ROM drive, sound card etc.
685. MS Excel is an application software for financial planning.
686. PowerPoint is a presentation software.
687. MS Access is a DBMS software.
688. MS Outlook is a software for DTP.
689. The number of computer viruses is some thousands.
690. Computer virus can hide itself in a diskette but not in a hard disk.
691. IC 7485 cannot be cascaded.
692. IC 74181 is a 4 bit device.
693. To add 8 bit words, two IC 74181 can be cascaded.
694. In IC 74181 a sum greater than 15 leads to low carry out.
695. IC 74181 performs arithmetic operations when M = 1.
696. IC 74181 has two select lines.
697. A computer memory is an ordered sequence of storage cells.
698. A compiler converts a high level computer program into machine language program.

699. A high level language is less efficient than machine language.
700. The character set of Fortran 77 includes lower case alphabets *a* to *z*.
701. – 56 is not a valid real constant in Fortran.
702. In Fortran 77 a variable name can contain special characters.
703. Some words like READ, WRITE, STOP etc. cannot be used as variable names in Fortran 77.
704. Fortran 77 has six arithmetic operators.
705. In Fortran 77 negation is a unary operation.

D. Fill in the Blanks with suitable words.

706. ....bus is unidirectional.
707. ....bus is bidirectional.
708. The results of arithmetic operations are stored in .... or ....
709. In a microprocessor the .... unit provides the .... and .... signals.
710. In 8085 accumulator has .... bits.
711. The arithmetic and logic operations are done by ....
712. In 8085 carry flag is .... when an arithmetic operation results in carry.
713. The fetch cycle has .... machine cycle.
714. .... allows direct access to memory.
715. ALE stands for ...., ..., ...
716. In 8085 $\overline{WR}$ and $\overline{RD}$ are active.................. .
717. In a microprocessor, oscillator frequency is .... the clock frequency.
718. 8088 has .... bit data bus.
719. EU stands for ...., ...
720. BIU stands for ...., ...
721. EU does the job of .... and ....
722. The instruction MOV r_1, r_2 causes transfer of data from register .... to register ....
723. In instruction MVI *r*, data the addressing mode is ....
724. The instruction STA addr causes movement of data from .... to the .... whose address is specified in the bytes 2 and 3 of instruction.
725. The instruction ADI data causes addition of .... second byte of instruction to the contents of ....
726. The instruction ADC *r* causes contents of .... and .... to be added to ....
727. The instruction SBB *r* means ...., ..., ..., ...
728. The instruction INR *r* causes contents of .... to be .... by ....
729. In instruction ANA *r*, the contents of register *r* are .... with the contents of ....
730. The addressing mode in instruction XRI data is ....
731. The instruction RCL causes contents of .... to be .... one position.
732. The number of logic or arithmetic operations which IC 74181 can carry out is ....
733. IC 74181 can accept two words each of .... bits.
734. To add two 8 bit words two IC 74181 have to be ....
735. In IC 74181 addition and subtraction is in ....
736. When two IC 74181 are cascaded the ....of first stage is connected to ....of second stage.
737. In Pascal the largest integer constant which can be accepted depends on the ....
738. A real constant may be expressed in .... or ....
739. In Pascal the word .... is used as a general term for names given to variables, functions, arrays etc.
740. In Pascal a valid declaration for real variables ABC, BETA, GAMMA is ....
741. If A, B, C are integers then the integer expression A/B + C in Pascal is ....
742. If K M are integers, then the integer expression K div-M in Pascal is ....
743. If X, Y, Z, P are real, then the expression (X + Y) (P + Z) in Pascal is ....
744. In Pascal no .... are allowed in a variable name.
745. In Pascal an integer .... be compared with a real quantity.
746. In Pascal the number of operators which operate on sets is ....
747. In Fortran 77 the common log (*i.e.*, to the base 10) is given the name of ....
748. In Fortran 77 rounding to the nearest whole number is done by the function named ....
749. In Fortran 77 assignment statement is used to .... values to variables.
750. Hyperbolic trigonometric functions are .... in Fortran 77.
751. The number of arithmetic operators in Fortran 77 is ....
752. A collection of values or data which are related in some way is called an ....
753. In Java the number of basic variable types for storage of integers, floating point numbers, characters and Boolean values are ....
754. The three kinds of variables in Java are called ....
755. In Java the number of data types which can be used to store integers are ....
756. In Java any number, text or other information that directly represent a value is called ....
757. The number of operators to accomplish basic arithmetic in Java is ....
758. In Java the modulus operator is written as ....
759. In Java the logical AND is written as ....
760. In Java if one assigns an integer value to a variable that is too large for that variable to hold then .... occurs.
761. The two categories of constants are called ....
762. In C a primary constant can be ....
763. In C the mantissa and exponential parts of a real constant are separated by letter ....

764. In C the maximum number of characters in a variable name can be ....

765. In C the number of key words is ....

766. In C the instruction used to declare the type of variables used in the program is called ....

767. The primary memory in most PCs is about .... or more.

768. The 3.5" floppy is usually called .... floppy.

769. The capacity of a zip floppy is about ....

770. The DOS command to compare two diskettes is ....

771. In Word Pad the default left and right hand page margins are .... each.

772. HTML stands for ....

773. The name of the window program used to create and edit web pages is ....

774. Dr. Solomon's is the name of a .... program.

775. The program which converts source code into object code is called ....

776. In banking a commonly used direct entry input device is ....

777. In 16 bit computer the .... is 16 bit wide.

778. WYSIWYG stands for ....

779. In a computer the .... bus is unidirectional.

780. In ALU the status register is called ....

ANSWERS WITH HINTS AND EXPLANATIONS

1. (*a*) **2.** (*b*) **3.** (*a*)

4. (*a*) See section 11.4.9.

5. (*b*) See Fig. 11.3. **6.** (*a*) See Section 11.4.3.

7. (*a*) See Section 11.4.5. **8.** (*b*) See Fig. 11.3.

9. (*b*) See Section 11.5. **10.** (*a*) See Section 11.4.9.

11. (*a*) See Section 11.4.10. **12.** (*a*) See Section 11.4.3.

13. (*b*) **14.** (*c*) **15.** (*c*)

16. (*b*) ALU performs arithmetic and logical computations.

17. (*a*) See Section 11.3.3.

18. (*a*) Use decimal to binary conversion method.

19. (*a*) **20.** (*b*) **21.** (*c*) **22.** (*a*)

23. (*c*) Complete real number has all the three, *i.e.*, sign, integer part and fraction part.

24. (*c*) A real number has a decimal point.

25. (*b*) Convert to decimal.

26. (*c*) Binary .101 equals 0.625 in decimal.

27. (*c*) **28.** (*a*) **29.** (*c*) **30.** (*a*)

31. (*a*) **32.** (*d*)

33. (*c*) One byte has 8 bits.

34. (*c*) **35.** (*a*) **36.** (*a*) **37.** (*b*)

38. (*c*) See Section 11.4.

39. (*a*) **40.** (*a*)

41. (*c*) See Section 11.4.4.

42. (*a*) See Fig. 11.18.

43. (*c*) See Section 11.10.3.

44. (*a*) See Section 11.10.3.

45. (*a*) See Section 11.10.3.

46. (*c*) See Section 11.10.3.

47. (*b*) See Section 11.10.3.

48. (*a*) See Section 11.10.3.

49. (*c*)

50. (*d*) See Section 11.10.3.

51. (*c*) See Section 11.10.3.

52. (*a*)

53. (*a*)

54. (*c*) See Section 11.10.3.

55. (*a*) See Section 11.10.3.

56. (*b*) See Section 11.10.3.

57. (*a*) See Section 11.10.3.

58. (*c*) **59.** (*c*) **60.** (*c*) **61.** (*c*)

62. (*b*) **63.** (*d*)

64. (*c*) See 8085 instruction sheet.

65. (*b*) **66.** (*c*) **67.** (*a*) **68.** (*a*)

69. (*b*) RAL means 'rotate left through carry'.

70. (*b*)

71. (*c*) When RAL instruction is used once the number is doubled.

72. (*b*) When RRC is used once the number is divided by 2.

73. (*b*) Adding X to PROD, Y times evidently gives the product XY.

74. (*b*) **75.** (*c*) **76.** (*c*) **77.** (*c*)

78. (*c*) **79.** (*a*) **80.** (*a*) **81.** (*d*)

82. (*a*) Cache memory has a speed of about four times the speed of main memory. It is a small high speed memory.

83. (*b*) Size of Cache memory varies from about 16 K bytes to about 256 K bytes.

84. (*a*) Time required to access different cells is the same.

85. (*d*) All these characteristics are important.

86. (*c*) **87.** (*d*) **88.** (*a*) **89.** (*a*)

90. (*b*) The starting word address is 0.

91. (*a*)

92. (*d*) See Section 11.8.

93. (*d*) **94.** (*c*) **95.** (*b*)

96. (*c*) Digital chips need 5 V supply.

97. (*c*) **98.** (*b*) **99.** (*b*) **100.** (*a*)

101. (*b*)

102. (*b*) Nibble has 4 bits.

103. (*a*) Power requirement in stand by mode is very low.

104. (*c*) **105.** (*a*) **106.** (*b*)

107. (*b*) $\text{MTBF} = \dfrac{1000}{0.01}$

108. (*c*) It is LSI chip.

109. (*b*) Instruction cycle consists of fetch and execute cycles.

110. (*c*) See Fig. 11.4.

111. (*b*) Some instructions do need more than one machine cycle.

112. (*c*) See Fig. 11.4.

113. (*c*) See Fig. 11.6.

114. (*c*) **115.** (*c*)

116. (*d*) See Section 11.4.8.

117. (*b*) See Section 11.4.8.

118. (*a*) See Section 11.4.8.

119. (*d*) See Section 11.4.8.

120. (*b*) See Section 11.4.5.

121. (*d*) **122.** (*c*) **123.** (*b*) **124.** (*d*)

125. (*a*) **126.** (*d*) **127.** (*b*) **128.** (*c*)

129. (*a*) **130.** (*a*) **131.** (*b*) **132.** (*b*)

133. (*a*) See handbook of digital chips.

134. (*b*) **135.** (*b*) **136.** (*d*) **137.** (*d*)

138. (*b*) See Section 11.7.

139. (*c*) See Section 11.7.

140. (*d*)

141. (*a*) See Section 11.7.

142. (*a*)

143. (*b*) See Section 11.7.

144. (*c*) See Section 11.7.2.

145. (*a*) See Section 11.7.

146. (*c*) See Section 11.7.2.

147. (*a*) See Section 11.7.2.

148. (*a*) See Section 11.7.2.

149. (*a*) See Section 11.8.

150. (*b*) See Section 11.8.

151. (*c*) See Section 11.7.

152. (*a*) See Section 11.13.

153. (*a*) See Fig. 11.13.

154. (*b*) See Fig. 11.13.

155. (*a*) See Fig. 11.7.

156. (*c*)

157. (*a*) See Section 11.4.5.

158. (*b*) See Section 11.4.5.

159. (*c*) See Section 11.11.

160. (*d*) **161.** (*a*)

162. (*a*) See 8085 instruction sheet.

163. (*a*) See Section 11.4.5.

164. (*a*) See Section 11.4.1.

165. (*d*) See 8085 instruction sheet.

166. (*a*) See 8085 instruction sheet.

167. (*c*) **168.** (*a*)

169. (*a*) See 8085 instruction sheet.

170. (*a*) See 8085 instruction sheet.

171. (*a*) See 8085 instruction sheet.

172. (*c*) See 8085 instruction sheet.

173. (*d*) See 8085 instruction sheet.

174. (*c*) See 8085 instruction sheet.

175. (*b*) See 8085 instruction sheet.

176. (*d*) See 8085 instruction sheet.

177. (*d*) **178.** (*d*)

179. (*a*) See Fig. 11.19.

180. (*a*) See Section 11.10.3.

181. (*c*)

182. (*b*) See Section 11.11.

183. (*d*) **184.** (*a*) **185.** (*a*) **186.** (*b*)

187. (*c*) If instruction RAL is executed once the number gets doubled.

188. (*d*) See Section 11.3.

189. (*c*) **190.** (*c*)

191. (*c*) See Fig. 11.19.

192. (*a*)

193. (*a*) See 8085 instruction sheet.

194. (*b*) **195.** (*b*)

196. (*a*) 2's complement is universally used in computers.

197. (*a*) In cascading carry out of first stage is connected to carry in of second stage.

198. (*c*)

199. (*c*) $\dfrac{1}{3 \times 10^6} = 0.333 \times 10^{-6}$ s.

200. (*a*) See Section 11.4.3.

201. (*d*) Integer constant should not have decimal point.

202. (*d*) – 810 is the only one without decimal point.

203. (*b*) Comma is not allowed in real constant.

204. (*a*) It is the only one with decimal and without comma.

205. (*a*) **206.** (*b*) **207.** (*c*) **208.** (*c*)

209. (*a*) See Section 11.17.

210. (*d*) **211.** (*a*)

212. (*b*) Integer variable must start with I, J, K, L, M, N.

213. (*d*) The first letter is not I, J, K, L, M, N.

214. (*d*) A real variable should not start with I, J, K, L, M, N.

215. (*b*) It starts with M and hence it is not a real variable.

216. (*a*) It is the only expression with correct symbols for exponential and product.

217. (*a*)

218. (*b*) It is the only one with decimal point.

219. (*a*) **220.** (*d*)

221. (*c*) All other operations require two operands.

222. (*b*) **223.** (*a*) **224.** (*a*)

225. (*c*) Two operators should not occur together.

226. (*c*)

227. (*b*) Only integer part appears in result.

228. (*a*) It is the only expression with correct symbols.

229. (*a*) It has a decimal point.

230. (*a*) **231.** (*d*)

232. (*b*) Others do not start withs I, J, K, L, M, N.

233. (*c*)

234. (*a*) The result should not contain decimal.

235. (*b*) The result should be sum of 50 and 30 and decimal must be added.

236. (*a*) $(95-32)\,\frac{5}{9} = 35$ and decimal must be added.

237. (*b*)

238. (*a*) Calculate and add decimal at proper place.

239. (*b*)

240. (*a*) $J = 12, K = 12^2 = 144, Z = 3 \times 24.4 = 73.2$. In the result J and K will be without decimal point.

241. (*b*)

242. (*a*) The given statement is logical IF statement. The control goes to statement 11 if SALT = PEPPER otherwise control goes to statement 13. Same is true in arithmetic statement (*a*).

243. (*c*) The given statement is logical IF statement. If SALT greater than or equal to PEPPER control goes to statement 11 otherwise control goes to statement 13. Same is true of arithmetic IF statement (*c*). If (SALT – PEPPFR) is negative, control goes to statement 13 and otherwise control goes to statement 11.

244. (*a*) The logical IF statement means that If SALT is less than or equal to PEPPER, control goes to statement 11 otherwise (*i.e.*, SALT greater than PEPPER), control goes to 13. Same is true of arithmetic IF statement (*a*).

245. (*a*) The logical IF statement means that IF (3 × J COKE) is equal to (L PEPSI), J COKE = 3 + 2 = 5 otherwise J COKE = 3 + 5 = 8. Since (3 × J COKE) is more than (L PEPSI), the result is 8.

246. (*b*) Since 2 × J COKE = 40 and (LPEPSI) is 15 control goes to statement J COKE = J COKE + 3 and result is 23.

247. (*a*) (J COKE – L PEPSI) is 11 – 19 = – 8. Since (J COKE – L PEPSI) is negative statement 5 is executed and result is 11 + 2 = 13.

248. (*a*) Since (J COKE –3 × L PEPSI) is negative, statement 5 is executed and the result is 98 + 5 = 103.

249. (*b*) It is a computed GO TO statement. Since J COKE = 3 + 1 = 4. the control is transferred to statement 11.

250. (*c*) Since L PEPSI = 2 × 2 + 1 = 5 the control goes to the fifth statement, *i.e.*, 13.

251. (*d*) All are valid.

252. (*a*) DO statement is executed when J = 1, 2, 3, 4, 5, 6. 7, *i.e.*, a total of 7 times.

253. (*d*) DO statement is executed when N = 1. 4, 7, 10. *i.e.*, a total of 4 times.

254. (*c*) DO loop executations are for A = 1.0, 1.5, 2.0, 2.5, 3.0, 3.5, 4.0, 4.5, 5.0, 5.5 or a total of 10 times.

255. (*d*) DO loop will not be executed since 0 is less than 10.

256. (*c*) $SUM = 1^2 + 2^2 + 3^2 + 4^2 + 5^2 + 6^2 + 7^2 + 8^2 + 9^2 + 10^2 = 385$.

257. (*c*) SUM = 2 + 4 + 6 + 8 + 10 + 12 + 14 + 16 + 18 = 90.

258. (*c*) 6! = 6 × 5 × 4 × 3 × 2 × 1 = 720.

259. (*b*) Domain of one DO loop is crossing the domain of another DO loop. This is not permissible.

260. (*c*) LI X = 3 and MIX = 2 + 3 = 5. Then LIX = 3 + 2 – 5 and MIX = 5 + 5 = 10. Since MIX is greater than 6, control goes to statement 25 according to which MIX = 3 × 10 = 30.

261. (*b*) Subscript R is not proper.

262. (*c*) Subscript not proper.

263. (*c*) **264.** (*b*) **265.** (*b*) **266.** (*b*)

267. (*a*) Both 1 and 2 are correct but 3 and 4 are wrong.

268. (*b*) **269.** (*d*) **270.** (*d*)

271. (*d*) See Section 11.16.

272. (*a*) **273.** (*c*) **274.** (*a*)

275. (*b*) See Section 11.16.

276. (*a*)

277. (*b*) Two operators cannot be together.

278. (*b*) Operator is missing.

279. (*a*) All others are incorrect.

280. (*a*) In Pascal unary has highest priority.

281. (*a*)

282. (*c*) Operators must be written.

283. (*b*) Operators must be written. They are not implied.

284. (*d*) In Pascal div and mod are not allowed with one real and the other integer operator.

285. (*a*) All are allowed in Pascal.

286. (*d*)

287. (*a*) Operator * must be written explicitly in Pascal.

288. (*a*) **289.** (*c*)

290. (*b*) $(2.57 + 0.06)\,10 = 26.3 \rightarrow \frac{26}{10} = 2.6$.

291. (*b*) $\frac{45}{10} = 4.5 \rightarrow 4 \times 10 = 40$.

292. (*a*) In Pascal multiplication has higher precedence than division so $10 \times 55 = \frac{550}{10} = 55$.

293. (*d*) **294.** (*a*) **295.** (*c*) **296.** (*d*)

297. (*d*) 'then' is missing.

298. (*c*) **299.** (*a*) **300.** (*c*) **301.** (*a*)

302. (*d*)

303. (*c*) All boolean operators are not scanned together. The order is NOT, AND, OR.

304. (*a*) All are correct.

305. (*a*)

306. (*a*) All are correct.

307. (*a*) All statements are true.

308. (*a*) All are true.

309. (*b*) Only 1, 2, 3 are correct and 4 is wrong.

310. (*c*)

311. (*a*) See Section 11.19.2.

312. (*b*) No comma or blank is allowed in integer constant in C.

313. (*c*) The range is from –32768 to +32767.

314. (*a*) Real constant must have a decimal point.

315. (*c*) See Section 11.19.2.

316. (*b*) E is not allowed and default sign of exponent is positive.

317. (*d*) In C real constant expressed in exponential form, mantissia and exponent should be separated by *e* (and not E).

318. (*a*) See Section 11.19.2.

319. (*d*) Character constant must be enclosed within single inverted commas.

320. (*b*) See Section 11.19.2.

321. (*a*) First character must be alphabet.

322. (*b*) First character must be alphabet only.

323. (*b*) First character is 1 which is not allowed.

324. (*a*) Comma is not allowed.

325. (*a*) else is not allowed as variable name in C.

326. (*a*) pop is not a keyword in C.

327. (*d*) See Section 11.19.2.

328. (*b*) See Section 11.19.2.

329. (*d*) See Section 11.19.2.

330. (*d*) **331.** (*a*)

332. (*d*) See Section 11.19.2.

333. (*c*) Left hand side must be a variable name.

334. (*b*) 0.562 is not an integer constant.

335. (*a*) Since both 5 and 2 are integers, the result will be 2.

336. (*b*) Mixed mode statement is allowed in C.

337. (*c*) $\frac{2.0}{5} = 0.4$.

338. (*c*) All operators must be written explicity in C.

339. (*a*) In C arithmetic operations can be performed on ints, floats and chars.

340. (*d*)

341. (*c*) $\frac{3}{10.1} = 0.297$. Since *k* is int hence answer is 0.

342. (*b*) $k = 0 + 1 - 2 = -1$ because *k* is an integer.

343. (*a*) See Section 11.19.5.

344. (*a*) See Section 11.19.5.

345. (*b*) $k = 1 + 1 + 8 - 2 - 0 = 8$ because *k* is an integer.

346. (*b*) See Section 11.19.5.

347. (*d*) **348.** (*b*) **349.** (*a*) **350.** (*a*)

351. (*c*) 7000.0 + 500.0 = 7500.0.

352. (*a*) $9000 + \frac{9000 \times 10}{100} = 9900$ and $14000 + 1500 = 15500$.

353. (*c*) Since pr = 43, the result will be printed as third.

354. (*b*) Since pr = 53, the result will be printed as second.

355. (*c*) **356.** (*d*)

357. (*c*) Whenever keyword 'break' is encountered inside any C loop control passes to the first statement after the loop.

358. (*c*) They are all used in C language.

359. (*d*) **360.** (*b*) **361.** (*b*)

362. (*b*) The two orders need not be the same in C.

363. (*b*)

364. (*c*) C has four storage classes viz. automatic storage class, register storage class, static storage class and external storage class.

365. (*c*) Preprocessor processes the source program before it is passed to compiler.

366. (*d*)

367. (*b*) An array can have all int or all floats or all chars.

368. (*d*)

369. (*a*) All are correct.

370. (*d*)

371. (*a*) All these statements are correct for Java.

372. (*b*) See Section 11.20.1.

373. (*a*) All these are correct for Java variables.

374. (*a*) See Section 11.20.2.

375. (*a*) See Section 11.20.2.

376. (*b*) See Section 11.20.4.

377. (*b*) Modulus operator gives the remainder of division as answer.

378. (*d*) **379.** (*a*)

380. (*b*) Java syntax is based on C++.

381. (*a*) The steps are : *a* is given the value 40, *b* is given the value of *a* before it is incremented (*i.e.*, 40) and *a* is incremented to 41, *a* is incremented to 42 and *c* is given this value. Thus the result is 42, 40, 42.

382. (*d*)
383. (*b*) See Section 11.20.7.
384. (*b*) An array on integers and strings is not possible.
385. (*a*) All are correct.
386. (*d*) The first letter of variable name should be lower case.
387. (*c*) In all others first letter is not lower case.
388. (*b*) Digit as first character is not allowed.
389. (*b*)
390. (*d*) The operator + = means add and assign.
391. (*b*)
392. (*b*) See Section 11.15.1.
393. (*c*) Statement number in basic should have only 5 digits.
394. (*d*) See Section 11.15.1.
395. (*a*)
396. (*d*) All are allowed as discussed in Section 11.15.1.
397. (*d*)
398. (*a*) Two operators cannot exist together.
399. (*d*) See Section 11.15.1.
400. (*b*) $3.1416 \times \frac{6^2}{4} = 28.27.$
401. (*a*) $S = 20 + 30 + 10 = 60,\ A = \frac{S}{3} = 20.$ Hence 60, 20, 20, 30, 10.
402. (*b*) **403.** (*a*)
404. (*a*) In this case Y > X, so the control goes to 25. The output is Y, Z, X which are 200, 250, 150.
405. (*b*) **406.** (*c*)
407. (*a*) It is called 'title bar'.
408. (*a*)
409. (*b*) Magnetic ink character recognition is used in banking.
410. (*a*) **411.** (*a*) **412.** (*d*)
413. (*d*) Zip floppy has high memory 100 MB or 250 MB.
414. (*a*) **415.** (*a*) **416.** (*b*) **417.** (*c*)
418. (*a*) Mouse is connected to serial port.
419. (*d*) Hard disc is a memory storage device.
420. (*d*) **421.** (*c*) **422.** (*d*) **423.** (*b*)
424. (*d*) **425.** (*d*)
426. (*a*) Portrait and worksheet are two orientations of worksheet page.
427. (*a*) MS Excel has all these features.
428. (*b*) MAX is not data base function.
429. (*b*) **430.** (*b*)
431. (*d*) Coral key is not DTP progr::n.
432. (*d*) dB III plus is DBMS and not for financial accounting.
433. (*b*) LISP is a computer language. The remaining are software packages.
434. (*c*) Primary memory holds these during processing of instructions.
435. (*c*) Magnetic tape can be accessed only sequentially.
436. (*c*) Diskette has volatile memory and therefore the matter has to be saved frequently.
437. (*d*) **438.** (*d*)
439. (*d*) Windows is an operating system.
440. (*a*) Modem is connected at serial port.
441. (*a*) **442.** (*c*) **443.** (*d*)
444. (*d*) LISP is a computer language.
445. (*a*) It gives remainder in division operation.
446. (*d*) **447.** (*b*)
448. (*a*) Both are true.
449. (*d*)
450. (*b*) *i* is incremented by 1.
451. (*a*)
452. (*c*) See Section 11.7.
453. (*b*) 8255 is interface chip for 8086 and ADC.
454. (*a*) **455.** (*b*)
456. (*b*) Assemblers do provide macro facilities.
457. (*d*) See Section 11.7.
458. (*a*) **459.** (*a*) **460.** (*a*) **461.** (*a*)
462. (*a*) Each SRAM cell has 4 MOS transistors.
463. (*a*)
464. (*a*) $2^8 = 256.$
465. (*b*) **466.** (*b*) **467.** (*b*) **468.** (*d*)
469. (*b*) **470.** (*c*)
471. (*d*) All the four are permitted.
472. (*b*) **473.** (*a*) **474.** (*b*)
475. (*a*) In normalised form, the integer part is zero.
476. (*a*)
477. (*c*) All are correct.
478. (*c*) Integer cannot start with A.
479. (*d*) Operator is not allowed.
480. (*c*) It is an integer. Hence N as first character.
481. (*c*) **482.** (*b*) **483.** (*b*) **484.** (*a*)
485. (*b*) **486.** (*a*)
487. (*b*) Exponents are made equal and integer part is made zero.
488. (*b*) **489.** (*c*) **490.** (*c*)
491. (*a*) All are correct.
492. (*a*) All are correct.
493. (*a*)
494. (*a*) Parenthesis has first priority.
495. (*c*) **496.** (*a*) **497.** (*c*) **498.** (*c*)
499. (*c*) **500.** (*c*) **501.** (*c*) **502.** (*b*)
503. (*c*) **504.** (*d*) **505.** (*a*) **506.** (*b*)

507. (*d*) **508.** (*c*)

509. (*b*) Average memory access time = (Hit ratio) (access time in cache memory) + (1 – hit ratio) (access time in main memory)

or 120 = (hit ratio) (100) + (1 – hit ratio) (1200)

or 12 = (10'– 120) hit ratio + 120

or hit ratio = $\frac{108}{110}$ = 0.98 or 98%

510. (*b*) **511.** (*b*) See section 11.4.8.

512. (*c*) See Fig. 11.5.

513. (*d*) MOV B, C 1 byte ORI 2 bytes STA 3 bytes

514. (*b*) **515.** (*b*) **516.** (*c*)

517. (*a*) See Section 11.5, Pin 35

518. (*c*)

519. (*a*) Software enables different functions

520. (*a*) **521.** (*a*) **522.** (*a*)

523. (*c*) Computer is not for real time video viewing

524. (*b*) **525.** (*d*) **526.** (*c*) **527.** (*b*)

528. (*d*) All statements are correct.

529. (*a*) Statement 3 is not a part of architecture.

530. (*c*) Cache memory uses DRAM, main memory uses SRAM.

531. (*d*)

532. (*a*) 8086 has 20 address lines. Hence address space = 2^{20} bytes = 1 Megabyte

533. (*d*) Mod 5 counter divides by 5.

534. (*b*) **535.** (*b*)

536. (*a*) % Resolution = $\frac{1}{2^N - 1} \times 100$ where N is the number of bits.

537. (*c*) See Section 11.4.9.

538. (*b*) Weather forecasting requires super computer.

539. (*c*) **540.** (*b*)

541. (*c*) Storage capacity

= π × diameter × tracks × storage density

= π × 8 × 100 × 200 = 502400 bits

542. (*d*)

```
  87 H =  1 0 0 0 0 1 1 1
 +79 H =  0 1 1 1 1 0 0 1
         1 0 0 0 0 0 0 0 0
       Carry     SUM
```

Both CARRY and ZERO FLAGS Set to 1.

543. (*a*) I/O is not involved in CALL

544. (*b*) *f* = 5 MHz, T = $\frac{1}{f} = \frac{1}{5} \times 10^{-6}$ s = $\frac{1}{5}$ μs

For 18 T starts time = $\frac{18}{5}$ μs = 3.6 μs

545. (*c*) TRAP is edge sensitive and level sensitive.

546. (*b*) Cluster is the name for such work.

547. (*c*) Number of chips = $\frac{32 \times 8 \times 1024}{2^{12} \times 4} = 16$

548. (*a*) **549.** (*b*) **550.** (*d*) **551.** (*d*)

552. (*b*) Data bus size = $\frac{\text{Memory chip size}}{\text{Address bus size}}$

for 1 kB memory = $\frac{2^{10} \times 8}{2^{10}}$ = 8 bits

2 kB memory = $\frac{2 \times 2^{10} \times 8}{2^{10}}$ = 16 bits

4 kB memory = $\frac{4 \times 2^{10} \times 8}{2^{10}}$ = 32 bits.

553. (*c*) Number of chips = $\frac{16 \times 2 \times 8}{2^{12} \times 4} = 8$.

554. (*b*) **555.** (*b*) **556.** (*d*)

557. (*b*) fetch, write, write

558. (*a*) **559.** (*b*) **560.** (*d*) **561.** (*a*)

562. (*b*)

563. (*c*)

Instruction	**Contents**
MVI, A, 45 H	A = 45 H = 0 1 0 0 0 1 0 1
MOV B, A	B = 45 H
STC	CS = 1
CMC	CS = 0
RAR	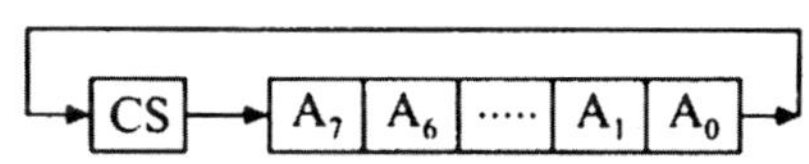 A = 0 0 1 0 0 0 1 0
XRA B	A ← 0 0 1 0 0 0 1 0 ⊕ 0 1 0 0 0 1 0 1 or A ← 0 1 1 0 0 1 1 1 or 67 H

564. (*d*)

565. (*c*)

Instructions	**Contents**
L X I, 30 AOH	HL = 30 A 0
DADH	HL = 6140
PCHL	PC = 6140

Hence PC = 6140, HL = 6140

566. (*b*) V_c can vary between $\frac{V_{CC}}{3}$ to $\frac{2V_{CC}}{3}$ or $\frac{9}{3}$ to $\frac{18}{3}$.

or 3 to 6 V.

567. (*b*) After line 1 A ← 1 0 1 1 0 1 0 1

line 2 B ← 0 0 0 0 1 1 1 0

line 3 ← 0 1 1 0 1 0 0 1

A ← 1 1 0 1 1 1 0 0

line 4 B ← 0 0 0 0 1 1 1 0

A ← E A

A = EAH

568. (*c*) After line 5 A ← 11101010

9B 10011011

ANDING 10001010 → A

After line 6 9F 10011111

As A < 9F, CY = 1, Z = 0

569. (*c*)

```
0100 LXI SP, 00FF
0103 LXI H, 0107
0106 MVI A, 20 H
0108 SUB M → A ← A – M
0109
```

M contains the data of memory whose address is in HL Pair, *i.e.*, 0107. Thus 0.107.

Corresponds to 20 H

A – M = 20 H – 20 H = 00H

570. (*c*) 0109 H ORI 40 H

ADD M

A ← 00 H = 00000000

40 H = 01000000

ORI → 01000000

A → 40 H

A → 40 H + 20 H = 60 H.

571. (*a*) In 8085 signal lines $AD_7 - AD_0$ are used for address bits in some T states and for data in some other T states.

572. (*a*)

573. (*a*) Dynamic RAM consists of a capacitor which has to be recharged.

574. (*b*) Both are correct but independent.

575. (*c*) **576.** (*a*) **577.** (*c*) **578.** (*c*)

579. (*b*) Both are correct but independent.

580. (*a*)

581. (*d*) A is wrong.

582. (*b*) Both are correct but independent.

583. (*b*)

584. (*c*) A is correct but R is wrong. Two IC 74181 can be cascaded to add two 8 bit words.

585. (*b*) Both are correct but independent.

586. (*b*) See Section 11.4.3.

587. (*a*) See Section 11.4.3.

588. (*b*) Both are correct but independent.

589. (*a*) $2^{16} = 65536$.

590. (*b*) See Section 11.4.4.

591. (*d*) SP and PC are 16 bit each in 8085.

592. (*b*) See Section 11.4.5.

593. (*b*) See 8085 instruction sheet.

594. (*b*) See Section 11.4.3.

595. (*c*) See Section 11.4.8.

596. (*d*) Crystal has to be connected externally to pins 1 and 2 of 8085 as discussed in Section 11.6.

597. (*b*) See Section 11.4.9.

598. (*b*) See Section 11.4.9.

599. (*b*) See Section 11.19.

600. (*d*) $\overline{WR}$ and $\overline{RD}$ are active low.

601. (*b*) See Fig. 11.19.

602. (*c*) The number of addressing modes is five.

603. (*b*) See Section 11.10.3.

604. (*b*) See Section 11.10.3.

605. (*b*) See Section 11.11.

606. (*b*) Both are correct but independent.

607. (*b*)

608. (*b*) Both are correct but independent. Computers work in binary only.

609. (*d*) A is wrong.

610. (*b*) Both are correct but independent.

611. (*b*) **612.** (*a*) **613.** (*b*)

614. (*b*) See Section 11.19.

615. (*c*) See section 11.19. A character constant is a single alphabet or digit or special character enclosed within single inverted commas.

616. (*d*) See Section 11.19.3.

617. (*b*) See Section 11.19.

618. (*b*) **619.** (*b*) **620.** (*b*)

621. (*c*) Mircoprocessor does need memory. Hence R is wrong.

622. (*a*)

623. (*b*) Both are correct but independent

624. (*a*) **625.** (*a*)

626. (*c*) Program counter is needed, R is wrong.

627. (*c*) R is wrong.

628. (*b*) Both are correct but independent.

629. (*d*) Processor level design is not heuristic. Hence A is wrong.

630. (*c*) **631.** T **632.** T

633. F, It is read and write memory. Its contents can be changed.

634. T **635.** T

636. F, Accumulator is register A.

637. T

638. F, It is 8 bit-register.

639. T

640. F, It has one sign flag S.

641. T

642. F, Z flag is set if ALU operation results in zero.

643. T

644. F, See Section 11.4.3.

645. T **646.** T

647. F, See Section 11.4.4.

648. F, See Fig. 11.4.
649. T **650.** T **651.** T **652.** T
653. T
654. F, 8085 needs +5 V supply.
655. F, Secondary memory is slow.
656. T **657.** T **658.** F **659.** T
660. T **661.** T **662.** F **663.** F
664. T **665.** T **666.** T **667.** T
668. T **669.** T **670.** T **671.** T
672. T **673.** T **674.** T
675. F, Diskette is a secondary storage.
676. T **677.** F
678. F, These are not held in ALU. Only computations are done by ALU.
679. T **680.** T **681.** T **682.** T
683. T **684.** T **685.** T **686.** T
687. T
688. F, See Section 11.21.8.
689. T
690. F, Computer virus can hide in harddisk also.
691. F **692.** T **693.** T **694.** T
695. F, When M = 1, IC 74181 performs logic operations.
696. F **697.** T **698.** T **699.** T
700. F, See Section 11.17.
701. T **702.** F, See Section 11.17.
703. T **704.** T **705.** T
706. address **707.** Data **708.** registers, memory
709. timing, control **710.** 8
711. ALU **712.** set **713.** one
714. DMA **715.** address latch enable
716. low **717.** double
718. 8 **719.** execution unit
720. bus interface unit
721. decoding, execution **722.** r_2, r_1 **723.** immediate
724. accumulators, memory location
725. contents, accumulator
726. register r, carry bit, accumulator
727. subtract register with borrow
728. register r, incremented, one
729. ANDed, accumulator
730. immediate **731.** accumulator, rotated left
732. 16
733. 4 **734.** cascaded
735. 2's complement **736.** carry out, carry in
737. computer
738. fractional form, exponential form
739. identifier
740. var ABC, BETA, GAMMA : real
741. incorrect **742.** incorrect
743. incorrect **744.** blanks
745. cannot **746.** seven
747. ALOGIO **748.** ANINT
749. assign **750.** permissible
751. six **752.** array
753. eight
754. instance variables, class variables and local variables
755. four **756.** literal
757. five **758.** %
759. && **760.** overflow
761. primary constants and secondary constants
762. integer constant, real constant and character constant
763. e **764.** eight
765. thirty two
766. type decleraction instruction
767. 1 MB **768.** micro
769. 100 MB **770.** DISKCOM
771. 1.25 inch
772. Hyper text markup Language
773. Front Page Express **774.** antivirus
775. compiler **776.** MICR
777. word length
778. What You See Is What You Get
779. address
780. stack pointer

COMMUNICATION SYSTEMS

CHAPTER 12

12.1. INFORMATION THEORY AND CODING

If a transmitter selects a message m_k with a probability p_k, the amount of information I_k is

$$I_k = \log_2 \frac{1}{p_k}$$

If two independent messages have probabilities p_1 and p_2 then probability of composite message is $p_1 p_2$ and amount of information is

$$I_{1,2} = \log_2 \frac{1}{p_1 p_2} = \log_2 \frac{1}{p_1} + \log_2 \frac{1}{p_2} = I_1 + I_2$$

Let M different and independent messages m_1, m_2, m_3.... have probabilities of occurrence p_1, p_2, p_3..... If a sequence of L messages have been generated during a long time interval, the total information is

$$I_{Total} = p_1 L \log_2 \frac{1}{p_1} + p_2 L \log_2 \frac{1}{p_2} + p_3 L \log_2 \frac{1}{p_3} + ...$$

Average information per message is called entropy and is denoted by H. Thus

$$H = \frac{I_{total}}{L} = p_1 \log_2 \frac{1}{p_1} + p_2 \log_2 \frac{1}{p_2} + p_3 \log_2 \frac{1}{p_3} + ...$$

The average information associated with an extremely likely message as well as that associated with an extremely unlikely message is zero.

H becomes maximum when all messages are equally likely

$$H_{max} = \Sigma \frac{1}{M} \log_2 M = \log_2 M$$

If a source generates messages at the rate of r messages per second, the information rate R is $R = rH$ = average number of bits of information/second.

If $M = 2$ the entropy is maximum when $p_1 = p_2 = \frac{1}{2}$.

$$H_{max} = \log_2 2 = 1$$

Shannon's Theorem, Channel Capacity

If a source of M equally likely messages ($M \geq 1$) is generating information at a rate R and a channel with channel capacity C exists such that $R \leq C$, there exists a coding technique such that the output of source may be transmitted over the channel with a probability of error in the received message which may be made arbitrarily small.

The negative statement associated with Shannon's theorem is : if the information rate R exceeds the channel capacity C, the error probability will increase towards unity as M increases. If noise is absent, the channel capacity C is

$$C = 2(\delta f) \log_2 (S)$$

where δf is bandwidth and S is number of coding levels

Shannon-Hartley Theorem

The channel capacity of a white band limited gaussian channel is

$$C = (\delta f) \log_2 \left(1 + \frac{S}{N}\right) \text{bits/s}$$

where δf is channel bandwidth, S is signal power and N is total noise within the channel bandwidth.

If $\frac{S}{N}$ is infinite, channel capacity is infinite. However, if δf becomes infinite, C does not become infinite because as δf increases, noise power also increases.

Some Algebraic Codes are: Single parity check bit code, Repeated codes, Hadamard code, Hamming code. Extended codes, cyclic codes, Golay code, BCH code, Reed-Solomon code etc.

Two techniques for controlling transmission errors are: Forward error correction (FEC) and automatic repeat request (ARQ).

12.2. NOISE

Any unwanted form of energy which tends to interfere with the reception and reproduction of wanted signals is called noise.

Noise can be external or internal. Sources of external noise are: atmospheric noise, solar noise, cosmic noise and industrial noise.

Internal Noise

Thermal agitation noise is the noise generated in a resistance. It is also called white noise or Johnson noise. Its *rms* value V_n is given by

$$V_n = \sqrt{4kT(\delta f)R}$$

where k is Boltzmann's constant Jouls/K, T is absolute temperature, δf is bandwidth and R is resistance.

For two resistances R_1 and R_2 in series, total noise voltage is

$$V_{n,total} = \sqrt{4kT(\delta f)(R_1 + R_2)}$$

Figure 12.1 shows two amplifiers with gains A_1 and A_2 in cascade. The first stage has a total input noise resistance of R_1, the second R_2 and the output resistance is R_3. The equivalent noise resistance at the input of first stage is

$$R_{eq} = R_1 + \frac{R_2}{A_1^2} + \frac{R_3}{A_1^2 A_2^2}$$

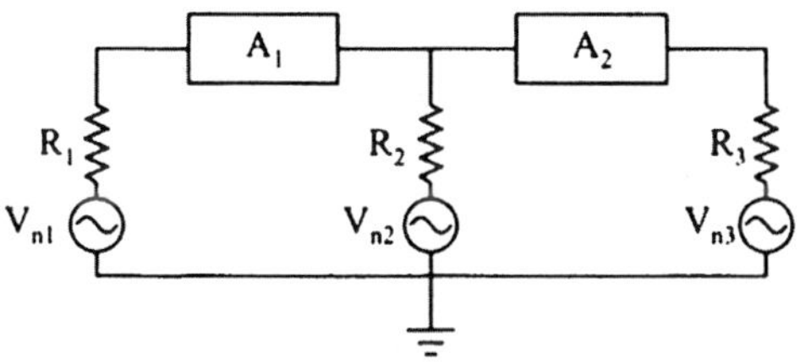

Fig. 12.1.

$$\text{Noise figure } F = \frac{\text{input S/N}}{\text{output S/N}}$$

Equivalent noise temperature $T_{eq} = T_0 (F - 1)$

where T_0 is standard temperature (= 290 K) and F is noise figure.

Shot Noise: It is caused by random variations in arrival of holes and electrons at the output electrode of amplifying device.

$$\text{For a diode } i_n = \sqrt{2ei_p(\delta f)}$$

where i_n is *rms* shot noise current, e is charge on electron, i_p is diode current and δf bandwidth.

Transit Time Noise

The transit time effect occurs when time taken by an electron to travel from one type of material to other in a transistor becomes comparable with the period of signal being amplified. This noise increases with frequency.

12.3. AMPLITUDE MODULATION

Let carrier and modulating voltage be

$$v_c = V_c \sin \omega_c t$$

$$v_m = V_m \sin \omega_m t$$

$$\text{Modulation index } m = \frac{V_m}{V_c}$$

The amplitude of AM wave is

$$A = V_c + v_m = V_c (1 + m \sin \omega_m t)$$

The instantaneous value of AM wave is

$$v = A \sin \omega_c t = V_c (1 + m \sin \omega_m t) \sin \omega_c t$$

$$= V_c \sin \omega_c t + \frac{mV_c}{2} \cos(\omega_c + \omega_m)t - \frac{mV_c}{2} \cos(\omega_c - \omega_m)t$$

$$\text{Also} \quad m = \frac{V_{max} - V_{min}}{V_{max} + V_{min}}$$

$$V_C = 0.5 (V_{max} + V_{min})$$

The frequency of lower side band (LSB) is $(f_c - f_m)$ and that of upper side band (USB) is $(f_c + f_m)$.

$$\text{Total power } P_t = P_c \left(1 + \frac{m^2}{2}\right)$$

where P_c is the power of carrier.

When $m = 1$, $P_t = 1.5P_c$

$$P_{LSB} = P_{USB} = \frac{m^2}{4} P_c$$

When several modulating waves V_1, V_2, V_3, ... are present, total modulating voltage v_t is

$$V_t = (V_1^2 + V_2^2 + V_3^2 + ...)^{0.5}$$

$$m_t = \left(\frac{V_1^2}{V_c^2} + \frac{V_2^2}{V_c^2} + \frac{V_3^2}{V_c^2} + ...\right)^{0.5}$$

$$= (m_1^2 + m_2^2 + m_3^2 + ...)^{0.5}$$

The carrier does not convey any information. Moreover each side band is image of other. Therefore it is sufficient to transmit only single side band (SSB). SSB technique requires less transmitted power and less bandwidth. The methods used for generation of SSB are: filter method, phase cancellation method and third method. All the these three methods use balanced modulator.

12.4. FREQUENCY MODULATION

The general equation of a carrier wave is

$$v = V_C \sin (\omega_c t + \phi)$$

If the amplitude V_C is varied by modulating wave, the process is amplitude modulation. If ω_c is varied it is called frequency modulation and if phase angle ϕ is varied it is called phase modulation. In frequency modulation the amount by which ω_c is varied from its unmodulated value is called deviation. The deviation is proportional to instantaneous value of modulating voltage. The rate at which this frequency variation takes place is equal to modulating frequency. Thus the instantaneous frequency f of FM wave is

$$f = f_c (1 + kV_m \cos \omega_m t)$$

The deviation is maximum when $\cos \omega_m t = \pm 1$. Therefore the instantaneous frequency f and maximum deviation δ are

$$f = f_c (1 \pm k V_m)$$

$$\delta = k V_m f_c$$

Modulation index m_f is defined as

$$m_f = \frac{\text{maximum frequency deviation}}{\text{modulating frequency}} = \frac{\delta}{f_m}$$

$$v = V_c \sin (\omega_c t + m_f \sin \omega_m t)$$

In frequency modulation, the number of side bands is infinite but higher side bands have insignificant relative amplitudes. A thumb rule states that bandwidth required for FM wave is twice the sum of deviation and highest modulating frequency.

Advantages of FM

1. FM reception is more immune to noise than AM.
2. In FM there is less adjacent channel interference than in AM.
3. FM stations operate in VHF and UHF bands at which the noise level is low.
4. At FM broadcast frequencies space wave is used for propagation. Thus the radius of reception is limited to line of sight distance. Therefore, several independent transmitters on the same frequency can be operated.

Disadvantages of FM

1. More bandwidth.
2. Complex transmitting and receiving equipment.
3. Limited area of reception.

Generation of FM wave is done by reactance modulator.

12.5. RADIO RECEIVERS

12.5.1 TRF Receivers

The simplest radio receiver is Tuned Radio Frequency (TRF) receiver shown in Fig. 12.2.

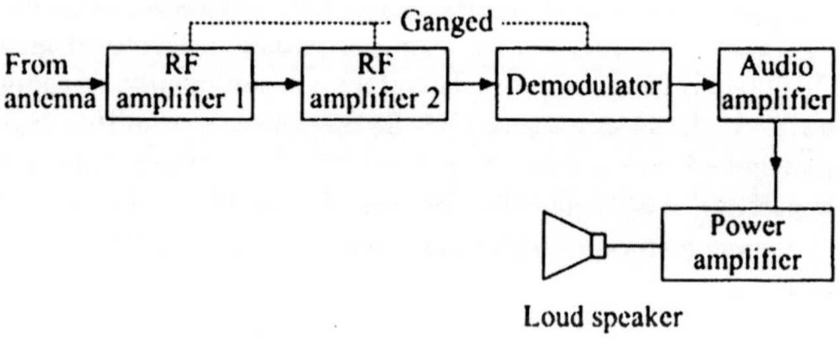

Fig. 12.2. TRF Receiver.

TRF receiver has the problems of instability, insufficient adjacent frequency rejection and bandwidth variation. In view of this TRF receiver is used only as fixed frequency receiver in special applications. Superheterodyne receiver does not have above problems.

12.5.2. Superheterodyne Receiver

In this receiver the signal voltage is combined with the local oscillator voltage and converted into a signal of a lower intermediate frequency (IF). The signal at IF is amplified and demodulated to reproduce the original information. Thus it has the same components as TRF receiver and has in addition mixer, local oscillator and IF amplifier. Figure 12.3 shows its simplified block diagram.

The frequency f_0 is always the sum of signal frequency f_s and intermediate frequency (IF). If a frequency f_{si} is able to reach the receiver such that $f_{si} = f_0 + \text{IF}$, *i.e.*, $f_{si} = f_s + 2\text{IF}$, then this frequency f_{si} will also produce IF when mixed with f_0. This unwanted signal will also be amplified and thus two stations will be tuned together. This frequency f_{si} is called image frequency. Additional circuitry is used to reject image frequency.

IF has a standard value of 455 kHz. Automatic gain control (AGC) helps in changing the overall gain automatically with the changing strength of received signal so as to keep the output constant.

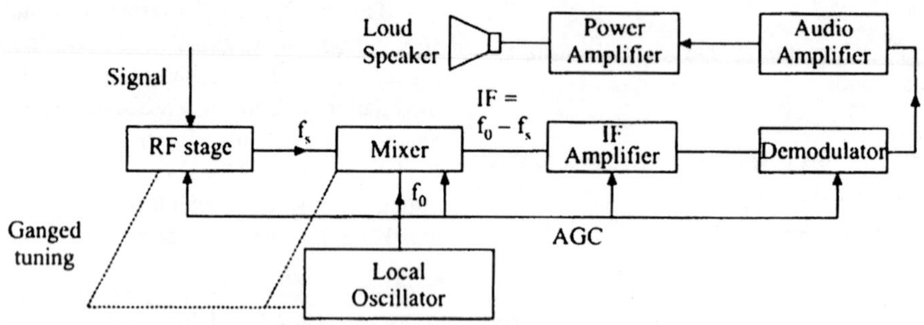

Fig. 12.3. Superheterodyne receiver.

12.5.3. Communication Receiver

The function of a communication receiver is reception of signals for communication. It is used for wireless communication (by police, army etc.) and is also used as detector in high frequency impedance bridges. It has very fine tuning, two or more IF, one or two RF stages, Delayed AGC, variable selectivity and sensitivity etc.

12.6. PROPAGATION OF WAVES

Electromagnetic waves are oscillations which propagate through free space at velocity of light $c = 3 \times 10^8$ m/s. These waves are transverse (*i.e.*, oscillations are perpendicular to direction of propagation). Moreover Electric field, Magnetic field and direction of propagation are mutually perpendicular to each other.

If an isotropic transmitter has power P_t, the power density P at a distance r from the transmitter is

$$P = \frac{P_t}{4\pi r^2}$$

Power density P is related to electric field intensity E by the equation

$$P = E^2/Z_0$$

When Z_0 is characteristic impedance of free space.

$$\text{For free space } Z_0 = \sqrt{\frac{\mu_0}{\epsilon_0}} = 377 \text{ ohms.}$$

P_t and E are related as

$$E = \frac{1}{r}(30P_t)^{0.5}$$

A plane parallel to E and H fields is called wave front. The direction of electric field is called direction of polarization. As waves travel through free space they undergo reflection, refraction and diffraction.

The mode of propagation classifies the radio waves into three categories viz. ground waves, sky waves and space waves.

Ground wave (also called surface wave) travels along the surface of ground. The ground behaves as a leaky capacitor. Therefore, the ground wave gets attenuated as it travels on the ground surface. Frequencies used in this mode are up to 1600 kHz.

$$\text{Field strength of ground wave} = \frac{AE_0}{r}$$

where E_0 is the field strength of wave at earth's surface at unit distance from transmitting antenna, r is distance from transmitting antenna and factor A takes into account ground losses.

Some radio waves reach the upper part of atmosphere known as ionosphere which acts like a mirror and reflects the waves back to earth. These waves are called sky waves. Frequencies up to 30 MHz use this mode of propagation.

For frequencies above 30 MHz (generally used for television) the attenuation of ground wave is very high. Moreover these waves penetrate the ionosphere and are not reflected by it. Communication at frequencies above 30 MHz is through space waves. The reception of signals through space waves is limited to line of sight distance.

The radio horizon of an antenna is given by

$$\text{Radio horizon} = 4.13\,(h)^{0.5}\text{ km}$$

where h is height of antenna in metres.

Total distance d is given by

$$d = 4.13\,(h_t)^{0.5} + 4.13\,(h_r)^{0.5}$$

where h_t and h_r are heights of transmitting and receiving antennas.

The classification of radio waves according to frequencies and application is given in Table 12.1.

Table 12.1 Classification of Radio Waves

Frequency	*Wave length, m*	*Class*	*Propagation characteristics and application*
10–30 kHz	3×10^4 to 10^4	VLF (Very low frequency)	Low attenuation at all times of day. Used in long distance point-to-point communication (ship communication).
30–300 kHz	10^4 to 10^3	LF (Low frequency)	During day time attenuation is more than that in VLF. Used in marine communication and navigation.
300 to 3000 kHz	10^3 to 10^2	MF (Medium frequency)	Day time attenuation high but night time attenuation low. Used for broadcasting and ship to shore communication.
3 to 30 MHz	10^2 to 10	HF (High frequency)	Attenuation depends on time of day, season and frequency. Used in broadcasting, telephones, telegraph communication and navigation.
30 to 300 MHz	10 to 1	VHF (Very high frequency)	Line of sight propagation. Used in TV, FM broadcast, radars and short distance communication.
300–1000 MHz	1 to 0.1	UHF (Ultra high frequency)	Line of sight propagation. Used in television.
1000 to 300000 MHz	0.1 to 0.001	SHF (Super high frequency)	Same as for UHF

A further sub classification is

UHF	300–1000 MHz
L band	1000–2000 MHz (1–2 GHz)
S band	2000–4000 MHz (2–4 GHz)
C band	4000–8000 MHz (4–8 GHz)
X band	8000–12000 MHz (8–12 GHz)
Ku band	12000–18000 MHz (12–18 GHz)
K band	18000–27000 MHz (18–27GHz)
Ka band	27000–40000 MHz (27–40 GHz)
Millimeter	40000–300000 MHz (40–300 GHz)
Sub millimeter	> 300000 MHz (> 300 GHz)

12.7. PULSE CODE MODULATION

In PCM the total amplitude range which the signal may occupy is divided into a number of standard levels. The actual number of these levels is a power of 2 since transmission is through binary code. By a process called quantizing the level actually sent at any sampling time is the nearest standard level. The quantizing process introduces some distortion called quantizing noise. By increasing the number of standard levels, the quantizing noise can be brought down within acceptable limits. However, more levels require more bits per second and, therefore, more bandwidth. Generally 128 levels for audio signals is considered satisfactory.

Generation of PCM involves sampling, conversion to PAM, quantization, encoding and addition of supervisory signals. Since PCM is immune to noise, amplitude modulation can be used. Then the signal is transmitted. At the receiver the signal is extracted, PCM translated to PAM pulses and demodulated. An integrating RC circuit is used for reconversion in the receiver. It is important to note that PCM requires very complex encoding and quantizing circuitry as also a large bandwidth as compared to analog systems.

Differential PCM is similar to PCM in some ways. However in differential PCM each word indicates difference in amplitude (positive or negative) between this sample and the previous sample. Thus in differential PCM, the relative value of each sample is indicated whereas in ordinary PCM the absolute value of each sample is indicated.

Two main applications of PCM are telegraphy and telemetry.

12.8. DIGITAL COMMUNICATION

Digital communication deals with digital signals, *i.e.*, signals which appear in discrete steps. Digital codes are the languages of computers and other digital equipment. The ability of computer to handle many input-output devices simultaneously has made data communications essential and important.

In most cases the data consists of pulse type energy. The data stream is similar to square wave signal with very fast transitions from one voltage level to other (*i.e.*, high to low and low to high). The exact duration of each half of square depends on the data word itself.

Most data transmissions use telephone channels. This channel has the frequency range of 300 Hz to 3400 Hz and is called 4 kHz channel. Sometimes 3 kHz channel is used. The data transmission speed is expressed in baud. One cycle of transmission can contain a maximum of two bauds. Thus in an ideal channel the maximum signalling speed in bauds (called Nyquist rate) is equal to twice the bandwidth of channel. For an ideal 3 kHz channel, the data speed is 6000 bits per second. The required signal to noise ratio is calculated as under

$$\frac{S}{N} = 2^{NR/\delta f} - 1$$

where S/N is signal noise ratio, NR is Nyquist rate and δf is channel width. If $\delta f = 3000$ and NR = 6000

$$\frac{S}{N} = 2^2 - 1 = 3 \quad \text{or} \quad 3:1 = 4.8 \text{ dB}$$

Baudot code (which is a 5 bit code) ASCII code (which is a 7 bit code), EBCDIC code (which is an 8 bit code) are some digital codes. Since ASCII is a 7 bit code and 8 bit word is easily accommodated in computers, the 7 bit code can be converted to 8 bit code by adding a parity bit. Many error detection and correction codes also exist.

Modems (Modulator-Demodulator) is required as an interface between the actual system (which is always analog) and digital communication channel. A data set which provides transmission only in one direction is said to be operating in simplex mode. Half duplex means data flow in one direction at one time and in the opposite direction at a second time. Full duplex operation permits transmission in both directions at the same time. Modems are classified as : Low speed (up to 600 bits/s), medium speed (600 to 2400 bits/s) and high speed (2400 to about 10800 bits/s).

12.9. BROAD BAND COMMUNICATION SYSTEMS

12.9.1. Frequency Division Multiplexing

Multiplexing means transmission of a number of separate signals together on the same cable simultaneously. It can be classified as Frequency Division Multiplexing and Time Division Multiplexing.

In frequency division multiplexing a bandwidth adequate for the required number of channels is selected. Each channel is alloted a frequency slot adjacent to previous channel. Standardised groupings of channels are used. The basic group consists of 12 adjacent 4 kHz channels occupying the frequency range 60 to 108 kHz. A low level pilot is transmitted at 104.08 kHz. In many submarine cables the basic group has 16 adjacent 3 kHz channels, the pilot being at 84 kHz. The next up from the basic group is basic super group consisting of five groups and occupying the frequency range of 312 to 552 kHz. Super groups are combined to form master groups, super master groups etc. The super groups belonging to opposite direction of transmission are allocated different frequencies, different coaxial tubes or different optical fibre cables to avoid interference.

12.9.2. Time Division Multiplexing

In this method use is made of the fact that in any pulse modulation system narrow pulses with wide spaces between them are generated. These spaces are used by signal from other sources. Slow speed TDM uses rotating mechanical switches while high speed TDM uses electronic switching and delay lines.

12.9.3. Fibre Optice Link

An optic fibre is a piece of very thin highly pure glass with an outside cladding of glass having a different refractive index. The core diameter is in the range of 2 to 200 μm. The advantages of optical fibres are (1) small size, (2) immunity to cross talk and electromagnetic interference, (3) Laser and LED modulation methods lend themselves ideally to digital operation.

12.9.4. Satellite Communication

A communication satellite is a microwave link repeater. The earth station sends the signals to the satellite which amplifies the signals and sends them back at a frequency about 2 GHz away. This difference in frequency prevents interference between uplink and downlink. Communication satellites are geostationary, *i.e.*, they have the same angular velocity as earth (one complete cycle in 24 hours). Thus it appears to be stationed at one spot in the globe. The main satellite systems include Intel sat (used for global point to point communication), Inmarsat (marine, *i.e.*, ship-to-ship communication), and many regional and domestic satellites used by different countries.

12.10. RADAR SYSTEMS

12.10.1 Range of Radar

Radar means Radio Aid to detection and Ranging. It is a system of gathering information about distant objects or targets by sending electromagnetic waves to them and analysing the echos (response).

A radar can be a Pulsed radar or Continuous Wave (CW) radar. Generally pulsed radar is used.

A radar consists of a transmitter and receiver with a common antenna. The transmitter produces high power ultra high frequency periodic pulses (of short duration) and feeds them to a directional antenna. The receiver collects as much energy as possible from the echos and feeds its output to a cathode ray tube so that the response is seen on the screen. A duplexer enables the use of a common directional antenna for transmitting and receiving pulses. The maximum range of a radar is given by

$$r_{max} = 48\left(\frac{P_t D^4 S}{(\delta f)\lambda^2 (F-1)}\right)^{0.25}$$

where r_{max} = maximum radar range, km

P_t = Peak pulse power, W

D = antenna diameter, m

S = effective cross-sectional area of target, m^2

δf = receiver bandwidth, Hz

λ = wavelength, m

F = noise Figure (expressed as a ratio)

An alternative expression for range of radar is

$$r_{max} = \left(\frac{P_t A_0^2 S}{4\pi\lambda^2 P_{min}}\right)^{0.25}$$

where A_0 = capture area of receiving antenna, m^2

P_{min} = Minimum receivable power of receiver, W

The minimum receivable power P_{min} is given by

$$P_{min} = k T_0 (\delta f)(F - 1)$$

where T_0 = Standard ambient temperature, K.

k = Boltzmann constant J/K

12.10.2. Moving Target Indication (MTI)

MTI system compares a set of received echos with those received during the previous sweep. Those echos whose phase has remained constant are, then, cancelled out. Echos corresponding to stationary targets are called 'clutter'. To eliminate clutter, MTI uses Doppler effect.

As per Doppler effect the apparent frequency of electromagnetic or sound waves depends on the relative radial motion of the source and observer. If the source and observer are moving away from each other, the apparent frequency decreases whereas if they are moving towards each other the apparent frequency increases. The Doppler frequency for radar is

$$f_d = \frac{2v_r}{\lambda}$$

where f_d = Doppler frequency for radar

λ = transmitted wavelength

v_r = relative velocity

When a target has a velocity whose radial component results in a phase difference of 2π rad between successive pulses, the situation is the same as having no phase difference. Then the target appears stationary. The radial velocity corresponding to this situation is called blind speed and is given by

$$v_b = 0.5\,(\text{PRF})\,(n\lambda)$$

where v_b = blind speed, m/s

n = any integer (*i.e.*, 1, 2, 3 etc.)

λ = wavelength of transmitted signal

PRF = Pulse repetition frequency (pulses per second)

12.10.3. Radar Beacon

A radar beacon is a small radar set having a receiver, a separate transmitter and an antenna (which is generally omni directional). When another radar sends a coded set of pulses at the beacon, the beacon responds by sending back its specific pulse code. Beacon is also called Transponder. A radar beacon is used in air traffic control. military operations etc.

12.11. MICROPHONES

A microphone is a transducer which converts variations in sound pressure into electrical signals of the same frequency. Its qualities are Sensitivity, Frequency Response, Signal to noise ratio. Output impedance, Distortion, Directivity. Table 12.2 gives a comparison of different microphones.

Table 12.2. Comparison of Microphones

Feature	*Moving coil*	*Ribbon*	*Condenser*	*Crystal*	*Carbon*
1. Sensitivity (μ V output for sound pressure of 0.1 Pa)	30 (without transformer) 90 (with transformer)	3	3000 (with amplifier)	50,000	1,00,000
2. Sensitivity (dB below IV)	90	110	50	26	20
3. Signal to noise ratio dB	25	50	40	40	20
4. Frequency response, Hz	50 to 8000	20 to 12000	40 to 15000	80 to 8000	200 to 5000
5. Distortion	5%	1%	1%	1%	10%
6. Directivity	Omni directional	bidirectional	Omni directional	Omni directional	Omni directional
7. Output impedance	30 Ω	0.5 Ω	50 MΩ	1 MΩ	100 Ω
8. Size	Big	Big	Big	Small	Small
9. Cost	Medium	High	High	Low	Very low
10. Applications	Public address system, broadcast	Drama, music, broadcast	Calibration	Home recording Mobile Communication	Telephone

12.12. LOUDSPEAKERS

A loudspeaker converts electrical audio frequency signal into sound waves of the same frequency. Its qualities are: sensitivity, frequency response, signal to noise ratio, distortion, directivity and input impedance.

12.12.1. Moving Coil Loudspeakers

Its principle is same as that of motor. A coil called voice coil is placed in the magnetic field of a permanent magnet. When electric current passes though the coil, a force is created and causes the coil to vibrate. A paper diaphragm of the shape of a cone is attached to the coil. Thus the diaphragm vibrates, causes pressure variations in air and thus sound waves are produced. It is used in radio receivers, TV receivers, cassette players etc.

Features: Poor efficiency (5–10%), signal to noise ratio about 30 dB, frequency response 200 to 5000 Hz non-linear distortion high (about 10%), omni directional.

12.12.2. Electrodynamic Loudspeaker

Its principle is the same as that of moving coil loudspeaker but it uses an electromagnet (instead of permanent magnet of moving coil loudspeaker).

12.12.3. Horn Loudspeaker

Its principle is also the same as that of moving coil loudspeaker. But it does not use a cone. Instead it delivers sound to air trapped in a horn which radiates sound to open space. It is used for public address systems, auditoriums and music concerts.

Features:

1. High efficiency (about 40%)
2. Signal to noise ratio about 40 dB.
3. Frequency response 25 to 10000 Hz.
4. Non-linear distortion less than 5%.
5. Omni directional at low frequencies but unidirectional at high frequencies.
6. Input impedance about 15 ohms.

12.13. TELEVISION

12.13.1. Scanning

For converting a scene or a picture into electrical signal, the electron beam has to explore the picture point-by-point. This process is called scanning. The scanning process is done at fast speed thus creating an illusion of continuity due to persistence of vision.

The complete picture is called a frame. Each frame is scanned 50 times. Moreover each frame is divided into 625 lines in India (in USA this number is 525). The present day scanning method is called interlaced scanning. In this process the 625 lines are divided into two sets of 312.5 lines each called the odd and even fields. Each of odd and even fields is scanned 25 times alternately one after the other. The scanning is done from left to right and from top to bottom.

12.13.2. CCIR-B System

CCIR stands for Committee Consultative for International Radio. In India we use CCIR-B standard. This system has 625 scanning lines per frame with 2 : 1 interlace (*i.e.*, two fields scanned sequentially but interlaced to form the complete picture). The field frequency is 50 Hz and picture frequency is 25 Hz. The horizontal line frequency is 625 × 25 = 15625 Hz and the maximum possible deviation is 0.1%. The ratio of width to height of picture is called aspect ratio and is 4 : 3. Total scanning period of 1 line including retrace period is 64 ms (52 μs for the scanning and 12 μs for blanking or retrace). The number of lines lost in vertical retrace is 20 per field.

12.13.3. Composite Video Signal

The composite video signal has three components, *i.e.*,

1. Video signal which contains the picture information.
2. Synchronising pulses to synchronise the horizontal and vertical scanning at transmitter and receiver.
3. Blanking pulses to make the retrace invisible.

12.13.4. Bandwidth of TV Signals

The required bandwidth depends on the total number of picture elements scanned. One pair of white and dark elements produces one cycle of electric signal. Therefore frequency is

$$\text{Frequency} = \frac{\text{Total number of picture elements scanned per second}}{2}$$

Since the aspect ratio is 4 : 3, the number of picture elements in horizontal direction is $\frac{4}{3}$ times the number of picture elements in vertical direction. Hence

$$\text{Frequency} = \frac{4}{3}\frac{(625)^2(25)}{2} = 6.66 \times 10^6 \text{ Hz} = 6.66 \text{ MHz}.$$

The effective number of lines is less than 625 because of lines lost during vertical blanking period. Thus the required bandwidth is 5 MHz.

TV systems use vestigial side band transmission. In this method one side band is transmitted fully and the other side band only partially. The actual bandwidth is 1.25 MHz on the lower side band. The video carrier is located at 1.25 MHz. The upper side band extends at full amplitude to 6.25 MHz with reduced amplitude upto 6.75 MHz.

12.13.5. Audio Signal Modulation

Every video signal has its own associated audio signal. Audio signal is transmitted through FM to give noise free operation. The two signals, *i.e.*, audio and video are sent to a common antenna. The diplexer combines them but maintains sufficient isolation between them. The difference between video carrier and audio carrier is called inter carrier frequency and is 5.5 MHz in CCIR-B standard. Since video signal is AM and audio signal is FM there is no chance of cross modulation. The sound carrier has a centre frequency of 6.75 MHz and has two side bands extending to a maximum of 100 kHz.

12.13.6. TV Channels

Each TV channel has a bandwidth of about 6.75 MHz. Therefore 7 MHz is allowed for each channel.

The alloted frequency range is divided into four bands I, III, IV and V. This is shown in Table 12.3.

Table 12.3. TV Channels in CCIR - B System

Band	*Channel*	*Channel range MHz*	*Video carrier MHz*	*Audio carrier MHz*
I (41 to 68	1	41 – 47	Not used for TV	
MHz	2	47 – 54	48.25	53.75
Lower VHF	3	54 – 61	55.25	60.75
	4	61 – 68	62.25	67.75
II	(68 to 174 MHz)	used for other services		
III	5	174 – 181	175.25	180.75
(Upper VHF)	6	181 – 188	182.25	187.75
	7	188 – 195	189.25	194.75
	8	195 – 202	196.25	201.75
	9	202 – 209	203.25	208.75
	10	209 – 216	210.25	215.75
	11	216 – 223	217.75	222.25
Additional Channel	12	223 – 230	224.25	222.25

12.13.7. Block Diagram of TV Transmitter

Figure 12.4 shows a simplified block diagram of TV transmitter.

The camera tube focuses the picture on to a photo sensitive plate to form an image on it. The picture elements are converted into electrical signals through scanning. The function of synchronising circuit is to supply synchronising pulses which identify the retrace so that scanning rate at receiver can be controlled and kept in synchronism with that at transmitter. The video amplifiers strengthens the video signal. The RF oscillator provides a high frequency radio wave for amplitude modulation of video signal. The sound waves are converted into electrical signal by microphone. The audio amplifier strengthens these signals. The RF oscillator provides carrier frequency for frequency modulation of audio signal. Both video and audio signals are radiated into atmosphere by transmitting antenna.

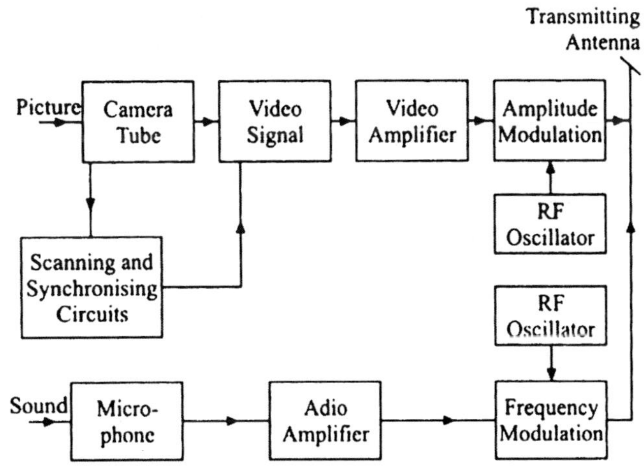

Fig. 12.4. Block diagram of TV transmission.

12.13.8. Black and White TV Receiver

Figure 12.5 shows a black and white TV receiver.

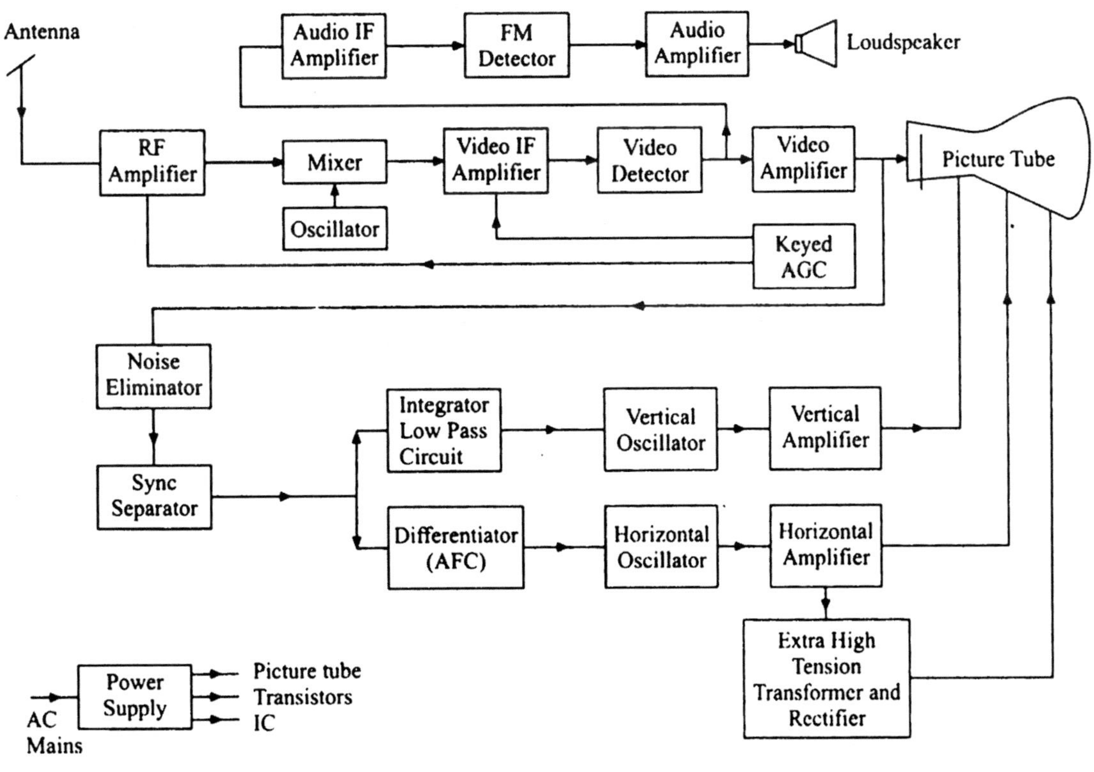

Fig. 12.5. Block diagram of B&W television.

12.13.9. Colour TV Signal

A colour is completely specified, viz., three terms, viz., Hue, Luminance and Saturation. Hue is the actual colour seen by eye. Red, green, blue, etc. represent different hues. Luminance or Brightness is the total amount of light intensity or light energy received by eye. Saturation indicates purity of colour.

The three primary colours are blue, green and red. The luminance signal is obtained by mixing three colours red, blue, green in proportion of 30%, 59% and 11%,

$$Y = 0.3\,R + 0.59\,G + 0.11\,B$$

where Y is luminance, R is red, G is green and B is blue.

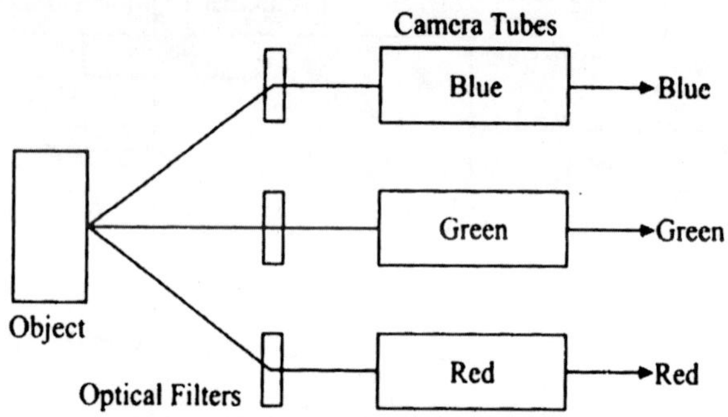

Fig. 12.6. Principle of Colour TV Camera.

Chrominance signal is also called C-signal. It indicates hue and saturation of colour. The chrominance information is transmitted as R-Y and B-Y colour differences.

PAL means phase alternation by line. In this method signal R-Y and B-Y are weighted by factors 0.877 and 0.493 to produce V and U chrome signals. Weighting reduces the chances of over modulation. They are both allowed a bandwidth of 1.3 MHz. The colour difference signals carry all the colour information and are transmitted as modulation of subcarrier, *i.e.*, 4.43 MHz. At the receiver the process is reversed to recover R, G, B signals Fig. 12.6 shows principle of colour TV camera.

12.13.10. Colour TV Receiver

Figure 12.7 shows a simple block diagram of colour TV receiver.

12.13.11. Video Monitor

A video monitor is a display device in computers, CCTV systems. It is similar to TV receiver but has some differences. A video monitor is fed directly by the base band signal. Therefore it does not have tuner and I F amplifier stages. The bandwidth is 8 or 10 MHz. The input signal is $IV_{p\text{-}p}$. Both internal and external synchronisation is provided. Figure 12.8. shows a simple block diagram of video monitor.

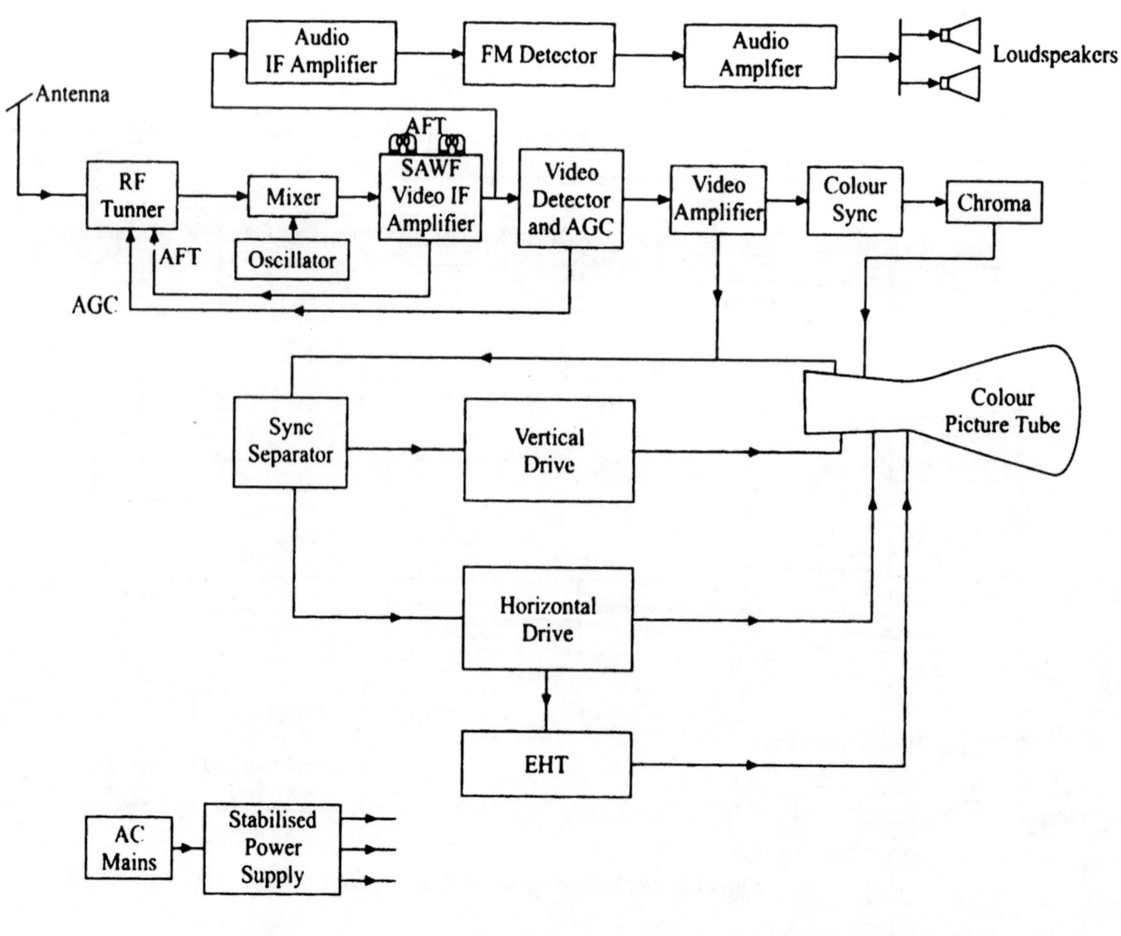

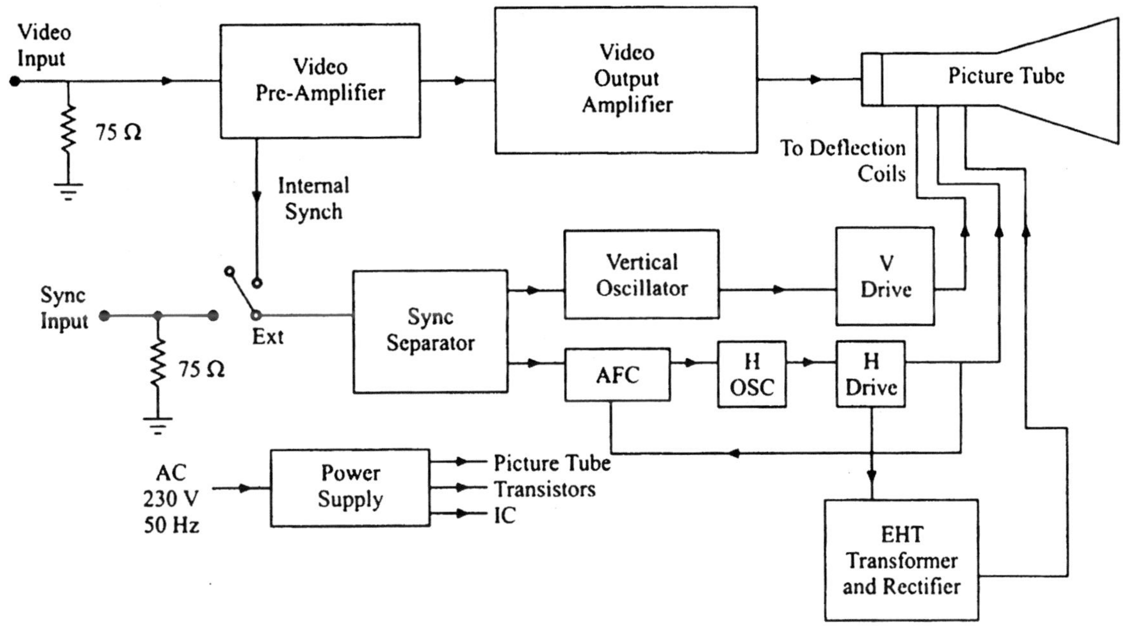

Fig. 12.8. Block Diagram of Black and White Video Monitor.

OBJECTIVE TYPE QUESTIONS

A. Select the most appropriate alternative

1. Sound waves travel as
   (*a*) longitudinal waves
   (*b*) transverse waves
   (*c*) both longitudinal and transverse waves
   (*d*) either longitudinal or transverse wave
2. If the frequency of a wave 20 Hz, the time period is
   (*a*) 20 secs (*b*) 2 secs
   (*c*) 0.2 secs (*d*) 0.05 secs
3. If a sound wave has a frequency of 50 Hz, the wavelength is
   (*a*) 66 m (*b*) 6.6 m
   (*c*) 0.66 m (*d*) 0.066 m
4. The audio frequency range is
   (*a*) 5 Hz to 1 kHz (*b*) 50Hz to 20 kHz
   (*c*) 16 Hz to 20 kHz (*d*) 0 Hz to 20 kHz
5. The velocity of sound waves in air
   (*a*) is constant at all temperatures
   (*b*) varies directly as temperature
   (*c*) varies inversely as absolute temperature
   (*d*) varies directly as square root of absolute temperature
6. If velocity of sound waves in air at 0°C is 330 m/s, the velocity at 40°C is about
   (*a*) 330 m/s (*b*) 334 m/s
   (*c*) 353 m/s (*d*) 380 m/s
7. If sound intensity is 8 W/m^2, the dB level of this sound is about
   (*a*) 5 dB (*b*) 50 dB
   (*c*) 129 dB (*d*) 1290 dB
8. A sound wave has an intensity of 10 Pa. The dB level is about
   (*a*) 1.139 dB (*b*) 11.39 dB
   (*c*) 113.9 dB (*d*) 1139 dB
9. A pure tone at 1000 Hz has a dB level of 50 dB. If two such tones exist simultaneously the dB level is about
   (*a*) 50 dB (*b*) 53 dB
   (*c*) 100 dB (*d*) 200 dB
10. A pure tone at 1000 Hz has a dB level of 40 dB. If the required dB level is 100, the number of such tones is
    (*a*) 2.5 (*b*) 60
    (*c*) 10^4 (*d*) 10^6
11. In a wave the electric field lines are vertical. Then the wave is
    (*a*) horizontally polarised (*b*) vertically polarised
    (*c*) either (*a*) or (*b*) (*d*) neither (*a*) nor (*b*)
12. The frequency range of 300 kHz to 3000 kHz is known as
    (*a*) low frequency (*b*) medium frequency
    (*c*) high frequency (*d*) very high frequency
13. UHF range is
    (*a*) 30 MHz is 300 MHz (*b*) 300 MHz to 3000 MHz
    (*c*) 3000 to 30000 MHz (*d*) above 30000 MHz
14. Ground wave can be used for communication up to
    (*a*) 5 kHz (*b*) 12 kHz
    (*c*) 16 kHz (*d*) 25 kHz

15. During day time ionosphere can be considered to be consisting of
(*a*) D, E, F_1 and F_2 layers (*b*) D, E, F_1 layers
(*c*) D, E, F_2 layers (*d*) D and E layers

16. At night the ionosphere can be considered to be consisting of
(*a*) D, E, F_1 and F_2 layers (*b*) D, E, F_1 layers
(*c*) D, E, F_2 layers (*d*) E and F_2 layers

17. Television transmission is by
(*a*) ground waves (*b*) sky waves
(*c*) space waves (*d*) all the above

18. An amplitude modulated wave is
$e = 50\,(1 + 0.4 \cos 1000\,t + 0.3 \cos 10000\,t) \cos 10^7 t$
The modulation index is
(*a*) 0.5 (*b*) 0.3
(*c*) 0.7 (*d*) 0.4

19. The maximum distance between television transmission and receiving antennas for reasonably good reception should be
(*a*) less than 20 km (*b*) less than 100 km
(*c*) less than 500 km (*d*) less than 1000 km

20. Radio broadcasts are generally
(*a*) amplitude modulated
(*b*) frequency modulated
(*c*) both amplitude and frequency modulated
(*d*) neither amplitude nor frequency modulated

21. Video signals are transmitted through
(*a*) amplitude modulation
(*b*) frequency modulation
(*c*) either amplitude or frequency modulation
(*d*) neither amplitude nor frequency modulation

22. The positive peak of an AM wave is 16 V and the minimum value is 4 V. Assuming single tone modulation, the modulation index is
(*a*) 0.6 (*b*) 0.25
(*c*) 4 (*d*) 0.36

23. A 200 Hz sound wave is to be transmitted by a 1000 kHz radio wave using frequency modulation. The instantaneous frequency variation will be
(*a*) 100 times a second (*b*) 200 times a second
(*c*) 400 times a second (*d*) 1000 times a second

24. A 300 Hz sound wave of intensity 20 dB is transmitted by a 1500 kHz radio wave by frequency modulation. It is seen that instantaneous frequency varies between 1500.2 kHz and 1499.8 kHz, 300 times a second. If the intensity of sound wave is 100 dB
(*a*) The instantaneous frequency variations will be between 1500.2 and 1499.8 kHz, 1500 times a second
(*b*) The instantaneous frequency variations will be between 1501 and 1499 kHz, 1500 times a second
(*c*) The instantaneous frequency variations will be between 1501 and 1499 kHz, 300 times a second
(*d*) The instantaneous frequency variations will be between 1501 and 1499 kHz, 60 times a second

25. The frequency response of a microphone means
(*a*) bandwidth of frequencies having output within ±1 dB of the output at 1000 Hz
(*b*) bandwidth of frequencies having output within ±3 dB of the output at 1000 Hz
(*c*) bandwidth of frequencies having output within ±1 dB of the output at 500 Hz
(*d*) bandwidth of frequencies having output within ±3 dB of the output at 500 Hz

26. Signal to noise ratio of microphone is defined as the ratio of output when the sound pressure is ..... to the output in the absence of any sound signal
(*a*) 0.5 Pa (*b*) 0.2 Pa
(*c*) 0.1 Pa (*d*) 0.01 Pa

27. The output resistance of a microphone is 20 ohms. It is to feed a line of impedance 320 ohms. To ensure maximum power transfer, the turn ratio of transformer is
(*a*) 16 (*b*) 256
(*c*) 2 (*d*) 4

28. A directional microphone gives an output of 0.5 mV whereas an omni directional one gives an output of 0.5 μV in the same position for the same sound pressure. The directivity of the directional microphone is
(*a*) 120 dB (*b*) 90 dB
(*c*) 60 dB (*d*) 30 dB

29. Which of the following microphones is the cheapest?
(*a*) Moving coil (*b*) Condenser
(*c*) Ribbon (*d*) Carbon

30. Which of the following microphones is used as a standard for calibration?
(*a*) Moving coil (*b*) Carbon
(*c*) Condenser (*d*) Ribbon

31. Which of the following microphones has maximum distortion?
(*a*) Ribbon (*b*) Condenser
(*c*) Crystal (*d*) Carbon

32. Which of the following microphones has highest output impedance?
(*a*) Carbon (*b*) Ribbon
(*c*) Crystal (*d*) Condenser

33. Which of the following microphones has lowest output impedance?
(*a*) Moving coil (*b*) Ribbon
(*c*) Crystal (*d*) Condenser

34. Which of the following microphones has bidirectional feature?
(*a*) Moving coil (*b*) Ribbon
(*c*) Condenser (*d*) Carbon

35. Which of the following microphones has an in-built amplifier?
(*a*) Condenser (*b*) Moving coil
(*c*) Ribbon (*d*) Carbon

36. Which of the following microphones works on the principle of Faraday's law?

(a) Carbon (b) Condenser
(c) Moving coil (d) Electret

37. Which of the following microphones works on the piezoelectric effect?
(a) Carbon (b) Condenser
(c) Moving coil (d) Crystal

38. A single tone modulating signal $V_m \cos \omega t$ modulates a carrier $A \cos \omega_c t$. The carrier power (modulated) and side band powers are respectively
(a) $\frac{A^2}{2}$ and $\frac{V_m^2}{4}$ (b) $\frac{A^2}{2}$ and $\frac{V_m^2}{2}$
(c) A^2 and $\frac{V_m^2}{4}$ (d) $\frac{A^2}{4}$ and $\frac{V_m^2}{4}$

39. The efficiency of a direct radiating cone type loudspeaker is about
(a) 5% (b) 20%
(c) 50% (d) 80%

40. In single tone AM modulation the transmission efficiency for $m = 1$ is
(a) 33.3% (b) 50%
(c) 100% (d) 16.65%

41. The material used for the magnet of moving coil cone type loudspeaker is
(a) steel alloy (b) soft iron
(c) hard steel (d) Alnico

42. In the equivalent circuit of a moving coil cone type loudspeaker, the mass of the moving system is represented by
(a) a resistance (b) a capacitance
(c) an inductance (d) a conductance

43. A direct radiating cone type loudspeaker attenuates :
(a) medium range frequencies
(b) very low frequencies
(c) high frequencies
(d) very low and high frequencies

44. In a two tone AM system the two modulating frequencies are 2000 π and 4000 π rad/sec. If carrier frequency is $2\pi \times 10^6$ rad/sec the frequencies of upper side bands are
(a) 1.002 MHz and 1.004 MHz
(b) 1.001 MHz and 1.002 MHz
(c) 1.001 MHz and 1.004 MHz
(d) 1.002 MHz and 1.008 MHz

45. The efficiency of a horn speaker is about
(a) 10% (b) 40%
(c) 60% (d) 80%

46. In a single tone AM the power of carrier is P and modulation index is m. The powers of each of the two side bands are
(a) $\frac{Pm^2}{4}$ (b) $\frac{Pm^2}{2}$
(c) $\frac{Pm^2}{8}$ (d) $\frac{Pm^2}{16}$

47. A woofer should be fed from the input through a
(a) low pass filter (b) high pass filter
(c) band pass filter (d) band stop filter

48. A tweeter should be fed from the input through a
(a) low pass filter (b) high pass filter
(c) band pass filter (d) band stop filter

49. The filter circuit for feeding a woffer has
(a) an inductance in series and capacitance in parallel
(b) a capacitance in series and an inductance in parallel
(c) a resistance in series and an inductance in parallel
(d) an inductance in series and a resistance in parallel

50. The filter circuit for feeding tweeter has
(a) an inductance in series and a capacitance in parallel
(b) a capacitance in series and an inductance in parallel
(c) a resistance in series and a capacitance in parallel
(d) a capacitance in series and a resistance in parallel

51. The most commonly used microphone for public address systems is
(a) carbon (b) crystal
(c) moving coil (d) condenser

52. A carrier $A \cos \omega_c t$ is modulated by a signal $m_1 A \cos \omega_1 t + m_2 A \cos \omega_2 t + m_3 A \cos \omega_3 t$. The total modulated power is
(a) $\frac{A^2}{2}(1 + 0.5m_1^2 + 0.5m_2^2 + 0.5m_3^2)$
(b) $A^2(1 + 0.5m_1^2 + 0.5m_2^2 + 0.5m_3^2)$
(c) $\frac{A^2}{2}(1 + m_1^2 + m_2^2 + m_3^2)$
(d) $\frac{A^2}{2}(m_1^2 + m_2^2 + m_3^2)$

53. The simplest mixer circuit for public address system uses
(a) fixed resistances (b) variable resistances
(c) inductances (d) capacitances

54. For a good public address system the sound intensity at the farthest point should be at least
(a) 20 dB over threshold of hearing
(b) 40 dB over threshold of hearing
(c) 80 dB over threshold of hearing
(d) 120 dB over threshold of hearing

55. If $F(\omega)$ is Fourier transform of $f(t)$, then the Fourier transform of $f(t) \cos \omega_c t$ is
(a) $F(\omega + \omega_c) + F(\omega - \omega_c)$
(b) $\frac{1}{2}[F(\omega + \omega_c) + F(\omega - \omega_c)]$
(c) $F(\omega + \omega_c) - F(\omega - \omega_c)$
(d) $\frac{1}{2}[F(\omega + \omega_c) - F(\omega - \omega_c)]$

56. If $F(\omega)$ is Fourier transform of $f(t)$, then Fourier transform of $f(t) \cos^2 \omega_c t$ is
(a) $\frac{1}{2}F(\omega) + \frac{1}{4}[F(\omega + 2\omega_c) + F(\omega - 2\omega_c)]$
(b) $F(\omega) + \frac{1}{4}[F(\omega + 2\omega_c) + F(\omega - 2\omega_c)]$

(c) $F(\omega) - \frac{1}{4}[F(\omega + 2\omega_c) + F(\omega - 2\omega_c)]$

(d) $\frac{1}{2}F(\omega) - \frac{1}{4}[F(\omega + 2\omega_c) + F(\omega - 2\omega_c)]$

57. The most commonly used transistor amplifier circuit is
(a) CB (b) CC
(c) CE (d) CE or CC

58. Which of the following has highest power?
(a) CE (b) CC
(c) CB (d) CB or CC

59. Which of the following amplifiers has a voltage gain of less than 1?
(a) CE (b) CC
(c) CB (d) CE or CB

60. Which of the following amplifiers is used for impedance matching?
(a) CE (b) CC
(c) CB (d) Push pull

61. Which of the following has the highest input impedance?
(a) CE (b) CB
(c) CC (d) Push pull

62. Which of the following amplifiers has highest output impedance?
(a) CE (b) CB
(c) CC (d) Push pull

63. AM-SC stands for
(a) Amplitude modulated-suppressed carrier
(b) Amplitude modulated-signal carrier
(c) Amplitude modulated-synchronous carrier
(d) Amplitude modulated-self carrier

64. Homodyne detection means
(a) coherent detection (b) non-coherent detection
(c) asyrchronous detection (d) none of the above

65. The reactance of coupling capacitors, for DC quantities, is
(a) zero (b) low
(c) infinite (d) zero or low

66. The reactance of coupling capacitor for AC signals in an amplifier, should be
(a) as low possible (b) as high as possible
(c) neither low nor high (d) either low or high

67. In a CE amplifier connection, the emitter is at ground potential for
(a) AC quantities (b) DC quantities
(c) all quantities (d) none of the quantities

68. The reactance of bypass capacitor in a CE amplifier circuit should be
(a) equal to the resistance in parallel with it
(b) equal to 10% or less of the resistance in parallel with it
(c) equal to 50% of the resistance in parallel with it
(d) independent of the resistance in parallel with it

69. Costa's receiver is used for
(a) coherent detection of AM-SC signal
(b) asynchronous detection of AM-SC signal
(c) amplitude modulation of signal
(d) frequency modulation of signal

70. SSB-SC stands for
(a) single side band-suppressed carrier
(b) single side band-synchronous carrier
(c) synchronous side band-suppressed carrier
(d) synchronous side band-self carrier

71. If each frequency component of $f(t)$ is shifted by $\pi/2$, the resulting signal is
(a) Hilbert transform of $f(t)$
(b) Inverse Hilbert transform of $f(t)$
(c) Laplace transform of $f(t)$
(d) Fourier transfer of $f(t)$

72. Audio amplifiers need
(a) Bass control
(b) Treble control
(c) Both Bass and Treble control
(d) Either Bass or Treble control

73. Push pull amplifier is
(a) voltage amplifier (b) current amplifier
(c) power amplifier (d) none of the above

74. In push pull amplifier each transistor conducts for
(a) 180° (b) 360°
(c) 90° (d) 45°

75. Push pull amplifier is
(a) class B (b) class A
(c) class C (d) class A or class C

76. The open loop gain of an amplifier is 50 but is likely to change by 20%. If negative feedback with $\beta = 0.1$ is used, the change in gain will be about
(a) 20% (b) 10%
(c) 4% (d) 0.05%

77. The open loop gain of an amplifier is 200. If negative feedback with $\beta = 0.2$ is used, the closed loop gain will be
(a) 200 (b) 40
(c) 4.878 (d) 1.212

78. Consider the following
(a) generation of SSB signals
(b) design of minimum phase type filters
(c) representation of band pass signals
In which of the above applications is Hilbert transform used
(a) 1 only (b) 1, 2 and 3
(c) 1, 2 only (d) 1 and 3

79. Filter method and phasing method are the methods for generating
(a) SSB signal
(b) Vestigial side band signal
(c) both SSB and vestigial side band signals
(d) none of the above

80. Resonant circuits are used in
(a) Audio amplifiers
(b) RF amplifiers

(c) Both audio and RF amplifiers
(d) none of the above

81. CSSB stands for
(a) compatible single side band
(b) compressed single side band
(c) carrier single side band
(d) coded single side band

82. For attenuation of low frequencies we should use
(a) shunt capacitance (b) series capacitance
(c) shunt inductance (d) series inductance

83. For attenuation of high frequencies we should use
(a) shunt capacitance (b) series capacitance
(c) inductance (d) resistance

84. Consider the following
1. A signal $f(t)$ and its Hilbert transform $f_h(t)$ have the same energy density spectrum
2. A signal $f(t)$ and its Hilbert transform $f_h(t)$ have the same auto correlation function
3. A signal $f(t)$ and its Hilbert transform are mutually orthogonal

Which of the above are true?
(a) 1, 2 and 3 (b) 1 and 2 only
(c) 1 only (d) 2 and 3 only

85. In major diatonic musical scale, an octave is divided into
(a) 12 intervals (b) 10 intervals
(c) 7 intervals (d) 5 intervals

86. In equally tempered musical scale, an octave is divided in to
(a) 12 intervals (b) 10 intervals
(c) 7 intervals (d) 5 intervals

87. In equally tempered scale, frequency of each tone equals the frequency of lower adjacent tone multiplied by
(a) $(2)^{\frac{1}{7}}$ (b) $(2)^{\frac{1}{12}}$
(c) $(2)^{\frac{1}{6}}$ (d) $(2)^{\frac{1}{5}}$

88. Which of the following is a better recording medium?
(a) Ferric oxide (b) Ferrous oxide
(c) Chromium dioxide (d) (a) or (b)

89. If $f_h(t)$ Hilbert transform of $f(t)$, then Hilbert transform of $f_h(t)$ is
(a) $-f(t)$ (b) $+f(t)$
(c) $f^2(t)$ (d) $\sqrt{f(t)}$

90. If $f(t)$ is the Hilbert transform of $f(t)$, and $f(t)$ is a real valued signal, the pre-envelope of $f(t)$ is
(a) $f(t)-jf_h(t)$ (b) $f(t)+jf_h(t)$
(c) $f_h(t)+jf(t)$ (d) $f_h(t)-jf(t)$

91. For maximum output the width of air gap of audio head should be
(a) equal to wavelength
(b) half the wavelength
(c) one quarter the wavelength
(d) very-very small as compared to wavelength

92. The speeds used in recording audio programmes in house hold recorders are
(a) 38 cm/s and 9.5 cm/s (b) 38 cm/s and 19 cm/s
(c) 9.5 cm/s and 4.75 cm/s (d) 38 cm/s and 4.75 cm/s

93. Which of the following is correct (f_m is maximum usable frequency, S is speed and G is width of air gap)
(a) $f_m = (\text{S})(2\text{G})$ (b) $f_m = \dfrac{\text{S}}{2\text{G}}$
(c) $f_m = \dfrac{\text{S}}{\text{G}}$ (d) $\text{G} = (f_m)(\text{S})$

94. Audio cassettes used in household tape recorders have a width of
(a) 1″ (b) 1/2″
(c) 1/4″ (d) 1/8″

95. In audio cassettes the width of each recording track is
(a) 5 mm (b) 2.5 mm
(c) 1.5 mm (d) 1 mm

96. An audio frequency of 5 kHz is translated to a radio frequency carrier of frequency 1 MHz and transmitted in to space through an antenna. The required height of antenna is
(a) 150 m (b) 300 m
(c) 75 m (d) 600 m

97. The tape recorders used in homes have
(a) common head and common amplifier system for recording and playback
(b) different heads and amplifiers for recording and playback
(c) same head but different amplifiers for recording and playback
(d) different heads but same amplifier for recording and playback

98. In a tape recorder the erase head is
(a) engaged during playback but disengaged during recording
(b) engaged both during recording and playback
(c) disengaged both during recording and playback
(d) engaged during recording but disengaged during playback

99. The width of air gap on recording head in a tape recorder is about
(a) 6 micron (b) 3 micron
(c) 1 micron (d) 25 micron

100. In a tape recorder
(a) the width of air gaps of record and erase heads is almost equal
(b) the air gap of record head is wider than the air gap of erase head
(c) the air gap of erase head is wider than the air gap of record head
(d) any of the two air gaps can be wider than the other

101. In a tape recorder
(a) both record and erase heads have high permeability core

(*b*) the record head has high permeability core but the erase head has low permeability core
(*c*) the record head has low permeability core but the erase head has high permeability core
(*d*) both record and erase heads have low permeability core

102. Tape recorders use
(*a*) cone type loudspeakers
(*b*) horn type loudspeakers
(*c*) either cone type or horn type loudspeakers
(*d*) none of the above

103. Household tape recorders generally use
(*a*) crystal microphone
(*b*) moving coil microphone
(*c*) condenser microphone
(*d*) cardioid microphone

104. Which of the following is not correct?
(*a*) In SSB amplitude modulated system, the transmitter circuit has to be more stable leading to better reception
(*b*) In SSB amplitude modulated system more channel space is available as compared to double side band amplitude modulated system
(*c*) In SSB amplitude modulated system the signal is more noise resistant as compared to double side band system
(*d*) An SSB-AM system requires lesser power (for the same signal strength) as compared to DSB-AM system

105. Household tape recorders use
(*a*) separate motor for each operation
(*b*) only one motor for all operations
(*c*) one motor for capstan and another for take up reel
(*d*) a total of 3 motors

106. During playback the take up reel is
(*a*) in clockwise motion
(*b*) in anticlockwise motion
(*c*) either clockwise or anticlockwise motion
(*d*) any of the above

107. In a tape recorder the output from RF oscillator is given to
(*a*) only erase head coil
(*b*) only record head coil
(*c*) coils of both erase head and record head
(*d*) none of the above

108. During fast forward operation of a tape recorder
(*a*) all heads are withdrawn
(*b*) all amplifiers are shut off
(*c*) tape is not pressed against capstan
(*d*) all the above

109. An SSB-AM signal cannot be demodulated by
(*a*) BJT balanced modulator
(*b*) Complete phase shift generator
(*c*) Diode balanced modulator
(*d*) Product detector

110. Which of the following is true in a tape recorder?
(*a*) Vow is due to variation in motor speed but flutter is due to some other reason
(*b*) both vow and flutter are due to variations in motor speed
(*c*) Flutter is due to variation in motor speed but vow is due to some other reason
(*d*) neither vow nor flutter is due to variation in motor speed

111. Satellite communication uses frequencies in the range of
(*a*) 3 to 6 GHz (*b*) 3–10 MHz
(*c*) 10–50 GHz (*d*) 100–200 GHz

112. Dolby A system of noise reduction divides the total audio spectrum into
(*a*) 8 frequency bands (*b*) 4 frequency bands
(*c*) 2 frequency bands (*d*) none of the above

113. For AM receivers the standard IF frequency is
(*a*) 106 kHz (*b*) 455 kHz
(*c*) 1.07 MHz (*d*) 10.7 MHz

114. A carrier is modulated by two modulating waves A and B having modulation indices of 0.6 and 0.8 respectively. The overall modulation index is
(*a*) 1 (*b*) 0.7
(*c*) 0.2 (*d*) 1.4

115. Compact discs use
(*a*) magnetic recording and retrieval
(*b*) optical recording and retrieval
(*c*) optical recording but magnetic retrieval
(*d*) magnetic recording but optical retrieval

116. In a CD, the depth and width of each pit is
(*a*) 0.5 μm each
(*b*) 0.5 μm and 1 μm respectively
(*c*) 1 μm and 0.5 μm respectively
(*d*) 1 μm each

117. In a CD the channel separation is
(*a*) 1 μm (*b*) 1.6 μm
(*c*) 2.5 μm (*d*) 4 μm

118. In a CD the sampling frequency is nearly
(*a*) half the highest frequency in audio signal
(*b*) equal to highest frequency in audio signal
(*c*) double the highest frequency in audio signal
(*d*) four times the highest frequency in audio signal

119. The direction of rotation of a CD is
(*a*) clockwise
(*b*) anticlockwise
(*c*) clockwise or anticlockwise depending on frequency of data stored
(*d*) mostly anticlockwise but some times clockwise

120. The diameter and thickness of CD are
(*a*) 6 cm and 1.2 mm (*b*) 6 cm and 0.6 mm
(*c*) 12 cm and 0.6 mm (*d*) 12 cm and 1.2 mm

121. In a CD player the speed of CD is
(*a*) constant at 500 rpm
(*b*) constant at 200 rpm
(*c*) varies from 200 to 500 rpm
(*d*) none of the above

122. The direction of pick up in a CD is
(*a*) linear from centre towards edge
(*b*) linear from edge to centre
(*c*) non-linear from centre to edge
(*d*) non-linear from edge to centre

123. In a CD player the laser uses
(*a*) silicon semiconductor
(*b*) silicon/germanium semiconductor
(*c*) aluminimum gallium arsenide semiconductor
(*d*) any of the above

124. The storage capacity of a CD is
(*a*) 360 k bytes per side (*b*) 360 M bytes per side
(*c*) 540 k bytes per side (*d*) 540 M bytes per side

125. A telephone exchange has 9000 subscribers. If the number of calls originating at peak time is 10,000 in one hour, the calling rate is
(*a*) 0.9 (*b*) 10/9
(*c*) 0.81 (*d*) 0.1

126. For a system to be designated as hi-fi, the signal to noise ratio should be at least
(*a*) 25 dB (*b*) 50 dB
(*c*) 75 dB (*d*) 100 dB

127. In a hi-fi system the non-linear distortion should be less than
(*a*) 10% (*b*) 5%
(*c*) 2% (*d*) 1%

128. Dynamic range of hi-fi system is at least
(*a*) 150 dB (*b*) 120 dB
(*c*) 80 dB (*d*) 40 dB

129. In a hi-fi system the frequency response over the audio frequency range should be within
(*a*) ± 10 dB (*b*) 5 dB
(*c*) 2 dB (*d*) 1 dB

130. If a signal voltage V_s is accompanied by a noise voltage V_n, the signal to noise ratio is
(*a*) $\frac{V_s^2}{V_n^2}$ (*b*) $\frac{V_s}{V_n}$
(*c*) $\sqrt{\frac{V_s}{V_n}}$ (*d*) $1 - \frac{V_s}{V_n}$

131. Brownian motion is
(*a*) random emission of electrons in vacuum tubes
(*b*) thermal motion of electrons in a resistance
(*c*) random diffusion of holes and electrons in semi-conductors
(*d*) both (*b*) and (*c*)

132. In shot noise
(*a*) the current wiggles around mean value
(*b*) the current rises to peak value, decreases to zero and then again rises
(*c*) the current may become negative also
(*d*) the current is always less than the mean value

133. To ensure good transient response in audio systems the natural frequencies should be
(*a*) above 40 Hz and below 15 kHz
(*b*) above 40 Hz
(*c*) above 15 kHz
(*d*) below 40 Hz and above 15 kHz

134. The number of channels in a stereophonic system is
(*a*) 8 (*b*) 4
(*c*) 2 (*d*) 1

135. In a stereo tape recorder the speed of the tape is
(*a*) 38 m/s (*b*) 19 m/s
(*c*) 9.5 m/s (*d*) 4.75 m/s

136. The power density spectrum of noise voltage of resistor R can be expressed as
(*a*) 2 kTR (*b*) kTR
(*c*) 0.5 kTR (*d*) 0.25 kTR

137. Loudness control gives
(*a*) high boost to bass and little boost to treble
(*b*) equal boost to bass and treble
(*c*) less boost to bass and more boost to treble
(*d*) less boost to both bass and treble

138. The function of blend control in stereo system is
(*a*) to dilute left channel
(*b*) to dilute right channel
(*c*) to dilute any of the channles
(*d*) none of the above

139. A resistor R is in parallel with capacitance C. The rms noise voltage of resistor in the equivalent circuit can be written as
(*a*) $\frac{kT}{C}$ (*b*) kTC
(*c*) $\frac{kC}{T}$ (*d*) $\frac{kT}{C^2}$

140. Present day stereophonic systems have
(*a*) only CD playback facility
(*b*) only tape playback facility
(*c*) both CD and tape playback facilities along with ratio
(*d*) none of the above

141. Graphic equaliser system in stereophonic systems generally have
(*a*) 3 or 4 bands (*b*) 1 band
(*c*) more than 8 bands (*d*) large number of bands

142. In telecast
(*a*) video and audio signals are frequency modulated
(*b*) video signals are frequency modulated but audio signals are amplitude modulated
(*c*) both audio and video signals are amplitude modulated
(*d*) video signals are amplitude modulated and audio signals are frequency modulated

143. In a Vidicon camera tube
(*a*) both sides of the target plate have the coating of tin oxide
(*b*) both sides of the target plate have the coating of antimony tri-sulphide
(*c*) the side of target plate facing the light has a coating of tin oxide and the side facing electron gun has a coating of antimony tri-sulphide

(*d*) the side of target plate facing the light has a coating of antimony tiisulphide while the side facing the electron gun has a coating of tin oxide

144. In a Vidicon camera tube, the beam current is controlled by
(*a*) grid G_1 (*b*) grid G_2
(*c*) grid G_1, G_2, G_3 (*d*) grid G_2, G_3

145. In a Vidicon camera tude
(*a*) grid G_1 accelerates electrons
(*b*) grid G_2 accelerates electrons and grid G_3 slows down electrons
(*c*) grid G_3 accelerates electrons and grid G_2 slows down electrons
(*d*) both grids G_2 and G_3 accelerate electrons

146. In Vidicon camera tube, the resistance of photoconductive material on the target plate
(*a*) is constant
(*b*) is more in darkness and less in brightness
(*c*) is more in brightness but less in darkness and brightness
(*d*) may be more or less in darkness and brightness

147. In the human eye, an image persists for a time of about
(*a*) 1 second (*b*) 0.1 second
(*c*) 0.05 second (*d*) 0.02 second

148. In a television camera tube, scanning of a picture is done
(*a*) line by line from left to right
(*b*) line by line from right to left
(*c*) line by line from top to bottom
(*d*) line by line from bottom to top

149. In the television system in India, each frame is scanned
(*a*) 100 times (*b*) 50 times
(*c*) 20 times (*d*) 10 times

150. In TV system in India, each frame is divided into
(*a*) 625 horizontal lines (*b*) 525 horizontal lines
(*c*) 425 horizontal lines (*d*) 325 horizontal lines

151. In the television system in India, the odd and even fields are scanned
(*a*) 25 times each
(*b*) 25 times each but alternately
(*c*) 50 times each
(*d*) 50 times each but alternately

152. In the television system in India, the aspect ratio is
(*a*) 4 : 3 (*b*) 3 : 4
(*c*) 5 : 4 (*d*) 4 : 5

153. For an RC low pass filter, the relation between noise bandwidth B_N and 3–dB bandwidth B_{3dB} is

(*a*) $B_N = \frac{\pi}{2} B_{3\text{-dB}}$ (*b*) $B_N = B_{3\text{-dB}}$

(*c*) $B_{3\text{-dB}} = \frac{\pi}{2} B_N$ (*d*) $B_N = \pi\, B_{3\text{-dB}}$

154. In the television system in India, each channel is allowed a bandwidth of
(*a*) 20 MHz (*b*) 12 MHz
(*c*) 7 MHz (*d*) 3 MHz

155. TV signal strength when expressed in dB μ has the reference of
(*a*) 10 μ V/m (*b*) 5 μ V/m
(*c*) 2 μ V/m (*d*) 1 μ V/m

156. The power density spectrum of noise voltage at the input of a network is S_{ni} and that at the output of the network is S_{no}. If transfer function of network is H(ω), the corresponding noise Figure is

(*a*) $\frac{S_{no}}{S_{ni}\,|H(\omega)|^2}$ (*b*) $\frac{S_{no}}{S_{ni}\,|H(\omega)|}$

(*c*) $\frac{S_{ni}}{S_{no}\,|H(\omega)|^2}$ (*d*) $\frac{S_{ni}}{S_{no}\,|H(\omega)|}$

157. In a TV receiver antenna the length of reflector rcd
(*a*) is the same as that of dipole
(*b*) is less than that of dipole
(*c*) is more than that of dipole
(*d*) may be equal, more or less than that of dipole

158. Which one of the following is true?
(*a*) characteristic impedance of twin wire feeder and coaxial cable are nearly the same
(*b*) the characteristic impedance of coaxial cable is more than that of twin wire cable
(*c*) the characteristics impedance of twin wire cable is more than that of coaxial cable
(*d*) the characteristic impedance of twin wire feeder may be more or less than that of coaxial cable

159. The characteristic impedance of a twin wire feeder used for TV signals is about
(*a*) 1000 ohm (*b*) 500 ohm
(*c*) 300 ohm (*d*) 100 ohm

160. The characteristic impedance of coaxial cable is about
(*a*) 1000 ohm (*b*) 500 ohm
(*c*) 250 ohm (*d*) 75 ohm

161. A balun transformer gives an impedance transformation of
(*a*) 4 : 1 (*b*) 2 : 1
(*c*) 1 : 4 (*d*) 1 : 2

162. The IF frequencies in a TV receiver
(*a*) 38.9 MHz for video and 33.4 MHz for audio
(*b*) 38.9 MHz for both video and audio
(*c*) 33.4 MHz for both video and audio
(*d*) 33.4 MHz for video and 38.9 MHz for audio

163. The number of noise sources in a BJT are
(*a*) 3 (*b*) 2
(*c*) 1 (*d*) 4

164. In a picture tube, the video signal is applied to
(*a*) cathode (*b*) grid
(*c*) any anode (*d*) either cathode or grid

165. Just when the video signal is applied to the picture in a TV receiver, the strength of video signal is about
(*a*) 1000 V (*b*) 200 V
(*c*) 50 V (*d*) 1 V

166. The EHT voltage in a picture tube of TV is
(*a*) about 1000 kV (*b*) about 100 kV
(*c*) about 15 kV (*d*) about 1 kV

167. The primary colours are
(*a*) blue and green (*b*) blue and red
(*c*) red and green (*d*) blue, red and green

168. Y signal has maximum percentage of
(*a*) red colour (*b*) blue colour
(*c*) green colour (*d*) red and blue colours

169. The subscarrier frequency for transmission of colour difference signals in TV systems in India is
(*a*) 9.47 MHz (*b*) 4.43 MHz
(*c*) 2.06 MHz (*d*) 1.3 MHz

170. The noise sources shot noise, partition noise and thermal noise are associated with which component
(*a*) BJT (*b*) diode
(*c*) resistor (*d*) capacitor

171. In which frequency range is the noise figure in a transistor minimum
(*a*) Intermediate frequencies (*b*) low frequencies
(*c*) high frequencies (*d*) very high frequencies

172. Generally a VCR has
(*a*) four stationary video heads
(*b*) two stationary video heads
(*c*) four rotating video heads
(*d*) two rotating video heads

173. In VCR the luminance signal is recorded by
(*a*) frequency modulation (*b*) phase modulation
(*c*) amplitude modulation (*d*) direct recording

174. In VCR, the chrominance signal's frequency is
(*a*) up converted during recording and down converted during playback
(*b*) down converted during recording and up converted during playback
(*c*) not changed during recording and playback
(*d*) any of above depending on VCR system

175. In which frequency range is the noise figure in a transistor independent of frequency
(*a*) Intermediate frequencies
(*b*) high frequency
(*c*) low frequencies
(*d*) both low and high frequencies

176. The video heads, in VCR, feed the other circuit
(*a*) directly
(*b*) through a transformer having two rotating windings
(*c*) through a transformer having two stationary windings
(*d*) through a transformer having one stationary and one rotating winding

177. In VHS format the width of tape is
(*a*) 2″ (*b*) 1″
(*c*) 1/2″ (*d*) 1/4″

178. In VHS format diameter of head drum is about
(*a*) 1 m (*b*) 20 cm
(*c*) 6 cm (*d*) 1 cm

179. In VHS format width of audio track is
(*a*) 5 mm (*b*) 2 mm
(*c*) 1 mm (*d*) 0.1 mm

180. In VHS format each video head is tilted from azimuth by
(*a*) 20° (*b*) 10°
(*c*) 6° (*d*) 1°

181. The intermediate frequencies in VCR are
(*a*) video 38.9 MHz and audio 33.4 MHz
(*b*) 38.9 MHz for both video and audio
(*c*) 33.4 MHz for both video and audio
(*d*) 33.4 MHz for video and 38.9 MHz for audio

182. DVD uses
(*a*) Laser beam for both recording and playback
(*b*) Laser beam for recording and video head for playback
(*c*) Video head for recording and laser beam for playback
(*d*) none of the above

183. A remote control of TV uses a 10 bit word. The maximum number of total commands it can transmit is
(*a*) 64 (*b*) 256
(*c*) 512 (*d*) 1024

184. A remote control for TV uses
(*a*) LED in receiver and photodiode in transmitter
(*b*) LED in transmitter and photodiode in receiver
(*c*) LED in both transmitter and receiver
(*d*) Photodiode in both transmitter and receiver

185. A TV remote control works on the principle of
(*a*) pulse code modulated ultra violet light
(*b*) pulse code modulated infra red light
(*c*) demodulation
(*d*) either (*a*) or (*b*)

186. The first pulse in a TV remote control is
(*a*) start pulse
(*b*) preliminary pulse
(*c*) either start pulse or preliminary pulse
(*d*) erase pulse

187. The total number of pulses in a TV remote control is
(*a*) 14 (*b*) 12
(*c*) 10 (*d*) 8

188. The seven segment display used in watches and clocks generally uses
(*a*) LED (*b*) LCD
(*c*) Photodiode (*d*) FET

189. The range of a cordless telephone is about
(*a*) 100 m (*b*) 500 m
(*c*) 100 m (*d*) 10 m

190. A cordless telephone using separate frequencies for transmission in base and portable units is known as
(*a*) duplex arrangement (*b*) half duplex arrangement
(*c*) either (*a*) or (*b*) (*d*) neither (*a*) nor (*b*)

191. The power to the base unit of a cordless telephone is supplied by
(*a*) battery (*b*) AC mains
(*c*) both battery and AC mains
(*d*) either battery or AC mains

192. The power to the portable unit of a cordless telephone is provided by
(*a*) pencil cells (*b*) dry cells
(*c*) rechargeable cells (*d*) AC mains

193. A mobile telephone has a range of about
(*a*) 50 km (*b*) 10 km
(*c*) 1 km (*d*) 0.1 km

194. The mobile unit of a mobile telephone is generally in
(*a*) vehicles (*b*) pocket
(*c*) office (*d*) vehicle or pocket

195. In cellular telephone each cell is designed to handle
(*a*) 45 two way communications
(*b*) 90 two way communications
(*c*) 45 one way communications
(*d*) 180 two way communications

196. The total number of channels alloted for cellular telephone is
(*a*) 666 (*b*) 333
(*c*) 221 (*d*) 111

197. The output of a battery eliminator is about
(*a*) 110 V DC (*b*) 110 V AC
(*c*) 45 V DC (*d*) 6 V DC

198. A UPS contains
(*a*) rectifier
(*b*) inverter
(*c*) either rectifier or inverter
(*d*) both rectifier and inverter

199. The decorative lighting works on the principle of
(*a*) periodic switching on and off
(*b*) special inverter
(*c*) rectifier
(*d*) motor

200. A digital watch contains
(*a*) quartz crystal oscillator
(*b*) RC oscillator
(*c*) LC oscillator
(*d*) either quartz or RC oscillator

201. Vidicon and plumbicon operate on the principle of
(*a*) photo conduction (*b*) photo emission
(*c*) photo volatic (*d*) none of above

202. To eliminate ghosts in the picture
(*a*) use a longer transmission line
(*b*) connect a booster
(*c*) change the antenna orientation of location
(*d*) twist the transmission line

203. Satellites for international communication are called
(*a*) cossort (*b*) domsat
(*c*) marisat (*d*) intelsat

204. Leak type bias is used in plate modulated class C amplifier to
(*a*) increase the bandwidth
(*b*) prevent over modulation
(*c*) prevent excessive grid current
(*d*) prevent tuned circuit damping

205. Frequency shift keying is used mostly in
(*a*) telegraphy
(*b*) telephony
(*c*) satellite communication
(*d*) radio transmission

206. In ionosphere incidence angle is 60°. The maximum usable frequency f_m is equal to
(*a*) $2f_c$ (*b*) f_c
(*c*) $0.5f_c$ (*d*) $\sqrt{3}f_c$

207. If transmission path, in PCM, is very long then
(*a*) repeater stations are used
(*b*) pulse width may be increased
(*c*) pulse amplitude is increased
(*d*) pulse spacing is reduced

208. When large amount of power is required, a push pull amplifier is designed for operation as
(*a*) class A (*b*) class B
(*c*) class C (*d*) class AB

209. In broadcast television, the number of lines per frame is chosen to be odd number to
(*a*) reduce the required bandwidth
(*b*) improve resolution
(*c*) reduce peak power requirement
(*d*) make inter lacing easier

210. A signal having uniformly distributed amplitude in the interval (–V to +V) is to be encoded using PCM with uniform quantization. The signal to quantization noise ratio is determined by
(*a*) dynamic range of signal
(*b*) sampling rate
(*c*) number of quantizing levels
(*d*) power spectrum of signal

211. In TV systems, equalising pulses are sent during
(*a*) horizontal blanking (*b*) vertical blanking
(*c*) serrations (*d*) horizontal retrace

212. In PCM, the biggest disadvantage as compared to AM is
(*a*) larger bandwidth
(*b*) larger noise
(*c*) inability to handle analog signals
(*d*) incompatibility with time division multiplex systems

213. Consider the drawbacks
1. Slope overload
2. Serration noise
3. Granular noise
Which of these drawbacks does Delta modulation have?
(*a*) 1 and 2 only (*b*) 2 and 3 only
(*c*) 1 and 3 only (*d*) 1, 2, 3

214. In a satellite system circular polarization is to be obtained. The antenna used is
(*a*) parabolic (*b*) horn
(*c*) log periodic (*d*) helical

215. Match List I with List II and select the answer using the codes

List I	List II
A. IF noise	1. Ganged tuning
B. Image frequency	2. Spurious response
C. Station selector	3. Leaky filter capacitor
D. Excessive Hum	4. Mixer stage

Codes:

	A	B	C	D
(a)	3	2	1	4
(b)	2	3	1	4
(c)	1	2	3	4
(d)	4	2	1	3

216. For signal amplitude modulated to a depth of 100% by a sinusoidal signal, power is
(a) same as that of unmodulated carrier
(b) twice the power of unmodulated carrier
(c) $\frac{3}{2}$ times the power of unmodulated carrier
(d) $\frac{2}{3}$ the power of unmodulated carrier

217. Frequency frogging is used in a carrier system to
(a) conserve frequencies (b) reduce distortion
(c) reduce cross talk (d) none of the above

218. A 1000 kHz carrier is simultaneously modulated by 300 Hz and 2 kHz audio sine wave. Which of the following frequencies will not be present in the output?
(a) 998 kHz (b) 999.7 kHz
(c) 1000.3 kHz (d) 700 kHz

219. Line of sight transmission is used in
(a) radio frequencies below 1 MHz
(b) VHF only
(c) audio signals only
(d) VHF and UHF

220. An image uses 512 × 512 picture elements. Each element can take any of eight distinguishable levels. The maximum entropy in the above image will be
(a) 2097152 bits (b) 786432 bits
(c) 648 bits (d) 144 bits

221. Match List I with List II and select correct answer using the given codes

List I	List II
A TV	1. 1 kHz
B Radio	2. 3 kHz
C Telephony	3. 9 kHz
D Telegraphy	4. 5 kHz

Codes:

	A	B	C	D
(a)	4	1	2	3
(b)	4	3	2	1
(c)	2	1	3	4
(d)	1	2	3	4

222. The saving in power in DSBSC modulated at 80% is
(a) NIL (b) 80%
(c) 75.76% (d) 50%

223. In PCM the amplitude levels are transmitted in a 7 unit code. Sampling frequency is 10 kHz. The bandwidth should be
(a) 5 kHz (b) 35 kHz
(c) 70 kHz (d) 5 MHz

224. An RF carrier of 12 kV at 1 MHz is amplitude modulated by 1 kHz signal having peak value of 6 kV. The modulation pattern is seen on CRO screen. The voltage indicated will be
(a) 18 kV (b) 1.001 kV
(c) 6 kV (d) 36 kV

225. The dc voltage on plate modulated class C amplifier is 8 kV. If the peak modulating voltage is 4 kV, the peak RF voltage delivered to load is
(a) 32 kV (b) 12 kV
(c) 0.5 kV (d) 24 kV

226. A communication channel disturbed by additive white, Gaussion noise has a bandwidth of 4 kHz and SNR of 15. The highest transmission rate that this channel can support (in k – bits/sec) is
(a) 16 (b) 1.6
(c) 3.2 (d) 60

227. A discrete zero memory information source has 40 symbols and each symbol is equally likely. The minimum number of bits required to code the source are
(a) 5 and 5.03 (b) 6 and 5.83
(c) 5 and 6.64 (d) 6 and 5.32

228. In a two stage cascade amplifier, each stage has a gain of 10 dB and noise figure of 10 dB. The overall noise figure is
(a) 10 (b) 1.09
(c) 1 (d) 10.9

229. The noise figure of individual stages of a two stage amplifier is 2.03 and 1.54 respectively. If gain of first stage is 62, the overall noise figure is
(a) 2.03 (b) 2.055
(c) 3.05 (d) 6.05

230. For a three stage cascaded amplifier, having gains of G_1, G_2 and G_3 and noise figures of F_1, F_2, F_3 the overall noise figure F =
(a) $F = F_1 + \frac{F_2 - 1}{G_1} + \frac{F_3 - 1}{G_1 G_2}$
(b) $F = F_3 + \frac{F_1 - 1}{G_1} + \frac{F_2 - 1}{G_1 G_2}$
(c) $F = F_3 + \frac{F_1 - 1}{G_1} + \frac{F_2 - 1}{G_1 G_2}$ (d) none of the above

231. In a two stage cascade amplifier the equivalent noise temperatures are T_{e1} and T_{e2}. If the gains of the two stages are G_1 and G_2, the overall noise temperature T_e is
(a) $T_e = T_{e2} + \frac{T_{e1}}{G_1}$ (b) $T_e = \frac{T_{e1}}{G_1} + \frac{T_{e2}}{G_2}$
(c) $T_e = T_{e1} + \frac{T_{e2}}{G_1}$ (d) $T_e = T_{e1} + \frac{T_{e2}}{G_2}$

232. The overall noise temperature T_e, absolute temperature T and noise figure F are related as
(*a*) $T_e = T(F-1)$ (*b*) $T_e = F(T-1)$
(*c*) $T = T_e(F-1)$ (*d*) $T_e = (T)(F)$

233. Which of the following are useful in comparing the noise performance of receivers
1. input noise voltage
2. noise figure
3. noise resistance
4. noise temperature

Select answer as per the following codes :
(*a*) 1, 2 and 3 (*b*) 1, 3 and 4
(*c*) 2, 3, 4 (*d*) 1, 2, 4

234. Which of the following is correct (FDM means frequency division multiplexing and TDM means time division multiplexing)
(*a*) FDM requires lower bandwidth but TDM has more noise immunity
(*b*) FDM has noise immunity and requires lower bandwidth than TDM
(*c*) FDM requires channel synchronisation while TDM has more noise immunity
(*d*) FDM requires more multiplexing while TDM requires band pass filter

235. If transmission bandwidth is doubled in FM, SNR is
(*a*) doubled (*b*) raised four times
(*c*) decreased four times (*d*) halved

236. Which statement is incorrect for the signal
$x(t) = 5 \sin(2\pi \times 10^3 t) \sin(2\pi \times 10^6 t)$
(*a*) upper side band frequency is 1001000
(*b*) carrier amplitude is 5
(*c*) lower side band frequency is 990000
(*d*) $x(t)$ is DSB-SC

237. Match List I with List II and select the answer as per given codes

	List I		List II
A.	Detection of a periodic signal in noise	1.	Increase in bandwidth
B.	Recovery of a band limited signal from its uniformly sampled values	2.	Slope overload error
C.	Finer quantization of signal	3.	Nyquist rate
D.	Delta modulation	4.	Cross correlation

Codes:

	A	B	C	D
(*a*)	4	3	1	2
(*b*)	3	4	2	1
(*c*)	4	3	2	1
(*d*)	4	2	1	3

238. In PCM, the information lies
(*a*) in one or two properties of the pulse
(*b*) in absence of pulse
(*c*) in presence of pulse
(*d*) in presence or absence of pulse

239. To couple a coaxial line to a parallel line wire it is best to use
(*a*) slotted line (*b*) balun
(*c*) directional coupler (*d*) $\frac{\lambda}{4}$ transformer

240. SWR for a transmission line is infinity when the load is
1. short-circuit 2. complex impedance
3. open circuit 4. pure reactance

Of the above statements
(*a*) 1, 3, 4 are correct (*b*) 1 and 3 are correct
(*c*) 2, 3, 4 are correct (*d*) 1, 2, 4 are correct

241. Match List I with List II and select the answer using the codes

	List I (Antenna)		List II (Communication system)
A	Turnstile	1.	Radar
B	Yagi	2.	TV transmission
C	Dish	3.	MW broadcasting
D	Vertical mast	4.	TV reception

Codes:

	A	B	C	D
(*a*)	4	2	3	1
(*b*)	2	4	1	3
(*c*)	2	4	3	1
(*d*)	4	2	1	3

242. Microwave link repeaters are typically 50 km apart in TV transmission due to
(*a*) atmosphere attenuation
(*b*) output power tube limitations
(*c*) attenuation of surface wave
(*d*) earth's curvature

243. In TV signals colour burst is used
(*a*) to maintain colour sequence
(*b*) to ensure I and Q phase correctly
(*c*) to synchronise colours
(*d*) for interlacing of horizontal lines

244. The minimum viewing angle of resolution of human eye is (D is iris opening and λ is wavelength of light)
(*a*) $\frac{\lambda}{D}$ radians (*b*) $\frac{1.22\lambda}{D}$ radians
(*c*) $\frac{5\lambda}{D}$ radians (*d*) $\frac{\lambda^2}{D}$ radians

245. In a TV studio the illuminance is about
(*a*) 200 lux (*b*) 1000 lux
(*c*) 10000 lux (*d*) 10 lux

246. The unit for luminous energy is
(*a*) lumens/sec (*b*) lumens/m^2
(*c*) lumen-sec (*d*) lumen/m

247. Kell factor is about
(*a*) 5 (*b*) 3
(*c*) 0.7 (*d*) 0.2

248. CCIR system B employs
(*a*) 625 lines
(*b*) 625 lines with 2 : 1 interlace
(*c*) 625 lines with 4 : 1 interlace
(*d*) 500 lines with 2 : 1 interlace

249. The complete video signal consists of
(*a*) video signal and synchronising pulses
(*b*) video signal and synchronising pulses and blanking pulses
(*c*) video signal corresponding to picture information only
(*d*) video signal and blanking pulses

250. VSB modulation is preferred in TV because
(*a*) it reduces the bandwidth requirement to half
(*b*) it avoids phase distortion at low frequencies
(*c*) it results in better reception
(*d*) none of the above

251. In TV broadcast, UHF range is
(*a*) 606 – 790 MHz (*b*) 100– 150 MHz
(*c*) 41– 68 MHz (*d*) 174 – 230 MHz

252. In TV scanning sequence of lines is
(*a*) left to right (*b*) right to left
(*c*) left to right for odd fields and right to left for even field
(*d*) left to right for even fields and right to left for odd field

253. Which of the following statements is correct?
(*a*) Convex lens has positive focal power and concave lens have negative focal power
(*b*) Convex lens has negative focal power and concave lens have positive focal power
(*c*) All lens have positive focal power
(*d*) All lens have negative focal power

254. If D is diagonal and F is focal length, lens angle θ is
(*a*) $\theta = \tan^{-1}(D/F)$ (*b*) $\theta = 2\tan^{-1}(D/F)$
(*c*) $\theta = \tan^{-1}(D/2F)$ (*d*) $\theta = 2\tan^{-1}(D/2F)$

255. A lens has a Hyperfocal distance of 7.5 m. It is focussed at 7.5 m. Then
(*a*) all objects from 7.5 m to infinity will appear in focus
(*b*) all objects from 3.75 m to infinity will appear in focus
(*c*) all objects from 0 to 7.5 m will appear in focus
(*d*) all objects from 0 to 3.75 m will appear in focus

256. Automatic sensitivity control by target voltage control is easily possible in
(*a*) Vidicon (*b*) Plumbicon
(*c*) New vicon (*d*) both (*b*) and (*c*)

257. Which camera tube has highest sensitivity and very wide spectral response?
(*a*) Vidicon (*b*) Plumbicon
(*c*) New vicon (*d*) Saticon

258. Which camera tube has gamma of almost unity?
(*a*) Vidicon (*b*) Plumbicon
(*c*) New vicon (*d*) Saticon

259. Which camera tube has highest dark current?
(*a*) Plumbicon (*b*) Vidicon
(*c*) Saticon (*d*) CCD imager

260. The colour of an object is decided by
(*a*) the reflected colour
(*b*) the wavelength transmitted through it
(*c*) reflected colour for opaque object and wavelength transmitted through it for transparent objects
(*d*) none of the above

261. Hue of a colour means
(*a*) amount of light energy received by eye
(*b*) predominant spectral purity of colour light
(*c*) amount of other colours present
(*d*) none of the above

262. In colour television V and U signals are
(*a*) same as I and Q signal (*b*) same as R–Y and B–Y
(*c*) modified I and Q signal (*d*) modified R–Y and B–Y

263. Chrome signals in PAL colour television system consist of
(*a*) R–Y and B–Y signal (*b*) V and U signals
(*c*) I and Q signals (*d*) R–Y and G–Y signals

264. The phase of colour subscarrier burst in PAL is along the axis at
(*a*) + 180° (*b*) –180°
(*c*) ± 180° (*d*) ± 45°

265. The decoder for PAL system is
(*a*) similar to NTSC decoder
(*b*) similar to NTSC decoder supplemented by 64 μs delay line
(*c*) similar to NTSC decoder supplemented by 64 μs delay line and an electronic commutator
(*d*) either (*b*) or (*c*)

266. PAL system was proposed first in
(*a*) USA (*b*) Japan
(*c*) England (*d*) Germany

267. In PAL system
(*a*) Phase of U signal is reversed on alternating lines
(*b*) Phase of V signal is reversed on alternating lines
(*c*) Phase of both U and V signal is reversed on alternating lines
(*d*) none of the above

268. In PAL system weighting of R–Y and B–Y signals is done to
(*a*) prevent over modulation on saturated colours
(*b*) prevent under modulation of saturated colours
(*c*) allow over modulation of saturated colours
(*d*) allow under modulation on saturated colours

269. PAL stands for
(*a*) Phase attenuation by line
(*b*) Phase alternative by line
(*c*) Parabolic alternation by line
(*d*) Picture alternation by line

270. The colour sub carrier and side bands produced by its modulation with the chrominance signals are accommodated in the standard channel width by the process of
(*a*) frequency adjustment

(*b*) frequency interleaving
(*c*) frequency changing
(*d*) frequency amalgamation

271. Match List I with List II and select the answer using the given codes

	List I (Camera pickup tubes)		List II (Special feature)
A.	Vidicon	1.	Highest sensitivity
B.	Plumbicon	2.	Temperature sensitive
C.	Saticon	3.	Large lag
D.	New vicon	4.	Special target for red response

Codes:

	A	B	C	D
(*a*)	4	3	2	1
(*b*)	1	2	3	4
(*c*)	3	4	2	1
(*d*)	4	3	1	2

272. Match List I with List I and select the answer using the given codes

	List I (Camera pick up tube)		List II (Developed by)
A.	CCD charger	1.	Toshiba
B.	New vicon	2.	Phillips
C.	Plumbicon	3.	Matushita
D.	Chalnicon	4.	RCA

Codes:

	A	B	C	D
(*a*)	3	4	1	2
(*b*)	4	3	2	1
(*c*)	4	3	1	2
(*d*)	1	2	3	4

273. Consider the following statements as regards advantages of CCD sensors as camera pick up tube

1. High sensitivity
2. Good spectral response
3. Image lag absent

Out of the above which are correct?

(*a*) 1, 2, 3 (*b*) 1 and 2 only
(*c*) 1 and 3 only (*d*) 2 and 3 only

274. Consider the following features of monochrome TV

1. Number of lines per picture 625
2. Interlace ratio 2 : 1
3. Aspect ratio 4 : 3
4. Channel bandwidth 8 MHz

As regards CCIR system B, which of the above are correct

(*a*) 1, 2, 3, 4 (*b*) 1, 2, 3
(*c*) 1, 2, 4 (*d*) 2, 3, 4

275. Some sources of TV programs are

1. TV camera **2.** telecine
3. VTR **4.** external signal

Which of the above are actually present in a TV studio?

(*a*) 1, 2, 3 and 4 (*b*) 1, 2, 3
(*c*) 1 and 3 (*d*) 1, 2 and 4

276. Match the List I with List II and select the correct answer using the codes

	List I (VCR system)		List II (Azimuth angle)
A.	VHS	1.	7°
B.	Video 2000	2.	15°
C.	Betamax	3.	6°

Codes:

	A	B	C
(*a*)	1	2	3
(*b*)	1	3	2
(*c*)	3	1	2
(*d*)	3	2	1

277. Some possible functions of AGC in a TV receiver are

1. to increase gain for weak signals
2. to decrease gain for strong signals
3. to change the contrast when we switch from one channel to other
4. to reduce flutter

Which of the above are correct?

(*a*) 1, 2, 3, 4 (*b*) 1, 2, 4
(*c*) 1, 2, 3 (*d*) 2, 3, 4

278. Telecine equipment in a TV studio consists of

(*a*) Projector telecine (*b*) Flying spot telecine
(*c*) Digital CCD telecine (*d*) any of above

279. The generally used value of EHV supply in TV receivers is

(*a*) 1 kV per diagonal cm
(*b*) 1 kV per diagonal inch
(*c*) 0.5 kV per diagonal cm
(*d*) 0.5 kV per diagonal inch

280. A computer monitor uses 180 W in normal operation and 20 W in sleep mode. The computer is on all the 24 hours but is in sleep mode for 18 hours a days. The energy units saved in 30 days are

(*a*) 0.54 units (*b*) 86.4 units
(*c*) 2.88 units (*d*) 4.8 units

281. Some possible disadvantages of high level modulation in TV transmitters are

1. Video monitor has to supply high video voltage upto several hundred volts into highly capacitive loads drawing high current at video frequencies
2. The grid modulated stage presents a varying load to its driver
3. The linearity and high video driving power present problems

Out of above which are true?

(*a*) 1 only (*b*) 1 and 2 only
(*c*) 1, 2 and 3 (*d*) 2 and 3 only

282. In TV transmitter, some possible advantages of IF modulation are

1. The modulator section and visual exciter can be solid state devices
2. VSB filter can be introduced at low power level

3. The subsequent stages following VSB filter can be tuned permanently for wide band linear operation
4. It is economical

Which of the above are true?

(a) 1 and 2 (b) 1, 2 and 3
(c) 2, 3 and 4 (d) 1, 2, 3 and 4

283. In TV transmitter, visual exciter may consist of
1. Video processing unit
2. Visual modulator
3. VSB filter
4. Delay equaliser
5. Frequency converter

Which of the above are actually a part of visual exciter?

(a) 1, 2 and 3 (b) 2, 3 and 4
(c) 1, 2, 3 and 4 (d) 1, 2, 3, 4 and 5

284. In TV transmitter, the function of diplexer is
(a) to amplify video signal
(b) to remove distortions in audio signal
(c) to combine outputs of visual and aural transmitters
(d) to amplify audio signal

285. Typical value of power consumption of TV transmitter is
(a) 0.5 kVA (b) 2 kVA
(c) 50 kVA (d) 1000 kVA

286. To relay outdoor remotely located live programs, TV transmitter use
(a) microwave links (b) coaxial cable link
(c) open wire link (d) either (b) or (c)

287. On a smooth surface, the line of sight distance between transmitting antenna (height h_t metres) and receiving antenna (height h_r metres) is about
(a) $(\sqrt{h_t} + \sqrt{h_r})$ km (b) $(\sqrt{h_t + h_r})$ km
(c) $(3.7\sqrt{h_t + h_r})$ (d) $3.7(\sqrt{h_r} + \sqrt{h_r})$ km

288. To avoid objectionable noise in a microwave TV link, the SNR should be about
(a) 50 or more (b) 10 or more
(c) 25 or more (d) 200 or more

289. If f is frequency and d is distance, path attenuation α between isotropic antennas is given by
(a) $\alpha = 33 + 20 \log f + 20 \log d$
(b) $\alpha = 100 + 33 \log f + 33 \log d$
(c) $\alpha = 33 + \dfrac{20}{\log f} + \dfrac{20}{\log d}$
(d) $\alpha = 100 + \dfrac{33}{\log f} + \dfrac{33}{\log d}$

290. At f = 7 GHz., a 120 cm dish antenna gives a gain of about
(a) 50 (b) 500
(c) 5000 (d) 50000

291. If an isotropic source is radiating power P in all directions, the electric field at a distance d from the source is proportional to
(a) $(P/d)^{0.5}$ (b) $\dfrac{P}{d}$
(c) $\dfrac{(P)^{0.5}}{d}$ (d) $\dfrac{P}{\sqrt{d}}$

292. A source is radiating power equally in all directions. If source power is P, the power density at distance d is
(a) $\dfrac{P}{4\pi d^2}$ (b) $\dfrac{P}{2\pi d^2}$
(c) $\dfrac{P}{d^2}$ (d) $\dfrac{P}{4\pi d}$

293. Each kilometer of travel of electromagnetic radiation means a time delay of
(a) 330 µs (b) 33 µs
(c) 3.3 µs (d) 0.33 µs

294. Circular polarization
(a) is useful in reducing depolarization effect on received wave
(b) involves critical alignment of transmitting and receiving antenna
(c) is useful in discriminating between reception of adjacent beams
(d) none of the above

295. RF choke used for suppressing interference in TV is generally
(a) honey comb wound coil (b) iron core coil
(c) single layer coil (d) multi layer coil

296. In a 50 cm TV tube, a path difference of 2 km between direct wave and interferring wave would mean a displacement between main image and ghost image equal to
(a) 0.3 cm (b) 3 cm
(c) 6 cm (d) 9 cm

297. In band III of TV, distant reception (beyond line of sight range) is possible due to
(a) tropospheric bending by refraction
(b) reflection of waves
(c) ionospheric scattering
(d) none of the above

298. Consider the following:
1. A TV receiver antenna should pick up the TV signal and develop maximum voltage from the available strength of signal
2. It should reject unwanted signals
3. It should not suffer from co-channel interference
4. It should have a very narrow band

The correct statements out of above are

(a) 1 and 2 only (b) 1, 2 and 3 only
(c) 1, 2, 3, 4 (d) 2, 3 only

299. Which SNR denotes excellent picture quality in TV?
(a) 50 dB (b) 30 dB
(c) 15 dB (d) 5 dB

300. A balun provides an impedance transformation of
(a) 16 : 1 (b) 8 : 1
(c) 4 : 1 (d) 2 : 1

301. In a TV receiver the video IF amplifier provides a gain of about
(*a*) 5 dB (*b*) 20 dB
(*c*) 60 dB (*d*) 200 dB

302. In TV system in India, IF video and IF audio frequencies are
(*a*) 33.4 and 28.4 MHz respectively
(*b*) 38.9 and 33.4 MHz respectively
(*c*) 43.8 and 38.9 MHz respectively
(*d*) 33.4 and 38.9 MHz respectively

303. In a TV the output from RF tuner is about
(*a*) 1 μV (*b*) 1 mV
(*c*) 100 mV (*d*) 0.5 V

304. Consider the following functions
1. It detects the AM video carrier IF so as to produce the video signal
2. It produces and separates out the inter carrier sound IF of 5.5 MHz
3. It amplifies the video signal
4. It amplifies the audio signal

The video detector in a colour TV performs which of the above functions
(*a*) 1, 2 and 3 (*b*) 1, 2, 3 and 4
(*c*) 1 and 2 (*d*) 1, 2 and 4

305. In colour TV receiver, varactor diode is used for
(*a*) detection (*b*) rectification
(*c*) tuning (*d*) both (*a*) and (*b*)

306. As the magnitude of reverse bias is increased, the capacitance of varactor diode
(*a*) increases (*b*) decreases
(*c*) first increase, then decreases
(*d*) first decreases then increases

307. The characteristics of H and V sync. pulses in a colour TV receiver are
1. H sync. pulses are narrow and have frequency of 15625 Hz
2. V sync. pulses are much wider and have frequency of 50 Hz
3. H and V pulses have almost the same size but frequencies of 15625 and 50 Hz respectively
4. V sync. pulses are very narrow but H sync. pulses are wide and their frequencies are 15625 Hz and 50 Hz respectively

Which of the above are correct?
(*a*) 1 and 2 only (*b*) 1 only
(*c*) 3 only (*d*) 4 only

308. The contrast control in colour TV receiver is done by
(*a*) chroma amplifier (*b*) SAW filter
(*c*) video detector (*d*) Luminance amplifier

309. In a colour TV receiver, which of the following incorporates colour saturation control
(*a*) chroma detector (*b*) colour killer circuit
(*c*) SAW filter (*d*) chroma amplifier

310. In a colour TV receiver, PAL Dematrix consists of
(*a*) chroma detector to separate the three primary colours
(*b*) delay line and PAL matrix to give U and V signals
(*c*) delay line and PAL matrix to give RGB colours
(*d*) none of the above

311. In a colour TV receiver the function of comb filter is
(*a*) to pass only the clusters of monochrome signal energy while blocking chroma signals
(*b*) to pass only chroma signals and block monochrome signals
(*c*) to block both monochrome and chrome signals
(*d*) none of the above

312. In a colour TV receiver AFT stands for
(*a*) automatic fine tuning
(*b*) automatic frequency tuning
(*c*) automatic field tuning
(*d*) alignment frequency tuning

313. In a colour TV receiver, AGC adjusts the gain of
(*a*) RF stage only
(*b*) IF stage only
(*c*) RF stage and one or two IF stages
(*d*) none of the above

314. In a colour TV receiver the bandwidth of RF amplifier is
(*a*) exactly equal to channel width
(*b*) a little less than channel width
(*c*) a little more than channel width
(*d*) either (*a*) or (*b*)

315. In a colour TV receiver SAW filter consists of
(*a*) piezo electric subtrate on which are deposited two inter digital transducers
(*b*) photo conductive subtrate on which are deposited two inter digital transducers
(*c*) photo voltaic subtrate on which are deposited two inter digital transducers
(*d*) photo resistive subtrate on which are deposited two inter digital transducers

316. A typical value of insertion loss of SAW filter is
(*a*) 15 dB (*b*) 2 dB
(*c*) 40 dB (*d*) 60 dB

317. For low impedance SAW filter the most suitable material for substrate of SAW filter is
(*a*) PZT ceramics (*b*) ZnO
(*c*) Lithium titanate (*d*) Lithium neobate

318. Consider the following statements as regards SAW filter :
1. The centre frequency is determined by the pitch of IDT comb shaped electrodes
2. The response can be modified shaping the overlap of comb electrodes
3. The response is maximum when acoustic wave has its wavelength equal to pitch
4. When a number of identical electrodes are arranged at the same pitch, the frequency characteristics of the filter is expressed by (sin x)/x

Which of the above statements are correct?
(*a*) 1, 2, 3, 4 (*b*) 1, 2, 3 only
(*c*) 1, 2, 4 only (*d*) 2, 3, 4 only

319. Trap circuits used in video IF amplifier of colour TV are
(a) High Q, LC resonant circuit
(b) Low Q, LC resonant circuit
(c) High Q, RLC resonant circuit
(d) Low Q, RLC resonant circuit

320. Some types of traps used in video IF amplifier of colour TV are
(a) shunt traps (b) series traps
(c) absorption traps (d) all the above

321. In a colour TV receiver, the sound intercarrier IF is generated in
(a) video detector (b) video IF amplifier
(c) tuner mixer stage (d) limiter stage

322. In TV studios, flying spot scanner is a device for
(a) mixing video and audio signals
(b) telecasting feature films
(c) live telecast of outdoor programs
(d) telecast of indoor music programs

323. Consider the following test signal patterns
1. Ramp 2. Grill
3. Grey shades 4. $\sin^2$ pulse
5. Picture line up generating euipment
Which of the above are used in TV studios?
(a) all (b) 1, 2 and 3 only
(c) 1, 2 and 5 only (d) 2, 3, 4 and 5 only

324. In a TV studio the function of vision switcher is to
(a) edit television programs
(b) relay television programs
(c) control zoom lens
(d) mix audio and video signals

325. If aspect ratio is $\frac{4}{3}$ and number of lines in each frame is 625, the number of picture elements in one frame is
(a) $\frac{4}{3} \times 625$ (b) $\frac{4}{3}(625)^2$
(c) 625^2 (d) $\frac{3}{4}(625)^2$

326. A line has standing waves with a voltage maximum of 35 mV and minimum of 17.5 mV. VSWR is
(a) 2 (b) 4
(c) 0.5 (d) 0.25

327. A line has L = 0.8 μH/m and C = 32 pF/m. Z_0 =
(a) 6000 (b) 600
(c) 158 (d) 15.8

328. The time required for horizontal blanking is 16% of that for each horizontal line. If horizontal time is 63.5 μs, the horizontal blanking time for each line is about
(a) 1 μs (b) 10 μs
(c) 100 μs (d) none of the above

329. If frames are repeated at 60 Hz and each frame has 525 lines, the time for scannig one horizontal line (for progressive scanning and no interlacing) is about
(a) 320 μs (b) 32 μs
(c) 3.2 μs (d) 0.32 μs

330. A + 1.9 lens has a focal length of 25 mm. The diameter of lens is about
(a) 13 mm (b) 1.3 mm
(c) 130 mm (d) none of the above

331. In CCIR-B system the time between start of one H syn. pulse and next is
(a) 64 μs (b) 6.4 μs
(c) 640 μs (d) 0.64 μs

332. A fascimile reproduction has a specification of 200 lines per frame, progressive scanning and 5 framer per second. The time to scan one line is
(a) 5 ms (b) 2.5 ms
(c) 1 ms (d) 0.1 ms

333. In a fascimile reproduction time to scan one line is 2 ms. If it has 4% blanking, the visible trace time is
(a) 1.92 ms (b) 2.08 ms
(c) 50 ms (d) 0.08 ms

334. The number of possible colours with a 4-line digital input monitor are
(a) 4 (b) 16
(c) 64 (d) 2

335. A 5000 W signal is attenuated by – 196 dB. The resultant signal strength is
(a) 1.26×10^{6} W (b) 1.26×10^{-2} W
(c) 1.26×10^{-8} W (d) 1.26×10^{-16} W

336. A transmitter power of 10 W is increased by 30 dB. The effective radiated power is
(a) 100 W (b) 1000 W
(c) 10000 W (d) 100000 W

337. The HV supply to a TV receiver is 25 kV and the total beam current for the three electron guns is 2000 μA. The power handled by 1HVT is
(a) 5 W (b) 0.5 W
(c) 50 W (d) 500 W

338. A TV receiver has a width of 60 cm. The aspected ratio is 4/3. The required EHT supply is about
(a) 29 kV (b) 2.9 kV
(c) 45 kV (d) 4.5 kV

339. A horizontal output stage is cut off for retrace and 40% of trace. If time for each horizontal line is 64 μs and retrace time is 12 μs, the transistor is conducting for about
(a) 64 μs (b) 6.4 μs
(c) 31μs (d) 3.1 μs

340. A transmitter power of 5 W is increased by 20 dB. The effective radiated power is
(a) 5 W (b) 100 W
(c) 25 W (d) 500 W

341. A TV receiver is 40 cm wide and 30 cm high. The EHV supply required is about
(a) 2 kV (b) 20 kV
(c) 53 kV (d) 10 kV

342. A TV receiver has 20 kV EHT supply and total beam current (for 3 electron guns) of 1500 μA. The power handled by IHVT is

(*a*) 3 W (*b*) 30 W
(*c*) 300 W (*d*) 0.3 W

343. The HV and other voltage supplies for a TV receiver are obtained from IHVT. The full name for this device is
(*a*) Internal high voltage transformer
(*b*) Integrated high voltage transformer
(*c*) Instant high voltage transformer
(*d*) Inter leaved high voltage transformer

344. If luminance signal Y = 0.3 R + 0.59 G + 0.11 B, the hue of R–Y colour is
(*a*) 0.7 R – 0.59 G – 0.11B (*b*) –0.7 R + 0.59 G + 0.11 B
(*c*) 1.3 R – 0.59 G – 0.11 B (*d*) none of the above

345. In digital TV digital imaging standard comprises
(*a*) six profiles and four levels
(*b*) four levels and six profiles
(*c*) six profiles and six levels
(*d*) four profiles and four levels

346. Digital television is also called
(*a*) High digital television
(*b*) High definition television
(*c*) High deflection television
(*d*) High display television

347. The power consumption of a present day colour TV receiver is about
(*a*) 8 W (*b*) 80 W
(*c*) 200 W (*d*) 400 W

348. Colour picture tube needs higher anode voltage and needs larger beam and cathode currents than monochrome tube because
(*a*) it is bigger in size
(*b*) it has higher level of illumination
(*c*) the tube has to illuminate three phosphors simultaneously
(*d*) the tube has lower efficiency because of aperture mask

349. A trinitron colour picture tube uses
(*a*) three electron guns
(*b*) one electron gun
(*c*) one electron gun with three separate in line cathodes
(*d*) none of the above

350. PIL tube means
(*a*) picture in line tube (*b*) precision in line tube
(*c*) pairing in line tube (*d*) pattern in line tube

351. Which colour picture tube uses slotted aperture mask?
(*a*) Trinitron (*b*) Shadow mask tube
(*c*) PIL (*d*) none of the above

352. PIL colour picture tube has the main advantage that
(*a*) it is cheap
(*b*) it ensures purity in colour
(*c*) it is self-convergent
(*d*) both (*a*) and (*b*)

353. Degaussing in a picture tube means
(*a*) Adding magnetism in mask and mounting frame
(*b*) Removing residual magnetism in mask, shield etc.
(*c*) Increasing resistance of coils to decrease time constant
(*d*) Decreasing resistance of coils to increase time constant

354. LCD uses
(*a*) nematic crystals (*b*) sematic crystals
(*c*) cholesteric crystals (*d*) twisted nematic crystal

355. TV receivers use flyback switched mode power supply because
(*a*) it can supply many voltage outputs
(*b*) it does not require a filter choke and is cheap
(*c*) it has very good regulation at all loads
(*d*) both (*a*) and (*c*)

356. In a TV receiver there is no raster, no picture and no sound. The most likely defective stage is
(*a*) power supply unit (*b*) video output circuit
(*c*) horizontal oscillator (*d*) balun

357. In a TV receiver the raster is normal but there is no sound and quality of picture is poor. The defective part is
(*a*) power supply (*b*) loudspeaker
(*c*) AGC, video IF amplifier, RF amplifier
(*d*) horizontal oscillator

358. CATV stands for
(*a*) Common antenna television
(*b*) Community antenna television
(*c*) Cable antenna television
(*d*) Channel antenna television

359. When satellite radio beam is aimed towards earth, it illuminates on the earth an oval shaped area called
(*a*) foot print (*b*) beam area
(*c*) satellite service area (*d*) radiated area

360. In a satellite transponder the power amplifier feeds power P to the transmitting antenna of gain G. If distance of earth is *d*, the power flux density is
(*a*) PG/*d* (*b*) $PG/4\pi d^2$
(*c*) $P/G\,(4\pi d^2)$ (*d*) $\frac{P(4\pi d^2)}{G}$

361. The distance of a direct broadcasting satellite from earth's surface is about
(*a*) 1000 km (*b*) 10000 km
(*c*) 36000 km (*d*) 55000 km

362. The orbital period of a geostationary communication satellite is
(*a*) 12 hours (*b*) 24 hours
(*c*) 36 hours (*d*) 48 hours

363. Fields are said to be circularly polarised if their magnitudes are
(*a*) equal and they are in phase
(*b*) equal but they differ in phase by ± 90°
(*c*) unequal and differ in phase by ± 90°
(*d*) unequal but are in phase

364. Skin depth at 1000 MHz in comparison with that at 500 MHz is
(*a*) 2 (*b*) $\sqrt{2}$
(*c*) $\frac{1}{\sqrt{2}}$ (*d*) $\frac{1}{2}$

365. Match List I (Unit of information measure) with List II (Base of logarithm) and select the correct answer using the codes

List I	List II
A. bit	1. 10
B. nat	2. e
C. decit	3. 2

Codes:

	A	B	C
(*a*)	1	2	3
(*b*)	1	3	2
(*c*)	3	2	1
(*d*)	2	3	1

366. The rate at which information can be carried through a communication channel depends on
(*a*) carrier frequency (*b*) bandwidth
(*c*) transmission loss (*d*) transmitted power

367. A source has two symbols x_1 and x_2 with probabilities p_1 and p_2. If a new source with four symbols is generated from the above source, the new symbol probabilities will be ($\in$ is a number such that $0 < \in < 1$)
(*a*) p_1, p_2, p_1, p_2 (*b*) p_1, p_1, p_2, p_2
(*c*) $(1 - \in)p_1, (1 - \in)p_2, p_1, p_2$
(*d*) $\frac{p_1}{4}, \frac{p_2}{4}, \frac{p_1}{4}, \frac{p_2}{4}$

368. The purpose of source coding is to
(*a*) increase the information transmission rate
(*b*) decrease $\frac{S}{N}$ ratio
(*c*) decrease the information transmission rate
(*d*) decrease probability of error

369. Match List I with List II and select the correct answer using the codes

List I	List II
A. Resistance	1. Current noise
B. Diode	2. Partition noise
C. Triode	3. Shot noise
D. *p-n* junction	4. Atmospheric noise
	5. Johnson noise

Codes:

	A	B	C	D
(*a*)	5	3	2	1
(*b*)	1	2	3	5
(*c*)	3	5	1	2
(*d*)	3	5	2	1

370. An amplifier operating over a frequency range of 18 to 20 MHz has 10 kΩ input resistance. At a temperature of 27°C, the noise voltage at input of amplifier is (Boltzman constant k = 1.38×10^{-23} J/K)
(*a*) 18.2 μV (*b*) 182 μV
(*c*) 16.2 μV (*d*) 162 μV

371. 24 telephone channels are frequency division multiflexed using an SSB modulation. Assuming 3 kHz per channel, the required bandwidth is
(*a*) 8 kHz (*b*) 27 kHz
(*c*) 21 kHz (*d*) 72 kHz

372. 24 telephone channels, each band limited to 3.4 kHz are to be time division multiplexed using PCM. If sampling frequency is 10 kHz and number of quantization levels is 128, the required bandwidth of PCM is
(*a*) 1.68 MHz (*b*) 240 kHz
(*c*) 81.6 kHz (*d*) 3.072 MHz

373. If carrier is modulated by a digital bit stream having one of the possible phases of 0°, 90°, 180°, and 270°, then modulation is called
(*a*) BPSK (*b*) QPSK
(*c*) QAM (*d*) MSK

374. BPSK stands for
(*a*) Binary phase shifting key
(*b*) Broad phase shifting key
(*c*) Bit phase shifting key
(*d*) Binary pulse shifting key

375. The relation between bandwidth B, modulating frequency f_m and frequency deviation Δf for a sinusoidally modulated FM signal is
(*a*) $B = \Delta f + f_m$ (*b*) $B = 2(\Delta f + f_m)$
(*c*) $B = f_m - \Delta f$ (*d*) $B = 2(f_m - \Delta f)$

376. PAM stands for
(*a*) Pulse analogue modulation
(*b*) Phase analogue modulation
(*c*) Pulse amplitude modulation
(*d*) Phase amplitude modulation

377. UHF range is
(*a*) 3 to 30 MHz (*b*) 30 to 300 MHz
(*c*) 300 to 3000 MHz (*d*) 3000 to 30000 MHz

378. The highest modulating frequency typically used in AM systems is
(*a*) 1 kHz (*b*) 5 kHz
(*c*) 20 kHz (*d*) 50 kHz

379. A telephone channel requires a bandwidth of about
(*a*) 1 kHz (*b*) 3 kHz
(*c*) 10 kHz (*d*) 50 kHz

380. In a system an input resistance of 20 kΩ causes a noise voltage of 15 μV. If two such input resistances each of 20 kΩ are used in parallel, the total noise voltage will be about
(*a*) 15 μV (*b*) 30 μV
(*c*) 7.5 μV (*d*) 10.6 μV

381. An amplifier has an input resistance of 10 kΩ and the noise voltage is 10 μV. If two such resistances of 10 kΩ each are connected in series at the input, the total noise voltage is about
(*a*) 20 μV (*b*) 5 μV
(*c*) 14.14 μV (*d*) 10 μV

382. Noise Figure is defined as
(*a*) $\frac{\text{S/N ratio at output}}{\text{S/N ratio at input}}$

(*b*) $\dfrac{\text{Noise voltage at output}}{\text{Noise voltage at input}}$

(*c*) $\dfrac{\text{S/N ratio at input}}{\text{S/N ratio at output}}$

(*d*) $\dfrac{\text{Noise voltage at input}}{\text{Noise voltage at output}}$

383. If V_s is signal voltage, V_n is noise voltage, signal to noise ratio S/N is defined as

(*a*) $\dfrac{S}{N} = \dfrac{V_s}{V_n}$ (*b*) $\dfrac{S}{N} = \left(\dfrac{V_s}{V_n}\right)^2$

(*c*) $\dfrac{S}{N} = \sqrt{\dfrac{V_s}{V_n}}$ (*d*) none of the above

384. Energy content of atmospheric noise

(*a*) does not depend on frequency
(*b*) decreases as frequency is increased
(*c*) increases as frequency is increased
(*c*) either (*a*) or (*c*) depending on the temperature

385. If B is the bandwidth, the thermal noise voltage is proportional to

(*a*) B (*b*) $B^{0.5}$
(*c*) $B^{1.5}$ (*d*) B^2

386. A sinusoidal voltage of 2 kV peak value is amplitude modulated to give 20% modulation. The peak value of each side band term is

(*a*) 400 V (*b*) 200 V
(*c*) 100 V (*d*) 800 V

387. A sinusoidal voltage amplitude modulates a carrier of peak value 5 kV. Each side band has an amplitude of 500 V. The modulation index is

(*a*) 0.1 (*b*) 0.2
(*c*) 0.4 (*d*) 0.8

388. In amplitude modulation the rms value of radio frequency voltage after 40% modulation is 60 mV. If modulation is 80% the rms value after modulation will be about

(*a*) 50 mV (*b*) 60 mV
(*c*) 66 mV (*d*) 80 mV

389. If ω is the frequency of modulating voltage in FM, the modulation index is proportional to

(*a*) ω (*b*) 1/ω
(*c*) $\sqrt{\omega}$ (*d*) $1/\sqrt{\omega}$

390. If carrier power is P and amplitude modulation index is *m*, the total power after modulation is

(*a*) P (*b*) *m*P
(*c*) P (1 + *m*) (*d*) $P\left(1 + \dfrac{m^2}{2}\right)$

391. Linear diode detector utilizes

(*a*) linear portion of static characteristics of diode
(*b*) linear portion of dynamic characteristic of diode
(*c*) square law portion of dynamic characteristics of diode
(*d*) rectification property of diode

392. A linear diode detector uses RC filter. As modulation index increases the maximum permissible value of time constant RC

(*a*) increases (*b*) decreases
(*c*) remains constant (*d*) either (*a*) or (*c*)

393. FM broadcast is used in

(*a*) VHF range (*b*) UHF range
(*c*) MF
(*d*) both VHF and UHF range

394. The requirements of carrier frequency are

(*a*) precise frequency value (*b*) low frequency drift
(*c*) both (*a*) and (*b*) (*d*) neither (*a*) nor (*b*)

395. SSB stands for

(*a*) single side band (*b*) suppressed side band
(*c*) single sound band (*d*) single sky band

396. FSK is a system involving

(*a*) pulse modulation (*b*) frequency modulation
(*c*) amplitude modulation (*d*) phase modulation

397. Armstrong FM transmitter uses

(*a*) LC oscillator (*b*) crystal oscillator
(*c*) RC oscillator (*d*) either (*a*) or (*c*)

398. In FM modulation pre emphasis is done for

(*a*) high frequency components
(*b*) low frequency components
(*c*) middle frequency components
(*d*) both (*a*) and (*b*)

399. If transmitted power is 75 kW, the field strength at 30 km distance will be about

(*a*) 10 V/m (*b*) 1 V/m
(*c*) 50 V/m (*d*) 2 V/m

400. If transmitted power is 10 kW, the field strength at a distance *d* is 50 mV/m. If transmitted power is reduced to 2.5 kW, the field strength at the same point will be

(*a*) 25 mV/m (*b*) 12.5 mV/m
(*c*) 6.25 mV/m (*d*) 3.125 mV/m

401. An antenna is 30 metre high. The frequency is 1 MHz. The radiation resistance is about

(*a*) 1500 Ω (*b*) 150 Ω
(*c*) 15 Ω (*d*) 1.5 Ω

402. A RADAR can be used to

(*a*) aim a gun at an aircraft
(*b*) to direct guided missiles
(*c*) to search submarines
(*d*) all the above

403. The power requirement of a RADAR used in military operations is

(*a*) a few watts (*b*) about 500 W
(*c*) about 2000 W (*d*) a few megawatts

404. In a pulsed RADAR, the transmitter is connected to the antenna

(*a*) during pulse interval
(*b*) during interval between pulses
(*c*) all the time
(*d*) only during a fraction of pulse interval

405. Consider the following statements :
1. In a pulsed radar, the transmitted pulses are of high frequency
2. In a pulsed radar, the transmitted pulses are of high power
3. In a pulsed radar, the transmitted pulses are of brief duration
4. In a pulsed radar, the transmitted pulses are of square wave shape

Of the above statements the correct statements are
(*a*) 1 only (*b*) 1,2 and 3 only
(*c*) 1, 2, 3, 4 (*d*) 1 and 3 only

406. The receiver of a pulsed RADAR can detect signals of power of the order of
(*a*) 1 W (*b*) 0.1 W
(*c*) 10^{-5} W (*d*) 10^{-12} W

407. In a pulsed RADAR typical value of pulse duration and pulse frequency respectively are
(*a*) 1 μs and 50 Hz (*b*) 1 μs and 500 Hz
(*c*) 4 μs and 50 Hz (*d*) 4 μs and 500 Hz

408. The function of duplexer in a RADAR is
(*a*) to permit the use of same antenna for transmission and reception
(*b*) to provide gain to the transmitted power
(*c*) to amplify the received signal
(*d*) both (*b*) and (*c*)

409. The currents fed to the two dipoles of turnstile antenna are
(*a*) in phase (*b*) in quadrature
(*c*) in phase opposition (*d*) either (*a*) or (*c*)

410. Which wave is also called surface wave?
(*a*) ground wave (*b*) sky wave
(*c*) space wave (*d*) both (*b*) and (*c*)

411. As the frequency increases, the absorption of ground wave by earth's surface
(*a*) decreases (*b*) increases
(*c*) remains the same (*d*) either (*a*) or (*c*)

412. Which of the following radio waves are not reflected by ionosphere
(*a*) waves above 1 kHz (*b*) waves above 1 MHz
(*c*) waves above 10 MHz (*d*) waves above 30 MHz

413. If a dipole is meant for a frequency of 1 MHz, the length of each rod is about
(*a*) 5 m (*b*) 20 m
(*c*) 100 m (*d*) 150 m

414. An antenna having a length of $\frac{\lambda}{4}$ and grounded at one end is called
(*a*) Marconi antenna (*b*) Yagi antenna
(*c*) turnstile antenna (*d*) none of the above

415. The radio receivers mostly used now a days are
(*a*) TRF receivers
(*b*) super hetrodyne receivers
(*c*) CW receivers
(*d*) pulsed receivers

416. In radio receivers the value of IF is
(*a*) 100 kHz (*b*) 225 kHz
(*c*) 455 kHz (*d*) 850 kHz

417. Some features of RF stage in good radio receivers is
1. It increases sensitivity by providing amplification to RF signal
2. it eliminates image frequency interference
3. it increases selectivity
4. it reduces noise level

Which of the above are true?
(*a*) 1, 2 and 3 only (*b*) 1, 2 and 4 only
(*c*) 2, 3, 4 only (*d*) 1, 2, 3, 4

418. In a radio receiver the input from local oscillator is fed to
(*a*) RF amplifier (*b*) mixer
(*c*) IF amplifier (*d*) detector

419. In transistor radio receivers the number of IF amplifier stages are
(*a*) 1 (*b*) 2
(*c*) 4 (*d*) 6

420. In push button type telephone set, pressing a button generates
(*a*) one frequency
(*b*) two high freqencies
(*c*) two low frequencies
(*d*) one low frequency and one high frequency

421. In a telephone system, the grade of service means
(*a*) the quality of sound reception at the receiver telephone
(*b*) the probability that a called for connection is established
(*c*) the type of wires cables used for the system
(*d*) none of the above

422. In telephone systems the relation between load ρ, number of calls initiated during a specified interval λ and average duration of a call (*h*) is
(*a*) $\rho = \lambda h$ (*b*) $\rho = \lambda h^2$
(*c*) $\lambda = \rho h$ (*d*) $h = \lambda \rho$

423. In telephone systems, load is expressed in the unit of
(*a*) erlangs (*b*) decibels
(*c*) bells (*d*) watts

424. Which of the following is true as regards telephone systems?
(*a*) a larger facility carrying a larger load operates less efficiently than a smaller capacity
(*b*) a larger facility carrying a lesser load operates more efficiently than a smaller capacity
(*c*) a larger facility carrying a larger load operates more efficiently than a smaller capacity
(*d*) a smaller capacity carrying a larger load operates more efficiently than a larger capacity

425. In a telephone system number of services is *s*. The calls are initiated at an average rate of λ calls/*s* and average time for which each call lasts is *h*. If $\rho = \lambda h$, the probability that call is blocked is B (*s*, ρ), then B (*s*, ρ) =

(a) $\dfrac{(\rho^s)/s!}{\sum_{j=0}^{s}\rho^j/j!}$ (b) $\dfrac{\sum_{j=0}^{s}\rho^j/j!}{(\rho^s)/s!}$

(c) $\sum_{j=0}^{s}\dfrac{\rho^s/j!}{\rho^j/s!}$ (d) none of the above

426. An antenna operates effectively when its dimensions are
(a) of the order of magnitude of wavelength to be transmitted
(b) very small than the magnitude of wavelength to be transmitted
(c) very large than the magnitude of wavelength to be transmitted
(d) either (a) or (b)

427. An audio signal (say from 50 Hz to 10000 Hz) is frequency translated by a carrier having a frequency of 10^6 Hz. The values of initial (without frequency translation) and final (after frequency translation) fractional change in frequency from one band edge to the other are
(a) 200 and 1.01 (b) 200 and 10.01
(c) 200 and 100.1 (d) 200 and 200

428. If $M(j\omega) = \mathfrak{F}[m(t)$ where $\mathfrak{F}$ denotes Fourier transform, then $\mathfrak{F}[m(t)\cos\omega_c t] =$
(a) $M(j\omega + j\omega_c)$
(b) $M(j\omega - j\omega_c)$
(c) $M(j\omega + j\omega_c) + M(j\omega - j\omega_c)$
(d) $\dfrac{1}{2}[M(j\omega + j\omega_c) + M(j\omega - j\omega_c)]$

429. In communication systems, the modulating signal is also called
(a) base band signal (b) side bands
(c) auxiliary signal (d) none of the above

430. A signal $m(t)$ amplitude modulates a carrier $A_c \cos\omega_c t$. The final signal will be
(a) $[A_c\, m(t)]\cos\omega_c t]$ (b) $A_c[1 + m(t)]\cos\omega_c t]$
(c) $A_c + [m(t)]\cos\omega_c t$ (d) $A_c \cos t\,\omega_c t + m(t)$

431. Figure 12.9 shows a simple detector circuit for amplitude modulated wave. $v(t) = (1 + m\cos\omega_m t)$ cost $\omega_c t$. If the detector is to follow the envelope of modulated wave at all times the extent of modulation m must be less than or equal to m_0 where m_0 is given by equation

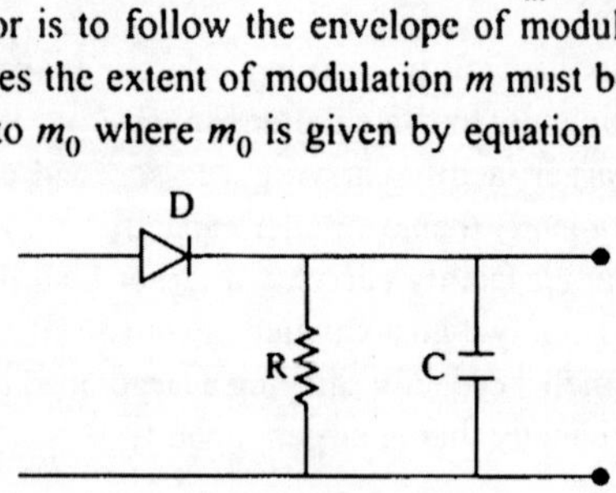

Fig. 12.9.

(a) $RC = \dfrac{1}{\omega_m}\dfrac{\sqrt{1-m_0^2}}{m_0}$ (b) $RC = \dfrac{\omega_m\sqrt{1-m_0^2}}{m_0}$

(c) $RC = \dfrac{m_0}{\omega_m}\sqrt{1-m_0^2}$ (d) $RC >> \dfrac{1}{\omega_m}\dfrac{\sqrt{1-m_0^2}}{m_0}$

432. The equation $v(t) = A\cos(\omega_c t + \phi(t))$ where A and ω_c are constant and $\phi(t)$ is a function of basis signal represents
1. Phase modulation 2. Amplitude modulation
3. Angle modulation 4. Frequency modulation
Which of the above are true?
(a) 3 only (b) 1, 3 and 4
(c) 1, 2, 3, 4 (d) 3 and 4 only

433. In Fig. 12.10, the output $v(t)$ is

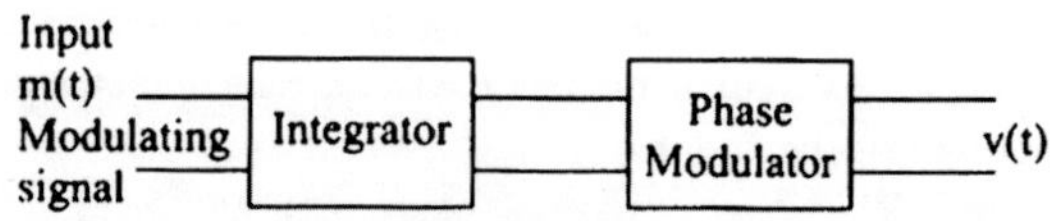

Fig. 12.10.

(a) phase modulated signal
(b) amplitude modulated signal
(c) frequency modulated signal
(d) none of the above

434. In Fig. 12.11, the output $v(t)$ is

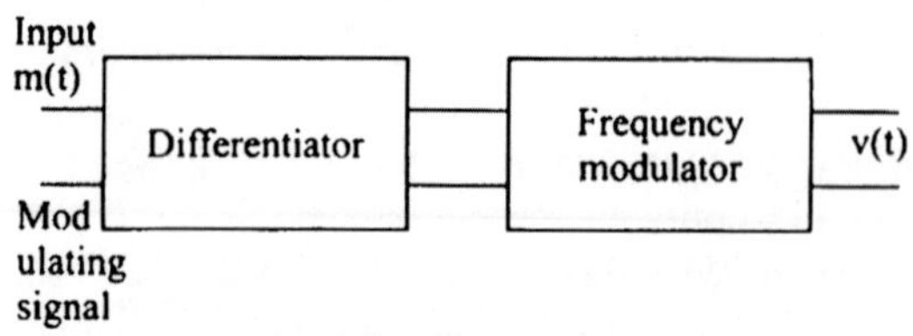

Fig. 12.11.

(a) an amplitude modulated signal
(b) a frequency modulated signal
(c) a phase modulated signal
(d) none of the above

435. The circuit of Fig. 12.12 can given an output which is

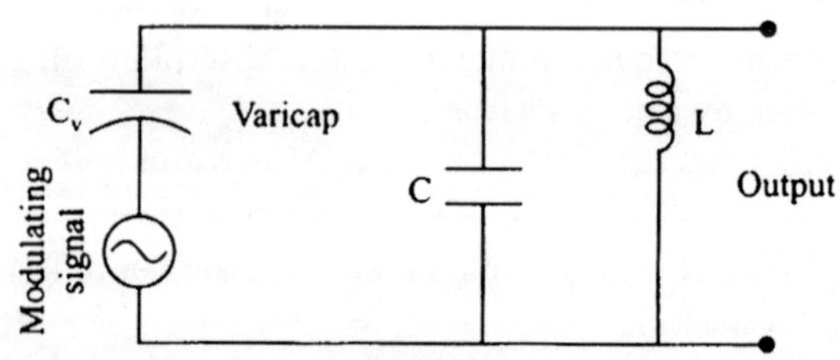

Fig. 12.12.

(a) a frequency modulated signal
(b) an amplitude modulated signal
(c) a phase modulated signal
(d) none of the above

436. PAM stands for
(a) phase angle modulation
(b) phase amplitude modulation
(c) pulse amplitude modulation
(d) pulse angle modulation

437. When a signal is quantized such that the step size S is small in comparison with the peak-to-peak range of the signal, the mean square quantization error is equal to
(a) S^2 (b) $\dfrac{S^2}{2}$

(c) $\frac{S^2}{6}$ (d) $\frac{S^2}{12}$

438. ADM stands for
(a) angle delta modulation
(b) angle digital modulation
(c) adaptive delta modulation
(d) adaptive digital modulation

439. In a channel voice coder, the outputs of sixteen 20 Hz low pass filters are sampled, multiplexed and A/D converted. If sampling is at 40 samples/second and 3 bits/sample represents each voltage sample, the bit rate is about
(a) 1.9 k bits/sec (b) 0.95 k bits/sec
(c) 3.8 k bits/sec (d) 7.6 k bits/sec

440. QAM stands for
(a) quasi amplitude modulation
(b) quadrature angle modulation
(c) quadrature amplitude modulation
(d) none of the above

441. Consider the following statements as regards DPSK and DEPSK
1. DPSK demodulator requires a device which operates at carrier frequency and provides a delay of bit time T_b.
2. In DEPSK, synchronous demodulation recovers the signal $b(t)$ and the decoding of $b(t)$ to generate $d(t)$ is done at base band
3. In both DPSK and DEPSK errors always occur in pairs

Which of the above statements are correct?
(a) 1 and 3 (b) 1and 2
(c) 1 and 3 (d) 1, 2 and 3

442. Consider the following statements as regards BPSK and QPSK
1. In BPSK we deal individually with each bit of durationT_b
2. In QPSK we lump two bits to form a symbol
3. In both BPSK and QPSK signal changes occur at bit rate
4. In QPSK the symbol can have any one of four possible values

Which of the above statements are correct?
(a) 1, 2, 3, 4 (b) 1, 2, 3
(c) 1, 2, 4 (d) 2, 3, 4

443. If N bits are lumped together to get an N-bit symbol, the possible number of symbols is
(a) 2^N (b) 2^{N-1}
(c) 2^{N-2} (d) 2^{N-4}

444. In a system using QPSK and 4 bits to form a symbol, the required bandwidth is
(a) Same as BPSK (b) half of that in BPSK
(c) One-fourth of that in BPSK
(d) none of the above

445. In orthogonal BFSK, the energy of information bearing term is
(a) the same as total transmitted energy
(b) one-half of total transmitted energy
(c) one-fourth of total transmitted energy
(d) none of the above

446. Consider the following differences between QPSK and MSK
1. The side in MSK are relatively much smaller in comparison to the main lobe
2. The spectrum of MSK has a main centre lobe which is 1.5 times as wide as the main lobe of QPSK
3. The waveforms of both MSK and QPSK exhibit phase continuity

Which of the above statements are correct?
(a) 1 only (b) 1 and 2 only
(c) 1, 2 and 3 (d) 2 and 3 only

447. Tamed FM is named so because
(a) it restrains the frequency excursions of FM
(b) it increases the frequency excursions of FM
(c) it decreases the amplitude of FM signal
(d) it increases the amplitude of FM signal

448. In the equation $d = \sqrt{4E_y \sin^2(\pi/M)}$ of M-ary PSK signal (where E_y is symbol energy), M becomes large. Then d will approach
(a) 2 (b) infinity
(c) zero (d) 1

449. In a 4-ary FSK, the frequencies are separated by f_s. Then they are
(a) in phase (b) in phase opposition
(c) orthogonal (d) none of the above

450. In a perfect conductor the resistor noise is
(a) maximum (b) zero
(c) very high (d) almost infinite

451. Consider the following statements as regards noise
1. If gaussian noise is applied at input a rectifier, the output is not gaussian
2. The noise encountered on a telephone line consists of short, pulse type disturbances whose amplitude distribution is not gaussian
3. Short noise results from random emission of electrons from a cathode
4. Short noise results from phenomenon associated with the flow of current across semiconductor junctions

Which of the above are correct?
(a) 1, 2, 3 only (b) 1, 2, 4 only
(c) 1, 2, 3, 4 (d) 2, 3 only

452. The spectral component of noise associated with kth frequency is given, in the limit as $\Delta f \to 0$, by $n_k(t)$ written as

$$n_k(t) = C_k \cos(2\pi k \Delta f t + \theta_k)$$

Consider the following statements as regards C_k and θ_k :
1. C_k and θ_k are independent of one another
2. C_k and θ_k are independent of the amplitude of a spectral component at a different frequency
3. C_k and θ_k are independent of the phase of a spectral component at a different frequency

Which of the above statements are correct?
(*a*) 1, 2 only (*b*) 1, 3 only
(*c*) 1, 2, 3 (*d*) 2, 3 only

453. A noise can be represented as superposition of spectral components. When this noise is passed through a narrow band filter, the output of filter should look like
(*a*) sinusoidal (*b*) rectangular
(*c*) triangular (*d*) none of the above

454. Thermal noise has a power spectral density which is quite uniform up to frequencies of the order of
(*a*) 100 Hz (*b*) 10^6 Hz
(*c*) 10^9 Hz (*d*) 10^{13} Hz

455. A maser is a
(*a*) low noise device
(*b*) high noise device
(*c*) medium noise device
(*d*) may be either high or medium noise device

456. If $\gamma = (S_0/N_0) / (S_i/N_m)$ where S_0 and S_i are signal power output and signal power input and N_0 and N_m are noise power output and noise power input, the value of γ for SSB-SC and DSB-SC amplitude modulated system respectively are
(*a*) 1 and 1 (*b*) 1 and 2
(*c*) 2 and 1 (*d*) 0.5 and 0.5

457. If the bandwidth is increased by 2, the $\gamma_{FM}\ \gamma_{AM}$ (where γ is the ratio of SNR of output to SNR at input, FM is frequency modulation and AM is amplitude modulation) is increased by a factor of
(*a*) 2 (*b*) 3
(*c*) 4 (*d*) 6

458. A bit synchronizer regenerates at the receiver a clock waveform which is
(*a*) synchronous with original clock waveform
(*b*) out of phase by 180° with original clock waveform
(*c*) out of phase by 90° with the original clock waveform
(*d*) none of above

459. In phase shift keying the input signal is
(*a*) $s_1(t) = A \cos \omega_0 t$ and $s_2(t) = -A \cos \omega_0 t$
(*b*) $s_1(t) = s_2(t) = A \cos \omega_0 t$
(*c*) $s_1(t) = A \cos \omega_0 t$ and $s_2(t) = A \cos (\omega_0 t + \pi/2)$
(*d*) $s_1(t) = A \cos \omega_0 t$ and $s_2(t) = -A (\cos \omega_0 t + \pi/2)$

460. If N is the number of binary digits needed to assign individual binary code designations to the M quantization levels, the signal to quantization noise ratio S_0/N_q in PCM is
(*a*) 2^N (*b*) 2^{2N}
(*c*) $\sqrt{2^N}$ (*d*) 2^{4N}

461. If delta modulation, the output signal to quantization noise ratio S_0/N_q is (N is the number of bits)
(*a*) $\frac{3}{\pi^2}N^2$ (*b*) $\frac{3}{\pi^2}N^3$
(*c*) $\frac{3}{\pi}N^2$ (*d*) $\frac{3}{\pi}N^3$

462. If there are M equally likely and independent message and $M = 2^N$ where N is an integer, the information in each message is equal to
(*a*) N bits (*b*) 0.5 N bits
(*c*) 2 N bits (*d*) 4 N bits

463. If there are M different and independent message m_1, m_2 etc. with probabilities of occurrence p_1, p_2... etc. and during a long period of transmission a sequence of L message has been generated, the entropy is
(*a*) $\sum_{k=1}^{M} p_k \log_2 \frac{1}{p_k}$ (*b*) $\sum_{k=1}^{M} \frac{1}{p_k} \log_2 p_k$
(*c*) $\sum_{k=1}^{M} p_k \log_2 p_k$ (*d*) $\sum_{k=1}^{M} \frac{1}{p_k} \log_2 \frac{1}{p_k}$

464. Entropies associated with an extremely unlikely message and that with an etremely likely message are respectively
(*a*) 0 and 0 (*b*) 1 and 1
(*c*) 1 and 0 (*d*) 0 and 1

465. Two messages m_1 and m_2 have a probability of 0.5 each. The entropy is
(*a*) 0.25 (*b*) 0.5
(*c*) 0.75 (*d*) 1

466. If there are M messages and each message has probability $p = \frac{1}{m}$, the entropy is
(*a*) 0 (*b*) 1
(*c*) $\log_2 M$ (*d*) $M \log_2 M$

467. If a source of information generates messages at the rate of r messages per second and H is entropy, the information rate is
(*a*) $\frac{H}{r}$ (*b*) $\frac{1}{rH}$
(*c*) rH (*d*) $\frac{r}{H}$

468. An analog signal is band limited to B Hz sampled at Nyquist rate and samples are quantized into 4 levels each with probability $\frac{1}{4}$. The information rate is
(*a*) 4 bits/s (*b*) B bits/s
(*c*) 4 B bits/s (*d*) 0.25 B bits/s

469. If B is channel bandwidth, S is signal power and N is the total noise within the channel bandwidth, the channel capacity C of a white band limited gaussian channel is
(*a*) $B \log_2 \left(1 + \frac{S}{N}\right)$ bits/s (*b*) $B \frac{S}{N}$ bits/s
(*c*) $B \log_2 \left(1 + \frac{N}{S}\right)$ bits/s (*d*) $B \log_2 \left(\frac{S}{N}\right)$ bits/s

470. As per Shannon-Hartley theorem, a noise less gaussian channel has
(*a*) zero capacity (*b*) infinite capacity
(*c*) small capacity (*d*) none of the above

471. As regards BCH code, Hamming code and Reed-Solomon code
(*a*) BCH and Hamming code are special cases of Reed-Solomon code
(*b*) Reed-Solomon and Hamming code are special cases of BCH code
(*c*) BCH and Reed-Solomon code are special case of Hamming code
(*d*) none of the above

472. ARQ stands for
(*a*) automatic repeat quest (*b*) analog repeat quest
(*c*) automatic repeat request (*d*) none of the above

473. A resistor of 10 kΩ has an rms noise voltage of 1 mV under a certain set of conditions. A resistor of 20 kΩ under the same conditions will have an rms noise voltage of
(*a*) 1 mV (*b*) 2 mV
(*c*) 1.414 mV (*d*) 4 mV

474. It is found that in a telephone system 60 subscribers initiate calls during a 30 minute interval and the total duration of all the calls is 100 minutes. The load is
(*a*) 2 (*b*) 3 erlang
(*c*) 3.333 erlang (*d*) 0.333 erlang

475. Pressing a key on telephone set generates two frequencies. The basis for selection of these frequencies is
(*a*) the harmonics of one frequency should not be mistaken for the other frequency
(*b*) the audio input to the handset transmitter should not be mistaken for a key pressing
(*c*) both (*a*) and (*b*) (*d*) neither (*a*) nor (*b*)

476. A three stage telephone switching structure is to have 128 input and 128 output terminals. There are 16 first stage and 16 third stage matrices. To avoid blocking the number of intermediate paths required is
(*a*) 16 (*b*) 15
(*c*) 8 (*d*) 7

477. A three stage telephone switching structure is to have 128 input and 128 output terminals. It has been found that at peak time the utilization probability is 0.1. The number of intermediate links required for non-blocking configuration is 15. If the actual number of intermediate links is 5, the probability of blocking of call is
(*a*) 1% (*b*) 0.5%
(*c*) 0.2% (*d*) 0.1%

478. Some multiplexing arrangements for telephone system are
1. Space multiplexing
2. Time division multiplexing
3. Frequency division multiplexing

Which of the above allow transmission of many independent signals over a single channel
(*a*) 1 only (*b*) 2 and 3 only
(*c*) 1, 2, 3 (*d*) 1 and 3 only

479. Which of the following sampling frequencies is considered adequate in voice telephony
(*a*) 4 kHz (*b*) 8 kHz
(*c*) 16 kHz (*d*) 32 kHz

480. Some features of advanced mobile phone service are
1. To provide telephone communication to a large number of telephone users within the area of a metropolitan only
2. To use minimum amount of frequency spectrum
3. To permit hand held as well as vehicle operation of telephone

Which of the above are true?
(*a*) 1 and 2 only (*b*) 1, 2, 3
(*c*) 1 and 3 only (*d*) 2 and 3 only

481. MTSO stands for
(*a*) Metropolitan telecommunication subscribers office
(*b*) Metropolitan telecommunication switching office
(*c*) Mobile telecommunication switching office
(*d*) Mobile telecommunication subscriber office

482. Some desirable features of a computer communication network are
1. If should accept an intended transmission without excessive delay
2. The data arriving at a receiving point should be processed rapidly
3. When a transmission calls for a response, the delay of response should be minimum,

Which of the above are true?
(*a*) 1 and 2 only (*b*) 1 and 3 only
(*c*) 1, 2, 3 (*d*) 2 and 3 only

483. Circuit switching, Message switching and Packet Switched Network belong to which of the following systems
(*a*) Computer communication system
(*b*) Electro-mechanical communication system
(*c*) PBX system
(*d*) EPBX system

484. Computer communication networks have
(*a*) star configuration (*b*) tree configuration
(*c*) loop/mesh configuration (*d*) any of the above

485. ISDN stands for
(*a*) Integrated services digital network
(*b*) Inter services digital network
(*c*) Inter subscriber digital network
(*d*) Integrated subscriber digital network

486. In spread spectrum technique
(*a*) a modulated signal is modulated again
(*b*) a modulated signal is modulated twice again
(*c*) the power of a modulated signal is increased
(*d*) the noise component of a modulated signal is decreased

487. In a communication system, noise is most likely to affect the signal
(*a*) at the receiver (*b*) at the transmitter
(*c*) at the information source
(*d*) in the channel

488. Fourier analysis indicate that a square wave can be represented as
(*a*) a fundamental sine wave and odd harmonics
(*b*) a fundamental sine wave and even harmonics

(c) a fundamental sine wave and harmonics
(d) fundamental and subharmonic sine waves

489. BFO stands for
(a) Beat frequency oscillator
(b) Broad frequency oscillator
(c) Blocking frequency oscillator
(d) Ballistic frequency oscillator

490. MEWS stands for
(a) Missile early warning system
(b) Moving target early warning system
(c) Missile early warning signal
(d) Moving target early warning signal

491. Which of the following types of noise assumes greater importance at high frequencies
(a) transit time noise (b) shot noise
(c) impulse noise (d) random noise

492. Point out the false statement
(a) impulse noise voltage does not depend on bandwidth
(b) thermal noise is independent of frequency at which it is measured
(c) industrial noise is mostly of impulse type
(d) space noise is observable at frequencies in the range of 8 MHz and higher

493. In a circuit the value of resistor is made four times. The noise voltage generated by this resistance will
(a) be doubled
(b) be increased four times
(c) remain the same
(d) become half

494. If k denotes Boltzmann's constant, the rms value of thermal noise voltage is proportional to
(a) k^2 (b) k
(c) $\sqrt{k}$ (d) $(k)^{1/3}$

495. Industrial noise has the frequency range of
(a) 1 to 100 Hz (b) 200 to 5000 Hz
(c) above 200 MHz (d) 1 to 600 MHz

496. For a resistance of about 15 kΩ operating in a frequency range of a few MHz, the noise voltage is in the range of
(a) a few μV (b) about 100 μV
(c) a few mV (d) a few volts

497. Which of the following is not considered as a useful quantity in comparing noise performance of receivers?
(a) noise temperature (b) noise figure
(c) noise resistance (d) input noise voltage

498. Which of the following noise sources is different from others?
(a) cosmic (b) solar
(c) atmospheric (d) galactic

499. A device has two resistances 300 Ω and 200 Ω respectively. The noise voltages due to these two resistances calculated separately are 5.38 μV and 4.38 μV respectively. When both of them are used together, the noise voltage will be
(a) 9.76 V (b) 1 μV
(c) 23.56 μV (d) 6.94 μV

500. An antenna has a resistance of 50 Ω and an equipment noise resistance of 30 Ω. The noise figure in decibels is
(a) 1.6 dB (b) 2 dB
(c) 3.6 dB (d) 6 dB

501. In a plate modulated class C amplifier, the peak supply voltage is V. The maximum plate-cathode voltage can be
(a) V (b) 1.5 V
(c) 2.5 V (d) 4 V

502. In a low level amplitude modulated system, the amplifier following the modulated stage must be
(a) class C (b) linear
(c) non-linear (d) harmonic

503. In an AM wave with $m = 1$, the carrier is suppressed. The resulting power saving will be
(a) $\frac{1}{3}$ (b) $\frac{1}{2}$
(c) $\frac{2}{3}$ (d) $\frac{3}{4}$

504. When modulation index of an AM wave is increased from 0.5 to 1, the transmitted power
(a) remains the same (b) increases by 25%
(c) increases by 33.3% (d) increases by 50%

505. One of the advantage of base modulation over collector modulation in a class C transistor amplifier is
(a) higher efficiency
(b) more linear operation
(c) more power output per transistor
(d) lesser requirement of modulating power

506. In an AM system, doubling of modulation index doubles the antenna current. The system is
(a) single side band full carrier
(b) vestigial side band
(c) single side band suppressed carrier
(d) double side band full carrier

507. In SSB generation the most commonly used filter is
(a) low pass (b) RC
(c) LC (d) mechanical

508. The device most likely to be found in an SSB transmitter is
(a) class C audio amplifier (b) tuned modulator
(c) class B RF amplifier (d) class A RF amplifier

509. In an AM wave the carrier and one of the side bands is suppressed. If $m = 1$, the percentage saving in power is
(a) 25% (b) 50%
(c) 66.6% (d) 83.3%

510. In a AM wave the carrier and one of the side bands is suppressed. If $m = 0.5$, the percentage saving in power is
(a) 50% (b) 83.3%
(c) 94.4% (d) 100%

511. The use of SSB
(a) halves the bandwidth required for transmission
(b) does not affect the bandwidth for transmission
(c) decreases the bandwidth required for transmission by 25%

(*d*) decreases the bandwidth required for transmission by 66.6%

512. The main systems used for generation of SSB are
(*a*) filter method
(*b*) phase cancellation method
(*c*) third method
(*d*) all the above

513. Consider the following statements in amplitude modulation
1. The harmonic and inter modulation distortion may occur in audio and RF amplifiers
2. The sum and difference frequencies are present in the output of mixer
3. The diode detector has audio frequencies in its output
4. A balanced modulator cannot produce an AM with suppressed carrier

Which of the above are true?
(*a*) 1, 2, 3, 4 (*b*) 1, 2, 3
(*c*) 1, 2 (*d*) 3, 4

514. The simplest method of suppression of unwanted side band in AM is
(*a*) filter method (*b*) phase shift method
(*c*) third method (*d*) both (*a*) and (*b*)

515. Which of the following refers to SSB suppressed carrier amplitude modulated system?
(*a*) A 3 E (*b*) H 3 E
(*c*) J 3 E (*d*) C 3 F

516. A 400 W carrier is amplitude modulated with $m = 0.75$. The total power in AM is
(*a*) 400 W (*b*) 512 W
(*c*) 588 W (*d*) 650 W

517. A 400 W carrier is amplitude modulated with $m = 0.75$. If single side band suppressed carrier is used the total power in modulated wave is
(*a*) 400 W (*b*) 512 W
(*c*) 112 W (*d*) 56 W

518. Which of the following can provide two or more voice circuits with the same carrier?
(*a*) SSB (*b*) Carrier reinsertion
(*c*) Lincompex
(*d*) Independent side band system

519. In the FM wave described by equation $v = 15 \sin (4 \times 10^8 t + 3 \sin 1100\, t)$, the carrier frequency and signal frequency respectively are
(*a*) 63.66 MHz and 175 Hz
(*b*) 4×10^8 Hz and 1100 Hz
(*c*) 175 Hz and 63.66 MHz
(*d*) 1100 Hz and 4×10^8 Hz

520. In the FM wave described by equation $v = 15 \sin (4 \times 10^8 t + 3 \sin 1100\, t)$, the maximum frequency deviation is
(*a*) 175 Hz (*b*) 525 Hz
(*c*) 3 Hz (*d*) 58.33 Hz

521. In an FM signal, the modulating frequency is 2 kHz and maximum deviation is 10 kHz. The bandwidth requirement is
(*a*) 2 kHz (*b*) 10 kHz
(*c*) 5 kHz (*d*) 32 kHz

522. Consider the following statements
1. The amplitude of an FM wave is constant
2. FM is more immune to noise than AM
3. FM broadcasts operate in upper VHF and UHF frequency ranges
4. FM transmitting and receiving equipments are simpler as compared to AM transmitting and receiving equipments

Which of the above are correct?
(*a*) 1, 2, 3, 4 (*b*) 1, 2, 3
(*c*) 2, 3, 4 (*d*) 1, 3, 4

523. In narrow band FM, the modulation index is
(*a*) near unity (*b*) about 100
(*c*) about 0.1 (*d*) about 300

524. In a reactance FET, $g_m = 9 \times 10^{-3}$ siemens, gate to source resistance is one-ninth of reactance of gate to drain capacitor and frequency is 4 MHz. The capacitive reactance is
(*a*) 1000 Ω (*b*) 1/1000 Ω
(*c*) 36×10^3 Ω (*d*) $\frac{4}{9} \times 10^3 \Omega$

525. In the stabilized reactance modulator AFC system
(*a*) phase modulation is converted into amplitude modulation by the equalizer circuit
(*b*) the discriminator frequency must not be too low or the system will fail
(*c*) the discriminator must have a fast time constant to prevent demodulation
(*d*) the higher the discriminator frequency, better the oscillator frequency stability

526. In the spectrum of an FM wave
(*a*) When modulation index is large, carrier frequency disappears
(*b*) carrier frequency cannot disappear
(*c*) amplitude of any side band depends on modulation index
(*d*) the total number of side bands depend on modulation index

527. An FM wave is made to pass through a frequency tripler. If original modulation index is m_f, the output of tripler has a modulation index equal to
(*a*) m_f (*b*) $\frac{1}{3} m_f$
(*c*) $3\, m_f$ (*d*) $\frac{1}{9} m_f$

528. Which of the following is the indirect way of FM generation?
(*a*) reactance bipolar transistor modulator
(*b*) armstrong modulator
(*c*) varactor diode modulator
(*d*) reactance FM modulator

529. A pre empasis circuit
(*a*) boosts bass frequencies
(*b*) pre amplifies the complete audio band
(*c*) amplifies higher audio frequencies
(*d*) converts FM into phase modulation

530. An FM signal with deviation δ is passed through mixer. Its frequency is reduced five fold. The deviation in output of mixer is
(*a*) $0.2\,\delta$ (*b*) δ
(*c*) $5\,\delta$ (*d*) very very high

531. Consider the following statements about advantages of phase discriminator over slope detector
1. Fewer tuned circuits 2. Greater limiting
3. Better linearity 4. Easy alignment
Which of the above are true?
(*a*) 1, 2, 3, 4 (*b*) 2, 3, 4
(*c*) 1, 2, 3 (*d*) 1, 3, 4

532. In a superheterodyne receiver having no RF amplifier, the IF is 455 kHz. The image frequency at 1000 kHz is
(*a*) 1455 kHz (*b*) 545 kHz
(*c*) 1910 kHz (*d*) 90 kHz

533. In a radio receiver with simple AGC
(*a*) the highest AGC voltage is produced between stations
(*b*) an increase in signal strength produces more AGC
(*c*) the audio stage gain is normally controlled by AGC
(*d*) the faster the time constant of AGC, more accurate the output

534. Consider the following types of distortions in diode detectors
1. Distortion is caused due to the reason that dc and ac load impedances are unequal
2. Distortion is caused because ac load impedance has a reactive component at high audio frequencies
3. Audio frequency diode current is less than dc current
Which of the above are true?
(*a*) 1, 2 and 3 (*b*) 1 and 2
(*c*) 1 and 3 (*d*) 2 and 3

535. Which of the following is present in both TRF receiver and superheterodyne receiver?
(*a*) detector (*b*) mixer
(*c*) IF amplifier (*d*) local oscillator

536. In a superheterodyne radio receiver
(*a*) RF amplifier normally operates at 455 kHz above carrier frequency
(*b*) mixer input must be tuned to the signal frequency
(*c*) local oscillator operates below signal frequency
(*d*) local oscillator frequency is double the IF

537. In a communication receiver a squelch circuit is normally inserted
(*a*) before detector (*b*) after detector
(*c*) before mixer (*d*) after power amplifier

538. Which of the following is not true as regards communication receiver?
(*a*) it is exactly same as radio receiver used in households
(*b*) it can be used, besides communication, for detection in high frequency bridges
(*c*) it is designed to perform the task of low and high frequency reception better than in household receiver.
(*d*) its tuning and operation are more complicated than those in household receiver

539. A communication receiver is
(*a*) TRF receiver
(*b*) a superheterodyne receiver
(*c*) either (*a*) or (*b*)
(*d*) neither (*a*) nor (*b*)

540. To prevent overloading in the last IF amplifier one should use
(*a*) variable sensitivity (*b*) squelch
(*c*) double conversion (*d*) variable selectivity

541. In a radio detector
(*a*) stabilization against signal strength variation is provided
(*b*) linearity is worse than in a phase discriminator
(*c*) the circuit is the same as in discriminator except that diodes are reversed
(*d*) output is twice that obtainable from a similar phase discriminator

542. The typical squelch circuit
(*a*) cuts off IF amplifier when AGC is minimum
(*b*) cuts off IF amplifier when AGC is maximum
(*c*) cuts off audio amplifier when carrier is absent
(*d*) eliminates RF interference when signal is weak

543. Consider the following statements as regards communication receiver
1. Variable sensitivity is used to eliminate selective fading
2. A product demodulator can be used to receive Morse code
3. The noise limiter cuts off receiver output during a noise pulse
4. Double conversion is used to improve image rejection
Which of the above statements are correct?
(*a*) 1, 2, 3, 4 (*b*) 1, 2, 3
(*c*) 2, 3, 4 (*d*) 1, 3, 4

544. A receiver has poor selectivity. It will also have poor
(*a*) sensitivity (*b*) double spotting
(*c*) blocking (*d*) diversity reception

545. A circuit which can not demodulate SSB is
(*a*) Balanced modulator (*b*) Product detector
(*c*) BFO (*d*) Phase discriminator

546. Consider the following statements as regards IF in radio receivers
1. If IF is too high poor selectivity and poor adjacent channel rejection could result
2. If IF is high, tracking difficulties increase
3. A very low IF makes selectivity too sharp
4. As IF is decreased, image frequency rejection becomes better

Which of the above are correct?
(a) 1, 2, 3, 4 (b) 1, 2, 3
(c) 2, 3, 4 (d) 1, 3, 4

547. In a phase discriminator f_c is the IF to which primary circuit is tuned. Consider the following statements: The primary and secondary voltage are
1. exactly 90° out of phase when input frequency is f_c
2. more than 90° out of phase when input frequency is higher than f_c
3. less than 90° out of phase when input frequency is lower than f_c

Which of the above statements are correct?
(a) 1 only (b) 1 and 2
(c) 1, 2 and 3 (d) 2 and 3

548. The image frequency of a superheterodyne receiver is
(a) created within the receiver itself
(b) not rejected by IF tuned circuits
(c) independent of the frequency of tuning
(d) due to insufficient adjacent channel rejection

549. Three point tracking is achieved with
(a) double conversion (b) padder capacitor
(c) double spotting (d) variable selectivity

550. If the ratio of ac to dc load impedance of a diode detector is low, the result will be
(a) poor AGC operation (b) poor AF response
(c) negative peak clipping (d) diagonal clipping

551. Which of the following cannot be used to demodulate SSB?
(a) complete phase shift generator
(b) product detector
(c) diode balanced modulator
(d) bipolar transistor balanced modulator

552. Indicate the false statement: If SWR on a transmission line is infinite, the line is terminated in
(a) open circuit (b) short circuit
(c) pure reactance (d) complex impedance

553. If k is dielectric constant of a medium, the velocity factor in that medium is
(a) $\sqrt{k}$ (b) $\frac{1}{\sqrt{k}}$
(c) $(k)^{1/3}$ (d) $\frac{1}{(k)^{1/3}}$

554. A $(75 - j50)\ \Omega$ load is connected to a coaxial line of $Z_0 = 75\ \Omega$ at 10 GHz. The best method of matching consists of using
(a) a capacitance at some specific distance from load
(b) a short circuited stub at some specific distance form load
(c) an inductance at load
(d) an short circuited at load

555. A $100 + j\ 100\ \Omega$ load is to be matched to a line with $Z_0 = 300\ \Omega$ to give SWR = 1. The reactance of stub is
(a) $j\,200\ \Omega$ (b) $-j\,200\ \Omega$
(c) $j\,100\ \Omega$ (d) $-j\,100\ \Omega$

556. Impedance inversion may be obtained with
(a) half wave line
(b) a short circuited stub
(c) complex impedance Z connected in series with line
(d) quarter wave line

557. If a line is terminated in characteristic impedance Z_0, the input impedance measured at the input will be
(a) Z_0 (b) $Z_0/2$
(c) Zero (d) infinite

558. Short circuited stubs are preferred to O.C. stubs because
(a) O.C. stubs are incapable of giving full range of reactances
(b) O.C. stubs are liable to radiate
(c) O.C. stubs are difficult to make and connect
(d) none of the above

559. For a parallel two-wire line with both conductors touching each other, the characteristic impedance is about
(a) $300\ \Omega$ (b) $200\ \Omega$
(c) $80\ \Omega$ (d) $24\ \Omega$

560. When energy is fed into open wire transmission line the energy may get dissipated due to
(a) radiation (b) conductor heating
(c) dielectric heating (d) all the above

561. It is desired that transmission line and load should match for a range of frequencies. The best method is
(a) to use a balun
(b) to use a single stub of adjustable position
(c) to use a double stub
(d) to use a broad band amplifier

562. In a transmission line, losses due to dielectric heating are
(a) proportional to voltage and inversely proportional to Z_0
(b) proportional to voltage and proportional to Z_0
(c) proportional to current and proportional to Z_0
(d) proportional to current and inversely proportional to Z_0

563. It is desired to couple a coaxial line to a two-parallel wire line. It is best to use
(a) quarter wave transformer (b) balun
(c) slotted line (d) directional coupler

564. A generator is feeding power to a transmission line of characteristic impedance Z_0. The line is terminated in impedance Z_L. Standing waves will be set up if
(a) $Z_L = Z_0$ (b) $Z_L < Z_0$
(c) $Z_L > Z_0$ (d) both (b) and (c)

565. In standing wave pattern on a transmission line
(a) voltage and current nodes coincide
(b) voltage and current anti-nodes coincide
(c) voltage nodes and current anti-nodes as well as current nodes and voltage anti-nodes coincide
(d) both (a) and (b)

566. It is found that a ship-to-ship communication suffers from fading. This can be avoided by using
(a) space diversity (b) frequency diversity
(c) broad band antenna (d) directional antenna

567. High frequency waves are
(*a*) absorbed by F_2 layer (*b*) affected by solar cycle
(*c*) reflected by D layer (*d*) none of the above

568. As *em* waves travel in free space they get
(*a*) reflected (*b*) refracted
(*c*) attenuated (*d*) absorbed

569. If transmitting antenna current is I and height h_t, the field strength for a wave of wavelength λ at a distance *d* from the antenna is given by
(*a*) $\frac{120\pi h_t I}{\lambda d}$ (*b*) $\frac{120\pi/\lambda}{h_t d}$
(*c*) $\frac{120\pi d I}{h_t \lambda}$ (*d*) $\frac{120\pi h_t d}{\lambda I}$

570. For F_2 layer the critical frequecy is about
(*a*) 500 MHz (*b*) 100 Mhz
(*c*) 25 MHz (*d*) 10 MHz

571. VLF waves are sometimes used because
(*a*) they penetrate the ionosphere easily
(*b*) they are very reliable
(*c*) they require low power
(*d*) they require transmitting antenna of convenient sizes

572. Which of the following terms are applicable to troposcatter propagation?
(*a*) Faradays rotation (*b*) SIOs
(*c*) Fading (*d*) atmospheric storms

573. In *em* waves, polarization
(*a*) is always vertical in an isotropic medium
(*b*) is caused by reflection
(*c*) is due to transverse nature of waves
(*d*) results from longitudinal nature of waves

574. The atmospheric absorption of radio waves depends on
(*a*) frequency of wave
(*b*) the distance from transmitter
(*c*) wave polarisation
(*d*) atmospheric polarisation

575. Diffraction of *em* waves
(*a*) arises only with spherical wave front
(*b*) may occur around the edge of an obstracle
(*c*) occurs when waves pass through a big slot
(*d*) none of the above

576. Tropospheric scatter is used with frequencies in the range of
(*a*) VLF (*b*) HF
(*c*) UHF (*d*) VHF

577. The maximum range of a transmitter depends on
(*a*) its frequency
(*b*) it power
(*c*) both its frequency and power
(*d*) none of the above

578. In MF range, increasing the transmitter power
(*a*) increases the range of transmitter
(*b*) does not increase the range of transmitter because propagation is limited by tilt
(*c*) decreases the range of transmitter because propagation is limited by tilt
(*d*) increases the range of transmitter because propagation is through space waves

579. Troposphere is the portion of atmosphere
(*a*) nearest to ground
(*b*) between 50 to 90 km above ground
(*c*) between 110 to 220 km above ground
(*d*) between 250 to 350 km above ground

580. The most frequent users of long distance VLF transmission is
(*a*) ship communication
(*b*) radio broadcasting
(*c*) TV broadcasting
(*d*) telephone and telegraph

581. For antenna of height h_t metres, the radio horizon is equal to
(*a*) $4\sqrt{h_t}$ metres (*b*) $4\sqrt{h_t}$ km
(*c*) $2.5\sqrt{h_t}$ km (*d*) $2.5\sqrt{h_t}$ metres

582. Which of the following is very useful as multiband HF receiving antenna
(*a*) square loop (*b*) conical horn
(*c*) folded dipole (*d*) log-periodic

583. Which of the following antennas is not wide band?
(*a*) Marconi (*b*) helical
(*c*) folded dipole (*d*) discone

584. When a wave reaches the open circuited end of a line
(*a*) all the forward energy is reflected by the open circuit
(*b*) all the forward energy is absorbed by the open circuit
(*c*) a small portion of energy is radiated into atmosphere and the remaining energy is reflected
(*d*) either (*a*) or (*b*)

585. Waves travel on each of the line configurations in Fig. 12.13. In which case would the radiation be maximum.

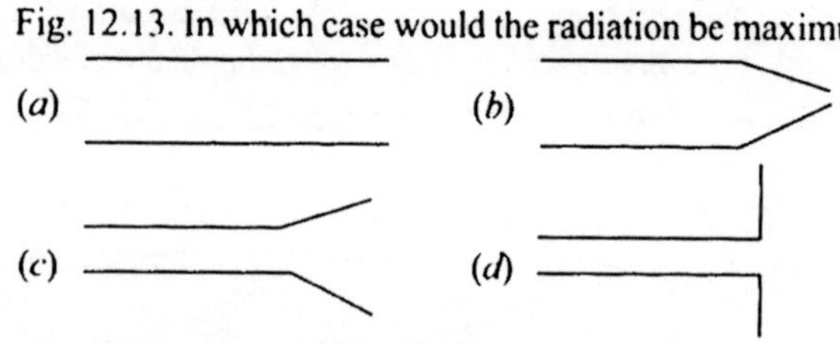

Fig. 12.13.

586. An ungrounded antenna near the ground
(*a*) is horizontally polarized
(*b*) acts as an antenna array
(*c*) acts as a single antenna of twice the height
(*d*) is not likely to need an earth mat

587. Which one of the following is non-resonant antenna?
(*a*) folded dipole (*b*) broad side array
(*c*) end fire array (*d*) rhombic antenna

588. When the length of antenna is a whole wave length
(*a*) the radiation at right angles is zero
(*b*) the radiation at right angles is maximum
(*c*) the radiation is zero in all directions
(*d*) the radiation is maximum in all directions

589. Directive gain is
(*a*) a power ratio (*b*) a voltage ratio
(*c*) a current ratio (*d*) a distance ratio

590. Directivity means
(*a*) maximum directive gain
(*b*) zero directive gain
(*c*) minimum directive gain
(*d*) none of the above

591. Radiation resistance of an antenna is a
(*a*) dc resistance (*b*) ac resistance
(*c*) either (*a*) or (*b*) (*d*) none of the above

592. Which of the following terms is not applicable to Yagi-Uda array?
(*a*) high gain (*b*) folded dipole
(*c*) parasitic elements (*d*) good bandwidth

593. Which of the following antennas is circularly polarized?
(*a*) Yagi-Uda (*b*) parabolic reflector
(*c*) small circular loop (*d*) helical

594. Which of the following is taken as reference antenna for directive gain
(*a*) half wave dipole (*b*) elementary doublet
(*c*) isotropic (*d*) infinitesimal dipole

595. For a short dipole having an effective length less than half the wavelength the radiation resistance is
(*a*) proportional to length
(*b*) proportional to square root of length
(*c*) proportional to square of length
(*d*) none of the above

596. Beam width of an antenna is expressed in
(*a*) metres (*b*) degrees
(*c*) ratio of two powers (*d*) ratio of two angles

597. The VLF, LF and MF antennas are
(*a*) horizotally polarized (*b*) vertically polarized
(*c*) non-linearly polarized (*d*) either (*a*) or (*b*)

598. A Marconi antenna
(*a*) is a grounded antenna with a length equal to $\lambda/4$
(*b*) is an ungrounded antenna with a length equal to $\lambda/4$
(*c*) has a length equal to λ
(*d*) has a length equal to $\lambda/2$

599. An earth mat for a communication tower consist of
(*a*) One or two radials extending from base of tower at a depth of about 5 cm below ground
(*b*) a large number of radials extending from base of tower at a depth of about 3 m
(*c*) a large number of radials extending from base of tower at a depth of about 30 cm
(*d*) none of the above

600. Counterpoise is used in soils having
(*a*) high resistivity (*b*) low resistivity
(*c*) medium resistivity (*d*) both (*b*) and (*c*)

601. The effective length of antenna is different from actual length due to
(*a*) top loading (*b*) end effects
(*c*) both (*a*) and (*b*) (*d*) none of the above

602. Top loading in an antenna increases
(*a*) effective height (*b*) beam width
(*c*) input capacitance (*d*) band width

603. A helical antenna is used for satellite tracking due to its
(*a*) good front to back ratio (*b*) broad bandwidth
(*c*) circular polarization (*d*) both (*a*) and (*b*)

604. The discone antenna is
(*a*) used in UHF range
(*b*) circularly polarized
(*c*) used in radar
(*d*) used as a direction finding antenna

605. Consider the following reasons for using antenna coupler
1. To prevent illegal radiation of spurious frequencies
2. To tune out reactive components of antenna impedance
3. To have correct value of load impedance for transmitter.

Which of the above are correct?
(*a*) 1, 2, 3 (*b*) 1, 2
(*c*) 1 only (*d*) 2, 3

606. In an end fire array, the radiation pattern
(*a*) is in the plane of array
(*b*) is at right angles to array
(*c*) is at an angle to the plane of array
(*d*) either (*b*) or (*c*)

607. Consider the following statements about Yagi-Uda antenna
1. It is unidirectional
2. Its gain is about 7 db
3. It is used at VLF
4. The optimum separation is 0.1 λ

Which of the above are correct?
(*a*) 1 and 2 only (*b*) 1, 2, 4 only
(*c*) 1, 2, 3, 4 (*d*) 1 and 4 only

608. Consider the following statements about Rhombic antenna
1. It is non-resonant
2. It is a broad band antenna
3. It is ideally suited for HF transmission

Which of the above are correct?
(*a*) 1, 2, 3 (*b*) 1, 3 only
(*c*) 2, 3 only (*d*) 1 only

609. A paraboloid reflector has mouth diameter D meters, wavelength λ meters. Its beam width ϕ is equal to
(*a*) $\frac{70\lambda}{D}$ (*b*) $\frac{70D}{\lambda}$
(*c*) $\frac{\lambda}{10D}$ (*d*) $\frac{D}{70\lambda}$

610. If D is mouth diameter of a paraboloid reflector, its direcetivity is proportional to
(*a*) D (*b*) D^2
(*c*) $\sqrt{D}$ (*d*) D^3

611. A Discone antenna is
(*a*) a combination of disk and cone in close proximity
(*b*) a combination of disk and cone at $\lambda/4$ spacing

(*c*) a combination of disk and cone far apart
(*d*) a combination of disk and cone at $\lambda/2$ spacing

612. Consider the following statements about loop antenna
1. It is suitable for direction finding application
2. It is a single turn or multiturn coil carrying RF current
3. The loop is surrounded by a magnetic field everywhere perpendicular to the loop
Which of the above statements are correct?
(*a*) 1 and 2 only (*b*) 1 only
(*c*) 1, 2, 3 (*d*) 2 and 3 only

613. A total of 16 equally probable events exist. If it is required to select 1 out of them, the number of required bits is
(*a*) 2 (*b*) 4
(*c*) 3 (*d*) 5

614. Indicate the false statement: To combat noise
(*a*) reduce signalling rate
(*b*) increase transmitted power
(*c*) use redundancy
(*d*) increase channel bandwidth

615. As per Hartley law
(*a*) redundancy is essential
(*b*) it is necessary to use only binary codes
(*c*) maximum rate of information transmission depends on the channel bandwidth
(*d*) the maximum rate of information depends on depth of modulation

616. Which one of the following is digital
(*a*) pulse code modulation
(*b*) pulse width modulation
(*c*) pulse frequency modulation
(*d*) pulse position modulation

617. Which one of the following is analog
(*a*) PCM (*b*) PWM
(*c*) Delta modulation (*d*) Differential PCM

618. To select one out of N equally probable events, the number of bits required is
(*a*) $\log_2 N$ (*b*) $\log_{10} N$
(*c*) $\log_e N$ (*d*) $\sqrt{N}$

619. The efficiency of using a binary system for selection of 1 out of 13 equally probable events is
(*a*) 100% (*b*) about 92%
(*c*) about 83% (*d*) about 71%

620. In the absence of noise, if C is channel capacity in bits/s, δf is channel bandwidth in Hz and N is number of coding levels. Then
(*a*) $C = \log_2 N$ (*b*) $C = (\delta f) \log_2 N$
(*c*) $C = 2(\delta f) \log_2 N$ (*d*) $C = 0.5(\delta f) \log_2 N$

621. If C is capacity of noisy channel, bits/s, δf is bandwidth, Hz and S/N is signal to noise ratio, then
(*a*) $C = (\delta f) \log_2 \left(1 + \frac{S}{N}\right)$
(*b*) $C = 2(\delta f) \log_2 \left(1 + \frac{S}{N}\right)$
(*c*) $C = \log_2 \left(1 + \frac{S}{N}\right)$
(*d*) $C = (\delta f) \log_{10} \left(1 + \frac{S}{N}\right)$

622. For telegraphy the most commonly used modulation system is
(*a*) FSK (*b*) two tone modulation
(*c*) PCM (*d*) single tone modulation

623. PWM can be generated
(*a*) by a free running multi vibrator
(*b*) by integrating the signal
(*c*) by a monostable multi vibrator
(*d*) by differentiating pulse position modulation

624. The most noise resistant modulation system is
(*a*) PCM
(*b*) SSB suppressed carrier
(*c*) FM
(*d*) pulse position modulation

625. Quantizing error occurs in
(*a*) TDM (*b*) FDM
(*c*) PCM (*d*) PWD

626. Consider the following statements about PCM
1. PCM is not noise resistant
2. PCM requires complex encoding and quantizing circuitry
3. PCM requires large bandwidth
Which of the above are correct?
(*a*) 1 only (*b*) 1, 2, 3
(*c*) 2, 3 (*d*) 1 and 3

627. If each short element of signal occupies 20 ms, the speed is
(*a*) 20 bauds (*b*) 50 bauds
(*c*) $\frac{1}{50}$ bauds (*d*) 0.05 bauds

628. The number of bits of information required to indicate the correct selection of 3 independent consecutive events out of 75 equally probable events is
(*a*) 6.23 (*b*) 18.69
(*c*) 12.46 (*d*) 24.92

629. Which of the following codes provides parity checks
(*a*) CCITT-2 (*b*) EBCDIC
(*c*) ASC II (*d*) Baudot

630. The start of modern computer age was started by the event
(*a*) development of disc drive
(*b*) development of transistor
(*c*) development of Hollerith code
(*d*) design of EVIAC computer

631. The Shannon Hartley law
(*a*) refers to noise (*b*) defines bandwidth
(*c*) describes signalling rate (*d*) refers to distortion

632. An S/N ratio of 3 expressed in dB is
(*a*) 4.8 dB (*b*) 2.4 dB
(*c*) 0.48 dB (*d*) 0.24 dB

633. For an ideal 3000 Hz channel, Nyquist rate is
(*a*) 3000 bps (*b*) 6000 bps
(*c*) 9000 bps (*d*) 12000 bps

634. For an ideal 3000 Hz channel, S/N ratio is
(*a*) 3 (*b*) 1.5
(*c*) 4 (*d*) 2

635. The baud rate is
(*a*) not equal to signalling rate
(*b*) always equal to bit transfer rate
(*c*) equal to twice the bandwidth of an ideal channel
(*d*) euqal to half the bandwidth of ideal channel

636. Full duplex operation
(*a*) requires modems at both ends of circuit
(*b*) requires two pairs of cables
(*c*) can transfer data in both directions simultaneously
(*d*) all the above

637. Switching systems
(*a*) are limited to small data networks
(*b*) require additional lines
(*c*) are not used in data system
(*d*) improve the efficiency of data transfer

638. Which of the following codes is based on BCD format
(*a*) EBCDIC (*b*) ASC II
(*c*) Hollerith (*d*) both (*a*) and (*b*)

639. The 2 out of 5 codes has
(*a*) two 1s and three 0s in each code group
(*b*) one 1s and four 0s in each code group
(*c*) four 1s and one 0s in each code group
(*d*) three 1s and two 0s in each code group

640. A modem is classified as low speed if data rate handled is
(*a*) upto 100 bps (*b*) upto 250 bps
(*c*) upto 400 bps (*d*) upto 600 bps

641. In a high speed modem the data rate handled is about
(*a*) 1000 bps (*b*) 5000 bps
(*c*) 20000 bps (*d*) 50000 bps

642. In which of the following schemes are several channels interleaved and transmitted together
(*a*) TDM (*b*) FDM
(*c*) group (*d*) sub group

643. The satellites used for intercontinental communication are called
(*a*) Comsat (*b*) Domast
(*c*) Marisat (*d*) Intelsat

644. Armoured submarine cable is used
(*a*) to prevent insulation breakdown due to high voltage
(*b*) to protect cable at great depths
(*c*) to prevent inadevertent damage due to ploughing in
(*d*) for the shallow shore ends of cable

645. The number of repeaters in a coaxial cable link depends on
(*a*) system bandwidth
(*b*) separation of equalizers
(*c*) the number of coaxical cables in the tube
(*d*) none of the above

646. The first geostationary satellite was named
(*a*) Intelsat I (*b*) Intelsat II
(*c*) Intelsat III (*d*) Early bird

647. Some advantages of optical fibre cables are
1. Small diameter
2. Immunity to cross talk and *em* interference
3. Laser and LED modulation methods lend themselves ideally to digital operation

Which of the above are correct?
(*a*) 1 only (*b*) 1 and 2
(*c*) 1, 2, 3 (*d*) 2, 3

648. A geostationary satellite
(*a*) is located at a height of 35800 km
(*b*) appears stationary over earth's magnetic pole
(*c*) is motionless in space but keeps spinning
(*d*) orbits the earth within a 24-hour period

649. TDM
(*a*) can be used with PCM only
(*b*) interleaves pulses belonging to different transmissions
(*c*) combines five groups into a super group
(*d*) stacks 24 channels in adjacent frequency slots

650. A basic group B
(*a*) consists of five super group
(*b*) has a frequency range of 60 to 108 kHz
(*c*) consists of erect channels only
(*d*) is formed at group translating equipment

651. Identical telephone numbers in different parts of a country are distinguished by their
(*a*) area codes (*b*) office codes
(*c*) access digits (*d*) language codes

652. Which of the following does not cause losses in optical fibre cables
(*a*) stepped index operation (*b*) impurities
(*c*) microbending (*d*) attenuation in glass

653. To separate channels in FDM receiver we have to use
(*a*) Integration (*b*) AND gate
(*c*) band pass filter (*d*) differentiation

654. Higher order TDM levels are obtained by
(*a*) forming super master groups
(*b*) using A law
(*c*) using μ law
(*d*) dividing pulse width

655. The different channels in a TDM receiver are separated by
(*a*) integration (*b*) differentiation
(*c*) use of AND gate (*d*) use of OR gate

656. The maximum power output of a standard A earth station over the total band allocated to satellite communication is about
(*a*) 0.5 kW (*b*) 8 kW
(*c*) 20 kW (*d*) 50 kW

657. If the transmitted pulse power of a radar is P_t (peak value) and the antenna is isotropic, the power density at a distance r is (*a*) $\frac{P_t}{r}$ (*b*) $\frac{P_t}{4\pi r^2}$

(*c*) $\frac{P_t}{4\pi r^3}$ (*d*) $\frac{P_t}{2\pi r^2}$

658. The maximum ranges of a radar is proportional to
(*a*) square root of peak transmitted pulse power
(*b*) fourth root of peak transmitted pulse power
(*c*) peak transmitted pulse power
(*d*) none of the above

659. If k is Boltzmann's constant, T_0 is standard ambient temperature, δf is bandwidth of receiver, F is noise figure, the minimum receivable signal in a radar receiver is given by
(*a*) $kT_0(\delta f)(F)$ (*b*) $kT_0(\delta f)(F-1)$
(*c*) $\frac{kT_0(\delta f)}{(F-1)}$ (*d*) none of the above

660. It is required to increase the maximum range of a radar by a factor of two. The peak power transmitted in the radar system has to be increased by a factor of
(*a*) 2 (*b*) 4
(*c*) 8 (*d*) 16

661. The antenna diameter of a radar system is doubled. The maximum range will
(*a*) be doubled (*b*) become four times
(*c*) be halved (*d*) decrease to one fourth

662. Consider the following statements: In a radar system having a high value of the ratio (antenna diameter)/wavelength
1. the maximum range is large
2. good target discrimination is achieved
3. capture area is increased
4. target acquisition is difficult

Which of the above are correct?
(*a*) 1, 2, 3, 4 (*b*) 1, 2, 3
(*c*) 1, 2 (*d*) 1, 3, 4

663. If the target area is small compared to wavelength of radar the situation is called
(*a*) resonance region (*b*) Rayleigh region
(*c*) optical region (*d*) binomial region

664. If the circumfrence of a spherical target is between 1 to 10 wavelength in a radar system the situation is called
(*a*) resonance region (*b*) Rayleigh region
(*c*) optical region (*d*) binomial region

665. The radar cross-section of a target depends an
(*a*) frequency
(*b*) polarization of incident wave
(*c*) aspect of target
(*d*) all the above

666. Which one of the following statements is wrong? The radar cross-section of a target
(*a*) depends on frequency
(*b*) may be reduced by special coating of non-spherical target
(*c*) depends on the aspect of target if it is non-spherical
(*d*) is equal to actual cross-sectional area for small targets

667. In high power radars, which of the follow'ng are used for switching
(*a*) Power diode (*b*) Power BJT
(*c*) Power MOSFET (*d*) Thyristor

668. Consider the following statements. The radar pulse must be flat topped to
1. allow accurate range measurement
2. allow good minimum range
3. prevent frequency changes in magnetron
4. make the returned echoes easier to distinguish from noise.

Which of the above are correct?
(*a*) 1, 2, 3, 4 (*b*) 1, 2, 3
(*c*) 1, 2 (*d*) 1, 3, 4

669. In which of the following is Doppler effect not used
(*a*) CW radar (*b*) FM radar
(*c*) MTI system
(*d*) moving target plotting on PPI

670. In a radar system the solution to blind speed problem is to
(*a*) use MTI
(*b*) use monopulse
(*c*) vary PRF
(*d*) change Doppler frequency

671. In radar systems PRF stands for
(*a*) Power return factor
(*b*) Pulse return factor
(*c*) Pulse repitition frequency
(*d*) Pulse response factor

672. Consider the following statements as regards pulse repitition frequency
1. Pulse interval should be as large as possible so that a pulse returns before next pulse is transmitted
2. Greater the number of pulse reflected from a target, greater is the probability of ditinguishing it from the noise

Which of the above are correct?
(*a*) 1 only (*b*) 2 only
(*c*) both 1 and 2 (*d*) neither 1 nor 2

673. A high pulse repitition frequency in a radar
(*a*) increases the maximum range
(*b*) decreases the maximum range
(*c*) has no effect on the maximum range
(*d*) either (*a*) or (*b*) depending on other factors

674. The IF bandwidth of a radar receiver is inversely proportional to
(*a*) pulse interval (*b*) pulse width
(*c*) PRF (*d*) none of the above

675. If the target cross-section is changing, the best scanning system for accurate tracking is
(*a*) lobe switching (*b*) monopulse
(*c*) sequential lobing (*d*) conical scanning

676. In a radar, the return echo arrives after the allocated pulse interval. Then
(*a*) the receiver may be over loaded
(*b*) the target will appear closer than its actual position
(*c*) the return echo may interfere with transmitter operation
(*d*) none of the above

677. MTI stands for
(*a*) moving target indicator
(*b*) microprocessor tracking index
(*c*) multiplexed telemetry index
(*d*) maximum transmission index

678. In radar system which of the following is called A scope
(*a*) deflection modulation of CRT
(*b*) intensity modulation of CRT
(*c*) computer display
(*d*) none of the above

679. In radar display system PPI stands for
(*a*) pulse position indicator (*b*) plan position indicator
(*c*) phase position indicator (*d*) point position indicator

680. An MTI radar operating at 5 GHz has a PRF of 800 pps. The lowest blind speed is
(*a*) 48 m/s (*b*) 96 m/s
(*c*) 192 m/s (*d*) 24 m/s

681. PA radar means
(*a*) phased array radar (*b*) pulsed array radar
(*c*) pulse amplitude radar (*d*) none of the above

682. In MTI radar, clutter means
(*a*) echos from moving targets
(*b*) echos from stationary targets
(*c*) atmospheric noise
(*d*) thermal noise

683. Tracking of extra terrestrial objects requires
(*a*) high transmitting power
(*b*) very sensitive receiver
(*c*) fully steerable antenna
(*d*) all the above

684. Consider the following applications/advantages of radar beacons
1. navigation
2. target identification
3. accurate tracking of enemy targets
4. large increase in maximum range

Which of the above are correct?
(*a*) 1, 2, 3, 4 (*b*) 1, 2, 4
(*c*) 1, 2 (*d*) 2, 3, 4

685. Consider the following advantages of phased array radar
1. simple circuitry
2. ability to track and scan simultaneously
3. ability to track many targets together
4. very fast scanning

Which of the above are correct?
(*a*) 1, 2, 3 (*b*) 1, 3, 4
(*c*) 2, 3, 4 (*d*) 1, 4 only

686. A 12 bit PCM system performance is better than 8 bit PCM because
(*a*) noise is lower and is proportional to reciprocal of bandwidth
(*b*) bandwidth is larger and detection is easier
(*c*) quantization noise is lesser other things equal
(*d*) quantization noise is nearly zero since 2^{12} levels are available **[IES 2010]**

687. The spectrum of a band pass signal spans 20 kHz to 30 kHz. The signal can be recovered ideally from the sampled values when the sapling rate is at least
(*a*) 20 kHz (*b*) 60 kHz
(*c*) 50 kHz (*d*) 10 kHz **[IES 2010]**

688. The Nyquist rate for the signal $x(t) = 2\cos(2000\,\pi t)\cos(5000\,\pi t)$ is
(*a*) 7 kHz (*b*) 5 kHz
(*c*) 14 kHz (*d*) 10 kHz **[IES 2010]**

689. Consider the following statements regarding PLL
1. Capture range of PLL primarily depends on the slope of transition band edge of low pass filter
2. Lock range is not affected by slope of transition band of low pass filter

Which of the above are correct?
(*a*) both 1 and 2 (*b*) 1 only
(*c*) 2 only (*d*) neither 1 nor 2
[IES 2010]

690. Quantization noise of PCM depends on
(*a*) number of quantization levels
(*b*) step size
(*c*) both number of quantization levels and step size
(*d*) sampling rate **[IES 2010]**

691. A TDM system is to be designed to multiplex the following two signals
$x_1 = 5\cos(2000\,\pi t)$
$x_2 = 2\cos(2000\,\pi t)\cos(3000\,\pi t)$
The minimum sampling rate is
(*a*) 4 kHz (*b*) 5 kHz
(*c*) 10 kHz (*d*) 6 kHz **[IES 2010]**

692. Consider the following statements
1. A multiplexer is analogus to rotary switch
2. A decoder is a combinational logic circuit which converts binary information from *n* input lines to a maximum of 2*n* distinct elements at output
3. Boolean expression of the output difference '*d*' from a full subtracter is the same as the output sum '*s*' from a full adder

Which of the above are correct?
(*a*) 2 and 3 (*b*) 3 only
(*c*) 1 and 3 only (*d*) 1, 2 and 3 **[IES 2010]**

693. In a typical satellite communication system which of the following could be uplink and downlink frequencies
(*a*) 40 GHz and 60 GHz (*b*) 60 GHz and 40 Ghz
(*c*) 6 GHz and 4 GHz (*d*) 4 GHz and 6 GHz
[IES 2010]

694. The output stage of a transponder on board a satellite has a maximum power output of 10 W. However it is not operated at maximum power output to
(*a*) conserve available limited battery power
(*b*) reduce noise due to device
(*c*) avoid inter modulation distortion
(*d*) avoid heating of satellite **[IES 2010]**

695. Encoder
(*a*) assigns quantized values
(*b*) changes quantized values to binary values
(*c*) changes quantized values to numerical values
(*d*) changes numerical values to binary values **[IES 2010]**

696. A single mode fibre does not suffer from which type of dispersion?
(*a*) wave guide dispersion (*b*) material dispersion
(*c*) intermodel dispersion
(*d*) polarization mode dispersion **[IES 2010]**

697. A balanced modulator is used in the generation of
(*a*) DSB–SC signal (*b*) FM signal
(*c*) PM signal (*d*) PAM signal **[IES 2010]**

698. An amplitude modulated signal occupies a frequency range of 395 kHz to 405 kHz. It can be demodulated by which of the following :
(*a*) Using an envelope detector and filter
(*b*) Multiplying with a 395 kHz local signal
(*c*) Multiplying with a 405 kHz local signal
(*d*) Low pass filter with cut off at 400 kHz **[IES 2009]**

699. A signal occupies the band 5 kHz to 10 kHz. For proper error free reconstruction the sampling should be at
(*a*) 10 kHz (*b*) 20 kHz
(*c*) 5 kHz (*d*) $(10 + 5) \times 2$ kHz **[IES 2009]**

700. The special range of a band pass signal extends from 10 MHz to 10.4 MHz. What is the minimum sampling frequency required for reconstruction
(*a*) 20 MHz (*b*) 20.8 MHz
(*c*) 20.4 MHz (*d*) 10.8 MHz **[IES 2009]**

701. For good quality signal transmission all frequency components should have the same transmission delay t_d and same phase shift ϕ_s. What can be said about this statement?
(*a*) correct
(*b*) true for t_d but not for ϕ_s
(*c*) Not true for t_d but true for ϕ_s
(*d*) Both t_d and ϕ_s are not involved in quality **[IES 2009]**

702. Which of the following introduces mode partition noise?
(*a*) coaxial line (*b*) wave guide
(*c*) fibre transmission line
(*d*) both coaxical line and wave guide **[IES 2009]**

703. Which of the following does not causes losses in optical fibre cables?
(*a*) impurities (*b*) microbending
(*c*) attenuation in glass
(*d*) stepped index operation **[IES 2009]**

704. On which bands do optical fibre cables operate?
1. ultra violet **2.** ultra high frequency
3. visible light **4.** infra red
Select your answer using following codes
(*a*) 1 only (*b*) 1 and 2 only
(*c*) 1, 2, 3 only (*d*) 1, 3, 4 only **[IES 2009]**

705. Which of the following photodetectors does not provide gain?
(*a*) photo transistor (*b*) photo conductor
(*c*) Avalanche photodiode (*d*) *p-i-n* photodiode **[IES 2009]**

706. Consider the following statement :
In space wave propagation, the signal strengths at receivers is
1. directly proportional to transmitter and receiver heights
2. inversely proportional to square of distance between transmitter and receiver
3. directly proportional to frequency
Which of the above are correct?
(*a*) 1, 2 and 3 (*b*) 1 and 3
(*c*) 2 and 3 (*d*) 3 only **[IES 2009]**

707. Which was the first commercial geostationary communication satellite?
(*a*) INTELSAT I (*b*) ECHO
(*c*) INSAT I–A (*d*) SPUTNIK **[IES 2009]**

708. What does a link budget for satellite communication include?
(*a*) total cost of satellite
(*b*) cost of satellite plus launch service
(*c*) signal and noise levels in dB
(*d*) margin of error permitted **[IES 2009]**

709. In optical fibre the light beam propagates due to
(*a*) simple reflection of light at a boundary between two media
(*b*) refraction of light in the medium
(*c*) total internal reflection at the boundary of the fibre
(*d*) scattering of hight in the medium **[IES 2008]**

710. In optical fibre used in a communication link, dispersion is of which type
(*a*) angular (*b*) modal
(*c*) chromatic
(*d*) dispersion due to standard irregularities in fibre **[IES 2008]**

711. Source S_1 produces 4 discrete symbols with equal probability
Source S_2 produces 6 discrete symbols with equal probability
If H_1 and H_2 are entropies of source S_1 and S_2, then which of the following is correct?
(*a*) H_1 is always less than H_2
(*b*) H_1 is always greater than H_2
(*c*) H_1 is always equal to H_2
(*d*) H_2 is 1.5 times H_1 **[IES 2009]**

712. The entropy of a digital source is 2.7 bit/symbol. It is producing 100 symbols per second. The source is
(*a*) binary source (*b*) quatenmary source
(*c*) octal source (*d*) hexadecimal source
[IES 2008]

713. Match List I and List II and give correct answer using the given codes

	List I		List II
A.	Shot noise	1.	Noise generated in resistor
B.	Thermal noise	2.	Power spectral density is independent of frequency
C.	White noise	3.	Temperature limited diode
D.	Narrow band noise	4.	Noise at the output of filter

Codes:

	A	B	C	D
(*a*)	2	4	3	1
(*b*)	3	1	2	4
(*c*)	2	1	3	4
(*d*)	3	4	2	1

714. If the value of resistor creating thermal noise is doubled, the noise generated is
(*a*) halved (*b*) doubled
(*c*) unchanged (*d*) slightly changed
[IES 2008]

715. Which of the following pulse modulations are digital?
1. PCM **2.** Differential PCM
3. PWM
Select your answer using the following codes :
(*a*) 1 and 2 only (*b*) 2 and 3 only
(*c*) 1 and 3 only (*d*) 1, 2 and 3 **[IES 2008]**

716. Which one is the indirect way of generating FM?
(*a*) Reactance FET modulator
(*b*) Varactor diode modulator
(*c*) Armstrong modulator
(*d*) Reactance tube modulator **[IES 2008]**

717. Multiplexing is possible if two signals are sampled. Two signals have bandwidth 0 to 4 kHz and 0 to 8 kHz. The sampling frequency is chosen to be 12 kHz.
The choice of sampling frequency
(*a*) is correct since the two signals have an integral relationship of 2
(*b*) will not lead to aliasing
(*c*) does not obey sampling theorem
(*d*) can never lead to multiplexing **[IES 2008]**

718. In TDM each signal is allotted a fixed
(*a*) frequency slot (*b*) time slot
(*c*) amplitude slot (*d*) phase slot **[IES 2008]**

719. Match List I and List II and select the correct answer using the codes :

	List I		List II
A.	Open wire	1.	100 kHz
B.	Twisted pair	2.	100 MHz
C.	Coaxical pair	3.	100 GHz
D.	Wave guide	4.	5 kHz

Codes:

	A	B	C	D
(*a*)	4	3	2	1
(*b*)	2	1	4	3
(*c*)	2	3	4	1
(*d*)	4	1	2	3

[IES 2008]

720. Match List I and List II and select the answer using the codes :

	List I		List II
A.	IF amplifier	1.	Demodulator
B.	Mixer	2.	Narrow band tuned amplifier
C.	Detector	3.	Variable gain amplifier
D.	AGC	4.	Frequency conversion

Codes:

	A	B	C	D
(*a*)	1	4	2	3
(*b*)	2	3	1	4
(*c*)	1	3	2	4
(*d*)	2	4	1	3

[IES 2008]

721. FM broadcast standards specify a maximum deviation of frequency to be equal to 75 kHz and a maximum modulating frequency of 15 kHz. What is modulation index?
(*a*) $\frac{1}{5}$ (*b*) 5
(*c*) 60 (*d*) 1125 **[IES 2007]**

722. In order to permit the selection of 1 out of 16 probable events. What is the number of bits required?
(*a*) 8 (*b*) 4
(*c*) $\log_{10} 16$ (*d*) 2 **[IES 2007]**

723. What does FM radio station perform better than AM station radiating the same power?
(*a*) FM is immune to noise
(*b*) AM has only two side bands while FM has more
(*c*) FM uses larger bandwidth for larger modulation depth
(*d*) Capture effect appears in FM **[IES 2007]**

724. What is the typical bandwidth of a 2-wire telephone line?
(*a*) 300 Hz to 3.4 kHz (*b*) 20 Hz to 4 kHz
(*c*) dc to 4 kHz (*d*) 20 Hz to 20 kHz
[IES 2007]

725. On modulating a carrier of frequency f_c by an audio signal f_s the following components have resulted
What is the type of modulation?
$f_c, f_c + f_s, f_c - f_s$
(*a*) Amplitude modulation
(*b*) Single side band modulation
(*c*) Frequency modulation
(*d*) (*a*) or (*c*) **[IES 2007]**

726. Modulation index of an AM wave is changed from 0 to 1. How is transmitted power changed?
(*a*) Gets halved (*b*) Gets doubled
(*c*) Gets increased by 50% (*d*) Remains uncharged
[IES 2007]

727. Which one is correct?
(*a*) Coding reduces the noise in the signal
(*b*) Coding introduces redundancy into messages

(*c*) Coding increases information
(*d*) Coding increases channel bandwidth **[IES 2007]**

728. In PCM system if we increase quantization level from 2 to 8 how does relative bandwidth requirement change?
(*a*) gets doubled (*b*) gets tripled
(*c*) remains same (*d*) becomes 8 times
[IES 2007]

729. How can quantizing noise be reduced
(*a*) by using de-emphasis circuit
(*b*) by using RF amplifier in receiver
(*c*) by increasing number of quantization levels
(*d*) by sending sloping pulses **[IES 2007]**

730. Match List I (type) units List II (application) and select the correct answer using the codes :

List I	List II
A. Speech signal	1. Received signal of a radar system monitoring variation in prevalent weather condition
B. Non-stationary	2. One dimensional signal where amplitude varies with time
C. Random signal	3. Signals of coupled system of non-linear difference
D. Chaotic signal	4. Ensemble of unpredictable waveform

Codes:

	A	B	C	D
(*a*)	2	1	4	3
(*b*)	4	3	2	1
(*c*)	2	3	4	1
(*d*)	4	1	2	3

[IES 2006]

731. Match List I (application of signals) with List II (definition) and select the correct answer using the codes :

List I	List II
A. Reconstruction	1. Sampling rate is chosen significantly greater than Nyquist rate
B. Over sampling	2. A mixture of continuous and discrete time signals
C. Interpolation	3. To convert discrete time sequence back to a continuous time signal and then resample
D. Decimation	4. Assign values between samples and signals

Codes:

	A	B	C	D
(*a*)	3	4	1	2
(*b*)	2	1	4	3
(*c*)	3	1	4	2
(*d*)	2	4	1	3

[IES 2006]

732. A communication channel has a bandwidth of 100 MHz. Channel is very noisy and signal power is very less than noise power. What is channel capacity?
(*a*) 100 Mbps (*b*) 50 Mbps
(*c*) 2400 bps (*d*) nearly zero
[IES 2006]

733. In general attenuation per unit length in coaxial cable
(*a*) increases with frequency
(*b*) decreases with frequency
(*c*) remains constant irrespective of frequency
(*d*) depends on type of cable **[IES 2006]**

734. What is half power beam width of a paraboloid of 1 metre diameter radiating 0.1 m wavelength
(*a*) 70° (*b*) 7°
(*c*) 35° (*d*) 78° **[IES 2006]**

735. What is directivity of isotropic antenna
(*a*) 1 dB (*b*) 0 dB
(*c*) 4π dB (*d*) 0.5π dB **[IES 2006]**

736. A paramidal horn antenna is best suited to be excited from
(*a*) microstrip line (*b*) microwave waveguide
(*c*) coaxial line (*d*) parallel twin line
[IES 2006]

737. A paraboloid of revolution is to have a power gain of 1000 at λ = 10 cm. What is the diameter of antenna?
(*a*) 129.1 m (*b*) 12.91 m
(*c*) 1.291 m (*d*) 0.1291 m **[IES 2006]**

738. A continuous signal has a voltage range of –2 V to 2 V. If this is quantized to 8 bits what does the resulting signal have
(*a*) 256 levels of step size 4/256
(*b*) 25 levels of step size 4/8
(*c*) 8 bits plus 2 levels at –2 V and + 2 V
(*d*) any of above depending on method of quantization
[IES 2005]

739. Maxwell's theorems are obeyed by EM wave
(*a*) only in free space
(*b*) only in water and free space
(*c*) only in ionosphere, water and free space
(*d*) in all solids, liquids, gases and all modes **[IES 2005]**

740. In a BPSK signal detector the local oscillator has a fixed phase error of 20°. By what factor does the phase error deteriorate SNR at the output?
(*a*) $\cos^2 20°$ (*b*) cos 20°
(*c*) cos 40° (*d*) cos 70° **[IES 2004]**

741. Consider an angle modulated signal
$x(t) = 6\cos[2\pi \times 10^6 t + 2\sin(8000\pi t) + 4\cos 8000\pi t]$ V. The average power of $x(t)$ is
(*a*) 10 W (*b*) 18 W
(*c*) 20 W (*d*) 28 W **[Gate 2010]**

742. Nyquist sampling rate for signal

$$s(t) = \frac{\sin(500\pi t)}{\pi t} \times \frac{\sin(700\pi t)}{\pi t} \text{ is}$$

(*a*) 400 Hz (*b*) 600 Hz
(*c*) 1200 Hz (*d*) 1400 Hz **[Gate 2010]**

743. A fully charged mobile phone with 12 V battery is good for 10 minutes talk time. During talk time battery delivers

2 A and its voltage drops linearly from 12 V to 12 V. How much energy does the battery deliver?

(*a*) 220 J (*b*) 12 kJ
(*c*) 13.2 kJ (*d*) 14.4 kJ **[Gate 2009]**

744. For a message signal $m(t) = \cos(2\pi f_m t)$ and carrier of frequency f_c which of the following represents single side band (SSB) signal

(*a*) $\cos(2\pi f_m t)\cos(2\pi f_c t)$
(*b*) $\cos(2\pi f_c t)$
(*c*) $\cos 2\pi (f_c + f_m) t$
(*d*) $[1 + \cos(2\pi f_m t)]\cos(2\pi f_c t)$ **[Gate 2009]**

745. A discrete random variable takes value from 1 to 5 with probabilities as shown below

K	1	2	3	4	5
P $(x = k)$	0.1	0.2	0.4	0.2	0.1

A student calculates mean as 3.5 and teacher calculates variance as 1.5. Which of the following is correct?

(*a*) both student and teacher are right
(*b*) both student and teacher are wrong
(*c*) student is wrong, teacher is wrong
(*d*) student is right, teacher is wrong **[Gate 2009]**

746. A communication channel with AWGN operating at a SNR >>1 and bandwidth B has capacity C_1. If SNR is doubled keeping B constant, capacity C_2 is given by

(*a*) $C_2 = 2C_1$ (*b*) $C_2 = C_1 + B$
(*c*) $C_2 = C_1 + 2B$ (*d*) $C_2 = C_1 + 0.3B$ **[Gate 2009]**

747. If signal to quantization noise ratio required in uniformly quantizing the signal is 43.5 dB, the step size of quantization is approximately.

[The amplitude of signal is uniformly distributed between − 5 V and +5 V.]

(*a*) 0.0333 V (*b*) 0.05 V
(*c*) 0.0667 V (*d*) 0.1 V **[Gate 2009]**

748. The amplitude of a signal is uniformly distributed between −5 V and 5 V. If the positive values of signal are uniformly distributed with step size of 0.05 V and negative values are uniformly distributed with step size of 0.1 V, the resulting signal to noise quantization ratio is about

(*a*) 46 dB (*b*) 43.8 dB
(*c*) 42 dB (*d*) 40 dB **[Gate 2009]**

749. Consider amplitude modulated signal $A_c \cos\omega_c t + 2\cos\omega_m t\cos\omega_c t$. For demodulating the signal using envelope detector the minimum value of A_c is

(*a*) 2 (*b*) 1
(*c*) 0.5 (*d*) 0 **[Gate 2008]**

750. A memoryless source emits n symbols each with probability p. The entropy of source as a function of n

(*a*) increases as log n (*b*) decreases as log n
(*c*) increases as n (*d*) increases as n log n **[Gate 2008]**

751. Four messages band limited to W, W, 2 W and 3 W respectively are to be multiplexed using TDM. The minimum bandwidth required is

(*a*) W (*b*) 3 W
(*c*) 6 W (*d*) 7 W **[Gate 2008]**

752. The signal $\cos\omega_c t + 0.5\cos\omega_m t\cos\omega_c t$ is

(*a*) FM only (*b*) AM only
(*c*) both FM and AM (*d*) neither AM nor FM **[Gate 2008]**

753. During transmission over a certain binary communication channel, bit error occur independently with probability p. The probability of at most one bit in error in a block of n bits is

(*a*) p^n
(*b*) $1 - p^n$
(*c*) $np(1-p)^{n-1} + (1-p)^n$
(*d*) $1-(1-p)^n$ **[Gate 2007]**

754. In a GSM 8 channels coexist in 200 kHz bandwidth using TDMA. AGSM operator is allotted 5 MHz bandwidth. Assuming frequency reuse factor of 1/5, a five cell repeat pattern the maximum number of simultaneous channels which can coexist in one cell is

(*a*) 20 (*b*) 40
(*c*) 25 (*d*) 5 **[Gate 2007]**

755. A zero mean white Gaussian noise is passed through an ideal low pass filter of bandwidth 10 kHz. The output is then uniformly sampled with sampling period $t_s = 0.03$ m sec. The samples so obtained would be

(*a*) corelated
(*b*) statistically independent
(*c*) uncorelated (*d*) orthogonal **[Gate 2006]**

756. A source generates three symbols with probabilities 0.25, 0.25 and 0.5 at a rate of 3000 symbols per second. Assuming independent generation of symbols the most efficient source encoder would have average bit rate

(*a*) 6000 bits/sec (*b*) 4500 bits/sec
(*c*) 3000 bits/sec (*d*) 1500 bits/sec **[Gate 2006]**

757. Refractive index of glass is 1.5. Find wavelength of a beam of light with frequency 10^{14} Hz in glass. Speed of hight 3×10^8 m/s in vacuum

(*a*) 3 μm (*b*) 2.5 μm
(*c*) 2 μm (*d*) 1 μm **[Gate 2005]**

758. A device with input $x(t)$ and output $y(t)$ is characterised by $y(t) = x^2(t)$. An FM signal with frequency deviation of 90 kHz and modulating signal bandwidth of 5 kHz is applied to this device. The bandwidth of output signal is

(*a*) 370 kHz (*b*) 190 kHz
(*c*) 380 kHz (*d*) 95 kHz **[Gate 2005]**

759. Noise with uniform power spectral density of N_0 W/Hz is passed through a filter with $H(\omega) = 2\exp(-j\omega t_d)$ followed by an ideal low pass filter of bandwidth B Hz. The output noise power in water is

(*a*) $2N_0B$ (*b*) $4N_0B$
(*c*) $8N_0B$ (*d*) $16N_0B$ **[Gate 2005]**

760. A 1 mW video signal having a bandwidth of 100 MHz is transmitted to a receiver through a cable which has 40 dB loss. If effective one sided noise spectral denisity at the receiver is 10^{-20} watts/Hz. What is SNR at receiver?

(*a*) 50 dB (*b*) 30 dB
(*c*) 40 dB (*d*) 60 dB **[Gate 2009]**

B. The following questions have two statements Assertion (A) and Reason (R). Examine them and answer as per following codes

(*a*) Both A and R are correct and R is correct explanation of A
(*b*) Both A and R are correct but R is not correct explanation of A
(*c*) A is correct but R is wrong
(*d*) A is wrong but R is correct

761. **A.** In AM the amplitude of carrier is made proportional to the instantaneous amplitude of modulating wave.
R. In AM, frequency of side band = frequency carrier ± frequency of modulating wave.

762. **A.** If a carrier is modulated by three sine waves and m_1, m_2, m_3 are the modulation indices, the total modulation index m_t is
$m_t = (m_1^2 + m_2^2 + m_3^2)^{0.5}$
R. An AM, higher the level of modulation, lower is the audio power required to produce modulation.

763. **A.** In AM carrier component remains constant in amplitude and frequency, no matter what the modulating voltage does.
R. In AM the two side bands are images of each other.

764. **A.** In AM, use of SSB reduces the power and bandwidth requirement
R. A balanced modulator produces AM with carrier suppressed.

765. **A.** In SSB, carrier is suppressed by about 45 dB.
R. SSB is standard form for radio communication.

766. **A.** SSB transmitter is generally smaller than AM transmitter.
R. SSB systems mostly use combined transmitter receiver.

767. **A.** SSB systems generally have duplex operation.
R. SSB systems are generally used in two-way communication.

768. **A.** In FM the frequency of the carrier is varied by the modulating voltage.
R. FM and PM are two forms of angle modulation.

769. **A.** In angle modulation the transmitted amplitude is constant.
R. Angle modulation has good immunity to noise and interference.

770. **A.** FM requires more bandwidth than AM.
R. Wide band FM is used for broadcast transmissions.

771. **A.** In FM all power amplifiers can be operated in class C.
R. FM is constant amplitude modulation system.

772. **A.** SSB and ISB receivers have good S/N ratio.
R. SSB and ISB receivers have good suppression of adjacent channels.

773. **A.** Demodulation of SSB and AM signals is exactly similar.
R. The basic SSB demodulation device is product detector.

774. **A.** If frequency is high, the distance between transmitter and antenna may be an appreciable fraction of wavelength.
R. Wavelength is inversely proportional to frequency.

775. **A.** Stubs, directional couplers and baluns are very commonly used in communication systems.
R. Characteristic impedance of a line is an important parameter.

776. **A.** In standing wave pattern, the nodes and antinodes are separated by half wavelength.
R. When line is terminated in resistance R_L, the standing wave ratio is SWR = Z_0/R_L or R_L/Z_0.

777. **A.** A smith chart can be used for a number of calculations of a transmission line.
R. Smith chart is basically a polar impedance diagram.

778. **A.** Free space does not interfere with normai radiation and propagation of radio waves.
R. Free space has no magnetic or graviational fields.

779. **A.** Water waves are longitudinal but *em* waves are transverse.
R. In an *em* wave electric field, magnetic field and propagation are mutually perpendicular.

780. **A.** Electromagnetic waves in free space spread uniformly in all directions from a point source.
R. Free space has no interference or obstacles.

781. **A.** EM waves are attenuated as they travel outward from a source.
R. Both field intensity and power density follow inverse square law.

782. **A.** Radio transmission is by sky waves.
R. Waves having frequencies higher than 30 MHz penetrate the atmosphere and do not return to earth.

783. **A.** In India we use CCI R–B standard.
R. Video signals are frequency modulated.

784. **A.** The colour signal consists of only luminance signal.
R. In India PAL colour signal system is used.

785. **A.** A video monitor does not need RF tuner and IF amplifier.
R. The picture tube in TV is basically a CRT.

786. **A.** In TV scanning is done from left to right and top to bottom.
R. Interlaced scanning is useful in eliminating flicker.

787. **A.** Data communication using digital data must be very accurate.
R. In analog signals redundancy is available but in digital signal redundancy is not available.

788. **A.** In digital communication data is coded prior to being transferred.
R. A number of codes for digital data processing are available.

789. A. The unit of telephone traffic measurement is erlang.
R. International circuits for communication are fitted with echo suppressors.

790. A. Radar gave birth to microwave technology.
R. In radar systems very high frequencies give the most accurate results.

791. A. The maximum range of a radar is proportional to fourth root of the peak transmitted pulse power.
R. Radar signals are subjected twice to inverse square law.

792. A. Many times a radar cannot detect 'grounding hopping' military aircraft.
R. Ground interference limits the range of radar.

793. A. For satellites and space probes, PRF of a radar is pretty low.
R. PRF of a radar depends on the range of a radar.

794. A. A colour TV uses a SAW filter to give stable band pass response and high gain.
R. Devices using SAW principle are used in many systems.

795. A. A composite video signal has synchronising pulses.
R. It is necessary to synchronise the scanning process in the transmitter and receiver.

796. A. CCIR-B system has 625 scanning lines per frame and 2 : 1 interlace.
R. The aspect ratio is 4 : 3.

797. A. Bandwidth of a TV signal is 5 MHz.
R. TV systems use vestigial side band transmission.

798. A. A balun is an RF impedance matching transformer.
R. Sometimes a booster amplifier is used with a TV set.

799. A. RF tuner is called front end of TV receiver.
R. RF tuner receives the TV signal and selects the required channel.

800. A. An electronic tuner uses varactor diode
R. The capacitance of a varactor diode depends on the magnitude of reverse bias.

801. A. Frequency modulation and phase modulation both produce different set of frequency bands for the same modulation depth.
R. Frequency modulation and phase modulation vary the carrier phase angle or its rate. [IES 2010]

802. A. In amplitude modulation system the value of modulation index should be about 1.
R. Power carried in the intelligence carrying side bands increases with modulation index. [IES 2010]

803. A. When coding signals like speech signals a-law or μ-law quantizers are used.
R. A-law and μ-law quantizers occupy smaller bandwidth than uniform quantizers. [IES 2010]

804. A. A PCM/FM systems transmit PCM pulses by modulating a high frequency carrier and hence occupy large bandwidth and eliminate distortion.
R. Large bandwidth ensures SNR tide off and hence distortion-less transmission is ensured. [IES 2010]

805. A. In satellite communication technique frequency reuse effectively doubles the bandwidth and information capacity of satellite.
R. EM waves radiated from a transmitting antenna could be received by a distant receiving antenna by tunning the receiver in terms of frequency and polarization. [IES 2010]

806. A. Entropy of a binary source is maximum if probabilities of occurrence of both events are equal.
R. Average amount of information per source symbol is called entropy for a memory less source. [IES 2008]

807. A. Isolation transformers should always be used when servicing TV receivers without power transformers.
R. TV receivers which use semiconductor rectifiers do not always use power transformers. [IES 2008]

808. A. Pulse broadening is a phenomenon associated with multimode fibres.
R. Multimode fibres have core diameter much larger than in single mode fibres. [IES 2008]

809. A. TDM can be employed to transmit channels having unequal bandwidth.
R. If sampling theorem is strictly followed an analog signal can be reconstructed from its samples. [IES 2006]

810. A. An 8 bit PCM performs better than 6 bit PCM.
R. 8 bit PCM produces smaller quantization noise than 6 bit PCM. [IES 2006]

811. A. The capture range of a PLL is always less than or equal to the look range and is related to cut off frequency of low pass filter in the loop.
R. The capture range is range of frequencies the VCO can produce. [IES 2006]

812. A. In a step index fibre, for single mode operation, the fibre parameters should be tailored such that V- parameter is less than 2.40 dB.
R. Modal fields of step index fibres are represented by Bessel function $J_n(x)$ in the cone and modified Bessel function $k_n(x)$ in the cladding. The first zero for $J_1(x)$ occurs at $x = 2.40$ dB. [IES 2005]

813. A. For optical communication three communication wavelength windows are decided and they are located near 850 nm, 1300 nm and 1550 nm.
R. At 850 nm first semiconductor laser for optical communication was fabricated, at 1300 nm fibre dispersion is minimum and at 1550 nm wavelength multiplexing is possible due to optical amplifiers. [IES 2005]

814. A. AM stereo broadcasting uses quadrature carrier multiplexing (QAM).
R. QAM involves more stringent synchronisation than an FDM system with SSB subcarrier modulation. [IES 2004]

815. A. In free space transmission between a transmitting and receiving antenna a high frequency wave attenuates more rapidly in magnitude than a low frequency wave.
R. For a given gain aperture of a high frequency antenna is smaller than that of low frequency antenna. [IES 2004]

C. Indicate whether the following are True (T) or False (F)

816. Audio and video signals are called modulating waves.
817. A carrier wave carries information.
818. Each frequency in the modulating wave gives rise to four side bands.
819. Sparking affects amplitude modulation.
820. Modulation is also called detection.
821. Sound waves are electromagnetic waves.
822. E and H fields are perpendicular to each other.
823. Speeds of sound and radio waves are equal.
824. A ground wave travels along surface of ground.
825. Waves of frequencies higher than 30 MHz penetrate the atmosphere.
826. Skip distance depends on time of day and angle of incidence.
827. The distance at which a sky wave is received back on earth decreases with angle of incidence for all angles of incidence.
828. Space wave may be a direct wave or wave reflected from earth.
829. Strength of space wave is not dependent on distance.
830. A shadow zone does not receive any space wave.
831. Reception through space wave is limited to line of sight distance.
832. Earth's curvature affects the effective height of antenna.
833. At night D layer of ionosphere is absent.
834. Super high frequency means more than 3000 MHz.
835. Direct and reflected sound waves may interfere with each other.
836. Public address systems generally use electric microphones.
837. As the area of a public meeting increases the amplifier power output required increases.
838. AC hum is picked up from power cables.
839. Audio systems use transverse recording.
840. The gap width in audio heads is about 0.01 mm.
841. In home tape recorders the tape width is 2 inches.
842. Audio tapes have two tracks each 2.5 mm wide.
843. A tape recorder and deck are entirely similar.
844. In tape recorder erase head is active during recording.
845. Pre-emphasis and De-emphasis is done by using R-C circuit.
846. In tape recorders the hissing noise is due to irregularities in magnetic coating on tape.
847. In sound recording the degree of pre-emphasis is about 6 dB/octave.
848. The sampling frequency in CD is 10 kHz.
849. The storage capacity of audio CD is 540 Mb per side.
850. A photo diode converts light signals to binary.
851. The signal plate of vidicon tube is coated with tin oxide.
852. The service area of a TV transmitter is a little beyond radio horizon.
853. In colour TV system the colour subcarrier frequency is 4.43 MHz.
854. A video monitor is exactly similar to TV receiver.
855. Trintron is a monochrome picture tube.
856. In cable TV local programmes can be telecast at any frequency.
857. Full duplex operation permits transmission in both directions at the same time.
858. The name modem is a contraction of terms modulator demodulator.
859. Baudot code is a six bit code.
860. Cross talk is due to reception of portions of a signal from one channel into another.
861. A planar array radar uses a high gain planar array antenna.
862. CW Doppler radar is capable of giving accurate measurements of relative velocities.
863. An MTI compares a set of received echos with those received during the previous sweep.
864. In radar PRF depends on maximum range.
865. A radar beacon is a small radar set having a receiver, a separate transmitter and an antenna.

D. Fill in the Blanks with suitable words.

866. The lowest audible frequency for human beings is ....
867. The wavelength of radio waves for line of sight propagation is .... m to .... m.
868. Frequencies used in ground waves are upto ....
869. The frequencies used in space waves are above ....
870. An AM wave can be expressed in the form of an equation as sum of .... waves.
871. For a fully amplitude modulated wave (m = 1) each side band carries .... of the energy of carrier.
872. In AM waves the energy in side bands .... as m decreases.
873. The .... in frequency modulated wave is less that in AM wave.
874. In TV transmission the picture transmission uses .... modulation and sound transmission uses .... modulation.
875. For depth in sound it is necessary to boost ....
876. For sharpness in sound it is necessary to boost ....
877. When light falls on a photo conductive plate its resistance ....
878. In TV terminology the isolated picture elements are called ....
879. In Indian standards each frame is divided into .... lines.
880. The two sets of 312.5 lines in each TV frame are called .... field and .... field.
881. If the picture on a TV screen is absent, the pattern of lines is called ....
882. In TV each frame is scanned .... times.
883. CCIR stands for ....
884. SAWF stands for ....
885. A resonant antenna has length in multiples of ....
886. In TV the ratio of width to height of picture is called ....

887. The proportion of red, green and blue in luminance signal is .... is and ....

888. PAL stands for ....

889. Video recording uses .... heads.

890. A remote control signal TV has ....

891. If μ and ∈ are permeability and permittivity of a medium, the quantity $\sqrt{\mu/\in}$ is equal to ....

892. The upper portion of atmosphere is called ....

893. The highest frequency which will be returned down to earth by a layer after having been beamed striaght at it is called .... of that layer.

894. The portion of atmosphere within about 15 km of ground is called ....

895. A short dipole is also called ....

896. The purpose of using a coupling network is ....

897. A helical antenna has .... characteristics.

898. An antenna which is a combination of disk and cone is called ....

899. The angle between half power points of the main lobe of radiation pattern of an antenna is called ....

900. When the bandwidth and noise level of a channel are known, the channel capacity is given by .... theorem.

901. PCM requires .... bandwidth compared to analog systems.

902. The process of sending a number of signals together over the same cable simultaneously and without interference is called ....

903. A radar uses a .... antenna.

904. Moving target indicator uses .... effect.

905. A radar beacon is also called ....

906. A radar which sends out continuous sine waves is called .... radar.

907. On a transmission line two sets of waves, *i.e.*, incident and reflected set up an interference pattern called ....

908. The ratio of maximum voltage to minimum voltage on a transmission line is called ....

909. When the load impedance of a transmission line is perfectly matching, SWR is equal to ....

910. The most important use of VLF is ....

ANSWERS WITH HINTS AND EXPLANATIONS

1. (*d*)

2. (*d*) $T = \frac{1}{f} = \frac{1}{20} = 0.05$ s.

3. (*b*) $\lambda = \frac{330}{50} = 6.6$ m.

4. (*c*)

5. (*d*) $v \propto \sqrt{T}$.

6. (*c*) $\frac{v}{330} = \sqrt{\frac{273+40}{273+0}}$ or $v = 353$ m/s.

7. (*c*) **8.** (*c*)

9. (*b*) **10.** (*d*) **11.** (*b*)

12. (*b*) See Table 12.1.

13. (*b*) See Table 12.1 **14.** (*c*) **15.** (*a*)

16. (*d*) During night D layer is almost absent. Moreover F_1 and F_2 layers merge to form single F_2 layer. Hence only E and F_2 layers at night.

17. (*c*) The reception of signals by space waves is governed by line of sight range limited to about 75 kms or so. For frequencies above 30 MHz (generally used for television) communication is through space waves.

18. (*a*)

19. (*b*) Television transmission is governed by line of sight distance which is about 100 km or less.

20. (*b*) Frequency modulation gives noise-free reception and is used for radio broadcasts.

21. (*a*) Because of large bandwidth requirement, modulation of video signals is possible only by amplitude modulation.

22. (*a*) Average value = 10 V and $m = \frac{16-10}{10} = 0.6$.

23. (*b*) The instantaneous frequency variation is equal to frequency of audio signal.

24. (*c*) Since the intensity of audio wave has been increased five times, the instantaneous frequency variation is also increased five times.

25. (*a*) 100 Hz is taken as standard for specifying the output.

26. (*c*) 0.1 Pa ($= 0.5 \times 10^{-4}$ W/m^2) is taken as standard sound pressure for defining S/N ratio of microphones.

27. (*d*) $(20)(4)^2 = 320$ ohms.

28. (*c*) Directivity $= 20 \log D = 20 \log \frac{0.5 \times 10^{-3}}{0.5 \times 10^{-6}}$

$= 20 \log 1000 = 60$ dB.

29. (*d*) Carbon microphone has the least cost.

30. (*c*) Condenser microphone is taken as standard for calibration (see Table 12.2).

31. (*d*) See Table 12.2.

32. (*d*) See Table 12.2.

33. (*b*) See Table 12.2.

34. (*b*) See Table 12.2.

35. (*a*) See Table 12.2.

36. (*c*) In moving coil microphone variations of sound pressure cause an emf to be induced in the moving coil.

37. (*d*) As per piezoelectric effect application of pressure across the faces of a crystal produces an emf across the faces.

38. (*a*) See Section 12.3.

39. (*a*) See Section 12.12.1.

40. (*a*) Since signal power = 0.5 P_C and total power is 1.5 P_C, efficiency is 33.3%.

41. (*d*) Alnico produces a strong magnetic field.

42. (*c*) Mass has inertia and is analogous to inductance.

43. (*d*)

44. (*b*) $1 + \frac{1000}{10^6} = 1.001$ MHz and $1 + \frac{2000}{10^6} = 1.002$ MHz.

45. (*b*) See section 12.12.3.

46. (*a*) See Section 12.3.

47. (*a*) Woofer is a low frequency loudspeaker covering the range 16 Hz to 500 Hz.

48. (*b*) Tweeter is high frequency loudspeaker covering the range above 5000 Hz.

49. (*a*) Series inductance and shunt capacitance is a low pass circuit.

50. (*b*) Series capacitance and shunt inductance is high pass circuit.

51. (*c*) See Table 12.2.

52. (*a*) See Section 12.3.

53. (*b*) A mixer circuit has variable resistance in series with each microphone to control intensity of sound.

54. (*c*) **55.** (*b*) **56.** (*a*)

57. (*c*) Common emitter connection has high power gain, and good current and voltage gains.

58. (*a*) CE connection has good current and voltage gains. Therefore power gain is very high.

59. (*b*) CC circuit has voltage gain of less than 1 and is used for impedance matching.

60. (*b*) CC connection has low output impedance and is suitable for impedance matching.

61. (*c*) CC connection has very high input impedance and low output impedance.

62. (*b*) CB connection has high output impedance and is used for radio frequency isolation.

63. (*a*) **64.** (*a*)

65. (*c*) For zero frequency capacitive reactance is infinite.

66. (*a*) X_C for ac signal should be low so that the attenuation is minimum.

67 (*a*) X_C for ac frequencies is very low and therefore emitter is nearly at ground potential.

68. (*b*) **69.** (*a*) **70.** (*a*) **71.** (*a*)

72. (*c*) Bass means low frequency tones and treble means high frequency tones.

73. (*c*) Push pull is a power amplifier.

74. (*a*) In push pull circuit each transistor conducts for 180° so that complete conduction for 360° occurs. Hence minimum distortion and maximum efficiency.

75. (*a*) It is a class B amplifier.

76. (*c*) $\frac{50}{1 + 50 \times 0.1} = 8.3333$ and $\frac{40}{1 + 40 \times 0.1} = 8$. Change in gain in about 4%.

77. (*c*) $\frac{200}{1 + 200 \times 0.2}$.

78. (*b*) Hilbert transform is used in all the applications.

79. (*a*)

80. (*b*) RF amplifiers use resonant loads (L and C in parallel).

81. (*a*)

82. (*b*) Series capacitance has high reactance for low frequencies.

83. (*a*) Since X_C for high frequencies is low, the high frequencies are shunted to ground and are not transmitted.

84. (*a*) All are correct.

85. (*c*) A major diatonic scale has 7 intervals between one octave.

86. (*a*)

87. (*b*) Since there are 12 intervals in one actave, hence $(2)^{1/12}$.

88. (*c*) Chromium dioxide has very good retentivity.

89. (*a*) **90.** (*b*)

91. (*b*) When width of air gap is 0.5 λ, output is maximum.

92. (*c*) **93.** (*c*)

94. (*c*) Home tape recorders use $\frac{1}{4}$ in, (*i.e.*, 6.3 mm) tape having two tracks of 2.5 mm width each with an in between separation of 1.3 mm.

95. (*b*) Each track is 2.5 mm wide.

96. (*a*)

97. (*a*) Common head and common amplifier is used.

98. (*d*) It is engaged during recording to erase the previous matter.

99. (*a*) **100.** (*c*) **101.** (*a*) **102.** (*a*)

103. (*a*) **104.** (*a*)

105. (*b*) For reasons of economy only one motor is used.

106. (*b*)

107. (*a*) RF waves (about 75 kHz frequency) are fed to head for erasing.

108. (*d*) **109.** (*a*)

110. (*b*) Small variation (less than 6%) in motor speed causes vow and large variation (more than 6%) in motor speed causes flutter.

111. (*a*)

112. (*b*) Dolby A system divides the frequency in 4 bands viz. $f < 80$ Hz, 80 Hz $< f <$ 2999 Hz, $f >$ 3000 Hz, $f >$ 9000 Hz.

113. (*b*)

114. (*a*) $\sqrt{0.6^2 + 0.8^2} = 1$.

115. (*b*) CD uses optical recording and retrieval.

116. (*c*) **117.** (*b*)

118. (*c*) For proper sampling and faithful reproduction sampling frequency must be at least twice the highest frequency in the signal.

119. (*b*) **120.** (*d*)

121. (*c*) Since the circumference of outer tracks is more than that of tracks near the centre, the speed of disc is varied from 200 rpm to 500 rpm.

122. (*a*) It is linear from centre to edge.

123. (*c*) **124.** (*d*)

125. (*b*) Calling rate is the number of calls per subscriber.

126. (*b*) Hi-Fi system has a $\frac{S}{N}$ ratio at least equal to 50 dB.

127. (*d*) **128.** (*c*) **129.** (*d*)

130. (*a*) Power is proportional to V^2.

131. (*b*) **132.** (*a*)

133. (*d*) The useful audio range is 40 Hz to 15 kHz. Therefore natural frequency should be outside this range.

134. (*c*) Left and right channels.

135. (*d*)

136. (*a*) See Section 12.2.

137. (*a*) For depth in sound the low frequency notes, (*i.e.*, bass) need more amplification.

138. (*c*) Dilution of the effect of any channel to give balanced output from left and right channels is called blend control.

139. (*a*) **140.** (*c*)

141. (*a*) Stereo systems have 3 or 4 bands at 300 Hz, 1000 Hz, 10000 Hz etc. in graphic equaliser system.

142. (*d*) **143.** (*c*) **144.** (*a*)

145. (*b*) Grid G_2 accelerates the electrons. Grid G_3 is focus grid and slows the electrons so that they fall perpendicularly on the target at a low speed without causing secondary emission.

146. (*b*) Light dependent resistor (LDR) has high resistance in darkness and low resistance in brightness.

147. (*d*) **148.** (*a*)

149. (*b*) Each frame is divided into odd and even fields. Each field is scanned 25 times so that each frame is scanned 50 times.

150. (*a*)

151. (*b*) Each field is scanned 25 times but alternately.

152. (*a*) Aspect ratio is the width/height ratio of TV receiver. It is $\frac{4}{3}$.

153. (*a*)

154. (*c*) Each channel is allowed a bandwidth of 7 MHz.

155. (*d*) TV signal strength has a reference of 1 $\frac{\mu V}{m}$.

156. (*a*) **157.** (*c*)

158. (*c*) Z_0 for twin wire cable is about 300 ohms while that for coaxial cable is about 75 ohms.

159. (*c*) **160.** (*d*)

161. (*a*) Balun gives 4 : 1 impedance transformation and can be used to interconnect twin wire feeder cable to coaxial cable.

162. (*a*)

163. (*a*) Shot noise, partition noise and thermal noise.

164. (*a*) Video signal is applied to cathode.

165. (*c*)

166. (*c*) EHT voltage is 1 kV for each diagonal inch of CRT screen.

167. (*d*)

168. (*c*) $Y = 0.3\,R + 0.59\,G + 0.11\,B$.

169. (*d*) **170.** (*a*) **171.** (*a*)

172. (*d*) VCR has two rotating leads about 180° apart.

173. (*a*) **174.** (*b*) **175.** (*a*)

176. (*d*) Rotary transformer, in VCR, has one stationary and one rotating winding.

177. (*c*) **178.** (*c*) **179.** (*c*)

180. (*c*) Tilting of heads is done to reduce cross-talk.

181. (*a*) **182.** (*a*)

183. (*d*) $2^{10} = 1024$.

184. (*b*) Infra red LED transits the pulse. At receiver photo diode converts infra red signals into electrical signals.

185. (*b*) Infra red light is used in TV remote control.

186. (*b*) The first pulse is preliminary pulse followed by start pulse.

187. (*a*) The 14 pulses are preliminary pulse, start pulse, stop pulse and 11 other pulses.

188. (*a*)

189. (*c*) The range depends on quality of instrument and is about 100 m.

190. (*a*) Separate frequencies for tramission from base and portable units allows two-way transmission and is called duplex arrangement.

191. (*b*) Base unit is fed from ac mains and portable unit from battery.

192. (*c*) Rechargable cells feed the portable unit.

193. (*a*) Mobile telephone has about 50 km range.

194. (*a*)

195. (*a*) Each cell is designed for 45 two-way conversations.

196. (*a*) Cellular telephones have been assigned 666 channels (826 to 845 MHz and 870 to 890 MHz). Each channel has width of 30 kHz.

197. (*d*)

198. (*d*) Rectifier is needed to charge the battery. Inverter is needed to feed the loads during power failure.

199. (*a*) **200.** (*a*)

201. (*a*) Photo conduction uses the fact that resistivity of some materials depends on intensity of light falling on their surface.

202. (*c*)

203. (*d*) See section 12.9.4.

204. (*c*)

205. (*a*) Frequency shift keying (FSK) is a system of frequency modulation used in telegraphy.

206. (*a*) $f_m = \dfrac{f_c}{\cos\theta} = \dfrac{f_c}{\cos 60^o} = 2f_c$.

207. (*a*) **208.** (*b*)

209. (*d*) Odd number of lines per frame makes interlacing easier.

210. (*b*) **211.** (*b*)

212. (*a*) PCM requires large bandwidth.

213. (*c*) In delta modulation the recovered waveform may be dc whereas the original signal is not dc. This noise is called granular noise.

214. (*d*) Helical antenna produces a circularly polarized radiation if its two components are at right angles and are equal. It is very commonly used in radio telemetry.

215. (*d*)

216. (*c*) When $m = 1$, $P_t = 1.5\ P_c$ as discussed in Section 12.3.

217. (*c*)

218. (*d*) 1000 ± 0.3 and 1000 ± 2 kHz will be present.

219. (*d*) See Table 12.1.

220. (*b*)

221. (*b*) TV uses bandwidth of 5 MHz and Telegraphy has bandwidth of 1 kHz.

222. (*c*) $P_t = P_c\left(1 + \dfrac{0.8^2}{2}\right) = 1.32\ P_c$. Power in side bands $= 0.32\ P_c$. Saving $= \dfrac{1.32 - 0.32}{1.32} \times 100 = 75.76\%$.

223. (*b*) Bandwidth $= \dfrac{1}{2}(10 \times 7) = 35$ kHz.

224. (*d*) Modulation pattern on CRO will show peak-to-peak value $= 2(12 + 6) = 36$ kV.

225. (*d*) Peak RF voltage delivered to load $= 2(8 + 4) = 24$ kV.

226. (*a*) Channel capacity $= 4000 \log_2(1 + 15) = 16000$ bits/s.

227. (*d*) See Section 12.1.

228. (*d*) $10 + \dfrac{10-1}{10} = 10.9$.

229. (*b*) See section 12.2.

230. (*a*)

231. (*c*) See Section 12.2.

232. (*a*)

233. (*c*) Noise figure, noise resistance and noise temperature are useful quantities. Input noise voltage alone is not sufficient unless gain of amplifier stages is also known.

234. (*c*)

235. (*c*) SNR changes in the ratio of square of change in bandwidth.

236. (*c*) **237.** (*a*)

238. (*d*) Depending on which logic is used.

239. (*b*) Balun provides a 4 : 1 Impedance transformation.

240. (*b*) SWR is infinite if $|\rho_v| = 1$.

241. (*c*)

242. (*d*) Earth's curvature limits line of sight distance. Therefore repeaters are needed.

243. (*b*) Colour burst consists of a short train of about 10 cycles sent to the receiver along with synchronous signals. It ensures I and Q phase correctly.

244. (*b*) **245.** (*b*) **246.** (*c*)

247. (*c*) Kell factor indicates reduction in vertical resolution and varies from 0.65 to 0.85.

248. (*b*) 625 lines divided into odd and even fields, *i.e.*, 2 : 1 interlace.

249. (*b*)

250. (*a*) VSB (vestigial side band) transmission transmits one side band fully and the other side band partially thus reducing the bandwidth requirement.

251. (*a*)

252. (*a*) Scanning is always from left to right.

253. (*a*) **254.** (*d*) **255.** (*b*) **256.** (*a*)

257. (*c*) Newvicon has very high sensitivity and very wide spectral response which extends well into infra red region.

258. (*d*)

259. (*a*) Dark current is the small signal current even when there is no illumination on the face plate of camera tube.

260. (*c*) **261.** (*b*)

262. (*d*) $U = 0.493\,(B - Y)$ and $V = 0.877\,(R - Y)$.

263. (*b*) **264.** (*c*) **265.** (*c*) **266.** (*d*)

267. (*b*)

268. (*a*) Weighting reduces the chances of over-modulation of the transmitter carrier in the presence of highly saturated colours when picture has very light or very dark parts.

269. (*b*)

270. (*b*) In frequency interleaving the space between bundles is used for colour signal.

271. (*c*) **272.** (*b*)

273. (*c*) CCD (charged coupled device) sensors have all these advantages.

274. (*b*) Channel bandwidth is not 8 MHz.

275. (*a*) All these are present.

276. (*c*) **277.** (*b*)

278. (*d*) It may be any one of these.

279. (*b*) It is 1 kV per diagonal inch.

280. (*b*) $\dfrac{(180 - 20)(18)(30)}{1000} = 86.4$ kWh.

281. (*c*) All these are true.

282. (*d*)

283. (*d*) All these are part of visual exciter.

284. (*c*) Diplexer combines video and audio signals so that they can be sent out by a common antenna.

285. (*c*)

286. (*a*) Microwave links are used.
287. (*d*) See Section 12.6.
288. (*c*) **289.** (*a*) **290.** (*d*)
291. (*c*) See Section 12.6.
292. (*a*) See Section 12.6.
293. (*c*) Since velocity of em waves is 300 m/μs, each kilometre means time delay of 3.3 μs.
294. (*a*) **295.** (*a*) **296.** (*c*) **297.** (*a*)
298. (*b*) The band should not be narrow.
299. (*a*)
300. (*c*) Balun has turn ratio of 2 :1. Therefore impedance transformation is 4 : 1.
301. (*c*) **302.** (*b*)
303. (*b*) It is about 1 mV.
304. (*c*) Video detector does not amplify any signal.
305. (*c*) In varactor diode the applied reverse bias controls the width and therefore capacitance of depletion layer. This capacitance is used for tuning.
306. (*b*) When magnitude of reverse bias is increased, thickness of depletion layer increases. Capacitance is inversely proportional to thickness.
307. (*a*)
308. (*d*) Luminance amplifier changes the luminance levels and provides contrast control.
309. (*d*) **310.** (*b*) **311.** (*a*) **312.** (*a*)
313. (*c*)
314. (*c*) It is kept a little more than channel width so that reception is not restricted.
315. (*a*) **316.** (*a*) **317.** (*a*) **318.** (*c*)
319. (*a*) High QLC resonant circuit provides high selectivity.
320. (*d*) All these are used.
321. (*a*)
322. (*b*) It is used to telecast feature films. Flying spot scanner uses a special low persistence high brightness cathode ray tube.
323. (*a*) All these are standard test signals used in TV studios.
324. (*a*) Vision switcher has a provision to select any one or more of a large number of inputs and switching them on to outgoing circuits.
325. (*b*) Since there are 625 lines in one frame, the number of picture elements is $(625)^2\left(\frac{4}{3}\right)$.
326. (*a*) $\text{VSWR} = \frac{\text{max.voltage}}{\text{min.voltage}} = 2.$
327. (*c*) $Z_0 = \sqrt{\frac{0.8\times10^{-6}}{32\times10^{-12}}}.$
328. (*b*) $\frac{63.5\times16}{100} \simeq 10\,\mu\text{s}.$
329. (*b*) $\frac{10^6}{60\times525} \simeq 32\,\mu\text{s}.$
330. (*a*) $\frac{25}{1.9} \simeq 13\text{ mm}.$
331. (*a*) The frequency of H pulses is 15625 Hz. Hence time = $\frac{10^6}{15625} \simeq 64\,\mu\text{s}.$
332. (*c*) $\frac{1}{200\times5}\text{ s} = 1\text{ ms}.$
333. (*a*) (2 ms) (1 – 0.04) = 1.92 ms.
334. (*b*) $2^4 = 16.$
335. (*d*) $10\log_{10}\frac{P}{5000} = -196$ or $P = 1.26\times10^{-16}$ W.
336. (*c*) $10\log_{10}\frac{P}{10} = 30$ or P = 10000 W.
337. (*c*) $25\times10^3\times2000\times10^{-6} = 50$ W.
338. (*a*) EHT supply is 1kV per diagonal inch.

Height $= 60\times\frac{3}{4} = 45$ cm. Diagonal $= \frac{\sqrt{60^2+40^2}}{2.54} \simeq$ 29 inches.

339. (*c*) Total time for each horizontal line $= \frac{10^6}{15625} = 64\,\mu$s. Time for trace = 64 – 12 = 52 μs. Time for which transistor is not conducting = 12 + 0.4 × 52 = 32.8 μs. Time for which transistor is conducting = 64 – 32.8 ≃ 31 μs.
340. (*d*) $10\log_{10}\frac{P}{5} = 20$ or P = 500 W.
341. (*b*) Diagonal $= \frac{\sqrt{40^2+30^2}}{2.54} \simeq 20$ inches and EHT supply is 1 kV per diagonal inch.
342. (*b*) $P = 20\times10^3\times1500\times10^{-6} = 30$ W.
343. (*b*)
344. (*a*) R – Y = R – 0.3R – 0.3R – 0.59C – 0.11B
= 0.7R – 0.59G – 0.11B.
345. (*a*) **346.** (*b*) **347.** (*b*) **348.** (*d*)
349. (*c*) **350.** (*b*) **351.** (*c*) **352.** (*c*)
353. (*b*) Since gauss is a unit of magnetic flux, degaussing means removal of magnetism.
354. (*d*) **355.** (*b*)
356. (*a*) If power supply is working, raster will exist.
357. (*c*)
358. (*b*) In USA the first cable service was called community antenna television.
359. (*a*)
360. (*b*) Available power is PG and is spread over $4\pi d^2$.
361. (*c*) **362.** (*b*)
363. (*b*) Magnitudes should be equal and phase angle ± 90°.
364. (*c*) Skin depth decreases as square root of increase in frequency.
365. (*c*)

366. (*b*) Rate of information depends on bandwidth.

367. (*c*) **368.** (*b*) **369.** (*a*)

370. (*a*) Using Eq. in Section 12.2, $V_a = (4 \times 1.38 \times 10^{-23} \times 300 \times 2 \times 10^6 \times 10^4)^{0.5} = 18.2 \times 10^{-6}$ V.

371. (*d*) Since SSB modulation is used, bandwidth = 24 × 3 kHz.

372. (*a*) Bandwidth $\simeq$ 24 ($\ln_2$ 128) 10 kHz = 1.68 MHz.

373. (*b*) **374.** (*a*) **375.** (*b*) **376.** (*c*)

377. (*c*) See Table 12.1.

378. (*b*) **379.** (*b*)

380. (*d*) Overall resistance = 10 kΩ. Noise voltage = $(15)\sqrt{\dfrac{10}{20}} = 10.6\,\mu$V.

381. (*c*) Noise voltage = $10\sqrt{\dfrac{20}{10}} = 14.14\,\mu$V.

382. (*c*) **383.** (*b*)

384. (*b*) Atmospheric noise decreases as frequency is increased.

385. (*b*)

386. (*b*) $\dfrac{0.2 \times 2000}{2} = 200$ V as discussed in Section 12.3.

387. (*b*) $\dfrac{mV_C}{2} = 500$ W or $\dfrac{m(5000)}{2} = 500$ or $m = 0.2$.

388. (*c*) **389.** (*b*)

390. (*d*) See Section 12.3.

391. (*d*) Detection (also called demodulation) is due to rectification property of diode.

392. (*b*) **393.** (*d*)

394. (*c*) Both, should have a precise value and low drift.

395. (*a*)

396. (*b*) FSK, *i.e.*, frequency shift keying is a system of frequency modulation.

397. (*b*) **398.** (*a*)

399. (*c*) Using the derivation in Section 12.6,

$$E = \frac{1}{r}\sqrt{30P_t} = \frac{\sqrt{30 \times 75000}}{30000} = 0.05\ \frac{V}{m} = 50\ \text{m V/m}.$$

400. (*a*) $E \propto \sqrt{P_t}$.

401. (*c*)

402. (*d*) Radar is used for all the three purposes.

403. (*d*) **404.** (*a*) **405.** (*b*)

406. (*d*) It can detect very very low power signals.

407. (*b*) **408.** (*a*) **409.** (*b*)

410. (*a*) See Section 12.6.

411. (*b*) Because of this reason ground waves can be used for frequencies up to about 1600 kHz only.

412. (*d*) See section 12.6.

413. (*d*) **414.** (*a*)

415. (*b*) Nowadays only super heteordyne radio receiver are used.

416. (*c*) **417.** (*d*) **418.** (*b*)

419. (*b*) Two IF amplifier stages are very commonly used.

420. (*d*) **421.** (*b*)

422. (*a*) Load = (number of calls) (duration of call).

423. (*a*) **424.** (*c*) **425.** (*a*)

426. (*a*) **427.** (*a*) **428.** (*d*)

429. (*a*)

430. (*b*) See section 12.3.

431. (*a*)

432. (*b*) This equation can represent frequency, angle and phase modulation.

433. (*c*) **434.** (*b*) **435.** (*a*)

436. (*c*) PAM or pulse amplitude modulation is a system in which signal is sampled at regular intervals and each sample is made proportional to amplitude of the signal at the instant of sampling.

437. (*d*) **438.** (*c*) **439.** (*a*) **440.** (*c*)

441. (*b*) **442.** (*c*)

443. (*a*) Each bit can have two levels. Hence 2^N.

444. (*c*) **445.** (*b*) **446.** (*b*) **447.** (*a*)

448. (*c*) When M is large, sine term will tend to zero.

449. (*c*)

450. (*b*) A perfect conductor does not generate noise.

451. (*c*) **452.** (*c*) **453.** (*a*) **454.** (*d*)

455. (*a*)

456. (*a*) SSB–SC and SSB–DB have $\gamma = 1$ and are similar from noise point of view.

457. (*c*) Doubling of frequency increase SNR in frequency modulation by 4. It has no effect in AM.

458. (*a*) **459.** (*a*) **460.** (*b*) **461.** (*b*)

462. (*a*)

463. (*a*) See section 12.1. **464.** (*a*)

465. (*d*) Entropy = $0.5 \log_2 2 + 0.5 \log_2 2 = 1$.

466. (*c*) **467.** (*c*)

468. (*c*) Entropy = $4\left(\dfrac{1}{4}\log_2 4\right) = 2$. Signal should be sampled at frequency 2B Hz. Hence information rate = 2 (2B) = 4B bits/s.

469. (*a*) See section 12.1.

470. (*b*) See section 12.1.

471. (*b*) **472.** (*c*)

473. (*c*) Noise $\propto \sqrt{R}$.

474. (*c*) Load = $\dfrac{\text{Total duration of calls}}{\text{Time}} = \dfrac{100}{3} = 3.333$ erlang.

475. (*c*) **476.** (*b*) **477.** (*c*) **478.** (*b*)

479. (*b*) Generally bandwidth of voice telephony is 4 kHz. Sampling frequency should be twice this value.

480. (*a*) All the three features are necessary.

481. (*c*)

482. (*c*) All the three features are necessary.

483. (*a*)

484. (*d*) All the three configurations are used.

485. (*a*) **486.** (*a*)

487. (*d*) Noise is mostly generated in channel.

488. (*a*) **489.** (*a*) **490.** (*a*)

491. (*a*) At high frequencies time period of wave and transit time become comparable.

492. (*a*)

493. (*a*) Noise voltage $\propto \sqrt{R}$.

494. (*c*) See Section 12.2.

495. (*d*)

496. (*a*) See Section 12.2.

497. (*d*) See Section 12.2.

498. (*c*)

499. (*d*) $5.38\sqrt{\frac{500}{300}} = 6.94\,\mu V$.

500. (*b*) $10\log_{10}\left(\frac{3}{5}\right)$.

501. (*a*) **502.** (*b*)

503. (*c*) $\frac{1.5-0.5}{1.5} = \frac{2}{3}$.

504. (*c*) Initial transmitted power $= P_c\left(1+\frac{0.5^2}{2}\right) = 1.125\,P_c$.

Final transmitted power $P_c\left(1+\frac{1^2}{2}\right) = 1.5P_c$.

$$\text{Increase} = \frac{1.5P_c - 1.125P_c}{1.125P_c} = 0.33.$$

505. (*d*) **506.** (*c*) **507.** (*d*) **508.** (*c*)

509. (*d*) Saving $= \frac{1.25}{1.5}$

510. (*c*) Saving $= \frac{1.0625}{1.125}$

511. (*a*) One side band is eliminated in SSB. Therefore bandwidth is halved.

512. (*d*) **513.** (*b*)

514. (*a*) Unwanted frequency is removed by filtering.

515. (*c*)

516. (*b*) $400\left(1+\frac{0.75^2}{2}\right) = 512$ W.

517. (*d*) $400 \times 0.5 \times \frac{0.75^2}{2} = 56$ W.

518. (*d*)

519. (*a*) $\frac{4\times10^8}{2\pi}$ and $\frac{1100}{2\pi}$.

520. (*b*) $f_m = \frac{1100}{2\pi} = 175$ Hz. Maximum frequency deviation $= 3 \times 175 = 525$ Hz.

521. (*d*) Modulation index $= \frac{10}{2} = 5$, Bandwidth $= 2(10+5) =$ 30 kHz. This is an approximate value. Exact value is 32 kHz.

522. (*b*) FM equipments are complex as compared to AM equipments.

523. (*a*)

524. (*a*) $X_c = \frac{9}{9\times10^{-3}} = 1000\ \Omega$.

525. (*b*) **526.** (*c*)

527. (*c*) Frequency tripler makes modulation index three times.

528. (*b*) It generates FM through phase modulation.

529. (*c*)

530. (*b*) Deviation is not affected by mixer.

531. (*d*)

532. (*b*) Image frequency = 1000 – 455 = 545 kHz.

533. (*b*) **534.** (*b*)

535. (*a*) Detection (also called demodulation) is an essential function.

536. (*b*)

537. (*b*) Squelch, *i.e.*, muting is added after detector to suppress noise when no carrier is present at input.

538. (*a*) **539.** (*b*)

540. (*a*) Variable sensitivity prevents overloading.

541. (*b*)

542. (*c*) To suppress noise when carrier is absent, squelch circuit cuts off audio amplifier.

543. (*c*) **544.** (*c*) **545.** (*d*) **546.** (*b*)

547. (*a*) **548.** (*b*)

549. (*b*) To obtain three point tracking a capacitor is connected in series with local oscillator coil. This capacitor is called padder capacitor.

550. (*c*) Negative peak will be clipped.

551. (*d*)

552. (*d*) If line is OC or SC or terminated is pure reactance, SWR is 1.

553. (*b*) Actual velocity $= \frac{3\times10^8}{\sqrt{k}}$.

Velocity factor $= \frac{\text{Actual velocity}}{3\times10^8} = \frac{1}{\sqrt{k}}$.

554. (*d*) A short-circuited stub placed at a suitable distance from load provides best matching.

555. (*b*) $Z_L = 100 + j\,100$,

$Y_L = \frac{1}{100+j100} = 0.005 - j\,0.005\ \mho$. Susceptance of stub $= j\,0.005\ \mho$.

Reactance of stub $= \frac{1}{j0.005} = -\,j200\ \Omega$.

556. (*d*)

557. (*a*) If line is terminated in Z_0, the input impedance is Z_0.

558. (*b*) **559.** (*c*)

560. (*d*) All the three cause dissipation of energy as heat.

561. (*c*) When matching over a range of frequencies is desired, double stub is necessary.

562. (*a*)

563. (*b*) Balun gives 4 : 1 impedance transformation and is suited for coupling a coaxial line to two wire line.

564. (*d*) Both the case, *i.e.*, $Z_1 > Z_0 < Z_0$ give rise to standing waves.

565. (*c*) When standing wave is set up voltage maxima and current minima and voltage minima and current maxima conicide.

566. (*b*) **567.** (*b*) **568.** (*c*) **569.** (*a*)

570. (*d*) **571.** (*b*)

572. (*c*) Tropospheric scatter propagation is subject to two types of fading called Rayleigh fading (due to multipath propagation) and fading due to variations in atmospheric conditions along the propagation path.

573. (*c*) **574.** (*a*) **575.** (*c*) **576.** (*c*)

577. (*c*) Both power and frequency determine the maximum range.

578. (*b*) **579.** (*a*)

580. (*a*) See Table 12.1.

581. (*b*) See Section 12.6.

582. (*d*) **583.** (*a*) **584.** (*c*) **585.** (*d*)

586. (*c*)

587. (*d*) Rhombic antenna is a non-resonant antenna capable of operating over 3 to 30 MHz range.

588. (*a*)

589. (*a*) Directive gain is the ratio of power density radiated in a particular direction by the antenna to the power density that would be radiated by isotropic antenna.

590. (*a*) Directivity means maximum directive gain.

591. (*b*) **592.** (*a*)

593. (*d*) The helical antenna is circularly polarized, *i.e.*, its polarization is evenly divided between vertical and horizontal components.

594. (*c*) Isotropic antenna is one which radiates equally in all directions. It is a standard reference antenna.

595. (*a*)

596. (*b*) Beam width is the angular separation between the two half power points on the power density radiation pattern. It is expressed in degrees.

597. (*b*) **598.** (*a*)

599. (*c*) The number of radials is large and depth is about 30 cm.

600. (*a*) Counterpoise is a horizontally buried wire used to provide low grounding resistance in high resistivity soils.

601. (*c*) The effective length is more than actual length due to the effect of top loading and end effects.

602. (*a*) **603.** (*a*)

604. (*a*) Discone antenna is a combination of disc and cone. It is omnidirectional and is used in VHF and UHF range especially at air ports.

605. (*a*) All the three are correct.

606. (*a*)

607. (*b*) This antenna is for UHF and VHF bands.

608. (*a*) All the three features are correct.

609. (*a*)

610. (*b*) Directivity of parabolic reflector $= 6\left(\frac{D}{\lambda}\right)^2$.

611. (*a*) It is a combination of disc and cone in close proximity.

612. (*c*) All the three are correct.

613. (*b*) $2^4 = 16$.

614. (*d*) **615.** (*c*) **616.** (*a*) **617.** (*b*)

618. (*a*)

619. (a) $\log_2 13 = 3.7$. Therefore number of bits = 4. Efficiency $= \frac{3.7}{4} \times 100 \simeq 92\%$.

620. (*c*) Hartley law $C = 2\,\delta f \log_2 N$.

621. (*a*) This is Shanon–Hartley theorem as discussed in section 12.1.

622. (*a*) **623.** (*c*) **624.** (*a*)

625. (*c*) An analog number cannot be converted into an exact digital number. This is called quantizing error.

626. (*c*)

627. (*b*) Speed $= \frac{1}{20 \times 10^{-3}} = 50$ bauds.

628. (*b*) **629.** (*c*) **630.** (*b*)

631. (*a*) See Section 12.1.

632. (*a*) $10 \log_{10} 3 = 4.8$.

633. (*b*) For 3000 Hz. Nyquist rate is twice, i.e, 6000 bps.

634. (*a*) $\frac{S}{N} = 2^{6000/3000} - 1 = 3$.

635. (*c*)

636. (*d*) Full duplex system has all these features.

637. (*d*) **638.** (*d*) **639.** (*a*)

640. (*d*) When data rate in bits per second is upto 600, modem is low speed.

641. (*b*) When data rate is in the range of 2400 to about 10800 bits per second, modem is designated as high speed.

642. (*b*) **643.** (*d*) **644.** (*d*) **645.** (*a*)

646. (*d*)

647. (*c*) Fibre optic cables have all these advantages.

648. (*d*) It orbits the earth in a 24 hour period. Thus it appears to be stationed over one spot of the globe.

649. (*b*) **650.** (*b*) **651.** (*a*) **652.** (*a*)

653. (*c*) Band pass filter allows a band of frequencies to pass through and attenuates other frequencies.

654. (*d*) **655.** (*c*) **656.** (*a*) **657.** (*b*)

658. (*b*) See Section 12.10.1.

659. (*b*) See Section 12.10.1.

660. (*d*) Range $\propto$ (Power)$^{0.25}$ as discussed in Section 12.10.1.

661. (*a*) Range is proportional to antenna diameter as discussed in Section 12.10.1.

662. (*b*) **663.** (*b*) **664.** (*a*)

665. (*d*) Effective area of target is called radar cross-section. It depends on frequency polarization of incident wave and aspect of target.

666. (*d*)

667. (*d*) For high power switching Thyristor is the most suitable device.

668. (*b*) **669.** (*d*)

670. (*c*)

671. (*c*) PRF in radar systems means pulse repitition frequency.

672. (*c*) Both are correct.

673. (*c*) PRF does not have effect on range of radar as seen in Section 12.10.1.

674. (*b*) **675.** (*b*) **676.** (*b*)

677. (*a*) See Section 12.10.2.

678. (*a*) **679.** (*b*)

680. (*a*) $\lambda = \dfrac{3\times10^8}{5\times10^9} = 0.06$ m. Lowest blind speed = 800 × 0.06 = 48 m/s.

681. (*a*) **682.** (*b*)

683. (*d*) All the three are required.

684. (*b*) See Section 12.10.3.

685. (*c*)

686. (*d*) Quantization level = 2^N where N is number of bits.

687. (*a*) Sampling rate should be at least twice the bandwidth, *i.e.*, 2 (30 – 20) = 20 kHz.

688. (*a*) Nyquist rate = $2\,[f_1 + f_2 + f_3 + ...]$

$\omega_1 = 2000\,\pi$ or $f_1 = 1000$ Hz

$\omega_2 = 5000\,\pi$ or $f_2 = 2500$ Hz

Nyquist rate = 2 (1000 + 2500) = 7000 Hz or 7 kHz.

689. (*b*) LPF controls capture range and VCO generates loading range.

690. (*c*) Quantization noise power $= \dfrac{\Delta^2}{12} = P$

Hence RMS value of noise $= \sqrt{P} = \sqrt{\dfrac{\Delta^2}{12}} = \dfrac{\Delta}{\sqrt{12}}$

$$\Delta = \frac{V_H - V_L}{M}$$

where $V_H - V_L$ = Difference of highest and lowest level of voltage

M = Number of quantization level

Hence both number of quantization levels and step size govern quantization noise.

691. (*b*) $f_1 = 2000\,\pi/2\pi = 1000$ Hz

$$f_2 = \frac{2000\pi}{2\pi} + \frac{3000\pi}{2\pi} = 2500 \text{ Hz}$$

Minimum sampling rate = 2 (max. of f_1 and f_2)

= 2 (2500) = 5000 Hz = 5 kHz

692. (*c*) Decoder converts binary words to alphanumeric characters. Hence statement 2 is wrong.

693. (*c*) Uplink frequency is about 2 GHz more than downlink frequency.

694. (*a*)

695. (*b*) Encoder converts quantized values to binary values.

696. (*c*) Intermodal dispersion occurs in multimode fibre.

697. (*a*) Balanced modulator is used to suppress carrier.

698. (*d*) Low pass filter having cut off frequency 400 kHz is suitable as shown below.

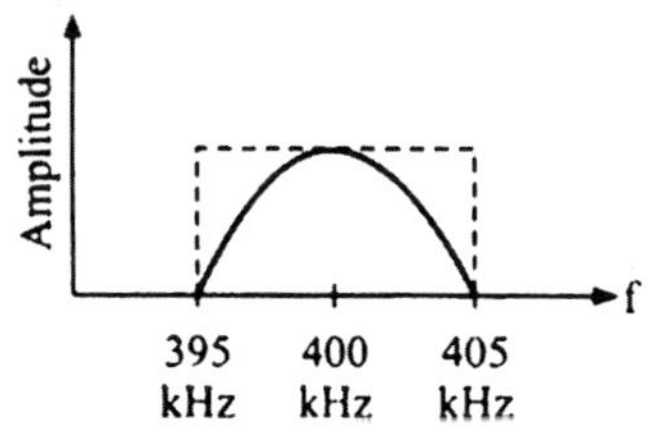

699. (*a*) 2 (10 – 5) = 10 kHz

700. (*d*) 2 (10.4 – 10) = 0.8 MHz

701. (*b*) Even same phase shift for different frequencies may distort the signal

702. (*c*) Fibre transmission line introduces mode partition noise. It is attributed to LASER

703. (*d*)

704. (*d*)

705. (*b*) Photoconductor does not provide gain.

706. (*a*) Fold strength at receivers $\propto \dfrac{\sqrt{P}h_t h_r}{d^2\lambda}$.

707. (*a*)

708. (*c*) Link budget pertains to dB loss in RF link.

709. (*c*) **710** (*b*)

711. (*a*) Entropy of $H_1 = \log 4$

Entropy of $H_2 = \log 6$

Hence $H_1 < H_2$

712. (*d*) Number of symbol = 100/sec.

Probability $= \dfrac{1}{100}$

If r is the number of digital signals, then

Entropy $= \log r\left(\dfrac{1}{P}\right) = \log_r 100 = \dfrac{\log_{10} 100}{\log_{10} r} = 2.7$

Hence $r = 5.5$

Therefore it is likely to be hexadecimal source.

713. (*b*)

714. (*c*) Noise does not depend on value of resistor.

715. (*a*) **716.** (*c*)

717. (*c*) Sampling frequency should be 2 × 8 = 16 kHz. Hence the choice does not obey sampling theorem.

718. (*b*) In TDM each signal has a fixed time slot.

719. (*d*) **720.** (*d*)

721. (*b*) Modulation index = $\dfrac{\text{Max. frequency devation}}{\text{Max. modulating frequency}}$

$= \dfrac{75}{15} = 5.$

722. (*b*) $n = \log_2 16 = 4.$

723. (*a*) FM is immune to noise because FM receivers are fitted with amplitude limiters to remove the amplitude variations due to noise.

724. (*c*) **725.** (*d*)

726. (*c*) To be power = $P_C\left(1+\dfrac{m^2}{2}\right)$

when $m = 0$, to be power $= P_C$

when $m = 1$ to be power $= 1.5\ P_C$

727. (*c*) Information per bit is increased

728. (*b*) Since $2^3 = 8$, bandwidth is tripled

729. (*c*) Increasing quantization levels reduce quantizing noise.

730. (*c*) **731.** (*c*)

732. (*d*) Channel capacity = Bandwidth $\log_2\left(1\times\dfrac{S}{N}\right)$. Since

$\dfrac{S}{N}$ is nearly zero, $\log_2(1) = 0$

733. (*a*) Attenuation increases with frequency.

734. (*b*) Half power bandwidth = $\dfrac{70\lambda}{D} = \dfrac{70\times0.1}{1} = 7°$

735. (*b*) Directivity = 1

Hence 1. log 1 = zero. dB

736. (*b*)

737. (*c*) Gain $= 6\left(\dfrac{\Delta}{\lambda}\right)^2$ or $1000 = 6\left(\dfrac{\Delta}{0.1}\right)^2$

or $\Delta = 1.291$ m

738. (*a*) Number of levels $= 2^8 = 256$

Step size $= \dfrac{+2-(-2)}{256} = \dfrac{4}{256}.$

739. (*d*) In all media.

740. (*a*) If phase error, θ, SNR deteriorates by $\cos^2\theta$.

741. (*b*) Average power of angle modulated signal

$x(t) = \dfrac{A_c^2}{2} = \dfrac{6^2}{2} = 18$ W.

742. (*c*) $s(t) = \dfrac{1}{2}\left[\dfrac{2\sin(500\pi t)\sin(700\pi t)}{\pi^2 t^2}\right]$

$= \dfrac{1}{2\pi^2 t^2}[\cos(700\pi t - 500\pi t) - \cos(700\pi t + 500\pi]$

$= \dfrac{1}{2\pi^2 t^2}[\cos 200\pi t - \cos 1200\pi t]$

Highest frequency component $= \dfrac{1200\pi}{2\pi} = 600$ Hz

Nyquist sampling rate = 2 × 600 = 1200 Hz

743. (*c*) Average voltage = 11 V

Energy = 11× 2 × 10 × 60 = 13200 J = 13.2 kJ

(1 Joule = 1 watt sec)

744. (*c*) See section 12.3.

745. (*b*) Mean = 1× 0.1 + 2 × 0.2 + 3 × 0.4 + 4 × 0.2 + 5 × 0.1

= 0.1 + 0.4 + 1.2 + .8 + .5 = 3

Variance $= E(x-m)^2 = (1-3)^2(0.01) + (2-3)^2(0.2) + (4-3)^2(0.2) + (5-3)^2(0.1)$

= 0.4 + 0.2 + 0.2 + 0.4 = 1.2

Hence both are wrong.

746. (*b*) Since SNR >> 1

$C_1 = B\log_2\left(1+\dfrac{S}{N}\right) = B\log_2\left(\dfrac{S}{N}\right)$

$C_2 = B\log_2\left(\dfrac{2S}{N}\right) = B\log_2\dfrac{S}{N} + B\log_2 2$

$= C_1 + B$

747. (*c*) SNR = 1.761 + 6.02 *n* or 43.5 = 1.761 + 6.02*n* or *n* = 7

Step size $= \dfrac{5-(-5)}{2n} = \dfrac{10}{14} \simeq 0.0667.$

748. (*b*) $Q = 2^n = \dfrac{5}{0.05} = 100$ or $n = 7$ for pisitive values

$\dfrac{S}{N} = 1.76 + 6n = 1.76 + 6\times7 = 43.76$ dB

For negative values $Q = 2^n = \dfrac{5}{0.1} = 50$ or $n = 5$

$\dfrac{S}{N} = 1.76 + 6\times5 = 31.76$

Best SNR = 43.76 ≃ 43.8 dB

749. (*a*) $(A_c + 2\cos\omega_m t]\cos\omega_c t$

For envelope detection of AM signal

$A_c + 2\cos\omega_m t \geq 0$

For $\cos\omega_m t = -1$, $A_c - 2 \geq 0$ or $A_c \geq 2$

750. (*a*) Entropy $= -\sum_{n=1}^{n} P_i \log P_i$ bits

$P_1 = P_2 = ... = P_n = \dfrac{1}{n}$

Hence entropy $= -\sum_{n=1}^{n}\dfrac{1}{n}\log n = \log n$

751. (*c*) Minimum bandwidth

= LCM of (W, W, 2 W, 3 W) = 6 W.

752. (*c*) Either AM or narrow band FM

753. (*c*) Probability of zero error in *n* bits $= (1-p)^n$

Probability of one error in *n* bits $= n.p(1-p)^{n-1}$

Total probability $= (1-p)^n + np(1-p)^{n-1}$

754. (*b*) Allotted Bandwidth 5 MHz

Frequency reuse factor $= \dfrac{1}{5}$

Bandwidth for 1 cell $= 5\times\dfrac{1}{5} = 1$ MHz

Number of simultaneous channels = $\frac{1 \text{ MHz}}{20 \text{ kHz}} \times 8 = 40$

755. (*a*) $f_s = \frac{1}{t_s} = 33$ kHz

f_s greater than $2f_m$. Hence corelated.

756. (*b*) $H = 0.25 \log \frac{1}{0.25} + 0.25 \log \frac{1}{0.25} + 0.5 \log \frac{1}{0.25}$

$= \frac{1}{2} + \frac{1}{2} + \frac{1}{2} = 1.5$

Average bit rate = 1.5 × 3000 = 4500 bits/sec.

757. (*c*) $\lambda = \frac{3 \times 10^8}{10^{14}} = 3 \times 10^{-6}$

In glass $\lambda_g = \frac{\lambda}{\mu} = \frac{3 \times 10^{-6}}{1.5} = 2\ \mu\text{m}$

758. (*a*) Bandwidth = $2(\Delta F + f_m) = 2(180 + 5) = 370$ kHz

759. (*b*) Noise output power = (Bandwidth of LP filter) (SN_{out})

$SN_{out} = |H(\omega)|^2 SN_i = 4 N_0$

Hence Noise output power = $4 B N_0$.

760. (*a*) $\text{SNR} = \frac{\text{Power}}{\text{Noise spectral density} \times \text{bandwidth}}$

$= \frac{10^{-3}}{10^{-20}(100 \times 10^6)} = 10^9$

SNR in dB = $10 \log 10^9 = 90$ dB

Cable loss = 40 dB

SNR at receiver = (90 – 40) = 50 dB

761. (*b*) Both are correct but independent, see Section 12.3.

762. (*c*) R is wrong.

763. (*b*) See Section 12.3.

764. (*b*) See Section 12.3.

765. (*b*) **766.** (*b*)

767. (*d*) A is wrong.

768. (*b*) Both are correct but independent.

769. (*b*) **770.** (*b*) **771.** (*a*)

772. (*b*) Both are correct but independent.

773. (*d*) **774.** (*a*)

775. (*b*) Both are correct but independent.

776. (*b*) $\text{VSWR} = \frac{1 - |\rho_v|}{1 + |\rho_v|}$ and $\rho_v = \frac{Z_L - Z_0}{Z_L + Z_0}$.

777. (*b*) Smith chart is a convenient tool for transmission line design and analysis.

778. (*a*) Since free space does not have magnetic and other fields, waves can propagate without any interference.

779. (*b*) Both are correct but independent.

780. (*a*) Since free space has no interference or obstacles, em waves spread uniformly.

781. (*a*) As per inverse square law, power is inversely proportional to square of distance.

782. (*b*) Both are correct but independent.

783. (*c*) Video signals are amplitude modulated.

784. (*d*) Colour signal consists of luminance and chrominance.

785. (*b*) Both are correct but independent.

786. (*b*) See Section 12.13.1.

787. (*a*) See Section 12.8.

788. (*b*) See Section 12.8.

789. (*b*) Both are correct but independent.

790. (*a*) See Section 12.10.

791. (*a*) See Section 12.10.1.

792. (*b*)

793. (*b*) PRF means pulse repetition frequency.

794. (*b*) SAW means surface acoustic wave. SAW filter is very commonly used in TV sets.

795. (*a*) Synchronisation of scanning process at transmitter and receiver is necessary for proper reproduction at TV receiver. For this purpose synchronising pulses are sent as a part of video signal.

796. (*b*) See Section 12.13.1. Aspect ratio = $\frac{\text{Width}}{\text{Height}}$ of TV receiver.

797. (*b*) See Section 12.13.

798. (*b*) Balun gives a 4 : 1 impedance transformation. In areas where TV reception is poor, booster is used to strengthen a TV signal.

799. (*a*) See Fig. 12.7.

800. (*a*) A varactor diode has a variable capacitance which depends on reverse bias. it is a very convenient device for tuning.

801. (*d*) A is wrong but R is correct.

802. (*a*) See section 12.3.

803. (*c*) **804.** (*a*) **805.** (*d*) **806.** (*b*)

807. (*b*) **808.** (*b*) **809.** (*b*) **810.** (*a*)

811. (*c*) **812.** (*a*) **813.** (*a*) **814.** (*b*)

815. (*d*)

816. T

817. F, Carrier does not have any information.

818. F, Each frequency gives two side bands.

819. T

820. F, Demodulation is called detection.

821. F, Sound waves are not electromagnetic in nature.

822. T

823. F, Radio waves have very high speed equal to that of light but speed of sound waves is very low.

824. T **825.** T **826.** T

827. F, It is not true.

828. T

829. F, Strength of space wave depends on distance as it suffers attenuation.

830. T **831.** T **832.** T **833.** T

834. T **835.** T

836. F, PAS use moving coil microphones.
837. T **838.** T
839. F, It is not true. See Reference 10.
840. T **841.** F **842.** T
843. F, Deck is much superior to tape recorder.
844. T, It erases the matter already existing on the tape.
845. T **846.** T **847.** T
848. F, See Reference 10.
849. T **850.** T **851.** T **852.** T
853. T
854. F, Video monitor does not have RF tuner and IF amplifier stages. See Section 12.13.11.
855. F, Trinitron is a colour picture tube.
856. F, Frequency band for local programs is fixed.
857. T **858.** T
859. F, Baudot code is 5 bit code.
860. T **861.** T **862.** T **863.** T
864. T **865.** T
866. 16 Hz **867.** 10, 0.1
868. 1600 kHz
869. 30 MHz
870. three **871.** 25% **872.** decreases
873. noise **874.** amplitude, frequency
875. bass **876.** treble **877.** decreases
878. pixels **879.** 625 **880.** odd, even
881. raster **882.** fifty
883. Committe consultative for international radio
884. Surface acoustic wave filter
885. $\lambda/4$
886. aspect ratio
887. 30%, 59%, 11%
888. Phase alternation by line
889. rotary **890.** Ten bits
891. characteristic impedance
892. ionosphere **893.** critical frequency
894. troposphere **895.** elementary doublet
896. Impedance matching **897.** circular polarization
898. Discone antenna **899.** beam width
900. Shannon-Hartley **901.** large
902. multiplexing **903.** directional
904. Doppler **905.** transponder
906. CW **907.** standing wave
908. standing wave ratio **909.** unity
910. ship communication

MICROWAVE ENGINEERING

13

CHAPTER

13.1. TRANSMISSION LINES

A transmission line is the medium through which electrical energy is sent from one place to another. The energy may be for lighting, heating etc. or for information (speech, music, data pictures etc.). Some commonly used configurations of transmission lines are shown Fig. 13.1.

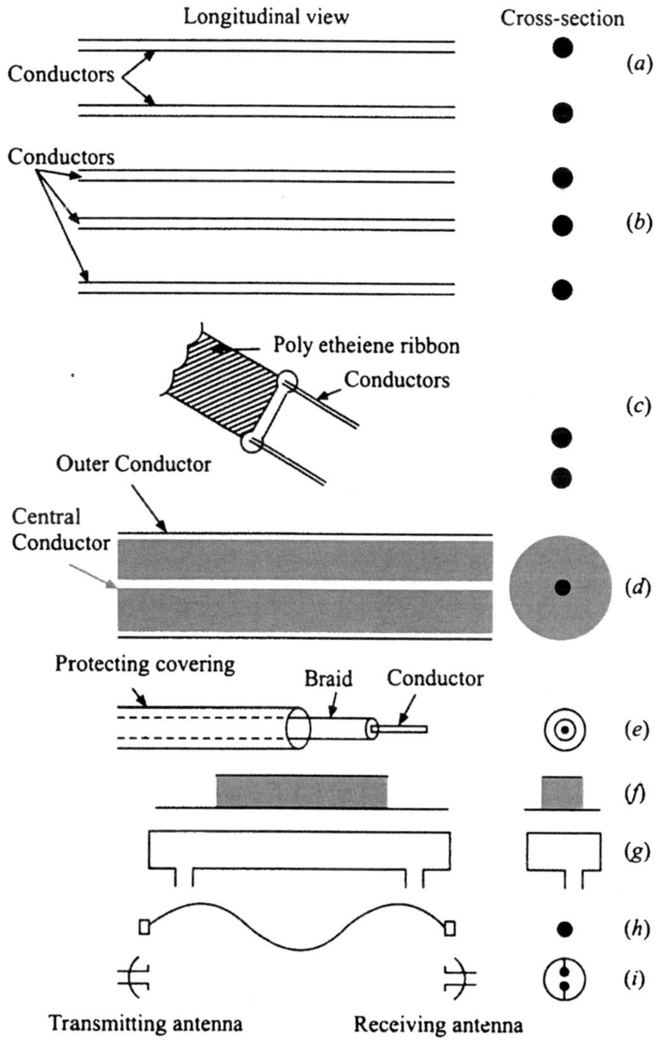

Fig. 13.1. Transmission lines:
(a) 2 wire line, (b) Multi wire line,
(c) Twin wire line, (d) Co-axial line, (e) Shielded cable,
(f) Microstrip line, (g) Waveguide,
(h) Optical fiber line, (i) Radio link.

13.1.1. Wave Equations

A transmission line is a distributed parameter circuit having R (*i.e.*, resistance per unit length ohms/metre), L (*i.e.*, inductance per unit length, H /m) C (*i.e.*, capacitance per unit length. F/m) and G (*i.e.*, conductance per unit length, siemens/m). Line equations in terms of above parameters are

$$v = V_1 e^{j\omega t} e^{\gamma x} + V_2 e^{j\omega t} e^{-\gamma x} \quad ...(13.1)$$

$$i = -\frac{V_1}{\sqrt{z/y}} e^{j\omega t} e^{\gamma x} + -\frac{V_2}{\sqrt{z/y}} e^{j\omega t} e^{-\gamma x} \quad ...(13.2)$$

Where $z = R + j\omega L$ and $y = G + j\omega C$

$$\gamma = \alpha + j\beta = \sqrt{zy}$$

$$\alpha = \text{Re}(\gamma) = \text{Re}\sqrt{zy} \text{ neper s/m}$$

$$\beta = \text{Im}(\gamma) = \text{Im}\sqrt{zy} \text{ rad/m}$$

For loss less line R = 0 and G = 0 and $\gamma = j\omega\sqrt{LC}$

$$\text{velocity of propagation} = v = \frac{1}{\sqrt{LC}} \text{ for forward wave} \quad ...(13.3)$$

$$\text{velocity of propagation} = -v = \frac{-1}{\sqrt{LC}} \text{ for backward wave} \quad ...(13.4)$$

$$\text{Characteristic impedance } Z_0 = \frac{e}{i} = \sqrt{\frac{z}{y}} = \sqrt{\frac{L}{C}} \quad ...(13.5)$$

For overhead lines $v = \frac{1}{\sqrt{LC}} = 3 \times 10^8$ m/s = 300 μ/s

For free space $v = \frac{1}{\sqrt{LC}} = \frac{1}{\sqrt{\mu_0 \epsilon_0}} = 3 \times 10^8$ m/s

For cables $v = \frac{1}{\sqrt{LC}} = \frac{3 \times 10^8}{\sqrt{\epsilon_r}}$ m/s

The ratio of velocity to velocity of light is called relative phase velocity or velocity factor. Characteristic impedance for various types of lines/media are given in Table 13.1.

Table 13.1. Characteristic Impedance/Intrinsic Impedance Z_0

Type of Line or medium	Z_0 *ohms*
Single overhead conductor in air	$60 \ln \frac{2h}{r}$
Two wire line, relative permittivity ε_r	$\frac{119.945}{\sqrt{\epsilon_r}} \ln\left(\frac{D}{r}\right)$
Two wire line in air	$119.945 \ln\left(\frac{D}{r}\right)$
Cable	$\frac{60}{\sqrt{\epsilon_r}} \ln\left(\frac{R}{r}\right)$
Free space	376.82
Lossless dielectric medium	$\frac{376.82}{\sqrt{\epsilon_r}}$
Conducting medium	$\sqrt{\frac{j\omega\mu}{\sigma + j\omega\epsilon}}$

13.1.2. Distortion

Types of distortion are frequency distortion and delay distortion.

Frequency distortion : Some frequencies suffer more attenuation than other frequencies.

Delay distortion : Different frequency components of a signal suffer different phase delays.

$$\text{Condition for minimum distortion } L = \frac{RC}{G} \quad \text{...(13.6)}$$

The process of adding inductance to satisfy Eq. (13.6) is called line loading.

13.1.3. Reflection and Refraction

Input impedance of line

$$= \frac{V_s}{I_s} = Z_0 \left[\frac{Z_L \cosh \gamma l + Z_0 \sinh \gamma l}{Z_0 \cosh \gamma l + Z_L \sinh \gamma l}\right] \quad \text{...(13.7)}$$

For open-circuited line, input impedance

$$= \frac{Z_0 \cosh \gamma l}{\sinh \gamma l} \quad \text{...(13.8)}$$

For short-circuited line, input impedance

$$= \frac{Z_0 \sinh \gamma l}{\cosh \gamma l} \quad \text{...(13.9)}$$

$$\rho_v = \text{Reflection coefficient} = \frac{Z_L - Z_0}{Z_L + Z_0} \quad \text{...(13.10)}$$

$$\rho_t = \text{Refraction coefficient} = \frac{2Z_L}{Z_L + Z_0} \quad \text{...(13.11)}$$

Two waves travelling in opposite directions on the line give rise to standing wave (Fig. 13.2)

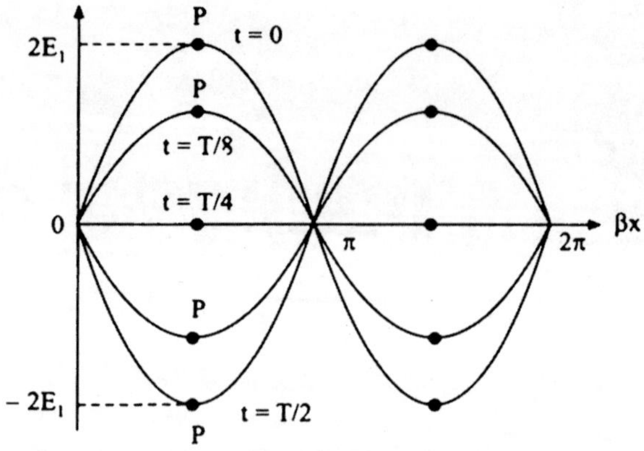

Fig 13.2. Two equal and opposite travelling waves resulting in a standing wave.

$$\text{Voltage Standing Wave Ratio (VSWR)} = \frac{1 + |\rho_v|}{1 - |\rho_v|} \quad \text{...(13.12)}$$

13.1.4. High Frequency Lines

$$\text{Depth of penetration } \delta = \sqrt{\frac{2}{\omega \mu \sigma}} \quad \text{...(13.13)}$$

where δ = depth of prenetration, m

ω = angular frequency, rad/sec.

μ = permeability, H/m

σ = conductivity, siemens/m

$$\text{For copper} \quad \delta = \frac{6.6 \times 10^{-2}}{\sqrt{f}} \quad \text{...(13.14)}$$

For copper if $f = 50$ Hz, $\delta = 9.33 \times 10^{-3}$ m

$f = 1$ MHz, $\delta = 6.6 \times 10^{-5}$ m

$f = 30$ GHz, $\delta = 3.8 \times 10^{-7}$ m

In transmission lines shunt conductance G is always negligible. In high frequency lines $X_L >> R$. Therefore high frequency lines are treated as loss less lines.

13.1.5. Quarter Wave Length Line

A line of length $\frac{\lambda}{4}$ is called quarter wave length line. Such a line is used for impedance matching.

If $l = \lambda/4$

$$Z_{in} = \text{input impedance} = \frac{Z_0^2}{Z_L} \quad \text{...(13.15)}$$

A quarter wave line can match a source impedance Z_{in} with load impedance Z_L by selecting a proper value of Z_0 so as to satisfy Eq. (13.15). Such a line is also called $\lambda/4$ transformer.

13.1.6. Single Stub for Impedance Matching

A short-circuited stub connected at suitable distance from load (Fig. 13.3) can be used for impedance matching.

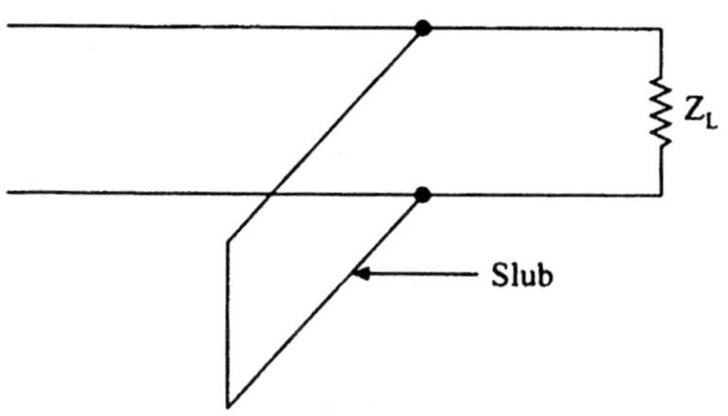

Fig 13.3. Impedance matching using short-circuited stub.

13.1.7. Reactance Properties of Transmission Lines

Many configurations of lines can represent series or shunt inductive or capacitive reactances. These are shown in Fig. 13.4. Such lines are also called transmission line resonators.

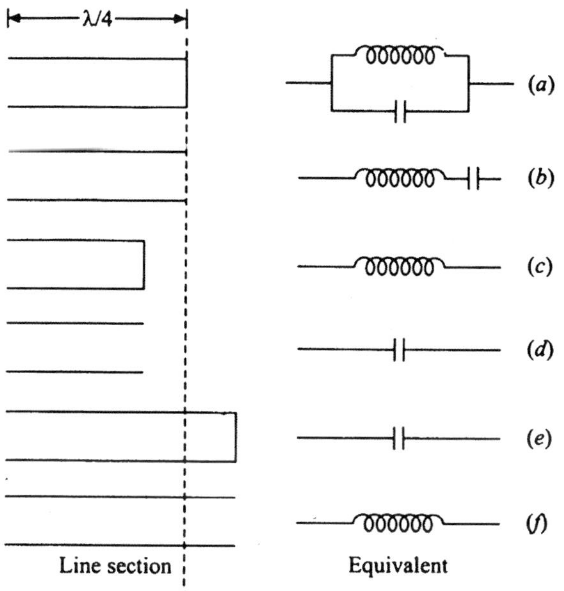

Fig. 13.4.

13.2. WAVE GUIDES

At frequencies above 5 GHz it is more convenient to use waveguides as transmission lines. The advantages of wave guide over coaxial line include higher power capacity, lesser attenuation and lesser cost. Mostly rectangular and circular wave guides are used.

Two main modes of electromagnetic wave propagation in wave guides are Transverse electric (TE) mode and Transverse magnetic (TM) mode.

For TE mode $E_z = 0$, *i.e.*, electric field component in the directions *z* (which is the direction of propagation) is zero.

For TM mode $H_z = 0$, *i.e.*, magnetic field component in *z* direction is zero.

In both TE and TM modes two subscripts *m* and *n* are used for sub classification. The subscript m denotes the number of half sine wave variations of electric (or magnetic field) across the wider dimension '*a*' of the wave guide. Similarly subscript *n* denotes the number of half sine wave variations of electric (or magnetic field) across the narrow dimension '*b*' of the wave guide. If in a TE mode $m = 1$ and $n = 0$, the wave is designated as TE_{10}. If in a TM wave $m = 1$ and $n = 1$ the wave is designated as TM_{11}. These subscripts *m* and *n* can take all integer values from 0 to infinity.

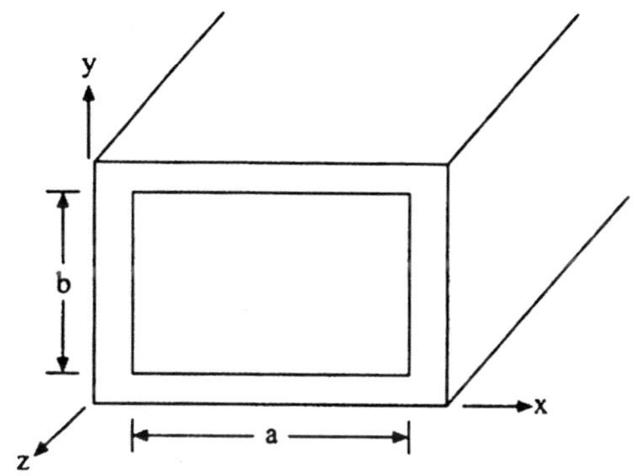

Fig. 13.5. Rectangular wave guide.

In each TE and TM mode, there is one mode which provides low loss and minimum distortion. This mode is called dominant mode. In TE category TE_{10} is the dominant mode. TE_{01} mode is merely TE_{10} mode rotated by 90°. Mostly air is used as the dielectric in all wave guides. Increase of air pressure increases the power handling capacity of the wave guide.

13.2.1. TE Modes of Rectangular Wave Guides

In this case $E_z = 0$ and $H_z \neq 0$.

For propagation in *z* direction H_z is a function of *x* and *y* only (Fig. 13.5).

$$H_z = H_0 e^{-j\beta z} \cos \frac{m\pi}{a} x \cos \frac{n\pi}{b} y \qquad ...(13.16)$$

where H_0 is a constant which indicates the amplitude of the electromagnetic wave and β is phase constant. The transverse field components are

$$E_x = j H_0 \frac{\omega\mu}{k_c^2} \frac{n\pi}{b} e^{-j\beta z} \cos \frac{m\pi}{a} x \sin \frac{n\pi}{b} y \qquad ...(13.17)$$

$$E_y = -jH_0 \frac{\omega\mu}{k_c^2} \frac{m\pi}{b} e^{-j\beta z} \sin \frac{m\pi}{a} x \cos \frac{n\pi}{b} y \qquad ...(13.18)$$

$$H_x = -E_y/Z_{TE} \qquad ...(13.20)$$

$$H_y = E_x/Z_{TE}$$

where
$$k_c^2 = \left(\frac{m\pi}{a}\right)^2 + \left(\frac{n\pi}{b}\right)^2$$

$$Z_{TE} = \text{wave impedance} = \frac{\eta}{\sqrt{1-(\lambda/\lambda_c)^2}}$$

$$\lambda_c = \text{cutoff wavelength}$$

$$= \frac{2\pi}{k_c} = \frac{1}{\sqrt{\left(\frac{m}{2a}\right)^2 + \left(\frac{n}{2b}\right)^2}}$$

$$\beta = \text{phase constant} = \sqrt{\omega^2\mu\epsilon - \left(\frac{2\pi}{\lambda_c}\right)^2}$$

For a particular mode the propagation can occur only when β is real, *i.e.*, $\omega^2 \mu\epsilon > (2\pi/\lambda_c)^2$.

Therefore $\lambda < \lambda_c$ and $f > f_c$.

The cutoff frequency f_c is given by

$$f_c = \frac{v}{\lambda_c} = \frac{c}{\lambda_c\sqrt{\mu_r \epsilon_r}} \qquad ....(13.21)$$

where c is velocity of light.

For a particular mode, phase velocity v_p and wave length in the guide λ_g are related by phase constant β

$$\beta = \frac{\omega}{v_p} = \frac{2\pi}{\lambda_g} \text{ rad/unit length} \qquad ...(13.22)$$

$$\lambda_g = \frac{\lambda_0}{\sqrt{\mu_r \epsilon_r - \left(\frac{\lambda_0}{\lambda_c}\right)^2}} = \frac{\lambda}{\sqrt{1-(f_c/f)^2}}$$

λ_0 is free space wave length and $\lambda = \dfrac{\lambda_0}{\sqrt{\mu_r \epsilon_r}}$

For TE_{10} mode

$$\lambda_c = 2a \quad \text{and} \quad f_c = \frac{c}{2a\sqrt{\mu_r \epsilon_r}}$$

For TE_{20} mode

$$\lambda_c = a \quad \text{and} \quad f_c = \frac{c}{a\sqrt{\mu_r \epsilon_r}}$$

TE_{10} mode has the lowest cutoff frequency and is called the dominant mode

$$\text{Attenuation } \alpha = \frac{54.6}{\lambda_c}\sqrt{1-\left(\frac{f}{f_c}\right)^2} \text{ dB/unit length} \qquad ...(13.24)$$

Characteristic Impedance

$$Z_0 = \frac{\pi}{2}\frac{b}{a} Z_{TE} = \frac{592\frac{b}{a}\sqrt{\mu_r/\varepsilon_r}}{\sqrt{1-(f_c/f)^2}} \text{ ohms} \qquad ...(13.25)$$

Also $$Z_0 = \frac{P}{I_z^2} \qquad ...(13.26)$$

where P is the power and I_z is longitudinal current.

13.2.2. TM Modes in Rectangular Wave Guides

$$H_z = 0$$

$$E_z = E_0 e^{-j\beta z} \sin\frac{m\pi}{a}x \sin\frac{n\pi}{b}y \qquad ...(13.27)$$

$$H_x = jE_0 \frac{\omega\epsilon}{k_c^2}\frac{n\pi}{b} e^{-j\beta z} \sin\frac{m\pi}{a}x \cos\frac{n\pi}{b}y \qquad ...(13.28)$$

$$H_y = -jE_0 \frac{\omega\epsilon}{k_c^2}\frac{m\pi}{a} e^{-j\beta z} \cos\frac{m\pi}{a}x \sin\frac{n\pi}{b}y \qquad ...(13.29)$$

$$E_x = \pm Z_{TM} H_y \qquad ...(13.30)$$

$$E_y = \mp Z_{TM} H_x \qquad ...(13.31)$$

where E_0 is the amplitude of the wave.

The expressions for β, λ_c, f_c, λ_g, α are the same as for TE waves.

For TM wave the lowest out off frequency in rectangular wave guide is for TM_{11} mode. If $b = \frac{a}{2}$, the cut-off frequency for TM_{11} mode is about 12% more than that for TE_{20} mode.

13.2.3. Circular Wave Guide

Some applications require dual polarization capability. Circular wave guide has this capability. Their analysis uses cylindrical coordinates. In circular wave guide TE_{11} mode has the lowest cutoff frequency and is the dominant mode. If D is diameter of wave guide

$\lambda_c = 1.706$ D for TE_{11} mode

$\lambda_c = 1.029$ D for TE_{21} mode

$\lambda_c = 0.82$ D for TE_{01} mode

$\lambda_c = 1.306$ D for TM_{01} mode

13.2.4. Special Wave Guide Configurations

These are shown in Fig. 13.6.

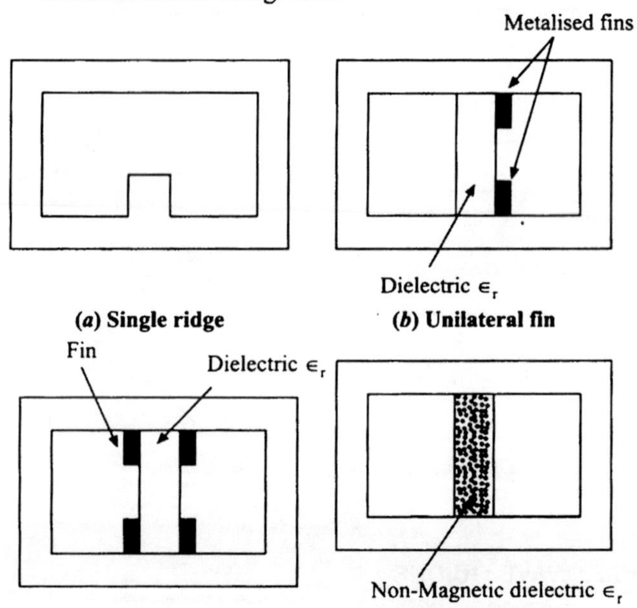

(*a*) **Single ridge** (*b*) **Unilateral fin** (*c*) **Bilateral fin** (*d*) **Dielectric loaded**

Fig. 13.6. Special wave guide configurations.

13.3. CAVITY RESONATOR

Hollow metallic enclosures exhibit resonance behaviour when excited by electromagnetic field. These enclosures are called cavity resonators.

13.3.1. Rectangular Cavity

It is like a rectangular wave guide shorted at both ends. The resonant wave length λ_r is given by

$$\lambda_r = \frac{1}{\sqrt{\left(\frac{m}{2a}\right)^2 + \left(\frac{n}{2b}\right)^2 + \left(\frac{p}{2l}\right)^2}} \qquad ...(13.32)$$

where a, b, l are the width, height and length of cavity, m and n refer to the modes, $p = 1, 2, 3....$, *i.e.*, positive integers. The modes are called TE_{mnp} and TM_{mnp}

Thus the resonant frequencies are different for different modes. The mode with lowest resonant frequency is called primary mode. If a is bigger dimension and b is smaller dimension of cavity the TE_{101} mode is the primary mode. For a given set of mode numbers the resonant frequency for TE and TM modes are the same. Such modes (*i.e.*, with same resonant frequency) are said to be degenerate.

13.3.2. Cylindrical Cavity

It is a cylindrical wave guide shorted at both ends. The modes are called TE_{mnp} and TM_{mnp} modes.

$$\lambda_r = \frac{1}{\sqrt{\left(\frac{1}{K_{nm}D}\right)^2 + \left(\frac{p}{2l}\right)^2}} \qquad ...(13.33)$$

where D is cavity diameter, l is length and $p = 1, 2, 3$...........

Q-factor of resonator is

$$Q = \omega_r \left[\frac{\text{Energy stored in resonator}}{\text{Power loss in resonator}}\right]_{\text{at } \omega_r} \qquad ...(13.34)$$

13.4. MICROWAVE JUNCTIONS

13.4.1. E-plane Tee Junction (Fig. 13.7 *a*)

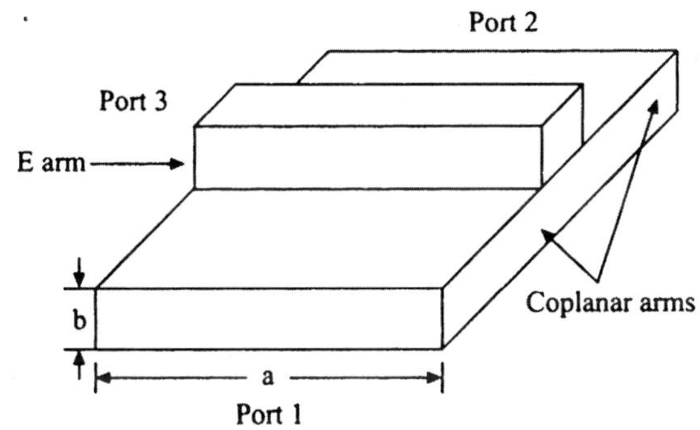

Fig. 13.7 (*a*) E-plane tee junction.

Ports 1 and 2 are called coplanar arms and port 3 is called E arm. The widths of the three guides are equal. If the tee is fed at port 3 and lengths of port I and port 2 arms are equal, the electric fields at the two outputs are 180° out of phase. The power delivered at port 1 and port 2 are equal.

13.4.2. H-plane Tee Junction (Fig 13.7*b*)

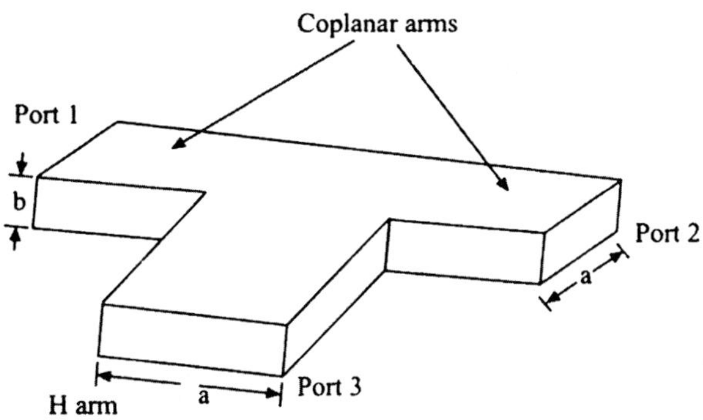

Fig. 13.7 (*b*) H-plane tee junction.

It is also called shunt tee. Port 3 is called *H* arm while ports 1 and 2 are called coplanar arms. A signal in *H* arm splits equally into coplanar arms. If lengths of coplanar arms are equal, the output electric fields are in phase.

13.4.3. Hybrid Tee or Magic Tee (Fig. 13.7c)

It is a four port device and is a combination of E plane tee and H plane tee. Ports 1 and 2 are called coplanar arms. Port 3 is E arm and port 4 is H arm. The features of magic tee are :

1. All ports are perfectly matched.
2. If a signal is fed at coplanar arms 1 (or 2) it splits equally between E and H arms. At each output port $P_{out} = 0.5\ P_{in}$ and $V_{out} = 0.707\ (V_{in})$. There will be no output at the other coplanar arm 2 (or 1).

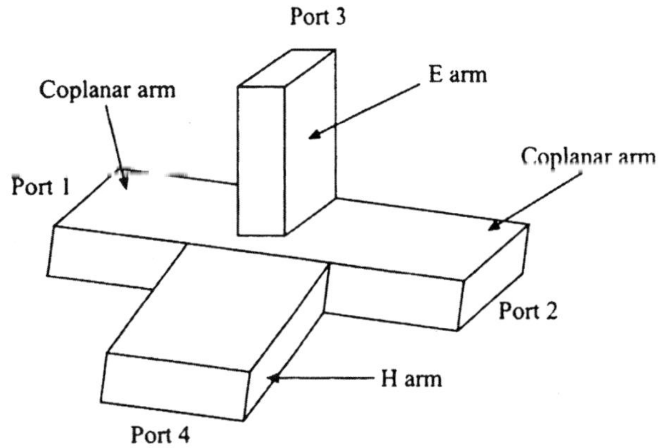

Fig. 13.7. (*c*) Hybrid tee.

3. If signal is fed in H arm it splits equally between coplanar arms, the outputs in coplanar arms being in phase equidistant from junction. There will be no output at E part.
4. If signal is fed at E arm it splits equally between coplanar arms the outputs being 180° out of phase equidistant from junction. There will be no output at H port.
5. If signals are fed into both coplanar arms then

(*a*) the signal output from E arm is equal to 0.707 (phasor difference of input signals).

(*b*) the signal output from H arm is equal to 0.707 (phasor sum of input signals).

Therefore E arm is called difference arm and H arm is called sum arm.

The scattering matrix for an ideal hybrid tee having equal coplanar arms and lengths of E and H arms chosen to make S_{13} and S_{14} positive real is as under

$$[S] = \begin{bmatrix} 0 & 0 & 0.707 & 0.707 \\ 0 & 0 & -0.707 & 0.707 \\ 0.707 & -0.707 & 0 & 0 \\ 0.707 & 0.707 & & \end{bmatrix}$$

13.4.4. Hybrid Ring

It is also called 'rat race' and is shown in Fig. 13.8. It has an annular ring to which are connected four ports. The ports 1 and 2 are decoupled. Ports 3 and 4 are also decoupled. Its features are

1. A signal entering port 1 or port 2 is split equally between ports 3 and 4.
2. A signal entering port 3 or port 4 is split equally between ports 1 and 2.

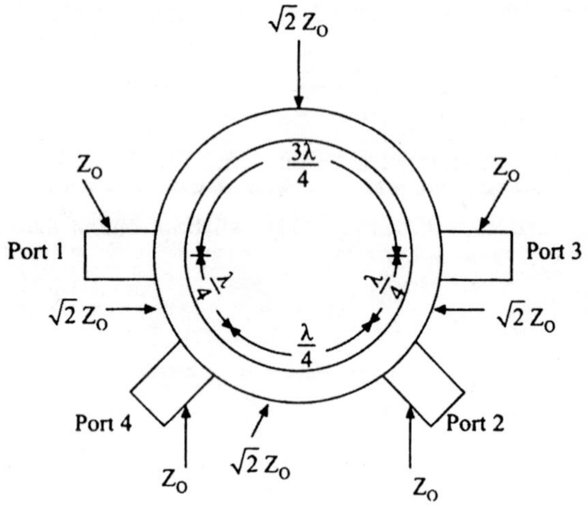

Fig. 13.8. Hybrid ring.

The scattering matrix for hybrid ring is as under

$$[S] = \begin{bmatrix} 0 & 0 & j0.707 & -j0.707 \\ 0 & 0 & -j0.707 & -j0.707 \\ j0.707 & -j0.707 & 0 & 0 \\ -j0.707 & -j0.707 & & \end{bmatrix}$$

The isolation between ports 1 and 2 as also between 3 and 4 is due to $\lambda/2$ difference in path lengths for the two waves in the ring.

13.5. DIRECTIONAL COUPLER (FIG. 13.9)

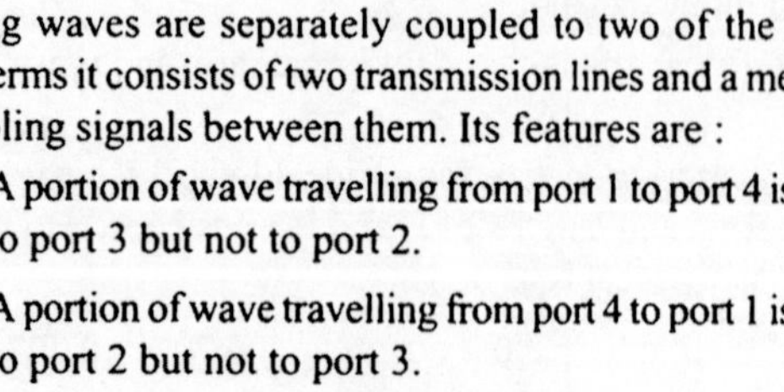

Fig. 13.9. Directional coupler.

It is a four-port device in which portions of forward and reverse travelling waves are separately coupled to two of the ports. In simple terms it consists of two transmission lines and a mechanism for coupling signals between them. Its features are :

1. A portion of wave travelling from port 1 to port 4 is coupled to port 3 but not to port 2.
2. A portion of wave travelling from port 4 to port 1 is coupled to port 2 but not to port 3.
3. A portion of wave travelling from port 2 to port 3 is coupled to port 4 but not to port 1.
4. A portion of wave travelling from port 3 to port 2 is coupled to port 1 and not to port 4.

A directional coupler is used for microwave measurements (measurement of reflected/refracted power, voltage standing wave ratio etc.)

13.6. MICROWAVE TUBES

Oscillator circuits using vacuum tubes (triodes, pentodes) have the following limitations at very high frequencies (microwave region).

1. The stray capacitances and inductances become important and affect the operation of the circuit.
2. At low frequencies the transit time between cathode and anode is a small fraction of period of oscillation. However at microwave frequencies this transit time becomes comparable to time period of oscillations.

13.6.1. Multicavity Klystron Amplifier

A Klystron is a vacuum tube used for generation/amplification of microwaves. An electron beam is produced by oxide coated indirectly heated cathode and is focussed and accelerated by focussing electrode. This beam is transmitted through a glass tube. The input cavity where the beam enters the glass tube is called buncher. As electrons move ahead they see an accelerating field for half cycle and retarding field for the other half cycle. Therefore some electrons are accelerated and some are retarded. This process is called velocity modulation. The velocity modulation causes bunching of electrons. This bunching effect converts velocity modulation into density modulation of beam. The input is fed at buncher cavity and output is taken at catcher cavity. In a two cavity klystron only buncher and catcher cavity are used. In multi cavity klystron one or more intermediate cavities are also used. The features of a multicavity klystron are :

1. Frequency range	0.25 GHz to 100 GHz
2. Power output	10 kW to several hundred kW
3. Power gain	60 dB (nominal value)
4. Efficiency	about 40%

A multicavity klystron is used in UHF TV transmitters, Radar transmitter and Satellite communication.

13.6.2. Reflex Klystron Oscillator

It uses a single cavity resonator for generating microwave oscillations. Its parts are electron gun, resonator, repeller and output coupling. It operates on the principle of positive feedback. The repeller electrode is at negative potential and sends the partially bunched electron beam back to resonator cavity. This positive feedback supports oscillations. Its features are :

1. Frequency range	2 to 100 GHz
2. Power output	10 mW to about 2W
3. Efficiency	10–20%

Its applications include radar receivers, local oscillator in microwave devices, oscillator for microwave measurements in laboratories etc.

13.6.3. Travelling Wave Tube (TWT)

In a klystron the resonant structure limits the bandwidth. A TWT is a broad band device. Its main components are electron gun (to produce the electron beam) and a structure supporting the slow electromagnetic wave. The velocity of wave propagation along the helix structure is less than velocity of light. The beam and wave travel along the structure at the same speed. Thus interaction occurs between beam and wave and the beam delivers energy to the RF wave. Therefore the signal gets strengthened and amplified output is delivered at the other end of tube. The main features of TWT are :

1. Frequency range	0.5 GHz to 90 GHz
2. Power output	5 mW at low frequencies (less than 20 GHz) 250 kW (continuous wave) at 3 GHz 10 MW (pulsed) at 3 GHz
3. Efficiency	about 5 to 20%
4. Noise	about 5 dB for low power TWT 25 dB for high power TWT

TWT is used as RF amplifier in broad band microwave receivers, repeater amplifier in broad band communication systems, communication satellites etc.

13.6.4. Backward Wave Oscillator

It is somewhat similar to TWT and can deliver microwave power over a wide frequency band. It has an electron gun and a helix structure. However the interaction between electron beam and RF wave is different than in TWT. The growing RF wave travels in opposite direction to the electron beam. The frequency of wave can be changed by changing the voltage which controls the beam velocity. Moreover the amplitude of oscillations can be decreased continuously to zero by changing the beam current. Its features are :

1. Frequency range	1 GHz to 1000 GHz
2. Power output	10 mW to 150 mW (continuous wave) 250 kW (pulsed)

It is used as signal source in transmitters and instruments.

13.6.5. Magnetron

The operation of magnetron is based on interaction of electromagnetic fields with electrons moving in static electric and magnetic fields oriented at 90° with respect to each other. In a magnetron three forces act on electron viz. force due to electric field (equal to $-eE$), force due to magnetic field [equal to $-e(v \times B)$] and centrifugal force (equal to mv^2/r). The path of the electron can be found by balancing these three forces. Magnetron can be of three types, *i.e.*, Negative resistance device, cyclotron frequency device and travelling wave or cavity device.

In negative resistance magnetron use is made of the negative resistance between two anode segments. It has low efficiency and is used at frequencies less than 0.5 GHz.

In cyclotron magnetron a synchronism exists between ac component of electric field and periodic oscillation of electrons in a direction parallel to the field. It is used for frequencies higher than 100 MHz.

A cavity magnetron has a number of cylindrical cavities in the interaction region.

The cavity magnetron is the most common type of magneton. Its features are :

1. Frequency range	500 MHz to 10 GHz
2. Power output	250 kW (pulses)
3. Efficiency	about 50%

It is used in radar systems, industrial heating systems and microwave ovens.

13.7. SOLID STATE MICROWAVE DEVICES

The limitation of solid state devices at high frequencies include those associated with transit time and junction capacitances. The devices used are : Transferred electron oscillators (Gunn diode), Avalanche diode oscillators (Impatt diode, Trapatt diode, Masers, Lasers, Tunnel diode, Varactor etc.).

13.7.1. Gunn Diode Oscillator

A Gunn diode uses Ga As which has a negative differential mobility, *i.e.*, a decrease in carrier velocity with increase in electric field. This effect is called transferred electron effect. The impedance of a Gunn diode is tens of ohms.

A Gunn diode oscillator has a resonant cavity, an arrangement to couple Gunn diode to cavity, biasing arrangement for Gunn diode and arrangement to couple RF power to load.

Applications of Gunn diode oscillator include continuous wave radar, pulsed radar and microwave receivers.

13.7.2. Impatt Diode Oscillator

An Impatt diode has $n^+ - p - i - p^+$ structure and is used with reverse bias. It exhibits negative resistance and operates on the principle of avalanche breakdown. Impatt diode circuits are classified as broadly tunable circuit, low Q circuit and high Q circuit. The impedance of Impatt diode is a few ohms. The word Impatt stands for Impact Avalanche Transit Time diode. The features of Impatt diode oscillator are : frequency 1 to 300 GHz, Power output (0.5 W to 5 W for single diode circuit and upto 40 W for combination of several diodes), efficiency about 20%. Its applications include police radar systems, low power microwave transmitter etc.

13.7.3. Trapatt Diode Oscillator

Avalanche diode can also be operated in large signal high efficiency mode called Trapped Avalanche Transit Time mode. The Trapatt oscillations depend on the delay in the current caused by avalanche process. The avalanche delay makes it possible to increase the diode voltage well above the breakdown voltage. Therefore a very rapid multiplication of charge carriers occurs. A Trapatt diode is also a negative resistance device. The features of Trapatt diode oscillator are : Frequency 3 to 50 GHz, Power output 1–3 W, efficiency about 25%. Its applications are low power doppler radar, microwave beacon landing system etc.

13.7.4. MASERS (Microwave Amplification by Stimulated Emission of Radiation)

Materials suitable for maser application include ammonia, hydrogen, cesium, ruby etc. Ruby with slight doping of chromium (to make it paramagnetic) is the most commonly used one. By supplying additional energy the electrons are stimulated from their initial energy level. It is a low noise level amplifier. Its features are : Frequency 1.6 GHz, Gain 25 dB, Bandwidth 25 MHz. It is suitable for radio astronomy, radio telescope etc.

13.7.5. Varactor Diode

It is a reverse biased *p-n* junction. The application of reverse bias increases the thickness of depletion layer. This can be considered as a capacitive effect. The incremental capacitance C_T is defined as :

$$C_T = \left|\frac{dQ}{dV}\right|$$

where dQ is the increase in charge due to increase dV in voltage. This voltage variable capacitance C_T is used in tuning of LC resonant circuits, self balancing bridge circuits etc. A varactor diode is also called varicap or voltacap.

13.7.6. Tunnel Diode

A tunnel diode uses heavily doped *p* and *n* materials. Another feature is that the junction barrier is very small (about 10^{-6} cm). Due to this thin barrier there is a large probability that an electron will penetrate through this barrier. This effect is called tunnelling. A tunnel diode has negative resistance in a portion of its *i–v* characteristics. It is used in tunnel diode amplifier, pulse and digital circuits.

13.7.7. PIN Diode

A PIN diode has an intrinsic (*i*) layer between *p* and *n* layers. When reverse bias is applied depletion layers are formed at *p-i* and *i-n* junctions. The effective width of depletion layer increases by the width of layer. It can be used as a voltage controlled attenuator. At high frequencies the rectification effect ceases and impedance of diode is effectively that of *i* layer. This impedance varies with the applied bias. It is used in high frequency switching circuits, limiters, modulators etc.

13.7.8. Mesfet

Mesfet is a field effect transistor for microwave frequencies. It is fabricated in Ga As and uses a metal-semiconductor Schottky junction for the gate contact. It has a moderately doped *n*-type epitaxial layer on a high resistance semiconductor substrate. On the top surface of epitaxial layer are the source and drain contacts. Mesfets are used in high frequency logic circuits.

13.7.9. Microwave Integrated Circuits (MIC)

An MIC has a hybrid structure. The active and passive devices in chip form are mounted on substrates. Thin film technique is used in fabrication. Advantages of MIC include reduced cross-section for transmission lines, reduced wavelengths of guide, elimination of interconnections, flexibility of layout, use of lumped elements, reproducibility, reliability and lower cost.

13.7.10. Microstrip Line

A Microstrip line has a single dielectric substrate with ground plane on one side and a strip on the other face. Its special feature is easy access to top surface so that minor adjustments can be made after circuit fabrication. The high dielectric constant of the substrate reduces guide wavelength and circuit dimensions. A microstrip line, is the most commonly used transmission structure for microwave integrated circuits.

It has high Q and high gain.

13.8. MICROWAVE MEASUREMENTS

The quantities required to be measured in microwave circuits are frequency, power and impedance.

Microwave frequency measurement is done by slotted line, resonant cavities and transfer oscillator. Microwave power measurement uses bolometers and microwave power meters. A self balancing bridge working on the principle of power substitution is commonly used. Microwave impedance measurement is done by measurement of reflection coefficient and VSWR. A slotted line and probe is a basic tool for these measurements. Network analyser enables rapid impedance measurement over a broad frequency range.

13.9. FERRITE DEVICES

A ferrite is a ceramic like material with dielectric constant of about 10, relative permeability of several thousand and specific resistance about 10^{14} times that of metals. The chemical composition is manganese, oxygen, ferrous etc. ($MOFe_2O_3$). Ferrites are used in fabrication of microwave isolator, circulator, resonator etc.

A microwave isolator allows transmission of power without any attenuation in one direction but power flowing in reverse direction is completely absorbed by isolator. A microwave circulator has 3 or more ports. All the three ports are matched. A signal incident at port 1 appears at port 2 but not at port 3. A signal at port 2 appears only at port 3 (and not at 1). A signal at port 3 appears only at port 1 (and not at port 2).

Yttrium iron garnet (YIG) is a rare earth material having ferrimagnetic properties but very low losses. It is used in fabrication of microwave resonator. Its resonance frequency can be changed by changing magnetic field strength.

13.10. MISCELLANEOUS APPLICATIONS OF MICROWAVES

Microwaves are widely used in radar and communication systems. They are also used in industrial, scientific and medical field (ISM field). The frequencies specified in USA for ISM applications are 915 MHz, 2450 MHz, 5800 MHz, 10525 MHz and 20125 MHz.

Industrial applications of microwaves include measurement of thickness of metal sheets in rolling mills, continuous measurement of diameter of wires, measurement and monitoring of moisture content, heating, cooking and processing of food. Medical applications include diathermy, cancer treatment etc. Doppler motion sensor uses the Doppler effect *i.e.*, change of frequency of wave when reflected by moving object. Doppler motion sensor is used in vehicle speed monitors for traffic control and intrusion alarms.

OBJECTIVE TYPE QUESTIONS

A. Select the most appropriate alternative

1. If VSWR is infinite, the transmission line is terminated in
   (*a*) short circuit (*b*) complex impedance
   (*c*) open circuit (*d*) either (*a*) or (*c*)
2. A (75 – *j*50) ohm load is connected to a coaxial line of $Z_0 = 75$ ohm at 10 GHz. The best method of matching is to connect
   (*a*) an inductance at load
   (*b*) a short-circuited stub at load
   (*c*) a short-circuited stub at specific distance from load
   (*d*) a capacitance at specific distance from load
3. The velocity factor of a transmission line depends on
   (*a*) temperature
   (*b*) skin effect
   (*c*) relative permittivity of dielectric
   (*d*) none of the above
4. Impedance inversion may be obtained with
   (*a*) a short-circuited stub (*b*) a quarter wave line
   (*c*) an open-circuited stub (*d*) a half wave line
5. A short-circuited stub is preferred to an open-circuited stub because an open-circuited stub
   (*a*) is more difficult to fabricate and construct
   (*b*) is made of transmission line of different Z_0
   (*c*) is liable to radiate
   (*d*) is incapable of giving full range of reactance
6. For matching over a range of frequencies in a transmission line it is best to use
   (*a*) a balun
   (*b*) a broad band directional coupler
   (*c*) double stub
   (*d*) a single stub of adjustable position
7. To couple a coaxial line to a parallel wire line it is best to use
   (*a*) slotted line (*b*) balun
   (*c*) directional coupler (*d*) $\lambda/4$ transformer
8. As the frequency is increased, the charging MVAR in a cable
   (*a*) decreases (*b*) increases
   (*c*) remains the same
   (*d*) decreases or remains the same
9. As frequency is increased, the skin effect
   (*a*) increases (*b*) decreases
   (*c*) remains the same (*d*) either (*b*) or (c)
10. In a cable with $\epsilon_r = 4$, the wave velocity is
    (*a*) 3×10^8 m/s (*b*) 0.75×10^8 m/s
    (*c*) 1.5×10^8 m/s (*d*) 10^8 m/s
11. The characteristic impedance of a cable is about
    (*a*) 300 Ω (*b*) 50 Ω
    (*c*) 5 Ω (*d*) 2 Ω
12. A 10 km long line has a characteristic impedance of 400 ohms. If line length is 100 km, the characteristic impedance is
    (*a*) 4000 Ω (*b*) 400 Ω
    (*c*) 40 Ω (*d*) 4 Ω
13. When a wave reaches an open circuit
    (*a*) the voltage at termination is double the incident voltage
    (*b*) the current at termination is double the incident voltage
    (*c*) the voltage at termination is zero
    (*d*) both V and I at termination are double the incident voltage and current
14. A line of characteristic impedance 400 Ω is terminated by a resistance of 400 Ω. The reflected quantities are
    (*a*) zero
    (*b*) equal to incident quantities
    (*c*) half the incident quantities
    (*d*) double the incident quantities
15. A line of characteristic impedance Z_0 branches into two lines each with characteristic impedance Z_0. The refracted voltage is
    (*a*) zero
    (*b*) infinite
    (*c*) equal to incident voltage
    (*d*) two-thirds of incident voltage
16. Which of the following is used in integrated circuits
    (*a*) microstrip line (*b*) coaxial line
    (*c*) twin wire line (*d*) shielded cable
17. The primary constants of a transmission line are
    (*a*) R and L (*b*) R, L and C
    (*c*) R and C (*d*) R, L, G and C
18. The secondary constants of a line are
    (*a*) Z_0 (*b*) Z_0 and γ
    (*c*) Z_0, γ and wave velocity (*d*) Z_0, γ and L
19. In terms of R, L, G and C, the propagation constant γ of a transmission line is
    (*a*) $\sqrt{R + j\omega L}$
    (*b*) $\sqrt{(R + j\omega L)(G + j\omega C)}$
    (*c*) $\sqrt{G + j\omega C}$
    (*d*) $\sqrt{(R + j\omega L)/(G + j\omega C)}$
20. Which of the following parameters is negligible in transmission lines?
    (*a*) R (*b*) L
    (*c*) C (*d*) G
21. In a loss less line the propagation constant is equal to
    (*a*) $j\omega\sqrt{LC}$ (*b*) $\sqrt{LC}$
    (*c*) $j\omega LC$ (*d*) LC
22. The condition of minimum distortion in a transmission line is
    (*a*) RL = 1 (*b*) R = 0
    (*c*) $RG = \frac{L}{C}$ (*d*) $L = \frac{RC}{G}$
23. Loading of a line means
    (*a*) increasing inductance of line
    (*b*) increasing distributed inductance of a line

(*c*) decreasing distributed inductance of a line
(*d*) decreasing inductance of a line

24. The attenuation characteristics of a loaded line is similar to that of
(*a*) low pass filter (*b*) high pass filter
(*c*) band pass filter (*d*) band stop filter

25. Effect of loading of line is
(*a*) increase in Z_0, α and β
(*b*) increases in Z_0 and β but decrease in α
(*c*) decrease in Z_0, α and β
(*d*) decrease in Z_0 and β but increase in α

26. Which of the following terminations makes the input impedance of a line equal to characteristic impedance Z_0
(*a*) line open-circuited at far end
(*b*) line terminated in Z_0
(*c*) in short-circuited at far end
(*d*) line terminated in inductance

27. When a line is short-circuited at far end, the minimum voltage occurs at
(*a*) far end
(*b*) source end
(*c*) midway between source and far end
(*d*) none of the above

28. Which of the following lines behaves as infinite line
(*a*) a line terminated in inductance
(*b*) a line terminated in capacitance
(*c*) a short line
(*d*) a line terminated in Z_0

29. The inductance of a twin feeder used to connect antenna to TV receiver is about
(*a*) 0.2 μ H/m (*b*) 0.6 μ H/m
(*c*) 1.6 μ H/m (*d*) 10 μ H/m

30. When a line is loaded the characteristic impedance is equal to
(*a*) $\sqrt{(R + j\omega L)(G + j\omega C)}$ (*b*) $\sqrt{\frac{L}{C}}$
(*c*) $\sqrt{\frac{C}{L}}$ (*d*) $\sqrt{LC}$

31. Which of the following is not a travelling wave
(*a*) $e = E_m \sin(\beta x - \omega t)$ (*b*) $e = E_m \cos(\beta x - \omega t)$
(*c*) $e = E_m \sin(\omega t - \beta x)$ (*d*) $e = E_m \sin(\beta x)$

32. If ρ_v is reflection coefficient and VSWR is voltage standing wave ratio, then
(*a*) $\rho_v = \frac{VSWR - 1}{VSWR + 1}$ (*b*) $|\rho_v| = \frac{VSWR - 1}{VSWR + 1}$
(*c*) $\rho_v = \frac{VSWR + 1}{VSWR - 1}$ (*d*) $|\rho_v| = \frac{VSWR + 1}{VSWR - 1}$

33. A line has $Z_0 = 300 \angle 0\ \Omega$. If $Z_L = 150 \angle 0\ \Omega$, reflection coefficient is
(*a*) 0.5 (*b*) 0.3333
(*c*) –0.3333 (*d*) – 0.5

34. If a line having $Z_0 = 300 \angle 0\ \Omega$ is open-circuited at far end, VSWR is
(*a*) 0 (*b*) 1
(*c*) ∞ (*d*) 2

35. If a line having $Z_0 = 300 \angle 0\ \Omega$ is short-circuited at far end, VSWR is
(*a*) 0 (*b*) 1
(*c*) ∞ (*d*) 2

36. If load impedance $Z_L >> Z_0$, then
(*a*) $Z_0 = Z_L(VSWR)$ (*b*) $Z_L = Z_0(VSWR)$
(*c*) $Z_L = Z_0(VSWR - 1)$ (*d*) $Z_L = Z_0\sqrt{VSWR}$

37. Loss angle of a good quality cable is about
(*a*) 1° (*b*) 30°
(*c*) 70° (*d*) 90°

38. A quarter wave line short-circuited at load end behaves as
(*a*) an inductance
(*b*) an inductance and capacitance in series
(*c*) a capacitance
(*d*) an inductance and capacitance in parallel

39. A line slightly longer than $\frac{\lambda}{4}$ length and short-circuited at far end behaves as
(*a*) an inductance (*b*) a capacitance
(*c*) L and C in series (*d*) L and C in parallel

40. A quarter wave line open-circuited at far and behaves as
(*a*) inductance (*b*) L and C in parallel
(*c*) capacitance (*d*) L and C in series

41. A line of length slightly less than $\frac{\lambda}{4}$ and open-circuited at far end behaves as
(*a*) inductance L (*b*) capacitance C
(*c*) L and C in series (*d*) L and C in parallel

42. A line of length slightly less than $\frac{\lambda}{4}$ and short-circuited at far end behaves as
(*a*) inductance L (*b*) capacitance C
(*c*) L and C in series (*d*) L and C in parallel

43. A line of length slightly longer than $\frac{\lambda}{4}$ and open-circuited at far end behaves as
(*a*) inductance L (*b*) capacitance C
(*c*) L and C in series (*d*) L and C in parallel

44. A directional coupler is used to determine
(*a*) ρ_v (*b*) ρ_v and VSWR
(*c*) L and C (*d*) wave velocity

45. Reflection from the object to be measured is significant only when the wave length is
(*a*) much bigger than the size of object
(*b*) much smaller than the size of object
(*c*) almost of the size of object
(*d*) either (*a*) or (*c*)

46. High speed logic circuits use
(*a*) pulses with very small widths
(*b*) pulses with very large width
(*c*) pulses whose width is neither small nor large
(*d*) either (*b*) or (c)

47. The main advantage of using microwaves for communications is
(*a*) large bandwidth (*b*) small bandwidth
(*c*) low power (*d*) high power

48. Which part of radio spectrum is more crowded?
(*a*) lower frequency part (*b*) higher frequency part
(*c*) both (*a*) and (*b*) (*d*) none of the above

49. The number of FM radio channels which can be accommodated in a spectrum of 300 MHz?
(*a*) 10 (*b*) 100
(*c*) 500 (*d*) 1500

50. The number of TV channels which can be accommodated in a spectrum of 300 MHz is about
(*a*) 5 (*b*) 10
(*c*) 50 (*d*) 200

51. Microwaves are used for
(*a*) telephony (*b*) radio broadcast
(*c*) TV systems (*d*) all the above

52. As wavelength decreases the size of high directivity antenna
(*a*) decreases (*b*) increases
(*c*) is not affected (*d*) either (*b*) or (*c*)

53. Microwaves
(*a*) get reflected by ionosphere
(*b*) are absorbed by ionosphere
(*c*) are neither reflected nor absorbed by ionosphere
(*d*) are both reflected and absorbed by ionosphere

54. To study electromagnetic radiations from astronomical objects, radio astronomers use
(*a*) low frequencies (*b*) medium frequencies
(*c*) microwaves (*d*) either (*a*) or (*b*)

55. The earliest uses of electromagnetic waves were
(*a*) point-to-point communication
(*b*) radar
(*c*) both (*a*) and (*b*)
(*d*) neither (*a*) nor (*b*)

56. The width of a radio beam from a 1 m diameter parabolic antenna at 1 GHz is about
(*a*) 100° (*b*) 50°
(*c*) 10° (*d*) 2°

57. The width of a radio beam from a 1 m diameter parabolic antenna at 10 GHz is about
(*a*) 100° (*b*) 50°
(*c*) 5° (*d*) 1°

58. The AM, FM and TV broadcast frequencies are
(*a*) of the same order as microwave range
(*b*) much lower than microwave range
(*c*) much higher than microwave range
(*d*) either (*a*) or (*c*)

59. Satellite communication systems use the
(*a*) 10 to 20 MHz (*b*) 50 to 70 MHz
(*c*) 3 to 6 GHz (*d*) 100 to 120 GHz

60. The beam of a radar should be
(*a*) very narrow
(*b*) very wide
(*c*) neither too narrow nor very wide
(*d*) either (*b*) or (*c*)

61. Radar principle is used in
(*a*) detection of aircraft (*b*) burglar alarms
(*c*) garage door openers (*d*) all the above

62. The directive gain of a transmitting antenna is proportional to
(*a*) its cross-sectional area
(*b*) square of cross-sectional area
(*c*) square root of cross-sectional area
(*d*) cube root of cross-sectional area

63. The directive gain of a transmitting antenna is
(*a*) proportional to wavelength
(*b*) inversely proportional to wavelength
(*c*) proportional to square of wavelength
(*d*) inversely proportional to square of wavelength

64. As frequency increases, the transmission efficiency
(*a*) decreases (*b*) increases
(*c*) is not affected (*d*) either (*a*) or (*c*)

65. If other parameters are constant, transmission loss at 5 GHz as compared to that at 0.5 GHz is
(*a*) 20 dB more (*b*) 20 dB less
(*c*) 700 dB more (*d*) 100 dB less

66. Roughly the time required for microwave cooking as compared to conventional cooking is
(*a*) about twice (*b*) about half
(*c*) about 10 times (*d*) about one-tenth

67. As compared to dimensions of most foods, the skin depth at microwave frequencies is
(*a*) of the same magnitude (*b*) much larger
(*c*) much smaller (*d*) either (*b*) or (c)

68. Atomic and molecular resonance is observed in many substances
(*a*) at microwave frequencies
(*b*) at low frequencies
(*c*) at frequencies used in AM broadcast
(*d*) at both (*b*) and (*c*)

69. In microwave detector used in laboratory measurements of relative power levels, standing wave ratio is about
(*a*) 0.1 (*b*) 1.25
(*c*) 5 (*d*) 10

70. Which of the following lines is non-radiating
(*a*) open two wire (*b*) coaxial
(*c*) both (*d*) none of these

71. As compared to their values in free space
(*a*) velocity of electric wave in dielectric is smaller but wavelength of electric wave is more
(*b*) both velocity and wavelength of electric wave in dielectric are higher
(*c*) both velocity and wavelength of electric wave in dielectric are smaller
(*d*) wavelength of electric wave is less but velocity is more in dielectric

72. If wavelength in free space is λ_0 wavelength in insulator is λ, μ_r is relative permittivity of insulator and ϵ_r is relative permittivity of insulator, then
(*a*) $\lambda = \frac{\lambda_0}{\mu_r \epsilon_r}$ (*b*) $\lambda = \frac{\lambda_0}{\sqrt{\mu_r \epsilon_r}}$
(*c*) $\lambda = \frac{\lambda_0}{(\mu_r \epsilon_r)^2}$ (*d*) $\lambda = \frac{\lambda_0}{(\mu_r \epsilon_r)^{1.5}}$

73. Hollow wave guides are used as transmission lines at frequencies
(*a*) above 1 GHz (*b*) above 5 GHz
(*c*) above 10 MHz (*d*) above 1 MHz

74. Compared to coaxial lines, the advantages of wave guides are
(*a*) higher power handling capacity
(*b*) lower loss per unit length
(*c*) cheaper and simple mechanical structure
(*d*) all the above

75. For a wave guide which of the following statements is wrong
(*a*) It is usually made of brass or aluminium
(*b*) It is usually rectangular or circular in shape
(*c*) The dielectric region is usually air
(*d*) The wall thickness is very small as compared to skin depth

76. In a rectangular wave guide
(*a*) the broad width is generally labelled '*a*' and height '*b*'
(*b*) the broad width is generally labelled '*b*' and height '*a*'
(*c*) any of the dimensions may be labelled a and the other *b*
(*a*) the length is labelled as '*a*'

77. If '*a*' is the width of rectangular wave guide and 'λ' is the wavelength, then
(*a*) $a > \frac{\lambda}{4}$ (*b*) $a > \frac{\lambda}{2}$
(*c*) $a < \frac{\lambda}{4}$ (*d*) $a < \frac{\lambda}{2}$

78. A rectangular wave guide has two stubs each length $\frac{\lambda}{4}$ connected at the source end. The stubs are short-circuited. Then
(*a*) the longitudinal current is power current and current in stubs is reactive
(*b*) the longitudinal current is reactive current and current in stubs is power current
(*c*) both power and reactive currents exist in both longitudinal direction and in stubs
(*d*) the current in stubs is zero

79. A rectangular wave guide has two stubs each of length $\frac{\lambda}{4}$ connected at the source end. The stubs are short-circuited. Then the stubs present
(*a*) an infinite impedance in shunt with line
(*b*) an infinite impedance in series with line
(*c*) a zero impedance in shunt with line
(*d*) a finite impedance in shunt with line

80. A rectangular wave guide has width a. Then at the cut-off frequency (with corresponding wave length λ_c)
(*a*) $a = \frac{\lambda_c}{2}$ (*b*) $a = \lambda_c$
(*c*) $a = \frac{\lambda_c}{4}$ (*d*) $a = 2\lambda_c$

81. In a wave guide the dielectric region is generally
(*a*) air (*b*) magnetic material
(*c*) brass (*d*) mica

82. Most electromagnetic transmission in rectangular wave guide uses
(*a*) TE_{10} mode (*b*) TE_{01} mode
(*c*) TE_{20} mode (*d*) TE_{11} mode

83. Which mode is called dominant mode in rectangular wave guide
(*a*) TE_{01} (*b*) TE_{11}
(*c*) TE_{10} (*d*) TE_{20}

84. In TE_{10} mode of rectangular wave guide
(*a*) there is one half sine wave variation along *x* direction and none along *y* direction
(*b*) there is one half sine wave variation along *y* direction and none along *x* direction
(*c*) there is one full sine wave variation along *x* direction and none along y direction
(*d*) none of the above

85. The electrical requirement on wall thickness of rectangular wave guide is
(*a*) wall thickness is more than skin depth
(*b*) wall thickness is less than skin depth
(*c*) wall thickness is equal to skin depth
(*d*) either (*b*) or (*c*)

86. Generally the wall thickness of rectangular wave guide is
(*a*) about 1–5 mils (*b*) about 40–100 mils
(*c*) about 0.1 to 0.5 mils (*d*) about 500 mils

87. Which mode has the minimum cut off frequency in rectangular wave guides?
(*a*) TE_{11} (*b*) TE_{10}
(*c*) TE_{01} (*d*) TE_{20}

88. For a rectangular wave guide having width *a* and height *b*, the cut off wavelength for TM_{11} mode is equal to
(*a*) $\frac{ab}{\sqrt{a^2+b^2}}$ (*b*) $\frac{2ab}{\sqrt{a^2+b^2}}$
(*c*) $\frac{ab}{\sqrt{a^2-b^2}}$ (*d*) $\frac{2ab}{\sqrt{a^2-b^2}}$

89. Which of the following statements is not correct regarding modes of electromagnetic energy transmission?
(*a*) In TE mode the direction of electric field is always and everywhere transverse to the direction of propagation
(*b*) In TM mode the magnetic field pattern is always and everywhere transverse to the direction of propagation
(*c*) The subscripts is TE and TM designations denote the number of half sine wave variations of the field components in *x* and *y* directions respectively
(*d*) TM_{11} is called the dominant mode

90. TE_{01} mode of propagation of electromagnetic energy
(*a*) is merely the TE_{10} mode rotated 90°
(*b*) is merely the TE_{10} mode rotated 180°
(*c*) is the same as TE_{10} mode
(*d*) either (*b*) or (*c*)

91. If *a* and *b* are the width and height of rectangular wave guide, the cut off wavelength λ_c for TE_{20} mode is equal to
(*a*) *a* (*b*) $a + b$
(*c*) $a - b$ (*d*) $\sqrt{a^2+b^2}$

92. Out of modes TE_{20} and TE_{30} of electromagnetic energy
(*a*) both have the same cutoff frequency
(*b*) TE_{20} has lower cut-off frequency as compared to TE_{30}
(*c*) TE_{30} has lower cut-off frequency as compared to TE_{20}
(*d*) either (*a*) or (*c*)

93. TE_{10} mode of propagation of electromagnetic energy in wave guides can be thought of as
(*a*) three TEM waves travelling along length of wave guide
(*b*) one TEM wave travelling along length of wave guide and one TEM wave propagation in the transverse plane
(*c*) one TEM wave travelling along the length of wave guide and two TEM waves propagating in the transverse plane
(*d*) none of the above

94. At a frequency equal to cut-off frequency in a rectangular wave guide
(*a*) the flow of electromagnetic energy is maximum
(*b*) the flow of electromagnetic energy is zero
(*c*) the flow of electromagnetic energy is 50% of maximum
(*d*) the flow of electromagnetic energy is 10% of maximum

95. If μ_r is relative permeability ϵ_r is relative permittivity, c is velocity of light, f_c is cutoff frequency, phase velocity v_p (*i.e.*, speed with which a particular phase of the wave travels down a rectangular wave guide at frequency f) is given by

(*a*) $v_p = \dfrac{c/\sqrt{\mu_r \epsilon_r}}{\sqrt{1-(f_{c/f})^2}}$ (*b*) $v_p = \dfrac{c/\sqrt{\mu_r \epsilon_r}}{\sqrt{1-(f/f_c)^2}}$

(*c*) $v_p = \dfrac{c}{\sqrt{1-\left(\dfrac{f_c}{f}\right)^2}}$ (*d*) $v_p = \dfrac{c}{\sqrt{1-\left(\dfrac{f}{f_c}\right)^2}}$

96. If f is the frequency of electromagnetic wave, f_c is cutoff frequency, then in a rectangular wave guide
(*a*) attenuation is low when $f > f_c$
(*b*) attenuation is low when $f < f_c$
(*c*) attenuation is high when $f < f_c$
(*d*) either (*b*) or (*c*)

97. In a rectangular wave guide cutoff wavelength for TE_{10} mode is 8 cm. Then cutoff wave length for TE_{20} mode is
(*a*) 8 cm (*b*) 6 cm
(*c*) 4 cm (*d*) 2 cm

98. In TE_{10} mode the largest electric field, in a rectangular wave guide, occurs
(*a*) along centre line of broad wall (*i.e.*, $x = a/2$)
(*b*) along the broad wall (*i.e.*, $x = 0$ and $x = a$)
(*c*) either (*a*) or (*b*)
(*d*) neither (*a*) nor (*b*)

99. If Z_{TE} is wave impedance for TE waves, E_d is the dielectric strength of insulating material, a and b are the width and height of a rectangular wave guide, the maximum power handling capability P_{max} for TE_{10} mode is

(*a*) $\dfrac{ab}{Z_{TE}} E_d^2$ (*b*) $\dfrac{ab}{4Z_{TE}} E_d^2$

(*c*) $\dfrac{ab}{Z_{TE}} (E_d)^3$ (*d*) $\dfrac{ab}{4Z_{TE}} (E_d)^3$

100. For a l cm × 2 cm air filled rectangular wave guide the maximum frequency should be less than about
(*a*) 1000 MHz (*b*) 5000 MHz
(*c*) 14000 MHz (*c*) 140000 MHz

101. In rectangular wave guides the ratio of width to height is about
(*a*) 0.5 (*b*) 1
(*c*) 2 (*d*) 4

102. If air pressure in a rectangular wave guide is increased the maximum power handling capacity
(*a*) is decreased (*b*) is increased
(*c*) remains the same
(*d*) may increase or decrease depending on pressure

103. To ensure that only dominant mode TE_{10} is allowed to propagate in an air-filled rectangular wave guide, the lower frequency limit and upper frequency limit are
(*a*) about 25% above f_c for TE_{10} and 5% below f_c for TE_{20} mode respectively
(*b*) about 50% above f_c for TE_{10} and 25% below f_c for TE_{20} mode respectively
(*c*) about 25% below f_c for TE_{10} mode and 25% above f_c for TE_{20} mode respectively
(*d*) none of the above

104. For a circular wave guide
(*a*) cutoff frequency for TE_{10} and TE_{01} modes are the same
(*b*) cutoff frequency for TE_{10} and TE_{01} mode are somewhat different
(*c*) cutoff frequency for TE_{10} mode = twice the cut-off frequency for TE_{20} mode
(*d*) cutoff frequency for TE_{10} mode = half the cut-off frequency for TE_{20} mode

105. As regards rectangular and circular wave guides
(*a*) both have infinite set of TE and TM modes
(*b*) rectangular wave guide has infinite TE and TM modes but not the circular wave guide
(*c*) circular wave guide has infinite TE and TM modes but not the rectangular wave guide
(*d*) none of the above

106. In a circular wave guide the dominant mode is
(*a*) TE_{01} (*b*) TE_{11}
(*c*) TE_{20} (*c*) TE_{21}

107. Which mode has the lowest cutoff frequency in circular wave guides
(*a*) TE_{01} (*b*) TE_{11}
(*c*) TE_{20} (*d*) TE_{21}

108. If D is diameter of circular wave guide the cutoff wave length for TE_{11} mode is equal to
(*a*) 1.706 D (*b*) 2.11 D
(*c*) 0.82 D (*c*) 0.41 D

109. At millimeter wavelength the favoured mode in circular wave guide is
(*a*) TE_{01} (*b*) TE_{11}
(*c*) TE_{20} (*d*) TE_{21}

110. In which mode of circular wave guide does the attenuation continuously decrease as frequency is increased
(*a*) TE_{01} (*b*) TE_{11}
(*c*) TM_{01} (*d*) TE_{21}

111. Which modes of circular wave guides have circular symmetry?
(*a*) TE_{11} (*b*) TE_{11} and TE_{21}
(*c*) TE_{01} (*d*) TE_{01} and TM_{01}

112. The first and second subscripts in modes of circular wave guides depict
(*a*) the number of full sine wave variations in circumferential direction and the Bessel function variations in radial direction respectively
(*b*) the Bessel function variations in radial direction and the number of full sine wave variations in circumferential direction respectively
(*c*) the Bessel function variations in circumferential direction and number of full sine wave variations in radial direction respectively
(*d*) none of the above

113. For air filled circular wave guides
(*a*) the attenuation due to imperfect conducting walls is negligible
(*b*) the attenuation due to imperfect dielectric is negligible
(*c*) both the attenuation, *i.e.*, due to imperfect conducting walls and due to imperfect dielectric are significant
(*d*) the total attenuation is almost zero

114. In TE and TM modes of rectangular wave guides having propagation in *z* direction
(*a*) E_z and H_z are both zero
(*b*) In TE mode E_z is zero and in TM mode H_z is zero
(*c*) In TE mode H_z is zero and in TM mode E_z is zero
(*d*) In both TE and TM modes both E_z and H_z are non-zero

115. Which TM mode in rectangular wave guide has lowest cutoff frequency?
(*a*) TM_{11} (*b*) TM_{01}
(*c*) TM_{10} (*d*) TM_{21}

116. In which mode of rectangular waveguide are all field components zero
(*a*) TM_{00} (*b*) TM_{11}
(*c*) TM_{01} (*d*) TM_{10}

117. The usable bandwidth of ridge guide :
(*a*) is lower than that of rectangular wave guide
(*b*) is higher than that of rectangular wave guide
(*c*) is almost the same as that of rectangular wave guide
(*d*) is very very small as compared to that of rectangular wave guide

118. Consider the following statements
1. Usable bandwidth of ridge guide is more than that of rectangular wave guide
2. For a given value of TM_{10} cutoff frequency the cross section of ridge wave guide is lesser than that of rectangular wave guide
3. Ridge wave guide has higher conductor losser than a rectangular wave guide
4. Ridge guide has higher power handling capacity than rectangular wave guide

Which of the above statements are correct?
(*a*) 1 and 2 only (*b*) 1, 2 and 3 only
(*c*) 2 and 3 only (*d*) all

119. Match List I and List II and give the answer using the codes

List I / **List II**

A Integrated fin line — 1. 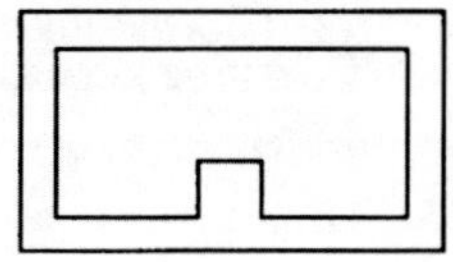

B Dielectric loaded wave guide — 2. 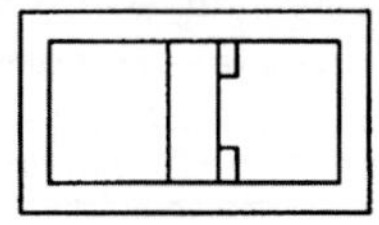

C Ridge waveguide — 3. 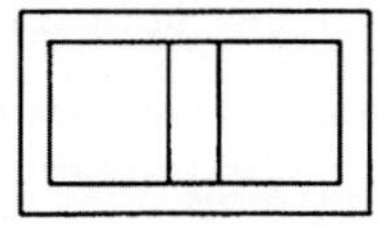

Codes :

	A	B	C
(*a*)	1	2	3
(*b*)	2	3	1
(*c*)	3	2	1
(*d*)	2	1	3

120. Which of these statements is correct?
(*a*) Ridge guide is essentially a form of capacitively loaded guide
(*b*) Ridge guide is essentially a form of inductively loaded guide
(*c*) Ridge guide is essentially a form of resistively loaded guide
(*d*) Ridge guide has higher power handling capacity than rectangular wave guide

121. The Q factor of a wave guide resonator is given by (ω_0 is resonant frequency, U is energy storage and W_L is the power loss)
(*a*) $Q = \frac{\omega_0 U}{W_L}$ (*b*) $Q = \frac{\omega_0 W_L}{U}$
(*c*) $Q = \omega_0 U W_L$ (*d*) $Q = \frac{U W_L}{\omega_0}$

122. The dominant mode in rectangular wave guide is
(*a*) TE_{01} (*b*) TM_{01}
(*c*) TE_{10} (*d*) TM_{11}

123. Match List I and List II and select the answer using the given code

List I	List II
A Reflex Klystron	1. High efficiency
B Magnetron	2. Slow wave structure
C TWT	3. Cross field device
D Multi Cavity Klystron	4. Oscillator

Codes :

	A	B	C	D
(*a*)	1	2	3	4
(*b*)	4	3	2	1
(*c*)	2	4	1	3
(*d*)	3	1	4	2

124. In Gunn diode electrons are transferred from
(*a*) high to low mobility energy bands
(*b*) low to high mobility energy bands
(*c*) valley to domain formation
(*d*) domain to valley formation

125. For a directional coupler the power is in the ratio of
(*a*) 40 dB (*b*) 30 dB
(*c*) 20 dB (*d*) 10 dB

126. In an optical fibre the refractive index of cladding material should be
(*a*) nearly 1 (*b*) very low
(*c*) less than that of core (*d*) more than that of core

127. Fibres have numerical aperture (NA) in the range of 0.15 to 0.4. Fibres with higher NA value would have
(*a*) reduced losses
(*b*) high bandwidth
(*c*) reduced losses and low bandwidth
(*d*) more losses and low bandwidth

128. Microwave links are typically 50 km apart
(*a*) because of atmospheric attenuation
(*b*) because of output tube power limitation
(*c*) because of earth's curvature
(*d*) to ensure that applied dc voltage is not excessive

129. Match List I and List II and give the answer using the given codes.

	List I		List II
A	T (reflection coefficient) = 0	1.	$\frac{Z_L - Z_0}{Z_L + Z_0}$
B	T = – 1	2.	$Z_L = Z_0$
C	T = + 1	3.	$Z_L = 0$
D	– 1 < T < + 1	4.	$Z_L = \infty$

Codes :

	A	B	C	D
(*a*)	2	3	4	1
(*b*)	1	2	3	4
(*c*)	2	4	1	3
(*d*)	3	2	4	1

130. A transmission line has $Z_0 = 300\ \Omega$ and $Z_L = (300 - j300)$ ohm. The transmission coefficient is
(*a*) $1.265\angle{-18.43°}$ (*b*) $1.01\angle{-10°}$
(*c*) $1.14\angle{66.68°}$ (*d*) $1.09\angle{66.68°}$

131. What is the numerical aperture in an optical fiber when critical angle is 30°?
(*a*) 0.5 (*b*) 0.704
(*c*) 0.866 (*d*) 0.2

132. If antenna diameter is increased four times, the maximum range is increased by a factor of
(*a*) $\sqrt{2}$ (*b*) 2
(*c*) 4 (*d*) 8

133. To couple a coaxial line to a parallel wire line it is best to use
(*a*) slotted line (*b*) balun
(*c*) directional coupler (*d*) λ/4 transformer

134. Consider the following statements :
SWR of a transmission line is infinite when the load is
1. a short circuit
2. a complex impedance
3. an open circuit
4. a pure reactance

Which of the above statements are correct?
(*a*) 1 and 3 only (*b*) 1, 2, 3
(*c*) 2, 3, 4 (*d*) 1, 2, 4

135. If the peak transmitted power in a radar is increased by a factor 81, the maximum range is increased by a factor of
(*a*) 3 (*b*) 9
(*c*) 27 (*d*) 81

136. A radar has a maximum range of 120 km. The maximum allowable pulse repetition frequency for unambiguous reception is
(*a*) 1250 (*b*) 330
(c) 2500 (*d*) 8330

137. Which of the following is not used as a microwave mixer or detector?
(*a*) PIN diode (*b*) Crystal diode
(*c*) Schottky barrier diode (*d*) Backward diode

138. A microwave junction is matched at all ports if the S matrix
(*a*) has all diagonal elements zero
(*b*) has all diagonal elements equal but not zero
(*c*) has all diagonal elements complex
(*d*) is Hermitian

139. Waveguides are pressurised above normal atmospheric pressure for
(*a*) increasing the power handling capacity
(*b*) improving conductivity of walls
(*c*) preventing higher modes from propagation
(*d*) varying the wave impedance

140. A duplexer is used to
(*a*) couple two antennas to a transmitter without interference
(*b*) isolate the antenna from the local oscillator
(*c*) prevent interference between two antennas connected to receiver
(*d*) use an antenna for reception or transmission without interference

141. In laboratory experiment the output from reflex Klystron is modulated by square wave because
(a) it is easy to generate a square wave
(*b*) crystal diode operates in square law region of *i-v* characteristics
(*c*) it prevents frequency modulation
(*d*) detector circuit is less complicated

142. Fields are said to be circularly polarized if
(*a*) their magnitudes are equal and they are in phase
(*b*) their magnitudes are equal and they differ in phase by ±90°
(*c*) their magnitudes are unequal and they differ in phase ±90°
(*d*) their magnitudes are unequal and they are in phase

143. Match List I and List II and give correct answer using the given codes

	List I (Microwave device)		List II (Application)
A	Reflex Klystron	1.	Wide band amplification
B	TWT	2.	Very low power amplification
C	2 cavity Klystron	3.	Low power FM generation
D	Maser	4.	Frequency multiplication

Codes :

	A	B	C	D
(a)	1	2	3	4
(b)	4	3	2	1
(c)	3	4	1	2
(d)	3	1	4	2

144. The semicondutor diode which can be used in switching circuit in microwave range is

(a) PIN diode (b) Varactor diode
(c) Tunnel diode (d) Gunn diode

145. Which of the following modes of transmission will not be supported by a rectangular wave guide

(a) TE_{10} (b) TE_{11}
(c) TM_{11} (d) TM_{10}

146. The skin depth at 1000 MHz as compared to that at 500 M Hz is

(a) 2 (b) $\sqrt{2}$
(c) 0.707 (d) 0.5

147. In a TWT the phase velocity of the axial component of the field on the slow wave structure is kept

(a) equal to velocity of electron
(b) slightly less than velocity of electron
(c) slightly more than velocity of electron
(d) equal to velocity of light in free space

148. A square wave guide carries TE_{11} mode whose axial magnetic field is given by $H_z = H_0 \cos \frac{\pi x}{\sqrt{8}}, \cos \frac{\pi x}{\sqrt{8}}$ where the guide dimensions are in cms. The cutoff frequency of the mode is

(a) 5 GHz (b) 7.5 GHz
(c) 6.5 GHz (d) 8 GHz

149. The reflection coefficient on a line is $0.2\angle 45°$. The SWR is

(a) 0.8 (b) 1.1
(c) 1.2 (d) 1.5

150. If the minimum range of a radar is to be doubled, the peak power has to be increased by a factor of

(a) 2 (b) 4
(c) 8 (d) 16

151. Match List I and List II and give the correct result using the given codes

List I (Solid state device)	List II (Application)
A PIN diode	1. Microwave ampiification
B Ga As Mosfet	2. Low noise microwave generation
C Transferred electron device	3. Electronic tuning of MW oscillator
D Varactor diode	4. Light wave detection

Codes :

	A	B	C	D
(a)	1	2	4	3
(b)	4	1	2	3
(c)	2	3	1	4
(d)	3	4	2	1

152. In π mode of operation of magnetron the spokes due to phase focussing effect rotate at an angular velocity corresponding to

(a) one pole/cycle (b) two pole/cycle
(c) four pole/cycle (d) six pole/cycle

153. In a rectangular wave guide there is one half wave variation of electric field across the narrow dimension and two half wave variations of electric field across wider dimension. The dominant mode is

(a) TM_{12} (b) TE_{12}
(c) TE_{21} (d) TM_{21}

154. Which one of the following is true for *em* wave propagation through rectangular wave guide of cross-section $2a \times a$

(a) critical wavelength for TE_{10} mode is the same as that for TE_{01} mode
(b) critical wavelength for TE_{10} mode is one half of that for TE_{01} mode
(c) critical wavelength for TE_{10} mode is the same as that for TM_{10} mode
(d) critical wavelength for TE_{10} mode is double that for TE_{01} mode

155. A TE_{10} rectangular wave guide is to be designed for operation over 25 -35 GHz and the band centre is 1.5 times the cutoff frequency. The dimension of broad side is

(a) 15 mm (b) 10 mm
(c) 9 mm (d) 7.5 mm

156. A disadvantage of microstrip line as compared to strip line is that former

(a) do not lend themselves to printed circuit technique
(b) are more likely to radiate
(c) are bulky
(d) are more expensive and complex

157. As compared to LED, LASER has

1. higher emission efficiency
2. no tuning arrangement
3. narrow spectral width
4. provision for confinement

Of these statements

(a) 1, 2, 3 are correct (b) 1, 2, 4 are correct
(c) 1 and 3 are correct (d) 1, 3, 4 are correct

158. Which one of the following is transferred electron device

(a) BARITT diode (b) IMPATT diode
(c) Gunn diode (d) Step recovery diode

159. The scattering matrix of magic tee is

(a) $\frac{1}{\sqrt{2}}\begin{bmatrix} 0 & 0 & 1 & 1 \\ 0 & 0 & -1 & 1 \\ 1 & -1 & 0 & 0 \\ 1 & 1 & 0 & 0 \end{bmatrix}$ (b) $\frac{1}{\sqrt{2}}\begin{bmatrix} 0 & 0 & 1 & 1 \\ 0 & 0 & 1 & 1 \\ 1 & 1 & 0 & 0 \\ 1 & 1 & 0 & 0 \end{bmatrix}$

(c) $\frac{1}{2}\begin{bmatrix} 0 & 0 & 1 & 1 \\ 0 & 0 & -1 & 1 \\ 1 & -1 & 0 & 0 \\ 1 & 1 & 0 & 0 \end{bmatrix}$ (d) $\frac{1}{2}\begin{bmatrix} 0 & 0 & 1 & 1 \\ 0 & 0 & 1 & 1 \\ 1 & 1 & 0 & 0 \\ 1 & 1 & 0 & 0 \end{bmatrix}$

160. A wave guide section in a microwave circuit acts as

(a) LP filter (b) band pass filter
(c) HP filter (d) band stop filter

161. Figure 13.10 shows an ideal 4 port circulator. The connections of

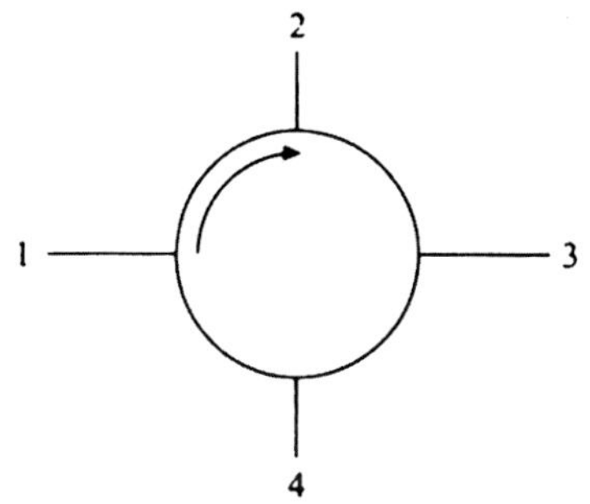

Fig. 13.10.

A microwave generator
B sychronising master source for injection locking
C antenna
D receiver are

	A	B	C	D
(*a*)	4	3	2	1
(*b*)	1	2	3	4
(*c*)	1	4	2	3
(*d*)	2	4	1	3

162. Two identical hybrid tees are joined as shown in Fig. 13.11. If 16 mW power is fed into port 1, the power output at ports 2, 3, 1′, 2′, 3′ respectively is

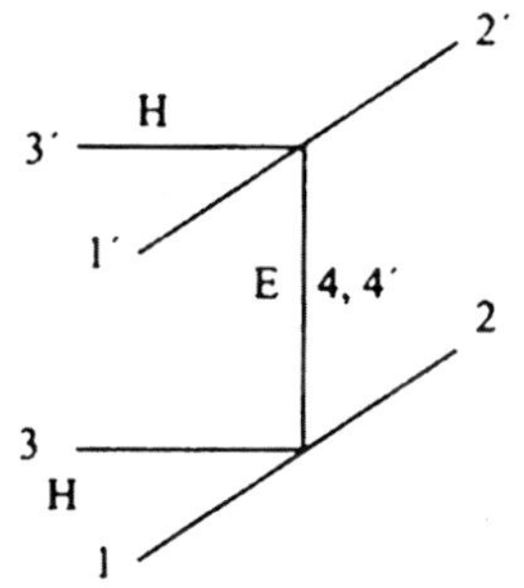

Fig. 13.11.

(*a*) 8, 4, 2, 2, 0 mW (*b*) 0, 8, 4, 4, 0 mW
(*c*) 0, 8, 2, 2, 4 mW (*d*) 8, 2, 4, 2, 0 mW

163. A 75 ohm line is first short terminated and minima locations are noted. Then the short is replaced by resistive load and minima locations are again noted. If minima location are not altered and VSWR is 3, the value of resistive load is
(*a*) 25 Ω (*b*) 50 Ω
(*c*) 225 Ω (*d*) 250 Ω

164. The transit time (in cycles) for electrons in repeller space of Reflex Klystron oscillator for sustaining oscillations is (n is any integer)
(*a*) $2(n-1)$ (*b*) $2n-1$
(*c*) $n+\frac{1}{2}$ (*d*) $n+\frac{3}{4}$

165. Which one of the following can be used for amplification of microwave energy
(*a*) TWT (*b*) magnetron
(*c*) Reflex klystron (*d*) Gunn diode

166. Guide wavelength (λ_g), cutoff wavelength (λ_c) and free space wavelength (λ_0) of a wave guide are related as
(*a*) $\frac{1}{\lambda_g^2}=\frac{1}{\lambda_0^2}-\frac{1}{\lambda_c^2}$ (*b*) $\frac{1}{\lambda_0^2}=\frac{1}{\lambda_g^2}-\frac{1}{\lambda_c^2}$
(*c*) $\frac{1}{\lambda_c^2}=\frac{1}{\lambda_0^2}-\frac{1}{\lambda_g^2}$ (*d*) $\frac{1}{\lambda_g}=\frac{1}{\lambda_0}-\frac{1}{\lambda_c}$

167. A loss less line of characteristic impedance Z_0 is terminated in pure reactance of $-j\,Z_0$ value. VSWR is
(*a*) 10 (*b*) 2
(*c*) 1 (*d*) infinity

168. In a circular wave guide with radius r, the dominant mode is
(*a*) TM_{01} (*b*) TE_{01}
(*c*) TM_{11} (*d*) TE11

169. Consider the following statements about a magic tee
1. The collinear arms arc isolated from each other
2. One of the collinear arms is isolated from E arm
3. One of the collinear arms is isolated from H arm
4. E and H arms are isolated from each other
Of the above statements
(*a*) 1 and 2 are correct (*b*) 1 and 3 are correct
(*c*) 1 and 4 are correct (*d*) 2 and 3 are correct

170. In a microwave power measurement using bolometer, the principle of working is variation of
(*a*) inductance with absorption of power
(*b*) resistance with absorption of power
(*c*) capacitance with absorption of power
(*d*) cavity dimension with heat generated by power

171. Match List I (component) with List II (dominant TE mode) and select the correct answer using the codes

	List I (Component)	**List II (Dominant TE mode)**
A	Rectangular wave guide	1. TE_{100}
B	Circular wave guide	2. TE_{10}
C	Rectangular cavity	3. TE_{111}
		4. TE_{11}

Codes:

	A	B	C
(*a*)	4	2	1
(*b*)	2	1	3
(*c*)	3	4	1
(*d*)	2	3	1

172. The main advantage of TWT over Klystron is
(*a*) higher bandwidth (*b*) higher gain
(*c*) higher frequency (*d*) higher output

173. The arrangement in Fig. 13.12 is a hybrid T. If 100 mW is fed at port 1, the power reflected back at port 1 is

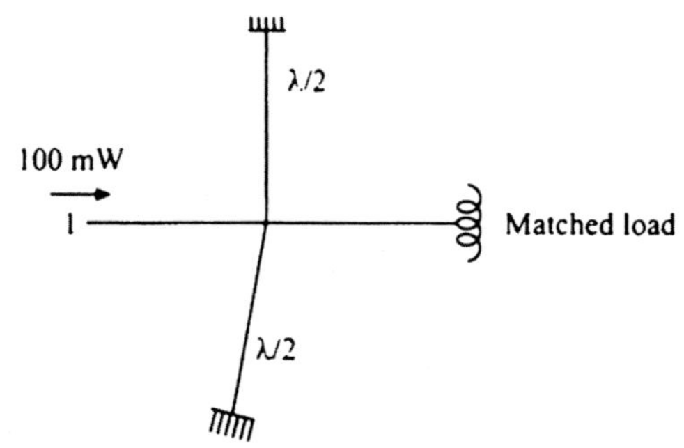

Fig. 13.12.

(*a*) 0 (*b*) 25 mW
(*c*) 50 mW (*d*) 100 mW

174. A 3-port circulator is shown in Fig. 13.13. Its scattering matrix is

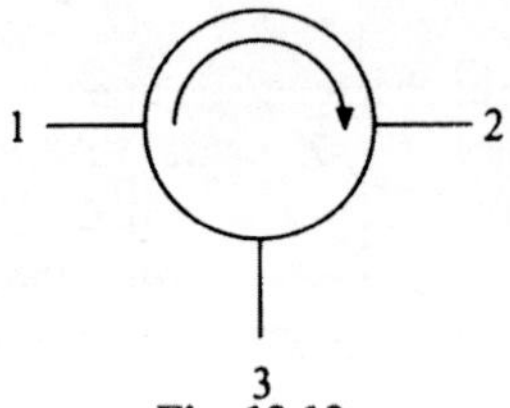

Fig. 13.13.

(*a*) $\begin{bmatrix} 0 & 0 & 1 \\ 1 & 0 & 0 \\ 0 & 1 & 0 \end{bmatrix}$ (*b*) $\begin{bmatrix} 1 & 0 & 0 \\ 0 & 1 & 0 \\ 0 & 0 & 1 \end{bmatrix}$

(*c*) $\begin{bmatrix} 1 & 0 & 1 \\ 0 & 0 & 0 \\ 1 & 0 & 1 \end{bmatrix}$ (*d*) $\begin{bmatrix} 1 & 1 & 1 \\ 1 & 0 & 1 \\ 1 & 1 & 1 \end{bmatrix}$

175. Figure 13.14 shows a 10 dB directional coupler. If power input at port 1 is 100 mW, power output at port 3 is

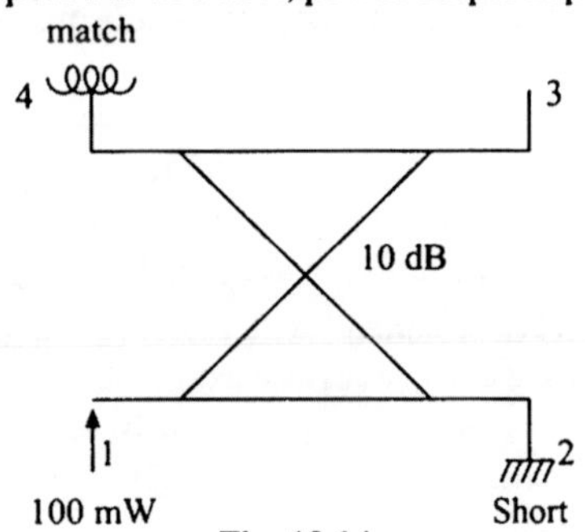

Fig. 13.14.

(*a*) 10 mW (*b*) 15 mW
(*c*) 20 mW (*d*) 25 mW

176. Which of the following are true for magic tee in Fig. 13.15. Select the answer as per given codes

1. $S_{13} = S_{23}$
2. $S_{14} = S_{24}$
3. $S_{12} = 0$
4. $S_{34} = 0$

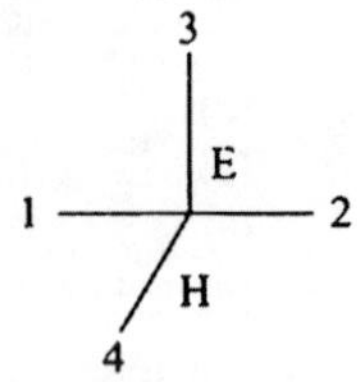

Fig. 13.15.

Codes:

(*a*) 1, 2 and 3 (*b*) 1, 2 and 4
(*c*) 2, 3 and 4 (*d*) 1, 3 and 4

177. A circular wave guide carries TE_{11} mode whole radial electric field is given by

$$E_r = E_0 J_1 (r) \sin \phi \text{ V}/m$$

where *r* is radial distance in cm from the axis. The cutoff wavelength is

(*a*) 10 cm (*b*) 3π cm
(*c*) 2π cm (*d*) 8 cm

178. The following components are used to measure power output of a 2 kW TWT amplifier:

1. TWTA
2. Low pass/High pass filter
3. Low power 20 dB attenuator
4. 40 dB directional coupler with matched load
5. Power meter

The correct sequence of connections of these components is:

(*a*) 1, 4, 2, 3, 5 (*b*) 1, 3, 4, 2, 5
(*c*) 1, 2, 4, 3, 5 (*d*) 1, 4, 3, 2, 5

179. In Fig. 13.16 power meters at A and B indicate 1 mW and 110 μW respectively. VSWR of line is

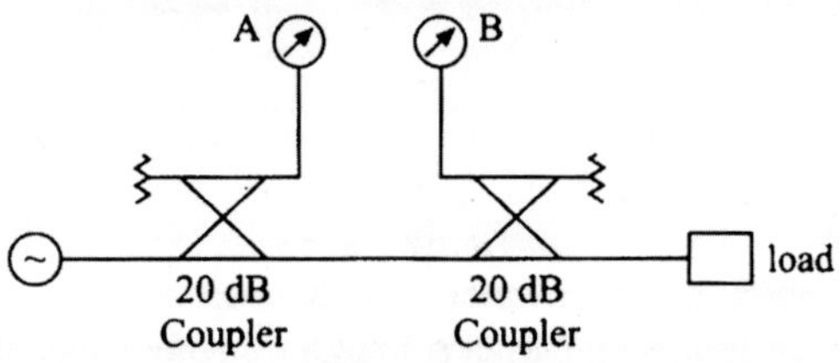

Fig. 13.16.

(*a*) 1.5 (*b*) 2
(*c*) 2.5 (*d*) 3

180. Match List I and List II and select the answer using the given codes :

List I (Transmission system)	List II (Mode)
A Rectangular wave guide	1. TE/TM
B Circular wave guide	2. TEM
C Coaxial line	3. Quasi TEM
D Microstrip line	

Codes:

	A	B	C	D
(*a*)	1	1	2	3
(*b*)	1	2	2	3
(*c*)	2	2	3	3
(*d*)	3	1	2	2

181. In microwave communication links the fading due to rain occurs. The methods to solve this problem include

(*a*) antenna replacement and feed correction
(*b*) amplitude trimming and phase correction
(*c*) polarization shifting and code diversity
(*d*) path diversity and frequency diversity

182. A coaxial RF cable has a characteristic impedance of 50 Ω and C equal to 40 pF/m. The inductance is

(*a*) 1 μ H/m (*b*) 10 μ H/m
(*c*) 0.1 μ H/m (*d*) 0.01 μ H/m

183. To ensure that only dominant TE_{10} mode propagates in a rectangular wave guide, the operating frequency

(*a*) must be more than cutoff frequency of TE_{10} mode
(*b*) must be less than cutoff frequency of TE_{20} mode
(*c*) must be more than cutoff frequency of TE_{10} mode and less than cutoff frequency of TE_{20} mode
(*d*) must be less than cutoff frequency of TE_{10} mode

184. TM_{01} mode in circular wave guide is analogous to

(*a*) TM_{10} mode in square wave guide
(*b*) TM_{11} mode in square wave guide
(*c*) TM_{20} mode in square wave guide
(*d*) TM_{02} mode in square wave guide

185. In a Gunn diode oscillator drift velocity of electron is 10^7 cm/s and active region length is 10×10^{-4} cm. The natural frequency of oscillation is

(*a*) 1 MHz (*b*) 10 MHz
(*c*) 1 GHz (*d*) 10 GHz

186. The correct sequence of parts in klystron amplifier are
(*a*) anode, catcher cavity, cathode, puncher cavity
(*b*) cathode, buncher cavity, catcher cavity, anode
(*c*) anode, buncher cavity, catcher cavity, cathode
(*d*) cathode, catcher cavity, anode, buncher cavity

187. Match List I and List II and select the answer using the codes

	List I		**List II**
A	Ratio of maximum energy stored to energy dissipated per cycle	1.	Propagation constant
B	TEM mode in loss less medium	2.	Cutoff frequency is zero
C	Ratio of frequency in radian to phase velocity of EM wave	3.	Quality factor of cavity
D	TEM is the mode of lowest cutoff frequency	4.	Cylindrical wave guide

Codes :

	A	**B**	**C**	**D**
(*a*)	3	2	4	1
(*b*)	2	3	1	4
(*c*)	3	2	1	4
(*d*)	2	3	4	1

188. The magnitudes of *oc* and *sc* input impedances of a transmission line an 100 Ω and 25 Ω. The characteristic impedance is
(*a*) 25 Ω (*b*) 50 Ω
(*c*) 75 Ω (*d*) 100 Ω

189. A TEM wave is incident normally on a perfect conductor. The E & H fields at boundary will respectively be
(*a*) minimum and maximum (*b*) minimum and minimum
(*c*) maximum and minimum (*d*) maximum and maximum

190. Frequency range for satellite communication is
(*a*) 1 kHz to 100 kHz (*b*) 100 kHz to 10 MHz
(*c*) 10 MHz to 100 MHz (*d*) 1 GHz to 30 GHz

191. Which of the following modes cannot exist in rectangular resonant cavity
(*a*) TE_{110} (*b*) TE_{011}
(*c*) TM_{110} (*d*) TM_{111}

192. The radiation resistance of a circular loop of one turn is 0.01 ohm. For 5 turn loop the radiation resistance is
(*a*) 0.002 Ω (*b*) 0.01 Ω
(*c*) 0.05 Ω (*d*) 0.25 Ω

193. In Fig. 13.17, reflection coefficient at load is

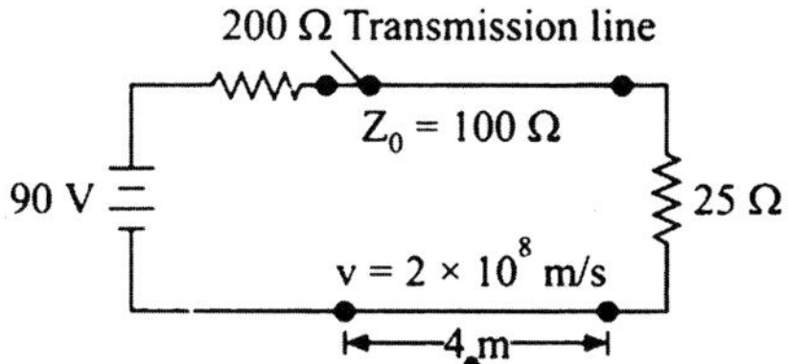

Fig. 13.17.

(*a*) 0.6 (*b*) – 0.6
(*c*) 0.4 (*d*) – 0.4

194. In Fig. 13.17 forward voltage wave has a magnitude of
(*a*) 90 V (*b*) 60 V
(*c*) 30 V (*d*) 10 V

195. In Fig. 13.17 forward current wave has a magnitude of
(*a*) 0.45 A (*b*) 0.3 A
(*c*) 0.2 A (*d*) 0.1 A

196. In Fig. 13.17 the time taken by the wave to make one round trip is
(*a*) 10 ns (*b*) 20 ns
(*c*) 30 ns (*d*) 40 ns

197. In Fig. 13.17 the reflection coefficient at source end is
(*a*) 1 (*b*) –1
(*c*) 1/3 (*d*) – 1/3

198. In Fig. 13.17 the reflected voltage wave after first reflection is
(*a*) 18 V (*b*) – 18 V
(*c*) 30 V (*d*) – 30 V

199. In Fig. 13.17 the reflected current wave after first reflection is
(*a*) 0.18A (*b*) – 0.18A
(*c*) 0.3 A (*d*) – 0.3 A

200. In Fig. 13.17 the line voltage after a number of reflections will be
(*a*) 30 V (*b*) – 30 V
(*c*) 10 V (*d*) – 10V

201. A line is excited by a 100 V, 200 ohm source and is terminated by 100 Ω resistance. If the voltage on line reaches steady value in 80 ns, the natural resonant frequency is
(*a*) 100 MHz (*b*) 50 MHz
(*c*) 12.5 MHz (*d*) 6.25 MHz

202. A line is excited by a 100 V dc source. If reflection coefficients at both ends are 1 each then
(*a*) there will be no oscillations on line
(*b*) there will be only 1 or 2 oscillations on line
(*c*) there will be a finite number of oscillations on line
(*d*) the oscillations will continue indefinitely

203. A coaxial line has L = 500 n H/m and C = 50 p F/m. The characteristic impedance is
(*a*) 500 Ω (*b*) 250 Ω
(*c*) 100 Ω (*d*) 50 Ω

204. A line has an attenuation of 0.054 Np/m. The attenuation in decibels is
(*a*) 4.7 dB/m (*b*) 0.47 dB/m
(*c*) 0.54 dB/m (*d*) 5.4 dB/m

205. A line has a phase constant of 29.8 rad/m. At 1000 MHz the wavelength is
(*a*) 29.8 m (*b*) 2.98 m
(*c*) 2.1 m (*d*) 0.21 m

206. In a high frequency coaxial line the characteristic impedance is a function of
(*a*) properties of insulating material only
(*b*) inner and outer conductor radius only
(*c*) both (*a*) and (*b*) (*d*) neither (*a*) nor (*b*)

207. In a coaxial line
(*a*) the characteristic impedance is less than that of two wire line
(*b*) the wave velocity is less than that of two wire line
(*c*) both (*a*) and (*b*)
(*d*) neither (*a*) nor (*b*)

208. Figure 13.18 shows the E & H lines in a coaxial cable
(*a*) The radial lines are E lines and circular lines are H lines
(*b*) Radial lines are H lines and circular lines are E lines
(*c*) The directions of E & H lines are wrong
(*d*) Radial lines may be E or H lines depending an direction of current

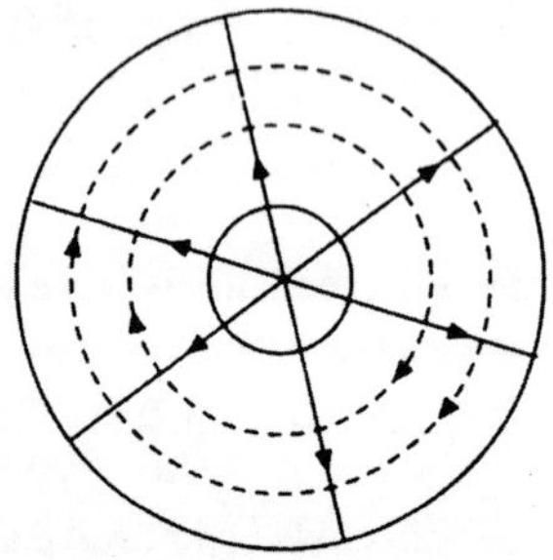

Fig. 13.18.

209. A loss less line is terminated in a finite value of R_L. Then
(*a*) SWR = 0 (*b*) SWR = ∞
(*c*) SWR is finite (*d*) SWR = 0.5

210. In a loss less line $R_L < Z_0$, then
(*a*) Reflection coefficient is zero
(*b*) Reflection coefficient is ∞
(*c*) Reflection coefficient is negative
(*d*) Reflection coefficient is positive

211. In a line with finite attenuation
(*a*) power transmission is least efficient when there are no standing waves on the line
(*b*) power transmission is most efficient when there are no standing waves on line
(*c*) load power depends on phase constant and length of line
(*d*) standing waves will always exist on the line

212. In a microwave coaxial cable the inner conductor radius is *a* and inner radius of outside conductor is *b*. For TE_{11} mode the cutoff wavelengths is about
(*a*) $a + b$ (*b*) $\pi(a + b)$
(*c*) πa (*d*) πb

213. The use of coaxial microwave line should be restricted to frequencies
(*a*) below cutoff frequency of TE_{11} mode
(*b*) above cutoff frequency of TE_{11} mode
(*c*) at frequencies much lower than cutoff frequency of TE_{11} mode
(*d*) at frequencies much higher than cutoff frequency of TE_{11} mode

214. In a microwave coaxial line, the maximum operating frequency is f_{max} and breakdown strength of dielectric is E_d. Then maximum power handling capacity is proportional to
(*a*) $\frac{E_d}{f_{max}}$ (*b*) $\left[\frac{E_u}{f_{max}}\right]^2$
(*c*) $\frac{E_d}{f_{max}^2}$ (*d*) $\frac{E_d^2}{f_{max}}$

215. The manufacturing technique used to manufacture strip type transmission line is
(*a*) photoetching technique (*b*) oxidation technique
(*c*) cladding (*d*) epitaxial

216. For the strip line shown in Fig. 13.19, the capacitance per unit length =

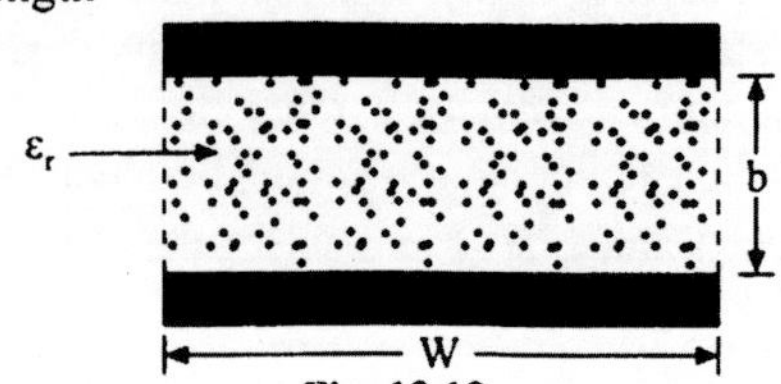

Fig. 13.19.

(*a*) $\frac{W}{b}\epsilon_0\epsilon_r$ (*b*) $\frac{b}{W}\epsilon_0\epsilon_r$
(c) $\frac{W}{b\epsilon_0\epsilon_r}$ (*d*) $\frac{b}{W\epsilon_0\epsilon_r}$

217. The fabrication of microstrip line is done by
(*a*) photo etching (*b*) printed circuit technique
(*c*) oxidation (*d*) cladding

218. If a microstrip is enclosed in a metal enclosure
(*a*) the radiation losses are increased
(*b*) the radiation losses are reduced
(*c*) the capacitance becomes zero
(*d*) the inductance becomes zero

219. In microstrip circuits, the dielectric material most commonly used is
(*a*) silicon (*b*) rubber
(*c*) plastic (*d*) bakelite

220. A microstrip line has dielectric having relative permittivity ϵ_r. The effective dielectric constant is
(*a*) more than ϵ_r (b) equal to 1
(*c*) less than ϵ_r (*d*) either (*a*) or (*b*)

221. A microstrip line has $\epsilon_r = 9$. The normalised wavelength $\left(\frac{\lambda}{\lambda_0}\right)$ is about
(*a*) 10 (*b*) 2
(*c*) 1 (*d*) 0.4

222. Match List I and List II and select the correct answer using the codes

	List I	List II
A	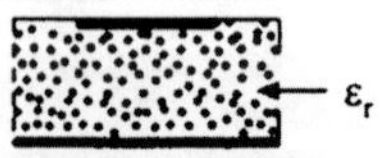	1. Slot line
B	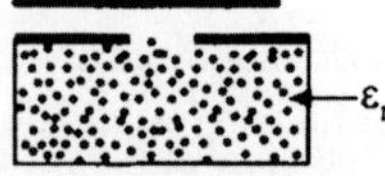	2. Coplanar line
C	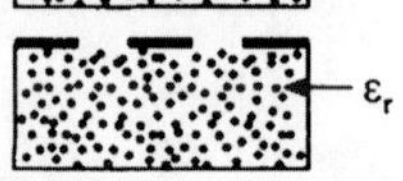	3. Microstrip line

Codes:

	A	B	C
(*a*)	1	2	3
(*b*)	3	2	1
(*c*)	2	3	1
(*d*)	3	1	2

223. For $l << \lambda$, the equivalent circuit of a microwave resistor is

(*a*)

(*b*) 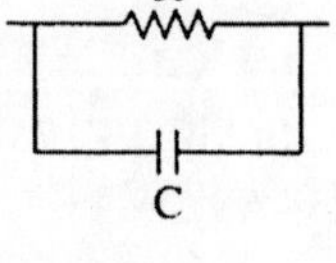

(c) 

(d) 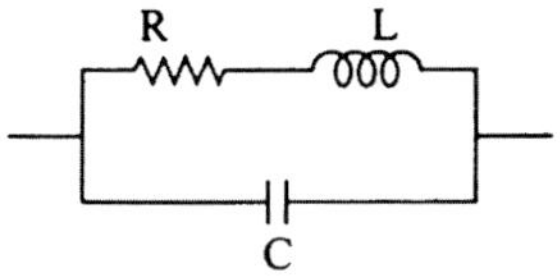

224. In a microwave resistor the wall thickness is made as small as possible to
(a) have high value of R (b) have low value of R
(c) minimize capacitive effect (d) reduce inductance

225. For a 50 Ω resistor for 3 GHz application, the stray capacitance should be less than
(a) 1 μF (b) 1nF
(c) 1 pF (d) 0.1 pF

226. A balun should have
(a) low SWR at both ports
(b) high SWR at both ports
(c) high SWR at input port and low SWR at output port
(d) low SWR at input port and high SWR at output port

227. Which of the following is a balanced line?
(a) coaxial line (b) strip line
(c) microstrip line (d) TV twin lead

228. Consider the following statements
1. Dissipative attenuator has a fixed value of attennuation
2. Reflective attenuator has a fixed value of attenuation
3. Both dissipative and reflective attenuators are available only with fixed attenuation
4. Both dissipative and reflective attenuators are available with either fixed or variable attenuation

Which of the above are correct?
(a) 1 only (b) 1 and 2
(c) 1, 2 and 3 (d) 4 only

229. Figure 13.20 shows the equivalent circuit of a T attenuator. If $\overline{R}_1 = \frac{R_1}{Z_0}$ and $\overline{R}_2 = \frac{R_2}{Z_0}$ the condition for matched attenuator is

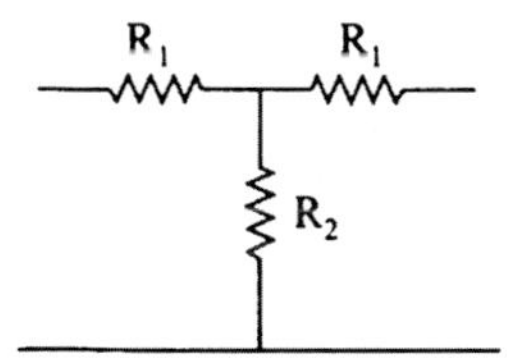

Fig. 13.20.

(a) $\overline{R}_2 = \frac{1-\overline{R}_1^2}{2\overline{R}_1}$ (b) $\overline{R}_1 = \frac{1-\overline{R}_2^2}{2\overline{R}_2}$
(c) $\overline{R}_2 = 1-\overline{R}_1^2$ (d) $\overline{R}_2 = \overline{R}_1^2$

230. A resistive microwave load with $Z_L = 150\ \Omega$ is connected to 50 Ω coaxial line. SWR is
(a) 1 (b) 2
(c) 3 (d) 3.5

231. If a matched attenuator is connected between a 150 Ω resistive load and 50 Ω coaxial line the SWR will be
(a) more than 3 (b) less than 3
(c) equal to 3 (d) either (a) or (c)

232. The component in Fig. 13.21 is

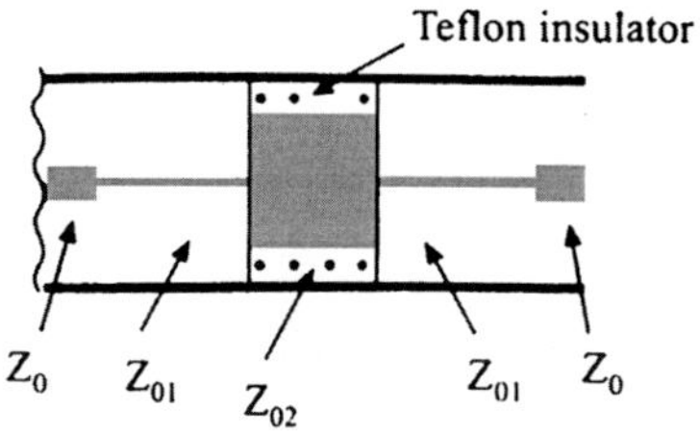

Fig. 13.21.

(a) attenuator (b) T type low pass filter
(c) T type high pass filter (d) phase shifter

233. The microwave circuits a mode transducer is used to
(a) change mode of wave transmission
(b) change direction of wave transmission
(c) change polarization of wave transmission
(d) change mode, direction or polarization of wave transmission

234. In microwave systems the function of mode filter is
(a) to suppress modes with lower cutoff frequencies
(b) to suppress modes with higher cutoff frequencies
(c) to change mode of wave transmission
(d) both (b) and (c)

235. Which of the following modes in circular wave guides are circularly symmetric?
(a) TE_{01} (b) TE_{02}
(c) both (a) and (b) (d) neither (a) nor (b)

236. In mode filter of Fig. 13.22, which of the following modes is unaffected

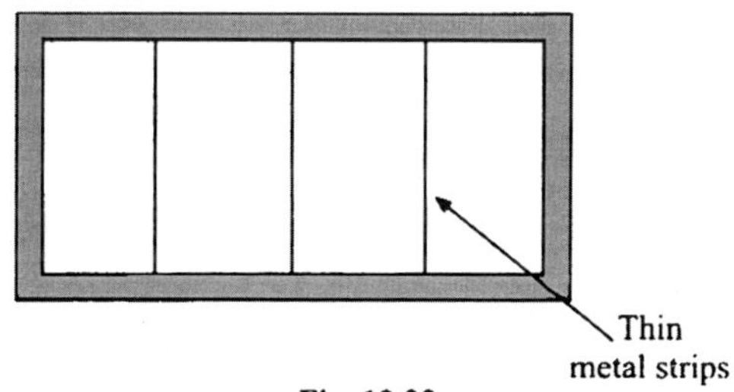

Fig. 13.22.

(a) TE_{01} (b) TE_{10}
(c) TE_{11} (d) all TM

237. In microwave systems, waveguide twists
(a) change the plane of polarization of a propagating wave
(b) change the direction of a propagating wave
(c) change the mode of a propagating wave
(d) both (b) and (c)

238. A rectangular wave guide has a discontinuity in the form of reduced narrow dimension. In the equivalent circuit this discontinuity can be represented by
(a) a shunt capacitance at the discontinuity
(b) a shunt inductance at the discontinuity
(c) a shunt resistance at the discontinuity
(d) none of the above

239. Which of the following is wrong for a magic tee
(a) E and H arms are decoupled
(b) Coplanar arms are coupled
(c) all ports are perfectly matched
(d) A signal into coplanar arm splits equally between E and H arms

240. Figure 13.23 shows the equivalent circuit of a magic tee. If all ports are matched

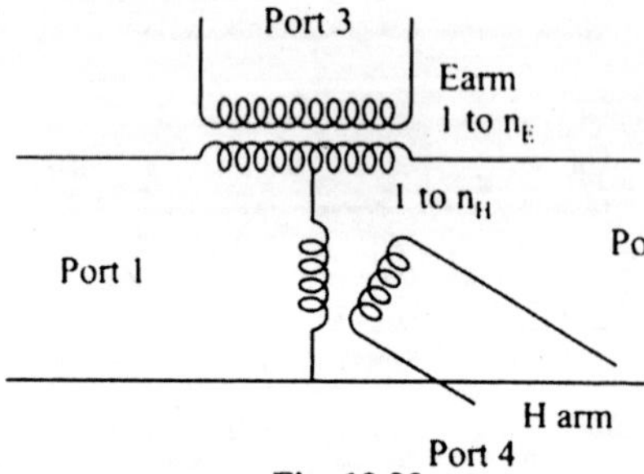

Fig. 13.23.

(*a*) $n_E = n_H = \sqrt{2}$ (*b*) $n_E = n_H = 0.707$
(*c*) $n_E = \sqrt{2}, n_H = 0.707$ (*d*) $n_E = 0.707, n_H = \sqrt{2}$

241. A matched generator ($Z_g = Z_0$) with 1W available power is connected to H arm of a magic tee. If all ports are matched, the power delivered at ports 1, 2, 3 respectively are
(*a*) 0.5 W, 0.5 W, 0 W (*b*) 0 W, 0 W, 1 W
(*c*) 0.33 W each (*d*) 0 W, 0.25 W, 0.25 W

242. A 'rat race' is a
(*a*) hybrid junction for microwaves
(*b*) a microwave oscillator
(*c*) a microwave amplifier
(*d*) a microwave antenna

243. Which one of the following is also called 'rat race'?
(*a*) E plane tee (*b*) H plane tee
(*c*) magic tee (*d*) hybrid ring

244. Consider the following statements about hybrid ring
1. Ports 1 and 2 are decoupled
2. Port 3 and 4 are coupled
3. A signal entering either port 1 or 2 splits equally between ports 3 and 4
4. A signal entering either port 3 or 4 splits equally between ports 1 and 2

Which of the above are correct?
(*a*) 1, 2, 3 and 4 (*b*) 1, 3 and 4
(*c*) 1 and 2 (*d*) 2, 3 and 4

245. Figure 13.24 shows the circuit for input impedance of a microwave device with output ports match terminated. The device is

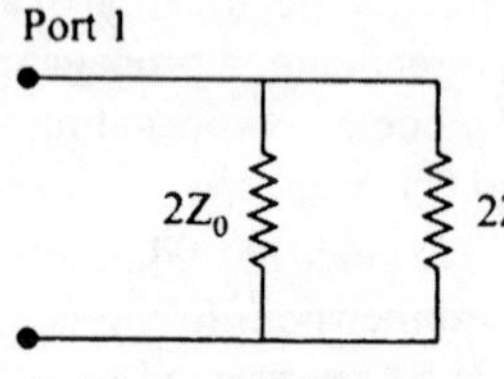

Fig. 13.24.

(*a*) E plane tee (*b*) H plane tee
(*c*) magic tee (*d*) hybrid ring

246. The path length between ports 1 and 2 for the two waves in a hybrid ring is
(*a*) λ (*b*) $\frac{3\lambda}{4}$
(*c*) $\frac{\lambda}{2}$ (*d*) $\frac{\lambda}{4}$

247. A directional coupler is used
(*a*) to transmit microwave signals
(*b*) to generate microwave signals
(*c*) to measure amplitude and phase of a travelling wave
(*d*) both (*a*) and (*b*)

248. Consider the following statements about directional coupler shown in Fig. 13.25
1. Ports 1 and 2 are decoupled
2. Ports 3 and 4 are decoupled
3. Ports 1 and 3 are decoupled
4. Ports 1 and 4 are decoupled

Which of the above statements are correct?

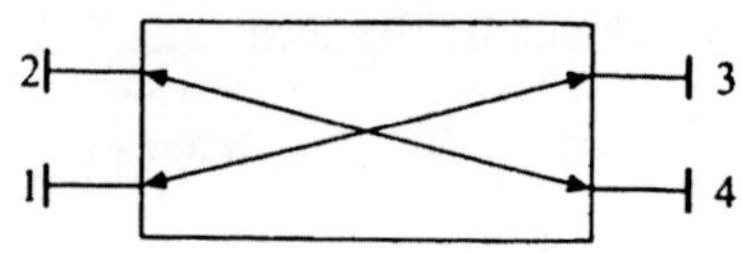

Fig. 13.25.

(*a*) 1 and 2 only (*b*) 3 and 4 only
(*c*) 1, 2 and 3 (*d*) 2, 3 and 4

249. In directional coupler of Fig. 13.25, the terms $|S_{14}|$ and $|S_{23}|$ of scattering matrix are nearly
(*a*) 1 each (*b*) 0.5 each
(*c*) 0.1 each (*d*) 1 and 0.5 respectively

250. The two terms used to describe performance of a directional coupler are
(*a*) coupling and directivity (*b*) gain and coupling
(*c*) gain and directivity (*d*) gain and isolation

251. In a directional coupler
(*a*) isolation (dB) equals coupling plus directivity
(*b*) coupling (dB) equals isolation plus directivity
(*c*) directivity (dB) equals isolation plus coupling
(*d*) isolation (dB) equals (coupling) (directivity)

252. Power monitor and reflectometer use the principle of
(*a*) directional coupler (*b*) klystron amplifier
(*c*) Gunn diode (*d*) TWT

253. If Q_L is loaded Q, Q_E is external Q and Q_U is unloaded Q of a microwave resonator, then
(*a*) $\frac{1}{Q_L} = \frac{1}{Q_E} + \frac{1}{Q_U}$ (*b*) $\frac{1}{Q_E} = \frac{1}{Q_L} + \frac{1}{Q_U}$
(*c*) $\frac{1}{Q_U} = \frac{1}{Q_L} + \frac{1}{Q_E}$ (*d*) $Q_L = Q_E + Q_U$

254. Figure 13.26 shows a short-circuited transmission line resonator. If $n = 1, 2, 3...$

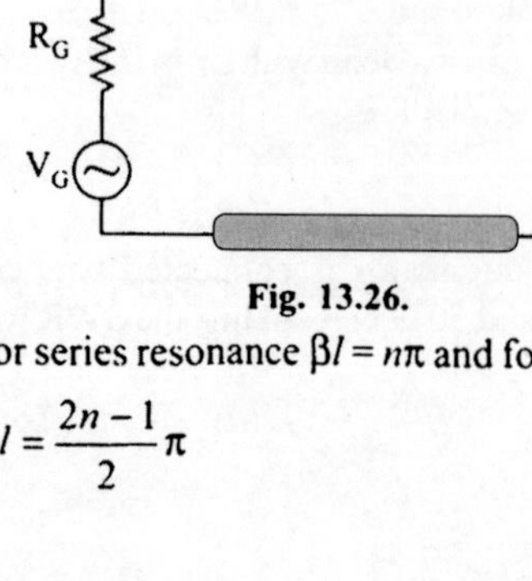

Fig. 13.26.

(*a*) For series resonance $\beta l = n\pi$ and for parallel resonance $\beta l = \frac{2n-1}{2}\pi$

(b) For series resonance $\beta l = \frac{2n-1}{2}\pi$ and for parallel resonance $\beta l = n\pi$
(c) $\beta l = n\pi$ for both series and parallel resonance
(d) $\beta l = \frac{2n-1}{2}\pi$ for both series and parallel resonance

255. A cavity resonator is
(a) a hollow metallic enclosure
(b) a hollow enclosure having magnetic material as its walls
(c) a hollow enclosure having dielectric material as its walls
(d) either (b) or (c)

256. Microwave resonators are used in
(a) microwave oscillators
(b) microwave narrow band amplifier
(c) microwave frequency metres
(d) all the above

257. A magic tee is to be used as CW duplexer. Then port 1, port 2, port 3 (Earm), port 4 (H arm) respectively should be connected to
(a) matched load, antenna, receiver and CW transmitter
(b) CW transmitter, antenna, receiver and matched load
(c) CW transmitter, receiver, antenna and matched load
(d) CW transmitter, matched load, receiver and antenna

258. A TR (transmitter receiver) tube has
(a) low loss for low power signals and high reflection loss for high load signals
(b) high loss for low power signals and low loss for high level signal
(c) low loss for both low level and high level signals
(d) high loss for both low level and high level signals

259. A branched duplexer requires
(a) TR tubes (b) ATR tube
(c) both TR and ATR tube (d) none of the above

260. The cutoff wave length in circular wave guide is proportional to
(a) guide diameter
(b) square of guide diameter
(c) cube of guide diameter
(d) square root of guide diameter

261. The expression
$H_z = C\cos(Bx)\cos(Ay)e^{-\gamma z}$ is for
(a) TE wave (b) TM waves
(c) both TE and TM waves
(d) some TE and some TM waves

262. In a vacuum tube, the transit time of electron between cathode and anode is important at
(a) low frequencies (b) high frequencies
(c) both (a) and (b)
(d) frequencies which are neither very low nor very high

263. In a three cavity klystron amplifier, the oscillations are excited in
(a) input cavity (b) output cavity
(c) intermediate cavity (d) both (a) and (b)

264. In a klystron amplifier the bunching effect
(a) converts velocity modulation into current modulation of beam
(b) converts current modulation into velocity modulation of beam
(c) both (a) and (b)
(d) neither (a) nor (b)

265. A reflex klystron oscillator uses
(a) one cavity resonator (b) two cavity resonators
(c) three cavity resonators (d) none of the above

266. In a reflex klystron oscillator, repeller electrode is at
(a) low positive potential (b) high positive potential
(c) negative potential (d) zero potential

267. Which device has internal positive feedback
(a) two cavity klystron amplifier
(b) multi cavity klystron amplifier
(c) Reflex klystron oscillator
(d) all the above

268. Consider the following statements :
1. Bunching of electrons occurs in two cavity klystron amplifier
2. Bunching of electrons occurs in multi cavity klystron amplifier
3. Bunching of electrons occurs in reflex klystron oscillator

Which of the above statements are correct?
(a) 1 and 2 only (b) 1, 2 and 3
(c) 1 and 3 only (d) 2 and 3 only

269. In a klystron amplifier the transit time of electron in passing the gap in the cavity should be
(a) less than one cycle (b) less than half cycle
(c) less than quarter cycle (d) more than quarter cycle

270. In a klystron amplifier the input cavity is called
(a) buncher (b) catcher
(c) Pierce gun (d) collector

271. In a klystron amplifier the force on all electrons
(a) causes acceleration of all electrons
(b) causes retardation of all electrons
(c) causes acceleration of some electrons and retardation of others
(d) none of the above

272. Power gain of a multicavity klystron amplifier is
(a) more than 30 dB (b) about 3 dB
(c) less than 3 dB (d) – 3 dB

273. The noise figure of klystron amplifier is about
(a) 25 dB (b) 3 dB
(c) zero (d) 1 dB

274. The diagram drawn to show distance time history of electrons in klystron amplifier is called
(a) apple gate diagram
(b) synchronous diagram
(c) bunching diagram
(d) velocity modulation diagram

275. Klystron amplifier is not used to amplify very weak microwave signals due to
(a) low gain for weak signals
(b) low noise for weak signals
(c) high noise
(d) both (a) and (b)

276. A reflex klystron oscillator is a
(*a*) low power device (*b*) high power device
(*c*) high efficiency device (*d*) both (*b*) and (*c*)

277. If the repeller voltage in a klystron oscillator is increased the transit time of electron
(*a*) will be increased (*b*) will be decreased
(*c*) will remain the same (*d*) either (*a*) or (*c*)

278. The frequency of oscillations obtained from a reflex klystron depends mainly on
(*a*) resonant frequency of cavity
(*b*) repeller voltage
(*c*) voltage of focussing electrode
(*d*) characteristics of cathode

279. Which of the following devices uses a helix
(*a*) klystron amplifier (*b*) klystron oscillator
(*c*) TWT (*d*) both (*a*) and (*c*)

280. In a TWT the loss in attenuator is
(*a*) almost equal to forward gain of tube
(*b*) much less than forward gain of tube
(*c*) much higher than forward gain of tube
(*d*) almost infinite

281. Read the following statements about TWT
1. It uses thermionic emission
2. It uses an attenuator
3. It is inherently a resonant device
4. It has broad bandwidth
Which of the above are correct?
(*a*) 1, 2, 3 and 4 (*b*) 1,2, end 3
(*c*) 1, 3 and 4 (*d*) 1, 2 and 4

282. In a TWT, the amplitude of resultant wave travelling down the helix
(*a*) increases exponentially (*b*) increases linearly
(*c*) decreases exponentially (*d*) is almost constant

283. The electric field, in a TWT, due to applied signal
(*a*) is directed along the helix axis
(*b*) is directed radially from halix axis
(*c*) is inclined to the helix axis by about 60°
(*d*) is inclined to the helix axis by about 45°

284. In a TWT the axial component of electric field advances along the tube
(*a*) at a velocity that is almost equal to speed of light
(*b*) at a velocity that is a small fraction of speed of light
(*c*) at a velocity that is about 50% of speed of light
(*d*) at a velocity that may be even more than speed of light

285. In a backward wave oscillator the wave
(*a*) travelling along the line winds itself back and forth
(*b*) progresses only in forward direction
(*c*) progresses only in backward direction
(*d*) either (*a*) or (*c*)

286. In a backward wave oscillator the efficiency
(*a*) is much higher than in TWT
(*b*) is much lower than in TWT
(*c*) is almost the same as in TWT
(*d*) either (*a*) or (*b*)

287. The action of backward wave oscillator is somewhat similar to that of
(*a*) klystron amplifier
(*b*) Reflex klystron oscillator
(*c*) TWT
(*d*) magnetron

288. Which of the following devices uses a slow wave structure?
(*a*) klystron two cavity amplifier
(*b*) klystron multi cavity amplifier
(*c*) Reflex klystron oscillator
(*d*) TWT

289. In a sheath helix
(*a*) the spacing between adjacent turns is very large
(*b*) the spacing between adjacent turns approaches zero
(*c*) the spacing between adjacent turns approaches zero and wire thickness approaches zero
(*d*) the spacing between adjacent turns is very large but wire thickness approaches zero

290. In which of the following devices can the operation be visiualised in terms of a series of feedback loops
(*a*) TWT
(*b*) Reflex klystron
(*c*) klystron amplifier
(*d*) backward wave oscillator

291. The frequency of oscillation in a backward wave oscillator can be changed by
(*a*) varying the voltage which controls beam velocity
(*b*) varying the beam current
(*c*) both by varying the beam current and by varying the voltage which controls beam velocity
(*d*) changing the rate of thermionic emission

292. The external magnetic field in a magnetron is such that flux lines are
(*a*) parallel to the axis of cathode
(*b*) perpendicular to the axis of cathode
(*c*) inclined to the axis of cathode
(*d*) either (*b*) or (*c*)

293. When *n* resonant cavities are coupled together the result is
(*a*) one resonant frequency (*b*) *n* resonant frequencies
(*c*) $\frac{n}{2}$ resonant frequencies (*d*) either (*a*) or (*c*)

294. The most commonly used magnetron is
(*a*) cylindrical (*b*) parallel plate
(*c*) inclined plate (*d*) flat plate

295. In 8 cavity magnetron the π mode corresponds to a total phase shift equal to
(*a*) $\pm 8\pi$ radians around the periphery
(*b*) $\pm 4\pi$ radians around the periphery
(*c*) $\pm 2\pi$ radians around the periphery
(*d*) $\pm 16\pi$ radians around the periphery

296. The π mode of operation of magnetron signifies that
(*a*) the phase difference between adjacent anode poles is π radians
(*b*) the phase difference between adjacent anode poles is a multiple of π radians
(*c*) the total phase shift around the periphery is π radians
(*d*) none of the above

297. In a klystron amplifier dc electron velocity is v_0 and dc electron charge density is ρ_0. The dc beam current density is
(*a*) $\rho_0 v_0$ (*b*) ρ_0/v_0
(*c*) v_0/ρ_0 (*d*) $(v_0/\rho_0)^2$

298. In a multicavity klystron amplifier the electron velocity is of the order of
(*a*) 10 m/s (*b*) 10^3 m/s
(*c*) 10^6 m/s (*d*) 10^{12} m/s

299. In a magnetron the magnetic flux density is of the order of
(*a*) m Wb/m^2 (*b*) Wb/m^2
(*c*) 10^6 Wb/m^2 (*d*) 10^3 Wb/m^2

300. The speed at which axial electric field due to signal advances in a TWT is equal to
(*a*) speed of light
(*b*) (speed of light) (helix pitch/helix circumference)
(*c*) (speed of light) (helix circumference/helix pitch)
(*d*) (speed of light) (helix pitch) (helix circumference)

301. Which one of the following is not a negative resistance device
(*a*) Gunn diode (*b*) Tunnel diode
(*c*) Impatt diode (*d*) varactor diode

302. Consider the following applications
1. TV tunning 2. Active filter
3. Microwave frequency multiplication
In which of above can a varactor diode be used?
(*a*) 1, 2 and 3 (*b*) 1 and 2 only
(*c*) 1 and 3 only (*d*) 2 and 3 only

303. The reverse voltage applied to a varactor diode
(*a*) is less than avalanche breakdown voltage
(*b*) is more than avalanche breakdown voltage
(*c*) may be more or less than avalnche breakdown voltage
(*d*) is very high as compared to avalanche breakdown voltage

304. Varactor diodes use
(*a*) silicon (*b*) GaAs
(*c*) either (*a*) or (*b*) (*d*) neither (*a*) nor (*b*)

305. A parametric amplifier uses
(*a*) non-linear resistance (*b*) non-linear reactance
(*c*) either (*a*) or (*b*) (*d*) neither (*a*) nor (*b*)

306. The main feature of a parametric amplifier is
(*a*) low noise (*b*) very high gain
(*c*) Broad bandwidth (*d*) both (*b*) and (*c*)

307. The resistance values of a PIN diode under positive bias and negative bias respectively are about
(*a*) a few ohms and a few kilo ohms
(*b*) about 100 ohms and 1000 ohms
(*c*) about 0.1 ohm and 100 ohms
(*d*) about 1000 ohms and 10000 ohms

308. PIN diode has
(*a*) *p* and *n* layers separated by *i* layer
(*b*) p^+ and n^+ layers separated by *i* layer
(*c*) p^- and n^- layers separated by *i* layer
(*d*) either (*a*) or (*b*) or (*c*)

309. Consider the following applications :
1. Switch 2. Attenuator
3. Phase shifter 4. Oscillator
In which of the above can a PIN diode be used?
(*a*) 1, 2, 3 and 4 (*b*) 1, 2 and 3 only
(*c*) 1 and 2 only (*d*) 2, 3 and 4 only

310. A PIN diode can be used
(*a*) as a shunt mounted switch
(*b*) as a series mounted switch
(*c*) both as series or shunt mounted switch
(*d*) none of the above

311. Performance of a PIN diode switch is represented by
(*a*) insertion loss when switch is off
(*b*) isolation loss when switch is on
(*c*) insertion loss when switch is on and isolation loss when switch is off
(*d*) insertion loss when switch is off and isolation loss when switch is on

312. Consider the following statements :
1. Gunn diode can be used in combinational logic circuits
2. Gunn diode can be used in sequential logic circuits
3. Both Gunn diode and GaAs Mcsfet can be used in logic circuits
Which of the above statements are correct?
(*a*) 1, 2 and 3 (*b*) 1 and 2 only
(*c*) 2 and 3 only (*d*) 1 and 3 only

313. Consider the following time parameters :
1. Domain growth time constant
2. Transit time
3. Dielectric relaxation time in positive mobility regime
Which of the above are used in connection width transferred electron device?
(*a*) 1, 2, and 3 (*b*) 1 and 2 only
(*c*) 1 and 3 only (*d*) 2 and 3 only

314. Consider the following statements :
1. Impedance of Gunn diode is about tens of ohms
2. Impedance of Impatt diode is a few ohms
3. Impedance of Gunn diode and Impatt diode are of the same order
4. Impedance of Impatt diode is more than that of Gunn diode
Which of the above statements are correct?
(*a*) all (*b*) 1 and 2 only
(*c*) 1, 2 and 3 (*d*) 1, 2 and 4

315. In high frequency circuits impedance matching can be done by
1. open-circuited stubs
2. short-circuited stubs
3. $\frac{\lambda}{4}$ transformer
Which of the above are correct?
(*a*) 1, 2 and 3 (*b*) 1 and 2 only
(*c*) 1 and 3 only (*d*) 2 and 3 only

316. The rotation of direction of E field of a linearly polarized wave passing through a magnetised ferrite medium is called
(*a*) Lorentz rotation (*b*) Faraday rotation
(*c*) Steinmitz rotation (*d*) Fleming's rotation

317. At frequencies of 10 MHz and higher the impedance represented by ground is primarily

(*a*) resistive
(*b*) inductive
(*c*) capacitive
(*d*) partly resistive and partly capacitive

318. Dielectric constant and conductivity of rocky soil and flat-sandy soil are around
(*a*) 10 and 20 mho /cm respectively
(*b*) 20 and 10 mho /cm respectively
(*c*) 2 and 5 mho/cm respectively
(*d*) 0.5 and 2 mho/cm respectively

319. Interference generated by electrical equipment and automobile ignition systems when observed close to earth is
(*a*) horizontally polarized
(*b*) vertically polarized
(*c*) polarized both horizontally and vertically
(*d*) not polarized

320. When duct propagation exists
(*a*) the line of sight and diffraction zone concepts are applicable
(*b*) the line of sight and diffraction zone concepts are not applicable
(*c*) line of sight concept is applicable but diffraction zone concept is not applicable
(*d*) diffraction zone concept is applicable but line of sight concept is not applicable

321. Consider the following statements
1. Fading of signals is generally greatest near radio horizon
2. Fading of signals is generally greatest in shadow zone
3. Fading is small when a good optical path is present
Which of the above statements are correct?
(*a*) 1, 2 and 3 (*b*) 1 and 2 only
(*c*) 1 and 3 only (*d*) 2 and 3 only

322. The effect of earth's magnetic field on radio waves is minimum when
(*a*) the electric field lines of radio waves and flux line of earth's magnetic field are in the same direction
(*b*) the electric field lines of radio waves and flux lines of earth's magnetic field are at right angles
(*c*) the electric field lines of radio waves and flux lines of earth's magnetic field oppose each other
(*d*) both (*b*) and (*c*)

323. Directive gain of elementary doublet is
(*a*) 100 (*b*) 10
(*c*) 1.5 (*d*) 0.5

324. The total field developed by an antenna array at a distant point is
(*a*) phasor sum of fields produced by individual antennas of the array
(*b*) algebraic sum of fields produced by individual antennas of the array
(*c*) either (*a*) or (*b*) depending on type of array
(*d*) neither (*a*) nor (*b*)

325. Consider the following statements (L is the length of array between centres of end antennas and λ is wavelength) :
1. Width of major lobe between nulls of broad side array and end fire array are different
2. Directive gain of broad-side array and end fire array are different
3. Directive gain of broad-side array is $4\frac{L}{\lambda}$
Which of the above statements are correct?
(*a*) 1, 2 and 3 (*b*) 1 and 2 only
(*c*) 1 only (*d*) 1 and 3 only

326. The radiation resistance of a doublet antenna of length δl is proportional to
(*a*) δl (*b*) $(\delta l)^2$
(*c*) $(\delta l)^3$ (*d*) $(\delta l)^{05}$

327. The total power radiated from an antenna system is determined by imagining that the antenna is at the centre of a very large
(*a*) sphere (*b*) rectangle
(*c*) square (*d*) ellipse

328. In a turnstile antenna two half wave resonant wire radiators are placed at 90° to each other in the same plane and are excited
(*a*) in phase
(*b*) in phase opposition
(*c*) 90° out of phase with each other
(*d*) 45° out of phase with each other

329. Match List I and List II as regards microwave frequency directivity of anennas and select the answer using the codes

List I	List II
A Horn antenna	**1.** High directivity
B Parabolic antenna	**2.** Very high directivity
C Lens antenna	**3.** moderate directivity

Codes :

	A	B	C
(*a*)	1	2	3
(*b*)	3	2	1
(*c*)	3	1	2
(*d*)	1	3	2

330. The directivity in a receiving antenna
1. increases the intercept area in favoured direction
2. reduces the noise picked up from other sources
3. provides a means of discriminating against undesired signals originating in directions other than in which the desired transmitter lies
Which of the above statements are correct?
(*a*) 1 only (*b*) 1 and 2 only
(*c*) 1, 2 and 3 (*d*) 2 and 3 only

331. The duty cycle of a radar transmitter is equal to
(*a*) (PRF) (pulse width) (*b*) (PRF)/(pulse width)
(*c*) (Pulse width)/(PRF) (*d*) (Pulse width) + (PRF)

332. A pulsed radar produces 1 μ sec pulses at a rate of 1000 per second. The duty cycle is
(*a*) 0.001 (*b*) 0.01
(*c*) 0.1 (*d*) 1

333. Consider the following statements about radar Beacon
1. It is also called transponder
2. It includes pulse transmitter, receiver and antenna
3. It replies with full power even when the signal it receives is weak

Which of the above statements are correct?
(*a*) all (*b*) 1 and 2 only
(*c*) 2 and 3 only (*d*) 2 only

334. The word 'LORAN' means
(*a*) long range navigation
(*b*) long range TV transmission
(*c*) long range cable transmission
(*d*) either (*b*) or (*c*)

335. Consider the following statements about instrument landing system of aircrafts
1. It ensures correct landing under conditions of low visibility
2. It consists of runaway localizer, glide path equipment and marker beacons
3. It is basically a blind landing system

Which of above statements are correct?
(*a*) all (*b*) 1 and 2 only
(*c*) 1 and 3 only (*d*) 2 only

336. Match List I and List II and select the answer using the codes :

	List I		List II
A.	Transferred electron devices	1.	Microwave amplifier
B.	Two cavity klystron	2.	Microwave oscillator
C.	PIN Diode	3.	Microwave low power measurement
D.	Bolometer	4.	Modulator

Codes :

	A	B	C	D
(*a*)	3	1	4	2
(*b*)	2	1	4	3
(*c*)	3	4	1	2
(*d*)	2	4	1	3

[IES 2010]

337. The main disadvantage of using coaxial cable for microwave signals is
(*a*) low selectivity (*b*) low distortion
(*c*) high attenuation (*d*) high sensitivity **[IES 2010]**

338. Match list I and list II and select the answer using the codes

	List I		List II
A.	Cavity wave meter	1.	Impedance measurement
B.	VSWR meter	2.	Frequency measurement
C.	Bolometer	3.	Antenna measurement
D.	Fraunhofer region	4.	Microwave power measurement

Codes :

	A	B	C	D
(*a*)	3	1	4	2
(*b*)	2	1	4	3
(*c*)	3	4	1	2
(*d*)	2	4	1	3

[IES 2010]

339. A ship to ship communication system is affected by fading. A useful solution for this is
(*a*) A more directional antenna
(*b*) A broad band antenna
(*c*) Use of frequency diversity
(*d*) Use of space diversity **[IES 2010]**

340. A standard waveguide WR 90 has inside wall dimensions $a = 2.286$ cm and $b = 1.016$ cm. What is cutoff wavelength for TE_{01} mode?
(*a*) 4.572 cm (*b*) 2.286 cm
(*c*) 2.032 cm (*d*) 1.857 cm **[IES 2009]**

341. When a particular mode is excited in a waveguide, there appears an extra electric component in the direction of propagation. In what mode is the wave propagating?
(*a*) Transverse electric (*b*) Transverse magnetic
(*c*) Transverse electromagnetic
(*d*) Longitudinal **[IES 2009]**

342. Which is dominant mode in rectangular waveguide?
(*a*) TE_{10} (*b*) TE_{11}
(*c*) TM_{01} (*d*) TM_{11} **[IES 2009]**

343. Consider the following :
For a square waveguide 3 m × 3 m, it has been found
(*a*) at 6 GHz dominant mode propagates
(*b*) at 4 GHz modes are evanescent
(*c*) at 11 GHz only dominant mode and no higher order mode will propagate
(*d*) at 7 GHz degenerate modes will propagate

Which of the above are correct?
(*a*) 1 and 2 only (*b*) 1, 2, and 4 only
(*c*) 2 and 3 only (*d*) 2, 3, 4 only **[IES 2009]**

344. Match List I (Modes) and List II (Characteristics) and give the answer using the codes

	List I		List II
A.	Evanescent mode	1.	Rectangular waveguide does not support
B.	Dominant mode	2.	No wave propagation
C.	TM_{10}, TM_{01}	3.	Lowest cutoff frequency

Codes :

	A	B	C
(*a*)	1	2	3
(*b*)	2	3	1
(*c*)	1	3	2
(*d*)	2	1	3

[IES 2009]

345. Which is transferred electron device?
(*a*) Barrit diode (*b*) IMPATT diode
(*c*) TRAPATT diode (*d*) Gunn diode **[IES 2009]**

346. Consider a line of length 37.5 cm terminated in zero resistance. Line is excited by 1 GHz source having internal impedance of 50 ohm. Find input impedance as seen by source.
(*a*) 50 ohm (*b*) zero ohm
(*c*) 100 ohm (*d*) infinite **[IES 2009]**

347. Which of the following is a microwave source with a crossfield structure?
(*a*) Double cavity klystron (*b*) Reflex klystron
(*c*) Magnetron (*d*) TWT **[IES 2009]**

348. In Reflex klystron the adjustment of volume of anode cavity will result in change of
(*a*) output power (*b*) output pulse width
(*c*) output frequency (*d*) output SNR **[IES 2008]**

349. The main purpose of inserting a microwave isolator in a microwave circuit is for
(*a*) offering minimum forward impedance

(*b*) offering maximum power output from source
(*c*) maintaining frequency stability of source
(*d*) maintaining amplitude stability of output power **[IES 2008]**

350. In measuring pulsed microwave power the pulse duration is 1.5 μs with PRF of 1 kHz. If CW power sensor reads 15 mW. What is the average signal power during pulse?
(*a*) 5 W (*b*) 10 W
(*c*) 15 W (*d*) 20 W **[IES 2008]**

351. For obtaining a beam of em wave with circular polarization, which is ideal antenna?
(*a*) Double conical reflector (*b*) helical antenna
(*c*) pyramidal horn antenna (*d*) hog-horn antenna **[IES 2008]**

352. Gunn diode is a negative resistance device used as a source of microwaves. What is the number of *p-n* junction?
(*a*) 1 (*b*) 2
(*c*) 3 (*d*) zero **[IES 2008]**

353. X band frequencies are in the range
(*a*) 3.5 to 5.5 GHz (*b*) 5.5 to 8 GHz
(*c*) 8 to 12 GHz (*d*) 12.4 to 16.4 GHz **[IES 2007]**

354. Match List I and List II and give correct answer using the code :

List I	List II
A. Microstrip line	**1.** Surface wave
B. Earth crust	**2.** Guided wave
C. Troposphere	**3.** Sky wave
D. Ionosphere	**4.** Space wave

Codes :

	A	B	C	D
(*a*)	1	2	3	4
(*b*)	1	2	4	2
(*c*)	2	1	3	4
(*d*)	2	1	4	3

[IES 2007]

355. Why is reflex klystron preferred as an ideal microwave source?
(*a*) It provides high power output with wide frequency range
(*b*) It provides high power output at one frequency
(*c*) It provides low power output with highly stable single frequency
(*d*) It provides low power output with multiple frequency components **[IES 2007]**

356. Which has dielectric, ohmic and radiation losses?
(*a*) Coaxial line (*b*) Twin parallel line
(*c*) Microstrip line (*d*) Waveguide **[IES 2007]**

357. A waveguide (a = 1.5 cm, b = 1 cm) is loaded with dielectric ($\epsilon_r = 4$). Which is correct?
8 GHz signal will
(*a*) pass through waveguide
(*b*) not pass through waveguide
(*c*) be absorbed in the waveguide
(*d*) none of the above **[IES 2007]**

358. Two microwave signals travelling in free space have path length difference of 3 cm when operating at 10 GHz. What is phase difference between signals?
(*a*) 2π (*b*) π
(*c*) 3π (*d*) 4π **[IES 2007]**

359. In dielectric medium ϵ_r (given $n = \sqrt{\epsilon_r}$). What is the expression for wavelength.
(*a*) $\frac{c}{n^2 f}$ (*b*) $\frac{nc}{f}$
(*c*) $\frac{c}{nf}$ (*d*) none of above **[IES 2007]**

360. An amplifier has a gain of 200. What is the gain in *dB* ($\log_{10}2$) = 0.3?
(*a*) 14 (*b*) 17
(*c*) 20 (*d*) 23 *dB* **[IES 2007]**

361. In parametric amplifiers used in microwave communication system the pump energy is the deciding factor for which of the following?
(*a*) Frequency stability (*b*) Max. output
(*c*) Amplification factor (*d*) Thermal noise level **[IES 2006]**

362. The working of a travelling wave tube is based on interaction between
(*a*) moving electrons and transverse component of progressive em wave
(*b*) moving electrons and axial component of a progressive em wave
(*c*) moving electrons and transverse component of rotating magnetic field
(*d*) moving electrons and axial component of steady magnetic field **[IES 2006]**

363. To prevent oscillations in TWT which of the following is resorted to
(*a*) Bunching defocussing mechanism is used
(*b*) Attenuator is used
(*c*) External magnetic field is provided
(*d*) Helix is used **[IES 2006]**

364. Which of the following express the phase velocity in microstrip line?
(*a*) $\frac{c}{\epsilon_r}$ (*b*) $\frac{c}{\sqrt{\epsilon_r}}$
(*c*) $c\epsilon_r$ (*d*) $c\sqrt{\epsilon_r}$ **[IES 2006]**

365. For a parabolic reflector of 5 m diameter the far field pattern measurement at 6 GHz should be carried out at a distance of at least
(*a*) 0.5 km (*b*) 1 km
(*c*) 1.5 km (*d*) 2 km **[IES 2005]**

366. A microstrip line of 50 ohm is terminated in $Z_L = 40 + j\,30\ \Omega$. What is VSWR of the load?
(*a*) 2 (*b*) 1.8
(*c*) 1.5 (*d*) 1.3 **[IES 2005]**

367. In an air-filled waveguide of dimensions $a \times b$ cm, at a given frequency, the longitudinal component of electric field of TM_{32} mode is of the form

$$E_z = 20 \sin (60\,\pi x) \sin (10\,xy)$$

Which form would E_z have for the lowest order TM mode?
(*a*) $E_z = 20 \sin (20\,\pi x)$ (*b*) $E_z = 20 \sin (20\,\pi y)$

(c) $E_z = 20 \sin(20\pi x) \sin(50\pi y)$
(d) $E_z = 20 \sin(20\pi x) \sin(100\pi y)$ **[IES 2005]**

368. For a rectangular waveguide of dimensions $a\sqrt{3}$ cm × a cm the cutoff frequency for TE_{10} mode is 2 GHz. What is the cutoff frequency for TM_{11} mode?
(a) 1 GHz (b) 3.46 GHz
(c) 4 GHz (d) 6 GHz **[IES 2005]**

369. Match list I and list II and give correct answer using the codes

List I	List II
A. S band	1. 18–26.5 GHz
B. X band	2. 2–4 GHz
C. KU band	3. 8–12.4 GHz
D. K band	4. 12.4–18 GHz

Codes:

	A	B	C	D
(a)	1	3	2	4
(b)	1	4	2	3
(c)	2	3	4	1
(d)	2	4	1	3

[IES 2004]

370. Consider the following statements
A TWT is a
1. broad band amplifier
2. linear beam amplifier
3. cross field amplifier
4. solid state amplifier
Which of the above are correct?
(a) 1 and 2 (b) 2 and 4
(c) 2 and 3 (d) 1, 3 and 4 **[IES 2004]**

371. In a GaAs Gunn diode the drift velocity is 10^7 cm/s and active region length is 10^{-3} cm. What is natural frequency of oscillation?
(a) 8 GHz (b) 9 GHz
(c) 10 GHz (d) 11 GHz **[IES 2004]**

372. A rectangular waveguide operating in dominant mode is deformed into circular cross-section in the direction of propagation. What is the mode at output?
(a) TM_{11} (b) TE_{11}
(c) TE_{01} (d) TM_{01} **[IES 2004]**

373. A 50 ohm coaxial line is connected to a 72 ohm load through a 60 ohm coaxial section of length 7.5 mm. What is the input VSWR on line if fed by 20 GHz source?
(a) 1 (b) 1.25
(c) 1.33 (d) 1.44 **[IES 2004]**

374. For a Gunn diode the drift velocity of electron through active drift region is 10^7 cm/s and active region is 10×10^{-4} cm. The critical voltage of the diode (critical field = 3.2 kV/cm) is
(a) 0.032 V (b) 0.32 V
(c) 3.2 V (d) 32 V **[IES 2003]**

375. Which of the following can be used for amplification of microwave signals?
(a) Gunn diode (b) Strapped magnetron
(c) Reflex klystron (d) Double cavity klystron **[IES 2003]**

376. If the height of a waveguide is halved, its cutoff frequency will
(a) be halved
(b) be doubled
(c) remain unchanged
(d) be one-fourth of previous value **[IES 2003]**

377. In a rectangular waveguide with broader dimension a and narrow dimension b the dominant mode of microwave propagation would be
(a) TE_{10} (b) TM_{10}
(c) TE_{01} (d) TM_{01} **[IES 2003]**

378. A parabolic dish antenna has a diameter of 1 m. The maximum possible (ideal) gain of the antenna at a wavelength of 3.14 cm is
(a) 20 dB (b) 30 dB
(c) 40 dB (d) 50 dB **[IES 2003]**

379. An IMPATT diode has a drift length of 4 μm. The drift velocity of Si is 10^5 m/s. The operating frequency of IMPATT diode is
(a) 25 GHz (b) 20 GHz
(c) 12.5 GHz (d) 125 GHz **[IES 2002]**

380. A rectangular waveguide measures 3 × 4.5 cm internally. It has a 9 GHz signal propagated in it. The cutoff wavelength for TE_{10} mode is
(a) 5 cm (b) 10 cm
(c) 15 cm (d) 9 cm **[IES 2002]**

B. The following statements have two parts : Assertion (A) and Reason (R). Read them and give correct answer as per following codes
(a) Both A and R are correct and R is correct explanation of A
(b) Both A and R are correct but R is not correct explanation of A
(c) A is correct but R is wrong
(d) A is wrong but R is correct

381. A. The condition of minimum distortion in a transmission line is L = RC/G.
R. Line loading means addition of inductance to satisfy the condition L = RC/G.

382. A. Artificial transmission lines are frequently used in laboratories.
R. An artificial transmission line can be used to represent an actual line and can also be used as a delay circuit, as attenuator, as filter network etc.

383. A. An artificial transmission line can be used as a pulse generator.
R. A pulse function is used very frequently in network analysis.

384. A. The velocity of electromagnetic waves on overhead lines and coaxial cables is the same.
R. Free space has an intrinsic impedance of 377 ohms.

385. A. A line of length $\frac{\lambda}{4}$ and short-circuited at far end has an input impedance of infinity.
R. A line of length $\frac{\lambda}{4}$ and short-circuited at far end behaves as a parallel resonant circuit.

386. **A.** A line of length slightly more than $\frac{\lambda}{4}$ and open-circuited at far end is equivalent to an inductive reactance.
R. A line of length slightly more than $\frac{\lambda}{4}$ and short-circuited at far end behaves as a capacitive reactance.

387. **A.** Impedance matching can be done by using stubs.
R. A double stub is used for impedance matching when frequency of signal varies.

388. **A.** A half wavelength line can be used as a 1 : 1 transformer.
R. The input impedance of a half wavelength line is equal to load impedance.

389. **A.** For high frequency lines inductive reactance is very high as compared to a c resistance.
R. Due to skin effect ac resistance of line is higher than dc resistance.

390. **A.** Two equal and opposite travelling waves result in a standing wave.
R. The ratio of maximum to minimum value of a standing wave is called SWR.

391. **A.** A backward wave oscillator can be used as a sweep generator.
R. The frequency of oscillation of a backward wave oscillator can be changed by varying the voltage which controls the beam velocity.

392. **A.** A backward wave oscillator has an internal positive feedback.
R. A positive feedback is necessary for sustained oscillations.

393. **A.** TWT uses a focussing mechanism to prevent the electron beam from spreading.
R. In a TWT the electron beam has to travel a much longer distance than in klystron.

394. **A.** Magnetron is generally used in π mode.
R. Frequency for π mode can be easily separated from adjacent modes.

395. **A.** In the interaction region of magnetron an electron is subjected to three forces : force due to electric field, force due to magnetic field and centrifugal force.
R. The mechanism of generation of microwaves, in a magnetron, involves interaction of electromagnetic fields with electrons moving in static electric and magnetic fields oriented at right angles to each other.

396. **A.** Klystron amplifiers use one or more intermediate cavities in addition to buncher and catcher cavity.
R. When one or more intermediate cavities are used the bandwidth can be increased.

397. **A.** Inter digitated geometry is widely used for microwave transistors.
R. Tuning at input and output of transistor uses MOS chip capacitors.

398. **A.** The phenomenon of differential mobility is called transferred electron effect.
R. Ga As exhibits transferred electron effect.

399. **A.** Gunn diode is a transferred electron device.
R. A Gunn oscillator uses the phenomenon of transferred electron effect.

400. **A.** IMPATT diode is an avalanche diode.
R. Avalanche breakdown phenomenon occurs when a *p-n* junction is reverse biased.

401. **A.** Impatt diode can be used in both amplifiers and oscillators.
R. Impatt diode has a low resistance.

402. **A.** For a reciprocal microwave junction the S matrix is symmetrical.
R. In a symmetrical matrix $S_{ij} = S_{ji}$.

403. **A.** In a power divider $S_{11} = 0$, $S_{21} = 0.707 = S_{31}$.
R. In a power divider the input port (*i.e.*, port 1) is matched and incident power is equally divided in output ports 2 and 3.

404. **A.** Bolometer is frequently used in microwave power measurement.
R. In a bolometer the input microwave power causes a change in resistance.

405. **A.** Impedance measurement at microwave frequencies is done by finding SWR.
R. SWR and reflection coefficient depend on the characteristic impedance and load impedance.

406. **A.** Slotted line is frequently used for microwave measurements.
R. The unknown impedance is mounted at the end of slotted line.

407. **A.** PIN diode is commonly used for microwave control.
R. A PIN diode uses heavily doped *p* and *n* materials.

408. **A.** PIN diode is used as a fast switch.
R. PIN diode has very high resistance when reverse biased and very low resistance when forward biased.

409. **A.** PIN diode can be used as attenuator and limiter.
R. PIN diode has a thin intrinsic layer.

410. **A.** Microstrip is very commonly used in microwave integrated circuits.
R. Microstrip has an easy access to the top-surface so that active and passive discrete components can be easily mounted.

411. **A.** The impedance of a matched load is equal to characteristic impedance of line.
R. A matched termination absorbs all the power incident on it.

412. **A.** A small section of high impedance line is equivalent to a series inductance.
R. A small section of a low impedance line is equivalent to a series resistance.

413. **A.** A quarter wave transformer is used to match a resistive load to a transmission line.
R. A quarter wave transformer is a transmission line of quarter wavelength.

414. A. The velocity factor of a line is the ratio of wave velocity on the line to speed of light.
R. If the conductors of a line are immersed in a non-magnetic insulating liquid, the wave velocity increases

415. A. A coaxial line is a non-radiating line.
R. In a coaxial line the electric and magnetic fields are confined to the region between the concentric conductors.

416. A. Hollow waveguides are frequently used as transmission lines at microwave frequencies.
R. A waveguide has high power handling capacity, low loss and simple structure.

417. A. In TE_{10} mode the cutoff frequency of rectangular wave guide is $\frac{c}{2a}$ where a is the broad dimension and c is velocity of light.
R. TE_{10} mode is the dominant mode in rectangular waveguide.

418. A. In rectangular waveguide TE_{01} mode is just TE_{10} mode rotated by 90°.
R. When a rectangular waveguide is rotated by 90°, the broad and narrow dimensions get interchanged.

419. A. In circular waveguide the attenuation of TE_{01} mode decreases as the operating frequency increases.
R. The wall currents in a circular waveguide in TE_{01} mode are completely circumferential.

420. A. In TE mode for rectangular waveguide $E_z = 0$.
R. In TM mode for rectangular waveguide $H_z = 0$.

421. A. A magnetron is not an amplifier but an oscillator producing microwaves.
R. In making their circular passes in the interaction space of a magnetron, the electrons excite the resonant cavities into oscillations. **[IES 2010]**

422. A. The slow wave structure of TWT is provided with an attenuator.
R. This is done for preventing oscillations in the device. **[IES 2010]**

423. A. Generally magnetrons are operated in π modes.
R. Frequency of a π mode can be easily separated from adjacent modes. **[IES 2010]**

424. A. Substrates chosen for fabrication of microstrip lines are usually thin and have high dielectric constant.
R. To minimize radiation loss from microstrip lines fringing of fields must be reduced as much as possible. **[IES 2007]**

425. A. In microwave communication links rain causes fading and this is a great concern in communication systems.
R. Water drops in the path of em wave will scatter the energy in the wave and this collective scattering will weaken the incident wave in forward direction. **[IES 2006]**

426. A. Microstrip lines are preferred for antenna application.
R. They have higher Q and higher gain. **[IES 2005]**

427. A. For a slow wave structure helix is a broad band low power device while coupled cavity chain is a high power narrow band device.
R. A helix is a good dissipater of power while a coupled cavity structure is poor for high power dissipation. **[IES 2004]**

428. A. TE_{10} mode is the dominant mode in a rectangular waveguide.
R. TE_{10} mode has lowest cutoff frequency for which the wavelength is double the broad dimension. **[IES 2002]**

429. A. GaAs is preferred to silicon for fabrication of Gunn diode.
R. GaAs has better frequency stability than silicon. **[IES 2002]**

430. A. One problem with microstrip circuits (or other planar circuit) is that of inevitable discontinuities at bends, step changes in width and junctions which can cause degradation of circuit performance.
R. Bends, step changes in width and junction discontinuities introduce parasitic reactances that can lead to phase and amplitude errors, input and output mismatch and spurious coupling. **[IES 2001]**

C. Indicate whether the following statements are True (T) or False (F)

431. Skin effect is more pronounced at high frequencies.
432. In an infinite line the input impedance at every point is equal to characteristic impedance.
433. If a short line is terminated in characteristic impedance it behaves as an infinite line.
434. If in a line $\frac{G}{C} = \frac{R}{L}$, the characteristic impedance becomes real.
435. Characteristic impedance of a coaxial cable is inversely proportional to $\sqrt{\epsilon_r}$.
436. A wave $E_m \cos(\beta x - \omega t)$ is a backward wave.
437. $VSWR = E_{max}/E_{min}$.
438. Depth of penetration does not depend on frequency.
439. If a line is open-circuited $Z_{in} = Z_0 \tanh(\gamma l)$.
440. $Z_0 = (Z_{oc} Z_{sc})^{0.5}$.
441. Ridge waveguide is essentially a form of capacitively loaded guide.
442. An electromagnetic wave can be completely guided by a dielectric rod or slab.
443. Integrated fin line can be utilized as a medium for constructing millimeter wave circuits.
444. In fibre optics the use is made of thin glass fibres for efficient transmission of light.
445. Standard mismatches are used to calibrate SWR and impedance measuring equipments.
446. A microwave connector should have high SWR.
447. A transmission line discontinuity is defined as any interruption in the uniformity of the line.
448. In a microstrip transmission the signal can be easily redirected by changing the centre strip.

449. For a given operating frequency a waveguide has more power handling capacity than a coaxial line.

450. A waveguide twist is used to change the plane of polarization.

451. A circular polarizer converts a linearly polarized wave into a circularly polarized wave.

452. A waveguide connected to a circularly polarized antenna must be able to efficiently propagate both vertically and horizontally polarized waves.

453. A circular waveguide has an infinite set of TE and TM modes.

454. In a circular waveguide TE_{21} mode has lowest cutoff frequency.

455. In TE mode of waveguide $E_z = 0$.

456. In a multicavity klystron amplifier the signal to be amplified develops an a c voltage of signal frequency across the gap in buncher cavity.

457. The bunching action which occurs in multicavity klystron amplifier can be represented by Applegate diagram.

458. The noise figure of multicavity klystron amplifier is very low.

459. Reflex Klystron oscillator is essentially a low power device.

460. A klystron amplifier generally uses Pierce gun.

461. In Reflex Klystron oscillator the focussing electrode is at a high potential.

462. Oscillations are obtained from a reflex klystron only for combinations of anode voltage and repeller voltage that give a favourable transit time.

463. In a reflex klystron the output frequency is determined primarily by resonant frequency of cavity.

464. In a travelling wave tube distributed interaction between an electron beam and a travelling wave takes place.

465. In a TWT the amplitude of resultant wave travelling down the helix remains constant.

466. In a backward wave oscillator the wave travelling along the line winds itself back and forth.

467. A magnetron has a cylindrical cathode surrounded by an anode structure having cavities opening into interaction space by means of slots.

468. A magnetron requires an external magnetic field with flux lines parallel to axis of cathode.

469. Radar is essentially an echo ranging system.

470. Radar receiver has high sensitivity and low noise figure.

471. Ga As exhibits negative differential mobility.

472. VTM stands for voltage tunable magnetron.

473. Impedance level of Impatt diodes is generally lower than that of Gunn diodes.

474. Both Impatt and Trapatt devices use avalanche effect.

475. Slotted line can be used for measurement of high frequencies.

D. Fill in the Blanks with suitable words

476. For free space the characteristic impedance is called ....

477. A $\frac{\lambda}{4}$ line is also called ....

478. A line longer than $\frac{\lambda}{4}$ and open-circuited at far end acts like ....

479. If a line is terminated in characteristic impedance VSWR = ...

480. Input impedance of a line of length $\frac{\lambda}{2}$ is equal to ....

481. Stubs used for impedance matching are generally ....

482. The impedance of a matched load is equal to .... of line.

483. In a microwave equipment it is essential to control the .... and .... of signal.

484. An abrupt change in '*a*' dimension of a wave-guide can be represented by ....

485. A microwave attenuator may be .... or ....

486. E plane tee is also called .... tee.

487. H plane tee is also called .... tee.

488. Hybrid tee is also called .... tee.

489. Hybrid ring is also called ....

490. Transmission lines at microwave frequencies use .... waveguides.

491. In a rectangular waveguide the power is delivered to the load via .... current flow along the length of guide.

492. In TE mode the direction of electric field is always and everywhere .... to the direction of propagation.

493. In rectangular waveguide the dominant mode is ....

494. In a waveguide the electromagnetic wave at a frequency less than cutoff frequency is ....

495. In TE_{10} mode of rectangular waveguide the largest electric field is along .... of broad wall.

496. Circular waveguide is the most common form of .... transmission line.

497. In a circular waveguide the dominant mode is ....

498. For analysis the guide walls of a waveguide are assumed to be .... conductors.

499. The negative differential mobility of Ga As is called ....

500. The electron gunn in multicavity klystron amplifier is generally ....

501. In reflex klystron oscillator the repeller is at .... potential.

502. In magnetron the static electric and magnetic fields are at .... to each other.

503. In a magnetron the number of forces acting on an electron in interaction region are ....

504. The most important mode of magnetron is .... mode.

505. A TWA is a .... band device.

ANSWERS WITH HINTS AND EXPLANATION

1. (*d*) $VSWR = \frac{1+|\rho_v|}{1-|\rho_v|}$ and $\rho_v = \frac{Z_0 - Z_L}{Z_0 + Z_L}$.
Since $VSWR = \infty$, $|\rho_v| = 1$. This can happen if $Z_L = 0$ or $Z_L = \infty$, *i.e.*, line is o.c. or s.c.
2. (*c*) A short-circuited stub is a suitable method of matching.
3. (*c*) ϵ_r of a dielectric changes with humidity.
4. (*b*) Impedance inversion occurs when load impedance is not matched with characteristic impedance of line.
5. (*c*)
6. (*c*) Double stub is useful for matching over a range of frequencies.
7. (*b*) A balun gives 4 : 1 impedance transformation.
8. (*b*) Charging current and MVAR are proportional to frequency.
9. (*a*) 10. (*c*) See Section 13.1.1.
11. (*b*)
12. (*b*) $Z_0 = \sqrt{\frac{L}{C}}$ and is independent of line length.
13. (*a*) For oc line substitute $Z_L = \infty$ in Eq. (13.10). Then $e_r = 2e_f$.
14. (*a*) Use Eq. (13.10).
15. (*d*) Substitute $Z_L = \frac{Z_0}{Z}$ in Eq. (13.11).
16. (*a*) 17. (*d*) 18. (*b*) 19. (*b*)
20. (*d*) Shunt conductance can be neglected in most of calculations.
21. (*a*) See Section 13.11.
22. (*d*) The condition for minimum distortion is found by equating $\frac{d\alpha}{dL} = 0$. The result is $\frac{L}{C} = \frac{R}{G}$.
23. (*b*) 24. (*a*) 25. (*b*)
26. (*b*) If Z_L is load impedance and Z_0 is characteristic impedance.
Input impedance $= Z_0 \left[\frac{Z_L \cosh \gamma l + Z_0 \sinh \gamma l}{Z_0 \cosh \gamma l + Z_L \sinh \lambda l}\right]$. If $Z_L = Z_0$, the input impedance $= Z_0$.
27. (*a*) When a line is short-circuited at far end, the voltage at far end is zero.
28. (*d*) 29. (*c*)
30. (*b*) $Z_0 = \sqrt{\frac{R + j\omega L}{G + j\omega C}}$

$$\text{or } Z_0 = \sqrt{\frac{R + j\omega R\frac{C}{G}}{G + j\omega C}}$$

$$= \sqrt{\frac{R(G + j\omega C)}{G(G + j\omega C)}} = \sqrt{\frac{R}{G}} = \sqrt{\frac{L}{C}}.$$

31. (*d*) In a travelling wave both *x* and *t* increase simultaneously so that a constant phase point moves in the direction of positive (or negative) *x*.
32. (*b*)
33. (*c*) Reflection coefficient

$$= \frac{Z_L - Z_0}{Z_L + Z_0} = \frac{150 - 300}{150 + 300} = -0.333$$

34. (*c*) If line is o.c. $|\rho_v| = 1$ and $VSWR = \frac{1+|\rho_v|}{1-|\rho_v|} = \infty$.
35. (*c*) If line is s.c. $|\rho_v| = 1$ and $VSWR = \infty$.
36. (*b*) If $Z_L \gg Z_0$, $|\rho_v| \simeq \frac{Z_L}{Z_0}$
and $VSWR = \frac{1+|\rho_v|}{1-|\rho_v|}$. Therefore, $Z_L = Z_0(VSWR)$.
37. (*a*) For ideal dielectric loss angle is zero. For good quality cables, loss angle is very small, about 1° only.
38. (*d*) A quarter wave line short-circuited at far end behaves as a parallel tuned circuit.
39. (*b*) See Fig. 13.4.
40. (*d*) A quarter wave line o.c. at far end behaves as a series tuned circuit.
41. (*b*) 42. (*a*) See Fig. 13.4.
43. (*a*) 44. (*b*) See Fig. 13.4.
45. (*b*)
46. (*a*) High speed means time period is small. Therefore, pulses should also have small width.
47. (*a*) 48. (*a*) 49. (*d*)
50. (*c*) A TV channel has a bandwidth about 5 MHz. Hence 300 MHz spectrum can accomodate about 50 channels.
51. (*d*) 52. (*a*) 53. (*c*) 54. (*c*)
55. (*c*)
56. (*b*) Beam width between nulls of a parabolic antenna

$$= \frac{70\times \text{Wavelength}}{\text{Diameter}} = \frac{140\times 3\times 10^8}{10^9\times 1} = 42°.$$ Therefore, 50° is the nearest value.

57. (*c*) $\frac{140\times 3\times 10^8}{10\times 10^9\times 1} = 4.2°$.
58. (*b*) See Table 12.1.
59. (*c*) 60. (*a*) 61. (*d*)
62. (*a*) Directive gain $= 6\left(\frac{D}{\lambda}\right)^2$ and is thus proportional to cross-sectional area.
63. (*d*) Directive gain $= 6\left(\frac{D}{\lambda}\right)^2$.
64. (*b*)
65. (*b*) As frequency increases, transmission loss increases slightly.

66. (*d*) Time required for microwave cooking is much less than the time required for conventional cooking.

67. (*a*)

68. (*a*) Atomic and molecular distances are very small. Therefore, resonance can occur only at microwave frequencies.

69. (*b*)

70. (*b*) Since fields are confined within a coaxial cable it is non-radiating.

71. (*c*) **72.** (*b*) See Fig. 13.21.

73. (*b*) **74.** (*d*) See Section 13.2.

75. (*d*) **76.** (*a*) See Fig. 13.5.

77. (*b*) **78.** (*a*) **79.** (*a*)

80. (*a*) See Section 13.2.1. **81.** (*a*) See Section 13.2.

82. (*a*) **83.** (*c*) See Section 13.2.

84. (*a*) See Section 13.2.

85. (*a*) **86.** (*b*)

87. (*b*) See Section 13.2.1.

88. (*b*) **89.** (*d*)

90. (*a*) See Section 13.2. **91.** (*a*) See Section 13.2.

92. (*b*) **93.** (*c*) **94.** (*b*)

95. (*a*) See Section 13.2.

96. (*a*) Wavelength should be less than cutoff wavelengths. Therefore, frequency should be higher than cutoff frequency for minimum attenuation.

97. (*c*) **98.** (*a*)

99. (*b*) See Section 13.2.

100. (*c*) **101.** (*c*)

102. (*b*) See Section 13.2.

103. (*a*)

104. (*a*) When a circular waveguide is rotated by 90°, the configuration remain the same.

105. (*a*)

106. (*b*) See Section 13.2.3.

107. (*b*) See Section 13.2.3.

108. (*a*) See Section 13.2.3.

109. (*a*) **110.** (*a*) **111.** (*d*) **112.** (*a*)

113. (*b*)

114. (*b*) See Section 13.2.1.

115. (*a*) See Section 13.2.2.

116. (*a*) **117.** (*b*) **118.** (*b*)

119. (*b*) See Section 13.6.

120. (*a*) **121.** (*a*)

122. (*c*) See Section 13.2.

123. (*b*) See Section 13.6.

124. (*a*) **125.** (*c*) **126.** (*c*) **127.** (*d*)

128. (*c*) Earth's curvature limits the distance between microwave links.

129. (*a*) $T = \dfrac{Z_L - Z_0}{Z_L + Z_0}$ and lies in the range of −1 to +1.

If $Z_L = Z_0$, $T = 0$.

130. (*a*) Transmission coefficient

$$= \frac{2Z_L}{Z_0 + Z_L} = \frac{600 - j600}{600 - j300} = 1.265\angle -18.43°.$$

131. (*a*)

132. (*c*) Range of radar is proportional to antenna diameter.

133. (*b*) A balun gives 4 : 1 impedance transformation.

134. (*a*)

135. (*a*) Range = $(\text{Power})^{0.25}$. See Section 12.10.1.

136. (*a*) **137.** (*a*) **138.** (*a*)

139. (*a*) See Section 13.2.

140. (*d*) Same antenna is used for transmission and reception.

141. (*b*) **142.** (b)

143. (*d*) See Section 13.6 and Section 13.7.4.

144. (*a*) See Section 13.7.7.

145. (*d*)

146. (*c*) See Eq. (13.13). **147.** (*a*) See Section 13.6.3.

148. (*b*)

149. (*d*) $\text{SWR} = \dfrac{1 + |\rho_v|}{1 - |\rho_v|} = \dfrac{1.2}{0.8} = 1.5.$

150. (*d*) Ranges $(\text{Power})^{0.25}$.

151. (*b*) See Section 13.7. **152.** (*b*)

153. (*c*) See Section 13.2.

154. (*c*) **155.** (*d*) **156.** (*b*) **157.** (*d*)

158. (*c*) See Section 13.7.1.

159. (*a*) See Section 13.4.3.

160. (*c*) It allows high frequencies.

161. (*b*)

162. (*b*) See Section 13.4.3.

163. (*c*) $\rho_v = \dfrac{Z_L - Z_0}{Z_L + Z_0}$ and $\text{VSWR} = \dfrac{1 + |\rho_v|}{1 - |\rho_v|}$.

164. (*d*) **165.** (*a*) See Section 13.6.3.

166. (*a*)

167. (*d*) A pure reactance does not absorb any power. Therefore, VSWR is infinite because $|\rho_v| = 1$.

168. (*d*) See Section 13.2.3.

169. (*c*) See Section 13.4.3.

170. (*b*)

171. (*d*) See Section 13.2. and Section 13.3.1.

172. (*a*) See Section 13.6.3.

173. (*c*) See Section 13.4.3.

174. (*a*) **175.** (*a*)

176. (*c*) See Section 13.4.3.

177. (*c*) **178.** (*b*) **179.** (*b*)

180. (*d*) See Section 13.2.

181. (*d*)

182. (*c*) $50 = \sqrt{\dfrac{L}{40 \times 10^{-12}}} = L = 0.1 \times 10^{-6}$ H/m.

183. (*c*) **184.** (*b*) **185.** (*d*)

186. (*b*) Cathode is the first part and anode is the last.

187. (*c*)

188. (*b*) $Z_0 = \sqrt{25 \times 100} = 50$ ohm.

189. (*a*) **190.** (*d*) **191.** (*a*) **192.** (*d*)

193. (*b*) $\rho_v = \dfrac{25-100}{25+100} = -0.6.$

194. (*c*)

195. (*b*) $\dfrac{30}{100} = 0.3$ A.

196. (*d*) $t = \dfrac{2 \times 4}{2 \times 10^8} = 40$ ns.

197. (*c*) $\dfrac{200-100}{200+100} = \dfrac{1}{3}.$

198. (*b*) $(30)\left(\dfrac{25-100}{25+100}\right) = -18$ V.

199. (*b*) $\dfrac{-18}{100} = -0.18$ A.

200. (*c*) **201.** (*c*)

202. (*d*) Because $\rho_v = 1$ at each end, the line voltage will not reach a steady value and oscillations will continue indefinitely.

203. (*c*) $Z_0 = \sqrt{\dfrac{L}{C}} = \sqrt{\dfrac{500 \times 10^{-9}}{50 \times 10^{-12}}} = 100\ \Omega.$

204. (*b*) $0.054 \times 8.68 = 0.47$ dB/m.

205. (*d*) $\lambda = \dfrac{2\pi}{\beta} = \dfrac{2\pi}{29.8} = 0.21$ m.

206. (*c*) **207.** (*c*)

208. (*a*) H lines are concentric circles around conductor and E lines are radial.

209. (*c*) $\rho_v = \dfrac{R_L - Z_0}{R_L + Z_0}$ = finite number and

$\text{VSWR} = \dfrac{1 + |\rho_v|}{1 - |\rho_v|}$ = finite number.

210. (*c*) $\rho_v = \dfrac{R_L - Z_0}{R_L + Z_0}$. Since $R_L < Z_0$, ρ_v is negative.

211. (*b*) If standing waves are not there, whole of the power is absorbed by load.

212. (*b*) **213.** (*a*)

214. (*b*) Power handling capacity ∝ $(\text{voltage})^2$ and is inversely proportional to f_{max}^2.

215. (*a*)

216. (*a*) $C = \epsilon_0 \epsilon_r \dfrac{\text{area}}{\text{thickness}} = \epsilon_0 \epsilon_r \dfrac{W}{h}$ farads/unit length.

217. (*b*) Printed circuit technique is used for microstrip line.

218. (*b*) **219.** (*a*) **220.** (*c*) **221.** (*d*)

222. (*d*) **223.** (*b*) **224.** (*c*)

225. (*d*) X_c should be at least ten times greater than R. Therefore for 50 Ω resistor at 3 GHz, C should be less than 0.1 pF.

226. (*a*) Turn ratio is so selected as to give low SWR at both ports.

227. (*d*)

228. (*d*) Both dissipative and reflective attenuators are available with either fixed or variable attenuation .

229. (*a*)

230. (*c*) $\rho_v = \dfrac{150-50}{150+50} = 0.5.$ $\text{SWR} = \dfrac{1+0.5}{1-0.5} = 3.$

231. (*b*) Use of matched attenuator reduces SWR.

232. (*b*) Z_{01} provides series inductances and teflon filled low impedance line Z_{02} provides shunt capacitance.

233. (*d*)

234. (*a*) Mode filter suppresses modes with lower cutoff frequencies.

235. (*c*)

236. (*a*) TE_{01} mode is unaffected because its horizontally oriented electric field is everywhere perpendicular to the conducting strips.

237. (*a*) **238.** (*a*)

239. (*b*) Coplanar ams are decoupled.

240. (*a*)

241. (*a*) Since all ports are matched, input SWR = 1. No power is reflected and $P_{in} = P_4 = 1$ W. This power splits between ports 1 and 2. Therefore $P_1 = P_2 = 0.5$ W, $P_3 = 0$.

242. (*a*) See Section 13.4.4.

243. (*d*) See Section 13.4.4.

244. (*b*) See Section 13.4.4.

245. (*d*) **246.** (*c*) See Fig. 13.8.

247. (*c*) **248.** (*a*) See Section 13.5.

249. (*a*) **250.** (*a*)

251. (*a*) Isolation equals coupling plus directivity.

252. (*a*) **253.** (*a*) **254.** (*a*)

255. (*a*) See Section 13.3.

256. (*d*) They are used in all the devices.

257. (*a*) **258.** (*a*)

259. (*c*) A branched duplexer finds use in narrow bandwidth applications. TR tube stands for transmit-receive tube and ATR tube stands for antitransmit receive tube. Both these tubes are used in branched duplexer.

260. (*a*) See Section 13.2.3.

261. (*a*) In TM mode $H_z = 0$.

262. (*b*) At high frequencies transit time is large as compared to the period of microwave signal.

263. (*c*) See Section 13.6.1.

264. (*a*) See Section 13.6.1.

265. (*a*) See Section 13.6.2.

266. (*c*) See Section 13.6.2.

267. (*c*) See Section 13.6.1.

268. (*b*) See Section 13.6.1.

269. (*c*)

270. (*a*) See Section 13.6.1.

271. (*c*) See Section 13.6.1.

272. (*a*) See Section 13.6.1.

273. (*a*)

274. (*a*) Applegate diagram is distance time plot.

275. (*c*)

276. (*a*) See Section 13.6.2.

277. (*c*) **278.** (*a*)

279. (*c*) See Section 13.6.3.

280. (*a*)

281. (*d*) See Section 13.6.3.

282. (*a*) See Section 13.6.3.

283. (*a*) See Section 13.6.3.

284. (*b*) See Section 13.6.3.

285. (*a*) See Section 13.6.4.

286. (*c*)

287. (*c*) See Section 13.6.4.

288. (*d*) See Section 13.6.3.

289. (*c*) **290.** (*d*)

291. (*a*) See Section 13.6.4.

292. (*a*) See Section 13.6.5.

293. (*b*) **294.** (*a*) **295.** (*a*) **296.** (*a*)

297. (*a*) $\left(\frac{\text{metres}}{\text{sec.}}\right)\left(\frac{\text{coulomb}}{\text{m}^3}\right)$ = coulomb | sec | m^2 = current density.

298. (*c*) **299.** (*a*) **300.** (*b*)

301. (*d*) Gunn diode, Tunnel diode and Impatt diodes are negative resistance devices as discussed in Section 13.7.

302. (*d*) Varactor diode is used in all the three applications.

303. (*a*) The action of varactor diode is due to the capacitance of depletion layer. The thickness of depletion layer depends on the reverse bias which should not cause breakdown.

304. (*c*)

305. (*b*) In a parametric amplifier the non-linear element is varactor diode or inductor.

306. (*a*) Since reactance does not contribute thermal noise to the circuit, it is a low noise device.

307. (*a*) **308.** (*b*)

309. (*b*) PIN diode is not used in oscillators.

310. (*c*) **311.** (*c*) **312.** (*a*) **313.** (*a*)

314. (*b*) See Section 13.7.1. and Section 13.7.2.

315. (*a*) All the three can be used for impedance matching in high frequency circuits.

316. (*b*) **317.** (*c*) **318.** (*a*) **319.** (*b*)

320. (*b*)

321. (*a*) All the three are correct.

322. (*a*) **323.** (*c*)

324. (*a*) Since field is a phasor quantity we have to take phasor sum.

325. (*d*) **326.** (*b*) **327.** (*a*)

328. (*c*) Since the two wire radiators are at 90° to each other, they have to be excited 90° out of phase with each other.

329. (*c*)

330. (*c*) All the three are correct

331. (*a*)

332. (*a*) $1 \times 10^{-6} \times 1000 = 0.001$.

333. (*a*) All the statements are correct.

334. (*a*) **335.** (*a*)

336. (*b*) Klystron is amplifier, Bolometer is a meter.

337. (*c*) At high frequency attenuation is high.

338. (*b*) Cavity meter is used to measure frequency, Bolometer is used for power measurement.

339. (*c*) Use of frequency diversity can solve fading problem.

340. (*c*) Cutoff wavelength for TE_{01} mode $= 2b = 2.032$ cm.

341. (*b*) Propagating waves in a waveguide can be TE or TM. Since an extra electric component is created, it is TM mode.

342. (*a*) For dominant mode cutoff wavelength is maximum.

343. (*a*) For dominant mode TE_{10}

$$\text{Cutoff frequency} = \frac{c}{2a} = \frac{3\times10^8}{2\times3\times10^{-2}} = 0.5\times10^{10}$$

$$= 5 \times 10^9 \text{ or } 5 \text{ GHz.}$$

344. (*b*) In evanescent mode, waveguide is excited under cutoff frequency. Hence no wave propagation Dominant mode has lower cutoff frequency. In TM_{01}, TM_{10} all field components (E_x, E_y, E_z) varnish. Hence these modes do not exist.

345. (*d*) See Section 13.7.

346. (*d*) $\lambda = \frac{3\times10^8}{1\times10^9} = 0.3 \text{ m} = 30 \text{ cm}$

$l = 37.5 \text{ cm} = 1.25\,\lambda$

$$Z_{in} = Z_0\left[\frac{Z_L + jZ_0 \tan\beta l}{Z_0 + jZ_L \tan\beta l}\right]$$

$$= Z_0\left[\frac{0 + jZ_0 \tan\beta l}{Z_0 + 0}\right]$$

$$= j\,Z_0 \tan\frac{2\pi}{\lambda}(1.25\lambda) = jZ_0 \tan(2.5\pi) = \infty$$

347. (*c*) In magnetron dc electric field and magnetic field are perpendicular to each other.

348. (*a*)

349. (*c*) Isolator eliminates reflections which may cause frequency instability.

350. (*b*) Average signal power $= \frac{15\times10^{-3}}{\text{Pulse duration}\times\text{PRF}}$

$$= \frac{15\times10^{-3}}{1.5\times10^{-6}\times10^3} = 10 \text{ W}$$

351. (*b*) Helical antenna provides circular polarization.

352. (*d*) Gun diode does not have *p-n* junction. It works on transferred electron effect.

353. (*c*) In super high frequency range bands are :

L band	1–2 GHz
S band	2–4 GHz
C band	4–8 GHz
X band	8–12 GHz
Ku band	12–18 GHz
K band	18–27 GHz
Ka band	27–40 GHz
Millimeter	40–300 GHz
Submillimeter	More than 300 GHz

354. (*d*) See Section 12.6.

355. (*c*) It provides low power output (10–500 mW) with high stable single frequency (1–25 GHz).

356. (*c*) Microstrip has all the three losses.

357. (*a*) $f_c = \dfrac{c}{(\sqrt{\epsilon_r})(2a)} = \dfrac{3\times10^8}{\sqrt{4}(2\times1.5\times10^{-2})} = 5\text{ GHz}$

Hence 8 GHz signal will pass through the waveguide.

358. (*a*) Phase difference $= \dfrac{2\pi}{\lambda}$(Path difference)

$$\lambda = \frac{c}{f} = \frac{3\times10^8}{10\times10^9} = 3\times10^{-2}\text{ m}$$

Phase difference $\dfrac{2\pi}{3\times10^{-2}}(3\times10^{-2}) = 2\pi$

359. (*c*) Velocity $= \dfrac{c}{\sqrt{\epsilon_r}} = \dfrac{c}{n}$

$$\lambda = \frac{\text{velocity}}{f} = \frac{c}{nf}$$

360. (*d*) Gain $= 10\log_{10}200 = 10(\log_{10}2 + \log_{10}100)$

$= 10(0.3+2) = 23$ dB

361. (*c*) Amplification factor

362. (*b*) See Section 13.6.3.

363. (*b*) Attenuator is used to prevent oscillations.

364. (*b*) Phase velocity $= \dfrac{1}{\sqrt{\mu_0\,\epsilon_0\,\mu_r\,\epsilon_r}}$

Since $\mu_r = 1$

Phase Velocity $= \dfrac{1}{\sqrt{\mu_0\,\epsilon_0}}\cdot\dfrac{1}{\sqrt{\epsilon_r}} = \dfrac{c}{\sqrt{\epsilon_r}}$.

365. (*b*) Distance $= \dfrac{2D^2}{\lambda} = \dfrac{2D^2 f}{c}$

$$= \frac{2(5^2)(6\times10^9)}{3\times10^8} = 1000\text{ m} = 1\text{ km}$$

366. (*a*) Reflection coefficient

$$= \frac{Z_L - Z_0}{Z_L + Z_0} = \frac{|40 + j30 - 50|}{|40 + j30 + 50|} = \frac{|-10 + j30|}{|90 + j30|}$$

or $|\rho| = \dfrac{\sqrt{1000}}{\sqrt{9000}} = \dfrac{1}{3}$

$$\text{VSWR} = \frac{1+|\rho|}{1-|\rho|} = \frac{1+1/3}{1-1/3} = \frac{4/3}{2/3} = 2$$

367. (*c*) Lowest order TM mode is TM_{11}.

$$E_z = 20\sin\left(\frac{69\pi x}{3}\right)\sin\left(\frac{100\pi y}{2}\right)$$

$= 20\sin(20\pi x)\sin(50\pi y)$

368. (*c*) For TE_{10} mode

$$f_c = 2\times10^9 = \frac{3\times10^{10}}{2}\left[\left(\frac{1}{a\sqrt{3}}\right)^2 + 0\right]^{0.5}$$

or $a = \dfrac{5\sqrt{3}}{2}$ cm

For TM_{11} mode

$$f_c = \frac{3\times10^{10}}{2}\sqrt{\left(\frac{1}{a\sqrt{3}}\right)^2 + \left(\frac{1}{a}\right)^2}$$

$$= \frac{3\times10^{10}}{2}\cdot\frac{2}{a\sqrt{3}}$$

$$= 1.5\times10^{10}\,\frac{2\times2}{\sqrt{3}\,(5\sqrt{3})}$$

$$= \frac{6\times10^{10}}{15} = 4\times10^9\text{ Hz}.$$

369. (*c*) See Section 12.6 (Table 12.1)

370. (*a*) See Section 13.6.3. TWT is a tube and a broad band amplifier.

371. (*c*) $f = \dfrac{\text{Drift velocity}}{\text{Active region length}} = \dfrac{10^7}{10^{-3}} = 10\times10^9$ Hz.

372. (*b*) The dominant mode in circular waveguide is TE_{11}.

373. (*d*) Input VSWR on line $= \dfrac{Z_L}{Z_0} = \dfrac{72}{50} = 1.44$.

374. (*c*) Voltage = Critical field × distance $= 3.2\times10\times10^{-4}$

$= 3.2\times10^{-3}$ kV $= 3.2$ V

375. (*d*) Only Double cavity klystron is amplifier. Other devices are oscillators.

376. (*c*) Cutoff wavelengths for dominant mode does not depend on height (see Section 5.8).

377. (*a*) Dominant mode in rectangular waveguide is TE_{10}.

378. (*c*) Maximum possible gain

$$= \pi^2\left(\frac{D}{\lambda}\right)^2 = \pi^2\left(\frac{1}{3.1[illegible]\times10^{-[illegible]}}\right)^2 = 10^4$$

Gain in *d*B $= 10\log_{10}10^4 = 40$ *d*B.

379. (*c*) Drift Velocity $= 10^5$ m/s

$$\text{Frequency} = \frac{10^5}{2\text{ (Drift length)}} = \frac{10^5}{2\times4\times10^{-6}}$$

$= 12.5\times10^9$ Hz

380. (*d*) Cut off wavelength for dominant mode

$= 2a = 2\times4.5 = 9$ cm.

381. (*b*) See Eq. (13.6).

382. (*a*) Artificial transmission lines have many applications. One of the applications is to simulate an actual line in the laboratory.

383. (*b*) Both are correct but independent.

384. (*d*) Velocity of em waves on coaxial cables is less than that on overhead lines.

385. (*a*) A parallel tuned circuit has an infinite impedance if R is zero.

386. (*b*) See Fig. 13.4. **387.** (*a*) See Fig. 13.3.

388. (*a*) Since a half wavelength line has an input impedance equal to load impedance, the impedance transformation ratio is 1 : 1.

389. (*b*) $X_L = \omega L$. As frequency increases line resistance increases slightly but X_L increases directly as per frequency.

390. (*b*) See Fig. 13.2. **391.** (*a*) See Section 13.6.4.

392. (*b*) See Section 13.6.4.

393. (*a*) See Section 13.6.3.

394. (*a*) See Section 13.6.4.

395. (*a*) See Section 13.6.4.

396. (*a*) See Section 13.6.1.

397. (*b*) Both are correct but independent.

398. (*b*) See Section 13.7.1.

399. (*b*) See Section 13.7.1.

400. (*b*) See Section 13.7.2.

401. (*b*) See Section 13.7.2.

402. (*b*) Both are correct but independent.

403. (*a*)

404. (*b*)

405. (*b*) $\rho_v = \frac{Z_L - Z_0}{Z_L + Z_0}$ and VSWR $\frac{1+|\rho_v|}{1-|\rho_v|}$.

406. (*b*) Both are correct but independent.

407. (*b*) See Section 13.7.7.

408. (*a*) See Section 13.7.7.

409. (*b*) See Section 13.7.7.

410. (*a*) See Section 13.7.10.

411. (*a*) Matched load means no reflections.

412. (*c*) See Fig. 13.4. **413.** (*b*) See Section 13.1.5.

414. (*c*) R is wrong because wave velocity decreases.

415. (*a*) Since the fields are confined, there is no radiation.

416. (*a*) See Section 13.2. **417.** (*b*) See Section 13.2.1.

418. (*a*) See Section 13.2.1.

419. (*b*) See Section 13.2.5.

420. (*b*) Both are correct but independent. See Section 13.2.1.

421. (*a*) See Section 13.6.5.

422. (*a*) **423.** (*b*) **424.** (*a*)

425. (*b*) **426.** (*a*) See Section 13.7.10.

427. (*c*) **428.** (*a*) See Section 13.2.1.

429. (*c*) See Section 13.7.1.

430. (*a*)

431. T, Skin effect increases as frequency increases.

432. T

433. T, In a line terminated by Z_0, there are no reflections.

434. T **435.** T, See Table 13.1.

436. F, it is a forward wave because as *t* increases, *x* increases.

437. T

438. F, Depth of penetration depends on frequency. See Eq. (13.14).

439. F, See Eq. 13.8. **440.** T

441. T, See Fig. 13.6. **442.** T

443. T **444.** T **445.** T **446.** F

447. T **448.** T, See Section 13.7.10.

449. T, See Section 13.2.

450. T **451.** T **452.** T **453.** T

454. F, See Section 13.2.3.

455. T, See Section 13.2.1.

456. T, See Section 13.6.1.

457. T, See Section 13.6.1.

458. F, See Section 13.6.1.

459. T, See Section 13.6.2.

460. T, See Section 13.6.2.

461. F, See Section 13.6.2.

462. T **463.** T

464. T, See Section 13.6.3.

465. F, The signal gets strengthened.

466. T, See Section 13.6.4.

467. T, See Section 13.6.4.

468. T, See Section 13.6.4.

469. T **470.** T

471. T, See Section 13.7.1.

472. T **473.** T, See Section 13.7.2.

474. T, See Section 13.7.2 and Section 13.7.3.

475. T

476. intrinsic impedance

477. quarter wave transformer

478. inductance **479.** one

480. load impedance **481.** short-circuited

482. characteristic impedance

483. amplitude, phase **484.** inductance

485. dissipative, reflective

486. series **487.** shunt

488. magic **489.** rat race

490. hollow **491.** longitudinal

492. transverse **493.** TE_{10}

494. attenuated **495.** centre line

496. dual polarization **497.** TE_{11}

498. perfect

499. transferred electron effect

500. Pierce gun **501.** negative

502. right angle **503.** three

504. π **505.** broad

Appendix

Model Test Paper-I
(Chapters 1 – 6)

Time : 2 Hours | *Max. Marks : 100*

A. Choose the most appropriate alternative.

1. Two resistances $R_1 = 18.7\ \Omega$ and $R_2 = 3.624\ \Omega$ in series. The total resistance to the appropriate number of significant figures is
   (*a*) 22.324 Ω (*b*) 22.32 Ω
   (*c*) 22.3 Ω (*d*) 22 Ω
2. Four capacitors are in parallel. Their values are 36.3 μF, 3.85 μF, 34.002 μF and 850 μF with an uncertainty of digit in the last place. The total capacitance in significant figures is
   (*a*) 75.0 μF (*b*) 75.6 ± 0.1 μF
   (*c*) 75.1 ± 0.1 μF (*d*) 75.002 ± 0.1 μF
3. A permanent magnet moving coil meter can be used for
   (*a*) ac only (*b*) dc only
   (*c*) both ac and dc
   (*d*) low frequency ac and dc
4. A moving coil meter has a resistance of 50 ohms and a full scale deflection current of 20 mA. To convert it into a voltmeter of 0–100 V range, the series resistance should be
   (*a*) 5000 Ω (*b*) 5050 Ω
   (*c*) 4950 Ω (*d*) 505 Ω
5. The coil of a moving iron instrument has a resistance of 500 Ω and an inductance of 1 H. It reads 250 V when a 250 V dc is applied. If series resistance is 2000 Ω, its reading when fed by 250 V, 50 Hz ac will be
   (*a*) 260 V (*b*) 252 V
   (*c*) 250 V (*d*) 248 V
6. A sinusoidal ac voltage of amplitude 100 V is applied to a rectifying device which offers 10 Ω resistance in forward direction and infinite resistance in backward direction. A moving coil ammeter is also connected in the circuit. The reading of ammeter will be
   (*a*) 10 A (*b*) 5 A
   (*c*) 3.185 A (*d*) 1.414 A
7. The method of frequency compensation in moving iron voltmeter is
   (*a*) to use a capacitor in series with coil
   (*b*) to connect a capacitor in parallel with coil
   (*c*) to connect a capacitor in parallel with series resistance
   (*d*) none of the above
8. The effect of lead resistance is important in measurement of
   (*a*) resistance of shunt
   (*b*) resistance of shunt winding
   (*c*) resistance of ammeter
   (*d*) both (*a*) and (*c*)
9. It is required to measure the true open circuit emf of a battery. The best device is
   (*a*) dc voltmeter
   (*b*) ammeter and known resistance
   (*c*) dc potentiometer
   (*d*) either (*a*) or (*c*)
10. For measurements of high voltage capacitors the suitable bridge is
    (*a*) Wien bridge
    (*b*) modified De-Sauty bridge
    (*c*) Schering bridge
    (*d*) either (*a*) or (*b*)
11. In a CRO the Y plates are oriented
    (*a*) horizontally
    (*b*) vertically
    (*c*) at an angle of 20° of horizontal
    (*d*) at an angle of 70° the horizontal
12. The wave applied to X plates, for obtaining trace of voltage wave, is
    (*a*) sawtooth wave (*b*) sinusoidal wave
    (*c*) rectangular wave (*d*) (*a*) or (*c*)
13. For measurement of surface temperature of solids the element of RTD is usually in the shape of
    (*a*) bar (*b*) coil
    (*c*) flat grid (*d*) either (*a*) or (*b*)
14. The dimensions of force are
    (*a*) MLT^{-2} (*b*) $ML^{-1}T^{-2}$
    (*c*) MLT^{-1} (*d*) ML^2T^{-2}
15. Figure 1 shows an RC potentiometer to measure ac voltage. It is desired that V_0/V_i should be independent of frequency. The value of C should be

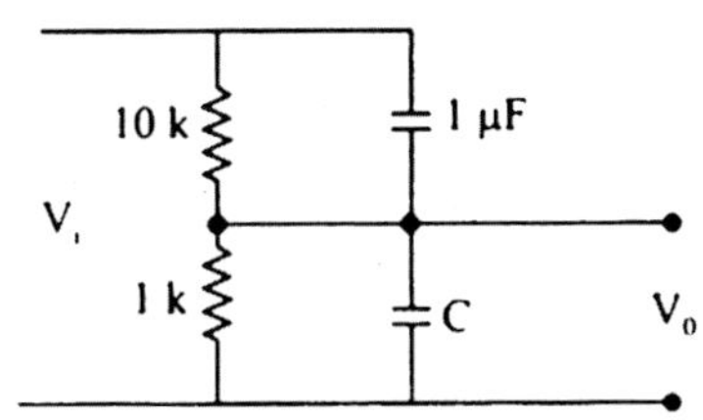

Fig. 1.

    (*a*) 10 μF (*b*) 11 μF
    (*c*) 0.1 μF (*d*) 0.09 μF

16. The output of a Piezo electric crystal has
(*a*) low amplitude and low impedance
(*b*) high amplitude and low impedance
(*c*) low amplitude and high impedance
(*d*) high amplitude and high impedance

17. If Fig. 2 the output V_0 is

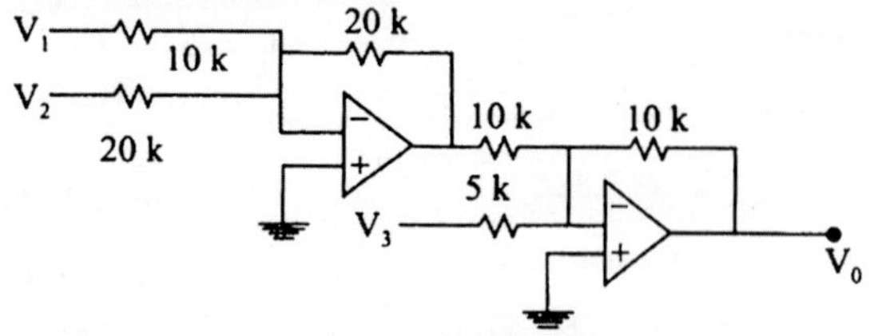

Fig. 2.

(*a*) $2V_1 + V_2 - 2V_3$ (*b*) $2(V_1 + V_2) - V_3$
(*c*) $V_1 + V_2 + V_3$ (*d*) $V_1 + 2V_2 - V_3$

18. A full wave rectifier type meter is used to measure the wave shape shown in Fig. 3. The voltmeter is calibrated to read rms value of sine wave. The reading of voltmeter will be

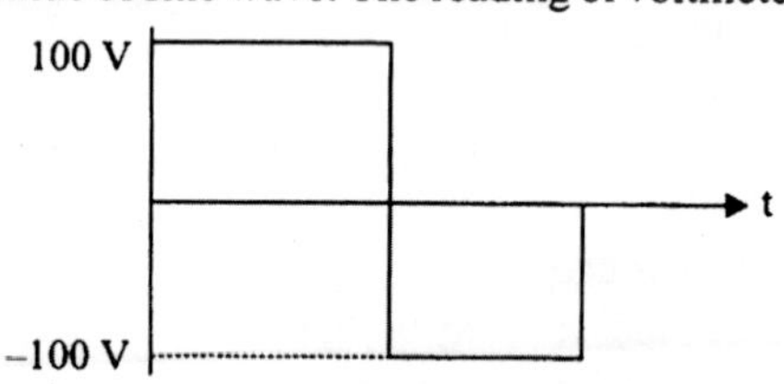

Fig. 3.

(*a*) $\frac{50}{\sqrt{2}}$ V (*b*) $\frac{100}{\sqrt{2}}$ V
(*c*) 100V (*d*) 111V

19. A digital frequency counter can be converted to DVM by addition of a suitable
(*a*) voltage controlled oscillator
(*b*) D/A converter
(*c*) power amplifier
(*d*) operational amplifier

20. A 0 – 200 V voltmeter has an accuracy of 0.75% of full scale reading. If voltage measured is 100 V, the error is
(*a*) 3% (*b*) 2%
(*c*) 1.5% (*d*) 0.75%

21. If $V = 100\angle 16°$, the $\sqrt{V}$ is
(*a*) $10\angle 32°$ (*b*) $10\angle 8°$
(*c*) $10\angle 16°$ (*d*) $10\angle 4°$

22. Two voltages are $v_1 = 100 \sin(\omega t + 15°)$ and $v_2 = 60 \cos \omega t$, then
(*a*) v_1 is leading v_2 by 15° (*b*) v_1 is leading v_2 by 75°
(*c*) v_2 is leading v_1 by 75° (*d*) v_2 is lagging v_1 by 15°

23. The resistance of the aluminium conductor ($\rho = 2.83 \times 10^{-8}$ Ω-m) in Fig. 4 is

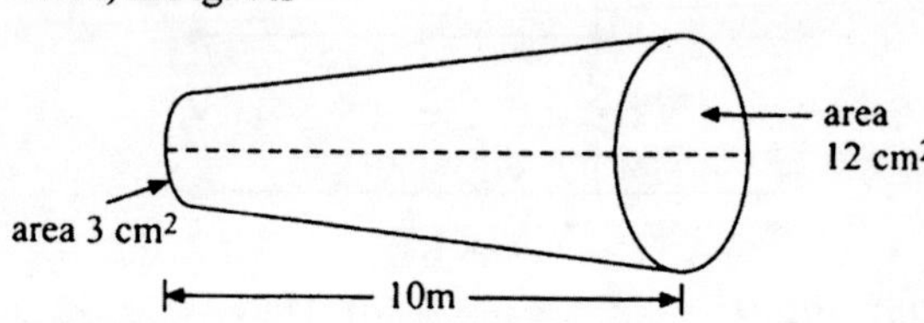

Fig. 4.

(*a*) 0.001 Ω (*b*) less than 0.001 Ω
(*c*) more than 0.001 Ω (*d*) more than 0.002 Ω

24. The input resistance of the network in Fig. 5 is

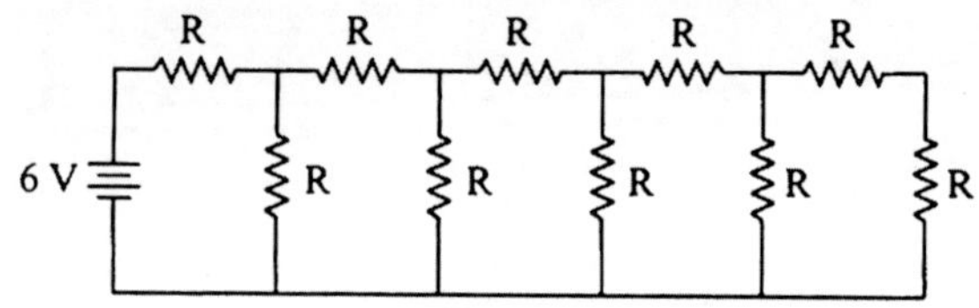

Fig. 5.

(*a*) 1.5R (*b*) $\frac{13}{8}$R
(*c*) $\frac{34}{21}$R (*d*) $\frac{89}{55}$R

25. In Fig. 6 the switch is closed at $t = 0$. At $t = 0^+$. The value of current supplied by battery is

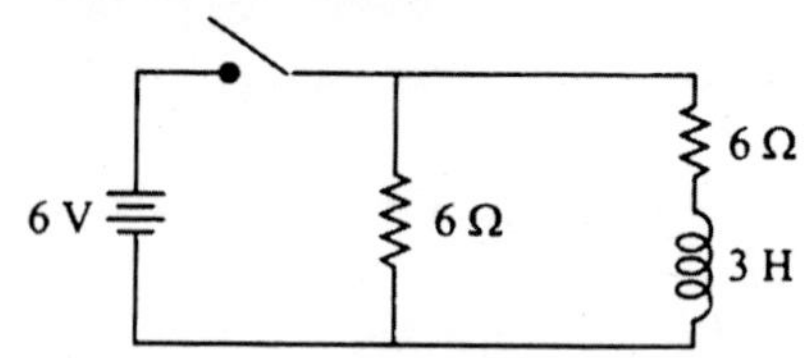

Fig. 6.

(*a*) 1 A (*b*) 2 A
(*c*) zero (*d*) infinity

26. In the circuit of Fig. 7 the switch is closed at $t = 0$. At $t = 0^+$ the current through inductance is

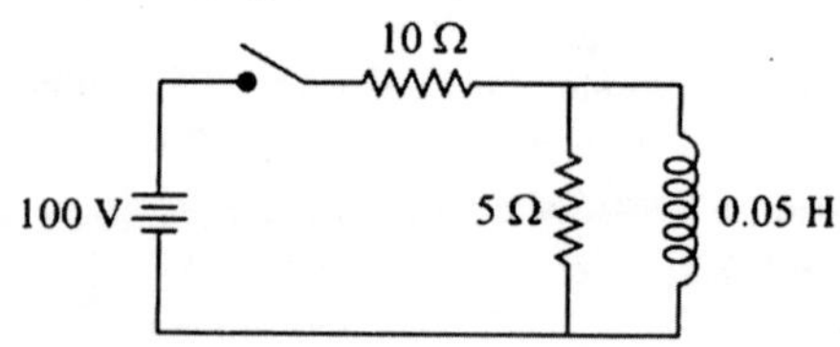

Fig. 7.

(*a*) 0 (*b*) 10 A
(*c*) 6.67 A (*d*) 5.1 A

27. The Q factor of a series RLC circuit is
(*a*) $\frac{1}{R}\sqrt{\frac{C}{L}}$ (*b*) $R\sqrt{\frac{C}{L}}$
(*c*) $R\sqrt{\frac{L}{C}}$ (*d*) $\frac{1}{R}\sqrt{\frac{L}{C}}$

28. In an ac network the sum of currents entering a node is $10\angle 45°$. The sum of currents leaving the node is
(*a*) $10\angle 45°$A (*b*) $10\angle 225°$A
(*c*) $10\angle -45°$ A (*d*) $10\angle -225°$ A

29. In the network of Fig. 8 it is desired that maximum power be supplied to Z_2. The values of Z_1 and Z_2 are fixed. However the transformation ratio *a* of transformer can be changed. The value of *a* for maximum power transfer is

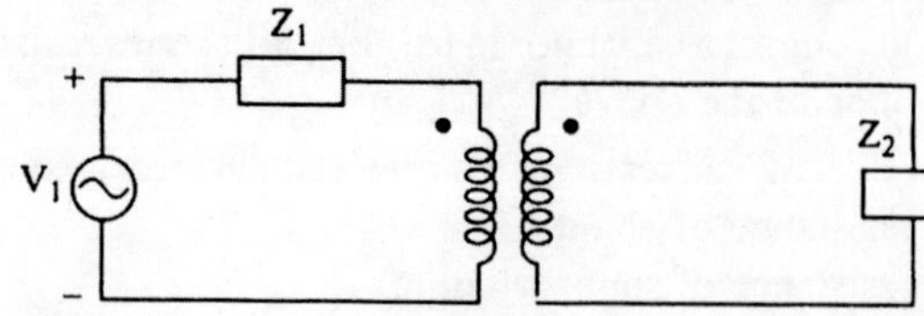

Fig. 8.

(a) $|Z_2|/|Z_1|$ (b) $[|Z_2|/|Z_1|]^{0.5}$
(c) $[|Z_2|/|Z_1|]^2$ (d) $[|Z_2|/|Z_1|]^{0.75}$

30. For the two-port of Fig. 9, z_{11} is

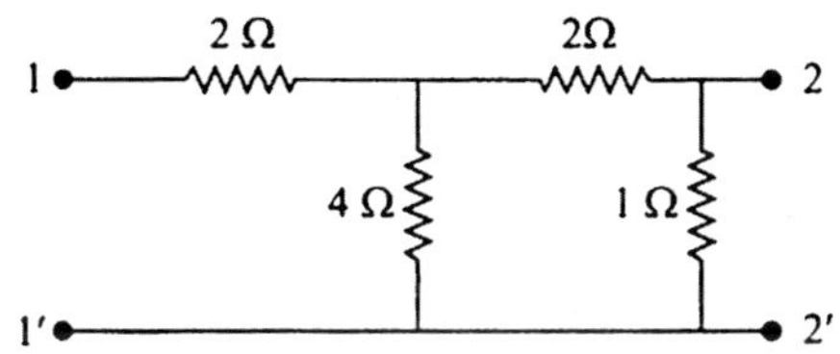

Fig. 9.

(a) 26/7 Ω (b) 10/3 Ω
(c) 6 Ω (d) 2 Ω

31. For the ladder network of Fig. 10, open circuit driving point impedance at port 1 =

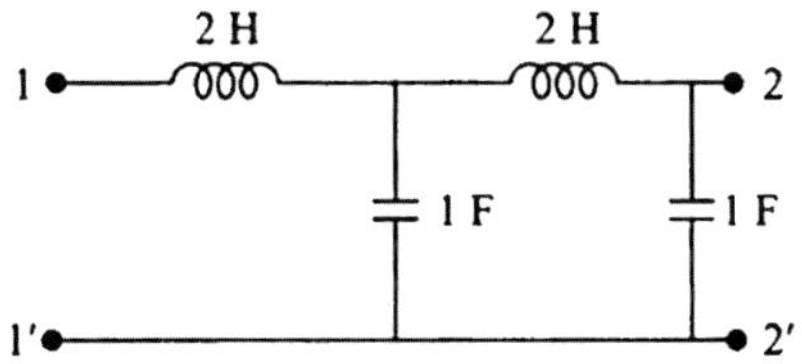

Fig. 10.

(a) $2s+\cfrac{1}{s+\cfrac{1}{2s+\cfrac{1}{s}}}$ (b) $2s+\cfrac{1}{2s+\cfrac{1}{s+\cfrac{1}{2s}}}$

(c) $s+\cfrac{1}{2s+\cfrac{1}{s+\cfrac{1}{2s}}}$ (d) $2s+\cfrac{1}{s+\cfrac{1}{s+\cfrac{1}{2s}}}$

32. For the two-port network of Fig. 11, the voltage transfer ratio in s domain is

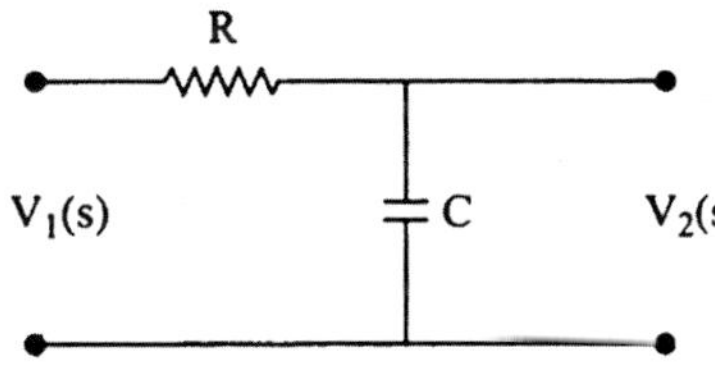

Fig. 11.

(a) $\dfrac{s}{1+RCs}$ (b) $\dfrac{1}{1+RCs}$

(c) $\dfrac{s^2}{1+RCs}$ (d) $\dfrac{1}{1+RCs^2}$

33. In Fig. 12 the current I

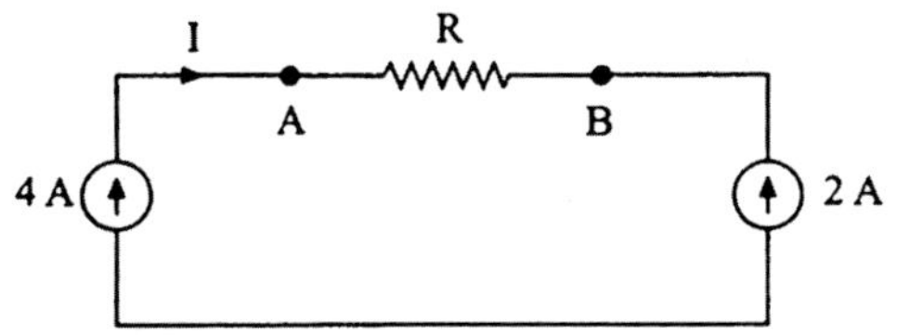

Fig. 12.

(a) is 2 A from A to B
(b) is 2 A from B to A
(c) depends on the value of R
(d) none of the above

34. In Fig. 13, the short circuit current through terminal AB will be

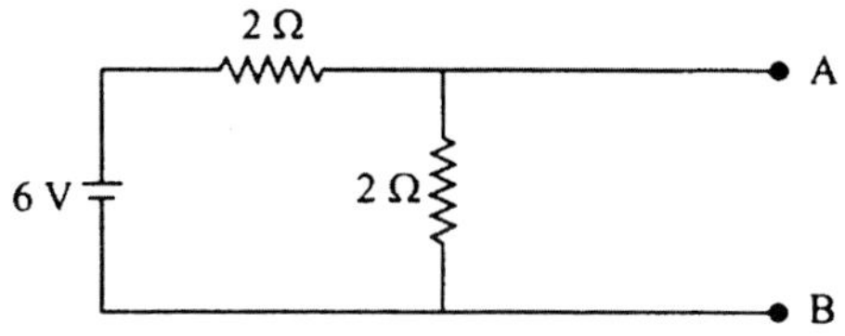

Fig. 13.

(a) 3 A (b) 1.5 A
(c) 2 A (d) 1 A

35. A series RLC circuit is supplied from a variable frequency ac source. The resonant frequency is ω_0. Voltage across C is maximum at a frequency

(a) equal to ω_0 (b) slightly less than ω_0
(c) slightly more than ω_0 (d) none of the above

36. A 10 ohm resistance and a 10 ohm capacitor are connected in series across a 100 V ac supply. The rms voltage across resistance is

(a) $100\sqrt{2}$ V (b) 100V
(c) $\dfrac{100}{\sqrt{2}}$ V (d) none of the above

37. The frame of an electric motor is earthed through three earthing plates resistances 30 Ω, 20 Ω and 10 Ω respectively. During fault the energy dissipated by three plates are W_1, W_2 and W_3. Then

(a) $W_1 > W_2 > W_3$ (b) $W_1 < W_2 < W_3$
(c) $W_1 = W_2 = W_3$ (d) $W_1 + W_3 < W_2$

38. A cut set schedule shows

(a) incidence of branches to loops
(b) incidence or branches 10 nodes
(c) incidence of nodes to loops
(d) none of the above

39. In Fig. 14

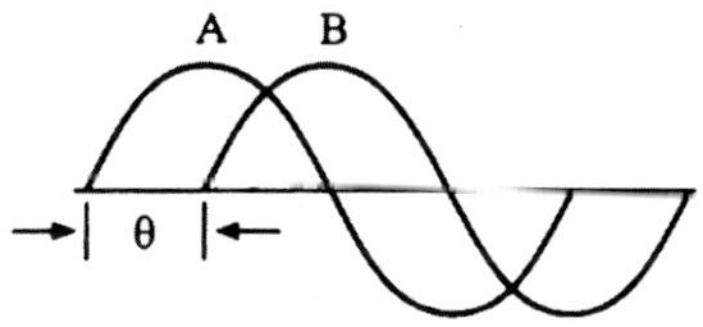

Fig. 14.

(a) Wave A is lagging wave B by $\angle\theta$
(b) Wave A is leading wave B by $\angle\theta$
(c) Wave A is leading wave B by $\angle\theta$ and the two waves have same rms values
(d) wave A is lagging wave B by $\angle\theta$ and the two waves have different rms values.

40. Henry is equivalent to

(a) Volts/ampere (b) Weber/volt
(c) Weber/ampere (d) Weber/ampere^2

41. A current increasing at the rate of 1 A/s is passed through a 1 H inductor. At $t = 0$, $i = 0$. The flux linkages after 1 sec are

(a) 0.5 Wb-turn (b) 1 Wb turn
(c) 2 Wb turn (d) 4 Wb turn

42. An electric dipole consists of a charge q and another charge $-q$ separated by a distance a. The electric field intensity at

a point P at perpendicular distance r from the centre O of the line joining the two charges
(*a*) is directed along line OP towards O
(*b*) is directed along line OP towards P
(*c*) is directed along a line parallel to the line joining $+q$ and $-q$ and directed in the direction of $+q$
(*d*) is directed along a line parallel to the line joining $+q$ and $-q$ and directed in the direction of $-q$

43. A charged particle having mass m and charge e is moved through a potential difference V. If the initial velocity of particle is zero, the final velocity is
(*a*) $(2eV/m)^{0.5}$ (*b*) $(eV/m)^{0.5}$
(*c*) $(eV/2m)^{0.5}$ (*d*) $V(e/m)^{0.5}$

44. A parallel plate capacitor has plates of area A. The dielectric between the plates is in two parts with thickness d_1 and d_2 and relative permittivities ϵ_1 and ϵ_2. The total capacitance is
(*a*) $\frac{\epsilon_1\epsilon_0 A}{d_1}+\frac{\epsilon_1\epsilon_0 A}{d_2}$ (*b*) $\frac{\epsilon_1\epsilon_0 A}{d_1}\left(1+\frac{d_2}{d_1}\right)$
(*c*) $\frac{\left(\frac{\epsilon_1\epsilon_0 A}{d_1}\right)\left(\frac{\epsilon_2\epsilon_0 A}{d_2}\right)}{\frac{\epsilon_1\epsilon_0 A}{d_1}+\frac{\epsilon_2\epsilon_0 A}{d_2}}$ (*d*) $\frac{\frac{\epsilon_1\epsilon_0 A}{d_1}+\frac{\epsilon_2\epsilon_0 A}{d_2}}{\left(\frac{\epsilon_1\epsilon_0 A}{d_1}\right)\left(\frac{\epsilon_2\epsilon_0 A}{d_2}\right)}$

45. Six capacitors of different capacitances $C_1, C_2, C_3, C_4, C_5, C_6$ are connected in series. $C_1>C_2>C_3>C_4>C_5>C_6$. The total cpacitance is almost equal to
(*a*) C_1 (*b*) C_3
(*c*) C_4 (*d*) C_6

46. A cylindrical conductor has an inner conductor and an outer conductor witn insulation between. The currents in the inner and outer conductors are in opposite directions. If E_t and E_n denote the fields tangential and normal to the boundary, then in the conductor
(*a*) $E_n=0$ (*b*) $E_t=0$
(*c*) both E_n and E_t are finite (*d*) either E_t or E_n is zero

47. In terms of current density J, KCL is
(*a*) $\oint_s J.ds = 0$ (*b*) $\oint_s J.ds =$ constant
(*c*) $\oint_s J.ds =$ infinity (*d*) $\oint_s J.ds =$ 0 or constant

48. Magnetic field at a point P at perpendicular distance r from a long wire carrying current I is
(*a*) $\frac{\mu I}{2\pi r}$ (*b*) $\frac{\mu I}{4\pi r}$
(*c*) $\frac{\mu I}{8\pi r}$ (*d*) $\frac{\mu I r}{4\pi}$

49. If H is magnetic field intensity, I is current and J is current density
(*a*) $\nabla . H = J$ (*b*) $\nabla \times H = J$
(*c*) $\nabla \times H = I$ (*d*) none of the above

50. If l is the distance between two points, a is the cross-sectional area of magnetic circuit and μ is the permeability, the reluctance between these points =
(*a*) $l/\mu a$ (*b*) μ/la
(*c*) $\frac{\mu a}{l}$ (*d*) $\frac{1}{\mu}al$

51. A non-conducting medium has $\epsilon = 4$ and loss tangent = 0.001. The power dissipated per unit volume at 9 MHz and with E = 1 kVm^{-1} is likely to be about
(*a*) 200 W/m^3 (*b*) 20 W/m^3
(*c*) 2 W/m^3 (*d*) 0.2 W/m^3

52. In free space the velocity of electromagnetic wave is about
(*a*) 3000 m/μs (*b*) 300 m/μs
(*c*) 30 m/μs (*d*) 3 m/μs

53. In a loss less line with ρ_v as the reflection coefficient for the voltage, voltage standing wave ratio (VSWR) =
(*a*) $\frac{1+|\rho_v|}{1-|\rho_v|}$ (*b*) $\frac{1-|\rho_v|}{1+|\rho_v|}$
(*c*) $\left(\frac{1+\rho_v}{1-\rho_v}\right)^2$ (*d*) $\sqrt{\frac{1+\rho_v}{1-\rho_v}}$

54. Retina of a human eye contains a bundle of about
(*a*) 1000 optical fibres (*b*) million optical fibres
(*c*) 10 million optical fibres (*d*) 100 million optical fibres

55. Poynting vector (instantaneous value) S =
(*a*) $E \times H$ (*b*) $E . H$
(*c*) $\sqrt{E \times H}$ (*d*) $\sqrt{E . H}$

56. Lorentz's force law for a point charge q is
(*a*) $\bar{F} = q(\bar{E} + v . \bar{B})$ (*b*) $\bar{F} = q(\bar{E} + v \times \bar{B})$
(*c*) $\bar{F} = q(\bar{B} + v \times \bar{E})$ (*d*) $\bar{F} = q(\bar{B} + v . \bar{E})$

57. Two concentric spherical conducting shells are held at differential potentials. Their centre coincides with the origin of spherical coordinate system. Which one of the following statements is true for E field in annular gap
(*a*) is purely radial and independent of (r, θ, ϕ)
(*b*) is purely radial, independent of (θ, ϕ) and varies as $1/r$
(*c*) the field lines depend on (θ, ϕ) but not on r
(*d*) none of the above

58. When a dielectric is placed in an electric field, E, then D, E and P (polarization) are related as
(*a*) $\bar{D} = \epsilon_0 \bar{E} \times \bar{P}$ (*b*) $\bar{D} = \epsilon_0 \bar{E} - \bar{P}$
(*c*) $\bar{D} = \epsilon_0 (\bar{E} + \bar{P})$ (*d*) $\bar{D} = \epsilon_0 \bar{E} + \bar{P}$

59. What are the directions of E and H in TEM mode of transmission?
(*a*) Both E and H are entirely transverse to direction of propagation
(*b*) E is entirely transverse but H has a component in the direction of propagation
(*c*) H is entirely transverse but E has a component in the direction of propagation
(*d*) Both E and H have components in the direction of propagation

60. A circular wire loop of radius R carries a current I. The magnetic flux density at centre of wire is
(*a*) $\frac{\mu I}{4\pi R}$ (*b*) $\frac{\mu I}{2\pi R}$

(*c*) $\frac{\mu I}{4R}$ (*d*) $\frac{\mu I}{2R}$

61. An atom of a rare gas is placed is an electric field E. Then
(*a*) the nucleus will be shifted in the direction of E by an amount which is proportional to E
(*b*) the nucleus will not be shifted
(*c*) the nucleus will be shifted in the direction of E by an amount which is independent of E.
(*d*) The nucleus will be shifted in a direction opposite to that of E

62. For a polyatomic gas, the equation relating dielectric constant $\in_r$ with molecular properties is

(*a*) $\in_0 (\in_r - 1) = N\left(\alpha_e + \alpha_i + \frac{\mu_p^2}{3kT}\right)$

(*b*) $\in_0 (\in_r - 1) = N\left(\alpha_e + \alpha_i + \frac{\mu_p^2}{3kT}\right)T^2$

(*c*) $\in_0 (\in_r - 1) = N\left(\frac{\alpha_e}{T} + \alpha_i + \mu_p^2\right)$

(*d*) $\in_0 (\in_r - 1) = N\left(\alpha_e + \frac{\alpha_i}{T} + \frac{\mu_p^2}{3kT}\right)$

63. An electron (charge $-e$) is rotating in a circular orbit of radius R, around a proton with charge $+e$ at a constant angular velocity ω. The orbital magnetic dipole moment is
(*a*) $0.5 e\omega R^2$ (*b*) $e\omega R^2$
(*c*) $2e\omega R^2$ (*d*) $0.5 e\omega R^3$

64. An electron is moving around a proton, in a circle of radius R. If flux density in the region is increased from zero to B, the force on the electron is

(*a*) $\frac{eR}{2}(dB/dt)$ (*b*) $eR(dE/dt)$

(*c*) eRB (*d*) $\frac{dR}{2}B$

65. In anti-ferromagnetic materials, the graph of susceptibility versus temperature shows a sharp maximum at a certain temperature. This temperature is known as
(*a*) Curie temperature (*b*) Neel temperature
(*c*) Weiss temperature (*d*) Bitter temperature

66. A metal has a large number of free electrons. If e is electronic charge and m is mass of electron and an electric field E is applied along (say) x direction, the electrons acquire a velocity having a component v_x is x direction. Then

(*a*) $\left(\frac{\partial}{\partial t}v_x\right) = \frac{e}{m}E$ (*b*) $\left(\frac{\partial}{\partial t}v_x\right) = \frac{m}{e}E$

(*c*) $\left(\frac{\partial}{\partial t}v_x\right) = \frac{-e}{m}E$ (*d*) $\left(\frac{\partial}{\partial t}v_x\right) = \frac{e}{Em}$

67. An electron in a metal suffers collision and if the probability of it being scattered over an angle $d\theta$ is $2n$ P(θ) sin θ $d\theta$ where P(θ) is an arbitrary function of scattering angle θ, then

(*a*) $\int_{\theta=0}^{\pi} 2\pi P(\theta) \sin\theta \, d\theta = 0$

(*b*) $\int_{\theta=0}^{\pi} 2\pi P(\theta) \sin\theta \, d\theta = 2$

(*c*) $\int_{\theta=0}^{\pi} 2\pi P(\theta) \sin\theta \, d\theta = 1$

(*d*) $\int_{\theta=0}^{\pi} 2\pi P(\theta) \sin\theta \, d\theta = -1$

68. The number of tightly bound electrons in silicon atom are
(*a*) 14 (*b*) 10
(*c*) 6 (*d*) 4

69. In which of the following semiconductor are holes minority carriers
(*a*) extrinsic (*b*) *p* type
(*c*) *n* type (*d*) intrinsic

70. The colour code on a carbon resistor is red-red-black-silver. The value of resistor is
(*a*) 22000 Ω (*b*) 2200 Ω
(*c*) 22 ± 5% Ω (*d*) 22 ± 10% Ω

71. Which capacitor cum store higher amount of energy?
(*a*) air capacitor (*b*) paper capacitor
(*c*) mica capacitor (*d*) plastic film capacitor

72. For a multiplate capacitor having *n* plates, A as area of each plate, the capacitance is proportional to
(*a*) nA (*b*) $(n - l)A$
(*c*) $(n + 1)A$ (*d*) $\frac{nA}{2}$

73. The temperature below which certain materials are antiferromagnetic and above which they are paramagnetic is called
(*a*) Weiss temperature (*b*) Curie temperature
(*c*) Neel temperature (*d*) None of above

74. In Fermi Dirac statistics the probability of electron occupation of an energy level equal to Fermilevel is
(*a*) 0 (*b*) 0.25
(*c*) 0.5 (*d*) 1

75. Match List I and List II and give correct answer using the given codes

List I	**List II**
A. Conductor	1. 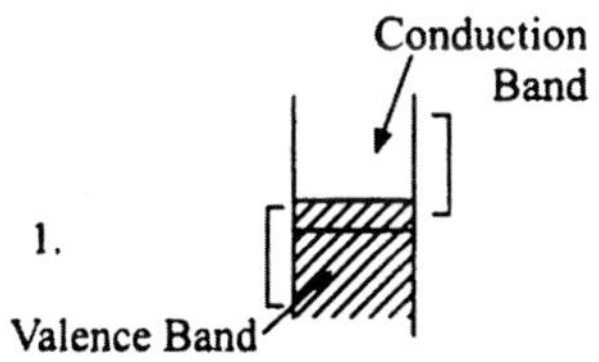
B. Semiconductor	2. 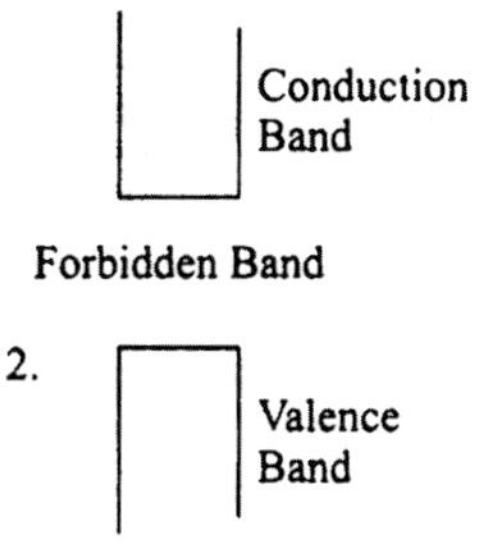

C. Insulator 3.

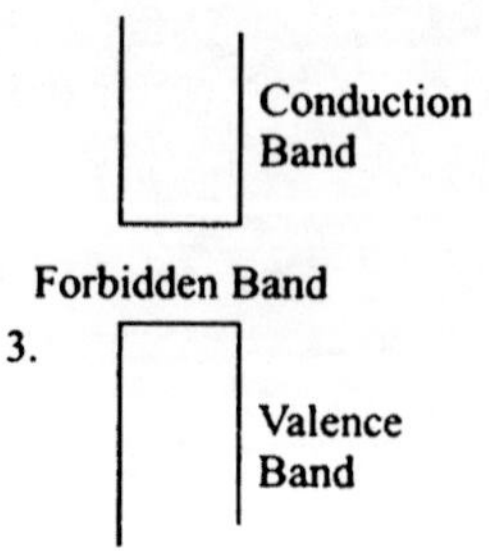

Codes:

	A	B	C
(*a*)	1	3	2
(*b*)	3	1	2
(*c*)	1	2	3
(*d*)	2	3	1

76. An air-cored inductance is a
(*a*) linear circuit element because its reactance varies linearly with voltage at a fixed frequency
(*b*) linear circuit element because its current varies linearly with voltage at a fixed frequency
(*c*) non-linear circuit element due to saturation of core
(*d*) non-linear circuit element because $v = \frac{L\,di}{dt}$

77. In a piezo electric crystal, the application of mechanical force will cause
(*a*) plastic deformation of crystal
(*b*) magnetic dipoles in the crystal
(*c*) electrical polarization in the crystal
(*d*) shift in fermi level

78. The real part of complex dielectric constant and tan δ for a dielectric are 2.1 and 5×10^{-4} at 100 Hz respectively. The imaginary part of dielectric constant at 100 Hz is
(*a*) 1.05×10^{-3} (*b*) 2.1×10^{-3}
(*c*) 5×10^{-3} (*d*) 1.05×10^{-2}

79. For a superconductor, magnetisation M in a field H is
(*a*) extremely small (*b*) – H
(*c*) – 1 (*d*) zero

80. If a small amount of Cu is added to Ni conductor, then
(*a*) resistivity of Ni will decrease at all temperatures because Cu is a better conductor than Ni
(*b*) residual resistivity of Ni at low temperature will increase as Cu atoms act as defect centres
(*c*) resistivity of Ni will increase at all temperatures as Cu destroys the periodicity of Ni and acts as defects
(*d*) resistivity of Ni remains the same

81. For most metals, fermi level E_F is less than
(*a*) 0.1 eV (*b*) 2 eV
(*c*) 5 eV (*d*) 10 eV

82. In *n* type semiconductor, the free electron concentration
(*a*) is nearly equal to density of donor atoms .
(*b*) is much greater than density of donor atoms
(*c*) is much less than density of donor atoms
(*d*) may be equal or more or less than density of donor atoms

83. Hall effect can be used
(*a*) to find type of semiconductor (whether *p* or *n*)
(*b*) to find carrier concentration
(*c*) to measure conductivity (*d*) all of the above

84. Which statement is false as regards holes?
(*a*) Holes exist in conductors as well as semiconductors
(*b*) Holes constitute positive charges
(*c*) Holes exist only in semiconductors
(*d*) Holes and electrons recombine

85. The concentration of minority carriers in a semiconductor depends mainly on
(*a*) the extent of doping (*b*) temperature
(*c*) the applied bias (*d*) none of above

86. Which of the following is basically a voltage controlled capacitance?
(*a*) zener diode (*b*) diode
(*c*) varactor diode (*d*) LED

87. A photo diode is
(*a*) forward biased (*b*) reverse biased
(*c*) either forward or reverse biased
(*d*) unbiased

88. Match the diodes in List I with symbol in List II using the given codes

List I		List II
A. Zener diode	1.	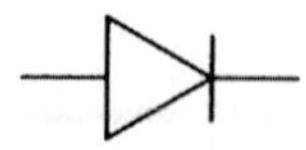
B. Tunnel diode	2.	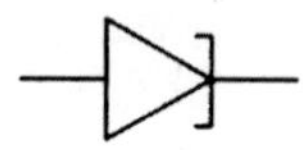
C. Schottky diode	3.	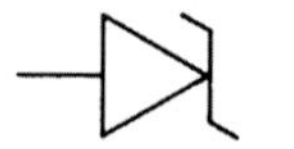
D. Diode	4.	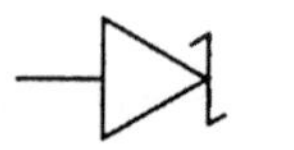

Codes:

	A	B	C	D
(*a*)	1	3	2	4
(*b*)	2	3	1	4
(*c*)	4	3	2	1
(*d*)	3	2	4	1

89. The value of α in a transistor
(*a*) is always equal 1
(*b*) is less than 1 but more than 0.9
(*c*) is about 0.4
(*d*) is about 0.1

90. For bipolar transistor
(*a*) $\beta_{dc} = \frac{\alpha_{dc}}{1-\alpha_{dc}}$ (*b*) $\beta_{dc} = \frac{\alpha_{dc}}{1+\alpha_{dc}}$
(*c*) $\beta_{dc} = \frac{1-\alpha_{dc}}{\alpha_{dc}}$ (*d*) $\beta_{dc} = \frac{1+\alpha_{dc}}{\alpha_{dc}}$

91. In forward active region of operation of a BJT
(*a*) both junctions are forward biased
(*b*) both junctions are reverse biased
(*c*) emitter base junction is forward biased and base collector junction is reverse biased
(*d*) emitter base junction is reverse biased and base collector junction is reverse biased

92. Between which regions does BJT act like switch
(*a*) cut off and saturation
(*b*) cut off and forward active
(*c*) forward active and cut off
(*d*) saturation and active

93. In *n* type MOSFET, the substrate
(*a*) is *p* type (*b*) is *n* type
(*c*) is metallic (*d*) may be *p* or *n* type

94. The normal operation of JFET is
(*a*) in constant voltage region
(*b*) constant current region
(*c*) both constant voltage and constant current regions
(*d*) either constant voltage or constant current region.

95. Match List I with List II using the given codes

	List I		**List II**
A	*n*-channel enhancement MOSFET	1.	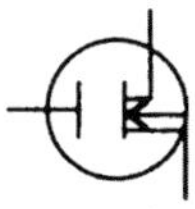
B	*p*-channel enhancement MOSFET	2.	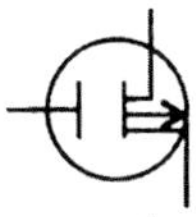
C	*n*-channel Depletion MOSFET	3.	
D	*p*-channel Depletion MOSFET	4.	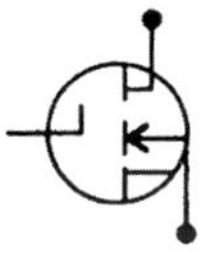

Codes:

	A	B	C	D
(*a*)	1	4	3	2
(*b*)	2	3	4	1
(*c*)	4	3	1	2
(*d*)	1	3	2	4

96. The collector to emitter cut off current (I_{CEO}) of a transistor is related to collector to base cut off current (I_{CBO}) as
(*a*) $I_{CEO} = I_{CBO}$ (*b*) $I_{CEO} = \alpha I_{CBO}$
(*c*) $I_{CEO} = \dfrac{I_{CBO}}{1-\alpha}$ (*d*) $I_{CEO} = \dfrac{I_{CBO}}{1+\alpha}$

97. For the transistor equivalent circuit of Fig. 15 match List I with List II using the given codes

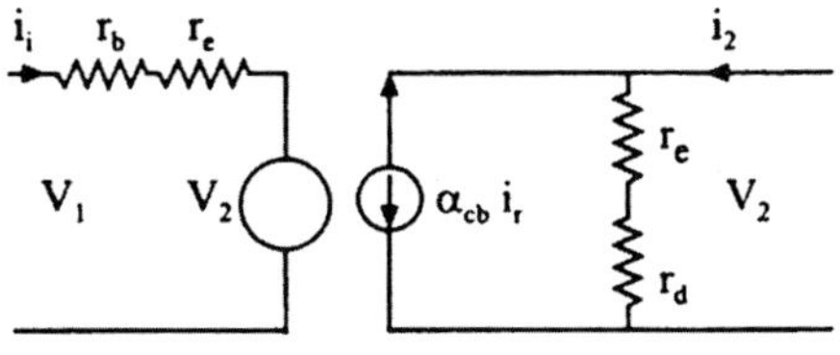

Fig. 15.

List I (Parameters)	**List II (Values)**
A. h_{22}	1. $r_b + r_e$
B. h_{11}	2. α_{cb}
C. h_{21}	3. $\dfrac{1}{r_e + r_d}$

Codes:

	A	B	C
(*a*)	3	1	2
(*b*)	1	3	2
(*c*)	2	3	1
(*d*)	3	2	1

98. If the drift velocity of holes under a field gradient of 100 V/m is 5 m/s, their mobility (in SI units) is
(*a*) 0.05 (*b*) 0.5
(*c*) 50 (*d*) 500

99. The conductivity of an intrinsic semi conductor is
(*a*) generally less than that of a doped semi conductor
(*b*) given by $\sigma_1 = en_i(\mu_n - \mu_p)$
(*c*) given by $\sigma_1 = en_i(\mu_p + \mu_n)$
(*d*) given by $\sigma_1 = n_i(\mu_n + \mu_p)$

100. Hall effect is observed in a specimen when it is carrying current and is placed in a magnetic field. The resultant electric field inside the specimen is
(*a*) normal to both current and magnetic field
(*b*) in the direction of current
(*c*) anti parallel to magnetic field
(*d*) in arbitrary direction

101. For an ac sinusoidal wave, the rms value is 10 A. For the same wave delayed by 60° in each half cycle, the rms value is likely to be
(*a*) 10 A (*b*) 7.07 A
(*c*) 6.35 A (*d*) 3.5 A

102. $\cos(n\omega_1 t) =$
(*a*) $0.5\,(e^{jn\omega_1 t} + e^{-jn\omega_1 t})$ (*b*) $0.5\,(e^{jn\omega_1 t} - e^{-jn\omega_1 t})$
(*c*) $\dfrac{0.5}{j}(e^{jn\omega_1 t} + e^{-jn\omega_1 t})$ (*d*) $\dfrac{0.5}{j}(e^{jn\omega_1 t} - e^{-jn\omega_1 t})$

103. A voltage $v = 100 \sin \omega t + 10 \sin 5\omega t$ is applied to a pure capacitor having capacitance of 1 μF. If ω = 314 rad/sec, the current through the capacitor is
(*a*) $0.0314 \cos 314t + 0.0157 \cos 1570t$
(*b*) $0.0314 \sin 314t + 0.0157 \sin 1570t$
(*c*) $0.0314 \cos 314t + 0.0314 \cos 1570t$
(*d*) $0.0157 \cos 314t + 0.0157 \cos 1570t$

104. If Laplace transform of $f(t)$ is F(*s*), then $\mathcal{L}\left[\dfrac{d}{dt}f(t)\right] =$
(*a*) $sF(s) - f(0^-)$ (*b*) $\dfrac{1}{s}F(s) - f(0^-)$
(*c*) $sF(s) + f(0^-)$ (*d*) $\dfrac{1}{s}F(s) + f(0^-)$

105. Out of the three transforms, viz., Z-transform, Laplace transform and Fourier transform
(*a*) all the three are used in continuous time domain
(*b*) all the three are used in both continuous time domain and discrete time domain
(*c*) Z transform is used in continuous time domain while Laplace and Fourier transforms are used in discrete time domain
(*d*) Z transform is used is discrete time domain while Laplace and Fourier transforms are used in continuous time domain

106. In Laplace transforms $s = \sigma + j\omega$. Which of the following represents true nature of σ.
Select your answer using the given codes
1. σ has a damping effect
2. σ is responsible for convergence of integral

$$\int_0^\infty f(t)e^{-st}\,dt$$

3. σ has a value less than zero

Codes:

(*a*) 1, 2 and 3 (*b*) 1 and 2
(*c*) 2 and 3 (*d*) 1 and 3

107. Match List I (source of signal) with List II (type of signal) and select the correct answer using the given codes

	List I	**List II**
A.	Output of signal generator	1. Modulated
B.	Error signal from a synchro	2. Digital
C.	Output of a JK flip-flop	3. Analog
D.	Signal received by radar	4. Stochastic

Codes:

	A	B	C	D
(*a*)	1	3	2	4
(*b*)	3	1	2	4
(*c*)	3	1	4	2
(*d*)	1	3	4	2

108. Given that $h(t) = 10e^{-10t}\, u(t)$ and $e(t) = \sin 10\, t\, u(t)$. The Laplace transform of

$$f(t) = \int_0^\tau h(t-\tau)\, e(\tau)\, d\tau \text{ is}$$

(*a*) $\dfrac{10}{(s+10)(s^2+100)}$ (*b*) $\dfrac{10(s+10)}{s^2+100}$

(*c*) $\dfrac{100}{(s+10)(s^2+100)}$ (*d*) $\dfrac{1}{(s+10)(s^2+100)}$

109. Match List I (Fourier transform) with List II (functions of time) and select the correct answer using the given codes

	List I	**List II**
A.	$\dfrac{\sin k\omega}{\omega}$	1. A constant
B.	$e^{-j\omega}$	2. Exponential function
C.	$1/(j\omega+2)^2$	3. t multiplied exponential function
D.	$k\,\delta(\omega)$	4. rectangular pulse
		5. impulse function

Codes:

	A	B	C	D
(*a*)	4	5	3	1
(*b*)	4	5	3	2
(*c*)	3	4	2	1
(*d*)	3	4	2	5

110. Which of the following is not true for impulse function $\delta(t)$?

(*a*) $\delta(t) = 0$ for $t \neq 0$ (*b*) $\delta(t)$ is undefined for $t = 0$

(*c*) $\int_{-\infty}^{\infty} \delta(t)dt = 1$ (*d*) Laplace transform is $\dfrac{1}{s}$

B. Questions 111 to 120 have two statements, viz., Assertion (A) and Reason (R). Read them and give correct answer as per the following codes.

(*a*) both A and R are true and R is correct explanation of A
(*b*) both A and R are true but R is not correct explanation of A
(*c*) A is true but R is false
(*d*) A is false but R is true

111. **A.** The line integral of $\bar{B}$ around a closed contour, in a region containing steady current, is μ_0 times the total net current, flowing through the contour of integration.
R. The curl of $\bar{B}$ is equal to μ_0 times current density vector J.

112. **A.** As per Maxwell's theorem $\nabla \cdot B$ is zero but $\nabla \cdot D$ is not necessarily zero.
R. An electric charge can be isolated but a magnetic pole cannot be isolated.

113. **A.** An electromagnetic compatibility chamber is a large completely enclosed metal room to provide shielding against outside sources of radio frequency interference.
R. Inside a metal cage the field is zero.

114. **A.** Many instruments use eddy current damping.
R. Eddy currents set up a magnetic flux which oppose the original movement which generates them.

115. **A.** If the resistance of shunt of an ammeter is increased, the range of ammeter is reduced.
R. If the series resistance of a voltmeter is increased, the range of voltmeter is increased.

116. **A.** Thermistors are commonly used to measure hot spot temperatures in electric machines.
R. Thermistor has the advantages of high temperature coefficient, small size and high speed of response.

117. **A.** In Fig. 16 the coefficients

$$a_n = \frac{4}{\pi(1-4n^2)}$$

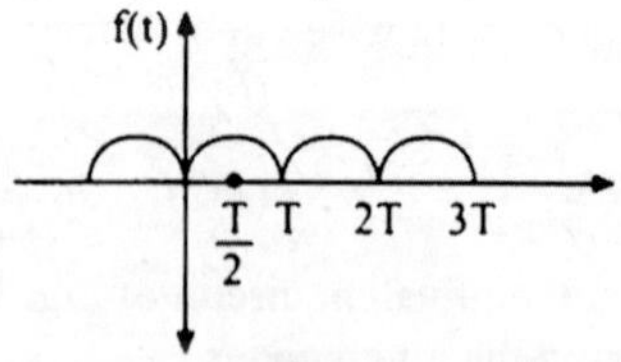

Fig. 16.

R. The periodic function is symmetrical about t = T and t = T/2.

118. **A.** For a physically realisable driving point function, the degree of numerator and denominator should be equal.
R. The sum of positive real functions is real.

119. **A.** If $Z_1(s)$ and $Z_2(s)$ are positive real then $Z_1(s) + Z_2(s)$ as well as $1/Z_1(s)$ and $1/Z_2(s)$ are positive real.
R. The poles of a positive real function are real or occur in conjugate pairs.

120. **A.** A complicated waveform can be replaced by sum or difference of two or more waveforms.
R. The method of superposition is applicable only to linear systems.

ANSWERS

1. (*c*)	**2.** (*b*)	**3.** (*b*)	**4.** (*c*)	**5.** (*d*)	**6.** (*c*)	**7.** (*c*)
8. (*d*)	**9.** (*c*)	**10.** (*c*)	**11.** (*a*)	**12.** (*a*)	**13.** (*c*)	**14.** (*a*)
15. (*a*)	**16.** (*c*)	**17.** (*a*)	**18.** (*d*)	**19.** (*b*)	**20.** (*c*)	**21.** (*b*)
22. (*c*)	**23.** (*b*)	**24.** (*d*)	**25.** (*a*)	**26.** (*a*)	**27.** (*d*)	**28.** (*a*)
29. (*b*)	**30.** (*a*)	**31.** (*a*)	**32.** (*b*)	**33.** (*a*)	**34.** (*a*)	**35.** (*b*)
36. (*c*)	**37.** (*b*)	**38.** (*b*)	**39.** (*b*)	**40.** (*c*)	**41.** (*b*)	**42.** (*d*)
43. (*a*)	**44.** (*c*)	**45.** (*d*)	**46.** (*a*)	**47.** (*a*)	**48.** (*a*)	**49.** (*b*)
50. (*a*)	**51.** (*c*)	**52.** (*b*)	**53.** (*a*)	**54.** (*d*)	**55.** (*a*)	**56.** (*b*)
57. (*b*)	**58.** (*d*)	**59.** (*a*)	**60.** (*d*)	**61.** (*a*)	**62.** (*a*)	**63.** (*a*)
64. (*a*)	**65.** (*b*)	**66.** (*c*)	**67.** (*c*)	**68.** (*b*)	**69.** (*c*)	**70.** (*d*)
71. (*c*)	**72.** (*b*)	**73.** (*c*)	**74.** (*c*)	**75.** (*a*)	**76.** (*a*)	**77.** (*c*)
78. (*a*)	**79.** (*b*)	**80.** (*c*)	**81.** (*d*)	**82.** (*a*)	**83.** (*d*)	**84.** (*a*)
85. (*b*)	**86.** (*c*)	**87.** (*b*)	**88.** (*d*)	**89.** (*b*)	**90.** (*a*)	**91.** (*c*)
92. (*a*)	**93.** (*a*)	**94.** (*b*)	**95.** (*c*)	**96.** (*c*)	**97.** (*a*)	**98.** (*a*)
99. (*c*)	**100.** (*a*)	**101.** (*c*)	**102.** (*a*)	**103.** (*a*)	**104.** (*a*)	**105.** (*d*)
106. (*b*)	**107.** (*b*)	**108.** (*c*)	**109.** (*a*)	**110.** (*d*)	**111.** (*b*)	**112.** (*a*)
113. (*a*)	**114.** (*a*)	**115.** (*b*)	**116.** (*a*)	**117.** (*a*)	**118.** (*d*)	**119.** (*b*)
120. (*a*)						

❑❑❑

Appendix

Model Test Paper-II
(Chapters 7 – 13)

Time : 2 Hours *Max. Marks : 100*

1. An RC coupled amplifier has an open loop gain of 200 and a lower cutoff frequency of 50 Hz. If negative feedback with ($\beta = 0.1$ is used the lower cut off frequency will be
(*a*) about 50 Hz (*b*) about 5 Hz
(*c*) about 2.38 Hz (*d*) about 70.5 Hz

2. A Wien bridge oscillator uses Wien bridge for
(*a*) tuning and phase shift
(*b*) tuning
(*c*) phase shift
(*d*) either tuning or phase shift

3. For an op-amp source follower circuit
(*a*) input and output voltages have same magnitude
(*b*) output and input voltages are in phase
(*c*) output and input voltage are in phase but have different magnitudes
(*d*) output and input voltages have same magnitude and are in phase

4. The element in the feedback circuit of a logarithmic amplifier using op-amp is
(*a*) a resistor (*b*) a capacitor
(*c*) an inductor (*d*) a diode

5. In a self-bias circuit for CE amplifier, an increase in emitter resistance R_E results in
(*a*) increase in emitter voltage
(*b*) decrease in emitter voltage
(*c*) increase in emitter current
(*d*) decrease in emitter current

6. In a CE amplifier circuit the emitter by pass capacitor is removed. The ac output voltage
(*a*) increases (*b*) remains the same
(*c*) increases or decreases (*d*) decreases

7. The phase angle of a lag circuit is
(*a*) between 0 and – 90° (*b*) between 0 and + 90°
(*c*) 90° (*d*) more than 90°

8. Equal amplitude ac signals are given as inputs to both inverting and non-inverting terminals of an op-amp. The output will be zero when the phase difference between inputs is
(*a*) 0° (*b*) 90°
(*c*) 180° (*d*) 270°

9. In the CE circuit shown in Fig. 1, the slope of ac load line will be

(*a*) $-\frac{1}{500}$ (*b*) $-\frac{1}{250}$

(*c*) $-\frac{1}{750}$ (*d*) $-\frac{1}{1000}$

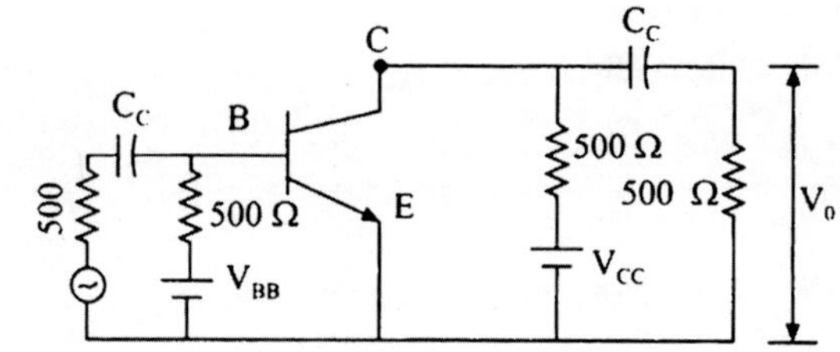

Fig. 1.

10. Match List I and List II and give correct answer using the given codes

List I (BJT amplifier)		List II (Characteristics)
A.	CE	1. good voltage gain, good current gain
B.	CC	2. voltage gain less than 1, high current gain
C.	CB	3. current gain less than 1, voltage gain high

Codes:

	A	B	C
(*a*)	2	1	3
(*b*)	1	2	3
(*c*)	3	1	2
(*d*)	3	2	1

11. In the r_e model of a BJT amplifier, the ac resistance of diode at room temperature is about (I_E is quiescent emitter current)

(*a*) $\frac{0.025}{I_E}$ (*b*) $\frac{0.25}{I_E}$

(*c*) $\frac{25}{I_E}$ (*d*) $\frac{250}{I_E}$

12. Figure 2 shows a self-bias circuit for FET amplifier, $I_D = 4$ mA

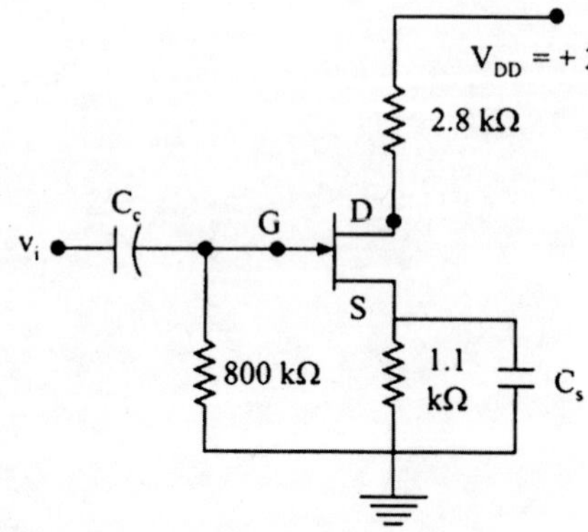

Fig. 2.

Then I_s =
(*a*) 4 mA (*b*) 2 mA
(*c*) 0.4 mA (*d*) 6.2 mA

13. In the circuit of Fig. 3 both diodes are ideal. If v_1 = 10 V and v_2 = 10 V which diode will conduct?

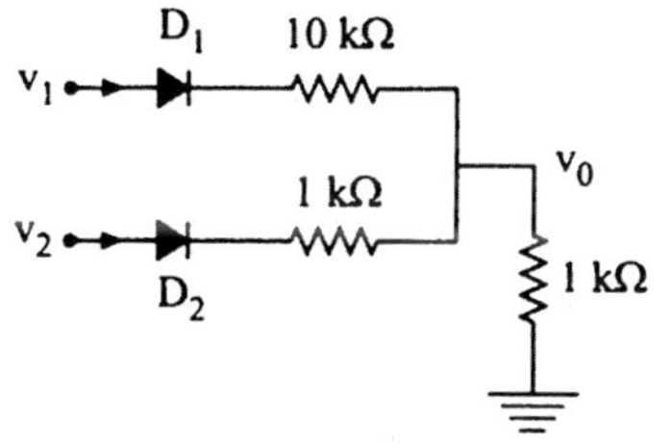

Fig. 3.

(*a*) D_1 only (*b*) D_2 only
(*c*) both D_1 and D_2 (*d*) neither D_1 nor D_2

14. In the circuit of Fig. 4, v_0 =

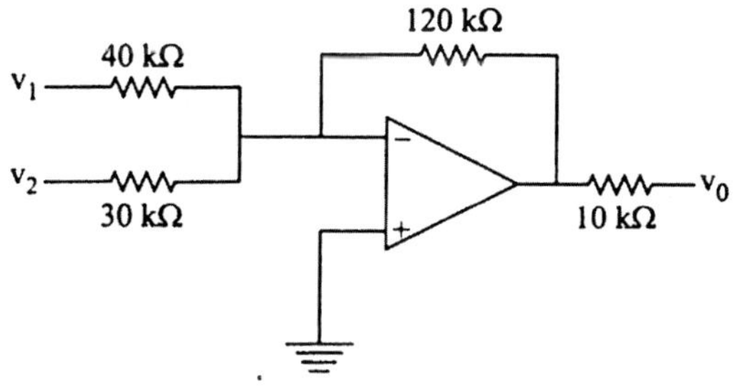

Fig. 4.

(*a*) $3v_1 + 4v_2$ (*b*) $-3v_1 - 4v_2$
(*c*) $4v_1 + 3v_2$ (*d*) $-4v_1 - 3v_2$

15. A two stage amplifier has an upper cutoff frequency of 2 MHz and a lower cutoff frequency of 30 Hz. If both stages are similar, their individual upper and lower cutoff frequencies respectively are
(*a*) 4 MHz and 60 MHz (*b*) 3 MHz and 20 Hz
(*c*) 3 MHz and 60 Hz (*d*) 4 MHz and 20 Hz

16. When an SCR is reverse biased
(*a*) two junctions are reverse biased and one junction is forward biased
(*b*) all the three junctions are reverse biased
(*c*) one junction is reverse biased and two junctions are forward biased
(*d*) any of above depending on the magnitude of reverse bias

17. In rectifier circuits, ripple factor RF and form factor FF are related as
(*a*) $RF = (FF^2 - 1)^{05}$ (*b*) RF = FF
(*c*) RF = FF – 1 (*d*) $RF = (FF - 1)^{05}$

18. In a series inverter supplying a load resistance R, the commutating elements L and C should be such that
(*a*) $R^2 = \frac{4L}{C}$ (*b*) $R^2 < \frac{4L}{C}$
(*c*) $R^2 > \frac{4L}{C}$ (*d*) $R^2 < \frac{2L}{C}$

19. In a bridge inverter
(*a*) an output transformer is essential
(*b*) an output transformer is not essential
(*c*) an output transformer cannot be connected
(*d*) an output transformer, if used, can improve the efficiency of the circuit

20. A step down dc chopper has an input voltage V. If duty cycle is α and the load is a resistance R, the rms value of output voltage is
(*a*) αV (*b*) $\sqrt{\alpha}\, V$
(*c*) $\alpha^2 V$ (*d*) $\frac{\alpha V}{R}$

21. In Jone chopper, the commutating element
(*a*) is a capacitor
(*b*) is a capacitor and an inductor
(*c*) is a capacitor and an auto transformer
(*d*) none of the above

22. In a 3-phase full wave regulator feeding a star connected resistance load the input voltage is 400 V line to line. The firing angle is 160°. The line to line output voltage would be
(*a*) 400 V (*b*) about 100 V
(*c*) about 20 V (*d*) zero

23. Which of these devices can conduct in both directions?
(*a*) SCR (*b*) SUS
(*c*) SBS (*d*) UJT

24. The stud in a stud mounted SCR acts as
(*a*) anode (*b*) cathode
(*c*) gate (*d*) anode or cathode

25. Which of the following devices has metal-silicon junction?
(*a*) general purpose power diode
(*b*) schottky diode
(*c*) SCR
(*d*) MOSFET

26. For Fig. 5, C(*s*)/R(*s*) =

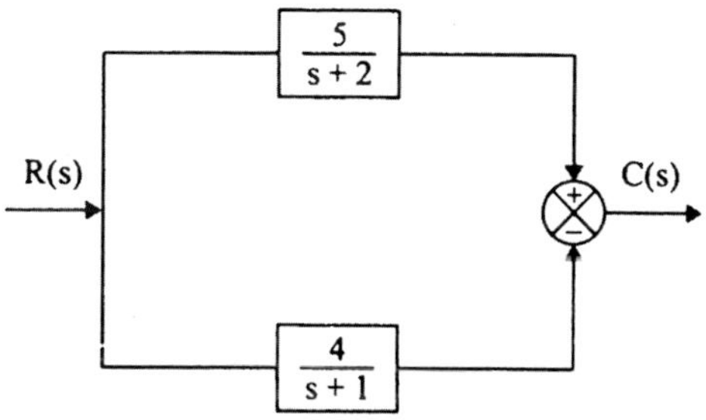

Fig. 5.

(*a*) $\frac{4(s+2)}{(s+2)(s+1)}$ (*b*) $\frac{s-3}{(s+2)(s+1)}$
(*c*) $\frac{9s+13}{(s+2)(s+1)}$ (*d*) $\frac{1}{(s+2)(s+1)}$

27. In some on-off controllers a differential gap is provided
(*a*) to make the manufacture process easier
(*b*) to prevent too often frequent operation of controller
(*c*) both (*a*) and (*b*) (*d*) neither (*a*) nor (*b*)

28. Examine the following statements
1. Pneumatic systems are fire proof
2. Pneumatic systems are explosion proof
3. Hydraulic systems are explosion proof
4. Hydraulic systems are fire proof

Which of the above statements are correct?
(*a*) 1 only (*b*) 1 and 2 only
(*c*) 1, 2, 3, 4 (*d*) 2, 3, 4 only

29. Gas flow resistance is defined as
(*a*) $\frac{\text{Change in gas pressure difference}}{\text{Change in gas flow rate}}$
(*b*) $\frac{\text{Change in gas flow rate}}{\text{Change in gas pressure difference}}$
(*c*) (change in gas pressure difference) (change in gas flow rate)
(*d*) none of the above

30. For the op-amp circuit of Fig. 6, $E_o(s)/E_i(s)$ =

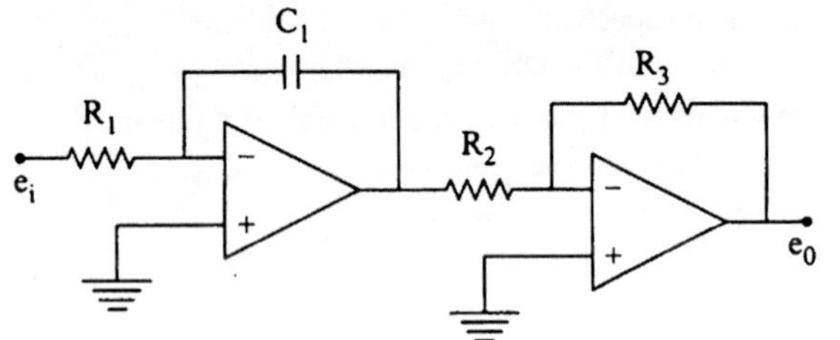

Fig. 6.

(*a*) $-\frac{R_3}{R_2}\left(\frac{1}{R_1C_1s}\right)$ (*b*) $\frac{R_3}{R_2}\left(\frac{1}{R_1C_1s}\right)$
(*c*) $\frac{R_3}{R_2}\left(\frac{C_1s}{R_1}\right)$ (*d*) $-\frac{R_3}{R_2}\left(\frac{C_1s}{R_1}\right)$

31. A network has $G(s) = \frac{s+1/T_1}{s+1/T_2}$. Then
(*a*) it is lead network
(*b*) it is lag network
(*c*) it is lead network if $T_1 > T_2$ and lag network $T_1 < T_2$
(*d*) it is lead network if $T_1 < T_2$ and lag network if $T_1 > T_2$

32. Transport lag usually exists in
(*a*) thermal systems
(*b*) hydraulic systems and thermal systems
(*c*) pneumatic systems
(*d*) thermal, hydraulic and pneumatic systems

33. The distance of poles from origin depends on
(*a*) undamped natural frequency
(*b*) damping ratio
(*c*) both (*a*) and (*b*)
(*d*) neither (*a*) nor (*b*)

34. The transfer function $G_c = 5\left(\frac{s+2.9}{s+5.4}\right)$ can be for
(*a*) lead compensator (*b*) lag compensator
(*c*) either (*a*) or (*b*) (*d*) neither (*a*) nor (*b*)

35. In Bode diagram (log magnitude plot) the factor $\frac{1}{j\omega}$ in the transfer function gives a line having slope
(*a*) – 20 *dB* per octave (*b*) – 10 *dB* per octave
(*c*) – 6 *dB* per octave (*d*) – 2 *dB* per octave

36. If $G(s) = \frac{1}{s^3+s^2+1}$, there are
(*a*) two poles of G(*s*) in the left half plane
(*b*) two poles of G(*s*) in the right half plane
(*c*) three poles of G(*s*) in the right half plane
(*d*) four poles of G(*s*) in the right half plane

37. Figure 7 shows a root locus plot. Consider the following statements for this plot

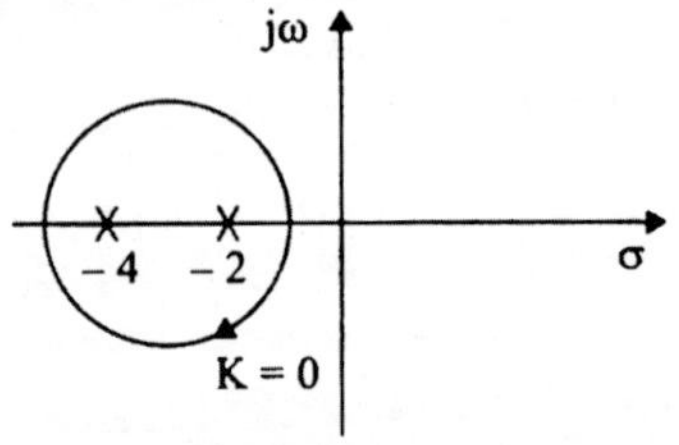

Fig. 7.

1. The open loop system is a second order system
2. The system is overdamped for K > 1
3. The system is stable for all values of K

Which of the above statements are correct?
(*a*) 1, 2, 3 (*b*) 1, 3
(*c*) 2, 3 (*d*) 1, 2

38. The effect of adding poles and zeros can be determined quickly by
(*a*) Nichols chart (*b*) Nyquist plot
(*c*) Bode plot (*d*) Root locus

39. Match List I and List II and select the answer using the codes

	List I (Location of characteristic roots)		**List II (System characteristic)**
A.	$(-1+j), (-1-j)$	1.	Marginally stable
B.	$(-2+j), (-2-j), (2j)$ $(-2j)$	2.	Unstable
C.	$-j, j, -1, 1$	3.	Stable

Codes:

	A	B	C
(*a*)	1	2	3
(*b*)	2	3	1
(*c*)	3	1	2
(*d*)	1	3	2

40. The system matrix of continuous time system described in state variable form is

$$A = \begin{bmatrix} x & 0 & 0 \\ 0 & y & -1 \\ 0 & 1 & -2 \end{bmatrix}$$

The systems is stable for all values of *x* and *y* satisfying
(*a*) $x < \frac{1}{2}, y < \frac{1}{2}$ (*b*) $x > \frac{1}{2}, y > 0$
(*c*) $x < 0, y < 2$ (*d*) $x < 0, y < \frac{1}{2}$

41. A compensating network has $G_c(s) = \frac{1+\alpha Ts}{1+Ts}$. If it is a phase lag circuit the value of α should be
(*a*) 0 (*b*) between 0 and 1
(*c*) 1 (*d*) more than 1

42. Consider the systems with following open loop transfer functions

1. $\dfrac{36}{s(s+3.6)}$ 2. $\dfrac{100}{s(s+5)}$

3. $\dfrac{6.25}{s(s+4)}$

If unity feedback is used, the increasing order of time taken for unit step response to settle is

(*a*) 1, 2, 3 (*b*) 3, 1, 2
(*c*) 2, 3, 1 (*d*) 3, 2, 1

43. Which of the following can act as error detecting devices?

1. a pair of potentiometers
2. a pair of synchros
3. a differential transformer
4. a metadyne
5. a control transformer

Select the answer using the following codes :

(*a*) 1, 2, 5 (*b*) 2, 3, 4, 5
(*c*) 1, 3, 4, 5 (*d*) 1, 2, 3, 4

44. Figure 8 shows a hybrid servo system. The components 1, 2, 3, 4 are respectively

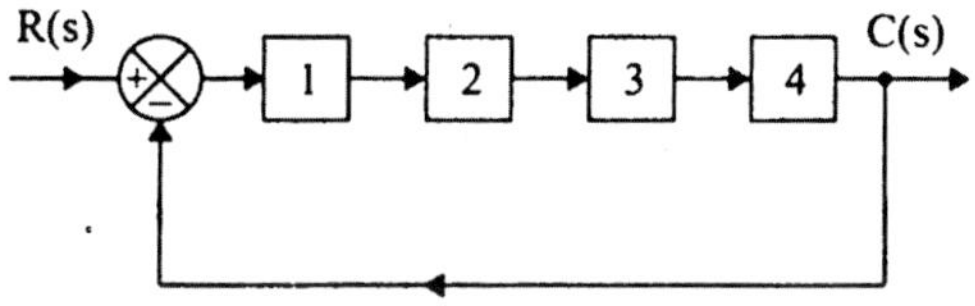

Fig. 8.

(*a*) amplifier, demodulator, dc servo motor and load
(*b*) demodulator, amplifier, dc servo motor and load
(*c*) demodulator, dc servo motor, amplifier and load
(*d*) amplifier, dc servo motor, demodulator and load

45. While forming Routh's array, the situation of a row of zeros indicates that the system

(*a*) has symmetrically located roots
(*b*) is not sensitive to variations in gain
(*c*) is stable
(*d*) is unstable

46. For a feedback system

$$\frac{C(s)}{R(s)} = \frac{G(s)}{1+G(s)H(s)}$$

Then the velocity error constant K_v is given by

(*a*) $K_v = \lim_{s\to 0} G(s)\,H(s)$

(*b*) $K_v = \lim_{s\to 0} \dfrac{sG(s)}{1+G(s)\,H(s)}$

(*c*) $K_v = \lim_{s\to 0} sG(s)$

(*d*) $K_v = \lim_{s\to 0} s[1+G(s)\,H(s)]$

47. M circles are symmetrical with respect to

(*a*) real and imaginary axis
(*b*) M = 1 straight line and real axis
(*c*) M = 1 straight line and imaginary axis
(*d*) M = 1 straight line

48. Symbol in Fig. 9 is IEEE symbol for

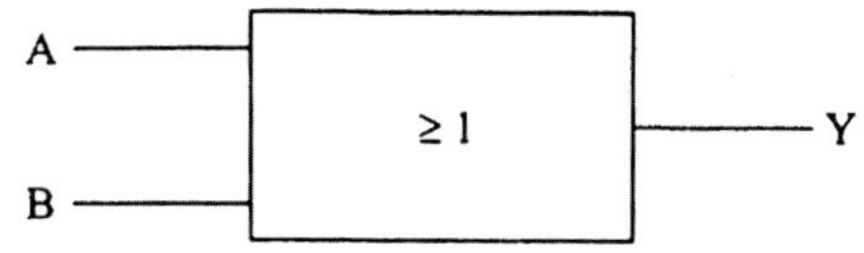

Fig. 9.

(*a*) AND (*b*) OR
(*c*) NAND (*d*) NOR

49. Decimal – 90 equals .... in 8 bit 2s complement

(*a*) 1000 1000 (*b*) 1010 0110
(*c*) 1100 1100 (*d*) 0101 0101

50. The number of inputs and outputs in a full adder are

(*a*) 2 and 1 (*b*) 2 and 2
(*c*) 3 and 3 (*d*) 3 and 2

51. In a positive edge triggered D flip-flop

(*a*) D input is called direct set
(*b*) Preset is called direct reset
(*c*) Preset and clear are called direct set and reset respectively
(*d*) D input overrides other inputs

52. A counter has 4 flip-flops. It divides the input frequency by

(*a*) 4 (*b*) 2
(*c*) 8 (*d*) 16

53. The basic storage element in a digital system is

(*a*) flip-flop (*b*) counter
(*c*) multiplexer (*d*) encoder

54. Which type of gate is in Fig. 10?

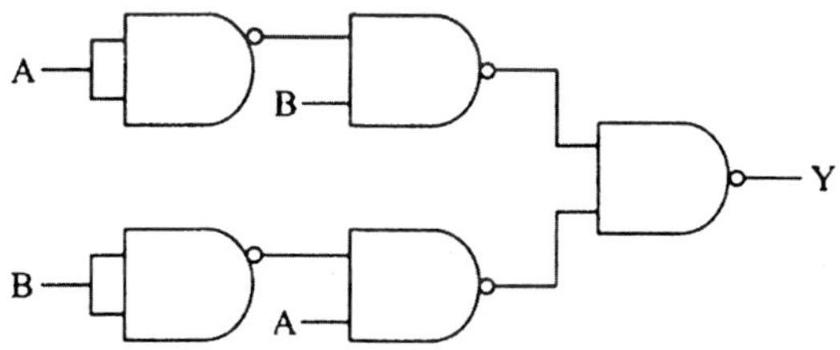

Fig. 10.

(*a*) NOR (*b*) OR
(*c*) NAND (*d*) EX-OR

55. The open wired circuit in Fig. 11 works as a

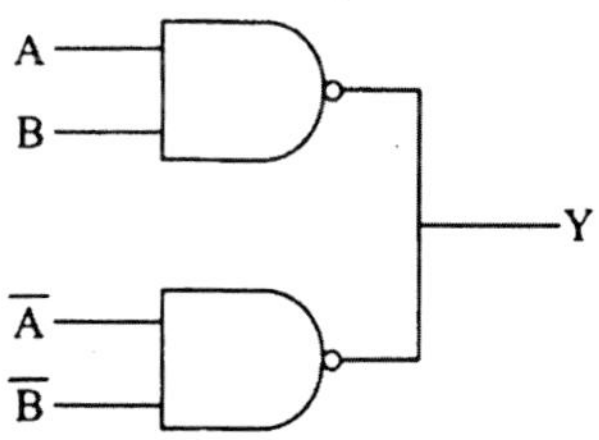

Fig. 11.

(*a*) Ex - NOR gate (*b*) AND gate
(*c*) XOR gate (*d*) NOR gate

56. ECL has high switching speed because the transistors are

(*a*) switching between cutoff and saturation regions
(*b*) switching between cutoff and active regions
(*c*) switching between active and saturation regions
(*d*) any of above

57. Examine the following truth table. The product of sum form for this truth table is

X	Y	Z
0	0	1
0	1	0
1	0	1
1	1	0

(*a*) $(\bar{X}+\bar{Y})(X+Y)$ (*b*) $(X+\bar{Y})(\bar{X}+\bar{Y})$

(*c*) $(\bar{X}+Y)(\bar{X}+\bar{Y})$ (*d*) $(X+\bar{Y})(X+Y)$

58. Which of the following is the truth table for AND gate?

(*a*)

Input		Output
0	0	1
0	1	0
1	0	0
1	1	1

(*b*)

Input		Output
0	0	0
0	1	1
1	0	1
1	1	0

(*c*)

Input		Output
0	0	0
0	1	0
1	0	0
1	1	1

(*d*)

Input		Output
0	0	1
0	1	1
1	0	1
1	1	0

59. The boolean expression for shaded area in Fig. 12 is

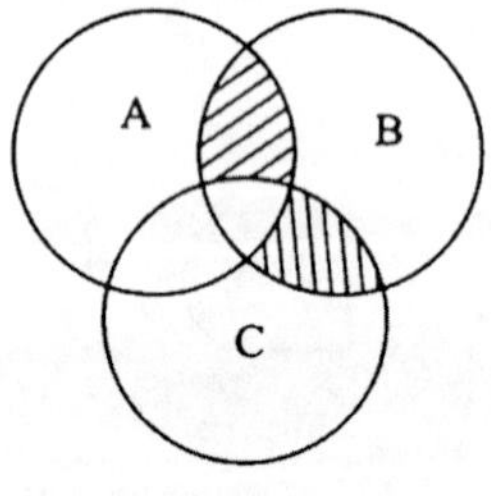

Fig. 12.

(*a*) $AB+AC$ (*b*) $ABC+\overline{A}\overline{B}\overline{C}$

(*c*) $AB\overline{C}+\overline{A}BC$ (*d*) None of the above

60. In Fig. 13, A = B = 1 and C = D = 0. Then Y =

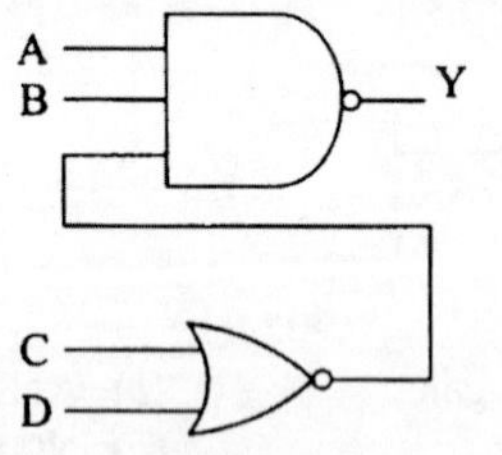

Fig. 13.

(*a*) 1 (*b*) 0
(*c*) either 1 or 0 (*d*) indeterminate

61. Which of the following finds application in pocket calculators?
(*a*) TTL (*b*) CMOS
(*c*) ECL (*d*) both (*a*) and (*c*)

62. The function of interface circuit is
(*a*) to match output conditions of driver with input conditions of load device
(*b*) to match output conditions of load device with input conditions of driver
(*c*) both (*a*) and (*b*)
(*d*) neither (*a*) nor (*b*)

63. A parity detector can detect
(*a*) single error only
(*b*) single error or odd number of errors
(*c*) two errors
(*d*) single or two errors

64. Which of them radiates emission?
(*a*) LED only (*b*) LCD only
(*c*) both LED and LCD
(*d*) neither LED nor LCD

65. In which counter does the maximum frequency depend on the modulus
(*a*) synchronous
(*b*) ripple
(*c*) both synchronous and ripple
(*d*) neither synchronous nor ripple

66. Match List I and List II and select the correct answer using the given codes

	List I (Logic)		List II (Characteristic)
A.	RTL	1.	High fan out
B.	CMOS	2.	Highest speed of operation
C.	I²L	3.	High noise immunity
D.	ECL	4.	Lowest product of power and delay

Codes:

	A	B	C	D
(*a*)	4	3	2	1
(*b*)	4	1	2	3
(*c*)	3	1	4	2
(*d*)	3	4	1	2

67. Which of the following assembler directives in 8085 does not cause any memory location to be used up?
(*a*) DS (*b*) DB
(*c*) DW (*d*) EQU

68. The memory hierarchy in microprocessor based system, is
(*a*) register, cache, primary, mass, off line back up
(*b*) cache, register, primary, mass, off line back up
(*c*) cache, register, mass, primary, off line back up
(*d*) register, cache, mass, primary, off line back up

69. Most of the memory chips in static RAMS
(*a*) do not require any supply
(*b*) need 2 V supply
(*c*) need 5 V supply
(*d*) need 12 V supply

70. Match List I and List II as regards register pairs in microprocessor 8085 and give answer using the codes

	List I (Bit Pattern)		List II (Register Pairs)
A.	00	1.	SP
B.	01	2.	B-C

C. 10 3. D-E
D. 11 4. H-L

Codes:

	A	B	C	D
(a)	1	2	3	4
(b)	2	3	4	1
(c)	2	3	1	4
(d)	4	3	2	1

71. Initially the number decimal 8 is stored. If instruction RAL is executed twice, the final number stored will be
(a) decimal 8 (b) decimal 16
(c) decimal 32 (d) decimal 2

72. Which of the following is a valid variable in Fortran 77?
(a) 3 A (b) ALPHA – 3
(c) ALPHA 3 (d) BALPHA C

73. The correct form of expression $B^3 - 2AC$ in Fortran 77 is
(a) B ** 3 – 2 * A * C (b) B * 3 – 2 * A AC
(c) B ** 3 – 2 ** A * C (d) B *** 3 – 2 * A * C

74. Which of the following is a valid integer variable in Fortran 77?
(a) ALPHA (b) NEXT
(c) BANGLORE (d) ABC

75. Consider the following logical IF statement in Fortran 77
IF (SALT. LE. PEPPER) 11
GO TO 13
The above statement using arithmetic IF statement would be
(d) IF (SALT – PEPPER) 11, 11, 13
(b) IF (SALT – PEPPER) 13, 11, 11
(c) IF (SALT – PEPPER) 13, 13, 11
(d) IF (SALT – PEPPER) 11, 13, 13

76. A computer program is used to read N and print the sum $1^2 + 2^2 + 3^2 + ... + N^2$. If N = 10, the print out will show the number
(a) 100 (b) 10
(c) 385 (d) 285

77. Which of the following statements are executable in Fortran 77?
(a) arithmetic IF (b) assign
(c) call (d) all the above

78. Which of the following real expressions is incorrect in Pascal?
(a) Y – X (b) (A + B) (C + D)
(c) X/Y – Z (d) P * Z – 5.0

79. The expression $A^2 + B^2 - 3AB$ when written is Pascal should be written as
(a) A * A + B * B – 3 A * B (b) A * A + B * B – 3.0 A * B
(c) A * A + B * B – 3.0 * A * B
(d) A * A + B * A – 3. * A * B

80. In Pascal the repeat and while loops will terminate
(a) only when the logical expression controlling the loop becomes true for both repeat and while loops
(b) only when the logical expression controlling the loop becomes false for both repeat and while loops
(c) only when the logical expression controlling the loop becomes true for repeat loop and false for while loop
(d) only when the logical expression controlling the loop becomes false for repeat loop and true for while loop

81. Which of the following operations will give the result 2 in C?
(a) 5/2 (b) 4.0/2 .
(c) 4.0/2.0 (d) 4/2.0

82. Which of the following is not correct in C?
(a) 6/4 = 1 (b) 2/5 = 0
(c) 2.0/5 = 0 (d) 2.0/5.0 = 0.4

83. Consider the following program in C
int *k* ;
float *b* ;
k = 3
b = 10.1
k = *k*/*b*
The final result would be *k* =
(a) 0.297 (b) 0.2
(c) 0 (d) 3

84. In the expression a = 5 * 2/3.2 + 4/8 – 2 + 5/9 which operation will have the first priority in C
(a) 5 * 2 (b) 2/3.2
(c) 5 * 2/3.2 (d) 4/8

85. Consider the following about 'arrays' in C
1. An array is a collection of similar elements
2. An array cannot have 10 numbers 5 of which are ints and 5 are floats
3. An array can have ints and chars in the same array
Which of the above are correct?
(a) all (b) 1 and 2 only
(c) 1 and 3 only (c) 2 and 3 only

86. Java's syntax is based on
(a) C (b) C++
(c) Fortran 77 (d) Pascal

87. In the human eye, an image persists for a time of about
(a) 1 second (b) 0.1 second
(c) 0.05 second (d) 0.02 second

88. The power density spectrum of noise voltage at the input of a network is S_{ni} and that at the output of the network is S_{no}. If transfer function of network is H(ω), the corresponding noise Figure is
(a) $\frac{S_{no}}{S_{ni}|H(\omega)|^2}$ (b) $\frac{S_{no}}{S_{ni}|H(\omega)|}$
(c) $\frac{S_{ni}}{S_{no}|H(\omega)|^2}$ (d) $\frac{S_{ni}}{S_{no}|H(\omega)|}$

89. In VCR, the chrominance signal's frequency is
(a) up converted during recording and down converted during playback
(b) down converted during recording and up converted during playback
(c) not changed during recording and playback
(d) any of above depending on VCR system

90. The power to the portable unit of a cordless telephone is provided by
(a) pencil cells (b) dry cells
(c) rechargeable cells (d) AC mains

91. In TV systems, equalising pulses are sent during
(*a*) horizontal blanking (*b*) vertical blanking
(*c*) serrations (*d*) horizontal retrace

92. SWR for a transmission line is infinity when the load is
1. short circuit
2. complex impedance
3. open circuit
4. pure reactance

Of the above statements
(*a*) 1, 3, 4 are correct (*b*) 1 and 3 are correct
(*c*) 2, 3, 4 are correct (*d*) 1, 2, 4 are correct

93. In TV broadcast, UHF range is
(*a*) 606 – 790 MHz (*b*) 100 – 150 MHz
(*c*) 41 – 68 MHz (*d*) 174 – 230 MHz

94. The generally used value of EHV supply in TV receivers is
(*a*) 1 kV per diagonal cm (*b*) 1 kV per diagonal inch
(*c*) 0.5 kV per diagonal cm (*d*) 0.5 kV per diagonal inch

95. Each kilometer of travel of electromagnetic radiation means a time delay of
(*a*) 330 μs (*b*) 33 μs
(*c*) 3.3 μs (*d*) 0.33 μs

96. The contrast control in colour TV receiver is done by
(*a*) chroma amplifier (*b*) SAW filter
(*c*) video detector (*d*) Luminance amplifier

97. Colour picture tube needs higher anode voltage and needs larger beam and cathode currents than monochrome tube because
(*a*) it is bigger in size
(*b*) it has higher level of illumination
(*c*) the tube has to illuminate three phosphors simultaneously
(*d*) the tube has lower efficiency because of aperture mask

98. In amplitude modulation the rms value of radio frequency voltage after 40% modulation is 60 mV. If modulation is 80% the rms value after modulation will be about
(*a*) 50 mV (*b*) 60 mV
(*c*) 66 mV (*d*) 80 mV

99. Consider the following statements
1. In a pulsed radar, the transmitted pulses are of high frequency
2. In a pulsed radar, the transmitted pulses are of high power
3. In a pulsed radar, the transmitted pulses are of brief duration
4. In a pulsed radar, the transmitted pulses are of square wave shape

Of the above statements the correct statements are
(*a*) 1 only (*b*) 1, 2 and 3 only
(*c*) 1, 2, 3, 4 (*d*) 1 and 3 only

100. In telephone systems, load is expressed in the unit of
(*a*) erlangs (*b*) decibels
(*c*) bells (*d*) watts

101. The directive gain of a transmitting antenna is proportional to
(*a*) its cross-sectional area
(*b*) square of cross-sectional area
(*c*) square root of cross-sectional area
(*d*) cube root of cross-sectional area

102. If '*a*' is the width of rectangular wave guide and 'λ' is the wavelength, then
(*a*) $a > \frac{\lambda}{4}$ (*b*) $a > \frac{\lambda}{2}$
(*c*) $a < \frac{\lambda}{4}$ (*d*) $a < \frac{\lambda}{2}$

103. TE_{01} mode of propagation of electromagnetic energy
(*a*) is merely the TE_{10} mode rotated 90°
(*b*) is merely the TE_{10} mode rotated 180°
(*c*) is the same as TE_{10} mode
(*d*) either (*b*) or (*c*)

104. In TE and TM modes of rectangular wave guides having propagation in *z* direction
(*a*) E_z and H_z are both zero
(*b*) In TE mode E_z is zero and in TM mode H_z is zero
(*c*) In TE mode H_z is zero and in TM mode E_z is zero
(*d*) In both TE and TM modes both E_z and H_z are non-zero

105. Which of the following is not used as a microwave mixer or detector?
(*a*) PIN diode (*b*) Crystal diode
(*c*) Schottky barrier diode (*d*) Backward diode

106. Fields are said to be circularly polarized if
(*a*) their magnitudes are equal and they are in phase
(*b*) their magnitudes are equal and they differ in phase by ± 90°
(*c*) their magnitudes are unequal and they differ in phase ± 90°
(*d*) their magnitudes are unequal and they are in phase

107. Match List I and List II and give correct answer using the given codes

	List I (Microwave device)		List II (Application)
A.	Reflex Klystron	1.	Wide band amplification
B.	TWT	2.	Very low power amplification
C.	2 cavity Klystron	3.	Low power FM generation
D.	Maser	4.	Frequency multiplication

Codes:

	A	B	C	D
(*a*)	1	2	3	4
(*b*)	4	3	2	1
(*c*)	3	4	1	2
(*d*)	3	1	4	2

108. Match List I and List II and give the correct result using the given codes

	List I (Solid state device)		List II (Application)
A.	PIN diode	1.	Microwave amplification
B.	Ga As Mosfet	2.	Low noise microwave generation
C.	Transferred	3.	Electronic tuning of MW oscillator
D.	Varactor diode	4.	Light wave detection

Codes:

	A	B	C	D
(*a*)	1	2	4	3
(*b*)	4	1	2	3
(*c*)	2	3	1	4
(*d*)	3	4	2	1

109. Consider the following statements about a magic tee

1. The collinear arms are isolated from each other
2. One of the collinear arms is isolated from E arm
3. One of the collinear arms is isolated from H arm
4. E and H arms are isolated from each other

Of the above statements

(*a*) 1 and 2 are correct (*b*) 1 and 3 are correct
(*c*) 1 and 4 are correct (*d*) 2 and 3 are correct

110. Match List I (component) with List II (dominant TE mode) and select the correct answer using the codes

	List I (Component)		List II (Dominant TE mode)
A.	Rectangular wave-guide	1.	TE_{101}
B.	Circular waveguide	2.	TE_{10}
C.	Rectangular cavity	3.	TE_{111}
		4.	TE_{11}

Codes:

	A	B	C
(*a*)	4	2	1
(*b*)	2	1	3
(*c*)	3	4	1
(*d*)	2	3	1

B. Questions 111 to 120 have two statements, viz, Assertion (A) and Resson (R). Read them and give correct answer as per the following codes.

(a) both A and R are true and R is correct explanation of A

(b) both A and R are true and R is not correct explanation of A

(c) A is true but R is false

(d) A is false but R is true

111. **A.** A fully controlled bridge converter can operate in first and fourth quadrant.
R. A semiconverter is cheaper than a full converter.

112. **A.** If the duty cycle of chopper is low the current would not become discontinuous.
R. If the time constant L/R of the load in chopper is low the current may become discontinuous.

113. **A.** A cycloconverter may be line commutated or forced commutated.
R. A stepup cycloconverter requires forced commutation.

114. **A.** Manufacturers quote *h* parameters of transistor.
R. *h* parameters of a transistor are easiest to measure.

115. **A.** In CE amplifier the emitter is at ground potential for ac signals.
R. A CE amplifier has near unity voltage gain.

116. **A.** In class C amplifier the efficiency is more than that for other classes of amplifiers.
R. In class C amplifier we can get high efficiency over a wide band.

117. **A.** In liquid system the ratio of change in liquid level to change in flow rate is called resistance of liquid flow.
R. In liquid level systems the ratio of change in liquid stored in a tank to the change in head is called capacitance of tank.

118. **A.** Increasing the system gain improves steady state behaviour but may result in instability.
R. A compensator is inserted into a system to ensure specified performance.

119. **A.** TTL is a very popular logic in SSI and MSI category.
R. In Schottky TTL the power dissipation is less than in ordinary TTL.

120. **A.** The advantages of totem pole output are: fast switching and low power consumption.
R. IC packages available are DIP, surface mount and J lead surface mount.

ANSWERS

1. (*c*)	**2.** (*a*)	**3.** (*d*)	**4.** (*d*)	**5.** (*d*)	**6.** (*d*)	**7.** (*a*)
8. (*a*)	**9.** (*b*)	**10.** (*b*)	**11.** (*a*)	**12.** (*a*)	**13.** (*c*)	**14.** (*b*)
15. (*b*)	**16.** (*a*)	**17.** (*a*)	**18.** (*b*)	**19.** (*b*)	**20.** (*b*)	**21.** (*c*)
22. (*d*)	**23.** (*c*)	**24.** (*a*)	**25.** (*b*)	**26.** (*b*)	**27.** (*b*)	**28.** (*b*)
29. (*a*)	**30.** (*b*)	**31.** (*c*)	**32.** (*d*)	**33.** (*a*)	**34.** (*a*)	**35.** (*c*)
36. (*b*)	**37.** (*b*)	**38.** (*c*)	**39.** (*c*)	**40.** (*c*)	**41.** (*b*)	**42.** (*c*)
43. (*a*)	**44.** (*a*)	**45.** (*a*)	**46.** (*a*)	**47.** (*b*)	**48.** (*d*)	**49.** (*b*)
50. (*d*)	**51.** (*c*)	**52.** (*d*)	**53.** (*a*)	**54.** (*d*)	**55.** (*c*)	**56.** (*b*)
57. (*d*)	**58.** (*c*)	**59.** (*c*)	**60.** (*b*)	**61.** (*b*)	**62.** (*a*)	**63.** (*b*)
64. (*a*)	**65.** (*b*)	**66.** (*c*)	**67.** (*a*)	**68.** (*a*)	**69.** (*c*)	**70.** (*b*)
71. (*c*)	**72.** (*c*)	**73.** (*a*)	**74.** (*b*)	**75.** (*a*)	**76.** (*c*)	**77.** (*b*)
78. (*b*)	**79.** (*c*)	**80.** (*c*)	**81.** (*a*)	**82.** (*c*)	**83.** (*c*)	**84.** (*a*)

85. (*b*)	**86.** (*b*)	**87.** (*d*)	**88.** (*a*)	**89.** (*b*)	**90.** (*c*)	**91.** (*b*)
92. (*b*)	**93.** (*a*)	**94.** (*b*)	**95.** (*c*)	**96.** (*d*)	**97.** (*d*)	**98.** (*c*)
99. (*b*)	**100.** (*a*)	**101.** (*a*)	**102.** (*b*)	**103.** (*a*)	**104.** (*b*)	**105.** (*a*)
106. (*b*)	**107.** (*d*)	**108.** (*b*)	**109.** (*c*)	**110.** (*d*)	**111.** (*b*)	**112.** (*d*)
113. (*a*)	**114.** (*a*)	**115.** (*c*)	**116.** (*c*)	**117.** (*b*)	**118.** (*b*)	**119.** (*c*)
120. (*b*)						

Bibliography

1. Boylestad RL; 'Electronic Devices and Circuits Theory' PHI, N. Delhi.
2. Collin, R.E. 'Fundamentals of Microwave Engineering' McGraw Hill Inc.
3. DekkerA.J. 'Electrical Engg. Materials' PHI, New Delhi.
4. Gabel. RA, et.al, 'Signals and Linear System' John Wiley and Sons.
5. Gaonkar, RS, 'Microprocessor Architecture, Programming and Applications' Wiley Eastern, N. Delhi.
6. Golding EW, 'Electrical Measurements and Measuring Instruments'.
7. Gottfried BS, 'Schaum Outline Series in Theory and Problems of Programming in C, TMH' N. Delhi.
8. Grob Bernard et.al, 'Basic TV and Video Systems' McGra Hill International,
9. Gulati, RR, 'Monochrome and Colour TV' New Age, New Delhi.
10. Gupta BR, 'Consumer Electronics' S.K. Kataria & Sons N. Delhi.
11. Gupta BR and Singhal V, 'Digital Electronics' S.K. Kataria & Sons N. Delhi.
12. Gupta BR and Singhal V, 'Electrical Science' S. Chand and Co., N. Delhi.
13. Gupta BR 'Electronics and Instrumentation' S. Chand and Co., N. Delhi.
14. Gupta BR and Singhal V, 'Fundamentals of Electric Networks' S. Chand and Co., N. Delhi.
15. Gupta BR 'Network Systems and Analysis' S. Chand and Co., N. Delhi.
16. Gupta BR and Singhal V, 'Power Electronics' S.K. Kataria & Sons, N. Delhi.
17. Gupta BR, 'Principles of Electrical Engg.' S. Chand and Co., N. Delhi.
18. Helfrick AD, et. al, 'Modern Instrumentation and Measurement Technique' PHI N. Delhi.
19. Jordon EC, 'Electromagnetic Waves and Radiating Systems' PHI, N. Delhi.
20. Kennedy G, 'Electronic Communication Systems' TMH, N. Delhi.
21. Kraus J.D., 'Electromagnetics' McGraw Hill Inc.
22. Lemay Laura, et. al, 'Java Platform 2' Techmedia, N. Delhi.
23. Malvino AP, 'Electronic Principles' TMH, N. Delhi.
24. Malvino AP, 'Digital Computer Electronics' TMH, N. Delhi.
25. Mathur AP, 'Introduction to microprocessors' TMH, N, Delhi.
26. Mayo WE, 'Schaum Outline Series on Theory and Problems in Programming with Fortran 77' McGraw Hill Inc.
27. Millman J, et. al, 'Electronic Devices and Circuits' TMH, N. Delhi
28. Nagrath IJ, et.al, 'Control System Engg,' New Age N. Delhi.
29. Ogata K, 'Modern Control Engg.' PHI, N. Delhi.
30. Oppenheim AV, 'Signals and Systems' PHI, N. Delhi.
31. Rajaraman V, 'Computer Programming in Pascal' PHI, N. Delhi.
32. Rashid M.H., 'Power Electronics Handbook' Academic Press N. Y.
33. Roddy D, et.al, 'Electronic Communication' PHI, N. Delhi.
34. Taub H, et. al, 'Principles of Communication Systems' TMH, N. Delhi.
35. Tocci RJ, 'Digital Systems' PHI, N. Delhi.
36. Valkenberg, VME, 'Network Analysis' PHI N. Delhi.

Win Prizes

ATTENTION READERS

Win Prizes

We request for your frank assessment regarding various aspects of the book as given below:

QUESTION BANK
IN
ELECTRONICS & COMMUNICATION ENGINEERING

By: Dr. B.R. Gupta & Vandana Singhal

Please Fill up the Sheet in Neat and Clean Writing

(i) What topics of your syllabus, important for your examination point of view, are not covered in this book ?

....

....

....

....

(ii) What are the chapters wherein the treatment of the subject matter is not systematic or organised or updated?

....

....

....

....

(iii) Have you come across misprints/mistakes/factual inaccuracies in the book? Please specify the chapters and the page numbers.

....

....

....

....

(iv) What other books on the same subject have you found/heard better than the present book? Please specify in terms of price and quality

....

....

....

....

(v) Further suggestions and comments for the improvement of this book.

....

....

....

....

OTHER DETAILS

(i) Who recommended you this book? (Please tick (✓) in the box near the option relevant to you.

☐ Teacher ☐ Friends ☐ Bookseller

....

....

....

....

(ii) Name of the recommending teacher, his designation and address:

....

....

....

....

(iii) Name and address of the bookseller you purchased the book from :

....

....

....

....

(iv) Name and address of your institution (Please mention the University or Board as the case may be)

....

....

....

....

(v) Your name and complete postal address:

....

....

....

....

(vi) Course you are enrolled in:

....

....

....

....

The best assessment will be awarded each month: The award will be in the form of our publications as decided by the Editorial Board.

Please mail the filled up coupon at your earliest to:

S.K. KATARIA & SONS®

4885/109, Prakash Mahal, Dr. Subhash Bhargav Lane,

Opposite Delhi Medical Association, Daryaganj, New Delhi–110002 (INDIA)

Phone: +91-11-43551243; *Mobile:* +91-9871775858

e-mail: katariabook@gmail.com; katariabooks@yahoo.com

Website: www.skkatariaandsons.com